Polymer Biomaterials in Solution, as Interfaces and as Solids

Professor Allan S. Hoffman

Polymer Biomaterials
in Solution, as Interfaces and as Solids

Festschrift Honoring the
60th Birthday of Dr. Allan S. Hoffman

Editors:
S.L. Cooper, C.H. Bamford and T. Tsuruta

CRC Press
Taylor & Francis Group
Boca Raton London New York

CRC Press is an imprint of the
Taylor & Francis Group, an **informa** business

First published 1995 by VSP Publishing

Published 2018 by CRC Press
Taylor & Francis Group
6000 Broken Sound Parkway NW, Suite 300
Boca Raton, FL 33487-2742

First issued in paperback 2019

ISBN-13: 978-0-367-44919-3 (pbk)
ISBN-13: 978-90-6764-180-7 (hbk)

CIP-DATA KONINKLIJKE BIBLIOTHEEK, DEN HAAG

Polymer

Polymer biomaterials in solution, as interfaces an as solids:
a festschrift honoring the 60th birthday of Dr. Allan S. Hoffman /
ed: S.L. Cooper, C.H. Bamford, T. Tsuruta. - Utrecht : VSP
Orig. publ. in Journal of Biomaterials Science, polymer
edition, ISSN 0920-5063 ; v. 4, no. 3 and 5, v. 5, no.
1/2, 1993 and v. 5, no. 4-6, v.6, no. 1 and 4, 1994
ISBN 90-6764-180-4 bound
NUGI 821/841
Subject headings: polymer chemistry / biochemistry

CONTENTS

VII. New Polymers and Applications

VIII. Biodegradable Polymers and Drug Delivery

A. Degradable Polymers

FOREWORD

This festschrift celebrates the 60th birthday of Professor Allan S. Hoffman. The broad range of the field of biomaterials that Allan Hoffman has influenced is evident in these manuscripts on original research as well as some review papers.

These articles were first published during 1993 - 1994 in 8 issues of the *Journal of Biomaterials Science, Polymer Edition.* We are most pleased now to make the entire festschrift available in book form.

The editors, publisher and the many colleagues who have participated in this project extend their best wishes to Allan on this happy occasion. We look forward to Allan Hoffman giving us many more years of continued vitality and leadership in biomaterials research.

S.L. Cooper
C.H. Bamford
T. Tsuruta

Personal Statement

When Professor Stuart Cooper, the North American/South American Editor of the *Journal of Biomaterials Science, Polymer Edition*, conceived the idea of a Festschrift in honor of my 60th birthday, and Professor Teiji Tsuruta and Professor "Bam" Bamford (the other Editors) enthusiastically agreed with him, non of us - at least of all myself - had any idea that it would turn out to be such a success. And indeed, it has been an *immense* success. This is due both to the large number and - especially - the high quality of the articles published in the eight issues of the Festschrift. Another important feature of this impressive collection of articles is their great diversity. They truly represent the *state of the art* in polymeric biomaterials today.

In preparing for this book, it has been both a joy and a challenge for me personally to organize the Festschrift journal articles into a logical order of sections, each having a distinct focus. This very diverse and eclectic "Interpenetrating network" of papers ressembles in may ways my own scientific wanderings over the past 30 years, through the subjects of natural tissue biomaterials, hydrogels, surface-modified polymers, and most recently, water-soluble polymers. It has been especially enjoyable because, as I "visited" each article, it felt like I was visiting with colleagues, renewing old friendships and making new friendships. Some contributors even included nice personal comments about me, which especially warmed my heart.

I was also inspired when I realized that this wonderful collection of research papers came from scientists of all ages, including many who have worked, as I have, for over 25 years in the field, working together with and mentoring their young and bright students and postdocs, who represent the future of our field. Indeed, the Festschrift includes a number of academic fathers, sons and grandsons, showing that we are essentially one large family. The *whole* is clearly much greater than the sum of the parts. I am so very honored to be the *raison-d'etre* of this great "family collection" of current biomaterials research of the 1990s.

I was further pleased to note the great geographic diversity of the authors. I would like to think that this reflects my own extensive and dedicated efforts over the past 25 years to stimulate interest in polymeric biomaterials around the world.

I am very grateful and indebted to all those who have contributed their research papers for this Festschrift, which has been such a very special event in my life. I am very grateful to the Editors, Teiji Tsuruta, "Bam" Bamford, and, especially, Stuart Cooper, of the excellent *Journal of Biomaterials Science, Polymer Edition*, who not only sponsored the Festschrift but who also put in many long and strenuous hours reviewing and sending out articles for review, editing and organizing. Of course, an essential component of the Festschrift has been the high standards, superb workmanship and generous assistance of the staff of VSP, International Science Publishers.

Finally, I want to recognize the tremendous contributions that so many special people have made to me personally and professionally over my lifetime. They have been inexorably involved in my life, and this recognition of me has to be for them as well.

Who are these people? They are, first and foremost, my first family, beginning with my parents, Saul and Frances, who gave me life and who instilled me in the love of people and the love of knowledge. And there is my brother, Mike, who excells in everything he does and who has always inspired me and stimulated me to achieve. And lastly there is my younger sister Marilyn, who arrived when I was 12, and who has inspired me so much by her huge love and compassion for others.

Then there are my many wonderful teachers in my schools over the years, who have had such a major impact onmy knowledge, my ability to think rationally and my joy in probing the unknown. I am grateful for the eduction I got at MIT, where I learned how to "talk to and travel with molecules" from many great teachers, including Ed Gilliland, Ed Merrill and Alan Michaels. I was also given a rare and valuable opportunity to teach classes at the Chemical Engineering Department at MIT while I was still a graduate student working on my Sc.D. It had a huge impact on my life. I taught my first class, a sophomore course in stolciometry, in 1955, when I had just turned 23 years old. I remember it as a thrilling experience, and it implanted in me a lifelong love for teaching. (Another kind of "biocompatible implant")!

And there is the lifeblood of my academic career - my students, post-doctoral fellows, visiting scientists, and collaborators - with whom I have worked over the past 25 years. Our very close and personal interactions, as well as our great and exciting adventures in watching our ideas about molecules come to life in the laboratory, have been an integral and indispensable part of my personal biomaterials life. I have learned so much from them, and I honestly feel that *I* have been the *lucky* one who has so often had the opportunity to "spread *their* word" in our joint publications and presentations. How can I truly express my deep gratitude to them all? It's impossible.

I do want to single out two of my former postdocs, Tom Horbett, who came with me at the University of Washington in 1971, and Buddy Ratner, who arrived in 1972. Tom and Buddy have been such close and indispensable collaborators and friends over more than 20 years, that it is very hard for me to adequately describe their huge and invaluable contributions to my own biomaterials career. Indeed, our three careers have been inexorably interlinked for the past 23 years. I have tremendous feelings of gratitude (for their collaboration) and pride (like that of an "academic" parent as well as a "talent scout" who has discovered new "stars") as I watched each of them grow to become individually recognized and established as outstanding researchers in the biomaterials field.

Last, but most certainly *not least*, I want to talk about my own family. My wife Susan has been an indispensable and invaluable "silent partner" in my career. She has been such a loyal and supportive wife to me and mother to our children over so many years, that I could not have achieved what I have without her help. Add to this our wonderful children, David and Lisa, who have been such a joy to come home to. Watching them grow up from loveable and exciting children into intelligent and compassionate adults has been *the greatest joy of my life.* Their patience and tolerance with a husband/father who is such a workaholic, as well as their sound advice to me when I most need it, have been invaluable sources of support and comfort over the years.

I am a very lucky man. Thank you all, so very much.

Allan S. Hoffman
Seattle, Washington
May 1, 1994

Highlights of Allan Hoffman's Biography

Education

MIT Chemical Engineering, Cambridge, MA
BS (1953), MS (1955), ScD (1957)
(Dissertation: 'Mechanisms of Radiation Induced Graft Polymerization'
Supervisors: E. W. Merrill, E. R. Gilliland, W. H. Stockmayer).

Post-Doctoral Fellowship

Fulbright Fellow (1957–1958)
Laboratoire de Chimie Physique, Paris, France
(A. Chapiro, M. Magat)

Professional Career

Academic:

MIT Chemical Engineering Department

Instructor	(1954–1956)
Assistant professor	(1958–1960)
Associate professor	(1964–1970)

University of Washington, Center for Bioengineering and Chemical Engineering Department
Professor (1970–present)
Assistant director, Center for Bioengineering (1973–1983)

Student/Post Doc Research Supervision

Supervised research of over 130 students (BS, MS, PhD) and Post Doctoral Fellows in 32 years at MIT and University of Washington.

Industrial

California Research Corp., Richmond, CA
(1960–1963)

Amicon Corp., Cambridge, MA
Associate Director of Research (1963–1964)

Some Professional Activities

Chairman, Gordon Conference on Biomaterials (1977)

President, Society for Biomaterials (1984)

Trustee and Member, Board of Directors, International Society for Artificial Organs (ISAO) (1987–1990)

Member, Board of Governors, Controlled Release Society (1991–1994)

Member of Editorial Boards of 7 professional journals (present)

Member, Scientific Advisory Boards of 4 companies (present)

Some Honors, Awards

Visking Fellow (1954–1955)

Kimberly-Clark Fellow (1956–1957)

Battelle Fellow (1970–1972)

Clemson Award from Society for Biomaterials for 'Outstanding Contributions to the Biomaterials Literature' (1984)

Recognition Award from Controlled Release Society for 'Excellence in Guiding Graduate Research' (1989)

'Biomaterials Science Prize' from the Japanese Society for Biomaterials (1990)

Founding Fellow, American Institute of Medical and Biological Engineering (1992)

Awards to Hoffman's Students and Postdoctoral Fellows

L. C. Dong (PhD): 'Outstanding Student Paper Presentation' Controlled Release Society (1989)

T. G. Park (PhD): Honorable Mention, 'Outstanding PhD Paper' from the Society for Biomaterials (1990)

K. P. Antonsen (Post-Doc): N. I. H. Bioengineering Training Fellowship (1989–1991)

J. R. Bain (PhD): N. I. H. Biotechnology Training Fellowship (1991–1992) N. I. H. Bioengineering Training Fellowship (1992–1994)

D. K. Pettit (PhD): 'Outstanding PhD Paper' from Society for Biomaterials (1992)

R. P. Mumper (Post-Doc): 'Outstanding Research Paper' from Sigma Xi, University of Kentucky (1991)

Publications

Approximately 210 papers or chapters in books, twelve patents, and co-authored or co-edited 5 books.

Major topics of research covered in these publications:

Basic Research Areas

Synthesis of inert, reactive, or responsive polymer systems in the form of soluble polymers, hydrogels, polyelectrolyte coacervates, composites with liposomes, IPNs and surface coatings on membranes, fibers, fabrics, particulates, and porous supports.

Surface modification of hydrophobic polymer supports using gas discharge, radiation grafting or ozone grafting to render them inert or chemically reactive or repellant or retentive to proteins and cells.

Bulk and surface characterization of these polymer systems (water uptake/loss; free vs. 'bound' water; thermodynamics of protein elasticity, pore structure, solute partitioning and transport mechanisms in hydrogels; ESCA; contact angles, surface energetics and wettability).

Physical immobilization or chemical conjugation of biologically active or surface active (amphipathic) molecules, on or within these polymer systems.

Biologic interactions (proteins and cells adsorption, adhesion and blood and tissue responses) of these polymer systems.

Bioactivity of the immobilized biomolecules.

Physical and biomechanical properties of natural structural tissues, using enzymes to probe the contributions of specific components such as collagen, elastin and GAGs.

Applied Research Areas

Drug delivery systems.

Affinity separations.

Immobilized, reactive enzymes and cells.

Desalination by reverse osmosis membranes.

Immunodiagnostics and biosensors.

Surface modification for improved dyability, static dissipation, wettability, stain and soil release, protein retention, protein and cell 'repulsion', cell culture, and chemical reactivity.

Small diameter vascular graft.

Allan S. Hoffman—Overview and appreciation

BUDDY D. RATNER and THOMAS A. HORBETT
Center for Bioengineering and Department of Chemical Engineering, BF-10, University of Washington, Seattle, WA 98195, USA

INTRODUCTION

In the field of biomaterials, Allan S. Hoffman has been a pioneer, a leader, and a mentor. Each of these contributions reflects well upon the man, and there is much that can be said about each. The pioneer opens new territories to exploration. An eye toward the horizon and unremitting curiosity are the characteristics of a pioneer. The leader not only opens those new territories, but guides others to them. The leader builds roads so others can follow. Finally, the mentor (in contrast to simply a teacher), as well as communicating facts, also imparts the creativity and free spirit of the pioneer and the art of leadership. Allan Hoffman's long, ongoing, and illustrious career prominently displays his contributions as pioneer, leader, and mentor. It is with pleasure and honor that we sketch the career of our mentor, colleague, and friend, Allan Hoffman.

Allan Hoffman was born in Chicago on October 27, 1932. He received B.S., M.S., and Sc.D. degrees from Massachusetts Institute of Technology, completing his doctoral thesis in 1957 under the guidance of three illustrious professors, E. R. Gilliland, E. W. Merrill, and W. H. Stockmayer. Why three professors? Perhaps this is an early manifestation of the interdisciplinary flexibility that is so evident throughout his career. The three advisors were needed to bring together the diverse skills and ideas that he proposed for his thesis research in the then new field of the radiation chemistry of polymers.

His first positions were at Massachusetts Institute of Technology, as the Assistant Director of the Oak Ridge Engineering Practice School (where he learned radiation chemistry) and then as an instructor. After a few industrial positions, including a stint with California Research Corporation in Richmond, CA, and Associate Director of Research for Amicon Corporation, Allan returned to MIT, first as Assistant Professor (1958–1960) and then as Associate Professor (1965–1970). A glimpse at the list of courses he taught during this period is revealing: surfaces, polymers, colloids, materials science, intermolecular forces, thermodynamics—many themes, to this day, associated with Allan Hoffman—but no biology or biomaterials. Still, this was an important time in the development of the biomaterialist we know today. Hoffman's supervision of an MIT thesis project on elastin, his discussions with Alan Rembaum on hydrogels, and his appreciation of the ideas communicated in seminal publications by Robert Leininger on biomolecule immobilization were influential to him.

In 1970, after a short stay at the Centre d'Etudes Nucleaires in Saclay, France, where he taught courses, in French, in polymers and surfaces, Allan Hoffman moved to the University of Washington. He has been Professor of Bioengineering and Chemical Engineering there since 1970. The Center for Bioengineering at the University of Washington, at that time under the inspired leadership of Dr. Robert Rushmer, provided a fertile environment for Allan to find appropriate collaborators and develop a new program in biomaterials. His list of courses at the University of Washington now includes biomaterials, controlled release, and biotechnology. Indeed, in 1974 he was invited to present a one week course back at Saclay, and this time the title was 'Biomaterials'.

Since 1970, the biomaterials interest, resting upon the foundation of polymer and surface science, has continued to grow, as has Allan Hoffman's stature and honors. His many invitations and commissions are too numerous to list, but a few choice items are worthy of mention: the invited main lecture at the 27th IUPAC Symposium on Macromolecules, Strasbourg, France (1981); a 6-day course on biomaterials at the invitation of the Ukrainian Academy of Sciences (1983); the keynote lecture at the First Kyoto International Symposium on Biomedical Materials (1983); invited plenary lecture at the 2nd World Congress on Biomaterials, Washington, DC; invited lecturer by the Chinese Chemical Society at Guangzhou University (1986); organizer of a three-day US/China Polymeric Biomaterials Research meeting (1987)—Allan Hoffman's CV lists well over 70 such prestigious lectureships and chairmanships.

Hoffman, throughout his career, has been a strong player in the international community. As well as the international honors reflected in the activities described above, he has been invited for courses and symposia in Mexico, Australia, Switzerland, Uzbekistan, Korea, Finland, Sweden, Czechoslovakia, Turkey, Israel, Italy, Taiwan, and a number of other countries as well. He speaks excellent French and an inspired smattering of about five other languages. With an ecumenical spirit and an ambassador's personality, Allan Hoffman has spread the gospel of biomaterials throughout the world, and it is probable that he has 'seeded' (intellectually) the biomaterials community in a number of countries.

In addition to his activities as a lecturer and communicator, Allan Hoffman has held many prestigious leadership positions in our field. These include advisor to the International Atomic Energy Agency, consultant to the World Health Organization, President of the Society For Biomaterials (1983–1984), chair of the Science and Technology of Biomaterials Gordon Conference (1977), member of the Board of Governors of the Controlled Release Society (1991–1993), a trustee and member of the Board of Directors of the International Society For Artificial Internal Organs (1986–1988), and an elected Fellow of the American Institute of Medical and Biological Engineering (1992).

Also, there are many awards that Allan Hoffman has received. These include the Clemson Award for Outstanding Contributions to the Literature on Biomaterials (1984) and the Japanese Biomaterials Science Prize (1990). Incidentally, Hoffman is the first non-Japanese to receive this prestigious award.

Of course, the essence of Allan Hoffman, the researcher, is contained in the body of knowledge he has generated and shared with the community. Starting with his first paper in 1955 ('Graphical Method Speeds Production Scheduling of Radioisotopes'), [1] Allan has, over 37 years, steadily contributed to the technical literature. Because of the large number of publications (well over 200), it is appropriate to divide his accomplishments into subject areas and describe some key papers and contributions in each.

RADIATION GRAFTING

Among the earliest polymer studies published by Allan Hoffman were those on radiation grafting [2, 3]. His position at Oak Ridge, coupled with an appreciation of the novel work published by Vivian Stannett and Adolphe Chapiro, led to this research theme that resurfaces throughout his career. Some of his first papers from the University of Washington were on radiation grafting [4, 5]. Both of the authors of this article came to the University of Washington to work with Allan on this then new class of materials, leading to many of the early papers from this group [6–10]. The radiation grafted hydrogels became the foundation for studies on blood compatibility [10, 11] enzyme immobilization [12], protein adsorption [9], surface characterization [13, 14], and non-fouling surface preparation [15].

SYNTHETIC MEMBRANES

Membrane technology is another theme evident throughout Allan Hoffman's career. Certainly, his work at Amicon Corporation as associate director of research (1963–1965) must have been influential. At that time, the first wave of interest in membranes as low energy consumption methods for separations and purification was building. However, even before that, Hoffman had published on gas permeation membranes [3]. A number of papers appeared in 1969 and 1970 on the preparation and characterization of polyacrylic membranes for reverse osmosis purification of water [16–18]. It is interesting to note that one of these papers [17] discussed the application of 'molecular engineering' for optimizing membrane performance.

MECHANICAL PROPERTIES OF CONNECTIVE TISSUE

Because of his interest in the properties of natural as well as synthetic biomaterials, Allan Hoffman and his young colleagues produced a number of studies on collagen and elastin, focusing on the contribution of these macromolecules to the stress–strain behavior of natural tissues, including ligaments and the aorta [19, 20]. A novel technique, mechanical testing after sequential enzymolysis of the tissue with specific enzymes, was used in this work [21–25]. Joon Bu Park, now a professor in biomaterials at the University of Iowa, performed many of these studies while a postdoctoral fellow in Allan's lab.

HYDROGELS

Synthetic hydrogels for biomedical applications were first proposed by Wichterle and Lim in 1960 [26]. With Hoffman's first paper on this subject dating back to 1969 [16], he is clearly a pioneer in this still expanding field. To this day, his steady output of creative, influential works on this subject continue to swell the literature. In particular, Hoffman has applied hydrogels for desalination [16–18], as radiation grafts [4, 27, 28], for blood interaction [29, 30], as enzyme immobilization supports [5, 31–34], for cell interaction [8, 35, 36], for protein separations [37–39], as non-fouling surfaces [40, 41], as stimuli-sensitive biomaterials [34, 42], and for controlled drug release [34, 43]. His review articles on hydrogels have also been influential [44–46].

BIOMOLECULE IMMOBILIZATION

The discovery of the special properties of immobilized enzymes in the late 1960s prompted much interest in the applications of this methodology, interest that continues to this day. Allan Hoffman made a seminal contribution to the extension of this idea to the biomaterials field when he began his work on the immobilization of various biomolecules to radiation grafted hydrogels [5, 47, 48]. The idea was to try to retain the good mechanical properties of a substrate polymer such as silicone rubber, but make it suitable for enzyme immobilization by converting the surface to a functionalized hydrogel. The biocompatibility of hydrogels was also a major reason for their choice, since Allan recognized the need to develop methods for biomolecule immobilization that might be transferable to devices that could be used in the treatment of human disease. Thus, the immobilization of proteolytic enzymes and heparin to either dissolve or prevent clots on surfaces won Allan and his colleague, Gottfried Schmer, justified praise for the novel extension of biomolecule immobilization to the field of biomaterials. Subsequent work on the immobilization of enzymes on ionic hydrogels focused on the mechanistic aspects of enzyme catalysis [49, 50]. More recent work with asparaginase on blood plasma exchangers [51, 52] or in thermally reversible gels [33], the immobilization of enzymes [53, 54], and even whole cells in thermally reversible gels [55, 56], all stem from Allan Hoffman's early recognition of the unique possibilities inherent in the immobilization of biomolecules on biocompatible surfaces in a variety of applications.

SURFACE MODIFICATION

An appreciation of the importance of surface composition as a modulator of bioreaction is clearly evident in Allan Hoffman's early realization that only the outermost region of material need be modified. Thus, many of his works emphasize surface modification by radiation grafting [14], RF-plasma glow discharge treatment [57, 58], biomolecule immobilization [5, 32], poly(ethylene oxide) immobilization [59], and surfactant immobilization [60]. His ideas on this subject were summarized in a comprehensive 1987 review article [61].

BLOOD COMPATIBILITY

The problem of blood clotting on medical device surfaces has been a central theme in Allan Hoffman's biomaterials research, and it is really this problem that attracted him, in the beginning, to biomaterials. At the time of his entry into the field in the early 1970s, a major revolution in the medical device field was underway in the form of the artificial kidney. This device was seeing a sharp increase in clinical use, along with laboratory study and further development. The success with the artificial kidney and with the heart–lung machine also prompted a major national effort directed toward the development of artificial hearts. Researchers were well aware of the great need for better blood compatibility for the materials used in these devices. Allan Hoffman realized that a good solution to the problem was to focus on making the surface of the devices more benign, and his background in radiation processing [2] naturally led him to pursue radiation grafting of hydrogels as a route to improved blood compatibility [4, 7, 62]. From this background stemmed much of the subsequent research in his laboratory.

This research included the unexpected finding that, while these hydrogels were, in fact, thromboresistant by the vena cava test [10, 11, 63]—the state-of-the-art test of that day—they caused enhanced destruction of platelets when tested in a then novel baboon *ex vivo* shunt model [29, 64–66]. This new test, developed by his student Steve Hanson and his colleague Laurie Harker, led to the new concept that some materials, including the hydrogels, only appear to be thromboresistant, while, in fact, they are constantly shedding thromboemboli. Thus, we now recognize that proper biomaterial blood compatibility evaluation must include an assessment of thromboemboli potential, as well as occlusion.

Allan Hoffman's more recent contributions to the field of blood compatibility include the development of plasma discharge treated fluorocarbon surfaces that may achieve their improved blood compatibility [67] because of their unique ability to bind adsorbed proteins much more tightly than conventional polymers, including ordinary PTFE Teflon [68]. In addition, Hoffman is actively working on improving blood compatibility by producing poly(ethylene oxide) surfaces that are non-fouling because of their extremely low retention of adsorbed proteins [40, 59, 69]. These surfaces may also prove to be particularly useful in preventing bacterial infection on implants, a new application that has generated much interest in Hoffman's laboratory.

DRUG DELIVERY SYSTEMS

In the early 1970s, Allan Hoffman was a consultant for an exciting, new, San Francisco Bay area start-up company, Alza. The imaginative coupling of polymer chemistry and pharmacy took place in an environment designed to stimulate innovation and excitement. This environment was peopled by individuals who would play important roles throughout his career, including Jorge Heller, Alan Michaels, Richard Baker, Nam Choi, Keith Gilding, George Green, Ed Schmidt, Harry Lonsdale, Richard Buckles, and others. The interest in drug release and controlled drug delivery was strongly stimulated by the Alza experience. Many of his publications, particularly starting in the late 1980s, featured novel drug delivery ideas [34, 70, 71]. In 1989, Hoffman's Ph.D. student Liang-chang Dong was honored by the Controlled Release Society with the Outstanding Student Presentation Award.

STIMULI SENSITIVE POLYMERS

Some of the original ideas on responsive polymers were published by Katchalsky in 1955 [72]. Hoffman was a leader in bringing these ideas to biotechnology and medicine. A conversation in August 1982 with Robert Nowinski, founder of Genetic Systems and Icos, triggered new ideas about the application for such polymer systems in biotechnology. In another conversation a year later with the polymer chemist, James Guillet, Allan Hoffman learned of poly(*N*-isopropyl acrylamide). This discussion precipitated a large and growing body of work on the synthesis and physical chemistry of this and related polymers that exhibit large molecular contractions upon gentle heating [33, 39, 42, 73, 74].

MATERIALS FOR BIOTECHNOLOGY APPLICATIONS

Many products of the biotechnology revolution are best used or made with a carrier polymer [75]. Thus, for example, the application of monoclonal antibodies for ELISA assays is often done with the antibody immobilized on a surface. Similarly, there are advantages in using immobilized cells in the production of biomolecules. In all of these applications, special properties of the material are needed to achieve full realization of the approach. Allan Hoffman, taking advantage of his rich experience in biomolecule immobilization, has therefore collaborated in a number of projects with biotechnological applications in mind.

Hoffman's career took a distinctly different turn at this juncture because he was able to establish productive collaborations with colleagues in the biotechnology industry, including John Priest, Robert Nowinski, and Nobuo Monji, that resulted in the development of some novel phase separation immunoassays employing thermally sensitive hydrogels [76, 77] and active esters [78] for separation and immobilization, respectively. Patentable inventions, as well as publications, resulted from this work. The application of the tight protein binding properties of plasma discharge deposited fluorocarbon surfaces [56] to enhance the activity and retention of immobilized antibodies used in solid phase immunoassays has been one of Allan Hoffman's more recent contributions to the biotechnology area.

The use of thermally reversible hydrogels for biotechnological applications also included the entrapment of cells [55] and enzymes [53, 54]. These gels could be shrunk or swollen at will to enhance or control production rates of these systems. The prospect of using such systems as improved bioreactors for applications such as the production of monoclonal antibodies from immobilized cells was based on the known biocompatibility of the hydrogels combined with the unusual ability to shrink or expand the gels around the cells to enhance transport as well as control the reactions.

ALLAN HOFFMAN—TEACHER AND MENTOR

Concurrent with the conduct of research activities at a university is the turnover of personnel as students enter and then graduate, postdoctorals come and go, and visiting academic and industrial scientists participate in the research for a year or so. In this environment, Allan Hoffman excelled at training researchers to achieve the level of productivity noted above. While many of us who work in this research environment concentrate on the scientific results and publications that mark our progress, the implied and explicit training of people in research is, in fact, a large part of the useful 'product' of research activities at a university. In this way, Allan has been unusually productive, particularly because of his ability to create relatively large research groups. His research groups have achieved the critical mass wherein many of the training activities are self-sustaining via the mentorship of individuals in the lab with particular expertise and various levels of training. Allan Hoffman has worked diligently to raise the funds to support this scale of enterprise. The nature of the Hoffman research groups has had a special impact on advancing the skills of the many, many people who have gone through his laboratory over the years.

The special contribution of Allan Hoffman to research training is evident in that three of his former postdoctoral fellows, Tom Horbett, Joon Bu Park, and Buddy Ratner, are now full professors in biomaterials. Because they have continued in biomaterials and established their own research groups, Allan's role in the advancement of their careers has leveraged a sizable impact on the field of polymeric biomaterials.

Some insight into the way that Allan Hoffman motivated Horbett and Ratner into committing to the biomaterials field seems in order, given the authorship of this article. How, exactly, did these researchers come into their current situation, and what did Allan do to facilitate this?

In the case of Tom Horbett, Allan Hoffman actively sought to recruit a biochemist into his group, almost from the day of his arrival at the University of Washington. He had established a good rapport with Hans Neurath, professor and chairman of biochemistry at the University of Washington, since each of them was, at the time, associated with the Seattle branch of Battelle Labs. Neurath knew Horbett well, since he had served on his doctoral committee, and Horbett had participated in Neurath's research meetings throughout his graduate school days. So, when Allan asked Neurath for the name of a physical biochemist to do a postdoctoral fellowship, he subsequently met Horbett and aggressively recruited Horbett into his lab. Allan's special insight in this case was to see the important role a biochemist could play in his new biomaterials venture, a recognition that few had made till then. A key contribution to Horbett's career was also made when Allan Hoffman decided, partly at the suggestion of Bob Leininger, that Horbett should concentrate his postdoctoral research on protein adsorption to hydrogels. In Allan's active interest and support of the research, and his clear admiration, respect, and enthusiasm for a biochemist's special contributions to his lab, he gave the necessary encouragement to get Horbett going and, eventually, committed and established in a career in biomaterials.

Buddy Ratner was also recruited to Allan Hoffman's laboratory as a postdoctoral fellow, about six months after Horbett arrived. Hoffman was already an expert in radiation chemistry, but desired to further develop his lab's capabilities in hydrogels. Ratner had just completed a thesis on the structure of poly(HEMA) hydrogels. But, Allan Hoffman had ideas beyond having Ratner simply make and characterize more hydrogels; he was able to get him, very early on, engaged in evaluating the blood compatibility of hydrogels via collaboration with James Whiffen of the University of Wisconsin [63] and Lester Sauvage at Providence Hospital in Seattle. Thus, Ratner gained perspective on the proper evaluation of blood compatibility. Later, Steve Hanson and Laurie Harker, working with Hoffman and Ratner, identified the need to pay attention to non-adhesive, non-occluding aspects of blood-biomaterials interactions. In another instance, it was Hoffman's interest in the hydrophilicity of the radiation grafted hydrogels (did they rearrange in water?) that led Ratner to his early work on ESCA with frozen, hydrated gels [13]. Ratner's recognition of the power of the ESCA method led him to concentrate an appreciable portion of his career on characterization of biomaterial surface chemistry, a contribution stimulated by Allan Hoffman.

CONCLUDING REMARKS

Allan Hoffman shares a birthday with the French scientist, Pierre Eugène Berthelot (October 27, 1827). Berthelot is considered the first person to synthesize organic compounds that did not exist in nature (combining glycerol with fatty acids), and is therefore associated with the transition from chemistry to biochemistry. Perhaps we can say that what Berthelot did for molecules, Allan Hoffman did for materials, leading the transition from materials to biomaterials. Maybe this analogy between these two denizens of Paris (Allan has lived in Paris for extended periods), born 105 years apart, is a bit stretched? In any event, it is clear that we in the biomaterials field are enriched by the body of work and the ideas generated by Allan Hoffman. It has been truly special knowing him, and his friendship and professional relationship are deeply valued. Happy birthday, Allan. We hope this joyous event will some day be viewed as a small celebration *mid course* in a career that continues to stimulate and educate those around you.

REFERENCES

1. A. D. Rossin, C. J. Billerbeck, W. S. Delicate, A. W. Wendling, A. S. Hoffman and R. C. Reid, *Nucleonics* **13**, 10 (1955).
2. A. S. Hoffman, E. R. Gilliland, E. W. Merrill and W. H. Stockmayer, *J. Polym. Sci.* **34**, 461–480 (1959).
3. A. W. Meyers, C. E. Rogers, V. Stannet, M. Szwarc, G. S. Patterson Jr., A. S. Hoffman and E. W. Merrill, *J. Appl. Polym. Sci.* **4**, 159–165 (1960).
4. A. S. Hoffman and C. Harris, *ACS Polym. Prepr.* **13**, 740–746 (1972).
5. A. S. Hoffman, G. Schmer, C. Harris and W. G. Kraft, *Trans. Am. Soc. Artif. Int. Organs* **18**, 10–17 (1972).
6. A. S. Hoffman, G. Schmer, T. A. Horbett, B. D. Ratner, L. N. Teng, C. Harris, W. G. Kraft, B. N. L. Khaw, T. T. Ling and T. P. Mate, *ACS Org. Coat. Plast. Chem. Prepr.* **34**, 568–573 (1974).
7. B. D. Ratner and A. S. Hoffman, *J. Appl. Polym. Sci.* **18**, 3183–3204 (1974).
8. B. D. Ratner, T. A. Horbett, A. S. Hoffman and S. D. Hauschka, *J. Biomed. Mater. Res.* **9**, 407–422 (1975).
9. T. A. Horbett and A. S. Hoffman, in: *Applied Chemistry at Protein Interfaces, Advances in Chemistry Series*, Vol 145, pp. 230–254, R. E. Baier (Ed.). American Chemical Society, Washington, DC (1975).
10. B. D. Ratner, A. S. Hoffman and J. D. Whiffen, *Biomat., Med. Dev., Art. Org.* **3**, 115–120 (1975).
11. B. D. Ratner, A. S. Hoffman, S. R. Hanson, L. A. Harker and J. D. Whiffen, *J. Polym. Sci., Polym. Symp.* **66**, 363–375 (1979).
12. T. P. Mate, T. A. Horbett, A. S. Hoffman and B. D. Ratner, in: *Enzyme Engineering*, Vol 2, pp. 137–139, E. K. Pye and L. B. Wingard Jr. (Eds). Plenum Press, New York (1974).
13. B. D. Ratner, P. K. Weathersby, A. S. Hoffman, M. A. Kelly and L. H. Scharpen, *J. Appl. Polym. Sci.* **22**, 643–664 (1978).
14. Y. C. Ko, B. D. Ratner and A. S. Hoffman, *J. Coll. Interf. Sci.* **82**, 25–37 (1981).
15. Y. H. Sun, W. R. Gombotz and A. S. Hoffman, *J. Bioactive Compatible Polym.* **1**, 316–334 (1986).
16. A. S. Hoffman, M. Modell and P. Pan, *J. Appl. Polym. Sci.* **13**, 2223–2234 (1969).
17. A. S. Hoffman, M. Modell and P. Pan, *J. Appl. Polym. Sci.* **14**, 285–301 (1970).
18. T. A. Jadwin, A. S. Hoffman and W. R. Vieth, *J. Appl. Polym. Sci.* **14**, 1339–1359 (1970).
19. D. P. Mukherjee and A. S. Hoffman, in: *Physical Properties of Skin*, Part 1, p. 219, H. R. Elden (Ed.). John Wiley & Sons, New York (1971).
20. D. P. Mukherjee, A. S. Hoffman and C. Franzblau, *Biopolymers* **13**, 2447 (1974).
21. A. S. Hoffman, J. B. Park and J. Abrahamson, *Biomat., Med. Dev., Art. Org.* **1**, 453 (1973).
22. J. B. Park, C. H. Daly and A. S. Hoffman, *Front. Matrix Biol.* **3**, 218 (1976).
23. A. S. Hoffman and J. B. Park, *Biomat., Med. Dev., Art. Org.* **5**, 121–145 (1977).
24. J. B. Park and A. S. Hoffman, *Ann. Biomed. Eng.* **6**, 167 (1978).
25. A. S. Hoffman, *Biorheology* **17**, 45 (1980).
26. O. Wichterle and D. Lim, *Nature* **185**, 117–118 (1960).
27. T. Sasaki, B. D. Ratner and A. S. Hoffman, *ACS Symp. Ser.* **31**, 283–294 (1976).
28. D. Cohn, A. S. Hoffman and B. D. Ratner, *J. Appl. Polym. Sci.* **29**, 2645–2663 (1984).
29. S. R. Hanson, L. A. Harker, B. D. Ratner and A. S. Hoffman, *J. Lab. Clin. Med.* **95**, 289–304 (1980).
30. A. S. Hoffman, T. A. Horbett, B. D. Ratner, S. R. Hanson, L. A. Harker and L. O. Reynolds, *ACS Adv. Chem. Ser.* **199**, 59–80 (1982).
31. S. Venkataraman, T. A. Horbett and A. S. Hoffman, *J. Biomed. Mater. Res.* **11**, 111–123 (1977).
32. L. C. Dong and A. S. Hoffman, *Radiat. Phys. Chem.* **28**, 177–182 (1986).

33. L. C. Dong and A. S. Hoffman, in: *Reversible Polymeric Gels and Related Systems*, Vol 350, pp. 236-244, P. S. Russo (Ed.). American Chemical Society, Washington, DC (1987).
34. L. C. Dong, A. S. Hoffman and P. Sadurni, *Proceed. Intern. Symp. Control. Rel. Bioact. Mater.* **16**, 95-96 (1989).
35. T. A. Horbett, J. J. Waldburger, B. D. Ratner and A. S. Hoffman, *J. Biomed. Mater. Res.* **22**, 383-404 (1988).
36. D. K. Pettit, T. A. Horbett, A. S. Hoffman and K. Y. Chan, *Invest. Ophthalmol. Vis. Sci.* **31**, 2269-2277 (1990).
37. A. S. Hoffman, *Pure Appl. Chem.* **56**, 1329-1334 (1984).
38. J. P. Chen and A. S. Hoffman, *Biomaterials* **11**, 631-634 (1990).
39. H. J. Yang and A. S. Hoffman, in: *Preprints of the IUPAC International Symposium on Molecular Design of Functional Polymers*, IUPAC. The Polymer Society of Korea, Seoul (1989).
40. W. R. Gombotz, W. Guanghui, T. A. Horbett and A. S. Hoffman, *J. Biomed. Mater. Res.* **25**, 1547-1562 (1991).
41. K. Bergstrom, K. Holmberg, A. Safranj, A. S. Hoffman, M. J. Edgell, A. Kozlowski, B. A. Hovanes and J. M. Harris, *J. Biomed. Mater. Res.* **26**, 779-790 (1992).
42. A. S. Hoffman, *Artif. Organs* **1**, 498-499 (1988).
43. A. S. Hoffman, A. Afrassiabi and L. C. Dong, *J. Controlled Release* **4**, 213-222 (1986).
44. A. S. Hoffman, in: *Polymer Gels*, pp. 289-297, D. DeRossi, K. Kajiwara, Y. Osada and A. Yamauchi (Eds). Plenum Press, New York (1991).
45. A. S. Hoffman, in: *Polymers in Medicine and Surgery*, pp. 33-44, R. L. Kronenthal, Z. Oser and E. Martin (Eds). Plenum Press, New York (1975).
46. B. D. Ratner and A. S. Hoffman, in: *Hydrogels for Medical and Related Applications, ACS Symposium Series*, Vol 31, pp. 1-36, J. D. Andrade (Ed.). American Chemical Society, Washington, DC (1976).
47. A. S. Hoffman and G. Schmer, in: *Coagulation: Current Research and Clinical Applications*, pp. 201-226. Academic Press, New York (1973).
48. A. S. Hoffman, G. Schmer, T. A. Horbett, B. D. Ratner, L. N. Teng and C. Harris, in: *Permeability of Plastic Films and Coatings*, pp. 441-451, H. B. Hopfenberg (Ed.). Plenum Press, New York (1975).
49. S. Venkataraman, T. A. Horbett and A. S. Hoffman, *J. Biomed. Mater. Res. Symp.* **8**, 111-123 (1977).
50. S. V. Raman, T. A. Horbett and A. S. Hoffman, *J. Mol. Catal.* **2**, 275-291 (1977).
51. W. R. Gombotz, A. S. Hoffman, G. Schmer and S. Uenoyama, *Radiat. Phys. Chem.* **25**, 549-556 (1985).
52. W. R. Gombotz, A. S. Hoffman, G. Schmer and S. Uenoyama, *J. Controlled Release* **2**, 375-383 (1985).
53. T. G. Park and A. S. Hoffman, *Appl. Biochem. Biotech.* **19**, 1-9 (1988).
54. T. G. Park and A. S. Hoffman, *J. Biomed. Mater. Res.* **24**, 21-38 (1990).
55. T. G. Park and A. S. Hoffman, *Biotech. Lett.* **11**, 17-22 (1989).
56. J. L. Bohnert, B. C. Fowler, T. A. Horbett and A. S. Hoffman, *J. Biomat. Sci. Polym. Edn.* **1**, 279-297 (1990).
57. W. R. Gombotz and A. S. Hoffman, *ACS Polym. Mater. Sci. Eng.* **56**, 720-724 (1987).
58. A. S. Hoffman, *Appl. Polym. Sci. Appl. Polym. Symp.* **42**, 251-267 (1988).
59. W. R. Gombotz, W. Guanghul and A. S. Hoffman, *J. Appl. Polym. Sci.* **37**, 91-107 (1989).
60. M.-S. Sheu, A. S. Hoffman, J. G. A. Terlingen and J. Feijen, *J. Clin. Mat.* (submitted).
61. A. S. Hoffman, *Ann. N. Y. Acad. Sci.* **516**, 96-101 (1987).
62. B. D. Ratner and A. S. Hoffman, *ACS Org. Coat. Plast. Chem. Prepr.* **33**, 386-392 (1973).
63. B. D. Ratner, A. S. Hoffman and J. D. Whiffen, *J. Bioeng.* **2**, 313-323 (1978).
64. L. A. Harker, S. R. Hanson and A. S. Hoffman, *Ann. N. Y. Acad. Sci.* **283**, 317-329 (1977).
65. S. R. Hanson, L. A. Harker, B. D. Ratner and A. S. Hoffman, *Ann. Biomed. Eng.* **7**, 357-367 (1979).
66. L. O. Reynolds, T. A. Horbett, B. D. Ratner and A. S. Hoffman, *Trans. Soc. Biomat.* **5**, 37 (1982).
67. A. M. Garfinkle, A. S. Hoffman, B. D. Ratner, L. O. Reynolds and S. R. Hanson, *Trans. Am. Soc. Artif. Int. Organs* **30**, 432-439 (1984).
68. D. Kiaei, A. S. Hoffman, B. D. Ratner and T. A. Horbett, *J. Appl. Polym. Sci. Appl. Polym. Symp.* **42**, 269-283 (1988).

69. W. R. Gombotz, W. Guanghui and A. S. Hoffman, *J. Appl. Polym. Sci.* **35**, 1-17 (1988).
70. A. S. Hoffman, in: *Polymers in Medicine III*, pp. 161-167, C. Migliaresi (Ed.). Elsevier, Amsterdam (1991).
71. K. P. Antonsen, J. L. Bohnert, Y. Nabeshima, M.-S. Sheu, X. S. Wu and A. S. Hoffman, *Biomat., Art. Cells Immob. Biotech.* (in press).
72. W. Kuhn, B. Hargitay, A. Katchalsky and H. Eisenberg, *Nature* **165**, 514-516 (1950).
73. C. A. Cole, S. M. Schreiner, J. H. Priest, N. Monji and A. S. Hoffman, in: *Reversible Polymeric Gels and Related Systems*, Vol 350, pp. 245-254, P. S. Russo (Ed.). American Chemical Society, Washington, DC (1987).
74. A. S. Hoffman, *J. Controlled Release* **6**, 297-305 (1987).
75. A. S. Hoffman, *Chemtech* **16**, 426-432 (1986).
76. K. Auditore-Hargreaves, R. L. Houghton, N. Monji, J. H. Priest, A. S. Hoffman and R. C. Nowinski, *Clin. Chem.* **33**, 1509-1516 (1987).
77. N. Monji and A. S. Hoffman, *Appl. Biochem. Biotech.* **14**, 107 (1987).
78. H. J. Yang, C. A. Cole, N. Monji and A. S. Hoffman, *J. Polym. Sci. A* **28**, 219-220 (1990).

Part I.A.

Surface Modification, Characterization, and Properties

RF Plasma Gas Discharge

Molecular surface tailoring of biomaterials via pulsed RF plasma discharges

V. PANCHALINGAM[1], BRYAN POON[1], HSIAO-HWEI HUO[2], CHARLES R. SAVAGE[1], RICHARD B. TIMMONS[1]* and ROBERT C. EBERHART[2]

[1] *Department of Chemistry, University of Texas at Arlington, TX 76019, USA*
[2] *Biomedical Engineering Program, University of Texas at Arlington and University of Texas Southwestern Medical Center at Dallas, Dallas, TX 75235-9031, USA*

Received 10 July, 1992; accepted 15 December, 1992

Abstract—A pulsed RF plasma glow discharge is employed to demonstrate molecular level controllability of surface film deposits. Molecular composition of plasma deposited films is shown to vary in a significant manner with the RF duty cycle. Three fluorocarbon monomers are used to illustrate the process. All three exhibit a trend towards increased surface CF_2 content with decreasing pulsed RF duty cycle, including exclusion of oxygen. Significant variations in carbon–fluorine surface functionalities are obtained over a controllable range of film thickness. Film growth rate measurements reveal the occurrence of surface reactions during significant portions of the *off* portion of the duty cycle. Albumin adsorption on fluorocarbon-treated PET films is unchanged from PET controls for a 100-fold range of bulk concentrations and 60-fold range of adsorption times. However, increased retention of albumin is observed following incubation with protein-denaturing sodium dodecyl sulfate solution, the retention decreasing with increasing bulk concentration of albumin. The increased retention of albumin suggests the treated surfaces may have promise as biocompatible materials.

Keywords: Pulsed plasma; surface modifications; fluorocarbons; albumin binding.

INTRODUCTION

The radiofrequency glow discharge (RFGD) technique employs a flowing gas system in which an ionized gas plasma is generated to provide thin film coatings on targeted substrates via polymerization of an appropriate monomer. The method has attractive features for surface modification which have prompted several applications for medical polymers. Three general categories of surface treatment have emerged: (1) removal of unsatisfied surface binding capacity by exposure to an RFGD-excited gas; (2) deposition of an ultrathin inert polymer film from an RFGD-created excited monomer population; and (3) deposition of a functionalized intermediate which permits immobilization of biological molecules. In each case the surface properties can be favorably altered without changing bulk properties, thus providing physical durability and chemical stability. Excellent general reviews of the use of plasma gas discharges in biomaterials applications are available [1–4].

There are several advantages associated with the use of gas discharges for surface modification. A simple, one-step process is involved which provides coatings which are both conformal and pinhole-free in nature. In some cases these coatings consist of unique molecular structures, resulting from the fact that under plasma conditions

* To whom correspondence should be addressed.

extensive, indiscriminate bond-breaking and bond-forming reactions occur. It is important to note that the plasma energetic conditions provide relatively little control of surface modification obtained with a given monomer. To a certain extent, the modulation of film compositions obtained from a given monomer can be effected by such tactics as substrate repositioning and control of substrate temperatures [5, 6].

This study was undertaken to develop an RFGD approach with better control of the molecular structure of the deposited film, with the possibility of 'surface tailoring' in mind. Our approach is based on the use of *pulsed* RF plasma discharges in lieu of the continuous (CW) process. Specifically, we are exploring the relationship between the molecular structure of the deposited film and the RF duty cycle, all other RFGD parameters being held constant [7]. Our rationale is as follows. Solids immersed in ionized gas plasmas acquire a high negative potential [8]; thus we reason that the surface reactions are biased by the selective attraction of positive ions. Since the negative potential disappears upon quenching the plasma, the competitive balance of surface reactions involving the residual free radicals consumed from the gas phase will be altered. Changes in film composition with the RF duty cycle are observed, reflecting changes in the competitive film-forming reactions and documenting that marked changes in film structure do indeed occur with RF duty cycle variations. Significant amounts of material are deposited during the 'off' portion of the duty cycle, confirming the occurrence of surface chemistry during this phase. In addition to variations in substrate bias voltages, a number of other factors must be considered in comparing the chemistry during plasma on and off periods, as noted in the discussion section.

We focus on three fluorocarbon monomers in this report, demonstrating similar film structural changes with duty cycle variations, thus suggesting the generality of the approach. Preliminary albumin binding studies are included to indicate modifications in protein behavior associated with the fluorocarbon plasma deposition process.

MATERIALS AND METHODS

Plasma film deposition and characterization

A 10-cm diameter, 30.5-cm long Pyrex glass cylindrical reactor was employed. RF power was provided by external concentric metal rings located at both ends of the reactor vessel. The system with associated electronics is shown in Fig. 1. A function generator (Wavetech model 166); pulse generator (Tetronik model 2101); RF amplifier (ENI model A300); frequency counter (HP model 5381A); wattmeter (Bird) and capacitor/inductor matching network were used. An oscilloscope calibrated against the wattmeter under CW conditions was used to tune the matching circuit to minimize reflected power during RF pulsed operation. All runs were carried out at 13.56 MHz. A butterfly valve (MKS Baratron model 252A), connected in a feedback loop with a pressure transducer, was used to monitor and control pressure in the reactor. Substrates were either polished Si or KCl disks or PET coverslips (Nunc, Urbana, IL). The PET coverslips were coated on both sides. Samples were placed on top of an inverted 20 ml beaker located at the center of the reaction chamber. The system was evacuated to a pressure of approximately 5 mTorr, following which a CW Argon plasma discharge was employed to clean the substrate. After 5 min the discharge was quenched, the Ar flow stopped and the chamber re-evacuated to background pressure.

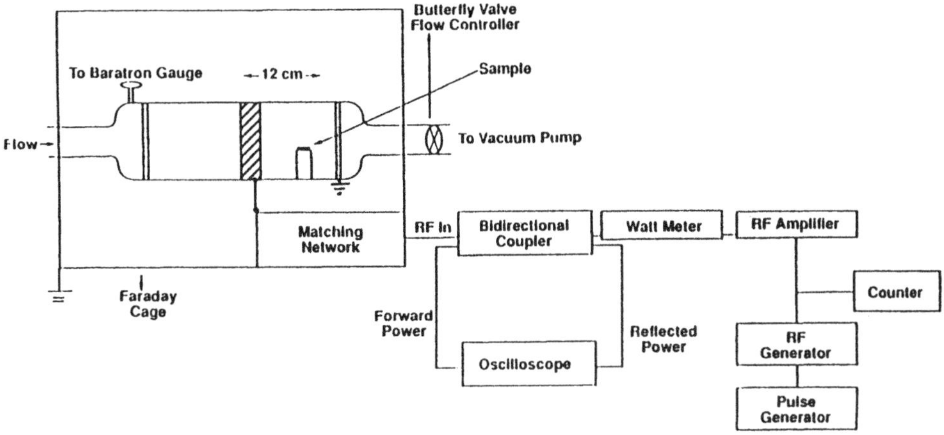

Butterfly Valve
Flow Controller
To Baratron Gauge
12 cm
Sample
Flow
To Vacuum Pump
Matching
Network
RF In
Bidirectional
Coupler
Watt Meter
RF Amplifier
Faraday
Cage
Counter
Forward
Power
Reflected
Power
RF
Generator
Oscilloscope
Pulse
Generator

Figure 1. Schematic of the pulsed RF plasma apparatus.

The fluorocarbon monomer was then introduced and the reactions commenced. We studied three fluorocarbon monomers: perfluoropropylene (C_3F_6); perfluoro-2-butyltetrahydrofuran ($C_8F_{16}O$); and hexafluoropropylene oxide (C_3F_6O), all obtained from PCR, Inc. (Gainesville, FL). They were outgassed before use, but subjected to no additional purification steps. Following each run the samples were removed, the chamber resealed, evacuated and subjected to an intense O_2 plasma to remove polymer deposits from chamber walls. A standard monomer pressure (0.43 Torr) and gas flow rate (9.6 cm^3 (STP)/min) were used for all film depositions. The applied RF power was either 200 W (C_3F_6) or 300 W (C_3F_6O and $C_8F_{18}O$). This RF power, relatively high compared with that employed in typical CW plasma polymerizations, was used in order to produce higher concentrations of reactive intermediates during the plasma on periods and thus to promote reactions during the plasma off periods. Other experiments have revealed relatively slow film growth rates for lower applied RF power (e.g., 50 or 100 W) under pulsing conditions. While RF pulse on- and off-times could be varied over a wide range, we chose on/off times in the intervals from 10 to 1000 ms. Preliminary studies yielded maximal changes in resulting films for this range of intervals for the selected fluorocarbon monomers.

Pulsed plasma-deposited films were characterized by ESCA and FTIR spectroscopies. ESCA spectra were obtained with a Perkin–Elmer PHI-5000 unit equipped with X-ray source and monochromator. Charge compensation was employed in the recording of all ESCA spectra. Binding energies were assigned by centering the F1*s* peak at 689 eV. Binding energy assignments for C1*s* high resolution spectra were based particularly on the work of Clark and Shuttleworth [9]: CF_3 = 294; CF_2 = 292; $\underline{C}F{-}CF_n$ = 290; CF = 289; $\underline{C}{-}CF_n$ = 287.5 eV. Although five distinct binding energies are noted above, plasma polymerized fluorocarbon films reveal only four distinct ESCA peaks in that CF-substituted (290 eV) and non-fluorine substituted (289 eV) peaks are not completely resolved. No peak was observed at a binding energy of 285 eV, indicating the absence of any CH_x or graphitic carbons. Since we are interested in trends in group functionality concentrations with varying RF duty cycle, we will consider the spectra to be simply combinations of CF_3, CF_2, CF and $\underline{C}$, with the latter referring to C bound to CF_3 and CF_2. Fourier Transform Infrared (FTIR) spectra were obtained with a Bio-Rad instrument with 2 cm^{-1} resolution. Transmission mode FTIR spectra were obtain for the Si and KBr substrates. Film thickness was obtained with a Tencor AlphaStep 200 profilometer. The ESCA and FTIR analyses confirm that the plasma-deposited films are independent of substrate at a set duty cycle. Furthermore, these spectral analyses reveal that the film compositions are independent of film thickness in the range 0.1–3.0 μm.

The samples prepared for the protein adsorption studies were fluorinated films deposited on PET with the C_3F_6O starting material. An RF duty cycle of 10 ms on and 200 ms off was employed to create these films. Advancing water contact angle measurements were made with a Ramé–Hart goniometer, using distilled deionized water at ambient temperature and humidity.

Albumin adsorption studies

Bovine serum albumin (fatty acid free, fraction V, Sigma, St. Louis, MO) was labeled with ^{125}I by a standard procedure [10], employing Iodogen (Pierce). The radiolabeled protein yield was typically greater than 95%. The radiolabeled albumin solution was diluted with unlabeled BSA (0.15, 1.50 and 15.00 mg/ml) to give a final radioactivity of 5 μCi ^{125}I/ml solution for all bulk concentrations. The polymer sample was secured in a frame, placed in a 0.6 × 2.5 × 20.3 cm flow chamber and equilibrated for 3 min with degassed PBS. The ^{125}I-BSA albumin solution was transferred at room temperature by slow volume replacement (10×) followed by static incubation for 1, 10, or 60 min. Samples were then washed with degassed PBS at a wall shear rate of 800 s^{-1} for 30 s. Care was taken in these steps to avoid the introduction of bubbles. Samples were then removed, placed in vials and counted in a well-type gamma counter (Abbott Autologic 100). A 0.3% sodium dodecyl sulfate (SDS, Aldrich) solution was infused into the vials: samples were incubed for 3 min, then transferred to fresh vials and recounted.

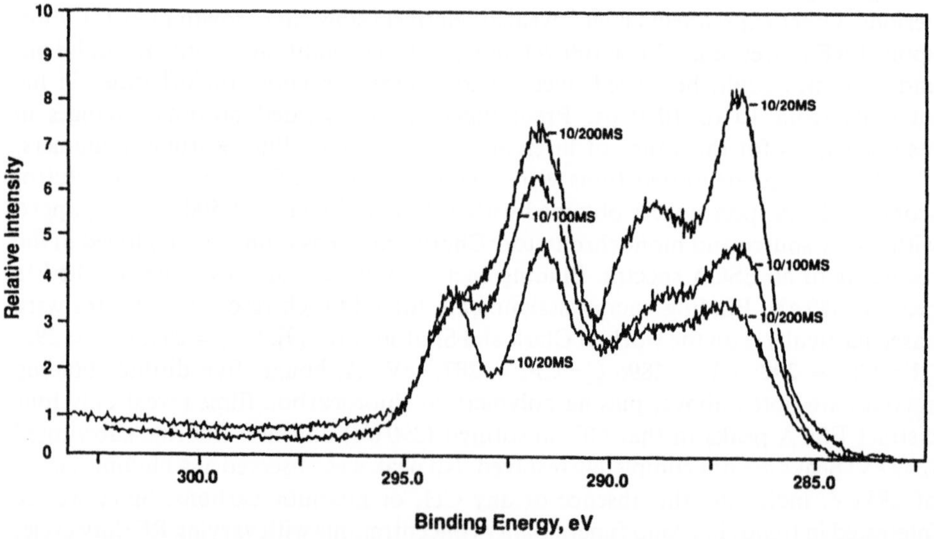
Relative Intensity
Binding Energy, eV
10
9
8
7
6
5
4
3
2
1
0
300.0
295.0
290.0
285.0
10/20MS
10/200MS
10/100MS
10/100MS
10/200MS
10/20MS

Figure 2. The variation in C_{1s} ESCA spectra obtained from C_3F_6 plasma deposited films for pulsed RF duty cycles of 10/20, 10/100, 10/200 ms on/off as shown.

RESULTS

C_3F_6

High resolution C1*s* ESCA spectra for the C_3F_6 starting materials indicate progressive changes were obtained in molecular composition with the extent of plasma discharge. C1*s* ESCA spectra for a constant *on* time (10 ms) and increasing *off* times (decreasing duty cycle, Fig. 2) indicate a shift in the relative concentrations of carbon–fluorine group functionalies favoring highly fluorine-substituted groups, i.e. CF_2 and CF_3 at 292 and 294 eV, respectively, at the expense of less highly substituted groups. The same phenomenon is observed in Fig. 3, in which *off* times is

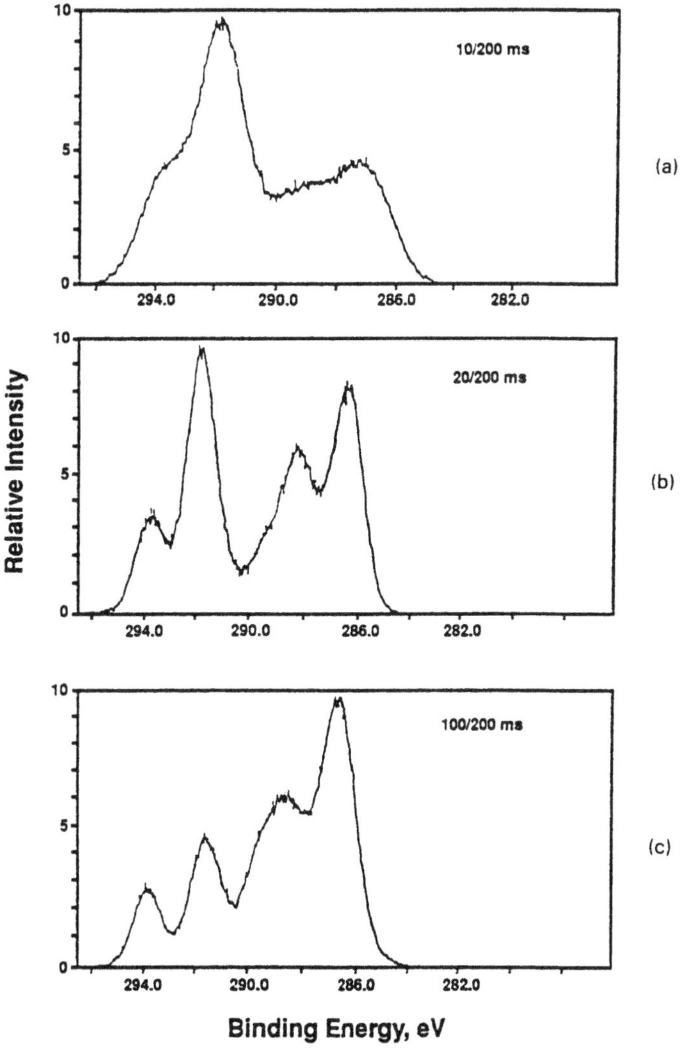
10/200 ms
(a)
20/200 ms
(b)
100/200 ms
(c)
Relative Intensity
Binding Energy, eV
0
5
10
294.0
290.0
286.0
282.0

Figure 3. Variation in fluorocarbon film composition obtained from pulsed plasma deposition of C_3F_6 at a constant plasma off time of 200 ms and variable plasma on time: (a) 10 ms; (b) 20 ms; (c) 100 ms.

Table 1.
RF Duty cycle, surface group and atom concentrations and F : C ratio for various starting monomers

RF duty cycle on/off (ms)	Relative functional group concentrations (%)				Surface atom concentrations (%)			
	CF_3	CF_2	CF	C	F	C	O	F : C ratio
$\underline{C_3F_6}$								
10/20	11	17	36	36	51	49	<0.3	1.04
10/100	14	34	31	21	59	41	0	1.44
10/200	18	42	24	16	62	38	0	1.63
20/200	11	34	27	28	56	44	0	1.27
100/200	9	17	34	40	49	51	<0.3	0.96
$\underline{C_8F_{18}O}$								
10/50	15	21	32	32	54	45	~0.6	1.19
10/100	14	34	27	25	58	42	~0.5	1.37
10/200	15	45	22	18	61	39	~0.4	1.57
10/400	17	61	13	9	65	35	~0.4	1.85
$\underline{C_3F_6O}$								
10/200	9	75	12	4	65	35	<0.3	1.86
20/200	12	55	20	13	62	38	0	1.63
30/200	16	40	29	15	61	39	0	1.56
40/200	19	40	24	17	62	38	0	1.63
70/200	16	27	23	34	56	44	<0.2	1.27

held constant (200 ms) while *on* time is decreased from 100 to 20 to 10 ms. Both figures demonstrate that a progressive increase in the surface atom ratio of fluorine to carbon is obtained as the RF duty cycle is decreased. Table 1 provides quantitative data on the concentrations of each of the four carbon atom functionalities (CF_3, CF_2, CF, $\underline{C}$) in each C1*s* core level spectral envelope as well as surface atom concentrations obtained from the ESCA scans. These data reveal a significant increase in the F : C ratio with decreasing RF duty cycle employed during the deposition. This increase in F : C ratio arises mainly from the increase in the concentration of CF_2 groups relative to the more highly substituted carbon atoms.

$C_8F_{18}O$

The C1*s* ESCA spectra for films generated from $C_8F_{18}O$ also indicate systematic changes in composition with RF duty cycle. The trend in the fluorine to carbon atom ratio (Fig. 4, Table 1) is similar to that observed with the C_3F_6 monomer, in that the relative importance of the more heavily fluorine-substituted groups increases with decreasing duty cycle. It is interesting to note that CW film deposition processes yield little, if any variation in film composition [11], in contrast with the present results for low duty cycle pulse discharge processes. ESCA surface analysis in reference 11 revealed little oxygen in the RFGD film (less than 1% oxygen atom concentration), despite the presence of oxygen in the starting material. As shown in Table 1, relatively little oxygen atom incorporation is observed under the pulsed RF plasma conditions employed in the present study. The low oxygen atom incorporation obtained under pulsing conditions (0.4–0.6%) is in contrast with the 3.7% concentration in the starting monomer.

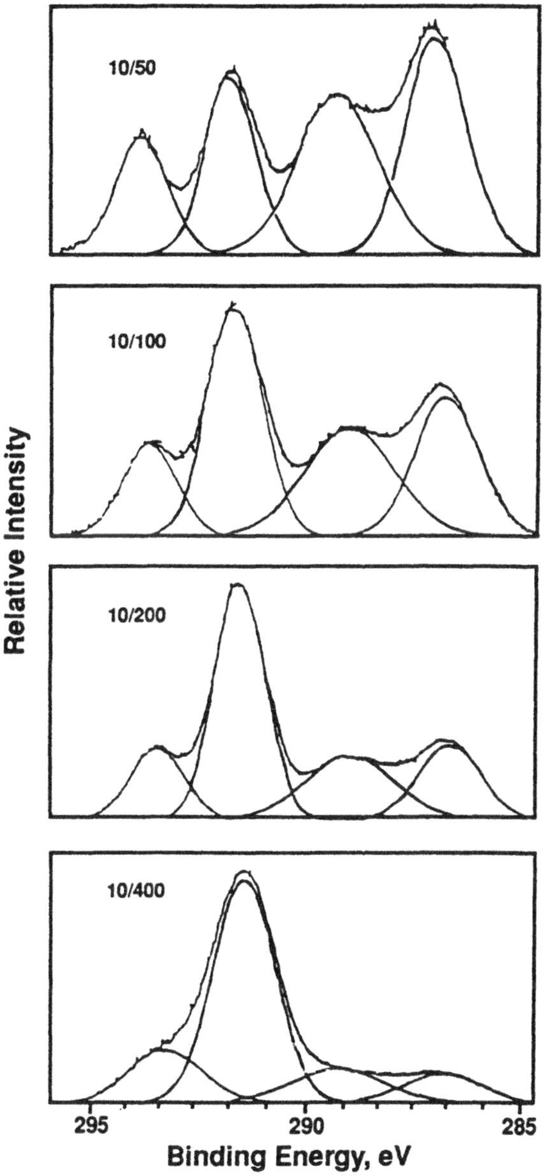
10/50
10/100
10/200
10/400
Relative Intensity
295
290
285
Binding Energy, eV

Figure 4. C1*s* ESCA spectra obtained from pulsed plasma deposition of $C_8F_{18}O$. RF duty cycles in ms on/ms off were 10/50, 10/100, 10/200, 10/400, reading from top to bottom.

FTIR analysis confirms the progressive change in film chemistry with the RF duty cycle. Results for films based on the $C_8F_{16}O$ starting monomer are shown in Fig. 5, stacked in order of increasing duty cycle for samples deposited on KBr disks. Included for comparison is the spectrum for a film generated with a 100 W CW discharge. These results are representative of those obtained with the other two monomers. The major component, a band in the 1200 cm^{-1} region, is characteristic of C—F stretching bonds [12]. A number of C—F stretching frequencies are

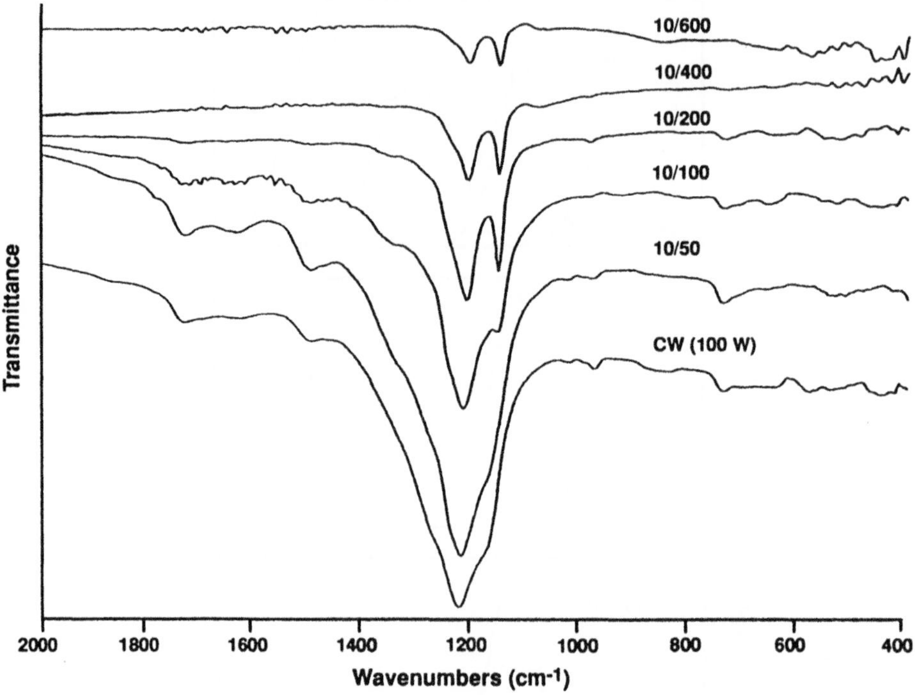
10/600
10/400
10/200
10/100
10/50
CW (100 W)
Transmittance
2000
1800
1600
1400
1200
1000
800
600
400
Wavenumbers (cm-1)

Figure 5. FTIR spectra of films obtained in the pulsed plasma polymerization of $C_8F_{18}O$ monomer. Spectra are stacked in order of increasing RF duty cycle, reading from top to bottom with on/off times (in ms) as shown. The bottom spectrum is from a film obtained under CW conditions using 100 W RF power.

apparent in the spectra for the CW and highest duty cycles (10 ms-on/50 ms-off); these are similar to results obtained in other plasma polymerizations of fluorocarbons [13]. However, as the duty cycle is shortened one can see, with increasing resolution, a doublet which is attributed to the dominance of CF_2 groups. The doublet is similar to that observed for polytetrafluoroethylene (Teflon) [14]. Thus the effect of decreasing the RF duty cycle is to convert a heterogeneous, low order, highly crosslinked fluorocarbon film to a more homogeneous ordered structure. Our experience with plasma-induced processes suggests that the appearance of the Teflon-like spectrum at low RF duty cycle is remarkable.

C_3F_6O

The ESCA spectra for the films originating with C_3F_6O monomer parallel those obtained with the other starting monomers (Fig. 6). The influence of a progressively decreasing *on* time (70–10 ms) set against a constant *off* time (200 ms) is to enhance the appearance of CF_2 and CF_3 groups at the expense of less highly fluorinated groups. This increase is shown quantitatively in Table 1. The controllability of the pulsed RF deposition process is suggested by the regular progression of the spectra. As shown in Table 1, virtually no oxygen is incorporated in these films despite the 10% oxygen atom concentration present in the starting monomer.

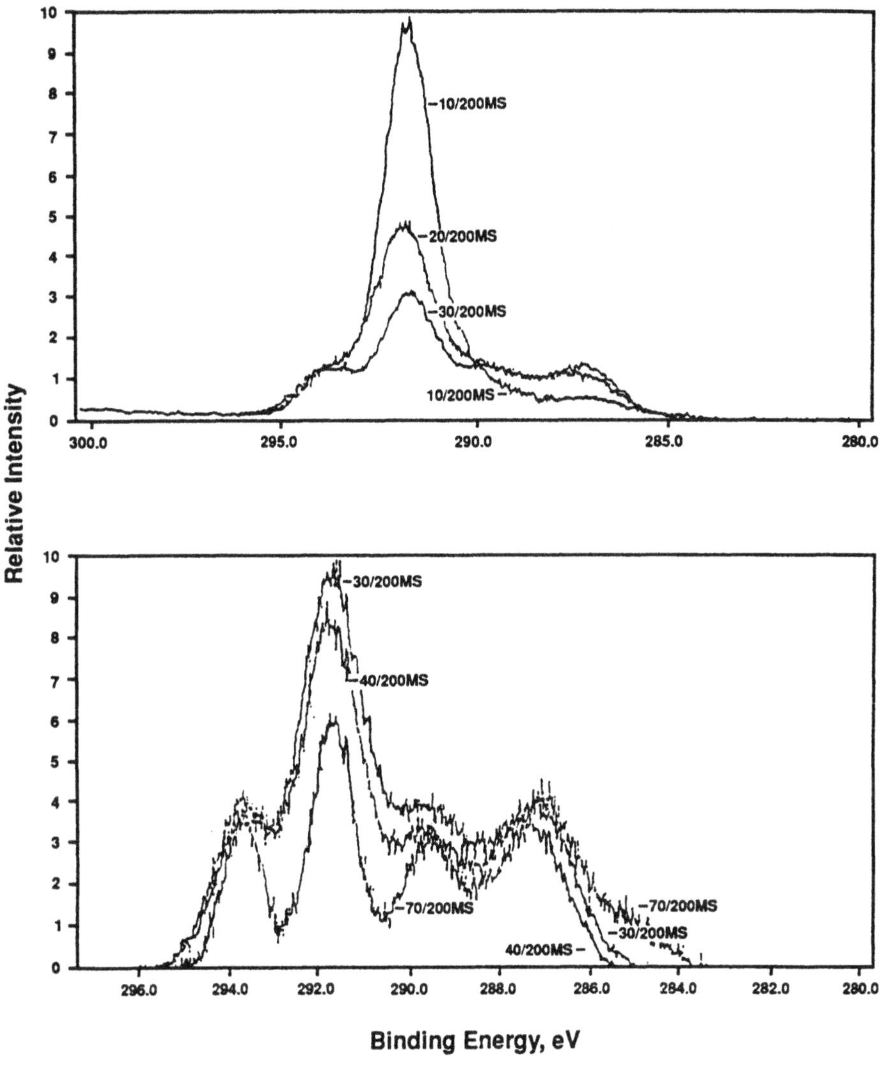
Relative Intensity
10
9
8
7
6
5
4
3
2
1
0
300.0
295.0
290.0
285.0
280.0
–10/200MS
–20/200MS
–30/200MS
10/200MS –
–30/200MS
–40/200MS
–70/200MS
–70/200MS
–30/200MS
40/200MS –
296.0
294.0
292.0
290.0
288.0
286.0
284.0
282.0
280.0
Binding Energy, eV

Figure 6. C1*s* ESCA spectra obtained from pulsed plasma deposition of C_3F_6O, illustrating the progressive uniform change in film composition for surface modification at a constant plasma off time of 200 ms and variable on times ranging from 10 to 70 ms (as shown).

Surface profilometry for the RFGD treatment of samples with hexafluoropropylene oxide (C_3F_6O) on silicon (Fig. 7) demonstrates that film deposition rate per unit of absorbed RF energy increases with plasma off time. In other words, the film thickness is inversely related to the duty cycle. Thus significant chemistry and film growth occurs during the off period. For the conditions listed in Fig. 7, with a 10 ms plasma discharge time, film growth occurs for as long as 600 ms following switchoff. Plasma chemistries following plasma quenching are expected to differ significantly from those obtained under plasma on conditions; these conditions must strongly favor film growth in the 1–5% duty cycle range, all other conditions held constant.

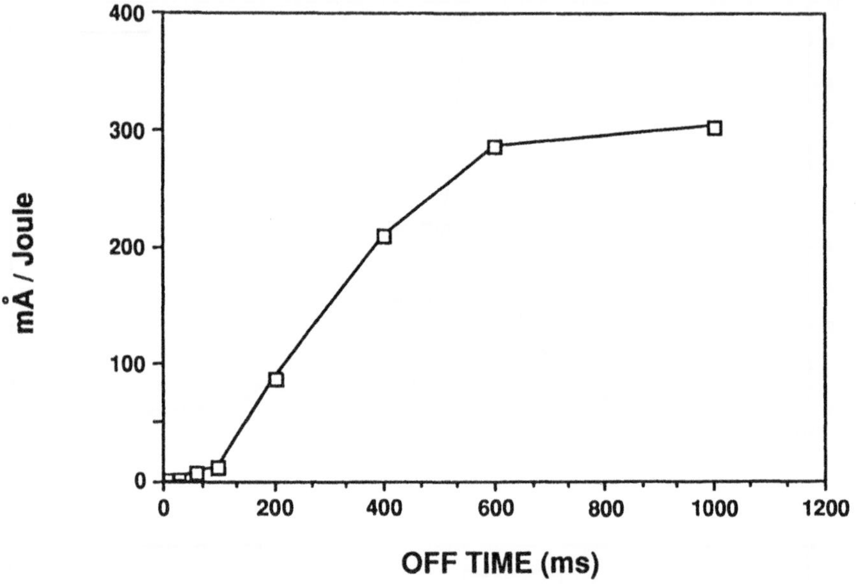

400
300
200
100
0
mÅ / Joule
0
200
400
600
800
1000
1200
OFF TIME (ms)

Figure 7. Film growth rate obtained during the pulsed RF plasma deposition of C_3F_6O as a function of the plasma off time and for a constant plasma on time of 10 ms. The film deposition rate is expressed in terms of milliAngstroms per Joule absorbed RF energy.

ESCA analysis of films slated for albumin adosorption experiments indicated that the film consists predominantly of CF_2 (Fig. 8). These fluorocarbon-coated PET surfaces are extremely hydrophobic as shown by advancing water contact angles of 125–130 deg, compared with measured values of 75–80 deg for uncoated PET substrates.

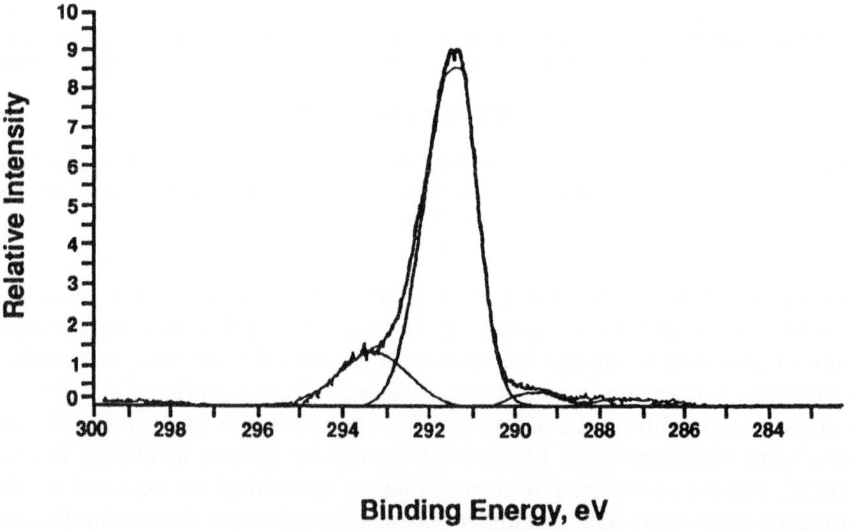

10
9
8
7
6
5
4
3
2
1
0
Relative Intensity
300
298
296
294
292
290
288
286
284
Binding Energy, eV

Figure 8. C1*s* ESCA spectrum of a typical fluorocarbon-coated PET sample used in the albumin binding studies.

Table 2.
Albumin adsorption and retention ($\mu g/cm^2$) on fluorinated and control PET surface

	Adsorption time					
	1 min		10 min		60 min	
Albumin solution	Adsorb	Retain	Adsorb	Retain	Adsorb	Retain
0.15 mg/ml						
Control	0.226	0.032	0.321	0.054	0.412	0.036
σ	±0.033	0.004	0.049	0.015	0.024	0.004
CF_2	0.175	0.085[a]	0.281	0.159[a]	0.426	0.274[a]
σ	±0.005	0.043	0.021	0.035	0.035	0.019
1.5 mg/ml						
Control	1.152	0.126	2.100	0.256	1.574	0.135
σ	±0.199	0.006	0.279	0.169	0.485	0.039
CF_2	1.116	0.319[a]	1.337	0.445[a]	2.874	0.421[a]
σ	±1.365	0.049	0.343	0.013	1.035	0.072
15 mg/ml						
Control	7.624	1.028	10.37	0.913	17.28	2.111
σ	±2.103	0.136	2.324	0.024	1.099	0.452
CF_2			10.34	2.597[a]	11.49	4.105[a]
σ			±4.820	0.165	4.856	0.410

$n = 3$ for all groups
[a] $p < 0.05$ wrt PET control

Albumin adsorption and retention results for the C_3F_6O RFGD-treated surfaces are given in Table 2. In general, albumin adsorption increased with solution concentration and with time. There were not significant differences in albumin adsorption for the fluorinated PET films versus PET controls. However, albumin retention following treatment in dilute SDS solution was significantly enhanced for the fluorinated PET films compared with controls. In other words, fluorination of the PET surface reduced the elution of albumin. The greatest improvements in retention were observed for the most dilute solutions (Fig. 9).

DISCUSSION

The focus of this study is to illustrate that pulsed RF plasmas can provide a substantial degree of controllability of film composition and thickness by regulation of the *off* portion of the duty cycle, more so than that achieved during the plasma process, the *on* portion of the cycle. There are a number of interesting applications for such technology. Of relevance to this forum is the ability to tailor the surface so that more precise knowledge of biological interactions with the polymer surface can be obtained. These interactions can be studied as a function of orderly progressive changes in surface functionality. Earlier studies involving pulsed RF plasmas failed to note the inherent film structural control available for pulsed RF plasma discharge film generation with a regulated duty cycle [15, 16]. While this study was restricted to fluorocarbon monomers, we have observed comparable levels of surface tailoring with a wide range of other monomers [17].

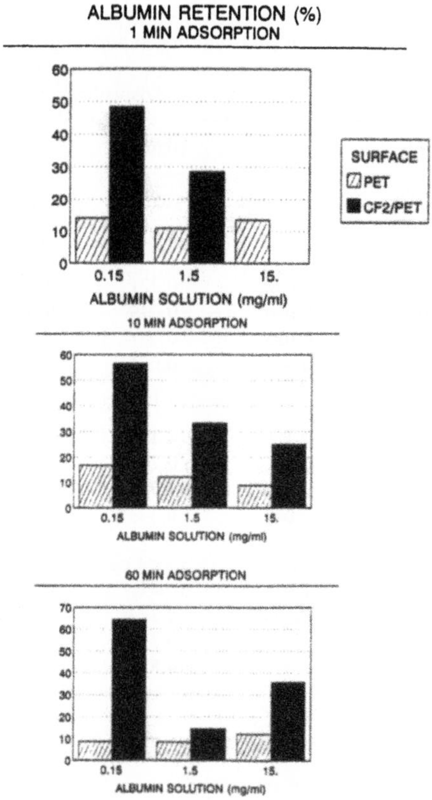
ALBUMIN RETENTION (%)
1 MIN ADSORPTION
60
50
40
30
20
10
0
0.15
1.5
15.
ALBUMIN SOLUTION (mg/ml)
SURFACE
PET
CF2/PET
10 MIN ADSORPTION
60
50
40
30
20
10
0
0.15
1.5
15.
ALBUMIN SOLUTION (mg/ml)
60 MIN ADSORPTION
70
60
50
40
30
20
10
0
0.15
1.5
15.
ALBUMIN SOLUTION (mg/ml)

Figure 9. Albumin retention, following short incubation in 0.3% sodium dodecyl sulfate solution. Results are shown for various initial albumin incubation times and solution concentrations.

We have not provided a kinetic scheme to rationalize these variations in film structure. This may be extremely difficult, given the variety and complexity of the reactive species generated in the plasma. However, the observation of film composition variations with prolonged *off* periods of the duty cycle must reflect differences in chemistry between the plasma *on* and *off* periods. The appearance and disappearance, respectively, of substrate bias during the *on/off* periods may heavily influence the film-forming conditions. However, other considerations come into play. For example, high concentrations of reactive species during discharge conditions will favor second order processes in the reactive intermediates. Thus the competitive balance between processes such as $R + R \rightarrow$ products and $R + M \rightarrow$ products (where R is a reactive intermediate and M is the monomer) will shift towards the monomer reactions during plasma quenching periods. Another factor which may exercise influence is surface temperature. Under set plasma conditions, as employed in these studies, substrate temperature is higher during processes at high RF duty cycle that at lower ones. Substrate temperatures were typically around 60°C for 10 ms on/20 ms off conditions, decreasing to about 35°C for 10 ms on/100 ms off values. Although temperature effects must be considered, significant changes in film composition were observed during low duty cycles, 10 on/100 off and less, when negligible changes in substrate temperature occurred.

Finally, plasma deposition processes involve a competition between ablation and film growth at the substrate surface. It is reasonable to anticipate that the relative importance of these competitive processes will shift with RF duty cycle changes. Any one, or combination thereof, of the factors noted above could be responsibe for the compositional changes with RF duty cycle observed in this work.

Our observation of increased albumin retention for hydrophobic (RFGD fluorocarbon treated) surfaces is consistent with others' findings. For example, Uyen *et al.* studied the influence of water wettability on albumin adsorption for polished PTFE, PMMA and glass in a parallel-plate flow cell, observing increasing albumin retention with decreasing wettability [18]. Bohnert *et al.* studied the elutability of albumin and fibrinogen which had been adsorbed from 37°C dilute baboon plasma onto RFGD fluorocarbon-treated PET and TFE substrates [19]. The SDS-elution of protein was significantly reduced for treated surfaces compared with untreated PET and TFE controls. Kiaie *et al.* obtained significantly increased albumin retention, independent of water wettability and ionic strength [20]. Recently Kiaie *et al.* have demonstrated a strong inverse correlation between albumin retention and surface free energy [21]. The surface with lowest free energy was obtained by RFGD fluorocarbon treatment (TFE/PET II in their terminology). This surface was rich in both CF_2 and CF_3 groups and was dominated by dispersion forces: it retained albumin perfectly in their dilute albumin solution incubation and SDS elution protocol. Our results were obtained for surfaces with much higher proportions of CF_2 than CF_3, with higher water contact angles than reported by Kiaie *et al.* for the TFE/PET II surface. Albumin retention in our study, while markedly increased over that for PET controls, was not perfect. Differences in albumin concentration, wash and SDS elution conditions between our study and that of Kiaie *et al.* [21] can explain the discrepancies.

Other hydrophobic surface functionalities, C_{16} and C_{18} saturated chains, substantially increase the early binding of albumin to treated surfaces *in vitro* [22–24] and modestly improved albumin retention *in vivo* [25]. This approach seeks to mimic circulating saturated fatty acids for which albumin has high affinity. Improved short-term biocompatibility has been oberved for such surfaces [10, 26]; longer term biocompatibility has been suggested but not proved. It is also worth noting that appropriate surface hydrophilic moieties also have resulted in improved short-term biocompatibility [27, 28]. However, no data yet exist to permit one to predict that an arbitrary hydrophobic or hydrophilic surface will improve long term biocompatibility of implanted blood-contacting devices.

Fluorocarbon treatments have also been developed for practical devices. Fluorocarbon coating was used to improve the patency of experimental Dacron small diameter grafts by Garfinkle *et al.* [29]. A noncrimped woven Dacron vascular graft prosthesis treated on the luminal surface by RFGD fluorocarbon was introduced into clinical practice but have since been discontinued. The luminal surface was shown to contain a nonuniform plasma coating and a high level of extractable contaminants [30], which might bear on the disappointing results. Other types of fluorocarbon treatments are under development. Schakenraad *et al.* have recently reported improved thromboresistance and short-term patency of a 'super-hydrophobic' e-PTFE small diameter vascular graft [31].

Our results may also be relevant to RFGD treatments for the surface immobilization of biological agents. Gombotz and Hoffman have used RFGD treatment with allyl alcohol to produce surface hydroxy groups with which to immobilize various biological agents [32]. Ertel *et al.* have extended the method, utilizing RF plasmas to introduce surface hydroxyl, carboxyl and carbonyl groups onto polystyrene to enhance the growth of endothelial cells [33]. Recently Marchant *et al.* have modified polyethylene surfaces by RFGD treatment with poly *N*-vinyl pyrrolidone and allyl alcohols, coupling to either amino-terminated polyethylene oxide or 3-aminopropyltriethoxysilane in order to bind heparin by reductive amination [34].

We have demonstrated the ability to more precisely define surface functionalities by film formation with pulsed RF plasmas of low duty cycle. The technique may permit improvement of film properties by a relatively simple means. It may also be relevant for RFGD treatments of functional intermediates for the immobilization of biological agents. Improved definition of surface functionality should also permit advances in understanding of the relationship between surface structure and adsorption/desorption phenomena. Finally, the use of pulse discharge technique, particularly for low duty cycles, may be useful for practical applications.

CONCLUSIONS

Pulsed RFGD plasmas provide substantial control of fluorocarbon film deposits on PET substrates. Significant film growth and surface chemistry occur during the plasma *off* period. Selective deposition of CF_2 groups from a variety of fluorocarbon monomers is observed for very low RF duty cycles. Adsorption of albumin from solution on CF_2 surfaces does not differ from PET controls, but albumin retention is significantly increased, especially for dilute solutions. Overall it is felt that the pulsed RFGD plasma approach may be useful for studying the interactions of biological molecules with polymers by offering means to more precisely control the molecular composition of the polymer surface.

Acknowledgements

Supported in part by Texas Advanced Research Program grant 003656-130 and USPHS grant HL 28690. The assistance of Dr. A. Constantinescu in protein labeling studies is gratefully acknowledged.

REFERENCES

1. H. K. Yasuda and M. Gazicki, *Biomaterials* **3**, 68 (1982).
2. W. R. Gombotz and A. S. Hoffman, In: *Critical Reviews of Biocompatibility*, Vol. 4, p. 1, D. F. Williams (Ed.). CRC Press, Boca Raton, FL (1987).
3. B. D. Ratner, A. Chilkoti and G. P. Lopez, In: *Plasma Deposition, Treatment and Etching of Polymers*, Plasma-Materials Interactions Series, p. 3, R. D'Agostino (Ed.). Academic Press, San Diego (1990).
4. R. Barbuci (Ed.), *J. Biomater. Sci., Polymer Edn* **4**, 1 (1992).
5. D. F. O'Kane and D. W. Rice, *J. Macromol. Sci. Chem.* **A10**, 567 (1976).
6. G. P. Lopez and B. D. Ratner, *J. Appl. Polym. Sci. Appl. Polym. Symp.* **46**, 493 (1990).
7. C. R. Savage, R. B. TImmons and J. W. Lin, *Chem. Mater.* **3**, 575 (1991).
8. M. Venugopalan and R. Avni, In: *Thin Films from Free Atoms and Particles*, ch. 3, p. 49, K. J. Klakunde (Ed.). Academic Press, Orlando FL (1985).
9. D. T. Clark and D. Shuttleworth, *J. Polym. Sci., Poly. Chem. Ed.* **18**, 27 (1980).
10. M. S. Munro, R. C. Eberhart, N. J. Maki, B. E. Brink and W. F. Fry, *J. ASAIO* **65**, 6 (1983).
11. D. T. Clark and M. Z. Abrahaman, *J. Polym. Sci., Polym. Chem. Edn* **20**, 291 (1982).
12. *Aldrich Library of Infrared Spectra*, 2nd Edn, Aldrich, Milwaukee, p.39 (1975).

13. H. Yasuda, *Plasma Polymerization*, p. 182. Academic Press, New York (1986).
14. D. O. Hamond, *Infrared Analysis of Polymers and Additives: An Atlas*, p. 154. Wiley Interscience, New York (1971).
15. H. Yasuda and T. Hsu, *J. Polym. Sci., Polym. Chem. Edn* **15**, 81 (1977).
16. K. Nakajima and A. K. Bell, *J. Appl. Polym. Sci.* **23**, 2627 (1979).
17. R. B. Timmons and C. R. Savage. Unpublished results.
18. H. M. W. Uyen, J. M. Schakenraad, J. Sjollema, J. Noordmans, W. L. Jongebloed, I. Stokroos and H. J. Busschert, *J. Biomed. Mater. Res.* **24**, 1599 (1990).
19. J. L. Bohnert, B. C. Fowler, T. A. Horbett and A. S. Hoffman, *J. Biomater. Sci., Polymer Edn* **1**, 279 (1990).
20. D. Kiaie, A. S. Hoffman and T. A. Horbett, *Trans. Biomater. Soc.* **12**, 110 (1989).
21. D. Kiaie, A. S. Hoffman and T. A. Horbett, *J. Biomater. Sci., Polymer Edn* **4**, 35 (1992).
22. N. A. Platé and M. N. Matrosovich, *Dokl. Akad. Nauk SSSR* **229**, 496 (1977).
23. M. S. Munro, A. J. Quattrone, S. R. Ellsworth, P. Kulkarni and R. C. Eberhart, *Trans. Am. Soc. Artif. Intern. Organs* **27**, 499 (1981).
24. W. G. Pitt and S. L. Cooper, *J. Biomed. Mater. Res.* **22**, 359 (1988).
25. R. C. Eberhart, M. S. Munro, G. B. Williams, W. A. Shannon, B. E. Brink and W. J. Fry, *Artif. Organs* **11**, 375 (1987).
26. T. G. Grasel, J. A. Pierce and S. L. Cooper, *J. Biomed. Mater. Res.* **21**, 815 (1987).
27. ?. Mori, S. Nagaoka, H. Takiuchi, T. KIkuchi, N. Noguichi, H. Tanzawa and Y. Noishiki, *Trans. Am. Soc. Artif. Intern. Organs* **28**, 459 (1982).
28. C. C. Tsai, M. L. Dollar, A. Constantinescu, P. Kulkarni and R. C. Eberhart, *Trans. Am. Soc. Artif. Intern. Organs* **37**, M192 (1991).
29. A. M. Garfinkle, A. S. Hoffman, B. D. Ratner, L. O. Reynolds and S. R. Hanson, *Trans. Am. Soc. Artif. Intern. Organs* **30**, 432 (1984).
30. R. Guidoin, M. King, M. Therrien, R. Paynter, S. SImoneau, E. Debille, L. Tremblay, D. Boyer and F. Gill. In: *High Performance Biomaterials, A Comprehensive Guide to Medical and Pharmaceutical Applications*, M. Szycher, Ed. Technomic Publishing Co., Lancaster, PA, 1991, pp. 381–399.
31. J. M. Schakenraad, I. Stokroos, H. Bartels and H. J. Busscher. *Letter and Materials*, in press (1993)
32. W. R. Gombotz and A. S. Hoffman, *Polym. Mater. Sci. Eng.* **56**, 720 (1987).
33. S. I. Ertel, A. Chilkoti, T. A. Horbett and B. D. Ratner, *J. Biomater. Sci., Polymer Edn* **2**, 163 (1991).
34. R. E. Marchant, G. Szakalas-Gratzl, S. Yuan, N. Ziats and K. Kottke-Marchant, *Trans. Soc. Biomater. (4th World Biomater. Cong.)* **15**, 202 (1992).

Introduction of amine groups on poly(ethylene) by plasma immobilization of a preadsorbed layer of decylamine hydrochloride

JOHANNES G. A. TERLINGEN[1], LAURA M. BRENNEISEN[1]*, HENDRIKA T. J. SUPER[1], A. P. PIJPERS[2], ALLAN S. HOFFMAN[3] and JAN FEIJEN[†]

[1]*Department of Chemical Technology, University of Twente, P.O. Box 217, 7500 AE Enschede, The Netherlands*
[2]*DSM Research, P.O. Box 18, 6160 MD Geleen, The Netherlands*
[3]*Center for Bioengineering FL-20, University of Washington, Seattle, WA 98195, USA*

Received 24 June 1992; accepted 8 September 1992

Abstract—In order to introduce amine groups on poly(ethylene) (PE) surfaces, PE surfaces were preadsorbed with decylamine hydrochloride (DA.HCl) and subsequently treated with an argon plasma. It was shown by XPS (X-ray Photoelectron Spectroscopy), that approximately half of the preadsorbed (mono)layer was immobilized and that a substantial part (60–70%) of the incorporated nitrogen containing groups were amine groups. The availability of the surface amine groups for reactions was investigated by applying a gas phase reaction with 4-trifluoromethylbenzaldehyde and by a reductive methylation reaction in aqueous solution with ^{14}C formaldehyde. A maximal number of reactive amine groups was found after a plasma treatment time of 2 s. The reductive methylation reaction was used to estimate the surface concentration of amine groups resulting in a typical surface concentration of 1×10^{-6} mol/m^2 after a plasma treatment time of 2 s.

Key words: Plasma immobilization of surfactants; functionalization of polymer surfaces; surface amine groups; reactivity of surface amine groups; immobilization of biomolecules.

INTRODUCTION

The success of a polymeric material in biomedical applications is not only determined by its mechanical properties, but also to a large extent by its surface properties [1]. Therefore an increasing demand for polymers with tailored surface properties exists. In particular the introduction of functional groups, which can be used for the improvement of adhesion in composite materials or for the covalent attachment of bioactive molecules [2, 3] for biomedical [4] and biosensor [5] applications has evolved into a major research field.

A promising strategy to modify polymer surfaces is the use of gas plasma technology [6–8]. In comparison to other surface modification techniques plasma techniques have some distinct characteristics [1]. First, only the surface is modified, leaving the bulk of the material unaffected. Furthermore, these techniques are usually fast, single step processes. Plasma techniques are also dry processes, which in combination with the usually mild chemicals used, yield an industrially and environmentally attractive process. Due to the fact that the gas is the reactive phase,

---

* Present address: Tebodin, P.O. Box 233, 7550 AE Hengelo, The Netherlands.
† To whom correspondence should be addressed.

the shape of the substrate and in the case of plasma polymerization also the chemical nature of the substrate are of minor importance. Some disadvantages should also be mentioned. Firstly, most plasma techniques are vacuum processes, which require relatively expensive equipment. Moreover, these processes are usually batch processes and the chemistry of the final modified surface is rather complex and generally cannot be precisely predicted.

By the selection of the gas used during the discharge, one can control the overall type of process. Generally, application of hydrocarbon gases/vapors results in a deposition process (plasma polymerization [9]) and by using inert or oxidizing gases etching of the substrate occurs [10]. The surface chemistry depends on the type of gas and the discharge conditions used. By applying an amine-containing hydrocarbon gas, a polymer deposit is obtained which not only contains a (small) amount of amine groups, but also imine and nitrile groups [11]. By applying an ammonia [12] or a nitrogen [13] plasma to a polymer surface, a similar spectrum of nitrogen-containing groups is introduced on the surface. Therefore several approaches have been developed to obtain a better control of the surface chemistry. Reduction of the power input per molecule generally reduces the fragmentation in the plasma phase and a chemistry which more closely resembles the precursor chemistry is obtained [9]. If, in addition, low substrate temperatures [14] are used an even better control of the surface chemistry can be obtained.

Recently we have developed a new approach, which leads to a good control of the surface chemistry [15]. In this approach, a hydrophobic substrate is preadsorbed with a thin layer—typically a monolayer—of a functional group-containing compound. The substrate with the preadsorbed layer of e.g., a surfactant is then plasma treated with an inert gas to covalently couple this layer to the substrate surface. The principle of this method is given in Fig. 1. Using this method, sulfate groups have

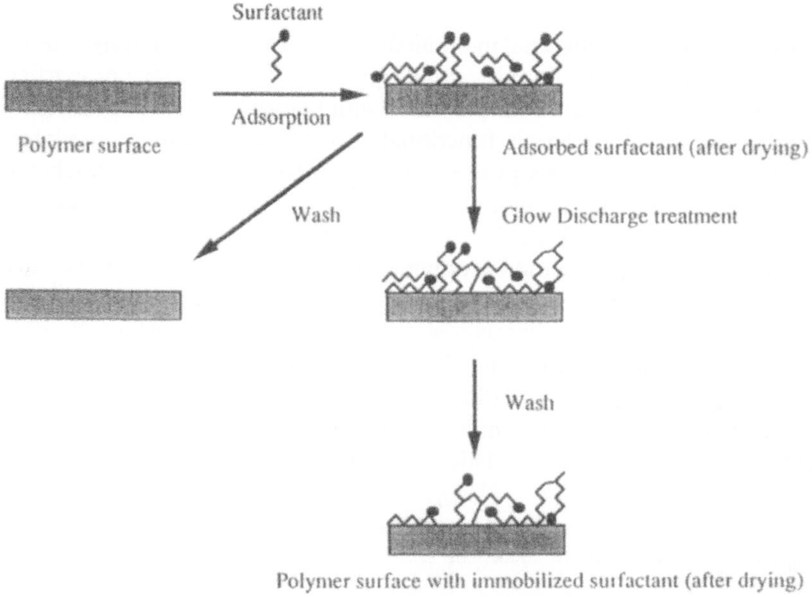
Surfactant
Adsorption
Polymer surface
Adsorbed surfactant (after drying)
Wash
Glow Discharge treatment
Wash
Polymer surface with immobilized surfactant (after drying)

Figure 1. Schematic representation of the plasma immobilization of a surfactant layer on a hydrophobic polymer surface.

been introduced on poly(propylene) (PP) surfaces by immobilization of a preadsorbed layer of a sulfate group-containing surfactant (sodium dodecylsulfate) [15]. Furthermore the immobilization of poly(ethylene oxide)-containing surfactants results in protein repellant surfaces, indicating that the poly(ethylene oxide) chains were immobilized in a reasonable intact state [16]. On the other hand the immobilization of preadsorbed layers of poly(acrylic acid) on PE was not successful due to rapid decarboxylation of the poly(acrylic acid) layer [17–19]. Therefore, although the immobilization of preadsorbed compounds seems generally applicable, it is necessary to check whether the specific functional groups selected are stable in the applied plasma environment.

Because amine groups can be readily utilized for the covalent coupling of bioactive molecules [20], the introduction of these groups by the plasma immobilization of an amine-containing surfactant has been studied. Therefore an inert hydrophobic polymer substrate (PE) was preadsorbed with decylamine hydrochloride (DA.HCl) and then treated with an argon plasma. After the plasma treatment the films were washed to remove any non-bound DA.HCl from the surface. Appropriate adsorption and washing conditions were established by XPS and water contact angle measurements. Subsequently, preadsorbed films have been plasma treated, washed and analyzed with water contact angle measurements, XPS and 'cold stage' XPS, in which the samples are cooled to −150°C prior to the XPS analysis. Furthermore the reactivity of the amine groups introduced on the surface has been investigated by applying a gas phase reaction with 4-trifluoromethylbenzaldehyde (TFBA) followed by XPS analysis and by a reductive methylation reaction [3, 21] with an aqueous solution ^{14}C formaldehyde (see Fig. 2). The reductive methylation reaction with ^{14}C formaldehyde was also used to estimate the concentration of amine groups introduced on the PE surface by the plasma immobilization of DA.HCl.

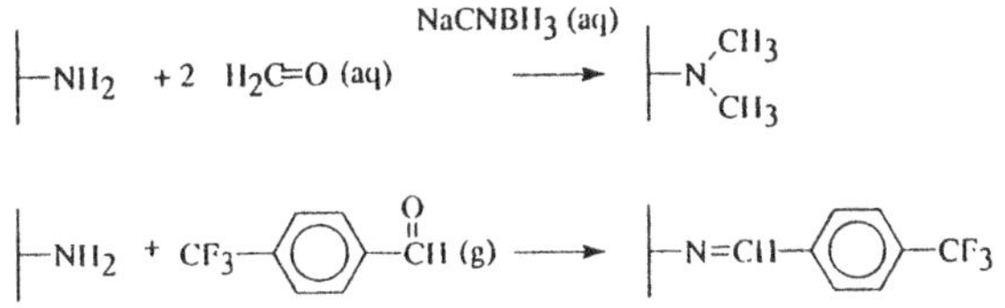
NaCNBH3 (aq)
—NH2 + 2 H2C=O (aq)
—N
CH3
CH3
—NH2 + CF3
CH (g)
—N=CH
CF3

Figure 2. Derivatization reactions used to tag amine groups on polymer surfaces. After reaction the surfaces are either analyzed with XPS (derivatization with 4-trifluoromethylbenzaldehyde) or by liquid scintillation counting of the surface radioactivity (reaction with ^{14}C formaldehyde).

MATERIALS AND METHODS

Materials

PE film: low density poly(ethylene), type 2300, thickness 0.2 mm, DSM, Geleen, The Netherlands. This LDPE does not contain additives. Dichloromethane (GR), toluene (GR), acetone (GR), ethanol (GR), diethylether (GR), decylamine (DA, purity ~95%), sodium chloride (NaCl, GR), aqueous sodium hydroxide solution (NaOH, 1.00 M), aqueous hydrochloric acid solution (HCl, 1.00 M), 4-trifluoromethylbenzaldehyde (TFBA, purity >98%), formaldehyde (37%, GR), sodium cyanoborohydride ($NaCNBH_3$, purity ~95%), sodium dihydrogenphosphate monohydrate (NaH_2PO_4, GR), di-sodium hydrogenphosphate dihydrate (Na_2HPO_4, GR), concentrated hydrochloric acid (37%, GR): Merck, Darmstadt, Germany. The GR grade chemicals have a purity of at least 99.5%. ^{14}C formaldehyde: Amersham Nederland B.V.'s Hertogen Bosch, The Netherlands, specific activity 8×10^{11} Bq/mol. Aqualuma: Hicol bv, Oud-Beijerland, The Netherlands. All chemicals were used as received. Water: Doubly deionized water. Argon (purity ≥ 99.999%): Hoekloos, Amsterdam, The Netherlands.

Pure decylamine hydrochloride (DA.HCl) was prepared as follows. DA was contacted with the vapor of concentrated HCl for 48 h. The white/yellow precipitate was dissolved in ethanol and precipitated in cold diethylether [22]. The precipitate (DA.HCl) was dried *in vacuo* for several hours. The formation of DA.HCl was confirmed by IR spectroscopy.

Methods

All films were stored in the dark in glass vials at atmospheric pressure and room temperature (RT).

Cleaning of PE films. PE films (13 × 25 mm) were cleaned ultrasonically in dichloromethane for 10 min. Fresh dichloromethane was added and the procedure was repeated. The total cleaning procedure consisted of 4 times treatment with dichloromethane, 4 times with acetone and 4 times with water. Subsequently the films were dried *in vacuo* at room temperature and stored.

Adsorption of DA.HCl on PE films. Each film was immersed in 5 ml of a solution of DA (0.07 M) in aqueous HCl (1 M)/ethanol (5% w/v) at room temperature. Ethanol was used to increase the solubility of the *in situ* formed DA.HCl. The surface tension of the DA.HCl solution measured at 22°C with a Krüss Tensiometer (Hamburg, Germany) was 25.3 ± 0.2 mN/m whereas the hydrochloric acid/ethanol solution without DA had a surface tension of 55.2 ± 0.1 mN/m. After 30 min the solution was removed and the films were dried *in vacuo* at room temperature.

Desorption of pre-adsorbed DA.HCl. Films with a preadsorbed DA.HCl layer were immersed in an aqueous HCl solution (1 M) at room temperature. After 60 min the solution was removed and the films were dried *in vacuo* at room temperature. In a control experiment the same procedure was used to treat clean PE.

Plasma treatment. An extensive description of the plasma treatment system has been given previously [18]. In brief the plasma system consists of a tubular reactor (internal diameter 6.5 cm) with three externally placed capacitively coupled electrodes spaced at 10 cm distances. One side of the tubular reactor was connected to a turbomolecular pump and to a two stage rotary vane pump. The other side was attached to a gas inlet controlled by mass flow controllers. The electrodes were powered through a matching network by a 13.56 MHz radio frequency generator, which was controlled by a timer (Apple IIe computer with a time control program). This system enables an optimal control of the plasma treatment time.

In order to test whether DA.HCl was stable at reduced pressures, pure DA.HCl was exposed to a 10^{-3} and to a 10^{-6} mbar vacuum for several hours at room temperature. In both cases no weight decrease could be measured (with a Mettler AE200 balance, accuracy 0.1 mg), indicating that under the vacuum conditions applied during the plasma treatment (0.07 mbar) DA.HCl was stable.

The following plasma treatment procedure was used. Twelve polymer films which were either clean or preadsorbed with DA.HCl, were placed in the center region of the reactor on a glass plate and the reactor was evacuated to a pressure of 0.01 mbar. Subsequently an argon flow of 10 cm^3/min (standard temperature and pressure) was established through the reactor. After 15 min the reactor was isolated at a pressure of 0.07 mbar, and the films were plasma treated (44 W, 0.07 mbar, no gas flow through the reactor). After the plasma treatment an argon flow of 10 cm^3/min was established through the reactor for 2 min. The reactor was then brought to atmospheric pressure with air, the films were taken out, turned and subsequently the other side of the films was treated according to the same procedure.

Washing after plasma treatment. Plasma treated films were washed according to the same procedure as was used for the desorption of preadsorbed DA.HCl. This wash step was necessary to remove any unbound DA.HCl from the surface.

Reaction of surface amine groups with TFBA. In order to convert the amine hydrochloride groups at the surface into free amine groups, plasma treated films which were washed afterwards, were rinsed once with an aqueous NaOH solution (1 M), three times with water and once with acetone. Subsequently these films were placed in glass vials on glass beads and 0.25 ml of TFBA was injected under the beads. The glass vial was closed and after 60 min, the film was taken out and immediately loaded for XPS analysis.

Reductive methylation of surface amine groups. The surface concentration of amine groups can be (semi) quantitatively determined by a reductive methylation reaction of the surface amine groups with ^{14}C formaldehyde followed by determination of the surface radioactivity [3, 21]. Each film was immersed in 5 ml of an aqueous solution of formaldehyde (1.8×10^{-4} M, 3×10^{10} CPM (counts per minute) in a $NaCNBH_3$ (1 g/l) containing phosphate buffer (pH 7.0, 0.028 M NaH_2PO_4, 0.049 M Na_2HPO_4) for 1 h. After the reaction the films were taken out, rinsed 3 times with an aqueous sodium chloride solution (4 M) and then three times with water. The films were placed in a scintillation vial and 20 ml Aqualuma was added. The ^{14}C activity on the surface was determined by liquid scintillation counting (1219 Rackbeta, LKB-Wallac, Turku, Finland). The measured activities were corrected for background activity.

Characterization

Contact angle measurements. Wilhelmy plate contact angle measurements [23, 24] using water were performed with all modified and control surfaces. When multiple measurements were performed on the same films the water was also refreshed after each dip.

X-ray photoelectron spectroscopy (XPS) measurements. XPS measurements were performed with a Kratos XSAM-800 (Manchester, United Kingdom) apparatus using a Mg K_α source (15 kV, 15 mA). The samples were analyzed with the analyzer placed perpendicular to the sample surface. A spot size with a diameter of 3 mm was analyzed. The pressure during measurements was typically 1×10^{-8} mbar. The spectra were recorded in the low resolution mode (pass energy 40 eV, FWHM $Ag3d^{5/2}$: 1.2 eV). Survey scans (0–1000 eV BE) were used to qualitatively determine the elements present at the surface. Quantification of the surface elemental composition was performed by recording detail scans (20 eV windows). Standard sensitivity factors delivered by Kratos were used to convert the measured peak areas into atom percentages.

Cold stage XPS measurements were performed with a Leybold MAX-200 Spectrometer (Leybold, Cologne, Germany). The sample holder on which films were mounted was cooled to −150°C before the films were evacuated. This procedure was necessary in order to keep the preadsorbed DA.HCl layer on the surface under ultra high vacuum conditions. The films were analyzed using a Mg K_α source (13 kV, 20 mA) at a pressure of typically 3×10^{-9} mbar. The analyzer was placed at a position perpendicular to the polymer sample surface and a spot of 4 × 7 mm was analyzed. Survey spectra (0–1300 eV KE) were recorded at a constant relative resolution (retardation factor (FRR) 4) and detail scans were recorded with a pass energy of 48 eV (FWHM Ag3$d^{5/2}$: 0.9 eV). The measured peak areas were normalized and converted into atom percentages by calculated sensitivity factors.

RESULTS

Adsorption and desorption of DA.HCl on PE

The adsorption of DA.HCl on and the desorption from PE surfaces were studied using water contact angle measurements and XPS. Table 1 shows that the advancing and receding contact angles for PE, PE treated with an ethanol/HCl solution, washed PE, and washed PE/DA.HCl films are identical. Only for PE/DA.HCl films a difference in advancing contact angles between the first and second dip is observed. During the second dip an advancing angle comparable to that of clean PE films is found.

When PE films preadsorbed with DA.HCl are analyzed with XPS, no nitrogen or chlorine is found (see Table 1). Although it has been reported that HCl salts of amine groups on surfaces can be detected with XPS [13, 25], it seems possible that under ultra high vacuum (UHV) conditions an ultra-thin DA.HCl layer on a polymer surface might not be stable. Therefore the preadsorbed films were cooled

Table 1.
Adsorption and desorption of decylamine hydrochloride on PE. The results of the XPS and Wilhelmy plate water contact angle measurements for PE and preadsorbed PE (PE/DA.HCl) films before and after washing with an aqueous hydrochloric acid solution (1 M) are given. The results of the XPS measurements in which the films were cooled (−150°C) prior to the XPS analysis are given in brackets.

	Atom percentages				
Sample	O	N	Cl	Θ_{adv}	Θ_{rec}
PE	0.3	<0.1	<0.1	109 ± 2	90 ± 3
PE HCl/ethanol	n.d.	n.d.	n.d.	112 ± 1	89 ± 2
PE washed	0.8	<0.1	<0.1	110 ± 2	92 ± 1
PE/DA.HCl	0.1 (0.2)	<0.1 (2.2)	<0.1 (3.0)	99 ± 1[a]	91 ± 1
PE/DA.HCl washed	<0.1 (<0.1)	<0.1 (≤0.1)	<0.1 (<0.1)	108 ± 2	92 ± 2

[a] The second dip deviates significantly: 110 ± 1. n.d.: not determined.

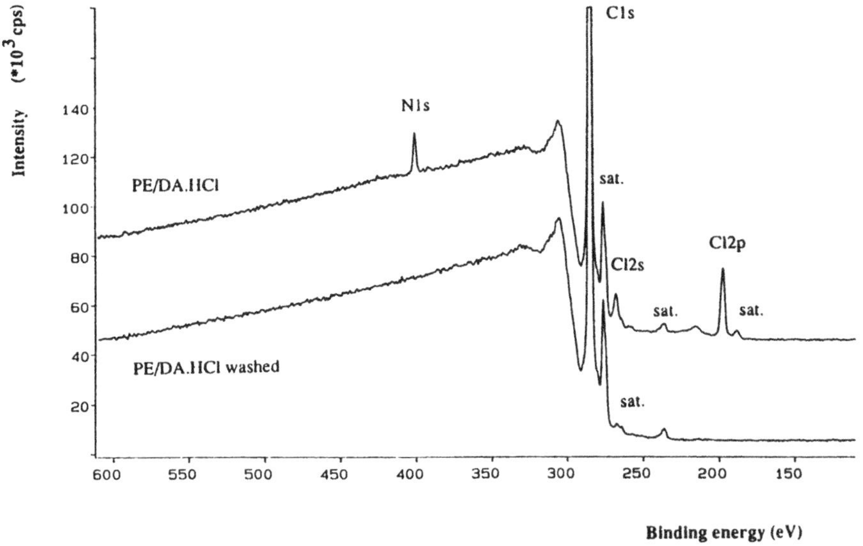
Intensity (*10^3 cps)
140
120
100
80
60
40
20
PE/DA.HCl
PE/DA.HCl washed
N1s
C1s
sat.
Cl2s
sat.
Cl2p
sat.
sat.
600
550
500
450
400
350
300
250
200
150
Binding energy (eV)

Figure 3. 'Cold stage' XPS survey spectra of a PE film which was preadsorbed with DA.HCl and of a preadsorbed PE film which was washed with an aqueous hydrochloric acid solution. Prior to the evacuation and XPS analysis the films were cooled to a temperature of −150°C. For the spectrum of PE/DA.HCl an offset of 40.000 CPS was used.

to −150°C prior to XPS analysis. The results of these 'cold stage' XPS measurements are qualitatively given in Fig. 3 and quantitatively in Table 1. In this case indeed nitrogen and chlorine are detected on the preadsorbed PE films. After washing with an aqueous HCl solution the amounts of nitrogen and chlorine on the surface have decreased to a level comparable or lower than the XPS detection limit.

Pure DA.HCl has also been analyzed with 'cold stage' XPS. As expected this compound only contained carbon, nitrogen and chlorine. Also a small amount of oxygen, probably due to residual water is observed. The elemental composition of pure DA.HCl is close to the theoretically expected composition. Theoretically an atom percentage of 8.3% for both nitrogen and chlorine is expected. The atom percentages observed are 6.9% nitrogen and 9.3% chlorine. Deviations from the theoretical values can be explained by the inherent inaccuracy of the method and by the use of calculated sensitivity factors, which might not be optimal for this particular system. The difference between the nitrogen and chlorine contents found for pure DA.HCl is also reflected in the atom percentages of nitrogen and chlorine of the preadsorbed DA.HCl layer on the PE surface.

Argon plasma treatment of PE/DA.HCl and PE films

PE films with or without a preadsorbed DA.HCl layer have been treated with a static argon plasma for varying time periods. The plasma treated films were washed afterwards with an aqueous HCl solution to remove any unbound DA.HCl from the surface. The results of the XPS analysis of these films are given in Fig. 4. From this figure it can be seen that the preadsorbed films after plasma treatment and washing contain more nitrogen and possibly less oxygen than the plasma treated and washed PE films. Furthermore no chlorine is found on both plasma treated films. The

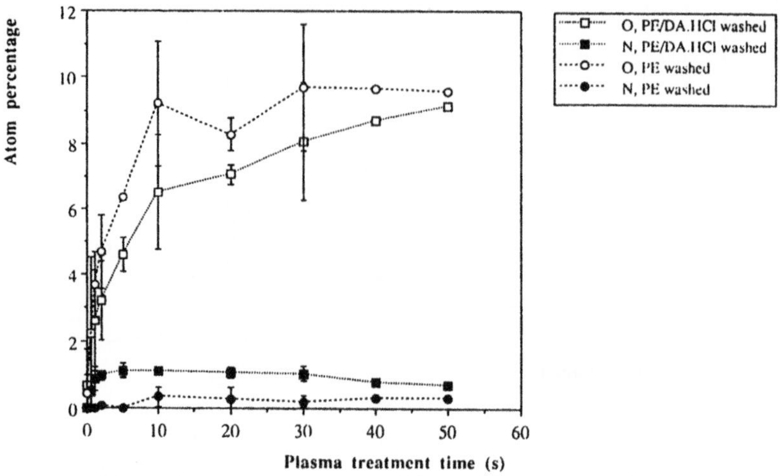
Atom percentage
12
10
8
6
4
2
0
0
10
20
30
40
50
60
Plasma treatment time (s)
O, PE/DA.HCl washed
N, PE/DA.HCl washed
O, PE washed
N, PE washed

Figure 4. XPS analysis of PE/DA.HCl and PE films which were first treated with an argon plasma and then washed. The atom percentages oxygen and nitrogen are given as a function of the plasma treatment time. The error bars indicate the standard deviation found for three independent experiments.

oxygen levels of PE/DA.HCl and PE films increase with increasing plasma treatment time. Wilhelmy plate water contact angles of these films (Fig. 5) show that the water contact angles of plasma treated and washed PE/DA.HCl and PE films are similar.

Because no chlorine was found on argon plasma treated and washed PE/DA.HCl films, these films were also analyzed with 'cold stage' XPS. Some typical examples of nitrogen (N1*s*) and chlorine Cl2*p*) spectra obtained during these analyses are given in Fig. 6. The argon plasma treated and washed PE/DA.HCl films not only

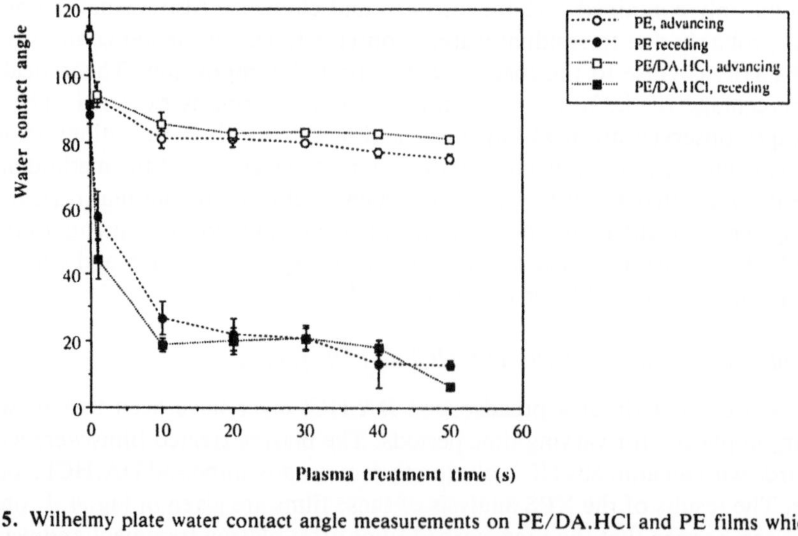
Water contact angle
120
100
80
60
40
20
0
0
10
20
30
40
50
60
Plasma treatment time (s)
PE, advancing
PE receding
PE/DA.HCl, advancing
PE/DA.HCl, receding

Figure 5. Wilhelmy plate water contact angle measurements on PE/DA.HCl and PE films which were first treated with an argon plasma and then washed. The advancing and receding water contact angles are given as a function of the plasma treatment time. The averages and standard deviations of three films within a batch are given.

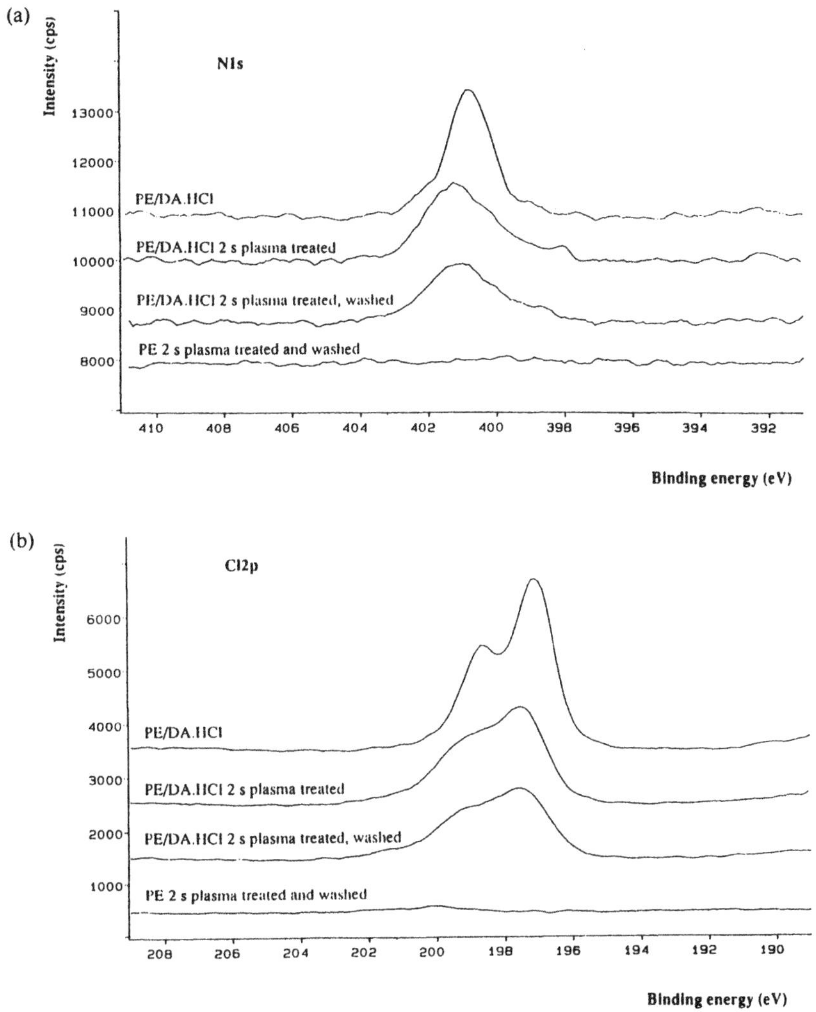
(a)
Intensity (cps)
N1s
13000
12000
11000
10000
9000
8000
PE/DA.HCl
PE/DA.HCl 2 s plasma treated
PE/DA.HCl 2 s plasma treated, washed
PE 2 s plasma treated and washed
410 408 406 404 402 400 398 396 394 392
Binding energy (eV)
(b)
Intensity (cps)
Cl2p
6000
5000
4000
3000
2000
1000
PE/DA.HCl
PE/DA.HCl 2 s plasma treated
PE/DA.HCl 2 s plasma treated, washed
PE 2 s plasma treated and washed
208 206 204 202 200 198 196 194 192 190
Binding energy (eV)

Figure 6. 'Cold stage' XPS measurements on argon plasma treated PE/DA.HCl and PE films. The films were either washed or not washed after plasma treatment. In Fig. 6(a) the N1*s* peaks and in Fig. 6(b) the Cl2*p* peaks are given. The C1*s* peak was set on 284.6 eV. An offset of 1000 CPS in between the different spectra was used.

contained carbon, oxygen and nitrogen, but also some chlorine. The plasma treated and washed PE films contained carbon, oxygen, very small amounts of nitrogen and negligible amounts of chlorine. The binding energies for the nitrogen (N1*s*) and chlorine (Cl2*p*) peaks of plasma treated PE/DA.HCl films before and after washing are similar to those observed for pure DA.HCl (N1*s*: 401.1, Cl$2p^{3/2}$ 197.6 eV). When plasma treated and washed PE/DA.HCl films are analyzed with XPS at room temperature a typical binding energy of the N1*s* peak of 399.0 eV is observed.

Figure 7 shows the quantitative results of the 'cold stage' XPS analysis of plasma treated and washed PE and PE/DA.HCl films. The amounts of chlorine and nitrogen on plasma treated and washed PE/DA.HCl films are similar and decrease with increasing plasma treatment time.

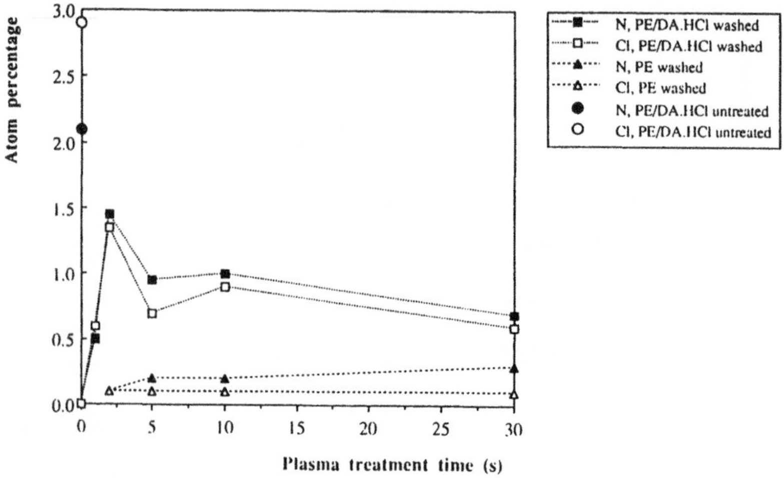
Atom percentage
3.0
2.5
2.0
1.5
1.0
0.5
0.0
0
5
10
15
20
25
30
Plasma treatment time (s)
N, PE/DA.HCl washed
Cl, PE/DA.HCl washed
N, PE washed
Cl, PE washed
N, PE/DA.HCl untreated
Cl, PE/DA.HCl untreated

Figure 7. 'Cold stage' XPS measurements on PE/DA.HCl and PE films which were first treated with an argon plasma and then washed. The observed amounts of nitrogen and chlorine are given as a function of the plasma treatment time. The amounts of oxygen found on the surface (not given in this figure) are comparable with those given in Fig. 4. Single measurements.

In order to investigate the chemical state of the nitrogen containing groups introduced on the surface, a gas phase derivatization reaction with TFBA has been performed on plasma treated and washed PE/DA.HCl and PE films. The results of the XPS analysis after the reaction with TFBA are given in Fig. 8. It is shown that substantial amounts of fluorine are found on plasma treated and washed PE/DA.HCl films compared to similarly treated PE films. Futhermore in contrast to the amount of nitrogen found on the surface, which is nearly constant after

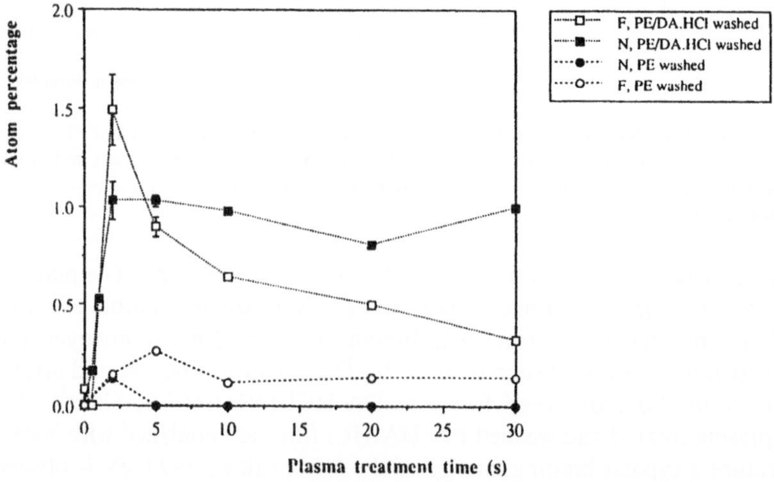
Atom percentage
2.0
1.5
1.0
0.5
0.0
0
5
10
15
20
25
30
Plasma treatment time (s)
F, PE/DA.HCl washed
N, PE/DA.HCl washed
N, PE washed
F, PE washed

Figure 8. Gas phase reaction of 4-trifluoromethylbenzaldehyde with PE/DA.HCl and PE films which were first treated with an argon plasma and then washed. The films have been analyzed with XPS after reaction. The atom percentages fluorine and nitrogen are given as a function of the plasma treatment time. The error bars indicate the standard deviation found for three samples within one batch.

applying a plasma of more than 2 s, a clear maximum in the amount of fluorine bound to the surface is found at a plasma treatment time of 2 s.

The reactivity of the surface amine groups in aqueous environments was checked by applying a reductive methylation with radioactive (^{14}C) formaldehyde. Because an unknown part of the β-radiation is absorbed by the polymer substrate and is thus not detected by scintillation counting, the measured surface concentration can only be used as a relative indication of the concentration of amine groups on the surface. An example of the amount of formaldehyde reacted to plasma treated and washed PE films with or without a preadsorbed DA.HCl layer is given in Fig. 9. From this figure it can be seen that again a maximum in the amount of amine groups available for reactions is found at a plasma treatment time of 2 s, and that the plasma treated and washed PE/DA.HCl films contain substantially more amine groups that the plasma treated and washed PE films. Although similar plots are obtained as a function of the plasma treatment time, each time this experiment is repeated, the absolute values vary. A large day to day variation has been found. The average amount of formaldehyde reacted to a plasma treated (2 s) and washed PE/DA.HCl film is: $(1.0 \pm 0.7) \times 10^{-6}$ mol/m^2 ($n = 7$) and to plasma treated (2 s) and washed PE films: $(0.07 \pm 0.03) \times 10^{-6}$ mol/m^2 ($n = 7$).

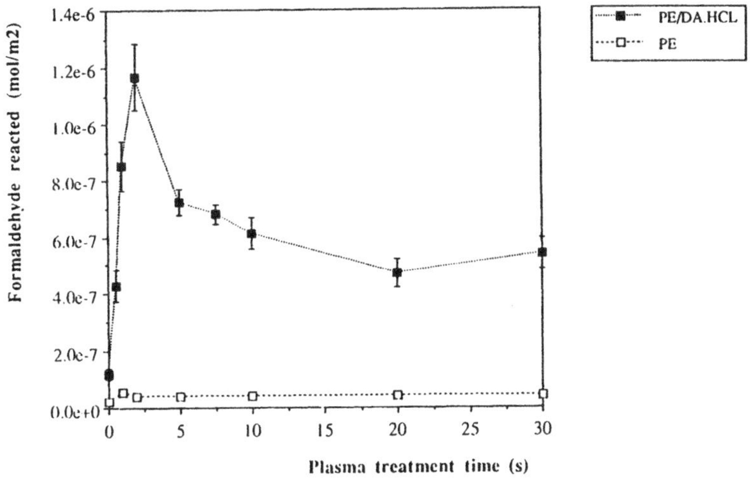

Formaldehyde reacted (mol/m2)
1.4e-6
1.2e-6
1.0e-6
8.0e-7
6.0e-7
4.0e-7
2.0e-7
0.0e+0
0
5
10
15
20
25
30
Plasma treatment time (s)
PE/DA.HCL
PE

Figure 9. Reaction of ^{14}C formaldehyde with PE/DA.HCl and PE films which were first treated with an argon plasma and then washed. The amount of formaldehyde reacted with the surface is given as a function of the plasma treatment time applied and is determined by measuring the surface radioactivity. The amount of formaldehyde reacted can be used as an indication of the amount of amine groups on the surface. The averages and standard deviations of four films within one batch are given.

DISCUSSION

Adsorption and desorption of DA.HCl on PE

Because the adsorption of DA.HCl on hydrophobic substrates has only been described for graphite [26], the adsorption/desorption behavior of DA.HCl on PE had to be studied in detail. From the water contact angle measurements and the cold stage XPS measurements given in Table 1, it can be clearly seen that DA.HCl adsorbs on PE and that by washing with a HCl solution the preadsorbed DA.HCl layer can be removed from the PE surface. Furthermore the water contact angle measurements show that by adsorbing a layer of DA.HCl on the surface a 'water unstable' surface is obtained. Especially the advancing water contact angles during the first and second dip differ significantly. Although water contact angle measurements on 'water unstable' surfaces will give erroneous absolute values, the double dip method is a valuable way to test whether a water unstable layer is present at the surface.

The quantitative XPS data for the layer of DA.HCl preadsorbed on PE can also be used to estimate its thickness. This thickness is calculated by using an overlayer model [27], in which it is assumed that the overlayer is homogeneously present on the surface. In the case of the adsorption of a surfactant on a hydrophobic surface this seems a reasonable assumption. Furthermore it was assumed that the layer adsorbed on the surface is not oriented. Probably the adsorbed DA.HCl on PE is initially oriented, with the alkyl tails facing the hydrophobic substrate, but after crossing the air-solution interface and drying *in vacuo*, this orientation is likely to be lost. The calculations were made using the elemental composition of the pure DA.HCl. The necessary inelastic mean free paths of nitrogen (N1*s*), chlorine (Cl2*p*) and carbon (C1*s*) were calculated according to the method of Seah and Dench [28]. A density of 1 g/cm^3 was assumed. Using these assumptions a layer thickness of 13 Å was calculated, which is in agreement with the length of a decylamine hydrochloride molecule. This indicates that the layer thickness of the preadsorbed DA.HCl is typically in the order of a monolayer.

Argon plasma treatment of PE/DA.HCl and PE films

A preadsorbed DA.HCl layer on a PE substrate can be immobilized by applying a static argon plasma. This is shown qualitatively in Fig. 6. After plasma treating PE/DA.HCl films and washing with an aqueous HCl solution, which removes any unbound DA.HCl from the surface, substantial levels of nitrogen and chlorine are found on the surface with 'cold stage' XPS. On plasma treated and washed PE surfaces without a preadsorbed DA.HCl layer hardly any nitrogen and chlorine are found. When plasma treated and washed PE/DA.HCl films are analyzed by XPS at room temperature, some nitrogen but no chlorine is found on the surface. Previously it was shown that when PE/DA.HCl films are analyzed with XPS at room temperature, the preadsorbed DA.HCl is removed prior or during the XPS analysis. Most likely this is caused by a two step process, consisting of a dehydrochlorination of the preadsorbed DA.HCl followed by a quick removal of the volatile amine from the surface. The lack of chlorine on plasma immobilized layers of DA.HCl during XPS analysis at room temperature, can be explained by assuming that a similar dehydrochlorination process occurs for the immobilized layers of DA.HCl. Furthermore the lack of chlorine shows that no chlorine was covalently incorporated in the immobilized layer during the plasma treatment. The fact that a substantially higher nitrogen level for plasma treated and washed PE/DA.HCl films compared to plasma treated and washed PE films is found during XPS measurements at room temperature, confirms the conclusion that the DA.HCl layer is covalently bound to the PE surface, because the UHV during the XPS measurements serves as a second 'desorption step'.

The N1*s* and Cl2*p* peaks of untreated PE/DA.HCl films measured with 'cold stage' XPS (given in Fig. 6) are shifted 0.5 eV downwards to a lower binding energy compared to pure DA.HCl. This is caused by differences in surface charge between the preadsorbed layer and the PE substrate. On the other hand the binding energies of the N1*s* and Cl2*p* peaks of plasma treated and of plasma treated and washed PE/DA.HCl films are identical to those observed for pure DA.HCl. These binding energies are in good agreement with those reported for amine hydrochloride salts [7, 13, 29]. This indicates that the amine hydrochloride groups are stable during the argon plasma treatment and that they are intactly immobilized on the surface. The broadening of the N1*s* at lower binding energies shows that possibly a second nitrogen containing group might be formed during the plasma treatment process.

When plasma treated and washed PE/DA.HCl films are analyzed with XPS at room temperature, no chlorine is detected on the surface and the N1*s* peak shifts to a binding energy of 399.0 eV. This is the expected binding energy for free amine groups [7, 29]. The immobilized DA.HCl layer looses its ionically bound hydrogen chloride in the UHV at room temperature, which renders the amine hydrochloride groups into free amine groups and explains the observed shift in binding energy of the N1*s* peak. The fact that the amine hydrochloride groups are not oxidized during the argon plasma treatment can be deduced from the observed binding energy of the N1*s* peak. For oxidized nitrogen containing groups ($-NO$, $-NO_2$, $-NO_3$) a N1*s* binding energy of at least 402 eV would be expected [29].

Generally an argon plasma treatment of a polymer film leads to surface oxidation [30, 31] and thus to an increase in the wettability of that surface. The surface oxidation can be explained by two mechanisms. Firstly the argon plasma introduces radicals at the polymer surface, which rapidly react with oxygen or nitrogen when the plasma treated surfaces are exposed to air [32]. It is also possible that, although all possible precautions are taken, an oxygen source like a small air leak or adsorbed water on the reactor walls might be present during the plasma treatment. In this case not only radical formation but also surface oxidation will occur during the plasma treatment. From Fig. 4 it can be seen that during or after the plasma treatment oxygen is incorporated on PE/DA.HCl and PE films. At longer plasma treatment times more oxygen is incorporated. Furthermore the amount of oxygen on the plasma treated and washed PE/DA.HCl films seems slightly lower than on plasma treated and washed PE films. This might be due to the fact that part of the preadsorbed layer which is oxidized by the plasma treatment, but not covalently bound to the surface, is washed off during the washing step applied afterwards.

Because the plasma treated and washed PE/DA.HCl and PE films contain slightly different amounts of oxygen and nitrogen on their surfaces, and because the exact contributions of the different oxygen and nitrogen groups to the wettability of the surface are not known, the differences in contact angles found between PE/DA.HCl and PE films in Fig. 5 cannot be unambiguously explained. The water contact angle measurements are thus not informative about the success or failure of the plasma immobilization of the preadsorbed DA.HCl layer on PE.

'Cold stage' XPS measurements on plasma treated and washed PE/DA.HCl films have shown that nearly identical levels of nitrogen and chlorine are present on the surface regardless of the plasma treatment time. From the analysis of the binding energies during the XPS measurements it was concluded that the immobilized nitrogen containing groups on the surface are mainly amine groups, which are able to bind hydrogen chloride. The fraction of nitrogen containing groups which can bind hydrogen chloride can be estimated from the chlorine levels found in Fig. 7 for plasma treated and washed PE/DA.HCl films. If a similar N/Cl ratio (1.35) is assumed for the immobilized DA.HCl layer as for the pure DA.HCl, it can be calculated that 60–70% of the immobilized nitrogen containing groups—presumably amine groups—can bind hydrogen chloride.

In order to confirm that the nitrogen groups created on the surface are amine groups, a gas phase reaction with TFBA was carried out. The reaction of an aldehyde group containing compound with surface amine groups, yielding a Schiff base, has been reported before as a possible derivatization reaction for surface amine groups. In this case surface amine groups were reacted with a solution of pentafluorobenzaldehyde in *n*-pentane [7, 13]. In order to prevent surface reorganization due to the solvent used in this reaction, we have modified this reaction to a gas phase reaction in which a more volatile fluorine containing benzaldehyde (TFBA) is used. Because the extent of this reaction on model surfaces with amine groups was not extensively checked or optimized, the amount of fluorine found on the surface, should be used as a relative indication of the surface amine group concentration. Furthermore it is assumed that this reaction only proceeds with primary and secondary amine groups (and hydrazines) although we have not checked the selectivity of this reaction. The fact that the reaction proceeds with a reasonable selectivity, can be derived from the low fluorine content of plasma treated and washed PE surfaces after reaction with TFBA. These plasma treated films contain a spectrum of oxygen containing groups. From Fig. 8 it can be seen that at a plasma treatment time of 2 s, a maximum in the amount of fluorine bound to the surface of plasma treated and washed PE/DA.HCl is found. This indicates that a maximum in the amount of amine groups available for reaction on the surface is formed at a plasma treatment time of 2 s. Although the amount of nitrogen containing groups remains constant at longer plasma treatment times, the amount of fluorine of TFBA reacted to the surfaces diminishes. This might be due to a decreasing amount of primary and secondary amine groups on the surface, a decreasing availability or decreasing reactivity of the surface amine groups for reaction with TFBA. It may be expected that at longer plasma treatment times the DA.HCl layer immobilized on the surface becomes more densely crosslinked. The surface amine groups would then become less available for reaction. It seems also possible that due to an increasing surface oxidation the amine groups become less reactive, because they can participate in hydrogen and/or ionic bonding to neighboring groups like e.g., ketone or carboxylic acid groups. Furthermore it cannot be ruled out that the amine groups introduced at the surface can be chemically modified by the plasma treatment. Based on the binding energy of the nitrogen peak (N1*s*) during the 'cold stage' and room temperature XPS measurements, and on the ability of a major part of the nitrogen groups on the surface to bind hydrogen chloride, the only possible side reaction seems to be the conversion of free amines to tertiary amine groups. Although dehydrogenation reactions frequently occur during plasma processes [33] a possible mechanism to convert free amines to tertiary amine groups is unclear.

The reaction of ^{14}C formaldehyde with plasma treated and washed PE/DA.HCl films also shows that a maximal surface concentration of amine groups available for reaction is found after applying an argon plasma for 2 s. This is in close agreement with the results of the TFBA coupling. At longer plasma treatment times the amount of reactive amine groups on the surface decreases. A similar explanation for the observed maximum can be given as for the gas phase reaction with TFBA. It is quite remarkable that both types of reactions—gas phase and aqueous solution—show the same maximum in the amount of reactive amine groups as a function of the plasma treatment time.

The large day to day variation observed during the reductive methylation with ^{14}C formaldehyde seems to be caused by an unknown parameter during the plasma treatment procedure. When films from the same batch are reacted with ^{14}C formaldehyde on different days, similar results are obtained. The reductive methylation itself seems thus reproducible. Although XPS measurements (given in Fig. 4) show that the amount of nitrogen is quite reproducible, the concentration of reactive amine groups on the surface shows a large fluctuation. Allred *et al.* have reported that due to residual oxygen during an ammonia discharge also large fluctuations in the surface concentration of amine groups have been observed [34]. A similar explanation might be appropriate here.

The exact mechanism by which the preadsorbed DA.HCl layer is immobilized on the PE surface is not yet clear. It seems likely that the preadsorbed layer of DA.HCl is immobilized on the PE substrate by a radical mechanism. Radicals could be formed by hydrogen abstraction from both the alkyl tails of the adsorbed DA.HCl and from the substrate. These radicals can recombine to give covalent bonds between the substrate molecules and the preadsorbed surfactant molecules. Based on the amount of nitrogen measured with 'cold stage' XPS approximately half of the preadsorbed monolayer is immobilized on the surface at plasma treatment times between 2 and 10 s. At longer plasma treatment times the immobilization efficiency decreases due to the fact that part of the immobilized layer is etched off. Furthermore it is shown in Fig. 7 that a substantial fraction of the immobilized groups can bind hydrogen chloride, and that this fraction does not significantly decrease with increasing plasma treatment time. This indicates that the amine hydrochloride group is reasonably stable during the plasma treatment and that the immobilization probably occurs through the alkyl tails of the preadsorbed DA.HCl molecules. This is in good agreement with the model system we have studied previously [15]. When a sodium dodecylsulfate (SDS) layer preadsorbed on poly(propylene) is plasma treated with a static argon plasma the sulfate groups are not chemically modified by the plasma and therefore the immobilization is also probably occurring through the alkyl tails of the SDS molecules.

By applying an argon plasma for 2 s to a preadsorbed layer of DA.HCl on PE a maximal amount of amine groups available for reaction is observed. Furthermore at this time the surface oxidation due to the plasma treatment is still very low (around 3% oxygen) indicating that the damage induced by the plasma on the original surface is very low. Furthermore it is possible to use the reaction with TFBA and the reductive methylation reaction to estimate the surface concentration of amine groups. At a plasma treatment time of 2 s a maximum fluorine concentration of 1.5% is found for plasma treated (2 s) and washed PE/DA.HCl films. Assuming that the reaction with TFBA is specific for amine groups, and that every amine group results in 3 fluorine atoms on the surface, around 50% of the nitrogen groups present on the surface are able to react with TFBA. On the other hand the amount of formaldehyde reacted with plasma treated (2 s) and washed PE/DA.HCl films can be used to make a rough estimation of the absolute surface concentration of amine groups. When it is assumed that each amine group reacts with two formaldehyde molecules and that only half of the β-radiation has been counted, the surface amine concentration is approximately 1×10^{-6} mol/m^2. When this surface concentration is converted to a surface area per amine group a value of 167 Å^2 is calculated. This value can be compared to the surface area of a DA.HCl molecule adsorbed at the air–water surface [35] (30 Å^2 per molecule). If it is assumed that DA.HCl is as closely packed on PE as at the air water interface, it can be estimated that one fifth of a monolayer is immobilized at the surface. This agrees well with the values obtained from the XPS measurements, which show that half of the preadsorbed monolayer is immobilized and that half of the immobilized nitrogen containing groups can react with TFBA.

Because the amine groups introduced by the plasma immobilization of DA.HCl on PE show a good reactivity towards aldehyde containing compounds, the coupling of bioactive molecules to these surface amine groups using dialdehyde chemistry is presently under investigation.

CONCLUSIONS

In order to introduce amine groups on PE surfaces, the feasibility of the plasma immobilization of a preadsorbed layer of decylamine hydrochloride (DA.HCl) on PE has been investigated. It was shown that after plasma treatment and washing of PE/DA.HCl films approximately half of the preadsorbed monolayer is immobilized and that a substantial fraction (60–70%) of the immobilized nitrogen containing groups are amine groups. These amine groups were reacted with aldehyde containing compounds both in the gas phase (with 4-trifluoromethylbenzaldehyde) and in aqueous solutions (with ^{14}C formaldehyde). A maximum in the amount of surface amine groups available for reaction with aldehydes is found for a plasma treatment time of 2 s. The surface concentration of amine groups available for reaction is in the order of 1×10^{-6} mol/m^2.

Acknowledgements

The authors would like to thank Dr G. Engbers for his valuable contributions during discussions, Mr A. van den Berg for assisting with the XPS measurements and DSM (Geleen, The Netherlands) for the financial support for this research.

REFERENCES

1. B. D. Ratner, A. Chilkoti and G. Lopez, in: *Plasma Deposition, Treatment and Etching of Polymers*, p. 463, R. d'Agostino (Ed.). Academic Press Inc., Boston (1990).
2. W. R. Gombotz, W. Guanghui and A. S. Hoffman, *J. Appl. Polym. Sci.* **37**, 91 (1989).
3. G. H. M. Engbers, PhD Thesis, University of Twente, Enschede, The Netherlands (1991).
4. H. Yasuda and M. Gazicki, *Biomaterials* **3**, 68 (1982).
5. G. Kampfrath and R. Hintsche, *Anal. Lett.* **22**, 2423 (1989).
6. P. W. Rose and E. M. Liston, *Plastic Eng.* **41**, 41 (1985).
7. C. N. Reilley, D. S. Everhart and F. F.-L. Ho, in: *Applied Electron Spectroscopy for Chemical Analysis*, p. 105, H. Windawi and F. F.-L. Ho (Eds). John Wiley, New York (1982).
8. W. R. Gombotz and A. S. Hoffman, *CRC Critical Rev. in Biocompatability* **4**, 1 (1987).
9. H. Yasuda, *Plasma Polymerization*. Academic Press Inc., Orlando (1985).
10. F. D. Egitto, V. Vukanovic, G. N. Taylor, in: *Plasma Deposition Treatment, and Etching of Polymers*, p. 321, R. d'Agostino (Ed.). Academic Press Inc., Boston (1990).
11. V. Krishnamurthy, I. L. Kamel and Y. Wei, *J. Polym. Sci.: Part A: Polym. Chem.* **27**, 1211 (1989).
12. J. Friedrich, J. Gähde, H. Frommelt and H. Wittrich, *Faserforsch. Textiltech.* **27**, 604 (1976).
13. D. S. Everhart and C. N. Reilley, *Anal. Chem.* **53**, 665 (1981).

14. G. P. López, B. D. Ratner, *Langmuir* **7**, 766 (1991).
15. J. G. A. Terlingen, J. Feijen and A. S. Hoffman, *J. Colloid Interface Sci.*, accepted for publication (1991).
16. M.-S. Sheu, A. S. Hoffman, J. G. A. Terlingen and J. Feijen, *J. Clinical Mater.* in press (1991).
17. J. G. A. Terlingen, A. S. Hoffman and J. Feijen, *Congress Proceedings ISPC-10*, Bochum, August 4–9 1991, 2.5-3, 1-6 (1991).
18. J. G. A. Terlingen, A. S. Hoffman and J. Feijen, *J. Appl. Polym. Sci.* submitted for publication (1992).
19. J. G. A. Terlingen, F. J. van der Gaag, A. S. Hoffman and J. Feijen, *J. Appl. Polym. Sci.* submitted for publication (1992).
20. Y. Inaki, in: *Functional Monomers and Polymers, Procedures, Synthesis, Applications*, p. 461, K. Takemoto, Y. Inaki and R. M. Ottenbrite (Eds). Marcel Dekker Inc., New York (1987).
21. H. van Damme, PhD Thesis, University of Twente, Enschede, The Netherlands (1990).
22. A. W. Ralston, E. J. Hoffman, C. W. Hoerr and W. M. Selby, *J. Am. Chem. Soc.* **63**, 1598 (1941).
23. H. S. van Damme, A. H. Hogt and J. Feijen, *J. Colloid Interface Sci.* **114**, 167–172 (1986).
24. H. S. van Damme, A. H. Hogt and J. Feijen, in: *Polymer Surface Dynamics*, p. 89–100, J. D. Andrade (Ed.). Plenum Press, New York (1988).
25. P. R. Moses, L. M. Wier, J. C. Lennox, H. O. Finklea, J. R. Lenhard and R. W. Murray, *Anal. Chem.* **50**, 576 (1978).
26. L. D. Skrylev and E. A. Strel'tsova, *Khim. Teknol. (Kiev)* **6**, 47 (1976) CA 86(20): 145179 a.
27. M. P. Seah, in: *Practical Surface Analysis, Volume 1, Auger and X-Ray Photoelectron Spectroscopy*, p. 437, D. Briggs and M. P. Seah (Eds). John Wiley, Chichester (1991).
28. M. P. Seah and W. A. Dench, *Surf. Interface Anal.* **1**, 2 (1979).
29. C. D. Wagner, W. M. Riggs, L. E. Davis and J. F. Moulder, in: *Handbook of X-ray Photoelectron Spectroscopy*, G. E. Muilinberg (Ed.). Perkin Elmer Corporation (1979).
30. H. Yasuda, H. C. Marsh, S. Brandt and C. N. Reilley, *J. Polym. Sci.: Polym. Chem. Ed.* **15**, 991 (1977).
31. J. L. Grant, D. S. Dunn and D. J. McClure, *J. Vac. Sci. Technol.* **A6**, 2213 (1988).
32. L. J. Gerenser, *J. Adhesion Sci. Technol.* **1**, 303 (1987).
33. H. Suhr, in: *Techniques and Applications of Plasma Chemistry*, p. 57, J. Hollahan and A. T. Bell (Eds). John Wiley, New York (1974).
34. R. E. Allred, E. W. Merrill and D. K. Roylance, *Polym. Prepr.* **24**, 223 (1983).
35. R. Perea-Cárpio, F. González-Caballero, J. M. Bruque and C. F. González-Fernández, *J. Colloid Interface Sci.* **110**, 96 (1986).

A wettability gradient as a tool to study protein adsorption and cell adhesion on polymer surfaces

JIN HO LEE and HAI BANG LEE

Biomaterials Laboratory, Korea Research Institute of Chemical Technology, P.O. Box 9, Daedeog Danji, Daejeon 305-606, Korea

Received 31 July 1992; accepted 2 October 1992

Abstract—A new method for preparing a wettability gradient on polymer surfaces was developed. Low density polyethylene sheets were treated in air with corona from a knife-type electrode whose power gradually increases along the sample length. The polymer surfaces oxidized gradually with the increasing power and the wettability gradient was created on the surfaces as evidenced by the measurement of water contact angles, Fourier-transform infrared spectroscopy in the attenuated total reflectance mode, and electron spectroscopy for chemical analysis. The wettability gradient surfaces prepared were used to investigate the interactions of model protein and cells in terms of the surface hydrophilicity/hydrophobicity of polymeric materials.

Key words: Wettability gradients; low density polyethylene; protein adsorption; cell adhesion; albumin; Chinese hamster ovary cells.

INTRODUCTION

The behavior of the adsorption and desorption of blood proteins or the adhesion and proliferation of different types of mammalian cells on polymeric materials depend on the surface characteristics such as wettability (hydrophilicity/hydrophobicity or surface free energy), chemistry, charge, roughness, and rigidity. A large number of research groups have studied the effect of the surface wettability on the interactions of biological species with polymeric materials since the wettability is one of the main parameters affecting the interactions. Some of them have studied the interactions of different types of cultured cells [1–4] or blood proteins [5–7] with various polymers with different wettabilities to correlate the relationship between surface wettability and blood- or tissue-compatibility. One problem derived from the study using different kinds of polymers is that the surfaces are heterogeneous both chemically and physically (different surface chemistry, roughness, rigidity, crystallinity, etc.), which may result in considerable variation. Some others have studied the interactions of different types of cells or proteins with a range of copolymers with different wettabilities like methacrylate copolymers [8–12]; the surfaces have the same kind of chemistry but are still physically heterogeneous. Another methodological problem is that such studies are often tedious, laborious, and time-consuming because a large number of samples must be prepared to characterize the complete range of the desired surface property. It may also involve the strong possibility of methodological error because the experiment for each sample is carried out separately.

Many studies have recently been focused on the preparation of surfaces whose properties are changed gradually along the material length. Such gradient surfaces are of particular interest for basic studies of the interactions between biological species and surfaces since the effect of a selected property can be examined in a single experiment on one surface. The preparation of surfaces with a wettability gradient was first described by Elwing *et al.* [13]. A gradient of methyl groups was formed by diffusion of dimethyl dichloro silane through xylene on flat hydrophilic silicon dioxide surfaces. The surfaces so formed had methyl groups at one end (hydrophobic) and hydroxyl groups at the other (hydrophilic) with a gradient of increasing wettability in between. Elwing and co-workers [13–19] and Hlady and co-workers [20–23] have used the wettability gradient surfaces to investigate surface hydrophilicity-induced changes of adsorbed proteins. The wettability gradient surfaces prepared by the above method seem to be useful as basic research tools, but have a limitation; they can be applied to only hydrophilic inorganic substrates such as silicon, silica, quartz, or glass.

Recently, a method for preparing wettability gradients on various polymer surfaces was developed by our group [24–26] and by Pitt *et al.* [27, 28]. The wettability gradients were produced via radio-frequency plasma discharge treatment by exposing the polymer sheets continuously to plasma. The polymer surfaces oxidized gradually along the sample length with increasing plasma exposure time and thus the wettability gradients were created.

We also developed a new method for preparing a wettability gradient on polymer surfaces using corona discharge treatment [29]. The wettability gradient was produced by treating the polymer sheets with corona from a knife-type electrode whose power was changed gradually along the sample length. The polymer surfaces oxidized gradually with the increasing power and the wettability gradient was created on the sample surfaces. This method of preparing a wettability gradient on polymer surfaces is simpler and more practical than the plasma treatment method because the samples are discharged in air at atmospheric pressure in this method, whereas they are discharged under vacuum in the plasma treatment method.

In this study, the wettability gradient prepared by the corona discharge treatment was used as a tool to investigate protein or cell interactions continuously related to the surface wettability of polymeric materials.

MATERIALS AND METHODS

Substrate

An additive-free low density polyethylene (PE) sheet, 250–300 μm thick (Hanyang Chemical Co., Korea), was used as the polymeric substrate for the preparation of wettability gradient surfaces. The PE sheet was cut into 5×7 cm pieces, which were ultrasonically cleaned twice in ethanol for 30 min and then dried at room temperature. The pieces were stored in a vacuum oven until use. The cleanliness of the surfaces were verified by electron spectroscopy for chemical analysis (ESCA).

Preparation of wettability gradient surfaces

The PE sheet was treated with a radio-frequency (RF) corona discharge apparatus made by our laboratory for the preparation of gradient surfaces (Fig. 1). The knife-type electrode is connected to the RF generator and the power is increased gradually by a motorized drive. The sample bed translates at a controlled speed.

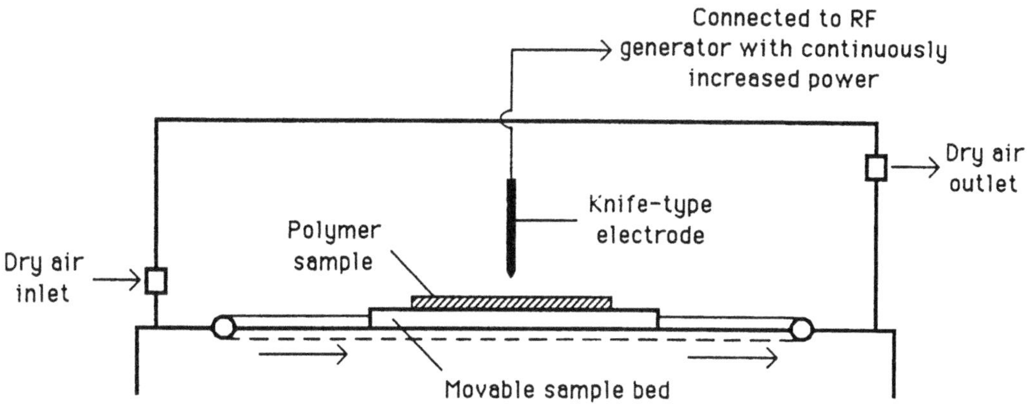
Connected to RF
generator with continuously
increased power
Dry air
outlet
Knife-type
electrode
Polymer
sample
Dry air
inlet
Movable sample bed

Figure 1. Schematic diagram showing the corona discharge apparatus for the preparation of wettability gradient surfaces.

The cleaned PE sheet was placed on the sample bed and dry air was purged into the apparatus at a flow rate of 20 l/min. The knife-type electrode was 1.0 mm away from the sample surface. At the same time that the sample bed was translated at a constant speed of 1.0 cm/s, the corona was discharged from the knife-type electrode onto the sample with increasing power (from 10 to 35 W at 100 kHz). The sample sheet (5 × 5 cm) was treated for 5 s. By this treatment, the sample surface was continuously exposed to the corona with increasing power. Thus, the surface oxidized gradually along the sample length and the wettability gradient was created. More details in the corona discharge apparatus and the characterization of the wettability gradient surfaces prepared were described in a previous paper [29].

Protein adsorption and desorption

Human albumin (crystallized powder, Sigma), the major constituent of plasma, as a model protein was used to investigate the effect of surface wettability on protein adsorption and desorption. The corona-treated PE surfaces were immersed in 1 mg/ml albumin solution (prepared with phosphate buffered saline (PBS) at pH 7.3–7.4) for 1 h and washed with PBS three times, followed by washing with purified water twice and vacuum drying. The albumin-adsorbed surfaces were analyzed by ESCA and Fourier-transform infrared spectroscopy in the attenuated total reflectance mode (FTIR-ATR). The ESCA (ESCA LAB MK II, V.G. Scientific Co., UK) was equipped with A1K_α radiation source at 1487 eV and 300 W power at the anode. The FTIR-ATR spectra were obtained using a Digilab FTS-80 (Bio-Rad) spectrophotometer equipped with KRS-5 reflection element (incidence angle, 45 deg). The corona-treated sheet was cut into five sections (1 × 5 cm each) in the direction perpendicular to the wettability gradient. For the ESCA analysis, survey scan and carbon 1*s* core level scan spectra of each section were obtained. For the FTIR-ATR analysis, each section was cut again in half (0.5 × 5 cm) and held against both faces of the reflection element. The absorption bands from the spectra after 2000 scans at 8.0 cm^{-1} resolution were compared between the sections of the gradient surface.

To study the desorption behavior of the protein preadsorbed on the wettability gradient surfaces, some corona-treated surfaces after the albumin adsorption and following washing were immersed in a nonionic polymeric surfactant solution, 1 mg/ml Tetronic 1504 (BASF-Wyandotte) solution which is an effective protein removal agent [30, 31], for 30 min and washed with PBS three times, followed by washing with purified water twice and vacuum drying. The amount of albumin remaining on the surfaces were also analyzed by ESCA.

Cell adhesion and growth

Chinese hamster ovary (CHO) cells (CHO-KI-BH_4, Oak Ridge National Laboratory, USA) were used to study the effect of surface wettability on cell adhesion and growth. The CHO cells are popularly used as a model system because they exist as reasonably stable single cells and are not unreasonably fastidious in terms of culture requirements [32]. They are grown in monolayer with fast generation time (about 12 h).

The CHO cells routinely cultured in tissue culture polystyrene (PS) flasks (Corning) at 37 °C under 5% CO_2 atmosphere were harvested after the treatment with trypsin (0.05% trypsin/0.02% EDTA (Gibco Laboratories)). The cells (4×10^4 cells/cm^2) were seeded to the corona-treated wettability gradient surfaces mounted in similar test chambers to those described by van Wachem *et al.* [33]. The culture medium used was Ham's F-12 nutrient mixture (Gibco Laboratories) containing 5% newborn calf serum, 100 units/ml penicillin, and 100 μg/ml streptomycin. After 2 h incubation at 37 °C under 5% CO_2 atmosphere, the surfaces were washed with Dulbecco's PBS (pH 7.2–7.3) free of Ca^{2+} and Mg^{2+} [34]. The surfaces were then cut into five sections along the wettability gradient and each section of the gradient surface was separately trypsinized. The cell adhesion to the sections of the gradient surface was determined by counting the number of detached cells with an electronic cell counter (Coulter Model ZM). The results were expressed as the percentage of cells seeded to the surfaces. The cell growth on the wettability gradient surfaces was carried out at intervals of 24 and 48 h in a same way to the cell adhesion experiment. The culture medium was changed once after 24 h. The cell growth to the surfaces was expressed in terms of the number of cells attached per cm^2. Further detailed procedures for the cell culture were described in previous papers [34, 35].

The cells adhered and grown on the wettability gradient surfaces were also examined by a scanning electron microscope (SEM, JSM-840A, Jeol Co., Japan). For the SEM observation, the cells attached on the gradient surfaces were washed with Dulbecco's PBS and fixed with 2.5% glutaraldehyde in PBS for 15 min at room temperature. After thorough washing with PBS, the cells were hydrated in ethanol graded series, 50, 60, 70, 80, 90 and 100%, for 10 min each and allowed to dry at room temperature. Then the cell-attached surfaces were cut into five sections along the wettability gradient, gold deposited in vacuum and examined by SEM with a tilt angle of 45 deg.

RESULTS AND DISCUSSION

Characterization of wettability gradient surfaces

The corona-treated PE surfaces did not show any visible changes, but the water contact angles of the surfaces gradually decreased (from 98 to 48 deg) along the

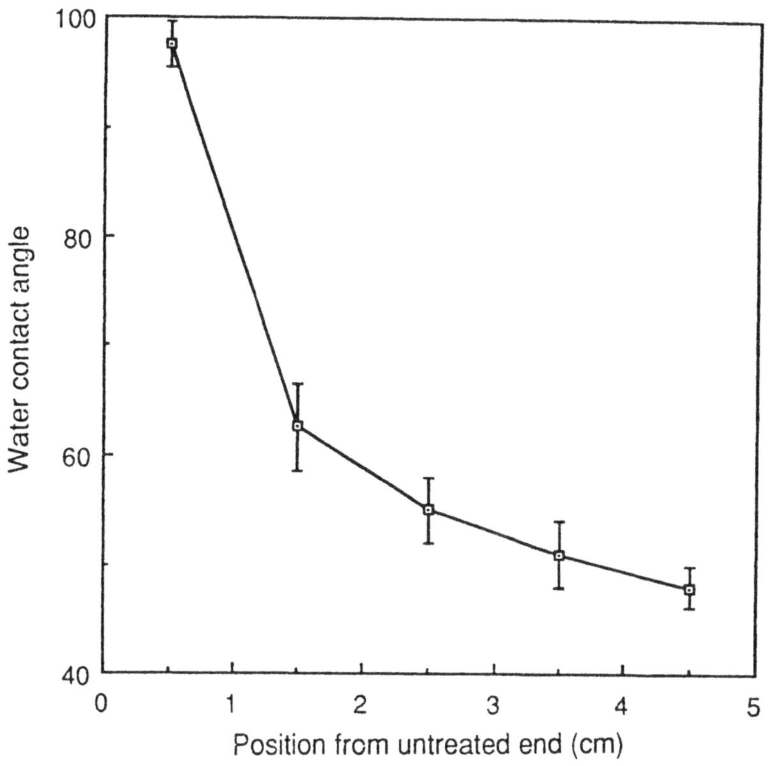
100
80
60
40
Water contact angle
0
1
2
3
4
5
Position from untreated end (cm)

Figure 2. Changes in water contact angle of corona-treated PE surface.

sample length with increasing corona power (Fig. 2). The contact angles were measured at room temperature at the static condition using a contact angle goniometer (Model 100-0, Rame-Hart, Inc.) by dropping 3 μl of purified water onto each section of the corona-treated surfaces. The decrease in the contact angles (and thus the increase in wettability) along the sample length was due to the oxygen-based polar functionalities (—C—O—, —C=O, and —COO—) incorporated on the surface by the corona treatment as evidenced by ESCA and FTIR-ATR analyses [29]. The oxygen-based functionalities gradually increased along the sample length and the relative distribution of them was almost invariant with respect to the position in the gradient (approximate ratio of —C—O— : —C=O : —COO— was 2:1:1 as determined by ESCA). The surfaces after the corona treatment were mildly roughened as the corona power increased, but the roughening was negligible in our experiment condition (the corona power used, up to 35 W at 100 kHz) as the surfaces were observed by SEM with a magnification of 10 000 [29].

Protein adsorption and desorption on wettability gradient surfaces

Human albumin as a model protein was adsorbed onto the PE gradient surfaces from PBS solution. The general structure and binding properties of the albumin are well-known [36–39]. The shape of this protein is a prolate ellipsoid with a size of about 140 × 40 × 40 Å, within which 17 disulfide bridges act to stabilize the conformation. It contains a comparatively large number of polar and charged residues and thus, it is highly soluble in water and negatively charged at pH 7.4 (the isoelectric point is 4.7–5.5 depending on the amount of bound fatty acid). It also has big hydrophobic patches on its surface.

The albumin-adsorbed gradient surfaces were analyzed by ESCA. The nitrogen signal from the surface was used as an indicator of the protein adsorption since we could see that little nitrogen is incorporated onto the surface by corona treatment in air [29]. The nitrogen content of human albumin is about 14–15 at % as determined by both theoretical calculation [40] and ESCA analysis [41, 42]. It is mainly derived from peptide bonds in the structure of the albumin. The albumin is adsorbed rapidly

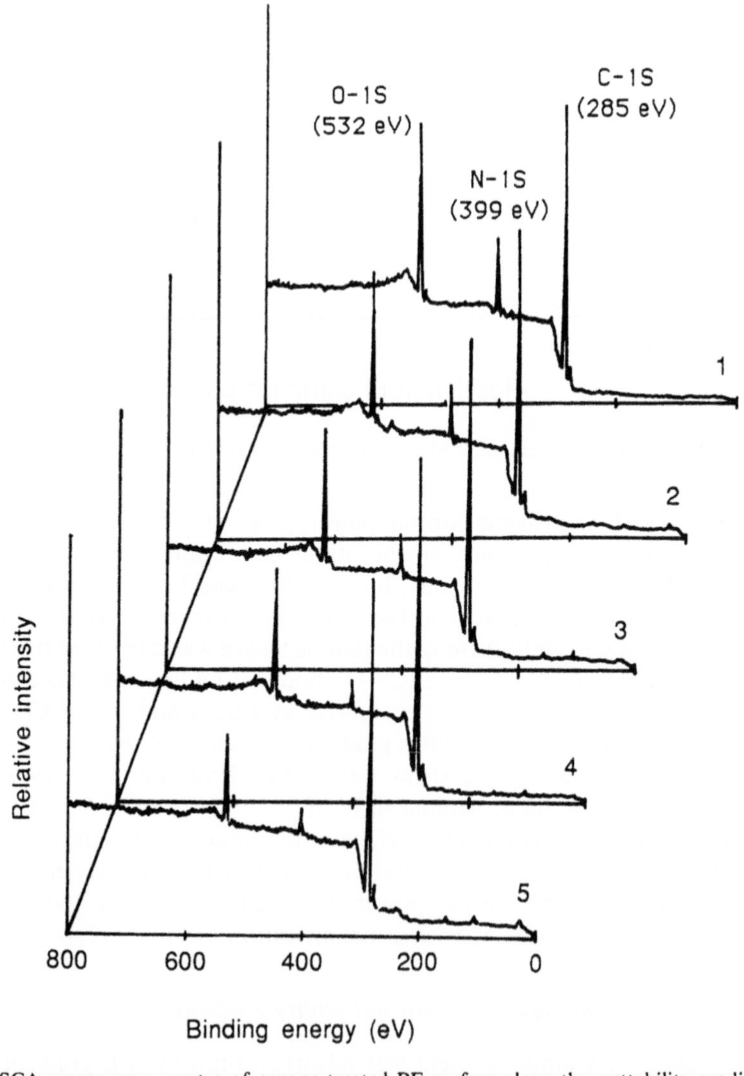
O-1S
(532 eV)
C-1S
(285 eV)
N-1S
(399 eV)
1
2
3
4
5
Relative intensity
800
600
400
200
0
Binding energy (eV)

Figure 3. ESCA survey scan spectra of corona-treated PE surface along the wettability gradient after albumin adsorption (1 h adsorption in 1 mg/ml albumin solution). Numbers labeled on the spectra (1 to 5) represent the positions from the untreated hydrophobic end of the gradient surface, 0.5, 1.5, 2.5, 3.5 and 4.5 cm, respectively.

on polymer surfaces and approaches a plateau within 30 min [42]. In this study the albumin (1 mg/ml) was adsorbed onto the PE gradient surfaces for 1 h. From Fig. 3, we can see that the nitrogen peaks (binding energy of ~399 eV) from the survey scan spectra of ESCA decreased gradually along the wettability gradient. This means that the albumin adsorption increased gradually with the increasing hydrophobicity of the surface. This is probably due to the increased hydrophobic interactions of the protein molecules with the hydrophobic sections of the gradient surface. The nitrogen content determined by ESCA was 9.3 ± 0.7% on the section 1 (contact angle, 98 deg) of the gradient surface and 5.0 ± 0.5% on the section 5 (contact angle, 48 deg) for more than three samples. The nitrogen content of 9.3% on the section 1 is nearly the value of the monolayer coverage of albumin [42]. Even though we consider an albumin monolayer randomly packed with side-on and

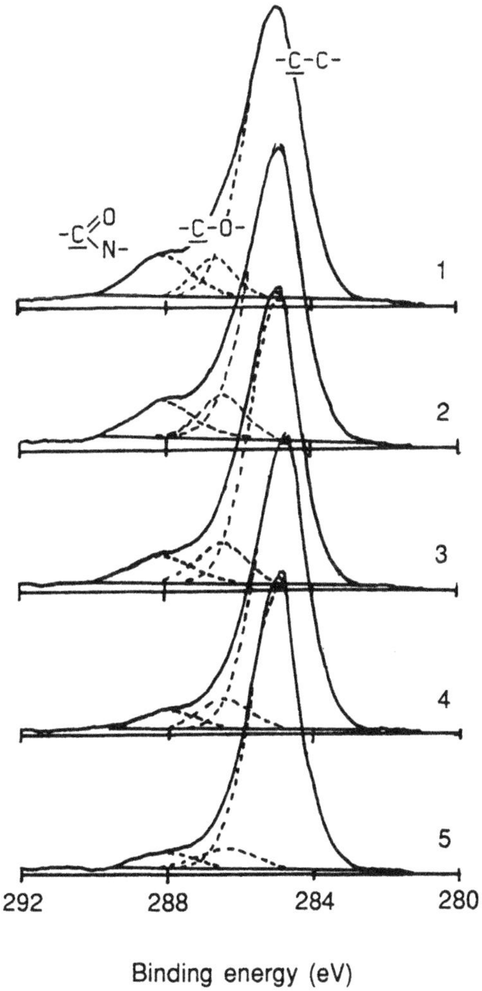
-C-C-
-C-O-
N-
1
2
3
4
5
292
288
284
280
Binding energy (eV)

Figure 4. ESCA C1*s* core level spectra of corona-treated PE surface along the wettability gradient after albumin adsorption (1 h adsorption in 1 mg/ml albumin solution). Numbers labeled on the spectra (1 to 5) represent the positions from the untreated hydrophobic end of the gradient surface, 0.5, 1.5, 2.5, 3.5, and 4.5 cm, respectively.

end-on ellipsoids, ESCA is still a good tool to determine the relative amount of the protein adsorbed on the surface since the photoelectron escape depths in the ESCA is more than 100 Å. Figure 4 shows the C1*s* core level spectra of the gradient surface. It has a similar trend to the survey scan spectra; the carbon peaks derived from peptide bonds of the protein (O=C̲—NH, binding energy of ~288.1 eV) decreased gradually along the the wettability gradient. As the albumin-adsorbed gradient surfaces were analyzed by FTIR-ATR, we could see that the amide II (1500–1600 cm^{-1}) and amide I (1600–1680 cm^{-1}) absorption bands from the protein adsorbed on the surface also decrease gradually along the wettability gradient (Fig. 5).

For the desorption study, the albumin-adsorbed gradient surfaces were exposed to a non-ionic polymeric surfactant, Tetronic 1504, which contains star-like 4 polyethylene glycol (ethylene glycol unit, 26)-polypropylene glycol (propylene glycol unit, 29) blocks in its structure [30]. The exposure of the albumin-adsorbed gradient surface to 1 mg/ml Tetronic 1504 solution for 30 min resulted in partial displacement of the protein (Fig. 6). This displacement was much greater on the hydrophobic sections of the gradient surface than the hydrophilic ones. The results of increasing protein adsorption and desorption toward the hydrophobic sections of wettability gradient surfaces in this study agree with those of others'

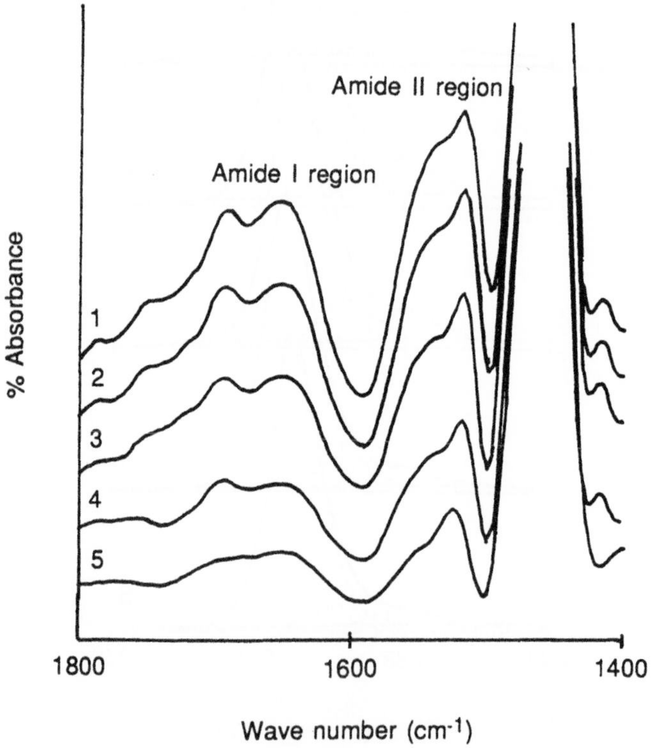
Amide II region
Amide I region
% Absorbance
1
2
3
4
5
1800
1600
1400
Wave number (cm-1)

Figure 5. FTIR-ATR spectra of corona-treated PE surface along the wettability gradient after albumin adsorption (1 h adsorption in 1 mg/ml albumin solution). Numbers labeled on the spectra (1 to 5) represent the positions from the untreated hydrophobic end of the gradient surface, 0.5, 1.5, 2.5, 3.5 and 4.5 cm, respectively.

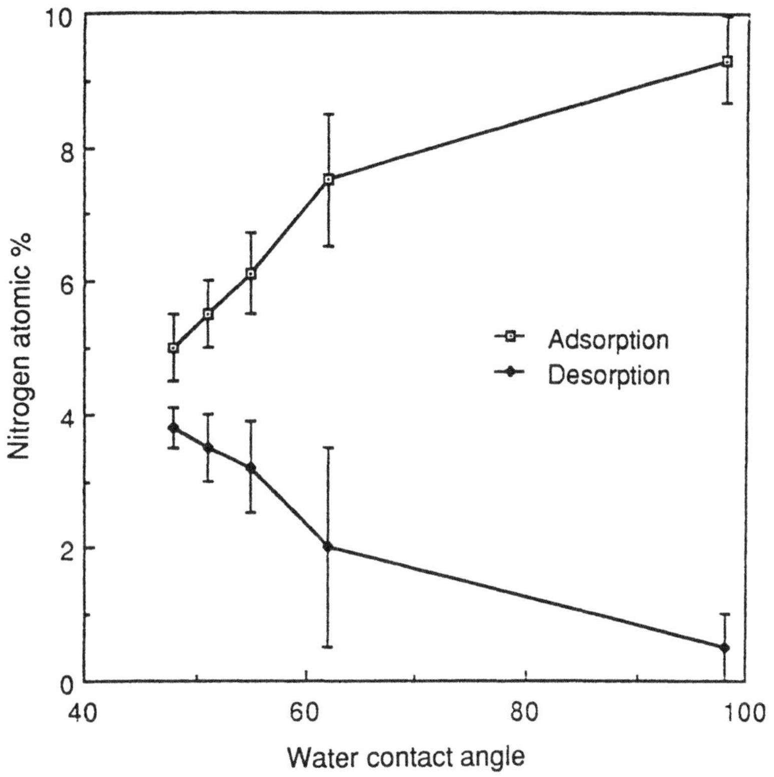
10
8
6
4
2
0
40
60
80
100
Nitrogen atomic %
Water contact angle
Adsorption
Desorption

Figure 6. Adsorption and desorption behavior of albumin on corona-treated PE surface. Nitrogen at % represents the amount of the protein adsorbed (for adsorption) and remaining (for desorption) on the surface (adsorption; 1 h adsorption in 1 mg/ml albumin solution: desorption; 30 min desorption in 1 mg/ml Tetronic 1504 solution after 1 h adsorption in 1 mg/ml albumin solution). Sample numbers, $n = 3$.

works [13, 14, 17, 21, 28], even though the functional groups introduced on the substrates are different.

Cell adhesion and growth on wettability gradient surfaces

The corona-treated PE surfaces were also used to study the effect of polymer surface wettability on cell adhesion and growth. CHO cells as a model system were cultured for 2 h on the PE gradient surfaces and the number of cells adhered onto each section of the gradient surfaces was determined by a electronic cell counter. As seen in Fig. 7, the cells were adhered more onto the sections with moderate hydrophilicity of the wettability gradient surface. The maximum adhesion of the cells appeared at around a water contact angle of 50–55 deg. SEM observation also verified that the cells are adhered more onto the sections with moderate hydrophilicity (Fig. 8). The section 3 (contact angle, 55 deg) shows more adhesion of the CHO cells than the section 2 (contact angle, 62 deg) or section 5 (contact angle, 48 deg). The SEM observation also revealed that the cells are spread better on the hydrophilic sections than the hydrophobic ones. On the section 1 (contact angle, 98 deg), the cells were

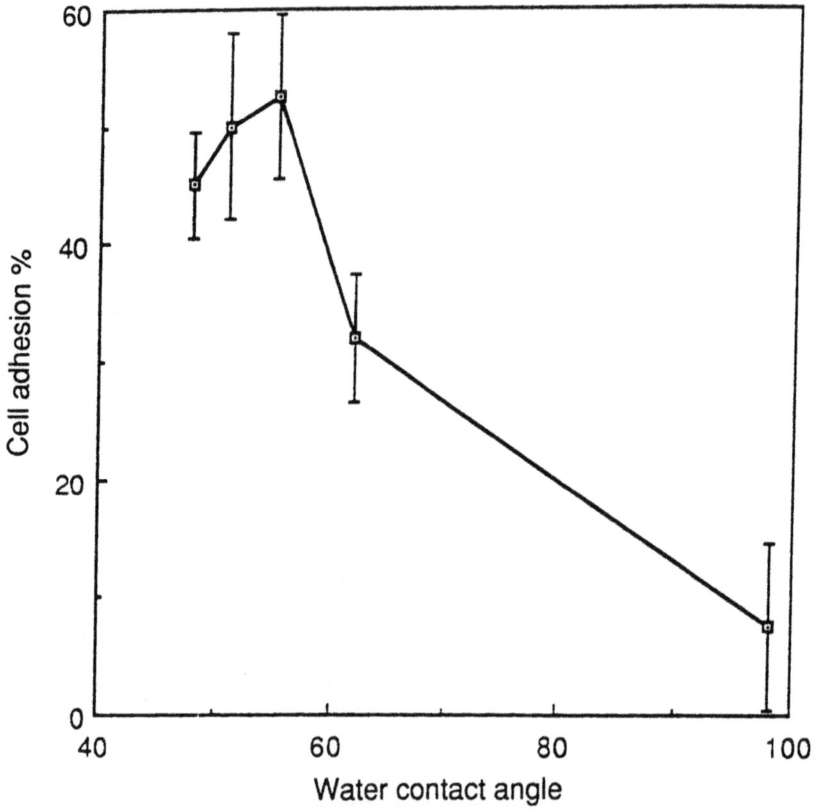
60
40
20
0
Cell adhesion %
40
60
80
100
Water contact angle

Figure 7. CHO cell adhesion on corona-treated PE surface (number of seeded cells, $4 \times 10^4/cm^2$; culture time, 2 h). $n = 3$.

not spread at all and have round shapes after 2 h culture. Cells attached on surfaces are spread only when they are compatible on the surfaces. It seems that surface wettability plays an important role for cell adhesion and spreading.

Figure 9 shows the SEM pictures of the CHO cells grown on the wettability gradient surface after 24 h culture. The cells were still not spread well on the section 1, while they were completely spread and flatten on the section 3 after 24 h culture. Figure 10 shows the number of CHO cells grown on the wettability gradient surface after 24 and 48 h culture. After 48 h culture the cells were grown more than twice compared to those after 24 h culture. The cells were grown better on the hydrophilic sections than the hydrophobic ones. The maximum growth of the cells were appeared on the sections with moderate hydrophilicity at around a water contact angle of 50–55 deg. This trend was similar to that of cell adhesion after 2 h culture as seen in Fig. 7. The fact that cells are more adhered, spread, and grown on the moderately wettable surfaces was also observed by van Wachem *et al.* [11, 33] and Kang [43] as they cultured endothelial cells, platelets, or fibroblast onto various polymer substrates with different surface wettability. They explained that preferential adsorption of some serum proteins like fibronectin and vitronection from culture medium onto the moderately wettable surfaces may be a reason for better cell adhesion, spreading, and growth. We do not have clear evidence for

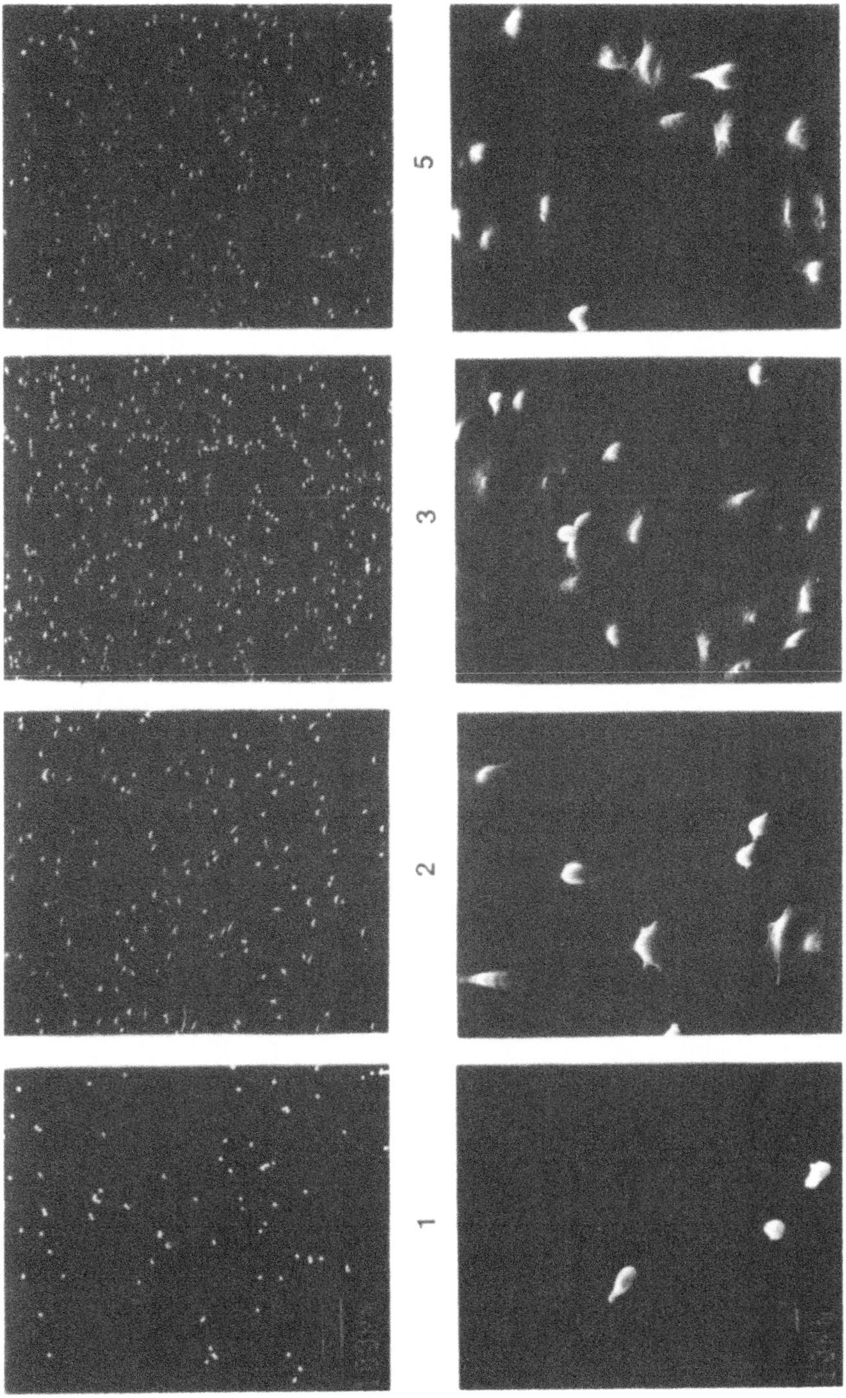
1
2
3
5

Figure 8. SEM pictures of CHO cells adhered on corona-treated PE surface along the wettability gradient after 2 h culture (Original magnification; top pictures, × 100 and bottom pictures, × 500). Numbers labeled (1 to 5) represent the positions from the untreated hydrophobic end of the gradient surface, 0.5, 1.5, 2.5 and 4.5 cm, respectively.

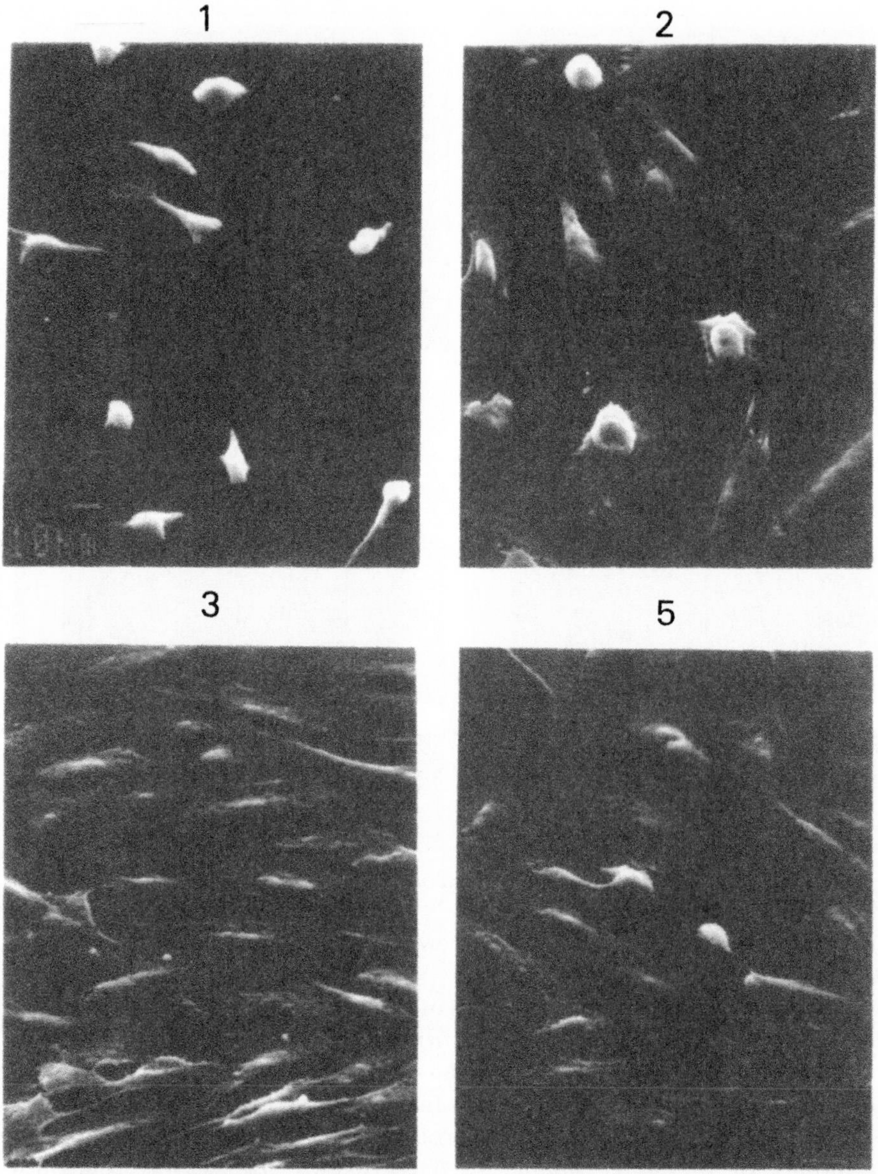
1
2
3
5

Figure 9. SEM pictures of CHO cells grown on corona-treated PE surface along the wettability gradient after 24 h culture (Original magnification, × 500). Numbers labeled (1 to 5) represent the positions from the untreated hydrophobic end of the gradient surface, 0.5, 1.5, 2.5 and 4.5 cm, respectively.

this yet. The work to study the adsorption behavior of the cell attachment proteins on the wettability gradient surfaces is in progress.

CONCLUSIONS

The interactions of different types of proteins or cells with various solid substrates depend mainly on the surface characteristics such as wettability (hydrophilicity/

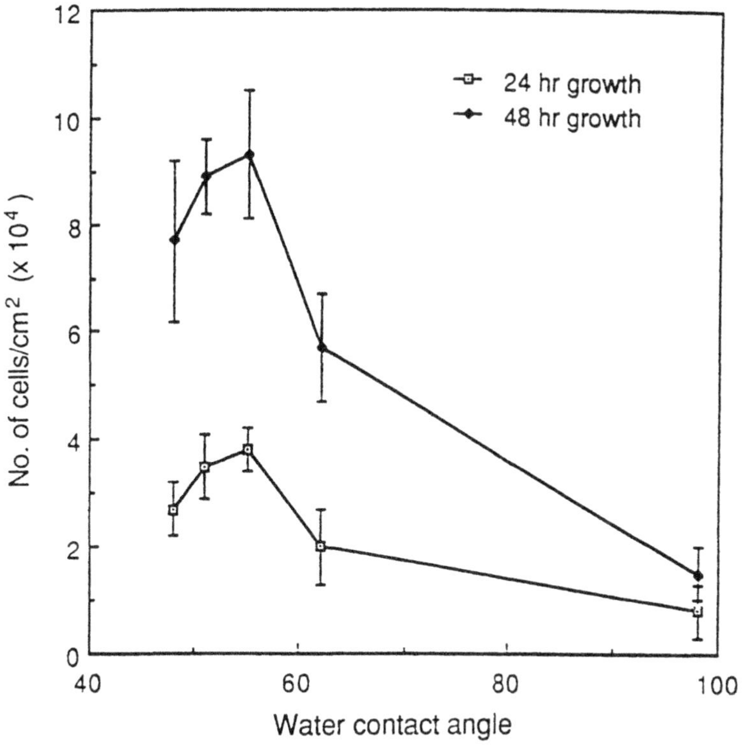

12
10
8
6
4
2
0
40
60
80
100
No. of cells/cm² (x 10⁴)
Water contact angle
24 hr growth
48 hr growth

Figure 10. CHO cell growth on corona-treated PE surface (number of seeded cells, $4 \times 10^4/cm^2$). $n = 3$.

hydrophobicity or surface free energy), chemistry, charge, roughness, and rigidity. We focused on the surface wettability in this study. We prepared wettability gradient surfaces by treating polymer sheets with corona whose power is changed gradually along the sample length. The major advantage to use wettability gradient surfaces as tools to study the interactions of biological species with solid surfaces is that several wettabilities are effectively compared at the same time and on the same surface. We studied the behaviors of the adsorption and desorption of albumin and the adhesion and proliferation of CHO cells on the wettability gradient surfaces prepared. The results clearly demonstrated that surface wettability is an important factor for protein adsorption/desorption and cell adhesion/proliferation.

The corona treatment with increasing power is a simpler and more practical method to produce wettability gradients than the diffusion or plasma treatment methods because polymer sheets are directly treated in air at atmospheric pressure within a few seconds; the diffusion method can be applied to only hydrophilic inorganic substrates with a short range (1–2 cm) of gradient and the plasma treatment method is carried out under vacuum. The wettability gradients prepared on various polymer surfaces by this corona treatment method can be used to systematically investigate the interactions of the different types of biological species in terms of the surface wettability of polymeric materials. The charge gradient surfaces or functional group gradient surfaces can be produced using this corona treatment and the following graft reaction, which may give us useful information on the interaction behaviors of biological species continuously related to the surface chemistry, density, or positive and negative charge of polymeric materials.

REFERENCES

1. D. H. Kaeble and J. Moacanin, *Polymer* **18,** 475 (1977).
2. P. van der Valk, A. W. J. van Pelt, H. J. Busscher, H. P. de Jong, C. R. H. Wildevuur and J. Arends, *J. Biomed. Mater. Res.* **17,** 807 (1983).
3. R. E. Baier, A. E. Meyer, J. R. Natiella, R. B. Nateiella and J. M. Carter, *J. Biomed. Mater. Res.* **18,** 337 (1984).
4. J. M. Schakenraad, H. J. Busscher, C. R. H. Wildevuur and J. Arends, *J. Biomed. Mater. Res.* **20,** 773 (1986).
5. J. Lyklema, *Colloids Surf.* **10,** 33 (1984).
6. H. M. W. Uyen, J. M. Schakenraad, J. Sjollema, J. Nordmans, W. L. Jongebloed, I. Stokros and H. J. Busscher, *J. Biomed. Mater. Res.* **24,** 1599 (1984).
7. D. R. Lu and K. Park, *J. Colloid Interface Sci.* **144,** 271 (1991).
8. D. L. Coleman, D. E. Gregonis and J. D. Andrade, *J. Biomed. Mater. Res.* **16,** 381 (1982).
9. T. A. Horbett, M. S. Schway and B. D. Ratner, *J. Colloid Interface Sci.* **104,** 28 (1985).
10. M. J. Lydon, T. W. Minett and B. J. Tighe, *Biomaterials* **6,** 396 (1985).
11. P. B. van Wachem, A. H. Hogt, T. Beugeling, J. Feijen, A. Bantjes, J. P. Detmers and W. G. van Aken, *Biomaterials* **8,** 323 (1987).
12. G. Harkes, J. Feijen and J. Dankert, *Biomaterials* **12,** 853 (1991).
13. H. Elwing, S. Welin, A. Askendal, U. Nilsson and I. Lundstrom, *J. Colloid Interface Sci.* **119,** 203 (1987).
14. H. Elwing, A. Askendal and I. Lundstrom, *Prog. Colloid Polym. Sci.* **74,** 103 (1987).
15. H. Elwing, A. Askendal and I. Lundstrom, *J. Biomed. Mater. Res.* **21,** 1023 (1987).
16. H. Elwing, B. Nilsson, K. E. Svensson, A. Askendal, U. R. Nilsson and I. Lundstrom, *J. Colloid Interface Sci.* **125,** 139 (1988).
17. H. Elwing, A. Askendal and I. Lundstrom, *J. Colloid Interface Sci.* **128,** 296 (1989).
18. S. W. Klintstrom, M. Wikstrom, A. Askendal, H. Elwing, I. Lundstrom, J. O. Karlsson and S. Renvert, *Colloids Surf.* **44,** 51 (1990).
19. H. Elwing and C. G. Golander, *Adv. Colloid Interface Sci.* **32,** 317 (1990).
20. V. Hlady, C. Golander and J. D. Andrade, *Colloids Surf.* **33,** 185 (1988).
21. C. G. Golander, Y. S. Lin, V. Hlady and J. D. Andrade, *Colloids Surf.* **49,** 289 (1990).
22. V. Hlady, *Appl. Spectrosc.* **45,** 246 (1991).
23. Y. S. Lin, V. Hlady and J. Janatova, *Biomaterials*, in press.
24. H. B. Lee and J. D. Andrade, *Trans. 3rd World Biomaterials Congr.*, p. 43 (1988).
25. H. B. Lee, in: *Frontiers of Macromolecular Science*, p. 579, T. Saegusa, T. Higashimura and A. Abe (Eds). Blackwell Scientific Publications, Oxford (1989).
26. J. H. Lee, J. W. Park and H. B. Lee, *Polymer (Korea)* **14,** 646 (1990).
27. W. G. Pitt, *J. Colloid Interface Sci.* **133,** 223 (1989).
28. C. G. Golander and W. G. Pitt, *Biomaterials* **11,** 32 (1990).
29. J. H. Lee, H. G. Kim, G. S. Khang, H. B. Lee and M. S. Jhon, *J. Colloid Interface Sci.* **151,** 563 (1992).
30. J. H. Lee, J. Kopecek and J. D. Andrade, *J. Biomed. Mater. Res.* **23,** 351 (1989).
31. J. H. Lee, P. Kopeckova, J. Kopecek and J. D. Andrade, *Biomaterials* **11,** 455 (1990).
32. W. B. Jakoby and I. H. Pastan (Eds). *Methods in Enzymology: Cell Culture*, Academic Press, New York (1979).
33. P. V. van Wachem, T. Beugeling, J. Feijen, A. Bantjes, J. P. Detmers and W. G. van Aken, *Biomaterials* **6,** 403 (1985).
34. J. H. Lee, J. W. Park and H. B. Lee, *Biomaterials* **12,** 443 (1991).
35. J. H. Lee, G. S. Khang, K. H. Park, H. B. Lee and J. D. Andrade, *J. Korea Soc. Med. Biol. Eng.* **10,** 195 (1989).
36. J. D. Andrade and V. Hlady, *Ann. N.Y. Acad. Sci.* **516,** 158 (1987).
37. J. R. Brown and P. Shockley, in: *Lipid–Protein Interaction*, Vol. 1, P. C. Jost and O. H. Grif (Eds). Wiley-Interscience, New York (1982).

38. V. M. Rosener, M. Oratz and M. A. Rothschild (Eds), in: *Albumin Structure, Function, and Use.* Pergamon Press, New York (1977).
39. T. Peters, *Adv. Clinical Chem.* **13,** 37 (1970).
40. H. Neurath, in: *The Proteins: Composition, Structure and Function*, 2nd Ed., Vol. 3, Ch. 14, F. W. Putnam (Ed.), Academic Press, New York (1965).
41. R. W. Paynter and B. D. Ratner, in: *Surface and Interfacial Aspects of Biomedical Polymers:* Vol. 2. Protein Adsorption, Ch. 5, J. D. Andrade (Ed.). Plenum Press, New York (1985).
42. J. H. Lee, Ph.D. Thesis, Ch. 6, University of Utah (1988).
43. I. K. Kang, Ph.D. Thesis, Chs. 6 and 7, Kyoto University, Japan (1987).

Activity of horseradish peroxide adsorbed on radio frequency glow discharge-treated polymers

JING PING CHEN, DAVID KIAEI and ALLAN S. HOFFMAN*
Center for Bioengineering, FL-20, University of Washington, Seattle, WA 98195, USA

Received 10 August 1992; accepted 4 February 1993

Abstract—Horseradish peroxidase (HRP) has been used as a model enzyme in this study of its physical adsorption and residual enzyme activity on radio frequency glow discharge (RFGD)-treated polymers. The specific enzymatic activity of HRP adsorbed on different surfaces was assumed to be an indication of the extent of its conformational alterations on the surfaces. The surfaces studied were poly (ethylene terephthalate) (PET), polytetrafluoroethylene (PTFE), and tetrachloroethylene and tetrafluoroethylene glow discharge-treated PET, abbreviated as TCE/PET and TFE/PET. All surfaces were characterized by electron spectroscopy for chemical analysis (ESCA) and liquid contact angles in air. HRP adsorbs more strongly onto the two discharge-treated surfaces than onto the untreated polymers, as evidenced by the lower amount of HRP eluted by sodium dodecyl sulfate (SDS) from the treated polymers. For example, seventy percent of the HRP adsorbed on TCE/PET or TFE/PET remains on the surface after overnight elution with a 1% solution of SDS. In contrast, untreated PET and PTFE each retains only *c.* 20% of the absorbed enzyme. The enzymatic activity of HRP adsorbed on the different surfaces was studied using hydrogen peroxide (H_2O_2) as the substrate. HRP adsorbed on the higher energy surfaces, PET and TCE/PET, retains significantly more activity than the HRP adsorbed on the lower energy surfaces, PTFE and TFE/PET which appear to destroy rapidly almost all of the activity of HRP after it adsorbs. HRP adsorbed on TCE/PET is relatively more stable over time than HRP adsorbed on PET or free HRP in solution. (For example, only 45% of the specific enzymatic activity of HRP adsorbed on TCE/PET was lost after 3 h of storage in phosphate buffer at 37°C, while 70% of that adsorbed on PET was lost.) In summary, when HRP is adsorbed on TCE/PET, it is very tightly bound, and yet it maintains a significant fraction of its initial specific activity and also retains this activity for 3 h in phosphate buffer at 37°C. Thus, tenacious physical adsorption of proteins such as enzymes on TCE glow discharge-treated surfaces may have potential as a new method of immobilization of such molecules, for uses in biosensors, diagnostics, bioseparations, cell culture and bioreactors.

Key words: Radio frequency glow discharge; horseradish peroxidase; enzyme immobilization; surface modification; adsorbed proteins.

INTRODUCTION

Protein adsorption at the solid–liquid interface is the subject of intense investigations in numerous areas, such as tissue culture, biosensors and biomaterial design [1–3]. The adsorption of proteins on hydrophobic polymeric surfaces is considered by many scientists to be an irreversible process since simple washes, such as with buffer, remove only a small fraction of the adsorbed protein [4, 5]. The use of a detergent, such as sodium dodecyl sulfate (SDS) is usually necessary in order to

* To whom correspondence should be addressed.

remove a large fraction of the proteins adsorbed on such polymer surfaces [6]. In our recent studies, we have shown that the adsorption of albumin and fibrinogen onto tetrafluoroethylene (TFE) and tetrachloroethylene (TCE) radio frequency glow discharge (RFGD)-treated polymers is so tenacious that the elution of protein, even with SDS, is surprisingly ineffective [7–9]. For example, over 95% of the albumin adsorbed on TFE/PET remains on the surface after overnight equilibration with a 1% solution of SDS [8, 10] in contrast to untreated PTFE or PET, which retain only 50 and 26%, respectively.

In this study we utilize the tight binding which proteins exhibit following adsorption on the TFE and TCE-treated polymers as a special technique for their immobilization on solid surfaces. The advantages of this approach over most other immobilization methods include simplicity and versatility. However, a possible drawback of this technique may be loss of activity of the protein upon such strong adsorption on surfaces. The concept of conformational changes of proteins upon adsorption on polymeric surfaces is well established. Techniques such as circular dichroism [11], Fourier transform infrared-attentuated total reflectance (FTIR–ATR) [12, 13] and antibody recognition [14, 15] have been utilized to study this process. The approach selected in this study is to measure the change in activity of an adsorbed enzyme as quantitative indicator of its conformational change on RFGD-treated surfaces, since enzymes have been known to lose activity following conformational changes [16–18].

We chose the enzyme horseradish peroxidase (HRP) as our model for several reasons. HRP exhibits high specificity for its substrate, and its enzymatic activity is observed only with hydrogen peroxide (H_2O_2), methyl peroxide and ethyl peroxide [19]. HRP also has high sensitivity and its enzymatic activity has been detected at concentrations as low as 5×10^{-4} M [20]. This enzyme is also very stable and a 1 mg/ml solution in buffer can be stored for more than one year at 4°C without a measurable loss in activity [19]. Furthermore, the availability of chromogenic substrates for HRP has made it widely useful as an enzyme marker for ELISA (enzyme linked immunosorbent assay) [21–23].

Our first aim was to determine whether HRP is also strongly retained on TFE and TCE treated surfaces, similar to albumin and fibrinogen. Another goal was to investigate the extent of conformational changes of adsorbed HRP by measuring its residual activity on the surfaces. We hypothesize that the tight binding observed previously for various proteins adsorbed on RFGD-treated surfaces is a direct consequence of protein unfolding on these surfaces. Measurement of the activity of HRP adsorbed on various surfaces should provide a direct connection between the degree of conformational changes of HRP adsorbed on a surface and the strength of binding of HRP to the same surfaces.

EXPERIMENTAL

Materials

HRP (EC 1.11.1.7, Type VI, Cat. No. P-8375), *O*-phenylenediamine (OPD) (Cat. No. P-1526) and hydrogen peroxide (30%) (H_2O_2) (Cat. No. H-1009) were purchased from Sigma Chemical Company (St. Louis, MO, USA). Sodium dodecyl sulfate (SDS, ultra pure) was obtained from ICN Biomedicals Inc. (Cleveland, OH, USA). $Na^{125}I$ was purchased from Amersham Corp. (Arlington Heights, IL, USA).

All salts used in preparation of buffers were reagent grade. Sulfuric acid (Cat. No. 9681-01) was obtained from J. T. Baker (Phillipsburg, NJ, USA).

PET coverslips were obtained from Nunc Inc. (Naperville, IL, USA). Polytetrafluoroethylene (PTFE) films were purchased from Berghof/America, Inc. (Raymond, NH, USA). Neither of the polymers had any type of surface modification treatments by the manufacturers. Polymers were cut to a convenient sample size (11 × 16 mm) and cleaned via ultrasonic treatments for 15 min in dichloromethane, followed by acetone and distilled water.

Clear polystyrene AutoAnalyzer sample cups (4 ml capacity) (VWR scientific) were used for film storage, enzyme adsorption and enzyme-substrate reaction without further cleaning procedure.

Argon (Ar) was obtained from Air Products and Chemicals Inc. (Allentown, PA, USA). TFE was received from PCR Inc. (Gainesville, FL, USA). TCE was purchased from Aldrich Chemical Co. (Milwaukee, WI, USA).

Radio frequency glow discharge treatment

RFGD treatments of the cleaned PET coverslips was carried out in a 135-cm-long 16-mm i.d. Pyrex reactor. PET coverslips, 11 × 16 mm, were placed horizontally in the reactor in order to get uniform treatment of both sides of the films. The RFGD reactor system has been described in detail elsewhere [24].

PET films were initially RFGD-treated with argon. The argon glow discharge, generated at 2.5 W, 3 cm^3 STP/min (sccm) and 0.1 Torr, was moved along the reactor at a speed of 3.3 mm/s. After the argon treatment, the reactor was evacuated to a pressure of 0.01 Torr. TFE gas was introduced and the TFE treatment was carried out under the same conditions used for the argon treatment (2.5 W, 3 sccm, 0.1 Torr, 3.3 mm/s). After that, the reactor was evacuated and then filled to atmospheric pressure with argon and the treated films were removed. Tetrachloroethylene, a liquid, was degassed by freeze–thawing once under vacuum prior to the treatment. TCE vapor was introduced into the evacuated reactor via a micrometering valve. The RFGD conditions for TCE were similar to those used for argon and TFE (2.5 W, 0.1 Torr, 3.3 mm/s). The treated films were stored under normal atmosphere at room temperature in capped polystyrene AutoAnalyzer sample cups and can be used months after preparation. Storage under these conditions for several weeks did not influence the measured adsorption, retention, or activity of the adsorbed HRP.

Surface characterization

Equilibrium contact angles of several test liquids were measured on surfaces using a Ramé-Hart goniometer (Ramé-Hart Inc., Mountain Lakes, NJ). The dispersion force contribution (γ_s^d) and the polar force contribution (γ_s^p) to the total surface free energy (γ_s) of each surface was estimated from the contact angle data according to techniques described previously [10, 25]. All surfaces were also characterized by electron spectroscopy for chemical analysis (ESCA) using a Surface Science Laboratories SSX-100 ESCA Spectrometer at the National ESCA and Surface Analysis Center for Biomedical Problems (NESAC/BIO) at the University of Washington. Both survey scans and high resolution scans of the C1*s* region were measured in order to determine the elemental composition of the surfaces.

Iodination of horseradish peroxidase

HRP was labeled by the iodine monochloride method of McFarlane [26] as modified by Helmkamp [27] and Horbett [28] using an equimolar ratio of iodine monochloride (ICI) to HRP. Briefly, 1 mCi of $Na^{125}I$ was added to 0.5 ml of borate buffer (0.4 M borate, 0.32 M sodium chloride, pH 7.75). This solution was mixed with 0.5 ml of cold ICI in 2 M sodium chloride. The ^{125}ICI solution was then added to 0.5 ml of the HRP solution (0.5 mg HRP in 0.05 M potassium phosphate buffer, pH 7.0) and mixed by gentle repipetting. The mixture was incubated for 30 min at 0°C. Unincorporated ^{125}I was separated from the labeled HRP by gel chromatography (Bio-Gel, P-4, Bio-Rad, Richmond, CA) at room temperature using potassium phosphate buffer (0.05 M, pH 7.0) as the mobile phase. The amount of free ^{125}I remaining in the labeled HRP solution following gel chromatography was determined by instant thin layer chromatography (ITLC) (Gelman Instrument Company, Ann Arbor, MI) to be approximately 2% of the total ^{125}I present in the protein solution. The labeled HRP was collected, stored at −70°C and used within 2 weeks of preparation.

HRP adsorption and retention

HRP was adsorbed on polymer surfaces from potassium phosphate buffer (0.05 M, pH 7.0) at 37°C (static adsorption). Adsorption was initiated by adding 0.5 ml of HRP at six times the desired final concentration to samples immersed in 2.5 ml of potassium phosphate buffer. Mixing was achieved by gentle repipetting in and out of the polystyrene AutoAnalyser cups. After a period of 2 h without further stirring, adsorption was terminated by displacing the protein solution with flowing potassium phosphate buffer by simultaneous filling and removing buffer (dilution-displacement technique) thereby avoiding exposure of the surfaces to air. Immediately after rinsing, samples were placed in 2.5 ml surfactant solutions (0.01 M Tris, 0.003 M phosphoric acid, 1% w/v SDS, pH 7.0) and the radioactivity of samples in the SDS solution was measured with a γ counter (T.M. Analytic, Elk Grove Village, IL). Following overnight incubation in SDS, samples were removed from the surfactant solutions, dip-rinsed nine times in potassium potassium phosphate buffer and the radioactivity retained on each sample was measured. The amount of HRP adsorbed initially and that remaining after elution, i.e. adsorbed and retained HRP, were calculated (in ng/cm^2) by correcting the radioactivity of each sample for background radioactivity and dividing by the specific activity of the HRP solution and the planar surface area of the sample. Percent retention was calculated by dividing the amount of HRP retained by the amount adsorbed multiplied by 100%. All error bars represent standard deviations of three replicate samples from the same experiment. The control surfaces were PET for the TCE/PET and PTFE for the TFE/PET.

Activity of the adsorbed HRP

HRP was adsorbed on all surfaces at 37°C using the adsorption protocol just described. Following adsorption, surfaces were rinsed with potassium phosphate buffer (0.05 M, pH 7.0) using the dilution-displacement technique and transferred to a new vial containing 2 ml of the substrate solution, OPD plus H_2O_2 in citrate-phosphate buffer (48.5 mM citric acid, 103 mM sodium phosphate dibasic, pH 5.0).

It was found that there was no further desorption of the HRP after this wash step. The enzyme-substrate reaction was allowed to proceed at room temperature (*c.* 21°C) with continuous shaking at 300 rpm (determined to be the optimum rpm for maximum HRP activity) for a specific length of time, after which the reaction was stopped by adding 1 ml of 2 M sulfuric acid. 100 μl of the final solution was placed in a Nunc-Immuno plate (Naperville, IL) and the optical density (O.D.) of the solution at 490 nm [29] was recorded using V_{max} Kinetic Microplate Reader (Molecular Devices Corp., Palo Alto, CA). The initial velocity of the reaction of HRP with OPD and H_2O_2 was calculated by dividing the change in the optical density at 490 nm by the enzyme reaction time. The specific activity of the adsorbed HRP was calculated by dividing the change in optical density at 490 nm by the enzyme–substrate reaction time, and the amount of HRP on the surface as determined from the result of ^{125}I-HRP adsorption study. The error bars represent standard deviations of three replicate samples from the same experiment.

Stability of the adsorbed HRP

HRP was adsorbed on PET and TCE/PET for 2 h, and on PTFE and TFE/PET for 1 min at 37°C. After being rinsed in the potassium phosphate buffer, the surfaces were transferred into new vials containing fresh buffer. The activity of the absorbed HRP was measured after storage in the potassium phosphate buffer at 37°C for various times. The specific activity of the absorbed HRP right after the adsorption, at zero storage time, was termed $(\text{Activity})_0$. The specific activity of the adsorbed HRP after a certain storage time was termed $(\text{Activity})_t$. The stability of the adsorbed HRP was characterized by the relative specific activity defined as follows:

$$\text{Relative specific activity (\%)} = \frac{(\text{Activity})_t}{(\text{Activity})_0} \times 100.$$

RESULTS AND DISCUSSION

Surface characterization

The ESCA C1*s* spectra of the untreated and RFGD-treated surfaces are compared in Fig. 1. The C1*s* spectrum of the untreated PET is comprised of three major peaks at binding energies of 285, 286.5, and 289 eV which have been assigned to the functional groups $\underline{C}-C$, $\underline{C}-O$, and $O-\underline{C}=O$, respectively (Fig. 1a). The C1*s* spectrum of PTFE exhibits a single peak at 292 eV corresponding to the $\underline{C}F_2$ repeating unit (Fig. 1b). The C1*s* spectrum of TCE/PET (Fig. 1c) may be resolved into three peaks corresponding to $\underline{C}-C$, $\underline{C}-Cl$, and $Cl-\underline{C}-Cl$ groups [30]. In contrast to PTFE, RFGD-deposited polymer of TFE has a complex C1*s* spectrum which is quite different from that of PTFE. The C1*s* spectrum of TFE/PET can be resolved into five peaks centered at 287.5, 289, 290.5, 292, and 294 eV corresponding to $\underline{C}-CF_n$, $\underline{C}F-C$, $\underline{C}F-CF_n$, $\underline{C}F_2$, and $\underline{C}F_3$ groups, respectively (Fig. 1d) [31]. The complexity of the C1*s* spectra of RFGD-treated surfaces results from the considerable molecular fragmentation and rearrangement of the monomer which occurs during the RFGD treatment [31]. The elemental compositions of all surfaces are shown in Table 1. TFE/PET and TCE/PET show little or no oxygen on the surface, attesting to the complete coverage of PET by the RFGD-deposited polymers. The Cl/C ratio of

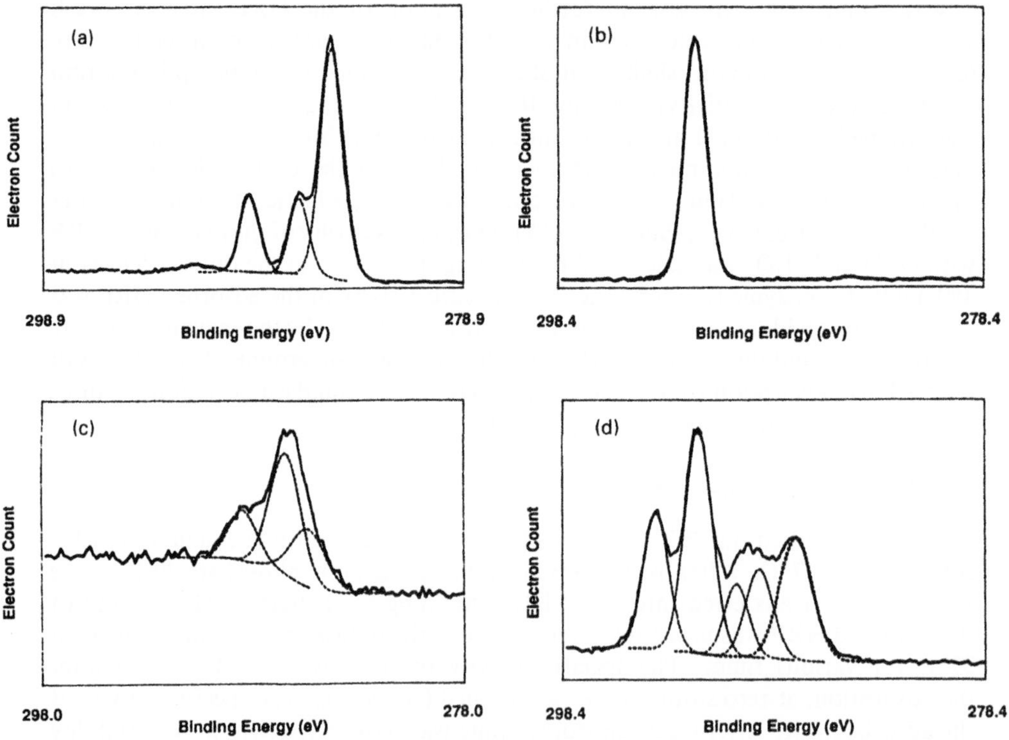
(a)
Electron Count
298.9
Binding Energy (eV)
278.9
(b)
Electron Count
298.4
Binding Energy (eV)
278.4
(c)
Electron Count
298.0
Binding Energy (eV)
278.0
(d)
Electron Count
298.4
Binding Energy (eV)
278.4

Figure 1. High resolution ESCA C1*s* spectra of untreated and RFGD treated surfaces. (a) PET, (b) PTFE, (c) TCE/PET, (d) TFE/PET.

TCE/PET, 0.9, and F/C ratio of TFE/PET surface, 1.6, are lower than the monomer theoretical value, 2.0, indicating the loss of Cl and F during the glow discharge process. The Cl/C or F/C ratios did not change significantly in storage during the time period of these studies.

Table 2 shows the contact angles measured for the various liquids on the untreated and RFGD-treated polymers. The dispersion (γ_s^d) and the polar (γ_s^p) components of the total surface free energy (γ_s) of the polymers, as determined from the contact angle measurements, are shown in Table 3. The surface free energy of polymers ranges from 46.5 dyn/cm for PET down to 11.2 dyn/cm for TFE/PET. TFE/PET

Table 1.
ESCA elemental compositions of the surfaces

	Element (Atomic %)				Elemental ratios			
Surface	C	O	F	Cl	Cl/C	(Theor.)	F/C	(Theor.)
PET	72	28	—	—	—	—	—	—
TCE/PET	52	—	—	48	0.9	2.0	—	—
PTFE	33	—	67	—	—	—	2.0	2.0
TFE/PET	38	1	61	—	—	—	2.6	2.0

Table 2.
Contact angles of various liquids on treated and untreated polymers[a]

Surface	Water	Glycerol	Formamide	Thiodiglycol	Methylene iodide	Ethylene glycol
PET	77 ± 1	63 ± 2	56 ± 1	38 ± 2	22 ± 1	50 ± 1
TCE/PET	81 ± 1	74 ± 1	67 ± 1	56 ± 1	—	56 ± 2
PTFE	109 ± 2	100 ± 1	87 ± 2	96 ± 2	79 ± 2	88 ± 2
TFE/PET	107 ± 1	104 ± 3	94 ± 1	100 ± 1	94 ± 1	95 ± 2

[a] Numbers listed in the table are the means and standard deviation of 15 measurements.

Table 3.
Surface free energy of RFGD treated polymers

Surface	γ_s^d (dyn/cm)	γ_s^p (dyn/cm)	γ_s (dyn/cm)
PET	44.9 ± 1.7	1.6 ± 0.7	46.5 ± 1.8
TCE/PET	23.5 ± 4.2	6.5 ± 2.8	30 ± 5.0
PTFE	18.2 ± 1.1	0.1 ± 0.1	18.3 ± 1.1
TFE/PET	9.5 ± 0.5	1.7 ± 0.5	11.2 ± 0.7

has a much lower surface free energy (11.2 erg/cm^2) than PTFE (which is well known to be *c.* 29 erg/cm^2). The lower surface free energy of TFE/PET may be attributed to the presence of $-CF_3$ groups, which have an inherently lower energy/surface area than the $-CF_2-$ groups mainly present in PTFE [32]. The estimated surface energy of TCE/PET ranges from 25 to 35 erg/cm^2, which is more like hydrocarbon surface energies than chlorocarbon polymers. Nevertheless, the unusually high polar component estimated for the TCE/PET (6.5 ± 2.8 erg/cm^2) clearly reflects its high Cl content (Cl/C ratio = 0.9). It should also be emphasized that gas discharge-deposited polymers are probably very different in molecular structure and composition from conventional polymers, and the composition of the uppermost layer of the plasma polymer, which has most influence on contact angle values, may be different from the layers underneath which were deposited earlier.

Horseradish peroxidase adsorption and retention

HRP adsorption isotherms are shown in Fig. 2. The amounts of HRP adsorbed on PTFE and TFE/PET are similar for the concentration range of 1×10^{-3} to 1×10^{-2} mg/ml. There is also no significant difference in the amount of HRP adsorbed on PET and TCE/PET in the concentration range of 5×10^{-5} to 7.5×10^{-4} mg/ml. (The explanation for using different HRP concentration ranges for PET and TCE/PET vs. PTFE and TFE/PET is provided in the specific activity section.) Figure 3 shows that the percent HRP retained on these surfaces after elution with a 1% solution of SDS varies widely. The two RFGD-treated surfaces, TFE/PET and TCE/PET, retain much higher percentages of adsorbed HRP than the untreated surfaces, PTFE and PET. This observation is consistent with our previous findings for albumin, fibrinogen and IgG on these same surfaces (Table 4) [7, 8, 10].

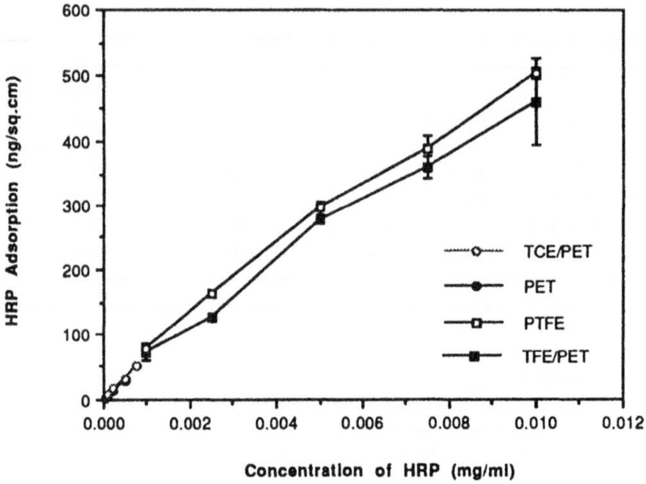
HRP Adsorption (ng/sq.cm)
600
500
400
300
200
100
0
0.000
0.002
0.004
0.006
0.008
0.010
0.012
Concentration of HRP (mg/ml)
TCE/PET
PET
PTFE
TFE/PET

Figure 2. HRP adsorption isotherms on RFGD-treated surfaces. HRP was adsorbed from potassium phosphate buffer (0.05 M, pH 7.0) for 2 h at 37°C. The concentration range used for PET and TCE/PET surfaces was from 10^{-4} mg/ml to 10^{-3} mg/ml. The concentration range used for PTFE and TFE/PET was from 10^{-3} (0.001) mg/ml to 10^{-2} (0.01) mg/ml. The PET control data points are essentially identical to, and hidden by the TCE/PET data points.

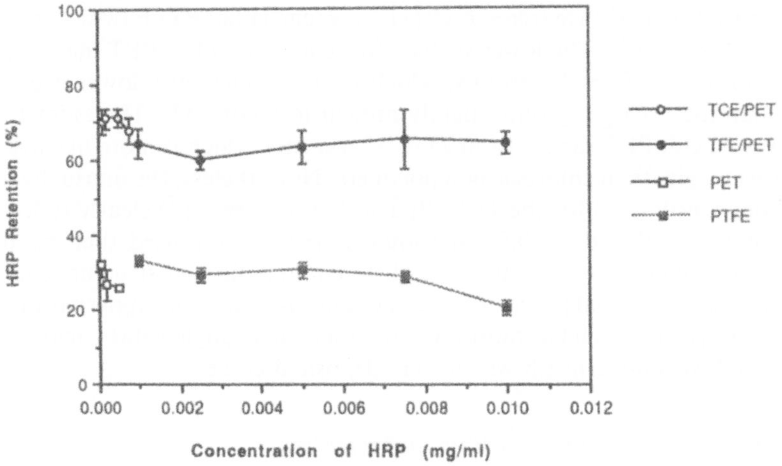
HRP Retention (%)
100
80
60
40
20
0
0.000
0.002
0.004
0.006
0.008
0.010
0.012
Concentration of HRP (mg/ml)
TCE/PET
TFE/PET
PET
PTFE

Figure 3. Percent HRP retention by various surfaces. HRP was adsorbed for 2 h at 37°C (see data in Fig. 2) and was eluted overnight with 1% SDS solution.

Table 4.
The percent retention of proteins on surfaces

	PET	TCE/PET	PTFE	TFE/PET
Albumin	26 ± 3	95 ± 1	41 ± 2	95 ± 5
Fibrinogen	37 ± 1	99 ± 1	54 ± 8	98 ± 1
IgG	14 ± 1	—	13.3 ± 3	50 ± 7
HRP	28 ± 2	70 ± 1	28 ± 5	68 ± 2

Properties of adsorbed horseradish peroxidase

To assure that the enzymatic activity observed for HRP adsorbed on surfaces is not due in part to the desorption of the enzyme into the substrate solution during the enzymatic reaction, the following study was performed: HRP was adsorbed on PET, PTFE, and TFE/PET at a concentration of 1×10^{-2} mg/ml, rinsed and placed in vials containing the citrate–phosphate buffer used for the enzymatic reaction, but without the addition of H_2O_2 or OPD. Following 5 min of incubation, 1 ml of this buffer was added to 1 ml of citrate–phosphate buffer containing H_2O_2 and OPD. No chromogenic product was observed during the 15 min incubation period suggesting that no active HRP molecules desorbed from the PET, PTFE, and TFE/PET surfaces. Since HRP is held more tenaciously on TCE/PET than on PET or PTFE, it was assumed that during the enzymatic reaction, no active HRP molecules had desorbed from TCE/PET either. Furthermore, the HRP molecules adsorbed on walls of the adsorption vial did not contribute to the enzymatic reaction because the samples were transferred to new vials after adsorption and prior to the enzymatic reaction. Therefore, it was concluded that active HRP molecules did not desorb from the surface during the enzymatic reaction and the enzymatic activity observed is due only to the surface-bound enzyme.

Adsorption of enzymes on surfaces may cause a shift in the optimum conditions for the enzyme–substrate reaction. Thus, our initial studies focussed on determining the catalytic characteristics (optimum pH and K_m) of HRP adsorbed on the different surfaces. The optimum pH of an enzyme can be dependent on the particular substrate. For H_2O_2 and OPD, free HRP in solution shows optimum enzymatic activity at pH 5 [20]. As Fig. 4 shows, the optimum pH of HRP adsorbed onto the untreated or RFGD-treated surfaces remains unchanged at pH 5. This suggests that the adsorption of HRP or its tight binding to RFGD-treated surfaces does not significantly perturb the chemical environment of the HRP, or the charge distribution in its active site, both of which should relate to the optimum pH of HRP.

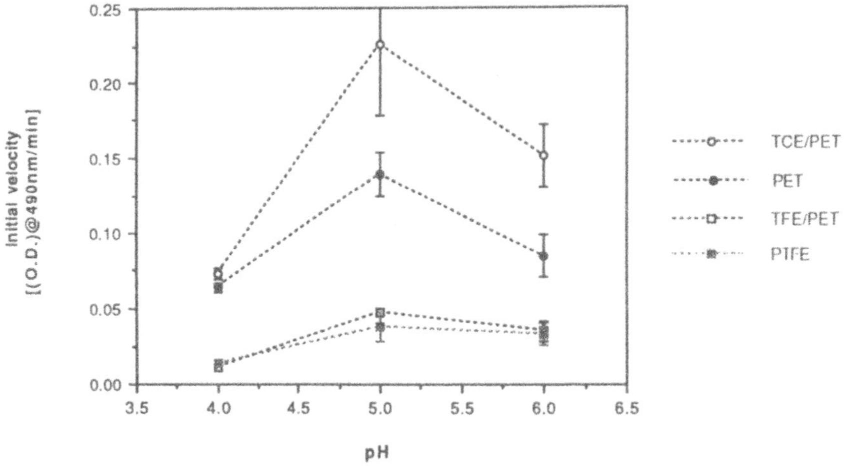
Initial velocity
[(O.D.)@490nm/min]
0.25
0.20
0.15
0.10
0.05
0.00
3.5
4.0
4.5
5.0
5.5
6.0
6.5
pH
TCE/PET
PET
TFE/PET
PTFE

Figure 4. The initial velocity [Δ(O.D.)/min] of HRP adsorbed on various surfaces. HRP was adsorbed from potassium phosphate buffer (0.05 M, pH 7.0) for 2 h at 37°C. The enzymatic reaction was carried out at room temperature with H_2O_2 (3.5 mM) and OPD (5.5 mM) (see text for details).

Table 5.
The Michaelis constants of adsorbed HRP[a]

Enzyme	K_m (mM)
on TCE/PET	0.63
on PTFE	0.82
on PET	0.83
on TFE/PET	1.06
Free HRP, in buffer	2.60

[a] HRP, 0.01 mg/ml, was adsorbed on surfaces from potassium phosphate buffer (0.05 M, pH 7.0) for 2 h at 37°C. The initial velocity of adsorbed HRP was measured by the reaction with H_2O_2 and OPD (5.5 mM) at pH 5 and room temperature for 10 min.

This result might also suggest that if there are HRP molecules which are denatured upon adsorption, then they completely lose their activity, while other adsorbed HRP molecules retain their native conformation and activity.

It is also reasonable to expect that whenever the conformation of microenvironment of an enzyme changes, its characteristic Michaelis constant (K_m) should also be affected. The Michaelis constants of HRP adsorbed on surfaces and free HRP in solution, determined from the Lineweaver–Burk plot, are listed in Table 5. The Michaelis constants of HRP adsorbed on all surfaces are all lower than that of free HRP in solution. One possibility suggested by these values is that the adsorbed HRP might have a higher specificity for the substrate than the free HRP in solution. One explanation could be that the adsorption process induces an allosteric enhancement of HRP activity on each of the surfaces studied.

Specific activity of adsorbed horseradish peroxidase

The specific activity of the free enzyme in phosphate buffer is approximately 5.2×10^3 Δ(O.D.)/mg HRP/min. The specific activity of HRP on the different surfaces was calculated from three slopes of the plots of the Δ(O.D.) @ 490 nm divided by the amount of HRP adsorbed (based on radiolabeled HRP) versus the reaction time. The results are shown graphically in Figs 5 and 6. HRP adsorbed on PET and TCE/PET have remarkably high specific activities (Table 6) in comparison with the HRP adsorbed on PTFE and TFE/PET. This is the primary reason why HRP adsorption was performed at lower concentrations and for longer time on PET and TCE/PET compared to PTFE and TFE/PET. In addition, HRP possesses similar specific activity whether adsorbed on PTFE or TFE/PET.

Table 6.
Specific activity of HRP on surfaces

Enzyme	Specific activity [Δ(O.D.) mg^{-1} min^{-1}]
Free HRP, in buffer	5201.7 ± 547.4
on PET	3234.8 ± 378.9
on TCE/PET	1655.5 ± 332.8
on TFE/PET	17.9 ± 1.5
on PTFE	17.9 ± 1.0

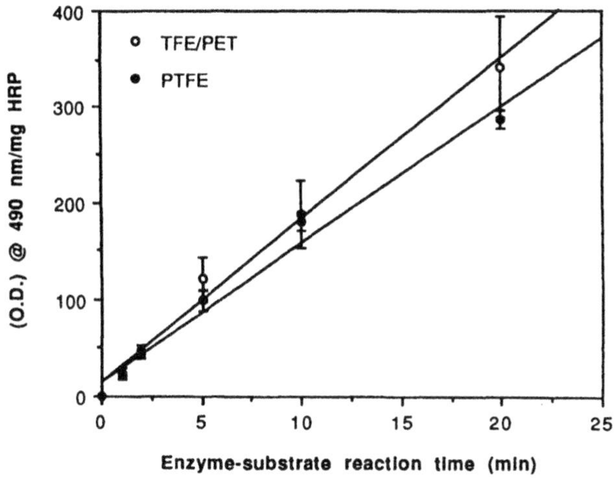
400
300
200
100
0
(O.D.) @ 490 nm/mg HRP
TFE/PET
PTFE
0
5
10
15
20
25
Enzyme-substrate reaction time (min)

Figure 5. The specific activity (measured at R.T.) of HRP adsorbed on PTFE and TFE/PET HRP, 0.01 mg/ml, was adsorbed for 2 h at 37°C. The enzymatic reaction was then carried out at room temperature with H_2O_2 (3.5 mM) and OPD (5.5 mM) and pH 5 (see text for details).

The hypothesis we had formulated prior to the initiation of this study was that the tight binding of proteins on TFE/PET relative to PTFE is a direct consequence of greater unfolding of proteins on that surface. If the specific activity of HRP was lower on TFE/PET than on PTFE, then that would be in support of our hypothesis. This is clearly not the case as there is no significant difference in the specific activity of HRP on these two surfaces. Furthermore, it can be seen that the specific activities of HRP adsorbed on both PET and TCE/PET are not only very high but are within the same order of magnitude, despite the tight binding of HRP on TCE/PET and lack of tight binding on PET. These observations clearly argue against our aforementioned hypothesis. It is possible that despite the close retention levels, HRP

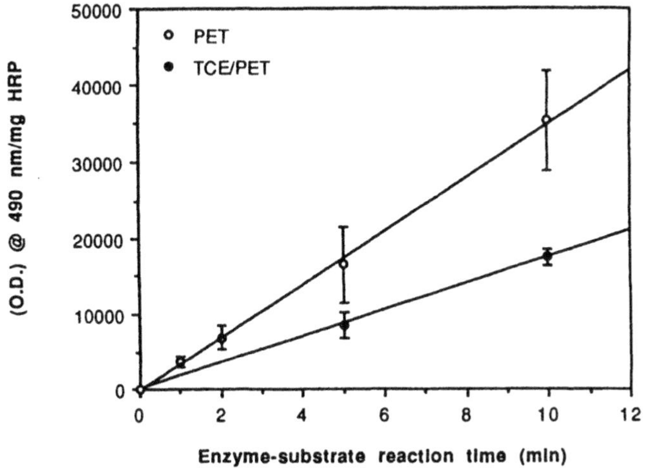
50000
40000
30000
20000
10000
0
(O.D.) @ 490 nm/mg HRP
PET
TCE/PET
0
2
4
6
8
10
12
Enzyme-substrate reaction time (min)

Figure 6. The specific activity (measured at R.T.) of HRP adsorbed on PET and TCE/PET. HRP, 0.0002 mg/ml, was adsorbed for 2 h at 37°C. The enzymatic reaction was then carried out at room temperature with H_2O_2 (3.5 mM) and OPD (5.5 mM) and pH 5 (see text for details).

adsorbs very differently on TFE/PET than on TCE/PET, such that the active site remains exposed and active on TCE/PET while it is either hidden or inactivated on TFE/PET. The high enzymatic activity of HRP adsorbed on the PET might be explained similarly to the TCE/PET.

The large difference in the HRP activity for the two groups, fluorocarbon surfaces (PTFE and TFE/PET) and nonfluorinated surfaces (PET and TCE/PET), is possibly due to the differences in the surface free energy of the polymers. The fluorocarbon polymers have lower surface energies (in air) than PET and TCE/PET. This means that the fluorocarbons have a higher interfacial tension when placed in water. HRP may unfold to a greater extent on the fluorocarbon surfaces in order to reduce the large interfacial tension which exists at the fluorocarbon/water interface.

To study the stability of immobilized enzyme, HRP was adsorbed on the four surfaces. Next, surfaces were rinsed and placed in phosphate buffer at 37°C. After various times, the residual activity of HRP on each surface was determined. Results of these studies are shown in Figs 7 and 8, and Fig. 9 replots all of the data together for comparison on the same time scale. As Fig. 7 shows, 95% of the activity of HRP is lost within 4 min on PTFE and TFE/PET. HRP adsorbed on PET loses 60% of its catalytic capability after 1 h of residence time (Fig. 8). In contrast, the loss in activity of HRP adsorbed on TCE/PET after 1 h of residence time is only 15% (Fig. 8). We conclude from these studies that HRP binds to TCE/PET and PET with relatively minor alterations to its conformation compared to the extent of conformational changes induced by the two fluorocarbon surfaces studied. HRP immobilized on TCE/PET also maintains its activity over longer times than on the other surfaces studied. One may speculate that the tight binding of HRP on TCE/PET is mainly via polar interactions and this may prevent the enzyme from unfolding and as such, is responsible for the longevity of the enzyme activity on that surface.

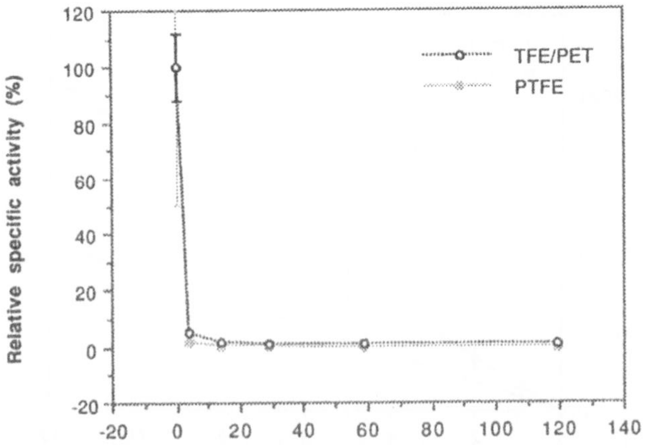
120
100
80
60
40
20
0
-20
-20 0 20 40 60 80 100 120 140
Relative specific activity (%)
TFE/PET
PTFE

Figure 7. The relative specific activity of HRP as a function of residual time. HRP, 0.01 mg/ml, was adsorbed on PTFE and TFE/PET at 37°C for 1 min, and then reacted with H_2O_2 (3.5 mM) and OPD (5.5 mM) at room temperature and pH 5 after various residence times in potassium phosphate buffer at 37°C.

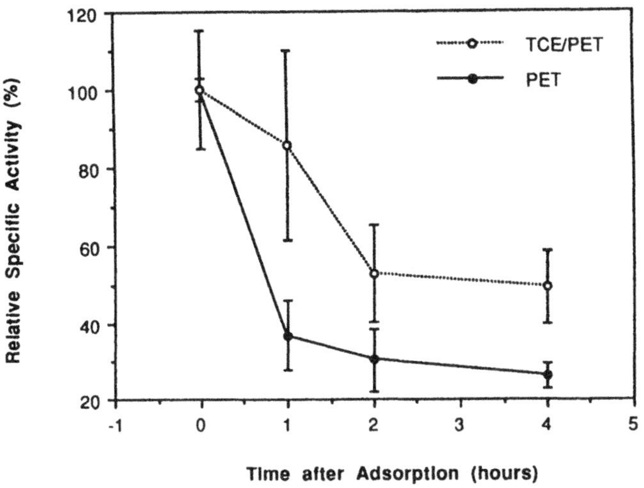

Relative Specific Activity (%)
120
100
80
60
40
20
-1
0
1
2
3
4
5
Time after Adsorption (hours)
TCE/PET
PET

Figure 8. The relative specific activity of HRP as a function of residence time. HRP, 0.0002 mg/ml, was adsorbed on PET and TCE/PET at 37°C for 2 h, and then reacted with H_2O_2 (3.5 mM) and OPD (5.5 mM) at room temperature and pH 5 after various residence times in potassium phosphate buffer at 37°C.

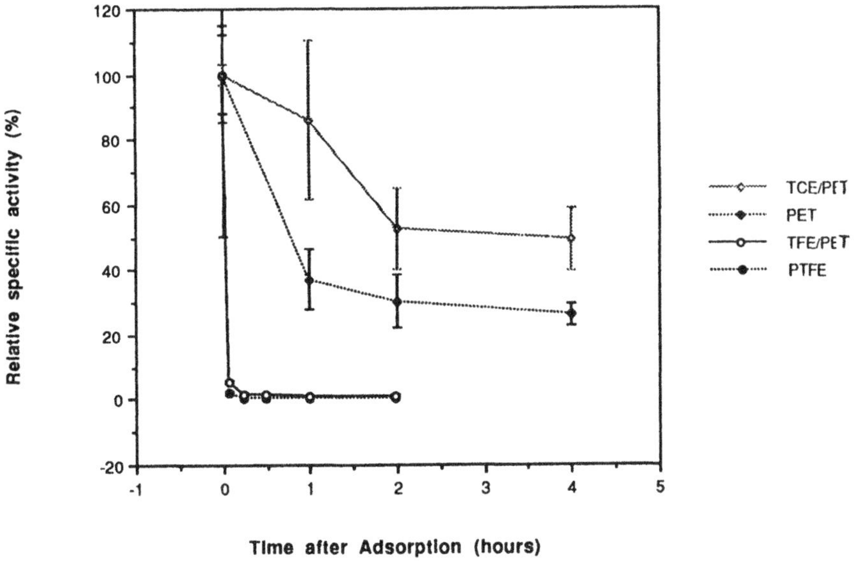

Relative specific activity (%)
120
100
80
60
40
20
0
-20
-1
0
1
2
3
4
5
Time after Adsorption (hours)
TCE/PET
PET
TFE/PET
PTFE

Figure 9. Replot of the data on Figs 7 and 8, using the same time scale.

CONCLUSIONS

HRP has been absorbed on untreated and RFGD-treated surfaces and we have studied the strength of enzyme/surface interactions and the residual enzyme activity at various times after adsorption. TFE/PET and TCE/PET surfaces retain approximately 70% of the adsorbed HRP following overnight elution with SDS. In contrast, untreated surfaces, PET and PTFE, retain only slightly more than 20% of the adsorbed HRP. The optimum pH for the enzymatic activity of the adsorbed HRP is the same as that for free HRP in solution. However, adsorbed HRP unexpectedly has a lower K_m than free HRP, for all the surfaces studied. A significant loss of enzymatic activity of HRP is found upon adsorption onto PTFE and TFE/PET when compared to PET and TCE/PET. This difference may be simply due to the much lower surface energy (in air) or much higher interfacial energy (in water) of PTFE and TFE/PET surfaces compared to PET and TCE/PET. HRP adsorbed on the TCE/PET surface resists not only elution by SDS but also loss of activity with time at 37°C in a phosphate buffer solution. More than 55% of the enzymatic activity of HRP on TCE/PET is retained after 3 h in phosphate buffer at 37°C. Under the same conditions, HRP adsorbed on PET retained less than 30% of its original activity, while HRP on the two fluorinated surfaces lost 100% of its original activity in a few minutes. The adsorption of HRP onto TCE glow discharge-treated surfaces may represent a new and simple method for immobilizing proteins while maintaining a large fraction of their activity.

Acknowledgements

The authors wish to acknowledge the financial support provided by the NIH (grant No. GM 40111-03 and 04) and the Washington Technology Centers. The ESCA spectra were obtained at the National ESCA and Surface Analysis Center for Biomedical Problems (NESAC/BIO, NIH grant RR 01296) at the University of Washington.

REFERENCES

1. P. B. van Wachen, C. M. Vreriks, T. Beugeling, J. Feijen, A. Bantjes, J. P. Detmers and W. G. van Aken, *J. Biomed. Mater. Res.* **21,** 701 (1987).
2. K. Kottke-Marchant, J. M. Anderson, Y. Umemura and R. E. Marchant, *Biomaterials* **10,** 147 (1989).
3. E. J. Guilbeau, L. C. Clark, U. B. Pizziconi, J. S. Schultz and B. C. Towe, *Trans. Am. Soc. Artif. Intern. Organs* **33,** 834 (1987).
4. J. L. Brash and Q. M. Samak, *J. Colloid Interface Sci.* **65,** 495 (1978).
5. J. L. Brash, S. Uniyal and Q. M. Samak, *Trans. Am. Soc. Artif. Intern. Organs* **20,** 69 (1974).
6. J. L. Bohnert and T. A. Horbett, *J. Colloid Interface Sci.* **111,** 363 (1986).
7. A. S. Hoffman, T. A. Horbett, J. L. Bohnert, B. C. Fowler and D. Kiaei, US patent No. 5,055,316 (1991).
8. J. L. Bohnert, B. C. Fowler, T. A. Horbett and A. S. Hoffman, *J. Biomater. Sci. Polymer Edn.* **1,** 279 (1990).
9. A. Safranj, D. Kiaei and A. S. Hoffman, *Biotechnol. Prog.* **7,** 173 (1991).
10. D. Kiaei, A. S. Hoffman and T. A. Horbett, *J. Biomater. Sci. Polymer Edn.*, in press.
11. T. Akaike, T. Tsuruta, K. Kosuge, S. Miyata, K. Kataoka and Y. Sakurai, *Jpn Bull. Heart Inst.* **00,** 4 (1979).
12. T. J. Lenk, B. D. Ratner, R. M. Gendreau and K. K. Chittur, *J. Biomed. Mater. Res.* **23,** 549 (1989).
13. T. Sanada, Y. Ito, M. Sissido and Y. Imanishi, *J. Biomed. Mater. Res.* **20,** 1179 (1986).
14. J. N. Lindon, G. McManama, L. Kushner, E. W. Merrill and E. W. Salzman, *Blood* **68,** 355 (1986).
15. H. Elwing, B. Nilsson, K. E. Svensson, A. Askendahl, U. R. Nilsson and I. Lundstrom, *J. Colloid Interface Sci.* **125,** 139 (1988).
16. D. Lewis and T. L. Whateley, *Biomaterials* **9,** 71 (1988).
17. R. K. Sandwick and K. J. Schray, *J. Colloid Interface Sci.* **115,** 130 (1987).
18. R. K. Sandwick and K. J. Schray, *J. Colloid Interface Sci.* **121,** 1 (1988).
19. M. Morrison and G. Bayes, in: *Oxidases and Related Redox Systems*, Vol. 1, p. 375, T. E. King, H. S. Mason and M. Morrison (Eds). University Park Press, Baltimore, MD (1973).
20. V. H. Gallati and H. Brodbeck, *J. Clin. Chem. Clin. Biochem.* **20,** 221 (1982).

21. T. J. Greewalt, E. M. Swierk and E. A. Steane, *J. Immunol. Methods* **8**, 351 (1975).
22. B. K. van Weemen and A. H. W. M. Schuurs, *FEBS Lett.* **43**, 215 (1974).
23. H. V. Malmatadt and T. P. Hadjioannou, *Anal. Chem.* **35**, 14 (1963).
24. D. Kiaei, A. S. Hoffman, B. D. Ratner, T. A. Horbett and L. O. Reynolds, *J. Appl. Polym. Sci.: Appl. Polym. Sym.* **42**, 269 (1988).
25. D. H. Kaelble and J. Moacanin, *Polymer* **18**, 475 (1977).
26. A. S. McFarlane, *Nature* **182**, 53 (1958).
27. R. W. Helmkamp, R. L. Goodland, W. F. Bale, I. L. Spar and L. E. Mutschler, *Cancer Res.* **20**, 1495 (1960).
28. T. A. Horbett, *J. Biomed. Mater. Res.* **15**, 673 (1981).
29. D. Richards, in: *ELISA and other Solid Phase Immunoassays*, p. 343, D. M. Kemeny and S. J. Challacombe (Eds). John Wiley & Sons Ltd., New York (1988).
30. A. Dilks, in: *Electron Spectroscopy: Theory, Techniques and Applications*, Vol. 4, p. 289, A. D. Baky and C. R. Brundle (Eds). Academic Press, London (1981).
31. D. T. Clark and D. Shuttleworth, *J. Polym. Sci. Polym. Chem. Ed.* **18**, 407 (1980).
32. F. M. Fowkes, in: *Contact Angle, Wettability, and Adhesion*, R. F. Gould (Ed.). American Chemical Society, Washington, DC (1964).

Patterned neuronal attachment and outgrowth on surface modified, electrically charged fluoropolymer substrates

ROBERT F. VALENTINI[1], TERRENCE G. VARGO[2], JOSEPH A. GARDELLA Jr.[2] and PATRICK AEBISCHER[1]*

[1]*Section of Artificial Organs, Biomaterials and Cellular Technology, Brown University, Providence, RI 02912, USA*

[2]*Department of Chemistry, SUNY at Buffalo, Buffalo, NY 14214, USA*

Received 23 October 1992; accepted 28 January 1993

Abstract—Fluorinated ethylenepropylene copolymer (FEP) and polyvinylidene fluoride (PVDF) can generate static and transient electrical charges, respectively, after bulk molecular rearrangements induced by electrical charging techniques. Neurons cultured on electrically active FEP and PVDF show increased levels of nerve fiber outgrowth compared to electrically neutral material. The purpose of the present study was to determine if the addition of charged surface groups to the surfaces of FEP and PVDF would modify the influence of bulk electrical charges on cultured neurons. Mouse neuroblastoma (Nb2a) cells were cultured on electrically charged and uncharged FEP and PVDF substrates with covalently modified surfaces containing hydroxyl (OH) and amine (NH_2) groups. Surface chemical modification was performed on the entire surface or in discrete striped regions. Nb2a cells cultured on electrically active FEP and PVDF showed greater levels of differentiation than cells on electrically neutral substrates. The presence of NH_2 groups attenuated these responses in serum-containing media. Cells attached to NH_2 rich surfaces generally displayed a flatter morphology and tended to remain attached for longer time periods. Cells cultured on stripe-modified substrates in serum-containing media showed a strong preferential attachment to modified regions, especially on NH_2 stripes. In summary, bulk electrical charges are more important than surface charges in stimulating Nb2a cell differentiation. Surface groups serve to modulate neuronal morphology and confer specific attachment promoting properties in serum-containing media. The development of an optimal neuronal regeneration template may require the incorporation of specific bulk and surface properties.

Key words: Surface modification; nerve regeneration; electrical field; polymer; neuron; tissue culture; piezoelectric; electret.

INTRODUCTION

A better understanding of the mechanisms underlying neuronal development, outgrowth and regeneration may provide clues toward optimizing the treatment of nerve injuries and neurodegenerative disorders. Due to the complexity of the nervous system, tissue culture systems are used to assess growth regulation *in vitro*. Neurons in culture extend long processes; typically a single axon, which transmits signals and several dendrites, which receive signals. *In vitro*, it is difficult to distinguish axons from dendrites and the neutral terms, 'neurite' or 'nerve fiber' are used. Three extrinsic factors are known to control neurite extension *in vitro*; diffusible trophic factors, cell/substrate interactions, and electric fields [1]. The

*To whom correspondence should be addressed.

roles of trophic substances such as nerve growth factor, ciliary neurotrophic factor and brain derived neurotrophic factor are being evaluated extensively and will not be addressed. In the present study, a relationship between substrate chemistry and local electrical charges is evaluated *in vitro* using electrically charged/chemically modified growth substrates.

The surface chemistry of synthetic substrates is known to influence the attachment and morphology of cultured cells [2]. For example, polystyrene substrates widely used in tissue culture must first be modified in order to enhance cell attachment [3]. Modification of polystyrene is usually accomplished through chemical treatment or glow discharge techniques which result in numerous charged surface groups and changes in surface energy. Synthetic coating factors with charged functional groups also influence cell attachment and morphology. In general, coating materials with net positive charges enhance attachment while negatively charged materials discourage it [4]. Specifically, positively charged polyamino acids, such as polylysine and polyornithine favor the attachment of neuronal cells and processes [5–7]. In contrast, negatively charged polyamino acids such as polygluconate and acetylated polyamines inhibit attachment [5, 7]. Even greater complexity is encountered when studying the influence of highly charged, high molecular weight extracellular matrix glycoproteins such as fibronectin and laminin on neuronal morphology [8]. Negatively charged extracellular matrix materials like sulfated proteoglycans inhibit neurite outgrowth *in vivo* and *in vitro* [9, 10]. Increasing the number of sialic acid residues on ganglisoides results in reduced neurite outgrowth [11]. Extracellular matrix molecules are believed to act primarily via receptor mediated events although electrostatic interaction may also be significant [12]. Due to the complexity of tissue culture substrates and synthetic and biological coating factors, it has not been possible to determine how surface properties optimize cell attachment and promote cellular differentiation. Cell/substratum interactions could be mediated through interrelated changes in surface morphology, chemistry and charge.

It has been suggested that time-varying charges generated by elements of the extracellular matrix play a role in tissue development *in vivo* [13, 14]. Exogenously applied electric fields affect cultured neurons by altering the rate, direction and number of neurites extended [15–17]. The interaction between neurons and local electric charges has been evaluated by culturing cells on electrically charged fluoropolymers [18, 19]. The simple, stable chemistry of fluoropolymers facilitates surface characterization and the absence of ionizable functionalities precludes any interaction between cells and charged surface groups. Fluorinated ethylenepropylene copolymer (FEP) can be fabricated into an electret which stores positive or negative monopolar charges that generate an external static electric field [20]. Polyvinylidene fluoride (PVDF) can be fabricated as a piezoelectric material that contains oriented molecular dipoles which generate transient electric charges upon mechanical deformation [21]. Charged injection and dipole orientation are achieved by exposing the polymer film to a non-breakdown, high voltage field. Mouse neuroblastoma (Nb2a) cells grown on piezoelectric PVDF showed greater levels of differentiation and process outgrowth than cells grown on non-piezoelectric PVDF [18]. Cells cultured on positively charged FEP displayed significantly greater levels of outgrowth than cells on negatively charged and uncharged FEP electrets [19]. No differences in the degree of cell attachment between charged and uncharged materials was observed in either study. Since exposure to a high voltage field could result in surface modification, all fluoropolymer surfaces were analyzed using several surface sensitive techniques. Scanning electron microscopy (SEM), electron spectroscopy for chemical analysis (ESCA) and contact angle measurements demonstrated that the surface chemistry and surface energy of charged and uncharged FEP and PVDF are identical [18, 19]. Enhanced neuronal outgrowth is, therefore, related to static or dynamic charge interactions between cells and charged substances.

Recently, a technique to modify fluoropolymers via a radiofrequency glow discharge (RFGD) plasma process using hydrogen gas and methanol vapor has been described [22]. OH groups can be added covalently to fluoropolymer surfaces in a controlled process that causes only a small increase in surface energy [22, 23]. Neither the surface topography nor the bulk properties of the polymer are altered. The OH groups can be used to further functionalize the surface through the addition of silane coupling agents. The covalent attachment of aminopropyl triethoxy silane (APTES), an amine-terminated coupling agent, is achieved by immersion in a dilute solution of the agent. The resulting APTES functionalized surfaces are stable for up to 6 months or longer when stored under ambient conditions [23]. It is also possible to localize surface modification to prescribed areas using lithographically patterned metallic overlays [24]. The ability to covalently refunctionalize materials which are capable of generating local charges could prove useful in the development of biosensors or tissue regeneration scaffolds. The first objective of the present study was to develop electrically active PVDF and FEP with covalently modified surfaces featuring OH and NH_2 groups.

The aforementioned properties of fluoropolymers make them ideal for studies investigating the simultaneous influence of substrate charge and chemistry on neuronal outgrowth. The second objective of the study was to determine how electrically charged/surface modified substrates influence neuronal morphology and neurite outgrowth. As in previous studies, the mouse neuroblastoma (Nb2a) cell line was used. Differentiation of neuroblastoma cells is recognized when a nerve fiber is elicited by the cell body [25, 26]. This provides a simple, objective way to assess differentiation. These cells also attach to fluoropolymer substrates without the need for other attachment factors which, themselves, are known to influence cell morphology. In addition, Nb2a cells can be cultured in serum-containing and serum-free media [27]. Cells differentiate more readily in serum-free conditions due to the absence of serum-derived inhibitory factors [28]. The ability to use serum free media, which contains no attachment factors and a minimal amount of protein, further simplifies the experimental environment.

Cells were plated directly on FEP and PVDF substrates with fully modified or stripe modified surfaces containing OH or NH_2 groups. Substrates with patterned (i.e. striped) surface modification makes it possible to evaluate adjacent areas of modified and virgin fluoropolymer with an underlying bulk charge. We investigated piezoelectric vs non-piezoelectric PVDF substrates and positively charged vs uncharged FEP substrates. Negatively charged FEP substrates were not evaluated since previous work established no difference between uncharged and negatively charged substrates regarding Nb2a differentiation [19]. This approach allows us to address the issue of how hydroxylated and amine-containing surfaces influence neurons and whether they modify the response of neurons to bulk electric charges.

MATERIALS AND METHODS

Surface modification of PVDF and FEP substrates

Fluoropolymer discs, 30 mm in diameter, were lathe-cut from commercially available sheets. Electrically poled (i.e. piezoelectric) and unpoled (i.e. non-piezoelectric) 40 μm thick sheets of biaxially stretched PVDF film were obtained from Solvay & Cie (solef® piezo film; Brussels, Belgium). Twenty five μm thick fluorinated ethylene propylene (FEP) copolymer sheets were a gift from DuPont (Teflon® FEP Type 200A; Wilmington, DE). FEP is a fully fluorinated ethylene propylene copolymer. All discs were cleaned initially by washing in 2% Alconox detergent solution, rinsing copiously with distilled water and air-drying. Surface contaminants were removed by ultrasonication in hexane and absolute methanol for one minute each. Hydroxyl (OH) groups were covalently bound to the fluoropolymer surfaces using an RFGD plasma process using hydrogen gas and methanol vapor [22]. On selected discs, patterned surface modification was achieved using a photolithographically etched nickel mask with 25 adjacent slots (15 mm long $\times$ 300 μm wide) spaced 500 μm apart. Plasma modifications were performed for the following periods of time: 1 min for fully modified PVDF, 10 s for PVDF modified with mask, 2 min for fully modified FEP, and 20 s for FEP modified with mask. Selected RFGD-treated materials were further functionalized with a silane coupling agent, aminopropyltriethoxysilane (APTES) which contains a primary amine (NH_2). APTES ethoxy groups react with surface OH groups resulting in an overlayer containing primary amines tethered to the substrate via propyl spacers [23]. APTES was reacted by immersing the OH-modified discs in a solution containing 40 ml hexane and 2 ml APTES for about 2 s. Addition of APTES to the hexane was carried out below the hexane/air interface to minimize prepolymerization. All discs were ultrasonicated in hexane and absolute methanol for 30 s each after modification and then stored in foil packets.

Preparation of electrically charged substrates

Unmodified, OH-modified and NH_2-modified PVDF and FEP discs were mounted onto custom made polycarbonate rings and held in position with a threaded insert to form tissue culture dishes. The exposed area of the discs was 5 cm^2. A silicone elastomer O-ring positioned between the discs and the insert held the discs taut and created a water-tight seal to prevent leakage of culture media.

Piezoelectric PVDF

Piezoelectric and non-piezoelectric PVDF membranes were used directly after surface modification and cleaning. The electrical output of all PVDF substrates was assessed to see if the surface modification processing caused any changes. The undersides of modified and unmodified PVDF (piezoelectric and non-piezoelectric) films were coated with a conductive silver paint and the dishes filled with 1 ml of culture media [18]. A grounding lead attached to the conductive undercoating and a stainless steel recording electrode immersed in the conducting fluid were connected to a Tektronix storage oscilloscope. Voltage outputs from three samples of each film type placed on the same incubator were tested.

FEP electrets

Since FEP is only available in uncharged form, FEP discs were charge injected (i.e. fabricated into electrets) using a corona-charging apparatus [19]. Briefly, mounted FEP discs were placed on a grounded aluminum block 4 cm below a single needle-point brass electrode connected to a low current, high voltage DC reversible polarity power supply (Bertan Associates model 205A-50R, Syosset, NY). A copper mesh assembly connected to a low voltage DC reversible polarity power supply (Bertan Associates model 205B-03R, Syosset, NY) was centered below the needle electrode, 2.5 cm above the FEP. The upper electrode was biased at 12 kV against the grounded FEP substrate and the copper mesh was biased to the desired surface voltage (+ 1000 V for fully modified and + 1500 V for stripe-modified FEP). FEP dishes were exposed to this corona field for 20 min. Electret charge density analysis was performed to determine how chemical modification and exposure to tissue culture media influenced charge decay. Partial charge neutralization is known to occur in ion containing media [29]. A charge decay study using charged dishes filled with tissue culture media and placed in a 37°C incubator was performed at 24 h intervals for 4 days. The projected surface voltage was measured with an electrostatic voltmeter (model 244, Monroe Electronics, Lyndonville, NY) equipped with a non-contact, vibrating probe (model 1017, Monroe Electronics, Lyndonville, NY). The surface voltages for all charged discs was calculated by sampling seven uniformly spaced points over the surface of the film. The numerical average was taken to be the average surface voltage of the film.

Surface characterization

The surfaces of charged and uncharged PVDF and FEP substrates were assessed for potential modifications induced by the surface modification procedures and, in the case of FEP, corona charging technique. The surface microtexture of representative samples was analyzed using scanning electron microscopy (SEM) performed on a S-2700 Scanning Electron Microscope (Hitachi), equipped with a 3 nm resolution LaB6 electron gun operated at 2 kV at a magnification of 1000 ×.

The surface chemistry of representative samples of fully modified PVDF and FEP samples was assessed using electron spectroscopy for chemical analysis (ESCA). ESCA was performed on: (1) a Physical Electronics model 5300 spectrometer with a Mg anode X-ray source at 300 W; and (2) a Surface Science Instruments model 206 spectrometer with a monochromatized Al X-ray source at 100 W. For both instruments, low resolution survey spectra were analyzed and all principal signals were analyzed under high resolution for qualitative and quantitative analysis. Three samples of each fully modified material were analyzed.

Tissue culture

All FEP and PVDF tissue culture dishes were sterilized by transferring them into sterile petri dishes and placing them in a dry, 56°C oven for 12 h. Sterilizing fluids were not used to avoid possible surface chemistry changes. The mouse neuroblastoma subclone Nb2a (American Tissue Type Collection) was maintained in 75 cm^2 flasks in Dulbecco's modified Eagle's medium (DMEM) supplemented with 10% fetal calf serum (FCS), penicillin G (100 U/ml) and streptomycin (100 μg/ml).

Cells were plated onto experimental dishes in either DMEM/10% FCS media or serum-free Opti-MEM media (GIBCO, Grand Island, NY), which contains trace amounts of insulin, transferrin and selenium and contains less than 15 μg/ml of protein. Nb2a cells were collected and pelleted after trypsinization in a solution of trypsin in Ca^{2+}-free, Mg^{2+}-free Hank's Balanced Salt Solution (GIBCO). After rinsing and centrifugation with DMEM/FCS or Opti-MEM media, cell counts and viability were assessed using a hemocytometer after trypan blue exclusion staining.

Neurite outgrowth analysis

A 1-ml aliquot of Nb2a cells was plated onto substrates in DMEM/FCS or serum-free Opti-MEM at a density of 1×10^5 cells/ml and brought up to a final volume of 2 ml. Experimental dishes were maintained at 37°C in a humidified 5%CO_2/95% air mixture. The extent of Nb2a differentiation and process outgrowth was quantified at 24 hour intervals after plating using 200 × Hoffman modulation optics on a Zeiss IM-35 inverted microscope. All neurons and neurons with neurites at least as long as one cell diameter were counted on random, successive fields until a total of 300–350 had been counted. The percentage of neuronal differentiation was calculated by dividing the number of differentiated neurons by the total population of neurons counted. Neurite lengths were quantified using 100 × phase contrast optics interfaced with an image processing system (CUE-2, Olympus, NJ). The longest neurites in successive fields were analyzed until a total of 30 neurites were counted. Cell attachment assays were performed 24 h after plating cells in serum-containing and serum-free media. Cell attachment densities were calculated by dividing the total number of cells counted by the total area of culture substrate analyzed. All data were analyzed using the unpaired Student's *t*-test.

RESULTS

Surface texture analysis

THE RFGD plasma conditions used to prepare OH containing surfaces did not change the surface texture or light transmittance properties of PVDF and FEP as analyzed by gross observation, with 400 × Hoffman modulation optics, or SEM. All surfaces were smooth and similar to unmodified surfaces. Surfaces undergoing further APTES modification were also similar to unmodified samples. Leaving the discs in the APTES solution for longer than 2 s or RFGD modifying the discs for longer periods of time than prescribed in the Materials and Methods section resulted in APTES overlayers, which caused a sputtered appearing surface. No surface differences between electrically charged and uncharged PVDF and FEP were observed with SEM.

ESCA analysis of OH and APTES modification

The RFGD plasma process resulted in a homogeneous deposition of OH and APTES on the surface as verified by ESCA (Table 1). For RFGD plasma modified PVDF, the surface atomic percentage of oxygen was 8%. APTES modified PVDF showed surface atomic percentages of 15% for oxygen and 10% for nitrogen. For RFGD plasma modified FEP, the surface atomic percentage of oxygen was 6%. APTES modified FEP showed surface atomic percentages of 16% for oxygen and

Table 1.
Surface elemental composition of RFGD and APTES treated PVDF and FEP material as ascertained via ESCA analysis. No other surface elements were observed. See text for details. Data ± 5% RSD.

	Carbon	Fluorine	Oxygen	Nitrogen	Silicone
PVDF-OH	67	25	8	—	—
PVDF-NH_2	50	15	15	10	10
FEP-OH	57	37	6	—	—
FEP-NH_2	54	12	16	9	9

9% for nitrogen. For all materials, the only other surface elements observed were carbon, fluorine and silicone (Table 1). Unmodified PVDF and FEP showed only surface carbon and fluorine in the expected bulk atomic ratios. Due to the spot beam resolution of the instruments used, ESCA was not used to evaluate stripe-modified specimens, which had stripes only 300 μm wide.

Electrical output

The piezoelectric output of surface modified, poled PVDF films was similar to that seen with unmodified, poled PVDF. Acting as sensitive mechanoelectrical transducers, poled PVDF dishes generated about 2–3 mV at 1200 Hz when placed in the tissue culture incubators used in the study [18]. Unpoled PVDF did not demonstrate any electrical output. Piezoelectric and non-piezoelectric PVDF substrates subjected to the OH and APTES modification steps did not exhibit any changes in their electrical properties; i.e. modified piezoelectric PVDF generated similar outputs as unmodified, piezoelectric PVDF, while non-piezoelectric PVDF showed no electrical output.

FEP electrets were prepared after OH and APTES modification. The modification steps did not result in any charge trapping in FEP. Selected FEP discs were charge-injected using conditions identical for unmodified FEP. The charge retention characteristics of OH and NH_2 modified, positively charged films were similar to

Table 2.
Cell Attachment to PVDF Substrates (cells/cm^2) after 24 h in culture. The left and middle columns display data for unmodified and fully modified OH- and NH_2-containing PVDF in serum-containing (FCS) and serum-free (Opti-MEM) media. The right column represents data for stripe-modified PVDF with FCS media.

PVDF type	Media Type FCS	Opti-MEM	Stripe-modified
Unpoled	3166	3783	—
Poled	2960	3145	—
OH-unpoled	4390	4079	16 113
OH-poled	3778	3451	16 393
NH_2-unpoled	4124	1666	21 484
NH_2-poled	4120	1992	19 985

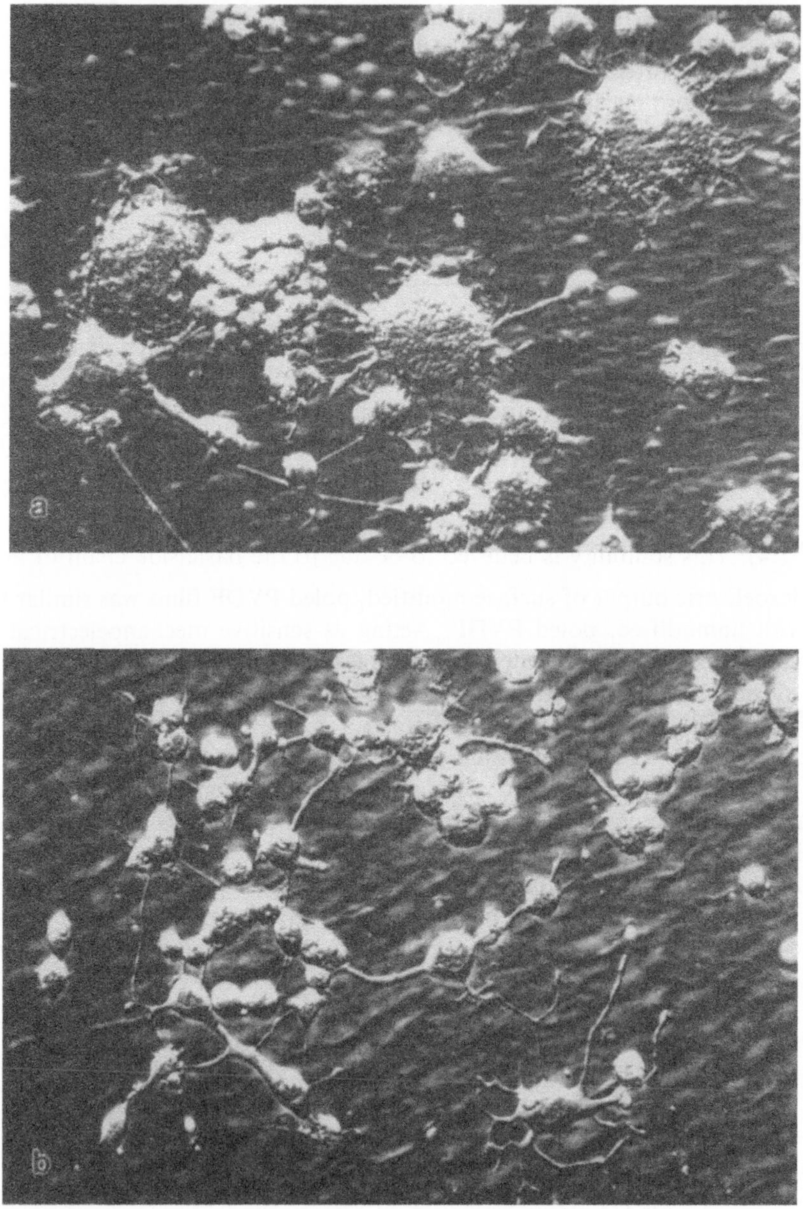
a
b

Figure 1. Light photomicrographs of mouse neuroblastoma (Nb2a) cells cultured on piezoelectric, fully NH_2 modified polyvinylidene fluoride substrates after 48 h in culture. (A) Cells in serum-containing media. Note the flattened appearance of the cell bodies and short neurites. (B) Cells in serum-free media. Note the predominance of rounded cell bodies and diminished cell spreading. Note also the greater number of nerve fibers extending (original magnification × 63).

that seen with unmodified, charged FEP. FEP discs initially charged to $+1023 \pm 44$ V showed a rapid decay to 357 ± 42 V after exposure to tissue culture media; discs initially charged to $+1500 \pm 54$ V showed a rapid decay to a level of $+493 \pm 31$ V. The voltage decay plateaued by 2 h and remained unchanged for the next 96 h [19].

Neuronal attachment, morphology and outgrowth

Piezoelectric PVDF substrates

1. Fully modified PVDF. (a) Serum-containing media. Cells cultured on amine-rich PVDF in serum-containing media often showed a flatter morphology and shorter, broader neurites than cells cultured on hydroxyl-rich and unmodified PVDF (Fig. 1A). Occasional cells on OH-rich and unmodified PVDF also showed flattening. No differences in cell morphology or the orientation of neurite outgrowth were observed between cells on piezoelectric vs non-piezoelectric PVDF. Substrates with OH surfaces supported similar levels of cell attachment as NH_2 surfaces (Table 2). The level of attachment was 25% higher on modified vs virgin PVDF surfaces. No differences in attachment between non-piezoelectric and piezoelectric materials were observed. All surfaces supported attachment throughout the 96-h study.

Nb2a cells cultured on piezoelectric unmodified, OH- and NH_2-rich PVDF showed significantly greater levels of neuronal differentiation than cells cultured on corresponding, unpoled substrates (Fig. 2A). The degree of differentiation increased gradually over time for all materials (data not shown). Differentiation levels on OH-modified PVDF were similar to those seen on unmodified PVDF. Differentiation levels on NH_2-modified PVDF, however, were lower than seen on OH-modified or unmodified PVDF. The degree of differentiation on the piezoelectric amine-rich surface was similar to that seen on the unmodified, non-piezoelectric surface. Piezoelectric materials supported greater neurite lengths than non-piezoelectric ones (Fig. 2B). Neurites extended on virgin and OH-modified piezoelectric PVDF were significantly longer than those extended on NH_2-modified PVDF.

(b) Serum-free media. Cells cultured in serum-free media were less flattened and showed a rounder morphology and extended thinner, straighter neurites than cells cultured in FCS (Fig. 1B). Cells grown on NH_2 surfaces did occasionally show a more flattened morphology than cells on OH or virgin PVDF surfaces, although the degree of spread was not as great as observed in serum-containing conditions. Attachment levels were similar for virgin and OH-modified substrates (Table 2). Cell attachment densities on NH_2-modified surfaces were about half that seen on the other surfaces. There were no differences in attachment between non-piezoelectric and piezoelectric materials. Quantitative analysis was not performed beyond 48 h since many cells cultured on virgin and OH-rich PVDF began detaching by 72 h.

As expected, serum-free conditions supported greater levels of differentiation and outgrowth than in the presence of serum. For all three substrate types, differentiation was significantly greater on piezoelectric vs non-piezoelectric materials (Fig. 3A). Unmodified and OH-containing surfaces supported similar levels of

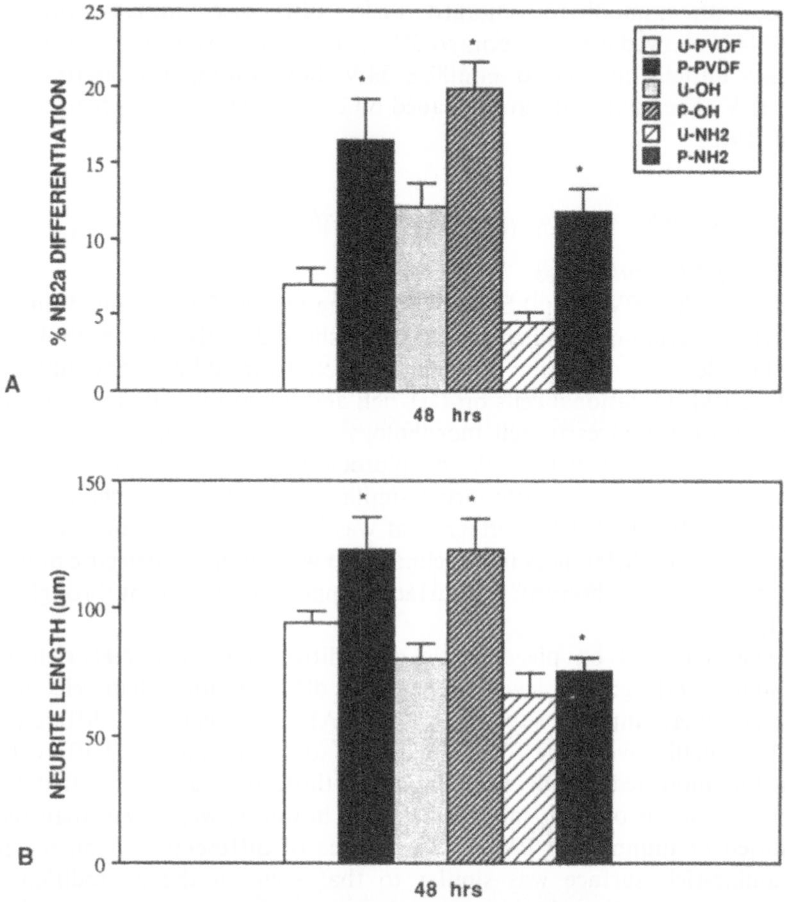
U-PVDF
P-PVDF
U-OH
P-OH
U-NH2
P-NH2
% NB2a DIFFERENTIATION
25
20
15
10
5
0
48 hrs
A
NEURITE LENGTH (um)
150
100
50
0
48 hrs
B

Figure 2. Histogram showing Nb2a differentiation and neurite outgrowth on unmodified and fully modified polyvinylidene fluoride (PVDF) in serum-containing media at 48 h. (A) The percentage of Nb2a cell differentiation. (B) The length of neurites extended by cells in A (U = non-piezoelectric and P = piezoelectric PVDF, OH = fully modified surfaces containing hydroxyl groups and NH_2 = fully modified surfaces containing amine groups). Scale bar = standard deviation, * = $p < 0.01$.

outgrowth which were somewhat greater than that seen on NH_2-rich PVDF, although the difference was not statistically significant. All of the non-piezoelectric substrates showed similar levels of outgrowth. In the absence of serum, neurite lengths were greater on piezoelectric vs non-piezoelectric PVDF on all surfaces (Fig. 3B).

2. Stripe modified PVDF. (a) Serum-containing media. In the presence of serum, Nb2a cells preferentially attached to NH_2-rich (Fig. 4A, B) and OH-rich stripes. This attachment pattern was observed after 6–8 h in culture and was observed through at least 96 h. Occasionally, cells attached between stripes and neurites projected from a striped region out to an unmodified region (Fig. 4B). This was especially true after 48 h when cells began clumping and overgrowing the striped regions. Again, cells on NH_2 surfaces were flatter morphologically than cells on OH regions. Attachment levels on NH_2 stripes were 25% higher than that seen on OH

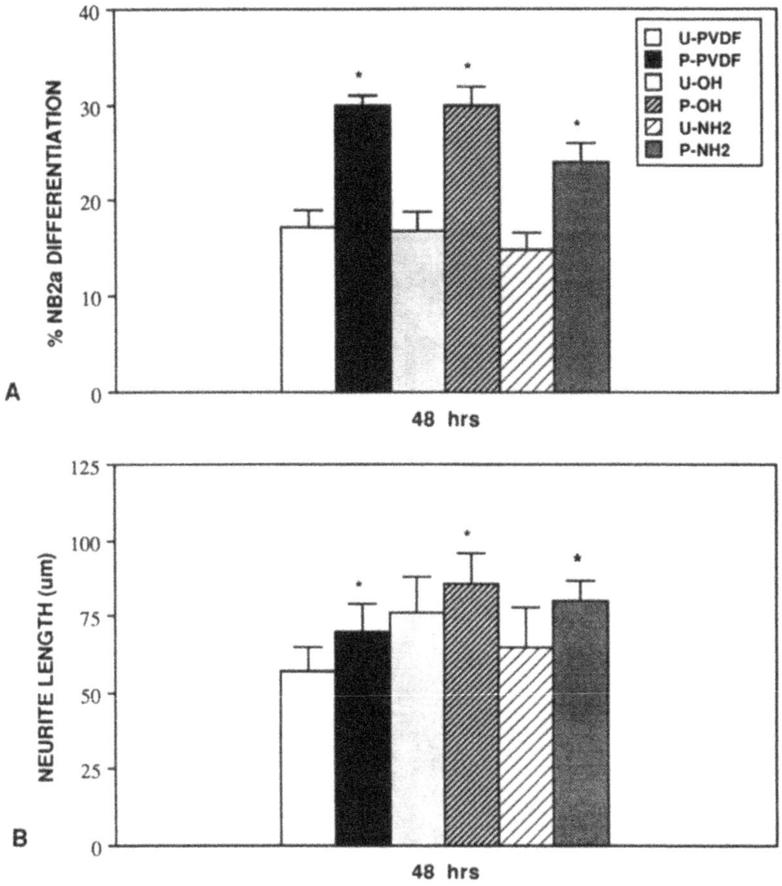
A
% NB2a DIFFERENTIATION
40
30
20
10
0
48 hrs
U-PVDF
P-PVDF
U-OH
P-OH
U-NH2
P-NH2
B
NEURITE LENGTH (um)
125
100
75
50
25
0
48 hrs

Figure 3. Histogram showing Nb2a differentiation and neurite outgrowth on unmodified and fully modified polyvinylidene fluoride (PVDF) in serum-free media at 48 h. (A) The percentage of Nb2a cell differentiation. The length of neurites extended by cells in A (U = non-piezoelectric and P = piezoelectric PVDF; OH = fully modified surfaces containing hydroxyl groups and NH_2 = fully modified surfaces containing amine groups). Scale bar = standard deviation, * = $p < 0.01$.

stripes (Table 2). Attachment on virgin PVDF were similar to results for the first experiment and were only 25% of that seen on striped regions. Increased attachment density was due to the fact that the cells only adhered to striped regions, whose total surface area was only about 25% that of the entire substrate.

The degree of differentiation was greater on piezoelectric vs non-piezoelectric substrates for NH_2, OH stripes, and unmodified PVDF (Fig. 5A). The degree of differentiation was similar for all surfaces. Differentiation levels were similar on all non-piezoelectric substrates also. Quantitative analysis was not performed after 48 h in culture due to cell crowding and overgrowth on the striped regions. Neurite lengths were greater on piezoelectric vs non-piezoelectric substrates for all chemistries (Fig. 5B). Neurite lengths on NH_2-modified PVDF were shorter than on OH-modified and unmodified PVDF.

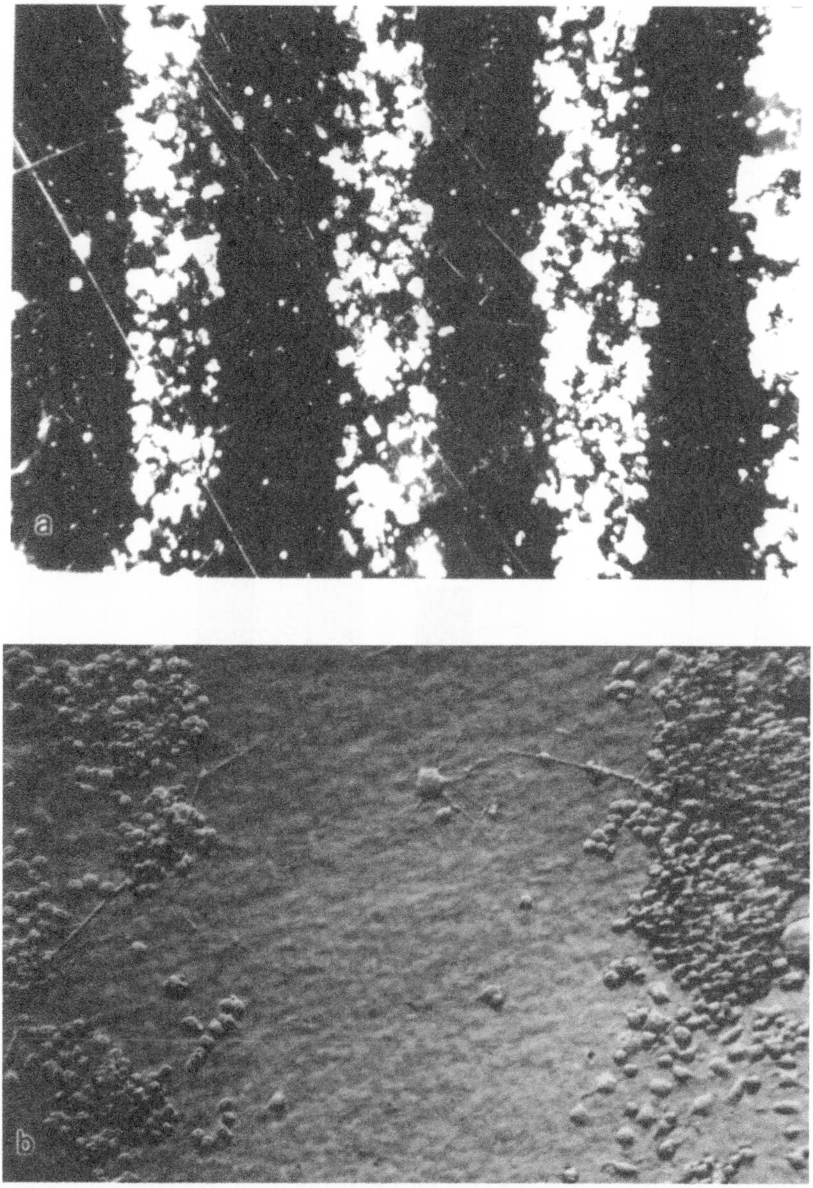

Figure 4. Light photomicrographs showing Nb2a cells cultured on NH_2 stripe-modified polyvinylidene fluoride substrates after 72 h in culture in serum-containing media. (A) Low power view showing Nb2a cells adhering preferentially to NH_2 stripes. Note the presence of numerous cell clumps indicating cell overgrowth (original magnification × 10). (B) Higher power view showing unmodified PVDF region surrounded by two NH_2 modified stripes. Note the large neuron attached on the unmodified region and extending a long neurite toward an NH_2-containing stripe. On the lefthand NH_2-modified stripe, a neurite is extending outward to an unmodified region (original magnification × 32).

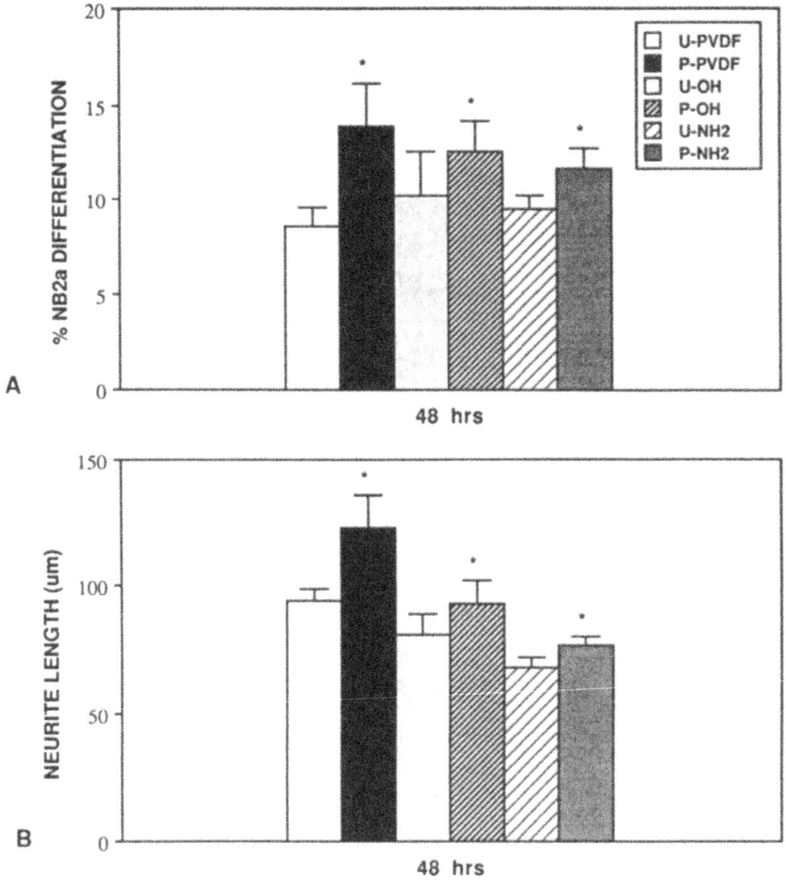

20
15
10
5
0
% NB2a DIFFERENTIATION
A
48 hrs
U-PVDF
P-PVDF
U-OH
P-OH
U-NH2
P-NH2
150
100
50
0
NEURITE LENGTH (um)
B
48 hrs

Figure 5. Histogram showing Nb2a differentiation and neurite outgrowth on stripe-modified polyvinylidene fluoride (PVDF) in serum-containing media at 48 h. (A) The percentage of Nb2a cell differentiation. (B) The length of neurites extended by cells in A (U = non-piezoelectric and P = piezoelectric PVDF; OH = stripe-modified surfaces containing hydroxyl groups and NH_2 = stripe-modified surfaces containing amine groups). Scale bar = standard deviation, * = $p < 0.01$.

(b) Serum-free media. In serum-free media, cells attached randomly and no striping was observed at any point in the study. No detectable differences in cell morphology or neurite outgrowth were observed in striped and non-striped regions. Differentiation levels were generally greater on piezoelectric vs non-piezoelectric PVDF. Since the precise location of stripe-modification was not apparent, it was not possible to accurately quantify differentiation on modified vs non-modified regions. Therefore, no quantitative analysis was performed.

FEP electret substrates

1. Fully modified FEP. (a) Serum-containing media. As was observed in the modified PVDF study, cells cultured on amine-rich FEP showed a flatter morphology and broader neurites than cells cultured on hydroxyl-rich or unmodified FEP (Fig. 6A). No differences in cell body shape or the orientation of neurite outgrowth were observed between charged and uncharged substrates. Attachment

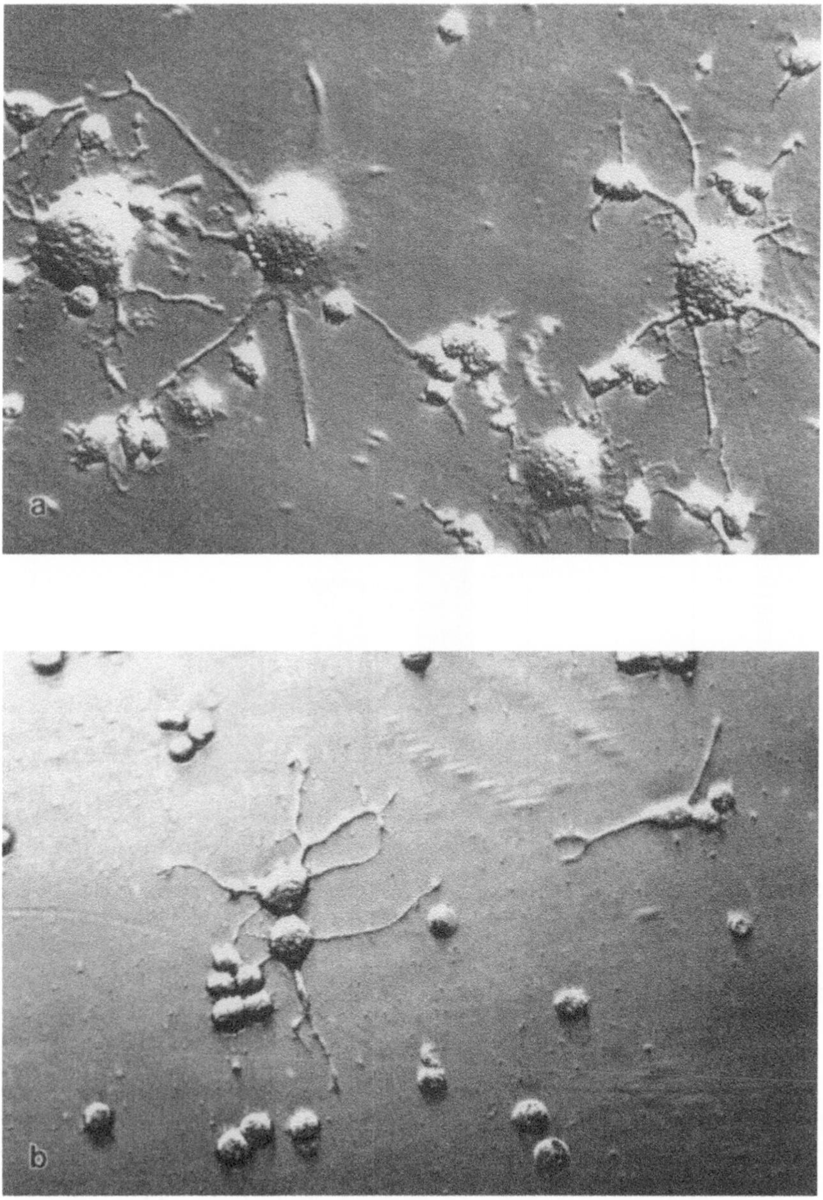

Figure 6. Light photomicrographs of mouse neuroblastoma cells cultured on positively charged, fully NH_2 modified fluorinated ethylene propylene substrates after 48 h in culture. (A) Cells in serum-containing media. Note the flattened appearance of the cell bodies and complex neurites. (B) Cells in serum-free media. Note the predominance of rounded cell bodies and diminished cell spreading (original magnification × 63).

Table 3.
Cell Attachment to FEP Substrates (cells/cm^2) after 24 h in culture. The left and middle columns display data for unmodified and fully modified OH- and NH_2-containing FEP in serum-containing (FCS) and serum-free (Opti-MEM) media. The right column represents data for stripe-modified FEP with FCS media.

FEP type	Media Type FCS	Opti-MEM	Stripe-modified
Uncharged	4445	5970	—
(+) charged	4776	5107	—
OH-uncharged	7074	6840	26 195
OH-(+) charged	6928	7119	22 995
NH_2-uncharged	6692	6346	25 599
NH_2-(+) charged	8107	6140	27 714

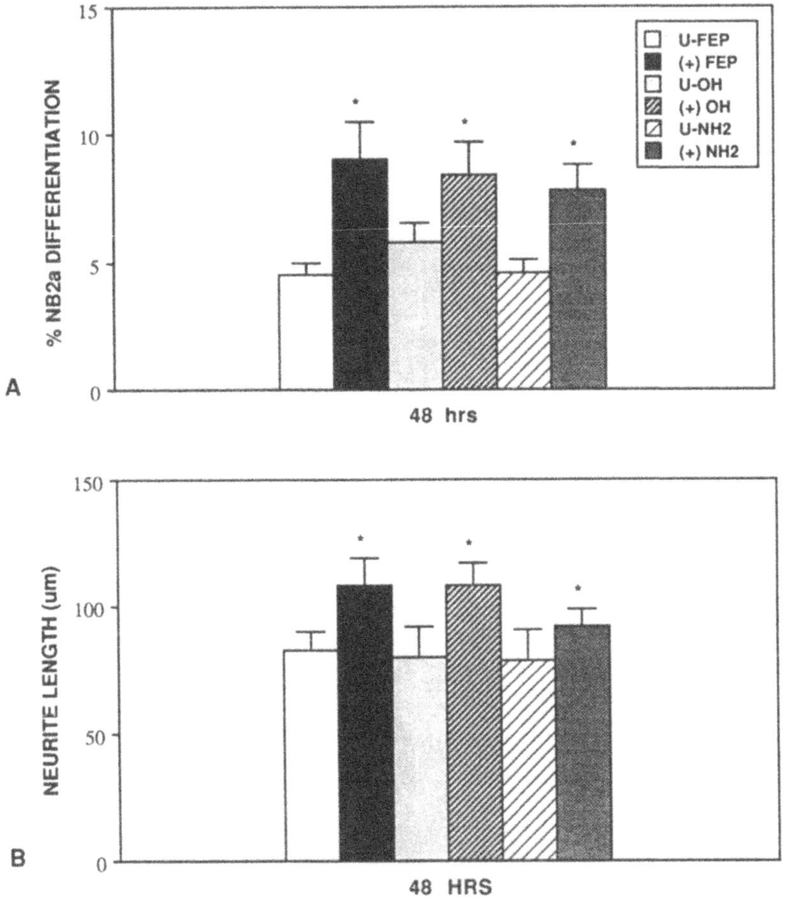

A
15
10
5
0
% NB2a DIFFERENTIATION
48 hrs
U-FEP
(+) FEP
U-OH
(+) OH
U-NH2
(+) NH2
B
150
100
50
0
NEURITE LENGTH (um)
48 HRS

Figure 7. Histogram showing Nb2a differentiation and neurite outgrowth on unmodified and fully modified fluorinated ethylene propylene (FEP) in serum-containing media at 48 h. (A) The percentage of Nb2a cell differentiation. (B) The length of neurites extended by cells in A (U = uncharged and (+) = positively charged FEP; OH = fully modified surfaces containing hydroxyl groups and NH_2 = fully modified surfaces containing amine groups). Scale bar = standard deviation, * = $p < 0.01$.

levels were greater on OH and NH_2 modified vs unmodified FEP by about 25% (Table 3).

Nb2a cells cultured on positively charged OH- and NH_2-rich and unmodified FEP showed greater levels of neuronal differentiation than cells cultured on corresponding, uncharged substrates (Fig. 7A). The degree of differentiation was similar for all groups. Neurite lengths were greater on positively charged vs uncharged surfaces for all chemistries (Fig. 7B). The longest neurites were observed on virgin and OH-rich, positively charged FEP.

(b) Serum-free media. Cells cultured on NH_2 surfaces often showed a somewhat more flattened morphology than cells on other substrates, although cells were not as flattened as observed in serum-containing media (Fig. 6B). Attachment levels were similar on all surfaces, although OH-modified surfaces supported a somewhat greater level of attachment (Table 3). By 48–72 h many cells on unmodified and OH-rich substrates began to detach.

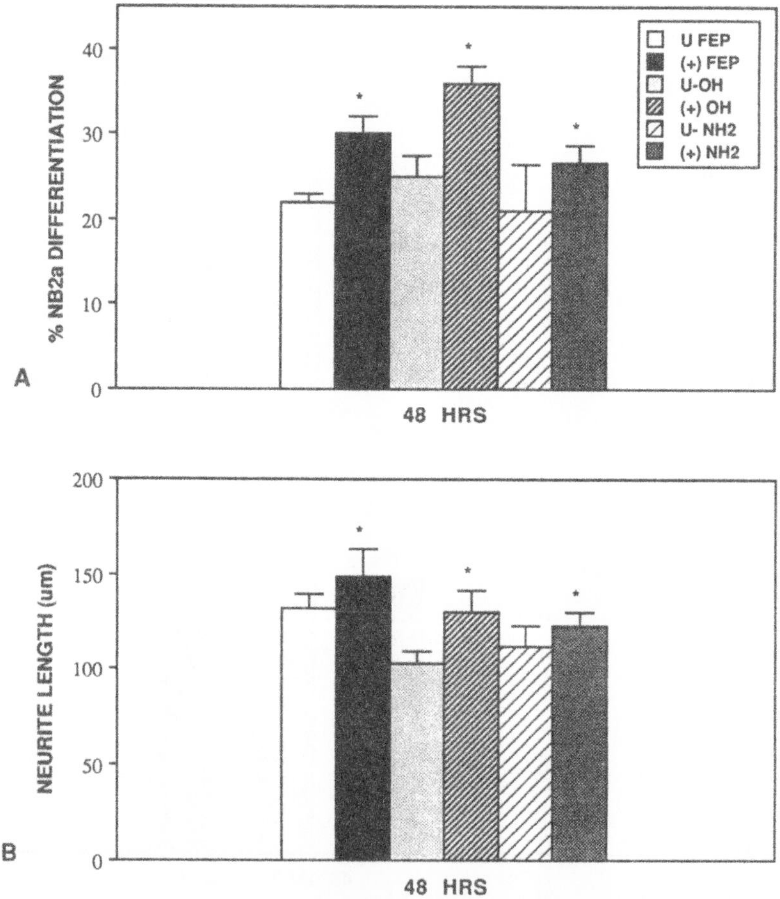
A
% NB2a DIFFERENTIATION
0
10
20
30
40
48 HRS
U FEP
(+) FEP
U-OH
(+) OH
U- NH2
(+) NH2
B
NEURITE LENGTH (um)
0
50
100
150
200
48 HRS

Figure 8. Histogram showing Nb2a differentiation and neurite outgrowth on unmodified and fully modified fluorinated ethylene propylene (FEP) in serum-free media at 48 h. (A) The percentage of Nb2a cell differentiation. (B) The length of neurites extended by cells in A (U = uncharged and (+) = positively charged FEP; OH = fully modified surfaces containing hydroxyl groups and NH_2 = fully modified surfaces containing amine groups). Scale bar = standard deviation, * = $p < 0.01$.

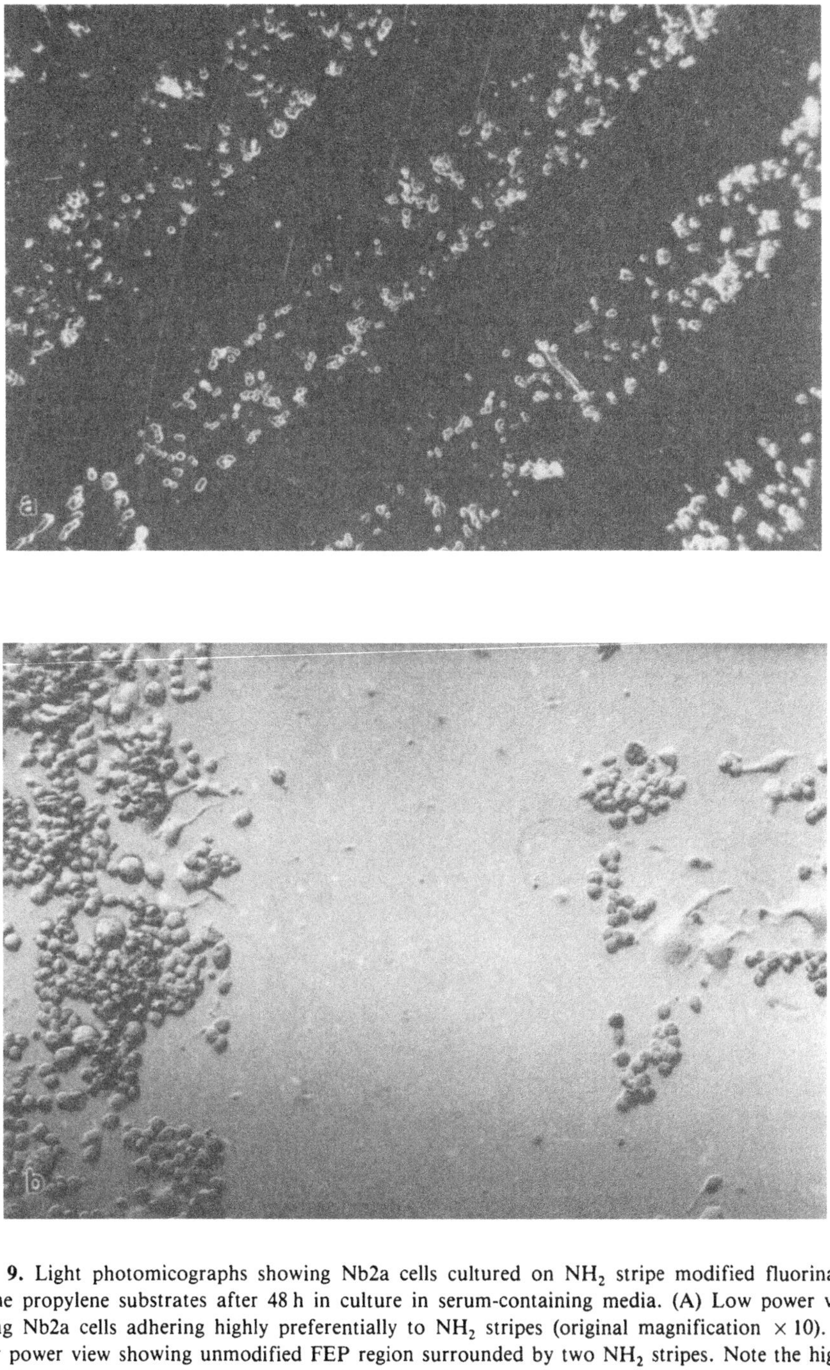
a
b

Figure 9. Light photomicrographs showing Nb2a cells cultured on NH_2 stripe modified fluorinated ethylene propylene substrates after 48 h in culture in serum-containing media. (A) Low power view showing Nb2a cells adhering highly preferentially to NH_2 stripes (original magnification $\times$ 10). (B) Higher power view showing unmodified FEP region surrounded by two NH_2 stripes. Note the highly preferential alignment of cells to the NH_2 stripes. No cells have attached to the unmodified region and no neurites are extending from a modified to an unmodified region (original magnification $\times$ 32).

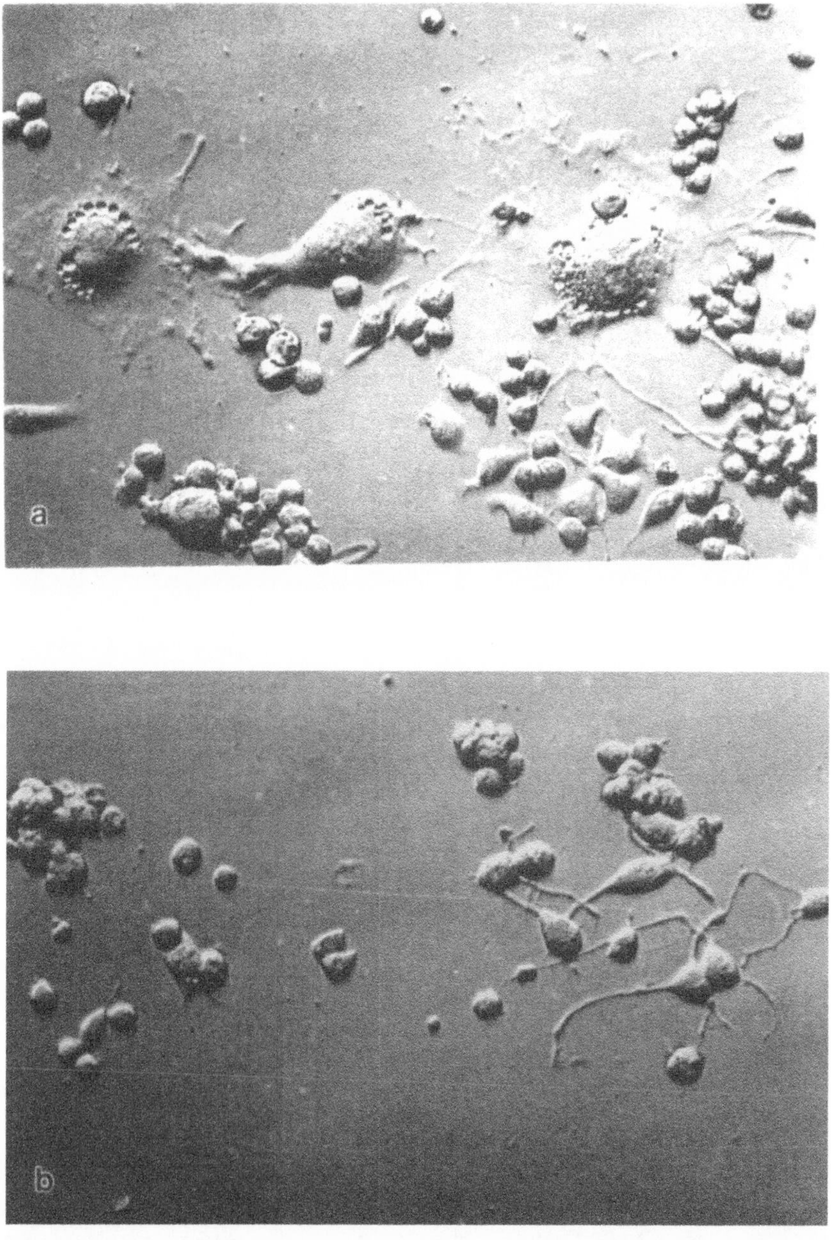
a
b

Figure 10. High power photomicrographs showing Nb2a cells cultured on stripe-modified FEP substrates in the presence of serum after 48 h. (A) Cells confined to an NH_2-containing region. Note the flattened appearance and veil-like lamellopodial structure of the cell bodies. (B) Cells confined to an OH-containing region. Note that cells are rounder and less spread and that neurites are thinner (original magnification × 63).

Serum-free conditions supported greater levels of differentiation and outgrowth than seen with serum present. By 48 h, cells cultured on positively charged FEP showed greater levels of differentiation than cells on uncharged FEP for all substrate chemistries (Fig. 8A). Differentiation was significantly greater on positively charged OH-vs NH_2-containing FEP. By 72 h, most cells on unmodified FEP had detached and no quantitative analysis was performed. Neurite lengths were significantly greater on positively charged FEP for all groups (Fig. 8B). Neurite lengths were somewhat greater on unmodified FEP than on OH- and NH_2-modified FEP, although the difference was not statistically significant.

2. Stripe modified FEP. (a) Serum-containing media. In the presence of serum, Nb2a cells showed a highly preferential attachment to NH_2-rich (Fig. 9A, B) and OH-rich stripes which was more striking than that observed with modified PVDF. This attachment pattern was apparent after 4–6 h in culture and was observed up to and beyond 96 h for NH_2 stripes. In contrast, cell detachment was observed starting

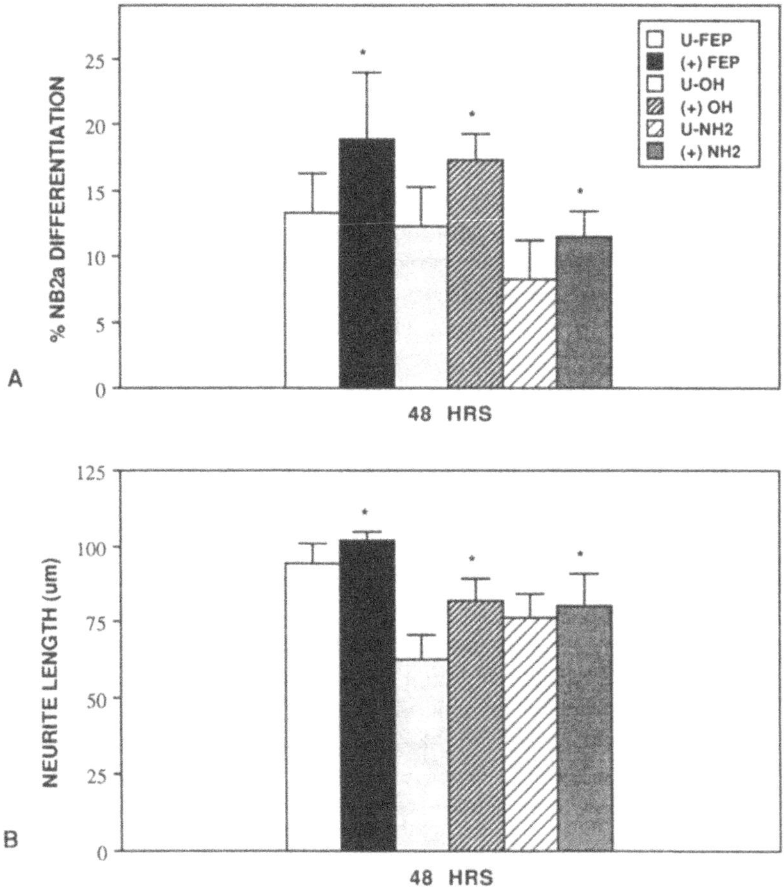

A
% NB2a DIFFERENTIATION
0
5
10
15
20
25
U-FEP
(+) FEP
U-OH
(+) OH
U-NH2
(+) NH2
48 HRS
B
NEURITE LENGTH (um)
0
25
50
75
100
125
48 HRS

Figure 11. Histogram showing Nb2a differentiation and neurite outgrowth on stripe-modified fluorinated ethylene propylene (FEP) in serum-containing media at 48 h. (A) The percentage of Nb2a cell differentiation. (B) The length of neurites extended by cells in A (U = uncharged and (+) = positively charged FEP; OH = stripe-modified surfaces containing hydroxyl groups and NH_2 = stripe-modified surfaces containing amine groups). Scale bar = standard deviation, * = $p < 0.01$.

at 48 h on OH stripes. Cells rarely attached between stripes and neurites never projected from a striped region out to an unmodified region (Fig. 9B). Cells confined to NH_2 stripes showed greater spreading than cells on OH stripes (Fig. 10A, B). Attachment levels were somewhat greater on NH_2 vs OH stripes independent of bulk charging (Table 3). It should be noted that cell density was much greater on modified vs unmodified substrates since cells attached only to striped regions.

The degree of differentiation was significantly greater on +1500 V positively charged vs uncharged substrates for NH_2, OH and unmodified material (Fig. 11A). The greatest degree of enhancement was seen on unmodified and OH-modified FEP, although charged substrates always supported greater levels of outgrowth than uncharged ones. Neurite lengths were also significantly greater on positively charged vs uncharged FEP (Fig. 11B). The longest neurites were observed on charged, unmodified FEP.

(b) Serum-free media. As observed with PVDF substrates, cells plated in serum-free media attached randomly and no striping was observed. This was true for positively charged and uncharged FEP. No quantitative analysis was performed.

DISCUSSION

The purpose of the present study was to culture neurons on electrically charged substrates with covalently bound amine and hydroxyl surface groups. Covalent modification was performed on the entire substrate surface or in discrete striped regions. Quantitative electrical charge analysis and surface characterization showed that the substrates were predictably and reproducibly fabricated. Surface and electrical characterization results can be summarized as follows: (1) covalent OH and NH_2 modification of PVDF and FEP can be accomplished without altering surface texture or bulk properties; (2) both patterned and total surface modification is possible; (3) chemically modified, piezoelectric PVDF and electrically charged FEP generate electrical outputs similar to those observed with unmodified, charged materials; and (4) chemically modified, non-piezoelectric PVDF and uncharged FEP do not show any electrical output. In the present study, neither the charging nor the surface modification processes altered the topographic, optical or bulk properties of the fluoropolymer substrates.

Limits of spatial resolution for the ESCA instrumentation used prevented evaluation of stripe-modified substrates in the present study. As previously reported, however, time of flight secondary ion mass spectroscopy (ToF-SIMS) can be used to analyze masked, RFGD-modified fluoropolymer modification [24]. ToF-SIMS analysis of RFGD treated FEP with or without APTES modification revealed well demarcated regions of OH and APTES modification adjacent to virgin fluoropolymer [24].

Both OH and NH_2 containing surfaces increased cell attachment in serum containing media, although virgin FEP and PVDF support significant levels of attachment. There were no differences in cell attachment between electrically active and electrically neutral substrates featuring similar surface chemistries. These results suggest that the adsorption of serum-derived attachment factors, such as fibronectin, is enhanced on chemically modified material. Enhanced attachment could be due to increased adsorption of attachment factors or to protein/substrate conformations which are more favorable for attachment. It has been shown that the binding and conformational structure of fibronectin is influenced by surface properties of substrates [30]. It has been reported that increasing the number of amine groups favorably influences cell attachment on surface modified silicon [5]. For example, spinal and hippocampal neurons attach more readily and remain viable longer on synthetic diamines and triamines than on monoamines [5]. In contrast, the mouse neuroblastoma subclone Nb2a shows poorer attachment levels and decreased viability on substrates coated with diamines or the polyamine, polylysine (unpublished results). These differences could be related to differences between the cells used or differences in the underlying substrates. Thus, it is important to individualize the choice of coating factors and charge densities when optimizing *in vitro* conditions for particular culture substrates and cell types. In the absence of serum, cells generally displayed a round morphology and displayed little or no spreading. Early attachment levels were high in serum-free conditions for all substrates but varied significantly with time, depending on surface chemistry. It is not known why NH_2 containing PVDF supported the lowest early level of attachment and may be due to changes in protein adsorption. Detachment on virgin and OH surfaces occurred after 48 h while many cells on NH_2-containing PVDF and FEP surfaces were still attached after 96 h. The enhanced attachment and viability of cells on NH_2 surfaces may be due to substrate mediated cell signals enhancing cellular function as well as early events optimizing cell attachment.

In serum free media, neurite lengths on amine-rich surfaces were more similar to those on hydroxyl-rich and virgin fluoropolymer surfaces. This suggests that in suboptimal attachment environments (i.e. the absence of serum-derived factors), amine surface groups may enhance neurite outgrowth by providing a more permissive surface. In the presence of serum, however, neurite lengths were generally lowest on NH_2 containing surfaces and greatest on virgin fluoropolymers. In concert with serum-derived factors, NH_2 groups promote cell adhesion and flattening and reduced neurite outgrowth. These results concur with recent reports demonstrating that extracellular matrix materials promoting the most vigorous neurite outgrowth show less neurite adhesion than materials supporting significantly less outgrowth [31, 32]. More adhesive substrates may enhance cell attachment but may hinder process outgrowth. This fact must be considered when designing substrates used to promote neuronal outgrowth *in vitro* and *in vivo*.

The presence of hydroxyl and amine surface groups do not increase the percentage of neurons which differentiate (i.e. extend neurites) on PVDF and FEP. This was true for cells cultured on electrically neutral and electrically active OH- and NH_2-containing fluoropolymers. For similar surface chemistries, neuronal differentiation was always greater on electrically active vs electrically neutral ones. The levels of differentiation were lowest on NH_2 containing substrates under serum-containing conditions. The percentage of differentiating neurons was similar on virgin and OH-containing substrates. It is possible that for the mouse neuroblastoma cells used, the NH_2 group provides a signal causing the neuron to attach and flatten but not to extend neurites. This signal is probably mediated by serum-derived factors since levels of differentiation were similar for all surface chemistries in the absence of serum. A recent study demonstrated that neurite outgrowth of a neuroblastoma cell line was influenced by chemically derived substrata in the presence of fibronectin [33]. As observed in the present study, amine containing surfaces supported less neurite outgrowth than other surface groups although attachment levels were similar [33].

Previous studies demonstrated that neuroblastoma cells cultured on electrically active PVDF and FEP substrates showed significantly higher levels of differentiation than cells on uncharged substrates [18, 19]. Enhanced differentiation was not related to differences in cell attachment or proliferation. The surface chemistry of charged and uncharged materials was identical, indicating that enhanced differentiation was due to the bulk charges. The present results suggest that charged surface groups are more important in influencing cell morphology than in generating signals to differentiate. The fact that a positively charged substrate (i.e. FEP electret) enhances neurite outgrowth, while a positively charged surface group, NH_2, does not, provides further evidence that bulk rather than surface charges are the main factor in eliciting Nb2a cell differentiation. While positively charged coating materials are known to enhance the attachment of several neuronal subclasses [11, 12], additional factors are needed in order to promote neurite outgrowth [34]. Electrical activity plays a similar role for the Nb2a cell line, since cells on NH_2 substrates demonstrate enhanced attachment and show greater differentiation levels on electrically active vs electrically neutral materials.

Results with the stripe-modified substrates confirmed results with fully modified substrates; electrically active material enhanced neuronal differentiation and NH_2 containing substrates showed greater flattening, improved long-term attachment and less neurite outgrowth. Enhanced differentiation on charged substrates was independent of attachment-related mechanisms since the degree of cell attachment was similar for charged and uncharged stripe-modified substrates. In an earlier study it was shown that FEP selectively bulk charged in striped regions did not show preferential attachment to the charged areas [19]. Thus, electrically charged material does not enhance differentiation through simple substratum/attachment mechanisms. Neurite lengths were shorter on striped vs fully modified substrates in general, probably because of the higher degree of cell packing.

Neurons showed a highly preferential attachment onto OH and especially NH_2 rich regions on stripe-modified materials. This was observed in serum-containing but not serum-free conditions and occurred within several hours after plating. This result is especially interesting since unmodified PVDF and FEP surfaces serve as suitable attachment substrates. Work in our laboratory suggests that serum-derived factors such as albumin may mediate this preferential attachment [35]. Preferential attachment was more striking on FEP vs PVDF substrates. Neurons on stripe-modified FEP rarely attached on unmodified regions and neurites were virtually never observed traversing onto virgin FEP. In contrast, cells on stripe-modified PVDF were more frequently observed in virgin regions and neurites did travel from modified to unmodified regions. The stricter adherence of cells to stripe-modified FEP vs PVDF may be due to the very low surface energy of FEP (water contact angle ~ 105 deg) contrasting with the relatively greater surface energy of PVDF (water contact angle ~ 85 deg). It is also possible that low levels of surface modification between stripes did occur on masked PVDF.

The increased attachment of cells onto NH_2 regions could be due to specific, favorable interactions of cells and attachment factors with positively charged groups. It is not possible to rule out more general interactions with the APTES overlayer, since the chemistry of APTES surface in aqueous environments has not been fully evaluated. The fact that cells did not show preferential attachment to chemically modified regions in the absence of serum indicates that charges or simple surface groups alone are not able to dictate attachment. In serum-free media, longer amine tethering groups or greater numbers of amines have also proven ineffective in promoting specific attachment of neurons [5]. Reactive OH or NH_2 surface groups may be used to couple more potent biological active molecules to fluoropolymers. The isolation of short amino acid sequences with biological activity have been described for several extracellular matrix attachment factors [36, 37]. For example, RGD and YIGSR are active sequences in fibronectin (and related proteins) and laminin, respectively. These small fragments show biological activity when non-specifically adsorbed to substrates, although activity levels are significantly less than seen with intact molecules. The covalent addition of short amino acid sequences to substrates has been described [38]. Current work in our laboratory is focusing on the covalent attachment of neurite-promoting and neuronal attachment sequences of laminin to fluoropolymer substrates.

Numerous biological tissues have been shown to possess piezoelectric and electret properties and to respond to electrical fields/charges [39, 40]. Electrically charged biomaterials have been reported to enhance tissue regeneration *in vivo*. For example, osteotomized bone in adult rabbits displays enhanced osteogenesis when repaired with piezoelectric foils [41]. Nerve regeneration is significantly enhanced when electrically active fluoropolymer tubes (electret and piezoelectric tubes) are used to repair severed nerves in adult rodents [42–44]. Since regenerating axons grow down the center of the tube lumens, enhanced outgrowth is due to electric field effects rather than surface properties, a finding substantiated in the present study. Advances in polymer surface modification have led to the development and improvement of many biomedical devices, including tissue implants and biosensors [45, 46]. Chemically engineered surfaces may be used to target cell attachment or to direct regenerating tissue. The results of the present study suggest that the bulk and surface properties of biomaterials play important roles in cell regulation. It may be possible to optimize tissue repair and enhance biosensor sensitivity by developing electrically charged templates featuring covalently attached growth peptides.

REFERENCES

1. P. C. Letourneau, in: *Molecular Bases of Neuronal Development*, pp. 269-293, G. M. Edelman, W. E. Gall, W. M. Cowan (Eds). John Wiley & Sons, New York (1985).
2. J. E. Davies, in: *Surface Characterization of Biomaterials*, p. 219, B. D. Ratner (Ed.). Elsevier, Amsterdam (1988).
3. A. S. G. Curtis, J. V. Forrester, C. McInnes and F. Lawrie, *J. Cell Biol.* **97**, 1500 (1983).
4. W. L. McKeen and R. G. Ham, *J. Cell Biol.* **71**, 727 (1976).
5. D. Kleinfeld, K. H. Kahler and P. E. Hockberger, *J. Neurosci.* **11**, 4098 (1988).
6. P. C. Letourneau, *Dev. Biol.* **44**, 92 (1975).
7. E. Yavin and Z. Yavin, *J. Cell Biol.* **62**, 540 (1974).
8. R. Timpl and M. Diadzek, Int. Rev. Exp. Pathol. **29**, 1 (1986).
9. D. S. Snow, V. Lemmon, D. A. Carrino, A. I. Caplan and J. Silver, *Exp. Neurol.* **109**, 111 (1990).
10. D. S. Snow, D. A. Steindler and J. Silver, *Dev. Biol.* **138**, 359 (1990).
11. H. Rauvala, *J. Cell Biol.* **98**, 1010 (1984).
12. V. W. Yong, H. Horie and S. U. Kim, *Dev. Neurosci.* **10**, 222 (1988).
13. C. T. Brighton, J. Black and S. R. Pollack (Eds). *Electrical Properties of Bone and Cartilage*. Grune and Stratton, New York (1979).
14. K. J. McLeod, R. C. Lee and H. P. Ehrlich, *Science* **236**, 1465 (1987).
15. L. F. Jaffe and M.-M. Poo, *J. Exp. Zool.* **209**, 115 (1979).
16. N. B. Patel and M.-M. Poo, *J. Neurosci.* **4**, 483 (1982).

17. N. B. Patel and M.-M. Poo, *J. Neurosci.* **4**, 2939 (1984).
18. R. F. Valentini, T. G. Vargo, J. A. Gardella and P. Aebischer, *Biomaterials* **13**, 183 (1992).
19. S. A. Makohliso, R. F. Valentini and P. Aebischer, *J. Biomed. Mat. Res.* in press (1993).
20. M. G. Broadhurst and G. T. Davies, in: *Electrets, Topics in Applied Physics 2nd edn*, p. 285, G. M. Sessler (Ed.). Springer Verlag, Berlin (1987).
21. A. J. Lovinger, *Science* **220**, 1115 (1983).
22. T. G. Vargo, J. A. Gardella, A. E. Meyer and R. E. Baier, *J. Polym. Sci. Part A: Polym. Chem.* **29**, 555 (1991).
23. D. J. Hook, T. G. Vargo, K. S. Litwiler, F. V. Bright and J. A. Gardella, *Langmuir* **7**, 142 (1991).
24. T. G. Vargo, P. M. Thompson, L. J. Gerenser, R. F. Valentini, P. Aebischer, D. J. Hook and J. A. Gardella, *Langmuir* **8**, 130 (1992).
25. D. Schubert, S. Humphreys, F. DeVitryt and F. Jacob, *Dev. Biol.* **25**, 514 (1971).
26. P. G. Nelson, *Physiol. Rev.* **55**, 1 (1975).
27. J. E. Bottenstein and G. H. Sato, *Proc. Natl Acad. Sci.* **76**, 514 (1979).
28. D. Gurwitz and D. D. Cunningham, *Proc. Natl Acad. Sci.* **85**, 3440 (1988).
29. E. W. Anderson, L. L. Blyer, G. E. Johnson and G. L. Link, in: *Electrets, Charge Storage and Transport in Dielectrics*, p. 424, M.-M. Perlman (Ed.). Electrochemical Soc., Princeton, NJ (1973).
30. F. Grinnell, *Ann. N.Y. Acad. Sci.* **516**, 280 (1987).
31. A. L. Caloff and A. D. Lander, *J. Cell Biol.* **115**, 779 (1991).
32. V. Lemmon, S. M. Burden, H. R. Payne, G. J. Elmslie and M. L. Hlavin, *J. Neurosci.* **12**, 818 (1992).
33. K. Lewandowska, E. Pergament, C. N. Sukenik and L. A. Culp, *J. Biomed. Mater. Res.* **26**, 1343 (1992).
34. S. Carbonnetto, D. Evans and P. Cochard, *J. Neurosci.* **2**, 610 (1987).
35. J. P. Ranieri, R. Bellamkonda, J. Jacob, T. G. Vargo, J. A. Gardella and P. Aebischer, *J. Biomed. Mat. Res.* in press (1993).
36. M. D. Pierschbacher and E. Ruoslahti, *Nature* **309**, 30 (1984).
37. M. Jucker, H. K. Kleinman and D. K. Ingram, *J. Neurosci. Res.* **28**, 507 (1992).
38. S. P. Massia and J. A. Hubbell, *Anal. Biochem.* **187**, 292 (1990).
39. R. Lipinski (Ed.), *Electronic Conduction and Mechanoelectrical Transduction in Biological Materials.* Marcel Dekker, New York (1982).
40. R. B. Borgens, K. R. Robinson, J. W. Vanable and M. E. McGinnis (Eds), *Electric Fields in Vertebrate Repair.* A. R. Liss, New York (1990).
41. E. Fukada, in: *Mechanisms of Growth Control*, p. 192, R. O. Becker (Ed.). C. C. Thomas, Springfield, MD (1982).
42. P. Aebischer, R. F. Valentini, P. Dario, C. Domenici and P. M. Galletti, *Brain Res.* **436**, 165 (1987).
43. R. F. Valentini, A. M. Sabatini, P. Dario and P. Aebischer, *Brain Res.* **480**, 300 (1989).
44. E. G. Fine, R. F. Valentini and P. Aebischer, *Biomaterials* **12**, 775 (1991).
45. A. S. Hoffman, *Clin. Mater.* **11**, 13 (1992).
46. A. S. Hoffman, *Art. Organs* **16**, 43 (1992).

Part I.B.

Surface Modification, Characterization, and Properties

Physico-Chemical Modification

New biomaterials through surface segregation phenomenon: New quaternary ammonium compounds as antibacterial agents

RONALD S. NOHR* and J. GAVIN MACDONALD

Kimberly-Clark Corporation, 1400 Holcomb Bridge Road, Roswell, GA 30076, USA

Received 5 May 1993; accepted 25 August 1993

Abstract—Five new trisiloxane quaternary ammonium compounds were synthesized from hydrotrisiloxane with allyl glycidyl ether to yield the epoxy function. Various amines were then reacted to yield trisiloxane amines which were further reacted to methyl substitute or oxidize the β-carbons in order to provide thermal stability. These new compounds were employed as melt additives in a nonwoven polypropylene fiber extrusion process to produce, through surface segregation, a new biomaterial with antibacterial properties.

Key words: Antimicrobial compounds; surface segregation; polydimethylsiloxane; quaternary ammonium compounds; polypropylene nonwovens.

INTRODUCTION

Surface modification of polypropylene nonwoven webs has proven to be a versatile approach to providing new and novel biomaterials [1–3]. However, most modifications have been confined to surface coatings. Attempts to increase properties such as coating durability and enhanced fluid transport have been limited for the most part to chemical grafting through the use of non-ionizing and ionizing radiation tools [4]. However, application of such tools to a complex fiber matrix, such as found in nonwoven fabrics, has found serious restrictions. These have included energy absorption phenomena which result in the outer fibers blocking the UV radiation from the inner fibers in the complex matrix [5], and in the case of ionizing radiation, in severe polymer degradation.

A promising new surface modification method for creating non-woven biomaterials for non-evasive applications has been developed. This innovative approach uses melt additives that spontaneously surface segregate during fiber formation thereby producing uniformly coated surfaces and eliminating the need for secondary process treatments.

This surface segregation approach was first demonstrated in the hydrophilic surface modification of polypropylene non-woven webs [6–8]. Using a low molecular weight silicone polyether copolymer as the melt additive, highly water wettable webs were produced which resulted from the surface active effects of the newly surface segregated compound. In addition to water wettability, these fabrics showed sustained and superior water adsorption and transport properties.

* To whom correspondence should be addressed.

The work described in this study extends this novel approach to producing antimicrobial surface attributes to polypropylene non-woven webs. Such a benefit may find non-evasive application in promoting skin wellness in such devices as incontinence pads. This may occur by reducing the microbial population in the restricted fluid volume between the skin and the material liner.

This paper centers on the synthetic methods for producing five new heptamethyltrisiloxane quaternary ammonium melt additives. Preliminary work examining their relative thermal process stability, proficiency in surface segregation during fiber formation, and their antibacterial efficacy as newly segregated surface active agents will be presented.

EXPERIMENTAL

Elemental analyses were performed by Schwarzkopf Microanalytical Laboratories (Woodside, NY). ^{1}H and ^{13}C NMR Spectra were run on 270 MHz and 360 MHz instruments by Spectral Data Services (Champaign, IL). ESCA analysis was performed by Surface Science Corporation (Mountain View, CA).

Samples for elemental analysis were Kogelruhr distilled. Surface tension measurements were performed using a Surface Tensiomat Model 21 (Fisher-Scientific) using the method P. L. du Douy.

Synthesis of 3-[3-(2,3-epoxypropoxy) propyl]-1,1,1,3,5,5,5,-hepta methyltrisiloxane **1** [9–12]

A 500 ml three-necked flash was equipped with a stirrer, addition funnel, and condenser and was continuously flushed with argon. The reaction flask was charged with 22.5 g (0.22 mol) allyl glycidyl ether (Aldrich), 50.0 g (0.22 mol) hydrotrisiloxane (Huls Americas) in 150 ml xylene. To the reaction mixture was dripped in a suspension of 2.8 g (0.03 mol) hexachloroplatinic acid (Aldrich) in 140 ml *n*-octyl alcohol. After the addition was complete the reaction mixture was heated at 100°C overnight. At the end of the reaction, the xylene was removed by rotary evaporation under vacuum. Selective extraction with hexane yielded the epoxy trisiloxane after solvent removal. Yield 68.4 g (94%). Analysis: theoretical %C 44.7, %H 9.3, %Si 26.8; found %C 44.3, %H 9.0, %Si 26.5; ^{1}H NMR ($CDCl_3$) 0.01 [*m*, $SiCH_3$], 0.60 [*m*, Si-CH_2-], and 3.60 [*m*, -CHO].

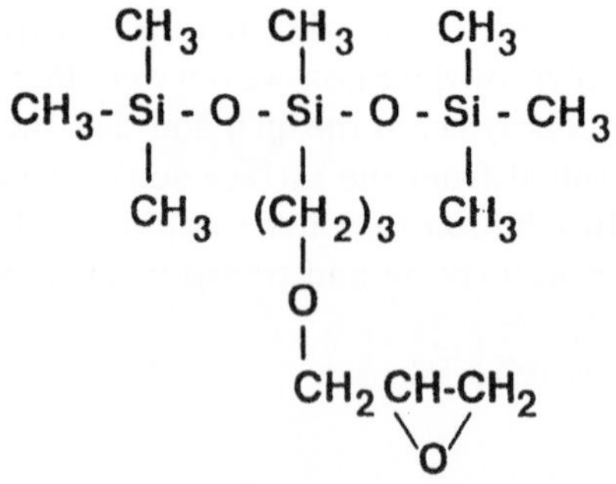

Scheme 1.

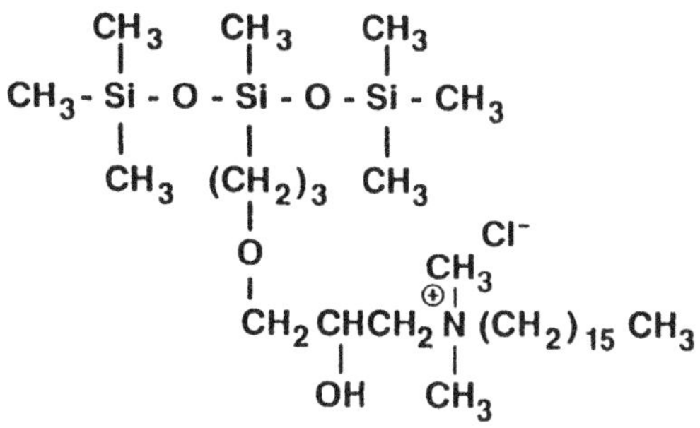

Scheme 2.

Synthesis of hexadecyl (2-hydroxy-3-[3-[1,3,3,3,-tetramethyl-1-(trimethylsiloxy) disiloxanyl] propoxy] propyl] dimethyl ammonium chloride **2** [13]

A 500 ml three-necked flask was equipped with a stirrer, addition funnel, dry ice/acetone condenser, thermometer, and electric heating mantle. The flask was continuously flushed with argon. The reaction flask was charged with 167.2 g (0.55 equivalent) of dimethyl hexadecyl ammonium chloride (Sartomer Chemical Company). 0.028 g (0.27 milliequivalent) of triethylamine (Aldrich) was transferred to the reaction vessel along with 250 g of isopropanol. To the stirred reactor contents 65.2 g of the epoxy trisiloxane was added over a 10-min period. The reaction mixture was stirred and heated to 80°C for 5 h to form a clear solution. The reaction mixture was cooled to ambient temperature and flushed overnight with dry argon. The solvent and other low-boiling materials were removed on a rota-evaporator at 45°C. 118 g (87%) yield of the trisiloxane quaternary ammonium salt was obtained as a light yellow oil. Analysis: theoretical %C 55.1, %H 10.7, %Si 12.6, %N 2.1; found %C 54.6, %H 10.4, %Si 12.1, %N 1.8; ^{1}H NMR ($CDCl_3$) 0.01 [*m*, $SiCH_3$], 0.60 [*m*, Si–CH_2–], 3.60 [*m*, –CH–O], and 2.86 [*m*, –N–CH_3].

Preparation of hexadecyldimethyl [3-[3-[1,3,3,3-tetramethyl-1-(trimethylsiloxy) disiloxanyl] propoxy] acetonyl] ammonium chloride **3**

A 250 ml three-necked flash was equipped with a stirrer, addition funnel, and condenser. The flask was charged with 14.5 g of chromium trioxide in 100 ml of water. To the solution was dripped in 50 g (0.07 mol) hexadecyl(2-hydroxy-3-[3-[1,3,3,3-tetramethyl-1-(trimethylsiloxy) disiloxanyl] propoxy] propyl] dimethyl

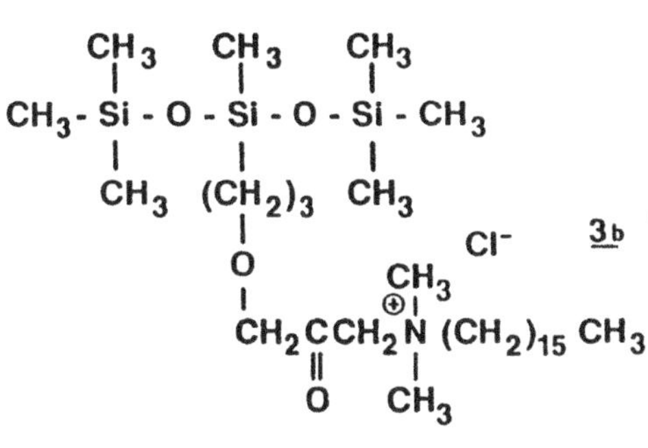

3b Tosylate ion

Scheme 3.

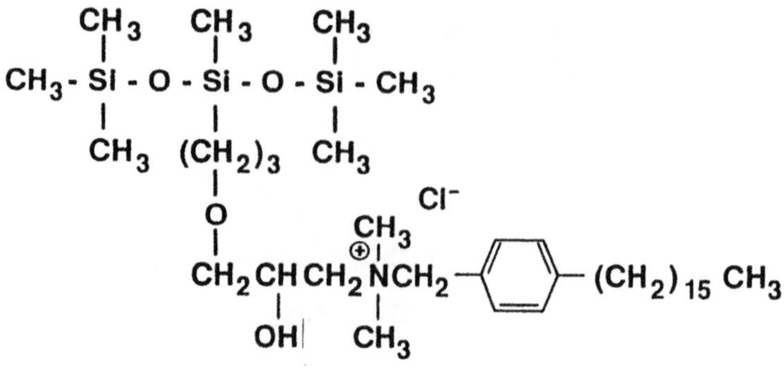

Scheme 4.

ammonium chloride (**2**) in 50 ml of tetrahydrofuran. The reaction mixture was stirred overnight and then poured into 200 ml of ice water and extracted with diethylether. The ether extract was dried, and the solvent removed on a rotary evaporator under vacuum to yield 47 g (94%) of a light yellow oil. Analysis: theoretical %C 55.3, %H 10.4, %Si 12.7, %N 2.1; found %C 54.7, %H 10.0, %Si 12.0, %N 1.7; IR (neat)-max 1740 (C=0), 1063 (N–C); ^{1}H NMR ($CDCl_3$) 0.01 [*m*, $SiCH_3$], 0.60 [*m*, Si–CH_2–], and 3.6 [*m*, –CH–0–).

Preparation of hexadecyldimethyl [3-[3-[1,3,3,3-tetramethyl-1-(trimethylsiloxy) disiloxanyl] propoxy] acetonyl] ammonium p-*toluenesulfonic acid sale* **3b**

A 250 ml three-necked flask equipped with a stirrer, addition funnel, and condenser was charged with 50.0 g (74 mmol) hexadecyldimethyl [3-[3-[1,3,3,3-tetramethyl-1-(trimethylsiloxy) disiloxanyl] propoxy] acetonyl] ammonium chloride (**3**) in 150 ml of isopropanol at ambient temperature. To the solution was added 57.5 g (0.30 mol) *p*-toluenesulfonic acid, sodium salt (Aldrich). The reaction mixture was stirred at ambient temperature of 8 h, after which 50 ml of water was added and the reaction mixture extracted with diethyl ether. Removal of the dried ether extract by rotary evaporation gave 53.4 g (94%). Analysis: theoretical %C 57.9, %H 10.3, %Si 11.2, %S 4.2, %N 1.8; found %C 57.6, %H 10.6, %Si 11.6, %s 3.9, %N 2.1; IR (neat) max 1740 (C=0), 1063 (N–C); ^{1}H NMR ($CDCl_3$) 0.01[*m*, Si–CH_3–], 0.60 [*m*, Si–CH_2–], and 3.60 [*m*, –CH–O–].

Preparation of trisiloxane **5** *from the Friedel–Crafts alkylation of* **8**

To a 500 ml three-necked flask fitted with a stirrer, addition funnel, and condenser being continuously flushed with argon was placed 20.0 g (0.03 mol) of **8**, 15.6 g (0.06 mol) 1-chlorohexadecane (Aldrich) and 200 ml hexane. The reaction mixture was cooled to 0°C using a crushed ice/salt bath and 2.0 g of anhydrous aluminium chloride added to the stirred reaction mixture. After 30 min the remaining 6.0 g of aluminium chloride (0.06 mol total) was added and the reaction slowly heated to 60°C over 4 h. After cooling, 100 g of crushed ice and 100 ml of water was slowly added and the organic layer washed with dilute hydrochloric acid, dried and the solvent removed under vacuum on a rotary evaporator. The oil was run through a short silica gel column using 30% ethyl acetate in hexane as eluent. Removal of the solvent gave 21.42 g (81%) of a colorless oil. Analysis: theoretical %C 62.5, %H

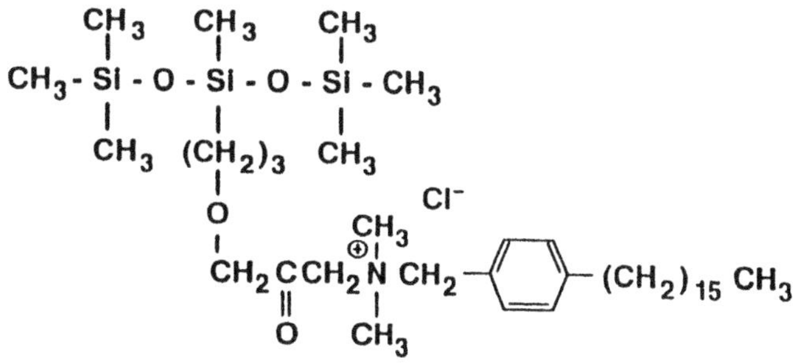

Scheme 5.

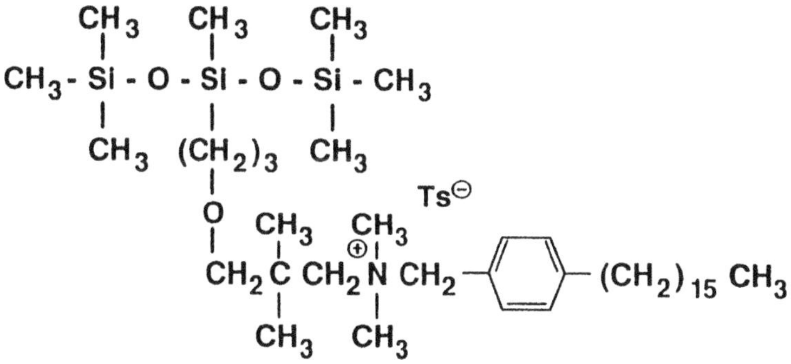

Scheme 6.

,10.4, %N 1.9, %Si 11.5; found %C 62.3, %H 10.6, %N 1.7, %Si 11.2; ^{1}H NMR ($CDCL_3$) 0.01 [*m*, Si–CH_3], 0.60 [*m*, Si–CH_2–], 2.85 [*m*, –N –CH_3], 3.56 [*m*, Ar –CH_2 –N], and 6.94 [*m*, *p*-substituted benzene].

Preparation of (p-*hexadecylbenzyl*) [*2-hydroxy-3-*[3-[*1,3,3,3-tetrameth l-1-(trimethylsiloxy) disiloxanyl*] *propoxy*) *propyl*) *dimethyl ammonium* p-*toluene sulfonic acid salt* **6**

Trisiloxane **6** was prepared using the synthetic pathway shown in Scheme 7. Experimental details of the synthetic steps are described below.

Oxirane ring opening with benzyl amine to prepare trisiloxane **7**

Into a 500 ml three-necked flask equipped with a stirrer, addition funnel, and condenser was placed 100 g (0.30 mol) epoxy trisiloxane **1** in 200 ml isopropanol. 42.8 g (0.4 mol) benzyl amine (Aldrich) in 100 ml isopropanol was slowly dripped into the reaction mixture at room temperature. The resultant solution was heated at 80°C for 8 h after which time the solvent was removed under vacuum using a rotary evaporator. The resultant oil was run through a short silica column using 10% ethyl accetate in hexane as eluent. 127.8 g (97%) yield of a colorless oil was obtained. Analysis: theoretical %C 48.7, %H 9.2, %Si 18.9, %N 3.1; found %C 48.4, %H 9.5, %Si 18.4, %N 3.0; IR (neat) max 3300 (OH), 1063 (C-N); ^{1}H NMR ($CDCl_3$) 0.01 [*m*, $SiCH_3$], 0.60 [*m*, Si–CH_2–], 3.60 [*m*, –CH–O], and 3.56 [*m*, Ar–CH_2–N].

Oxidation followed by methylation of **7**

The oxidatin of the alcohol function to the ketone was similar to that described earlier in the preparation of **3**.

Methylation of the amine was carried out as follows. In a 500 ml three-necked flask fitted with a stirrer, addition funnel, and condenser was added 80.0 g (0.18 mol) of the above ketone product in 200 ml dimethyl sulfate (Aldrich) and the resultant mixture heated at reflux for 8 h after which the solvent and other volatiles were removed under vacuum on a rotary evaporator. The oil was run through a short silica gel column using 30% ethyl acetate in hexane as eluent. 78.4 g (92%) yield of a colorless oil was obtained. IR (neat) max 1730 (C=0); ^{1}H NMR ($CDCl_3$) 0.01 [*m*, $SiCH_3$], 0.60 [*m*, Si–CH_2–], 2.85 [*m*, –N–CH_3], and 3.56 [*m*, Ar–CH_2–N].

Preparation of **9** *by exhaustive methylation of* **8** *by trimethyl aluminium*

Exhaustive methylation of **8** was carried using a method described by Meisterg and Mole [14].

In a thick walled glass tube was placed 20.0 g (0.04 mol) of the trisiloxane ketone **8** in 40 ml of benzene (Aldrich), 0.4 ml water, and 8.6 g (0.12 mol) trimethyl aluminium (Aldrich). The glass tube was sealed, placed in a steel bomb which was then heated at 140°C for 8 h. After cooling to ambient temperature the glass tube was carefully opened and the contents dripped into a mixture of 200 ml diethyl ether and 50 ml 0.5 *N* hydrochloric acid chilled in an ice bath. Separation of the organic layer, drying, and removal of the solvent under vacuum using a rotary evaporator gave 9.5 g (46%) of the dimethyl trisiloxane **9**. ^{1}H NMR ($CDCl_3$) 0.01 [*m*, Si–CH_3], 0.60 [*m*, Si–CH_2–], 0.81 [*m*, –$(CH_3)_2$], 2.85 [*m*, –N–CH_3], and 3.56 [*m*, Ar–CH_2–N].

Friedel–Crafts alkylation of **9**

To a 500 ml three-necked round bottom flask fitted with a stirrer, addition funnel, and condenser being continuously flushed with argon was placed 20.0 g (0.04 mol) of **9**, 15.6 (0.06 mol) 1-chlorohexadecane (Aldrich) and 200 ml hexane. The reaction mixture was cooled to 0°C using crushed ice/salt bath and 2.0 g of anhydrous aluminum chloride added to the stirred reaction mixture. After 30 min the remaining 5.98 g of aluminium chloride (0.06 mol total) was added and the reaction mixture slowly heated to 60°C over 4 h. After cooling, 100 g crushed ice and 100 ml water were slowly added and the organic layer washed with dilute hydrochloric acid, dried and the solvent removed under vacuum on a rotary evaporator. The oil was run through a short silica gel column using 30% ethyl acetate in hexane as eluent. Removal of the solvent gave 23.9 g (82%) of a colorless oil. Analysis: theoretical %C 62.7, %H 11.6, %Si 12.1, %N 2.0; found %C 62.1, %H 11.2, %Si 12.4, %N 2.4; ^{1}H NMR ($CDCl_3$) 0.01 [*m*, Si–CH_3], 0.60 [*m*, Si–CH_2–], 0.81 [*m*,–$(CH_3)_2$] 2.85 [*m*, –N–CH_3], 3.56 [*m*, Ar–CH_2–N], and 6.94 [*m*, *p*-substituted benzene).

Preparation of (p-*hexadecylbenzyl*) *dimethyl* [*3-*[*3-*[*1,3,3,3-tetramethyl-1-(trimethylsiloxy) disiloxanyl*] *propoxy*] *2,2,-dimethyl propyl*] *ammonium* p-*toluene sulfonic acid salt* **6**

The method described for the preparation of trisiloxane **3b** was used and a colorless oil obtained in 94% yield. Analysis: theoretical %C 59.8, %H 10.5, %Si 10.2, %N

1.5; found %C 59.5, %H 10.7, %Si 10.6, %N 1.4; ^{1}H NMR ($CDCl_3$) 0.01 [*m*, Si–CH_3], 0.60 [*m*, Si–CH_2–], 0.81 [*m*, –$(CH_3)_2$], 2.85 [*m*, –N–CH_3], 3.56 [*m*, Ar–CH_2–N], and 6.94 [*m*, *p*-substituted benzene).

Fabric preparation

The polypropylene non-woven spunbond fabric was prepared on pilot scale equipment essentially as described in US Patent No. 4,360,563 [15]. The fibers were thermally bonded and the basis weight was 27 gm^{-2}.

The heptamethyltrisiloxane quaternary ammonium salts were compounded into polypropylene pellets prior to extrusion. The additive concentration was calculated as a straight wt/wt ratio. On extrusion the fabric was formed and received no further post-extrusion treatment. X-ray photoelectron spectroscopy (ESCA) analysis was used together with methanol extractions to evaluate surface concentrations.

Methanol surface extraction

The non-woven fabrics were surface analyzed by methanol extractions for determination of additive concentrations and composition. The fabrics were soaked in methanol for 1 min and the resulting solution was chromatographed using a HPLC procedure.

Samples were run on an HPLC system comprising ISCO model 2350 pump, Waters RCM pack unit containing a Waters C18 (5 μm column), a Waters 410 differential refractometer, and a Waters 745 data module integrator. The solvent used was deairated 10% water in methanol. Relative concentrations were determined from the areas under the peaks.

Biological preparations

The bacteria were selected as examples of gram negative and gram positive strains commonly found in non-evasive applications. The biological evaluation protocol was designed as a preliminary screen to check on the efficacy of the surface segregated melt additive. The 4 h ambient temperature test was designed to closely mimic that found in applications related to disposable materials.

The bacterial strains *Escherichia coli* (ARCC # 13706) and *Staphylococcus epidermidis* (ATCC # 1859) were used to evaluate the bacterial activity of the coatings. Bacterial suspensions containing about 10^8 colony forming units (CFUs) per ml were obtained by collecting overnight growth from tryptic soy agar (Difco, USA) in saline.

All samples of fabric were cut into 1 in × 1 in squares. To the individual fabric sample placed in a 50 ml centrifuge tube was added 100 μl of 2.8×10^8 cfu ml^{-1} of each stock solution of bacteria. Samples were left at ambient for 4 h. At the end of the 4 h period, 30 ml of Letheen broth (Difco, USA) was added to each centrifuge tube. The tubes were vortexed at a setting of 4 *g* for 1 min. The survival of bacteria was determined by plating suitable dilutions on Letheen agar (Difco, USA) and counting the number of colony forming units (CFUs) after 18 h of incubation at 37°C. Survival was considered as the ratio between the number of CFUs per ml observed in bacterial suspensions after four hours of incubation in the presence of fabric and the number of CFUs per ml of the same bacterial suspensions in the control tubes [16, 17]. The bacteria kill is presented below in Tables 4 and 5 as log drop [16, 17]. The trisiloxane quaternary ammonium compounds were tested against the bacteria using 10^{-2} g l^{-1} solutions and the results are shown in Table 5.

RESULTS AND DISCUSSION

Chemistry

The heptamethyltrisiloxane quaternary ammonium salts described in this paper were all synthesized from the vacuum distilled heptamethylhydrotrisiloxane. Reaction of the heptamethylhydrotrisiloxane with allyl glycidyl ether yields the epoxy function

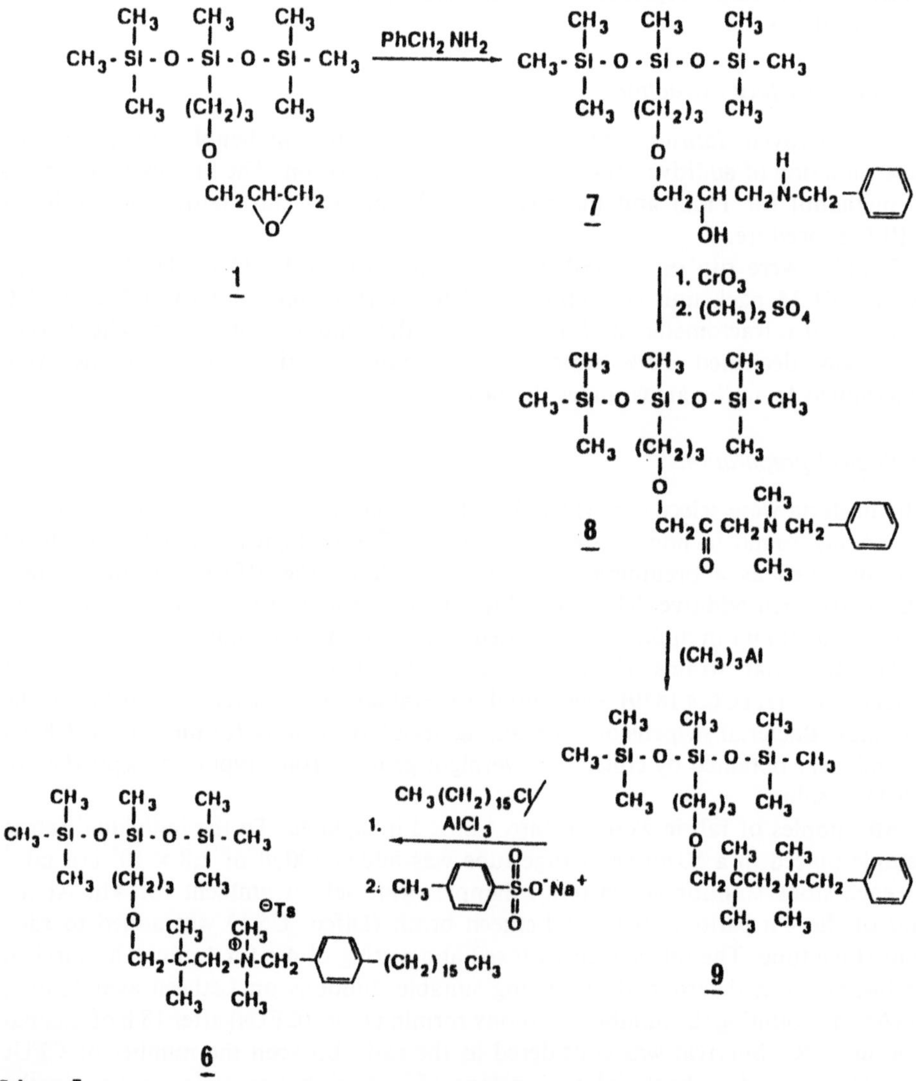
PhCH₂NH₂
1
7
1. CrO₃
2. (CH₃)₂SO₄
8
(CH₃)₃Al
9
1. CH₃(CH₂)₁₅Cl / AlCl₃
2.
6

Scheme 7.

which in turn can be reacted with various amines to yield the heptamethyltrisiloxane amines. In order to avoid thermal decomposition of the heptamethyltrisiloxane ammonium salt via a Hoffman Elimination Pathway [18, 19], the β-hydrogens were substituted by methyl groups in the case of compounds **6** and **9** by oxidation of the β-hydroxy group followed by exhaustive methylation by trimethyl aluminum. The availability of the two other β-hydrogens on the other side of the nitrogen in compounds **2** and **3** were replaced by an aryl group as shown in compounds **5** and **6** (see Scheme 7 for the synthetic pathway for compound **6**).

Surface segregation properties

All five of the heptamethyltrisiloxane quaternary ammonium compounds synthesized showed spontaneous surface segregation when added to polypropylene and melt extruded by a spunbond process into a fibrous non-woven web. The X-ray photoelectron spectroscopy (ESCA) data (Table 2) strongly suggest that the fiber surfaces of the non-woven web have a heptamethyltrisiloxane ammonium salt composition of greater than 90%. In addition, it was observed that the newly formed non-woven webs were hydrophilic, i.e. water wettable. This additional surface property is in marked contrast to unmodified hydrophobic polypropylene.

It has been recently suggested that polydimethylsiloxane copolymers when dispersed in polypropylene may form micellular structures [20, 21]. It has also been reported that the stability of these silicone copolymer molecular aggregates in the melt are in part a function of structure, process temperature, shear, and molecular weight [22]. Therefore, polydimethylsiloxane–copolymer melt additives with molecular weights of less than 1000 are more susceptible to surface segregation than to phase separation during fiber extrustion [23]. The five newly synthesized silicone quaternary ammonium compounds all have molecular weights appreciably less than 1000 (Table 1).

The spontaneous surface segregation during fiber formation may in part be explained by their micellular instability at high process temperatures (*c.* 232°C) and their relatively low surface tensions (Table 1) [15]. That is, the thermodynamic driving force of the melt additives, in their non-aggregate state, is to form an interfacial fiber surface of minimal free energy.

Table 1.
Physical properties of trisiloxane ammonium salts

Compound and % additive	Molecular weight	Surface tension dyn cm^{-1}	Fabric wettability
2 0.5	641.5	22	W
2 1.0	641.5	22	W
3 0.5	639.5	23	W
3 1.0	639.5	23	W
3b 0.5	775.0	22	W
3b 1.0	775.0	22	W
5 0.5	705.5	23	W
5 1.0	705.5	23	W
6 0.5	843.0	24	W
6 1.0	843.0	24	W

Table 2.
ESCA analysis of non-woven fabric

Compound and % additive	Found (At. %) Si	C	N	Theoretical (% element) Si	C	N
2 0.5	10.0	68.2	1.8	12.6	55.1	2.1
2 1.0	10.4	67.8	1.9	12.6	55.1	2.1
3 0.5	11.0	68.0	1.9	12.7	55.3	2.1
3 1.0	12.0	66.8	1.9	12.7	55.3	2.1
3b 0.5	10.0	68.0	1.5	11.2	57.9	1.8
3b 1.0	10.5	67.8	1.6	11.2	57.9	1.8
5 0.5	11.5	64.4	1.8	12.1	61.2	2.0
5 1.0	11.8	65.2	1.9	12.1	61.2	2.0
6 0.5	10.0	62.4	1.5	10.2	59.8	1.7
6 1.0	10.0	62.6	1.6	10.2	59.8	1.7

Process stability of new compounds

Thermal stability of the new melt additive compounds during fiber processing is critical in achieving surface segregation and antimicrobial surface activity. The spunbond oxygen free process residence times at 232°C did not exceed 15 min. The thermal stability of these compounds was studied under these anaerobic and temperature conditions and the percent decompositions are presented in Table 3. The products of the decomposition include chiefly those of the expected Hoffman elimination pathway, namely hexamethyl trisiloxane unsaturated alkyls, and aryl and alkyl amines [18, 19]. Those compounds, **3b**, **5**, and **6**, in which the β-hydrogens were substituted and/or a weak counter ion base was used showed higher thermal stability under the spunbond fiber temperature conditions. However, those with one or more unsubstituted β-hydrogens and/or strong counter ion base were more suceptable to elimination and therefore showed up to 36% thermal decomposition by weight.

Methanol surface extraction studies of the non-woven materials made with compounds **2** and **3** did not show quantifiable decomposition products. However, the studies did reveal corresponding lower surface concentrations than **3b**, **5** and **6**. These preliminary results suggest that the elimination products produced during fiber processing phase separated in the polymer bulk.

Table 3.
Thermal stability of neat trisiloxane quarternary ammonium salts via HPLC analysis

Compound	% Decomposition
2	36
3	32
3b	8
5	3
6	1

Fluid interactions with new surface segregated additive fabric

Surface tension experiments were performed on treated fabric materials in order to understand the sustenance of the heptamethyltrisiloxane extended surface coatings during exposure to water.

The 3 in. × 4 in. fabric samples containing 0.5% heptamethyltrisiloxane **6** were soaked in 100 ml of deionized water for 4 h, then dried in a convection oven at 80°C for 30 min. The exposure/dry cycle was repeated five times and the exposed fluid was collected and its surface tension measured by the du Douy method [25]. From the results shown in Table 4, the surface segregated additive on the fabric reduced in each of the above cases, the surface tension of the water to approximately 26 dyne cm^{-1}. It can be deduced that the heptamethyltrisiloxane additive on the fabric surface is present in sufficient amount to survive multiple exposures to a relatively small volumes of water.

The above results suggest that the melt additives are not entrapped in the fabric, but rather, act as replenishable surface active agents.

Table 4.
Surface tension reduction of water by fabric

Number of exposure/dry cycles	Surface tension (dyn cm^{-1})
1	26
2	25
3	26
4	27
5	26

Antibacterial surface property

As described earlier, these newly formed additive segregated surfaces provide for the surface tension reduction of the contacting fluids. Whilst in solution, these quaternary ammonium salts also have the potential for reducing the antimicrobial population [16, 17] at the fluid–fabric interfacial surface.

The antibacterial surface property of the fabric materials produced by the five new melt additive compounds is presented in Table 5. Compounds **3b**, **6**, and **7** showed the greatest antibacterial activity for both the gram negative and gram positive strains tested. These superior killing rates of greater than 99.9%, when compared to less than 90.0% for **2**, and **3** may be explained in part by their effective surface concentration differences as shown in the methanol extraction studies. Melt additives **3b**, **5**, and **6** have substituted β-hydrogens and/or low reactive anions for minimal thermal decomposition and subsequent phase separation. It should be noted as shown in Table 6 that the relative antibacterial properties of the five new compounds as preliminarily studied are apparently independent of their small structural differences.

Table 5.
Antimicrobial activity of non-woven fabric containing trisiloxane quaternary ammonium salts

	Log drop of bacterial strain	
Compound and % additive	*Escherichia coli* (ATCC # 13706)	*Staphylococcus epidermidis* (ATCC # 1895)
2 0.5	1.2	1.9
2 1.0	1.8	2.2
3 0.5	1.6	2.2
3 1.0	1.8	2.3
3b 0.5	3.1	3.8
3b 1.0	3.5	4.0
5 0.5	3.8	4.4
5 1.0	4.1	4.5
6 0.5	3.8	4.4
6 1.0	4.0	4.5
Control fabric	No drop	No drop

Table 6.
Antibacterial activity of trisiloxane quaternary ammonium salts

	Log drop of bacterial strains	
Compounds (10^{-2} g l^{-1})	*Escherichia coli*	*Staphylococcus epidermidis*
2	3.5	4.0
3	3.5	4.1
3b	3.7	4.2
5	3.8	4.4
6	3.8	4.4

CONCLUSION

Surface segregation phenomenon of specially designed melt additives in polypropylene has proven to be a versatile approach in providing a new non-evasive biomaterial with antibacterial properties. It has been demonstrated that thermal decomposition of these new melt additives during high temperature processing have been reduced by introducing substituents on the β-carbons next to the ammonium group. The high surface area of the non-woven treated fabrics provide for sustainable release in restricted volume interfacial applications. Further communications will describe additional work in progress on introducing other important surface attributes through surface segregation.

REFERENCES

1. S. M. Gendel and R. S. Nohr, *Appl. Microbiol. Biotechnol.* **31,** 138 (1989).
2. J. G. MacDonald and R. S. Nohr, US Patent 4,879, 232.
3. J. G. MacDonald and R. S. Nohr, US Patent 4,950,601.
4. S. P. Pappas (Ed.) *Radiation Curing Science and Technology* 1–448 (1992).
5. W. E. Maycock, R. S. Nohr and J. G. MacDonald, US Patent 4,859,759.
6. R. S. Nohr and J. G. MacDonald, US Patent 4,923,914.

7. R. S. Nohr and J. G. MacDonald, US Patent 5,057,262.
8. R. S. Nohr and J. G. MacDonald, US Patent 5,120,888.
9. A. J. Chalk and J. F. Harrod, *J. Am. Chem. Soc.* **87**, 6 (1965).
10. J. S. Riffle, I. Yilgor, C. Tran, G. L. Wilkes, J. E. McGrath and A. K. Banthis, Epoxy Resin Chemistry II, R. S. Bauer ACS Symp. Ser., No. 221, Washington, DC (1983).
11. E. Lukevics, *Russ. Chem. Revs.* **46**(3), 264 (1977).
12. M. Galin and A. Mathis, *Macromolecules* **14**, 677 (1981).
13. A. J. Margida, US Patent 4,895,964 (1990).
14. A. Meisterg and T. Mole, *J. Chem. Soc.* 595 (1972).
15. D. W. Appel and M. T. Morman, US Patent 4,360,563.
16. R. A. Robinson, H. L. Bodily and R. P. Christesen, *Appl. Environ. Microbiol.* **54**, 158 (1988).
17. E. Nester, *Microbiology,* 2nd Edition (1978).
18. T. W. Bentley and A. J. Kirby, *Elucidation of Organic Structures by Physical and Chemical Methods,* pt. 2, p. 255, Wiley, New York (1973).
19. A. C. Cope and E. R. Trumbull, *Org. React.* **11**, 317 (1960).
20. J. V. Dawkins and G. Taylor, *Makromol. Chem.* **180**, 1737 (1979).
21. D. S. Brown, J. V. Dawkins, A. S. Farnell and G. Taylor, *Eur. Polym. J.* **23**, 463 (1987).
22. V. P. Bolko, *Zh. Prikl. Khim.* **51**, 2362 (1978).
23. J. E. McGrath, D. W. Dwight, J. S. Riffle, T. F. Davidson, D. C. Webster and R. Viswanathan, *Polym. Prepr. Am. Chem. Soc.* **20**, 528 (1979).
24. T. C. Kendrick, B. M. Kingston, N. C. Lloyd and M. J. Owen, *J. Colloid. Interface Sci.* **24**, 135 (1967).
25. Fisher Surface Tensiomat Manual, Catalog No. 14-814, 9th issue (April, 1988).

Biomaterials with permanent hydrophilic surfaces and low protein adsorption properties

B. E. RABINOW[1]*, Y.S. DING[1], C. QIN[1], M. L. McHALSKY[1], J. H. SCHNEIDER[1], K. A. ASHLINE[1], T. L. SHELBOURN[1] and R. M. ALBRECHT[2]

[1] *I. V. Systems Division, Scientific Affairs, Baxter Healthcare Corporation, Round Lake, IL 60073, USA*

[2] *Animal Health and Biomedical Sciences, University of Wisconsin, Madison, WI 53706, USA*

Received 17 May 1993; accepted 9 December 1993

Abstract—Low protein adsorbing polymer films have been prepared with which to fabricate intravenous containers, designed for compatibility with low concentrations of protein drugs. The material is economically manufactured utilizing physical melt blending of water-soluble surface-modifying polymers (PEO, PEOX, PVA, and PNVP) with a base polymer (EVA, PP, PETG, PMMA, SB, and nylon). Permanency of the hydrophilic surfaces so generated was confirmed by surface contact angle experiments and total organic carbon leachables analysis of the aqueous contacting solutions. Binding of IgG, albumin and insulin was studied. A sixfold reduction of protein adsorption was obtained by adding 5% PVA13K to EVA, for IgG at a bulk concentration of 2.5 ppm. Surface bound protein measured by micro-BCA colorimetry, agreed with the solution protein lost, as determined by the Fluoraldehyde procedure. Imaging of the protein exposed plastic surfaces by silver enhanced protein conjugated gold staining agreed with the quantitative assay determinations.

Key words: Protein; adsorption; surface; polymer; pharmaceutical; container.

INTRODUCTION

When artificial surfaces (constructed from organic, inorganic or metallic materials) are exposed to protein-containing solutions, protein adsorption usually occurs. The protein adsorption may lead to serious problems. In the pharmaceutical industry, for example, drug adsorption to materials used in the manufacture of intravenous administration sets has previously been documented [1]. Measurable reduction in potency has been demonstrated for digitoxin, insulin, mithramycin, and vincristine sulfate when passed through in-line I.V. filters containing cellulose ester membranes [2]. The National Coordinating Committee on Large Volume Parenterals recommended limits of 5 $\mu g\ ml^{-1}$ for drug concentration and 5 mg 24 h^{-1} for total amount of drug, below which drug binding should be tested [3]. Additionally, significant adsorption of proteins such as porcine insulin, human chorionic gonadotrophin, bovine serum albumin and sheep immunoglobulin G can occur, depending upon the materials used to sterile filter such preparations [4].

Adsorption of insulin to I.V. containers and administration set tubing, in addition to filter materials, has been demonstrated repeatedly [5–7]. This problem has been circumvented clinically by the process of 'titrating' the patient; that is, sufficient

* To whom correspondence should be addressed.

insulin is administered to produce the appropriate level of blood glucose. This method of compensation for inaccurately known drug amounts has been accepted because a biological endpoint is readily determined and the drug is relatively inexpensive. This permits administration of an intentional excess of drug, solely to compensate for losses due to adsorption.

However, genetically engineered drugs at the present time are inherently more expensive than insulin, making such waste unacceptable. Indeed, price has become such an issue that regulatory approval requires, in some cases, a detailed protocol specifying which patient subgroups should receive these therapies, in an attempt to limit health care cost escalation. These new drug therapies are represented by the classes of colony stimulating factors, growth factors and hormones, interferons, interleukins, and monoclonal antibodies [8]. Because of the biological specificity of these reagents, they are administered in amounts much smaller than those of many more conventional drugs. This smaller dose size renders these drugs more susceptible to adsorptive potency losses. Furthermore, these classes of drugs are accounting for an exponentially larger proportion of the number of approved drugs with each passing year. There is clearly a need in the pharmaceutical industry to decrease protein adsorption to artificial surfaces with the goal of maintaining drug potency.

The study of protein-resistant materials has been a very active field of research during the past two decades. There has been much effort to minimize protein adsorption due to its importance in the areas of blood-contacting devices, chromatographic supports, coatings to minimize biofouling, separation membranes, contact lenses, immunoassays, protein drug-contacting materials, etc.

Most recently, increased attention has been paid to the water soluble polymer, poly(ethylene oxide) (PEO), believed to be the most effective protein-resistant material due to the steric stabilization effect, unique solution properties and its molecular conformation in aqueous solution. The steric repulsion forces can be attributed to osmotic pressure and elastic restoring forces. The competition between this steric exclusion force and van der Waals attraction between PEO and protein dominates the protein adsorption process.

However, PEO or other water soluble polymers cannot be directly used as a biomaterial due to their complete water solubility and poor mechanical properties. Thus, many researchers, both in academia and in industry, have focused on how to immobilize PEO or PEO-containing molecules on artificial surfaces. Most of the work has been related to blood-contacting materials and almost all involved chemical or chemical-physical modifications e.g., covalent grafting [9–16], coating of PEO-containing block copolymeric surfactants [17, 18], block copolymerization [19–21], surface physical interpenetrating network [22, 23] and synthesis of PEO-containing interpenetrating networks [24, 25]. Most of the above research emphasized PEO homopolymers and PEO-containing copolymeric surfactants with few being commercialized.

While covalent bonding is, theoretically, the most attractive way to immobilize PEO macromolecules onto surfaces, such surface modification methods are difficult to commercialize for economic reasons. More economically attractive material preparation techniques, described previously [26], involve physical blending processes only and do not require small reactive organic molecules as in the surface modification techniques described above. This contribution will focus on the analytical characterizations of these new surface modified, protein-compatible, biopolymers.

The permanent hydrophilic properties of the different polymer surfaces generated by the blending technique have been confirmed by surface contact angle experiments and total organic carbon leachables analysis. Experimental protein adsorption measurements were made using the Micro-BCA™ (BCA—bicinchoninic acid), Fluoraldehyde™ and silver enhanced protein conjugated gold staining techniques. Uniformity of the water soluble polymer covered surface was confirmed by attenuated total reflectance Fourier transform infrared spectroscopy (ATR-FT-IR).

DESIGN PRINCIPLES

Melt blending approach

Simple and low cost manufacturing processes are always desirable for industrial applications. Conventional methods of preparing a protein compatible surface typically involve surface adsorption, surface grafting or solvent swelling induced physical interpenetrating networks on the surface. Adsorption of the surface modifying agent is normally not permanent and desorption will eventually occur. Additionally, surface grafting and swelling require multiple treatment steps including the painstaking removal of the solvent or residual reactive ingredients. While there do exist commercially feasible surface modification techniques such as gamma and UV irradiation, these require discrete processing steps subsequent to the manufacturing of the film. In this study, a simple melt blending process was utilized. A small amount of protein compatible additives, such as PEO, polyvinyl alcohol (PVA), poly(ethyl oxazoline) (PEOX), and poly(vinyl pyrrolidone) (PNVP) materials were used to modify the substrate polymer. It is well known that in the molten state and under high shear, the low viscosity component in the mixture will tend to move to the surface. In this work, low viscosity additives or surface modifiers were thus driven onto the surface through shear, employing extrusion and injection molding processes. Figure 1 schematically illustrates this hypothetical approach.

Analytical screening strategy

Four analytical methods were employed to confirm the surface properties as well as the permanence of the modified surfaces. Contact angle methodology, which assesses the wettability, was used as the first level test to screen a large number of potential formulations for the degree of surface modification. Materials which exhibited low contact angle characteristics were then tested for protein adsorption with a modification of the Micro-BCA assay [27]. The permanence of the surface modification was confirmed by time dependent contact angle measurements. The migration of the surface modifier additives was monitored by measuring the total organic carbon (TOC) leachable levels of the materials following long term aqueous storage. Finally, ATR-FT-IR Spectroscopy was used to characterize the surface chemistry as to its uniformity as well as surface vs bulk concentration.

The methods chosen were considered for their universal applicability in that, for the most part, they could be used regardless of the protein or surface studied. This

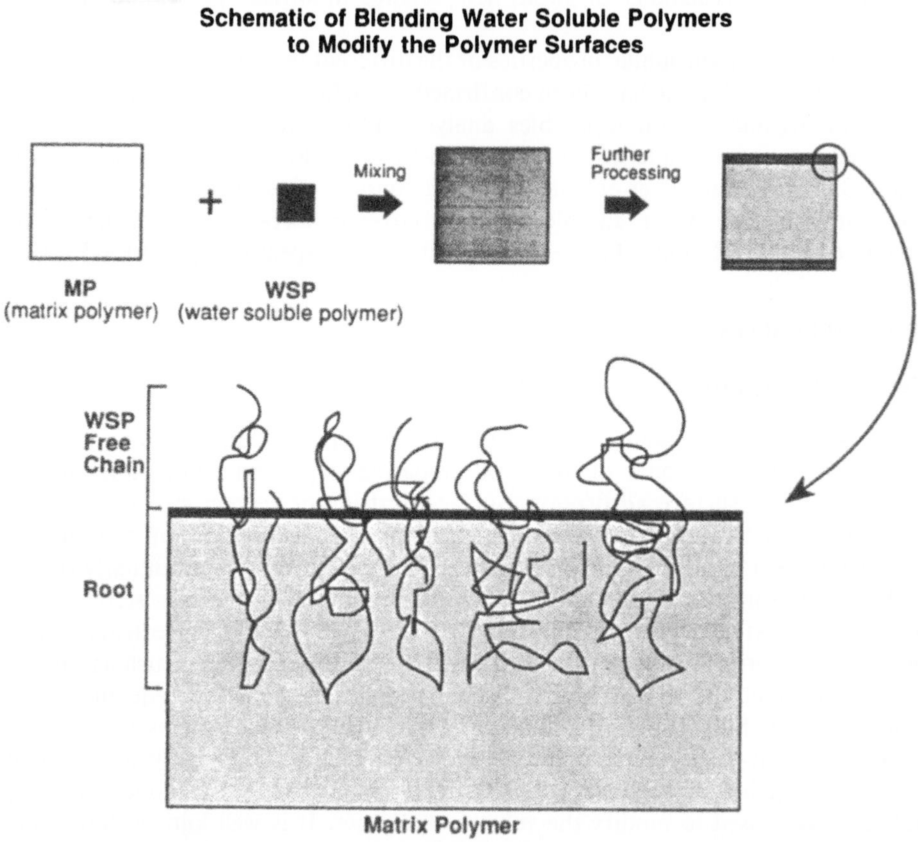
Schematic of Blending Water Soluble Polymers
to Modify the Polymer Surfaces
Mixing
Further
Processing
+
MP
(matrix polymer)
WSP
(water soluble polymer)
WSP
Free
Chain
Root
Matrix Polymer

Figure 1. Schematic representation of blended water soluble polymers to modify the polymer surfaces. Low viscosity water soluble polymers were melt blended into matrix polymers and extruded into films. It is theorized that the water soluble polymer (WSP) hydrophilic components are present on the surface while the root of the WSP is bound in the polymer matrix.

would preclude the use of radiolabeling methods for protein adsorption measurement, since radiolabeled analogs may be unavailable and/or difficult to prepare for the more exotic protein drugs. In addition, radiolabeling a protein may result in its conformational change which could alter the adsorption characteristics of that protein.

EXPERIMENTAL

Material description

Samples were prepared by the melt-blending process. The surface modifiers and substrate polymers were blended and pelletized using a 1.5 in. single screw extruder equipped with a Maddox mixing section or a 1 in. intermeshing twin-screw extruder and then cut into small cylindrical pellets. The pellets were then extruded into 10-mm thick films using standard film extrusion equipment with a 1.5 in. single screw extruder and chromium plated stainless steel chill rolls.

The surface modifiers used in this study are water soluble polymers such as polyethylene oxide (PEO from Aldrich Chemicals, M_w of about 100 000 and 600 000), poly(ethyl oxazoline) (PEOX-500 from Dow Chemicals, M_w of about 500 000), polyvinyl alcohol (PVA from Aldrich Chemicals, 89% hydrolyzed, M_w of 13 000–23 000), and poly(vinyl pyrrolidone) (PNVP, from Aldrich Chemicals, M_w of about 40 000).

The substrate polymers used in this study include ethylene-vinyl acetate (EVA, UE648 resin from Quantum Chemicals), polypropylene (PP, 23M2C from Rexene Corporation), glycol modified poly(ethylene terephthalate (PETG, Kodar 6763 resin from Eastman Chemicals), poly(methyl methacrylate) (PMMA, CP-82 resin from ICI Acrylic, Inc.), styrene–butadiene copolymer (KR-03 resin from Phillips), polyamide 6.6/6.10 copolymer (XE3303 resin from EMS America Grilon, Inc.), and polyamide 12 (L-16 resin from EMS America Grilon, Inc.).

Bovine serum albumin, Sigma Chemicals, fraction V 98–99% pure; immunoglobulin (IgG) human, 95%, Sigma; insulin, bovine pancreas, 25.7 U mg^{-1}, Sigma and insulin injection USP, beef and pork, 25.7 U mg^{-1}, Eli Lilly were the proteins used in this study. Other common chemicals used, such as sodium chloride, potassium phosphate etc., were AR grade.

IgG was chosen as the primary test protein because it adsorbs at concentrations higher than that of the other common proteins tested as well as serving as a model for a general class of monoclonal antibodies. Except where noted the protein adsorption experiments were typically performed in phosphate buffered saline solutions of KH_2PO_4/Na_2HPO_4, pH 7.4 buffer (0.01 M total phosphate) and 0.15 M NaCl.

Surface hydrophilicity measurements

The surface contact angle measurement was used to characterize the wettability of the modified materials. The Tensiometric (Wilhelmy Plate) method with a Cahn Electrobalance and a LVDT was used to measure the contact angle between water (Aldrich HPLC grade) and sample films. Dynamic advancing and receding contact angles were calculated from the force-depth curves. Contact angles were measured from fresh films as well as from films soaked in water at room temperature for a period of time ranging from several minutes to 1 month to monitor possible changes in surface dynamics. The receding and advancing rates of 25 mm min^{-1} were used for all tests. The instrument was calibrated using clean glass slides. All glass slides used in this experiment were cleaned by chromic acid, HPLC-grade water and/or plasma treatment followed by a vacuum drying procedure. Sample sizes of 3 cm × 2 cm were used in the testing. The receding contact angles were calculated using the following equation:

$$F_0 = p\gamma \cos\theta \tag{1}$$

where θ is the contact angle, F_0 the force at immersion depth (d) = 0, p the plate perimeter, and γ is the liquid surface tension which for purified water is 72.8 dyn cm^{-1}.

Surface protein adsorption analysis

Direct analysis of surface protein adsorption was performed by using a modification of the Micro-BCA method. Essentially the method involves the interaction of protein with Cu^{2+} with the subsequent reduction of Cu^{2+} to Cu^{1+}. The Cu^{1+} produced is specifically complexed with bicinchoninic acid (BCA) and measured using colorimetry. Containers of 250–350 cm^2 surface area prepared from the polymer films of interest were filled with 50 ml of solution either with or without protein. Air was removed and the containers were sealed and allowed to equilibrate for at least 18 h with the solutions. Following equilibration, the containers were opened, drained of contents and rinsed with distilled water for up to 2 min to remove loosely bound protein. Triton X-100, phosphate buffered saline and the BCA reagent to a total of 5 ml were added to the containers. All air was removed and the containers resealed. The containers with reagents were placed in a water bath held at room temperature and allowed to incubate overnight. Protein standards were also prepared, combined with reagents and incubated for the same time as the samples. After incubation the containers were opened and the contents analyzed by a Hewlett Packard Model 8451A diode array UV-visible Absorption Spectrometer at the 562 nm wavelength using 1-cm disposable polystyrene cuvettes. Absorption measurements were converted to concentrations using a calibration curve produced from similarly measured standards. The concentrations were blank corrected and converted to protein surface coverage in $\mu g\ cm^{-2}$ using the measured surface area of the container.

Solution protein analysis

Solution protein concentration was determined using the Fluoraldehyde Protein assay [28]. The fluoraldehyde reagent, *o*-phthalaldehyde along with mercaptoethanol were added to the protein solution to be quantitated. The *o*-phthalaldehyde reagent reacts with any primary amine groups present on a protein. In the presence of mercaptoethanol this reaction produces an intense blue fluorescent compound with a maximum emission wavelength at 430 nm for an excitation wavelength at 360 nm. The fluorescence measurements were made with a Perkin Elmer model LS-50 fluorescence Spectrophotometer using 1-cm quartz cells.

To determine surface protein adsorption using the Fluoraldehyde assay, the protein solutions were measured before and after contact with the container material. The protein lost due to adsorption is the difference between the two measurements.

Total organic carbon leachables analysis

In order to maintain a stable polymer surface, the loss of additives from the film must be held to a minimum. Total organic carbon (TOC) analysis provides a rapid and easy method to screen the amount of container leachables. The TOC method involves the conversion of all solution organic leachables to CO_2 using thermally promoted acidic persufate oxidation. The CO_2 produced is collected and measured by an infrared detector. The TOC analysis is performed on water stored for various periods of time within containers composed of the films being studied. The analysis is performed using a commercially available instrument, the Model OI-700 TOC analyzer (OI Instruments).

Initial experiments were performed by immersing weighed film strips into capped glass BOD bottles containing 40 ml of pure water. The bottles were capped and placed at room temperature, 40, and 60°C. The samples were analyzed for TOC on a periodic basis replacing with fresh water following each analysis. The total organic leachables for each film is a sum of all TOC analyses.

Subsequent leachables analyses were performed using containers prepared from the test films. The samples were stored at room temperature for 1 month. The container contents were periodically removed and analyzed for TOC. The containers were subsequently refilled with fresh purified water.

Silver enhanced protein conjugated gold staining

The 18 nm colloidal gold was prepared by the sodium citrate procedure for the reduction of gold chloride ($HAuCl_4$, Sigma) [29]. Briefly, 4% gold chloride was added to filtered distilled water. This mixture was heated to boiling at which time sodium citrate was added. The resulting solution was boiled for 30 min and then allowed to cool. The amount of protein used to stabilize the gold was determined by the salt flocculation test using NaCl as previously described [30]. Both the amount of protein and the pH were varied to determine the least amount of protein and the appropriate pH required to stabilize the colloidal gold.

Bovine insulin (Sigma) was dissolved in 0.01 M HCl at a concentration of 10 mg ml^{-1} and then diluted with filtered distilled water to 0.1 mg ml^{-1}. Two micrograms of insulin were used to stabilize each ml of 18 nm colloidal gold at pH 6.0. Sixty micrograms of BSA (Sigma) were required to stabilize each ml of 18 nm colloidal gold at pH 6.7. Sixteen micrograms of IgG (anti-fibrinogen) (Sigma) were required to stabilize each ml of 18 nm colloidal gold at pH 7.1. After concentrating by centrifugation, each of the gold conjugates was suspended in 0.05 M Tris buffer at pH 7.3. Additionally, all gold conjugates were adjusted to an adsorbance of 4.5–5 at a wavelength of 525 nm, which is equal to 4.5×10^{12} to 5×10^{12} particles per ml[1]. Disks were cut out of the various polymer films and marked to differentiate the internal and external surfaces. After a brief rinse in distilled water, polymer disks were floated (inner surface down) on a 2-ml pool of the appropriate gold conjugate. The samples were placed in a moist chamber and incubated 18 h at room temperature. Polymer samples were thoroughly rinsed with distilled water after labeling and allowed to air dry. Polymer disks were cut in half and one half was silver enhanced (Sigma kit # SE-100) for 8 min in the dark and then rinsed well in distilled water and air dried. Disks could be examined visually to determine the extent of binding and scored. Non-silver enhanced material appeared as a reddish-pink color with the intensity proportional to the extent of binding. Silver enhancement served to intensify the staining such that qualitative differences in binding were readily seen and trends immediately apparent. Scanning electron microscopy examination of the polymer films was performed to ensure all silver enhancement was due to the presence of gold particles and to determine numbers of individual particles and patterns of distribution of the protein-gold on the different polymer surfaces [33].

Surface content and uniformity

Films of EVA with 0, 1, 2, 3, and 5% PVA13K were analyzed by ATR-FT-IR spectroscopy. The attenuated total reflectance spectra were collected using a Horizontal Analyzer (Spectra Tech) equipped with a 45 deg ZnSe, multiple pass internal reflectance element of 72 mm^2 surface area. Infrared spectra were obtained on a Bio-Rad Digilab FTS-60 Fourier transform infrared spectrometer equipped with a liquid nitrogen cooled MCT detector. Spectra were recorded at a resolution of 8 cm^{-1} and 200 scans were co-added per spectrum.

Five areas of each film were randomly selected and the infrared absorbance measurements taken. The ratio of PVA13K to EVA was determined from the peak height or the peak area of the PVA hydroxyl absorption at ~3292 cm^{-1} ratioed to the peak height or area of the EVA methylene symmetric stretch absorption at ~2847 cm^{-1}.

RESULTS AND DISCUSSION

Surface hydrophilicity and dynamics

It is well understood that some polymer surfaces, unlike more rigid surfaces, such as inorganic glasses or metals, are mobile and can modify their molecular compositions in response to contacting media. Information on this changing surface molecular architecture becomes important if polymeric materials are used for long term solution storage. To obtain surface dynamics information of the various polymer material systems used in this study, long-term contact angle experiments were carried out.

Figure 2 shows the long-term contact angle results for pure EVA films. A strong time dependence of contact angle is shown. This may possibly indicate either that the polar vinyl acetate groups are rearranging on the EVA surface or a migration of these same groups to the surface. In either case, minimization of the interfacial tension between the surface and polar water results.

Long-term contact angle studies of water-soluble-polymer PEOX-modified EVA surfaces indicate much less of a time dependence of contact angle (Fig. 3). This is not surprising since these surfaces are quite hydrophilic to begin with and, while rearrangements may occur, little, if any, net difference in surface polarity is likely to be detected.

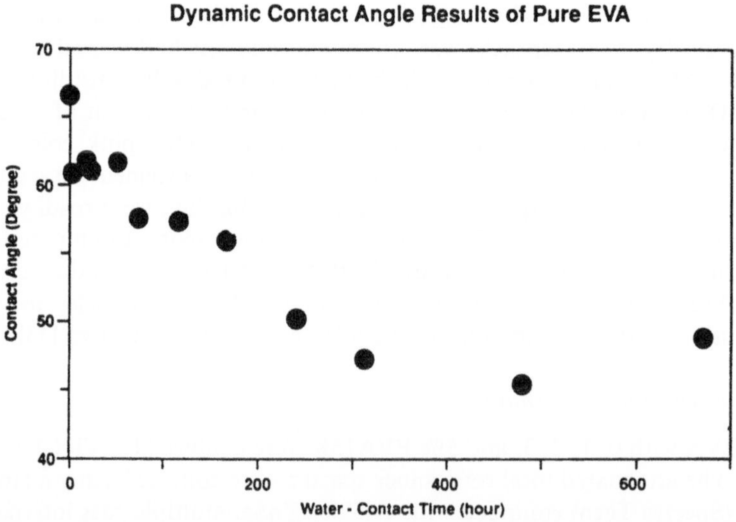
Dynamic Contact Angle Results of Pure EVA
Contact Angle (Degree)
70
60
50
40
0
200
400
600
Water - Contact Time (hour)

Figure 2. Dynamic contact angle results of pure EVA film as a function of water contact time. Films were stored in contact with HPLC water at room temperature. Measurements are receeding angle.

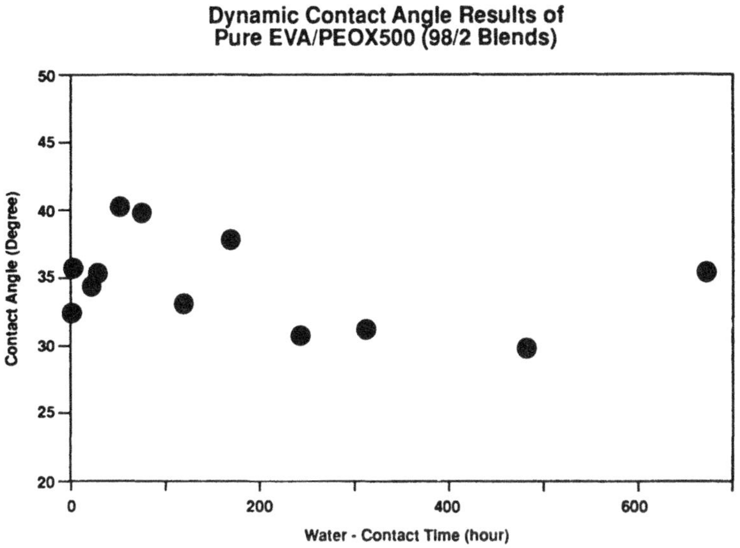
Dynamic Contact Angle Results of
Pure EVA/PEOX500 (98/2 Blends)
Contact Angle (Degree)
50
45
40
35
30
25
20
0
200
400
600
Water - Contact Time (hour)

Figure 3. Dynamic contact angle results of EVA/PEOX500 film (98/2Blend) as a function of water contact time. Films were stored in contact with HPLC water at room temperature. Measurements are receeding angle.

The PVA-modified EVA hydrophilic surface is clearly different in architecture from the pure EVA surface. It is likely that long exposed arms of the PVA molecules may protect the more hydrophobic surface of the EVA film from protein adsorption.

Figure 4 shows the effectiveness of increasing the polymer's hydrophilic character with increasing PVA additive level. It is worth noting that low contact angles can be achieved with only a small amount of added water-soluble polymer. This gives the materials specialist a large composition window in which to optimize other film properties.

Protein adsorption

Hydrophilic character usually but not always coincides with low protein adsorption. Direct measurement of surface protein was used to confirm the preliminary findings of the contact angle measurements.

Table 1 shows the surface concentrations in $\mu g\,cm^{-2}$ of various proteins for common surfaces. The surface protein coverage seems quite similar, in most cases, for the same protein indicating a similar hydrophobic character for these surfaces. The differences detected between proteins show the relative effects caused by differences in protein character.

Table 2 shows the effect of film modification on protein adsorption. IgG was chosen as the test protein because it adsorbs at higher levels than that of the other common proteins tested. It also serves as a model for the general class of monoclonal antibodies. In the cases presented, PEO was added to various base polymers. Dramatic decreases in protein adsorption were noted in several of the polymers so modified. The addition of various amounts of PVA13K additive to EVA

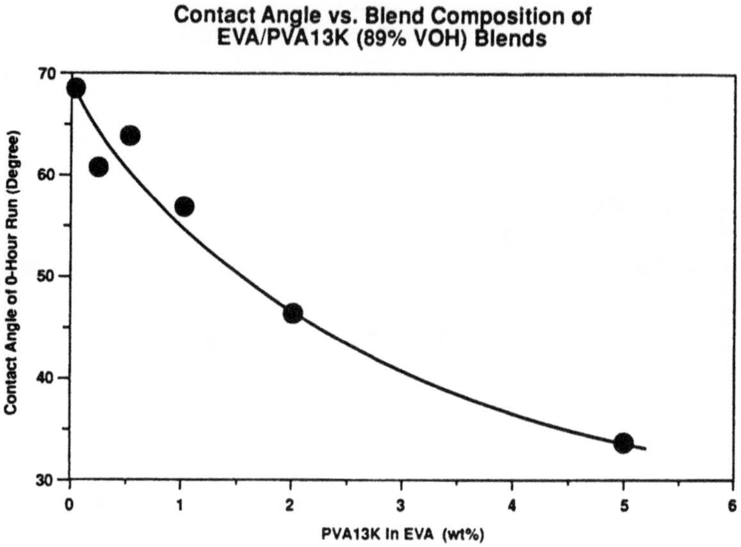
Contact Angle vs. Blend Composition of
EVA/PVA13K (89% VOH) Blends
Contact Angle of 0-Hour Run (Degree)
70
60
50
40
30
0
1
2
3
4
5
6
PVA13K In EVA (wt%)

Figure 4. Effect of EVA/PVA blend composition on surface contact angle. The EVA is Quantum Chemical's UE648 with 18% VA content. PVA, from Aldrich Chemical, has a 13-23K molecular weight range and is 89% hydrolyzed. The contact angle was measured on the extruded film.

Table 1.
Protein surface adsorption on common surfaces

Film type	Surface protein concentration ($\mu g\, cm^{-2}$) IgG	Albumin	Insulin
LLDPE (production)	0.24	0.11	0.09
	0.22	0.11	0.10
LLDPE (laboratory)	0.22		
	0.20		
EVA (production)	0.20	0.09	0.13
	0.21	0.08	0.12
EVA (laboratory)	0.25	0.09	0.12
	0.25	0.09	0.10
PVC (DOP plasticizer)	0.20	0.07	0.07
	0.20	0.07	0.06
PP	0.24		
	0.24		
PCCE[a]	0.15		
	0.15		
	0.14		
Glass	0.11		
	0.13		
	0.10		

[a]PCCE Copolyester (Eastman Ecdel)

Solutions of 2.5 ppm protein in 70 ml PBS, pH = 7.4 were stored in containers made from the various films with surface areas of 200–250 cm^2. The estimated standard deviation of the pooled data in this table is 0.01 $\mu g\, cm^{-2}$. The analysis was performed on duplicate containers (triplicate for glass and PCCE) by the modified BCA surface analysis method.

Table 2.
Immunoglobulin adsorption on different films and different concentrations

Film type	Surface adsorption ($\mu g\ cm^{-2}$) Solution concentration 2.5 ppm	10 ppm	50 ppm
EVA/5% PVA13K	0.02	0.09	0.18
EVA/2% PVA13K	0.05	0.10	0.27
EVA control	0.23	0.26	0.45
XE3303/2% PEO100K	0.21	0.36	0.55
XE3303 control	0.21	0.32	0.48
EMS-L16/5% PEO100K	0.12		
EMS-L16/2% PEO100K	0.21	0.36	0.52
EMS-L16 control	0.30	0.40	0.53
PETG/5% PEO600K	0.07		
PETG/2% PEO600K	0.22	0.32	0.50
PETG control	0.25	0.37	0.47

Solutions of IgG in 70 ml PBS, pH = 7.4 were stored in containers made from the various films with surface areas of 200–250 cm^2. Only single determinations by the modified BCA surface analysis method were made. An estimated standard deviation of 0.01 $\mu g\ cm^{-2}$ would be expected for this method, however.

polymer show a dramatic improvement in minimizing surface adsorption (Fig. 5). The change in surface adsorption as a factor of changing PVA13K additive concentration in EVA copolymer matches the contact angle results shown in Fig. 4 and confirms the hypothesis that increasing surface hydrophobic character decreases protein adsorption.

Table 3, in addition to displaying adsorption changes with film modification, also compares the surface protein assay (modified Micro-BCA) with the solution protein assay (Fluoraldehyde). The results, showing surface gain versus solution loss, compare quite favorably demonstrating the effectiveness of both assays.

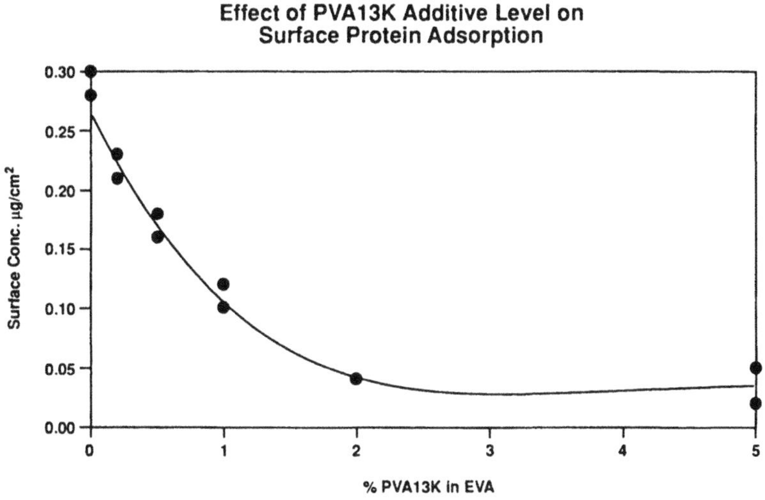
Effect of PVA13K Additive Level on
Surface Protein Adsorption
Surface Conc. µg/cm²
0.30
0.25
0.20
0.15
0.10
0.05
0.00
0
1
2
3
4
5
% PVA13K in EVA

Figure 5. Effect of PVA13K additive level on surface protein adsorption as measured by the modified BCA method. Solutions of 2.5 ppm IgG in PBS, pH = 7.4, (70 ml) were stored for 18 h in film containers of approximately 250 cm^2 surface area.

Table 3.
Immunoglobulin adsorption on different films—assay comparison

Film/additive	Surface adsorption ($\mu g\ cm^{-2}$) BCA	Fluoraldehyde
PMMA/5% PEO600K	0.06	0.07
	0.04	0.06
PMMA/5% PEO100K	0.07	0.07
	0.06	0.07
PMMA control	0.09	0.09
KR03/5% PEO100K	0.05	0.06
	0.03	0.06
KR03 control	0.21	0.23
	0.20	0.25
PETG/5% PEO600K	0.07	0.08
	0.07	0.02
PETG control	0.24	0.24
	0.24	0.22

Solutions of 2.5 ppm protein in 70 ml PBS, pH = 7.4 were stored in containers made from the various films with surface areas of 200–250 cm^2. The modified BCA surface method was used for the determination of adsorbed protein while the Fluoraldehyde method measured solution protein. The analysis was performed on duplicate containers for both methods. No differences in the two methods were detected ($p = 0.31$). The estimated standard deviation of all the pooled data in this table is 0.01 $\mu g\ cm^{-2}$.

Table 4 shows the surface adsorption results of different proteins using different additive/substrate components. As shown, some matrix/additive combinations appear to be more successful than others.

Protein adsorption isotherms were obtained for both unmodified and modified EVA films for an IgG concentration range up to 1 mg/ml. The adsorption profile of IgG on both films was determined for different solution concentrations as shown in Fig. 6. Although the same protein adsorption maximum was reached with both films, differences in adsorption were determined at the lower IgG concentrations. Similar adsorption maxima for different films has been observed previously, for example, albumin on PVC, PE, PU, and silicone rubber [34]. The constancy in value has been attributed to a common geometric packing of the adsorbed protein molecule, regardless of the nature of the surface [35]. The maximum adsorption for IgG of 0.70–0.75 $\mu g\ cm^{-2}$ is based on a Langmuir type adsorption calculation. This adsorption limit lies within the range of 0.27–1.37 $\mu g\ cm^{-2}$ calculated for deposition of a monolayer of IgG, assuming adsorption by either the side or end of the molecule and neglecting its possible spreading once adsorbed [36]. Molecular dimensions of 235 Å × 44 Å were assumed.

As shown in Fig. 6, there appears to be little change in the maximum protein adsorption with polymer surface modification. Improvements in adsorption at the protein maximum plateau are not necessary, however, because the starting protein concentration is so high that the percent loss of protein due to adsorption will be minimal. Of more importance is the effect surface modification will have on adsorption from solutions of low protein concentration.

Table 4.
Protein adsorption on different films

Film/additive	Surface concentration ($\mu g\ cm^{-2}$) IgG	Albumin	Insulin
EVA/5% PEOX500K	0.04	0.03	0.07
	0.05	0.02	0.07
EVA/5% PEOX50K	0.12	0.04	0.07
	0.12	0.03	0.07
EVA/2% PEO100K	0.24	0.07	0.09
	0.25	0.07	0.11
EVA/5% PNVP40K	0.16		
	0.16		
EVA/2% PVA13K	0.06	<0.01	0.05
	0.06	0.02	0.06
EVA control	0.25	0.09	0.10
	0.25	0.09	0.12
PP 5% PVA13K	0.09		
	0.07		

Solutions of 2.5 ppm protein in PBS, pH = 7.4 (70 ml) were stored in containers made from the various films with surface areas of 200–250 cm^2. The estimated standard deviation of the pooled data in this table is 0.01 $\mu g\ cm^{-2}$. The analysis was performed on duplicate containers by the modified BCA surface method. Significant differences ($p < 0.05$) were found for each film when compared with its base polymer with the exception of the 2% PEO100K/EVA formulation.

Figure 7 replots the same data in Fig. 6 in terms of a percent of total remaining solution protein. The addition of PVA13K to the EVA polymer shows a relative improvement in decreased surface adsorption that is most apparent at lower IgG concentrations. This is fortunate because relative protein losses due to adsorption are much more important at the lower concentrations. As an example, an adsorption level of 0.20 $\mu g\ cm^{-2}$ for a 2.5 ppm solution IgG concentration stored in a 50 ml EVA container with a 200 cm^2 surface area would result in a 32% loss of protein from solution. On the other hand, a 50 ppm protein solution in the same EVA container showing a 0.50 $\mu g\ cm^{-2}$ surface adsorption would lose only 4% of its solution protein. Modifying the surface for a container used to store a 50 ppm protein solution would, therefore, make little sense. However, modifying a container used for a 2.5 ppm protein solution to decrease the surface adsorption from 0.2 to 0.02 $\mu g\ cm^{-2}$ will decrease the adsorption loss from 32 to 3.2%, a much more acceptable loss.

Leachables analysis

One of the more important issues concerning surface-modified biomaterials is long-term surface stability. Since the water-soluble polymers are physically trapped within the matrix polymer rather than chemically bonded with the matrix, the possibility exists for loss of the additive present on the surface to the solution. Although Fig. 3 shows little change in hydrophilic character of the surface over time, it is not conclusive of minimal additive loss. Time dependent total organic carbon analysis of

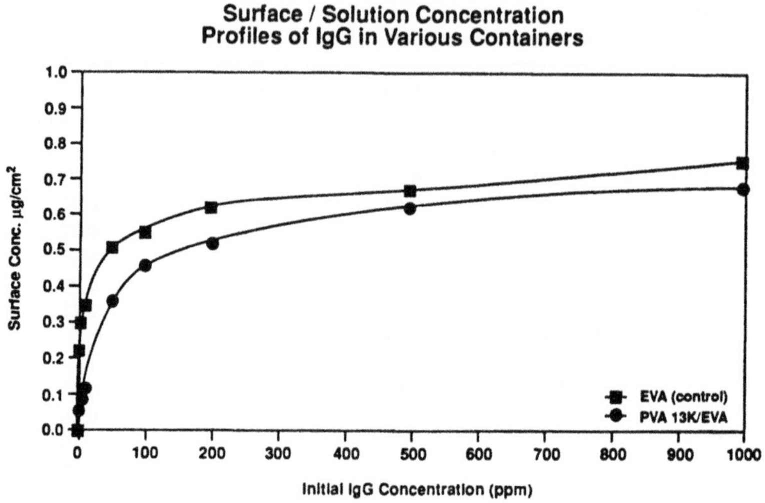
Surface / Solution Concentration
Profiles of IgG in Various Containers
Surface Conc. µg/cm²
1.0
0.9
0.8
0.7
0.6
0.5
0.4
0.3
0.2
0.1
0.0
0
100
200
300
400
500
600
700
800
900
1000
Initial IgG Concentration (ppm)
EVA (control)
PVA 13K/EVA

Figure 6. Solutions of PBS at pH = 7.4 with varying concentrations of IgG (70 ml) were stored for 18 h in containers made with base EVA (■) and EVA modified with 5% PVA13K (●) films of approximately 200 cm^2 surface area.

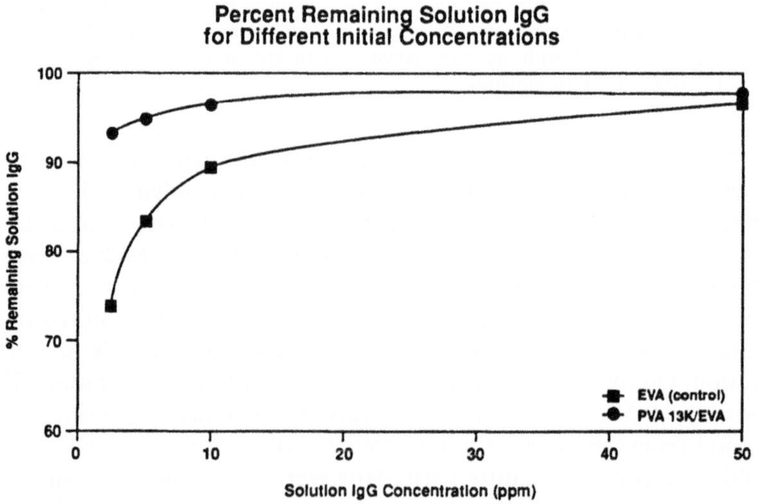
Percent Remaining Solution IgG
for Different Initial Concentrations
% Remaining Solution IgG
100
90
80
70
60
0
10
20
30
40
50
Solution IgG Concentration (ppm)
EVA (control)
PVA 13K/EVA

Figure 7. Same conditions as Fig. 6. Differences in the percent remaining protein were greatest at the lowest initial IgG concentrations.

pure water, stored in contact with the film surface, would confirm that the water-soluble polymer was not being lost to the solution.

Cumulative leachables for each film/additive type were compared with the same film type without additive to determine whether its addition had an adverse effect on the film. Some examples of these comparisons are shown in Table 5. One example of an unsuccessful film modification was the EVA containing 2% PEO100K, which resulted in a 3000 $\mu g\, g^{-1}$ carbon leachable burden versus the unmodified EVA control which released only 140 $\mu g\, g^{-1}$ carbon over the same test period and

Table 5.
Leachables from different films at various temperatures

Film/additive	Organic leachables (μg carbon g^{-1} film) RT	40°C	60°C
EVA 5% PEOX500K	160	600	750
EVA 2% PEO100K	490	3000	3400
EVA 5% PNVP40K		660	780
EVA 5% PVA13K	210	400	480
EVA 2% PVA13K		560	1000
EVA	140	140	180
PP 5% PVA13K		200	330

Estimated total organic leachables, in μg carbon g^{-1} film, were obtained by TOC analysis. Strips of film were weighed and placed in water stored in BOD bottles. The room temperature samples were stored for 29 days while the 40 and 60°C samples were stored for 48 days. Only single determinations were made for each film/temperature condition.

temperature of 40°C. The modified film showed what was presumed to be a loss of the additive to solution, thus mitigating its effectiveness as a low protein adsorbing film type. Most of the other modified films showed only modest gains in leachables over their unmodified controls.

Surface imaging by protein staining

Figure 8a–d present photographs depicting the results of silver enhanced protein conjugated gold nanobead staining of PVC, EVA and EVA modified with PVA13K. Split circular pieces of plastic 2 cm in diameter are shown. For each experiment, the left hemicircle shows the effect of exposure of the plastic to the gold conjugated protein only. The right hemicircle depicts subsequent treatment by the silver enhancement.

Gold conjugated to anti-fibrinogen was studied in Fig. 8a. While slight shading is apparent on the left side, dense silver coating characterizes the right side of the plastic samples of PVC and EVA. By comparison, very little density is apparent for the PVA13K modified EVA sample, in qualitative agreement with the micro-BCA protein adsorption data for IgG.

Density for the silver enhanced gold conjugated to BSA experiment, shown in Fig. 8b, was much lighter than that for anti-fibrinogen. This again is in agreement with the micro-BCA data and reflects the effects of experimental pH and isoelectric point of the proteins studied. The reduced adsorption of BSA on the PVA13K modified EVA is nevertheless apparent, in agreement with the micro-BCA quantitative determinations.

Heavy staining of the silver enhanced gold conjugated to insulin adsorption to PVC and EVA is visible in Fig. 8c. Reduced staining on the PVA modified EVA is clear. The gold, when used by itself in Fig. 8d selectively coats the plastics that have been modified. A possible reason for this may be electrostatic attraction between the positively charged gold nanospheres and the surface.

The imaging approach will be used to refine the surface coating process, inasmuch as a degree of surface inhomogeneity in the adsorption characteristic is apparent for

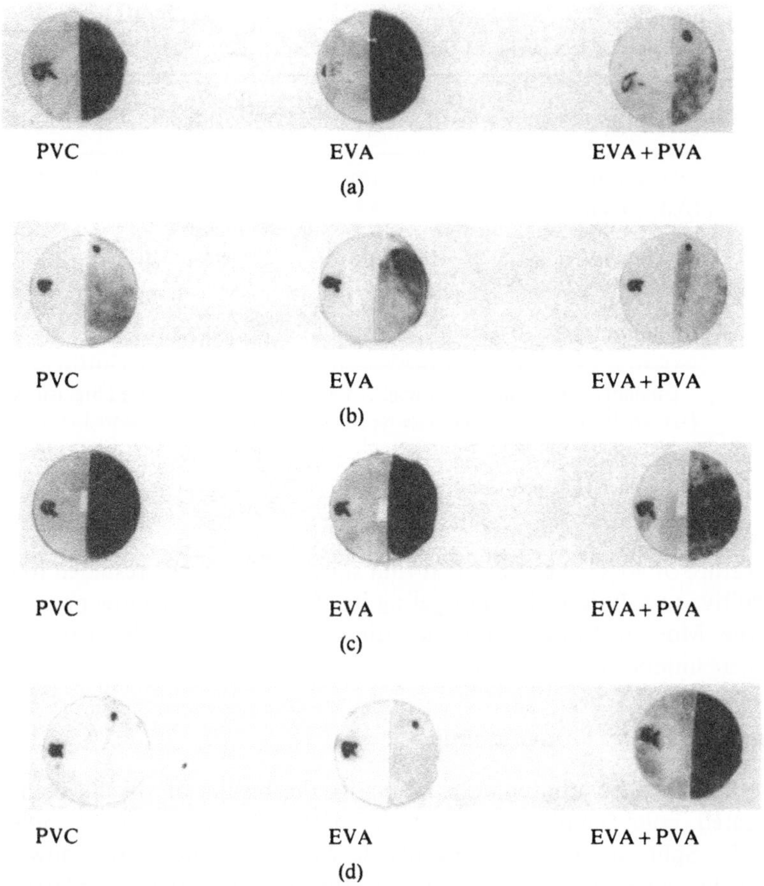
PVC
EVA
EVA + PVA
(a)
PVC
EVA
EVA + PVA
(b)
PVC
EVA
EVA + PVA
(c)
PVC
EVA
EVA + PVA
(d)

Figure 8. Silver enhanced protein conjugated gold nanobead staining of PVC, EVA and EVA modified with PVA13K. Split circular pieces of plastic 2 cm in diameter are shown. The left hemicircle shows the effect of exposure of the plastic to the gold conjugated protein only. The right hemicircle depicts subsequent treatment by the silver enhancement. Increased protein adsorption is demonstrated by a darkening of the film by the stain. The proteins used in the study were: (a) Anti-fibrinogen; (b) bovine serum albumin; (c) insulin; and (d) none.

the EVA modified with PVA. Causes for the inhomogeneity found by these tests will have to be further investigated.

In order to provide a rapid survey procedure sufficient for this study and to eliminate problems associated with varying antibody affinities, direct conjugates of 18 nm colloidal gold were used to label surfaces. All gold particles are of the same size and, hence, gold-protein conjugates are similarly of nearly equal diameters. Thus, differences in binding due to varying sizes of conformations of the protein molecule are not apparent and maximal labeling on a per-molecule basis may vary from studies using soluble proteins.

Studies, where quantitation is of primary importance or where a high degree of spatial resolution is desirable, can employ the use of very small, 2–3 nm gold particles directly conjugated to the protein. In this case, the gold is much smaller than the protein molecule and a constant number of one or more gold particles can be conjugated to each protein molecule [37, 38].

Alternatively, the soluble, unlabeled protein can be incubated with the surface of interest and attached molecules subsequently labeled with 2–3 nm gold–Fab high affinity anti-protein antibody. In this case, problems associated with antibody affinity can exist and care must be taken [37, 38].

In both cases, the size of the 'probe', gold–protein, or protein–Fab–gold, is nearly equal to the size of the soluble protein molecule and a relatively constant number of gold particles per molecule is achieved. Silver enhancement is readily performed on the small gold particles; however, direct observation of the unenhanced small gold particles requires high resolution SEM or TEM to view and count the particles [39, 40].

In addition to the possibilities of rapid screening, the gold labeling procedure is most useful in determining patterns of labeling relative to the substrate structure. This can be evaluated at all levels from macroscale to microscale, tens of angstroms, depending on the method of observation [41].

Surface content and uniformity

Peak absorbance ratios of PVA13K measured at ~3292 cm^{-1} to EVA measured at ~2847 cm^{-1} were determined for EVA films with varying amounts of PVA13K. The average peak area and height ratios of $n = 5$ measurements of 72 mm^2 surface area for each film along with the measurement variations are given in Table 6. The variation in surface uniformity was shown to be 10% or less for the films containing 2% or more of the PVA13K additive. The surface vs bulk % PVA13K shows reasonable linearity with correlations of R^2 of 0.968 for the peak height and 0.975 for the peak area measurements, respectively.

Table 6.
Surface uniformity measurements

	Average ratios			
PVA13K (%)	Peak area	(S.D.)	Peak Ht.	(S.D.)
0%	0.002	(0.004)	0.001	(0.001)
1%	0.065	(0.023)	0.009	(0.003)
2%	0.154	(0.016)	0.016	(0.001)
3%	0.208	(0.012)	0.021	(0.001)
5%	0.426	(0.017)	0.043	(0.002)

(S.D.) Standard deviation

Linear Regression of peak ratios (y) vs bulk % PVA13K (x)

Peak Area Ratios	$y = 0.0841x - 0.0142$	$R^2 = 0.975$
Peak Height Ratios	$y = 8.06 \times 10^{-3}x + 9.72 \times 10^{-5}$	$R^2 = 0.968$

The peak area or height values are ratios of absorbance measurements taken at ~3292 and ~2847 cm^{-1}, respectively, for the PVA (OH) and EVA (CH_2) values. Each film was randomly sampled with a measurement area of 72 mm^2 and the absorbance readings taken. The ratios in the table are averages of $n = 5$ measurements. The linear regressions were calculated using all individual measurements.

CONCLUSIONS

Preparation of low protein adsorbing polymer films can readily be accomplished by the physical blending of water-soluble polymers with a base polymer. This process is much more economical than the various surface modification technologies currently used. Once prepared, polymer films can be characterized using contact angle, protein adsorption, involving both quantitation and imaging, and leachables determinations.

Development and delivery of low concentration protein drug solutions requires more than the material optimization discussed in this paper. Additionally, the drug solution filling process must be optimized to avoid shear and bubbles which can denature proteins, and the use of filters which would adsorb proteins. The final packaging must be able to withstand distribution-induced denaturing of the product. The I.V. administration set must also minimize adsorbing surfaces, shear and bubble formation which induce denaturation.

NOTE

[1] Absolute protein concentrations were 10 ppm insulin, 300 ppm BSA and 80 ppm Anti-fibrinogen; however, since the suspension behaves as a colloid the number of individual colloidal particles, not absolute protein concentration governs binding [31, 32].

REFERENCES

1. S. J. Turco, *Am. J. IV Ther., Clin. Nutr.* **9**, 6 (1982).
2. L. D. Butler, J. M. Munson and P. P. DeLuca, *Am. J. Hosp. Pharmacol.* **37**, 935 (1980).
3. National Coordinating Committee on Large Volume Parenterals, *Recommendations of the NCCLVP for the Compounding and Administration of Intravenous Solutions*, K. N. Barker (Ed.). Amer. Soc. Hosp. Phar., Inc. (1981).
4. A. M. Pitt, *J. Parent. Sci. Technol.* **41**, 110 (1987).
5. M. V. Sefton and G. M. Antonacci, *Diabetes* **33**, 674 (1984).
6. J. Hirsch, M. Fratkin, J. Wood and R. Thomas, *Am J. Hosp. Pharmacol.* **34**, 583 (1977).
7. S. Weisenfeld, S. Podolsky, L. Goldsmith and L. Zeff, *Diabetes* **17**, 766 (1968).
8. Biotechnology Medicines in Development, 1991 Survey Report, Pharmaceutical Manufacturers Association, Washington D.C.
9. E. Brinkman, A. Poot, L. van der Dose and A. Bantjes, *Biomaterials* **11**, 200 (1990).
10. N. P. Desai and J. A. Hubbell, *J. Biomed. Mater. Res.* **25**, 829 (1991).
11. N. P. Desai and J. A. Hubbell, in: *Proceedings of the ACS Division of Polymeric Materials: Science and Engineering* **62**, 731 (1990).
12. Y. Mori, S. Nagaoka, H. Takiuchi, T. Kikuchi, N. Hoguchi, H. Tanzawa and Y. Noishiki, *Trans. Am. Soc. Artif. Intern. Organs* **28**, 459 (1982).
13. N. Chisato, K. D. Park, T. Okano and S. W. Kim, *Trans. Am. Soc. Artif. Intern. Organs* **35**, 357 (1989).
14. S. Nagaoka and A. Nakao, *Biomaterials* **11**, 119 (1990).
15. W. R. Gombotz, W. Guanghui and A. S. Hoffman, *J. Appl. Polym. Sci.* **37**, 91 (1989).
16. M. V. Sefton, G. Llanos and W. F. Ip, *Proceedings of the ACS Division of Polymeric Materials: Science and Engineering* **62**, 741 (1990).
17. J. H. Lee, J. Kopecek and J. D. Andrade, *J. Biomed. Mater. Res.* **23**, 351 (1989).
18. C. Maechling-Strasser, P. Dejardin, J. C. Galin and A. Schmitt, *J. Biomed. Mater. Res.* **23**, 1385 (1989).
19. E. W. Merrill, E. W. Salzman, S. Wan, N. Mahmud, L. Kushner, J. N. Lindon and F. Curme, *Trans. Am. Soc. Artif. Intern. Organs* **28**, 482 (1982).
20. S. K. Hunter, D. E. Gregonis, D. L. Coleman, B. Hanover, R. L. Steohen and S. C. Jacobsen, *Trans. Am. Soc. Artif. Intern. Organs* **29**, 250 (1983).
21. D. W. Grainger, C. Nojiri, T. Okano and S. W. Kim, *J. Biomed. Mater. Res.* **23**, 979 (1989).

22. N. P. Desai and J. A. Hubble, *Biomaterials* **12**, 144 (1991).
23. N. P. Desai and J. A. Hubble, *Macromolecules* **25**, 226 (1992).
24. K. Mukae, Y. H. Bae, T. Okano and S. W. Kim, *Polym. J.* **22**, 250 (1990).
25. C. Sung, M. R. Sobarzo and E. W. Merrill, *Polymer* **31**, 556 (1990).
26. S. Ding, C. Qin and B. Rabinow, "Biomaterials With Hydrophilic Surfaces" patent—pending.
27. Pierce Bulletin No. 23235, *Micro-BCA Protein Assay Reagent*. Pierce Chemicals, Rockford, IL (1989).
28. Pierce Bulletin No. 23255X, *Fluoraldehyde Protein/Peptide Assay Reagent*. Pierce Chemicals, Rockford IL (1990).
29. G. Frens, *Nature Phys. Sci.* **241**, 20–22 (1973).
30. R. M. Albrecht, J. A. Oliver and J. C. Loftus, *Science of Biological Specimen Preparation*, p. 185 (1986).
31. K. Park, S. R. Simmons and R. M. Albrecht, *Scan, Microsc.* **1**, 339 (1987).
32. K. Park, H. Park and R. M. Albrecht, in *Colloidal Gold: Principles, Methods and Applications*, Vol. 1, p. 490, M. A. Hayat (Ed.). Academic Press (1989).
33. K. D. Murthy, A. R. Diwan, S. R. Simmons, R. M. Albrecht and S. L. Cooper, *Scan Microsc.* **1**, 765 (1987).
34. B. R. Young, *Protein adsorption on polymeric biomaterials and its role in thrombogenesis*, Ph.D. Dissertation, University of Wisconsin, Madison (1984).
35. B. R. Young, W. G. Pitt and S. L. Cooper, *J. Colloid Interface Sci.* **124**, 28 (1988).
36. C.-G. Gölander and E. Kiss, *J. Colloid Interface Sci.* **121**, 240 (1988).
37. S. R. Simmons and R. M. Albrecht, *Proceedings Scanning 91, Scanning* **13**, 142 (1991).
38. R. M. Albrecht, Q. J. Lai, S. R. Simmons, S. L. Goodman and S. L. Cooper, Proc. Cardiovascular Sci. and Tech. Conf., Assoc. for Adv. of Med. Inst., Arlington, VA, 3 (1991).
39. J. Pawley and R. M. Albrecht, *Scanning* **10**, 184 (1988).
40. S. R. Simmons and R. M. Albrecht, in *The Science of Biological Specimen Preparation*, R. M. Albrecht and R. L. Ornberg (Eds). *Scan. Microsc. Suppl.* **3**, 27–34 (1989).
41. S. L. Goodman, S. R. Simmons, S. L. Cooper and R. M. Albrecht, *J. Colloid. Interface Sci.* **139**, 561 (1990).

Surface properties of RGD-peptide grafted polyurethane block copolymers: Variable take-off angle and cold-stage ESCA studies

HORNG-BAN LIN[1], KENNETH B. LEWIS[2], DEBORAH LEACH-SCAMPAVIA[2], BUDDY D. RATNER[2] and STUART L. COOPER[1]*

[1]*Department of Chemical Engineering, University of Wisconsin-Madison, Madison, WI 53706, USA*
[2]*National ESCA and Surface Analysis Center for Biomedical Problems and Department of Chemical Engineering, University of Washington, Seattle, WA 98195, USA*

Received 12 August 1992; accepted 6 October 1992

Abstract—Variable take-off angle and cold-stage ESCA measurements were utilized to analyze the surface composition of five polyurethane block copolymers. The polymers studied included a PTMO-polyurethane control, a carboxylated version of the control polyurethane, and three different peptide grafted (GRGESY, GRGDSY, and GRGDVY) polyurethanes. On dry samples the nitrogen signal detected using ESCA decreased with increasing take-off angle (i.e. as the specimen was probed closer to the surface) for all five polymers. This was believed to be due to the depletion of nitrogen-containing urethane hard segments at the surface. For all five polymers, the surface nitrogen concentration, associated with the hard segment, increased upon hydration. A greater increase of nitrogen concentration was observed for the peptide grafted polymers which suggests that grafting of the hydrophilic peptides to the polyurethane augments the hard segment enrichment at the surface upon hydration. Upon dehydration, the nitrogen concentration decreased for all five polymers suggesting migration of the more hydrophobic PTMO soft segment to the surface. *In vitro* endothelial cell adhesion showed an increase of cell attachment on prehydrated RGD-containing peptide grafted polyurethanes, but not on the other polymers. This result suggests an enhancement of peptide density at the aqueous interface, in good agreement with the ESCA studies.

Key words: Polyurethane copolymer; RGD-peptide; surface dynamics; ESCA; endothelial cell adhesion.

INTRODUCTION

Immobilization of cell adhesive peptides containing the YIGSR (Tyr–Ile–Gly–Ser–Arg) and RGD (Arg–Gly–Asp) sequences on polymer surfaces has been under study recently where it has been shown that these peptide grafted substrates significantly increase cell adhesion [1–3]. This suggests that surface immobilization of cell adhesive peptides may provide a promising route for the preparation of hybrid biomaterials. In our previous studies, a polyurethane block copolymer, based on methylene diphenylene diisocyanate (MDI), butanediol (BD), and poly(tetramethylene oxide) (PTMO), was modified by covalently grafting RGD-containing peptides onto the polymer backbone. Dramatic enhancement of endothelial cell adhesion was observed on films prepared from these solution-cast, bulk modified RGD-peptide grafted polyurethane block copolymers [4, 5].

Polyurethane block copolymers are typically composed of alternating blocks of hard and soft segment units. The bulk microphase separation of polyurethane block

* To whom correspondence should be addressed. Present address: College of Engineering, University of Delaware, Newark, DE 19716.

copolymers has been well established with the driving force for the phase separation believed to be due to the thermodynamic incompatibility of the hard and soft segment blocks [6]. Several studies have shown that the chemical composition at the polyurethane surface is different from that in the bulk, an observation which has been suggested to be due to the difference in interfacial energy between either the hard or soft segment blocks and the surface or interface to which the polymer is exposed [7, 8]. Polymer surfaces may restructure or reorient in response to different environments. For example, it is believed that a surface will restructure to orient its more non-polar or lower surface energy components to the air–polymer interface in order to minimize its interfacial free energy. In an aqueous environment, the same surface is expected to restructure due to the high driving force for its more polar components to migrate to the aqueous interface [9].

Polyurethane based on PTMO soft segments have been shown to exhibit soft segment enrichment at the air–polymer interface. The enrichment of PTMO soft segment at the air interface has been ascribed to the lower air–polymer interfacial energy of the soft segment relative to that of the hard segment [10, 11]. Studies [12] using dynamic contact angle techniques revealed an enhancement of the hard segment at the aqueous interface for a series of PTMO-based polyurethanes. Using cold-stage ESCA on freeze-dried samples, Takahara and coworkers [13], studied the surface reorganization of a polyurethane composed of a fluorine-containing diol chain extender in the hydrated and dehydrated states. They concluded that the hard and soft segment composition at the surface of hydrated and dehydrated samples was different due to molecular reorganization at the surface.

Since only the surface of a biomaterial has contact with a biological system of interest, a better understanding of surface dynamics is important. We are interested in studying how the RGD-peptide grafted polyurethane surfaces reorient or restructure when placed in contact with different environments. The orientation and availability of the cell adhesive RGD-containing peptide at the interface of peptide grafted polymers is particularly important for applications where cell adhesion is being attempted. In this study, variable take-off angle and cold-stage ESCA techniques were utilized to examine the surface composition of the base and peptide grafted polyurethanes in the as-cast, hydrated, and dehydrated states. Additionally, in order to investigate the effect of hydration on the cell-polymer interactions, *in vitro* endothelial cell adhesion studies were conducted on the polyurethane substrates with and without prehydration.

MATERIALS AND METHODS

Materials

The synthesis of a poly(tetramethylene oxide) (PTMO) based polyurethane as well as versions which are carboxylated and then subsequently peptide-grafted have been reported previously in detail [4] and are only summarized here. A two-step solution polymerization was used to synthesize a PTMO-based polyurethane [14]. The polyurethane was based on a 3:2:1 molar ratio of methylene diphenylene diisocyanate (MDI), 1,4-butanediol (BD), and poly(tetramethylene oxide) (PTMO, Mn = 1000). A bimolecular nucleophilic substitution reaction was used to graft ethyl carboxylate groups to the urethane nitrogen (i.e. creating a carboxylated polyurethane) using sodium hydride (NaH; Aldrich, 60% dispersion in mineral oil) to abstract the urethane hydrogen, followed by reaction with β-propiolactone (Aldrich, 90%) [15, 16]. The extent of ethyl carboxylate group substitution was calculated from the weight fraction of sodium as determined by elemental analysis (Galbraith Laboratories). The carboxylated polyurethane was then passed over a strong acid cation-exchange resin (Amberlyst-15, Aldrich) to protonate the carboxylate groups prior to the peptide coupling reaction. A carboxylated polyurethane with approximately 4% substitution of urethane hydrogen with ethyl carboxylate groups was used in this study.

Three protected hexapeptides including Fmoc-Gly-Arg(Pmc)-Gly-Asp(OtBu)-Ser(OtBu)-Tyr(OtBu)-OtBu (Fmoc-GRGDSY*), Fmoc-Gly-Arg(Pmc)-Gly-Asp(OtBu)-Val-Tyr(OtBu)-OtBu (Fmoc-GRGDVY*), and Fmoc-Gly-Arg(Pmc)-Gly-Glu(OtBu)-Ser(OtBu)-Tyr(OtBu)-OtBu (Fmoc-GRGESY*) were synthesized by a standard solution phase method using N^{α}-9-Fluorenylmethyloxycarbonyl (Fmoc)-protected amino acids (Bachem Bioscience Inc.). The guanidino functionality of arginine was protected with the N_G-2,2,5,7,8-pentamethylchroman-6-suphonyl (Pmc) derivative group [17], and the *tert*-butyl ether or ester (OtBu) group protected the side chains of serine, tyrosine, aspartic acid, glutamic acid, and the C-terminus of the peptide. Prior to coupling to the carboxylated polyurethane, the N-terminal Fmoc protecting group of each hexapeptide was removed to provide free amine groups. This was accomplished by reacting the peptide with 10% diethylamine in dimethylformamide (DMF) at room temperature for 1.5 h.

Each of these peptides was grafted onto the polyurethane backbone via the formation of an amide linkage between the N-terminal amino group of the protected peptide and the carboxyl group on the polymer using 1-(3-dimethylaminopropyl)-3-ethylcarbodiimide hydrochloride (EDCI) as the coupling agent. The protecting groups on the grafted peptide were removed by freshly prepared Reagent K[18]. The peptide grafted polyurethanes were then precipitated, thoroughly washed with anhydrous ethyl ether, and dried under vacuum.

The amount of peptide incorporation on the grafted polyurethanes was quantified by amino acid analysis (Department of Chemistry, University of Minnesota). Table 1 summarizes the nomenclature of samples which were used in this study. The starting polymer and the carboxylated polyurethane are denoted as PEU and PEU-COOH, respectively. The peptide grafted polyurethanes are referred as PEU-GRGDSY, PEU-GRGDVY, and PEU-GRGESY.

The chemical structures of the base, carboxylated, and peptide-grafted polyurethanes are illustrated in Fig. 1. It should be noted that only approximately 4% of the urethane hydrogen has been substituted to produce the carboxylated polyurethane, and that separately each of the hexapeptides was grafted onto the urethane hard segment via an amidation reaction utilizing the carboxyl groups.

Table 1.
Sample nomenclature

Sample	Description
PEU	Base polyurethane
PEU-COOH	Carboxylated polyurethane
PEU-GRGDSY	Free peptide GRGDSY grafted polyurethane
PEU-GRGDVY	Free peptide GRGDVY grafted polyurethane
PEU-GRGESY	Free peptide GRGESY grafted polyurethane

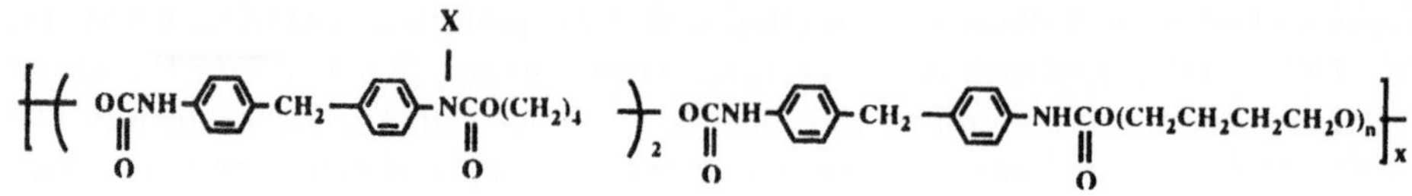

HARD SEGMENT **(MDI + BD)**

SOFT SEGMENT **(PTMO)**

SIDE CHAIN	**POLYMER NOTATION**
X = H	PEU
$(CH_2)_2COOH$	PEU-COOH
$(CH_2)_2CONH$(GRGDSY)	PEU-GRGDSY
$(CH_2)_2CONH$(GRGDVY)	PEU-GRGDVY
$(CH_2)_2CONH$(GRGESY)	PEU-GRGESY

Figure 1. Schematic of the chemical structures of the base, carboxylated, and peptide grafted polyurethanes.

ESCA measurements

ESCA experiments were performed on samples prepared by spin casting 3% polymer/DMF solutions onto clean glass coverslips (9 mm dia.) followed by drying in a laminar flow hood (SEMIFAB Inc.). A Surface Science Instruments SSX-100 spectrometer with a monochromatized Al K_α X-ray source was utilized for the variable take-off angle and cold-stage ESCA experiments. Elemental composition was based on peak area data collected at a pass energy of 150 eV. Binding environment information for carbon was determined from spectra collected at a pass energy of 25 eV.

Variable take-off angle ESCA. Samples were mounted on a sample stage which can be accurately tilted to the desired angles. The ESCA spectra were collected at 10, 39, 55, 68 and 80° take-off angles at ambient temperature. The take-off angle was defined as the angle between the normal to the sample surface and the axis of the analyzer lens system. In order to improve the spectrum resolution at each angle, an aperture was placed over the analyzer lens to decrease the acceptance angle of the lens from the normal 30 to 6° [18]. A conductive screen was placed above the sample surface and a low energy (5–20 eV) electron flood gun was used to minimize surface charging. The elemental compositions measured at each take-off angle were used to determine a depth profile using the procedure described by Paynter [20] and Ratner *et al.* [21]. The mean free paths used in this procedure were calculated from the equations given by Seah and Dench [22]. These calculations predict the sampling depth (three times the mean free path) decreasing from *c*. 110 to 20 Å as the take-off angle increasing from 10 to 80°.

Cold-stage ESCA. Samples were hydrated with deionized water overnight in covered petri dishes. Prior to mounting a sample onto the stage, the sample surface was rinsed off with deionized water. A single drop of deionized water was then placed on each sample. The sample was then rapidly cooled to −160°C in the preparation chamber, such that an ice layer was formed on the surface. The ice layer was removed by sublimation at 10^{-6} Torr and −60°C for approximately 1 h. It was assumed that the polymer surface restructuring was hindered during the freeze-drying since the T_g of the control polyurethane is approximately −45°C as determined by differential scanning calorimetry. After the ice layer was completely removed from the surface, the sample was introduced to the analytical chamber, where the ESCA spectra were collected at 0 and 72° take-off angles. After analysis in the freeze-dried state, the sample was removed to the preparation chamber and vacuum dried at 30°C for aproximately 3 h. It was assumed that polymer surface would restructure itself to minimize its interfacial energy due to the vacuum-drying. The surface was then again analyzed at 0 and 72° take-off angles.

Cell adhesion

Human umbilical vein endothelial cells (HUVECs) (generously provided by Dr. Shinji Asakura, University of Wisconsin) were used for the *in vitro* cell adhesion experiments. Cells were harvested for experiments by incubating in 0.5 mg/ml trypsin and 0.2 mg/ml EDTA in Hank's balanced salt solution (Ca^{2+} and Mg^{2+} free, pH = 7.4; SIGMA) for approximately 2 min at 37°C. Trypsinization was stopped by the addition of soybean trypsin inhibitor (SBTI) to a final concentration of 0.5 mg/ml. Cells were collected by centrifugation and resuspended in DMEM supplement with 0.1% heat inactivated bovine serum albumin (BSA).

Cell attachment was measured using the method of Grinnell [23] with modifications. Polymer coated coverslips with and without prehydration were placed in 24-well tissue culture plates (Corning). Prehydration of polymer coated coverslips was done by incubating coverslips in Tris buffered saline (TBS, pH = 7.4) at 37°C for 3 h, then TBS was discarded prior to the experiment. 2×10^4 cells/cm^2 were seeded on each polymer substrate and allowed to attach for 80 min. The wells were subsequently rinsed with TBS to remove non-adherent cells, and the adherent cells were fixed with a solution of 3% paraformaldehyde for 2 h. The number of attached cells was determined visually using 100 × magnification on a microscope equipped with phase contrast objectives (Nikon, Japan). Adherent cells were counted in seven or eight areas randomly chosen in the central and peripheral regions of each polymer substrate. The morphology of attached cells was examined by scanning electron microscopy (SEM) using a JEOL JSM-35C SEM. Samples were serially dehydrated in 30, 50, 75, 85, 90, 95, and 100% ethanol and critical point dried using CO_2 as the transitional fluid. Samples were then sputter-coated with gold and examined with the scanning electron microscope at an accelerating voltage of 15 kV.

RESULTS AND DISCUSSION

Peptide grafted polyurethanes

The amount of grafted peptide on PEU-GRGESY, PEU-GRGDSY, and PEU-GRGDVY was quantified by amino acid analysis. The amount of peptide incorporated in PEU-GRGESY, PEU-GRGDSY, and PEU-GRGDVY was approximately 80, 100, and 100 μmol/g polymer, respectively. As indicated in Fig. 1, each of these peptides was grafted onto the urethane hard segment.

Variable take-off angle ESCA analysis

Table 2 shows the surface atomic composition determined from variable take-off angle ESCA measurements and calculated from polyurethane stoichiometric chemistry. In addition, the atomic percentages of carbonyl carbon (C=O) resulting

Table 2.
Variable take-off angle ESCA measurements

Sample	take-off angle (deg)	Atomic concentration (%)							Atomic ratio (%)
		C	O	N	Si	F	Na	Cl	$C{=}O/C_{total}$
PEU	80	79.2	18.3	2.5	—	—	—	—	—
	68	79.2	18.4	2.4	—	—	—	—	1.5
	55	78.9	18.0	3.0	—	—	—	—	2.0
	39	78.2	18.3	3.5	—	—	—	—	2.3
	10	78.1	18.4	3.5	—	—	—	—	2.5
	S[a]	77.9	17.9	4.3	—	—	—	—	5.5
PEU-COOH	80	67.6	24.3	2.5	5.5	—	—	—	—
	68	67.7	24.5	1.8	6.0	—	—	—	1.5
	55	72.3	21.7	2.9	3.2	—	—	—	2.0
	39	76.4	19.2	3.0	1.4	—	—	—	2.3
	10	74.5	20.0	3.5	1.9	—	—	—	2.8
	S[a]	77.7	18.0	4.3	—	—	—	—	5.7
PEU-GRGESY	80	78.3	18.8	3.0	—	—	—	—	—
	68	78.6	18.5	3.0	—	—	—	—	1.6
	55	78.9	17.9	3.3	—	—	—	—	2.6
	39	78.4	18.0	3.6	—	—	—	—	2.4
	10	78.8	17.4	3.9	—	—	—	—	3.5
	S[a]	76.6	18.2	5.2	—	—	—	—	6.4
PEU-GRGDSY	80	76.0	20.7	3.3	—	—	—	—	—
	68	76.8	19.6	3.6	—	—	—	—	2.7
	55	78.2	18.2	3.6	—	—	—	—	2.2
	39	77.7	18.1	4.1	—	—	—	—	2.8
	10	76.8	18.4	4.8	—	—	—	—	2.8
	S[a]	76.3	18.3	5.4	—	—	—	—	6.6
PEU-GRGDVY	80	78.6	19.0	2.5	—	—	—	—	—
	68	78.3	18.5	3.2	—	—	—	—	1.6
	55	74.4	18.5	4.6	—	1.2	0.9	0.4	3.0
	39	71.0	18.5	6.2	—	2.0	1.5	0.9	5.9
	10	72.6	19.2	5.4	—	1.4	1.0	1.4	5.6
	S[a]	76.4	18.2	5.4	—	—	—	—	6.6

[a] Stoichiometric

from curve-fitting of the high resolution, C1*s* spectra are listed. For the polyurethanes studied, nitrogen (N) is only associated with the urethane hard segment component (i.e. methylene diphenylene diisocyanate, MDI) for PEU and PEU-COOH, while it is associated with both the urethane group and the grafted peptide in the polymers designated as PEU-GRGESY, PEU-GRGDSY, and PEU-GRGDVY. The ether carbon (C—O) is primarily associated with the PTMO soft segment, whereas the carbonyl carbon (C=O) is related to the urethane hard segment and the amide linkages of the grafted peptides.

At each take-off angle, the amount of nitrogen on PEU-GRGESY, PEU-GRGDSY, and PEU-GRGDVY was higher than that detected on PEU and PEU-COOH. Similarly, the carbonyl carbon (C=O) content (at approximately 289.5 eV) was higher for PEU-GRGESY, PEU-GRGDVY, and PEU-GRGDSY compared to PEU and PEU-COOH. Since the N and C=O can be only attributed to the urethane hard segment for PEU and PEU-COOH, and to the urethane hard segment and the grafted peptide for PEU-GRGESY, PEU-GRGDSY, and PEU-GRGDVY, these data suggest the presence of the grafted peptide at and near the air-polymer interface *in vacuo*.

Silicon (Si) was detected on PEU-COOH. The amount of Si increased with increasing take-off angle which suggests Si was concentrated near or on the surface. The Si contamination may have occurred during the sample preparation process, since Si has been documented as one of the most commonly encountered contaminants in the laboratory environment [18]. Fluorine (F), sodium (Na), and chlorine (Cl) were detected on PEU-GRGDVY at take-off angles less than 55 deg. These contaminants may be due to residual trifluoroacetic acid (TFA) and NaCl which were used in the polymer synthesis procedure [4], and they could be eliminated by hydrating and rinsing the sample surface as suggested by the cold-stage ESCA study (see below).

Figure 2 shows the compositional depth profiles of the polyurethanes studied except PEU-GRGDVY. The experimentally observed composition at low take-off angles and the values used in the depth profiles at depth >20 Å show good agreement with the expected bulk elemental compositions. The depth profile models also show a decrease of N concentration and an increase of O concentration at depth ≤ 20 Å for the polyurethanes studied. This indicates a depletion of the urethane hard segment or an enrichment of the PTMO soft segment at the surface.

The Si impurity on PEU-COOH is enriched at a depth <20 Å (Fig. 2B). This is consistent with the finding of Si impurity on polymer surfaces reported by other researchers [16, 19]. The PEU, PEU-GRGESY, and PEU- GRGDSY samples which are free of any impurity deserve further discussion (Fig. 2A, C and D). The difference between the detected and stoichiometric N atomic percentage was greatest in the outer 15 Å of PEU, PEU-GRGESY, and PEU-GRGDSY. This suggests that the depletion of the hard segment and the grafted peptide is greatest at the air-polymer interface.

The depletion of hard segment at the PEU and PEU-COOH surfaces may be ascribed to migration of the hydrophobic PTMO soft segment to the surface. This is consistent with the studies reported by other researchers [16, 25]. For the peptide grafted polyurethanes, the hexapeptides GRGDSY, GRGESY, and GRGDVY are very hydrophilic and are highly soluble in phosphate buffered saline (PBS, pH = 7.4). This solubility is believed to be due to the basic side chain of arginine

(R; pK = 12.5) and the acidic side chains of aspartic acid (D; pK = 3.9) and glutamic acid (E; pK = 4.3). Like the urethane hard segment, these hydrophilic peptides may have a tendency to migrate away from the air–polymer interface *in vacuo*. Looking at 80° take-off angle (the top 20 Å layer), the differences between

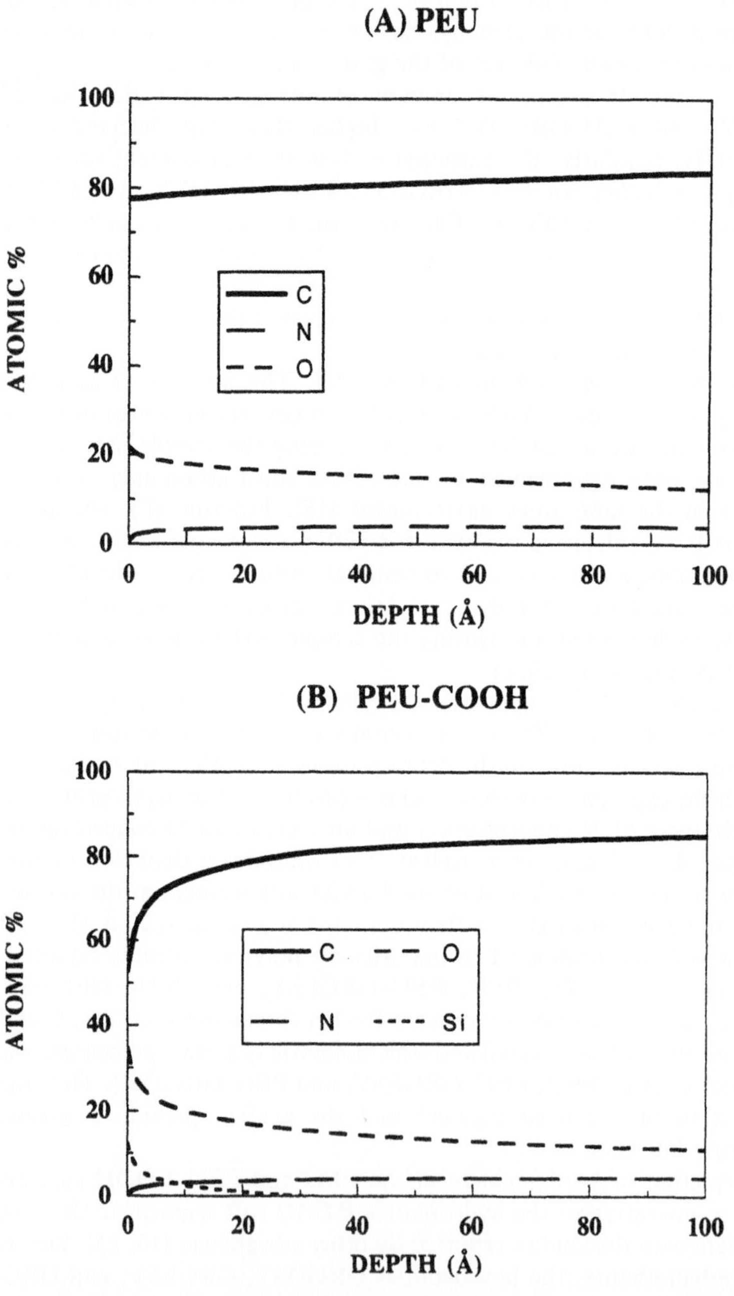
(A) PEU
ATOMIC %
DEPTH (Å)
C
N
O
0
20
40
60
80
100
(B) PEU-COOH
ATOMIC %
DEPTH (Å)
C
O
N
Si
0
20
40
60
80
100

Figure 2. Compositional depth profiles determined for (A) PEU; (B) PEU–COOH; (C) PEU–GRGESY; and (D) PEU–GRGDSY.

the detected and stoichiometric N atomic percentage on the PEU-GRGDSY, PEU-GRGESY, and PEU-GRGDVY samples were greater than those for PEU and PEU-COOH. This suggests that the grafted peptide on the hard segment enhances the migration of hard segment away from the air-polymer interface.

(C) PEU-GRGESY

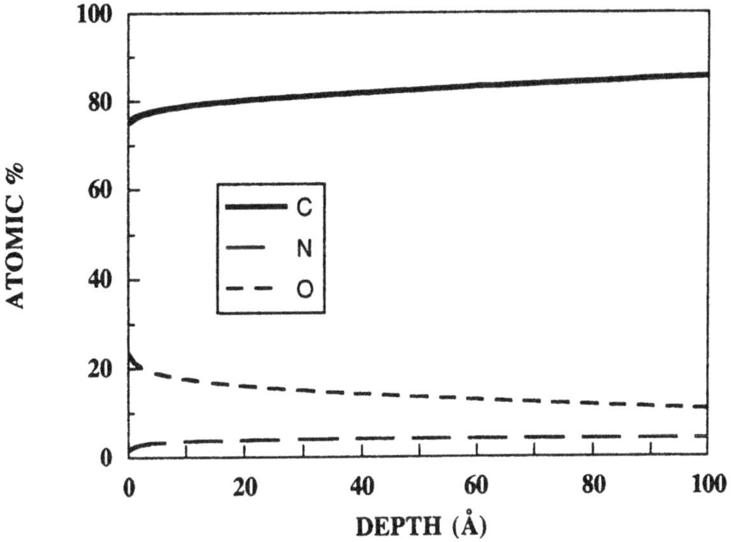
100
80
60
40
20
0
ATOMIC %
C
N
O
0
20
40
60
80
100
DEPTH (Å)

(D) PEU-GRGDSY

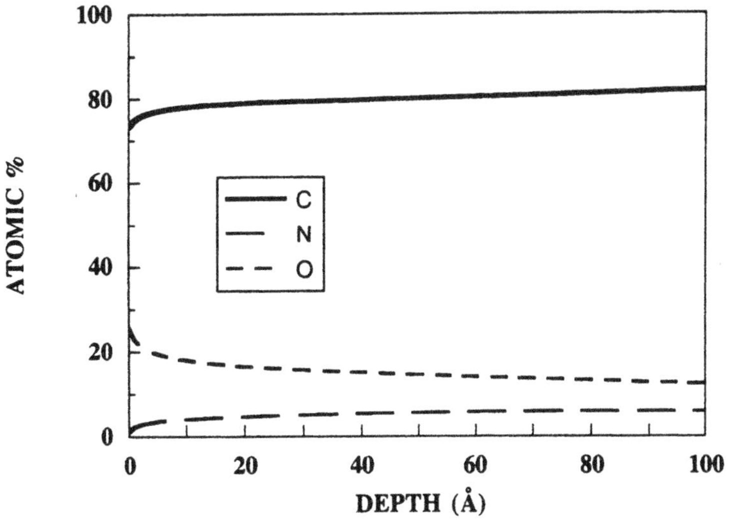
100
80
60
40
20
0
ATOMIC %
C
N
O
0
20
40
60
80
100
DEPTH (Å)

Figure 2. Continued

Table 3.
Cold-stage ESCA measurements

Sample	State	take-off angle (deg)	Atomic concentration (%)			
			C	O	N	Si
PEU	FD	72	79.0	18.3	2.7	—
	FD	0	78.5	17.8	3.7	—
	VD	72	78.7	18.7	2.6	—
	VD	0	79.7	17.0	3.3	—
PEU-COOH	FD	72	77.8	16.4	1.2	4.6
	FD	0	68.5	23.4	2.2	5.9
	VD	72	—	—	—	—
	VD	0	—	—	—	—
PEU-GRGESY	FD	72	79.7	16.9	3.4	—
	FD	0	78.7	17.1	4.2	—
	VD	72	81.3	16.5	2.2	—
	VD	0	78.7	17.7	3.7	—
PEU-GRGDSY	FD	72	77.6	18.6	3.8	—
	FD	0	77.0	18.6	4.4	—
	VD	72	77.9	18.7	3.4	—
	VD	0	77.2	18.4	4.4	—
PEU-GRGDVY	FD	72	76.7	18.7	4.6	—
	FD	0	76.7	18.2	5.1	—
	VD	72	75.8	20.5	3.7	—
	VD	0	76.9	18.2	4.9	—

FD: freeze-dried; VD: vacuum-dried.

Cold-stage ESCA analysis

The surface atomic compositions determined from the cold-stage ESCA experiments are summarized in Table 3. Freeze-dried and vacuum-dried states refer to spectra that were collected from the frozen, hydrated samples at −60°C and the dehydrated samples at 30°C, respectively. It was assumed that the surface of hydrated samples would not rearrange during the freeze-drying and cold ESCA data acquisition period. This assumption was based on the fact that the large scale chain translational molecular motion does not take place at −60°C, since this temperature is well below the glass transition temperature (approximately −45°C) of the polyurethanes studied. At 30°C in the vacuum-dried state, large scale of molecular motion in the bulk would be expected so that polymer surface restructuring would be facilitated.

At 72° take-off angle (i.e. the top 25 Å layer), the amount of nitrogen found in the freeze-dried state was greater than that found in the vacuum-dried state for all polyurethanes studied. The increase of the nitrogen signal for PEU in the hydrated state can be explained by the migration of the hard segment to the water–polymer interface upon hydration. This is consistent with the studies of polyurethane surfaces by Andrade [12] and Takahara [13] who concluded that concentration of the more hydrophilic hard segment indeed occurred at the water-polymer interface. For PEU-GRGESY, PEU-GRGDSY, and PEU-GRGDVY, the increase in nitrogen

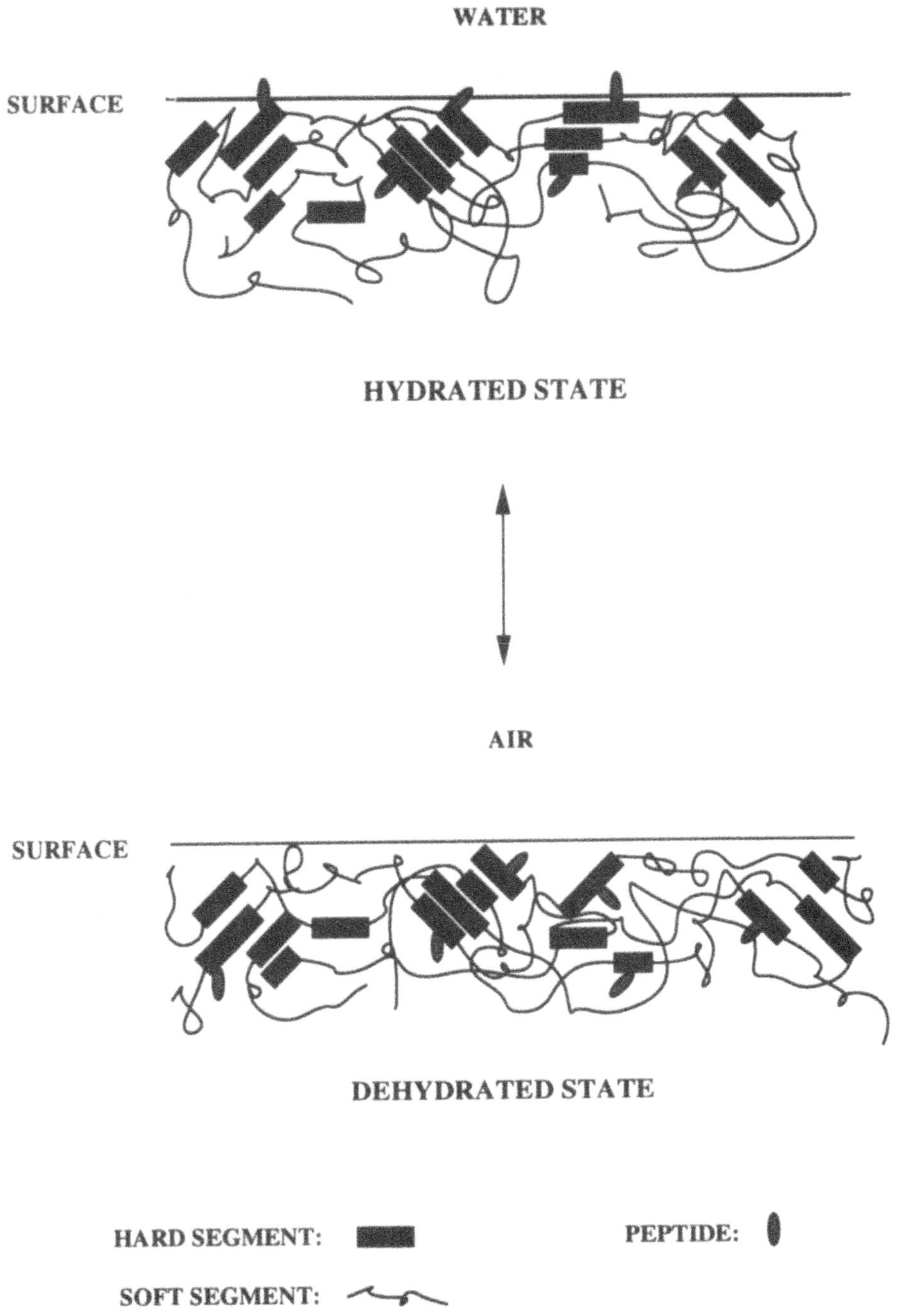
WATER
SURFACE
HYDRATED STATE
AIR
SURFACE
DEHYDRATED STATE
HARD SEGMENT:
PEPTIDE:
SOFT SEGMENT:

Figure 3. Schematic of the surface reorientation of the peptide grafted polyurethanes in the hydrated and dehydrated states.

signal was greater than that for PEU. This was a good indication of the enhancement of grafted peptide at the water–polymer interface, since these hydrophilic, grafted peptides were also contributing to the nitrogen signal. At 0° take-off angle (i.e. the top 110 Å layer), all of the polymers studied except PEU exhibited smaller differences between the detected nitrogen signal of the freeze-dried and the vacuum-dried samples. This implies that changes of the hard segment concentration due to the equilibrating interface was concentrated near the surface of these polymers. A schematic representation of the surface reorientation which is occurring in the peptide grafted polyurethanes due to hydration and dehydration is illustrated in Fig. 3.

It was interesting that fluorine, sodium, and chlorine contaminants in PEU-GRGDVY were not detected on the freeze-dried samples. These contaminants were believed to be residual trifluoroacetic acid (TFA) and NaCl from the polymer synthesis process. Since both TFA and NaCl are highly water soluble, they may be dissolved in the deionized water during hydration, and any residual may be rinsed off before mounting the sample onto the ESCA stage. The freeze-dried sample of PEU-COOH exhibited a decrease of Si with increasing take-off angle. This suggests a low concentration of Si at the hydrated surface. The variable angle study indicated that Si was concentrated at the surface *in vacuo*. These results are in agreement with the findings that Si is localized at the air–polymer interface but migrates away from the surface in aqueous environments [24].

Cell adhesion

Endothelial cell attachment on the polyurethane substrates was measured in culture medium containing 0.1% heat inactivated bovine albumin. Bovine albumin was used to minimize non-specific cell adhesion. The polyurethane substrates were either

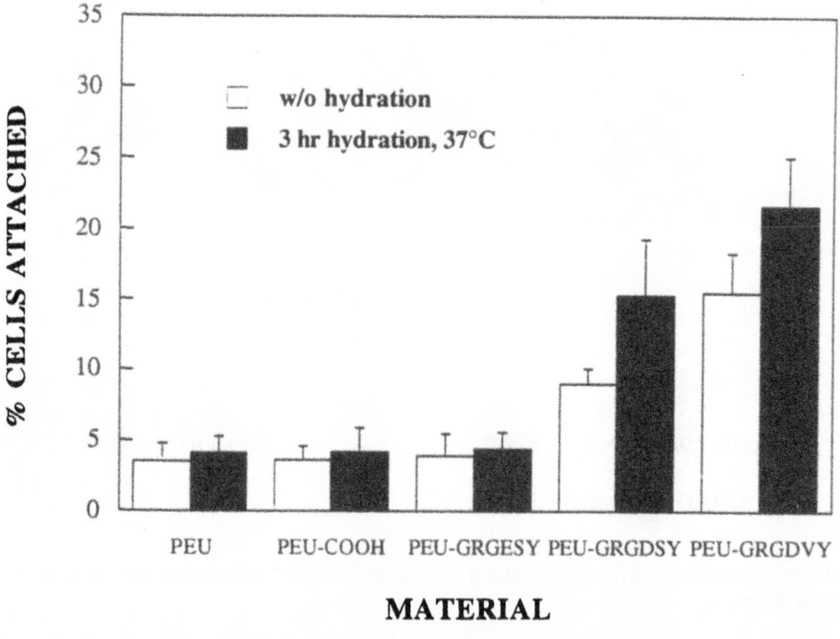
% CELLS ATTACHED
35
30
25
20
15
10
5
0
w/o hydration
3 hr hydration, 37°C
PEU
PEU-COOH
PEU-GRGESY
PEU-GRGDSY
PEU-GRGDVY
MATERIAL

Figure 4. Effect of substrate prehydration on human umbilical vein endothelial cell attachment.

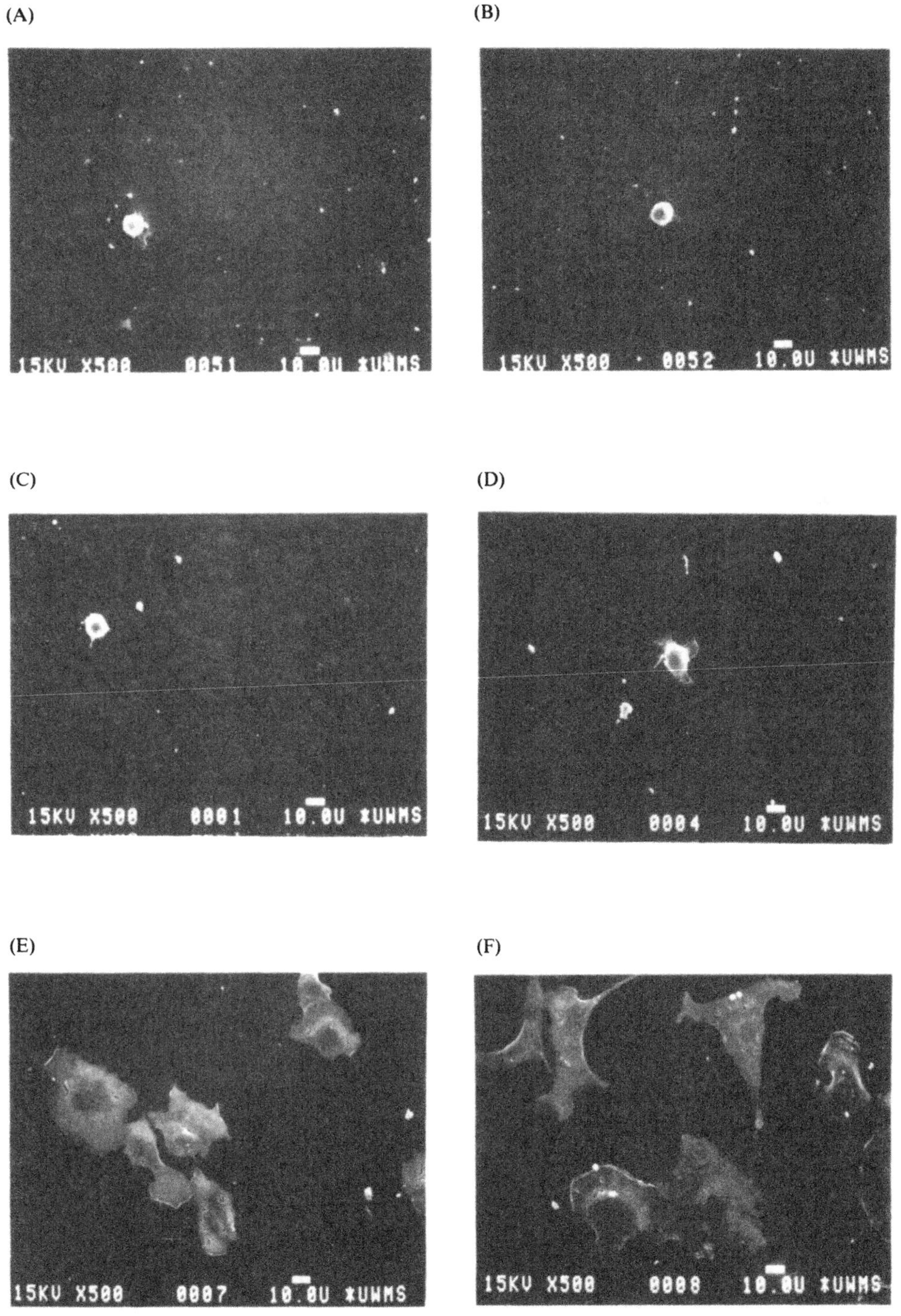
(A)
15KU X500 0051 10.0U *UWMS
(B)
15KU X500 0052 10.0U *UWMS
(C)
15KU X500 0001 10.0U *UWMS
(D)
15KU X500 0004 10.0U *UWMS
(E)
15KU X500 0007 10.0U *UWMS
(F)
15KU X500 0008 10.0U *UWMS

Figure 5. Scanning electron micrographs showing human umbilical vein endothelial cells adherent on: (A) and (B) PEU; (C) and (D) PEU-GRGESY; and (E) and (F) PEU-GRGDVY after 80 min incubation in DMEM containing 0.1% BSA. (A), (C), and (E) are substrates without prehydration, whereas (B), (D), and (F) are substrates with prehydration treatment (scale bar = 10 μm).

prehydrated at 37°C for 3 h or were studied without prehydration prior to the experiment. The goal of this experiment was to determine the effect that substrate prehydration had on cell attachment and to correlate the results to the cold-stage ESCA experiments. Figure 4 shows the percentages of seeded cells attached onto the polyurethane substrates after 80 min incubation. Both with and without prehydration, the ranking of the substrates for supporting cell adhesion was PEU-GRGDVY > PEU-GRGDSY > PEU-GRGESY ≈ PEU-COOH ≈ PEU. This is consistent with our previous experiments [5]. Without prehydration, only approximately 3-4% of cells attached to PEU, PEU- COOH, and PEU-GRGESY, whereas 9 and 16% of cells attached to PEU-GRGDSY and PEU-GRGDVY. With prehydration of substrates at 37°C for 3 h, no significant increase ($p > 0.05$) of cell attachment was observed on PEU, PEU-COOH, and PEU-GRGESY. Although the cold-stage ESCA data suggest that the urethane hard segment and possibly the grafted GRGESY peptide may migrate to the water-polymer interface upon hydration, the increase of the hard segment and the GRGESY peptide at the interface did not increase cell attachment. On the other hand, an increase ($p < 0.05$) in cell attachment was observed on the prehydrated PEU- GRGDSY and PEU-GRGDVY substrates. Since the increase of the hard segment and grafted inactive peptide did not increase cell attachment, the increase of cell attachment on the prehydrated PEU-GRGDSY and PEU-GRGDVY substrates was likely due to the increase of the active peptide density and availability at the interface.

The morphology of attached cells after 80 min incubation in 0.1% BSA was examined using scanning electron microscopy. Figures 5A-F show the attached cells on the PEU, PEU-GRGESY, and PEU-GRGDVY substrates with and without prehydration. The attached cells on PEU and PEU-GRGESY (Figs 5A-D) were round in shape and no extended pseudopodia were observed. This indicates a minimal extent of cell spreading or no spreading. Prehydration of these substrates did not improve cell attachment and spreading. This may be attributed to the fact that bovine albumin was used to block or minimize the non-specific cell adhesion and neither the urethane hard segment nor the inactive peptide GRGESY provided proper ligands for cell receptor-mediated (i.e. integrin-mediated) cell adhesion. The attached cells on PEU-GRGDVY showed extended pseudopodia and became flat in shape (Figs 5E and F). Improvement of cell attachment and spreading on the RGD-containing peptide grafted polyurethanes and other polymer substrates has been reported by our laboratory [4] and other research groups [1-3]. Furthermore, more attached cells were observed on the prehydrated PEU-GRGDVY substrate. This supports the hypothesis that there is an increase of the GRGDVY peptide density or availability at the polymer-water interface upon hydration which increases the probability for integrin-mediated cell adhesion [26].

CONCLUSIONS

Five polyurethane block copolymers, including a PTMO-based (PEU), a carboxylated (PEU-COOH), and three peptide grafted (PEU-GRGESY, PEU-GRGDSY, and PEU-GRGDVY) polyurethanes, were examined using variable take-off angle and cold-stage ESCA. Peptides were grafted onto a carboxylated polyurethane via formation of amide linkages. In order to determine the polymer surface composition and its reorientation in response to different environments, samples were analyzed in the as-cast, hydrated, and dehydrated states.

The variable angle ESCA analysis showed a decrease in nitrogen concentration with increasing take-off angle (or decreasing sample depth) for all five polyurethanes. This indicates a depletion of hard segment component at the surface. The cold-stage ESCA analysis of hydrated samples showed higher surface N atomic percentage compared to the dehydrated state. This suggests a surface enrichment of the hard segment and the grafted peptide for the hydrated polyurethanes. Upon dehydration, a decrease in the surface nitrogen concentration was observed. This surface restructuring may be attributed to the migration of the more hydrophobic PTMO soft segment to the surface.

The effect of substrate prehydration on endothelial cell adhesion was investigated. An increase of cell adhesion was observed on prehydrated samples of PEU-GRGDSY and PEU-GRGDVY, but not on prehydrated PEU, PEU-COOH, or PEU-GRGESY. The results suggest that there is an increase in the concentration of RGD-containing peptide at the aqueous interface which is in agreement with the cold-stage ESCA data.

Acknowledgements

This work was supported by the National Institutes of Health through grants HL-24046 and HL-47179. The ESCA experiments were performed at the National ESCA and Surface Analysis Center for Biomedical Problems (NESAC/BIO), funded by a grant from the National Institute of Health, Division of Research Resources (RR-01296). The assistance of Dr W. Sun in carrying out the cell culture experiments is gratefully acknowledged.

REFERENCES

1. B. K. Brandley and R. L. Schnaar, *Anal. Biochem.* **172,** 270–278 (1988).
2. S. P. Massia and J. A. Hubbell, *Anal. Biochem.* **187,** 292–301 (1990).
3. T. Matsuda, A. Kondo, K. Makino and T. Akutsu, *Trans. Am. Soc. Artif. Intern. Organs* **35,** 677–679 (1989).
4. H.-B. Lin, Z.-C. Zhao, C. García-Echeverría, D. H. Rich and S. L. Cooper, *J. Biomater. Sci., Polymer Edn* **3,** 217–227 (1992).
5. H.-B. Lin, C. García-Echeverría, S. Asakura, W. Sun, D. F. Mosher and S. L. Cooper, *Biomaterials* in press (1992).
6. M. D. Lelah and S. L. Cooper, *Polyurethanes in Medicine*. CRC Press, Boca Raton, Florida (1986).
7. K. Knutson and D. J. Lyman, in: *Biomaterials: Interfacial Phenomena and Applications*, pp. 109–132, S. L. Cooper and N. A. Peppas (Eds). Advances in Chemistry Series 199, Washington DC (1982).
8. D. B. Ratner and R. W. Paynter, in: *Polyurethanes in Biomedical Engineering*, pp. 41–68, H. Plank, G. Egbers and I. Syre (Eds). Elsevier Science Publishers, Amsterdam (1984).
9. J. D. Andrade, *Surface and Interfacial Aspects of Biomedical Polymers, Vol. 1, Surface Chemistry and Physics*. Plenum Press, New York (1985).
10. A. Takahara, J.-C. Tashita, T. Kajiyama, M. Takayanagi and W. J. MacKnight, *Polymer* **26,** 978–986 (1985).
11. A. Takahara, J.-C. Tashita, T. Kajiyama, M. Takayanagi and W. J. MacKnight, *Polymer* **26,** 987–996 (1985).
12. K. G. Tingey and J. D. Andrade, *Langmuir* **7,** 2471–2478 (1991).
13. A. Takahara, N.-J. Jo, K. Takamori and T. Kajiyama, in: *Progress in Biomedical Polymers*, pp. 217–228, G. Gebelein and R. Dunn (Eds). Plenum Press, New York (1990).
14. J. H. Saunders and K. C. Frisch, *Polyurethane Chemistry and Technology, Part I*. Interscience Publishers, New York (1965).
15. K. K. Hwang, T. A. Speckhard and S. L. Cooper, *J. Macromol. Sci.-Phys.* **B23,** 153–174 (1984).

16. A. Z. Okkema, *The Physical and Biointerfacial Properties of Polyurethane Ionomers*. Ph.D. Dissertation, University of Wisconsin-Madison (1990).
17. R. Ramage and J. Green, *Tetrahedron Lett.* **28**, 2287–2290 (1987).
18. D. S. King, C. G. Fields and G. B. Fields, *Int. J. Peptide Protein Res.* **36**, 255–266 (1990).
19. T. G. Grasel, D. G. Castner, B. D. Ratner and S. L. Cooper. *J. Biomed. Mater. Res.* **24**, 605–620 (1990).
20. R. W. Paynter, *Surf. Interface Anal.* **3**, 186–187 (1981).
21. B. J. Tyler, D. G. Castner and B. D. Ratner, *Surf. Interface Anal.* **14**, 443–450 (1989).
22. M. P. Seah and W. A. Dench, *Surf. Interface Anal.* **1**, 2–11 (1979).
23. F. Grinnell, *Exp. Cell. Res.* **102**, 51–62 (1976).
24. B. D. Ratner, P. K. Weathersby and A. S. Hoffman, *J. Appl. Polymer Sci.* **22**, 643–664 (1978).
25. M. J. Hearn, B. D. Ratner and D. Briggs, *Macromolecules* **21**, 2950–2959 (1988).
26. E. Ruoslahti and M. D. Pierschbacher, *Science* **238** 491–497 (1987).

Effect of polyurethane surface chemistry on its lipid sorption behavior

ATSUSHI TAKAHARA, KOHZO TAKAHASHI and TISATO KAJIYAMA*

Department of Chemical Science & Technology, Faculty of Engineering, Kyushu University, Hakozaki, Higashi-ku, Fukuoka 812, Japan

Received 25 September 1992; accepted 14 October 1992

Abstract—The relationships among surface, bulk properties and lipid sorption behaviors of segmented polyurethanes (SPUs) with various polyol soft segments were investigated. The polyols used in this study were poly(ethylene oxide) (PEO), poly(tetramethylene oxide) (PTMO), and poly(dimethylsiloxane) (PDMS). The hard segment of these segmented polyurethanes was composed of 4,4′-diphenylmethane diisocyanate and 1,4-butanediol, present at 50 wt %. X-ray photoelectron spectroscopic (XPS) and dynamic contact angle measurements were carried out in order to analyze the surface chemical structure in the air- and water-equilibrated states. XPS revealed that in the air-equilibrated state, lower surface free energy components were enriched at the air–solid interface, whereas in the water-equilibrated state, higher surface free energy components were enriched at the water–solid interface. The change in environment from air to water induced the surface reorganization in order to minimize interfacial free energy. Lipid sorption behaviors of SPUs were investigated by means of infrared spectroscopy. Even after extensive rinsing of the surface, the amount of lipid present on the SPU surface was more than that calculated on the assumption that a monolayer covers the SPU surface. Therefore, the lipid was not only adsorbed on the surface of SPU but absorbed into SPU. The SPU with hydrophilic PEO sorbed larger amount of phospholipid compared with that with hydrophobic polyol such as PTMO and PDMS. Also, the competitive sorption behaviors of phospholipid and cholesterol from their mixed liposome solution were studied. The ratio of sorbed cholesterol to phospholipid increased with an increase in surface hydrophobicity owing to the hydrophobic nature of cholesterol.

Key words: Segmented polyurethane; soft segment; hydrophilicity; hydrophobicity; X-ray photoelectron spectroscopy; dynamic contact angle; lipid sorption; selective sorption.

INTRODUCTION

Since the surface layer plays an important role in the practical application of polymeric materials, the surface properties of polymers have recently received great attention. Multiphase polymeric systems which usually consist of more than two components with different cohesive energies, show a unique character due to their amphiphilic nature. In the bulk state, they show the microphase separated structure due to the incompatibility of each component. On the other hand, the enrichment of lower surface free energy components are often observed at the air–solid interface [1]. This is due to a thermodynamic driving force to minimize the interfacial free energy.

Segmented polyurethanes (SPUs) showed both excellent blood compatibility and mechanical properties, and therefore, have been used for clinical applications. The study of the surface properties of SPUs has been of primary importance for their application in the biomedical field. Surface analyses of a variety of SPUs and

* To whom correspondence should be addressed.

segmented poly(urethaneureas) (SPUs) have been performed by using X-ray photoelectron spectroscopy (XPS) [2–4], secondary ion mass spectroscopy (SIMS) [5], and attentuated total reflection Fourier-transform infrared spectroscopy (ATR FT-IR) [6, 7]. Since most of the measurements have been carried out in a high-vacuum or an air-equilibrated state, the surface under these conditions was enriched with lower surface free energy components. The depth profile of a SPU containing a hydrophobic soft segment has been analyzed using angular dependent XPS technique. The results of angular dependent XPS were analyzed using the Paynter algorithm [8] or inverse theory developed by Ratner and Yih [9]. The depth profiles indicate that within the top 2 nm from the air–solid interface, there is a significant excess of hydrophobic polyether component compared with the average bulk composition, in conjunction with the relative exclusion of the nitrogen-containing hard segment domains. For the biomedical purpose, it is necessary to obtain the surface structural information at the water–solid interface. However, little investigation has been made to characterize the surface structure in the water-equilibrated state.

It has been reported that lipid components are involved in blood coagulation process [10] and biodegradation phenomena [11]. The surface with phosphorylcholine group has been developed by Chapman and coworkers [12, 13]. They intended to prepare the polymer surface which mimicked the thromboresistent surfaces of blood cell membrane. Also, Ishihara and Nakabayashi prepared various copolymers containing methacrylate with phosphorylcholine group (MPC) [14, 15]. Copolymers of MPC exhibited excellent blood compatibility due to the protein resistance behavior of its surface. Also, the selective adsorption of phospholipid was proposed for the explanation of its attainment of excellent blood compatibility. Calcification [16, 17] and environmental stress cracking (ESC) [18] are biodegradation phenomena with which lipid are concerned. Calcification involves the deposition of calcium phosphate on the surface in a complex interaction between the surface and biological components. On the other hand, ESC occurs as the result of the cooperative interaction of stress and biological components. ESC is commonly observed for plastics such as polyethylene under the action of detergent and stress [19]. Calcification phenomena and ESC might be strongly related to the absorption of biological components such as lipids and fatty acids. The surface cracking of SPUs after lipid sorption [20] and subsequent decrease in fatigue strength have been reported [11, 21, 22]. Though it is important to study the interaction between lipid and SPUs, little attention has been paid on the interaction between the lipid component and polyurethane surface.

In this study, the surface property and structure of SPUs with various polyol components in the water-equilibrated state were investigated based on X-ray photoelectron spectroscopic and dynamic contact angle measurements. Also, the interaction between SPU surface and lipid has been studied based on infra-red spectroscopic measurement.

EXPERIMENTAL

Materials

A series of segmented polyurethanes (SPUs) containing various polyol soft segments was prepared by a conventional two-step solution polymerization procedure under a nitrogen atmosphere [23]. The hard segment content was *c.* 50 wt%. Figure 1

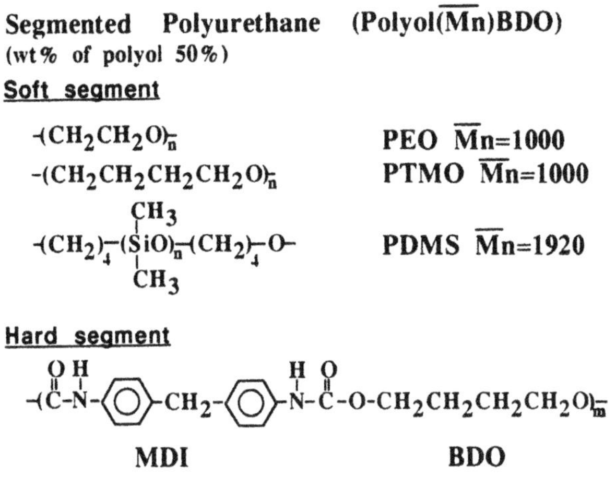
Segmented Polyurethane (Polyol(Mn)BDO)
(wt% of polyol 50%)
Soft segment
PEO Mn=1000
PTMO Mn=1000
PDMS Mn=1920
Hard segment
MDI
BDO

Figure 1. Chemical structures of segmented polyurethanes (SPUs).

shows the chemical structures of the SPUs used in this study. The details of the synthesis of these polymers have been previously described [24]. In these polymers, the hard segment consisted of 4,4′-diphenylmethane diisocyanate (MDI) and 1,4-butanediol (BDO). The polyols used in this study were poly(ethylene oxide) (PEO), poly(tetramethylene oxide) (PTMO), and poly(dimethylsiloxane) (PDMS). These SPUs are designated as $X(\overline{M_n})$BDO, where X, $\overline{M_n}$, and BDO represent polyol component, number average molecular weight of polyol, and butanediol chain extender. The proceeding of the synthesis of SPU was confirmed by infrared spectroscopy. In order to eliminate the effect of any residual antioxidants which may influence the surface characteristics of SPU, all of the additives to polymers were extracted with suitable solvents. After removal of reaction solvent under vacuum, polymer films were cast from 1–10 wt% DMAc-based solutions on clean substratum at 338 K. A glass plate was used as a substratum for bulk property and surface characterization, whereas a calcium fluoride (CaF_2) plate was used for lipid sorption experiment. After air drying the film, residual solvent was removed under vacuum at 348 K for at least 48 h. Gel permeation chromatographic (GPC) measurement of SPU showed that the polymer synthesized in this study has a sufficiently high molecular weight for most biomedical applications (M_n = 20 000–100 000).

Bulk property characterization

Bulk characterization of SPU was carried out by means of differential scanning calorimetry (DSC). DSC thermograms were recorded at a heating rate of 10 K min^{-1} using a Rigaku DSC 8230B under a dry nitrogen purge.

Surface characterization

X-ray photoelectron spectra (XPS) were obtained on a Shimadzu ESCA750 using MgK_α radiation at 1253.0 eV with emission angle of 90 deg. The emission angle is defined as the angle between the electron path to the analyzer and the specimen surface. All aliphatic C1*s* peaks were assigned a binding energy of 285.0 eV to correct for the charging energy shift. Since XPS is usually measured under high vacuum condition, the surface structure might be quite different from that in water. In order to characterize the surface chemical composition at the water interface, the XPS spectra of SPU in the water-equilibrated state were examined in the frozen state [25]. SPU films were placed soaking wet onto the sample stage, rapidly frozen to 220–240 K and then advanced into the sample chamber. Water was evaporated from the surface at 220–240 K and XPS spectra were obtained *in situ*. Sample was heated up to the room temperature *in vacuo* and the XPS spectra under the vacuum-(air) equilibrated state were obtained.

The dynamic contact angle of SPU was measured with a dynamic Wilhelmy plate technique [26–30]. Water for the dynamic contact angle measurement was purified with a Milli-Q system (Millipore Co. Ltd.). A motor controls the cross head for advancing or receding the water level at a speed of 20 mm min^{-1}. The surface tension was measured with a high-sensitivity strain gauge. The SPU was coated on a glass cover-slip ($24 \times 50\ mm^2$) which had been cleaned with a sodium-free alcoholic potassium hydroxide solution. The polymer coated glass cover-slip was suspended to a strain-gauge by a fine thread. The contact angle hysteresis curve was obtained at 293 K by plotting the observed force against the immersion depth.

Analysis of interaction between polyurethane surface and lipid

Figure 2 shows the general scheme for analytical method of lipid sorption. The SPU film was coated on a CaF_2 plate by the solvent evaporation method. Thickness of the film was *c.* 10 μm. The SPU-coated CaF_2 was immersed in 0.25% (w/v) L-α-phosphatidylcholine (PC, from egg yolk, Nakarai Chem. Co., Ltd.) solution or a mixture solution of 0.25% (w/v) PC and 0.1% (W/V) cholesterol (Ch, ash-free, Sigma) for a certain period at 310 K. Lipid liposome solutions were prepared by the sonication method. After immersing in lipid solution, each sample was rinsed with distilled water and vacuum-dried at 343 K for at least 2 days. The IR spectra was collected with a Nicolet system 510 with a resolution of 2 cm^{-1}. The spectra of

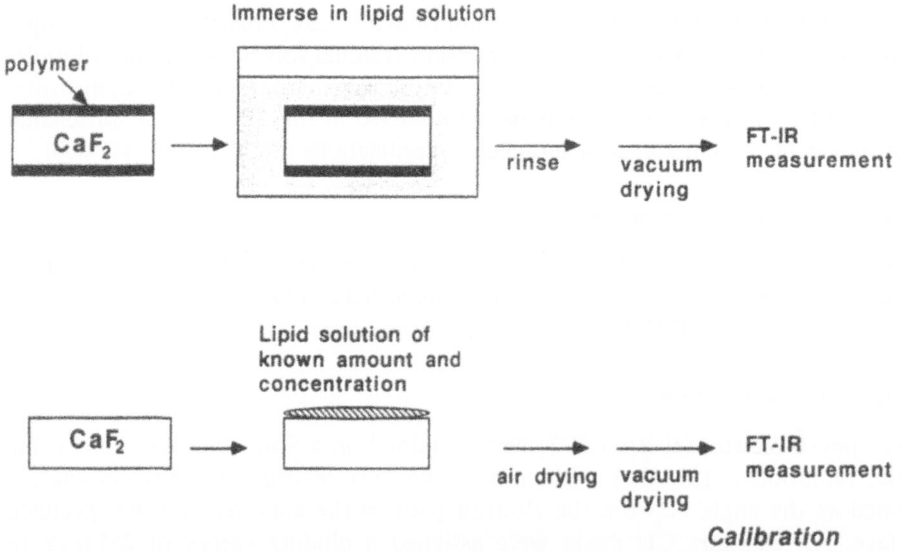
Immerse in lipid solution
polymer
CaF2
rinse
vacuum drying
FT-IR measurement
Lipid solution of known amount and concentration
CaF2
air drying
vacuum drying
FT-IR measurement
Calibration

Figure 2. Schematic representation of the lipid sorption measurement based on infra-red spectroscopy.

sorbed lipid was obtained after subtracting the spectra of polymer + CaF_2. The strong characteristic absorption bands at *c.* 1230 and 3300 cm^{-1} was used for the quantitative analysis of PC and Ch, respectively. The calibration curve for the quantitative analysis was obtained by placing a known amount of lipid solution with a known concentration on the CaF_2 plate and measured IR absorbance as a function of the surface concentration of lipid.

RESULTS AND DISCUSSION

Bulk property characterization

DSC measurement revealed the state of microphase separation of SPU in the bulk as shown in Fig. 3. The glass transition of polyol component was observed in a temperature range of 180–250 K, whereas endotherm related to the hard segment transition was observed at around 420–470 K. The transitions observed between 400 and 470 K could be attributed either to a melting of the microcrystalline hard segment domain or to endotherms due to the destruction of long range order in the hard segment region [31]. This transition shifted to a higher temperature region with an increase in cohesive energy of polyol. The PDMS(1920)BDO had the lowest glass transition temperature due to the low T_g of PDMS homopolymer and small degree of phase mixing. The higher T_g for the PEO(1000)BDO compared with the PTMO(1000)BDO was due to the enhanced phase mixing between the hard segment and polar PEO segments [24]. Therefore, the results indicated that the larger the difference in the cohesive energy between the hard and soft segments, the better in degree of microphase separation.

Surface characterization

The magnitude of surface free energy evaluated from the contact angle measurement increased in the order of PDMS, PTMO, hard segment (MDI-BDO) and PEO [32]). The XPS spectra of SPUs in the water- and air(vacuum)-equilibrated states were obtained by means of freeze-etch XPS. Figure 4 shows the typical C1*s* and N1*s* spectra of PTMO(1000)BDO in the air-equilibrated and in the water-equilibrated

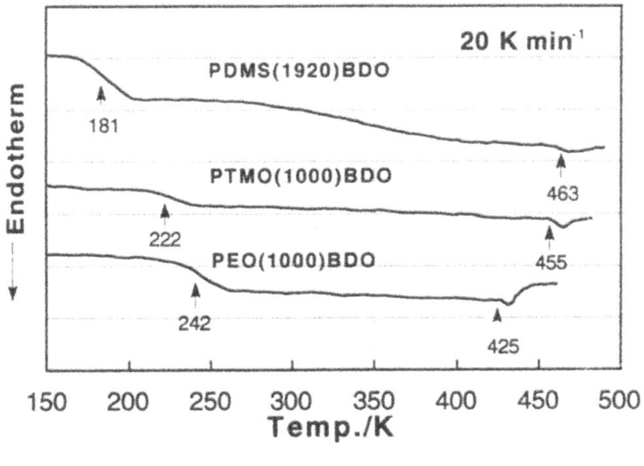
20 K min⁻¹
PDMS(1920)BDO
181
463
PTMO(1000)BDO
222
455
PEO(1000)BDO
242
425
Endotherm
150
200
250
300
350
400
450
500
Temp./K

Figure 3. DSC thermograms for SPUs.

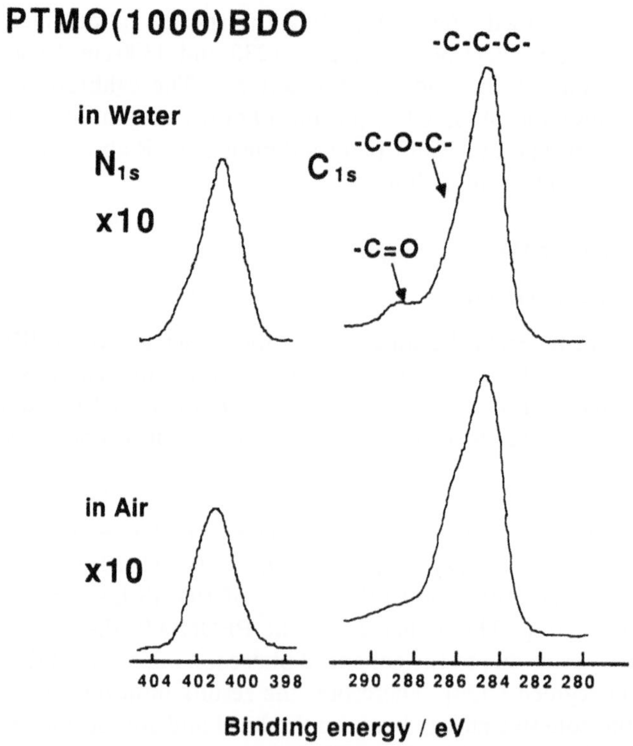
PTMO(1000)BDO
in Water
N_{1s}
x10
C_{1s}
-C-C-C-
-C-O-C-
-C=O
in Air
x10
404 402 400 398
290 288 286 284 282 280
Binding energy / eV

Figure 4. X-ray photoelectron spectra of PTMO(1000)BDO in air- and water-equilibrated states.

states. The C1*s* core level spectra consisted of well-resolved peaks as previously reported [24]. The main peak, referenced to 285 eV, was due to unsubstituted aliphatic and aromatic hydrocarbons. The peak shifted approximately 1.5 eV toward the higher binding energy side of the main peak which corresponded to carbon singly bonded to oxygen. The small peak observed at around 288.8 eV is ascribed to the carbonyl carbon. The N1*s* core level peak observed at around 400 eV is attributed to the urethane nitrogen in the hard segment. The relative intensity ratio of N1*s* and C1*s* (N/C) in the water-equilibrated state was greater than that in the air-equilibrated states. These results apparently indicate the surface reorganization upon environmental change [30].

Figure 5 summarized the magnitude of the ratio of number of nitrogen to carbon atoms (N/C) on the surface of SPUs in the air- and water-equilibrated states. N/C was calculated from the intensity ratio of N1*s* and C1*s* after correction of photoionization cross-section and photoelectron mean free path of each photoelectron. The standard error of N/C was *c*. 10%. The large magnitude of N/C indicates the large concentration of hard segment at the surface. In the water-equilibrated state, a large concentration of the hard segment at interface was observed for PTMO(1000)BDO and PDMS(1920)BDO. Even though XPS measurements was made at 220 K, the temperature was slightly above T_g of soft segment of PDMS(1920)BDO. Thus, the water-equilibrated surface of PDMS(1920)BDO might rearrange to some extent during the XPS measurement. On the other hand, PEO(1000)BDO showed an enrichment of PEO at solid–water interface. In the

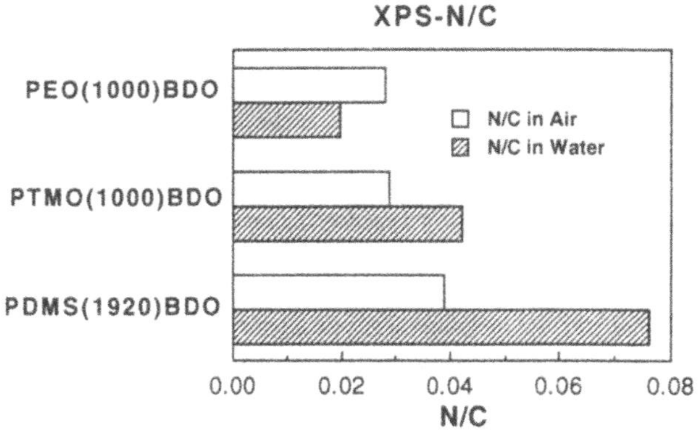
XPS-N/C
PEO(1000)BDO
PTMO(1000)BDO
PDMS(1920)BDO
N/C in Air
N/C in Water
0.00
0.02
0.04
0.06
0.08
N/C

Figure 5. Ratio of nitrogen to carbon atoms (N/C) for SPUs in the air- and water-equilibrated states.

air-equilibrated state, large concentration of the component with lower surface free energy was observed at the air–solid interface of those SPUs. These results indicated that the surface structure was reorganized upon environmental change in order to minimize interfacial free energy between solid surface and its environment.

The dynamic contact angle data against water reflect aspects of surface molecular mobility, surface reorganization, and surface heterogeneity at the air–water–solid interface [27]. The advancing and receding contact angles of SPU surface were obtained from the interfacial tension at the point of sample immersion and removal from the water. The detailed procedure of determining contact angles has been reported previously [27–30]. Figure 6 summarizes the magnitudes of advancing and receding contact angles and also contact angle hysteresis for SPUs at 293 K. The advancing contact angle decreased with an increase in the surface free energy of the polyol. On the other hand, the receding contact angle for the SPUs studied showed almost the same magnitude. The small magnitude of hysteresis observed for PEO(1000)BDO suggested the small difference in the surface free energy between PEO phase and the hard segment phase. This behavior agreed well with the fact that the hard segment melting endotherm for PEO(1000)BDO was located at lower temperature due to the phase mixing between hard and soft segments. The immersion

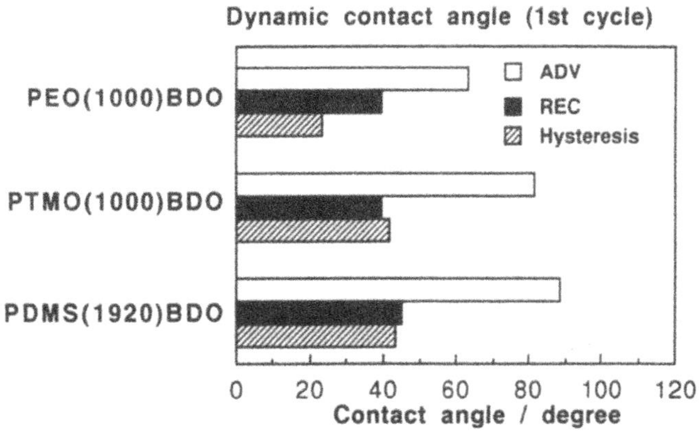
Dynamic contact angle (1st cycle)
PEO(1000)BDO
PTMO(1000)BDO
PDMS(1920)BDO
ADV
REC
Hysteresis
0
20
40
60
80
100
120
Contact angle / degree

Figure 6. Magnitudes of advancing and receding angles against water for SPUs at 293 K.

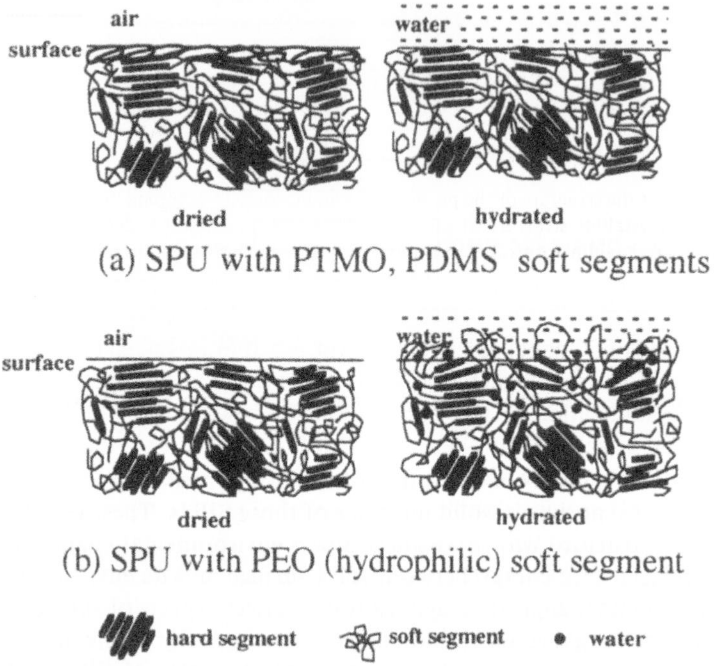
air
surface
water
dried
hydrated
(a) SPU with PTMO, PDMS soft segments
air
surface
water
dried
hydrated
(b) SPU with PEO (hydrophilic) soft segment
hard segment
soft segment
water

Figure 7. Schematic representation of surface structure of SPUs in the air- and water-equilibrated states. (a) SPU with hydrophobic poly(PTMO, PDMS), (b) SPU with hydrophilic polyol (PEO).

cycle dependence of the magnitude of advancing and receding contact angles was not prominent. This indicates that relatively fast response to reorganize the surface chemical composition occurs upon the environmental change.

Based on the results of both dynamic contact angle and XPS measurements, the surface structural models of SPUs were proposed as shown in Fig. 7. Figure 7(a) and (b) shows the surface structural models for SPU with a hydrophobic polyether soft segment (Fig. 7(a)) and a hydrophilic one (Fig. 7(b)), in the air- (dried) and water-equilibrated states. The hydrophilic component is dominant at the water–solid interface, whereas the hydrophobic component is at the air–solid interface. Since the glass transition temperature of the polyol in SPU employed in this study is lower than room temperature, the surface layer may be mobile enough to reorganize its chemical composition upon immersion in water at room temperature. Also, the sorption process of water enhancing plasticization can induce molecular rearrangement at the surface. Thus, the surface chemical composition of SPU is easily organized to minimize the interfacial free energy between the surface and its environment at room temperature, if thermal molecular motion of the components is sufficient.

Interaction between lipid liposome and SPU surface

Lipid sorption behavior was investigated by a difference IR spectra, before and after immersing SPU coated CaF_2 in a lipid solution. The SPU sample for lipid sorption experiment was equilibrated with a distilled water for 1 h. Figure 8 shows the IR

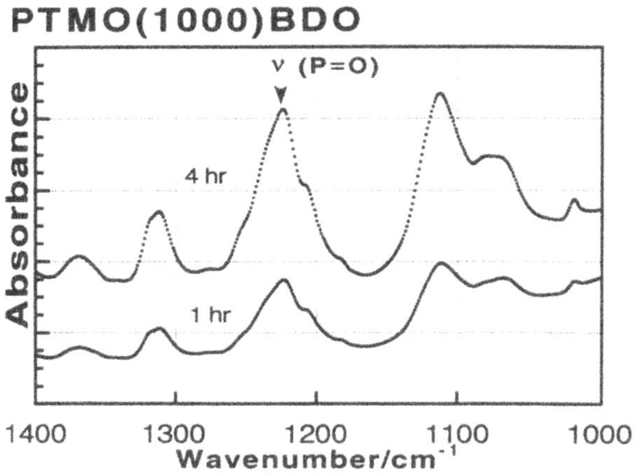
PTMO(1000)BDO
ν (P=O)
4 hr
1 hr
Absorbance
1400
1300
1200
1100
1000
Wavenumber/cm⁻¹

Figure 8. Infrared spectra of L-α-phosphatidylcholine (PC) sorbed in PTMO(1000)BDO after immersing in PC solution for 1 and 4 h at 310 K.

spectra of PC after immersing PTMO(1000)BDO in a PC solution for 1 and 4 h. The IR absorption intensity increased with an increase in immersion time in a PC solution. The characteristic strong IR absorption bands assigned to phosphatidyl moiety has been observed at around 1000–1300 cm^{-1}. The strong absorption at *c.* 1230 cm^{-1} is due to the stretching vibration of P=O in a phosphate moiety [33]. This band was used for the determination of lipid concentration at the surface.

Figure 9 shows the variation of amount of sorbed PC on SPUs with immersion time in a 0.25% (w/v) PC solution at 310 K. Even though the surface of SPU was subsequently rinsed with buffer solution before IR measurements, the amount of PC at the SPU surface was much greater in comparison with the amount of PC for monolayer adsorption of PC. The amount of PC for the monolayer absorption was calculated to be *c.* 0.7 $\mu g\ cm^{-2}$ on the assumption that the alkyl chain with 16 methylene and one methyl carbons was completely oriented perpendicular to the SPU surface. This result indicates that PC is not only adsorbed on the surface of SPUs, but also diffused into bulk SPUs. The absorption of lipid into SPU has been

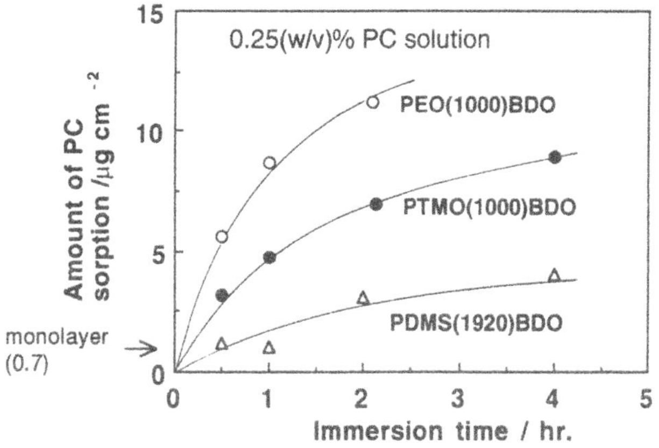
0.25(w/v)% PC solution
PEO(1000)BDO
PTMO(1000)BDO
PDMS(1920)BDO
Amount of PC sorption /μg cm⁻²
monolayer (0.7)
15
10
5
0
0
1
2
3
4
5
Immersion time / hr.

Figure 9. Sorption behavior of PC on the surface of SPUs from PC solution at 310 K.

previously confirmed from the change in modulus and shift in mechanical absorption peak after immersion of the SPU in lipid solution [11]. The amount of lipid sorption of SPU increased with an increase in hydrophilicity of polycol component as shown in Fig. 9, that is, PEO(1000)BDO showed the largest PC sorption rate and also, the large equilibrium PC sorption among SPUs studied. Lipid molecules in water forms liposome in which the hydrophilic group orients to the water interface. Since the surface of PEO(1000)BDO was enriched with PEO in an aqueous lipid solution, the rate of lipid sorption was faster and the equilibrium amount of lipid sorption was larger than other SPU with rather hydrophobic soft segment, such as PTMO(1000)BDO and PDMS(1920)BDO. Therefore, it seems apparent that the results in Fig. 9 can be attributed to the magnitudes of interfacial free energy between the outer surface of phospholipid liposome and hydrated PEO-rich surface. On the other hand, in the hydrated-state, the surface of SPUs with hydrophobic polyol such as PTMO and PDMS exhibited larger concentration of crystalline hard segment in comparison with the case of PEO(1000)BDO. Since it is difficult for lipid liposome with a hydrophilic surface characteristic to penetrate into either hydrophobic polyol phase or the crystalline hard segment region, the lipid sorption rate for SPUs with hydrophobic polyol is smaller than that for SPU with hydrophilic polyol.

Since lipid in the biomembrane contains more than two kinds of lipid component, it may be useful to study lipid sorption behaviors in an aqueous solution of lipid mixture. Cholesterol (Ch) is present in almost all biological membrane and shows hydrophobic character compared with that of phospholipid. Also, the phase transition behavior of phospholipid membrane was strongly influenced by the Ch concentration. Since the SPU specimen studied here has both hydrophobic and hydrophilic segments, the selective absorption(sorption) of lipid onto each phase of SPU would be expected. The SPU was immersed in a mixture solution of PC and Ch and the amount of lipid sorption was measured with transmission IR spectroscopy. Since the characteristic absorption of PC and Ch are located at a different wavenumber region, the amount of PC and Ch can be measured independently with fairly good precision. The amount of Ch was estimated by using the absorption band of OH stretching at around 3300 cm^{-1}. Figure 10 shows the sorption behavior of PC

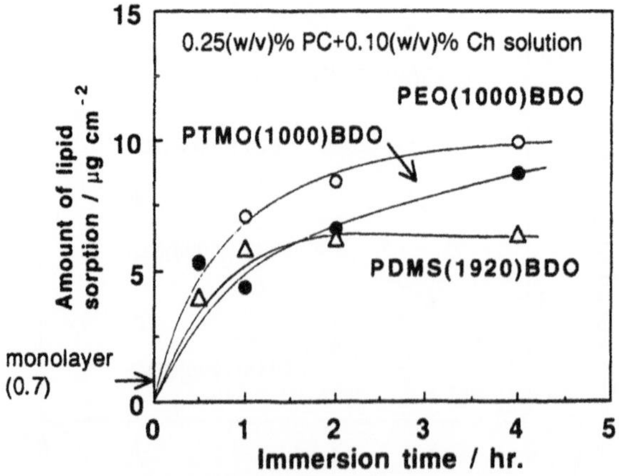
0.25(w/v)% PC+0.10(w/v)% Ch solution
PEO(1000)BDO
PTMO(1000)BDO
PDMS(1920)BDO
Amount of lipid sorption / μg cm -2
monolayer (0.7)
Immersion time / hr.
15
10
5
0
0
1
2
3
4
5

Figure 10. Sorption behavior of PC on the surface of SPU from PC and cholesterol (Ch) mixture solution at 310 K.

from a mixture solution of 0.25% (w/v) PC and 0.1% (w/v) Ch at 310 K even after extensive rinsing of the surface. The amount of lipid present on the SPUs was more than the monolayer adsorption, suggesting the diffusion of lipid into the SPU bulk. The amount of PC sorption from a PC/Ch mixture solution of PEO(1000)DBO was smaller than that from a PC solution as shown in Fig. 9. This can be ascribed to the decrease in hydrophilicity of lipid liposome due to the presence of Ch. However, in the cases of the hydrophobic SPU such as PTMO(1000)BDO and PDMS(1920)BDO, the lipid sorption rate and the amount of equilibrium sorption were increased due to the increase in hydrophobic character of lipid liposome.

The competitive sorption behavior of PC and Ch from a mixture solution has been obtained for each SPUs. Figures 11(a), (b) and (c) show the competitive sorption behavior of PC and Ch for PEO(1000)BDO, PTMO(1000)BDO, and PDMS(1920)BDO at 310 K, respectively. The total amount of PC and Ch sorbed by PTMO(1000)BDO and PDMS(1920)BDO from a mixture solution was greatly increased compared with that from PC liposome solution. The extent of increase was largest for PDMS(1920)BDO. This can be ascribed to the hydrophobic character of both Ch containing liposome and PDMS-based SPU surface. In the case of PEO(1000)BDO, the amount of PC sorption was larger than that of Ch. However, PDMS(1920)BDO and PTMO(1000)BDO showed a larger sorption of Ch compared with that of PC even though the PC and Ch weight ratio in an aqueous solution was 5 : 2. Though lipid sorption might occur mainly at both the polyol phase and the phase mixed region of polyol and hard segment, lipid cannot penetrate the hard segment phase unless it breaks intermolecular hydrogen-bonding between the hard segment. Thus, the hydrophobic Ch tended to concentrate at the hydrophobic phase. The ratio of Ch to PC sorbed into hydrophobic SPU was larger than that for SPU with hydrophilic PEO. Also, it should be mentioned that the Ch/PC ratio in PEO(1000)BDO is larger than that in an aqueous solution. Even if the PEO phase in PEO(1000)BDO is hydrated, the environment among sorbed lipid in PEO(1000)BDO is still more hydrophobic compared with aqueous solution, thus the Ch/PC ratio in the polymer matrix became larger than that in a solution. These results indicate that the polymer surface and lipid molecule in liposome recognize the hydrophobic and hydrophilic characters in each other. It has been reported that the silicone rubber heart valve prosthesis did not show the sorption of phospholipid, but showed the sorption of hydrophobic sterol esters, triglycerides and cholesterol [34]. The presence of Ch in liposome solution increased the hydrophobicity of liposome and increased the amount of lipid sorption from a mixture solution into SPU. It has been supposed that the charged amphiphilic species such as long chain fatty acid salt and anionically charged phospholipid may provided an achoring site for calcium deposition [17]. Thus phospholipid sorption might be ultimately lead to calcification. This fact suggests that either the molecular design or surface structural control of SPU is necessary in order to reduce phospholipid sorption. One approach is to improve microphase separation of SPU with hydrophobic polyol. Another approach might be the surface cross-linking of SPU in order to inhibit penetration of lipid [35]. Since lipid sorption has been accelerated in the flexing SPU elastomers [21], further study, especially analysis of lipid sorption behavior under cyclic straining, is necessary to understand the lipid sorption mechanism of SPUs for cardiovascular application.

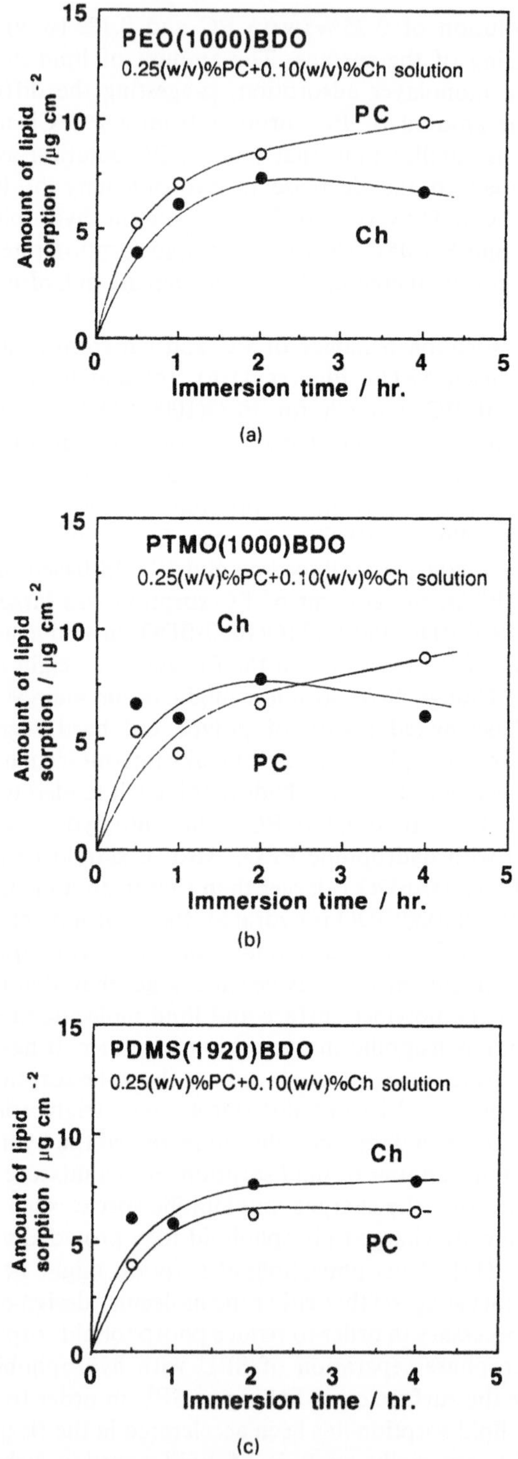
PEO(1000)BDO
0.25(w/v)%PC+0.10(w/v)%Ch solution
PC
Ch
Amount of lipid sorption /μg cm^-2
Immersion time / hr.
(a)
PTMO(1000)BDO
0.25(w/v)%PC+0.10(w/v)%Ch solution
Ch
PC
Amount of lipid sorption / μg cm^-2
Immersion time / hr.
(b)
PDMS(1920)BDO
0.25(w/v)%PC+0.10(w/v)%Ch solution
Ch
PC
Amount of lipid sorption / μg cm^-2
Immersion time / hr.
(c)

Figure 11. Competitive sorption of PC and Ch from mixture solution of PC and Ch at 310 K. (a) PEO(1000)BDO, (b) PTMO(1000)BDO, (c) PDMS(1920)BDO.

CONCLUSION

The surface characterization of segmented polyurethanes revealed that the component with lower surface free energy was enriched at the air-solid interface, whereas the component with larger surface free energy was dominant at water-solid interface. The change in environment from air to water induced the surface reorganization in order to minimize interfacial free energy.

Lipid sorption behavior of SPUs was investigated based on infrared spectra. Lipid was not only adsorbed on the surface but absorbed into bulk SPU. The SPU with hydrophilic PEO showed a larger amount of sorbed phospholipid compared with that with hydrophobilic polyol such as PTMO and PDMS. This suggests that the hydrated liposome surface has preferential interaction with hydrated PEO of PEO(1000)BDO. The competitive sorption behavior of cholesterol and phospholipid into SPU was also studied. The ratio of sorbed cholesterol to phospholipid in the SPU changed depending on the magnitude of surface free energy of SPU. It was obvious that hydrophobic cholesterol in lipid liposome was selectively sorbed by hydrophobic component of SPU.

REFERENCES

1. A. Takahara, in: *Modern Approaches to Wettability—Theory and Applications*, p. 179, G. Leob and M. Schroeder (Eds.) Plenum, New York (1992).
2. T. Kajiyama and A. Takahara, *J. Biomater. Appl.* **6**, 42 (1991).
3. B. D. Ratner and B. J. McElroy, in: *Spectroscopy in the Biomedical Sciences*, p. 107, R. M. Gendreau (Ed.). CRC Press, Boca Raton, FL (1986).
4. A. Takahara, J. Tashita, T. Kajiyama, M. Takayanagi and W. J. MacKnight, *Polymer* **26**, 978 (1985).
5. M. Heara, B. D. Ratner and D. Briggs, *Macromolecules* **21**, 2950 (1988).
6. K. Knutson and D. J. Lyman, in: *Surface and Interfacial Aspects of Biomedical Polymers, Vol. 1, Surface Chemistry and Physics*, J. D. Andrade. (Ed.). Plenum, New York (1985).
7. P. M. Bummer and K. Knutson, *Macromolecules* **23**, 435 (1990).
8. R. W. Paynter, *Surf. Interface. Anal.* **3**, 186 (1981).
9. R. S. Yih and B. D. Ratner, *J. Electron Spectrosc. Relat. Phenom.* **43**, 61 (1987).
10. H. C. Hemker, M. J. Lindhut and C. Vermeer, *Ann. N.Y. Acad. Sci.* **283**, 104 (1977).
11. A. Takahara, J. Tashita, T. Kajiyama and M. Takayanagi, *J. Biomed. Mater. Res.* **19**, 13 (1985).
12. J. A. Hayward and D. Chapman, *Biomaterials* **5**, 135 (1984).
13. O. Albrecht, S. S. Johnston, C. Villaverde and D. Chapman, *Biochim. Biophys. Acta* **687**, 165 (1982).
14. K. Ishihara, T. Ueda and N. Nakabayashi, *Polym J.* **22**, 355 (1990).
15. M. Kojima, K. Ishihara, A. Watanabe and N. Nakabayashi, *Biomaterials* **12**, 121 (1991).
16. D. R. Owen and R. M. Zone, *Trans. Am. Soc. Artif. Intern. Organs* **27**, 528 (1981).
17. E. Dalas, P. V. Ibanhou and P. G. Koutsoukons, *Langmuir* **6**, 535 (1990).
18. E. F. Cuddihy, J. Moacanin, E. J. Roschke and E. C. Harison, *J. Biomed. Mater. Res.* **10**, 471 (1971).
19. S. Bardyopadhay and H. R. Brown, *Polymer* **19**, 489 (1978).
20. A. Takahara, R. W. Hergenrother, A. J. Coury and S. L. Cooper, *J. Biomed. Mater. Res.* **26**, 801 (1992).
21. A. Takahara, K. Takamori and T. Kajiyama, in: *Artificial Heart* 2, p. 19, T. Akutsu (Ed.). Springer, Tokyo (1989).
22. A. Takahara and T. Kajiyama, *Kobunshi Ronbunshu* **42**, 793 (1985).
23. J. H. Saunders and K. C. Frisch, *Polyurethane Chemistry and Technology, Part 1: Chemistry*, Interscience, New York (1962).
24. A. Takahara, A. Z. Okkema, A. J. Coury and S. L. Cooper, *Biomaterials* **12**, 324 (1992).
25. A. Takahara, K. Korehisa, K. Takahashi and T. Kajiyama, *Koubunshi Ronbunshu* **49**, 275 (1992).
26. L. Wilhelmy, *Ann. Phys.* **199**, 177 (1963).

27. J. D. Andrade, L. M. Smith and D. E. Gregonis, in: *Surface and Interfacial Aspects of Biomedical Polymers, Vol. 1, Surface Chemistry and Physics*, p. 249, J. D. Andrade (Ed.). Plenum, New York (1985).
28. T. Kajiyama, T. Teraya and A. Takahara, *Polym. Bull.* **24,** 333 (1990).
29. T. Teraya, A. Takahara and T. Kajiyama, *Polymer* **31,** 114 (1990).
30. A. Takahara, N. J. Jo and T. Kajiyama, *J. Biomater Sci. Polymer Edn* **1,** 19 (1989).
31. T. R. Hesketh, J. W. C. Van Bogart and S. L. Cooper, *Polym. Eng. Sci.* **20,** 190 (1980).
32. A. Takahara and T. Kajiyama, in preparation.
33. L. J. Bellamy, *The Infrared Spectra of Complex Molecules*, Third Ed., Vol. 1, Ch. 18. Chapman & Hall, London (1975).
34. H. P. Chin, E. C. Harrison, D. H. Blankenhorn and J. Moacanin, *Circulation, Suppl. I* **43/44,** 51 (1970).
35. Shouren Ge, A. Takahara and T. Kajiyama, *Rept. Progr. Polym. Phys. Japan* **34,** 649 (1991).

Part II

Protein Adsorption

Residence time effects on monoclonal antibody binding to adsorbed fibrinogen

THOMAS A. HORBETT* and KENNETH R. LEW

Department of Chemical Engineering and Bioengineering, BF-10, University of Washington Seattle, WA 98195, USA

Received 4 January 1993; accepted 15 July 1993

Abstract—Fibrinogen adsorbed to polymeric surfaces and then allowed to reside on the surface while it is kept in a buffer solution for a period of time (the 'residence time') undergoes postadsorptive changes that decrease its SDS elutability, displaceability by plasma, polyclonal antifibrinogen binding, and ability to support platelet adhesion (summarized in Chinn *et al. J. Biomed. Mater. Res.* **26,** 757 (1992)). In order to better understand the nature of the changes in adsorbed fibrinogen, the binding of ten different monoclonal antifibrinogen molecules to fibrinogen adsorbed from plasma to Biomer and several other surfaces has been measured after increasing residence time in buffer. Three of the monoclonal antibodies used bind to sequences that have been implicated in platelet binding to fibrinogen. One of these (M1) binds to the C-terminal region of the gamma chain (402–411), another (R1) binds to the N-terminal region of the A alpha chain containing an RGDF sequence (95–98), and the third (R2) binds to the C-terminal region of the A alpha chain containing an RGDS sequence (572–575). Two other antibodies (P1 and K4) also bind to the C-terminal region of the gamma chain (373–385 and 392–406, respectively). Five other antibodies that bind to other regions in fibrinogen were also used. Two of the antibodies (K4 and P1) are also known to be sensitive to conformational changes in the fibrinogen molecule.

The binding of the various antibodies changed with residence time in ways that were highly dependent on the particular antibody. The binding of some antibodies was very stable with respect to residence time, others rose with time, some declined with residence time and one appears to pass through a maximum. However, none of the changes in antibody binding were nearly as fast as has been observed for the changes in platelet binding reported previously. Binding to the platelet binding region near the gamma chain C-terminal region either did not change with residence time (M1), increased with residence time (K4), or else decreased more slowly than observed for platelets (P1). Binding of the antibodies to the RGD sequences near the N-terminus of the A alpha chain (95–98) was very low initially but increased with residence time, while the binding to the RGD sequence near the C-terminus of the A alpha chain (572–575) increased slightly at short residence times but then declined substantially after longer residence times. Thus, the changes in the expression of the putative platelet binding domains do not correlate with the declines in platelet binding to plasma preadsorbed Biomer. We therefore suggest that other features of the adsorbed fibrinogen, such as how tightly it is held to the surface, may influence the ability of the platelets to bind to it.

Key words: Monoclonal antibody; fibrinogen; protein adsorption; Biomer; platelets.

INTRODUCTION

Fibrinogen adsorption to biomaterials surfaces probably plays a key role in mediating cellular interactions that are central events in determining the biocompatibility of these materials. Fibrinogen is well known to be a major mediator of platelet adhesion and aggregation because of its binding to the platelet IIb/IIIa receptor [1–7] and it is, therefore, thought to play an important role in blood compatibility of

* To whom correspondence should be addressed.

materials because of its propensity to adsorb to surfaces. Fibrinogen could also be important in bacterial infection and in the foreign body reaction. Thus, for example, preadsorption of fibrinogen has been shown to cause increased adhesion of bacteria [8] and common staphylococcus cause clumping and clotting of fibrinogen [9]. More recently, several different bacteria have been shown to bind to fibrinogen [10, 11] as well as to fibronectin [12] through specific receptors. The ability of bacteria to adhere to fibrinogen or fibrin or fibronectin in blood clots provides a likely way for these organisms to overcome normal clearance mechanisms. A role for fibrinogen in macrophage or neutrophil adhesion in the foreign body reaction has not been shown as yet, but the extracellular fluid in peritoneal exudate includes some fibrinogen. For all these reasons, the behavior of fibrinogen on surfaces has been the subject of many studies in this and several other laboratories.

Fibrinogen adsorption from plasma to surfaces has been shown to be a major mediator of platelet adhesion to Biomer exposed to plasma *in vitro*; this process was also shown to involve the platelet's IIb/IIIa receptor [13]. However, the ability of fibrinogen to mediate adhesion to Biomer declined greatly if the platelets were not introduced immediately after the preadsorption step [14, 15]. In the latter study, the adsorbed fibrinogen was allowed to undergo a 'residence time' by leaving it in buffer after the adsorption step, during which it evidently undergoes what we have called a 'post-adsorptive transition' that results in decreased platelet adhesion [13, 16, 17]. The residence time effect can be seen not only on platelet adhesion, but also in changes in SDS elutability [16, 18, 19], displaceability by plasma [20, 21], FTIR spectra [22], and the binding of polyclonal antibodies [14, 17]. However, the post-adsorptive transitions affecting platelet adhesion occurred much more rapidly than the changes in polyclonal antibody binding, SDS elutability, or FTIR spectra, but at about the same rate as the changes in plasma displaceability.

Platelet binding to fibrinogen has been reported to be mediated by only a few small peptide segments out of the entire fibrinogen molecule. The regions implicated are the two RGD sequences in the A-alpha chain, RGDF at 95–98 and RGDS at 572–575 [7, 23] and the C-terminal dodecapeptide of the gamma chain [24–26], as illustrated in Fig. 1. It therefore seemed possible to us that the discordance between

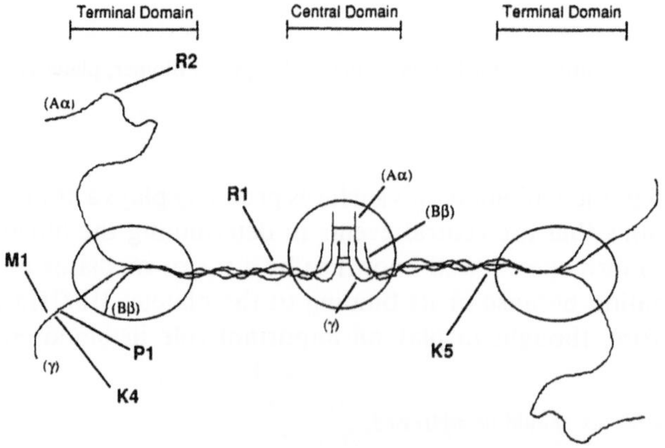
Terminal Domain
Central Domain
Terminal Domain
R2
(Aα)
R1
(Aα)
(Bβ)
M1
(Bβ)
(γ)
P1
K5
(γ)
K4

Figure 1. A schematic illustration of the fibrinogen molecule and the location of some of the binding sites for the antibodies used in this study. The central domain contains the NH_2 terminus of the three subunits. The antibodies M1, P1, K4, R2, R1, and K5 are identified in Table 1.

changes in platelet binding and the other, more global measures of the state of fibrinogen arose from highly localized changes in the fibrinogen molecule after it adsorbed. In this study, we have used a series of monoclonal antibodies directed to the platelet binding regions as well as to several other epitopes in the fibrinogen molecule to determine in more precise detail the regions of the molecule undergoing changes after adsorption. In addition, we hoped the antibody binding measurements would reveal whether any of the antibody binding domains were changing at the rate observed for the changes in platelet adhesion and thus identify those regions undergoing changes affecting platelet adhesion.

MATERIALS AND METHODS

Substrate preparation

Poly(ethylene terephthalate) (PET) samples (Lux Thermanox, Nunc, Inc., Naperville, IL) were cut into 6.7-mm squares and then cleaned in methanol using an ultrasonic bath (model T-21, L&R Manufacturing Company, Kearny, NJ). Glass coverslips (9.0 mm diameter, #2, VWR Scientific Inc., San Francisco, CA) were also cleaned using an ultrasonic bath with a 4% Isopanasol solution (C.R. Callen Corporation, Seattle, WA).

Polymer preparation

Biomer (segmented polyetherurethane urea, code E0205, non pigmented, Ethicon, Inc., Somerville, NJ) was purified by repeated precipitation in methanol. The urethane-urea hard segment (HS-PEU) was synthesized in our laboratory based on the conventional two-step polyetherurethane condensation reaction [27]. HS-PEU was synthesized using 4,4′-methylenebis (phenylene isocyanate) (MDI), ethylene diamine (ED), and dipropylene glycol (DPG). The reactant ratio was (MDI : ED : DPG) (2 : 1 : 1) [14]. Biomer and HS-PEU were dissolved in 1-methyl-2-pyrrolidinone (m-pyrol) for 12 h at a 2% concentration (w/w). The polymer solutions were then filtered through a 0.5 μm Teflon filter (Millipore, Bedford, MA). Uniform, thin films of Biomer and HS-PEU were centrifugally cast onto the 6.7 mm square PET substrate using a photoresist spinner (Model EC101, Headway Research, Inc., Garland, TX). The coated samples were then air dried in a laminar flow hood for 24 hours. Immulon I is a treated polystyrene microplate commercially available through Dynatech Laboratories, Inc.

Radiolabeled fibrinogen

Human fibrinogen (KABI Vitrum) was radiolabeled using ^{125}I (Na^{125}I, catalog number IMS-30, Amersham, Arlington Heights, IL) and the iodine monochloride (ICL) technique [28] as modified by Horbett [29]. An equimolar ratio of ICL to fibrinogen was used. The unincorporated ^{125}I was separated from labeled fibrinogen using gel filtration with P-4 BioGel (Bio-Rad, Richmond, CA). Prior to use the labeled fibrinogen was dialyzed overnight in citrate phosphate-buffered saline with the addition of sodium azide (CPBSz: 0.01 M citrate; 0.01 M phosphate; 0.120 M sodium chloride; 0.02% sodium azide; pH = 7.4). Residual free ^{125}I was 2–3% of the total ^{125}I fibrinogen solution as determined using instant thin-layer chromatography (ITLC) [30]. ^{125}I fibrinogen was then added to 1% human plasma prior to adsorption to Biomer. After plasma adsorption and residence time the samples were counted on a gamma radiation counter (model 1185R, TM Analytic, Elk Grove, IL) and fibrinogen was calculated by dividing the retained radioactivity of the sample less background by the sample surface area and the specific activity of the plasma.

Plasma adsorption

Normal and factor 1 (fibrinogen) deficient human plasma were obtained from George King Bio-Medical Inc., Overland Park, KS. Citrated baboon plasma was prepared from 10% ACD anticoagulated blood obtained from the Regional Primate Center at the University of Washington. Prior to human plasma or baboon plasma adsorption the polymer samples (typically quadruplicates) were incubated in beaker cups (Evergreen Scientific, Los Angeles, CA) containing CPBSz at 37°C for 2 h. The buffer was then removed and 1% human plasma (in CPBSz) or 1% baboon plasma (in CPBSz) was then added (0.75 ml). The 6-min plasma adsorption was terminated by six successive CPBS (no sodium azide) buffer rinses. The rinsing step consisted of aspirating the solution and immediately adding an equal volume of buffer. The polymer samples were placed in new beaker cups containing a monoclonal antibody solution or placed in beaker cups containing fresh 37°C CPBSz for a specific residence time. After a specific residence time the polymer samples were then placed in new beaker cups containing the monoclonal antibody solution.

We have also examined antibody binding to fibrinogen adsorbed from many other plasma concentrations, including 100% plasma, and found large effects of loading on the binding of the antibody. These results will be reported elsewhere (Horbett and Lew, manuscript in preparation).

Buffers

The buffer (antibody buffer) used to dilute the monoclonal antibodies and the goat anti-mouse IgG-horseradish peroxidase conjugate consisted of 0.02 M Trizma base (no. T-1503, Sigma Chemical Company, St Louis, MO) and 0.02 M sodium chloride (J.T. Baker Inc., Phillipsburg, NJ). The addition of 1% polyethylene glycol (PEG) (catalog number 6102, M_w 4000, Polysciences, Inc., Warrington, PA), 0.5% gelatin from swine skin (Type I, No. G-2500, Sigma Chemical Company, St Louis, MO) and 0.5% bovine albumin (Fraction V, ICN ImmunoBiologicals, Costa Mesa, CA) were used as blocking agents to prevent non-specific uptake of antibodies. These blocking agents were found to be effective in our previous studies with polyclonal antibodies [14] and generally were equally effective with the monoclonal antibodies used here, except as noted (see Results). The NaCl, Trizma base, PEG, buffer (pH = 7.4) was degassed for 5 min, then gelatin was dissolved in this solution by heating for 5 min to 60°C. This solution was then cooled to 37°C before adding the albumin. The rinsing buffer was CPBS (no azide). For antibodies R1 and R2 only, 1% bovine serum was also added (see Results).

The substrate solution was hydrogen peroxide/3,3′,5,5′-tetramethylbenzidene (catalog number 98-060) from ICN ImmunoBiologicals, Costa Mesa, CA. The stopping solution was 2 M sulfuric acid.

Table 1.
Monoclonal antifibrinogens

No.	Abb.	Antibody, Lot No., Conc., Isotype	Cross-reacts with	Dilution used ×1000	Plasma type
1	K1	Fd4-7B3, RC1136 (1) 1–2 mg ml^{-1}, IgG 1, κ	Fibrinogen, γ chain, FgD, FgDD	140	human
2	K2	44-3 (xl-f), RC1154 (1 + 2) 3.65 mg ml^{-1}	γ, $\gamma\gamma$, not fibrinogen	6	human
3	K3	1C2-2 (xl-f) RC1116 (1) 1.44 mg ml^{-1}, IgG 1, κ	Aα chain, fibrinogen, fibrin, plasmin digest of both (before/after heat precipitation)	120	human
4	K4	4-2 (xl-f), (6)Bk20 (5) 3.4 mg ml^{-1}, IgG 1, κ	Fibrinogen, fragment D, fragment DD, γ 392–406	7	human/ baboon
5	K5	1D4 (xl-f), RC1193 (1) 3.9 mg ml^{-1}, IgG 1, κ	Aα chain, Aα 241–476, fbgn., fibrin, plasmin digest of both (before/after heat ppt.)	450	human
6	K6	2N3H10, RC1135 (1) 3.1 mg ml^{-1}, IgG 1, κ	Fibrinogen, fibrin, plasmin-derived fragment E from both, N-DSK but not reduced/alkylated N-DSK	250	human
7	P1	Fg-RIBS-I, IgG, γ_1, κ	C-terminus of γ chains (RIBS); 373–385 of γ chain	400	human
8	M1	Matsueda, F37 4A5 1E5 f97-103, 3.4 mg ml^{-1}, γ-1, κ	C-terminus of γ chains (402–411) a platelet binding region	100/1600	human/ baboon
9	R1	Ruggeri-1, 155 B1616.1 OSC IgG_1	Sequence of Aα chain (87–100) near N-terminal including the RGDF at 95–98, a platelet binding region	10	human
10	R2	Ruggeri-2, 34B 29.11 OSC IgG_1	Sequence of Aα chain (566–580) near C-terminal including the RGDS at 572–575, a platelet binding region	10	human

Antibodies

The mouse anti-human fibrinogen antibodies were generous gifts from B. Kudryk, E. Plow and G. R. Matsueda. Goat anti-mouse IgG-HRP conjugate (H + L) came from Bio-Rad Laboratories, Richmond, CA. The antibodies are listed in Table 1.

Indirect ELISA

After the plasma adsorption and rinsing the polymer samples were placed in new beaker cups containing a particular monoclonal antibody (0.75 ml). The antibody dilutions are listed in Table 1.

The plasma adsorbed polymer samples were allowed to react with the monoclonals for 1 h at 37°C and then rinsed five times with CPBS. The samples were then placed in new beaker cups containing the anti-IgG-HRP conjugate (1 : 3000 dilution,

0.75 ml) and allowed to react for 1 h at 37°C. Samples were then rinsed five times with CPBS and placed in new beaker cups. The substrate solution was added (0.6 ml) and allowed to react for 20 min on a shaker (Gyrotory Shaker-model G2, New Brunswick Scientific Co., Inc., Edison, NJ) at room temperature. Stopping solution (0.3 ml) was added to stop the reaction. The solution (0.2 ml) was transferred to a microplate (Immulon I, Dynatch, Alexandria, VA) and the optical density read at 450 nm on a V_{max} Kinetic Microplate Reader (Molecular Devices, Palo Alto, CA). The O.D. cm^{-2} was calculated by dividing the final optical density of each sample by the polymer sample area (0.898 cm^2). The O.D. cm^{-2} data in the tables and figures are averages of data obtained from three or more separate polymer samples, with the exception of the kinetics. The errors listed in the Tables are standard deviations. The error bars in the Figures are two standard deviations in length; however, only the top half of the error bar is visible in the bar charts.

Anti-IgG-HRP kinetics

Various dilutions (0.04 ml) of anti-IgG-HRP in duplicates were placed in beaker cups and 0.6 ml substrate solution was added. Samples were then mixed using the shaker. At one minute intervals for 20 min the reactions were stopped using 0.3 ml stopping solution. The solutions (0.2 ml) were then transferred to a microplate and the optical density read at 450 nm on the microplate reader.

RESULTS

Kinetics of ELISA assay

The kinetics of color generation in the HRP reaction were found to be approximately linear with time to an optical density value greater than 2.0 (see Fig. 2). In all our

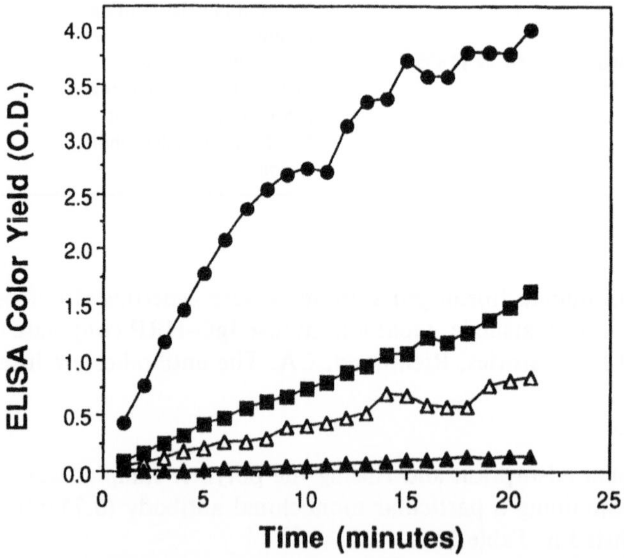
ELISA Color Yield (O.D.)
4.0
3.5
3.0
2.5
2.0
1.5
1.0
0.5
0.0
0
5
10
15
20
25
Time (minutes)

Figure 2. Kinetics of ELISA assay color formation catalyzed using various HRP-anti-IgG conjugate dilutions. 1:4000 (circles), 1:20000 (squares), 1:40000 (open triangles), 1:200000 (solid triangles). Duplicate samples for each time point.

studies we therefore adjusted the antibody concentrations to try to given an O.D. of less than 2.0 in the standard substrate incubation period of 20 min. This should assure that the substrate had not been depleted and therefore the color generated would be proportional to the amount of bound antibody. Thus, this experiment assured that the experiments were run in the zero order regime for this enzyme/ substrate pair.

ELISA titration

For each primary monoclonal antibody in our studies, a preliminary experiment was done in which the dilution of the primary, monoclonal antibody was varied over a wide range. Similarly, the dilution of the secondary antibody, the polyclonal goat antimouse IgG/HRP conjugate, was also varied. Typical titration curves are shown in Fig. 3. Based on this experiment, the residence time studies with the R2 antibody were then done at primary antibody dilutions of 1/10 000 together with a secondary antibody dilution of 1/3000 since this gave a high color value in a region in which color increased approximately linearly with primary antibody dilution, while still maintaining relatively low values for non-specific adsorption (see next section). The choice of primary antibody and secondary antibody dilutions (see Table 1) for the other antibodies was based on similar experiments with each of them, the object in each case being the achievement of an O.D. of 1–2 and low background values.

Controls for non-specific uptake of antibody

In all experiments, samples of Biomer that had not been preadsorbed with plasma were included as controls for the effectiveness of the blocking agents in the buffers

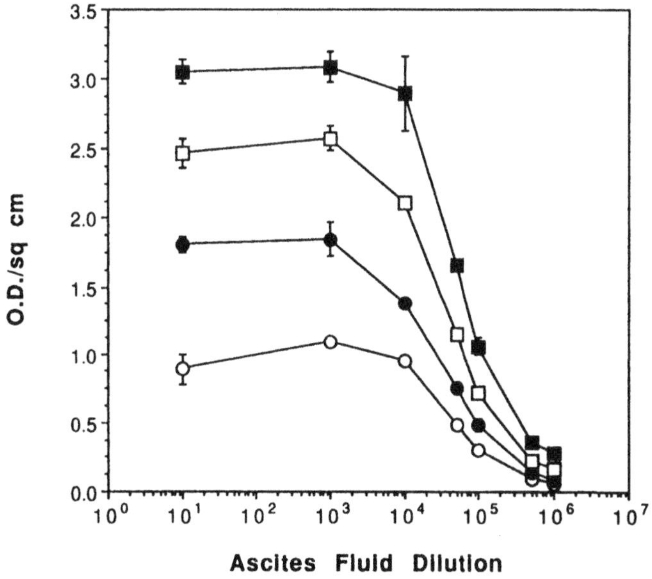
O.D./sq cm
3.5
3.0
2.5
2.0
1.5
1.0
0.5
0.0
10^0
10^1
10^2
10^3
10^4
10^5
10^6
10^7
Ascites Fluid Dilution

Figure 3. R2 antibody binding to fibrinogen adsorbed to biomer from 1% human plasma using anti-IgG-HRP. Anti-IgG-HRP dilutions: 1:1000 (solid squares), 1:3000 (open squares), 1:6000 (solid circles), 1:12 000 (open circles). Triplicate samples for each point.

Table 2.
Non-specific and specific uptake of antibody on Biomer samples

	Amount bound (O.D. cm^{-2}) after exposure to		
Antibody	Buffer	Plasma with 0 residence	Plasma and 3 day residence
K1	0.029[a]	0.640[b]	0.272 ± 0.032[c]
K2	0.740	0.480	0.484 ± 0.035
K3	0.0336	0.769	0.701 ± 0.084
K4	0.186	0.589	1.82 ± 0.12
K5	0.031	1.19	1.21 ± 0.16
K6	0.041	0.695	0.726 ± 0.058
P1[d]	0.049 ± 0.003	2.23 ± 0.31	0.117 ± 0.046
M1	0.033 ± 0.004	0.905 ± 0.120	1.205 ± 0.228
R1	0.070 ± 0.021	0.030 ± 0.016	0.530 ± 0.141
R2	0.012 ± 0.004	1.30 ± 0.17	0.470 ± 0.238

[a] In the preliminary study with K1 through K6 antibodies from which these data were obtained, only one buffer blank was measured. In the subsequent studies with these antibodies that are represented in the Figures, we obtained similar buffer blank values and in addition always obtained triplicates or more for the blanks.
[b] The average of duplicate samples for K1 through K6 after zero residence time is listed.
[c] The average of triplicate samples for K1 through K6 after three day residence time is listed.
[d] The average of quadruplicate buffer blank and plasma preadsorbed samples of P1, M1, R1, and R2 is listed.

in reducing non-specific uptake of the antibodies. Table 2 shows optical density for these controls on all ten antibodies as well as the O.D. for the samples that had been preadsorbed with plasma. It can be seen that except for antibody K2, the blocking agents appeared to be effective, since the control O.D. was substantially less than observed on the plasma treated samples.

In the case of antibody R1, a high O.D. was observed for the Biomer sample that had not been preadsorbed. In this case, it was found necessary to add 1% bovine serum to the buffer as an additional blocking agent. For the sake of consistency, we also included 1% serum in the buffer for the R2 studies even though the non-specific uptake of the antibody was low in our normal buffer.

In the case of R1 and R2, a further control was also run by incubating non-preadsorbed Biomer samples in buffer for the full 3-day residence time, but the O.D. for these samples was not significantly different from Biomer control samples that had not been incubated, as expected. In all subsequent data in this paper, the reported O.D. has been corrected for the control O.D. of Biomer that has not been preadsorbed with plasma.

A further control for non-specific uptake of antibody was performed using afibrinogenemic plasma to preabsorb Biomer. The results are shown in Fig. 4, from which it can be seen that the binding of M1 and R2 by the Biomer samples exposed to deficient plasma is far less than for Biomer exposed to normal plasma. The data also illustrate that there was a slight increase in M1 uptake on Biomer samples preadsorbed with deficient plasma in comparison to Biomer that had no plasma exposure.

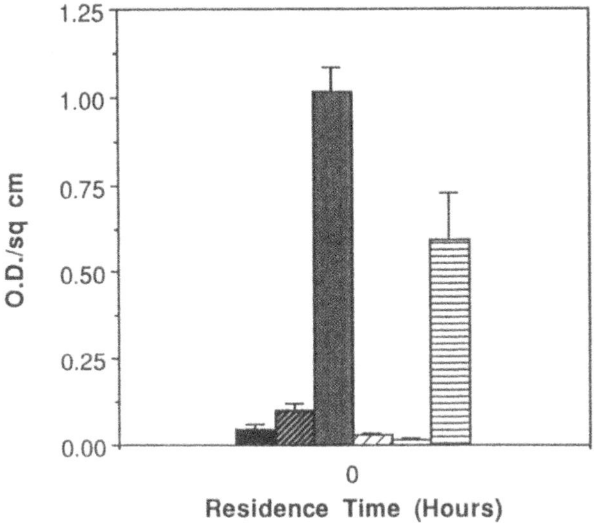
1.25
1.00
0.75
0.50
0.25
0.00
O.D./sq cm
0
Residence Time (Hours)

Figure 4. M1 and R2 antifibrinogen binding after exposure of Biomer to various solutions. The three bars to the left are for M1 binding to surfaces treated with buffer only (solid, leftmost), factor I deficient plasma (dark diagonal, second from left), or normal human plasma (cross-hatched, third from left). The three bars to the right are for R2 binding to surfaces treated with buffer only (light diagonal, third from right), factor I deficient plasma (open, second from right), or normal human plasma (horizontal lines, rightmost bar). Quadruplicate samples.

Retention of adsorbed fibrinogen during the residence time

The amount of fibrinogen adsorbed from plasma to Biomer samples immediately after the adsorption and rinsing step, as well as after periods of increasing residence time in buffer for periods up to 125 h, is shown in Fig. 5. The amount adsorbed from the human plasma used in this study is similar to the amount previously reported for adsorption from baboon plasma for similar plasma dilutions and adsorption times [17]. In addition, the amount of adsorbed fibrinogen remains almost unchanged during the residence time in CPBSz buffer. Similarly high

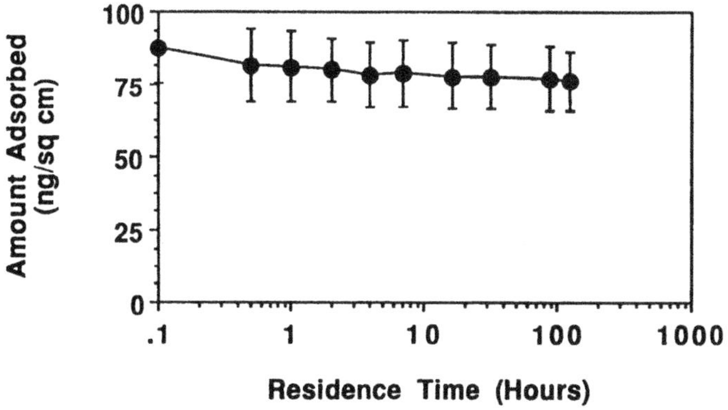
100
75
50
25
0
Amount Adsorbed (ng/sq cm)
.1
1
10
100
1000
Residence Time (Hours)

Figure 5. Retention of human fibrinogen adsorbed to Biomer from 1% human plasma after residence time in buffer. Quintuplicate samples.

retention of fibrinogen adsorbed to other polymers during the residence time has been previously observed [15].

Residence time effects

Since variations in the color yield in the antibody binding assays can occur as a result of aging of the substrate solution (expecially as it reaches the end of its shelf life) or due to a decline in the enzymatic activity of the antimouse IgG/HRP conjugate, the antibody binding studies were designed so that antibody binding to all the samples from a residence time study were assayed at the same time. To accomplish this, the samples were preadsorbed at several times prior to the day of the antibody binding study. On the day of the antibody binding experiment, other samples were preadsorbed with fibrinogen, rinsed and then immediately placed in solutions containing the primary monoclonal antibody. The latter type of samples are considered to have had 'zero' residence time. At the same time, the samples that had been previously adsorbed with fibrinogen were placed into an antibody solution and from then on all samples were treated the same.

In the residence time studies, a preliminary experiment was done to see whether the binding of the antibody to adsorbed fibrinogen changed at all with residence time by examining antibody binding at only two time points, namely 'zero' and three days of residence time. Antibodies that showed significant change were then selected for further experiments done with more frequent residence times. The results of the preliminary screening of the ten antibodies in the 'zero' vs 3-day residence time are shown in Table 2, while Figs 6 and 7 show some of the subsequent experiments done with more residence times.

As shown in Table 2, the binding of five of the monoclonal antibodies did not change significantly between 'zero' and three day residence times (K2, K3, K5, K6,

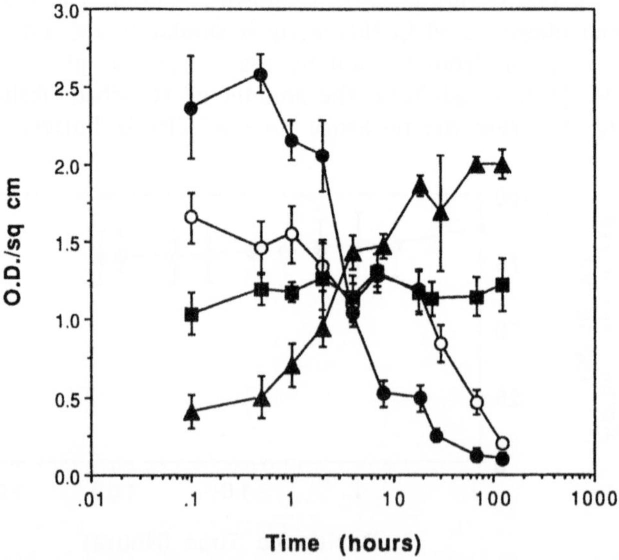
3.0
2.5
2.0
1.5
1.0
0.5
0.0
O.D./sq cm
.01
.1
1
10
100
1000
Time (hours)

Figure 6. Monoclonal antifibrinogen binding to Biomer adsorbed with human plasma: effect of residence time. P1 (solid circles), K1 (open circles), K6 (squares), K4 (triangles). Quadruplicate samples.

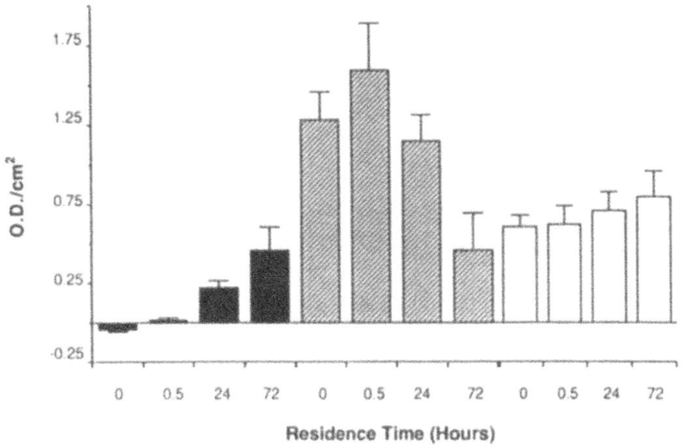
1.75
1.25
0.75
0.25
-0.25
O.D./cm²
0
0.5
24
72
0
0.5
24
72
0
0.5
24
72
Residence Time (Hours)

Figure 7. Monoclonal antifibrinogen binding to human plasma adsorbed Biomer. Effects of residence time. R1 (solid), R2 (diagonals), M1 (open). Quadruplicate samples.

and M1). The binding of three antibodies (K1, P1 and R2) was less after three days of residence time, with an especially large decrease in the binding of P1. The binding of antibodies K4 and R1 appeared to increase with residence time. These variations in the patterns of change in antibody binding to fibrinogen suggest that fairly complex rearrangements occur in the fibrinogen molecule during the residence time. Complex changes are indicated because multiple regions of the molecule appear to be involved and because the changes are not all in one direction such as might occur via complete denaturation of the molecule (see Discussion).

When the binding of antibodies K1, K4, K6, and P1 was measured again after more frequent residence times, the patterns of change observed in the preliminary studies were reproduced in that there was little change in the binding of K1 and K6, while the binding of K4 increased and the binding of P1 decreased with residence time (Fig. 6). It is interesting to note that the largest changes in the binding of antibody K4 and P1 occur at approximately the same residence time (*c.* 1–10 h) but that the changes are in opposite directions. Both of these antibodies bind to the C-terminal domain of the gamma chain, although the epitopes are thought to be located at slightly different positions in the sequence (gamma 373–385 for P1 vs gamma 392–402 for K4; see Discussion). Thus, the differences in the effect of residence time on binding of K4 and P1 also suggest that complex and possibly interrelated rearrangements of this region of fibrinogen may be occurring.

The changes in the binding of three antibodies that bind to the putative platelet binding domains in fibrinogen are compared in Fig. 7. The qualitative nature of the changes are the same as observed in the first experiments with these antibodies in that the binding of M1 does not display a significant change, the binding of R1 increases with residence time, while the binding of R2 at 72 hours is less than after 'zero' residence times. However, the higher time resolution studies in Fig. 6 reveal that the binding of R2 appears to increase slightly after 0.5 h before decreasing after longer times and that the major changes in the binding of R1 do not occur until 1 day or more of residence time. All of these changes seem to occur more slowly than the changes in platelet adhesion to plasma preadsorbed Biomer [14]. Since our initial

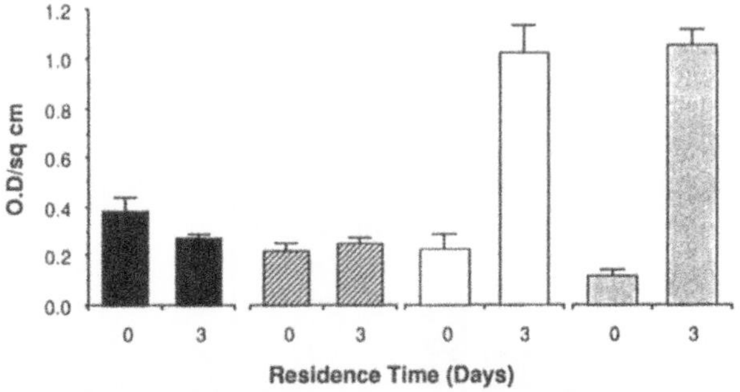
1.2
1.0
0.8
0.6
0.4
0.2
0.0
O.D/sq cm
0
3
0
3
0
3
0
3
Residence Time (Days)

Figure 8. M1 and K4 antifibrinogen binding to fibrinogen adsorbed to Biomer and HS-PEU from baboon plasma. M1 on Biomer (solid), M1 on HS-PEU (diagonals), K4 on Biomer (open), K4 on HS-PEU (dotted). Quadruplicate samples.

platelet experiments were performed using baboon fibrinogen, residence time effects on monoclonal antibody M1 and K4 binding to baboon fibrinogen were also examined and found to be similar to those for human fibrinogen in that the M1 binding did not change much after three days and the K4 binding increased by a large amount (Fig. 8).

Reproducibility of residence time effects

Most of the studies of the binding of antibodies to adsorbed fibrinogen have been done at least twice, separated in time by 1 month or more. In the separate experiments, the changes in antibody binding as well as the O.D. of the samples agreed quite well. An example of the reproducibility of the separate experiments is shown in Fig. 9 in which the binding of R1 and R2 to fibrinogen adsorbed from plasma to Biomer (Fig. 9A) and Immulon (Fig. 9B) is shown as a function of residence time. The two bars at each residence time represent data taken in two separate experiments. As can be seen, the changes in antibody binding with residence time were qualitatively very similar in the two separate experiments. In addition, with one exception (R1 at 72 h), there was good quantitative agreement in the O.D. values in the two experiments.

Polymer effects

The effect of residence time of the binding of some of the antibodies to fibrinogen adsorbed to several other polymers was also examined. As shown in Fig. 9, the binding of R1 to fibrinogen adsorbed to both Biomer and Immulon increased with residence time, although the initial antibody binding was greater on Immulon. However, the binding of R2 decreased with residence time on Biomer but increased on Immulon.

The binding of the M1 antibody did not change with residence time on either Biomer or the hard segment analog polyurethane (HS-PEU) (see Fig. 10). Binding of the antibody K4 to fibrinogen adsorbed on either Biomer of HS-PEU increased substantially with residence time. These experiments show that the changes in the binding of the antibodies are often similar on quite different polymers, showing that the results are not exclusively associated with the polymer Biomer used for most of our studies.

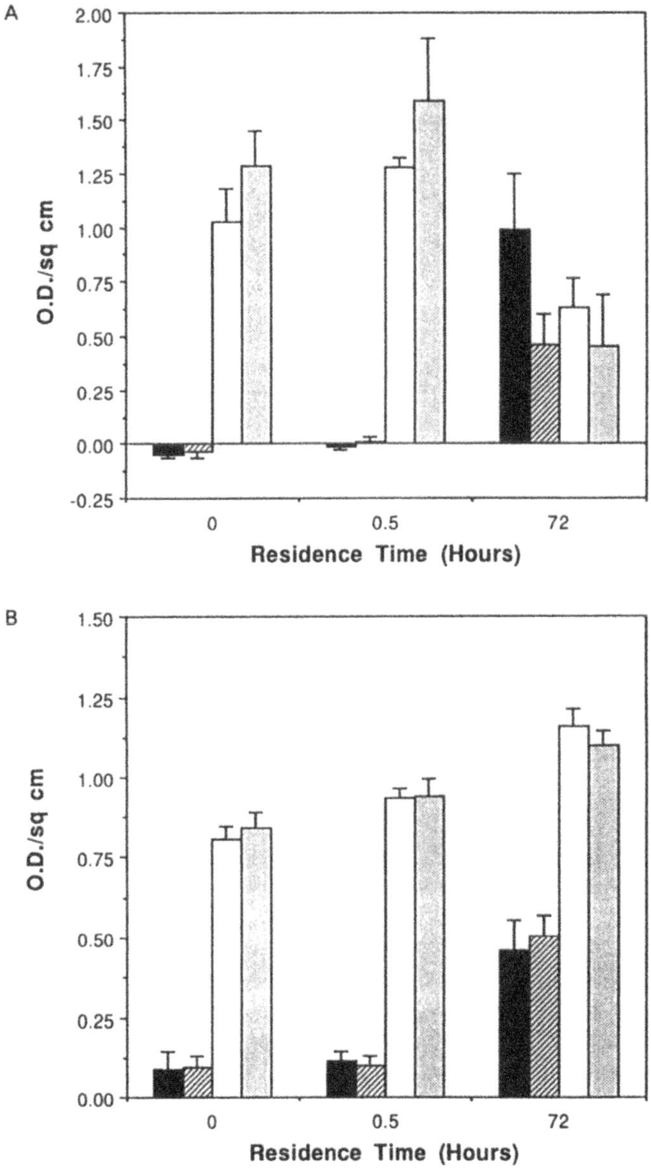
A
2.00
1.75
1.50
1.25
1.00
0.75
0.50
0.25
0.00
-0.25
O.D./sq cm
0
0.5
72
Residence Time (Hours)
B
1.50
1.25
1.00
0.75
0.50
0.25
0.00
O.D./sq cm
0
0.5
72
Residence Time (Hours)

Figure 9. R1 and R2 antifibrinogen binding to fibrinogen adsorbed to Biomer or Immulon I from human plasma: reproducibility of the effects of residence time. A: Two sets of data for Biomer with R1 (solid and hatched) and R2 (open and dotted). B: Two sets of data for Immulon I with R1 (solid and hatched) and R2 (open and dotted). Quadruplicate samples.

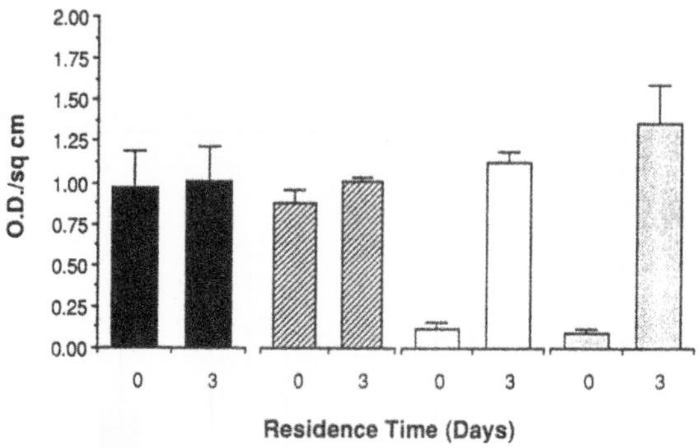
2.00
1.75
1.50
1.25
1.00
0.75
0.50
0.25
0.00
O.D./sq cm
0
3
0
3
0
3
0
3
Residence Time (Days)

Figure 10. M1 and K4 antifibrinogen binding to fibrinogen adsorbed to Biomer or HS-PEU from human plasma. M1 on Biomer (solid), M1 on HS-PEU (diagonals), K4 on Biomer (open), K4 on HS-PEU (dotted). Quadruplicate samples.

DISCUSSION

Properties of the monoclonal antifibrinogens

Some of the antifibrinogens used here were also used previously in studies of platelet adhesion. Antibodies K4 and M1 (numbers 4 and 8 in Table 1) were reported to partially inhibit platelet retention on poly(butylmethacrylate) bead columns that had been preadsorbed with plasma and then exposed to a platelet suspension containing the Fab fragment of these antibodies [31]. We have evaluated the ability of the M1 antibody to block platelet adhesion to plasma preadsorbed Biomer and found that it does inhibit adhesion by approximately 50% when present in the platelet suspension. We also found the antibody inhibited the ADP induced aggregation of PRP by approximately the same amount (unpublished observations). Since the M1 and K4 antibodies are known to have epitopes in the C-terminal region of the gamma chain of fibrinogen, these results support the involvement of that region of fibrinogen in mediating platelet adhesion to adsorbed fibrinogen. The C-terminal dodecapeptide of the gamma chain had previously been shown to inhibit fibrinogen binding to ADP-treated suspension of platelets [25].

Cheresh *et al.* [23] have shown that the monoclonal antibodies to the RGDS containing sequence near the C-terminal of the A alpha chain (572–575) partially blocks platelet adhesion to fibrinogen adsorbed to microtitre wells, while a monoclonal to the RGDF containing sequence near the N-terminal (95–98) was much less effective in blocking platelet adhesion. These authors also showed that platelet adhesion to plasmin-digested fibrinogen lacking the C-terminus of the A alpha chain is reduced (but not eliminated) in comparison to intact fibrinogen. More recently, Savage and Ruggeri have shown that platelet adhesion to fragments X and D from fibrinogen was considerably reduced, but not eliminated, in comparison to fibrinogen. Both X and D contain the gamma chain 400–411 dodecapeptide sequence but lack the A alpha 572–575 RGDS sequence. Adhesion to fragments X and D was about the same, even though the A alpha 95–98 RGDF sequence is present in X but absent in D. Savage and Ruggeri concluded that both the gamma chain dodecapeptide 400–411 and the A alpha chain RGDS at 572–575, but not RGDF 95–98, may play a role in the interaction of surface bound fibrinogen with glycoprotein IIb/IIIa in platelets.

Several earlier studies showing the inhibition of fibrinogen binding to platelets with peptides containing an RGD sequence [6, 32] led to the belief that these sequences in fibrinogen were important in platelet interactions. The importance of the RGD sequences has been emphasized in earlier reviews [5] as well as in more recent ones [33]. However, as has been shown, there is now some doubt about the importance of the RGD sequences in the A alpha chain. More recent studies have emphasized the possibly dominant role of the C-terminal dodecapeptide of the gamma chain. The question of the importance of the RGD sequences is currently considered an open issue [34]. The RGDS containing peptides and the gamma chain peptides compete for the same binding site of GPIIb/IIIa [35].

Antibody P1 (no. 7 in Table 1) has previously been shown to be specific for a conformationally sensitive epitope on fibrinogen. Solution phase fibrinogen does not inhibit the binding of the P1 antibody to adsorbed fibrinogen [36]. This antibody also binds to fibrinogen only after fibrinogen has bound to the IIb/IIIa receptor in platelets. The epitope on fibrinogen to which P1 binds has therefore been called the receptor induced binding site, 'Fg-RIBS-I' [37]. The epitope is in the C-terminal region of the gamma chain. A key element of the epitope was localized to residues 373–385 of the gamma chain [37]. The Fab fragment of antibody P1 also inhibits ADP induced platelet aggregation. The region of fibrinogen to which this antibody binds seems to undergo conformational changes upon its binding to surfaces or to the platelet IIb/IIIa receptor. Thus, the changes in the expression of the epitope in fibrinogen to which antibody P1 binds in our studies of the residence time effect are consistent with the previous work with this antibody.

Antibody K4 (no. 4 in Table 1) has also been reported to be conformationally sensitive in its binding to fibrinogen in that its binding to adsorbed fibrinogen was not inhibited by bulk phase fibrinogen [31, 38]. The changes in K4 binding reported here are consistent with previous work. The epitope to which this antibody binds has been localized to the sequence 393–406 of the gamma chain of fibrinogen through experiments with peptide fragments of fibrinogen [38]. This antibody does not cross react with the peptide (397–411) that was used to raise antibody M1. Also, the epitope to which it binds is not expressed in fibrinogen or fibrin, whereas M1 binds both of these molecules (B. Kudryk, pers. commun.). Thus, while the epitope for antibodies K4 and M1 are adjacent in the fibrinogen gamma chain C-terminal region, it appears that the key binding region for K4 (392–406) is close to but not the same as M1 (402–411). The K4 binding region is evidently in a more conformationally sensitive region slightly to N-terminal side of the site for M1 (402–411).

The other antibodies in Table 1 (nos 1, 2, 3, and 6) are not known to bind to epitopes in fibrinogen that involve platelets but instead were raised in studies attempting to produce fibrin specific antibodies. In this study, therefore, they serve as 'controls' for the effect of adsorption and residence time on changes in the entire fibrinogen molecule.

Postadsorptive changes in fibrinogen

The observation that platelet adhesion to adsorbed fibrinogen declines rapidly when a delay occurs between the preadsorption step and the introduction of the platelets was originally made in this laboratory [14]. The platelet adhesion residence time studies were done because of a desire to understand whether any of the changes in the physicochemical (SDS elutability, displaceability with plasma, infrared spectra) or polyclonal antibody (pAb) binding properties of adsorbed fibrinogen might have an effect on the ability of fibrinogen to mediate platelet adhesion. In collaboration with another lab, the observations of decreased reactivity with increasing residence time originally made with Biomer have now been extended to poly(ethylmethacrylate) [15]. The changes in most of the physicochemical measures and pAb binding took a long time to occur (on the order of days), while platelet adhesion decreased by more than 2/3 after a residence time of less than 25 min [14]. We therefore thought the much more rapid changes in platelet binding might have reflected changes in one or more of the platelet binding domains of fibrinogen, rather than more global changes measured by SDS elution or infrared spectroscopy. However, the results presented in this paper show that the changes in the binding of the monoclonal antibodies do not occur as fast as the changes in the platelet binding. Furthermore, the antibodies to the platelet binding domains either increase in binding (R1), go through a slight maximum and then decrease (R2), or do not exhibit any significant changes in binding at all (M1), with residence time. The potential platelet binding regions in fibrinogen undergo changes at a seemingly much slower rate than previously seen for platelets. Those changes in binding which do occur would seem to be opposite from those seen for platelet adhesion. These results suggest that other mechanisms are needed to explain the changes in platelet binding to adsorbed fibrinogen with increasing residence time. The availability of the platelet binding regions of fibrinogen is evidently not the only factor involved in mediating adhesion.

The stability of binding of the antibody M1 to adsorbed fibrinogen is particularly important in view of good evidence suggesting that the gamma chain C-terminus to which M1 binds may be the most important in mediating platelet adhesion to fibrinogen. For this reason, the effect of residence time on the binding of M1 to fibrinogen adsorbed to Biomer and other surfaces has been studied most extensively in our work, but we have not been able to observe significant changes in the binding of this antibody. Our results indicate that the gamma chain C-terminal dodecapeptide region of the molecule is available for binding at all residence times. We therefore measured the changes in other regions of fibrinogen that may play a role in platelet binding, namely the RGD sequences, using R1 and R2. As shown in Figs 7 and 9, the binding of antibodies R1 and R2 increased only slightly after short (30 min) residence times, while longer residence times result in further increases in R1 binding but decreases in R2 binding. Thus, there is no clear evidence from these measurements of rapid or large scale changes in the RGD sequences that could explain the large decreases in platelet binding.

It is possible that the binding of M1 and of the IIb/IIIa receptor to the C-terminal dodecapeptide are not equally sensitive to changes in the accessibility or conformation of this region of the fibrinogen molecule. In this case, there might not be an equally rapid change in the binding of the antibody and in the binding of platelets to adsorbed fibrinogen. Thus, for example, the binding of IIb/IIIa to the fibrinogen molecule may require cooperative changes in both of the interacting molecules to achieve the final liganded state, while any changes in the antibody or fibrinogen that occur in the process of their binding may be quite different. Very recent evidence is available to suggest that these ideas might have some validity. Thus, Blumentein *et al.* recently reported that while M1 binding to the 392–411 gamma chain segment of fibrinogen is inhibited by D- or L-alanine 409 analogs of this peptide, these peptides do not inhibit fibrinogen binding to platelets [39]. The authors concluded that the recognition requirements for the platelet receptor differ considerably from those for antibody M1.

Other factors that may influence platelet adhesion to adsorbed fibrinogen

Changes in adsorbed fibrinogen have so far been studied with SDS elution [16, 18, 19], infrared spectroscopy [22], polyclonal antibody binding [17], platelet adhesion [13], and the displaceability of adsorbed fibrinogen with plasma [20, 21]. We have previously noted that the changes measured using these different methods do not occur on the same time scale and attributed this to differences in the properties measured. For example, because SDS is a powerful surfactant, the SDS elutability test is probably a severe test for the occurrence of increased bonding of fibrinogen to surfaces. The changes in fibrinogen needed to become more resistant to SDS elution may be extensive and therefore take a long time, as has been observed in studies to date. In contrast, displacement of adsorbed fibrinogen by plasma is a milder test. Changes in plasma displaceability occur within a few minutes of residence time [20, 21]. Thus, platelet adhesion is not the only measure of fibrinogen's surface state displaying changes after short residence times. Is the change in fibrinogen causing decreased platelet adhesion somehow related to the changes in plasma displaceability?

The reduction in plasma displaceability of adsorbed fibrinogen with residence time is thought to involve further contact formation between the fibrinogen molecule and the substrate polymer. This results in a more tightly held molecule that is less subject to competitive displacement by other proteins in plasma that bind in the place of fibrinogen. Essentially, this explanation postulates that the fibrinogen molecule becomes more tenaciously held by the substrate as more contact time occurs. Thus the similar time scale for changes in platelet adhesion and plasma displaceability could indicate that platelet adhesion is very sensitive to how tightly the fibrinogen molecule is held to the substrate, rather than having much to do with changes in the availability of the platelet binding domains in fibrinogen. At this time there is no direct evidence for this idea, but there are indications in other studies supporting a role for the tightness of binding of a protein on a surface with its ability to influence cell interactions.

Substrate adherent platelets have been shown to undergo a centripetal redistribution of fibrinogen occupied receptors on their dorsal surface that is related to cytoskeletal recorganization [41, 42]. Similarly, movement of colloidal gold labeled fibrinogen on the ventral surfaces of adherent platelets has also been observed when platelets were placed in contact with surfaces preadsorbed with fibrinogen conjugated with colloidal gold [43]. In these studies, no movement of colloidal gold labeled albumin was observed, suggesting that movement of bound fibrinogen was receptor mediated. More recently, adherent platelets have been shown to cause redistribution of adsorbed fibrinogen [44]. Fluorescently labeled fibrinogen on a surface was removed from areas adjacent to the platelets and became concentrated centripetally [44]. The movement of receptor bound and surface bound fibrinogen in these studies provides evidence that platelet interaction with adsorbed fibrinogen is not an entirely passive process. The interaction may involve removal of the fibrinogen from the substrate. These observations indicate that the tightness with which the fibrinogen is held could influence platelet behavior.

Other indications in support of a role for tightness of binding of adsorbed fibrinogen in platelet adhesion come from a study of the effects of sequential adsorption of albumin and fibrinogen on surfaces. It has been found that surfaces loaded equally with albumin and fibrinogen behave quite differently in regard to platelet adhesion [45]. Surfaces loaded partially first with fibrinogen and then filled subsequently with albumin induce more platelet adhesion than when the order is reversed. Placing surfaces preadsorbed with fibrinogen in albumin containing buffers prevents most of the residence time dependent decrease in platelet binding to fibrinogen [14] as well as preventing the decreases in polyclonal antifibrinogen binding and SDS elutability [17]. The albumin in the buffer probably occupies some of the empty sites on the surface and prevents further contact formation between the fibrinogen molecule and the surface. Thus, the effects of sequential albumin adsorption on platelets may indicate that more tightly bound fibrinogen is less adhesive for platelets. This idea is also supported by the finding that certain types of plasma polymerized TFE surfaces, which bind fibrinogen more tightly than most surfaces, also exhibit reduced adhesion of platelets compared to other fibrinogen treated surfaces [46, 47]. These surfaces are considered to bind proteins tightly because SDS does not remove as much of the protein as on other surfaces, apparently reflecting a 'molecular spreading' phenomenon in which many more bonds to the adsorbing substrate are formed per protein molecule. However, fibrinogen adsorbed to the tight binding plasma TFE surfaces binds antibody M1 at least as well as fibrinogen adsorbed on PTFE (unpublished observations of T. Horbett and D. Kiaei). Finally, the migration rate of corneal epithelial cells seems to correlate well with the tightness of binding of fibronectin [48], an observation that also indicates a possible role for the tightness of binding of proteins to substrates in affecting cell interactions.

In summary, the evident tendency for platelets to remove fibrinogen from surfaces, as well as the effects of albumin in preventing losses in fibrinogen elutability and platelet adhesion, together with the finding in this work of no correlated changes in the binding of antibodies to the platelet binding regions of fibrinogen, all are consistent with the possible role of tightness of binding of fibrinogen to the substrate in platelet binding. Clearly, direct evidence for a cause-and-effect relationship between changes in the tightness of fibrinogen binding and platelet adhesion will require further experimentation. Since we have been systematically studying the adhesion of platelets to a series of other polyurethane surfaces and have observed large differences in the adhesion of the platelets, we expect to subject this hypothesis to experimental testing in the near future.

Acknowledgement

The financial support of the NHLBI through grant HL19419 is gratefully acknowledged.

REFERENCES

1. J. Hawiger, *Methods of Enzymology, Platelets: Receptors, Adhesion, Secretions, Part A*. Harcourt Brace Jovanovich, San Diego, CA (1989).
2. M. A. Packham, *Proc. Soc. Exp. Biol. Med.* **189**, 261 (1988).
3. E. A. Wayner, W. G. Carter, R. S. Piotrowicz and T. J. Kunicki, *J. Cell Biol.* **107**, 1881 (1988).

4. P. Thiagarajan and K. Kelly, *Thomb. Haemostas.* **60,** 514 (1988).
5. D. R. Phillips, I. F. Charo, L. V. Parise and L. A. Fitzgerald, *Blood* **71,** 831 (1988).
6. D. M. Haverstick, J. F. Cowan, K. M. Yamada and S. A. Santoro, *Blood* **66,** 946 (1985).
7. B. Savage and Z. M. Ruggeri, *J. Biol. Chem.* **266,** 11227 (1991).
8. S. F. Mohammad, N. S. Topham, G. L. Burns and D. B. Olsen, *Trans. Am. Soc. Artif. Int. Organs* **34,** 573 (1988).
9. A. White, P. Handler and E. L. Smith, *Principles of Biochemistry*, 3rd Edn, p. 648. McGraw-Hill, New York (1964).
10. M. S. Lantz, R. D. Allen, T. A. Vail, L. M. Switalski and M. Hook, *J. Bacteriol.* **173,** 495 (1991).
11. M. S. Lantz, R. W. Rowland, L. M. Switalksi and M. Hook, *Infect. Immun.* **54,** 654 (1986).
12. P.-E. Lindgren, P. Speziale, M. McGavin, H.-J. Monstein, M. Hook, L. Visai, T. Kostiainen, S. Bozzini and M. Lindberg, *J. Biol. Chem.* **267,** 1924 (1992).
13. J. A. Chinn, T. A. Horbett and B. D. Ratner, *Thromb. Haemostas.* **65,** 608 (1991).
14. J. A. Chinn, S. E. Posso, T. A. Horbett and B. D. Ratner, *J. Biomed. Mater. Res.* **25,** 535 (1991).
15. I. A. Feuerstein, W. G. McClung and T. A. Horbett, *J. Biomed. Mater. Res.* **26,** 221 (1992).
16. R. Rapoza and T. A. Horbett, *J. Biomed. Mater. Res.* **24,** 1263 (1990).
17. J. A. Chinn, S. E. Posso, T. A. Horbett and B. D. Ratner, *J. Biomed. Mater. Res.* **26,** 757 (1992).
18. R. J. Rapoza and T. A. Horbett, *J. Biomat. Sci. Polym. Edn.* **1,** 99 (1989).
19. R. J. Rapoza and T. A. Horbett, *J. Coll. Interf. Sci.* **136,** 480 (1990).
20. S. M. Slack and T. A. Horbett, *J. Coll. Interf. Sci.* **133,** 148 (1989).
21. S. M. Slack and T. A. Horbett, *J. Biomed. Mater. Res.* **26,** 1633 (1992).
22. T. J. Lenk, T. A. Horbett, B. D. Ratner and K. K. Chittur, *Langmuir* **7,** 1755 (1991).
23. D. A. Cheresh, S. A. Berliner, V. Vicente and Z. M. Ruggeri, *Cell* **58,** 945 (1989).
24. M. Kloczewiak, S. Timmons and J. Hawiger, *Biochem. Biophys. Res. Commun.* **107,** 181 (1982).
25. M. Kloczewiak, S. Timmons, T. J. Lukas and J. Hawiger, *Biochemistry* **23,** 1767 (1984).
26. I. F. Chatro, L. Nannizzi, J. W. Smith and D. A. Cheresh, *J. Cell Biol.* **111,** 2795 (1990).
27. J. H. Saunders and K. C. Frisch, *Polyurethanes; Chemistry and Technology, Part 1. Chemistry.* Interscience, New York (1962).
28. R. W. Helmkamp, R. L. Goodland, W. F. Bale, I. L. Spar and L. E. Mutschler, *Cancer Res.* **20,** 1495 (1960).
29. T. A. Horbett, *J. Biomed. Mater. Res.* **15,** 673 (1981).
30. A. Rosenberg and F. W. Teare, *Anal. Chem.* **77,** 289 (1977).
31. E. Shiba, J. N. Lindon, L. Kushner, G. R. Matsueda, J. Hawiger, M. Kloczewiak, B. Kudryk and E. W. Salzman, *Am. J. Physiol.* **260,** C965 (1991).
32. A. Andrieux, G. Hudry-Clergeon, J. J. Ryckewaert, A. Chapel, M. H. Ginsberg, E. F. Plow and G. Marguerie, *J. Biol. Chem.* **264,** 9258 (1989).
33. T. J. Kunicki and P. J. Newman, *Blood* **80,** 1386 (1992).
34. E. F. Plow, S. E. D'Souza and M. H. Ginsberg, *J. Lab. Clin. Med.* **120,** 198 (1992).
35. S. C.-T. Lam, E. F. Plow, M. A. Smith, A. Andrieux, J.-J. Ryckwaert, G. Marguerie and M. H. Ginsberg, *J. Biol. Chem.* **262,** 947 (1987).
36. C. Zamarron, M. H. Ginsberg and E. F. Plow, *Thromb. Haemostas.* **64,** 41 (1990).
37. C. Zamarron, M. H. Ginsberg and E. F. Plow, *J. Biol. Chem.* **266,** 161193 (1991).
38. R. Procyk, B. Kudryk, S. Callender and B. Blomback, *Blood* **77,** 1469 (1991).
39. M. Blumentein, G. R. Matsueda, S. Timmons and J. Hawiger, *Biochemistry* **31,** 10692 (1992).
40. W. G. Laver, G. M. Air, R. G. Webster and S. J. Smith-Gill, *Cell* **61,** 553 (1990).
41. J. C. Loftus and R. M. Albrecht, *J. Cell Biol.* **99,** 822 (1984).
42. R. M. Albrecht, O. E. Olorundare and S. R. Simmons, in: *Fibrinogen 3. Biochemistry, Biological Functions, Gene Regulation and Expression*, p. 211, M. W. Mosesson (Ed.). Elsevier, Amsterdam (1988).
43. S. L. Goodman, Q. J. Lai, K. Park and R. M. ALbrecht, in: *Proceedings of the XII International Congress for Electron Microscopy*, p. 22. San Francisco Press, San Francisco, CA (1990).
44. K. Gaebel and I. A. Feuerstein, *Biomaterials* **12,** 597 (1991).
45. W. G. Pitt, K. Park and S. L. Cooper, *J. Coll. Interf. Sci.* **111,** 343 (1986).
46. D. Kiaei, A. S. Hoffman, B. D. Ratner and T. A. Horbett, *J. Appl. Polym. Sci. Appl. Polym. Symp.* **42,** 269 (1988).
47. D. Kiaei, A. S. Hoffman and S. R. Hanson, *J. Biomed. Mater. Res.* **26,** 357 (1992).
48. D. K. Pettit, T. A. Horbett and A. S. Hoffman, *J. Biomed. Mater. Res.* **26,** 1259 (1992).

Adsorption behavior of fibrinogen to sulfonated polyethyleneoxide-grafted polyurethane surfaces

DONG KEUN HAN[1], GYU HA RYU[2], KI DONG PARK[1], SEO YOUNG JEONG[1], YOUNG HA KIM[1,*] and BYOUNG GOO MIN[2]

[1]*Polymer Chemistry Laboratory, Korea Institute of Science and Technology, P.O. Box 131, Cheongryang, Seoul 130-650, Korea*

[2]*Department of Biomedical Engineering, College of Medicine, Seoul National University, Seoul 110-744, Korea*

Received 4 June 1992; accepted 1 September 1992

Abstract—Fibrinogen adsorptions to surface modified polyurethanes (PU, PU-PEO, and PU-PEO-SO_3) were studied from plasma *in vitro*. PU and PU-PEO surfaces demonstrated that initial adsorption increases with increasing plasma concentration in kinetic profiles and adsorption time in adsorption profiles as a function of plasma concentration, but after the plateau is reached, its adsorption amount decreases as plasma concentration (0.2–2.0%) and adsorption time (1–120 min) increase, respectively. In contrast, PU-PEO-SO_3 showed that initial adsorption is almost same regardless of plasma concentration and adsorption time, which is due to the high affinity of surface sulfonate group to fibrinogen. All the surfaces indicated the Vroman effect at about 0.6% plasma concentration; however, the displacement was relatively low. Adsorbed amount of fibrinogen at steady state decreased in the order: PU > PU-PEO-SO_3 > PU-PEO, regardless of adsorption time and plasma concentration. The adsorption behavior of PU-PEO-SO_3 is attributed to both effect of low binding affinity of PEO chain and high affinity of pendant sulfonate group toward fibrinogen.

Key words: Polyurethanes; surface modification; polyethyleneoxide; sulfonate group; sulfonated PEO; protein adsorption; fibrinogen.

INTRODUCTION

Polyurethane (PU) has been widely used for biomedical polymers because of its excellent mechanical properties. However, the inherent thrombogenicity of PU remains a problem and limits greater widespread blood-contacting devices.

A variety of approaches have been taken to enhance the blood compatibility of polymer surfaces. Among them, of interest to us is the surface modification of existing polymers without changing bulk properties. The modifications are based on factors such as surface free energy, protein adsorption, platelet adhesion and other blood coagulation factors. Practically every physical and chemical material property has been suggested as being important in thrombosis [1, 2].

A study of protein adsorption is valuable in understanding blood response to materials and developing blood compatible polymers. In particular, fibrinogen is known to be a protein of high surface activity, which can play a leading part resulting in thrombus formation. However, Cooper and coworkers [3, 4] reported that the sulfonated PUs exhibit more fibrinogen adsorption and less platelet adhesion than untreated PU, but enhanced thromboresistance. Santerre *et al.* [5]

---

* To whom correspondence should be addressed.

also described high affinity of fibrinogen to the sulfonated PUs. In addition, from time dependent patterns of fibrinogen adsorption, the 'Vroman effect' suggests that after rapid initial adsorption, fibrinogen was displaced from some surfaces, including glass by other proteins, such as high molecular weight kininogen (HMWK) and coagulation factor XII [6-8].

In our previous report [9], *in situ* polyethyleneoxide (PEO) and sulfonated PEO immobilization onto PU surface (PU-PEO and especially, PU-PEO-SO_3) were shown to be effective in improving the blood compatibility of PU surfaces *in vitro* and *ex vivo*. The rationale for this research suggests that the hydrophilic PEO and PEO-SO_3 are expected to reduce protein adsorption and platelet adhesion (PEO hypothesis). As shown in the 'negative cilia' model in Fig. 1, the advantage of using hydrophilic PEO and PEO-SO_3 is due to low interfacial free energy, nonadhesive property and the high dynamic motion of PEO chains to diminish surface thrombus formation [10-12]. Moreover, pendant negative charge of the sulfonate group is expected to repel protein adsorption and cell adhesion as well as suppress the procoagulant factors, possibly by electrostatic repulsion and the heparin-like anticoagulant activity of surface bound sulfonate groups [13].

The modification and surface characterization have been described in detail previously [14, 15]. *In vitro* and *ex vivo* biological responses of modified PU surfaces were also reported previously [9, 16]. Less platelet was adhered on modified PUs (both PEO and PEO-SO_3 grafted surfaces) as compared to control PU *in vitro*. PEO grafted PU surface demonstrated a more prolonged occlusion time than PU control, and PEO-SO_3 grafted PU surface illustrated a much longer occlusion time than PU-PEO surface in *ex vivo* rabbit A-A shunt experiment.

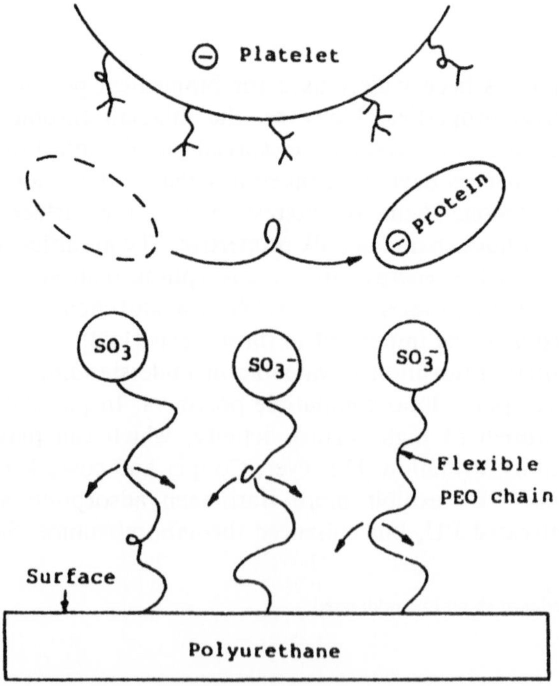

Platelet
Protein
SO3-
SO3-
SO3-
Flexible
PEO chain
Surface
Polyurethane

Figure 1. 'Negative cilia' model on PU-PEO-SO_3 system.

In this study, fibrinogen adsorption in plasma onto modified PU surfaces were performed to elucidate the relationship between fibrinogen adsorption and the degree of surface-induced thrombosis. Also, to examine the existence of the Vroman effect in plasma, studies of kinetics as well as adsorption as a function of plasma concentration on fibrinogen adsorption to polymer surfaces were performed. Since fibrinogen displacement from the surface is considered too rapid to be observed [17], diluted plasma solutions were used to cause adsorbed fibrinogen to remain on the surface longer than the 1 min needed for detection.

MATERIALS AND METHODS

Surfaces

The polyethyleneoxide (PEO)-grafted polyurethane (PU) (PU–PEO) and further sulfonated PEO-grafted PU (PU–PEO–SO_3) beads were surface-modified and characterized as previously described [14, 15]. The surface modification scheme as shown in Fig. 2 involved the coupling of hexamethylene diisocyanate (HMDI) to PU (Pellethane®, Dow Chem. Co.) through an allophanate reaction. The free isocyanate groups attached to PU were then coupled to PEO (M_W: 1000) through a condensation reaction to terminal hydroxyl end groups on PEO (PU–PEO). The OH group remaining on the PEO was then treated with propane sultone (PST) to produce the PU–PEO–SO_3.

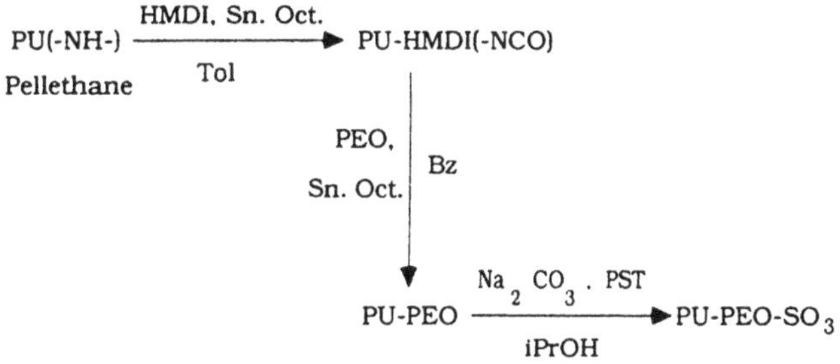
HMDI, Sn. Oct.
PU(-NH-)
PU-HMDI(-NCO)
Pellethane
Tol
PEO,
Sn. Oct.
Bz
Na2 CO3 . PST
PU-PEO
PU-PEO-SO3
iPrOH

Figure 2. Modification scheme of polyurethane (PU) surfaces: HMDI = $OCN(CH_2)_6NCO$; PEO = $HO(CH_2CH_2O)_nH$; PST = $(CH_2)_3SO_3$.

Fibrinogen and plasma

Bovine fibrinogen (Sigma Chem., 95% clottable) was dialyzed against 0.04 M potassium phosphate buffer (pH 7.0), lyophilized and stored at −70°C. Bovine fibrinogen was radiolabeled using reductive alkylation with ^{14}C-formaldehyde (Amersham Int., 1 mCi) and sodium cynanoborohydride using the method of Dottavio-Martin and Ravel [18]. Unreacted radioactive label and salts were removed by dialysis against 0.04 M potassium phosphate buffer until no further radioactivity was detected in the dialysate. Radiolabeled fibrinogen was then lyophilized and weighed into appropriate solutions and assayed for radioactivity using a liquid scintillation counter (Beckman DP 5500). The radiolabeling procedure produced specific activities of fibrinogen ranging from 1.12×10^4 cpm/μg. Competitive ^{14}C-labeled fibrinogen adsorption studies were performed with diluted bovine plasma.

Fresh bovine whole blood was collected in 3.8% sodium citrate anticoagulant solution (final dilution 9 : 1, v/v). The blood was then carefully transferred to polystyrene tubes, centrifuged at 2500 *g* at 4°C to prepare platelet-poor plasma (PPP). Concentrations of fibrinogen in bovine plasma were determined by standard methods. The control plasma was found to have contained 2.66 mg/ml fibrinogen.

Fibrinogen solutions for the adsorption experiment from plasma were prepared by adding tracer amounts of radiolabeled fibrinogen, in amounts corresponding to approximately 0.2–10% of protein concentration, to freshly prepared bovine plasma. For experiments, labeled fibrinogen in plasma was diluted with phosphate buffered saline (PBS, pH 7.4).

Fibrinogen adsorption

Quantities of surface modified PU beads (800 mg, surface area 13.80 cm^2) were carefully weighed into plastic disposable 10 ml syringes and equilibrated with 5 ml of PBS overnight. Prior to adsorption studies, the buffer was removed and 1.6 ml of protein solution introduced into the syringe system.

Sets of syringes were arranged for varied adsorption time intervals (1–120 min) as well as different concentrations (0.2–10%) of protein solution to produce the profiles of adsorption kinetics and adsorption as a function of plasma concentration. Adsorption kinetics were constructed for different adsorption times at a fixed plasma concentration, and adsorption as a function of plasma concentration for each applied concentration at a fixed adsorption time.

After adsorbed at each time and concentration, the plasma was expelled from the syringe, leaving the beads inside. The syringe was filled and washed extensively with PBS until no further activity was detected in the eluent. Finally, a 2% (w/v) solution of sodium dodecyl sulfate (SDS) in PBS was added to the syringes and agitated for 48 h to dissolve any bound proteins. Aliquots of these washings were assayed for radioactivity and compared to the eluent and stock solutions for quantifiable depletion of radioactive species.

To compare each adsorption behavior of ^{14}C-labeled and unlabeled fibrinogen onto modified PU surfaces, the amount of adsorbed fibrinogen was measured as a function of the percentage (5–100% of labeled protein) of fibrinogen present in the plasma. No preferential adsorption of labeled fibrinogen was found. A control sample of plasma adsorbed without beads was used as a reference for each experiment, and all data were taken from values measured in at least three separate adsorption experiments. Each point represented an average with standard deviations ranging from ± 2.0 to $\pm 3.5\%$.

RESULTS

The surface characteristics of modified PUs was previously reported elsewhere [15, 16]. As shown in Table 1, PU-PEO and particularly, PU-PEO-SO_3 surfaces demonstrated high smoothness and homogeneity. Dynamic contact angle measurements in water revealed that surface hydrophilicity increased in the following order: PU < PU-PEO < PU-PEO-SO_3, resulting from the surface coupling of PEO and PEO-SO_3.

Figures 3–8 show the adsorption behaviors of fibrinogen to respective PU, PU-PEO, and PU-PEO-SO_3 surfaces. On the whole, PU-PEO and PU-PEO-SO_3 surfaces demonstrated less fibrinogen adsorption than PU. The kinetic profiles

Table 1.
Surface characteristics of modified PUs

Material	Surface morphology[a]	Surface wettability[b]	
		θ_{adv}	θ_{rec}
PU	relatively smooth	86	41
PU-PEO1000	smoother	40	15
PU-PEO1000-SO_3	smoothest	45	wetting

[a] By SEM (X3000).
[b] By Wilhelmy plate method.

show that as the plasma concentration increases, the amount of fibrinogen initially present on the surface increases, but is then later displaced, and decreases with increasing plasma concentration. Also, the adsorption profiles as a function of plasma concentration show that as the adsorption time increases, the adsorption amount of fibrinogen increases. At higher plasma concentrations, more fibrinogen is then adsorbed at shorter times, and that the amount of fibrinogen decreases with increasing adsorption time. This result means that as adsorption time and plasma concentration increase, fibrinogen adsorbed was displaced or competed with other plasma proteins, which indicates that the Vroman effect occurred in all surfaces.

Figures 3 and 4 show the adsorption kinetics and adsorption as a function of plasma concentration of fibrinogen on control PU surface, respectively. Initial fibrinogen adsorption increased as plasma concentration increases as shown in Fig. 3. At very diluted plasma concentration (0.2 and 0.6% plasma), it showed a conventional time dependence, i.e. rapid adsorption to steady value. In contrast, at 1 and 2% plasma concentration, initial fibrinogen adsorption at adsorption time

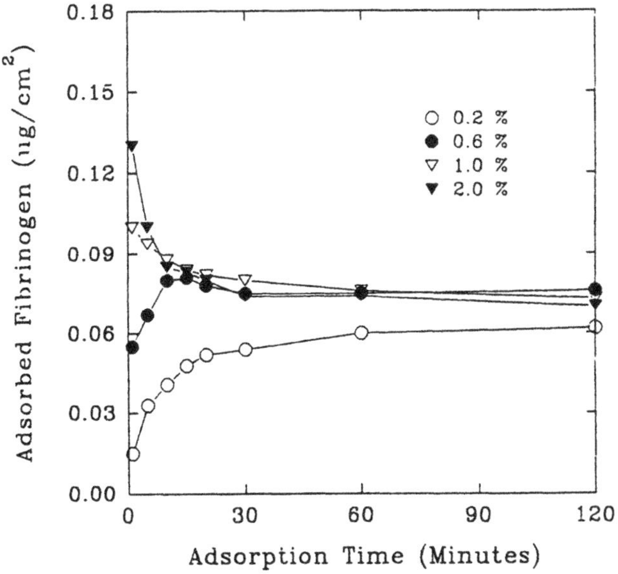

0.18
0.15
0.12
0.09
0.06
0.03
0.00
0
30
60
90
120
Adsorbed Fibrinogen (ug/cm²)
Adsorption Time (Minutes)
0.2 %
0.6 %
1.0 %
2.0 %

Figure 3. Adsorption kinetics for fibrinogen of PU from plasma at various plasma concentrations.

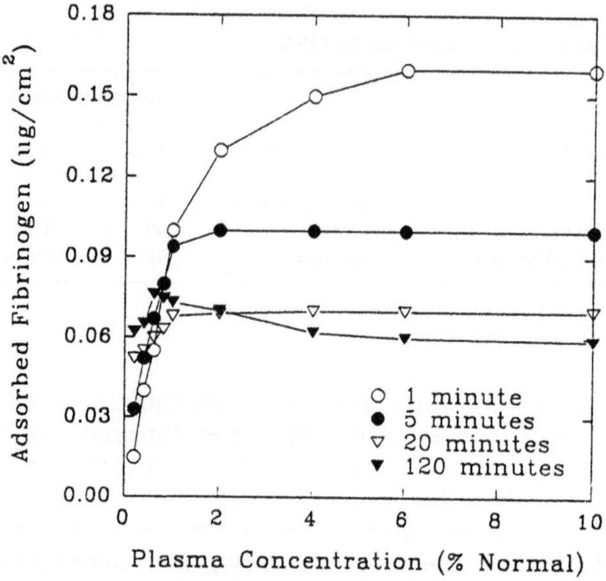
0.18
0.15
0.12
0.09
0.06
0.03
0.00
0
2
4
6
8
10
Adsorbed Fibrinogen (ug/cm²)
Plasma Concentration (% Normal)
1 minute
5 minutes
20 minutes
120 minutes

Figure 4. Adsorption as a function of plasma concentration for fibrinogen on PU from plasma at various times.

below 1 min cannot be observed since displacement is too rapid, whereas at concentration less than 0.6% there is no displacement and a plateau is reached. However, the adsorption as a function of plasma concentration, as shown in Fig. 4, exhibited significant maximum peak at 120 min plasma adsorption.

Figures 5 and 6 demonstrate, respectively, the adsorption kinetics and adsorption as a function of plasma concentration of fibrinogen on PU-PEO surface. Although

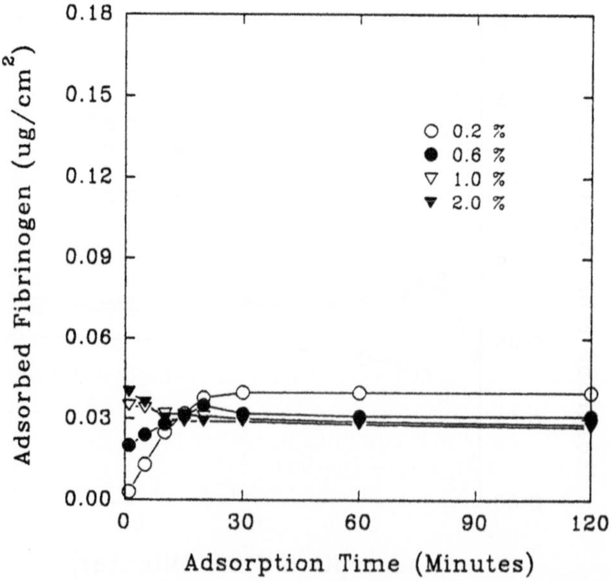
0.18
0.15
0.12
0.09
0.06
0.03
0.00
0
30
60
90
120
Adsorbed Fibrinogen (ug/cm²)
Adsorption Time (Minutes)
0.2 %
0.6 %
1.0 %
2.0 %

Figure 5. Adsorption kinetics for fibrinogen of PU-PEO from plasma at various plasma concentrations.

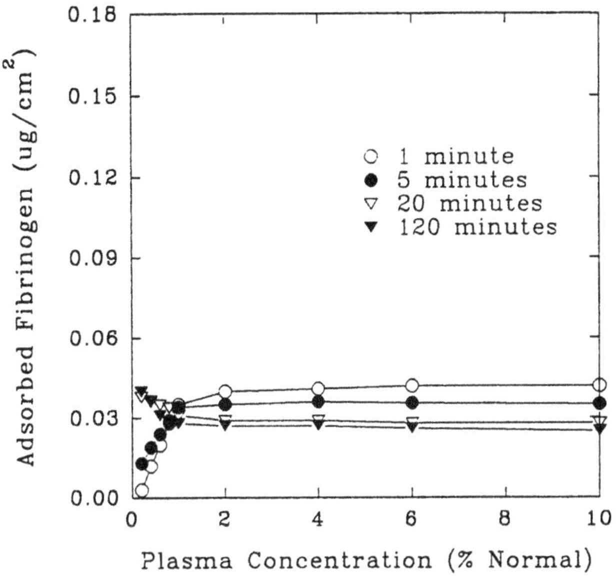
Adsorbed Fibrinogen (ug/cm^2)
0.18
0.15
0.12
0.09
0.06
0.03
0.00
0
2
4
6
8
10
Plasma Concentration (% Normal)
○ 1 minute
● 5 minutes
▽ 20 minutes
▼ 120 minutes

Figure 6. Adsorption as a function of plasma concentration for fibrinogen on PU-PEO from plasma at various times.

PU-PEO surface showed much less fibrinogen adsorption than control PU, PU-PEO surface showed an adsorption pattern similar to control PU surface, i.e. conventional adsorption to steady value at a concentration of less than 0.6%, and initially high adsorption followed by displacement of adsorbed fibrinogen at higher than 0.6%. In addition, adsorption as a function of plasma concentration demonstrated a similar trend to the adsorption kinetics. Fibrinogen adsorption from plasma shows a maximum below 1% plasma for 20 and 120 min fibrinogen adsorption (conventional Langmuir type isotherm after 1 and 5 min adsorption).

The decrease in fibrinogen adsorption on hydrophilic PEO surface as compared to the unmodified PU surface is as expected. It is well known that PEO surface decreases protein adsorption and cell adhesion due to low interfacial free energy, highly dynamic motion and extended chain conformation of PEO chain, and its nonadhesive property [10-12].

The adsorption kinetics and adsorption as a function of plasma concentration of fibrinogen on PU-PEO-SO_3 surface are shown in Figs 7 and 8, respectively. It is of interest to note the different adsorption behavior as compared to PU and PU-PEO. The initially adsorbed fibrinogen amount of PU-PEO-SO_3 is almost the same regardless of plasma dilution concentration (0.2-2.0%) and adsorption time (1-120 min). Although the adsorption pattern is similar to that of PU and PU-PEO at steady state, the initial adsorption rate of fibrinogen on PU-PEO-SO_3 is much faster when compared to that of PU and PU-PEO at low plasma concentrations and short adsorption times, suggesting that the negatively charged SO_3 group, attached to one end of the PEO chain, has a high affinity to fibrinogen.

Figure 9 demonstrates the fibrinogen adsorption profiles of adsorbed fibrinogen versus adsorption times at various plasma concentrations, in order to compare three different PU surfaces. Overall, it seems that at high plasma concentrations,

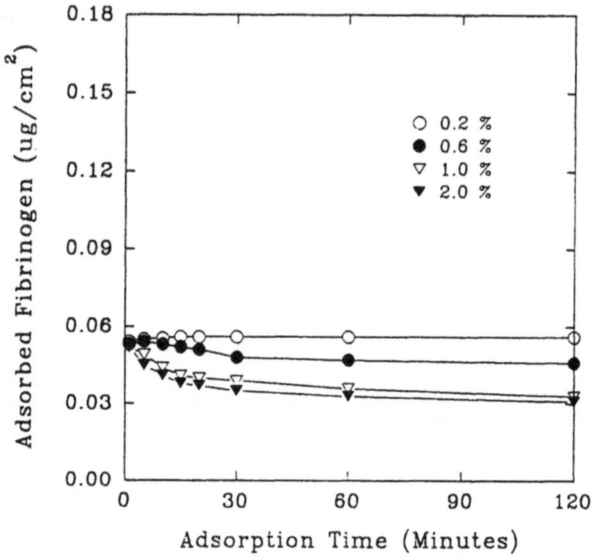
0.18
0.15
0.12
0.09
0.06
0.03
0.00
0
30
60
90
120
Adsorbed Fibrinogen (ug/cm2)
Adsorption Time (Minutes)
0.2 %
0.6 %
1.0 %
2.0 %

Figure 7. Adsorption kinetics for fibrinogen of PU-PEO-SO_3 from plasma at various plasma concentrations.

fibrinogen adsorption is limited by the displacement, whereas at low concentrations, it is limited by the fibrinogen concentration. Also, all the surfaces indicated the Vroman effect at about 0.6% plasma concentration, but displacement by other plasma proteins such as contact phase clotting factors was relatively low. The amount of adsorbed fibrinogen at steady state decreased in the order: PU > PU-PEO-SO_3 > PU-PEO, regardless of plasma dilution.

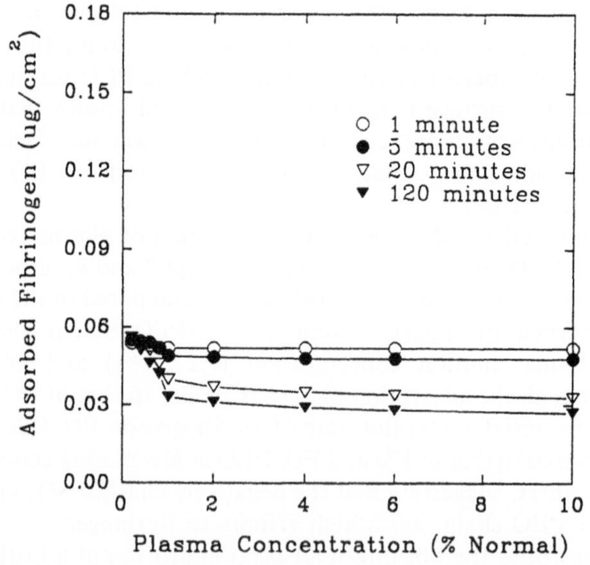
0.18
0.15
0.12
0.09
0.06
0.03
0.00
0
2
4
6
8
10
Adsorbed Fibrinogen (ug/cm2)
Plasma Concentration (% Normal)
1 minute
5 minutes
20 minutes
120 minutes

Figure 8. Adsorption as a function of plasma concentration for fibrinogen on PU-PEO-SO_3 from plasma at various times.

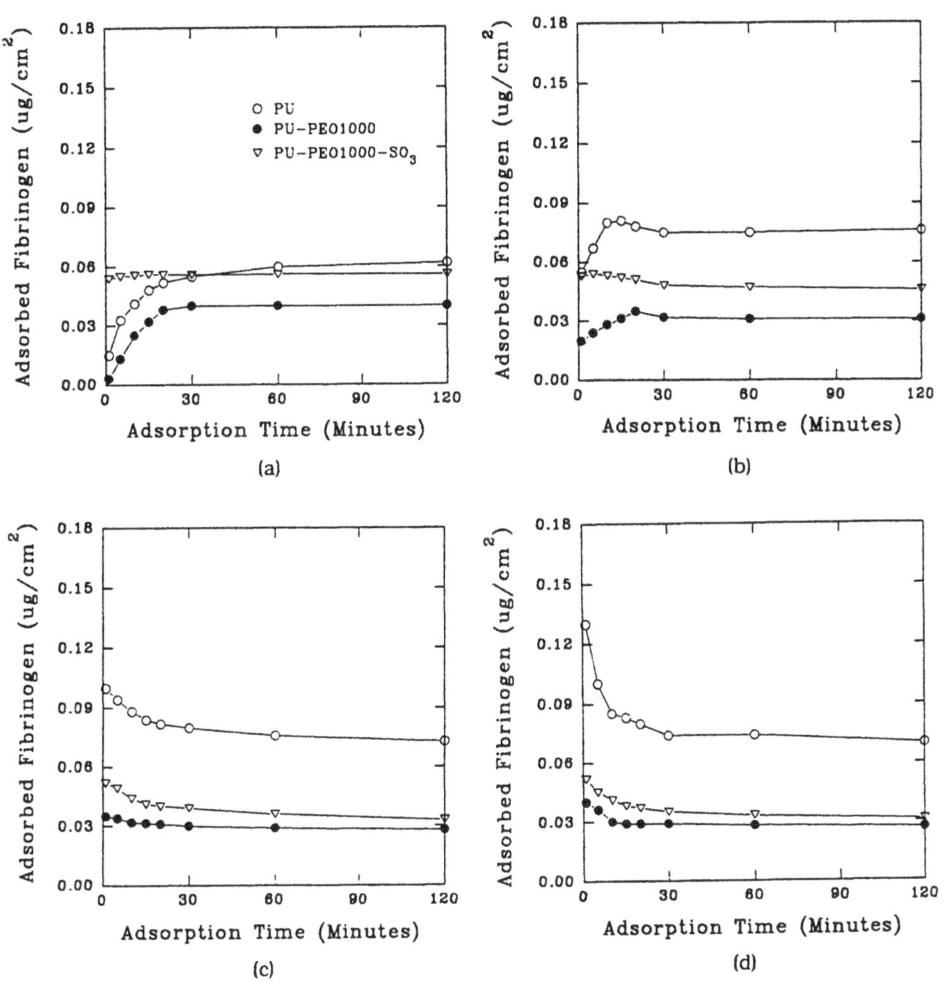

Adsorbed Fibrinogen (ug/cm^2)
Adsorption Time (Minutes)
0.18
0.15
0.12
0.09
0.06
0.03
0.00
0
30
60
90
120
PU
PU-PEO1000
PU-PEO1000-SO3
(a)
(b)
(c)
(d)

Figure 9. Adsorption kinetics for fibrinogen on different surfaces: (a) plasma concentration: 0.2%; (b) 0.6%; (c) 1.0%; (d) 2.0%.

Figure 10 displays the fibrinogen adsorption profiles of adsorbed fibrinogen versus plasma concentrations at various adsorption times, in order to compare three different PU surfaces. It is also evident that at concentrations of low normal plasma, initial fibrinogen adsorption rate is highest in the case of PU-PEO-SO_3. The amount of adsorbed fibrinogen at steady state decreased in the same order (PU > PU-PEO-SO_3 > PU-PEO), regardless of adsorption time.

DISCUSSION

Although a variety of research on protein-surface interaction has been done, the relationship between the protein adsorption and surface properties is not yet conclusive. It was known that many factors play an important role in protein adsorption onto solid surfaces. Among them, of importance are hydrophobic interaction, electrostatic interaction and acceptor-donor interaction [19-21]. It is well agreed

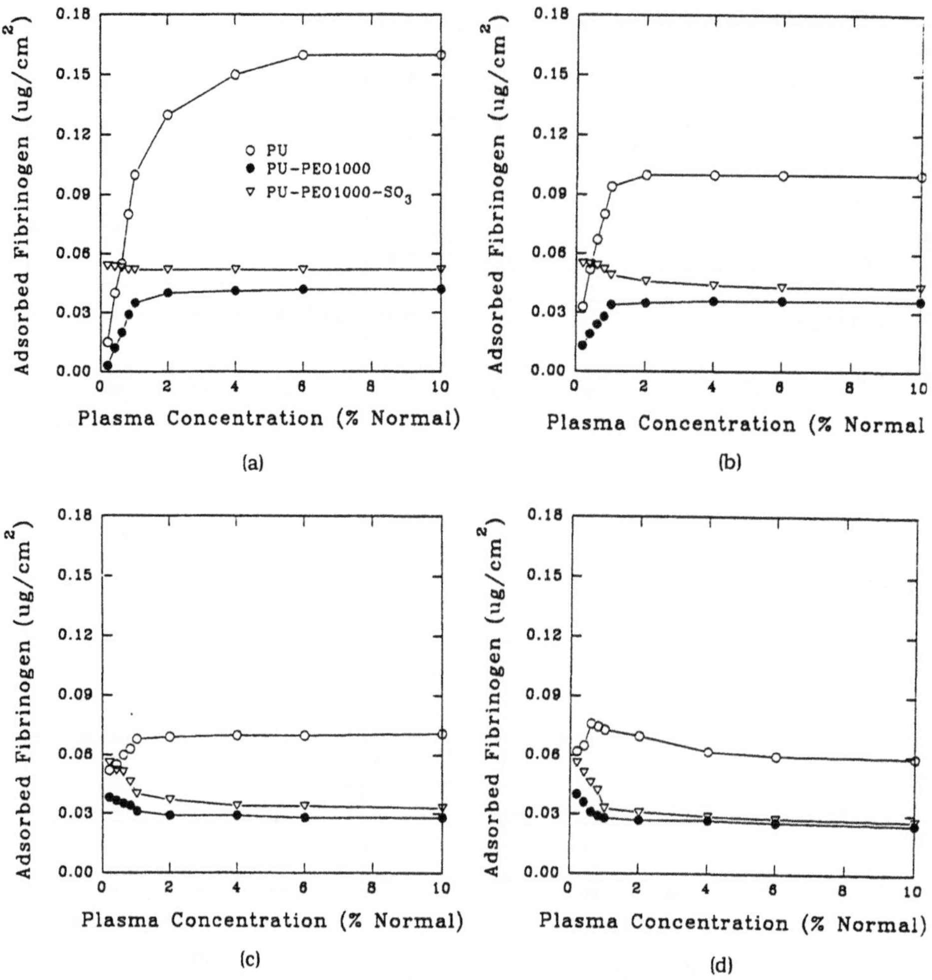

Adsorbed Fibrinogen (ug/cm^2)
Plasma Concentration (% Normal)
PU
PU-PEO1000
PU-PEO1000-SO$_3$
(a)
(b)
(c)
(d)

Figure 10. Adsorptions as a function of plasma concentration for fibrinogen from plasma on different surfaces: (a) adsorption time: 1 min; (b) 5 min; (c) 20 min; (d) 120 min.

that more protein generally adsorbs on hydrophobic surfaces compared to hydrophilic surfaces, which is mainly attributed to hydrophobic interaction between protein and surface.

This is consistent with the results for relatively hydrophobic PU and hydrophilic PU-PEO surfaces. Comparing PU-PEO with PU-PEO-SO_3, it is shown that both hydrophobic interaction and electrostatic interaction contribute to fibrinogen adsorption. Based upon the more hydrophilic, negatively charged surface properties of PU-PEO-SO_3 as compared to PU-PEO, it can be expected that PU-PEO-SO_3 surface adsorbs less fibrinogen than PU-PEO due to the charge repulsion and greater hydrophilicity. In fact, it was observed that PU-PEO adsorbs slightly less fibrinogen than PU-PEO-SO_3, which might be due to the pendant negatively charged SO_3 groups. This result suggests that the negatively charged SO_3 group has a binding affinity toward fibrinogen, which is consistent with other reports [3-5].

Cooper and coworkers [3, 4] studied the biological interaction of ionic polyurethanes. The strategies in their ionic polyurethanes involve the incorporation of sulfonate or carboxylic groups into a polyurethane copolymer. Blood compatibility studies with sulfonated polyurethane demonstrated that as the level of sulfonate incorporation increases, fibrinogen adsorption increases and platelet interaction decreases (i.e., decreased platelet attachment, spreading and activation). Brash's research group also reported the effect of polyurethane sulfonation on fibrinogen adsorption from plasma *in vitro* [5]. It was found that sulfonated polyurethane adsorbs much more fibrinogen than unsulfonated PU and fibrinogen surface concentration increases as sulfonate content increases. Our previous platelet adhesion studies revealed that less platelet adhesion on PU–PEO–SO_3 was observed than on PU–PEO and PU surfaces, although PU–PEO adsorbs less fibrinogen than PU–PEO–SO_3. In addition, PU–PEO–SO_3 surface showed less platelet activation than PU–PEO and PU surfaces [9].

The reason why PU–PEO–SO_3 surface adsorbs much less fibrinogen than PU control can be explained by the use of PEO chain grafted to the PU surface. The protein resistant characteristics of the grafted PEO chains due to the many factors including dynamic motion of PEO chain, overcomes the strong affinity of fibrinogen to negatively charged SO_3 group, resulting in the order of fibrinogen adsorption as described above. To clarify this, PU–SO_3 (without use of PEO chain) surface is presently in progress.

Therefore, resultant fibrinogen adsorption on PU–PEO–SO_3 can be explained by the contribution of counterforces between the low protein binding affinity of PEO chain and high fibrinogen binding affinity of the sulfonate group. As known, surface hydrophilicity is a determinant of protein adsorption and it is also clear that the presence of a surface sulfonate group influences protein–surface interaction.

On the other hand, the significance of the fibrinogen displacement phenomenon (i.e., the Vroman effect) to thrombogenicity is still not conclusive. As postulated by Brash [22], if the critical surface property is inertness toward coagulation, then fibrinogen retention, which probably prevents access of coagulation factors (contact activation) to the surface, may be desirable. However, if less interaction with platelets is the most important property for nonthrombogenicity, rapid and complete displacement of fibrinogen would be desirable since adsorbed fibrinogen has been shown to be highly reactive to platelets [23, 24].

Regardless of which aspect is correct, fibrinogen retention and/or fibrinogen displacement is expected to contribute to the improvement of blood compatibility of PU–PEO–SO_3 surface. PU–PEO–SO_3 showed a more reduced fibrinogen adsorption than PU. Reduced fibrinogen adsorption might be less interactive to platelets. In addition to the fibrinogen retention hypothesis mentioned above (i.e., fibrinogen retention can prevent access of contact activation coagulation factors to the surface), based upon the affinity of sulfonate group toward fibrinogen, it can be suggested that the interaction between the sulfonate group and fibrinogen subsequently provide its conformational change such that the binding site of fibrinogen for glycoprotein IIb/IIIa of platelets is not recognizable, resulting in less platelet interaction [4].

In order to elucidate further the correlation between protein adsorption and blood compatibility in blood–material interaction, the effect of sulfonate density (number of PEO–SO_3 chain) and PEO length on fibrinogen as well as adsorption behaviors of other proteins such as albumin and gamma globulin are under investigation.

CONCLUSIONS

A new surface modification of polyurethane (PU) to improve the blood compatibility has been developed. Fibrinogen adsorption (kinetics and adsorption as a function of plasma concentration) to surface modified PUs (PU–PEO and PU–PEO–SO_3) was evaluated from plasma *in vitro*.

PU and PU–PEO surfaces demonstrated that initial adsorption increases with increasing plasma concentration in kinetic profiles and adsorption time in adsorption profiles as a function of plasma concentration, but after the plateau is reached, its adsorption amount decreases as plasma concentration (0.2–2.0%) and adsorption time (1–120 min) increase, respectively. In contrast, PU–PEO–SO_3 showed that initial adsorption is almost the same regardless of plasma concentration and adsorption time, which is due to the high affinity of the surface sulfonate group to fibrinogen.

All the surfaces indicated the Vroman effect at about 0.6% plasma concentration. However, the displacement by other plasma proteins such as contact coagulation factors was relatively low.

Adsorbed amount of fibrinogen at steady state decreased in the order: PU > PU–PEO–SO_3 > PU–PEO, regardless of adsorption time and plasma concentration. The adsorption behavior of PU–PEO–SO_3 is attributed to both the effect of low binding affinity of PEO chain and high affinity of pendant sulfonate group toward fibrinogen. Therefore, although PU–PEO–SO_3 showed more enhanced fibrinogen adsorption than PU–PEO, it exhibited decreased platelet adhesion and activation, and then excellent blood compatibility.

Acknowledgment

This work was supported by the Korean Ministry of Science and Technology (MOST), Grant BSN 733(1)-4203-6.

REFERENCES

1. S. L. Cooper and N. A. Peppas (Eds), in: *Advances in Chemistry Series*, Vol. 199, ACS Press, Washington, DC (1982).
2. E. F. Leonard, V. T. Truitto and L. Vroman (Eds), in: *Ann. N. Y. Acad. Sci.*, Vol. 516, New York Academy of Science, New York (1987).
3. T. G. Grasel and S. L. Cooper, *J. Biomed. Mater. Res.* **23**, 311 (1989).
4. A. Z. Okkema, X-H. Yu and S. L. Cooper, *Biomaterials* **12**, 3 (1991).
5. J. P. Santerre, P. Hove, N. H. VanderKamp and J. L. Brash, *J. Biomed. Mater. Res.* **26**, 39 (1992).
6. L. Vroman, A. L. Adams, G. C. Fischer and P. C. Munoz, *Blood* **55**, 156 (1980).
7. J. L. Brash and P. Hove, *Thromb. Haemostas.* **51**, 326 (1984).
8. T. A. Horbett, *Thromb. Haemostas.* **51**, 174 (1984).
9. D. K. Han, S. Y. Jeong, Y. H. Kim, B. G. Min and H. I. Cho, *J. Biomed. Mater. Res.* **25**, 561 (1991).
10. Y. Mori, S. Nagaoka, H. Takiuchi, T. Kikuchi, N. Noguchi, H. Tanzawa and Y. Noishiki, *Trans. Am. Soc. Artif. Intern. Organs* **28**, 459 (1982).
11. D. L. Coleman, D. E. Gregonis and J. D. Andrade, *J. Biomed. Mater. Res.* **16**, 381 (1982).
12. E. W. Merrill and E. W. Salzman, *ASAIO J.* **6**, 60 (1983).
13. D. K. Han, K. D. Park, S. Y. Jeong, Y. H. Kim, N. Y. Lee, H. I. Cho and B. G. Min, submitted to *Biomaterials* (1992).
14. D. K. Han, K. D. Park, K.-D. Ahn, S. Y. Jeong and Y. H. Kim, *J. Biomed. Mater. Res.: Appl. Biomater.* **23**(A1), 87 (1989).
15. D. K. Han, S. Y. Jeong, K.-D. Ahn, Y. H. Kim and B. G. Min, *J. Biomater. Sci. Polymer Edn*, in press (1993).

16. D. K. Han, S. Y. Jeong and Y. H. Kim, *J. Biomed. Mater. Res.: Appl. Biomater.* **23**(A2), 211 (1989).
17. L. Vroman and A. L. Adams, *J. Biomed. Mater. Res.* **3**, 43 (1969).
18. D. Dottavio-Martin and J. M. Ravel, *Anal. Biochem.* **87**, 562 (1978).
19. J. D. Andrade (Ed.), *Surface and Interfacial Aspects of Biomedical Polymers: Protein Adsorption.* Plenum Press, New York (1985).
20. W. Norde, *Adv. Colloid Interface Sci.* **25**, 267 (1986).
21. J. D. Andrade and V. Hlady, *Ann. N. Y. Acad. Sci.* **516**, 158 (1987).
22. J. L. Brash, *Ann. N. Y. Acad. Sci.* **516**, 206 (1987).
23. S. J. Whicher and J. L. Brash, *J. Biomed. Mater. Res.* **12**, 181 (1978).
24. J. N. Lindon, G. McNamana, L. Kushner, E. W. Merrill and E. W. Salzman, *Blood* **68**, 355 (1986).

Effects of branching and molecular weight of surface-bound poly(ethylene oxide) on protein rejection

KARIN BERGSTRÖM[1], EVA ÖSTERBERG[1], KRISTER HOLMBERG[1*], ALLAN S. HOFFMAN[2], THOMAS P. SCHUMAN[3], ANTONI KOZLOWSKI[3] and J. MILTON HARRIS[3†]

[1] *Berol Nobel, 444 85 Stenungsund, Sweden*
[2] *Department of Bioengineering, FL-20, University of Washington, Seattle, WA 98149, USA*
[3] *Department of Chemistry, University of Alabama in Huntsville, Huntsville, AL 35899, USA*

Received 6 August 1992; accepted 27 May 1993

Abstract—To understand better the origin of protein rejection observed with surface-bound poly(ethylene oxide) (or PEO), we have measured fibrinogen adsorption for a series of linear and branched, low-molecular-weight PEOs bound to solid polystyrene surfaces. The results show that a dependence on molecular weight is found below 1500 g mol^{-1} for linear PEO. Branched PEOs are less effective at protein rejection than linear PEOs. The branched PEOs have smaller exclusion volumes (from GPC) than the corresponding linear PEOs, consistent with restriction in conformational freedom for the branched compounds. The protein rejection results are interpreted in terms of entropy changes that result upon protein adsorption. In addition, some practical problems in preparation of PEO glycidyl ethers have been clarified, thus making these PEO derivatives more useful for surface modification.

Key words: PEO; surfaces; protein adsorption.

INTRODUCTION

The ability of surface-bound poly(ethylene oxide) (PEO, equivalent to polyethylene glycol or PEG) to reduce protein adsorption continues to be a subject of much interest in the biomaterials community [1–4]. We recently completed a study in which branched (four arms) and linear PEOs were covalently bound to polystyrene (PS) and the resulting reduction in fibrinogen adsorption was measured [5]. Interestingly, little effect of polymer branching was observed. We had reasoned that branched PEOs should be less effective at rejecting protein for two reasons: first, a branched PEO should have fewer degrees of freedom, and thus compression of the polymer upon protein adsorption should not be as unfavorable entropically as for the more randomly oriented linear chain. Secondly, multipoint attachment of the branched chain could also further reduce motion and entropy. As noted, these effects were not observed. Also, no effect of PEO molecular weight (M_w) (from 1500 to 20 000 g mol^{-1}) on the extent of protein adsorption was found, in contrast to several previous works indicating such a dependence [1, 21–24].

---

* Current address: Institute for Surface Chemistry, P.O. Box 5607, 114 86 Stockholm, Sweden.
† To whom correspondence should be addressed.

To elucidate the effects of PEO M_w and branching on protein rejection, we have extended our previous study to include a series of branched and linear PEOs of M_ws as low as 250 g mol^{-1}. Covalent binding of PEOs to the PS surface was accomplished by reaction of PEO glycidyl ethers to aminated PS.

In the course of this work we have identified an epoxide impurity that at least in part accounts for our previous observations of greater than 100% yield in preparation of PEO glycidyl ethers. In addition, ^{1}H NMR studies were conducted that give a better understanding of the structure of the branched PEOs used in this and our earlier study [5].

METHODS AND MATERIALS

Protein adsorption measurement

PEG-coated test tubes were rehydrated overnight at 4°C in CPBS (0.01 M citric acid, 0.01 M sodium phosphate monobasic, 0.12 M sodium chloride, 0.02% sodium azide, pH 7.4). The buffer was replaced with 2 ml fresh, degased CPBS, and the samples were equilibrated for 2 h at 37°C. Human fibrinogen (0.4 mg) was added and mixed by gentle repipetting. After 2 h at 37°C, each sample was rinsed by dilution-displacement with 100 ml CPBS. Two ml of a 0.1 M phosphate buffer, pH 7.0 (containing 1.0 M NaCl and 0.2% Tween 20) containing 0.01 mg ml^{-1} peroxidase-conjugated goat immunoglobulin to human fibrinogen and 0.05 mg ml^{-1} rabbit immunoglobulin to human fibrinogen (Dakopatts, Denmark) was added to each test tube. After 1 h at room temperature the test tubes were rinsed extensively with water. Colour was developed by adding 2 ml of a solution containing 0.67 mg ml^{-1} 1,2-phenylene-diamine dihydrochloride (Dakopatts) in 0.01% hydrogen peroxide (Merck). The reaction was terminated after 5 min with 2 ml of 1 M sulphuric acid and the absorbance read at 490 nm. The color produced is proportional to the amount of adsorbed fibrinogen. Adsorption of fibrinogen on the microplates was measured in the same way but with a total amount of 250 μl of solution per well.

The method described here provides relative amounts of protein adsorbed. We showed in an earlier publication that 100% fibrinogen adsorption on the same polystyrene tubes used here corresponds to fibrinogen adsorption of 100 ng cm^{-2} [5].

PEO immobilization

As described previously [5], the polystyrene (PS) surfaces (test tubes or microtiter plates) were first oxidized by $KMnO_4/H_2SO_4$ to generate charged and polar groups on the surface. Branched poly(ethylene imine) (PEI–Polymin SN-BASF—containing a 1:2:1 ratio of primary to secondary to tertiary amino groups, $M_w = 1.5 \times 10^6$) was then adsorbed to produce a surface containing many primary amine groups (Eqn 1) [5]. PEO glycidyl ethers (**1**), prepared as described previously (Eqn 2) were then coupled to this surface (Eqn 3) [5]

$$\text{PS} \xrightarrow{[O]} \text{PS}(-\text{COOH}, -\text{CHO, etc.}) \xrightarrow{\text{PEI}} \text{PS}-\text{NH}_2 \quad (1)$$

$$\text{PEO}-\text{OH} + \underset{\diagdown\ \text{O}\ \diagup}{\text{H}_2\text{C}-\text{CH}}-\text{CH}_2\text{Cl} \xrightarrow{\text{NaOH}} \text{PEO}-\text{O}-\text{CH}_2-\underset{\diagdown\ \text{O}\ \diagup}{\text{CH}-\text{CH}_2}\ \ \mathbf{1} \quad (2)$$

$$\text{PS}-\text{NH}_2 + \mathbf{1} \longrightarrow \text{PS}-\text{NH}-\text{CH}_2-\text{CHOH}-\text{CH}_2\text{O}-\text{PEO} \quad (3)$$

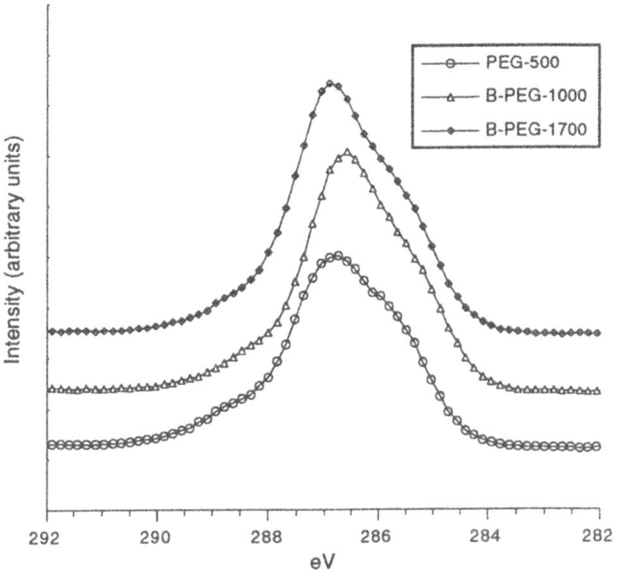
Intensity (arbitrary units)
PEG-500
B-PEG-1000
B-PEG-1700
292
290
288
286
284
282
eV

Figure 1. XPS C1*s* of PEG-500, B-PEG-1000, and B-PEG-1700.

Binding of PEO was confirmed by increase in surface oxygen content by XPS analysis (Surface Science Laboratories SSX-100) as well as by appearance of ether-type carbon in the XPS spectra (Fig. 1). The XPS spectra of the different PEO surfaces are very similar, presumably because the variation in PEO molecular weights is so small. Exhaustive washing of the surfaces with warm water (50°C) failed to remove PEO or alter XPS spectra, thus indicating that the coatings are durable. Binding of PEO to the surfaces was also reflected by the large reduction in fibrinogen adsorption (results below).

PEO molecular weights (M_n) were determined by gel permeation chromatography using ten PEO standards (from 106 to 23 000 g mol^{-1}) for calibration (Polymer Laboratories) (Fig. 2). Polydispersities (M_w/M_n) for the PEOs used in this work were all less than 1.05, so M_w and M_n values are similar. Gel permeation chromatography of PEOs was performed with a 30-cm Waters Ultrahydrogel 250 column using a Biorad pump and a refractive index monitor. Molecular weights were also determined using NMR spectroscopy for measurement of the $-OH/(CH_2CH_2O)$ ratio [9], and from the amount of ethylene oxide consumed relative to the amount of base initiator. NMR spectra were obtained with a Bruker 200 MHz instrument.

PEO molecular weights (M_n) as given in the text are round numbers obtained from ethylene oxide consumption. The values given for the linear PEOs are, within experimental error, the same as those obtained by NMR and GPC. Similarly, M_n values from ethylene oxide consumption and NMR are the same within experimental error. GPC values for the branched PEOs are discussed below.

Impurity identification in PEO glycidyl ethers

As shown in Eqn 2, PEO glycidyl ethers were synthesized by reaction of PEO with epichlorohydrin and solid sodium hydroxide. We have observed with this synthetic method that yields of greater than 100% (from ^{1}H-NMR and epoxide titration) [5]

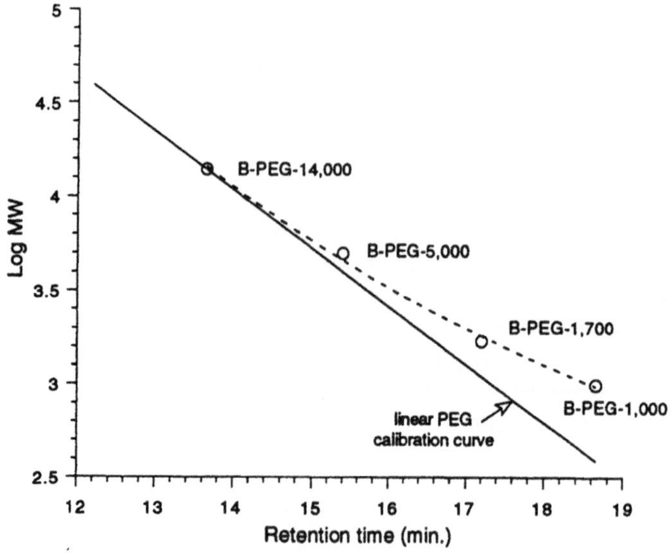
Log MW
5
4.5
4
3.5
3
2.5
12
13
14
15
16
17
18
19
Retention time (min.)
B-PEG-14,000
B-PEG-5,000
B-PEG-1,700
B-PEG-1,000
linear PEG
calibration curve

Figure 2. GPC analysis of branched vs linear PEGs.

are often obtained. During the course of the present work the proton NMR spectra of the glycidyl ethers revealed two small peaks in the region expected for alkenes (multiplet at 6.05 ppm and doublet at 6.49 ppm). Solid extraction on a silica gel column permitted purification of the PEO-glycidyl ether and isolation of the impurity (**2**). Ten grams of silica (Merck grade 60, 230–400 mesh) was used per gram PEO. The first eluent, dichloromethane, removed impurities, and the second eluent, methanol, removed product. After removal of the impurity, the glycidyl ethers were found to be almost exactly 100% pure (by NMR and titration) [5]. A proposed mechanism for formation of **2** is given in the Discussion section.

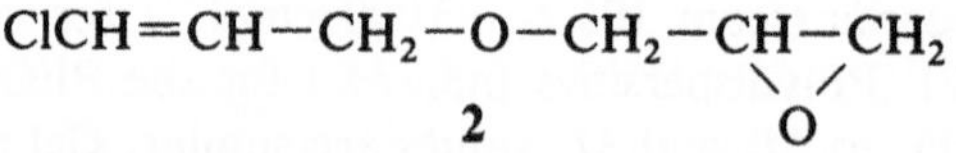
ClCH=CH−CH2−O−CH2−CH−CH2
O

2

Proton NMR of impurity **2** showed the following spectrum in DMSO-d_6: 2.55 ppm (*m*, CH_2 epoxy ring), 2.73 ppm (*m*, CH_2 epoxy ring), 3.08 ppm (*m*, CH epoxy ring), 3.26 ppm (*dd*, CH_2 next to ring), 3.70 ppm (*dd*, CH_2 next to ring), 3.99 ppm (*dd*, CH_2 next to double bond), 6.05 ppm (*m*, =CH−), 6.49 ppm (*d*, =CHCl). The structure of **2** was confirmed by mass spectroscopy: (*m/z*) 113 (M^+-Cl), 91 (M^+-CH_2-epoxide), 75 (M^+-−OCH_2-epoxide), 57 (epoxide-CH_2^+), 49 (CH_2Cl^+), 43 (CH_2OCH^+), 39 ($C_3H_3^+$), 29 (CHO^+). Mass spectroscopy was performed with a Hewlett-Packard GC/MS 5970 spectrometer.

Although the NMR spectrum for impure PEO glycidyl ether showed only a small peak in the alkene region due to **2**, the great difference in M_w between **2** and the glycidyl ether means that this small amount of impurity by weight is a much larger percent on a molar scale. In some cases we have found the molar amount of **2** in unpurified glycidyl ether to be as high as 20%. Also, it is important to note that **2** is chemically active since it contains an epoxide ring, and thus it is important that it be removed from preparations of PEO glycidyl ether to avoid surface contamination.

NMR analysis of branched PEOs

Branched PEOs [5] (B-PEOs **3**) were obtained from Berol Nobel and were prepared by base-catalyzed ethoxylation of **4**. The ^{1}H-NMR spectra of the small, branched PEOs reveal an important practical point regarding their structure. A typical spectrum for a large B-PEO (i.e. $M_w > 5000$) shows the following peaks (DMSO-d_6): 0.80 ppm (t, CH_3-), 1.29 ppm (q, $CH_3-\underline{CH_2}-$), 3.23 ppm (s, $C-CH_2-O$), 3.51 ppm (s, polymer backbone), 4.57 ppm (t, $\overline{-OH}$). The small B-PEOs ($M_w < 1700$) show an additional triplet at 4.19 ppm that can be assigned to an unethoxylated hydroxyl as in structure **5**; the area of this hydroxyl peak plus that of the 4.57 ppm hydroxyl peak always add up to four hydrogens when compared to the PEO backbone and ethyl groups. Compound **4** also has this absorption at 4.19 ppm.

It is apparent from these spectra that the four arms of the smaller B-PEOs are not equivalent in length since at least one arm or branch is not ethoxylated (i.e. structure **5**). The NMR spectra give the following amounts of unethoxylated hydroxyls (from comparison of the 4.19 ppm peak to the 4.57 ppm peak): B-PEO-1700, 18%; B-PEO-1000, 24%; and B-PEO-500, 41%. Presumably, this unethoxylated hydroxyl of **5** is sterically hindered and thus not readily available to ethoxylation; similarly, we have found that it is necessary to use 'forcing' reaction conditions (as described in ref. [5]) to convert these hindered hydroxyls to glycidyl ethers. Also, it is known from studies of alcohol ethoxylation that ethoxylated chains are more reactive toward ethylene oxide than an unethoxylated alcohol group (the chain is more acidic), so there is also a kinetic reason for an arm of **4** remaining unethoxylated despite extensive overall ethoxylation [10].

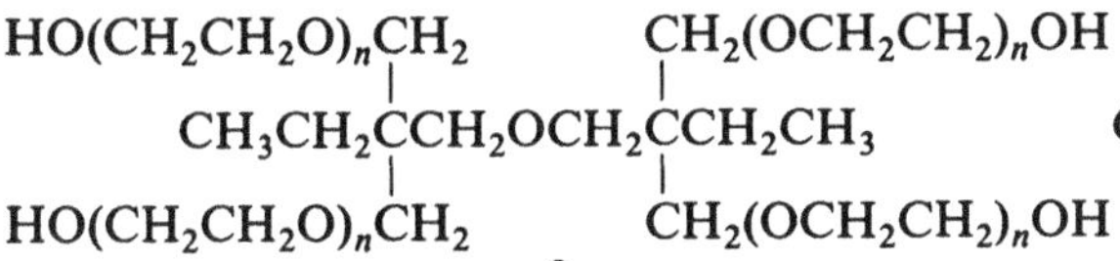

3

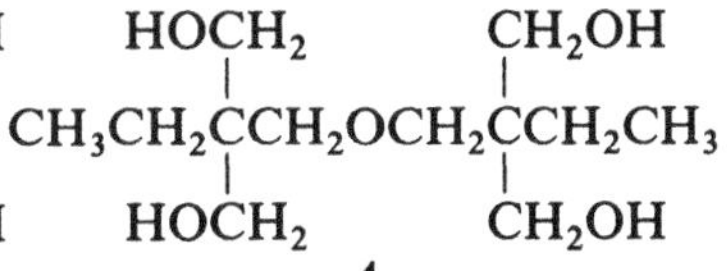

4

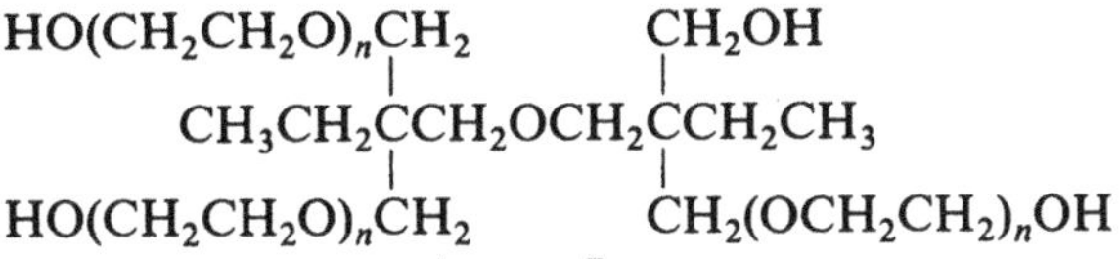

5

RESULTS AND DISCUSSION

Protein adsorption measurements

The results of this study are shown in Table 1. Note that two sets of measurements are listed. The first is the original study with large M_ws [5], and the second is from the present work. The general agreement among these data is reassuring. Note that the ELISA technique for protein adsorption provides relative values rather than absolute amounts of protein adsorbed. We showed in an earlier publication that 100% fibrinogen adsorption under the exact conditions used here corresponds to adsorption of 100 ng cm^{-2} [5].

Table 1.
Effects of linear (L) and branched (B) PEOs bound to polystyrene on fibrinogen adsorption[a]

	Fibrinogen adsorption (%)	
PEO	Previous study [5]	Current study
untreated	100 ± 11	100 ± 6
L-19000	5 ± 1	—
L-4000	3 ± 2	5 ± 2
L-1500	5 ± 1	9 ± 1
L-1000	—	17 ± 3
L-500	—	52 ± 5
L-250	—	100 ± 1
B-6000	6 ± 5	—
B-1700	11 ± 3	14 ± 10
B-1000	—	46 ± 17

[a] B-500 proved to be insoluble in water. All measurements in triplicate. Error limits are standard deviations.

As can be seen from Table 1, reduction in linear PEO M_w below 1500 g mol^{-1} is accompanied by an increase in fibrinogen adsorption; this effect becomes especially dramatic below 1000 g mol^{-1}, with the 250 g mol^{-1} material showing no protein rejecting ability at all. Similarly, the branched 1700 is seen to be marginally less effective than the linear 1500, while the branched 1000 is shown to be dramatically less effective than the linear 1000. Interestingly, other workers have noted that there are some distinct, general differences in properties (other than protein rejection on surfaces) between low M_w and high M_w PEOs [11–13]. In the following discussion, the overall dependence of protein adsorption of PEO M_w is considered, following by examination of the effect of PEO branching.

Several factors could contribute to the observed increase in fibrinogen adsorption for the linear low-M_w PEOs. First, the small PEOs may ineffectively hide the exposed surface from approaching protein; i.e. the PEO layer produced by small PEOs may be so thin that the protein can sense the surface through the PEO layer. Alternatively, PEO packing density could be lower for the small PEGs so that protein is able to find an exposed, unprotected portion of the surface (in future work we will directly measure the effect of PEO packing density on protein adsorption) (Fig. 3). Evidence against this mechanism is that the XPS spectra for all the PEO-coated surfaces, regardless of PEO M_w or branching, are virtually identical (Fig. 1). In every case the ratio of —C—O— to —C—C— carbon in the C1*s* peak is approximately eight (within experimental error). It is noteworthy that these spectra are similar to those obtained for other PEO-coated surfaces in which it is known that the surface PEOs occupy areas less than their solution exclusion volumes [14–17]. While the adsorbed protein may be directly contacting the surface it must, in our view, also be compressing surface-bound PEO (Fig. 2).

A second contributing factor derives from the theory that PEOs reject proteins because of the unfavorable entropy change that results upon compression of these conformationally-random, heavily-hydrated PEO chains by protein adsorption (Fig. 3) [5, 6, 18–21]. Adsorption could either be by interpenetration to provide

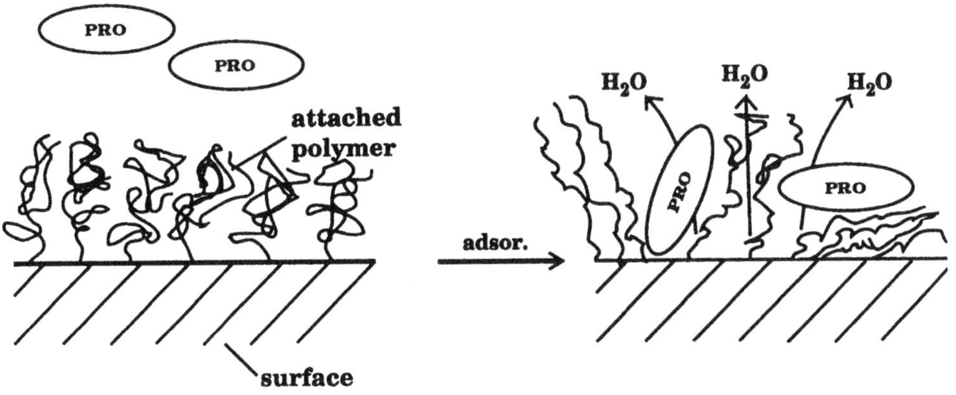
PRO
PRO
attached polymer
surface
adsor.
H_2O
H_2O
H_2O
PRO
PRO

Figure 3. Compression of polymer layer by protein adsorption. A negative entropy change accompanies compression of random, surface-bound polymer. May be opposed by release of water of hydration, but this produces a positive enthalpy change.

contact with the surface or it could be directly on top of the bound PEO layer. Although the chains are randomly oriented, there is considerable ordering of water around these chains, with two to three waters closely associated with each ethylene-oxide unit [11, 12]. The negative entropy change upon PEO compression ($-\Delta S_{\mathrm{PEO\ comp}}$) can be compensated by a positive entropy change associated with loss of waters of hydration from the PEO chain ($-\Delta S_{\mathrm{H_2O\ loss}}$), although there would be a large endothermic enthalpy change opposing this loss ($+\Delta H_{\mathrm{H_2O\ loss}}$). Thus if water is lost, protein adsorption is enthalpically unfavorable, and if water is not lost, protein adsorption is entropically unfavorable; in either case the overall free energy change (ΔG_{PEO}) is positive and unfavorable (Eqn 4). For adsorption to occur, the free energy for interaction of protein with the surface (ΔG_{ads}) must offset ΔG_{PEO}.

$$\Delta G_{\mathrm{PEO}} = +\Delta H_{\mathrm{H_2O\ loss}} - T(\Delta S_{\mathrm{H_2O\ loss}} - \Delta S_{\mathrm{PEO\ comp}}). \tag{4}$$

This thermodynamic explanation of reduction in protein adsorption by the PEO coating can be related to the observed loss of effectiveness with small PEOs. Possibly the $+\Delta G_{\mathrm{PEO}}$ is simply too small in the case of the smaller PEOs to offset the $-\Delta G_{\mathrm{ads}}$.

A third factor has been proposed by Nagaoka *et al.* who presented NMR evidence that small PEO chains are less mobile than the larger ones [6]. These workers propose that PEOs repel proteins because an adsorbed protein does not have sufficient contact time with the mobile PEO chain, and the variation in mobility with M_w can then be seen to be the origin of the M_w effects on protein adsorption.

A fourth possible contributing factor to our observed results derives from the proposal that PEO coatings are effective at protein rejection because the PEO-water interface has a low interfacial free energy [22, 23]. Earlier work by Gombotz *et al.* has shown an expected reduction in water contact angle for PEO-coated surfaces [4]. Possibly the smaller PEOs do not provide an effective lowering of the interfacial free energy.

Now we turn to discussion of the effect of PEO branching. The observed reduction in effectiveness of protein rejection for the small, branched PEOs, relative to the small, linear PEOs (Table 1), lends credence to the theory that PEOs reject proteins because of the unfavorable change in entropy accompanying protein adsorption. As was suggested above, the branched PEOs are more constricted in their motion than the linear PEOs, with the result that the unfavorable entropy change produced by protein adsorption is less than that for the linear polymers. Earlier work indicates that multipoint attachment is not occurring with these branched PEOs [5].

If the branched PEOs are more constricted in their motion than the linear PEOs then one might expect the branched PEOs to have smaller exclusion volumes and to produce thinner layers when surface bound. We reported earlier that branched and linear PEOs fall on the same plot of log M_n vs. retention volume, indicating similar molecular volumes for branched and linear PEOs of the same M_w. However, the small branched-PEO 1700 and 1000, which were not examined previously, do not fit this plot and appear as if their M_ws were 1100 and 400 g mol^{-1}, respectively (Fig. 2). This result shows that placing a branching point in the polymer backbone does restrict conformational randomness and extension of the polymer chains for the small branched PEOs. For the larger branched PEOs this effect is apparently negligible.

Thus we can see that the branched PEOs would, when compressed by protein adsorption, produce a smaller entropy change than that for a corresponding linear PEO. Also since the branched PEO has a lower hydrated volume than a corresponding linear PEO of the same M_w, the branched PEO will produce a thinner surface layer.

Mechanism of formation of impurity 2

Reaction of PEO with epichlorohydrin and solid sodium hydroxide, analogous to reaction of other alcohols with epichlorohydrin, proceeds by addition and dehydrochlorination (Eqns 5 and 6).

$$\text{PEG}-\text{OH} + \underset{\text{O}}{\text{H}_2\text{C}-\text{CH}}-\text{CH}_2\text{Cl} \longrightarrow \underset{\mathbf{6}}{\text{PEG}-\text{OCH}_2-\underset{|}{\overset{}{\text{CH}}}}\!\!\!\!\!\!\underset{\text{OH}}{}-\text{CH}_2\text{Cl} \qquad (5)$$

$$\mathbf{6} + \text{NaOH} \longrightarrow \underset{\mathbf{1}}{\text{PEG}-\text{O}-\text{CH}_2-\underset{\text{O}}{\text{CH}-\text{CH}_2}} + \text{NaCl} + \text{H}_2\text{O}. \qquad (6)$$

It is important to note, however, that PEO is not acting as a simple alcohol in this two-phase reaction between solid sodium hydroxide and liquid PEO and epichlorohydrin. Typically, alcohols react efficiently under these conditions only if a phase transfer agent (such as tetralkylammonium salt or crown ether) is added [24–26], and presumably PEO is acting here as a phase transfer agent (a property for which PEO is well known [27, 28]) as well as a reactant.

In addition to the desired reactions 5 and 6, it has long been known that epichlorohydrin is itself an excellent dehydrochlorinating agent, so that certain side reactions can occur during glycidyl ether formation. In these side reactions the initial reaction intermediate **6** is dehydrochlorinated by excess epichlorohydrin to produce by-products such as **7** and **8** [24–26].

$$\underset{\mathbf{7}}{\text{ClCH}_2-\text{CH(OH)}-\text{CH}_2\text{Cl}} \qquad \underset{\mathbf{8}}{\text{ClCH}_2-\text{CH(Cl)}-\text{CH}_2\text{O}-\text{CH}_2-\underset{\text{O}}{\text{CH}-\text{CH}_2}}$$

We suggest that the troublesome impurity **2**, formed during preparation of **1** (see Methods and Materials), is derived from the previously undocumented dehydrochlorination reaction of Eqn 7. Product **9**, which might have been expected, is not found.

$$8 + H_2C-CH-CH_2Cl \longrightarrow ClCH=CH-CH_2-O-CH_2-CH-CH_2 \quad (7)$$
$$\underset{O}{\diagdown \diagup} \qquad\qquad\qquad\qquad\qquad\qquad \mathbf{2} \qquad\qquad \underset{O}{\diagdown \diagup}$$

$$+ \; CH_2=\underset{\underset{\mathbf{9}}{Cl}}{\overset{|}{C}}-CH_2O-CH_2-CH-CH_2$$
$$\underset{O}{\diagdown \diagup}$$

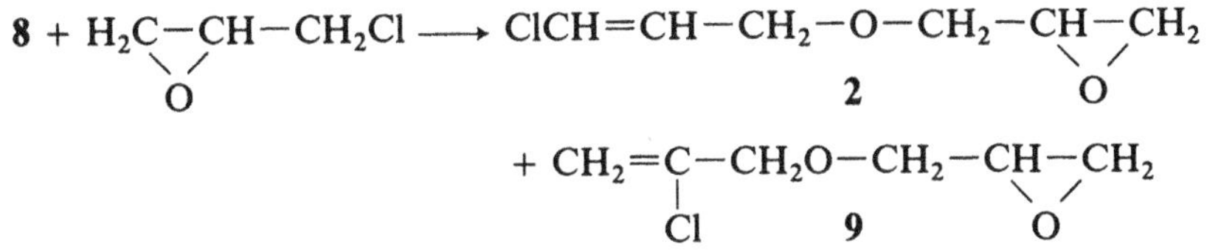

CONCLUSIONS

To summarize, the present results show that large linear and branched PEOs are effective at preventing fibrinogen adsorption on polystyrene surfaces. Small linear PEOs are ineffective, and small branched PEOs are even less effective than linear PEOs of similar M_w. The branched PEOs are constricted in their conformational randomness and they have smaller exclusion volumes than corresponding linear PEOs; this effect is clearly evident below molecular weight of $1700\,g\,mol^{-1}$. In our view these results support the entropic theory of reduction in protein adsorption (Fig. 3).

Acknowledgement

The authors gratefully acknowledge the financial support of this work by the National Institutes of Health (GM40111).

REFERENCES

1. J. M. Harris (Ed). *Poly(Ethylene Glycol) Chemistry, Biotechnical and Biomedical Applications.* Plenum, New York (1992).
2. N. P. Desai and J. A. Hubbell, *Macromolecules* **25**, 226 (1992).
3. D. W. Grainger, T. Okano and S. W. Kim, *J. Colloid Interface Sci.* **132**, 161 (1989).
4. W. R. Gombotz, W. Guanghui, T. A. Horbett and A. S. Hoffman, *J. Biomed. Mater. Res.* **25**, 1547 (1991).
5. K. Bergström, K. Holmberg, A. Safranj, A. S. Hoffman, M. J. Edgell, B. A. Hovanes and J. M. Harris, *J. Biomed. Maters. Res.* **26**, 779 (1992).
6. S. Nagaoka, Y. Mori, H. Takiuchi, K. Yokota, H. Tanzawa and S. Nishiumi, in: *Polymers as Biomaterials*, p. 361, S. W. Shalaby, A. S. Hoffman, B. D. Ratner and T. A. Horbett (Eds). Plenum Press, New York (1985).
7. Y. Sun, A. S. Hoffman and W. R. Gombotz, *ACS Polymer Prepr.* **28**, 292 (1987).
8. V. Hlady, R. A. Van Wagenen and J. D. Andrade, in: *Surface and Interfacial Aspects of Biomedical Polymers*, Vol. 2, Ch. 2, J. D. Andrade (Ed.). Plenum, New York (1985).
9. J. M. Dust, Z.-H. Fang and J. M. Harris, *Macromolecules* **23**, 3742 (1990).
10. B. Weibull, in: *Grenzflächenactive Äthylenoxid-Addukte*, Ch. 2, N. Schönfeldt (Ed.). Wissenschaftliche Verlagsgessellschaft, Stuttgart, Germany (1984).
11. K. P. Antonsen and A. S. Hoffman, in: *Poly(Ethylene Glycol) Chemistry. Biotechnical and Biomedical Applications*, Ch. 2, J. M. Harris (Ed.), Plenum, New York (1992).
12. N. B. Graham, In: *Poly(Ethylene Glycol) Chemistry. Biotechnical and Biomedical Applications*, Ch. 17, J. M. Harris (Ed.). Plenum, New York (1992).
13. T. Arakawa and S. N. Timasheff, *Biochemistry* **24**, 6756 (1985).
14. E. Kiss, C.-G. Gölander and J. C. Eriksson, *Prog. Colloid Polym. Sci.* **74**, 113 (1987).
15. J. C. Eriksson, C.-G. Gölander, A. Baszkin and L. Ter-Minassian-Saraga, *J. Colloid Interface Sci.* **100**, 1 (1984).

16. C.-G. Gölander and J. C. Eriksson, *J. Colloid Interface Sci.* **119**, 38 (1987).
17. B. J. Herren, S. G. Shafer, J. M. Van Alstine, J. M. Harris and R. S. Snyder, *J. Colloid Interface Sci.* **115**, 46 (1987).
18. S. I. Jeon, J. H. Lee, J. D. Andrade and P. G. de Gennes, *J. Colloid Interface Sci.* **142**, 149 (1991).
19. Y. Mori, S. Nagaoka, H. Takiuchi, T. Kikuchi, N. Noguchi, H. Tanzawa and Y. Noishiki, *Trans. Am. Soc. Artif. Internal Organs* **28**, 459 (1982).
20. C.-G. Gölander, J. Herron, K. Lim, P. Claesson, P. Stenius and J. D. Andrade, in: *Poly(Ethylene Glycol) Chemistry. Biotechnical and Biomedical Applications*, Ch. 15, J. M. Harris (Ed.). Plenum, New York (1992).
21. W. R. Gombotz, W. Guanghui, T. A. Horbett and A. S. Hoffman, in: *Poly(Ethylene Glycol) Chemistry. Biotechnical and Biomedical Applications*, Ch. 16, J. M. Harris (Ed.). Plenum, New York (1992).
22. J. H. Lee, P. Kopeckova, J. Kopecek and J. D. Andrade, *Biomaterials* **11**, 455 (1990).
23. H. Elwing, S. Welin, A. Askendal, U. Nilsson and I. Lundström, *J. Colloid Interfaces Sci.* **119**, 203 (1992).
24. W. Bradley, J. Forrest and O. Stephenson, *J. Chem. Soc.* 1589 (1951).
25. J. Buddrus and W. Kimpenhaus, *Chem. Ber.* **106**, 1648 (1973).
26. M. Kamel, W. Kimpenhaus and J. Buddrus, *Chem. Ber.* **109**, 2351 (1976).
27. J. M. Harris, N. H. Hundley, T. G. Shannon and E. C. Struck, in: *Crown Ethers and Phase Transfer Catalysis in Polymer Science*, p. 371, L. Mathias and C. E. Carreher (Eds.). Plenum, New York (1984).
28. Y. Kimura and S. L. Regen, *J. Org. Chem.* **47**, 2493 (1982).

Review
Formation of proteinmultilayers and their competitive replacement based on self-assembled biotinylated phospholipids

W. MÜLLER[1], H. RINGSDORF[1*], E. RUMP[1], X. ZHANG[1], L. ANGERMAIER[2], W. KNOLL[2] and J. SPINKE[2]

[1]*Institute of Organic Chemistry, Johannes Gutenberg-University, J.J. Becher Weg 18-20, 55099 Mainz, Germany*
[2]*MPI for Polymer Chemistry, Ackermannweg 10, 55128 Mainz, Germany*

Received 23 September 1993; accepted 28 January 1994

Abstract—Based on specific recognition processes the build-up of protein multilayers was achieved using streptavidin layers as a docking matrix. For this purpose, streptavidin was organized at biotin-containing monolayers, liposomes, and self-assembled layers on gold. Thus, mixed double and triple layers of streptavidin, Con A, Fab fragments, and hormones were prepared and characterized by fluorescence microscopy and plasmon spectroscopy. Using biotin analogues with lower binding constants several cycles of multilayer formation followed by competitive replacement could be achieved.

Key words: Bioreactive surfaces; streptavidin matrix; protein mono- and multilayers; molecular recognition; competitive replacement.

INTRODUCTION

The classic pathway of the immune cascade [1] nicely shows the perfect interplay of molecular interaction and self-organization to build up oriented and highly specific protein structures. After activation of the immune cascade by an antigen, a number of specific protein recognition processes take place, leading to a complex of proteins. This finally leads to the lysis of the antigen carrying cell. In order to avoid the affection of cells of the host, the stepwise build up of this complex via docking of proteins onto proteins has to be reversible. 'Mother Nature' insures this by different controlling mechanisms like competitive replacement. Looking at this fascinating molecular architecture of the system one may ask, if—after transfer to model systems—it is possible to use this principle of repeated docking of proteins onto other proteins for the build up and variation of bioreactive surfaces. In the following review it will be discussed, how—based on the streptavidin/biotinlipid system—such a model system can be developed and studied at the air water interface or on solid supports.

2-D CRYSTALLIZATION OF STREPTAVIDIN ON BIOTIN-LIPID MONOLAYERS

The structurally analogous proteins streptavidin (M_w = 68 kDa) and avidin (M_w = 60 kDa) have become a model system for biotechnological related research

---

* To whom correspondence should be addressed.

and fundamental protein/ligand binding studies [2–4]. Both are water-soluble proteins with four identical subunits, each having a specific binding site for biotin.

A unique feature of the recognition reaction of streptavidin/avidin with biotin is the extremely high binding constant of $10^{-15}\,M^{-1}$. Other advantages of the system are that the protein is very stable and that biotin can be functionalized on its free carboxylic group without impairing its binding properties. Thus, biotin can be introduced in the hydrophilic headgroup of lipids to investigate the recognition between avidin/streptavidin and biotin in model membrane systems e.g., monolayers and liposomes [5, 6].

The specific recognition process occurring at the gas–water interface was examined by fluorescence microscopy [5, 6]. Combination of a Langmuir balance with a fluorescence microscope allows processes in or at the monolayer to be observed by using fluorescent dyes as probes. It is advantageous when investigating the interaction of a protein with a lipid monolayer to be able to observe the lipid layer and the protein selectively. This can be achieved by using a suitable filter system and two fluorescence dyes whose absorption and emission spectra do not overlap (e.g., sulforhodamine: $\lambda_{Ex} = 584$ nm, $\lambda_{Em} = 607$ nm and fluorescein: $\lambda_{Ex} = 497$ nm, $\lambda_{Em} = 521$ nm). One of the two dyes is mixed with the lipid monolayer while the other one is covalently bound to the protein. Thus for example it is possible by using the sulforhodamine filter to visualize the changes in the morphology of the lipid layer doped with sulforhodamine and to observe only the protein, being labelled with fluorescein, by use of the fluorescein filter.

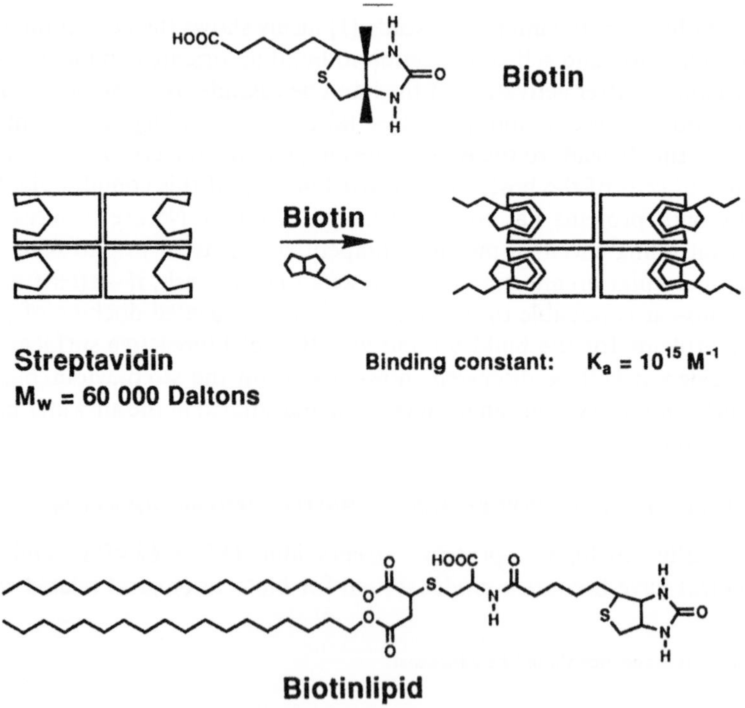
HOOC
Biotin
Biotin
Streptavidin
M$_w$ = 60 000 Daltons
Binding constant: K$_a$ = 10^{15} M^{-1}
HOOC
Biotinlipid

Figure 1. Schematic representation of the binding of biotin (vitamin H) to the tetrameric proteins avidin and streptavidin.

If fluorescence-labelled streptavidin is injected under a monolayer of biotinlipids in the gas analogous state the formation of regularly shaped streptavidin domains can be observed after a certain time (Fig. 1) [5–7]. These domains show a strong optical anisotropy, if linearly polarized light is used for the excitation of fluorescence. This indicates a highly ordered arrangement and a low mobility of the streptavidin molecules. Control experiments with biotin-saturated streptavidin and with lipid layers incapable of binding (biotin-free e.g., DPPC) produced no domains in any instance. Thus the formation of these domains is a consequence of the specific recognition of biotin by streptavidin.

The structural investigation of the streptavidin domains with electron crystallography have shown them to be 2-D single crystals. These 2-D crystals could be used to determine the 3-D structure of the streptavidin bound to the biotin lipid layer with a resolution of 15 Å [8]. Hendrickson *et al.* determined the 3-D structure of the protein, including the biotin binding sites by X-ray structural analysis with a resolution of 3 Å [9]. When the structure found by Hendrickson and that obtained from the 2-D crystals are superimposed, there is found to be a very good agreement. This permits the position of the biotin binding sites to be transferred to the reconstruction of the streptavidin molecule obtained from the 2-D crystals [6]: two of its four binding sites are bound to the biotinlipid monolayer, while the two remaining binding sites are facing the subphase [8, 9]. These results allow to visualize the orientation of the streptavidin molecules bound to the biotin-containing monolayer.

There are two points worthy of note here: the distance between the binding site and the monolayer corresponds to the length of the spacer (see Fig. 1) of the biotin lipid used. This favours the rapid interaction of the protein with this biotin lipid. In addition, each streptavidin molecule contains two other free binding sites for biotin which point into the aqueous subphase. Thus in principle every biotinylated molecule can be docked onto the highly ordered streptavidin matrix.

A number of docking experiments at this protein matrix have proved these free binding sites to be available for biotin-containing molecules dissolved in the subphase. This opens the possibility of the above-mentioned build-up and variation of bioreactive surfaces based on the streptavidin matrix.

STREPTAVIDIN AS A PROTEIN MATRIX FOR THE BUILD-UP OF BIOREACTIVE SURFACES

The highly specific and extremely strong binding of biotin to avidin/streptavidin has found application in affinity chromatography for some time now [10]. There are also a number of immunoassays which utilize this protein-ligand interaction. However, no use has yet been made of the properties of highly ordered monomolecular protein layers which can be prepared with streptavidin at biotin-lipid monolayers. As shown in Fig. 5, a number of possible applications, mainly based on the interplay between molecular recognition and molecular selforganization, can be realized on the basis of the streptavidin matrix.

Bifunctional biotin derivatives can, for example, serve for the build-up of very thin and well defined streptavidin multilayer structures. The primary streptavidin layer can also be used as a highly ordered, protein-compatible matrix for the docking of other biotinylated proteins (as e.g., enzymes, antibodies) or nucleic acids (RNA, DNA) [11–15]. Points of interest are the function of the proteins being docked onto

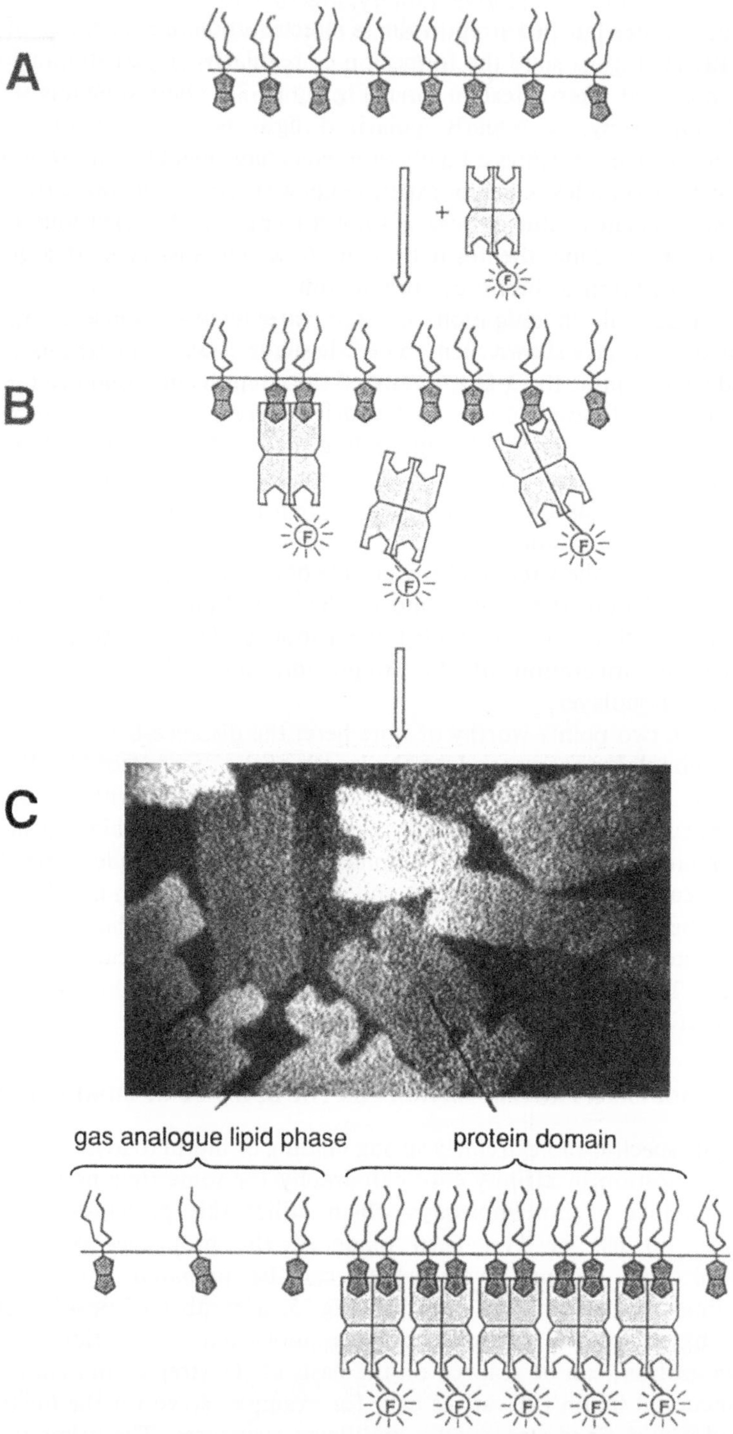
A
B
C
gas analogue lipid phase
protein domain

Figure 2. Diagram representing the binding of streptavidin to monolayers of biotinlipids: the fluorescent labelled streptavidin dissolved in the subphase binds specifically to biotin on the membrane (A, B); formation of streptavidin domains can be observed under the fluorescence microscope (bright areas through the fluorescein filter; (C) $P = 0\,\mathrm{mN\,m^{-1}}$ (10 mm correspond to 23 μm)).

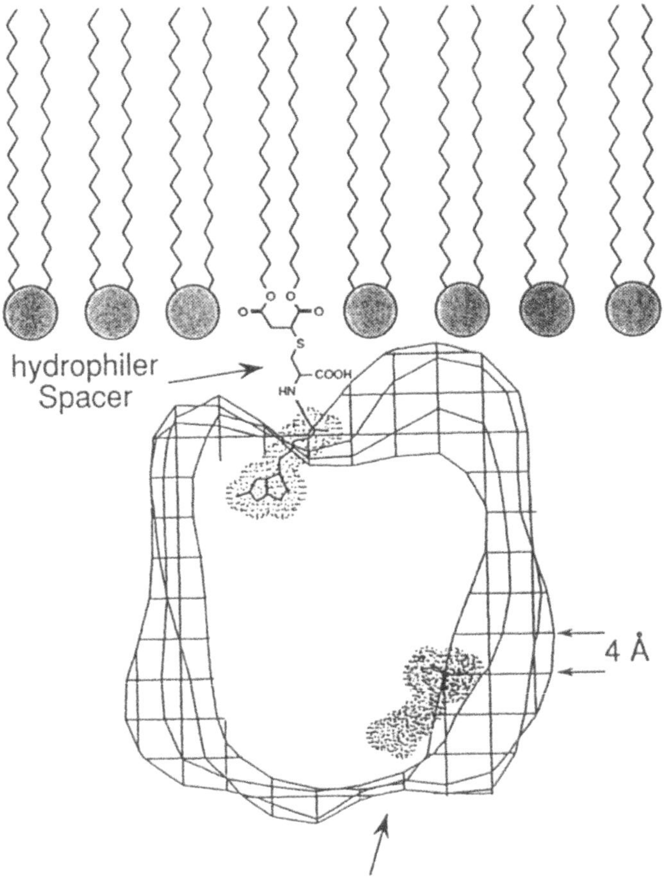
hydrophiler
Spacer
COOH
HN
S
4 Å

Figure 3. Section through a monolayer containing biotinlipids with bound streptavidin: the distance of the upper biotin binding site from the monolayer approximately corresponds to the length of the spacer in the biotinlipid; the lower free binding site is accessible from the subphase (arrow).

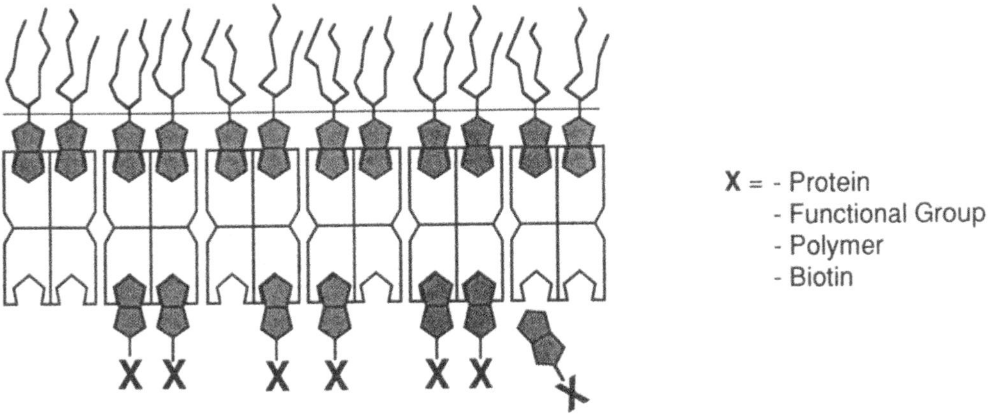
X = - Protein
- Functional Group
- Polymer
- Biotin

Figure 4. 2-D crystalline streptavidin layer functionalized by binding of biotinylated molecules X to free biotin binding sites: X = proteins, polymers, and functional groups like e.g., biotin (bis-biotin compound) or sugar.

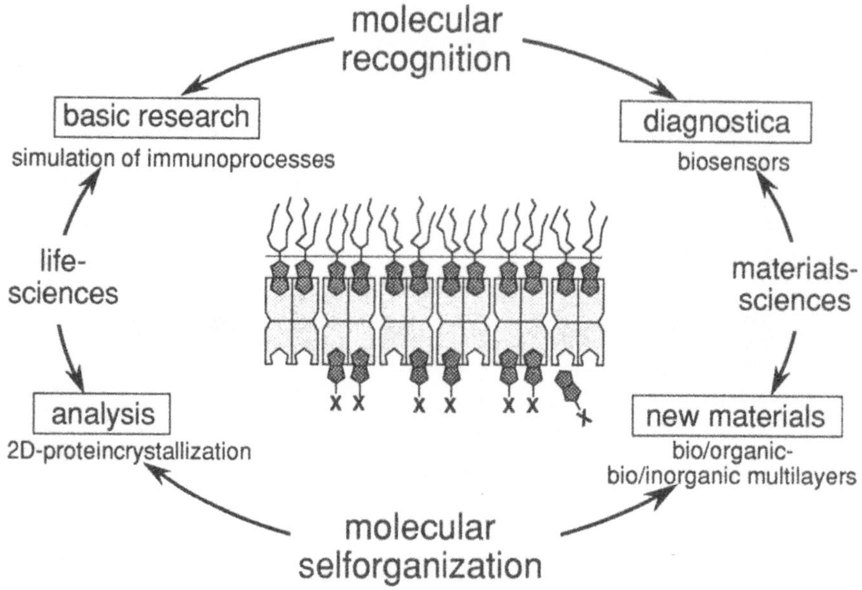
molecular
recognition
basic research
simulation of immunoprocesses
diagnostica
biosensors
life-
sciences
materials-
sciences
analysis
2D-proteincrystallization
new materials
bio/organic-
bio/inorganic multilayers
molecular
selforganization

Figure 5. Bioreactive surfaces based on streptavidin matrices open the possibility to investigate protein/protein interactions and to build up biosensors as well as bio/inorganic multilayers.

the matrix (simulation of immune processes, biosensors) on the one hand, on the other hand the tendency of these proteins to crystallize (2-D crystallization). As shown in Fig. 5 even mixed organic/inorganic multilayers are possible, for example, by use of biotinylated ligands which are able to complex inorganic salts.

A large number of bifunctional ligands (Fig. 6) on one side and variation of the primary protein matrices on the other side open many different possibilities to build up defined protein multilayers. In Fig. 7 two principal examples are given.

Using e.g., a biotin/sugar linker (see Fig. 6) combined streptavidin/lectin layers can be obtained (Fig. 7a); use of bisbiotin linkers allows us to obtain pure streptavidin or streptavidin/avidin multilayers (Fig. 7b) [16]. The influence of the primary protein matrix is also of some interest. The lectin concanavalin A (Con A), docked as a second, non-crystalline layer onto a crystalline streptavidin matrix, only traces the structures of the primary layer (Fig. 7a). If streptavidin is bound via a bisbiotin linker to an amorphous avidin layer as primary matrix 2-D crystalline streptavidin domains are obtained (Fig. 7b). In this case the high crystallization tendency of streptavidin leads to a structure in the second layer, which is independent of the one of the primary matrix. These two examples are demonstrated in Figs 8 and 9.

If Con A is docked onto a streptavidin layer as a primary matrix the domain structure of the streptavidin layer is traced by the Con A layer. To carry out this experiment, the Con A, labelled with sulforhodamine, was first saturated with four equivalents of the biotin sugar linker and then injected underneath a 2-D streptavidin matrix at the air–water interface. The fluorescence labelled streptavidin crystals can be observed before addition of Con A in the fluorescein filter mode. After docking of the sulforhodamine-labelled Con A, the same domains can now be seen in the sulforhodamine filter. The docking process does not influence the structure of the streptavidin crystals in a noticeable way [16].

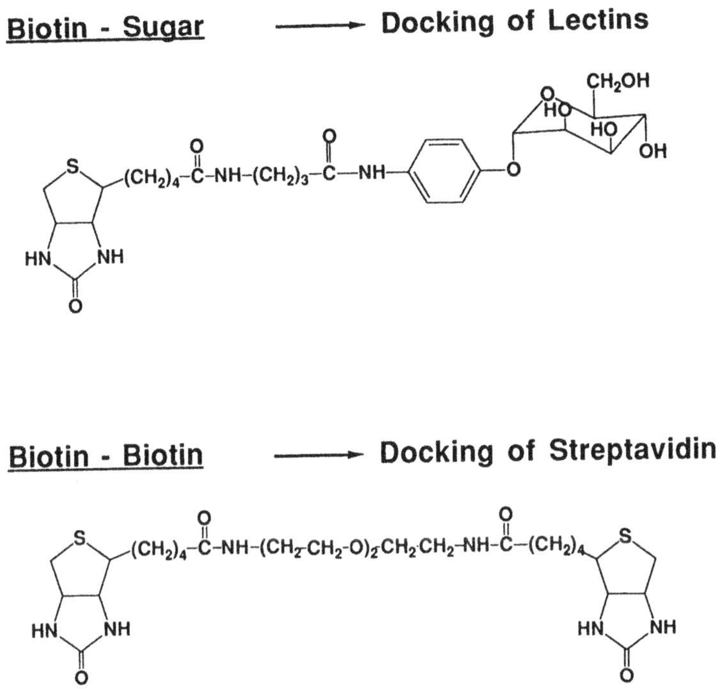
Biotin - Sugar
Docking of Lectins
Biotin - Biotin
Docking of Streptavidin

Figure 6. Examples for biotin/sugar and bisbiotin derivatives to dock the lectin concanavalin A or streptavidin respectively onto a primary streptavidin matrix.

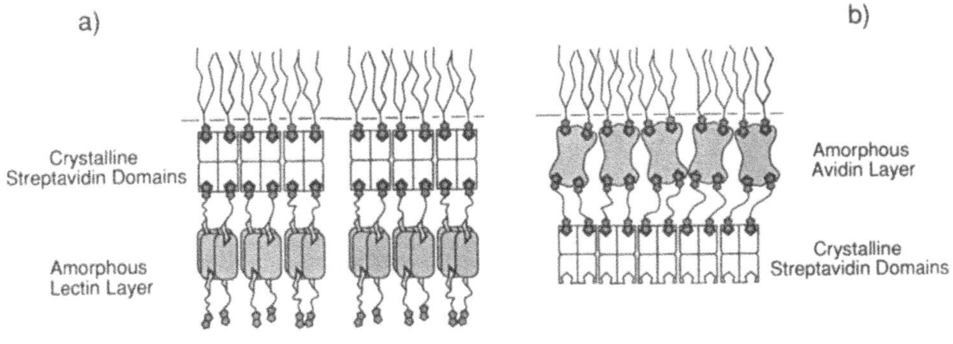
a)
b)
Crystalline
Streptavidin Domains
Amorphous
Lectin Layer
Amorphous
Avidin Layer
Crystalline
Streptavidin Domains

Figure 7. Influence of the structure of the primary protein matrix on the second protein layer: (a) Con A is only tracing the crystalline structure of the streptavidin matrix, when it is docked as second protein layer onto it; (b) streptavidin forms 2-D crystalline domains underneath a primary amorphous avidin matrix.

If streptavidin (FITC-labelled) is bound via a bisbiotin linker (see Fig. 6) to an amorphous avidin matrix the different structure of the primary avidin layer does not suppress the crystallization of the streptavidin layer (see Fig. 7). The fluorescence micrographs and a scheme of the process are shown in Fig. 9. The appearing, brightly fluorescing streptavidin crystals can clearly be seen [16].

The build up of protein double layers based on the streptavidin matrix can perfectly be demonstrated with a quartz crystal micro-balance (QCM) [12, 15]. The resonance frequency of the quartz crystal of the QCM is sensitive to the mass bound

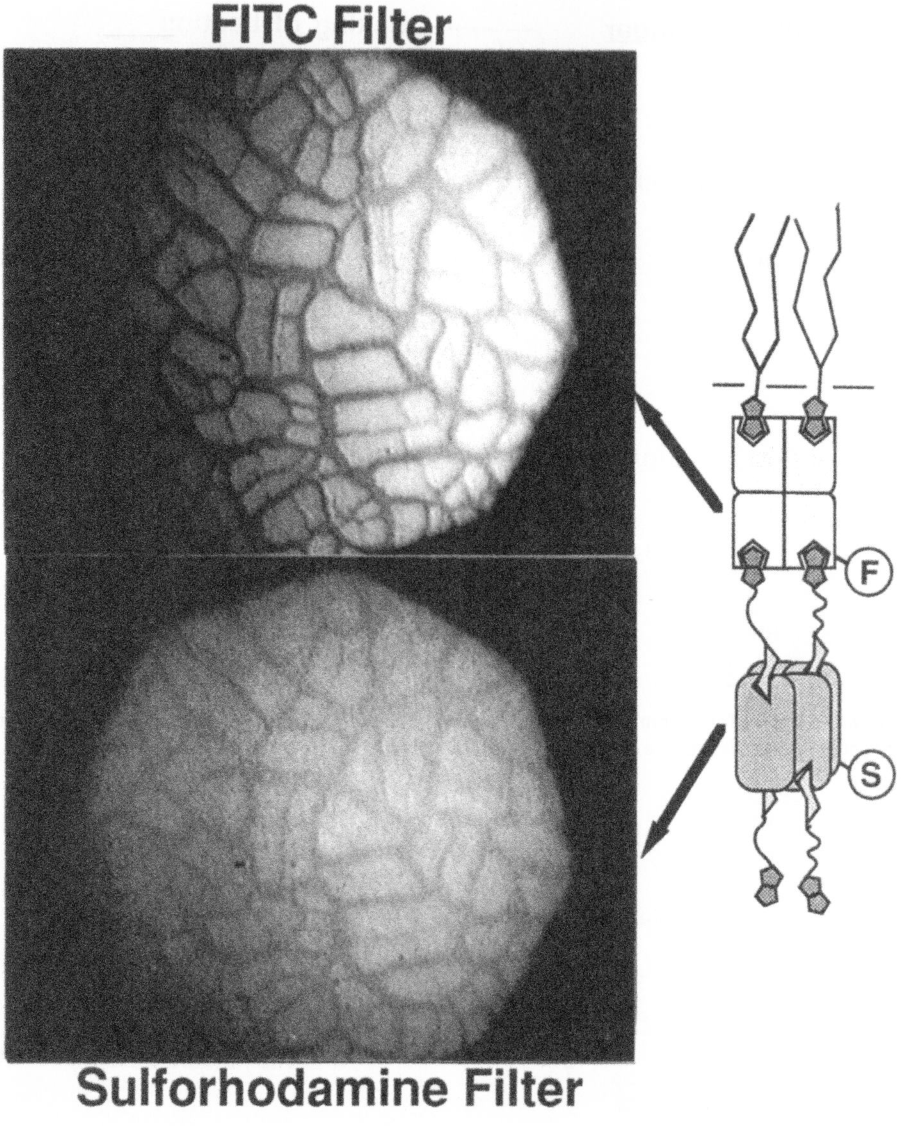
FITC Filter
F
S
Sulforhodamine Filter

Figure 8. Fluorescence micrograph of a combined streptavidin/concanavalin A double layer. Con A (sulforhodamine labelled) is functionalized with a biotin/mannose linker and docked as a second protein layer onto a primary streptavidin matrix (FITC-labelled) (10 mm correspond to 20 μm).

to the surface of the gold electrode. Thus, in principle, the QCM technique allows one to quantify the binding process of proteins to various membranes. Furthermore the course of the interaction with time can be followed in real time. DPPE/biotinlipid mixtures were deposited on the gold electrode of a QCM by means of the Langmuir–Blodgett (LB) technique. The quartz crystals (9 MHz) were immersed in a buffer solution (50 mmol phosphate buffer, pH 7.5) and streptavidin was added. Experiments with different concentrations of streptavidin and biotinlipid allowed to fix the optimum concentrations to obtain a densely-packed streptavidin layer as a template for docking of a second functional protein layer.

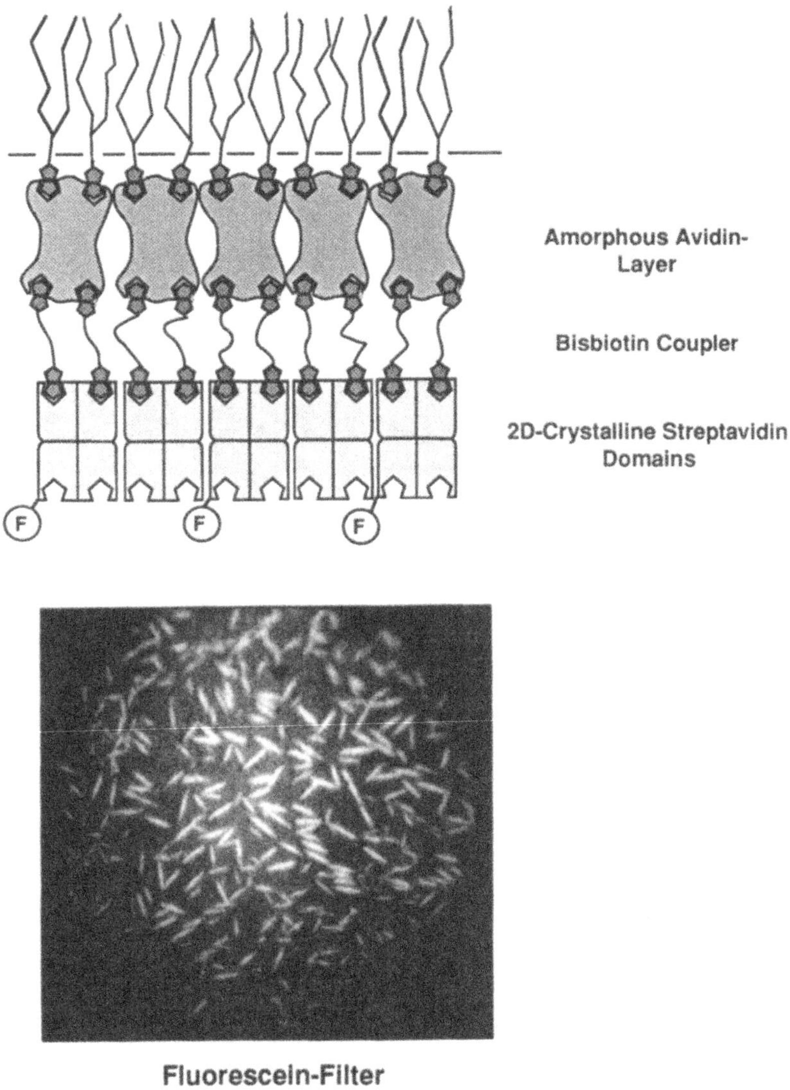
Amorphous Avidin-Layer
Bisbiotin Coupler
2D-Crystalline Streptavidin Domains
F
F
F
Fluorescein-Filter

Figure 9. Fluorescence micrograph of a protein double layer: Streptavidin is docked as a second layer onto a primary avidin matrix by use of a bisbiotin linker. Streptavidin forms 2-D crystalline domains (bright needles) underneath the amorphous avidin layer (10 mm correspond to 21.5 μm).

Anti-fluorescyl-antibody fragments (Fab fragments) were specifically biotinylated at their hinge region to prepare a streptavidin/Fab-fragment double layer on the quartz support.

A mixed biotinlipid/DPPE layer (5 mol% of biotinlipid) on the QCM electrode was incubated with streptavidin (5×10^{-7} M) up to saturation, the quartz was rinsed and transferred into a fresh buffer solution. Then the biotinylated Fab was added to the buffer resulting in a slow but significant frequency change, with a saturation frequency change indicating a densely packed Fab-layer (Fig. 11). This Fab layer can serve as a matrix for further binding of fluorescein-containing molecules (see Fig. 10).

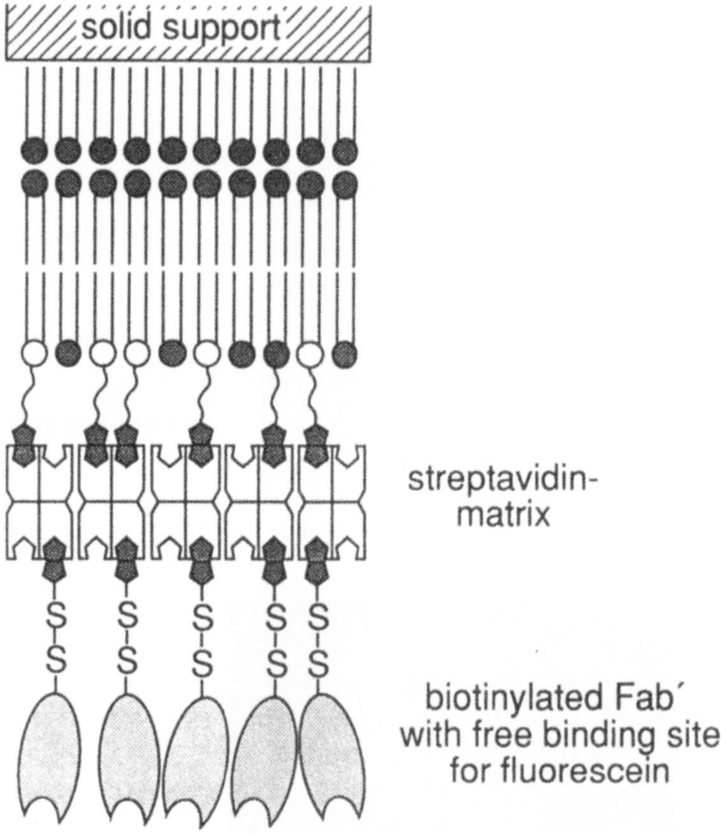
solid support
streptavidin-matrix
biotinylated Fab′
with free binding site
for fluorescein

Figure 10. Schematic representation of the docking of a Fab fragment which was specifically biotinylated in the hinge region to a two-dimensional streptavidin layer. The Fab layer can serve as a matrix for further binding of fluorescein containing molecules.

Besides the build-up of these protein double layers the same principle can be used to obtain triple or, in general, multilayers. The schematic of the build up of a triple layer consisting of streptavidin and Con A and the corresponding fluorescence micrographs is shown in Fig. 12. In this case the primary streptavidin layer was not fluorescently labelled. The tetra-biotinylated Con A was labelled with sulforhodamine. Injection of this tetra-functional Con A underneath the primary streptavidin matrix followed by further addition of streptavidin (FITC labelled) led to a protein triple layer, where the Con A layer is embedded between two streptavidin layers. Even in this case the ternary protein layer (streptavidin) is exactly tracing the structure of the initial crystalline streptavidin matrix [16].

These results show that, based on a streptavidin or avidin matrix, the controlled build-up of a number of protein multilayer systems is possible. In this context the possibility of using the streptavidin layers on solid supports as a basis for biosensors is discussed [11–17]. A problem in this context is that, due to the strong interaction of biotin/streptavidin, the sensor surface cannot be regenerated. A solution to this problem lies in the use of biotin analogues, which—having a lower binding constant to streptavidin than biotin—allow regeneration via competitive replacement of the

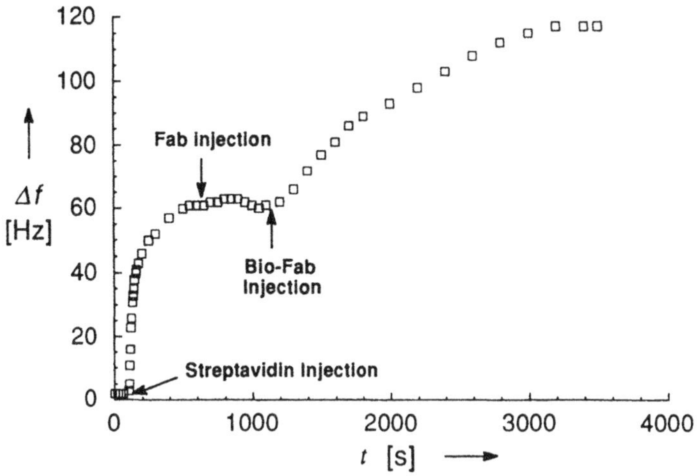
120
100
80
60
40
20
0
Δf
[Hz]
Fab injection
Bio-Fab
injection
Streptavidin injection
0
1000
2000
3000
4000
t [s]

Figure 11. Course of streptavidin binding with time to a lipid membrane containing 5 mol% of biotinlipid and interaction of non-biotinylated antifluorescyl Fab and biotinylated antifluorescyl Fab with a streptavidin matrix.

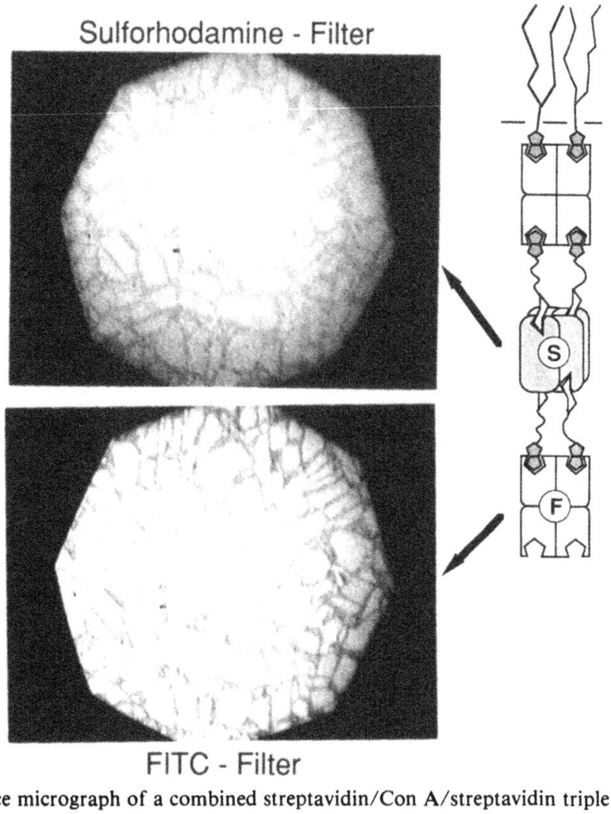
Sulforhodamine - Filter
S
F
FITC - Filter

Figure 12. Fluorescence micrograph of a combined streptavidin/Con A/streptavidin triple layer. (A) A second layer of Con A (sulforhodamine labelled) is bound to the first layer of crystalline streptavidin (non-labelled), tracing the structure of the primary streptavidin matrix. (B) The third layer of streptavidin (fluorescein labelled) is docked onto the second Con A layer and again images the shape of the streptavidin domains in the first layer (10 mm correspond to 30 μm).

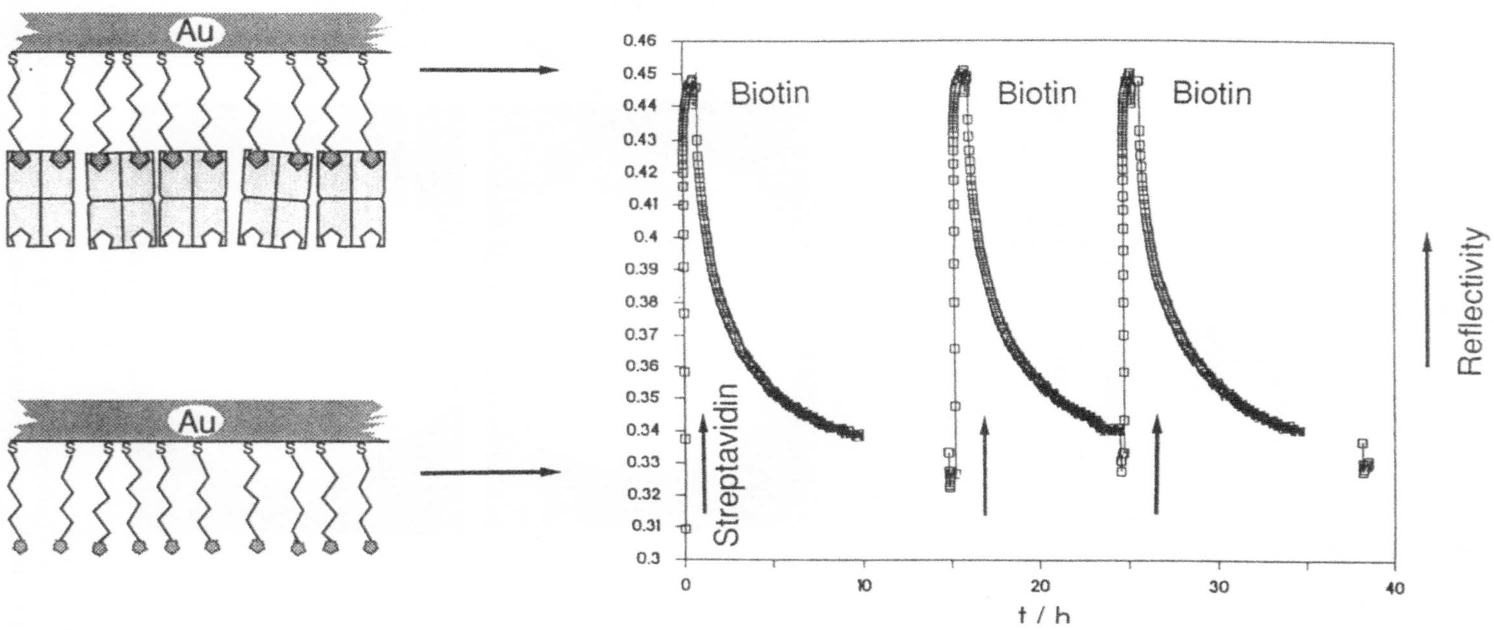
Au
Au
0.46
0.45
0.44
0.43
0.42
0.41
0.4
0.39
0.38
0.37
0.36
0.35
0.34
0.33
0.32
0.31
0.3
0
10
20
30
40
t / h
Biotin
Biotin
Biotin
Streptavidin
Reflectivity

Figure 13. Plasmon spectroscopical measurement of the docking of streptavidin onto the desthiobiotin-(DTB) mixed layer and of the competitive replacement of the streptavidin layer by biotin (three cycles) (DTB-mixed layer: 10% DTB-SH, 90% HS-C11H22-OH).

streptavidin-covered surfaces for a further application. One of these biotin-analogues is desthiobiotin (DTB) with a binding constant to streptavidin of 10^{-13} $M^{-1.}$ Thiol group containing DTB derivatives have, besides other analogues, been synthesized [18, 19], which can be chemisorbed to gold surfaces to give a self-organizing layer. Binding of streptavidin to these DTB-functionalized gold surfaces and its competitive replacement with biotin have mainly been investigated with plasmon spectroscopy [20–21]. This surface-sensitive method allows the measurement of the thickness of a medium, which is in contact with a gold surface, by the change in reflectivity of a gold-evaporated prism. This medium is in our case the protein layer. It was possible to show that a streptavidin layer docked onto a DTB-functionalized gold surface can completely be removed via competitive replacement with biotin, to give the initial DTB-surface. Several binding/replacement cycles can be effected without affecting the initial DTB-thiol layer (Fig. 13).

These supported layer systems allow the build up of protein multilayers like those obtained at the air–water interface. This has been shown with a biotinylated Fab fragment of an antibody against HCG (human choreonic gonadotropin), which has been docked as a second protein layer onto a primary streptavidin layer at the gold surface. As the third layer, the antigen HCG can now be bound to the immobilized Fab fragment. As shown in Fig. 14, the point of interest now became: Could such a protein triple layer be bound reversibly via streptavidin and replaced with biotin. Figure 14 shows the results of the plasmon spectroscopy. The docking of the different proteins can be followed by the increase of the reflectivity, and the competitive replacement process after addition of biotin is expressed by the decrease of the reflectivity. After approximately 30 min the free DTB-surface is reobtained, which again can be used to dock streptavidin onto it.

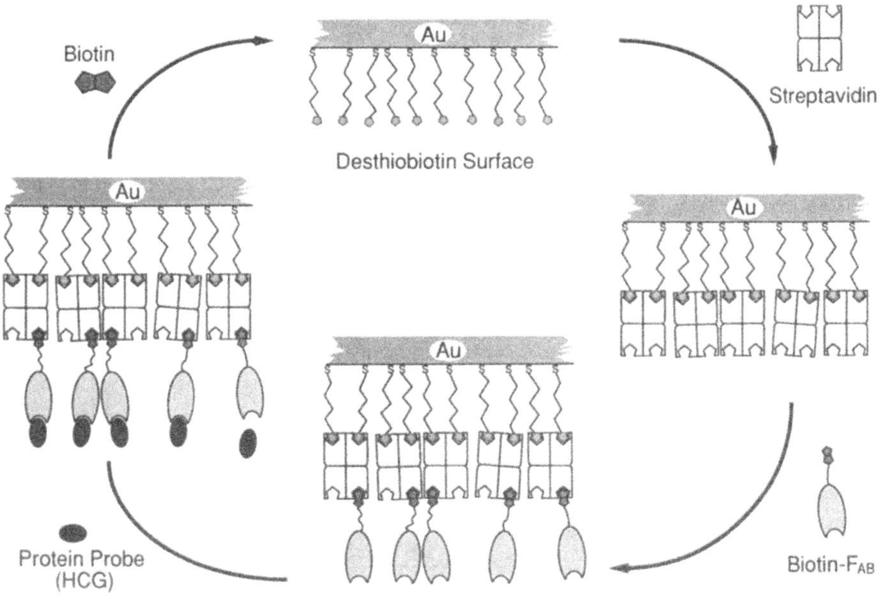
Biotin
Au
Desthiobiotin Surface
Streptavidin
Au
Au
Au
Protein Probe
(HCG)
Biotin-FAB

Figure 14. Schematic of the reversible build up of a protein triple layer based on a streptavidin matrix: (a) formation of the streptavidin matrix; (b) addition of the biotinylated Fab-fragment; (c) binding of HCG to the Fab fragment; and (d) competitive replacement of the protein triple layer with biotin.

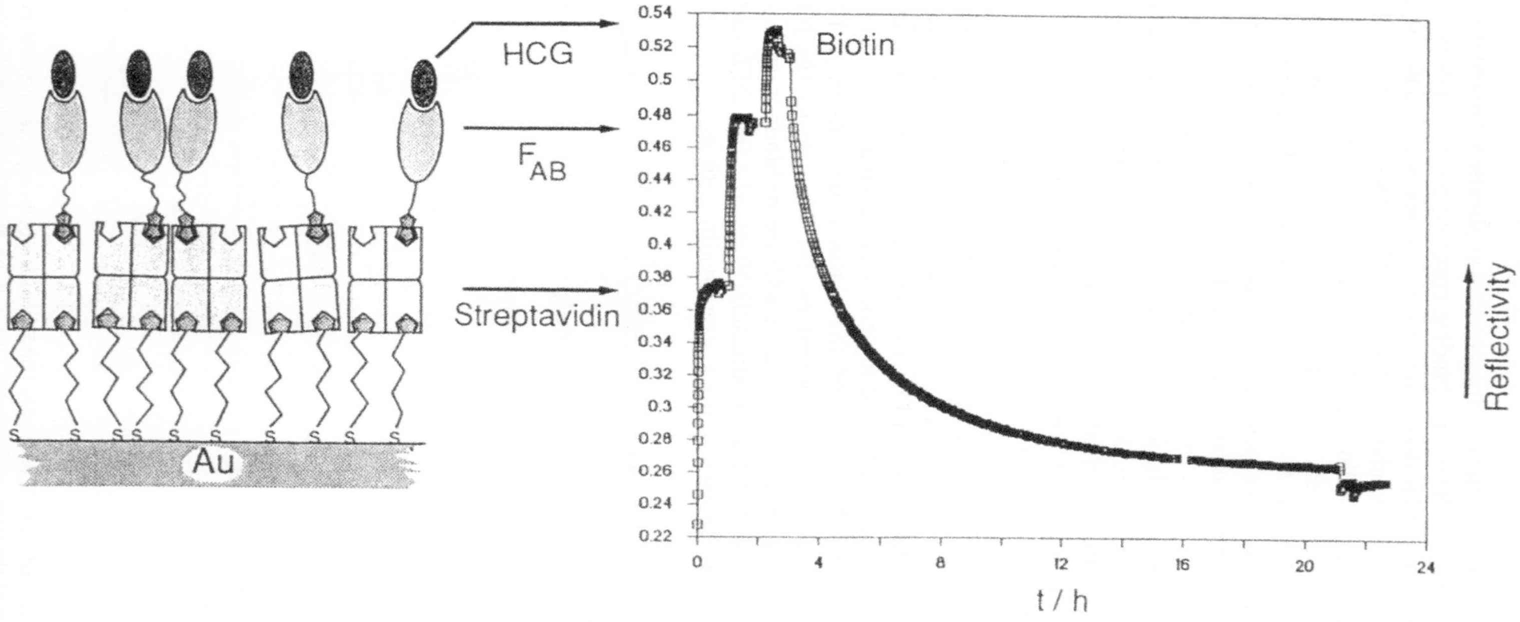

HCG
F_AB
Streptavidin
Au
Biotin
Reflectivity
t / h
0.54
0.52
0.5
0.48
0.46
0.44
0.42
0.4
0.38
0.36
0.34
0.32
0.3
0.28
0.26
0.24
0.22
0
4
8
12
16
20
24

Figure 15. The stepwise build-up of the protein triple layer (streptavidin/Fab fragment/HCG, see Fig. 14) as well as the following competitive replacement by biotin can be followed using surface plasmon spectroscopy.

These results have shown the streptavidin matrix to be an interesting and most versatile base for recognition and docking processes of proteins. Furthermore the possibility of defined replacement with biotin derivatives, having different binding constants, gives an additional chance for a defined reversible build up of proteinateous multilayers.

REFERENCES

1. B. Alberts, D. Bray, J. Lewis, M. Raff, K. Roberts and J. D. Watson in: *Molecular Biology of the Cell*, Garland, New York (1983); *Molekularbiologie der Zelle*, VCH Verlagsgesellschaft, Weinheim (1986).
2. M. Wilchek and E. A. Bayer, *Anal. Biochem.* **171**, 1 (1988).
3. G. Gitlin, I. Khait, E. A. Bayer, M. Wilchek and K. A. Muzkat, *Biochem. J.* **259**, 493 (1989).
4. R. M. Buckland, *Nature* **320**, 557 (1986).
5. R. Blankenburg, P. Meller, H. Ringsdorff and C. Salesse, *Biochemistry* **28**, 8214 (1989).
6. M. Ahlers, W. Müller, A. Reichert, H. Ringsdorf and J. Venzmer, *Angew. Chem.* **102**, 1310 (1990); *Angew. Chem. Int. Ed. Engl.* **29**, 1269 (1990).
7. M. Ahlers, R. Blankenburg, D. W. Grainger, P. Meller, H. Ringsdorf and C. Salesse, *Thin Solid Films* **180**, 93 (1989).
8. S. A. Darst, M. Ahlers, P. H. Meller, E. W. Kubalek, R. Blankenburg, H. O. Ribi, H. Ringsdorf and R. D. Kornberg, *Biophys. J.* **59**, 387 (1991).
9. (a) P. C. Weber, D. H. Ohlendorf, J. J. Wendoloski and F. R. Salemme, *Science* **243**, 85 (1989); (b) W. A. Hendrickson, A. Pähler, J. L. Smith, Y. Satow, E. A. Merrit and R. P. Phizackerley, *Proc. Natl. Acad. Sci. USA* **86**, 2190 (1989).
10. M. Wilchek and E. A. Bayer, *Analyt. Biochem.* **171**, 1 (1988).
11. H. Ebato, J. N. Herron, W. Müller, Y. Okahata, H. Ringsdorf and P. Suci, *Angew. Chem.* **104**, 1064 (1992); *Angew. Chem. Int. Ed. Engl.* **31**, 1078 (1992).
12. J. N. Herron, W. Müller, M. Paudler, H. Riegler, H. Ringsdorf and P. A. Suci, *Langmuir* **8**, 1413 (1992).
13. L. Häussling, W. Knoll, H. Ringsdorf and F.-J. Schmitt, *Langmuir* **7**, 1837 (1991).
14. H. Morgan and D. M. Taylor, *Biosensors* **7**, 405 (1992).
15. R. C. Ebersole, J. A. Miller, J. R. Moran and M. D. Ward, *J. Am. Chem. Soc.* **112**, 3239 (1990).
16. (a) W. Müller, H. Ringsdorf and E. Rump in: *Horizonte: Wie weit reicht unsere Erkenntnis heute?, Verhandlungen d. Gesellschaft Deutscher Naturforscher und Ärzte, 117. Versammlung, Aachen 1992*, Wiss. Verlagsges.mbH, Stuttgart; (b) W. Müller, H. Ringsdorf, E. Rump, G. Wildburg, X. Zhang, L. Angermaier, W. Knoll, M. Liley and J. Spinke, *Science* **262**, 1706 (1993).
17. (a) M. Ahlers, M. Hoffmann, H. Ringsdorf, A. M. Rourke and E. Rump, *Makromol. Chem., Macromol. Symp.* **46**, 307 (1991); (b) M. Hoffmann, W. Müller, H. Ringsdorf, A. M. Rourke, E. Rump and P. A. Suci, *Thin Solid Films* **210/211**, 780 (1992).
18. L. Häußling, B. Michel, H. Ringsdorf and H. Rohrer, *Angew. Chem.* **103**, 568 (1991).
19. F.-J. Schmitt, L. Häußling, H. Ringsdorf and W. Knoll, *Thin Solid Films* **210/211**, 815 (1992).
20. E. Burstein, W. P. Chen, Y. J. Chen and A. J. Hartstein, *Vac. Sci. Technol.* **11**, 1004 (1974).
21. H. Raether in: *Physics of Thin Films*, p. 145, Vol. 9, G. Hass, M. H. Francombe and R. W. Hoffmann (Eds). Academic Press, New York (1977).
22. J. Gordon and J. D. Swalen, *Opt. Commun.* **22**, 374 (1977).
23. W. Hickel and W. Knoll, *J. Appl. Phys.* **67**, 3572 (1990).

Identification of proteins adsorbed to hemodialyser membranes from heparinized plasma

RENA M. CORNELIUS and JOHN L. BRASH*

Departments of Pathology and Chemical Engineering, McMaster University Hamilton, Ontario L8S 4L7, Canada

Received 13 July 1992; accepted 6 October 1992

Abstract—The protein layers formed during contact of plasma with hemodialysis membranes were studied. Dialysers having membranes of cellulose acetate (CA), saponified cellulose ester (SCE), cuprophane (CUP), polymethylmethacrylate (PMMA), and polyacrylonitrile (PAN) were used. Heparinized human plasma was recirculated through the dialysers for four hours. They were then rinsed and the proteins adsorbed to the membranes were eluted with 2% SDS. The yields of protein from the different membranes increased in the order: PMMA < CA < SCE < CUP < PAN. This is the probable order of increasing hydrophilicity. SDS-PAGE and Western blots were performed on the dialyser eluates. The blots were positive for most of the twenty proteins tested for. There were some interesting differences in adsorption patterns among the different membrane materials, notably for high molecular weight kininogen (HMWK), plasminogen and the C3 component of complement. HMWK was intact in the eluates from CA, CUP and SCE, whereas on PMMA and PAN there was evidence of cleavage, suggesting that activation of the contact phase of coagulation was more extensive on the latter two materials. Intact plasminogen was visible on all the blots. However, low molecular weight fragments were visible in the PAN eluates, suggesting activation of the fibrinolytic pathway. Low molecular weight fibrinogen fragments eluted from PAN membranes support this conclusion. C3 was visible in the blots obtained for all membrane materials, and the data suggest that complement is activated by all the membranes. A C3 fragment at about 30 kD (possibly C3d) was seen in the blots for the cellulosic membranes but not for PMMA or PAN.

Key words: Blood compatibility; protein adsorption; hemodialysers; SDS-PAGE; immunoblotting.

INTRODUCTION

Dialysis is an established and effective form of treatment for patients suffering from chronic renal failure, although a number of adverse side effects such as chest pain, dyspnea, hypotension, and neutropenia are sometimes encountered [1–3].

Protein adsorption is one of the first observable events that occurs when blood comes into contact with an artificial surface such as a dialyser membrane [4–6]. The adsorbed protein layer is believed to affect all subsequent interactions, and may be at least partially responsible for some of the adverse reactions suffered by patients on hemodialysis. It has been suggested that some of these problems may be linked to activation of the complement pathway [7, 8], the coagulation system and the kinin system [9]. Furthermore these systems may be activated by adsorption of the relevant proteins to the dialyser surfaces.

Since protein adsorption plays an important role in the events occurring at the blood/dialyser membrane interface, the identification of adsorbed proteins is of considerable interest. The main objective of this work was to identify the plasma

* To whom correspondence should be addressed.

proteins adsorbed to dialyser membranes, and to compare the adsorption profiles of different membrane materials. In a previous study [10] we investigated protein adsorption to dialyser membranes during clinical hemodialysis. The adsorbed layers were extremely complex and many of the proteins were found to be degraded. The work reported here represents a simplification relative to the clinical study. *In vitro* experiments on adsorption to dialyser membranes using normal human plasma were carried out. The possible effects of cellular proteases in degrading adsorbed proteins were thus avoided.

MATERIALS AND METHODS

Dialysers

The following five types of hollow fiber dialysers were utilized in this study: CDAK 4000 with cellulose acetate (CA) fibers (Cordis Dow, Miami Lakes, FL, USA); SCE 135 with saponified cellulose ester (SCE) fibers (Cordis Dow); CF 1511 with cuprophane (CUP) fibers (Baxter Healthcare, Deerfield, IL, USA); B2-1.5H with polymethylmethacrylate (PMMA) fibers (Toray Industries, Japan); Filtral model 12 with polyacrylonitrile (PAN) fibers (Hospal, Meyzieu, France). The physical characteristics of these dialysers are shown in Table 1.

Two dialysers of each type were examined. In addition, two preliminary studies to establish experimental protocol were performed using the SCE model. The CA, SCE, CUP, and PMMA dialysers were used in our previous study of adsorption under clinical use conditions [10].

Table 1.
Physical data on dialysers.

Dialyser model	Membrane type	Blood volume (ml)	Membrane area (m^2)
CDAK 4000	CA	74	1.4
SCE 135	SCE	98	1.5
CF 1511	CUP	74	1.1
B2-1.5H	PMMA	96	1.5
Filtral 12	PAN	99	1.3

Primary antibodies

Antibodies to 20 different human plasma proteins were used in the Western blotting experiments. These primary antibodies, as well as their source, are detailed below. Except as indicated, they were in the form of fractionated antisera developed in goat.

Antibodies to Factor XI, Factor XII, prekallikrein, HMWK, plasminogen, albumin, and prothrombin were from Nordic Immunology, Tilberg, The Netherlands. Antibodies to fibrinogen (raised in rabbit), antithrombin III, C3, α_1-antitrypsin, and fibronectin, were from Cappel Laboratories, Cochraneville, PA, USA. Antibodies to α_2-macroglobulin, transferrin, β_2-microglobulin (raised in rabbit), and hemoglobin (raised in rabbit) were from Sigma Chemical Company, St. Louis, MO, USA. Anti-IgG and anti-β-lipoprotein were from Miles Scientific,

Rexdale, Ontario, Canada. Anti-vitronectin (raised in mouse) and anti-protein C (raised in rabbit) were from Calbiochem, Behring Diagnostics, La Jolla, CA, USA.

Enzyme conjugated second antibodies

Affinity purified second antibodies were as follows: rabbit anti-goat IgG-alkaline phosphatase conjugate (Sigma); goat anti-rabbit IgG-alkaline phosphatase conjugate (Bio-Rad Laboratories, Richmond, CA); and goat anti-mouse IgG-alkaline phosphatase conjugate (Bio-Rad).

Dialysate

Acetate concentrate was from BDH Chemical, Toronto, Ontario, Canada. When diluted one part concentrate to 34 parts purified water, the composition was as follows: sodium, 142 mM; potassium, 1.0 mM; calcium, 1.55 mM; magnesium, 0.5 mM; chloride, 105.1 mM; acetate, 42 mM.

Heparinized plasma

Human platelet poor plasma was prepared from heparinized blood (3 IU/ml). The blood was taken from nonmedicated, normal healthy donors and centrifuged to prepare plasma (3 min at 2000 g, 3 min at 2000 g, 15 min at 2000 g, 20 °C). Single donors were used in order that plasma would be fresh for each dialyser experiment.

Dialyser experiment

A schematic of the experimental setup is shown in Fig. 1. Before an experiment the dialysate compartment was filled with dialysate fluid. Dialysate circulation was initiated, and was continued during the preparation of the blood compartment. The dialysate fluid that was used during preparation of the blood compartment was discarded, and fresh dialysate was added to the system and circulated just prior to the initiation of plasma flow in the blood compartment. Normal saline solution (0.9% NaCl) was circulated for 10 min at 300 ml/min through the blood compartment

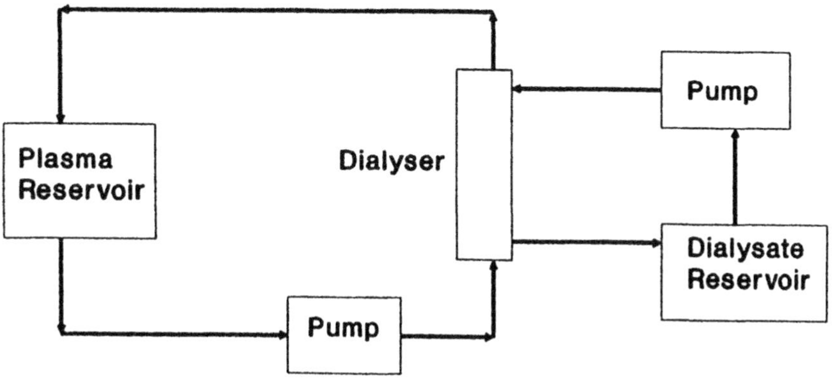
Pump
Plasma Reservoir
Dialyser
Dialysate Reservoir
Pump

Figure 1. Schematic of the experimental setup. Dialysate flowrate, 500 ml/min. Plasma flowrate, 200 ml/min. Temperature, 37 °C. Plasma recirculation time, 4 h.

which was then flushed thoroughly with tris buffered saline (TBS) containing 20 mM ethylene diamine tetraacetic acid (EDTA), pH 7.4. The TBS-EDTA buffer was then displaced with heparinized plasma diluted with TBS, and the initial fluid discarded. Diluted heparinized plasma was recirculated through the blood compartment at a flowrate of 200 ml/min, and dialysate through the dialysate compartment at a flowrate of 500 ml/min. These flowrates were considered to be representative of clinical conditions. Plasma–dialyser contact time was defined from completion of displacement of the buffer in the blood compartment to the beginning of the wash step. A contact time of 4 h, representative of clinical dialysis times, was used in all experiments, and both dialysate and plasma were maintained at 37 °C. The total volume of plasma used for an experiment was about 180 ml, with the plasma reservoir containing 80–120 ml depending on the dialyser type.

Following plasma exposure, the blood compartment was rinsed extensively with TBS-EDTA (six to eight dialyser volumes). The effluent was monitored continuously by absorbance at 280 nm, and washing was continued until the A-280 reached a value of about 0.01. The TBS-EDTA solution was then displaced with 2% sodium dodecyl sulfate (SDS) in TBS, the inlet and outlet lines clamped, and the dialyser maintained overnight at a temperature of 4 °C. Elution was then continued at room temperature by gravity flow of SDS through the dialyser. The A-280 of the eluted protein was monitored continuously, and fractions collected. The fractions having the highest A-280 values were used for analysis. The samples collected from the PAN, SCE, and CUP dialysers were analyzed with no further processing. The PMMA and CA eluates were concentrated by ultrafiltration using a membrane with a molecular weight cut-off of 5 kD.

Polyacrylamide gel electrophoresis and immunoblotting

All electrophoresis reagents were obtained from Bio-Rad, Richmond, CA, USA. The polyacrylamide gel electrophoresis and immunoblotting procedures were performed in the manner that has been detailed in previous reports from this laboratory [10, 11]. Briefly, the eluates were treated by conventional SDS-PAGE to separate the proteins according to molecular weight. The proteins were then transferred electrophoretically (blotted) from the gel onto an Immobilon PVDF membrane (Millipore Co., Bedford, MA, USA). The blots were then cut into 3 mm strips and, following blocking with 5% nonfat dry milk, the strips were incubated with the primary antibodies to the different proteins and then with the appropriate alkaline phosphatase-conjugated second antibody. The substrate system used to develop a color reaction for alkaline phosphatase was 5-bromo-4-chloro-3-indolyl phosphate (BCIP) and nitroblue tetrazolium (NBT) (both from Bio Rad), prepared as described by the supplier.

RESULTS AND DISCUSSION

The proteins adsorbed to the dialyser membranes were eluted with 2% SDS. Plots of A-280 versus fraction number indicated that most of the protein emerged as a sharp peak. Based on peak areas, the relative yields were found to increase in the order: PMMA < CA < SCE < CUP < PAN. This trend is similar to that reported in our clinical study [10] and shows that the relative amounts of eluted protein for the different membranes increase in the probable order of increasing hydrophilicity.

Kuwahara [12] found similarly that the relative amounts of proteins recovered from dialyser membranes increased in the order: PMMA < CA < CUP < PAN. Recent work by Gachon *et al.* [13] showed that the amount of protein eluted from PAN, following clinical use, was substantially higher than from CUP or CA. However, most of the proteins were obtained by application of a reverse transmembrane pressure applied to the dialysate compartment. This suggests that the porous nature of the PAN membranes results in considerable trapping of proteins in membrane pores. Thus, the simple SDS incubation that was performed in the present study may result in only a portion of the sorbed proteins being eluted. Nonetheless it is possible that the high yield of protein from the PAN dialyser is due to some release of trapped, as opposed to adsorbed protein.

Samples of normal heparinized platelet poor plasma were obtained prior to each dialyser experiment and subjected to SDS-PAGE and immunoblotting techniques.

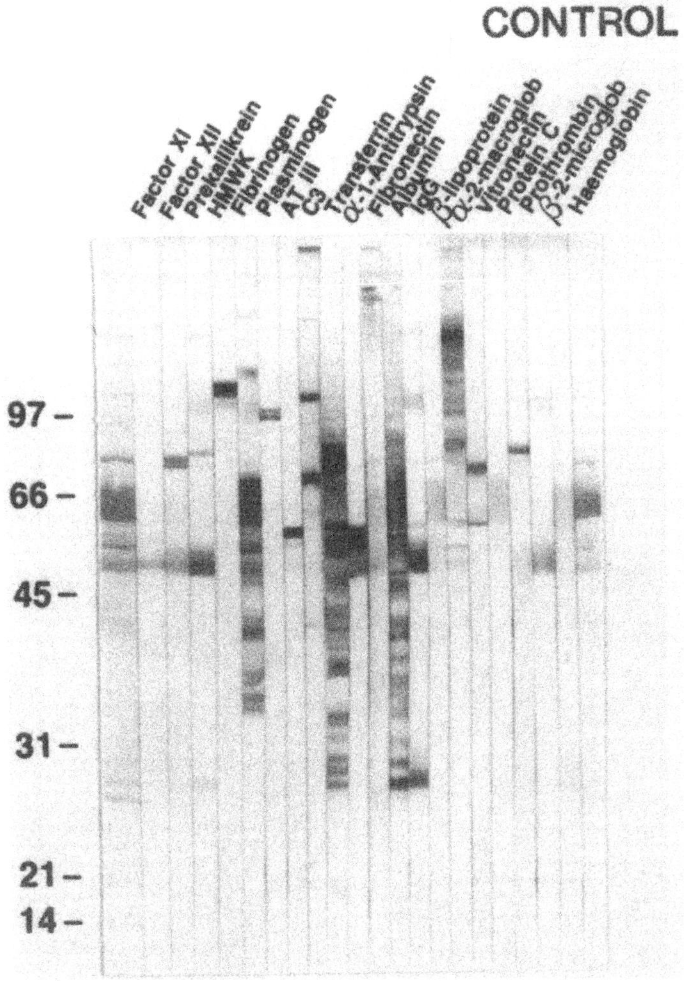
CONTROL
Factor XI
Factor XII
Prekallikrein
HMWK
Fibrinogen
Plasminogen
AT III
C3
Transferrin
α-1-Antitrypsin
Fibronectin
Albumin
IgG
β-lipoprotein
α-2-macroglob
Vitronectin
Protein C
Prothrombin
β-2-microglob
Haemoglobin
97 -
66 -
45 -
31 -
21 -
14 -

Figure 2. Immunoblot from SDS-PAGE (reduced) of normal heparinized plasma. Leftmost lane and rightmost lanes: amido black stained plasma sample. Remaining 20 lanes: immunostaining patterns for specific antisera as indicated. Molecular weight scale in kD.

The resulting 'control' blots showed no marked differences between different donors. Figure 2 shows a typical control blot. The lanes using antibodies to the twenty proteins tested for show bands at the expected molecular weights [10]. Some degradation is apparent in fibrinogen, transferrin and albumin; Factor XII and prekallikrein are somewhat activated (bands less than 80 kD); and Factor XI is strongly activated (most of material at ~50 kD).

Four plasma exposure experiments were carried out with the SCE135 dialyser, and two with each of the other four types. Blots of the eluates from dialysers having the same membrane type were similar, and thus only one blot for each type is presented (Figs 3–7). The extreme right and left lanes show the blots after staining with amido black and give a rough indication of the molecular weights of the more abundant proteins in the eluates. From these lanes it is difficult to make positive identifications of adsorbed proteins. It seems reasonable to conclude, however, that the strong band at about 70 kD seen in all the eluates indicates adsorption of substantial

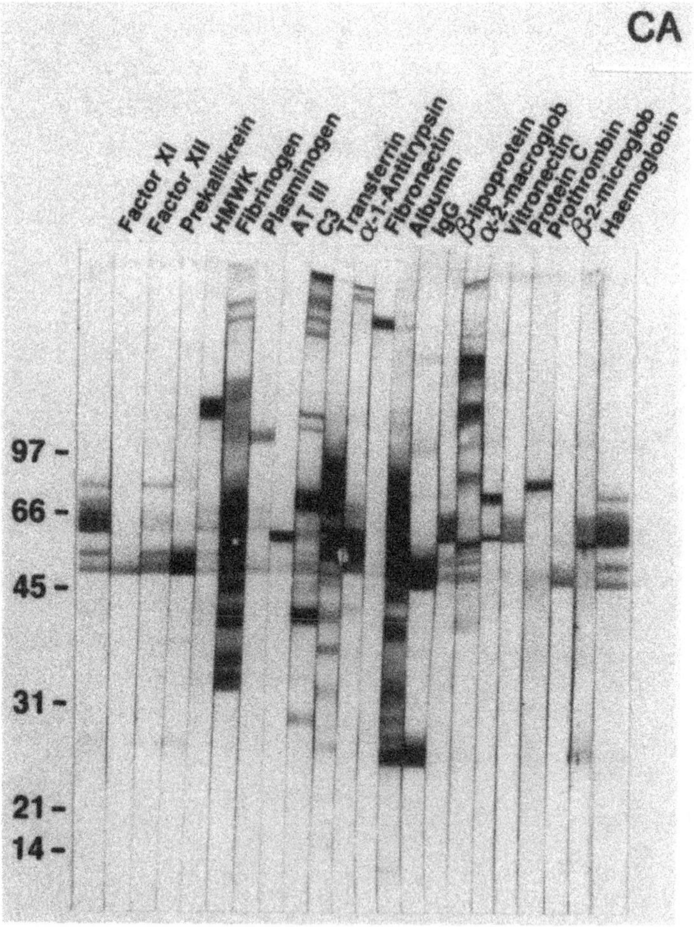
CA
Factor XI
Factor XII
Prekallikrein
HMWK
Fibrinogen
Plasminogen
AT III
C3
Transferrin
α-1-Antitrypsin
Fibronectin
Albumin
IgG
β-lipoprotein
α-2-macroglob
Vitronectin
Protein C
Prothrombin
β-2-microglob
Haemoglobin
97 -
66 -
45 -
31 -
21 -
14 -

Figure 3. Immunoblot from SDS-PAGE (reduced) of 2% SDS eluate from CA dialyser after 4 h exposure to heparinized plasma. Leftmost and rightmost lanes: amido black stained eluate. Remaining 20 lanes: immunostaining patterns for specific antisera as indicated. Molecular weight scale in kD.

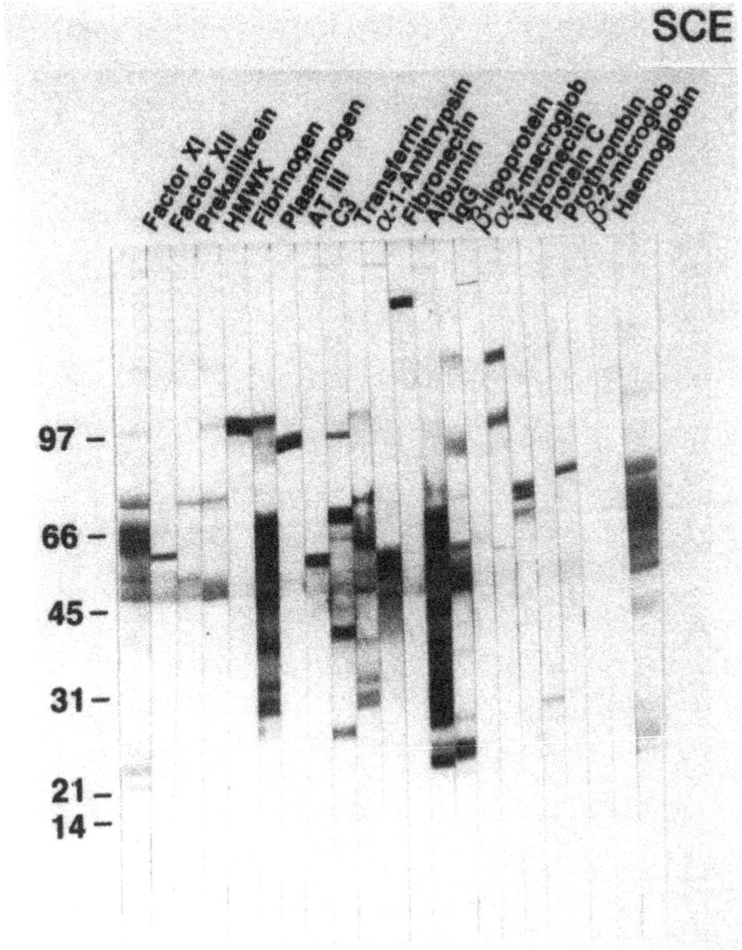
SCE
Factor XI
Factor XII
Prekallikrein
HMWK
Fibrinogen
Plasminogen
AT III
C3
Transferrin
α-1-Antitrypsin
Fibronectin
Albumin
IgG
β-lipoprotein
α-2-macroglob
Vitronectin
Protein C
Prothrombin
β-2-microglob
Haemoglobin
97 -
66 -
45 -
31 -
21 -
14 -

Figure 4. Immunoblot from SDS-PAGE (reduced) of 2% SDS eluate from SCE dialyser after 4 h exposure to heparinized plasma. Leftmost and rightmost lanes: amido black stained eluate. Remaining 20 lanes: immunostaining patterns for specific antisera as indicated. Molecular weight scale in kD.

amounts of albumin to all membrane types. The amido black-stained lanes were similar for all dialyser types with the exception of PAN which showed a markedly different banding pattern, perhaps indicating that different proteins are adsorbed, and/or different biochemical pathways are activated on this membrane.

It should be acknowledged that, while SDS is generally found to be a powerful eluent, it is possible that not all the adsorbed proteins are eluted, so that the eluate composition may not correspond exactly to the adsorbed layer composition. However since the antibody tests were positive for most of the proteins studied, this consideration is probably moot for the present discussion. There were some interesting differences from one membrane to another in the blot banding patterns of a few of the proteins, most notably HMWK, plasminogen, and C3. The discussion that follows emphasizes these three proteins. A few other proteins, where clear differences among the dialysers were seen, are also discussed.

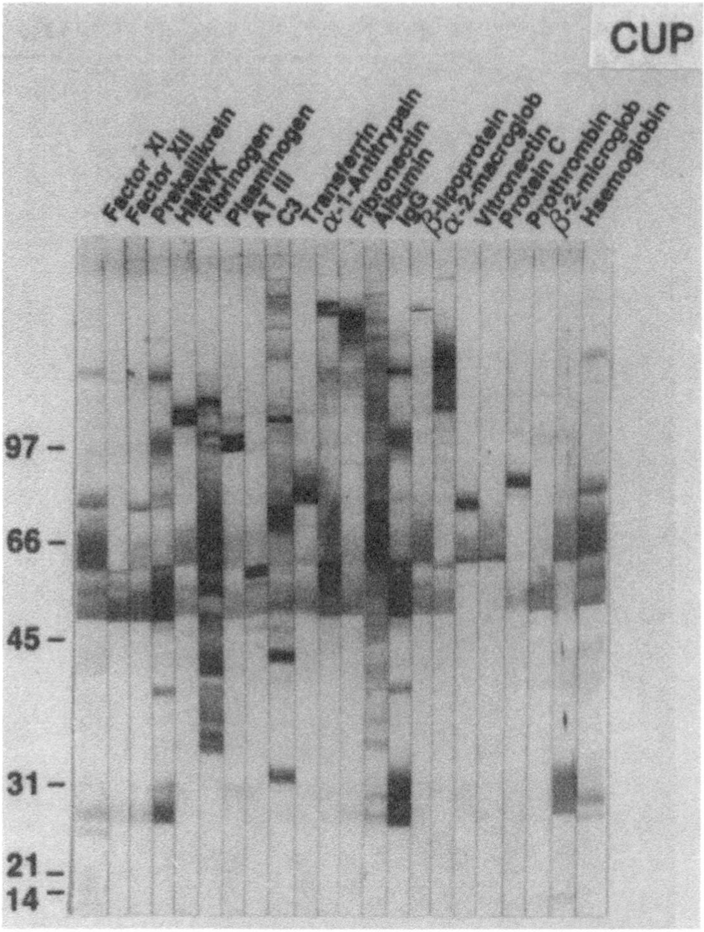
CUP
Factor XI
Factor XII
Prekallikrein
HMWK
Fibrinogen
Plasminogen
AT III
C3
Transferrin
α-1-Antitrypsin
Fibronectin
Albumin
IgG
β-lipoprotein
α-2-macroglob
Vitronectin
Protein C
Prothrombin
β-2-microglob
Haemoglobin
97 -
66 -
45 -
31 -
21 -
14 -

Figure 5. Immunoblot from SDS-PAGE (reduced) of 2% SDS eluate from CUP dialyser after 4 h exposure to heparinized plasma. Leftmost and rightmost lanes: amido black stained eluate. Remaining 20 lanes: immunostaining patterns for specific antisera as indicated. Molecular weight scale in kD.

Contact system coagulation proteins

Factor XI, Factor XII, prekallikrein, and HMWK are of interest because they are involved in the contact phase of the intrinsic blood coagulation pathway. In addition these proteins play an important role in the kallikrein/kinin system [14, 15]. HMWK is known to accelerate the activation of Factor XII, prekallikrein, and Factor XI [16]. Factor XII is activated by kallikrein or activated Factor XI in the presence of a foreign surface [14]. It was hypothesized that some if not all of the eluted contact phase coagulation proteins would exhibit cleavage, indicating that they had been activated by contact with the dialyser membranes.

As can be seen from the control blot, Factor XI appears at a molecular weight of ~50 kD and is thus extensively cleaved, probably during the plasma preparation procedure. Factor XII is essentially intact at a molecular weight of 80 kD, although some cleavage products can be seen in the control blots. The blots of the dialyser eluates for both Factor XI and XII are similar to the control blots, and to the blots obtained in the clinical study [10].

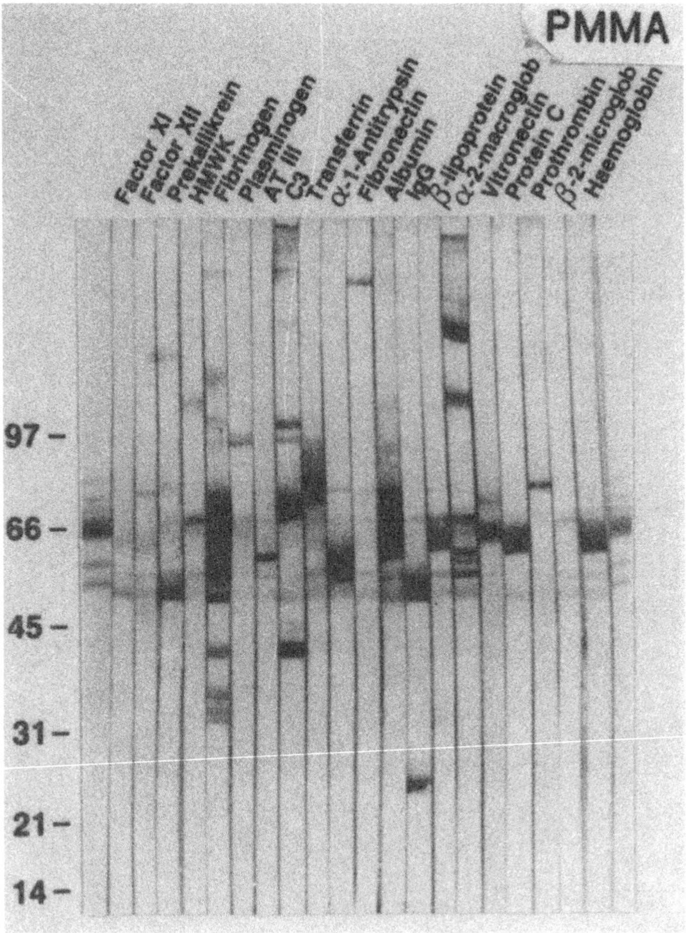
PMMA
Factor XI
Factor XII
Prekallikrein
HMWK
Fibrinogen
Plasminogen
AT III
C3
Transferrin
α-1-Antitrypsin
Fibronectin
Albumin
IgG
β-lipoprotein
α-2-macroglob
Vitronectin
Protein C
Prothrombin
β-2-microglob
Haemoglobin
97-
66-
45-
31-
21-
14-

Figure 6. Immunoblot from SDS-PAGE (reduced) of 2% SDS eluate from PMMA dialyser after 4 h exposure to heparinized plasma. Leftmost and rightmost lanes: amido black stained eluate. Remaining 20 lanes: immunostaining patterns for specific antisera as indicated. Molecular weight scale in kD.

Prekallikrein is substantially intact in the control blots (molecular weight 85 kD), but again some cleavage products are visible. The dialyser eluates show only a faint band at 85 kD, and a stronger band at 50 kD, presumably an activation fragment. The cuprophane blot is unique in that several lower molecular weight fragments (36 and 25 kD) are visible. The data from the clinical study also showed extensive degradation of prekallikrein in the eluates from cuprophane [10].

HMWK is essentially intact in the control blots at a molecular weight of 120 kD. This protein is known to be highly susceptible to plasma kallikrein, and a 'nicked' kininogen can be produced by traces of kallikrein during plasma preparation [17]. HMWK is intact in the eluates from CA, CUP, and SCE and no fragments are visible. PMMA and PAN, on the other hand, show only a faint band at 120 kD, along with numerous lower molecular weight fragments, again suggesting activation of the contact system. This conclusion is further supported by blots of plasma after exposure to PAN (data not shown), which clearly showed cleavage of HMWK. The HMWK dialyser blots in the present work differ considerably from those of our

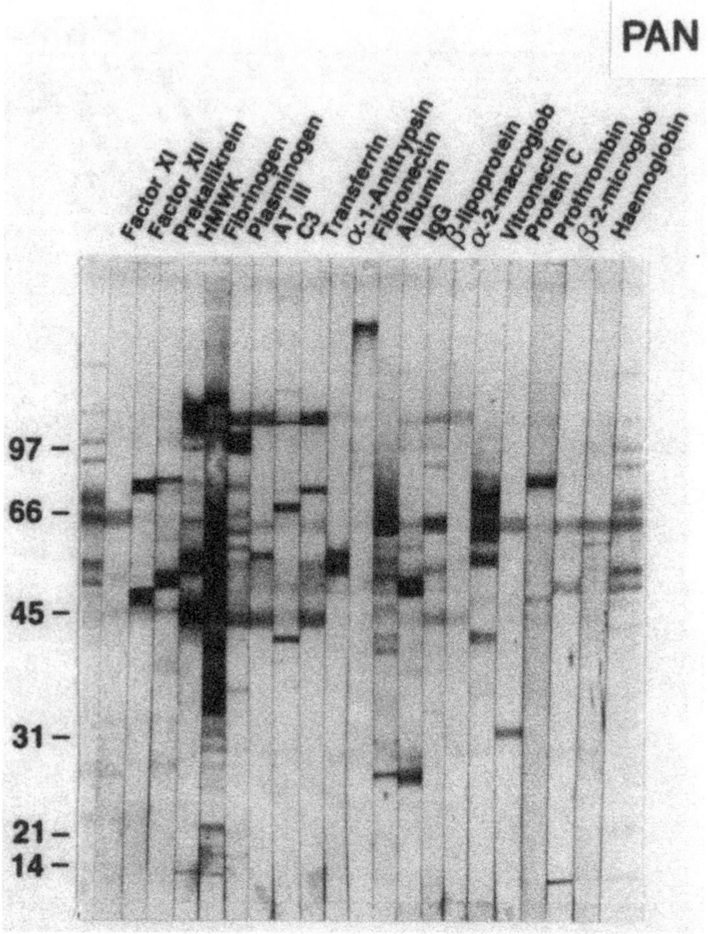
PAN
Factor XI
Factor XII
Prekallikrein
HMWK
Fibrinogen
Plasminogen
AT III
C3
Transferrin
α-1-Antitrypsin
Fibronectin
Albumin
IgG
β-lipoprotein
α-2-macroglob
Vitronectin
Protein C
Prothrombin
β-2-microglob
Haemoglobin
97 -
66 -
45 -
31 -
21 -
14 -

Figure 7. Immunoblot from SDS-PAGE (reduced) of 2% SDS eluate from PAN dialyser after 4 h exposure to heparinized plasma. Leftmost and rightmost lanes: amido black stained eluate sample. Remaining 20 lanes: immunostaining patterns for specific antisera as indicated. Molecular weight scale in kD

clinical study in which both intact HMWK and numerous fragments were visible in the eluates from all dialyser types [10].

Plasminogen

Intact plasminogen (glu-plasminogen) has a molecular weight of 94 kD, although it exists in a low molecular weight form (lys-plasminogen, ~84 kD) as well. Activation of plasminogen to plasmin involves scission of the Arg 560–Val 561 peptide bond, yielding a molecule having a heavy chain (60 kD) and a light chain [18]. Reduced gels of plasmin are therefore expected to exhibit bands at 60, 24, and 10 kD. The control plasma blot (Fig. 2) indicates that plasminogen is present in intact form. Both glu- and lys-plasminogen were visible on most of the control blots obtained, but no plasmin was detected. Intact plasminogen is visible on all the dialyser blots (Figs 3–7), but numerous low molecular weight fragments are also detected on the PAN blots suggesting possible activation of the fibrinolytic pathway. Plasmin degradation of fibrinogen may explain the low molecular weight fragment at about 14 kD that is visible in the fibrinogen blot of the PAN eluate (Fig. 7). Activation of the fibrinolytic system on PAN may be seen as secondary to activation of intrinsic coagulation which is also strong on PAN. Activated Factor XII is a known activator of plasminogen [19].

The plasminogen blots differed considerably from those in the clinical study, where plasmin was detected in the eluates from all the dialysers except CA.

C3 component of complement

C3 is the most abundant of the complement proteins and it plays an important role in the complement pathway, which may be activated during hemodialysis [20]. Complement activation associated with hemodialysis has been reviewed extensively [1, 7, 21–23]. It has been reported that the extent of complement activation depends on the type of dialysis membrane [1, 24–26] and, in the case of cellulosic membranes, whether the dialyser is new or has been used previously [27, 28]. The extent of complement activation seems to correlate well with the severity of symptoms of patients undergoing routine dialysis. In a clinical situation, the activation of C3 results in the α chain being cleaved giving the products C3a and C3b. C3a is an anaphylatoxin, high concentrations of which can result in severe allergic reaction.

Figure 2 shows that in the control heparinized plasma the C3 molecule was intact, with a heavy (α) chain of molecular weight 110 kD, and a light (β) chain of molecular weight 75 kD [29]. In previous work [10] we reported that a C3 blot of plasma treated with cobra venom factor or zymosan to activate complement, showed cleavage of the α chain of C3 with formation of a fragment at about 40 kD, probably related to C3bi. Evidence for cleaved C3 is visible in the blots for the dialysers (Figs 3–7). There is a decrease in intensity of the 110 kD band relative to the control blot, and the appearance of a strong band at about 40 kD. The complement system thus appears to have been activated on all the membrane materials. A C3 fragment at about 30 kD (possibly C3d) is seen in the blots for CA, CUP and SCE, but not for PMMA or PAN. The significance, if any, of this fragment in relation to possible differences in complement activation on the cellulosic versus non-cellulosic membranes is unknown.

A more common approach to assessing complement activation associated with dialysis membranes is to measure fluid phase C3a [25, 26]. This type of measurement has led to the conclusion that cuprophan is a more potent complement activator than polyacrylonitrile. Recently it has been found that significant amounts of C3a generated by complement activation may be bound to the membrane, suggesting that membrane-bound as well as fluid phase C3a should be measured [30]. Indeed it has been concluded by Cheung *et al.* [30] that total C3a generated on PAN (i.e. fluid phase plus surface bound) is more than on cuprophan. The criterion of complement activation used in the present work relies on the detection of membrane-bound fragments and neglects fluid phase activation products. Deppisch *et al.* [31] have suggested an alternative index of complement activation, namely the generation of fluid phase terminal complement component. Clearly the question of assessing complement activation by membranes is an area of some controversy and awaits a satisfactory resolution.

The data in the present study differ significantly from those in our clinical study [10] which showed a very complex gel banding pattern of eluted C3 indicating extensive, nonspecific degradation for all membrane types.

Fibronectin and vitronectin

Fibronectin functions as an adhesive protein that binds cells to other cells as well as to solid surfaces [32–34] and may act as a 'glue' for platelet adhesion to blood contacting surfaces. It is a dimeric glycoprotein, with subunit polypeptides of 200 kD. As can be seen in Figs 2–7 this protein is intact in the control plasma and in the eluates. It is of interest to note that the bands are stronger in the dialyser eluates than in the control blots. This may indicate that fibronectin is enriched in the adsorbed layer.

Vitronectin, like fibronectin, is an adhesive protein that promotes the attachment and spreading of a variety of cells [34, 35]. Vitronectin was visible in the control blots as two subunits with molecular weights of approximately 70 and 60 kD, close to the reported values of 75 and 65 kD [35]. The vitronectin eluted from the membranes was for the most part similar to the control blots, although again the bands are much stronger than in the control blots suggesting that this protein may be selectively adsorbed.

Albumin

Adsorption of albumin to artificial surfaces is known to have a passivating effect on blood response, probably due to a reduction in platelet adhesion [36–38]. In Figs 2–7 it can be seen that the control plasma and eluates all contain degraded as well as intact albumin. The albumin concentrations in the eluates from PMMA and PAN appear to be relatively low. These two materials show a strong band at 66 kD representative of the intact molecule and some faint bands indicating cleavage products.

Similar observations were made in the clinical study which showed that albumin was present in the eluates of all dialyser types, and that it was extensively and nonspecifically degraded. Since degradation also occurs in plasma it is clear that not all of the effect can be attributed to cell-derived proteases.

β_2-microglobulin

Changes in the plasma concentration of β_2-microglobulin have been reported during routine hemodialysis. Whether the concentration is elevated or reduced appears to depend on the type of membrane used [39]. The decrease that is sometimes observed, for example on PAN membranes, may be due to transport across the membrane since this protein when uncomplexed has a molecular weight of 14 kD. An increase in plasma β_2-microglobulin (for example on CUP membranes), which may be caused by leakage from damaged leukocytes, can result in a number of clinical problems because it deposits in joints and synovia where it may induce amyloidosis [40]. The banding pattern in Fig. 2 shows only a faint band at the expected molecular weight of 14 kD, and a stronger band at approximately 60 kD representing presumably a complexed form. The eluate blots show a clear band at 14 kD for PAN but only very faint bands for the other materials. It may be that PAN membranes selectively adsorb (or entrap) β_2-microglobulin resulting in the reported decrease in plasma levels associated with hemodialysis using PAN [39].

Comparison to clinical study

A key conclusion from our earlier clinical study [10] was that many of the eluted proteins were extensively and nonspecifically degraded. By 'nonspecifically' is meant that the fragment molecular weight patterns could not be explained by known processes such as complement activation, blood coagulation, and fibrinolysis. Complement component C3, in particular, showed a large number of fragments which could not be related to complement activation. The protein damage noted in the clinical experiments was tentatively attributed to proteases released from damaged cells. This explanation is supported by the observation in the present experiments using plasma that protein degradation is much less extensive. When it does occur it can usually be explained in terms of known and expected biological processes. Besides those proteins already discussed in this context (C3, albumin, and HMWK), one should also draw attention to α_1-antitrypsin, transferrin, fibronectin, and IgG as proteins which showed serious damage in the clinical experiments.

Acknowledgments

Financial Support of this work by the Medical Research Council of Canada and the Heart and Stroke Foundation of Ontario is gratefully acknowledged.

REFERENCES

1. C. Woffindin and N. A. Hoenich, *Biomaterials* **9**, 53 (1988).
2. R. M. Hakim and E. G. Lowrie, *Nephron* **32**, 32 (1982).
3. M. Rancourt, K. Senger and P. DeOreo, *Trans. Am. Soc. Artif. Int. Organs* **30**, 49 (1984).
4. L. Vroman, A. L. Adams, M. Klings, G. C. Fischer, P. C. Munoz and R. P. Solensky, *Ann. N.Y. Acad. Sci.* **283**, 65 (1977).
5. J. L. Brash and D. J. Lyman, *J. Biomed. Mater. Res.* **3**, 175 (1969).
6. R. E. Baier, *Ann. N.Y. Acad. Sci.* **283**, 17 (1977).
7. A. K. Cheung, C. J. Parker and J. Janatova, *Kidney Int.* **35**, 576 (1989).
8. R. M. Hakim, J. Breillatt, J. M. Lazarus and F. K. Port, *New Eng. J. Med.* **14**, 878 (1984).
9. R. M. Hakim, *Clin. Nephrol.* **26**, S9 (1986).
10. S. R. Mulzer and J. L. Brash, *J. Biomed. Mater. Res.* **23**, 1483 (1989).
11. S. R. Mulzer and J. L. Brash, *J. Biomater. Sci. Polymer Edn.* **1**, 173 (1990).
12. T. Kuwahara, M. Markert and J. P. Wauters, *Artif. Organs* **11**, 325 (1987).
13. A. M. F. Gachon, J. Mallet, A. Tridon and P. Deteix, *J. Biomater. Sci. Polymer Edn* **2**, 263 (1991).
14. R. C. Wiggins, B. N. Bouma, C. G. Cochrane and J. H. Griffin, *Proc. Natl. Acad. Sci.* **74**, 4636 (1977).
15. C. F. Scott, L. D. Silver, A. D. Purdon and R. W. Colman, *J. Biol. Chem.* **5**, 10856 (1985).
16. R. W. Colman, *J. Clin. Invest.* **73**, 1249 (1984).
17. H. Kato, S. Nagasawa and S. Iwanaga, *Methods Enzymol.* **80(C)**, 172 (1981).
18. H. L. Lijnen and D. Collen, *Sem. Thromb. Hemostas.* **8**, 2 (1982).
19. D. Ogston and B. Bennett, *Br. Med. Bull.* **34**, 107 (1978).
20. R. B. Sim, T.M. Twose, D. S. Paterson and E. Sim, *J. Biochem.* **193**, 115 (1981).
21. M. D. Kazatchkine and M. P. Carreno, *Biomaterials* **9**, 30 (1988).
22. W. H. Horl, W. Riegel, P. Schollmeyer, W. Rautenberg and S. Neumann, *Clin. Neph.* **25**, 304 (1986).
23. P. R. Craddock, J. Fehr, K. L. Brigham, R. S. Kronenberg and H. S. Jacob, *New Engl. J. Med.* **296**, 769 (1977).
24. D. E. Chenoweth, *Artif. Organs* **8**, 281 (1984).
25. D. E. Chenoweth, A. K. Cheung and L. W. Henderson, *Kidney Int.* **24**, 764 (1983).
26. A. I. Jacob, G. Gavellis, R. Zarco, G. Perez and J. J. Bourgoignie. *Kidney Int.* **18**, 505 (1980).
27. K. S. Kant, V. E. Pollack, M. Chathey, D. Goetz and R. Berlin, *Kidney Int.* **19**, 728 (1981).

28. E. Savdie, L. Bruce and P. C. Vincent, *Clin. Nephrol.* **8,** 422 (1977).
29. E. Sim, A. B. Wood, L. Hsuing and R. B. Sim, *FEBS Lett.* **132,** 55 (1981).
30. A. K. Cheung, C. J. Parker, L. A. Wilcox and J. Janatova, *Kidney Int.* **37,** 1055 (1990).
31. R. Deppisch, V. Schmitt, J. Bommer, G. M. Hansch, E. Ritz and E. W. Rauterberg, *Kidney Int.* **37,** 696 (1990).
32. K. Sekiguchi and S. Hakomori, *J. Biol. Chem.* **258,** 3967 (1983).
33. R. O. Hynes, *Sci. Am.* **254(6),** 42 (1986).
34. M. H. Ginsberg, J. C. Loftus and E. F. Plow, *Thrombos. Haemostas.* **59,** 1 (1988).
35. C. R. Ill and E. Ruoslahti, *J. Biol. Chem.* **260,** 15610 (1985).
36. H. V. Roohk, S. Pick, R. Hill, E. Hung and R. H. Barrett, *Trans. Am. Soc. Artif. Int. Organs* **22,** 1 (1976).
37. E. W. Salzman, in: *Chemistry of Biosurfaces*, M. L. Hair (Ed.). Marcel Dekker, New York (1972).
38. T. M. S. Chang, *Can. J. Physiol. Pharmacol.* **52,** 275 (1984).
39. M. E. De Broe, J. Nouwen and J. P. Waelegham, *Nephrol. Dial. Transplant* **2,** 124 (1978).
40. C. Basile and T. Drueke, *Nephron* **52,** 113 (1989).

Part III

Blood Interactions

Mechanism of cytoplasmic calcium changes in platelets in contact with polystyrene and poly(acrylamide-co-methacrylic acid) surfaces

NOBUHIKO YUI[1]*, KEN SUZUKI[1], TERUO OKANO[1], YASUHISA SAKURAI[1], CHIKAKO ISHIKAWA[2], KEIJI FUJIMOTO[2] and HARUMA KAWAGUCHI[2]

[1]*Institute of Biomedical Engineering, Tokyo Women's Medical College, 8-1 Kawada-cho, Shinjuku, Tokyo 162, Japan*

[2]*Department of Applied Chemistry, Keio University, 3-14-1 Hiyoshi, Kohoku, Yokohama 223, Japan*

Received 3 April 1992; accepted 4 August 1992

Abstract—Changes in cytoplasmic free calcium levels ($[Ca^{2+}]_i$) in platelets in contact with polystyrene (PSt) and poly(acrylamide-co-methacrylic acid) (PAAmMAc) particles were evaluated and results were compared with those from two representative biological calcium agonists; thrombin and calcium ionophore A23187. PSt particles stimulated a steep increase in cytoplasmic calcium levels in platelets as much as thrombin and A23187. Serratia protease-treated platelets showed a steep increase in $[Ca^{2+}]_i$ by PSt particles, suggesting that PSt surfaces can initiate platelet activation independent of a glycoprotein Ib (GPIb)-mediated pathway. By contrast, dibucaine-treated platelets showed little increase in $[Ca^{2+}]_i$ by PSt particles, indicating that microfilament assembly, including binding of GPIb with actin binding protein, should be required for platelet activation in contact with PSt surfaces. PAAmMAc particles induced little increase in cytoplasmic calcium levels in platelets. However, PAAmMAc particle-treated platelets demonstrated little response to thrombin in terms of an increase in $[Ca^{2+}]_i$ and ATP release, suggesting the possibility that PAAmMAc surfaces may regulate $[Ca^{2+}]_i$ by influencing platelet metabolism. Furthermore, sodium azide-treated platelets showed an increase in $[Ca^{2+}]_i$ in platelets when contacting PAAmMAc particles, supporting the suggestion that PAAmMAc surfaces could regulate platelet functions. Fluorescence polarization measurements using 1,6-diphenyl-1,3,5-hexatriene-loaded platelets revealed that PAAmMAc particles increased membrane fluidity in platelets, which may be due to physicochemical interaction with PAAmMAc surfaces.

Key words: Polystyrene; poly(acrylamide-co-methacrylic acid); cytoplasmic free calcium level; membrane fluidity; ATP release; glycoprotein.

INTRODUCTION

Development of improved blood-contacting devices requires design of non-thrombogenic polymers with high reliability. Interaction of platelets with polymer surfaces has been extensively studied in terms of platelet adhesion, plasma protein adsorption, and their dynamics on polymer surfaces. We have performed systematic studies on non-thrombogenicity of various types of block copolymers in terms of platelet interaction with these surfaces both *in vitro* and *in vivo*. Throughout these studies, we have suggested the importance of controlling polymer surface microstructure in order to prevent platelets from activating on polymer surfaces [1, 2]. We have proposed that prevention of contact-induced activation of platelets on polymer surfaces is a critical step in the design of non-thrombogenic polymers.

* To whom correspondence should be addressed.

Platelet activation involves several interacting systems that ultimately result in thrombus formation. Platelet activation by stimuli such as thrombin and collagen is well known to be mediated by an elevation of cytoplasmic free calcium levels [3, 4]. Therefore, the elucidation of platelet activation in relation to intracellular calcium levels forms an important investigative basis for progress in non-thrombogenic polymers. We have previously demonstrated that polystyrene (PSt) latex particles stimulate a drastic increase in cytoplasmic free calcium levels in platelets, indicating the importance of intracellular calcium concentration as a secondary messenger involved in polymer–platelet interaction [5]. Furthermore, we have suggested the possibility of platelet activation occurring on PSt surfaces via a different pathway from thrombin-stimulated activation [6]. Recently, we reported that increases in cytoplasmic free calcium levels, in platelets contacting our designed microstructured block-copolymer surfaces, are strongly reduced by both modifying the surface microstructure of the polymer itself and adsorbing plasma protein onto the polymer surface [7]. This result suggests that platelets contacting a particularly microstructured surface are blocked from contact-induced activation in terms of a cytoplasmic calcium-mediated activation pathway.

Hydrophilic surfaces such as poly(2-hydroxyethyl methacrylate) (PHEMA), poly(ethylene oxide) (PEO), and polyacrylamide (PAAm) surfaces have been promoted as blood compatible, non-adherent biomaterials because they minimize interfacial free energy in an aqueous solution [8, 9]. Candidates for this type of polymer were proposed by Andrade *et al.* [8]. However, blood compatibility of these polymers is probably limited because of the possibility of embolization and chronic damage to the circulating platelets *in vivo*. For example, Ratner and coworkers examined the blood compatibility of hydrogel surfaces PAAm and PHEMA in a renal embolus test *in vivo* and suggested the hypothesis that these hydrogel surfaces were platelet consumptive, but they were insufficiently broad to prove such a hypothesis [10, 11]. Furthermore, Okano and colleagues evaluated thrombogenicity of PEO-grafted surfaces using a rabbit arterio–arterial (A–A) shunt model. In this report, there were definite decreases in occlusion time of coated tubing and circulating platelet count during A–A shunt experiments, indicating surface-induced platelet activation and systemic thromboembolization linked to the presence of PEO [12]. Hydrophilic surfaces are now thought to interact with platelets and other blood components, leading to thrombosis *in vivo*. Even so, quantifying the interaction of platelets with such hydrophilic surfaces *in vitro* is actually more difficult than other hydrophobic surfaces [13], and this situation has made it difficult to evaluate the interactions between platelets and hydrophilic polymer surfaces *in vitro*. Thus, little information is available to clarify the mechanism of platelet interaction with hydrophilic surfaces. Recently, Feuerstein *et al.* studied the adherence and detachment of platelets on HEMA-coating glass plates using epifluorescent video microscopy. They demonstrated that such hydrophilic surfaces favored platelet adhesion with detachment, over long-term adhesion [14]. They pointed out the possibility of changes in platelet plasma membrane as a result of transient adhesion. However, no specific evidence for supporting this conclusion was presented.

Hydrophilic microspheres containing acrylamide and/or other comonomers have been prepared by a precipitation copolymerization technique in ethanol [15]. This polymerization technique allowed preparation of poly(acrylamide-co-methacrylic acid) (PAAmMAc) particles with mono-disperse diameters. Moreover, PAAmMAc and its related particles exhibited interesting features during the phagocytosis and oxygen consumption by polymorphonuclear neutrophils in contact with these particles [16].

The objective of this paper is to clarify the mechanism of platelet metabolism in contact with PSt and PAAmMAc particles. Changes in platelet cytoplasmic calcium levels by polymer surfaces are compared with two representative biological calcium agonists, thrombin and calcium ionophore A23187. Adenosine triphosphate (ATP) release from platelets exposed to PSt or PAAmMAc particles was examined using thrombin. Furthermore, a fluorescence polarization measurement was performed in order to estimate changes in membrane fluidity of platelets in contact with PAAmMAc particles.

EXPERIMENTAL

Preparation of biochemicals and polymer particles

Fura 2-AM [1-(2-(5′-carboxyoxazol-2′-yl)-6-aminobenzofuran-5-oxy)-2-(2′-amino-5′-methylphenoxy)ethane-N,N,N′,N′-tetraacetic acid pentaacetoxymethyl ester] and a Chrono-Lume® reagent kit were purchased from Dojin Chemical Co., Japan, and Chrono-Log Co., USA, respectively. Unless otherwise noted, all biochemicals were purchased from Sigma Chemical Co., USA. Polystyrene (PSt) latex particles with average diameter of 1.0 μm were supplied from Sekisul Chemical Co., Japan, as an aqueous suspension. ζ potential of PSt particles was −55.23 mV at 37°C in 0.1 M phosphate buffer solution, which was determined by an electropholetic mobility measurement. Poly(acrylamide-co-methacrylic acid) (PAAmMAc) particles were prepared from a precipitation polymerization technique using ethanol as solvent, in which 10 mol % of methacrylic acid and 10 mol % methylenebis-acrylamide were added to stabilize suspension of the particles and to crosslink the particles, respectively. The polymerization was initiated by 0.05 wt % azobis-isobutyronitrile at 60°C, and the resulting particles were washed twice each with ethanol and distilled water, and finally resuspended with distilled water. The average diameter of PAAmMAc particles was 1.0 μm at 37°C in distilled water as estimated by photon correlation spectroscopy (Ohtsuka Electronics Co., Japan, LPA-3000/3100). ζ potential of PAAmMAc particles was −13.28 mV which was determined by the same manner as that of PSt particles. These particles were diluted with Ca^{2+}- and Mg^{2+}-free Hanks' balanced salt solution (HBSS), so that final concentrations were 6×10^8, 6×10^9, 1.8×10^{10}, or 1.8×10^{11}/ml.

Preparation of Fura 2-loaded platelet suspension

The loading of Fura 2 into platelets was carried out according to the method previously reported in detail [5]. Platelet suspensions in HBSS (platelet concentration: 3×10^8/ml) were prepared from the citrated blood of male Japanese white rabbits, weighing from 2.0 to 2.5 kg. Fura 2-AM was loaded into platelets by incubating the platelet suspension with Fura 2-AM solution at 37°C for 45 min at a Fura 2-AM concentration of 5 μM. Platelets were washed with HBSS and were finally resuspended in HBSS, so that final platelet concentration was 3×10^8/ml. The platelet suspension was recalcified with $CaCl_2$, so that the external calcium concentration was 1 mM just prior to use in particle measurements.

Pretreatment of Fura 2-loaded platelets

Fura 2-loaded platelet suspensions were incubated with benzyl alcohol (BA)–HBSS solution at 37°C for 5 min at a BA concentration of 40 mM. BA-treated platelet suspensions were used for the following calcium measurement without further purification. Pretreatment of platelets by serratia protease (SP) was performed by incubating the suspension at 37°C for 60 min at a SP concentration of 0.05 unit/ml. Dibucaine (DB) treatment was performed by incubating the suspension at room temperature for 25 min at a DB concentration of 1 mM. Both pretreated suspensions were centrifuged once at 1500 rpm for 10 min and resuspended in HBSS for the following measurement.

Measurement of cytoplasmic calcium levels in platelets

PSt or PAAmMAc particles (100 μl) were mixed with 400 μl of the Fura 2-loaded platelet suspension in a fluorescence cuvette in a fluorimeter (Japan Spectroscopic Co., model CAF-100) at 37°C with magnetic stirring (1000 rpm) [5]. The mixture was excited at both 340 and 380 nm and emission was measured at 500 nm. Fluorescence intensities at each of the two wavelengths were used to determine the fluorescence dichroic ratio 340/380 (R). The cytoplasmic free calcium concentration ($[Ca^{2+}]_i$) was calculated based on the R value as described by Tsien and coworkers [17].

Scanning electron microscopic (SEM) observation of platelets in contact with polymer particles

SEM observation of platelets mixed with polymer particles was made. After 1 min of contact with polymer particles, a saline solution containing 1.25% glutaraldehyde were poured into the platelet suspension in order to fix platelets. These mixtures of platelets and polymer particles were rinsed with distilled water, freeze-dried, then coated with gold. The morphology of platelets mixed with polymer particles was observed by SEM (JEOL, Japan, JSM-5300LV).

Measurement of adenosine triphosphate (ATP) release from platelets

ATP release from activated platelets was evaluated using a Lumiaggregometer (Chrono-Log Co., USA, C-400). The procedure followed used the Chrono-Lume® Reagents for platelet function testing, which is based on a luciferin–luciferase reaction [18]. Fura 2-loaded platelet suspensions (400 μl) were pipetted into glass cuvettes, followed by addition of 50 μl of reconstituted Chrono-Lume® Reagent. This cuvette was then transferred into the Lumi-aggregometer to incubate 37°C for 1 min with stirring. Volumes of calcium agonists such as thrombin, PSt, and PAAm particles (100 μl), were added to the cuvette to record luminescence. After getting a peak luminescence, 5 μM of ATP standard solution was added to calibrate the amount of released ATP in this cuvette.

Fluorescence polarization measurement of 1.6-diphenyl-1,3,5-hexatriene (DPH) in platelets

Fluorescence anisotropy of DPH in platelet membranes was measured to assess the membrane fluidity of platelets [19]. DPH solution in tetrahydrofuran (2 mM) was diluted into platelet suspensions to 2 μM and incubated with gentle agitation at 37°C for 60 min [20]. Fluorescence polarization of DPH-loaded platelets was examined in a spectrofluorimeter (Japan Spectroscopic Co., FP-770) equipped with a fluorescence polarization accessory (Japan Spectroscopic Co., ADP-300) at 37°C with magnetic stirring. DPH was excited at 360 nm and the fluorescence was detected at 430 nm. Fluorescence intensities were measured with polarizers inserted into the excitation and emission light paths. Two relative intensities were measured, I_{VV} and I_{VH}, where the subscripts indicate the position of the polarizers, the first one being that in the excitation, the second being that in the emission path. I_{HV} and I_{HH} were also measured to determine the instrumental factor, G. From these measurements, the fluorescence anisotropy, $\langle r \rangle$, was calculated as follows [21]:

$$\langle r \rangle = (I_{VV} - G \cdot I_{VH})/(I_{VV} + 2G \cdot I_{VH}) \qquad G = I_{HV}/I_{HH}$$

RESULTS AND DISCUSSION

Cytoplasmic calcium changes in platelets by polymer particles

Interaction of platelets with PSt or PAAmMAc surfaces was estimated in terms of evaluating intracellular calcium change in platelets. In this paper, all the experiments were examined in protein-free suspensions, which may be different from those in protein-containing suspensions. Our previous paper demonstrated the eliminating effect of intracellular calcium change in platelets in contact with PSt particles in the presence of protein [5]. When extracellular protein are present, they could rapidly absorb to the particles, resulting in eliminating an intracellular calcium change. Figure 1 shows a representative change in fluorescence ratio of Fura 2 in platelets in contact with PSt and PAAmMAc particles. Changes in the fluorescence ratios of Fura 2 in platelets stimulated by 0.1 U/ml thrombin and 0.1 μM A23187 were examined and are also shown in Fig. 1. PSt particles exhibited a steep increase in the fluorescence ratio as much as thrombin and A23187, whereas PAAmMAc showed little increase. Based on the fluorescence ratio, $[Ca^{2+}]_i$ in platelets was calculated. Results of the changes in cytoplasmic calcium levels in platelets in contact with

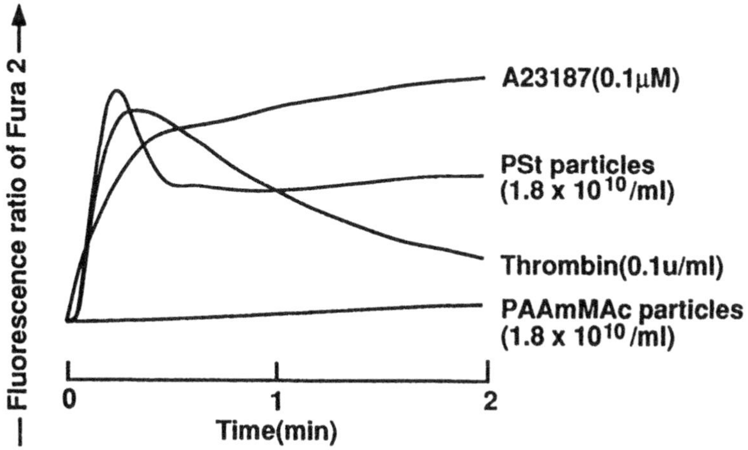
Fluorescence ratio of Fura 2
A23187(0.1μM)
PSt particles
(1.8 x 10¹⁰/ml)
Thrombin(0.1u/ml)
PAAmMAc particles
(1.8 x 10¹⁰/ml)
0
1
2
Time(min)

Figure 1. Changes in the fluorescence ratio of Fura 2 in platelets stimulated by thrombin, A23187, and polymer particles.

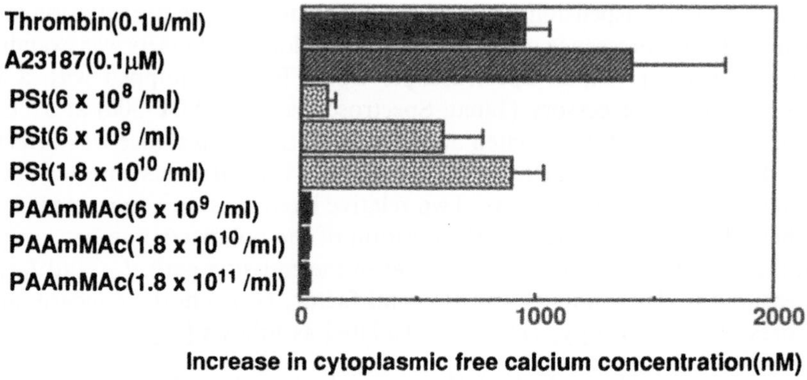
Thrombin(0.1u/ml)
A23187(0.1μM)
PSt(6 x 10^8 /ml)
PSt(6 x 10^9 /ml)
PSt(1.8 x 10^10 /ml)
PAAmMAc(6 x 10^9 /ml)
PAAmMAc(1.8 x 10^10 /ml)
PAAmMAc(1.8 x 10^11 /ml)
0
1000
2000
Increase in cytoplasmic free calcium concentration(nM)

Figure 2. Changes in cytoplasmic free calcium levels in platelets stimulated by thrombin, A23187, and polymer particles (± S.E.M.).

PSt and PAAmMAc particles are summarized in Fig. 2, together with those from thombin and A23187. Addition of HBSS solution (100 μl) to Fura 2-loaded platelet suspension (400 μl) showed few increase in $[Ca^{2+}]_i$. The possibility of fluorescence quenching by PAAmMAc particles was discounted because the fluorescence intensity of PAAmMAc contacting platelets was almost the same as that of native platelets when all the platelets were solubilized by the addition of Triton X detergent. The result of Fig. 2 indicates that cytoplasmic calcium acts as the important secondary messenger for platelet activation by PSt surfaces, although PAAmMAc surfaces do not initiate the cytoplasmic calcium-mediated activation process. In this system, platelets were mixed with PSt particles with ratios of

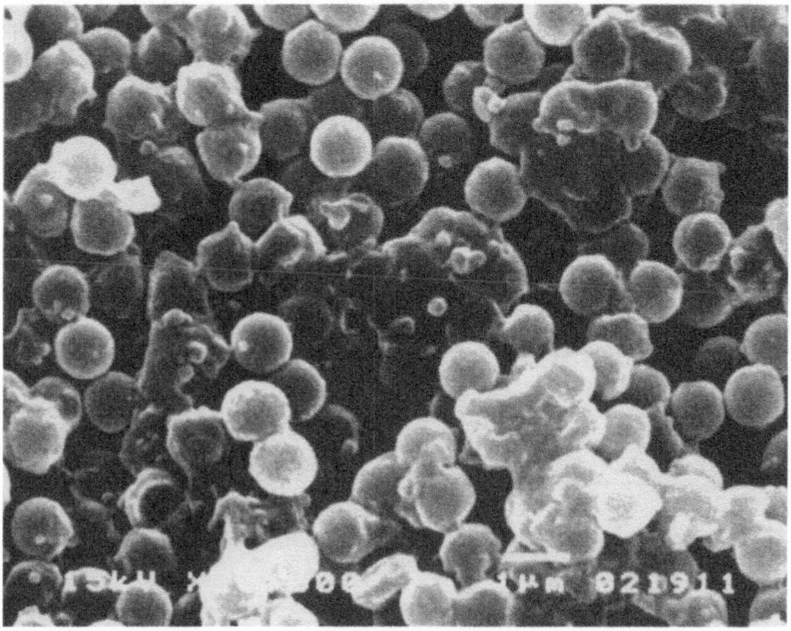

Figure 3. SEM view of platelets mixed with PSt particles.

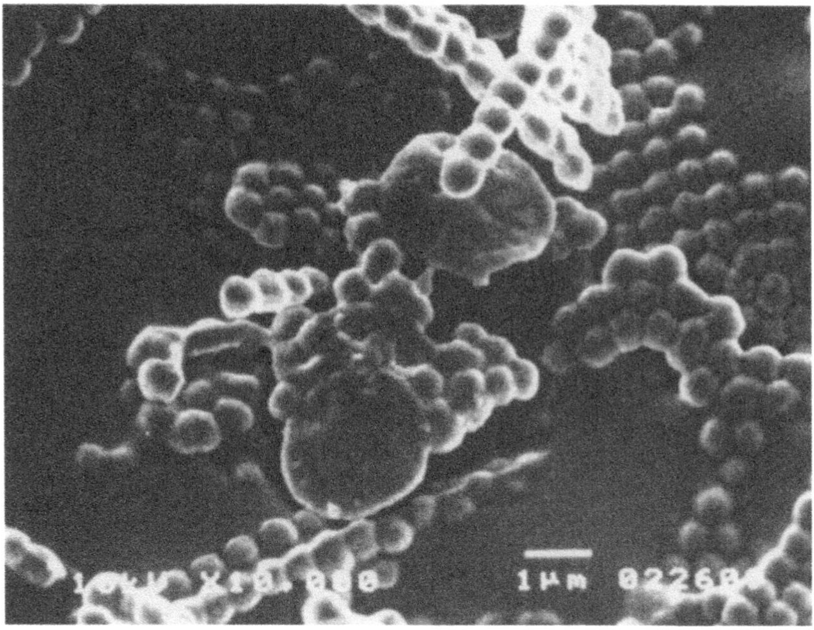
1µm

Figure 4. SEM view of platelets mixed with PAAmMAc particles.

0.5–15 particles per platelet. SEM observations of platelets mixed with polymer particles after 1 min of contact with the particles were also examined, the typical results of which are shown in Figs 3 and 4. Platelets mixed with PSt particles showed major shape change, spread over some PSt particles (Fig. 3). Therefore, an increase in $[Ca^{2+}]_i$ in platelets by PSt particles is considered as a process of platelet adhesion and activation on PSt particles. In contrast, PAAmMAc particles were mixed with platelets with ratios of 5–150 particles per platelet. SEM observation revealed that majority of platelets kept their discoid shape and showed few adhesion onto PAAmMAc particles (Fig. 4). Thus, few increase in $[Ca^{2+}]_i$ in platelets by PAAmMAc particles is considered as a result of platelets in contact with PAAmMAc particles.

Mechanism of cytoplasmic calcium change in platelets by PSt particles

One question arising from the above results is how PSt particles induce increases in $[Ca^{2+}]_i$ in platelets. It has already been reported that the majority of the calcium increase induced by PSt particles was due to calcium influx across the plasma membrane [5], indicating a possible physicochemical change in the plasma membrane by its contact with PSt particles. In this paper, the increase in $[Ca^{2+}]_i$ in platelets by contact with PSt particles was investigated by focusing on glycoprotein Ib (GPIb) and its network with the cytoskeleton. Figure 5 shows an illustration of GPIb on the platelet membrane and its network with the cytoskeleton. In thrombin-stimulated platelet activation, glycocalicin has been well defined as a thrombin-binding domain on GPIb [22]. SP is known to preferentially hydrolyse glycocalicine from GPIb, which indicates that SP is a suitable tool to examine GPIb-mediated activation pathway [23]. Figures 6 and 7 show changes in the fluorescence ratio of Fura 2 in

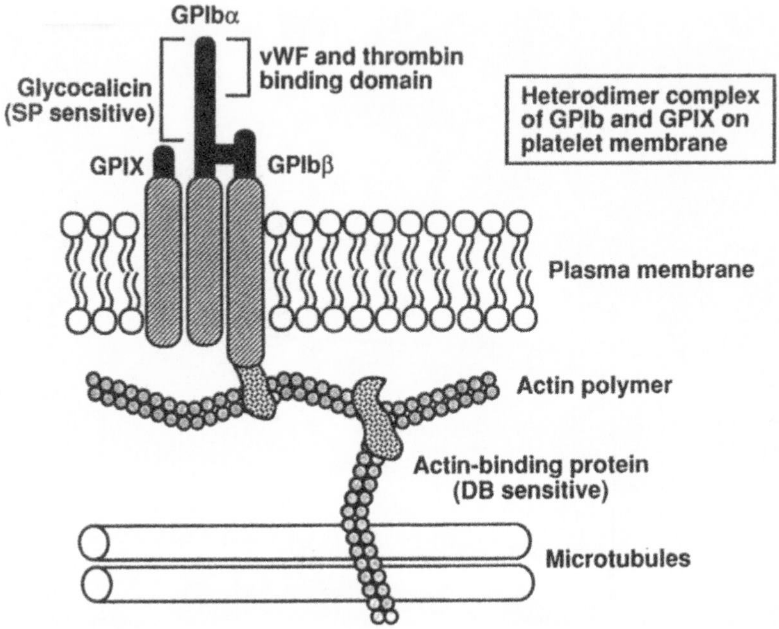
GPIbα
vWF and thrombin binding domain
Glycocalicin (SP sensitive)
GPIX
GPIbβ
Heterodimer complex of GPIb and GPIX on platelet membrane
Plasma membrane
Actin polymer
Actin-binding protein (DB sensitive)
Microtubules

Figure 5. GPIb on platelet membrane and its network with the cytoskeleton.

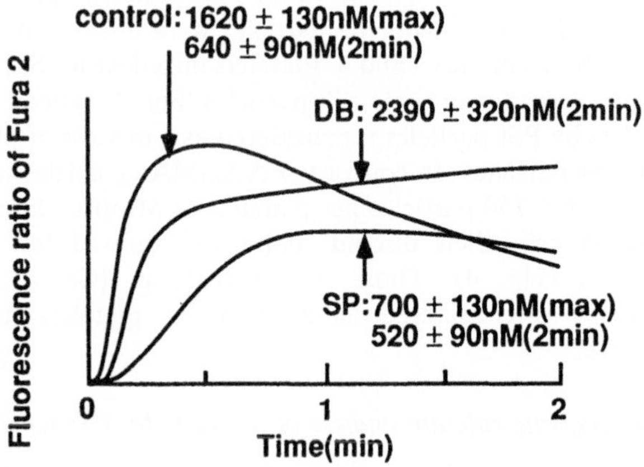
control: 1620 ± 130nM(max)
640 ± 90nM(2min)
DB: 2390 ± 320nM(2min)
SP:700 ± 130nM(max)
520 ± 90nM(2min)
Fluorescence ratio of Fura 2
0
1
2
Time(min)

Figure 6. Changes in the cytoplasmic free calcium levels in platelets stimulated by 0.1 U/ml thrombin (±S.E.M.).

SP-treated platelets by thrombin and PSt particles, respectively. SP-treated platelets show a drastic increase in $[Ca^{2+}]_i$ by PSt particles, but only gradual increases by thrombin. Reduced activation of SP-treated platelets by thrombin is consistent with the previously reported results [24]. Since there is no prothrombin in the extracellular fluids, thrombin generation will be very low or zero. Therefore, this result suggests that PSt surfaces can initiate cytoplasmic calcium-mediated platelet activation either independent of a GPIb-mediated pathway or via direct activation of GPIb without glycocalicin.

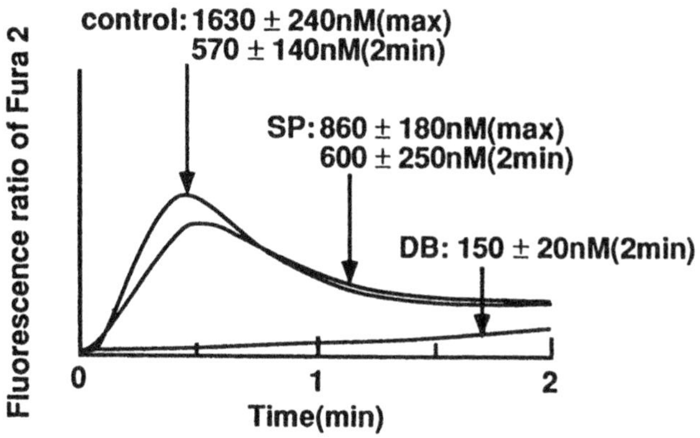
control: 1630 ± 240nM(max)
570 ± 140nM(2min)
SP: 860 ± 180nM(max)
600 ± 250nM(2min)
DB: 150 ± 20nM(2min)
Fluorescence ratio of Fura 2
0
1
2
Time(min)

Figure 7. Changes in the cytoplasmic free calcium levels in platelets stimulated by PSt particles (1.8×10^{10}/ml) (± S.E.M.).

It has been reported that GPIb is associated with cytoskeletal structures in intact platelets and that a Ca^{2+}-dependent protease is able to cleave the actin-binding protein (ABP) of platelets in their activation process [25, 26]. DB is a local anesthetic which promotes the presentation of glycocalicin extracellularly at the membrane surface by initiating the activation of a Ca^{2+}-dependent protease [27, 28]. This is related to an irreversible disruption of microfilament assembly in the platelet cytoskeleton. Changes in the fluorescence ratio of Fura 2 in DB-treated platelets by thrombin and PSt particles are also shown in Figures 6 and 7, respectively. DB-treated platelets showed gradual increases in $[Ca^{2+}]_i$ by thrombin but little increase by PSt particles. This result indicates that microfilament assembly, including the binding of GPIb with ABP, should be required for the activation of platelets by contact with PSt surface. Another interesting feature in Fig. 6 is that the increase in $[Ca^{2+}]_i$ by thrombin in DB-treated platelets was not transient. As normal platelets show a transient increase of $[Ca^{2+}]_i$ by thrombin followed by Ca^{2+} uptake and recovery, this unusual lack of recovery is considered to be due to disruption of the microfilament assembly by DB. Furthermore, PSt particles-contacting platelets also exhibited the same phenomenon in $[Ca^{2+}]_i$ change by thrombin as this DB-treated ones (data not shown). It is thus proposed that a change in membrane glycoprotein-cytoskeleton network may be important in initiating the activation of platelets on PSt surfaces.

Interaction of platelets with PAAmMAc particles

In order to examine the effect of PAAmMAc particles on platelet function, 0.1 U/ml thrombin or 0.1 μM A23187 solution was added to PSt or PAAmMAc particle-contacting platelet suspensions (after 2 min contact time, unless otherwise mentioned). The results are summarized in Figs 8–11. Platelets incubated with PSt particles showed almost the same response to thrombin after 10 min as normal platelets, although these platelets showed a smaller response after 2 min (Fig. 8). Platelets contacting PAAmMAc particles showed smaller response to thrombin whenever thrombin was added (Fig. 9). First, low levels of activation for platelets contacting PAAmMAc particles were initially thought to result from the interaction of

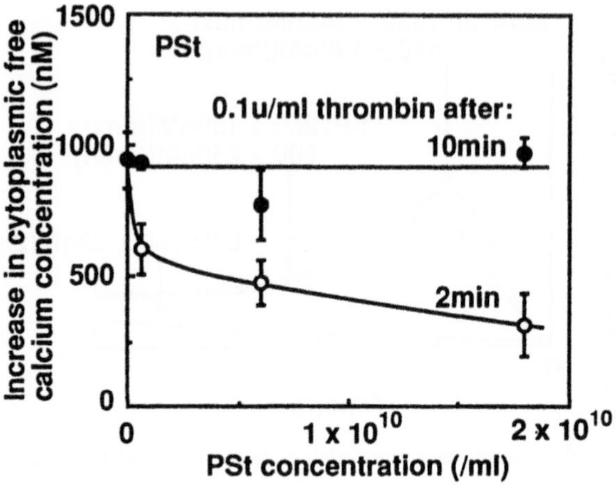
1500
PSt
0.1u/ml thrombin after:
1000
10min
500
2min
0
0
1 x 10^10
2 x 10^10
PSt concentration (/ml)
Increase in cytoplasmic free calcium concentration (nM)

Figure 8. Change in the cytoplasmic free calcium levels in PSt-contacting platelets by 0.1 U/ml thrombin (± S.E.M.).

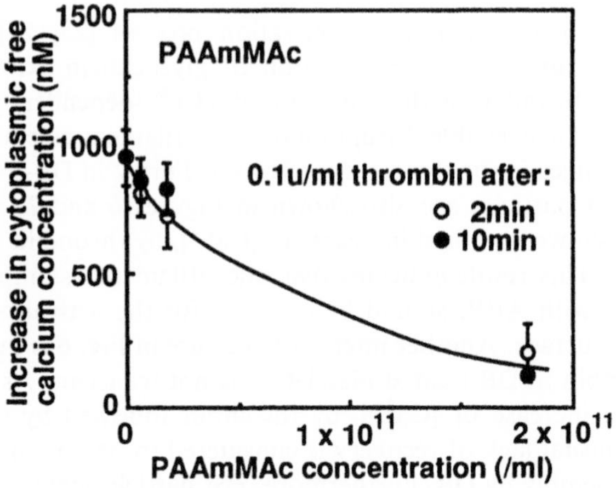
1500
PAAmMAc
1000
0.1u/ml thrombin after:
o 2min
● 10min
500
0
0
1 x 10^11
2 x 10^11
PAAmMAc concentration (/ml)
Increase in cytoplasmic free calcium concentration (nM)

Figure 9. Change in the cytoplasmic free calcium levels in PAAmMAc-contacting platelets by 0.1 U/ml thrombin (± S.E.M.).

PAAmMAc particles directly with thrombin, particularly thrombin ab- or adsorption in- or onto PAAmMAc particles. However, this hypothesis was discounted because the supernatant of the mixture of PAAmMAc particles and thrombin consistently showed the same activity in cytoplasmic Ca^{2+} increase as thrombin control (data not shown). At least, three possibilities might be considered to explain the inhibition of thrombin-stimulated calcium increase with PAAmMAc particles: (1) PAAmMAc particles competitively inhibit thrombin-induced activation in terms of thrombin-binding domain of membrane glycoproteins such as GPIb; (2) PAAmMAc particles can accelerate the biochemical mechanism regulating $[Ca^{2+}]_i$ in platelets; or (3) PAAmMAc particles adsorb or absorb Ca^{2+}.

Platelet contact with PSt particles caused a decrease in cytoplasmic calcium changes by A23187, and the rate of this change is quite low (Fig. 10). PAAmMAc particles also showed a diminished change in $[Ca^{2+}]_i$ by A23187, however, the rate

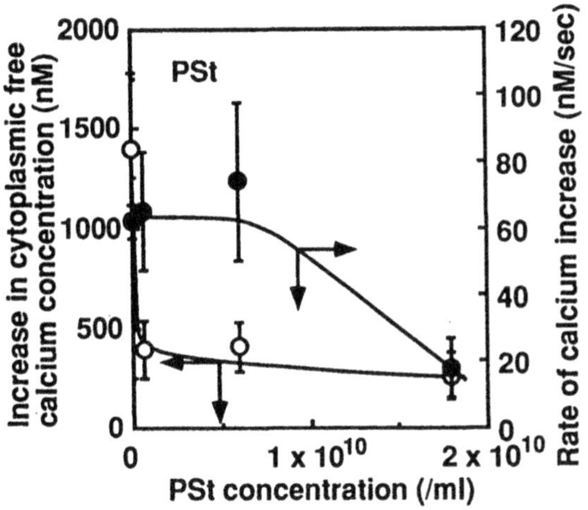
PSt
Increase in cytoplasmic free calcium concentration (nM)
Rate of calcium increase (nM/sec)
PSt concentration (/ml)
2000
1500
1000
500
0
120
100
80
60
40
20
0
1 x 10^10
2 x 10^10

Figure 10. Changes in the cytoplasmic free calcium levels and the rate of increasing calcium levels in PSt-contacting platelets by 0.1 μM A23187 ($\pm$ S.E.M.).

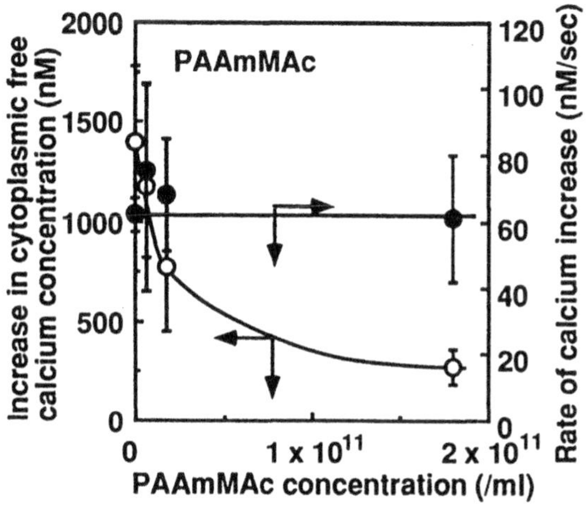
PAAmMAc
Increase in cytoplasmic free calcium concentration (nM)
Rate of calcium increase (nM/sec)
PAAmMAc concentration (/ml)
2000
1500
1000
500
0
120
100
80
60
40
20
0
1 x 10^11
2 x 10^11

Figure 11. Changes in the cytoplasmic free calcium levels and the rate of increasing calcium levels in PAAmMAc-contacting platelets by 0.1 μM A23187 ($\pm$ S.E.M.).

of increasing $[Ca^{2+}]_i$ in the platelets with PAAmMAc particles was almost the same as for normal platelets (Fig. 11). A further interesting feature of these experiments was that A23187-induced intracellular calcium increase in platelets contacting PAAmMAc particles was transient especially at the very high particle dose, as shown in Fig. 12. It has been reported that A23187 transports Ca^{2+} across the plasma membrane into the cytoplasm and that this efficiency as a divalent cation carrier depends upon the physicochemical characteristics of the plasma membrane [29]. These results thus suggest the possibility of change in both physicochemical characteristics and biochemical functions of the plasma membrane in platelets by contact with PAAmMAc particles.

Changes in $[Ca^{2+}]_i$ in sodium azide (NaN_3)-treated platelets by the addition of polymer particles or calcium agonists was examined in order to clarify the inhibition of the thrombin-induced activation process by PAAmMAc particles. The effect of

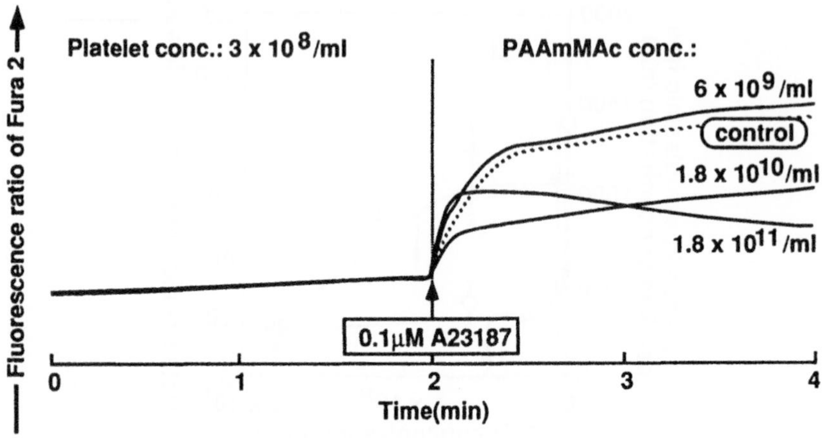
Fluorescence ratio of Fura 2
Platelet conc.: 3 x 10⁸/ml
PAAmMAc conc.:
6 x 10⁹/ml
control
1.8 x 10¹⁰/ml
1.8 x 10¹¹/ml
0.1μM A23187
0
1
2
3
4
Time(min)

Figure 12. Change in the fluorescence ratio of Fura 2 in PAAmMAc particles-contacting platelets by 0.1 μM A23187.

NaN_3 on intracellular calcium change by polymer particles or calcium agonists is shown in Fig. 13. With increasing NaN_3 concentration, increases in $[Ca^{2+}]_i$ in platelets diminishes when platelets were stimulated by thrombin and PSt particles. As NaN_3 acts as an inhibitor of energy metabolism in platelets [30], this result indicates that thrombin and PSt particles induce activation of platelets by affecting platelet metabolism. A23187 induced increases in $[Ca^{2+}]_i$, independent of NaN_3 concentration. As A23187 transports Ca^{2+} across the plasma membrane, evidence suggests that NaN_3 was ineffective in altering the physicochemical nature of the plasma membrane. In contrast, PAAmMAc particles exhibited an increase in $[Ca^{2+}]_i$ in platelets with increasing NaN_3 concentration. For example, an increase in $[Ca^{2+}]_i$ in 40 mM NaN_3-treated platelets by PAAmMAc particles was 290 nM, while no treatment of NaN_3 induced an increase of $[Ca^{2+}]_i$ of 60 mM in platelets by PAAmMAc particles. Because NaN_3 inhibits the platelet metabolism via regulation of ATP production

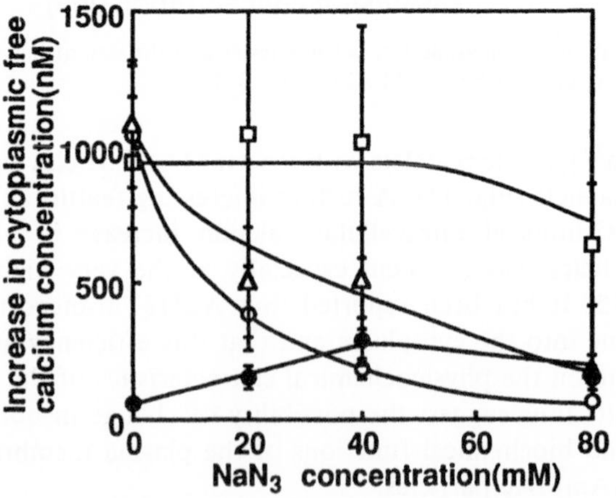
Increase in cytoplasmic free calcium concentration(nM)
1500
1000
500
0
0
20
40
60
80
NaN₃ concentration(mM)

Figure 13. Effect of NaN_3 on the change in cytoplasmic free calcium levels in platelets stimulated by: (○), 0.1 U/ml thrombin; (□), 0.1 μM A23187, and (△), PSt particles (1.8×10^{10}/ml); (●), PAAmMAc particles (1.8×10^{11}/ml) (± S.E.M.).

in mitochondria of the cytoplasm, the increase in $[Ca^{2+}]_i$ in NaN_3-treated platelets by PAAmMAc particles must not be due to metabolic processes. Thus it is suggested that regulation of $[Ca^{2+}]_i$ in platelets is enhanced by contact with PAAmMAc particles, the mechanism of which is metabolically directed. One representative example of the regulation mechanism of $[Ca^{2+}]_i$ in platelets is the calcium inhibition by adenosine cyclic monophosphate (cAMP)-dependent protein kinase (A kinase) via the activation of adenylate cyclase [31]. These cAMP-dependent processes have been well characterized to exert two types of effects on $[Ca^{2+}]_i$: (1) they inhibit the rise of Ca^{2+} that would normally occur in response to platelet stimulation by a calcium agonist; and (2) they stimulate resequestration of Ca^{2+} that has already been released into the cytoplasm by a calcium agonist [32].

ATP release from platelets

In order to confirm the possibility that PAAmMAc particles can regulate platelet functions involving intracellular calcium levels, ATP release from platelets in contact with PAAmMAc by thrombin was evaluated by a lumi-aggregometer using the luciferin-luciferase reaction. This reaction provides luminescence in the presence of ATP secreted during dense granule release [18]. Figure 14 shows the relation between the intracellular calcium change and the release of ATP from polymer-contacting platelets by 0.1 U/ml thrombin. The relation between the intracellular calcium change and the ATP release from native platelets by thrombin with different concentrations was also shown as a dark curve. As indicated by the dark curve, it is clear that thrombin initiates the platelet activation process (ATP release) in relation to the increase in $[Ca^{2+}]_i$. PSt-contacting platelets responded to thrombin, and the relation between intracellular calcium change and ATP release in PSt-contacting platelets is close to that seen in thrombin-stimulated native platelets.

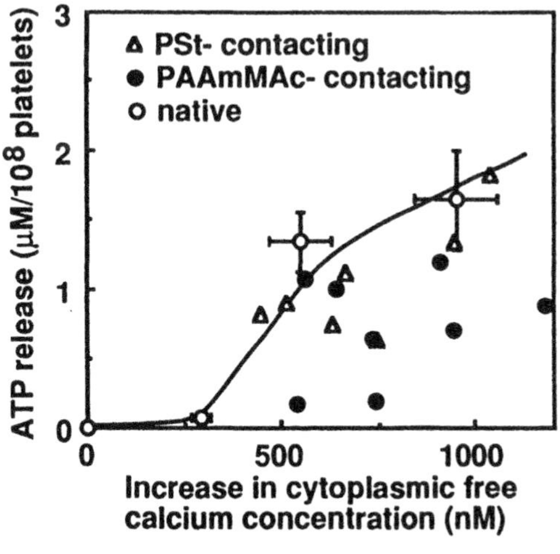

ATP release (µM/10⁸ platelets)
3
2
1
0
Δ PSt- contacting
● PAAmMAc- contacting
o native
0
500
1000
Increase in cytoplasmic free calcium concentration (nM)

Figure 14. Relation between cytoplasmic free calcium levels and ATP release from PAAmMAc particles-contacting platelets by 0.1 U/ml thrombin. Before adding thrombin, PSt (100 μl, 6×10^8 or 6×10^9/ml) or PAAmMAc (100 μl, 6×10^9 or 1.8×10^{10}/ml) suspension was mixed with the platelet suspension. Relation between an increase in cytoplasmic free calcium levels and ATP release from native platelets was carried out by using 0.01, 0.05, and 0.1 U/ml thrombin ($\pm$ S.E.M.).

PAAmMAc-contacting platelets showed much lower ATP release by thrombin, thus, the relationship between calcium change and ATP release deviates substantially from the standard curve made by thrombin-induced results with both native and PSt-contacting platelets. Such a deviation between $[Ca^{2+}]_i$ and ATP release has also been reported in platelet disfunction, which include storage pool defects, thrombasthenia, and functional disorders by drugs [18]. Therefore, results in Fig. 14 suggest that PAAmMAc particles can regulate platelets in terms of not only intracellular calcium changes but also the sequential activation process after increasing $[Ca^{2+}]_i$.

Membrane fluidity in platelets

Any physicochemical changes in the plasma membrane, including fluidity and perturbation, might be related to platelet activation during contact with polymer surfaces. Benzyl alcohol (BA) is well known to bind the headgroups of membrane lipids and/or displace cholesterol from the lipid bilayer, resulting in an increase in membrane fluidity [33]. Figure 15 shows the effect of BA on intracellular calcium change by thrombin, A23187, PSt and PAAmMAc particles. BA-treated platelets showed higher $[Ca^{2+}]_i$ by A23187, but lower by thrombin and PSt particles. BA-treated platelets showed little response to PAAmMAc particles. This result suggests that elevation of $[Ca^{2+}]_i$ by PSt surfaces is not due to an increase in membrane fluidity. Previously, thrombin-, ADP-, and arachidonic acid-stimulated platelets exhibited a transient decrease in the membrane fluidity in their activation processes [34]. Reduced increases in $[Ca^{2+}]_i$ in BA-treated platelets by PSt particles may indicate a decrease in the membrane fluidity.

In order to examine changes in platelet membrane fluidity by contact with polymer particles, fluorescence polarization of DPH was measured in platelets mixed with polymer particles. Among probes used for fluorescence polarization, DPH has been the most extensively used [35]. The precise location of DPH incubated with living cells has not been accurately determined. However, it can be

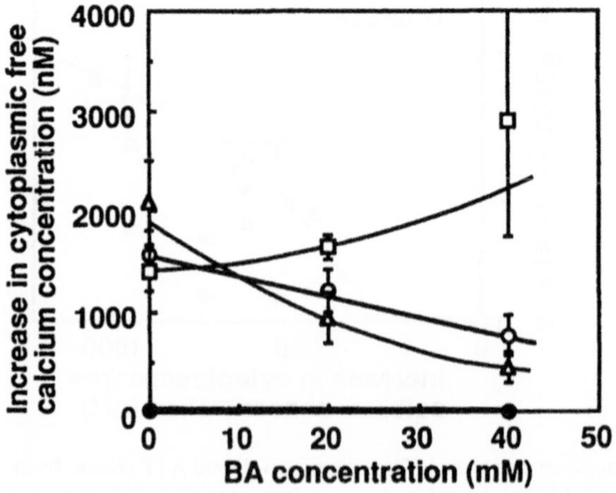
Increase in cytoplasmic free calcium concentration (nM)
4000
3000
2000
1000
0
0
10
20
30
40
50
BA concentration (mM)

Figure 15. Effect of BA on the change in cytoplasmic free calcium levels in platelets stimulated by: (○), 0.1 U/ml thrombin; (□), 0.1 μM A23187, and (△), PSt particles (1.8×10^{10}/ml); (●), PAAmMAc particles (1.8×10^{11}/ml) (± S.E.M.).

assumed that DPH is situated in the apolar hydrocarbon zone of the membrane bilayer after incubation [36]. Fluorescence results indicate that PSt particles enhance platelet-incubated DPH intensity with increasing concentration of PSt particles. We confirmed that fluorescence of DPH in platelets mixed with either PSt or PAAmMAc particles have the same emission maxima (430 nm) as DPH in native platelets. Changes in hydrocarbon packing in the platelet plasma membrane in contact with PSt surfaces are proposed to cause this change because such increases in DPH intensity were not observed in PSt particle suspensions containing DPH without platelets (data not shown). Our previous paper demonstrated that PSt-contacting platelets showed low cytoplasmic Ca^{2+} change by A23187 [7], suggesting the possibility of decreased membrane fluidity for platelets in contact with PSt particles. Furthermore, results in Fig. 15 support this contention. Therefore, the increased DPH intensity from platelets by contact with PSt particles could indicate that PSt surfaces decrease membrane fluidity in platelets.

Figure 16 shows a time course of fluorescence polarization of DPH in platelets after the addition of PAAmMAc particles to platelet suspensions. This change was too fast to estimate the rate of decreasing $\langle r \rangle$ value. Figure 17 summarizes the results of the changes in DPH fluorescence anisotropy in platelets as a function of PAAmMAc particle concentrations. PAAmMAc particles induce a significant decrease in anisotropy at the very high PAAmMAc concentration. Furthermore, decreases in $\langle r \rangle$ values upon PAAmMAc particle addition were reduced by both addition of 1% bovine serum albumin (BSA) in platelet suspension and precoating of PAAmMAc particles with 10% BSA solution. Addition of NaN_3 (40 mM) into platelet suspensions had no influence on changes in anisotropy (data not shown). From these results, we propose that the increase in membrane fluidity of platelets by PAAmMAc particles is due to physicochemical interaction rather than platelet metabolism.

Recently, non-crosslinked PAAmMAc, which has the same PAAm/PMAc ratio but is water soluble, was prepared by similar synthetic techniques, and the

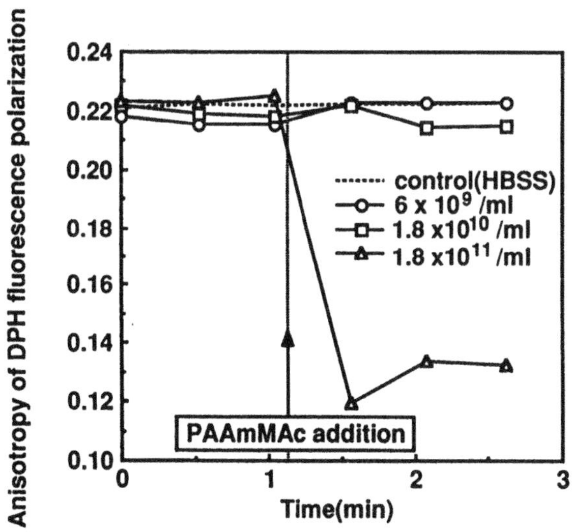

Anisotropy of DPH fluorescence polarization
0.24
0.22
0.20
0.18
0.16
0.14
0.12
0.10
0
1
2
3
Time(min)
control(HBSS)
6 x 10⁹ /ml
1.8 x10¹⁰ /ml
1.8 x10¹¹ /ml
PAAmMAc addition

Figure 16. Absolute change in the anisotropy of DPH fluorescence polarization in platelets by contact with PAAmMAc particles.

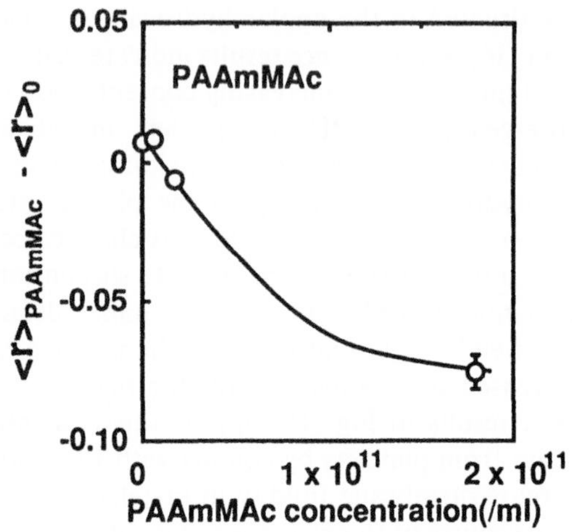
PAAmMAc
0.05
0
-0.05
-0.10
$<r>_{PAAmMAc} - <r>_0$
0
1 x 10^11
2 x 10^11
PAAmMAc concentration(/ml)

Figure 17. Differences ($\langle r \rangle_{PAAmMAc} - \langle r \rangle_0$) in the anisotropy of DPH fluorescence polarization in platelets by contact with PAAmMAc particles compared to DPH in native platelets (± S.E.M.).

interaction of platelets with this water soluble PAAmMAc was investigated in inducing changes in $[Ca^{2+}]_i$ and platelet membrane fluidity. Water soluble PAAmMAc exhibited the same result on the regulation of thrombin-induced $[Ca^{2+}]_i$ change as the PAAmMAc particles, but showed no effect on platelet membrane fluidity, the detailed results of which will be reported in our forthcoming paper [37]. Therefore, we suggest that aggregated macromolecular surface assemblies of PAAmMAc dominate the increase in membrane fluidity of platelets, but Ca^{2+} changes are induced by discrete molecular level interactions between PAAmMAc and platelet membranes.

CONCLUSIONS

The interaction of platelets with PSt and PAAmMAc particles was investigated in terms of the changes in $[Ca^{2+}]_i$ in platelets, ATP release from platelets, and membrane fluidity of platelets. Based on the above mentioned results and discussion, the following conclusions are presented: (1) Polymer surfaces induced different platelet metabolic processes from thrombin and A23187. (2) PSt surfaces can initiate platelet activation via glycoprotein-cytoskeleton networks, and this process is metabolism-dependent. (3) Results of cytoplasmic Ca^{2+} increases from PSt surface-stimulated platelets are consistent with a decrease in membrane fluidity during their activation process. (4) PAAmMAc surfaces can regulate platelet functions involving cytoplasmic free calcium levels and calcium-mediated sequential activation processes apparently in terms of cell metabolism. From our experience with platelets and soluble PAAmMAc, this effect is consistent with biochemical interaction of plasma membrane components with discrete PAAmMAc molecules or monomeric groups rather than PAAmMAc aggregated surfaces. (5) PAAmMAc surfaces act to increase the membrane fluidity of platelets, and this phenomenon is independent of platelet metabolism. Because soluble PAAmMAc shows no influence on membrane fluidity, macromolecular surface aggregates of PAAmMAc surface may dominate the increase in membrane fluidity of platelets.

Acknowledgements

The authors are grateful to Dr David W. Grainger, Oregon Graduate Institute of Science and Technology, USA, for his valuable discussion, and to Sekisul Chemical Co., Japan, for their supply of polystyrene microsphere particles. This research was financially supported by a Grant-in-Aid from the Ministry of Education, Science and Culture, Japan.

REFERENCES

1. T. Okano, M. Uruno, N. Sugiyama, M. Shimada, I. Shinohara, K. Kataoka and Y. Sakurai, *J. Biomed. Mater. Res.* **20,** 1035 (1986).
2. N. Yui, K. Kataoka, Y. Sakurai, T. Aoki, K. Sanui and N. Ogata, *Biomaterials* **9,** 225 (1988).
3. M. B. Feinstein, J. J. Egan, R. I. Shaafi and J. White, *Biochem. Biophys. Res. Commun.* **113,** 598 (1983).
4. J. A. Ware, P. C. Johnson, M. Smith and E. W. Salzman, *J. Clin. Invest.* **77,** 878 (1986).
5. N. Yui, K. Kataoka, Y. Sakurai, Y. Fujishima, T. Aoki, A. Maruyama, K. Sanui and N. Ogata, *Biomaterials* **10,** 309 (1989).
6. N. Yui, K. Suzuki, T. Okano and Y. Sakurai, *Jpn. J. Artif. Organs* **21,** 222 (1992).
7. N. Yui, K. Kataoka, T. Okano and Y. Sakurai, in: *Artificial Heart 3*, p. 23, T. Akutsu and H. Koyanagi (Eds). Springer, Tokyo (1991).
8. J. D. Andrade, H. B. Lee, M. S. Jhon, S. W. Kim and J. B. Hibbs Jr., *Trans. Am. Soc. Artif. Intern. Organs* **19,** 1 (1973).
9. S. D. Bruck, *J. Biomed. Mater. Res.* **7,** 387 (1973).
10. B. D. Ratner, A. S. Hoffman, S. R. Hanson, L. A. Harker and J. D. Whiffen, *J. Polym. Sci. Polym. Symp.* **66,** 363 (1979).
11. S. R. Hanson, L. A. Harker, B. D. Ratner and A. S. Hoffman, *J. Lab. Clin. Med.* **95,** 289 (1980).
12. C. Nojiri, T. Okano, K. D. Park and S. W. Kim, *Trans. Am. Soc. Artif. Organs* **34,** 386 (1988).
13. J. L. Brash, *Ann. N. Y. Acad. Sci.* **283,** 356 (1977).
14. I. A. Feuerstein, W. G. McClung and T. A. Horbett, *J. Biomed. Mater. Res.* **26,** 221 (1992).
15. H. Kawaguchi, Y. Yamada, S. Kataoka, Y. Morita and Y. Ohtsuka, *Polym. J.* **23,** 955 (1991).
16. Y. Uragami, Y. Kasuya, K. Fujimoto, H. Kawaguchi, M. Miyamoto and T. Juhji, *Polym. Prepr. Jpn.* **40,** 2509 (1991).
17. G. Grynkiewicz, M. Poenie and R. Y. Tsien. *J. Biol. Chem.* **260,** 3440 (1985).
18. C. M. Ingerman-Wojenski, *J. Med. Technol.* **1,** 697 (1984).
19. M. Shinitzky and M. Inbar, *Biochim. Biophys. Acta* **433,** 133 (1976).
20. M. Steiner, *Biochim. Biophys. Acta* **640,** 100 (1981).
21. L. Davenport, R. E. Dale, R. H. Bisby and R. B. Cundall, *Biochemistry* **24,** 4097 (1985).
22. T. Okumura, M. Hasitz and G. A. Janieson, *J. Biol. Chem.* **253,** 3435 (1978).
23. Y. Ikeda, *Jpn. J. Artif. Organs* **15,** 1887 (1986).
24. Y. Ikeda, K. Satoh, M. Handa, Y. Yoshii, M. Imai, K. Toyama, M. Yamamoto, K. Watanabe and Y. Ando, *Acta Haematol. Jpn.* **45,** 173 (1982).
25. J. R. Okita, D. Pidard, P. J. Newman, R. R. Montgomery and T. J. Kunicki, *J. Cell Biol.* **100,** 317 (1985).
26. J. E. B. Fox, *J. Biol. Chem.* **260,** 11970 (1985).
27. B. S. Coller, *Blood* **60,** 731 (1983).
28. N. O. Solumn, T. M. Gogstad, I. Hagen and F. Brosstad, *Biochim. Biophys. Acta* **729,** 53 (1983).
29. J. G. White, G. H. Rao and J. M. Gerrard, *Am. J. Pathol.* **77,** 135 (1974).
30. D. F. Wilson and B. Chance, *Biochem. Biophy. Res. Commun.* **23,** 751 (1966).
31. L. E. Limbird, *Biochem. J.* **195,** 1 (1981).
32. M. B. Feinstein, G. B. Zavoico and S. P. Halenda, in: *The Platelets: Physiology and Pharmacology*, p. 237, G. L. Longenecker (Ed.). Academic, London (1985).
33. C. M. Colley and J. C. Metcalfe, *FEBS Lett.* **24,** 241 (1972).
34. M. Steiner and E. F. Luscher, *Biochemistry* **23,** 247 (1984).
35. J. F. Stoltz and M. Donner, *Biorheology* **22,** 227 (1985).
36. M. Donner and J. F. Stoltz, *Biorheology* **22,** 385 (1985).
37. N. Yui, K. Suzuki, T. Okano, Y. Sakurai, C. Ishikawa, K. Fujimoto and H. Kawaguchi, in prep.

The synthesis of a water soluble complement activating polyacrylic acid–IgG polymer

GOTTFRIED SCHMER[1]*, RUTH A. HENDERSON[1] and
WOLFGANG MÜLLER[2]

[1]*Department of Laboratory Medicine, University of Washington, Seattle, WA 98195, USA*
[2]*Rheumatologische Universitatsklinik des Felix-Platter Spitals, 4012 Basel, Switzerland*

Received 29 June 1992; accepted 21 September 1993

Abstract—Polyacrylic acid–IgG polymer (PAA–IgG) activates the classical and alternative pathway of complement as shown by specific C1q-(IgG–PAA) interaction and the generation of C4d, Bb, C3b, C3a and C5a. This water soluble and stable PAA–IgG polymer represents a model substance for the study of humoral and cellular inflammatory mechanisms, mediated by the complement system.

Key words: Polyacrylic acid–IgG polymer; complement activation; immune complex.

INTRODUCTION

The formation of immune complexes (IC) is normally an important protective mechanism for the elimination of foreign material through formation of antibodies, mostly of the IgG type, which complex with the antigen and facilitate its elimination via the reticulo-endothelial system [1–5]. In immune complex diseases like systemic lupus erythematosus, rheumatoid arthritis, and rapidly progressing glomerulonephritis, auto-antibodies are formed which react with autoantigens to form IC. It is thought that IC cause tissue injury mediated by the activation of the complement system [6] and by the activation of macrophages, which lead to the release of inflammatory mediators [7]. There is evidence to suggest that the molecular basis of these events is a conformational change in the antibody molecule associated with the IC to expose its Fc region [8]: it was found that an IgM antibody, a pentamer of IgG, against the hapten phenyl-β-lactoside exhibited a lower K_a for the interaction of IgM with the monohapten substituted antigen (the hapten carrier) compared to the free hapten. While the former leads to C1q fixation the latter does not. Since the affinity constant at equilibrium is a mixed function which represents net binding energy and conformational change it is suggested that some of the binding energy was used to induce a conformational change in the Fc region with subsequent complement fixation. The interaction of the antigen with the Fab region of the N-terminal region of the antibody triggers a signal, which is transmitted to the Fc region at the C-terminal part of the molecule leading to an exposure of the C1q binding site also referred to as CH2 Fc domain. It is postulated that the signal transmission requires a stabilized immunoglobulin structure achieved either by disulfide bonds or antigen aggregation of monomeric immunoglobulins. The Fc region is then capable of reacting with C1q, the first component of the classical

* To whom correspondence should be addressed.

complement pathway, leading to the sequential activation of complement factors with the generation of the powerful inflammatory mediators C3a and C5a. In addition, the exposed Fc region of the antibody is capable of reacting with macrophages via the Fc receptor thereby activating phagocytosis of IC. However, it is difficult to predict any tissue damage by antigen–antibody complexes, since the formation of the antigen–antibody complexes is a dynamic process and the steady change of the antigen–antibody ratio influences the biological effect of the resulting IC. The pathogenic significance of IC in immune complex diseases is therefore not fully understood. Opinions range from disease causing to disease maintaining importance of IC and even to being simply an 'epiimmunological phenomenon' without major importance in the pathogenesis of immune complex diseases. To define more clearly the role of structurally altered IgG in the kinetics of inflammation we decided to bind IgG covalently to a water soluble matrix, polyacrylic acid polymers, in the hope of inducing a conformational change in IgG leading to complement activation and the generation of powerful inflammatory mediators. The covalent linkage would prevent the constantly changing complex ratio as seen in IC and would allow a better prediction and understanding of inflammatory mechanisms on a molecular basis. Two requirements must be met before IgG will bind C1q. First, the IgG must undergo a conformational change to expose the Fc binding region. Second, the density of the IgG binding regions must be such that two closely spaced IgG molecules are available to interact with one C1q molecule [9]. We previously described the synthesis and purification of polyacrylic acid–IgG polymer (PAA–IgG) capable of activating human granulocytes by specifically binding to the Fc receptors [10]. In this study we show that the polyacrylic acid polymer containing covalently bound IgG molecules is a powerful activator of the complement system by specifically binding C1q, the first component of complement and leading to the formation of C3a and C5a through the classical, as well as alternative pathway, thereby imitating the pathophysiological effect of naturally occurring IC.

MATERIALS AND METHODS

p-Nitro phenyl phosphate, anti-human IgG (Fab specific) alkaline phosphatase conjugate from goat, rabbit anit-human C_3, human C1q, anti-human C1q from goat, goat anti-human IgG (Fc specific), and bovine albumin were purchased from Sigma, St. Louis, MO, USA. 99% pure human IgG derived from Cohn fraction II and III (Sigma) was routinely subjected to immune electrophoresis to exclude the presence of IgA and in particular IgM and was dialyzed against 100 volumes of imidazole buffered saline (IBS) prior to use. All reagents used were reagent grade.

IgG starting material for the synthesis was also checked routinely by gel filtration over Sepharose 6B CL to exclude high molecular weight aggregates. Five ml of an IgG solution (10 mg ml^{-1}) were passed over a Sepharose 6B CL column (2.5 × 42 cm) equilibrated in IBS and 5 ml fractions were collected. Protein throughout the gel filtration profile was measured by the extinction at 280 nm and by a color assay (Bio-Rad Protein assay, BIORAD, Richmond, CA, USA).

The synthesis of PAA–IgG

One-and-a-half grams of polyacrylic anhydride (Polysciences, Warrington, PA) was dissolved in 100 ml dry dimethylformamide (DMF) with stirring. The clear solution was stored at −80°C in tightly closed polyethylene plastic tubes. Five milliliters of cold solution of polyacrylic anhydride was added dropwise under stirring within 5 min to 100 ml of imidazole buffered saline (0.05 M imidazole, 0.15 M NaCl, pH 7.4) containing 1 g human IgG. The IgG solution was dialyzed extensively against the same buffer prior to use and was cooled to 0°C before the addition of polyacrylic anhydride solution. The mixture was stirred for 24 h at 0°C. The pH of the solution drops rapidly during the first 30 min of the reaction. 1 M NaOH was added dropwise to maintain the pH at 7.4. Ten milliliters of the reaction mixture were passed at room temperature over a column (2.5 × 100 cm) containing Sepharose CL-6B (Pharmacia, Uppsala, Sweden) equilibrated with imidazole-buffered saline and fractions of 7 ml were collected. Samples containing the PAA–IgG (first protein peak) were pooled and kept at 4°C prior to use.

The binding of C1q to PAA-IgG polymer

The specific reaction of PAA–IgG with C1q was determined by 3 different methods: (1) enzyme linked C1q solid phase essay (ELISA); (2) adsorption of radiolabeled PAA–IgG to solid phase C1q; and (3) liquid phase C1q binding assay.

Method 1 ELISA. The ELISA was carried out in polystyrene tubes and followed essentially the procedure described for the determination of IC in serum [11]. Polystyrene tubes (11 × 55 mm Falcon) were coated at room temperature for 2 h under horizontal shaking (Lab-Line Shaker, Melrose Park, IL) with 4 μg C1q in 0.5 ml phosphate buffered saline (PBS, 0.15 M NaCl, 0.02 M phosphate, pH 7.4). The tubes were washed three times with 4 ml of 0.15 M NaCl containing 1% Tween (Sigma Chem. Co., St. Louis, MO). Next, 280 μg PAA–IgG (stock solution = 3 mg ml^{-1} IBS) were incubated with 1 ml fresh human serum for 15 min at 37°C to imitate the *in vivo* conditions of natural IC. One tenth milliliter of the serum-PAA–IgG mixture was mixed with 0.4 ml of 0.2 M EDTA, pH 7.4. Forty μl of this serum dilution was added to 1 ml of PBS containing 1% Tween giving an overall dilution of 1:125 or 2.3 μg PAA–IgG per ml. Eight more serial dilutions were carried out in PBS–Tween. A 0.5 ml aliquot of each dilution sample was incubated at room temperature for 2 h with C1q coated polystyrene tubes under horizontal shaking (100 times min^{-1}). The tubes were then washed with PBS–Tween as described and incubated under horizontal shaking for 2 h with 0.5 ml PBS–Tween containing 5 μl of a 1:300 dilution (PBS–Tween) of phosphatase conjugated goat-anti-human IgG. After a final wash as described above, 2 ml 0.5 M sodium carbonate, 1 mM $MgCl_2$, pH 9.8 containing 2 mg *p*-nitro-phenyl-phosphate were added to the tubes and incubated for 30 min at 37°C. The absorbance at 405 nm was measured in a 8451A Diode Array Spectrophotometer (Hewlett Packard) after stopping the reaction with 0.1 ml 1 M NaOH. Comparable serum dilutions minus the PAA–IgG conjugate in C1q coated polystyrene tubes, were used as a control. To prove the specific interaction of PAA–IgG with the solid phase C1q the following experiments were carried out: PAA–IgG was incubated in a final concentration of 0.5, 1, 2, and 3 μg ml^{-1} PBS–Tween solution containing 0.2% bovine albumin with the C1q solid phase system described above. The absorbance curve of C1q bound PAA–IgG was compared to the curves obtained after the simultaneous addition of 2 and 4 μl of an anti-human C1q solution (44 mg fractionated total IgG ml^{-1}, Lot #99F8828) to PAA–IgG. In addition a control experiment was carried out adding polyacrylic acid

(M_w250 × 10^3, pH adjusted to 7.5) to PAA–IgG in a ratio of 1:5 to assess possible nonspecific interactions between the PAA residues of PAA–IgG with solid phase C1q. The resulting binding of PAA–IgG was measured exactly as described in Method 1.

Method 2. The specific interaction between solid phase C1q and PAA–IgG was also checked by the use of radiolabeled PAA–IgG: PAA–IgG was radiolabeled with ^{125}I by the lactoperoxidase method as described for C1q [12]. Four parallel experiments were carried out: (a) Radiolabeled PAA–IgG (2.4 μg, specific activity = 17.3 × 10^6 cpm mg^{-1} in 0.5 ml PBS containing 0.2% bovine albumin was incubated overnight at 4°C in C1q coated polystyrene tubes [11]. (b) The same experiment was carried out with free neutralized polyacrylic acid (PAA) added to PAA–IgG in a weight ratio of 1:5. (c) C1q coated tubes were first incubated overnight at 4°C with 10 μl goat anti-human C1q in PBS–Tween, then washed with the same buffer and incubated with radiolabeled PAA–IgG as described. (d) The iodinated PAA–IgG polymer in PBS/0.2% albumin was first mixed with 10 μg goat anti-human IgG (Fc specific) and incubated overnight at 4°C in uncoated polystyrene tubes. The reaction mixture was then transferred to C1q coated tubes with the assay carried out as described in (a). (e) A parallel control experiment to (d) used the same conditions described in (d) in the absence of Fc specific antibody. The blank consisted of iodinated PAA–IgG in PBS/0.2% bovine albumin using uncoated polystyrene tubes. The tubes were washed free of nonbound iodinated PAA–IgG as described in Method 1 and surface bound radioactivity was counted in a Packard Model PGD gamma counter.

Method 3. C1q binding to PAA–IgG was finally determined by a liquid phase C1q binding assay [12]. This assay is carried out in IC containing sera treated with EDTA and consists of a competitive binding of radiolabeled C1q with IC and subsequent polyethylene glycol precipitation of the radioactive C1q–IC complex.

The determination of C3a and C5a

A commercially available RIA kit (Amersham, Arlington Heights, IL) was used for the determination of C3a and C5a. Fifty, 100 and 200 μg PAA–IgG (stock solution = 4 mg ml^{-1} IBS) were added to 1 ml aliquots of fresh human serum in polystyrene tubes and incubated for 20 min at 37°C. In a parallel experiment 100, 250, 500 and 1000 μg of the starting IgG material (stock solution = 10 mg ml^{-1} IBS) were incubated with 1 ml aliquots of human serum as a control. The reaction was stopped after 20 min by the addition of 50 μl 0.2 M EDTA, pH 7.4. Sera exceeding 10 μg C3a per ml were diluted with bovine EDTA plasma and reassayed. Kinetic experiments of C3a and C5a generation were performed by incubating 25 μg PAA–IgG with 1 ml fresh serum at 37°C and removing 0.1 ml samples every 5 min for 20 min to determine the rate of C3a and C5a formation. A serum blank containing 25 μg IgG with identical serial determinations was used as a control. All assays were done in duplicate.

Demonstration of C5a formation by the granulocyte aggregate method

Human granulocytes were isolated by a simple one step Percoll density gradient [13]. Aggregation assays were carried out on a Payton dual channel aggregation module as described by Craddock *et al.* [14]. In brief, 50 μg PAA–IgG were incubated with 1 ml fresh human serum for 30 min at 37°C. Next, 50 μl of the activated serum were added to 0.4 ml of the granulocyte suspension (8×10^6 cells) in Hank's buffer containing 1% human albumin. Zymosan activated serum served as the positive control [14]. Two negative controls consisted of 50 μl IgG added to 1 ml human serum and 50 μg PAA-IgG incubated with 1 ml EDTA plasma with the granulocyte aggregation assay carried out as described.

Immune electrophoresis of C3 activation products

This was carried out as described [15] with the following modifications: 1 ml aliquots of fresh human serum were incubated for 15 min at 37°C with 100 μg PAA-IgG (stock solution—1 mg ml^{-1} IBS). The reaction was stopped by the addition of 50 μl 0.2 M EDTA pH 7.4. A negative control consisted of serum incubated with 100 μg IgG (stock solution—1 mg ml^{-1} IBS) under identical conditions. Zymosan activated serum served as a positive control [15].

Classical v alternative complement pathway involvement

To measure the respective contribution of both complement pathways C4d and Bb levels were measured in fresh human serum incubated with PAA-IgG using an ELISA Kit (Quidel, San Diego, CA). (1) Aliquots of 1 ml fresh human sera were incubated with 25 μg PAA-IgG for 30 min at 37°C. A total of ten different PAA-IgG preparations were used for the determinations of Bb and C4d generated. Serum without PAA-IgG served as the control. All determinations were done in duplicate. (2) Kinetic studies of Bb and C4d generation by PAA-IgG were carried out by incubating 1 ml serum samples with 25 μg PAA-IgG for 5, 10, 15, 20, 25, and 30 min. The assay was carried out exactly as described by the company. Only single determinations were done in this case. V_{max} kinetic microplate reader (Molecular Devices) was used to determine the optical density.

RESULTS

The binding of C1q to PAA-IgG

Method 1. Figure 1a shows the typical binding curve of PAA-IgG to C1q coated tubes using the ELISA technique. The detection limit of the conjugate was 0.035 μg ml^{-1} sample. The blank using serum dilutions containing the IgG starting material gave the same low background values throughout the dilution series.

Figure 1b shows the decrease of the absorbance at 405 nm as a measure of PAA-IgG adsorbed to the C1q solid phase in the absence and presence of anti C1q. There is about a 40% decrease of PAA-IgG binding after the addition of 2 μl anti C1q and a 60% decrease after the addition of 4 μl anti C1q. The values shown are not corrected for the blank consisting of C1q solid phase in the absence of PAA-IgG (blank—OD280 = 0.08). The addition of polyacrylic acid did not decrease the adsorbance of 405 nm indicating a specific C1q-(PAA-IgG) interaction through the IgG component (results not shown).

Method 2. The specific interaction of PAA-IgG with C1q is shown in Table 1. The addition of 0.5 μg PAA does not cause a significant decrease of PAA-IgG binding to C1q. However, the coating of C1q with anti C1q, as well as the incubation of PAA-IgG with Fc specific anti-human IgG, causes a dramatic decrease of PAA-IgG binding to solid phase C1q of 96.7 and 92%, respectively.

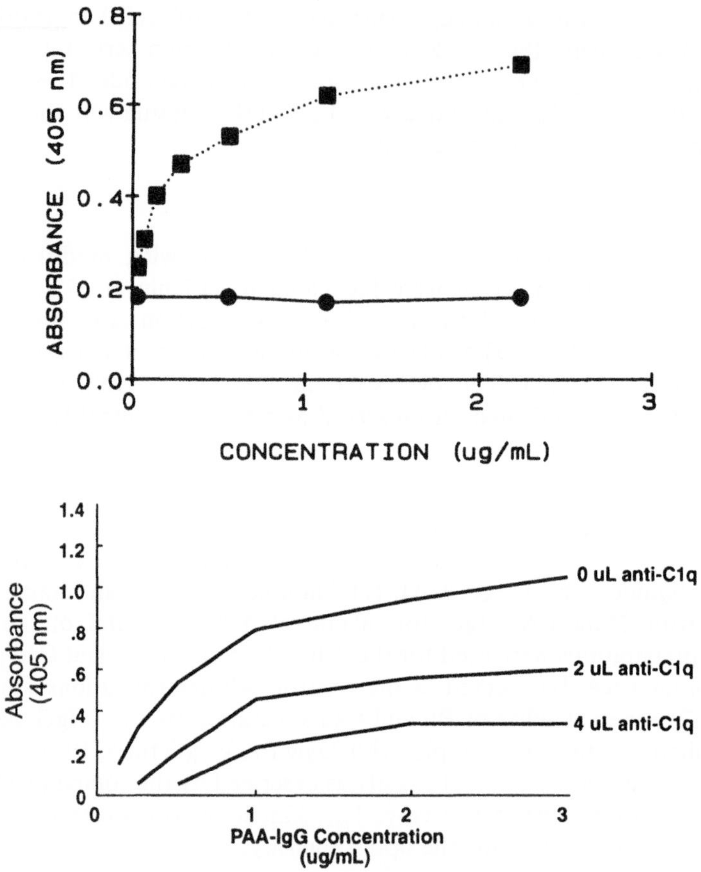
0.8
0.6
0.4
0.2
0.0
ABSORBANCE (405 nm)
0
1
2
3
CONCENTRATION (ug/mL)
1.4
1.2
1.0
.8
.6
.4
.2
0
Absorbance (405 nm)
0 uL anti-C1q
2 uL anti-C1q
4 uL anti-C1q
PAA-IgG Concentration (ug/mL)

Figure 1. (a) Binding of PAA–IgG (■···■) and IgG control (●–●) to C1q coated polystyrene tubes (ELIZA). For details see Methods. (b) Binding of PAA–IgG to C1q coated polystyrene tubes in the presence and absence of anti C1q.

Table 1.
The Binding of polyacrylic acid–IgG polymers to C1q coated tubes. Values reflect the average of two determinations

Samples	c.p.m.	Percent adsorption
(a) PAA–IgG	17 200	100
(b) PAA–IgG + PAA	16 684	97
(c) PAA–IgG + anti C1q	568	3.3
Blank	590	3.4
(d) PAA–IgG + anti Fc	1150	8
(e) PAA–IgG	14 380	100

The baseline value of radioactivity bound to the C1q solid phase in the absence of free PAA, anti C1q, and anti Fc = 100%. Each adsorption experiment reflects the average of 2 experiments. For experimental details see Methods.

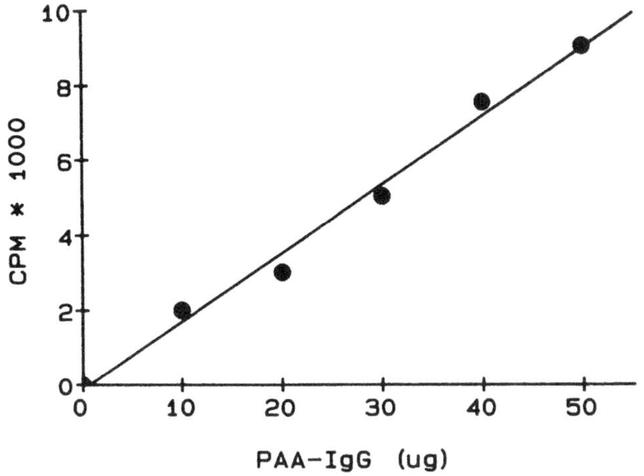
CPM * 1000
10
8
6
4
2
0
0
10
20
30
40
50
PAA-IgG (ug)

Figure 2. Dose response curve of PAA–IgG as measured by the liquid phase C1q binding assay. Blank Values (av. 300 cpm) have been subtracted.

Method 3. The high affinity of the conjugate for C1q was also shown by the linear increase of radiolabeled C1q binding to PAA–IgG as a function of conjugate concentration using a liquid phase C1q-binding assay (Fig. 2) which measures IC by its binding to radiolabeled C1q [12].

The generation of C3a and C5a by PAA-IgG

Figure 3a and 3b summarize the results of the generation of C3a and C5a in serum by PAA–IgG. The addition of as little as 50 μg PAA–IgG to 1 ml of human serum generated an average of 19.7 μg C3a in five experiments, while the addition of 100 μg IgG to 1 ml serum generated only 4.3 μg C3a, slightly above the serum blank of 3.3 μg. With 500 μg IgG added to 1 ml serum, an increase in C3a generation to 7.4 μg ml^{-1} could be observed indicating that a small amount of complement activating IgG fraction was present in the preparation. The generation of C5a by PAA–IgG was comparable to the results of C3a formation. The addition of 100, 250, and 500 μg IgG to serum did not generate C5a above the serum blank (27 ng ml^{-1}). Only with 1000 μg IgG added to 1 ml serum could an increase in C5a generation be seen (43 ng). In contrast a significant increase in C5a (83 ng ml^{-1}) could be observed with only 50 μg PAA–IgG. With 1000 μg PAA–IgG an average C5a generation of 520 ng ml^{-1} was seen. Figure 4a and b show the time course of C3a and C5a generation in 1 ml of serum by 25 μg PAA–IgG. The serum blank of C3a started out relatively high (4.0 μg ml^{-1} at '0' min incubation) rising slowly to 5.0 μg ml^{-1} after 5 min and 8.0 μg ml^{-1} after 20 min. C3a in serum incubated with PAA–IgG rose sharply and reached 20 μg ml^{-1} after only 5 min with a slower increase to 25 μg ml^{-1} after 20 min. C5a increased to 55 ng ml^{-1} after 10 min with no further significant increase. The serum blank for C5a remained around 9 ng ml^{-1} throughout the 20 min incubation period.

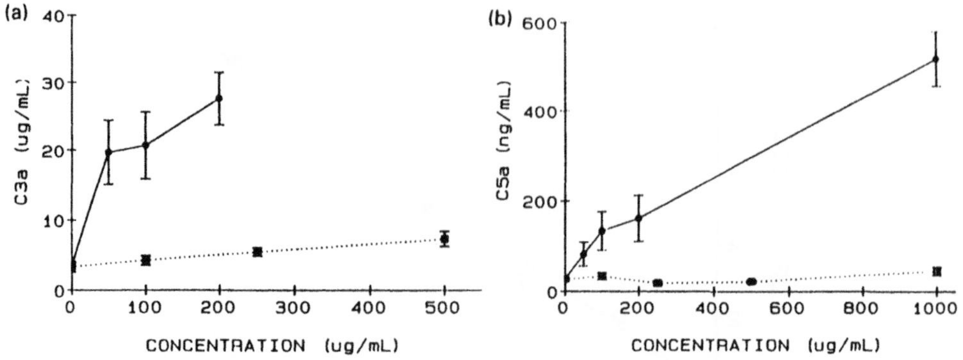
(a)
C3a (ug/mL)
CONCENTRATION (ug/mL)
(b)
C5a (ng/mL)
CONCENTRATION (ug/mL)

Figure 3. (a) Generation of C3a measured as C3a desARG by competitive radioimmune assay as a function of PAA–IgG concentration. Increasing amounts of PAA–IgG were added to one ml fresh human serum. Endpoint of C3a generation was 20 min at 37°C. C3a generation by PAA–IgG = (●–●–●). C3a generation by IgG = (■···■). Values reflect an average of five determinations. Error bars indicate ±SEM. (b) Generation of C5a measured as C5a desARG by competitive radioimmune assay as a function of PAA–IgG concentration. Reaction conditions are as described in (a). C5a generation by PAA–IgG = (●–●–●). C5a generation by IgG = (■···■). Values reflect an average of five determinations. Error bars indicate ±SEM.

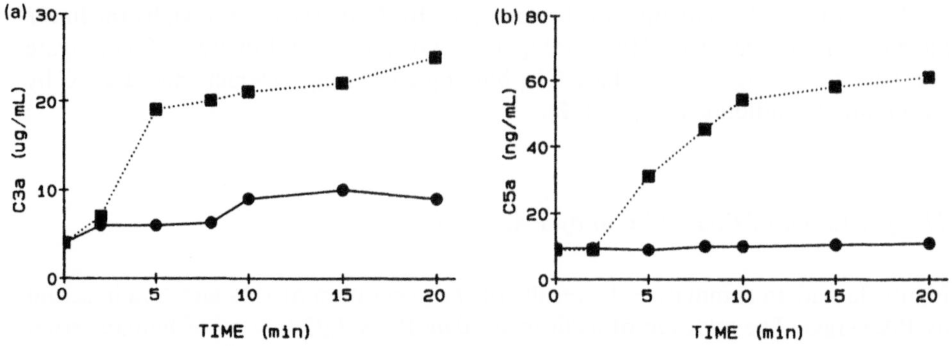
(a)
C3a (ug/mL)
TIME (min)
(b)
C5a (ng/mL)
TIME (min)

Figure 4. (a) Generation of C3a measured as C3a desARG by competitive radioimmune assay as a function of time. (■···■) = C3a activity generated by 25 μl PAA–IgG in 1 ml fresh serum. (●–●–●) = C3a activity generated by IgG. Values reflect the average of two determinations. (b) Generation of C5a measured as C5a desARG by competitive radioimmune assay as a function of time. (■···■) = C5a generated by 25 μl PAA–IgG in 1 ml fresh human serum. (●–●–●) = C5a activity generated by IgG. Values reflect the average of two determinations.

The generation of C3 activation products

Figure 5 shows the qualitative generation of C3 activation products by immune electrophoresis as described [15]. Serum incubated with PAA–IgG (100 μg ml^{-1} serum) showed the elongated precipitin line indicative of C3b activation products. In contrast, serum incubated with 100 μg IgG showed only non-activated C3. Zymosan activated serum was used as a positive control [15] to demonstrate C3 activation.

Measurement of PAA–IgG mediated C5a generation by granulocyte aggregation

The addition of 50 μl of PAA–IgG activated serum as described above to isolated human granulocytes (8 × 10^6 cells 0.4 ml^{-1} cuvette) caused a pronounced

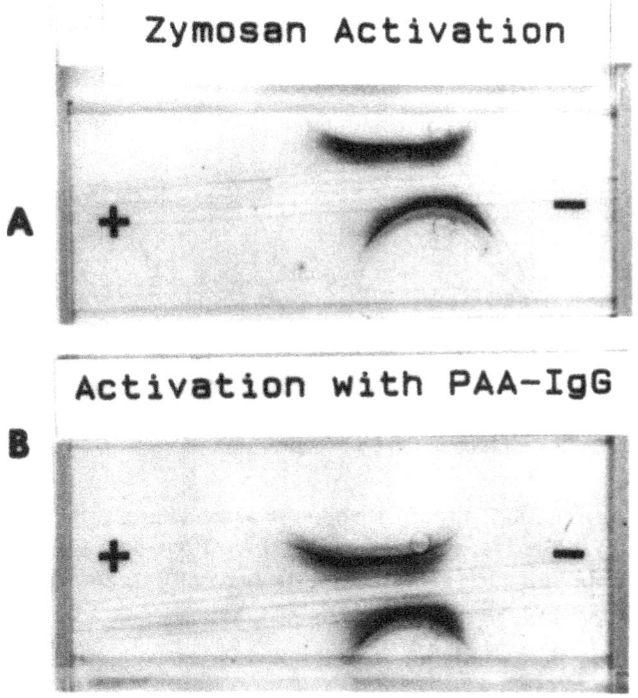
Zymosan Activation
A
+
-
Activation with PAA-IgG
B
+
-

Figure 5. Immune electrophoresis of fresh human serum incubated with PAA-IgG or Zymosan. Upper half of the slides represent the pattern obtained by Zymosan activation (A) and by PAA-IgG activation (b). Lower half of the slides show the IgG control. Rabbit antihuman C3 was used. The increased electrophoretic mobility of the precipitin lines reflect smaller C3 activation products [15].

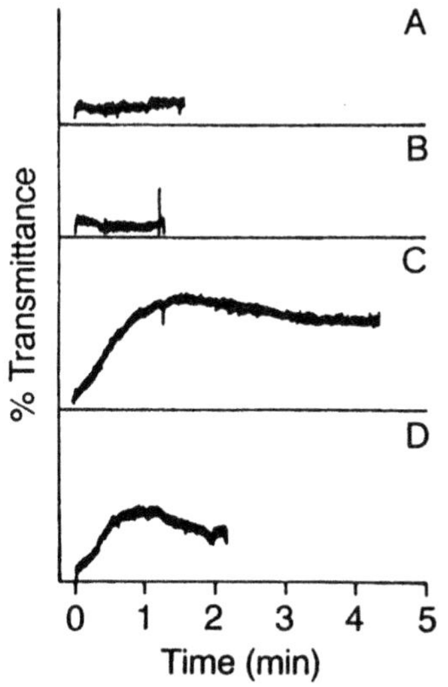
A
B
C
D
% Transmittance
0
1
2
3
4
5
Time (min)

Figure 6. C5a mediated granulocyte aggregation with human sera incubated with Zymosan or PAA-IgG. 8×10^6 granulocytes in Hank's buffer + 1% human albumin were mixed with 50 μl of the controls (IgG + serum and PAA-IgG + EDTA plasma), and PAA-IgG of Zymosan activated fresh human serum. A = blank (IgG + serum), B = blank (EDTA plasma), C = PAA-IgG activated serum, D = Zymosan activated serum. Transmission was measured in a Payton dual channel aggregation module.

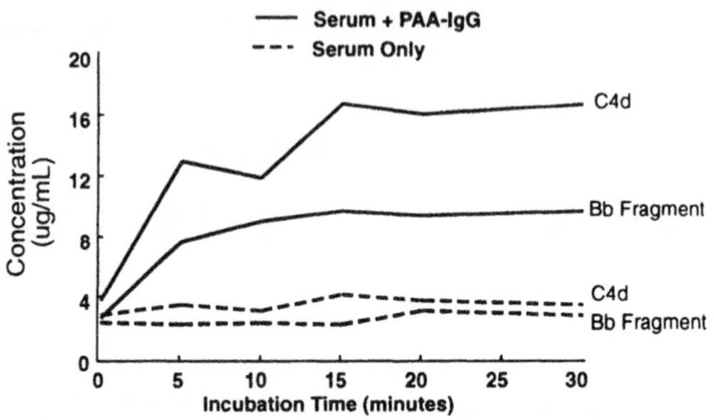
Serum + PAA-IgG
Serum Only
Concentration (ug/mL)
20
16
12
8
4
0
C4d
Bb Fragment
C4d
Bb Fragment
0
5
10
15
20
25
30
Incubation Time (minutes)

Figure 7. Generation of C4d and Bb in fresh human serum incubated with 25 μg PAA-IgG ml^{-1} measured by ELISA as a function of time.

granulocyte aggregation (Fig. 6C) while the same amount of non activated serum (IgG + serum) showed no aggregation (Fig. 6A). PAA–IgG added to EDTA plasma (Fig. 6B) also did not elicit any granulocyte aggregation. Zymosan activated serum served as the positive control (Fig. 6D).

C4d and Bb generation by PAA–IgG

The incubation of 1 ml fresh human serum with 25 μg PAA–IgG for 30 min at 37°C resulted in a C4d generation of 13.3 ± 2 μg (corrected for serum blank) with a serum blank of 4.0 ± 0.6 μg. Bb levels rose to 8.1 ± 2.04 μg (corrected for serum blank) with a serum blank of 2.6 ± 0.58 μg.

Figure 7 shows the increase in Bb and C4d levels in serum by addition PAA–IgG as a function of time. While the serum blank for C4d rose from 2.8 μg ml^{-1} to only 3.5 μg ml^{-1} within 30 min, an increase of C4d levels were observed from 3.5 to 16.8 μg ml^{-1}.

Bb levels in the serum blank showed no significant increase within 30 min (from 2.3 to 2.66 μg ml^{-1}) with a significant increase in PAA–IgG activated serum (from 2.66 to 9.8 μg ml^{-1}). The reactions reached two thirds of its maximum activation after 5 min and were complete after 15 min with no further increase in Bb and C4d levels.

DISCUSSION

Of primary importance was the assessment of a specific interaction of PAA–IgG and C1q with the possibility of nonspecific binding between the residual negative groups of PAA in the PAA–IgG complex and the highly positively charged solid phase C1q. The competitive experiments carried out with the addition of free PAA demonstrated that free PAA did not decrease the C1q–(PAA–IgG) interaction demonstrating that the binding of PAA–IgG to C1q cannot be explained by a charge effect between the highly positively charged C1q and the negative charges of PAA residues.

The dramatic decrease of the binding of the polymer to C1q coated tubes pretreated with an anti C1q antibody, as well as the decrease of binding after preincubation of PAA–IgG with Fc specific anti IgG, reflects the specific binding between C1q and the Fc region of PAA–IgG, a prerequisite for the activation of the classical complement pathway.

We further demonstrated the generation of C3a and C5a as well as of the C3 activation product C3b. Although the amount of C3a and C5a generated by as little as 50 μg PAA–IgG was considerable in the light of its potent inflammatory capability, we might have underestimated the amount generated by a possible tight binding of the highly positively charged anaphylatoxins to the highly negatively charged PAA residues. C5a formation was also shown conclusively by a biological assay system (granulocyte aggregation assay). C5a, and in particular C3a generation, was not linear as a function of PAA–IgG added to serum. Also the formation of C3a and C5a by a fixed amount of PAA–IgG was complete after 5 and 10 min respectively, with no further increase in C3a and C5a levels.

This raised the question of an activation of the alternative pathway with the generation of C3a and C5a with subsequent control of C3 and C5 convertase by factor H and I [16, 17]. Our data indeed show an activation of the alternative complement pathway as measured by the generation of Bb fragment. In addition, however, we found clear signs of activation of the classical complement pathway as shown by the specific interaction of C1q and PAA–IgG with the subsequent generation of C4d, which is derived from the C4 activation product C4b and represents an end product of C4 activation. C4b is strictly regulated and its activity is inhibited by Factor I [17] which can further degrade C4b to C4d. The flattening of the C4d generation curve after 5–10 min again implies the action of an inhibitor, possibly C1s esterase inhibitor.

While the data on specific C1q–(PAA–IgG) interaction with subsequent C4d formation clearly demonstrate the activation of the classical complement pathway, less certainty exists as to the mechanism of complement activation through the alternative pathway. However, it is known that autoimmune diseases with immune complexes activate the classical pathway, leading to the formation of C3b, which is capable of binding Factor B and initiating the formation of Bb fragments. A special example of an autoantibody inducing alternative pathway activation is the C3-nephritic factor [18]. Thus, alternative pathway can be activated in autoantibody mediated diseases leading to an increase in complement activation with subsequent tissue destruction. The importance of PAA–IgG therefore, lies in its similarity with certain features of immune complexes in autoimmune diseases and in the possibility of using this compound as a specific inflammatory model substance to elicit humoral and cellular inflammatory mechanisms. In contrast to natural IC, PAA–IgG is non-dissociable under physiological conditions and would, therefore, allow a clearer interpretation of results under biological conditions.

Acknowledgement

We are greatly indebted to the Stanley Thomas Johnson Foundation for the generous support of this project.

REFERENCES

1. B. Albini, E. Penner and P. Knoflach, *Wiener Klinische Wochenschrift* **8**, 363 (1985).
2. A. O. Haakenstad and M. Mannik, *J. Immunol.* **112**, 1939 (1974).

3. M. I. Hamburger, E. H. Gerardi, T. S. Fields and M. L. Bernstein, *Arthritis Rheum.* **24**, (Suppl. 5), 98 (1981).
4. M. Mannik, A. O. Haakenstad and W. P. Arend, in: *Progress in Immunology II*, Vol. 5, p. 91, L. Brent and J. Holborow (Eds). North Holland, Amsterdam (1974).
5. A. O. Haakenstad and M. Mannik, *Lab. Invest.* **35**, 283 (1976).
6. T. E. Hugli and H. Muller-Eberhard, in: *Advances in Immunology*, Vol. 26, pp. 1–53, F. J. Dixon and E. J. Kunkel (Eds). Academic Press, New York (1976).
7. D. T. Fearson and K. F. Austen, *Biol. Pathobiol. Essays Med. Biochem.* **2**, 1 (1976).
8. J. C. Brown and M. E. Koshland, *Proc. Natl. Acad. Sci. USA* **72**, 5111 (1975).
9. H. J. Muller-Eberhard and M. A. Calcott, *Immuno-chemistry* **3**, 500 (1966).
10. R. J. Klauser, G. Schmer, W. L. Chandler and W. Muller, *Biochem. Biophys. Acta* **1052**, 408 (1990).
11. S. Ahlstedt, L. A. Hanson and C. Wadsworth, *Scand. J. Immunol.* **5**, 293 (1976).
12. P. J. Spaeth, A. Corvetta, V. E. Nydegger, M. Montroni and R. Buetler, *Scand. J. Immunol.* **18**, 319 (1983).
13. R. J. Harbeck, A. A. Hoffman, S. Redecker, T. Biundo and J. Kurnick, *Clin. Immunol. Immunopathol.* **23**, 682 (1982).
14. P. R. Craddock, D. Hammerschmidt, J. G. White, A. P. Dalmasso and H. S. Jacob, *J. Clin. Invest.* **60**, 260 (1977).
15. P. R. Craddock, J. Fehr, A. P. Dalmasso, K. L. Brigham and H. S. Jacob, *J. Clin. Invest.* **59**, 879 (1977).
16. J. M. Weiler, M. R. Daha, K. F. Austen and D. T. Fearon, *Proc. Natl Acad. Sci. USA* **73**, 3268 (1976).
17. M. K. Pangburn, R. P. Schreiber and H. J. Eberhard, *J. Exp. Med.* **146**, 257 (1977).
18. W. E. Ratnoff, D. T. Fearon and K. F. Austen, in: *Complement*, p. 215, H. J. Muller-Eberhard and P. A. Miescher (Eds). Springer Verlag, Berlin (1984).

A novel biomaterial: Poly(dimethylsiloxane)-polyamide multiblock copolymer I. Synthesis and evaluation of blood compatibility

TSUTOMU FURUZONO[1], EIJI YASHIMA[1*], AKIO KISHIDA[1], IKURO MARUYAMA[2], TAKEO MATSUMOTO[3] and MITSURU AKASHI[1†]

[1]*Department of Applied Chemistry and Chemical Engineering, Faculty of Engineering, Kagoshima University, 1-21-40 Korimoto, Kagoshima 890, Japan*

[2]*Department of Clinical Laboratory Medicine, Faculty of Medicine, Kagoshima University, 8-35-1 Sakuragaoka, Kagoshima 890, Japan*

[3]*Tsukuba Research Laboratory, NOF Corporation, Ltd., 5-10 Tokodai, Tsukuba 300-26, Japan*

Received 30 June 1992; accepted 8 December 1992

Abstract—Aramid-silicone resins (PASs) consisting of aromatic polyamide (aramid) and poly(dimethylsiloxane) (PDMS) segments were synthesized by low temperature solution polycondensation. For the evaluation of blood compatibility *in vitro*, two kinds of experiments were carried out. One was the thromboxane B_2(TXB_2) release test from platelets attaching to PAS and Biomer®. The other was the observation of the platelet adhesion on the surfaces of PAS by scanning electron microscopy (SEM). The results indicated that PAS was bio-inert *in vitro*. The surface chemical composition of PAS films was investigated by means of electron probe micro analysis (EPMA), X-ray photoelectron spectroscopy (XPS), and dynamic contact angle measurements. The relationship between blood compatibility and surface composition of PAS is discussed.

Key words: Poly(dimethylsiloxane); aramid; multiblock copolymer; blood compatibility; electron probe micro analysis; X-ray photoelectron spectroscopy; dynamic contact angle.

INTRODUCTION

Silicone rubber is widely used in medical fields, in many forms, such as blocks, tubes, sutures and films [1–4]. Generally, the desirable properties of the silicone are high thermal stability, oxidative stability, low surface energy, water repellency, good dielectric properties, high gas permeability and good biocompatibility [1–4]. Especially in the medical field, the last two characteristics were important. On the other hand, the mechanical properties of silicone rubber, especially low tensile strength, sometimes limit their medical application. To improve the mechanical properties of silicone rubber, many researchers have reported the syntheses of high-modulus resin including silicone unit which possess both good mechanical properties and the characteristics of silicone [5–12]. For instance, poly(dimethylsiloxane-co-carbonate) [6–11] and poly(dimethylsiloxane-co-hydroxystyrene) [12], of which both mechanical properties and oxygen permeability were excellent, were synthesized and applied to the membrane for preparing the oxygen-enriched air in the industrial field.

Recently, Imai *et al.* reported the synthesis of PDMS and aromatic polyamide (aramid) copolymer (PAS) [13, 14]. PAS is a multiblock copolymer consisting of

* Present address: Nagoya University, Nagoya 464-01, Japan.

† To whom correspondence should be addressed.

hard segment (aramid) and soft segment (PDMS). The aramid is known to have very high-mechanical properties, such as high-thermal stability and high modulus of elasticity [15]. It is expected that the incorporation of PDMS domain into an aramid matrix allows the system to exhibit many of the often desirable properties of both polymers. However, there has so far been only a few studies reported on characteristics of PAS [13]. In this paper, we report the synthesis, evaluation of *in vitro* blood compatibility and investigation of the surface composition of PAS from the point of view of novel biomaterial. This is the first part of our research on developing a novel biomaterial, PAS.

MATERIALS AND METHODS

Materials

In the following, numbers such as (1) refer to the specific molecules shown in Scheme 1. PDMS-diamine (3) of number-average molecular weight of 1680 g/mol was obtained from Shin-Etsu Chemical Co. (Tokyo, Japan), and dried at 100°C for 3 h under vacuum. 3,4′-Diaminodiphenylether (1) (3,4′-DAPE: Wakayama Seika Industry Co., Wakayama, Japan), isophtaloyl chloride (IPC) (2), triethylamine (TEA) and chloroform were purchased from Wako Pure Chemicals (Osaka, Japan) and were purified by distillation. Triethylamine hydrochloride (TEA·HCl: Wako Pure Chemicals Co., Osaka, Japan) was purified by recrystallization from ethanol. All other solvents and chemicals were purified by distillation. Biomer® was purchased from Ethicon Inc. (Somerville, New Jersey, USA). Silastic® 500-1 was kindly donated by Dow Corning Corp. (Tokyo, Japan).

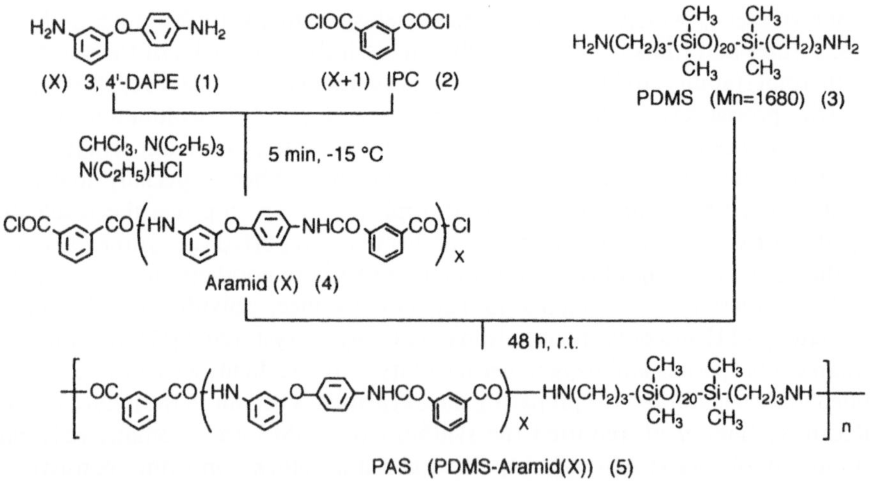

(X) 3, 4'-DAPE (1)
(X+1) IPC (2)
PDMS (Mn=1680) (3)
CHCl3, N(C2H5)3
N(C2H5)HCl
5 min, -15 °C
Aramid (X) (4)
48 h, r.t.
PAS (PDMS-Aramid(X)) (5)

Scheme 1. Synthesis of PAS

Synthesis of PAS

PAS (5) was synthesized by low temperature solution polycondensation through a two step procedure according to the method developed by Imai *et al.* with a little modification [13] (Scheme 1). First, α, ω-dichloroformyl-terminated aramid oligomers (4) were prepared by the reaction of IPC (2) with 3,4′-DAPE (1) in a chloroform-TEA·HCl system at −15°C for 5 min in the presence of TEA as hydrogen chloride acceptor under nitrogen. Next, the preformed aramid oligomers were then reacted with PDMS (3) at −15°C for 1 h. The reaction was then continued at room temperature for another 48 h under nitrogen. The polymer was isolated by pouring the reaction mixture into methanol. After the product was washed successively with excess amounts of methanol, a low molecular weight fraction enriched in PDMS was removed by washing the product with a good deal of *n*-hexane three times, then dried at 60°C for 48 h under vacuum. PAS films for surface analysis and platelet adhesion were cast from 10 wt% *N,N′*-dimethylacetamide (DMAc) solution in stainless steel petri dishes.

Tensile properties measurement

Tensile properties were determined by the stress–strain curves obtained with Tensilon UTM-I (Toyo Baldwin Co., Tokyo, Japan) at an elongation rate of 1 mm/min. Measurements were performed at room temperature with film specimens (3 mm wide, 15 mm long and 0.1 mm thick) [16] and four individual determinations were averaged.

Molecular composition characterization

^1H NMR spectra were recorded on JEOL EX-90 Fourier transform spectrometer (JEOL, Tokyo, Japan). The polymers were dissolved in DMSO-d_6 and D_2SO_4.

Surface characterization

The electron probe micro analysis (EPMA) spectra were obtained with Shimadzu EMX-SM (Shimadzu Co., Kyoto, Japan). EPMA was operated under an excitation voltage of 15 kV and a sample current of 0.003 mA. A spectrometer ESCA 1000 (Shimadzu Co., Kyoto, Japan) was employed to carry out X-ray photoelectron spectroscopy (XPS) [17–19]. The dynamic contact angle of PASs was measured using dynamic Wilhelmy plate technique [20] by means of a Orientic DCA-20 (Orientic Co., Tokyo, Japan).

Thromboxane B_2 releasing test

The 10 wt% of PASs and Biomer® solutions were poured into the glass tubes and coated the inside of glass tubes (7 mm I.D. × 50 mm) respectively, the solvent evaporated under vacuum at 60°C for 48 h. One ml human whole blood was poured into the tubes and allowed to stand for 7 min. After centrifuging (4000 rpm, 30 min), 300 μl samples of separated plasma were collected and the thromboxane B_2(TXB_2) concentrations were measured by RIA method [21, 22].

Platelet adhesion experiment

Platelet rich plasma (PRP) and platelet poor plasma (PPP) were prepared from normal human blood using sodium citrate as anti-coagulant [23]. Platelet suspensions (1 ml) with a concentration of 200 000 platelets/mm^3, prepared by mixing the

PRP with the PPP, were placed on the sample films and glass plate of 1.8 cm^2, then incubated at 37°C for 15 min. After being rinsed with a phosphate buffered saline, the sample films were fixed with glutaraldehyde, freeze-dried overnight and coated with gold. Then the platelets on the films were observed using a scanning electron microscope (SEM) (EMX-SM, Shimadzu Co., Kyoto, Japan).

RESULTS AND DISCUSSION

Preparation of PAS

Table 1 summarizes the results of the preparation of PAS used in this study. The structure of the resulting copolymers was confirmed to be the proposed block copolymers by means of ^{1}H NMR spectroscopy. In the ^{1}H NMR spectra, two remarkable peaks at 0 ($SiCH_3$) and 6.7–8.5 ppm (aromatic H) were observed. The observed PDMS contents of PASs were calculated from the $SiCH_3$/aromatic H ratio on the ^{1}H NMR spectra. For molecular weight determination, gel permeation chromatographic (GPC) measurements were attempted. However, we could not obtain reliable results about the molecular weight of PASs because of the poor solubility of PDMS unit of PAS for *N,N'*-dimethylformamide (DMF) as an eluent. Imai *et al.* also reported the abnormal solubility of PAS in the viscosity measurements of PAS/DMAc solution [13]. From the results of the mechanical properties which are discussed below, we confirm that the PASs synthesized in this study have a high molecular weight.

Table 1.
Preparation of PAS

No.	Aramid oligomer (x) in feed	in polymer[b]	PDMS[a] content (wt%) Calcd[c]	Found[b]	Yield (%)
1	—	—	0	0	100
2	7	14.9	41	25	83
3	3	4.1	60	53	77
4	1	1.3	78	75	56

[a] M_n (PDMS) = 1680 g/mol.
[b] Calculated from the $SiCH_3$/aromatic H ratio in the ^{1}H NMR spectrum.
[c] Weight (PDMS)/[weight (PDMS) + weight (aramid)] in the feed.

Mechanical property of PAS

The stress–strain curves of PASs are shown in Fig. 1. It is obvious that the tensile properties of the films are dependent on the PDMS content in the copolymers. A comparison of the curves shows a significant decrease in tensile strength and an increase in elongation at break with increasing PDMS content. Over the whole composition range, PAS can afford films between a rubber-toughened plastic and an elastomer. Therefore, we could infer that these block-copolymers had a high molecular weight.

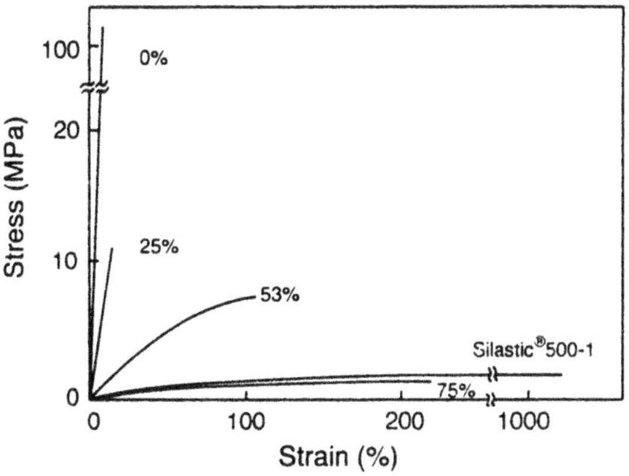
Stress (MPa)
100
20
10
0
0%
25%
53%
Silastic®500-1
75%
0
100
200
1000
Strain (%)

Figure 1. Stress–strain curves for PAS and Silastic® 500-1. PDMS contents are shown.

Surface analysis

Figure 2 shows an SEM of the silicon distribution on PAS films by EPMA. The silicon units are indicated as white areas when the X-ray detector is set to the wave length of silicon. It was obvious that the white areas increased with increasing PDMS content in PAS. From this result, it seemed that the silicone and aramid blocks underwent microphase-separation. To determine the microphase-separation

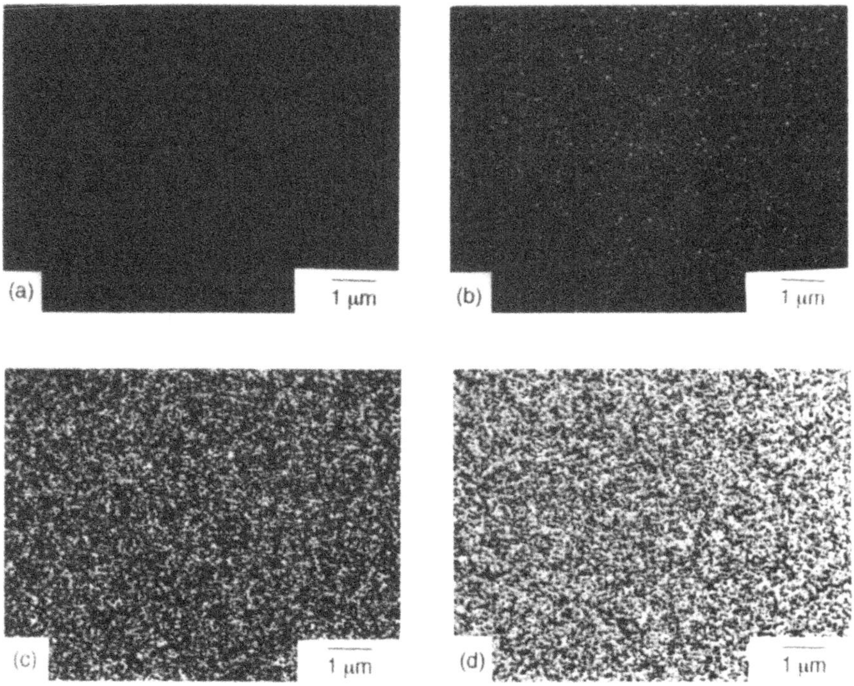
(a)
1 μm
(b)
1 μm
(c)
1 μm
(d)
1 μm

Figure 2. Scanning pictures of silicon distribution of PAS films by EPMA. (a)–(d) PAS: PDMS content is (a) 0% (aramid homopolymer), (b) 25%, (c) 53%, (d) 75%.

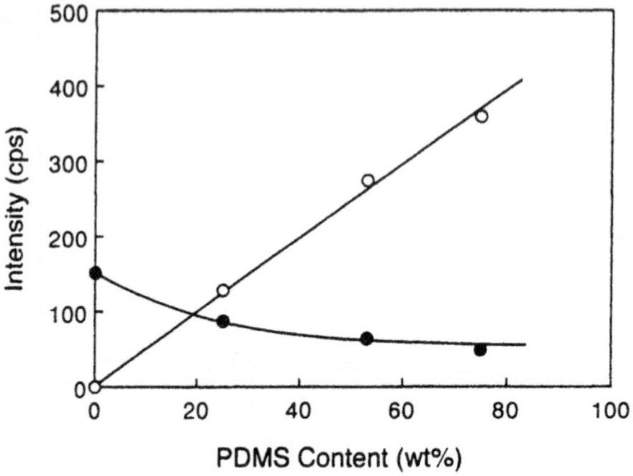
Intensity (cps)
500
400
300
200
100
0
0
20
40
60
80
100
PDMS Content (wt%)

Figure 3. Elemental analysis of PAS surface by EPMA: (○) silicon; (●) carbon.

of PAS, more data are needed. The results of the elemental surface analysis of silicon and carbon of PAS by means of EPMA are shown in Fig. 3. The signal intensity from silicon increased almost linearly with increasing PDMS content in PAS, whereas the signal intensity from carbon gradually decreased to a value of about 30 wt% and subsequently became constant. The EPMA measurement gives the information on the atomic composition in the range of a few micrometers depth from surface. In this case, the stoichiometries determined by EPMA spectra are in good agreement with the bulk composition determined by ^{1}H NMR spectra (Table 1).

Figure 4 shows Si/C stoichiometries of the PAS films measured by XPS. It is apparent that the Si/C ratio increases with the PDMS content in the PAS films. The Si/C ratio is approximated at 0.44 when the surface of the film is fully covered with PDMS unit. The Si/C ratio in bulk phase was calculated from $SiCH_3$/aromatic H ratio from ^{1}H NMR spectra. The Si/C ratio calculated from XPS spectra increased logarithmically with PDMS content, whereas the Si/C of bulk phase increased exponentially. From the difference between the surface and bulk composition, it

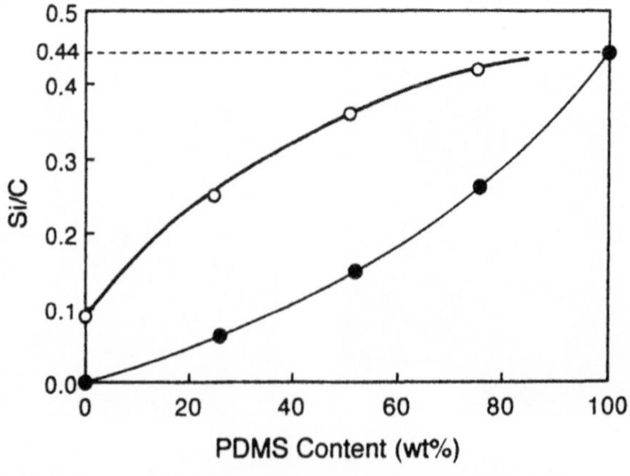
Si/C
0.5
0.44
0.4
0.3
0.2
0.1
0.0
0
20
40
60
80
100
PDMS Content (wt%)

Figure 4. Elemental analysis of PAS surface by XPS: (○) Si/C calculated by XPS; (●) Bulk value obtained by ^{1}H NMR.

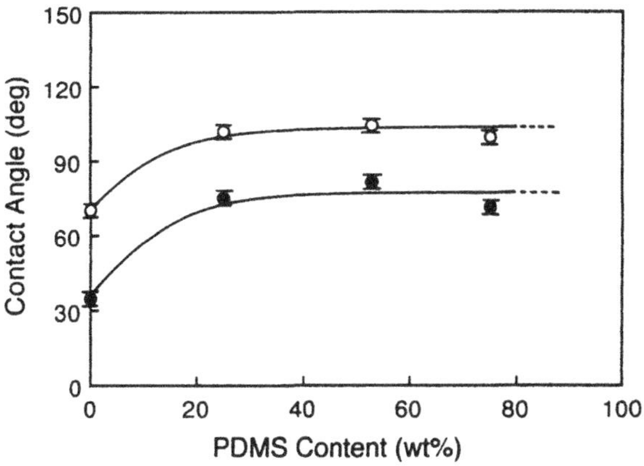

Contact Angle (deg)
150
120
90
60
30
0
0
20
40
60
80
100
PDMS Content (wt%)

Figure 5. Advancing and receding contact angle for PAS: (○) advancing contact angle; (●) receding contact angle. Standard deviations are shown.

seemed that the silicone block was separated or condensed at the surface. As we have no apparatus for varying the photoelectron take-off angle in ESCA 1000, we did not attempt the depth profiling of compositional variation.

Figure 5 shows the relationship between the advancing or receding contact angle and PDMS content of PAS. It was apparent that both advancing and receding contact angles of PASs did not change regardless of the PDMS content. This result indicates that the outermost surface of PAS film in all ranges of PDMS content is almost fully covered with silicone unit in air or water. These results are acceptable because PDMS has a low surface energy and it seemed that PDMS unit migrated to the surface when the PAS film was casting.

Blood compatibility

The TXB_2 releasing test was carried out in order to investigate interaction platelets with PAS. The results are shown in Fig. 6. Biomer®, which is a well-known polymer in bio-medical use, was adopted as a standard. TXB_2 is the prostaglandin derivative, and causes the platelet aggregation and release of contents of platelets [21, 22]. The release of TXB_2 is due to the adhered platelets or bulk platelets after surface activation. It is apparent that the aramid homopolymer surface induced severe release of TXB_2 from the attached platelets. This high value was almost in the same range as that obtained from glass-attached platelets (4×10^4 pg/ml, data not shown). On the other hand, the TXB_2 release from platelets attached to the other PASs were significantly lower regardless of the PDMS content. Biomer® induced slightly higher release of TXB_2 than PAS, but the level was lower than that of glass. We could not compare the release of TXB_2 with Silastic® because Silastic® could not be applied to the coating in the glass tube successfully.

Figure 7 shows the SEM micrograph of platelets adhering to the surfaces of PASs, Biomer®, Silastic® 500-1 and glass plate. On the surface of the aramid homopolymer and glass, many platelets adhered, deployed pseudopods and aggregated. On the other hand, on the surfaces of PAS films (25%, 53% and 75% PDMS), only a few platelets were found without any aggregation and morphological deformation.

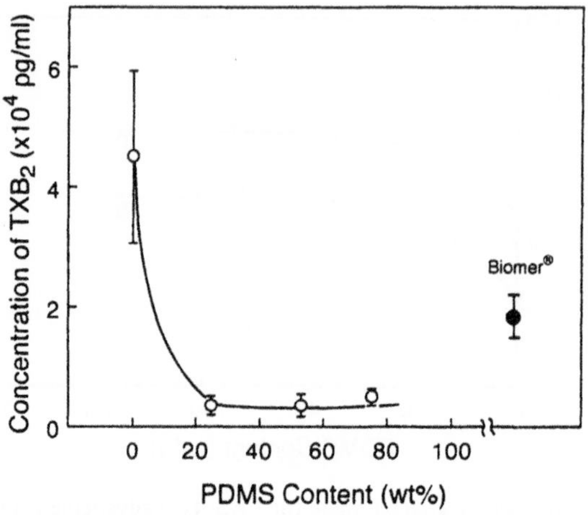
Concentration of TXB2 (x10⁴ pg/ml)
6
4
2
0
Biomer®
0
20
40
60
80
100
PDMS Content (wt%)

Figure 6. *In vitro* thromboxane B_2 released from platelets on PAS and Biomer®. Standard errors are shown (n = 4): (○) PAS; (●) Biomer®.

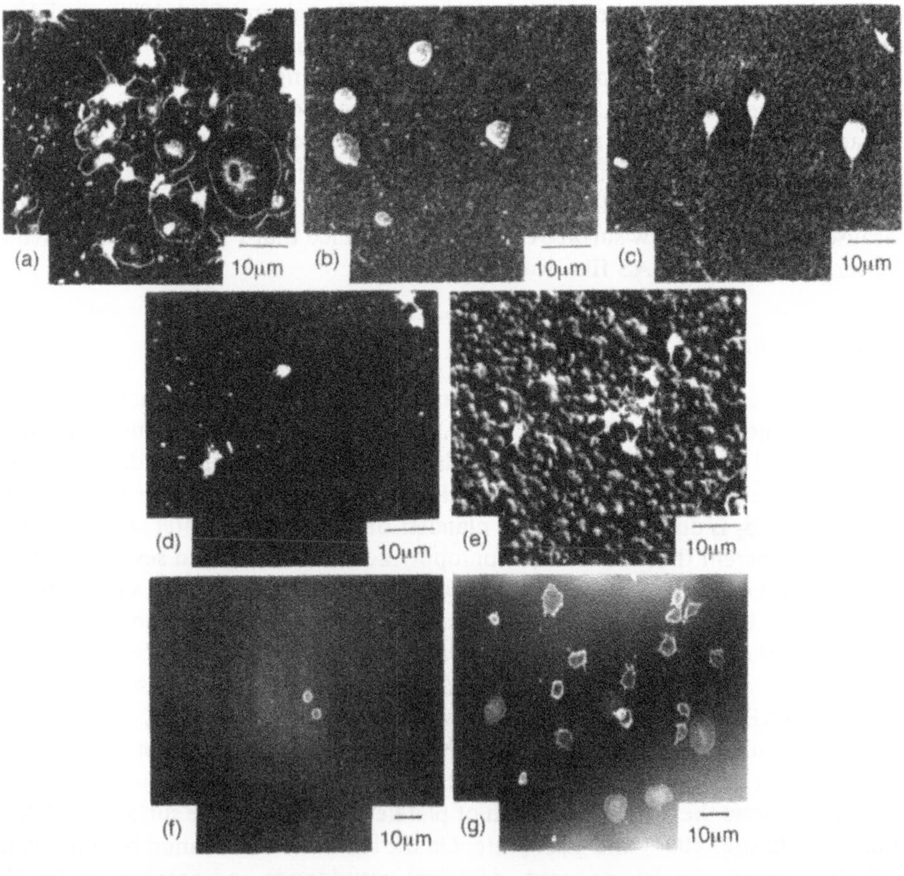
(a)
10μm
(b)
10μm
(c)
10μm
(d)
10μm
(e)
10μm
(f)
10μm
(g)
10μm

Figure 7. Scanning electron micrographs of adhered platelets using Shimadzu EMX-SM (original magnification = ×1000). (a)–(d) PAS; PDMS content is (a) 0% (aramid homopolymer), (b) 25%, (c) 53%, (d) 75%, (e) Biomer®, (f) Silastic®, (g) Glass.

The numbers of adherent platelets on the Biomer® surface were similarly few, but they deployed pseudopods slightly and aggregated, compared with those on the PASs surfaces.

From the results of surface analysis, it might be concluded that the surface of PAS was fully covered with PDMS blocks and that the *in vitro* bio-inertness of PASs resulted from the outermost PDMS layer similar to PDMS (Silastic® 500–1). Microphase-separation-like structures were observed for PASs by EPMA analysis. In this study, this phase separation seems not to improve blood compatibility of aramid, but to improve the mechanical strength. The XPS and dynamic contact angle studies of PAS surfaces verify that hypothesis. Therefore, the PASs show ideally desirable properties for both bulk phase and surfaces.

CONCLUSION

Poly(dimethylsiloxane)-polyamide copolymers (PASs) with different silicone contents were prepared, and the relationships between the surface chemical characteristics and blood compatibility were studied. EPMA measurements suggested that the surface layer (i.e. a few micrometers in depth) of cast films contained PDMS units proportional to its concentration. Microphase-separation-like structures were observed. However, XPS and dynamic contact angle measurements suggested that PDMS units were condensed on the outermost surface of PAS compared to that of the calculated bulk. *In vitro* platelet activation was studied by measuring TXB_2 release and platelet adhesion. SEM showed an ability of these PASs not to activate platelets significantly compared with aramid homopolymer, glass and Biomer®. PAS has a very weak interaction with platelets, compared to Silastic® (PDMS), due to the surface layer of PDMS. It has superior mechanical properties which are widely variable from tough plastic to rubber-like elastomer, which is a function of the PDMS content. From the point of view of gas permeability, we have already reported this earlier [24]. PAS had enough oxygen permeability to be used as a membrane for an artificial lung. Because of its unique properties, we confirm that PAS is an important novel biomaterial.

Acknowledgements

The authors would like to express thanks to Prof. Yoshio Imai, Tokyo Institute of Technology, for his continuing interest and encouragement. We also thank Yoshihisa Ozono, Terumi Kakoi and Toru Shinoyama of Kagoshima University for their assistance in obtaining surface property data and scanning electron micrographs of platelet adhesion.

REFERENCES

1. W. Lynch, *Handbook of Silicone Rubber Fabrication*, Van Nostrand Reinhold Company, New York (1978).
2. R. Rudoluph, J. Abraham, T. Vechione, S. Guber and M. Woodward, *Plast. Reconstr. Surg.* **62,** 185 (1978).
3. L. C. Hartman, R. W. Bessette, R. E. Baier, A. E. Meyer and J. Wirth, *J. Biomed. Mater. Res.* **22,** 475 (1988).
4. E. E. Frisch, in: *Silicones in Artificial Organs,* p. 63, C. G. Gebelein (Ed.), ACS Symposium Series, 256 (1984).
5. E. Nyilas, U.S. Patent 3 562 352 (1971).

6. E. P. Goldberg, U.S. Patent 2 999 845 (1961).
7. W. D. Merritt Jr. and J. H. Vestergard, U.S. Patent 3 821 325 (1974).
8. D. G. Laurin, U.S. Patent 3 994 988 (1976).
9. D. G. LeGrand, *J. Polym. Sci.* **B7**, 597 (1969).
10. D. G. LeGrand and Magila, *J. Eng. Sci.* **10**, 349 (1970).
11. W. J. Ward III, W. R. Browall and R. M. Salemme, *J. Membr. Sci.* **1**, 99 (1976).
12. S. Asakawa, Y. Saito, M. Kawahito, Y. Ito, S. Tsuchiya and K. Sugata, *Natl. Tech. Rep.* **29**, 93 (1983).
13. M. Kajiyama, M. Kakimoto and Y. Imai, *Macromolecules* **22**, 4143 (1989).
14. Y. Imai and M. Kakimoto, *Polymer Appls.* **39**, 438 (1990).
15. K. Koga and M. Takayanagi, *Hyomen (Surface)* **30**, 288 (1991).
16. Y. Imai and Y. Kadoma, *Polymer Preprints, Japan* **39**, 1850 (1990).
17. J. D. Andrade, in: *Surface and Interfacial Aspects of Biomedical Polymers,* Vol. 1 *X-ray Photoelectron Spectroscopy* pp. 105–195, J. D. Andrade (Ed.). Plenum Press, New York (1985).
18. A. Takahara, N. J. Jo, T. Kajiyama, *J. Biomater. Sci. Polymer Edn.* **1**, 29 (1989).
19. A. Takahara, H. Wabers, A. T. Okkema, S. L. Cooper, *Biomaterials* **12**, 324 (1991).
20. A. Takahara, *Hyomen (Surface)* **26**, 233 (1988).
21. F. Kurimoto, *Nippon Rinsho* **48**, 167 (1990).
22. W. S. Powell, *Prostaglandins* **20**, 947 (1980).
23. A. Kishida, Y. Tamada and Y. Ikada *Biomaterials* **13**, 113 (1992).
24. Y. Koinuma, T. Miyazaki, T. Matsumoto, T. Furuzono, M. Akashi and I. Maruyama, *Polym. Prep. Japan* **41**, 489 (1992).

Synthesis and nonthrombogenicity of fluoroalkyl polyetherurethanes

TAKASHI KASHIWAGI, YOSHIHIRO ITO, and YUKIO IMANISHI

Department of Polymer Chemistry, Kyoto University, Yoshida Honmachi, Sakyo-ku, Kyoto 606-01, Japan

Received 4 September 1992; accepted 15 January 1993

Abstract—New polyetherurethanes carrying fluoroalkyl substituents in the side chains were synthesized from *N,N*-di(hydroxyethyl)heptadecafluorooctyl-sulfonamide (a chain extender), 4,4′-disocyanatodiphenylmethane, and poly(tetramethylene glycol). Various kinds of polyetherurethanes having different tensile properties were prepared by changing the content of fluoroalkyl chain extender or the molecular weight of poly(tetramethylene glycol). The surface of a film made from the fluoroalkyl polyetherurethane was strongly water-repulsive. The *in vitro* thrombus formation on the fluoroalkyl polyetherurethanes was reduced by increasing the content of chain extender for the same molecular weight of poly(tetramethylene glycol). Protein adsorption, platelet adhesion, and platelet activation on the fluoroalkyl polyetherurethanes were also investigated.

Key words: Polyetherurethanes; fluoroalkylation; nonthrombogenicity; platelet; inactive surface; water repulsion.

INTRODUCTION

To reduce thrombogenicity of synthetic materials, many investigations have been carried out by the design and synthesis of material surface e.g., hydrophilization, hydrophobilization, blending or immobilization of antithrombogenic substances, or endothelialization of material surface as reviewed by us [1–5].

Polyetherurethanes have been used for artificial organ materials that necessitate nonthrombogenicity, like Biomer®. On the other hand, expanded polytetrafluoroethylene with its microporous structure is often used for vascular grafts and patches. Therefore, some researchers modified polymer surfaces with fluoro chemicals to prepare nonthrombogenic materials [6, 7]. The present authors synthesized new polyetherurethanes containing fluoroalkyl groups in side chains of hard segments and characterized their mechanical properties and *in vitro* interactions with blood components.

EXPERIMENTAL

Materials

Polyetherurethanes carrying fluoroalkyl substituents in the side chains were synthesized by the method shown in Fig. 1 by using a fluoroalkyl-substituted dihydroxyl compound as a chain extender, which was kindly supplied by Dainihon Ink Chem. Ind. (F-528 type, Osaka) and Dupont Japan Co. (Tokyo).

Poly(tetramethylene glycol)s (PTMG) having molecular weights of 870, 1280, and 2080 were donated by Hodogaya Chem. Ind. (Tokyo) and used without any further

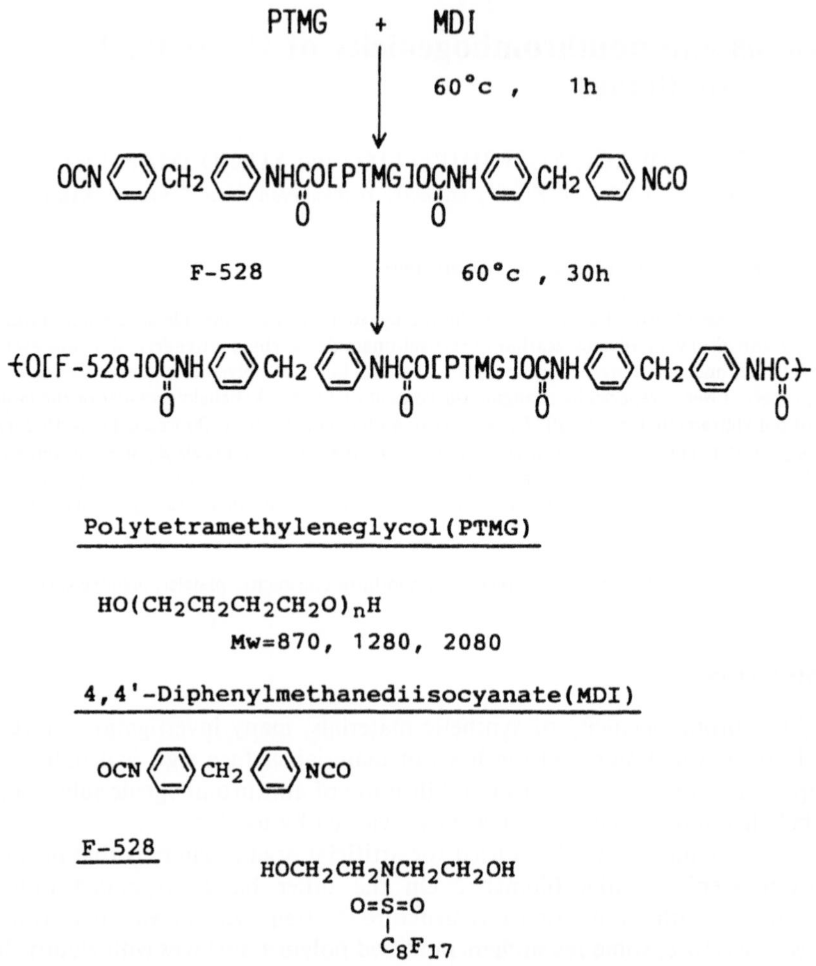
PTMG + MDI
60°c , 1h
OCN CH2 NHCO[PTMG]OCNH CH2 NCO
O O
F-528
60°c , 30h
O[F-528]OCNH CH2 NHCO[PTMG]OCNH CH2 NHC
O O O O
Polytetramethyleneglycol(PTMG)
HO(CH2CH2CH2CH2O)nH
Mw=870, 1280, 2080
4,4'-Diphenylmethanediisocyanate(MDI)
OCN CH2 NCO
F-528
HOCH2CH2NCH2CH2OH
O=S=O
C8F17

Figure 1. Synthetic scheme of polyetherurethane containing fluoroalkyl substituents in hard segments.

purification. 4,4′-Diisocyanatodiphenylmethane (MDI) and *N,N*-dimethylformamide (DMF) were purchased from Wako Pure Chem Ind. (Osaka) and both were distilled before utilization. *n*-Butyltin laurate was purchased from Nacalai Tesque, Inc. (Kyoto) and used without any further purification. Glasswares were predried at 100°C. PTMG and MDI (2–6 molar equivalent to PTMG) were dissolved in DMF (10 wt.%), and *n*-butyltin laurate (8×10^{-4} molar equivalent to PTMG) as a polymerization catalyst was added to the solution. The mixture was stirred under a nitrogen atmosphere at 60°C for 1 h to yield prepolymers terminated with isocyanate groups at both ends of a chain. The reaction solution was diluted with DMF and mixed with a DMF solution of F-528 (1–5 molar equivalent to PTMG). The chain-extension reaction was continued for 30 h at 60°C under stirring. After the reaction, the solution was poured into distilled water to precipitate polymers. The polymers were put in distilled water for 24 h, kept in methanol for 24 h, and dried in a vacuum. The chain-extension reaction was carried out also by using 1,5-dihydroxypentane in place of F-528 under otherwise similar conditions to obtain a control sample of polyetherurethane. The composition of the fluoroalkyl polyetherurethanes was determined by elemental analysis (F analysis). The intrinsic viscosities of the polyetherurethane produced were determined in a DMF solution at 30°C. Polyetherurethane films were cast on a glass plate by eluting a 10 wt.% DMF solution of the polymer, and used for various measurements.

Measurement of contact angle

Contact angles of the surface that was cast to the air (opposed to the casting mold surface) of the fluoroalkyl polyetherurethane film were measured by the liquid-droplet method or by the air-bubble method. In the liquid-droplet method, a water droplet was put on a polymer film and the contact angle was measured after keeping the film at 30°C for 30 s with an Erma Optical Instrument G-1 type apparatus (Tokyo). More than ten measurements were repeated for a sample and the average value was calculated. In the air-bubble method, a polymer film cast on a glass was immersed in distilled water for 1 h and an air bubbble was put on the film in water. Contact angles were measured according to the method proposed by Adamson [8]. More than ten measurements were repeated for a sample and the average value was calculated.

Measurement of mechanical properties

The stretching test of the fluoroalkyl polyetherurethane film was performed using a KES-G1 stretching apparatus (Katotech, Kyoto), and the tensile strength, the elongation at break, and the Young's modulus of a polymer film were determined.

In vitro *thrombus formation experiment*

The *in vitro* thrombus formation on the fluoroalkyl polyetherurethane film was measured by the method previously reported by us [9–11]. Polymer films having diameters of 4–5 cm were cast from a DMF solution of the polymer on a watch glass having a diameter of 4.5 cm. Fresh blood (9 parts) was collected from an adult dog into a bag (Japan Medical Supply, Hiroshima) containing the ACD solution (1 part), which contains anhydrous D-glucose (2.45 g), sodium citrate dihydrate (2.20 g), and citric acid monohydrate (0.8 g) in distilled water (100 ml). The experiments using the ACD blood were carried out within several hours after the collection. A $CaCl_2$ solution (0.1 M, 20 μl) was added to the ACD blood (200 μl) on the sample film kept at 37°C to initiate the blood clotting. After 20 min, distilled water (2 ml) was added to stop the clotting reaction. After standing the mixture for 5 min, the thrombus formed was put in 37% formalin (5 ml) for 10 min under shaking, and washed by shaking in distilled water for 10 min. The wet thrombus was dehydrated with tissue paper and weighed. The sample was divided into four pieces and tested for thrombus formation. Three to four experiments were repeated with a test piece. The amount of thrombus formed on the polymer samples relative to that on a glass plate was determined and compared. The statistical significance of the results has been reported previously by us [9–11] and Imai and Nose [12].

Protein adsorption experiment

The fluoroalkyl polyetherurethane film was cast from a DMF solution (0.1 wt%) on CaF_2 (diameter 2.5 cm and thickness 2 mm). The polymer-cast CaF_2 plate was immersed in an aqueous solution of a plasma protein as reported previously [13, 14]. The plasma proteins used were bovine serum albumin (BSA, A6003), bovine γ-globulin (BγG, G3500), and bovine plasma fibrinogen (BPF, F4753). These proteins were purchased from Sigma (St. Louis, MO). The protein concentrations in a buffer solution (0.01 M Tris-HCl, pH 7.4) containing NaCl (0.9 wt.%) were 4.5 g/dl, 1.6 g/dl, and 0.3 g/dl, respectively. After immersing in the protein solution for 1 h at 37°C, the polymer-cast CaF_2 plate was transferred to the Tris-buffer solution for washing for 10 s. After drying in vacuum for 24 h, the plasma protein adsorbed was investigated by Fourier-transform (FT) infrared (IR) spectroscopy using Digilab FTS-15E/D apparatus (Cambridge, MA).

Platelet adhesion experiment

Adhesion of platelets onto the surface of the fluoroalkyl polyetherurethane film was investigated by the method previously reported [15, 16]. The membrane proteins of platelets were labelled with ^{51}Cr and the number of platelets adhered was determined by counting the radio activity. Serotonin in the dense granules of platelets was replaced with ^{3}H-labelled serotonin, and the amount of serotonin remaining in adhered platelets was determined by the β-counting. The statistical significance of the results has been reported previously [15, 16].

RESULTS AND DISCUSSION

Synthesis and properties of fluoroalkyl polyetherurethanes

The experimental conditions for preparation and the properties of polyetherurethanes synthesized are summarized in Table 1. With increasing concentration of F-528 in the feed of the polymerization, the fluorine content of the polymer increased and intrinsic viscosity of the polymer decreased. The decrease of intrinsic viscosity should not simply be related with a decrease of molecular weight, because polyetherurethanes with high contents of fluoroalkyl groups may not expand well in a DMF solution. The molecular weight of PTMG influenced the intrinsic viscosity of the polymer very little, under otherwise similar conditions.

Contact angles of fluoroalkyl polyetherurethanes

Water contact angles in air and air contact angles in water of the fluoroalkyl polyetherurethanes are summarized in Table 2 along with those of the reference materials. The contact angle of fluoroalkyl polyetherurethane surface was larger than that of the control polyetherurethane, indicating that the film surface became more water-repulsive. Incorporation of small amounts of fluoroalkyl groups strongly increased the water contact angle of the film, but further incorporation increased it very slightly. In addition, the contact angles of fluoroalkyl polyetherurethanes are relatively close to that of polytetrafluoroethylene. This change of water contact angle with the content of fluoroalkyl groups implies that fluoroalkyl groups tend to appear on the air-side surface of a film upon casting on the glass plate from a DMF solution, and that a small amount of fluoroalkyl groups is enough to cover almost completely the film surface.

Although the contact angles increased with increasing content of fluoroalkyl chain extenders for the same molecular weight of PTMG, there was no statistical

Table 1.
Feed composition, fluorine content and intrinsic viscosity of polymer

Polymer	Feed composition			Polyurethane	
	PTMG[a]	MDI	Chain extender	F(wt%)	[η](dl/g)[b]
Control	1 (1280)	2	1[c]		1.07
			F-528		
870A	1 (870)	2	1	14.3	0.20
870B	1 (870)	3.2	2.2	20.2	0.13
870C	1 (870)	6	5	27.2	0.11
1280A	1 (1280)	2	1	12.8	0.21
1280B	1 (1280)	4	3	21.9	0.13
1280C	1 (1280)	6	5	26.2	0.11
2080A	1 (2080)	2	1	7.8	0.25
2080B	1 (2080)	4	3	16.5	0.18
2080C	1 (2080)	6	5	21.5	0.13

[a]Values in the parentheses show the molecular weight of PTMG.
[b]Measured in DMF solution at 30°C.
[c]1,4-Dihydroxpenane was used.

difference between polymers having different PTMG molecular weights or different fluorine contents.

The water contact angles of fluoroalkyl polyetherurethanes were slightly smaller in water than in air. In water, chain motion becomes active on the surface of polymer film and hydrophilic segments appear on the water/film interface to decrease the water contact angle. A similar phenomenon has been reported by Kang *et al.* [17].

Table 2.
Water contact angle of polymers at 25°C

Polymer	F(wt.%)	In air (deg)	In water (deg)
Glass	—	30 ± 2	35 ± 2
Polytetrafluoroethylene	76.0	121 ± 3	107 ± 2
Polydimethylsiloxane	—	100 ± 2	93 ± 2
Control	—	71 ± 1	55 ± 2
870A	14.3	106 ± 2	94 ± 1
870B	20.2	108 ± 2	95 ± 1
870C	27.2	109 ± 2	99 ± 1
1280A	12.8	106 ± 2	82 ± 1
1280B	21.9	107 ± 2	85 ± 1
1280C	26.2	108 ± 2	105 ± 2
2080A	7.8	106 ± 2	90 ± 1
2080B	16.5	107 ± 2	93 ± 3
2080C	21.5	108 ± 2	106 ± 2

Table 3.
Tensile properties of polymers at 25°C

Polymer	Tensile strength (MPa)	Per cent elongation at break	Young's modulus (MPa)
Control	1.90 ± 0.21	61 ± 4	20.6 ± 1.0
870A	5.87 ± 0.26	475 ± 66	28.4 ± 1.0
870B	7.15 ± 0.40	67 ± 3	176.5 ± 14.7
1280A	3.50 ± 0.14	218 ± 26	30.4 ± 6.9
1280B	5.32 ± 0.11	21 ± 4	177.5 ± 14.7
2080A	2.40 ± 0.11	346 ± 47	7.9 ± 1.0
2080B	2.81 ± 0.28	87 ± 2	52.0 ± 4.9
2080C	4.67 ± 0.33	5 ± 1	156.9 ± 8.8

Tensile properties of fluoroalkyl polyetherurethanes

The results of stretching experiments of the fluoroalkyl polyetherurethane are shown in Table 3. The chain extender F-528 enhanced the tensile strength, the elongation at break, and Young's modulus of polyetherurethane more than the control chain extender1,5-dihydroxypentane. The increasing content of fluoroalkyl chain extender led to a significant increase of the tensile strength and Young's modulus and to a decrease of the elongation at break of the polyetherurethanes. These changes indicate that the polyetherurethane becomes stronger and stiffer with the introduction of fluoroalkyl chain extenders. The increase of fluoroalkyl content necessarily increases the content of urethane bondings and hence interpolymer

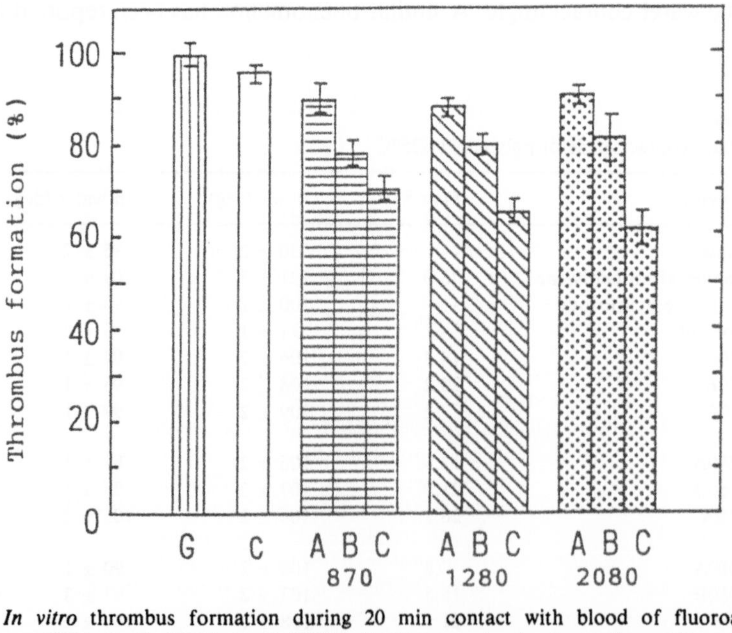

Thrombus formation (%)
100
80
60
40
20
0
G
C
A B C
870
A B C
1280
A B C
2080

Figure 2. *In vitro* thrombus formation during 20 min contact with blood of fluoroalkyl polyetherurethanes (870A, B, C; 1280A, B, C; 2080A, B, C), control polyetherurethane (C), and glass (G).

crosslinking by hydrogen bondings. The interpolymer hydrogen bondings strengthen the cohesivity of polymers to make the polymer stronger and more rigid. On the other hand, increasing molecular weight of PTMG segment weakens the cohesivity of polymers to make the polyetherurethanes weaker and softer.

In vitro *thrombus formation*

The amounts of thrombus formed on various polymer samples during 20 min contact with blood are shown in Fig. 2. The histograms show that the increasing content of fluorine in the polyetherurethanes decreased the amount of thrombus for the same molecular weight of PTMG used.

Protein adsorption

FT-IR spectra of albumin, γ-globulin, and fibrinogen adsorbed onto the fluoroalkyl polyetherurethanes are shown in Fig. 3 after the spectral subtraction of the

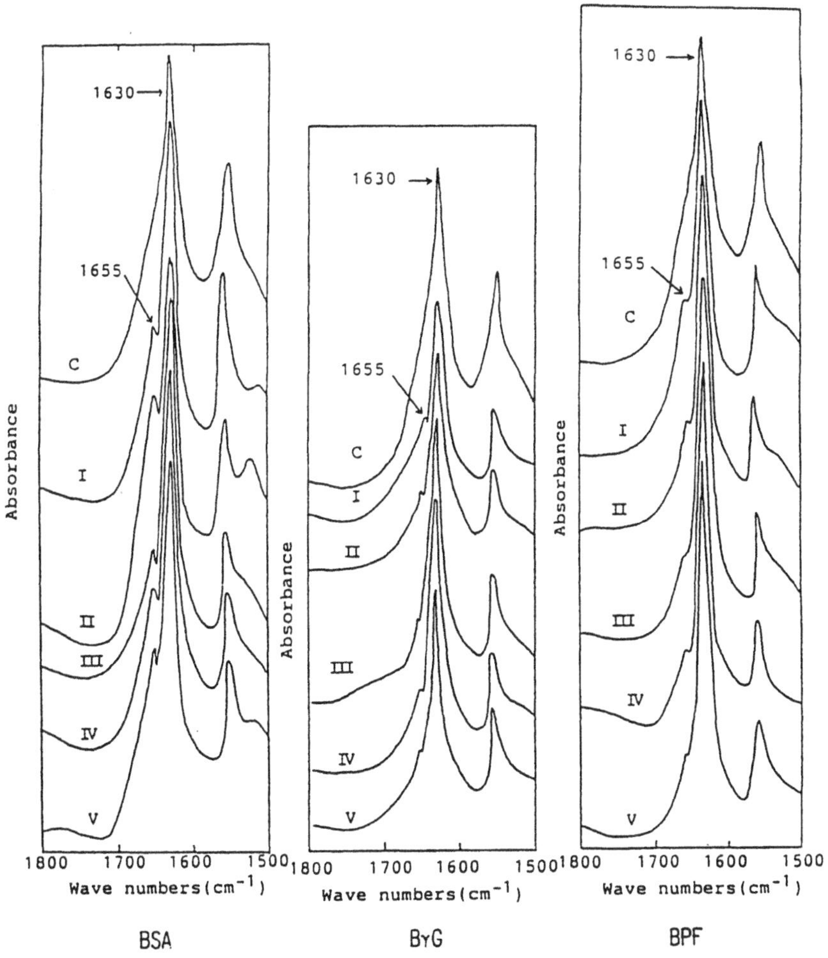
1630
1655
C
I
II
III
IV
V
Absorbance
1800 1700 1600 1500
Wave numbers(cm^{-1})
BSA
BγG
BPF

Figure 3. FT-IR spectra of proteins adsorbed on to fluoroalkyl polyetherurethanes 870A (I), 1280A (II), 1280B (III), 1280C (IV), and 2080A (V), and control polyetherurethane (C).

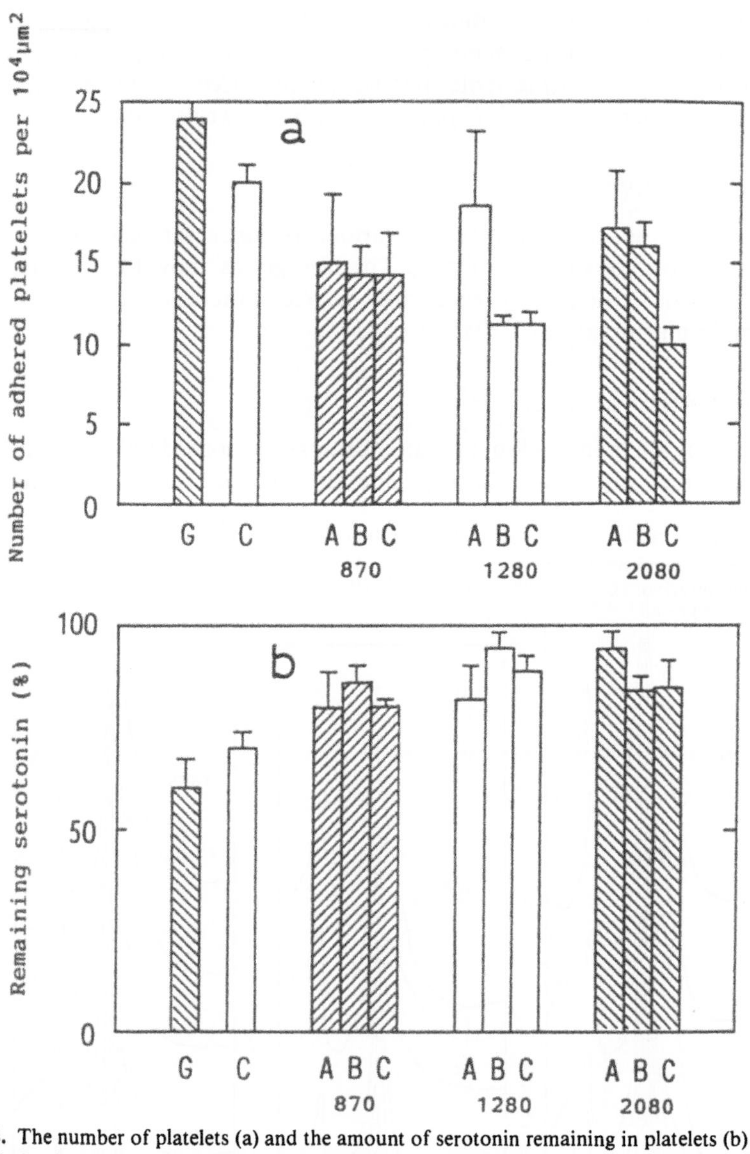
Number of adhered platelets per $10^4 \mu m^2$
a
25
20
15
10
5
0
G C A B C A B C A B C
870 1280 2080
Remaining serotonin (%)
b
100
50
0
G C A B C A B C A B C
870 1280 2080

Figure 4. The number of platelets (a) and the amount of serotonin remaining in platelets (b) adhered on fluoroalkyl polyetherurethanes (870A, B, C; 1280A, B, C; 2080A, B, C), control polyetherurethane (C), and glass (G).

contribution of base polymers. Ito *et al.* reported that the amide I adsorptions of native albumin, γ-globulin, and fibrinogen appeared at 1659, 1643, and 1655 cm^{-1}, respectively, and that of completely denatured ones at 1632 cm^{-1}, indicating that β-sheet conformation was evoked by the denaturation upon adsorption, and that the degree of denaturation of adsorbed proteins depended on the nature of polymers [14]. In this investigation, the 1655-cm^{-1} absorption was absent in the spectra of proteins adsorbed on the control polyetherurethane. However, the IR spectra of all proteins adsorbed on the fluoroalkyl polyetherurethanes showed a shoulder absorption at 1655 cm^{-1}, although the proportion of denaturation were not clearly

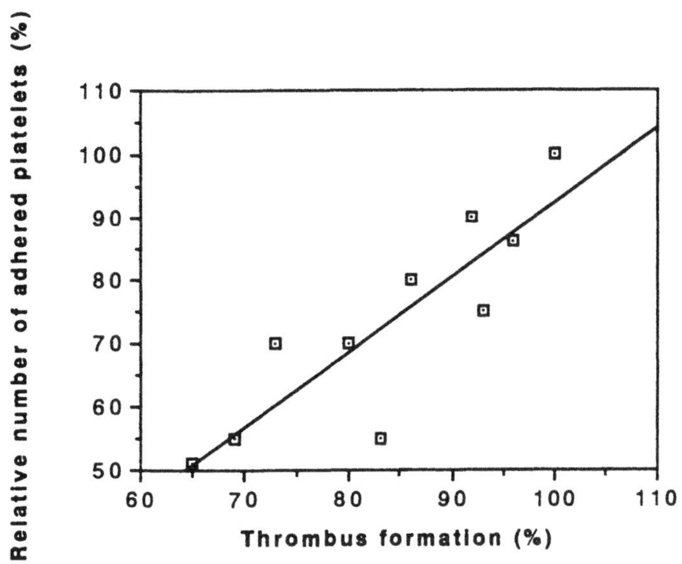
Relative number of adhered platelets (%)
110
100
90
80
70
60
50
60
70
80
90
100
110
Thrombus formation (%)

Figure 5. Correlation between thrombus formation and number of platelets adhered on fluoroalkyl polyetherurethanes and control polyetherurethane. The line was drawn according to the equation $y = -25.995 \pm 1.1851x$, $r = 0.748$, which was calculated by the least-squares method.

related with the composition of fluoroalkyl polyetherurethanes. It is considered that the interaction between the polymer surface and the proteins was reduced by the low surface free energy of the fluoroalkyl polyetherurethanes.

Platelet adhesion

The number of platelets adhered on the polymer films and the amount of serotonin remaining in the adhered platelets are shown in Fig. 4. The fluoroalkyl polyetherurethanes reduced the number of adhered platelets and increased the amount of serotonin remaining in the adhered platelets more than the control polyetherurethane. For the same molecular weight of PTMG, the increase of fluoroalkyl chain extenders in the polyetherurethane significantly decreased the platelet adhesion, although the amount of serotonin remaining in adhered platelets was not significantly dependent on the composition of the polymers.

We previously demonstrated that the protein adsorption on the polymers significantly affected platelet adhesion on the polymers, and proposed that the blood compatibility of materials should be evaluated by a number of parameters including both platelet and coagulation systems [1–4, 18]. In the present investigation, the *in vitro* reduction of thrombogenicity was well correlated with the suppression of platelet adhesion as shown in Fig. 5.

In conclusion, the introduction of fluoroalkyl groups into polyetherurethane affected both the surface and bulk properties. However, because of the property of fluoroalkyl groups tending to be concentrated on the air-side surface, the presence of fluoroalkyl chain extender beyond a certain amount did not influence the surface property, although the bulk property, similar to tensile properties, was affected very significantly. Similarly, because the surfaces of the fluoroalkyl polymers were almost completely covered with fluoroalkyl groups, the polymer compositions more significantly affected the mechanical properties of the polymers than the interactions of the polymers with blood components.

Acknowledgements

The authors acknowledge Dainihon Ink Chem. Ind. and Dupont Japan Co. for the supply of *N*,*N*-di(hydroxyethyl)heptadecafluorooctylsulfonamide (F-528).

REFERENCES

1. Y. Ito and Y. Imanishi, *CRC Crit. Rev. Biocomp.* **5**, 45–104 (1989).
2. Y. Ito, *J. Biomater. Appl.* **2**, 235–265 (1987).
3. Y. Ito, in: *Synthesis of Biocomposite Materials*, pp. 15–84, Y. Imanishi (Ed.). CRC Press, Boca Raton (1992).
4. Y. Ito and Y. Imanishi, *J. Biomat. Appl.* **6**, 293–318 (1992).
5. S. Q. Liu, Y. Ito and Y. Imanishi, *J. Biomed. Mater. Res.*; accepted.
6. Y. S. Yeh, Y. Iriyama, Y. Matsuzawa, S. R. Hanson and H. Yasuda, *J. Biomed. Mater. Res.* **22**, 7905–7919 (1988).
7. J. L. Bohnert, B. C. Fowler, T. A. Horbett and A. S. Hoffman, *J. Biomater. Sci., Polym. Edn.* **1**, 279–292 (1990).
8. A. W. Adamson, *Physical Chemistry of Surface*, 3rd edn. John Wiley & Sons, Inc., New York (1976).
9. Y. Ito, M. Sisido and Y. Imanishi, *J. Biomed. Mater. Res.* **20**, 1017–1033 (1986).
10. Y. Ito, M. Sisido and Y. Imanishi, *J. Biomed. Mater. Res.* **20**, 1157–1177 (1986).
11. S. Q. Liu, Y. Ito and Y. Imanishi, *J. Biomater. Sci., Polym. Ed.* **1**, 111–122 (1989).
12. Y. Imai and Y. Nose, *J. Biomed. Mater. Res.* **6**, 165–172 (1972).
13. Y. Ito, M. Sisido and Y. Imanishi, *J. Biomed. Mater. Res.* **20**, 1139–1155 (1986).
14. T. Sanada, Y. Ito, M. Sisido and Y. Imanishi, *J. Biomed. Mater. Res.* **20**, 1179–1196 (1986).
15. A. Mori, Y. Ito, M. Sisido and Y. Imanishi, *Biomaterials* **7**, 386–392 (1986).
16. Y. Ito, M. Sisido and Y. Imanishi, *Biomaterials* **8**, 458–463 (1987).
17. I.-K. Kang, Y. Ito, M. Sisido and Y. Imanishi, *Biomaterials* **9**, 138–144 (1988).
18. Y. Ito, M. Sisido and Y. Imanishi, *J. Biomed. Mater. Res.* **24**, 227–242 (1990).

Surface and bulk effects on platelet adhesion and aggregation during simple (laminar) shear flow of whole blood

T. M. ALKHAMIS* and R. L. BEISSINGER**

**Department of Chemical Engineering, Mu'tah University, Jordan*
***Department of Chemical Engineering, Illinois Institute of Technology, Chicago, IL 60616, USA*

Received 14 September 1992; accepted 15 February 1994

Abstract—This study attempts to clarify the role of the artificial surface and the fluid bulk on platelet adhesion and aggregation events during simple shear flow of whole blood. The experimental approach involved the shearing of fresh whole blood samples over the shear rate range of 720–5680 s^{-1}, which corresponded to a shear stress maximum of about 150 dyn cm^{-2}. Results on platelet adhesion, measured as surface coverage by platelets, and platelet aggregation, measured in terms of reduction in platelet count and adenosine diphosphate (ADP) release, were determined as a function of the surface to volume ratio (*S/V*); and *artificial* surface used. Both shear-induced platelet adhesion and platelet count reduction showed significant variation over the range of *S/V* employed. The ratios between the three different *S/V* values used in this system (10:6:4) were about the same as the ratio of the shear rate-averaged results obtained. Also, for shear-induced hemolysis, an increase in the release of hemoglobin from red blood cells was found as *S/V* was increased, again with ratios between the shear rate-averaged values similar to the ratio of *S/V* values employed. The shear-induced release of ADP, presumably from platelets and from red blood cells indicated a different dependence of ADP release on *S/V* than was observed for the other parameters reported. Irreversible platelet aggregation was expected to occur because the amount of ADP that was released as a result of the shear was substantial. Models proposed to explain the experimental results were found to support a surface-controlled mechanism.

Key words: Shear-induced; platelet surface coverage; platelet count reduction; surface-to-volume-ratio; artificial surface; red blood cells; ADP release.

INTRODUCTION

The use of artificial organs, prosthetic devices, and extracorporeal flow systems in medical treatment is complicated by the chemical (materials of construction of the device) and mechanical (hydrodynamic and geometric parameters of the device) interactions that occur at the artificial surface with blood's plasma solutes and cellular components. Therefore, an improved understanding of these interactions is needed before better systems that use artificial surfaces can be designed. Under normal circumstances platelets in the circulation do not appear to adhere to the endothelial-lined blood vessel walls nor do they usually aggregate with other platelets. However, systems involving artificial surfaces were found to activate platelets [5]. Also, shear-induced red blood cell (RBC) and platelet-surface interactions were found to occur. These include RBC-augmentation of platelet transport to the surface [1–3] with subsequent platelet deposition/adhesion [3–5], release of adenosine diphosphate (ADP) [4, 6] and hemoglobin (Hb) [4], and platelet aggregation [4, 6]. Also, shear may induce formation of thrombin [7] and generation of thromboxane A_2 at the artificial surface, both of which can cause platelet activation followed by release reaction and aggregation (secondary) [8, 9]. Platelet aggregation in the bulk fluid, as in primary aggregation, can occur as a result of stimulation by various platelet aggregating agents, such as ADP, without direct interaction of the platelets with the device surface [4, 6, 10]. Separating the surface and bulk interaction mechanisms from each other has proved to be difficult.

Low-stress, shear induced hemolysis (without the occurrence of rupture of RBCs wherein Hb is released via deformation of the cell membrane) appeared to be sensitive to wall material chemistry, to be proportional to device/volume ratio (S/V), to have slow kinetics, and to be controlled by fluid shear rate, while being essentially independent of shear stress [11–13]. Data obtained for freshly out-dated blood in a cone-and-plate viscometer, demonstrated that hemolysis correlated with shear rate, S/V, and hematocrit [14]. From that study it was concluded that for blood (or at least out-dated blood) a dual mechanism, that combines the effect of the surface and the fluid bulk, could explain the hemolytic results obtained. In a more recent study Hb release for freshly out-dated RBCs sheared in a capillary viscometer was modeled based on the dual mechanism in which the release of Hb was initiated at the artificial surface due to shear-induced RBC interactions there, followed by diffusion of the RBCs into the fluid bulk [15].

In this study, we investigated as a function of shear rate for fresh samples of human blood the relative contributions of the surface and the bulk to cell component release (Hb from RBCs and ADP from RBCs and platelets), percent surface coverage by platelets and reduction in platelet count. The relative importance of the surface to the bulk for each of these parameters was obtained by variation of the S/V of the test system. The results obtained served as a basis for attempting through simple correlations to better elucidate and explain mechanistically (using the dual mechanism approach) the artificial surface and bulk effects on RBCs and platelets in simple (laminar) shear flow.

MATERIALS AND METHODS

Apparatus

A model R16/R18 Weissenberg rheogoniometer (Sangamo Weston Controls Ltd., Sussex, UK) was used with a cone-and-plate (CP) arrangement to generate viscometric flow [14]. Three different CP arrangements were used: a 10.0 cm diameter cone of 0.3 deg angle and a 10.0 cm flat plate; a 10.0 cm diameter cone of 0.5 deg angle and a 10.0 cm flat plate; and a 7.5 cm diameter cone of 1.0 deg angle along with a 7.5 cm flat plate. The cone of the CP arrangement does not make contact with the plate surface; its apex was chopped. The distance between the chopped cone apex and flat plate was 10 μm for the 0.3 deg cone, 25 μm for the 0.5 deg cone, and 41 μm for the 1 deg cone. Blood samples were contained within the conical gap while the cone was rotated at various speeds giving shear rates to a maximum of 5680 s^{-1}. The CP geometry is useful at small cone angles because a good approximation to uniform shear is achieved if rotational speed is slow enough [14]. Thin Teflon™ films (50 μm) of fluorinated ethylene propylene copolymer (FEP, DuPont, Wilmington, DE, USA) were used as the primary test surface in these studies for their relevance to artificial organs [7]. The primary test surface was compared with results obtained using the unfinished side of the FEP film, i.e. the sticky[R] FEP surface. Also, the primary test surface was compared with results obtained using thin films (50 μm) of silicone rubber (SR) and polyethylene (PE).

Surface-to-volume-ratio

The CP arrangements described above were used to evaluate the relative contribution of the surface and the bulk on the phenomena investigated. These CP arrangements corresponded to three different S/V: 114.6, 68.8, and 46.9 cm^{-1}. Data were obtained on several parameters: reduction in platelet count as an approximate measure of platelet aggregation in the bulk, platelet surface coverage as an index of the extent of platelet adhesion, and both Hb release from RBCs and ADP release from RBCs and platelets as indicators of cell leakage.

Solutions: acid citrate dextrose (ACD), Formula A

This anticoagulant/preservative solution was prepared by dissolving 22.0 g of trisodium citrate, 8.0 g of citric acid, and 24.5 g of dextrose in sufficient deionized water to make 1 l of pH 5.5 solution [16]. For studies on platelets and platelet function, ACD is known to stabilize the reactivity of platelets [17]. Experiments were begun 15 min after collection of blood samples and involved up to a maximum time of 5 h after collection during which the cells were studied.

Blood collection

Blood was obtained from male or female donors, who were made aware that their blood was being used for scientific research. They appeared healthy, had fasted overnight and were asked if they had taken any medications within the 14 days prior to donation that might affect platelet activation (such as aspirin). Volunteers on such medications were excluded from the study. Whole blood samples of 10–20 ml were withdrawn at either the IIT clinic or the Michael Reese Special Coagulation Laboratory (Michael Reese Hospital, Chicago, IL). Samples were collected in plastic tubes with ACD anticoagulant at a ratio of 15 ml of ACD per 100 ml of blood which resulted in a final pH value of 6.9 at which the experiments of this study were conducted; at this pH it is known that the reactivity of platelets was stabilized [17]. Blood samples of 45 $\pm$ 3 hematocrit and 200 000 $\pm$ 20 000 platelet count ranges were transferred to our laboratory within 15 min for further testing.

Experimental procedure

Thin films of biomaterial were mounted on the viscometer platen surfaces using either double faced adhesive sheets (Sun Process Company, Elk Grove Village, IL, USA) or an adhesive material (Glue Stick, Faber Castell Dist., Lewisburg, TN, USA) [4]. Unless otherwise stated we adhered the unfinished side of the FEP film to the stainless steel platen surfaces of the CP arrangement. In this configuration we refer to the FEP film surface that faces the blood as the normal side. However, when we exposed the unfinished side of the FEP film to the blood, we refer to this side of the FEP film as the sticky[R] surface. The modified platen surfaces were checked for a maximum allowable roughness of 2.5 μm using a sensitive dial clock. These virgin surfaces were cleaned using phosphate detergent solution (Sparkleen, Fisher Scientific, Itasca, IL, USA), rinsed with deionized water, and dried with a cool stream of filtered air.

The gap of the viscometer platens was set following standard procedures. Then, the blood sample was placed on the apex of the cone. An excess volume of sample was used in order to fill all of the space between the platens and to assure uniform distribution; e.g., 2 ml was used for the 0.3 deg CP arrangement (which requires only 1.38 ml sample); excess sample was carefully wiped from the rim of the platen system. Next, the upper platen was lowered until the required gap setting was obtained. The lower platen was rotated at the required rotational speed while the upper one was held stationary. The duration of rotation for each experimental run was 2 min. (For runs at zero shear rate samples were loaded into the CP arrangement, both plates were held stationary for 2 min, and then the samples were removed). After the required time, the motion was stopped quickly (within a fraction of a second) using an electric brake unit.

A plastic pipet was used to collect the sheared or unsheared (load/unload control) sample, which was transferred to polyethylene tubes for further processing. As described in a previous report 0.02 ml of the sample was mixed with ammonium oxalate solution for platelet counting [4]. The remaining portion of the sample was centrifuged at 14 000 *g* and 4°C for 5 min to remove the platelets from the supernatant. Then the supernatant was divided into portions to allow determination of the concentration of ADP and Hb, also as described in the previous report [4].

In the last step the biomaterial films were peeled from both platen surfaces, rinsed thoroughly with phosphate buffered saline to remove non-adherent material, fixed by incubation with 3% glutaraldehyde solution for 1 h, and air dried. Then, a 1-cm wide piece along the radius was cut and divided into five subregions for microscopic evaluation of surface area occupied by the platelets. Pictures of every subregion were taken and the area occupied by platelets or platelet material was quantified using a digitizing pad system as described elsewhere [4]. The average value of the five subregions was taken as a representation of the whole surface.

Statistical methods

The blood damage data for the normal FEP surface were analyzed using the Student's *t*-test. The experimental results obtained (as represented in the figures below) include interdonor variability (for about three to five donors) along with the associated experimental error and are represented by the mean along with the standard deviation.

RESULTS

Surface and bulk solution effects

Viscosity of whole blood. The steady shear viscosity of whole blood (for the hematocrit range of 45 ± 3) was measured at room temperature (25°C) in the CP geometry over the shear rate range of 720–5680 s^{-1}. The blood was seen to behave as a Newtonian fluid with an average viscosity of 3.3 ± 0.25 centipoise over the entire shear rate range studied.

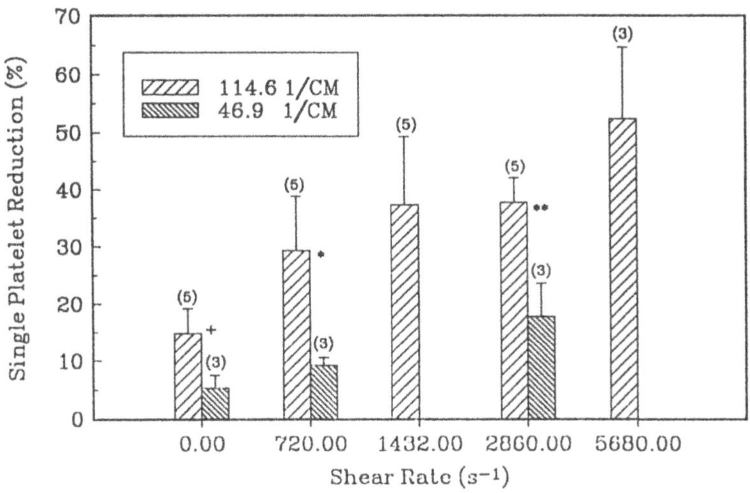

Single Platelet Reduction (%)
70
60
50
40
30
20
10
0
114.6 1/CM
46.9 1/CM
(5)
(3)
(5)
(3)
(5)
(5)
(3)
(3)
+
*
**
0.00
720.00
1432.00
2860.00
5680.00
Shear Rate (s⁻¹)

Figure 1. Reduction in platelet count (%) as a function of surface-to-volume ratio and system shear rate using fluorinated ethylene propylene copolymer normal (FEP–NOR) surfaces. The data at each shear rate value tested are represented by the corresponding mean along with the standard deviation. $+P < 0.05$ vs the other group at the same shear rate. $^*P < 0.02$ vs the other group at the same shear rate. $^{**}P < 0.002$ vs the other group at the same shear rate.

Percent decrease in platelet count. Reduction in platelet count was measured as a function of shear rate and *S/V*; results are reported in Fig. 1. These results show a strong dependence on the *S/V* employed. The ratios between the two different *S/V* values in this system are about 10:4, and the respective ratios of the resulting shear-averaged values (defined as the average values of all points over the entire shear rate range) of reduction in platelet count (based on Fig. 1) are also about 10:4. These values are approximately the same as the *S/V* and indicate that the surface has an important effect on reduction in platelet count measured in the fluid bulk, which is due to platelet adhesion at the surface and platelet aggregation in the bulk. Results from an appropriate statistical comparison at some of the different shear rates tested showed significant differences. In addition to determining shear-induced platelet count reduction, in future studies platelet activation will be evaluated using flow cytometry by a fluorescently labeled antibody for p-selectin.

Percent surface coverage by platelets. Platelet adhesion was measured in terms of percent surface coverage by platelets as a function of shear rate and *S/V*; results are reported in Fig. 2. These results also show a strong dependence on the *S/V* employed. For the *S/V* values used in our system of 10:4, the respective ratios between the shear-averaged results (based on Fig. 2) are about 10:5. These values are not too dissimilar as compared to the *S/V*. The results indicate that percent surface coverage (hence platelet adhesion) increases as the *S/V* increases. As indicated in Fig. 2, at the higher shear rate levels (i.e. higher than 1500 s^{-1}) the ratios obtained between the adhesion results for the different *S/V* values were about 10:4. Results from an appropriate statistical comparison at a shear rate of 2860 s^{-1} showed a significant difference.

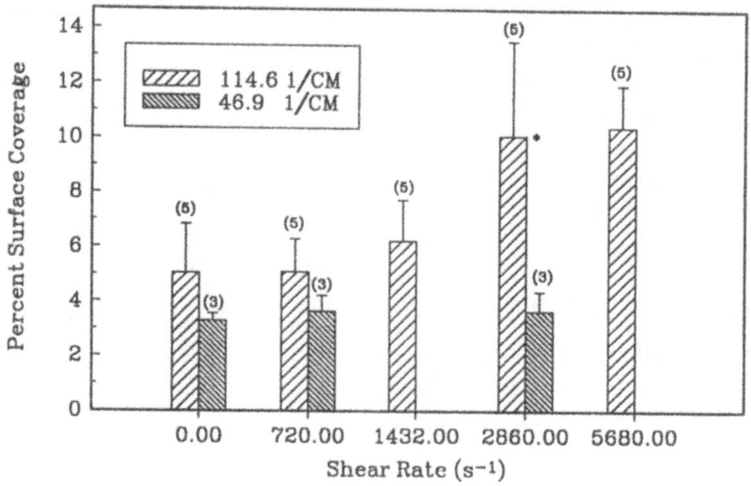

14
12
10
8
6
4
2
0
Percent Surface Coverage
114.6 1/CM
46.9 1/CM
(5)
(3)
*
0.00
720.00
1432.00
2860.00
5680.00
Shear Rate (s-1)

Figure 2. Platelet surface coverage (%) as a function of surface-to-volume ratio and system shear rate using fluorinated ethylene propylene copolymer normal (FEP–NOR) surfaces. The data at each shear rate value tested are represented by the corresponding mean along with the standard deviation. $^{*}P < 0.02$ vs the other group at the same shear rate.

Release of Hb and ADP. Hb release from RBCs as a function of shear rate vs S/V employed is reported in Fig. 3. These results showed the expected trend: an increase in the release of Hb as S/V increases. The ratios between the shear-averaged values were 10:4, which is the same as the ratios of S/V values employed. Results from an appropriate statistical comparison at a shear rate of 2860 s^{-1} showed a significant difference.

The shear-induced release of ADP, presumably from platelets and from RBCs (as reported recently) [1, 4], was also measured in this study. The results are shown in

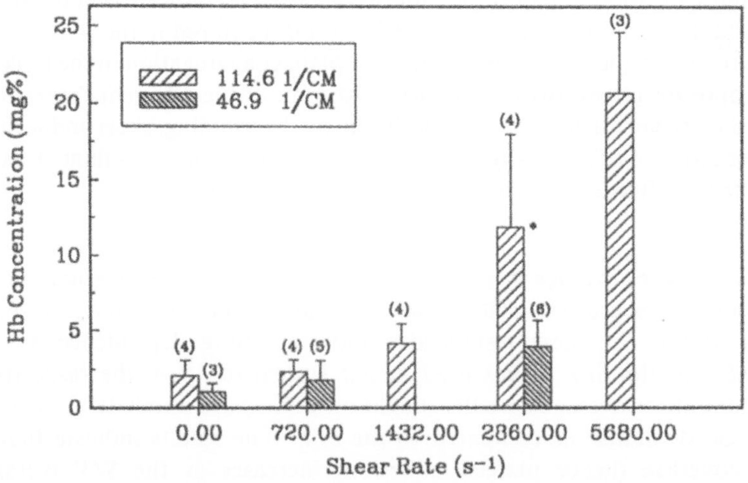

25
20
15
10
5
0
Hb Concentration (mg%)
114.6 1/CM
46.9 1/CM
(3)
(4)
(5)
(6)
*
0.00
720.00
1432.00
2860.00
5680.00
Shear Rate (s-1)

Figure 3. Plasma hemoglobin (Hb) concentration (mg%) as a function of surface-to-volume ratio and system shear rate using fluorinated ethylene propylene copolymer normal (FEP–NOR) surfaces. The data at each shear rate value tested are represented by the corresponding mean along with the standard deviation. $^{*}P < 0.02$ vs the other group at the same shear rate.

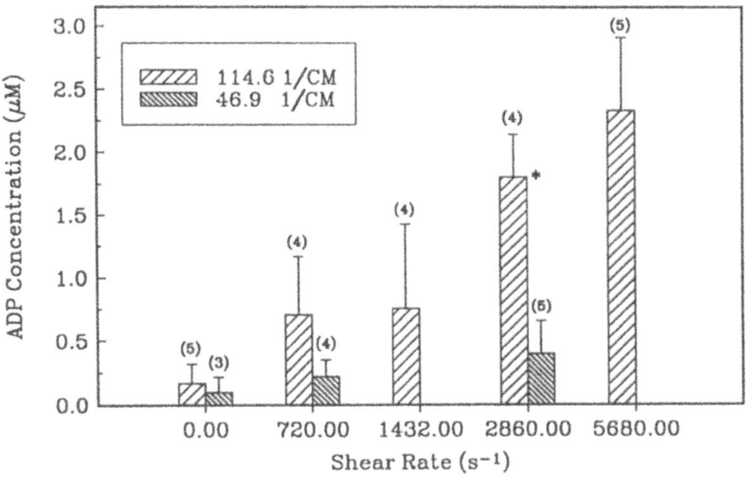

ADP Concentration (μM)
3.0
2.5
2.0
1.5
1.0
0.5
0.0
114.6 1/CM
46.9 1/CM
(5)
(3)
(4)
(4)
(4)
(4)
*
(6)
(5)
0.00
720.00
1432.00
2860.00
5680.00
Shear Rate (s⁻¹)

Figure 4. Plasma adenosine diphosphate (ADP) concentration (μM) as a function of surface-to-volume ratio and system shear rate using fluorinated ethylene propylene copolymer normal (FEP-NOR) surfaces. The data at each shear rate value tested are represented by the corresponding mean along with the standard deviation. $^*P < 0.001$ vs the other group at the same shear rate.

Fig. 4. The ratios for the shear-averaged values are 10:2.5 and indicate a different dependence of ADP release on shear and S/V than was observed for the other parameters reported above. Results from an appropriate statistical comparison at a shear rate of 2860 s^{-1} showed a significant difference.

Artificial surface effects

Different synthetic surfaces, i.e. FEP (normal), PE, and SR were compared to each other with respect to the blood damage parameters noted above. The unfinished FEP surface (sticky^R) was used (as the results bear out as shown below) to emphasize the effect of a more thrombogenic surface on platelet adhesion and aggregation results. The idea behind using this surface was that if surface was controlling blood damage, then a thrombogenic surface such as the FEP sticky^R surface would show a significant increase in the blood damage when compared to results obtained with the primary test surface (FEP-NOR). Both SR and PE gave blood damage results very similar to those shown for FEP-NOR in Figs 5–8. Therefore in Figs 5–8 artificial surface effects were reported and compared for just two surfaces, FEP-NOR and FEP-STI, which showed significant differences in blood damage results.

Percent decrease in platelet count and percent surface coverage. Typical results on reduction in platelet count as a measure of platelet loss in the whole blood sample due to adhesion to artificial surface and aggregation in the bulk (suspension) phase were obtained (see Fig. 5). Obvious differences were observed between the data representing the FEP-NOR and FEP-STI surfaces. As shown in Fig. 5 the FEP-NOR surface demonstrated significantly less reduction in platelet count compared to the FEP-STI surface. Typical results on platelet adhesion as represented by the percent surface coverage were also obtained (see Fig. 6). Again the FEP-STI surface showed significantly higher values.

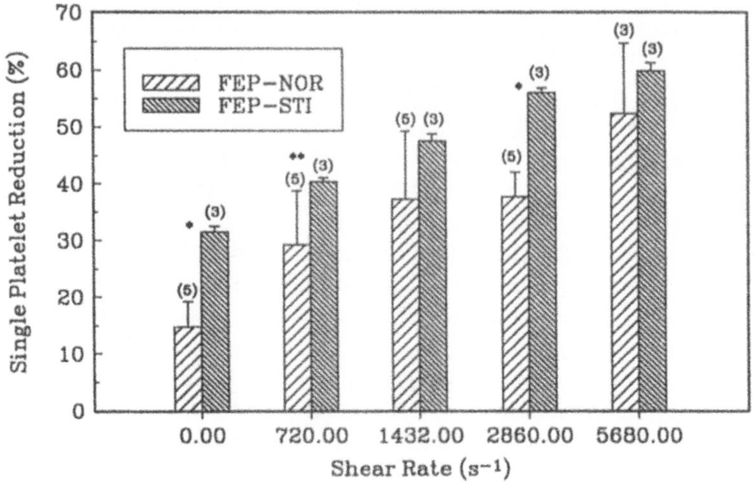
Single Platelet Reduction (%)
70
60
50
40
30
20
10
0
FEP-NOR
FEP-STI
0.00
720.00
1432.00
2860.00
5680.00
Shear Rate (s-1)

Figure 5. Reduction in platelet count (%) for NOR-FEP and sticky FEP (FEP-STI) surfaces as a function of system shear rate. The data at each shear rate value tested are represented by the corresponding mean along with the standard deviation. *$P < 0.001$ vs the other group at the same shear rate. **$P < 0.1$ of the other group at the same shear rate.

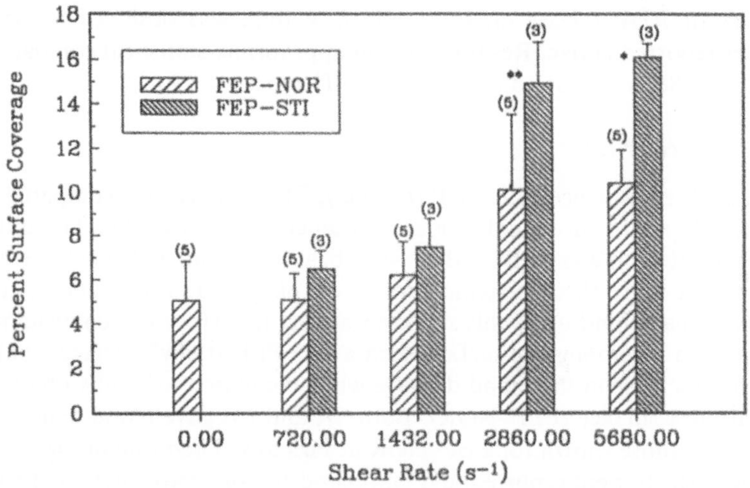
Percent Surface Coverage
18
16
14
12
10
8
6
4
2
0
FEP-NOR
FEP-STI
0.00
720.00
1432.00
2860.00
5680.00
Shear Rate (s-1)

Figure 6. Platelet surface coverage (%) for NOR-FEP and sticky FEP (FEP-STI) surfaces as a function of system shear rate. The data at each shear rate value tested are represented by the corresponding mean along with the standard deviation. *$P < 0.001$ vs the other group at the same shear rate. **$P < 0.1$ of the other group at the same shear rate.

Hb and ADP release. The release of Hb and ADP was measured for the two synthetic surfaces as a function of shear rate; typical results are shown in Figs 7 and 8. Hb release data were represented by plasma Hb concentration. The FEP-STI surface showed significantly higher plasma Hb concentrations compared to the results obtained for the FEP-NOR surfaces. The shear-induced release of ADP

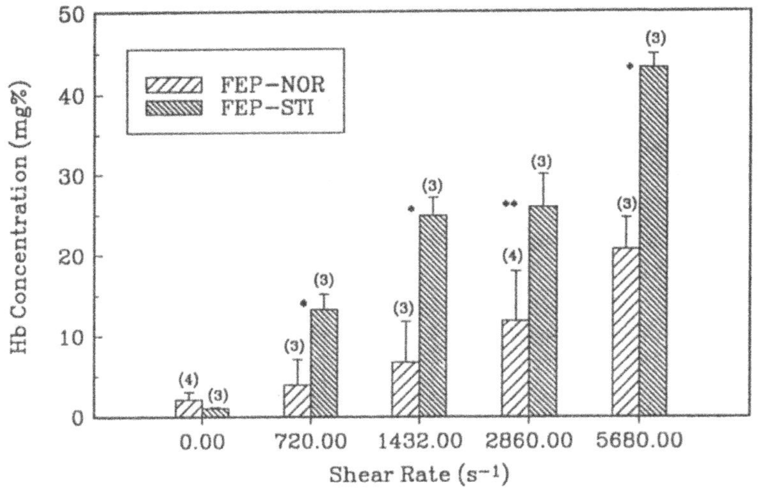
Hb Concentration (mg%)
FEP-NOR
FEP-STI
0.00
720.00
1432.00
2860.00
5680.00
Shear Rate (s-1)

Figure 7. Plasma hemoglobin (Hb) concentration (mg%) for NOR-FEP sticky FEP (FEP-STI) surfaces as a function of system shear rate. The data at each shear rate value tested are represented by the corresponding mean along with the standard deviation. *$P < 0.001$ vs the other group at the same shear rate. **$P < 0.02$ vs the other group at the same shear rate.

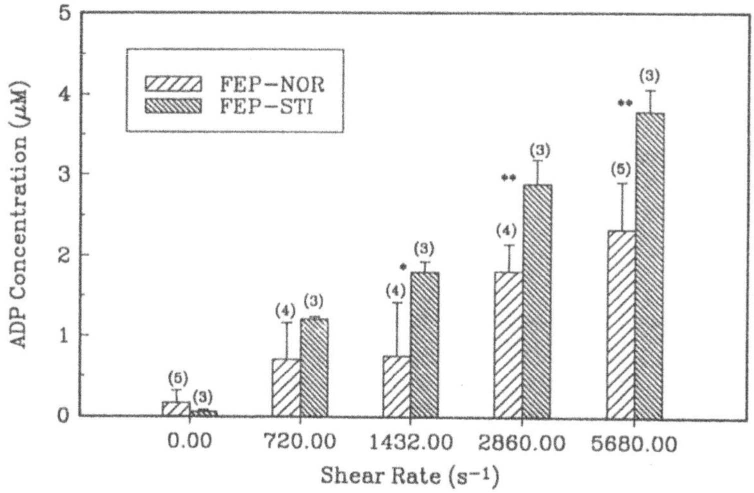
ADP Concentration (μM)
FEP-NOR
FEP-STI
0.00
720.00
1432.00
2860.00
5680.00
Shear Rate (s-1)

Figure 8. Plasma adenosine diphosphate (ADP) concentration (μM) for NOR-FEP sticky FEP (FEP-STI) surfaces as a function of system shear rate. The data at each shear rate value tested are represented by the corresponding mean along with the standard deviation. *$P < 0.05$ vs the other group at the same shear rate. **$P < 0.01$ vs the other group tested at the same shear rate.

from RBCs and platelets was represented by its release into the surrounding plasma. The FEP-STI surface showed significantly higher ADP release results compared to the FEP-NOR surfaces.

Blood damage model: surface and bulk effects

The presentation of the various blood damage parameters—which in this study include reduction in platelet count, surface coverage by platelets, ADP release from platelets and RBCs, and Hb release from RBCs, all as a function of shear rate—requires some concept of the blood damage mechanism. An approach analogous to the one used previously for describing shear-induced low stress Hb release from outdated RBCs [14] was adopted in this study. Models were developed from which to evaluate surface-controlled and bulk-controlled mechanisms for the blood damage parameters measured in this study.

Thus, with respect to surface-controlled effects, we assume that low-stress blood damage occurs at the cone-and-plate viscometer walls and is controlled by the magnitude of shear rate at the wall. If the blood damage results are a surface-dominated phenomenon, then they should be represented in terms of an integral over all device surfaces: therefore, the surface-averaged shear rate was chosen as the correlational parameter. For viscometric flow, which is closely approximated in this study, the shear rate is constant throughout the bulk of the CP device as well as at the wall and, therefore, for the CP system the volume-averaged and surface-averaged shear rates are equal. The effects of secondary flow on hemolytic blood damage were evaluated in a previous study [14]. The surface is of major importance because it may act to damage platelets through two different pathways: (1) directly, through activation of platelets coming into contact with the surface; and (2) indirectly, by which RBCs are traumatized at the surface leading to the partial release of their ADP (as well as their participation in the formation of other platelet agonists) which then diffuses back to activate platelets in the bulk for subsequent adhesion to the surface and/or aggregation of platelets in the bulk.

This mechanism assumes that ADP is released from RBCs and platelets only at the surface. The released ADP then activates platelets passing near the surface as well as diffusing back into the bulk to activate platelets there. Consider that Hb and ADP release rates and platelet adhesion and platelet count reduction rates are all controlled by wall-surface effects such that the flux rates (j_i) are proportional to shear rate ($\dot{\gamma}$) at the platen surface:

$$j_i = K_{si}\,\dot{\gamma}. \tag{1}$$

Also for more complicated cases, other than linear, the same argument basically holds. Then, the ADP and Hb release rates and the platelet surface coverage and platelet count reduction rates for both top and bottom platen surfaces (w_i) are:

$$w_i = \int j_i \, ds = K_{si}\,\dot{\gamma} \int ds = K_{si}\,\dot{\gamma}\, S, \tag{2}$$

where K_{si} includes surface-related kinetic factors and S is total available solid wall surface area. For example, the change, i.e. the increase in plasma ADP concentration (ΔC_i) due to shear is:

$$\Delta C_i = \frac{\int_0^{t_0} w_i \, dt}{V(1 - H)} = \left(\frac{S}{V}\right) \dot{\gamma}\, \bar{K}_{si} \frac{t_0}{(1 - H)}, \tag{3}$$

where H is the hematocrit, V is the total blood volume in the CP system, t_0 is the experimental time, and $\bar{K}_{si}$ is the surface-related time-averaged rate constant, where

$$\bar{K}_{si} = \int_0^{t_0} K_{si} \, dt. \tag{4}$$

Note the dependence of ΔC_i on S/V and $\dot{\gamma}$ in the surface-controlled mechanism. The other blood damage parameters are obtained in an analogous fashion.

An alternative to the surface-controlled model for the various blood damage results would be the bulk-flow one, which depends on homogeneous rates in the fluid bulk on a unit volume basis. For example, for ADP release the bulk-controlled model gives:

$$\Delta C_i = \bar{K}_{bi}\,\dot{\gamma}\,\frac{t_0}{(1-H)} \tag{5}$$

where $\bar{K}_{bi}$ is the bulk-related time-averaged rate constant; note there is no S/V dependence.

The surface and bulk time-averaged rate constants were determined using the above equations. Hb and ADP release data and platelet count reduction and platelet surface coverage were fitted by the least square methods of Eqn (3) to get K_{si} and of Eqn (5) to get K_{bi} and are presented in Tables 1–4, respectively. Although these results do not give a clear picture of whether the release mechanism is surface- or bulk-controlled, the surface-related rate constant is more nearly of constant value, i.e. the variation in the rate constant values for a surface-controlled mechanism is about 50% of the variation in rate constant values for a bulk-controlled one and therefore suggests that the mechanism of release is more likely to be surface-controlled. Surface and bulk time-averaged rate constants were determined by least squares fits of the reduction in platelet count and percent platelet surface coverage data, which are reported in Tables 3 and 4. For these results the variation in $\bar{K}_{bi}$ values is much more than that for the variation in $\bar{K}_{si}$ values, which again seems to indicate that the surface mechanism is controlling. However, these results show that

Table 1.
Values for surface related time-averaged rate constant for Hb and ADP using Eqn (3)

Surface-to-volume ratio (cm^{-1})	Hemoglobin $\bar{K}_s \times 10^5$ ($mg \cdot cm^{-2}$)	ADP $\bar{K}_s \times 10^6$ ($mg \cdot cm^{-2}$)
114.6	2.8	4.8
68.8	4.04	2.44
46.9	2.97	2.59

Table 2.
Values for bulk related time-averaged rate constant for Hb and ADP using Eqn (5)

Surface-to-volume ratio (cm^{-1})	Hemoglobin $\bar{K}_b \times 10^3$ ($mg \cdot ml^{-1}$)	ADP $\bar{K}_b \times 10^4$ ($mg \cdot ml^{-1}$)
114.6	3.22	6.07
68.8	2.78	1.68
46.9	1.39	1.20

Table 3.
Values for surface related time-averaged rate constant for reduction in platelet count and platelet surface coverage using Eqn (3)

Surface-to-volume ratio (cm^{-1})	Platelet count reduction $\bar{K}_s \times 10^5$ (cm^{-2})	Platelet surface coverage $\bar{K}_s \times 10^6$ (cm^{-2})
114.6	8.02	9.49
68.8	9.25	5.09
46.9	10.47	4.27

Table 4.
Values for bulk related time average rate constant of platelet count reduction and platelet surface coverage using Eqn (5)

Surface-to-volume ratio (cm^{-1})	Platelet count reduction $\bar{K}_b \times 10^3$ (ml^{-1})	Platelet surface coverage $\bar{K}_b \times 10^4$ (ml^{-1})
114.6	9.18	10.89
68.8	6.36	3.60
46.9	4.91	3.64

the blood damage mechanism is probably more complicated than the above linear shear rate models suggest. Therefore, more comprehensive global models describing blood damage either in series or in parallel are discussed below.

A series approach, previously developed for hemolysis [14] that can be used for correlating the blood damage results involves coupling the surface and bulk mechanisms via a simple two-parameter ($\lambda, \bar{K}_i$) model. For example, for ADP release

$$\Delta C_i = \left[\left(\frac{S}{V} \right)^{\lambda} \dot{\gamma} \bar{K}_i \right] \frac{t_0}{(1 - H)}. \tag{6}$$

The other blood-damage relations were developed in an analogous fashion. Note that the value of the exponent of the S/V ratio accommodates both the surface-controlled case ($\lambda = 1$) and the bulk-controlled case ($\lambda = 0$). The global least squares fits of the experimental results for the various blood damage parameters were obtained by using either program 3R and AR (or both) from the BMDP statistical software [18] and are reported in Table 5. The results indicate a strong dependence on the surface for ADP and Hb release as well as for the platelet adhesion because the value of the exponent is either 1 (high dependence on the surface) or very close to 1, as is the case with Hb ($\lambda = 0.92$). However, for reduction in platelet count the exponent is about 0.5, which implies a dependence on both surface and the bulk effects.

Equation (6) above is only correlative, not mechanistic. Therefore, more insight into the surface vs bulk-controlled blood damage process may be obtained through a simple mechanistic approach. One approach that models two independent generators of blood damage operating in parallel gives [14]:

$$\Delta C_i = [(S/V)\, \dot{\gamma} \bar{K}_{si} + \dot{\gamma} \bar{K}_{bi}] \frac{t_0}{(1 - H)}. \tag{7}$$

Table 5.
Values for λ and $\bar{K}_b$ in Eqn (6) for the different blood damage parameters

Parameter	λ	$\bar{K}$
ADP	1.0 ± 0^a	$1.0 \times 10^{-6} \pm 0^a$
Hb	0.92 ± 0.22	$(1.15 \pm 1.2) \times 10^{-5}$
Platelet count reduction	1.0 ± 0^a	$(2.48 \pm 0.34) \times 10^{-6}$
Platelet surface coverage	0.52 ± 0.40	$(1.76 \pm 3.26) \times 10^{-4}$

[a] Value is of an order of 4×10^{-7} or less

Table 6.
Values for $\bar{K}_s$ and $\bar{K}_b$ in Eqn (7) for the different blood damage parameters

Parameter	$\bar{K}_s$	$\bar{K}_b$
ADP	$(9.80 \pm 0.83) \times 10^{-7}$	0^a
Hb	$(7.66 \pm 1.39) \times 10^{-6}$	$(3.24 \pm 14.27) \times 10^{-5}$
Platelet count reduction	$(2.48 \pm 0.34) \times 10^{-6}$	0^a
Platelet surface coverage	$(1.11 \pm 0.91) \times 10^{-6}$	$(7.96 \pm 9.31) \times 10^{-4}$

[a] Value is of an order of 3×10^{-20} or less

Hemolytic damage in the above equation is assumed to have two independent generators (surface and bulk) operating in parallel. Simple hemolytic blood damage is assumed to depend on a surface and bulk mechanism and they are coupled in one equation in order to see their importance with respect to the surface and bulk blood damage parameters. Dependence of ΔC_i on S/V can range from first to zeroth order, depending on the relative magnitude of $\bar{K}_{si}$ to $\bar{K}_{bi}$. In theory all the blood damage data can be fit by this type of an equation. $\bar{K}_{si}$ and $\bar{K}_{bi}$ were computed for the various blood damage parameters from the data presented in our study using the BMDP statistical software programs mentioned above and are reported in Table 6. The results again support the surface-controlled mechanism and suggest that (S/V) $\bar{K}_{si}$ is considerably larger than $\bar{K}_{bi}$ for all cases, which is an interpretation similar to that obtained in the previous correlation (see Eqn (6)). A time-dependent analysis of low stress hemolysis (with respect to RBCs) in terms of a series model in partial differential equation form has been presented elsewhere and was used to rationalize results as well as give predictions of the relative contributions of the surface and bulk based on the low-stress flow conditions imposed [15].

DISCUSSION

The experimental results obtained in this study using normal FEP surfaces showed that platelet counts of sheared whole blood samples were reduced compared to unsheared samples and that the reduction appeared to depend on the shear rate and S/V of the test system. Shear-induced platelet count reduction, as noted above, is due to the combined effects of platelet adhesion at the surface and aggregation in the bulk, where aggregation can be considered to be on top of the first adhered monocellular platelet layer as well as in the bulk. Platelet adhesion was also shown to be a function of shear rate and S/V ratio. Although platelet aggregation in the bulk was not quantified, its occurrence was expected because the amount of ADP that was released (especially at the higher shear rate levels used in this study) is known to cause irreversible aggregation of platelets [10]. Moritz *et al.* [19] suggested that ADP was the major, if not the sole mediator of shear-induced platelet aggregation. Adenosine diphosphate is likely to act by binding with ADP-platelet membrane receptors to activate platelets [4, 5, 10, 20, 21].

The results obtained in this study for Hb suggest that the mechanism by which RBCs release their Hb is controlled by shear rate and surface. Comparison with another hemolytic study using recently outdated whole blood (i.e. about 21 days of storage) in a capillary viscometer with a residence time range of 10 s to 2 min, showed that hemolytic damage may also have been surface-induced [15] and that the storage age of the blood up to 21 days did not appear to be a factor. However, the present findings are different from those found in other investigations [12, 14]. In one study recently outdated whole blood was sheared for 10 min in a torsional viscometer [14]. Results showed that the hemolytic damage had little dependence on the S/V values tested. These results may suggest that the hemolysis measured was a bulk-induced phenomenon. A possible explanation is that the longer residence time in the viscometer resulted in damage to all the fragile outdated RBCs at the viscometer plater surface (outdated blood is expected to have more fragile RBCs than fresh blood) and low-stress hemolysis would continue in the bulk, but only as a function of bulk flow conditions. In the other study [12] artificial surface effects were found to be less pronounced. This was also probably due to the longer residence times used and the shear sensitivities of the fragile outdated RBCs tested.

Based on the results for shear-induced Hb release, it seemed reasonable to suggest that shear-induced release of ADP (a much smaller molecule than that of Hb) from RBCs in fresh whole blood samples may also (like Hb) have been controlled by the surface. Other studies, using PRP, found that blood cell interactions with the viscometer surface have little influence on platelet aggregation [22–24]. In such systems much fewer interactions would be expected with the surface, which may be due to the absence of RBCs. Blood damage as measured in this study was found to significantly depend on the S/V ratio found in this study, which may be due to the effect of RBCs present in the system. A recently published study [4] showed that of the total ADP release, about 67% originated from RBCs, which caused six times as much reduction in platelet count and about twice as much platelet adhesion as did the ADP released by the platelets themselves.

Low-stress, shear-induced reduction in platelet count may only represent normal platelet phenomena (i.e. without platelet damage) such as platelet clumping and thus may reflect platelet adhesion without extensive spreading and loss of subcellular granular components like ADP [25–27]. However, for sheared whole blood samples with substantial release of ADP from RBCs as in this study, primary aggregation and secondary aggregation (with platelet adhesion) and spreading (pseudopod formation) was expected to follow. Platelet-rich plasma samples exposed to shear stresses above 50 dyn cm^{-2} exhibited platelet damage as indicated by loss of lactic dehydrogenase (LDH) from the platelets [19]. The low-stress, shear-induced release from RBCs of the cytoplasmic components ADP and LDH as well as Hb, as was found in a recent study [14], represents RBC damage. It will be appropriate in future work to study the release of not only ADP, but also other alpha and dense granule components as expressions of stronger platelet activation and blood damage. For example, as was mentioned above, platelet activation will be evaluated using flow cytometry by a fluorescently labeled antibody for p-selectin.

It has been believed by many that surface effects (at least initially) control blood damage. Of the artificial surfaces tested many display to various extents incompatibilities with whole blood. This study supports that belief as evidenced by the success seen with the surface-controlled correlational blood damage models used. The experimental results presented suggest that the surface is of primary importance in terms of damage to blood as it is exposed to shear rate levels to $5680\ s^{-1}$. However, this conclusion needs to be qualified since there are interactions between the bulk flow and surface effects. Along with the release of ADP from the RBCs the rotational and translational motion of the RBCs physically augment platelet transport in the bulk to the surface and therefore dramatically increase the platelet concentration in the wall boundary layers [1-3]. Subsequently the platelets increase their adhesive interaction with artificial surfaces as well as their aggregation with other platelets at the surface and in the bulk, which leads to more aggregation [1, 6, 28]. Recent studies on artificial surfaces concluded that these surfaces activate platelets requiring GPIIb/IIIa (and probably fibrinogen) participation for adhesion and GPIb (and probably von Willibrand factor) for paradhesion [5].

The effect of the artificial surface on blood damage was also seen in a recent study concerning thrombin activation of platelets [7]. That study's results suggested that thrombin was being generated at the platelet-artificial surface as a function of shear rate for ACD-anticoagulated whole blood samples because of the increased platelet surface coverage seen compared to coverage found for the whole blood samples containing hirudin. It was suggested in that study that shear at the artificial surface caused the release of enough intracellular calcium [29], most likely due to ADP (from RBCs) [4] induced platelet activation, so that thrombin could be generated; without calcium, prothrombin conversion to thrombin via the prothrombinase complex would not be expected [30]. Also the work of Wagner and Hubbell [31] demonstrated, even under the conditions of heparin anticoagulation, a role for the local action of thrombin at the surfaces of thrombi through platelet recruitment and thrombi stabilization.

Of the different artificial surfaces investigated in this study results with the primary test surface (FEP–NOR) (data shown), a well as with the PE and SR surfaces (data not shown), gave similar blood damage results, i.e. platelet adhesion and aggregation and Hb and ADP release, that were significantly less than those obtained with the unfinished FEP surfaces (sticky^{R}, data shown). Therefore, this set of results for the two groupings of artificial surfaces further supports the hypothesis that surface events were controlling the blood damage phenomena measured.

Acknowledgement

The authors wish to thank Margaret Winters of the IIT Clinic for the collection of donor blood. The authors also want to thank Woonou Cha for making the illustrations. This work was supported by a grant-in-aid from the American Heart Association (Chicago Chapter).

REFERENCES

1. V. T. Turitto and H. J. Weiss, *Science* **207**, 541 (1980).
2. A. W. Tilles and E. C. Eckstein, *Microvascular Res.* **33**, 211 (1987).

3. P. A. Aarts, S. A. van den Broek, G. W. Prins, G. D. Juiken, J. J. Sixma and R. M. Heethaar, *Aterioscleroris* **8**, 819 (1988).
4. T. M. Alkhamis, R. L. Beissinger and J. R. Chediak, *Blood* **75**, 1568 (1990).
5. R. A. Sheppeck, M. Bentz, C. Dickson, S. Hribar, J. White, J. Janosky, S. Berceli, H. S. Borovetz and P. C. Johnson, *Blood* **78**, 673 (1991).
6. R. C. Reimers, S. P. Sutera and J. H. Joist, *Blood* **64**, 1200 (1984).
7. T. M. Alkhamis, R. L. Beissinger and J. R. Chediak, *Biomaterials* **14**, 865 (1993).
8. Report of the National Heart, Lung, and Bood Institute Working Group. 1985. Concepts of thrombus formation, dissolution, and antithrombotic therapy. In Guidelines For Blood-Material Interactions (NIH publication no. 85-2185).
9. J. M. Anderson and K. Kottke-Marchant, *CRC Crit. Rev. Biocompatibility* **1**, 111 (1985).
10. G. V. R. Born, *Nature* **194**, 927 (1962).
11. J. M. Monroe, R. C. Lijana and M. C. Williams, *Biomater. Med. Dev. Artif. Organs* **8**, 103 (1980).
12. R. D. Offeman and M. C. Williams, *Biomater. Med. Dev. Artif. Organs* **7**, 359 (1979).
13. S. I. Shapiro and M. C. Williams, *Am. Inst. Chem. Eng. J.* **16**, 575 (1970).
14. R. L. Beissinger and M. C. Williams, *Am. Inst. Chem. Eng. J.* **30**, 569 (1984).
15. R. L. Beissinger and J.-F. Laugel, *Am. Inst. Chem. Eng. J.* **33**, 99 (1987).
16. W. V. Miller (Ed.), *Technical Manual of Blood Banks*, 7th ed., Lippincott Company, Philadelphia (1977).
17. H. Holmsen, *Clin. Hematol.* **1**, 235 (1972).
18. The BMDP Statistical Software, Los Angeles (1987).
19. M. W. Moritz, R. C. Reimers, R. K. Baker, S. P. Sutera and J. H. Joist, *J. Lab. Clin. Med.* **101**, 537 (1983).
20. R. W. Coleman, *Semin. Hematol.* **23**, 119 (1986).
21. S. A. Kalambakas, F. M. Robertson, S. M. O'Connell, S. Sinha, K. Vishnupad and G. I. Karp, *Blood* **81**, 2652 (1993).
22. G. H. Anderson, J. D. Hellums, J. L. Moake and C. P. Alfrey, *Thromb. Res.* **13**, 1039 (1978).
23. T. K. Belval, J. D. Hellums and R. T. Solis, *Microvasc. Res.* **28**, 279 (1984).
24. J. H. Joist, E. J. Bauman, M. Speer and S. P. Sutera, *Thromb. Haemostasis* **54**, 109 (1985).
25. H. R. Baumgartner, R. Muggli, T. B. Tschopp and V. T. Turitto, *Thromb. Haemost.* **35**, 124 (1976).
26. E. F. Grabowski, K. K. Herther and P. Didisheim, *J. Lab. Clin. Med.* **88**, 368 (1976).
27. E. F. Grabowski, P. Didisheim, J. C. Lewis, J. T. Franta and J. Q. Stropp, *Trans. Am. Soc. Artif. Intern. Organs* **23**, 141 (1977).
28. Y. A. Butruille, S. R. Savitz, E. F. Leorard and R. S. Litwak, *J. Biomed. Mater. Res.* **10**, 145 (1976).
29. M. H. Kroll and A. I. Schafer, *Blood* **74**, 1181 (1989).
30. K. G. Mann, M. E. Nesheim, W. R. Church, P. Haley and S. Krishnaswamy, *Blood* **76**, 1 (1990).
31. W. R. Wagner and J. A. Hubbell, *J. Lab. Clin. Med.* **116**, 636 (1990).

A model for thromboembolization on biomaterials

L. O. REYNOLDS♣, W. H. NEWREN, Jr., J. F. SCOLIO and I. F. MILLER*

Bioengineering Program, University of Illinois at Chicago, Box 4348 (Mail Code 063), Chicago, IL 60680, USA

Received 1 July 1992; accepted 2 October 1992

Abstract—A model was developed to describe the kinetics of protein and platelet deposition and embolization on biomaterials. The model assumes that proteins can be adequately represented by fibrinogen, albumin, and Factor XII, that protein adsorption is Langmuir-type, that surfaces are homogeneous, and that all adsorption and deposition steps are first order. Eleven model parameters were determined from literature experimental data from *ex vivo* experiments utilizing canine and baboon blood on Silastic, one parameter came from adsorption of Factor XII on glass, and three parameters were obtained by minimizing differences between experimental and predicted fibrinogen adsorption, and platelet deposition and embolization behavior. The model well predicted observed behavior for fibrinogen adsorption, platelet deposition, and platelet embolization on Silastic, and platelet embolization from both polyacrylamide and HEMA-MAAC.

Key words: Thrombosis; biomaterials; platelets; embolization; protein adsorption; models.

INTRODUCTION

Thrombogenesis is one of the major problems associated with the use of biomaterials in blood transport applications. Since the growth and subsequent embolization of blood aggregates can lead to vascular occlusion and downstream organ damage, fully blood compatible artificial organs cannot be developed until embolization can be controlled or prevented. The process of thrombogenesis on a synthetic biomaterial involves a large number of separate events that occur simultaneously or in series. Although a substantial amount of effort has been expended to develop an understanding of the details of the process, its complexity has made the goal of a complete understanding of the process elusive.

The major events that occur during thrombogenesis can be grouped into protein adsorption and activation steps, platelet attachment and activation steps, and embolism growth and detachment. Since investigation of these individual steps is difficult, workers have typically relied on models of the steps or of the overall phenomena to predict behavior which can then be checked against experiment. Data for these models come from a variety of sources.

Most *in vitro* experiments on plasma protein adsorption indicate that a Langmuir isotherm is followed, though with complexities not present in the original Langmuir model [1]. For example, Dillman and Miller [2] showed that for a single protein on a single surface, two independent types of adsorption can occur simultaneously: a hydrophilic, exothermic, easily reversible type; and a hydrophobic, endothermic, and largely irreversible type. Other workers have shown that protein adsorption

---

♣ Deceased.

* To whom correspondence should be addressed.

depends on the diffusivity and concentration of the protein [3, 4], that multilayers of protein build up upon the initial Langmuir monolayer [5], that blood flow rate has important effects [3, 6–8], and that surface charge is an important mediating variable [9, 10]. Andrade [11] and Beissinger and Leonard [12] proposed that a hydrophobic surface may produce a conformational change in the bound protein, while Matsuda *et al.* [5] proposed that only the initial Langmuir-type protein layer in direct surface contact is denatured. Proteins also appear to compete for adsorption sites [12–15].

Activation of adsorbed coagulation factors appears to depend on surface properties [16, 17]. Surface activation of Factor XII initiates the intrinsic coagulation cascade. Part of the Factor XII desorbs from the surface while the remaining portion is slowly inactivated by a histidine-rich peptide [18]. Factor XII is not thought to replace surface fibrinogen in any significant amount [19], although fibrinogen appears to be easily replaced by other proteins. It is reasonable, therefore, to assume that Factor XII does not replace other proteins on the surface.

Platelet contact with surfaces appears to be convection/diffusion-controlled, and initial adhesion of platelets to surfaces appears to be independent of the material [20, 21]. It has been reported that platelets appear to adhere only to surfaces to which fibrinogen has been adsorbed [22], although others report that platelets may adhere to albumin [23] or to other serum proteins [24]. Platelet adhesion also depends on flow rate and exposure time [25]. Platelet activation and aggregation appear to be a function of platelet concentration, flow rate, and time [26–28].

The amount of thrombus deposited and removed from a surface has been found to depend on the physical dimensions of the surface, its shape, and the surface morphology. Linear correlations between the rate of production of microemboli volume and shunt surface area have been found [29, 30]. Although DePalma *et al.* [31] found that irregularities of 1 μm or less did not alter surface thrombogenic properties, Eberhart *et al.* [32] found that fibrinogen preferentially adsorbed at microscopic cracks on the surface, while albumin did not.

Attempts to understand the contribution of these various phenomena to thromboembolization has led to the development of models for various aspects of the process, and for the process as a whole. For example, Bagnall [33] developed a competitive surface spreading hypothesis for protein adsorption to account for protein denaturation and competition on the surface, while Cuypers *et al.* [34] developed a simulation model for competitive protein adsorption/desorption kinetics. Models for platelet aggregation and disaggregation have been developed by Nguyen and O'Rear [35].

Attempts to integrate all the steps of protein and platelet adsorption and activation, and thrombus formation and removal into a single model have been rarer. Schultz *et al.* [36] developed equations to describe both platelet and fibrinogen diffusion and deposition, but did not model thrombus removal. Marmur and Cooper [37] developed a model that included protein adsorption, platelet deposition, and thrombus removal with co-embolization of proteins. More recently, Wilson *et al.* [38] developed a model that included protein and platelet deposition and detachment, and which considered hydrodynamic forces in determining embolization rates. This model, the most comprehensive to date, assumed a single protein type, a characteristic time between adsorption of protein and detachment of protein and associated platelet aggregates, and a distribution of thrombus size that was determined by hydrodynamic forces.

Herein, we present an integrated model for protein, platelet, and embolization kinetics on biomaterial surfaces that improves on the Wilson *et al.* model in the following ways. Rather than assuming a single protein type, the model considers three types of protein: platelet-adherent proteins exemplified by fibrinogen, 'other proteins' exemplified by albumin, and active coagulation factors exemplified by Factor XII. The production of fibrin is explicit and considers the active coagulation factors. Platelet deposition is considered to be directly related to adsorbed 'fibrinogen'. Embolization from the surface depends on the area covered by the platelet aggregate, and on fibrin-related platelet–platelet bridges. Most of the parameters used in the model equations have values derived from the literature, while the few remaining parameters for which literature values are unavailable have been evaluated to minimize differences between predicted and observed rates of fibrinogen deposition and platelet deposition and embolization.

THEORY

Protein adsorption kinetics are modeled as first order rate equations, and are based on the Langmuir isotherm. Each protein subset ('fibrinogen', 'other proteins', and 'active coagulation factors') is assumed to have its own characteristic rate constants. Fibrinogen adsorption is assumed to occur onto a bare surface, or by replacement of other proteins. Fibrinogen loss is assumed to occur by either desorption from an uncovered fibrinogen-covered surface, by replacement by other proteins, or when adhered platelets embolize. Deposition of all substances is treated in terms of area covered and, initially, no protein is adsorbed to the surface. Platelets are assumed to adhere only to a fibrinogen-coated surface.

$$\frac{dF_s(t)}{dt} = k_{fa}A(t) - k_{fr}[F_s(t) - P_f(t)] + k_{pr}O_s(t) - E_{fs}(t) \tag{1}$$

where F_s = surface area covered by fibrinogen, P_f = surface area covered by platelets, O_s = surface area covered by 'other proteins'. E_{fs} = rate of fibrinogen area co-embolizing from surface with embolizing platelets (area/time), k_{fa} = fibrinogen adsorption rate constant (time^{-1}), k_{fr} = fibrinogen replacement rate constant (time^{-1}), and k_{pr} = 'other protein' replacement rate constant (time^{-1}).

The bare surface area fraction is:

$$A(t) = 1 - F_s(t) - O_s(t) - C_s(t) \tag{2}$$

where C_s = surface area covered by active coagulation factors.

Fibrinogen that is adherent to platelets is not available for replacement by other proteins. Since it is assumed that platelets adhere only to a fibrinogen-coated surface, and not to other proteins, the rate equation for 'other proteins' does not include a co-embolization term:

$$\frac{dO_s(t)}{dt} = k_{pa}A(t) + k_{fr}[F_s(t) - P_f(t)] - k_{pr}O_s(t) \tag{3}$$

where k_{pa} = 'other proteins' adsorption rate constant (time^{-1}).

Assuming that active coagulation factors do not compete with other proteins, the adsorption rate of the active coagulation factors is described by:

$$\frac{dC_s(t)}{dt} = k_{ca}A(t) - k_{cr}C_s(t) \tag{4}$$

where k_{ca} = coagulation factor adsorption rate constant (time^{-1}), and k_{cr} = coagulation factor removal rate constant (time^{-1}).

The coagulation factors are assumed to be adequately modeled by Factor XII, which undergoes contact activation on the surface and initiates the intrinsic pathway of the coagulation cascade. Removal of this factor from the surface is achieved by replacement with other proteins, or deactivation via conformational changes, proteolysis, or interactions with other proteins.

The kinetics of platelet deposition onto a fibrinogen layer is described by:

$$\frac{dP_f(t)}{dt} = k_{pf}[F_s(t) - P_f(t)] - E_{pf}(t) \tag{5}$$

where E_{pf} = rate of platelet surface area loss by embolization from a fibrinogen surface (area/time), and k_{pf} = platelet loss rate constant (time^{-1}).

The model assumes that platelets do not adhere either to a bare surface or to 'other proteins'.

The rate of platelet deposition onto a previously deposited platelet layer is described by:

$$\frac{dP_p(t)}{dt} = k_{pp}[P_p(t) + P_f(t)][1 + W] - E_{pp}(t) \tag{6}$$

where P_p = area of platelets deposited onto previously deposited platelets, W = fractional enhancement of effective adhesive platelet surface area caused by the presence of 'fibrinogen receptors', E_{pp} = rate of platelet surface area loss by embolization from a platelet surface (area/time), and k_{pp} = platelet–platelet adsorption rate constant (time^{-1}).

Thrombin kinetics are not directly characterized in the model, but are considered tied to fibrin kinetics.

The rate of change of fibrin content on the platelets is modeled by:

$$\frac{dB_p(t)}{dt} = k_f\frac{[C_s(t - T_1)][P_f(t) + P_p(t)]}{F_s(t)} - E_{bp}(t) \tag{7}$$

where B_p = platelet surface area covered by fibrin (for $t < T_1$, $B_p = 0$), E_{bp} = rate of fibrin surface area loss by embolization (area/time), k_f = fibrin formation rate constant (time^{-1}), and T_1 = intrinsic clotting activation (prothrombin) time.

The equation states that the rate of change of fibrin content is proportional to the concentration of active coagulation factors at an appropriate previous time, to account for clotting activation time. Since the equation relates to fibrin on the platelets only, the proportion is modified by the fraction of fibrinogen-coated surface that is covered by platelets.

In the absence of fibrin, the platelet aggregate can be stabilized by platelet–platelet bridges formed by apposition of 'fibrinogen receptors' on adjacent platelets.

This bridging area, P_{rb}. is described by:

$$P_{rb}(t) = W\,[P_f(t) + P_p(t)]. \tag{8}$$

Although the effect of platelet activation is not explicitly included in the equations, it is implicit in the definition of 'fibrinogen receptors'.

Embolization is modeled by a set of algebraic equations as follows. The embolization of platelets from platelets is described by:

$$E_{pp}(t) = \frac{[K_{epp}][E_{bp}(t)][P_p(t)]}{B_p(t) + P_{rb}(t)} \tag{9}$$

where K_{epp} = platelet-platelet embolization coefficient (dimensionless), and E_{bp} = rate of release of fibrin area by platelet embolization (area/time). This is assumed to be proportional to the total area covered by fibrin. Thus,

$$E_{bp}(t) = k_{bp}[B_p(t)] \tag{10}$$

where k_{bp} = platelet-fibrin release rate constant (time^{-1}).

The embolization of platelets from fibrinogen is described by:

$$E_{pf}(t) = \frac{[K_{epf}][E_{pp}(t)][P_f(t)]}{P_p(t)} \tag{11}$$

where E_{pf} = rate of area release by embolization of platelets from fibrinogen (area/time), and K_{epf} = platelet-fibrinogen rate embolization coefficient (dimensionless), with a value between 0 and 1. If platelets are tenaciously bound to the fibrinogen layer, K_{epf} will be near 0. If attachments are weak, it will be near 1.

Embolization of fibrinogen from the surface is described by:

$$E_{fs}(t) = [K_{efp}][E_{pf}(t)] \tag{12}$$

where K_{efp} = platelet-surface embolization coefficient (dimensionless).

If appropriate values for the 15 coefficients in this set of equations can be determined from experimental data or reasonable physical considerations, then the model can be used to predict protein, platelet, and embolization kinetics on biomaterials, and to estimate the biocompatibility of candidate biomaterials.

MATERIALS AND METHODS

The model was tested with data for the interaction of blood with Silastic, polyacrylamide, and HEMA-MAAC, using literature sources for the various coefficients. This approach has a number of limitations. First of all, experimental conditions vary among the various sources of the data, and many of the experimental details necessary to determine the reliability of the data are missing. Second, no two samples of biomaterial can be assumed to be exactly alike, making comparisons problematic. Third, many of the coefficients are not explicitly given in the references, but must be calculated from other data. Finally, literature values could be determined for only 12 of the 15 coefficients and even then, not under all conditions. As a consequence, model testing was limited to qualitative behavior and comparisons with experimental observations and no *a priori* predictions of biomaterial behavior were possible.

Values for the 12 coefficients that were found in the literature all had some uncertainty. Some of these were adjusted somewhat to improve the fits to experimental fibrinogen adsorption, platelet deposition, and embolization data, or to make them more realistically fit the experimental conditions. Values for the three remaining coefficients that could not be found in the literature were also fitted to minimize differences between the predicted and experimental values for fibrinogen and platelet deposition and embolization.

The experimental values for platelet embolization came from data of Reynolds that remained unpublished at the time of his death. From his notes, we have reconstructed the protocols as follows. Two cuvettes were constructed from 1.0 mm diameter medical grade Silastic tubing, 5.0 mm in length. The cuvettes were encased in an aluminum housing that held three optical fibers in contact with and orthogonal to the Silastic tube. Light from a He–Ne laser was focused on one of the optical fibers and the intensity of the scattered light detected by the other two orthogonal fibers was analyzed to determine the size and number density of the microemboli within the cuvette. One cuvette was placed upstream and the other downstream of the tubular biomaterial to be tested. The entire system was connected with 3.0 mm diameter Teflon tubes to form an arteriovenous shunt in an *ex vivo* baboon model similar to that described by Horbett *et al.* [30]. The simultaneous signals from the two cuvettes were multiplexed to eliminate the need for baseline thromboembolization measurements and, thus, to provide an accurate measure of the size and number of microemboli produced during blood/material contact within the test specimen.

A computer algorithm was constructed to solve the system of equations numerically. Simulations were run on an IBM 370 mainframe computer, using the FORTRAN VS language. Fifth- and sixth-order Runge–Kutta–Verner differential equation interpolations were carried out using International Mathematical Statistical Routines Library (IMSL) software. Graphs were produced using Display Integrated Software System and Plotting Language (DISSPLA) software routines. Double precision floating point variables were used in all iterative procedures.

The simulations were performed by first evaluating the model parameters at or near values found in the literature; values were assumed for those parameters for which literature values could not be found, using expected qualitative behavior as criteria for initial estimates. The unknown parameters were adjusted in each simulation to minimize the differences between experimental and predicted acute responses (peak amplitude and width) for fibrinogen adsorption and platelet deposition and embolization. Adjustments to parameter values were made only to those parameters for which exact literature values were not available. These adjustments were made only in directions consistent with expected qualitative behavior.

RESULTS

Figure 1 compares predicted rates of fibrinogen adsorption on Silastic with *ex vivo* clottable and anticoagulated baboon blood data of Horbett and co-workers [30]. With the exception of the fibrinogen rate constants (k_{fa} and k_{fr}), which were smaller than the literature values, the fibrin formation rate constant k_f for anticoagulated blood, which was 20% of the literature value for clottable blood, the platelet–platelet embolization coefficient (K_{epp}) and intrinsic clotting time (T_1), which were higher

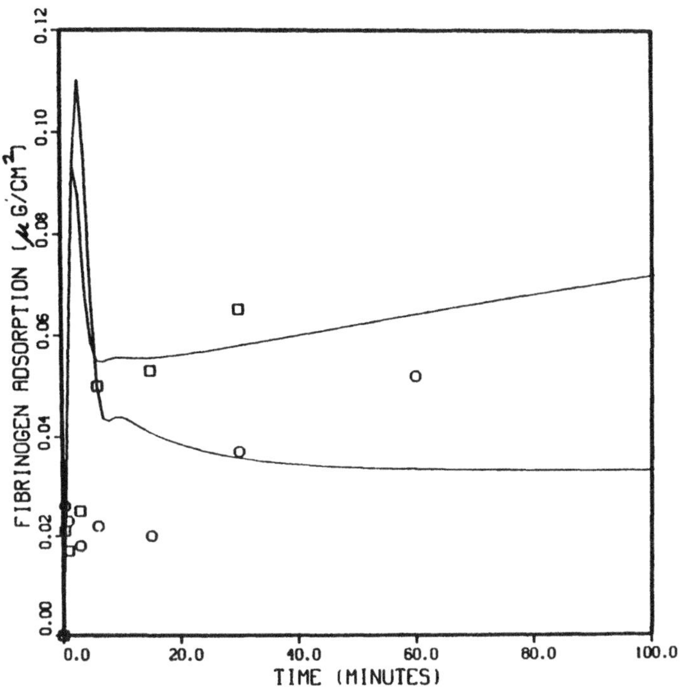
0.12
0.10
0.08
0.06
0.04
0.02
0.00
FIBRINOGEN ADSORPTION (μG/CM²)
0.0
20.0
40.0
60.0
80.0
100.0
TIME (MINUTES)

Figure 1. Fibrinogen adsorption onto Silastic as a function of time, for clottable and anticoagulated baboon blood. Experimental data for clottable (□) and anticoagulated (○) blood come from Horbett *et al.* [30]. Predicted adsorptions for clottable and anticoagulated blood are the bottom and top curves, respectively.

than the literature values, and W, which was about 3% of its literature value, all the values for the coefficients used in the model were taken directly from the literature.

The adjustments from literature values were made to qualitatively account for known differences from the literature (e.g., the difference between clottable and anticoagulated blood), or to improve the fit. With the exception of W, all the adjustments from literature values were within the uncertainty of the literature values. The reduction in k_{fa} of 57% caused a reduction in peak height of about 48%, while the reduction in k_{fr} of 50% had no significant effect on the results. The reduction in k_f for anticoagulated blood was necessitated by the fact that the literature value was for clottable blood, and fibrin should not form in anticoagulated blood. This change had no effect on the peak response, but reduced fibrinogen adsorption at later times. The changes in K_{epp} and T_1 made very little difference in peak response, but decreased the post-peak plateau by about 48%. The reduction of W to 3% of its literature value reduced the peak response by 64%.

The figure demonstrates that the model predicts behavior that follows experiment fairly closely. An initial spike in fibrinogen adsorption was predicted, although it was not apparent in the experimental data used herein. Such an initial spike in fibrinogen adsorption was seen, however, by Horbett *et al.* [30] in a different experiment, and by others [39]. There is a reasonable quantitative fit between predicted and experimental fibrinogen adsorption at times later than about five minutes.

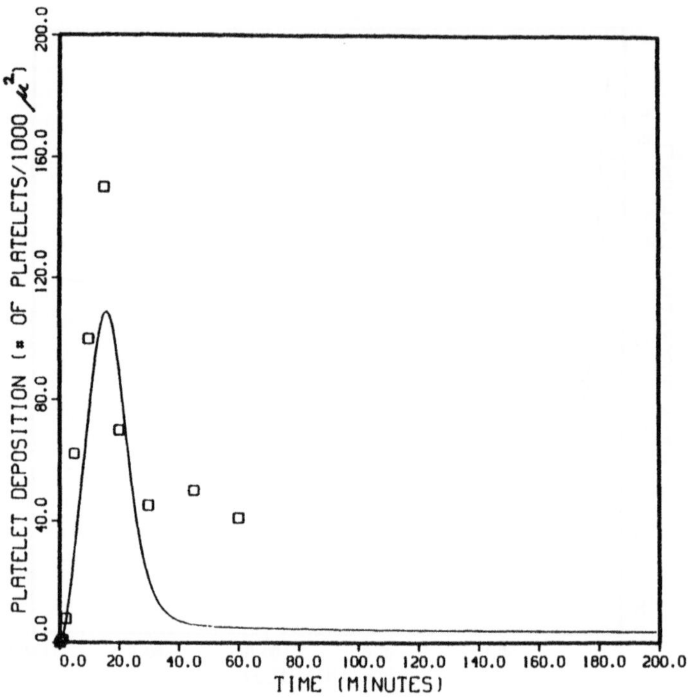
PLATELET DEPOSITION (# OF PLATELETS/1000 μ²)
0.0
40.0
80.0
120.0
160.0
200.0
0.0 20.0 40.0 60.0 80.0 100.0 120.0 140.0 160.0 180.0 200.0
TIME (MINUTES)

Figure 2. Platelet deposition from canine blood onto Silastic as a function of time. Experimental data (□) come from Lelah *et al.* [39]. The curve shows the predicted deposition.

Figure 2 compares predicted rates of platelet deposition on Silastic with data from a canine model of Lelah and co-workers [39]. With the exception of k_{fr}, which was 50% of its literature value, all the parameters were set to their literature values. The change in k_{fr} was motivated by the desire to set the same value as for the fibrinogen adsorption simulation (Fig. 1). This change had no discernible effect on the results. The model predicts platelet deposition amounts that closely follow experimental values.

Figure 3 compares predicted rates of platelet embolization from Silastic with unpublished data of Reynolds from an *ex vivo* baboon model similar to the Horbett *et al.* model [30]. The fit was obtained by using a value for k_{pf} that was about two-thirds the value used in Fig. 2, a value for k_{pp} about half the value used in Fig. 2, and a value for W as used in Fig. 1 (about 3% of its literature value). All other parameters retained their literature values. The reduction of k_{fr} to 50% of its literature value reduced the peak response by about 12%. The reduction of k_{pf} by 35% from its literature value had no significant effect on the results. The reduction in k_{pp} to 49% of its literature value reduced the peak response about 19%, while the reduction of W to 3% of its literature value reduced the peak response by about 32%. For the first 30 min, the model predictions closely follow the data, while at later times, the predictions do not fall off as rapidly as the data.

Figures 4 and 5 compare predicted rates of platelet embolization from polyacrylamide and HEMA–MAAC with unpublished data of Reynolds using the identical protocols used in Fig. 3. The coefficients changed from the literature values

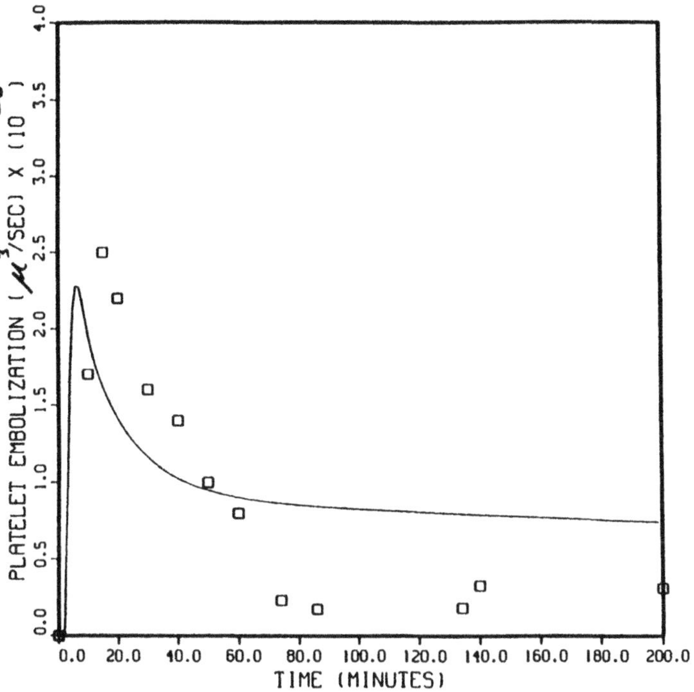
PLATELET EMBOLIZATION (μ^3/SEC) X (10^{-6})
0.0
0.5
1.0
1.5
2.0
2.5
3.0
3.5
4.0
0.0 20.0 40.0 60.0 80.0 100.0 120.0 140.0 160.0 180.0 200.0
TIME (MINUTES)

Figure 3. Platelet embolization from Silastic as a function of time. Predicted values (curve) are compared with unpublished data of Reynolds (□) using an *ex vivo* model similar to Horbett *et al.* [30].

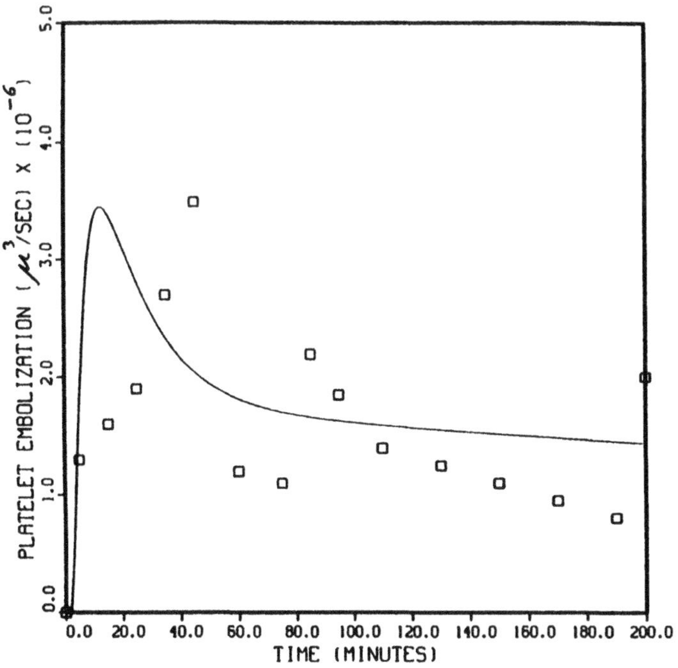
PLATELET EMBOLIZATION (μ^3/SEC) X (10^{-6})
0.0
1.0
2.0
3.0
4.0
5.0
0.0 20.0 40.0 60.0 80.0 100.0 120.0 140.0 160.0 180.0 200.0
TIME (MINUTES)

Figure 4. Platelet embolization from polyacrylamide as a function of time. Predicted values (curve) are compared with unpublished data of Reynolds (□) using an *ex vivo* model similar to Horbett *et al.* [30].

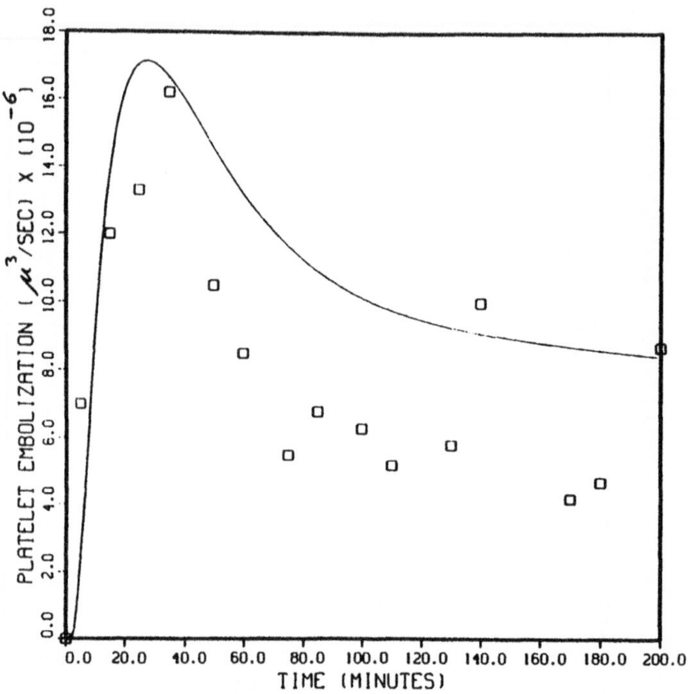
PLATELET EMBOLIZATION (μ³/SEC) X (10⁻⁶)
0.0
2.0
4.0
6.0
8.0
10.0
12.0
14.0
16.0
18.0
0.0 20.0 40.0 60.0 80.0 100.0 120.0 140.0 160.0 180.0 200.0
TIME (MINUTES)

Figure 5. Platelet embolization from HEMA-MAAC as a function of time. Predicted values (curve) are compared with unpublished data of Reynolds (□) using an *ex vivo* model similar to Horbett *et al.* [30].

Table 1.
Model and literature values of model coefficients for Silastic (k has units of min^{-1}, K and W are dimensionless, T_1 has units of min.)

	Literature		Fibrinogen Adsorption		
Coefficient	Value	(Ref.)	Clottable	Anticoagulated	Platelet deposition
k_{fa}	0.07	[6]	0.03	0.03	0.07
k_{pa}	0.05	[6]	0.05	0.05	0.05
k_{ca}	0.0066[a]	[40]	0.007	0.007	0.007
k_{fr}	0.03	[41]	0.015	0.015	0.015
k_{pr}	0.015	[6]	0.015	0.015	0.015
k_{cr}	—	—	0.01	0.01	0.01
k_f	0.05	[7]	0.05	0.01	0.05
k_{pf}	0.65	[42]	0.65	0.65	0.65
k_{pp}	0.21	[42]	0.21	0.21	0.21
K_{epp}	0.19	[42]	0.30	0.35	0.19
K_{epf}	—	—	1.0	1.0	1.0
K_{efp}	—	—	1.5	1.5	1.5
k_{bp}	0.017	[7]	0.02	0.02	0.02
W	0.03	[43]	0.001	0.001	0.03
T_1	0.4–0.7	[44]	1.0	1.0	1.0

[a] Adsorption of active Factor XII onto glass

Table 2.
Coefficients for platelet embolization from Silastic, polyacrylamide, and HEMA-MAAC (k has units of min^{-1}, K and W are dimensionless. T_1 has units of min.)

Coefficient	Silastic	Polyacrylamide	HEMA-MAAC
k_{fa}	0.07	0.07	0.2
k_{pa}	0.05	0.05	0.05
k_{ca}	0.007	0.007	0.007
k_{fr}	0.015	0.015	0.015
k_{pr}	0.015	0.015	0.015
k_{cr}	0.01	0.01	0.01
k_f	0.05	0.05	0.05
k_{pf}	0.42	0.65	0.65
k_{pp}	0.10	0.21	0.21
K_{epp}	0.19	0.15	0.12
K_{epf}	1.0	0.23	0.15
K_{efp}	1.5	3.0	3.5
k_{bp}	0.02	0.02	0.02
W	0.001	0.001	0.001
T_1	1.0	1.0	1.0

to simulate these curves were k_{fa} for HEMA-MAAC, which was higher, k_{fr} (which had the same value as for Silastic), and K_{epp}, which was lower. K_{epf} for both materials was higher than for Silastic, while K_{efp} was lower. Again, the simulated embolization rates follow the experimental values reasonably closely.

Table 1 contains the literature values of the 12 coefficients used in the simulation of activity on Silastic, their sources, and the values used in the simulations. Since three different experiments (fibrinogen adsorption from clottable and anticoagulated blood, and platelet deposition) were modeled, a total of 36 separate values of the coefficients were needed. Of these, 10 were adjusted as described above. In nearly all of these cases, the adjustments were well within the uncertainty limits of the literature values. The table also contains best fit values for the three parameters for which literature data were not available: k_{cr}, the coagulation factor removal rate constant, was best fit with a value of 0.01 min^{-1}, close to the value for k_{pr}, 'other protein' replacement rate constant; K_{epf}, the ratio of embolization of platelets from platelets to embolization of platelets from a fibrinogen layer, was best fit by a value of 1.0; K_{efp}, the ratio of fibrinogen embolization to platelet embolization from fibrinogen, was best fit by a value of 1.5.

Table 2 compares the coefficients used in the simulations for platelet embolization from Silastic, polyacrylamide, and HEMA-MAAC.

DISCUSSION

Blood coagulation on synthetic surfaces is a very complex series of processes. It involves a large number of different proteins, many of whose structures are still unknown, ionic catalysts (e.g., Ca^{2+}), diverse surface sites, multiple pathways, platelet activation factors, etc. Considering the complexity of the processes involved, it is noteworthy that a model as simple as the one proposed herein does so well in simulating behavior. It is especially noteworthy when one considers the known limitations in the simulations.

The many proteins involved in the process were modeled by just three, namely, fibrinogen (representing 'platelet-adherent proteins'), albumin (representing 'other proteins') and Factor XII (representing 'coagulation factors'). The surface was modeled as homogeneous, with Langmuir adsorption the only mechanism, and all adsorption and deposition phenomena were assumed to be first order. All adsorption processes were treated in terms of area covered only, with no reorganization on the surfaces considered. Emboli were treated as homogeneous entities.

Of the 15 coefficients needed by the model, 12 came from experimental data in the literature that involved different experimental protocols, different animal models, and different conditions which, because of a lack of details in the literature, could not be carefully compared. Of the twelve parameters determined from the literature, eight rate constants and one coefficient were derived from adsorption/deposition experiments on Silastic. The rate constant for coagulation factor activation, k_{ca}, was derived from an experiment on glass. W, the fractional enhancement of effective platelet area, was determined from experiments with sedimented washed platelets. The characteristic clotting time, T_1, came from measurements of human partial thromboplastin time. In a few cases, some adjustments of the values of these literature-derived coefficients were made either to improve the fit or to qualitatively express a difference between the model and the literature. Except in the case of W for fibrinogen adsorption, the adjustments were well within the uncertainty range of the literature values.

The three remaining coefficients were determined by minimizing the differences between predicted and experimental values of fibrinogen adsorption, platelet deposition, and platelet embolization on Silastic, shown in Figs 1–3. Even this optimization process shows the strength of the model, since the coefficients for which literature values were unavailable seem to fit qualitative conceptions of how the phenomena actually occur [45]. These best fit values imply that the rate of removal of coagulation factors from the surface is indistinguishable from the rate of removal of other proteins from the surface, that embolization from a platelet surface is indistinguishable from embolization from a fibrinogen surface, and that the amount of fibrinogen that embolizes with a platelet is about 50% higher than the ratio of fibrinogen to platelets present on the surface.

The best fit values for all the coefficients given in Tables 1 and 2 lead to a number of other implications as well. For fitting the fibrinogen adsorption experiments, the only significant difference among the coefficients for clottable and anticoagulated blood is for k_f, the fibrin formation rate constant, where the anticoagulated blood value is 20% of the normal blood value. Although this makes qualitative sense, it is interesting that no other coefficient seems to be affected.

For fitting the platelet embolization experiments on the three biomaterials tested, nearly all the protein interaction rate constants turned out to be the same for all three surfaces. The single exception was k_{fa}, the fibrinogen adsorption rate constant, which was three times higher on HEMA–MAAC than on either Silastic or polyacrylamide. The differences in observed behavior among these three materials appear to be a result only of changes in the three embolization coefficients, K_{epp}, K_{epf}, and K_{efp}. If these results are a measure of how these phenomena actually occur, then the observed differences in thromboembolization behavior among biomaterials cannot be attributed to differences in protein adsorption/desorption behavior.

In the Wilson *et al.* [38] model, embolization was considered to be strongly influenced by local hydrodynamics, expressed in terms of the 'flow coefficient' and the 'critical height' of the thrombus. Such hydrodynamic control can explain the apparent lack of an effect of differences in protein adsorption/desorption behavior on thromboembolization. Although hydrodynamic considerations were not explicitly considered in the model presented herein, such considerations are implicit in the way the embolization, Eqs (9)–(12), were formulated. For example, Eq (9) states that the rate at which platelet–platelet emboli break free increases with the number of platelets deposited onto other platelets, and decreases with the number of platelet–platelet bridges. Since the 'height' of the thrombus is measured by the number of platelets within the thrombus while the 'strength' of the thrombus is measured by the number of bridges, the concept of a 'critical height' emerges naturally from this formulation.

The success of the simulations are likely due to the fact that the model contains the principal steps that actually occur in the experiments, while the many phenomena that are known to occur but which are not explicitly included in the model either are not central to the measurements, or are implicitly included within the pseudo-steps of the model. Correctly predicted behavior includes early rapid Silastic fibrinogen adsorption rates followed by a decline and a slower rise phase, a tenaciously bound fibrinogen layer on Silastic, and higher rates of platelet embolization and co-embolized fibrinogen from polyacrylamide and HEMA–MAAC than from Silastic. These higher rates of embolization are probably a result of the higher degree of fibrinogen and albumin adsorption on these surfaces than on Silastic, which causes these surfaces to be more thrombotic but less thrombo-adherent [45].

Although the model presented herein is successful insofar as its ability to predict measured fibrinogen and platelet interactions with several synthetic surfaces is concerned, its value as a general tool for investigating blood-surface interactions must await the availability of more kinetic data on individual steps in the process, taken under conditions in which different experiments can be compared. The availability of such data will also allow the model to be refined to allow other steps in the process to be investigated.

Acknowledgements

Larry O. Reynolds came to UIC from the University of Washington, where he worked with Allan Hoffman. He passed away in January, 1992. This work is partly based on an incomplete study of Prof. Reynolds and an M.S. thesis by John F. Scolio, one of Reynolds' graduate students. It also is part of an M.S. thesis currently being written by William H. Newren, Jr. under the direction of Irving F. Miller, in partial fulfillment of the requirements for the M.S. degree in Bioengineering at UIC.

REFERENCES

1. Langmuir, *Phenomena, Atoms, and Molecules*. Philosophical Library, New York (1950).
2. W. J. Dillman and I. F. Miller, *J. Colloid Interface Sci.* **44**, 221 (1973).
3. D. Basmadjian, *Ann. Biomed. Eng.* **18**, 685 (1990).
4. J. D. Andrade and V. Hlady, *Ann. N.Y. Acad. Sci.* **516**, 158 (1987).

5. T. Matsuda, H. Takano, K. Hayashi, Y. Taenaka, S. Takaichi, M. Umezu, T. Nakamura, H. Iwata, T. Nakatani, T. Tanaka, S. Takatani and T. Akutsu, *Trans. ASAIO* **30**, 353 (1984).
6. S. W. Kim and R. G. Lee, in: *Applied Chemistry at Protein Interfaces*, p. 218, R. E. Baier (Ed.). American Chemical Society, Washington (1975).
7. B. R. Young, L. K. Lambrecht, S. L. Cooper and D. F. Mosher, in: *Interfacial Phenomena and Applications*, p. 317, S. L. Cooper and N. A. Peppas (Eds). American Chemical Society, Washington (1982).
8. J. L. Brash and Q. M. Samak, *J. Colloid Interface Sci.* **65**, 495 (1978).
9. W. Norde and J. Lyklema, *J. Colloid Interface Sci.* **66**, 257 (1978).
10. B. W. Morrissey, L. E. Smith, R. R. Stromberg and C. A. Fenstermaker, *J. Colloid Interface Sci.* **56**, 557 (1976).
11. J. D. Andrade, in: *Protein Adsorption*, p. 1, J. D. Andrade (Ed.). Plenum, New York (1985).
12. R. L. Beissinger and E. F. Leonard, *J. Colloid Interface Sci.* **85**, 521 (1982).
13. G. A. Adams and I. A. Feuerstein, *Trans. ASAIO* **27**, 219 (1981).
14. L. Vroman, A. L. Adams, G. C. Fischer and P. L. Munoz, *Blood* **55**, 156 (1980).
15. J. L. Brash, S. Uniyal and Q. Samak, *Trans. ASAIO* **20**, 69 (1974).
16. J. H. Griffin, in: *Interaction of the Blood with Natural and Artificial Surfaces*, p. 139, E. W. Salzman (Ed.). Marcel Dekker, New York (1981).
17. M. Verstraete and J. Vermylen, *Thrombosis*. Pergamon Press, New York (1984).
18. H. L. Meier, C. F. Scott, R. Mandle, Jr., M. E. Webster, J. V. Pierce, R. W. Colman and A. P. Kaplan, in: *The Behavior of Blood and its Components at Interfaces*, p. 65, L. Vroman and E. F. Leonard (Eds). Ann. N.Y. Acad. Sci., 283, New York (1977).
19. L. Vroman, A. L. Adams, M. Klings, G. C. Fischer, P. C. Munoz and R. P. Solensky, in: *The Behavior of Blood and its Components at Interfaces*, p. 93, L. Vroman and E. F. Leonard (Eds). Ann. N.Y. Acad. Sci., 283, New York (1977).
20. L. I. Friedman, H. Liem, E. F. Grabowski, E. F. Leonard and C. W. McCord, *Trans. ASAIO* **26**, 63 (1970).
21. H. R. Baumgartner, *Microvasc. Res.* **5**, 167 (1973).
22. L. Vroman, A. L. Adams, G. C. Fischer, P. C. Munoz and M. Stanford, in: *Interfacial Phenomena and Applications*, p. 265, S. L. Cooper and N. A. Peppas (Eds). American Chemical Society, Washington (1982).
23. I. A. Feuerstein and J. Kush, *Thrombosis Haemostasis* **55**, 184 (1986).
24. D. J. Fabrizius-Homan, S. L. Cooper and D. F. Mosher, *Thrombosis Haemostasis* **68**, 194 (1992).
25. L. I. Friedman and E. F. Leonard, *Fed. Proc.* **30**, 1641 (1971).
26. P. D. Richardson, *Nature* **245**, 103 (1973).
27. E. Ruckenstein, A. Marmur and W. N. Gill, *J. Theor. Biol.* **66**, 147 (1977).
28. T. K. Belval and J. D. Hellums, *Biophys. J.* **50**, 479 (1986).
29. L. O. Reynolds, T. A. Horbett, B. D. Ratner and A. S. Hoffman, *Trans. Soc. Biomater.* **5**, 37 (1982).
30. T. A. Horbett, C. M. Cheng, B. D. Ratner and A. S. Hoffman, *J. Biomed. Mater. Res.* **20**, 739 (1986).
31. V. A. DePalma, R. E. Baier, J. W. Ford, V. L. Gott and A. Furuse, in: *Biomaterials for Skeletal and Cardiovascular Applications*, p. 143, C. Homsy and C. D. Armeniades (Eds). Interscience, New York (1972).
32. Eberhart, M. E. Lynch, F. H. Bilge, J. F. Wissinger, M. S. Munro, S. R. Elsworth and A. J. Quattrone, in: *Interfacial Phenomena and Applications*, p. 293, S. L. Cooper and N. A. Peppas (Eds). American Chemical Society, Washington (1982).
33. R. D. Bagnall, *J. Biomed. Mater. Res.* **12**, 203 (1978).
34. P. A. Cuypers, G. M. Willems, H. C. Hemker and W. T. Hermans, *Ann. N.Y. Acad. Sci.* **516**, 244 (1987).
35. P. D. Nguyen and E. A. O'Rear, *Ann. Biomed. Eng.* **18**, 427 (1990).
36. J. S. Schultz, A. Ciarkowski, J. D. Goddard, S. M. Lindenauer and J. A. Penner, *Trans. ASAIO* **21**, 269 (1976).
37. A. Marmur and S. L. Cooper, *J. Colloid Interface Sci.* **89**, 458 (1982).
38. R. S. Wilson, A. Marmur and S. L. Cooper, *Ann. Biomed. Eng.* **14**, 383 (1986).
39. M. D. Lelah, L. K. Lambrecht and S. L. Cooper, *J. Biomed. Mater. Res.* **18**, 475 (1984).
40. S. D. Revak, C. G. Cochrane and J. H. Griffin, *J. Clin. Invest.* **59**, 1167 (1977).

41. S. M. Slack, J. L. Bohnert and T. A. Horbett, *Ann. N.Y. Acad. Sci.* **516,** 223 (1987).
42. L. K. Lambrecht, M. D. Lelah, C. A. Jordan, M. E. Pariso, R. M. Albrecht and S. L. Cooper, *Trans. ASAIO* **29,** 194 (1983).
43. G. A. Marguerie, T. S. Edgington and E. F. Plow, *J. Biol. Chem.* **255,** 154 (1980).
44. R. W. Colman, J. Hirsh, V. J. Marder and E. W. Salzman (Eds), *Hemostasis and Thrombosis, Basic Principles and Clinical Practice*, 2nd Ed., p. 1052, J. B. Lippincott, Philadelphia (1987).
45. A. S. Hoffman, T. A. Horbett, B. D. Ratner, S. R. Hanson, L. A. Harker and L. O. Reynolds, in: *Interfacial Phenomena and Applications*, p. 59, S. L. Cooper and N. A. Peppas (Eds). American Chemical Society, Washington (1982).

A canine *ex vivo* shunt for isotopic hemocompatibility evaluation of a NHLBI DTB primary reference material and of a IUPAC reference material

JOSSELINE CAIX, GERARD JANVIER, BENOIT LEGAULT, LAURENCE BORDENAVE, FRANCOIS ROUAIS, BERNARD BASSE-CATHALINAT and CHARLES BAQUEY

Inserm U. 306, Université de Bordeaux II, 33076 Bordeaux, France

Received 4 December 1991; accepted 17 September 1992

Abstract—Factors determining the thrombogenic response to particular artificial surfaces were investigated *ex vivo* in a canine shunt model. Methods using radioisotopic tracers made it possible to dynamically monitor the deposition of labelled blood cells and proteins on a NHLBI.DTB primary reference material polydimethylsiloxane (PRM.PDMS) and on a IUPAC reference material polyvinyl chloride (IUPAC.PVC). On the one hand, leukocyte affinity $\tau_{s_{(leu)}}$ (number of deposited leukocytes $mm^{-2}\,s^{-1}$) was not significantly different between IUPAC.PVC ($\tau_{s_{(leu)}} = 1.2$–2.5) and PRM.PDMS ($\tau_{s_{(leu)}} = 1.5$–3.4) and the fibrinogen adsorption rate varied from 33 to $48.10^{-5}\,\mu g\,mm^{-2}\,s^{-1}$ for both these materials. On the other hand, platelet affinity $\tau_{s_{(plat)}}$ (number of deposited platelets $mm^{-2}\,s^{-1}$) was significantly different ($p < 0.05$) for IUPAC.PVC and PRM.PDMS ($\tau_{s_{(plat)PVC}} = 683 \pm 200 > \tau_{s_{(plat)PDMS}} = 327 \pm 80$). Scanning electron micrographs of adherent platelets, red cells and leukocytes after blood contact *ex vivo* were performed after each experiment. This preliminary work contributes not only to quantify the adsorption of different radiotracers, but also to evaluate the superficial distribution of the labelled biological species on the inner surface of the tested biomaterials.

Key words: Blood cell–polymer interactions; canine *ex vivo* shunt; hemocompatibility; reference materials.

INTRODUCTION

A number of organic polymers or biological grafts are used clinically in vascular surgery. For a long time, it has been known that platelets and fibrin are involved in the failure of small caliber vascular grafts [1]. In order to develop methods to predict the thrombogenicity of vascular grafts, a number of laboratories have studied the early interactions of platelets and proteins with a variety of biomaterials. To compare results obtained with various testing systems available, the Devices and Technology Branch (DTB) of the National Heart, Lung and Blood Institute (NHLBI) selected two different materials: low density polyethylene (LDPE) and polydimethyl siloxane (PDMS) as primary reference materials (PRM) and IUPAC Working Party selected two other different materials: Polyvinylchloride (PVC) and Cellulose (Cell).

The experimental model used in our laboratory is based on an *ex vivo* arterio-arterial by-pass surgically performed in a dog. The surgical procedure applied needs: flexibility properties of the material in order to build an extracorporeal loop, and a sufficient diameter of the material compatible with the diameter of dog arteries. Thus, the two most suitable materials were PDMS and PVC. Indeed, the diameter of cellulose tubing was too small and LDPE tubing was too rigid for this particular surgical use.

The present methodology [2, 3] was designed to allow continuous monitoring of the behaviour of blood elements at the interface between the inner surface of the tubing and blood flow. By means of suitable radiotracers corresponding to blood cells and proteins usually involved in thrombogenic phenomena which were intravenously injected to the animal on the one hand, and by means of powerful detecting devices on the other hand, we were able to continuously compare the respective concentrations of these tracers in the whole circulation and in a given length of the extracorporeal by-pass. If a difference between the measured values was observed, we were able to attribute this difference to the amount of tracers absorbed on the inner surface of the tested biomaterial.

In this paper, we describe the behaviour of labelled red cells, platelets, neutrophils and fibrinogen when blood flow is in contact with two different polymers PRM.PDMS and IUPAC.PVC, connected to in the canine femoral artery.

The aim of this work is to show by means of isotopic methods the important role of leukocytes neutrophils in thrombogenicity. Currently, we are trying to understand and demonstrate the many functions of neutrophils in the coagulation phenomena like Spilizewski *et al.* [4]. We hope that the results obtained will be complementary to those published by other teams that have tested the same materials applying different techniques and models like Cooper's team [5]

MATERIALS AND METHODS

Materials

The NHLBI.DTB primary reference silica free polydimethylsiloxane (PRM.PDMS) was provided as 4.0 mm ID tubing from Mercor, a division of Thoratec Laboratories Corporation, Berkeley, CA, USA sterilized by exposure to ^{60}Co γ-radiation up to absorption of 25 kGy. The polyvinylchloride selected by IUPAC Working Party was obtained as 4.0 mm ID unsterilized tubing, from Norton Company, Akron, OH, USA (bath 0.9 formulation S-50-HL) and corresponded to a surgical grade resin. The medical grade PVC tubing used as control was provided as 3.17 mm, ID tubing by Bentley, Plaisir, France.

Labelling

Red blood cells collected immediately after the first centrifugation of 50 ml of freshly drawn canine whole blood were washed twice in a Tyrode buffer with glucose at pH: 9.2. One milliliter of the red cell concentrate was labelled with 74 MBq of 99^{m} technetium in the presence of stannous pyrophosphate [6]. Autologous platelets were harvested from the canine platelet rich plasma (PRP) and then labelled with 37 MBq ^{111}indium oxinate (Mallinckrodt, Evry, France) according to the method originally described by Thakur *et al.* [7], and in our laboratory [8]. From 50 ml of autologous canine blood collected on heparin (10 IU per ml blood), a sedimentation-flotation technique could be performed and lead to the isolation of granulocytes for labelling with 37 MBq ^{111}indium oxinate (Mallinckrodt, Evry, France), according to a method derived from Thakur and improved by McAfee [9] and by Moisan [10]. Unlike with Anglo-American teams, the lymphocytes and the other mononuclear cells were not labelled. Human fibrinogen (Kabi, Oslo, Norway) was made up to 1 ml with phosphate buffered saline (PBS) and labelled with ^{123}iodine in the presence of chloramine T [11].

Equipment

Two detectors based on junctions of semi-conductors, high purity germanium (HP.Ge) and germanium lithium (Ge.Li) were used. They had an energy resolution capacity (≅1.5%) allowing simultaneous detection of several proximal energies such as 143, 159 and 173 keV photons emitted respectively by ^{99m}Tc, ^{123}I and ^{111}In. Both detectors were connected to a spectrometric analysing system and to a recording device. Data were processed by a computer. One detector was placed above the extracorporeal loop made of the tested biomaterial via the extracorporeal circuit (ECC), and the other one was set above one forelimb, in order to monitor the different radiotracer concentrations in the systemic circulation. Blood samples were periodically collected by venous punction and their volumic radioactivity was measured and analyzed in order to determine their relative content for each of the tracers.

Hematological check up

The dogs used were always female Beagle, between 8 and 15 months old, weighing between 15 and 20 kg, without specific diet. Given the differences in the coagulation patterns between humans and dogs, the evaluation of canine hemostasis parameters requires some modifications to the techniques used in human bioclinics, particularly the dilution factors [12].

In Table 1, we can see the different values recorded for the dogs used in this work and the average values of the same canine parameters, usually found in our own experience. Dog 3 could not be taken into account due to too great a difference

Table 1.
Canine hematological parameters

	Dog 1	Dog 2	Dog 3	Average values usually obtained
WBC	7.110	8.800	10.830	10.3 ± 5.3
RBC	5.9	4.6	5.2	5.7 ± 1.1
Hb	12.1	10.9	11.4	12.9 ± 2.5
Ht	46	41	43	40.8 ± 8.2
Plat	342	266	520	336 ± 127
PL	100	100	100	90 ± 10
APTT	19	17	10	23.5 ± 13.5
TT	20	19	16	22.3 ± 6.9
CT	1.8	1.5	1.3	2
Fib	2.5	1.6	3	1.83 ± 0.73

WBC = white blood cells per $mm^3 \times 10^{-3}$
RBC = red blood cells per $mm^3 \times 10^{-6}$
Hb = hemoglobin in g/100 ml
Ht = hematocrit in %
Plat = platelets per $mm^3 \times 10^3$
Fib = fibrinogen in g/l
PL = prothrombin level (%)
APTT = activated partial thromboplasmin time (s)
TT = thrombin time(s)
CT = Clotting time (min).

Table 2.
Radiotracers used and biomaterials tested

		Dog 1	Dog 2
Radiotracers used	RBC ^{99m}Tc	+	+
	WBC ^{111}In	−	+
	Platelets ^{111}In	+	−
	Fibrinogen ^{123}I	+	+
Biomaterials tested (number of samples)	PRM.PDMS	1	2
	IUPAC.PVC	2	2
	PVC control	1	1

between its hematological values and the corresponding mean values measured for dogs in our laboratory.

Experimental schedule

Data reported here concern two dogs which received respectively radiotracers and biomaterials as summarized in Table 2.

Surgical procedure

Dogs were tranquilized with a subcutaneous ketamine chlorhydrate (250 mg) and acepromazine (10 mg) cocktail before being anaesthetized by IV injection of pentobarbital (150 mg). A perfusion line fed with Plasmion® was then set up. The animal was kept asleep for the duration of the experiment by breathing a cocktail of oxygen and halothane (1.5%) through an endotracheal duct.

Before surgery the dog received the three radiotracers through the perfusion line. Then the canine blood was by-passed through an extracorporeal loop made of the tubing tested from an artery to downstream the same artery. One end of the sample, which was 35 cm long and filled with saline, was inserted through a longitudinal arteriotomy into the femoral artery which was clamped 3 or 4 cm upstream from the arteriotomy. The artery was then tightly bound around the tubing. The other end of the sample was inserted downstream the same artery by a similar procedure.

In addition, blood samples were collected periodically from a catheter inserted in a vein of a forelimb and their radioactivity was measured and analyzed over time. When the blood volumic radioactivity related to labelled blood cells became constant or when the blood volumic radioactivity related to radioiodinated fibrinogen decreased according to a constant slope (which is proportional to the reverse of the plasmatic turn over for this protein) the blood was allowed to flow through the ECC which was placed on the bottom of a lead-walled container. The radioactivity of the ECC was then measured and analyzed every 30 s for a total period of 15 min by a detector. At the end of this period, measurements were stopped to let the surgeon disconnect the ECC and rinse it with saline. The ECC residual radioactivity could then be carefully measured and analyzed. A second ECC was then set with another tube at the same site or on the opposite femoral site.

Scanning electron microscopy (SEM)

Small pieces from the mid-portion of each material tested (0.5 cm^2 large) were fixed by immersion in a mixture (1 : 1) of a glutaraldehyde solution (2%) and a cacodylate buffer (pH 7.2). Surfaces were then washed with 0.15 M cacodylate for 10 min. Samples were desiccated at room temperature, and finally metallized with metal using a gold target.

Data representation modalities

Two kinds of plots have been chosen: for Figs 1 and 3, the recorded radioactivities for the volume seen by the detector and respectively related to platelets, and white blood cells (WBC), were plotted versus time, the initial time corresponding to the complete filling of the ECC; for Figs 2, 4 and 5, the so-called ECC radioactivity ratios related to the red blood cells radioactivity recorded at a given time and divided by the radioactivity respectively related to platelets, WBC or fibrinogen and recorded at the same time, is plotted versus time.

This second type of plot offers the following advantage: at any moment, the radioactivity value related to red blood cells is proportional to the volume of blood taken into account by the detector, being stated that no red blood cell adhere to the tubing wall. This seems to be an acceptable assumption since red blood cells are known to not be involved in the initial steps of thrombogenic phenomenon. In other words, red blood cells may not be entrapped by the wall before the constitution of a minimal fibrin network.

Should any variation of the ECC length under the detector or of the ECC position occur (in the first case the volume of blood seen by the detector changes, and in the second one, the detection efficacy changes), the recorded value for the radioactivity related to RBC changes in both cases, without any variation of the haematocrit, and this value provides a permanent index of the volume of blood, i.e. the ECC length, taken into account by the detector; and the ratios Q_{RBC}/Q_{plat}, Q_{RBC}/Q_{leu} and Q_{RBC}/Q_{F} are directly proportional to the reverse of the concentrations of the related blood species.

Expression of data

From the recorded data, one can calculate the kinetics of several phenomena i.e. platelet adhesion, leukocyte adhesion and fibrinogen adsorption. These kinetics are expressed by the parameters τ_s or μF which represent respectively the number of cells adhering every second by square millimeter and the amount of fibrinogen in μg which is deposited every second by square millimeter.

τ_s is given by the following expression:

For platelets:

$$\tau_{s(\text{plat})} = \frac{N_p \cdot r}{2(t - t_0)} \times \left[\frac{(Q_p)_{t_0}}{(Q_{RBC})_{t_0}} \cdot \frac{C_{RBC}}{C_p} \right] \left[\frac{(Q_p)_t}{(Q_p)_{t_0}} - 1 \right], \tag{1}$$

$\tau_{s(\text{plat})}$ = number of deposited platelets $mm^{-2}\ s^{-1}$, N_p = platelet count mm^{-3}, r = biomaterial tube radius expressed in millimeter (mm), t_0 = time at which the ECC is completely filled with flowing blood, t = time at which measurements are stopped, $t - t_0$ = is expressed in seconds (s), $Q_{pt_0 \text{ or } t}$ = radioactivity due to platelets of the extracorporeal circulation at t or t_0, Q_{RBCt_0} = radioactivity of the ECC related to the red blood cells, C_{RBC} = volumic radioactivity of the systemic circulation related to red blood cells and C_p = volumic radioactivity of the systemic circulation platelets.

For leukocytes:

$$\tau_{s(\text{leu})} = \frac{N_{(\text{leu})} \cdot r}{2(t - t_0)} \times \left[\frac{Q_{(\text{leu})t_0}}{Q_{(\text{RBC})t_0}} \cdot \frac{C_{\text{RBC}}}{C_{(\text{leu})}} \right] \left[\frac{Q_{(\text{leu})t}}{Q_{(\text{leu})t_0}} - 1 \right], \tag{2}$$

where $\tau_{s(\text{leu})}$ is the number of deposited leukocytes $\text{mm}^{-2}\,\text{s}^{-1}$, $N_{(\text{leu})}$ leukocyte count mm^{-3}, $Q_{(\text{leu})t_0 \text{ or } t}$ radioactivity due to leukocytes of the extracorporeal circulation at t or t_0 and $C_{(\text{leu})}$ volumic radioactivity of the systemic circulation related to leukocytes.

For fibrinogen:

$$\mu F_{(t)} = \frac{r.M_{\text{F}}}{2} \cdot \left[\frac{Q_{\text{F}t}}{Q_{\text{F}t_0}} \cdot e^{\lambda(t-t_0)} - 1 \right], \tag{3}$$

where r = biomaterial tube radius (mm), M_{F} = fibrinogen concentration in the systemic circulation $\mu\text{g.mm}^{-3}$ $(Q_{\text{F}})_t \cdot e^{\lambda(t-t_0)}$ = is used instead of $Q_{\text{F}(t)}$ to take into account the leakage of radioactive fibrinogen from the vascular compartment $(t - t_0)$ is expressed in seconds with $\lambda = 0.693/t_{1/2}$ and $t_{1/2}$ = is the plasmatic half-life of fibrinogen for the dog used in the study and expressed in seconds.

$t_{1/2}$ is assessible from the following procedure: after intravenous injection of radiotracers, to the dogs, blood is periodically collected in order to monitor its radioactivity; after a first increase, a decrease of the recorded value for fibrinogen can be observed and the corresponding values, plotted on a semi-logarithmic scale versus time, fall on a straight line of which the slope λ is related to $t_{1/2}$ by the following expression $t_{1/2} = 0.693/\lambda$.

A 95% confidency interval was determined from these above expressions and from the statistical fluctuations concerning the different radioactivity measurements involved in these expressions.

RESULTS

Firstly, we shall report results corresponding to the experiment lead with dog # 1. During a 15 min exposure to blood, we observed, as shown in Fig. 1, a very steep

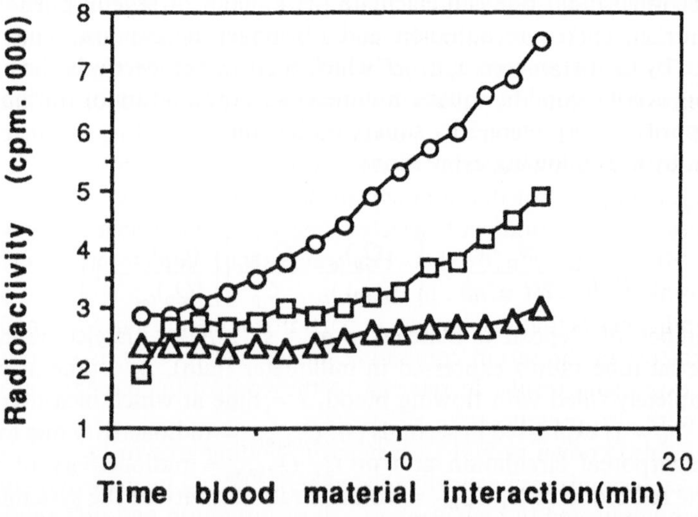
Radioactivity (cpm.1000)
1
2
3
4
5
6
7
8
0
10
20
Time blood material interaction(min)

Figure 1. Platelet radioactivity profiles on PVC.IUPAC (circles), PRM.PDMS (squares) and PVC control (triangles).

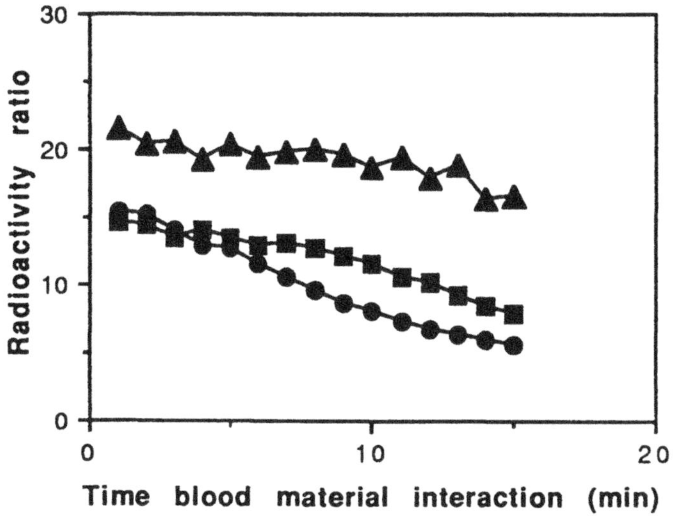

Radioactivity ratio
30
20
10
0
0
10
20
Time blood material interaction (min)

Figure 2. Red cell/platelet radioactivity ratio evolution for PVC.IUPAC (circles), PRM.PDMS (squares), PVC control (triangles).

increase of platelet concentration for PVC.IUPAC and a low increase for PVC control. For PRM.PDMS, the number of platelets seen by the detector increased moderately between the second and the fifteenth minute. Between the first and the second minute, this number appeared to increase by a factor of 2. But if we look at Fig. 2 where the same results are demonstrated differently and where the respective evolution of the radioactivity ratios RBC/plat for each type of material are given, the behaviour of platelets when blood was exposed to PRM.PDMS does not feature any sudden variation during the same period of time. Accordingly, the sudden increase reported in Fig. 1 must be due to a transient variation of geometrical conditions for the radioactivity measurements.

Using the expression previously given (1), the kinetics parameters related to platelet adhesion on the various types of biomaterial under test have been computed and are given in Table 3.

According to these data, PVC.IUPAC appears to feature the greatest affinity, $\tau_{s(plat)}$ being for this material twice as great as $\tau_{s(plat)}$ for PRM.PDMS and six times as great for PVC control although data were more scattered in this last case. These differences must be smoothed if one considers the scattering of the data which is larger for the PVC control than for the two other materials.

Secondly, we shall address the behaviour of leukocytes. Comparing the three biomaterials tested with the same dog (dog 2), a stable level of the radioactivity

Table 3.
Platelet adhesion on biomaterials tested
PVC.IUPAC > PRM.PDMS > PVC control

Tested samples	$\tau_{s(plat)} \pm 2$ SD (number of deposited leukocytes mm^{-2} s^{-1})
PVC control	110 ± 75
PVC.IUPAC	680 ± 200
PRM.PDMS	330 ± 80

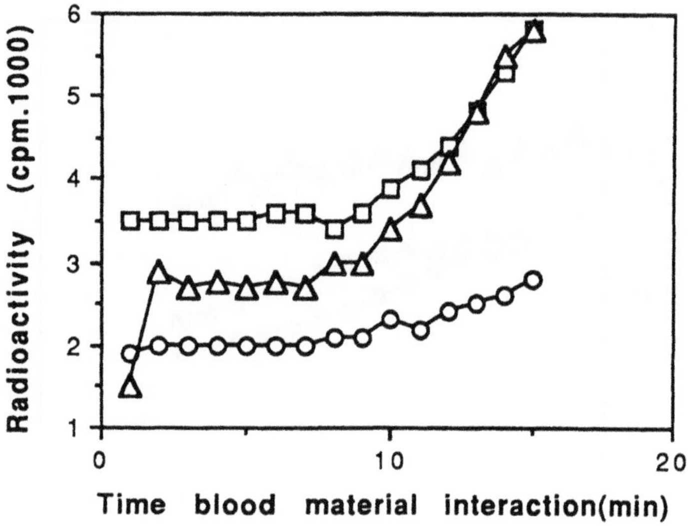

Radioactivity (cpm.1000)
6
5
4
3
2
1
0
10
20
Time blood material interaction(min)

Figure 3. White blood cell radioactivity profiles on PVC.IUPAC (circles), PRM.PDMS (squares) and PVC control (triangles).

related to these blood elements can be observed for the first eight minutes whichever is the biomaterial under consideration in Fig. 3. The sudden initial increase observed for PVC control can be explained as made before in the previous section for platelets and PRM.PDMS taking into account the quasi-absence of variation of the radioactivity ratio $Q_{RBC}/Q_{(leu)}$ at the same time in Fig. 4. After the eighth minute, a very important increase of the radioactivity related to leukocytes for both PRM.PDMS and PVC control can be noticed while the accumulation of leukocytes was much less important for PVC.IUPAC. Using the expression previously given (2), the kinetic parameters related to leukocytes adhesion on the various types of biomaterial are computed and listed in Table 4. According to these data, PVC control appears to

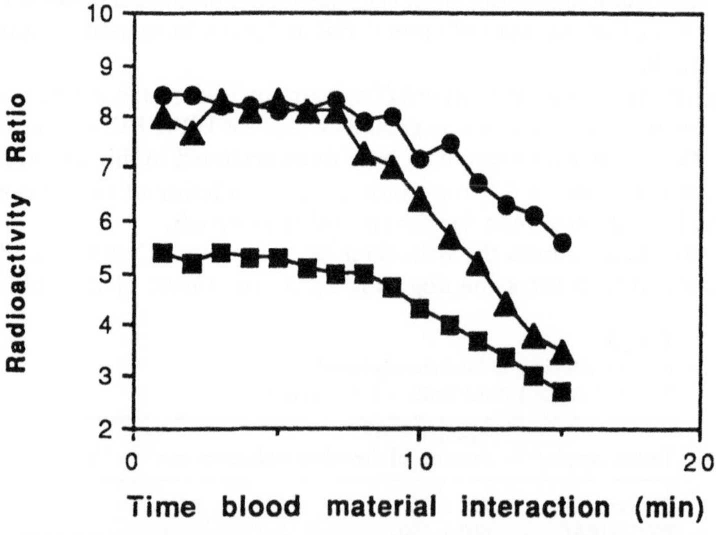

Radioactivity Ratio
10
9
8
7
6
5
4
3
2
0
10
20
Time blood material interaction (min)

Figure 4. Red blood cell/white blood cell radioactivity ratio on PVC.IUPAC (circles), PRM.PDMS (squares) and PVC control (triangles).

Table 4.
Leukocyte adhesion on biomaterials tested
PVC control ≫ PRM.PDMS ≥ PVC.IUPAC

Samples $n = 5$	$\tau_{s(leu)} \pm 2$ SD (number of deposited leukocytes mm^{-2} s^{-1})
PVC control $n = 1$	5.3 ± 1.1
PVC.IUPAC $n = 2$	1.85 ± 0.70
PRM.PDMS $n = 2$	2.45 ± 0.60

exhibit the greatest affinity, $\tau_{s(leu)}$ being for this material three times as great as $\tau_{s(leu)}$ for PVC.IUPAC and twice as great as $\tau_{s(leu)}$ for PRM.PDMS.

Thirdly, data related to fibrinogen behaviour will be discussed. Qualitatively, the three materials under test behave similarly as far as fibrinogen accumulation on their surface is concerned. The amount of radioactive fibrinogen seen by the detector above the ECC decreases slightly with a slope which is however less steep than expected if the fibrinogen seen was entirely belonging to the flowing blood; the observed difference must be clearly attributed to fibrinogen which accumulated onto the material surface.

Data analysis from the expression (3) leads to the kinetics parameters of this phenomenon and the corresponding values are given in Table 5. From these values, one may calculate approximately the amount of fibrinogen which has been deposited between the first and the fifteenth minute. These amounts were respectively equal to 13.5, 29.7–43.2, and 30–43 $\mu g.cm^{-2}$ for PVC control, PVC.IUPAC and PRM.PDMS.

Now a question must be raised: Is there no fibrinogen which would adsorb on the biomaterials surface during the first minute of exposure? The corresponding amount can be assessed through a comparison of the values respectively taken by the radioactivity ratio Q_{RBC}/Q_F in the systemic circulation on the one hand, and in the ECC on the other, for the time at which blood is allowed to flow through the ECC.

These values are computed from the intercept of the curves with the y-axis in Fig. 5 and from the radioactivity measurements made on blood samples which have been collected at the beginning of each extracorporeal circulation. It appears that Q_{RBC}/Q_F is systematically smaller for the ECC than for the systemic circulation.

Table 5.
Fibrinogen adsorption on biomaterials
PVC.IUPAC = PRM.PDMS > PVC control

	Fibrinogen adsorption $\mu g.cm^2.s^{-1}$	
Tested samples	(dog 1) $M_F = 2.3\ \mu g\ mm^{-3}$ $t_{1/2} = 115$ min	(dog 2) $M_F = 1.63\ \mu g\ mm^{-3}$ $t_{1/2} = 145$ min
PVC Control	(15 ± 10) 10^{-5}	(14.9 ± 8) 10^{-5}
PVC.IUPAC	(48 ± 12) 10^{-5}	(33 ± 12.5) 10^{-5}
PRM.PDMS	(34 ± 11) 10^{-5}	(47 ± 9.5) 10^{-5}

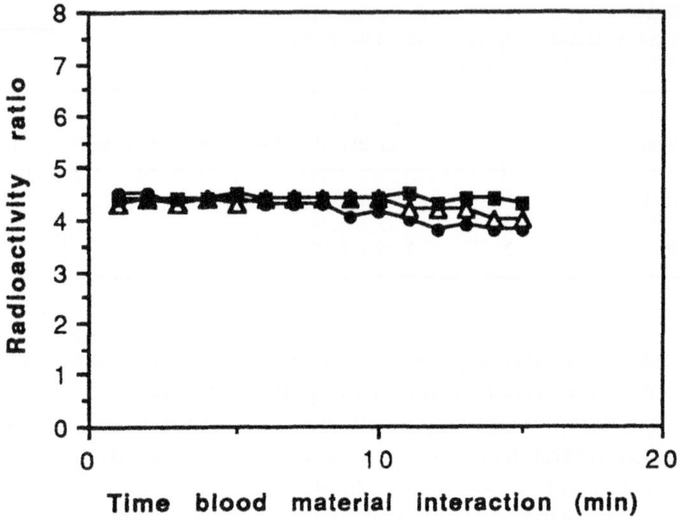
Radioactivity ratio
0
1
2
3
4
5
6
7
8
0
10
20
Time blood material interaction (min)

Figure 5. Radioactivity ratio red blood cell/fibrinogen for PVC.IUPAC (circles), PRM.PDMS (squares) and PVC control (triangles).

(a)

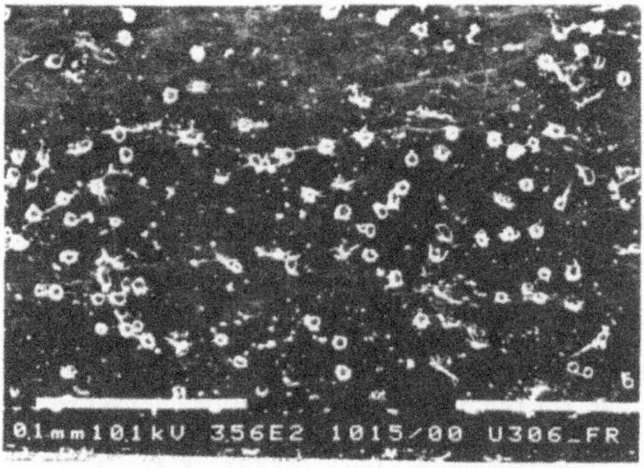
0.1mm 10.1kV 3.56E2 1015/00 U306_FR

(b)

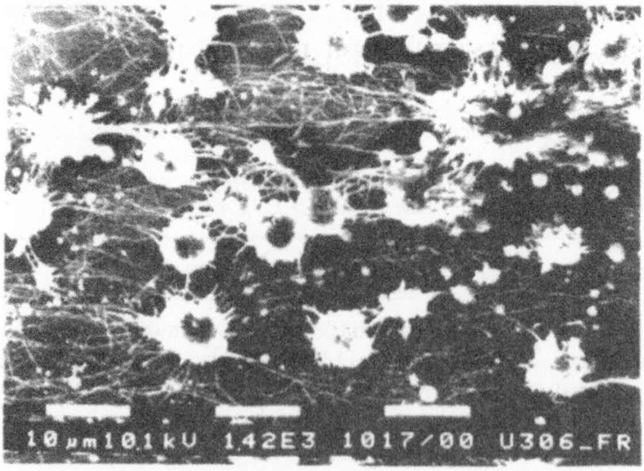
10µm 10.1kV 1.42E3 1017/00 U306_FR

Figure 6. Isotopic hemocompatibility evaluation in a canine *ex vivo* shunt: scanning electron micrographs of adherent platelets, red cells and debris after blood contact *ex vivo* (PVC.IUPAC). (a) ×356, (b) ×1420.

(a)

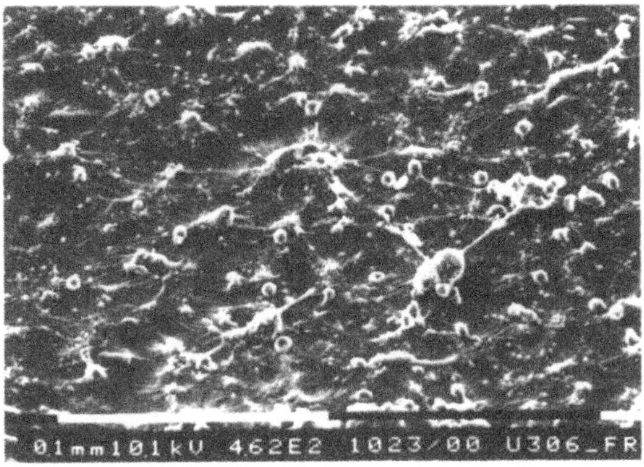
01mm101kV 462E2 1023/00 U306_FR

(b)

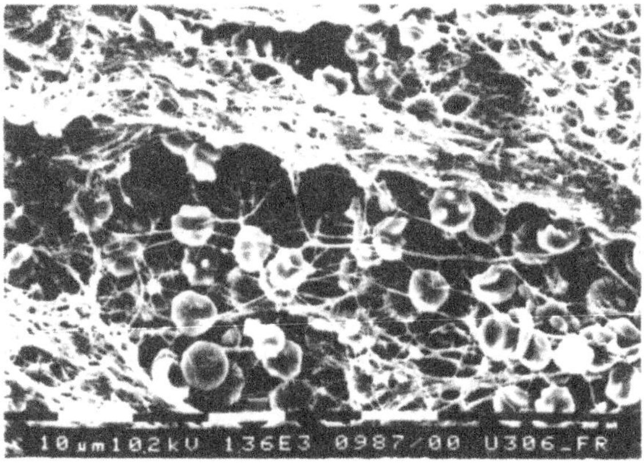
10um102kV 136E3 0987/00 U306_FR

Figure 7. Isotopic hemocompatibility evaluation in a canine *ex vivo* shunt: scanning electron micrographs of adherent platelets, red cells and after blood contact *ex vivo* (PRM.PDMS). (a) ×462, (b) ×1360.

This observation implies that fibrinogen is relatively more abundant inside the ECC than in systemic circulation, the difference being due to a fibrinogen layer which coats the surface as soon as the latter comes into contact with blood.

Scanning electron microscopy (SEM)

The internal surfaces of the PRM.PVC (Fig. 6a, b) and PRM.PDMS (Fig. 7a, b) tubing were not significantly different after blood exposure. Uniform dense platelet adhesion was observed for both materials at a higher magnification. Platelet pseudopodes, fibrin deposits, a few leukocytes and some entrapped erythrocytes were visible on the surfaces of the mid-portion of the tubes. Fig. 7b shows the different cells and proteins constituting a spectacular thrombus; this phenomenon occurred not earlier than the fifteenth minute after the first contact with blood. Red cells, leukocytes and platelets were stuck together by a fibrin network.

DISCUSSION

Recently, Cooper and coworkers [5] examined acute and chronic canine *ex vivo* blood interactions with PRM.PE and PRM.PDMS in very specific conditions. The two materials elicit a relatively benign response to platelets which results in low level for subsequent thrombus formation. The comparison of acute and chronic experiments showed that these materials exhibit similar platelet deposition patterns. The only difference concerned the absolute numbers of platelets adhering to these materials which was explained by a lower shear rate in the chronic experiment, [670–800 compared to 1300–1600 s^{-1} for the acute experiment].

Our study involves a canine arterio-arterial by-pass which cannot allow a flow higher than 100 ml min^{-1} which corresponds to a shear rate of 66 s^{-1} for PVC.IUPAC and PRM.PDMS (ID: 4 mm). Obviously shear rates were slightly higher for PVC control tubing, as its diameter is narrower. That may explain the greater number of adhering platelets detected on biomaterials tested in this study. Moreover, the presence of heavy amounts of fibrinogen on these materials may account for the observed platelet affinity.

The leukocyte affinity may be due to two factors: the possible activation of polymorphonuclear neutrophils, which would be responsible for an accumulation; and the possible trapping in a fibrin network. But this phenomenon would concern RBC as well. This does not seem to occur since the ratio $Q_{RBC}/Q_{(leu)}$ decreases which would not be the case if both kinds of blood cells were entrapped.

The presence of fibrinogen or fibrin on the biomaterial surfaces as well as of fibrinopeptides and fibrin degradation products, all these substances being chemiotactic for the neutrophils trigger the recruitment of these cells on these surfaces. This phenomenon is followed by activation of neutrophils, leading to the expression of integrins which promote their adhesion to the fibrinogen layer and a further recruitment of circulating neutrophils through aggregation processes.

It must also be said that initial activation of neutrophils can be triggered as well by inflammatory mediatiors which are generated by the arteriotomy or for rheological reasons associated with flow disturbances which appear near to the connections of the ECC with the natural vessels. A last remark must be made: as the recruitment of leukocytes by the biomaterial surfaces is delayed regarding the platelet retention, which starts very early after blood is allowed to flow through the ECC, we cannot exclude the potential role of thrombocytes in leukocyte adhesion. Nevertheless, this hypothesis seems difficult to support is we remember that PVC control exhibits the greatest affinity for leukocytes and the lowest affinity for platelets.

CONCLUSION

Considering the overall data (Table 6), we attempted to establish a hierarchy between the different biomaterials studied here. From platelet affinity and fibrinogen adsorption data, the best material appears to be the PVC control followed by PRM.PDMS and PVC.IUPAC. From leukocyte affinity data, PRM.PDMS

Table 6.
Summary of results

Platelet affinity	PVC.IUPAC > PRM.PDMS > PVC control
Leukocyte affinity	PVC Control > PRM.PDMS ≥ PVC.IUPAC
Fibrinogen adsorption	PVC.IUPAC ≥ PRM.PDMS > PVC control

behaves worse than PVC.IUPAC but better than PVC control. It is not so certain that in terms of thrombogenicity, these hierarchies are relevant. Long term testing would probably be more appropriate than the acute *ex vivo* experiment reported here and in a precedent work (13).

All synthetic surfaces may activate or change platelet and leukocyte functions. Marois *et al.* [14] reported their *in vivo* investigations on four chemically processed grafts developed in 1980 and concluded that if the ideal vascular prosthesis has not yet been achieved, nevertheless these grafts present marginal improvements compared to the previously chemically processed biological grafts commercially available.

In the future, it may be possible to reduce the thrombogenicity of conventional prostheses by inducing endothelial cells to grow to confluence and cover the synthetic surface prior to implantation. Preliminary studies in vitro in our laboratory show that PRM.PDMS and PVC.IUPAC meet the requirements of an initial cytotoxicity screening test [15].

Acknowledgements

The authors wish to thank Xavier Barthe and Marie-Odile Rubio for their photographic work, Josette Gautreau of her technical assistance concerning the animals. Monique Rouais and Véronique Silvério for typing this manuscript.

REFERENCES

1. M. D. Lelah, L. K. Lambrecht and S. L. Cooper, *J. Biomed. Mater. Res.* **18**, 475 (1984).
2. B. Basse-Cathalinat, Ch. Baquey, Y. Llabador and A. Fleury, *Int. J. Appl. Radiat. Isot.* **31**, 747 (1980).
3. Ch. Baquey, B. Basse-Cathalinat, L. Bordenave, J. Caix, A. J. Brendel and D. Ducassou, *Artif. Organs.* **10** (6), 481 (1986).
4. K. L. Spilizewski, R. E. Marchant, J. A. Anderson and A. Hiltner, *Biomaterials* **8**, 12 (1987).
5. T. J. McCoy, T. G. Grasel, A. Z. Okkema and S. L. Cooper, *Biomaterials,* **10**, 243 (1989).
6. D. Ducassou, D. Arnaud, A. Bardy, J. Beydon, M. Hegesippe and Ch. Baquey, *Br. J. Radiol.* **49**, 344 (1976).
7. M. L. Thakur, M. J. Welsh, J. H. Joist and R. E. Coleman, *Thromb. Res.* **9**, 345 (1976).
8. L. Bordenave, J. Caix, B. Basse-Cathalinat, Ch. Baquey and D. Ducassou, *R.B.M.* **9** (6), 283 (1987).
9. J. G. McAfee and A. Samin, *Radiol.* **155**, 221 (1985).
10. O. Loreal, A. Moisan, J. F. Bretagne, J. Le Cloirec, J. L. Raoul, J. Gastard and J. Y. Herry, *J. Nucl. Med.* **31**, 1470 (1990).
11. M. Rabaud, F. Lefebvre, Y. Piquet, F. Bellocq, J. Chevaleyre, M. F. Roudault, and H. Bricaud, *Thromb. Res.* **43**, 205 (1986).
12. A. C. P. Zondag, A. M. Kolb and N. M. A. Bax, *Haemostasis* **15**, 318 (1985).
13. L. Bordenave, J. Caix, B. Basse-Cathalinat, Ch. Baquey, D. Midy, J. C. Baste and H. Constans *Biomaterials* **10**, 235 (1989).
14. Y. Marois, D. Boyer, R. Guidoin, Y. Douville, M. Marois, F. J. Teijeira and P. E. Roy, *Biomaterials* **10** (6), 369 (1989).
15. L. Bordenave, R. Bareille, F. Lefebvre, J. Caix and Ch. Baquey, *J. Biomater. Sci. Polym. Ed.* **3**, 509 (1992).

Part IV

Cell Interactions

Effects of environmental parameters and composition of poly(2-hydroxyethyl methacrylate)-*graft*-polyamine copolymers on the retention of rat lymphocyte subpopulations (B- and T-cells)

AKIHIKO KIKUCHI[1,2], TEIJI TSURUTA[2] and KAZUNORI KATAOKA[1,2]*

[1]*Department of Materials Science, Science University of Tokyo, 2641 Yamazaki, Noda, Chiba 278, Japan*

[2]*Research Institute for Biosciences, Science University of Tokyo, 2669 Yamazaki, Noda, Chiba 278, Japan*

Received 16 November 1992; accepted 3 May 1993

Abstract—Retention behavior of rat lymphocyte subpopulations (B- and T-cells) was investigated on poly(2-hydroxyethyl methacrylate)-*graft*-polyamine (HA) copolymers with various copolymer compositions. Separation mechanism of B- and T-cells was then evaluated by focusing on several parameters, such as pH, temperature, and ionic strength. The interaction of lymphocytes with HA surfaces was mainly through the electrostatic force from their retention profile at varying ionic strengths of the medium. Temperature also has a crucial effect on the response of lymphocytes toward pH-induced phase transition of polyamine grafts at the polymer interface with aqueous milieu. At 4°C, both B- and T-cells showed minimal retention on HA surfaces at pH 8. At this pH, polyamine grafts existed in a compact conformation with a low degree of protonation. However, at pHs below 8, at which polyamine grafts existed in extended conformation, the resolution of B- and T-cells was achieved. In contrast, at 23°C, the phase transition of polyamine grafts significantly influenced T-cell retention, resulting in a decrease in the retention of T-cells on HA with polyamine in a compact conformation. Consequently, preferential retention of B-cells was achieved under this condition. The polyamine content was found to be another important factor affecting the retention behavior of lymphocyte subpopulations. On HA copolymers with low polyamine content (HA7, HA10), conformational transition of polyamine grafts showed a significant influence for B-cell retention, although the influence decreased with increasing polyamine content. From the study estimating the effect of neuraminidase treatment of lymphocytes on their retention to HA surfaces, sialic acid residues on the plasma membrane surface of lymphocytes are suggested as feasible anionic sites showing electrostatic interaction with polyamine grafts.

Key words: PHEMA-*graft*-polyamine copolymer; electrostatic interaction; lymphocyte subpopulations; separation; pH-induced phase transition.

INTRODUCTION

The development of novel materials with specific affinities toward particular cell populations has an important basis in applied and basic biosciences [1, 2]. In particular, cell-specific materials can be applied for the separation of a particular cell population from a mixture of other cell populations. The recovery of high purity cells with high viability can be utilized in cell biology, biotechnology, and medicine [3–5]. In response to growing demands for cell separation, novel materials having specific affinities toward particular cell populations have been developed, and advanced methods for cell separation utilizing these materials have been proposed [6, 7].

* To whom correspondence should be addressed.

Among the many methods for cell separation, we have focused on cellular adsorption chromatography for a convenient method with high efficiency to separate a desired cell population. Placing special emphasis on the separation of lymphocyte subpopulations, we have carried out a series of studies applying amine-containing polymers for cellular adsorption chromatography [6–15]. Our interest has been on the separation of two major subpopulations of lymphocytes: B-cells which are derived from bone marrow and differentiate to antibody-secreting cells, and T-cells which differentiated in the thymus and play important roles in cell-mediated immune response. Separation of these two subpopulations has been recognized as a basic requirement in the fields of immunology as well as biotechnology. Separated lymphocytes can be used for mass production of monoclonal antibodies, as well as bioactive compounds like lymphokines. In the course of our studies, it was elucidated that rat B-cells derived from mesenteric lymph nodes were selectively retained on the surface of poly(2-hydroxyethyl methacrylate)-*graft*-polyamine (HA copolymer). Successful resolution of B- and T-cells was achieved by using HA13 copolymer, which contains 13 wt % of polyamine grafts in the copolymer, as column matrices [12, 13]. An optimum chain length of HA13 polyamine grafts was determined for the successful resolution of B- and T-cells [14]. It was further noted that lymphocyte retention on HA13 surface was strongly influenced by the conformational transition of polyamine grafts accompanied with the deprotonation of amino groups in the chain from water-soluble extended state to water-insoluble compact conformation [11]. Another parameter is the effect of polyamine content in HA copolymer on the resolution of lymphocyte subpopulations. Thus, in this paper, we focused on several parameters (pH, temperature, and ionic strength) which may crucially affect lymphocyte retention. Furthermore, the contribution of sialic acid residues existing on the plasma membrane surfaces on the electrostatic interaction of lymphocytes with protonated amino groups of HA was discussed.

MATERIALS AND METHODS

Preparation of polymers

Poly(2-hydroxyethyl methacrylate) (PHEMA) was prepared by radical polymerization of 2-hydroxyethyl methacrylate (HEMA) in a sealed glass ampoule with stirring. To avoid gel formation, the homopolymerization of HEMA was carried out in ethanol at a low monomer concentration (less than 1 mol dm^{-3}) at 45°C using 2,2′-azobis(2,4-dimethylpentanonitrile) (V-65) as an initiator [16]. Polyamine macromonomer[1] was prepared according to previous reports [17, 18]. A typical procedure is as follows: 1,4-divinylbenzene (1,4-DVB) was reacted with *N,N′*-diethylethylenediamine (DEDA) through an anionic addition reaction to form a 1 to 1 adduct, *N,N′*-diethyl-*N*-(4-vinylphenethyl)ethylenediamine (EDAS, b.p.; 70°C/0.006 mmHg, yield; 60%). Polyamine macromonomer was then prepared by a self-polyaddition reaction of EDAS (1 mol dm^{-3}) in THF using lithium diisopropylamide (0.06 mol dm^{-3}) as a catalyst for 48 h at room temperature. After evaporation of the solvent, a residuum containing polyamine macromonomer was washed three times with dimethylformamide to remove lower molecular weight fraction (yield; 55%).

---

[1] Systematic name: α-(4-vinylphenethyl)-ω-[(*N*-ethyl)-2-ethylaminoethylamino)polyethyliminoethylene-ethyliminoethylene-1,4-phenyleneethylene.

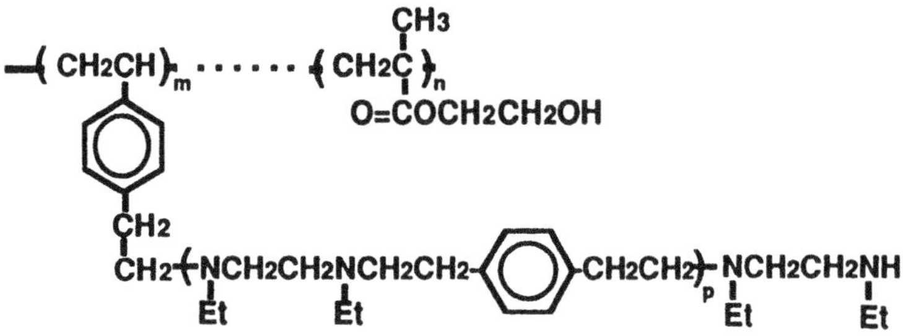
CH3
CH2CH
m
CH2C
n
O=COCH2CH2OH
CH2
CH2
NCH2CH2NCH2CH2-
Et
Et
CH2CH2
p
NCH2CH2NH
Et
Et

Figure 1. Structural formula of PHEMA-*graft*-polyamine copolymer.

Table 1.
Copolymerization of polyamine macromonomer with HEMA[a]

Code	Feed composition HEMA mmol	Polyamine mmol	Macromonomer[b] wt %	Yield %	Polyamine content in the copolymer[c] mol %	wt %
PHEMA	38.4	—	—	20.4	—	—
HA2	38.0	0.014	1	24.5	0.09	2.3
HA7	37.3	0.043	3	24.2	0.27	7.0
HA10	36.5	0.071	5	19.8	0.43	10.4
HA13	35.8	0.100	7	21.2	0.56	13.2
HA20	34.6	0.143	10	18.3	0.89	19.7
HA23	32.7	0.214	15	13.2	1.09	22.8

[a] Solvent—ethanol; initiator—V-65 (5 mM); react. temp.—45°C; react. time—1.5 h, total vol.—40 ml

[b] Number average molecular weight = 3500, determined by UV method [18].

[c] Copolymer composition determined by elemental analysis (C/N).

PHEMA-*graft*-polyamine copolymer (HA), of which the structural formula is shown in Fig. 1, was prepared by radical copolymerization of polyamine macromonomer with HEMA in ethanol at 45°C for 1.5 h using V-65 as an initiator [16, 18]. A number average molecular weight of polyamine macromonomer used in this study was in the range of 2500–3500. Polymer was recovered as a precipitate from an excess of diethyl ether at room temperature, followed by thorough drying *in vacuo*. The polyamine content of the HA copolymers was calculated from C/N ratio which was determined by elemental analysis of the copolymer. The results of copolymerization were summarized in Table 1.

Acid–base titration of polyamine and diamine compounds

Approximately 0.5 mmol equivalent of diamine unit of polyamine macromonomer, or *N,N'*-diethyl-*N*-(4-vinylphenethyl)ethylenediamine (EDAS) as a model compound for repeating unit of polyamine macromonomer, was dissolved into 15 ml of 0.1 M HCl solution. The solution was titrated with standardized 0.1 M NaOH solution. Change in the pH of the solution was measured by using a pH-meter (Horiba, M-8s). The protonation degree of amino groups (α) was then estimated from the acid–base titration curve of polyamine macromonomer. The point at which an excess of HCl was neutralized by NaOH was defined as $\alpha = 1$, and the point at which protonated amino groups of polyamine macromonomer were fully neutralized was defined as $\alpha = 0$. To determine the solubility of these compounds in an aqueous medium, solution turbidity was also measured at 500 nm using a spectrophotometer (Hitachi, Type 228) at the same time as the pH measurement.

Fluorescence study of polyamine macromonomer using pyrene probe

Polyamine macromonomer was dissolved in 5 mM HEPES-buffered solution at different pH, at a concentration of 1 mg ml^{-1}. Pyrene was dissolved into ethanol at a concentration of 10^{-2} M. 1.2 μl of pyrene solution was added to a solution of polyamine macromonomer (20 ml) to give a final pyrene concentration of 6×10^{-7} M. Sample solutions were degassed by bubbling with argon.

Fluorescence emission spectra of pyrene in polyamine solution at different pH were recorded by fluorophotometer (Jasco Co., FP 770) in the range of 350–550 nm at an excitation wavelength of 339 nm. The fluorescence intensity of the first and the third vibronic bands were measured at approximately 372 and 383 nm, respectively.

Polymer coating and column preparation [9, 10, 19]

Glass beads (average diameter: 250 μm) (Toshiba Ballotini Co.) were coated with each polymer sample using a solvent evaporating technique. Coating procedures are as follows: 40 g of glass beads were immersed into 40 ml of ethanol solution of HA copolymer (0.2 w/v %) at room temperature for 1 h. Polymer-coated glass beads were recovered by filtration, dried under a purified nitrogen atmosphere for 1 h at room temperature, followed by thorough drying *in vacuo* over $CaCl_2$. PHEMA coating was performed in a similar manner using ethanol as solvent. The polymer coating procedure of glass bead surface was nearly complete based on ESCA measurement [12].

The column was prepared by packing 1 g of polymer-coated glass beads into poly(vinyl chloride) tubing (inner diameter: 0.3 cm, column length: 9 cm) fitted with nylon mesh column supports and a stopcock. The column, packed with polymer coated glass beads, was primed with physiological saline for at least 12 h. The primed saline was replaced with Ca^{2+}- and Mg^{2+}-free Hanks' balanced salt solution (HBSS) at a definite pH prior to lymphocyte loading.

Preparation of lymphocyte suspension

Rat lymphocytes were obtained from the mesenteric lymph nodes of Wistar male rats (5 weeks old, Japan Rat Co.) by squeezing the mesenteric lymph nodes with a pair of clean glass slides into HBSS (pH 7.2) [20]. This lymphocyte suspension was centrifuged at 200 *g* for 5 min at room temperature to sediment lymphocytes. After discarding the supernatant, the lymphocyte pellets were resuspended in HBSS, and the number of lymphocytes was counted by a Coulter Counter (Coulter Electronics, Model ZBI). The lymphocyte suspension was adjusted to a desired concentration by adding an appropriate amount of HBSS. Lymphocytes viability was evaluated by dye exclusion method using trypan blue [21] and revealed above 90% of cell viability.

Ionic strength variation of HBSS

Ionic strength of HBSS was varied by substituting NaCl with corresponding amount of sucrose to sustain physiological osmotic pressure of the solution at pH 7.2.

pH adjustment of HBSS

N-2-hydroxyethyl-*N'*-2-sulfoethylpiperazine (HEPES), as a buffer agent, was added to the HBSS to a final concentration of 5 mM. An appropriate amount of 0.1 M HCl or 0.1 M NaOH solution was added to adjust the pH.

Estimation of cell/materials interaction by time-dependent fractionation of effused suspension [12, 13, 20]

The lymphocyte suspension was continuously loaded into the column using an infusion pump (Precidol Type 5003) at a flow rate of 0.4 ml min^{-1} for 3.5 min. The effluent from the column was collected as a single aliquot. The lymphocyte concentration in the aliquot was determined by using Coulter Counter. The percentages of lymphocyte effusion from the column and retention in the column were estimated according to Eqs. (1) and (2), respectively [20].

$$\text{Effusion (\%)} = [L]/[L]_0 \times 100 \quad (1)$$

$$\text{Retention (\%)} = (1 - [L]/[L]_0) \times 100 \quad (2)$$

where $[L]_0$ (cells ml^{-1}) is the concentration of lymphocytes prior to loading into the column, and $[L]$ (cells ml^{-1}) is the concentration of lymphocytes in the aliquot (correcting for the priming volume of the column). Data are expressed as the mean of triplicate experiments with standard error of mean (S.E.M.).

Identification of the lymphocyte subpopulations (B- and T-cells) by immunofluorescence methods [22, 23]

Immunoglobulin (Ig) molecules, presented on the plasma membrane surface of B-cells, were used as markers for B-cells. The Ig molecules were detected by immunofluorescence method using fluorescein isothiocyanate (FITC) labeled rabbit anti-rat immunoglobulins (Dako Co., Denmark). Approximately 1×10^7 lymphocytes were resuspended in 0.1 ml of HBSS (pH 7.2) containing FITC-labeled antibodies (0.51 mg ml^{-1}) with 0.02 M sodium azide. After 1 h incubation in an ice bath, the lymphocytes were washed three times with HBSS (pH 7.2) containing 0.02 M sodium azide to remove excess FITC-labeled antibodies at 4°C. The percentage of stained cells (B-cells) was determined by counting more than 800 cells using fluorescence microscope (Nikon Fluorophoto VFD-TR). T-cells were also stained by FITC labeled anti-T-cell monoclonal antibody (Serotec, W3/13). It was confirmed that the lymphocyte population in mesenteric lymph nodes consists of B- and T-cells with a negligible amount of other cell populations. Thus, unstained cells with FITC labeled anti-rat immunoglobulins were regarded as T-cells [24].

Preparation of neuraminidase-treated lymphocytes and estimation of their interaction with HA surfaces

Approximately 10^8 lymphocytes were treated with 0.5 ml of *Arthrobacter ureafaciens* neuraminidase (NACALAI TESQUE INC., Japan) solution in HBSS (concentration; 0.5 U ml^{-1}) at pH 7.2 at 37°C for 90 min to digest sialic acid residues. Following digestion, the cell suspension was centrifuged at 200 *g* and supernatant was collected in a glass tube with a stopcock. This supernatant was used for quantification of removed sialic acid concentration by thiobarbituric acid assay as described below. After washing twice with HBSS, neuraminidase-treated lymphocytes were resuspended in HBSS at 1×10^7 cells ml^{-1}. This lymphocyte suspension was used for the estimation of cell/materials interaction as described in the former section. Cell viability before and after neuraminidase treatment was determined by a dye exclusion method [21].

Removed sialic acid from the plasma membrane of lymphocytes in the supernatant was quantified by using thiobarbituric acid assay [25, 26]. The procedure is as follows: 0.25 ml of 25 mM sodium periodate in 0.125 M H_2SO_4 was added to 0.5 ml of the supernatant of cell suspension and incubated for 30 min at 37°C. 0.2 ml of 2% sodium arsenite in 0.5 M HCl was added and the tube was shaken gently. After the disappearance of the yellow-brown color, 2 ml of 0.1 M 2-thiobarbituric acid solution (pH 9.0) was added. The tube was capped with a cock and heated in boiling water for 7.5 min. The colored solution was then cooled in ice–water bath and shaken vigorously with 5 ml of 1-butanol/HCl (95/5 v/v). The solution was centrifuged at 900 *g* for 10 min to separate the butanol layer from the water layer. The absorbance at 551 nm was measured by a spectrophotometer (Hitachi, Type 557) using 3-ml glass cell. The concentration of removed sialic acid from the plasma membrane of lymphocytes was calculated by using a previously prepared calibration curve for sialic acid.

RESULTS AND DISCUSSION

Protonation and solubility of polyamine macromonomer in aqueous media

It was previously reported that amino groups in polyamine grafts of HA copolymers were partially protonated over a wide pH range [11, 14]. The protonation degree of amino groups significantly influenced the retention behavior of rat lymphocyte subpopulations on HA13 at ambient temperatures [11, 14]. Temperature, as well as pH, is considered to affect the ionization of polyamine macromonomer, which may lead to the variation in retention behavior of B- and T-cells. Acid–base titration of polyamine macromonomer was carried out at varying temperature to determine protonation degree of amino groups. In every case, a two-stage titration curve was obtained, which was attributed to two step deprotonation of amino groups of ethylenediamine repeating units of polyamine macromonomer (Scheme 1). This was also confirmed from the similar two-stage titration curve obtained for EDAS, a model compound of repeating unit of polyamine macromonomer.

The protonation degree (α) was determined from the acid–base titration curve of polyamine macromonomer (α = [protonated amino groups]/[total amino groups]) [14]. Table 2 summarized the apparent p*K* values and the turbidity points of diamine derivatives. Protonation degree at the turbidity point is also shown in the

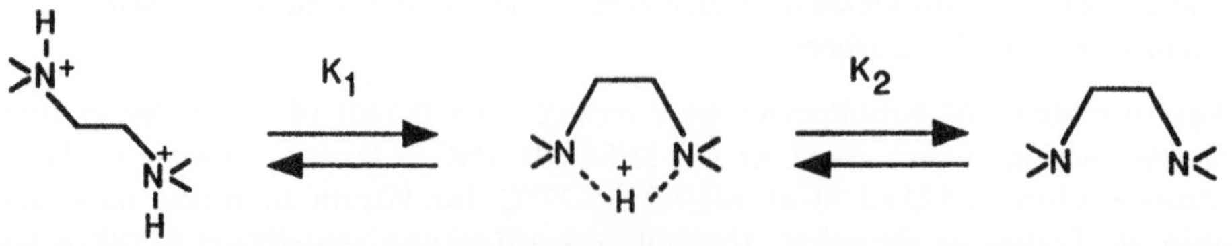
K1
K2

Scheme 1. Two-step deprotonation of diamine unit.

Table 2.
Apparent pK values and solubility of diamine derivatives[a]

Samples	Added NaCl[b]	pK_1 4°C	pK_1 23°C	pK_2 4°C	pK_2 23°C	$pH_{\alpha=0.5}$[c] 4°C	$pH_{\alpha=0.5}$[c] 23°C	pH at turbidity point[d] 4°C(α)	pH at turbidity point[d] 23°C(α)
Polyamine	yes	5.8	5.4	8.5	7.7	7.4	6.7	8.0 (0.45)	6.8 (0.49)
macromonomer	no	5.3	5.0	8.3	7.5	7.0	6.4	8.0 (0.40)	7.0 (0.48)
EDAS[e]	no	n.d.[f]	5.8	n.d.	10.2	n.d.	8.0	n.d.	10.0 (0.29)

[a] Estimated from acid-base titration curve. pK_1 and pK_2 are the pH at $\alpha = 0.75$ and 0.25, respectively.
[b] 0.9 wt % of NaCl was added to the solution.
[c] Half equivalent point (protonation degree: $\alpha = 0.5$).
[d] The number in parenthesis is the protonation degree at turbidity point.
[e] *N,N'*-diethyl-*N*-(4-vinylphenethyl)ethylenediamine.
[f] Not determined.

parenthesis of the last column of Table 2. As shown in Table 2, deprotonation of amino groups in polyamine macromonomer took place at a lower pH range than EDAS. At 23°C, the pK_1 value of the first deprotonation stage of polyamine macromonomer was slightly lower than that of EDAS ($\Delta pK_1 = pK_{1EDAS} - pK_{1polyamine} = \sim 1$), while the p$K_2$ value of the second deprotonation stage of polyamine macromonomer was lower than that of EDAS ($\Delta pK_2 = pK_{2EDAS} - pK_{2polyamine} = \sim 2.5$). This is due to the transition of polyamine chain from water-soluble extended conformation to water-insoluble compact conformation. The turbidity point occurred during the second deprotonation stage, indicating the increased hydrophobicity through the deprotonation of ethylenediamine units further facilitates the deprotonation process to provoke phase transition from water-soluble state to water-insoluble state. Consequently, this transition causes a shift in apparent pK_2 to a lower value than expected. Polyamine macromonomer consists of alternately repeating units of apolar diethylenephenylene portion and ethylenediamine portion, the latter of which becomes polar through the protonation of amino groups. This apolar–polar balance in the polyamine macromonomer collapsed at $\alpha < 0.5$, resulting in the phase transition of polyamine macromonomer to a dehydrated compact conformation.

Both ionic strength and the surrounding temperature are influential factors on pK values of ionizable groups. The change in pK values of polyamine macromonomer in the presence of sodium chloride at physiological concentration (0.9 wt %) or by changing the surrounding temperature were also summarized in Table 2. The pK values of polyamine macromonomer increased in the presence of NaCl at physiological concentration (0.9 wt %; *c*. 0.15 M), indicating the increased affinity of proton to amino groups due to the salt effect. The electrostatic shielding effect of the protonated amino groups by counter ions such as Cl^- caused the decreased dissociation of protonated amino groups resulting in the increase in pK values of polyamine macromonomer.

Remarkable increase in the apparent pK values, especially pK_2, of polyamine macromonomer was observed by lowering the surrounding temperature to 4°C. Furthermore, the turbidity point increased from pH 6.8 ($\alpha = 0.49$) at 23°C to pH 8.0 ($\alpha = 0.45$) at 4°C in the presence of 0.9% of NaCl. By lowering the temperature, there may be an increased stabilization of ionic hydration around protonated amino groups. Polar ethylenediamine units then contribute to the solubilization of polyamine macromonomer due to the stabilization of the protonation of amino groups.

The water of hydrophobic hydration has a negative enthalpy ($\Delta H < 0$), and a negative entropy ($\Delta S < 0$) of hydration [27]. By decreasing the surrounding temperature, the hydrophobic hydration around the apolar portion becomes less unstable, allowing increased water-solubility of polyamine macromonomer. This synergistic effect of hydration around apolar and polar portions leads to an increase in pK_2 value at 4°C. The effect of these protonation features of polyamine grafts on retention behavior of rat lymphocytes will be discussed.

Lymphocyte retention to HA surfaces with varying polyamine content at different pH

pH is one of the most influential factors on the retention behavior of lymphocytes on HA copolymer surfaces. Lymphocyte retention was investigated by varying the pH as well as polyamine content in HA copolymer. pH-Dependency of lymphocyte retention on polymer surfaces with varying polyamine content is shown in Fig. 2. The lymphocyte retention on HA copolymer surfaces was remarkably influenced with the surrounding pH, increasing with a decreasing pH. Even HA7, which retained only 5% of loaded lymphocytes under the physiological pH condition (pH 7.4), retained approximately 90% of loaded lymphocytes at pH 5.6. At a certain pH, lymphocyte retention increased with an increased amount of polyamine portions in HA. This result indicates that the interaction force of lymphocytes with HA surfaces was increased with increasing polyamine content in HA copolymer.

In contrast, approximately 50% of the lymphocyte were retained on PHEMA surface and retention was constant over the range of pH 5.6–8.0. It was reported

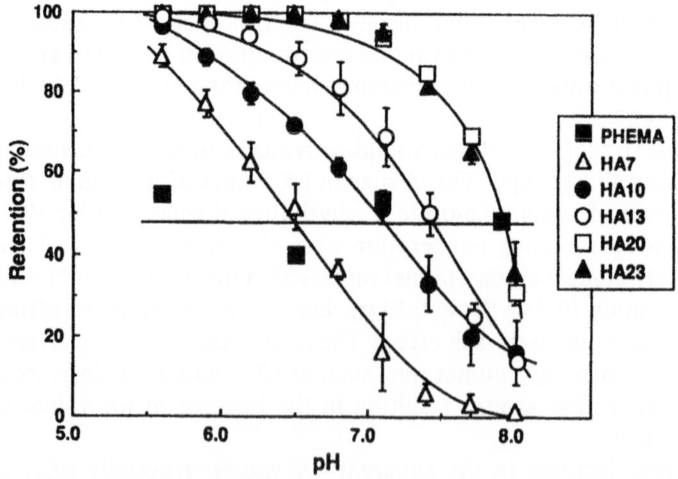
100
80
60
40
20
0
Retention (%)
5.0
6.0
7.0
8.0
pH
PHEMA
HA7
HA10
HA13
HA20
HA23

Figure 2. pH-Dependency of lymphocyte retention on PHEMA and HA copolymers at 23°C. Initial cell concentration: $[L]_0 = (1.00 \pm 0.03) \times 10^7$ cells ml^{-1}.

that the electrophoretic mobilities of lymphocyte subpopulations derived from the mesenteric lymph nodes of Wistar rats were unchanged in this experimental pH range [28, 29]. Consequently, the change in lymphocyte retention to HA with pH is considered to be due to the variation in the electrostatic characteristic of HA copolymer surfaces through protonation. Indeed, pH-dependent change in the protonation degree of amino groups correlated with retention behavior of lymphocytes, as will be described in the following section.

Effects of protonation degree of amino groups and temperatures on lymphocyte retention to HA surfaces

As reported, the protonation of amino groups in polyamine portions is the most influential factor on retention behavior of rat lymph node lymphocytes on the HA13 surface [11, 14]. However, the contribution of this factor has not been fully investigated for lymphocyte retention on HA with varying composition of polyamine. Thus, the effect of the protonation of amino groups on lymphocyte retention to HA surfaces with different polyamine composition was estimated. Figure 3 shows the semilogarithmic plots of inverse-effusion of lymphocytes from the HA column with varying polyamine composition at 23°C against the protonation degree of amino groups (α). An increase in the logarithm of inverse-effusion of lymphocytes indicates increased lymphocyte retention in the column. As shown in Fig. 3, the logarithmic value increased with increasing the protonation degree, as well as increasing

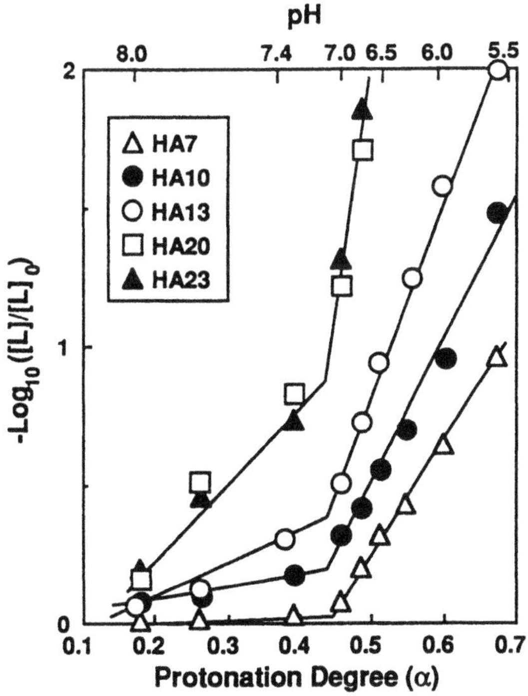
pH
8.0
7.4
7.0
6.5
6.0
5.5
2
1
0
Δ HA7
● HA10
○ HA13
□ HA20
▲ HA23
-Log10([L]/[L]0)
0.1
0.2
0.3
0.4
0.5
0.6
0.7
Protonation Degree (α)

Figure 3. Logarithmic plots of inverse-effusion of lymphocytes from HA copolymer columns as a function of the protonation degree of amino groups at 23°C. Upper abscissa represents the corresponding pH. Initial cell concentration: $[L]_0 = (1.00 \pm 0.03) \times 10^7$ cells ml^{-1}.

polyamine content in HA copolymer. As data shown in Fig. 3 were obtained from the various HA columns with the same bed volume, their differences are considered to reflect the differences in the apparent number of adsorption sites in the adsorbent, i.e. the number of protonated amino groups. Thus, pronounced cell–materials interaction is observed by increasing the number of protonated amino groups.

Of interest, break points were observed in the logarithmic plots for each HA copolymer, which are located in the range of $0.4 < \alpha < 0.5$. This range corresponds to the phase transition of polyamine macromonomer as described in the foregoing section. The slopes increased steeply above this range of the protonation degree, indicating a change in the mode of interaction. These results suggest that the interaction of HA copolymer surfaces with lymphocytes is closely coupled with the phase transition of polyamine grafts, and increased through conformational change from the compact chain state to the extended chain state.

The number of protonated amino groups in HA copolymers at a given pH should change by varying the surrounding temperature due to the change in p*K* value of amino groups as cited in Table 2. Figure 4 shows the semilogarithmic plots of inverse-effusion of lymphocytes from HA copolymer columns at 4°C. Logarithmic values were linearly decreased with decreasing the protonation degree of amino groups. Worth noting is that all lines obtained for HA copolymer with different composition converged to the abscissa at $\alpha = 0.45$. As shown in Table 2, polyamine grafts undergo phase change at around $\alpha = 0.45$ at 4°C in physiological saline

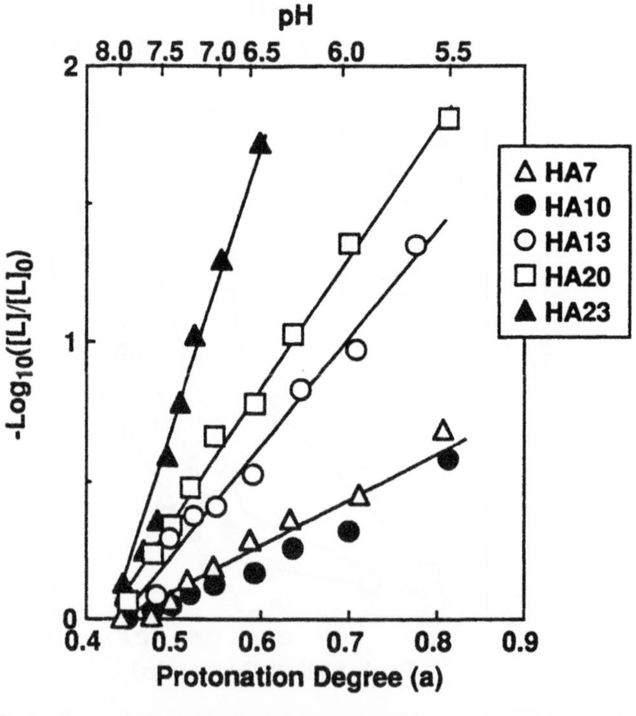
pH
8.0 7.5 7.0 6.5 6.0 5.5
2
1
0
-Log10([L]/[L]0)
0.4 0.5 0.6 0.7 0.8 0.9
Protonation Degree (a)
HA7
HA10
HA13
HA20
HA23

Figure 4. Logarithmic plots of inverse-effusion of lymphocytes from HA copolymer columns as a function of the protonation degree of amino groups at 4°C. Upper abscissa represents the corresponding pH. Initial cell concentration: $[L]_0 = (0.96 \pm 0.04) \times 10^7$ cells ml^{-1}.

buffer solution. Consequently, at 4°C, lymphocyte retention on HA surfaces was significantly influenced by the phase transition of polyamine grafts, and occurred only when polyamine grafts exist in water-compatible state.

As shown in Figs 3 and 4, the phase transition of polyamine grafts from the extended state to a compact conformation, as a function of decreasing protonation degree, was found to be a crucial factor for the lymphocyte retention on HA surfaces at both 4 and 23°C. Furthermore, significant difference in lymphocyte retention profiles between these temperatures were observed at the range where polyamine grafts exist in water-incompatible state with compact conformation; no retention of lymphocytes occurred at 4°C.

Given that electrostatic interaction is the main attractive force for lymphocyte retention, the total amount of protonated amino groups of HA copolymer can be correlated with lymphocyte retention on HA surfaces. The total amount of the protonated amino groups is represented by the amount of protonated nitrogen (N^+ content in wt %) which is calculated by multiplying the nitrogen content (N) in the copolymer (in wt %) by the protonation degree (α). The relation between the N^+ content of HA copolymers and lymphocyte retention at 23°C is shown in Fig. 5. Introducing a small amount of protonated nitrogen in the copolymer resulted in a

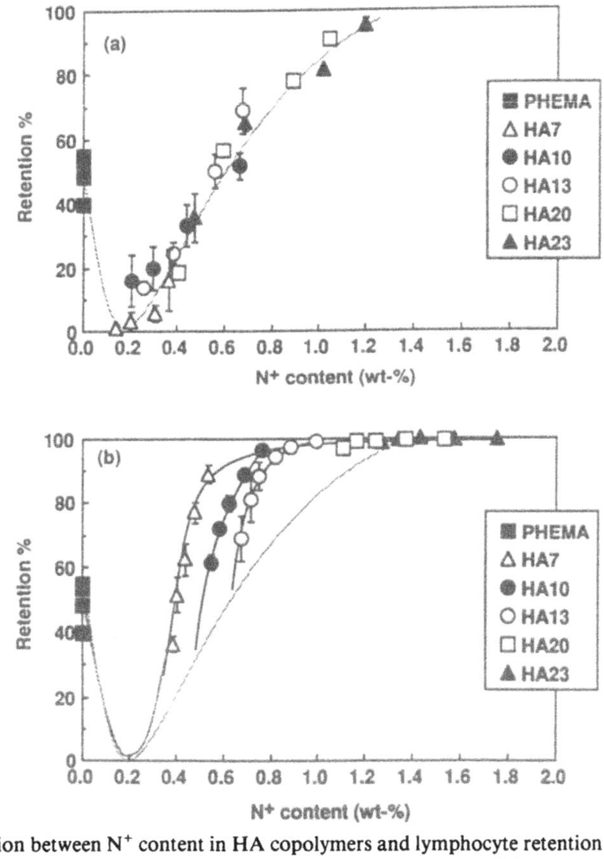
(a)
Retention %
N+ content (wt-%)
PHEMA
HA7
HA10
HA13
HA20
HA23
(b)
Retention %
N+ content (wt-%)
PHEMA
HA7
HA10
HA13
HA20
HA23

Figure 5. Relation between N^+ content in HA copolymers and lymphocyte retention at 23°C (a) $\alpha < 0.5$, (b) $\alpha > 0.5$. Gray curve in (b) corresponds to the retention curve at $\alpha < 0.5$, represented in (a). Initial cell concentration: $[L]_0 = (1.00 \pm 0.03) \times 10^7$ cells ml^{-1}.

minimum in lymphocyte retention (N^+ content = ~0.2 wt %). Increased protonation of nitrogen caused a monotonous increase in lymphocyte retention. This phenomenon of significant elimination of cellular retention is a unique characteristic of amine-containing copolymers, and a similar result was obtained for platelet retention as previously reported [30, 31]. This might be due to the steeply decreased contribution of non-electrostatic force by introducing a small amount of the protonated amino groups, which was featured by an increase in free water content in polymer films as reported elsewhere [24].

The retention of lymphocytes showed different profiles depending on whether the polyamine existed in a water-compatible state or in a phase-separated state. When polyamine grafts exist in phase-separated state (compact conformation), lymphocyte retention followed the same master curve, regardless of the polyamine content in the HA copolymer (Fig. 5(a)). This result indicates that the amount of N^+ is the primary factor for lymphocyte retention at this state.

Figure 5(b) represents data when the polyamine grafts exist in a water-compatible state (considerably extended conformation) ($\alpha > 0.5$). HA copolymer with these extended polyamine grafts followed different retention curves depending on polyamine content. Comparing the same N^+ content in Fig. 5(b), higher retention was observed for HA copolymer with less polyamine content. Since the number of protonated amino groups per polyamine chain is higher for HA copolymer with less content of polyamine at given N^+ content, this result suggests that the number of N^+ sites per chain plays a substantial role in lymphocyte retention in the condition at which polyamine takes extended conformation. It is likely that the polyamine chain with a higher protonation degree may lead to stronger binding with lymphocytes through multi-site interaction.

At 4°C, lymphocyte retention on HA surfaces took place only when polyamine exists in an extended conformation, as described. Lymphocyte retention at 4°C, as a function of N^+ content, is summarized in Fig. 6 for HA copolymers with different polyamine content. Comparing similar N^+ contents, higher retention took place for HA copolymer with lower polyamine content. This may reflect the difference in the

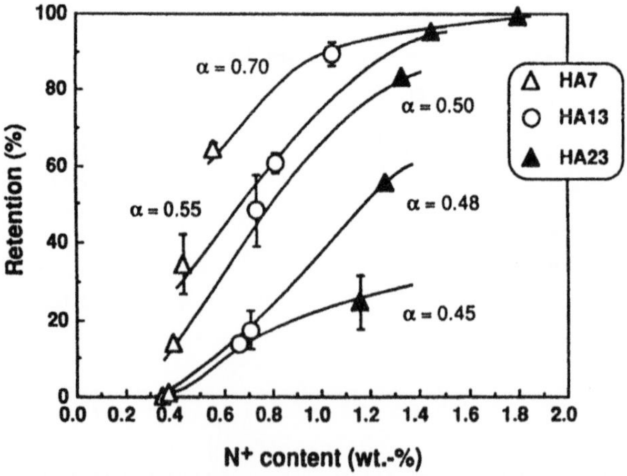

Retention (%)
100
80
60
40
20
0
0.0
0.2
0.4
0.6
0.8
1.0
1.2
1.4
1.6
1.8
2.0
N+ content (wt.-%)
α = 0.70
α = 0.50
α = 0.55
α = 0.48
α = 0.45
HA7
HA13
HA23

Figure 6. Relation between N^+ content in HA copolymers and lymphocyte retention at 4°C. Initial cell concentration: $[L]_0 = (0.96 \pm 0.04) \times 10^7$ cells ml^{-1}.

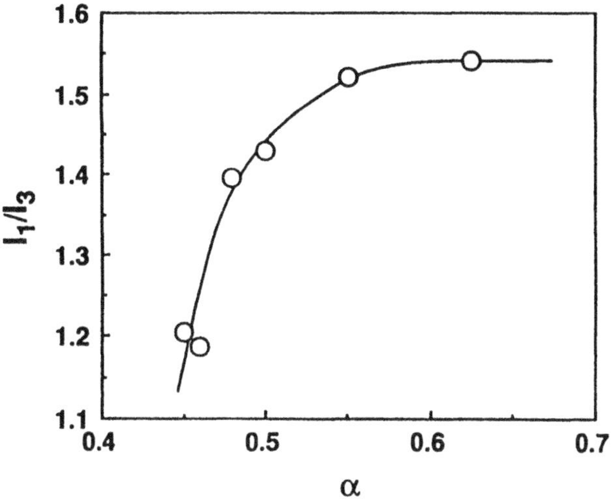
1.6
1.5
1.4
1.3
1.2
1.1
I_1/I_3
0.4
0.5
0.6
0.7
α

Figure 7. The ratio of the first and the third vibronic bands of pyrene fluorescence in polyamine solution at 4°C.

number of the protonated amino groups per polyamine chain (e.g., cationic sites per polyamine chain). Although polyamine chains with this range of protonation degree are in the water-compatible state, it may be assumed that the gradual extension of polyamine graft from the surface may occur with increased N^+ number per polyamine chain due to intramolecular electrostatic repulsion. Thus, lymphocytes may be captured to cationic sites more readily in a polyamine chain with highly extended conformation. As shown in Fig. 6, indeed, lymphocyte retention drastically increased in the range of $0.45 < \alpha < 0.50$.

This consideration was confirmed by the study using pyrene probe. The fluorescence of pyrene is known to change with the polarity of the surroundings, and is applied as a probe molecule to explore the micro-environment of the polymer and its associates [32]. The ratio of the first and the third vibronic bands of fluorescence spectrum is especially sensitive to the polarity of the surroundings. Thus, pyrene was used as a fluorescent probe for the estimation of the pH-dependent conformational change in polyamine macromonomer. The fluorescence spectrum of pyrene in the solution of polyamine macromonomer was measured by changing the pH at 4°C. Fig. 7 shows the ratio (I_1/I_3) of the first (I_1) and the third (I_3) vibronic bands of fluorescence, plotted against the protonation degree of amino groups. Although the macromonomer is soluble in this region of protonation degree at 4°C, the I_1/I_3 values steeply decreased at $\alpha < 0.5$ with a decrease in the protonation degree, indicating that pyrene is incorporated into the apolar micro-environment of macromonomer formed through decreasing protonation degree of amino groups. This result suggests the conformational change in polyamine in the range of $0.45 < \alpha < 0.50$, which is consistent with the drastic increase in lymphocyte retention at this region, as observed in Fig. 6.

Differential retention of lymphocyte subpopulations to HA surfaces with different polyamine content

Lymphocytes consist of two major subpopulations, B- and T-cells. These cells have different electrostatic features of plasma membrane surfaces as represented by lower

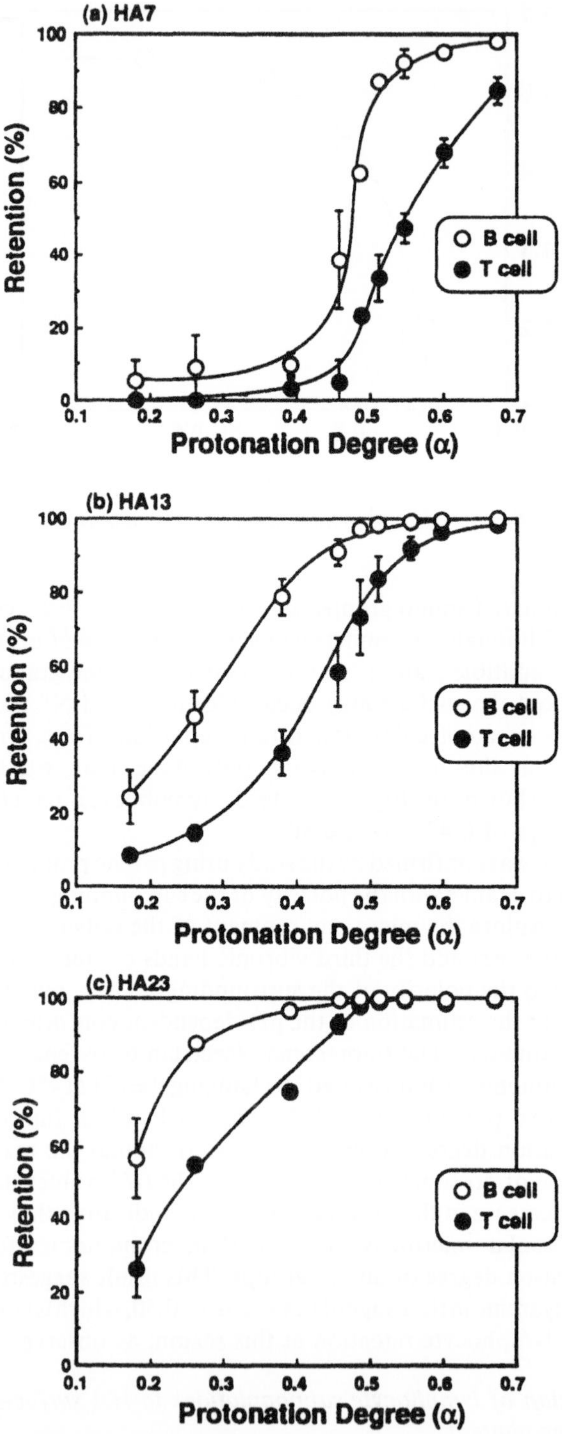
(a) HA7
100
80
60
40
20
0
0.1
0.2
0.3
0.4
0.5
0.6
0.7
Retention (%)
Protonation Degree (α)
B cell
T cell
(b) HA13
100
80
60
40
20
0
0.1
0.2
0.3
0.4
0.5
0.6
0.7
Retention (%)
Protonation Degree (α)
B cell
T cell
(c) HA23
100
80
60
40
20
0
0.1
0.2
0.3
0.4
0.5
0.6
0.7
Retention (%)
Protonation Degree (α)
B cell
T cell

Figure 8. Relation between protonation degree and retention of B- and T-cells on HA copolymers at 23°C. (a) HA7, (b) HA13, (c) HA23. Initial cell concentration: $[L]_0 = (1.00 \pm 0.03) \times 10^7$ cells ml^{-1}. Initial B- cell fraction: $[B]_0 = 32.7 \pm 0.9\%$.

isoelectric point of B- than T-cells [28, 29, 33]. In this section, the effects of polyamine content and the surrounding temperature on the resolution of B- from T-cells were evaluated based on electrostatic interaction. Retention of B- and T-cells on HA surfaces was evaluated for mixed lymphocytes. In Fig. 8, the retention profiles of B- and T-cells at 23°C were plotted against the protonation degree of amino groups. On HA7 surface, as shown in Fig. 8(a), retention of B- and T-cells steeply decreased at around $\alpha = 0.5$, the range corresponding to the phase transition of polyamine grafts. On HA13 (Fig. 8(b)), T-cell retention decreased at around $\alpha = 0.5$, whereas B-cell retention was still significant in this range. On HA23, as shown in Fig. 8(c), both B- and T-cells showed considerable retention in the region below $\alpha = 0.5$. These results indicate that retention of both B- and T-cells is influenced by the conformational transition of polyamine grafts, while as the polyamine content increases, interaction of lymphocytes with HA surfaces becomes strong enough to obscure the effect of phase transition of polyamine grafts.

Data of B- and T-cell retention on HA with different composition was summarized as the function of N^+ content (Fig. 9). As shown in Fig. 9(a), when the

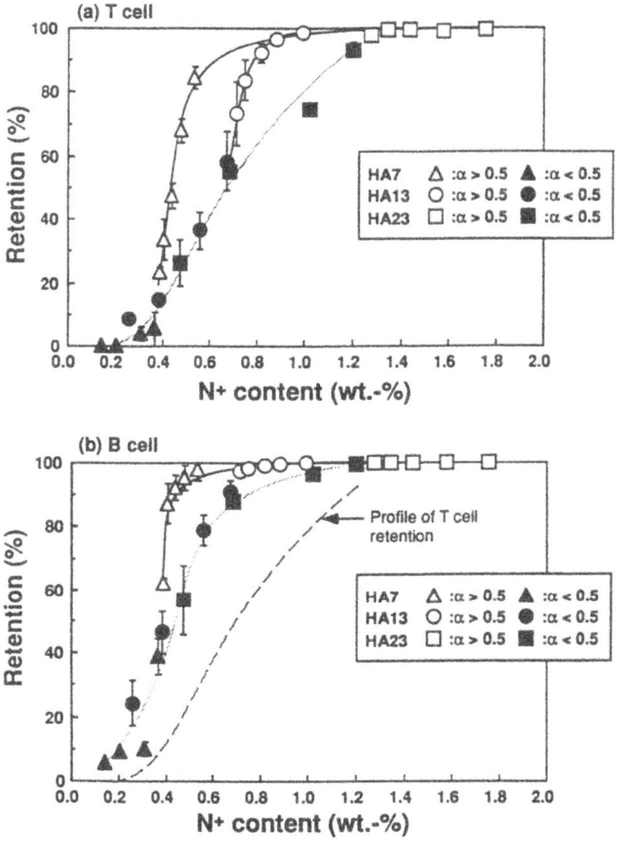
(a) T cell
Retention (%)
N+ content (wt.-%)
HA7 △ :α > 0.5 ▲ :α < 0.5
HA13 ○ :α > 0.5 ● :α < 0.5
HA23 □ :α > 0.5 ■ :α < 0.5
(b) B cell
Profile of T cell retention

Figure 9. Relation between N^+ content in HA copolymers and retention of B- and T-cells at 23°C. (a) T-cell, (b) B-cell, open plot; $\alpha > 0.5$, closed plot; $\alpha < 0.5$. Initial cell concentration: $[L]_0 = (1.00 \pm 0.03) \times 10^7$ cells ml^{-1}. Initial B-cell fraction: $[B]_0 = 32.7 \pm 0.9\%$. Dashed line without any plots in (b) represents the retention curve of T-cells at $\alpha < 0.5$.

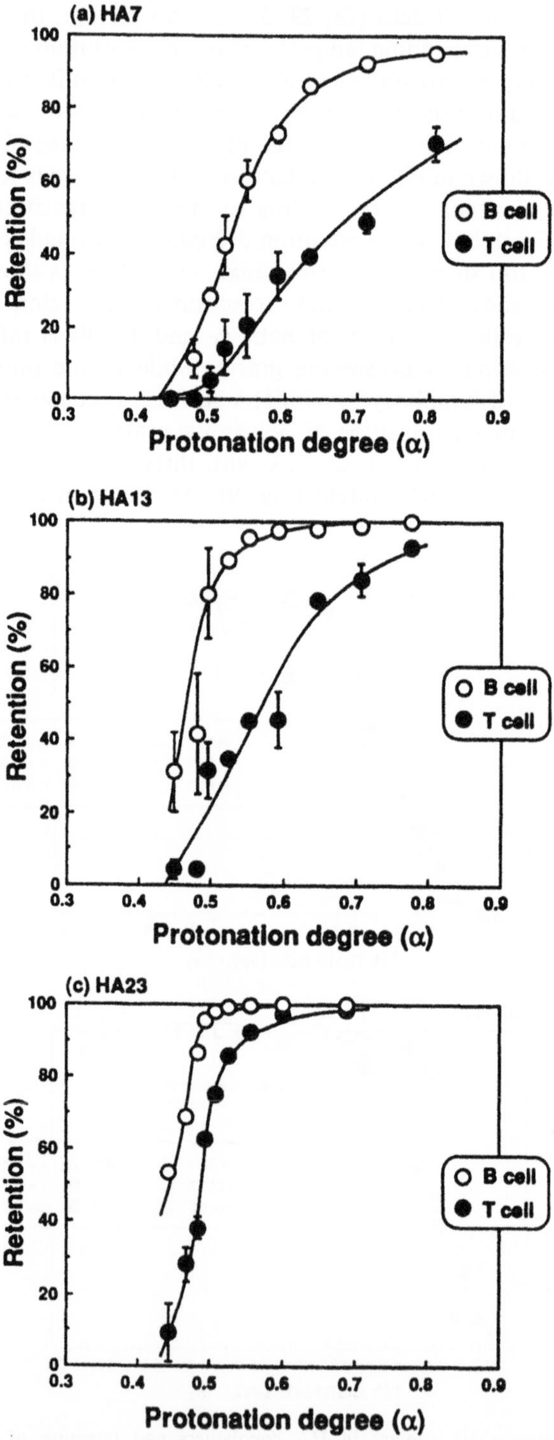
(a) HA7
100
80
60
40
20
0
Retention (%)
0.3
0.4
0.5
0.6
0.7
0.8
0.9
Protonation degree (α)
B cell
T cell
(b) HA13
100
80
60
40
20
0
Retention (%)
0.3
0.4
0.5
0.6
0.7
0.8
0.9
Protonation degree (α)
B cell
T cell
(c) HA23
100
80
60
40
20
0
Retention (%)
0.3
0.4
0.5
0.6
0.7
0.8
0.9
Protonation degree (α)
B cell
T cell

Figure 10. Relation between protonation degree and retention of B- and T-cells on HA copolymer surfaces at 4°C. (a) HA7, (b) HA13, (c) HA23. Initial cell concentration: $[L]_0 = (0.96 \pm 0.04) \times 10^7$ cells ml^{-1}. Initial B-cell fraction: $[B]_0 = 34.7 \pm 0.2\%$.

polyamine exists in a compact conformation, T-cell retention followed the same curve, regardless of the polyamine content in HA copolymers (closed plots). The amount of N^+ sites is a dominant factor for T-cell retention on HA surface with polyamine in compact conformation. However, retention of T-cells followed different curves depending on the composition, while conformation of polyamine grafts changed from a compact state to an extended state. As B-cell retention steeply increased with N^+ content, even in the range where polyamine takes compact conformation, the effect of phase transition became obscure.

In Fig. 9(b), the retention curve of T-cells on HA with compact conformation was superimposed as dotted curve without any plots. B-cell retention was always higher than T-cell retention on HA surfaces with the polyamine in a compact conformation. This difference in retention is not solely explained by the different affinity between B- and T-cells toward protonated amino groups, since the difference is not significant for the random copolymers of HEMA and EDAS with varying composition [14]. It is likely that the microdomained distribution of polyamine grafts as well as the amount of protonated amino groups are important factors for successful resolution of B- and T-cells by HA copolymers.

Effect of temperature on differential retention of B- and T-cells

Figure 10 shows the retention profiles of B- and T-cells at 4°C, plotted against the protonation degree of amino groups. Retention of B- and T-cells was only observed above $\alpha = 0.45$, and increased with increasing the protonation degree. At 4°C, the affinity of both B- and T-cells was extremely weak for HA surfaces with polyamine in compact conformation. It is obvious that B-cell, as well as T-cell retention, correlates to the phase transition.

Figure 11 shows the relation between N^+ content and the retention of B- and T-cells at 4°C. Although both B- and T-cell retention drastically increased in the range of $0.45 < \alpha < 0.50$, steeper increase in B-cell retention was obvious in this region. This α range corresponds to the region where the small amount (<10%) of deprotonated ethylenediamine unit still exist in polyamine grafts, and thus, polar-apolar balance in polyamine chain might take considerable change in this region, as suggested from the results using the molecular probe, pyrene (see Fig. 7). Consequently, the gradual and considerable extension in the chain conformation may occur accompanied with the increased protonation degree in this α region due to the compensative contribution of electrostatic repulsion and hydrophobic interaction. This conformational change may affect the retention of B- and T-cells in this α range.

Effect of ionic strength of the surrounding medium on lymphocyte retention

The ionic strength of the medium is considered to be another crucial factor influencing the retention behavior of lymphocytes with HA copolymer surfaces. Figure 12 shows the effect of the ionic strength on lymphocyte retention to HA surfaces at pH 7.2. The ionic strength of the surrounding medium is varied by substituting sodium chloride of HBSS with the corresponding amount of sucrose to maintain physiological osmotic pressure. Quite different retention profiles of lymphocytes were observed between PHEMA and HA copolymers by decreasing ionic strength of the medium.

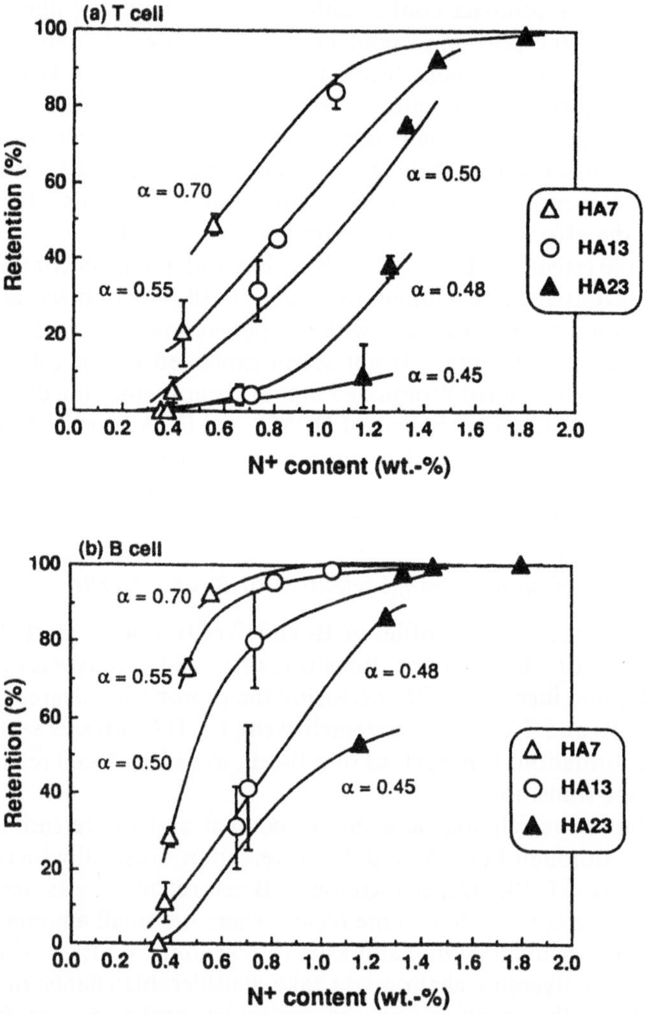

(a) T cell
Retention (%)
N+ content (wt.-%)
α = 0.70
α = 0.50
α = 0.55
α = 0.48
α = 0.45
HA7
HA13
HA23
(b) B cell
Retention (%)
N+ content (wt.-%)
α = 0.70
α = 0.55
α = 0.48
α = 0.50
α = 0.45
HA7
HA13
HA23

Figure 11. Relation between N^+ content in HA copolymers and retention of B- and T-cells at 4°C. (a) T-cells, (b) B-cells. Initial cell concentration: $[L]_0 = (0.96 \pm 0.04) \times 10^7$ cells ml^{-1}. Initial B-cell fraction: $[B]_0 = 34.7 \pm 0.02\%$.

On PHEMA surface, lymphocyte retention decreased with decreasing ionic strength. Retention decreased from 50 to 20% by decreasing NaCl from 8 to $0\,g\,dm^{-3}$. The mode of lymphocyte retention on HA surfaces was in marked contrast with that on PHEMA; increasing retention with decreasing ionic strength. The shielding effect of the positively charged polyamine with the counter ions like Cl^- ions was considered to decrease with decreasing ionic strength. The shielding effect would also influence the negative charge expressed on the plasma membrane of lymphocytes. Decreasing the shielding effect of the surface charge increases attractive force of lymphocytes toward HA surfaces based on electrostatic interaction. Consequently, lymphocyte retention on HA surfaces was increased with decreasing ionic strength. These results are consistent in assuming that lymphocyte-HA interaction is mainly due to electrostatic forces.

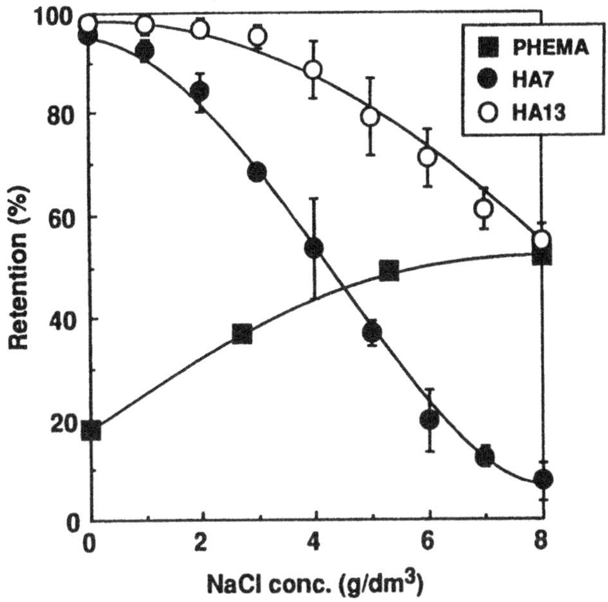
PHEMA
HA7
HA13
100
80
60
40
20
0
Retention (%)
0
2
4
6
8
NaCl conc. (g/dm3)

Figure 12. The effect of ionic strength on lymphocyte retention to PHEMA and HA surfaces at 23°C (pH 7.2). Initial cell concentration: $[L]_0 = (1.00 \pm 0.01) \times 10^7$ cells ml^{-1}.

Further, T-cell retention drastically increased with decreasing ionic strength of the medium. The extent of both B- and T-cell retention reached approximately 95% in the absence of NaCl in the medium, and no selectivity to B- over T-cells was observed. This result demonstrates that the effective resolution of lymphocyte subpopulations by HA was achieved under the physiological ionic strength at pH 7.2. It seems likely that ion exchange process between cationic sites in the polyamine chain and anionic sites in cellular membrane might be important for the resolution of B- and T-cells.

Estimation of interaction sites of lymphocytes with HA copolymer surfaces

From the above results, it can be assumed that the interaction of lymphocyte subpopulations with HA copolymer surfaces is mainly influenced by the electrostatic interaction between the negatively charged plasma membrane surface of lymphocytes and the protonated amino groups in HA copolymers.

It is known that sialic acid residues exist at chain ends of gangliosides and major glycoproteins in plasma membrane surface of cells [34]. Negative charges of the plasma membrane of lymphocytes originate from carboxylate anions of these sialic acid residues. It is plausible that the sialic acid residues are the feasible interaction sites with HA surfaces. To gain insight into this mechanism, sialic acid residues were partially removed from plasma membrane of lymphocytes by neuraminidase treatment, followed by retention studies of the neuraminidase-treated lymphocytes on HA13. As over 2 h treatment of lymphocytes with 0.5 U ml^{-1} of neuraminidase caused a considerable decrease in cell viability (<50%), lymphocytes were treated with 0.5 U ml^{-1} of neuraminidase for 90 min at 37°C. Figure 13 shows the lymphocyte retention on HA13 before and after neuraminidase treatment by changing the

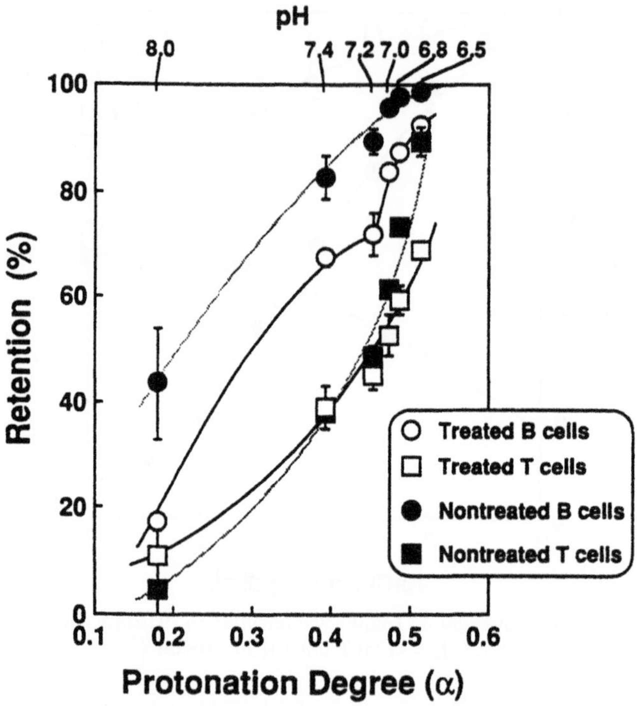
pH
8.0
7.4 7.2 7.0 6.8 6.5
100
80
60
40
20
0
0.1
0.2
0.3
0.4
0.5
0.6
Retention (%)
Protonation Degree (α)
Treated B cells
Treated T cells
Nontreated B cells
Nontreated T cells

Figure 13. Neuraminidase treated lymphocyte retention on HA13 surface at 23°C. Retention of B- and T-cells are shown before and after neuraminidase treatment for 90 min at 37°C. Viability of the neuraminidase treated cells = 86.9 ± 4.0%. Viability of the non-treated cells = 90.5 ± 1.0%. Initial cell concentration: $[L]_0 = (0.95 \pm 0.03) \times 10^7$ cells ml^{-1}. Initial B-cell fraction: $[B]_0 = 42.5 \pm 0.7\%$. Removed sialic acid concentration was 0.24 ± 0.01 $\mu g/10^7$ cells.

surrounding pH. Viability of neuraminidase treated lymphocytes was estimated by dye exclusion method using trypan blue [21], and determined to be approximately 87%, indicating minimal decrease in viability from non-treated lymphocytes (*c.* 90% viability).

Removed sialic acid concentration was determined by thiobarbituric acid assay [25, 26], and evaluated to be approximately 0.24 $\mu g/10^7$ cells. Assuming that the sialic acid residues existing on the plasma membrane of rat lymphocytes to be similar with human lymphocytes [35], approximately 10% of sialic acid residues was removed. As can be seen in Fig. 13, B-cell retention on HA13 decreased after the neuraminidase treatment, while only a small change in T-cell retention was seen. Anderson *et al.* pointed out that the amount of sialic acid residues which exist on the plasma membrane surface of B-cells was 1.6 times greater than that of T-cells [36]. By neuraminidase treatment, sialic acid residues existing on the membrane surface of B-cells were considered to be sufficiently removed at least under this experimental condition, which may result in the decreased retention of B-cells as shown in Fig. 13.

This result indicates that sialic acid residues which exist on plasma membrane would contribute to B-cell retention on protonated HA surfaces. Difference in the retention of B- and T-cells on HA surfaces may originate at least partially from differences in the contribution of sialic acid residues existing on the surfaces of B- and T-cells.

CONCLUSIONS

Rat lymphocyte retention on polyHEMA-*graft*-polyamine (HA) copolymers was evaluated by microsphere column method. Lymphocyte retention was determined to be influenced through the electrostatic interaction between negatively charged cell surface and positively charged amino groups in HA surface. Polyamine grafts of HA copolymer underwent conformational transition due to deprotonation of amino groups as well as intrachain cohesive force at higher pH ($\alpha < 0.5$). This conformational transition of polyamine grafts, accompanied with electrostatic interaction, influences lymphocyte retention, as well as resolution of B- and T-cells. Since the retention of cells on HA copolymer surfaces is mainly based on the physiochemical interaction, but not on biospecific interaction, the results described in this paper suggest that many kinds of cells can be separated by using polymers with an appropriate chemical composition of polymers, as well as operating condition.

Acknowledgement

This work was supported by Grant-in-Aid for Scientific Research (Priority Area Research Program: New Functionality Materials-Design, Preparation, and Control), the Ministry of Education, Science, and Culture, Japan, and Tokyo Ohka Foundation for the Promotion of Science and Technology. We are grateful to Professor Yasuhisa Sakurai, Tokyo Women's Medical College, and Dr. Atsushi Maruyama, Tokyo Institute of Technology, and the late Dr. Yoshikuni Nabeshima, for their valuable collaboration throughout this work. We are grateful to Dr. Harvey A. Jacobs, University of Utah, for his valuable suggestions and to Ms. Mami Karasawa for her assistance in the preparation of this manuscript.

REFERENCES

1. J. Jozefonvicz and M. Jozefowicz, *J. Biomater. Sci., Polymer Edn* **1,** 147 (1990).
2. K. Kataoka, T. Okano, Y. Sakurai, A. Maruyama and T. Tsuruta, in: *Multiphase Biomedical Materials*. p. 1, T. Tsuruta and A. Nakajima (Eds). VSP, Utrecht, The Netherlands (1989).
3. N. Catsimpoolas (Ed.), *Methods of Cell Separation*, Vols 1–3. Plenum Press, New York (1977, 1980).
4. T. G. Pretlow II and T. P. Pretlow (Eds), *Cell Separation—Methods and Selected Applications*, Vols 1–5. Academic Press, New York (1982, 1983, 1984, 1987, 1987).
5. D. S. Kompala and P. Todd (Eds), *Cell Separation Science and Technology*. ACS Symposium Series **464,** ACS, Washington D.C. (1991).
6. K. Kataoka, *CRC Crit. Rev. in Biocompatibility* **4,** 341 (1988).
7. K. Kataoka, in: *Cell Separation Science and Technology*, ACS Symposium Series **464.** p. 159, D. S. Kompala and P. Todd (Eds). ACS, Washington D.C. (1991).
8. K. Katoaka, T. Okano, Y. Sakurai, T. Nishimura, M. Maeda, S. Inoue, T. Watanabe and T. Tsuruta, *Makromol. Chem. Rapid Commun.* **3,** 275 (1982).
9. K. Kataoka, T. Okano, Y. Sakurai, T. Nishimura, S. Inoue, T. Watanabe, A. Maruyama and T. Tsuruta, *Eur. Polym. J.* **19,** 979 (1983).
10. K. Kataoka, Y. Sakurai and T. Tsuruta, *Makromol. Chem. Suppl.* **9,** 53 (1985).
11. A. Maruyama, T. Tsuruta, K. Kataoka and Y. Sakurai, *Makromol. Chem. Rapid Commun.* **8,** 27 (1987).
12. A. Maruyama, T. Tsuruta, K. Kataoka and Y. Sakurai, *J. Biomed. Mater. Res.* **22,** 555 (1988).
13. A. Maruyama, T. Tsuruta, K. Kataoka and Y. Sakurai, *Biomaterials* **10,** 393 (1989).
14. Y. Nabeshima, T. Tsuruta, K. Kataoka and Y. Sakurai, *J. Biomater. Sci. Polymer Edn* **1,** 85 (1989).
15. A. Kikuchi, S. Mizutani, K. Kataoka and T. Tsuruta, *Polym. Adv. Tech.* **2,** 245 (1991).
16. A. Maruyama, E. Senda, T. Tsuruta and K. Kataoka, *Makromol. Chem.* **187,** 1895 (1986).

17. Y. Nitadori and T. Tsuruta, *Makromol. Chem.* **180,** 1877 (1979).
18. Y. Nabeshima, A. Maruyama, T. Tsuruta and K. Kataoka, *Polym. J.* **19,** 593 (1987).
19. K. Kataoka, M. Maeda, T. Nishimura, Y. Nitadori, T. Tsuruta, T. Akaike and Y. Sakurai, *J. Biomed. Mater. Res.* **14,** 817 (1980).
20. A. Maruyama, T. Tsuruta, K. Kataoka and Y. Sakurai, *Biomaterials* **9,** 471 (1988).
21. L. Hudson and F. C. Hay, *Practical Immunology*, p. 30. Blackwell Scientific Publications, Oxford (1976).
22. L. Hudson and F. C. Hay, *Practical Immunology*, p. 26. Blackwell Scientific Publications, Oxford (1976).
23. W. L. Ford, in: *Handbook of Experimental Immunology*, Vol. 2. Ch. 23. D. M. Weir (Ed.). Blackwell Scientific Publications, Oxford (1978).
24. A. Kikuchi, M. Karasawa, T. Tsuruta and K. Kataoka, *J. Colloid Interface Sci.* **158,** 10 (1993).
25. L. Warren, *J. Biol. Chem.* **234,** 1971 (1959).
26. D. Aminoff, *Biochem. J.* **81,** 384 (1961).
27. C. Tanford, *The Hydrophobic Effect.* Wiley, New York (1980).
28. N. Muramatsu, Y. Yoshida, Y. Katayama, H. Ohshima and T. Kondo, *J. Biomater. Sci. Polymer Edn* **2,** 139 (1991).
29. K. Morita, N. Muramatsu, H. Ohshima and T. Kondo, *J. Colloid Interface Sci.* **147,** 457 (1991).
30. K. Kataoka, Y. Sakurai, A. Maruyama and T. Tsuruta, *Proc. ACS Polym. Mater. Sci. Eng.* **53,** 37 (1985).
31. A. Kikuchi, K. Kataoka and T. Tsuruta, *J. Biomater. Sci. Polymer Edn* **3,** 355 (1992).
32. M. A. Winnik, in: *Polymer Surfaces and Interfaces*, p. 1. W. J. Feast and H. S. Munro (Eds). John Wiley & Sons, Ltd., New York (1987).
33. R. L. Hirsch, I. Gray and J. A. Bellanti, *J. Immunol. Methods* **18,** 95 (1977).
34. B. Alberts, D. Bray, J. Lewis, M. Raff, K. Roberts and J. D. Watson (Eds). *Molecular Biology of the Cell*, 2nd Edn, Ch. 6, Garland Publishing, New York (1989).
35. S. Kataoka, T. Kikuchi and T. Toyota, *Tohoku J. Exp. Med.* **145,** 73 (1985).
36. R. E. Anderson, J. C. Standefer and J. V. Scaletti, *Lab. Invest.* **37,** 329 (1977).

In vitro leukocyte adhesion to modified polyurethane surfaces: III. Effect of flow, fluid medium, and platelets on PMN adhesion

A. BRUIL[1], J. I. SHEPPARD[2], J. FEIJEN[1] and I. A. FEUERSTEIN[2]*

[1] *Department of Chemical Technology, Biomaterials Section, University of Twente, P.O. Box 217, 7500 AE Enschede, The Netherlands*

[2] *Departments of Chemical Engineering and Pathology, McMaster University, Hamilton, Ontario, Canada L8S 4L7*

Received 27 July 1992; accepted 4 February 1993

Abstract—The operation of filters used to remove leukocytes from red cell concentrates may depend on the adhesion and mechanical trapping of leukocytes. If adhesion is a major component of filtration then filter materials which augment leukocyte adhesion will be useful. In previous leukocyte adhesion studies, done without flow, poly(ethyleneimene) (PEI) modified polyurethane (PU) films were shown to have greater adhesion when compared with unmodified PU. Since filtration is done under flow conditions, it was decided to study PMN adhesion at a number of flow rates using an established parallel plate flow cell. The influence of divalent cations, plasma and platelets were investigated in the presence of red cells, 40% Hematocrit.

The number of adherent PMNs to the PEI modified films was always substantially higher than that for the unmodified ones when the shear rate was set at $30\,s^{-1}$. When using Tyrode's solution containing albumin, with or without divalent cations, a maximum in PMN adhesion was found between the shear rates of 10 and $100\,s^{-1}$. With Tyrode's solution containing albumin and with 10% (v/v) plasma in saline, the addition of platelets increased PMN adhesion when divalent cations were absent. Adhesion levels with 10% (v/v) plasma in saline were reduced when compared to Tyrode's solution containing albumin without divalent cations. These results support the use of filtration conditions where the concentration of plasma is reduced and the concentration of divalent cations is increased. Detailed evaluation of filter function with flow rate is also recommended. A cell adhesion promoting polymer coating, such as PEI, may be useful in improving filter efficiency.

Key words: Polyurethane; poly(ethyleneimine); polymorphonuclear leukocyte adhesion; platelet; leukocyte filtration.

INTRODUCTION

Leukocyte filtration is applied at blood banks and hospitals to remove leukocytes from blood components. The clinical relevance of this process has been reviewed by several authors [1, 2]. Although the leukocyte removal of currently available leukocyte filters is acceptably high [3], the mechanism of the filtration process is still not clearly understood. This may obstruct the development of improved filter types with respect to leukocyte depletion capacity, red cell recovery and filtration time. In our previous research on the mechanism of leukocyte filtration, in which we used polyurethane (PU) membranes as model filters, we have found that adhesion plays an important role in the depletion of leukocytes by the filter [4]. Similar results have recently been reported for commercially available leukocyte filters composed of cellulose acetate [5] or polyester [6].

* To whom correspondence should be addressed.

Leukocyte adhesion to the filter material may be substantially influenced by the physico-chemical properties of the filter surface. In order to optimize leukocyte adhesion to PU membrane filters, we have previously studied the effect of ionizable groups [7] and wettability [8] on the *in vitro* adhesion of leukocytes, suspended in synthetic buffer, to modified PU surfaces under static conditions. Of all surfaces tested, the number of adherent leukocytes was highest on PU surfaces which were modified with poly(ethyleneimine) (PEI). PEI is a highly branched macromolecule composed of ethyleneimine units, and is generally regarded as a polycation since the primary, secondary and tertiary amine groups in the polymer may be partially protonated at physiological pH [9]. It is assumed that short range interactions, such as hydrogen bonding between the amine groups of the PEI and carboxylic acid groups on the leukocyte membrane account for the enhanced extent of leukocyte adhesion to PEI modified surfaces [7]. The cell adhesive properties of PEI are well known in cell biology, and have been explored in chemotherapy [10], cell diagnostic procedures [11], biotechnology [12], and water purification [13]. Up to now, the use of PEI to improve leukocyte adhesion in leukocyte filters has not been reported. It is likely, however, that modification with PEI offers opportunities for the development of improved leukocyte filters. In this respect it is essential to evaluate the efficacy of the PEI coating when routine blood filtration conditions are applied.

The conditions during blood filtration differ largely from the conditions used in our previously reported experiments. Leukocyte adhesion to PEI modified film surfaces was studied under static conditions using a synthetic medium to suspend the leukocytes [8] whereas leukocyte adhesion to the filter surface during blood filtration occurs under flow conditions with leukocytes suspended in blood. Blood flow during filtration may induce substantial fluid forces in the filter, thus influencing the extent of leukocyte adhesion to the filter surface. The effect of flow rate on cell adhesion has been frequently reported in the literature, and generally an increased number of adherent cells to solid surfaces was found when the flow rate of the fluid cell suspension was increased [14, 15]. Truskey and Pirone reported that the number of preattached fibroblasts to fibronectin coated glass surfaces was exponentially decreased as the shear rate increased [16]. Morley and Feuerstein have studied leukocyte adhesion to fibrinogen coated glass surfaces in the presence of adherent platelets and found a maximum number of adherent leukocytes at an optimal shear rate of $100\,s^{-1}$ [17].

Various components present in the blood may have an effect on the adhesion of leukocytes to solid surfaces. The presence of divalent cations is generally known to enhance the cell adhesion process [18–20]. Plasma proteins in the cell suspension medium will generally reduce the number of adherent cells [19]. Forrester and Lackie have studied these effects on the adhesion of leukocytes to glass surfaces under conditions of flow and found that the presence of Ca^{2+} and Mg^{2+} was essential for adhesion whereas adhesion was substantially inhibited when the cells were suspended in 25% autologous plasma [15]. The presence of platelets in the medium may also be involved in the adhesion of leukocytes. Several authors have reported an increased extent of leukocyte adhesion to adherent platelets [17, 21] probably due to expression of PADGEM (platelet activation dependent granule-external membrane protein, GMP-140) on the surface of activated platelets [22, 23]. Considering the number of possible effects on the adhesion of leukocytes to solid surfaces, the question arises whether the enhancing effect of the PEI coating on leukocyte adhesion *in vitro* can be extended to a system which mimics the conditions applied during blood filtration.

The aim of the present study was to investigate the effects of flow rate and specific components in the medium on the extent of leukocyte adhesion to PEI modified PU films as compared to untreated PU films. The results of this study may be important for both the design of new leukocyte filter materials and the optimization of the filtration process. In order to quantify leukocyte adhesion, we have used purified polymorphonuclear leukocytes (PMNs) which were labeled with mepacrine, a fluorescent dye. The use of this label allows continuous observation of cell-surface phenomena by epifluorescent microscopy [17]. PMNs were resuspended in buffered salt solution containing 40% (v/v) of red cells, to which divalent cations, platelets, and plasma were systematically added. PMN adhesion to film surfaces was studied at three shear rates, using a previously described flow system [17].

MATERIALS AND METHODS

Film surfaces

Polyurethane films (2.5 cm × 6.0 cm, thickness 0.25 mm) were prepared from a solution of Pellethane-2363-80AE (Dow Chemical Nederland BV, Delfzijl, The Netherlands) as described previously [7]. Briefly, the polymer was first purified by adding a 5% (w/v) solution of PU in dimethylformamide to water while stirring at high speed. The precipitate was dried *in vacuo* at 60°C. To prepare films, a 10% (w/v) solution of the purified PU in tetrahydrofuran was poured into large petri dishes, after which the solvent was allowed to slowly evaporate at room temperature. The smooth side of the PU film, originating from contact with the glass surface of the dishes, was used for PMN adhesion studies. The films were washed with analytical grade organic solvents to remove contaminating surface active agents. This washing procedure involved two-fold sonication for 15 min in cyclohexane, followed by two-fold sonication for 15 min in ethanol. The washed films were dried *in vacuo*.

Clean PU films were coated with poly(ethyleneimine) (PEI, M_w 40 000–60 000; Sigma Chemicals Co., St. Louis, MO, USA) by immersion of the films for 1 h in a solution of 1% (w/v) PEI in water. The mechanism for PEI attachment to polymer surfaces has been attributed to electrostatic interactions between the positively charged PEI and the negatively charged substrate surface [12]. Weakly adsorbed PEI was removed by rinsing with distilled water, followed by threefold sonication in a phosphate buffered sodium chloride solution (PBS, pH 7.4) and threefold sonication in water successively (15 min each). The modified films were dried *in vacuo*.

Before use in adhesion experiments, the surfaces of PEI modified and untreated PU films were tested by dynamic water contact angle measurements (Wilhemy plate method) and X-ray spectroscopy (XPS), according to standard surface characterization procedures [7]. We have recently demonstrated, with ^{14}C-labeled PEI, that PEI forms a monolayer, 0.1 mg/m^2, at the surface of PU films and does not desorb from the surface upon exposure to flowing media, including 10% (v/v) plasma in saline [24]. Surface properties of the films are summarized in Table 1. The increase of nitrogen at the surface of PEI modified PU films, as compared to unmodified films, is consistent with the presence of PEI at the surface of modified films.

Table 1.
Surface properties of untreated and PEI modified PU films used in this study

	Surface-elemental composition[a] (at.%)				Contact angle[b] (deg)	
Film sample	C	O	N	Si	$\theta_{adv.}$	$\theta_{rec.}$
PU-untreated	77.8 ± 1.6	19.4 ± 1.6	2.7 ± 0.1	0.1 ± 0.1	91 ± 1	44 ± 3
PU-PEI modified	77.3 ± 2.7	19.4 ± 2.4	3.6 ± 0.3	0.1 ± 0.1	92 ± 1	39 ± 2

[a] Relative surface elemental composition ±S.D. was determined by means of XPS, using four independently prepared samples.
[b] Advancing ($\theta_{adv.}$) and receding ($\theta_{rec.}$) water contact angles ±S.D. were determined by means of the Wilhelmy plate method, using four independently prepared samples.

Cell preparation

Venous blood collected from healthy adult donors was anticoagulated with acid-citrate-dextrose (ACD) at an ACD to blood ratio of 1 : 6 (v/v). Polymorphonuclear leukocytes (PMNs), platelets, plasma, and red blood cells were extracted from the whole blood and purified through various centrifugation and washing protocols. Initial centrifugation of the anticoagulated blood produced a plasma/platelet supernatant (PRP, platelet rich plasma) and a leukocyte/red cell concentrate.

Preparation of PMNs. Polymorphonuclear leukocytes were isolated from the packed red cell concentrate, according to a method previously described [17]. This involved a sedimentation of the packed cell suspension mixed at a 1 : 1 (v/v) with a solution of 6% (w/v) dextran (Dextran D-7265, Sigma Chemical Co.) in Dulbecco's phosphate buffered saline (DPBS), for 1 h at 37°C. The leukocyte rich supernatant was collected and centrifuged and the resultant pellet was resuspended in DPBS containing 0.35% (w/v) albumin (Bovine albumin fraction V, Sigma Chemical Co.). This suspension was layered on ice cold Ficoll (Ficoll-Paque, density 1.077 g/ml, Pharmacia (Canada) Ltd., Dorval, Quebec, Canada) and then centrifuged. The PMN rich pellet was resuspended in DPBS containing 0.1% (w/v) D-glucose and 0.35% (w/v) albumin. The PMNs in this suspension were labeled with mepacrine (Quinacrine dihydrochloride, Sigma Chemical Co.) at a final concentration of 25 μM, for 20 min at room temperature. The cells were washed three more times, by centrifugation/resuspension in DPBS with albumin and glucose. The final resuspension was made in Tyrode's solution with 0.35% (w/v) of bovine albumin added (T.A. solution). Final PMN suspensions had a purity of at least 96%, determined microscopically after staining with May–Grünwald dye.

A control experiment comparing adhesion of unlabeled PMNs and labeled PMNs to PU and PEI-modified PU was performed at 30 s^{-1} to determine the effect of mepacrine labeling on PMN adhesion. There was no significant difference between the number of labeled and unlabeled adherent cells after the 20-min study. The viability of these cells was greater than 99% (Trypan blue dye exclusion). The PMNs were kept at 4°C until use in adhesion experiments, not exceeding 4 h. The viability of the PMN suspension did not change during the course of the adhesion experiments.

Preparation of platelets. A platelet suspension was prepared from the PRP, according to the method described by Mustard *et al.* [25]. After centrifugation, the plasma was removed from the platelet pellet and kept on ice for later use in the adhesion studies. The platelet pellet underwent three resuspension/centrifugation steps in T.A. solution with apyrase to remove residual plasma. The platelets were finally resuspended in T.A. solution with apyrase. These platelet suspensions were responsive to ADP, thrombin and collagen in a platelet aggregation test (Aggregation Module 300B-5, Payton Associates Ltd., Scarborough, Ontario, Canada).

Preparation of red blood cells. Red blood cells, collected after dextran sedimentation, were washed by mixing with Tyrode's solution (1 : 1 v/v) followed by centrifugation. This washing procedure was repeated three times. A final wash was done in T.A. solution with apyrase. After the final centrifugation, the washed red cells were collected and kept at room temperature until needed.

For flow studies, blood cell suspensions were prepared by mixing aliquots of the purified cell suspensions described above. A typical cell suspension contained 5×10^6 PMNs/ml and 40% (v/v) of red blood cells in T.A. solution with apyrase, either with or without Ca^{2+} (2 mM) and Mg^{2+} (1 mM). When platelets were required, a final concentration of 3×10^8 platelets/ml was prepared. In the case of studies with plasma, cells were suspended in 10% (v/v) autologous plasma in saline (0.9% sodium chloride in distilled water). All suspensions were incubated at 37°C, 20 min prior to exposure to the polymer surfaces.

Cell adhesion

A PU or PEI-modified polyurethane film was placed in a flow cell, which is separated from a Perspex block by a 254 μm thick sheet of Silastic having a flow channel of 30 mm × 2.5 mm, as described previously [17]. Approximately 20 cm of polyethylene tubing (inner diameter 1.6 mm) was used to connect the flow cell to the reservoir. Using tracer particles in Tyrode's solution, it was determined that the flow paths in the vicinity of the inlet port were parallel under the experimental conditions. Reynolds numbers for the flow in the channel were always lower than 0.3 in the present study, indicating that the flow was laminar with a short development length.

The flow cells were primed with T.A. solution (either with or without Ca^{2+} and Mg^{2+}) or saline (in plasma studies only) by drawing the solution through the cell at a shear reate of 100 s^{-1} (0.2 ml/min) for 5 min at 37°C, using an infusion/withdrawal pump (Harvard Apparatus Co. Inc., Millis, MA, USA). The adhesion experiment was started by drawing the suspension of cells, held at 37°C, over the film's surface.

The shear rate for these experiments was between 10 and 100 s^{-1} (0.02–0.20 ml/min). This range was chosen to match average operating conditions in the membrane filters of our earlier study [4]. Since filter dimensions, flow rates and pore sizes, are comparable to those of commercially available units, we assume that our applied shear rate range is close to the practical situation in leukocyte filters.

For the studies using 30 and 100 s^{-1} shear rates, the number of PMNs adherent to the film's surface was determined by counting images of the adherent PMNs present on 35 mm negatives. These were prepared with the aid of a photomicrographic unit (Wild MPS46 photoautomat, Leica Canada Inc., Willowdale, Ontario, Canada) attached to a microscope (Laborlux-S, Leica Canada Inc.) having blue-excitation epi-fluorescent optics. Photographs of six different areas, 0.0504 mm^2, in the centre of the flow channel, less than 1 cm from the inlet, were taken at 5, 10, 15, and

20 min after initial blood cell exposure to the surface. A single microscope field was photographed only once, with less than 5 s of exposure to the fluorescent light. Adhesion data for the lower shear rate, $10\,s^{-1}$, was determined manually through visual records of six different areas at the four different times. Accurate numbers of adherent cells could not be produced from photography due to difficulty in distinguishing moving from stationary cells. The six individual numbers of adherent cells at each time interval were averaged and expressed as the mean number of adherent PMNs/mm^2. Each adhesion experiment was repeated at least four times.

Scanning electron microscopy (SEM)

Immediately after use in the adhesion experiment, small pieces of the films were prepared for scanning electron microscopy (SEM). Following the 20-min adhesion study a 5-min rinse with Tyrode's buffer ($30\,s^{-1}$) and subsequently a 5-min rinse in 2% glutaraldehyde in PBS ($30\,s^{-1}$) was performed. The films were then disconnected from the flow cell and further fixed by immersion in 2% glutaraldehyde for at least 10 min. After rinsing with distilled water, the samples were dehydrated by successive immersions of 10 min duration in 25, 50, 75% (v/v) ethanol in water and twice in 98% ethanol. After drying in a vacuum desiccator the samples were sputter-coated with gold and examined by means of a JSM-35 CF scanning electron microscope (Japan Optics Laboratory) using a 15 kV accelerating voltage.

Statistical analysis

Analysis of variance (ANOVA) [26] was carried out to evaluate the significance of differences between PMN adhesion as a function of time with a number of experimental variables using a commercial statistical package (Minitab Release 7, Minitab Inc., State College, PA, USA). A p-value less than 0.05 was required for statistical significance.

RESULTS

The experiments described here deal with PMN and platelet interactions with a PU surface and a surface of poly(ethyleneimine) modified PEI. The results to follow will be in the form of either surface concentrations of adherent PMNs or SEM photomicrographs of adherent platelets and PMNs. The principal variables of this investigation are the two surface types, the suspending medium for the cells, including the presence or absence of Ca^{2+} and Mg^{2+}, and the surface shear rate.

A series of experiments was performed with the two surfaces at three shear rates using Tyrode's solution with albumin (T.A. solution) in the presence or absence of Ca^{2+} and Mg^{2+} (Fig. 1). The PMN adhesion levels are vanishingly small for all of the measurements with PU. The data with PEI were grouped together and an Analysis of Variance (ANOVA) was performed with time, shear rate, and the presence or absence of Ca^{2+} and Mg^{2+} as factors. The analysis revealed that a significant difference in PMN adhesion between the three shear rates tested existed ($p < 0.001$) with a maximum in adhesion occurring between 10 and $100\,s^{-1}$. The number of adherent PMNs at a shear rate of $30\,s^{-1}$ was significantly larger than the number of adherent PMNs at the shear rates of 10 and $100\,s^{-1}$, both in the presence or absence of Ca^{2+} and Mg^{2+} ($p < 0.01$). These data also showed that, at

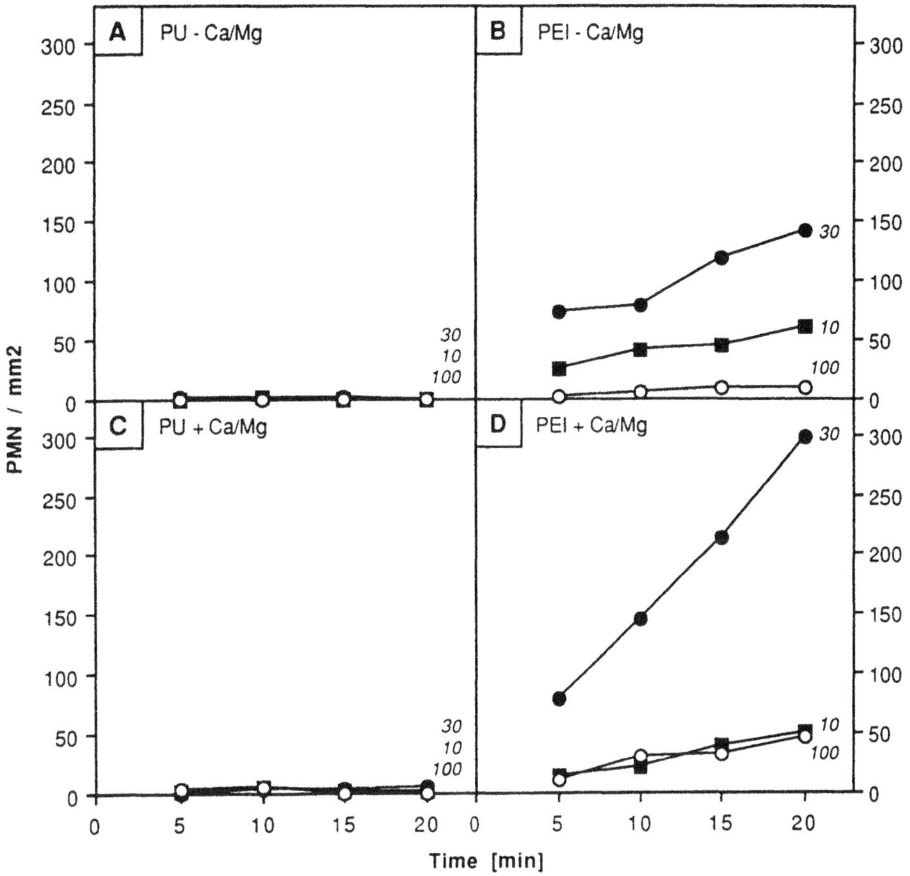
A
PU - Ca/Mg
B
PEI - Ca/Mg
C
PU + Ca/Mg
D
PEI + Ca/Mg
PMN / mm2
Time [min]
30
10
100
0
5
10
15
20
0
50
100
150
200
250
300

Figure 1. Effect of shear rate on the adhesion of PMNs to untreated (left: A and C) and PEI modified (right: B and D) PU surfaces, in the absence (top: A and B) or presence (bottom: C and D) of Ca^{2+} and Mg^{2+} in T.A. solution. PMN adhesion as a function of time at shear rate $10\,s^{-1}$ (■), $30\,s^{-1}$ (●), and $100\,s^{-1}$ (○). The standard deviation (S.D.) associated with the ANOVA for the PEI data is ±43.3.

the shear rates of 30 and $100\,s^{-1}$, the extent of PMN adhesion to PEI is significantly greater when Ca^{2+} and Mg^{2+} were present, as compared to the absence of these ions ($p < 0.001$).

A second set of comparisons between PEI and PU was made as before, except for the addition of platelets to the cell suspension and the restriction to one shear rate, $30\,s^{-1}$ (Fig. 2). Here the numbers of adherent PMNs on PU were also vanishingly small. For measurements with PEI, PMN adhesion was greater in the presence of than in the absence of Ca^{2+} and Mg^{2+}, $p < 0.002$. Using ANOVA, a comparison between the presence of absence of platelets in T.A. solution, with and without Ca^{2+} and Mg^{2+} (data from Figs 1B and 2B and Figs 1D and 2D), indicated that the platelets significantly increased the extent of PMN adhesion to the PEI-surface in the absence of divalent cations, $p < 0.05$, but not in their presence.

In a third set of experiments, PMN adhesion measurements were made in the presence of plasma and platelets in the cell suspending medium (Fig. 3). For this purpose, the cells were suspended in 10% (v/v) autologous plasma in saline. The composition of this fluid medium is comparable to the composition of the

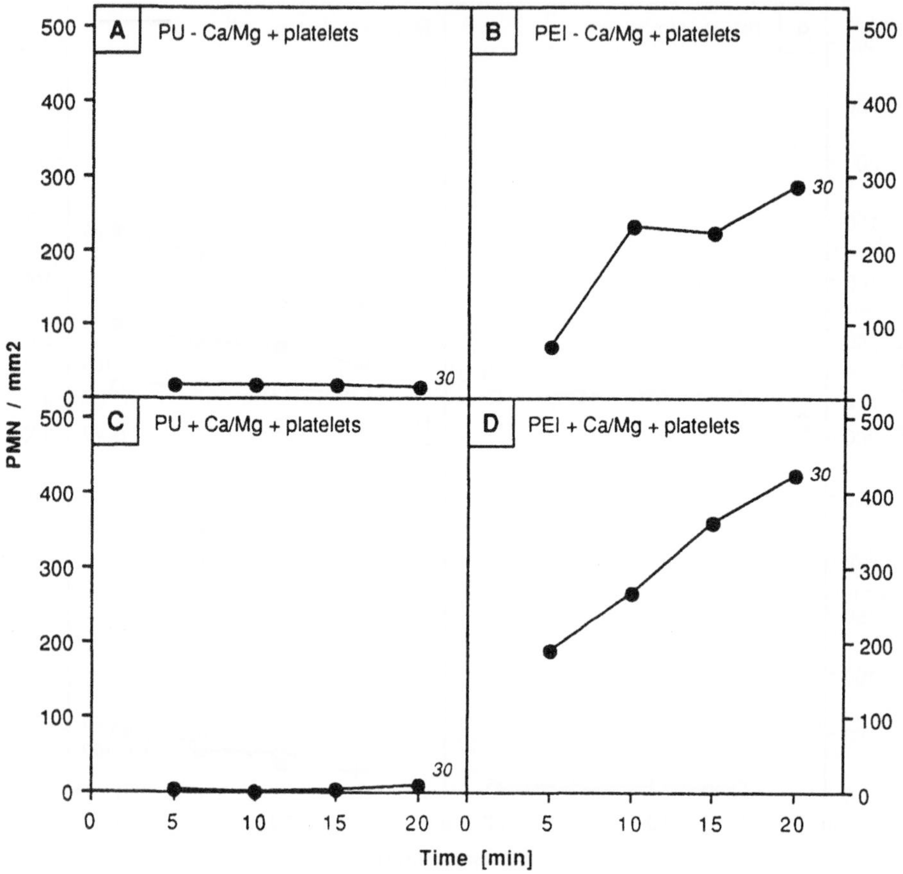
A
PU - Ca/Mg + platelets
B
PEI - Ca/Mg + platelets
C
PU + Ca/Mg + platelets
D
PEI + Ca/Mg + platelets
PMN / mm2
Time [min]
30
0
100
200
300
400
500
5
10
15
20

Figure 2. Effect of the presence of platelets on the adhesion of PMNs to untreated (left: A and C) and PEI modified (right: B and D) PU surfaces, at a shear rate of $30\ s^{-1}$, in the absence (top: A and B) or presence (bottom: C and D) of Ca^{2+} and Mg^{2+} in T.A. solution. The S.D. associated with the ANOVA for the PEI data is ± 68.2.

saline-diluted, packed blood cell concentrate generally used during routine blood filtration at blood banks [6]. Dilution is necessary to reduce the viscosity of the suspension for filtration. Note that the Ca^{2+} and Mg^{2+} concentrations were much lower here than in the above cases where Ca^{2+} and Mg^{2+} were present. Only the amount remaining unsequestered by citrate from the plasma portion of the suspending medium could be present. Again, the PMN adhesion levels for PU were much less than those for PEI. Comparisons between a medium not containing plasma and divalent cations (T.A. solution without Ca^{2+} and Mg^{2+}) and 10% autologous plasma in saline were made. It was found that the presence of plasma led to a significant decrease in the number of adherent cells on PEI at a shear rate of $30\ s^{-1}$ as compared to the cases in which plasma was absent (in the absence of platelets Figs 1B and 3B, $p < 0.01$; and in the presence of platelets Figs 2B and 3D, $p < 0.001$). It could also be shown for PEI at a shear rate of $30\ s^{-1}$ with 10% (v/v) plasma in saline that the presence of platelets in the suspension again, at a low concentration of divalent cations, significantly augmented PMN adhesion ($p < 0.002$).

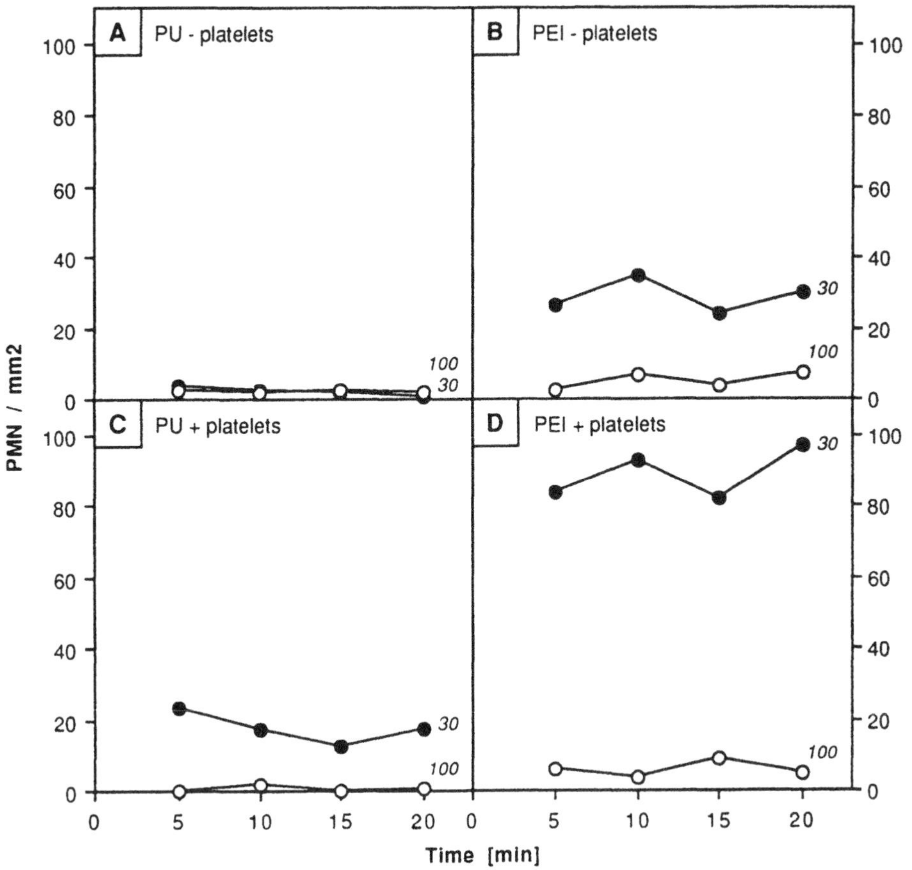
A
PU - platelets
B
PEI - platelets
C
PU + platelets
D
PEI + platelets
PMN / mm2
Time [min]
100
30

Figure 3. Effect of the presence of 10% (v/v) plasma in saline on the adhesion of PMNs to untreated (left: A and C) and PEI-modified (right: B and D) PU surfaces, at a shear rate of 30 s^{-1} (closed circles) and 100 s^{-1} (open circles) in the absence (top) or presence (bottom) of platelets. The S.D. associated with the ANOVA for the combined PEI and PU data is ±26.3.

Photomicrographs from samples of PEI were prepared from experiments with suspensions containing red cells, PMNs and platelets for two suspending media: (1) T.A. solution (with Ca^{2+} and Mg^{2+}) and (2) 10% (v/v) plasma in saline (Fig. 4). In both cases activated platelets were present but the degree of spreading and pseudopod formation was considerably larger for the PEI sample coming from T.A. solution (Fig. 4A). Samples for both cases showed evidence for PMN adhesion to adherent platelets and PMN adhesion to regions where platelets were not present.

DISCUSSION

Although established methods and filters are available for the removal of leukocytes from red cell concentrates, the mechanism for filtration has not been fully elucidated. Knowledge of the mechanism for filtration will be useful for optimizing the filtration process and for the design of new filter materials. The operation of a blood filter will depend potentially on two processes: (1) adhesion of the leukocyte to the filter surface [4–6]; and (2) mechanical trapping of the leukocyte by the filter.

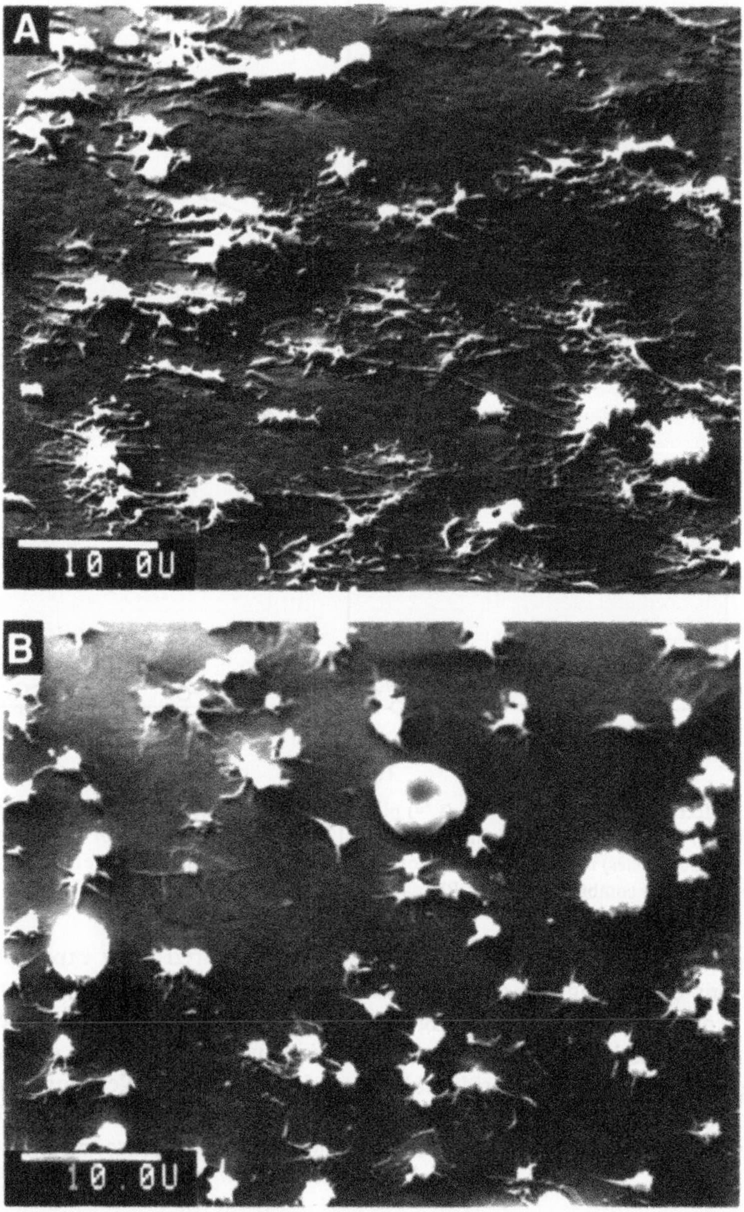
A
10.0U
B
10.0U

Figure 4. Typical SEM micrographs of adherent PMNs and platelets after flow (shear rate $30\,s^{-1}$) for 20 min at 37°C over PEI modified surfaces. Cells were suspended in: (A) T.A. solution containing Ca^{2+} and Mg^{2+} or (B) 10% (v/v) plasma in saline. Original magnification ×2000, stage tilt 60 deg.

There are several factors which lead one to feel that leukocyte adhesion will be an important contributor to blood filter performance and that a study of leukocyte adhesion can be useful in understanding the filtration mechanism. This approach is justified because: (1) there is a large surface to volume ratio in blood filters so that adhesion may be an important event; (2) there are specific receptors on the leukocyte for a number of adhesive plasma proteins e.g., fibrinogen, fibronectin, and von Willebrand Factor, likely to be adsorbed to the filter material [27, 28]; and (3) it is possible to design materials which promote cell adhesion. At the moment there is little information about the absolute role of adhesion in blood filters.

This work has focused on the adhesion of PMNs within a flow cell under well defined conditions of flow and medium which allows us to study PMN adhesion in the absence of mechanical trapping. Assuming that PMN adhesion will be an important feature in PMN removal with filters, we have proceeded with a study which emphasizes two key variables which may influence cell adhesion to our test surface: shear rate and the composition of suspending medium. Platelets can be present in red cell concentrates and may adhere to polymer surfaces. Since PMNs can adhere to platelets in suspension [22] and to platelets adherent to surfaces under conditions of flow [23], we have done studies in the presence and absence of platelets. The protocol for this study has then provided a varied milieu to evaluate filter materials as well as an opportunity to study other variables.

In a previous study, we have found that PEI modified PU surfaces can potentially promote leukocyte adhesion under static conditions [7]. The results of the present study clearly demonstrate, under flow conditions, that PMN adhesion to PEI modified PU surfaces is enhanced when compared to unmodified PU surfaces. Although the presence of divalent cations (Fig. 1), platelets (Fig. 2), or plasma (Fig. 3) in the cell suspension seriously affects the absolute extent of adhesion, the number of adherent PMNs to PEI modified surfaces was always substantially higher when compared to unmodified surfaces. It is not completely understood how this enhanced adhesion to the PEI modified surface is accomplished. We assume that amine groups of the PEI bind to negatively charged groups at the cell surface, such as sialic acid [29]. In addition to specific ionic attractions, other interactions, such as hydrogen bonding, may occur. It is, however, not clear if such a direct interaction between the PEI modified surface and the cells is possible when plasma proteins are present. In this case, it is more likely that cells bind to specific proteins, which are selectively adsorbed to the modified surface. The PEI may substantially change the conformation of adsorbed proteins as compared to untreated surfaces in such a way as to favour adhesion.

Using the PEI modified surfaces it was shown that flow can have a large effect on PMN adhesion. It has been found that there is a maximum in adhesion between 10 and 100 s^{-1} for the PEI surface where cells are suspended in Tyrode's solution with albumin (T.A. solution) either without (Fig. 1B) or with Ca^{2+} and Mg^{2+} (Fig. 1D). Because of the maximum in adhesion there are some conditions of flow where adhesion levels for both surfaces are low and differences between the two surfaces become difficult to assess. Amongst the conditions studied there was a shear rate, 30 s^{-1}, where the PEI surface was consistently more adhesive than the PU surface.

A maximum in PMN adhesion may be attainable with counteracting factors. On the one hand, adhesion of PMNs in the presence of red cells may be *promoted* by increasing shear rate as a result of an increase in the translation and rotation of red cells on PMN transport to the surface [30]. A similar effect of red cells on cell transport has also been demonstrated for platelets [31–33]. This would lead to an increase in the number of cell surface contacts with a consequent increase in the number of successful adhesion events. On the other hand, adhesion, at increased shear rate, can also be *reduced* by the associated increase in shear stress. Others have demonstrated that PMN adhesion, in the absence of red cells decreases with increasing shear stress [34, 35]. Such a reduction could be due to the requirement for more PMN-substrate bonds with the higher shear force on adherent cells [36]. Sufficient bonds for stable cell attachment may not be available because of insufficient time for ligand–receptor or receptor–receptor bonding to occur. Since the effects of these two factors on PMN adhesion will act counter to one another and with different magnitudes, the possibility for a maximum in adhesion with shear rate exists. Such a maximum in PMN adhesion has been previously demonstrated with platelets [37] adherent to subendothelium and with PMNs adherent to spread platelets [17].

It is of interest to note that PMN adhesion may vary as much as five-fold over the shear rate range of 10–100 s^{-1}. For the data with T.A. solution, in the presence of Ca^{2+} and Mg^{2+}, the adhesion levels at 10 and 100 s^{-1} on PEI were less than 20% of those at 30 s^{-1} (Fig. 1D). In a previous study of PMN adhesion to adherent platelets, done over a shear rate range of 10–500 s^{-1} [17] the peak in adhesion, at 100 s^{-1}, was sharp and the tails of the adhesion curve were long. This information, in conjunction with the results of this study, supports the idea that there is a small range of shear rate over which the contribution of PMN adhesion will be optimal in blood filter function. The removal of leukocytes by filtration may be improved considerably by optimization of the flow rate. Since the effect of flow on adhesion is likely to be quite sensitive to operating conditions, each application would need an independent determination for the optimal flow rate.

Divalent cations are known to influence cell adhesion [18–20, 23, 38] and cohesion [39]. The concentration of these ions is normally very low when citrate anti-coagulated red cell concentrates are filtered. To test for the effect of Ca^{2+} and Mg^{2+} on PMN adhesion, we worked with T.A. solution with physiological levels of these ions and without them. Significant comparisons could only be made with the PEI surface since the levels of PMN adhesion on PU were very small in all cases. It was found that there was an effect of the presence of these ions but that adhesion could be supported in their absence. This probably indicates that mechanisms other than those requiring specific interactions between ligands and receptors, which require divalent cations [23, 38], are operating. The addition of divalent cations increased adhesion by about 100% in the absence of platelets (Fig. 1B and 1D) and by about 33% in their presence (Fig. 2B and 2D). Binding facilitated by Ca^{2+} and Mg^{2+}, acting as bridging molecules between negatively charged groups on the cell and the albumin-coated substrate, and between PMN and adherent platelet are possible mechanisms.

Since suspended platelets are commonly found in red cell concentrates with leukocytes, a set of experiments was done in the presence and absence of platelets. It was found that the presence of platelets augmented the adhesion of PMNs to PEI: (1) from T.A. solution in the absence of divalent cations; and (2) from 10% (v/v) plasma, anticoagulated with citrate, in saline having a reduced concentration of divalent cations. This could occur through the adhesion of platelets to the substrate followed by their activation by the substrate. Once activated, the platelets could provide a surface for PMN adhesion possibly through PADGEM, a receptor for PMNs found on the plasma membrane of activated platelets [22]. Alternatively, the platelets may secrete adhesive proteins which can adsorb to the PEI. Thus, fibrinogen, fibronectin, and von Willebrand Factor from platelets [40, 41] may promote the adhesion of PMNs. These proteins have been shown to promote platelet adhesion [42]. Morphological information from scanning electron microscopy, SEM, of adherent PMNs in the presence of platelets showed that both mechanisms were possible.

In the presence of divalent cations, platelets did not augment PMN adhesion in our experiments. This may be due to a high binding efficiency with Ca^{2+} and Mg^{2+} through bridging between surface bound proteins and PMNs. The secretion and adsorption of adhesive proteins from platelets would likely be part of such a situation.

In order to study PMN adhesion in a medium close to actual blood bank filter conditions, we used saline diluted plasma as the fluid phase and worked with and without platelets added to the red cell/PMN suspension. Studies done at the shear rates of 30 and $100\ s^{-1}$ were consistent with the maximum in PMN adhesion with shear rate found with T.A. solution as the fluid phase. At these shear rates, the surface concentration of PMNs with 10% (v/v) plasma in saline was approximately 25% of those with T.A. solution. Others have also presented sizeable reductions in leukocyte adhesion when comparing a buffer with plasma as a suspending medium [15]. This information, together with our results with and without divalent cations, leads us to recommend that means should be found to decrease the concentration of plasma and increase the concentration of divalent cations under filtration conditions.

Using conditions representative of those during routine blood filtration, i.e. in the presence of platelets and plasma in the fluid medium, we have found substantial differences between the extent of PMN adhesion on PEI modified PU surfaces (Fig. 3D) and unmodified PU surfaces (Fig. 3C). At a shear rate of $30\ s^{-1}$, the number of adherent cells on the modified surface is about four times higher than on unmodified surface. Although the present experiments relate to planar film surfaces composed of PU, whereas commercially available leukocyte filters are generally composed of fibers of different materials, there is some evidence that the PEI modification procedure may offer opportunities for the development of improved leukocyte filters. It was shown by Ostrovidova *et al.* that PEI irreversibly binds to poly(ethylene terephthalate) (PET) surfaces [43]. PET is currently used for the preparation of various commercially available leukocyte filters [6]. Specific reports on leukocyte adhesion to PEI modified PET surfaces are not known to us, but in general the adhesion of cells to PEI modifed surfaces is higher as compared to unmodified surfaces [10–13]. It should be noted, however, that the influence of surface roughness [44, 45] and filter porosity [46–48] on leukocyte adhesion have to be further evaluated.

Acknowledgements

Part of the contribution of A.B. and J.F. to this work was financially supported by the Nederlands Produktielaboratorium voor Bloedtransfusieapparatuur en Infusievloeistoffen (NPBI), Emmer-Compascuum, The Netherlands. Financial support for the contributions of J.I.S. and I.A.F. was from the Heart and Stroke

Foundation of Ontario, the Natural Sciences and Engineering Research Council of Canada and the Ontario Centre for Materials Research. The authors wish to thank Mr W. G. McClung for performing the statistical analysis.

REFERENCES

1. A. Brand, in: *Transfusion Medicine in the 1990's*, p. 35, S. T. Nance (Ed). American Association of Blood Banks, Arlington (1990).
2. P. Rebulla, F. Bertolini, A. Parravicini and G. Sirchia, *Transfustion Med. Rev.* **4,** (Suppl.), 19 (1990).
3. R. N. I. Pietersz, I. Steneker, H. W. Reesink, W. J. A. Dekker, E. J. M. Al, J. G. Huisman and J. Biewenga, *Vox Sang.* **62,** 76 (1992).
4. A. Bruil, W. G. van Aken, T. Beugeling, J. Feijen, I. Steneker, J. G. Huisman and H. K. Prins, *J. Biomed. Mater. Res.* **25,** 1459 (1991).
5. I. Steneker and J. Biewenga, *Vox Sang.* **58,** 192 (1990).
6. I. Steneker and J. Biewenga, *Transfusion* **31,** 40 (1991).
7. A. Bruil, J. G. A. Terlingen, T. Beugeling, W. G. van Aken and J. Feijen, *Biomaterials* **13,** 915 (1992).
8. A. Bruil, L. M. Brenneisen, J. G. A. Terlingen, T. Beugeling, W. G. van Aken and J. Feijen, *J. Colloid Interface Sci.* Submitted.
9. D. Horn, in: *Polyamines and Polyquarternary Ammonium Salts*, p. 333, E. J. Goethals (Ed.). Pergamon Press, Oxford (1980).
10. H. Moroson and M. Rotman in: *Polyelectrolytes and Their Applications*, p. 187, A. Rembaum and E. Selegny (Eds). Reidel Publishing Company, Dordrecht, The Netherlands (1976).
11. K. C. Watts, O. A. N. Husain, J. H. Tucker, M. Stark, P. Eason, G. Shippey, D. Rutovitz and G. T. B. Frost, *Anal. Quant. Cytol.* **6,** 272 (1984).
12. S. F. D'Souza and N. Kamath, *Appl. Microbiol. Biotechnol.* **29,** 136 (1988).
13. T. Tashiro, *J. Appl. Polym. Sci.* **43,** 1369 (1991).
14. D. M. Lederman, R. D. Cumming, H. E. Petschek, P. H. Levine and N. I. Krinskey, *Trans. Am. Soc. Artif. Intern. Organs* **24,** 557 (1978).
15. J. V. Forrester and J. M. Lackie, *J. Cell Sci.* **70,** 93 (1984).
16. G. A. Truskey and J. S. Pirone, *J. Biomed. Mater. Res.* **24,** 1333 (1990).
17. D. J. Morley and I. A. Feuerstein, *Thromb. Haemostas.* **62,** 1023 (1989).
18. B. Kvarstein, *Scand. J. Clin. Lab. Invest.* **24,** 41 (1969).
19. F. Grinnell, *Int. Rev. Cytol.* **53,** 65 (1978).
20. R. L. Hoover, R. Folger, W. A. Haering, B. R. Ware and M. J. Karnovsky, *J. Cell. Sci.* **45,** 73 (1980).
21. F. L. Rasp, C. C. Clawson and J. E. Repine, *J. Lab. Clin. Med.* **97,** 812 (1981).
22. E. Larsen, E. Celi, G. E. Gilbert, B. C. Furie, J. K. Erban, R. Bonfanti, D. D. Wagner and B. Furie, *Cell* **59,** 305 (1989).
23. E. L. Yeo, J. Kennington and I. Feuerstein, *Blood* **78** (Suppl. 1), 278 (1991).
24. A. Bruil, H. A. Oosterom, I. Steneker, E. J. M. Al, T. Beugeling, W. G. van Aken and J. Feijen, *J. Biomed. Mater. Res.* Submitted.
25. J. F. Mustard, D. W. Perry, N. G. Arlie and M. A. Packham, *Br. J. Haematol.* **22,** 193 (1972).
26. G. E. P. Box, W. G. Hunter and J. S. Hunter, *Statistics for Experimenters*. John Wiley and Sons, New York (1978).
27. G. F. Burns, L. Cosgrove, T. Triglia, J. A. Beall, A. F. Lopez, J. A. Werkmeister, C. G. Begley, A. P. Haddad, A. J. F. d'Apice, M. A. Vadas and J. C. Cawley, *Cell* **45,** 269 (1986).
28. C. G. Pommier, J. O'Shea, T. Chused, K. Yancey, M. M. Frank, T. Takahashi and E. J. Brown, *J. Exp. Med.* **159,** 137 (1984).
29. P. S. Vassar, J. M. Hards and G. V. F. Seaman, *Biochim. Biophys. Acta* **291,** 07 (1973).
30. H. L. Goldsmith and V. T. Turitto, *Thromb. Haemostas.* **55,** 415 (1986).
31. V. T. Turitto and H. R. Baumgartner, *Microvasc. Res.* **17,** 38 (1979).
32. E. F. Grabowski, L. I. Friedman and E. F. Leonard, *Ind. Eng. Chem. Fundam.* **11,** 224 (1972).
33. I. A. Feuerstein, J. M. Brophy and J. L. Brash, *Trans. Am. Soc. Artif. Int. Organs* **21,** 427 (1975).
34. M. B. Lawrence, L. V. McIntire and S. G. Eskin, *Blood* **70,** 1284 (1987).

35. M. B. Lawrence and T. A. Springer, *Cell* **65**, 859 (1991).
36. D. A. Hammer and D. A. Lauffenburger, *Biophys. J.* **52**, 475 (1987).
37. V. T. Turitto, H. J. Weiss and H. R. Baumgartner, *Microvasc. Res.* **19**, 352 (1980).
38. J.-P. Cazenave, M. A. Packham and J. F. Mustard, *J. Lab. Clin. Med.* **82**, 978 (1973).
39. R. L. Kinlough-Rathbone, J. F. Mustard, M. A. Packham, D. W. Perry, H.-J. Reimers and J.-P. Cazenave, *Thromb. Haemostas.* **37**, 291 (1977).
40. K. L. Kaplan, M. J. Broekman, A. Chernoff, G. L. Lesznik and M. Drillings, *Blood* **53**, 604 (1979).
41. J. C. Giddings, L. R. Brookes, F. Piovella and A. L. Bloom, *Br. J. Haematol.* **52**, 79 (1982).
42. J. N. Mulvihill, J. A. Davies, F. Toti, J.-M. Freyssinet and J.-P. Cazenave, *Thromb. Haemostas.* **62**, 989 (1989).
43. G. U. Ostrovidova and V. B. Kopylov, *Chem. Abst.* **106**, 103193b (1987).
44. P. Predecki, L. Life, P. A. Russell and M. M. Neumann, *J. Biomed. Mater. Res.* **14**, 405 (1980).
45. D. R. Clarke and J. B. Park, *Biomaterials* **2**, 9 (1981).
46. A. Rich and A. K. Harris, *J. Cell Sci.* **50**, 1 (1981).
47. W. Zingg, A. W. Neumann, A. B. Stong, O. S. Hum and D. R. Absolom, *Biomaterials* **2**, 156 (1981).
48. C. A. Ward, A. Koheil, W. R. Johnson and P. N. Madras, *J. Biomed. Mater. Res.* **18**, 255 (1984).

35. M. B. Lawrence and T. A. Springer, Cell 65, 859 (1991).
36. D. A. Hammer and D. A. Lauffenburger, Biophys. J. 52, 475 (1987).
37. V. T. Turitto, H. J. Weiss and H. R. Baumgartner, Microvasc. Res. 19, 352 (1980).
38. J.-P. Cazenave, M. A. Packham and J. F. Mustard, J. Lab. Clin. Med. 82, 978 (1973).
39. R. L. Kinlough-Rathbone, J. F. Mustard, M. A. Packham, D. W. Perry, H.-J. Reimers and J.-P. Cazenave, Thromb. Haemostas. 37, 291 (1977).
40. K. L. Kaplan, M. J. Broekman, A. Chernoff, G. R. Lesznik and M. Drillings, Blood 53, 604 (1979).
41. J. C. Giddings, L. R. Brookes, F. Piovella and A. L. Bloom, Br. J. Haematol. 52, 79 (1982).
42. L. N. Sandifield, J. A. Davies, T. Tait, J-M. Freyssinet and J.-P. Cazenave, Thromb. Haemostas. 61, 95 (1989).
43. G. U. Ostrovskova and V. B. Koptelov, Chem. Abst. 106, 107158 (1987).
44. F. Fredrick, L. Lie, F. A. Russell and M. M. Neumann, J. Biomed. Mater. Res. 14, 403 (1980).
45. D. R. Clarke and J. L. Rich, Biomaterials 2, 9 (1981).
46. A. Rich and A. K. Harris, J. Cell Sci. 50, 1 (1981).
47. W. Zingg, A. W. Neumann, A. B. Strong, O. S. Hum and D. R. Absolom, Biomaterials 2, 156 (1981).
48. C. A. Ward, A. Koheil, W. R. Johnston and D. N. Madras, J. Biomed. Mater. Res. 18, 255 (1984).

Basic fibroblast growth factor production *in vitro* by macrophages exposed to Dacron and polyglactin 910

HOWARD P. GREISLER[1,2,*], SCOTT C. HENDERSON[2] and TINA M. LAM[1]

[1]*Department of Surgery, Loyola University Medical Center, Maywood, IL 60153, USA*
[2]*Hines VA Hospital, Hines, USA*

Received 2 July 1992; accepted 18 September 1992

Abstract—Macrophage activation by implanted bood-contacting biomaterials modulates smooth muscle cell and endothelial cell ingrowth. The present study evaluates the *in vitro* interactions between Dacron or polyglactin 910 with macrophages derived from rabbits fed either normal or atherogenic diets. Peritoneal macrophages were cultured in the presence or absence (negative controls) of either biomaterial for 7 weeks. Conditioned media was evaluated for mitogenic activity using a rabbit aortic smooth muscle cell bioassay with or without preincubation with neutralizing anti-basic-FGF antibody. Results demonstrated increased mitogen release from macrophages harvested from the atherosclerotic rabbits. Only macrophages harvested from normal diet fed rabbits increased their mitogen release following exposure to either polyglactin 910 ($p < 0.05$) or to Dacron ($p < 0.005$) over controls. The stimulation of mitogen release by polyglactin 910 did not significantly exceed that in response to Dacron. In rabbits fed normal diets neutralization with the anti-basic-FGF antibody inhibited 100% of the Dacron induced mitogen release as compared to 36% of the polyglactin 910 induced mitogen release ($p < 0.01$). These results demonstrate significant induced mitogen release from macrophages exposed to biomaterials *in vitro*, much of the smooth muscle cell mitogen represented by basic-FGF.

Key words: Macrophage; Dacron; polyglactin; vascular grafts; basic fibroblast growth factor; smooth muscle cells.

INTRODUCTION

Arterial smooth muscle cell proliferation occurs following most therapeutic interventions including percutaneous transluminal angioplasty and the implantation of both biologic and synthetic vascular graft materials. The extent of the proliferative response to graft implantation is a function of both the biochemical characteristics of the prosthetic material and the hemodynamic and biomechanical characteristics of the anastomotic region and the biomaterial respectively. While the healing of bioresorbable lactide/glycolide copolymeric vascular grafts may be dependent upon an extensive smooth muscle cell and/or fibroblast proliferation, an abundant smooth muscle cell proliferative response to Dacron, ePTFE, and other nonresorbable vascular grafts may significantly contribute to the clinical entity of anastomotic pseudointimal hyperplasia often resulting in graft occlusion. Most if not all prosthetic materials elicit a monocyte/macrophage/giant cell rich foreign body response. Activated macrophages produce a variety of smooth muscle cell mitogens including platelet derived growth factor (PDGF) [1] and basic fibroblast growth factor (bFGF) [2]. We have postulated in the past that macrophages may be differentially activated by specific properties of different biomaterials, resulting in a

* To whom correspondence should be addressed.

differential secretion of both endothelial cell and smooth muscle cell mitogens. We have published in previous work [3–5] that macrophages cultured in the presence of polyglactin 910, a lactide/glycolide bioresorbable material, release into the culture media significantly more mitogenic activity for rabbit aortic smooth muscle cells, mouse BALB/c3T3 embryonic fibroblasts, and murine capillary lung LE-II endothelial cells than do macrophages cultured in the presence of either Dacron or neither biomaterial. Conversely we have previously reported [6] that macrophages cultured in the presence of Dacron secrete into the media more TGF-β_1 than do macrophages cultured in the presence of polyglactin 910 or neither biomaterial.

The current study utilizes a neutralizing anti-bFGF antibody to document the presence of a mitogen with immunoreactivity to this antibody in the media in which macrophages are cultured in the presence or absence of either Dacron or polyglactin 910. The hypothesis of this investigation was that a portion of the mitogenic activity released by macrophages following exposure to either Dacron or polyglactin 910 is caused by the release of basic-FGF into the media and that the different biomaterials differentially induce basic-FGF production by macrophages with which they interact.

METHODS

Macrophage harvest

Peritoneal macrophages were harvested from adult female New Zealand White (NZW) rabbits weighing 3–4 kg. Animals were housed in an American Association for Accreditation of Laboratory Animal Care (AAALAC) accredited facility and fed either standard Purina rabbit chow or Purina rabbit chow formulated specifically to provide a 2% cholesterol, 6% peanut oil diet. Baseline serum cholesterol and triglyceride levels were measured at the time of institution of the diet and again two months later at the time of macrophage harvest.

Peritoneal macrophages were harvested from six rabbits of each group using a peritoneal lavage technique with Hanks Balanced Salt Solution containing calcium and magnesium and 10 U/ml of heparin by our previously published technique [3, 4]. The dialysate (peritoneal lavage fluid returned by gravity drainage) was centrifuged at 500 *g* for 10 min and the cell pellet washed three times with phosphate buffered saline. Cells from the six rabbits per group were combined and suspended in Minimum Essential Medium (MEM) with 10% equine platelet poor plasma derived serum, penicillin/streptomycin (100 U/ml and 100 μg/ml respectively), L-glutamine (100X, 10 μl/ml), pyruvate (100X, 10 μl/ml), and non-essential amino acids (100X, 10 μl/ml). These suspended cells, 1×10^6 cells/ml, were then seeded into three T-25 Falcon (Becton-Dickinson Labware, Lincoln Park, NJ) tissue culture flasks containing 3 ml of MEM. Cell viability and cell counts were determined by Trypan blue dye exclusion and a hemocytometer. Following a 2 h adherence separation, adherent cells were washed with MEM and 5 ml of complete media were added. Cells were cultured at 37°C in a humidified 5% CO_2 atmosphere. Macrophages were then passaged twice resulting in nine T-75 flasks of macrophages of both dietary groups.

Macrophage identification and culture purity were determined by morphologic characteristics under phase contrast microscopy, non-specific esterase staining by the method of Yam *et al.* [7], identification of F_c (immunoglobulin G) receptors on cell membranes by the method of Bianco and Pytowski [8], and by immunoperoxidase staining using the monoclonal anti-Ram 11 antibody to a macrophage specific cytoplasmic antigen, the antibody kindly provided by Dr. Alan Gown of the University of Washington.

Dacron (polyethylene terephthalate) and polyglactin 910 (both grafts provided by Ethicon Inc, Somerville, NJ) woven vascular grafts were finely shredded by scalpel techniques previously published [3–6]. The biomaterials were then each added to one-third of the nine T-75 flasks of both of the two groups of macrophages yielding a concentration of 1.6 mg of biomaterial per cm^2.

Flasks were routinely fed three times weekly and examined for macrophage/biomaterial interactions under phase contrast microscopy.

Collection of conditioned media

Macrophages of both groups remained viable in all three conditions (Dacron, polyglactin 910, and control) at subconfluent levels with relatively infrequent observed mitotic figures and minimal polyploidy. Macrophages of the rabbits fed the atherogenic diet contained more lipid droplets within their cytoplasm. After 5 weeks in culture, intracytoplasmic polyglactin 910 inclusions were seen within the macrophages of both dietary groups (Fig. 1). Cells cultured in the presence of Dacron

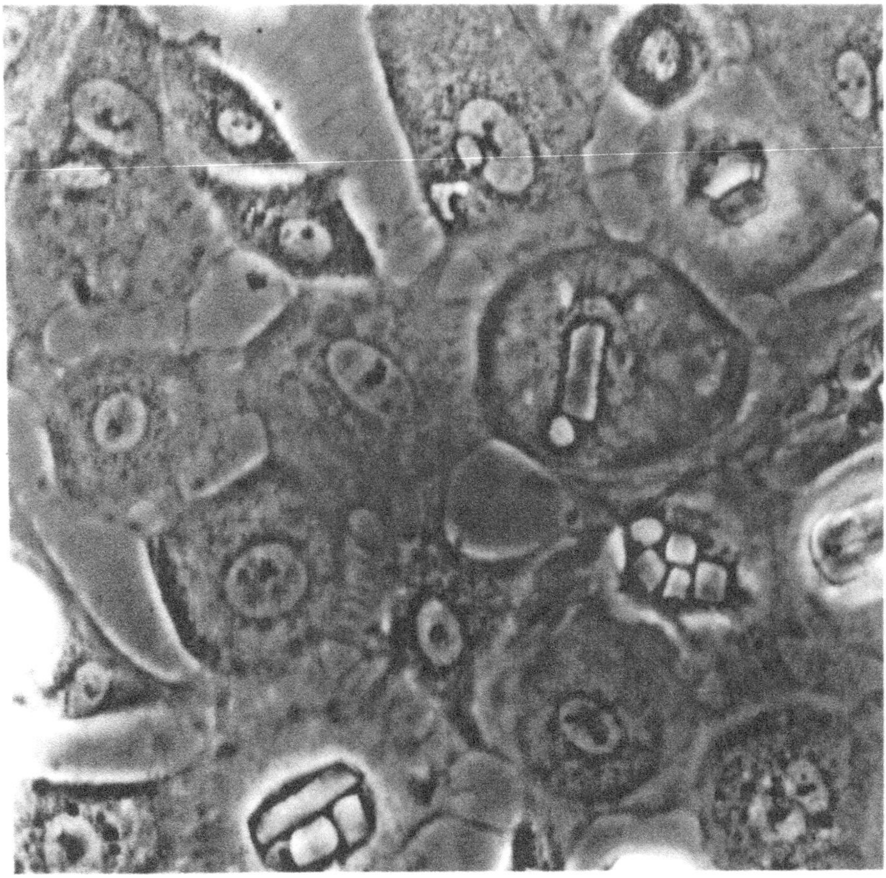

Figure 1. Phase contrast photomicrograph of peritoneal macrophages following six weeks in culture in the presence of the bioresorbable polyglactin 910. Intracytoplasmic polyglactin 910 inclusions can be seen (original magnification ×400).

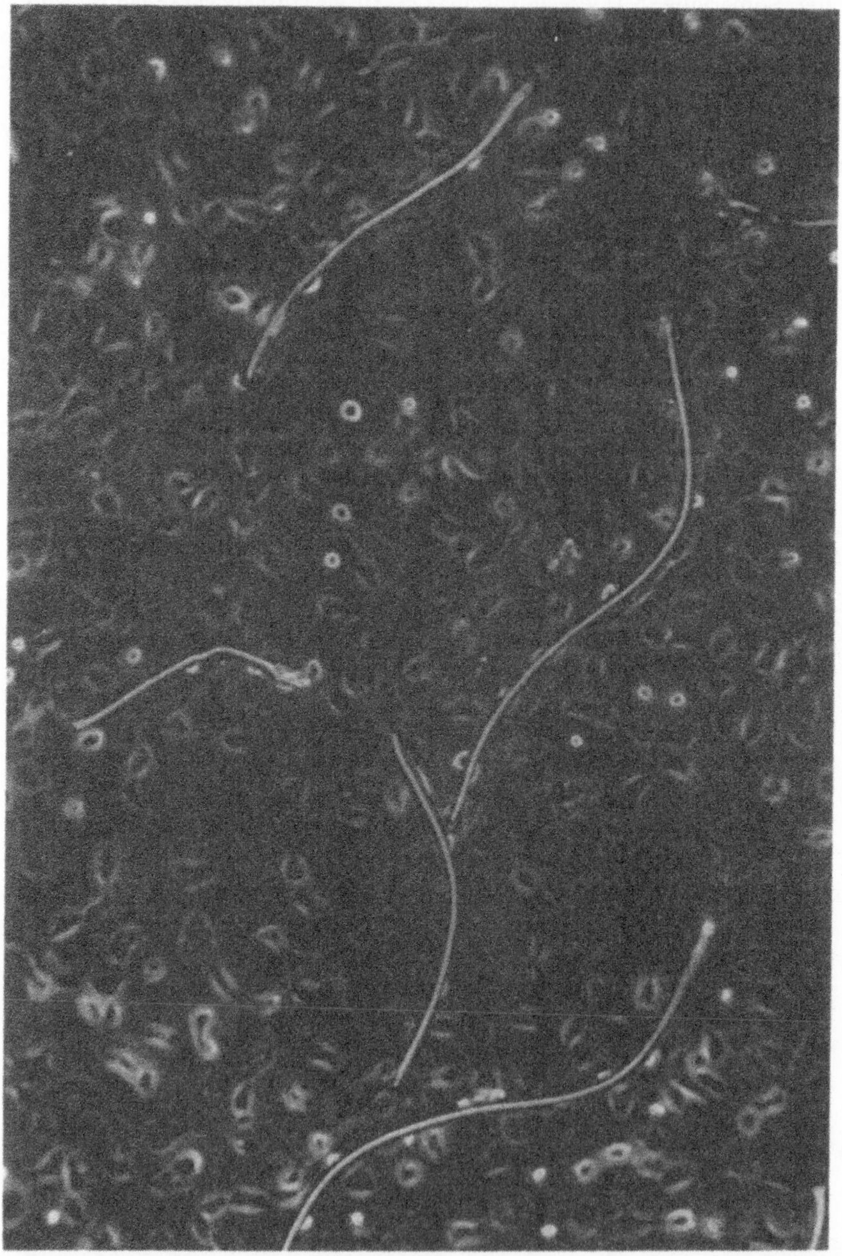

Figure 2. Phase contrast photomicrograph of peritoneal macrophages following five weeks in culture in the presence of Dacron. Macrophages can be seen adherent to Dacron particles but no intracytoplasmic inclusions are observed (original magnification ×100).

adhered to the biomaterial but no phagocytosis was observed (Fig. 2). Conditioned media from each flask was collected at the time of refeeding three times weekly beginning at the time of addition of prosthetic materials. The three media collections per week were pooled and frozen to −70°C resulting in three groups of conditioned media for each of the two groups of harvested macrophages per week in culture. Media were later thawed for analysis of growth promoting activity against test quiescent cells in the presence and absence of a specific neutralizing anti-bFGF antibody.

Rabbit aortic smooth muscle cell harvest

One adult New Zealand White rabbit fed the standard Purina rabbit chow diet was anesthetized with intravenous 2% methohexital and the thoracic and abdominal aorta from aortic arch to bifurcation was resected using sterile technique. The specimen was washed vigorously with Hanks Balanced Salt Solution to remove adherent blood. The aorta was then opened longitudinally and the endothelium removed by gentle passage of a number 10 scalpel blade. The adventitia was removed by careful sharp dissection and the remaining media was minced into 1-mm^3 fragments which were suspended in 1 ml of complete media. The complete media consisted of Dulbecco's Modified Eagles Medium (DMEM—glucose 4.5 g/l in 10% fetal bovine serum [FBS]) supplemented with non-essential amino acids, sodium pyruvate, pencillin/streptomycin, and gentamicin. This suspension was transferred into a T-25 Falcon tissue culture flask and incubated for 4 h at 37°C in humidified 5% CO_2 to allow adherence followed by the addition of 1 ml of complete media. Cells were refed three times/week with 5 ml of complete media.

Aortic smooth muscle cells migrated off the explants after one week and following confluence were subcultured twice. Smooth muscle cell identification was documented by immunofluorescence staining using a smooth muscle specific anti-alpha actin monoclonal antibody (Sigma Chemical Company, St. Louis, MO).

Rabbit aortic smooth muscle cell bioassay

The rabbit aortic smooth muscle cells were trypsinized and seeded into 96 well microtiter plates at a density of 5000 cells/well (in a volume of 200 μl) and were cultured in complete media at 37°C in humidified 5% CO_2 for 3 days until confluent. The media was then aspirated and cells washed and refed with 200 μl serum free media for 48 h. Serum free media consisted of 1 : 1 mixture of DMEM and HAM'S F-10 media supplemented with transferrin (5 μg/ml), penicillin (100 U/ml), streptomycin (100 μg/ml), insulin (1 μM [6 μg/ml]), and ascorbate (0.2 μM [35.2 μg/ml]) [4, 9].

After 48 h in serum free media, 50 μl of phosphate buffered saline (PBS) was added to the negative control wells and 50 μl of fetal bovine serum (FBS) was added to the positive control wells. Fifty μl of either undiluted or 1 : 10 diluted conditioned media was added to the remaining test microtiter wells, each in replicates in four. Tritiated thymidine (^{3}H-TdR, 1 μCi/well) was then added to each well and allowed to incubate for 48 h. Wells were then processed for trichloroacetic acid (TCA) precipitable material for scintillation counting of ^{3}H-TdR incorporation into newly synthesized DNA. Cells were vigorously washed with normal saline three times and then twice fixed with 250 μl of 100% methanol at 4°C for 5 min followed by vigorous washing with distilled water three times. The wells were then incubated twice in 5% TCA for 10 min, again vigorously washed and the acid precipitable material was solubilized with 150 μl of 0.3 N NaOH at room temperature for 20 min. Each 150-μl sample was then transferred to 10 ml of Ready Protein scintillation fluid and counted in a Beckman LS6800 (Beckman Instruments, Inc., Arlington Heights, IL) scintillation counter.

Rabbit aortic smooth muscle cell bioassay utilizing a neutralizing anti-basic-FGF antibody

The presence of a mitogen with immunoreactivity to a neutralizing anti-bFGF antibody was documented by preincubation of conditioned media with a mouse monoclonal anti-bFGF antibody (Upstate Biotechnologies Inc, Lake Placid, NY). The concentration of antibody was determined by dose response studies utilizing the same rabbit aortic smooth cells.

Smooth muscle cells were seeded at 5000 cells/well into 96 well microtiter plates and cultured at 37°C in humidified 5% CO_2 ambient conditions. At confluence (3 days) cells were washed with PBS and refed with 200 μl of serum free media and incubated for 48 h. Wells were then fed either 50 μl of FBS (positive control), PBS (negative control), or serial dilutions of bFGF (Upstate Biotechnologies, Inc., Lake Placid, NY) covering a range from 0.025 ng/well to 100 ng/well, each group in replicates of four. Wells were then processed for ^{3}H-TdR incorporation into TCA precipitable material as described above. Results demonstrated a maximum smooth muscle cell mitogenic effect of this bFGF at 2.5 ng/well.

The neutralizing capacity of the anti-bFGF antibody was then determined. Rabbit aortic smooth muscle cells were seeded 5000 cells/well into 96 well microtiter plates, grown to confluence, and quiesced for 48 h in serum free media. Wells were then given 2.5 ng of basic-FGF plus a range of anti-bFGF antibody concentrations varying from 0.125 μg/ml to 10 μg/ml. Again ^{3}H-TdR incorporation into TCA precipitable material was quantitated and percent neutralization of the maximal stimulating bFGF dose was calculated. From these data the 50% neutralizing concentration (ND 50) of this anti-basic-FGF antibody was calculated to be 0.7175 μg/ml (Fig. 3).

The presence of mitogen with immunoreactivity to this anti-bFGF antibody within the conditioned media groups was documented by the preincubation of 50 μl aliquots of each group of conditioned media with a dose of anti-bFGF antibody corresponding to a five times ND 50 concentration in a 10 μl aliquot for 2 h at 37°C. Concomitant controls of 50 μl of conditioned media plus 10 μl of PBS were similarly incubated. Conditioned media preincubated with the anti-basic-FGF antibody were then added to cultures of rabbit aortic smooth muscle cells prepared as described above and ^{3}H-TdR incorporation into TCA precipitable material similarly quantitated. To optimize the neutralization of bFGF in the conditioned media, the media groups from macrophages of the atherosclerotic rabbits, which had greater bFGF concentrations, were diluted 1:1 with PBS. This dilution of conditioned media was added to the smooth muscle cell cultures both in the presence and absence of the neutralizing antibody. Thymidine incorporation into smooth muscle cells in response to a conditioned media group was then compared with the simultaneously assayed thymidine incorporation into identical smooth muscle cells in response to

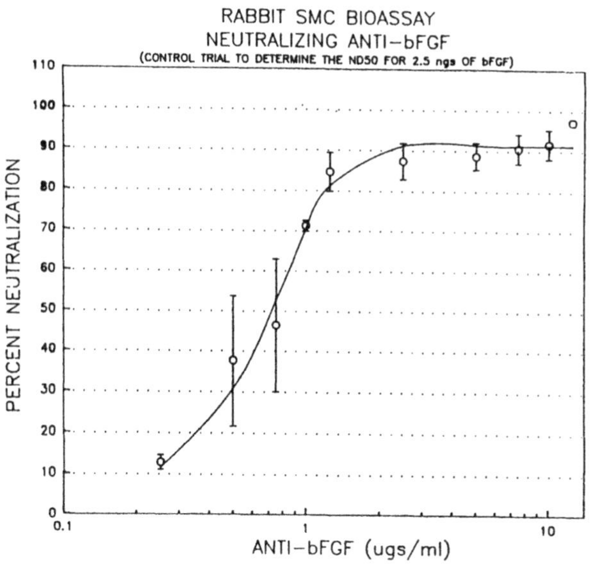
RABBIT SMC BIOASSAY
NEUTRALIZING ANTI-bFGF
(CONTROL TRIAL TO DETERMINE THE ND50 FOR 2.5 ngs OF bFGF)
PERCENT NEUTRALIZATION
110
100
90
80
70
60
50
40
30
20
10
0
0.1
1
10
ANTI-bFGF (ugs/ml)

Figure 3. Anti-bFGF Ab dose response curve showing percent neutralization of 2.5 ng of basic-FGF by serial dilutions of anti-basic-FGF antibody.

that conditioned media group following preincubation with the anti-bFGF antibody and results thus expressed as percent inhibition by immunoprecipitation with the antibody.

Statistical analysis

Mitogenic activity in a conditioned media sample was calculated as percent increase above quiescence which was defined as the tritiated thymidine incorporation into smooth muscle cells in response to the PBS negative control. Percent inhibition of smooth muscle cell proliferation in response to the addition of the neutralizing anti-bFGF antibody was calculated as ^{3}H-TdR incorporation without the antibody minus ^{3}H-TdR incorporation in the presence of the antibody divided by ^{3}H-TdR incorporation in the absence of the antibody. Biomaterial induced mitogen release was calculated as thymidine incorporation due to conditioned media of macrophages cultured in the presence of a biomaterial minus thymidine incorporation due to conditioned media in the absence of any biomaterial (polystyrene control). Percent inhibition of biomaterial induced mitogen release was calculated as thymidine incorporation due to conditioned media of macrophages in the presence of biomaterial but in the absence of antibody minus that corresponding group in the presence of antibody divided by the calculated biomaterial induced mitogen release as described in the previous sentence.

Mean values with standard deviations were calculated over the 7 week periods and corresponding values compared using Student *t*-tests. Statistical significance was defined as $p < 0.05$. Statistical analyses were not performed to compare basic-FGF released from macrophages from normal versus atherosclerotic rabbits because of the 1:1 dilution of the media from the macrophages of the atherosclerotic rabbits.

RESULTS

The 2 months feeding of the atherogenic diet resulted in a significant increase ($p < 0.001$) in serum cholesterol levels with cholesterol and triglyceride levels measuring 2840 ± 339 mg% and 167.5 ± 86.1 mg% respectively compared to 32.3 ± 23.9 mg% and 67.3 ± 29.5 mg% prior to institution of the diet. Under phase contrast microscopy, the macrophages harvested from the atherosclerotic rabbits were seen to contain greater numbers of lipid vacuoles. Examination of the aortas of the atherogenic diet fed rabbits revealed gross atherosclerotic plaques (Fig. 4) primarily in the aortic arch and in the periosteal areas throughout the aortas. Histologically these areas showed extensive intimal thickening with foam cells and cholesterol clefts.

The addition of macrophage conditioned media from all groups of macrophages with or without the exposure to either biomaterial led to an increased ^{3}H-TdR incorporation into smooth muscle cells (Tables 1–6). Greater mitogenic activity was encountered in the groups of macrophages harvested from atherosclerotic rabbits and the observed suboptimal neutralization with the antibody necessitated the 1:1 dilution of these groups of media with PBS as described above. The addition of either biomaterial to the macrophages harvested from normal rabbits resulted in the release of more mitogen into the media, seen particularly in weeks 3 to 5 of culture as compared to the corresponding cultures of macrophages in the absence of biomaterial (Fig. 5). These differences over the entire 7 week period were significant in the polyglactin 910 group ($p < 0.05$) and in the Dacron group ($p < 0.005$). The mitogenic activity within the media of macrophages cultured in the presence of polyglactin 910 did not statistically significantly exceed that in the corresponding group of macrophages cultured in the presence of Dacron. The mitogenic activity in the

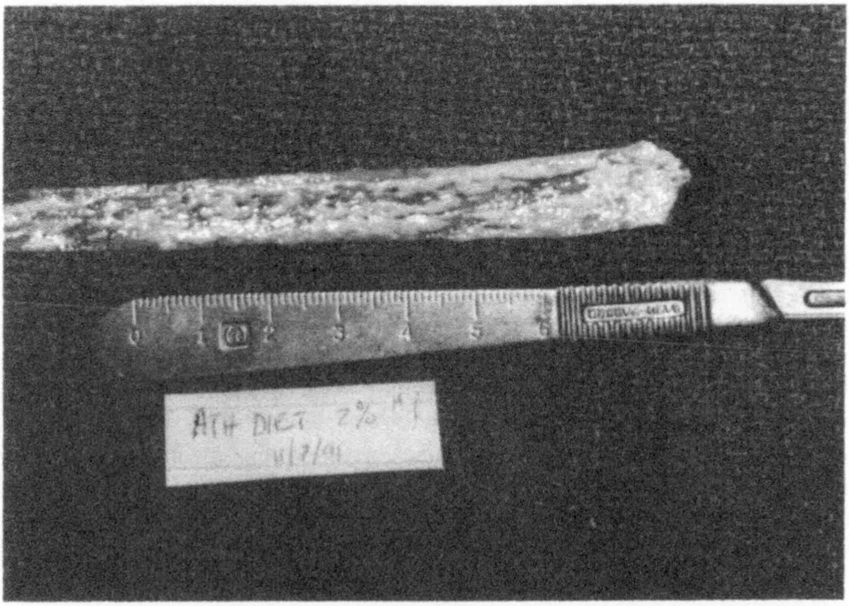
ATH DIET 2%

Figure 4. Gross photograph of the thoracic aorta explanted from a New Zealand White rabbit following 2 months of feeding of the 2% cholesterol/6% peanut oil diet. Gross atherosclerotic plaques are seen throughout the luminal surface.

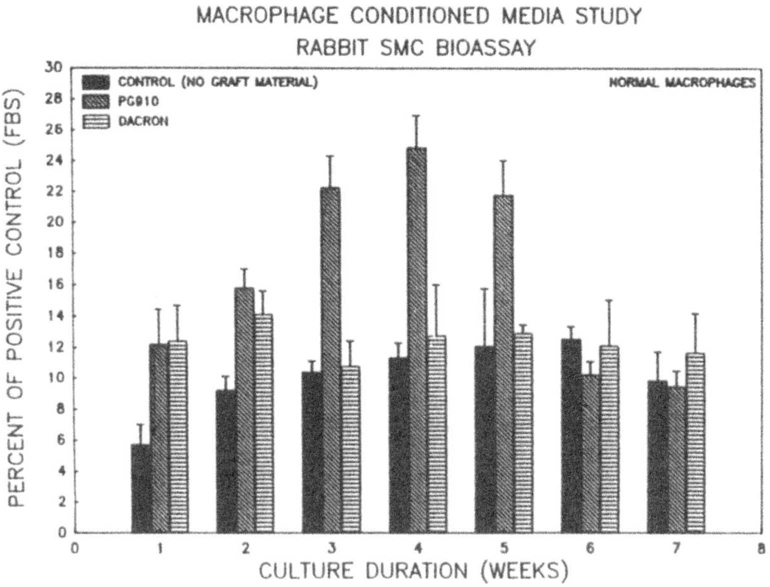
MACROPHAGE CONDITIONED MEDIA STUDY
RABBIT SMC BIOASSAY
CONTROL (NO GRAFT MATERIAL)
PG910
DACRON
NORMAL MACROPHAGES
PERCENT OF POSITIVE CONTROL (FBS)
CULTURE DURATION (WEEKS)

Figure 5. Rabbit aortic smooth muscle cell bioassay results. Conditioned media from macrophages harvested from New Zealand White rabbits fed a normal diet and cultured from one through seven weeks in the presence of either no biomaterial (first bar), polyglactin 910 (second bar), or Dacron (third bar). Results are expressed as counts per minute of the sample divided by counts per minute of the fetal bovine serum positive control $\times 100$ (mean $\pm$ SD).

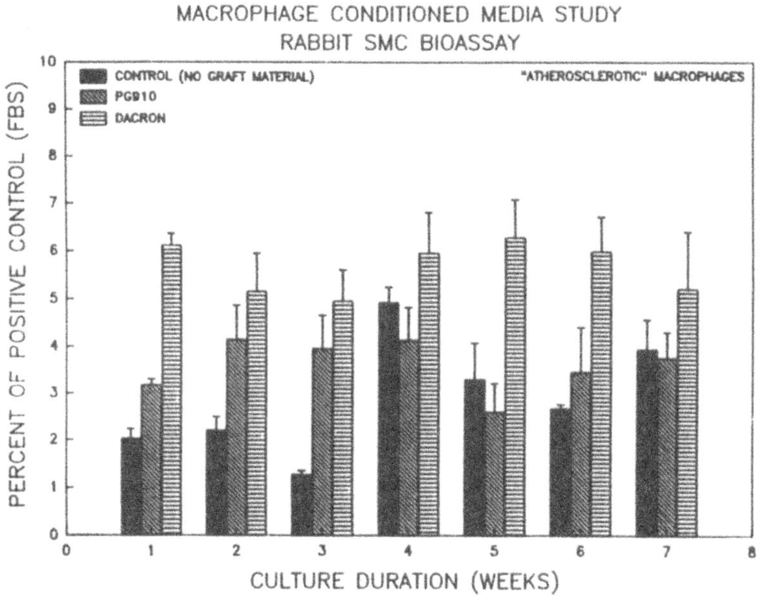
MACROPHAGE CONDITIONED MEDIA STUDY
RABBIT SMC BIOASSAY
CONTROL (NO GRAFT MATERIAL)
PG910
DACRON
"ATHEROSCLEROTIC" MACROPHAGES
PERCENT OF POSITIVE CONTROL (FBS)
CULTURE DURATION (WEEKS)

Figure 6. Rabbit aortic smooth muscle cell bioassay results. Conditioned media from macrophages harvested from New Zealand White rabbits fed a 2% cholesterol/6% peanut oil diet and cultured from 1 through 7 weeks in the presence of either no biomaterial (first bar), polyglactin 910 (second bar), or Dacron (third bar). Results are expressed as counts per minute of the sample divided by counts per minute of the fetal bovine serum positive control $\times 100$ (mean $\pm$ SD).

weekly collections of conditioned media of macrophages not exposed to biomaterials was near constant.

Mitogenic activity in the media of macrophages from atherosclerotic rabbits (Fig. 6), in the absence of biomaterials, was more variable among the weekly collections. The exposure of these macrophages to either biomaterial tended to increase mitogen release but with less consistency and with less predominance of the polyglactin 910 over the Dacron as a stimulus for increased mitogen release.

The pre-incubation of conditioned media groups with neutralizing anti-basic-FGF antibody resulted in a diminution of ^{3}H-TdR incorporation by smooth muscle cells in most groups (Figs. 7 and 8 and Tables 1 and 2). In the absence of exposure to biomaterial however the conditioned media of macrophages of normal diet fed rabbits showed no immunoneutralization of mitogenic activity with the anti-basic-FGF antibody. By contrast the corresponding group of macrophages from atherosclerotic rabbits, not exposed to biomaterial, did demonstrate immunoneutralization

Table 1.
Results—^{3}H-TdR incorporation—rabbit aortic SMCs macrophages of normal diet rabbits

No dilution	No biomaterial % Increase above quiescence ± anti-bFGF Ab		
	A_1 Control	B_1 Control-Ab	A_1–B_1/A_1 % inhibition
Week 1	472.2	484.9	0
Week 2	530.5	471.2	11.2
Week 3	553.4	481.0	13.1
Week 4	832.0	747.3	10.2
Week 5	742.5	668.8	9.9
Week 6	900.4	718.7	20.2
Week 7	698.3	957.2	0
	675 ± 162	647 ± 181	9.2 ± 7.2

Table 2.
Results—^{3}H-TdR incorporation—rabbit aortic SMCs macrophages of normal diet rabbits

No dilution	Polyglactin 910 % increase above quiescent ± anti-bFGF Ab				
	A_2 PG910	B_2 PG910-Ab	A_2–B_2/A_2 % inhibition PG910	A_2–A_1 (C) PG910 induced	A_2–B_2/C % inhibition of PG910 induced
Week 1	906.1	815.5	10.0	433.9	20.9
Week 2	727.7	688.0	5.4	196.8	20.0
Week 3	1402.8	1026.4	26.8	849.4	44.3
Week 4	1681.3	1407.8	16.3	849.3	32.2
Week 5	2181.4	1476.8	32.3	1438.9	49.0
Week 6	679.0	694.7	0	0	...
Week 7	1474.9	1071.6	27.3	776.6	51.9
	1293 ± 593	1026 ± 322	16.9 ± 12.3	649 ± 482	36.4 ± 14.1

Table 3.
Results—^{3}H-TdR incorporation—rabbit aortic SMCs macrophages of normal diet rabbits

	No dilution	Dacron % increase above quiescence ± anti-bFGF Ab			
	A_3 Dacron	B_3 Dacron-Ab	A_3–B_3/A_3 % inhibition Dacron	A_3–A_1 (D) Dacron induced	A_3–B_3/D % inhibition of Dacron induced
Week 1	728.7	609.8	16.3	256.5	46.4
Week 2	761.1	431.9	43.2	230.6	142.8
Week 3	1269.6	593.9	53.2	716.2	94.3
Week 4	1085.7	624.5	42.5	253.7	181.8
Week 5	1129.7	742.3	34.3	387.2	100.1
Week 6	1299.6	882.7	32.1	399.2	104.4
Week 7	1532.8	639.8	58.3	834.5	107.0
	1115 ± 291	646 ± 139	40.0 ± 14.0	440 ± 241	111.0 ± 42.1

Table 4.
Results—^{3}H-TdR incorporation—rabbit aortic SMCs macrophages of atherosclerotic diet rabbits

1:1 dilution	No biomaterial % increase above quiescence ± anti-bFGF Ab		
	A_4 Control	B_4 Control-Ab	A_4–B_4/A_4 % inhibition
Week 1	467.7	242.2	48.2
Week 2	268.4	195.2	27.3
Week 3	576.4	377.7	34.5
Week 4	837.1	641.5	23.4
Week 5	543.8	524.9	3.5
Week 6	617.7	431.2	30.2
Week 7	995.1	602.9	39.4
	615 ± 238	430 ± 171	29.5 ± 14.1

Table 5.
Results—^{3}H-TdR incorporation—rabbit aortic SMCs macrophages of atherosclerotic diet rabbits

	1:1 Dilution	Polyglactin 910 % increase above quiescence ± anti-bFGF Ab			
	A_5 PG910	B_5 PG910-Ab	A_5–B_5/A_5 % inhibition PG910	A_5–A_5 (E) PG910 induced	A_4–B_5/E % inhibition of PG910 induced
Week 1	391.9	321.2	18.1	0	—
Week 2	358.8	297.3	17.1	90.4	68.3
Week 3	1222.0	505.2	58.7	645.6	111.0
Week 4	568.1	310.6	45.3	0	—
Week 5	521.6	434.6	16.7	0	—
Week 6	613.1	624.9	0	0	—
Week 7	1086.2	915.9	15.7	91.1	186.9
	680 ± 338	487 ± 223	24.5 ± 20.2	118 ± 236	122.1 ± 60.1

Table 6.
Results—^{3}H-TdR incorporation—rabbit aortic SMCs macrophages of atherosclerotic diet rabbits

	1:1 Dilution		Dacron % increase above quiescence ± anti-bFGF Ab		
	A_6	B_6	A_6-B_6/A_6	A_6-A_4 (F)	A_6-B_6/F % inhibition
	Dacron	Dacron-Ab	% inhibition Dacron	Dacron induced	of Dacron induced
Week 1	664.8	465.3	30.0	197.1	101.2
Week 2	506.0	366.0	27.7	237.6	58.9
Week 3	906.7	539.7	40.5	330.3	111.1
Week 4	1092.9	982.2	10.1	255.8	43.3
Week 5	1055.1	795.8	24.6	511.3	50.7
Week 6	647.0	741.5	0	29.3	—
Week 7	1173.0	927.0	21.0	177.9	138.3
	863 ± 258	608 ± 236	22.02 ± 13.4	248 ± 148	76.7 ± 38.6

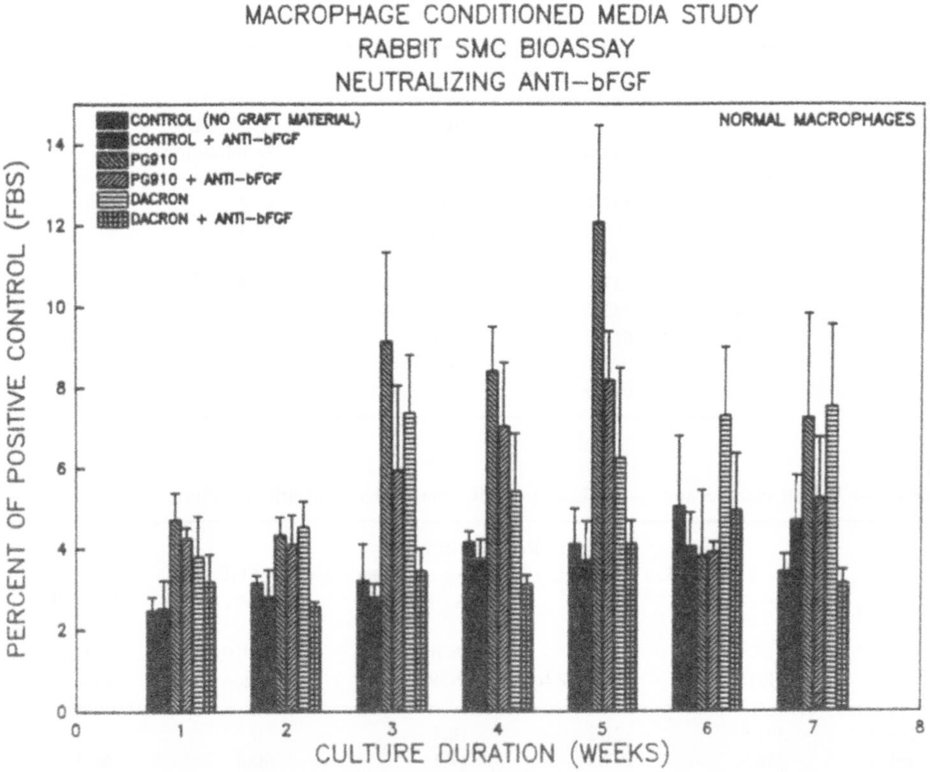
MACROPHAGE CONDITIONED MEDIA STUDY
RABBIT SMC BIOASSAY
NEUTRALIZING ANTI-bFGF
CONTROL (NO GRAFT MATERIAL)
CONTROL + ANTI-bFGF
PG910
PG910 + ANTI-bFGF
DACRON
DACRON + ANTI-bFGF
NORMAL MACROPHAGES
PERCENT OF POSITIVE CONTROL (FBS)
CULTURE DURATION (WEEKS)

Figure 7. New Zealand White rabbit aortic smooth muscle cell bioassay using conditioned media from macrophages harvested from rabbits fed a normal diet and exposed for 1 through 7 weeks to either no biomaterial, polyglactin 910, or Dacron. The solid bars represent the results in the absence of the neutralizing anti-basic-FGF antibody and the patched bars represent the results following the pre-incubation of conditioned media with a concentration of neutralizing anti-basic-FGF antibody equal to 5 times the ND50 (mean ± SD).

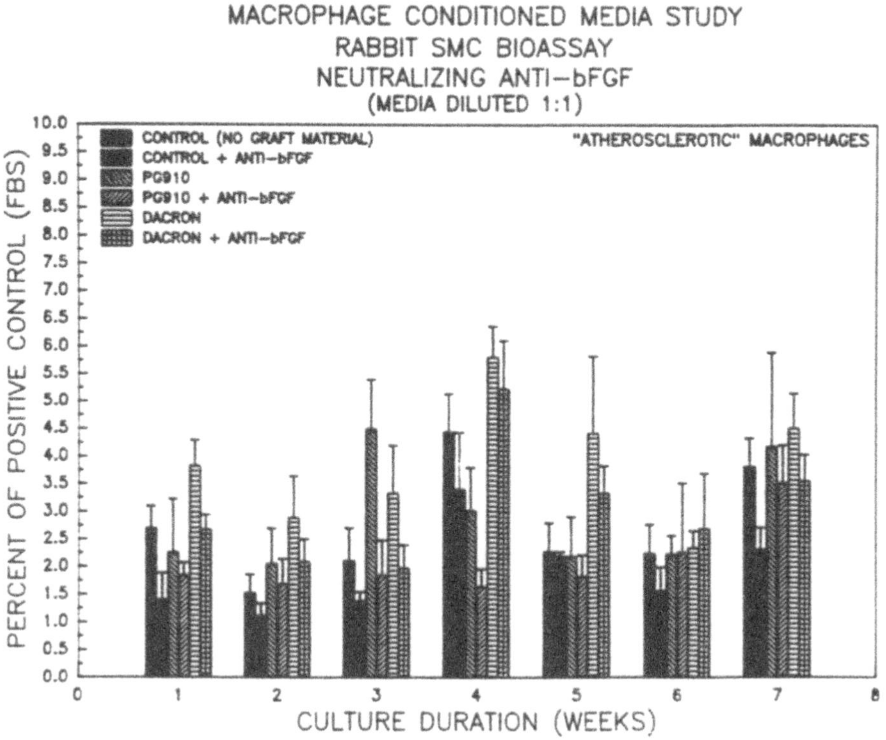
MACROPHAGE CONDITIONED MEDIA STUDY
RABBIT SMC BIOASSAY
NEUTRALIZING ANTI-bFGF
(MEDIA DILUTED 1:1)
CONTROL (NO GRAFT MATERIAL)
CONTROL + ANTI-bFGF
PG910
PG910 + ANTI-bFGF
DACRON
DACRON + ANTI-bFGF
"ATHEROSCLEROTIC" MACROPHAGES
PERCENT OF POSITIVE CONTROL (FBS)
10.0
9.5
9.0
8.5
8.0
7.5
7.0
6.5
6.0
5.5
5.0
4.5
4.0
3.5
3.0
2.5
2.0
1.5
1.0
0.5
0.0
0
1
2
3
4
5
6
7
8
CULTURE DURATION (WEEKS)

Figure 8. New Zealand White rabbit aortic smooth muscle cell bioassay using conditioned media from macrophages harvested from rabbits fed a 2% cholesterol/6% peanut oil diet and exposed for 1 through 7 weeks to either no biomaterial, polyglactin 910, or Dacron. The solid bars represent the results in the absence of the neutralizing anti-basic-FGF antibody and the hatched bars represent the results following the preincubation of conditioned media with a concentration of neutralizing anti-basic-FGF antibody equal to 5 times the ND50 (mean ± SD).

by the anti-basic-FGF antibody. This is consistent from our previous report [4] showing immunoreactivity to basic-FGF antibody on Western blot analysis comparing the media of macrophages of normal diet fed rabbits when exposed to either biomaterial but not in the absence of that biomaterial exposure.

Statistical analysis of the mean values and their standard deviations of each group over the seven week period reveals a significantly greater percent inhibition by the antibody of Dacron induced mitogenic activity (111 ± 42%), as compared to the percent inhibition of polyglactin 910 induced mitogenic activity (36 ± 14%), $p < 0.02$. Similarly the percent inhibition by the antibody of the total mitogenic activity in the conditioned media of these macrophages exposed to Dacron (40 ± 14%) was significantly greater than the percent inhibition in the media of macrophages exposed to polyglactin 910 (17 ± 12%, $p < 0.005$) and similarly greater than the percent inhibition of total mitogenic activity in the media of macrophages exposed to neither biomaterial (9 ± 7%, $p < 0.002$).

Among the macrophages harvested from atherosclerotic rabbits, such statistical significance was not reached. This is a reflection of the greater percent inhibition by the antibody of the control groups of macrophages, those not exposed to either biomaterial (29.5 ± 14% inhibition). The percent inhibition by the antibody of mitogenic activity released by this group of macrophages when exposed to either Dacron or polyglactin 910 did not significantly differ from each other or from the polystyrene control macrophage group.

DISCUSSION

The implantation of small diameter vascular grafts constructed of Dacron and of ePTFE into atherosclerotic patients frequently fails due to the development of anastomotic pseudointimal hyperplasia. This lesion is characterized by smooth muscle cell and fibroblast proliferation and collagen deposition in the juxta-anastomotic zone. The regulation of this proliferative activity is certainly multifactorial involving the modulation of growth factor release by all of the cells present in the local microenvironment. The regulation of this growth factor secretion by endothelial cells, smooth muscle cells, fibroblasts, macrophages, leukocytes, and platelets is a function of both the chemical composition of the prosthetic material and the hemodynamic and biomechanical characteristics of the anastomotic region. Previous reports from our laboratory have demonstrated that the implantation of bioresorbable lactide/glycolide copolymeric vascular prostheses results in a significantly more extensive ingrowth of myofibroblasts into the inner capsules as compared to Dacron graft implantation into the same rabbit and canine models. This ingrowth of myofibroblasts is a function of both cell migration and cell proliferation with a significantly greater rate of myofibroblast proliferation seen in response to the bioresorbable polymers [10]. Following the implantation of these bioresorbable polymers, monocyte derived macrophages phagocytize the material and shortly thereafter the migration and proliferation of myofibroblasts occurs. Similarly, monocyte derived macrophages are seen following Dacron implantation although in the latter situation the macrophages adhere to and surround the biomaterial without observable phagocytosis occurring.

The current study demonstrates that rabbit peritoneal macrophages exposed to either Dacron or to polyglactin 910 *in vitro* release mitogenic activity into the media which stimulates DNA synthesis in rabbit aortic smooth muscle cells. Included amongst the mitogens released is bFGF. Interestingly there appears to be no bFGF released from macrophages harvested from normal diet fed rabbits without exposure of these macrophages to biomaterials. By contrast macrophages harvested from rabbits fed atherogenic diets release into the media a mitogen which immunoprecipitates with the neutralizing anti-bFGF antibody, even in the absence of biomaterial exposure. When the macrophages of either group are exposed to Dacron or to polyglactin 910 they release greater amounts of smooth muscle cell mitogen into the media. Although macrophages harvested from rabbits fed a normal diet are stimulated to release more mitogen when exposed to polyglactin 910 as compared to Dacron, the proportion of this mitogen represented by bFGF is significantly greater following Dacron exposure as compared to polyglactin 910 exposure. In addition, the feeding of the atherogenic diet to the rabbits resulted in their macrophages releasing more of the smooth muscle cell mitogenic activity into the media as compared to macrophages harvested from rabbits fed a normal diet. It is thus tempting to speculate that atherosclerotic patients in whom Dacron grafts are implanted may have a significantly increased propensity toward macrophage bFGF release, this growth factor perhaps playing a role in the stimulation of the proliferative lesion seen in the region of the anastomosis.

While differences in the reactions to the two biomaterial types may reflect a difference in their chemical composition, surface texture and shape may similarly play a role. These differences will be investigated in future studies. One cannot conclude necessarily that the results encountered are a reflection of the issue of resorbability of the biomaterial. However, an understanding of macrophage/ interactions with any biomaterial is important to the potential utility of that biomaterial for purposes of development of new bioprostheses.

The 7-week study period was selected because macrophage phagocytosis of polyglactin 910 occurs *in vitro* between weeks four and seven as evidence by intracytoplasmic inclusions of the material.

The possibility exists that results may in part reflect the presence of small amounts of growth factors in the serum containing media. However, this effect should be negated by expressing results in relation to the control no biomaterial groups in which sera were also present.

Basic-FGF, also known as FGF-2 or heparin binding growth factor-2 is one of a family of seven recognized chemically related heparin binding growth factors. Basic-FGF has been implicated in regulation of cell proliferation and differentiation in a large spectrum of cells derived from all three embryonic germ layers [10]. FGF has been immunolocalized to the basement membrane in the arterial wall [11] where it is apparently inactive in the absence of arterial wall injury. Basic-FGF is produced by macrophages, smooth muscle cells, and endothelial cells among others. In the case of polyglactin 910 implantation, it is tempting to speculate that polyglactin 910 activation of macrophages results in their release of smooth muscle cell mitogens among which is included bFGF which results in the increased mitotic index of the ingrowing myofibroblasts previously reported. In the situation of Dacron, the Dacron may activate the macrophage to release bFGF which may modulate the process of anastomotic pseudointimal hyperplasia. However, the localization of this proliferative activity to the peri-anastomotic regions and not to the central portions of the implanted prosthesis suggests that other factors contribute to either further stimulate cell growth in the area of the anastomosis or to block much of the bFGF affect on cell proliferation in the regions of the graft removed from the anastomotic areas.

Basic-FGF has not been found to contain a signal peptide sequence. It is thus widely believed that the release of this growth factor from cells is dependent upon injury or death of the cell. In the current studies, no evidence of cell death over the period of culture was seen. No differences were encountered in cell number/flask among the groups at the conclusion of the seven weeks of culture. Similarly no differences among the groups were seen in pH or desquamation. The current studies may reflect an increased synthesis of bFGF and their release by a cell toxicity not resulting in cell death or by a previously unrecognized mechanism for bFGF secretion. Additionally the current studies do not discriminate between release of stored pools of bFGF and up regulation of gene transcription following exposure of the cells to the biomaterials. These studies are currently being repeated with the addition of RNA analysis to define the transcriptional versus translational regulation of bFGF release following biomaterial exposure of macrophages.

Recent studies by Lindner *et al.* [12, 13] have demonstrated that following rat carotid artery balloon de-endothelialization injury, the intravenous infusion of an anti-bFGF antibody results in a significant decrease in the first wave of smooth muscle cell proliferation, i.e. smooth muscle cell proliferation with the media prior to migration of the cells into the intima, and conversely the infusion of a recombinant bFGF resulted in significantly increased smooth muscle cell proliferation. The current study suggests a possible role for bFGF in similarly regulating pseudo-intimal cell proliferation following the implantation of vascular grafts, which represent another form of arterial injury.

Acknowledgement

This work was supported by a grant from the National Institutes of Health, RO1 HL41272.

REFERENCES

1. K. Shimokado, E. W. Raines, D. K. Madtes, T. B. Barrett, E. P. Benditt and R. Ross, *Cell* **43**, 277 (1985).
2. A. Baird, P. Mormede and P. Bohlen, *Biochem. Biophys. Res. Commun.* **126**, 358 (1985).
3. H. P. Greisler, J. W. Dennis, E. D. Endean, J. Ellinger, R. Friesel and W. H. Burgess, *J. Vasc. Surg.* **4**, 588–593 (1989).
4. H. P. Greisler, J. Ellinger, S. C. Henderson, A. M. Shaheen, W. H. Burgess, D. U. Kim and T. M. Lam, *J. Vasc. Surg.* **14**, 10–23 (1991).
5. T. M. Lam, N. E. Whereat, S. C. Henderson, W. H. Burgess, A. Shaheen and H. P. Greisler, in: *Angiogenesis Key Principles—Science—Technology—Medicine*, pp. 346–356, R. Steiner, P. B. Weisz and R. Langer (Eds). Birkhauser Verlag, Basel (1992).
6. D. Petsikas, D. J. Cziperle, T. M. Lam, P. M. Murchan, S. C. Henderson and H. P. Greisler, *Surgical Forum* **42**, 326–328 (1991).
7. L. T. Yam, C. Y. Li and W. H. Crosby, *Am. J. Clinical Pathol.* **55**, 283–290 (1971).
8. C. Bianco and B. Pytowski, in: *Methods for Studying Mononuclear Phagocytes*, pp. 273–280, D. O. Adams, P. J. Edelson and H. S. Koren (Eds). Academic Press, New York (1981).
9. P. Libby and K. V. O'Brien, *J. Cell Physiol.* **115**, 217–223 (1983).
10. D. Gospodarowicz, in: *The Fibroblast Growth Factor Family*, Vol. 638, pp. 1–8, A. Baird and M. Klagsbrun (Eds). Ann. NY Acad. Sci. (1991).
11. A. M. Gonzalez, M. Buscaglia, M. Ong and A. Baird, *J. Cell Biol.* **109**, 203–215 (1990).
12. V. Lindner, D. A. Lappi, A. Baird, R. A. Majack and M. A. Reidy, *Cir. Res.* **68**, 106–113 (1991).
13. V. Lindner and M. A. Reidy, *NAS* **88**, 3739–3743 (1991).

Part V

Immobilized Cell Receptor Ligands and Immobilized Cells

Fibroblast attachment to Arg–Gly–Asp peptide-immobilized poly(γ-methyl L-glutamate)

K. KUGO[1]*, M. OKUNO[1], K. MASUDA[1], J. NISHINO[1], H. MASUDA[1] and M. IWATSUKI[2]

[1]*Department of Applied Chemistry, Konan University, Okamoto 8-9-1, Higashinada-ku, Kobe 658, Japan*

[2]*Ajinomoto Co. Inc., Planning & Development Department, Chuo-ku, Kyobashi 1-5-8, Tokyo 104, Japan*

Received 30 July 1992; accepted 28 January 1993

Abstract—The attachment of MRC-5 human fibroblasts was investigated on poly(γ-methyl L-glutamate) (PMLG), and upon cell adhesion peptides Arg–Gly–Asp–Ser (RGDS)- and Gly–Arg–Gly–Asp–Ser (GRGDS)-immobilized PMLG (RGDS-PMLG and GRGDS-PMLG). The peptides were immobilized by their N-terminal amine to activated PMLG surfaces. Prior to peptide immobilization, the aminolysis of PMLG surfaces was performed with hydrazine hydrate (HA), ethylenediamine (EDA), and hexamethylenediamine (HMDA) and was followed by the activation with hexamethylene diisocyanate. Surface characterization of these films was carried out by means of a Fourier transform IR (FT-IR) spectrometer equipped with an attenuated total reflectance (ATR) attachment. The amount of immobilized RGDS could be controlled by the reaction time of the aminolysis. The effects of HA, EDA, and HMDA as a spacer on the cell attachment were also investigated, and it was suggested that a longer spacer promoted the cell attachment via specific receptor-ligand interaction.

Key words: Cell attachment; Arg–Gly–Asp (RGD) peptide; immobilization; poly(γ-methyl L-glutamate).

INTRODUCTION

It has been proved that the peptide sequence Arg–Gly–Asp (RGD) in cell adhesion proteins such as fibronectin and vitronectin is crucial for the attachment of many cells to these proteins [1, 2]. Fibronectin is the best characterized multifunctional glycoprotein mediating cell–extracellular matrix interactions, and plays an important role in cell adhesion [3, 4]. Recent research on vitronectin in blood plasma revealed that it exists in two forms of molecular weight 75 000 and 65 000 Da, and plays a major role in the cell adhesion and spreading activity of serum [5–7]. It was also pointed out that for an oxygen-containing fluoropolymer film, the attachment and spreading of human vein endothelial cells in serum-containing medium was very dependent upon the vitronectin content of the serum but was substantially independent of the fibronectin content of the culture medium [8]. For nitrogen-containing films, however, the attachment of human vein endothelial cells was not simply dependent upon either the vitronectin content of the serum, or the fibronectin content [8]. The only homology between the two glycoproteins is the RGD sequence, as described above.

RGDS bound to albumin adsorbed to a substrate was found to promote fibroblast adhesion [9]. Furthermore, an RGD sequence peptide covalently immobilized to the

* To whom correspondence should be addressed.

surface of polyacrylamide gel [10] and glycophase glass [11] has been found to support the adhesion of fibroblasts. Also, synthetic elastomeric polypeptide matrices containing the covalently incorporated RGDS sequence were ascertained to support the attachment and growth of fibroblasts [12]. We have recently investigated the effect of the conformational transition [13] of polypeptide such as poly(γ-methyl L-glutamate) (PMLG) surfaces on cell attachment and cell growth properties. It is considered that polypeptides may mimic native cell surfaces to a certain extent since amino acids are components of proteins. Moreover, a wide range of surface properties of polypeptides can be provided by various chemical reactions, such as transformation of the hydrophobic moiety into the hydrophilic one by saponification. It is also known that some polypeptides and proteins exist in various secondary structures. Consequently, it should be of interest from functional and morphological points of view to investigate to what extent functional groups and conformations of polypeptide surfaces exert an influence on cell–substrate interactions such as cell attachment. PMLG is one of the synthetic polypeptides which are readily available and are inexpensive. As experimental results on FT-IR attenuated total reflectance spectra measurements, it becomes clear that the α-helical form in the surface layer transformed readily into the β-structure by the formic acid treatment, and the β-structure content increased with increasing time of the formic acid treatment and/or approaching the outermost surface [13]. It was also proved that cells on PMLG with a higher β-structure content were spread and flattened more than those on untreated PMLG existing in α-helical form [14]. The increase in cell spreading depending upon the secondary structure of PMLG should be due to non-biospecific interaction. Compared with this nonspecific effect, the biospecific interaction by immobilizing RGD-peptide to PMLG is also interesting, because it should improve the cell attachment, as mentioned above.

In this study, we accordingly examined the fibroblast attachment activity toward Arg–Gly–Asp–Ser (RGDS)- and Gly–Arg–Gly–Asp–Ser (GRGDS)-immobilized PMLG (hereafter designated simply as RGDS-PMLG and GRGDS-PMLG, respectively). The aminolysis of PMLG surfaces was first performed with hydrazine hydrate (HA), ethylenediamine (EDA), and hexamethylenediamine (HMDA) and was followed by the activation with hexamethylene diisocyanate. The RGDS- and GRGDS-peptides were immobilized by their N-terminal amine to these activated PMLG surfaces. The effects of HA, EDA, and HMDA as a spacer on the cell attachment were also investigated.

MATERIALS AND METHODS

Materials

PMLG, which is AJICOAT A-2000 (10% w/v in ethylene dichloride-tetrachloroethylene (70/30 v/v) mixture, $[\eta] = 1.00$ in dichloroacetic acid), was supplied by Ajinomoto Co. Inc. and purified by a precipitation into *c.* 5 times cold methanol. The PMLG films of *c.* 50 μm in thickness were cast onto petri dishes (27.6 mm diameter) from chloroform-2,2,2-trifluoroethanol (95/5 v/v) solution at 25°C and relative humidity less than 60% using a Tabai Model LHL-111 Humidity Cabinet.

The RGDS and GRGDS peptides used in this study were supplied by Peptide Institute Inc. According to the analytical data, these peptides were synthesized by conventional solution procedures, and purified to show a single TLC spot in two different solvent systems upon application of 50 μg. Impurities detectable by ordinary gradient HPLC systems are less than 1%. These peptides were dissolved in calcium, magnesium-free phosphate-buffered saline (PBS (–), pH 7.4) each time just before use.

Immobilization of RGD-peptides

The aminolysis of PMLG film surfaces was first performed at 25°C with 1 ml hydrazine hydrate (HA), ethylenediamine (EDA), and hexamethylenediamine-water (5/1 v/v) mixture (HMDA). The reaction time was varied to produce PMLG films with different degrees of aminolysis. After the aminolysis reaction, PMLG films were washed three times with water and dried under vacuum overnight. These films were then activated at 25°C for 18 h with 1 ml hexamethylene diisocyanate, and after three times washing with toluene, were dried at 150°C under vacuum overnight.

Surface characterization of these films was carried out by means of FT-IR spectra obtained on a Nicolet 20DXB FTIR spectrometer equipped with an ATR attachment (Spectra-Tech Inc.) using ZnSe (60 deg facecut angle), which is hereafter called simply ZeSe 60. All spectra were collected by co-adding 1000 scans at 2 cm^{-1} resolution with Happ–Genzel apodization with a long-range HgCdTe (MCT) detector.

In a vial, 0.56 mg (1.3 μmol) RGDS and 0.54 mg (1.1 μmol) GRGDS peptides (Peptide Institute Inc.) were dissolved in 3.9 ml and 3.3 ml of PBS (–), respectively. 1.2 ml of peptide solution was then reacted with the activated PMLG films at 25°C for 18 h. RGDS-PMLG and GRGDS-PMLG films were finally rinsed with PBS (–). The amount of immobilized peptides on the PMLG surfaces was determined by gel permeation chromatography (GPC) analysis for the amount of free peptides in the reaction mixture. The calibration curve was prepared by using 0.7 ml of the same peptide solution before the measurements. A Shimadzu GPC Model LC-6A, SPD-2A detector (220 nm), and GS-320 column (7.6 mm inner diameter and 500 mm length from Asahi Chemical Industry Co., Ltd) were used for the GPC measurements. The carrier solvent was PBS (–) and its flow rate was 1.0 ml/min at 35°C.

Cell culture

MRC-5 human fibroblasts, which were obtained commercially (ATCC CCL, 171, Flow Laboratories Inc.), were used for the cell culture experiments. The cells were maintained in a 5% CO_2 atmosphere at 37°C in Eagle's minimum essential medium (MEM: Nissui Pharmaceutical Co., Ltd.) supplemented with 10% fetal bovine serum (FBS). MRC-5 cell stock cultures were detached with 0.25% trypsin/0.02% EDTA (50/50 v/v; GIBCO Laboratories). After centrifugation, cells were resuspended at 1×10^5 cells/ml in Eagle's MEM containing 10% FBS, and 1.8 ml was innoculated into petri dishes (27.6 mm diameter) coated with untreated PMLG and peptide-immobilized PMLG polymers. After the desired cell–substrate contact time had elapsed, the petri dishes were washed three times with PBS (–). The number of attached cells was determined by the total number of innoculated cells and the number of cells in the washings counted using the hemacytometer. For the cell attachment, duplicate samples were incubated at 37°C in a CO_2 incubator. The experiments were usually done at least three times to yield a total of six estimates or more.

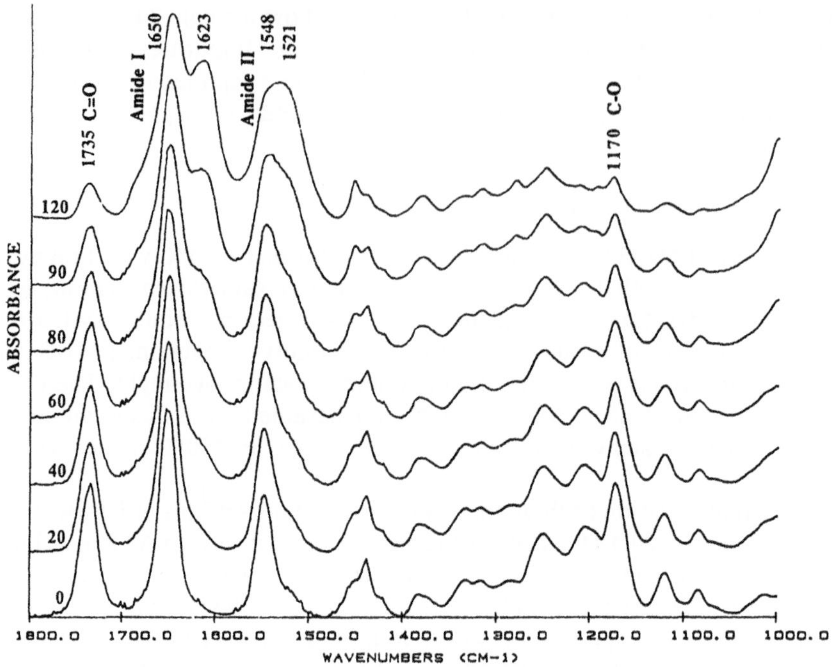
1735 C=O
Amide I
1650
1623
Amide II
1548
1521
1170 C-O
ABSORBANCE
120
90
80
60
40
20
0
1800.0
1700.0
1600.0
1500.0
1400.0
1300.0
1200.0
1100.0
1000.0
WAVENUMBERS (CM-1)

Figure 1. FT-IR ATR spectra from 1800 to 1000 cm^{-1} of PMLG and PMLG treated with HA surfaces. The treatment time in minutes is indicated on each spectrum. The ATR crystal of ZnSe 60 was used.

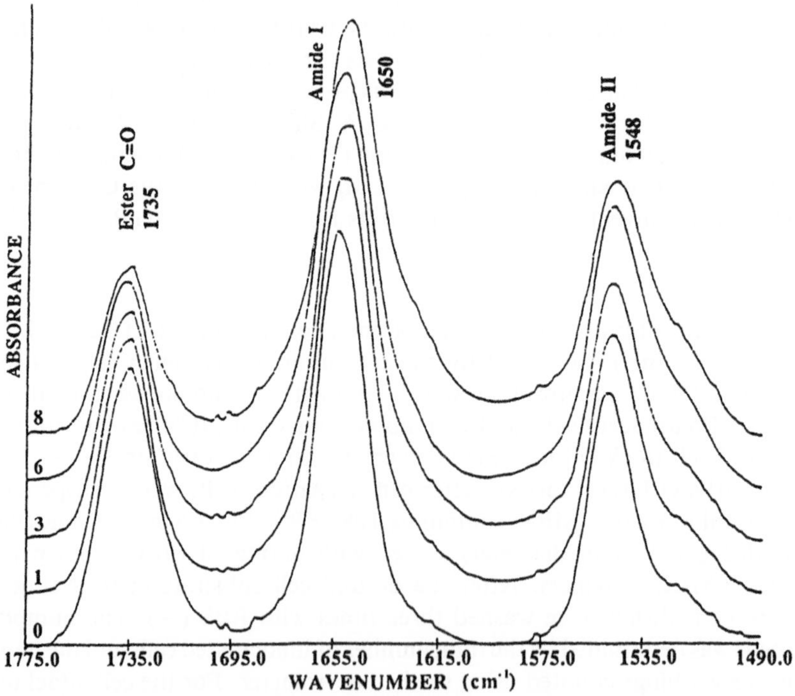
Ester C=O
1735
Amide I
1650
Amide II
1548
ABSORBANCE
8
6
3
1
0
1775.0
1735.0
1695.0
1655.0
1615.0
1575.0
1535.0
1490.0
WAVENUMBER (cm⁻¹)

Figure 2. Expanded FT-IR ATR spectra in the 1775–1490 cm^{-1} region of PMLG and PMLG treated with EDA surfaces. The treatment time in hours is indicated on each spectrum.

RESULTS AND DISCUSSION

Immobilization of RGD-peptides

Figure 1 shows representative attenuated total reflectance (ATR) spectra from 1800 to 1000 cm^{-1} of PMLG untreated and PMLG treated with HA surfaces. The treatment time in minutes is indicated on each spectrum. The bands of the C=O stretching vibration of ester, amide I, amide II, and C—O stretching vibration of ester of PMLG were observed at 1735, 1650, 1548, and 1170 cm^{-1}, respectively, and this means that the surface of PMLG exists in α-helical form [15]. For HA-treated PMLG surfaces a new absorption band and a discernible shoulder appear around 1623 and 1521 cm^{-1}, which are assigned to the β-structure [16]. The relative intensity of these bands increased with increasing time of HA treatment. It is obvious that a transition of the α-helical form to the β-structure took place in the PMLG surface layer by the HA treatment.

The expanded ATR spectra of PMLG untreated and PMLG treated with EDA surfaces in the 1775–1490 cm^{-1} region are shown in Fig. 2. The treatment time in hours is indicated on each spectrum. The bands of the C=O stretching vibration of ester, amide I, and amide II of EDA-treated PMLG were observed at 1735, 1650, and 1548 cm^{-1} respectively, but the relative intensity of a shoulder due to the β-structure was very small and/or negligible. For HMDA-treated PMLG a new absorption band or a shoulder assigned to the β-structure was not observed (spectra not shown).

Some models were proposed by Nakajima *et al.* to account for the formation of hydrogen bonds between monomeric or dimeric formic acid and PMLG chains [17].

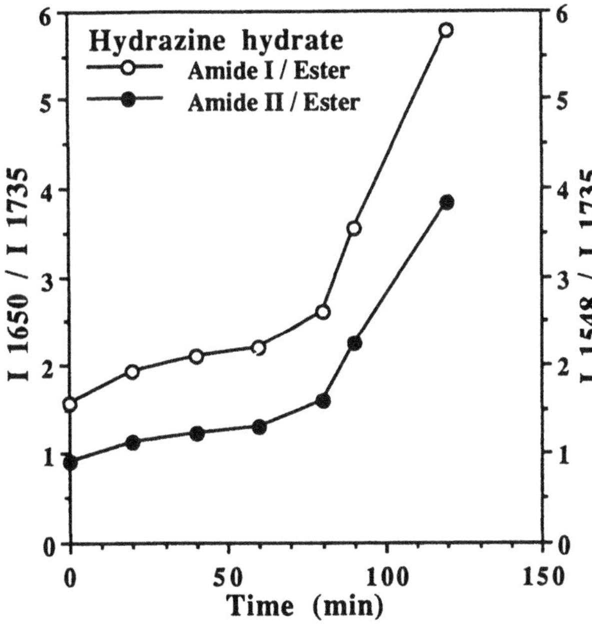

Hydrazine hydrate
Amide I / Ester
Amide II / Ester
I 1650 / I 1735
I 1548 / I 1735
0
1
2
3
4
5
6
0
50
100
150
Time (min)

Figure 3. Variation of the peak intensity ratios of amide I and amide II bands to ester band for HA-treated PMLG surfaces with time of the HA treatment. Open circles: amide I/ester (left ordinate); solid circles: amide II/ester (right ordinate).

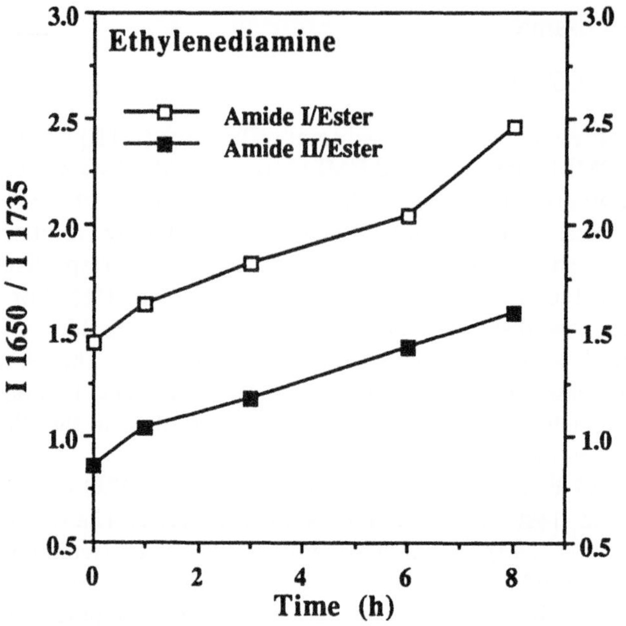
Ethylenediamine
Amide I/Ester
Amide II/Ester
I 1650 / I 1735
I 1548 / I 1735
Time (h)
3.0
2.5
2.0
1.5
1.0
0.5
0
2
4
6
8

Figure 4. Variation of the peak intensity ratios of amide I and amide II bands to ester band for EDA-treated PMLG surfaces with time of the EDA treatment. Open squares: amide I/ester (left ordinate); solid squares: amide II/ester (right ordinate).

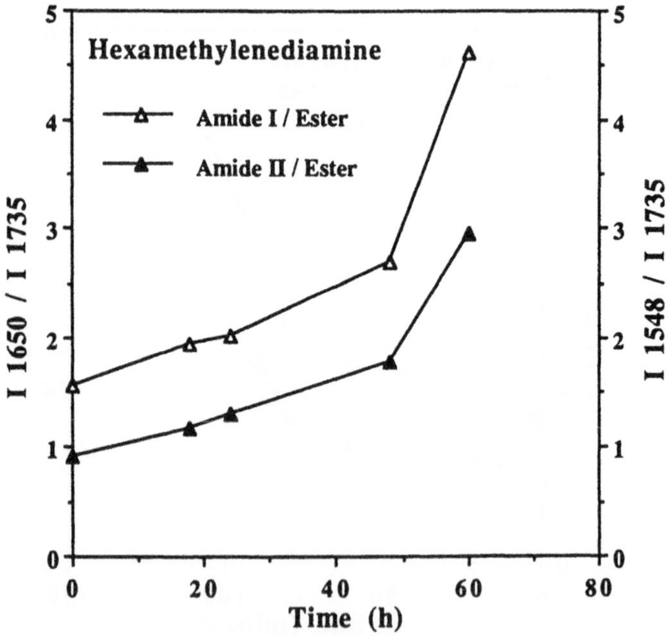
Hexamethylenediamine
Amide I / Ester
Amide II / Ester
I 1650 / I 1735
I 1548 / I 1735
Time (h)
5
4
3
2
1
0
0
20
40
60
80

Figure 5. Variation of the peak intensity ratios of amide I and amide II bands to ester band for HMDA-treated PMLG surfaces with time of the HMDA treatment. Open triangles: amide I/ester (left ordinate); solid triangles: amide II/ester (right ordinate).

According to their results, dimeric formic acid should form hydrogen bonds with PMLG chains and after evaporation of formic acid molecules the interchain hydrogen bonds (the β-structure) should be formed. When the aminolysis reaction is performed with the HA treatment in this study, the crosslinking may partly occur between two PMLG chains [18]. This crosslinking should cause the intrachain hydrogen bonds in the α-helices to compete with the interchain hydrogen bonds (the β-structure). Nakajima *et al.* also concluded that the size and geometry of the solvent are responsible for the formation of the β-chain conformation. The chain length of EDA and HMDA is longer than that of HA. For these reasons, a new absorption band or a shoulder assigned to the β-structure should be negligible and/or not observed for EDA- and HMDA-treated PMLG surfaces.

For all spectra of HA-, EDA- and HMDA-treated PMLG surfaces, the relative intensity of ester band at 1735 cm^{-1} decreased with increasing time of the aminolysis. Figures 3–5 show the changes of the peak intensity ratios of amide I and amide II bands to ester band for HA-, EDA- and HMDA-treated PMLG surfaces, respectively. As is obvious in these figures, the peak intensity ratios increased with increasing time of the aminolysis.

Figure 6 shows FT-IR ATR spectra from 4000 to 700 cm^{-1} of PMLG untreated and PMLG activated with hexamethylene diisocyanate after HA treatment for 60 and 90 min. As shown in Fig. 6 a new peak of around 2266 cm^{-1} due to the N=C=O stretching mode appeared, and the peak intensity increased with increasing time of HA treatment. The same tendency was also observed in FT-IR ATR

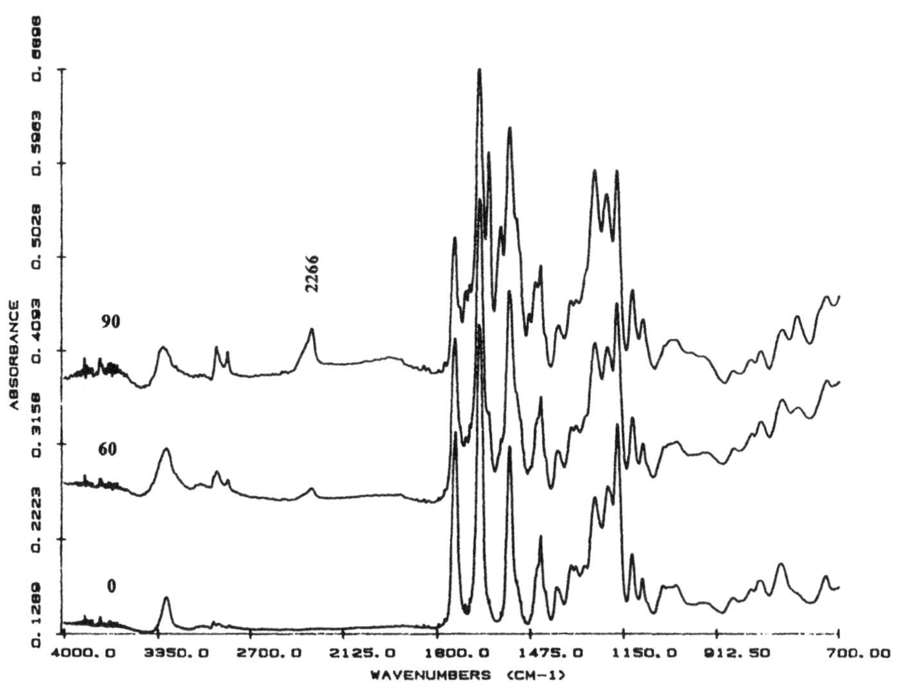
ABSORBANCE
0.1289
0.2223
0.3158
0.4093
0.5028
0.5963
0.6898
90
60
0
2266
4000.0
3350.0
2700.0
2125.0
1800.0
1475.0
1150.0
912.50
700.00
WAVENUMBERS (CM-1)

Figure 6. FT-IR ATR spectra from 4000 to 700 cm^{-1} of PMLG untreated and PMLG activated with hexamethylene diisocyanate after the HA treatment for 60 and 90 min. The treatment time in minutes is indicated on each spectrum.

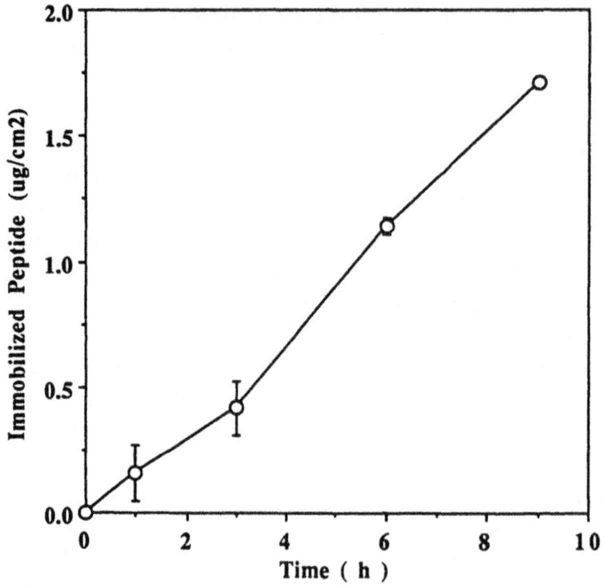
Immobilized Peptide (ug/cm2)
2.0
1.5
1.0
0.5
0.0
0
2
4
6
8
10
Time (h)

Figure 7. Relationship between the amount of RGDS immobilized to PMLG and time of the aminolysis reaction by the EDA treatment.

spectra of PMLG activated with hexamethylene diisocyanate after the EDA and HMDA treatments. Figure 7 shows the variation of the amount of RGDS immobilized to PMLG. As shown in the figure, the amount of RGDS immobilized to PMLG increased linearly with time of the aminolysis reaction by EDA. The amount of RGDS immobilized was *c.* 0.2 μg/cm^2 for PMLG treated with EDA for 1 h

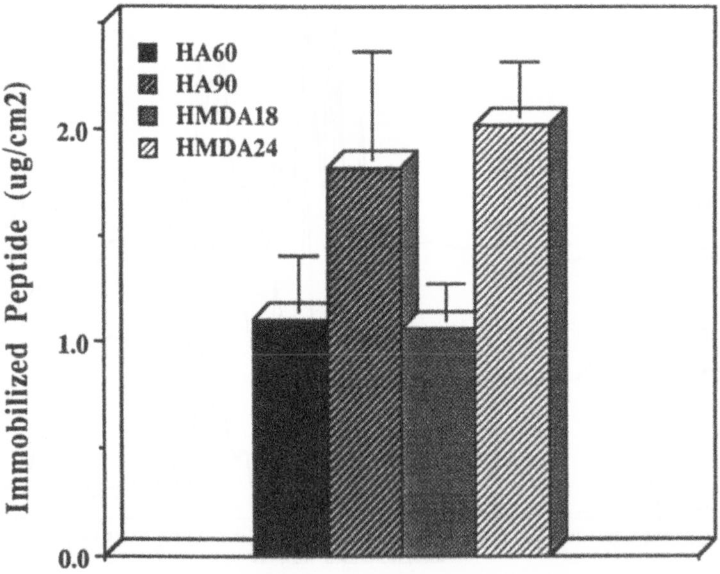
HA60
HA90
HMDA18
HMDA24
Immobilized Peptide (ug/cm2)
2.0
1.0
0.0

Figure 8. Variation of the amount of RGDS immobilized to PMLG by the HA treatment and the HMDA treatment. The treatment time in minutes by HA and in hours by HMDA is indicated behind each abbreviation.

(EDA1) and *c.* 0.4 μg/cm² for PMLG treated with EDA for 3 h (EDA3). The amount of GRGDS immobilized was *c.* 0.45μg/cm² for EDA3 (data not shown). Figure 8 illustrates the variation of the amount of immobilized RGDS peptides to PMLG by the HA- and HMDA-treatments. The treatment time in minutes by HA and in hours by HMDA is indicated in the Fig. 8. These results indicate that the amount of immobilized RGDS could be controlled by the reaction time of the aminolysis. As Fig. 8 shows, the amount of immobilized RGDS by the HA treatment for 60 min (HA60) corresponded to that by the HMDA treatment for 18 h (HMDA18), and the amount by the HA treatment for 90 min (HA90) corresponded to that by the HMDA treatment for 24 h (HMDA24).

Cell attachment

RGDS-PMLG surfaces, which were prepared by using PMLG treated with EDA for 3 h, were tested for their ability to support the attachment of cells after 24 h of incubation, as shown in Fig. 9. The amount of RGDS immobilized was *c.* 0.4 μg/cm², as mentioned above. Cell attachment was measured with cells suspended in Eagle's MEM with 10% FBS. The attachment of MRC-5 cells increased sharply up to 70% between 0 and *c.* 0.4 μg/cm² immobilized RGDS, and gradually attained to *c.* 75% at *c.* 1.7 μg/cm² immobilized RGDS. Since these experiments were performed in the presence of 10% FBS, the increase of cell attachment could be partially due to the adsorption of serum proteins such as fibronectin and vitronectin.

Figure 10 shows the results of initial attachment of MRC-5 cells to untreated PMLG, RGDS-PMLG, GRGDS-PMLG, and glass surfaces after 60 and 180 min of

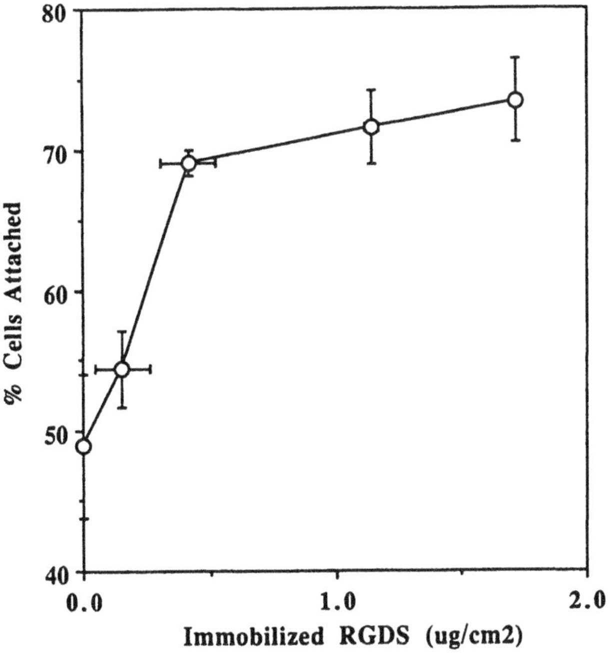

80
70
60
50
40
% Cells Attached
0.0
1.0
2.0
Immobilized RGDS (ug/cm2)

Figure 9. Cell attachment after 24 h of incubation on PMLG and RGDS-PMLG prepared by using PMLG treated with EDA for 3 h. The amount of RGDS immobilized was *c.* 0.4 μg/cm².

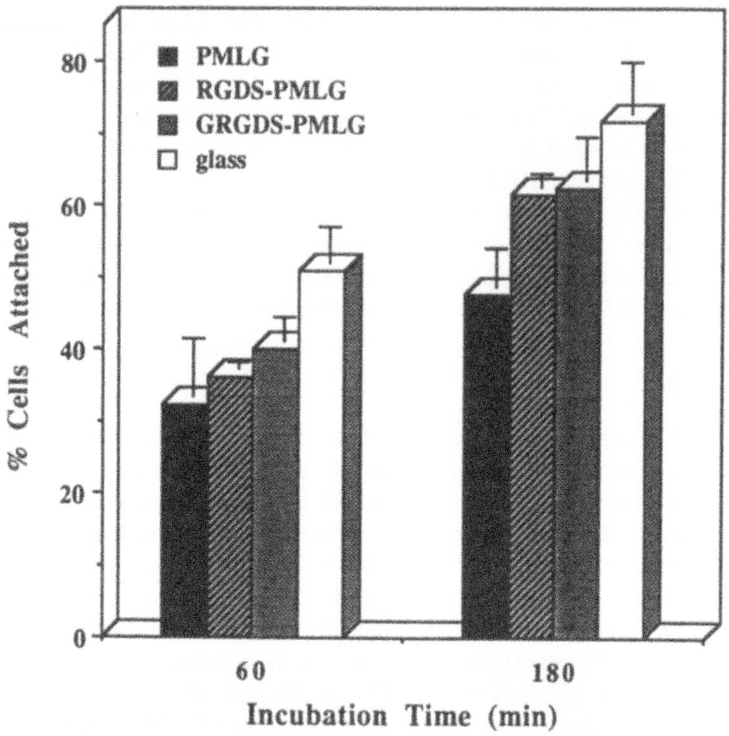
80
60
40
20
0
% Cells Attached
PMLG
RGDS-PMLG
GRGDS-PMLG
glass
60
180
Incubation Time (min)

Figure 10. MRC-5 cell attachment to untreated PMLG, RGDS-PMLG, GRGDS-PMLG, and glass surfaces after 60 and 180 min of incubation. The RGDS-PMLG and GRGDS-PMLG substrates were prepared by using PMLG treated with EDA for 3 h, and the amount of immobilized RGDS and GRGDS was *c*. 0.4 and *c*. 0.45 $\mu g/cm^2$, respectively.

incubation. The RGDS-PMLG and GRGDS-PMLG substrates were prepared by using PMLG treated with EDA for 3 h, and the amount of immobilized RGDS and GRGDS was *c*. 0.4 and *c*. 0.45 $\mu g/cm^2$, respectively, as described above. Cell attachment on RGDS- and GRGDS-PMLG was greater than on untreated PMLG even after 60 min of incubation. After 180 min of incubation attachment on untreated PMLG was 47%, but that on RGDS- and GRGDS-PMLG reached more than 62%. It may be somewhat surprising that the cell attachment is better on the glass substrate than it is on the RGDS- and GRGDS-PMLG. One of the causes may be the paucity of the amount of immobilized RGDS and GRGDS. It can be also assumed that EDA and hexamethylene diisocyanate as a spacer inhibited the biospecifc interaction of the RGD peptides with the cells.

Figure 11 shows the results of MRC-5 cell attachment to various substrates after 60 and 180 min of incubation. The cell attachment to HA60, HA90, HMDA18, and HMDA24 as the RGDS-PMLG surfaces is illustrated to examine the effect of spacers. The cell attachment to PMLG activated with hexamethylene diisocyanate after the aminolysis using HMDA for 24 h (HD-HMDA24) is also shown in Fig. 11. The RGDS peptide is not immobilized on HD-HMDA24. As experimental results on water contact angles at 25°C for these samples, we obtain 66, 76, and 64 deg for untreated PMLG, HD-HMDA24, and HMDA-24, respectively. Thus untreated PMLG is relatively hydrophobic, but by treating the surface via hexamethylene

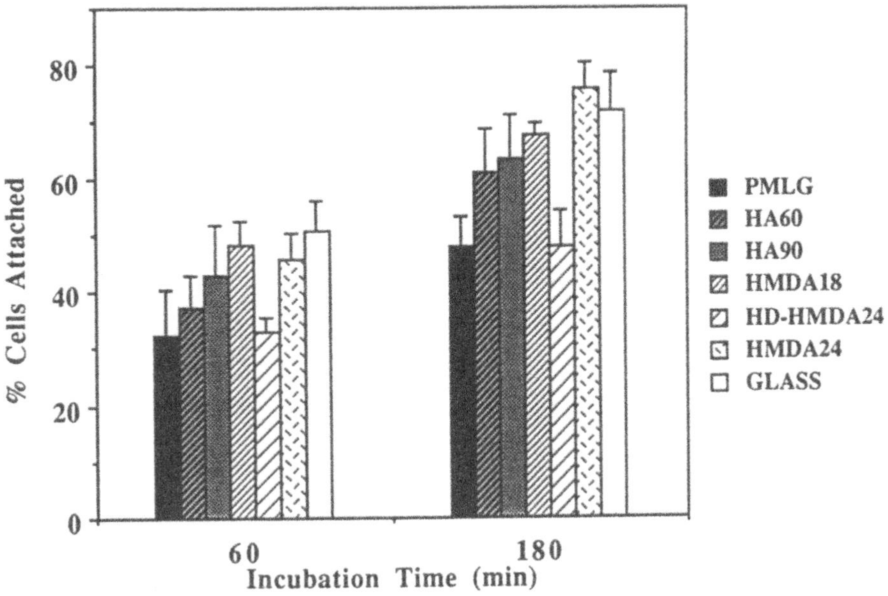
80
60
40
20
0
% Cells Attached
60
180
Incubation Time (min)
PMLG
HA60
HA90
HMDA18
HD-HMDA24
HMDA24
GLASS

Figure 11. MRC-5 cell attachment to various substrates: effect of spacers. Cell attachment to untreated PMLG, HA60, HA90, HMDA18, HMDA24, and glass surfaces is illustrated. Cell attachment to PMLG activated with hexamethylene diisocyanate after the aminolysis using HMDA for 24 h (HD-HMDA24) is also shown.

diisocyanate after the aminolysis using HMDA, HD-HMDA24 becomes more hydrophobic. HMDA24 then becomes hydrophilic compared to HD-HMDA24 and PMLG by peptide immobilization. Surface energy and wettability have been related to the cell attachment and cell growth on various substrates [19–22]. Although the influence of hydrophobic/hydrophilic balance on the cell attachment is poorly understood, it was proved by Ikada *et al.* [22] that the maximum attachment of fibroblasts was obtained with polymer materials having water contact angles of around 70 deg. According to their conclusions, it can be assumed that the cell attachment to HMDA24 is almost the same as or less than that of HD-HMDA24 and PMLG. As is shown in Fig. 11, however, the cell attachment to HMDA24 increased greatly, while the cell attachment to HD-HMDA24 was almost the same as that of PMLG. These results should imply that the increase in the cell attachment to HMDA24 is due to the biospecific interaction of immobilized RGDS peptides with the cells. The inactive control peptides, such as RDGS and RGES, were not immobilized in this study. To immobilize such peptides might be useful in demonstrating that immobilized RGDS peptides are indeed biologically active. It is, however, considered that experimental results mentioned above should support their biological activity. Unfortunately there is the possibility that the increase of the cell attachment to HMDA18, HA90, and HA60 could just as well be due to changes in the surface charge or structure by the aminolysis step or by some other step in the process, as it could be due to the biospecific interaction since we do not have data about that. In Fig. 11, the cell attachment to the glass substrate was usually better than that of peptide-immobilized surfaces. This result seems to be due to the paucity of the amount of immobilized RGDS peptides. It is clear that

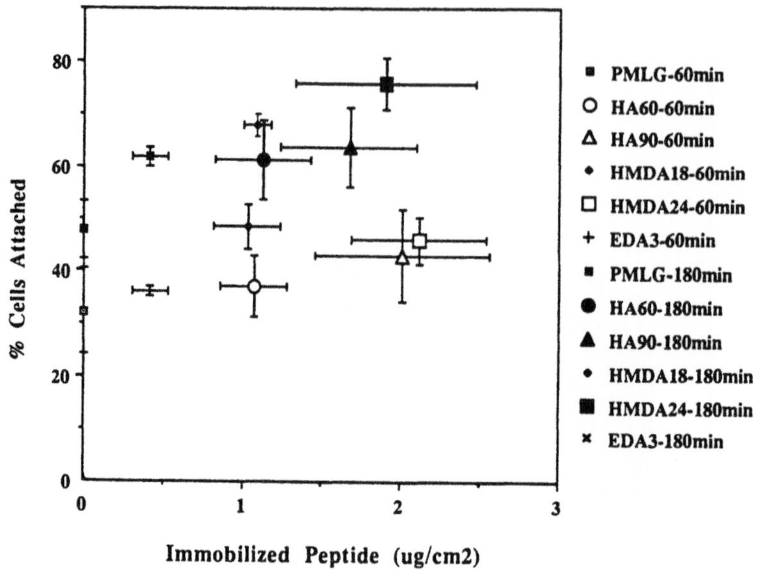
80
60
40
20
0
0
1
2
3
% Cells Attached
Immobilized Peptide (ug/cm2)
PMLG-60min
HA60-60min
HA90-60min
HMDA18-60min
HMDA24-60min
EDA3-60min
PMLG-180min
HA60-180min
HA90-180min
HMDA18-180min
HMDA24-180min
EDA3-180min

Figure 12. Relationship between the cell attachment and the amount of immobilized RGDS. Data for attachment to untreated PMLG, HA60, HA90, EDA3, HMDA18, and HMDA24 after 60 (open symbols) and 180 min (solid symbols) of incubation are shown. The incubation time is indicated behind each abbreviation.

RGDS-PMLG surfaces promoted the cell attachment compared to untreated PMLG even after 60 min of incubation, as expected. After 180 min of incubation for cell attachment to HMDA24 attained up to *c*. 76%, which was higher than the 72% of glass. The amount of immobilized RGDS of HA60 corresponded to that of HMDA18, and the amount of HA90 corresponded to that of HMDA24, as shown in Fig. 8.

To clarify the spacer effect, the relationship between the cell attachment and the amount of immobilized RGDS is illustrated in Fig. 12. Data for attachment to untreated PMLG and RGDS-PMLG surfaces after 60 (open symbols) and 180 min (solid symbols) of incubation are shown. The cell attachment to RGDS-PMLG surfaces increases with increasing amount of immobilized RGDS after 180 min of incubation. Although the amount of immobilized RGDS of EDA3 is less than half the amount of RGDS of HA60, the attachment to EDA3 was almost the same as that of HA60 or rather higher. Furthermore, compared with the attachment to HA60 and HA90, the attachment to HMDA18 and HMDA24 was higher even though the amount of immobilized RGDS was almost the same. It can be also assumed that the hydrophobicity of the spacer promoted the cell attachment. As mentioned above, it was actually proved that cells on hydrophobic PMLG with a higher β-structure content were spread and flattened more than those on untreated PMLG existing in α-helical form [14]. However, significant differences in the attachment to RGDS-PMLG surfaces compared to the differences due to the hydrophobicity should suggest that a longer spacer promoted cell attachment via specific receptor–ligand interaction by increasing proximity.

REFERENCES

1. M. D. Pierschbacher and E. Ruoslahti, *Nature* **309**, 30 (1984).
2. M. D. Pierschbacher and E. Ruoslahti, *Proc. Natl Acad. Sci. USA* **81**, 5985 (1984).
3. E. Ruoslahti, *Ann. Rev. Biochem.* **57**, 375 (1988).
4. K. M. Yamada, in: *Fibronectin,* p. 47, D. F. Mosher (Ed.). Academic Press, Inc., New York (1989).
5. E. G. Hyman, M. D. Pierschbacher and E. Ruoslahti, *J. Cell Biol.* **100**, 1948 (1985).
6. E. G. Hyman, M. D. Pierschbacher, S. Suzuki and E. Ruoslahti, *Exp. Cell Res.* **160**, 245 (1985).
7. D. W. Barnes and J. Silnutzer, *J. Biol. Chem.* **258**, 12548 (1983).
8. J. G. Steele, G. Johnson, C. McFarland, Z. R. Vasic, R. C. Chatelier, P. A. Underwood and H. J. Griesser, *Transactions of Fourth World Biomaterials Congress,* April 24–28, 218 (1992).
9. H. B. Streeter and D. A. Rees, *J. Cell Biol.* **105**, 507 (1987).
10. B. K. Brandley and R. L. Schnaar, *Anal. Biochem.* **172**, 270 (1988).
11. S. P. Massia and J. A. Hubbell, *Anal. Biochem.* **187**, 292 (1990).
12. A. Nicol, D. C. Gowda and D. W. Urry, *J. Biomed. Mater. Res.* **26**, 393 (1992).
13. K. Kugo, M. Okuno, K. Kitayama, T. Kitaura, J. Nishino, N. Ikuta, E. Nishio and M. Iwatsuki, *Biopolymers* **32**, 197 (1992).
14. K. Kugo, M. Okuno, J. Nishino, H. Masuda, M. Kayama and M. Iwatsuki, *Biopolymers,* to be submitted.
15. K. Kugo, M. Murashima, T. Hayashi and A. Nakajima, *Polym. J.* **15**, 267 (1983).
16. T. Miyazawa and E. R. Blout, *J. Am. Chem. Soc.* **83**, 712 (1961).
17. A. Nakajima, T. Fujiwara, T. Hayashi and K. Kaji, *Biopolymers* **12**, 2681 (1973).
18. T. Kinoshita, A. Takizawa, Y. Tsujita and M. Ishikawa, *Kobunshi Ronbunshu* **43**, 827 (1986).
19. H. Yasuda, B. S. Yamanashi and D. P. Devito, *J. Biomed. Mater. Res.* **12**, 701 (1978).
20. D. Gingell and S. Vince, *J. Cell Sci.* **54**, 255 (1982).
21. T. A. Horbett, M. B. Schway and B. D. Ratner, *J. Colloid Interface Sci.* **104**, 28 (1985).
22. Y. Tamada and Y. Ikada, in: *Polymers in Medicine II,* p. 101, E. Chiellini, P. Giusti, C. Migliaresi and L. Nicolais (Eds). Plenum Press, New York (1986).

Cell-attachment activities of surface immobilized oligopeptides RGD, RGDS, RGDV, RGDT, and YIGSR toward five cell lines

YOSHIAKI HIRANO[1]*, MOTOYA OKUNO[1], TOSHIO HAYASHI[2], KUNIO GOTO[1] and AKIO NAKAJIMA[1]

[1]*Department of Applied Chemistry, Osaka Institute of Technology, 5-16-1 Ohmiya, Asahi-ku, Osaka 535, Japan*

[2]*Research Center for Biomedical Engineering, Kyoto University, 53 Kawahara-cho, Shogoin, Sakyo-ku, Kyoto 606, Japan*

Received 22 May 1992; accepted 28 August 1992

Abstract—Tetrapeptides, Arg-Gly-Asp-Ser (RGDS), Arg-Gly-Asp-Val (RGDV), and Arg-Gly-Asp-Thr (RGDT), respectively, appearing in the cell-attachment domains of fibronectin, vitronectin, and collagen, and pentapeptide Tyr-Ile-Gly-Ser-Arg (YIGSR) appearing in B1 chain of laminin, were synthesized by liquid-phase procedure. Bioactivities of RGD, RGDX (X=S, V and T), YIGSR, and YIGSR-NH_2 as cell recognition determinants were investigated by cell-attachment test using these oligopeptides immobilized to ethylene-acrylic acid copolymer (PEA) film. The cell lines used were A431, NRK, CHO-K1, HeLa.S3, and RLC-16 cells. It was found that the residue X in RGDX plays an important role for cell-attachment activity of RGDX, and, regarding YIGSR, introduction of NH_2 residue at the C-terminal of the pentapeptide enhances the cell-attachment activity.

Key words: Cell-attachment activity; oligopeptides, RGD, RGDS, RGDV, and RGDT; oligopeptide YIGSR; oligopeptides immobilized to ethylene-acrylic acid copolymer films; A431, NRK, CHO-K1, HeLa.S3, and RLC-16 cells.

INTRODUCTION

Investigations on the cell-attachment activities of various proteins, such as fibronectin, vitronectin, collagen and fibrinogen, have revealed that oligopeptide sequences including Arg-Gly-Asp (RGD) may act as ligands toward the cell-attachment receptor (integrin super family) distributing in the cell membranes. Synthetic oligopeptides containing RGD sequences have been employed to study cell attachment activity on the molecular level. It has been demonstrated that these oligopeptides strongly interact with some kinds of cells such as fibroblast [1-5], epithelial [2, 5-7], and platelet [8-10]. In addition, these oligopeptides were used for disruption of cell migration *in vivo* [11], in tissue culture [11, 12], and in experimental metastasis.

Laminin, the major non-collagenous glycoprotein distributing only in the basement membrane—an extracellular matrix between epithelial cells and stroma—has various biological activities such as adhesion, migration, differentiation, and growth of cells [13, 14]. Laminin is composed of three chains designated A, B1 and B2, which are arranged in a cross-shaped structure [15]. Iwamoto *et al.* [16] and

* To whom correspondence should be addressed.

Graf *et al.* [17, 18] suggested that the peptide sequence Tyr–Ile–Gly–Ser–Arg (YIGSR) in the B1 chain is concerned with cell-attachment activity, and, in particular, plays important roles for cell-attachment, cell-spreading and chemotaxis. In addition, Iwamoto and co-workers [16] have demonstrated that the synthetic pentapeptide YIGSR inhibited metastasis of malignant melanoma cells. The YIGSR sequence, however, does not exist in fibronectin, vitronectin, and collagen.

Oligopeptides containing RGD and YIGSR sequences may be used as a new field of drugs owing to their apparently selective inhibition effect in certain important biological processes [19]. For example, in 1989, Murata, Saiki, and co-workers [20, 21] reported that synthetic sequential polypeptides poly(RGDS), and poly(YIGSR) drastically inhibited the metastatic formation of B16-BL16 melanoma cells. However, only few papers have been published on cell-attachment behavior of RGDX, and YIGSR-immobilized films [2, 22–25].

We have synthesized RGD, RGDX (X=S, V, and T), and YIGSR, and evaluated their cell-attachment activities toward L-929 fibroblast cell, using RGDS-immobilized polyvinyl alcohol film [26], RGDX-immobilized ethylene-acrylic acid copolymer (PEA) film [27] and YIGSR-immobilized PEA film [23]. The aim of our research is to accumulate the basic information required for synthetic materials to be used as biomedical materials. In this report, the cell-attachment activities of PEA films, to which RGD, RGDX, YIGSR or YIGSR-NH_2 was immobilized at their N-terminals, were determined by measuring the number of adhered cells to the film in the absence of added serum. The cell lines used were ovary CHO-K1 cell originating in Chinese hamster, kidney NRK cell originating in rat, epidermoid carcinoma A431 cell originating in human, cervix HeLa.S3 cell originating in human, and liver RLC-16 cell originating in rat.

EXPERIMENTAL

Synthesis of RGDX, RGD, YIGSR, and YIGSR-NH_2 oligopeptides

RGDX (X=S, V, and T), RGD, and YIGSR were synthesized by liquid-phase procedures reported previously [27, 28]. Boc–Tyr–Ile–Gly–Ser–Arg(Tos)–NH_2 was synthesized from Boc–Tyr–Ile–Gly–Ser–Arg(Tos)–OMe, and the deblocked YIGSR-NH_2 was obtained from Box–Tyr–Ile–Gly–Ser–Arg(Tos)–NH_2 using trifluoromethanesulfonic acid (TFMSA) as a deblocking agent.

Immobilization of oligopeptides onto ethylene-acrylic acid copolymer film and surface characterization

The schema of surface activation of ethylene-acrylic acid copolymer (PEA) and oligopeptide immobilization are shown in Fig. 1.

The ethylene–acrylic acid copolymer pellet (acrylic acid content: 20 wt%, lot. no. EAA-A510W) contributed from Teijin Co. Ltd. was purified in a methanol Soxhlet extractor system. PEA films were prepared by casting from 5 wt% THF solution on glass plates. Finally the films were dried *in vacuo* for 5 days at 40 °C. The —COOH residues locating on the PEA film surface were activated at 4 °C for 30 min with 4 g water-soluble carbodiimide (1-ethyl-3-(3-dimethylaminopropyl)-carbodiimide) dissolved in 400 ml phosphate buffer solution (PBS). Then 80 mg oligopeptide was dissolved in 80 ml PBS and reacted with the activated PEA films for 24 h at 4 °C.

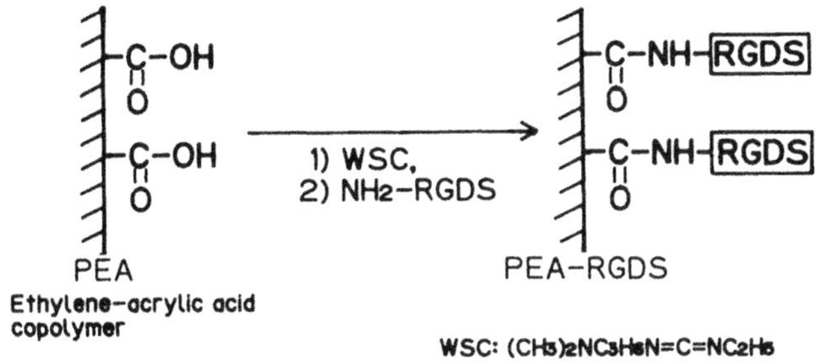
C–OH
O
C–OH
O
PEA
Ethylene–acrylic acid copolymer
1) WSC,
2) NH2–RGDS
C–NH–RGDS
O
C–NH–RGDS
O
PEA–RGDS
WSC: (CH3)2NC3H6N=C=NC2H5

Figure 1. Schema of surface activation and oligopeptide immobilization.

N-terminals of oligopeptide were immobilized to —COOH residues to the PEA surface. The immobilized films were rinsed with PBS and dodecyl sulfate sodium salt solution (SDS) by ultrasonic cleaner and washed with pure water for 72 h. The surface characterization of the oligopeptide-immobilized films was carried out by means of C1*s* and N1*s* spectra measured with a Shimadzu 750 electron spectrometer for chemical analysis (ESCA) using $MgK\alpha_{1,2}$ exciting radiation. The surface composition of the immobilized film was tested by the N/C ratio estimated from ESCA. From the N/C ratio obtained, it was confirmed that the oligopeptide was really immobilized onto PEA film surface. For cell-attachment tests, the films were stirred in 70% ethanol overnight, and ethanol was displaced with PBS for 1 day. The test films were placed in a 24-well cell culture dish (Falcon).

Cell culture on, and cell-attachment activity of RGDX, RGD, YIGSR, and YIGSR-NH_2 immobilized films

All the cell lines used were supplied from Riken Gene Bank (Tsukuba, Japan). CHO-K1 [29], NRK [30], HeLa.S3 [31, 32], A431 [33], and RLC-16 [34] were cultured at 37 °C, under 5% CO_2, and 95% air atmosphere, respectively, in Ham's F12 (Nissui Seiyaku Co.), Eagle's MEM (Nissui Seiyaku Co.), Dulbecco's Modified Eagle Medium (DMEM) (Nissui Seiyaku Co.), and DM-160 AU medium (Kyokuto Seiyaku Co.), each containing 10% fetal bovine serum (FBS) (M. A. Bioproducts, Maryland, USA). The cultured cells were detached from the culture dish with 0.25% trypsin-EDTA (Difco Laboratory, Trypsin 1:125), and washed once in FBS containing medium and once in serum-free medium. The cell density was adjusted to 1.9×10^5 cells/ml (1.0×10^5 cells/cm^2) for cell adhesion tests in serum-free medium. Each 1 ml cell suspension was added to the oligopeptide immobilized polymer films ($n = 5$) placed in each well of the cell culture dish, and kept for predetermined periods (30 min, 1 and 3 h) in a humidified incubator conditioned to 37 °C, 5% CO_2 and 95% air atmosphere. After incubation, unattached cells were removed from the film surface by washing with PBS. Adhered cells were fixed using glutaraldehyde, stained by Giemsa, and finally treated with glycerin jelly. The number of cells attached to the film was counted under the microscope at 30 positions for each film. Five films ($n = 5$) were used for each cell species, so the total number of test positions was 150. Student's t-test was used to determine if significant differences existed. For test, the difference was regarded as significant for $p < 0.05$.

In addition, cell adhesion efficiency was measured. Subcultures of cell were carried out in a medium containing 10% serum for 8 h, in a humidified incubator conditioned to 37 °C, 5% CO_2, and 95% air atmosphere. At 8 h incubation, all cells were stable to cell growth, cell extension, and cell division in the culture dish so the value of cell adhesion efficiency at 8 h was referred to the cell attachment tests on immobilized films. After incubation, the number of cells attached and unattached to the cell culture dish were counted. The cell adhesion efficiencies obtained (Table 2) were given by the percentage of cells attached to the cell culture dish.

RESULTS AND DISCUSSION

Surface characterization of oligopeptides immobilization PEA films

N/C and O/C ratios of PEA and oligopeptide immobilized PEA films estimated from ESCA spectra were shown in Table 1. As mentioned earlier, the PEA sample is a copolymer composed of 91.1 mol% ethylene and 8.9 mol% acrylic acid. If we assume that the surface composition is almost the same as the bulk composition, then the elemental ratio O/C (%) of PEA surface is calculated as 8.5%. Comparing this value with the experimental value 7.7%, we may conclude that the surface composition is almost the same as the bulk composition.

Table 1.
Surface composition of oligopeptide-immobilized PEA films and % of reacted acrylic acid units

Designation	Elemental ratio O/C (%)	Elemental ratio N/C (%)	% of reacted acrylic acid unit
PEA	8.0	0	—
PEA-RGD	13.1	4.62	19.5%
PEA-RGDS	14.9	5.20	19.2%
PEA-RGDV	13.4	4.24	19.7%
PEA-RGDT	14.8	5.30	20.1%
PEA-YIGSR	13.6	5.63	19.8%
PEA-YIGSR-NH_2	12.4	6.38	20.0%

For oligopeptide immobilized PEA, the experimental values of the N/C ratio were used for calculating the immobilized oligopeptide densities, but O/C ratios were not used for such purpose because of possible adsorption of water on the film surface. If we designate the mol% immobilization of oligopeptide to the PEA molecules on the film surface as x, then the experimental N/C values lead to x. As obvious from Table 1, it was confirmed that 19–20% of the whole acrylic acid units reacted with every oligopeptide.

Cell culture on, and cell-attachment activity of RGDX–, RGD–, YIGSR–, and YIGSR–NH_2-immobilized films

Cell anchorage dependency and cell adhesion efficiency of CHO-K1, NRK, A431, HeLa.S3, and RLC-16 cells are shown in Table 2. These data may suggest that the percentage cell attachment for oligopeptide-immobilized films is appreciably affected by both the cell anchorage dependency and cell adhesion efficiency.

Table 2.
Cell anchorage dependency and cell adhesion efficiency of various cells

Cells	Anchorage dependency[a]	Adhesion efficiency
CHO-K1	Yes	98 ± 5%
NRK	Yes	90 ± 5%
A431	Yes	65 ± 5%
HeLa.S3	No	50 ± 5%
RLC-16	Yes	30 ± 5%

[a] Data from catalogue (No. 4 1991) prepared by Riken gene bank.

Figure 2 shows the percentage of attachment of CHO-K1 cells to PEA, PEA-RGD, PEA-RGDS, PEA-RGDV, PEA-RGDT, PEA-YIGSR, and PEA-YIGSR-NH_2 films at incubation times of 30 min, 1 and 3 h. The initial cell concentration was 1.0×10^5 cells/cm^2. The error bars represent the standard errors for five different experiments ($n = 5$). With CHO-K1 cells, on the whole many cells adhered to PEA, PEA-RGD, PEA-RGDX (X=S, V, and T), PEA-YIGSR and PEA-YIGSR-NH_2 films because the cell anchorage dependency of CHO-K1 is high. The surface condition of films is so important to adhere to the cell attachment oligopeptides that oligopeptide attached to the cell may accelerate the interaction of cell-attachment receptor of the cell with oligopeptides. At 3 h incubation, about 90% of the cells were adhered to PEA-RGDS films, and about 80% to PEA-RGDV and PEA-RGDT. The percentage of cell adhesion depends on the incubation time. At each incubation time, the percentage of cell adhesion for PEA-RGDX (X=S, V, and T) films was about twice to three times that for PEA (control). However, the percentage of cell adhesion for PEA-RGD film was almost the same as that of PEA. For YIGSR and YIGSR-NH_2, the percentage of cell adhesion for PEA-YIGSR-NH_2 films was higher than for PEA-YIGSR films ($p < 0.05$), and moreover, the percentage of cell adhesion for PEA-YIGSR and PEA-YIGSR-NH_2 films was always lower than that for PEA-RGDS film ($p < 0.05$), but higher than those for PEA-RGD and PEA

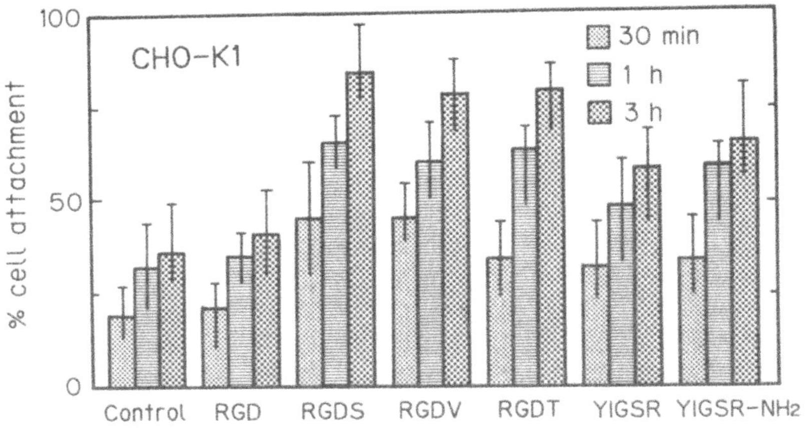
CHO-K1
30 min
1 h
3 h
% cell attachment
100
50
0
Control
RGD
RGDS
RGDV
RGDT
YIGSR
YIGSR-NH2

Figure 2. Percentage of CHO-K1 cells adhered onto PEA, PEA-RGD, PEA-RGDS, PEA-RGDV, PEA-RGDT, PEA-YIGSR, and PEA-YIGSR-NH_2 films, in the absence of serum, at 30 min, 1 and 3 h.

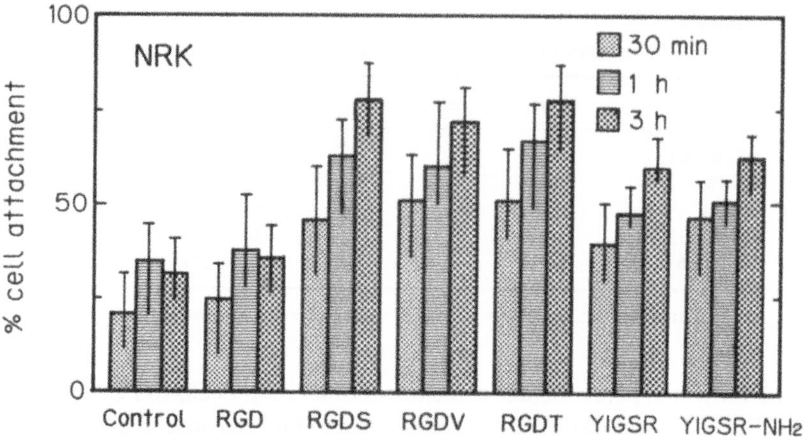
NRK
30 min
1 h
3 h
100
50
0
% cell attachment
Control
RGD
RGDS
RGDV
RGDT
YIGSR
YIGSR-NH2

Figure 3. Percentage of NRK cells adhered onto PEA, PEA-RGD, PEA-RGDS, PEA-RGDV, PEA-RGDT, PEA-YIGSR, and PEA-YIGSR-NH_2 films, in the absence of serum, at 30 min, 1 and 3 h.

(control) films ($p < 0.05$). At 3 h incubation, cell spreading was observed on RGDX, YIGSR, and YIGSR-NH_2 immobilized films. The percentage of cell spreading on these immobilized films was about 30% as a whole.

Cell attachments for NRK and A431 are shown in Figs 3 and 4, respectively. The percentages of cell attachment on RGDX and YIGSR series films for NRK and A431 are almost the same as for CHO-K1. But, the number of cells adhered for NRK is lower than for CHO-K1 ($p < 0.05$), and the number of cells adhered for A431 is the lowest among three kinds of cells ($p < 0.05$). Such an order of cell attachment agrees with the order of cell adhesion efficiency shown in Table 2. In addition, for NRK and A431, the percentages of cell adhesion to PEA-RGD are almost the same as to PEA, similar to the case of CHO-K1. In this case, the percentages of cell attachment to PEA-YIGSR and PEA-YIGSR-NH_2 were lower than those for PEA-RGDX ($p < 0.05$), however, that for PEA-YIGSR series was higher than that for RGD tripeptide-immobilized film ($p < 0.05$).

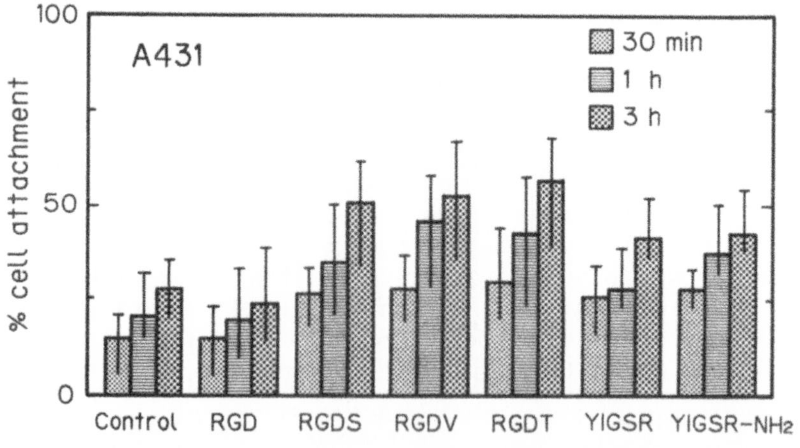
A431
30 min
1 h
3 h
100
50
0
% cell attachment
Control
RGD
RGDS
RGDV
RGDT
YIGSR
YIGSR-NH2

Figure 4. Percentage of A431 cells adhered onto PEA, PEA-RGD, PEA-RGDS, PEA-RGDV, PEA-RGDT, PEA-YIGSR, and PEA-YIGSR-NH_2 films, in the absence of serum, at 30 min, 1 and 3 h.

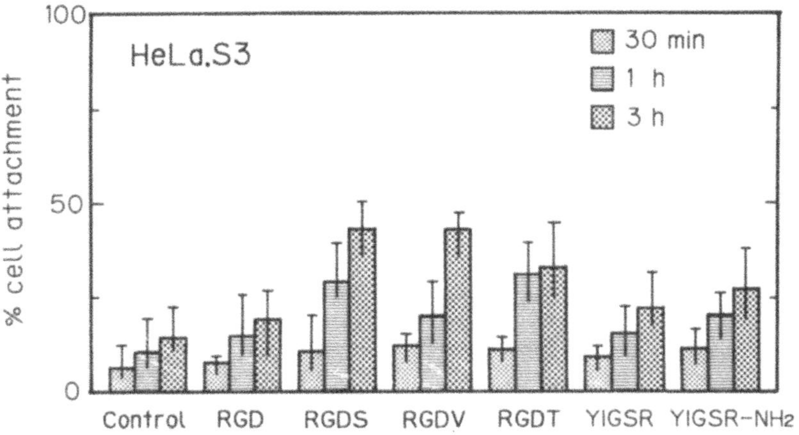
100
50
0
% cell attachment
HeLa.S3
30 min
1 h
3 h
Control
RGD
RGDS
RGDV
RGDT
YIGSR
YIGSR-NH2

Figure 5. Percentages of HeLa.S3 cells adhered onto PEA, PEA-RGD, PEA-RGDS, PEA-RGDV, PEA-RGDT, PEA-YIGSR, and PEA-YIGSR-NH_2 films, in the absence of serum, at 30 min, 1 and 3 h.

Figure 5 shows the percentage of HeLa.S3 cells adhered to PEA, PEA-RGD, and PEA-RGDX (X=S, V, and T) films at incubation times of 30 min, 1 and 3 h. About 10% of the cells were adhered to every film at 30 min, however, 40–50% of HeLa.S3 cells were adhered to PEA-RGDX after 1 h. Concerning the YIGSR series, 20–30% of cells were adhered to PEA-YIGSR and PEA-YIGSR-NH_2 at 3 h. The percentage of cell adhesion to PEA-RGDX was about twice that to PEA at 1-3 h. Such rather low values of percentage of cell adhesion for PEA-RGDX may originate from the fact that the adhesion efficiency of HeLa.S3 is rather low. Cell-attachment for HeLa.S3 does not depent on anchorage as shown in Table 2, however, the value of about 40% for PEA-RGDX films is compared with 15% for PEA film (control). This fact may suggest that the RGDX oligopeptides immobilized to the films accelerate the cell attachment by the interaction of the ligand oligopeptide with the receptor distributing on the cell surface. The interaction or recognition of receptor for the YIGSR series seems weaker than for the RGDX series ($p < 0.05$).

Figure 6 shows the percentage of RLC-16 cells adhered to PEA, PEA-RGD, PEA-RGDX (X=S, V, and T), PEA-YIGSR and PEA-YIGSR-NH_2 films at incubation times of 30 min, 1 and 3 h. On the whole, the percentage of cell adhesion for RLC-16 is fairly low compared with other types of cells. The cell adhesion efficiency of RLC-16 cell is as low as about 30%. The percentage of cell adhesion, as shown in Fig. 6, is slowly increasing for every film at incubation times from 30 min to 1 h. In particular, for PEA-RGDS, the percentage of cell adhesion considerably increases from 1 to 3 h and reaches a value equal to the cell adhesion efficiency at 3 h. These results may indicate that RGDX-immobilized PEA surface strongly interacts with the cell attachment receptors on the RLC-16 cells, because cell attachment of RLC-16 depends on the anchorage. In this case, the percentage of cell adhesion for YIGSR and YIGSR-NH_2 immobilized films is almost the same as those for other types of cells.

Thus, it was elucidated that tetrapeptide RGDX is more effective than tripeptide RGD, ($p < 0.05$) and YIGSR-NH_2 is more effective than YIGSR ($p < 0.05$). The behavior of tripeptide RGD is almost the same as the control (PEA). Though not shown here (our unpublished data), a non-biologically active oligopeptide

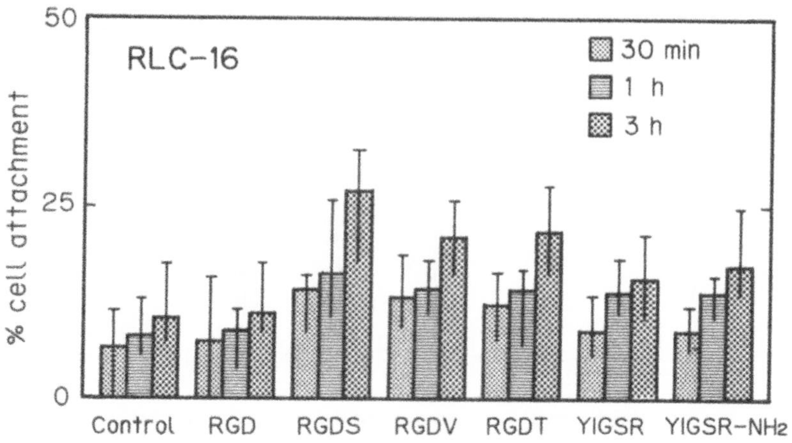
RLC-16
30 min
1 h
3 h
% cell attachment
50
25
0
Control RGD RGDS RGDV RGDT YIGSR YIGSR-NH2

Figure 6. Percentage of RLC-16 cells adhered onto PEA, PEA-RGD, PEA-RGDS, PEA-RGDV, PEA-RGDT, PEA-YIGSR, and PEA-YIGSR-NH_2 films, in the absence of serum, at 30 min, 1 and 3 h.

Arg-Gly-Glu-Ser (RGES) was examined by cell attachment test and cell inhibition test ($n = 2$). It was confirmed that RGES is almost the same as PEA in affinity toward cells.

It seems that the fourth residue X in RGDX is important, and X may contribute to hold the RGDX conformation to fit the cell attachment receptor. YIGSR-NH_2 is effective, because the minus charge $-COO^-$ at the C-terminal of YIGSR disappeared. Such a result agreed with Graf and co-worker's findings [17]. The percentage of cell attachment to PEA, PEA-RGD, PEA-RGDX, PEA-YIGSR, and PEA-YIGSR-NH_2 are dependent on the cell adhesion efficiency and anchorage dependency of the used cells. Among various films used in the experiments, PEA-RGDS seems to be the most remarkable in cell-attachment activity.

The peptide sequences which include the RGDX in cell-adhesive proteins, fibronectin, vitronectin and collagen are known as Thr-Gly-Arg-Gly-Asp-Ser-Pro-Ala (TGRGDSPA) [1], Val-Thr-Arg-Gly-Asp-Val-Phe-Thr (VTRGDVFT) [35], and Gly-Leu-Arg-Gly-Asp-Thr-Gly-Ala (GLRGDTGA) [3, 36], respectively. RGDS in fibronectin is emphasized with Gly and Pro locating before and after the RGDS sequence. But, in vitronectin and collagen, the amino acid residues locating before and after the RGDX sequence are those residues having bulky side chains and may weaken the interaction of RGDX with the receptor of the cells. The interaction of YIGSR series with cell receptors was found to be weaker than that of RGDX. YIGSR-NH_2 oligopeptide seems to accelerate the interaction of the oligopeptide with the receptor because the minus charge of the C-terminal of YIGSR was eliminated, and instead a plus charge appeared on the side chain of Arg residue. In other words, the receptors existing in the cell surface may properly interact with various oligopeptides as ligands.

Acknowledgments

The authors wish to thank Messrs Masaaki Shirai, Masayuki Morimoto, Kouji Hadani, and Katsuhiko Isota of the Osaka Institute of Technology, for their assistance in the experiments. This work was supported by the Grant-in-Aid

#01604026 to A.N. for Scientific Research on Priority Area, the Ministry of Education, Science and Culture, Japan.

REFERENCES

1. M. D. Pierschbacher and E. Ruoslahti, *Nature* **309**, 30 (1984).
2. D. F. Mosher (Ed.), *Fibronectin*. Academic Press Inc., New York (1989).
3. M. D. Pierschbacher and E. Ruoslahti, *Proc. Natl. Acad. Sci, USA* **81**, 5985 (1984).
4. K. M. Yamada and D. W. Kennedy, *J. Cell Biol.* **99**, 29 (1984).
5. J. E. Silnutzer and D. W. Barnes, In vitro *Cell Dev. Biol.* **21**, 73 (1985).
6. K. M. Yamada and D. W. Kennedy, *J. Cell. Biochem.* **28**, 99 (1985).
7. E. G. Hayman, M. D. Pierschbacher and E. Ruoslahti, *J. Cell Biol.* **100**, 1948 (1985).
8. M. Ginsberg, M. D. Pierschbacher, E. Ruoslahti, G. Marguerie and E. Plow, *J. Biol. Chem.* **260**, 3931 (1985).
9. Y. Takemoto, T. Matsuda, E. Alfaro, H. Fukumura, T. Kishimoto, M. Maekawa and T. Akutsu, *Jpn. J. Artif. Organs* **18**, 381 (1989).
10. E. Ozeki, T. Matsuda and T. Akutsu, *Jpn. J. Artif. Organs* **19**, 1078 (1990).
11. J. C. Boucaut, T. Darribere, T. J. Poole, H. Aoyama, K. M. Yamada and J. P. Thiery, *J. Cell Biol.* **99**, 1822 (1984).
12. J. B. McCarthy, S. T. Hagen and L. T. Furcht, *J. Cell Biol.* **102**, 179 (1986).
13. J. R. Couchman, M. Höök, D. A. Rees and R. Timpl, *J. Cell Biol.* **96**, 177 (1983).
14. H. K. Kleinman, F. B. Cannon, G. W. Laurie, J. R. Hassell, M. Aumailley, V. P. Terranova, G. R. Martin and M. Dubois-Dalcq, *J. Cell. Biochem.* **27**, 317 (1985).
15. J. E. Engel, E. Odermatt, A. Engel, J. Madri, H. Rhode and R. Timpl, *J. Mol. Biol.* **150**, 97 (1981).
16. Y. Iwamoto, F. A. Robey, J. Graf, M. Sasaki, H. K. Kleinman, Y. Yamada and G. R. Martin, *Science* **238**, 1132 (1987).
17. J. Graf, R. C. Ogel, F. A. Robey, M. Sasaki, G. R. Martin, Y. Yamada and H. K. Kleinman, *Biochemistry* **26**, 6896 (1987).
18. J. Graf, Y. Iwamoto, M. Sasaki, G. R. Martin, H. K. Kleinman, F. A. Robey and Y. Yamada, *Cell* **48**, 989 (1987).
19. M. J. Humphries, K. Olden and K. M. Yamada, *Science* **233**, 467 (1986).
20. I. Saiki, J. Iida, I. Azuma, N. Nishi and K. Matsuno, *Int. J. Biol. Macromol.* **11**, 23 (1989).
21. J. Murata, I. Saiki, I. Azuma and N. Nishi, *Int. J. Biol. Macromol.* **11**, 97 (1989).
22. B. K. Brandley and R. L. Schnaar, *Anal. Biochem.* **172**, 270 (1988).
23. H.-B. Lin, Z.-C. Zhao, C. García-Echeverría, D. H. Rich and S. L. Cooper, *J. Biomater. Sci. Polymer Edn.* **3**, 217 (1992).
24. S. P. Massia and J. A. Hubbell, *Biochemistry* **187**, 292 (1990).
25. S. P. Massia and J. A. Hubbell, *J. Biomed. Mater. Res.* **25**, 223 (1991).
26. Y. Hirano, T. Hayashi, K. Goto and A. Nakajima *Polymer Bull.* **26**, 363 (1991).
27. Y. Hirano, Y. Kando, T. Hayashi, K. Goto and A. Nakajima, *J. Biomed. Mater. Res.* **25**, 1523 (1991).
28. Y. Hirano, M. Okuno, T. Hayashi and A. Nakajima, *Polymer J.* **24**, 465 (1992).
29. F. T. Kao and T. T. Puck, *Proc. Natl. Acad. Sci. USA* **64**, 872 (1968).
30. J. E. De Larco and G. J. Todaro, *J. Cell. Physiol.* **94**, 335 (1978).
31. T. T. Puck, P. I. Marcus and S. J. Cieciura, *J. Exp. Med.* **103**, 273 (1956).
32. D. M. Prescott, *Methods in Cell Biology* IV, 1 (1973).
33. D. J. Giard, S. A. Aaronson, G. J. Todaro, P. Arnstein, J. H. Kersey, H. Dosik and W. P. Parks, *J. Natl. Cancer Inst.* **51**, 1417 (1973).
34. T. Takaoka, S. Yasumoto and H. Katsuta, *Jap. J. Exp. Med.* **50**, 329 (1980).
35. S. Suzuki, A. Oldberg, E. G. Hayman, M. D. Pierschbacher and E. Ruoslahti, *EMBO J.* **4**, 2529 (1985).
36. S. Dedhar, E. Ruoslahti and M. D. Pierschbacher, *J. Cell Biol.* **104**, 585 (1987).

Receptor-mediated regulation of differentiation and proliferation of hepatocytes by synthetic polymer model of asialoglycoprotein

A. KOBAYASHI[1], M. GOTO[1], K. KOBAYASHI[2] and T. AKAIKE[3]*

[1]*Kanagawa Academy of Science and Technology, Takatsu-ku, Kawasaki 213, Japan*
[2]*Faculty of Agriculture, Nagoya University, Chikusa-ku, Nagoya 464-01, Japan*
[3]*Faculty of Bioscience and Biotechnology, Tokyo Institute of Technology, Midori-ku, Yokohama 227, Japan*

Received 29 January 1993; accepted 7 January 1994

Abstract—Morphology and responses of hepatocytes are investigated using an artificial asialoglycoprotein model polymer—lactose-carrying polystyrene (PVLA) as a culture substratum, especially in focusing on the effect of the surface density of the PVLA substratum. The surface density of PVLA on polystyrene dishes was determined using fluorescein-labeled PVLA as a probe under a fluorescence laser microscope. PVLA-coated surfaces were observed by scanning electron microscope and atomic force microscopies under air and water, which showed that PVLA molecules were adsorbed patchily on low density surfaces and uniformly concentrated all over the dish on high density surfaces. It is suggested from the requirement of the Ca^{2+} ion, inhibition of galactosyl substances, and localization of receptors that the adhesion of hepatocytes to both low and high PVLA-density surfaces is mediated by galactose-specific interactions between PVLA and asialoglycoprotein receptors. At low PVLA densities (0.07 $\mu g\ cm^{-2}$), the hepatocytes were flat and expressed high levels of ^3H-thymidine uptake and low levels of bile acid secretion. Contrastingly, at high PVLA densities (1.08 $\mu g\ cm^{-2}$), they were round and expressed a low level of ^3H-thymidine uptake and a high level of bile acid secretion. The shapes, proliferation, and differentiation of hepatocytes could be regulated by varying the densities of PVLA adsorbed to polystyrene dishes. We assume that there are two recognition mechanisms operating between PVLA and hepatocytes: (1) adhesion through highly concentrated or clustered galactose-specific interaction; and (2) responses in shape, proliferation, and differentiation by PVLA-coating densities.

Key words: Lactose-carrying polystyrene (PVLA); hepatocytes; control of proliferation and differentiation; cell-shape; asialoglycoprotein receptors; bioartificial liver.

INTRODUCTION

Proliferation and differentiation of hepatocytes are regulated not only by soluble factors but also extracellular matrices. For example, DNA synthesis in cultured hepatocytes is induced [1–4] by several soluble proteins such as insulin, epidermal growth factor (EGF), and hepatocyte growth factor (HGF), and inhibited [5, 6] by transforming growth factor (TGF-β). Extracellular matrices such as collagen, laminin, and fibronectin have been recently revealed to control spreading, migration, adhesion, and proliferation of hepatocytes since they have several globular domains which are specialized for binding to a particular molecule or cell [7–10]. Cell–cell contact is also important in controling proliferation and differentiation of hepatocytes [11]. When hepatocytes on collagen were not in contact with each other, they were stimulated to grow but their differentiated

* To whom correspondence should be addressed.

functions were depressed. When they were in contact with each other, they did not grow but expressed highly differentiated functions. In these interactions, cell–matrix interactions are especially important. The surroundings of cells are arranged through these interactions which may determine control functions and cell-shapes.

We are interested in the regulation of the proliferation, differentiation, and shapes of hepatocytes using an artificial cellular matrix, PVLA, as a substratum (containing an ideal ligand to asialoglycoprotein receptors on hepatocytes [12–15]). We reported that a lactose-carrying polystyrene, poly[*N*-*p*-vinylbenzyl-4-*O*-*β*-D-galactopyranosyl-D-gluconamide] (PVLA, Fig. 1) [16] is a useful substratum for hepatocyte culture [17–21]. The behavior of hepatocytes on PVLA were distinct from those on naturally occurring substrata such as collagen, fibronectin, laminin, and proteoglycan. It was reported that hepatocytes on PVLA showed a round morphology and expressed highly differentiated functions. They could be detached easily by treating the culture with EDTA to remove Ca^{2+} ions from the medium [20]. Epidermal growth factors and insulin stimulated the formation of multicellular aggregates of hepatocytes [22]. The co-culture of hepatocytes with nonparenchymal liver cells induced the formation of tissue-like structures composed of hepatocytes, Kupffer cells, and endothelial cells. It was demonstrated that the adhesion is mediated by the galactose-specific interactions between hepatocytes and PVLA which carries highly concentrated *β*-galactose residues along the polymer chain.

It was found that regulation of the proliferation, differentiation, and hepatocyte-shape could be achieved when various concentrations of aqueous PVLA solutions were used to coat the culture dishes. These findings have been extended in this paper. First, an estimation method of the PVLA-coating density on polystyrene dishes was established and the surfaces were examined by various types of microscopes. Second, the adhesion behavior of hepatocytes depending on the PVLA-coating density were investigated in order to confirm the participation of asialoglycoprotein receptors in adhesion. Third, it has been found that morphology, functions, and proliferation of hepatocytes are closely correlated with the PVLA-coating density and microscopic surface structures on the artificial extracellular matrix. The culture system using the asialoglycoprotein model polymer (PVLA) as a substratum is useful to investigate regulations of hepatocytes.

MATERIALS AND METHODS

Preparation of polymer-coated dishes

PVLA and its reference substance PVMA (maltose-carrying polystyrene) (Fig. 1) were prepared via homopolymerization of *N*-*p*-vinylbenzyl-*O*-*β*-D-galactopyranosyl-(1-4)-D-gluconamide and *N*-*p*-vinylbenzyl-*O*-*α*-D-glucopyranosyl-(1-4)-D-gluconamide respectively [16]. Aqueous solutions of PVLA (100, 10, 5, and 1 $\mu g\ ml^{-1}$) were prepared by dissolving the required amounts of PVLA in deionized water. Collagen (3 $mg\ ml^{-1}$ solution, Type 1 collagen, Koken Co., Tokyo, Japan) was diluted with 0.001 *N* HC to 300 $\mu g\ ml^{-1}$. An aliquot (0.5 ml) of PVLA or collagen solution was placed in each well (15.6 mm diameter) of a polystyrene bacteriological multiwell plate (Sumilon MS 80240, Sumitomo Bakelite Co., Japan) at room temperature for 12 h. The solution was then decanted and each well was rinsed with Dulbecco's phosphate buffer solution (PBS, Nissui, Co., Tokyo, Japan) or distilled water three times.

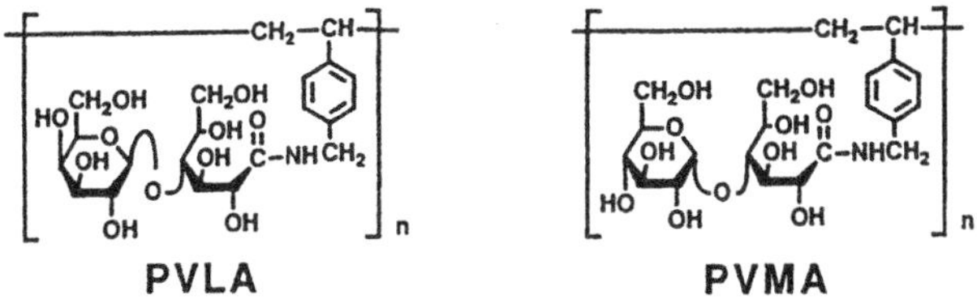
PVLA
PVMA

Figure 1. The chemical structures of PVLA and PVMA.

Determination of the amount of adsorbed FITC-PVLA

PVLA labeled with fluorescein isothiocyanate (FITC) was prepared by treating PVLA (1 g) with FITC (50 mg, Wako Pure Chem. Co., Japan) dibutyltin dilaurate (15 mg), and one drop of pyridine in dimethyl sulfoxide (5 ml) at 60°C for 12 h, followed by precipitation into ethanol and dialysis through a cellulose tube (Sanwa Seiyaku, Japan) against distilled water. About one FITC molecule was introduced to 100 structural units in PVLA.

The calibration curve was prepared between the amounts of FITC-PVLA and fluorescence color values. Aqueous solutions containing known amounts of fluorescein-labeled PVLA (400 μl) were poured into polystyrene dishes (Falcon 1008 bacteriological petri dishes, 35 mm diameter), the solvent was dried uniformly at 40°C, and the average fluorescence intensities of the dried surfaces were determined with a confocal fluorescence laser microscope. The dish surface was excited with a 488-nm argon laser beam irradiated from the bottom and the fluorescence was detected at 530 nm with a confocal laser microscope (ACAS 570, Meridian Co., USA). The scan area was 18 mm × 18 mm, which was enough to detect the amount of FITC-PVLA on the polystyrene dish.

Various concentrations (1, 5, 10, and 100 $\mu g\ ml^{-1}$) of the FITC-PVLA solution were placed in polystyrene dishes for a designated time. The solutions were decanted, and the surfaces were rinsed with distilled water three times and dried at 40°C. The fluorescence color values of the dish surfaces were determined and the amount of FITC-PVLA adsorption on polystyrene dishes was calculated from the calibration curve.

Microscopic observation of PVLA-coated polystyrene surface

Polystyrene dishes (15.6 mm diameter) were treated with 0.5 ml of PVLA solution for 12 h. After decanting, the surface was rinsed with distilled water five times, and dried under vacuum at room temperature for 3 h. The dried samples were sputtered with 10 Å of gold and observed with an SEM (S-900, Hitachi, Japan) at 1×10^5 magnification. The AFM images were obtained with an atomic force microscope (Digital Instruments, USA) under air by Nanoscope 2 (the scan area, 1 μm × 1 μm), and under distilled water by Nanoscope 3 (the scan area, 2 μm × 2 μm).

Preparation and culture of cells

Hepatocytes were isolated from the livers of female Sprague Dawley rats (150–200 g) by the modified collagenase perfusion technique of Seglen [23]. The viability of the isolated hepatocytes was determined by the trypan blue staining method. Hepatocytes with higher than 90% viability were used for the following experiments. The cell density was adjusted to 4×10^4 cells cm^{-2} in Williams' E medium containing 10^{-6}M insulin, 10^{-6}M dexamethasone, 100 units ml^{-1} of penicillin, and 100 mg ml^{-1} streptomycin, with or without 2.5% fetal calf serum (FCS). An aliquot was seeded onto a polymer-coated culture dish and maintained at 37°C in a humidified air/CO_2 incubator (95/5 vol.%) for a prescribed time. The number of unattached cells was counted with a hemocytometer under a phase-contrast microscope (Olympus IM).

Effect of divalent cations and inhibitors

The effect of divalent cations was investigated using Hanks' balanced salt solution (HBSS) in the presence and absence of 1.8 mM of Ca^{2+} and Mg^{2+} ions. Inhibitors were dissolved in the Williams' E medium containing bovine serum albumin (0.03 mM). The hepatocytes were treated with the medium in a humidified air/CO_2 incubator at 37°C for 60 min and then plated and cultured on PVLA-coated dishes of different PVLA densities for 30 min. Non-adhering cells were collected and the number of cells was counted with a hemocytometer.

Movement of asialoglycoprotein receptors along hepatocyte membranes

Hepatocytes were preincubated on PVLA-coated dishes of different PVLA densities (0.07 and 1.08 $\mu g\ cm^{-2}$) at 37°C for a prescribed time. The medium was removed and the hepatocytes were treated with fluorescein-labeled PVLA (10 $\mu g\ ml^{-1}$) in Williams' E medium containing bovine serum albumin (0.2% w/v) at 4°C for 60 min. The medium was removed and the hepatocytes were rinsed with PBS containing Ca^{2+} and Mg^{2+} ions at 4°C three times, and then fixed with paraformaldehyde (Wako Pure Chem. Co., Japan) (4% w/v) at 4°C. Fluorescence images of FITC–PVLA bound to hepatocytes were observed with a confocal laser microscope and the average fluorescence color values were calculated. Control experiments were carried out using PVMA-coated dishes.

Measurement of the projected area of cultured hepatocytes

After the hepatocytes were cultured for 2 days, the cells were treated with 0.01% (w/v) fluorescein isothiocyanate (FITC, Wako Pure Chem. Co., Japan) in PBS at room temperature for 20 min and then rinsed with PBS. The cells in each well were fixed by incubation in 2 ml glutaraldehyde (Wako Pure Chem. Co., Japan) (2.5% w/v) in PBS and rinsed three times with PBS. The cells were excited with a 488-nm argon laser emission and the fluorescence was monitored at 530 nm (ACAS 570, Meridian Co., USA). The average projected area of each cell was calculated from the fluorescence image of FITC labeled hepatocytes at the same magnification using the standard algorithm provided by Meridian Co.

Assay of the incorporation of ^{3}H-thymidine and growth of hepatocytes

After preincubation for 24 h, hepatocytes were cultured for 2 days, and then ^{3}H-thymidine (0.5 μCi per well) was added and incubated for 17 h. The hepatocytes were rinsed with cold Hanks' solution three times, with 10% trichloroacetic acid and 70% ethanol twice, and with 100% ethanol twice. After drying, they were treated with 0.3 ml of 1 M NaOH for 1 h to allow complete digestion of the cells. The sample was neutralized with 0.3 ml of 1 M HCl, and 0.3 ml of the resulting solution in each sample was counted in 3 ml Atomlight (NEN Research Products, Boston).

Hepatocyte growth was assayed by counting the number of cells with a phase contrast microscope after 3 days culture including preincubation for 24 h. Each well was treated with PBS (1 ml) containing EDTA (0.02% w/v) and trypsin (0.25% w/v) and incubated at 37°C for 20 min. The hepatocytes detached from the substratum were collected by pipette, centrifuged at 50 *g* for 90 s, resuspended in Williams' E medium, and the number of cells then counted with a hemocytometer.

Assay for bile acid secretion

Hepatocytes (5.2×10^5 cells ml^{-1}) were incubated in 4.5 ml of Williams' E medium containing 5 K IU ml^{-1} of aprotinin, 10^{-9} M of insulin, and dexamethasone for 24 h. The dishes were rinsed with the same medium three times and then incubated with 4.5 ml fresh medium for an additional 48 h. The conditioned mediums (4.5 ml) were assayed using a 3α-hydroxysteroid dehydrogenase–diaphorase method [24].

RESULTS

Determination of PVLA-coating densities

The fluorescent color values (y) on polystyrene dishes increased linearly with the amounts (x) of FITC-labeled PVLA (Fig. 2). $y = (1.18x + 0.5) \times 10^3$ (correlation coefficient: $r = 0.99$). Once given the calibration, we can estimate the PVLA-coating density of the dish whose surface was equilibrated with aqueous FITC–PVLA solutions. FITC–PVLA solutions of 1, 5, 10, and 100 $\mu g\ ml^{-1}$ gave PVLA-coating densities of 0.07, 0.76, 0.90, and 1.08 $\mu g\ cm^{-2}$, respectively.

Microscopic observation of PVLA-coated surfaces

The surfaces of the dishes were examined with a confocal laser microscope and an SEM and AFM. Different surface appearances were observed on PVLA-coated polystyrene dishes having low (0.07 $\mu g\ cm^{-2}$) and high (1.08 $\mu g\ cm^{-2}$) PVLA-coating

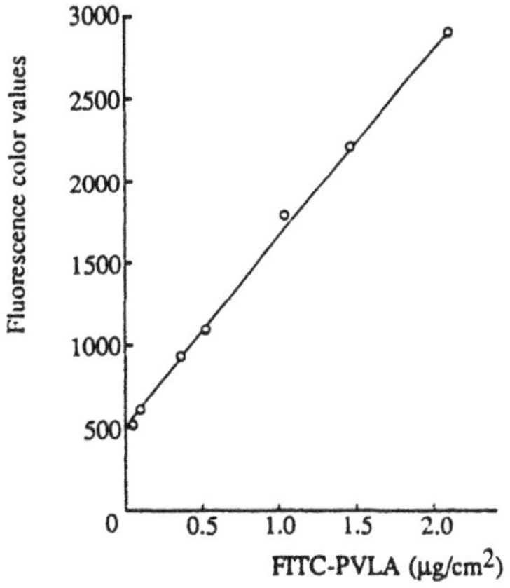

Fluorescence color values
3000
2500
2000
1500
1000
500
0
0.5
1.0
1.5
2.0
FITC-PVLA (μg/cm²)

Figure 2. Calibration of fluorescence color values and amounts of fluorescein isothiocyanate-labeled PVLA (FITC–PVLA) on polystyrene dishes.

densities and non-coated polystyrene dishes (Figs 3–5). According to the observation with a confocal laser microscope, strong and homogeneous fluorescence color values appeared over the whole surface of the dish coated with high density FITC–PVLA, and faint fluorescent color values appeared on the surface coated with low density FITC–PVLA (data not shown).

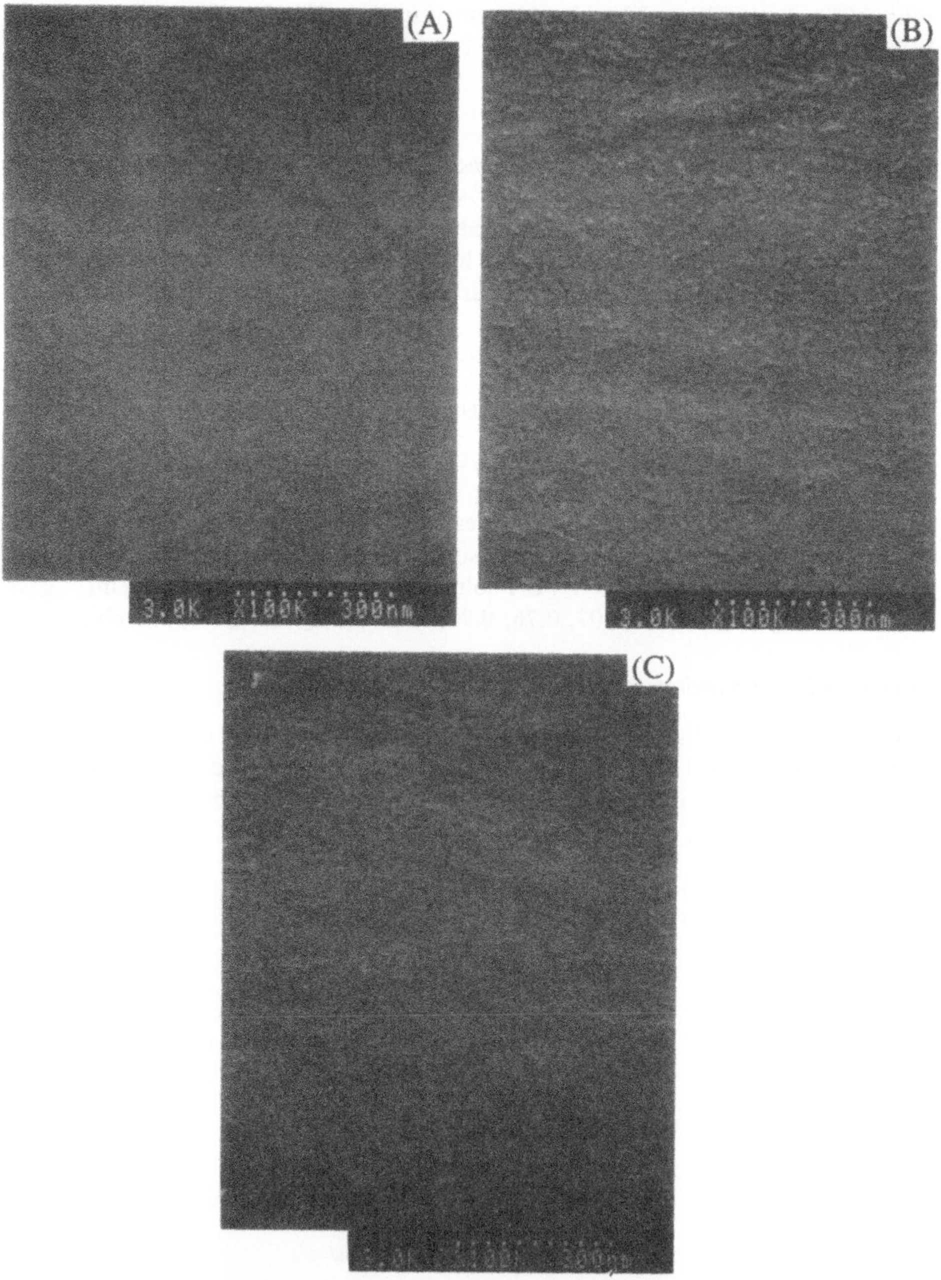
(A)
3.0K X100K 300nm
(B)
3.0K X100K 300nm
(C)

Figure 3. SEMs of the surface of PVLA-coated dishes. PVLA-coating densities: (A) $0\,\mu g\,cm^{-2}$; (B) $0.07\,\mu g\,cm^{-2}$; and (C) $1.08\,\mu g\,cm^{-2}$.

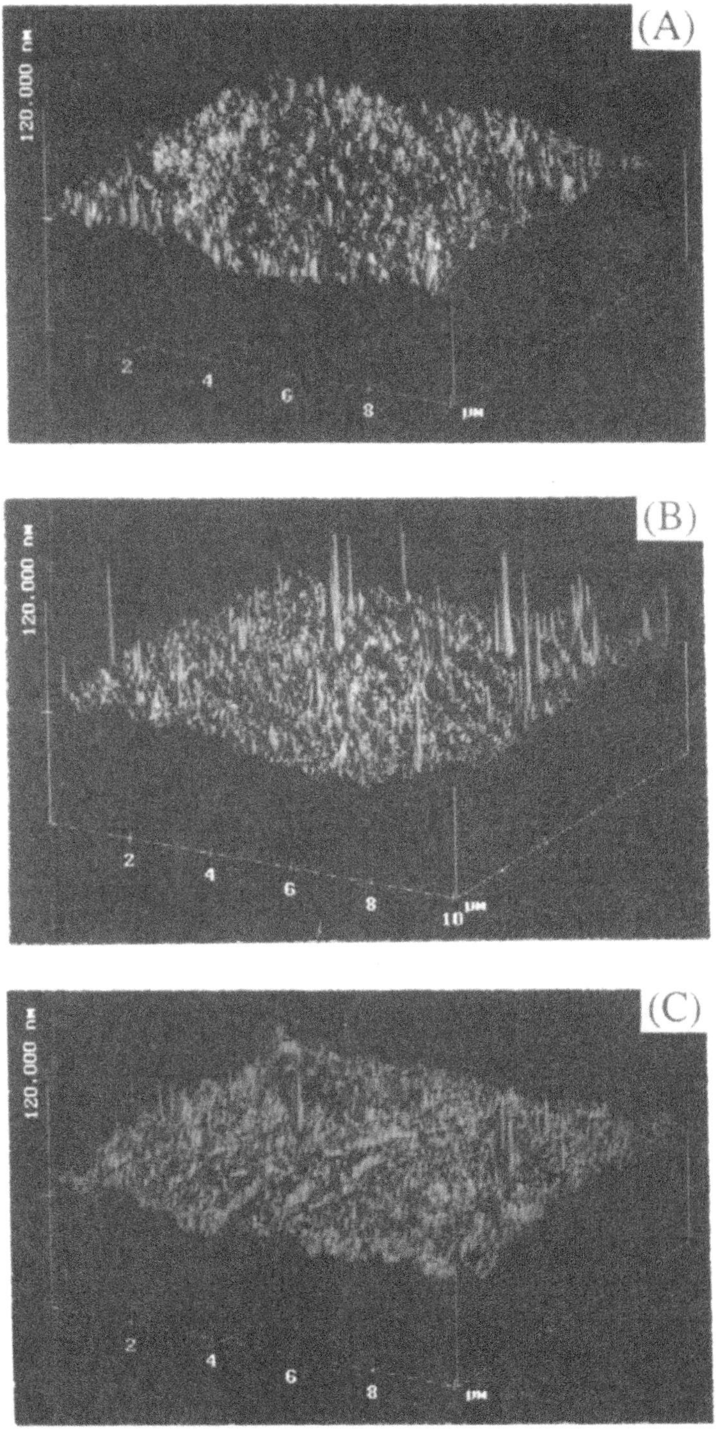
(A)
120.000 nm
2
4
6
8
μm
(B)
120.000 nm
2
4
6
8
10
μm
(C)
120.000 nm
2
4
6
8
μm

Figure 4. AFM images (under air) of PVLA adsorbed to polystyrene dishes. PVLA-coating densities: (A) 0 $\mu g\ cm^{-2}$: (B) 0.07 $\mu g\ cm^{-2}$; and (C) 1.08 $\mu g\ cm^{-2}$.

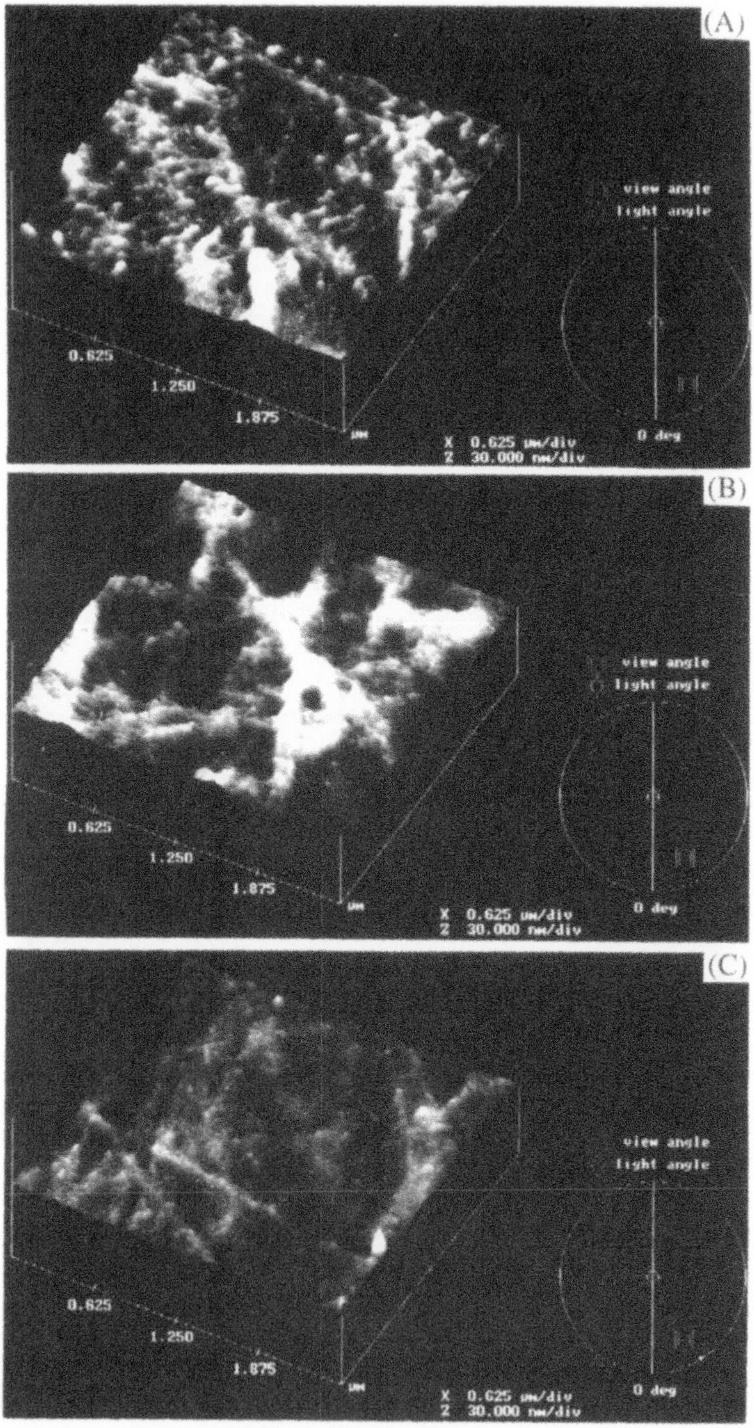
(A)
view angle
light angle
0.625
1.250
1.875
µm
X 0.625 µm/div
Z 30.000 nm/div
0 deg
(B)
view angle
light angle
0.625
1.250
1.875
µm
X 0.625 µm/div
Z 30.000 nm/div
0 deg
(C)
view angle
light angle
0.625
1.250
1.875
µm
X 0.625 µm/div
Z 30.000 nm/div
0 deg

Figure 5. AFM images (under the distilled water) of PVLA adsorbed to polystyrene dishes. PVLA-coating densities: (A) 0 $\mu g\ cm^{-2}$; (B) 0.07 $\mu g\ cm^{-2}$; and (C) 1.08 $\mu g\ cm^{-2}$.

Observation of these surfaces was difficult with conventional SEMs, but successful with a high performance SEM (S-900 Hitach) operated under low acceleration voltage. Patch-like patterns, probably resulting from the formation of aggregates of PVLA molecules, were scattered here and there on the SEM photograph (Fig. 3B) of the low PVLA-density surface, which is in contrast to the rather smooth surface of the non-coated surface (Fig. 3A). On high PVLA-density surfaces (Fig. 3C), a lot of small pockmarks similar to moon-craters were observed here and there, but the surface became rather smooth, indicating a homogeneous adsorption of PVLA all over the dish.

Figure 4 is an AFM image of a 1 μm × 1 μm area under air. On the low-PVLA density surface, some sharp, high, and mountainous patterns were superimposed onto small, rather uniform mountains (Fig. 4B), the latter of which were similar to the patterns observed on non-coated dishes (Fig. 4A). On the high density-PVLA surfaces, the number of sharp, high mountains decreased and the mountainous patterns became dispersed (Fig. 4C). Under wet conditions (Fig. 5), smooth, roundish patterns different from those under air conditions are observed on the AFM images (an area of 2 μm × 2 μm) on the polystyrene dishes (Fig. 5A). A lot of small irregularities appeared on the low PVLA-density surfaces (Fig. 5B). High PVLA-density surfaces became flat and less irregular (Fig. 5C).

In summary, the surfaces of non-coated, low and high PVLA-density dishes were different from each other. PVLA molecules were adsorbed in dispersed clusters on low density surfaces and uniformly concentrated over the dish on high density surface.

Adhesion behavior of hepatocytes on PVLA having different coating-densities

Figure 6 shows time courses of hepatocyte adhesion to PVLA-coated dishes. Hepatocyte adhesion became as high as 80% after 120 min incubation, whether or not serum was present in the culture medium. On the other hand, hepatocyte adhesion to PVMA-coated dishes with various coating densities was low (about 60%), especially in the presence of serum (about 20% data not shown).

The interaction between asialoglycoprotein receptors and galactoses requires Ca^{2+} ions. Therefore, we examined the effect of divalent cations on hepatocyte attachment to dishes having different PVLA-coating densities, which is summarized in Table 1. Adhesion to non-coated dishes was low and not dependent on the presence or absence of these cations. On PVLA-coated dishes, however, the effect of Ca^{2+} ions was remarkable, with more than 70% attachment when Ca^{2+} ions were present in the medium. Effective adhesion could not be induced by Mg^{2+} ions alone.

The effect of inhibitors on hepatocyte adhesion is summarized in Table 2. No inhibition was detected with glucose, PVMA, and fetuin. Galactose, lactose, and VLA (the monomeric substance leading to PVLA) were found to inhibit adhesion. However, rather large amounts of these low molecular-weight carbohydrates were required for effective inhibition. Contrastingly, the attachment was remarkably inhibited by quite small amounts of PVLA and asialofetuin. It is suggested from the galactose specific interactions, the effect of Ca^{2+} ions and inhibitors, that adhesion of hepatocytes to PVLA-coated dishes having low and high PVLA densities is mediated by asialoglycoprotein receptors.

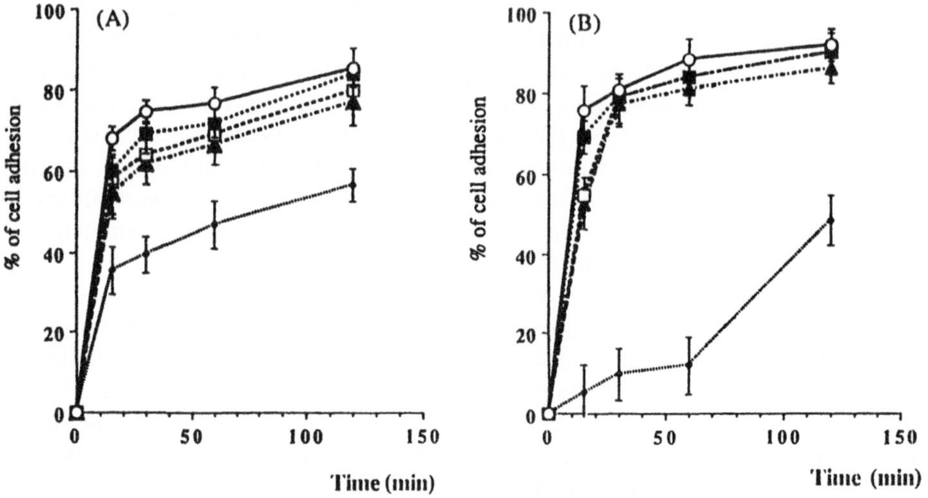
(A)
(B)
% of cell adhesion
Time (min)
0
20
40
60
80
100
0
50
100
150

Figure 6. Time course of hepatocyte attachment to PVLA-coated dishes. (A) in serum-free Williams' E medium and (B) in serum-supplemented Williams' E medium. PVLA-coating densities: ◇ 0 $\mu g\ cm^{-2}$; ○ 0.07 $\mu g\ cm^{-2}$; ■ 0.76 $\mu g\ cm^{-2}$; □ 0.90 $\mu g\ cm^{-2}$; and ▲ 1.08 $\mu g\ cm^{-2}$. Cell density, 4×10^4 cells cm^{-2} in Williams' E medium containing 10^{-6} M insulin and 10^{-6} M dexamethasone.

The experiments in Fig. 7 were carried out in order to clarify the behavior of asialoglycoprotein receptors. Hepatocytes were preincubated on PVLA- and PVMA-coated dishes and treated with the medium-containing FITC–PVLA probe. The fluorescence intensity on the hepatocyte surfaces was determined with a confocal microscope and the amount of bound FITC–PVLA plotted as fluorescent color values against preculture times. When the hepatocytes were preincubated on PVMA-coated dishes (Fig. 7A), high fluorescence values were detected for any preincubation times for both high and low PMVA densities. However, preincubation of hepatocytes on PVLA-coated dishes with low and high densities brought about a decrease of fluorescence values. The decrease was remarkable on high PVLA-density dishes (1.08 $\mu g\ cm^{-2}$) at early preincubation times (Fig. 7B).

The phenomenon can be elucidated on the basis of the movement of asialoglycoprotein receptors along the cell surface. When hepatocytes adhere to a PVLA-coated dish, an interaction occurs firstly between the asialoglycoprotein

Table 1.
The effect of divalent cations (Ca^{2+} and Mg^{2+} ions, 1.8 mM) on hepatocyte attachment to PVLA-coated dishes various coating densities

	PVLA-coating density ($\mu g\ cm^{-2}$)				
	Cell attachment (%)				
Conditions	0	0.07	0.76	0.90	1.08
Ca^{2+}, Mg^{2+} (1.8 mM)	40 ± 6	77 ± 2	73 ± 3	75 ± 5	71 ± 5
Ca^{2+} (1.8 mM)	38 ± 5	76 ± 4	72 ± 4	71 ± 2	69 ± 5
Mg^{2+} (1.8 mM)	43 ± 6	33 ± 3	12 ± 7	10 ± 4	8 ± 7
Without Ca^{2+}, Mg^{2+}	32 ± 5	30 ± 3	11 ± 4	10 ± 5	9 ± 5

Table 2.
The effect of inhibitors on hepatocyte attachment to PVLA-coated dishes at various coating densities

		PVLA-coating density ($\mu g\ cm^{-2}$)			
		Inhibition rate of cell attachment (%)			
Inhibitor conc.	(mM)	0.07	0.76	0.90	1.08
Glucose	30	8 ± 4	5 ± 8	7 ± 6	9 ± 7
	240	5 ± 3	3 ± 5	1 ± 2	14 ± 4
Galactose	30	18 ± 8	14 ± 2	7 ± 3	12 ± 6
	240	45 ± 7	28 ± 5	25 ± 7	25 ± 4
Lactose	30	15 ± 5	8 ± 6	1 ± 3	6 ± 9
	240	80 ± 7	88 ± 9	78 ± 9	86 ± 8
PVMA	2.1	12 ± 3	2 ± 5	7 ± 7	7 ± 4
VLA	30	95 ± 7	92 ± 6	96 ± 2	97 ± 5
PVLA	2.1	97 ± 10	97 ± 5	94 ± 7	98 ± 8
Fetuin	0.5	4 ± 5	14 ± 4	19 ± 7	28 ± 6
Asialofetuin	0.5	96 ± 8	94 ± 9	96 ± 5	97 ± 2

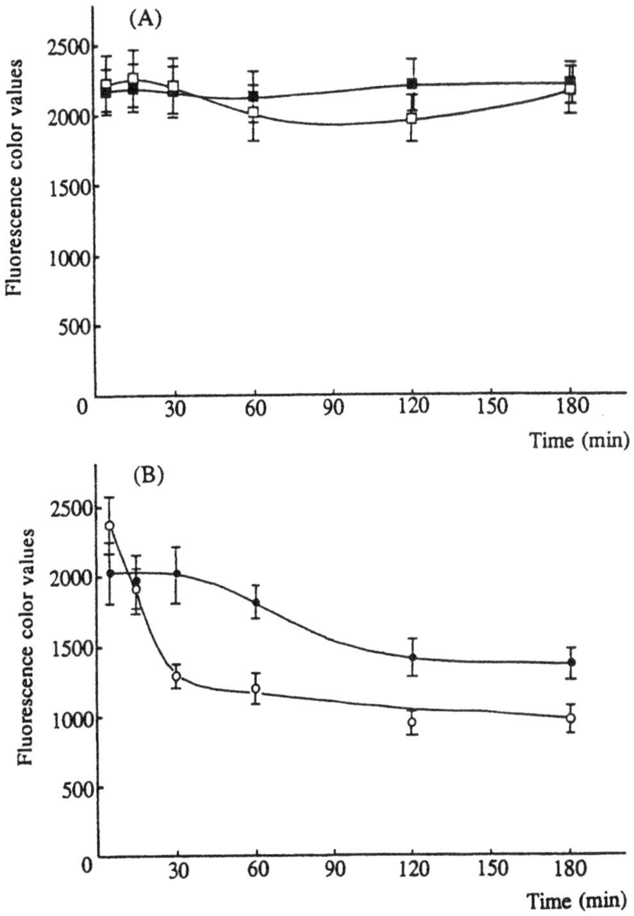
(A)
Fluorescence color values
2500
2000
1500
1000
500
0
30
60
90
120
150
180
Time (min)
(B)
Fluorescence color values
2500
2000
1500
1000
500
0
30
60
90
120
150
180
Time (min)

Figure 7. Dynamic behavior of hepatocyte membrane on (A) PVMA- and (B) PVLA-coated polystyrene dishes: □ and ○ high density; ■ and ● low density.

receptors on the cell surface in contact with the dish and the PVLA substratum on the dish. The receptors in contact with the medium move around the cell surface, and when they come into contact with the PVLA-coated surface, they are bound to the PVLA on the surface. As a result, receptors disappeared from the upper sides of cells over time—the disappearance time depending on the surface density of the PVLA.

Effects of PVLA-coating densities on morphology, differentiation, and proliferation of hepatocytes

The hepatocytes exhibited quite different morphologies depending on the PVLA-coating density. As reported previously, one of the most important characteristics of hepatocytes on PVLA is their round, tight and contracted morphology [17, 20]. When hepatocytes were incubated on the polystyrene dish, they showed a little dispersion (Fig. 8A), but on the dishes of high PVLA-coating density ($1.08\,\mu g\,cm^{-2}$), a round morphology was observed, as shown in Fig. 8C. Hepatocytes did not disperse even after 48 h incubation. On the dishes of low PVLA density ($0.07\,\mu g\,cm^{-2}$), however, they were not round but disperse as shown in Fig. 8B, which was similar to the commonly observed morphology on collagen-coated dishes (Fig. 8D).

The extent of round and dispersed morphologies can be estimated qualitatively by the projected areas. The average projected areas determined by a confocal laser microscope are plotted against PVLA-coating density in Fig. 9. The round-shaped hepatocytes with a high density of $1.08\,\mu g\,cm^{-2}$ had $(6.3 \pm 2.8) \times 10^2\,\mu m^2$. The dispersed morphology of the lower density of $0.07\,\mu g\,cm^{-2}$ had $(102 \pm 18) \times 10^2\,\mu m^2$. Hepatocytes on PVLA having a intermediate density were a mixture of dispersed and round shapes.

The relationship between the shape of hepatocyte and proliferation was examined by measuring the ^{3}H-thymidine uptake of hepatocytes as an index of proliferation. The uptake of ^{3}H-thymidine by hepatocytes are plotted against the PVLA density is shown in Fig. 10. Hepatocytes cultured on the highest PVLA density exhibited the lowest ^{3}H-thymidine uptake. However, the uptake increased with a decrease in PVLA density. At the lowest PVLA density, the highest ^{3}H-thymidine uptake was attained [$(2.9 \pm 0.3) \times 10^3$ dpm/(8×10^3) cells]. This value was similar to that of hepatocytes cultured on collagen where the hepatocytes were dispersed [$(3.0 \pm 1.3) \times 10^3$ dpm/(8×10^4) cells]. Similar tendencies were observed when the experiments were carried out in the presence and absence of EGF and serum and even after the surfaces of PVLA-coated dishes were rinsed with distilled water. It is reasonable to assume that the ^{3}H-thymidine uptake, as a measure of proliferation, is dependent on the shape of the hepatocytes. We assume that the hepatocytes exhibited the lower ^{3}H-thymidine uptake under the round morphology on the higher PVLA density, and the higher ^{3}H-thymidine uptake under the dispersed morphology on the lower PVLA density.

Hepatocytes were cultured in Williams' E medium containing serum for 1 day and then in the medium containing EGF instead of serum for an additional 2 days. Figure 11 shows phase-contrast microphotographs of the resulting hepatocytes. When hepatocytes were incubated on the polystyrene dish, there were many unpopulated spaces (Fig. 11A). With low-density PVLA, they could disperse and

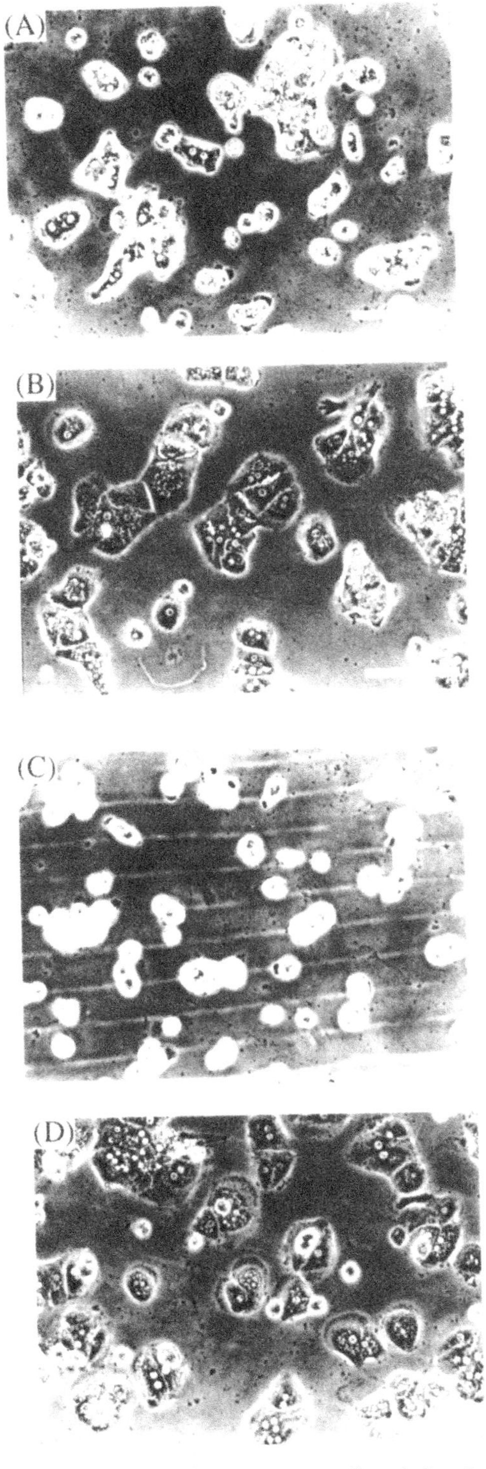
(A)
(B)
(C)
(D)

Figure 8. Phase-contrast microphotographs of hepatocytes cultured for 2 days (bar = 75 μm): (A) uncoated polystyrene dish; (B) on PVLA at 0.07 $\mu g\ cm^{-2}$ density; (C) at 1.08 $\mu g\ cm^{-2}$ density; and (D) on collagen coated with 300 $\mu g\ ml^{-1}$ solution.

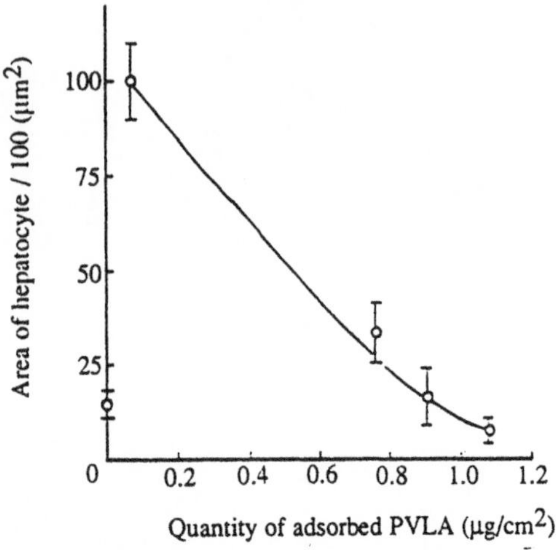
Area of hepatocyte / 100 (μm2)
100
75
50
25
0
0.2 0.4 0.6 0.8 1.0 1.2
Quantity of adsorbed PVLA (μg/cm2)

Figure 9. Average projected areas of hepatocytes on PVLA-coated dishes after incubation for 2 days.

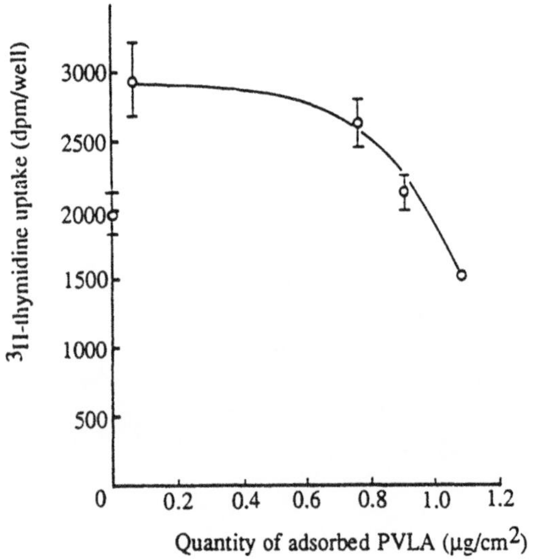
3H-thymidine uptake (dpm/well)
3000
2500
2000
1500
1000
500
0
0.2 0.4 0.6 0.8 1.0 1.2
Quantity of adsorbed PVLA (μg/cm2)

Figure 10. ^{3}H-Thymidine uptake by hepatocytes cultured on PVLA-coated dishes.

grow, but there were some unpopulated spaces (Fig. 11B), because it took about 24 h to disperse. With high-density PVLA, hepatocytes could hardly grow and had a round morphology (Fig. 11C). On collagen they grew and dispersed over the whole surface of the dish (Fig. 11D). Thereafter, the hepatocytes were detached by treating the dish with a solution of EDTA (0.02% w/v) and trypsin (0.25% w/v) at 37°C for 20 min to count the number of hepatocytes. Figure 12 shows that the number of cells detached from the low PVLA density is larger than that from the high PVLA density, the latter of which is comparable to that detached from the collagen-coated dish. It is assumed that the hepatocytes on low PVLA density showed growth similar to those on collagen.

Figure 13 shows secretion of bile acid as typical of the differentiated functions of hepatocytes which were cultured on PVLA-coated dishes with various coating

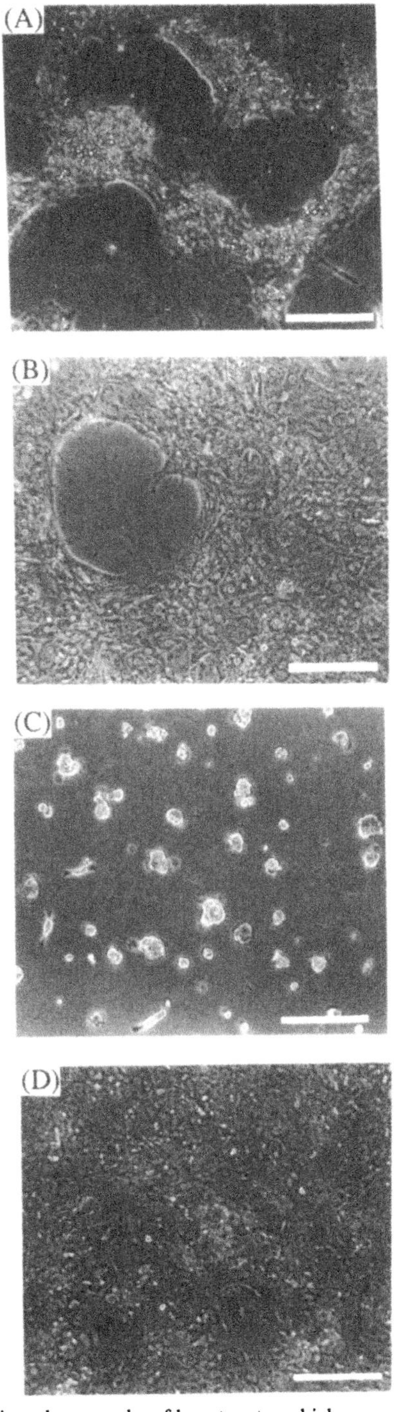
(A)
(B)
(C)
(D)

Figure 11. Phase-contrast microphotographs of hepatocytes which were cultured in Williams' E medium containing serum for 1 day and then in the medium containing EGF instead of serum for 2 days (bar = 100 μm): (A) uncoated polystyrene dish; (B) on PVLA at 0.07 $\mu g\,cm^{-2}$ density; (C) at 1.08 $\mu g\,cm^{-2}$ density; and (D) on collagen coated with 300 $\mu g\,ml^{-1}$ solution.

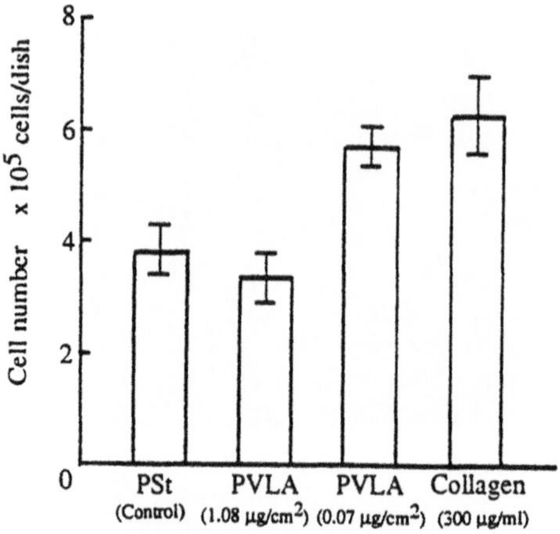
Cell number x 10^5 cells/dish
8
6
4
2
0
PSt
(Control)
PVLA
(1.08 μg/cm²)
PVLA
(0.07 μg/cm²)
Collagen
(300 μg/ml)

Figure 12. Growth of hepatocytes on various substrata. Hepatocytes were cultured on the substrata for 3 days, detached with a solution of EDTA (0.02% w/v) and trypsin (0.25% w/v) at 37°C for 20 min and counted with a hemocytometer.

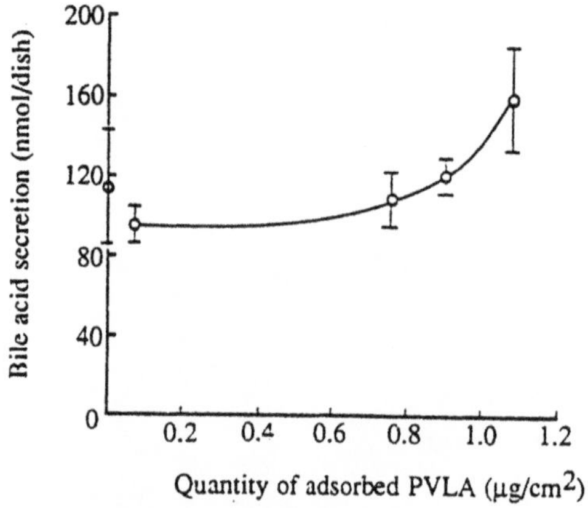
Bile acid secretion (nmol/dish)
200
160
120
80
40
0
0.2
0.4
0.6
0.8
1.0
1.2
Quantity of adsorbed PVLA (μg/cm²)

Figure 13. Secretion of bile acid of hepatocytes on PVLA.

densities. At high PVLA densities (1.08 μg cm^{-2}), on which attached hepatocytes were round in shape, high bile acid secretion (160 ± 30 nmol/2.3 × 10^6 cells) was observed. On the other hand, at low PVLA densities (0.07 μg cm^{-2}), on which the attached hepatocytes were dispersed, low bile acid secretion (95 ± 10 nmol/2.3 × 10^6 cells) was observed. A correlation between the shape of hepatocyte and differentiated function (bile acid release) was also observed. Namely, hepatocytes cultured at high PVLA densities are round with low ^{3}H-thymidine uptake and high bile acid release, whereas at low densities, the hepatocytes are dispersed with high ^{3}H-thymidine uptake, low bile acid release, and a similar growth rate to hepatocytes on collagen coated dishes.

DISCUSSION

PVLA has an amphiphilic structural unit composed of hydrophilic carbohydrate and hydrophobic vinylbenzyl moieties. Owing to these structural properties, the polymer molecules are strongly adsorbed onto the surface of polystyrene dishes as confirmed by several methods including XPS [19]. The polymer also forms multimolecular assemblies in water as suggested by dynamic light scattering. We suppose that the PVLA characteristic is reflected on the microscopic observations. The molecular assemblies were adsorbed patchily on low density surfaces, and uniformly concentrated on high density surfaces. We assume that the difference in surface structures, depending on the coating density, causes some effect on the shape of the hepatocytes.

It was reported that proliferation, differentiation, and shape of the hepatocytes were regulated by changing the coating densities of the extracellular matrices (ECM) such as laminin, fibronectin, and collagen [10]. At lower coating densities, the hepatocytes had a round morphology and maintained highly differentiated functions such as secretion of albumin, transferrin, and fibronectin. At their higher coating densities, both dispersion and proliferation were enhanced. Hepatocytes trapped in collagen gels retained a round morphology owing to the shrinkage of gels and they expressed differentiated functions. It is possible that the kind of ECM and its 3D structure regulate expression of differentiated functions and proliferation of hepatocytes.

The inherent role of asialoglycoprotein receptors *in vivo* is the clearance of asialoglycoproteins from blood through endocytosis. Nevertheless, shape, proliferation, and differentiation of hepatocytes were regulated directly through the PVLA substratum. The receptors may not be related in signal transduction necessary for the regulation of differentiation and proliferation, but may play an indirect role by interactions through cytoskeletons. We assume that differentiation and proliferation of hepatocytes could be regulated by the following two different types of recognition mechanisms operating on hepatocytes cultured in PVLA-coated dishes. The first is the galactose-specific recognition between PVLA molecules and the asialoglyprotein receptors of hepatocytes. The second recognition is based on the different densities of adsorbed PVLA on polystyrene dishes, which brought about a change of cell shape and regulation of differentiation and proliferation. We would like to conclude that control of differentiation and proliferation in hepatocytes could also be achieved through the artificial matrix designed to interact specifically with asialoglycoprotein receptors on the cell surface.

REFERENCES

1. T. Nakamura, Y. Tomita and A. Ichihara, *J. Biochem.* **94,** 1029 (1983).
2. G. Michalopoulos, H. D. Cianciulli, A. R. Novotony, A. D. Kligerman, S. C. Strom and R. L. Jirtle, *Cancer Res.* **42,** 4673 (1982).
3. J. A. McGowan and N. R. L. Bucher, *In Vitro* **19,** 159 (1983).
4. G. K. Michalopoulos, *FASEB J.* **4,** 176 (1990).
5. T. Nakamura, Y. Tomita and R. Hirai, *Biochem. Biophys. Res. Commun.* **133,** 1042 (1985).
6. W. E. Russel, R. J. Coffey, Jr., A. J. Ouellette and H. L. Moses, *Proc. Natl. Acad. Sci. USA* **85,** 5126 (1988).
7. A. Ben-Ze'ev, S. L. Farmer and S. Penman, *Cell* **21,** 365 (1980).
8. J. Folkman and A. Moscona, *Nature* **273,** 345 (1978).
9. N. Sawada, A. Tomomura, C. A. Sattler, G. L. Sattler, H. K. Kleinman and H. C. Pitot, *Exp. Cell Res.* **167,** 458 (1986).
10. D. Mooney, L. Hansen, J. Vacanti, R. Langer, S. Farmer and D. Ingber, *J. Cell Physiol.* **151,** 497 (1992).
11. T. Nakamura, K. Yoshimoto, Y. Nakayama, Y. Tomita and A. Ichihara, *Proc. Natl. Acad. Sci. USA* **80,** 7229 (1983).

12. T. Tanabe, W. E. Pricer, Jr. and G. Ashwell, *J. Biol. Chem.* **254**, 1038 (1979).
13. L. Vanlenten and G. Ashwell, *J. Biol. Chem.* **247**, 4633 (1972).
14. T. Tanabe, W. E. Pricer, Jr. and G. Ashwell, *J. Biol. Chem.* **254**, 1038 (1979).
15. T. Sawamura, H. Nakada, Y. Fujii-Kuriyama and Y. Tashiro, *Cell Structure Function* **5**, 133 (1980).
16. K. Kobayashi, H. Sumitomo and Y. Ina, *Polym. J.* **17**, 567 (1985).
17. A. Kobayashi, T. Akaike, K. Kobayashi and H. Sumitomo, *Makromol. Chem., Rapid Commun.* **7**, 645 (1986).
18. K. Kobayashi, H. Sumitomo, A. Kobayashi and T. Akaike, *J. Macromol. Sci. Chem.* **A25**, 655 (1988).
19. T. Kugumiya, A. Yagawa, A. Maeda, H. Nomoto, T. Tobe, K. Kobayashi, T. Matsuda, T. Onishi and T. Akaike, *J. Bioactive Compatible Polym.* **7**, 337 (1992).
20. A. Kobayashi, K. Kobayashi and T. Akaike, *J. Biomater. Sci. Polym. Edn* **3**, 499 (1992).
21. T. Akaike, *Proceedings of the 1st Japan International SAMPE Symposium*, Nov. 28–Dec. 1 (1989).
22. S. Tobe, Y. Takei, K. Kobayashi and T. Akaike, *Biochem. Biophys. Res. Commun.* **184**, 225 (1992).
23. P. O. Seglen, *Methods Cell Biol.* **13**, 29 (1976).
24. S. Skrede and H. E. Solberg, *Clin. Chem.* **24**, 1095 (1978).

Growth of human cells on plasma polymers: Putative role of amine and amide groups

HANS J. GRIESSER[1]*, RONALD C. CHATELIER[1], THOMAS R. GENGENBACH[1], GRAHAM JOHNSON[2] and JOHN G. STEELE[2]

[1]*Division of Chemicals and Polymers, CSIRO, Private Bag 10, Clayton 3168, Australia*
[2]*Division of Biomolecular Engineering, CSIRO, PO Box 184, North Ryde 2113, Australia*

Received 16 September 1992; accepted 31 March 1993

Abstract—The attachment and growth of human endothelial cells and fibroblasts was studied on polymer surfaces fabricated by the polymerization of volatile amine and amide compounds in a low pressure gas plasma, and by the treatment of various surfaces in ammonia plasmas, which served to increase the nitrogen content of the surface layers. Infrared spectra showed the presence of amide groups, including those cases where the volatile compound ('monomer') did not contain oxygen. The performance of the surfaces in cell attachment correlated with the surface hydrophilicity and the nitrogen content, although for the latter a fair degree of scatter indicated that a more complex relationship applies. All these surfaces supported the attachment and growth of human cells. Generally, amide plasma polymers were best but the individual monomer and the plasma parameters also played a role. From comparisons of the various surfaces, it is suggested that the amide group is the main promoter of cell attachment in nitrogen-containing plasma surfaces.

Key words: Endothelial cell; fibroblast; cell growth; plasma polymerization; plasma surface treatment; amine groups; amide groups.

INTRODUCTION

The ability of polymer surfaces to support the attachment and growth of anchorage-dependent cells varies widely. Polystyrene, polyethylene terephthalate, polytetrafluoroethylene, and perfluorinated ethylene propylene copolymer, for instance, are poor supports [1-6], while other polymeric materials suitable for the attachment and proliferation of cells are now commercially available. The attachment of cells is mediated by adhesive glycoproteins such as fibronectin and vitronectin [7-10], which compete with other proteins for adsorption on to polymer surfaces. The composition of the protein layer is influenced by the properties of the polymer surface [11-14]. On some surfaces, proteins are bound very tightly and resist elution [15, 16]. Much attention has been directed towards identifying the role of particular chemical groups on polymer surfaces in the attachment of proteins and cells, and surface modification by gas plasma (glow discharge) methods has become popular as a versatile means of fabricating various surface chemistries for evaluation in biological tests [17-21].

In vitro cell growth assays are particularly convenient for the study of the effects of varying surface composition on biological responses. Most studies have focussed on surfaces with oxygen-containing groups, often produced by oxidative plasma

* To whom correspondence should be addressed.

surface treatments or by the plasma polymerization of oxygen-containing monomers, and the role of various oxygen-containing functional groups in promoting the attachment of cells has been discussed. The involvement of surface sulfonate [22], hydroxyl [2, 23–27], or carboxyl [27–30] groups has been suggested, but a recent study found little correlation between the growth of bovine endothelial cells and the surface density of hydroxyl and carboxyl groups; instead, cell growth correlated with the density of carbonyl groups [31].

Little information is available on the role of nitrogen-containing groups in cell attachment. When nitrogen gas was admixed to acetone vapor used for plasma polymerization, the resultant incorporation of nitrogen-containing groups into the plasma polymer coating promoted only moderate growth of bovine endothelial cells [5]. The treatment of poor cell growth supports with ammonia plasma, on the other hand, led to improved performance [32, 33] for reasons which are not understood. A possible advantage of placing amine groups on surfaces is that a fraction of them may be positively charged at physiological pH: since most proteins and cells carry a net negative charge, their interaction with the surface will be enhanced. A disadvantage of treatment in an ammonia plasma is, however, that the modified surfaces undergo surface restructuring which results in a partial loss of the treatment effects [34].

In the light of the relative paucity of data regarding the role of amine and amide groups in supporting cell growth, we fabricated plasma polymers from amine and amide monomer vapors [35] and tested their ability to support colonization by human endothelial cells and fibroblasts. Other samples were fabricated by the plasma polymerization of oxygen containing monomers, and by two-step treatments involving the exposure of plasma polymer coatings to ammonia plasma. The latter provided additional degrees of freedom in fabricating various surface chemistries. In a separate study [36], experiments involving the removal of fibronectin and/or vitronectin from the culture medium were performed in order to study the relative involvement of these two adhesive glycoproteins in the promotion of cell attachment. The discovery that the requirement of serum proteins for cell attachment to nitrogen-containing surfaces differed from oxygen-containing surfaces [36] led us to surface analytical studies aimed at elucidating the relative contribution of the various possible nitrogen-containing groups in supporting cell attachment and growth.

This paper discusses currently available data on the chemical composition of plasma polymer surfaces from amine and amide monomers, describes the evaluation of their ability to support cell colonisation, and speculates about the chemical basis of the attachment mechanism. Although at this stage, the surface chemistry of these plasma polymers has not been determined in detail, we nevertheless are able to present a preliminary interpretation and to discuss some related issues in the study of the relationship between polymer surface compositions and cell colonization.

EXPERIMENTAL

Materials

Fluorinated ethylene propylene copolymer (FEP) was selected as the main polymeric substrate to be modified for cell culture assessment. It had several advantages over more commonly used materials such as polystyrene; namely, its perfluorinated composition which provides fluorine to be used as a marker for X-ray photoelectron spectroscopy (XPS) analysis, its poor cell attachment characteristics in an unmodified state [6] and its availability as a tape with a relatively smooth surface topography. PTFE tape was also investigated but was found by STM to possess a surface topography much rougher than FEP and unsuitable for angle resolved XPS work. FEP tape of 25 μm thickness (Du Pont FEP 100 Type A) was used as received following assessment of the cleanliness of its surface by XPS [37]. On the FEP batch used in this study, no contamination was observed, the C1*s* region showing a signal characteristic of fluorocarbon only, at 292.2 eV, with no hydrocarbon contribution at 285.0 eV. Surface cleanliness of fluorocarbon polymers is important as small amounts of hydrocarbon contamination can markedly affect the results of non-depositing plasma treatments [38]. As an additional check, several samples were subjected to ultrasonic washing in ethanol, and this produced no detectable changes in the contact angles and XPS elemental ratio F/C. Moreover, XPS was used periodically to check the cleanliness of samples from various locations along the roll of tape.

Another substrate used for plasma polymerizations was polyimide (Du Pont Kapton 100HN) tape which was custom aluminized, by Dr R. Spahn of Eastman-Kodak, Rochester, USA. This substrate provided a highly reflective layer onto which plasma polymer coatings were deposited. This multilayer structure enabled reflection infrared spectroscopic analysis of the plasma polymers, optimization of plasma conditions on-line, and estimation of the coating thickness under a given set of conditions.

The 'monomer' liquids heptylamine (HA), butylamine (BA), allylamine (AA), dimethyl formamide (DMF), dimethyl acetamide (DMAc), dimethyl propionamide (DMP), *n*-hexane (*n*H), methyl salicylate (MS), and hexamethyldisilazane (HMDSA), all from Ajax Chemicals (Auborn, Australia), were of purity $\geq$99% and were used as received; purification prior to plasma experimentation was unnecessary by adopting the following procedure. A fresh batch of the monomer liquid was placed in a round bottom flask and connected to the reactor chamber by a stainless steel line and a manual flow control valve. We removed any impurities which may have been present and which were more volatile than the monomer by pumping on the liquid for a few minutes prior to ignition of the plasma. This procedure is akin to a vacuum distillation as used for the purification of organic liquids, a process in which the initial and final fractions also are discarded. Ammonia gas (Matheson) was supplied from a cylinder via a stainless steel line and a mass flow controller (MKS).

Plasma equipment

Plasma experiments were performed using a custom built reactor [39] shown schematically in Fig. 1. The reactor chamber consisted of a vertical glass cylinder of diameter 175 mm and height 350 mm. The glow discharge was established between two vertical electrodes, mounted on PTFE blocks, of dimensions 90 $\times$ 18 mm and with a spacing of 16 mm. Two glass plates were held against the sides of the electrodes and the mounting blocks; they extended to the gas inlet which was situated immediately below the interelectrode space. Confined by the blocks and the glass plates, the incoming process gas flowed along a controlled path (a rectangular

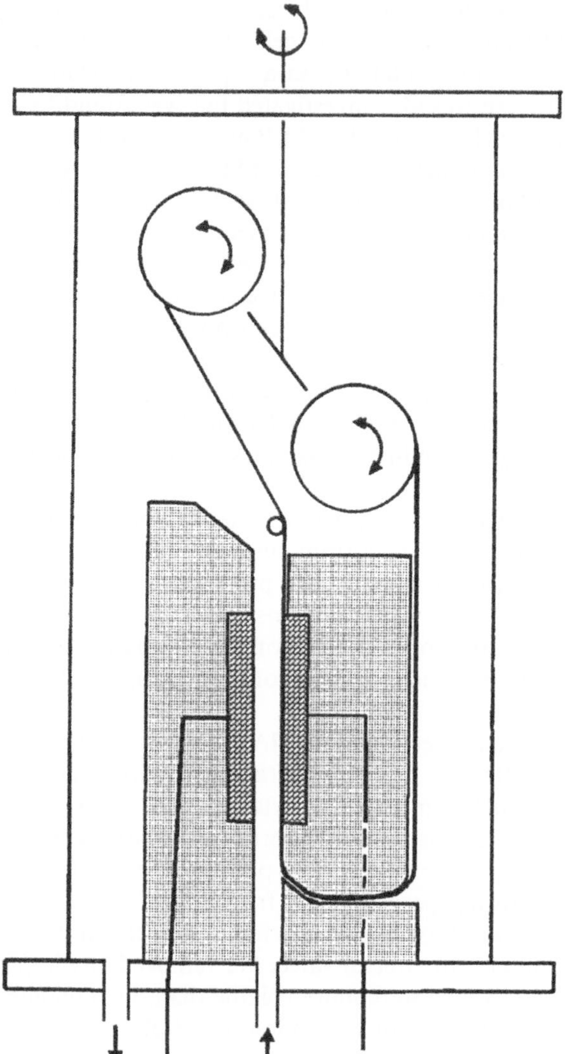
RF
Gas in
RF

Figure 1. Schematic diagram of the plasma reactor.

channel of 16 × 18 mm) to beyond the discharge region. Without losses to the sides upstream of the glow region, accurate determination of the amount of gas provided to the discharge zone and of the power/gas ratio was achieved, as opposed to configurations consisting of unconfined electrodes placed in a bell jar where much of the gas may bypass the glow discharge zone.

The glow discharge was powered by commercial generators, ENI models ACG-3 and HPG-2, operating at 13.56 MHz, and at variable frequency in the range 125–375 kHz, respectively. The other ancillary fittings, such as the pressure sensor (MKS Baratron) and the pumping line, were standard commercial items.

The reactor also incorporated a tape transport system designed for semi-continuous treatment of $\frac{1}{2}$ inch wide tape which can be moved through the plasma at a controlled speed. Up to 120 m of tape can be loaded per run, enabling the fabrication of multiple samples of several metres length each, thus providing numerous identical specimens for the various analyses. In addition, the tape drive was reversible, a feature which allowed liberation of air entrapped in the roll prior to treatments, and multiple plasma treatments without breaking vacuum. The latter feature was used for the subsequent surface treatment of plasma polymers. Plasma polymer coated tapes were partly re-wound immediately after plasma deposition and exposed to an ammonia plasma; this scheme was designed to provide additional variations in the range of amine/amide surface chemistries.

Treatment of extended lengths of tape presented several advantages. Glow discharges are of non-steady state composition for several seconds on startup and shutdown, whereas in our unit the coatings were produced from a stable, time-independent glow discharge. Also, it is more difficult to obtain uniform plasma polymer thickness when coating substrates attached to electrodes because the electric field strength varies with the position on the electrode. Often we found that the coating thickness diminished along the gas flow path; transporting the tape substrate eliminated the effect of spatial variation of the deposition rate. In our reactor, the surface chemistry of plasma polymers is determined by the chemistry in the upper (trailing) end of the plasma zone.

Plasma experimentation

Sections of aluminized Kapton tape were spliced in-between sections of FEP tape and the rolls loaded into the cleaned reactor. After extended pumpdown of the entire system, the exit valve to the pump was closed and the rate of pressure increase monitored. This enabled assessment of the sum of any leakage which may be present and the residual outgassing of air and water vapour from the substrate tape roll and the plastic reactor fittings. If this was satisfactorily low, the reactor was re-evacuated and the monomer vapor feed started.

The monomer vapor was evaporated at ambient temperature (22 ± 1°C) from the liquid contained in a round bottom flask, thermostatted by partial immersion in a water bath. The vapor was fed to the reactor chamber via a manually operated flow control/shutoff valve and the flow rate determined prior to deposition by recording the rate of pressure increase in the reactor with the exit valve to the pumping line closed. The exit valve was then partially opened such that a stable pressure reading was obtained at the set flow rate.

Performing plasma polymerizations onto sections of aluminized tape resulted in the production of interference coloring at a plasma polymer thickness > ~50 nm, enabling visual on-line assessment of the uniformity of the coating, and an approximate estimate of the thickness, and hence the deposition rate. Once adequate uniformity of the interference color over most of the electrode area was achieved, the tape transport was started and an extended section of aluminized tape coated; in stable plasmas, constant thickness, as shown by constant interference color, resulted. The deposition was continued on to the following section, of transparent FEP tape, on which no interference color could be observed on-line on account of the small difference between the refractive indices of plasma polymers and FEP (off-line observation at low angle usually showed a weak color).

Several sets of the experimental plasma parameters (power, pressure, flow rate, frequency) were used for each monomer. The range of parameters conducive to obtaining good coatings from each monomer depends markedly on the geometry of the electrodes and the gas flow. As our reactor differs substantially from reactors employed in other laboratories, it is not instructive to detail plasma parameters. Three DMAc and two HA samples fabricated under different conditions were evaluated in detail, in order to ascertain that the results were qualitatively independent of the particular set of fabrication parameters.

After shutdown of the plasma, pumping was maintained for a few minutes with monomer flow, then pumping continued with the monomer flow shut off, and the reactor finally brought up to atmospheric pressure by venting with nitrogen gas. The plasma modified tapes were then removed and cut into sections, thus providing a large number of identical specimens. The start and end sections were discarded. For a few samples, XPS analyses were performed on a number of specimens cut from various positions along the treated tape; no compositional changes were detected within experimental error. The surface composition of plasma polymers, and also of plasma treated FEP, did, however, change on post-deposition storage due to oxidative processes as shown by XPS [34, 40]. These compositional changes were reflected in changes to contact angles with time. Samples intended for cell culture experimentation were stored in clean tissue culture polystyrene dishes at ambient temperature until contact angles had been stable for at least two weeks (see Results).

Contact angle determinations

Air-water contact angles were measured with triply distilled water using a modified Kernco G-II Contact Angle Meter equipped with a syringe comprising a reversible plunger driven by a micrometer, to allow determination of advancing (ACA), sessile (SCA), and receding (RCA) contact angles. The SCA was determined by placing a drop of water on the surface and recording the angle between the horizontal plane and the tangent to the drop at the point of contact with the substrate. The ACA was measured by inserting the needle of the syringe into the drop, adding water to the drop and recording the maximum angle achieved. The RCA was measured by removing water from the drop and recording the minimum angle achieved. Five independent determinations were averaged. Ultrasonically cleaned FEP was used as a reference material to check reproducibility of measurements and absence of artefacts, such as contamination of the water. The values determined on clean FEP were ACA = 117, SCA = 107, and RCA = 98 deg. The uncertainty in the contact angle data was generally ± 3 deg, except at low CAs where tangential optical adjustment became much more difficult.

Spectroscopic analyses

Compositional surface analysis by XPS was performed on a VG Escalab V spectrometer using non-monochromatic AlK_α radiation at a power of 200 W. The pressure in the analysis chamber typically was 2×10^{-9} mb. The binding energy scale was calibrated using sputter cleaned foils of nickel, silver, gold, and copper. The charging of the polymer samples was corrected by referencing to the CF_2 contribution as an internal standard with a binding energy of 292.2 eV. Elements present were identified by survey spectra. High resolution spectra were acquired to determine elemental surface compositions; these were calculated from numerically integrated peak areas, using a Shirley background correction, according to a first principles method [41]. A new specimen cut from the same tape region was used for each XPS analysis in order to avoid accumulation of effects arising from the inevitable slow degradation under the non-monochromatic X-ray irradiation. The effects of sample decomposition during analysis were minimized by exposing specimens to radiation for less than 1 min before the start of data acquisition and limiting the total analysis time to less than 30 min. Changes to the elemental composition due to sample degradation were thus kept below 5% as assessed by separate experiments involving the monitoring of compositional changes over extended periods of irradiation. In addition, analysis of several specimens on a monochromatic SSX-100 spectrometer at NESAC/BIO, University of Washington, Seattle, showed that the non-monochromatic analyses were sufficiently accurate. Angle resolved XPS was performed using two emission angles, θ, of 0 and 70 deg as measured from the surface normal.

Infrared spectra of the plasma polymer films coated onto aluminized Kapton tape were measured on a BOMEM MB-100 FTIR spectrometer fitted with a SpectraTech FT80 Specular Reflectance attachment at a fixed grazing angle of 80 deg; the incident beam was plane-polarized parallel to the surface of the tape. Spectra were accumulated with a resolution of 4 cm^{-1} over 64 scans or with a resolution of 8 cm^{-1} over 32 scans. As the IR bands of plasma polymers tend to be broader than bands of conventional polymers, higher spectral resolution is not beneficial. Prior to each analysis a spectrum of a clean mirror was recorded to compensate for the presence of small vibrational bands due to water vapor and CO_2.

Plasma polymer samples were surface analyzed both before and after cleaning by ultrasonication in ethanol. Results were identical in most cases. For a few samples, however, partial delamination appeared to occur, as indicated by the presence of fluorine in the XPS spectra after ultrasonication; these samples were not used in subsequent evaluations.

Cell culture

Human umbilical vein endothelial (HUVE) cells were isolated and cultured according to the method of Jaffe *et al.* [42]. The cells were routinely grown on tissue-culture polystyrene (TCPS) coated with 40 $\mu g\ ml^{-1}$ bovine plasma fibronectin and maintained in a culture medium containing Medium 199 (ICN, Sydney) supplemented with 20% (v/v) fetal calf serum (FCS; CSL, Sydney), 100 $\mu g\ ml^{-1}$ heparin, 60 $\mu g\ ml^{-1}$ endothelial cell growth supplement (both Sigma Chemicals), 60 $\mu g\ ml^{-1}$ penicillin and 100 $\mu g\ ml^{-1}$ streptomycin (both CSL, Sydney). The human dermal fibroblast cell strain (HDFb) was a gift from Dr Robin Holliday and was maintained in a culture medium containing Dulbecco's Modified Eagles Medium (ICN, Sydney) supplemented with 10% (v/v) FCS (PA Biologicals, Sydney), 60 $\mu g\ ml^{-1}$ penicillin and 100 $\mu g\ ml^{-1}$ streptomycin (CSL, Sydney). All cells were maintained at 37°C in an atmosphere of 5% CO_2 in air.

Cell attachment

In order to assess cell attachment, sample materials were cut into pieces 10 mm square and ultrasonically cleaned in 70% ethanol prior to cell seeding. Cell attachment assays were performed by placing the samples, modified side up, in 16 mm diameter tissue culture trays (TCPS, Corning 24-well, Cat. No. 3847) and seeding with metabolically labelled cells as previously described [10]. Briefly, on the day before the attachment assay, confluent cell cultures were passaged, then metabolically labelled by culture for 18 h in methionine-free Medium 199 (Cytosystems, Sydney) containing either 15% (v/v) FCS for the HUVE cells, or 10% (v/v) FCS for the HDFb cells, both supplemented with ^{35}S-methionine (Amersham, Australia) at a concentration of 5 μCi ml^{-1}. Following labelling, the cells were treated with 0.1% trypsin/0.02% EDTA solution, centrifuged and re-suspended in fresh serum-free medium. Cells were then seeded in the desired FCS/medium combinations to give a final density of 3–5 $\times$ 10^4 cells per 16 mm well and incubated for 90 min at 37°C in a mixture of 5% CO_2 in air. Following incubation, the wells were washed twice with PBS to remove non-attached cells. The remaining cells were removed by overnight digestion with a solution of 0.1% trypsin and 0.02% EDTA, transferred to counting vials and the amount of ^{35}S determined by liquid scintillation counting. Cell attachment was expressed as a proportion of the number of cells which attached to TCPS in medium containing either 15% (v/v) or 10% (v/v) intact FCS for HUVE or HDFb cells, respectively.

Cell growth

For cell growth studies, the sample preparation was the same as for the attachment assay. Non-labelled cells were seeded at a density of 3–5 $\times$ 10^4 per 16 mm well in 10% (v/v) intact FCS in DMEM for the HDFb, or 20% (v/v) intact FCS in medium 199 containing 100 μg ml^{-1} heparin and 60 μg ml^{-1} endothelial cell growth supplement for the HUVE cells. The culture medium was replenished every second day. Nine randomly selected fields of each culture were photographed per day for up to 7 days, or until confluent. Cell numbers were determined from the photographic negatives and the mean value and the standard error of the mean of each set of fields calculated. A commercial, polystyrene based material with a nitrogen-containing surface (Primaria, Becton Dickinson) was also evaluated for comparison with the plasma modified FEP surfaces.

RESULTS AND INTERPRETATION

When coating onto moving metallized Kapton substrate tape, uniform interference colors were obtained for the central 8–10 mm of plasma polymers coated onto 12.7 mm wide tape. The outermost 2 mm of the tape were thus not used for evaluation. Although faint, the interference colors of plasma polymers coated onto moving FEP tape likewise appeared uniform over a width of at least 8 mm and tape lengths of several meters.

Following deposition, HA, *n*H, and MS plasma polymers were also surface treated in ammonia plasmas; HA with the intention of increasing the density of amine groups, in the case of *n*H as a means of producing an analog for the HA plasma polymer but with amine groups only on the surface rather than throughout the coating, and in the case of MS for study of whether attaching nitrogen-containing groups onto an oxygen-containing surface might affect the mechanism of cell attachment. For comparison, MS was also polymerized with some nitrogen gas admixed. The effects of these treatments were readily observable by contact angle measurements, the results of which are listed in Table 1.

Table 1.
Sessile (SCA), advancing (ACA), and receding (RCA) air–water contact angles, N/C and O/C ratios determined by XPS at an emission angle of 70 deg, and HUVE cell attachment (relative to TCPS) of selected plasma polymer coatings on FEP.

Sample	SCA (deg)	ACA (deg)	RCA (deg)	N/C	O/C	Cell attachment (%)
HA (1)	80	89	18	0.072	0.12	35.1
HA (2)	74	82	≤10	0.075	0.12	44.2
BA	65	72	≤10	0.059	0.14	61.9
AA	67	72	≤10	0.093	0.12	60.2
DMAc (1)	61	69	≤10	0.12	0.24	59.1
DMAc (2)	35	43	≤10	0.19	0.17	85.1
DMAc (3)	33	60	≤10	0.18	0.21	90.7
DMF	51	66	≤10	0.12	0.19	70.4
HMDSA	96	103	51	0.065	0.34	41.3
HA + NH_3	53	62	≤10	0.20	0.33	81.8
*n*H + NH_3	60	50	≤10	0.18	0.29	63.0
MS/N_2	53	63	≤10	0.052	0.30	58.6
MS + NH_3	44	59	≤10	0.17	0.26	72.5

Contact angles

Air–water contact angles (CAs) were used to screen plasma polymer coatings prior to cell culture studies. CA analysis using water as the test liquid can provide an indication of whether the polymeric surface changes significantly over a time scale of minutes, while cells attempt to settle and grow. Swelling and penetration of the probe liquid, dissolution of polymeric material, and reorientation of polymer chains can produce time dependent effects in CA determinations [43]. These processes may also give rise to complicated, time dependent interfacial interactions with adsorbing proteins when cell growth medium is brought into contact with the polymer surface. Time dependence of the cell assay substrate surface obviously causes additional complexities in attempts to correlate cell growth performance with particular surface compositions. Moreover, the analysis of mobile surfaces, by the XPS freeze-hydration technique [44], is much more involved; we have chosen to utilize surfaces which do not respond rapidly to a change in the environment from air to aqueous media. The criterion used was that CAs should decrease only slightly over observation periods of 10 min, consistent with only evaporation of some of the water occurring. A few samples whose CAs were not stable, or variable with position on the tape, were not used in further tests. The CAs of samples used in cell culture assays are compiled in Table 1.

Soluble low molecular weight fragments may possibly be present in some plasma coatings produced at low power levels; acting as surfactants, such fragments might interfere with CA determinations and the cell colonization assay, but at low levels may not be readily detectable in CA analysis. However, samples were also washed with 70% EtOH prior to cell seeding; hence, such interference is unlikely to have occurred.

Even when stable over periods of 10 min, however, CAs were observed to change, over periods of days, on storage in air. For plasma polymers the changes typically were relatively slow and small but for ammonia plasma treated FEP, a substantial

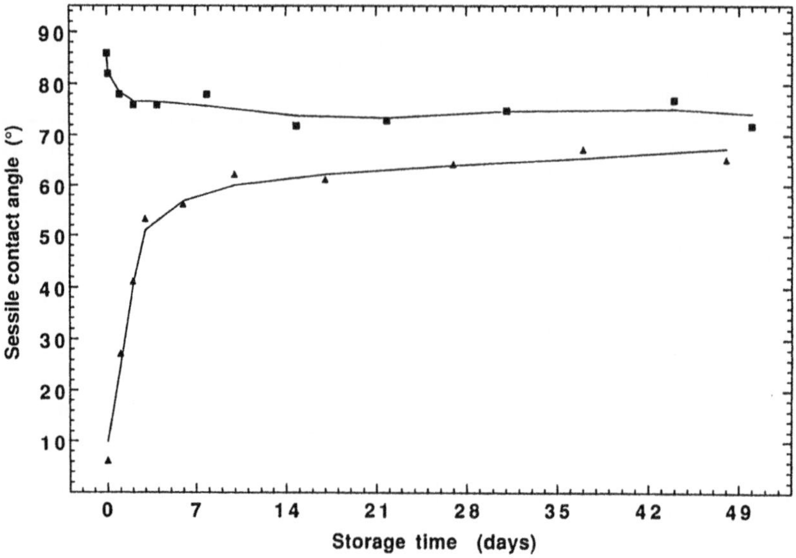
Sessile contact angle (°)
90
80
70
60
50
40
30
20
10
0
7
14
21
28
35
42
49
Storage time (days)

Figure 2. Air–water sessile contact angles of heptylamine plasma polymer on FEP (■) and ammonia plasma treated FEP (▲).

increase in the CAs was observed within days after fabrication. Representative results are reproduced in Fig. 2. Plots of CAs as a function of time of storage were used to ascertain the period of storage required for obtaining equilibrated, stable surface compositions prior to cell culture studies. The values listed in Table 1 are those determined after such equilibration.

IR spectroscopic analyses

The IR bands of all plasma polymers were rather broad; this is believed to result from the heterogeneous structure of plasma polymers, with random bonding patterns. For all amine and amide plasma polymers, strong absorption bands were observed around 3300 cm^{-1} (very broad) and around 1700 cm^{-1}; representative spectra are shown in Fig. 3. Interpretation of the spectra was complicated by the fact that all coatings contained oxygen-containing groups, as shown both by FTIR and XPS, even in plasma polymers deposited from monomers that do not contain oxygen. With leaks and significant residual outgassing absent, the only likely source was oxygen incorporation after deposition, by reaction of remaining radicals with atmospheric oxygen. This phenomenon, which is not instantaneous on venting, but continues on for several weeks, is discussed elsewhere [40]. As a consequence, the IR band at ~3300 cm^{-1} may contain contributions not only from amine and amide groups but also from hydroxyl and hydroperoxide groups, making it impossible to use this band for assessment of the amine and amide density.

Similarly, the C=O stretching mode at ~1700 cm^{-1} can originate from amides as well as carbonyls, carboxyls and other groups not directly associated with the precursor monomer. The C=O stretch spectral region is shown in Fig. 4 for plasma polymers fabricated from *n*H (without subsequent ammonia plasma treatment in this instance), HA, and DMAc. For the *n*H coating, the main group appears to be

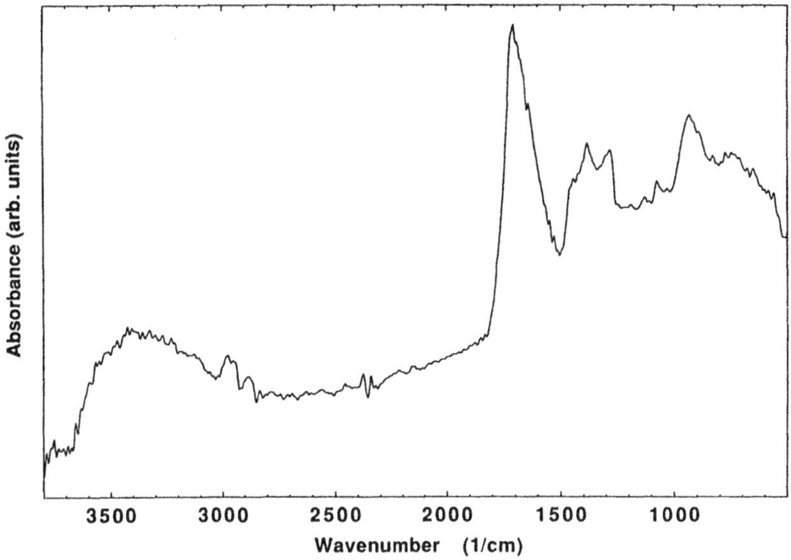
Absorbance (arb. units)
3500
3000
2500
2000
1500
1000
Wavenumber (1/cm)

Figure 3. Infrared spectra of a plasma polymer deposited from heptylamine.

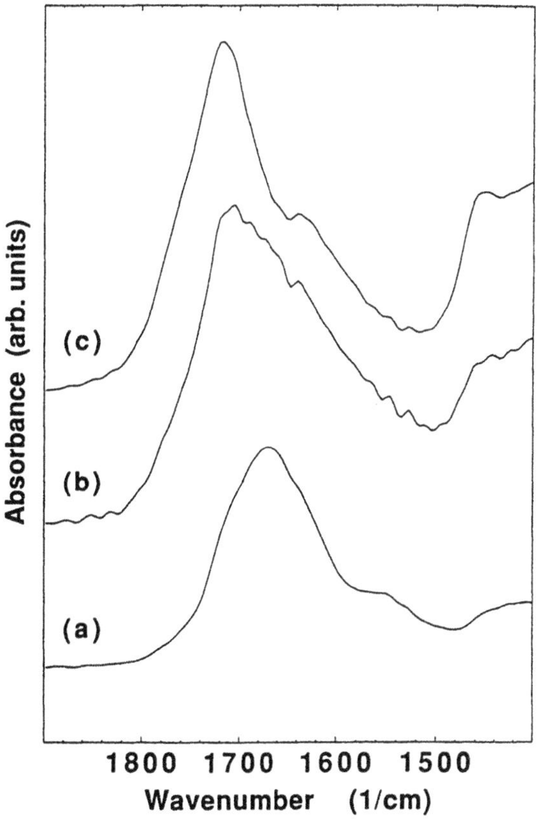
Absorbance (arb. units)
(c)
(b)
(a)
1800
1700
1600
1500
Wavenumber (1/cm)

Figure 4. Expanded part of the infrared spectra of plasma polymers from DMAc (a), HA (b), and *n*H (c).

carbonyl as judged from the peak position of 1717 cm^{-1}. For the DMAc plasma polymer, the main intensity is located at 1670 cm^{-1}, a frequency which is indicative of amide. In addition, a high intensity shoulder is also evident and suggests that some carbonyl groups are also present in the DMAc coating. For the HA plasma polymer, the C=O band evidently is a superposition of carbonyl and amide, the latter as a result of post-deposition oxidation adjacent to an amine group. Assuming that the extinction coefficients are the same for these two groups, more carbonyls are present.

It would be of interest to compare the amounts of amide groups in the various samples from IR spectra but quantitative comparison of samples by reflection IR is unreliable at present; firstly, there is considerable overlap of bands, and secondly, a reference band for calibration and compensation for possible polarization effects does not exist.

XPS analysis

XPS survey spectra (Fig. 5) invariably contained peaks assignable to C, N, and O. Table 1 lists elemental composition data obtained at a photoelectron emission angle of 70 deg. On our unit, elemental ratios generally are reproducible to ±10% or better for the elements and range of ratios applicable here. Few comparative

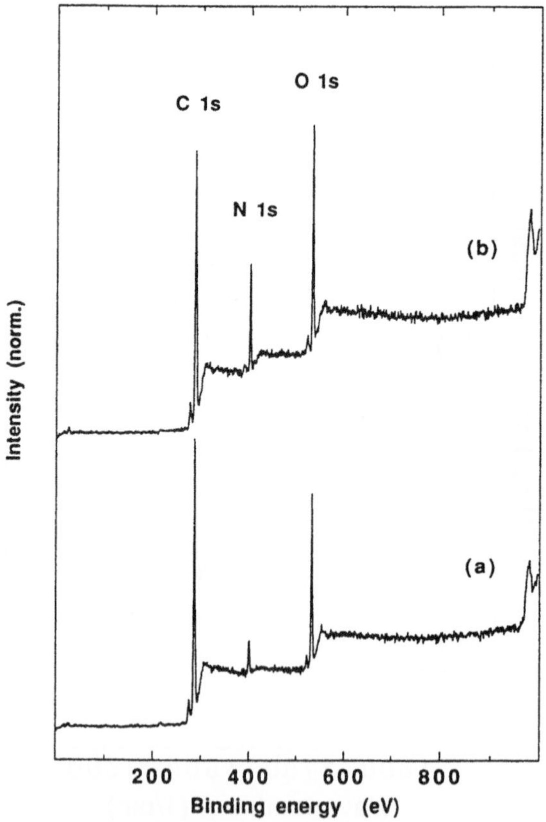
O 1s
C 1s
N 1s
(b)
(a)
Intensity (norm.)
200
400
600
800
Binding energy (eV)

Figure 5. XPS survey spectra of HA (a) and DMAc (b) plasma polymers.

literature data exist; Gombotz and Hoffman have previously fabricated and analyzed AA plasma polymers, and they also found oxygen in addition to nitrogen [45, 46]. With the different reactor geometries and operating parameters, however, one might expect the compositions of their AA coatings to differ somewhat from ours.

The incorporation of nitrogen into the *n*H, HA and MS plasma polymer surfaces by ammonia plasma treatment was quite efficient, leading to substantial amounts of nitrogen in the treated surfaces (Table 1). In fact, the combination of *n*H plasma polymerization followed by ammonia plasma surface treatment provided a higher density of incorporated nitrogen than the direct one-step HA polymerization, as well as a more convenient process alternative: the deposition of *n*H could be done faster because relatively low power levels had to be used in order to retain a substantial amount of nitrogen in HA plasma polymers. We also note in Table 1 that the subsequent ammonia plasma treatment of an HA plasma polymer led to a considerable increase in the nitrogen content.

Angle dependent XPS (ADXPS) was used to determine whether the composition of the samples varied with depth. Prior to ADXPS, however, identical specimens were analyzed by STM to ascertain whether the surface topography was smooth enough for meaningful work at a 70 deg emission angle. Untreated FEP tape was found to have a surface topography described by relatively smooth, regular undulations. Over a scan area of 1000 nm square the vertical excursion (difference between the highest and lowest point in the scan area) was between 21 and 45 nm, and over a scan area of 100 nm it was 7 to 11 nm. By contrast, PTFE tape showed a less regular and coarser surface topography described by a vertical excursion of 230 nm over a scan area of 1000 nm; this corresponded to slopes whose steepness prohibited angle resolved XPS studies. Hence, only plasma modified FEP tapes were used for ADXPS, at emission angles of 0 and 70 deg. While most plasma modified FEP samples showed a topography not much rougher than untreated FEP, the AA plasma polymer listed had a very rough surface topography, presumably due to rapid gas phase polymerization by double bond radical additions [47].

ADXPS showed little variation of the nitrogen content with depth of analysis for most samples; selected examples are listed in Table 2. One clear exception was the commercial reference material Primaria, whose fabrication method is not known to us. The oxygen content also varies with depth in Primaria.

The C1*s* region of two representative plasma polymers is reproduced in Fig. 6. With several functional groups and a variety of secondary shifts, curve fitting

Table 2.
Angle dependence of atomic ratios by XPS of ammonia plasma treated samples, selected plasma polymers on FEP, and Primaria.

Sample	N/C (70 deg)	N/C (0 deg)	O/C (70 deg)	O/C (0 deg)
FEP + NH_3	0.17	0.16	0.26	0.28
*n*H + NH_3	0.18	0.17	0.29	0.30
HA + NH_3	0.20	0.17	0.33	0.33
MS + NH_3	0.17	0.20	0.26	0.30
MS/N_2	0.052	0.065	0.30	0.30
HMDSA	0.065	0.067	0.34	0.34
DMAc (1)	0.12	0.12	0.24	0.25
Primaria	0.091	0.044	0.18	0.099

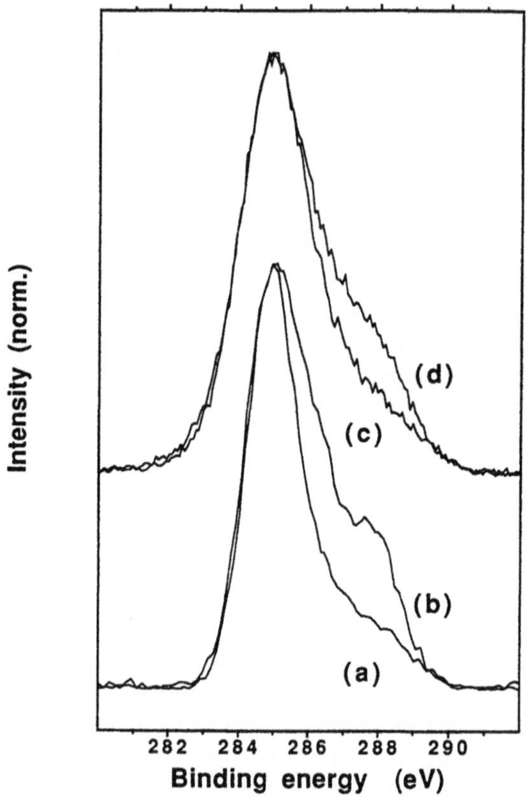
Intensity (norm.)
(d)
(c)
(b)
(a)
282
284
286
288
290
Binding energy (eV)

Figure 6. XPS C1*s* spectra of HA (a, c) and DMAc (b, d) plasma polymers recorded on the SSI X-Probe (a, b) and VG Escalab (c, d) spectrometers.

appeared not meaningful. The larger high energy tail of the DMAc plasma polymer was, however, consistent with the higher N and O content. The VG Escalab spectrometer showed a linewidth of 1.8 eV for the C1*s* signal of clean polyethylene compared to 2.1 eV for that of plasma polymers. On an SSI X-Probe 100 spectrometer with monochromatized AlK_α radiation, the linewidth was 1.6 eV for plasma polymers, confirming their inherently large linewidths. The C1*s* spectra of a DMAc and a HA plasma polymer obtained on the latter instrument are shown in Fig. 6 in addition to the corresponding spectra recorded from the same samples on the VG Escalab unit. The contributions from amides and carbonyls at approximately 288 eV are better resolved in the SSI X-Probe spectra, but there still is excessive overlap of components. In addition to multifunctionality, the relatively broad nature of these signals also indicates considerable variations in binding energies, consistent with varying secondary shift contributions within a heterogeneous matrix provided by the random structure of plasma polymers.

The N1*s* peak (not shown) of amine and amide plasma polymers typically was asymmetric and broad, particularly so for the DMAc coatings, suggesting a multiple signal. However, the difference in the N1*s* binding energy of amine and amide groups is far too small for reliable quantification by curve fitting. The same applied to the O1*s* signal.

Cell attachment and growth

The comparative performance of the various materials was tested in a short term (90 min) cell attachment assay using HUVE cells. Initial evaluation of samples was done in a number of runs at various times, following which a direct 'head-to-head' comparison of the most relevant samples was performed to obtain the data plotted here without the uncertainties involved in the calibration procedures between different cell runs. The data from the 'head-to-head' comparison are listed in Table 1, as percentages relative to TCPS, onto which close to a 100% of the total number of cells applied became attached. Note that all of the samples appeared to give less cell attachment than TCPS. This was a result of the assay format, in which pieces of modified FEP tape were placed at the bottom of the culture wells, as in preliminary experiments we demonstrated that when 15 mm diameter PET (Thermanox) coverslips were cut into squares the same size as the plasma modified samples and coated with fibronectin, the number of cells attaching to these PET pieces was consistently 75–80% of the number that attached to an intact, fibronectin coated PET coverslip.

The cell attachment data are plotted in Fig. 7 as a function of the air–water SCA of the surfaces. The ACA values paralleled the SCA values; hence, a plot of cell attachment versus ACA was very similar and is not reproduced. The RCA values of these samples were mostly so low (Table 1) that an analogous plot was not possible.

Figure 7 shows quite a good correlation between initial cell attachment and SCA. This suggests that cell attachment is promoted by the presence of polar surface groups. Both oxygen-containing (carbonyl, amide, and perhaps others) and nitrogen-containing (amine, amide) groups are present on these surfaces. Cell attachment did not correlate at all with the oxygen content of the surface (Fig. 8) whereas there was some correlation with nitrogen content (Fig. 9), with scatter indicating that another factor is also involved. A high polarity of the surface is evidently of most benefit for cell attachment (Fig. 7), suggesting that the chemical nature of the groups involved (i.e. their polarities) is as important as the absolute density of groups/heteroatoms. In other words, among the chemical groups which contain N and/or O, those that are less polar than others may be less effective in supporting cell attachment.

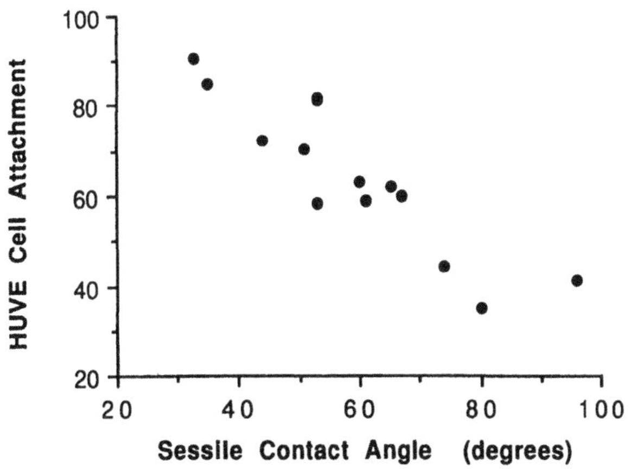
100
80
60
40
20
HUVE Cell Attachment
20
40
60
80
100
Sessile Contact Angle (degrees)

Figure 7. HUVE cell attachment (% relative to TCPS) as a function of the air–water SCA of the surfaces.

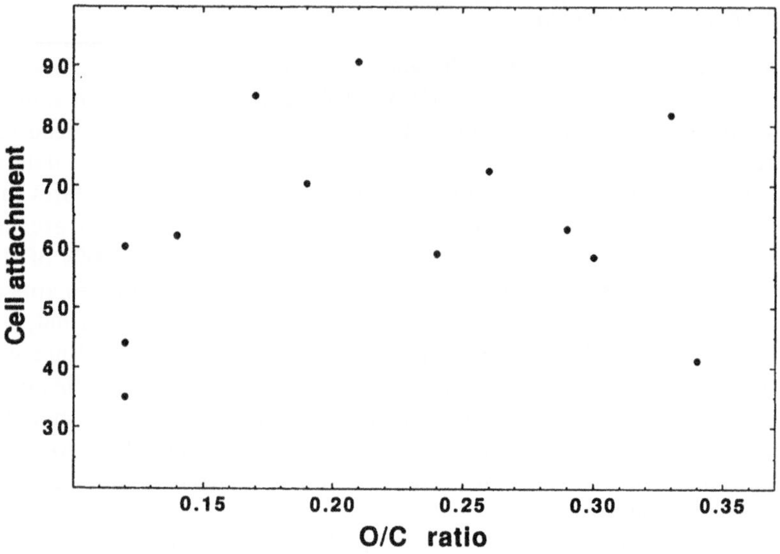
Cell attachment
90
80
70
60
50
40
30
0.15
0.20
0.25
0.30
0.35
O/C ratio

Figure 8. HUVE cell attachment (% relative to TCPS) as a function of the O/C ratio determined by XPS at an emission angle of 70 deg.

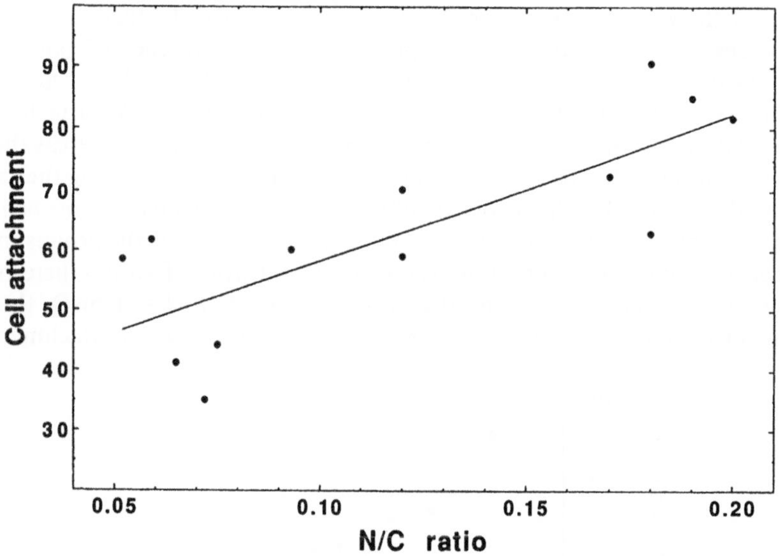
Cell attachment
90
80
70
60
50
40
30
0.05
0.10
0.15
0.20
N/C ratio

Figure 9. HUVE cell attachment (% relative to TCPS) as a function of the N/C ratio determined by XPS at an emission angle of 70 deg.

Data for Primaria are not included in Figs 8 and 9 as, on account of the pronounced variation of its composition with depth, we could not be confident of the accuracy of an estimate for the nitrogen and oxygen contents within the depth probed by the interfacial interactions with proteins. The probe depth of contact angle measurements has been stated to be ~0.5 nm [48], and the interfacial forces involved in protein-polymer interactions are likely to have a similar range. Very smooth surfaces and high quality ADXPS data are required for the elucidation of

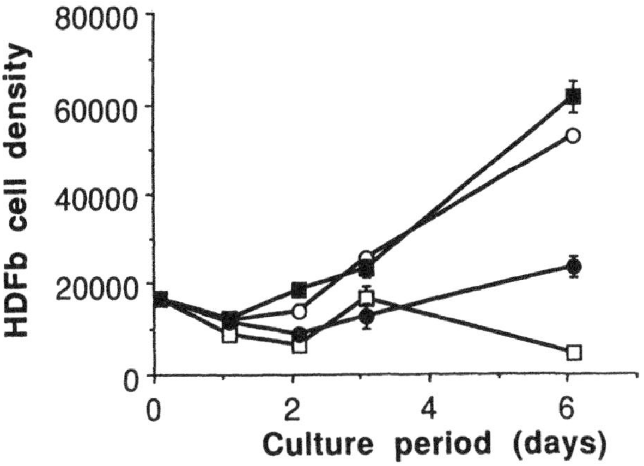
80000
60000
40000
20000
0
HDFb cell density
0
2
4
6
Culture period (days)

Figure 10. Density of human dermal fibroblasts, in cells cm^{-2}, on untreated FEP (□), and plasma polymers from DMF (■), DMAc (○), and DMP (●).

the composition of the top 0.5 nm by construction of depth profiles; for these samples this is not feasible. When searching for correlations such as in Figs 8 and 9 it is important to use samples whose compositional depth profiles are well defined; plasma polymers are much better suited to this than are plasma treated surfaces.

Whilst short-term attachment assays provide an extremely useful means of investigating the early events involved in interactions between cells and surfaces, they are not necessarily predictive of the long-term cell support potential of a given material [49, 50]. However, such an assay does identify those surfaces that are capable of supporting the initial attachment of cells, an essential prerequisite for long-term colonization. The capability of amine and amide plasma surfaces to support longer term cell growth was studied as well, using both human dermal fibroblasts and human endothelial cells.

Figures 10 and 11 show the relative rate of growth of fibroblasts on various surfaces over a period of 6 days. While the data within each figure are consistent, comparison between the data of the two figures is not entirely reliable as they were

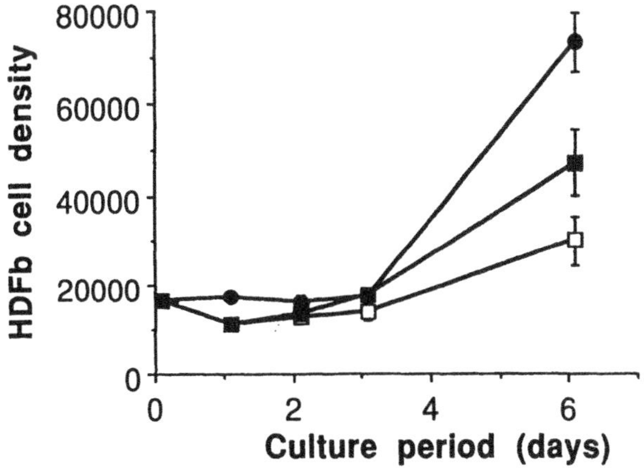
80000
60000
40000
20000
0
HDFb cell density
0
2
4
6
Culture period (days)

Figure 11. Density of human dermal fibroblasts, in cells cm^{-2}, on plasma polymers from AA (■), BA (●), and HA (□).

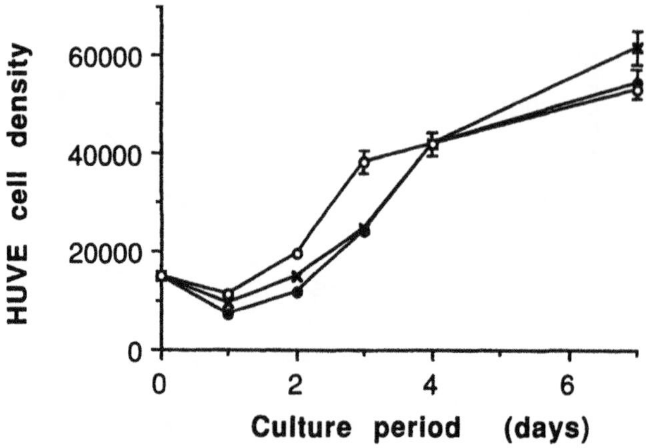
HUVE cell density
60000
40000
20000
0
0
2
4
6
Culture period (days)

Figure 12. Density of human endothelial cells, in cells cm^{-2}, on a DMAc plasma polymer (●), ammonia plasma treated FEP (○), and primaria (×).

obtained on two different cell growth runs; quantitative comparison of samples requires direct 'head-to-head' evaluation in order to eliminate differences in the growth rate between runs. On untreated FEP, cells which attached could not proliferate adequately and the total number of cells decreased (Fig. 10).

HUVE cells gave similar results; some representative data are reproduced in Fig. 12 which also shows the comparison with the commercial reference material Primaria. The mode of fabrication of Primaria is not known to us, and the shallowness of its surface modification (Table 2) prevents identification of the main functional groups by FTIR analysis. The performance of Primaria is very similar to that of our plasma fabricated, nitrogen-containing surfaces.

DISCUSSION

FEP and PTFE are very poor substrata for cell colonization; few cells attach, and the cells that do attach do not proliferate well. This makes PTFE and FEP ideal substrates for qualitative and quantitative studies of the effects of surface treatments and coatings on cell colonization. The cell growth curves of Figs 10–12 clearly demonstrate the benefits arising from surface modification of fluoropolymers in ammonia, amine and amide plasmas. Our results agree with, and extend, earlier studies which reported improved growth of cells on ammonia plasma modified substrates [32, 33]. Pratt *et al.* have previously described the use of plasma modification for endothelialization [51]. Our amine and amide plasma treatments likewise appear very promising for enabling the endothelialization of GoreTex vascular grafts.

Cell colonization and surface composition

In which way do these plasma modifications provide support for cell growth? The role of polymer surface properties and/or composition in promoting the attachment and growth of cells has been studied by a number of workers; attention has, however, focused on oxygen-containing surfaces, and very little is known about the cell culture performance of nitrogen-containing surfaces. Various hypotheses have been proposed for the surface properties which allow efficient cell colonization; a recent review by Ratner *et al.* has summarized the main theories [52]. Previous studies using other types of polymer surfaces have found that the hydrophilicity of the surfaces plays a role [1, 46, 53–56], and Fig. 7 indicates that for the nitrogen-containing materials of the present study likewise, the more hydrophilic surfaces tend to be better cell attachment substrata. Polar groups are therefore involved in the mediation of cell attachment. It appears unlikely that the various polar groups present on these multifunctional surfaces all contribute equally, and it seems reasonable to test the hypothesis developed in previous studies using other types of surfaces, which have sought to attribute support of cell attachment to specific surface groups [2, 22–31].

Although a detailed chemical description of modified polymer surfaces is a formidable task, as plasma processes do not provide chemical specificity, the surface analyses described in this paper do identify the major chemical groups present on the various surfaces produced by ammonia, amine and amide plasmas. IR spectra indicate that the principal functional groups in the amine and amide plasma polymers invariably are amine, amide and carbonyl groups, and the latter two are found even in plasma polymers fabricated from vapors which do not contain oxygen. IR probes the total thickness of these thin coatings, but as ADXPS showed little compositional variation with depth, these groups are also likely to dominate at the surface.

The ability of plasma polymers fabricated from oxygen-containing monomers (but not containing nitrogen) to support good cell attachment and growth [2, 5, 31] has been assigned to carbonyl groups [31]. Our amine and amide plasma polymers also contain carbonyl groups, and it is tempting to ascribe cell attachment to their presence. Figure 4 indicates, however, that the content of carbonyl groups was relatively low in the dimethylacetamide plasma polymers, which were superior surfaces for cell attachment (Table 1). In addition, the combination of Figs 7–9 suggests that there is a better correlation between cell attachment and the surface *nitrogen* content than there is with the oxygen content. Furthermore, cell culture experimentation using a range of plasma polymer coatings and sera selectively depleted of fibronectin or vitronectin have shown distinct mechanistic differences in the modes of cell attachment on nitrogen-containing surfaces and oxygen-containing surfaces, in that cell attachment on nitrogen-containing surfaces took place when either one of the two serum adhesive glycoproteins was present, whereas on oxygen-containing surfaces the presence of vitronectin was required and fibronectin alone was not able to effectively stimulate cell attachment [36].

There is thus a component or property of nitrogen-containing surfaces which evokes a distinctly different interfacial interaction, compared with oxygen-containing surfaces, in the sequence of events leading to cell attachment. This distinct factor is common to the nitrogen-containing surfaces studied, and it appears reasonable to assign it to a chemical group which spectroscopy indicates to be present on all these surfaces and which contains nitrogen. IR spectra attest to the presence of amines or amides; there is no spectroscopic evidence for other nitrogen-containing groups. It seems unlikely that a group present in minor amounts could dominate interfacial interactions with serum adhesive glycoproteins when amine and amide groups, which are capable of efficient interaction with proteins by hydrogen bonding, are present at substantial surface densities.

The samples with higher cell attachment for a given N/C ratio are plasma polymers fabricated from *amide* monomers. Hence, we speculate that it is the *amide* fraction of the N content which is primarily responsible for cell attachment. To support this contention, quantification is required of the surface density of amide groups. However, direct XPS analysis is not feasible because of the small difference in the N1*s* peak position of amine and amide groups. XPS analysis following derivatization of a selected group can in principle be used to overcome such resolution problems but we have not been able to date to assay surface amide groups by a derivatization reaction, as they are relatively unreactive and thus very difficult targets for selective labeling (especially in the presence of other, more reactive groups such as amines and carbonyls). In IR analysis, band overlaps, polarization effects, and the difficulties in referencing band intensities in different polymeric environments prevent accurate quantitation. Qualitatively, both XPS and IR data indicate a higher density of amide groups in amide plasma polymers compared with amine plasma polymers; the XPS C1*s* signal recorded on the former materials shows a broader high energy shoulder (Fig. 6), consistent with the higher concentration of amide groups inferred from the IR spectra (Fig. 4) when assuming similar polarization effects for these samples in the C=O stretch region. Thus, substantial involvement of surface amide groups in the attachment of human cells on these nitrogen-containing surfaces appears to be the most consistent interpretation of available data.

Comparison of FEP/NH_3 surfaces and plasma polymers is of interest here in that the cell responses on FEP/NH_3 surfaces were similar to those on plasma polymerized coatings (Table 1 and Figs 11 and 12) although the surface compositions differed: the plasma polymers formed a *hydrocarbon* network structure with attached amine and amide groups whereas the ammonia plasma treated FEP surface contained N and O attached to a *fluorocarbon* polymeric backbone. On a fluorocarbon surface, amine groups should be considerably more acidic than amine groups on the surface of alkylamine plasma polymers because of the electron withdrawing effect of adjacent fluorine atoms. The degree of protonation of amine groups in these buffered media may thus vary considerably between the FEP/NH_3 and the other surfaces used; we are currently measuring the surface p*K* values. The similarity of cell culture performance on amine-containing hydrocarbon and fluorocarbon surfaces would suggest that amine groups do not have an overriding influence in the initial protein adhesion events or, alternatively, that any role they play is substantially independent of the degree of protonation.

Our working hypothesis is that on the nitrogen-containing plasma surfaces described here, amide groups play the principal role in the support of cell attachment. This hypothesis is based upon correlative evidence: surfaces rich in amide groups tend to be more supportive of cell growth but the involvement of amine groups cannot be ruled out, as amide plasma polymer surfaces are also more hydrophilic, and there exists a correlation between contact angles and cell attachment (Fig. 7). The enabling role in the initial cell attachment onto amide plasma polymer surfaces of the adsorption of serum adhesive glycoproteins, and particularly fibronectin, is of interest here. Studies with depleted sera [36] have shown that the nitrogen-containing surfaces differ from plasma polymers which do not contain nitrogen, in that cell attachment could occur as a result of adsorption of fibronectin from fetal bovine serum. On oxygen-containing plasma polymer surfaces, this mechanism of initial cell attachment was not significant [36]. The reason for this difference is not clear at present. The *presence* of particular groups, such as amides or carbonyls, may not be the sole criterion for efficient adsorption of a particular protein; the *density* of such groups may well also be of relevance, a contention which is supported by the correlation of cell attachment with hydrophilicity (Fig. 7), which is an indirect measure of the density of polar groups.

The geometry of interfacial interactions between surface groups and cell-adhesive glycoproteins such as vitronectin and fibronectin is likely to be quite complex. To interpret the cell attachment results, it will be necessary to identify the mechanism for fibronectin adsorption onto nitrogen-containing surfaces *in competition with other serum proteins*, and contrast this with the lack of adsorption onto oxygen-containing surfaces. Detailed description of the interfacial interactions between well characterized surface compositions and proteins such as vitronectin and fibronectin, singly and in combination in cell growth media, will require much further work and advances in a number of techniques. Perhaps the attempted description of cell growth in terms of a specific group and its density is too simplistic even for a given class of plasma surfaces. However, protein interaction with surface amide groups should be of much interest for model studies.

Surface mobility

Plasma treatment with a simple process gas such as ammonia provides a more facile process than plasma polymerization for the fabrication of nitrogen-containing surfaces. However, as shown in Fig. 2, ammonia plasma treated FEP (and PTFE) surfaces change markedly on storage following fabrication whereas plasma polymer surfaces offer the advantage of varying less with time. The increase in the air/water contact angles of FEP/NH_3 surfaces on storage in air has been assigned to the capacity of the modified surfaces to undergo rearrangement motions towards a less polar surface composition when in contact with air [34]. Polymer surfaces generally are highly mobile [57–59], and surface restructuring has often been observed to lead to a considerable decrease of the surface density of polar groups attached by non-depositing surface treatments [34, 37, 60–63].

The surface mobility of polymers is a complication in the analysis of surface compositions and interfacial events, and has not been considered in earlier work using ammonia plasmas for improving the cell attachment capability of polymers, nor was the time elapsed between sample fabrication and evaluation specified. Even when using a mobile modified surface of known age, however, uncertainties arise because, as discussed recently by Andrade [57], mobile polymer surfaces may be able to change their conformation in response to adsorbing proteins. The surface composition assessed by XPS, whether 'dry' or freeze-hydrated, may not be identical to the composition which the polymer adopts for interfacial interactions with proteins. When polymers which have been stored in air are immersed in serum, buried polar groups may re-emerge at the surface; an adsorbing protein layer may adapt to the existing polymer surface; or, in the most complex case, restructuring both of the polymer and of the proteins may occur until a mutually satisfactory interfacial energy minimization is achieved. Very little is currently known of how polymer surface mobility events would interrelate with protein adsorption and denaturation, and the results of interactions between FEP/NH_3 surfaces and fibronectin and vitronectin cannot be predicted. The use of mobile polymer surfaces thus complicates the interpretation of cell culture performance in terms of surface compositions.

The response of biological media to mobile polymer surfaces is a topic of considerable interest but difficult to study. At present, the relationship between surface composition and cell culture performance may advantageously be studied using plasma polymers which can eliminate problems associated with surface restructuring; absorbing proteins are likely to be in contact with surfaces which are the same as those presented in air and the vacuum surface analysis environment. Plasma polymerization generally induces a substantial extent of crosslinking in the deposited thin coatings: Yasuda *et al.* found that of two oxidatively treated hydrocarbon plasma polymers, one possessed no mobility while the other slowly rearranged a little, although much less than oxidatively treated polypropylene [60].

In contact with air, the surface rearrangements of NH_3/FEP samples take place on a time scale of several days (Fig. 2 and ref. 34). On the time scale of the attachment assay (90 min), one may thus expect effects due to surface mobility to be small although it is not known whether one can extrapolate from air-induced surface restructuring to protein-induced restructuring. The surface mobility of Primaria is not known to us as we have no access to freshly treated material, but these considerations may also apply. By comparison with plasma polymers, it may be possible to glean whether the surface mobility of polymers treated in non-depositing plasmas is of relevance in short-term and long-term biological assays.

CONCLUSIONS

Surfaces fabricated by the plasma polymerization of volatile amine and amide monomers onto perfluorinated polymer substrates provided generally good support for colonization by human endothelial cells and fibroblasts. Plasma polymers from amide monomers tended to perform better than plasma polymers from amine monomers, although the performance in supporting cell attachment depended on the particular monomer and the plasma conditions. The efficiency of attachment of human endothelial cells correlated with the hydrophilicity of the surface and, to a lesser extent, with the nitrogen content. The available spectroscopic evidence is consistent with a major role played by surface amide groups in the adsorption of adhesive glycoproteins. The study of cell culture on amine and amide plasma polymers, which can avoid complications arising from surface mobility, is of relevance towards an understanding of cell attachment on nitrogen rich surfaces, and may lead to designed, biomimetic surface chemistries for improved synthetic biomaterials. The coating of fluoropolymers with amide plasma polymers appears to be a promising strategy for the efficient endothelialization of vascular grafts.

Acknowledgements

Part of this work was supported by DITAC under the Generic Technoloy component of the Industry Research and Development Act 1986, and by AMBRI Ltd., Cyanamid Australia Pty. Ltd., Telectronics Pty. Ltd., and Terumo (Australia) Pty. Ltd. We are indebted to Prof. Buddy Ratner, NESAC/BIO and the Division of Research Resources (N.I.H. Grant RRO1296) for support and access to the SSI X-Probe spectrometer. We thank Dr. Robert Spahn for supplying custom metallized tapes, and Zoran Vasic and Samantha Mayer for technical assistance.

REFERENCES

1. P. B. Van Wachem, C. M. Vreriks, T. Beugeling, J. Feijen, A. Bantjes, J. P. Detmers and W. G. van Aken, *J. Biomed. Mater. Res.* **21**, 701 (1987).
2. J. A. Chinn, T. A. Horbett, B. D. Ratner, M. B. Schway, Y. Haque and S. D. Hauschka, *J. Coll. Interf. Sci.* **127**, 67 (1989).
3. K. J. Pratt, S. K. Williams and B. E. Jarrell, *J. Biomed. Mater. Res.* **23**, 1131 (1989).
4. H. J. Griesser, J. H. Hodgkin and R. Schmidt, in: *Progress in Biomedical Polymers*, p. 205, C. G. Gebelein and R. L. Dunn (Eds.). Plenum Press, New York (1990).
5. S. I. Ertel, B. D. Ratner and T. A. Horbett, *J. Biomed. Mater. Res.* **24**, 1637 (1990).
6. H. J. Griesser, G. Johnson and J. G. Steele, *Polym. Mater. Sci. Eng.* **62**, 828 (1990).
7. M. D. Bale, L. A. Wohlfahrt, D. M. Mosher, B. Tomasini and R. C. Sutton, *Blood* **74**, 2698 (1989).
8. A. Dekker, T. Beugeling, H. Wind, A. Poot, A. Bantjes, J. Feijen and W. G. van Aken, *J. Mater. Sci.: Mater. Med.* **2**, 227 (1991).
9. J. G. Steele, G. Johnson, W. D. Norris and P. A. Underwood, *Biomaterials* **12**, 531 (1991).
10. J. G. Steele, G. Johnson and P. A. Underwood, *J. Biomed. Mater. Res.* **26**, 861 (1992).
11. A. S. Hoffman, in: *Biomaterials: Interfacial Phenomena and Applications*, p. 3, S. L. Cooper and N. A. Peppas (Eds.). Adv. Chem. Ser. **199**, ACS, Washington (1982).
12. T. A. Horbett, in: *Biomaterials: Interfacial Phenomena and Applications*, p. 233, S. L. Cooper and N. A. Peppas (Eds.). Adv. Chem. Ser. **199**, ACS, Washington (1982).
13. M. D. Lelah, L. K. Lambrecht, B. R. Young and S. L. Cooper, *J. Biomed. Mater. Res.* **17**, 1 (1983).
14. A. S. Hoffman, *Adv. Polym. Sci.* **57**, 141 (1984).
15. D. Kiaei, A. S. Hoffman, B. D. Ratner, T. A. Horbett and L. O. Reynolds, *J. Appl. Polym. Sci.* **42**, 269 (1988).
16. J. L. Bohnert, B. C. Fowler, T. A. Horbett and A. S. Hoffman, *J. Biomater. Sci., Polym. Ed.* **1**, 279 (1990).
17. W. R. Gombotz and A. S. Hoffman, in: *Critical Reviews in Biocompatibility*, Vol. 4, p. 1, D. F. Williams (Ed.). CRC Press, Boca Raton, FL (1987).
18. A. S. Hoffman, *J. Appl. Polym. Sci., Appl. Polym. Symp.* **42**, 251 (1988).
19. B. D. Ratner, A. Chilkoti and G. P. Lopez, in: *Plasma Deposition, Treatment, and Etching of Polymers*, p. 463, R. d'Agostino (Ed.). Academic Press, New York (1990).
20. T. A. Giroux and S. L. Cooper, *J. Appl. Polym. Sci.* **43**, 145 (1991).
21. S. D. Johnson, J. M. Anderson and R. E. Marchant, *J. Biomed. Mater. Res.* **26**, 915 (1992).
22. H. G. Klemperer and P. Knox, *Lab. Pract.* **28**, 179 (1977).
23. A. S. G. Curtis, J. V. Forrester, C. McInnes and F. Lawrie, *J. Cell Biol.* **97**, 1500 (1983).
24. N. F. Owens, D. Gingell and A. Trommler, *J. Cell Sci.* **91**, 269 (1988).
25. K. D. Thomas, B. J. Tighe and M. J. Lydon, in: *Biological and Biomechanical Performance of Biomaterials*, p. 379, P. Christel, A. Meunier and A. J. C. Lee (Eds.). Elsevier, Amsterdam (1986).
26. J. H. Lee, J. W. Park and H. B. Lee, *Biomaterials* **12**, 443 (1991).
27. A. S. G. Curtis, J. V. Forrester, and P. Clark, *J. Cell Sci.* **86**, 9 (1986).
28. W. S. Ramsey, W. Hertl, E. D. Nowlan and N. J. Binkowski, *In Vitro* **20**, 802 (1984).
29. B. R. McAuslan and G. Johnson, *J. Biomed. Mater. Res.* **21**, 921 (1987).
30. E. A. Vogler and R. W. Bussian, *J. Biomed. Mater. Res.* **21**, 1197 (1987).
31. S. I. Ertel, A. Chilkoti, T. A. Horbett and B. D. Ratner, *J. Biomater. Sci. Polymer Edn.* **3**, 163 (1991).
32. Y. Nakayama, T. Takahagi, F. Soeda, K. Hatada, S. Nagaoka, J. Suzuki and A. Ishitani, *J. Polym. Sci.: Pt. A: Polym. Chem.* **26**, 559 (1988).
33. R. Sipehia, *Biomat., Art. Cells Art. Org.* **18**, 437 (1990).
34. X. Ximing, T. R. Gengenbach and H. J. Griesser, *J. Adhes. Sci. Tech.* **6**, 1411 (1992).
35. H. J. Griesser and R. C. Chatelier, *J. Appl. Polym. Sci.: Appl. Polym. Symp.* **46**, 361 (1990).
36. J. G. Steele, G. Johnson, C. McFarland, Z. R. Vasic, R. C. Chatelier, P. A. Underwood and H. J. Griesser, *Trans. 4th World Biomater. Congr.*, 218 (1992); full paper in preparation.
37. H. J. Griesser, Da Youxian, A. E. Hughes, T. R. Gengenbach and A. W. H. Mau, *Langmuir* **7**, 2484 (1991).
38. M. A. Golub, T. Wydeven and R. D. Cormia, *Langmuir* **7**, 1026 (1991).
39. H. J. Griesser, *Vacuum* **39**, 485 (1989).
40. T. R. Gengenbach, R. C. Chatelier, Z. R. Vasic and H. J. Griesser, in preparation.
41. J. T. Grant, *Surf. Interf. Anal.* **14**, 271 (1989).

42. E. A. Jaffe, L. W. Hoyer and R. L. Nachman, *J. Clin. Invest.* **52,** 2757 (1973).
43. M. Morra, E. Occhiello and F. Garbassi, *Adv. Coll. Interf. Sci.* **32,** 79 (1990).
44. B. D. Ratner, P. K. Weathersby, A. S. Hoffman, M. A. Kelly and L. H. Scharpen, *J. Appl. Polym. Sci.* **22,** 643 (1978).
45. W. R. Gombotz, W. Guanghui and A. S. Hoffman, *J. Appl. Polym. Sci.* **37,** 91 (1989).
46. W. R. Gombotz and A. S. Hoffman, *J. Appl. Polym. Sci.: Appl. Polym. Symp.* **42,** 285 (1988).
47. H. J. Griesser, R. C. Chatelier, T. R. Gengenbach, Z. R. Vasic, G. Johnson and J. G. Steele, *Polym. Int.* **27,** 109 (1992).
48. C. D. Bain and G. M. Whitesides, *J. Am. Chem. Soc.* **110,** 5897 (1988).
49. F. Grinnell, *Int. Rev. Cytol.* **53,** 65 (1978).
50. T. A. Horbett, M. B. Schway and B. D. Ratner, *J. Coll. Interf. Sci.* **104,** 28 (1985).
51. K. J. Pratt, S. K. Williams and B. E. Jarrell, *J. Biomed. Mater. Res.* **23,** 1131 (1989).
52. B. D. Ratner, A. Chilkoti and G. P. Lopez, in: *Plasma Deposition, Treatment, and Etching of Polymers*, p. 463, R. d'Agostino (Ed.). Academic Press, San Diego (1990).
53. H. Yasuda, B. S. Yamanashi and D. P. Devito, *J. Biomed. Mater. Res.* **12,** 701 (1978).
54. P. B. van Wachem, T. Beugeling, J. Feijen, A. Bantjes, J. P. Detmers and W. G. van Aken, *Biomater.* **6,** 403 (1985).
55. J. M. Schakenraad, H. J. Busscher, C. R. H. Wildevuur and J. Arends, *J. Biomed. Mater. Res.* **20,** 773 (1986).
56. T. A. Horbett and M. B. Schway, *J. Biomed. Mater. Res.* **22,** 763 (1988).
57. J. D. Andrade, *Clin. Mater.* **11,** 19 (1992).
58. F. J. Holly and M. F. Refojo, *J. Biomed. Mater. Res.* **9,** 315 (1975).
59. J. D. Andrade (Ed.), *Polymer Surface Dynamics*. Plenum Press, New York (1988).
60. H. Yasuda, A. K. Sharma and T. Yasuda, *J. Polym. Sci., Polym. Phys. Ed.* **19,** 1285 (1981).
61. F. Garbassi, M. Morra, E. Occhiello, L. Barino and R. Scordamaglia, *Surf. Interf. Anal.* **14,** 585 (1989).
62. M. Morra, E. Occhiello and F. Garbassi, *J. Coll. Interf. Sci.* **132,** 504 (1989).
63. Da Youxian, H. J. Griesser, A. W. H. Mau, R. Schmidt and J. Liesegang, *Polymer* **32,** 1127 (1991).

Synthesis and application of new microcarriers for animal cell culture.
Part I: Design of polystyrene based microcarriers

ANITA ZÜHLKE, BETTINA RÖDER, HARTMUT WIDDECKE and JOACHIM KLEIN*

Gesellschaft für Biotechnologische Forschung, Mascheroder Weg 1, W-3300 Braunschweig, Germany

Received 14 July 1992; accepted 10 November 1992

Abstract—In this work (Part I), surface modified styrene polymers as new microcarrier material for animal cell culture were extensively investigated. The first synthesis steps—carried out by chloromethylation, sulphonation and nitration of the polystyrene matrix—resulted in precursors with a defined surface layer thickness. The obtained hydrophobic bulk phase showed a limited absorption of hydrophilic media components compared to polysaccharides matrices like dextran. By varying reaction conditions for microcarrier synthesis and/or by using similar styrene type polymer matrices like polyvinyltoluene, the specific density (1.028–1.05 g/cm^3) of the microcarrier matrix was adjusted without problems. Chemical varying of the microcarrier surface by reaction of the precursors with different amines, saccharides or proteins led to new microcarriers with optimal conditions for cell adhesion and cell growth. All biological investigations were carried out with a BHK 21 (c-13) cell line. Detailed results will be discussed and summarized in Part II of this work.

Key words: new microcarrier synthesis; surface functionalization—all growth interaction.

INTRODUCTION

In general, animal cell growth can be adherent or suspended. An efficient step forward in cultivation technology of anchorage-dependent cells [1–4] was the development of microcarriers, originally based on modified dextran matrices containing amino groups like DEAE (DEAE = *N,N*-diethyaminoethyl-) [5]. The main advantage obtainable by using such microcarrier suspensions is the higher cultivation surface area per unit volume compared to bioreactor configurations [6–10]. Commercially available microcarriers usually consist of hydrophilic polysaccharide or polypeptide matrices e.g., dextran [11, 12], cellulose [13, 14], gelatin [15, 16] or collagen [17].

A potential disadvantage of these materials is the absorption of proteins and nutrients from the cell culture medium. Many other features and properties may be considered for the design of an ideal microcarrier for animal cell culture [18–20]. Inorganic materials, like silica glass beads [21], typically have densities higher than 2.0 g/cm^3, whereas the cell culture media possess densities generally in the range of 1.02–1.05 g/cm^3. In addition, settling and aggregation of microcarriers in growth medium is usually avoided by vigorous agitation which may be very destructive to many shear sensitive cell lines [22].

* Offprint requests to: Prof. J. Klein, Scientific Director, GSF, W-8042 München-Neuherberg, Germany.

Microcarriers based on synthetic polymers like polystyrene offer the further advantage to avoid the absorption of hydrophilic medium components and growth factors due to their hydrophobicity. The preparation of these polymers in the required bead size by suspension polymerization is easy and inexpensive, their density range can be varied between 1.03 and 1.06 g/cm^3. In addition, they are stable in size and can be reused [23]. Finally, polystyrene beads do not collapse with the use of organic solvents during the necessary dehydration procedure required for ESR microscopy sampling [8]. Surface functionalization of polystyrene beads can be adapted to the cell line by a variety of functional groups via electrophilic surface substitution of the aromatic polymer [24]. However, until now the use of surface modified polystyrene for animal cell culture is only described for surface treated polystyrene dishes [25, 26] and some surface sulfonated polystyrene beads [27] but not in depth with regard to investigating the interaction between surface functionalization and cell growth behavior. Thus, commercially available polystyrene microcarriers—e.g., Cytosphere (Lux) and Biosilon (Nunc)—are not suitable for practical use [28].

In this work, the versatility of styrene polymers for the synthesis of tailor-made microcarriers will be studied in more detail. By variation of the chemical nature of the microcarrier surface, the interaction between carriers and cells will be influenced and optimal conditions for cell adhesion and growth can be determined. A first biological screening test to select group effectively supporting cell growth was performed with BHK 21 to (c-13) cells in petri-dishes. Afterwards, scale-up cultivation tests were carried out in spinner flasks. The methods used for microcarrier synthesis are described in this part I whereas the topic of cultivation results will be discussed in part II [29].

MATERIALS AND METHODS

Resins

The crosslinked nonporous polystyrene beads (6 or 12% divinylbenzene) were produced by suspension polymerization. Commercial resins were used from Bayer (Leverkusen) resin I (12% DVB, 0.5–1 mm), resin II (6% DVB, 0.5–1 mm) and resin III (6% DVB, 200–300 μm). Prior to use all resins were washed in a glass column with 1 N HCL, 1 N NaOH, H_2O, tetrahydrofuran (THF), methanol (MeOH). All surface modified resins were dialysed 3 days (distilled water) prior to use in cell culture.

Poly [1-(4-chloromethylphenyl)]ethylen (surface chloromethylation)

To an ice-cooled solution of 9 ml methylal in 5 ml CCl_4, 6.7 ml $ClSO_3H$ were added dropwise under nitrogen atmosphere and the reaction mixture was stirred for 30 min at 0°C. After addition of 5 g polystyrene resin the suspension was stirred for 15–30 min at 0°C. The reaction was stopped by dropwise addition of ice water. The beads were filtered off and successively washed with H_2O, 0.1 N NaOH, H_2O, H_2O/THF, THF, THF/acetone, acetone, MeOH and dried for 24 h at 74°C (vacuum: <0.1 h Pa). Direct determination by elemental analysis resulted in 0.27% chlorine (0.077 mmol/g). The chlorine amount could be determined indirectly as follows: 5 g of chloromethylated beads were suspended in 100 ml 33% methanolic solution of trimethylamin, shaken for 24 h at room temperature and worked off as described in the next instruction. The anion exchange capacity was found to be 0.079 mmol/g.

Microcarriers modified with amines (see Fig. 5)

One gram of surface chloromethylated polystyrene resin was suspended in 10 ml of an aqueous or methanolic 20% amine solution (depending on solubility) and shaken for 24 h at room temperature. The beads were filtered off, carefully washed with MeOH, acetone, 2 N HCl, H_2O, H_2O/THF, THF, THF/MeOH, MeOH and dried for 24 h at 75°C (vacuum: <0.1 h Pa). The anion exchange capacity was determined by Mohr titration with 0.1 N $AgNO_3$ and $K_2Cr_2O_7$ as indicator.

Poly [1-(4-nitrophenyl)]ethylen (surface nitration)

An ice-cooled suspension of 5 g polystyrene resin in 50 ml 75% aqueous nitric acid was stirred for 0.5–3 h at 0°C according to the desired degree of surface functionalization. The beads were filtered off, washed with ice water, H_2O, H_2O/THF, THF, THF/MeOH, MeOH and dried for 24 h at 75°C (vacuum: <0.1 h Pa). The N-content of the nitrated polystyrene was found to be 0.25–0.5% (elemental analysis) depending on reaction time.

Poly [1-(4-aminophenyl)]ethylen (reduction of nitropolystyrene)

A suspension of 5 g nitrated resin in 100 ml of dimethylformamide (DMF) was heated to 75°C. A solution of 42 g $SnCl_2 \cdot 2H_2O$ in 35 ml of slightly warm DMF (c. 40°C) was slowly added to the stirred suspension (exothermic reaction). The temperature was raised and kept at 140–150°C. After 15 min the suspension was cooled down to 75°C and 35 ml of conc. HCl were added. The mixture was heated for 1 h to 100°C. The resin was filtered off, washed with DMF, H_2O, H_2O/THF, THF, THF/MeOH, MeOH and dried for 24 h at 75°C (vacuum: <0.1 h Pa). The N-content was found to be 0.1–0.3% N (elemental analysis). Free amine groups on the polymer surface were determined qualitative by ninhydrine test (blue color) [30].

Poly [1-(4-sulphophenyl)]ethylen (surface sulphonation)

Fifty millilitres of conc. H_2SO_4 were added to a suspension of 5 g polystyrene resin and shaken for 2–4 h at 60°C in a waterbath. The reaction mixture was carefully added into 200 ml ice water, filtered off, washed with ice water, H_2O, 0.1 N NaOH, H_2O, H_2O/THF, THF, THF/MeOH, MeOH and dried for 24 h at 75°C (vacuum: <0.1 h Pa). The S-content was found to be 0.02–0.06% S (elemental analysis and titration).

Poly [1-(4-formylphenyl)]ethylen

A suspension of 1 g surface chloromethylated polystyrene resin in 16.7 ml dimethylsulphoxide (DMSO) and 0.1 g Na_2CO_3 was heated for 6 h to 50°C and shaken for 24 h at room temperature. The beads were filtered off, washed twice with DMSO, hot water, cold water, H_2O/dioxane (2:1), dioxan, acetone, ethanol (EtOH), MeOH and dried for 24 h at 75°C (vacuum: <0.1 h Pa). The CHO-contents were determined by a sensible colorimetric method [31]: 0.020–0.080 mmol CHO/g were found.

Poly [1-(4-hydroxymethylphenyl)]ethylen

To an ice-cooled solution of 201 mg $AlCl_3$ (1 mmol) in 10 ml diethylether (Et_2O), 180 mg (3 mmol) $LiAlH_4$ were added under nitrogen atmosphere. After stirring for 15 min, a suspension of 1 g formylpolystyrene resin in 10 ml abs. Et_2O was added to the solution and the reaction mixture refluxed for 1.5 h. The beads were filtered off after they were hydrolysed by dropwise adding of 1 N HCl, washed with H_2O, H_2O/THF, THF, THF/MeOH, MeOH and dried for 24 h at 75°C (vacuum: <0.1 h Pa).

Poly [1-(4-(3-methoxy-1,2-epoxypropane)phenyl)]ethylen

A suspension of 1 g hydroxypolystyrene resin in 10 ml epichlorhydrine was heated for 24 h under stirring up to 60°C. The beads were filtered off, washed with acetone, acetone/THF, THF, THF/H_2O, H_2O, MeOH and dried for 24 h at 75°C (vacuum: <0.1 h Pa). The epoxy-content was found to be 0.023 mmol/g (titration).

Poly [1-(-(4-carboxyphenyl)]ethylen

A suspension of 0.5 g formylpolystyrene (0.049 mmol CHO/g) in 30 ml saturated aq. $Na_2Cr_2O_7$ solution and 0.3 ml conc. H_2SO_4 was shaken for 48 h in a waterbath at 60°C. The beads were filtered off, washed with H_2O, hot acetic acid, H_2O, acetone, acetone/THF, THF, THF/MeOH, MeOH and dried for 24 h at 75°C (vacuum: <0.1 h Pa). The carboxy-content (titration) was found to be 0.042 mmol/g (86% conversion of the formyl groups).

Poly [1-(4-chloroformylphenyl)]ethylen

Under nitrogen atmosphere at room temperature, 0.42 ml SO_2Cl_2 were added dropwise to a suspension of 0.5 g carboxypolystyrene resin (RS 120, 0.05 mmol/g) in 6 ml abs. benzene and 1 ml abs. dimethylformamide (DMF). The reaction suspension was heated for 3 h at 75°C. The beads were filtered off, washed with abs. DMF, abs. dioxan, abs. THF, abs. MeOH and dried for 24 h at 75°C (vacuum: <0.1 h Pa).

Microcarriers surface-modified with proteins and monosaccharides

From surface-chloromethylated polystyrene. A suspension of 1 g surface chloromethylated polystyrene resin in 20 ml 10% solution of the desired saccharid (or protein) in 1 N NaOH was shaken for 1–3 h at 50°C and for 24 h at room temperature. The beads were filtered off, washed with H_2O, H_2O/THF, THF, THF/MeOH, MeOH and dried for 24 h at 75°C (vacuum: <0.1 h Pa).

From surface-sulphonated polystyrene. General procedure of the coupling method with 1-ethyl-3-(3-dimethyl-aminopropyl)carbodiimide (EDC): A suspension of 0.5 g surface-sulfonated polystyrene resin in 20 ml 10% saccharide (or protein, amino acid) solution in 2 mM pyridine (pH 4.5) was shaken for 24 h at room temperature.

The resin was filtered off, suspended in 30 ml 50 mM aq. EDC-solution and shaken for 24 h at room temperature. The beads were filtered off, washed with 3 M NH_4Cl, hot H_2O (80°C), H_2O, H_2O/THF, THF, THF/MeOH, MeOH and dried for 24 h at 75°C (vacuum: <0.1 h Pa).

From formylpolystyrene. A suspension of 1 g formylpolystyrene resin in 40 ml 1% aqueous solution of collagen or gelatin (pH 8) was heated for 1–5 h at 50°C. After cooling to room temperature 3 × 45 mg $NaBH_4$ were added and the reaction suspension shaken for 24 h at room temperature. The beads were filtered off, washed with MeOH, acetone, 2 N HCl, H_2O, H_2O/THF, THF, THF/MeOH, MeOH and dried for 24 h at 75°C (vacuum: <0.1 h Pa).

Microcarriers surface-modified with chitosan

From formylpolystyrene. See description above (1% chitosan solution in aq. acetic acid).

From epoxypolystyrene. A suspension of 0.76 g epoxypolystyrene in 50 ml chitosan solution was shaken for 1 h at 50°C and for 24 h at room temperature. The beads were filtered off, washed with hot H_2O, 0.01 N CH_3COOH, 0.1 N NaOH, H_2O, H_2O/THF, THF, THF/MeOH, MeOH and dried for 24 h at 75°C (vacuum: <0.1 h Pa). The anion exchange capacity of the microcarrier was found to be 0.022 mmol/g.

From chloroformylpolystyrene. A suspension of 0.5 g chloroformylpolystyrene in 50 ml chitosan solution was shaken for 2 h at 50°C and for 24 h at room temperature. The separation and purification of the beads was carried out as described above.

From surface-sulphonated polystyrene (with spacer). To a cooled (0–5°C) suspension of 0.5 g surface sulfonated polystyrene (0.005 mmol/g) in 20 ml aq. 0.1 N ε-aminocapronic acid (pH 6–7), 0.3 g EDC were added and the reaction mixture stirred 24 h at room temperature. The resin was filtered off and suspended in 50 ml chitosan solution. To the suspension, 0.3 g EDC was added and the mixture stirred for an additional 24 h room temperature. The separation and purification of the beads was carried out as described above.

RESULTS AND DISCUSSION

Concept for the synthesis of microcarriers based on polystyrene

The specific objective of this work was to develop suitable and flexible surface modification processes for polystyrene beads obtaining microcarriers with a hydrophobic bulk phase and a modified surface layer (Fig. 1).

All surface modified polystyrene beads obtained by the presented synthetic routes showed densities between 1.04 and 1.06 g/cm^3 which is slightly too high for an ideal microcarrier as mentioned before. But the required density of maximal 1.03 g/ml can be adjusted without problems by the use of polyvinyltoluene (density: 1.028 g/cm^3) or polystyrene microballoons. The modification techniques developed

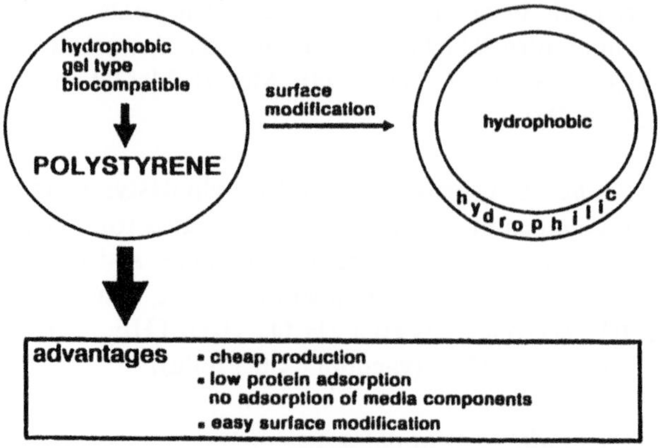
hydrophobic
gel type
biocompatible
POLYSTYRENE
surface
modification
hydrophobic
hydrophilic
advantages
• cheap production
• low protein adsorption
no adsorption of media components
• easy surface modification

Figure 1. Concept for the synthesis of microcarriers based on polystyrene.

for polystyrene beads—especially the first steps to prepare the microcarrier precursors—can be completely transferred to other styrene type polymer matrices as was demonstrated for polyvinyltoluene: surface sulphonation (2 h, 60°C) led to beads with 0.13% S*, by surface nitration (15 min, 0°C) microcarriers with 0.7% N* were obtained and chloromethylation (30 min, 0°C) resulted in a resin with 1.08% C*. The density could be varied between 1.028 and 1.037 g/ml depending on reaction time. These experiments showed that the use of polyvinyltoluene instead of the polystyrene model substance led to similar results, although the electrophilic substitution reactions are forced to take place on other positions of the aromatic ring system caused by the different molecular structure.

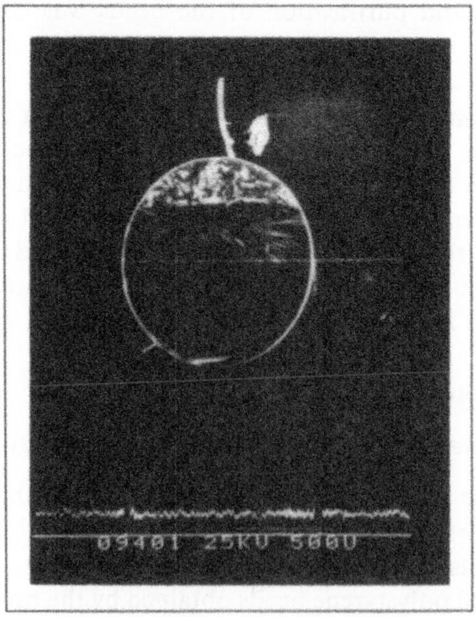
09401 25KU 500U

Figure 2. Surface distribution of chlorine after chloromethylation.

* Determined by elemental analysis.

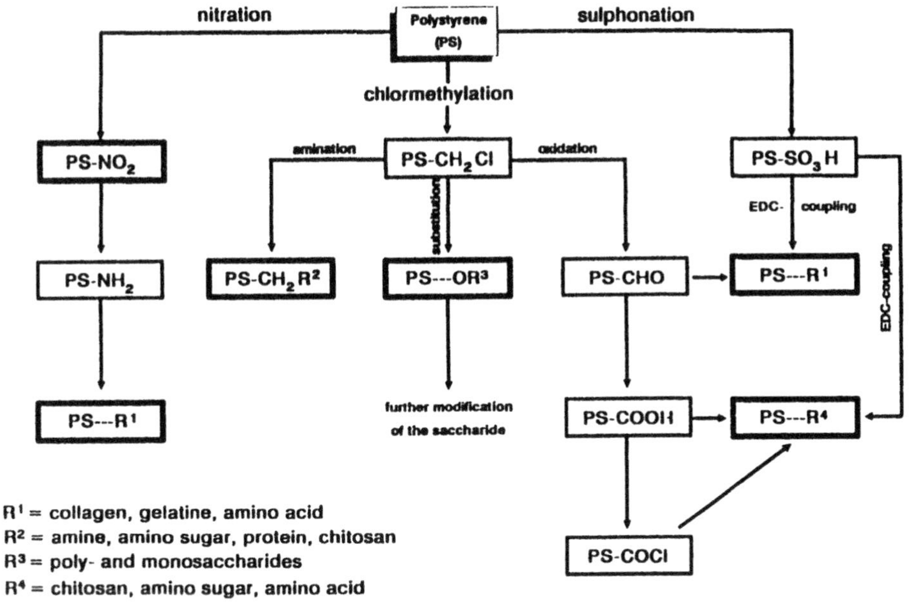
nitration
Polystyrene (PS)
sulphonation
chlormethylation
PS-NO2
amination
PS-CH2Cl
oxidation
PS-SO3H
substitution
EDC- coupling
PS-NH2
PS-CH2R2
PS---OR3
PS-CHO
PS---R1
EDC-coupling
PS---R1
further modification of the saccharide
PS-COOH
PS---R4
PS-COCl
R1 = collagen, gelatine, amino acid
R2 = amine, amino sugar, protein, chitosan
R3 = poly- and monosaccharides
R4 = chitosan, amino sugar, amino acid

Figure 3. Methods used for the synthesis of surface-modified microcarriers based on polystyrene (PS = polystyrene).

Microcarrier precursors and synthesis strategies

Three main routes were followed for the synthesis of new microcarriers for animal cell culture based on polystyrene. Polystyrene matrices with a chemically defined surface layer were obtained by chloromethylation, sulphonation and nitration. Figure 3 shows the overall chlorine distribution determined by EDX analysis (electron diffraction X-ray) across the whole diameter (750 μm) of a commercially available polystyrene bead (resin I, 12% crosslinked). Repeated EDX analysis carried out with self-made microcarriers (average diameters: 250 μm) showed that the chlorine layer thickness generally varies between 5 and 10 μm depending on the reaction time of the surface chloromethylation: 30 min resp. 105 min at a reaction temperature of 0°C.

The precursors obtained were successfully converted to new polystyrene microcarriers by reaction with several different agents (Fig. 3). Basic reaction conditions for the functionalization of polystyrene [32, 33] were partly changed in order to obtain only surface modification.

Surface modification of polystyrene with different amines

The main starting material of amine functionalized microcarriers were surface chloromethylated polystyrene beads with different degrees of crosslinking (6–12% DVB). Conversion takes place under similar conditions as described for the synthesis of anion exchange resins [34, 35]: new surface modified microcarriers were obtained showing typical capacities of between 0.013 and 0.112 mmol/g ± 0.002. The capacity values increased with increasing amine reactivity (Fig. 4).

Best cultivation results of BHK 21 (c-13) were obtained by the use of triethylamine modified polystyrene resins (Table 1 and part II): the cell concentration reached a

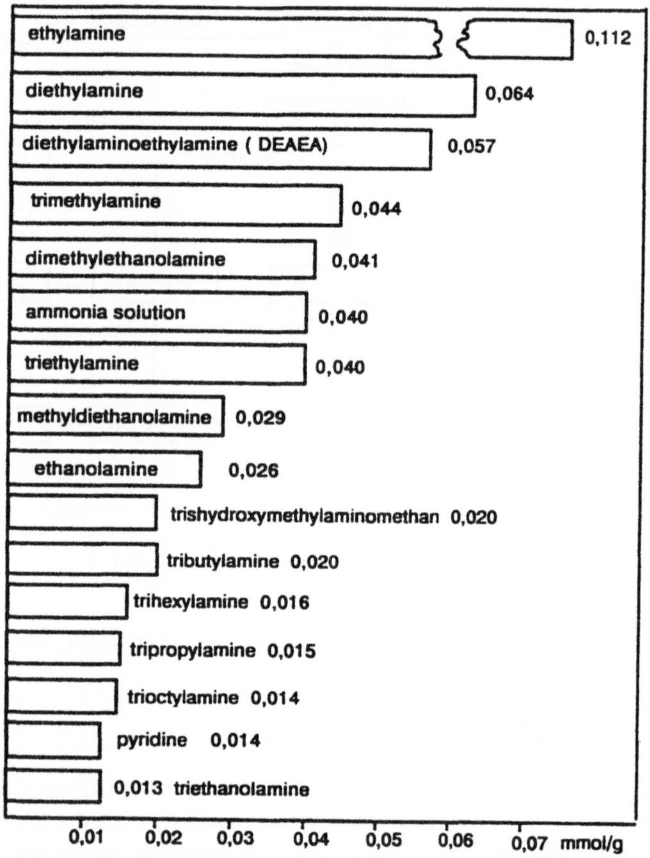
ethylamine
0,112
diethylamine
0,064
diethylaminoethylamine (DEAEA)
0,057
trimethylamine
0,044
dimethylethanolamine
0,041
ammonia solution
0,040
triethylamine
0,040
methyldiethanolamine
0,029
ethanolamine
0,026
trishydroxymethylaminomethan 0,020
tributylamine 0,020
trihexylamine 0,016
tripropylamine 0,015
trioctylamine 0,014
pyridine 0,014
0,013 triethanolamine
0,01
0,02
0,03
0,04
0,05
0,06
0,07 mmol/g

Figure 4. Capacities (mmol/g) obtained by the reaction of different amines with chloromethylated polystyrene resin (1 g resin, 24 h reaction time).

Table 1.
Polystyrene-microcarrier surface modified with triethylamine.

Reaction conditions	N-content [%,(mmol/g)]	Cl-content of the precursor [%,(mmol/g)]	Exp. capacity (mmol/g)	Percentage of[a] conversion (%)	Ratio Et_2N/NR_4^+
24 h rt	0.55 (0.393)	1.89 (0.542)	0.346	73	1:7.5
72 h rt	1.43 (1.020)	4.30 (1.230)	0.591	83	1:1.4
3 h 60°C/24 h rt	0.74 (0.530)	4.16 (1.190)	0.366	45	1:2.2
3 h 90°C	0.61 (0.436)	1.89 (0.542)	0.288	81	1:1.9
3 h 90°C	0.37 (0.264)	1.23 (0.350)	0.148	75	1:1.3
6 h 90°C	0.10 (0.071)	1.19 (0.340)	0.016	21	3.6:1

rt = room temperature (25°C)
Et_3N = triethylamine adsorption (calculated)
NR_4^+ = ammonium groups
exp. = experimentally determined exchange capacity
[a] With relation to the precursor capacity

level of 106–250 × 10^5 cells/cm^2 after 43 h cultivation time and was therefore comparable with the control and in some cases even better (Cytodex 3: 112–185 × 10^5 cells/cm^2).

Table 1 shows that the experimentally obtained capacity values (mmol/g) determined by titration slightly differed from those obtained by elemental analysis (N-content). This interesting observation can be explained by an additional adsorption of not covalently bound amine [36]. ESCA-investigations supported this assumption: in the ESCA-spectra two N*ls* peaks at 402 eV (quarternary ammonium nitrogen) and 400 eV (nitrogen without positive charge resp. 'free amine') were observed. The relative intensity ratio (charged : uncharged nitrogen) given in atom percentages of both N*ls* peaks are found to be 1 : 1 and 1 : 1.4 for higher surface modification. Microcarriers which were obtained after 24 h reaction time at room temperature have shown the lowest amount of adsorbed triethylamine and the best results in the cell culture screening test using petri-dishes. An increase of reaction time resulted in a decrease of 'charged nitrogen', followed by a reversed ratio of 'charged' (1 : 1.28 after 3 h reaction time) to 'uncharged' nitrogen (3.58 : 1 after 6 h). This can be explained by the so called 'Hofmann elimination of quarternary ammonium salts' taking place in alkaline medium as well as at higher temperatures and should result in an increase of 'free amines' content. After optimization with regard to surface charge density, exchange capacity and matrix density, microcarriers functionalized with triethylamine were also successfully tested in spinner flask experiments (detailed results: [29]).

Different microcarriers modified with amino groups were also obtained by nitration of the polystyrene surface (Figs 3 and 5). When used as a microcarrier, the nitrated polystyrene resin did show comparable results in the cell culture test with regard to growth behavior and final cell concentration.

Reduction of the nitro groups led to aminopolystyrene being used as a support matrix for solid-phase peptide synthesis [37]. The cell culture performance of those aminopolystyrene beads was less satisfactory than for the triethylamine carriers probably because triethylamine is a stronger base than the corresponding ammonia. From this it follows directly that the positive charge stimulates cell growth more strongly than the 'free amine' group.

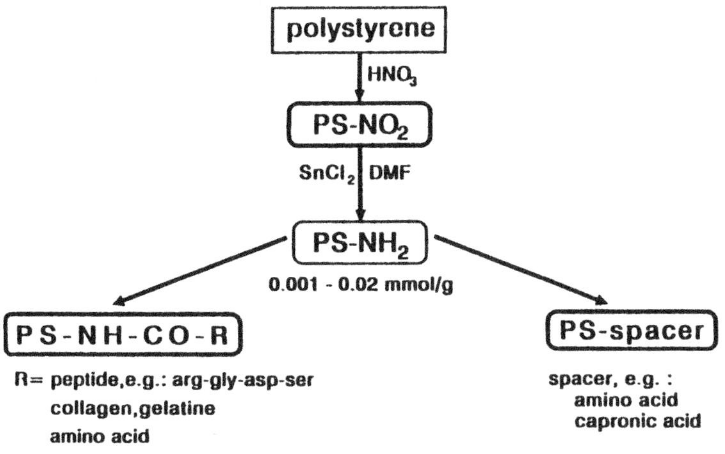
polystyrene
HNO3
PS-NO2
SnCl2 DMF
PS-NH2
0.001 - 0.02 mmol/g
PS-NH-CO-R
PS-spacer
R= peptide,e.g.: arg-gly-asp-ser
collagen,gelatine
amino acid
spacer, e.g. :
amino acid
capronic acid

Figure 5. Application of nitro- and aminopolystyrene microcarrier material.

Table 2.
Polystyrene modified with poly- and monosaccharides.

Precursor	Capacity of the precursor (mmol/g)[a]	Used saccharide	Coupling method	Amount of saccharide mg/g (mmol/g)	Percentage of conversion
PS-CH_2Cl	0.022 (Me_2N)	methyl-D-glycopyranoside	ether-	3.79 (0.019)[b]	89
	0.010 (Et_3N)	glucose	synthesis	2.63 (0.015)[b]	73
	0.021 (Et_3N)	*N*-methyl-D-glucamine		0.033 mmol/g[c]	78
	0.026 (Me_3N)	galactose		0.43[d]	—
	0.026 (Me_3N)	xylose		0.27[d]	—
	0.058 (Me_3N)	*N*-acetyl-glucosamine		0.015 mmol/g[c]	26
	0.021 (Et_3N)	dextran ($M = 6000$)		6.55[e]	—
	0.037 (Me_3N)	DEAE-dextran		8.22[e]/0.013[c]	49
PS-CH_2Cl	0.044 (Me_3N)	glucosamine	amination	0.022 mmol/g[c]	50
	0.026 (Et_3N)	lactosamine		0.032 mmol/g[c]	69
	0.058 (Me_3N)	chitosan		0.014 mmol/g[c]	24
PS-CHO	0.058	chitosan	EDC-coupling	0.015 mmol/g[c]	26
PS-epoxide	0.023	chitosan	epoxide cleavage	0.022 mmol/g[c]	96
PS-COCl	0.049 mmol CHO/g	chitosan	condensation	0.026 mmol/g[c]	54
PS-SO_2H	0.005–0.2	*N*-methyl-D-glucamine	EDC-coupling	0.107 mmol/g	—
		N-acetyl-D-glucosamine		0.186 mmol/g	—
		glucosamine		0.114 mmol/g	—
		chitosan		1.7×10^{-5} mmol/g[f]	—

rt = room temperature
Et_3N = triethylamine
Me_3N = trimethylamine
[a] Anion exchange capacity of a resin obtained by reaction from the chloromethylated polystyrene with Et_3N or Me_3N
[b] After ether cleavage determination by glucose analyzer
[c] Determination of the exchange capacity by Mohr-titration
[d] After ether cleavage determination by DNS-test (DNS = dinitrosalicylic acid)
[e] After ether and enzymatic cleavage by dextranase determination with the DNS-test
[f] Determination by dye adsorption

Surface modification of polystyrene with saccharides and proteins

Generally, cell surfaces contain a wide variety of polysaccharides, glycolipides, and glycoproteins [38]. Hence, the surface modification of polystyrene with saccharides and/or glycoproteins was a promising step in the synthesis of microcarriers for animal cell culture. Some similar methods described in solid-phase-oligosaccharide-synthesis [39–41] are usually laborious and many reaction steps are needed since the saccharide is fixed to the polymer, whereas the binding of proteins and amino acids to the polymeric surface can be obtained much more easily. Appropriate variation of the reaction conditions of many established methods of solid-phase-peptide-synthesis [42, 43] worked successfully for the surface modification of polystyrene. In this work, the binding of mono- and polysaccharides, glycoproteins and amino acids to polystyrene has been studied in detail. Mono- and polysaccharides were bound to the polystyrene surface via substitution reaction of the chloromethyl group or amide binding formation via sulpho- or carboxy groups as well as 'Schiff base' formation via formyl groups (Table 2).

Three different ways were used for the coupling of mono- and polysaccharides to the polystyrene surface: (1) The direct coupling of the desired saccharide: water soluble saccharides were bound to the chloromethylated polystyrene known for the methylation of sugars in alkaline solution (NaOH/MeOH or NaOH/H_2O). By the use of methylglycosides, uncontrolled aldol condensations of the saccharides can be prevented. Water unsoluble saccharides like chitosan [44] were coupled to chloromethyl-, chloroformyl-, carboxy-, epoxy- and sulpho groups respectively by reaction with amino functionalized saccharides. (2) Another method for the synthesis of polystyrene bound saccharides was the direct conversion of amino functionalized saccharides, e.g. glucosamine or lactosamine, with surface chloromethylated polystyrene. (3) As a third method surface sulphonated polystyrene can be used for the saccharide-coupling.

The following saccharides were bound successfully to the polystyrene surface: D-(+)glucose, *N*-methyl-D-glucamine, *N*-acetyl-glucosamine, D-(+)glucosamine as well as chitosan. The resulting microcarriers were tested for the cultivation of

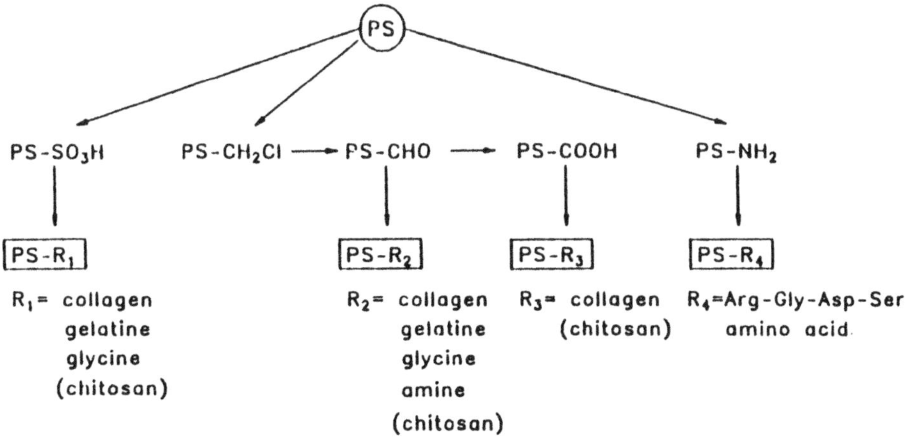
PS
PS-SO3H
PS-CH2Cl
PS-CHO
PS-COOH
PS-NH2
PS-R1
PS-R2
PS-R3
PS-R4
R1= collagen
gelatine
glycine
(chitosan)
R2= collagen
gelatine
glycine
amine
(chitosan)
R3= collagen
(chitosan)
R4=Arg-Gly-Asp-Ser
amino acid

Figure 6. Surface modification of polystyrene with proteins, peptide and/or amino acids (PS = polystyrene).

Table 3.
Polystyrene microcarrier modified with proteins.

Precursor	Capacity of the precursor (mmol/g)	Protein, peptide, amine, amino acid	Coupling method	Amount of bounded /'protein'	Percentage of conversion*
PS-CH_2Cl	0.022 (Me_3N)[a]	collagen	*amination:*	0.022 mmol/g[b]	96
	0.038 (Et_3N)[a]	glycine	pH = 8,	0.010 mmol/g[b]	26
	0.038 (Et_3N)[a]	glycinethylester	24 h rt	0.016 mmol/g[b]	42
PS-COOH	0.058	collagen	EDC-coupling	0.65 mg/g[c]	—
PS-CHO	0.025	collagen	*Schiff base*	0.066 mmol/g[d]	81
	0.060	gelatine	*formation*: 50°C/ + $NaBH_4$	5.83 mg/g[c]	—
	0.028	glycine	24 h rt	0.037 mg/g[d]	—
PS-CHO	0.019	2-diethylaminoethylamine	*coupling*:	0.0126 mmol/g[b]	65
	0.030	2-ethylaminoethylamine	MeOH/amine,	0.0075 mmol/g[b]	25
	0.030	diethylentriamine	24 h rt	0.0112 mmol/g[b]	41
	0.030	ethylendiamine		0.0093 mmol/g[b]	31
PS-NH_2	0.006[b]	H–Arg–Gly–Asp–Ser–OH	EDC/2 h rt	0.0020 mmol/g	33
PS-SO_3H	0.005–0.2	collagen	EDC-coupling	1–15 mg/g[e]	—
		gelatine		2–3 mg/g[e]	—
		glycine		0.063 mmol/g[f]	—

* = with regard to precursor capacity
rt = room temperature
Et_3N = triethylamine
Me_3N = trimethylamine
MeOH = methanol
[a] Anion exchange capacity of a resin obtained by reaction from the chloromethylated polystyrene with Et_3N or Me_3N
[b] Determination of the exchange capacity by Mohr titration
[c] After protein decoupling determination by Bradford test
[d] Determination by ninhydrine test
[e] After protein decoupling determination by BCA-test (Pierce)
[f] Elemental analysis

BHK 21 (c-13) cells [29]. Proteins, peptides and amino acids were bound to polystyrene according to similar enzyme immobilization methods [45] or via various precursors obtained from surface chloromethylated polystyrene (Fig. 6 and Table 3).

CONCLUSION

In this work, the synthesis of surface modified polystyrene resins as microcarriers for animal cell culture has been investigated in detail. The microcarriers are characterized by a hydrophilic surface layer with defined charge distribution and a hydrophobic bulk phase. The protein adsorption can be reduced as compared to commercial hydrophilic carriers based on dextran or chitosan.

The surface charge—determined by polyelectrolyte titration or dye adsorption (10^{-5}–10^{-4} mmol/g)—the density—measured by sedimentation (1.02–1.06 g/cm^3)—and the protein adsorption (0–0.4 mg BSA/g) can be adjusted without problems according to the reaction conditions used for the synthesis [46]*. A coating thickness of 300 Å to 1 μm has only small measurable effects on the carrier density. The wide variety of functionalization methods makes polystyrene especially suited as microcarrier core material to investigate the positive effect of differently charged or polar groups on growth of anchorage-dependent animal cells. By this route tailor made microcarriers based on polystyrene or similar polymers like polyvinyltoluene can be developed.

Acknowledgement

Financial support by Boehringer Mannheim AG, Germany is gratefully acknowledged.

REFERENCES

1. D. Schwengers and I. Keller, EP 0150796, Pfeiffer & Langen (1985).
2. D. Barngrover, in: *Mammalian Cell Technology*, p. 131, W. G. Thilly (Ed.). Butterworth, London (1986).
3. M. W. Glacken, R. J. Fleischaker and A. L. Sinskey, *Trends Biotechnol.* **1**, 102 (1983).
4. R. Fleischaker, in: *Large Scale Cell Culture Technology*, p. 59, B. Lydersen (Ed.). Carl Hanser, Stuttgart (1987).
5. A. L. van Wezel, *Nature* **216**, 64 (1967).
6. M. Hirtenstein, J. Clark, G. Lindgren and P. Vretblad, *Dev. Biol. Stand.* **46**, 109 (1980).
7. R. E. Spier, *Adv. Biochem. Eng.* **14**, 119 (1980).
8. A. O. A. Miller, F. D. Menozzi and D. Dubois, *Adv. Biochem. Eng./Biotechnol.* **39**, 73 (1989).
9. S. Reuveny, *Adv. Biotechnol. Proc.* **2**, 1 (1983).
10. D. W. Levin, D. I. C. Wang and W. G. Thilly, *Biotechnol. Bioeng.* **21**, 821 (1979).
11. A. David, E. Segard and G. Broun, *Ann. NY Acad. Sci.* **369**, 61 (1981).
12. D. W. Levin, J. S. Wong, D. I. C. Wang and W. G. Thilly, *Somatic Cell Genetics* **3**, 149 (1977).
13. S. Reuveny, T. Bino, H. Rosenberg and A. Mizrahi, *Dev. Biol. Stand.* **46**, 137 (1980).
14. S. Reuveny, L. Silberstein, A. Shahar, E. Freeman and A. Mizrahi, *In Vitro* **18**, 92 (1982).
15. K. Nilsson and K. Mosbach, *Pure & Appl. Biochem.*, paper presented at the 2nd European congress of biotechnology, p. 199 (1981).
16. K. Nilsson, F. Buzsaky and K. Mosbach, *Biotechnology* **4**, 989 (1986).

---

* For details of test methods used please contact the authors.

17. J. N. Vournakis and P. W. Runstadler, Technical Memorandum TM-250, Verax cooperation (1989).
18. A. L. van Wezel, *Dev. Biol. Stand.* **37**, 143 (1977).
19. M. Butler, *Adv. Biochem. Eng./Biotechnol.* **34**, 57 (1987).
20. M. Butler, *Dev. Biol. Stand.* **60**, 269 (1985).
21. J. Varani, M. Dame, T. F. Beals and J. A. Wass, *Biotechnol. Bioeng.* **25**, 1359 (1983).
22. R. S. Cherry and E. T. Papoutsakis, *Biotechnol. Bioeng.* **32**, 1001 (1988).
23. A. L. van Wezel, *J. Chem. Technol. Biotechnology* **25**, 1359 (1983).
24. P. Hodge, in: *Synthesis and Separations Using Functional Polymers*, p. 43, D. C. Sherrington and P. Hodge (Eds), John Wiley, New York (1988).
25. N. G. Maroudas, *J. Theor. Biol.* **49**, 417 (1975).
26. N. G. Maroudas, *J. Cell Physiol.* **90**, 511 (1977).
27. B. S. Jacobson and U. S. Ryan, *Tissue Cell* **14**, 69 (1982).
28. A. Johannson and V. Nielsen, *Dev. Biol. Stand.* **46**, 125 (1979).
29. B. Röder, A. Zühlke, H. Widdecke and J. Klein, *J. Biomater. Sci. Polymer Edn* **5**, 79 (1993).
30. V. K. Sarin, S. B. H. Kent, J. P. Tamm, R. B. Merrifield, *Anal. Biochem.* **117**, 147 (1981).
31. R. Ulbrich and A. Schellenberger, *Z. Chem.* **261**, 261 (1979).
32. J. M. L. Frechet and K. E. Haque, *Macromol.* **8**, 130 (1975).
33. H. W. Gibson and F. C. Bailey, *J. Polym. Sci. Chem. Ed.* **13**, 1951 (1971).
34. R. E. Barron and J. S. Fritz, *Reactive Polymers* **1**, 215 (1983).
35. R. E. Barron and J. S. Fritz, *J. Chromath.* **284**, 13 (1984).
36. V. Dourtoglou, W. E. E. Stone, M. Genet and P. G. Rouxhet, *Surface Interface Anal.* **9**, 446 (1986).
37. W. Machleidt and E. Wachter, in: *Methods Enzym.* **47**, p. 203, C. H. W. Hirs and S. N. Timasheff (Ed.). Academic Press, New York (1977).
38. G. M. W. Cook and R. W. Stoddard (Eds), in: *Surface Carbohydrates of the Eukaryotic Cell*, p. 98. Academic Press, New York (1973).
39. J. M. J. Frechet, in: *Polymer-Supported Reactions in Organic Synthesis*, p. 407, P. Hodge and D. C. Sherrington (Eds). John Wiley, New York (1980).
40. J. M. J. Frechet and C. Schuerch, *J. Am. Chem. Soc.* **93**, 492 (1971).
41. G. Excoffier, D. Gagnaire, J. P. Utille and M. Vignon, *Tetrahedron Lett.* **1972**, 5065 (1972).
42. M. S. Doscher, in: *Methods Enzym.* **47**, p. 578, C. H. W. Hirs and S. N. Timasheff (Eds). Academic Press, New York (1977).
43. J. M. Stewart, in: *Polymer-Supported Reactions in Organic Synthesis*, p. 343, P. Hodge and D. C. Sherrington (Eds). John Wiley, New York (1980).
44. G. A. F. Roberts and K. E. Taylor, *Macromol Chem., Rapid. Comm.* **64**, 478 (1975).
45. G. P. Royer, F. A. Liberatore and G. M. Green, *Biochem. Biophys. Res. Commun.* **64**, 478 (1975).
46. A Zühlke, Synthese neuer Microcarrier, Ph.D. Thesis, TU Braunschweig (1990).

Synthesis and application of new microcarriers for animal cell culture.
Part II: Application of polystyrene microcarriers

BETTINA RÖDER, ANITA ZÜHLKE, HARTMUT WIDDECKE and JOACHIM KLEIN*

Gesellschaft für Biotechnologische Forschung, Mascheroder Weg 1, W-3300 Braunschweig, Germany

Received 14 July 1992; accepted 20 November 1992

Abstract—In this work (Part II) the application of new polystyrene based microcarriers in cell culture technology is demonstrated. Carriers with a variety of surface modifications were tested as a growth support for cell line BHK 21. The growth behavior of the cells and cell to surface attachment were compared to Cytodex 3 (Pharmacia), which was used as a reference carrier. To select carriers with growth supporting surfaces, broad screening in petri dish experiments was carried out. Candidates with the highest growth rates were investigated in spinner flask experiments in further detail. Polystyrene carrier with a surface modification like triethylamine, maltamine or *N*-methylglucosamine were able to support growth as good or better as the reference carrier Cytodex 3. Economies of ingredients and ease in laboratory handling could make amine-modified polystyrenes a competitive alternative to currently commercially available microcarrier types.

Key words: cell culture technology; microcarrier; BHK.

INTRODUCTION

In vivo, most animal cells grow in cell tissues, where they are attached to other cells or support layers. In cell culture technology, especially in large scale cultivation, microcarriers can effectively play the role as support surfaces to provide an appropriate environment for the fragile cells. The microcarrier technique was first introduced by A. L. van Wezel [1] and has been established since as a common technique [2-9]. Many important pharmaceuticals are currently produced by anchorage dependent cells of, e.g., interferon [10-12], urokinase [13] or viral vaccines [14-17]. For industrial production, microcarriers are needed in large quantities but are quite expensive, require pretreatment and are usually not reuseable. Cheap inorganic materials, like silica glass beads [5], typically have densities higher than 2.0 g/cm^3, whereas the culture media possess densities generally in the range of 1.02-1.05 g/cm^3. To avoid settling and aggregation of microcarriers in growth medium, vigorous agitation is needed which could harm many shear sensitive cell lines [18]. The commercially available polystyrene microcarriers—e.g., Cytosphere (Lux) an Biosilon (Nunc)—are not very suitable for practical use [19].

Therefore, there is a need for inexpensive and easy to handle microcarriers with an appropriate surface structure sustainable for cell growth. In this work, the synthesis of new microcarriers based on modified hydrophobic polymers and their

*Offprint requests to: Prof. J. Klein, Scientific Director, GSF, W-8042 München-Neuherberg, Germany.

application in animal cell culture is described. In part I, the synthesis of the new microcarriers has been described. The present part II discusses their application, using BHK 21 as the cell line, and compares their performance with a state of the art commercial carrier as reference material.

MATERIAL AND METHODS

Chemicals

Dulbecco's Modified Eagle Medium (DMEM), fetal calf serum (FCS) and trypsin solution (0.125% trypsin in 3.3 mmol/l EDTA solution) were obtained from GIBCO (Paisley, UK), Bacto tryptose was purchased from Difco (Detroit, MI, USA), Cytodex 3 was purchased from Pharmacia (Uppsala, Sweden) and phosphate buffered saline (PBS) was bought from Boehringer Mannheim (Germany). All other chemicals were of analytical grade and obtained commercially.

Cell-line. All experiments were carried out with BHK 21 (c-13) cells (baby hamster kidney cells) [ATCC CCL 10] (American Type Culture Collection, Rockville, MD, USA).

Carrier concentration. In all experiments a carrier surface of 7900 cm^2/l was used. Estimation of the carrier surface is given in [20]. This surface corresponds to a carrier concentration of 3.0 g of dry Cytodex 3 per liter.

Conditioning of microcarriers. Cytodex 3 was prepared and sterilized as recommended by the manufacturer. Thus, the dry beads were swollen in PBS overnight for at least 6 h and washed twice with culture medium. All other carriers were only suspended in PBS, steam sterilized (20 min, 121°C) and washed twice with culture medium.

Culture medium. Cells were cultivated in all experiments in DMEM supplemented with 10% FCS and 10% tryptose-phosphate-broth (20 g/l Bacto tryptose, 2 g/l glucose, 5 g/l NaCl and 2.5 g/l Na_2HPO_4; pH 7.2 ± 0.2).

Seed culture. Cells were routinely cultured in 75 or 175 cm^2 culture flasks (Falcon, Oxnard, CA, USA) in an incubator at 37°C in a 7.5% CO_2/air atmosphere and a humidity of 95%. Cells were subcultured during the exponential growth phase with a seeding concentration of 1×10^5 cells/ml according to a concentration of 4×10^4 cells/cm^2.

Analytical procedures. Glucose and lactate concentrations in the culture supernatant were measured enzymatically. Glucose was determined with a glucose analyser (Yellow Springs Instruments, OH, USA) and lactate with a lactate analyser (Yellow Springs Instruments, OH, USA).

Experiments in spinner flasks. All experiments were carried out in 200 ml spinner flasks (Techne, Fernwald, Germany) with a working volume of 100 ml. The flasks were inoculated with a cell concentration of 2×10^5 cell/ml according to a concentration of 2.5×10^3 cells/cm^3. Cytodex 3 was used as reference carrier. Reference and test carrier were inoculated with the same culture suspension. The cells were cultivated in an incubator at 37°C in a 7.5% CO_2/air atmosphere and a humidity at 95%. Agitation speed was maintained at 40 rpm according to sufficient homogeneous mixing. Growth was measured by taking samples at certain intervals. Concentration of cells in free suspension was counted directly in the suspension on a haemocytometer. Total cell concentration was estimated by enumeration of released nuclei using crystal violet method [21]. Thus, a sample of 1.0 ml was centrifuged in an Eppendorf centrifuge. The supernatant was discarded and replaced by crystal violet solution (0.1% crystal violet in 0.1 mol/l citric acid solution). The mixture was incubated for 1 h at 37°C. The stained nuclei were counted on a haemocytometer. The growth quality (homogeneous distribution of the cells to the beads, performance of mono- and multilayers, etc.) was examined microscopically.

RESULTS AND DISCUSSION

Carriers

Cytodex 3 was chosen as reference carrier. Despite the fact that Cytodex 1 is the most frequently used carrier type in industry and inexpensive, we have chosen Cytodex 3 because it shows better cell attachment and cell growth than Cytodex 1. All carriers tested are based on polystyrene matrix and are of a nonporous type with a chemically modified surface. Several functional groups were introduced to the surface as described in part I. The strategy was, first, to introduce those functional groups to the surface, which results in a positive charge of the surface, like the amine or triethylamine group, and secondly to introduce those groups, which are part of the cell surface itself, like saccharides or aminosaccharides.

Experiments in petri dishes

The tests for adherent cell growth were carried out in petri dishes, where the cells grow statically without mixing on the beads. Due to the simplicity of the experiment and the possibility to test large numbers of different carriers simultaneously, this type of experiment was chosen as a qualitative assay to select those functional groups which lead to a good attachment of the cells as well as to a sufficient growth of the cells.

Petri dishes were chosen where the cells are not able to grow on the dish surface itself. Cells were seeded in the dishes using an initial concentration of 1.9×10^5 cells/ml according to a carrier concentration of 2.5×10^3 cells/cm^3 and harvested after about 48 h. The doubling time was estimated using the presumption that the cells are still in the exponential growth phase (this was tested previously). As a comparison, cells growing on the control carrier Cytodex 3 have a doubling time of 23 h under identical conditions. Thereafter, the cells were investigated microscopically. Carriers with suitable functional groups show a homogeneous cell distribution on the beads in monolayers and only a few unattached single cells in the supernatant. The results of the petri dish experiments are summarized in Tables 1 and 2.

It is obvious from these data (Table 1) that the polystyrene matrix itself did not support the growth of the BHK cells; in fact the cells tend more to cling to each other and form big aggregates in suspension than to stick to the plastic surface. The

Table 1.
Growth behavior of BHK 21 in petri dish experiments on unmodified polystyrene matrix.

Batch no.	Doubling time (h)	Microscopic view
1	34	only few cells on beads, all other
2	34	cells form aggregates in suspension
3	37	
control	23	nearly all cells on beads, homogeneous distribution

doubling time was more than 50% higher compared to the control Cytodex 3. The polystyrene beads did not show any harmful effects on cell growth but on the other hand in the pure unmodified form they are not suitable for attachment of the cells.

To get an optimal performance of the carrier surface for attachment of the cells, several functional groups were introduced to the surface of the polystyrene beads [22]. These groups include amines, amino acids, saccharides, aminosaccharides as well as polymers like collagen or chitosan which are commonly used for commercial carriers. The results using these surface modified carrier in petri dish experiments are shown in Table 2.

For the polystyrene carriers with a modified surface, results are quite different. There are carriers, like the triethylamine carrier, where the cells built monolayers on the carrier surfaces and the major part of the cells is located on the carrier beads. In addition to that, the cells grow with the same doubling time as on the control carrier and sometimes even more rapidly than on Cytodex 3. Because BHK cells did not grow on the pure unmodified matrix, these growth results are correlated to the surface modification; in case of the triethylamine carrier it is correlated to the charge of the surface (Table 1 of parts I and II). There are other candidates, like the lactosamine carrier, where the cells are able to attach to the carrier surface, but the doubling time is significantly elongated due to a toxic or inhibiting effect of the functional group. There is a third type of carrier, like the glycine–ethylesterhydrochlorid carrier, where no significant stimulation of the cell growth could be observed in comparison to the unmodified matrix. That might be the result of a lack of charge and/or a slightly toxic or inhibiting effect of the introduced functional group.

In general, all tested polar functional groups had a positive effect on cell attachment and growth behavior of the tested cell line BHK 21 (c-13). Several of the test carriers show comparable results to the reference material Cytodex 3; low values for the doubling time as well as nearly complete attachment and homogeneous distribution was observed. In addition to the polystyrene carriers modified with collagen and dextran, especially polystyrene beads modified with amines and aminosaccharides, leads to good results for growth and attachment. These groups were selected for further investigations.

Experiments in spinner flasks

It had to be considered that the conditions in petri dishes do not necessarily represent conditions in a fermenter, because in petri dishes no mixing of the culture suspension exists. Previous experiments showed that the growth rate of BHK on

Table 2.
Growth behavior of BHK 21 in petri dish experiments on modified polystyrene matrix

Functional group	No. of batches	Doubling time (h)	Microscopic view Cells on carriers	Cells in suspension
Amine	5	29	cells and beads form aggregates	only few cells; no aggregates
Triethylamine	11	23	homogeneous distribution	only few cells; no aggregates
Ethanolamine	2	24	homogeneous distribution	more cells; aggregate formation
Trimethylamine	4	52	inhomogeneous distribution	more cells; aggregate formation
Ethylamine	1	32	inhomogeneous distribution	more cells; aggregate formation
Diethylamine	3	44	slightly inhomogeneous	more cells; aggregate formation
Methionine	1	20	homogeneous distribution	more cells; aggregate formation
Glycine	2	27	slightly inhomogeneous	only few cells; no aggregates
Glycine-ethylesterhydrochlorid	1	36	slightly inhomogeneous	only few cells; no aggregates
N-Methyl-D-glucosamine	2	20	homogeneous distribution	more cells; aggregate formation
Glucosamine	2	25	slightly inhomogeneous	only few cells; no aggregates
N-Acetylglucosamine	2	21	homogeneous distribution	only few cells; no aggregates
D-(+)-Glucose	1	33	slightly inhomogeneous	only few cells; no aggregates
Galactose	2	26	homogeneous distribution	more cells; aggregate formation
D-(+)-Xylose	1	29	homogeneous distribution	only few cells; no aggregates
Lactosamine	1	62	homogeneous distribution	only few cells; no aggregates
Dextrane	1	26	homogeneous distribution	more cells; aggregate formation
Chitosan	5	33	homogeneous distribution	more cells; aggregate formation
Collagen	8	23	homogeneous distribution	more cells; aggregate formation
Gelatine	2	41	slightly inhomogeneous	more cells; aggregate formation
DEAE-Dextrane	2	33	cells and beads form aggregates	more cells; aggregate formation
DEAE-Glucose	1	36	homogeneous distribution	more cells; aggregate formation
Methyl-D-glucopyranoside	1	32	inhomogeneous distribution	more cells; aggregate formation
DEAE	1	25	homogeneous distribution	only few cells; no aggregates
DEAEA	1	31	homogeneous distribution	only few cells; no aggregates
Maltamine	1	20	homogeneous distribution	only few cells; no aggregates

Table 3.
Growth behavior of BHK 21 in spinner flask experiments on modified polystyrene matrix

Functional group	No. of batches	μ_{max} / θ_{max} (Cy3)	%C att. / %C att. (Cy3)	max. Cell. / max. Cell. (Cy3)	q_g / q_g (Cy3)	q_l / q (Cy3)	Microscopic view
Amine	2	0.77	0.93	0.54	1.13	1.01	slightly inhomogeneous
Triethylamine	4	0.96	0.96	0.87	1.00	0.94	homogeneous
Ethanolamine	1	0.59	1.00	0.75	0.84	0.70	cells and beads from aggregates
DEAE	1	0.90	1.01	1.00	0.82	0.85	cells and beads form aggregates
Maltamine	1	1.00	1.01	1.00	0.95	1.06	homogeneous
Glucosamine	1	0.59	Aggl.	0.40	1.54	1.63	inhomogeneous; most cells in big aggregates in suspension
N-Methyl-glucosamine	1	0.90	0.98	1.20	0.99	0.95	homogeneous
N-Acetyl-glucosamine	1	0.51	0.95	0.63	1.16	1.15	slightly inhomogeneous
Chitosan	1	0.59	Aggl.	0.47	1.21	1.02	cells and beads form aggregates; unattached cells in big aggregates
Collagen	1	0.73	0.84	1.21	0.91	1.09	slightly inhomogeneous

μ_{max} = specific growth rate
q_g = specific glucose consumption rate
q_l = specific lactate production rate
%C att. = percentage of cells attached to carrier beads
max. Cell. = maximum of cell concentration during batch cultivation
Cy3 = Cytodex 3

Cytodex 3 in spinner flasks is the same as in fermentation systems [23, 24]. In general, spinner flasks best represent the conditions in a fermenter and are therefore chosen for the scale-up tests. Parallel to the fermentation with the test carriers in spinner flasks, control runs using Cytodex 3 were carried out. Reference and test runs were inoculated with the same preculture. An important parameter for microcarrier cultures is the available carrier surface per volume to provide the cells with enough surface to build up a homogeneous monolayer on the beads. In previous experiments [23], BHK cells were grown on Cytodex 3. In a batch of culture cells grow up to a cell concentration of 2–3 × 10^6 per ml. For a carrier concentration of 3.0 g/l dry weight of Cytodex 3, the single beads are at a cell concentration of 3 × 10^6 cells per ml confluent populated. That means that a surface of 7900 cm^2/l provides enough space for the cells. The surface of Cytodex 3 was estimated according to Zühlke [20]. For all experiments in this article a surface ratio of 7900 cm^2/l was used.

The results of the spinner flask experiments are summarized in Table 3. Growth behavior of BHK 21 cells on test carriers is characterized by the specific growth rate, the specific lactate production rate and the specific glucose consumption rate. These values were determined from the exponential growth phase. In addition the

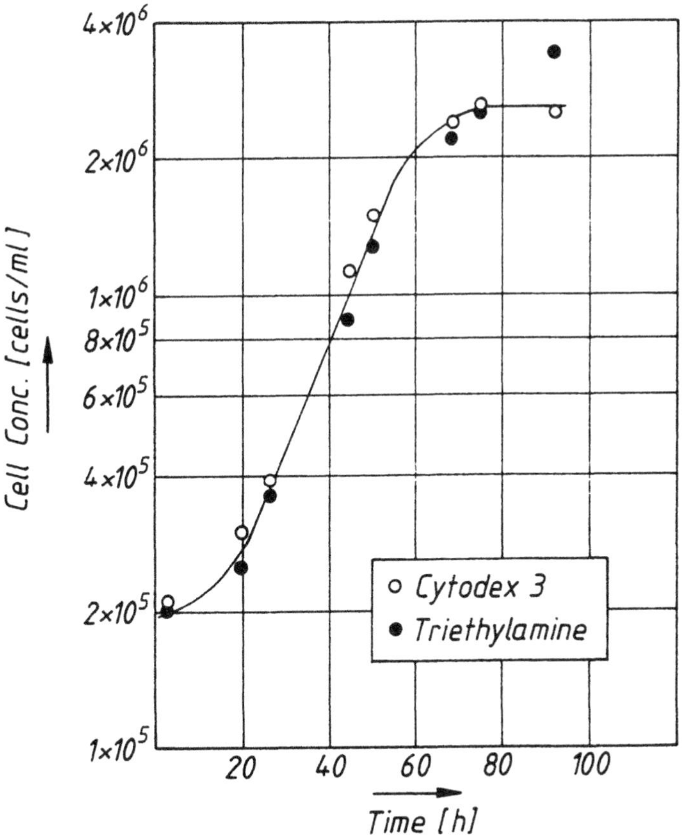
4×10^6
2×10^6
1×10^6
8×10^5
6×10^5
4×10^5
2×10^5
1×10^5
Cell Conc. [cells/ml]
20
40
60
80
100
Time [h]
o Cytodex 3
• Triethylamine

Figure 1. Time course of cultivation of BHK 21 cells on Cytodex 3 and triethylamine modified polystyrene carrier.

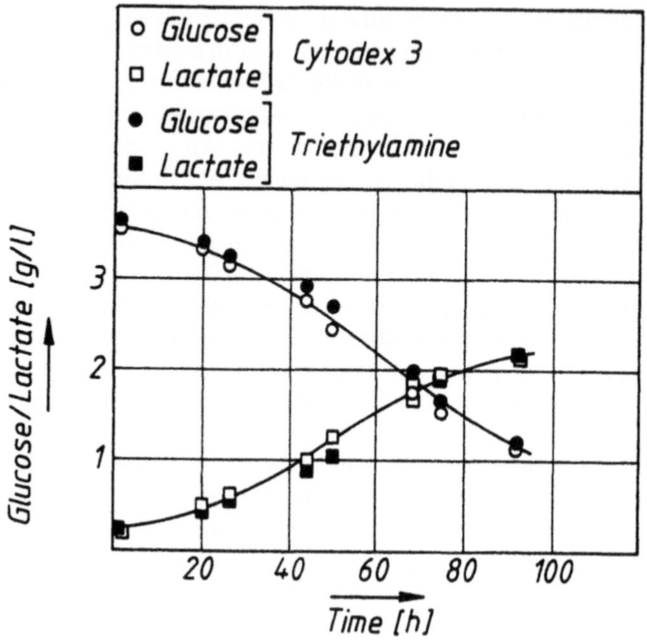

○ Glucose
□ Lactate
Cytodex 3
● Glucose
■ Lactate
Triethylamine
Glucose/Lactate [g/l]
3
2
1
20
40
60
80
100
Time [h]

Figure 2. Glucose consumption and lactate formation during cultivation of BHK 21 cells on Cytodex 3 and triethylamine modified polystyrene carrier.

maximum cell concentration and the percentage of attached cells on the carriers were estimated. All values were related to corresponding values obtained from growth experiments on Cytodex 3. Under the experimental conditions used, no significant difference in lag-phase was observed, when the cells were grown on Cytodex 3 or on the different test carriers, except for the collagen modified carrier where a slightly longer lag-phase was observed. In conclusion, polystyrene beads modified with triethylamine, maltamine and *N*-methylglucosamine seem to be

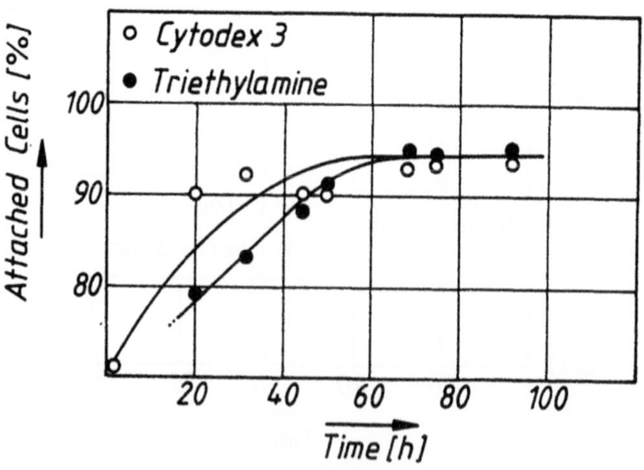

○ Cytodex 3
● Triethylamine
Attached Cells [%]
100
90
80
20
40
60
80
100
Time [h]

Figure 3. Percentage of cells attached to the beads during cultivation of BHK 21 on Cytodex 3 and triethylamine modified polystyrene.

especially good and inexpensive alternatives compared with dextrane-collagen based Cytodex 3 carrier.

In addition, the modified polystyrene carrier does not need any pretreatment, because they do not swell. As an example, growth of BHK 21 cells in spinner flasks on a triethylamine modified microcarrier is shown in Fig. 1. No significant difference between the test and the reference carrier was observed. In Figs 2 and 3, the amount of attached cells, as well as glucose consumption and lactate formation, is identical for both carrier types. Attachment of the cells to the Cytodex 3 beads is rapid. After 20 h about 90% of the cells are attached to the carrier surface. Attachment to the triethylamine carrier takes more time, but after 40 h the growth is as good as on Cytodex 3.

CONCLUSION

Polystyrene based microcarrier matrix as a growth support for cell line BHK 21 (c-13) in cell culture technique with chemical surface modification by functional groups (synthesis described in part I) were tested in petri dish experiments. Carrier modifications with promising properties were further investigated in spinner flasks experiments. All experiments were carried out using commercial Cytodex 3 as a reference. Triethylamine, maltamine and *N*-methylglucosamine modified carrier surfaces are able to support growth as well as the reference carrier. The new microcarrier material is inexpensive and easy to handle, and in addition to that the carriers are reusable. Further investigations will include use of different cells lines, production of the selected carriers in a larger scale and the final test in a fermenter.

Acknowledgement

This work was supported by Boehringer Mannheim Ag, Germany.

REFERENCES

1. A. L. van Wezel, *Nature* **216**, 64 (1967).
2. A. O. A. Miller, F. D. Menozi and D. Dubois, *Adv. Biochem. Eng./Biotechnol.* **39**, 73 (1989).
3. T. Y. Tao, *Biotechnol. Bioeng.* **32**, 1037 (1988).
4. S. Inooka, *J. Agricult. Res.* **20**, 19 (1969).
5. J. Varani, M. Dame, T. F. Beals and J. A. Wass, *Biotechnol. Bioeng.* **25**, 1359 (1983).
6. A. L. van Wezel, *Animal Cell Biotechnol.* **1**, 265 (1985).
7. W. R. Tolbert and J. Feder, *Ann. Rep. Fermentation Process* **6**, 35 (1983).
8. D. W. Levin, J. S. Wong, D. I. C. Wang and W. G. Thilly, *Somatic Cell Genetics* **3**, 149 (1977).
9. J. P. Whiteside and R. E. Spier, *Biotechnol. Bioeng.* **18**, 659 (1976).
10. A. L. Smeley, W. S. Hu and D. I. C. Wang, *Biotechnol. Bioeng.* **33**, 1182 (1989).
11. K. Nilsson, S. Birnbaum and K. Mosbach, *Appl. Microbiol. Biotechnol.* **27**, 366 (1988).
12. A. Mizrahi, *Process Biochem.* **1986**, 108 (1986).
13. R. Fleischaker, in: *Large Scale Culture Technology*, p. 59, B. Lydersen (Ed.). Carl Hanser, Stuttgart (1987).
14. D. I. C. Wang and M. Tyo, *Adv. Biotechnol.* **1**, 141 (1980).
15. B. Föhring, S. T. Tjia, W. M. Zenke, G. Sauer and W. Doerfler, *Proc. Soc. Exp. Bio. Med.* **164**, 222 (1980).
16. J. P. Whiteside, B. R. Whiting and R. E. Spier, *Dev. Biol. Stand.* **42**, 113 (1979).
17. M. Butler, *Adv. Biochem. Eng./Biotechnol.* **34**, 57 (1987).
18. R. S. Cherry and E. T. Papoutsakis, *Biotechnol. Bioeng.* **32**, 1001 (1988).

19. A. Johannson and V. Nielson, *Dev. Biol. Stand.* **46,** 125 (1979).
20. A. Zühlke, *Synthese neuer Microcarrier*, Ph.D. Thesis, TU Braunschweig (1990).
21. K. K. Sanford, W. R. Earle and V. J. Evans, *J. Natl. Cancer Inst.* **11,** 773 (1951).
22. A. Zühlke, B. Röder, H. Widdecke and J. Klein, *J. Biomater. Sci. Polymer Edn* **00,** 000 (1993).
23. J. Lehmann, G.-W. Piehl, J. Vorlop and B. Röder, *BAP Progr. Rep.* **3,** 893 (1988).
24. B. Röder, J. Lehmann, Bead-to-bead transfer. An important factor for scale-up of anchorage dependent mammalian cells, poster. BAP Final Meeting Animal Cell Technology, Braunschweig (1989).

Monosize microbeads based on polystyrene and their modified forms for some selected medical and biological applications

E. PISKIN*, A. TUNCEL, A. DENIZLI and H. AYHAN

Chemical Engineering Department and Bioengineering Division, Hacettepe University, 06420 Ankara, Turkey

Received 30 June 1992; accepted 1 March 1993

Abstract—Polymeric particles are produced by different polymerization techniques. Phase inversion (dispersion) polymerization is one of the recent techniques to obtain monosize polymeric microbeads in the size range of 1–50 μm. The size and monodispersity of these microbeads can be adjusted by using several solvent systems (e.g., alcohol-water mixtures) with different polarities and by changing the type and amount of monomer, initiator and stabilizer. Surfaces of these microbeads can be further modified by different techniques including coating with different copolymers. Monosize polymeric microbeads are widely used in medical and biological applications as carriers, such as in immunoassays and cell separation, in site-specific drug delivery systems, in nuclear medicine for diagnostic imaging, in studying the phagocytic process, in affinity separation of biological entities, etc. Here, some important aspects of the production of monosize microbeads based on polystyrene and their modified forms are briefly discussed, and some selected medical and biological applications are summarized.

Key words: Monosize polystyrene microbeads; dispersion polymerization; surface modification; GIT imaging; phagocytosis; protein adsorption; dye affinity.

INTRODUCTION

Recently, monosize polymeric microbeads have attracted much attention as carrier matrices in a wide variety of medical and biological applications [1]. These polymeric carriers have been found valuable in several immunoassays (e.g., latex agglutination tests for pregnancy, immunoenzymometric assay for placental alkaline phosphates, direct assay for carcinoembryonic antigen, etc.) [1–5]. Various cells (e.g., red cells, human B and T lymphocytes, bone marrow cells, etc.) have been separated successfully by using polymeric microbeads carrying specific ligands (mainly antibodies) on their surfaces [1, 6–9]. A variety of colloidal delivery systems in the form of microspheres and microcapsules have been developed for controlled release of bioactive substances and for targeting therapeutic and diagnostic agents to their site of action [10–13]. Therapeutic or diagnostic bioactive agents have been incorporated on or within these polymeric carriers for site-specific drug delivery. For diagnostic purposes, microbeads have been labeled with radioactive agents. Colloidal particles labeled with radionuclides have been used in nuclear medicine for *in vivo* visualizing of several diseases [14, 15]. Monosize polymeric microbeads with different size and surface properties have emerged as an important tool in the study of phagocytosis and the factors influencing the process [16–19].

* To whom all correspondence should be addressed.

Polymeric particles and their biologically modified forms (carrying a wide variety of ligands) have been widely used in affinity separation of many important biologicals e.g. proteins [20–22].

Suspension polymerization and emulsion polymerization are two classical polymerization techniques to produce spherical polymeric particles. Larger particles (usually larger than 50 μm) with an appreciable size distribution are produced by suspension polymerization. Submicron polymeric particles (usually smaller than 0.1 μm) with extremely uniform size are obtained by conventional emulsion polymerization processes. Recent techniques, such as swollen emulsion polymerization, dispersion polymerization, etc. give micron-size (usually between 1 and 50 μm) monosize polymeric particles. The first part of this paper briefly presents current polymerization techniques, and describes the phase inversion polymerization (i.e. dispersion polymerization) technique that we have used to produce polystyrene based monosize polymeric microbeads and their modified forms (carrying different surface functional groups). The second part discusses some selected medical and biological applications of these microbeads.

PRODUCTION OF MONOSIZE POLYMERIC MICROBEADS

Current techniques

Suspension polymerization is a conventional technique to produce polymers in bead form in large quantities for industrial applications. In this technique, the monomer phase is broken into droplets (a few microns in diameter) within a dispersion medium

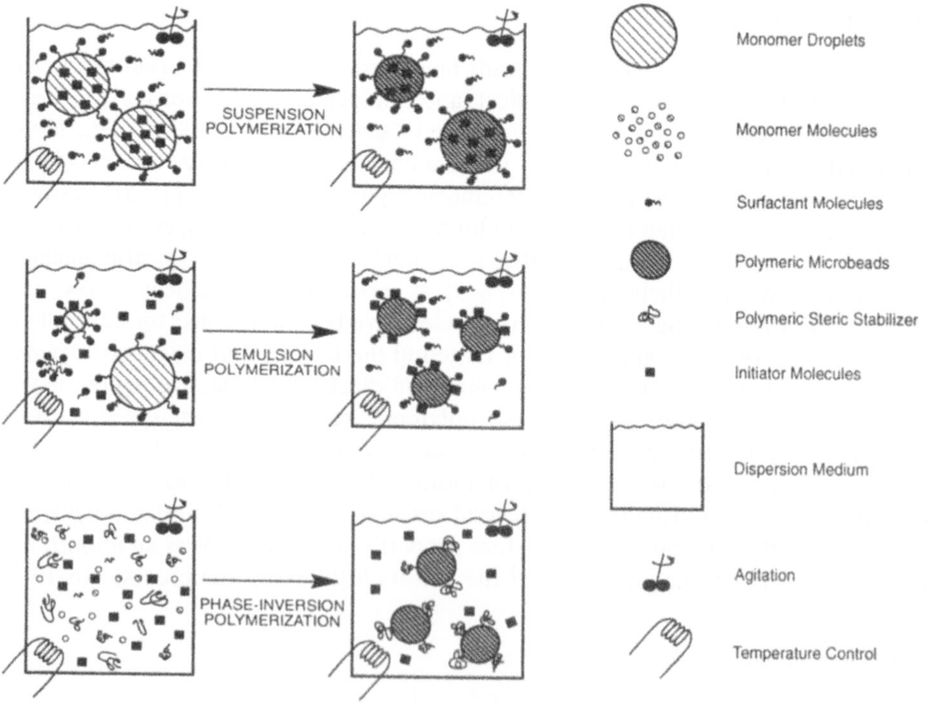
SUSPENSION
POLYMERIZATION
EMULSION
POLYMERIZATION
PHASE-INVERSION
POLYMERIZATION
Monomer Droplets
Monomer Molecules
Surfactant Molecules
Polymeric Microbeads
Polymeric Steric Stabilizer
Initiator Molecules
Dispersion Medium
Agitation
Temperature Control

Figure 1. Schematic description of current polymerization techniques for the production of polymeric microbeads.

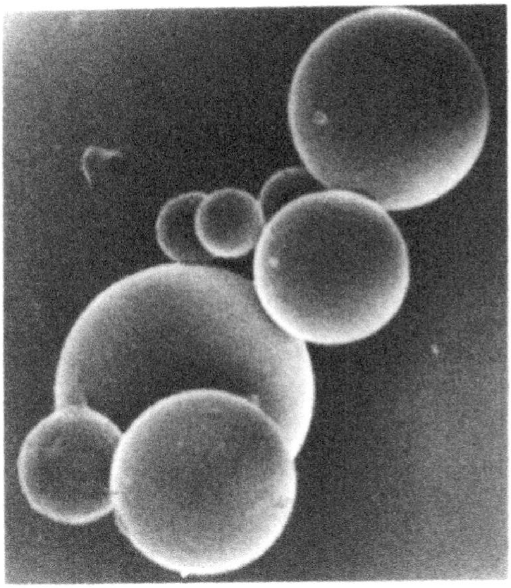

Figure 2. Polystyrene microbeads (50–250 μm) produced by suspension polymerization.

(usually an aqueous phase), and stabilized by a surfactant dissolved in the dispersion medium (Fig. 1). These monomer droplets containing a monomer phase soluble initiator are then individually polymerized by applying a temperature/agitation program. This technique is usually used for the production of spherical polymeric particles between about 50 and 1000 μm. A wide particle size distribution is usually observed because of inherent size distribution in the mechanical homogenization (agitation) step and because of a coalescence problem that arises in this type of polymerization. The representative picture given in Fig. 2 clearly shows the particle size distribution of polymeric particles obtained by a suspension polymerization.

Emulsion polymerization is another classical technique for manufacturing of polymers in latex at industrial level. In this process the monomer is present in the form of large droplets suspended in a dispersion medium (Fig. 1). Here, polymerization proceeds inside the micelles formed by emulsifier molecules in the existence of the initiator in the dispersion medium. The classical emulsion polymerization process is convenient for producing spherical polymeric particles in submicron size with extreme uniformity in a size up to about 0.1 μm, in a latex form. Figure 3a shows a representative picture of polystyrene particles produced by a classical emulsion polymerization protocol.

Many attempts have been made to fill the size-gap between these two traditional techniques. The so-called 'swollen emulsion polymerization' developed by Ugelstad and his coworkers was the first successful technique, where monomer and/or solvents were used to start the swelling of the seed polymer particles, which in turn caused a size increase after repolymerization [23–26]. Figure 3b shows a representative picture of polystyrene particles produced by a swollen emulsion polymerization. Note that in order to obtain latex particles larger than 1 μm, multistep swollen emulsion polymerizations must be performed [25–28].

By following a similar technique, Vanderhoff and coworkers were also able to produce polymeric particles in the micron-size range aboard the Space Shuttle in

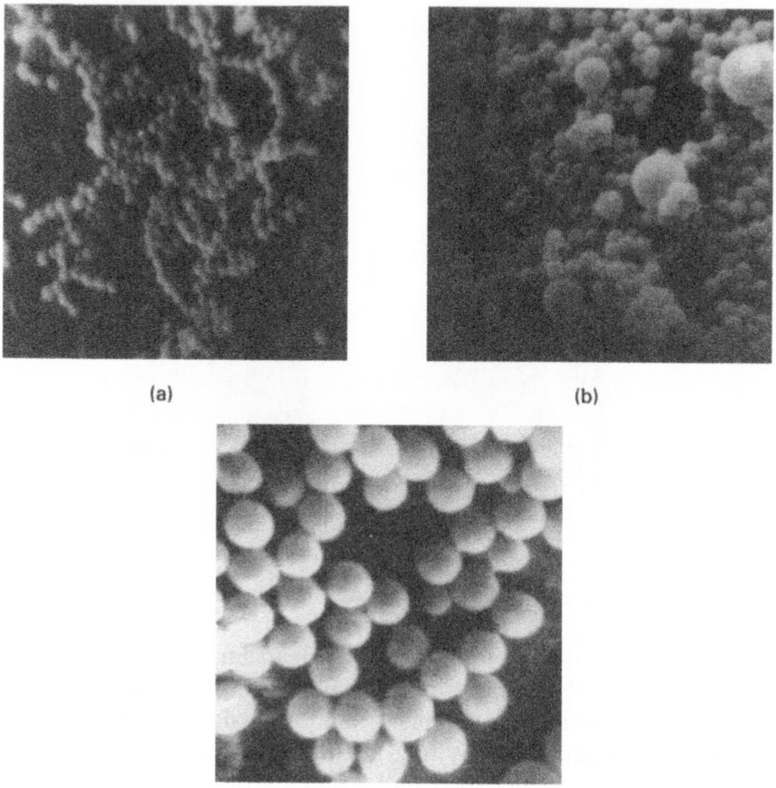

(a) (b) (c)

Figure 3. Polystyrene microbeads produced by different forms of emulsion polymerization: (a) by classical emulsion (0.15 μm); (b) by swollen emulsion (0.3 μm) (swelling agent; cetyl alcohol); and (c) by emulsion initiated from the monomer phase (0.65 μm) (swelling agent: lauryl alcohol and initiator: lauryl peroxide).

microgravity [29]. As discussed by Ugelstad and his coworkers, and as shown recently by us, if polymerization is initiated in the monomer phase, the resultant average size of the polymer particles can further be increased significantly [23, 24, 30–33]. Figure 3c shows a representative picture of polystyrene particles produced by using this latest approach with lauryl alcohol and lauryl peroxide as the swelling agent and the initiator, respectively.

Monosize polymeric microbeads were also prepared by dispersion polymerization [4, 34–50]. This process involves polymerization of a monomer dissolved in a solvent or solvent mixture in the presence of a steric stabilizer. There is initially only one phase where the monomer, initiator, and stabilizer molecules are dissolved (Fig. 1). The polymerization reaction starts out as a homogeneous solution containing monomer dissolved in an inert medium. When the growing polymeric chains reach a certain size, a phase inversion takes place. These polymer chains aggregate and are separated from the medium to form the polymeric particles which are stabilized by the steric stabilizers. Therefore, a two-phase system (i.e. a latex) is obtained at the end of the polymerization. The final particle size is determined by the inherent polymer aggregation behavior under a given set of conditions.

Barrett prepared polymethylmethacrylate monosize particles in hydrocarbon media [34]. Almog *et al.* reported the preparation of monosize polystyrene and polymethylmethacrylate particles up to 5 μm by dispersion polymerization [35]. Corner described polystyrene particles produced in aqueous ethanol media using polyacrylic acid as steric stabilizer [36]. Ober and his coworkers produced monosize polystyrene particles up to 9 μm by polymerization of styrene in a variety of solvent systems by using different non-ionic cellulosic polymers as steric stabilizers with no charged cosurfactants [40–42]. They were also successful in forming large monosize copolymer particles by dispersion polymerization in the presence of polyacrylic acid as a steric stabilizer [43]. Paine and co-workers studied the dispersion polymerization of styrene in alcoholic media in the presence of poly(*N*-vinyl pyrrolidone) as a steric stabilizer. They reported the production of monosize polymer particles up to 18 μm in size [44]. Hoffman *et al.* prepared temperature sensitive large, monosize hydrogel microspheres in micron size range and used this structure for immobilization of bioactive agents [45–46].

Monosize polystyrene microbeads by phase inversion

Recently, we were also able to produce monosize polystyrene microbeads by following a similar 'phase inversion polymerization' technique [47–51]. Note that instead of 'dispersion polymerization' we use the term of 'phase inversion polymerization' to define this polymerization technique which we believe better describes the phenomenon occurring during the polymerization process, as discussed above. In order to obtain polystyrene microbeads with different sizes, we studied a wide variety of solvent systems (i.e. ethanol/water, isopropanol/water, and ethanol/2-methoxyethanol). In order to control the polarity of the medium, we changed both the alcohol/water ratio (in the range of 90/10–60/40 ml ml^{-1}) and the type of alcohol. We used polyacrylic acid (PAA) as a steric stabilizer, and 2,2′-azobisisobutyronitrile (AIBN) as an initiator. We also investigated the effects of the monomer/dispersion medium ratio (i.e. 1/10–2.5/10 ml ml^{-1}), and the initial concentrations of initiator (i.e. 0.5–2.0 mol %) and stabilizer (i.e. 0.5–2.0 g dl^{-1}) on the size and the monodispersity of the polystyrene microbeads [50].

Table 1.
Polymerization conditions for producing monosize PS microbeads in the size range of 1–6 μm.

	Size of polymeric microbeads (μm)						
Compounds	1.0	1.5	2.0	2.5	3.0	4.0	6.0
Isopropanol (ml)	180	160	180	180	—	—	—
Water (ml)	20	40	20	20	—	—	—
Methoxyethanol (ml)	—	—	—	—	110	100	130
Ethanol (ml)	—	—	—	—	90	100	70
PAA (g)	2.00	2.00	2.00	3.00	3.50	2.50	2.50
Styrene (ml)	10.0	20.0	20.0	40.0	35.0	35.0	35.0
AIBN (g)	0.14	0.28	0.28	0.28	0.50	1.00	1.00
Conditions							
Temperature (°C)	75	80	80	80	75	75	70
Time (h)	10	10	10	24	24	24	24
Agitation rate (rpm)	250	250	250	250	250	250	250

In a typical polymerization system a glass, jacketed, magnetic drive, sealed cylindrical reactor was used. Dispersion medium was prepared by mixing of the relevant alcohol and water in the desired ratio, and the stabilizer was then dissolved in this medium. The initiator was dissolved within the monomer, and this phase was added to the dispersion medium. This polymerization medium was then charged to the reactor. Polymerizations were conducted for 24 h at 75°C with a stirring speed of 150 rpm. The experimental conditions for producing monosize PS microspheres are given in Table 1. Note that we were able to produce monosize polystyrene microbeads by following this simple phase inversion protocol in the size range of 1–6 μm, without any significant deformation in the monodispersity. Representative pictures of polystyrene microbeads obtained at different polymerization conditions are given in Fig. 4.

Details of our studies related the variation of monomer conversion, and particle size and monodispersity with the polymerization parameters were given elsewhere [50]. The following points taken from these studies should be noted to achieve monodispersity and to control the size of the particles.

Polarity. Polarity of the dispersion medium is one of the most important parameters which controls the average size and the monodispersity of the microbeads. We prepared different dispersion media having different polarities by changing the alcohol type and the water content. Note that in these types of mixtures, water should be soluble in the medium. Water content is important to achieve phase inversion and monodispersity. We found that average particle size increases with decreasing polarity of the dispersion medium. This can be explained as follows: the oligomer chains forming in the polymerization medium are apolar, and their apolar character

Figure 4. Monosize polystyrene microbeads with different sizes from 1.5 to 5 μm by phase inversion polymerization.

increases due to an increase in their molecular weight with polymerization time. If the polarity of the polymerization medium is low (i.e. the medium with low water content), the polymer chains have a chance to reach higher molecular weights before the phase inversion occurs. Therefore, fewer nuclei will be produced, leading to larger particles containing polymer chains with higher molecular weights.

Initiator concentration. The average particle size increased with increasing initiator concentration for all dispersion media that we have used. We explained this result as follows: The increase in the initiator concentration causes an increase in the number of free radicals for polymerization. In other words, polymerization starts with more radicals per unit volume in the case of high initiator concentration. This leads to lower molecular weight polymeric chains, which are more soluble in the medium. Note that in phase inversion polymerization, the nucleation or phase inversion occurs when the polymer chains reach a certain molecular weight when they become insoluble in the dispersion medium. At high initiator concentrations, due to low concentration of high molecular weight chains in the medium, fewer polymerization nuclei are produced which leads to a lower number of particles but with a larger size.

Stabilizer concentration. The average particle size decreased with increasing stabilizer concentration for all dispersion media that we have tested. It may be explained as follows: phase inversion polymerization begins and progresses around the stabilizer chains. During the nucleation period, the stabilizer chains form a structure which acts as a skeleton for particle growth. The number of forming nuclei increases with increasing stabilizer concentration which leads to more particles but with a smaller size.

Monomer concentration. The average particle size and size distribution increased with increasing monomer concentration for all dispersion media that we have tested.

Monosize polystyrene microbeads with surface functional groups

In addition to uniformity in diameter and in surface area, surface chemistry of polymeric microbeads is another main consideration for biomedical applications. Chemical modification of polystyrene based monosize microbeads is relatively difficult. For instance, one method of chemically modifying the benzene ring of cross-linked polystyrene particles which has been used commercially involves introduction of nitro groups on the ring with fuming nitric acid [52]. These can be further modified to amines and diazonium ion groups. In addition, styrene can be copolymerized with other monomers which contain specific functional groups, such as methacrylic acid. Bangs extensively reviewed some of the techniques used for copolymerization of polystyrene and further surface modification of these copolymeric microbeads for biological applications [2]. More recently, Okuho *et al.* reported preparation of monosize polymer microbeads having chloromethyl groups by following a two-step polymerization process under various conditions [53]. In the first step, they polymerized styrene by dispersion polymerization in the presence of polyacrylic acid, and obtained monosize particles about 2 μm. Then, by seeded copolymerization of styrene and chloromethyl styrene, the final structure was achieved. Monosize styrene–acrylamide copolymer lattices with different sizes were prepared by Kawaguchi *et al.* [39, 54].

Monosize microbeads having different surface chemical groups were also synthesized from various monomers. Polyacrylic microbeads were prepared by polymerizing the respective acrylic monomers (e.g., methylmethacrylate, 2-hydroxyethylmethacrylate, isopropylacrylamide) [34–35, 45, 55]. Monosize polyacrolein microspheres and their phenylated forms were investigated [56, 57]. Monosize polyvinylpyridine and copolymers were produced as large as 10–12 μm [58].

Recently, we were able to coat our monosize polymeric microbeads prepared by phase inversion polymerization [48–49, 59–60]. Monosize polystyrene microbeads prepared in the first step given above were coated with polystyrene/polyacrylate copolymers. As acrylate monomers, 2-hydroxyethylmethacrylate (HEMA), acrylic acid (AA), and dimethylaminoethylmethacrylate (DMAEMA) in order to have functional groups, namely hydroxyl (OH, uncharged), carboxyl (COOH, negatively charged), and dimethylamino ($N(CH_3)_2$, positively charged) on the surfaces of the polystyrene microspheres, respectively.

A short description of the copolymerization procedure is as follows: Prior to the copolymerization, the relevant acrylate monomer was mixed with styrene, and the initiator was dissolved in this mixture which was then added to polystyrene latex diluted with water. This medium was stirred at room temperature for 24 h in order to allow the adsorption of monomers on the polystyrene microbeads. The adsorbed monomer layer at the outer shelf of the microbeads was then polymerized. The polymerization conditions are given in Table 2.

The polystyrene/polyacrylate copolymer microbeads were examined by optical microscopy (Nikon Alphaphot, Japan). Representative pictures are given in Fig. 5. Note that the coating procedure did not change the size and monodispersity of the original polystyrene microbeads.

Table 2.
The polymerization conditions for coating of PS microbeads

Compounds	PS/PDMAEMA	PS/PAA	PS/PHEMA
Polystyrene latex[a] (ml)	100	100	100
Water (ml)	100	100	100
Styrene (ml)	2	2	2
AIBN (g)	0.12	0.12	0.12
HEMA (ml)	—	—	4
AA (ml)	—	4	—
DMAEMA (ml)	4	—	—
Conditions			
Adsorption temp. (°C)	20	20	20
Adsorption time (h)	24	24	24
Stirring rate in adsorption (rpm)	500	500	500
Copolymerization temp. (°C)	85	85	85
Copolymerization time (h)	24	24	24
Agitation rate (rpm)	200	200	200

[a] The latex suspension contains 20% w/v solid latex particles in a solution containing 90% ethyl alcohol and 10% water by volume.

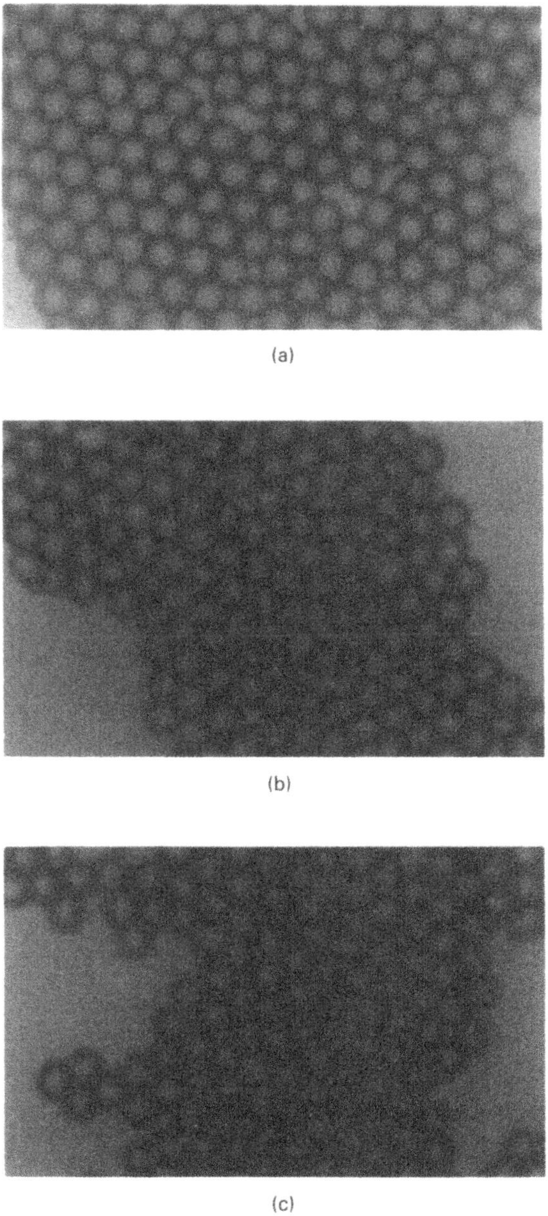

Figure 5. Monosize polystyrene microbeads (2.5 μm) with surface functional groups; (a) PS/PHEMA; (b) PS/PAA; and (c) PS/PDMAEMA.

The surface atomic compositions of the cleaned and dried polystyrene and polystyrene/polyacrylate copolymer microbeads were determined by X-ray photoelectron spectroscopy (ESCA, Hewlett Packard, 5950 B, USA). The data obtained with ESCA are given in Table 3. Higher surface oxygen contents of PHEMA and PAA coated microbeads relative to that of polystyrene clearly show the existence of these

Table 3.
The surface atomic compositions of monosize polystyrene/polyacrylate microbeads

Sample	C	N	O	Na
PS	89	0.1	10	0.4
PS/PAA	86	n.d.	14	n.d.
PS/PDMAEMA	89	1.3	9.5	0.1
PS/PHEMA	86	0.7	13	n.d.

n.d. indicates the element was not detected

respective coatings on the polystyrene microbeads. The increase in the surface nitrogen content relative to that of polystyrene confirm the existence of the PDMAEMA coating.

The average zeta potentials and electrophoretic mobilities of the microbeads were also measured. Electrophoretic-mobility measurements were done by laser-Doppler electrophoresis. The Zetasizer 3 (Malvern Instruments, Ltd., London, UK) with the AZ4 standard cell was used for measuring electrophoretic mobilities. The data obtained in these measurements are given in Table 4. As seen here, the zeta potential and electrophoretic mobility decreased due to the incorporation of more hydrophilic polymers (i.e. PAA, PHEMA, and PDMAEMA) on the hydrophobic polystyrene.

Table 4.
The electrophoretic mobilities and zeta potentials of polystyrene/polyacrylate microbeads

Sample	Electrophoretic mobility	Average zeta potential
PS	−37.73	−2.823
PS/PDMAEMA	−39.89	−2.985
PS/PHEMA	−41.91	−3.136
PS/PAA	−45.23	−3.384

MEDICAL AND BIOLOGICAL APPLICATIONS

Diagnostic imaging of GIT by radiopharmaceuticals

Currently available pharmaceuticals. Nuclear imaging techniques are rapid and effective methods for *in vivo* visualization of several diseases. The three radioactive isotopes most commonly used in clinical nuclear medicine are ^{67}Ga, ^{111}In, and ^{99m}Tc. Technetium radiopharmaceuticals make up more than 90% of all *in vivo* imaging studies in nuclear medicine today because of the optimal radionuclide properties of technetium (i.e. $t_{1/2}$ of 6 h; 140 keV gamma energy, and low radiation dosimeter) for imaging with gamma cameras [15, 61].

^{99m}Tc-pharmaceuticals, such as ^{99m}Tc-labeled diethylenetriamine pentaacetic acid, ^{99m}Tc-sulfur, antimony sulfide, tin, rhenium and phytate colloids, and ^{99m}Tc-albumin microspheres, and liposome have extensively been studied for scintigraphic visualization [15, 61–64]. All these radiopharmaceuticals proposed and tested so far for the gastrointestinal tract (GIT) imaging have been considered as non-ideal because of their following limitations: most of the carriers are not stable enough to survive the drastic pH values and changes, and enzymatic action prevailing in the gastrointestinal tract without either losing the radiolabel, getting metabolized, or decomposing. The free radiolabel released from the carrier during its passage through GIT is absorbed and causes an elevated blood background activity. The free label also accumulates in the thyroid, stomach, and urinary bladder, which then interferes with the precision of the colon studies. The labeling efficiency of the existing carriers is low, and they distribute unevenly (nonhomogeneously) within the GIT.

Polystyrene based pharmaceuticals. Recently, we proposed to use monosize polystyrene microbeads as a carrier matrix to study GIT transit and morphology. The polystyrene microbeads that we had used in our initial studies were not successful simply because of very low labeling efficiency and because of easy detachment of the label from the carrier during use. We then started to use dimethylaminoethylmethacrylate (DMAEMA) coated polystyrene microbeads. We extensively investigated these beads in *in vitro* and *in vivo* studies [59, 65–67].

The PS/PDAEMA microbeads (2.5 μm in diameter) carrying amino groups on their surfaces were labeled by the following very simple procedure: 0.5 ml of the latex containing 25 mg polymer particles were placed into a vial. 0.5–1.0 ml of pertechnate solution containing 10000 μCi ^{99m}Tc, which was obtained from a CIS generator, and 0.5 ml $SnCl_2 \cdot 2H_2O$ aqueous solution were added. The vial was shaken for a few seconds, and then left for 10 min at room temperature of about 20°C for the reaction to be completed. Note that the final pH of this solution was 5–6.

The labeling efficiency and the stability of the label were examined at different time intervals for 48 h after storage at room temperature by impregnated thin layer chromatography. The stability studies were also repeated at pH 1 and 8.

The labeling efficiency of the PS/PDMAEMA microbeads carrying amino groups was very high—more than 99%. As shown in Table 5, the labelled carriers were quite stable up to 48 h of storage at room temperature at all pH values which reflects the pH values and its change in GIT.

In vivo studies were performed by using adult Chinchilla rabbits. After overnight fast, 1000 μCi, ^{99m}Tc-PS/PDMAEMA microbeads in 1 ml of emulsion were given orally. Whole body scintigrams were obtained immediately and at selected intervals for about 48 h after administration, by a gamma camera. The scintigrams were drawn over the GIT and the whole body. The feces were also collected at the end of 24 and 48 h and counted.

Table 5.
In vitro stability of ^{99m}Tc-PS/PDMAEMA pharmaceuticals at three different pH (T = 20°C)

	^{99m}Tc-pertechnetate (%)[a]		
Time (h)	pH 1	pH 5–6	pH 8
1/2	0.48 ± 0.27	0.18 ± 0.18	1.42 ± 0.13
6	0.44 ± 0.52	0.41 ± 0.28	3.63 ± 0.22
24	0.71 ± 0.52	0.26 ± 0.29	2.75 ± 1.46
48	0.91 ± 0.51	0.33 ± 0.37	5.32 ± 1.30

[a] Mean ± S.D. of 5 tests

In scintigraphic studies in rabbits, the label stayed intact during 48 h of observation as there was no detectable radioactivity outside the GIT (>90% in GIT). The absence of thyroid, stomach (at late scintigrams), and urinary bladder images reinforced the *in vivo* stability of the labeled microbeads. The amount of activity

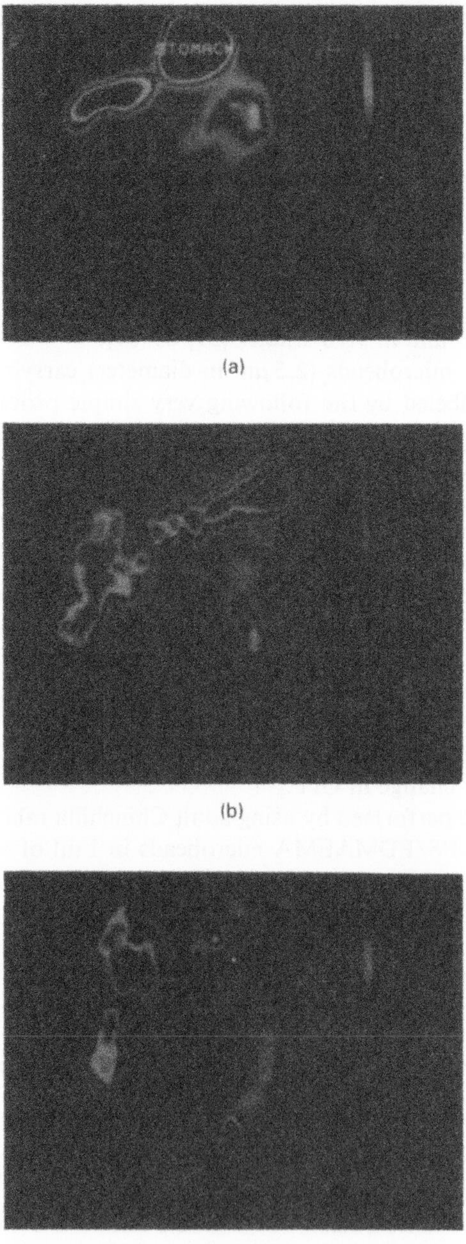
(a)
(b)
(c)

Figure 6. Representative scintigrams (anterior view) obtained in a normal subject after oral administration of ^{99m}Tc-PS/PDMAEMA latex: (a) immediately; (b) 4 h; and (c) 7.5 h.

recovered in feces was 8.6 ± 6.7 and 21.3 ± 13.8% (mean and standard deviations of 5 observations) after 24 and 48 h, respectively.

The ^{99m}Tc-PS/PDMAEMA pharmaceuticals were also tested clinically. A group of five healthy volunteers and six patients with different problems were studied in our initial clinical trials. These studies in man confirmed the findings obtained in rabbits. Figure 6 shows scintigrams obtained in a human. Notice that only the GIT was visualized. Almost no radioactivity accumulation was detected outside the GIT. The ratio of GIT whole body was 99.2 ± 0.05% (mean and standard deviations of six normals).

According to these encouraging results, we concluded that currently available pharmaceuticals may be replaced by these novel ^{99m}Tc-PS/PDMAEMA pharmaceuticals to study the colon transit time and morphology for the following reasons: (i) the labeling is very simple, effective, and reproducible; (ii) the labeled particles are very stable and remain in the GIT until excretion by feces without dissociation; and (iii) they allow monitoring of the passage of radioactivity through the GIT without any interference from background and other organs.

Phagocytosis of polymeric microbeads

One of the main defense mechanisms of living systems against foreign bodies (e.g., viruses and bacteria) is 'phagocytosis' which is the internalization of these micron-size solid foreign materials by cells. In early studies, the uptake of vital dyes was utilized to identify phagocytic cells [68]. Colloidal suspensions (i.e. carbon, iron oxide) were then used to assess phagocytic activity [69]. The use of these particles improved cell specificity, visualization, and quantitation of the phagocytic process [70]. However, low colloidal stability, lack of size uniformity, aggregation and other problems associated with these colloidals were noted as important limitations [70, 71]. More recently, synthetic polymeric particles, which can be readily manufactured in a desired size range and with different surface properties, have been utilized extensively in studying qualitatively and quantitatively the phagocytic process [16–19, 57, 69–72].

Monosize polystyrene particles are the most widely used polymeric particles to study phagocytosis because they are uniform in size, commercially available, non-toxic and highly stable. However, these particles may also have two important disadvantages related to surface properties. Firstly, they are produced by using different polymerization processes. Therefore, they may contain different chemical groups on their surfaces, which are coming from the initiator, surfactant or other ingredients used in the polymerization media. This should be one of the main concerns, especially when commercial products are used. The second disadvantage of the polystyrene particles is that it is very difficult to modify the surfaces of these particles to study the effects of different surface chemistries of the solid particles on the phagocytosis behavior of different cells.

Several other monosize polymeric microbeads have also been utilized in studying phagocytosis. Styrene–acrylamide copolymer latices with different sizes were prepared by Kawaguchi *et al.* and used for phagocytosis by leukocyte [18, 54]. Tabata and Ikada investigated monosize polyacrolein microspheres and their phenylated forms to study phagocytosis by macrophages [19, 57]. The same group proposed to use cellulose based microspheres and their chemically modified forms to study phagocytosis and studied phagocytosis of biologically modified cellulose microbeads (i.e. precoated with different proteins and pretreated with fetal calf serum). They also used biodegradable polylactic/glycolic acid copolymers in their studies. In contrast to tailor-made surface properties, unfortunately these microbeads exhibited a significant polydispersity in size, which may be considered as the main limitation in their studies and related results.

Recently, we investigated phagocytosis of our monosize polystyrene based microbeads by blood cells and also mouse peritoneal macrophages. The monosize polystyrene microbeads with different sizes (0.9–6.0 μm), their modified forms carrying functional groups on their surfaces (i.e. PS/PHEMA, PS/PAA, and PS/PDMAEMA), and the polystyrene microbeads pretreated within bovine serum albumin (BSA) and fibronectin (Fn) aqueous solutions were utilized in this study [73].

Blood samples from healthy volunteers were incubated in heparinized tubes with microbeads (10 μl microbead latex for 0.25 ml blood sample) for 20 min at 37°C. After fixing in methanol and dying with a proper dye (i.e. Giemsa or Jenner, Merck, Germany), leukocytes were observed by phase-contrast microscopy. In each test at least 200 leukocytes were examined and the average number of microbeads internalized by one leukocyte was counted. Experiments were repeated three times for each type of microbead. Phagocytosis behavior of the peritoneal macrophages freshly

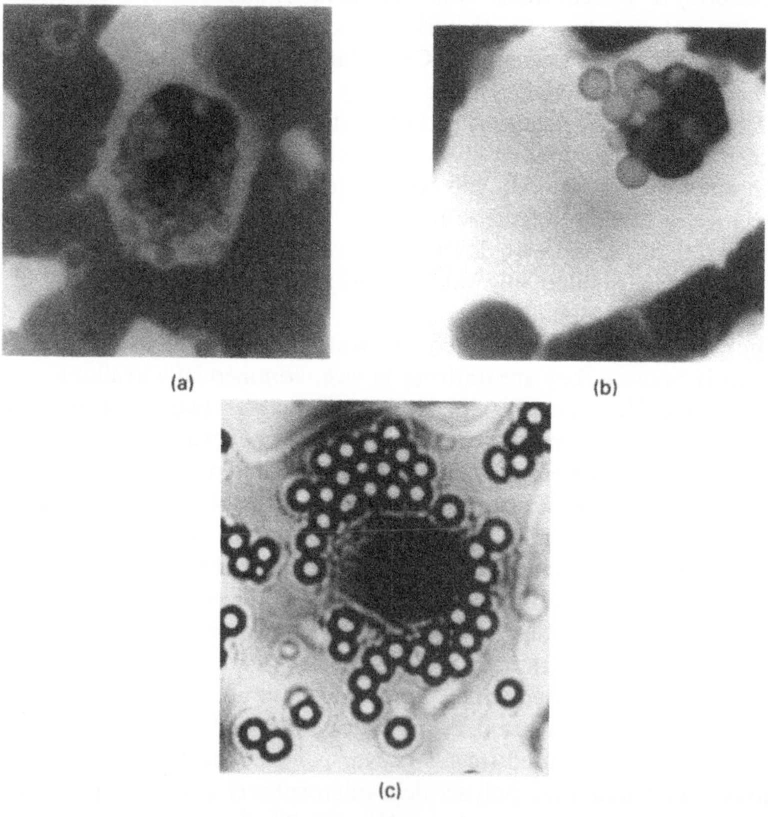

(a) (b) (c)

Figure 7. Phagocytosis of polystyrene microbeads: (a) a monocytes and PS microbeads; (b) a neutrophil and PS microbeads; and (c) a macrophage and BSA pretreated microbeads.

Table 6.
Effect of the microbead size on phagocytosis

Microbead size (μm)	Microbead no/cell Neutrophils	Monocytes	Macrophages
0.9	29 ± 7	36 ± 9	65 ± 10
1.5	13 ± 4	15 ± 5	42 ± 7
2.0	9 ± 3	11 ± 3	33 ± 5
2.5	7 ± 2	8 ± 3	26 ± 2
3.0	4 ± 2	5 ± 2	9 ± 3
4.0	1 ± 0	1 ± 0	3 ± 1
6.0	1 ± 0	1 ± 0	1 ± 0

isolated from mouse were also studied in Hanks' balanced salt solution and phosphate-buffered saline solution at pH 7.4, by following the same technique used for leukocytes.

Figure 7 shows some representative pictures of phagocytosis. The following important results should be noted:

(i) Size of microbeads: The number of microbeads phagocytosed per cell was the maximum for the smallest particles (0.9 μm) that we used in this study, for both leukocytes (i.e., neutrophils and monocytes) and peritoneal macrophages, as given in Table 6. The particle uptake dropped significantly with the particle size. However, both cell groups were still able to internalize one to three microbeads of 4–6 μm.

(ii) Surface properties: Table 7 shows the effects of surface chemistries on the number of particles phagocytosed by different cells. Notice that the uptake of the more hydrophobic PS particles was much higher than the less hydrophobic PS/PHEMA microbeads. Amino groups (i.e. positively charged) on PS/PDMAEMA microbeads significantly increased the particle uptake. In contrast to amino groups, carboxyl groups (i.e. negatively charged) on PS/PAA microbeads caused a pronounced drop in the number of microbeads phagocytosed. Biological modification of microbead surfaces dramatically changed the behavior of phagocytosis of polystyrene microbeads. The albumin molecules preadsorbed on the

Table 7.
Effect of the microbead surface properties on phagocytosis

Microbead[a]	Microbead no/cell Neutrophils	Monocytes	Macrophages
PS	13 ± 4	15 ± 5	42 ± 7
PS/PHEMA	4 ± 2	6 ± 1	10 ± 2
PS/PDMAEMA	20 ± 3	21 ± 4	55 ± 9
PS/PAA	9 ± 3	10 ± 3	18 ± 4
PS/BSA[b]	0	0	9 ± 6
PS/Fn[c]	22 ± 8	24 ± 7	54 ± 8

[a] Microbead size: 1.5 μm.
[b] PS microbeads were soaked in bovine serum albumin (BSA) solution (4 mg BSA ml^{-1}) for 24 h.
[c] PS microbeads were soaked in fibronectin (Fn) solution (1 mg Fn ml^{-1}) for 24 h.

polystyrene microbead surfaces drastically reduced the number of microbeads internalized by both cell groups while preadsorbed fibronectin, known as 'biological glue', significantly enhanced the phagocytosis.

Polystyrene microbeads as carriers for dye affinity chromatogaphy

The interest in and demand for proteins in biotechnology, biochemistry, and medicine have contributed to an increased exploitation of affinity chromatography. Unlike other forms of protein separation, affinity chromatography relies on the phenomenon of biological recognition, which enables biopolymers to recognize specifically, and bind reversibly, their complementary ligands (e.g. enzymes) and their substrates, hormones and their receptors, and antibodies and their antigens [20–22].

Unfortunately, preparation of sorbents carrying biological ligands is usually very expensive because the ligands themselves often require extensive purification and it is difficult to immobilize them on the carrier matrix with retention of their biological activity. As an alternative to their natural biological counterparts, the reactive triazinyl dyes, have been investigated as ligands for protein affinity separation [74–76]. These dyes are able to bind proteins in a remarkably specific manner. They are inexpensive, readily available, biologically and chemically inert, and are easily coupled to support materials. Cibacron Blue F3GA, and many other reactive dyes have been coupled to a variety of supports including agarose, cellulose, polyacrylamide, sephadex, silica, and glass [77–82]. Dye–ligand chromatography has now enabled the purification of a wide range of proteins (e.g. lactate hydrogenase, alcohol dehydrogenase, hexokinase, carboxyl peptidase, etc.) [74–82].

Recently we attempted to use our monosize polymeric microbeads as a carrier matrix for affinity purification of proteins. We selected albumin as a potential model protein. We studied both nonspecific albumin adsorption on the polystyrene based microbeads and also specific albumin adsorption on the dye-attached polystyrene microbeads [60, 83–84]. Some interesting results of these studies are briefly discussed below.

Nonspecific albumin adsorption. Bovine serum albumin (BSA) adsorption on various polymeric latices have been intensively investigated by several groups [85–93]. Note that the latices used in these previous studies had been produced either by conventional emulsion polymerization or seeded copolymerization recipes. Therefore, most of these latex particles contain charged groups on their surfaces coming from the water soluble initiator used in the polymerization recipes. These charged groups exhibit a strongly acidic character which, of course, significantly affects the protein–latex interactions.

Recently, we investigated the adsorption behavior of BSA onto our monosize PS based polymeric microbeads (2.5–4.0 μm), which have quite different surfaces than those produced with conventional recipes. We studied the effects of pH, ionic strength, and coexistent electrolytes on adsorption of BSA both at native and denatured states [60, 83–84]. Effects of pH and ionic strength of the adsorption medium on BSA adsorption capacity of the microbeads are briefly presented below.

For equilibrium adsorption studies, 10 mg of BSA was dissolved in 10 ml of buffer solution. The ionic strength of the medium was adjusted by using different

Table 8.
BSA adsorption on PS based microbeads at different pH and ionic strengths (adsorption conditions: temperature: 25°C; stirring rate: 40 r.p.m.; and equilibrium time: 2 h; BSA initial concentration: 1.0 mg ml^{-1})

Microbead	Adsorption capacity (mg BSA m^{-2}) Ionic Strength: 0.01					Ionic Strength: 0.1				
pH	3.5	5.0	6.0	7.0	8.0	3.5	5.0	6.0	7.0	8.0
PS[a]	1.3	9.4	8.5	0	0	1.9	8.5	3.5	0	0
PS[b]	2.3	13.3	8.0	0	0	2.3	11.6	2.0	0	0
PS/PHEMA[a]	1.3	9.4	4.2	0	0	2.3	8.5	2.7	0	0
PS/PDMAEMA[a]	3.8	13.1	10.0	1.7	0.2	2.3	9.8	4.2	0.8	0.4
PS/PAA[a]	0	10.4	2.3	0	0	0	7.5	0	0	0

The density of solid latex particles was determined as 1.03 g cm^{-3} by the picnometric method.
'0' denotes that the albumin adsorption capacities of the microbeads are less than 0.2 mg BSA ml^{-2}.
[a] Microbead size: 2.5 μm.
[b] Microbead size: 4.0 μm.

amounts of NaCl. PS microbeads (0.05–0.5 g) were added to the BSA solution. Adsorption experiments were conducted for 2 h at a constant temperature of 25°C and at a stirring rate of 40 r.p.m. BSA concentration was measured spectrophotometrically at 279 nm.

Table 8 gives the adsorption data. As seen here, the maximum adsorption capacities were observed near the isoelectric point (IEP) of BSA (i.e. about 5) in all cases. It has been shown that proteins have no net charge at their isoelectric points (IEP), therefore, the maximum amount of adsorption from aqueous solutions is observed at IEP [85–92]. Below or above their IEP values, proteins are charged positively or negatively, respectively. Therefore, they are more hydrated, which increases their stability and solubility in the aqueous phase (which means low adsorption). We also obtained significantly lower and nearly zero adsorptions with all our microbeads in both acidic and alkaline regions. Shirahama *et al.* investigated BSA adsorption onto different polymeric latices prepared by conventional processes [92]. They found an appreciable BSA adsorption capacity in the pH region higher than IEP. This difference may be explained by considering the chemical groups on their latex surfaces. Their latices contain strongly acidic sulfate groups on their surfaces coming from the initiator. These groups may change the degree of hydration of latex particles, and hence, may increase the adsorption of BSA molecules. In the phase inversion polymerization procedure, we used AIBN, which has a more hydrophobic character, to produce our PS and modified PS microbeads. As it is expected we did observe no adsorption of BSA, which is highly hydrated (high solubility in aqueous phase) in the alkaline region, on our relatively hydrophobic surfaces. The relatively higher adsorption capacities obtained with the PS/PDMAEMA microbeads may be due to electrostatic interaction between positively charged amino groups on these microbeads and the negatively charged BSA molecules.

The amount of BSA adsorbed on all of the microbeads tested in this study decreased with an increase in the ionic strength of the adsorption medium. A similar tendency was also observed for the polymeric latices studied by Shirahama and Suzawa [88]. This may be explained by the formation of more compact structures of BSA molecules at high ionic strength. More ions may be attached to BSA molecules at higher ionic strengths. This causes further stabilization of the protein molecules (higher solubility), which may lead to lower adsorption of BSA on the latex.

Specific albumin adsorption. Recently we investigated the feasibility of using monosize polystyrene based microbeads as a support matrix for dye affinity purification of proteins. We coated the surfaces of polystyrene microbeads with a thin layer of cross-linked polyvinylalcohol (PVAL) in order to attach the dye and also to eliminate nonspecific interactions between albumin and the hydrophobic PS surfaces.

Coating the PS microbeads with PVAL was achieved by a two step procedure. At the first adsorption step, the PS microbeads were treated in a magnetically stirred vessel containing PVAL with different molecular weights (i.e. 14 000–100 000) at different concentrations (100–1400 mg PVAL l^{-1}). Different amounts of Na_2SO_4 was used to adjust the ionic strength (0.05–0.2) of the adsorption medium. Adsorption was completed in 2 h. The stirring rate and temperature were 200 r.p.m. and 25°C, respectively. At the second step the PVAL coating on the PS microbeads chemically cross-linked by using terephatalaldehyde (TPA). After the adsorption process, the final acid concentration of the medium was adjusted to 0.1 M by adding concentrated HCl. The cross-linker, TPA was then added (10 mg TPA for 10 ml medium). The batch containing the microbeads and the cross-linker was stirred in a sealed reactor, first for 48 h at 500 r.p.m. at 25°C, and then for 4 h at 500 r.p.m. at 80°C. The latex was filtered. The microbeads were washed with distilled water several times and then resuspended in distilled water. No PVAL leakage was detected in the leakage tests performed at different conditions for about 24 h [83].

For the attachment of dye to the PVAL coated microbeads, 300 mg Cibacron Blue F3GA (Polyscience, USA) was dissolved in 10 ml of water. This solution was then mixed with 90 ml of an aqueous dispersion of microbeads containing 3 g of the dry polymeric microbeads and 4 g of NaOH. This mixture was then treated in a sealed

Table 9.
BSA adsorption on PS based microbeads (microbead size: 4.0 μm; adsorption conditions: pH: 5.0 with CH_3COONa/CH_3COOH; Ionic strength: 0.01 with NaCl; Temperature: 25°C; Stirring rate: 40 rpm; and Equilibrium time: 2 h)

BSA initial conc. ($mg\ ml^{-1}$)	BSA adsorption ($mg\ BSA\ m^{-2}$) PS/PVAL	Untreated PS	Dye-attached PS
0.5	0.3	5.0	15.7
1	0.7	13.3	24.3
2	0.9	16.1	28.9
3	1.3	17.9	35.5
4	1.7	18.6	36.5
5	1.8	17.7	38.1
6	1.9	17.7	38.9
7	1.9	17.9	39.6

The density of solid latex particles was determined as 1.03 g cm^{-3} by the picnometric method.

reactor for 4 h at 400 r.p.m. and a constant temperature of 80°C. The latex was filtered, and the microbeads were washed several times with water and methanol to remove the unbound dye, and resuspended in 30 ml of distilled water. No dye leakage was detected in the leakage tests performed at different conditions for about 24 h [83].

Effects of the initial BSA concentration, pH, and ionic strength on adsorption behavior of BSA on the untreated PS, the PVAL coated PS, and the dye-attached PS microbeads was investigated in batch adsorption-equilibrium studies. Table 9 gives typical BSA adsorption data obtained in this group of experiments. As was expected, adsorption increased with the initial concentration of BSA. Note that there was a pronounced BSA adsorption on PS microbeads (up to 20 mg BSA m^{-2} at the conditions given on the legend of Table 9) because of the hydrophobic interactions between albumin and PS. The PVAL coating significantly decreased BSA adsorption on these untreated hydrophobic PS microbeads, as was expected. Very high (up to 40 mg m^{-2}) BSA adsorption capacities were achieved with the dye-attached PS microbeads.

The desorption of the adsorbed BSA from the dye-attached PS microbeads was also studied in a batch experimental setup. We were able to desorb up to 92% of the adsorbed protein by using 0.5 M NaSCN [83]. The ultimate design of the dye affinity separation system for protein purification by using monosize PS based microbeads is still under investigation.

Acknowledgements

We would like to thank Prof. Joseph D. Andrade and Paul Dryden for ESCA measurements performed in the Department of Bioengineering, University of Utah, USA. The electrophoretic-mobility measurements were done in the Department of Metallurgical Engineering, University of Utah, USA. It is a pleasure to thank Prof. Jan D. Miller and Jaroslaw Drelich for their help and respective contributions in electrophoretic-mobility measurements.

REFERENCES

1. A. Rembaum and Z. A. Tokes (Eds.), *Microspheres: Medical and Biological Applications*. CRC Press, Boca Raton, FL (1988).
2. L. B. Bangs, *Uniform Latex Particles*, 3rd Edition. Seradyn Inc., Indianapolis (1987).
3. J. L. Millan, K. Nustad and B. N. Pederson, *Clin. Chem.* **31**, 54 (1985).
4. A. Rembaum, R. Yen, D. Kempner and J. Ugestad, *J. Immunol. Meth.* **52**, 341 (1982).
5. O. Bormer, *Clin. Biochem.* **15**, 128 (1982).
6. R. K. Kumar and A. W. J. Lykke, *Pathology* **16**, 53 (1984).
7. A. Rembaum and W. J. Dreyer, *Science* **208**, 364 (1990).
8. P. L. Kronick, G. L. Campbell and K. Joseph, *Science* **200**, 1074 (1978).
9. S. Margel, U. Beitler and M. Ofarim, *J. Cell. Sci.* **56**, 157 (1982).
10. E. P. Goldberg (Ed.), *Targeted Drugs*. John Wiley, New York (1983).
11. R. Juliano (Ed.), *Drug Delivery Systems*. Oxford Univ. Press, Oxford (1980).
12. J. J. Ariens (Ed.), *Drug Design*. Academic Press, New York (1980).
13. S. S. Davis, L. Illum, J. G. McVie and E. Tomlinson (Eds.), *Microspheres and Drug Therapy*. Elsevier, Amsterdam (1984).
14. S. E. Strand, L. Anderson and L. Bergqvist, in: *Microspheres: Medical and Biological Applications*, p. 193, A. Rembaum and Z. A. Tokes (Eds.). CRC Press, Boca Raton, FL (1988).
15. J. W. Froelich, in: *Nuclear Medicine Annual 1985*, p. 23, L. M. Freeman and H. S. Weismann (Eds.). Raven Press, New York (1985).

16. H. E. Walter, E. J. Krob and R. Garza, *Biochim, Biophys. Acta.* **165,** 507 (1968).
17. C. J. van Oss, *Ann. Rev. Microbial.* **32,** 19 (1978).
18. H. Kawaguchi, N. Koiwai, Y. Ohtsuka, M. Miyamoto and S. Sasakawa, *Biomaterials* **7,** 61 (1986).
19. Y. Tabata and Y. Ikada, *Biomaterials* **9,** 356 (1988).
20. P. D. G. Dean, W. S. Johnson and F. A. Middle (Eds.). *Affinity Chromatography,* IRL Press, Oxford (1985).
21. W. H. Scouten, *Affinity Chromatography.* John Wiley, New York (1981).
22. B. R. Dunlap (Ed.). *Immobilized Biochemicals and Affinity Chromatography.* Plenum Press, New York (1974).
23. J. Ugelstad, M. S. El-Aasser and J. W. Vanderhoff, *J. Polymer Sci.: Polymer Lett. Ed.* **11,** 503 (1973).
24. J. Ugelstad, F. K. Hansen and S. Lange, *Makromol. Chem.* **175,** 507 (1974).
25. F. H. Hansen and J. Ugelstad, *J. Polym. Sci. Polym. Chem. Ed.* **16,** 1953 (1978).
26. J. Ugelstad, K. H. Kaggerard, F. K. Hansen and A. Berge, *Makromol. Chem.* **180,** 737 (1979).
27. J. Ugelstad, P. C. Mork, K. H. Kaggerud, T. Ellingsen and A. Berge, *Adv. Coll. Int. Sci.* **13,** 101 (1980).
28. J. Ugelstad, P. C. Mork, A. Berge, T. Ellingsen and A. A. Khan, in: *Emulsion Polymerization,* I. Piirma, (Ed.). Academic Press, New York (1982).
29. A. M. Lovelace, J. W. Vanderhoff, F. J. Micale, M. S. El-Asser and D. M. Kornfeld, *J. Coating Technol.* **54,** 691 (1982).
30. A. Tuncel, Ph.D. Thesis. Hacettepe University, Ankara, Turkey (1989).
31. A. Tuncel and E. Piskin, *Polym. Plast. Technol. Eng.* **29,** 561 (1990).
32. A. Tuncel and E. Piskin, *Polym. Plast. Technol. Eng.* **31,** 787 (1992).
33. A. Tuncel and E. Piskin, *Polym. Plast. Technol. Eng.* **31,** 807 (1992).
34. K. E. J. Barrett (Ed.), *Dispersion Polymerization in Organic Media.* John Wiley, London (1975).
35. Y. Almog, S. Reich, S. and M. Levy, *Br. Polym. J.* **14,** 131 (1982).
36. T. Corner, *Coll. Surf.* **3,** 119 (1981).
37. J. W. Vanderhoff, M. S. El-Aasser, F. J. Micale, E. D. Sudol, C. M. Tseng, A. Silvanowicz, D. M. Kornfeld and F. A. Vincente, *J. Disp. Sci. Tech.* **5,** 231 (1984).
38. S. Margel and E. Weisel, *J. Polym. Sci. Polym. Chem. Ed.* **22,** 145 (1984).
39. H. Kawaguchi, H. M. Nakamura, M. Yanagisawa, F. Hishino and Y. Ohtsuka, *Makromol. Chem. Rapid Commun.* **6,** 315 (1985).
40. C. K. Ober, K. P. Lok and M. L. Hair, *J. Polym. Sci.: Polym. Lett. Ed.* **23,** 103 (1985).
41. K. P. Lok and C. K. Ober, *Can. J. Chem.* **63,** 209 (1985).
42. C. K. Ober and K. P. Lok, *Macromolecules* **20,** 268 (1987).
43. C. K. Ober and M. L. Hair, *J. Polym. Sci.* **25,** 1395 (1987).
44. A. J. Paine, W. Luymes, J. McNulty, *Macromolecules* **23,** 3104 (1990).
45. T. G. Park, A. S. Hoffman, *J. Polym. Sci., Part A: Polym. Chem.* **30,** 505 (1992).
46. T. G. Park, A. S. Hoffman, *Biotechnol. Bioeng.* **35,** 152 (1990).
47. A. Tuncel and E. Piskin, Turkish Patent, No: 24125 (1991).
48. A. Tuncel and E. Piskin, Turkish Patent, No: 24126 (1991).
49. R. Kahraman, M. Sc. Thesis, Hacettepe University, Ankara, Turkey (1991).
50. A. Tuncel, R. Kahraman and E. Piskin, *J. Appl. Polym. Sci.* (1993), (in press).
51. A. Tuncel, R. Kahraman, H. Ayhan and E. Piskin, *J. Appl. Polym. Sci.* (1993), (accepted).
52. H. J. Tenoso and D. B. Smith, NASA Tech. Brief B72-10006; NASA TSP 72-10006, Washington, D.C. (1972).
53. M. Okuho, K. Ikegami, Y. Yamamoto, *Coll. Polym. Sci.* **267** 193 (1989).
54. H. Kawaguchi, H. Hoshino, H. Amagasa and Y. Ohtsuka, *J. Colloid Interface Sci.* **97,** 465 (1984).
55. R. S. Molday, W. J. Dreyer, A. Rembaum and S. P. S. Yen, *J. Cell. Biol.* **64,** 75 (1975).
56. S. Margel, *Methods Enzymol.* **112,** 165 (1985).
57. Y. Tabata and Y. Ikada, in: *High Performance Biomaterials,* p. 621, M. Szycher (Ed.). Technomic Publ. Comp., Basel (1991).
58. R. van Furth, T. L. Zwet and P. C. J. Leijh, in: *Cellular Immunology.* Vol. 2, Ch. 32, D. M. Weir (Ed.), Blackwell Scientific, Oxford (1978).
59. M. T. Ercan, A. Tuncel, B. E. Caner, M. Mutlu and E. Piskin, *Nucl. Med. Biol.* **18,** 253 (1991).
60. A. Tuncel, A. Denizli, M. Abdelaziz, H. Ayhan and E. Piskin, *Clin. Mater.* **11,** 139 (1992).
61. G. Subramanian, B. A. Rhodes, J. F. Cooper and V. J. Sodd (Eds.), *Radiopharmaceuticals.* Society of Nuclear Medicine, New York (1975).

62. V. J. Claride, E. K. Prokop, F. J. Troncale, W. Buddoura, K. Winchenbach and R. W. McCallum, *Gastroenterology* **86**, 714 (1984).
63. J. Isolauri, M. O. Koskinen and H. Markula, *J. Thorac. Cardiovasc. Surg.* **94**, 521 (1987).
64. J. C. Maublant, M. Sournac, J.-M. Aiache and A. Veyre, *J. Nucl. Med.* **28**, 1199 (1987).
65. M. T. Ercan, A. Tuncel, B. E. Caner, M. Mutlu and E. Piskin, *Proc. Eur. Assoc. Nucl. Med. Cong.*, p. 136, Amsterdam, (1990).
66. B. E. Caner, M. T. Ercan, L. O. Kapucu, A. Gezici, A. Tuncel, C. F. Bekdik, G. F. Erbengi and E. Piskin, *Proc. Eur. Assoc. Nucl. Med. Cong.* p. 277, Amsterdam (1990).
67. B. E. Caner, M. T. Ercan, L. O. Kapucu, A. Tuncel, C. F. Bekdik, G. F. Erbengi and E. Piskin, *Nucl. Med. Commun.* **12**, 539 (1992).
68. H. W. Florey (Ed.), *General Pathology*. Lloyd-Luke Medical Books, London (1970).
69. R. van Furth (Ed.), *Mononuclear Phagocytes*. Blackwell Scientific, Oxford (1970).
70. R. J. Kanet and J. D. Brain, *RES J. Reticuloendothel. Soc.* **27**, 201 (1980).
71. C. J. van Oss (Ed.), *Phagocytic Engulfment and Cell. Adhesiveness*. Marcel Dekker, New York (1975).
72. E. F. Repollet and A. Schwartz, *Microspheres: Medical and Biological Applications*, p. 139, A. Rembaum and Z. A. Tokes (Eds.). CRC Press, Boca Raton, FL (1988).
73. H. Ayhan, M. Sc. Thesis, Hacettepe University, Ankara, Turkey (1992).
74. A. Fiechter (Ed.), *Advances in Biochemical Engineering*, Spring-Verlag, Berlin (1982).
75. Y. D. Clonis, A. A. Atkinson, C. J. Bruton and C. R. Lowe (Eds.), *Reactive Dyes in Protein and Enzyme Technology*. Macmillan, Basingstoke (1987).
76. Y. D. Clonis, *CRC Crit. Rev. Biotechnol.* **7**, 263 (1988).
77. C. R. Lowe, M. Hans, N. Spibey and W. T. Drabble, *Anal. Biochem.* **104**, 23 (1980).
78. S. Angal and P. D. G. Dean, *Biochem J.* **167**, 301 (1977).
79. M. F. Meldolesi, V. Macchia and P. Laccetti, *J. Biol. Chem.* **251**, 6 (1976).
80. R. L. Easterday and I. M. Easterday, *Adv. Exp. Med. Biol.* **42**, 123 (1974).
81. C. R. Lowe, M. Glad, P. O. Larsson, S. Ohlson, D. A. P. Small, T. Atkinson and K. Mosbach, *J. Chromatogr.* **299**, 175 (1975).
82. W. C. Thresher and H. E. Swaisgood, *Biochem. Biophys. Acta.* **749**, 214 (1983).
83. A. Denizli, Ph.D. Thesis, Hacettepe University, Ankara, Turkey (1992).
84. A. Tuncel, A. Denizli, D. Purvis, C. R. Lowe and E. Piskin, *J. Chromatogr.* **634**, 161 (1993).
85. T. Suzawa, H. Shirahama and T. Fujimoto, *J. Colloid Interface Sci.* **86**, 144 (1982).
86. T. Suzawa, H. Shirahama and T. Fujimoto, *J. Colloid Interface Sci.* **93**, 498 (1983).
87. H. Shirahama and T. Suzawa, *Colloid Polym. Sci.* **263**, 141 (1985).
88. H. Shirahama and T. Suzawa, *J. Colloid Interface Sci.* **104**, 416 (1985).
89. T. Suzawa and T. J. Murakami, *J. Colloid Interface Sci.* **78**, 266 (1980).
90. L. J. Zsom, *Colloid Interface Sci.* **111**, 434 (1986).
91. H. Shirahama, K. Takeda and T. Suzawa, *J. Colloid Interface Sci.* **109**, 552 (1986).
92. H. Shirahama, T. Shikawa and T. Suzawa, *Colloid Polym. Sci.* **267**, 587 (1989).
93. M. Okubo, T. Azume and Y. Yamamoto, *Colloid Polym. Sci.* **268**, 598 (1990).

Covalent immobilization of microorganisms in polymeric hydrogels

LEV I. VALUEV*, VLADIMIR V. CHUPOV and NICOLAI A. PLATE'
Institute of Petrochemical Synthesis of the Russian Academy of Sciences, Leninsky Prospect 29, GSP-1, 117912 Moscow, Russia

Received 27 February 1992; accepted 17 September 1992

Abstract—A method of covalent immobilization of microorganisms (marine luminescent bacteria and yeast) in polymeric hydrogels is described. It is shown that cell immobilization leads to the creation of materials having properties of both synthetic polymers and physiologically active systems. Application of systems containing covalent immobilized yeast and photobacteria in biotechnological and other processes is proposed.

Key words: Cells; covalent immobilization; polymeric hydrogels; bioluminescence.

INTRODUCTION

One of the important problems of modern biotechnology is to establish methods of cell immobilization, including microorganisms, without loss of their physiological or functional activity. Several methods of enzyme immobilization are suitable for cells: adsorption [1], inclusion in organic and inorganic gels [2], crosslinking of cells with bifunctional agents [3], and covalent bonding to insoluble matrices [4]. It should be noted that methods of covalent immobilization of cells have not been extensively developed, and that the number of publications on theoretical and practical problems of covalent immobilization is small [5].

The effective covalent coupling of cells requires minimal interaction between the cell and the insoluble matrix to allow for maximum availability to the cell of low-molecular weight substrates, coenzymes, and metabolytes.

The creation of covalent bonds between the cell and the support matrix forms a stable conjugate which is likely to dissociate during the normal lifetime of cell. The conditions under which such a conjugate is formed often subjects the cells to strong, disruptive physical and chemical stresses. Thus covalent bonding is only justified where the superior performance and stability of immobilized cells outweighs the irreversible loss of their activity which inevitably accompanies the covalent bonding chemistry. Some immobilization methods are based on the use of different chemical substances which are toxic to the cells. To study the interaction of the modifying agents with the cells during the course of covalent immobilization is very important for understanding how the cell functions and how the metabolic pathways might be altered.

On the other hand, if the immobilization process does not affect the physiological activity of the cells, this can lead to the creation of long-lived systems with

* To whom correspondence should be addressed.

applications as biocatalysts in technological processes. For example, immobilized yeast cells reactors could represent a less costly and less complex alternative to yeast recycle reactors [6].

The aim of this paper is to describe a new effective method for cell immobilization and to study the amounts and kinetics of the metabolites obtained.

MATERIALS AND METHODS

The cells of marine luminescent bacteria *Photobacterium fischeri*, cultures 6 and 137 [7] and yeast cells *Saccharomyces cerevisia* [8] were used. Acroylchloride was synthesized according to the procedure of Stempel [9]. ^{14}C-labelled acetylchloride with 81.4 MBk/ml of activity and ^{125}I-labelled NaI with 2.8 GBk/ml of activity were used without purification. The acylation of the cells was carried out according to the procedure described by Labonesse and Gervais [10]. The cells were acylated in 0.05 M phosphate buffer, pH = 7.8, containing 0.3 M NaCl, using acroyl- or acetylchloride for 30 min at 0–4°C. After this reaction the suspension of acylated cells was filtered and washed with phosphate buffer solution until the level of radioactivity of the washing solution reached the background level. The acylation of the destroyed cells was carried out by the same method. The radioactivity of fractions was measured using a 'Mark-3' counter (Tracor Analytic, USA).

Cell disruption was accomplished either by osmotic shock combined with freezing–thawing cycles (5 times) according to [11] (100% cell disruption), or by ultrasonic disintegration (35 kHz, 3 min) for complete disruption. The membrane fraction was separated by centrifugation. Centrifugation conditions after each disruption procedure were typically 3000 *g* for 10 min. Acylation of the cell fractions was carried out by the same method as for intact cell acylation. The membrane fraction was washed to remove free label with the above mentioned buffer solution and the intracellular extract was purified by dialysis. The protein content in the membrane fraction after disruption and in the intracellular extract was determined by the method of Lawry [12].

The immobilization of the acylated cells in polymeric hydrogels was carried out by means of copolymerization of the acylated acroylchloride cells with acrylamide (10.0 wt.%) and *N,N'*-methylenebisacrylamide (1.0 wt.%) in the presence of a redox initiating system consisting of ammonium persulphate and *N,N,N',N'*-tetramethylethylenediamine.

The luminescence of both intact and immobilized cells was measured with a special luminometer at 490–500 nm. High levels of bioluminescence of photobacteria do not require a nutrition-containing substrate for maximal response [7].

A polyethylene carrier film for cell immobilization was created by grafting of a copolymer of acrylamide and acroylchloride using γ-irradiation [13]. The graft copolymer had an acroylchloride content of 60%. The immobilization of cells onto the surface of this carrier film was studied by contacting a cell suspension with the carrier and measuring the intensity of bioluminescence of both the suspended and immobilized cells with time at the ambient temperature.

The activity of yeast cells was evaluated by following ethanol production in batch fermentations. Fermentations were carried out in a standard synthetic medium, pH = 4.7, containing 12% glucose, 0.3% yeast extract, 0.2% NH_4Cl, 0.2% KH_2PO_4, and 0.2% lactic acid. 10 ml of intact or immobilized yeast was added to

20 ml of medium in 100 ml flasks. The ethanol concentration was measured densitometrically.

The efficiency of yeast cells immobilization was calculated by equation (1):

$$w(\%) = (w_o - w_r)/w_o \times 100\%, \qquad (1)$$

where w_o and w_r are the numbers of cells in the original cell suspension, and in the rinse solution removed from the gel, respectively. The number of cells was measured using an optical microscope with a special Goryaev's camera [14].

Seed cultures for cell reproduction experiments were grown in 100 ml lots in peptone media consisting of 1% glucose and 0.3% of aliphatic aldehyde $C_{13}H_{27}CHO$ in marine water. Growth was carried out aerobically with agitation at 25°C for 24 h. The biomass growth was recorded spectrophotometrically using optical density measurements at $\lambda = 580$ nm.

RESULTS AND DISCUSSION

The covalent binding of cells to a support is usually achieved by activating a surface which has a high concentration of functional groups and contacting this surface with the cells under mild conditions between pH 4 and 9, near or below the ambient temperature. Thus, the reactions which can be considered are limited to those capable of derivatizing the free amino, carboxylic, phenolic, and thiol groups present on the cell surface or the carbohydrate substituents of the cell wall glycoproteins. Insoluble reagents, developed for cell immobilization and have proved a rich source of ideas for activation procedures, are rarely rewarded by significant improvements in the efficiency with which the cells finally bind to the matrix [15].

To carry out immobilization of microorganisms in polymer hydrogels we used a 'macromonomeric' method based on the introduction of reactive C=C double bonds onto the physiologically active compound. Copolymerization of this 'macromonomer' with a hydrophylic comonomer and crosslinking agent is described in Scheme 1.

Many physiologically active substances can be immobilized in the hydrogels by this method: aminoacids, peptides, proteins and enzymes, amino-containing polysaccharides, etc. [16].

To use this technique for cell immobilization, it is necessary to study the interaction of the cells with the acylating agent and properties of the modified cells. The introduction of C=C bonds as cell functional groups suggests that these products may be considered as 'supermacromonomers', and the copolymerization products as cell filled graft copolymers. As model cells having easily measurable and controlled activity, we used various cultures of marine luminescent bacteria having

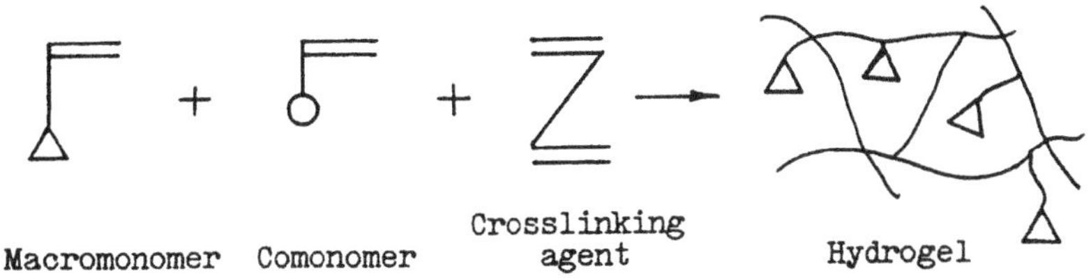
Macromonomer
Comonomer
Crosslinking agent
Hydrogel

Scheme 1.

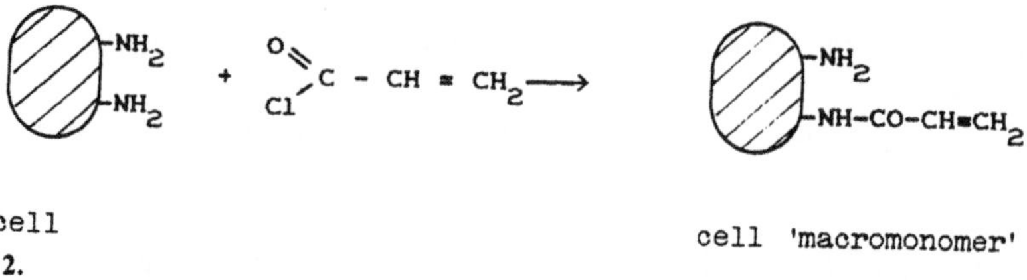
-NH2
-NH2
cell
+
O
C - CH = CH2
Cl
-NH2
-NH-CO-CH=CH2
cell 'macromonomer'
Scheme 2.

high levels of bioluminescence, and well known yeast cells which effectively ferment sugar into ethanol.

One of the most reactive compounds which permits the introduction of C=C double bonds into biological compounds containing free amino groups is acroyl- (or methacroyl)-chloride [17]. This reagent reacts with the above mentioned groups in aqueous media, leading to the formation of amidobonds between the reacting molecules (Scheme 2).

The presence of free amino groups both in the walls of the cells and in the intracellular space provides the opportunity to produce a cell bound reactive 'monomer'. The acylation reaction of *P. fischeri* with ^{14}C-labelled acetylchloride which reacts with amino groups in the same way as acroylchloride was studied. The suspension of intact cells was acylated by labelled acetylchloride and after disruption of the modified cells by osmotic shock or ultrasonic disintegration, the radioactivity of membrane fraction and intracellular extract was measured.

The results obtained (Table 1) show that for intact cells the radioactivity is distributed between membrane fraction and the intracellular extract: the ratio of radioactivities in these fractions is constant and equal to 3:1 on average. The main portion of the radioactivity (65–80%) resides in the membrane fraction. The rest of the radioactivity (20–35%) is found in the intracellular extract although the concentration of the protein in the membrane is smaller than in the extract (the experimentally determined protein ratio in these fractions is equal to 1:8). The radioactivity of the overall protein from the membrane fraction is 90–95% of the total activity of the membrane fraction and does not depend on either the amount of acetylchloride or the method of cell disruption. The acylation of the destroyed

Table 1.
Results of radiochemical experiments on the modification of photobacteria by ^{14}C-labelled acetylchloride

Initial number of –COCl groups onto one cell	% of radioactivity, ±0.5%: in membrane fraction	in intracellular extract	Ratio of radioactivities of fractions
	Intact cells		
9.49×10^3	72.7	27.3	2.66:1
1.88×10^4	62.8	37.2	1.68:1
1.89×10^5	81.8	18.2	4.50:1
2.03×10^5	71.3	28.6	2.49:1
2.36×10^6	63.5	36.5	1.74:1
	Destroyed cells		
3.67×10^5	20.3	79.7	1:3.92
1.03×10^6	31.0	69.0	1:2.22
5.11×10^6	30.0	70.0	1:2.33

cells leads to a different distribution of radioactivity: the main part of the radioactivity is in the extract, and the rest in the membrane fraction. The ratio of radioactivities is 1:3 and is proportional to the protein ratio in these fractions. Hence, the interaction of the amino-containing components of the cell membrane (mainly proteins) with acylchlorides is more effective than that with the intracellular components. The membrane prevents the penetration of the acylating agent into the cell although it does not totally exclude this process. This modification technique permits substitution of about 2 million amino groups onto a cell or to introduce one C=C double bond onto 200 $Å^2$ of cell surface on average.

The modification process and the chemical reagents used influence the physiological activity of the bound cells. Figure 1 shows the kinetics of luminescence inhibition of intact photobacteria with the gel-forming components (acrylamide and *N,N'*-methylenebisacrylamide) as well as with their saturated analog (acetamide). One can see that all reversible substances used inhibit the cell luminescence. After 3 h of washing the modified cells from these substances with a fresh portion of 0.05 M phosphate buffer solution at pH = 7.8 containing 0.3 M NaCl, the luminescence level is increased and reaches 60–70% of the initial one.

The reason for the luminescence inhibition is the reversible noncovalent interaction of amides with intracellular enzyme—luciferase [18]. In contrast, the interaction of photobacteria, even with a small amount of acylating agents containing reactive chloroanhydride groups, leads to a nonreversible deactivation of luciferase luminescence because of the chemical coupling of acylchlorides with amino groups of the enzyme. Figure 2 shows the relationship between the bioluminescence of photobacteria and the type of reactive modifier used in the system. One can see that the level of bioluminescence is not restored even after washing the cells for 3 h. The decreasing luminescent intensity in this case occurs due to the chemical interaction of acylchlorides with the amino-containing components of the intracellular enzymatic system. However, the effect of inhibition is absent

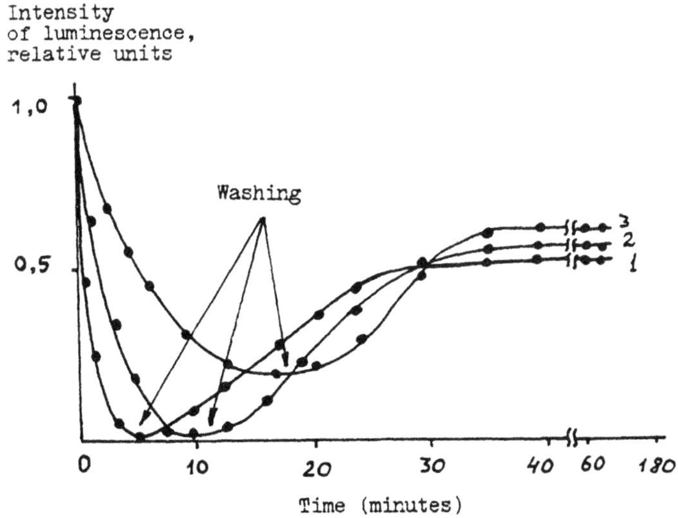
Intensity
of luminescence,
relative units
1,0
0,5
Washing
3
2
1
0
10
20
30
40
60
180
Time (minutes)

Figure 1. The inhibition of luminescence of intact photobacteria with 5 wt.% of acetamide (1), 5 wt.% of *N,N'*-methylenebisacrylamide (2) and 5 wt.% of acrylamide (3) at room temperature.

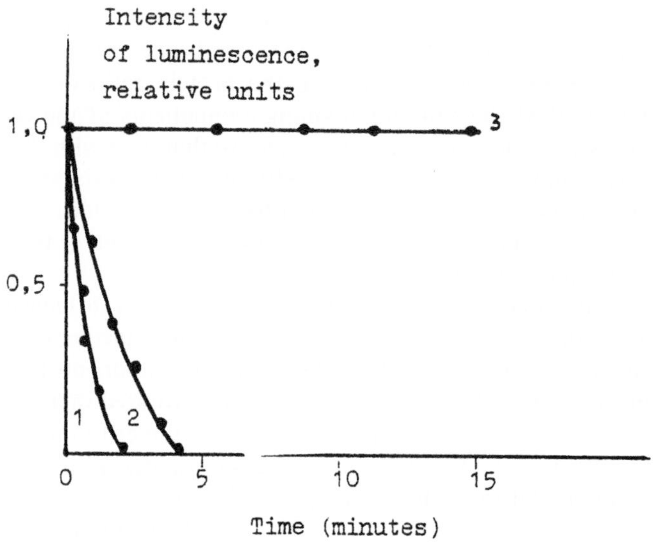
Intensity
of luminescence,
relative units
1,0
0,5
3
1
2
0
5
10
15
Time (minutes)

Figure 2. The influence of reactive modifier on luminescence intensity of intact photobacteria at room temperature: 2 mol % of acetylchloride (1), 2 mol % of acroylchloride (2) and 2 mol % of polyacroylchloride with M_w = 90 000 (3).

when polymeric modifying agents (for example, polyacroylchloride with molecular mass 90 000) is used for the acylation reaction at the same molar ratio: the high molecular components can not penetrate through the cell membrane of the photobacteria because its pore size excludes substances having molecular masses of more than 10 000 Da [19].

The behavior of yeast cells is quite different. The comparative results obtained for *S. cerevisiae* are presented in Fig. 3. It is seen that the treatment of yeast suspension with acroylchloride does not decrease its fermentational activity: the yield of ethanol for both intact and modified cells is the same.

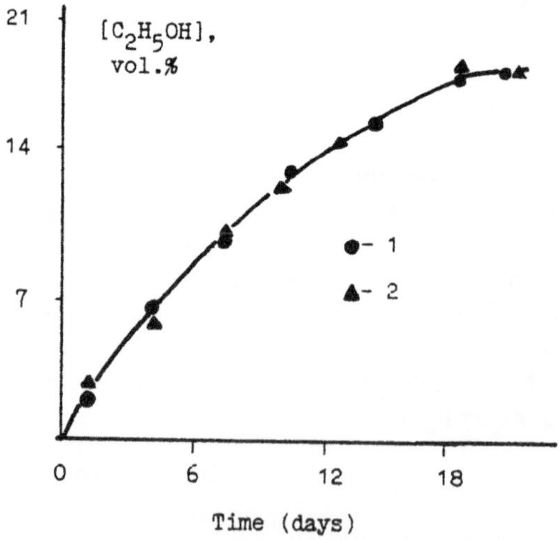
21
[C2H5OH],
vol.%
14
7
●- 1
▲- 2
0
6
12
18
Time (days)

Figure 3. The fermentation activity of intact (1) and modified with acroylchloride (2) yeast cells during the culture.

Thus the synthesis of 'supermacromonomers' on the bases of microorganisms has been accomplished along with subsequent copolymerization into polyacrylamide gels. The influence of the modification methods on the activity of the microorganisms is conditioned by the pecularities of the interaction between reactive modifiers with each cell system and by stability of these systems under the modification conditions.

Photobacteria labelled ^{125}I were acylated with acetylchloride and then incubated with an initiating redox system in the presence of acrylamide and a crosslinking agent. It was found that incubation of the cells, modified with acetylchloride, for 2–3 h results in an increase in optimal density of the system and subsequent gel formation. For cells acylated with acroylchloride incubation under the same conditions also leads to the formation of stable hydrogels. Figure 4 (a and b) shows the radioactivity of a cell containing hydrogels as a function of washing time, with and without the presence of nutritions. One can see that cells modified with acetylchloride are removed from the gel rather quickly while those cells modified with acroylchloride are removed much more slowly. Washing with nutritions in the medium gives similar results. It can be assumed that cells modified with acetylchloride are not chemically linked to the matrix of the gel and are easily removed. The acroylchloride modified cells were more effectively bound to the polymeric matrix.

Practically the same results were obtained for the yeast cells. Modification of the cells in this case and subsequent copoloymerization also results in the formation of hydrogels with immobilized cells (Table 2). The efficiency of immobilization of the yeast cells does not depend on the number of cells in the initial suspension if the ratio of modifier to cells remains constant. The high degree of immobilization in this case is a result of two processes: the mechanical entrapping of unmodified cells; and covalent binding of the yeast cells in the polymeric matrix. A comparison of the two processes shows that efficiency of covalent immobilization is higher than that of entrapment (Table 3).

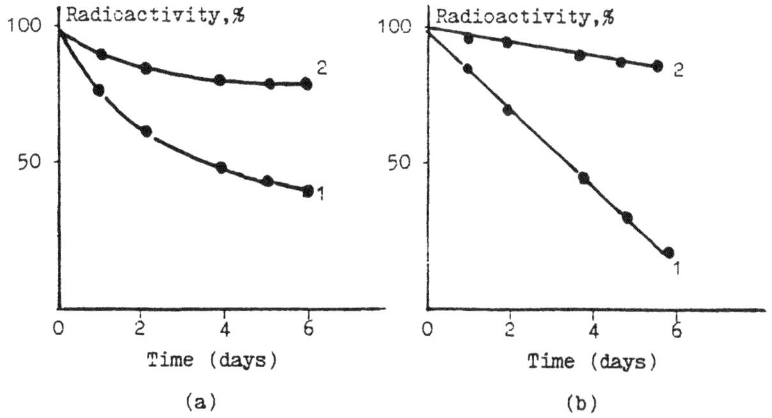
Radioactivity,%
100
50
2
1
0
2
4
6
Time (days)
(a)
Radioactivity,%
100
50
2
1
0
2
4
6
Time (days)
(b)

Figure 4. The radioactivity of hydrogels with entrapped (1) and covalently immobilized (2) photobacteria vs the washing time in nutrition-containing (a) and nutrition-free (b) media at 25°C. The initial reaction mixture consisted of 8.3×10^7 cells labeled with ^{125}I, 100 mg acrylamide, 10 mg *N,N'*-methylenebisacrylamide, 1 ml of water and 0.03 wt.% of redox initiating system. The gel blocks were cut into samples with a diameter of about 0.5 mm.

Table 2.
Efficiency of immobilization of yeast as a function of modification conditions

Number of cells in 1 ml of intact suspension	Quantity of acroylchloride on one cell, μmol ± 0.0002	Number of removed cells	Efficiency of immobilization, ±1%
1.0×10^7	0.0015	0.020×10^7	98
5.2×10^6	0.0016	0.156×10^6	97
1.1×10^6	0.0015	0.011×10^6	99
1.3×10^5	0.0014	0.013×10^5	99

Table 3.
The influence of immobilization technique on efficiency of yeast immobilization.

Method of immobilization	Efficiency, % ±2%
Physical entrapping of yeast in polyacrylamide hydrogel	35–40
Modification of yeast with acetylchloride and copolymerization with acrylamide and *N,N'*-methylenebisacrylamide	28–33
Modification of yeast with acroylchloride and copolymerization with acrylamide and *N,N'*-methylenebisacrylamide	87–90

Thus, the introduction of reactive double bonds onto a cell surface and copolymerization of the modified system with hydrophilic comonomers is a simple and useful method to immobilize cells in polymeric hydrogels.

An important property of a cell is its viability, which can be evaluated as its ability to reproduce. The question arises of how immobilization of photobacteria affects their ability to reproduce? The incubation of intact, modified, and immobilized cells in nutritions-containing media results in an increase of biomass. It is seen from Fig. 5 that the biomass produced from intact, modified, and immobilized cells is the

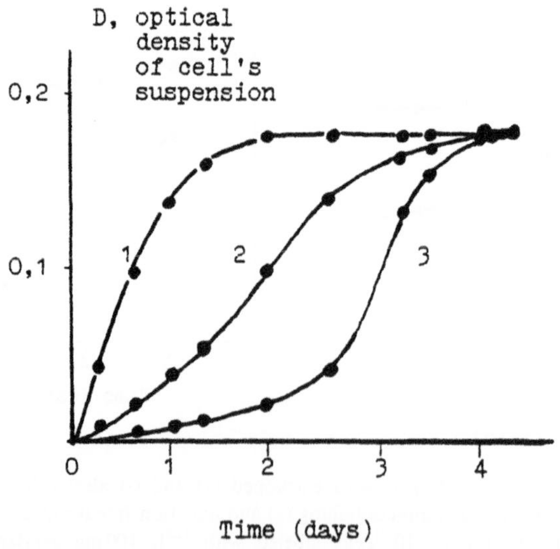
D, optical density of cell's suspension
0,2
0,1
0
1
2
3
4
Time (days)

Figure 5. The kinetics of biomass growth of photobacteria for intact cells (1), cells activated with acroylchloride (2) and cells covalently immobilized in a polyacrylamide hydrogel (3) at 25°C.

same but for the latter the growth rate is slower. The long lag-time in this case arises because of oxygen deficit and diffusion limitation nutrient components [20]. It should be noted that the daughter cells exhibit bioluminescence of about one third of that of the intact cells.

Covalently immobilized yeast can also reproduce themselves and exhibit high fermentation activity. Figure 6 shows comparative data on the fermentation activity of samples with intact, entrapped, and immobilized yeast cells. The yields of the final product (ethanol) for these samples are the same but the rates of the fermentation reaction are different: for intact cells the maximum concentration of ethanol is reached within 10 days; for entrapped and covalent immobilized cells the maximum ethanol concentration is reached within 6 and 3–4 days, respectively. In the latter case fermentation activity of immobilized yeast is not only maintained, but is increased. This means that the industrial processes of sugar fermentation or other natural raw material into alcohol can be accelerated due to the increased productivity of immobilized cells. The higher productivity can be explained by the fact that the microenvironments offered by the carrier are stabilizing to the microorganism or its enzymes, which generally show optimal activity only under very narrowly prescribed physical conditions.

Besides accelerating the fermentation, the immobilization of the yeast cells leads to an increased stability and intensification of the fermentation process. Repeat batch cultures with intact, entrapped, and immobilized yeast cells was undertaken to evaluate the durability of the immobilized cells. For a number of repeated fermentations the cultures gave a constant ethanol yield. The maximum number of repeated batch cultures without loss or decrease of fermentation activity was 5–6 for intact cells, 13–15 for entrapped and 35–45 for cells immobilized in the polyacrylamide hydrogel formed in the presence of acroylchloride acylated yeast cells. These results

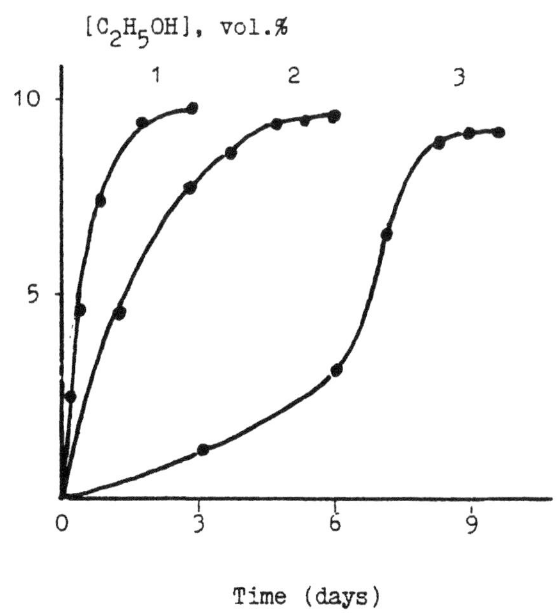
[C2H5OH], vol.%
1
2
3
10
5
0
3
6
9
Time (days)

Figure 6. Fermentational activity of immobilized (1), entrapped (6), and intact (3) yeast cells vs culture time.

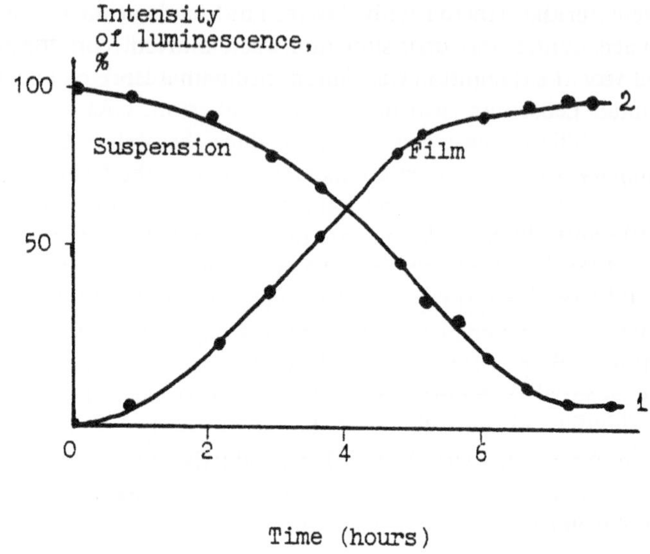
Intensity
of luminescence,
%
100
50
Suspension
Film
2
1
0
2
4
6
Time (hours)

Figure 7. The level of luminescence with time of intact cells in suspension (1) and upon contacting this suspension with modified reactive polyethylene film (2).

indicate that the immobilized yeast cells are suitable for use in continuous fermentations in packed beds or stirred tank reactors.

Thus covalent immobilization of cells in the volume of polyacrylamide hydrogel formed by radical copolymerization of 'supermacromonomers' on the base of the microorganisms with acrylamide and crosslinking agent, leads to highly active and stable systems for yeast cells. Photobacteria immobilized by similar chemistry are not as active as yeast cells.

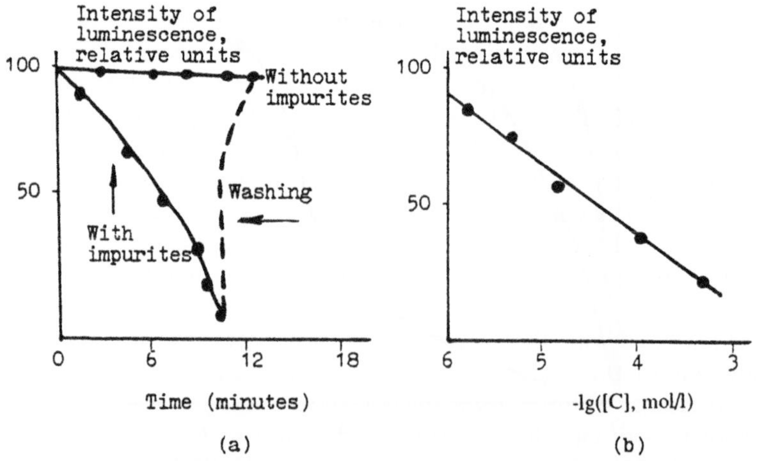
Intensity of
luminescence,
relative units
100
50
Without
impurites
Washing
With
impurites
0
6
12
18
Time (minutes)
(a)
Intensity of
luminescence,
relative units
100
50
6
5
4
3
-lg([C], mol/l)
(b)

Figure 8. The relationship between the intensity of luminescence of immobilized photobacteria and the concentration of organic molecules in flowing sea water at ambient temperature. The influence of 2×10^{-3} mol/l *m*-chlorcarbonylphenylhydrazone on the luminescence with time and its recovery on rinsing (a) and intensity of luminescence of film as a function of impurity concentration (b).

Figure 7 shows the results obtained for the immobilization of photobacteria onto a solid carrier with a grafted polyacrylamide hydrogel. During the contact of graft copolymer with the cell suspension, the luminescence intensity of the suspension decreases but the intensity of the film luminescence increases. Thus, this immobilization process proceeds without loss of cell activity. Similar to the immobilized yeast samples the immobilized photobacteria are more stable than the intact cells. The parameter which describes the stability is the half-life of the luminescence. For intact cells this parameter is 9 h, but for the immobilized cells the half-life is 33–35 h. Samples with photobacteria immobilized in grafted polyacrylamide hydrogels are sensitive to some organic substances in solution (Fig. 8). If the film with immobilized cells is put into sea water no inhibition of luminescence occurs. The presence of some organic impurities (for example, *m*-chlorcarbonylphenylhydrazone) leads to a reversible inhibition of luminescence which can be restored after washing with fresh sea water (Fig. (8a)). Inhibition–restoration cycles of luminescence on one and the same sample can be repeated more than 50 times. The linear relationship between the intensity of luminescence and concentration of the organic impurity in the media (Fig. (8b)) suggests a potential for this system to be employed as an analytical biosensor [21].

CONCLUSION

It may be concluded that the proposed approach of covalent immobilization of microorganisms in polymeric hydrogels has great promise. The activation of cells by low-molecular unsaturated chloroanhydrides is accompanied by the loss of their physiological activity only in individual cases when the intracellular systems are very sensitive to the chemical reagents employed. Copolymerization of activated 'macromonomers' with hydrophylic comonomers leads to preparation of highly active and stable systems in the use of immobilized yeast cells. Similarly, immobilized photobacteria lose their bioluminescence but not ability to reproduce. For the latter type of cell the creation of physiologically active immobilized systems can be realized using a high-molecular weight insoluble modifier as a reactive carrier.

Acknowledgements

We are grateful to Dr A. Usova and Dr N. Jakovenko for help in some of the experiments and for fruitful discussions.

REFERENCES

1. Ch. Divies, *Bull. O.I.V.* **54**, 843 (1981).
2. K. Jamamoto, T. Tosa, K. jamashita and J. Chibata, *Biotechnol. Bioeng.* **19**, 1101 (1977).
3. J. Kennedy, S. Darcer and J. Humphreys, *Nature* **265**, 242 (1976).
4. G. Durand and J. Navarro, *Process Biochem.* **13**, 14 (1978).
5. K. Koshcheenko and G. Suchodol'skaya, in: *Immobilized Cells in Biotechnology*, p. 4, Pushchino (1987) (in Russian).
6. G. Belfort, *Biotechnol. Bioeng.* **33**, 1047 (1989).
7. N. Baranova, N. Alexandrushking and N. Egorov, *Biol. Nauki* **12**, 77 (1980) (in Russian).
8. J. Campbell and J. H. Duffens (Eds), *Yeasts. A Practical Approach.* LKL Press, Oxford (1988).
9. G. Stempel, *J. Am. Chem. Soc.* **72**, 2299 (1950).
10. J. Labonesse and M. Gervais, *Eur. J. Biochem.* **2**, 215 (1967).
11. J. W. Hastings, T. O. Baldwin and M. Z. Nicoli, *Methods Enzymol.* **4**, 135 (1978).

12. G. Folin and V. Chiocalteu, *J. Biol. Chem.* **3,** 627 (1927).
13. I. Burdygina, V. Chupov, L. Valuev, A. Alexandushkina, V. Golubev and N. Plate', *Polym. Sci. USSR* **24A,** 862 (1982).
14. V. Chupov, A. Usova, V. Zemlyanskaya and N. Plate', *Vest. Mosk. Univ.* **32,** 293 (1991) (in Russian).
15. C. R. Phillips and Y. C. Poon, *Immobilization of Cells*. Springer-Verlag (1988).
16. N. Plate', L. Valuev and V. Chupov, *Pure Appl. Chem.* **56,** 1351 (1984).
17. I. Burdygina, L. Valuev, V. Chupov and N. Plate', *Polym. Sci. USSR* **24B,** 862 (1982).
18. H. Wang and D. Hettwer, *Biotechnol. Bioeng.* **24,** 1827 (1982).
19. I. Tysyachnaya, V. Fechina and V. Yakovleva, *Prik. Biochim. Microbiol.* **20,** 433 (1984) (in Russian).
20. S. Narvaha and J. Kennedy, *Enzyme Microd. Technol.* **6,** 18 (1984).
21. F. Scheller, F. Schubert, *Biosensors*. Elsevier, Amsterdam (1992).

Part VI.A.

Immobilized Biomolecules and Synthetic Derivatives of Biomolecules

Heparin

Immobilization of a lysine-terminated heparin to polyvinyl alcohol

J. S. TURNER and M. V. SEFTON*

Department of Chemical Engineering and Applied Chemistry, Centre for Biomaterials, University of Toronto, Toronto, Ontario M5S 1A4, Canada

Received 22 June 1992; accepted 28 January 1993

Abstract—Lysine terminated heparin, prepared by the nitrous acid partial depolymerization and reductive amination of heparin, failed to increase the active heparin content of a heparin-polyvinyl alcohol (heparin–PVA) hydrogel relative to the unmodified commercial heparin. The depolymerization of heparin resulted in a loss of biological activity which outweighed the increase in the terminal amine groups (produced by reductive amination), that were used for glutaraldehyde immobilization to the PVA. The loss in anti-thrombin activity (thrombin time or chromogenic substrate) paralleled the increase in anhydromannose end groups due to depolymerization making it necessary to optimize the loss of activity against the increase in terminal amine groups after amination. For example, depolymerization at a high sodium nitrite concentration (81 g/l) at pH4 and 25°C for 20 min, resulted in a loss of 22–40% of the biological activity but achieved an anhydromannose content of 600 nmoles/mg (~7 cleavage sites/molecule). After the anhydromannose groups were reductively aminated by lysine, the anhydromannose content was reduced to 190 nmol/mg indicating a terminal lysine content of 410 nmol/mg. This resulted in an increase in heparin content of the final hydrogel by 53% on mass terms. However, given the reduction in biological activity, it was not surprising that the modified heparin-PVA hydrogel coated on a polyethylene tube was no better than the hydrogel with unmodified heparin in inactivating thrombin in a flow circuit. These results point out the need for care in interpreting heparin immobilization results and for new strategies to increase the active heparin content of this hydrogel.

Keywords: Heparin-PVA; immobilized heparin; nitrous acid depolymerization; lysine.

INTRODUCTION

In 1972 Hoffman published [1] one of the first reports of the covalent immobilization of heparin in order to prepare a nonthrombogenic material. Working with G. Schmer, he used the then-new cyanogen bromide chemistry [2] to immobilize heparin to radiation grafted silicone rubber. Thus began a long interest in biomolecule immobilization which went from heparin to enzymes [3] to cells [4]. For one of us (M. V. Sefton) the 1972 paper was our first 'contact' with A. S. Hoffman. A few years later, Alan came to Toronto as a keynote lecturer of the Canadian Biomaterials Society. His words of encouragement were extremely helpful to my student (M. F. A. Goosen) and myself, as we began a research program on heparin–PVA hydrogels. This paper reports on one of the latest episodes in the heparin–PVA saga.

Considerable effort [6–10] has been placed on the immobilization of a commercially available heparin to polyvinyl alcohol for the twin purpose of evaluating the role that covalently immobolized heparin can play in the preparation of

* To whom correspondence should be addressed.

non-thrombogenic materials [10] and of preparing a coating that can be used to reduce the thrombogenicity of cardiovascular materials [7]. Heparin was immobilized using glutaraldehyde through a serine group at the end of many of the molecules in the commercial heparin we used. We presumed that such end-linked heparin was able to assume its native conformation and that binding did not interfere with the thrombin and antithrombin III binding sites, thereby accounting for the retention of biological activity despite covalent or irreversible immobilization.

Heparin exerts its anticoagulant activity by accelerating the inactivation of thrombin by antithrombin III. The active site of binding on the heparin molecule for antithrombin III is a particular pentasaccharide sequence contained in only about one-third of the chains of commercial heparin [11, 12]. One proposed mechanism for this reaction (the approximation-template mechanism) involves the interaction of heparin with both antithrombin III and thrombin in order to enhance the inhibition of the enzyme [13, 14]. This is supported by findings that octa- to dodecasaccharides with high affinity for antithrombin III have high anti-Factor Xa activity, but are virtually unable to potentiate the inhibition of thrombin. Oligosaccharides with 22 residues or greater, which are able to bind to both thrombin and antithrombin III, are required in order to potentiate the inhibition of thrombin by antithrombin III.

Reaction between glutaraldehyde and the terminal serine or more specifically the terminal amine, was presumed to occur through a Schiff base [15] which ultimately underwent rearrangement to a more stable linkage: reduction with a borohydride was not found to be necessary to stabilize the product. Because the serine was necessary for binding, the level of binding was limited in part by the fraction of heparin molecules that contained the terminal serine. Consequently the amount of bound heparin was sensitive to changes in the process for producing the heparin. For example, a few years ago Waitaki International (formerly Canada Packers) 'improved' the process with the unexpected (to us) consequence of reducing the serine content and the binding to PVA. The serine content was reduced by a factor of approximately two, necessitating a doubling of the heparin concentration (to 2%) in the heparin–PVA gel solution to produce a gel with the same bound heparin content as what we had previously produced with a 1% heparin solution. Unfortunately further increases in heparin concentration were not possible because of phase separation in the aqueous solution of the two polymers, PVA and heparin. Hence it was the objective of this study to increase the terminal amine content of the starting heparin and thereby increase the fraction of molecules that could be bound to the PVA. We chose to use lysine since one amino group could be used for binding to anhydromannose groups (generated by nitrous acid treatment of heparin), leaving the other amine free to react with glutaraldehyde (Fig. 1). Nitrous acid depolymerization of heparin has elsewhere been used to end-link heparin to biomaterial surfaçes [16] or to prepare a form of heparin suitable for incorporation in a block copolymer [17].

MATERIALS AND METHODS

Heparin

Heparin (porcine intestinal, Waitaki International, Toronto, Canada, 165 U/mg) was used as is or after modification by nitrous acid and reductive amination with lysine. A high serine content heparin from Diosynth (Chicago, IL, USA; intestinal mucosal, 164 U/mg courtesy of Dr. G. Van Dedem) was also used on occasion.

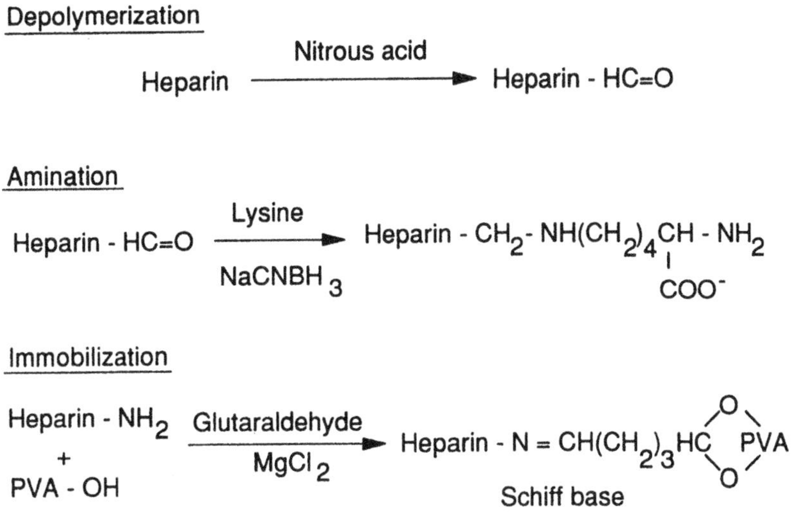
Depolymerization
Nitrous acid
Heparin
Heparin - HC=O
Amination
Heparin - HC=O
Lysine
NaCNBH 3
Heparin - CH2- NH(CH2)4CH - NH2
COO-
Immobilization
Heparin - NH2
+
PVA - OH
Glutaraldehyde
MgCl 2
Heparin - N = CH(CH2)3HC PVA
O
O
Schiff base

Figure 1. Schematic diagram of reactions used to prepare lysine terminated heparin.

Depolymerization

For nitrous acid depolymerization 6 g (~0.4 mmol) of heparin (Waitaki) was dissolved in 50 ml of 0.2 M sodium acetate buffer (pH 1.5 or 4). A specified amount of sodium nitrite (BDH, Toronto, Canada) was dissolved in 25 ml distilled water and mixed with 25 ml of the sodium acetate buffer. The sodium nitrite ($NaNO_2$) solution sat for 10 min to allow for the formation of nitrous anhydride (N_2O_3) and was then added to the heparin solution. At the end of each reaction 30 ml of 70% ammonium sulphamate (BDH Ltd.) was added to quench the reaction and the solution pH was readjusted to 7 using saturated sodium hydroxide.

The depolymerized heparin was precipitated using cold (4°C) ethanol (400 ml added to 100 ml H_2O) washed with another 100 ml of cold ethanol and filtered. A distilled water solution (100 ml) was dialysed against 2 l of distilled water for 12 h and lyophilized (Labconco Model 5, Kansas City, MO, USA) for future use. The yield was 50–65%.

Amination. To reductively aminate [18] the depolymerized heparin, 3 g (0.18 mmol) of depolymerized heparin or the starting Waitaki heparin was added to 100 ml of a 100 mM L-lysine solution (pH 9) and reacted for 30 min. 1 g of $NaCNBH_3$ (16 mmol) was added to the heparin–lysine mixture and the reaction continued for 12 h at which point 1.5 g of $NaCNBH_3$ was added. At 24 h the reaction was terminated by adjusting to pH 7 using 6 M hydrochloric acid.

The heparin–lysine mixture to which was added NaCl (1 M final concentration) was dialysed against distilled water for 12 h. The desalted, reductively aminated heparin solution was passed through a Sephadex SP-C50 cationic exchanger to remove all ionically bound amino acids and lyophilized for future use.

Heparin–PVA films

Heparin–PVA was prepared in the standard manner [8, 20] from an aqueous solution of PVA (7% w/v, 100% hydrolysed, Aldrich), glutaraldehyde (2% w/v,

Grade II, Sigma), glycerol (5%), $MgCl_2$, and 1% heparin. For films the gel solution was cast into a petri dish, dried overnight, and cured at 70°C for 2 h, and then washed for a two week period with distilled water to remove unreacted components. PVA gels were prepared without the addition of heparin solution.

Etched polyethylene (PE) tubing (1.14 mm i.d., PE 160, Intramedic, Clay Adams, Parsippany, NJ, USA) was coated with gel as before [8, 9] and washed with 500 ml PBS (phosphate buffered saline: 0.15 M NaCl, 0.05 M phosphate, pH 7.4) containing 0.7% PEG 8000 using a peristatic pump at 7 ml/min. Coating weights and hence average thickness were as before [8, 9].

Heparin characterization

Molecular size. The change in molecular size on depolymerization was estimated by measurement of the HPLC retention time with a micro-gel E-125 gel permeation column at ambient temperature connected to a refractive index detector (HP 1090 HPLC system). A 0.2 M sodium acetate solution was used as the mobile phase at 0.4 ml/min. Peak widths of modified heparins were similar to that of the unmodified heparin.

Anticoagulant activity. The anticoagulant activity of modified heparin was determined by thrombin time and chromogenic substrate assays. For the thrombin time, fresh frozen pooled human plasma (0.2 ml), buffer (0.2 ml, pH 8.4, 0.05M Tris-HCl, 0.15 M NaCl) and heparinized buffer (0.2 ml, 0–6 USP U/ml) were incubated in polystyrene tubes at 37°C for 1 min. Bovine thrombin (0.2 ml, 10 USP U/ml, Sigma) was added and mixed immediately. The endpoint was reached when the first strands of fibrin were detected visually.

For the chromogenic substrate assay, 0.5 ml of heparinized buffer (0.05 M Tris-HCl, pH 8.4) was mixed with 0.2 ml of diluted fresh frozen pooled human plasma (1 ml diluted to 10 ml using 0.15 M NaCl). After the addition of 0.05 ml of 10 USP U/ml bovine thrombin solution (zero time) and mixing for 30 s the cuvette was placed in the cell holder of the spectrophotometer. 0.2 ml of 0.75 mM Spectrozyme TH dissolved in 0.15 M NaCl, 0.3% polybrene solution was added at 1 min. The absorbance was monitored at a wavelength of 380 nm. The slope of the absorbance vs time curve (from 75 s on) indicated the rate of cleavage of the chromogenic substrate by thrombin and was plotted against the concentration of the heparinized buffer. A calibration curve was prepared with unmodified heparin.

Amino acid analysis. Heparins were analyzed for amino acid content by the Amino Acid Analyzer Facilities (University of Toronto) using Beckman 121 M amino acid analyzer with a purified Dowex 50 ion exchange column after hydrolysis in 6 M HCl at 100°C for 21 h. The samples were quantified spectrophotometrically following their reaction with ninhydrin.

Anhydromannose end groups. The method of Dische and Borenfreund [19] was used to quantify the number of anhydromannose groups created during the depolymerization reaction. The deamination step in the original procedure was omitted and the anhydromannose content of the depolymerized heparins was determined immediately after ammonium sulphamate was used to quench the depolymerization reaction.

For the calibration curve, 0.2 ml of a 5% $NaNO_2$ (w/v) and 0.2 ml of 33% acetic acid solution were added to a 0.2 ml glucosamine solution of known concentration ($0.2–10 \times 10^{-4}$ M) in a screw-top pyrex test tube. Deamination proceeded for 60 min at which point 0.2 ml of 12.5% ammonium sulphamate was added and the test tube was shaken intermittently for 30 min. 1 ml of 5% (v/v) hydrochloric acid and 0.2 ml of 5% (w/v) indole in ethanol were then added, the tubes were covered and heated at 100°C for 5 min. After allowing the tubes to cool, 1 ml of ethanol was added to remove any formed precipitate. Absorbances at 492 and 520 nm were measured and the difference was plotted against the concentration of the glucosamine solution.

The anhydromannose groups of depolymerized heparin was quantified using a similar procedure with the following exceptions. The $NaNO_2$, acetic acid and ammonium sulphamate solutions were added to the test tube and the mixture was shaken intermittently for 30 min. The appropriately diluted heparin solution (0.2 ml; in most cases, 0.1%) was added to the mixture. The hydrochloric acid and indole solutions were then added and the procedure was carried out as above.

Concentration determination

The heparin content of distilled water solutions was determined using the toluidine blue/metachromatic shift assay [20] at 600 nm (HP 8452). Lysine concentrations of aqueous solutions were assayed using picryl sulphonic acid (0.3 M).

Characterization of the heparin–PVA hydrogel

Heparin content. Elemental analysis (DRL laboratories, Toronto) was used to quantify the nitrogen content of the various heparins and the heparin content of the heparin–PVA films. Organic nitrogen after conversion to ammonia was analyzed using the Kjeldahl method [21]. The heparin content was also assayed indirectly by mass balance using the toluidine blue method [20] based on the difference between the heparin initially added to the film and the heparin assayed by toluidine blue removed from the film during washing of the film. Toluidine blue analysis was not used for the modified heparins because it was assumed that each heparin would interact differently with toluidine blue depending on the extent of the modification.

Thrombin inactivation assay

Human plasma (diluted 1 : 10 with PBS-PEG) preincubated with bovine thrombin (5 USP U/ml) for 60 s, was injected into a 30 cm coated tube, connected to a 13 cm long piece of Silastic® tubing (1 mm i.d.) [22]. The two ends of the tubing were connected to form a closed loop, the Silastic® tubing was mounted under the rollers of a peristaltic pump and the mixture was circulated at a constant flow rate (7 ml/min). After 2 min of circulation, the solution was pumped into a polystyrene test tube which contained 0.1 ml of a 1 mM Spectrozyme TH solution (American Diagnostica). After a further 2 min of incubation with the chromogenic substrate the reaction was quenched with 1 ml of 50% acetic acid solution. The optical density was measured at a wavelength of 405 nm. A calibration curve of the absorbance vs thrombin concentration was used to calculate the amount of thrombin inactivated.

RESULTS

Heparin modification

Depolymerization. Depolymerization was carried out in two series. In the first series, the temperature (0 and 25°C), pH conditions (1.5 or 4), sodium nitrite concentrations (0.03 or 2 or 81 g/l) and reaction times were varied. The reaction parameters from the first series (Table 1) were chosen based on experimental procedures carried out by others [16, 23]. Only one batch of heparins from this series was prepared. During the second series of depolymerizations the pH was maintained at 4 and the reactions were carried out at room temperature (Table 2). Only the sodium nitrite concentration (0.03 and 81 g/l) and reaction time (10, 20, 30 min, 2, 12, 24 h) were varied. Any further reference to these heparins were made using Hi10, Hi20, Hi30, and Lo2, Lo12, and Lo24. 'Hi' and 'Lo' refer to the relative sodium nitrite concentrations. The number following refers to the reaction time.

In general, as the HPLC based retention time of depolymerized heparin samples increased, indicating a decrease in the relative molecular weight of the sample, the fraction of retained anticoagulant activity decreased (Fig. 2). Extensive

Table 1.
Reaction parameters and retention time of the first series of depolymerized heparins

Sample	Sodium nitrite (g/l)	pH	Reaction time (h)	Reaction temperature (°C)	Retention time (min)
1	0.03	4	2	0	4.63
2	0.03	4	2	25	4.67
3	0.03	4	24	25	4.95
4	0.03	1.5	2	25	5.3
5	2.0	4	2	25	5.28
6	2.0	4	24	25	6.54
7	81	4	0.17	0	4.79
8	81	1.5	0.17	25	6.0
9	81	4	0.5	25	4.84
Waitaki heparin					4.64

Table 2.
Reaction parameters and characteristics of the second series of depolymerized heparins

Sample	Sodium nitrite (g/l)	Reaction time (h)	Retention time (min)	% Retained activity	
				Thrombin time	Chromogenic substrate
Lo2	0.03	2	4.99	100	88
Lo12	0.03	12	5.08	67	82
Lo24	0.03	24	5.23	40	60
Hi10	81	0.17	5.17	74	73
Hi20	81	0.33	5.31	60	78
Hi30	81	0.5	5.43	43	45
Waitaki heparin			4.99		

Reactions at pH 4, 25°C

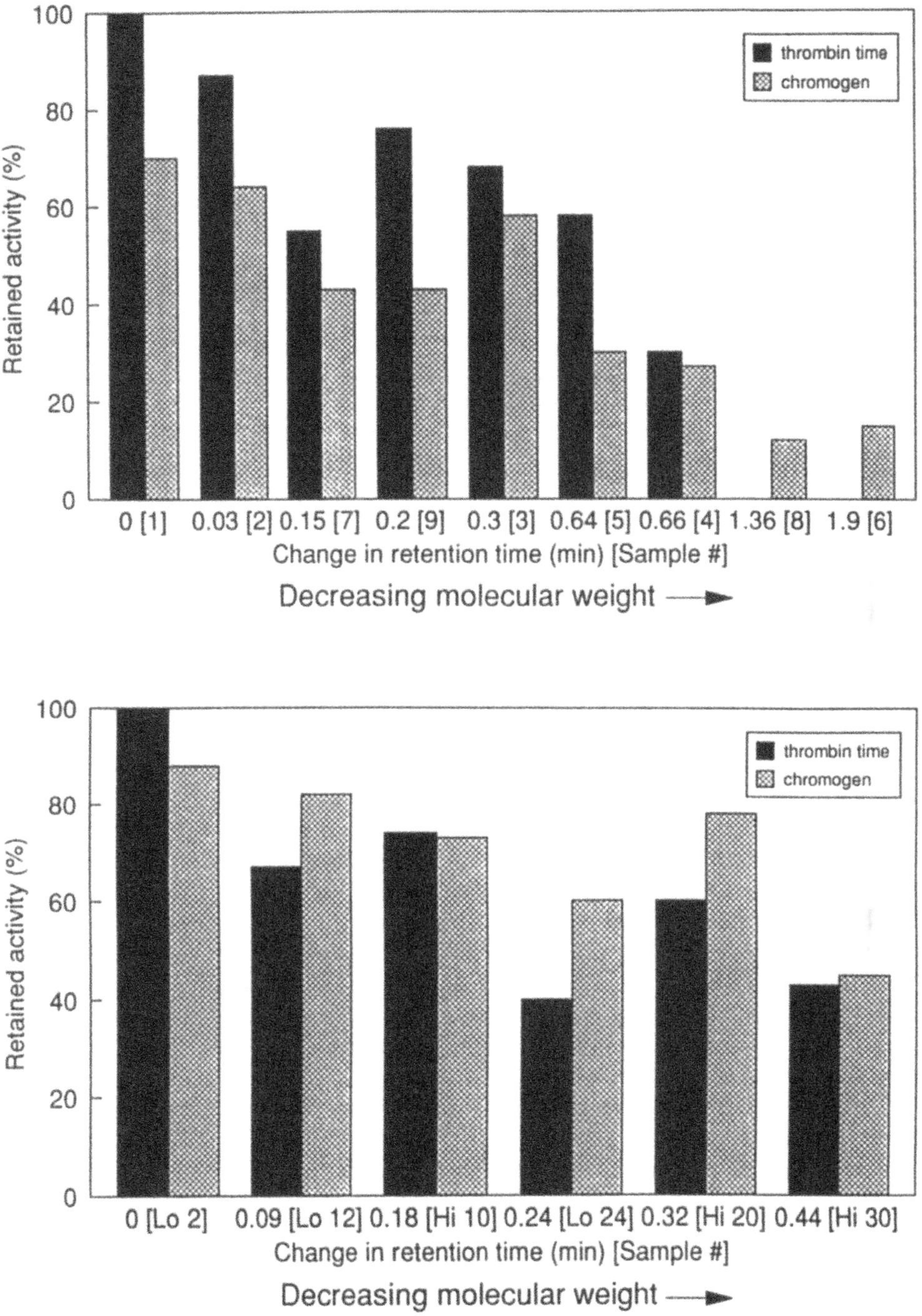
100
80
60
40
20
0
Retained activity (%)
thrombin time
chromogen
0 [1]
0.03 [2]
0.15 [7]
0.2 [9]
0.3 [3]
0.64 [5]
0.66 [4]
1.36 [8]
1.9 [6]
Change in retention time (min) [Sample #]
Decreasing molecular weight
100
80
60
40
20
0
Retained activity (%)
thrombin time
chromogen
0 [Lo 2]
0.09 [Lo 12]
0.18 [Hi 10]
0.24 [Lo 24]
0.32 [Hi 20]
0.44 [Hi 30]
Change in retention time (min) [Sample #]
Decreasing molecular weight

Figure 2. Effect of depolymerization reaction conditions on the anti-thrombin activity of the partially depolymerized heparin. The percent of the initial activity that is retained after depolymerization, measured using thrombin time ■ or chromogenic substrate ▩, plotted against the change in HPLC retention time. The latter reflects the reduction in molecular weight which is the consequence of depolymerization. Also shown is the sample number in [ ]: (a) series 1; (b) series 2.

Table 3.
Anhydromannose (AHM) content of modified heparins

	Before amination		After amination	
Sample	AHM (nmol/mg)	Cleavage sites[a] (#/molecule)	AHM (nmol/mg)	Lysine[b] (nmol/mg)
Lo2	42	0.5	0	42
Lo12	350	4.2	0	350
Lo24	580	7	4	580
Hi10	180	2.2	30	150
Hi20	600	7	190	410
Hi30	3000	36	1600	1400

[a] Estimated from anhydromannose content.
[b] Assumed equal to difference between anhydromannose contents before and after amination.

depolymerization and corresponding large losses in anticoagulant activity occurred when the pH of the depolymerization reaction was decreased from 4 to 1.5. For example, a heparin depolymerized at a pH of 4, had a retention time of 4.79 and a retained activity of 49%. Lowering the pH to 1.5 increased the retention time to 6.0 and decreased its retained activity to only 6%. At a pH of 4 the extent of heparin depolymerization increased with an increase in reaction time or sodium nitrite concentration. The reaction temperature had a relatively small effect on the extent of depolymerization.

Within each group of 'Hi-nitrite' and 'Lo-nitrite' reactions, an increase in the 'change in retention time' was accompanied by an increase in the anhydromannose content and a decrease in the retained activity (Tables 2 and 3). It was presumed that as the retention time increased, the extent of depolymerization increased and the relative molecular weight of the product decreased.

Within this series a smaller molecular weight did not always coincide with a smaller retained activity. For example, Hi20 experienced a greater amount of depolymerization than Lo24, but Hi20 also retained a larger amount of anticoagulant activity than Lo24. 'Hi-nitrite' reactions were able to depolymerize heparin to the same extent as 'Lo-nitrite' reactions, but without the same detrimental effect on the anticoagulant activity.

In terms of the anhydromannose content of the depolymerized heparin samples, a larger increase in retention time did not always coincide with a larger anhydromannose content. For example, Hi10 experienced a greater extent of depolymerization than Lo12 but Lo12 contained a larger number of anhydromannose units than Hi10. 'Lo-nitrite' reactions were able to create the same number of anhydromannose units as 'Hi-nitrite' reactions without decreasing the molecular weight by the same amount.

The organic nitrogen content of depolymerized heparin was less than the organic nitrogen content for Waitaki heparin (Table 4). The decrease in the nitrogen content ranged from 4.4 mg/g heparin (Hi30) to 10mg/g heparin (Lo12); it did not seem to follow any noticeable trend. The sulphur content of the 'Lo-nitrite' heparin was the same as the sulphur content of Waitaki heparin. The serine content of depolymerized heparin (Table 5) decreased during the nitrous acid depolymerization reaction.

Table 4.
Elemental analysis (mg/g) of depolymerized and aminated-depolymerized heparins

	Depolymerized heparin		Aminated-depolymerized heparin	
Sample	Nitrogen	Sulphur	Nitrogen	Sulphur
Lo2	10.9	81.3	9.2	38.5
Lo12	8.4	87.3	16.7	61.1
Lo24	11.6	83	12.6	39.7
Hi10	13.7	96.7	12.2	37.1
Hi20	ND	95.3	16.5	38.4
Hi30	14.1	92.7	17.3	42.7

Waitaki heparin: nitrogen, 18.5 mg/g; sulphur, 85.7 mg/g.

Table 5.
Amino acid analysis of heparin and modified heparins

Heparin	As-received		After ion exchange		After depolymerization		After amination	
sample	Lys	Ser	Lys	Ser	Lys	Ser	Lys	Ser
Waitaki	3.8	12.1	1.1	3.8				
Diosynth			0.6		18.2			
Hi10					7.6	3.3	2.3	
Hi30					3.8	1.4	1.9	1.0

Reductive amination. The number of terminal anhydromannose groups decreased after the depolymerized heparin was reductively aminated using lysine. This was attributed to the formation of a Schiff base between the aldehyde group of the terminal anhydromannose unit and the amine group of the lysine molecule. Assuming all of the loss in anhydromannose was due to covalently bound lysine, the latter was 3 (Lo2) to 100 (Hi30) times greater than the amount of serine found in unmodified heparin (Table 3). However, much lower lysine contents were estimated by amino acid analysis (Table 5).

The amino acid analysis of depolymerized heparin confirmed that unreacted, ionically bound lysine was removed during the ion exchange process (Table 5). Although excess lysine was added to depolymerized heparin solutions during the reductive amination reaction, the lysine content of Hi10 was 3 times lower after reduction and ion exchange than immediately after depolymerization.

Heparin–PVA films (unmodified heparin)

The heparin content of heparin–PVA films did not significantly change when the percentage of unmodified Waitaki heparin in the gel solution was increased from 0.5 to 4% heparin in a 7% PVA solution: 20.4 ± 6.5 mg/g wet gel ($n = 7$; toluidine blue assay) or 17.2 ± 2.8 mg/g wet gel ($n = 10$; organic nitrogen analysis). All subsequent work was with 1% modified heparin. The low solubility of heparin in PVA solutions limited the amount of heparin which was bound to the polymer film. Since heparin did not form a homogeneous solution with PVA, not all the heparin molecules were accessible to bind to PVA through glutaraldehyde. As a result, the heparin content of these films was independent of the concentration of heparin in the gel solution.

Table 6.
Estimated active heparin content of modified heparin–PVA films

Heparin	Heparin content[a] (mg/g gel)	%increase content[b] (mass)	Molecular weight[c] (Da)	Heparin content (nmol/mg)	% increase content (moles)	Active content[d] (U/mg)
Waitaki[e]	17		15000	1.1		2.8
Diosynth[e]	23		15000	1.5		
Lo2	14	−18	15000	0.9	−18	2.2
Lo12	23	+35	12750	1.8	+64	2.9
Lo24	27	+59	9000	3.0	+170	2.2
Hi10	16	−6	10500	1.4	+27	1.9
Hi20	26	+53	7000	3.6	+227	3.0
Hi30	26	+53	4000	6.4	+482	1.9

[a] Based on nitrogen analysis.
[b] Increase (or decrease) relative to unmodified Waitaki heparin–PVA.
[c] Estimated from change in retention time (see text).
[d] Estimated from heparin content of gel and fractional loss of activity of heparin in solution assuming no loss on immobilization. For modified heparin, activity was calculated by multiplying activity of Waitaki heparin (165 U/mg) by fraction retained activity (Table 2).
[e] After ion exchange.

The amino acid content of unmodified heparin decreased after it was ion exchanged from 12.1 nmol serine/mg to 13.8 nmol serine/mg. However, the heparin content of films prepared using this heparin had the same value as those prepared without ion exchanging the heparin (Table 6). Ion exchanged Diosynth heparin contained 18.2 nmol serine/mg heparin. Its serine content was larger than the Waitaki heparin and films prepared using it contained 30% more than films prepared from Waitaki heparin.

Modified heparin–PVA films

Heparin content. The heparin content of films prepared using modified heparins (Hi10 or Lo2) were similar (~15 mg/g wet gel) to the heparin content of films prepared using unmodified heparins (organic nitrogen analysis). The heparin content of the other modified heparin–PVA films was approximately 10 mg heparin/g wet gel higher and in proportion (up to a limit) to the decrease in anhydromannose content after amination (i.e. apparent lysine content, Fig. 3). The amount of heparin actually present at the surface is unknown; the problem of defining the 'surface' of a hydrogel and the accessibility of heparin molecules in the interior of the gel has been addressed elsewhere [29].

The theoretical anticoagulant activity of the modified heparin–PVA was estimated using the retained fraction of activity (Table 2) and the heparin content of the films assuming that no activity was lost on immobilization (Table 6). Loss of activity upon immobilization was fully expected so that these calculated activities are useful in a relative sense: if the loss of activity upon immobilization is assumed to be the same for modified and unmodified heparins. Hi20–PVA and Lo12–PVA had slightly larger theoretical anticoagulant activities than the unmodified heparin–PVA films. However, taking into account the changes in retention time and the corresponding estimated decrease in molecular weight upon depolymerization, some reaction conditions lead to an almost fivefold increase in heparin content in molar terms (Table 6).

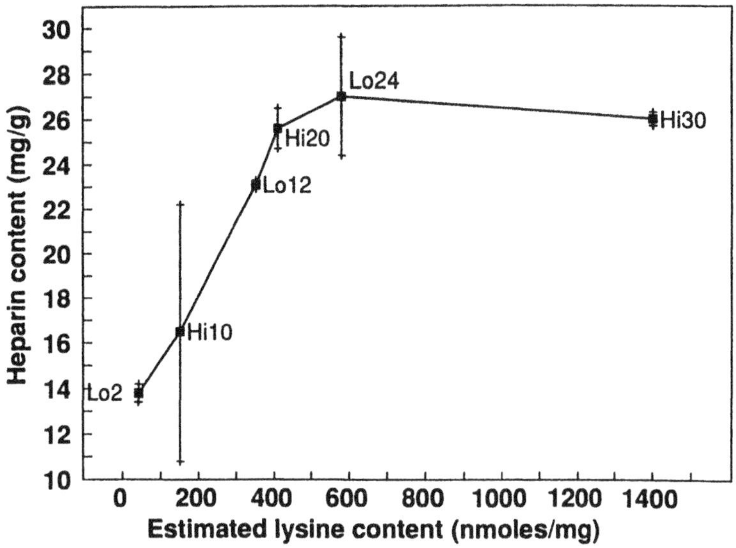
Heparin content (mg/g)
Estimated lysine content (nmoles/mg)
Lo2
Hi10
Lo12
Hi20
Lo24
Hi30
10
12
14
16
18
20
22
24
26
28
30
0
200
400
600
800
1000
1200
1400

Figure 3. Heparin content of heparin–PVA hydrogel films prepared with aminated heparins, plotted against the estimated lysine content. Lysine content calculated as difference in anhydromannose content before and after depolymerization. $n = 3$, $\pm$S.D.

Anticoagulant activity. Three tubes of each type of coating (PVA, Waitaki heparin–PVA, Hi20–PVA, Lo12–PVA, and a bare chromic acid etched polyethylene (PE) tube) were examined for their ability to inactivate thrombin (Fig. 4). PE tubes, which were chromic acid etched, inactivated 69% of the thrombin added to the system, whereas tubes coated with PVA (no heparin) inactivated 33% of the thrombin added to the system, in both cases presumably by adsorption only.

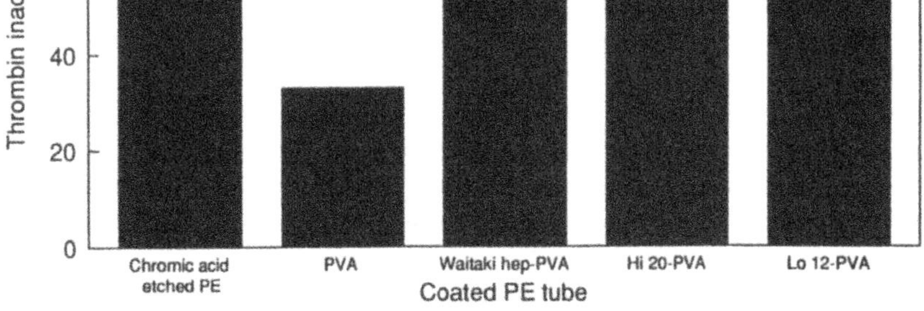
Thrombin inactivation (%)
0
20
40
60
80
100
Chromic acid etched PE
PVA
Waitaki hep-PVA
Hi 20-PVA
Lo 12-PVA
Coated PE tube

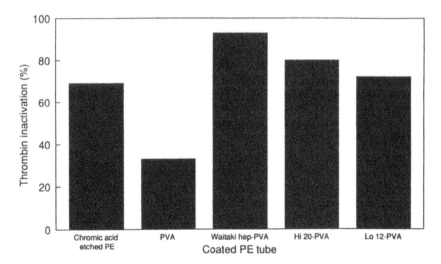

Figure 4. Biological activity of heparin–PVA coated polyethylene tubing or control tubings measured as the percent of thrombin that is inactivated (or lost by adsorption) in plasma after 2 min of circulation in a tube of one of the indicated materials. Mean of three values.

Polyethylene (PE) tubes coated with Waitaki heparin–PVA inactivated 92% of the thrombin added to the system. The percentage of thrombin inactivated using coatings prepared from Hi20 and Lo12 was less.

DISCUSSION

Heparin–PVA hydrogel–commerical heparin

Past work [11], which investigated the site of binding of the heparin molecule to the PVA hydrogel, supported the hypothesis that heparin binds to PVA via a Schiff base reaction between the terminal serine residue on the heparin molecule and the glutaraldehyde within the polymer film. In this study the serine content of Waitaki heparin, used to prepare the heparin–PVA hydrogel, was 68% less than the serine content of heparin used in past work. Waitaki International attributed [H. Ko, pers. commun] the decrease in serine content to new proteolytic methods which have been more efficient in removing the protein linkage from the heparin molecule. When Waitaki heparin was ion exchanged its serine content decreased a further 66%, indicating that a large portion of the serine which was measured by amino acid analysis was ionically bound to the heparin molecule.

The decreased serine content of Waitaki heparin was expected to directly affect the heparin content of the polymer films. Polymer films prepared using ion exchanged Diosynth heparin, with a serine content four times greater than that of ion exchanged Waitaki heparin, contained 50% more heparin than films prepared using ion exchanged Waitaki heparin.

However, as in past work [11], a discrepancy existed between the number of heparin molecules which contained a terminal serine residue and the number of heparin molecules which were found to bind to PVA. Assuming: (1) commercial heparin had a molecular weight of 15 000 Da; and (2) all serine molecules in ion exchanged Diosynth heparin were covalently bound to its terminal end, then 1 out of every 3.7 molecules or 27% of all heparin molecules contained a terminal serine residue. However, more than 56% of the heparin added to the polymer solution (1% ion exchanged Diosynth heparin–PVA gel formulation) was to bind to the PVA film. Using the same assumptions, films prepared using 0.5% Waitaki heparin-gel formulations contained 93% of the original heparin added. At best, the serine content of Waitaki heparin could only account for 18% of the heparin molecules binding. These observations suggested that either heparin was bound to the hydrogel at a point other than its terminal serine residue or that the amino acid analysis of heparin was inaccurate.

The precision of the amino acid analysis was good: triplicate serine analyses of Waitaki heparin had a coefficient of variation of ± 12%. However, the accuracy may not have been as good. Amino acid analysis is used mainly for the analysis of proteins, which are easily hydrolysed. Hydrolysis of glycosaminoglycans, such as heparin, often results in the formation of a carbonaceous precipitate, which is difficult to remove, and if present in large amounts may contaminate the column used for amino acid analysis. Therefore, heparin was analyzed in very dilute quantities, and the amount injected into the column would be very small and near the detection limit of the amino acid analyzer.

Another problem is that serine is progressively destroyed under conditions of acid hydrolysis. In a 24 h acid hydrolysis, only 90% of a 20% serine solution is recovered [24] and correction factors (based on protein hydrolysis) are used to compensate for the lost serine. However, the amount of serine destroyed changes greatly from one analysis to another because its destruction is dependent on the concentration of serine in the system, the concentration of high molecular weight proteins, and elements such as iron and calcium, which have a catalytic effect on the destruction of serine. The presence of heparin or the salts from the buffer solutions may have affected the amount of serine destroyed.

Modification of heparin

Depolymerization. Heparin was first depolymerized with nitrous acid to create terminal anhydromannose units on the heparin molecule while limiting the reduction in anticoagulant activity. In general, it was found that although 'Hi-nitrite' heparins had a smaller molecular weight (and, therefore, underwent a larger degree of depolymerization) they contained a smaller number of anhydromannose units (Table 3). Thus it was concluded that terminal functional groups other than anhydromannose units were created during the 'Hi-nitrite' depolymerization reaction. 'Lo-nitrite' heparins depolymerized using reaction times greater than 12 h contained the same number of anhydromannose units as 'Hi-nitrite' heparins, but had a smaller anticoagulant activity, regardless of the fact that the molecular weight of 'Lo-nitrite' heparins was larger (Table 2). Desulphation of heparin was probably more extensive during the longer reaction times.

For applications of modified heparin preparations which require a relatively lower anhydromannose content (less than 350 nmol/mg heparin) without a large loss of anticoagulant activity >75%), 'Lo-nitrile' depolymerization reactions with reaction times between 2 and 12 h may be used. 'Hi-nitrite' reactions with reaction times between 15 and 20 min should be used for depolymerized heparin preparations requiring a relatively larger amount of anhydromannose units (up to 600 anhydromannose U/mg heparin. At longer reactions times depolymerization was too extensive and the loss of activity was very large.

The loss in the anticoagulant activity of heparin was attributed primarily to the depolymerization reaction and not to the reductive amination reaction. Aminated heparin activity was approximately the same as the activity of depolymerized heparin. The loss of sulphate groups, cleavage of the antithrombin III binding site in the heparin molecule and the decrease in the molecular weight of heparin during the depolymerization reaction most likely contributed to the decrease in heparin's anticoagulant activity.

The relative molecular weight of the depolymerized heparin was estimated from increase in HPLC retention time in order to compare some results on a molar basis. Since heparin standards were not available, the retention time was 'calibrated' with a molecular weight value (~10000 or 35% decrease) from the literature [23] for a depolymerized heparin prepared under conditions similar to Hi10. The molecular weight of the remaining samples was estimated by assuming a linear relationship between retention time difference and molecular weight reduction. The difficulty of using the literature in this context is illustrated by the report [13] that the molecular weight of a depolymerized heparin similar to Lo2 also had a molecular weight of ~10000, despite Hi10 and Lo2 having very different retention times.

Reductive amination. It was assumed that the epsilon amino group of lysine, known for its relatively high degree of reactivity [18], formed a Schiff base with the terminal anhydromannose residues on heparin, leaving the alpha group on the lysine molecule free to react with glutaraldehyde when mixed in the heparin–PVA formulations. Because an excess of lysine was used in the reaction, it was also assumed that both amine groups on lysine would not react with two different heparin molecules to form a heparin-lysine-heparin complex. All reductive aminations were carried out a pH of 10 in order to provide a large source of the unprotonated amine.

It was assumed that the decrease in the number of anhydromannose units quantified using the indole colorimetric assay was a direct consquence of the reductive amination of these groups using lysine. However the discrepancy between amino acid analysis and the lysine estimates of Table 3 likely reflects the above noted problem in amino acid analysis coupled with the difficulty of detecting an anhydromannose–lysine conjugate in the amino acid chromatogram. The presence of side reactions, however, cannot be ruled out. Direct determination of covalent bound lysine was not possible because the presence of serine interfered with the picryl sulphonic acid assay and because a unique peak due to anhydromannose-lysine conjugate was not identifiable in the amino acid analysis chromatogram.

Modified heparin–PVA

The heparin content of the modified heparin-PVA films (except Hi10, Lo2) increased on both a mass and molar basis when compared to the heparin content of Waitaki heparin–PVA films. Using modified heparin Hi30, which contained 100 times more terminal amino acid residues than commercial heparin (based on anhydromannose assay) the heparin content increased from 15–20 mg/g wet gel by as much as 56% on a mass basis and as much as 482% on a molar basis.

The increase in the heparin content of modified heparin–PVA did not correspond completely with the terminal amino acid content of the modified heparin. For example, although modified heparin Hi10 and Lo2 contained a larger terminal lysine content than the serine content of Diosynth heparin, films prepared using the latter had a larger heparin content (mass and molar basis) than films prepared using the modified heparins. This implied that the terminal lysine group on modified heparin was less reactive than the terminal serine group found on Waitaki heparin or Diosynth heparin. On the other hand, the heparin content of the modified heparin–PVA films increased (Fig. 3) as the terminal amine content of the modified heparin increased and also as the molecular weight of the modified heparin decreased (Tables 2 and 6). Therefore, it was not certain if it was the increase in the terminal amine group of modified heparin which actually caused the heparin content to increase. The modification of heparin may have created another functional group on the heparin molecule which may have reacted with the glutaraldehyde–PVA solution. Also the lower molecular weight of the modified heparin may have increased the solubility of heparin in a 7% PVA solution.

It is interesting to note that even though the terminal amine content of the various modified heparin samples had substantially increased, the heparin content of the modified heparin–PVA films did not contain more than 60% (w/w) of the original heparin added to the gel formulation, perhaps reflecting the solubility limit of even modified heparin in the PVA solution.

Assuming that the expected anticoagulant activity of the polymer films could be estimated by using the heparin content of that film, along with the activity of heparin in solution, the 'expected' anticoagulant activity of the modified heparin films was estimated. This calculation did not account for the loss of anticoagulant activity which likely occurred during immobilization. Except for Hi20 and Lo12 the 'expected' anticoagulant activities of the other modified heparin-PVA films were less than Waitaki heparin-PVA either because a small amount of heparin was bound or because the modified heparin had lost a very large amount of activity during depolymerization. Based on this calculation only modified heparin Hi20 and Lo12 were selected for anticoagulant activity testing.

Anticoagulant activity of the hydrogel coating

Hydrogel coatings prepared using Waitaki heparin inactivated a larger amount of thrombin than hydrogel coatings prepared using modified heparin. Although the 'expected' anticoagulant activity of modified heparin Lo12-PVA and modified heparin Hi20-PVA was estimated to be equivalent to that of Waitaki heparin-PVA (Table 6), hydrogel coating with Hi20 inactivated only 80% of the thrombin in solution and hydrogel coating with Lo12 inactivated 72% of the thrombin in solution, whereas Waitaki heparin-PVA inactivated 93% of the thrombin in solution. Thus modified heparin, upon immobilization, experienced a larger decrease in anticoagulant activity than Waitaki heparin.

The decrease in molecular weight may have reduced the fraction of molecules with an active site. It has been shown that for heparin in solution a molecular weight of at least 8000 Da is required to catalyze the inactivation of thrombin by antithrombin III [26]. Perhaps for immobilized heparin, the minimum molecular weight of heparin required to catalyze the antithrombin III-thrombin inactivation is larger. Past studies have found that although low molecular weight heparins were unable to inactivate thrombin, they were able to interrupt the coagulation mechanism by inactivating Factor Xa [26]. Thus, although the hydrogel coatings prepared using modified heparin did not inactivate a larger amount of thrombin, they may be more successful in activating Factor Xa. Further studies are needed.

Another factor may have been that other sites of binding were created during their modification. The immobilization of heparin through these 'other' sites of binding may have prevented modified heparin Hi20 and modified heparin Lo12 from fully exerting their potential anticoagulant activity. Thus, non-aminated depolymerized heparins should be used to prepare heparin-PVA hydrogels to test whether another site of binding was created during the depolymerization reaction. A related factor may be steric hindrance since it is conceivable that immobilizing shorter chains (i.e. the modified heparins) may lead to greater loss of activity than that expected for unmodified heparin.

Tay [23] found that heparinized materials were more active when prepared using modified heparin (nitrous acid depolymerization followed by reductive amination) than when prepared using commercial heparin. This may be because in their case the commercial heparin was bound at sites along heparin's backbone, whereas, modified heparin was bound, predominantly, at its terminal end. In the case of heparin-PVA, the immobilization of commercial heparin occurred predominantly at its terminus. Therefore, the modification of heparin carried out here was not expected to change its site of binding, but rather to increase the number of molecules that could bind. However, nitrous acid depolymerization resulted in a substantial decrease in the anticoagulant activity of heparin. Unfortunately, the detrimental effects caused by the loss in anticogulant activity far outweighed the positive effects which an increased heparin content could bring about.

SUMMARY

Although the modification of heparin, using nitrous acid depolymerization followed by reductive amination, was successful in increasing the terminal amine content of heparin and in increasing the heparin content of PVA polymer films, it did not increase the active heparin content (in terms of thrombin inactivation) of the heparin–PVA hydrogel.

It was expected that heparin would lose a portion of its anticoagulant activity upon depolymerization with nitrous acid. Thus different reaction conditions were used in order to find the optimum conditions. Unfortunately, none of the six modified heparins was able to increase the active heparin content of heparin–PVA. The increase in the heparin content of heparin–PVA brought about by the increase in heparin's terminal amino acid content was not large enough to overcome the adverse affects which the nitrous acid depolymerization of heparin had on the anticoagulant activity of modified heparin. Unless milder methods are used to increase heparin's terminal amino acid content, it is recommended that future work focus on the modification of the PVA hydrogel, rather than the heparin molecule. 'Chemical amplification' techniques [21] or immobilized functional groups [29] may be useful approaches in future studies.

Acknowledgements

The authors acknowledge the financial support of the National Institutes of Health (HL24020) and the Medical Research Council and the technical advice of Gerry Llanos.

REFERENCES

1. A. S. Hoffman, G. Schmer, C. Harris and W. G. Kraft, *Trans. ASAIO* **18,** 10–16 (1972).
2. P. Cuatrecasas, M. Wichek and C. B. Anfinsen, *Proc. Natl. Acad. Sci.* **61,** 636 (1968).
3. S. Ventkataraman, T. A. Horbett and A. S. Hoffman, *J. Biomed. Mater. Res.* **8,** 111–123 (1977).
4. W. R. Gombotz and A. S. Hoffman, in: *Hydrogel in Medicine and Pharmacy*, Vol 1, pp. 95–125, N. A. Peppas (Ed.). CRC Press, Boca Raton, FL (1986).
5. M. F. A. Goosen and M. V. Sefton, *J. Biomed. Mater. Res.* **13,** 347 (1979).
6. M. V. Sefton, C. H. Cholakis and G. Llanos, in: *Blood Compatibility*. Vol. 1, pp. 151–198, D. F. Williams (Ed.). CRC Press, Boca Raton, FL (1987).
7. R. Evangelista and M. V. Sefton, *Biomaterials* **7,** 206–211 (1986).
8. W. F. Ip and M. V. Sefton, *Biomaterials* **10,** 313–317 (1989).
9. C. H. Cholakis, W. Zingg and M. V. Sefton, *J. Biomed. Mater. Res.* **23,** 417–441 (1989).
10. M. V. Sefton, W. F. Ip, C. H. Cholakis and W. Zingg, *ASAIO J.* **8,** 207–212 (1985).
11. I. Bjork and U. Lindahl, *Mol. Cell. Biochem.* **48,** 161 (1982).
12. R. D. Rosenberg, in: *Hemostasis and Thrombosis: Basic Principles and Clinical Practice*, pp. 1373–1392, 2nd Edition. R. W. Colman, J. Hirsh, V. J. Marder and E. W. Salzman (Eds.). J. B. Lippincott Company, Philadelphia, PA (1987).
13. A. Danielsson and I. Bjork, *Biochem. J.* **213,** 345 (1983).
14. B. Pasche, K. Kodama, O. Larm, P. Olsson and J. Swedenborg, *Thromb. Res.* **44,** 739 (1986).

15. C. H. Cholakis and M. V. Sefton, in: *Polymers as Biomaterials*, pp. 305–315, S. W. Shalaby, A. S. Hoffman, B. D. Ratner and T. A. Horbett (Eds). Plenum Press, New York (1984).
16. J. Hoffman, O. Larm and E. Scholander, *Carb. Res.* **117**, 328 (1983).
17. D. Grainger, S. W. Kim and J. Feijen, *J. Biomed. Mater. Res.* **22**, 231–249 (1988).
18. B. A. Schwartz and G. R. Gray, *Acta Biochem. Biophys.* **181**, 542 (1977).
19. Z. Dische and E. Borenfreund, *J. Biol. Chem.* **184**, 517 (1950).
20. M. F. A. Goosen and M. V. Sefton, *J. Biomed. Mater. Res.* **17**, 359–373 (1983).
21. C. C. Wang and L. B. Jacques, *Arzheim.-Forsch. (Drug Res.)* **24**, 1945 (1974).
22. G. Rollason and M. V. Sefton, *J. Biomater. Sci. Polym. Ed.* **1**, 31–41 (1989).
23. S. W. Tay, *Hydrogel by Covalent Bonding*, Sc.D. Thesis, Department of Chemical Engineering, M.I.T. (1986).
24. D. Lagunoff, *Biochemistry* **13**, 3982 (1974).
25. T. W. Barrowcliffe and D. P. Thomas, in: *Heparin and Related Polysaccharides: Structures and Activities*, F. A. Ofosu, I. Danishefsky and J. Hirsh (Eds). *Ann NY Acad. Sci.* **556**, 132–145 (1989).
26. D. A. Lane, I. R. McGregor, R. Michalski and V. V. Kakkar, *Thromb. Res.* **12**, 257–271 (1978).
27. S. C. Lin, H. A. Jacobs and S. W. Kim, *J. Biomed. Mater . Res.* **25**, 791–795 (1991).
28. X. Ma, S. F. Mohammad and S. W. Kim, *J. Coll. Interface Sci.* **147**, 251–261 (1991).
29. B. A. Smith and M. V. Sefton, *J. Biomed. Mater. Res.* **22**, 673–685 (1988).

Surface modification of polymeric biomaterials with poly(ethylene oxide), albumin, and heparin for reduced thrombogenicity

MANSOOR AMIJI and KINAM PARK*

Purdue University, School of Pharmacy, West Lafayette, IN 47907, USA

Received 13 April 1992; accepted 13 August 1992

Abstract—Appropriate surface modification has significantly improved the blood compatibility of polymeric biomaterials. This article reviews methods of surface modification with water-soluble polymers, such as polyethylene oxide (PEO), albumin, and heparin. PEO is a synthetic, neutral, water-soluble polymer, while albumin and heparin are a natural globular protein and an anionic polysaccharide, respectively. When grafted onto the surface, all three macromolecules share a common feature to reduce thrombogenicity of biomaterials. The reduced thrombogenicity is due to the unique hydrodynamic properties of the grafted macromolecules. In aqueous medium, surface-bound water-soluble polymers are expected to be highly flexible and extend into the bulk solution. Biomaterials grafted with either PEO, albumin, or heparin are able to resist plasma porotein adsorption and platelet adhesion predominantly by a steric repulsion mechanism.

Key words: Poly(ethylene oxide); albumin; heparin; surface modification; blood compatibility; steric repulsion.

INTRODUCTION

Many different materials have been used for biomedical applications including polymers, ceramics, metals, carbon, and composites [1]. Of all these materials, polymers offer ease in processing and control over the physical properties necessary for appropriate biomedical applications [2]. Like other blood-contacting biomaterials, long-term use of polymers in blood is limited in part due to surface-induced thrombosis, initiated by the adsorption of plasma proteins and activation of platelets [3, 4]. The adsorption of plasma proteins is known to produce a 'conditioning film' which determines the outcome of other processes, such as cell adhesion and activation of the complement system [5].

Following protein adsorption, the adhesion of platelets onto biomaterials contributes to surface-induced thrombosis. In irreversible interactions with the surface, adherent platelets change their shape by extending pseudopods and spread on the surface [6, 7]. When fully activated, platelets release the biochemical contents of their granules, which causes further activation of other platelets and simultaneous initiation of the coagulation cascade reaction to form thrombi [6]. Thrombi formed on the biomaterial surface could be either red or white depending on their content. Red thrombi, for instance, consist of platelet aggregates and red blood cells all trapped in a fibrin network [8]. Interactions at the biomaterial-blood interface leading to surface-induced thrombosis are illustrated in Fig. 1.

* To whom correspondence should be addressed.

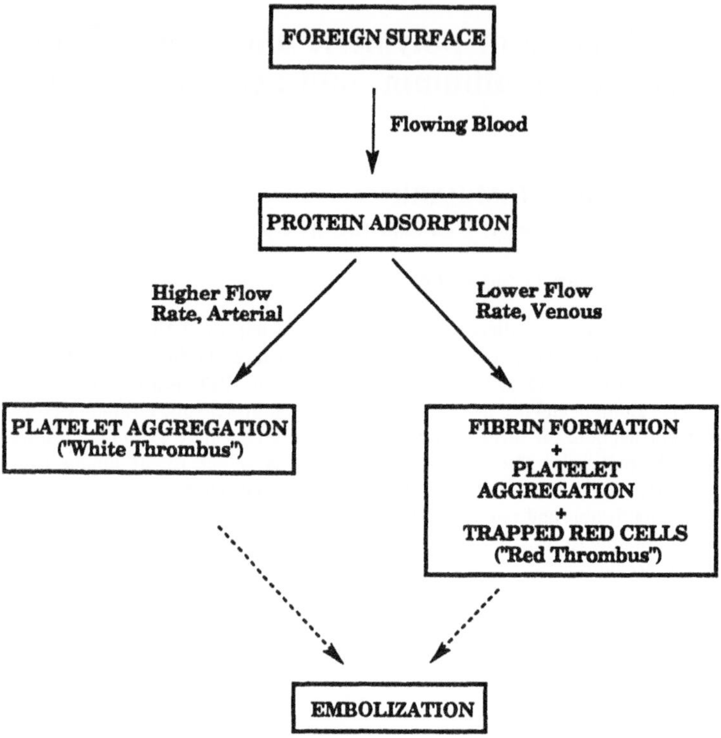
FOREIGN SURFACE
Flowing Blood
PROTEIN ADSORPTION
Higher Flow Rate, Arterial
Lower Flow Rate, Venous
PLATELET AGGREGATION ("White Thrombus")
FIBRIN FORMATION + PLATELET AGGREGATION + TRAPPED RED CELLS ("Red Thrombus")
EMBOLIZATION

Figure 1. Schematic illustration of surface-induced thrombosis on synthetic biomaterial surface (from ref. 1).

Since the interactions that lead to thrombus formation occur at the biomaterial-blood interface, appropriate surface modification methods will be beneficial in improving the blood compatibility of biomaterials without altering the bulk properties of the material necessary for biomedical applications. Hoffman [9] reviewed the surface modification methods and grouped them into two general categories: physicochemical and biological methods. Examples of physicochemical methods are coating deposition, chemical modification, graft copolymerization, and plasma treatment. The biological methods include pre-adsorption of proteins, drug or enzyme immobilization, cell seeding, and preclotting. Ikada [10] proposed that a diffuse hydrophilic biomaterial surface would be blood compatible. To prepare diffuse hydrophilic surfaces, various types of water-soluble monomers have been graft-polymerized to biomaterial surfaces [11]. Water-soluble polymers have also been grafted to create a diffuse hydrophilic surface. The hydrophilic polymers in the diffuse layer exert steric repulsion to proteins and cells that reach the surface.

Steric repulsion is due to a loss of configurational entropy resulting from volume restriction and/or osmotic repulsion between the two overlapping polymer layers [12, 13]. Steric repulsion of plasma proteins and platelets from a biomaterial surface by water-soluble polymers is illustrated in Fig. 2. For effective steric repulsion, water-soluble polymers of the diffuse layer must satisfy the following three requirements. First, the polymer molecules should have high affinity to the surface so that they can anchor tightly to the surface. Second, part of the polymer chain should also

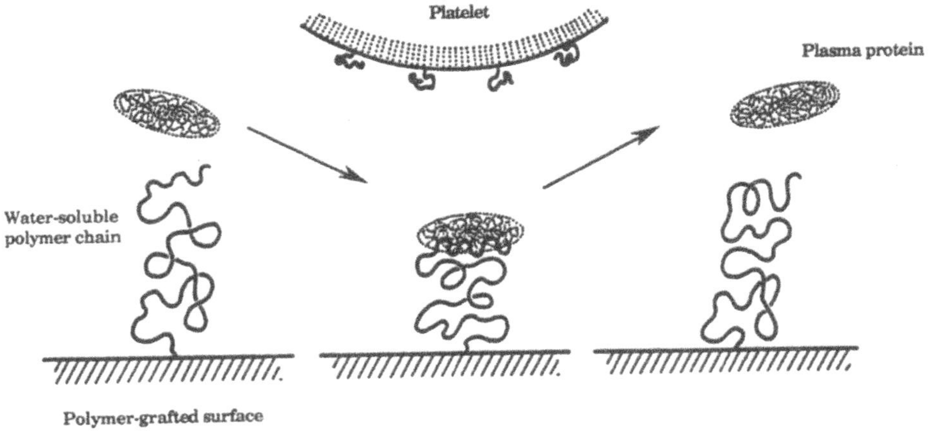
Platelet
Plasma protein
Water-soluble polymer chain
Polymer-grafted surface

Figure 2. Steric repulsion of plasma proteins and platelets by surfaces grafted with water-soluble polymers.

extend into the bulk solution. The domination of steric repulsion over the van der Waals attractive forces is dependent on the extension and flexibility of the polymer chain in the bulk solution. Block copolymers containing both hydrophilic and hydrophobic segments are more effective in steric repulsion than homopolymers [13, 14]. The hydrophobic segment can anchor the copolymer to the hydrophobic surface, while the water-soluble segment can extend into the bulk solution. If covalently grafted, homopolymers will be as effective as amphiphatic block copolymers. Finally, the polymer molecules should completely cover the surface. If a significant portion of the surface is exposed, steric repulsion is not effective.

In this review, we describe grafting of three water-soluble polymers—PEO, albumin, and heparin—which have been used for surface modification of biomaterials. PEO is a synthetic neutral polymer; albumin and heparin are natural macromolecules. Albumin is a globular plasma protein, while heparin is an anionic polysaccharide. Biomaterial surfaces modified with these three water-soluble polymers have shown a remarkable improvement in blood compatibility.

PEO-TREATED SURFACES

PEO in aqueous medium

As shown below, PEO is a polyether type of the water-soluble synthetic polymer.

$$—[CH_2CH_2O]_n—$$

When the molecular weight is less than 10000 Daltons, the polymer is called poly(ethylene glycol) (PEG). The polymers with higher molecular weight are known as poly(ethylene oxide) (PEO) or polyoxyethylene [15]. Compared to other polyethers, such as poly(propylene oxide), PEO is highly water soluble [16]. Kjellander and Florin [17] explained the high water solubility of PEO in terms of a good structural fit between water molecules and the polymer, resulting in hydrogen bonding between the ether oxygen of PEO and water molecules.

With PEO, aqueous solubility decreases with increasing temperature. The decreased solubility at elevated temperatures is due to a decrease in hydrogen bonding and corresponding increase in hydrophobic interactions between polymer chains [18].

Using ^{13}C-NMR, Bjorling *et al.* [19] reported that PEO adapts to a gauche conformation in the polar solvent such as water and a trans conformation in nonpolar medium. The gauche conformation would be more suitable for hydrogen bonding.

Steric repulsion with terminally attached PEO

At the solid–liquid interface, PEO anchored through terminal end will interact with water molecules and extend into the bulk aqueous medium [20]. Surface-bound PEO molecules are very effective in preventing adsorption of other macromolecules by the steric repulsion mechanism [21]. Using the surface force technique, Luckam [22, 23] observed that steric repulsion with PEO is mainly due to osmotic repulsion between interdigitated PEO chains. Jeon *et al.* [24, 25] have theoretically modeled protein–surface interactions in the presence of PEO and found that steric repulsion by surface-bound PEO chains is mainly responsible for the prevention of protein adsorption on PEO-rich surface.

Interfacial properties of PEO-treated biomaterials

PEO-rich surfaces have been prepared by physical adsorption of PEO [26, 27]. Only high molecular weight PEO ($M_w > 100\,000$) can be effectively adsorbed on hydrophobic surfaces [28]. Chromatographic supports used for the separation of proteins, cells, and viruses have been treated by physical adsorption of high molecular weight PEO [26, 27]. Physically adsorbed PEO homopolymers, however, can be displaced by other macromolecules which have higher affinity for the surface. Many proteins and cells in the blood can easily displace physically adsorbed PEO from the surface.

Adsorption of PEO-containing amphipathic block copolymers would be more stable than that of homopolymers as the hydrophobic segment can anchor the copolymer molecule on the surface, while the hydrophilic PEO chains can extend into the bulk aqueous medium. Adsorption of PEO/poly(propylene oxide)(PPO)/PEO triblock copolymers (Pluronics®) have been used for the prevention of protein adsorption and cell adhesion on hydrophobic surfaces. Lee *et al.* [29] did not observe a significant decrease in albumin adsorption on Pluronic-treated surfaces if Pluronics had 30 propylene oxide (PO) residues, since the copolymer was weakly bound to the surface. Albumin could easily displace weakly bound Pluronics and interact directly with the surface. To improve the stability of copolymers on the surface, Lee *et al.* [30] synthesized copolymers of PEO-methacrylates containing alkyl chains for tight binding to hydrophobic surface. Tight surface binding was achieved with Pluronics containing longer hydrophobic PPO segments. Pluronic F-108 which has 56 PO residues and 129 ethylene oxide (EO) residues, minimized protein adsorption on polystyrene latex particles [31]. The adhesion of cells onto octadecyldimethylsilane-treated glass was also inhibited, when the surface was coated with Pluronic F-108 [32]. Using ten different Pluronics with varying chain length of PEO and PPO, we have shown that Pluronics containing a minimum of 56 PO residues and 19 EO residues were sufficient to repel proteins and platelets [33, 34]. The adsorption of Pluronics onto hydrophobic surface presents the simplest method to modify the surface for prevention of protein adsorption and cell adhesion.

Desai and Hubbell [35] have entrapped PEO chains to the surface by partially dissolving the base polymer with a suitable solvent. Poly(ethylene terephthalate) (PET) with entrapped PEO (M_w 18 500) decreased albumin adsorption by 80% and platelet adhesion by more than 95% as compared to the control PET. Ruckenstein and coworkers [36, 37] have used a similar approach to entrap PEO-containing block copolymers to poly(methyl methacrylate), polystyrene, and poly(vinyl acetate) surfaces. The base polymer was swollen in an organic solution containing PEO block copolymer. This caused the hydrophobic segment of the copolymer to be entangled within the swollen surface and was secured by placing the system in a non-solvent (water). Physical entrapment of PEO homopolymers or block copolymers should consider the toxicity of the organic solvents used for swelling the base polymer and poor adaptability of the technique to other polymeric substrates.

Covalent grafting of PEO to surfaces is the most effective way of creating a permanent PEO layer. Merrill *et al.* [38] initially reported that PEO soft-segment polyurethanes are highly blood compatible. Ito and Imanishi [39] reviewed the work of many investigators who prepared polyurethanes with polyether soft-segment. In some cases, polyurethanes made of low molecular weight PEG ($M_w \sim 1000$) allow platelets to adhere and activate. The overall results, however, have proven that segmented polyurethanes containing high molecular weight PEO had improved blood compatibility as compared to other types of polyethers [39]. Yu *et al.* [40] developed PEO-containing polyurethaneurea hydrogels as coatings for biomedical products. Polyurethanes with PEO grafted to the side chains were found to be highly blood compatible [41, 42]. Figure 3 shows the reaction scheme for synthesis of polyurethanes with methoxy-PEG side chains. Chaikof *et al.* [43] recently developed interpenetrating polymer networks (IPN's) of PEO and polyether substituted polysiloxane. A significant decrease in platelet adhesion was observed when the PEO (M_w 8000) content in the IPN was increased up to 65%.

Allmer *et al.* [44] covalently coupled PEO chains to a glycidyl methacrylate-bound polyethylene surface. Akizawa *et al.* [45] coupled methoxy-PEG with a terminal carboxyl group to form an ester linkage with the hydroxyl groups of cellulose dialysis membranes. Improved dialysis efficiency and blood compatibility was observed with PEG grafting. Desai and Hubbell [46] grafted cyanuric chloride activated PEO to amine derivatized PET surfaces. Figure 4 illustrates the reaction for covalent coupling of cyanuric chloride activated PEO to amine-derivatized PET film. A 50% decrease in plasma protein adsorption and more than 90% decrease in platelet adhesion was observed using PEO (M_w 18 500 and 100 000)-grafted surfaces. Gombotz *et al.* [47] recently reported coupling of *bis*-amino PEO to cyanuric chloride activated PET films. The adsorption of albumin and fibrinogen was found to decrease with increasing molecular weight of immobilized PEO. Chemical coupling of PEO to polymeric surfaces, as described, is possible only if the surface has functional groups that can react with PEO derivatives. For inert polymers such as polyethylene, PEO coupling is possible only when the surface is pre-modified with reactive functional groups [44]. Grafting by use of UV or gamma irradiation, however, may not require premodification of the polymer surface.

Mori and Nagaoka [48] prepared PEO-rich surfaces by photoinduced grafting of methoxy poly(ethylene glycol) methacrylates to poly(vinyl chloride) surface in the presence of dithiocarbamate. With increasing PEO chain length up to 100 EO residues, plasma protein adsorption and platelet adhesion was significantly

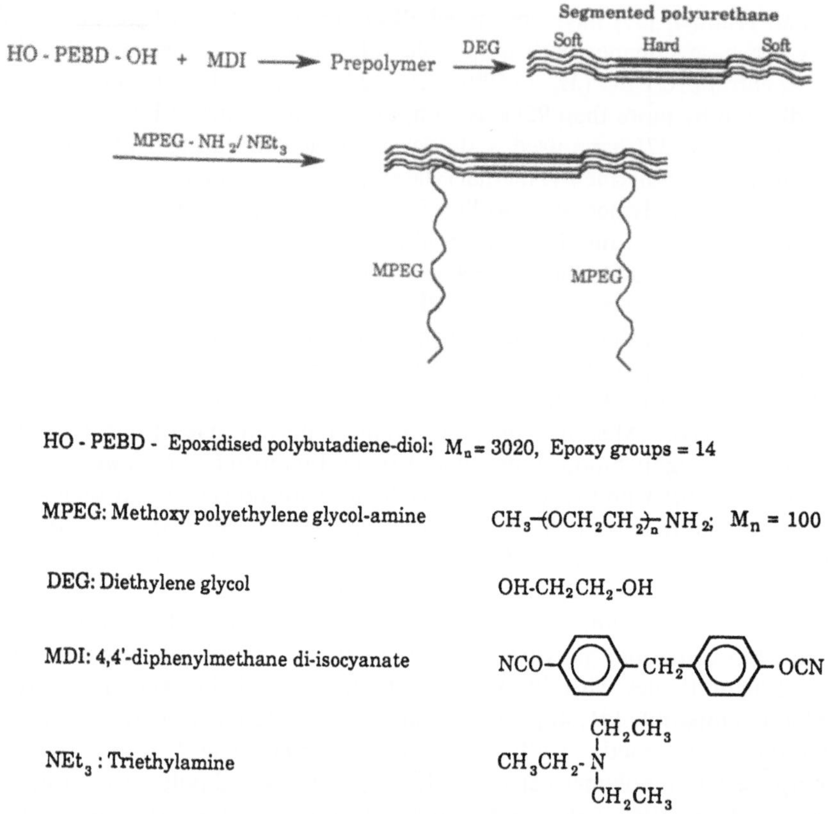
Segmented polyurethane
HO - PEBD - OH + MDI ⟶ Prepolymer
DEG
Soft
Hard
Soft
MPEG - NH$_2$/ NEt$_3$
MPEG
MPEG
HO - PEBD - Epoxidised polybutadiene-diol; M_n = 3020, Epoxy groups = 14
MPEG: Methoxy polyethylene glycol-amine
$CH_3(OCH_2CH_2)_n NH_2$; M_n = 100
DEG: Diethylene glycol
OH-CH_2CH_2-OH
MDI: 4,4'-diphenylmethane di-isocyanate
NCO
CH_2
OCN
NEt$_3$: Triethylamine
CH_2CH_3
CH_3CH_2- N
CH_2CH_3

Figure 3. Covalent coupling reaction of methoxy-PEG (MPEG) to segmented polyurethane (from ref. 39).

decreased on PEO-grafted surfaces. Tseng and Park [49] synthesized PEG-phenylazide for photoinduced grafting to various polymeric surfaces. In the presence of UV light, azide groups are converted into highly reactive nitrine groups. The reaction scheme for synthesis of PEG-phenylazide and photoinduced grafting onto dimethyldichlorosilane-treated glass (DDS-glass) is shown in Fig. 5. The number of platelets decreased by 95% on PEG-grafted DDS-glass as compared to that on control DDS-glass. The spread area of individual platelets also decreased from 45 μm^2 on control DDS-glass to less than 20 μm^2 on PEG-grafted DDS-glass. If platelets are in the contact adherent state, their spread area is less than 20 μm^2.

Sheu *et al.* [50] recently introduced a method of grafting PEO-containing block copolymers (Brij®) by exposing the adsorbed copolymers to glow discharge treatment. High energy gamma or electron beam irradiation can also be used to graft PEO to various surfaces. We have grafted Pluronic F-68 (76/30/76) copolymers to DDS-glass by γ-irradiation. Effective grafting was achieved when Pluronic-adsorbed DDS-glass was exposed to γ-irradiation in the presence of an aqueous buffer [51]. The number of adherent platelets decreased by 85% on Pluronic-grafted DDS-glass as compared to that on control DDS-glass. In addition, the spread area of individual platelets decreased from 45 μm^2 on control DDS-glass to 15 μm^2 on Pluronic-grafted DDS-glass. Sun *et al.* [52] grafted PEG-methacrylates to

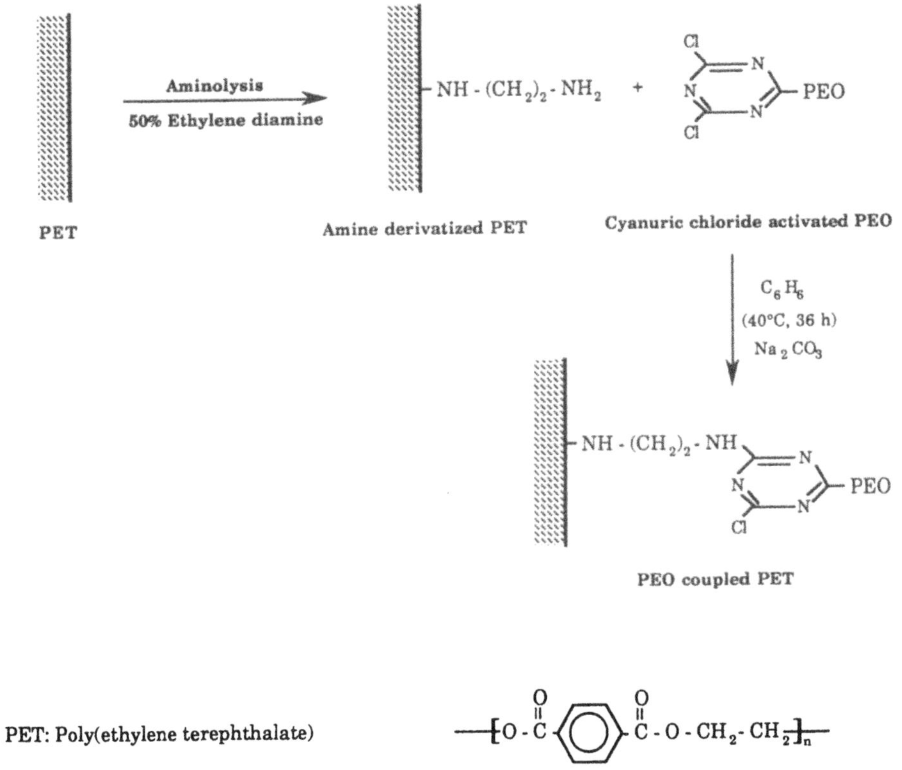

Aminolysis
50% Ethylene diamine
NH - (CH2)2 - NH2
+
Cl
N
N
N
PEO
Cl
PET
Amine derivatized PET
Cyanuric chloride activated PEO
C6H6
(40°C, 36 h)
Na2CO3
NH - (CH2)2 - NH
PEO
PEO coupled PET
PET: Poly(ethylene terephthalate)

Figure 4. Covalent coupling reaction of cyanuric chloride activated PEO to amine-derivatized poly(ethylene terephthalate) (PET) film (from ref. 46).

Silastic® films by mutual irradiation in the prescence of Cu^{2+} ions to prevent homopolymer gelation. A 72% decrease in fibrinogen adsorption was observed when the number of EO residues of the grafted PEG were 100. However, gamma irradiation at high doses may alter the bulk properties of some polymers such as polypropylene.

ALBUMIN-TREATED SURFACES

Properties and functions of albumin

Albumin is the most abundant protein found in blood. The concentration of albumin in healthy adults varies from 35 mg/ml to 50 mg/ml [53]. There are 585 amino acid residues in human serum albumin, 181 of which have either acidic or basic side-chains. Albumin has some unique properties which distinguishes it from other globular proteins in the blood. It is an acidic protein with high aqueous solubility and stablity. Stability of albumin against thermal denaturation and low pH conditions is due to 17 disulfide bonds [54]. It has high aqueous solubility, attributed to the polar surface of the molecule. At pH 7, albumin molecule has a net charge of −15 [55].

Albumin serves three important functions in the body [56]. First, a sufficient amount of albumin is synthesized by the liver to serve as a nutritional source in

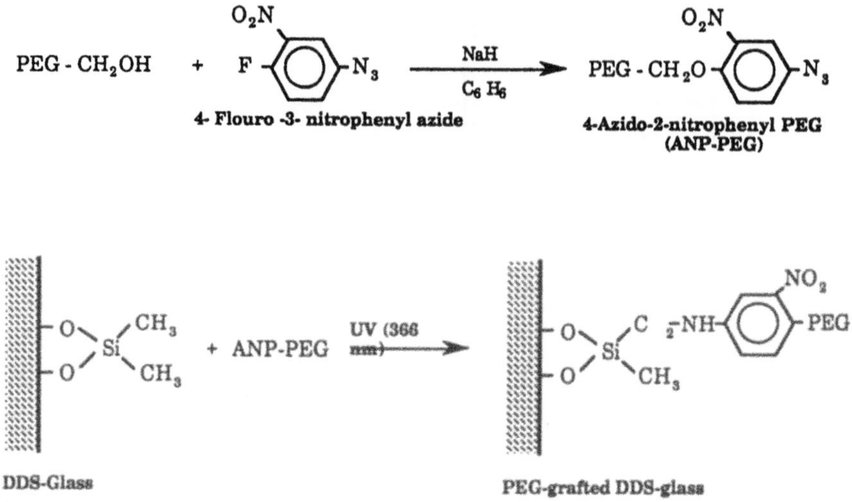
O2N
PEG - CH2OH
+
F
N3
NaH
C6 H6
4- Flouro -3- nitrophenyl azide
O2N
PEG - CH2O
N3
4-Azido-2-nitrophenyl PEG
(ANP-PEG)
O
Si
CH3
O
CH3
+ ANP-PEG
UV (366 nm)
O
Si
C
NH
NO2
PEG
O
CH3
DDS-Glass
PEG-grafted DDS-glass

Figure 5. Synthesis and photoinduced grafting of PEG-phenylazide to dimethyldichlorosilane-treated glass (DDS-glass) (from ref. 49).

cellular metabolism. Second, albumin serves to transport small molecules such a steroids, ions, and fatty acids. Finally, albumin maintains the osmotic pressure of the blood.

Mechanism of surface passivation by albumin

Plasma protein adsorption and platelet adhesion and activation is significantly reduced on albumin-coated biomaterial surfaces [57, 58]. Albumin has poor affinity for hydrophilic surfaces. Previously, we have found that platelets were able to adhere and activate on albumin-coated hydrophilic glass even when the bulk albumin concentration was increased up to 50 mg/ml [59]. Albumin adsorbs tightly onto hydrophobic surfaces and the adsorption is usually irreversible [60]. Since albumin molecules do not have receptors on the platelet membrane, platelets do not specifically interact with albumin or adhere to albumin-coated surface [61].

Platelets were able to adhere and activate completely on albumin-coated DDS-glass, if the adsorbed albumin was crosslinked with glutaraldehyde, dried and rehydrated, or digested with trypsin, a proteolytic enzyme [62]. Crosslinking with glutaraldehyde is expected to reduce the motional freedom or flexibility of the surface accessible segments of adsorbed albumin molecules. Drying of absorbed albumin resulted in the exposure of bare surface sites. Tryptic digestion may have resulted in the exposure of bare surface sites and cleavage of flexible albumin segments. The three requirements of steric repulsion by water-soluble polymers are met by albumin molecules which are adsorbed on DDS-glass. First, albumin binds tightly to DDS-glass [63]. Second, the flexibility of surface accessible fragments of albumin is important in preventing platelet adhesion. Finally, prevention of platelet adhesion by albumin is effective only when the surface is fully covered. This suggests that steric repulsion is the predominant mechanism for the prevention of protein adsorption and platelet adhesion by the absorbed albumin.

The relationship between colloidal stabilization by steric repulsion and prevention of platelet adhesion on the albumin layer was first proposed by Morrissey [64].

Recently, using the surface force technique, Blomberg *et al.* [65] concluded that short range repulsion between albumin-coated mica surfaces could be due to steric repulsion. It should also be noted that albumin can be used for the stabilization of colloidal gold particles [66] and negatively charged polystyrene latices [67]. Van der Scheer *et al.* [67] concluded that the optimum stability of polystyrene latices was obtained only when albumin molecules completely covered the surface of the latices. Further experimental evidence of colloidal stabilization with adsorbed albumin on polystyrene latices has been obtained by Tamai *et al.* [68]. In high ionic strength media, adsorbed albumin was effective in preventing coagulation of latices by steric repulsion.

Interfacial properties of albumin-treated biomaterials

Chang [69, 70] initially suggested the use of albumin coating as a method to prevent surface-induced thrombosis in clinical situations. Physically adsorbed albumin has been used for improving blood compatibility of clinical implants [71]. Physically adsorbed albumin, however, is effective in preventing platelet adhesion and activation only for a short period of time. For long-term use of blood-contacting biomaterials, adsorbed albumin is desorbed from the surface by other plasma proteins with high surface affinity.

In an attempt to improve the stability of albumin on the surface, Sigot-Luizard *et al.* [72] treated adsorbed albumin with glutaraldehyde. The results, however, showed that platelets were able to adhere and activate on the glutaraldehyde-crosslinked albumin layer. For permanent albumin immobilization, Hoffman *et al.* [73] covalently attached albumin to hydrogel-grafted polymer surface using ε-aminocaproic acid spacer group. Sharma and Kurian [74] attempted covalent grafting of albumin to polyurethane surface by γ-irradiation. After 1.0 Mrads of irradiation, the number of adherent platelets actually increased. It is unclear whether albumin was grafted at all or if adsorbed albumin was damaged by γ-irradiation. Matsuda and Inoue [75] synthesized photoreactive albumin for grafting to fabricated medical devices. Platelet reactivity on photografted albumin surface was found to decrease significantly. Unfortunately, no long-term stability studies have been done to suggest that albumin remains on the surface in the native, undenatured state.

Eberhart *et al.* [76–78] studied selective adsorption of albumin from blood onto polymeric surfaces. Long carbon chains (C-16 and C-18) grafted on polyurethane surfaces were found to have high affinity for albumin. Alkyl-grafted surfaces were found to have high affinity for albumin adsorbed from simple protein solution, from binary protein mixture, from plasma, and from whole blood [76]. Similar alkylated surfaces were later tested in canine *ex vivo* shunt model [79, 80]. Figure 6 shows the reaction scheme for synthesis of polyurethanes with alkyl chains for enhanced albumin affinity. Fibrin formation and platelet adhesion and aggregation were completely inhibited in short-term *in vivo* experiments. The alkyl-modified surfaces, however, could not achieve total specificity towards albumin [77]. In long-term applications, thrombogenic proteins in blood can displace albumin from the surface to create a favorable environment for platelet activation.

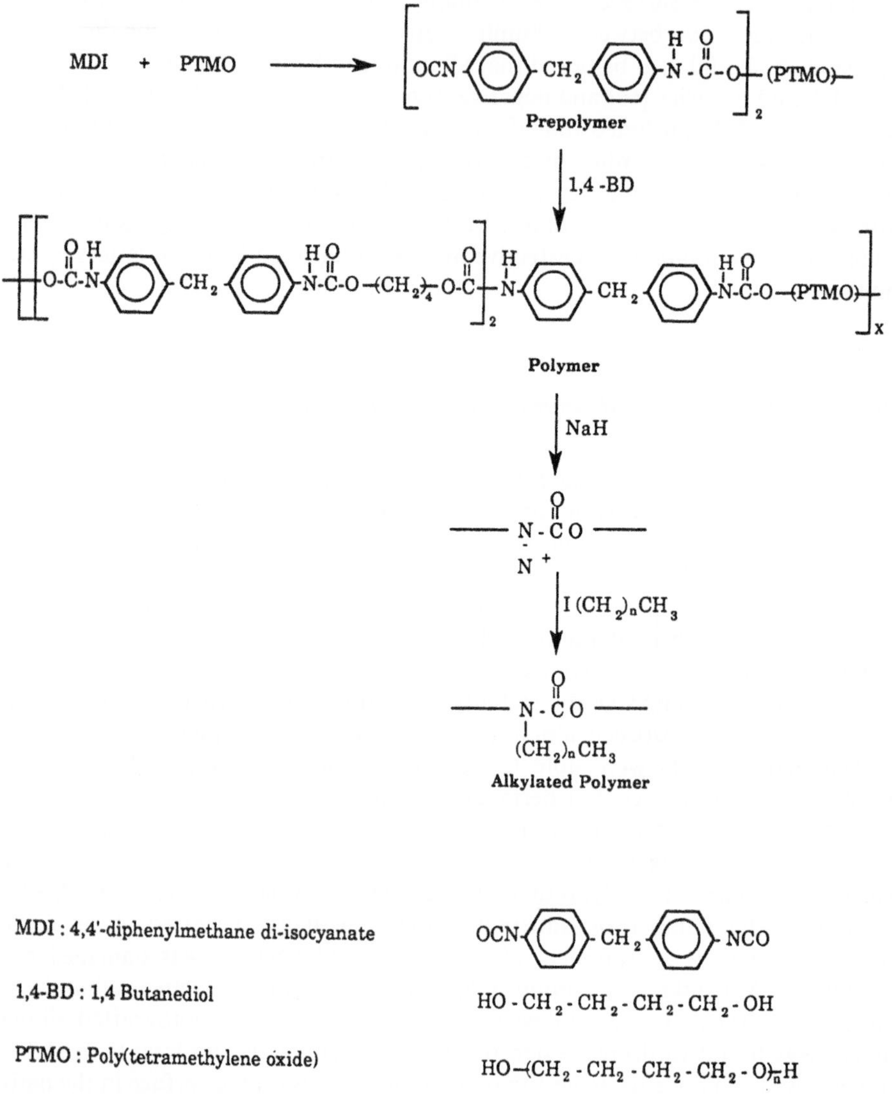
MDI + PTMO
OCN CH2 N-C-O (PTMO)
Prepolymer
1,4 -BD
O-C-N CH2 N-C-O-(CH2)4-O-C N CH2 N-C-O-(PTMO)
Polymer
NaH
N-CO
N+
I (CH2)nCH3
N-CO
(CH2)nCH3
Alkylated Polymer
MDI : 4,4'-diphenylmethane di-isocyanate
OCN CH2 NCO
1,4-BD : 1,4 Butanediol
HO - CH2 - CH2 - CH2 - CH2 - OH
PTMO : Poly(tetramethylene oxide)
HO-(CH2 - CH2 - CH2 - CH2 - O)n-H

Figure 6. Reaction scheme for synthesis of alkyl-derivatized polyurethanes with enhanced albumin affinity (from ref. 39).

HEPARIN-TREATED SURFACES

Properties and functions of heparin

Heparin is an anionic linear polysaccharide chemically known as glycosaminoglycans. The molecular weight of heparin ranges from 3000 to 40 000 Daltons with an average of 15 000 Daltons [81]. Commercially available heparin is isolated from bovine and porcine lung or intestinal mucosa. The chemical structure of heparin sodium is shown in Fig. 7, consisting of two repeating disaccharide units: D-glucosamine-L-iduronic acid and D-glucosamine-L-glucouronic acid. Heparin is strongly acidic because of its content of covalently bound sulfonate and carboxylic acid groups [82].

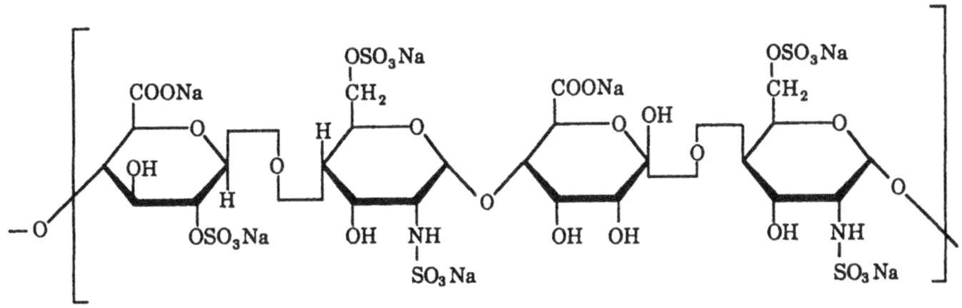
COONa
OSO3Na
CH2
COONa
OSO3Na
CH2
OH
OH
H
H
O
O
O
O
O
O
O
O
OSO3Na
OH
NH
SO3Na
OH
OH
OH
NH
SO3Na

Figure 7. Proposed structure of heparin sodium.

Endogenous or intravenously injected heparin is known to inhibit blood coagulation. The anticoagulant activity of heparin is mediated through a co-factor, antithrombin III [83]. Heparin–antithrombin III complex inactivates serine proteases, most importantly thrombin, which catalyzes the conversion of fibrinogen to fibrin in the final step of the coagulation cascade reaction [84]. Heparin markedly accelerates the rate, but not the extent, of thrombin inactivation. Heparin-antithrombin III complex may also bind with the activated clotting factors (XII_a, kallikrein, XI_a, IX_a, X_a, II_a, and XIII_a) resulting in inactivation of these factors [83].

Mechanism of surface passivation by heparin

Heparinized materials, first described by Gott *et al.* in 1963 [85], have been extensively used as blood-contacting biomaterials. Several detailed review articles on the subject of heparinized biomaterials are available [86, 87]. Although heparinized surfaces have been studied extensively, the exact mechanism of surface passivation by heparin is still not well understood. In addition, the effect of heparin seems to be different *in vitro* and *in vivo*. When heparin is ionically immobilized, there is a gradual release of heparin into the biological environment. The released heparin can then interact with antithrombin III and mediate the anticoagulant effect [88]. The exact mechanism of thrombo-resistance of heparin grafted to the biomaterial surface is still unclear. Ebert and Kim [89] have shown that the activity of heparin grafted to the surface through alkyl spacer groups was dependent on the length of the spacer, while prevention of platelet activation occurred independent of the spacer length. These results suggest that the ability of heparin to prevent platelet activation is independent from the anticoagulant activity of heparin.

Larsson *et al.* [90–92] proposed a mechanism of surface passivation of heparin based on steric repulsion. Heparinized materials were tested by measuring the activity of thrombin after exposure to heparin–antithrombin III complex. Heparin was attached to the surface either by end-point or multi-point attachment. End-point attached heparin completely inhibited thrombin activity, whereas multi-point attached partially inhibited thrombin activity. End-point grafted heparin showed inactivation of thrombin similar to that of the vascular endothelium, since heparin was flexible and may extend into the bulk aqueous medium. These studies have shown that the flexibility of immobilized heparin is important for optimum surface passivation.

Heparin flexibility is also improved by the use of a hydrophilic spacer group, such as PEO or albumin, when immobilizing heparin onto a surface [93–97]. The anticoagulant properties of heparin can be improved by end-point attachment or with the use of a hydrophilic spacer group. In addition, end-point grafted or the use of a hydrophilic spacer group would extend heparin into the bulk solution for effective steric repulsion. For effective surface passivation, heparin should be covalently or tightly bound to the surface, remain flexible and extend into the bulk solution, and cover the surface completely.

Interfacial properties of heparin-treated biomaterials

Gott *et al.* [85] ionically immobilized negatively charged heparin onto the positively charged benzalkonium chloride-treated graphite surface. Other more hydrophobic quaternary ammonium salts were also used, since they are soluble in organic solvents [98]. The organic solution can be spread on hydrophobic substrates and nonrigid hydrophobic polymers such as silicone rubber can be effectively coated. Although effective for short-term blood-compatibility, ionically bound heparin was easily displaced from the surface by an ion exchange mechanism when exposed to blood. In order to increase the stability of ionically-bound heparin, heparin-bound surfaces were treated with glutaraldehyde [99]. Treatment with glutaraldehyde significantly reduced the release of heparin from the surface.

Even with crosslinking by glutaraldehyde, surface-bound heparin is slowly released into the circulation upon exposure to blood. To further improve heparin stability on the surface, albumin-heparin conjugates were synthesized for preparing heparin-surfaces by simple coating on hydrophobic biomaterials [93, 94]. In contact with plasma, however, physically adsorbed albumin-heparin conjugates were gradually replaced by other plasma proteins.

Heparinized hydrogels combine two biologically important features: anti-coagulant properties of heparin and blood compatibility of hydrogels. Sefton [100] reviewed the work on heparinized hydrogels, especially the blood compatibility of heparin-poly(vinyl alcohol) (PVA). *In vitro* and short-term *in vivo* studies provided sufficient evidence of improved blood compatibility with heparinized-PVA [101, 102]. Recent studies by Cholakis *et al.* [103, 104], however, suggested that the number of adherent platelets as well as platelet reactivity remained similar to the control PVA surface. In addition, platelet consumption was found to increase on heparin-PVA surface when exposed to blood in an *ex vivo* shunt. On some hydrogel-grafted surfaces platelet adhesion is decreased, while platelet consumption increases [105]. The exact mechanism of increased platelet consumption on hydrogel-grafted surfaces is not understood [106]. Cholakis *et al.* [104] concluded that there was no correlation between the improved blood compatibilities of heparin-PVA surface obtained from *in vitro* and *ex vivo* tests.

Polyurethanes with covalently grafted heparin were synthesized and their blood compatibility reported [39]. Heparin linked with alkyl groups was reacted with the isocyanate-derivatized polyurethanes. Blood compatibility of heparin-grafted polyurethanes was much better compared to physically adsorbed or ionically bound heparin. Larsson *et al.* [90–92] immobilized heparin on polyethylene imine (PEI)-grafted polyethylene (PE) tubing. Heparin was immobilized either by end-point or by multi-point attachment procedure. Platelet adhesion was almost negligible when heparin was attached through end-point attachment.

In order to increase the flexibility of heparin, it was attached to surfaces through a hydrophilic PEO spacer. Grainger *et al.* [95, 96] prepared poly(dimethylsiloxane)–poly(ethylene oxide)–heparin (PDMS–PEO–heparin) block copolymers which were

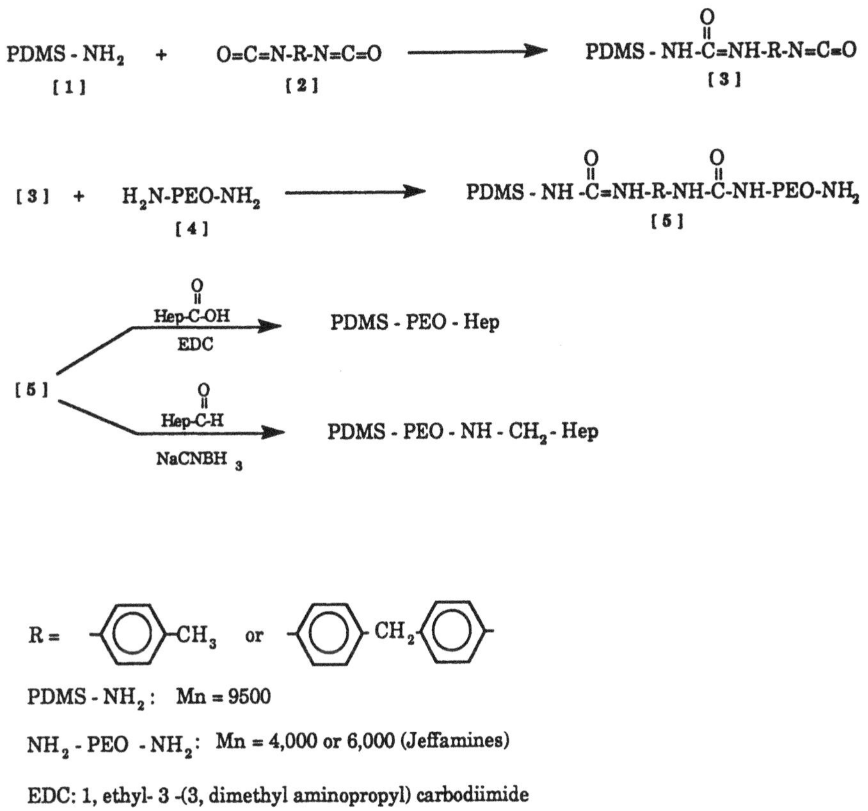
PDMS - NH₂ + O=C=N-R-N=C=O → PDMS - NH-C=NH-R-N=C=O
[1] [2] [3]
[3] + H₂N-PEO-NH₂ → PDMS - NH -C=NH-R-NH-C-NH-PEO-NH₂
[4] [5]
[5]
Hep-C-OH
EDC
PDMS - PEO - Hep
Hep-C-H
NaCNBH 3
PDMS - PEO - NH - CH₂ - Hep
R =
CH₃
or
CH₂
PDMS - NH₂ : Mn = 9500
NH₂ - PEO - NH₂: Mn = 4,000 or 6,000 (Jeffamines)
EDC: 1, ethyl- 3 -(3, dimethyl aminopropyl) carbodiimide

Figure 8. Covalent coupling of heparin to poly(dimethylsiloxane) (PDMS) through PEO spacer group (from ref. 95).

adsorbed onto hydrophobic surfaces. Figure 8 shows the reaction scheme for the synthesis of PDMS–PEO–heparin. Platelet adhesion and activation actually increased on surfaces treated with PDMS–PEO–heparin compared to PDMS alone. This is probably due to poor grafting efficiency of heparin to PDMS. Park *et al.* [97] synthesized segmented polyurethaneurea–PEO–heparin (Biomer®–PEO–Hep) copolymers using three different molecular weights of PEO. PEO was grafted to Biomer® through free isocyanate linkage and heparin was reacted with isocyanate-derivatized PEO. Figure 9 illustrates the reaction scheme for synthesis of Biomer®–PEO–Hep. Although the amount of heparin immobilized was highest with PEO-1000 ($0.47\ \mu g/cm^2$), the heparin bioactivity ratio was highest with PEO-3400 (22.17%).

Copolymers containing tertiary amine groups which can be protonated at the physiological pH such as *N*,*N*′-dimethylaminoethyl or *N*,*N*′-diethylaminoethyl have been synthesized [107]. This approach avoids the toxicity of using quaternary ammonium salts; however, heparin is still bound ionically to the surface. Ferruti and co-workers [108–111] have synthesized poly(amido-amine) (PAA) which can be grafted to various biomaterial surfaces. PAA forms a strong complex with heparin prior to blood–polymer contact [109]. Figure 10 illustrates the reaction scheme for grafting PAA to polyurethane surface. Studies with PAA-grafted surfaces have

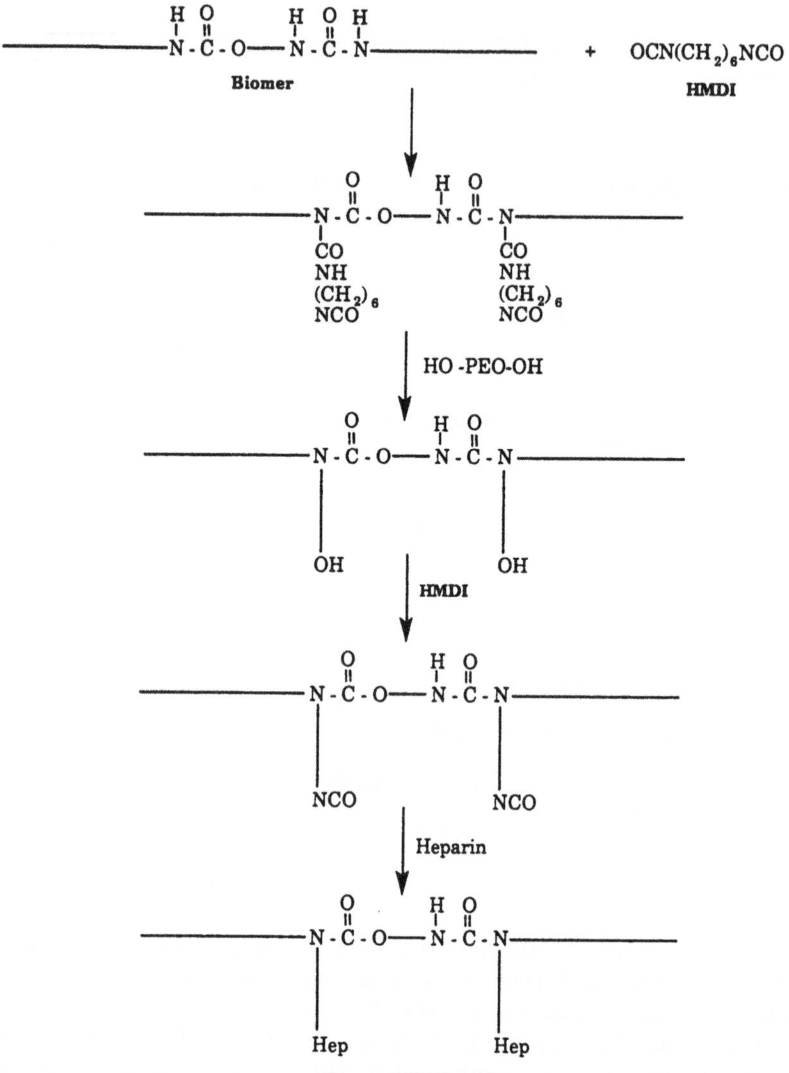
Biomer
OCN(CH2)6NCO
HMDI
HO-PEO-OH
HMDI
Heparin
Hep
Hep
Biomer-PEO-Heparin

Figure 9. Covalent coupling of heparin to Biomer® through PEO spacer group (from ref. 97).

shown tight heparin binding. Upon long-term exposure to blood, even tightly bound heparin can be displaced by ion-exchange mechanism with endogenous cations and proteins.

CONCLUDING REMARKS

This paper reviewed the effectiveness of PEO, albumin, and heparin in reducing the surface-induced thrombosis in terms of the steric repulsion mechanism. PEO, albumin, and heparin are effective in preventing surface-induced thrombosis only when they meet the three requirements: (1) tight binding to the surface; (2) complete surface coverage; and (3) flexibility of the grafted polymers. The tight binding of the hydrophilic polymers to the surface is important, since the polymers can be

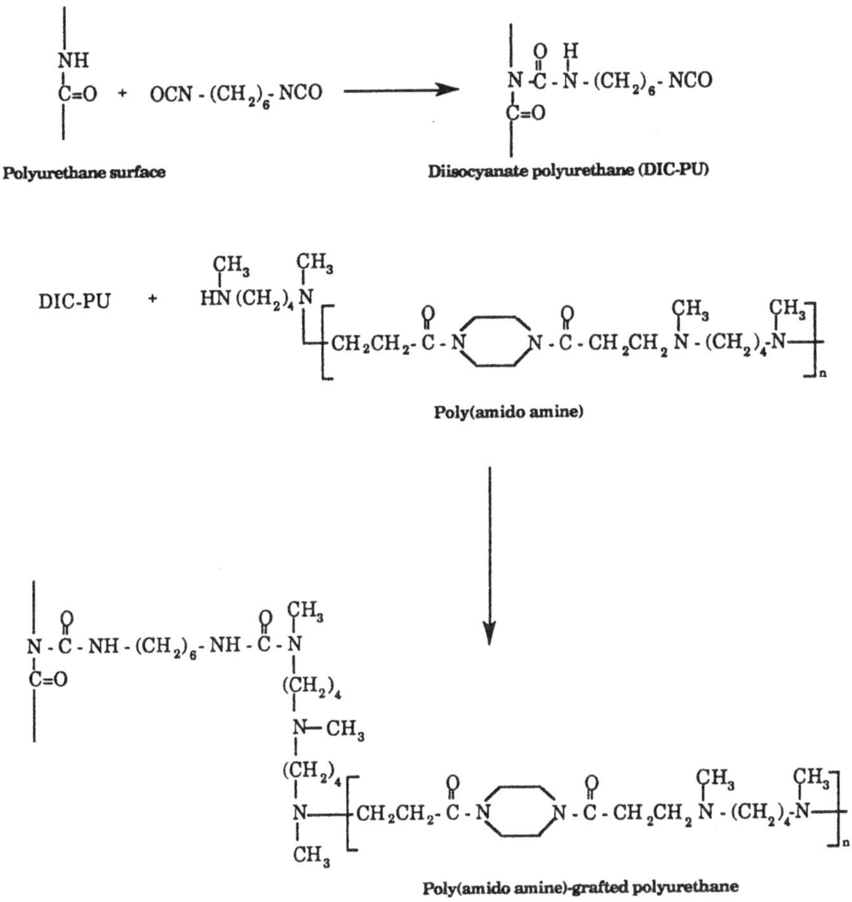
NH
C=O + OCN - (CH2)6 - NCO
Polyurethane surface
N - C - N - (CH2)6 - NCO
C=O
Diisocyanate polyurethane (DIC-PU)
DIC-PU +
Poly(amido amine)
Poly(amido amine)-grafted polyurethane

Figure 10. Reaction scheme for grafting poly(amido amide) to polyurethane (from ref. 111).

otherwise easily displaced from the surface by proteins and cells in blood. Thus, covalent grafting to the surface is most desirable. The grafted polymers have to cover the surface in such a way that no significant bare surface sites are exposed to blood proteins and cells. The beneficial effect of PEO, albumin, and heparin will not be observed if the surface coverage is not complete. The grafted hydrophilic polymers should remain flexible on the surface. Otherwise, it will simply create another surface on which protein adsorption and cell adhesion can occur. The steric repulsion mechanism can explain many of the observed phenomena described in the literature. The concept of the steric repulsion mechanism should be useful in the design of improved biomaterials.

Acknowledgements

This work was supported by the National Heart, Lung, and Blood Institute of the National Institute of Health through grant HL 39081.

REFERENCES

1. A. S. Hoffman, in: *Polymeric Materials and Artificial Organs. ACS Symposium Series*, Vol 256, pp. 13–29, C. G. Gebelein (Ed.). American Chemical Society, Washington, DC (1984).
2. E. P. Goldberg and A. Nakajima (Eds), *Biomedical Polymers, Polymeric Materials and Pharmaceuticals for Biomedical Use*. Academic Press, New York, NY (1980).
3. A. S. Hoffman, in: *Biomaterials: Interfacial Phenomena and Applications*, Vol 199, pp. 3–8, S. L. Copper and N. A. Peppas (Eds). American Chemical Society, Washington, DC (1982).
4. J. D. Andrade, S. Nagaoka, S. L. Cooper, T. Okano and S. W. Kim, *Trans. Am. Soc. Artif. Intern. Organs* **33**, 75–84 (1987).
5. J. D. Andrade and V. Hlady, *Adv. Polymer Sci.* **79**, 1–63 (1986).
6. J. M. Anderson and K. Kottke-Marchant, *CRC Crit. Revs. Biocomp.* **1**, 111–204 (1985).
7. E. W. Salzman, J. Lindon, D. Brier and E. W. Merrill, *Ann. NY. Acad. Sci.* **283**, 114–127 (1977).
8. J. S. Schultz, S. M. Lindenauer and J. A. Penner, in: *Biomaterials: Interfacial Phenomena and Applications*, Vol 199, pp. 43–58, S. L. Copper and N. A. Peppas (Eds). American Chemical Society, Washington, DC (1982).
9. A. S. Hoffman, *Ann. NY. Acad. Sci.* **516**, 96–101 (1987).
10. Y. Ikada, *Adv. Polymer Sci.* **57**, 103–140 (1984).
11. B. D. Ratner and A. S. Hoffman, in: *Synthetic Biomedical Polymers: Concepts and Applications*, pp. 133–151, M. Szycher and W. J. Robinson (Eds). Technomic Publishing Co., Westport, CT (1980).
12. F. Th. Hesselink, A. Vrij and J. Th. G. Overbeek, *J. Phys. Chem.* **75**, 2094–2103 (1971).
13. D. H. Napper, *J. Colloid Interface Sci.* **58**, 390–407 (1977).
14. D. H. Napper, in: *Colloidal Dispersions*, pp. 99–128, J. W. Goodwin (Ed.). The Royal Society of Chemistry, London, UK (1981).
15. F. E. Bailey and J. V. Koleske, *Poly(ethylene oxide)*. Academic Press, New York, NY (1976).
16. P. Molyneux, *Water-Soluble Synthetic Polymers: Properties and Behavior*, Vol I, Chapter 2, pp. 19–72. CRC Press, Inc. Boca Raton, FL (1983).
17. R. Kjellander and E. Florin, *J. Chem. Soc., Faraday Trans.* **77**, 2053–2077 (1981).
18. F. E. Bailey and J. V. Koleske, in: *Non-ionic Surfactants: Physical Chemistry*, pp. 927–969, M. J. Schick (Ed.). Marcel Dekker, Inc. New York, NY (1987).
19. M. Bjorling, G. Karlstrom and P. Linse, *J. Phys. Chem.* **95**, 6706–6709 (1991).
20. S. T. Milner, *Science* **251**, 905–914 (1991).
21. S. Nagaoka, Y. Mori, T. Tanzawa, Y. Kikuchi, F. Inagaki, Y. Yokota and Y. Nioshiki, *Trans. Am. Soc. Artif. Intern. Organs* **33**, 76–78 (1987).
22. P. F. Luckham, in: *Polymer Surfaces and Interfaces*, pp. 55–73, W. J. Feast and H. S. Munro (Eds). John Wiley and Sons, New York, NY (1987).
23. P. F. Luckham, *Adv. Colloid Interface Sci.* **34**, 191–215 (1991).
24. S. I. Jeon, J. H. Lee, J. D. Andrade and P. G. DeGennes, *J. Colloid Interface Sci.* **142**, 149–158 (1991).
25. S. I. Jeon and J. D. Andrade, *J. Colloid Interface Sci.* **142**, 159–166 (1991).
26. C. W. Hiatt, A. Shelokov, E. J. Rosenthal and J. M. Galimore, *J. Chromatography* **56**, 362–364 (1971).
27. G. L. Hawk, J. A. Cameron and L. B. Dufault, *Prep. Biochem.* **2**, 193–203 (1972).
28. T. Kato, K. nakamura, M. Kawaguchi and A. Takahashi, *Polymer J.* **13**, 1037–1043 (1981).
29. J. H. Lee, J. Kopecek and J. D. Andrade, *J. Biomed. Mater. Res.* **23**, 351–368 (1989).
30. J. H. Lee, P. Kopeckova, J. Kopecek and J. D. Andrade, *Biomaterials* **11**, 455–464 (1990).
31. J. Lee, P. A. Martic and J. S. Tan, *J. Colloid Interface Sci.* **131**, 252–266 (1989).
32. N. F. Owens, D. Gingell and P. R. Rutter, *J. Cell Sci.* **87**, 667–675 (1987).
33. M. Amiji, H. Park and K. Park, *Trans. Soc. for Biomaterials* **14**, 41 (1991).
34. M. Amiji and K. Park, *Biomaterials* **13**, 682–692 (1992).
35. N. P. Desai and J. A. Hubbell, *Biomaterials* **12**, 144–153 (1991).
36. E. Ruckenstein and D. B. Chung, *J. Colloid Interface Sci.* **123**, 170–185 (1988).
37. J. H. Chen and E. Ruckenstein, *J. Colloid Interface Sci.* **142**, 545–553 (1991).
38. E. W. Merrill, E. W. Salzman, S. Wan, N. Mahmud, L. Kushner, J. N. Lindon and J. Curme, *Trans. Am. Soc. Artif. Intern. Organs* **28**, 482–487 (1982).
39. Y. Ito and Y. Imanishi, *CRC Crit. Revs. Biocomp.* **5**, 45–104 (1989).

40. J. Yu, S. Sundaram, D. Weng, J. M. Courtney, C. R. Moran and N. B. Graham, *Biomaterials* **12**, 119-120 (1991).
41. E. Brinkman, A. Poot, T. Beugeling, L. Van der Does and A. Bantjes, *Int. J. Artif. Organs* **12**, 390-394 (1989).
42. E. Brinkman, A. Poot, L. Van der Does and A. Bantjes, *Biomaterials* **11**, 200-205 (1990).
43. E. L. Chaikov, E. W. Merrill, J. E. Coleman, K. Ramberg, R. J. Connolly and A. D. Callow, *AIChE J.* **36**, 994-1002 (1990).
44. K. Allmer, J. Hilborn, P. H. Larsson, A. Hult and B. Ranby, *J. Polym. Sci. Part A: Polym. Chem.* **28**, 173-183 (1990).
45. T. Akizawa, K. Kino, S. Koshikawa, Y. Ikada, M. Yamashita and K. Imamura, *Trans. Am. Soc. Artif. Intern. Organs* **35**, 333-335 (1989).
46. N. Desai and J. A. Hubbell, *J. Biomed. Mater. Res.* **25**, 829-843 (1991).
47. W. R. Gombotz, W. Guanghui, T. A. Horbett and A. S. Hoffman, *J. Biomed. Mater. Res.* **25**, 1547-1562 (1991).
48. Y. Mori and S. Nagaoka, *Trans. Am. Soc. Artif. Intern. Organs* **28** ,459-463 (1982).
49. Y. C. Tseng and K. Park, *J. Biomed. Mater. Res.* **26**, 373-391 (1992).
50. M. S. Sheu and A. S. Hoffman, *Polymer Preprints* **32**, 239-240 (1991).
51. M. Amiji and K. Park, *J. Colloid Interface Sci.* (in press).
52. Y. H. Sun, W. R. Gombotz and A. S. Hoffman, *J. Bioactive Compatible Polym.* **1**, 316-334 (1986).
53. W. F. Ganong, *Review of Medical Physiology*, 11th Edn, pp. 430. Lange Medical Publications, Los Altos, CA (1983).
54. T. Peters Jr., *Adv. Protein. Chem.* **37**, 161-245 (1985).
55. M. Meloun, L. Moravek and V. Lostka, *FEBS Lett.* **58**, 134-137 (1975).
56. Peters, T. Jr., in: *The Plasma Proteins*, Vol I, pp. 133-181, F. W. Putnam (Ed.). Academic Press, New York, NY (1975).
57. K. Park, D. F. Mosher and S. L. Cooper, *J. Biomed. Mater. Res.* **20**, 589-612 (1986).
58. S. L. Goodman, S. L. Cooper and R. M. Albrecht, *J. Biomater. Sci. Polym. Edn.* **2**, 147-159 (1991).
59. K. Park and H. Park, *Scanning Microscopy* **Supp 3**, 137-146 (1989).
60. S. H. Lee and E. Ruckenstein, *J. Colloid Interface Sci.* **125**, 365-379 (1988).
61. E. Brynda, M. Houska, Z. Pokorna, N. A. Cepalova, Y. V. Moiseev and J. Kalal, *J. Bioeng.* **2**, 411-418 (1978).
62. M. Amiji, H. Park and K. Park, *J. Biomater. Sci. Polym. Edn.* **3**, 375-388 (1992).
63. K. Park, F. W. Mao and H. Park, *Biomaterials* **11**, 24-31 (1990).
64. B. W. Morrissey, *Ann. NY. Acad. Sci.* **283**, 50-64 (1977).
65. E. Blomberg, P. M. Claesson and C. G. Golander, *J. Disp. Sci. Tech.* **12**, 179-200 (1991).
66. K. Park, *Scanning Microsc.* **Supp 3**, 15-25 (1989).
67. A. van der Scheer, M. A. Tanke and C. A. Smolders, *Chem. Soc. Faraday Discussion* **65**, 264-287 (1978).
68. H. Tamai, A. Fujii and T. Suzawa, *J. Colloid Interface Sci.* **118**, 176-181 (1987).
69. T. M. S. Chang, *Can. J. Physiol. Pharmacol.* **47**, 10-43 (1969).
70. T. M. S. Chang, *Can. J. Phys. Pharmacol.* **52**, 275-285 (1974).
71. J. N. Mulvihill, A. Faradji, F. Oberling and J. P. Cazenave, *J. Biomed. Mater. Res.* **24**, 155-163 (1990).
72. M. F. Sigot-Luizard, D. Domurado, M. Sigot, R. Guidoin, C. Gosselin, M. Marois, J. F. Girard, M. King and B. Badour, *J. Biomed. Mater. Res.* **18**, 895-909 (1984).
73. A. S. Hoffman, G. Schmer, C. Harris and W. G. Kraft, *Trans. Am. Soc. Artif. Intern. Organs* **18**, 10-17 (1972).
74. C. P. Sharma and G. Kurian, *J. Colloid Interface Sci.* **97**, 39-41 (1984).
75. T. Matsuda and K. Inoue, *Trans. Am. Soc. Artif. Intern. Organs* **36**, 161-164 (1990).
76. R. C. Eberhart, M. S. Munro, J. R. Frautschi and V. I. Sevastianov, in: *Proteins at Interfaces: Physicochemical and Biochemical Studies*, pp. 378-400, J. L. Brash and T. A. Horbett (Eds). American Chemical Society, Washington, DC (1987).
77. R. C. Eberhart, M. S. Munro, J. R. Frautschi, M. Lubin, F. J. Clubb Jr., C. W. Miller and V. I. Sevastianov, *Ann. NY. Acad. Sci.* **516**, 78-95 (1987).
78. M. S. Munro, A. J. Quattrone, S. R. Ellsworth, P. Kulkarni and R. C. Eberhart, *Trans. Am. Soc. Artif. Intern. Organs* **27**, 499-503 (1981).
79. T. G. Grasel, J. A. Pierce and S. L. Cooper, *J. Biomed. Mater. Res.* **21**, 815-842 (1987).

80. W. G. Pitt and S. L. Cooper, *J. Biomed. Mater. Res.* **22**, 359-382 (1988).
81. L. B. Jaques, *Trends Pharmacol. Sci.* **3**, 289-291 (1982).
82. I. Bjork and U. Lindahl, *Mol. Cell. Biochem.* **48**, 181-182 (1982).
83. R. A. O'Reilly, in: *The Pharmacological Basis of Therapeutics*, 7th Edn, pp. 1338–1359, A. G. Gilman, A. S. Goodman, T. W. Rall and F. Murad (Eds). Macmillan Publishing Co., New York, NY (1985).
84. H. A. Scheraga and M. Laskowski Jr., *Adv. Protein Chem.* **12**, 1-131 (1957).
85. V. L. Gott, J. D. Whiffen and R. C. Dutton, *Science* **142**, 1297-1298 (1963).
86. S. W. Kim and J. Feijen, *CRC Crit. Revs. Biocomp.* **1**, 229-260 (1985).
87. M. V. Sefton, C. H. Cholakis and G. Llanos, in: *Blood Compatibility*, Vol I, pp. 151–198, D. F. Williams (Ed.). CRC Press, Inc. Boca Raton, FL (1987).
88. Y. Idezuki, H. Watanabe, M. Hawgira, K. kanasygi, Y. Mori, S. Nagaoka, M. Hagio, K. Yamamoto and H. Tanzawa, *Trans. Am. Soc. Artif. Intern. Organs* **21**, 436–438 (1975).
89. C. D. Ebert and S. W. Kim, *Thromb. Res.* **26**, 43-57 (1982).
90. R. Larsson, O. Larm and P. Olsson, *Ann. NY. Acad. Sci.* **516**, 102–115 (1987).
91. R. Larsson, P. Olsson and U. Lindahl, *Thromb. Res.* **19**, 43-54 (1980).
92. O. Larm, R. Larsson and P. Olsson, *Biomat. Med. Devices Artif. Organs* **11**, 161-174 (1983).
93. W. E. Hennink, S. W. Kim and J. Feijen, *J. Biomed. Mater. Res.* **18**, 911-926 (1984).
94. W. E. Hennink, C. D. Ebert, S. W. Kim, W. Breemhaar, A. Bantjes and J. Feijen, *Biomaterials* **5**, 264-268 (1984).
95. D. W. Grainger, S. W. Kim and J. Feijen, *J. Biomed. Mater. Res.* **22**, 231–249 (1988).
96. D. W. Grainger, K. Knutson, S. W. Kim and J. Feijen, *J. Biomed. Mater. Res.* **24**, 403-431 (1990).
97. K. D. park, A. Z. Piao, H. Jacobs, T. Okano and S. W. Kim, *J. Polym. Sci. Part A: Polym. Chem.* **29**, 1725-1737 (1991).
98. J. C. Eriksson, G. Gillberg and H. Lagergren, *J. Biomed. Mater. Res.* **1**, 301-312 (1967).
99. P. Olsson, H. Lagergren, H. Larsson and K. Radegan, *Thromb. Hemostas.* **37**, 274–282 (1977).
100. M. V. Sefton, in: *Hydrogels in Medicine and Pharmacy, Vol III. Properties and Applications*, pp. 17-51, N. A. Peppas (Ed.). CRC Press, Inc., Boca Raton, FL (1987).
101. E. W. Merrill, E. W. Salzman, P. S. L. Wong, T. P. Ashford, A. H. Brown and W. G. Austen, *J. Appl. Physiol.* **29**, 723-730 (1970).
102. N. A. Peppas and E. W. Merrill, *J. Biomed. Mater. Res.* **11**, 423-434 (1977).
103. C. H. Cholakis and M. V. Sefton, *J. Biomed. Mater. Res.* **23**, 399-415 (1989).
104. C. H. Cholakis, W. Zingg and M. V. Sefton, *J. Biomed. Mater. Res.* **23**, 417-441 (1989).
105. S. R. Hanson, L. A. Harker, B. D. Ratner and A. S. Hoffman, *Ann. Biomed. Eng.* **7**, 357–367 (1979).
106. W. Zingg and A. W. Neumann, *Biofouling* **4**, 293-299 (1991).
107. H. Tanzawa, Y. Mori, N. Harumiya, H. Miyama, M. Hori, N. Ohshima and Y. Idezuki, *Trans. Am. Soc. Artif. Intern. Organs* **19**, 188-194 (1973).
108. P. Feruti and R. Ranucci, *Polymer J.* **23**, 541-550 (1991).
109. E. Ranucci and P. Ferruti, *macromolecules* **24**, 3747-3752 (1991).
110. P. Ferruti, G. Casini, F. Tempesti, R. Barbucci, R. Mastacchi and M. Sarret, *Biomaterials* **5**, 234-236 (1984).
111. R. Barbucci, M. Benvenuti, P. Ferrruti and M. Nocentini, in: *Advances in Biomedical Polymers*, pp. 259-276, G. C. Gebelien (Ed.). Plenum Press, New York, NY (1985).

Antiproliferative capacity of synthetic dextrans on smooth muscle cell growth: the model of derivatized dextrans as heparin-like polymers

DIDIER LETOURNEUR, D. LOGEART, T. AVRAMOGLOU and JACQUELINE JOZEFONVICZ

Laboratoire de Recherches sur les Macromolécules, CNRS URA 502, Université Paris-Nord, 93430 Villetaneuse, France

Received 17 July 1992; accepted 18 September 1992

Abstract—Proliferation of vascular smooth muscle cells (SMC) is postulated to be a key step in the pathogenesis of atherosclerosis or restenosis after vascular interventions such as angioplasty. Natural glycosaminoglycans, such as heparin and heparan sulfate, are known for their ability to inhibit SMC proliferation *in vivo* and *in vitro*. The antiproliferative activity of synthetic derivatized dextrans exhibiting heparin-like anticoagulant and anticomplement capacities have been investigated with rat aorta smooth muscle cells in culture.

We report here that some derivatized dextrans grafted with benzylamide sulfonate moieties are potent antiproliferative agents for rat smooth muscle cell (SMC) *in vitro*. These synthetic polymers inhibit the SMC proliferation as well as heparin. The SMC growth inhibition is dose dependent, reversible and non-toxic. Highly anionic carboxylic dextrans are not capable of inhibiting the SMC growth, excluding a simple charge effect mechanism. Using fluorescent (DTAF) probes, we demonstrated that the synthetic antiproliferative polymers and heparin are internalized into the SMC. No binding or internalization was observed with native dextran devoid of antiproliferative capacity. We conclude that a suitable distribution of functional groups on the dextran backbone can simulate heparin activity in terms of antiproliferative capacity on SMC growth.

Key words: Derivatized dextrans; smooth muscle cells; growth inhibition; antiproliferative; heparin-like.

INTRODUCTION

Biologically active natural products utilized in an increasing number of biological and medical applications are often extracted from animal and human tissues or fluids such as blood. There are numerous problems associated with the preparation and utilization of these natural products. These problems include reproducibility from batch to batch, contaminant-free (e.g., HIV) preparations, and the availability of products in quantities large enough for industrial scale preparation. In addition, the poor stability of natural products and their high costs are also critical points in the extensive uses of these products. To avoid these drawbacks, studies have focused on synthetic products for biomedical applications. In theory, these synthetic products should exhibit similar biological activities of their natural analogs and also limit side-effects. With the data available on chemical groups or sequences within natural products that confer biological properties, the first step in the research and development of synthetic products is the synthesis of analogs. This approach allows the study of the precise role of some substituents such as carboxyl groups, phosphate groups, or sulfate groups [1–3]. Because these functional groups are usually disposed on complex macromolecules, synthetic products could be polymers with suitable distributions of these functional groups.

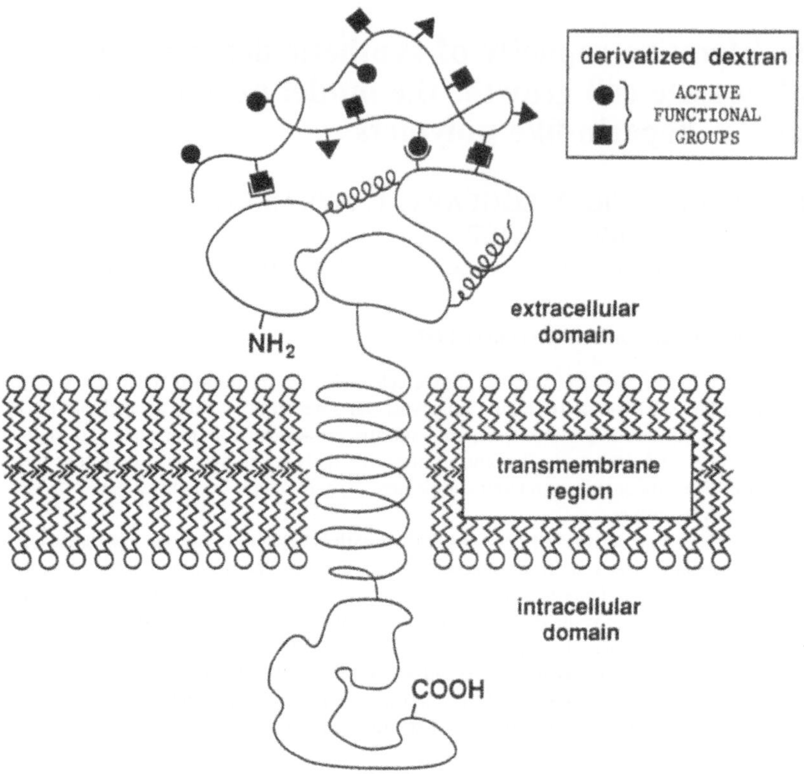
derivatized dextran
ACTIVE
FUNCTIONAL
GROUPS
extracellular
domain
NH_2
transmembrane
region
intracellular
domain
COOH

Figure 1. Macromolecular chains of biospecific polymers allow the creation of 'keys' matching to protein domains.

To investigate the structure–activity relationships between synthetic polymers and biologically active natural products, we have undertaken studies directed toward the hypothesis that a random distribution of particular chemical groups on polymer backbones can confer specific interactions [1]. Random substitution of macromolecular chains with a suitable flexibility would permit the creation of specific 'keys' which could match protein domains (Fig. 1).

In this work, we have studied the ability of a family of synthetic polysaccharides to inhibit smooth muscle cell proliferation. This family of synthetic polysaccharides is modeled after heparin, a natural glycosaminoglycan mainly used as an anticoagulant in clinical practice. However, the target biological activity in these studies is the well-known ability of heparin to inhibit SMC proliferation *in vivo* and *in vitro* [4–6].

Previous efforts in our laboratory have been directed toward the synthesis from water-soluble dextran, a poly(α, 1-6 glucose) [2]. On these macromolecular chains, a statistical distribution of chemical groups gave derivatized dextrans, heparin-like properties in terms of anticoagulation [7] and anti-complement activating properties [8]. Studies on the molecular weight of these polymers in relation to their biological activities have also been carried out [8].

Here, we have studied the anti-proliferative capacity of derivatized dextrans on smooth muscle cell (SMC) growth. Using fluorescent probes, we have investigated the internalization of the synthetic anti-proliferative polymers and heparin into the smooth muscle cells.

MATERIALS AND METHODS

Materials

Hog intestinal heparin (H 108) with specific anticoagulant activity of 173 IU/mg was a generous gift of Dr J. Choay (Sanofi-Recherche; Centre Choay, Gentilly, France). The chromatographic molecular weight was given by the manufacturer: 10 700 g/mol. Calciparine®, a heparin calcium salt, was commercially available by Sanofi. The two low molecular weight heparins (LMWH), Fraxiparine® (250 anti-Xa units/mg; $M = 4500$ g/mol) and Lovenox® (110 anti-Xa units/mg) were obtained from Sanofi and Pharmuka respectively. All chemicals used were analytical grade products purchased from Sigma (La Verpilliere, France).

Derivatized dextrans

Water-soluble dextran derivatives were prepared as previously described [2], from dextran T40 batch 32 202 ($M_W = 43\,900$ g/mol; $M_n = 26\,200$ g/mol) obtained from Pharmacia (St Quentin en Y., France). Carboxymethyl dextrans were synthesized from native dextran (D), by substituting glucosyl units with carboxymethyl groups (CM). In a second step, benzylamine was coupled to carboxylic groups to form benzylamide units (B). Finally, benzylamide aromatic rings were sulfonated (S). The chemical structure of these synthetic polysaccharides is shown on Fig. 2. Changes in chemical conditions resulted in the formation of various substituted macromolecules from native dextran. Thus, native dextran (T40), three carboxymethyl dextrans (CMD) and one carboxymethylbenzylamide sulfonated dextran named E9 were tested in cell culture. All the sodium salts of the derivatives were previously ultrafiltrated on a YM 2 Diaflo membrane (1000 M_W cut-off; Amicon Co., Danvers, MA, USA) and lyophilized.

Carboxymethyl dextran sulfate (CMD ~ S) was synthesized by direct reaction of chlorosulfonic acid on carboxymethyl dextran. It was characterized and purified as described above. The chemical compositions are characterized by acidimetric titrations and elemental analysis. They are reported in Table 1.

Fluorescent probes

Polysaccharides were labeled with 5-([4,6-dichlorotriazin-2-yl]amino) fluorescein (Sigma) as described [9, 10]. Briefly, the DTAF solutions (100 mg into 15 ml of 0.1 M borate buffer; pH = 9) were slowly added to heparin, T40 and E9 solutions (300 mg into 15 ml of borate buffer). After 18 h at room temperature, under constant stirring, a concentrated HCl solution was added until the pH reached 7. In order to remove all the free-DTAF, the solutions were extensively ultrafiltrated on YM2 membrane and then lyophilized. As a control, native dextran-T40 labeled with Rhodamine from Sigma was also used.

Cell culture

Eagle's Minimal Essential Medium (MEM) was purchased from Gibco BRL (Cergy Pontoise, France). Penicillin, streptomycin, L-glutamine and trypsin-EDTA were

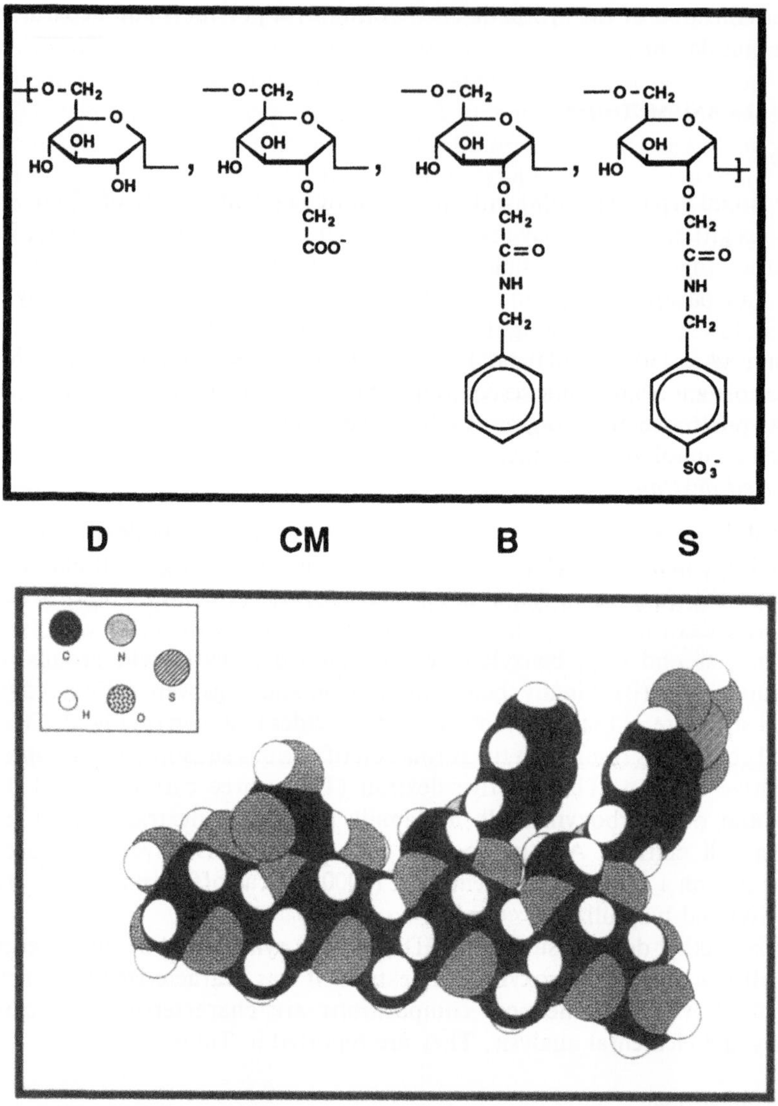
D
CM
B
S
C
N
S
H
O

Figure 2. Structure and space-filling representations of water-soluble derivatized dextrans. The conformations shown maximize the two-dimensional area but are not necessarily the most stable geometries. D: dextran unit; CM: carboxymethyl unit; B: benzylamide unit; S: sulfonate unit.

from Flow Laboratories (Les Ulis, France), collagenase from Seromed (Berlin, Germany). New Born Calf Serum (NBCS) was obtained from Eurobio (Les Ulis, France). Twenty-four well tissue culture dishes from Costar were used. Platelet-poor human citrated plasma was obtained from several blood healthy donors in our laboratory.

Thoracic aorta smooth muscle cells from male Sprague-Dawley rats (strain OFA: Iffa-Credo, L'arbresle, France) were isolated as previously described [11, 12]. Cells were cultured at 37°C in a humidified 5% CO_2 atmosphere. Cells between the second and fourth passage were used in this work. Confluent cells appeared to

Table 1.
Chemical characterization of polymers in % of glucosidic units. Their specific anticoagulant activities are expressed in International Units per mg. D: dextran unit; CM: carboxymethyl unit; B: benzylamide unit; S: sulfonate unit; TT: thrombin time; KCT: kaolin cloting time; T40: native dextran; CMD: carboxymethyl dextran; CMD ~ S: carboxymethyl dextran sulfate; E9: carboxymethyl benzylamide sulfonated dextran; H108: Heparin; n.d.: not determined.

	Composition (%)				Anticoagulant activity (IU/mg)	
Polymers	D	CM	B	S	TT	KCT
T40	100	0	0	0	0	0
CMD1	22	78	0	0	0	0
CMD2	20	80	0	0	n.d.	n.d.
CMD3	0	107	0	0	n.d.	n.d.
CMD ~ S	54	33	0	13[a]	n.d.	n.d.
E9	0	58	19	26	0.29	2.1
H108	—	—	—	—	173	173

[a] Sulfate instead of sulfonate.

present the characteristic 'hills and valleys' pattern. They are also identified as SMC by the presence of numerous vesicles near the plasma membrane. Indirect immunofluorescence revealed specific alpha actin filaments. Absence of mycoplasma contamination was routinely checked using fluorescent hoechst 33258 dye [12].

Growth inhibition studies

1.2×10^4 SMC were plated into 16 mm multiwell plates in MEM containing 10% NBCS. After 24 h, they were growth arrested by placing them in medium with 0.1% NBCS for 48 h (except when specified). The control cells were exposed to medium with 10% NBCS. Treated-cells were exposed to MEM with 10% NBCS containing heparin or heparin-like solutions. Cell numbers were measured in quadruplicate samples and counted with a Coulter counter. The degree of inhibition was determined after 5 days as described by Castellot *et al.* [13] according to the relationship:

$$I\% = \left(1 - \frac{\text{net growth with polysaccharide}}{\text{net growth in controls}}\right) \times 100\%$$

Internalization of DTAF-compounds

The binding and internalization of DTAF-polysaccharides were studied as follows: a suspension of 25 000 cells/ml in 10% NBCS was plated on 2 ml cell culture chamber containing a glass slide (Lab-tek®, Nunc). After 24 h, the cells were growth arrested by placing them in 0.1% NBCS for 48 h. The growth arrested SMC were then exposed to 10% NBCS containing DTAF-polysaccharides (100 μg/ml) for the desired incubation times at 37°C. The cells were washed three times with PBS and then fixed in 3.7% formaldehyde solution for 15 min. The cell chamber was removed and a drop of 50% glycerol–50% PBS solution was added to the slide. Glass coverslips were inverted on the preparations which were then observed under a Leitz epifluorescent microscope equipped with H3 filter.

Coagulation and complement assays

Anticoagulant activities of derivatized dextrans and heparin were assessed by measuring the thrombin clotting time (TT) and the activated-cephalin kaolin clotting time (KCT), recorded automatically at 37°C with a Dade KC1 coagulometer (Dade, Cergy-Pontoise, France). The thrombin clotting time was assessed using 100 μl of polysaccharide solutions in Michaelis buffer, incubated for 2 min at 37°C with 200 μl of human citrated plateled-poor plasma (PPP). Then 100 μl of thrombin (Roche) solution was added at 10 NIHu/ml and the clotting times were measured. KCT determinations were carried out by incubating at 37°C for 5 min of 100 μl of PPP with 100 μl of cephalin-kaolin suspension (Stago, Asnieres, France) and 100 μl of polysaccharides solutions. The kaolin clotting time started with the addition of 100 μl of 0.025 M $CaCl_2$.

The specific anticoagulant activities of polysaccharides were reported in Table 1. They were expressed in IU/mg as previously described [7, 8] by comparison with 173 IU/mg of H108-heparin.

The anticomplement activities of derivatized dextrans and heparin were carried out as previously described [7].

RESULTS

Anticoagulant and anticomplement activities of derivatized dextrans

Heparin, a glycosaminoglycan composed of repeating glucosamine and uronic acid sugar residues, is a potent anticoagulant [6]. Previous work in our laboratory has demonstrated that the grafting of carboxylic and benzylamide sulfonate units confered to the dextran anticoagulant activities [2, 7]. Structure-function experiments indicated that the anticoagulant capacity required at least 45% of glucosyl units bearing carboxylic groups. Moreover, the anticoagulant activity increases with the global amount of sulfonated units [7]. Results were obtained from dextran of 10 000 g/mol [2, 7] and 40 000 g/mol [8]. These works demonstrated that the anticoagulant activity is related to the molecular weight of the derivatized dextrans. Finally, these synthetic polysaccharides exert their anticoagulant activity, as heparin, by enhancing factor IIa inhibition in the presence of natural inhibitor antithrombin III [1].

The E9 derivative potentiates, as do heparin and heparan sulfate [6], the inhibition of thrombin (factor IIa). As reported in Table 1, native dextran (T40) and the carboxymethyl dextrans have, under the same conditions, no anticoagulant capacity. The anticoagulant activity of E9 compound is low but significantly higher than dextran samples devoid of benzylamide sulfonate units. From these classical coagulation assays and previous efforts [1], we have concluded that the mechanism of these anticoagulant polysaccarides involves the specific binding of thrombin-antithrombin III to the functional polymer. The biospecific sequences on the macromolecular chains were obtained by random substitution with suitable carboxylic and sulfonate groups.

Our previous studies have shown that these synthetic products also possess high anticomplement activity [7]. In the human complement system, they could inhibit the formation of amplification C3 convertase [7, 8]. With derivatized dextrans, the anticomplement activities are in the same order of magnitude as heparin, while the anticoagulant capacity is low. The random distribution of functional groups along the macromolecular chains allows the creation of domains interacting specifically with proteins of the complement system [1]. Interestingly, the heparin anticoagulant activity also requires structural determinants distinct from those responsible for the anticomplement activity.

Taken together, these results provided valuable clues of the presence of heparin-like sequences on the derivatized dextran chains. In view of these results, we checked whether various derivatized dextrans are able to interact with SMC [12]. In addition, Castellot and coworkers demonstrated that non-anticoagulant heparins also inhibit the SMC growth [14]. Thus, derivatized dextran with very low anticoagulant capacity was studied for its effect on SMC growth.

Heparins inhibit the in vitro *SMC growth*

Heparin was previously described to inhibit *in vivo* [15, 16] and *in vitro* [17–20] SMC growth. The SMC inhibition by heparin is not restricted to some species but demonstrated on human and numerous animal SMC [21, 22]. Heparin exhibits specific growth inhibitory properties for SMC as compared to endothelial and fibroblastic cells [21]. It was also demonstrated that the growth inhibition by heparin was reversible and increases with the dose [22].

We have isolated vascular SMC from rat aorta by the explant method after partial digestion with collagenase [11, 12]. SMC cultures were characterized as described above. SMC growth in the presence of various heparins was studied after growth arrest for 24 h without serum.

In these dose-response experiments (Fig. 3), SMC growth inhibition was observed with all the samples. The IC_{50}, defined as the concentration required to yield 50% of inhibition, are in the 0.2–7 μg/ml range. These results are in good agreement with those published in the literature [21, 22].

In comparing heparins in terms of molar concentration, no significant differences were found with the high and low molecular weight molecules (Fig. 3). Finally, no correlation was observed between the inhibition rates and the anticoagulant activities (i.e. anti-Xa units). As previously described by Castellot and coworkers [21], our experiments also show that the inhibitory capacity and the anticoagulant activity are clearly separate. The results reported here permit us to definitively validate our SMC isolation. Because of the need of a positive control in the next experiments, we chose the H108 heparin, already studied for its anticoagulant and anticomplement activity.

Effect of growth arrested conditions

The growth conditions affect the SMC response to antiproliferative compounds. We tested the inhibitory capacity of H108 heparin in various growth arrested conditions. The cells were seeded for 24 h in 10% serum and then were growth arrested with low serum content (0–0.5%) for 1 or 2 days before adding heparin.

As shown in Fig. 4, the growth arrested conditions modulate the inhibitory capacity of the same heparin sample. The more drastic the growth arrested conditions, the more efficient the heparin was. An increase in the growth arrested time gave higher inhibitory results. A decrease of the serum content lead to the same effect. Important differences were seen with the different conditions and for instance with H108 heparin, the IC_{50} were from 4 to 400 μg/ml. A possible explanation is that

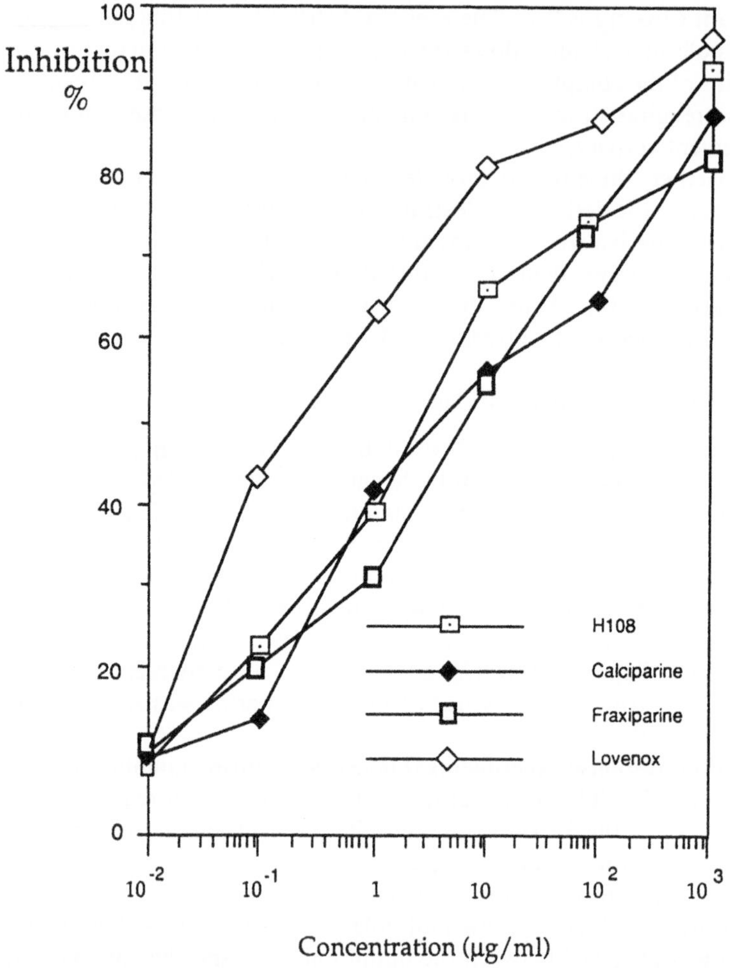
100
Inhibition
%
80
60
40
20
0
H108
Calciparine
Fraxiparine
Lovenox
10 -2
10 -1
1
10
10 2
10 3
Concentration (μg/ml)

Figure 3. Antiproliferative capacity of various heparins on SMC growth in culture. Growth arrested cells were released from G0 phase by addition of culture media containing 10% serum plus high molecular weight heparins (H108; Calciparine) or low molecular weight heparins (Fraxiparine: 250 anti-Xa units/mg; Lovenox: 110 anti-Xa units/mg). After 5 days the cell numbers were determined and the growth inhibition was calculated. Mean values are indicated; SD for all points was less than 5% of the mean value.

exponentially growing SMC have been described to be 50–100 times less sensitive to heparin than SMC which have been growth arrested [21].

In view of these results, we definitively chose one SMC inhibition protocol; all the following experiments were achieved by placing the SMC for 48 h in 0.1% FCS. Under these conditions, ^{3}H-thymidine uptake into DNA indicates that more than 90% of the cells were in G0/G1 phase.

Derivatized dextrans inhibit the SMC growth

Derivatized dextrans were assessed for *in vitro* SMC growth. In order to study the role of functional groups, carboxymethyl dextrans (CMD) and carboxymethyl-benzylamide sulfonated dextran were synthesized from native dextran (Table 1) and

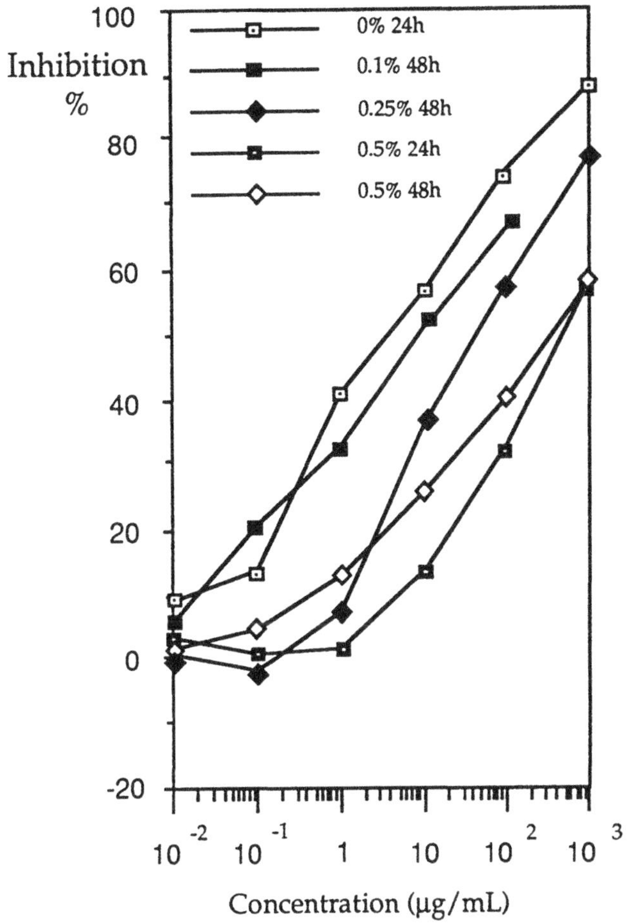
100
Inhibition
%
80
60
40
20
0
-20
0% 24h
0.1% 48h
0.25% 48h
0.5% 24h
0.5% 48h
10^-2 10^-1 1 10 10^2 10^3
Concentration (µg/mL)

Figure 4. Effect of growth arrested conditions. SMC were growth arrested by placing them in 0–0.5% of serum for the indicated times and then exposed to media containing 10% serum plus heparin for 5 days. Growth inhibition was determined by cell counts. Mean values are indicated; SD for all points was less than 5% of the mean value.

then assessed in cell culture. After 5 days in the presence of various polysaccharides, cell numbers were determined and the percentage inhibition was calculated. As shown in Fig. 5, E9 is a potent inhibitor of the SMC proliferation. The IC_{50} for E9 is of 20 μg/ml, i.e. 440 nM taking into account a molecular weight of 45 000 g/mol. Under the same conditions, the IC_{50} for H108 heparin is 9 μg/ml (840 nM). In comparison, native dextran had no effect on SMC growth even with doses as high as 1 mg/ml (Fig. 5). Moreover, when the same amount of T40 was added to the culture-wells containing E9, the native dextran did not change the inhibition values of E9 (data not shown).

To prove that the growth inhibitory effects observed with E9 and heparin are not due to direct cytotoxicity, classical trypan blue assays were performed. After 4 days in the presence of 1 mg/ml of polysaccharides, the higher dose used here, the viability of the SMC was always >95%. In another experiment, after incubation of SMC with E9 and heparin, the cells were rinsed with culture media without polysaccharides.

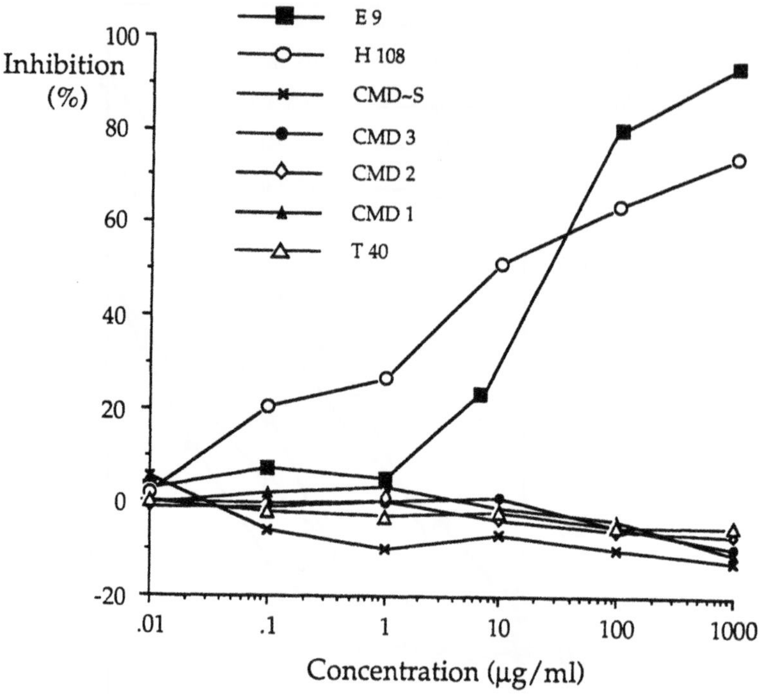
E 9
H 108
CMD~S
CMD 3
CMD 2
CMD 1
T 40
Inhibition (%)
100
80
60
40
20
0
-20
.01
.1
1
10
100
1000
Concentration (µg/ml)

Figure 5. SMC growth inhibition by derivatized dextrans and heparin. SMC were growth arrested and the antiproliferative activities were assessed as described in Materials and Methods. T40: native dextran; CMD: carboxymethyl dextran; CMD ~ S: carboxymethyl dextran sulfate; E9: carboxymethyl benzylamide sulfonated dextran; H108: Heparin. Mean values are indicated; SD for all points was less than 5% of the mean value.

The following of the growth curves during 4 days indicated for both compounds that the doubling times were the same as the control without polymers. The reversibility of the antiproliferative activity by rinsing the media is also a proof of the non-toxicity of the derivatized dextran.

To study the charge effect of the synthetic polymers, we checked various CMD bearing different percentages of carboxylic groups Even the highly substituted carboxymethyl dextran, which is more anionic than E9, had no effect on SMC growth (Fig. 5). In addition, a carboxymethyl dextran was directly sulfated (Table 1). We did not observe any effect on SMC growth with this compound grafted with carboxylic and sulfate units. As previously described, we confirmed that the inhibition effect is related to the benzylamide sulfonate content [12]. However, the precise structure-function relationship of the antiproliferative heparin-like polymers were not elucidated. Specific determinants with a given distribution of chemical groups were probably able to form sequences interacting with cells (Fig. 1).

Internalization of DTAF-antiproliferative polysaccharides

To understand the mechanism that controls the SMC growth inhibition and whether heparin-like molecules are related to the natural heparin, we have labeled both heparin and E9 with a fluorescein derivative (DTAF). To ensure that DTAF-polysaccharides retained their antiproliferative activities, we first tested them in the SMC inhibition assay. DTAF-E9 and DTAF-H108 heparin exhibited the same potency as unlabeled compounds. Therefore, these probes could be of use for studying the heparin and heparin-like processing by SMC.

Binding experiments were performed at 37°C by incubation DTAF-polysaccharides with SMC for different periods of time. The pattern of uptake is shown in Fig. 6. A diffuse fluorescence was first seen at short incubation times (Fig. 6A)

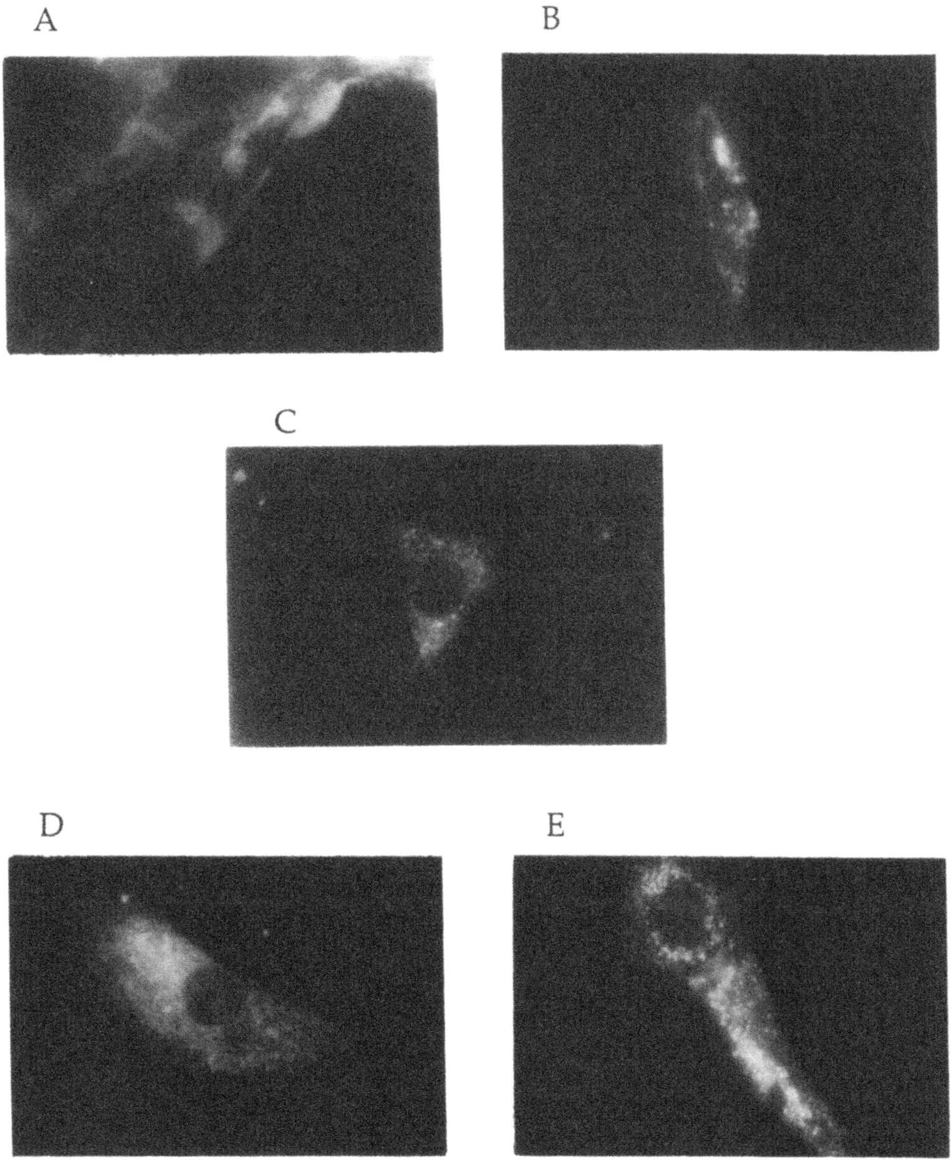
A
B
C
D
E

Figure 6. Uptake of DTAF-E9 (A, B, C, D) and DTAF-H108 (E). SMC were exposed to 100 μg/ml of DTAF-polysaccharides at 37°C for 45 min (A), 2 h (B), 4 h (C) and 3 days (D, E). The cells were fixed and examined with an epifluorescent microscope. Magnification: ×1250.

with E9 and heparin. In addition, no binding was observed with DTAF-T40 or Rhodamine-T40. The cell surface staining was then decreased and vesicles were observed (Fig. 6B). After 4 h, all the vesicles were clearly concentrated around the nucleus (Fig. 6C). The E9 uptake by SMC follows the internalization of heparin on SMC as described by Castellot *et al.* [23]. The pattern of DTAF-E9 uptake by a rapid internalization and the formation of surface clusters suggest a receptor-mediated endocytosis process, similar to that of heparin. These results also indicated that heparin-like polymer can act early in the cell cycle, leading to an inhibition of cell proliferation. We previously observed by ^{3}H-thymidine uptake into DNA that E9 as well as heparin block SMC in the early G0/G1 phase of the cell cycle [24]. In agreement with our results, a recent publication [25] indicated that the SMC inhibition was seen if heparin was present at least during the first hour following serum stimulation.

Upon warming the cells at 37°C for 3 days with DTAF-E9 and DTAF-heparin, the fluorescence in the vesicles was increased (Fig. 6D, 6E). Thus, the amount of polysaccharides internalized into the cells was increased with longer time exposure. Kinetic experiments performed by cell counting indicated that at least 75% of the maximal inhibitory effect was obtained after 72 h (data not shown). Under the same conditions, binding or internalization were never seen when using fluorescent-T40 suggesting that the uptake of heparin and heparin-like polymers (E9) is related to the antiproliferative activities.

DISCUSSION

The proliferation of vascular SMC plays a crucial role in the pathogenesis of atherosclerosis and in persistent pulmonary hypertension of the newborn [21]. The hyperplastic response of SMC to the vessel injury during vascular surgery is also another area of serious concern. A large number of vascular surgical procedures such as vein grafts, bypasses, angioplasties, arteriovenous shunts, endarterectomies and heart transplants fail due to SMC proliferation [14]. A highly antiproliferative molecule would be ideal to achieve good therapeutic control of the diseases and surgically-induced pathological states. To avoid hemorrhagic complications, the antiproliferative drug should be non-anticoagulant. Unfortunately, heparin and heparan sulfate, which are demonstrated inhibitors of *in vivo* and *in vitro* SMC proliferation, are more or less naturally anticoagulant. Moreover, the structure-function relationships are difficult to interpret because of the structural complexity and heterogeneity of heparin and heparan sulfate preparations.

The development of non-anticoagulant antiproliferative heparin may have clinical significance. Such studies from natural heparin preparations are now in different phases for clinical uses.

Another approach is to prepare new heparin-like molecules with SMC antiproliferative activity. These heparin-like biomaterials can be used to understand the mechanisms regulating SMC growth. By the control of the chemical determinants on these synthetic compounds, the structural requirements for the antiproliferative effect could be studied. These compounds were carried out by substitution of macromolecular chains with suitable chemical groups [1, 2]. We have shown that dextran grafted with carboxymethyl benzylamide sulfonated groups inhibit SMC proliferation. In comparison, native dextran or dextran substituted with carboxyl units (i.e. anionic charges) have no effect on SMC growth (Fig. 5).

We previously demonstrated [12, 24] that the antiproliferative and anticoagulant activities of substituted dextrans are distinct. With heparin, these two activities are also clearly separate (Fig. 3). Taking into account these results, we synthesized a polymer with a very low anticoagulant capacity but having SMC antiproliferative activity (Fig. 5) in the same magnitude as heparin (Fig. 3 and 4). By grafting a fluorescent probe (DTAF) on this derivatized dextran, we followed its binding and then the uptake into SMC (Fig. 6). A similar pattern was observed with DTAF-heparin. On the other hand, fluorescent unsubstituted-dextran exhibiting no effect on SMC growth was never internalized.

In conclusion, the synthesis of dextran derivatives, where the nature and distribution of chemical groups and the composition can be altered, lead to polymers which mimic the action of heparin in different biological systems [2, 8, 12, 26]. In addition to the anticoagulant and anticomplement activities [2, 8], derivatized dextrans can also mimic heparin in the protection, stabilization and potentiating effects with aFGF and bFGF [27, 28]. Here, we show that a low anticoagulant derivatized dextran has the capacity to inhibit SMC growth. This synthetic polymer binds to cell membrane and is internalized by SMC as heparin. Thus, derivatized dextrans may also express heparin-like properties in terms of SMC growth inhibition. Studies are now under investigation with these useful biomaterials as tools for elucidating the biological and molecular processes involved in SMC regulation.

Acknowledgements

This work was supported by G.I.P. Thérapeutiques Substitutives. We gratefully acknowledge K. A. Zinnack (Tufts University, MA, USA) for the preparation of the manuscript. We also appreciate the helpful discussions and critical reading by Jim Anderson. The authors wish to acknowledge M. C. Bourdillon and N. Blaes (Inserm, Lyon, France) for their cooperation in rat aorta smooth muscle cell isolation.

REFERENCES

1. J. Jozefonvicz and M. Jozefowicz, *J. Biomater. Sci. Polymer Edn* **1**, 147 (1990).
2. M. Mauzac and J. Jozefonvicz, *Biomaterials* **5**, 301 (1984).
3. D. Letourneur and M. Jozefowicz, *Biomerials* **13**, 59 (1992).
4. J. J. Castellot, D. L. Beeler, R. D. Rosenberg and M. J. Karnovsky, *J. Cell. Physiol.* **120**, 315 (1984).
5. J. J. Castellot, J. Choay, J. C. Lormeau, M. Petitou, E. Sache and M. J. Karnovsky, *J. Cell Biol.* **102**, 1979 (1986).
6. D. A. Lane and U. Lindahl, in: *Heparin—Chemical and Biological Properties, Clinical Applications.* Edward Arnold, London (1989).
7. M. Mauzac, F. Maillet, J. Jozefonvicz and M. D. Kazatchkine, *Biomaterials* **6**, 61 (1985).
8. B. Crepon, F. Maillet, M. D. Kazatchkine and J. Jozefonvicz, *Biomaterials* **8**, 248 (1987).
9. D. Blakeslee, *J. Immunol. Methods* **17**, 361 (1977).
10. D. Blakeslee and M. G. Baines, *J. Immunol. Methods* **13**, 305 (1976).
11. M. C. Bourdillon, J. P. Boissel and B. Crousset, *Prog. Biochem. Pharmacol.* **13**, 103 (1977).
12. T. Avramoglou and J. Jozefonvicz, *J. Biomater. Sci. Polymer Edn.* **3**, 149 (1992).
13. J. J. Castellot, L. V. Favreau, M. J. Karnovsky and R. D. Rosenberg, *J. Biol. Chem.* **257**, 11256 (1982).
14. J. J. Castellot, *Am. J. Respir. Cell. Mol. Biol.* **2**, 11 (1990).
15. A. W. Clowes and M. M. Clowes, *Lab. Invest.* **52**, 611 (1985).
16. J. R. Guyton, R. D. Rosenberg, A. W. Clowes and M. J. Karnovsky, *Circ. Res.* **46**, 625 (1980).
17. J. J. Castellot, M. L. Addonizio and R. D. Rosenberg, *J. Cell Biol.* **90**, 372 (1981).
18. L. M. S. Fritze, C. F. Reilly and R. D. Rosenberg, *J. Cell Biol.* **100**, 1041 (1985).

19. J. J. Castellot, D. L. Cochran and M. J. Karnovsky, *J. Cell. Physiol.* **124,** 21 (1985).
20. E. R. Edelman, D. H. Adams and M. J. Karnovsky, *Proc. Natl. Acad. Sci. USA* **87,** 3773 (1990).
21. J. J. Castellot and M. Karnovsky, in: *Vascular Smooth Muscle in Culture*, Vol. 1, p. 93, J. H. Campbell and G. R. Campbell (Eds). CRC Press, Boca Raton, Florida (1987).
22. M. J. Karnovsky, T. C. Wright, J. J. Castellot, J. Choay, J. C. Lormeau and M. Petitou, *Ann. N.Y. Acad. Sci.* **556,** 268 (1989).
23. J. J. Castellot, K. Wong, B. Herman, R. L. Hoover, D. F. Albertini, T. C. Wright, B. L. Caleb and M. J. Karnovsky, *J. Cell. Physiol.* **124,** 13 (1985).
24. D. Logeart, T. Avramoglou, D. Letourneur and J. Jozefonvicz, *Am. Soc. Cell Biol.* **A174,** 1611 (1991).
25. L A. Pukac, M. E. Ottlinger and M. J. Karnovsky, *J. Biol. Chem.* **26,** 3707 (1992).
26. P. Vaudaux, T. Avramoglou, D. Letourneur, D. P. Lew and J. Jozefonvicz, *J. Biomat. Sci. Polymer Edn*, in press.
27. M. Tardieu, F. Slaoui, J. Jozefonvicz, J. Courty, C. Gamby and D. Barritault, *J. Biomat. Sci. Polymer Edn* **1,** 63 (1989).
28. M. Tardieu, C. Gamby, T. Avramoglou, J. Jozefonvicz and D. Barritault, *J. Cell Physiol.* **150,** 194 (1992).

Heparin surface immobilization through hydrophilic spacers: Thrombin and antithrombin III binding kinetics

YOUNGRO BYUN, HARVEY A. JACOBS and SUNG WAN KIM*

Department of Pharmaceutics and Pharmaceutical Chemistry and Center for Controlled Chemical Delivery, 421 Wakara Way # 318, University of Utah, Salt Lake City, UT 84108, USA

Received 1 February 1992; accepted 7 July 1993

Abstract—The immobilization of heparin onto polymeric surfaces using hydrophilic spacer groups has been effective in curtailing surface induced thrombus formation. In this study, the effect of hydrophilic spacers (PEO) on the binding kinetics of immobilized heparin with antithrombin III (ATIII) and thrombin was investigated. Monodispersed, low molecular weight heparin was fractionated on an ATIII affinity column to isolate high-ATIII affinity heparin. This high-ATIII affinity fraction was immobilized onto a styrene/*p*-amino styrene random copolymer surface using hydrophilic poly(ethylene oxide) (PEO) spacer groups. Styrene/*p*-amino styrene random copolymer was chosen as the model surface to provide quantitative and reproducible surface concentrations of available amine groups, grafted PEO spacers, and immobilized heparin. The polymer substrate was coated onto glass beads, tolylene diisocyanate modified PEO was covalently coupled to the surface, followed by heparin immobilization.

The bioactivity of immobilized heparin was 16.2%, relative to free heparin, and a 1:1 binding ratio between heparin and PEO was achieved. The binding of ATIII and thrombin to control surfaces (no heparin), soluble heparin, heparin immobilized directly onto the surface, and heparin immobilized via spacer groups, were compared.

Soluble heparin bound both thrombin and ATIII, while heparin immobilized directly onto the surface bound only thrombin. Spacer-immobilized heparin bound both ATIII and thrombin, although to a lesser extent than soluble heparin. Thus, the enhanced bioactivity of spacer-immobilized heparin, compared to direct-immobilization, may be attributed to the retention of ATIII binding.

Key words: Fractionated low molecular weight heparin; spacer-immobilized heparin; antithrombin III; thrombin; binding kinetics.

INTRODUCTION

Heparin's role as an anticoagulant is to catalyze and enhance the binding of antithrombin III (ATIII) and thrombin, thus preventing blood coagulation. Heparin binding to ATIII changes the conformation of the protein, which increases the rate of thrombin inhibition. Several mechanisms of heparin binding to ATIII and thrombin have been proposed. Rosenberg and Damus suggested that heparin bound mainly to ATIII [1], while Griffith showed that heparin could bind thrombin [2]. Laurent and Lindahl suggested that heparin was able to bind to thrombin, as well as ATIII [3]. Lane *et al.* established that the minimum saccharide sequence of heparin which can simultaneously accommodate ATIII and thrombin was an octasaccharide [4]. Similarly, Goosen and Sefton showed that heparin immobilized onto a PVA matrix was able to bind thrombin, maintaining its anticoagulant properties [5].

* To whom correspondence should be addressed.

A variety of polymeric materials have been developed for improving the blood compatibility of artificial devices. Among them, heparin immobilized onto polymer surfaces using grafted hydrophilic spacer groups have been reported to improve the blood compatibility. Tanzawa *et al.* modified a PVC surface through surface grafting of hydrophilic cationic polymer containing quaternary ammonium groups. These surfaces bound heparin through ionic interactions, creating a nonthrombogenic surface [6]. The use of spacer groups was further investigated by Danishefsky and Tzeng [7]. They synthesized an aminoethyl derivatized agarose surface and covalently coupled heparin through carboxylic acid groups. Schmer *et al.* [8] coupled heparin to agarose by using the free amine groups of heparin, comparing a directly coupled system with heparin coupled via a spacer system. Their results showed enhanced ATIII binding to heparin by using the spacer system, compared to direct coupling. Kim *et al.* demonstrated that the activity of immobilized heparin increased with increasing length of the alkyl spacer group [9]. Nagaoka *et al.* reported that a PEO grafted surface reduced protein adsorption by means of water content, surface mobility, and volume restrictions of PEO spacer [10]. Gregonis *et al.* also showed that covalently bonded PEO decreased protein adsorption and this effect increased with an increase in the molecular weight of PEO [11]. Merrill *et al.* synthesized a PEO spacer system as a block copolymer, as well as by PEO grafting [12]. They reported that PEO was unreactive to the blood proteins and prevented blood coagulation.

In a previous manuscript [13], our laboratory reported the bioactivity of heparin immobilized through PEO spacer groups onto polyurethane surfaces. In *in vitro*, *ex vivo*, and *in vivo* experiments, heparin immobilized through PEO consistently demonstrated increased bioactivity, compared to direct-immobilized surfaces. For example, heparin immobilized through a PEO 4000 spacer groups retained ~20% of its bioactivity, compared to ~4% bioactivity when immobilized directly onto the surface. The increased bioactivity of immobilized PEO was also demonstrated in *ex vivo* rabbit A–A shunt occlusion experiments and *in vivo* canine vascular graft implants. These experiments demonstrated the effectiveness of PEO-immobilized heparin in reducing thrombus formation; however, no mechanistic interpretations, such as heparin binding to ATIII and/or thrombin were discussed.

Therefore, the objective of this current research is to investigate and compare the binding of ATIII and thrombin to soluble heparin, heparin immobilized directly onto the model surface, and heparin immobilized via hydrophilic spacer groups onto the model surface. Through these studies, the extent of heparin binding to ATIII and/or thrombin may help elucidate the enhanced bioactivity observed for heparin immobilized through hydrophilic spacers, compared to direct surface immobilization.

MATERIALS AND METHODS

Heparin fractionation

Low molecular weight (LMW) heparin (M_w 6000, anticlotting activity 95 IU mg^{-1}, Hepar Ind. Inc., Franklin, OH) was fractionated on an ATIII-sepharose column (5 × 30 cm) (Sigma Chemical Co., St. Louis, MO) to obtain high-ATIII affinity heparin. A 2% heparin solution (0.025 M Tris-HCl, pH 7.4) was loaded onto the ATIII column and initially eluted with a continuous ionic strength gradient

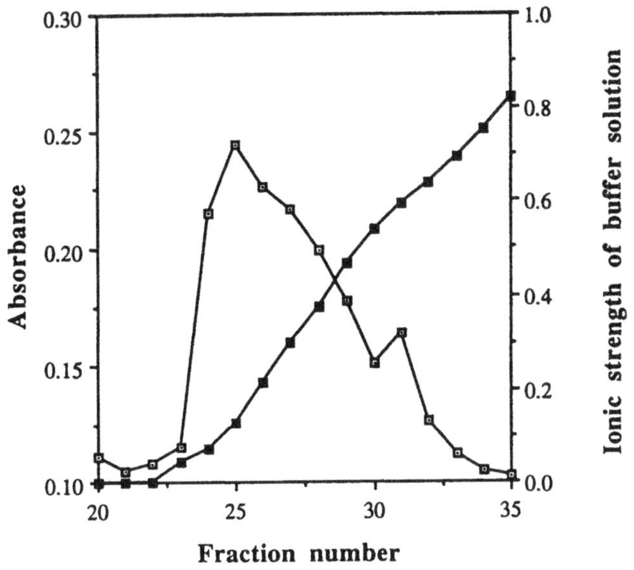
0.30
0.25
0.20
0.15
0.10
1.0
0.8
0.6
0.4
0.2
0.0
20
25
30
35
Absorbance
Ionic strength of buffer solution
Fraction number

Figure 1. Fractionation of low molecular weight heparin (6000 M_w) on an ATIII-sepharose affinity column. Elution of fractions with changing buffer ionic strength.

(0–0.8 M NaCl) to qualitatively determine its ATIII affinity, as shown in Fig. 1. For preparative fractionation, low-ATIII affinity heparin was eluted with a 0.15 ionic strength buffer solution and the high ATIII affinity heparin was eluted with a 1.0 ionic strength buffer solution at 4°C, as shown in Fig. 2. The amount of heparin in each volume fraction was determined using the Azure II assay [14] and the

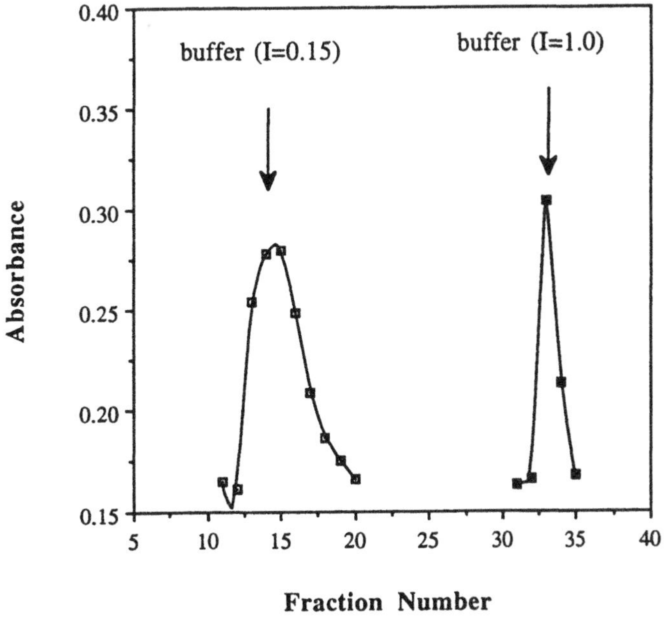
buffer (I=0.15)
buffer (I=1.0)
0.40
0.35
0.30
0.25
0.20
0.15
5
10
15
20
25
30
35
40
Absorbance
Fraction Number

Figure 2. Fractionation of heparin by step elution of $I = 0.15$ and $I = 1.0$ ionic strength solutions.

bioactivity of each fraction was measured by the Factor Xa assay [15]. The high ATIII affinity heparin fraction was dialyzed, freeze dried, and used exclusively in all subsequent experiments.

Preparation of polymer substrate

Random copolymers of styrene (STY. Aldrich Chemical Co., Milwaukee, WI) and *p*-amino styrene (PAS, Polysciences Inc., Warrington, PA) were synthesized as the model polymer substrate, as illustrated in Fig. 3. The two monomers (feed ratio

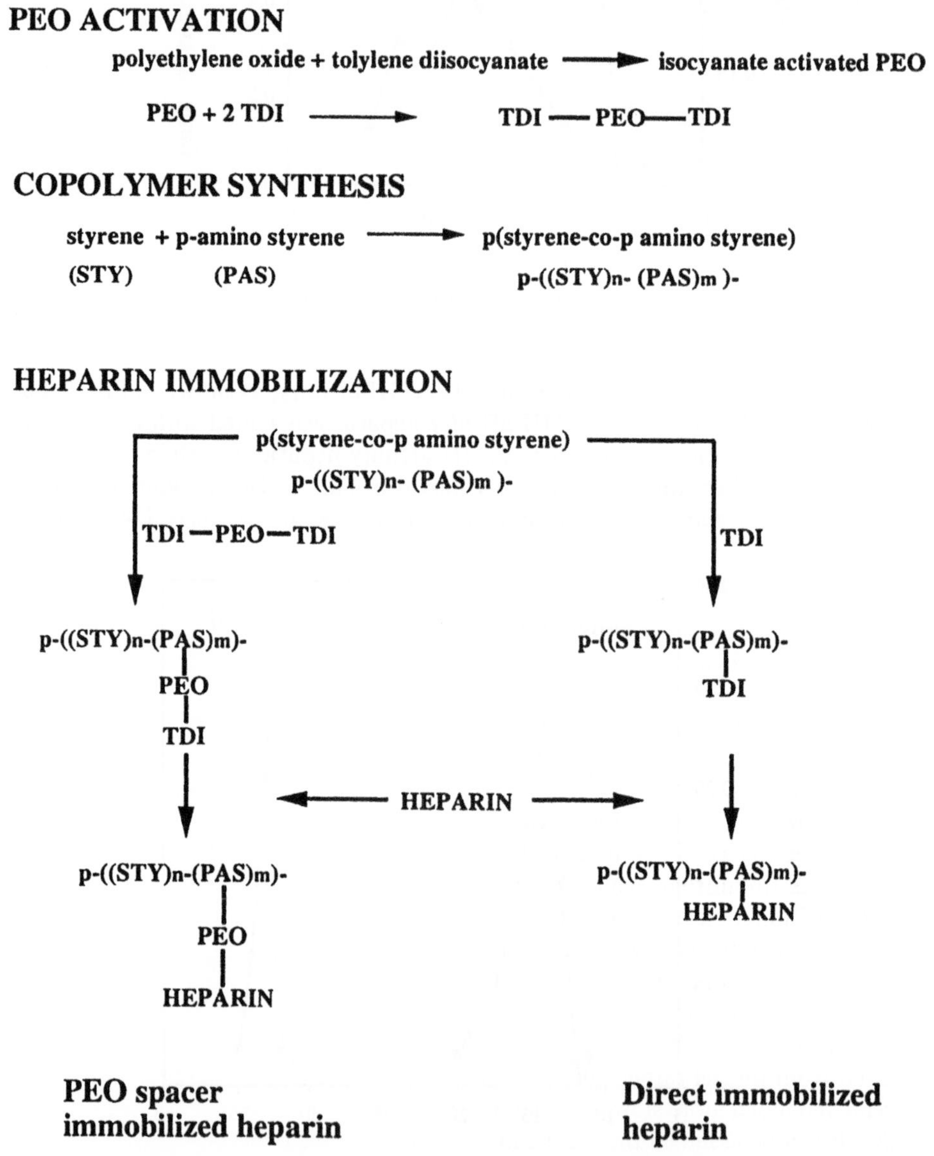
PEO ACTIVATION
polyethylene oxide + tolylene diisocyanate → isocyanate activated PEO
PEO + 2 TDI → TDI — PEO — TDI
COPOLYMER SYNTHESIS
styrene + p-amino styrene → p(styrene-co-p amino styrene)
(STY) (PAS)
p-((STY)n- (PAS)m)-
HEPARIN IMMOBILIZATION
p(styrene-co-p amino styrene)
p-((STY)n- (PAS)m)-
TDI — PEO — TDI
TDI
p-((STY)n-(PAS)m)-
PEO
TDI
p-((STY)n-(PAS)m)-
TDI
HEPARIN
p-((STY)n-(PAS)m)-
PEO
HEPARIN
p-((STY)n-(PAS)m)-
HEPARIN
PEO spacer immobilized heparin
Direct immobilized heparin

Figure 3. Synthetic scheme for the preparation of the *p*-(styrene-co-*p* aminostyrene) model surface, activation of PEO with TDI, and coupling heparin directly and through hydrophilic spacers to the surface.

0–1.0 mol mol^{-1}) were mixed in dioxane at 60°C for 24 h. As reported [16], the initiator was AIBN but no catalyst was used. The formed polymer (STY–PAS) was precipitated in methanol, followed by filtering and drying. The synthesized copolymer was qualitatively characterized by IR spectrometry and the composition ratio determined by element analysis.

Glass beads (mean diameter 100 μm, Ferro-Cataphote, Corp., Jackson, MS) were cleaned by soaking in chromic acid for 24 h and washed alternately with pure ethanol and distilled water. The cleaned glass beads were coated by flowing a 1% solution of STY–PAS in benzene over a sintered glass filter. Filtered glass beads were dried in vacuum for 24 h then sieved to remove aggregates. The surface of the coated glass beads was examined by SEM. The amount of polymer coated onto the surface was determined by dissolving the polymer from the glass beads in benzene, removing the beads, and drying and weighing the residual polymer.

PEO grafting onto polymer surfaces

Both hydroxyl groups of PEO (M_w 3400, Sigma Chemical Co., St. Louis, MO) were derivatized with tolylene diisocyanate (TDI, Kodak, Co., Rochester, NY). PEO in benzene (5 mM) was added drop-wise into 40 mM TDI in benzene and the reaction proceeded under nitrogen atmosphere at 60°C for 24 h [13]. The TDI derivatized PEO (TDI–PEO–TDI) was then precipitated in diethylether, filtered and dried under nitrogen.

The TDI–PEO–TDI activated hydrophilic spacer was grafted onto the STY–PAS surface through isocyanate groups of TDI–PEO–TDI and the amino groups of the STY–PAS surface, as shown in Fig. 3. The grafting reaction proceeded in formamide under nitrogen at 25°C for 24 h.

The amount of PEO spacers grafted onto the surface was quantitated by reductive alkylation using ^{14}C-formaldehyde (Sigma Chemical Co., St. Louis, MO). The isocyanate end group of the spacers was hydrolyzed in water to obtain amine groups, and were further reacted with ^{14}C-formaldehyde at 25°C for 24 h. The radiolabelled spacer grafted polymer substrate was dissolved from the glass beads, and the radioactivity of the solution was determined with a Beckman LS 1801 liquid scintillation counter (Beckman Instruments Inc., Fullerton, CA).

Heparin immobilization

Heparin was covalently bound to the PEO spacers-STY–PAS coated glass beads through a coupling reaction between the hydroxyl of amine group of heparin and the isocyanate group of the PEO spacer, as shown in Fig. 3. In addition, heparin was direct-immobilized onto the STY–PAS coated glass beads (Fig. 3). In this reaction, the amine groups of STY–PAS were modified with TDI and heparin was again coupled through hydroxyl or amine groups. In both reactions, the coupling reaction took place in formamide at 4°C for 3 days. Both the PEO spacer and direct-immobilized heparin systems were washed in methanol and distilled water successively.

Heparin characterization

Azure II assay. The determination of heparin concentrations eluted during the ATIII affinity fractionation was determined by an Azure II assay. A heparin solution (0.5 ml) was added to 4.5 ml of Azure II (Allied Chemical Co., NY) solution (0.01 mg ml^{-1}) and the absorbance was measured at 500 nm on a Perkin Elmer Lambda 19 UV/VIS/NIR spectrophotometer (Perkin Elmer, Norwalk, CT).

Toluidine blue assay. The surface concentration of immobilized heparin was determined by a modified toluidine blue colorimetric method of Smith *et al.* [17]. A solution heparin standard curve was obtained by vortexing 2 ml of toluidine blue (0.005% w/v, Sigma Chemical Co., St.Louis, MO) solution (25 mg toluidine blue dissolved in 500 ml 0.01 N HCl containing 0.2% NaCl) and 0.2 ml of heparin standard solutions for 30 s. Hexane (2 ml) was added to the solution, mixed and allowed to extract the heparin–toluidine blue complex. The absorbance of the aqueous layer was determined at 631 nm on a Perkin Elmer Lambda 19 UV/VIS/NIR spectrophotometer.

Heparin immobilized beads (150 mg, both spacer and direct-immobilized) were mixed with 2 ml of toluidine blue solution for 30 s. The toluidine blue solution was decanted from the beads, mixed with 2 ml hexane, and the absorbance of the toluidine blue solution layer was determined at 631 nm.

Factor Xa assay. The bioactivity of surface immobilized heparin was analyzed by a chromogenic assay which measures the potentiating effect of heparin on ATIII's binding and inactivation of Factor Xa (FXa). ATIII, S-2765 (Bz-Ile-Glu-Gly-Arg-*p*NA), and FXa were reconstituted from lyophilized powders with sterile water, as per supplier instructions (Kabi, Franklin, OH). Heparin standards, between 0 and 150 ng ml^{-1} (0.2 ml), were diluted with 0.2 ml ATIII. Aliquots of this solution (0.4 ml) were incubated at 37°C for 5 min and then 0.4 ml FXa (71 nkcat/40 ml) was added and incubated at 37°C for 3 min. S-2765 (25 mg/38.5 ml) was then added and incubated. The reaction rate of the substrate with free FXa was monitored spectrophotometrically at 405 nm. Heparin immobilized beads (40 mg, both spacer and direct-immobilization) were incubated with 0.4 ml ATIII solution (1 U ml^{-1}) for 5 min at 37°C, followed by the addition of FXa and S-2765. Samples were monitored for absorbance at 405 nm and compared to the heparin standards to determine the bioactivity of immobilized heparin.

Glass treatment

All glass vials used in the kinetic experiments were treated with dichlorodimethylsiloxane (Kodak, Co., Rochester, NY) to prevent non-specific protein adsorption onto the glass surface. The glass surface was soaked in 5% dichlorodimethylsiloxane in toluene at 25°C for 1 h, then washed in pure ethanol, followed by distilled water [18, 19].

Kinetic experiments

Binding kinetics between immobilized Heparin and ATIII. Heparin immobilized beads (250 mg, both spacer and direct-immobilized) were gently stirred (magnetic stirrer) in an ATIII solution (0–1 μM, Kabi, Franklin, OH) at 37°C for 10 min. The unbound ATIII in solution was sampled with a filtered syringe and assayed using the chromogenic substrate H-D-Phe-Pip-Arg-*p*NA (S-2238). Human ATIII (100 IU/mg) was dissolved in 50 mM Tris buffer solution (7.5 mM NaEDTA and 0.15 M NaCl, pH 8.4) and stored at −20°C. H-D-Phe-Pip-Arg-*p*NA (S-2238,

25 mg/38.5 ml) was dissolved in distilled water to 1.5 M and stored in the dark at 4°C. The sample solution (600 μl) was incubated with 100 μl of buffer solution containing heparin (0.03 IU ml^{-1}) at 37°C for 3 min. 100 μl thrombin solution (2 IU ml^{-1}, Sigma Chemical Co., St. Louis, MO) was added into the mixture. After mixing for 1 min, 300 μl S-2238 solution was added. The reaction rate of S-2238 with free thrombin was monitored by measuring the enzymatic cleavage of *p*-nitroanilide (*p*NA) spectrophotometrically at 405 nm.

Binding kinetics between immobilized heparin and thrombin. Heparin immobilized beads (250 mg, both spacer and direct-immobilized) were incubated in thrombin solution (0 to 5 nM) at 37°C for 5 min. The supernatant was sampled with a filtered syringe and the chromogenic substrate S-2238 was used to determine the amount of free, unbound thrombin in solution (i.e. that concentration not bound to the heparinized surfaces). Several concentrations of thrombin solution (700 μl Tris buffer) were mixed with 300 μl of S-2238 (25 mg/38.5 ml) solution. The rate of S-2238 reacting with unbound thrombin was monitored spectrophotometrically at 405 nm, again monitoring the cleavage of *p*NA.

In every kinetic experiment, the data obtained from heparin immobilized beads were compared with those of the control beads (STY–PAS surfaces and STY–PAS–PEO grafted surfaces). The non-specific binding of ATIII and thrombin to the control surfaces was used to determine a more accurate measurement of these proteins binding to the heparinized surfaces.

RESULTS AND DISCUSSION

Heparin fractionation

Heparin is a polydispersed, heterogeneous polysaccharide, varying in molecular weight and binding affinity for ATIII, as shown in Fig. 1. Heparin was fractionated on an ATIII-sepharose affinity column to separate high- and low-ATIII affinity heparin. Low-affinity heparin was eluted with an ionic strength buffer of 0.15, while high ATIII affinity heparin eluted with a buffer ionic strength of 1.0, as shown in Fig. 2.

The yield of high-ATIII affinity heparin was 12% of total heparin and had an absolute activity of 323 IU mg^{-1}; ~3.4 times the activity of unfractionated heparin (95 IU mg^{-1}). The yield of low-ATIII affinity heparin was 30% of the total compound and had an activity of 185 IU mg^{-1}; ~1.95 times the activity of unfractionated heparin. Nearly 50% of the heparin molecules had no affinity for ATIII. In ensuing kinetic experiments, high-ATIII affinity heparin (both direct and spacer-immobilized) was exclusively used to evaluate ATIII and thrombin binding.

The molecular weight distribution of heparin used in these experiments was monodispersed (M_w 6000 ± 100). This monodispersity was important to optimize heparin immobilization reactions and ATIII and thrombin binding kinetics. As the molecular weight of heparin increases, the number of spacers bound per one heparin molecule could increase, decreasing the mobility and possibly the binding affinity of heparin. The molecular weight of heparin was also important to thrombin binding. The binding ratio of heparin and thrombin depends on the molecular weight of heparin. Low molecular weight heparin (M_w 5000) has a thrombin binding ratio of one or two, while high molecular weight heparin ($M_w > 20\,000$) has a binding ratio of two to four [20–22]. In addition, heparin had no affinity for thrombin with a molecular weight lower than 5000.

Heparin immobilization

Two important aspects in the immobilization of heparin via hydrophilic spacer groups was to control the surface density of spacers and to control the binding ratio of heparin to the spacer groups. An ideal situation would be where 1:1 binding of heparin and spacer groups was achieved. As the surface density of spacer groups increases, the amount of spacer groups not coupled to heparin may increase. These hydrophilic spacer groups not coupled to heparin may affect the binding kinetics of ATIII and thrombin.

The density of spacer groups coupled to the surface may also affect the coupling efficiency of heparin. As the surface density of spacer groups increases, the coupling ratio of heparin to spacer may decrease, leading to heparin binding with several spacers. Multi-point binding of heparin to the spacer groups may decrease the bioactivity of heparin and may result in decreased binding properties with the ATIII and thrombin. Therefore, 1:1 binding system may be viewed as an optimum condition to demonstrate the interaction of immobilized heparin with ATIII and thrombin. It is recognized that by manipulating the reaction conditions to obtain 1:1 binding (heparin : spacer group), maximum *in vitro* nonthrombogenicity may not be achieved, due to sub-maximum heparin surface density. However, accurate and reproducible binding interactions between immobilized heparin and ATIII/thrombin may be realized by establishing a fixed surface density of immobilized heparin.

Random copolymers of styrene and *p*-amino styrene were synthesized as the model polymer substrate. This copolymer was chosen to insure quantitative determination of available grafting sites (—PAS) from which PEO coupling ratios could be

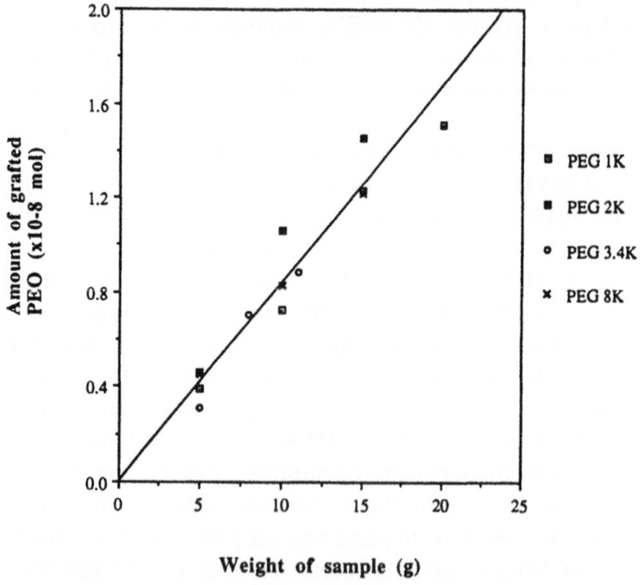
2.0
1.6
1.2
0.8
0.4
0.0
0
5
10
15
20
25
Amount of grafted PEO (x10-8 mol)
Weight of sample (g)
PEG 1K
PEG 2K
PEG 3.4K
PEG 8K

Figure 4. Quantitation of PEO as a function of spacer length.

determined. This was important to assure equivalent surface densities of PEO and heparin in the evaluation of binding kinetics. Initially, copolymers were synthesized with feed ratios (STY/PAS) of 100/0, 90/10, 80/20, 50/50, and 0/100, with final compositions of 100/0, 89/11, 74/26, 50/50, and 0/100, respectively, as determined by element analysis. The 74/26 STY/PAS copolymer achieved a 1:1 binding ratio of heparin to spacer group, and was therefore used throughout the kinetic experiments. The STY–PAS copolymer substrate was then coated onto glass beads, achieving a smooth surface as confirmed by SEM (data not shown). The amount of coated polymer substrate was 0.81 wt% (determined gravimetrically), relative to the glass beads.

PEO was modified with TDI to create reactive end groups for isocyanate-amine (biuret) coupling of the spacer to the polymer substrate. The amount of grafted spacers was measured by hydrolyzing the unreacted isocyanate groups to amine, and coupling with C^{14}-formaldehyde, resulting in ~90% coupling efficiency [23]. In previous research [24], PEO molecular weights of 1000, 2000, 3400, and 8000 were investigated as the hydrophilic spacer groups. For each molecular weight (Fig. 4), the amount of beads (surface area) was a dependent variable, and the surface density of spacer was calculated by the slope. The results showed that for each molecular weight spacer, ~0.91 nmol of PEO spacer was grafted per gram of polymer coated glass beads. In terms of surface density, 16.3 pmol PEO were coupled per cm^2 surface area, or one PEO spacer occupied $33 \times 33\,Å^2$ area.

The most effective hydrophilic chain length, in terms of binding efficiency and *ex vivo/in vitro* heparin bioactivity was PEO 3400 [13]. This chain length enhanced the dynamic motion of the immobilized heparin and presumably, minimized steric hindrance of immobilized heparin interacting with ATIII and thrombin. For this reason, detailed information as to the binding kinetics of ATIII and thrombin to heparin immobilized, specifically, with PEO 3400 was desired. Therefore, PEO 3400 was used in subsequent experiments to help explain previously determined results.

An important consideration in the immobilization of diisocyanate derivatized PEO spacer group was to prevent the coupling of both functional groups of one chain to the surface (intramolecular bridging). If the intramolecular bridging was significant, the amount of free end groups might decrease with increasing spacer length until steric effects prevent the coupling of PEO. As shown in Fig. 4, the amount of free end groups of the grafted PEO surfaces were similar for each PEO molecular weight evaluated; thus, intramolecular bridging of PEO was minimized in this procedure.

The amount of high ATIII affinity heparin immobilized onto the surface using PEO 3400 was 0.83 nmol of heparin per gram of polymer coated beads, resulting in a surface density of 13.7 pmol heparin per cm^2 surface area, or one heparin molecule occupied $35 \times 35\,Å^2$ area. The binding ratio of heparin to PEO spacer was 0.91:1, inferring that nearly every PEO group was coupled with one heparin molecule. In addition, the bioactivity of immobilized heparin, as determined by FXa assay, was 16.2% relative to heparin in solution.

A direct-immobilized (non-spacer) heparin surface was also synthesized to compare to the PEO spacer system. The amount of heparin direct-immobilized onto the STY–PAS surface (toluidine blue assay) was 0.99 nmol heparin/gm beads (16.3 pmol cm^{-2}), a value similar to the PEO spacer system. Therefore, the surface density of heparin immobilized directly onto the surface, and immobilized via a hydrophilic spacer group were nearly identical. Additionally, the bioactivity of direct-immobilized heparin (FXa assay) was 2.58% (based on solution heparin), giving an absolute bioactivity of 8.33 U mg^{-1}.

Binding kinetics of immobilized heparin and ATIII

The three heparin systems, soluble, direct-immobilized, and PEO 3400 spacer-immobilized heparin systems, were used to investigate the binding kinetics of heparin to ATIII and thrombin. In addition, control surfaces, STY–PAS and PEO grafted STY–PAS, were used to determine non-specific binding of ATIII and thrombin. Non-specific binding accounted for less than 5% of the total specific binding.

Literature values for soluble, low molecular weight, high-ATIII affinity heparin binding constants to ATIII ranged from 10^7 to 10^8 M^{-1} [25–27]. The binding constant of PEO 3400 spacer-immobilized heparin and ATIII, as determined by Scatchard plots (Fig. 5) was 0.958×10^7 M^{-1}, as listed in Table 1. The binding constant was not significantly altered by immobilization via spacer. However, the binding constant of direct-immobilized heparin for ATIII was below measurable values.

Table 1.
Binding constants of Antithrombin III to heparin

	K_a	r	activity
Soluble heparin	10^7–10^8 M^{-1} [a]	<1.0	323 U mg^{-1} (100%)
Direct immobilized heparin	N.D.	N.D.	8.33 U mg^{-1} (2.58%)
Heparin immobilized via PEO (M_w 3400) spacer	0.958×10^7 M^{-1}	0.31	52.33 U mg^{-1} (16.2%)

[a] Ref [25–27]. N.D.—not determined.

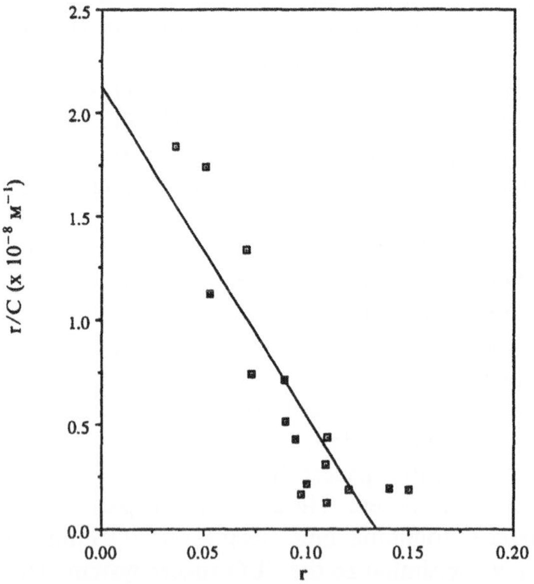
2.5
2.0
1.5
1.0
0.5
0.0
r/C (x 10^{-8} M^{-1})
0.00
0.05
0.10
0.15
0.20
r

Figure 5. Scatchard plot of ATIII binding to spacer-immobilized heparin.

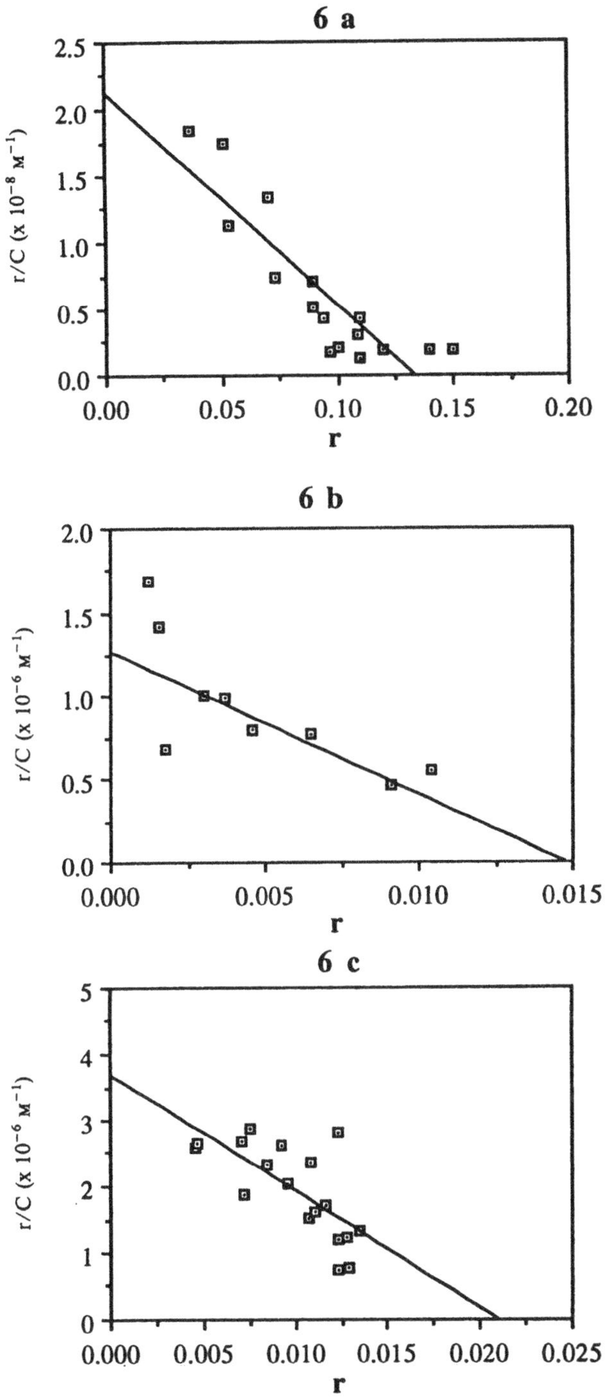
6 a
2.5
2.0
1.5
1.0
0.5
0.0
r/C (x 10⁻⁸ M⁻¹)
0.00
0.05
0.10
0.15
0.20
r
6 b
2.0
1.5
1.0
0.5
0.0
r/C (x 10⁻⁶ M⁻¹)
0.000
0.005
0.010
0.015
r
6 c
5
4
3
2
1
0
r/C (x 10⁻⁶ M⁻¹)
0.000
0.005
0.010
0.015
0.020
0.025
r

Figure 6. (a) Scatchard plot of thrombin binding to soluble heparin; (b) Scatchard plot of thrombin binding to the direct-immobilized heparin; and (c) Scatchard plot of thrombin binding to spacer-immobilized heparin.

Based on the FXa assay, the bioactivity of direct-immobilized heparin was estimated to be 2.58% of soluble heparin. Thus, the hydrophilic spacer effect was effective in maintaining the binding interaction between PEO 3400 immobilized heparin and ATIII. When heparin was immobilized onto the polymer substrate without spacer, the specific binding site of heparin may have been within the interface of the heparin and polymer surface, and the steric effect of heparin immobilized directly onto the surface may have prevented the binding of ATIII to heparin. On the other hand, when heparin was immobilized via the hydrophilic PEO 3400 spacer group, heparin maintained the dynamic mobility necessary to maintain specific binding with ATIII.

Binding kinetics of immobilized heparin and thrombin

Literature values for the binding of soluble, low molecular weight heparin to thrombin ranged from 10^5 to $10^9\ \text{M}^{-1}$ [2, 20, 28]. However, the binding constants of soluble, direct-immobilized, and spacer-immobilized heparin to thrombin were extrapolated from Scatchard plots (Fig. 6a–c) and calculated to be 1.59×10^9, 0.85×10^8, and $1.76 \times 10^8\ \text{M}^{-1}$, respectively.

Thus, the binding constant of PEO 3400 spacer-immobilized heparin and direct-immobilized heparin to thrombin, as listed in Table 2, were ten and twenty times less than soluble heparin. However, when compared to ATIII binding, thrombin bound to both direct- and spacer-immobilized heparin.

To summarize the results, soluble heparin bound to both thrombin and ATIII. For the PEO 3400 spacer-immobilized system, heparin also bound to thrombin, as well as ATIII, similar to soluble heparin. However, direct-immobilized heparin bound only thrombin, and not to ATIII. These binding kinetic results may help explain the enhanced bioactivity of PEO spacer-immobilized heparin, compared to direct-immobilized heparin, as demonstrated in previous *in vitro* and *in vivo* experiments [13].

Table 2.
Binding constants of thrombin to heparin

	K_a	r
Soluble heparin	$1.59 \times 10^9\ \text{M}^{-1}$	0.13
Direct immobilized heparin	$0.85 \times 10^8\ \text{M}^{-1}$	0.015
Heparin immobilized via PEO (M_w 3400) spacer	$1.76 \times 10^8\ \text{M}^{-1}$	0.02

CONCLUSIONS

The use of styrene/*p*-amino styrene random copolymer as the model surface provided a reproducible system for reliable and quantitative immobilization of PEO and heparin. Through exact determinations of available surface amine groups (PAS), coupling ratios of PEO and heparin were determined. The immobilization of heparin through hydrophilic PEO spacers enhanced its binding to ATIII and thrombin, compared to direct-immobilization. This was most significant for immobilized heparin binding to ATIII. Direct-immobilized heparin did not bind to

ATIII, while PEO spacer-immobilized heparin showed a ten-fold decrease in binding constants. The main difference between direct and spacer-immobilized heparin may be the dynamic motion of immobilized heparin coupled through the hydrophilic spacer. The increased mobility of the spacer group may allow increased interaction of heparin with ATIII, while direct-immobilization may bury or hide the specific binding site within the interface of the heparin and the polymer surface. However, direct and spacer-immobilized heparin showed similar binding to thrombin, perhaps due to non specific binding. Therefore, the surface immobilization of heparin via hydrophilic PEO spacers was shown to bind both thrombin and ATIII, similar to soluble heparin, in contrast to direct-immobilized heparin which was shown only to bind thrombin.

Acknowledgement

This work was supported by NIH grant HL 20251-15.

REFERENCES

1. R. D. Rosenberg and P. S. Damus, *J. Biol. Chem.* **248**, 6490 (1973).
2. M. J. Griffith, *J. Biol. Chem.* **254**, 12044 (1979).
3. T. C. Laurent and U. Lindahl, *Biochem. J.* **175**, 691 (1978).
4. D. A. Lane, J. Denton, A. M. Flynn, L. Thunberg and U. Lindahl, *Biochem. J.* **218**, 725 (1984).
5. M. F. A. Goosen and M. V. Sefton, *Thrombosis Res.* **20**, 543 (1980).
6. H. Tanzawa, Y. Mori, N. Harumiya, H. Miyama, M. Hori, N. Ohshima and Y. Idezuki, *Trans. Am. Soc. Art. Int. Org.* **19**, 188 (1973).
7. I. Danishefsky and F. Tzeng, *Thrombosis Res.* **4**, 237 (1974).
8. G. Schmer, L. M. L. Teng, J. J. Cole, H. E. Vizzo, M. M. Francisco and B. H. Scribner, *Trans. Am. Soc. Art. Int. Org.* **22**, 654 (1976).
9. C. D. Ebert, E. S. Lee and S. W. Kim, *J. Biomed. Mater. Res.* **16**, 629 (1982).
10. Y. Mori, S. Nagaoka, Y. Masubuchi, M. Itoga, H. Tanzawa, T. Kiucki, Y. Yamada, T. Yonaha, H. Watanabe and T. Idezuki, *Trans. Am. Soc. Art. Int. Org.* **24**, 736 (1978).
11. D. Gregonis, R. Van Wagonen and J. D. Andrade, *Trans. Sec. World Cong. Biomater.* **7**, 266 (1984).
12. E. W. Merrill, E. W. Salzman, K. A. Dennison, S. W. Tay and R. W. Pekala, *Prog. Art. Org.* **9**, 909 (1985).
13. K. D. Park and S. W. Kim, in: *Poly(ethylene glycol) Chemistry: Biotechnical and Biomedical Applications*, p. 283, M. Harris (Ed.). Plenum Press, New York (1992).
14. H. J. Conn, in: *Biological Stains*, p. 96, H. J. Conn (Ed.). The Williams and Wilkins Co., MD (1961).
15. A. N. Teien, M. Lie and N. Abildgaard, *Thrombosis Res.* **8**, 413 (1976).
16. A. P. Donya and G. N. Agafonova, *Igr. Vyssh, Uchebn, Zaved, Kjim. Khim. Tekhnol.* **18**, 796 (1975).
17. P. K. Smith, A. K. Mallia and G. T. Harmanson, *Anal. Biochem.* **109**, 466 (1980).
18. G. Elgue, B. Pasche, M. Blomback and P. Olsson, *Thrombosis Haemostasis* **63**, 435 (1990).
19. J. H. Lee, Ph.D dissertation, The University of Utah (1988).
20. R. E. Jordan, G. M. Oosta, W. T. Gardner and R. D. Rosenberg, *J. Biol. Chem.* **255**, 10073 (1980).
21. E. H. H. Li, C. Orton and R. D. Feinman, *Biochemistry* **13**, 5012 (1974).
22. G. Oshima, H. Uchiyama and K. Nagasawa, *Biopolymers* **25**, 527 (1986).
23. C. F. Lane, *Synthesis* 135 (1975).
24. Y. R. Byun, H. A. Jacobs, S. W. Kim, *ASAIO J.* **38**, M638 (1992).
25. B. Nordenman and L. Bjork, *Biochemistry* **17**, 3339 (1978).
26. R. Jordan, D. Beeler and R. Rosenberg, *J. Biol Chem.* **254**, 2902 (1979).
27. S. T. Olson, K. R. Srinivasan, I. Bjork and J. D. Shore, *J. Biol. Chem.* **256**, 11073 (1981).
28. M. O. Longas, W. S. Ferguson and T. H. Finlay, *Arch. Biochem. Biophys.* **200**, 595 (1980).

Part VI.B.

Immobilized Biomolecules and Synthetic Derivatives of Biomolecules

Collagen and Polysaccharides

Physico-chemical surface characterization of hyaluronic acid derivatives as a new class of biomaterials

R. BARBUCCI[1],A. MAGNANI[1], A. BASZKIN[2],
M. L. DA COSTA[2], H. BAUSER[3], G. HELLWIG[3],
E. MARTUSCELLI[4] and S. CIMMINO[4]
[1]*C.R.I.S.M.A. and Department of Chemistry, University of Siena, Siena, Italy*
[2]*C.N.R.S., URA 1218 University of Paris Sud, 92296 Chatenay-Malabry, France*
[3]*FhG-IGB, Fraunhofer Institute, Stuttgart, Germany*
[4]*Institute of Research on Polymer Technology, C.N.R., Napoli, Italy*

Received 10 June 1992; accepted 16 September 1992

Abstract—Three hyaluronic acid derivatives with different types and/or percentages of esterification, were analyzed by means of static contact angle measurements, SEM, ESCA, ATR/FT-IR, WAXS, DSC and TGA. The physico-chemical characterization of the three different samples, in both dry and wet state, was provided in terms of surface and bulk properties. ESCA and infrared analyses showed that the surface composition of all samples differs from that of the bulk. The hydrophilic–hydrophobic character of the samples changed according to the chemical composition as shown by ESCA and contact angle measurements. Both infrared and contact angle measurements reveal that surface restructuring occurred upon hydration for all the samples and the greater the hydrophilic character of the sample, the greater and faster the restructuring phenomenon. A clear picture of the different types of chemical groups has been established at different depth for the three materials.

Key words: Biomaterials; hyaluronic acid; chemical modification; infrared spectroscopy; electron spectroscopy; electron microscopy; contact angle; thermal analysis.

INTRODUCTION

Hyaluronic acid is a nonsulphonated glycosaminoglycan, consisting of alternate residues of *N*-acetyl-D-glucosamine and D-glucuronic acid.

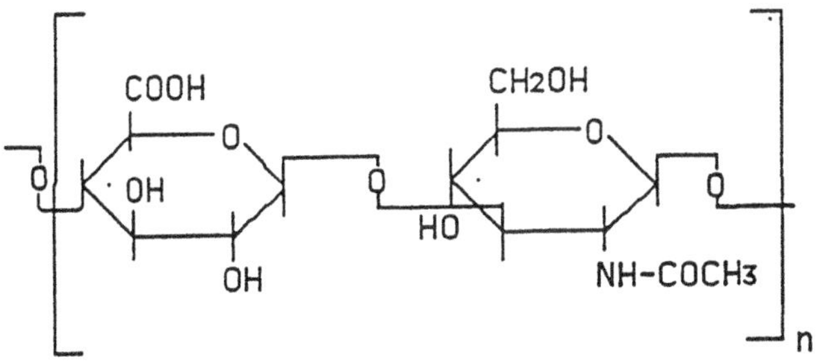
COOH
CH2OH
O
OH
HO
OH
NH-COCH3
n

It is a naturally occurring mucopolysaccharide present in all tissues in varying concentrations, with the highest concentrations occuring in soft connective tissue. By chemical modification of hyaluronic acid [1] it is possible to produce polymers with physico-chemical properties significantly different from those of the hyaluronic acid itself, which can be regarded as a new class of semisynthetic biopolymers. These can be processed leading to a great variety of products such as threads, films, fabrics and sponges of great interest in plastic surgery, wound dressing, orthopedics etc. Amongst them ethylhyaluronic acid ester (HYAFF-07) and benzyl hyaluronic acid ester (HYAFF-11) seem to be potentially very promising biomaterials capable of replacing short nerve segments [2] and ensuring the growth of human epidermal keratinocytes in wound healing [3].

The purpose of this paper is to analyse the surface properties of these materials by different techniques such as contact angle measurements, ATR/FT-IR, ESCA, SEM, together with some bulk characteristics, in order to obtain a better understanding of the interfacial interactions of these biomaterials in contact with the physiological liquid. Since those interactions occur in a very narrow interfacial zone (of the thickness <1 nm) their surface composition in dry and wet state needs to be precisely determined.

EXPERIMENTAL SECTION

Hyaluronic acid derivatives

Films of the ethyl ester of hyaluronic acid (HYAFF-07), the benzyl ester in which 100% of the carboxylate groups were esterified (HYAFF-11), and the benzyl ester with 75% of the carboxylate groups esterified by benzyl alcohol (HYAFF-11p75) 25% remaining in the form of sodium salt, were supplied by FIDIA S.p.A.

As previously described [1, 4] the films were obtained by extrusion of a solution of the three different derivatives into DMSO as the coagulant. Briefly, the polymers (HYAFF-07, HYAFF-11, HYAFF-11p75) were solubilized in DMSO at high concentrations (130–150 mg per ml of DMSO). The solutions were preliminarily filtered through a 22 mm filter and degassed for 2 h by vacuum, and then were pumped through an extruder with a slit (150 mm width, 200 mm thickness). The films coming out from the slit were immersed in the coagulant: a 20 : 80 ethanol : water mixture was used for the totally esterified derivatives HYAFF-07 and HYAFF-11; 100% ethanol was used for HYAFF-11p75. In both cases the coagulation baths were miscible with DMSO but they were not solvents for the respective polymers; therefore, as soon as the films were immersed in the bath, a coagulation process took place. The resulting films were washed several times in fresh ethanol in order to eliminate the polymer solvent. Subsequently, they were rolled up on reels and extensively washed in salt solution (9 g/l NaCl). $12 \times 12\ cm^2$ films were obtained from the original continuous films. They were dried under vacuum to get rid of any remaining ethanol and packed.

Physico-chemical characterization

SEM analysis. The surfaces of the HYAFF films were analyzed after coating by Au-Pd evaporization, by using a Philips 505 scanning electron microscope at 30 kV.

WAXS measurements. Wide angle X-ray scattering (WAXS) measurements were carried out on a Philips PW 1050 model. WAXS patterns were collected by a flat camera with a sample-film distance = 50 mm.

Thermal analysis. The thermal behaviour of the samples was investigated by a Mettler DSC-30. Two sets of experiments were performed; the first one, on the untreated samples; the second one on the samples which had been kept in a vacuum oven at 60°C for 2 days in order to remove eventual solvents or water adsorbed by the films. The experiments consisted of heating the samples (about 8 mg) from −120 to 300°C at a heating rate (HR) of 10°C/min.

TGA measurements. Thermogravimetric analysis was performed from 30 to 350°C at a heating rate of 20°C/min. in nitrogen gas by using a Mettler TG-50.

Contact angle measurements. The samples were analysed according to two distinct procedures: (a) drop-on-plate method and (b) air-under-water and octane under water contact angle measurements. The former yielded advancing and receding contact angles on dry surfaces, the latter the receding angles on hydrated surfaces. Dry samples were analysed after having been stored for 2 days under vacuum at room temperature, wetted samples were kept in tri-distilled water at room temperature for 1 day prior to contact angle measurements in a separate water bath. The method is described in detail in [5]. In addition, in order to enable the comparison of results from different surface analysis tests, the samples were labeled with an asymmetric identification mark to distinguish between the two sides of a given film. All glassware used in contact angle measurements (glass cells, pipettes, syringes) were cleaned with freshly prepared sulphochromide solution and rinsed well with tri-distilled water. The advancing and receding contact angles measured on dry surfaces were used to calculate water-polymer works of adhesion according to $W_{(slv)} = \gamma_{H_2O}(1 + \cos\theta)$ where $W_{(slv)}$ is the work of adhesion of a water vapour covered polymer surface. The hysteresis (H) of polymer-water works of adhesion were then found from $W_R - W_A = H = \gamma_{H_2O}(\cos\theta_R - \cos\theta_A)$ where θ_R and θ_A are the receding and advancing contact angles, respectively.

Advancing contact angles were also used to determine the surface energy of polymer films (γ_{sv}) adopting the equation of state approach [6].

$$\cos\theta_A = \frac{(0.015\gamma_{sv} - 2.00)(\gamma_{sv}\cdot\gamma_{lv})^{1/2} + \gamma_{lv}}{\gamma_{lv}[0.015(\gamma_{lv}\cdot\gamma_{sv})^{1/2} - 1]}.$$

The air-under-water contact angles (θ_A') were determined on hydrated surfaces. These values should be compared with (θ_R) values on the dry surface, at the same time the corresponding hysteresis $\theta_A - \theta_R'$ should be compared with that of $\theta_A - \theta_R$.

The polar acid-base components of the polymer-water work of adhesion I_p^{AB} were calculated using octane-under-water contact angles (θ_0) from $I_p^{AB} = 50.6(1 - \cos\theta_0)$. The I_p^{AB} values were then subtracted from the total values of polymer water works of adhesion (W_{slv}) yielding the contribution of Lifshitz van der Waals interactions to these works, W_{SL}^{LW} [7].

FT-IR spectroscopic measurements. The FT-IR spectra were obtained using a Perkin-Elmer M1800 spectrophotometer connected to a Data Station 7500 professional computer. Samples of 50 mm × 30 mm were cut from the films immediately after opening the transport packages and pressed against the 45 deg end-faces KRS-5 ATR crystal without touching the surface area to be analysed. Spectra of the hydrated samples were taken in a flow ATR cell. An MCT detector was used and the apparatus was purged with N_2. The frequency scale was internally calibrated with a helium-neon reference laser to an accuracy of 0.01 cm^{-1}. Spectra of 300 scans at a resolution of 2 cm^{-1} were collected on both sample faces (each experiment consists at least of two replica) and stored for further manipulation. Difference spectra were obtained according to the previously described procedure [8, 9] and the frequency values were calculated by the 'center-of-gravity' method.

The different penetration depths of the beam into the material were realized simply by changing the angle of incidence of the infrared radiation into the ATR crystal [10]. In ATR the depth of penetration for a non absorbing medium, defined as the distance required for the electric field amplitude to fall to e^{-1} of its value at the surface is given by Harrick [11]):

$$d_p = \frac{\lambda}{2\pi n_2[\sin^2\theta - (n_1/n_2)^2]^{1/2}},$$

where n_1 is the refractive index of the sample and n_2 is the refractive index of the ATR element. If we were analysing polymeric materials with refractive indices of approximately 1.5, the sampling depth of the bands around 1050 cm^{-1} would be about 2 μm for a sample analysed on a 45° KRS-5 element ($n_2 = 2.35$). The incident angle is a function of the two variable angles (endface and optics) and the index of refraction of the crystalline materials being used as the ATR element as described by the equation:

$$\theta = \beta - \sin^{-1}\frac{[\sin(\beta - \psi)]}{n_2}$$

where β is the endface angle and ψ is the optics angle.

In our experiment we used a 45° endface KRS-5 with optic angles of 60°, 45°, 35°. The incident angles θ and the corresponding sampling depth at 1050 cm^{-1} are reported below:

Incident angles and sampling depth for KRS-5 at different optics angles

Endface angles (β) at optics angle (ψ)	Incident angle (θ)	Sampling depth (μm)
(β/ψ)		
45/60	51.3	1.4
45/45	45	2.0
45/35	40.8	3.8

ESCA studies. ESCA analyses were done using a modified Leybold E10 electron spectrometer with an MCD attachment. The X-ray source was a double-anode tube. For the measurements reported, Mg $K\alpha$ radiation was applied throughout (excitation voltage 13 kV, emission current 20 mA).

Samples of 10 mm × 14 mm were cut from the films immediately after opening the transport packages without touching the surface area to be analyzed. Each sample was introduced into ESCA chamber via a load lock system without further handling. In order to investigate the other side of the film, a new sample was cut out each time. The pressure in the analysis chamber was below 10^{-8} mb. The sample was held by two gold screws; one of which reached slightly into the analysis in order to provide additional calibration lines for the spectra. Correction of the apparent energy shift due to the sample charging, however, was done by making use of the prominent carbon lines. Survey spectra for each side of the films were acquired at a constant retardation ratio of 4 in 2 scans, a step time of 20 ms, and a step energy of 500 meV. High resolution spectra for the elemental lines were taken with a pass energy of 50 eV, a step time of 100 ms, and a step energy of 500 meV. The numbers of scans was between 8 and 12. High resolution spectra were collected for each element indicated in the survey spectrum [12] or held to be of interest by other considerations. The detector voltage was always 2.15 kV. In order to resolve the different carbon subpeaks owing to chemical shift, a commercial fitting programme was applied. Input values for this fitting programme are the halfwidth (FWHM) of the spectrometer (1.3 eV), the number of the expected bands, and their expected binding energy. The programme calculates the best fit under the condition that the sum of the single peak areas is equal to the area under the 'envelope band'. The programme does small corrections in the binding energy.

RESULTS

Morphology and structure

The microscopic analysis of hyaluronic acid derivative films shows that their surface display regular depressions of two ellipsoidal forms: one of which is about 100 μm long and 20 μm wide, the other 30 μm long and 10 μm wide. These depressions are particularly obvious on the surface of the HYAFF-11p75 film (Fig. 1A 11p75) and less pronounced on the surfaces of the other two films (HYAFF-07 (Fig. 1B 07) and HYAFF-11 (Fig. 1A 11). These depressions seem to result from the offset of the steel holder structure used in the preparation of the films since the surface of the oposite side which is homogeneous and smooth. The presence of some small particles and cracks was observed on the surfaces of HYAFF-11 and HYAFF-07 films.

The wide angle X-ray scattering (Fig. 2) indicates that the three samples are amorphous. The three films are stable up to 220°C after which they degrade (Figs 3 and 4). The DSC thermograms of all three films studied show the presence of an endotherm with the maximum at 90°C for the HYAFF-07 and HYAFF-11p75 films, and at about 80°C for the HYAFF-11 film (Fig. 3) whilst the glass transition was not observed in the temperature range scanned. It is interesting to note that the thermograms of all films annealed in a vacuum oven at 60°C for 2 days, exhibited an endotherm peak with a maximum at about 95°C which was sharper than those annealed in the presence of air (Fig. 3).

WAXS measurements have shown that the three samples are not semicrystalline, so the endothermal peaks are not due to the melting of crystals. They are probably due to the dissociation of hydrogen bonds. This phenomenon is also reported for other materials whose molecules are able to form hydrogen bonds, for example some polyurethanes [13]. At the moment, it does not seem plausible to attribute the peaks to the loss of water molecules bound to the saccharide matrix of the HYAFF molecules [14], because the phenomenon is still observed after annealing the samples in a vacuum oven at 60°C for two days. Moreover there is no evidence of weight loss on the thermogravimetric curves (see Fig. 4).

ESCA analysis

Figure 5 displays, as an example, the survey spectrum of HYAFF-07 (side A). The prominent peaks are photoelectron bands of C, O and N, and Auger lines of C

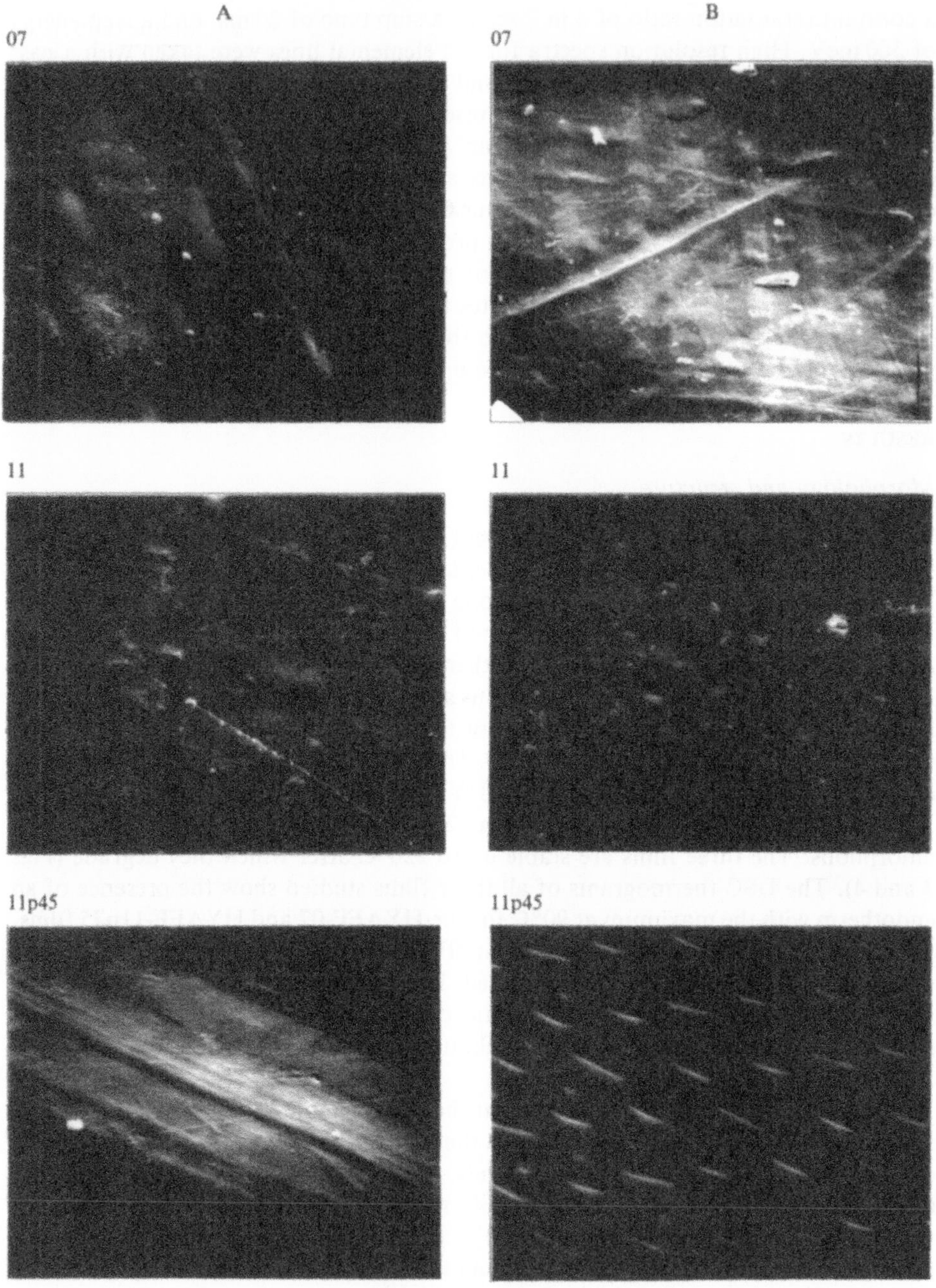
A
B
07
07
11
11
11p45
11p45

Figure 1. Electron micrographs of surface of HYAFF films.

and O. The Au lines belong to a gold screw used for additional calibration. High resolution spectra were taken for the elements found in the survey spectrum, which are nominally present in the HYAFF samples. No search for trace elements was done. Examples of high resolution spectra are shown in Figs 6–11. The energy scales (abscissa) are not corrected for shifts caused by sample charging.

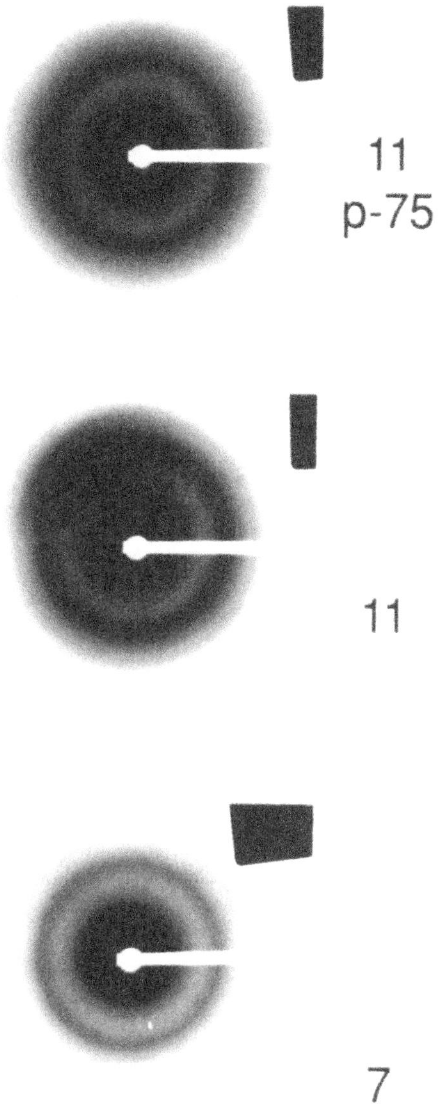
11
p-75
11
7

Figure 2. WAXS spectra of HYAFF films.

Carbon. Figures 6–8 show the carbon high resolution spectra of HYAFF-07, -11 and -11p75 (side A) respectively, together with the peaks obtained by application of the line fitting programme. The dominant C1*s* peak is attributed to the alcoholic (C—O—H) and ether (C—O—R) groups. Their chemical shifts are practically identical. The peak belonging to the second C atom of the glucosamine ring (to which the amide group is bound) is found in the same energy range. At a lower binding energy the aliphatic C line shows up as a shoulder in the HYAFF-07's C1 band and as a peak in the HYAFF-11 and -11p75 samples. At the high binding energy side two shoulders occur, one belongs to the amide groups, and the other one to the O—C—O groups; their C1 peaks are too close together to be resolved. The

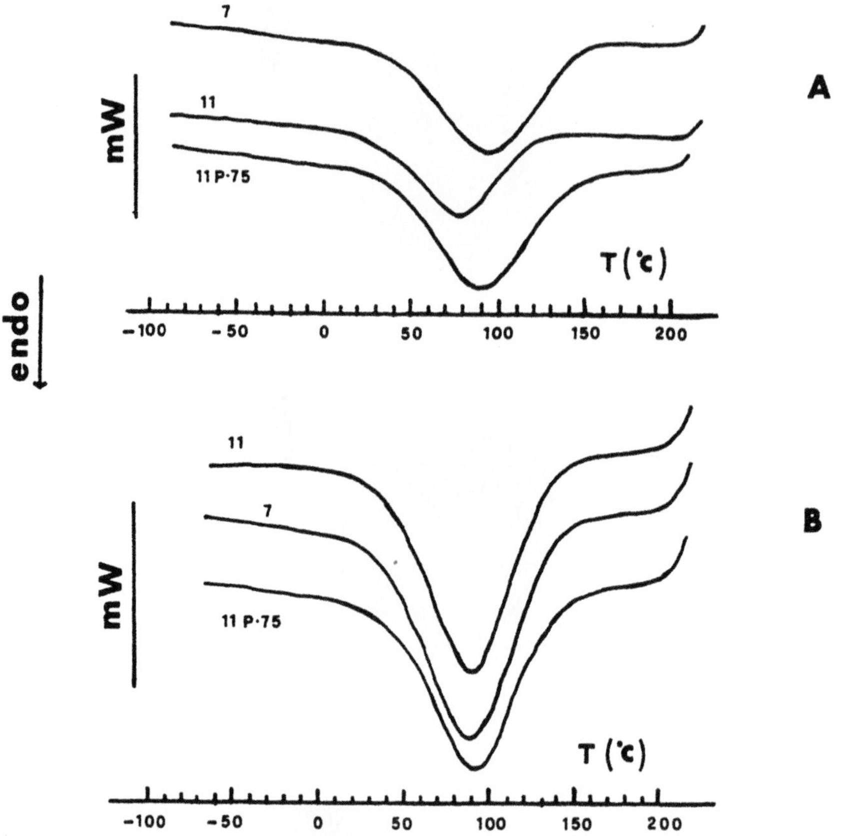
7
11
11 P·75
mW
A
endo
T(°c)
-100
-50
0
50
100
150
200
11
7
11 P·75
mW
B
T (°c)
-100
-50
0
50
100
150
200

Figure 3. DSC thermograms of HYAFF films (A) films as obtained from the synthesis; (B) samples annealed in the vacuum oven at 60°C for 2 days.

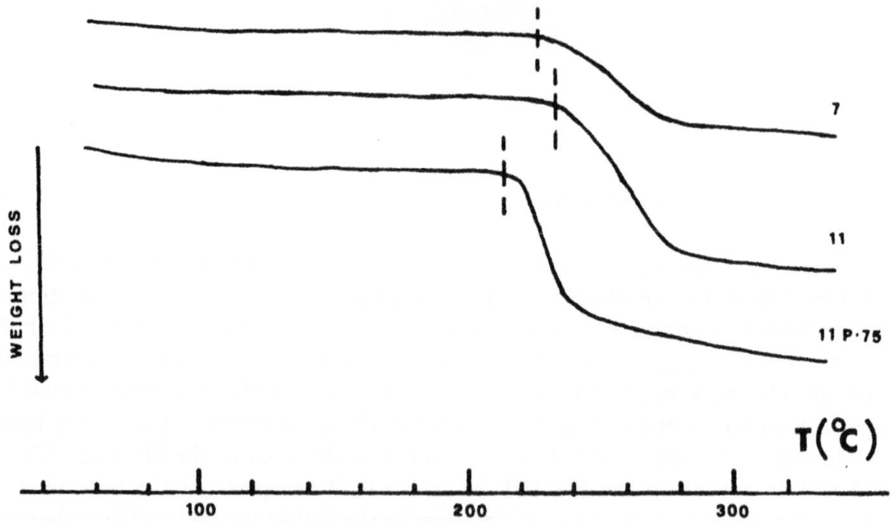
7
11
11 P·75
WEIGHT LOSS
T(°C)
100
200
300

Figure 4. Thermogravimetric curves of HYAFF films.

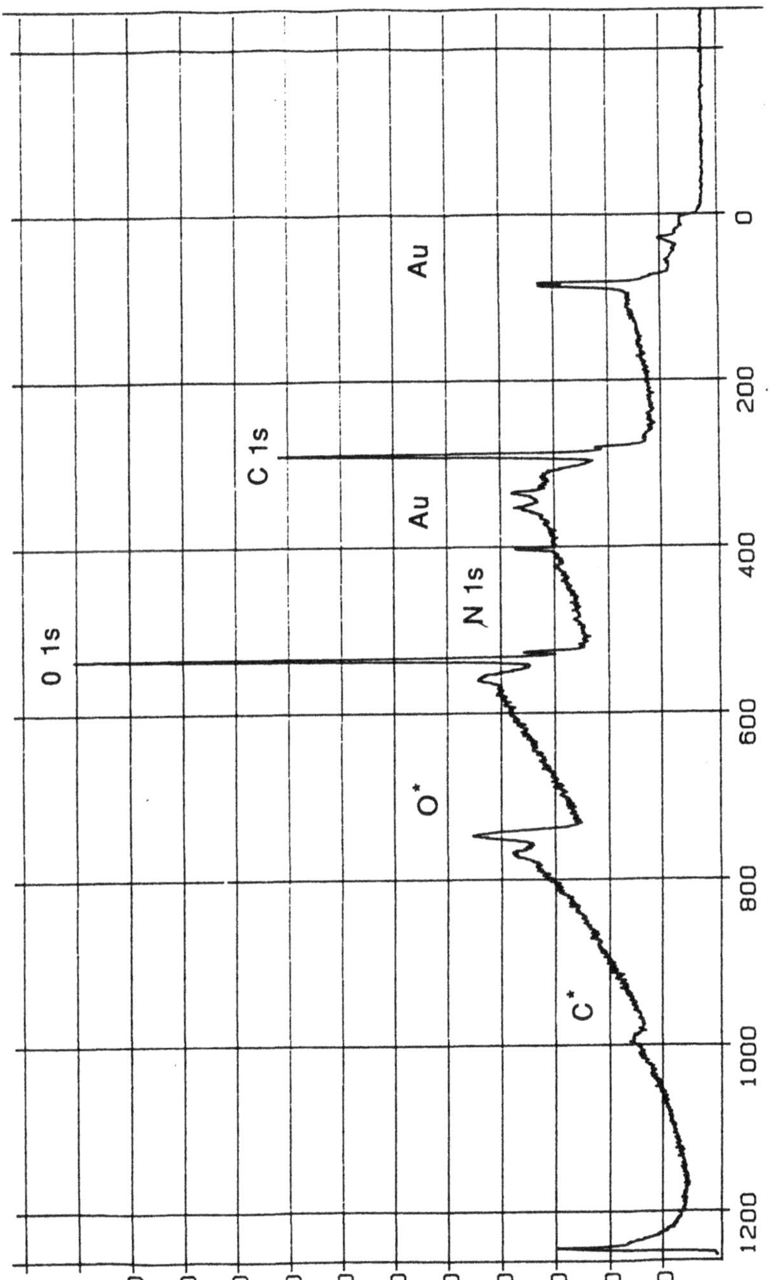

[*10³]
intensity
[cps]
lin
550
500
450
400
350
300
250
200
150
100
50
1200
1000
800
600
400
200
0
bind. energy (eV) lin
O 1s
C 1s
N 1s
Au
Au
O*
C*

Figure 5. ESCA survey spectrum; HYAFF-07, side A. Auger peaks are marked by an asterisk. Gold peaks are for calibration.

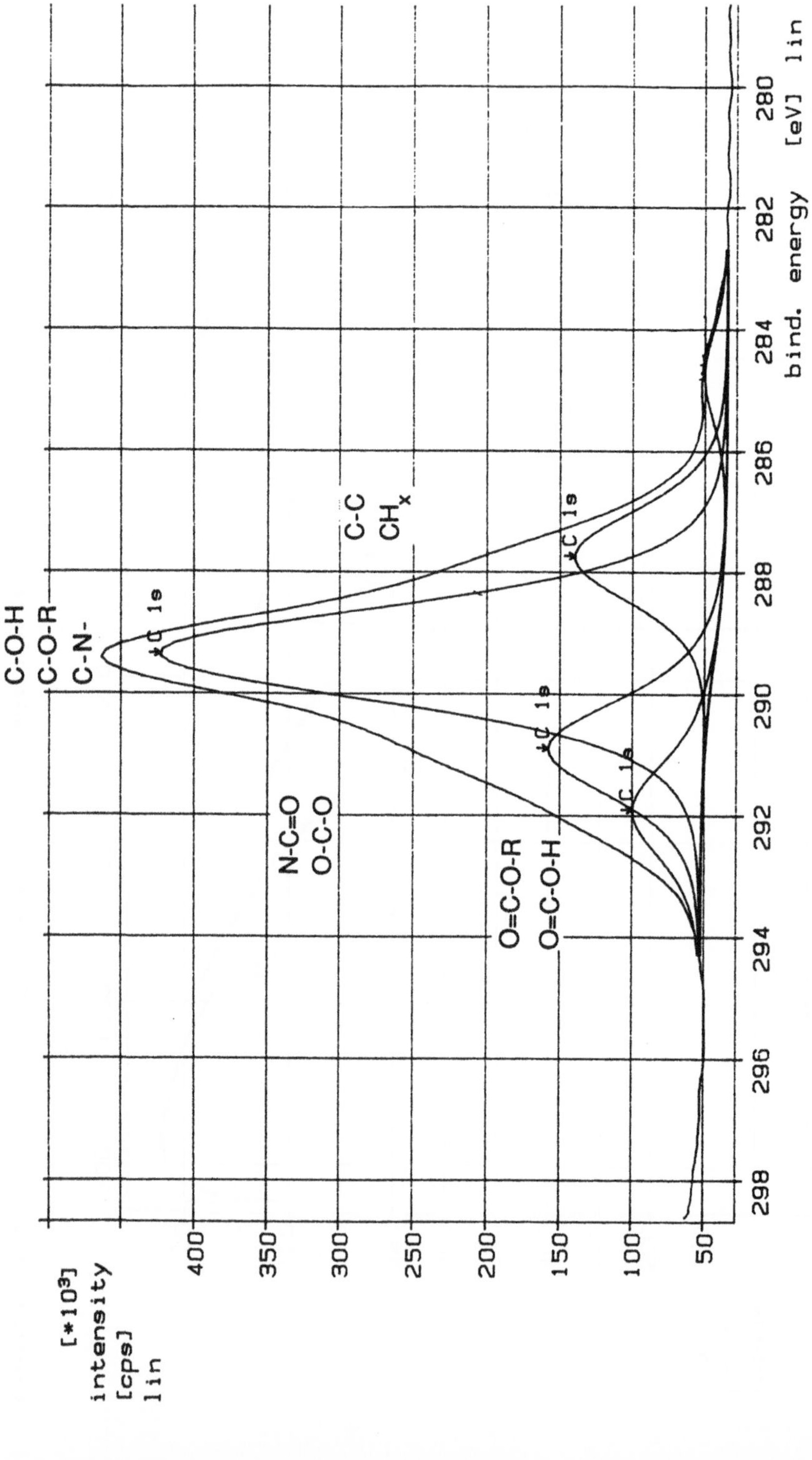
[*10³]
intensity
[cps]
lin
400
350
300
250
200
150
100
50
298
296
294
292
290
288
286
284
282
280
bind. energy [eV] lin
C-O-H
C-O-R
C-N-
C 1s
C-C
CHx
C 1s
N-C=O
O-C-O
C 1s
C 1s
O=C-O-R
O=C-O-H

Figure 6. Carbon spectrum of HYAFF-07, side A, with peak fitting.

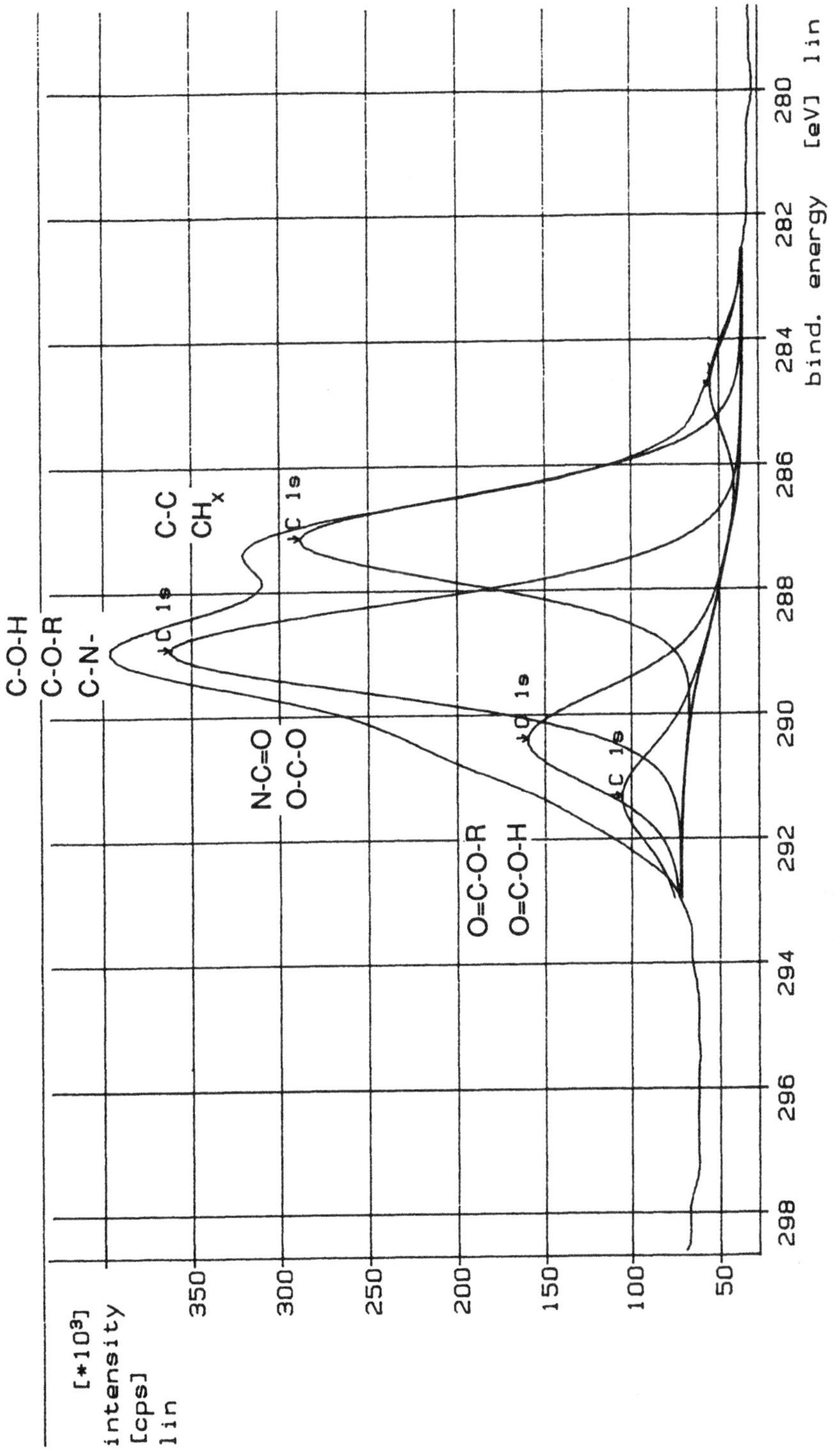
C-O-H
C-O-R
C-N-
C-C
CH$_x$
N-C=O
O-C-O
O=C-O-R
O=C-O-H
C 1s
[*10³]
intensity
[cps]
lin
350
300
250
200
150
100
50
298
296
294
292
290
288
286
284
282
280
bind. energy [eV] lin

Figure 7. Carbon spectrum of HYAFF-11, side A, with peak fitting.

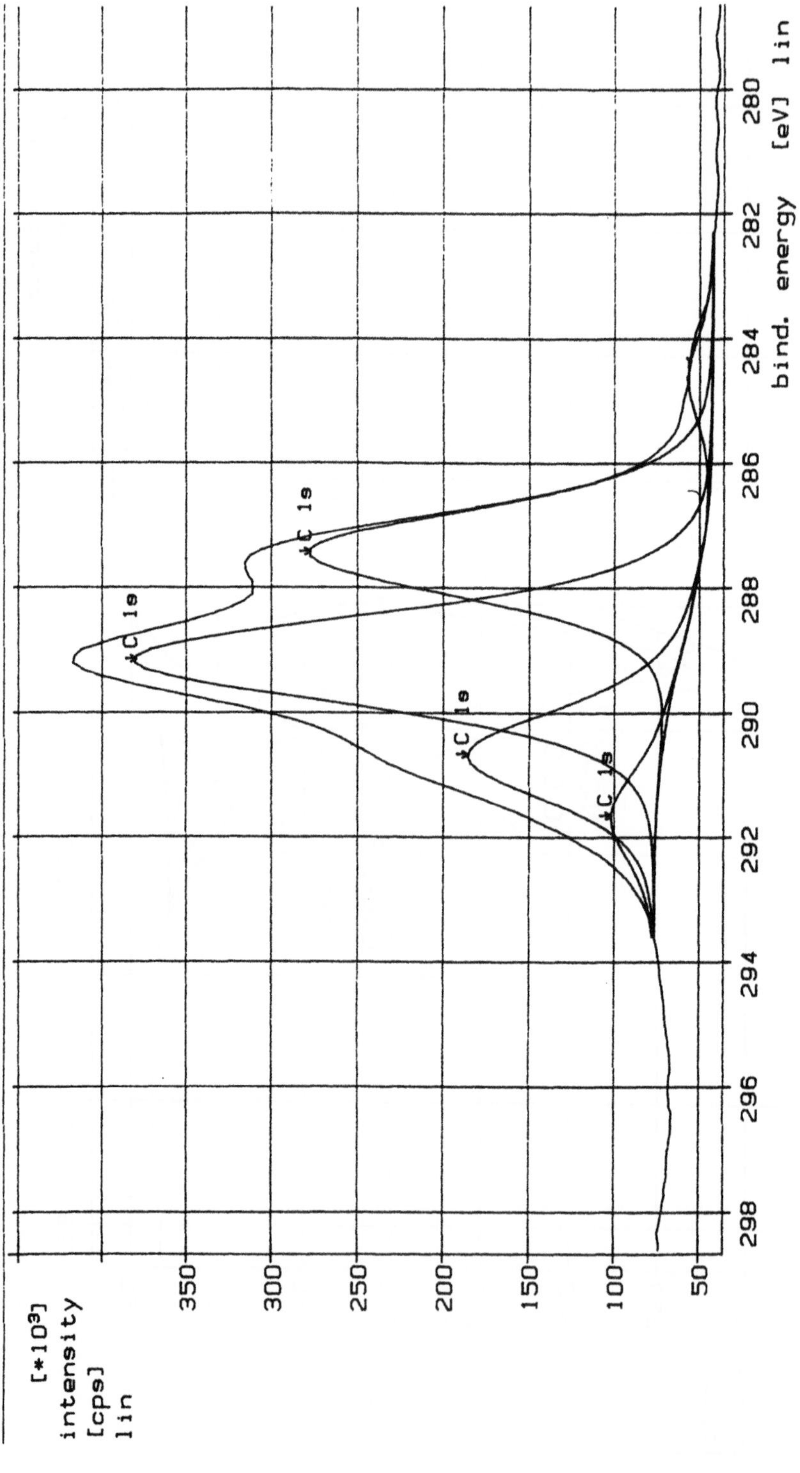
[*10³]
intensity
[cps]
lin
350
300
250
200
150
100
50
298
296
294
292
290
288
286
284
282
280
bind. energy [eV] lin
C 1s
C 1s
C 1s
C 1s

Figure 8. Carbon spectrum of HYAFF-11p75, side A, with peak fitting.

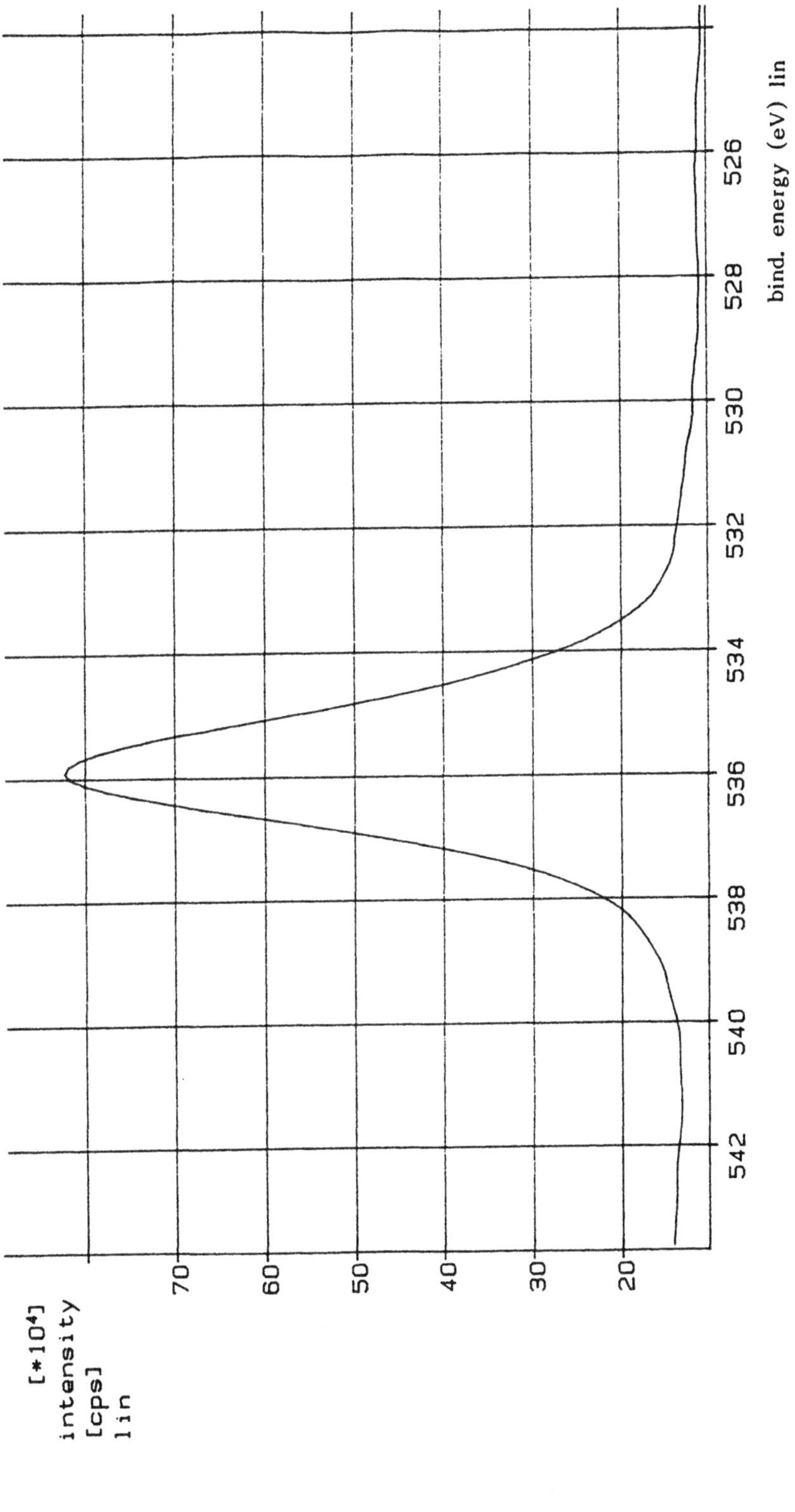
[*10^4]
intensity
[cps]
lin
70
60
50
40
30
20
526
528
530
532
534
536
538
540
542
bind. energy (eV) lin

Figure 9. Oxygen spectrum of HYAFF-07.

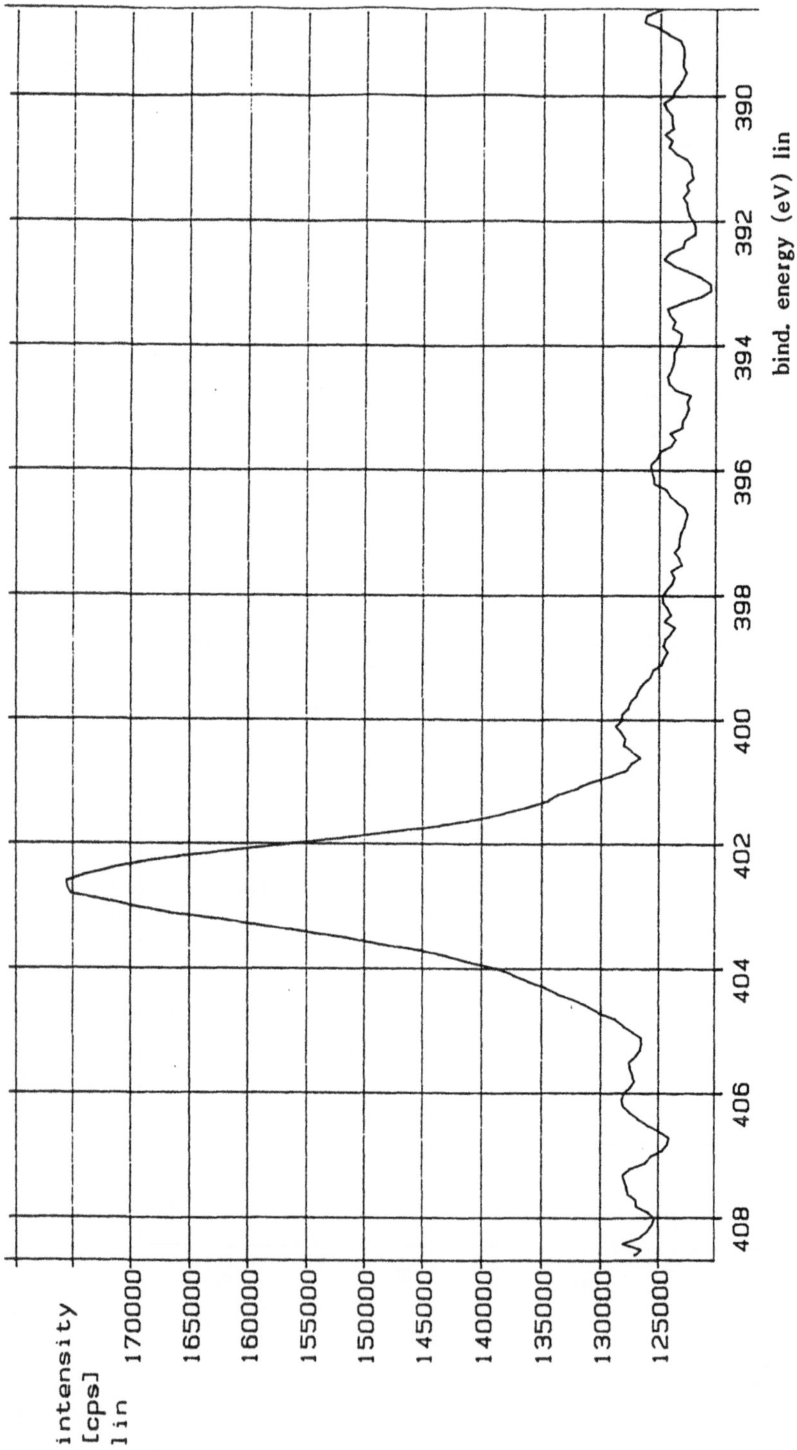
intensity
[cps]
lin
170000
165000
160000
155000
150000
145000
140000
135000
130000
125000
390
392
394
396
398
400
402
404
406
408
bind. energy (eV) lin

Figure 10. Nitrogen spectrum of HYAFF-11p75.

furthest shoulder on the high energy tail is the contribution of the protonated and esterified carboxyl C1 bands neither of which are discriminated.

Oxygen. The oxygen band with a halfwidth of 2.25 eV (Fig. 9) is expected, to contain, in principle, three contributions. The overlap of the O1*s* peak of O—C—O (2 atoms per repeat unit) and O—C=O (1 atom) is too high to resolve from the peak of C—O—C (2 or 3 atoms) and C—O—H (6 atoms). Under these circumstances application of a fitting programme does not seem appropriate. The smaller C=O peak (2 atoms) might be resolved, but with uncertainty. The oxygen of adsorbed water would virtually cause a peak shift of about 3–5 eV to lower binding energies; such a hypothetical peak, if present in the tail of the band, is not evaluated. Apparently in the ultrahigh-vacuum surrounding of ESCA the samples are rather dry. The energy position of the total oxygen peak is at a lower value than expected (by approximately 0.7–0.8 eV). A possible explanation could be a hydrogen interaction between neighbouring chains which could impose a partial double bond character on the O—H groups. Such effects may also be a reason for the excessive overlap between the oxygen bands.

Nitrogen. The N peak (example in Fig. 10) is somewhat noisy. Since only one N peak was to be expected and since the relative concentration is not too far from the theoretical bulk value, a considerably increased number of scans for further smoothing did not seem appropriate.

Sodium. The sodium peak is only seen with the HYAFF-11p75 samples (Fig. 11). This peak is also noisy, for by the same reason as for nitrogen, an excessive number of scans were not attempted.

Surface composition. Table 1 contains the elemental composition of the sample surfaces in terms of atomic percentages. The experimental values are calculated from the area under the ESCA peaks (taking into account atomic sensitivity factors). In addition, Table 1 contains the theoretical composition of the HYAFF films (in brackets). Carbon has been subdivided into four groups as discussed above.

Although the elemental composition is relatively close to the chemical formula of the polymer, the difference is big enough to demonstrate that the surface composition differs from that of the bulk. Carbon in C—C binding states is considerably more abundant at the surface. This is not surprising since air-borne carbohydrate molecules are usually adsorbed at the surfaces. This adsorbed layer reduces the surface free energy and, hence, puts the surface into a more stable state.

HYAFF-07 displays a clear compositional difference between side A and B, whereas for the two other films the differences observed are within the limits of error. Differences in the surface area may be partially masked by the contribution of the adsorption layer.

In spite of the superposition of the adsorbed hydrocarbon layer, the percentage of C atoms in the C—C environment reflects qualitatively in the correct order, the degree of esterification in the three samples. The esterification is expressed by one of six additional C-atoms contributing to the C—C peak for ethyl or benzyl ester, respectively, and by an additional C-atom contributing to the C—O peak.

Although most elemental contributions (except C—C) are relatively close to the theoretical bulk values, there are some differences in the carboxyl groups. These differences should not be overemphasized with regard to the problem of peak fitting

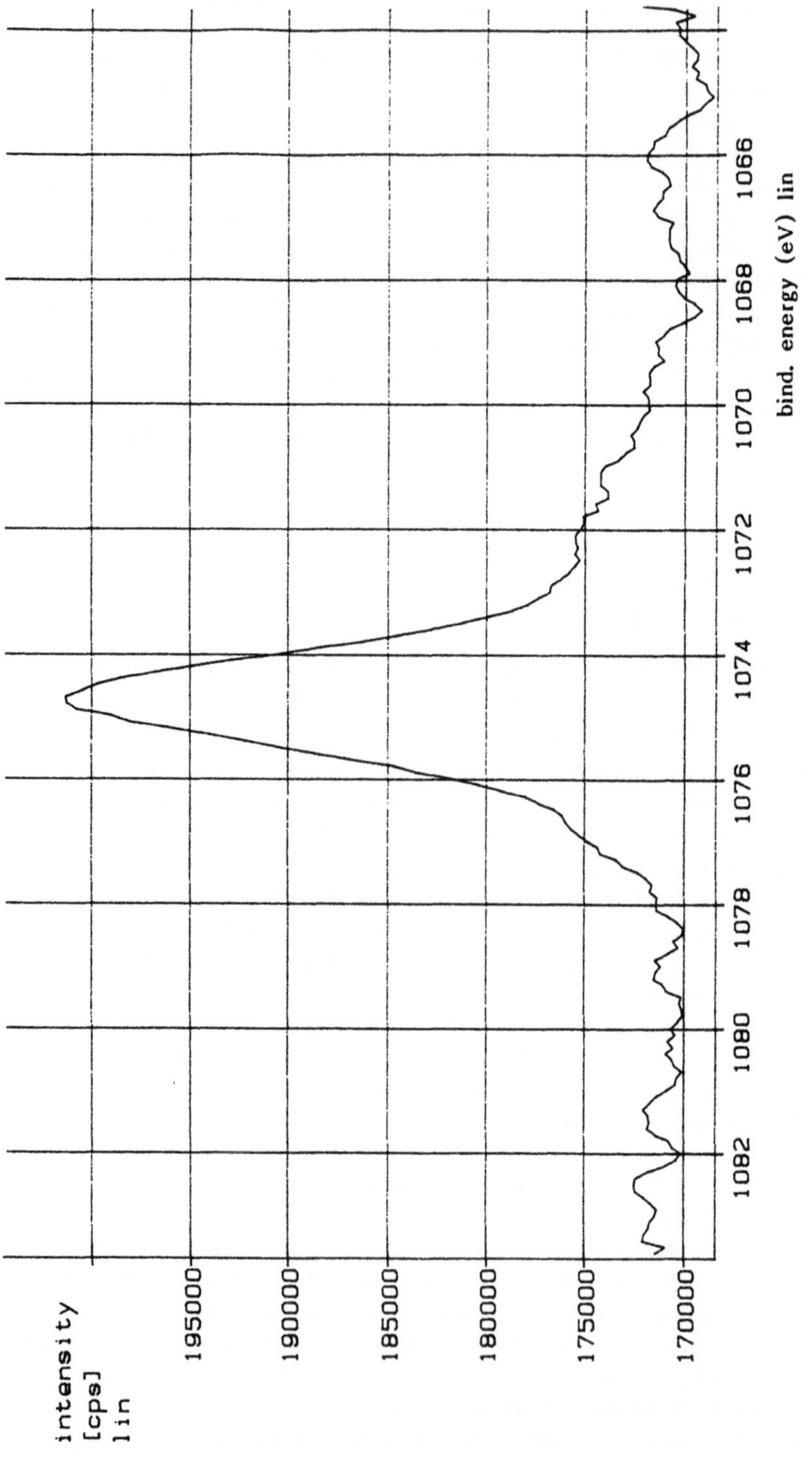
intensity
[cps]
lin
195000
190000
185000
180000
175000
170000
1066
1068
1070
1072
1074
1076
1078
1080
1082
bind. energy (eV) lin

Figure 11. Sodium spectrum of HYAFF-11p75.

Table 1.
Surface composition of HYAFF films according to ESCA analysis

Film No.	Side	Charging potential (V)	Element E_B (eV) Species	C 288.9 COO^- COOR	C 287.8 O—C—O N—C=O	C 286.3 COH COC C—N	C 284.6 C—C CH_x	O 532.7 O_{total} O—C [O=C]	N 399.8 NH	Na 1072 Na—O—
HYAFF-07	A	3.1	at. %, surface (exp)	4.7	10.6	37.1	10.1	34.1	3.5	0.0
			(theoretical at. %, bulk)	(3.57)	(10.71)	(35.71)	(7.14)	(39.29)	(3.57)	(0)
HYAFF-07	B	3.1	at. %, surface (exp)	4.9	10.8	35.3	12.6	33.1	3.4	0.0
			(theoretical at. %, bulk)	(3.57)	(10.71)	(35.71)	(7.14)	(39.29)	(3.57)	(0)
HYAFF-11	A	2.6	at. %, surface (exp)	3.4	9.1	29.8	23.9	30.7	3.0	0.0
			(theoretical at. %, bulk)	(3.03)	(9.09)	(30.3)	(21.21)	(33.33)	(3.03)	(0)
HYAFF-11	B	2.6	at. %, surface (exp)	3.8	9.3	29.2	24.9	30.0	2.9	0.0
			(theoretical at. %, bulk)	(3.03)	(9.09)	(30.3)	(21.21)	(33.33)	(3.03)	(0)
HYAFF-11p75	A	2.8	at. %, surface (exp)	2.5	10.8	30.2	21.7	31.4	2.7	0.7
			(theoretical at. %, bulk)	(3.17)	(9.52)	(30.95)	(17.46)	(34.92)	(3.17)	(0.79)
HYAFF-11p75	B	2.7	at. %, surface (exp)	2.3	9.8	30.0	22.0	32.1	2.9	0.8
			(theoretical at. %, bulk)	(3.17)	(9.52)	(30.95)	(17.46)	(34.92)	(3.17)	(0.79)

Table 2.
Ratio of atomic concentration (in different binding states) to carbon concentration in the C—O— binding state. ESCA results and theoretical values according to chemical formula (bulk)

Film No.	Side	Element E_B (eV) Feature		C 288.9 COO^- COOR	C 287.8 O—C—O N—C=O	C 286.3 COH COC C—N	C 284.6 C—C CH_x	CO 532.7 O_{total}	N 399.8 NHCO	Na 1071.8 Na—O—R
HYAFF-07	A	ratio to [C—O—]	exp (surface)	0.13	0.29	1.00	0.27	0.92	0.09	0.00
			(theor. bulk)	(0.10)	(0.30)	(1.00)	(0.20)	(1.10)	(0.10)	(0.00)
HYAFF-07	B	ratio to [C—O—]	exp (surface)	0.14	0.31	1.00	0.36	0.94	0.10	0.00
			(theor. bulk)	(0.10)	(0.30)	(1.00)	(0.20)	(1.10)	(0.10)	(0.00)
HYAFF-11	A	ratio to [C—O—]	exp (surface)	0.11	0.31	1.00	0.80	1.03	0.10	0.00
			(theor. bulk)	(0.10)	(0.30)	(1.00)	(0.70)	(1.10)	(0.10)	(0.00)
HYAFF-11	B	ratio to [C—O—]	exp (surface)	0.13	0.32	1.00	0.85	1.03	0.10	0.00
			(theor. bulk)	(0.10)	(0.30)	(1.00)	(0.70)	(1.10)	(0.10)	(0.00)
HYAFF-11p75	A	ratio to [C—O—]	exp (surface)	0.08	0.36	1.00	0.72	1.04	0.09	0.02
			(theor. bulk)	(0.10)	(0.31)	(1.00)	(0.56)	(1.13)	(0.10)	(0.03)
HYAFF-11p75	B	ratio to [C—O—]	exp (surface)	0.08	0.33	1.00	0.73	1.07	0.10	0.03
			(theor. bulk)	(0.10)	(0.31)	(1.00)	(0.56)	(1.13)	(0.10)	(0.03)

(the biggest errors are with the smallest peaks) but there seems to be a tendency for the HYAFF-07 to show higher values than the chemical formula, whereas HYAFF-11p75 gives lower values.

Since the deviation in the C—C groups owing to the adsorption layer distorts all other percentages, it is helpful to consider the ratios of all elemental groups to the most abundant C—O groups. These have been compiled in Table 2. Again the theoretical ratios according to the chemical formula have been included in brackets. This way of presenting the results reveals the most prominent difference between surface and bulk composition to be in the the hydrocarbon adsorption layer.

ATR/FT-IR analysis

Dry samples. Figure 12 depicts the spectra of the studied films together with that of Hyaluronic acid in the form of the Hyaluronate salt. These spectra are independent of the film side from which they were taken. The main frequencies observed in these spectra together with their assignment are presented in Tables 3 and 4.

Table 3.
Main frequencies observed in the spectra of HYAFF-07 and their assignments

Frequencies (cm^{-1})	Assignments
3550–3200	OH + NH stretching
2900–2800	CH_2 stretching
1745	C=O stretching of carboxyl group
1636	Amide I
1557	Amide II
1425	CH_2 bending
1370–1210	Amide III
1077	C=O stretching of carboxyl group
1050, 1022	C—O—C stretching
860	CH_2—CH_3 bending

Table 4.
Main frequencies observed in the spectra of HYAFF-11 and HYAFF-11p75 and their assignments

	Assignments	
Frequencies (cm^{-1})	HYAFF-11	HYAFF-11p75
3550–3200	OH + NH stretching	OH + NH stretching
2900–2800	CH_2 stretching	CH_2 stretching
1745	C=O stretching of carboxyl group	
1742		C=O stretching of carboxyl group
1655–1620	Amide I + aromatic ring	
1650–1615		Amide I + aromatic ring + asym. stretch. of carboxylate group
1550–1540	Amide II	Amide II
1430		CH_2 bending
1425	CH_2 bending	
1407		Sym. stretch. of carboxylate group
1380–1210	Amide III	Amide III
1077	C=O stretching of carboxyl group	C=O stretching of carboxyl group
1050, 1022	C—O—C stretching	C—O—C stretching
998	Aromatic ring	Aromatic ring

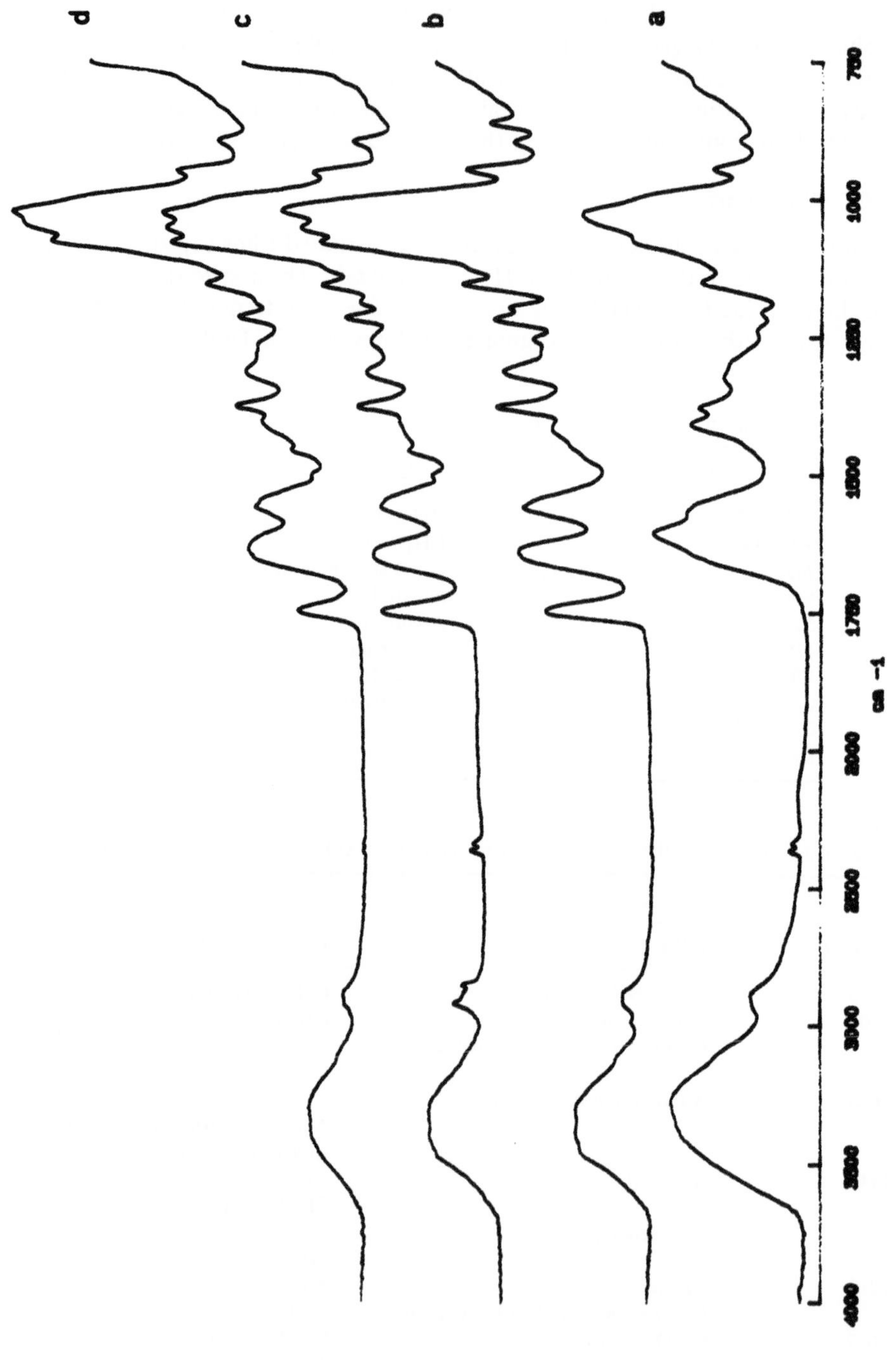
d
c
b
a
4000
3500
3000
2500
2000
1750
1500
1250
1000
750
cm -1

Figure 12. ATR/FT-IR absorbance spectra of: (a) HYALURONATE salt film; (b) HYAFF-07 film; (c) HYAFF-11 film; (d) HYAFF-11p75 film.

Hydrated samples. In Figs 13–15 the spectrum of the hydrated sample is compared to that of the dry sample of HYAFF-07, HYAFF-11, and HYAFF-11p75 respectively. As in the case of dry samples these spectra were independent of the film side from which they were taken.

HYAFF-07. The film swells in 0.9% NaCl and its IR spectrum in the hydrated state exhibits a small decrease in the frequencies (less than 10 cm^{-1}) of ester, amide and ether bands relative to the dry sample (Fig. 13). This may be explained by the possibility of hydrogen bond formation between these groups and water molecules, or with other ester, amide, or ether groups via intercalated water molecules. Moreover, in the hydrated state the band intensity at 1077 cm^{-1} is higher than that at 1022 cm^{-1}, whereas in the dry state the intensity at 1022 cm^{-1} is higher than that at 1077 cm^{-1}. This would suggest that the rearrangement of some chemical groups takes place during hydration of the film.

HYAFF-11. The film slightly swells in 0.9% NaCl. The Amide I, II and III spectral regions of this film depend very little on when the film is hydrated or dry (Fig. 14). However, the lower frequencies of amide groups are more pronounced for the sample in its hydrated state than in the dry state indicating that the H-bond interactions involving amide groups are favoured upon hydration.

HYAFF-11p75. The film swells in 0.9% NaCl solution. As for the HYAFF-07 film, the band intensity at 1077 cm^{-1} of this sample with respect to that at 1022 cm^{-1} increases after its hydration in 0.9% NaCl (Fig. 15). This again would indicate that some surface rearrangement occurs upon hydration.

Spectra at different depth. The comparison of the representative regions of the spectra of HYAFF-07 and HYAFF-11 (Fig. 16) shows a decrease in the band intensity at 1077 cm^{-1} (—COO stretching) compared to that at 1022 cm^{-1} (C—O—C stretching) with increasing penetration depth. This indicates that ester group concentration is higher at the surface than in the bulk of the material. In the case of HYAFF-11p75 the ester (and carboxylate) group concentration increases from the layer 1.4 μm deep to the layer 2.0 μm deep and then decreases to the layer 3.8 μm deep. Band intensity ratios at 1077 and 1022 cm^{-1} were also obtained at different depths for the hydrated samples. The results are summarized in Table 5. In the case of hydrated samples we observed different trends for the different kind of

Table 5.
Ester (1077 cm^{-1})/ether (1022 cm^{-1}) band intensity ratios at different depths

Compound	Dry sample	Depth (μm)	Wet sample
HYAFF-07	1.02	1.4	1.28
	0.89	2.0	1.25
	0.68	3.8	1.20
HYAFF-11	1.02	1.4	0.96
	1.00	2.0	1.02
	0.92	3.8	1.07
HYAFF-11p75	0.78	1.4	1.11
	0.86	2.0	1.14
	0.77	3.8	1.10

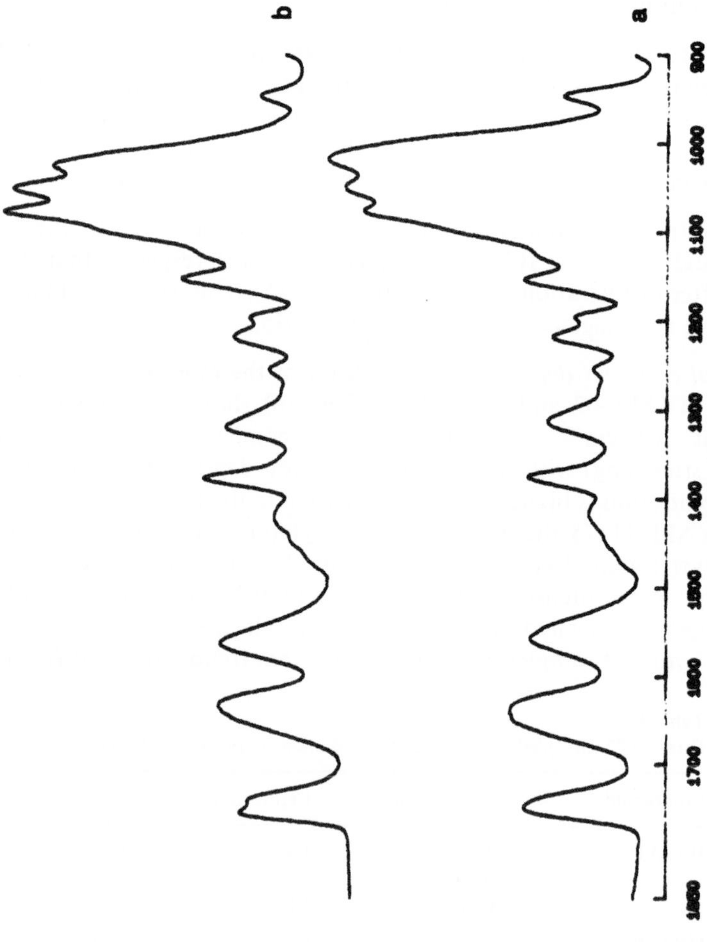
b
a
900
1000
1100
1200
1300
1400
1500
1600
1700
1800
cm −1

Figure 13. ATR/FT-IR absorbance spectra of HYAFF-07 film: (a) dry sample; (b) wet sample.

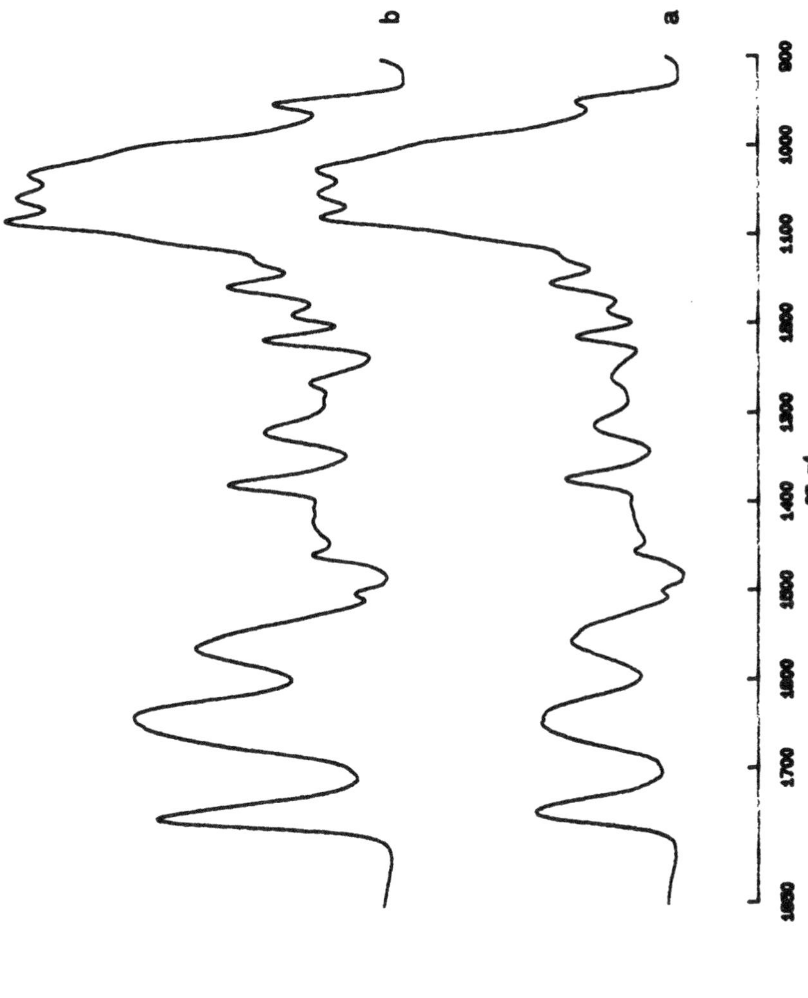
b
a
1800
1700
1600
1500
1400
1300
1200
1100
1000
900
cm -1

Figure 14. ATR/FT-IR absorbance spectra of HYAFF-11 film: (a) dry sample; (b) wet sample.

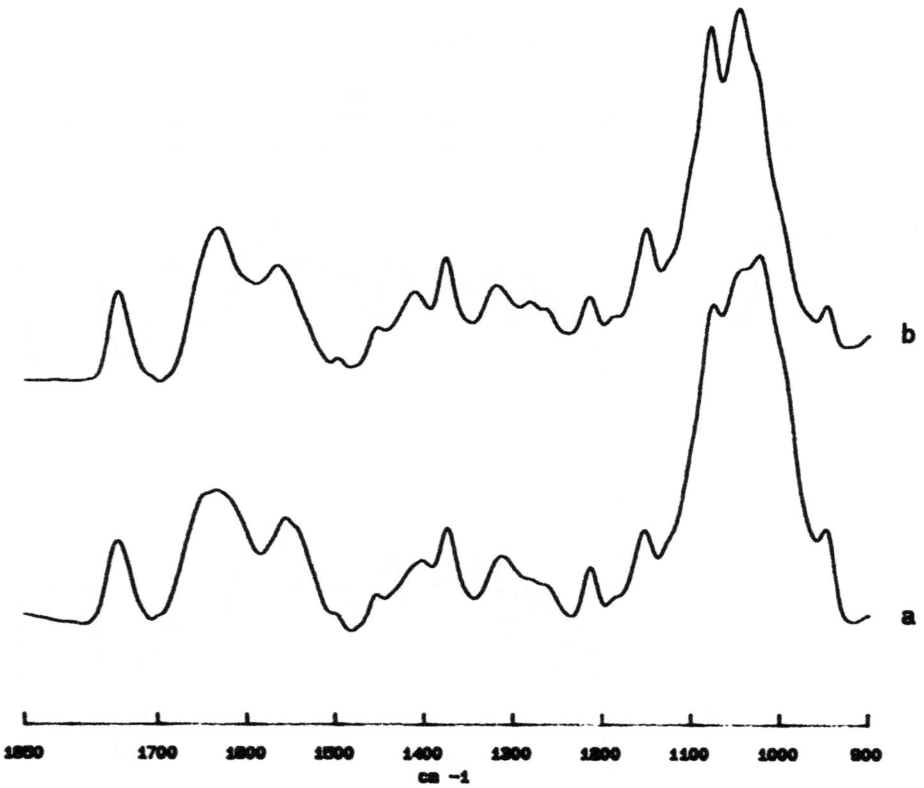
b
a
1850
1700
1600
1500
1400
1300
1200
1100
1000
900
cm -1

Figure 15. AFT/FT-IR absorbance spectra of HYAFF-11p75 film: (a) dry sample; (b) wet sample.

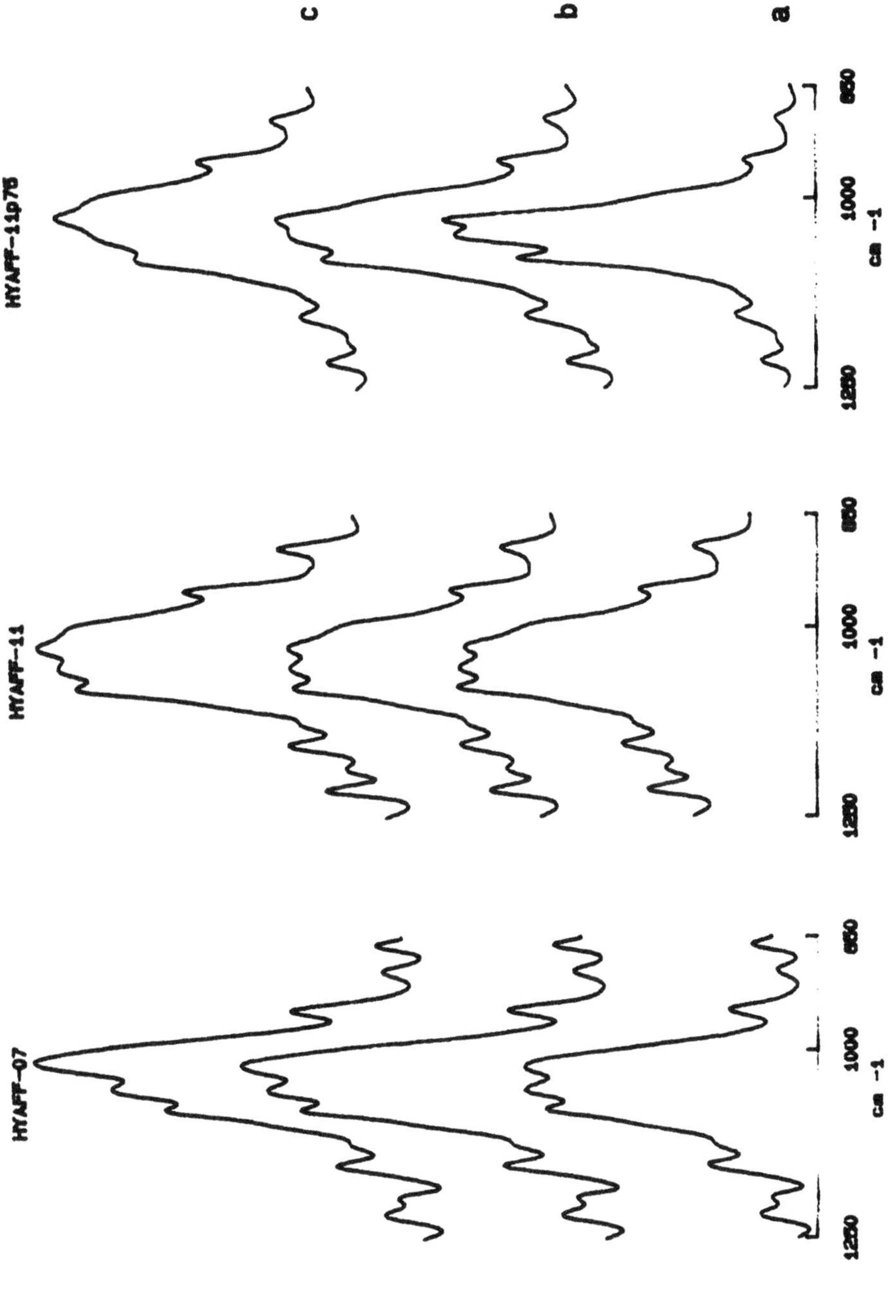
HYAFF-07
HYAFF-11
HYAFF-11p75
1250
1000
950
cm -1
a
b
c

Figure 16. ATR/FT-IR absorbance spectra of HYAFF-07, HYAFF-11 and HYAFF-11p75 at different depth: (a) 1.4 μm; (b) 2.0 μm; (c) 3.8 μm.

materials. In particular: hydrated HYAFF-07 sample behaves like the dry sample, i.e. the ester group concentration decreases with increasing depth; in the hydrated HYAFF-11 sample the ester group concentration increases from the surface to the bulk, contrary to what happened for the dry sample; and for hydrated HYAFF-11p75 we found an increase of the carboxyl band from the layer 1.4 μm deep to the layer 2.0 μm deep, and then a decrease to the layer 3.8 μm deep, as for the dry sample.

Contact angle measurements

Table 6 summarizes the results of contact angle measurements for HYAFF derivative films.

The main conclusions which one may draw from these results are as follows: (1) All films exhibit differences in wettability between their two faces. However these differences were less pronounced with HYAFF-11p75. (2) All films swell in water. This is clearly demonstrated when comparing the θ_R values with the values of θ'_R. The hysteresis of the contact angle which on dry surfaces ($\theta_A - \theta_R$) is relatively small, becomes very important on wetted films ($\theta_A - \theta'_R$). (3) From θ_A values the most hydrophobic out of the three HYAFF films is the HYAFF-11p75 film. The film is brittle and breaks when fixed on a sample holder in contact with water. (4) The comparison between HYAFF-07 and HYAFF-11 films clearly shows that HYAFF-11 is more hydrophobic (θ_A is close to 60 deg) than HYAFF-07 (θ_A is close to 50 deg). Consequently the γ_{SV} for HYAFF-11 was found to be in the 31–39 mN/m range while for HYAFF-07 this value was in the 42–50 mN/m range. (5) The

Table 6.

		Face A			Face B		
		HYAFF-7	HYAFF-11	HYAFF-11p75	HYAFF-7	HYAFF-11	HYAFF-11p75
Dry surfaces	θ_A	46.4	56.5	61.5	54.5	64.0	65.0
	θ_R	38.6	50.8	51.4	46.8	60.7	54.3
	$\theta_A - \theta_R$	7.8	5.8	10.1	7.7	3.3	11.7
	W_A	122.0	111.9	106.7	114.1	104.0	104.6
	W_R	128.7	117.8	117.4	121.6	107.5	114.3
	$W_R - W_A$	7.7	5.9	10.7	7.5	3.5	9.7
	γ_{SV}	49.8	39.7	34.5	41.9	31.78	29.36
Wetted surfaces	θ'_R	25.4	21.4	*a*	42.5	38.7	*a*
	$\theta_A - \theta'_R$	21.0	29.2	*a*	12.0	25.3	*a*
	W'_R	137.4	136.3	*a*	125.4	128.6	*a*
	$W'_R - W_A$	15.4	24.4	*a*	11.3	21.1	*a*
	θ_0	*b*	141.0	*a*	*b*	125.8	*a*
	I_p^{AB}	*b*	89.9	*a*	*b*	80.2	*a*
	W_{SL}^{LW}	*b*	46.4	*a*	*b*	48.4	*a*
	I_p^{AB}/W_{SL}^{LW}	*b*	1.94	*a*	*b*	1.96	*a*

The reported contact angle values are the mean values from the measurements on 12 drops of liquid or air bubbles. The scatter of the mean never exceeded ±3, except on HYAFF-11p75 Face B θ_R where the deviation from the mean was 73.5 deg and HYAFF-11 Face A for θ'_R ±4, 3.

[a] Impossible to measure, the film breaks in contact with water.

[b] The surface is too slippery. Octane drop escapes.

HYAFF-11 exhibits the lowest hysteresis of contact angles which would indicate that its roughness and its heterogeneity are relatively small.

DISCUSSION

Dry surfaces

All the three HYAFF materials are amorphous and thermostable up to 220°C. Although the surfaces are smooth and homgeneous on one side of the film, some regular depressions were observed on the other face and seem to be the offset of the surface texture of the holder which is used in their fabrication.

The surface composition of HYAFF samples differs from the elemental composition obtainable from the chemical formula. The percentage of ester groups present at the surface is higher than that which was calculated from the chemical formula of the materials for HYAFF-07 and -11. On the contrary, for HYAFF-11p75, a decrease in the amount of COOR (COO^-) groups on the surface with respect to the theoretical value was observed (Tables 1, 2 and 8). Thus, the carboxyl groups seem to be orientated towards the bulk, favouring the presence of other more hydrophobic groups at the surface, generally N—C=O or O—C—O.

Infrared analysis shows that the [ester]/[ether] ratio decreases with increasing sampling depth in HYAFF-07 and HYAFF-11 films, whereas in the HYAFF-11p75 sample an increase in the [ester(carboxylate)]/[ether] ratio is observed from the layer 1.4 μm deep to the layer 2.0 μm deep.

The wetting behaviour of a surface is basically determined by its surface chemical composition. The degree of ESCA analysis achieved in this study does not completely reflect the contributions of polar groups being more polar than others. An approximation of the polar groups, however, can be made from the intensity of total oxygen to aliphatic carbon. The 'hydrophilicity ratio' is entered in Table 7 together with the contact angle data. Both measurements are in agreement, with regard to HYAFF-07 being the most hydrophilic of all the films under investigation. Furthermore, both measurements confirm that side A is more hydrophilic. However, the hydrophilicity order of HYAFF-11 and HYAFF-11p75 is different, although the quantitative difference in the respective data is not large.

Table 7.
Contact angle Q, surface energy g, and ESCA hydrophilicity ratio.

Film no.	Side	Contact angle θ (degrees)	cos θ	γsv [mN/m]	ESCA hydrophilicity ratio
HYAFF-07	A	46.4	0.69	49.8	3.4
HYAFF-07	B	54.5	0.58	41.9	2.6
HYAFF-11	A	56.6	0.55	39.7	1.3
HYAFF-11	B	64.0	0.44	31.8	1.2
HYAFF-11 p 75	A	61.5	0.48	34.5	1.5
HYAFF-11 p 75	B	66.0	0.41	29.4	1.5

Hydrated surfaces

Hydration of the HYAFF films results in significant differences in film composition just below the surface; whilst for HYAFF-07 and HYAFF-11p75 the [ester and/or carboxylate]/[ether] ratio in region just below the surface increases with respect to the dry sample, a decrease in this ratio was observed for HYAFF-11 films (see infrared data). The benzyl group interacts less with water molecules than ethyl or COO^- group. HYAFF-11 also exhibits the lowest degree of swelling when in contact with NaCl solution.

Upon hydration the 1077 cm^{-1}/1022 cm^{-1} band intensity ratio of HYAFF-07 decreases with increasing depth, as occurred for the dry sample. The opposite trend is observed with HYAFF-11 with respect to the dry sample, showing that the hydrophobic benzyl group repels the water and prefers to be buried in the bulk. As for the dry sample, the 1077 cm^{-1}/1022 cm^{-1} ratio of the hydrated HYAFF-11p75 does not change linearly with increasing sampling depth. This nonlinear trend can be

Table 8.

			COOR		COR(H)
HYAFF-07 Theor.			COOR 3.6		COR(H) 35.7
DRY	Surface ↓ Bulk		4.7–4.9[b,1]	Hydrophilic[a]	37.1–35.3[b,1]
			decreases[c]		increases[c]
		↓(+ H_2O)			
WET	Surface ↓ Bulk			Swells[a,c]	
			decreases[c]		increases[c]
HYAFF-11 Theor.			COOR 3.0		COR(H) 30.3
DRY	Surface ↓ Bulk		3.4–3.8[b,1]	Hydrophobic[a]	29.8–29.2[b,1]
			decreases[c]		increases[c]
		↓(+ H_2O)			
WET	Surface ↓ Bulk			Slightly swells[a,c]	
			increases[c]		decreases[c]
HYAFF-11p75 Theor.			COOR(Na) 3.2		COR(H) 31.0
DRY	Surface ↓ Bulk		2.5–2.3[b,1]	Hydrophobic[a]	30.2–30.0[b,1]
			decreases[c]		increases[c]
		↓(+ H_2O)			
WET	Surface ↓ Bulk			Swells[a,c]	
			increases[c]		decreases[c]

explained in terms of the different character of the carboxylate (hydrophilic) and benzyl (hydrophobic) groups.

In Table 8 a picture of composition and properties of the three different HYAFF samples is given by gathering the information from the performed measurements both in the dry and wet state. All that allows us to know the behaviour of these materials when they come in contact with physiological liquids and with cells. It is worth mentioning that HYAFF-11 characterized by a high polar/apolar ratio of polymer–water work of adhesion ($I_p^{AB}/W_{SL}^{LV} \simeq 2$) for the reorientation of COOR/COR groups when in contact with water shows the lowest number of platelets adhering to the surface.

Acknowledgements

The authors would like to thank FIDIA S.p.A. for supplying the HYAFF samples.

REFERENCES

1. F. Della Valle and A. Romeo, Hyaluronic acid esters and their medical and cosmetic uses formulations. *Eur. Patent Appl. EP 216453*, April 1987, p. 129.
2. A. Pastorello, M. Stefani, L. Benedetti, F. Biviano and L. Callegaro, *Proceedings of the 9th European Conference of Biomaterials.*
3. L. Andreassi, L. Casini, E. Trabucchi, S. Diamantini, A. Rastrelli, L. Donati, M. L. Teuchini and M. Malcovati, *Wounds* **3**, 116–126 (1991).
4. A. Rastrelli, M. Beccaro, F. Biviano, G. Calderini and A. Pastorello, *Clinical Implant Materials, Advances in Biomaterials*, Vol. 9, G. Heimke, U. Soltesez and A. J. C. Lee (Eds). Elsevier, Amsterdam (1990).
5. R. Barbucci, A. Baszkin, M. Benvenuti, M. L. Da Costa and P. Ferruti, *J. Biomed. Mater. Res.* **21**, 443–47 (1987).
6. A. W. Newman, R. J. Good, C. J. Hope and M. Sejpal, *J. Colloid Interface Sci.* **49**, 291–302 (1974).
7. C. J. van Oss, M. K. Chandury and R. J. Good, *Chem. Rev.* **88**, 927–941 (1988).
8. R. Barbucci, M. Casolaro, A. Magnani and C. L. Roncolini, *Polymer* **32**, 897–904 (1991).
9. R. Barbucci, A. Albanese, A. Magnani and F. Tempesti, *J. Biomed. Mater. Res.* **25**, 1259–1274 (1991).
10. K. Knutson and D. J. Lyman, in: *Surface and Interfacial Aspects of Biomedical Polymers*, Vol. 1, Chapter 6, J. D. Andrade (Ed.). Plenum Press, New York (1985).
11. N. J. Harrick, in: *Internal Reflection Spectroscopy*, p. 30. John Wiley and Sons, New York (1979).
12. A. Dilks, in: *Electron Spectroscopy: Theory. Techniques and Applications*, Vol. 4, p. 227, D. R. Brundle and A. D. Baker (Eds). Academic Press, London (1981).
13. S. B. Clough and N. S. Schnideider, *J. Macromol. Sci. Phys.* **2**, 553 (1968).
14. A. Rastrelli, M. Beccaro, F. Biviano, G. Calderini and A. Pastorello, *Clinical Implant Materials, Advanced in Biomaterials*, Vol. 9, p. 199, G. Heimke, U. Soltz and A. J. C. Lee (Eds). Elsevier, Amsterdam (1990).

Fibroblast contraction of collagen matrices with and without covalently bound hyaluronan

LYNN L. H. HUANG-LEE and MARCEL E. NIMNI*

Departments of Biochemistry and Molecular Biology and Surgery, School of Medicine, University of Southern California and Children Hospital Los Angeles, Los Angeles, California, USA

Received 1 September 1992; accepted 9 December 1992

Abstract—Hyaluronan, which is found in high concentrations in fetal tissues, has been suggested to play a major role in preventing scar formation in fetal wounds. We have developed a floating collagen fibrillar matrix (CFM) made out of reconstituted type I collagen for the purpose of evaluating the ability of hyaluronan to inhibit the fibroblast induced contraction of the matrix. When hyaluronan is covalently bound to collagen it appears to better support fibroblast proliferation and matrices are less contractible by these cells than when hyaluronan interacts only ionically. When hyaluronan is bridged between collagen fibrils by a network of extensive covalent crosslinks, contractibility by fibroblasts is abolished. These modified collagen matrices may prove to be very useful in the development of bioprostheses and implants.

Key words: Hyaluronan; collagen; contraction; periodate; crosslink; fibroblast.

INTRODUCTION

Scarless fetal wound healing is a remarkable process which has been actively studied by many researchers [1-4]. Stern *et al.* [1] found that increased levels of hyaluronan appear early during the repair process of both fetal and adult wounds. The level of hyaluronan drops rapidly after day 3 in adult wounds. In fetal wounds before the third trimester, hyaluronan deposition remains elevated for an extended time and no scar formation is found. Siebert *et al.* [3] applied hyaluronan to postnatal wounds topically and found a reduced amount of scarring. The underlying mechanisms by which hyaluronan inhibits wound contracture to prevent scar formation during the healing process remain unknown.

Hyaluronan extracted from human skin and scar tissue is found to associate with collagen and other proteins [3, 5]. Therefore, we chose a collagen fibrillar matrix (CFM) as an *in vitro* model to study the effects of hyaluronan on wound contracture, since collagen is not only predominant in the extracellular matrix but also serves as a support for connective tissues. In addition, collagen and its derived matrices have been suggested as bioprostheses for burn wound dressings [6] and tissue templates [7] due to their biocompatibility [8]. The desirable properties of these matrices for facilitating wound healing processes include stimulation of cell migration and infiltration [9], support of cell proliferation [9], as well as matrices which are non-contractible [10].

Collagen matrices have been used as a model to study the interactions of fibroblasts with extracellular matrices [11-15]. The collagen matrices used in most

* To whom correspondence should be addressed. Children Hospital Los Angeles, Smith Research Tower, 8th Floor, 4650 Sunset Blvd, Los Angeles, CA 90027, USA.

experiments contain 1.5 mg/ml of collagen with fibroblasts distributed throughout the attached matrix and the contraction is measured by the changes of its thickness [11, 13]. At this low concentration (1.5 mg/ml), collagen matrices undergo significant self contraction under physiological conditions. Usually only the outermost surfaces of implanted matrices are initially in contact with cells. Therefore, we sought to establish a CFM using a higher concentration (3 mg/ml) of reconstituted type I collagen, and measure the surface area of the floating CFM as an indicator of contraction after fibroblasts seeding on one surface. This floating CFM model has allowed us to observe the effects of hyaluronan on the inhibition of CFM contraction by fibroblasts better [16]. In our earlier experiments, we found that the contraction of CFM by fibroblasts was greatly reduced when high concentrations (>1 mg/ml) of hyaluronan were present in the media. In the present work, we compared the effects of hyaluronan which was either ionically or covalently bound to collagen on the process of CFM contraction by fibroblasts.

MATERIALS AND METHODS

Preparation of reconstituted type I CFM

Pepsin soluble type I collagen was isolated and purified from bovine skin as described [16]. Collagen solutions were dialyzed against physiological phosphate buffer saline (PBS) at 4°C and sterilized by ultracentrifugation twice at 38,000 rpm, 4°C for 3 h. CFM was reconstituted from PBS in Plastek B non-stick cultureware (Mattek Co.) at 37°C for 2 h. Two sizes of CFM containing 3.0 ± 0.1 mg/ml of collagen were used. In the floating CFM contraction assay, each of 0.5 ml of CFM had a circle area of 2 cm^2 and 2.5 mm thickness. When examining fibroblast proliferation on attached CFM, each 0.25 ml of collagen was used to make disks of 1 $cm^2 \times 2.5$ mm.

Preparation of periodate pre-modified hyaluronan

Hyaluronan was a gift from MedChem Products, Inc. and had a molecular weight of $1 \sim 2 \times 10^6$. It was dialyzed against deionized water to remove salts and sterilized by ultracentrifugation as described above. After lyophilization into a solid, each milligram of hyaluronan contains 2.5 μmol of glucosamine which was determined by hexosamine assay (see below).

Sodium periodate was added to 3 μmol/ml glucosamine of hyaluronan at 4°C in the dark to obtain a final concentration of 20 mM [17, 18]. The reaction was allowed to continue for 4 h and periodate-modified hyaluronan was obtained after dialysis against deionized water followed by ultracentrifugation.

Preparation of hyaluronan-collagen matrices (HCM)

Table 1 is an outline listing the preparation of various HCM in two series of experiments. The major differences between the two series of experiments are: (1) the incubation time with hyaluronan or periodate-modified hyaluronan, and (2) the rinsing temperature to remove non-reacted molecules. Some details of the procedures are described as follows. CFM were incubated with 0 or 12.5 μmol/ml glucosamine of hyaluronan or 2.5 or 12.5 μmol/ml glucosamine of periodate-modified hyaluronan (1:3, v/v) in PBS at 37°C for the time indicated on a rotator

Table 1.
Preparation of various hyaluronan–collagen matrices (HCM)

	(I) Experiments in Figs 2–4						(ii) Experiment in Fig. 6								
	A	B	C	D	E	F	A	B	C	D	E	F	G	H	I
CFM (1 × vol.)															
+ 3 × vol. of hyaluronan or periodate-modified hyaluronan (μmol/ml of glucosamine)	0	12.5	2.5	12.5	12.5	12.5	0	12.5	2.5	2.5	2.5	2.5	0	12.5	12.5
Rotate at 37°C for certain hours	8	8	8	8	6	6	24	24	48	48	48	24	24	24	24
+ periodate, room temp., 2 h	−	−	−	−	+	+	−	−	−	−	−	−	+	+	+
Change to glycine(G), lysine(L) or $NaBH_3CN$(S), 2 h			G	G	G	S				L	S	L[a] S		G	S
			at room temperature								at 37°C				
Extensive rinse with PBS(P) or 4M NaCl–PBS(N)			at room temperature								at 37°C				
	N	P	N	N	N	N	N	P	N	N	N	N	N	N	N
Glucosamine (nmol/matrix[b])[c]	0	16.6	24.2	28.4	18.6	31.2									

[a] Change to lysine in PBS and incubate for 24 h.
[b] Here each matrix was formed from 0.5 ml of 3 mg/ml of collagen.
[c] Each datum is the average of duplicate samples.

which was adjusted to a 30 deg angle above the horizontal plane at 50 rpm. For periodate post-treatment, CFM and HCM which were in 50 ml conical tubes wrapped with foil to prevent light penetration, sodium periodate (final concentration 20 mM) was added to the gels and the reaction allowed to continue for 2 h. Certain groups of crosslinked HCM were changed to 1 mg/ml of glycine or 1 mg/ml of lysine or twice to fresh 20 mM sodium cyanoborohydride [19] in PBS for at least another 2 h of incubation in the rotator. Finally non-ionically bound hyaluronan was rinsed out extensively with at least five changes of PBS over 2 days at room temperature or 37°C depending on the experiments. Control CFM, periodate post-treated CFM, as well as crosslinked HCM were rinsed extensively with 4M NaCl–PBS to remove non-reacted molecules. All procedures were carried out under sterile conditions. PBS contained 50 μg/ml of gentamicine and 2.5 μg/ml of fungizone when used for overnight incubations.

Hexosamine determination

The amounts of hyaluronan either ionically or covalently bound to CFM were determined by a hexosamine assay [20] after hydrolysis with 4N HCl at 110°C for 1 h. D(+)-Glucosamine (Sigma G4875) was used as a standard and the data expressed as nmol glucosamine.

Cell culture

For the experiment in Fig. 1, CFM (0.5 ml/well) were formed *in situ* in 24-well Plastek B non-stick cultureware as described and were pre-equilibrated with two changes of Dulbecco's Modified Eagle Medium (DMEM) (1:2, v/v). Each concentration 0, 0.02, 0.4, 2 and 10 mg/ml of hyaluronan in DMEM (containing 50 μg/ml gentamicine and 2.5 μg/ml fungizone) was added in triplicate to the CFM in a volume of 0.5 ml/well. Subsequently DMEM-25% fetal calf serum (0.4 ml/well) was added and mixed with hyaluronan solutions by quickly rotating the cultureware for 15 s. The CFM soaked in media containing hyaluronan were allowed to stand at 37°C with 5% CO_2 overnight. Next day, 100 μl of 1×10^6 cells/ml of human foreskin fibroblasts (8th passage, Clonetics Co.) suspended in DMEM were added on each CFM. Fibroblasts were evenly distributed by quickly rotating the cultureware for 15 s. Final culture media were composed of various concentrations 0, 0.01, 0.2, 1, and 5 mg/ml of hyaluronan in 1 ml of DMEM containing 10% fetal calf serum, 50 μg/ml gentamicine and 2.5 μg/ml fungizone. The experimental CFM with fibroblasts in the presence of hyaluronan in the culture medium were incubated at 37°C with 5% CO_2.

For the experiments shown in Figs 2–6, in which preparations of CFM and HCM are outlined in Table 1, HCM with ionically and covalently bound hyaluronan as well as control CFM were pre-equilibrated with five changes of DMEM (1:3, v/v) and subsequently glued onto cultureware with 3 mg/ml of type I collagen solution in PBS (5–10 μl/matrix). Early passages (6–8) of human foreskin fibroblasts (Clonetics Co., 2×10^4 cells) in 0.5 ml of culture medium (DMEM containing 10% fetal calf serum, 50 μg/ml gentamicine and 2.5 μg/ml fungizone) were added onto each 0.25 ml of CFM or HCM fitted in the well of 48-well tissue culture plate. For 0.5 ml of CFM or HCM in 24-well Plastek B non-stick cultureware (Mattek Co.), 1 ml of culture medium containing 4×10^4 fibroblasts was added to each well. The CFM and HCM with fibroblasts were incubated at 37°C with 5% CO_2 and the culture media changed after culturing fibroblasts for 3 and 5 days.

Floating CFM and HCM contraction assay

After plating fibroblasts for two hours, CFM and HCM detached from the bottom of the wells and appeared floating in the culture media. At the end of the experiments, the culture medium was removed and the CFM and HCM flattened by surface tension. The close-up of each CFM or HCM was photographed together with a ruler using a 35 mm camera (Nikon N2000) with attached extension rings (PK 11–13). The surface area of each matrix was integrated by using a computer program of SigmaScan. The data are expressed as % of initial area.

Cell viability and total DNA determination

At the end of each experiment, each sample was harvested for collagenase (720 U/ml of original matrix) digestion at 37°C for 2 h and subsequent 0.25% trypsin-EDTA (200 μl/ml of original matrix, Gibco) together with 200 mM EDTA (40 μl/ml of original matrix) digestion for 20 min. Fibroblasts were pelleted by centrifugation (1000 rpm, 5 min) and suspended in 2 ml of DMEM. To an aliquot (50 μl) was added 10 μl of trypan blue (0.4% in saline, Gibco) and the total cell number as well as viable cell number were counted with a hemocytometer under a phase contrast light microscope. The rest of the samples were rinsed repeatedly with PBS and total DNA contents were determined as described by West *et al.* [21]. To the cell pellet, 1.5 ml of cold 10 mM EDTA, pH 12.3 was added to solubilize cells. After neutralization with 1 M KH_2PO_4 (about 140 μl) and mixing with 0.5 ml of 100 mM NaCl-10 mM Tris, pH 7.0, 10 μl of 66.7 μg/ml Hoescht 33258 (Sigma B-2883) was added to give a fluorescent reading at emission wavelengths of 350 and 455 nm. Cell numbers were obtained by comparing the readings to a standard curve of cellular DNA (1×10^4–1×10^6 fibroblasts).

RESULTS

Fibroblast contraction of CFM (3 mg/ml) started right after matrices were released from the bottom of the cultureware, and this contraction continued with time. CFM in the presence of high concentration (5 mg/ml) of hyaluronan in the culture medium were much less contracted by fibroblasts (Fig. 1). Some possible mechanisms in which high concentrations of hyaluronan inhibit CFM contraction by fibroblasts are: (1) direct interaction of hyaluronan with fibroblasts; (2) hyaluronan modification of collagen fibrils; and (3) special physical chemical properties of hyaluronan.

Various CFM and HCM were prepared as outlined in Table 1. Since periodate treatment on hyaluronan does not affect the moiety of glucosamine [22], the amounts of hyaluronan and periodate pre-modified hyaluronan actually bound to collagen were determined by a hexosamine assay and were expressed as nmol glucosamine/matrix. The data are presented in the last row of Table 1. Elevated amounts of hyaluronan were bound to collagen through the periodate crosslinking reaction. The contraction of these various CFM and HCM are seen in Fig. 2. Crosslinked HCM were less contractible than HCM with ionically bound hyaluronan and control CFM. With time, CFM containing either ionically or covalently bound hyaluronan continue to be less contractible.

Figure 2C and D is a comparison in which 2.5 and 12.5 μmol/ml glucosamine of periodate-modified hyaluronan were used in the crosslinking reaction. Although

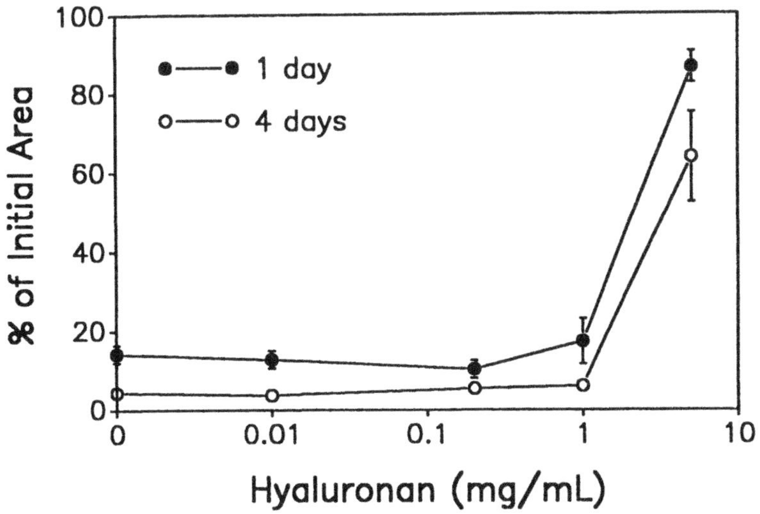
100
80
60
40
20
0
% of Initial Area
1 day
4 days
0
0.01
0.1
1
10
Hyaluronan (mg/mL)

Figure 1. Fibroblast contraction of floating CFM in the presence of hyaluronan in the culture medium. The data are expressed as mean ± S.E. ($n = 3$).

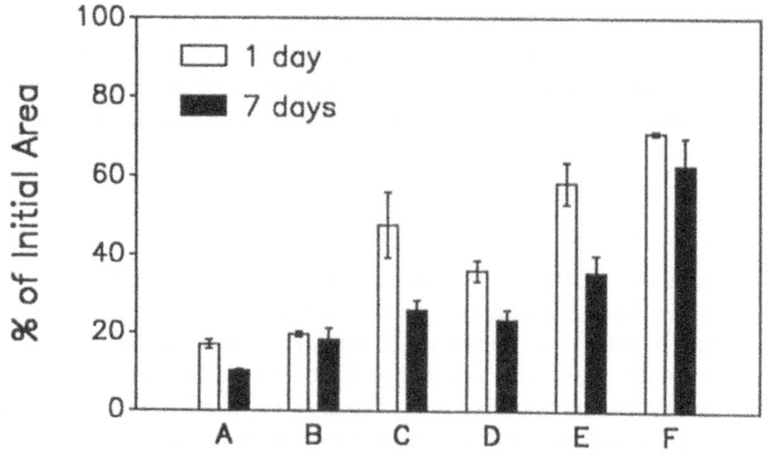
100
80
60
40
20
0
% of Initial Area
1 day
7 days
A
B
C
D
E
F

Figure 2. Fibroblast contraction of floating CFM and HCM. Each matrix was formed from 0.5 ml of 3.0 mg/ml of collagen. Group A: control CFM. Group B: HCM with ionically bound 16.6 ± 0.5 nmol glucosamine of hyaluronan per matrix. Group C: crosslinked HCM containing 24.2 ± 0.5 nmol glucosamine of periodate-modified hyaluronan per matrix. Group D: crosslinked HCM containing 28.4 ± 0.5 nmol glucosamine of periodate-modified hyaluronan per matrix. Group E: periodate and subsequent glycine post-treated HCM containing 18.6 ± 0.5 nmol of glucosamine per matrix. Group F: periodate and subsequent sodium cyanoborohydride post-treated HCM containing 31.2 ± 0.5 nmol of glucosamine per matrix. The data are expressed as mean ± S.E. ($n = 3$).

more periodate-modified hyaluronan was bound to CFM in Fig. 2D, the HCM were not less contractible by fibroblasts than those in Fig. 2C. It is possible that periodate-modified hyaluronan was bound differently in the two cases. Each molecule of periodate-modified hyaluronan may bind to collagen fibrils multivalently such as in the case of Fig. 2C. This may strengthen the CFM as well as block some fibroblasts which are in direct communication with the collagen fibrils, so as to cooperatively reduce forces to contract the HCM. If the concentration of periodate-modified hyaluronan is high, more molecules are in contact with collagen fibrils initially and simple point attachments may prevail such as in the case of Fig. 2D. In this situation, the strength of HCM may be smaller than Fig. 2C and more portions of collagen fibrils could be exposed to fibroblasts and allow contraction due to repulsion among highly negative charged hyaluronan molecules. In some extreme situation, these periodate-modified hyaluronan molecules may interfere with the integrity of CFM due to their highly negative charges and by their property to attract water. Besides such periodate-modified hyaluronan could be more susceptible to enzyme degradation.

In Fig. 2E and F, periodate post-treatment of HCM is shown to convert vicinal hydroxyl groups of both hyaluronan and collagen molecules into active aldehyde groups, which may lead to increased crosslinks via aldol condensation. The elevated crosslinks may confer the HCM a less contractible property. This is also evident by comparing the HCM with subsequent glycine (Fig. 2E) or sodium cyanoborohydride (Fig. 2F) treatment. Glycine not only blocks excess aldehyde groups but also competes with amino groups of collagen to interact with aldehyde groups of periodate modified hyaluronan, thus inhibiting crosslink formation. Sodium cyanoborohydride, on the other hand, further strengthens the crosslinks by reducing

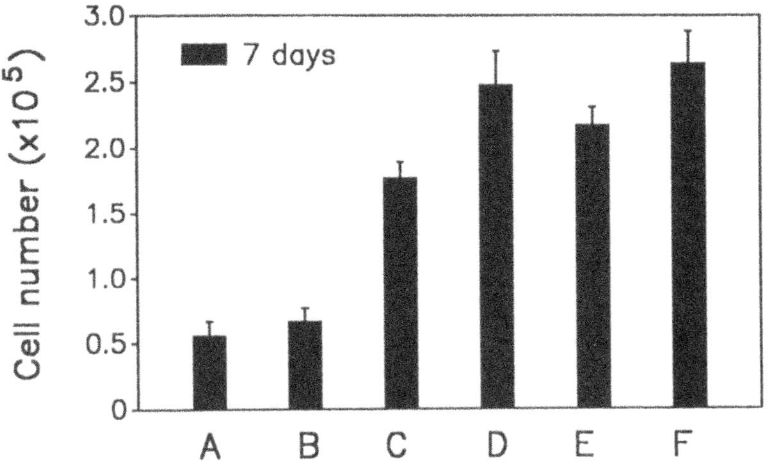

3.0
2.5
2.0
1.5
1.0
0.5
0
Cell number (x10^5)
7 days
A
B
C
D
E
F

Figure 3. Total numbers of fibroblasts cultured on floating 0.5 ml of CFM and HCM. The six groups of matrices are described in Fig. 2. The data are expressed as mean ± S.E. ($n = 3$).

selectively the Schiff's bases [23, 24] formed between modified hyaluronan and collagen molecules.

In order to confirm that reduced contractibility of the crosslinked HCM was not due to their cytotoxicity, cell viability and total cell numbers of floating CFM and HCM at various intervals were determined. The viabilities of fibroblasts which were released from different matrices after being cultured for 24 h were over 90% in all cases. After culturing fibroblasts on various matrices for 7 days, less cell numbers were obtained from smaller matrices such as the control CFM and HCM with ionically bound hyaluronan (Fig. 3). This indicates that limited space within the matrices may restrict cell growth and the crosslinked HCM are not cytotoxic.

To further understand how effective these crosslinked HCM were in supporting fibroblast proliferation, cell numbers of attached CFM and HCM at various

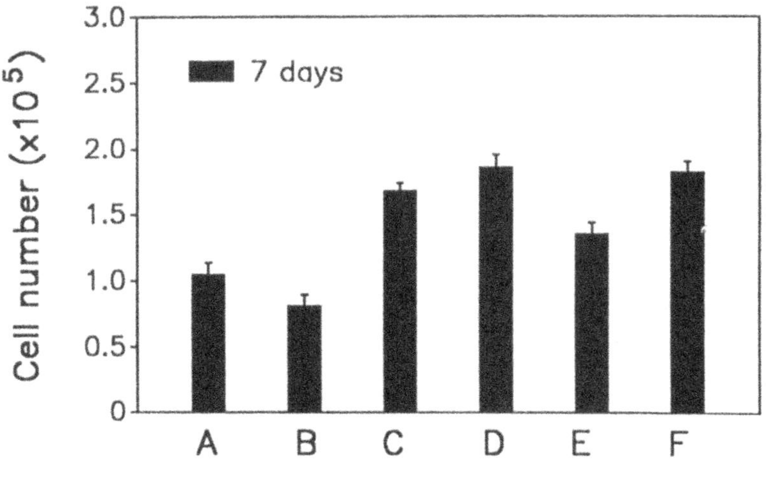

3.0
2.5
2.0
1.5
1.0
0.5
0
Cell number (x10^5)
7 days
A
B
C
D
E
F

Figure 4. Total numbers of fibroblasts cultured on attached 0.25 ml of CFM and HCM. See Fig. 2 for description of various matrices. The data are expressed as mean ± S.E. ($n = 3$).

intervals were counted. The results in Fig. 4 demonstrate that crosslinked HCM better support fibroblast proliferation.

Since crosslinked HCM were better in supporting cell proliferation and were less contractible by fibroblasts but not for prolonged periods, we further improved the crosslinking network. (1) The incubation of periodate-modified hyaluronan with collagen was prolonged to 48 h. From our previous experience [25], the reaction of aldehyde groups with lysine in collagen seems to be initially fast, but continues with time. (2) Lysine was added after the incubation of periodate-modified hyaluronan and collagen to introduce further crosslinks. Amino groups of lysine can form Schiff's base with aldehyde groups [25] and thus crosslink periodate-modified hyaluronan molecules to the collagen. (3) Sodium cyanoborohydride was added either after the incubation of periodate-modified hyaluronan with collagen or after further crosslinking the HCM with lysine. Sodium cyanoborohydride is known to reduce selectively the imminum salt formed between an aldehyde and an amine [23]. Figure 5 illustrates how periodate can modify the vicinal hydroxyl groups of hyaluronan to aldehyde groups, free amino groups in collagen can form Schiff's bases with these aldehyde groups, and the Schiff's bases can be reduced subsequently by addition of sodium cyanoborohydride to

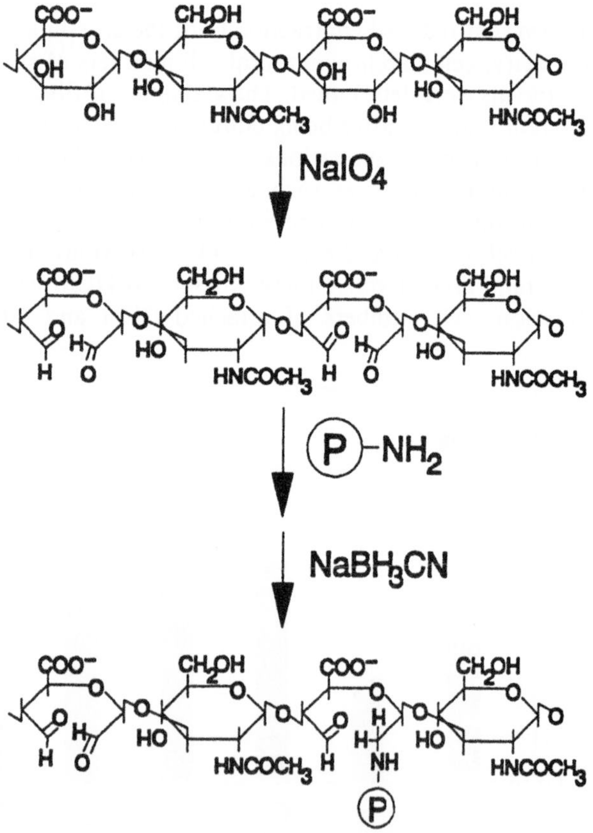
NaIO4
P-NH2
NaBH3CN

Figure 5. Illustration of the crosslinking reactions which occur between hyaluronan and collagen after treatment with sodium periodate. (P = collagen polypeptide.)

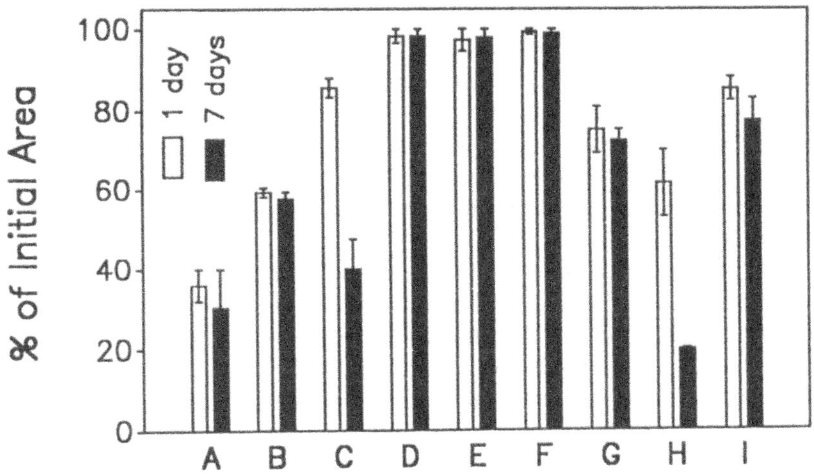

% of Initial Area
100
80
60
40
20
0
1 day
7 days
A
B
C
D
E
F
G
H
I

Figure 6. Fibroblast contraction of floating CFM and HCM. The conditions for preparation of the various matrices are different from those in Fig. 2. Refer to Table 1 for the differences between these experiments. Group A: control CFM. Group B: HCM with ionically bound hyaluronan. Groups C–F are crosslinked HCM with periodate-modified hyaluronan. Group C had no further treatment after the crosslinking reaction. Groups D–F were further treated with lysine, sodium cyanoborohydride, and with both lysine and sodium cyanoborohydride after crosslinking was completed. Group G: periodate post-treatment of the CFM. Group H and I: periodate post-treatment of the pre-incubated HCM with subsequent glycine (Group H) or sodium cyanoborohydride (Group I) treatment. The data are expressed as mean ± S.E. (n = 3).

stable amides. Periodate can also transform the hydroxyl group of hydroxylysine into aldehydes.

In addition, extensive rinsing procedures using PBS or 4M NaCl–PBS were carried out at 37°C in order to maintain the integrity of the matrices. The results (Fig. 6) show that rinsing at 37°C (rather than at room temperature, Figs 2–4) greatly reduced the contractibility of the matrices, especially those with ionically bound hyaluronan probably by stabilizing the endothermic assembly of collagen (Fig. 6B).

Periodate post-treated CFM (Fig. 6G) were less contractible than the control CFM (Fig. 6A). This further indicates that the rigidity of collagen can reduce contractibility since periodate can introduce crosslinks between collagen molecules [26]. In comparing the data of Fig. 6D–F and I to those of Fig. 6C and H respectively, it is apparent that fibroblasts fail to contract the highly crosslinked HCM and that immobilized hyaluronan can reduce matrix contractibility for a prolonged period. The results suggest that the presence of hyaluronan in CFM strengthens the matrices and/or somehow blocks direct communications between fibroblasts and collagen fibrils thus inhibiting fibroblasts to contract them. Besides, further treatment of the crosslinked HCM with lysine or sodium cyanoborohydride or both abolishes contractibility for an extended period of time (Fig. 6D–F). These treatments may not only generate a more stable network but may also prevent the biodegradation of hyaluronan.

DISCUSSION

In this study, an *in vitro* floating CFM was used to investigate the effects of hyaluronan on collagen contraction by fibroblasts. Hyaluronan can bind ionically to collagen fibrils and thus strengthen the matrices and interrupt the direct communications between fibroblasts and collagen fibrils. In consequence, the contractibility of the HCM is decreased. When hyaluronan is bridged between collagen fibrils by extensively covalent crosslinking (i.e. copolymerization with lysine or by reduction of the crosslinks with sodium cyanoborohydride), the contractibility of the HCM by fibroblasts was abolished. Whether this is due to the improvement of the strength of the matrices or due to prevention of hyaluronan degradation or both needs to be further studied.

The effects of hyaluronan on the inhibition of CFM contraction are specific, since we have demonstrated in Fig. 1 that hyaluronan at high concentrations without crosslinking greatly reduced CFM contraction. Some of our results (Fig. 6G) showed that crosslinked CFM were also less contractible. We know that CFM treated with formaldehyde or other bifunctional crosslinking reagents are also inhibited from contracting. In this case, the cytotoxic [25] effects are associated with matrix crosslinks. From the results of Fig. 4 in which crosslinked HCM support cell proliferation better, periodate-modified hyaluronan could be an alternative crosslinker for collagen.

Our results suggested that hyaluronan can strengthen collagen fibrils and/or interrupt direct communications between fibroblasts and collagen fibrils to prevent wound contracture. This could provide possible explanations of how hyaluronan affects wound healing in fetal and adult tissues. However, it is still possible that hyaluronan can affect cells directly through cellular mechanisms which have not yet been identified.

Crosslinking of hyaluronan to collagen allows hyaluronan to persist in the environment for extended periods and completely prevents CFM contraction. Such modified collagen matrices may prove to be useful in the manufacturing of bioprostheses.

Acknowledgement

This research is part of a dissertation by Lynn L. H. Huang-Lee, submitted to the University of Southern California in partial fulfillment of the requirements for a Ph.D. in biochemistry and molecular biology.

REFERENCES

1. M. G. Stern, M. T. Longaker and R. Stern, in: *Fetal Wound Healing*, p. 189, N. S. Adzick and M. T. Longaker (Eds). Elsevier Science Publishing Co., New York (1992).
2. B. A. Mast, R. F. Diegelmann, T. M. Krummel and I. K. Cohen, *Surg. Gynecol. Obstet.* **174**, 441 (1992).
3. J. W. Siebert, A. R. Burd, J. G. McCarthy, J. Weinzweig and H. P. Ehrlich, *Plast. Reconstr. Surg.* **85**, 495 (1990).
4. T. M. Krummel, J. M. Nelson, R. F. Diegelmann, W. J. Lindblad, A. M. Salzberg, L. J. Greenfield and I. K. Cohen, *J. Pediatr. Surg.* **22**, 640 (1987).
5. D. A. Burd, J. W. Siebert, H. P. Ehrlich and H. G. Garg, *Matrix* **9**, 322 (1989).
6. G. F. Murphy, D. P. Orgill and I. V. Yannas, *Lab. Invest.* **63**, 305 (1990).
7. I. V. Yannas, in: *Collagen, Vol. 3: Biotechnology*, p. 87, M. E. Nimni (Ed.). CRC Press, Inc., Florida (1988).
8. M. E. Nimni, D. T. Cheung, B. Strates, M. Kodama and K. Sheikh, in: *Collagen, Vol. 3: Biotechnology*, p. 1, M. E. Nimni (Ed.). CRC Press, Inc., Florida (1988).

9. R. A. F. Clark, in: *The Molecular and Cellular Biology of Wound Repair*, p. 3, R. A. F. Clark and P. M. Henson (Eds). Plenum Press, New York (1988).
10. J. M. McPherson and K. A. Piez, in: *The Molecular and Cellular Biology of Wound Repair*, p. 471, R. A. F. Clark and P. M. Henson (Eds). Plenum Press, New York (1988).
11. F. Grinnell and C. R. Lamke, *J. Cell Sci.* **66,** 51 (1984).
12. G. G. Reid, J. M. Lackie and S. D. Gorham, *Cell Biol. Int. Rep.* **14,** 1033 (1990).
13. C. Guidry and F. Grinnell, *J. Cell Biol.* **104,** 1097 (1987).
14. L. W. Adams and G. C. Priestley, *J. Invest. Dermatol.* **87,** 544 (1986).
15. G. Parry, E. Y. Lee, D. Farson, M. Koval and M. J. Bissell, *Exp. Cell Res.* **156,** 487 (1985).
16. L. L. H. Huang-Lee, Ph.D. Thesis, University of Southern California, Los Angeles, California (1993).
17. J. R. Dyer, in: *Methods of Biochemical Analysis Vol. III*, p. 111, D. Glick (Ed.). Wiley Interscience, New York (1956).
18. R. Montgomery and S. Nag, *Biochim. Biophys. Acta.* **74,** 300 (1963).
19. A. S. Acharya, L. G. Sussman and J. M. Manning, *J. Biol. Chem.* **258,** 2296 (1983).
20. E. A. Davidson, in: *Methods in Enzymology, Vol. VIII: Complex Carbohydrates*, p. 53, E. F. Neufeld and V. Ginsburg (Eds). Academic Press, New York and London (1966).
21. D. C. West, A. Sattar and S. Kumar, *Anal. Biochem.* **147,** 289 (1985).
22. G. J. Hooghwinkel and G. Smits, *J. Histochem. Cytochem.* **5,** 120 (1957).
23. R. F. Borch, M. D. Bernstein and H. D. Durst, *J. Amer. Chem. Soc.* **93,** 2897 (1971).
24. G. R. Gray, *Arch. Biochem. Biophys.* **163,** 426 (1974).
25. L. L. H. Huang-Lee, D. T. Cheung and M. E. Nimni, *J. Biomed. Mater. Res.* **24,** 1185 (1990).
26. M. Tardy and J.-L. Tayot, U.S. Patent 4931546 (1990).

Chemical modification of biopolymers—mechanism of model graft copolymerization of chitosan

WEI LI, ZHAOYANG LI, WENSHEN LIAO and XIN-DE FENG

Department of Chemistry, Peking University, Beijing 100871, China

Received 25 June 1992; accepted 29 October 1992

Abstract—Mechanism of graft copolymerization of vinyl monomers onto chitosan initiated by Ce(IV) ion, one of the important ways of chemical modification of chitosan, has been investigated by means of kinetics measurement and polymer chain structure analysis. It is found that when Ce(IV) ions reacts with adjacent hydroxylamine structure in chitosan, a chelate complex is first formed and then it disproportionates to radical for initiation of graft copolymerization. There exist two ways for such initiation depending upon the reaction temperature. For a reaction temperature lower than 40°C, a —CHO group and a —C=NH group are introduced, and the aldehyde group reacts with Ce(IV) ions to form an acyl radical which initiates a graft polymer chain in this saccharide unit. For temperatures higher than 90°C, the —C=NH group hydrolyzes to form an amine and an aldehyde group which also reacts with Ce(IV) ions. In this case there should be two aldehyde groups, i.e. two initiation sites, in one adjacent hydroxyl-amine structure. That means that the initial radical in the chitosan/Ce(IV) system is similar to that in the cellulose/Ce(IV) one, but the former usually gives a lower grafting reactivity than the latter due to the higher stability of the chelate.

Key words: Graft copolymerization; chitosan model/Ce(IV); adjacent hydroxylamine/Ce(IV); Ce(IV) initiation; reaction mechanism.

INTRODUCTION

Chitosan, having a similar chemical backbone to cellulose, is a linear polymer composed of a partially deacetylated material of chitin [(1-4)-2-acetamide-2-deoxy-β-D-glucan]. Unlike most commercial polysaccharides, chitosan is a basic polysaccharide. Chitosan has been shown to be a reactive polysaccharide [1] and been noted as a functional polymer which is applicable to various fields [2]. It can be used as a wound healing accelerator [3], blood anticoagulant [4], carrier for controlled release of drugs, functional membranes [5] and so on. It seems to fulfil a number of demands in our highly technological world.

Grafting copolymer chains onto chitosan can enhance or introduce desired properties. A lot of papers and patents related to chitosan graft copolymers have been reported [6–8]. The first paper about using Ce(IV) salt as an initiator was reported by Yang *et al.* [9]. No paper, however, considered the detailed mechanism of this initiation system and the structure of initial radical, until now.

The authors recently re-investigated the mechanism of grafting vinyl monomers onto carbohydrates such as starch and cellulose initiated by Ce salt with some new results [10, 11].

In this work, the mechanism of model graft copolymerization of vinyl monomers onto chitosan will be firstly revealed. Here we choose compounds containing the adjacent hydroxyl–amine structure, which is the reaction group in the chitosan unit, as the models of chitosan. Thus we can determine and understand the mechanism of initiation more easily.

EXPERIMENTAL

Materials

D-Glucosamine hydrochloride (GlA, Aldrich), trans-2-amino-cyclohexanol hydrochloride (ACH, Aldrich), (s)-(−)-2-amino-3-phenyl-1-propanol (APhP, Aldrich) and DL-1-amino-2-propanol (AP, Aldrich) were used without further purification. Amino alcohol (AA, Beijing Chemical Manufacturers) was distilled and collected at 97.5–98.5°C/40 mmHg.

Instruments

UV spectra were measured by Shimadzu UV-250 spectrophotometer, ESB spectra by Bruker ESR 300, IR spectra by NICOLET 7199B, and kinetics measurements were performed with a dilatometer.

Oxidization

A reaction was carried out by mixing $CeSO_4$ (as oxidant) and model compound (as reductant) aqueous solutions together, which had previously been heated to the reaction temperature. The mixed solution was analyzed when the reaction had finished.

RESULTS AND DISCUSSION

Properties of chelate complex

The amino groups of chitosan are powerful ligands for metals other than alkali, alkaline-earth metals, and thallium. Here we compare the chelate complex of chitosan/Ce(IV) with that of cellulose/Ce(IV).

The mechanism of formation of the chelate complex and its decomposition is usually described as follows:

$$Ce^{4+} + RH \xrightleftharpoons{K} complex \xrightarrow{k_d} R\cdot + Ce^{3+} + H^+$$

$$Ce^{4+} + R\cdot \xrightarrow{\text{fast}} \text{product} + Ce^{3+} + H^+.$$

In the case of the presence of excess reductants, one has

$$-d[\mathrm{Ce(IV)}]_T/\mathrm{d}t = k' \, [\mathrm{Ce(IV)}]_T$$

where $k' = k_d K [\mathrm{RH}]/(1 + K [\mathrm{RH}])$.

The equilibrium constant for complex formation (K) and the rate constant for complete decomposition (k_d) of trans-2-amino-cyclohexanol(ACH)/Ce(IV) and

Table 1.
Comparison of complex constant (K and k_d).

Reductant	ACH	APhP	CH
K, 1/mol.	24.6	16.0	18.0
k_d, 1/min.	3.45×10^{-3}	2.38×10^{-2}	3.6×10^{-1}

$[\mathrm{Ce(IV)}] = 2.5$ mmol/l, $[\mathrm{HNO_3}] = 0.1$ N, 26°C
ACH: trans-2-aminocyclohexanol; APhP: 2-amino-3-phenyl-1-propanol; CH: trans-1,2-cyclohexanediol.

2-amino-3-phenyl-1-propanol (APhP)/Ce(IV) were compared with those of trans-1,2-cyclohexanediol (CH)/Ce(IV) system [10] as shown in Table 1. It is found that the *K*-values for the adjacent hydroxyl-amine structure are greater than those of glycol, but the decomposition constants are smaller.

Hence, it can be expected that the polymerization rate for using the adjacent hydroxyl–amine compound/Ce(IV) as an initiator will be slower than that for the glycol/Ce(IV) system. Data of polymerization rate and induction time are shown in Table 2. It is concluded that high reactivity of forming chelate complex does not mean high polymerization rate.

Table 2.
Comparison of polymerization rate (R_p) and induction time (t_i).

Reductant	ACH	CH	GlA	Gl
R_p, % min.	0.222	0.468	0.247	0.569
R_r	1.00	2.11	1.11	2.56
t_i, min.	23.5	17.0	18.0	10.0

[Ce(IV)] = 2.0 mmol/l; [Reductant] = 2.5 mmol/l; [AN] = 0.5 mol/l; [HNO_3] = 0.1 N; 25°C; GlA: Glucosamine; Gl: Glucose; ACH and CH: the same as Table 1.

Rate of polymerization

Dependence of polymerization rate on Ce salt concentration was measured at fixed monomer concentration (0.50 mol/l), reductant concentration (2.50 mmol/l) and nitric acid concentration (0.10 mol/l). The Ce salt concentration varied from 1.0 to 22.0 mmol/l.

The relationships between R_p and Ce(IV) concentration for GlA/Ce(IV), AA/Ce(IV), AP/Ce(IV) and ACH/Ce(IV) systems are shown in Figs 1–3. A

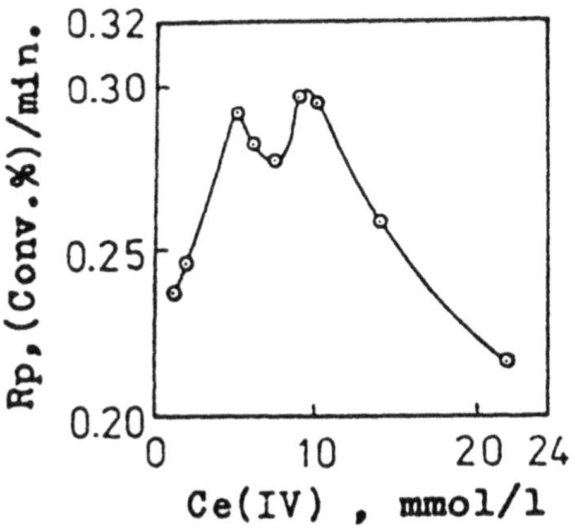
Rp,(Conv.%)/min.
0.32
0.30
0.25
0.20
0
10
20
24
Ce(IV) , mmol/l

Figure 1. Effect of [Ce(IV)] on R_p, [GlA] = 2.5 mmol/l; [AN] = 0.5 m; [HNO_3] = 0.1 N 25°C.

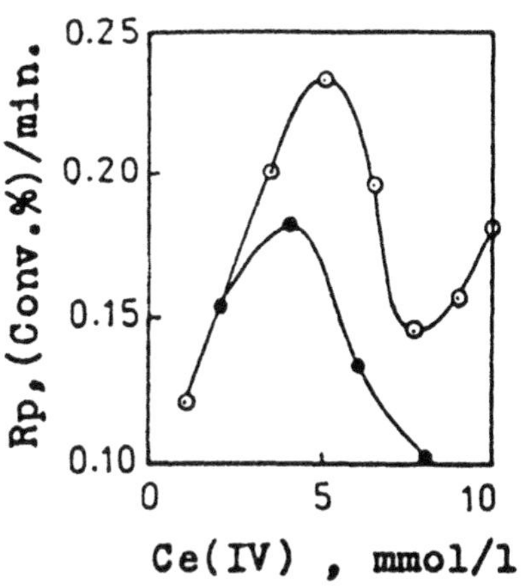
Rp,(Conv.%)/min.
0.25
0.20
0.15
0.10
0
5
10
Ce(IV) , mmol/l

Figure 2. Effect of [Ce(IV)] on R_p, [AA](⊙) = [AP](●) = 2.5 mmol/l, 25°C [AN] = 0.5 m; [HNO_3] = 0.1 N.

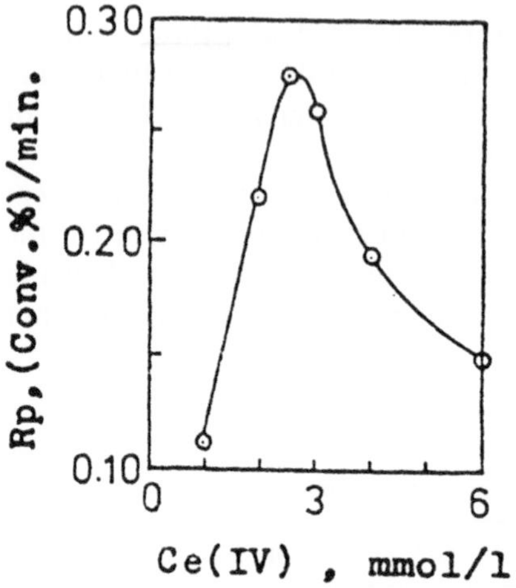
0.30
0.20
0.10
0
3
6
Rp, (Conv.%)/min.
Ce(IV) , mmol/l

Figure 3. Effect of [Ce(IV)] on R_p, [ACH] = 2.5 mmol/l; [AN] = 0.5 m; [HNO_3] = 0.1 N 25°C.

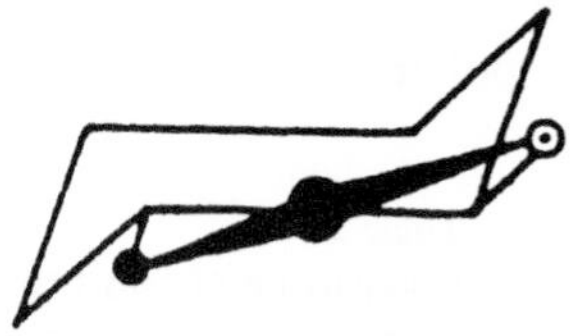

Figure 4. Chelate complex of trans-ACH/Ce(IV) system. ● = N; ⊙ = 0; ● = Ce(IV).

Table 3.
Comparison of R_p and E_a (activation energy).

Reductant	/	GlA	ACH	AP	AA
R_p, % min.	0.119	0.247	0.222	0.154	0.131
R_r	1.00	2.08	1.87	1.29	1.10
E_a, kg/mol	61.86	28.92	33.26	42.88	47.11

T = 15–40°C, AP: 1-amino-2-propanol, AA: amino alcohol
others: the same as Tables 1 and 2.

bimodal curve was observed in GlA/Ce(IV) system as in cyclic 1,2-diols/Ce(IV) system [10]. We predict that a bimodal curve should appear for AA/Ce(IV) system if the [Ce(IV)] range is wide enough.

E_a *of polymerization*

Total E_a values of polymerization of AN initiated by GlA/Ce(IV), ACH/Ce(IV), AA/Ce(IV) and AP/Ce(IV) systems were determined at temperatures ranging from 15 to 40°C. The results are shown in Table 3. It is found that all of these four adjacent hydroxyl and amine compounds can form an effective redox initiation system with Ce(IV), i.e. increasing R_p and decreasing E_a. What is more important is that the cyclic molecule seems more effective than the linear one. For example, R_r of cyclic systems is about 1.98 ± 0.10, and linear systems about 1.20 ± 0.10. Again, the decrease of E_a for cyclic systems is 30.77 ± 2.17 (kJ/mol) and 16.87 ± 2.11 (kJ/mol) for linear systems.

These results can be explained as the breakdown of the C—C band between the hydroxyl and amine groups. For 6-numbered ring adjacent hydroxyl-amine compounds, the chelate complex combines a 5- and a 6-numbered ring (as shown in Fig. 4) which is more unstable than a simple 5-numbered ring structure in linear compounds. As a result, the redox initiator with cyclic molecules gives high activity.

Table 4.
Results of Fehling test

Reductant	Phenomena	Conclusion
$CH_2-CH(OH)-CH_3$ with NH_2 on C_1	Brown, purplish red precipitate	exist $-CH{=}O$ and $HCH{=}O$
$CH_2(OH)-CH(NH_2)-CH_2Ph$	purplish red precipitate	exist $HCH{=}O$
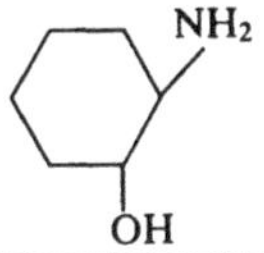	brown red precipitate	exist $-CH{=}O$

[Ce(IV)] = [Reductant] = 1.0×10^{-2} mol./l, $[H^+]$ = 0.1 N, 40°C.

Analysis of the products of oxidization

Evidence of aldehyde group. Fehling's solution [12] was added to the reacted solutions and then heated to 100°C. The results, as seen in Table 4, indicate that carbonyl group is present in the oxidization products, similar to the poly(vinyl alcohol)/Ce^{4+} ion system [13]. This suggests that the initial radical is an acyl one. Formaldehyde formed in 1-amino-2-propanol system (see Table 4) means that the C—C linkage between the amino and hydroxyl groups has been broken off.

Evidence of —C=NH group. The reaction solutions (40°C) were tested by the S—$Fe(NO_3)_3$ method [14] to detect —C=NH group directly. The reaction solution was dropped onto a piece of melting sulfur in a test tube. The test tube was covered with a test paper which had been moist with acidic $Fe(NO_3)_3$ aqueous solution. The cover paper was light red for mixed solutions while the control's was white. It is proved that —C=NH group is produced during oxidization. Thus the reactions of 40°C of the model compounds with Ce(IV) must proceed as follows:

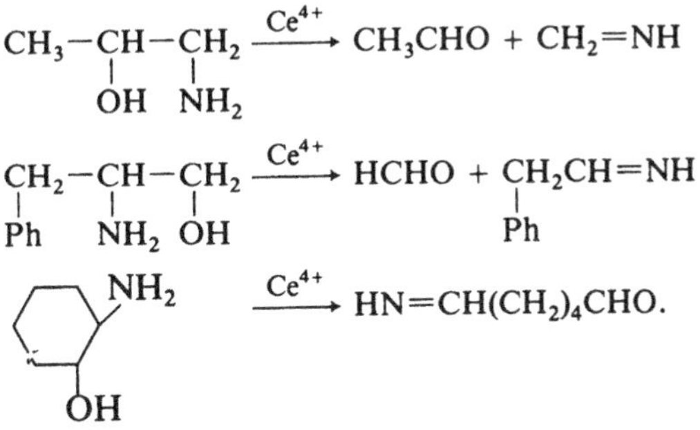

Now the same reaction solutions were tested by Nessler's reagent [15]. As the reductant contained amino group, its solution was used as control. The results are shown in Table 5. It is seen that at 40°C there were no amine groups present. So, in this case, there are aldehyde groups and imine groups in the reacted chitosan unit. As mentioned before [10, 11], the aldehyde group has high reactivity with Ce^{4+} to produce an acyl radical which initiates polymerization while the imine group does not react with Ce^{4+}. Under this condition, we can only have one polymer chain

Table 5.
Results of Nessler's reagent test

Reductant	40°C	90°C	Control[a]	Conclusion
$CH_2(NH_2)-CH(OH)-CH_3$	light yellow	brown red precipitate	light yellow	no NH_4^+ at 40°C, forming NH_4^+ at 90°C
$CH_2(OH)-CH(NH_2)-CH_2Ph$	milky white	brown red precipitate	light white	no NH_4^+ at 40°C, forming NH_4^+ at 90°C
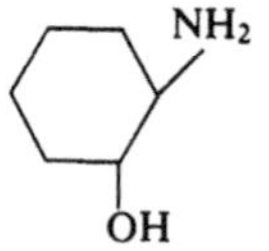	light yellow	brown red precipitate	light yellow	no NH_4^+ at 40°C, forming NH_4^+ at 90°C

[a] Control kept the same at 40 and 90°C.
$[Ce^{4+}] = [\text{Reductant}] = 1.0 \times 10^{-2}$ mol/l, $[H^+] = 0.1$ N.

grafted to a reacted chitosan unit. But at higher reaction temperatures (90°C), the imine group is no longer stable in acid aqueous solution and can hydrolyze to produce another aldehyde and amine. That is:

$$RC{=}NH \xrightarrow{H_2O} RCHO + NH_3 \xrightarrow{H^+} RCHO + NH_4^+.$$

Like the early one, the newly produced aldehyde further reacts with Ce^{4+} ion to initiate another polymer chain in the same chitosan unit. Mechanism of oxidization of adjacent amino-hydroxyl structure by Ce(IV) is therefore suggested as follows:

$$Ce^{4+} + H_2O \longrightarrow Ce(OH)^{3+} + H^+$$

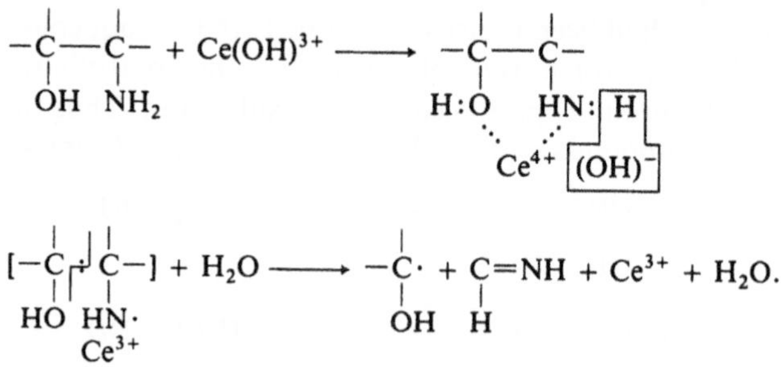

As the radical $-\dot{C}-OH$ reacts with Ce^{4+} to form a carbonyl group very quickly [11], we have

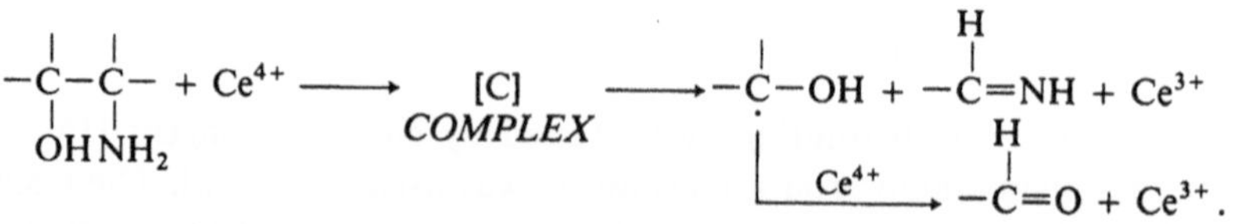

Macromolecular chain structure

ESR results. By using MNP(2-methyl-2-nitrol-propane) as a spin trapper, ESR spectra were measured (see Figs 5 and 6). Figure 5 was obtained when adding MNP

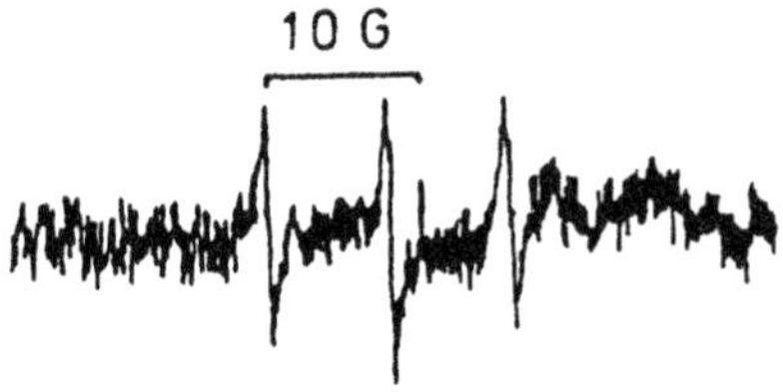
10 G

Figure 5. ESR spectrum, adding MNP immediately after mixing of Ce^{4+} and 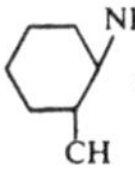 , a_N = 7.90 [G].

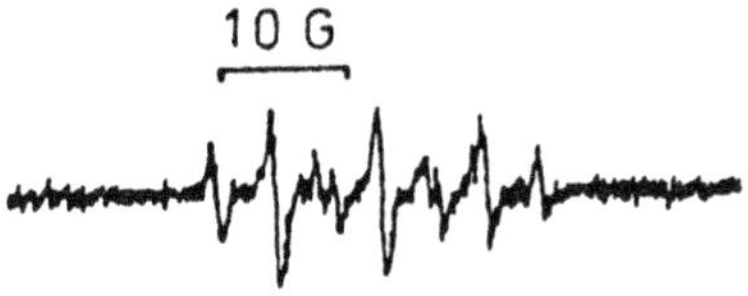
10 G

Figure 6. ESR spectrum, adding MNP immediately after mixing of Ce^{4+} and 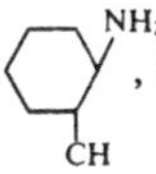 , 10 min later than in Fig. 5.

to the reaction solution immediately after model compound and Ce salt are mixed together and Figure 6 results when adding MNP 10 min later. These results indicate that the initial radical of grafting polymerization of vinyl monomer onto chitosan initiated by Ce(IV) ion is an acyl radical, such as in starch and cellulose/Ce(IV) systems.

IR spectra. The end groups of polyacrylonitrile (PAN) samples initiated by Ce(IV) only and model compounds/Ce(IV) at 40°C have been determined by FTIR spectrometry. Before IR measurement, the samples were first cleaned [2] and then detected by UV-spectrometer to ensure there was no —C=O group there. It can be seen from Figs 7 and 8 that all of these samples contained carbonyl end-groups (characteristic band of —CO— at 1716 cm^{-1}). The differential FTIR spectra of PAN initiated by model compounds/Ce(IV) and by Ce(IV) only are shown in Fig. 9.

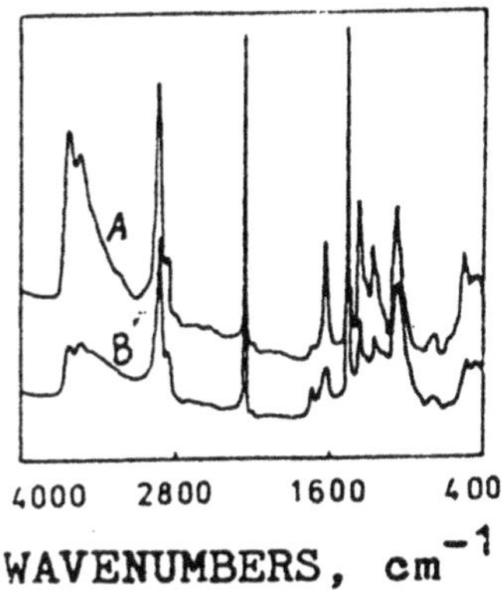
A
B
4000 2800 1600 400
WAVENUMBERS, cm⁻¹

Figure 7. FTIR spectrum of PAN initiated by (A): 1-amino-2-propanol/Ce^{4+} and (B): 2-amino-3-phenyl-1-propanol/Ce^{4+}, 40°C, $[HNO_3]$ = 0.1 N.

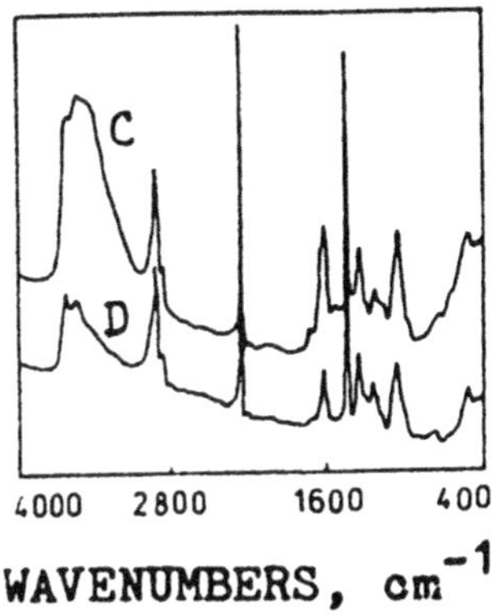
C
D
4000 2800 1600 400
WAVENUMBERS, cm⁻¹

Figure 8. FTIR spectrum of PAN initiated by (C): trans-2-amino-cyclohexanol/Ce^{4+} and (D): Ce^{4+} only, 40°C, $[HNO_3]$ = 0.1 N.

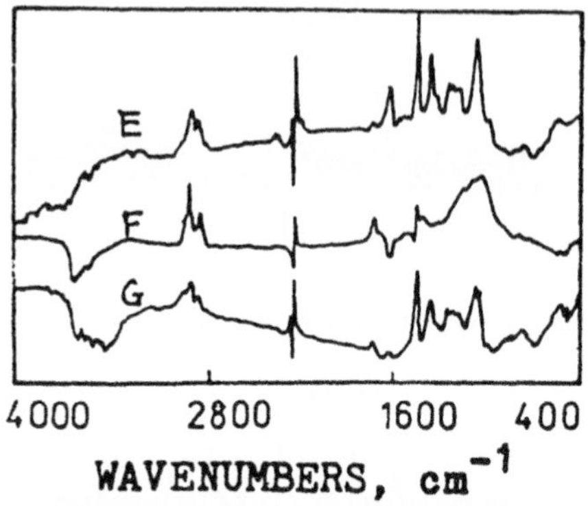
E
F
G
4000
2800
1600
400
WAVENUMBERS, cm^{-1}

Figure 9. FTIR different spectrum between PAN initiated by (E): 1-amino-2-propanol/Ce^{4+} and by Ce^{4+} only; (F): 2-amino-3-phenyl-1-propanol/Ce^{4+} and by Ce^{4+} only; (G): trans-2-amino-cyclohexanol/Ce^{4+} and by Ce^{4+} only, 40°C, $[HNO_3] = 0.1$ N.

It is seen that PAN samples initiated by model compound systems contain larger amounts of carbonyl group than PAN initiated by only Ce(IV). For PAN initiated by 2-amino-3-phenol-1-propanol/Ce(IV) and 1-amino-2-propanol/Ce(IV), there are no characteristic bands of —N—H and phenol in their spectra. But the —N—H bands (at 1520 and 1660 cm^{-1}) can be seen clearly in the differential IR spectrum of the trans-2-aminocyclohexanol/Ce(IV) initiated PAN sample. These results again indicate that the C—C linkage between the amino and hydroxyl group was broken in the initiation reaction (i.e. oxidization). The sole initial radical, at 40°C, is an acyl radical and the —C=NH group does not initiate polymerization.

GPC results. Because there was only one radical created in one adjacent amino-hydroxyl structure at 40°C, the molecule weight distribution of polymer initiated by such a system gives a good single peak curve in GPC measurement. GPC results of PAM initiated by 1-amino-2-propanol/Ce(IV) and trans-aminocyclohexanol/Ce(IV) are shown in Figs 10 and 11, respectively. Here we have

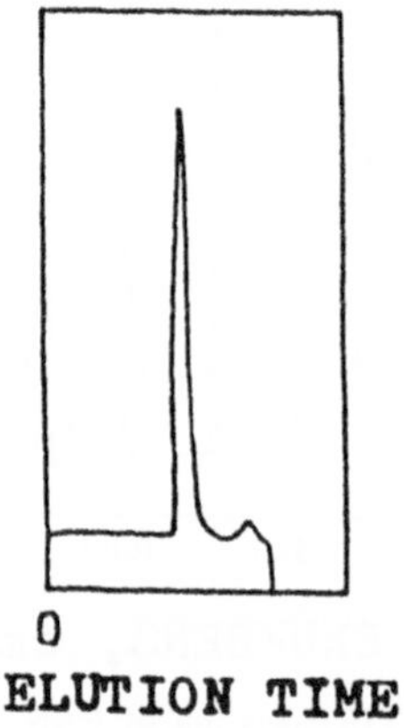
0
ELUTION TIME

Figure 10. GPC result of PMMA initiated by 1-amino-2-propanol/Ce^{4+} at 40°C, $[HNO_3] = 0.1$ N.

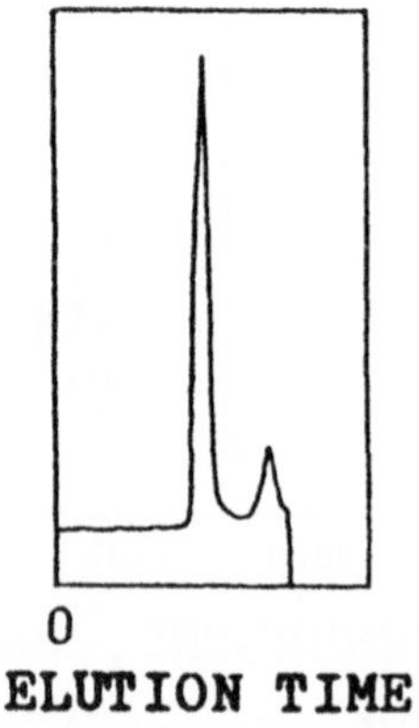
0
ELUTION TIME

Figure 11. GPC result of PMMA initiated by trans-2-amino-cyclohexanol/Ce^{4+} at 40°C. $[HNO_3] = 0.1$ N.

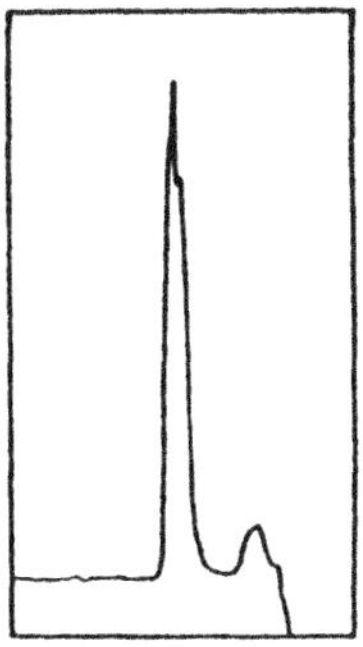
0
ELUTION TIME

Figure 12. GPC result of PMMA initiated by trans-2-amino-cyclohexanol/Ce^{4+} at 90°C, $[HNO_3] = 0.1$ N.

single peaks with a narrow molecule weight distribution. But in the case of reactions at 90°C there should be two radicals in a chitosan unit, similar to the cellulose and starch systems [11], and this is confirmed by the GPC spectrum shown in Fig. 12.

CONCLUSION

By combining the results of kinetics study and polymer chain structure analysis, we find that model compounds of chitosan form an efficient redox initiation system with Ce ion, and that the chelate complex is formed when Ce(IV) ion reacts with the adjacent hydroxyl–amino structure. There are two ways for this initiation reaction to take place, depending on the reaction temperature. At 40°C, the amino and hydroxyl groups are changed to aldehyde and —CH=NH groups, respectively. At 90°C, the —CH=NH group is hydrolyzed to form another aldehyde group. There are no hydroxyl groups in the long polyacrylonitrile (PAN) chain initiated by model compounds and Ce ion. It is suggested that one adjacent hydroxyl–amine structure creates the sole initial radical (acyl radical) at 40°C or two initial radicals at 90°C. We can therefore get two kinds of model graft copolymer chain where a chitosan unit is grafted, by a single chain for one kind and by double chains for the other. The mechanism of initiation of graft copolymerization of vinyl monomer onto chitosan is proposed as follows:

At 40°C:

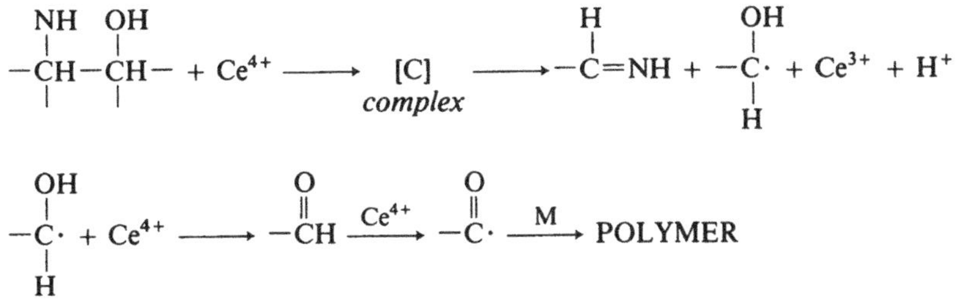
NH OH
—CH—CH— + Ce⁴⁺ ⟶ [C] ⟶ —C=NH + —C· + Ce³⁺ + H⁺
complex
H
OH
H
OH
—C· + Ce⁴⁺ ⟶ —CH ⟶ —C· ⟶ POLYMER
H
O
O
Ce⁴⁺
M

At 90°C:

$$\underset{\substack{|\\ \text{NH}}}{\underset{\substack{|\\ \text{OH}}}{-\text{CH}-\text{CH}-}} + Ce^{4+} \longrightarrow \underset{complex}{[C]} \longrightarrow \underset{\substack{|\\ \text{H}}}{-\text{C}}=\text{NH} + \underset{\substack{|\\ \text{H}}}{-\overset{\text{OH}}{\text{C}}\cdot} + Ce^{3+} + H^+$$

$$\underset{\substack{|\\ \text{H}}}{-\overset{\text{OH}}{\text{C}}\cdot} + Ce^{4+} \longrightarrow -\overset{\overset{\text{O}}{\|}}{\text{CH}} \xrightarrow{Ce^{4+}} -\overset{\overset{\text{O}}{\|}}{\text{C}}\cdot \xrightarrow{M} \text{POLYMER}$$

$$-\text{CH}=\text{NH} \xrightarrow[H^+]{H_2O} -\text{CHO} + NH_4^+$$
$$\xrightarrow{Ce^{4+}} \xrightarrow{M} \text{POLYMER}$$

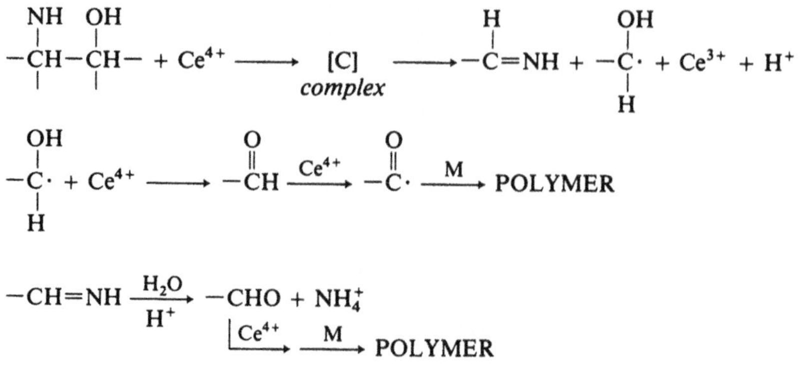

REFERENCES

1. R. A. A. Muzzarelli, Italian Patent 625-A-78 (1978).
2. R. A. A. Muzzarelli, *Chitin*, Pergamon Press, Oxford (1977).
3. K. Kifune, Y. Yamaguchi, K. Motosugi and Y. Oshima, *Proceedings of the 11th Annual Meeting of the Society for Biomaterials*, p. 135. San Diego, CA (1985).
4. R. A. A. Muzzarelli, in: *Polymers in Medicine*, E. Chiellini and E. Giusti (Eds). Plenum Press, New York (1984).
5. A. Domard, C. Gey, M. Rinaudo and C. Terrassin, *Int. J. Biol. Macromol.* **9**, 233 (1987).
6. R. C. Slagel and G. D. M. Sinkovitz, U.S. Patents 3,709,780 and 3,770,673 (1973).
7. K. Kojima, M. Yoshikuni and T. Suzuki, *J. Appl. Polym. Sci.* **24**, 1587 (1979).
8. Y. Shigeno, K. Kondo and K. Takemoto, *J. Macromol. Sci. -Chem.* **A17**(4), 571 (1982).
9. J. X. Yang, D. Q. He, J. Wu, H. Yang and Y. Gao, *J. Shandong College Oceanogr.* **14**(4), 58 (1984).
10. W. Li, C. Yu, J. J. Zhang and X. D. Feng, *The 6th China/Japan Symp. on Radical Polymer*, p. 27. Guilin, China (1991).
11. C. Yu, W. Li and X. D. Feng, *IUPAC Polymer. '91*, p. 55. Hangzhou, China (1991).
12. B. Herstein, *J. Am. Chem. Soc.* **22**, 779 (1910).
13. Y. Ikada, Y. Nishizaki, H. Iwata and I. Sakurada, *J. Polym. Sci: Polym. Chem. Edn* **15**, 451 (1977).
14. F. Feigl, V. Gentil and E. Jungreis, *Mikrochim. Acta* 47, (1959).
15. R. L. Shriner and R. C. Fuson, *The Systematic Indentification of Organic Compounds*, 3rd Edn, John Wiley & Sons, New York (1948).

Preparation of DNA-carrying affinity latex and purification of transcription factors with the latex

Y. INOMATA[1], T. WADA[2], H. HANDA[2], K. FUJIMOTO[1] and H. KAWAGUCHI[1]*

[1]*Department of Applied Chemistry, Faculty of Science and Technology, Keio University, 3-14-1 Hiyoshi, Yokohama 223, Japan*

[2]*Faculty of Bioscience and Biotechnology, Tokyo Institute of Technology, 4259 Nagatsuta-cho, Midori-ku, Yokohama 227, Japan*

Received 23 June 1992; accepted 19 September 1992

Abstract—We have developed DNA-carrying latex particles for the separation and purification of transcription factors. These particles consist of styrene (St), glycidyl methacrylate (GMA) and divinylbenzene (DVB). It was confirmed that the ethanolamine-treated surface of these particles suffered no nonspecific adsorption of proteins. To the latex particles sequence-specific DNA oligomers were immobilized via covalent coupling. A transcription factor, E4TF3, was efficiently purified to homogeneity using the latext particles. In contrast, the purification using DNA-carrying Sepharose gel yielded poor results. Compared to DNA-carrying Sepharose gel, the latex particles exhibited several times higher efficiency in the purification of E4TF3 from the crude nuclear extract.

Key words: DNA-carrying latex particles; purification; transcription factor; nonspecific adsorption; sequence-specific DNA; E4TF3; Sepharose gel.

INTRODUCTION

Sequence-specific DNA binding proteins directly affect the transcription initiation in higher organisms [1, 2]. The separation and purification of these transcription factors are very important in understanding the transcriptional control and the biological phenomena of generation, specialization and canceration. Proteins are most often fractionated by column chromatography. A more efficient procedure, known as affinity chromatography, takes advantage of the biologically important binding interactions that occur on protein surfaces [3–5]. The purification of transcription factor has been carried out with DNA affinity chromatography developed by Kadonaga and Tjian [6]. We have developed the separation and purification methodology using latex particles [7]. By using the particles, we were able to purify a DNA-binding transcription factor E4TF3 from an enriched protein fraction. However, the particles had some defects in the direct purification of E4TF3 from crude cell extracts. Some improvements were therefore necessary in terms of purification efficiency.

In this paper, new latex particles were prepared by soap-free emulsion copolymerization followed by seeded polymerization and their excellent features as the support of DNA were confirmed. Comparison between our latex batch system and Sepharose-gel column system indicated the overwhelming preference of the former in the purification of a transcription factor E4TF3 from the crude nuclear extracts.

* To whom correspondence should be addressed.

EXPERIMENTAL PROCEDURES

Preparation of latex particles

The monomers used here were glycidyl methacrylate (GMA), styrene (St) and divinylbenzene (DVB) purchased from Wako Pure Chemicals Co. To prepare latex particles, a mixture of 1.8 g GMA, 1.2 g St, 0.04 g DVB and 110 g distilled water was put into a 200-ml three-necked round-bottom flask equipped with a stirrer, a nitrogen gas inlet, and a condenser. Nitrogen gas was bubbled into the mixture to purge oxygen. The system was kept at 70 °C in a water bath. To the flask was added 10 g distilled water containing 0.06 g azobisamidinopropane dihydrochloride for the initiation of the soap-free emulsion polymerization. Since the resulting GMA–St copolymer particles partially have hydrophobic surface due to exposed polystyrene microdomains, 0.3 g GMA was supplemented to the reaction mixture 2 h after the initiation of polymerization and the polymerization was continued for 22 h in order to cover the whole surface of the particles with poly-GMA. The particles were collected by centrifugation, and washed three times with water. The particles before and after the seeded polymerization are referred to as (St–GMA) and (St–GMA)/GMA latex particles, respectively.

Adsorption of proteins to latex particles

Human γ-globulin was dissolved in 50 mM Tris–HCl buffer (0.1 M KCl, pH 8.0) to prepare the solutions of concentrations of 200, 400, 600, and 800 ppm. Then the particles which were masked with 1 M ethanolamine·HCl (pH 8.0) and washed with 50 mM Tris–HCl buffer were added to the protein solutions. The adsorption was performed for 30 min at 0 °C and terminated by centrifugation to separate the particles from the supernatant. After the centrifugation, the amount of γ-globulin in the supernatant was measured with a spectrophotometer (HITACHI U-2000, wavelength: 280 nm).

Determination of amount of epoxy groups [8]

The latex was mixed with 3.6 ml of 1 N HCl·$CaCl_2$ and 60 ml of distilled water. The mixture was shaken for 3 h at 40 °C to open the epoxy rings. After the reaction, 14.4 ml of 0.5 M KOH was added to the mixture, which was kept standing for 15 min at room temperature. Electric conductimetric titration of the latex was then carried out with 0.1 N HCl.

DNAs

Two complementary oligonucleotides, 5′-AAGTGACGTAACGTGGGG-3′ and 5′-ACGTTACGTCACTTCCCC-3′, were chemically synthesized, annealed, 5′-phosphorylated and ligated to give double stranded oligomers that mainly ranged from 150 to 250 bp as described previously [7]. 5′-Phosphorylation and ligation were carried out as follows. The annealed oligonucleotides (200 μg) were combined in 66 mM Tris-HCl buffer (pH 7.4) containing 10 mM $MgCl_2$, 15 mM dithiothreitol, 1 mM spermidine, 0.2 mg/ml BSA, 1 mM ATP, and [γ^{32}P] ATP in a total volume of 100 μl. T4 polynucleotide kinase (72 units) was added and the resulting solution was incubated at 37 °C for 1 h. Then T4 DNA ligase (2800 units) was added, and the mixture was incubated at 4 °C for 24 h. The ligated oligonucleotides were extracted with phenol/chloroform and ethanol-precipitated. The oligomer contained tandemly-repeated E4TF3-binding sites. The oligomers with 3′-protruding ends were treated with T4 DNA polymerase (0.5 units/μg of oligomers) in the absence of the substrates at 37 °C for 0, 15, 30, 60, and 120 s to give single-stranded 5′-ends with −5, approximately 0, +5, +15, and +35 nucleotide-lengths, respectively. After extraction with phenol/chloroform and precipitation with ethanol, the oligomers were dissolved in 10 mM potassium phosphate buffer (pH 8.0) and coupled to the latex particles. Complementary oligonucleotides shown below, in which part of base units were replaced, were also synthesized. The oligonucleotides were annealed, phosphorylated, and ligated to give oligomers. They were referred to as M1, M2, and M3, respectively.

M1: 5′-GGGGGAAGTGAC**T**TAACGT-3′
3′-TTCACTG**A**ATTGCACCCCC-5′

M2: 5′-GGGGGAAGTGACGT**GG**CGT-3′
3′-TTCACTGCA**CC**GCACCCCC-5′

M3: 5′-GGGGGAAGTGACG**C**AACGT-3′
3′-TTCACTGC**G**TTGCACCCCC-5′

(The thick lettered bases are different from those of wild type.)

Immobilization of DNA on the latex particles

(St–GMA)/GMA latex particles (2.5 mg) were washed twice with 500 μl of 10 mM potassium phosphate buffer (pH 8.0), and mixed with 30 μg of the DNA oligomers in 200 μl of 10 mM potassium phosphate buffer. The coupling reaction was carried out at 50 °C for 24 h. The particles were collected by brief centrifugation in a microcentrifuge at 15 000 rpm for 20 s, washed twice with 500 μl of 2.5 M NaCl. Because the DNA oligomers are labeled with ^{32}P, the efficiency of immobilization can be estimated by comparing the amount of radioactivity that is retained on the particles with that which remains in solution after the coupling reaction. The particles were then suspended in 1 ml of 1 M ethanolamine·HCl (pH 8.0). The suspension was incubated at room temperature for 24 h to inactivate unreacted epoxy groups on the surface of the particles. The particles were collected, washed once with distilled water (500 μl), washed three times with 500 μl of a storage buffer containing 10 mM Tris–HCl (pH 8.0), 0.3 M KCl, 1 mM EDTA and 0.02% (wt/vol) NaN_3, suspended in 200 μl of the storage buffer, and stored at 4 °C.

Purification of E4TF3

One ml of HeLa cell nuclear extracts was mixed with 100 μg of poly(dI–dC) and 1.0 mg single-stranded DNA (ssDNA) to sweep up strongly non-specifically binding proteins. After this pretreatment, the DNA-immobilizing particles (2.5 mg) were added to the nuclear extracts to bind the specific proteins to the target sequence for 30 min at 4 °C. The binding reaction was terminated by brief centrifugation to separate the particles from supernatant. The particles were then washed five times with 120 μl of TGEN buffer [50 mM Tris–HCl (pH 7.9), 20% (vol/vol) glycerol, 1 mM EDTA, 0.1% (vol/vol) NP-40, and 1 mM DTT] containing 0.1 M KCl. Then the particles were soaked in 100 μl of TGEN buffer containing 1.0 M KCl for 2 min. After brief centrifugation, the E4TF3 activity was recovered in the supernatant. This elution step was repeated three times.

When E4TF3 was purified using a DNA-immobilizing Sepharose gel, 1 ml of pretreated HeLa cell nuclear extracts was loaded on the affinity gel immobilizing 20 μg of DNA oligomers which was equilibrated with TGEN buffer containing 0.1 M KCl. The column was then washed with 5 ml of TGEN buffer containing 0.1 M KCl and the E4TF3 bound to the particles was eluted with 1 ml of TGEN buffer containing 1.0 M KCl.

The proteins purified were assayed by gel retardation assays and SDS-polyacrylamide gel electrophoresis (SDS-PAGE) [9].

Gel retardation assays were performed as described previously [10, 11]. DNA probes containing a single ATF/E4TF3 binding site were prepared from pUCTF3 which contained the sequence from −29 to −84 relative to the cap size of adenovirus E4 mRNA into Sma I − BamH I sites in the polylinker sites of pUC19 [11]. The plasmid was digested with Hind III and end-labeled with Klenow fragment and [α-^{32}P] dATP and then cut with EcoRI. The 96-bp fragments were purified by agarose gel electrophoresis and used for the assay.

RESULTS AND DISCUSSION

Adsorption of proteins on latex particle

The particles to be used as bioseparator must reject non-specific adsorption of proteins. To check whether this is the case or not on our particles, the adsorption of some proteins was examined. Figure 1 shows the results for the adsorption

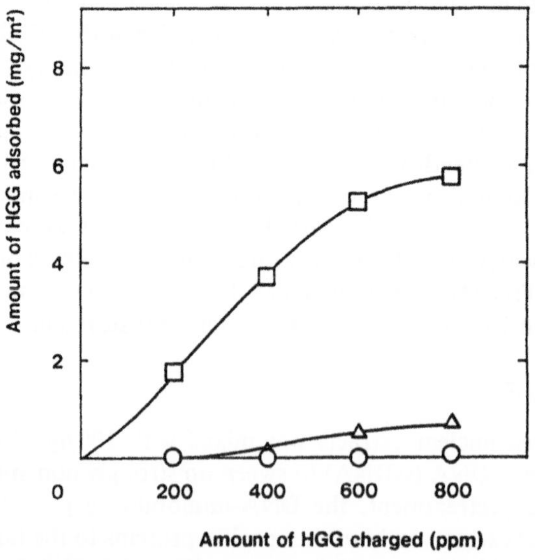
Amount of HGG adsorbed (mg/m²)
0
2
4
6
8
200
400
600
800
Amount of HGG charged (ppm)

Figure 1. Adsorption of γ-globulin on different latex particles. □: poly-St; △: (St–GMA) copolymer; ○: (St–GMA)/GMA copolymer.

of γ-globulin on the different particles. Hydrophobic poly-St particles adsorbed γ-globulin by approximately 5.8 mg/m^2 when 800 ppm γ-globulin was charged. Protein is generally liable to adsorb on the hydrophobic polymer surfaces as poly-St [12–14]. On (St–GMA) copolymer particles a small amount of γ-globulin was adsorbed. The surfaces of (St–GMA)/GMA particles themselves and those whose epoxy groups were converted to inactive groups by the treatment with 0.01 N HCl or 1 M ethanolamine·HCl adsorbed little γ-globulin. Because the coating of (St–GMA) particles with poly-GMA makes the particle surface hydrophilic and less ionic, non-specific adsorption of proteins is suppressed. Ethanolamine might not be the best masking reagent because the treatment of the particles with ethanolamine generated secondary amine. However, it was found that the amine gave no undesirable properties to the particles as a bioseparator.

Nuclear extracts from which transcription factors are purified contain many kinds of proteins and some of them might be adsorbed even on the (St–GMA)/GMA particle surface. To check this point, we examined the adsorption of the proteins in crude nuclear extracts on the particles. The results in Fig. 2 indicate that (St–GMA)/GMA copolymer particles and the ones treated with HCl did not adsorb the proteins in nuclear extracts but the particles treated with 1 M ethanolamine·HCl did. Even in this case, however, the proteins adsorbed were removed by only one washing with 200 μl of 50 mM Tris–HCl buffer (0.1 M KCl, pH 8.0) as shown in Fig. 2. Consequently, it was concluded that non-specific adsorption of proteins is negligible on the surface of our particles and that the proteins adsorbable on the DNA-immobilizing latex particles should be only DNA-binding proteins (still) to be purified.

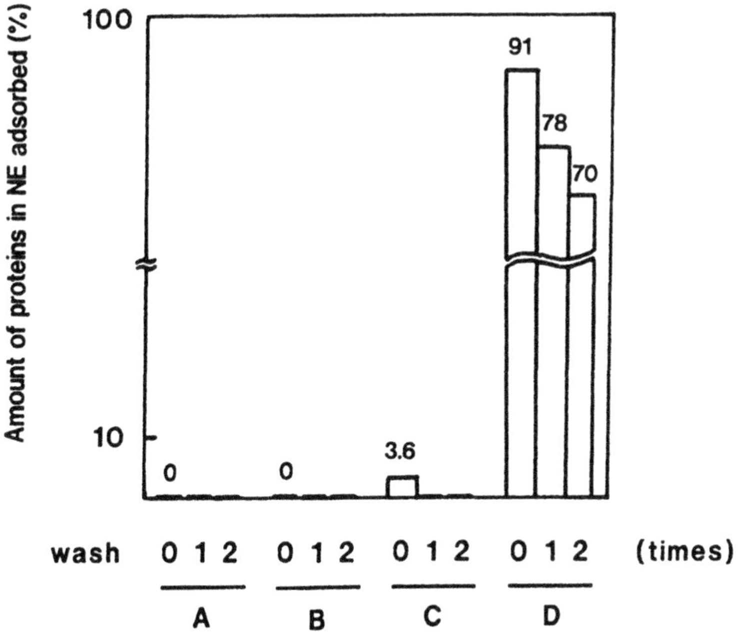
Amount of proteins in NE adsorbed (%)
100
10
91
78
70
0
0
3.6
wash 0 1 2 0 1 2 0 1 2 0 1 2 (times)
A
B
C
D

Figure 2. Adsorption of proteins in nuclear extracts (NE) on different particles. A: (St–GMA)/GMA particle; B: (St–GMA)/GMA particle treated with 0.01 M HCl; C: (St–GMA)/GMA particle treated with 1 M ethanolamine·HCl; D: poly–St particle.

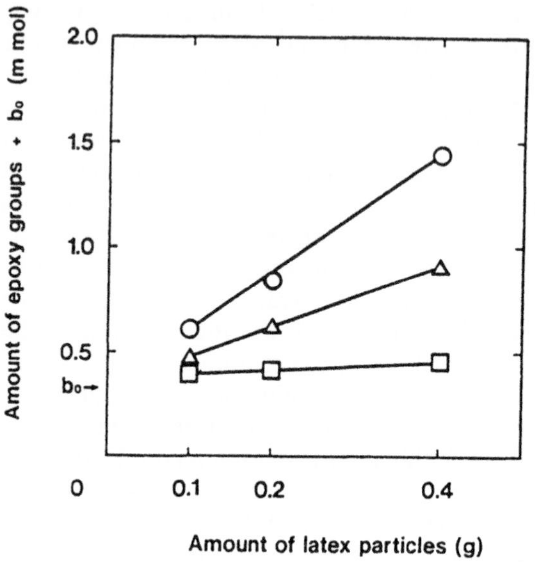
Amount of epoxy groups + b₀ (m mol)
2.0
1.5
1.0
0.5
b₀→
0
0.1
0.2
0.4
Amount of latex particles (g)

Figure 3. Determination of epoxy groups on latex particle. ○: (St–GMA)/GMA copolymer; △: (St–GMA) copolymer; □: GMA homopolymer, b_0: blank value.

Determination of epoxy groups on particle

We intended to immobilize sequence-specific DNA on the latex particle via covalent coupling using epoxy groups on (St–GMA)/GMA particles. Figure 3 shows the result of titration for the determination of epoxy groups on the particle surface. It indicates that the titre increases linearly with the amount of particles. The blank value is estimated by extrapolation to 0 g of the particles. The amount of epoxy groups on the surface of (St–GMA) and (St–GMA)/GMA copolymer particles were calculated to be approximately 1.4 and 2.8 mmol/g particles, respectively. When GMA was polymerized on (St–GMA) particles, the amount of epoxy groups on the surface has doubled and the particles are believed to be fully coated with poly-GMA. GMA homopolymer particles shown in Fig. 3 were prepared by persulfate-initiated polymerization in which most epoxy groups were converted to glycol units.

Effect of single-stranded lengths on DNA immobilization efficiency

DNA was immobilized on (St–GMA)/GMA particles by direct coupling between the amino groups of DNA and the epoxy groups on the particles. We examined the effect of single-stranded chain lengths at the 5′-protruding ends on the amounts of DNA oligomers immobilized to the particles (Fig. 4), in order to study the coupling mode. DNA oligomers with blunt ends at either strand hardly immobilized to the particles. Whereas, with increasing 5′-protruding ends DNA oligomers were more efficiently immobilized. But beyond the single-stranded lengths of 15 nucleotides, immobilization efficiency decreased. This is because DNA oligomers with a lot of free amino groups at the single stranded parts are immobilizable with a number of points. The DNA oligomers, which immobilized with a lot of points, are supposed to cover the surface of the particles and prevent free DNA oligomers from

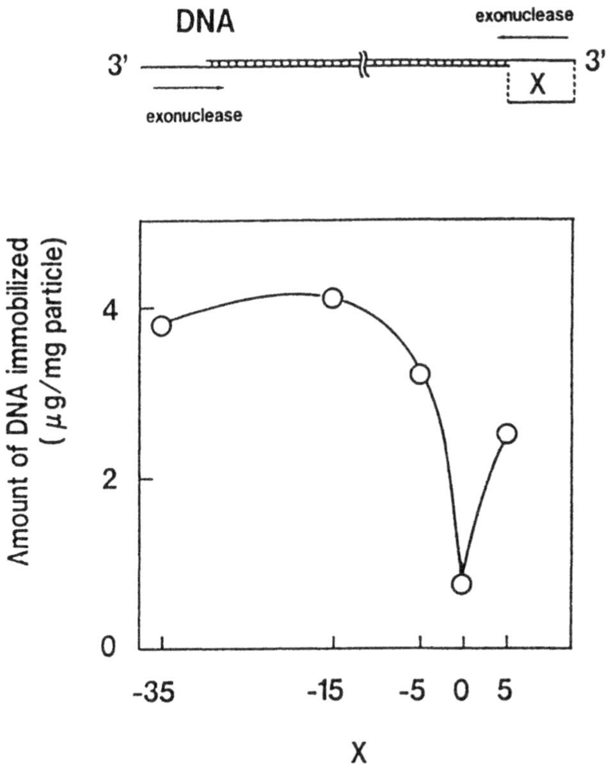
DNA
exonuclease
3'
3'
X
exonuclease
Amount of DNA immobilized
(μg/mg particle)
4
2
0
-35
-15
-5
0
5
X

Figure 4. Dependence of immobilization efficiency on DNA chain structure.

immobilizing to the particles. DNA oligomers with about 15 nucleotides at the 5′-protruding ends were bound to the particles about five times more efficiently than those with blunt ends. This suggests that the majority of DNA oligomers were bound to the particles through their protruding ends. The series of particles carrying DNA with different single-stranded chain lengths were used for the purification of E4TF3 and the effect of single-stranded chain lengths on the quality and quantity of purified E4TF3 was investigated. The DNA-binding activity of affinity-purified E4TF3 was examined by gel retardation assay. The particles carrying DNA oligomers with about 15 nucleotides at the protruding ends gave the most efficient purification of the E4TF3 (data not shown).

Comparison between DNA-carrying latex particles and Sepharose gel

Using the particles, we could purify a transcription factor, E4TF3, from crude nuclear extracts of HeLa cells within a few hours. We also tried to purify E4TF3 from the extracts by using DNA-carrying Sepharose gel, which was used in the DNA affinity chromatography developed by Kadonaga and Tjian [6], with batch-wise and column operations, Sepharose(B) and Sepharose(C), respectively. Judging from the thickness of arrowed bands in gel retardation assay lines for particles and Sepharose(C) in Fig. 5, it is obvious that the eluate from DNA-carrying Sepharose gel packed into column has a small but almost the same DNA-binding activity with that from DNA-carrying latex particles, whereas the eluate from DNA-carrying

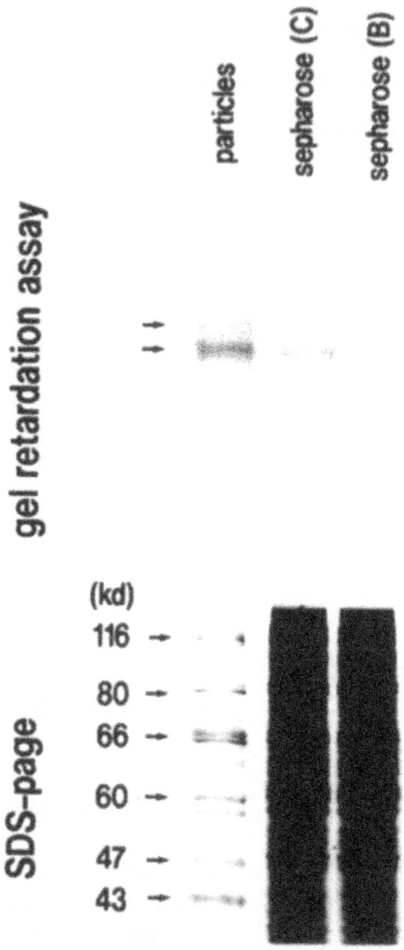
particles
sepharose (C)
sepharose (B)
gel retardation assay
SDS-page
(kd)
116
80
66
60
47
43

Figure 5. Purification of E4TF3 by the DNA-immobilizing latex particles and DNA-immobilizing Sepharose gel. Upper and lower columns present gel retardation assay and SDS-page, respectively, for proteins purified by three systems: particle/batch; sepharose/column(C); and sepharose/batch(B).

Sepharose gel in batch-wise operation has lower activity. E4TF3 eluted from DNA-carrying Sepharose gel, however, was contaminated by a large amount of non-specific proteins (shown by SDS-PAGE in Fig. 5). It is concluded that, as compared to DNA-carrying Sepharose gel, DNA-carrying latex particles were significantly high in efficiency in the purification of E4TF3 from crude nuclear extracts.

Effect of DNA-sequence on E4TF3 purification

We tried selective purification of E4TF3 using four kinds of particles which carried either the E4TF3-specific sequence [15–18] or the single-point substituted mutants (M1, M2, and M3) described above. Figure 6 indicates that E4TF3 family (80, 65/66, 60, 47, 45, and 43 kDa) could not efficiently bind to the M1 sequence on the surface of the particles but bind to the M2 and M3. The composition of purified

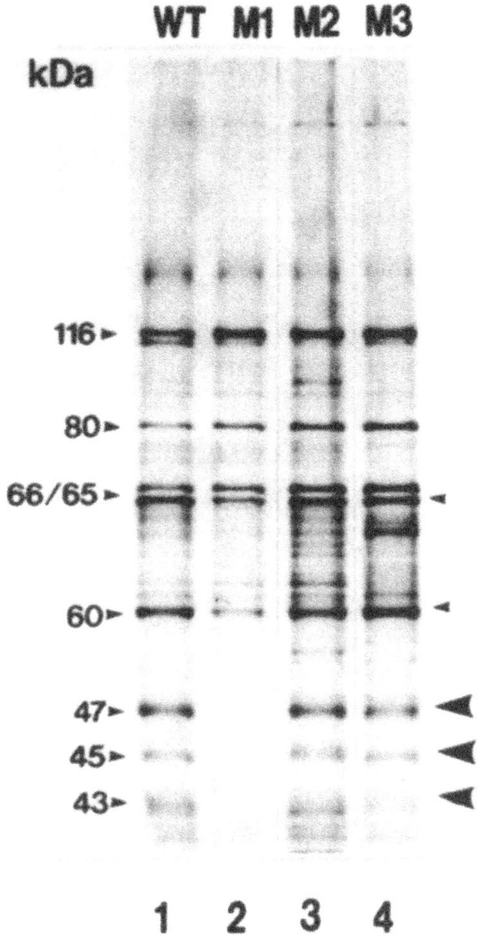
WT M1 M2 M3
kDa
116
80
66/65
60
47
45
43
1 2 3 4

Figure 6. Effect of DNA-sequence on E4TF3 purified. WT, M1, M2, and M3: see Experimental section.

proteins has obviously changed with the difference of only a base unit in DNA sequence on the surface of the particles. These results suggest that DNA-carrying affinity latex particles are useful for the investigation of the affinity of specific DNA sequences with DNA-binding proteins.

CONCLUSIONS

On the surface of (St–GMA)/GMA copolymer latex particles, γ-globulin and the proteins in the crude nuclear extracts were little adsorbed. This result indicates that the proteins adsorbable on the DNA-carrying latex particles are only DNA-binding proteins to be purified. A transcription factor, E4TF3, was efficiently purified to homogeneity using the latex particles. Compared to DNA-carrying Sepharose gel, the latex particles gave several-fold higher efficiency in the purification of E4TF3 from the crude nuclear extracts. The composition and purity of E4TF3 family purified changed obviously with the difference of only a base unit in DNA sequence on the surface of the particles.

Acknowledgement

This study was supported by a Grant-in-Aid for Scientific Research from Ministry of Education, Japan.

REFERENCES

1. P. J. Mitchell and R. Tjian, *Science* **245**, 371 (1989).
2. T. Hai, F. Liu, J. Coukos and M. R. Green, *Genes Dev.* **3**, 2083 (1986).
3. D. E. Krieger, B. W. Erickson and R. B. Merrifield, *Proc. Natl. Acad. Sci. USA* **73**, 3160 (1976).
4. B. P. Cuatrecasas, M. Wilchek and C. B. Anfinsen, *Biochemistry* **61**, 636 (1968).
5. D. J. Arndt-Jovin, T. M. Jovin, W. Bähr, A. M. Frischauf and M. Marquardt, *Eur. J. Biochem.* **54**, 411 (1975).
6. J. T. Kadonaga and R. Tjian, *Proc. Natl. Acad. Sci. USA* **83**, 5889 (1986).
7. H. Kawaguchi, A. Asai, Y. Ohtsuka, H. Watanabe, T. Wada and H. Handa, *Nucleic Acids Res.* **17**, 6229 (1989).
8. T. Matsumoto, M. Okubo and Y. Takahashi, *Koubunshi Ronbunshu* **34**, 571 (1977).
9. U. K. Laemmli, *Nature* **227**, 680 (1970).
10. H. Watanabe, T. Wada and H. Handa, *EMBO, J.* **9**, 841 (1990).
11. T. Wada, H. Watanabe, Y. Usuda and H. Handa, *J. Virol.* **65**, 557 (1991).
12. A. Baszkin and D. J. Lyman, *J. Biomed. Mater. Res.* **14**, 393 (1980).
13. J. L. Brash and S. Uniyal, *J. Polym. Sci. Polym. Symp.* **66**, 377 (1979).
14. M. D. Bale, D. F. Mosher, L. Wolfarht and R. C. Sutton, *J. Colloid Interface Sci.* **125**, 516 (1988).
15. K. A. W. Lee and M. R. Green, *EMBO J.* **6**, 1345 (1987).
16. K. A. W. Lee, T. Y. Hai, L. SivaRaman, R. Thimmappaya, H. C. Hurst, N. C. Jones and M. R. Green, *Proc. Natl. Acad. Sci. USA* **84**, 8355 (1987).
17. H. C. Hurst and N. C. Jones, *Genes Dev.* **1**, 1132 (1987).
18. Y. S. Lin and M. R. Green, *Proc. Natl. Acad. Sci. USA* **85**, 3396 (1988).

Part VII

New Polymers and Applications

In vivo evaluation of polyurethanes based on novel macrodiols and MDI

ARTHUR BRANDWOOD[1]*, GORDON F. MEIJS[2], PATHIRAJA A. GUNATILLAKE[2], KATHRYN R. NOBLE[1], KLAUS SCHINDHELM[1] and EZIO RIZZARDO[2]

[1]*Centre for Biomedical Engineering, University of New South Wales, PO Box 1, Kensington, NSW 2033, Australia*

[2]*CSIRO Division of Chemicals and Polymers, Clayton, Victoria, Australia*

Received 1 March 1993; accepted 16 July 1993

Abstract—A series of novel polyurethane elastomers based on methylenediphenyl diisocyanate, 1,4-butanediol and the macrodiols, poly(hexamethylene oxide), poly(octamethylene oxide), and poly(decamethylene oxide) were implanted subcutaneously in sheep for periods of 3 and 6 months. The specimens that were subjected to 3 months of implantation were strained to 250% of their resting length, while those implanted for 6 months had no applied external strain. SEM examination of the explanted specimens revealed that the novel materials displayed resistance to environmental stress cracking. Proprietary materials, Pellethane 2363-80A, Biomer and Tecoflex EG-80A, which had been implanted under identical conditions, showed evidence of significant stress cracking. The extent of stress cracking in the 3-month strained experiment was similar to that from the 6-month unstrained experiment. Stress cracking was also observed in Pellethane 2363-55D, when implanted for 6 months (unstrained). Neither changes in molecular weight nor in tensile properties provided a clear indication of early susceptibility to degradation by environmental stress cracking.

Key words: Polyurethanes; biostability; stress cracking; chronic implant.

INTRODUCTION

Polyurethane elastomers are widely used in the manufacture of implantable medical devices. They exhibit a high degree of haemocompatibility and excellent mechanical properties, such as softness, high strength, and resistance to fatigue and tear. However, in applications requiring long term implantation in tissues, many polyurethanes undergo a gradual deterioration, resulting in stress cracking at the material surface. This may ultimately lead to failure of the device.

Medical-grade thermoplastic polyurethane elastomers are microphase separated copolymers comprising a hard domain, derived from a diisocyanate (e.g., methylenediphenyl diisocyanate [MDI]) and a chain extender (e.g., 1,4-butanediol [BDO]), and a soft domain derived from a macrodiol (e.g., poly(tetramethylene oxide) [PTMO]). The soft domain is believed to be mainly responsible for imparting flexibility, while the hard domain imparts strength.

Stress cracking of polyether polyurethanes is believed to be associated with biologically induced oxidation [1]. The recent work of Zhao and coworkers [2] has elucidated the important role played by macrophages in the biodegradation of

* To whom correspondence should be addressed.

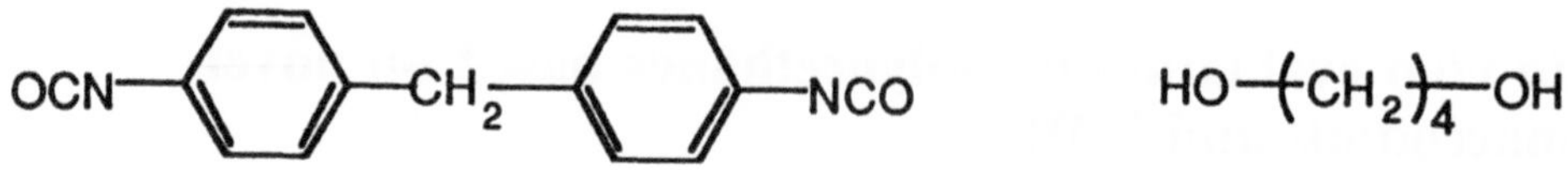

MDI BDO

Scheme 1.

polyurethanes. Most investigators believe that the soft domain is the likely site of oxidative attack [1, 3, 4] and rationalize this on the basis that the constituent macrodiol contains numerous CH_2 groups adjacent to oxygen. Such moieties are prone to attack by free radicals, especially by radicals that are oxygen centred.

We have previously reported [5, 6] the synthesis, mechanical properties and processability of a series of novel polyurethanes based on MDI, BDO, and the macrodiols poly(hexamethylene oxide) [PHMO], poly(octamethylene oxide) [POMO], and poly(decamethylene oxide) [PDMO].

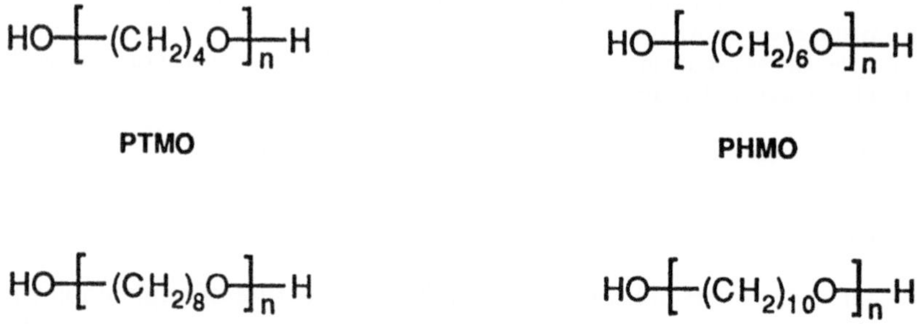

PTMO PHMO

POMO PDMO

Scheme 2.

Data relating to the stability of the polymers to hydrolytic and oxidative reagents were also reported [5]. We had prepared these polyurethanes to investigate whether they would exhibit enhanced stability over those prepared from PTMO, which is the conventional macrodiol used to prepare medical-grade polyurethanes, such as Dow's Pellethane™ 2363 series. The macrodiols PHMO, POMO, and PDMO have a lower proportion of labile CH_2 groups adjacent to oxygen (compared with PTMO). They are also less hydrophilic; the resultant polyurethanes would be expected to swell less *in vivo* and therefore, aqueous oxidative reagents will permeate these materials more slowly.

The process of polyurethane stress cracking is accelerated by the application of strain [1, 7]. We have developed an *in vivo* model, described elsewhere [8, 9], in which static strain is applied to polymer specimens. In this model, polyurethane dumbbells are stretched to a length of 250% of their resting length, over poly(methyl methacrylate) (PMMA) holders, and then implanted subcutaneously in sheep for 90 days. Following retrieval of the implant, the extent of any stress cracking is determined by scanning electron microscopy (SEM). We have used this model in the present study to investigate the biostability of the polyurethanes prepared from PHMO, POMO, and PDMO.

Degradation might also be detected by changes in molecular weights and mechanical properties [2, 10], although these are generally less sensitive to early degradation [1]. In additional studies, unstressed samples were implanted for a period of 6 months. In this series of experiments, changes in the mechanical properties and molecular weights following implantation were also monitored.

METHOD

Materials

Pellethane 2363-80A and Pellethane 2363-55D were obtained from Dow Chemicals. Tecoflex EG80A was obtained from Thermedics Corporation, and Biomer (lot no. BSZ 038) was obtained from Ethicon. These materials were used as received. The novel polyurethanes, listed in Table 1, were prepared in dimethylformamide solution using the general two stage polymerization method of Lyman [11]. The syntheses, including the preparation of the macrodiols PHMO, POMO, and PDMO, are described in detail elsewhere [5, 12]. A solution-polymerized 'equivalent' of Pellethane 2363-80A was also prepared [5].

Flat sheets with a nominal thickness of either 0.6 mm or 1 mm and with dimensions of 60 mm × 100 mm were prepared by either solvent casting or compression moulding at temperatures below 190°C. All samples were inspected under crossed

Table 1.
Formulations of polymers

Polymer	Macrodiol	Macrodiol molecular weight ($\bar{M}_n$)	% Macrodiol (w/w)	Diisocyanate	Chain extender	Shore hardness
Laboratory synthesized polymers (solution polymerized)						
P4	PTMO	1000	56	MDI	BDO	80A
P6	PHMO	650	47	MDI	BDO	90A
P8	POMO	1685	70	MDI	BDO	93A
P10	PDMO	1270	64	MDI	BDO	55D
Commercial polymers						
Biomer	PTMO	[a]	[a]	MDI	[b]	64A
Pellethane 2363-80A	PTMO	[a]	[a]	MDI	BDO	82A
Pellethane 2363-55D	PTMO	[a]	[a]	MDI	BDO	55D
Tecoflex EG80A	PTMO	[a]	[a]	$H_{12}MDI$[c]	BDO	80A

[a] Undisclosed—proprietary formulations
[b] Undisclosed, but primarily 1,2-diaminoethane
[c] $H_{12}MDI$ = hydrogenated MDI

polarizers and were considered free of significant residual internal processing stress because birefringence was absent. The strained *in vivo* experiments were carried out using material of 0.6 mm thickness. Sheet material of 1 mm thickness was used in all other experiments described in this paper. Following preparation of the sheets, all materials were conditioned at ambient temperatures for a minimum of 4 weeks prior to use.

In vivo *experiments*

Strained implant (3 months). Small dumbbells (Fig. 1(b)) were cut from the sheets using a specially manufactured steel punch. Poly(methyl methacrylate) (PMMA) holders (Fig. 1(a)) were prepared by laser machining from 1.5 mm thick PMMA sheet. Dumbbells and holders were washed by soaking in reverse-osmosis water for 48 h with a minimum of six changes of water in that period. Dumbbells were stretched over holders (Fig. 1(c)) taking care not to extend the polyurethane any further than was necessary to engage the sample onto the holder. A polypropylene suture (3–0 Prolene) was firmly tied around the centre of each strained dumbbell to act as a localized stress raiser (Fig. 1(c)). Each material was identified by a binary code marked on its PMMA holder using a 1 mm drill.

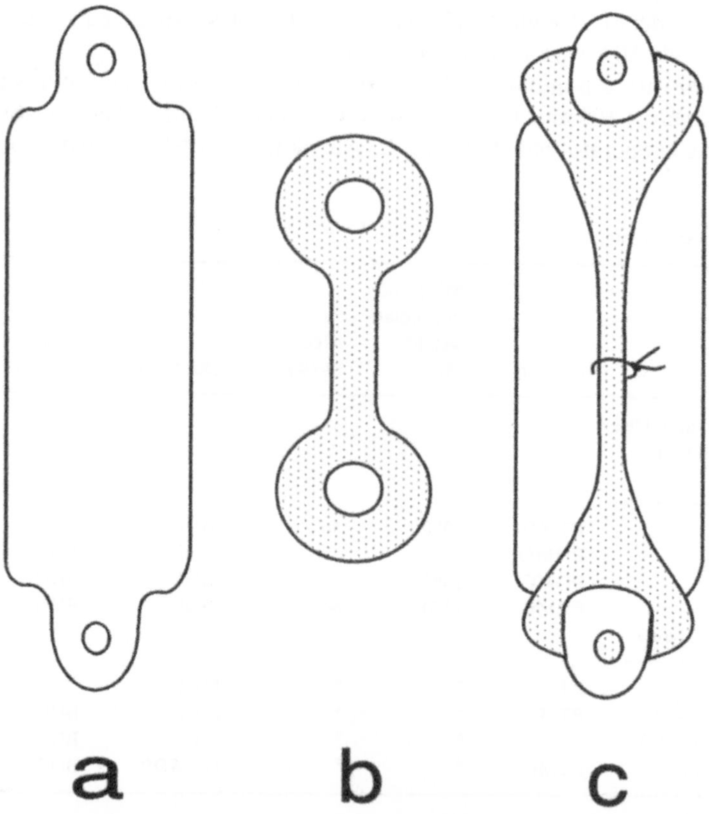
a
b
c

Figure 1. (a) PMMA holder. (b) Polyurethane dumbbell. (c) Dumbbell stretched over PMMA holder with the central section extended to 250% of its resting length. Note the use of a suture as a stress raiser.

The prepared implants were sterilized by ethylene oxide (EtO). The maximum temperature reached was 60°C. Residual EtO was removed by degassing *in vacuo* for a minimum of 48 h and then in air for 7 days.

The sterile samples were implanted subcutaneously in the dorsal thoraco-lumbar region of 1–2-year-old cross-bred wether sheep. The implants were placed via a skin incision in a position resting on the surface of the dorsal musculature and beneath the subcutaneous fat, oriented with the polyurethane facing upwards, in contact with the adipose tissue and with the PMMA holder resting on the muscle. Two replicates of each test material were implanted in each of two animals. Two sets of non-implanted controls were prepared. One set (referred to later as the 'wet controls') was stored in darkness in sterile phosphate buffered saline (PBS) at pH 7.3 and at a temperature of 37°C. The other set of controls (the 'dry controls') was stored in sealed containers under dry nitrogen in darkness at room temperature. The purpose of these controls was firstly to confirm that neither washing and sterilizing nor any non-specific degradation such as hydrolysis had contributed to any changes and secondly to provide a reference material for comparison of surface texture with the materials retrieved from animals.

After 90 days the implants were retrieved. Remaining adhered biological material was removed by soaking in 0.1M NaOH for 7 days followed by washing in copious reverse-osmosis water and air drying in a laminar flow cabinet. Non-implanted controls were washed in the same way.

All samples were sputter coated with gold palladium and examined in a Cambridge Stereoscan S360 scanning electron microscope at magnifications up to 2000

Table 2.
Surface degradation after implantation as assessed by SEM

	Stability ranking (median value[a])	
Polymer	Strained (3 months)[b]	Unstrained (6 months, $n = 4$)
P4	3.5	3.5
P6	1	2
P8	1	1
P10	1	1
Biomer	5	4
Pellethane 2363-80A	4	3
Pellethane 2363-55D	[c]	3
Tecoflex EG80A	2.5	3

[a] These data are ranks and the calculation of means is inappropriate; the appropriate statistic is the median [17]

[b] $n = 4$ except for Pellethane 2363-80A, where $n = 6$

[c] Pellethane 2363-55D was not examined in this experiment

Key to ranks: 1 = sample smooth on all surfaces; 2 = small amounts of cracking or pitting in high stress areas (e.g. near the suture ligature in stressed samples or adjacent to coding holes or corners of unstressed samples); 3 = larger patches of stress cracks in stressed areas; 4 = patches of stress cracking in all parts of sample; 5 = generalized stress cracking over most of surface; and 6 = extensive, deep stress cracking of entire surface.

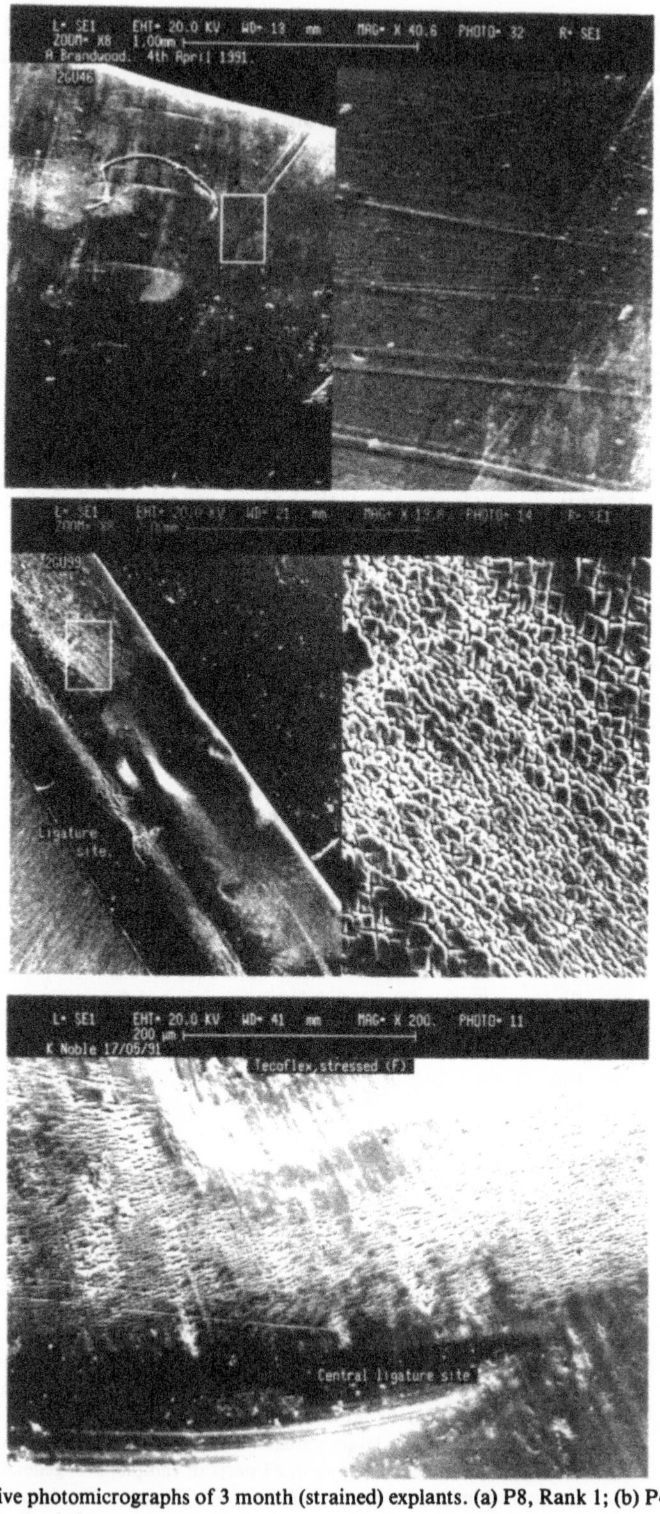
(a)
A Brandwood: 4th April 1991.
(b)
Ligature site.
(c)
K Noble 17/05/91
Tecoflex stressed (F)
Central ligature site

Figure 2. Representative photomicrographs of 3 month (strained) explants. (a) P8, Rank 1; (b) P4, Rank 4, (c) Tecoflex EG80A, Rank 3.

for evidence of surface degradation. Each sample was ranked on a scale of 1 (no evidence of biodegradation, smooth surface) to 6 (extensive deep stress cracking of the entire surface) as described in the footnote to Table 2. Representative photomicrographs are shown in Fig. 2.

Unstrained implant (6 months). Rectangular samples (30 × 10 mm) were cut from the sheets. One corner of each sample was cut off for orientation and each was coded by means of a unique binary code punched into one end of the sample with a modified 19-gauge hypodermic needle. All samples were washed, sterilized, and degassed using the same procedures as those described above for the strained 3 month implants.

The sterile samples were implanted subcutaneously in sheep, via a skin incision, using a specially manufactured trochar to standardize placement beneath the subcutaneous fat layer and resting on the surface of the dorsal musculature. Duplicate samples were implanted in each of two sheep. The samples were retrieved after 180 days by carefully cutting the end of the fibrous capsule around each sample and sliding out the polymer rectangle. The samples were then washed and dried as described above for the three month stressed samples.

A small dumbbell test piece was cut from each retrieved sample using a specially manufactured punch. An offcut from one end of each sample was examined in the scanning electron microscope for evidence of surface cracking. Molecular weights were determined from a portion of an offcut of each material.

Mechanical properties and molecular weight determinations were carried out using the methods described below. Specimens for the SEM examinations were prepared using the same techniques as for the strained 3-month *in vivo* samples (described above) and the extent of biodegradation of the different materials was ranked on a scale of 1 to 6. Representative photomicrographs are shown in Fig. 3.

Mechanical characterization

Tensile mechanical properties were determined using an Instron model 4302 tensometer. The mechanical properties of the wet controls and of the explanted polyurethanes were obtained after drying of the samples at ambient temperature and pressure. The sample preparation and testing protocol followed the recommendations of ASTM D412-87 [13], with the following exceptions. Firstly, because of limited availability of material and the constraints on sample size for the *in vivo* implant, non-standard dumbbells that were smaller than the recommended sizes were used. The dumbbells had external dimensions of 30 mm × 10 mm with a central reduced section of nominal dimensions 10 mm × 4 mm. These dumbbells were cut from the polymer sheets using a specially manufactured punch. Secondly, because of technical limitations imposed by small sample sizes, extensions were determined by cross-head displacement rather than by direct extensometry. For such highly extensible materials, the errors introduced by this indirect determination of extension would be small and the data would be suitable for the comparative purposes of this investigation. Replicate samples were tested to failure at an extension rate of 500 mm/min^{-1}. From the results, ultimate tensile stress (UTS), strain at failure and stress at an extension of 100% were calculated from the formulae given in ASTM D412-87. Hardness was determined using Shore A and D durometers (Shore Instrument Co., New York).

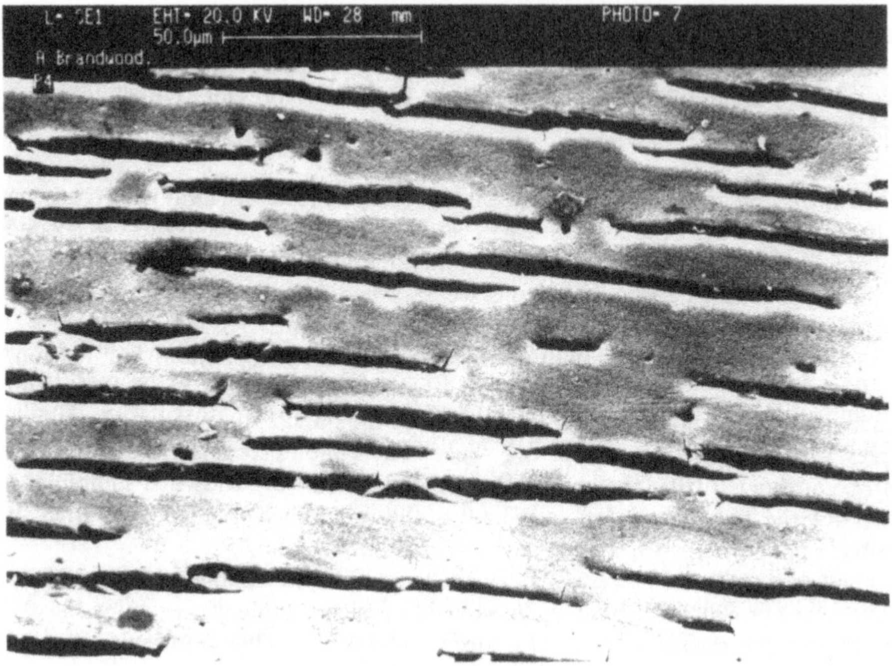
L= SE1 EHT= 20.0 KV WD= 28 mm
50.0µm
PHOTO= 7
A Brandwood.
P4

(a)

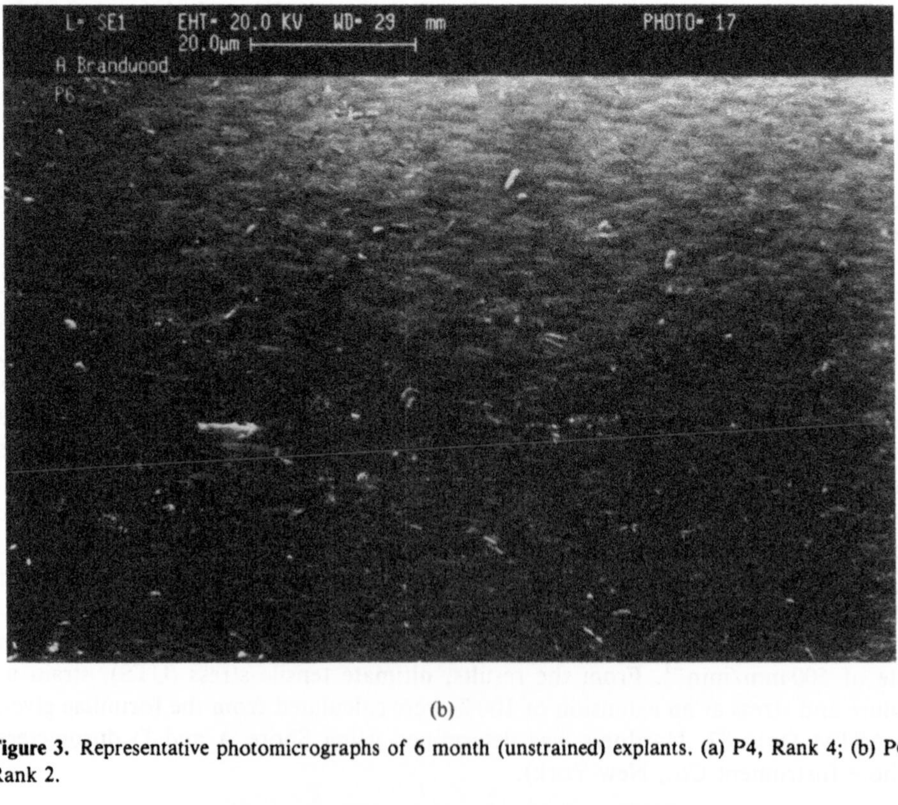
L= SE1 EHT= 20.0 KV WD= 29 mm
20.0µm
PHOTO= 17
A Brandwood
P6

(b)

Figure 3. Representative photomicrographs of 6 month (unstrained) explants. (a) P4, Rank 4; (b) P6, Rank 2.

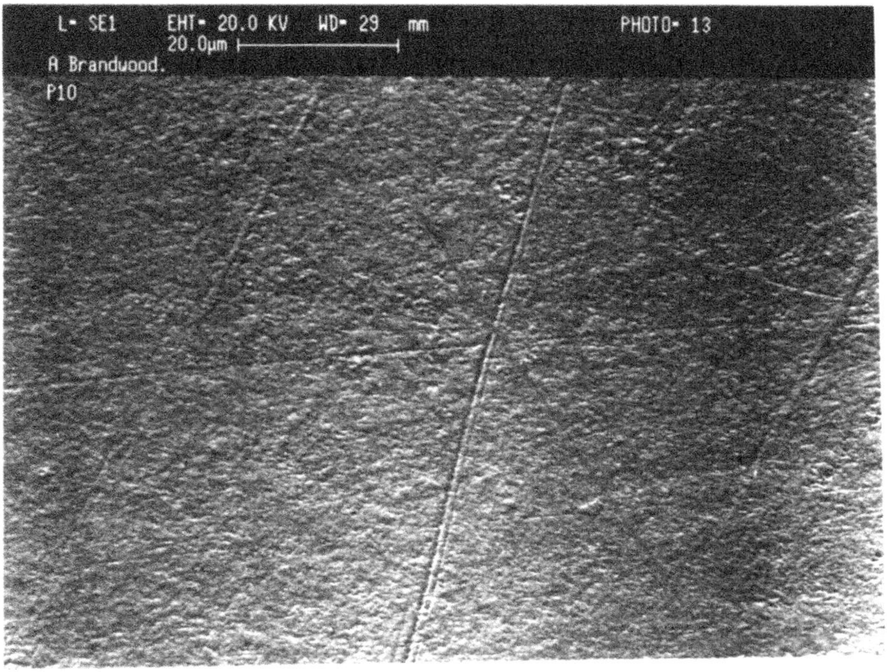
L= SE1 EHT= 20.0 KV WD= 29 mm PHOTO= 13
20.0μm
A Brandwood.
P10

(c)

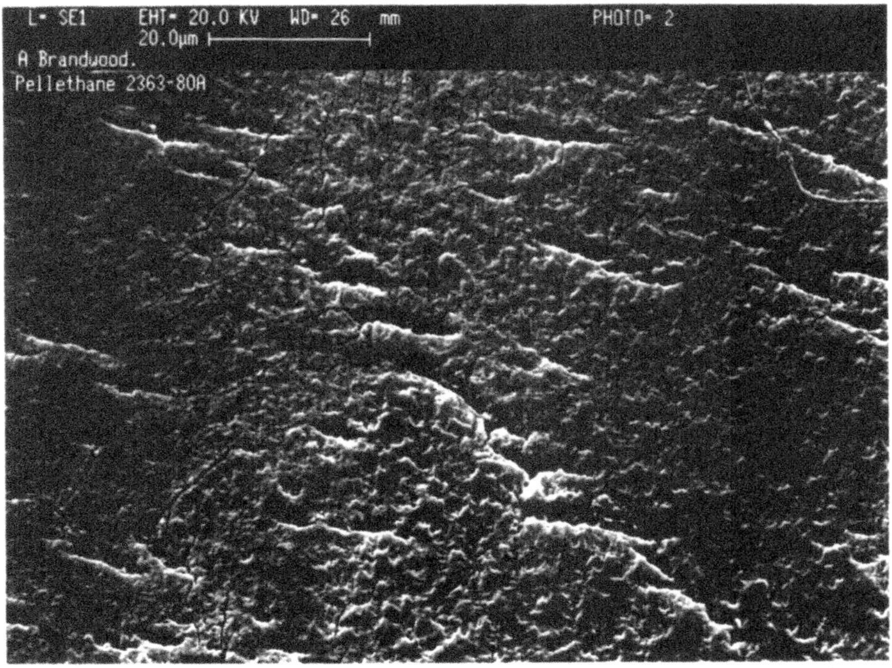
L= SE1 EHT= 20.0 KV WD= 26 mm PHOTO= 2
20.0μm
A Brandwood.
Pellethane 2363-80A

(d)

Figure 3. Representative photomicrographs of 6 month (unstrained) explants. (c) P10, Rank 1; (d) Pellethane 2363-80A, Rank 4.

Molecular weight determinations

Gel permeation chromatography was carried out at 80°C with 0.05 M lithium bromide in dimethylformamide as eluent on a Waters Associates chromatograph with 10^5, 10^3, and 50 Å μ-Styragel columns. The chromatograph was equipped with a refractive index detector and was calibrated with narrow distribution polystyrene standards. Results are expressed as polystyrene-equivalent molecular weights.

RESULTS AND DISCUSSION

The formulations of the materials that were examined are given in Table 1. Since the novel materials P6-P10 were prepared by solution polymerization and Pellethane 2363-80A, Pellethane 2363-55D and Tecoflex EG-80A are presumably prepared by bulk polymerization, a solution polymerized Pellethane 2363-80A 'equivalent' was prepared [5] and included in the testing protocol to isolate any effects due to the different modes of polymerization.

Surface degradation as assessed by SEM

The extent of surface degradation from both 6 months of implant without external static strain and from the 3-month *in vivo* experiment, in which static strain was applied, is summarized in Table 2. There were considerable differences between implanted materials, with some polyurethane formulations exhibiting relatively severe stress cracking, while other formulations were crack-free. In all cases, the non-implanted controls that were either stored wet in buffered saline or stored dry showed no evidence of surface degradation.

Polyurethanes P6-P10, prepared from PHMO, POMO, and PDMO, did not exhibit any stress cracking in the 3-month strained implant. A representative photomicrograph is shown in Fig. 2(a). The surface texture apparent in the micrograph was a consequence of the method of sample preparation and was present in unimplanted controls. The polymers, Biomer, Pellethane 2363-80A, and P4 (Fig. 2(b)) all of which were prepared from the macrodiol PTMO, showed significant evidence of cracking, with Biomer performing the worst. Tecoflex EG-80A (Fig. 2(c)), also a polyurethane based on PTMO, showed patches of stress cracking.

This situation was essentially paralleled in the case of the unstrained 6-month implant. Biomer, Pellethane 2363-80A (Fig. 3(d)), Pellethane 2363-55D, P4 (Fig. 3(a)), and Tecoflex EG80A all showed significant signs of degradation, while, except for P6 (Fig. 3(b)), the materials that were not based on PTMO, showed no signs of cracking (Fig. 3(c)). In the case of P6, the surface was generally clear, but there were small areas of fine surface cracking in stressed areas adjacent to the punched coding holes (Fig. 3(b)).

The solution-polymerized laboratory 'equivalent' of Pellethane 2363-80A, polyurethane P4, did not perform markedly differently to the commercial material in both of the implants. It is noteworthy that, overall, there were similarities in the extent of surface degradation in the 3-month strained implant and in the 6-month unstrained implant, despite the fact that the former implant was of shorter duration. This attests to the accelerating effect of externally applied strain on polyurethane biodegradation [7].

The results indicate that, under the experimental conditions and timeframes, the polyurethanes prepared from the macrodiols PHMO, POMO, and PDMO exhibit an increased resistance to biologically induced environmental stress cracking when compared with those prepared from the more conventional macrodiol PTMO. This is consistent with results from *in vitro* testing of chemical stability [5, 8], in which the polyurethanes P6–P10 showed enhanced resistance to hydrolysis and to oxidation with hydrogen peroxide compared with PTMO-based materials. Further studies will be required to characterize the long-term biostability of these materials. Such studies are in progress.

Mechanical properties after storage in PBS and after 6 months of implantation

Table 3 shows the mechanical properties of dry (darkness-stored) and wet (37°C PBS-stored) control materials and of materials retrieved from the 6-month implant, in which no static external strain was applied. The tensile properties of the dry controls were unchanged from those of the freshly prepared materials. Accordingly, only the dry control values are presented here. The mechanical properties differed slightly from those previously reported [5], since different batches were tested. Small variations in the mechanical properties between different batches are common in polyurethane manufacture.

Table 3.
Mechanical properties of control and explanted materials

Polymer	Treatment	UTS (MPa) $\bar{x}$	S.D.	Fail strain (%) $\bar{x}$	S.D.	Stress at 100% strain (MPa) $\bar{x}$	S.D.
P4	D	22.7	2.4	540	9	8.3	0.4
P4	W	24.3	4.5	601	59	8.6	0.4
P4	E	23.0	1.2	559	8	8.5	0.5
P6	D	23.9	1.1	283	45	18.4	1.4
P6	W	20.8	4.3	292	17	15.7	4.2
P6	E	24.1	1.5	300	12	19.0	1.0
P8	D	12.4	0.2	263	37	11.1	0.1
P8	W	12.8	0.1	244	43	11.9	0.1
P8	E	12.6	0.9	265	90	11.5	0.3
P10	D	19.0	2.2	110	65	20.3	0.0
P10	W	21.3	0.6	147	31	21.3	0.7
P10	E	20.1	0.1	160	12	20.2	0.2
Biomer	D	26.9	2.7	679	35	5.4	0.3
Biomer	W	27.7	4.3	678	54	5.3	0.3
Biomer	E	24.6	2.8	549	16	5.4	0.4
Pellethane 2363-80A	D	29.7	1.5	565	80	8.1	0.1
Pellethane 2363-80A	W	26.1	2.9	493	66	8.9	0.9
Pellethane 2363-80A	E	34.5	1.7	496	59	8.7	0.3
Pellethane 2363-55D	D	37.4	4.7	401	33	19.9	1.2
Pellethane 2363-55D	W	39.6	3.2	450	28	20.4	3.0
Pellethane 2363-55D	E	38.0	3.0	404	25	20.6	0.4
Tecoflex EG80A	D	15.2	1.01	661	100	4.3	0.2
Tecoflex EG80A	W	13.5	1.28	492	93	4.1	0.1
Tecoflex EG80A	E	12.0	0.19	462	51	4.5	0.0

All values are means ($\bar{x}$) ± standard deviation (S.D.). UTS, ultimate tensile stress; D 'dry control'; W, 'wet control'; E, explant retrieved after 6 months

In most cases, the observed differences between the mechanical properties of the dry and wet controls and the explanted materials were small and within experimental errors of measurement. The fail strains of implanted Biomer and Tecoflex were lower than those of the dry controls (81 and 79%, respectively, of the dry control value). The fail strain of the Tecoflex wet control was also lower than the dry control. However, the ultimate tensile stress of the explanted Pellethane 2363-80A increased by 16%. A small amount of crosslinking in the soft segment, thought to be an early manifestation of oxidation degradation [14], may be responsible for this. The fail strain of this polymer, however, remained essentially unchanged after implantation. There were no changes in the tensile properties of Pellethane 2363-55D after implantation.

There was no significant change in the mechanical properties of the polymers P4–P10 after immersion in PBS or after explant. In the case of P6–P10, this was consistent with the SEM results showing that little degradation *in vivo* had occurred; however, in the case of P4, this contrasts with the SEM examinations which revealed a significant degree of surface cracking.

Our conclusion is that compared with SEM examination of the surface, measurements of changes in tensile properties of the bulk material are insensitive probes of early biodegradation, which initially is a surface phenomenon. This conclusion has also been reached by other workers [15]. It has been suggested that fatigue properties are more sensitive indicators of degradation that static tensile properties [15].

Molecular weights after 6 months of implantation

Table 4 shows the molecular weights, expressed as polystyrene-equivalent molecular weights, of the materials explanted from the 6-month unstrained experiment, compared with those of the dry controls. Most noteworthy was the presence of significant quantities of insoluble material in the Pellethane 2363-80A and P8 after implant, consistent with some crosslinking. The molecular weight ($\bar{M}_n$) of Tecoflex EG80A decreased by 26%, while the $\bar{M}_n$ of P10, P6, and P4 decreased by 8, 19, and 12%, respectively. The polydispersity ($\bar{M}_w/\bar{M}_n$) of P10 increased somewhat. The molecular weight of Pellethane 2363-55D remained essentially unchanged after implant.

Table 4.
Molecular weights before and after 6 months of subcutaneous implantation

Polymer	Before implant $\bar{M}_n$	Before implant $\bar{M}_w/\bar{M}_n$	After implant $\bar{M}_n$	After implant $\bar{M}_w/\bar{M}_n$	Reduction in $\bar{M}_n$ (%)
P4	73 750	1.41	65 050	1.45	12
P6	89 000	1.46	72 050	1.56	19
P8	44 950	1.36	[b]	[b]	—
P10	48 550	1.44	44 500	1.71	8
Biomer	[b]	[b]	[b]	[b]	—
Pellethane 2363-80A	159 500	1.56	[b]	[b]	—
Pellethane 2363-55D	95 250	1.53	94 550	1.53	1
Tecoflex EG80A	146 000	1.44	107 600	1.44	26

[a] $\bar{M}_n$, $\bar{M}_w$ number and weight average molecular weights, respectively
[b] Not determined because of insolubility in dimethylformamide/LiBr; presumably cross-linked species are present

These results show that changes in molecular weight of the bulk polymer do not correlate well with the SEM observations of polymer stress cracking. There are likely to be several reasons for this. Firstly, degradation is often localized and confined to the surface. Accordingly, reliable sampling becomes a problem, especially when comparisons are made between polymers showing different stress-cracking patterns. Secondly, GPC molecular weight determinations have only modest reproducibility and it is difficult to ascertain whether small changes (e.g., 8%) are real. Finally, there may be differences in the extent of disruption of any residual allophanate branching during GPC analysis, or indeed after implantation[1], that contribute to the unreliability of molecular weight analysis as an indicator of propensity towards stress cracking.

CONCLUSIONS

Neither changes in molecular weight nor in tensile properties provided any clear indication of early susceptibility to degradation by environmental stress cracking. Implantation of polymers that were strained over poly(methyl methacrylate) holders to 250% of their resting length provided an acceleration of the degradation which closely correlated with the results of longer term *in vivo* experiments with unstrained samples. The extent of degradation by cracking or pitting could be readily assessed by SEM examination after removal of biological debris.

The polyurethanes produced with macrodiols PHMO, POMO, and PDMO were markedly more resistant in these experiments to biologically induced stress cracking compared with P4 and the commercial materials, Biomer, Pellethane 2363-80A, Pellethane 2363-55D and Tecoflex EG80A, all of which are based on the macrodiol PTMO. The polyurethanes P8 and P10 showed no evidence of stress cracking in both accelerated (3 month strained) and longer term (6 month unstrained) *in vivo* experiments.

Acknowledgements

Support was provided under the Generic Technology component of the Australian Government Industry Research and Development Act 1986, in conjunction with Telectronics Pacing Systems, Cyanamid (Australia) Pty. Ltd. and Terumo (Australia) Pty. Ltd. We are grateful to Veronica Tatarinoff, Donna McIntosh and Dr Ron Chatelier for assistance with aspects of the experimental work. We thank Dr. Mike Skalsky (Telectronics) for helpful discussions.

NOTE

[1]Small quantities of allophanate (or biuret) linkages are present in many polyurethanes and are relatively labile [16]. The proportion of allophanates present depends on the detailed conditions used to prepare the polymer, the stoichiometry of the reactants and the conditions of storage and processing.

REFERENCES

1. M. Szycher, *J. Biomat. Appl.* **3**, 297 (1988).
2. Q. Zhao, M. P. Agger, M. Fitzpatrick, J. M. Anderson, A. Hiltner, K. Stokes and P. Urbanski, *J. Biomed. Mat. Res.* **24**, 621 (1990).

3. K. Stokes, A. Coury and P. Urbanski, *J. Biomat. Appl.* **1**, 411 (1987).
4. G. F. Meijs, S. J. McCarthy, E. Rizzardo, Y. Chen, R. C. Chatelier, A. Brandwood and K. Schindhelm, *J. Biomed. Mat. Res.* **27**, 345 (1993).
5. P. A. Gunatillake, G. F. Meijs, E. Rizzardo, R. C. Chatelier, S. J. McCarthy, A. Brandwood and K. Schindhelm, *J. Appl. Polym. Sci.* **46**, 319 (1992).
6. G. F. Meijs, E. Rizzardo, P. A. Gunatillake, A. Brandwood and K. Schindhelm, *International Patent Application* PCT/AU91/00270 (1991).
7. K. Stokes, A. W. Frazer and E. A. Carter, *Proc. ANTEC* **84**, 1073 (1984).
8. A. Brandwood, K. R. Noble, K. Schindhelm, G. F. Meijs, P. A. Gunatillake, R. C. Chatelier, S. J. McCarthy and E. Rizzardo, in: *Degradation Phenomena in Polymeric Biomaterials*, H. Planck, M. Dauner and M. Renardy (Eds.). Springer-Verlag, Berlin (1992).
9. A. Brandwood, K. R. Noble, K. Schindhelm, G. F. Meijs, P. A. Gunatillake, R. C. Chatelier, S. J. McCarthy and E. Rizzardo, in: *Biomaterial-Tissue Interfaces (Proc. 9th European Soc. Conf. Biomat.)*, P. J. Doherty, R. L. Williams, D. F. Williams and A. J. C. Lee (Eds.). Elsevier, London (1991).
10. W. Lemm, in: *Polyurethanes in Biomedical Engineering*, p. 93, H. Plank, G. Egbers and J. Sym (Eds.). Elsevier, Amsterdam (1984).
11. D. J. Lyman, *J. Polym. Sci.* **45**, 49 (1960).
12. P. A. Gunatillake, G. F. Meijs, R. C. Chatelier, D. M. McIntosh and E. Rizzardo, *Polymer Int.* **27**, 275 (1992).
13. D412-87 Standard Test Methods for Rubber Properties in Tension, *Annual Book of ASTM Standards*, Vol. 9, ASTM, Philadelphia (1992).
14. A. Takahara, A. J. Coury, R. W. Hergenrother and S. L. Cooper, *J. Biomed. Mater. Res.* **25**, 341 (1991).
15. Q. Zhao, R. E. Marchant, J. M. Anderson and A. Hiltner, *Polymer* **28**, 2040 (1987).
16. R. Arshady and M. H. George, *Polym. Commun.* **31**, 448 (1990).
17. R. F. Sokal and F. J. Rohlf, *Biometry*, Freeman, San Francisco (1981).

Morphology of block copolyurethanes: V. The effect of $-CH_2CH_2-$ vs $-CH_2-$ spacers between aromatic rings

L. A. GOWER, T.-L. D. WANG* and D. J. LYMAN†

Department of Bioengineering and Department of Materials Science and Engineering, University of Utah, Salt Lake City, UT 84112, USA

Received 5 June 1992; Accepted 15 December 1993

Abstract—Copolyether-urethane-ureas based on 1,2-ethylene bis(4-phenylisocyanate), polypropylene glycol and ethylene diamine were synthesized by both the standard two-step and a multi-step procedure and their properties compared with analogous copolyurethanes based on methylene bis(4-phenylisocyanate). The infrared and dynamic mechanical properties indicate differences in the packing structure of the hard domains between these two copolymer systems. The infrared data suggest that the introduction of a $-CH_2CH_2-$ spacer between the two aromatic rings of the diisocyanate results in a coplanar packing in the hard domains for these copolyurethanes.

Key words: Copolyurethanes; segmented poly(ether-urethane-ureas); synthesis, morphology; infrared spectra; infrared dichroism; dynamical mechanical spectroscopy.

INTRODUCTION

Block copolyurethanes have been used in medical devices and implants involving blood contact. The relatively good blood compatibility of some of these urethane materials appears to be related to the heterogenous nature of their surfaces due to the microphase separation of the copolymer blocks [1, 2]. Data indicate that changes in the morphology of these heterogenous surfaces, whether by changes in block chemical structure, size of the soft block, or by changes in fabrication, do affect the initial protein adsorption and subsequent platelet adhesion [3].

Attempts to correlate the morphology of these block copolyurethanes with their infrared and electron spectroscopy data assumed that these materials had a simple alternating $(AB)_n$ chain structure. However, computer modeling studies [4] suggested that the chain structures are more complex, i.e. they contain multimetric segment lengths. Indeed, synthetic studies [5] have indicated that the block copolyether-urethane-ureas synthesized by the standard two-step procedure can contain up to 25% dimeric hard and soft segments in the copolymer chains. While the presence of these dimeric soft and hard segments in the material do not appear to greatly affect the domain-matrix phase separation, they do affect the nature and size of the interface between the hard and soft segments as well as the properties of the copolymers. Even with complete phase separation, these materials would have urethane linkages in the matrix from both the dimeric soft segments and the dimeric hard segments in the domains extending into the matrix. These dimeric hard segments also create a more diffuse interface and could introduce urea groups into

* Present address: Department of Chemical Engineering, National Kaohsiung Institute of Technology, Kaohsiung 80782, Taiwan, China.

† To whom correspondence should be sent: P.O. Box 5314, Lacey, WA 98503, USA.

the matrix. This diffuse interface has been partially demonstrated for other block copolyurethane materials [6, 7]. Thus, to better interrelate the effects of surface morphology to the infrared spectra of these urethane materials, it would be desirable to prepare the block copolymers under conditions which should eliminate these dimeric segments. In the absence of these dimeric segments, one would expect a phase-separated structure having a sharper interface with less urethane (or urea) linkages in the matrix.

Earlier studies on the effect of spacer groups (between the two aromatic rings of the diisocyanate) on the structure and properties of homopolyurethanes [8] showed that the chain packing of the homopolyurethane based on 1,2-ethylene bis(4-phenylisocyanate) was strikingly different than that for the analogous polymer based on the more commonly used diisocyanate, methylene bis(4-phenylisocyanate) (MDI). The WAXD crystalline diffraction patterns of drawn film strips showed a triclinic packing for the phenylene–ethylene–phenylene (P2P) material, while the crystalline diffraction patterns of drawn film strips for the phenylene–methylene–phenylene (P1P) material showed a hexagonal type of packing (see Fig. 1). The single methylene group present in the P1P structure sterically forces the phenylene rings into a fixed tetrahedral orientation (109 deg) with a canting of the rings 26 deg from coplanarity. This type of structure appeared to prevent the urethane groups from forming inter-urethane hydrogen-bonds when the urethane group was in a position perpendicular to the aromatic ring. The presence of the additional methylene group in the P2P structure allows the phenylene rings to rotate more freely with respect to each other. This allows a planar stacking of the aromatic rings (as would occur in triclinic packing) and inter-urethane hydrogen-bondings which would be perpendicular to the aromatic rings.

Later infrared studies on urethane model compounds associated with the P2P structure [7] showed a low frequency shoulder (1690 cm^{-1}) in the urethane Amide I band, which was assigned to carbonyl groups that are packed in the crystalline state in a coplanar structure. Coplanarity of similar structures has been shown to result in a lower frequency carbonyl stretching Amide I band [9]. Since the model compounds and the block copolyether–urethane-ureas based on MDI, a P1P type of structure, do not show this low wave number shoulder, it is possible that their domains exist in a hexagonal packing state. This would contradict much of the literature where a triclinic packing state is assumed for the MDI based block copolyurethanes [10].

This paper describes the syntheses of block copolyether–urethane-ureas based on 1,2-ethylene bis(4-phenylisocyanate), polypropylene glycol and ethylene diamine and compares their infrared spectra and other properties with the analogous copolymers based on MDI. In order to better determine the subtle spectral effects resulting from replacing the $-CH_2-$ spacer group with the $-CH_2CH_2-$ spacer group, the block copolyurethanes were prepared by both the two-step method to give the standard material as well as by the three-step method to give a purer material having little or no dimeric segments.

MATERIALS AND METHODS

Chemicals

4,4′-Diamino-1,2-diphenylethane was synthesized by adding dropwise, with slow stirring, a solution of 0.2 mol of 4,4′-ethylene dianiline in 1000 ml of ethyl acetate

(A)

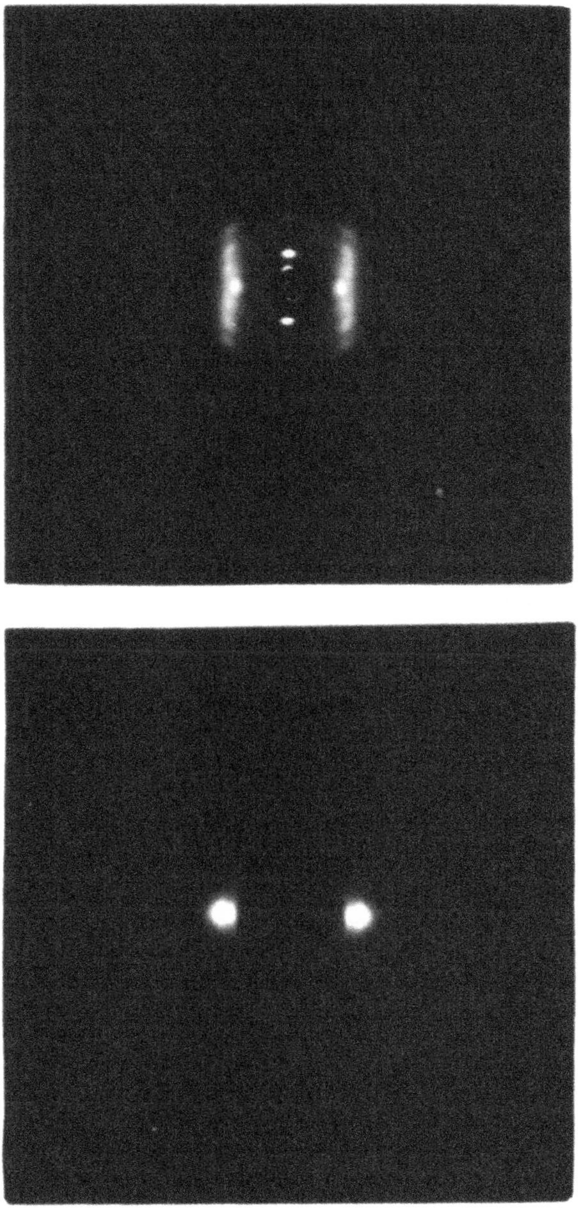

(B)

Figure 1. WAXD crystalline diffraction patterns for homopolyurethanes based on ethylene glycol and (A) methylene bis(4-phenylisocyanate); (B) 1,2-ethylene bis(4-phenylisocyanate) [8].

to a cooled (0°C) solution of 2 mol of phosgene in 400 ml of ethyl acetate. A white precipitate formed. After the addition was complete, the reaction mixture was heated at reflux, giving a clear, brownish solution. Excess solvent was then distilled leaving brownish white crystals (50.7 g, 96% yeld) of the ethylene bis(4-phenylisocyanate). Recrystallization from ethyl acetate gave white crystals, melting 89–89.5°C. Methylene bis(4-phenylisocyanate) (Multrathane, Mobay Chemical) was distilled under reduced pressure (141°C at 0.01 mm Hg).

Polypropylene glycol (NIAX Polyol PPG, Union Carbide) was degassed under vacuum at 55°C at 4.5 mm Hg for 3 h to remove any absorbed water, then stored over type 4A molecular sieves. The PPG 1025 hydroxy number was 113.6, giving a calculated molecular weight of 987.8. Ethylene diamine (EKC) was used as received. Dimethyl sulfoxide (J. T. Baker) was distilled under reduced pressure (68°C at 9.5 mm Hg). Methyl isobutyl ketone (J. T. Baker) was distilled under reduced pressure (45°C at 53 mm Hg).

Polymer synthesis

The standard two-step block copolymer synthesis [11] was conducted by reacting 0.2 mol of polypropylene glycol with 0.4 mol of 1,2-ethylene bis(4-phenylisocyanate) in 70 ml of a 50/50 mixture of dimethyl sulfoxide/methyl isobutyl ketone. The reaction was stirred and heated at 85°C for 3 h, then cooled to room temperature and 0.2 mol of ethylene diamine in 30 ml of dimethyl sulfoxide was added to the rapidly stirred solution. After 30 min of stirring, the viscous solution was poured into water to isolate the polymer. Inherent viscosity was 0.36 dl g^{-1} in dimethyl sulfoxide at 30°C.

The standard block copolymer based on methylene bis(4-phenylisocyanate) was prepared in a similar manner. Inherent viscosity was 0.53 dl g^{-1} in *N,N*-dimethylformamide at 30°C.

The pure $(AB)_n$ block copolymers were prepared in a multi-step process similar to that described earlier [5]. Polypropylene glycol (0.005 mol) was reacted with a solution of 0.05 mol of 1,2-ethylene bis(4-phenylisocyanate) in 20 ml of anhydrous ethyl acetate at 85°C for 2.5 h. At the reaction temperature (85–95°C), the ethyl acetate distilled off. The clear viscous solution was then cooled to room temperature, then triturated 18 times with 100-ml portions of hexane (or until the hexane rinse, upon evaporation, contained no diisocyanate). The dried, white residue was then weighed, redissolved in 35 ml of dimethyl sulfoxide and reacted with an equimolar amount of ethylene diamine. A gel formed initially, but became a viscous solution on standing. The solution was poured into water to precipitate the polymer. Inherent viscosity was 0.51 dl g^{-1} in dimethyl sulfoxide at 30°C.

The pure block copolymer based on methylene bis(4-phenylisocyanate) was prepared in a similar manner except in the prepolymer step no solvent was used. Inherent viscosity was 0.39 dl g^{-1} in dimethyl sulfoxide at 30°C.

Films of the copolyether–urethane-ureas were prepared by solvent casting filtered solutions in dimethyl sulfoxide (3–5% w/w solids for the thin films for infrared analysis and 10–20% w/w solids for the mechanical testing) on to clean glass plates. The films were dried in a forced draft oven at 70°C for 1 h, then vacuum dried at approximately 0.5 mm Hg for 24 h to ensure complete solvent removal.

It is interesting to note that while the P1P copolyurethanes were easily solvent cast into clear films, the P2P copolyurethanes tended to precipitate during the casting process, resulting in films that were less clear. This affect was also shown by the P2P homopolyurethanes, and was thought to be related to the ease in which the P2P structure could pack in a triclinic manner, forming insoluble molecular aggregates. The precipitated homopolymer was found to be nearly as crystalline as the drawn film strips [8].

Polymer characterization

Infrared spectra were obtained using a Digilab FTS 14B/D Fourier transform infrared spectrometer (1000 scans per film sample at a resolution of 1-cm^{-1} over the 4000–400 cm^{-1} region). The urethane Amide I band was deconvoluted manually to obtain 'free' (non-hydrogen bonded) and hydrogen bonded peak area ratios. It was assumed that the 'free' peak was a symmetrical peak, and since no bands overlap on the high frequency side, the low frequency side could be plotted and subtracted for the H-bonded peak. In actuality, these are probably not two distinct peaks, but rather a distribution of secondary bonds ranging from entirely 'free' to tightly H-bonded carbonyls.

The infrared dichroism spectra were obtained using a Nicolet MX-1 FTIR spectrometer with a 1200-S data terminal and a quadruple diamond polarizer (20 scans per film sample at a resolution of 8 cm^{-1}, peak heights were obtained from the peak finder program). A stretching device built in our laboratory was used to hold the center of the film stationary in the beam path. The films were relaxed for 3 min between each strain cycle in order to relate the orientation functions to the mechanical hysteresis. The strains were measured at 25, 50, 100, 200%, etc., until failure. Because the Nicolet is a single beam spectrometer, a baseline spectra had to be recorded initially for both polarizations.The baseline changed as the sample was elongated, so each baseline was subtracted from the peak height to give a relative measure of the peak absorbance (assuming that the absorbance is proportional to the peak height).

The dichroic ratio, D, calculated as the parallel absorbance divided by the perpendicular absorbance at each elongation (for the various vibrational bands that characterize the hard and soft phases), can be related to the orientation function, f, as follows:

$$f = [D_0 + 2)/(D_0 - 1)][D - 1)/(D + 2)],$$

where $D_0 = 2\cot^2\alpha$ (D_0 is the dichroic ratio for perfect alignment and α is the angle between the transition moment vector for the vibration and the local chain axis). The D_0 values were calculated using values of $\alpha = 90$ deg for N—H, C—H, and urea C=O vibrations, and $\alpha = 79$ deg for the urethane C=O vibrations [12–14]. The limits of this orientation function are $f = -0.5$ for a perfect perpendicular orientation of the chain axis, and $f = 1.0$ for a perfect parallel orientation, with $f = 0$ indicating a random chain orientation.

The dynamic mechanical properties were determined using a Rheovibron DDV-II-C (Toyo Rayon Co.) at a frequency of 3.5 Hz. The sample size was $3.0 \times 0.4 \times 0.03$ cm and the temperature ranged from −165° to 150°C. A heating rate of *c.* 1°C min^{-1} was used. The dissipation factor (tan δ) and storage modulus (E') as a function of temperature were calculated using our own computer program.

Dry film densities were measured using a Micromeritics Auto Pycnometer 1320. The density decreased a small amount (about 0.05) during five consecutive runs (possibly due to either a small amount of absorbed water being evaporated under the vacuum, or to a collapse of the film structure under the repeated evacuations). Therefore, the five consecutive trials were averaged and a standard deviation (N weighing) was used to indicate the variability of each value. The densities of P2P standard was 1.135 ± 0.029; P2P pure, 1.151 ± 0.088; P1P standard, 1.081 ± 0.021; and P1P pure, 1.107 ± 0.018.

RESULTS AND DISCUSSION

The unusual range of mechanical, physical and chemical properties associated with block copolyurethanes results from their separation into a domain-matrix morphology. Indeed, this microphase separation appears to be why a number of block copolyurethanes have also shown varying degrees of blood compatibility [1, 2]. Thus, these materials are widely studied for their potential use in a variety of medical implant applications.

Attempts to correlate the morphology of these block copolyurethanes with their infrared and electron spectroscopy data assumed that these materials (prepared by the standard two-step procedure) had a simple alternating $(AB)_n$ chain structure. However, it has been shown [4, 5] that these copolyurethanes contain dimeric hard and soft segments which would complicate the interpretation of the infrared spectra. Since the spectral effects resulting from replacing the $-CH_2-$ spacer group with the $-CH_2CH_2-$ spacer group may be subtle and thus masked in the copolymers prepared by this standard procedure, the analogous copolyurethanes were also prepared by a multi-step procedure [5] so as to eliminate or at least greatly reduce the presence of these dimeric segments.

Infrared analysis

The FT-IR spectra of the block copolymers based on 1,2-ethylene bis(4-phenylisocyanate), polypropylene glycol and ethylene diamine show differences between the standard preparation copolymer (P2P std) and the pure copolymer (P2P pure) (Figs 2 and 3). Of particular interest is the urethane Amide I band which is sensitive to secondary bonding [7]. In the standard copolymer, the hydrogen-bonded carbonyls form a shoulder at 1707 cm^{-1} with the nonhydrogen bonded or free carbonyls forming the major absorption at 1729 cm^{-1}. This latter peak is from

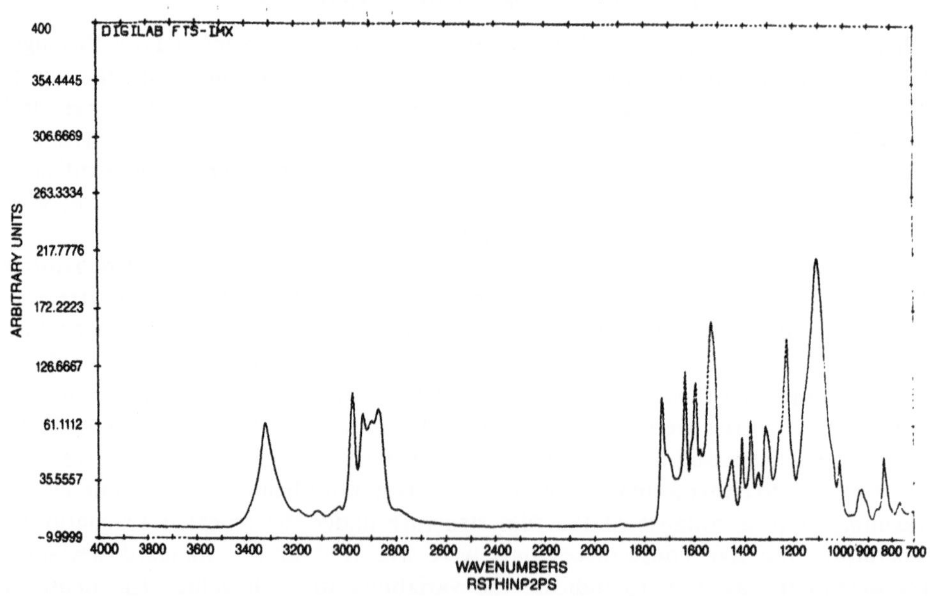
DIGILAB FTS-IMX
400
354.4445
306.6669
263.3334
217.7776
172.2223
126.6667
61.1112
35.5557
-9.9999
ARBITRARY UNITS
4000 3800 3600 3400 3200 3000 2800 2600 2400 2200 2000 1800 1600 1400 1200 1000 900 800 700
WAVENUMBERS
RSTHINP2PS

Figure 2. FT-IR spectrum of the standard copolyether-urethane-urea based on 1,2-ethylene bis(4-phenylisocyanate).

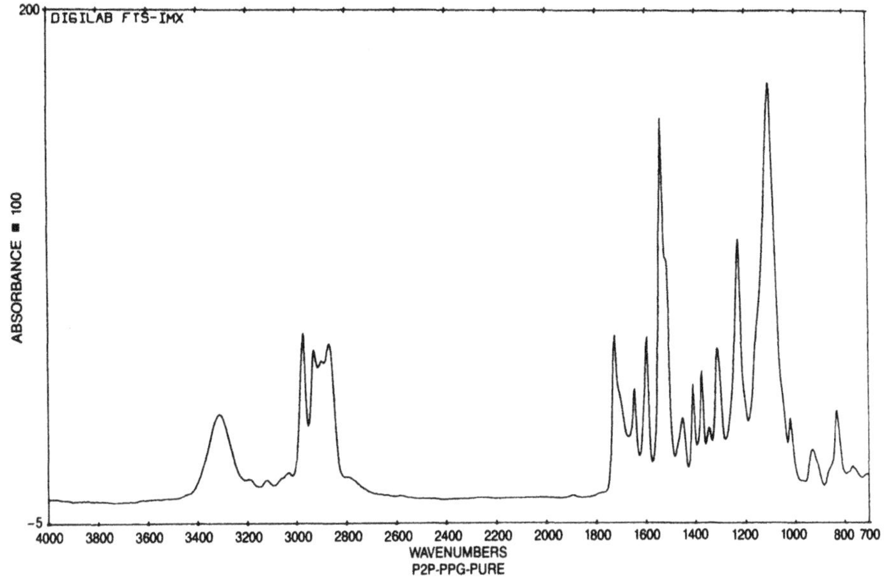
DIGILAB FTS-IMX
200
-5
ABSORBANCE * 100
4000 3800 3600 3400 3200 3000 2800 2600 2400 2200 2000 1800 1600 1400 1200 1000 800 700
WAVENUMBERS
P2P-PPG-PURE

Figure 3. FT-IR spectrum of the pure copolyether–urethane-urea based on 1,2-ethylene bis(4-phenylisocyanate).

the urethane linkages interconnecting the two soft segments of the dimer, from the penetration of the urethane linkage into the broadened interface due to the presence of dimeric hard segments, as well as from phase mixing. When these dimeric chain segments are eliminated (or reduced) in the pure copolymer, there is about a 15% decrease in the non-hydrogen bonded carbonyl (1729 cm^{-1}) absorption relative to the hydrogen bonded carbonyl (1707 cm^{-1}) absorption. Since the dynamic mechanical analyses indicate similarly high phase separation, the improvement in hydrogen bonded urethane Amide I band is mainly due to a sharper interface and a cleaner matrix phase. This is similar to that observed for the analogous P1P copolymers based on methylene bis(4-phenylisocyanate).

In addition, expansion of the urethane Amide I region comparing the P2P copolymers with the P1P copolymers (Fig. 4), does indicate a shoulder at 1690 cm^{-1} on the low wavenumber side of the hydrogen bonded carbonyl for the P2P copolymers. This had also been seen in urethane model compounds associated with the P2P structure [7]. For example, *N*-(4-tolyl) propyl carbamate showed a low frequency shoulder (1690 cm^{-1}) in the urethane Amide I band which was assigned to carbonyl groups that were packed in a coplanar structure in the crystalline state. Coplanarity of similar structures has been shown to result in a lower frequency carbonyl stretching Amide I band [9]. This suggests that differences in chain packing do result from the introduction of a $-CH_2CH_2-$ spacer between the aromatic rings of the diisocyanate.

The urea Amide I band shows several interesting features. The first is the decrease in intensity of the urea Amide I peak in the pure P1P and P2P copolymers as compared to the standard copolymers. However, other peaks, such as the aromatic ring breathing modes at 1597, 1413, and 1312 cm^{-1}, show similar intensities (or possibly slight increases) relative to the aliphatic $-CH_3$ peaks at 2972 and 1374 cm^{-1}. Since these bands arise from groups which are not sensitive to hydrogen

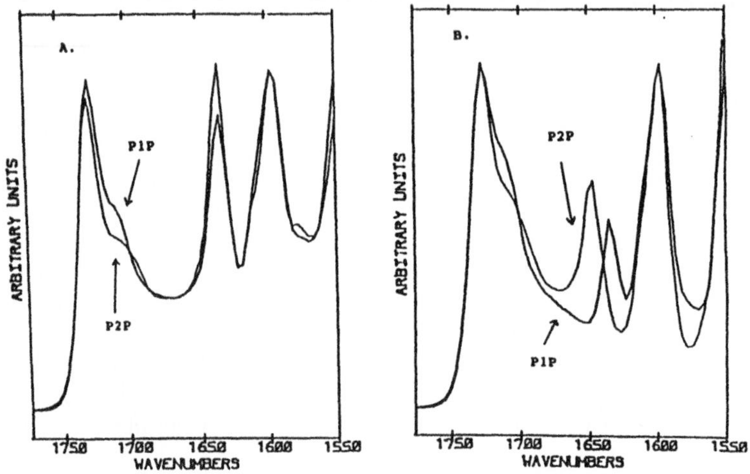
A.
P1P
P2P
ARBITRARY UNITS
1750 1700 1650 1600 1550
WAVENUMBERS
B.
P2P
P1P
ARBITRARY UNITS
1750 1700 1650 1600 1550
WAVENUMBERS

Figure 4. Expanded Amide I region of FT-IR spectra for (A) the standard P1P and P2P copolyurethanes and (B) the pure P1P and P2P copolyurethanes.

bonding, they would tend to indicate a similarity in hard/soft composition for the standard and pure copolymers. In annealing studies [15], it was observed that there were large differences in the extinction coefficients for the free vs hydrogen bonded urea carbonyls. Thus, we feel this is the reason for the decrease in the urea Amide I intensity.

The second feature is the frequency shift shown by the urea Amide I peak from 1636 cm^{-1} for the standard copolymer to 1647 cm^{-1} for the pure P2P copolymer. This shift is not shown by the analogous P1P copolymer spectra. The pure P2P copolymer also shows minor frequency shifts in the N—H stretch region, the aromatic ring breathing bands and the Amide II and III bands. We feel that the shift in the urea Amide I peak is due to a difference in the hydrogen bond structure related to a difference in chain packing for the pure P2P copolymer as compared to that for the pure P1P copolymer. In the standard P2P copolymer, the presence of large amounts of dimeric hard segments either prevents this type of packing and hydrogen bond interaction, or greatly reduces it (there could be a shoulder on the urea Amide I band at 1636 cm^{-1}).

In addition, in the pure P2P material, the Amide II (urea and urethane) band shows a splitting into peaks and the 1250 cm^{-1} shoulder of the Amide III (urea and urethane) shows a decrease in intensity. Interpretation of these urea regions is difficult because this group is capable of two different types of hydrogen bonds due to the double N—H groups. However, the data again suggests that differences in chain packing do result from the introduction of the $-CH_2CH_2-$ spacer between the aromatic rings of the diisocyanate in place of the $-CH_2-$ spacer.

Dynamic mechanical analysis

The storage modulus (E') and dissipation factor ($\tan \delta$) of the P2P copolymers (Figs 5 and 6) are generally characteristic of a system where a two-phase morphology

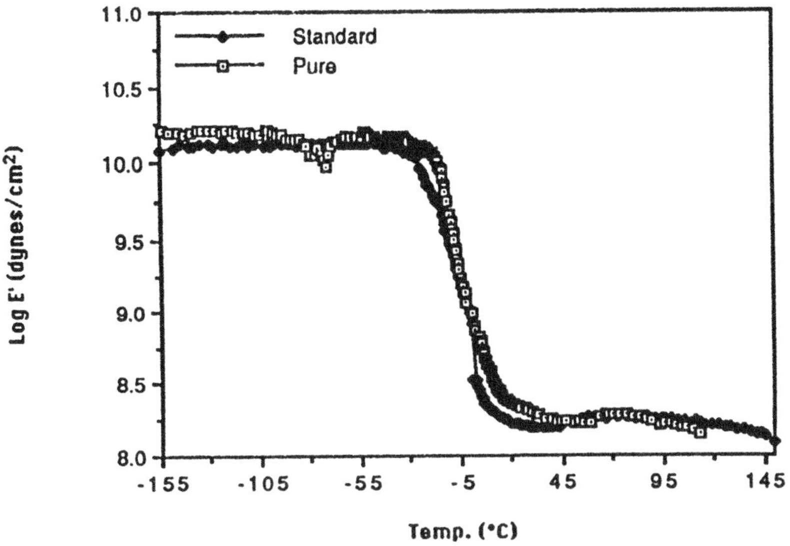
11.0
10.5
10.0
9.5
9.0
8.5
8.0
Log E' (dynes/cm²)
Standard
Pure
-155
-105
-55
-5
45
95
145
Temp. (°C)

Figure 5. Storage modulus (E') vs temperature for the standard and pure copolyether–urethane-ureas.

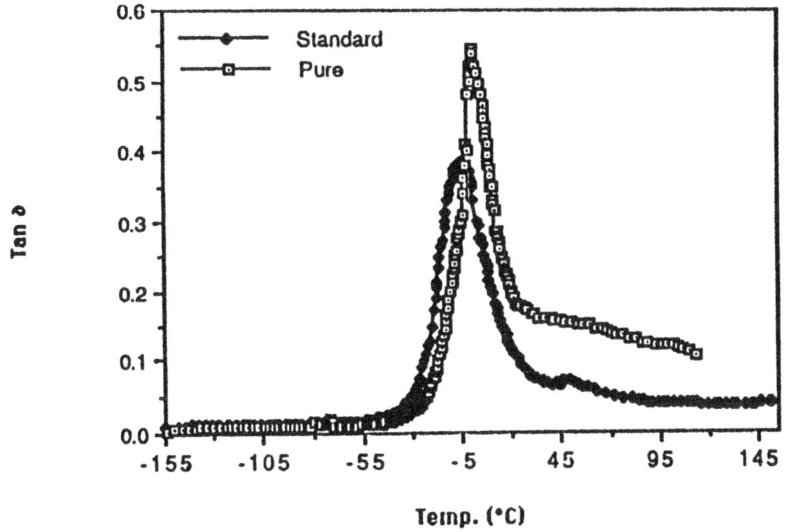
0.6
0.5
0.4
0.3
0.2
0.1
0.0
Tan δ
Standard
Pure
-155
-105
-55
-5
45
95
145
Temp. (°C)

Figure 6. Loss modulus vs temperature for the standard and pure copoloyether–urethane-ureas.

is observed. The soft segment glass transitions (the β_s relaxations) occur at 1°C for the pure copolymer and at −5°C for the standard copolymer. These soft segment glass transition temperatures are only slightly lower than that shown by the P1P copolymers (10°C for the pure and −1°C for the standard copolymer). The shift in the glass transition of the standard copolymers to a lower temperature than that shown by the pure copolymers could be attributed to the presence of dimeric soft segments (which would act as higher molecular weight soft segment additives) in the standard materials. An alternative explanation for the higher T_g's for the pure copolymers is that a decrease in soft segment flexibility resulted from a decrease in free volume. This explanation is offered because it was observed that in both systems, the pure copolymers had higher densities than the standard materials. If it is assumed that the density of the crystalline hard domains remains the same in each copolymer system, it seems reasonable to deduce that the enhanced purity of the phases and the cleaner interface would allow for tighter packing in the soft phase.

The tan δ peak of the pure P2P copolymer was higher than that for the standard material. The increase in peak height may result from a freeing up of more soft segments due to the elimination of the hydrogen bonding from the presence of dimeric soft segments [16] which were present in the standard materials. This again was similar to that observed for the P1P copolymers, except that the pure P1P copolymer had a shoulder on the higher temperature side of the tan δ curve. This lack of shoulder in the pure P2P copolymer may indicate a sharper distribution of phases in this material.

The slope of the storage moduli curves for the P2P copolymers (0.055 dyn cm^{-2}) were a little steeper than that shown for the P1P copolymers (0.083 dyn cm^{-2} for the standard copolymer and 0.063 dyn cm^{-2} for the pure copolymer), signifying a cleaner phase separation for the P2P materials. Also, the overall shape and the plateaus in the storage moduli were similar for both P2P copolymers. This is in contrast to the P1P copolymers in which the rubbery plateau for the analogous pure P1P material was lower than the plateau for the standard material.

Infrared dichroism

Infrared dichroism offers a quantitative way to study what is occurring at the molecular level during mechanical deformation. The relationship between the orientation functions and the elongation is shown in Figs 7 and 8. The standard materials in both copolymer systems (Fig. 7) are similar in their behavior. For example, the urea carbonyl bending vibrations show the hard segments orient transverse to the stretch direction up to an elongation of about 350%, and reach a maximum of transverse orientation (–0.3) at about 100% elongation. The pure P1P copolymer (Fig. 8), while also reaching a maximum of transverse orientation (of –0.4) at about 100%, is still orientating transverse to the stretch direction up to an elongation of about 400%. In contrast, the pure P2P copolymer (Fig. 8) shows orientation transverse to the stretch direction up to about 150% elongation, with a maximum of transverse orientation of –0.2 at about 50% elongation. The pure P2P copolymer's domains appear to stop orienting at relatively low elongations. If this is due to domain disintegration, it may relate to the larger elongation set (end %) shown by the pure P2P copolymer (142.7 at 600% elongation vs 69.1 for the pure P1P copolymer and 118.5 for the standard P2P copolymer). Thus, the disintegration of the domains may be the significant factor (rather than the large transverse orientations of the hard domains) that leads to worse mechanical hysteresis. These differences between the pure P1P and P2P copolymers would suggest differences in the packing of the hard segments [17].

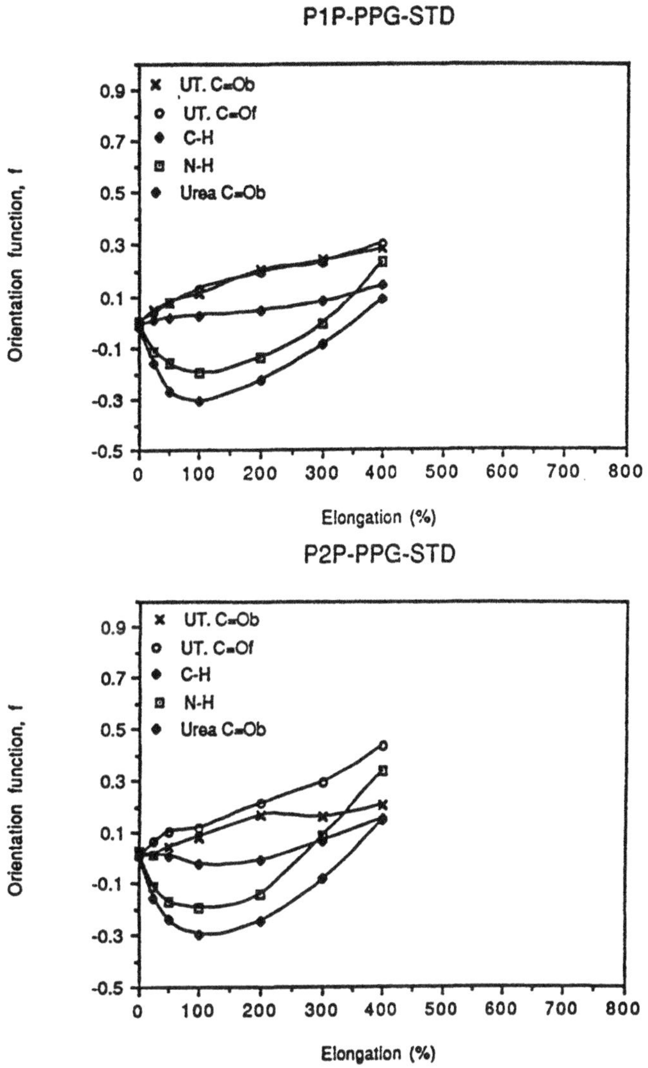
P1P-PPG-STD
UT. C=Ob
UT. C=Of
C-H
N-H
Urea C=Ob
Orientation function, f
Elongation (%)
P2P-PPG-STD
UT. C=Ob
UT. C=Of
C-H
N-H
Urea C=Ob
Orientation function, f
Elongation (%)

Figure 7. Orientation function vs elongation for the standard P1P and P2P copolyether-urethane-ureas.

SUMMARY

The infrared and dynamic mechanical data all indicate that replacement of the $-CH_2-$ spacer group between the two aromatic rings of the diisocyanate by a $-CH_2CH_2-$ spacer group results in a different, more dense packing structure in the hard domains of the copolyether-urethane-ureas. The infrared data would suggest that this packing structure is coplanar. These differences in the copolymers are best studied using materials prepared by the multiple step synthesis which eliminates or greatly reduces the presence of multimeric segment lengths.

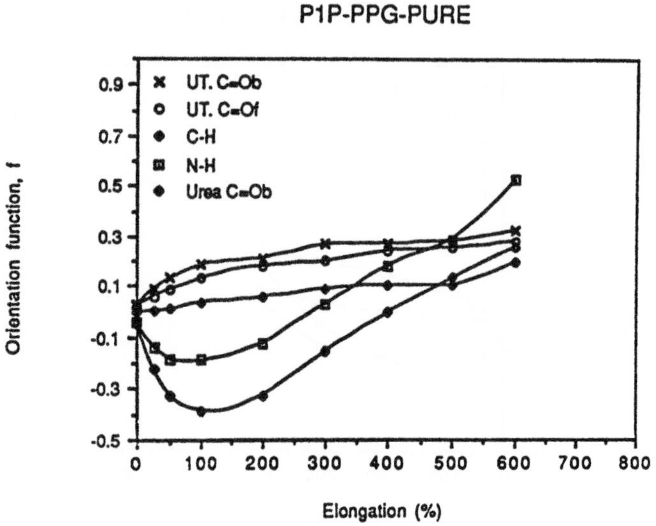
P1P-PPG-PURE
UT. C=Ob
UT. C=Of
C-H
N-H
Urea C=Ob
Orientation function, f
Elongation (%)

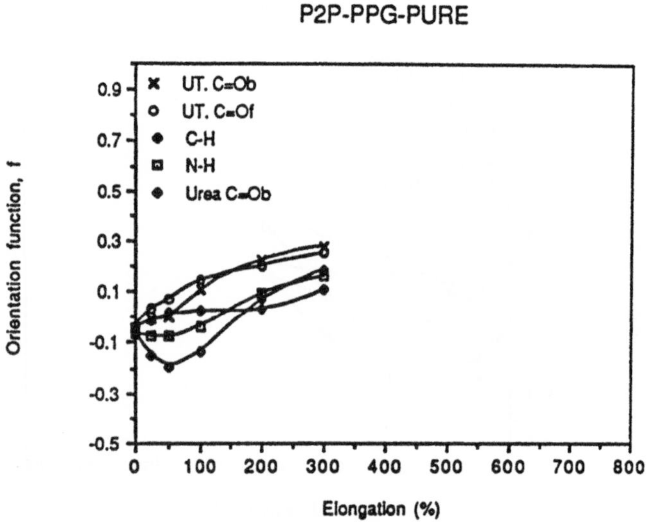
P2P-PPG-PURE
UT. C=Ob
UT. C=Of
C-H
N-H
Urea C=Ob
Orientation function, f
Elongation (%)

Figure 8. Orientation function vs elongation for the pure P1P and P2P copolyether–urethane-ureas.

Acknowledgement

This work was supported by the University of Utah Biomaterials Research and Development fund.

REFERENCES

1. D. J. Lyman, K. Knutson, B. McNeill and K. Shibatani, *Trans. Am. Soc. Artif. Int. Organs* **21**, 49 (1975).
2. K. Kataoka, T. Akaike, Y. Sakurai and T. Tsuruta, *Makromol. Chem.* **179**, 1121 (1978).
3. D. J. Lyman, L. C. Metcalf, D. Albo, Jr., K. F. Richards and J. Lamb, *Trans Am. Soc. Artif. Int. Organs* **20**, 474 (1974).
4. L. H. Peebles, Jr., *Macromolecules* **7**, 872 (1974).
5. T.-L. D. Wang and D. J. Lyman, *J. Poly. Sci., Part A: Poly. Chem.* **31**, 1983 (1993).

6. C. D. Eisenbach, M. Baumagartner and C. Gunter, in: *Advances in Elastomers and Rubber Elasticity*, p. 51, J. Lal and J. E. Mark (Eds). Plenum, New York (1986).
7. K. Knutson and D. J. Lyman, in: *Biomedical and Dental Application of Polymers*, p. 163, C. G. Gebelein (Ed.). Plenum, New York (1981).
8. D. J. Lyman, J. Heller and M. Barlow, *Die Makromol. Chemie* **84**, 64 (1965).
9. R. T. Conley, *Infrared Spectroscopy*, 2nd ed., Allyn and Bacon, Boston, MA (1972).
10. J. Blackwell, M. R. Nagarajan and T. B. Hoitink, *Polymer* **23**, 950 (1982).
11. D. J. Lyman, in: *Kinetics and Mechanisms of Polymerization, Vol. 3. Step-Growth Polymerizations*, Ch. 3, D. H. Solomon (Ed.). Dekker, New York (1972).
12. J. C. West and S. L. Cooper, *J. Poly. Sci.: Polym. Symp.* **60**, 127 (1977).
13. H. Ishihara, *J. Macromol. Sci.-Phys.* **B22**, 763 (1983).
14. H. Ishihara, I. Kimura, K. Saito and H. Ono, *J. Macromol. Sci.-Phys.* **B10**, 591 (1974).
15. L. A. Gower and D. J. Lyman, manuscript in preparation.
16. O. V. Starsev, A. L. Iordanski and G. E. Gaikov, *Polym. Degrad. Stab.* **17**, 273 (1987).
17. H. Ishihara, I. Kimura, K. Saito and H. Ono, *J. Macromol. Sci.-Phys.* **B10**, 591 (1974).

Polyisobutylene-toughened poly(methyl methacrylate): III. PMMA-*l*-PIB networks as bone cements

JOSEPH P. KENNEDY[1], MICHAEL J. ASKEW[2] and G. CAYWOOD RICHARD[1]

[1]*Institute of Polymer Science, The University of Akron, Akron, OH 44325-3909, USA*
[2]*Musculoskeletal Research, Akron City Hospital, Akron, OH 44325, USA*

Received 27 March 1992; accepted 17 September 1992

Abstract—A series of novel polyisobutylene (PIB)-toughened poly(methyl methacrylate) (PMMA) networks consist of rubbery PIB domains covalently bonded to a glassy PMMA matrix. Materials containing 8.5–17 wt.% PIB (M_n = 18 000 g/mol) in a PMMA matrix (PIB/PMMA) were evaluated to assess their feasibility as the powder component along with methyl methacrylate (MMA) as the liquid component in a standard powder/liquid bone cement formulation. A standard ISO four-point bend test, commonly used for testing bone cements, was employed to investigate flexural properties. The mixing time and powder/liquid (*P/L*) ratio were studied to formulate novel PIB/PMMA cements for optimum toughness. Appropriate formulations led to improved toughness while adequate flexural strength and modulus were maintained. An experimental PIB/PMMA-system exhibited ~57 MPa flexural strength and ~2000 MPa flexural modulus.

Key words: Polyisobutylene; carbocationic polymerization; methacrylate.

INTRODUCTION

Self-curing PMMA systems have been used as bone cements in joint replacement surgery since 1958 [1]. While this cement fixation system has performed well, there have been persistent problems with prosthesis loosening. The multifactorial causes of these problems are of both mechanical and biological origin. One is the brittleness of the current cement which results in cracking and mechanical failure as well as the liberation of particular debris that causes adverse biological reaction [2].

The goal of this research was the synthesis, characterization and mechanical testing of a series of new PIB-modified PMMA's suitable for use as rubber-toughened bone cements in orthopaedic surgery. The incorporation of the inert, saturated PIB rubber was anticipated to improve crack resistance and energy adsorption, while maintaining acceptability for biomedical applications.

The recent discovery of living carbocationic polymerization [3, 4] provides an opportunity for the synthesis of controlled molecular weight (M_n) PIB telechelics. The synthesis, characterization and tensile properties of networks composed of PMMA linked by methacrylate-capped PIB has been described in the first paper of this series [5]. Specifically, these new materials are composed of a two-component PMMA-*l*-PIB network [6] plus PMMA interwoven throughout the network. They are referred to as semi-simultaneous interpenetrating networks (semi-SINs) [7] because network and homopolymer are formed simultaneously during the synthesis. In work reported in detail elsewhere a series of twelve materials were prepared containing 5, 10, 20, and 30 wt.% of each of three different M_n PIBs [5]. These experimental materials were screened by uniaxial tensile testing to determine the effect of PIB M_n and wt.% on tensile properties and relative toughness (area under the stress–strain curve) [5]. As an initial step in the assessment of these materials on bone cements, the static mechanical properties of these materials, as measured in a standardized test, were compared with those of a currently used commercial cement.

EXPERIMENTAL

Bending test specimens of PIB-toughened PMMA and of commercial bone cement (Zimmer Regular, Zimmer Inc., Warsaw, Indiana) were fabricated and tested as specified by ISO standard ISO5833/1 (proposed revision 1986) [8]. For both materials, powder and liquid components were mixed and cast into plates which were then cut into the required test specimens. Table 1 compares the compositions of the commercial and the experimental, PIB-modified components. Standard commercial bone cement packages (Zimmer Regular) of 40 g of powder component and 20 ml of liquid component were used. Based on their superior performance in previous testing [5], PIB-toughened PMMA having 15 and 30 wt.% PIB (M_n 18 000) were synthesized as previously described [5] and ground to fine powder (0.1–0.3 mm) in an analytical mill (Model A-10, Tekmar, Inc.). Benzoyl peroxide (0.75 wt.%) was milled into powder. Forty grams of the milled powder were used as the PIB-modified powder component and were mixed with the commercial liquid component (Zimmer Regular MMA Monomer, Zimmer, Inc., Warsaw, Indiana). Table 2 provides sample codes and cement compositions, and indicates the powder/liquid (*P/L*) ratios employed to obtain the final compositions.

Table 1.
Comparison of commercial and PIB-toughened bone cement compositions

Liquid		Powder		
Zimmer regular radiopaque bone cement				
methyl methacrylate	97.25%	PMMA		89.25%
N,N-dimethyl-*p*-toluidine	2.75%	Barium sulfate		10.0%
hydroquinone	75 ppm	BPO		0.75%
PIB-toughened bone cement				
As above		PMMA	70.0%	85.0%
		PIB	30.0%	15.0%
		BPO	0.75%	0.75%

Table 2.
Bone cement formulation and final compositions

Powder (40 g)	*P/L*	Bone cement final composition
Zimmer PMMA	2/1	PMMA
18K15[a]	1.33/1	18K8.5
18K15	1.6/1	18K9.2
18K30	1/1	18K15
18K30	1.33/1	18K17.1

[a] PIB $\bar{M}_n$ = 18 000 g/mol, PIB wt.% = 15.

The components were mixed manually in an open disposable plastic mixing bowl with a plastic spatula by a clockwise–counterclockwise mixing cycle at ~30 cycles/min. The commercial cement was mixed for ~40 s. The experimental PIB/PMMA cement was mixed from 40 s to 1.5 min. The dough was then placed into a mold.

The mold was a rectangular hole, 90 mm × 75 mm, in a 3.5 mm thick Teflon plate. (One standard commercial package of bone cement, consisting of 40 g of powder component and 20 ml of liquid monomer filled the mold.) The mold was placed on a glass plate with a 0.2 mm thick polyester film between the glass plate and the mold. After the mixture had been poured or placed into the mold, a second polyester film and a glass plate were placed on top of the mold and the assembly secured by clamps. After 1 h, the cement plate was removed from the mold and cut into specimen strips 75 mm long, 10 mm wide and 3.5 mm thick, using a water-cooled saw. The sides of samples were wet-ground with 400-grit carborundum paper. The final dimensions of each specimen were measured with a micrometer and recorded. The specimens were then stored in a water bath at 37°C for 48 h.

Four-point bending tests of the prepared specimens were carried out in a water bath at 37°C on a materials testing machine (model 812, MTS Systems, Inc., Mineapolis, MN) at a support displacement rate of 15 mm/min. The bending load, measured by the load cell of the materials testing system, and the midspan deflection, measured by a linear variable displacement transducer (LVDT, model 7307-X2-AO, Pickering and Co.), were recorded on an *x–y* plotter. The bending modulus and the bending strength were calculated according to ISO standard 5833/1 [8].

RESULTS AND DISCUSSION

For the PIB-toughened PMMA it was not possible to maintain the commercially used 2/1 *P/L* composition, since the PIB-containing powder was more difficult to wet with MMA than the commercial powder. The PIB-toughened doughs did not become pourable like the commercial dough, and they had to be packed into the mold manually. The problem was lessened by reducing the amount of PIB in the initial powder from 30% to 15%, so that less MMA was required to reach maximum wetting. However, a *P/L* of 2/1 still could not be achieved, and the dough still could not be poured. The PIB-toughened PMMA cements containing a lower amount of PIB exhibited fewer voids indicating improved wetting.

A second approach for improving wetting was to increase the mixing time before transferring the dough to the mold; Table 3. For example, the 18K9.2 cement was prepared by using a slightly lower amount of MMA and increasing the mixing time from 45 s to 1.5 min. This protocol gave better wetting, and, after curing, the cements prepared with longer mixing time appeared to contain fewer voids than those prepared with extra monomer.

Contrary to expectations, the samples 18K15 exhibited a lower modulus (E) and strength (σ_b) than the 18K17 samples, most likely because the shorter mixing time used for the 18K15 samples gave incomplete wetting and more voids in the final test pieces. However, both test 18K17 and 18K15 samples were significantly weaker than the commercial samples probably because of the higher porosity resulting from poor wetting of the relatively high PIB content powder.

The 18K8.5 and 18K9.2 samples behaved similarly. The material with the lower PIB content was weaker due to incomplete wetting during the shorter mixing; the

Table 3.
Four point bend results [8, 9]

PIB content[a] (wt.%)	E (MPa)	σ_b (MPa)	Δd (mm)	Comments
0.0[b]	2190 ± 150	53.4 ± 4.2	4.90 ± 0.16	45 s mix. [10–12]
8.5	1680 ± 240	41.4 ± 6.2	5.10 ± 0.04	45 s mix. little s.w.
9.2	1980 ± 100	57.0 ± 2.3	6.40 ± 0.87	1.5 min mix. much s.w.
15.0	970 ± 180	21.2 ± 5.1	4.80 ± 0.60	45 s mix. s.w., inc. wet.
17.0	1260 ± 140	26.7 ± 7.2	3.80 ± 1.30	1.5 min mix. s.w., inc. wet.

[a] PIB $\bar{M}_n$ = 18 000 g/mol.
[b] Commercial product (Zimmer).
Abbreviations: E = flexural modulus, σ_b = ultimate flexural strength at break, Δd = maximum deflection before break, s.w. = stress whitening, inc. wet. = incomplete wetting.

dough was difficult to pack into the mold and the final composite appeared to contain more voids. However, by reducing the PIB content to 8–10%, both E and σ_b remained quite high.

The 18K9.2 material (less MMA, longer mixing) performed excellently: it exhibited ~90% of the modulus and a slightly greater breaking strength than that of the commercial cement. This result was surprising, since this material appeared to contain more voids than the commercial cement. The results were quite reproducible, with the lowest standard deviation among the PIB-toughened materials. The maximum deflection (Δd) of the 18K9.2 specimens was also higher than that of the commercial cement, indicating improved toughness; Table 3.

Based on these results, this class of materials, particularly the 18K9.2 composition, shows great promise as PIB-toughened bone cements. The results indicate that PIB joined covalently to PMMA increases the toughness of the material without adversely effecting its modulus or strength. However, the results also indicate that much further work is needed. The greater number of voids in the experimental materials, observed subjectively in this study, suggest that the optimal toughness has not yet been reached. A slight decrease in the P/L ratio appears to improve wetting, and an increase in mixing time may also be necessary. The satisfactory performance of the present experimental materials indicate that further work to improve powder wetting, dough rheology and to reduce the number of voids would be worthwhile and is currently under investigation.

Acknowledgement

Financial help by EPIC and the National Science Foundation (grant DMR-89-20826) are gratefully acknowledged.

REFERENCES

1. J. Charnley, *J. Bone Jt. Surg.* **42B**, 28 (1960).
2. L. C. Jones and D. S. Hungerford, *Clin. Orthop.* **225**, 192 (1987).
3. R. Faust and J. P. Kennedy, *J. Polym. Sci., Part A: Polym. Chem.* **25**, 1847 (1987).
4. G. Kaszas, J. Puskas and J. P. Kennedy, *Makromol. Chem., Makromol. Symp.* **13, 14,** 473 (1988).
5. J. P. Kennedy and G. C. Richard, *Macromolecules*, in press.
6. M. Weber and R. Stadler, *Polymer* **29**, 1071 (1988).

7. L. H. Sperling, *Interpenetrating Polymer Networks and Related Materials*, Plenum Press: New York (1981).
8. Implants for Surgery—Acrylic Resin Cements, Part I: Orthopaedic Application (Revised version of ISO 5833/1-1979), International Standards Organization, Geneva, Switzerland, 1986.
9. ASTM Specification F451-86, Acrylic Bone Cement, Vol. 13.01: Medical Devices Annual Book of ASTM Standards, Philadelphia (1987).
10. S. Saha and S. Pal, *J. Biomed. Mater. Res.* **18**, 435 (1984).
11. J. P. Davies, D. O. O'Connor, J. A. Greer and W. H. Harris, *J. Biomed. Mater. Res.* **21**, 719 (1987).
12. M. J. Askew, M. F. Kufel, P. R. Fleissner, Jr., I. A. Gradisar, Jr., S.-J. Salstrom, and J. S. Tan, *J. Biomed. Mater. Res.* **24**, 573 (1990).

Part VIII.A.

Biodegradable Polymers and Drug Delivery

Degradable Polymers

Synthesis and characterization of a new biodegradable semi-solid poly(ortho ester) for drug delivery systems

ALAIN MERKLI[1], JORGE HELLER[2], CYRUS TABATABAY[1] and ROBERT GURNY[1]

[1] *Department of Pharmaceutical Technology, University of Geneva, 1211 Geneva 4, Switzerland*
[2] *Controlled Release and Biomedical Polymers Department, SRI International, Menlo Park, CA 94025, USA*

Received 9 July 1992; accepted 23 October 1992

Abstract—Since the late 1970s, three families of poly(ortho esters) (POE) were synthesized to provide bioerodible carriers for drug delivery devices. The most recent POE is a semi-solid polymer with a viscous behavior at room temperature. Polymer synthesis by a transesterification reaction between a triol and a trialkyl orthoester is described. The structure of the polymer was confirmed by conventional methods such as ^{1}H-NMR, ^{13}C-NMR and FT-IR. Information concerning average molecular weight and intrinsic viscosity was obtained respectively by GPC and viscosimetry. Residual solvents in the polymer were determined using gas chromatography. The chromatographic conditions were optimized to enable the quantification of the solvents in concentrations of a few percent. The mechanical behavior of the semi-solid POE was determined by rheometric measurements. Hydrolysis of the polymer leads to the formation of the original triol and the carboxylic acid derived from the trialkyl orthoester used in the transesterification step. No toxicological problems associated with these compounds are anticipated.

Key words: Poly(ortho esters); bioerodible polymer; polymer synthesis; polymer characterization; polymer hydrolysis.

INTRODUCTION

In recent years, the development of new degradable polymers as possible biomaterials has attracted considerable attention. One widely studied application of these polymers are implantable drug delivery devices and currently a wide range of new applications for degradable polymers is being investigated. The release of active agents from bioerodible polymeric matrices has been classified by Heller [1, 2] into three groups, according to the breakdown mechanism involved: water-soluble polymers which are made insoluble by hydrolytically unstable cross-links (mechanism I); linear polymers which are initially water-insoluble and which become solubilized by ionization resulting from hydrolysis, or protonation of pendent groups, but without backbone cleavage (mechanism II); and polymers which are water-insoluble and break down to small soluble products by backbone cleavage (mechanism III). This last category includes poly(ortho esters) (POE), which are hydrophobic polymers that under certain conditions can undergo an erosion (heterogeneous) process that is confined to the polymer–water interface.

The first POE's were described in a series of patents [3–6] assigned to the Alza Corporation [7]. They were synthesized by a transesterification reaction between a 2,2-diethoxytetrahydrofuran (**I**) and a diol (**II**) (Fig. 1). The tradename Chronomer® was originally used for these polymers and later they were marketed as Alzamer®.

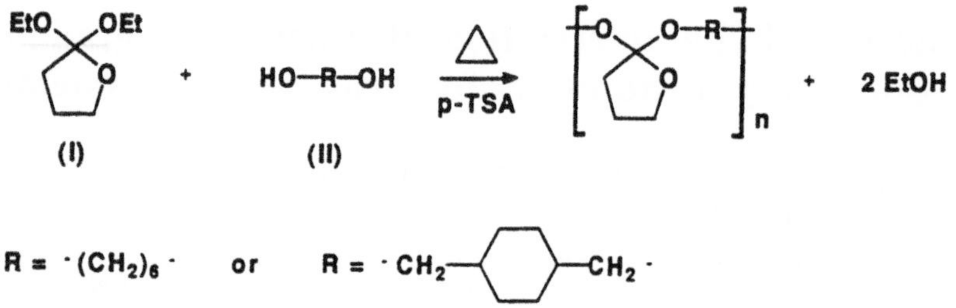

Figure 1. Synthesis of first family of POE.

Various potential applications were described by Choi and Heller in their patents [3–6] in particular for the delivery of contraceptive steroids such as norethindrone or levonorgestrel [8], for the delivery of the narcotic antagonist naltrexone [9, 10], and for the treatment of burns [11].

One of the problems with these polymers is the release of acidic by-products that autocatalyze the hydrolysis process, resulting in degradation rates which increase with time. For this reason, a second family of POE's, different from the Alzamer®, was subsequently developed by Heller *et al.* [12, 13]. The preparation is based on the addition of polyols to the diketene acetal 3,9-bis(ethylidene-2,4,8,10-tetraoxaspiro [5, 5] undecane) (DETOSU) (**III**). The reaction, catalyzed by a small amount of acid, is exothermic and produces a high molecular weight polymer almost instantaneously. The structure of this POE is shown in Fig. 2.

By using different mixtures of a flexible diol such as 1,6-hexanediol (1,6-HD) (**IV**) and a rigid diol such as *trans*-cyclohexanedimethanol (*t*-CDM) (**V**), a family of linear polymers can be produced, with glass transition temperatures ranging from 20°C for 100% 1,6-HD to 120°C for 100% *t*-CDM, and a range of different mechanical properties [14]. This polymer hydrolyzes to first form pentaerythritol dipropionate and the original diols. The pentaerythritol dipropionate later hydrolyzes to propionic acid and the corresponding alcohol [15].

The backbones of these polymers contain acid-labile linkages which are stable in alkaline media, hydrolyze slowly in a pH 7.4 phosphate buffer and hydrolyze at increasing rates as the acidity of the medium increases. In order to produce drug delivery devices with lifetimes ranging from hours to months, studies were carried out by different investigators, who incorporated acid or basic excipients into hot-molded samples, to modify erosion rates [15–18]. Recently, a new semi-solid POE has been described [19] and in this manuscript we present a detailed characterization of this polymer.

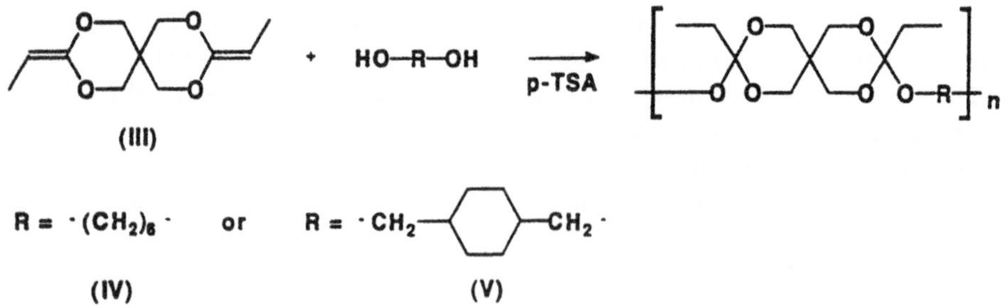

Figure 2. Synthesis of second family of POE.

EXPERIMENTAL

Polymer synthesis

The synthesis of the semi-solid POE is based on a transesterification reaction under anhydrous conditions, between 34.68 g (300 mmol) of trimethyl orthoacetate (99%) (**VI**) (Aldrich®-Chemie, Steinheim, Germany) and 40.25 g (300 mmol) of 1,2,6-hexanetriol (98%) (**VII**) (Aldrich®-Chemie, Steinheim, Germany) (Fig. 3). The mixture is introduced into a round bottom flask, placed on a magnetic stirrer and 400 ml of cyclohexane (Fluka® Chemie AG, Buchs, Switzerland) are added. The reaction is catalyzed by 25 mg of *p*-toluene sulfonic acid (*p*-TSA) (Fluka® Chemie AG, Buchs, Switzerland).

The reaction flask is adapted to a distillation equipment and heated under argon to 120°C with vigorous stirring. In the first step, the reaction by-product methanol is readily removed at 54°C during the first 4 h of the distillation. In the second step, the temperature of the column head climbs above 54°C. At this time, the distillation flow is decreased and the solution is heated for 6 additional hours until the boiling point of 81°C is reached. The solution is then cooled to room temperature, and 10 drops of triethylamine (TEA) (Fluka® Chemie AG, Buchs, Switzerland) are added to neutralize the acid catalyst and to stabilize the polymer. Excess solvent is poured off and the polymer is dried thoroughly overnight under vacuum at 40°C. The polymer is then purified by dissolution in 100 ml tetrahydrofurane (THF) (Romil® Chemicals, Leics, England) and precipitation in 500 ml anhydrous methanol containing 10 drops of TEA. After separation of the solvent the POE is dried in a rotavapor under high vacuum for 48 h.

Polymer hydrolysis

The exposure of POE to a pH 7.4 phosphate buffer induces an initial hydrolysis of the ortho ester bonds with the possible formation of three isomeric esters of the triol, followed by a slower hydrolysis to the original hexanetriol (**VII**) and a carboxylic acid (**VIII**). Because the second hydrolysis is much slower ($K_2 \ll K_1$), no autocatalysis is observed (Fig. 4).

The hydrolysis of the ortho ester bonds in the polymer can take place by two different mechanisms as shown in Fig. 5. Exocyclic protonation of the alkoxy group followed by the cleavage of the ortho ester bonds and reaction with water will produce two esters; the 1- and 2-isomers. The endocyclic protonation followed by the cleavage of the ortho ester bonds and reaction with water will produce two esters; the 1- and 6-isomers. A detailed study of the initial hydrolysis products using gas

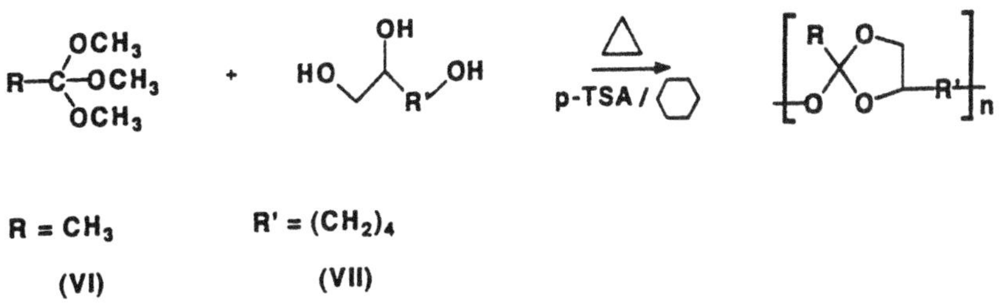
OCH₃
R—C—OCH₃
OCH₃
+
OH
HO
OH
R'
p-TSA /
R
O
O
O
R'
n
R = CH₃
R' = (CH₂)₄
(VI)
(VII)

Figure 3. Synthesis of third family of POE.

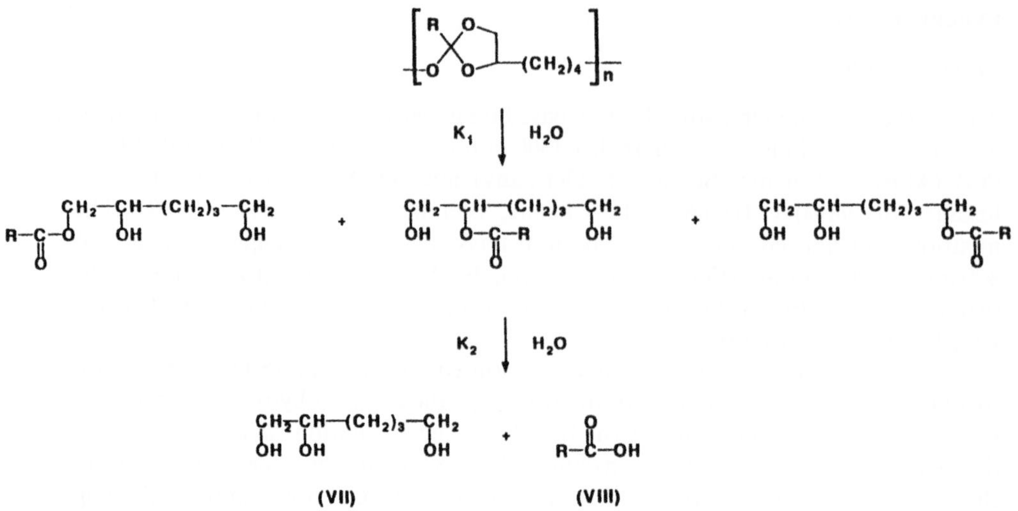

Figure 4. The two hydrolysis steps of the semi-solid POE.

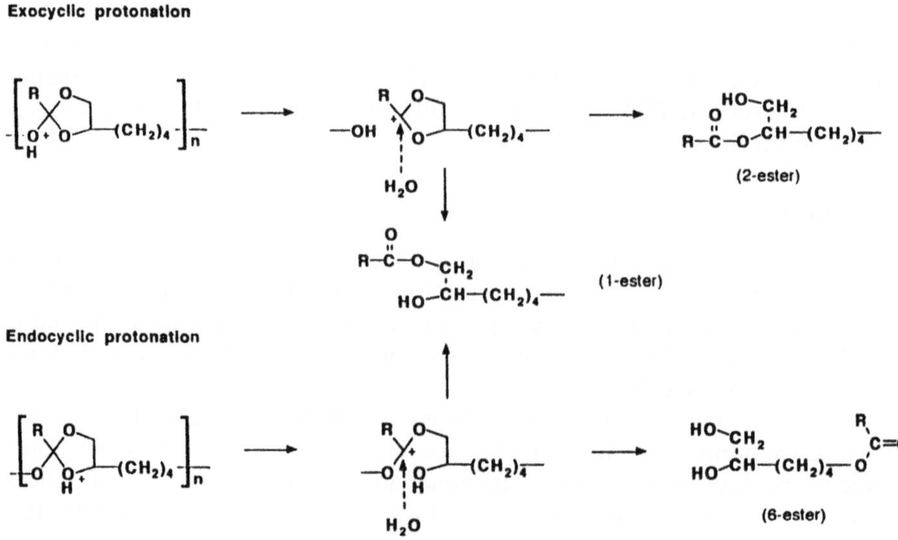
Exocyclic protonation
(2-ester)
(1-ester)
Endocyclic protonation
(6-ester)

Figure 5. The two possible mechanisms during the first step of the hydrolysis of the semi-solid POE.

chromatography has established that only the 1- and 2-isomer are produced so that hydrolysis of this particular polymer proceeds exclusively by the exocyclic cleavage path [20]. The two final breakdown products are well-known and thus no toxicological problems are anticipated [21, 22].

Polymer characterization

^{1}H and ^{13}C-NMR. The POE was characterized by recording its ^{1}H and ^{13}C-NMR spectra and by carrying out a two-dimensional NMR using a Hetcor Correlation between ^{1}H-^{13}C-NMR. The 200 MHz ^{1}H and 50 MHz ^{13}C spectra of POE as a

20% (w/v) solution in $CDCl_3$ (Dr. Glaser® AG, Basel, Switzerland) at room temperature was obtained with a XL 200 NMR Spectrometer (Varian®, Basel, Switzerland). A pulse sequence termed APT (Attach Proton Test) has been employed with parameters chosen so that differences between CH, CH_2, and CH_3 appear in the ^{13}C spectra [23].

Fourier transform infra-red (FT-IR). The molecular composition determination of the polymer by analyzing the characteristic vibrations of functional groups was obtained with a 1600 series FT-IR spectrometer (Perkin Elmer® AG, Küsnacht, Switzerland). Sample preparation of the semi-solid POE was achieved by directly dispersing the viscous paste on the surface of a sodium chloride disc.

Gel permeation chromatography (GPC). The average molecular weight of this semi-solid POE was determined by GPC on Microstyragel 10^3 and 10^4 Å (Waters®, Volketswil, Switzerland) at 15°C using a 600E apparatus coupled with a R401 differential refractometer. The solvent, THF, was stabilized with a small amount of triethylamine. Polystyrene was used as the narrow molecular weight standard. Each sample of polymer was diluted in THF (0.1% w/v), and filtered through a 0.45 μm Teflon® filter (Millipore®, Volketswil, Switzerland) before injection.

Gas chromatography (GC). The method was developed on a HP 5880 A series gas chromatograph (Hewlett Packard®, Meyrin, Switzerland) equipped with a flame ionization detector (FID). The column was a 30 m × 0.53 mm ID fused-silica capillary column with a 1 μm thick bonded stationary phase (Supelcowax 10®) (Supelco® SA, Gland, Switzerland). The pressure of the carrier gas nitrogen was set at 4.2 psi, while hydrogen and air pressure were 30 and 16 psi respectively. Injector and detector temperatures were set at 220°C. The GC determination of all residual solvents was carried out using a single step oven temperature at 40°C.

Samples were prepared by dissolution of the polymer in acetone (0.500 g/10 ml). The analysis of the solutions was performed by injecting in the direct mode 1 μl into the chromatographic system described above. The concentrations of the solvents used during synthesis and purification were calculated from calibration curves which had been plotted using a concentration range of 0.1–1 g/10 ml acetone (Fluka® Chemie AG, Buchs, Switzerland).

Rheological behavior. The rheological behavior was studied using a Bohlin® Controlled Stress Rheometer (Bohlin® Rheology GmbH, Mühlacker, Germany). A cone-plate with a 4 deg cone angle and a 40 mm cone diameter (CP 4/40) was used as the measuring device. The temperature of the sample was set at 20.0°C using the Bohlin® Extended Temperature Option (ETO) (Bohlin® Rheology GmbH, Mühlacker, Germany).

Two different tests were carried out; a stress viscometry test, which is the most commonly used measurement of dynamic viscosity, and a creep and creep recovery test, which gives, at a selected stress level, full information on the viscoelastic behavior of the material.

Intrinsic viscosity. The intrinsic viscosity was measured with a modified Ubbelohde® viscosimeter type 0c (Schott®, Hofheim, Germany). A known volume of the highly concentrated solution was placed in the viscosimeter and its flow time determined. A known volume of solvent was then added. After mixing and temperature equilibration, the flow time of the diluted solution was measured. This procedure was then repeated for several dilutions. The different dilutions were made with toluene stabilized with TEA and the temperature of the analysis was controlled at 37.0°C (±0.02°C) with a D8 thermostatic controller (Haake®, Karlsruhe, Germany). The flow time was recorded automatically by an AVS 310 (Schott®, Hofheim, Germany).

RESULTS AND DISCUSSION

Polymer characterization

^{1}H and ^{13}C-NMR. Nuclear magnetic resonance spectroscopy allows a study of the structure and dynamics of polymer chains both in solution and in the solid state. In the solid state, movements of polymer chains are relatively slow and the resonances are broad, owing to the local dipolar field at each observed nucleus. This phenomenon of dipolar broadening tends to abolish all structural information. However, in solution, where chain movement is fast and where this effect almost disappears, detailed structural information can be obtained [23, 24].

The total range of proton chemical shifts in organic compounds extends over 10 ppm. ^{13}C-NMR offers the advantage when studying polymers of a much broader range; over 200 ppm (Fig. 6). Table 1 shows chemical shifts and assignments for the ^{1}H and ^{13}C-NMR spectra.

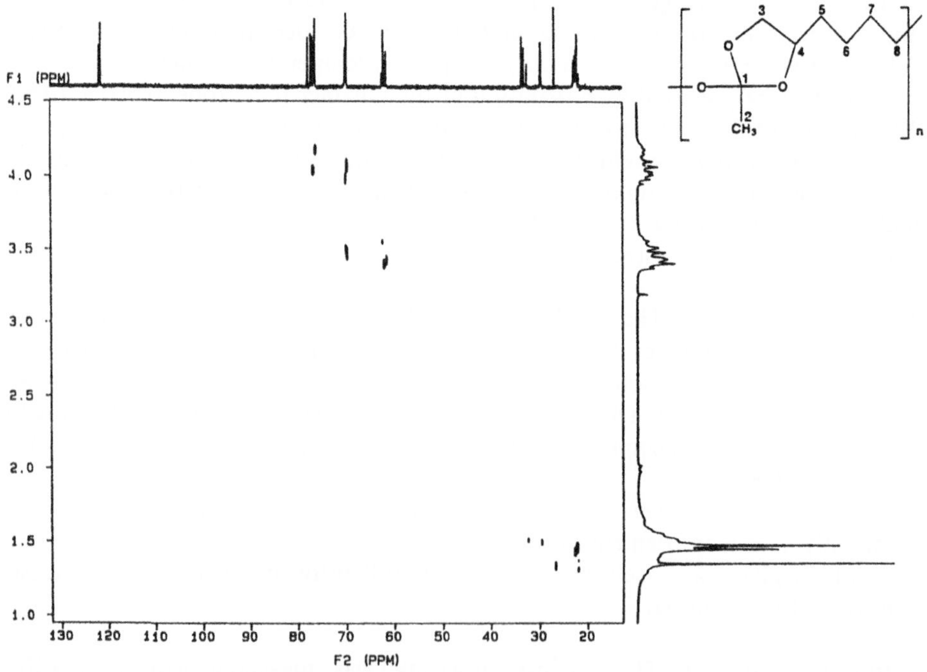
F1 (PPM)
4.5
4.0
3.5
3.0
2.5
2.0
1.5
1.0
130 120 110 100 90 80 70 60 50 40 30 20
F2 (PPM)
O
O
O
O
CH3
n

Figure 6. Two dimensional 200 MHz proton and 50 MHz carbon spectrum of the atactic semi-solid POE (20% (w/v) in $CDCl_3$ at room temperature).

Table 1.
Chemical shifts and assignment for the 200 MHz proton and the 50 MHz carbon spectrum of the POE

Carbon no.	Chemical shifts δ ^{1}H-NMR (ppm)	^{13}C-NMR (ppm)	Assignment
1	—	121.31 and 121.67	C (quaternary)
2	1.35	22.44 and 22.92	CH_3
3	3.42–3.55 and 3.95–4.10	69.75 and 70.03	CH_2 (cycle)
4	3.65–4.22	76.31 and 76.84	CH (cycle)
5	1.42–1.51	22.17 and 22.63	CH_2
6	1.42–1.55	29.60 and 29.67	CH_2
7	1.42–1.55	33.12 and 33.54	CH_2
8	3.37–3.47	61.50 and 62.04	CH_2 (terminal)

The ^{13}C-NMR spectrum shows that all peaks are double, a phenomenon which is often observed for atactic polymers [24, 25]. Two conformations are indeed obtained in the synthesis of the monomer. The methyl group (R) of the trimethyl orthoacetate can be oriented in the same direction as the alkyl chain of the triol or in the opposite direction. Accordingly, the ^{1}H-NMR spectrum shows a multiplicity of peaks and the coupling constants are difficult to determine.

Fourier transform infra-red (FT-IR). The utility of IR spectroscopy to determine the molecular composition of polymers by analyzing the characteristic vibrations of functional groups should not be underestimated. Fourier transform IR spectroscopy has proven to be a powerful tool in polymer characterization. FT-IR techniques make it possible to improve the signal to noise ratio and to increase the speed of data acquisition when studying time dependent phenomena. In this study, this method was used to identify the functional groups contained in the polymer chain. Table 2 shows the characteristic vibrations and the functional groups of the polymer.

The presence of a small peak is noted at 1736 cm^{-1} (Fig. 7). This is characteristic of the IR absorption band corresponding to the carbonyl group of an ester. Because it was very difficult to carry out these analyses under absolutely anhydrous

Table 2.
Characteristic vibrations and functional groups of the POE (w = weak, m = medium, s = sharp)

Wave numbers (cm^{-1})	Vibrations	Groups
3520 w	υ(O—H)	OH (terminal chain)
2995 s–2940 s	υ(C—H)	CH_3, CH_2
2880 s	υ(C—H)	OCH_3 (terminal chain)
1460 m	δ(C—H)	CH_2
1380 s	δ(C—H)	C—CH_3
1220 s–1160 s	υ(C—O)	C—O—C (cycle)
1040 s	υ(C—O)	C—OH (terminal chain)
740 w	δ(C—C)	$(CH_2)_4$

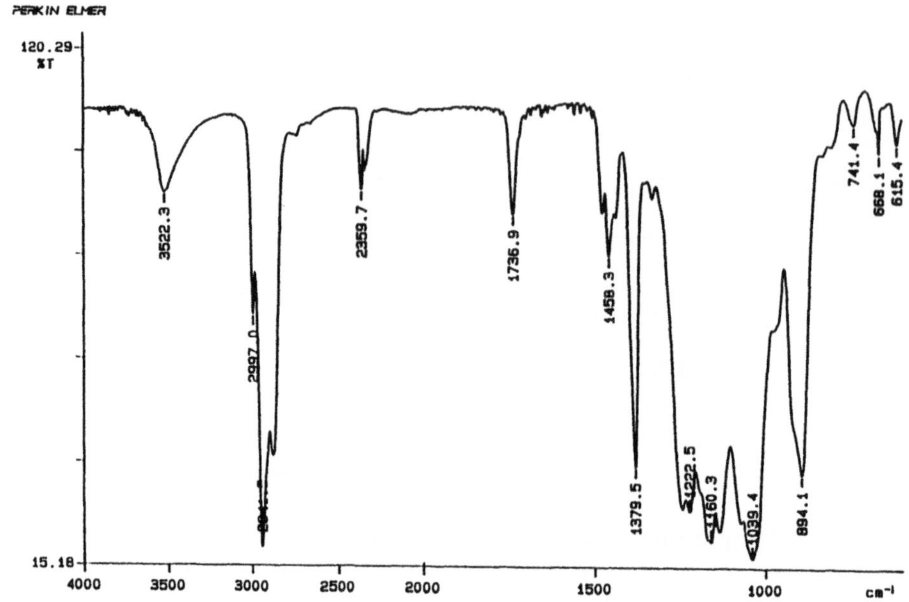
PERKIN ELMER
120.29
%T
15.18
3522.3
2997.0
2359.7
1736.9
1458.3
1379.5
1222.5
1160.3
1039.4
894.1
741.4
668.1
615.4
4000
3500
3000
2500
2000
1500
1000
cm-1

Figure 7. Infrared absorption spectrum of the semi-solid POE (NaCl disc).

conditions, the occurrence of this band is very likely due to the presence of small amounts of hydrolysis products. The quantification of the breakdown products is currently being investigated by GC-MS (gas chromatography—mass spectroscopy).

Gel permeation chromatography. The gel permeation chromatogram indicates a unimodal, polydisperse molecular weight distribution with a retention time (*Rt*) between 13 and 21 min. The average molecular weight $\overline{M_w}$ is 5.8×10^3 Da, $\overline{M_n}$ is 1.6×10^3 and the polydispersity ($\overline{M_w}/\overline{M_n}$) is 3.61. A broadening of the chromatogram for the low molecular weights (high Rt) is observed. The asymmetry factor As is 1.6 (tailing phenomena). In optimal conditions, good separation should give As-values between 0.9 and 1.2. A better As-value will be achieved by including a third 500 Å column for the separation of low molecular weights.

Gas chromatography. Assessing the amount of residual solvents and other volatile contaminants in bulk pharmaceuticals has become an important analytical problem, owing to the increasingly strict regulations [26]. Residual solvents can influence the quality of a polymer not only through their intrinsic toxicity but also as a result of certain characteristics such as crystal forming capacity [27], viscosity, which is very important in our case, or bioavailability. The residual solvents found after purification in POEs are given in Table 3.

Different limits are given in the European and United State Pharmacopoeia and by the Societa Italiana di Scienze Farmaceutiche (SISF) [26, 28, 29]. Some of them are listed in Table 3. It was then difficult to choose one or the other because these limits are largely dependent on the known toxicity of the solvent, the route of administration of the pharmaceutical substance, the dosage and administration and the likely duration of the treatment. In our case, these results give an indication of good manufacturing practice and provide the information needed for improving the elimination of the residual solvents used in the purification step.

Table 3.
Determination of residual solvents in the POE and generally acceptable limits

Solvents	Residual solvents (mg/g)	(%) (w/w)	Residual solvents limits Pharm. Eur.[a] (ppm)	Pharm. forum[b] (ppm)	SISF[c] (mg/j)
Cyclohexane	0.2	0.02	—	—	3.5 [29]
Tetrahydrofuran	1.5	0.15	—	100 [26]	2.0 [29]
Methanol	4.1	0.41	1000 [28]	1000 [26]	0.9 [29]

[a] European Pharmacopoeia.
[b] United State Pharmacopoeia.
[c] Societa Italiana di Scienze Farmaceutiche.

Rheological behavior. Figure 8 shows a stress viscometry test run on a POE sample. The experiment was carried out using a proportional delay time of 20 s and an integration time of 5 s. A Newtonian behavior is observed. The viscosity does not vary with shear rate or shear stress and a constant dynamic viscosity of 22.3 Pa·s is measured under these conditions.

The creep and creep recovery test gives information about the viscoelasticity of the sample. It is not unreasonable to assume that all materials are viscoelastic. In all materials, both viscous and elastic properties coexist [30, 31]. We can see in Fig. 9 that POE shows a Newtonian viscous flow behavior. No significant instantaneous or delayed elastic compliance was found. A constant strain was observed after $t = 63.1$ s, even if the stress was removed. Under these conditions, the recoverable compliance estimated from the creep J_C and the creep recovery J_r curves was 1.6 and 3.69×10^{-2} Pa^{-1} respectively.

Intrinsic viscosity. The intrinsic viscosity $[\eta]$ is defined as the limit value of the ratio η_{sp}/c at infinite dilution, $\lim_{c\text{-}o} \eta_{sp}/c = [\eta]$, where the specific viscosity (η_{sp}) is defined as $\eta_{sp} = \eta/\eta_o - 1 = \eta_{rel} - 1$. The term η is the viscosity of the polymer solution and η_o the viscosity of the pure solvent. Experience shows that solute

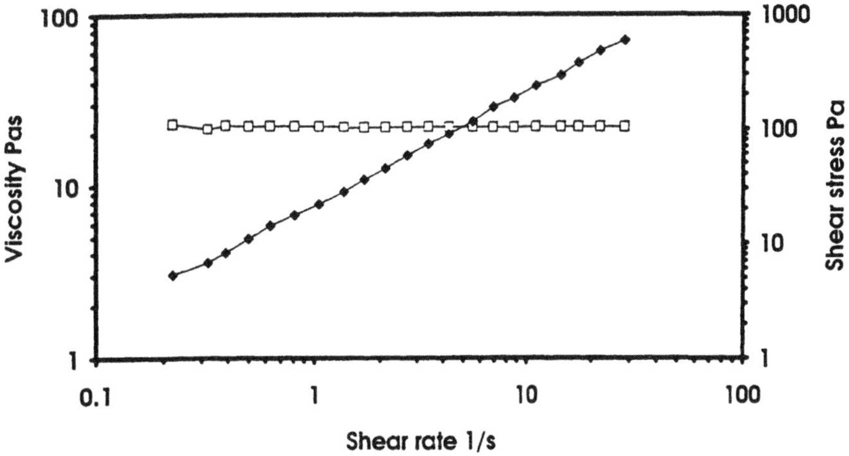
100
10
1
0.1
1
10
100
1000
100
10
1
Viscosity Pas
Shear stress Pa
Shear rate 1/s

Figure 8. Stress viscometry test on POE. (□: viscosity; ◆: shear stress).

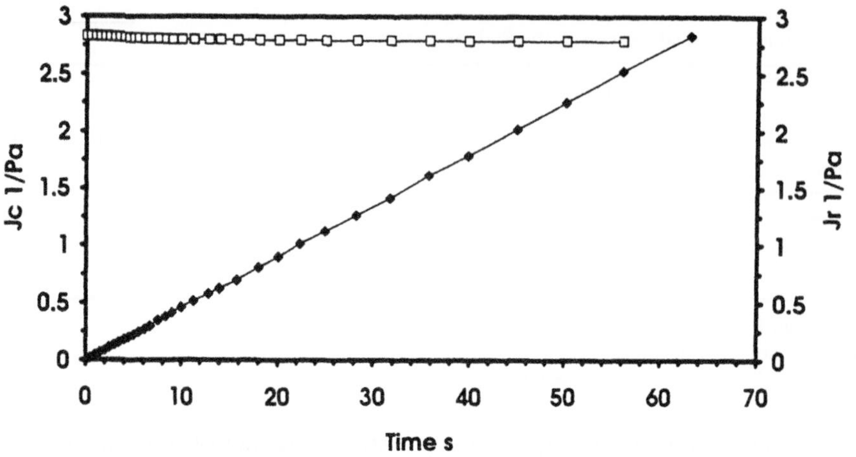

Jc 1/Pa
Jr 1/Pa
Time s

Figure 9. Creep and creep recovery test on POE. (◆: creep J_c; □: creep recovery J_r).

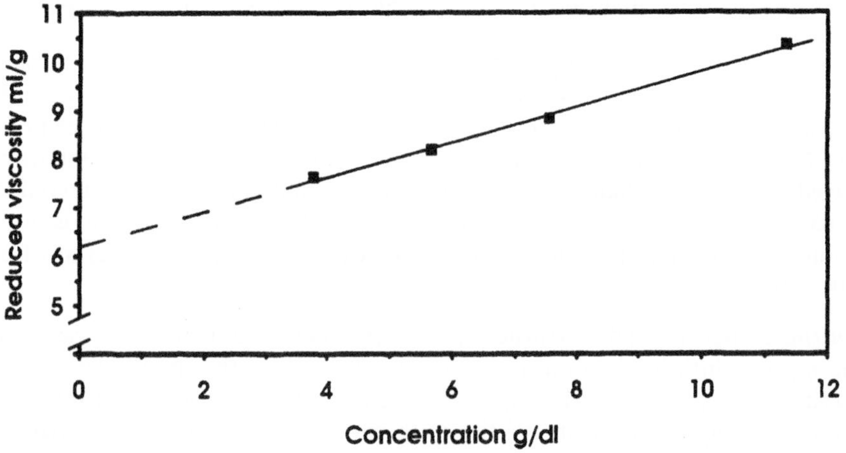

Reduced viscosity ml/g
Concentration g/dl

Figure 10. Determination of the intrinsic viscosity of POE in toluene at 37.0°C. (■: reduced viscosity; —: linear fit) (n = 3).

concentrations are optimum when η/η_0 lies between about 1.2 and 2.0. For values of $\eta_{rel} > 2.0$, the relationship between η_{sp}/c and c becomes nonlinear. For values of $\eta_{rel} < 1.2$, anomalies begin to appear at low concentrations. These anomalies are usually considered to be apparatus related and to result from the adsorption of macromolecules on the capillary walls [32]. For these reasons, the intrinsic viscosity of POE was measured by using an initial concentration of 11.25% (w/v) and by diluting progressively 1.5, 2, and 3 times (Fig. 10). Under these conditions, an intrinsic viscosity of 6.19 ml/g (r = 0.9981) was obtained for the POE.

CONCLUSION

The physical behavior studied and the structure of this new semi-solid POE show promising characteristics for potential application in drug delivery. This POE is being presently investigated in our laboratories as a bioerodible ocular implant containing the antimetabolite 5-fluorouracil for application in glaucoma filtration surgery [33]. The possibility of injecting this polymer under the conjunctiva using a syringe with an appropriate hypodermic needle without any other surgical intervention is a significant advantage when compared with solid bioerodible devices that must be placed with either a trocar or by a surgical procedure. Another advantage of this viscous paste-like material is the fact that drugs can be incorporated by simple mixing into the polymer at room temperature without the use of a solvent. This latter characteristic is of considerable interest for the delivery of peptides [20].

A detailed study of the biocompatibility of POE releasing 5-FU over an extended period of time (days up to several weeks) using sodium hyaluronate as a non-inflammatory control [34] is currently under way.

Acknowledgement

This study was supported by FNSRS grant #32.28009.89. We would like to acknowledge Prof. U. Burger, Dr. F. Barbalat and M. J. P. Saulnier for their technical support in the NMR studies, and Dr. C. Corvi for the use of his HP 5880 A gas chromatograph.

REFERENCES

1. J. Heller, *Biomaterials* **1,** 51 (1980).
2. J. Heller, in: *Recent Advances in Drug Delivery Systems*, p. 101, J. M. Anderson and S. W. Kim (Eds). Plenum Press, London (1984).
3. N. S. Choi and J. Heller, US Patent 4,093,709, June 6 (1978).
4. N. S. Choi and J. Heller, US Patent 4,131, 648, December 26 (1978).
5. N. S. Choi and J. Heller, US Patent 4,138,344, February 6 (1979).
6. N. S. Choi and J. Heller, US Patent 4,180,646, December 25 (1979).
7. J. Heller, *Biomaterials* **11,** 659 (1990).
8. B. B. Pharriss and L. Sendelbeck, *J. Reprod. Med.* **17**(2), 91 (1976).
9. R. C. Capozza, E. E. Schmitt and L. Sendelbeck, *Natl. Inst. Drug Abuse Res. Monogr. Ser.* **4,** 39 (1976).
10. R. C. Capozza, L. Sendelbeck and W. J. Balkenhol, in: *Polymeric Delivery Systems*, p. 59, R. J. Kostelnik (Ed.). Gordon & Breach, New York (1978).
11. L. M. Vistnes, E. E. Schmitt, G. A. Ksander, E. H. Rose, W. J. Balkenhol and C. L. Coleman, *Surgery* **79,** 690 (1976).
12. N. S. Choi and J. Heller, US Patent 4,304,767, December 8 (1981).
13. J. Heller, D. W. H. Penhale and R. F. Helwing, *J. Polymer Sci., Polymer Lett. Edn.* **18,** 82 (1980).
14. J. Heller, D. W. H. Penhale, B. K. Fritzinger, J. E. Rose and R. F. Helwing, *Contracept. Deliv. Syst*, **4,** 43 (1983).
15. J. Heller and K. J. Himmelstein, *Methods Enzymol.* **112,** 422 (1986).
16. C. Shih, T. Higuchi and K. J. Himmelstein, *Biomaterials* **5,** 237 (1984).
17. Y. F. Maa and J. Heller, *J. Contr. Rel.* **13,** 11 (1990).
18. C. Shih, S. Lucas and G. M. Zentner, *J. Contr. Rel.* **15,** 55 (1991).
19. J. Heller, S. Y. Ng, B. K. Fritzinger and K. V. Roskos, *Biomaterials* **11,** 235 (1990).
20. P. Wüthrich, S. Y. Ng, B. K. Fritzinger, K. V. Roskos and J. Heller, *J. Contr. Rel.* **21,** 191 (1992).
21. R. E. Lenga (Ed.), The Sigma-Aldrich Library of Chemical Safety Data, I and II Edition. Sigma-Aldrich Corp., Milwaukee (1988).
22. H. F. Smyth, U. C. Pozzani, C. S. Weil, M. J. Tallant and C. P. Carpenter, *Toxicol. Appl. Pharmacol.* **15** (2), 282 (1969).
23. V. D. Fedotov and H. Schneider, in: *NMR 21, Basic Principles and Progress*, P. Diehl, E. Fluck, H. Günther, R. Kosfeld, J. Seeling (Eds.). Springer Verlag, Berlin (1989).
24. F. W. Wehrli and T. Wirthlin (Eds.), in: *Interpretation of Carbon-13 NMR Spectra*, Heyden, London (1978).

25. H. O. Kalinowsky, S. Berger and S. Braun (Eds), in: *13C-NMR Spektroskopie*, Georg Thieme Verlag, Stuttgart (1984).
26. J. Rabiant, *STP Pharma Pratiques* **1**(3), 278–283 (1991).
27. A. M. Guyot-Hermann, *STP Pharma Pratiques* **1**(3), 258–266 (1991).
28. Anonymous, *Pharmeuropa* **2**(2), 142–146 (1990).
29. Societa Italiana di Scienze Farmaceutiche, *Cronache Farmaceutiche* **29**, 14 (1986).
30. J. D. Ferry (Ed.), in: *Viscoelastic Properties of Polymer*, 3rd Edition. J. Wiley and Sons, New York (1980).
31. G. Couarraze and J. L. Grossiord (Eds), in: *Initiation à la Rhéologie, Technique et Documentation*. Lavoisier, Paris (1983).
32. C. Booth and C. Price (Eds), in: *Comprehensive Polymer Science, Vol. 1, Polymer Characterization*. Pergamon Press, Oxford (1989).
33. A. Merkli, P. Wüthrich, J. Heller, C. Tabatabay and R. Gurny, *Invest. Ophtalmol. Vis. Sci.* (suppl.) **11**, 1392 (1992).
34. S. F. Bernatchez, C. Tabatabay, R. Gurny, T. Le Minh, T. Seemayer and J. M. Anderson, *Invest. Ophtalmol. Vis. Sci.* (suppl.) **11**, 1391 (1992).

Design, synthesis, and preliminary characterization of tyrosine-containing polyarylates: New biomaterials for medical applications

JAMES FIORDELISO, SAMUEL BRON and JOACHIM KOHN*

Department of Chemistry, Rutgers—The State University of New Jersey, New Brunswick, NJ 09803, USA

Received 22 January 1993; accepted 15 April 1993

Abstract—Five structurally related, aliphatic polyarylates were synthesized from tyrosine-derived diphenols and diacids. The diphenols were a homologous series of three desaminotyrosyl-tyrosine alkyl esters (ethyl, hexyl, octyl) which had previously been used in the synthesis of mechanically strong and tissue-compatible polycarbonates. The diacids (succinic acid, adipic acid, sebacic acid) were selected among compounds that were known to be of low systemic toxicity. By using different diacids as comonomers, the flexibility of the polymer backbone could be varied while the desaminotyrosyl-tyrosine alkyl esters provided pendent chains of various length. Some of the thermal and mechanical properties of the five polymers could be correlated to their chemical structure: the glass transition temperature decreased from 53 to 13°C, and the tensile modulus (measured at room temperature) decreased from 1500 to about 3 MPa when the length of the aliphatic diacid in the polymer backbone and/or the length of the alkyl ester pendent chain was increased. The presence of an arylate bond in the polymer backbone introduced a hydrolytically labile linkage into the polymer structure. Under physiological conditions *in vitro* all polymers degraded: thin films retained only about 30–40% of their initial molecular weight (M_w) after 26 weeks of storage in phosphate buffer solutions (pH 7.4) at 37°C. Release studies with *p*-nitroaniline as a model drug indicated that a diffusion controlled release process occurred. The rate of *p*-nitroaniline release could be correlated with the glass transition temperature of the polymer.

Key words: Degradable polymers; biomaterials; tyrosine; polyarylate; pseudo-poly(amino acid); thermomechanical properties; drug release profiles; degradation.

INTRODUCTION

Because of their hydrolytic instability, a wide range of aliphatic polyesters including poly(lactic acid), poly(glycolic acid), polydioxanone, poly(hydroxybutyrate), and poly(caprolactone) have been intensively explored as degradable biomaterials [1]. In addition, a number of degradable polymer systems based on other backbone linkages have become available such as polyanhydrides [2], poly(ortho esters) [3], polyphosphazenes [4], polyphosphoesters [5], and a series of pseudo-poly(amino acids) [6, 7].

Because the potential toxicity of the degradation products is an important concern, attempts have been made to identify polymers that degrade *in vivo* to natural metabolites. The widely used poly(lactic acid) represents an example of such a polymer. Likewise, several pseudo-poly(amino acids) have recently been developed [6–9]. In these polymers, natural amino acids or dipeptides are polymerized by bonds other than the conventional amide linkage found in poly(amino acids) or in

* To whom correspondence should be addressed at: Department of Chemistry, Rutgers University—Busch Campus, Piscataway, NJ 08855-0939, USA.

polypeptides. For example, polyesters have been obtained from *N*-protected derivatives of L-serine [10] and hydroxy-L-proline [11, 12], and polyiminocarbonates [13] or polycarbonates [9] have been prepared from derivatives of L-tyrosine. Those polymers appear to be nontoxic, tissue-compatible materials [14] that exhibit significantly improved engineering properties compared to most conventional poly(amino acids). In particular, the tyrosine-derived polycarbonates were found to be readily processible, mechanically strong materials that are currently being evaluated for use in degradable bone fixation devices [15] and in long-term drug delivery systems [16].

In this paper we present a preliminary investigation of new, tyrosine-containing polyarylates (Fig. 1). According to Imai and Kakimoto [17], aliphatic polyarylates are polymers derived from *aliphatic* diacids and *aromatic* dialcohols (diphenols). These polymers are distinctly different from aromatic polyesters (such as Dacron) which are derived from aromatic diacids and aliphatic dialcohols. While the nondegradable aromatic polyesters are used as synthetic fibers and have found biomedical applications (e.g., as artificial vascular grafts [18, 19]), the degradable, aliphatic polyarylates have heretofore not been systematically explored as biomaterials.

In the development of tyrosine-containing polyarylates, we applied a number of 'design guidelines' intended to facilitate the ultimate application of these polymers in medicine. First of all, we selected the monomeric starting materials to lead as much as possible to nontoxic degradation products. To increase the mechanical strength and stiffness of the materials, we included aromatic rings within the backbone structure. To facilitate the attachment of drugs or crosslinkers, and to allow for the modification of the polymer surface, we incorporated pendent carboxylic acid groups into the polymer structure. The presence of hydrolytically labile arylate bonds (in the backbone) and ester bonds (in the pendent chain) was intented to impart degradability to the polymers under physiological conditions. Finally, we attempted to create a family of structurally related polymers, making it possible to control the macroscopic polymer properties by means of small changes in chemical

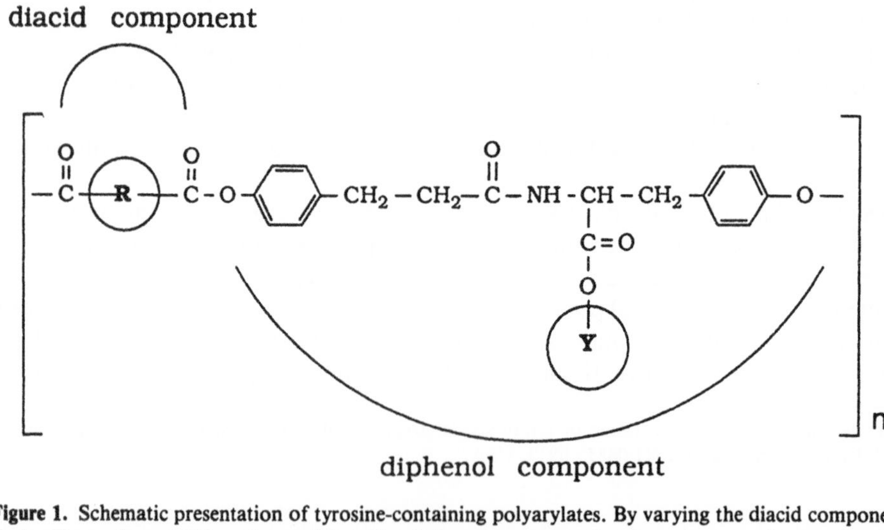
diacid component
O
C
R
O
C-O
CH2 – CH2 – C – NH - CH – CH2
O
C=O
O
Y
O
n
diphenol component

Figure 1. Schematic presentation of tyrosine-containing polyarylates. By varying the diacid component, the backbone element (R) can be modified. The diphenol component is a desaminotyrosyl-L-tyrosine alkyl ester. By varying the alkyl ester group, the pendent chain (Y) of the polymer can be modified.

diacid component	diphenol component		
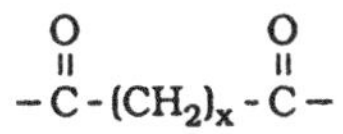	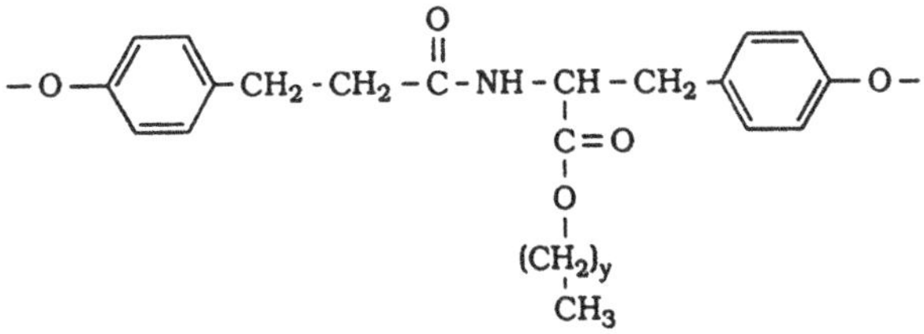		
	DTE $y + 1 = 2$	DTH $y + 1 = 6$	DTO $y + 1 = 8$
succinic $x = 2$		poly(DTH-succinate)	
adipic $x = 4$	poly(DTE-adipate)	poly(DTH-adipate)	poly(DTO-adipate)
sebacic $x = 8$		poly(DTH-sebacate)	

Figure 2. Chemical structures for the five polyarylates included in this study. Desaminotyrosyl-L-tyrosine ethyl ester (DTE), desaminotyrosyl-L-tyrosine hexyl ester (DTH), and desaminotyrosyl-L-tyrosine octyl ester (DTO) were used as the diphenol components, while succinic, adipic, and sebacic acid served as the diacid components.

structure such as the length and effective volume of the pendent chains and the flexibility of the polymer backbone. In this way, it may be possible to adapt the polymer properties to a wide range of possible applications.

The general structure of tyrosine-containing polyarylates and the structural relationship between the five polymers included in this preliminary study are described in Fig. 2. Although additional polyarylates can be designed and prepared by choosing different diacids or pendent alkyl ester chains, the particular polymers included in this study cover a 'representative selection' of backbone and pendent chain combinations, including pendent chains from ethyl to octyl and diacids providing from two to eight flexible CH_2-units in the polymer backbone. Here we report on the synthesis and preliminary thermomechanical characterization of these new polymers. A detailed study of their biological and toxicological properties will be published at a later stage.

MATERIALS AND METHODS

Materials

The aliphatic diacids (adipic, succinic, and sebacic), desaminotyrosine (3-*p*-hydroxyphenylpropionic acid), L-tyrosine, hexanol, octanol, 1-hydroxybenzotriazole hydrate, *p*-toluenesulfonic acid monohydrate (PTSA-H_2O), dicyclohexylcarbodiimide (DCC) and diisopropylcarbodiimide (DiPC) were purchased from Aldrich Chemicals. Thionyl chloride was from Fisher. Dimethylaminopyridine was from Eastman Kodak Co. All solvents were HPLC grade and were used as received.

Methods

Spectroscopy. FT-IR spectra were recorded on a Matson Cygnus 100 spectrometer. Polymer samples were dissolved in methylene chloride and films were cast directly onto NaCl plates. All spectra were collected after 16 scans at 2 cm^{-1} resolution. UV/Vis spectra were recorded on a Perkin-Elmer Lambda 3B spectrophotometer. NMR spectra of polymer solutions in deuterated dimethylsulfoxide (d_6-DMSO) were recorded on a Varian VXR-200 spectrometer (64 scans).

Wide angle X-ray scattering (WAXS). A Scintag PAD-V diffractometer in θ–2θ mode with Cu$K\alpha$ radiation was used. For measurement, 5 thin film samples (thickness about 0.1 mm) were stacked on top of each other.

Gel permeation chromatography (GPC). The chromatographic system consisted of a Perkin-Elmer Model 410 pump, a Waters Model 410 RI detector, and a PE-Nelson Model 2600 computerized data station. Two PL-gel GPC columns (pore size 10^5 and 10^3 Å) were operated in series at a flow rate of 1 ml min^{-1} in DMF containing 0.1% (w/v) of LiBr. Molecular weights were calculated relative to polystyrene standards without further correction.

Thermal analysis. The glass transition temperature (T_g) was determined by differential scanning calorimetry (DSC) on a DuPont 910 DSC instrument calibrated with indium. Each specimen was subjected to two consecutive DSC scans. After the first run the specimen was quenched with liquid nitrogen and the second scan was performed immediately thereafter. T_g was determined in the second DSC scan as the onset point, obtained at the intersection of the regression lines in the glassy and transition regions. The decomposition temperature (T_d) was determined by thermogravimetric analysis (TGA) on a DuPont 951 TGA instrument and was reported at 10% decrease in weight according to Manami *et al.* [20]. The heating rate for all polymers was 10°C min^{-1} and the average sample size was 10 mg.

Preparation of solvent cast films. 1.5 g of polymer was dissolved in 25 ml of methylene chloride. The solution was filtered through a teflon syringe filter (0.45 μm pore diameter) and poured into a leveled glass mold. The mold surfaces were treated with chlorotrimethylsilane to facilitate the removal of the films from the glass mold. The solvent was left to evaporate in a glove bag under dry nitrogen flow for 3 days. Films of about 0.1 mm thickness were obtained by this method. The films were dried in vacuum to constant weight before use.

Mechanical testing. Thin (approx. 0.1 mm) solvent cast polymer films were tested on a Sintech 5/D tensile tester according to ASTM standard D882-83 at room temperature. For each polymer, 10 individual specimens were used. Prior to testing, the films were examined under a polarizing light microscope to confirm the absence of crystallinity and/or orientation. For the first 20% of strain, the crosshead speed was 2 mm min^{-1}, allowing for a reliable calculation of the elastic modulus. Thereafter, the crosshead speed was increased to 100 mm min^{-1}. The yield point was determined based on the zero slope criterion.

Hydrolytic degradation studies. Samples were cut from solvent cast films. The samples were incubated at 37°C in phosphate buffer (0.1 M, pH 7.4) containing 200 mg l^{-1} of sodium azide to inhibit bacterial growth. The degradation process was followed by recording weekly the changes in weight and in the molecular weight of the polymer. Results are the average of five separate specimens per polymer.

Dye release studies. Accurately weighed amounts of *p*-nitroaniline and polymer were dissolved in methylene chloride such that a 20:1 ratio of polymer:dye was obtained (5% nominal loading). The volume of this solution was adjusted so that the final polymer concentration was about 7% (w/w). Any undissolved residues were removed by filtration using a Teflon membrane filter. The filtered solutions were then used for solvent casting as described above. Homogeneous and transparent films (thickness: 0.2–0.3 mm) were obtained. Circular disks (diameter: 1.0 cm, weight: 35–40 mg) were cut from the films. The disks were first placed into 50 ml of distilled water for 24 h to remove any drug on or near the surface. Thereafter the devices were placed into 20 ml of phosphate buffer (0.1 M, pH 7.4) and incubated at 37°C. The amount of dye released was determined spectrophotometrically at 380 nm (ε = 13 100 $l mol^{-1} cm^{-1}$). After each analysis, the device was transferred into fresh, sterile phosphate buffer.

Synthesis

The monomers used in this study are readily available. The required diacids were purchased in high purity and were used without further purification. For the synthesis of desaminotyrosyltyrosine alkyl esters suitable procedures had been developed previously [9, 13].

Preparation of monomers. The tyrosine-derived diphenols (DTE, DTH, and DTO) were synthesized by a dicyclohexylcarbodiimide mediated coupling reaction according to procedures reported earlier [9, 13]. The crude diphenols were purified by flash chromatography using a 100:2 mixture of methylene chloride:methanol as the mobile phase.

Preparation of dimethylaminopyridinium p-*toluenesulfonate (DPTS).* The procedure of Moore and Stupp [21] was modified as follows: A solution of *p*-toluenesulfonic acid monohydrate in toluene was azeotropically dehydrated in a Dean–Stark trap and allowed to cool to room temperature. An equimolar amount of dimethylaminopyridine in warm toluene was then added. White crystals of DPTS formed immediately. The crystals were retrieved by vacuum filtration and were purified by recrystallization from boiling dichloroethane. Colorless needles were obtained (m.p. = 174–176°C). The purity of DPTS was estimated by melting point depression and by ^{1}H-NMR.

Polymerization. The procedure of Moore and Stupp [21] was modified as follows: An exactly equimolar mixture of the column purified diphenol (for example, 8.2712 g of DTH) and the diacid (for example, 2.9228 g of adipic acid) was placed in a dry 150 ml round-bottomed flask. Dimethylaminopyridinium *p*-toluenesulfonate (2.3552 g, 0.4 equivalents) was then added to the flask which was fitted with a septum and flushed with dry nitrogen. Next, methylene chloride (150 ml) was added via syringe and the solution was stirred to suspend the reactants. Diisopropyl-carbodiimide (10 g, about 4 equivalents) was added via syringe. Within 10 min the reaction mixture first cleared up, followed by the formation of a precipitate of diisopropylurea. The reaction was allowed to proceed overnight and was then terminated by pouring the viscous reaction mixture into ten volumes of rapidly stirred methanol. Stirring was continued until the supernatant became clear and the polymer had completely precipitated. The supernatant was decanted and the polymer was dried under high vacuum to constant weight. For further purification, the polymer was redissolved in methylene chloride (10% w/w) and reprecipitated from methanol. The chemical structure of all polymers was confirmed by ^{1}H-NMR and elemental analysis (Table 1). For representative FT-IR and ^{1}H-NMR spectra (see Fig. 4).

Table 1.
Elemental analysis of new polymers

Polymer	Formula (repeat unit)	Calculated values (%)			Found values (%)		
		C	H	N	C	H	N
poly(DTE-adipate)	$C_{26}H_{29}NO_7$	66.80	6.25	3.00	66.16	6.29	2.91
poly(DTH-adipate)	$C_{30}H_{37}NO_7$	68.81	7.12	2.68	68.25	7.11	2.62
poly(DTO-adipate)	$C_{32}H_{41}NO_7$	69.67	7.49	2.54	69.36	7.46	2.52
poly(DTH-succinate)	$C_{28}H_{33}NO_7$	67.86	6.71	2.83	67.33	6.72	2.78
poly(DTH-sebacate)	$C_{34}H_{45}NO_7$	70.44	7.82	2.42	70.24	7.83	2.37

RESULTS AND DISCUSSION

Synthesis

The preparation of polyarylates is not straightforward. Since acid-catalyzed, direct esterification is not applicable to phenols, simple methods used for the preparation of aliphatic polyesters cannot be employed. Until very recently, only a few synthetic methods for the polyesterification of phenols were available. These procedures usually revolved around the reaction of acid chlorides with diphenolate salts and proceeded under conditions that were too harsh for the tyrosine-derived monomers employed in this study [22].

Major advances in the polyesterification of phenols were recently made by the use of condensing agents. A particularly noteworthy procedure was published by Moore and Stupp, who utilized diisopropylcarbodiimide (DiPC) in conjunction with a specially designed catalyst, dimethylaminopyridinium *p*-toluenesulfonate (DPTS) [21]. Moore and Stupp estimated the degree of polymerization (DP) of their polyarylates to be above 50 based on end group analysis but did not actually determine the molecular weight or molecular weight distribution of their products.

We tested this procedure for the synthesis of poly(DTH-sebacate), the polyarylate derived from DTH and sebacic acid. We realized that Moore and Stupp had under-estimated the amount of carbodiimide needed in the polymerization: by increasing the excess of carbodiimide in the reaction mixture and by optimizing the reaction time, the molecular weight (M_w) of poly(DTH-sebacate) could be increased from

37 000 to over 200 000. The synthesis of poly(DTH succinate), poly(DTH adipate), poly(DTE adipate), and poly(DTO adipate) was also accomplished.

Our studies revealed that the procedure of Moore and Stupp yields high polymers only if solvents with low dielectric constants are used. Such solvents suppress the undesirable rearrangement of the active *O*-acyl isourea intermediate to the unreactive *N*-acylurea derivative [23]. Since this rearrangement terminates further chain growth, solvent effects strongly influence the outcome of the polymerization reaction. Consequently, the need to use a nonpolar solvent with low dielectric constant has to be balanced with the requirement that the reactants and the polymeric product be soluble in the reaction mixture. In our experiments, the highest molecular weights were obtained when the polymerization was performed in pure methylene chloride. For example, the addition of as little as 0.5 ml of pyridine as a cosolvent into 15 ml of methylene chloride reduced the maximum molecular weight (M_w) of the resulting poly(DTH sebacate) to about 80 000. Likewise, polymers with molecular weights (M_w) of only about 50 000 were obtained when THF was the reaction medium.

The need to perform the polymerization in pure methylene chloride in order to obtain high polymers can be a limitation since not all monomers of interest are readily soluble in methylene chloride. For example, due to the reduced solubility of succinic acid in the reaction mixture, the preparation of poly(DTH succinate) required much longer reaction times (Fig. 3) and we were not able to obtain poly(DTH succinate) with weight average molecular weights significantly above 100 000 while all other polymers were obtained with molecular weights in excess of 200 000 (Table 2).

All crude polymers were purified by repeated precipitation into methanol from a 10% (w/w) solution in methylene chloride. Chemical structure and polymer purity

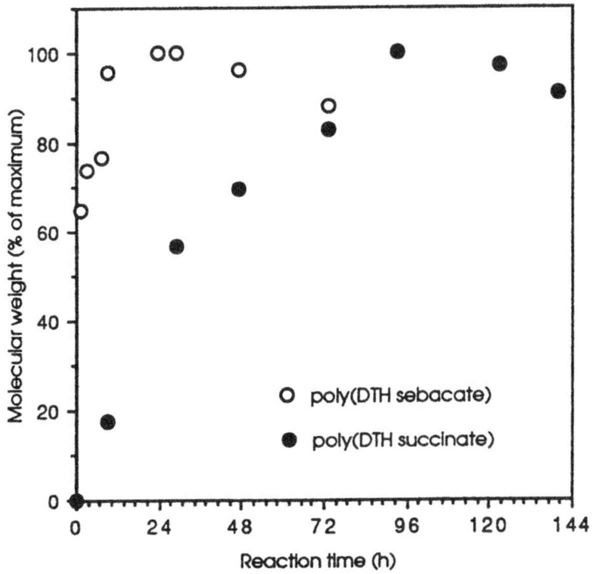
Molecular weight (% of maximum)
0
20
40
60
80
100
0
24
48
72
96
120
144
Reaction time (h)
poly(DTH sebacate)
poly(DTH succinate)

Figure 3. Molecular weight (M_w) as function of the reaction time. The polymerization of poly(DTH sebacate) was complete within about 24 h since all monomers were readily soluble in methylene chloride. In the case of the sparingly soluble succinic acid, the polymerization of poly(DTH succinate) required about 96 h for completion.

Table 2.
Representative molecular weights

Polymer[a]	Yield (%)	Weight average molecular weight[c]	Number average molecular weight[c]	Polydispersity (M_w/M_n)
Poly(DTE-adipate)	68	209 000	114 000	1.83
Poly(DTH-adipate)	53	232 000	132 000	1.76
Poly(DTO-adipate)	78	220 000	135 000	1.63
Poly(DTH-succinate)	66	102 000	68 000	1.50
Poly(DTH-sebacate)	76	212 000	133 000	1.59

[a] Representative polymer batches. Molecular weight determined after purification.

[b] Yield in % of theory after purification, based on the amount of diphenol added into the reaction mixture.

[c] Calculated by GPC relative to polystyrene standards in DMF containing 0.1% of LiBr, without further corrections.

were monitored by FT-IR and ^{1}H-NMR (Fig. 4). Only pure and well-characterized polymer samples with polydispersity below 2 were used in the characterization studies described below.

Thermomechanical properties

In this study, we focused on five polymers which represent a wide range of possible molecular structures (Fig. 2). One of our goals was to explore to which extent small variations of the molecular structure (such as changes in the length of the pendent chain or the length of the diacid comonomer in the polymer backbone) affect macroscopic polymer properties.

The most obvious changes were observed in the glass transition temperature (T_g). Tyrosine-containing polyarylates having short pendent chains and relatively rigid backbones (i.e. short diacid components) are glassy at room temperature. Increasing the length of the pendent chain and/or the length of the aliphatic diacid component, lowered T_g and the polymers became rubbery at room temperature. Specifically, in the adipate series, T_g decreased from 52 to 21°C when the length of the pendent chain was increased from two carbons (ethyl) to eight carbons (octyl). Likewise, at constant pendent chain length (hexyl series), T_g decreased from 43 to 13°C when the length of the backbone diacid was increased from two carbons in poly(DTH succinate) to eight carbons in poly(DTH sebacate) (Table 3).

Table 3.
Selected thermomechanical properties of tyrosine-containing polyarylates

Polymer[a]	Side chain length ($y + 1$)	Diacid chain length (x)	T_g (°C)	T_d (°C)	Elastic modulus (MPa)	Yield stress (MPa)	Break stress (MPa)	Strain at break (%)	Energy to break (Nm)
DTE-adipate	2	4	52	355	1525 ± 84	38.6 ± 1.8	34.8 ± 2	157 ± 154	0.40 ± 0.40
DTH-adipate	6	4	29	361	431 ± 58	9.6 ± 0.6	30.7 ± 5	418 ± 42	0.80 ± 0.25
DTO-adipate	8	4	21	368	15 ± 8.0	0.9 ± 0.1	29.6 ± 8	443 ± 80	0.48 ± 0.20
DTH-succinate	6	2	43	353	1202 ± 142	34.6 ± 5.2	30.7 ± 9	9 ± 6	0.012 ± 0.009
DTH-sebecate	6	8	13	364	2.6 ± 0.4	1.1 ± 0.1	23.4 ± 5	449 ± 82	0.35 ± 0.14

[a] See Fig. 2 for detailed chemical structure.

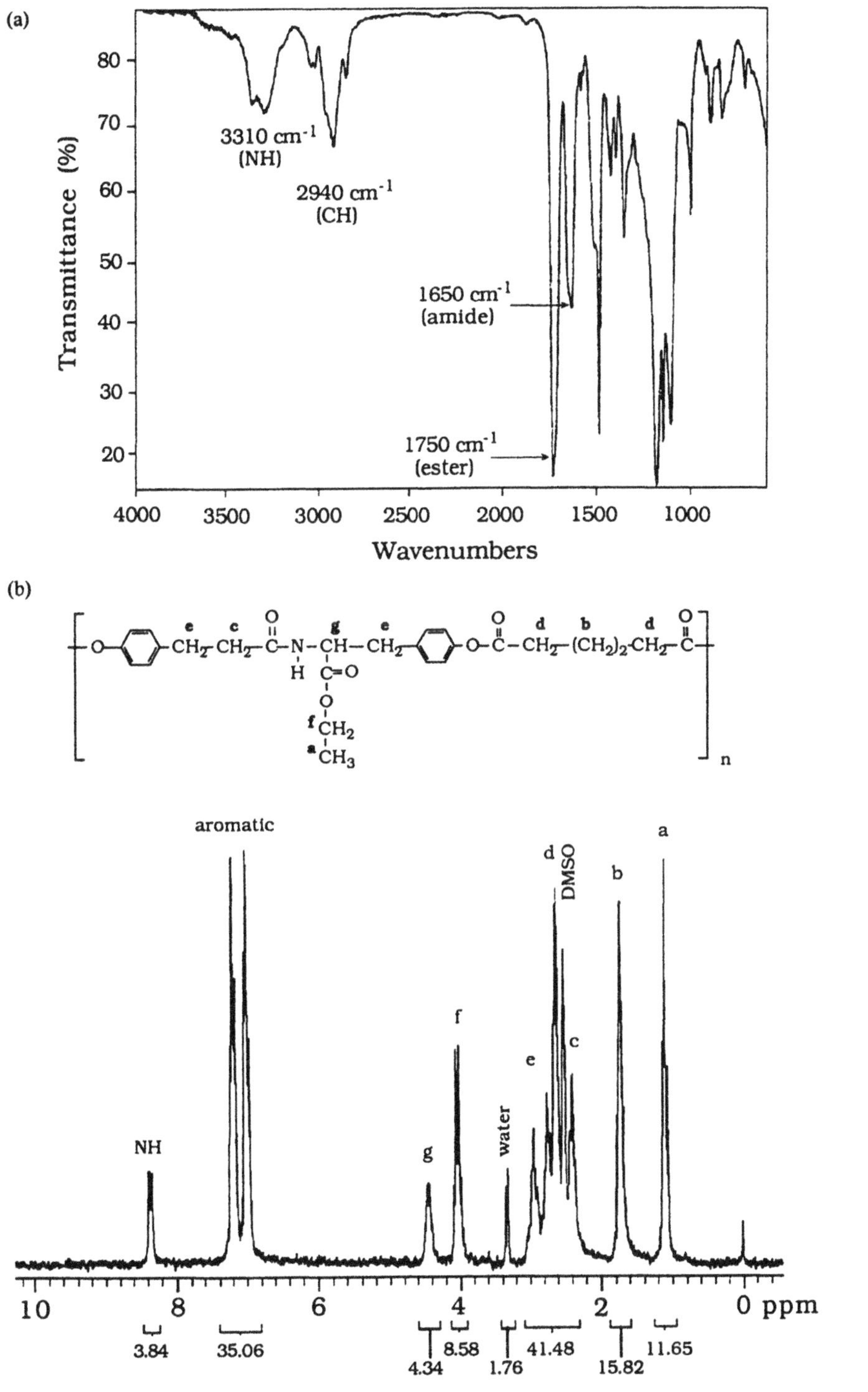
(a)
Transmittance (%)
80
70
60
50
40
30
20
3310 cm-1
(NH)
2940 cm-1
(CH)
1650 cm-1
(amide)
1750 cm-1
(ester)
4000
3500
3000
2500
2000
1500
1000
Wavenumbers
(b)
n
aromatic
NH
g
f
water
e
d
DMSO
c
b
a
10
8
6
4
2
0 ppm
3.84
35.06
4.34
8.58
1.76
41.48
15.82
11.65

Figure 4. Representative spectra for tyrosine-containing polyarylates. For details on spectral acquisition, see Methods. (a): FT-IR spectrum of poly(DTE-adipate). The expected ester and amide absorptions are at 1750 and 1650 cm^{-1} respectively. (b): ^{1}H-NMR spectrum of poly(DTE-adipate). See figure insert for the assignment of the peaks in the NMR spectrum.

The changes in the glass transition temperature are also reflected in the mechanical properties in tension, measured on polymer films obtained by solvent casting (Table 3). Polymers that are glassy at room temperature are strong, ductile materials whose tensile strength and Young's moduli are comparable to those of other nonreinforced degradable biomaterials such as poly(ortho esters), poly(D,L-lactic acid), or poly(hydroxybutyrate) [1]. Polymers that are rubbery at room temperature, are soft, elastomeric materials that undergo very large elongations before failure. Since currently only a small number of degradable elastomers are known, the combination of elastomeric behavior and large break strength, high ductility and toughness may provide unique applications for these polyarylates.

Our preliminary observations indicate that changes in the length of the diacid component in the polymer backbone had a stronger effect on the mechanical properties of the polymer than corresponding changes in the length of the pendent chain. Specifically, increasing the length of the diacid comonomer from two carbons in poly(DTH-succinate) to eight carbons in poly(DTH-sebacate) resulted in a reduction of the elastic modulus by almost three orders of magnitude, while a corresponding increase in the length of the pendent chain from two carbons in poly(DTE-adipate) to eight carbons in poly(DTO-adipate) led to a decrease in modulus of only two orders of magnitude. Those changes in modulus were closely paralleled by corresponding changes in the yield stress, while no significant effect on the break stress was observed (Table 3).

Since no endothermal peak in the expected range of melting ($T_m = 1.5 * T_g$) [24] was observed by DSC, all investigated polyarylates appeared to be amorphous. This observation was supported by wide angle X-ray scattering (WAXS): only a broad and relatively low intensity peak in the 10–25° 2θ range was observed, characteristic for amorphous polymers [25]. The absence of birefringence when solvent cast films were placed between crossed polarizers in an optical microscope also indicated a lack of orientation or crystallinity in nonelongated film specimens [26].

After solvent cast film specimens were highly elongated, the DSC thermogram showed a sharp endothermal peak at 20–30°C above the glass transition temperature of the polymer. This peak was observed only in the first scan and disappeared in subsequent DSC scans. Since the shape and location of the peak suggested a melting transition, the possibility of strain-induced crystallinity was tested by wide angle X-ray scattering, using the procedure detailed by Salem [27]. The diffractogram showed no crystallinity, but only an amorphous peak in the same 2θ range as for unstrained specimens. However, the significantly increased intensity of the amorphous peak in elongated specimens suggested that the elongation process resulted in a denser molecular packing [28]. This was confirmed by the observation of birefringence in elongated film specimens. When the elongated film specimens were heated to about 20–30°C above T_g on a hot stage diffractometer, the intensity of the amorphous peak dropped back to the value measured in the unstrained state.

The fact that film specimens became oriented during the tensile test can possibly explain the apparent independence of the break stress on the molecular structure: since break stress is determined mainly by long-range intermolecular interactions (which became similar in all polymers due to orientation), the effect of the variations in monomer structure (which are short-range in nature) could no longer be detected.

The thermal stability of tyrosine-containing polyarylates was investigated by thermogravimetric analysis (TGA). At a heating rate of 10°C min^{-1}, the temperature at which the specimen had lost 10% of its initial weight, was taken as the decomposition temperature (T_d) [20]. While we cannot yet explain the slight variations in the decomposition temperature among the tested polyarylates, it is obvious that all investigated polymers have decomposition temperatures above 300°C (Table 3). Hence we expect these polyarylates to be readily processible by conventional techniques such as extrusion or injection molding even if relatively high temperatures are applied.

Hydrolytic degradation

At physiological conditions (0.1 M phosphate buffer solution, pH 7.4, 37°C), the arylate linkage should be more susceptible to hydrolysis than the aromatic ester linkage. Thus, while aromatic polyesters such as Dacron are stable at physiological conditions, we expected the tyrosine-containing polyarylates to slowly degrade. To test this hypothesis and to explore the effect of the molecular structure on the rate of backbone degradation, thin films of poly(DTE adipate), poly(DTH adipate), and poly(DTO adipate) were exposed to physiological buffer solutions (pH 7.4, 0.1 M phosphate) at 37°C for up to 26 weeks. At predetermined times, the film specimens were inspected for any macroscopic changes, carefully weighed, and their residual molecular weight was determined by GPC analysis.

Over the course of 26 weeks, the films became increasingly opaque and the formation of small indentations at the polymer surface indicated that some erosion occurred. In addition, all film specimens became noticeably more brittle and lost much of their initial mechanical strength. At 26 weeks, the films were still physically intact but crumbled easily when handled. The degradation of the polymer backbone was unambiguously confirmed by GPC which revealed that the polymer molecular weight had significantly decreased for all tested polymers over the course of 26 weeks (Fig. 5). In spite of the degradation of the polymer backbone, however, the

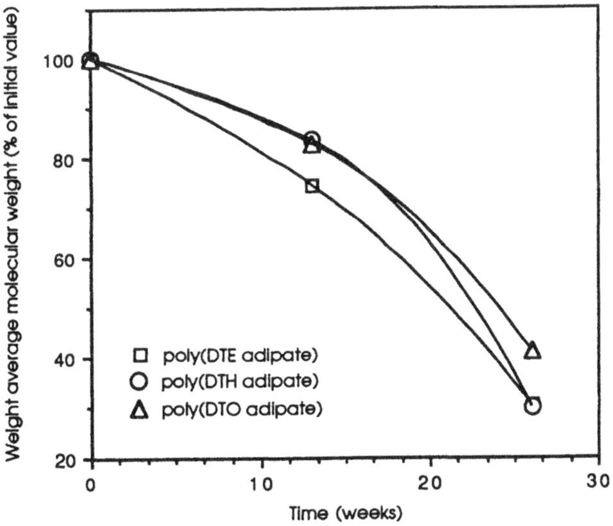
Weight average molecular weight (% of initial value)
100
80
60
40
20
0
10
20
30
Time (weeks)
poly(DTE adipate)
poly(DTH adipate)
poly(DTO adipate)

Figure 5. Changes in the *weight average* molecular weight of poly(DTE adipate), poly(DTH adipate), and poly(DTO adipate) while stored in physiological buffer solution at 37°C. All three polymers had an initial weight average molecular weight of about 200 000 (relative to polystyrene standards).

weight of all film specimens remained unchanged (±1%) throughout the study. This can be attributed to the highly hydrophobic nature of the polymer which prevented the imbibition of water and the insolubility of the expected polymer degradation products in aqueous buffer solutions.

Our degradation studies revealed no clear dependence of the degradation rate on the molecular structure of the tested polymers (Fig. 5). The fact that poly(DTE adipate), poly(DTH adipate), and poly(DTO adipate) degraded at about the same rate can possibly be attributed to the operation of two opposing effects: on one hand, the increasing length of the pendent chain increased the hydrophobicity of the polymer (which should slow degradation); on the other hand, the increasing length of the pendent chain decreased the glass transition temperature so that poly(DTE adipate) was glassy at 37°C, while the polymers with the longer pendent chains (poly(DTH adipate) and poly(DTO adipate)) were rubbery at 37°C. This should increase the rate of hydrolytic degradation due to the higher chain mobility and greater free volume fraction in the rubbery state. More detailed studies are required to establish quantitative correlations between the rate of degradation and the molecular structure of tyrosine-containing polyarylates.

Release studies

To evaluate the effect of the pendent chain length on the transport properties of tyrosine-containing polyarylates, the release of *p*-nitroaniline from thin films of the adipate series of polymers was investigated. Solvent cast films were prepared containing a nominal loading of 5% (w/w) of *p*-nitroaniline. Since *p*-nitroaniline is soluble in methylene chloride, clear yellow films were obtained. The dye appeared to be homogeneously dispersed throughout the polymer matrix (detailed studies on

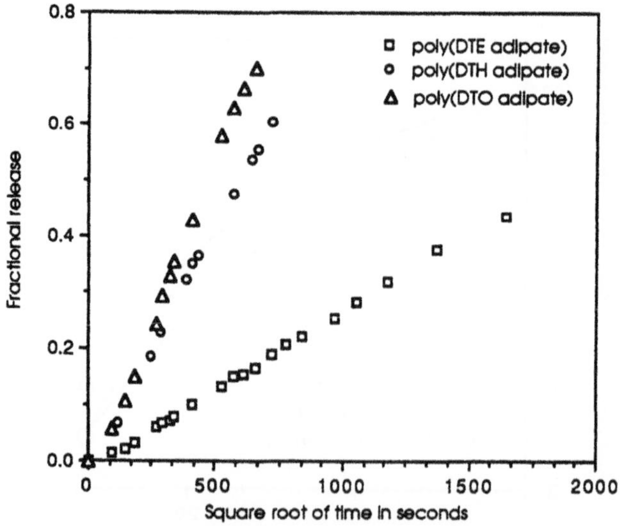
0.8
0.6
0.4
0.2
0.0
Fractional release
0
500
1000
1500
2000
Square root of time in seconds
poly(DTE adipate)
poly(DTH adipate)
poly(DTO adipate)

Figure 6. Plots of fractional release of *p*-nitroaniline from several polyarylates as function of the square root of time. The release rates were calculated from the slopes of the linear portions of the release curves. The release rates were 1.90×10^{-4}, 8.51×10^{-4} and 10.88×10^{-4} ($s^{-1/2}$) for poly(DTE-adipate), poly(DTH-adipate), and poly(DTO-adipate) respectively.

the physical state of the dye within the polymer matrix are currently in progress). During the initial washing step (see Methods), a negligible amount of *p*-nitroaniline was released from poly(DTE adipate), while 5 and 10% of the total dye content was released from the poly(DTH-adipate) and poly(DTO-adipate) respectively. Thereafter, the release of *p*-nitroaniline from the films was monitored spectrophotometrically in the usual fashion.

Plots of the fractional release versus the square root of time in seconds were linear throughout much of the release period (Fig. 6). This indicated that the release of *p*-nitroaniline from all three polymers was predominantly diffusion controlled [29]. A comparison of the release rates (calculated from the linear portions of the plots of Fig. 6) for the three polymers indicated that the rate of dye release was related to the morphological state of the polymers: poly(DTH adipate) and poly(DTO adipate) have glass transition temperatures below 37°C and were rubbery. These two polymers exhibited about a 10-fold faster release rate than the glassy poly(DTE adipate). This observation can be explained by the fact that rubbery polymers have a higher free volume fraction than glassy polymers and therefore facilitate the diffusion of small molecules through the polymer matrix. Since these results indicated that small changes in the molecular structure of the polymers can lead to widely differing release kinetics, we are currently evaluating the drug release characteristics of tyrosine-derived polyarylates in more detail.

CONCLUSION

Five aliphatic polyarylates were synthesized using a series of tyrosine-derived diphenols and a group of aliphatic diacids of low toxicity. The structural design of these polymers allowed the systematic variation of the polymer backbone (through the use of different diacids) while independently changing the length of the polymer pendent chain. Our most significant observation was that these polymers are readily processible materials with potentially useful engineering properties for biomedical applications. In particular, poly(DTH-adipate), poly(DTO-adipate), and poly(DTH sebacate) were highly ductile polymers that exhibited elastomeric behavior. All tyrosine-containing polyarylates degraded slowly in phosphate buffer solution due to hydrolysis of the arylate linkages in the polymer backbone. Although more data are necessary to identify quantitative relationships between polymer structure and a range of polymer properties, our preliminary studies revealed a number of useful structure–property correlations: generally, increasing the length of either the pendent chain or the diacid in the polymer backbone resulted in a decrease in glass transition temperature, modulus, and yield strength. However, lengthening the aliphatic diacid in the polymer backbone appears to be a more effective tool in controlling the elastic modulus and the yield strength than lengthening the pendent chain. This could be a useful observation for future attempts to design polymers for specific applications. While the release profiles of *p*-nitroaniline from thin polymer films indicated that the rate of release is related to the morphological state of the polymer, initial degradation studies *in vitro* did not reveal an obvious correlation between the rate of polymer degradation and the length of the pendent alkyl ester chain.

Acknowledgements

This work was supported by a Focused Giving Award from Johnson & Johnson and by a grant from the Biomaterials Research Fund program sponsored by the

University of Medicine and Dentistry of New Jersey. The technical assistance of Ms. Ligia Bron in performing the DSC and TGA analyses and the help of Dr. K. V. Ramanujachary with the WAXS testing and data interpretation are gratefully acknowledged.

REFERENCES

1. I. Engelberg and J. Kohn, *Biomaterials* **12,** 292 (1991).
2. M. Chasin, A. Domb, E. Ron, E. Mathiowitz, R. Langer, K. Leong, C. Laurencin, H. Brem and S. Grossman, in: *Biodegradable Polymers as Drug Delivery Systems*, p. 43, M. Chasin and R. Langer (Eds). Marcel Dekker, New York (1990).
3. J. Heller, R. V. Sparer and G. M. Zentner, in: *Biodegradable Polymers as Drug Delivery Systems*, p. 121, M. Chasin and R. Langer (Eds). Marcel Dekker, New York (1990).
4. H. R. Allcock, in: *Biodegradable Polymers as Drug Delivery Systems*, p. 163, M. Chasin and R. Langer (Eds). Marcel Dekker, New York (1990).
5. M. Richards, B. I. Dahiyat, D. M. Arm, P. R. Brown and K. W. Leong, *J. Biomed. Mater. Res.* **25,** 1151 (1991).
6. J. Kohn, in: *Biodegradable Polymers as Drug Delivery Systems*, p. 195, M. Chasin and R. Langer (Eds). Marcel Dekker, New York (1990).
7. J. Kohn, *Drug News Perspectives* **4,** 289 (1991).
8. S. Pulapura and J. Kohn, in: *Peptides—Chemistry and Biology: Proceedings of the 12th American Peptide Symposium*, p. 539, J. A. Smith and J. E. Rivier (Eds). Escom Science Publishers, Leiden, The Netherlands (1992).
9. S. Pulapura and J. Kohn, *Biopolymers* **32,** 411 (1992).
10. Q.-X. Zhou and J. Kohn, *Macromolecules* **23,** 3399 (1990).
11. J. Kohn and R. Langer, *J. Am. Chem. Soc.* **109,** 817 (1987).
12. H. Yu-Kwon and R. Langer, *Macromolecules* **22,** 3250 (1989).
13. S. Pulapura, C. Li and J. Kohn, *Biomaterials* **11,** 666 (1990).
14. F. H. Silver, M. Marks, Y. P. Kato, C. Li, S. Pulapura and J. Kohn, *J. Long-Term Effects Med. Implants* **1,** 329 (1992).
15. S. Lin, S. Krebs and J. Kohn, *Proceedings, The 17th Annual Meeting of the Society of Biomaterials*, p. 187, Scottsdale, AZ. Society for Biomaterials, Algonquin, IL (1991).
16. D. Coffey, Z. Dong, R. Goodman, A. Israni, J. Kohn and K. O. Schwarz, presented at the Symposium on Polymer Delivery Systems at the Spring Meeting of the American Chemical Society, San Francisco (1992).
17. Y. Imai and M. Kakimoto, in: *Handbook of Polymer Science and Technology*, Vol. 1, p. 177, N. P. Cheremisinoff (Ed.). Marcel Dekker, New York (1989).
18. T. J. Hunter, S. P. Schmidt, W. V. Sharp and G. S. Malindzak, *Trans. Am. Soc. Artif. Intern. Organs* **29,** 177 (1983).
19. S. R. Hanson, H. F. Kotze, B. Savage and L. A. Harker, *Arteriosclerosis* **5,** 595 (1985).
20. H. Manami, M. Nakazawa, Y. Oishi, M. Kakimoto and Y. Imai, *J. Polym. Sci., Part A: Polym. Chem.* **28,** 465 (1990).
21. J. S. Moore and S. I. Stupp, *Macromolecules* **23,** 65 (1990).
22. W. M. Eareckson, *J. Polym. Sci.* **40,** 399 (1959).
23. D. H. Rich and J. Singh, in: *The Peptides: Analysis, Synthesis, Biology*, Vol. 1, p. 241, E. Gross and J. Meienhofer (Eds). Academic Press, New York (1979).
24. L. E. Nielsen, *Mechanical Properties of Polymers*, Reinhold Publishing, New York (1962).
25. H. J. Biangardi, *Makromol. Chem.* **183,** 1785 (1982).
26. M. Pick, R. Lovell and A. H. Windle, *Polymer* **21,** 1017 (1980).
27. D. R. Salem, *Polymer* **33,** 3182 (1992).
28. H.-H. Song and R.-J. Roe, *Macromolecules* **20,** 2723 (1987).
29. N. M. Franson and N. A. Peppas, *J. Appl. Polym. Sci.* **28,** 1299 (1983).

A study on the *in vitro* degradation of poly(lactic acid)

C. MIGLIARESI[1], L. FAMBRI[1] and D. COHN[2]

[1]*Dipartimento di Ingegneria dei Materiali, Universitá di Trento, 38050 Mesiano di Povo, Trento, Italy*
[2]*Casali Institute of Applied Chemistry, The Hebrew University of Jerusalem, Jerusalem, Israel*

Received 25 January 1993; accepted 17 June 1993

Abstract—The *in vitro* degradation of samples of L- and D,L-lactic acid polymers, P(L)LA and P(DL)LA respectively, having different molecular weights, morphology and/or geometry, has been studied through the determination of viscometric molecular weight, mass and mechanical properties as function of the immersion time in Ringer solution at 37°C. In particular have been compared the degradation kinetics of P(L)LA, amorphous and crystalline, and of P(L)LA and P(DL)LA having different molecular weight and sample geometry.

From the molecular weight versus the degradation time data, a degradation rate has been defined, as the derivative of the function best fitting the data, normalized to the molecular weight of the polymer at each time. The behavior of the degradation rate curves, plotted against the degradation time, has been interpretated and compared with relation to the initial physical and geometrical characteristics of the PLA samples.

Key words: Poly(lactic acid); degradation; crystallinity; morphology.

INTRODUCTION

Starting from their initial use in the fabrication of sutures, biodegradable polymers have found numerous applications in various areas of surgery. The unique feature of this type of biomedical system, their biodegradability, not only eliminates the need of a second surgery for the removal of the implant, but also enables the gradually biodegradable system to interact with the surrounding tissue, eliciting an improved healing process [1–3].

The clinical use of biodegradable polymers can be categorized, as suggested by Gilding [4], into three main classes: when they perform as a transient scaffold for autologous tissue regeneration [5–10], as temporary barriers [11, 12], and as drug delivery systems [13–15].

Polylactic acid (PLA) is one of the most important biodegradable polymers, being used in a wide range of clinical applications, such as in orthopaedic surgery [16–24], in the cardiovascular system [1–3, 25–28], and as drug delivering implants [29–36]. Much work has been devoted also to investigating the synthesis [37–47], morphology [48–54], thermal [55–58] and mechanical properties of PLA [22, 28, 58–60], as well as its degradation kinetics under both *in vitro* and *in vivo* conditions [61–70].

Previous work by the authors [58] demonstrated that rigorously dry annealing can result in P(L)LA polymers displaying extremely high degrees of crystallinity, as determined by differential scanning calorimetry (DSC) and dynamic mechanical thermal analysis (DMTA). While temperature and time played a central role in enhancing the polymers crystallinity, it was also shown that thermal scission of the polymeric backbone during the annealing process, could not be avoided.

A more recent study conducted in our laboratory [56] shed some light on the effect of thermal history on the crystallinity of various P(L)LA samples having different molecular weights (from 18 000 to 425 000). Controlled thermal treatments conducted in a differential scanning calorimeter (DSC), at various heating or cooling rates, underscored the substantially different crystallizability of the various P(L)LA polymers, a kinetic phenomenon dependent on molecular mobility and chain length. It was apparent from the data presented that, as a function of molecular weight, a given thermal treatment resulted in different degrees of crystallinity. While the lower molecular weight polymers exhibited significant crystallizability during their cooling from the melt, higher molecular weight P(L)LAs were unable to crystallize, unless especially slow cooling rates (1 and 0.5°C min^{-1}) were used. On the other hand, when heated after being quenched from the melt, all polymers displayed a similar behaviour, with increasingly high degrees of crystallinity being attained, as the heating rate decreased. Also, for any given heating rate, the lower the molecular weight (within the broad range investigated), the higher the crystallinity gain.

The present work focused on the mechanical and morphological analyses of various PLA polymers, degrading under *in vitro* conditions. The first two sections of this study investigated the influence of the morphology of the polymer on its degradation profile. First, P(L)LA samples exhibiting different initial degrees of crystallinity, due to the annealing or quenching of the polymer, were investigated. Then, the degradation behavior of crystalline P(L)LA was compared with that of amorphous P(DL)LA. Aiming at gaining further insight into the effect of chain length on the degradation process, P(L)LA samples differing substantially in their molecular weight were studied. Light was also shed on the effect of specimens geometry on the behaviour of the degrading PLA polymer. Viscometric molecular weight results, differential scanning calorimetry findings, mechanical properties measurements as well as scanning electron microscopy analysis of the degrading samples, will be presented and discussed.

EXPERIMENTAL

Materials

Polymers of L-lactic (P(L)LA, molecular weight 425 000 and 156 000) and D,L-lactic (P(DL)LA, molecular weight 300 000 and 180 000) acid were supplied by Boehringer, Ingelheim, and were stored under vacuum in a desiccator until use.

Flat samples, cut from a sheet of 100 × 50 × 1.5 mm^3, or cylindrical samples (length: 70 mm, diameter: 3 mm) were prepared by using a Carver press, by compression moulding of the as received polymer flakes, then quenched by circulating cold water into the mold plates. Molding temperatures of 210, 200, 120 and 110°C were used, for the two types of P(L)LA and P(DL)LA, respectively, kept for about 5 min.

After molding, viscometric molecular weight of the different samples were measured. Among them, specimens of P(L)LA with molecular weights of 27 000 (cylindrical), 103 000 (flat), 153 000 (flat), and 177 000 (cylindrical), and of P(DL)LA with molecular weights of 156 000 (flat) and 180 000 (cylindrical) were selected. Flat samples had a slab geometry with an A/V (area to volume) ratio of 1.57, while the cylindrical samples had an A/V ratio of 1.36.

All quenched samples were amorphous or displayed a small amount of crystallinity (less than 5%), as revealed by DSC analysis. Only P(L)LA samples with a molecular weight of 153 000 annealed at 120°C for 30 min exhibited a substantial degree of crystallinity.

In the text all samples are indicated by the name of the polymer (P(L)LA or P(DL)LA) followed by a number which corresponds, in thousands, to its molecular weight after moulding and by the letter C (crystalline) or A (amorphous), for the annealed or non-annealed samples, respectively (e.g., P(L)LA103A indicates an amorphous (non-annealed) sample of P(L)LA, having a molecular weight of 103 000).

Thermal analysis

Differential scanning calorimetry was performed in a Mettler DSC 30 calorimeter, assisted by a computer. Samples weighing approximately 15 mg were used, heated at $+10°C\ min^{-1}$ from 0 to 230°C. Glass transition temperatures were calculated as the temperature of the inflection point of the curve, whereas relative crystallinity, χ, was assessed determining the difference between the normalized areas of the melting endotherm peak and the exotherm crystallization peak, developed during the DSC scan, and rating it to $93.6\ J\ g^{-1}$ relative to the 100% crystalline polymer [57].

Mechanical properties

Mechanical properties of dry polymer samples were measured in bending at room temperature with a span length of 40 mm and a cross rate of $1.27\ mm\ min^{-1}$, by using an Instron 4502 dynamometer. Flexural strength was defined as the maximum of the stress–strain curve.

Molecular weight analysis

Intrinsic viscosity of polymer was determined by using an Ubbelohde viscometer (type I) at 25.0°C for diluted chloroform polymer solutions at different concentrations. The viscometric average molecular weight, M_w, was calculated from the intrinsic viscosity η by using the Mark–Houwink equation:

$$\eta = K \times M_w^a$$

Values of K and a equal to 5.45×10^{-4} and 0.73 for P(L)LA samples and 6.06×10^{-4} and 0.64 for P(DL)LA polymers, respectively, were used [48].

Hydrolitic degradation

Ageing of PLA samples was performed in Ringer solution changing the degradation medium every month. The water sorption process and mass loss during degradation were also investigated.

Scanning electron microscopy

A Cambridge Stereoscan 200 scanning electron microscope (SEM), was used to investigate the fracture surface of different polymers during their hydrolitic degradation.

Table 1.
Molecular weight (M_w), degradation rate (M_w'/M_w), glass transition temperature (T_g), percent amount of crystallinity (χ) and melting temperature (T_m) versus degradation time for P(L)LA153C.

Time (days)	M_w (dalton)	M_w'/M_w (day^{-1})	T_g (°C)	χ (%)	T_m (°C)
0	153 000	0.004 14	65.5	45.2	181.5
10	142 000	0.004 19	69.4	46.3	182.0
44	—	0.004 37	71.3	47.3	182.0
59	90 000	0.004 45	73.2	51.2	181.8
107	85 000	0.004 73	74.4	46.3	183.1
148	80 500	0.004 98	73.1	47.6	182.3
189	64 500	0.005 24	75.0	49.0	185.3
268	51 200	0.005 75	73.2	47.4	185.0
383	31 000	0.006 18	68.7	60.0	180.7
576	7 200	0.003 84	61.5	87.3	178.3
693	—	—	61.9	78.3	176.9
779	—	—	62.4	69.3	183.1

RESULTS AND DISCUSSION

The effect of initial crystallinity on the degradation behaviour of P(L)LA

The polymers investigated in this section comprised exclusively the P(L)LA enantiomer, which, even when quenched to generate an essentially amorphous matrix, remained potentially crystallizable. Tables 1 and 2 summarize some of the data generated by these samples, as degradation took place.

The DSC revealed a rather similar behaviour of both the initially crystalline and amorphous P(L)LA polymers. Figures 1 and 2 present thermograms of four P(L)LA153C and P(L)LA103A samples, respectively, prior to degradation and after $\frac{1}{2}$, 1 and 2 years immersion in Ringer solution, at 37°C. The thermograms of the two

Table 2.
Molecular weight (M_w), degradation rate (M_w'/M_w), glass transition temperature (T_g), crystallization temperature (T_c), melting temperature (T_m) and percent amount of crystallinity (χ) versus degradation time for P(L)LA103A.

Time (days)	M_w (dalton)	M_w'/M_w (day^{-1})	T_g (°C)	T_c (°C)	T_m (°C)	χ (%)
0	103 000	0.005 56	70.6	128.8	184.6	0.7
10	73 000	0.005 65	66.5	125.7	179.9	3.3
44	—	0.005 97	66.9	111.4	181.3	1.7
59	50 000	0.006 12	70.6	124.3	181.6	0
107	49 000	0.006 62	70.9	121.5	181.6	0.9
148	43 000	0.007 06	71.2	109.9	183.9	0.7
189	40 300	0.007 47	72.7	111.5	184.2	0.4
268	18 000	0.007 91	67.8	101.0	180.8	13.5
383	11 500	0.005 74	67.2	98.7	173.7	27.5
576	4 900	—	62.3	95.7	162.6	39.6
693	—	—	50.0	96.0	138.6	47.7
779	—	—	46.7	91.6	140.6	63.2

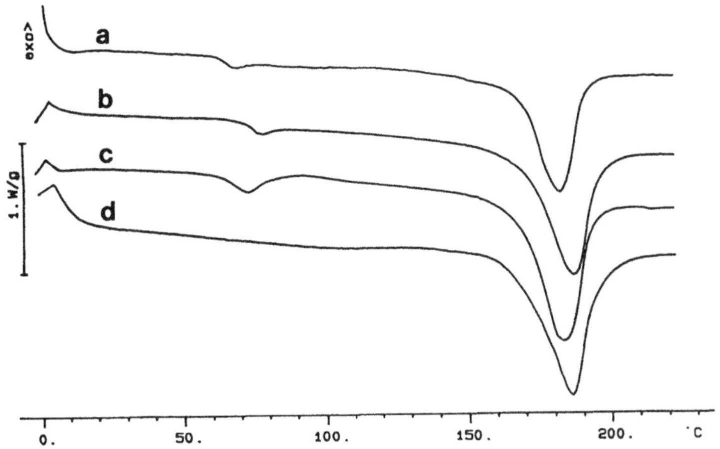
exo>
a
b
c
d
1. W/g
0.
50.
100.
150.
200.
°C

Figure 1. DSC thermograms of P(L)LA153C: (a) prior to degradation; (b) after half; (c) 1; and (d) 2 years immersion in Ringer solution at 37°C.

P(L)LA samples, prior to degradation, demonstrate the substantial degree of crystallinity of the annealed polymer ($\chi = 45\%$) (see Fig. 1, curve a), as well as the effectiveness of the quenching procedure in generating an amorphous matrix ($\chi \approx 1\%$) (see Fig. 2, curve a). It is important to realize that the melting endotherm shown by the latter developed as the result of a crystallization taking place in the initially amorphous matrix at temperatures between 90 and 160°C, during the DSC scanning. The sharp glass transition seen around 70°C also indicates the amorphous state of the P(L)LA103A starting polymer. For both P(L)LA polymers, T_m and T_g proved to be less sensitive to chain cleavage and molecular weight reduction than the degree of crystallinity (χ) of the system (see Tables 1 and 2). The melting temperature of the P(L)LA103A kept constant, around 180°C, until 270 days when it begun to gradually decrease, reaching values as low as 140°C after 2 years. The melting temperatures of the annealed polymer remained constant for the whole period

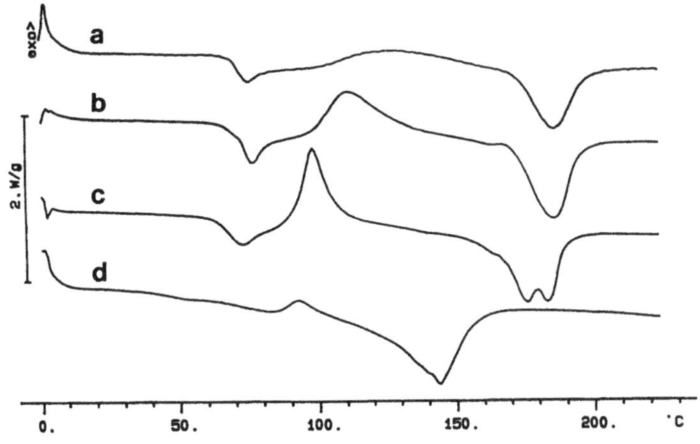
exo>
a
b
c
d
2. W/g
0.
50.
100.
150.
200.
°C

Figure 2. DSC thermograms of P(L)LA103A: (a) prior to degradation; (b) after half; (c) 1; (d) 2 years immersion in Ringer solution at 37°C.

studied (two years), even though the molecular weight was reduced to about 5% of its initial value. The glass transition temperature of both polymers remained fairly constant for approximately 1 year, decreasing gradually after this point. Expectedly, the effect on the quenched, essentially amorphous polymer was more pronounced, exhibiting a T_g decrease of about 25°C, from initial values of around 70°C. At this advanced stage of the degradation process only, the concentration of the chain ends became sufficiently high to have a measurable plasticizing effect on the amorphous phase of the matrix.

Since chain cleavage and molecular weight reduction represent the fundamental phenomena of polymer degradation, special attention was devoted to this process. Table 1 and 2 present the molecular weight decrease of the annealed and quenched P(L)LA samples, respectively, as immersion time in the Ringer solution increases. Expectedly, the annealed polymer degraded at a slower rate, due to the ordered, tightly packed spatial array of the crystalline domains. The weight vs time curves shown in Fig. 3, are based on measurements conducted on samples taken out periodically from the Ringer solution and weighed in their wet state. These data reflect, therefore, the combined effect of two processes: (a) water uptake, which should be predominant during early stages of the process, and whose increase during degradation is mainly due to the formation of voids or microcavities into the sample (see SEM micrographs reported in Figs 7 and 8); and (b) mass loss due to degradation, whose occurrence explains the decrease of the wet sample weight, as time elapses. The numbers reported on the curves are the corresponding molecular weight values, as determined by viscometry, for any given specimen. The behavior encountered is consistent with the different initial morphology of both P(L)LA samples. It should be stressed that viscometric analysis provides important though partial information on the molecular weight decrease of the system. In order to integrate this information, GPC molecular weight distribution studies will be conducted.

Separate gravimetric studies performed on dried samples, indicated that neither one of the two polymers showed any mass loss during the first 270 days of the degradation process. This, notwithstanding the fact that the molecular weight decreased drastically, by 83 and 67%, for the amorphous and crystalline P(L)LA samples, respectively. Since the degraded chains were still significantly long (51 200 and

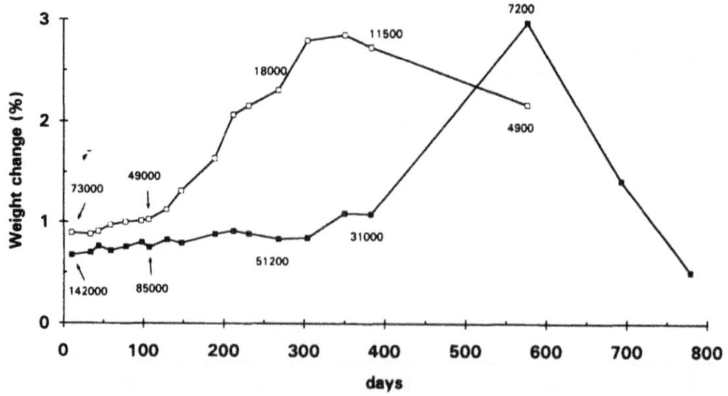
Weight change (%)
days
0
1
2
3
0
100
200
300
400
500
600
700
800
73000
49000
18000
11500
7200
4900
31000
51200
142000
85000

Figure 3. Percentage of weight change of P(L)LA103A (□) and P(L)LA153C (■) during degradation in Ringer solution at 37°C. Numbers reported on the curves are the corresponding molecular weight.

18 000 for the crystalline and amorphous samples, respectively), P(L)LA continued to show no water solubility, resulting in the observed weight constancy. Only when the molecular weight fell in the few thousands range, significant mass loss could be detected. This occurs first for the amorphous sample, losing around 15% of its initial mass after 580 days, when its molecular weight was 4900. At that point in time, the crystalline P(L)LA153C showed only very limited weight decrease, approximately 2%, when its molecular weight was around 7200.

It is apparent from the results presented in Fig. 4 (see also Tables 1 and 2), that the degree of crystallinity increases for both polymers, as degradation proceeds. The initial 45% χ level displayed by the annealed polymer P(L)LA153C, remained constant for approximately 270 days, followed by a period during which it increased steadily, reaching a high 87% value after approximately 580 days degradation. This behavior is attributed mainly to what could be called 'degradation induced crystallization', a process whereby the repeated cleavage of the polymeric backbone reduces the constraints imposed on mobility by chain entanglements, a phenomenon which renders the system with additional freedom for spatial rearrangement and enhanced crystallizability. Finally, after 580 days immersion in the Ringer solution, when the molecular weight was as low as 7200, the degree of crystallinity of the polymer started to drop, the final χ value measured being 70%, after 780 days degradation.

The behavior encountered could also be related to the faster degradation of the amorphous phase of the semicrystalline polymer, resulting in loss of amorphous material and a concomitant increase in χ. Nevertheless, gravimetric data which showed almost no mass loss during the first 580 days, seem to rule this factor out.

The plasticising effect of water molecules which diffused into the P(L)LA matrix, lowering T_g and allowing some degree of segmental motion, could be another factor to take into consideration. Water absorption was also ruled out as playing an important role, since it takes place early on in the process, a period during which χ remained constant. Furthermore, after 270 days χ increased drastically, while the water content of the materials remained constant.

It can be surmised, therefore, that the increase in the degree of crystallinity of the P(L)LA sample during this period, was not caused by the selective solubilization of

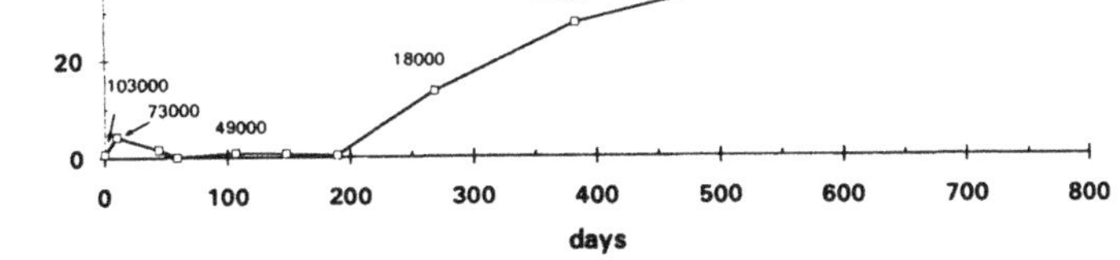

100
80
60
40
20
0
χ(%)
7200
31000
153000
142000
85000
51200
4900
11500
18000
103000
73000
49000
0
100
200
300
400
500
600
700
800
days

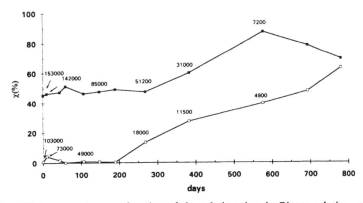

Figure 4. Crystallinity percent, χ, as function of degradation time in Ringer solution at 37°C, for P(L)LA153C (■) and P(L)LA103A (□). Numbers reported on the curves are the corresponding molecular weight.

rapidly degrading amorphous material, but rather the result of its gradual crystallization, as molecular weight decreases.

The fact that the degree of crystallinity of P(L)LA153C remained constant during the first 270 days, until the molecular weight dropped to about 50 000, seems to indicate that a threshold chain length must be reached to reduce their entanglement sufficiently, rendering the polymer with the mobility required for spatial rearrangement and enhanced crystallinity.

It is clear from the data presented in Fig. 4, that also the initially amorphous P(L)LA103A ($\chi \approx 1\%$), becomes increasingly crystalline, after an initial period of about 200 days where χ values remained constant and very small. This first stage can also be attributed to the fact that a critical molecular length must be attained during degradation, sufficiently low to allow for disentanglement of chains, providing sufficient mobility for the polymer to rearrange spatially. It is also apparent from these data, that the increase in χ is not the result of loss of amorphous material, since this is a process which would have no effect in the quenched P(L)LA103A samples, comprising a fully amorphous matrix.

It should be stressed that the 'degradation induced crystallization' behavior just described, was not found in other P(L)LA systems studied by the authors, a puzzling phenomenon now under further investigation.

The varying degradability of both P(L)LA samples was determined at different stages of the degradation process and expressed in terms of the degradation rate. The purpose of this study was to assess the rate of the chain cleavage process caused by the ester groups hydrolysis, as a function of the morphology and molecular weight of both polymers. The degradation rate, for each time t, M_w'/M_w (expressed in days^{-1}), is defined in this work as the absolute value of the derivative function of the best fit equation of the molecular weight decrease curve, dM_w/dt (or, M_w'), normalized to M_w, the molecular weight of the material at the time t.

Figures 5(a) and (b) illustrate the way the rate of degradation is calculated, for P(L)LA103A and P(L)LA153C samples. The process starts from the molecular weight profile based on the experimental data and the best fit equation, goes through the derivative function shown in Fig. 5(a), while the degradation rate (see its definition above) is plotted in Fig. 5(b). The data presented for P(L)LA153C reveal a gradually increasing rate of degradation, for about 360 days, followed by a decrease as degradation proceeds further. It is important to stress that at least up to 270 days the degree of crystallinity of the polymer remains essentially constant around 47% (see Fig. 4). Contrasting with the constant χ values measured during the initial stage of the degradation process, the degree of crystallinity is then seen to increase, up to 60% after 400 days, reaching a peak value of 87% after 580 days. This large increase of the crystallinity of the system can explain the decrease in the degradation rate observed, due to the increasingly tightly packed polymeric chains, consistent with previous data presented.

The quenched P(L)LA103A sample follows an essentially similar pattern, but displaying an initial larger increase in the degradation rate, followed by a more pronounced decrease. As shown in Fig. 4, the quenched polymer remained fully amorphous for the first 200 days, displaying extremely low χ values (around 11%) even after 270 days degradation. At this stage the degree of crystallinity of the polymer started to increase substantially, consistent with the sharp decrease in the rate of degradation seen after 280 days immersion in the Ringer solution.

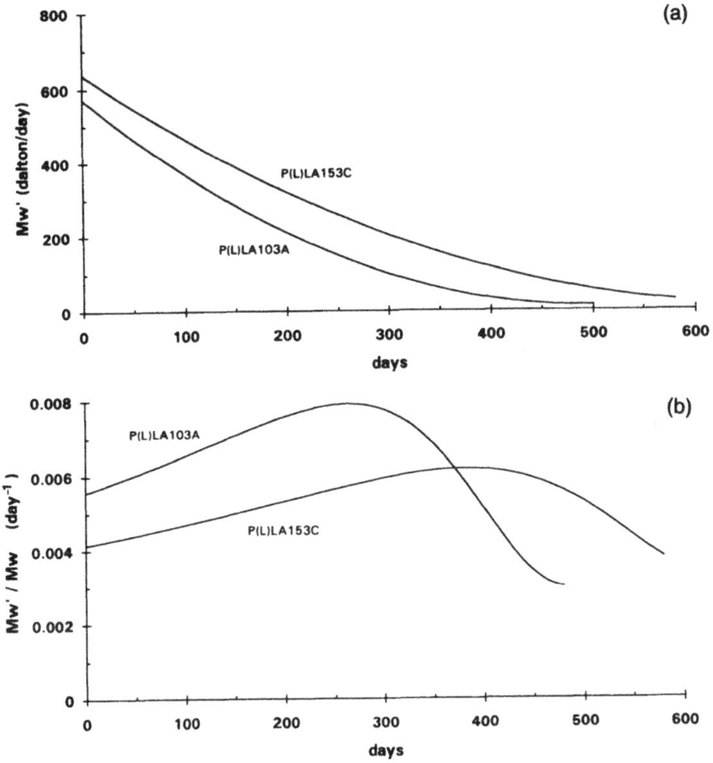
(a)
800
600
400
200
0
Mw' (dalton/day)
P(L)LA153C
P(L)LA103A
0
100
200
300
400
500
600
days
(b)
0.008
0.006
0.004
0.002
0
Mw' / Mw (day⁻¹)
P(L)LA103A
P(L)LA153C
0
100
200
300
400
500
600
days

Figure 5. (a) Time derivative (absolute value) of the best fit molecular weight versus degradation time function and (b) degradation rate (see text), versus degradation time, for P(L)LA103A and P(L)LA153C.

solution. Contrasting with the behavior displayed by the annealed P(L)LA153C, which showed very little mass loss, even after 580 days, the quenched P(L)LA103A sample lost 15% of its initial mass, at the same point in time, starting the mass loss process already after 280 days degradation. This considerable mass loss could also contribute to the sharp decrease observed in the degradation rate. Since the material leached into the Ringer solution represents the low molecular weight fragments generated by the polymer hydrolysis, the steadily decreasing molecular weights measured are relatively higher, resulting therefore, in artifactually lower degradation rates. It is also worth noting that, as one would expect, the degradation rate values obtained for the quenched P(L)LA103A are higher than those of the annealed P(L)LA153C. This can be illustrated by the peak values of both polymers, the former attaining a relatively high 0.0079 days^{-1} value, while the latter reached peak value of 0.0062 days^{-1}.

When the behavior of P(L)LA and P(DL)LA samples having similar molecular weights is compared (P(L)LA177A and P(DL)LA159), the effect of the crystallizability of the polymer in determining the degradation dependent properties of the polymer becomes apparent. While 173 days were enough to cause an 88% decrease in the initial molecular weight of the P(DL)LA159 polymer, 479 days were required to reduced the initial molecular weight of the P(L)LA177A, by the same percentage. These data clearly demonstrate the dramatic effect of crystallinity and enantiomeric purity on the degradation kinetics.

Table 3.
Molecular weight (M_w) and degradation rate (M_w'/M_w) versus degradation time for P(L)LA177A and P(L)LA27A.

	P(L)LA177A			P(L)LA27A	
Time (days)	M_w (dalton)	M_w'/M_w (day^{-1})	Time (days)	M_w (dalton)	M_w'/M_w (day^{-1})
0	177 000	0.005 55	0	27 000	0.005 79
17	162 500	0.005 68	—	—	—
24	150 300	0.005 74	21	23 000	0.005 99
32	139 000	0.005 80	28	22 000	0.006 06
47	131 500	0.005 92	46	—	0.006 24
63	121 400	0.006 04	60	17 500	0.006 39
95	94 000	0.006 29	98	15 000	0.006 8
173	62 800	0.006 70	175	—	0.007 61
327	32 000	0.004 68	321	3 500	0.006 92
446	21 300	0.001 26	440	—	0.002 39
479	18 500	0.001 12	473	1 500	0.002 09
593	12 400	—	—	—	—

The effect of molecular weight on the degradation behavior of P(L)LA

Table 3 presents the gradual molecular weight decrease of P(L)LA177A and P(L)LA27A, as a function of immersion time in a Ringer solution. The degradation rate data shown reveal that these samples follow the same basic pattern already described in the previous section, where the degradation rate shows a mild increase, followed by a more pronounced reduction.

One of the most important properties of biodegradable polymers which determine their possible clinical use, is the mode in which their mechanical properties change as a function of degradation. As shown in Fig. 6, P(L)LA177A and P(L)LA27A differ significantly, the higher molecular weight polymer retaining its mechanical properties for a longer period. It is apparent from Fig. 6, that the initial molecular weight plays a central role in determining the strength retention profile of the

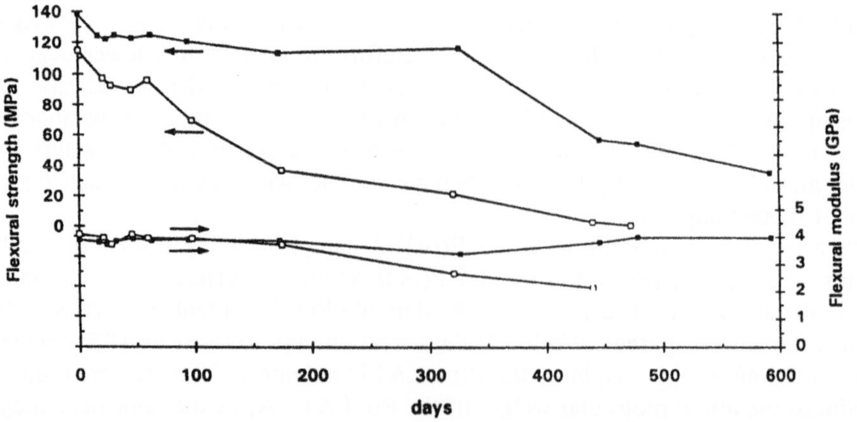
Flexural strength (MPa)
140
120
100
80
60
40
20
0
Flexural modulus (GPa)
5
4
3
2
1
0
0
100
200
300
400
500
600
days

Figure 6. Flexural strength and flexural modulus of P(L)LA177C (■) and P(L)LA27A (□) as function of degradation time in Ringer solution at 37°C.

polymer. P(L)LA177A retained most of its initial flexural strength during the first year of immersion in the Ringer solution (approximately 85% of prior to degradation value for 327 days), a substantial decrease being measured only after 446 days. The fact that after 327 days, the average molecular weight of P(L)LA177A was still relatively high (32 000), may explain this behavior. In striking contrast, P(L)LA27A showed a rapid and sharp decrease, losing more than 80% of its initial strength after the same degradation period (at this stage M_w = 3500). Furthermore, the stiffness of P(L)LA177A was not affected, as revealed by the fairly constant values displayed by the flexural modulus throughout the 600 days of the study. Once again in clear contrast, P(L)LA27A showed a significant loss in stiffness already after 170 days, the modulus dropping from its initial value of 4.1 to 2.7 and 2.1 GPa, after 321 and 440 days, respectively.

(a)

(b)

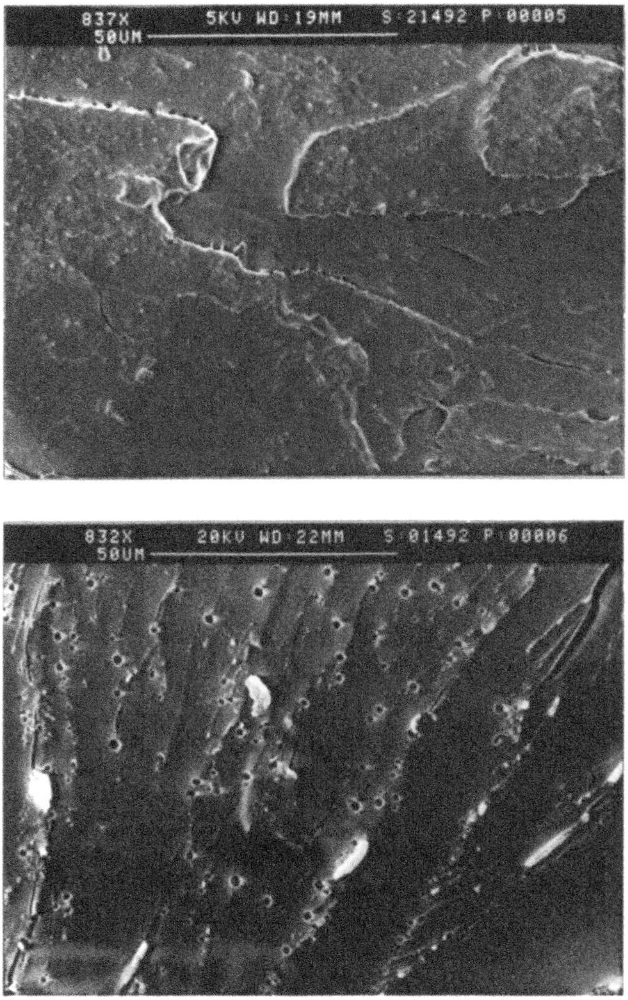
837X
5KU WD 19MM
S 21492 P 00005
50UM
832X
20KU WD 22MM
S 01492 P 00006
50UM

Figure 7. Scanning electron micrographs of P(L)LA27A after (a) 98 and (b) 321 days degradation in Ringer solution at 37°C.

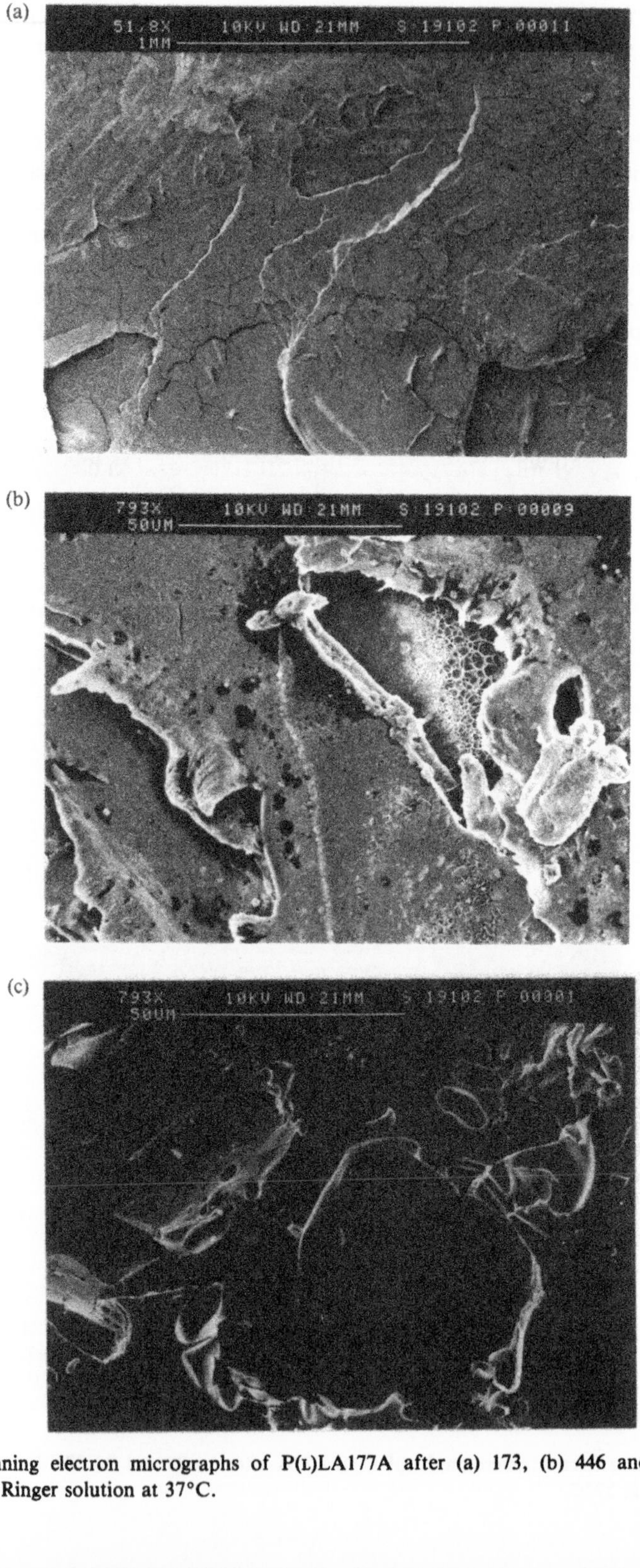
(a)
51.8X 10KV WD 21MM S 19102 P 00011
1MM
(b)
793X 10KV WD 21MM S 19102 P 00009
50UM
(c)
793X 10KV WD 21MM S 19102 P 00001
50UM

Figure 8. Scanning electron micrographs of P(L)LA177A after (a) 173, (b) 446 and (c) 597 days degradation in Ringer solution at 37°C.

Scanning electron micrographs of the fracture surface of P(L)LA27A samples after 98 and 321 days degradation are presented in Fig. 7(a) and (b), respectively. While the surface of the former shows no special features, the development of cracks and numerous microvoids 2–3 microns in diameter are apparent in the latter. These findings are consistent with the molecular weight of the two samples. A considerable 15 000 molecular weight value was measured for P(L)LA27A after 98 days, while a very low molecular weight was found for the polymer after 321 days degradation (M_w = 3500) which also comprised fragments able to be solubilized into the aqueous medium. The mechanical data, which reveal a large difference in both stiffness (3.7 and 2.5 GPa, after 98 and 321 days, respectively) and ultimate strength (37.4 and 22.4 MPa, for the two samples, accordingly) are also in accordance with the SEM findings.

The fracture surface developed by P(L)LA177A as a function of degradation is illustrated in Fig. 8 (a)–(c), for samples degrading for 173, 446, and 597 days, respectively. Even though the molecular weight of the polymer decreased to 63 000 after 173 days degradation, most of its initial mechanical properties were unaffected, displaying a modulus of 3.9 GPa and an ultimate strength of 114.0 MPa. Also after degrading for 446 days, when the polymer molecular weight was no more than 20 800, the stiffness of the sample did not change, even though a considerable drop in strength, down to 58.4 MPa, was measured. In the fracture surface, some limited areas could be found where degradation generated sponge-like localized structures, as shown in Fig. 8(b). After 597 days immersion in Ringer solution, while still retaining overall mass integrity, a few large voids could be seen in the fracture surface of P(L)LA177A presented in Fig. 8(c). In addition to these voids approximately 100 μm large, numerous small microvoids were apparent, not only on the surface for they also seem to be penetrating into the bulk of the sample. Even at this advanced stage of the degradation process (M_w = 12 400), P(L)LA177A samples retained mass integrity and unaffected stiffness (E = 4.0 GPa). Expectedly, though, the ultimate strength showed a pronounced decrease, to a low 37 MPa value, mainly due to the stress concentrating effect of the voids generated in the degrading matrix.

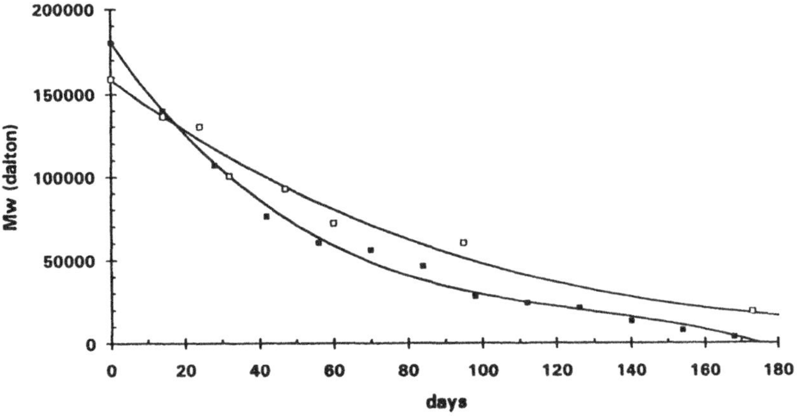
200000
150000
100000
50000
0
Mw (dalton)
0
20
40
60
80
100
120
140
160
180
days

Figure 9. Molecular weight profile during degradation in Ringer solution at 37°C of a flat P(DL)LA180 (■) and a cylindrical P(DL)LA159 (□) sample. The lines represent the best fit functions.

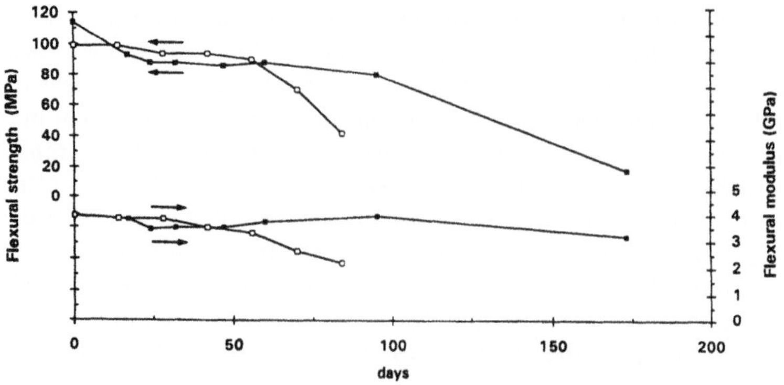
Flexural strength (MPa)
Flexural modulus (GPa)
days
120
100
80
60
40
20
0
5
4
3
2
1
0
0
50
100
150
200

Figure 10. Flexural strength and flexural modulus of a flat P(DL)LA180 sample (□) and a cylindrical P(DL)LA159 sample (■) as function of degradation time in Ringer solution at 37°C.

The effect of specimen geometry on the degradation process of PLA

Since transport phenomena largely affect the degradation process, the geometry of the specimen is seen as a major factor influencing the behavior of the degrading system. The effect of the sample geometry was investigated by comparing two P(DL)LA polymers having similar molecular weights (180 000 and 159 000) but markedly differing in the A/V ratio, namely, a P(DL)LA180 having a slab geometry with an A/V ratio of 1.57, and a P(DL)LA159 in the shape of a cylinder having an A/V ratio of 1.36.

It is apparent from the data plotted in Fig. 9, that the P(DL)LA180 flat specimen exhibits a faster molecular weight decrease. The higher A/V ratio of this sample results in a relative shorter diffusion path of water molecules into the bulk of the specimen, allowing also a more efficient extraction of the degraded fragments into the Ringer solution.

Figure 10 presents the mechanical properties of both P(DL)LA samples, as a function of time. After an initial period during which both specimens exhibited rather constant ultimate strength and modulus values, the flat specimen displayed a substantially faster loss of mechanical properties. Contrasting with the sharp reduction in both ultimate strength and modulus, shown by the flat specimen after approximately two months, the cylindrical sample exhibited a slower mechanical deterioration profile, following a much slower path.

CONCLUSIONS

The present study focused on the effect of molecular weight, crystallinity and enantiomeric structure on the degradation of lactic acid polymers. In agreement with previous findings reported in the literature, crystallization of P(L)LA polymers during degradation has been observed. Furthermore, the crystallinity which develops as degradation proceeds counterbalances and often even overcomes the initial faster degradation of initially less crystalline polymers.

If the molecular weight degradation is expressed in terms of degradation rate, as defined elsewhere in this paper, some noteworthy differences in the degradation pattern of the different investigated polymers can be observed. The data presented reveal a gradually increasing degradation rate, which is followed by a maximum and by a decrease of rate as degradation time proceeds further. Depending on the morphology of the polymer, such a behavior has been explained considering the effect of crystallization and of the loss of the low molecular weight polymer fragments, both resulting in a real or apparent decrease of the degradation rate.

REFERENCES

1. S. Bowald, C. Busch and I. Eriksson, *Surgery* **86,** 722 (1979).
2. S. Bowald, C. Busch, I. Eriksson and T. Aberg, *Scand. J. Thorac. Cardiovasc. Surg.* **9A,** 91 (1981).
3. E. Lommen, S. Gogolewski, A. J. Pennings, C. R. H. Wilderuur and P. Nienwenhuis, *Trans. ASAIO* **29,** 255 (1983).
4. D. K. Gilding, in: *Biocompatibility of Clinical Implant Materials*, p. 209, vol II, D. F. Williams (Ed.). CRC Press, Boca Raton, FL (1981).
5. D. F. Williams, *J. Mater. Sci.* **17,** 1233 (1982).
6. R. Duvivier, *Bull. NY Acad. Med.* **61,** 621 (1985).
7. T. H. Barrows, *Clin. Mater.* **1,** 233–257 (1986).
8. S. Vainionpaa, P. Rokkanen and P. Tormala, *Prog. Polym. Sci.* **14,** 679 (1989).
9. L. Claes, in: *Clinical Implant Materials, Advances in Biomaterials Vol 9*, p. 161, G. Heimke, U. Soltesz and A. J. C. Lee (Eds). Elsevier, Amsterdam (1990).
10. K. M. Rehm, in: *Degradation Phenomena on Polymeric Biomaterials*, p. 163, H. Planck, M. Dauner and M. Renardy (Eds). Springer-Verlag, Berlin, Heidelberg (1992).
11. D. E. Cutright and R. L. Reid, *Hand* **7,** 228 (1975).
12. J. Feijen, in: *Polymeric Biomaterials*, p. 62, E. Piskin and A. S. Hoffmann (Eds). Martinus Nijhoff Publ., Dordrecht, The Netherlands (1986).
13. T. H. Tice and D. R. Cowsar, *Pharm. Technol.*, Nov., 26 (1984).
14. E. H. Schacht, *Med. Dev. Techn.* **1,** 15 (1990).
15. D. H. Lewis, in: *Biodegradable Polymers as Drug Delivery Systems*, M. Chasin and R. Langer (Eds), Marcel Dekker, Inc., New York (1990).
16. L. Getter, D. E. Cutright, S. N. Bhaskar and J. K. Augsburg, *J. Oral Surg.* **30,** 344 (1972).
17. M. Vert and F. Chabot, *Makromol. Chem. Suppl.* **5,** (1981).
18. M. Vert, P. Christel, H. Garreau, M. Audion, M. Chavanaz and F. Chabot, in: *Polymers in Medicine II: Biomedical and Pharmaceutical Applications*, p. 263, E. Chiellini, P. Giusti, C. Migliaresi and L. Nicolais (Eds). Plenum Press, New York (1986).
19. J. W. Leenslag, A. J. Pennings, R. R. M. Bos, F. R. Rozema and G. Boering, *Biomaterials* **8,** 70 (1987).
20. D. C. Tunc, M. W. Rohovsky, B. Jadhav, W. B. Lehman, A. Strongwater and F. Kummer, in: *Advances in Biomedical Polymers; Polymer Science Technology Vol. 35*, p. 87, C. G. Gebelein (Ed.). Plenum Press, New York (1987).
21. R. R. M. Bos, F. R. Rozema, G. Boering, A. J. Nijenhuis, A. J. Pennings and H. W. B. Jansen, *Br. J. Oral Maxillofac. Surg.* **27,** 467 (1989).
22. D. G. Tunc, *Clin. Mater.* **8,** 119 (1991).
23. A. Maiola, S. Vainionpaa, P. Rokkanen, H. M. Mikkola and P. Tormala, *J. Mater. Sci., Mater. Med.* **3,** 43 (1992).
24. R. Suuronen, L. Wessman, M. Mero, P. Tormala, J. Vasenius, E. Partio, K. Vihtonen, S. Vainionpaa. *J. Mater. Sci., Mater. Med.* **3,** 288 (1992).
25. R. A. Olson, D. L. Roberts and D. B. Osbon, *Oral. Surg.* **53,** 441 (1982).
26. B. Van der Lei, H. L. Bartek, P. Nieuwenhuis and C. R. H. Wildevuur, *Surgery* **98,** 955 (1985).
27. J. W. Leenslag, M. T. Kroes, A. J. Pennings and B. Van der Lei, *New Polym. Mater.* **1,** 111 (1988).
28. C. M. Agrawal, K. F. Haas, D. A. Leopold and H. Clark, *Biomaterials* **13,** 176 (1992).
29. S. Yolles, US Patent 3 887 699 (1975).
30. T. M. Jakanicz, H. A. Nash, D. L. Wise and J. B. Gregory, *Contraception* **8,** 227 (1973).
31. L. R. Beck, D. R. Cowsar, D. H. Lewis, R. J. Coxgrove, C. T. Riddle, S. L. Lowry and T. Epperly, *Fert. Ster.* **31,** 545 (1979).
32. J. P. Kitchell and D. L. Wise, *Methods Enzymol.* **112,** 436 (1985).
33. K. Juni, J. Ogata, M. Nakano, T. Ichihara, K. Mori and M. Akagi, *Chem. Pharm. Bull.* **33,** 313 (1985).

34. R. L. Dunn, J. P. English, J. P. Strobel, D. R. Cowsar and T. R. Tice, *Polymers in Medicine III*, p. 149, C. Migliaresi *et al.* (Eds). Elsevier, Amsterdam (1988).
35. N. Leelarusamee, S. A. Howard, C. J. Malango and J. K. Ma, *J. Microencapsul.* **147,** 5 (1988).
36. D. H. Lewis, in: *Biodegradable Polymers as Drug Delivery Systems*, p. 1, M. Chasin and R. Langer (Eds). Marcel Dekker, Inc., New York (1990).
37. R. K. Kulkarni, E. G. Moore, A. F. Hegyeli and F. Leonard, *J. Biomed. Mater. Res.* **5,** 169 (1971).
38. V. W. Dittrich and R. C. Schulz, *Angew. Makromol. Chem.* **109,** 15 (1971).
39. T. A. Augurt, M. N. Rosenaft and V. A. Perciaccante, US Patent 4 033 038 (1977).
40. D. K. Gilding and A. M. Reed, *Polymer* **20,** 1459 (1979).
41. F. E. Kohn, J. W. A. van den Berg, G. van Ridden and J. Feijen, *J. Appl. Polym. Sci.* **29,** 4265 (1984).
42. H. R. Kricheldorf and A. Serra, *Polym. Bull.* **14,** 497 (1985).
43. D. S. Kaplan and R. R. Muth, US Patent 4 523 591 (1985).
44. H. R. Kricheldorf and R. Dunsing, *Makromol. Chemie* **187,** 1611 (1986).
45. J. W. Leenslag and A. J. Pennings, *Makromol. Chemie* **188,** 1809 (1987).
46. A. Duda and S. Penczek, *Macromolecules* **23,** 163 (1991).
47. Z. Jedlinski, W. Watach, P. Kurcok and G. Adamus, *Makromol. Chem.* **192,** 2051 (1991).
48. A. Schindler and D. Harper, *J. Polym. Sci., Polym. Chem. Ed.* **17,** 2593 (1979).
49. B. Kalbe and A. J. Pennings, *Polymer* **21,** 607 (1980).
50. R. Vasanthakumari and A. J. Pennings, *Polymer* **24,** 175 (1983).
51. K. Kishore and R. Vasanthakumari, *J. Polym. Sci., Polym. Phys. Ed.* **22,** 537 (1984).
52. W. Hoogsteen, A. R. Postema, A. J. Pennings, G. ten Brinke and P. Zugenmaier, *Macromolecules* **23,** 632 (1990).
53. A. Marigo, C. Marega, G. Paganetto and R. Zannetti, *Makromol. Chem.*, in press.
54. S. Mazzullo, G. Paganetto and A. Celli, *Progr. Coll. Polym. Sci.* **87,** 32 (1992).
55. K. Jamshidi, S. H. Hyon and Y. Ikada, *Polymer* **29,** 2229 (1988).
56. C. Migliaresi, A. De Lollis, L. Fambri and D. Cohn, *Clin. Mater.* **8,** 111 (1991).
57. A. Celli and M. Scandola, *Polymer* **33,** 2699 (1992).
58. C. Migliaresi, D. Cohn, A. De Lollis and L. Fambri, *J. Appl. Polym. Sci.* **43,** 83 (1991).
59. J. W. Leenslag, S. Gogolewski and A. J. Pennings, *J. Appl. Polym. Sci.* **29,** 2829 (1984).
60. I. Engelberg and J. Kohn, *Biomaterials* **12,** 292 (1991).
61. S. Li, H. Garreau and M. Vert, *J. Mater. Sci., Mater. Med.* **1,** 123 (1990).
62. S. Li, H. Garreau and M. Vert, *J. Mater. Sci., Mater. Med.* **1,** 131 (1990).
63. S. Li, H. Garreau and M. Vert, *J. Mater. Sci., Mater. Med.* **1,** 198 (1990).
64. B. Buchholz, *Degradation Phenomena on Polymeric Biomaterials*, p. 67, H. Planck, M. Dauner and M. Renardy (Eds). Springer-Verlag, Berlin, Heidelberg (1992).
65. M. Dauner, E. Muller, B. Wagner and H. Planck, in: *Degradation Phenomena on Polymeric Biomaterials*, p. 107, H. Planck, M. Dauner and M. Renardy (Eds). Springer-Verlag, Berlin, Heidelberg (1992).
66. C. G. Pitt, M. M. Gratzel, G. L. Kimmel, J. Surles and A. Schindler, *Biomaterials* **2,** 215 (1981).
67. A. S. Chawla and T. M. S. Chang, *Biomater. Med. Dev. Art. Org.* **13,** 153 (1985).
68. K. L. Gerlach and J. Eitenmuller, in: *Biomaterials and Clinical Applications*, p. 439, A. Pizzoferrato, P. G. Marchetti, A. Ravaglioli and A. J. C. Lee (Eds). Elsevier, Amsterdam (1987).
69. T. Nakamura, S. Hitomi, S. Watanabe, Y. Shimizu, S. H. Hyon and Y. Ikada, *J. Biomed. Mater. Res.* **23,** 1115 (1989).
70. F. R. Rozema, R. R. M. Bos, G. Boering, A. J. Nijenhuis, A. J. Pennings, H. W. B. Jansen and W. C. de Bruijin, in: *Degradation Phenomena on Polymeric Biomaterials*, p. 123, H. Planck, M. Dauner and M. Renardy (Eds). Springer-Verlag, Berlin, Heidelberg (1992).

Poly-DL-lactic acid: Polyethylene glycol block copolymers. The influence of polyethylene glycol on the degradation of poly-DL-lactic acid

S. S. SHAH, K. J. ZHU and C. G. PITT*

Amgen Inc., Mail Stop 8-1-A-215, 1840 Dehavilland Drive, Thousand Oaks, CA 91320, USA

Received 23 October 1992; accepted 19 January 1993

Abstract—ABA block copolymers of polyethylene glycol and poly-DL-lactic acid were prepared by ring-opening polymerization of DL-dilactide with α,ω-dihydroxy polyethylene glycol, M_n 1000 or 2000. The morphology of the resulting copolymers, with PEG:PLA ratios(mol/mol) of 1:2, 1:3 and 1:4, was characterized by DSC and ESR spectroscopy. The rate of water uptake was biphasic, reflecting the contribution of two processes: rapid diffusion of water into the initially miscible PEG and PLA blocks; then a slower rate of hydration possibly due to phase separation and hydrolytic cleavage of the PLA blocks. The rate of hydrolytic degradation of the block copolymers in DI water at 37°C was measured by two methods: weight loss and colorimetric analysis of the carboxy end group concentration resulting from chain scission of PLA blocks. As a result of phase separation, the rate of scission of PLA blocks in the copolymers was similar to that of the PLA homopolymer. The more rapid onset of weight loss of the copolymers, relative to PLA, is attributed to the greater water solubility of PEG–PLA oligomers and their greater diffusivity in the more highly hydrated copolymers.

Key words: Poly-DL-lactic acid; polyethylene glycol; block copolymers; morphology; hydration; chain scission.

INTRODUCTION

A number of the important properties of hydrogels used for biomedical purposes, including permeability to solutes, biocompatability, and rates of enzymatic or hydrolytic degradation, can be directly related to their water content [1]. For example, free volume theory has been used to establish a quantitative relationship between the water content of hydrogels and their permeability to a wide range of hydrophobic and hydrophilic drugs [1, 2]. Plasticization by water is also known to be important. It has been shown that absorption of water is responsible for large reductions in the glass transition temperature of hydrophilic polymers e.g., PVP, nylons, PEG, and PHEMA [3–7]. Even with more hydrophobic materials e.g., poly(glycolic acid-co-DL-lactic acid), absorption of water can convert the polymer from the glassy to the rubbery state [8]. The role of water in determining the rate of hydrolytic degradation of polymers is less well understood. Recent work has demonstrated that an increase in the water content can result in an exponential increase in the rate of hydrolytic chain scission of poly(glycolic acid-co-DL-lactic acid) in miscible blends with polyvinyl alcohol; comparable increases in hydrolysis rates are not observed with phase-separated blends [9]. We have now extended these correlations of hydrolysis rates versus water content to ABA block copolymers of poly-DL-

* To whom correspondence should be addressed.

lactic acid (PLA) and polyethylene glycol (PEG). Here the length of the hydrophilic PEG block and the PEG/PLA ratio provide means of varying the water content of the copolymers and studying its effect on the rate of ester hydrolysis of PLA blocks. In the following text, the compositions of the copolymers are abbreviated to n : m : n, where n and m refer to the number average molecular weights ($\times 10^3$) of the PLA and PEG blocks, respectively.

METHODS

D,L-Dilactide (Boehringer Ingelheim) was purified by recrystallization from ethyl acetate. PEG was dried *in vacuo* (0.1 Torr) at 100°C for 24 h. PLA was prepared by heating D,L-dilactide (0.25 mol) in an evacuated glass tube at 150°C for 16–18 h in the presence of 1,6-hexanediol (3.11 mmol) and stannous 2-ethylhexanoate (0.25 wt%). The resulting polymer was dissolved in chloroform, precipitated from methanol and dried at 60°C *in vacuo* for 2 days. Block copolymers of PEG and PLA were prepared by the same procedure, substituting PEG (M_n 1000 and 2000) for 1,6-hexanediol as the initiator. Each copolymer was extracted with a mixture of ethyl acetate and *n*-heptane (2:1 v/v) to remove PLA homopolymer, then water to remove PEG, before drying at 60°C *in vacuo* for 2 days. At this time there was no evidence of a DSC endotherm associated with residual solvent and DSC transitions were reproducible. Copolymer films, thickness 0.6–0.8 mm, were obtained by melt compression at 80–100°C. PLA films were obtained by casting a methylene chloride solution on to a glass plate, followed by melt compression at 140°C.

Thermal transitions of dry samples were measured using a differential scanning calorimeter (DSC, Dupont Model 910). Each sample was heated under nitrogen to 120°C and held at that temperature for 5 min to ensure removal of residual moisture and solvent. The annealed sample was then cooled to −100°C before heating to 200°C at a rate of 10°C/min. The T_g was obtained from the second heating cycle. Thermal transitions of hydrated samples were measured using a Microcal, MC-2, DSC instrument. Weighed samples were hydrated in DI water, introduced into the solid insert probe, cooled to 5°C, and then heated to 60°C at a rate of 1.5°C/min. ^{1}H-NMR spectra were measured in $CDCl_3$ using a GE 300 spectrometer operating at 300 MHz; the copolymer compositions were determined by comparing the integrations of the methylene protons of polyethylene glycol (3.66 ppm) and the methine proton of polylactide blocks (5.20 ppm). The carboxyl end group concentrations of partly hydrolyzed copolymers were determined by the colorimetric method of Palit and Ghosh [10]; UV-Vis spectra were obtained with a Hewlett-Packard 9000 spectrometer. The water content and weight loss of polymers were measured after immersion of the polymer films (200–230 mg) in DI water at 37°C. Samples were recovered at the indicated time intervals, and water uptake and weight loss were measured gravimetrically after drying the film to constant weight *in vacuo* at room temperature for 2 days, then at 40°C for 1 day.

RESULTS

Polymer synthesis and characterization

ABA block copolymers of PLA and PEG were prepared by stannous octoate catalyzed ring-opening polymerization of DL-dilactide using α, ω-dihydroxy-terminated

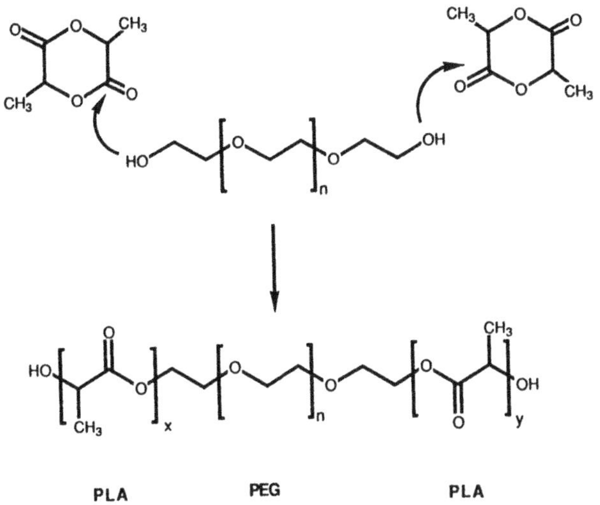

Figure 1. Synthesis of poly-DL-lactic acid-b-polyethylene glycol-b-poly-DL-lactic acid by ring opening polymerization of dilactide with polyethylene glycol.

PEG as the initiator (Fig. 1) [11–14]. The mole ratio of PEG (B block) to DL-dilactide was used to control the lengths of the PLA (A) blocks. Two PEG initiators $M_n = 1000$ and 2000, $M_w/M_n = 1.09$, were utilized and the ratios of the A:B block lengths were changed systematically to obtain a series of polymers with increasing PLA contents and hydrophobicities; PEG:PLA = 1:2, 1:3, and 1:4. The relative A and B block lengths of the six polymers were confirmed by ^{1}H-NMR spectroscopy (Table 1). There was good agreement between the observed PLA block lengths and the values calculated from dilactide:PEG ratio assuming each lactide chain is initiated by a terminal PEG hydroxyl group. A low intensity multiplet at 4.3 ppm, assigned to the $-CH_2CH_2OC(=O)-$ part structure, confirmed the presence of a covalent link between the PEG and PLA blocks.

Water uptake by films of the block copolymers was measured as a function of time in DI water at 37°C (Fig. 2A–C). The rate of water uptake was biphasic and during the first 24 h was dependent on both the PEG block length and the PEG:PLA ratio (Fig. 2A). There was no obvious relationship between the rate

Table 1.
Calculated and experimentally observed copolymer compositions, experimental and calculated T_g of the unhydrated copolymers and the extent of phase separation of PLA blocks after hydration determined by DSC

Composition (Calcd.)	Composition (^{1}H-NMR)	T_g (DSC)/dry	T_g (Fox Eq.)	PLA phase separation[a]
2000:1000:2000	2000:1000:2000	24.5	17.8	62%
1500:1000:1500	1650:1000:1650	19.6	14.2	40%
1000:1000:1000	1100:1000:1100	8.4	5.9	34%
4000:2000:4000	4120:2000:4120	21.2	18.3	67%
3000:2000:3000	3100:2000:3100	13.7	13.0	69%
2000:2000:2000	2100:2000:2100	3.0	4.8	73%

[a] DSC measured after 1 day (PEG 2000 series) or 5 days (PEG 1000 series).

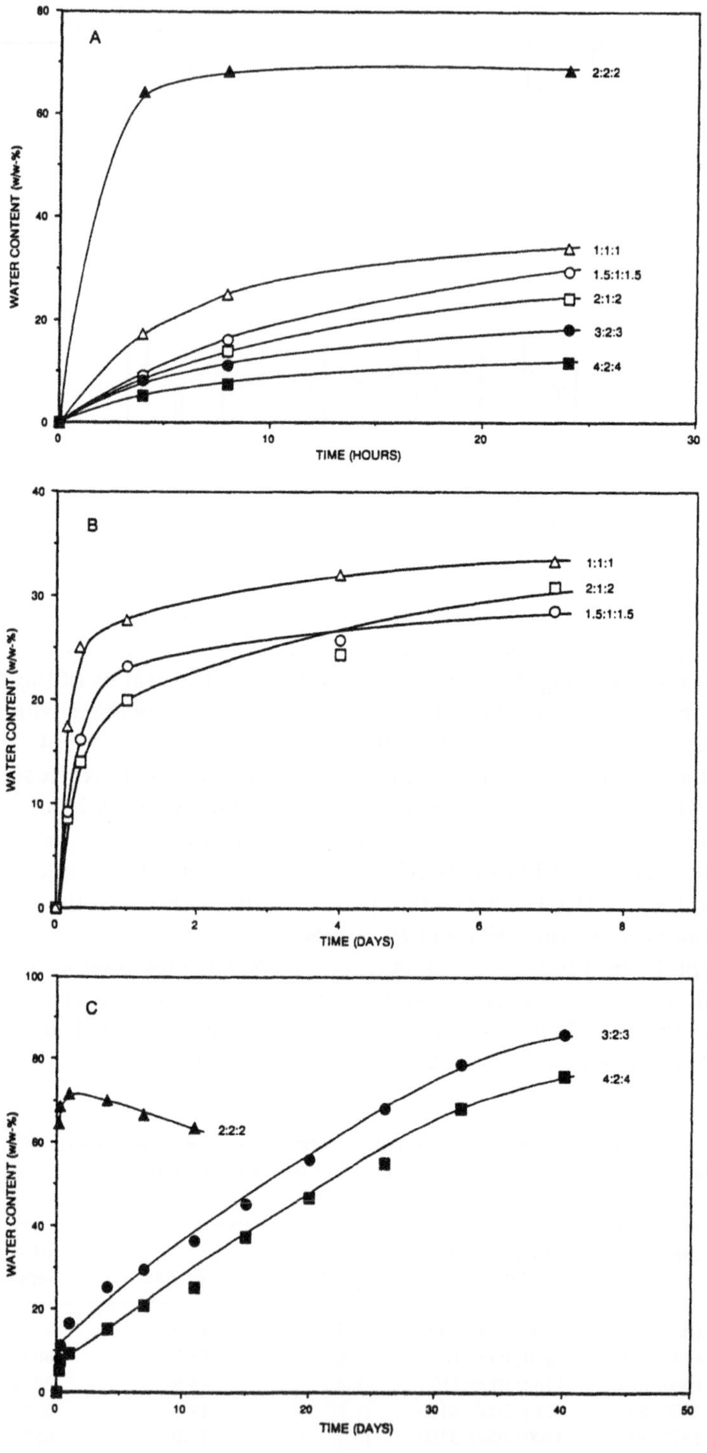
A
2:2:2
1:1:1
1.5:1:1.5
2:1:2
3:2:3
4:2:4
WATER CONTENT (w/w-%)
TIME (HOURS)
B
1:1:1
2:1:2
1.5:1:1.5
WATER CONTENT (w/w-%)
TIME (DAYS)
C
3:2:3
4:2:4
2:2:2
WATER CONTENT (w/w-%)
TIME (DAYS)

Figure 2. Rates of hydration of all PLA–PEG block copolymers during the first 24 h (A) and copolymers of PEG 1000 (B) and copolymers of PEG 2000 (C) after longer time periods.

or magnitude of water uptake and the copolymer composition. The initial water uptake was greatest and most rapid for the 2:2:2 copolymer derived from PEG 2000. Increasing the PLA:PEG ratio, but maintaining the same PEG 2000 center block, resulted in a substantial decrease in both the rate and extent of the initial hydration. During the subsequent 40 days, the water content of the copolymers with a PEG 2000 center block increased almost linearly to approximately 80%. The copolymers derived from PEG 1000 differed little in their initial degree of hydration, which decreased slightly with the increase in hydrophobicity (PLA:PEG ratio). All of the copolymers were hydrated to a greater degree than the homopolymer, PLA, which absorbed only 4.7% water over a 40 day period.

Phase separation of the PEG and PLA blocks of the copolymers was evaluated by DSC. In the dry state, single glass transition temperatures (T_g) were observed. The experimental values were in good agreement with those calculated from the weighted averages of the T_gs of the parent homopolymers (PLA 42°C, PEG −50°C) using the Fox equation (1):

$$(1/T_g)_{AB} = (\mathrm{Wt}/T_g)_A + (\mathrm{Wt}/T_g)_B. \tag{1}$$

This observation of a single T_g, predicted by the Fox equation, is consistent with the miscibility of the two phases in the dry state.

The films were optically transparent immediately after immersion in water. However, within 24 h, the films showed varying degrees of cloudiness attributable to light scattering resulting from phase separation of the PEG and PLA blocks. Phase separation was confirmed by DSC of the hydrated polymers (Fig: 3A–C). Upon hydration, the T_gs of the PEG 2000 copolymers increased from the values (3–25°C) listed in Table 1 for the dry polymers to 30°C, which was coincident with the T_g of a sample of PLA hydrated under the same conditions (Fig. 3B). An increase in the T_g cannot be explained by plasticization by water, which will generally cause a decrease in the T_g [3–7]. Assuming that the increase in the T_g is the result of phase separation of PLA blocks, the degree of separation may be calculated from the change in the specific heat of each copolymer at its T_g, relative to that of the PLA homopolymer. The magnitude of phase separation of each copolymer calculated by this method is listed in Table 1. The data indicate that substantial phase separation of PLA blocks had occurred.

The DSC scans of the hydrated PEG 1000 copolymers were more complex (Fig. 3A) and time dependent than those of the PEG 2000 copolymers. For example, 24 h after hydration of the 2:1:2 copolymer, no transition in the range of 10–50°C was evident (Fig. 3C). However, over a period of 5 days, a transition developed at a temperature slightly lower than that of hydrated PLA. The same behavior was observed for 1.5:1:1.5 and 1:1:1 copolymers. Evidently, phase separation of the PEG 1000 copolymers was a relatively slow process, occuring over a number of days and possibly facilitated by the hydrolytic chain scission of the PLA blocks.

Freezing of water prevented measurement of T_gs below 0°C, and it was not possible to determine whether the transition observed in the dry state associated with a miscible phase was still present to some degree, but shifted to a lower temperature by hydration.

The phase separation of two of the PEG–PLA copolymers has previously been studied by ESR spectroscopy [15]. The latter technique is based on the use of a spin

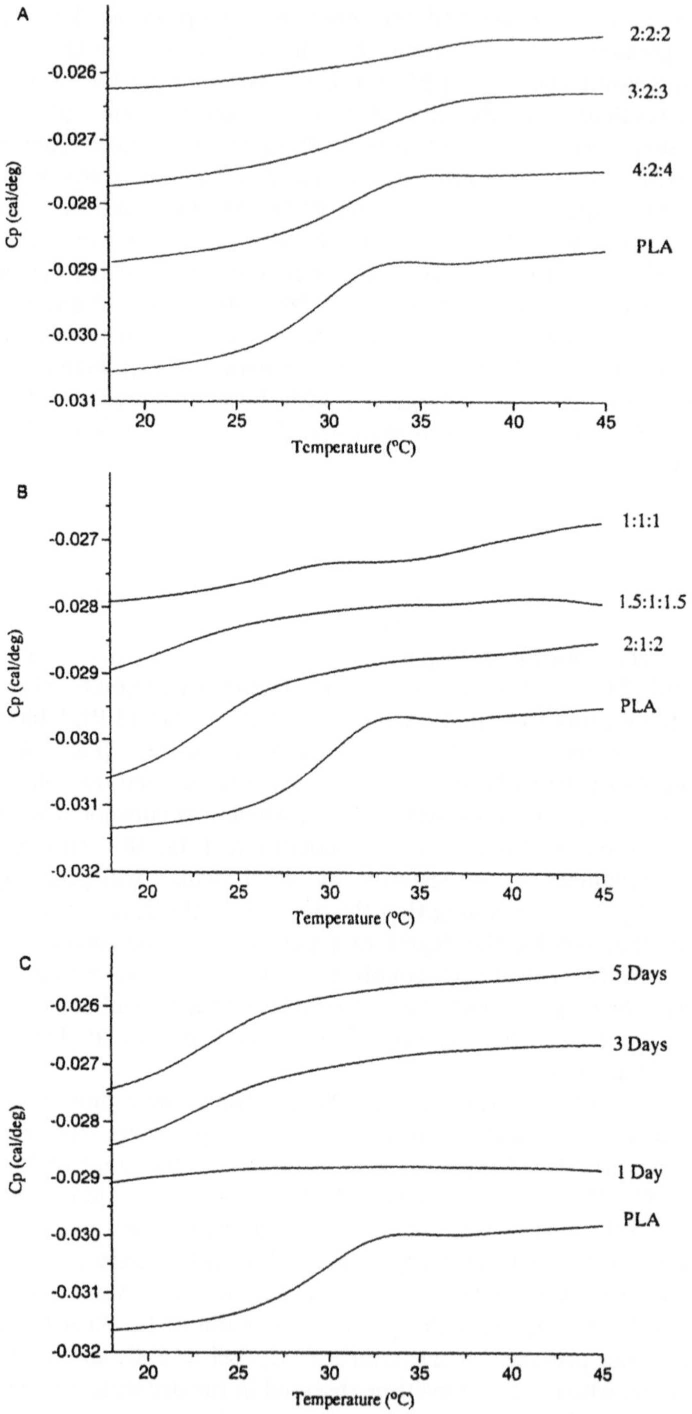
A
2:2:2
3:2:3
4:2:4
PLA
Cp (cal/deg)
-0.026
-0.027
-0.028
-0.029
-0.030
-0.031
20
25
30
35
40
45
Temperature (°C)
B
1:1:1
1.5:1:1.5
2:1:2
PLA
Cp (cal/deg)
-0.027
-0.028
-0.029
-0.030
-0.031
-0.032
20
25
30
35
40
45
Temperature (°C)
C
5 Days
3 Days
1 Day
PLA
Cp (cal/deg)
-0.026
-0.027
-0.028
-0.029
-0.030
-0.031
-0.032
20
25
30
35
40
45
Temperature (°C)

Figure 3. DSC scans of copolymers of PEG 2000 after hydration for 1 day (A), copolymers of PEG 1000 after hydration for 5 days (B), and changes in the DSC scans of the 2:1:2 copolymer after 1, 3 and 5 days (C).

probe, *N*-butylamino-TEMPO, the hydrophobicity of which may be controlled by changing the pH of the medium. Phase separation is indicated by a composite spectrum resulting from the protonated (hydrophilic) and non-protonated(hydrophobic) spin probe partitioning between the two phases; in particular, a sensitivity of the spectrum to pH changes is strong evidence of phase separation of PEG and PLA blocks. For the 2:2:2 copolymer, partial phase separation could be detected after hydration for 24 h, consistent with the DSC result. No phase separation was evident with the 1:1:1 copolymer after 24 h, again consistent with the DSC result at that time period. The signal-to-noise ratio of the spectrum of the 2:1:2 copolymer was too low to permit analysis.

Hydrolytic chain scission

The rate of hydrolysis of the block copolymers was measured in DI water at 37°C by two methods: weight loss and colorimetric analysis of the carboxy end group concentration resulting from cleavage of PLA blocks.

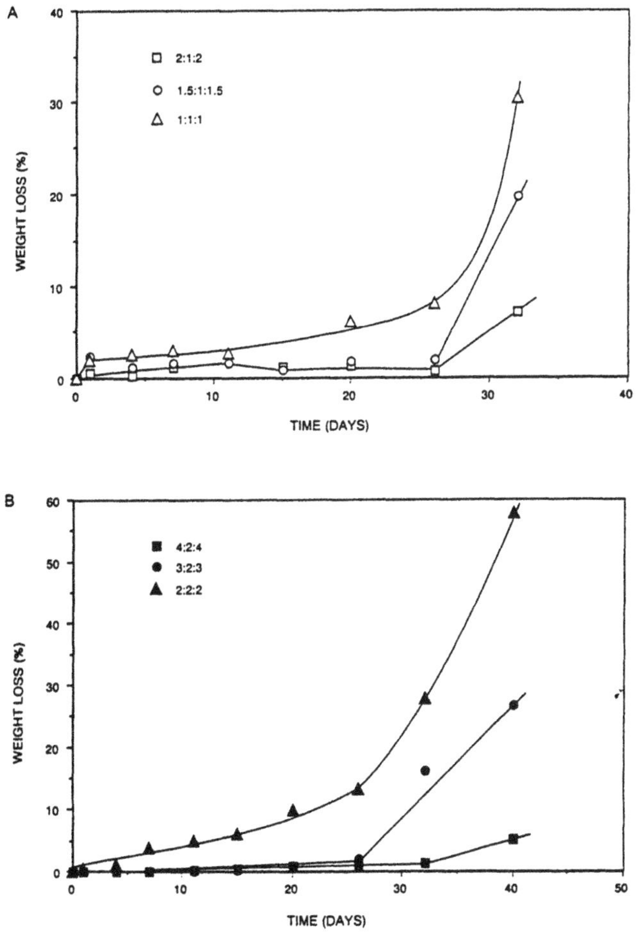
A
2:1:2
1.5:1:1.5
1:1:1
WEIGHT LOSS (%)
TIME (DAYS)
0
10
20
30
40
B
4:2:4
3:2:3
2:2:2
WEIGHT LOSS (%)
TIME (DAYS)
0
10
20
30
40
50
60

Figure 4. Weight loss of PLA-PEG block copolymers derived from PEG 1000(A) and PEG 2000(B) in water at 37°C.

Table 2.
Erosion and change in the composition of copolymers as a function of time in DI water at 37°C

Time (Days)	PLA:PEG Ratio	Weight Loss (w – %)	PLA:PEG Loss
1:1:1 Copolymer			
0	2.2:1	0	0
20	2.4:1	6	0.3:1
26	2.6:1	8	0.2:1
32	2.6:1	30	0.7:1
40	2.7:1	*a*	—
1.5:1:1.5 Copolymer			
0	3.3:1	0	0
26	3.3:1	2	1.7:1
32	3.3:1	20	1.1:1
40	4.4:1	*a*	—
2:1:2 Copolymer			
0	4.0:1	0	0
26	4.0:1	1	100% PLA
32	3.6:1	7	100% PLA
40	5.0:1	*a*	—
2:2:2 Copolymer			
0	2.2:1	0	0
26	4.1:1	13	100% PEG
32	5.7:1	28	0.2:1
40	6.9:1	58	0.6:1
3:2:3 Copolymer			
0	3.1:1	0	0
32	4.2:1	16	0.3:1
40	4.8:1	27	0.4:1
4:2:4 Copolymer			
0	4.1:1	0	0
32	4.7:1	1	100% PEG
40	5.1:1	5	0.1:1

[a] could not be quantitated because of difficulty of handling of copolymer

Table 3.
Rate constants of hydrolytic cleavage of poly-DL-lactide blocks

Copolymer composition (PLA:PEG:PLA)	Rate constant (k')	Normalized rate constant, k'/[Ester]
2000:1000:2000	8.03	10.04
1500:1000:1500	7.79	10.39
1000:1000:1000	6.45	9.68
4000:2000:4000	5.74	7.18
3000:2000:3000	5.82	7.76
2000:2000:2000	5.83	8.75
PLA	9.77	9.77

Weight loss of the polymer films was measured gravimetrically until the loss of mechanical strength of the films introduced handling errors. The 3:2:3 and 4:2:4 copolymers showed induction periods of 26 and 32 days, respectively, before significant weight loss was observed (Fig. 4B). In contrast, slow erosion of the more hydrophilic 2:2:2 copolymer was observed immediately, with a marked acceleration in the rate of 26 days. Essentially the same results were obtained with the block copolymers derived from PEG 1000, the 1:1:1 copolymer losing weight in a biphasic manner, and all of the copolymers showing a significant increase in weight loss at 26 days (Fig. 4A). In contrast, no loss in weight of a PLA sample was observed over a 40-day period under the same conditions.

^{1}H-NMR analysis was used to detect changes in the PEG:PLA ratio arising from selective erosion of the blocks. Calculation of the PLA:PEG ratio of the solubilized polymer from the weight and composition of the residual polymer showed that the initial weight loss was largely due to erosion of the PEG component (Table 2).

Molecular weights could not be determined by GPC because of the unknown hydrodynamic volumes and Mark–Houwink relationships of the copolymers and their changing compositions. Therefore, the extent of cleavage of the PLA block was determined by measurement of the carboxy end group concentration [COOH] in the polymer using the dye interaction technique of Palit and Ghosh [10]. This method has previously been applied successfully to measure polyester cleavage [16]. Data were collected for 26 days, after which time substantial weight loss of the PLA component of the copolymers was observed. The rates of chain scission were derived from the slopes of semilog plots of $[COOH]^{-1}$ vs time (Table 3). It is evident from Table 3 that the rates of chain scission of the different copolymers did not differ greatly with changes in composition, and were not significantly different from PLA homopolymer.

DISCUSSION

In principle, the water content of a polyester may contribute to the kinetics of hydrolysis by two mechanisms: indirectly by solvating the transition state and, more directly, as a concentration term in the kinetic rate law. There is good evidence that the bulk hydrolytic chain scission of PLA and related polyesters is autocatalyzed by the carboxylic acid end groups and follows the kinetic law (Eq. (3)),

$$-d[\mathrm{COOH}]/dt = k[\mathrm{H_2O}][\mathrm{Ester}][\mathrm{COOH}] \qquad (3)$$

where $[H_2O]$, [Ester], and [COOH] are the water, ester and carboxylic acid end group concentrations in the polymer bulk [17, 18]. For many systems, especially during the initial phase of hydrolysis, when the number of chain scissions is small, both the [Ester] and $[H_2O]$ terms can be treated as constants. Integrating leads to Eq. (4) where $[COOH]_0$ is the initial carboxy end group concentration.

$$[\mathrm{COOH}]_0/[\mathrm{COOH}] = \exp(-kt[\mathrm{H_2O}][\mathrm{Ester}]) \qquad (4)$$

or

$$\ln[\mathrm{COOH}] - \ln[\mathrm{COOH}]_0 = -k't. \qquad (5)$$

Thus a plot of ln[COOH] vs time is linear during the initial stages of degradation, with a slope equal to $-k[H_2O][Ester]$. Such kinetic behavour appears to be followed

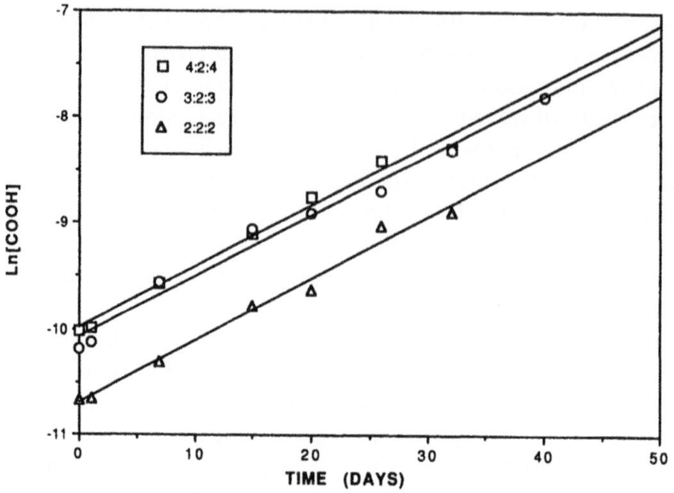
4:2:4
3:2:3
2:2:2
Ln[COOH]
-7
-8
-9
-10
-11
0
10
20
30
40
50
TIME (DAYS)

Figure 5. Semilog plot of the increase in the carboxy end group concentration, [COOH], resulting from chain scission of copolymer of PEG 2000 in DI water at 37°C.

by the block copolymers used in the present study (Fig. 5). This is despite the differing water contents of the copolymers and the increasing hydration with time, which would be predicted to result in a non-linear correlation of ln[COOH] and time.

Previous studies with the 1:1 copolymer of DL-lactic acid-glycolic acid (PGLA) have shown that an increase in the water content does not necessarily result in an increased rate of ester hydrolysis [8]. Polymer morphology has been shown to be important. For example, when PGLA is blended with polyvinyl alcohol, the kinetics of chain scission of PGLA follow the rate law (Eq. (4)) but the rate constant (k') does not increase in proportion to the measured water content until the blends become miscible and the water is distributed homogeneously throughout the blend [9].

The DSC and ESR results obtained with the present ABA block copolymers point to at least partial hydration-induced phase separation of the PLA and PEG blocks. Logically then, water uptake by the copolymers is largely associated with the PEG phase and chain scission of the PLA blocks will not be accelerated in proportion to the water content. The more rapid weight loss of the copolymers, relative to PLA, is easily understood because of the greater water solubility of PEG–PLA oligomers and their greater diffusivity in the more highly hydrated copolymers.

An earlier paper [14] reported the contact angles and the rate of weight loss of block copolymers of glycolic acid, lactic acid and ε-caprolactone with Pluronic F-68, an ethyleneoxy-propyleneoxy block copolymer. The contact angle measurements indicated phase separation at the film surface after hydration, while the half-life of erosion was slightly increased by the presence of the F-68 block.

CONCLUSION

ABA block copolymers of PLA and PEG are miscible in the dry state. Phase separation of the PEG and PLA blocks is induced by their water uptake and, as a result, the water content is primarily associated with the PEG phase. Consequently, the rate of hydrolytic chain scission of the PLA blocks is not accelerated, and is the same as observed for the PLA homopolymer. The shorter induction period prior to the onset of erosion of the PEG–PLA copolymers is not a reflection of a greater rate of chain scission, but arises because of the greater solubility of the PEG-PLA oligomers and their greater rate of diffusional loss in the hydrated copolymer films.

Acknowledgements

This work was supported in part by the National Institute on Drug Abuse, Grant No. DABB 5-RO1-DA-3616-05.

REFERENCES

1. N. A. Peppas, *Hydrogels in Medicine and Pharmacy*. CRC Press, Boca Raton, FL (1986).
2. H. Yasuda, A. Peterlin, C. K. Colton, K. A. Smith and E. W. Merrill, *Makromol. Chem.* **126**, 177 (1969).
3. C. A. Oksanen and G. Zografi, *Pharm. Res.* **7**, 654 (1990).
4. H. K. Reimschuessel, *J. Polym. Sci. Polym. Chem. Ed.* **16**, 1229 (1978).
5. Y. Y. Tan and G. Challa, *Polymer* **17**, 739 (1976).
6. D. T. Turner and A. Schwartz, *Polymer* **26**, 757 (1985).
7. W. E. Roorda, *Structure and application of hydrogels as drug delivery systems*, PhD Dissertation, University of Leiden (1988).
8. S. S. Shah, Y. Cha and C. G. Pitt, *J. Controlled Release* **18**, 261 (1992).
9. C. G. Pitt, Y. Cha, S. S. Shah and K. J. Zhu, *J. Controlled Release* **19**, 189 (1992).
10. S. R. Palit and P. Ghosh, *J. Polym. Sci.* **58**, 1225 (1962).
11. X. M. Deng, C. D. Xiong, L. M. Cheng and R. P. Xu, *J. Polym. Sci., Part C: Polym. Lett.* **28**, 411 (1990).
12. K. J. Zhu, L. Xiangzhou and Y. Shilin, *J. Appl. Polym. Sci.* **39**, 1 (1989).
13. D. Cohn, G. Marom and H. Younes, *Adv. Biomater.* **7**, 503 (1987).
14. A. S. Sawhney and J. A. Hubbell, *J. Biomed. Mater. Res.* **24**, 1397 (1990).
15. C. G. Pitt, S. S. Shah, R. Sik and C. F. Chignell, *Macromolecules*, submitted. (1992).
16. A. Schindler and D. Harper, *J. Polym. Sci., Polym. Chem. Ed.* **17**, 2593 (1979).
17. C. G. Pitt, in: *2nd International Scientific Workshop on Biodegradable Polymers and Plastics* M. Vert (Ed.). The Royal Society of Chemistry, Montpellier, France (1991).
18. C. Pitt and Z. W. Gu, *J. Controlled Release* **4**, 283 (1987).

Synthesis and characterization of putrescine-based poly(phosphoester-urethanes)

B. I. DAHIYAT, E. HOSTIN, E. M. POSADAS and K. W. LEONG*

Department of Biomedical Engineering, Johns Hopkins University, Charles and 34th Sts., Baltimore, MD 21218, USA

Received 21 August 1992; accepted 28 October 1992

Abstract—A novel set of putrescine-based segmented polyurethanes was synthesized using 1,4-butane-diisocyante and phosphoester diols, and was characterized for its potential as a degradable biomaterial. These poly(phosphoester-urethanes) (PPU) were flexible polymers with ultimate tensile strength (UTS) from 2 to 3 MPa, elongations up to 80% and tan δ near 0.15. The incorporation of phosphoester bonds in the backbone of the polymer by using bis(2-hydroxyethyl)phosphite (BGP) and bis(6-hydroxyhexyl)-phosphite (BHP) as chain extenders resulted in hydrolytic degradation which was evaluated *in vitro*. By varying the content of the phosphoester diol BGP, degradation rate, as followed by mass loss and GPC, could be modulated. Polymers based on the more hydrophobic monomer, BHP, showed slower degradation than corresponding BGP based polymers. Tensile properties of PPU-B2 after 22 days *in vitro* degradation show more than a 50% drop in UTS and ultimate elongation, likely caused by void spaces left behind in the polymer after mass loss and swelling. The attachment of a drug, PAS, pendant to the phosphoester group of the PPU was demonstrated. PAS was linked via the spacer 4-hydroxybenzaldehyde, and free, intact drug was released in about 5 h from a thin film.

Keywords: Degradable polymers; poly(phosphoesters); polyurethanes; drug delivery.

INTRODUCTION

Polymers are ubiquitous in medical devices because of the tremendous range of properties they can be designed to possess. A particularly attractive feature of polymers is the possibility of biodegradability. Surgical removal of the implant is not necessary, and there is the potential for cellular ingrowth and eventual replacement of the device by regenerated tissue. By careful material design and manipulation of degradation rate, a device can be programmed to function over a wide time scale. The rational development of biodegradable polymers for medical applications is the objective of this study.

Since structural versatility is of paramount importance in tuning the properties of polymers, polyurethanes are an ideal candidate. They have found use in a wide range of devices and are characterized by excellent mechanical properties and good biocompatibility [1, 2]. Owing to the high chemical reactivity of diisocyanates, a wide range of diol or diamine structures can be used to yield high molecular weight polyurethanes or polyureas. We exploit this flexibility by incorporating labile phosphoester bonds in the backbone, using bis(2-hydroxyethyl)phosphite (BGP) and bis(6-hydroxyhexyl)phosphite (BHP) as chain extenders (Fig. 1). These poly(phosphoester-urethanes) (PPU) then become hydrolytically degradable, breaking

* To whom correspondence should be addressed.

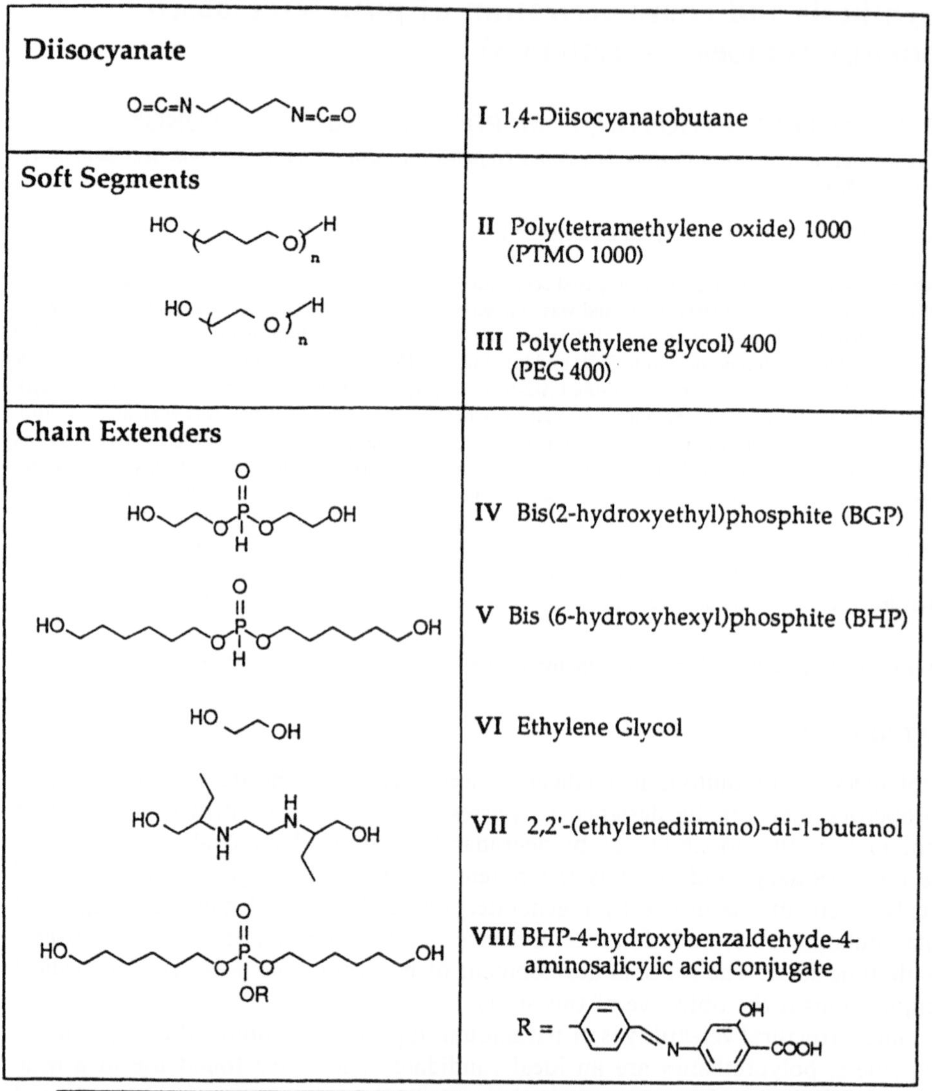
Diisocyanate
O=C=N N=C=O
I 1,4-Diisocyanatobutane
Soft Segments
HO H
O n
II Poly(tetramethylene oxide) 1000 (PTMO 1000)
III Poly(ethylene glycol) 400 (PEG 400)
Chain Extenders
IV Bis(2-hydroxyethyl)phosphite (BGP)
V Bis (6-hydroxyhexyl)phosphite (BHP)
VI Ethylene Glycol
VII 2,2'-(ethylenediimino)-di-1-butanol
VIII BHP-4-hydroxybenzaldehyde-4-aminosalicylic acid conjugate
R =
OH
COOH
OR

Polymer	BDI (mole ratio)	Diols						
		II	III	IV	V	VI	VII	VIII
PPU-B1	4	2		1			1	
PPU-B2	2	1		1				
PPU-B3	2	1				1		
PPU-B4	4	2		1		1		
PPU-B5	4		2	1			1	
PPU-H1	2	1			1			
PPU-H2	2	1						1

Figure 1. Chemical structures and names of the monomers used in PPU synthesis, and compositions and names of PPUs studied. Mole ratios of monomers are given.

down into phosphate, diols and possibly diamines, the last arising from the diisocyanate component [3, 4]. We have chosen 1,4-butane-diisocyanate (BDI) (Fig. 1) as the building block for our hard segment. BDI forms hard-segments composed of urethane groups linked by residues of 1,4-butanediamine, an intermediate in the urea cycle that is also called putrescine. Upon hydrolysis of the urethane bond, putrescine is released. Since putrescine is already present in the body, we hope that BDI can minimize the toxicity problems associated with the degradation products of many aromatic diisocyanates. The soft segment, which makes up most of the rest of the backbone, was usually poly(ethylene glycol) (PEG) or poly(tetramethylene oxide) (PTMO). In addition to erodibility, the pentavalency of phosphorus provides phosphoesters with a site for pendant linkage of bioactive compounds to the polymer, where they can function as a targeting moiety or as a therapeutic agent to be released in a sustained manner by hydrolytic cleavage.

We have explored a variety of structures for PPU (Fig. 1). The array of possible PPU is vast, however, and to gain a better understanding of how structure affects properties and degradation, a systematic investigation of the effect of phosphoester content on PPUs was performed. A study of bisphenol A based polyphosphoesters assessing the effect of different pendant groups has been reported [5]. The current work, however, examines the properties and relative hydrolytic stabilities of a set of PPU with different BGP contents: PPU-B2 with 25 mol%, PPU-B4 with 12.5 mol% and PPU-B3, a control, with 0 mol%. Changes in physical and chemical properties due to *in vitro* degradation were monitored by mass loss, swelling, molecular weight distribution, thermal profiles, dynamic mechanical moduli and tensile properties. PPU containing the more hydrophobic BHP were also studied. Further, to demonstrate the viability of drug loading via the pendant route, a spacer, 4-hydroxybenzaldehyde, was linked to the backbone of PPU-H2 and the antibiotic *p*-aminosalicylic acid was coupled to this spacer.

METHODS AND MATERIALS

Synthesis

Monomers. All reagents were obtained from Aldrich unless otherwise noted. Before use, all solvents were dried over 8 mesh pellets of 4 Å molecular sieves overnight. BGP, BHP and BDI were synthesized as described previously [6, 7]. 4-hydroxybenzaldehyde (HBA) was linked to BHP via a P—O bond by adapting the method of Penczek [8]. One gram (3.55 mmol) of BHP was dissolved in 50 ml of CH_2Cl_2 and Cl_2 was bubbled through the solution until a persistent yellow color was observed. After 1 h, the excess Cl_2 was removed under vacuum to give a clear solution. 0.495 g (7.10 mmol) of imidazole in 20 ml of CH_2Cl_2 was added dropwise to the BHP solution and a precipitate of imidazole hydrochloride rapidly formed. The reaction mixture was stirred for half an hour and then cooled to −20°C for 1 h to ensure complete precipitation of the excess imidazole. After vacuum filtration of the mixture to remove the imidazole, 0.433 g (3.55 mmol) of HBA in 80 ml of CH_2Cl_2 was added dropwise and allowed to react for 2 days. The solvent was removed by rotary evaporation and the resulting product, a tan colored paste, was washed exhaustively with ethyl ether and then vacuum dried.

The HBA–BHP conjugate was dissolved in 140 ml of absolute ethanol and 0.543 g of *p*-aminosalicylic acid (PAS) in 20 ml of absolute ethanol was added dropwise over 15 min. A yellow color rapidly appeared in the reaction solution and after 15 min the solution was filtered, rotary evaporated and vacuum dried. The product was a bright orange viscous liquid readily soluble in DMF, DMAc and alcohols.

Polymers. Polymerization was a two step reaction under nitrogen atmosphere in dry dimethylacetamide (DMAc). Fifty mole percent polyol as soft segment was added to 1 g of BDI and heated to 90–110°C for 3 h. Either poly(tetramethylene oxide) M_w 1000 (PTMO-1000) or poly(ethylene glycol) M_w 400 (PEG-400) was used as soft segment after vacuum drying at 60°C in a vacuum oven overnight before use. A solution of chain extenders was then added containing the balance of diol needed to react with the diisocyanate and heating was continued for 6 h. The reaction mixture was quenched into ethyl ether, and the product was isolated as a solid by filtration or as a syrup by centrifugation and then vacuum dried. These poly(phosphoester-urethanes) (PPU) are named PPU-B*x* for BGP based materials and PPU-H*x* for BHP based materials where the *x* is a number designating the particular polymer. Trials were run where the diisocyanate was in 20% excess of total diol to assess the effect of stoichiometry on the reaction.

Characterization

Chemical properties. Gel permeation chromatography (GPC) was performed on a Hewlett Packard (HP) 1090M liquid chromatograph. DMF at 1 ml/min flow rate and 60°C was used as the mobile phase to elute the sample through two mixed pore size polystyrene gel columns (Polymer Laboratories) in series. The system was calibrated with monodispersive polystyrene standards (Polymer Laboratories). A HP 1037A Refractive Index detector and HP Chemstation software acquired and analyzed the chromatographs. Fourier transform infrared spectroscopy (FTIR) was performed on a Perkin-Elmer 1600 Series machine and samples were prepared by film casting on NaCl plates or using KBr pellets. Proton nuclear magnetic resonance spectroscopy (H^1 NMR) was performed on a Bruker AMX300 in $CDCl_3$, DMSO or D_2O. Elemental analysis was performed by Galbraith Laboratories (Knoxville, TN).

Thermal properties. Differential scanning calorimetry (DSC) and thermogravimetric analysis (TGA) were done on a Seiko DSC 220 and a Seiko TG/DTA 220, respectively, under nitrogen at a heating rate of 20°C/min in open pans. Analysis was performed on a Seiko SDM 5500 Rheostation. T_g were measured from inflection points, and breakdown temperatures were taken at onset.

Mechanical properties. Dynamic moduli were measured on a Seiko TMA/SS 120C. Dynamic mechanical testing samples were solvent cast from 10 wt% chloroform on Teflon. The films were cut into rectangular samples measuring 12 mm × 3 mm × 300 μm and were measured with a micrometer prior to testing. Gauge length was 5 mm, the offset was 1% and the amplitude of oscillation was fixed at 0.3% at a frequency of 0.01 Hz. All experiments were conducted at room temperature. Moduli were calculated from displacement and load versus time curves.

Tensile tests were conducted following ASTM Method D638M and using the M-III sample specification. Specimens were molded at 70°C. Testing to failure was conducted on an Instron MTT-L at a crosshead speed of 1 cm/min. Ultimate strengths, elongations and moduli were calculated from load-deflection data.

In vitro *degradation and swelling*

Degradation and swelling samples were solvent cast from chloroform into films about 300 μm thick and were cut with a razor into rectangles weighing about 50 mg. Three samples per time point were placed in 5 ml each of 0.1M pH 7.4 phosphate buffer at 37°C. Solutions were changed periodically to better approximate perfect sink conditions. Mass loss was followed by removing film samples from buffer and drying to constant weight under vacuum. GPC samples were cut from the dried films and then dissolved in DMF and filtered through a 0.45 μm PTFE filter (Gelman Sciences) prior to injection. FTIR, thermal analysis, elemental analysis and dynamic mechanical testing samples were also cut from the dried films and then assayed as described above.

Swelling sample were removed from buffer and weighed after blotting the surface dry. Samples were then dried to constant weight in a desiccator containing P_2O_5 at 1 Torr and room temperature. The difference between wet and dry weight was normalized with respect to dry weight to give swelling values.

Tensile testing specimens identical to those described above were placed in 15 ml of buffer for *in vitro* degradation. Samples were vacuum dried before testing.

In vitro *drug release*

Films of PPU-H2 were cast from 10 wt% chloroform on to Teflon. Samples weighing about 50 mg were cut from the film and placed in 10 ml of 0.1M pH 7.4 phosphate buffer at 37°C. The buffer solutions were periodically changed and analyzed for drug concentration by HPLC. A HP 1090M liquid chromatograph was used and the column was a 15 cm × 7.5 mm Hypersil C8 (Alltech) running 70 : 30 acetonitrile : H_2O at 1 ml/min and 40°C. Detection was by a diode array operating from 200 to 400 nm, with the signal at 265 nm integrated by HP Chemstation software to give peak areas.

RESULTS AND DISCUSSION

Synthesis and general properties

The structure of BDI was confirmed by FTIR and ^1H NMR. BHP structure was verified by FTIR and elemental analysis, but the elemental analysis of BGP showed discrepancies with expected values. Specifically, phosphorus content was 20.2% while the calculated value was 18.2%, and carbon was 26.1% while the calculated value was 28.2%. Side reactions, such as cyclization and elimination, that consume the ethylene glycol are likely to be limiting conversion of dimethyl phosphite to BGP [9]. The BHP–HBA conjugate structure was confirmed by the presence of an aldehyde peak at 9.8 ppm, aromatic protons at 6.9 and 7.8 ppm and sets of resonances for the methylene protons of BHP at 3.9 and 1.3 ppm. Further, the distinctive phosphite P–H doublet, 6.8 ppm and $J_{P\text{-}H}$ = 694 Hz, in the BHP spectrum was absent in the conjugate's spectrum indicating coupling at the pendant position.

Polymerizations run at 110°C yielded PPUs with higher molecular weight, strength and more flexibility than PPUs synthesized at less than 100°C. This improvement in physical properties could have resulted from crosslinking due to allophanate or biuret bonds that only occurs at the higher temperature [2]. Moreover, the reaction of the P–H group with isocyanates has been reported and might have contributed to crosslinking [6]. Using 20% excess BDI in the synthesis improved yields but did not result in any noticeable changes in material properties. The remainder of this paper deals with polymers synthesized at 110°C with 20% excess BDI.

PTMO 1000 based PPUs were generally light colored, soft and flexible while PEG 400 based PPU were hard, brittle and tacky. All of the PEG 400 polymers dissolved readily in water, in addition to DMF, DMAc, alcohols and chloroform. The PTMO 1000 polymers generally dissolved in DMF, DMAc and chloroform but, as expected from the hydrophobic nature of the polyol, were insoluble in alcohols or water. The ready solubility of these PPU in chlorinated solvents can likely be attributed to the solubilizing effect of the phosphoester group in the backbone [10]. These polymers can be cast into clear resilient films, and can be molded readily at 70°C.

Chemical and physical characterization

PPUs typically have polydispersities over 20 and a bimodal molecular weight distribution, with one peak corresponding to chains in the 20000 Dalton range and the other consisting of very high molecular weight polymer that often extended beyond the 3×10^6 Dalton exclusion limit of the GPC columns used. Figure 2 is a cumulative molecular weight distribution plot of PPU-H1, both before and after repeated washes with ethyl ether. The weight fraction of polymer equal to or less than a given molecular weight is plotted against that molecular weight. The difference between the two curves shows that the fraction of polymer comprised of low molecular weight chains was reduced significantly by ether washes, which resulted in loss of about 40 wt% and in improved material properties for the remaining polymer. By comparison, polymers synthesized without extra BDI had greater than two thirds weight fraction of low molecular weight polymer and consequently lost far more mass during ether washes.

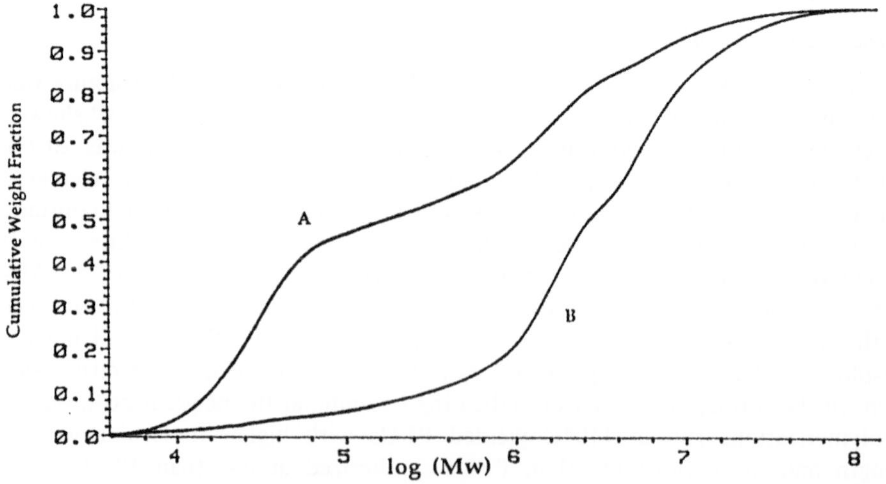
Cumulative Weight Fraction
1.0
0.9
0.8
0.7
0.6
0.5
0.4
0.3
0.2
0.1
0.0
A
B
4
5
6
7
8
log (Mw)

Figure 2. Molecular weight distribution overlay for PPU-H1 (A) before and (B) after repeated washes with ethyl ether. Cumulative weight fraction of polymer equal to or less than a given molecular weight is plotted against the logarithm of molecular weight.

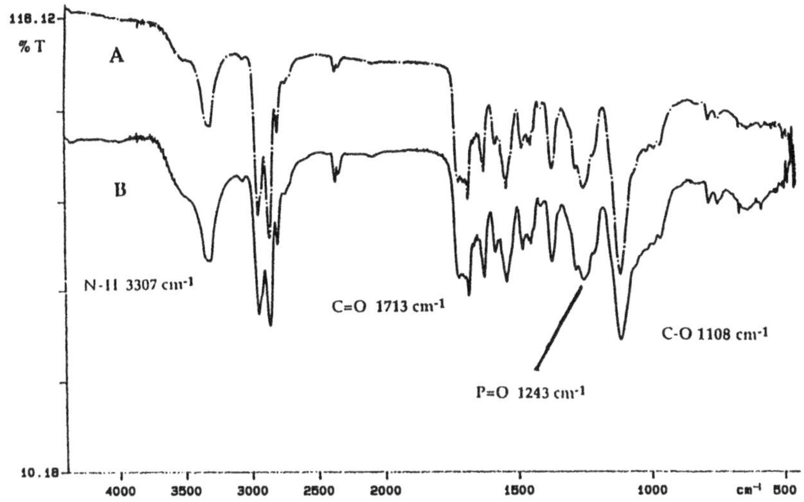
118.12
%T
A
B
N-H 3307 cm-1
C=O 1713 cm-1
P=O 1243 cm-1
C-O 1108 cm-1
10.18
4000
3500
3000
2500
2000
1500
1000
cm-1 500

Figure 3. FTIR spectra of PPU-B2 (A) before and (B) after 90 days of *in vitro* degradation. Films of samples were cast on NaCl from $CHCl_3$.

FTIR confirms the presence of the phosphoester bond in the polymers studied with a strong P=O stretch at 1240–1245 cm^{-1} (Fig. 3). This peak was absent from PPU-B3, the only polymer with no phosphoester component. Urethane structure is confirmed by a carbonyl stretch at 1710–1715 cm^{-1}, the amide II band at 1537–1540 cm^{-1}, N—H stretching at 3300–3315 cm^{-1} and a very strong C—O band at 1108–1112 cm^{-1} arising from the polyol component.

All of the PPU showed a soft segment T_g between −60 and −65°C due to relaxation of the soft segment (Table 1). This value is lower than those reported for aromatic diisocyanate based polyurethanes because of the greater flexibility of the aliphatic BDI [11]. PPUs also showed several transitions between 30 and 140°C that almost entirely disappeared if the samples were heated to 150°C for 10 min and rapidly quenched to −120°C immediately before DSC analysis. The annealing appeared to disrupt crystal structure that was slow to reorder. Letting annealed samples sit for several days resulted in reappearance of the affected peaks. Breakdown of PPUs as measured by TGA began at 280°C (Table 1 and Fig. 4) for all polymers tested with the exception of PPU-B3, which decomposed at 315°C. Thus, the presence of phosphoester decreased the thermal stability of the polymer.

Table 1.
Physical Properties of some PPUs

	% Phosphorus						% Swelling
Polymer	Calc.	Meas.	T_g (°C)	T decomp (°C)	E^* (MPa)	tan δ	at 10 days
PPU-B2	2.11	1.1	−66	280	10.2 ± 2.6	0.14 ± 0.01	48
PPU-B4	1.06	0.8	−63	280	8.8 ± 1.6	0.11 ± 0.01	44
PPU-B3	0.00	<0.06	−65	315	7.4 ± 1.6	0.08 ± 0.01	11
PPU-H1	—	—	−66	280	6.5 ± 0.5	0.12 ± 0.002	14

E^* and tan δ values are mean ± S.E.M. ($n \geq 4$.)

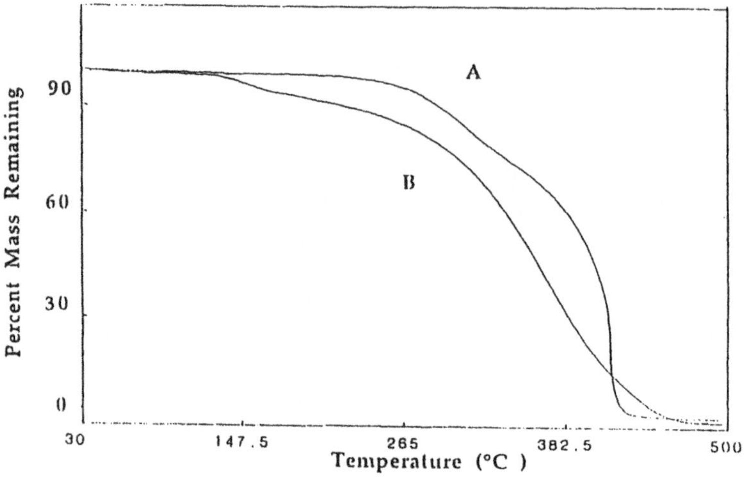

Percent Mass Remaining
90
60
30
0
A
B
30
147.5
265
382.5
500
Temperature (°C)

Figure 4. TGA profiles of PPU-B2 (A) before and (B) after 90 days of *in vitro* degradation. Samples were analyzed under N_2 at 20°C/min.

Dynamic moduli measurements allowed us to probe the stiffness of the PPUs in a quantitative manner while using small amounts of material. The apparatus used required less than 10 mg of polymer while a tensile test required about 1 g per test. Conserving material is an important consideration when conducting degradation experiments, and dynamic testing allowed us to study stiffness and viscoelasticity as a function of time *in vitro*. One percent offset and 0.3% strain amplitude were selected to stay within the linear ranges of the stress-strain curves for the materials and were the smallest values that gave reproducible results. Since varying the frequency from 0.001 Hz to 0.1 Hz changed the dynamic modulus only slightly, 0.01 Hz was used for all measurements because it simplified data analysis. For PTMO based PPUs, the tan δ ranged from 0.07 to 0.21, depending on chain extender composition, while the magnitude of the modulus, E^*, varied from 1.6 MPa for PPU-B1 to 10.2 MPa for PPU-B2. For the BGP based PPU, phosphoester tended to increase both E^* and tan δ (Table 1), while BHP based polymers were usually softer than their BGP analogues as can be expected from the longer hexamethylene chains in the monomer. The results of TMA testing on degraded PPU samples are discussed in the following section.

The ultimate tensile strengths (UTS) of PTMO based PPUs were in the 2–3 MPa range and were quite similar for all of the various compositions (Table 2). The elongations and elastic moduli were more varied, and the trends could be correlated to the chain extender structures. PPU-B3 with ethylene glycol as its only chain extender was stiffer than the BGP extended PPU-B2. PPU-B1, with half of its chain extender made up of the longer diol VII, is softer, and PPU-H1 is the softest yet. Elongations reflect the same tendency, with PPU-B1 and PPU-H1 stretching more than PPU-B2 and PPU-B3. An analogue to PPU-B2 was synthesized with hexamethylene diisocyanate (HDI) replacing BDI, in order to gauge the effect of changing the diisocyanate. Tensile properties were not significantly different. Other monomers, such as poly(caprolactone) diol and diamines based on lysine, are being explored for better mechanical properties.

Table 2.
Tensile properties of PPUs

Polymer	UTS (MPa)	Ultimate elongation (%)	Elastic modulus (MPa)
PPU-B2	2.27 ± 0.39	36 ± 11	7.84 ± 0.92
PPU-B3	3.07 ± 0.71	48 ± 13	10.37 ± 0.57
PPU-H1	1.89 ± 0.31	84 ± 18	4.03 ± 0.62
PPU-B1	2.50 ± 0.69	60 ± 18	6.03 ± 0.23
HDI analogue of PPU-B2	3.28 ± 0.81	36 ± 9	6.97 ± 0.46

Values are mean ± S.E.M. ($n = 4$.)

In vitro *degradation*

The effect of *in vitro* degradation on PPU-B2, PPU-B3 and PPU-B4, polymers with 25, 0 and 12.5 mol% BGP respectively, was assessed. The thin, 300 μm, samples reached swelling equilibrium in less than 3 days, with PPU-B2 and PPU-B4 swelling about 45% and remaining nearly constant for 110 days. PPU-B3, without the phosphoester chain extender, swelled only about 10% and remained clear and translucent throughout the study in contrast to the other polymers which turned opaque. In addition, PPU-B2 became appreciably softer and more difficult to handle until breaking up at around 100 days. Mass loss from the PPUs is shown in Fig. 5. While PPU-B3 predictably showed no significant degradation for over 3 months, a rapid initial weight loss was seen for PPU-B2 and PPU-B4, of approximately 20 and 5% respectively. The mass loss then leveled off for about 75 days

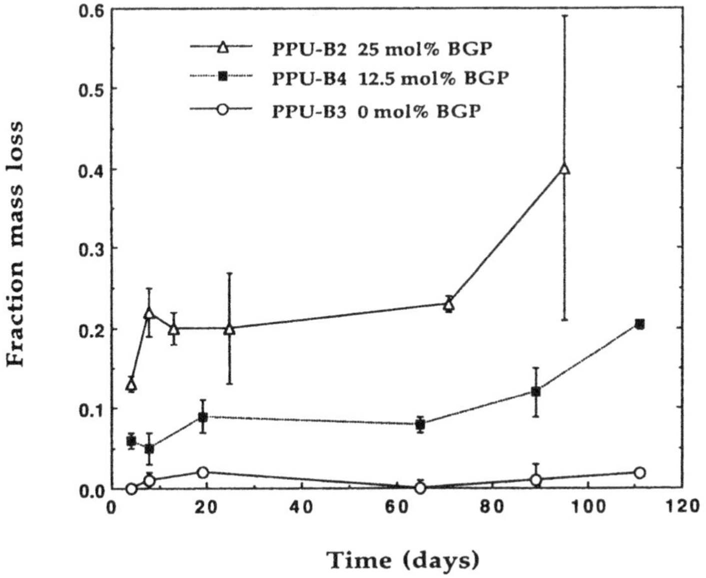
PPU-B2 25 mol% BGP
PPU-B4 12.5 mol% BGP
PPU-B3 0 mol% BGP
Fraction mass loss
0.0
0.1
0.2
0.3
0.4
0.5
0.6
0
20
40
60
80
100
120
Time (days)

Figure 5. *In vitro* mass loss of PPU-B2, PPU-B3 and PPU-B4 as a function of time. All points are averages ± S.E.M. ($n = 3$.)

before accelerating again. The complete breakup of PPU-B2 samples at late time points confirmed that extensive degradation had occurred but, unfortunately, also increased the scatter of the mass loss values. Degradation was still occurring during the plateau phase as verified by GPC. During the first 20 days of the study, molecular weight distributions apparently increased as low molecular weight oligomers leached out and accounted for most of the early mass loss. The distributions started to decrease at around 75 days as the longer chains were cleaved (Fig. 6), when the degradation showed up in the mass loss. No significant change occurred in the molecular weight distribution of PPU-B3.

Although considerable mass loss and chain cleavage was seen, FTIR spectra of degraded samples showed no noticeable change from pristine polymer (Fig. 3). DSC curves showed small shifts in the set of melting transitions above room temperature but not in any consistent or reproducible manner. Apparently, little change in chemical composition was occurring in the polymer samples during degradation. The identity of degradation products was, however, not accounted for from these data. TGA did show differences after degradation (Fig. 4). Mass loss begins at 150°C after 90 days *in vitro* compared to 280°C for fresh polymer.

Changes in dynamic moduli during degradation are shown in Fig. 7. Thicknesses of the films were measured immediately prior to testing, and hence, moduli values are normalized to degraded sample dimensions. PPU-B2 and PPU-B4 showed initial increases in E^* followed by declines at later time points. PPU-B2 dropped off steeply and was too soft for the testing after 75 days while PPU-B4 showed a more deliberate softening and PPU-B3 remained unchanged. The changes in E^* appear to correlate with the mass loss and chain cleavage discussed above. The initial loss of low molecular weight oligomers, effectively raising the molecular weight, would be expected to stiffen the polymer, then, as high molecular weight polymer degraded, the modulus would be expected to drop. PPU-B2 showed a greater effect than PPU-B4 for both E^* and mass loss, which can be attributed to the higher phosphoester content, while the control PPU-B3 remained unchanged. Tan δ showed similar trends, but the structural features responsible for energy dissipation are not as apparent as those effecting modulus. The results here show that dynamic modulus measurements can be a sensitive probe of degradation.

Tensile specimens of PPU-B2 were tested to failure after 22 days *in vitro* (Fig. 8). Both UTS and ultimate elongation dropped to less than 50% of initial values, while modulus remained unchanged. These declines are far more rapid than the changes in other properties discussed above, and, indeed, mass loss and swelling from tensile samples are even slower than for the thin films due to the 7 fold increase in thickness. Only 20% swelling and 4% mass loss occurred in 22 days. However, the cracks left behind by mass loss and drying, which were clearly visible, had a profound effect on failure properties such as UTS and elongation by providing sites for fracture and tearing to occur. Modulus and dynamic modulus were not affected as much because they are properties measured at low strain where defects do not play a large role. Tensile modulus was nearly constant, contrary to the dynamic modulus data, possibly due to the low mass loss not eliminating enough low molecular weight chains to cause an increase in stiffness. It should be noted that no cracks were seen in the dynamic testing samples, an effect that could be related to the different fabrication methods used, molding for tensile specimens versus solvent casting for dynamic testing specimens.

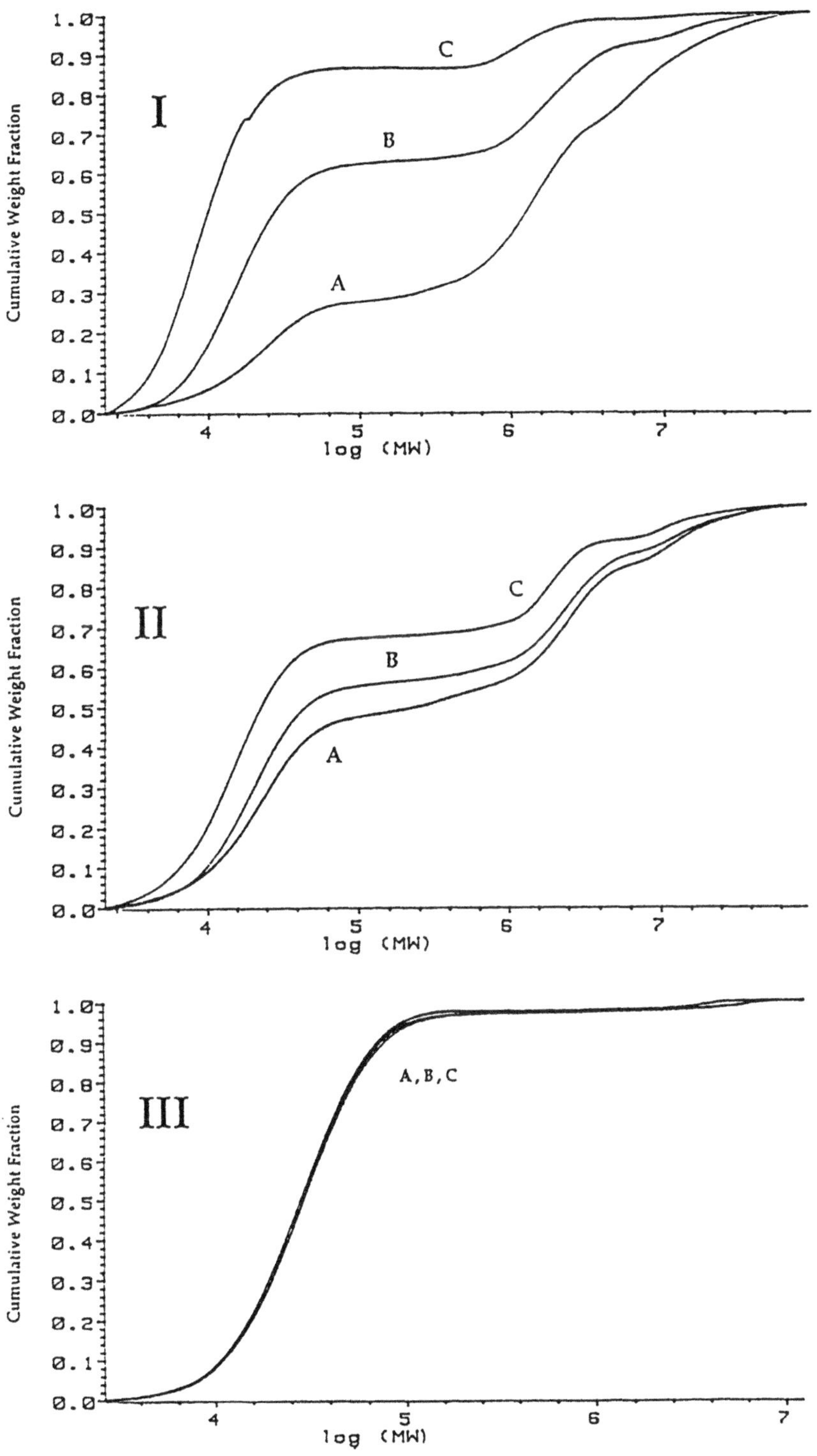
I
C
B
A
Cumulative Weight Fraction
1.0
0.9
0.8
0.7
0.6
0.5
0.4
0.3
0.2
0.1
0.0
4
5
6
7
log (MW)
II
C
B
A
Cumulative Weight Fraction
log (MW)
III
A, B, C
Cumulative Weight Fraction
log (MW)

Figure 6. Molecular weight distribution overlays for (I) PPU-B2, (II) PPU-B4 and (III) PPU-B3 after (A) zero days, (B) 75 days and (C) 90 days of *in vitro* degradation.

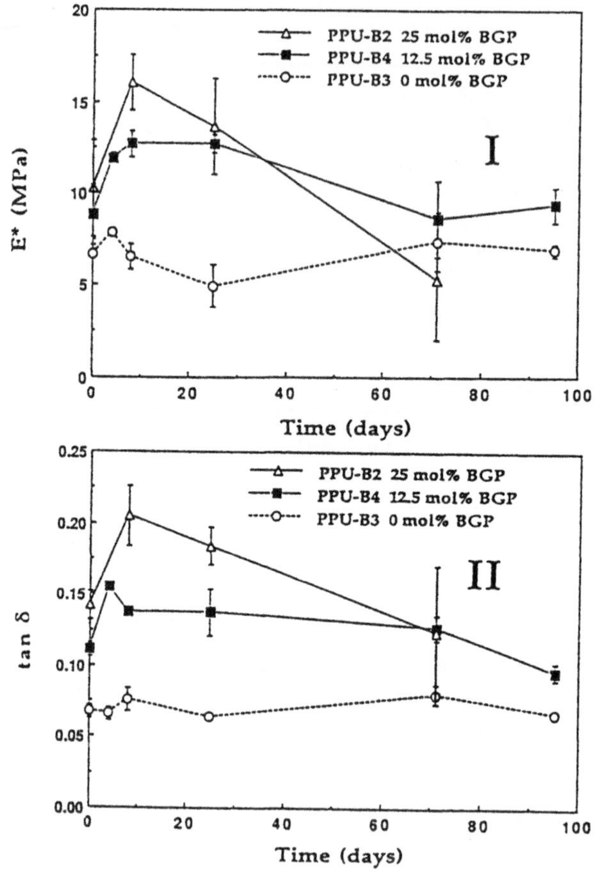
PPU-B2 25 mol% BGP
PPU-B4 12.5 mol% BGP
PPU-B3 0 mol% BGP
I
E* (MPa)
Time (days)
PPU-B2 25 mol% BGP
PPU-B4 12.5 mol% BGP
PPU-B3 0 mol% BGP
II
tan δ
Time (days)

Figure 7. (I) Magnitude, E^* and (II) tan δ of complex modulus for PPU-B2, PPU-B3 and PPU-B4 as a function of degradation time. All points are averages ± S.E.M. (n = 3.)

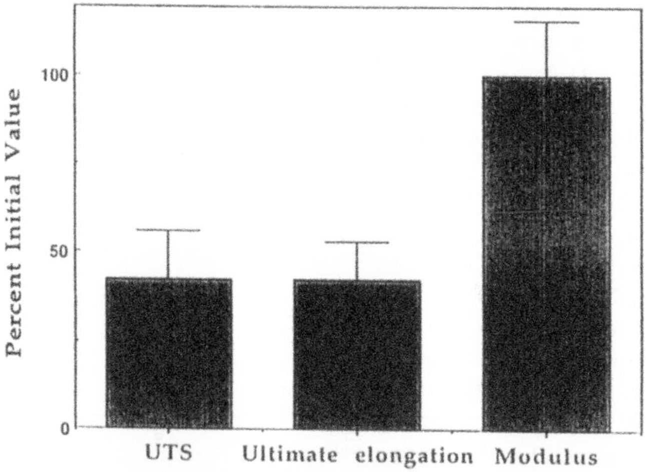
Percent Initial Value
UTS
Ultimate elongation
Modulus

Figure 8. Effect of 22 days *in vitro* degradation on tensile properties of PPU-B2. Percent of the initial values of UTS, elongation and modulus that remain are plotted. Values are averages ± S.E.M. (n = 6.)

PPU-H1, the BHP based PPU, swells only 14% at 10 days as compared to 48% for the BGP based PPU-B2 (Table 1). Also, after 11 days *in vitro* PPU-H1 lost 4% mass while PPU-B2 lost 13%. Even though they have the same mole ratio of phosphoester, BHPs more hydrophobic nature slows water penetration and retards degradation. Monomer structure variation, in addition to monomer ratios, therefore can be used to affect PPU degradation.

In vitro *drug release*

PPU-H2 was synthesized with 25 mol% *p*-aminosalicylic acid (PAS) conjugated monomer, diol VIII (Fig. 1). The polymer was still elastic and resilient, and cast dark orange translucent films from chloroform. The orange color and FTIR peaks at 1682 cm^{-1} for the PAS carboxylic acid and at 1620 cm^{-1} for the conjugated imine bond confirm that drug attachment survived the polymerization. Drug release from 100 μm thick film samples was assayed by HPLC, where free drug, drug-spacer conjugate and spacer were separated. Figure 9 shows release of free drug which is complete at about 5 h. Only about a quarter of the PAS introduced into the synthesis was recovered during release experiments. Both loss of PAS conjugated monomer during workup and side reactions during polymerization involving the free hydroxyl on PAS could account for the loss of drug. Several mechanisms of drug cleavage and release can be expected, such as backbone scission, polymer–spacer bond cleavage and drug–spacer bond cleavage (Fig. 10). However, no spacer or drug–spacer was found in the release medium, suggesting that the polymer–spacer bond is much more stable than the drug–spacer bond. Scission of high molecular weight chains was demonstrated by GPC after 11 days *in vitro*, but this mode of degradation was much slower than drug release.

The presence of the hydrophilic drug apparently facilitated the degradation as the mass loss at 11 days for PPU-H2 was 44% versus 4% for PPU-H1, which has no pendant drug linkage. PPU-H2 also imbibed more water but measurement was

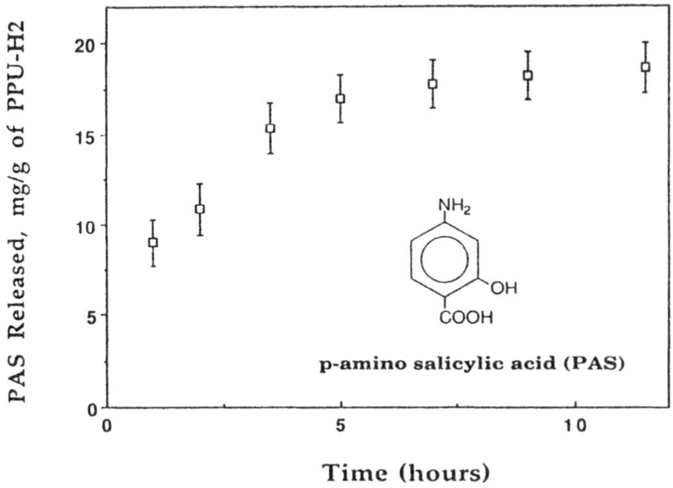
PAS Released, mg/g of PPU-H2
20
15
10
5
0
0
5
10
Time (hours)
NH2
OH
COOH
p-amino salicylic acid (PAS)

Figure 9. *In vitro* release of *p*-amino salicylic acid (PAS) from PPU-H2. Values are averages ± S.E.M. ($n = 3$.)

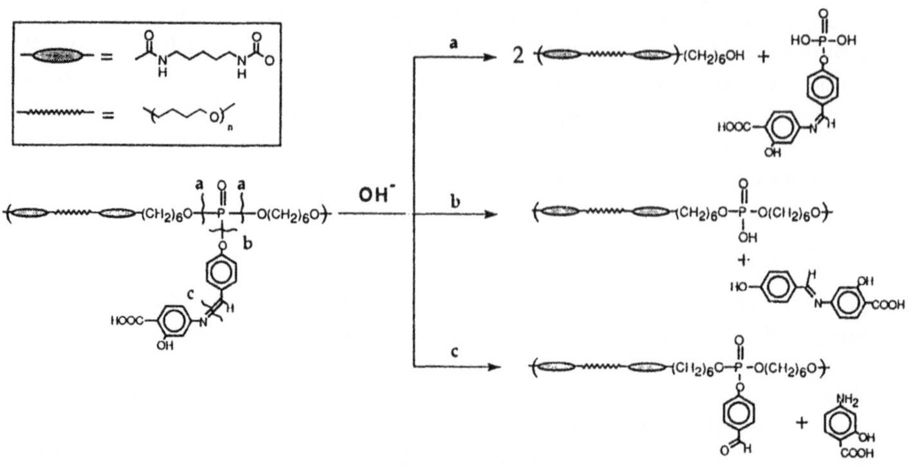

Figure 10. Degradation modes for PPU drug conjugates. Pathway (a) is backbone cleavage resulting in chain scission and phosphate-spacer-drug conjugate release. Pathway (b) is polymer-spacer bond cleavage resulting in spacer-drug conjugate release. Pathway (c) is drug-spacer bond cleavage resulting in free drug release.

impossible because of sample breakup during handling. A high fraction of low molecular weight oligomers in the polymer probably also led to the rapid weight loss. The high reactivity of the aldehyde group has allowed the linking of other compounds to the polymer, such as phenylalanine and cycloserine, an antibiotic. Currently, protection of the aldehyde group during polymerization is being explored in order to allow conjugation of delicate compounds to the PPU after polymer synthesis.

CONCLUSION

A set of putrescine based segmented polyurethanes was synthesized using 1,4-butanediisocyanate and phosphoester diols, and was characterized for its potential as a degradable biomaterial. These PPU were flexible polymers with UTS from 2 to 3 MPa, elongations up to 80% and tan δ near 0.15. The incorporation of phosphoester bonds in the backbone of the polymer resulted in hydrolytic degradation. By varying the content of the phosphoester diol BGP, degradation rate could be modulated. A PPU with 25 mol% BGP, PPU-B2, lost 40% mass by 3 months, and GPC showed almost total cleavage of high molecular weight chains by this time. PPU-B4, with 12.5 mol% BGP, only lost 20% mass after 3 months while a control polymer without any BGP was stable. In addition, dynamic mechanical modulus was monitored and, for the phosphoester containing polymers, showed an initial stiffening, probably due to loss of low molecular weight oligomers, followed by steady softening as long polymer chains were cleaved. The loss of E^* was more pronounced for PPU-B2. Tensile properties of PPU-B2 after 22 days *in vitro* degradation show more than a 50% drop in UTS and ultimate elongation, likely caused by void spaces left behind in the polymer after mass loss and swelling. Polymers based on a more hydrophobic monomer, BHP, showed slower degradation than corresponding BGP based polymers. The attachment of a drug, PAS, pendant to the phosphoester group of the PPU was demonstrated. PAS was linked via the spacer 4-hydroxybenzaldehyde, and free, intact drug was released in about 5 h from a thin film.

Acknowledgements

The authors would like to acknowledge support from the American Foundation for AIDS Research. BID was supported by a National Defense Science and Engineering Graduate Fellowship in Chemistry. We would also like to thank Steve Johnston and Dr. Norman Sheppard for technical assistance.

REFERENCES

1. M. D. Lelah and S. L. Cooper, *Polyurethanes in Medicine.* CRC Press, Boca Raton, FL (1986).
2. M. Szycher (Ed.), *High Performance Biomaterials: A Comprehensive Guide to Medical and Pharmaceutical Applications*, Ch. 3-5. Technomic Pub., Lancaster (1991).
3. B. Dahiyat, F. Shi, Z. Zhao and K. Leong, *Proc. ACS Div. PMSE.* **66,** 87 (1992).
4. B. I. Dahiyat, E. Hostin, E. M. Posadas, M. H. Cynamon and K. Leong, *Proc. ACS Div. PMSE.* **67,** 87 (1992).
5. M. Richards, B. I. Dahiyat, D. M. Arm, P. R. Brown and K. W. Leong, *J. Biomed. Mater. Res.* **25,** 1151 (1991).
6. G. Borisov and K. Troev, *Europ. Polym J.* **9,** 1077 (1973).
7. P. Eckert and E. Herr, *Chem. Abs.* 1564 (1944).
8. J. Pretula, K. Kaluzynski and S. Penczek, *Macromolecules* **19,** 1797 (1986).
9. W. Vogt and S. Balasubramanian, *Makromol. Chem.* **163,** 111 (1973).
10. M. Sander and E. Steininger, *J. Macromol. Sci. (Revs.)* **C1**(1), 91 (1967).
11. G. B. Guise and G. C. Smith, *J. Appl. Polym. Sci.* **25,** 149 (1980).

Oxidative degradation of Biomer^{TM} fractions prepared by using preparative-scale gel permeation chromatography

B. J. TYLER[1] and B. D. RATNER[2]*†

[1]*Chemical Engineering Department, Montana State University, Bozeman, MT 59717, USA*
[2]*Department of Chemical Engineering and Center for Bioengineering, University of Washington, Seattle, WA 98195, USA*

Received 6 July 1992; accepted 15 February 1994

Abstract—The possibility that some macromolecular chains within a chemically heterogeneous polyether-urethane (PEU) may be more susceptible to degradation than others has been investigated. Preparative scale gel permeation chromatography has been used to separate chemically different fractions of a sample of the commercial PEU, Biomer^{TM}. The fractions were characterized and then tested for susceptibility to oxidative degradation by exposing them to hydrogen peroxide. After exposure to hydrogen peroxide, the samples were analyzed using high pressure gel permeation chromatography (HPGPC), Fourier transform infrared spectroscopy (FTIR), and X-ray photoelectron spectroscopy (XPS). By using these methods, we were able to identify chemical changes in some of the Biomer^{TM} fractions but not in others. Clear differences in the chemistry and reactivity of the fractions were observed. Changes in the weight average molecular weight varied from a decrease of 55.8% to an increase of 3.9%. A decrease in hard segment content at the surface and in the bulk was observed in some samples, but opposite trends were observed in others. The evidence suggests that there may be a number of mechanisms by which hydrogen peroxide can react with PEUs. Some fractions separated from the Biomer^{TM} were not significantly affected by concentrated hydrogen peroxide solutions. This suggests an intrinsic stability in some PEUs and points the way to the development of more degradation-resistant PEUs.

Key words: Polyurethane; degradation; oxidation; gel permeation chromatography; ESCA; IR.

INTRODUCTION

During the past decade, the susceptibility of poly(ether urethanes) (PEUs) to biodegradation has aroused much interest in the biomaterials community. Although many researchers have investigated both *in vivo* and *in vitro* degradation of PEUs, widely varying results have been obtained and few generalizations on PEU degradation exist [1–14]. This has led to much controversy over the mechanisms and extent of degradation. One reason for the difficulties in identifying mechanisms involved in the biodegradation of PEUs lies in the heterogeneity of these polymers. Commercial PEUs commonly exhibit several levels of heterogeneity. Not only can the length of the polymer chains vary, but the lengths of urethane hard segments and ether soft segments within the chains differ [15, 16]. Also, side reactions such as allophanate and biuret formation take place to varying degrees further complicating the structure.

---

* To whom correspondence should be addressed.

† One of the authors (BDR) was encouraged by Allan S. Hoffman, years ago, to avoid getting involved in research on polyurethanes as they are unreasonably complicated systems. This work proves just how prescient Allan Hoffman was!

The results of several studies [6, 7, 10, 11] have suggested that some portions of a heterogeneous PEU may be highly susceptible to attack by components of the *in vivo* environment, while others may be degradation-resistant. To investigate this hypothesis, we have used preparative-scale gel-permeation chromatography (PSGPC) to take apart a common commercial medical grade PEU, Biomer^{TM}. Degradation experiments have been performed on each of the samples obtained in the chromatographic separation. The objective of this work was to determine if a clinically relevant PEU contains portions that are highly susceptible to degradation and others that are degradation-resistant. To accomplish this, the polymer fractions were degraded in a controlled *in vitro* environment to allow clear comparisons.

When PEUs are implanted in animals, the complex biological environment and the heterogeneous polymer both contribute to the degradation process. Because of the complexity of the biologic and polymeric systems, it can be difficult to identify which components of the host are reacting with a specific component of the PEU. In addition, experimental variability can result because of differences in implant sites, degrees of surgical trauma, animal-to-animal differences and other factors. To minimize interferences with our observations on the differences in reactivity between the polymer fractions, we chose here to use an *in vitro* assay, the hydrogen peroxide oxidation of the polymers. This system was selected because of its relative simplicity and because of its relevance to *in vivo* studies. Hydrogen peroxide is released in the normal inflammatory response to biomaterial implants and therefore, the reactions that hydrogen peroxide undergo with PEUs are pertinent to the long term biostability of implantable PEU devices. The objective of this assay was not to provide a rigorous model of the *in vivo* environment, but to emulate a key, reactive component of that environment.

Another reason behind many of the controversies regarding the biodegradation of polyurethanes is that consistent criteria for assessing the biostability of materials do not exist. Researchers have used a wide array of techniques to measure degradation including weight loss or release of low molecular weight species [1, 5, 10, 11], changes in mechanical properties [3, 13], changes in molecular weight distribution [7, 12, 13], changes in infrared spectrum [3, 4, 6, 12], changes in surface chemistry [4, 6, 8, 12], and surface cracking [9, 14]. Although it was not the intent of this study to develop consistent criteria for use in future biodegradation studies, it was necessary to establish criteria and protocols that could be used to consistently compare the biostability of the Biomer^{TM} fractions. For the purposes of this work, biodegradation will be defined as chemical modification of the polymer by chemicals produced by the host. Even if the extent of attack is insufficient to adversely affect mechanical properties, chemical modification may produce leachable materials or alter the interaction of the polymer with proteins and cells. In this study, three methods for measuring chemical differences between the Biomer^{TM} PSGPC fractions before and after exposure to hydrogen peroxide have been used: high pressure gel permeation chromatography (HPGPC), Fourier transform infrared spectroscopy (FTIR), and X-ray photoelectron spectroscopy (XPS).

MATERIALS PREPARATION AND CHARACTERIZATION

Biomer^{TM}

The PEU samples used in this study were prepared from Biomer^{TM} lot BSUA-001 using PSGPC. Biomer^{TM} is a commercially available medical grade PEU that is similar in composition to the Dupont polymer, Lycra Spandex™ [17]. Based on published studies of Biomer™, it is synthesized from a poly(tetramethylene oxide) (PTMO) with an average molecular weight of 1800, methylene bisphenyl diisocyanate (MDI), and an ethylene diamine chain extender [18]. A smaller amount of a second chain extender is also used in the polymer to inhibit crystallinity in the soft segment [19]. Biomer™ lot BSUA-001, which has been used in this study, has been found to contain a high molecular weight additive, poly(diisopropylaminoethyl methacrylate) (DPAEMA), that dominates the surface of the polymer [20]. DPAEMA forms an unfilterable gel when Biomer™ is dissolved in *n,n*-dimethyl acetamide (DMAC) and did not elute from the preparative chromatography columns.

The weight average molecular weight (M_w) of samples of Biomer™ lot BSUA-001 that were exposed to 30% hydrogen peroxide for 24 h decreased by 20.6%. This change was substantially greater than the change seen for samples of BSP-067 and PEU-PTMO-2000 (a model PEU similar to Biomer™) [21].

Chromatographic sample preparation

Ten fractions of Biomer™ lot BSUA-001 were prepared for oxidative degradation studies using PSGPC. Two Waters Microstyrogel™ columns with pore sizes of 10 000 and 1000 Å were used in series for the separation. The columns were each 1 in. in diameter and 4 ft in length and were packed with 25 μm beads. The solvent DMAC was used as the mobile phase. Separations were performed at a flow rate of 10 ml min^{-1}. Ten injections of 5 ml of 20 mg ml^{-1} PEU solution in DMAC were made. Samples were collected at 2-min intervals starting one minute after injection. Fractions from a given time point from the ten runs were combined and analyzed using analytical scale HPGPC (see procedure, below) to determine which samples contained sufficient polyurethane for further experimentation. The samples that eluted between 33 and 53 min were dried and used in degradation studies. The samples were identified with numbers that indicated the elution time at the beginning of sample collection. Each sample was dried under vacuum at 35°C to a volume of 5 ml. The polymer was then precipitated from the 5 ml of solution by the addition of 5 ml distilled-deionized water. The precipitate was collected and dried for 24 h in a vacuum desiccator at ambient temperature.

Characterization of the fractions

After drying, each sample was dissolved at a concentration of 1 mg ml^{-1} and analyzed with analytical HPGPC to determine if the drying procedure had altered the molecular weight distribution. Changes in the molecular weight distribution of the samples were minimal. The samples were then analyzed with both ESCA and FTIR to quantify differences in the bulk and surface chemistry. A summary of the results is found in Table 1. The MDI/PTMO ratio of the fractions which have been used in this study varied from 3.2 to 2.2. Two regions of the original polymer molecular weight distribution had low MDI content. Principal components analysis of the FTIR spectra suggest the differences between the fractions can be described adequately by three components: a hard segment content, an aliphatic hydrocarbon component, and a component suggesting differences in the urea hydrogen bonding. A more detailed description of the preparation and analysis of these fractions is presented elsewhere [22, 23].

Table 1.
Characteristics of samples before treatment

Sample	M_w	M_n	Bulk MDI/PTMO[a]	Surface N/C (×100)[b]
33	693653	574917	3.2	4.9
35	572830	428596	3.1	4.7
37	421035	323820	2.7	2.4
39	375187	252575	2.3	2.3
41	290678	184333	2.4	2.6
43	233622	144729	2.5	4
45	200159	96098	2.7	3.5
47	139150	53240	2.3	3.3
49	91990	35002	2.2	3.4
51	72234	18691	2.3	4.3

[a] Molar ratios determined from FTIR measurements
[b] Molar ratios determined from ESCA measurements

EXPERIMENTAL

Oxidative degradation studies

Dried samples of each Biomer™ fraction were dissolved at a concentration of 20 mg ml^{-1} in 1,1,1,3,3,3-hexafluoro-2-propanol (HFIP). Samples were prepared for the degradation experiments by evaporating 150 μl of solution onto clean 9-mm diameter glass cover slips. The cast films weighed between 2 and 3 mg and were 15–20 μm thick.

Samples of each Biomer™ fraction were exposed to 5% hydrogen peroxide, 30% hydrogen peroxide or a distilled water control for 24 h at 37°C. The concentration of the hydrogen peroxide solutions was assayed by titration with potassium permanganate. The Biomer™ samples delaminated from the glass coverslips within a few seconds of addition of the liquid so that both surfaces of all samples were exposed to the media. The samples were thick enough that they could easily be remounted on the coverslips as they were removed from the treatment solutions. Following treatment, the samples were dip rinsed in deionized water and then dried in a vacuum desiccator overnight. The samples were then examined with XPS for surface changes. Following XPS analysis, half of each sample was dissolved at 1.0 mg ml^{-1} in DMAC and filtered through 0.5-μm Teflon Millipore filters for HPGPC analysis. The other half was dissolved at 20 mg ml^{-1} in HFIP and cast on an NaCl crystal for FTIR analysis.

High pressure gel permeation chromatography

High pressure gel permeation chromatography analysis was performed to determine changes in the molecular weight distribution of the polymers. A 20-μl injection of each sample was introduced into a DMAC mobile phase pumped at 1.0 ml min^{-1}. Two Alltech™ C-18 derivatized Nucleosil™ columns, one containing 100 Å pore size packing and the other containing 1000 Å pore size packing, were used. Monodisperse polystyrene standards gave a linear calibration curve over a molecular weight range from $>10^6$ to 500. A Waters™ 410 refractive index detector was interfaced to a computer for data collection and calculation of the molecular weight distributions. Weight and number average molecular weights (M_w and M_n) were calculated using the methods outlined by Yau *et al.* [24]. Each sample was injected five times to permit statistical comparison. A Student's *t*-test at the 95% confidence level was used to determine the significance of changes in M_w, M_n, and P_d. A multivariate statistical test was used to determine the significance of changes in the peak shapes.

The multivariate statistical test was used to ask the question "Did the samples exposed to hydrogen peroxide have a different molecular weight distribution?" Since M_w, M_n, and P_d contain only a small amount of the information available in the chromatogram, *t*-tests of these values are only an indirect comparison of chromatographic peak shapes. A modification of the T^2 test for changes in profile was used for the multivariate statistical test. To test for changes in peak shape, the peaks were normalized to unit area and the derivative of each peak was taken. Too many variables were present in the derivatives of the peaks to use a T^2 test directly, so principal components analysis was used to decrease the number of variables. Principal components were calculated for data sets made up of ten samples of a Biomer™ fraction, five water treated samples, and either five 5% H_2O_2 treated samples or five 30% H_2O_2 treated samples. A discriminant function was then calculated using the scores for the first three principal components. The significance of the separation between the two groups was then determined using a T^2 statistic. A more detailed description of this multivariate peak shape comparison is presented elsewhere [25].

Infrared absorption spectroscopy

FTIR spectra were obtained using a Biorad™ FTS-80A spectrometer. Samples were solution cast onto NaCl crystals from 20 mg ml^{-1} HFIP solution and allowed to dry for a minimum of 2 h at ambient conditions. Spectra were collected for 256 scans at 80 Hz with 2 wavenumber resolution.

X-ray photoelectron spectroscopy

XPS data were obtained on a Surface Science Instruments Model SSX-100 XPS system. A 5-eV electron flood gun was used to offset charge accumulation on the samples. Elemental compositions were determined on the basis of peak areas from the C1*s*, N1*s*, and O1*s* orbitals using spectra collected at a pass energy of 150 eV. The binding environments for C, N, and O were determined using spectra collected at 25 eV pass energy with a 1000 μm diameter X-ray spot size. The binding energy was referenced by setting the C1*s* hydrocarbon peak to 285 eV. The C1*s* peak envelopes were resolved into sub-peaks using the standard SSX-100 software. Gaussian shape peaks were used for all peak fits.

RESULTS

Molecular weight distributions

Figure 1 shows the molecular weight distributions from HPGPC analysis of the samples. The samples treated with water showed less than 1% change in weight average molecular weight compared to untreated samples. This difference was not

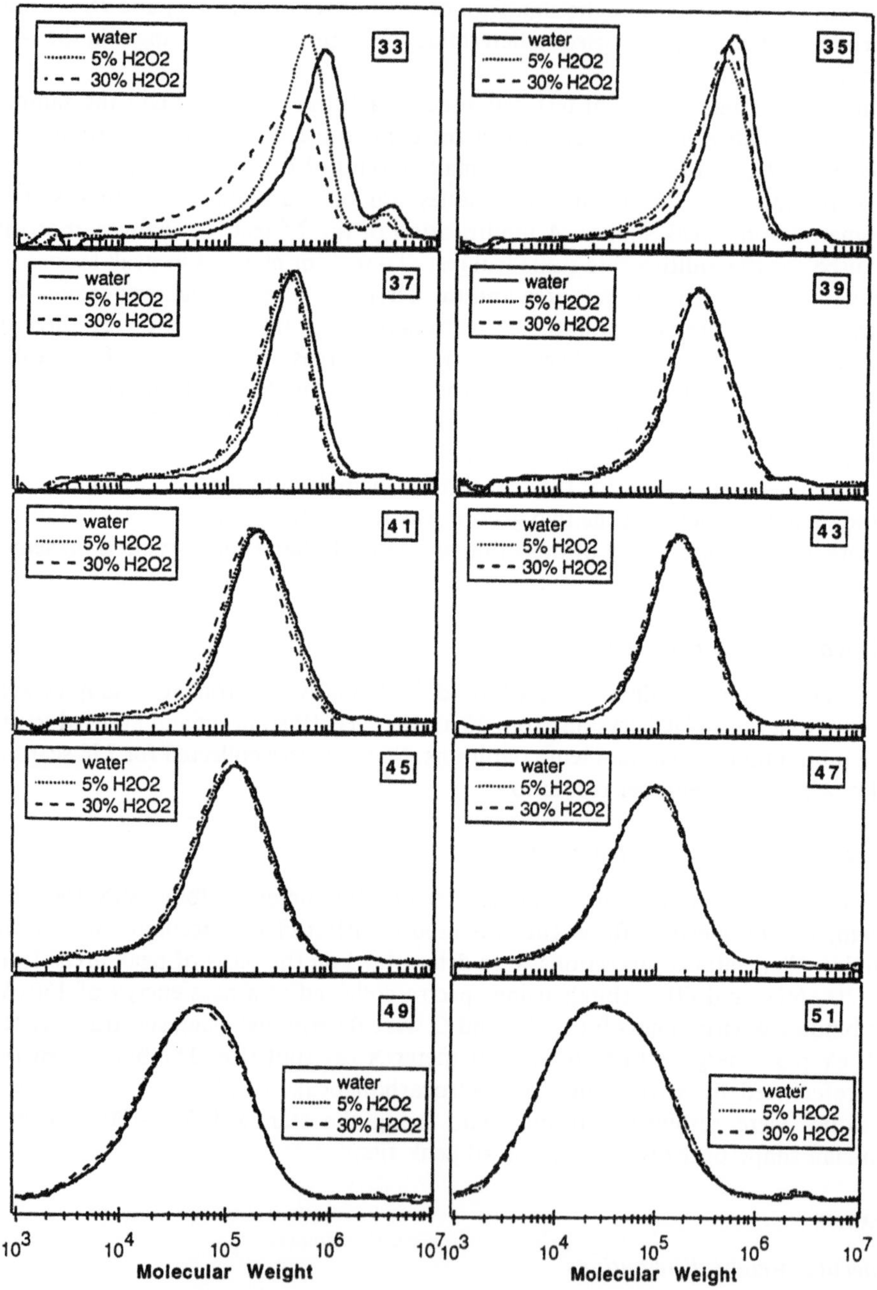
water
5% H2O2
30% H2O2
33
35
37
39
41
43
45
47
49
51
10^3
10^4
10^5
10^6
10^7
Molecular Weight

Figure 1. HPGPC chromatograms for the treated samples. The molecular weight distributions for some samples, for example 33 and 35, change dramatically when exposed to hydrogen peroxide. Other samples, for example 47 and 49, show no change.

found to be statistically significant. There are significant changes in many of the peak shapes for the samples treated with 5 and 30% hydrogen peroxide. The changes in molecular weight distributions for samples treated with 5% hydrogen peroxide are summarized in Table 2. Table 3 contains the same summary for samples treated with 30% hydrogen peroxide. All comparisons were made to samples treated with water, rather than untreated samples. The percent change in weight average molecular weight, number average molecular weight and polydispersity (P_d) are given along with the T^2 value calculated using the first three principal components (PCs) and the percentage of the total variance contained in the first three PCs. The T^2 value is a measure of the change in peak shape based on multivariate comparison of the chromatogram derivatives. The percentage variance in the first three PCs indicates the amount of the total information in the chromatograms that was used to calculate T^2. The statistical significance of changes in molecular weight, molecular number and polydispersity were determined by a Student's *t*-test.

Table 2.
Change in molecular weight distribution for samples treated with 5% hydrogen peroxide

Sample	% M_w change	% M_n change	% P_d change	T^2 PCs 1-3	% var in PCs 1-3
33	−34.4[b]	−33.2[b]	2.1	749.5[b]	83.0
35	−20.0[b]	−31.6[b]	17.2[b]	563.4[b]	81.4
37	−4.4[a]	−19.4[b]	18.8[b]	193.6[b]	68.5
39	−5.0	−5.5	0.3	5.8	52.5
41	−8.8[a]	−18.9[b]	12.2[a]	42.8[b]	55.6
43	−2.7	−6.1	3.5	147.5[b]	59.3
45	−4.7[b]	−5.9[b]	1.4	42.6[b]	55.7
47	−1.0	−6.3	7.2	10.4	56.3
49	1.9	−0.2	2.3	9.9	50.9
51	7.0[a]	9.1	−2.1	66.1[b]	55.7

[a] Indicates change was significant at 95% level
[b] Indicates change was significant at the 99% level

Table 3.
Change in molecular weight distribution for samples treated with 30% hydrogen peroxide

Sample	% M_w change	% M_n change	% P_d change	T^2 PCs 1-3	% var in PCs 1-3
33	−55.8[b]	−76.0[b]	84.8[b]	1489.2[b]	87.0
35	−13.5[b]	−22.7[b]	12.1[b]	411.9[b]	78.9
37	−10.5[b]	−16.5[b]	7.2[b]	291.3[b]	77.0
39	−11.1[b]	−7.4	3.2	100.3[b]	65.1
41	−21.0[b]	−25.9[b]	6.3[a]	703.0[b]	67.6
43	−9.3[b]	−8.7	0.5	213.3[b]	63.2
45	−10.0[b]	−6.0[a]	−4.0	132.1[b]	63.9
47	−2.7	−7.2	6.5	73.7[b]	56.1
49	−3.2	−9.3	7.5	20.4	48.8
51	3.9	8.6	−4.6	32.5	50.8

[a] Indicates change was significant at 95% level
[b] Indicates change was significant at the 99% level

A statistically significant change in the molecular weight distribution (based on the T^2 value) was found in the samples 33, 35, 37, 41, 43, 45, and 51 when they were treated with 5% hydrogen peroxide solution. The change in molecular weight distribution for samples 33, 35, 37, 39, 41, 43, 45, and 47 was statistically significant when they were treated with 30% hydrogen peroxide. A large difference in the extent of degradation of the samples was also observed. The largest changes in molecular weight distribution were observed for the two highest molecular weight samples (33 and 35). These two samples also have the highest hard segment content of the series. Sample 41 also showed a large change in the molecular weight distribution when it was treated with 30% hydrogen peroxide, although it has a below average hard segment content.

Bulk chemistry

FTIR revealed significant changes in the bulk chemistry for several of the peroxide treated samples. Table 4 gives the MDI/PTMO ratio calculated for the samples. The MDI/PTMO ratio was calculated using a PLS model that incorporated the absorbances of every fourth wave number from 3600 to 2550 cm^{-1} and from 1768 to 900 cm^{-1}. The hard segment content of samples 33, 35, and 37 decreased when the samples were exposed to 5 and 30% hydrogen peroxide. The hard segment content of sample 45 decreased only when exposed to 30% peroxide. Changes in the hard segment ratios of the other polymers were within the error bounds of the PLS model. Figures 2–4 show the difference between samples treated with water and hydrogen peroxide for samples 33, 35, and 45. The difference spectra were obtained by subtracting spectra that had been normalized to the sum of the absorbance at 1111, 2855, and 2941 cm^{-1}. The hard segment peaks both for urea and urethane have decreased after exposure to hydrogen peroxide. Changes in the C—H stretching region are also evident, particularly for the samples treated with 30% hydrogen peroxide.

Surface chemistry

XPS also revealed changes in surface chemistry for the peroxide treated samples. The water treated samples showed no change. Table 5 lists the surface N/C (at.%)

Table 4.
Molar MDI/PTMO ratio[a] of samples before and after treatments

Sample	Original	Water	5% H_2O_2	30% H_2O_2
33	3.2	3.2	2.8	2.2
35	3.1	3.1	2.3	2.1
37	2.7	2.6	2.4	2.2
39	2.3	2.3	2.2	2.3
41	2.4	2.4	2.2	2.1
43	2.5	2.5	2.5	2.5
45	2.7	2.7	2.7	2.2
47	2.3	2.3	2.4	2.3
49	2.2	2.2	2.5	2.3
51	2.3	2.4	2.3	2.5

[a] Determined by FTIR

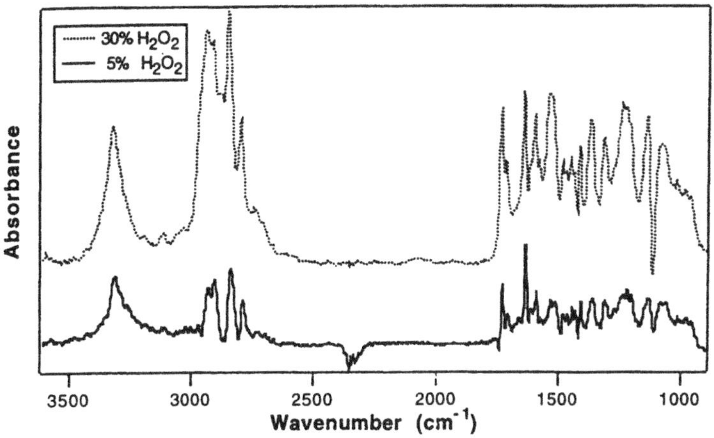

30% H_2O_2
5% H_2O_2
Absorbance
3500
3000
2500
2000
1500
1000
Wavenumber (cm^{-1})

Figure 2. Above are shown the differences between the FTIR spectra for sample 33, which has been soaked in water, and sample 33, which has been soaked in 30 or 5% hydrogen peroxide. The difference spectra show that urethane and urea components are lost when the samples are treated with hydrogen peroxide. The difference spectra also show a loss of aliphatic hydrocarbon. This loss is particularly large for the sample exposed to 30% hydrogen peroxide.

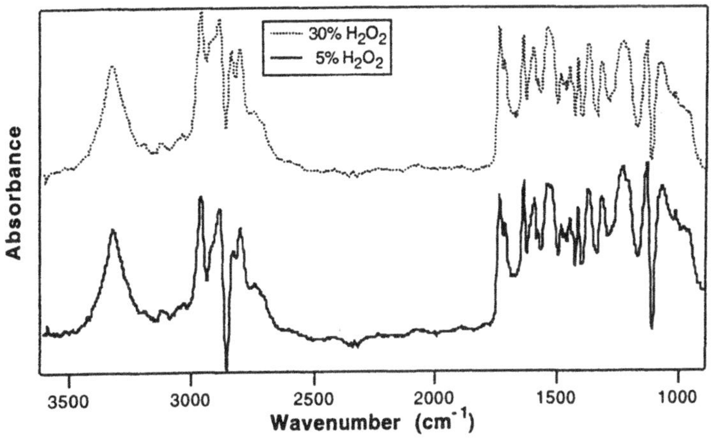

30% H_2O_2
5% H_2O_2
Absorbance
3500
3000
2500
2000
1500
1000
Wavenumber (cm^{-1})

Figure 3. Above are shown the differences between the FTIR spectra for sample 35, which has been soaked in water, and sample 35, which has been soaked in 30 or 5% hydrogen peroxide. The difference spectra show that urethane and urea components are lost when the samples are treated with hydrogen peroxide. The difference spectra also show a loss of aliphatic hydrocarbon.

ratio for the samples. The surface N/C of samples 33, 35, 43, and 51 decreases after exposure to hydrogen peroxide but the N/C ratio of the other samples increases. Changes in the C1*s* spectra for the samples correspond to the N/C ratios. Figures 5–7 show C1*s* spectra for samples 33, 35, and 45. For samples 33 and 35, which show a decrease in N/C ratio when treated with hydrogen peroxide, the peak from urethane and urea carbon at 289 eV decreased and the peak from ether carbon at 286.5 eV increased. For sample 45, which shows an increase in the surface N/C ratio when treated with peroxide, the peak for urethane and urea carbon increased. The decreases in the surface N/C ratio for samples 33 and 35 correspond with the

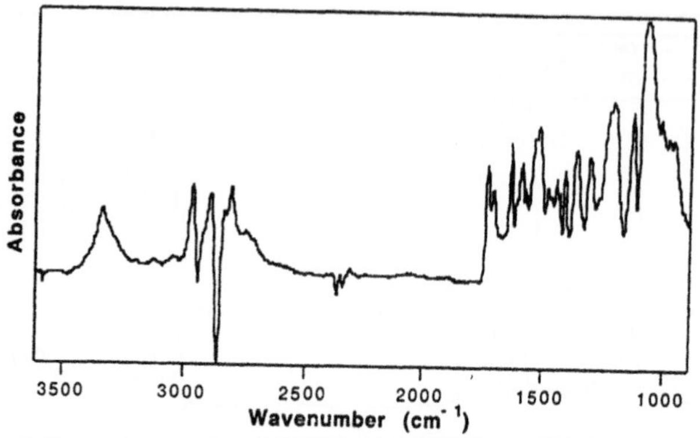

Absorbance
3500
3000
2500
2000
1500
1000
Wavenumber (cm⁻¹)

Figure 4. Above are shown the differences between the FTIR spectrum for sample 41, which has been soaked in water, and sample 41, which has been soaked in 30% hydrogen peroxide. The spectrum shows that urethane and urea components are lost when the samples are treated with hydrogen peroxide. The difference spectra also show a loss of aliphatic hydrocarbon.

Table 5.
Surface atomic N/C ratio[a] (×100) for treated samples

Sample	Water	5% H_2O_2	30% H_2O_2
33	4.9	3.2	2.5
35	4.7	2.9	2.3
37	2.4	2.7	2.7
39	2.3	2.3	2.6
41	2.6	3.3	2.9
43	4.0	2.5	3.2
45	3.5	3.9	4.2
47	3.3	4.7	4.3
49	3.4	3.9	4.5
51	4.3	3.7	3.2

[a] Determined by XPS

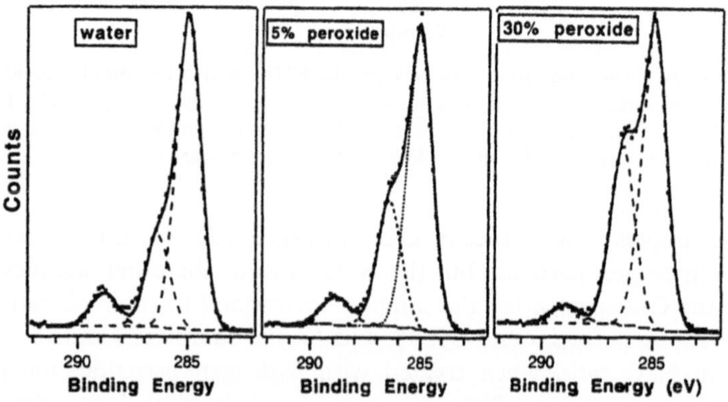

Counts
water
5% peroxide
30% peroxide
290
285
Binding Energy
290
285
Binding Energy
290
285
Binding Energy (eV)

Figure 5. The XPS C1*s* peaks for sample 33, which has been soaked for 24 h in water, 5% hydrogen peroxide or 30% hydrogen peroxide. The C1*s* peaks for the samples that have been exposed to hydrogen peroxide show an increase in the ether peak at 286.5 eV and a decrease in the urethane and urea peak at 289 eV.

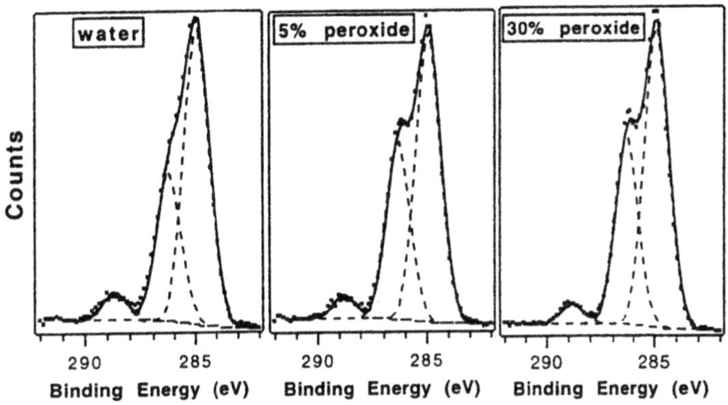
water
5% peroxide
30% peroxide
Counts
290
285
Binding Energy (eV)

Figure 6. The XPS C1*s* peaks for sample 35, which has been soaked for 24 h in water, 5% hydrogen peroxide or 30% hydrogen peroxide. The C1*s* peaks for the samples that have been exposed to hydrogen peroxide show an increase in the ether peak at 286.5 eV and a decrease in the urethane and urea peak at 289 eV.

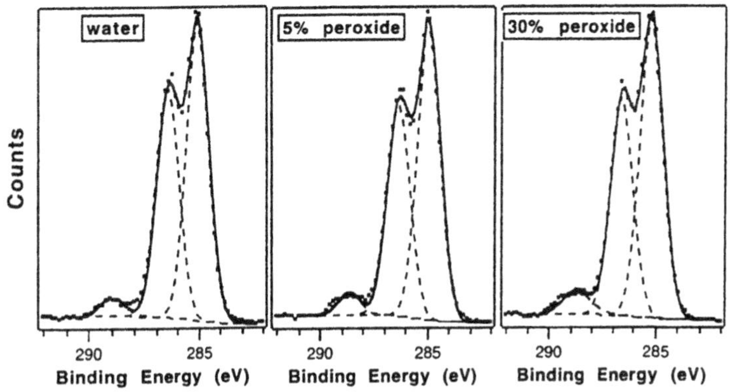
water
5% peroxide
30% peroxide
Counts
290
285
Binding Energy (eV)

Figure 7. The XPS C1*s* peaks for sample 45, which has been soaked for 24 h in water, 5% hydrogen peroxide or 30% hydrogen peroxide. The C1*s* peaks for the samples that have been exposed to hydrogen peroxide show an increase in the urethane and urea peak at 289 eV.

changes seen in the bulk. However, the changes in N/C ratio for the other samples do not parallel those in the bulk, which indicates that the reactions that dominate in the surface region are different from those in the bulk polymer.

When the samples of Biomer™ lot BSUA-001 were exposed to hydrogen peroxide, protonated amines and nitroso groups formed at the surface. These species were attributed to reactions of the DPAEMA with hydrogen peroxide. Because the DPAEMA was removed in the preparative scale GPC procedure, these species were not observed on the surface of any of the fractions either before or after exposure to hydrogen peroxide.

DISCUSSION

For both the 5 and 30% hydrogen peroxide treatments there is a correlation between the hard segment content and the extent of degradation (see Figs 8 and 9). The correlation between the hard segment content for 5% hydrogen peroxide is 0.932

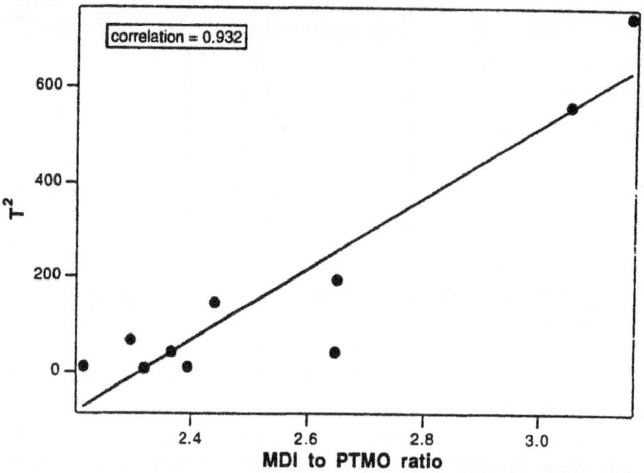
correlation = 0.932
600
400
200
0
T^2
2.4
2.6
2.8
3.0
MDI to PTMO ratio

Figure 8. For samples treated with 5% hydrogen peroxide, the correlation between the extent of degradation, indicated by the T^2 value and the MDI to PTMO ratio of the original samples, is 0.932.

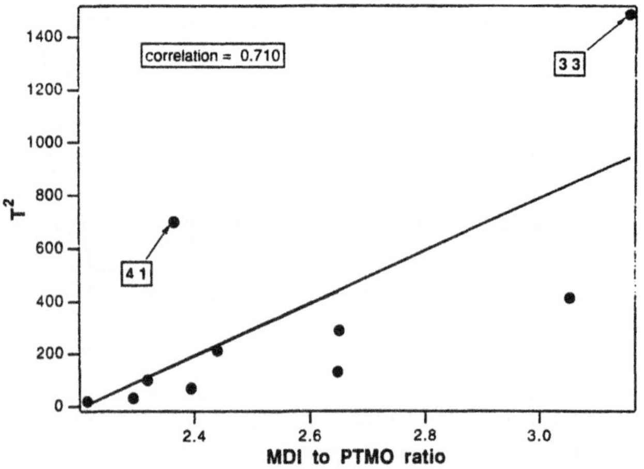
correlation = 0.710
33
41
1400
1200
1000
800
600
400
200
0
T^2
2.4
2.6
2.8
3.0
MDI to PTMO ratio

Figure 9. For samples treated with 30% hydrogen peroxide, the correlation between the extent of degradation, indicated by the T^2 value and the MDI to PTMO ratio of the original samples is 0.710. Samples 33 and 41 deviate noticibly from the correlation seen for the other samples.

and for 30% hydrogen peroxide is 0.710. The correlation is poorer for 30% hydrogen peroxide, largely due to samples 33 and 41. This correlation corresponds well with the data in Table 4 and Figs 2–4, which show decreases in the bulk MDI content for samples with a large extent of degradation. Combined, these two pieces of information suggest that the hard segment is being attacked in the degradation process. Because no evidence for alcohol or acid groups is observed in the spectra of the degraded samples, it is unlikely that the hard segment is being lost as the result of hydrolysis of the urethane or urea linkages.

Earlier studies have suggested that the polyether soft segment is susceptible to oxidative attack [5, 13, 26]. The hydrogen atoms adjacent to the ether linkage are more labile than those in standard aliphatic hydrocarbons; abstraction of these carbons can lead to cleavage of the bond between the α- and β-carbons of the ether.

Oxidative attack of the hard segment has not been previously reported. The attack site may actually be between the urethane and ether. The hydrogens adjacent to the urethane ether are more labile than the aliphatic ether linkages [6]. Abstraction of these hydrogens may lead to cleavage of the polyether chain at the interface between the hard and soft segments. The hydrophilic hard segment fragments would then be dissolved in the solution and lost from the PEU. In order to verify that this is occurring, further experiments measuring the composition of the solution phase would be required.

When hydrogen bonding between the urethane carbonyl and the hydrogen on the urethane nitrogen is not present, a five membered ring structure, caused by hydrogen bonding of the urethane carbonyl to the hydrogens on the carbon adjacent to the urethane ether, can form. Formation of this structure may facilitate cleavage of the PEU. The FTIR spectra of Biomer™ shows a large fraction of free urethane carbonyls. Hydrogen bonding of the urethane carbonyls is reduced when the hard segment length is not uniform. In a high hard segment containing fraction, the urethane carbonyls are frequently pushed outside the hard segment domains [27] so less hydrogen bonding of the urethane carbonyl occurs. In addition, non-crystalline high hard segment containing fragments will absorb more water in an aqueous environment. This water interferes with hydrogen bonding in the hard segment domains [28].

The extent of degradation for samples treated with 30% hydrogen peroxide is also correlated (0.762) to the aliphatic hydrocarbon component described by PC2 (see Fig. 10). Samples 33 and 41 follow the correlation with PC2 better than they follow the correlation with the MDI to PTMO ratio. For the samples treated with 5% hydrogen peroxide, the correlation between the extent of degradation and the scores for PC2 is 0.699. The difference between the FTIR spectra of sample 33, exposed to water, and sample 33, exposed to hydrogen peroxide, shows a large decrease in the aliphatic C—H stretching region. A comparable loss is not observed in the C—O stretching region. This corresponds well to the characteristics of PC2. PC2

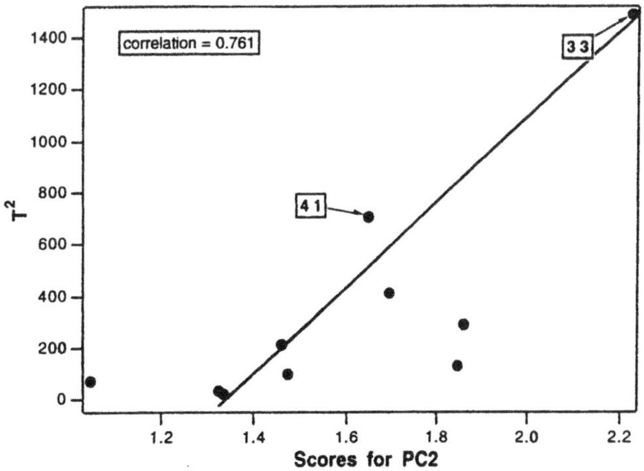

correlation = 0.761
33
41
1400
1200
1000
800
600
400
200
0
T^2
1.2
1.4
1.6
1.8
2.0
2.2
Scores for PC2

Figure 10. For samples treated with 30% hydrogen peroxide, the correlation between the extent of degradation, indicated by the T^2 value and the scores for PC2, is 0.762. Samples 33 and 41 are described better by this correlation than by the correlation between T^2 and the MDI to PTMO ratio.

suggests that an aliphatic diol, in addition to PTMO, was present in Biomer™ lot BSUA-001. The correlation between this component and the extent of degradation by 30% hydrogen peroxide, as well as the difference spectra for the degraded samples, indicate that this aliphatic portion of the PEU is lost in the degradation process.

The changes in the surface chemistry suggest that different reactions may dominate at the surface of some of the components. Samples 41, 45, 47, and 49 show an increase in the surface N/C ratio and an increase in the carbon signal due to urethane and urea type carbon. This change could reflect rearrangement of the polymer. Because the change is not seen in samples exposed to water only, it is more likely that the ether component at the surface has been degraded by the hydrogen peroxide. Earlier investigations indicate that this process is common in PEU based polyurethanes [13, 26]. SIMS studies of the oxidative degradation of Biomer™ lot BSP-067 suggested cleavage of the PTMO chains between the α- and β-carbons [26].

CONCLUSIONS

PSGPC fractions of the commercial medical grade PEU Biomer™ reveal variations in the PEU chemistry as a function of molecular weight. When exposed to 5 and 30% hydrogen peroxide solutions, the PSGPC fractions reacted in different ways. The molecular weight of the highest molecular weight fraction decreased 55.8%. The molecular weight of the lowest molecular weight fraction increased 3.9%. The extent of degradation of the PEU samples correlated strongly with the hard segment content of the PEUs and with the content of an aliphatic hydrocarbon moiety. The results suggest that the hydrogen peroxide may be initiating cleavage of the PEU chains at the urethane ether linkage as well as cleavage at the PTMO ether linkage.

PEUs of the type commonly used in medical applications are highly heterogeneous polymers. This heterogeneity arises from the kinetics of competing reactions in the synthesis of the polymers. If reaction conditions are not adequately reproduced from one PEU batch to another, the heterogeneity of the PEUs would be expected to show batch to batch variation as well. As a result of this heterogeneity, some fractions of the polymer are more susceptible to attack by hydrogen peroxide than other fractions. This work illustrates the importance of minor components of a PEU in the biostability of the polymer. The need for quality control in production of medical grade PEUs is clear.

Some fractions prepared from the Biomer™ were not significantly affected by concentrated hydrogen peroxide solutions. These fractions appear to be free of the moieties that are highly susceptible to oxidative attack. This result suggests possibilities for the synthesis of degradation resistant PEUs.

REFERENCES

1. S. K. Hunter, B. E. Gregonis, D. L. Coleman, J. D. Andrade and T. Kessler, *Trans. Am. Soc. Artif. Intern. Organs* **28**, 473 (1982).
2. R. E. Marchant, J. M. Anderson, K. Phua and A. Hiltner, *J. Biomed. Mater. Res.* **18**, 309 (1984).
3. Q. Zhao, R. E. Marchant, J. M. Anderson and A. Hiltner, *Polymer* **28**, 2040 (1987).
4. R. E. Marchant, Q. Zhao, J. M. Anderson and A. Hiltner, *Polymer* **28**, 2032 (1987).
5. K. Hayashi, T. Matsuda, H. Takano and M. Umezu, *J. Biomed. Mater. Res.* **18**, 939 (1984).
6. A. Takahara, A. J. Coury, R. W. Hergenrother and S. L. Cooper, *J. Biomed. Mater. Res.* **25**, 341 (1991).

7. B. D. Ratner, K. W. Gladhill and T. A. Horbett, *J. Biomed. Mater. Res.* **22**, 509 (1988).
8. R. W. Paynter, H. Martz and R. G. Guidoin, *Biomaterials* **8**, 93 (1987).
9. Y. Wu, Q. Zhao, J. M. Anderson, A. Hiltner, G. A. Lodoen and C. R. Payet, *J. Biomed. Mater. Res.* **25**, 725 (1991).
10. R. Smith, D. F. Williams and C. Oliver, *J. Biomed. Mater. Res.* **21**, 1149 (1987).
11. C. Batich, J. Williams and R. King, *J. Biomed. Mater. Res.: Appl. Biomater.* **23**(A3), 311 (1989).
12. M. Bouvier, A. S. Chawla and I. Hinberg, *J. Biomed. Mater. Res.* **25**, 773 (1991).
13. K. Stokes, P. Urbanski and J. Upton, *J. Biomater. Sci., Polymer Edn* **1**, 207 (1990).
14. Q. Zhao, N. Topham, J. M. Anderson, A. Hiltner, G. Lodoen and C. R. Payer, *J. Biomed. Mater. Res.* **25**, 177 (1991).
15. B. D. Ratner, in: *Physicochemical Aspects of Polymer Surfaces*, Vol. 2, p. 969, K. L. Mittal (Ed.). Plenum, New York (1983).
16. L. H. Peebles, *Macromolecules* **9**, 58 (1976).
17. J. M. Richard, W. H. McClennen and H. L. C. Meuzelaar, *J. Appl. Polym. Sci.* **40**, 1 (1990).
18. M. D. Lelah and S. L. Cooper, *Polyurethanes in Medicine*. CRC Press, Boca Raton, FL (1986).
19. J. Belisle, S. K. Maier and J. A. Tucker, *J. Biomed. Mater. Res.* **24**, 1585 (1990).
20. B. J. Tyler, B. D. Ratner, D. G. Castner and D. Briggs, *J. Biomed. Mater. Res.* **26**, 273 (1992).
21. B. J. Tyler and B. D. Ratner, *J. Biomed. Mater. Res.* **27**, 327 (1993).
22. B. J. Tyler, Doctoral Dissertation, University of Washington (1992).
23. B. J. Tyler and B. D. Ratner, submitted (1994).
24. W. W. Yau, J. J. Kirkland and D. D. Bly, *Modern Size Exclusion Liquid Chromatography*. John Wiley, New York (1979).
25. B. J. Tyler, submitted to *Anal. Chem.* (1992).
26. B. J. Tyler and B. D. Ratner, in: *Degradation Phenomena of Polymeric Biomaterials, Proceeding of the 4th International ITV Conference on Biomaterials*. Elsevier, Amsterdam, in press.
27. C. E. Miller, P. G. Edelman and B. D. Ratner, *Appl. Spectrosc.* **44**, 576 (1990).
28. K. G. Tingey, J. D. Andrade, C. W. McGary and R. J. Zdrahala, in: *Polymer Surface Dynamics*, p. 153, J. D. Andrade (Ed.). Plenum Press, New York (1988).

Characterization of extractable species from poly(etherurethane urea) (PEUU) elastomers

M. RENIER[1], Y. K. WU[1], J. M. ANDERSON[2]
A. HILTNER[1]*, G. A. LODOEN[3] and C. R. PAYET[4]

[1]*Department of Macromolecular Science and*
[2]*Institute of Pathology, Case Western Reserve University, Cleveland, OH 44106, USA*
[3]*E.I. du Pont de Nemours & Company, Inc., Waynesboro, VA 22980, USA*
[4]*E.I. du Pont de Nemours & Company, Inc., Wilmington, DE 19880, USA*

Received 29 June 1992; accepted 26 March 1993

Abstract—Methanol extracts of four poly(etherurethane urea) (PEUU) materials were analyzed using Gel Permeation Chromatography (GPC). The additives in the materials were Santowhite® powder at 1 wt% and Methacrol® 2138 F at 5 wt% loading levels. One-to-two wt% of the original PEUU films was extractable with methanol. The extractables consisted of a low molecular weight (*Mw*) PEUU polymer, an MDI-rich oligomer, the additives Santowhite® (SW) powder and Methacrol® 2138 F, and aniline. The low *Mw* PEUU polymer had a M_w of 12 000 relative to polystyrene, and the MDI-rich oligomer had a M_w of 1000 relative to polystyrene. Quantitation of all extracted species was achieved using GPC; the use of dual-detectors on the GPC made it possible to determine the soft-to-hard composition of the PEUU extracts as a function of molecular weight.

Key words: Poly(etherurethane urea) (PEUU); additives; extraction; gel permeation chromatography (GPC); solubility parameter.

INTRODUCTION

Poly(etherurethane ureas) (PEUUs) are materials that possess good blood compatibility and mechanical properties [1]. Their uses include: vascular prostheses, ventricular assist bladders, heart valves, pacemaker lead wire insulation, and catheters [2–5]. PEUUs used in implanted devices are complex formulations of polymer and possibly other ingredients: antioxidants, catalysts, solvents, processing aids, and trace impurities. Antioxidants are employed to stabilize PEUUs from oxidative degradation [6]. The extraction of these constituents and low molecular weight fractions of polymer may have considerable influence on the PEUU matrix and also the biocompatibility of the PEUU. The loss of material from the PEUU matrix may cause voids which allow for the penetration of biological media such as enzymes which are known to degrade PEUU materials [7–11]. The surface deposited fatty amide extrusion lubricant, which can be extracted by organic solvents, may affect the biocompatibility of the Pellethane series of poly(ether urethanes) (PEUs) [12, 13]. Another important consideration of the extractables is the fate of leachable materials in an *in vivo* environment. The leachability of possible degradation products from PEU-covered breast implants has been an area of wide discussion and controversy [14]. However, the other viable extractable constituents of the PEU have not been studied for their potential as *in vivo* irritants.

* To whom correspondence should be addressed.

The extraction of PEUUs may be influenced by kinetic and thermodynamic factors. The kinetic factors may include: extraction temperature, length of extraction period, amount of extraction media, and amount of polymer surface area. Absorption and permeation phenomena in an *in vivo* or *in vitro* environment may be controlled by thermodynamic factors [15, 16]. The main thermodynamic factor is the difference in the Hildebrand solubility parameter between the penetrant and the polymer. The biological constituents of human or animal plasma vary in solubility parameter from 16.2 MPa$^{0.5}$ for cholesterol esters to 32.0 MPa$^{0.5}$ for phospholipids [17]. The true overall solubility parameter of human or animal blood plasma is too complex to measure, since it contains various components such as electrolytes, proteins, and lipids in a dynamic environment. Previous reviews on the extraction of leachables from polymers have emphasized the use of the Hildebrand solubility parameter [16]. In this study, methanol was used to extract PEUUs because its solubility parameter 28.7 MPa$^{0.5}$ is similar to those of the PEUU 23.0 MPa$^{0.5}$ and the two additives, Santowhite® powder 25.9 MPa$^{0.5}$ and Methacrol® 2138 F 18.8 MPa$^{0.5}$. Results of the extraction are correlated with differences in the solubility parameters.

MATERIALS AND METHODS

Materials

Four materials based on a single poly(etherurethane urea) (PEUU) but differing in additives were provided by the E.I. du Pont Co. The base polymer was prepared from poly(tetramethylene glycol) (nominal Mw = 2000) (PTMEG) and 4,4′-methylene bis(phenyl isocyanate) (MDI) in an approximately 1.0–1.6 capping ratio. The polymer was chain extended primarily with ethylenediamine; in addition, a small quantity of aromatic monoamine was added as an end-capping agent for molecular weight control. The base polymer had the average molecular weights of M_n = 15 000–20 000, and M_w = 40 000–50 000 relative to polystyrene standard. The additives were blended into the PEUU solution after polymerization. The four materials were as follows: material A was the base polymer without additives, material B was the base polymer with 1 wt% Santowhite® (SW) powder, material C was the base polymer with 5 wt% Methacrol® 2138 F (MT), and material D was the base polymer with both 1 wt% SW and 5 wt% MT. Methacrol® 2138 F is a poly(diisopropylaminoethyl methacrylate-co-decyl methacrylate) (DIPAM-co-DM) copolymer that provides resistance to chloride and hypochlorite ion attack and ultraviolet (UV) degradation, while SW(4,4′-butylidene bis(6-tert-butyl-*m*-cresol)) is a phenolic antioxidant [18]. The chemical structures of the base PEUU and the additives are shown in Fig. 1. Two poly(etherurethane) prepolymers and a poly(tetramethylene glycol) (uncapped polyol) were also supplied by E.I. du Pont Co. The two prepolymers were an MDI-capped polyol (CG I) with a nominal molar ratio of 1.0 : 1.6 (polyol : MDI) and a polyol-capped MDI (CG II) with a nominal molar ratio of 1.6 : 1.0 (polyol : MDI).

Sample preparation and extraction method

The PEUU materials were solution cast on Mylar® polyester film from *N,N*-dimethylacetamide. The 0.2 mm thick films were prepared for extraction by cutting each into a 3 cm × 3 cm piece and then removing the Mylar® substrate. Each

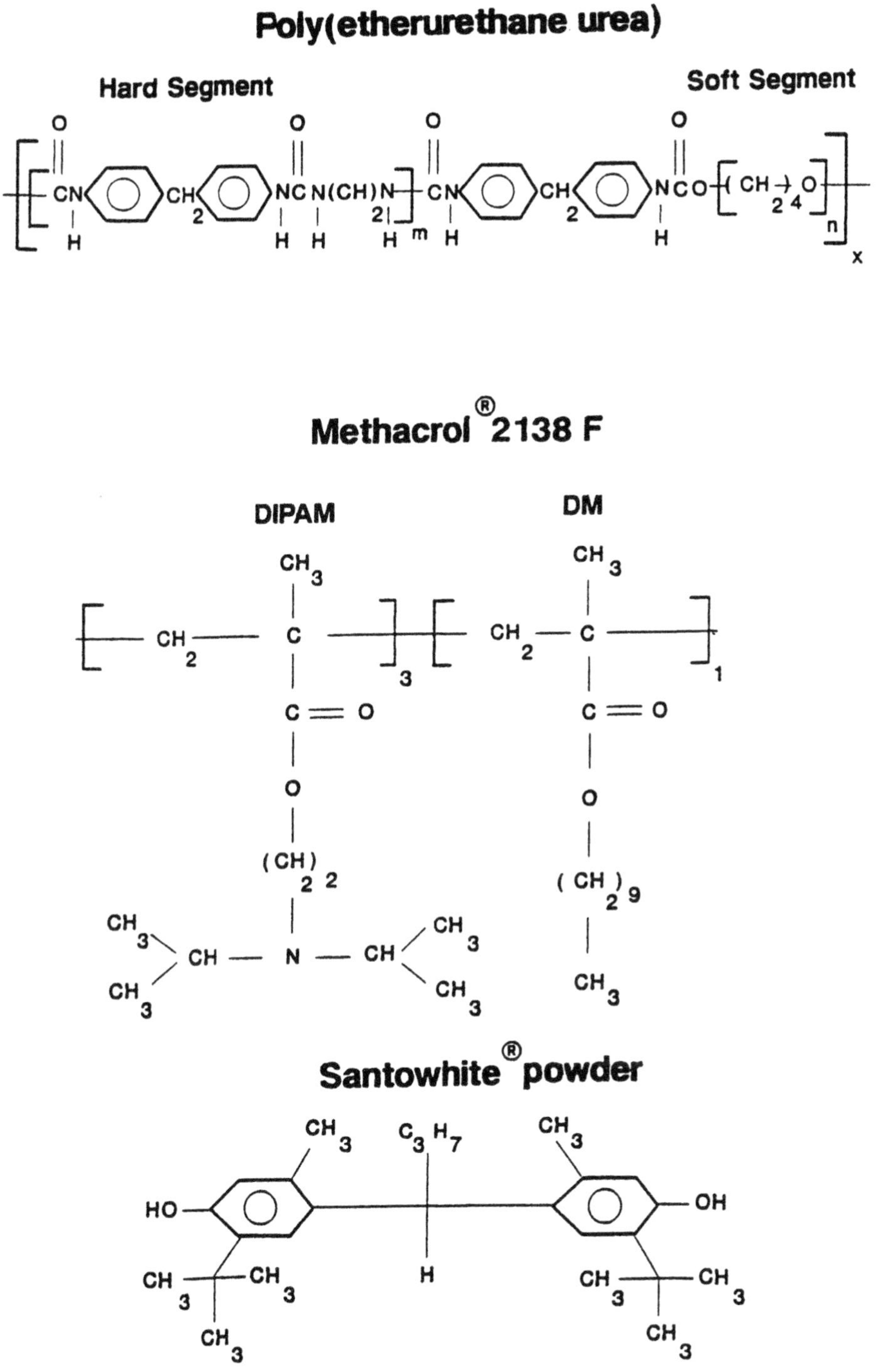
Poly(etherurethane urea)
Hard Segment
Soft Segment
Methacrol® 2138 F
DIPAM
DM
Santowhite® powder

Figure 1. Chemical structures of poly(etherurethane urea) (PEUU), Methacrol® 2138 F, and Santowhite® powder.

specimen was rinsed briefly with 95% ethanol followed by distilled water at room temperature. The materials were vacuum dried for 24 h at 50°C, and stored in a desiccator before extraction. Two specimens were weighed, placed in a flask with 100 ml of HPLC grade methanol, and left in a convection oven at 50°C for 1 week. After extraction, the PEUU films were retrieved, and the solution containing the extractables was concentrated to 1 ml using a rotovaporator. The concentrated extract solution was diluted with 10 ml of HPLC grade tetrahydrofuran (THF) and further concentrated to 1 ml for GPC analysis. The retrieved polymer films were vacuum dried for 24 h at 50°C and weighed.

Instrumental techniques and methods

The instrument employed for GPC analysis was a Varian DS-651 LC Star System with a 9010 Solvent Delivery System, a 9065 Polychrom® UV diode-array detector, and an RI-4 refractive index detector set at 40°C. The eluent was THF and the flow rate was 1.0 ml min^{-1}. Two sets of calibration standards were used to calibrate a series of PL-Gel columns of 500 and 10000 Å pore size. The PS standards had a polydispersity index of $(M_w/M_n) \leq 1.1$, while the PTHF standards had a polydispersity index of $(M_w/M_n) \leq 1.2$. The standards and columns were obtained from Polymer Laboratories, Inc.

The untreated PEUU films were dissolved in THF (0.5 g PEUU per 5 ml THF) with lithium chloride (0.2 g per 100 ml). The PEUU extracts, CG I and CG II were dissolved in THF and analyzed by GPC. The GPC analysis was repeated four times for each film, extract, and prepolymer. Methacrol® 2138 F (MT) was extracted with HPLC grade methanol because it was insoluble in THF, approximately 4 wt% of the MT was extractable with methanol. The extract was concentrated to 1 ml and then analyzed by GPC. Santowhite® powder was dissolved in THF and analyzed by GPC.

Quantitative analysis of the additives in the extracts was accomplished by resolving the GPC chromatograms with spectral curve-fitting functions. The chromatograms were deconvoluted using Gaussian and exponentially-modified Gaussian functions which are known to accurately represent chromatographic peaks [19, 20].

The trace amount of aromatic amine in the PEUU extracts was analyzed using a mass spectrometer equipped with a standard electron impact (EI) source and positive ion detection. The compositions of the prepolymers CG I and CG II were obtained by ^{1}H NMR using a 200 MHz Varian instrument.

RESULTS AND DISCUSSION

Qualitative analysis of extracts

The original PEUU films A–D had a 1–2% weight loss after methanol extraction. The mean weight loss and standard deviation for PEUUs A–D are presented in Table 1. The UV chromatograms of the A–D extracts in Fig. 2a had two major peaks at 13.0 and 16.2 min; the B and D extracts also had a shoulder at 17.0 min. The RI chromatogram of the A extract in Fig. 2b had two major peaks that corresponded to those in the UV chromatograms. The RI chromatograms of the B, C and D extracts also showed the two major peaks, and in addition, the B and D extracts had a very distinct peak that had the same retention time, 17.0 min, as the shoulder in the UV chromatograms. The C and D extracts had additional features

Table 1.
Gravimetric loss of PEUU films A–D after methanol extraction

Sample	Weight loss (%)[a]
A	1.53 ± 0.16
B	1.39 ± 0.32
C	1.79 ± 0.18
D	1.79 ± 0.14

[a] Mean values ± SD, n = 4 measurements.

in the 14.0–16.5 min region. Comparison of the RI chromatograms of C and D extracts with that of the MT extract in Fig. 2b suggested the presence of extractable MT (M_w about 1850) in the C and D extracts. A relatively small peak at 18.8 min for the A, B, C, and D extracts appeared in both the UV and RI chromatograms of all the PEUU extracts indicating the presence of a very low molecular weight extractable component.

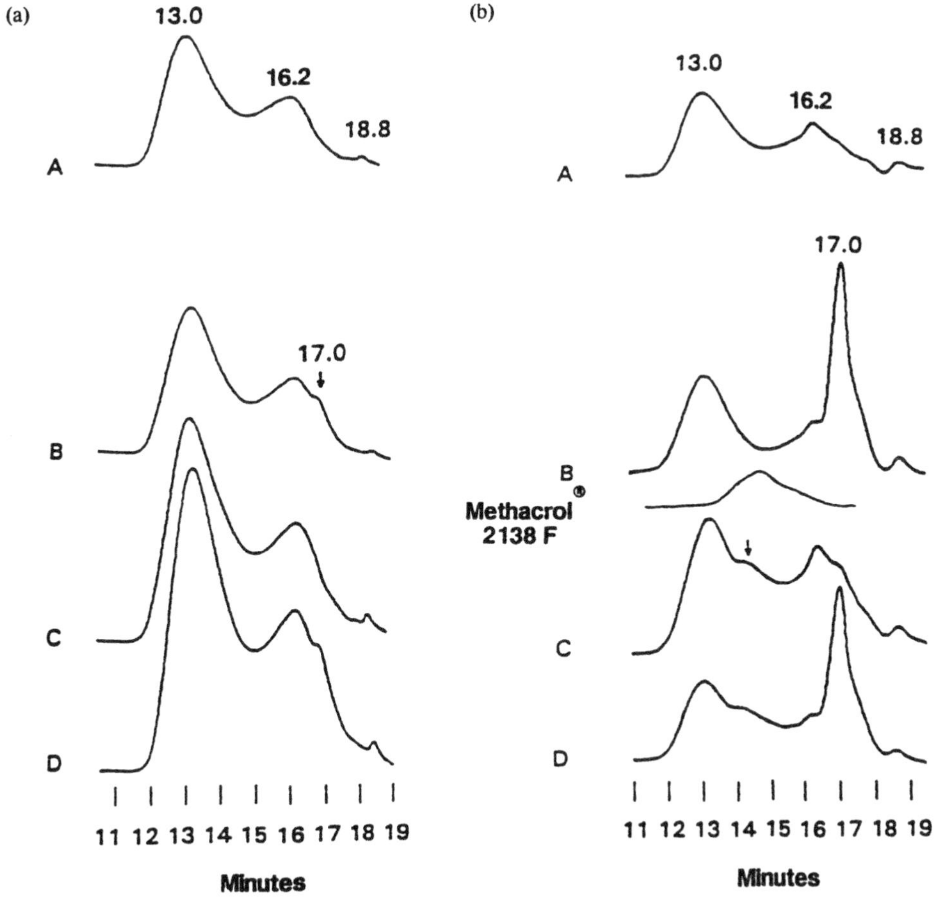
(a)
13.0
16.2
18.8
A
17.0
B
C
D
11 12 13 14 15 16 17 18 19
Minutes
(b)
13.0
16.2
18.8
A
17.0
B
Methacrol®
2138 F
C
D
11 12 13 14 15 16 17 18 19
Minutes

Figure 2. GPC chromatograms of PEUU extracts A–D. (a) UV chromatograms; and (b) RI chromatograms.

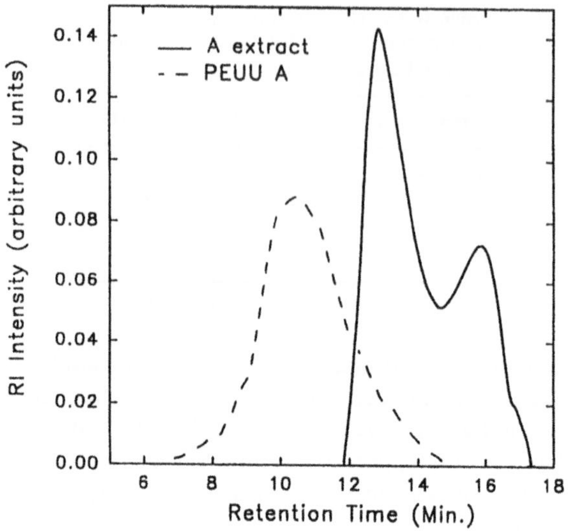
A extract
PEUU A
RI Intensity (arbitrary units)
0.14
0.12
0.10
0.08
0.06
0.04
0.02
0.00
6
8
10
12
14
16
18
Retention Time (Min.)

Figure 3. Comparison of the chromatograms of the first extract peak (——) and PEUU A (- - -) demonstrating overlap of retention times.

Previous investigators have suggested that the first major extract peak in the UV and RI chromatograms may come from a low molecular weight component of the PEUU, while the lower molecular weight extract peak is probably an oligomer of the PEUU [21]. Overlap in retention times of the low molecular weight tail of the PEUU A chromatogram and the first chromatographic peak in the A extract, shown in Fig. 3, supported the hypothesis that the first major extract peak originated from extractable low molecular weight polymer. The retention time of the lower molecular weight extract material overlapped that of the low molecular weight tail of the MDI-capped prepolymer (Fig. 4). Since the MDI-capped prepolymer was the

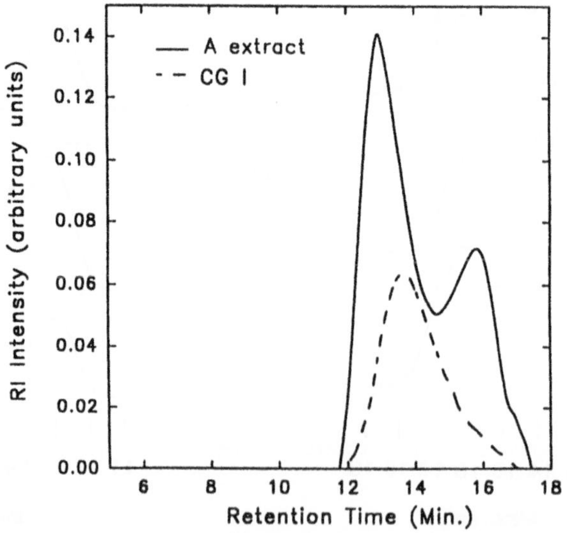
A extract
CG I
RI Intensity (arbitrary units)
0.14
0.12
0.10
0.08
0.06
0.04
0.02
0.00
6
8
10
12
14
16
18
Retention Time (Min.)

Figure 4. Comparison of the chromatograms of the second extract peak (——) and CG I (- - -) demonstrating overlap of retention times.

precursor to the PEUU polymer, the second major extract peak in the A extract chromatogram was probably an MDI-rich oligomer.

Identification of the 17.0 and 18.8 min peaks in the extracts was facilitated by using a UV diode-array detector. The UV detector incorporates a 38-channel linear photodiode array that yielded data in three domains: time, absorbance, and wavelength [22]. This detection system allowed the spectrum to be taken at any time point on the chromatogram. The UV spectrum at the 17.0-min peak in the chromatograms of the B and D extracts is compared with the UV spectrum of pure SW powder in Fig. 5a. When the two spectra were normalized and overlaid for a qualitative comparison, the similarity confirmed the presence of extractable SW powder in the

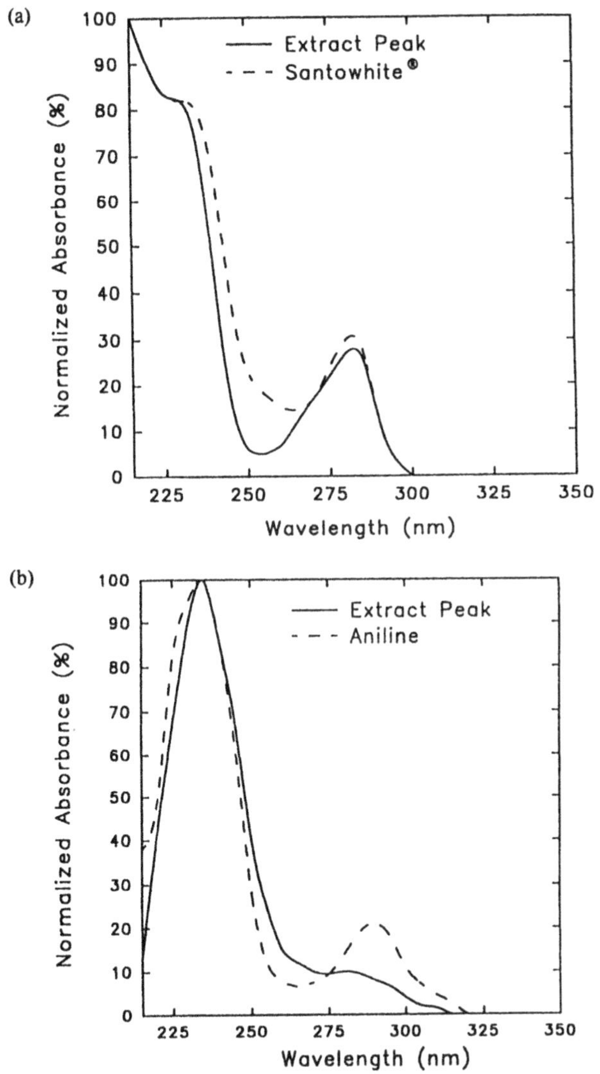
(a)
Extract Peak
Santowhite®
Normalized Absorbance (%)
Wavelength (nm)
(b)
Extract Peak
Aniline
Normalized Absorbance (%)
Wavelength (nm)

Figure 5. Identification of the extract components by an UV spectral comparison. (a) The 17.0-min extract peak in the B and D extracts (——) and pure Santowhite® powder (- - -); and (b) The 18.8-min extract peak in PEUUs A–D (——) and pure aniline (- - -).

B and D extracts. Phenolic antioxidants such as SW powder have strong UV absorption maxima at 235 nm (E_2 band) and 287 nm (B band) due to the aromatic ring, whereas the UV absorption at 254 nm is weak for SW [23].

Confirmation that a trace of aniline or another aromatic amine was present in the extracts was provided by a comparison of the normalized UV spectrum of the 18.8-min peak in the A–D extracts and pure aniline (Fig. 5b). Aniline and other aromatic amines have UV absorption maxima at 230 nm (E_2 band) and 280 nm (B band) due to the aromatic ring [24]. Analysis of the A extract by mass spectroscopy revealed a molecular ion peak at 93 m/z, the mass corresponding to the molecular weight of aniline [25]. Aniline could have originated as an impurity in the PEUU polymerization since it is a starting reagent for 4,4′-methylene bis(phenyl isocyanide) (MDI). In other preparations of the PEUU, aniline was not detected. Aniline might have survived the polymerization since the reaction of MDI with aromatic amines is much slower than reaction with aliphatic amines.

Quantitative analysis of extract components

The molecular weight (*Mw*) and molecular weight distribution (*MwD*) values for the two major GPC peaks in the PEUU extracts are summarized in Table 2. The first set of *Mw* and *MwD* data was calculated using polystyrene (PS) standards with UV detection. The first extract peak had an *Mw* of about 12 000 and the second extract peak had an *Mw* of about 1000. Molecular weights and molecular weight distributions of the A–D extracts were similar.

The PTHF standards, which have the same chemical structure as the PEUU soft segment, gave lower *Mw* values for the PEUU extracts with RI detection than the PS standards. Relative to PTHF, low molecular weight PEUU had an *Mw* of about

Table 2.
Molecular weight and molecular weight distribution values for PEUU extracts A–D[a]

Extract	Peak	M_n	M_w	M_z	PDI
		Polystyrene standards (UV detector)			
A	1	12 200 ± 300	16 300 ± 500	20 300 ± 700	1.34
	2	1030 ± 20	1130 ± 20	1240 ± 10	1.10
B	1	12 800 ± 200	17 200 ± 300	21 800 ± 400	1.34
	2	920 ± 20	1040 ± 20	1180 ± 10	1.13
C	1	11 700 ± 200	15 600 ± 300	19 700 ± 400	1.34
	2	1010 ± 20	1110 ± 20	1220 ± 20	1.10
D	1	11 900 ± 300	15 700 ± 600	19 700 ± 700	1.33
	2	890 ± 30	1010 ± 31	1160 ± 30	1.14
		Polytetrahydrofuran standards (RI detector)			
A	1	8800 ± 470	12 100 ± 700	15 300 ± 900	1.35 ± 0.02
	2	510 ± 10	592 ± 10	690 ± 20	1.16 ± 0.03
B	1	9300 ± 300	13 100 ± 400	17 500 ± 900	1.41 ± 0.06
	2	—	—	—	—
C	1	7700 ± 100	10 800 ± 200	13 700 ± 300	1.40 ± 0.01
	2	460 ± 10	520 ± 20	600 ± 10	1.14 ± 0.01
D	1	7600 ± 200	10 800 ± 300	13 900 ± 500	1.42 ± 0.01
	2	—	—	—	—

[a] Mean values ± SD, $n = 4$ measurements.

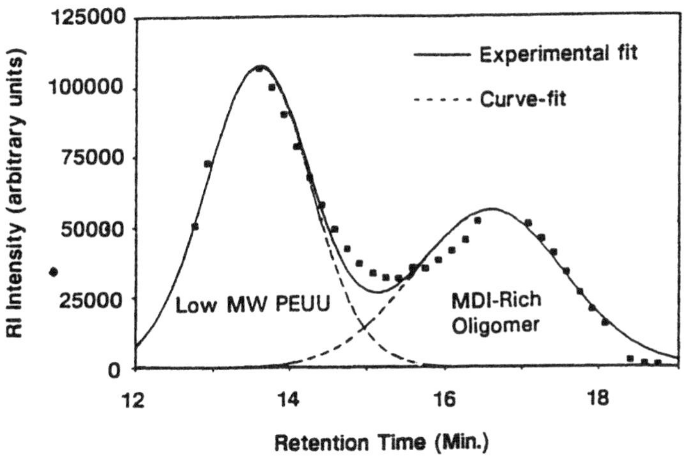

125000
100000
75000
50000
25000
0
12
14
16
18
RI Intensity (arbitrary units)
Retention Time (Min.)
Experimental fit
Curve-fit
Low MW PEUU
MDI-Rich
Oligomer

Figure 6. Deconvolution of the RI chromatogram of A extract into its two polymer components, low molecular weight PEU polymer and the MDI-rich oligomer, experimental line (——), curve-fitted line (- - -).

8800 for the A and B extracts and an approximate *Mw* of 7700 for the C and D extracts. This apparent difference between the A and B extracts compared to the C and D extracts was caused by overlap with extractable MT in the lower molecular weight portion of the RI chromatogram. The oligomer in the A and C extracts had an *Mw* of about 500 relative to PTHF. The *Mw* and *MwD* calculations for the oligomer peak were not attempted for the B and D extracts since the chromatograms contained an intense overlapping SW peak. The explanation for the substantially higher *Mw* values using the PS standards lies in the *Q* factor, the parameter which is used to relate *Mw* to molecular size. The value of *Q* for PS is about 41, while that for PTHF is approximately 18 at ambient temperature [26].

Quantitation of the extractable additives, SW powder and MT, along with low molecular weight PEUU, MDI-rich oligomer, and aniline, was accomplished using spectral curve-fitting functions to deconvolute the GPC RI chromatograms. Figure 6 shows the RI chromatogram of the A extract deconvoluted into the two polymer components with each peak fitted to a Gaussian distribution. The first step in deconvolution of the chromatograms of the B, C, and D extracts was to normalize the chromatograms to the chromatogram of the A extract using the relative weight loss on extraction. The next step was to subtract the polymer components by subtracting the RI chromatogram of the A extract. The multiplication factor that gave the best subtraction was used to obtain the relative amounts of total polymer components in the different extracts. Since the relative areas of the two polymer peaks were the same in all extracts, it was not necessary to subtract the two polymer components separately. After subtraction of the polymer components, the remaining additive peaks were each fitted with an exponentially-modified Gaussian function.

Figure 7a displays the normalization of the B extract to the A extract, and Fig. 7b provides the resolved SW peak after subtraction of the polymer components curve-fitted with an exponentially-modified Gaussian function. The same procedure was used for the C and D extracts, and the results are shown in Figs 8 and 9 respectively.

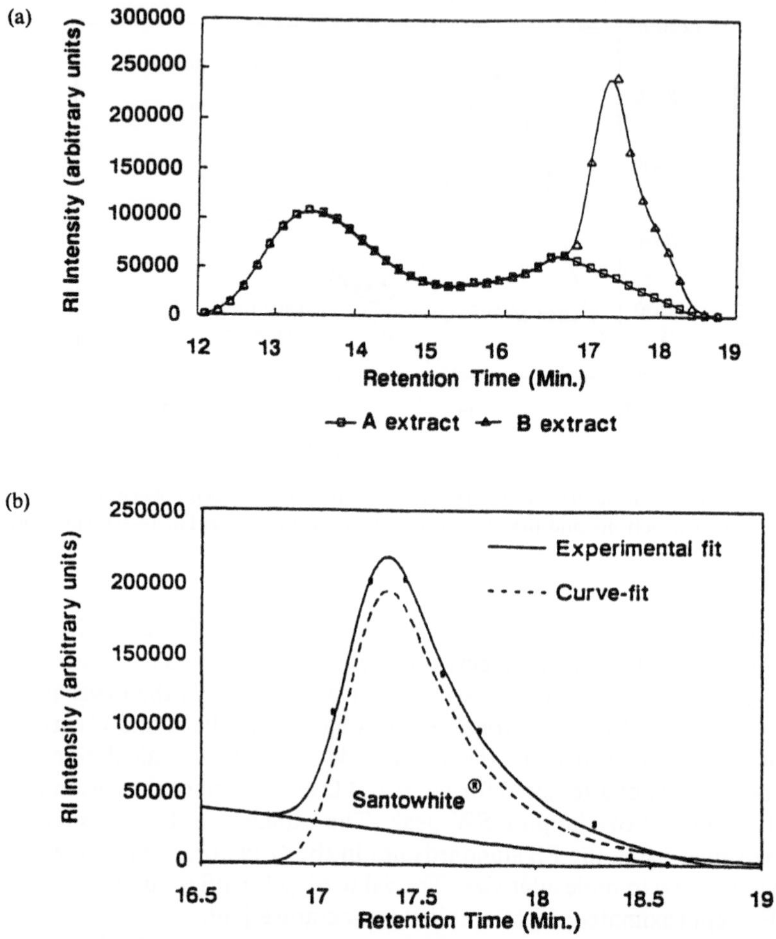
(a)
RI Intensity (arbitrary units)
300000
250000
200000
150000
100000
50000
0
12
13
14
15
16
17
18
19
Retention Time (Min.)
A extract
B extract
(b)
RI Intensity (arbitrary units)
250000
200000
150000
100000
50000
0
Experimental fit
Curve-fit
Santowhite®
16.5
17
17.5
18
18.5
19
Retention Time (Min.)

Figure 7. Deconvolution of the RI chromatogram of B extract. (a) Normalization to the RI chromatogram of A extract using relative weight loss after extraction; and (b) subtracted Santowhite® peak fitted to an exponentially-modified Gaussian function, experimental line (——), curve-fitted line (- - -).

The amounts of low molecular weight PEUU, MDI-rich oligomer, SW powder, MT, and aniline in the extracts determined by the curve-fitting procedures are summarized in Table 3. The weight of each component extracted, expressed as the percent of the initial weight of the PEUU film (wt%), was calculated from the total weight loss of the PEUU films. The wt% of SW extracted from PEUUs B and D was similar; of the initial 1 wt% of SW powder added to the films during processing, 0.30–0.50 wt% or almost half was extractable with methanol. Table 3 indicates that MT was much less easily extracted than SW. Extractable quantities of MT were similar for PEUUs C and D, of the 5 wt% present initially in the films, less than 0.2 wt% was extracted.

The relative methanol extractabilities of the polymer components and the two additives was ascribed to differences in the Hildebrand solubility parameters. Solubility parameters for methanol and PEUU were taken from the literature, while the solubility parameters of the additives were calculated using group contribution

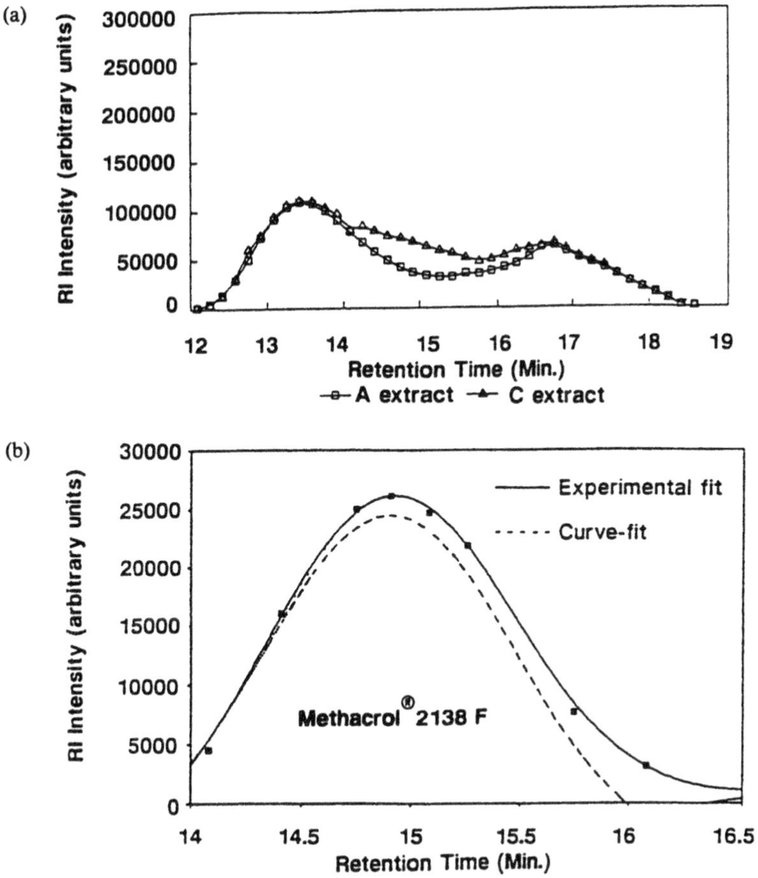
(a)
RI Intensity (arbitrary units)
300000
250000
200000
150000
100000
50000
0
12
13
14
15
16
17
18
19
Retention Time (Min.)
A extract
C extract
(b)
RI Intensity (arbitrary units)
30000
25000
20000
15000
10000
5000
0
Experimental fit
Curve-fit
Methacrol® 2138 F
14
14.5
15
15.5
16
16.5
Retention Time (Min.)

Figure 8. Deconvolution of the RI chromatogram of C extract. (a) Normalization to the RI chromatogram of extract A using the relative weight loss after extraction; and (b) subtracted Methacrol® 2138 F peak fitted to an exponentially-modified Gaussian function, experimental line (——), curve-fitted line (- - -).

Table 3.
Weight percent (wt%) of extractable components based on the total weight of PEUU films[a]

Extract	A	B	C	D
Low *Mw* PEUU	0.87 ± 0.07	0.58 ± 0.05	0.91 ± 0.08	0.66 ± 0.44
MDI-rich oligomer	0.65 ± 0.05	0.44 ± 0.04	0.68 ± 0.04	0.50 ± 0.03
Santowhite® powder	—	0.36 ± 0.04	—	0.46 ± 0.05
Methacrol®	—	—	0.18 ± 0.03	0.14 ± 0.02
Aniline	0.01 ± 0.00	0.01 ± 0.00	0.02 ± 0.00	0.02 ± 0.00
Total weight	1.53	1.39	1.79	1.79

[a] Mean values ± S.D., $n = 4$ measurements.

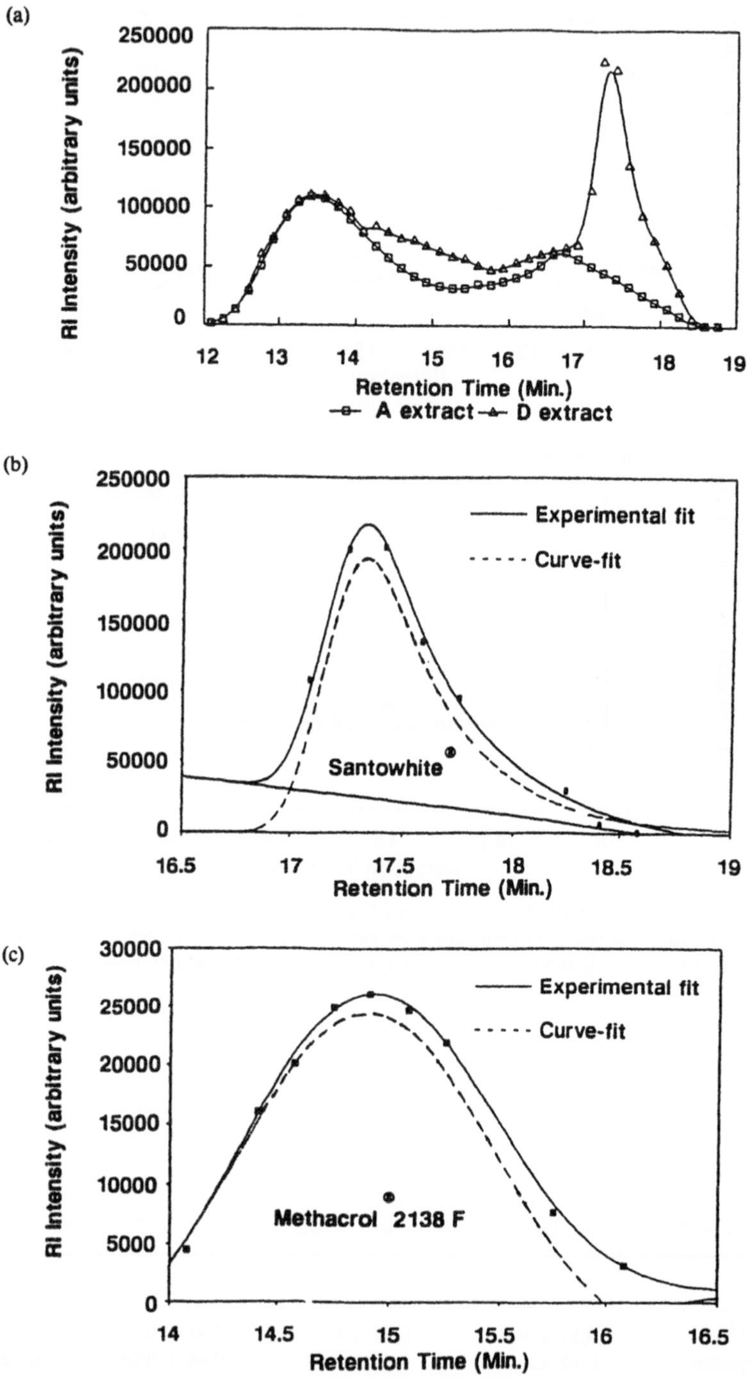
(a)
250000
200000
150000
100000
50000
0
RI Intensity (arbitrary units)
12 13 14 15 16 17 18 19
Retention Time (Min.)
A extract D extract
(b)
250000
200000
150000
100000
50000
0
RI Intensity (arbitrary units)
Experimental fit
Curve-fit
Santowhite®
16.5 17 17.5 18 18.5 19
Retention Time (Min.)
(c)
30000
25000
20000
15000
10000
5000
0
RI Intensity (arbitrary units)
Experimental fit
Curve-fit
Methacrol® 2138 F
14 14.5 15 15.5 16 16.5
Retention Time (Min.)

Figure 9. Deconvolution of the RI chromatogram of D extract. (a) Normalization to the RI chromatogram of extract A using the relative weight loss after extraction; (b) subtracted Santowhite® peak fitted to an exponentially-modified Gaussian function, experimental line (——), curve-fitted line (- - -); and (c) subtracted Methacrol® 2138 F peak fitted to an exponentially-modified Gaussian function, experimental line (——), curve-fitted line (- - -).

Table 4.
Hildebrand solubility parameters of the material, additives, and solvent [27].

Material	δ (MPa$^{0.5}$)	Additives	δ (MPa$^{0.5}$)	Solvent	δ (MPa$^{0.5}$)
PEUU	23.0	Methacrol® 2138 F	18.8	Methanol	29.7
		Santowhite® powder	25.9		

methods [27]. Table 4 summarizes the solubility parameters and shows that methanol should solvate SW powder preferentially to MT. This is consistent with the observation that the percentage of SW powder extracted was higher than the percentage of MT. The Hildebrand parameter for a typical PEUU is only slightly lower than that of the extraction solvent, methanol; and solubility would be further enhanced by the potential for intermolecular hydrogen bonds to form between methanol and the PEUU [28]. These results show that the PEUU contains low molecular weight components that potentially could be solubilized by the biological milieu. The results of *in vitro* extraction with methanol, however, cannot be directly correlated with an *in vivo* situation.

Dual detector method for composition

Identification of the soft-to-hard segment ratio in the two polymer peaks of the extracts was accomplished using the ratio of the RI and UV responses in the chromatograms of the PEUU extracts and two prepolymers (CG I and CG II). The RI response was taken to be proportional to the total polymer concentration, while the UV response at 254 nm was assumed to measure only the hard segment concentration. There were two fundamental assumptions in the use of the UV and RI detectors to determine PEUU composition. First, the absorption at 254 nm followed Beer's law and was proportional to the concentration of the aromatic hard segment

$$A_{UV} = K'_H C_H \tag{1}$$

where A_{UV} is the total area of the UV chromatogram; C_H, the concentration of the aromatic hard segment; and the hard segment response factor K'_H is the product of the molar extinction coefficient and the cell path length. The second assumption was that the total RI detector response for the polymer A_{RI}, measured as the total area under the RI chromatogram, was the sum of the soft and hard segment contributions

$$A_{RI} = K_S C_S + K_H C_H \tag{2}$$

where K_S and K_H are the RI response factors for the soft and hard segments of the PEUU respectively, C_S is the concentration of soft segment, and C_H, the concentration of hard segment, is the same as in Eqn (1).

Although recent studies have shown that under certain conditions the refractive index of polymer solutions may vary with molecular weight [29], the total RI response area A_{RI} for a series of PTHF standards was found to be a constant. Since the PEUU hard segment was insoluble in THF it was assumed that the RI response of the hard segment was also independent of molecular weight. It was then possible to obtain the soft-to-hard segment ratio at any time point on the chromatogram.

Combining Eqns (1) and (2), the ratio of the RI response to the UV response at time t, is given as

$$\left(\frac{A_{RI}}{A_{UV}}\right)_t = K_1\left(\frac{C_S}{C_H}\right)_t + K_2 \tag{3}$$

where $K_1 = K_S/K'_H$ and $K_2 = K_H/K'_H$ then rearranging Eqn (3)

$$\left(\frac{C_S}{C_H}\right)_t = \left(\frac{1}{K_1}\right)\left[\left(\frac{A_{RI}}{A_{UV}}\right)_t - K_2\right]. \tag{4}$$

The quantities of interest were the weight fractions of soft segment F_S and the hard segment F_H at time t, given by

$$F_S = \frac{(C_S/C_H)_t}{(C_S/C_H)_t + 1} \tag{5}$$

and $F_H = 1 - F_S$.

Determination of RI and UV response factors

Three model compounds, CG I, CG II, and uncapped polyol, were used to obtain the constants, K_1 and K_2, in Eqn (4). The soft-to-hard segment ratios of CG I and CG II, determined experimentally using ^{1}H NMR, were 1.0:1.7 (Polyol:MDI) and 1.1:1.0 (Polyol:MDI) respectively. The RI chromatograms of CG I and CG II solutions with total concentration, C_T, provided a value of A_{RI} for each model compound. From Eqn (2) and the values of C_S and C_H obtained from NMR, values of $K_S = 110 \pm 18$ RI units g^{-1} ml^{-1} and $K_H = 750 \pm 120$ RI units g^{-1} ml^{-1}, were obtained. In addition, K_S was determined independently from the RI response of the uncapped polyol. The plot of A_{RI} against polyol concentration was linear, and a value of $K_S = 128 \pm 12$ RI units g^{-1} ml^{-1} ($n = 3$) was obtained from the slope. The values of K_S obtained by the two different methods were within 1 S.D. of each other.

The UV response factor of the prepolymer hard segment was acquired from the 254 nm UV chromatograms of CG I and CG II, by plotting the total area A_{UV} vs the concentration of hard segment C_H for a series of prepolymer solutions. Values obtained for the UV response factor were 1.5×10^4 absorbance units g^{-1} ml^{-1} for CG I and 1.7×10^4 absorbance units g^{-1} ml^{-1} for CG II. It was assumed that the UV response factor of the PEUU chain extended hard segment would not differ significantly from that of the MDI-capped PEU prepolymers, and an average value of 1.6×10^4 absorbance units g^{-1} ml^{-1} was subsequently used for K'_H.

Compositional heterogeneity

The variation in soft-to-hard segment ratio with molecular weight of the PEUUs, PEUU extracts, and prepolymers was determined by analyzing the RI-to-UV ratio along the chromatographic time axis. Figure 10a shows the RI-to-UV ratio of the PEUU extracts compared with the two prepolymers CG I and CG II. Since the soft segment content of CG II was higher than that of CG I, in the plot the data for CG II was multiplied by a factor of 0.5 to normalize it to that of CG I. The soft-to-hard ratio of CG II gradually increased with decreasing molecular weight, while the

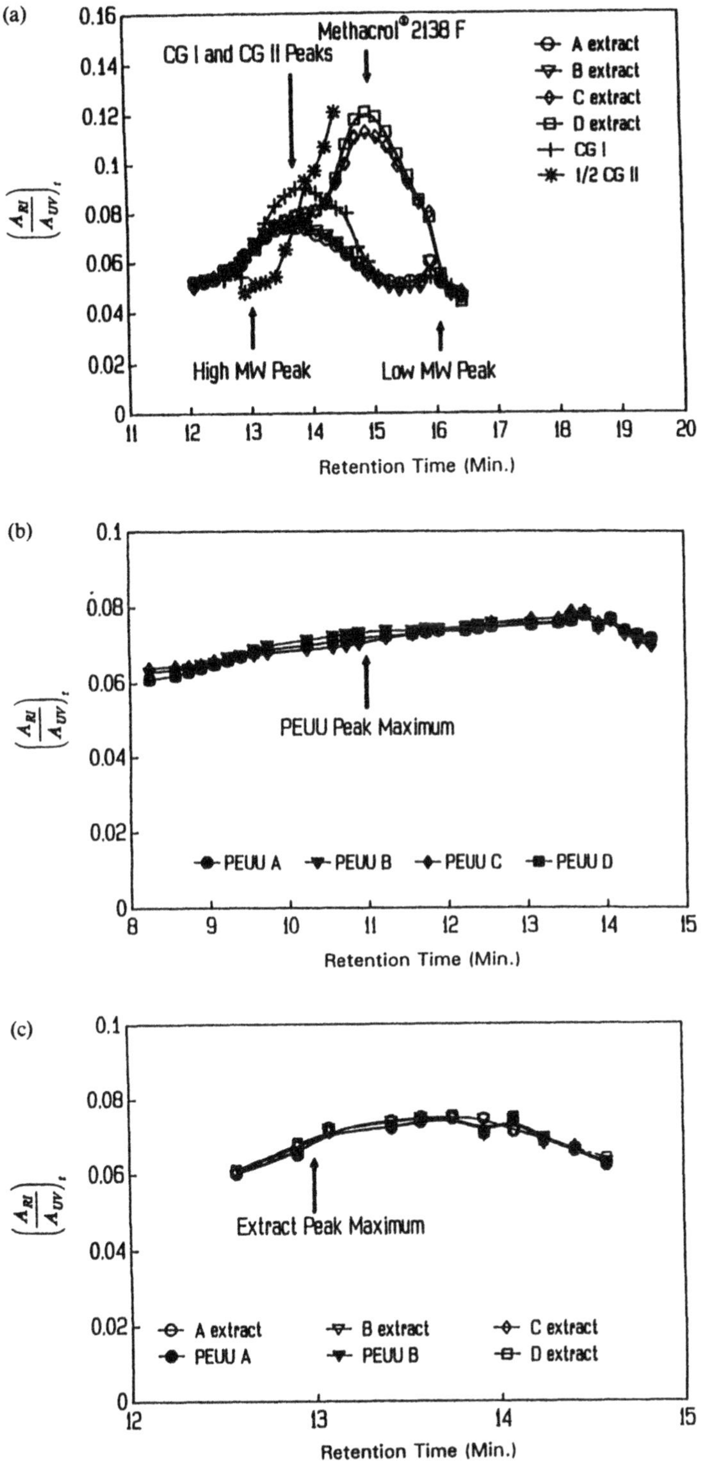
(a)
Methacrol® 2138 F
CG I and CG II Peaks
A extract
B extract
C extract
D extract
CG I
1/2 CG II
High MW Peak
Low MW Peak
Retention Time (Min.)
(b)
PEUU Peak Maximum
PEUU A
PEUU B
PEUU C
PEUU D
Retention Time (Min.)
(c)
Extract Peak Maximum
A extract
PEUU A
B extract
PEUU B
C extract
D extract
Retention Time (Min.)

Figure 10. Comparison of the A_{RI}/A_{UV} profiles: (a) PEUU extracts A–D with two prepolymers, CG I and CG II; (b) PEUUs A–D; and (c) PEUU extracts A–D with PEUUs A and B.

soft-to-hard ratio of CG I went through a maximum near the position of the peak in the chromatogram.

The RI-to-UV ratio of all the extracts was the same in the region of the first peak, about 13 min. The ratio was lower than that of CG I indicating that the low molecular weight extractable PEUU had a lower soft segment content than either of the prepolymers. The RI-to-UV profile of the C and D extracts was strongly affected in the 14.0–15.0-min region by the presence of MT. The substantially larger RI-to-UV response for the C and D extracts in this time range was consistent with the lack of a UV absorbing chromophore in MT at 254 nm. The A and B extracts had essentially the same RI-to-UV curve, the soft segment content gradually decreased with increasing retention time, then increased to another small maximum at the oligomer peak, about 16 min, and then decreased to 17.5 min. The RI-to-UV ratios of the A and B extracts and CG I in the 15.0–17.5-min time frame were similar. This provided evidence that the second extract peak was extractable low molecular weight oligomer.

The RI-to-UV ratios were essentially the same for PEUUs A–D. Figure 10b shows the soft segment content gradually increasing from the high molecular weight end to the low molecular weight tail of the polymer. In Fig. 10c, the RI-to-UV ratios of the A and B extracts are compared with those of the PEUUs in the 13.0–14.0-min region where the first extract peak appeared. The similarity in composition of the extracts and the PEUUs in this range strongly supported the hypothesis that the first extract peak was low molecular weight extractable PEUU.

The weight ratio of soft-to-hard segment at any time point on the chromatograms was calculated from the RI-to-UV ratio using Eqn (4) with constants $K_1 = 0.0069$ and $K_2 = 0.0469$. The weight fractions of soft segment were subsequently obtained from Eqn (5). The weight fractions of soft segment F_S in CG II, CG I, the A extract and PEUU A as a function of retention time or molecular weight are compared in Fig. 11. The other PEUUs and extracts were omitted because the weight fraction curves were essentially the same as those of PEUU A and A extract. The weight fraction of soft segment in the PEUU gradually increased with decreasing molecular weight and, on average, the low molecular weight tail contained the PEUU that was highest in soft segment. Both the low molecular weight and the higher soft segment content favored extraction by methanol. Figure 11 clearly shows the coincidence in the composition of the PEUU and extract at retention times in the region of 13.0 min. On the other hand, the 16-min extract peak had the same soft-to-hard ratio as the low molecular weight tail of CG I, which suggested that the MDI-rich starting material may have lost some of its functionality through side reactions leaving oligomers with a chemical composition similar to CG I.

The soft segment fractions determined from the RI-to-UV ratios made it possible to speculate on the molecular structures of species present in the various samples. The molecular weight of the polyol was determined from the peak in the RI chromatogram to be 3200 relative to PTHF and the molecular weight of the hard segment in the prepolymers was taken to be 268 based on chemical structure. The chain extended PEUU hard segment also had a molecular weight distribution with a repeat unit *Mw* of 310 based upon chemical structure. Table 5 shows possible chemical structures of the species present in CG II, CG I, A extract, and PEUU A at 13.8 min or a molecular weight of 6580 relative to PTHF. This retention time was chosen because it corresponded to the peak in the chromatograms of the

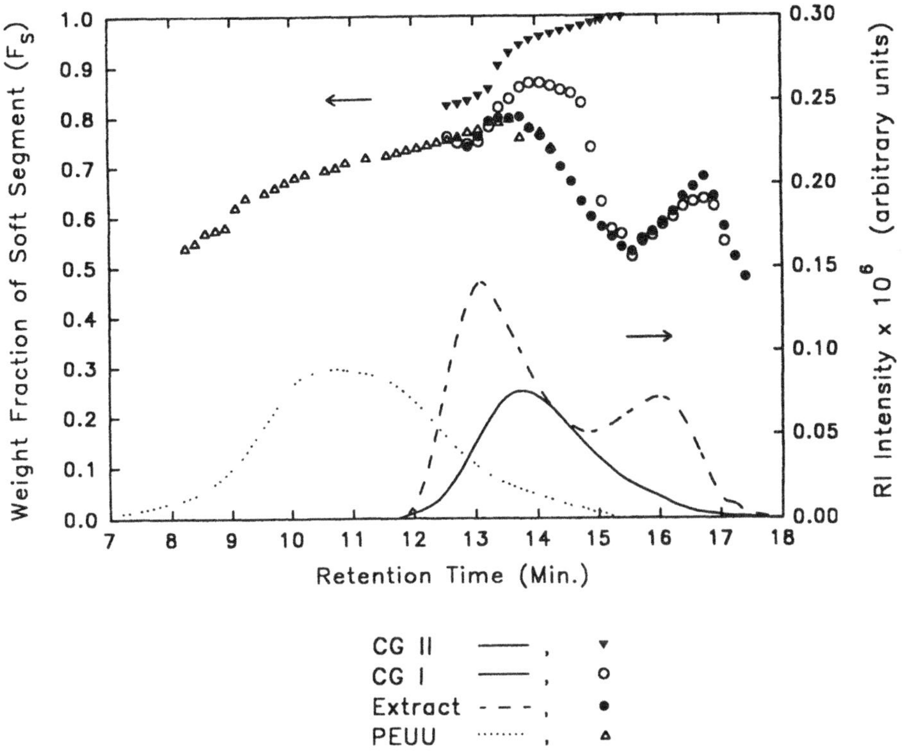
Weight Fraction of Soft Segment (F_S)
1.0
0.9
0.8
0.7
0.6
0.5
0.4
0.3
0.2
0.1
0.0
7
8
9
10
11
12
13
14
15
16
17
18
Retention Time (Min.)
0.30
0.25
0.20
0.15
0.10
0.05
0.00
RI Intensity x 10^6 (arbitrary units)
CG II
CG I
Extract
PEUU

Figure 11. Comparison of the weight fractions of soft segment F_S of CG I and CG II, A extract, and PEUU A in relation to their respective RI chromatograms.

Table 5.
Chemical composition of the capped prepolymers, A extract, and PEUU A at 13.8 min, Mw = 6580 relative to poly(tetrahydrofuran).

Sample	F_S	Weight ratio (S : H)	Mw ratio (w_S : w_H)	Chemical structure
CG II	0.96	24.0 : 1.0	6320 : 260	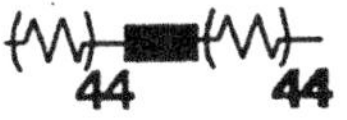
CG I	0.87	6.7 : 1.0	5730 : 860	
Extract	0.80	4.0 : 1.0	5260 : 1320	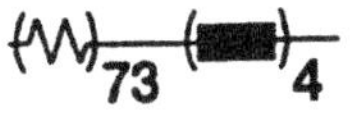
PEUU	0.75	3.1 : 1.0	4940 : 1650	

 : the soft segment unit, i.e. $(CH_2)_4O$, with a Mw of 72.

: the prepolymer hard segment unit with a Mw of 268.

 : the PEUU hard segment repeat unit with a Mw of 310.

prepolymers. There were differences in the soft-to-hard segment ratios for all of the samples at 13.8 min. The polyol-capped prepolymer CG II had the highest soft segment content. Its model structure consisted of one hard segment unit, $M_w = 268$, capped by two polyol oligomers. CG I was presumed to have an MDI-capped prepolymer structure, the soft-to-hard segment ratio was consistent with a structure of the polyol-capped prepolymer CG II capped by an additional two hard segment units, $M_w = 268$. The A extract had a hard segment content that was slightly lower than that of PEUU A at 13.8 min, again supporting the hypothesis that methanol preferentially extracted low molecular weight PEUU with higher soft segment content. The chemical structures of PEUU A and its extract were not put into the same model form as the prepolymers since there would have been a distribution in the structures of the soft and hard segments. Therefore, at this elution time, the average molecule in A extract had a total of 73 $(CH_2)_4O$ units combined with four hard segment units, while an average PEUU A molecule had 68 soft segment units and five hard segment units.

CONCLUSIONS

GPC analysis of four PEUU materials and their methanol extracts led to the following conclusions: (1) Only 1–2 wt% of the PEUU was extractable with methanol. Components of the PEUU extracts were low molecular weight PEUU (*Mw* about 12 000); an MDI-rich oligomer (*Mw* about 1000); the additives, Santowhite® powder and Methacrol® 2138 F; and trace amounts of aniline, an impurity in MDI. (2) Deconvolution of the GPC RI chromatograms of the four PEUU extracts revealed that most of the extractable material was low molecular weight PEUU and the MDI-rich oligomer. Santowhite® powder constituted about 25.0–29.0% of the total extractables from PEUUs B and D, equivalent to about 40% of that present initially. Methacrol® 2138 F comprised 8.0–10.0% of the total extractables from PEUUs C and D, equivalent to about 20% of that present initially. Aniline, found in all of the four PEUU extracts, composed 1.0% of the extracted material. The relative amount of extractables in the four PEUU extracts was related to the differences in solubility parameters of the PEUU and the two additives. (3) The two detector GPC system, RI and UV, provided a soft-to-hard segment composition measurement as a function of the molecular weight. The soft-to-hard segment ratio of the PEUU extracts varied with molecular weight. Chemical structures were proposed on the basis of the soft-to-hard segment measurements.

Acknowledgement

This work was generously supported by the E.I. du Pont Company and the National Institutes of Health, grant number of HL25239.

REFERENCES

1. M. D. Lelah and S. L. Cooper, *Polyurethanes in Medicine*. CRC Press, Boca Raton, FL (1986).
2. K. Stokes and B. Chem, in: *Polyurethanes in Biomedical Engineering*, p. 243, H. Planck, G. Egbers and I. Syre (Eds). Elsevier, Amsterdam (1984).
3. K. Stokes and B. Chem. *J. Biomed. Appl.* **3**, 228 (1988).
4. M. Szycher, V. Poirier and D. Dempsey, *Elastomerics* **115**(3), 11 (1983).

5. S. Gogolewski, *Colloid Poly. Sci.* **267**, 757 (1989).
6. Y. Wu, Q. Zhao, J. M. Anderson, A. Hiltner, G. A. Lodoen and C. R. Payet, *J. Biomed. Mater. Res.* **25**, 725 (1991).
7. T. G. Grasel, D. C. Lee, A. Z. Okkemea, T. J. Slowinski and S. L. Cooper, *Biomaterials* **9**, 383 (1988).
8. R. E. Marchant, J. M. Anderson, A. Hiltner, E. J. Castillo and J. Gleit, *J. Biomed. Mater. Res.* **20**, 799 (1986).
9. S. D. Bruck, *Properties of Biomaterials in the Physiological Environment*. CRC Press, Boca Raton, FL (1980).
10. Q. Zhao, R. E. Marchant, J. M. Anderson and A. Hiltner, *Polymer* **28**, 2040 (1987).
11. A. J. Coury, K. B. Stokes, P. T. Cahalan and P. C. Slaikeu, *Prog. Rubber Plast. Technol.* **3**, 24 (1987).
12. A. J. Coury, P. Slaikeu, P. T. Cahalan, K. B. Stokes and C. M. Hobot, *J. Biomater. Appl.* **3**, 130 (1988).
13. B. Ratner, Transactions, Polymer-IUPAC Symposium, Amherst, MA, 677 (1982).
14. C. Chan, D. C. Birdsell and Y. Clair Gradeen, *Clin. Chem.* **37**, 756 (1991).
15. A. Peterlin, *Controlled Drug Delivery*, Vol. I, pp. 15–28, S. D. Bruck (Ed.). CRC Press, Boca Raton, FL (1983).
16. S. D. Bruck, *Med. Prog. Technol.* **16**, 131 (1990).
17. J. Moacanin, D. D. Lawson, H. P. Chin, E. C. Harrison and D. H. Blankenhorn, *Biomater. Med. Dev., Artif. Organs* **6**, 327 (1975).
18. J. M. Richards, W. H. McClennen and H. C. Meuzelaar, *J. Appl. Poly. Sci.* **40**, 1 (1990).
19. J. P. Foley and M. S. Jeansonne, *J. Chromatogr. Sci.* **29**, 258 (1991).
20. R. S. Juvet, Jr. and J. T. Lundeen, *Anal. Chem.* **53**, 1369 (1981).
21. R. E. Marchant, Q. Zhao, J. M. Anderson, A. Hiltner and R. S. Ward, in: *Surface Characterization of Biomaterials*, p. 29, B. D. Ratner (Ed.). Elsevier, Amsterdam (1988).
22. Varian Associates, Inc. *Polyview™ Spectral Processing Application Operator's Manual* (1990).
23. M. G. Loudon, *Organic Chemistry*, Ch. 17. Addison-Wesley, Reading, MA (1984).
24. T. C. Morrill, D. Y. Curin and B. L. Shriner, *Systematic Identification of Organic Compounds*, 6th ed. John Wiley & Sons, New York (1980).
25. W. G. Stillwell, M. S. Bryant and J. S. Wishnok, *Biomed. Env. Mass Spectrom.* **14**, 221 (1987).
26. Varian Associates, Inc., *DSC-650 GPC Software Manual*, Appendix 2 (1989).
27. E. A. Grulke, in: *Polymer Handbook*, p. 519, 3rd ed., E. H. Immergut and J. Brandup (Eds). John Wiley & Sons, New York (1989).
28. G. L. Grobe III, A. S. Nagel, J. A. Gardella, Jr., R. L. Chin and L. Salvati, Jr., *Appl. Spectrosc.* **42**, 980 (1988).
29. H. Rubio-Garcia, A. E. Luis Hamielec and J. F. MacGregor, in: *Computer Applications in Applied Polymer Science*, p. 151, American Chemical Society Symposium Series 197, Theodore Provder (Ed.) (1982).

Part VIII.B.

Biodegradable Polymers and Drug Delivery

Drug Delivery

Controlled release of TGF-β_1 from a biodegradable matrix for bone regeneration

WAYNE R. GOMBOTZ*, SUSAN C. PANKEY, LISA S. BOUCHARD, JANE RANCHALIS and PAULI PUOLAKKAINEN

Bristol-Myers Squibb, Pharmaceutical Research Institute, 3005 First Avenue, Seattle, WA 98121, USA

Received 2 July 1992; accepted 6 November 1992

Abstract—Although bone has a remarkable capacity for regenerative growth, there are many clinical situations in which the bony repair process is impaired. TGF-β_1 is a 25 kD homodimeric protein which modulates the growth and differentiation of many cell types. The ability of TGF-β_1 to promote bone formation suggests that it may have potential as a therapeutic agent in disease of bone loss. However, there still exists a need for an effective method of delivering TGF-β_1 to the site of an osseous defect for the promotion of bone healing. This paper describes a novel biodegradable controlled release system for TGF-β_1 comprised of poly (DL-lactic-co-glycolic acid) (PLPG) and demineralized bone matrix (DBM). The amount and activity of TGF-β_1 released was determined using several methods including ^{125}I-labeled TGF-β_1 as a tracer, an enzyme linked immunosorbent assay (ELISA) and a growth inhibitory assay (GIA). Protein was released from the devices for time periods of more than 600 h. The amount of TGF-β_1 released was directly proportional to both the TGF-β_1 loading and the weight percent of DBM in the device. The release kinetics could be further controlled by applying polymeric coatings of varying porosity to the devices. The GIA indicated that between 80 and 90% of the TGF-β_1 released from the delivery system retained its bioactivity. The PLPG and DBM existed in phase separated domains within the device as determined by differential scanning calorimetry. Scanning electron microscopy suggested that the devices were sufficiently porous to allow bone ingrowth.

Keywords: Growth factor; TGF-β_1; osteoinduction; drug delivery; poly (DL-lactic-co-glycolic acid).

INTRODUCTION

Bone has a remarkable capacity for regenerative growth, however, there are many clinical indications in which the bony repair process is impaired. These include complications in which there is too much bone loss for the bone to regenerate such as skeletal deformations caused by trauma, malformation, cancer or reconstructive surgeries. There is a critical need to develop implant technologies to augment or promote bone healing.

Autogeneous cartilage and bone are the materials of choice for skeletal augmentation and reconstruction, but there are several drawbacks to using these materials including donor site morbidity, limited quantities of grafting materials and unpredictable implant resorption. As a result, a variety of different materials have been studied that could function as artificial implants for bone repair. These include ceramics, composites, bone derivatives and natural or synthetic polymers [1].

The ideal implant material for bone repair must possess certain basic properties. It should be biocompatible, sterilizable and exhibit osteoinduction (the stimulation

* To whom correspondence should be addressed.

of phenotypic conversion of mesenchymal cells into osteoblasts with bone formation). The implant should also be osteoconductive, acting as a trellis for new bone formation. Porosity and pore density of the implant play an important role in its osteoconductive properties. The materials should degrade in concert with new bone growth without interfering with the formation of new bone. The implant should neither induce adverse local tissue reaction nor be immunogenic or systemically toxic. Many of the materials presently under investigation only partially satisfy these requirements.

Poly(lactic acid) (PLA) poly(glycolic acid) PGA and poly(lactic-co-glycolic acid) (PLPG) polymers have been used experimentally for osseous repair. They are both biocompatible and biodegradable and in some cases have been shown to induce bony wound healing [2–3]. Another attractive feature of these polymers is their ability to be manufactured or sculpted into a variety of forms to fit a given defect. High molecular weight PLA and PGA possess adequate mechanical strength to be used in devices such as pins, screws and plates to augment the healing of load bearing bones [4].

Recent reports in the literature have indicated that the use of PLA and PGA pins for bone fixation may cause inflammatory foreign-body reactions due to chemical irritation from the polymer or the degradation products [4]. In addition, these polymers are not osteoinductive and in some cases their lack of porosity can obstruct bone penetration into the implant [5]. Clearly, these polymers would not be expected to act as ideal implant materials for bone repair by themselves. However, the addition of appropriate components to increase the porosity of the polymers and to stimulate bone growth could result in a viable system for enhancing bone healing.

Demineralized bone matrix (DBM) is a bone derivative that is prepared by demineralizing cadaver bone with HCl. It is comprised of more than 90% collagen along with small amounts of lipids, proteins and proteoglycans [6, 7]. Residual hydroxyapatite can also be found in some preparations. Various forms of DBM have been produced including powder, chips and blocks. The material has been shown to exhibit osteoinduction [8–10] and clinically it is used primarily for facial skeletal augmentation and reconstruction [11]. In many cases, however, bone resorption often occurs with DBM implants. A recent study concluded that demineralized bone had an unacceptably high resorption rate and should only be used in cases where the implant is positioned in sites rich in primitive mesenchymal cells or bone-forming cells [12]. In addition, DBM implants do not possess the cohesive properties to be used in applications where the implant is exposed to movement (when hydrated, DBM is the consistency of wet sand). Various natural [13, 14] and synthetic [15–17] materials have been combined with DBM to form composites that are capable of enhancing new bone formation while improving the mechanical properties of the implant. Only limited success has been achieved with these systems [18].

The incorporation of DBM into a PLPG matrix would enhance the structure forming properties and strength of the DBM while providing an osteoconductive surface. The only component lacking in this system is an osteoinductive catalyst that could abrogate the bone resorption which often occurs with DBM implants.

In an effort to enhance osteogenesis some focus has been directed toward understanding the underlying mechanisms that govern the regulation of healing.

Growth factors are known to regulate the cellular functions of many processes, particularly the healing of tissue. Transforming growth factor-β (TGF-β), basic fibroblast growth factor (bFGF) and platelet derived growth factor (PDGF) are several polypeptide growth factors that are important in the tissue repair process. Among these proteins, the TGF-β superfamily appears to play a critical role in bone healing [19].

TGF-β_1 is a 25 kD homodimeric protein that stimulates the migration, proliferation and matrix synthesis of mesenchymal cells. Bone is the one of the largest known reservoirs of TGF-β in the body [20] and its ability to promote bone formation suggests that it may have potential as a therapeutic agent in diseases of bone loss [21]. The subperiostol injection of TGF-β in rats has been shown to increase bone thickness and chondrogensis [22–24]. TGF-β_1 has also been shown to induce closure of large bony defects in the rabbit skull [25].

Despite these promising findings, there exists a need for an effective method of delivering TGF-β to osseous defects for the promotion of bone healing. The growth factor should be delivered in a stable form without denaturation or loss of bioactivity. Also, there exists a need for a system which predictably controls the amount and duration of TGF-β release.

In the past few years, considerable research has been devoted to the development of protein and peptide drug delivery systems [26, 27]. In this paper we describe a novel biodegradable controlled release system for TGF-β_1 comprised of PLPG and DBM. The goal of this research was to incorporate the TGF-β_1 into a delivery system and release it in an active form. Further, we sought to vary the release kinetics of the growth factor from the delivery device in a controlled and predictable manner.

MATERIALS AND METHODS

Preparation of the delivery systems

Demineralized bone matrix was prepared by the Northwest Tissue Center, Seattle, WA, USA. Long bones of mature New Zealand white rabbits were obtained under sterile conditions. The bones were crushed, water extracted, ground, HCl-demineralized and lyophilized. The particle size of the resulting demineralized bone powder ranged from 70 to 565 μm in diameter. The TGF-β_1 was colyophilized with the DBM in a 30 mM sodium citrate buffer containing 30 mg/ml mannitol, pH 2.5. A typical lyophilized preparation was prepared by suspending 54 mg of the DBM in 2 ml of buffer solution. To this suspension was added 25 μg of TGF-β_1 (Bristol-Myers Squibb Pharmaceutical Research Institute, Seattle, WA, USA). Similar preparations were made that contained varying amounts of TGF-β_1 while keeping the amount of DBM and buffer constant. The suspensions were allowed to equilibrate for 12 h at 4°C. The samples were then frozen in liquid nitrogen and lyophilized.

Solutions of DL-PLPG (50:50, M_w 40–100 kD, Birmingham Polymers, Inc., Birmingham, AL, USA) were prepared in methylene chloride which ranged in concentration from 20–30% (w/v). The PLPG solutions were added to the colyophilized DBM/TGF-β_1 powder which were then poured into polypropylene molds (12 mm in diameter) on a glass plate. The methylene chloride was allowed to evaporate at room temperature for 2 days under atmospheric pressure. Samples were then placed in a vacuum desiccator for 2 more days to remove more of the residual solvent. Devices were prepared which contained between 15 and 35% colyophilized DBM/TGF-β_1 by weight and were loaded with approximately 78, 100, 174, 440 or 1000 μg TGF-β_1/g of device. Cylindrical devices were excised from the dried matrices with a 2 mm diameter dermal punch. The dimensions of the devices used in release studies were 2 mm in diameter by 2 mm thick.

Some of the colyophilized DBM/TGF-β_1 was prepared which contained ^{125}I-labeled TGF-β_1 as a tracer in order to quantitate the incorporation and release kinetics of protein from the devices. The tyrosine residues of the TGF-β_1 molecule were radiolabeled by the chloramine-T method [28]. The ^{125}I-labeled material was added to the native TGF-β_1 resulting in preparations having specific activities ranging from 30 000 to 100 000 cpm/μg TGF-β_1. This preparation was incorporated into the delivery systems using the same method described above. To minimize degradation, the ^{125}I-TGF-β_1 was used within 1 week of iodination.

The devices were treated with various coatings to further control the release kinetics. Delivery systems were prepared which contained approximately 26% colyophilized DBM/TGF-β_1 by weight and 100 μg TGF-β_1/g device. The coatings included a 10% PLPG (50 : 50, M_w 40 000–100 000) solution, a 10% PLPG solution containing 20% mannitol by weight and a 10% oligo-L-lactide solution (Boehringer Ingelheim, Rhein, Germany, M_w 310), all in methylene chloride. Coatings were applied by first immersing the devices in liquid nitrogen and then dipping them into the coating solution. Samples were allowed to air dry.

Device characterization

Scanning electron microscopy. Scanning electron microscopy was used to analyze the DBM alone, and the PLPG/DBM devices. The internal structure of the devices was exposed by cutting samples in half with a razor blade. In addition, the surfaces of both PLPG coated and uncoated devices were analyzed. Studies were also done with devices that had been releasing TGF-β_1 in buffer at 37°C for 2 months. All samples were mounted on stubs with carbon paint and sputter coated with gold-palladium. A JOEL model 35C scanning electron microscope was used to visualize the samples.

Differential scanning calorimetry. Thermal transitions including glass transition temperatures and melting temperatures were determined using a Seiko model SSC/5200 differential scanning calorimeter (DSC). Temperature scans were run on the PLPG, the colyophilized DBM/TGF-β_1 and the completed delivery system. Samples were heated from 10 to 225°C at a rate of 5°C/min.

Evaluation of release kinetics

The release of TGF-β_1 from the PLPG devices ($\sim$10 mg) and from the colyophilized DBM/TGF-β_1 powder ($\sim$3 mg) was evaluated by placing samples, in triplicate, in 4 ml phosphate buffered saline (PBS), pH 7.4 containing 1% human serum albumin (HSA) (Sigma Chemical Co., St Louis, MO, USA). The HSA was necessary to prevent the TGF-β_1 from adsorbing to the surface of the pipette tips and vials used in the release study. The samples were incubated while shaking at 37°C and at various times the buffer was removed and replaced with new buffer. Control samples containing a known amount of TGF-β_1 in buffer were also incubated at 37°C. The amount of TGF-β_1 present in the release buffer was determined by counting on a Micromedic 4/600 gamma counter (ICN Micromedic Systems, Inc., Costamesa, CA, USA) for samples containing ^{125}I-TGF-β_1. An ELISA was also used to quantitate the amount of unlabeled TGF-β_1 present in the buffer. A cell growth inhibitory assay (GIA) was done to measure the bioactivity of the released TGF-β_1.

ELISA. The ELISA was based on the ability of a mouse anti-TGF-β monoclonal antibody, 1D11 (Bristol-Myers Squibb, Seattle, WA, USA), to bind to the TGF-β_1 molecule. When the 1D11 was coated on 96 well immunoassay plates it captured TGF-β_1 from the applied sample and standard solutions. Captured TGF-β_1 was then bound by rabbit anti-TGF-β_1 antibody which in turn was bound by a goat anti-rabbit IgG-horse radish peroxidase probe (Southern Biotech, Birmingham, AL, USA, cat #4040-05). A color reaction occurred by adding the chromophore/substrate solution of 3,3′,5,5′-tetramethylbenzidine in citrate/phosphate buffer containing hydrogen peroxide. The reaction was stopped with the addition of 1*N* sulfuric acid and the A_{450} determined by a plate reader. Concentrations of the unknown samples were quantified relative to a TGF-β_1 standard curve run on the same plate.

Growth inhibitory assay (GIA). The GIA measured the ability of TGF-β_1 to inhibit the growth of mink lung epithelial cells (ATCC #CCL64) [29]. The activity of the growth factor was determined by the inhibitory reponse of the cells to different concentrations of TGF-β_1. Cell viability was based on the enzymatic cleavage by metabolically active cells of a tetrazolium salt into an orange/red formazan product.

Prior to the assay cells were trypsinized and plated in a 96-well flat bottomed plate (Costar, Cambridge, MA, USA) at a concentration of 1000 cells/well. After allowing the cells to attach, samples containing the TGF-β_1 and a reference standard were diluted to concentrations ranging from 1000 to 1.95 pg/ml and added to the wells. The cells were incubated for four days after which time a 25 μl solution containing 25 μg of sodium 3″-[1-(phenylamino)-carbonyl]-3,4-tetrazolium]-bis(4-methoxy-6-nitro) benzene sulfonic acid hydrate (Diagnostic Chemicals, Ltd., Oxford, CT, USA) and 5 μM phenazine methosulfate (Aldrich, Milwaukee, WI, USA) in media was added to each well. The cells were then incubated for 7 h and the plates read on a microplate reader at an absorbance of 450 nm with a 630 nm reference filter. The specific activity of a sample was calculated relative to the reference material.

RESULTS

Scanning electron microscopy

The resulting PLPG/DBM devices had a macroporous morphology upon visual inspection. The samples could be easily cut with a scalpel or excised with a dermal punch to produce a desired shape. Representative SEM micrographs of the DBM alone and the PLPG/DBM devices are shown in Fig. 1(a)–(f). Figure 1(a) and (b) are low and high magnification pictures of the DBM alone. A longitudinal orientation of the DBM particles can be seen in Fig. 1(a). This is indicative of the structure of long bones from which the DBM was prepared. The DBM also has a porous structure. Figure 1(b) shows a higher magnification of the DBM surface. The small, white granular particles on the DBM are pieces of hydroxyapatite that were not

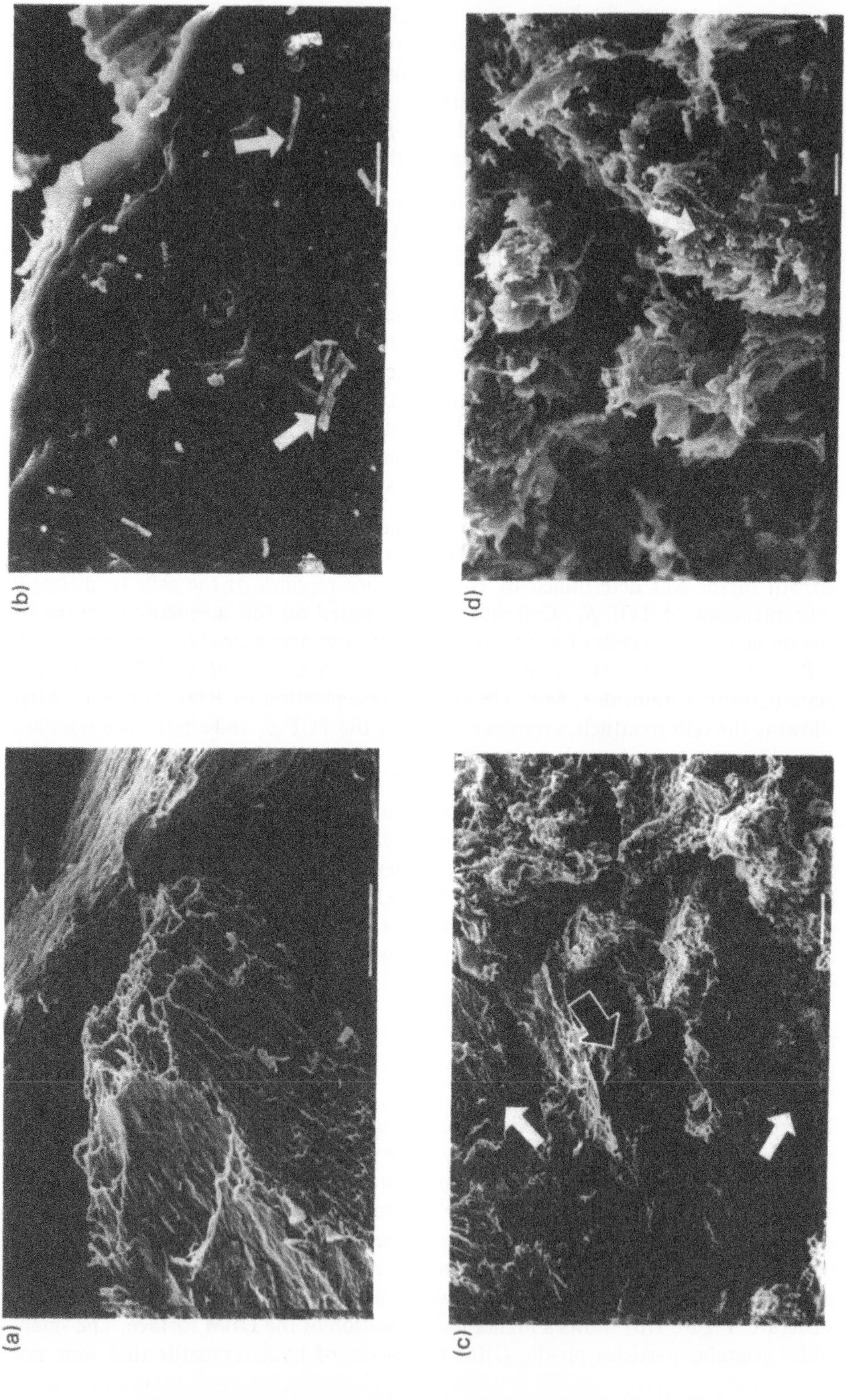
(a)
(b)
(c)
(d)

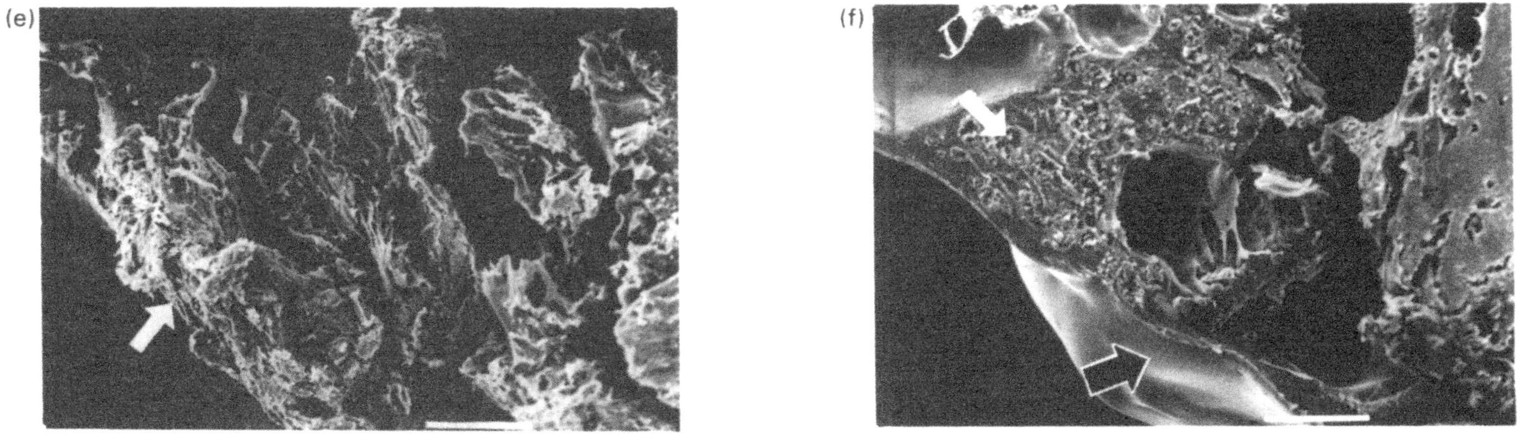
(e)
(f)

Figure 1. Scanning electron micrographs of: (a) DBM alone, bar = 100 μm; (b) DBM alone, bar = 10 μm; (c) PLPG/DBM cross-section, bar = 100 μm; (d) PLPG/DBM cross-section, bar = 10 μm; (e) Uncoated PLPG/DBM cross-section and surface, bar = 100 μm; and (f) PLPG/DBM device coated with 10% PLPG, cross-section and surface, bar = 100 μm. In (b) white arrows identify hydroxyapatite particles. In (c) white arrows identify DBM particles while black arrow shows PLPG. In (d) arrows show hydroxyapatite and/or buffer salts and mannitol. White arrow in (e) points to the surface of the uncoated device. Black arrow in (f) identifies 10% PLPG coating on the device and white arrow identifies buffer salts and mannitol incorporated in PLPG matrix.

completely removed by the demineralization process. Figure 1(c) and (d) are low and high magnification micrographs of PLPG/DBM device cross-sections. Many pores are present which range in diameter from a few microns up to several hundred microns. Pieces of DBM are interdispersed throughout the device and coated with PLPG. Figure 1(e) and (f) show cross-sections and some of the surface of uncoated and coated PLPG/DBM devices. The uncoated sample (Fig. 1(e)) exhibits a porous structure as seen in Fig. 1(c). The surface of this device is also quite rough and porous. In contrast, the same device coated with a 10% PLPG solution has a relatively smooth surface with some very small pores present (Fig. 1(f)). The thickness of the coating varies from approximately 50 μm down to less than 10 μm. Pieces of DBM, buffer salts and mannitol can also be seen dispersed throughout the device.

Differential scanning calorimetry

The thermal transitions of the delivery system and of the individual components in the system determined by DSC are shown in Fig. 2. The PLPG by itself exhibited a glass transition at about 43°C. Because the polymer was amorphous it did not have a defined melting point. The colyophilized DBM/TGF-β_1 had several thermal transitions between 105 and 160°C, with a large endotherm occurring at about 138°C. The multiple transitions are most likely due to the different components in the lyophilized material including the DBM, citric acid, and mannitol. It is unlikely that any of the transitions were due to the TGF-β_1 since the amount of TGF-β_1 in the colyophilized powder was insignificant compared to the DBM and buffer components. The DSC trace of the completed device had distinct thermal transitions corresponding to those seen in the PLPG and the colyophilized DBM/TGF-β_1 alone. Since there were no significant shifts in the dominant thermal transitions of the components when incorporated into the device, we can conclude that they exist in phase separated domains. There was an additional broad endothermic transition that was seen in the device between 50 and 120°C. This could be due to a slow vaporization of water or residual methylene chloride that was trapped in the device and escaped upon heating.

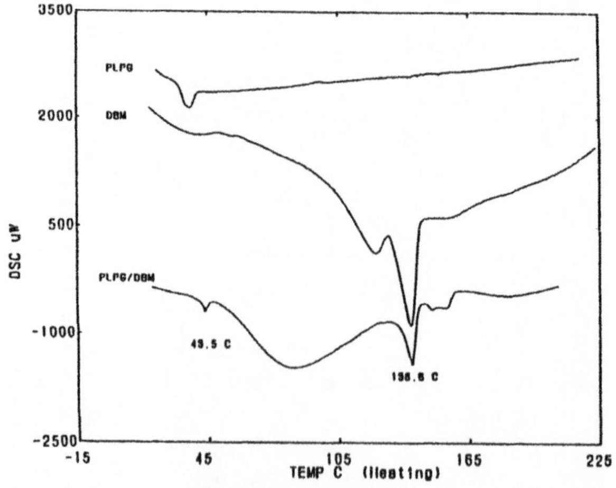
PLPG
DBM
PLPG/DBM
DSC uW
2000
500
-1000
-2500
-15
45
105
165
225
TEMP C (Heating)

Figure 2. DSC traces for the polymer (PLPG), the DBM/TGF-β_1 colyophilate (DBM) and the delivery system (PLPG/DBM).

Release kinetics

ELISA and GIA were used to determine the amount of TGF-β_1 that was released from a 3.0 mg PLPG/DBM device containing 1345 ng of TGF-β_1 (equivalent to 440 μg TGF-β_1/g device) and cast from a 20% PLPG solution (Table 1). After incubating the sample for 2 days at 37°C in PBS with 1% HSA, 35% of the initially incorporated TGF-β_1 was released from the device when assayed by ELISA. When the buffer containing the released TGF-β_1 was assayed by a GIA, 31% of the protein was shown to be released. These numbers are in relatively close agreement indicating that the released TGF-β_1 has a high specific activity. In order to determine if there was released TGF-β_1 that was completely inactive (i.e. not detected by ELISA or GIA) a release study was done with a similar device containing ^{125}I-labeled TGF-β_1. About 28% of the initially incorporated TGF-β_1 was released from this system which corresponds to the value determined by ELISA and GIA. More detailed release kinetics which are described below were done using devices containing the ^{125}I-labeled TGF-β_1 preparations.

Table 1.
Amount of TGF-β_1 released from a PLPG/DBM device that was incubated for 48 h at 37°C in PBS with 1% HSA, pH 7.4. Samples were cast from a 20% PLPG solution and contained 15% colyophilized DBM/TGF-β by weight. The devices weighed 3.0 mg and contained 1345 ng of TGF-β_1 (equivalent to 440 μg TGF-β_1/g device).

Theoretical loading (ng)	TGF-β_1 released (ng)		% TGF-β_1 released		
	ELISA	GIA	ELISA	GIA	^{125}I-TGF-β_1
1345	477 ± 71[a]	415 ± 83[a]	35	31	28[b]

[a] Mean of three measurements ± standard deviation taken from the same device.
[b] Measurement taken from a different device containing ^{125}I-TGF-β_1.

Figure 3 shows the cumulative percent of TGF-β_1 released as a function of time from a colyophilized DBM/TGF-β_1 sample (inset) and various PLPG/DBM delivery systems. Devices were prepared which contained approximately 26% colyophilized DBM/TGF-β_1 by weight and were loaded with 1000 μg TGF-β_1/g device. The inset shows that after 10 h all of the TGF-β_1 was released from the colyophilized DBM/TGF-β_1. By incorporating this material into a PLPG delivery system, the release of TGF-β_1 was sustained for over 600 h (top curve). The release kinetics were further controlled by applying various coatings to the devices. The most significant modification in release was achieved by coating the device with a 10% PLPG solution. While the uncoated device released over 40% of the TGF-β_1 within the first 50 h, less than 20% was released in this same time period from a device coated with PLPG. The device coated with PLPG and mannitol exhibited a slight increase in TGF-β_1 release when compared to the device coated with PLPG alone. The oligo-L-lactide coating had no effect on the TGF-β_1 release kinetics.

Another set of devices containing 26% colyophilized DBM/TGF-β_1 by weight was loaded with different amounts of TGF-β_1 (440, 174 and 78 μg/g device). The cumulative amount of growth factor released over time from these devices is

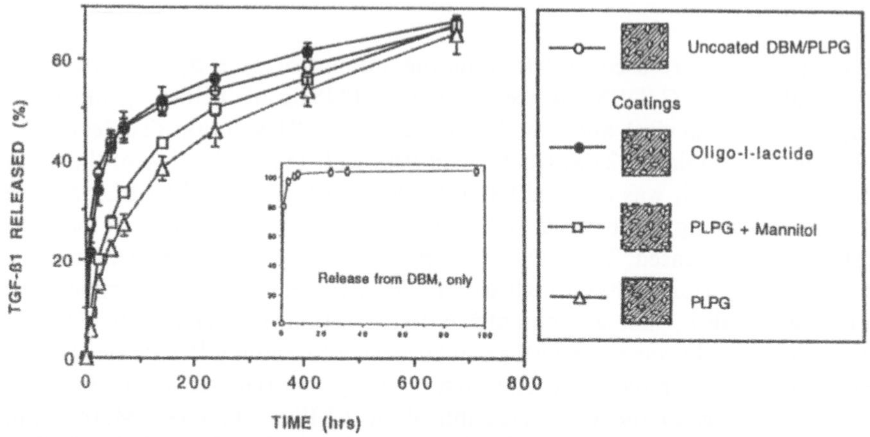

TGF-ß1 RELEASED (%)
TIME (hrs)
Release from DBM, only
Uncoated DBM/PLPG
Coatings
Oligo-l-lactide
PLPG + Mannitol
PLPG

Figure 3. Cumulative percent of TGF-β_1 released as a function of time from colyophilized DMB/TGF-β_1 (inset) and PLPG/DBM devices with no coating and with polymeric coatings of varying porosity. Devices which were cast from a 30% PLPG solution (w/v) contained 26% colyophilized DBM/TGF-β_1 by weight and 1000 μg TGF-β_1/g device. The coatings included a 10% PLPG (50 : 50) solution, a 10% PLPG solution containing 20% mannitol by weight and a 10% oligo-L-lactide solution. Each data point represents the mean of three samples ± standard deviation.

shown in Fig. 4. As the TGF-β_1 loading was increased, the total amount of growth factor released over time also increased.

Figure 5 shows the effects of increasing the weight percent of colyophilized DBM/TGF-β_1 in the device while keeping the amount of TGF-β_1 relatively constant at 190 and 174 μg TGF-β_1 g device. Increasing the amount of DBM in the device resulted in an increase in the percent of TGF-β_1 released and an increase in the initial release rate (as indicated by an increase in the slope of the release curve).

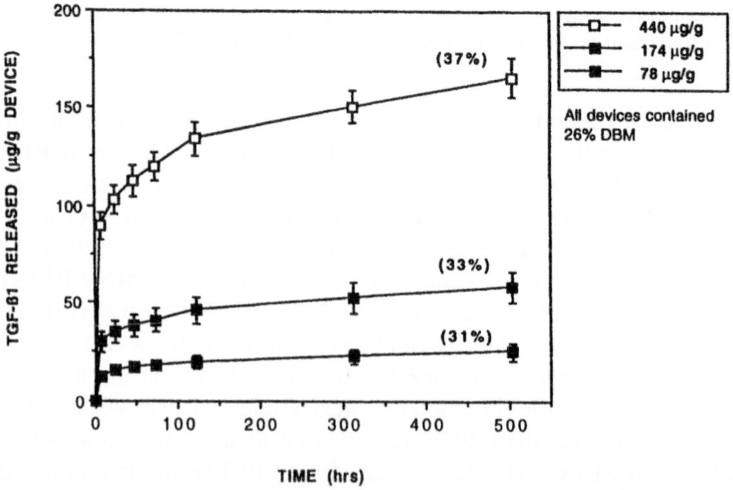

TGF-ß1 RELEASED (µg/g DEVICE)
TIME (hrs)
(37%)
(33%)
(31%)
440 µg/g
174 µg/g
78 µg/g
All devices contained
26% DBM

Figure 4. Cumulative amount of TGF-β_1 released (μg TGF-β_1/g device) as a function of time from PLPG/DBM devices loaded with 440, 174, and 78 μg TGF-β_1/g device. Devices which were cast from a 30% PLPG solution (w/v) contained 26% DBM/TGF-β_1 colyophilate. Each data point represents the mean of three samples ± standard deviation. The numbers in parentheses represent the cumulative percent of TGF-β_1 released at 502 h.

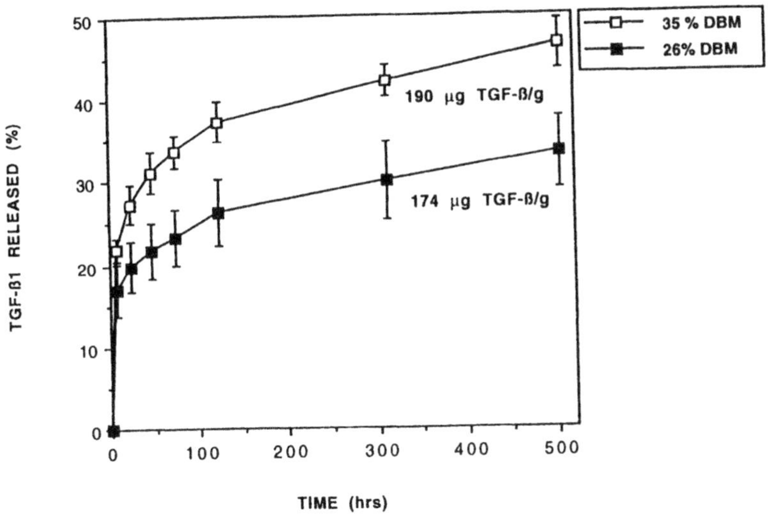
35 % DBM
26% DBM
190 µg TGF-ß/g
174 µg TGF-ß/g
TGF-ß1 RELEASED (%)
TIME (hrs)
0
10
20
30
40
50
0
100
200
300
400
500

Figure 5. Cumulative percent of TGF-β_1 released as a function of time from PLPG/DBM devices loaded with 35% and 26% DBM/TGF-β_1 colyphilate by weight. The devices contained 190 and 174 μg TGF-β_1/g device and were cast from a 30% PLPG solution (w/v). Each data point represents the mean of three samples ± standard deviation.

DISCUSSION

Whenever a protein or polypeptide is incorporated into a controlled release system it is essential that the stability and bioactivity of the active agent are maintained. Many biopharmaceuticals undergo a variety of physical and chemical reactions in solution which lead to their instability and denaturation [30, 31]. The problem of inactivation can be compounded when the protein is incorporated into a delivery system. Considerable effort was therefore taken in developing the DBM/TGF-β_1 colyophilized formulation comprised of 30 mM citrate and 30 mg/ml mannitol, pH 2.5. Stability assays using ELISA combined with a GIA showed that the TGF-β_1 in this colyophilized preparation retained 90–100% of its original activity after being stored for more than three months (Table 2). The use of this formulation in the PLPG delivery system is probably one of the reasons why the released TGF-β_1 retained a large amount of activity.

The optimal porosity of a bone implant material is an ongoing subject of debate and a range of pore diameters has been reported to permit bone ingrowth. In one study with ceramic materials a minimum pore diameter of 100 μm was required [32] while another recommends 500 μm as optimal [33] Based on these reports, the

Table 2.
Quantitation of TGF-β_1 using both ELISA and GIA before and 3.5 months after colyophilization in a buffer containing 30 mM citric acid and 30 mg/ml mannitol, pH 2.5.

Theoretical conc. (μg/ml)	Pre-colyophilized ELISA (μg/ml)	Post-colyophilized ELISA (μg/ml)	Post-colyophilized GIA (μg/ml)
100	100 ± 14[a]	97 ± 12[a]	99 ± 22[a]

[a] Mean of three samples ± standard deviation.

DBM/PLPG devices made in our study appear to be sufficiently porous to allow bone ingrowth. The formation of the pores in these devices is probably due to the drying process during device fabrication. At room temperature the methylene chloride is highly volatile. Small bubbles were observed forming in the devices upon drying and these bubbles resulted in pores.

The release of TGF-β_1 from the colyophilized formulation was relatively fast and essentially complete within 10 h. Probable mechanisms controlling the release from the colyophilized powder are the dissolution of the TGF-β_1 in the buffer and desorption of the growth factor from the DBM particles.

The incorporation of the colyophilized DBM/TGF-β_1 into a PLPG matrix resulted in a significant sustained release of the protein for more than 600 h. The uncoated devices exhibited an initial fast release followed by a slower more linear release profile. The reason that the TGF-β_1 release rate decreased with time from this type of system is probably because the protein released from the surface layer first has a shorter distance to travel than protein which is released at later times from deep within the matrix.

The uncoated devices had two different phases of TGF-β_1 release which are similar to that seen with insulin release from an ethylene-vinyl acetate delivery system [34]. The first phase is an initial faster release of TGF-β_1 resulting from dissolution of the TGF-β_1 that is near the surface of the pores throughout the device. This phase lasts for about 24 h. The second phase results from the continued dissolution and diffusion of TGF-β_1 from the pores and the DBM in the device. If the cumulative percent of protein released is plotted as a function of the square root of time (Fig. 6), a more linear release profile is observed, particularly for the PLPG coated samples and at later times (greater than 25 h) for the uncoated devices. The linear release with the square root of time resembles a first order diffusion controlled release mechanism originally described by Higuchi [35]. Due to the

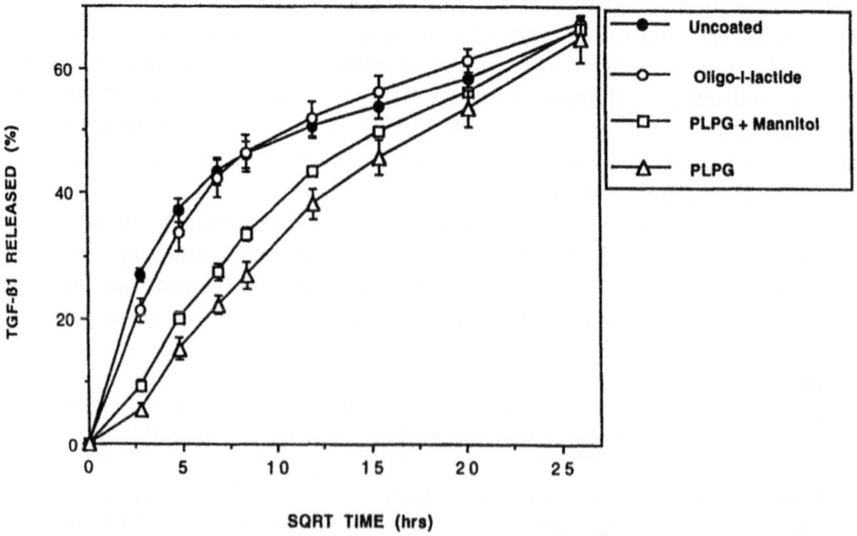
TGF-B1 RELEASED (%)
0
20
40
60
0
5
10
15
20
25
SQRT TIME (hrs)
Uncoated
Oligo-l-lactide
PLPG + Mannitol
PLPG

Figure 6. Cumulative percent of TGF-β_1 released as a function of the square root of time from the same PLPG/DBM devices shown in Fig. 3. Each data point represents the mean of three samples ± standard deviation.

physical structure of the PLPG/DBM device, however, the TGF-β_1 release is too complex to be described completely by a simple diffusion mechanism. The devices have considerable heterogeneity in pore sizes and in sizes of the dispersed DBM phase. In addition, at later times, hydrolytic degradation of the PLPG will also contribute to release of the TGF-β_1 from the device [36]. The release would then be controlled by both a diffusion and a polymer degradation mechanism. If the devices were allowed to completely degrade all of the TGF-β_1 would eventually be expected to be released.

Coating the device with PLPG significantly slowed down the initial release phase of TGF-β_1. The PLPG coating provides a barrier to diffusion of the TGF-β_1 from the device (Fig. 1(f)), thus reducing the initial amount of protein released. Over time the PLPG coating swells and begins to degrade. The barrier to protein diffusion is reduced and after 600 h the coated and uncoated devices have both released about 65% of the TGF-β_1. The PLPG coating containing mannitol exhibited a faster release than PLPG alone. The mannitol probably created pores in the PLPG coating which allowed for faster diffusion of TGF-β_1 from the device. The oligo-L-lactide coating had such a low molecular weight that it provided a negligible barrier to diffusion of the TGF-β_1 from the device and therefore exhibited the same release kinetics as the uncoated device. By treating the delivery systems with different coatings the same dose of TGF-β_1 can be delivered over a given time period but at different rates.

The total dose of TGF-β_1 delivered from the devices was also controlled by the initial TGF-β_1 loading and the amount of colyophilized DBM/TGF-β_1 in the device. Devices which contained different loadings of TGF-β_1, but the same weight percent of DBM, exhibited very similar cumulative percent release profiles (see numbers in parentheses in Fig. 4). The cumulative amount of TGF-β_1 released from these devices, however, was directly proportional to the TGF-β_1 loading (Fig. 4). Increasing the weight percent of colyophilized DBM/TGF-β_1 while keeping the TGF-β_1 loading relatively constant, resulted in a greater cumulative percent release of TGF-β_1 (Fig. 5). This can be explained by the higher swellability of the devices containing larger amounts of DBM. This, in turn, caused more water diffusion into the device, resulting in a higher initial release of TGF-β_1.

In order to further investigate the mechanism of TGF-β_1 release from these systems, a study was carried out to monitor the dissolution of DBM alone and in a device. Samples which did not contain TGF-β_1 were incubated in PBS at 37°C and the total amount of protein in the release buffer was determined after various times up to one month using a bicinchoninic acid assay (Pierce). No protein was detected in any of the samples indicating that the dissolution of DBM was not contributing to the release of TGF-β_1. In fact, the DBM particles showed no evidence of visual degradation after 2 months of incubation. In addition, when PLPG/DBM devices were incubated for 2 months they underwent a significant loss in mechanical strength and crumbled when handled with forceps. SEM analysis revealed that much of the PLPG had degraded, leaving behind particles of DBM. This is probably due to the composition of DBM which contains a relatively large amount of insoluble collagen.

The DBM/PLPG devices containing TGF-β_1 described in this work are anticipated to function well when implanted in a bone defect. The release of TGF-β_1 from the delivery system would produce an osteoinductive effect. As new bone growth occurred, the PLPG component of the device would degrade, leaving behind the DBM. The DBM is an osteoconductive material which would provide a trellis into which new bone could continue to grow. These devices can also be tailored to deliver different doses of TGF-β_1 to bone at different release rates which may be important factors in the optimization of bone regeneration.

CONCLUSIONS

TGF-β_1 can be incorporated into and released from a delivery system made of PLPG and DBM. The growth factor remains intact as determined by ELISA and retains a high specific activity when assayed by a GIA. The release kinetics can be controlled by coating the device with polymers of varying porosity and by varying the TGF-β_1 and DBM loading. Future experiments are planned to evaluate this drug delivery system *in vivo* in a rat critical calvarial defect model.

Acknowledgements

The authors wish to thank Dr. M. Strong at the Northwest Tissue Center, Seattle, WA, USA for supplying the demineralized bone powder and Dr. D. Kibblewhite of University Hospital Vancouver, BC, Canada for his helpful discussions.

REFERENCES

1. C. J. Damien and J. R. Parsons, *J. Appl. Biomater.* **2,** 187 (1991).
2. J. O. Hollinger, *J. Biomed. Mater. Res.* **17,** 71 (1983).
3. J. O. Hollinger and J. P. Schmitz, *J. Oral Maxillofac. Surg.* **45,** 594 (1987).
4. O. M. Bostman, *J. Bone Joint Surg.* **73,** 148 (1991).
5. A. Tencer, P. L. Woodard, J. Swenson and K. L. Brown, *J. Orthop. Res.* **5,** 275 (1987).
6. M. R. Urist, *Science* **150,** 893 (1965).
7. P. V. Hauschka, A. E. Mavrakos, M. D. Iafreti, S. E. Doleman and M. K. Klagsburn, *J. Biol. Chem.* **261,** 12665 (1986).
8. I. A. Guterman, T. E. Boman, G. Wang and G. Balian, *Collagen Rel. Res.* **8,** 419 (1988).
9. A. H. Reddi and W. A. Anderson, *J. Cell. Biol.* **69,** 557 (1976).
10. H. E. Firschein and M. R. Urist, *Clin. Orthop.* **84,** 263 (1972).
11. J. B. Mulliken, J. Glowacki, L. B. Kaban, J. Folkman and J. E. Murray, *Ann. Surg.* **3,** 366 (1981).
12. D. M. Toriumi, W. F. Larrabee, J. W. Walike, D. J. Millay and D. W. Gisele, *Arch. Otolaryngol Head Neck Surg.* **116,** 676 (1990).
13. E. Green, C. Hinton and J. T. Triffitt, *Clin. Orthop.* **205,** 292 (1986).
14. T. S. Lindholm and M. R. Urist, *Clin. Orthop.* **150,** 288 (1980).
15. P. Kohler and A. Kreicbergs, *Clin. Orthop.* **218,** 247 (1987).
16. J. J. Vamdersteenhoven and M. Spector, *J. Biomed. Mater. Res.* **17,** 1003 (1983).
17. S. G. Hopp, L. E. Dahners and J. A. Gilbert, *J. Orthop. Res.* **7,** 579 (1989).
18. C. J. Damien and J. R. Parsons, *J. Appl. Biomater. Res.* **2,** 187 (1991).
19. G. F. Pierce, T. A. Mustoe, T. Lingelbach, V. R. Masakowski, G. L. Griffis, R. M. Senior and T. F. Duel, *J. Cell Biol.* **109,** 429 (1989).
20. S. M. Seydin, P. M. Segarini, D. M. Rosen, A. Y. Thompson, H. Bentz and J. Graycar, *J. Biol. Chem.* **281,** 5693 (1986).
21. M. Noda and J. Camilliere, *Endocrinology* **124,** 2991 (1989).
22. M. E. Joyce, A. B. Roberts, M. B. Sporn and M. Bolander, *J. Cell. Biol.* **110,** 2195 (1990).
23. C. Marcelli, A. J. Yates and G. R. Mundy, *J. Bone Miner. Res.* **5,** 1087 (1990).
24. E. J. Mackie and U. Trechsel, *Bone* **11,** 295 (1990).
25. L. S. Beck, L. Deguzman, W. P. Lee, Y. Xu, C. A. McFatridge, N. A. Gillett and E. P. Amento, *J. Bone Miner. Res.* **6,** 1257 (1991).

26. R. Langer, *Science* **249**, 1527 (1990).
27. C. G. Pitt, *Int. J. Pharmaceut.* **59**, 173 (1990).
28. C. A. Frolik, L. M. Wakefield, D. M. Smith and M. B. Sporn, *J. Biol. Chem.* **10**, 10995 (1984).
29. T. Ikeda, M. N. Loubin and H. Marquardt, *Biochemistry* **26**, 2406 (1987).
30. M. C. Manning, K. Patel and R. T. Borchardt, *Pharmac. Res.* **6**, 903 (1989).
31. Y. J. Wang and M. A. Hanson, *J. Parent. Sci. Tech.* **Supp. S3**, 42 (1988).
32. J. Klawitter and S. Hubert, *J. Biomed. Mater. Res. Symp.* **2**, 161 (1983).
33. T. Flatley, K. Lynch and M. Benson, *Clin. Orthop. Rel. Res.* **179**, 246 (1983).
34. L. Brown, L. Siemer, C. Munoz and R. Langer, *Diabetes* **35**, 684 (1986).
35. T. Higuchi, *J. Pharm. Sci.* **50**, 874 (1961).
36. S. S. Shah, Y. Cha and C. G. Pitt, *J. Controlled Rel.* **18**, 261 (1992).

Controlled release of β-estradiol from biodegradable microparticles within a silicone matrix

L. BRANNON-PEPPAS

Biogel Technology, Inc., 9521 Valparaiso Ct., P.O. Box 681513, Indianapolis, IN 46278, USA

Received 18 November 1992; accepted 28 January 1993

Abstract—Novel, biodegradable controlled release systems were prepared from biodegradable microparticles of poly(lactic acid-co-glycolic acid) containing β-estradiol in the presence or absence of silicone. The release behavior of β-estradiol from free microparticles as well as from microparticles embedded within a silicone matrix was compared with the release behavior shown by nonencapsulated β-estradiol within a silicone matrix. It was found that incorporating biodegradable microparticles within a silicone matrix lessens the initial burst of release often seen with these types of formulations and provides a controlled rate of drug release. In addition, the release rate of β-estradiol from biodegradable microparticles within silicone is higher than for unencapsulated β-estradiol in silicone. This type of formulation may be useful in a number of instances such as release of drugs from implants for which a simple drug-silicone formulation does not yield desired release behavior, formulations which are currently developed for microparticles but which may need to be removed if necessary, and implant formulations containing drugs which will not diffuse through silicone.

Key words: Biodegradable microparticles; silicone; β-estradiol; drug release; PLA/PGA.

INTRODUCTION

Biodegradable microparticles have been widely studied in the area of controlled release. These microparticles are most often used as injectable controlled drug delivery systems and little work has been done in combining the microparticles with other polymers to produce other, novel, drug delivery systems. Previous work by this research group has included design and modeling of drug release from surface-eroding microparticles imbedded within a matrix of another biodegradable polymer [1]. The formulations described here include another hybrid system in which biodegradable microparticles containing drug are embedded within a silicone matrix. A unique characteristic of the formulation designed and evaluated here is that the erosion of the polymeric spheres within the matrix creates a porous polymeric structure, whose release characteristics will be distinctly different from the release behavior of drug from either free microparticles or polymeric matrices.

These hybrid, implantable drug delivery systems could be used when there might be a need to halt the drug release at some future date, which is impossible with injectable microparticle formulations. In addition, these implants may be used with a wide range of hydrophilic and hydrophobic drugs, with the only restriction being that they can be microencapsulated in biodegradable polymers. This is a significant improvement over current silicone-based implant technology which is only useful for hydrophobic drugs which can diffuse through silicone.

A few references to systems similar to this [2–4] are found in the patent literature and have been developed for a wide range of applications. The systems described by Langer and collaborators [2] include microparticles of poly(carboxyphenoxy propane) polymerized with sebacic acid within a polyurethane matrix, microparticles of polylactic acid within a polyurethane matrix, and microparticles of polylactic acid within a polystyrene matrix. These systems could be used as transdermal delivery systems, vascular grafts, and wound-healing films. Other systems designed by Suominen *et al.* [3] are based upon enzymatically modified starch particles within a synthetic polymer matrix. The formulation is biodegradable because the starch particles contain enzymes that will degrade the synthetic polymer matrix. In films prepared of this type of formulation, the starch particles dissolve upon exposure to water in a landfill, exposing the enzymes which then degrade the synthetic polymer, completely degrading the film. Oral drug delivery systems consisting of coated particles within a matrix of hydroxypropylmethyl cellulose have been patented by Urquhart and Theeuwes [4]. The cellulose matrix here serves two purposes: controlled release of the particles and prolonged gastric retention.

Even though biodegradable microparticles and silicone matrices have been extensively studied as controlled release systems, to date no other studies have combined the two and examined hybrid microparticle/matrix systems consisting of poly(lactic acid-co-glycolic acid) microparticles within a silicone matrix.

Steroid release from implants

Steroids have been used since 1964 in a number of silicone-based controlled release formulations [5, 6] including a contraceptive implant (containing levonorgestrel, Norplant) and an implant, administered in the ear, for increased growth rates in cattle (containing β-estradiol, Compudose) [7]. Estradiol has successfully been delivered from collagen matrices as a means of preventing allograft skin rejection in laboratory animals [8]. Currently, estradiol is also available in transdermal delivery systems [9, 10]. The transdermal patch Estraderm (developed by Alza for Ciba-Geigy) delivers hormone directly into the blood stream as a treatment to relieve menopausal symptoms.

Biodegradable microparticles

Biodegradable microparticles, especially those containing proteins and peptides, have been extensively studied in recent years. The possibility of using biodegradable polymers as drug carriers was brought to the attention of many scientists when bioresorable sutures entered the market 20 years ago. Since that time, researchers in pharmaceutical sciences, chemical engineering, and other disciplines have strived to design biodegradable polymers with desired degradation mechanisms and mechanical properties. The products of these biodegradable polymers may be completely broken down and removed from the body by normal metabolic pathways.

The most widely used and studied class of biodegradable polymers is the polyesters, including poly(lactic acid), poly(glycolic acid), and their copolymers. Poly(glycolic acid) (henceforth referred to as PGA) was first marketed in 1970 as a biodegradable suture and poly(lactic acid) (henceforth referred to as PLA) was investigated as a drug delivery material as early as 1971. By varying the monomer ratios in the polymer processing and by varying the processing conditions, the resulting polymer can exhibit drug release capabilities for months or even years [11]. The degree of crystallinity has a significant effect on the rate of degradation. These polymers, which have been prepared as films, microparticles, rods, and other forms, display a bulk erosion hydrolysis.

Biodegradable polymer technology is progressing rapidly enough that researchers have at their disposal a number of biodegradable polymers with a range of degradation rates. Not only may researchers use a single polymer (copolymer or blend) in their formulation, but they may also use a combination of polymers to yield the desired degradation profile. Many successful techniques for preparation of microparticles from biodegradable polymers such as PLA, PGA, and their copolymers have been both published [12–18] and patented [19–21].

Biodegradable microparticles of PLA/PGA copolymers have been used to deliver a wide variety of active agents. Microparticles containing norethisterone for 90 days of release have been produced by a solvent evaporation procedure (similar to that used in this research project). These microparticles have been successfully used in Phase II and Phase III clinical trials. Other active agents released from PLA/PGA microparticles include diphtheria toxoid [22], triptoreline (a luteinizing hormone-releasing hormone agonist) [23], staphylococcal enterotoxin [24], leuprolide acetate (an analog of luteinizing hormone-releasing hormone) [25], and insulin [26].

The last 10 years have seen an increase in new bioactive proteins and peptides produced using biotechnology. These molecules often have short biological half-lives and relatively high molecular weights, thereby necessitating formulations utilizing controlled delivery as opposed to conventional formulations [27]. Biodegradable microparticulate delivery of proteins and peptides allows for more controlled delivery, increased stability, and greater bioactivity than could ever be achieved through current delivery routes, especially oral delivery.

However, there are some applications where there might be a need or a desire to halt the drug delivery process some time after the microparticles have been administered. With injectable microparticle formulations, this is impossible. These same drugs, sometimes proteins and peptides, cannot be delivered using a traditional silicone implant because of their inability to diffuse through the silicone matrix. The formulations described in this paper have the potential to show improved controlled release profiles over microparticles alone and can be removed, to stop drug release entirely, if needed.

MATERIALS AND METHODS

Materials

Three grades of poly(lactic acid-co-glycolic acid) obtained from Birmingham Polymers, Inc. were used: 50/50 PLA/PGA (M_n = 3859, M_w = 5963), 65/35 PLA/PGA (M_n = 3901, M_w = 6041), and 75/25 PLA/PGA (M_n = 3701, M_w = 5900). Poly(vinyl alcohol) (13 000–23 000 M_w, 98% hydrolyzed) (PVA) and 1,3,5[10]-Estratriene-2,17β-diol ($C_{18}H_{24}O_2$, formula weight 242.4) (β-estradiol) were obtained from Sigma Chemical Co. Medical grade silastic MDX4-4210 and curing agent were purchased from Dow Corning. The methylene chloride and ethanol used were reagent grade.

Microparticle formation

Microparticles were prepared by a solvent evaporation technique. A 0.25 wt.% solution of PVA was prepared by heating appropriate amounts of PVA and distilled, deionized water until the PVA dissolved. The solution was then cooled to room temperature before use. This PVA solution was then stirred with a tri-blade teflon-coated stirrer at 350 r.p.m. An appropriate amount of PLA/PGA was dissolved in methylene chloride (i.e. 1.0 g PLA/PGA in 8 ml methylene chloride) and β-estradiol was added to this solution to give an overall composition of 40 wt.% β-estradiol and 60 wt.% PLA/PGA. The PLA/PGA solution was added dropwise into the stirring PVA solution. The stirring was continued overnight to allow for complete evaporation of the methylene chloride.

Once the microparticles had been formed, the solution was allowed to settle. The particles were then vacuum filtered using Whatman No. 1 filter paper and stored in a desiccator at 4°C. The microparticles were sieved and particles of diameter 300–425 μm were used in silicone film preparation and all drug release studies.

Preparation of silicone films

Silastic MDX4-4210 and its curing agent were weighed out in a 9 : 1 weight ratio onto glass plates. Appropriate amounts of β-estradiol or PLA/PGA microparticles containing β-estradiol were added and the materials were mixed together well. The silastic mixture was spread to approximately a 1 mm thickness using a casting apparatus and blade and was then placed in an oven at 100°C for 15 min to cure. All samples were cured within this time and were easily peeled from the glass plate. Individual samples were cut from these films for the drug release studies.

Drug release studies

The microparticles or silicone films to be studied were placed in glass vacutainer test tubes and the test tube caps were fitted with filter paper underneath and surrounding the cap so that samples could be withdrawn, using a syringe, from the space between the cap and the filter paper to avoid removing any microparticles during the sampling process. The weight of microparticles or film used for each sample was chosen so that the amount of β-estradiol in each sample would be 2–3 mg. The release solution used was a 30% (v/v) distilled, deionized water and ethanol mixture. This solution was chosen because of the low solubility of β-estradiol in water or buffered solutions [28]. It was important that the solubility of β-estradiol be high enough in the release solution that it would be the rate of drug release due to polymer degradation and release that was measured, not the rate of β-estradiol solubility. The samples were kept in a shaker bath at 37°C. At each sample time, the entire release solution (2.5 ml) was removed and replaced with fresh solution. The absorbance of the withdrawn solution was read at 280 nm and the corresponding β-estradiol concentration was calculated.

RESULTS AND DISCUSSION

The goal of this work was to examine and compare the release of β-estradiol from three types of controlled release formulations: (i) β-estradiol in silicone films; (ii) β-estradiol in biodegradable microparticles; and (iii) β-estradiol in biodegradable microparticles within silicone films. All microparticles formulations discussed have been prepared with 40 wt.% β-estradiol. The silicone film containing β-estradiol alone contained 4 wt.% β-estradiol. The silicone films containing microparticles were prepared at 10 wt.% microparticles (for an overall β-estradiol loading of 4 wt.%, the same as the films alone) and at 20 wt.% microparticles (for an overall β-estradiol loading of 8 wt.%). This gave a total of four specific formulations to be tested.

There were three different low molecular weight PLA/PGA polymers used: 50/50 PLA/PGA (M_n = 3859, M_w = 5963), 65/35 PLA/PGA (M_n = 3901, M_w = 6041), and 75/25 PLA/PGA (M_n = 3701, M_w = 5900). Each of the four formulation types (described above) was prepared with each of the three polymers, for a total of twelve formulations. The cumulative release and rate of release data will be discussed by polymer type so that the differences between the formulation types are most apparent. For consistency, the four formulations are compared on each graph for a specific polymer; (a) microparticles at 40 wt.% loading of β-estradiol; (b) microparticles at 40 wt.% loading of β-estradiol at a loading of 10% in silicone for a net β-estradiol loading of approximately 4 wt.%; (c) microparticles at 40 wt.% loading of β-estradiol at a loading of 20% in silicone for a net β-estradiol loading of approximately 8 wt.%; and (d) β-estradiol at a 4 wt% loading in silicone.

Cumulative release

It was found that all formulations of free microparticles tested released the β-estradiol quickly. In fact, over 70 wt.% of the drug was released within the first 24 h of the *in vitro* test in all cases. In contrast, microparticles within silicone released only 30 wt.% or less of their β-estradiol within the first 24 h. Figures 1–3

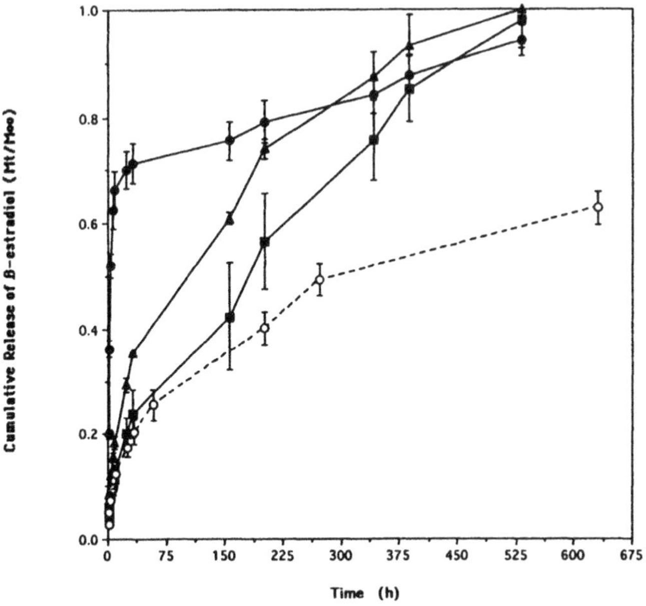

Cumulative Release of β-estradiol (Mt/M∞)
1.0
0.8
0.6
0.4
0.2
0.0
0
75
150
225
300
375
450
525
600
675
Time (h)

Figure 1. Cumulative release (M_t/M_∞) of β-estradiol from formulations containing 50/50 PLA/PGA (M_n = 3859, M_w = 5963); (●) microparticles alone (40 wt.% loaded); (■) 10 wt.% microparticles (40 wt.% loaded) in silicone; (▲) 20 wt.% microparticles (40 wt.% loaded) in silicone; and (○) 4 wt.% β-estradiol in silicone.

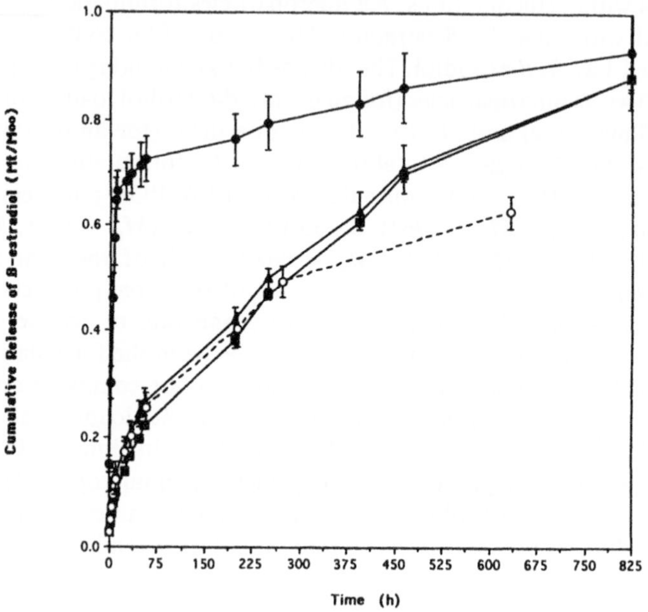
Cumulative Release of β-estradiol (Mt/Moo)
1.0
0.8
0.6
0.4
0.2
0.0
0
75
150
225
300
375
450
525
600
675
750
825
Time (h)

Figure 2. Cumulative release (M_t/M_∞) of β-estradiol from formulations containing 65/35 PLA/PGA (M_n = 3901, M_w = 6041); (●) microparticles alone (40 wt.% loaded); (■) 10 wt.% microparticles (40 wt.% loaded) in silicone; (▲) 20 wt.% microparticles (40 wt.% loaded) in silicone; and (○) 4 wt.% β-estradiol in silicone.

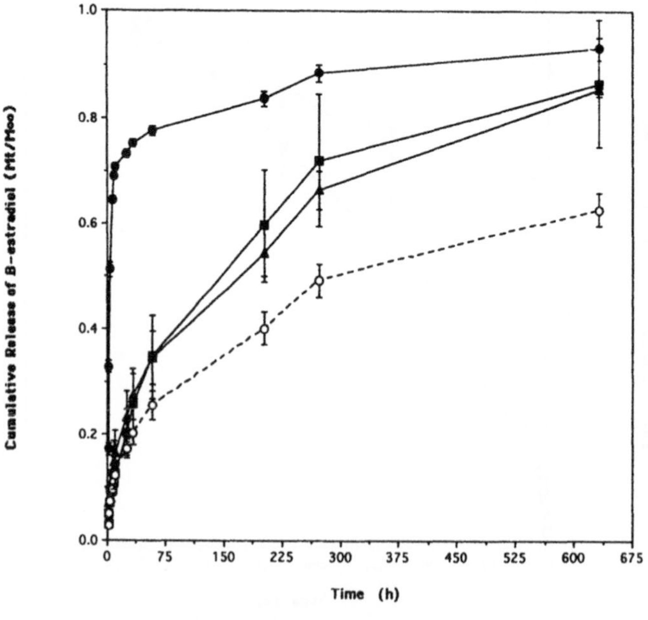
Cumulative Release of β-estradiol (Mt/Moo)
1.0
0.8
0.6
0.4
0.2
0.0
0
75
150
225
300
375
450
525
600
675
Time (h)

Figure 3. Cumulative release (M_t/M_∞) of β-estradiol from formulations containing 75/25 PLA/PGA (M_n = 3701, M_w = 5900): (●) microparticles alone (40 wt.% loaded); (■) 10 wt.% microparticles (40 wt.% loaded) in silicone; (▲) 20 wt.% microparticles (40 wt.% loaded) in silicone; and (○) 4 wt.% β-estradiol in silicone.

show the cumulative release of β-estradiol from 50/50 PLA/PGA (Fig. 1), 65/35 PLA/PGA (Fig. 2), and 75/25 PLA/PGA (Fig. 3). It can be seen, for all three polymers, that the cumulative amount of β-estradiol released from PLA/PGA microparticles within silicone is, at almost all times, lower than the corresponding release from microparticles alone but higher than that of a corresponding loading of β-estradiol (nonencapsulated) in silicone.

The cumulative release for the first 60% of drug release was analyzed according to the following equation [29]:

$$\frac{M_t}{M_\infty} = kt^n \tag{1}$$

In this equation, M_t is the total mass of drug released through time t, M_∞ is the total amount of drug released, k is a constant and n is the overall time dependence of the drug release. For this analysis, M_∞ was taken as the total initial amount of β-estradiol in the sample being tested. An analysis of cumulative release data (M_t/M_∞) versus time for all formulations yields the dependence on time of the drug release, n, as well as the constant, k. This value of the exponential dependence of the drug release on time indicates whether degradation, diffusion, anomalous transport, or other mechanisms are the predominant factors in the mechanism of drug release. An n of 0.5 is characteristic of Fickian diffusion whereas higher values of n indicate that other methods of drug transport may be present [29].

Table 1 shows the exponent n for β-estradiol release as calculated from the release data from free microparticles (40 wt.% loaded with drug), microparticles at a 10% loading in silicone (for a net 4 wt.% loading in silicone), and β-estradiol in silicone (at a 4 wt.% loading). It can be seen that the exponent n for β-estradiol release from all biodegradable microparticles is equal to or higher than the 0.5 that one would expect from Fickian diffusion. The exponential dependence on time, however, for release of β-estradiol from a silicone matrix is 0.422. The value of the exponent n for all formulations of microparticles within a silicone matrix is higher than that for release from silicone alone, but less than the corresponding value for release from microparticles alone. This gives an additional indication, beyond the physical observations of watching the microparticles disappear from within the silicone, that both diffusion and degradation affect the drug release from microparticles within a silicone film.

Table 1.
Exponential dependence of drug release on time (as calculated from Eq. (1))

Polymer and formulation	Exponent n
50/50 PLA/PGA (M_n = 3859, M_w = 5963)	
microparticles (40 wt.% β-estradiol)	0.504
10 wt.% microparticles (40 wt.% β-estradiol) in silicone	0.465
65/35 PLA/PGA (M_n = 3901, M_w = 6041)	
microparticles (40 wt.% β-estradiol)	0.669
10 wt.% microparticles (40 wt.% β-estradiol) in silicone	0.506
75/25 PLA/PGA (M_n = 3701, M_w = 5900)	
microparticles (40 wt.% β-estradiol)	0.675
10 wt.% microparticles (40 wt.% β-estradiol) in silicone	0.512
4 wt.% β-estradiol in silicone	0.422

Rates of release

The rates of β-estradiol release in terms of milligrams per hour are shown in Figs 4–9. Figures 4 and 5 show the rate of release from formulations containing 50/50 PLA/PGA, with Fig. 4 showing the first 24 h of release and Fig. 5 showing the β-estradiol release from 25 to 625 h. Similarly, Figs 6 and 7 show the rate of release from formulations containing 65/35 PLA/PGA, with Fig. 6 showing the first 24 h of release and Fig. 7 showing the β-estradiol release from 25 to 825 h. The final set of data plotted in Figs 8 and 9 show the rate of release from formulations containing 75/25 PLA/PGA, with Fig. 8 showing the first 24 h of release and Fig. 9 showing the β-estradiol release from 25 to 625 h. The initial release rates from the microparticles are four to five times the release rates from the particles within silicon or from the silicone matrices alone. However, after the first 24 h, the microparticle release rates have dropped to below that of the silicone matrix or the microparticles within silicone. By 25 h into the experiments, the microparticles within the silicone matrix give the highest release rates. These results show that β-estradiol release from biodegradable microparticles within a silicone matrix has a diminished burst over free microparticles and an extended period of release. In addition, the release rate is consistently higher than for β-estradiol released from silicone matrices, at the same net drug loading.

It is clear that the silicon matrix retards the β-estradiol release from the biodegradable microparticles. However, if the drug release were due only to diffusion through the silicone matrix after degradation of the microparticles, then the release of the encapsulated drug from the silicone matrices would have been slower than the release of the unencapsulated drug in silicone. This was not the case. Therefore, the drug release also occurs by diffusion through the pores formed by degradation of the microparticles as shown schematically in Fig. 10. In formulations where drugs cannot diffuse through the silicone proper, the release will be limited to diffusion through the pores formed as the biodegradable polymer degrades.

This phenomenon could be seen during the release experiments as the white microparticles in the silicone films were gradually replaced by clear holes. There may be some microparticles, depending upon the loading rate, that will not be exposed to a series of pores and drug release from these particles will be only by the slow process of diffusion through the silicone matrix. Processing conditions can also be designed to add or remove air bubbles from the silicone during processing. These air bubbles, when present, will increase the rate of drug delivery because they will increase the porosity of the silicone matrix, thereby not restricting the diffusion of drug through only those pores which previously contained biodegradable microparticles.

Not all drugs will readily diffuse through silicone and the formulations developed and studied in this work have the advantage that they can be used to deliver a wide variety of small molecular weight drugs as well as proteins and peptides, provided that they can be successfully encapsulated and are stable under dry conditions for the silicone curing process. Other studies have been conducted on these formulations by this research group, including variations of the drug and microparticle loading using the same PLA/PGA polymers and β-estradiol as well as the same PLA/PGA polymers and a variety of proteins and peptides.

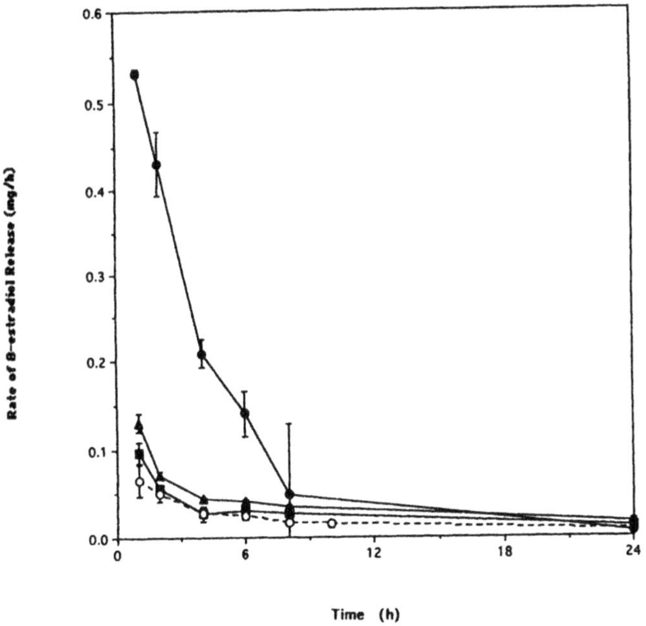
Rate of ß-estradiol Release (mg/h)
Time (h)
0.6
0.5
0.4
0.3
0.2
0.1
0.0
0
6
12
18
24

Figure 4. Rate of release (mg/h) for first 24 h of β-estradiol from formulations containing 50/50 PLA/PGA (M_n = 3859, M_w = 5963): (●) microparticles alone (40 wt.% loaded); (■) 10 wt.% microparticles (40 wt.% loaded) in silicone; (▲) 20 wt.% microparticles (40 wt.% loaded) in silicone; and (○) 4 wt.% β-estradiol in silicone.

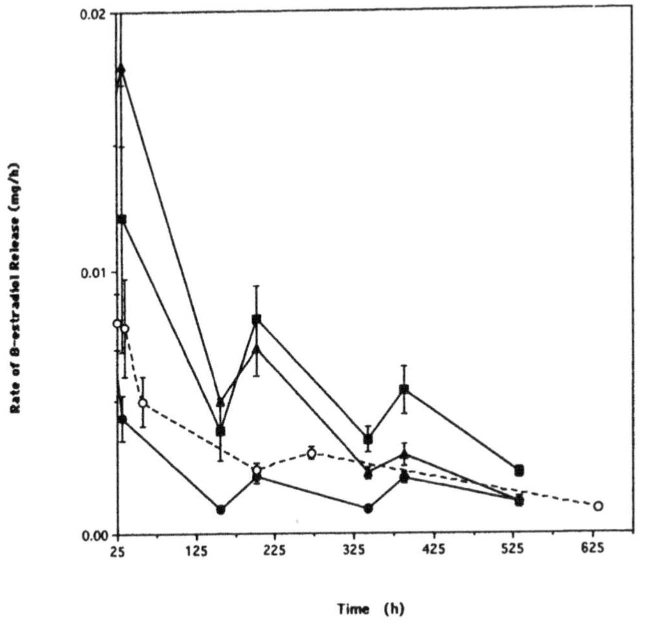
Rate of ß-estradiol Release (mg/h)
Time (h)
0.02
0.01
0.00
25
125
225
325
425
525
625

Figure 5. Rate of release (mg/h) after first 24 h of β-estradiol from formulations containing 50/50 PLA/PGA (M_n = 3859, M_w = 5963): (●) microparticles alone (40 wt.% loaded); (■) 10 wt.% microparticles (40 wt.% loaded) in silicone; (▲) 20 wt.% microparticles (40 wt.% loaded) in silicone; and (○) 4 wt.% β-estradiol in silicone.

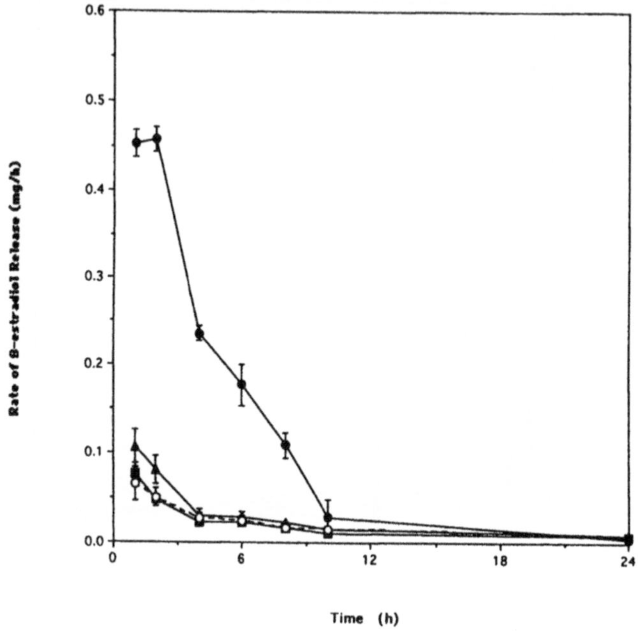
0.6
0.5
0.4
0.3
0.2
0.1
0.0
0
6
12
18
24
Rate of B-estradiol Release (mg/h)
Time (h)

Figure 6. Rate of release (mg/h) for first 24 h of *β*-estradiol from formulations containing 65/35 PLA/PGA (M_n = 3901, M_w = 6041): (●) microparticles alone (40 wt.% loaded); (■) 10 wt.% microparticles (40 wt.% loaded) in silicone; (▲) 20 wt.% microparticles (40 wt.% loaded) in silicone; and (○) 4 wt.% *β*-estradiol in silicone.

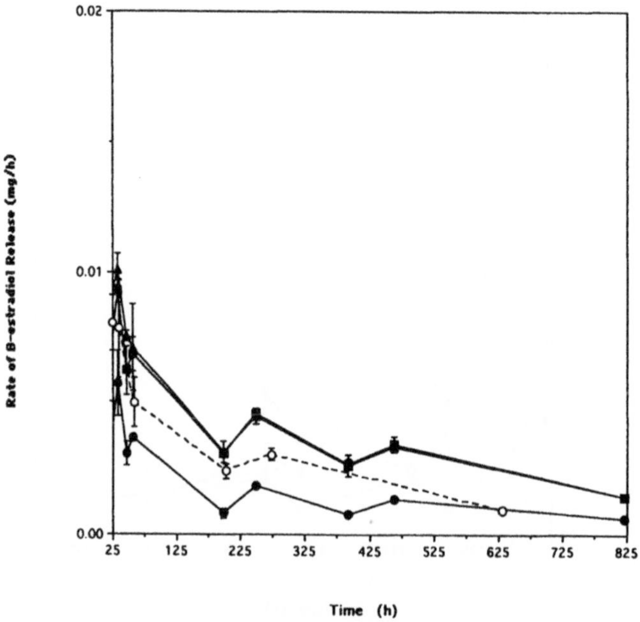
0.02
0.01
0.00
25
125
225
325
425
525
625
725
825
Rate of B-estradiol Release (mg/h)
Time (h)

Figure 7. Rate of release (mg/h) after first 24 h of *β*-estradiol from formulations containing 65/35 PLA/PGA (M_n = 3901, M_w = 6041): (●) microparticles alone (40 wt.% loaded); (■) 10 wt.% microparticles (40 wt.% loaded) in silicone; (▲) 20 wt.% microparticles (40 wt.% loaded) in silicone; and (○) 4 wt.% *β*-estradiol in silicone.

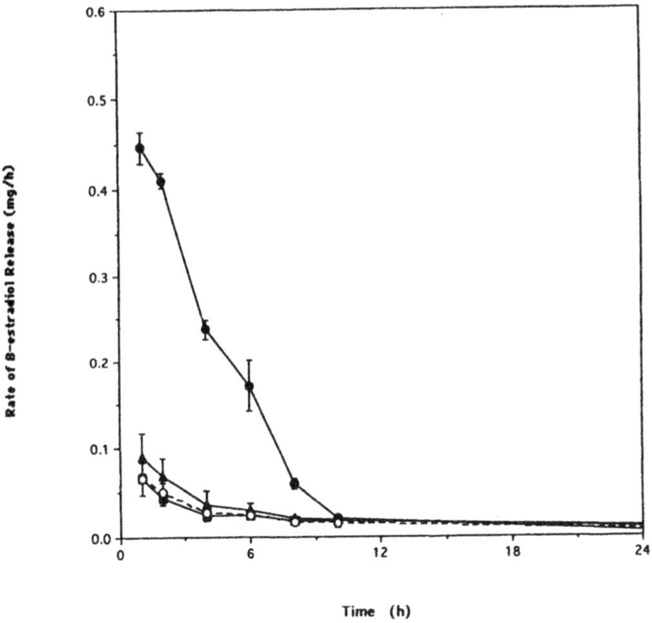
Rate of B-estradiol Release (mg/h)
0.6
0.5
0.4
0.3
0.2
0.1
0.0
0
6
12
18
24
Time (h)

Figure 8. Rate of release (mg/h) for first 24 h of β-estradiol from formulations containing 75/25 PLA/PGA (M_n = 3701, M_w = 5900): (●) microparticles alone (40 wt.% loaded); (■) 10 wt.% microparticles (40 wt.% loaded) in silicone; (▲) 20 wt.% microparticles (40 wt.% loaded) in silicone; and (○) 4 wt.% β-estradiol in silicone.

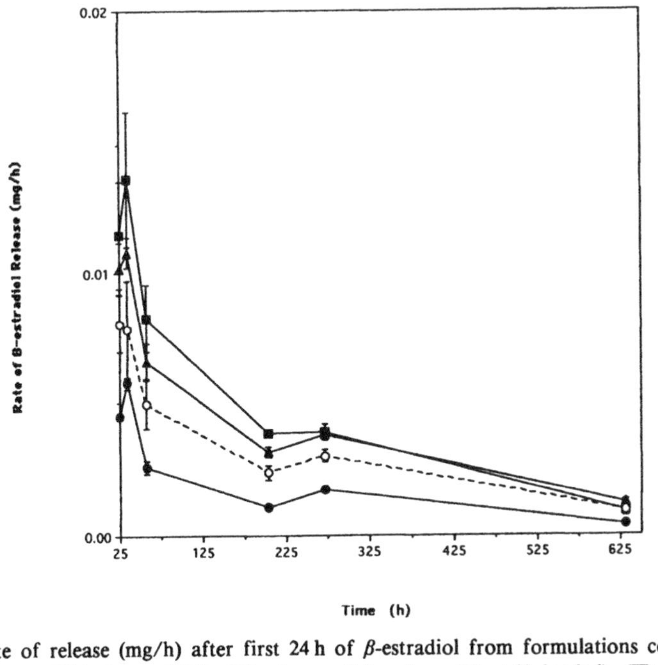
Rate of B-estradiol Release (mg/h)
0.02
0.01
0.00
25
125
225
325
425
525
625
Time (h)

Figure 9. Rate of release (mg/h) after first 24 h of β-estradiol from formulations containing 75/25 PLA/PGA (M_n = 3701, M_w = 5900): (●) microparticles alone (40 wt.% loaded); (■) 10 wt.% microparticles (40 wt.% loaded) in silicone; (▲) 20 wt.% microparticles (40 wt.% loaded) in silicone; and (○) 4 wt.% β-estradiol in silicone.

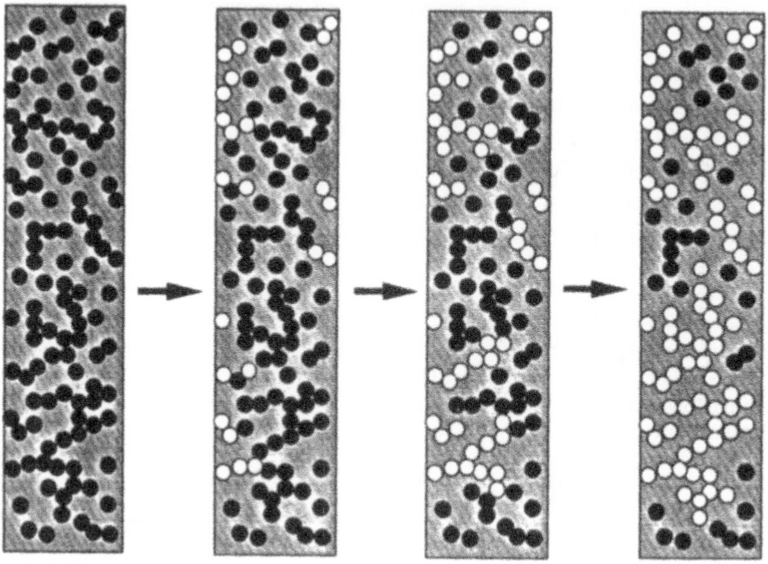

Figure 10. Schematic representation of erosion process of biodegradable polymer microparticles within a silicone matrix.

CONCLUSIONS

The hybrid formulations of biodegradable microparticles within silicone described in this paper exhibit a controlled drug delivery profile distinctly different from that of biodegradable microparticles alone or unencapsulated drug in silicone. These formulations do not show the burst of release found with some microparticles and show higher long-term rates of release than comparable silicone formulations. This type of system would be most useful in developing implants for delivery of drugs where removal of the delivery system might be necessary and for which implants of silicone alone would not yield the desired drug delivery profile.

Acknowledgements

The author wishes to thank Enzytech, Inc. for their generous gift of the poly(lactic acid-co-glycolic acid) polymers used in this study.

REFERENCES

1. L. Brannon-Peppas, *J. Controlled Release* **20**, 201 (1992).
2. E. Mathiowitz, R. S. Langer, A. Warshawsky and E. Edelman, US Patent 4,898,734, February 6, 1990.
3. H. L. Suominen, J. Melartin and K. Karimo, WO patent application 89/103381, November 2, 1989.
4. J. Urquhart and F. Theeuwes, US Patent 4,851,232, July 25, 1989.
5. M. Danckwerts and A. Fassihi, *Drug Dev. Ind. Pharm.* **17**, 1465 (1991).
6. Y. W. Chien, in: *Controlled Drug Delivery. Fundamentals and Applications*, p. 481, J. R. Robinson and V. H. L. Lee (Eds). Marcel Dekker, New York (1987).
7. T. H. Ferguson, G. F. Needam and J. F. Wagner, *J. Controlled Release* **8**, 45 (1988).
8. S. Bogdansky, in: *Biodegradable Polymers as Drug Delivery Systems*, p. 231, M. Chasin and R. Langer (Eds). Marcel Dekker, New York (1990).
9. R. Duncan and L. W. Seymour, *Controlled Release Technologies*. Elsevier Advanced Technology, Oxford (1989).

10. P. Liu, T. Kurihara-Bergstrom and W. R. Good, *Pharmmacol. Res.* **8**, 9388 (1991).
11. D. H. Lewis, in: *Biodegradable Polymers as Drug Delivery Systems*, p. 1, M. Chasin and R. Langer (Eds). Marcel Dekker, New York (1990).
12. R. Jali and J. R. Nixon, *J. Microencapsulation* **6**, 473 (1989).
13. R. Bodmeier and H. Chen, *J. Pharm. Pharmacol.* **40**, 754 (1988).
14. R. Wada, Y. Tabata, S.-H. Hyon and Y. Ikada, *Bull. Inst. Chem. Res., Kyoto Univ.*, **66**, 241 (1988).
15. G. Spenlehauer, M. Vert, J. P. Benoit and A. Boddaert, *Biomaterials* **10**, 557 (1989).
16. R. Arshady, *J. Controlled Release* **17**, 1 (1991).
17. B. Wichert and P. Rohdewald, *J. Controlled Release* **14**, 269 (1990).
18. S. Cohen, T. Yoshioka, M. Lucarelli, L. H. Hwang and R. Langer, *Pharmac. Res.* **8**, 713 (1991).
19. T. R. Tice and R. M. Gilley, European patent EP 0302582, March 31, 1988.
20. D. A. Epstein and B. B. Schryver, European patent EP 0251476, May 23, 1986.
21. C. T. Laurencin, G. T. Syftestad, J. Glowacki and R. S. Langer, Patent application WO 90/09873, February 22, 1990.
22. M. Singh, A. Singh and G. P. Talwar, *Pharmac. Res.* **8**, 958 (1991).
23. J. M. Ruiz and J. P. Benoit, *J. Controlled Release* **16**, 177 (1991).
24. J. H. Eldridge, J. K. Staas, J. A. Meulbroek, J. R. McGhee, T. R. Tice and R. M. Gilley, *Mol. Immunol.* **28**, 287 (1991).
25. H. Okada, T. Heya, Y. Igari, Y. Ogawa, H. Toguchi and T. Shimamoto, *Int. J. Pharmaceutics* **54**, 231 (1989).
26. A. K. Kwong, S. Chou, A. M. Sun, M. V. Sefton and M. F. A. Goosen, *J. Controlled Release* **4**, 47 (1986).
27. F. G. Hutchinson and B. J. A. Furr, in: *Drug Carrier Systems*, p. 111, F. H. Roerdink and A. M. Kroon (Eds). John Wiley & Sons, New York (1989).
28. W. I. Higuchi, U. D. Rohr, S. A. Burton, P. Liu, J. L. Fox, A. H. Ghanem, H. Mahmoud, S. Borsadia and W. R. Good, in: *Controlled-Release Technology. Pharmaceutical Applications*, p. 232, P. I. Lee and W. R. Good (Eds). ACS Symposium Series 348, American Chemical Society, Washington, D.C. (1987).
29. P. L. Ritger and N. A. Peppas, *J. Controlled Release* **5**, 37 (1987).

Ultrasonically enhanced transdermal drug delivery. Experimental approaches to elucidate the mechanism

M. MACHLUF and J. KOST*.

Department of Chemical Engineering, Ben-Gurion University, Beer-Sheva, 84105, Israel

Received 3 September 1992; accepted 13 January 1993

Abstract—The effect of therapeutic range ultrasound on skin permeability was studied *in vitro*. Permeating molecule ionization state, pH, ultrasound duration, reversibility of the enhancement phenomenon, and skin structural alterations were evaluated. It was found that ultrasound affects the permeability of both ionized and unionized molecules. No irreversible structural alterations due to the ultrasound exposure were detected in the stratum corneum. Ultrasound enhancing mechanism was discussed.

Key words: Transdermal drug delivery; ultrasound; skin permeability enhancers.

INTRODUCTION

The barrier functions of skin stratum corneum serve to prevent intrusion of materials from the environment and water loss. These properties have been attributed to the nature and structural state of lipids which are often important regulators in many biological membrane functions and properties. The stratum corneum, which is 15–20 mm thick over much of the human body, primarily consists of blocks of cytoplasmic protein matrices (keratins) embedded in extracellular lipid. The stratum corneum therefore, can offer two possible routes of penetration: one transcellular, the other via the tortous but continuous intercellular lipid [1]. The route through which permeation occurs is largely dependent on the penetrant's physiochemical characteristics, the most important being the relative ability to partition into each skin phase. There have been intensive studies to bypass, in a controlled and reversible way, the permeability barrier of the stratum corneum. Much effort has been focused upon the use of chemical penetration enhancers [2–4], which partition into and interact with the stratum corneum constituents, thereby decreasing its barrier properties. Iontophoresis, the permeation of substances across biological membranes under the influence of electrical current, has also attracted considerable interest [5–7].

Previously we reported on the effect of ultrasound on transdermal drug delivery to rats and guinea pigs [8]. We also suggested and demonstrated the feasibility of pulsatile drug delivery from implantable devices controlled by ultrasound [9]. Recently Bommannan *et al.* [10, 11] studied the mechanism of ultrasound enhancing effects on transdermal permeability and found that ultrasound can induce considerable and rapid transport via the intercellular route. Longer exposure (20 min) of the skin to 16 MHz resulted in structural alterations of the stratum granulosum and

* To whom correspondence should be addressed.

stratum basale cells. These results lead the authors to question the reversibility and the adverse effects of certain frequencies and exposure times.

Phonophoresis or sonophoresis, the movement of drugs through living intact skin and into soft tissue under the influence of an ultrasonic perturbation, [12] was reviewed by Tyle and Agrawala [13] who concluded that 'for most situations in phonophoretic drug delivery, the exact physical mechanisms are not known due to the complexity of the factors involved during the process of phonophoresis'.

The objective of the present research was to assess the effect of various experimental parameters (permeating molecules ionization state, ultrasound duration, skin pretreatment and pH) on the transdermal transport affected by ultrasound, in order to understand and characterize the permeability enhancement phenomenon.

EXPERIMENTAL METHODS

Materials

The membranes evaluated were female hairless mice skin (Balb-C, 6–8 weeks). The permeating molecules were [^{3}H] Hydrocortisone (NET-396, DuPont), [^{14}C] Salicylic acid, SA (NEC-263, DuPont) and Caffeine (Sigma Chemical Co.).

Instruments

The therapeutic ultrasound unit was Sonopuls 434 (Enraf Nonius Delft, The Netherlands), having a probe surface area of 1.4 cm^2 and an effective radiating area (ERA) of 0.8 cm^2. The frequency was 1.0 MHz ± 0.2%, power output 3.0W/cm^2, pulse mode 20% duty cycle.

Permeability studies

Permeability experiments were performed in a two compartment glass transdermal transport cell, adjusted to accommodate an ultrasound probe in the donor compartment (Fig. 1). The transport cell consisted of two compartments (3 ml donor and 7 ml receiving). The available area for diffusion between the two compartments was 4.9 cm^2. The receiving compartment was mixed by a magnetic stirrer (Fried Electric, Israel) at 560 ± 30 rpm. The temperature of the solution was controlled at 27 ± 1°C, pumping water at a constant temperature (Fried Electric thermostat, Israel) throughout the cell jacket.

All permeability experiments used full-thickness skin which was removed from the animal at sacrifice (a CO_2 chamber) and equilibrated for 90 min in the buffer of interest. The skin section was then mounted between the two compartments of the diffusion cell with the dermis side facing the receiving compartment. In order to eliminate ultrasound attenuation in the liquid the ultrasound probe was placed in contact with the skin. The donor solution was left in place after the ultrasound unit was turned off.

SA permeability experiments were performed at two pH values: pH = 2.65 (0.005 M phosphate buffer) and pH = 9.9 (0.0025 M borax buffer). Hydrocortisone permeability experiments were performed in a phosphate buffer pH 7.4 (0.005 M phosphate buffer). Caffeine permeability experiments were performed at three pH

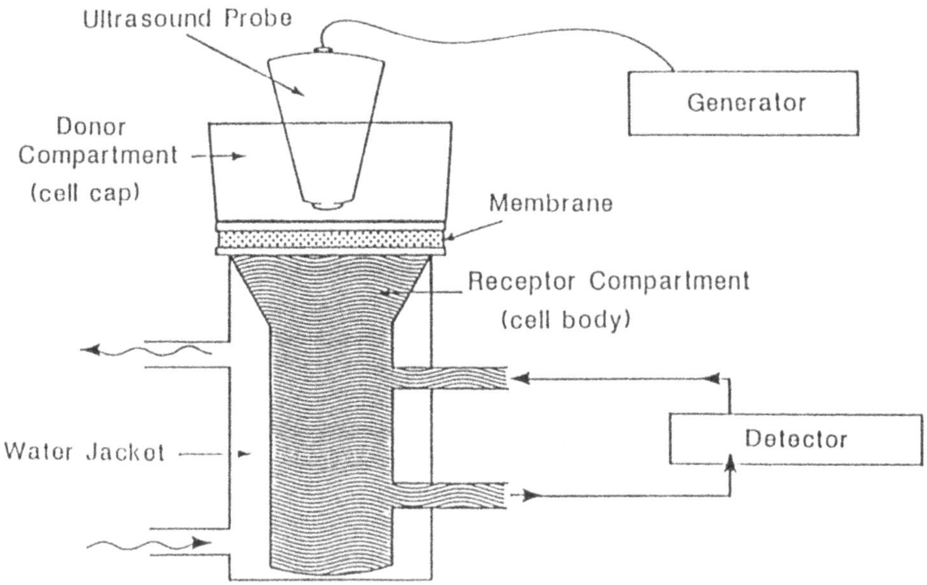
Ultrasound Probe
Generator
Donor
Compartment
(cell cap)
Membrane
Receptor Compartment
(cell body)
Detector
Water Jacket

Figure 1. Schematic diagram of the transdermal transport cell.

values: pH 4 (0.05 M phosphate buffer), pH 9 (0.0025 M borax bufer), and pH 7.4, (0.005 M phosphate buffer).

Permeability evaluation

Salicylic acid or hydrocortisone solution samples from the donor and receiving compartments (10 μl) were placed immediately into 4 ml of liquid scintillation counting solution consisting of 2,5-diphenyl-oxazole (15 g PPO), 1,4 bis-2-(4-methyl-5-phenyloxazole)-benzene (0.3 g POPOP), 2 l toluene, and 1 l triton. Radioactivity was determined in a β-counter (Beckman-LF 1800).

Caffeine samples from the donor and receiving compartments (50 μl) were detected by high precision liquid chromatography (HPLC), using a C-18 Millipore column at a flow rate of 1 ml/min of a mobile phase consisting of 28/72 methanol sodium acetate adjusted to pH of 3.6 with acetic acid. The retention time of caffeine was 2.7 min, detected by UV adsorption at 297 nm [14].

All the results are presented as a mean and standard deviation of five independent repetitions.

Stratum corneum preparation

Stratum corneum was separated from epidermis by placing the full-thickness skin dermis-side down on a filter paper saturated with 1% trypsin (Sigma Chemical Co., purified porcine pancreas, Type IX) solution at 37°C for 4 h following the procedure described by Knutson *et al.* [15]. The separated stratum corneums before and after exposure to ultrasound were evaluated by Fourier transform infrared spectroscopy (FTIR) from 4000 to 400 cm^{-1} (Nicolet, S-ZDX) and differential scanning calorimetry (DSC) (Mettler, TA-4000), 0–150°C at a rate of 10°C/min. The procedures were repeated at least five times for different stratum corneum samples.

RESULTS AND DISCUSSION

A series of experiments were performed in which the influence of applied ultrasound on *in vitro* transdermal transport of salicylic acid (SA) was assessed. As the route through which permeation occurs largely depends on the penetrant's physiochemical characteristics, the effect of ultrasound on the permeability of ionized and unionized states of SA was evaluated. Salicylic acid is predominantly unionized at pH 2.65, and therefore, presumably permeates the skin through the lipoidal pathway, while at pH 9.9 it is ionized, and therefore more aqueous routes are involved in its permeation through the skin [4, 16]. As lipoidal configuration changes were suggested to be the enhancing mechanism for some chemical transdermal enhancers [2–4] and also proposed [17] as a possible mechanism of improved percutaneous absorption by ultrasound, the effect of ultrasound on the possible different routes of permeation through the skin was evaluated.

Figure 2 (a and b) shows the effect of ultrasound on the permeability of SA through hairless mice skin at pH 9.9 and 2.65. The initial concentration of SA in the donor compartment was $1.8\ 10^{-5}$ M. The donor compartment was exposed for 5 h to ultrasound in a pulsed mode (20% duty cycle, 1 MHz, $3 W/cm^2$). As can be seen from these experiments, ultrasound affects the permeability of both, ionized and unionized SA. We hypothesize that if the main effect of ultrasound is on the lipoidal configuration, the effect of ultrasound on molecules which do not permeate through this route should be relatively small or none. The observed effect of ultrasound on the transport of the two molecular states and especially the very pronounced effect of ultrasound on permeability of the ionized state of SA, suggests that the enhancing mechanism of ultrasound might be more complex than just its effect on lipoidal configuration changes.

In order to evaluate the effect of pH on transdermal permeability, control experiments were performed with caffeine at pHs of 4, 7.4, and 9. In contrast to SA, caffeine is unionized in this pH span, and thus is not affected by pH changes. As can be seen in Fig. 3 (a–c) the effect of 3 h of ultrasound exposure, at comparable frequencies, pulse mode and intensities to the SA experiments, was very similar at the three different pH values of 4, 7.4, and 9. These results confirm that the effect of ultrasound on SA permeability at the ionized and unionized state is not just due to pH differences.

In experiments performed with hydrocortisone in an attempt to evaluate the effect of ultrasound exposure duration, it was found that 5 h exposure and 90 min exposure resulted in the same permeability enhancement (Fig. 4a, b). Shorter exposures such as 30 min (Fig. 4c) were not sufficient to cause any effect on hydrocortisone permeability when compared to controls which were not exposed to ultrasound. In previous *in vivo* experiments [8] performed on rats and guinea pigs we were able to demonstrate the effect of ultrasound on mannitol, inulin and physostigmine permeability with only 3–5 min of ultrasound irradiation at comparable frequencies, intensities, and ultrasound efficiency (probe placed in contact with the skin). The difference between the *in vivo* and *in vitro* results may be attributed to: the different permeating molecules; the differences between nude skin which was

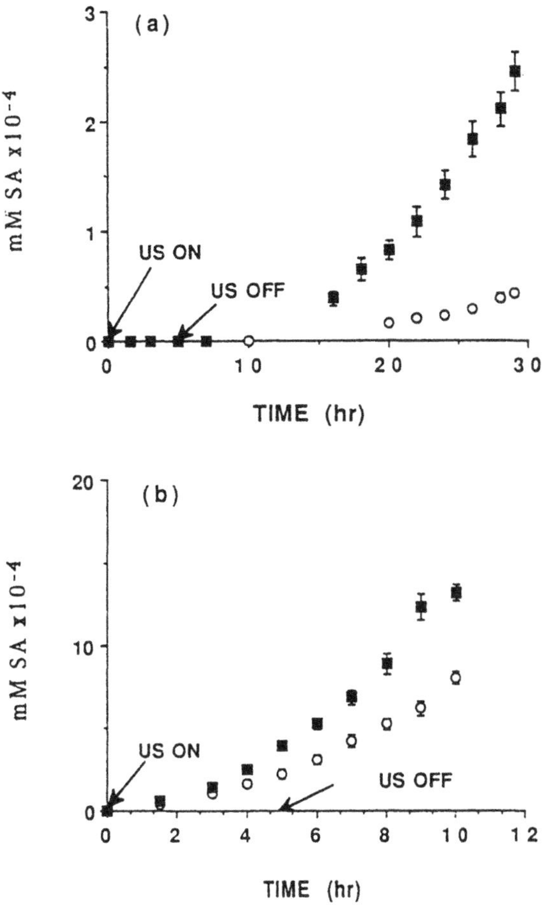
(a)
mM SA x10-4
3
2
1
0
US ON
US OFF
0
10
20
30
TIME (hr)
(b)
mM SA x10-4
20
10
0
US ON
US OFF
0
2
4
6
8
10
12
TIME (hr)

Figure 2. Effect of ultrasound (20% duty cycle, 1 MHz, $3W/cm^2$) on salicylic acid permeability through hairless mice skin (displayed as SA concentration in the receiving compartment vs time) at: (a) pH of 9.9; and (b) pH of 2.65. (■) ultrasound applied for 5 h, (O) controls, no ultrasound. The initial concentration of SA in the donor compartment was 1.8 to 10^{-5} M.

evaluated in the *in vitro* experiments and furry closely clipped skin in the *in vivo* applications; different ultrasound coupling agents, gel in the *in vivo* and water in the *in vitro* experiments; and the different drug distributions and metabolism profiles [18].

In an attempt to investigate the reversibility of the enhancement phenomenon, permeability experiments were performed on skin that was pretreated by exposing it to ultrasound for 2 h while mounted in the transport cell. After that exposure was complete, hydrocortisone (1.2 10^{-7} M) was added to the donor compartment and its concentration, in the receiving compartment, was monitored over a period of time. Figure 5 displays the results of these experiments and controls which were not subjected to ultrasound. No difference between the two can be detected, suggesting that there are no irreversible changes in the skin which would affect its permeability. These results were supported by differential scanning calorimetry (DSC), and

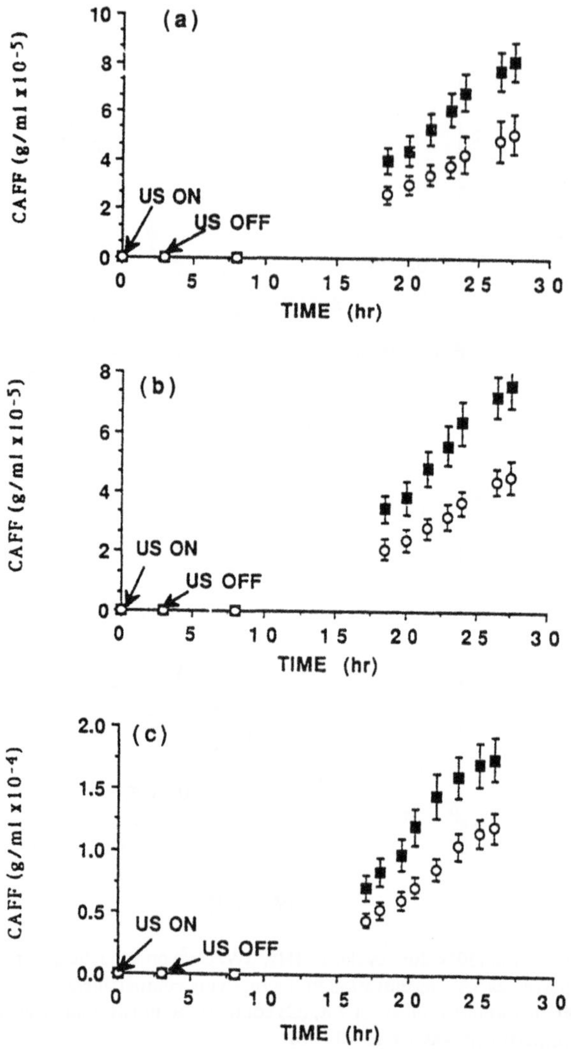

(a)
CAFF (g/ml x10-5)
10
8
6
4
2
0
US ON
US OFF
0
5
10
15
20
25
30
TIME (hr)
(b)
CAFF (g/ml x10-5)
8
6
4
2
0
US ON
US OFF
0
5
10
15
20
25
30
TIME (hr)
(c)
CAFF (g/ml x10-4)
2.0
1.5
1.0
0.5
0.0
US ON
US OFF
0
5
10
15
20
25
30
TIME (hr)

Figure 3. Effect of ultrasound (20% duty cycle, 1 MHz, 3W/cm^2) on caffeine permeability through hairless mice skin (displayed as caffeine concentration in the receiving compartment vs time) at: (a) pH of 4; (b) pH of 7.4; and (c) pH of 9. (■) ultrasound applied for 3 h, (O) controls, no ultrasound. The initial concentration of caffeine in the donor compartment was 4.6 10^{-2} M.

Fourier transform infra-red (FTIR) analysis of the ultrasound exposed and not exposed stratum corneum samples (Figs 6 and 7).

DCS thermograms of the ultrasound exposed and not exposed stratum corneum samples are presented in Fig. 6 for the 5–125°C temperature region. Temperature transitions below 80°C are primarily associated with lipid transitions [19]. No significant difference between the exposed and not exposed to ultrasound stratum corneum thermograms could be detected. The small differences in peak shapes or location are possibly due to the difference in skins origin.

The infrared bands evaluated were in the 2400–3000 and 1360–1840 cm^{-1} ranges (Fig. 7). The C—H asymmetric and symmetric stretching vibrations forming the

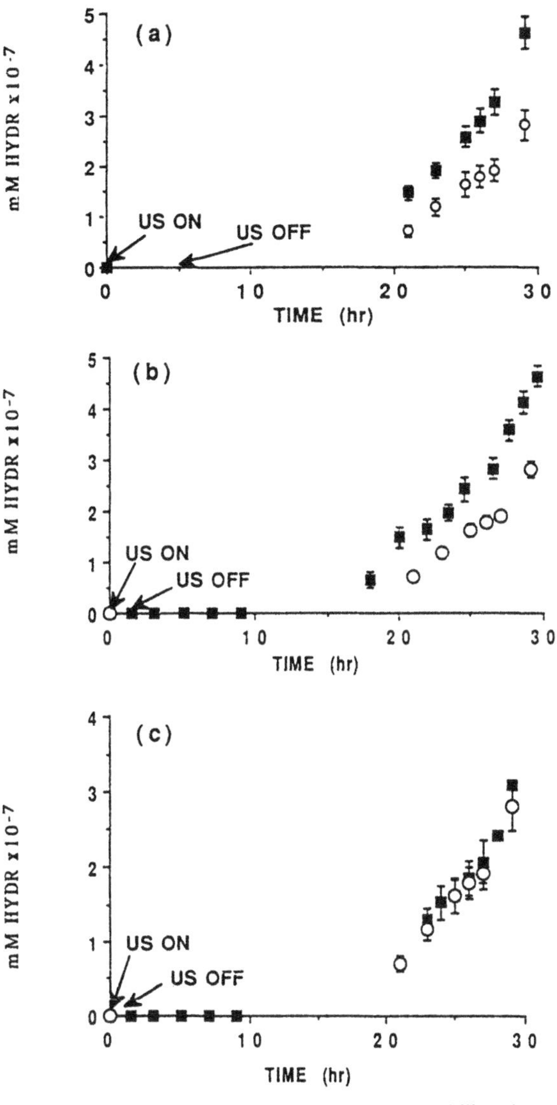
(a)
US ON
US OFF
mM HYDR x10-7
TIME (hr)
(b)
US ON
US OFF
mM HYDR x10-7
TIME (hr)
(c)
US ON
US OFF
mM HYDR x10-7
TIME (hr)

Figure 4. Effect of ultrasound duration on hydrocortisone permeability through hairless mice skin exposed to ultrasound (20% duty cycle, 1 MHz, $3W/cm^2$) (displayed as hydrocortisone concentration in the receiving compartment vs time) for: (a) 5 h; (b) 1.5 h; and (c) 30 min. (○) exposed to ultrasound, (■) controls, no ultrasound. The initial concentration of hydrocortisone in the donor compartment was $1.2\ 10^{-7}$ M.

series of bands in the 2700–3000 cm^{-1} range arose from the CH_2 and CH_3 molecular groups of the hydrocarbon groups tails of the lipids, and minor contribution of proteins. The C=O stretching vibrations of amide groups formed the 1640 cm^{-1} band (Amide I). The C—N stretching and N—H bending vibration of the protein amide group formed the band near 1550 cm^{-1} (Amide II). Amide I and II bands arose primarily from proteins, although sphingolipids were minor contributors to the bands due to the presence of amide groups within the polar head groups. The C—H bending vibrations of CH, CH_2, and CH_3 groups formed the bands between

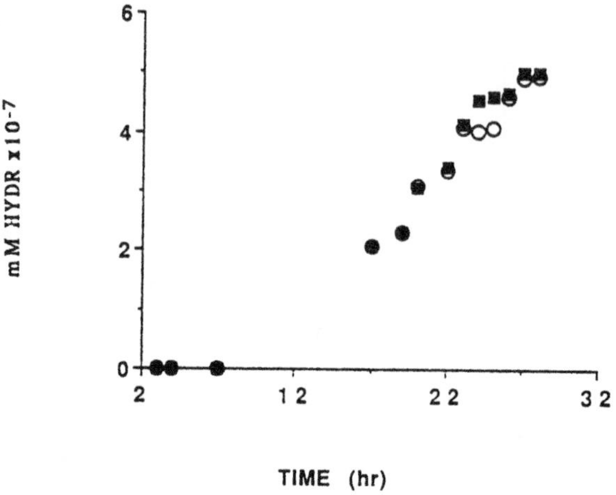
mM HYDR x10-7
6
4
2
0
2
12
22
32
TIME (hr)

Figure 5. Effect of skin pretreatment by ultrasound on hydrocortisone permeability (displayed as hydrocortisone concentration in the receiving compartment vs time) through hairless mice skins (■) which were pretreated for 2 h by ultrasound (20% duty cycle, 1 MHz, $3W/cm^2$), (O) control experiments with skins which were not pretreated by ultrasound.

1500 and 1360 cm^{-1} [15, 19]. No significant difference between the exposed and not exposed to ultrasound stratum corneum FTIR spectra could be detected.

The permeability results of the pretreated skins, and the DSC and FTIR analysis indicate that there are no irreversible structural changes in the stratum corneum caused by ultrasound exposure at therapeutic levels ($3W/cm^2$) that are sufficient to enhance transdermal drug permeability.

In conclusion, we have evaluated the effect of permeating molecule ionization state, pH, ultrasound duration, reversibility of the enhancement phenomenon and

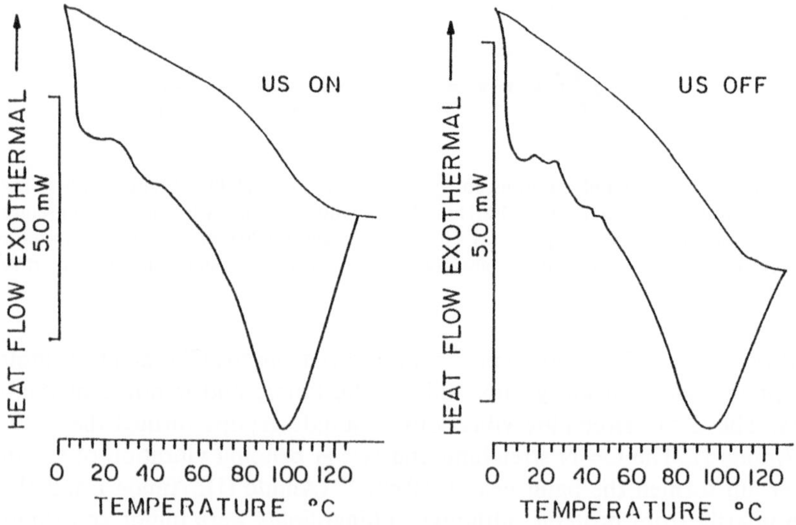
HEAT FLOW EXOTHERMAL
5.0 mW
US ON
0 20 40 60 80 100 120
TEMPERATURE °C
HEAT FLOW EXOTHERMAL
5.0 mW
US OFF
0 20 40 60 80 100 120
TEMPERATURE °C

Figure 6. DSC scans of hairless mice stratum corneum before (US OFF), and after exposure of 2 h (US ON) to ultrasound (20% duty cycle, 1 MHz, $3W/cm^2$).

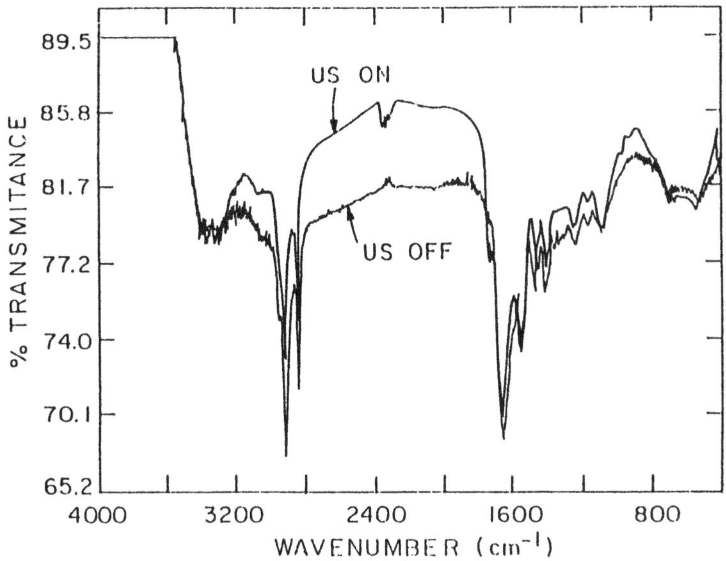

89.5
85.8
81.7
77.2
74.0
70.1
65.2
% TRANSMITANCE
US ON
US OFF
4000
3200
2400
1600
800
WAVENUMBER (cm⁻¹)

Figure 7. FTIR scans of hairless mice stratum corneum before (US OFF) and after exposure of 2 h (US ON) to ultrasound (20% duty cycle, 1 MHz, $3W/cm^2$).

skin structural changes on the enhanced skin permeability while exposed to ultrasound. The results show that ultrasound affects the permeability of both ionized and unionized molecules. The pronounced effect on the ionized SA, together with our and other previous mechanistic studies [8, 9, 20–22] which showed that ultrasound can reversibly enhance the permeability of synthetic membranes (phenomenon which is not attributed to mixing, temperature effects, or irreversible changes in membrane integrity), and the results of caffeine transport at different pHs, suggest that the effect of ultrasound on the lipoidal configuration might not be the only possible mechanism responsible for the enhanced transport through the skin when exposed to ultrasound. No irreversible structural changes due to the therapeutic level ultrasound exposure were detected in the stratum corneum. Further studies on skin reversible morphological changes are required in order to elucidate the transdermal permeability enhancing mechanism. Current studies are in progress to evaluate the contribution of ultrasonically induced convective transport (acoustic streaming) as a possible additional contributor to the enhancing phenomenon.

REFERENCES

1. P. M. Elias, *Int. J. Dermatol.* **20**, 1–19 (1981).
2. B. W. Barry, *J. Controlled Release* **15**, 237–248 (1991).
3. K. A. Walters, in: *Transdermal Drug Delivery*, p. 197, J. Hadgraft and R. H. Guy (Eds). Marcel Dekker, New York (1989).
4. R. H. Guy and J. Hadgraft, *J. Controlled Release* **5**, 43–51 (1987).
5. P. Tyle, *Pharmacol. Res.* **3**, 318–326 (1986).
6. S. Del Terzo, C. R. Behl and R. A. Nash, *Pharmacol. Res.* **6**, 85–90 (1989).
7. A. Banga and Y. W. Chien, *J. Controlled Release* **7**, 1–24 (1988).
8. D. Levy, J. Kost, Y. Meshulam and R. Langer, *J. Clin. Invest.* **83**, 2074–2078 (1989).
9. J. Kost, K. Leong and R. Langer, *Proc. Natl. Acad. Sci.* **86**, 7663–7666 (1989).
10. D. Bommannan, H. Okuyama, P. Stauffer and R. H. Guy, *Pharmacol. Res.* **9**, 559–564 (1992).

11. D. Bommannan, G. K. Menon, H. Okuyama, P. M. Elias and R. H. Guy, *Pharmacol. Res.* **9**, 1043–1047 (1992).
12. D. M. Skauen and G. M. Zentner, *Int. J. Pharmacol.* **20**, 235–245 (1984).
13. P. Tyle and P. Agrawala, *Pharmacol. Res.* **6**, 355–361 (1989).
14. T. Kurihara-Bergstrom and W. R. Good, *J. Controlled Release* **6**, 51–58 (1987).
15. K. Knutson, S. L. Krill, W. J. Lambert and W. I. Higuchi, *J. Controlled Release* **6**, 59–74 (1987).
16. E. R. Cooper, in: *Solution Behavior of Surfactants*, p. 1505, K. L. Mittal and E. J. Fendler (Eds). Plenum Press, New York (1982).
17. H. A. Benson, J. C. McElnay, R. Harland and J. Hadgraft, *Pharmacol Res.* **8**, 204–209 (1991).
18. D. B. Guzek, A. H. Kennedy, S. C. McNeill, E. Wakshull and R. O. Potts, *Pharmacol Res.* **6**, 33–39 (1989).
19. K. Knutson, R. O. Potts, D. B. Guzek, G. M. Golden, J. E. McKie, W. J. Lambert and W. I. Higuchi, *J. Controlled Release* **2**, 67–87 (1985).
20. T. N. Julian and G. M. Zentner, *J. Controlled Release* **12**, 77–85 (1990).
21. L. S, Liu, J. Kost, A. D'Emanuele and R. Langer, *Macromolecules* **25**, 123–128 (1992).
22. A. D'Emanuele, J. Kost, J. Hill. and R. Langer, *Macromolecules* **25**, 511–515 (1992).

Part IX

Water-Soluble Biomolecules, Synthetic Polymers and their Conjugates

Synthesis of carboxylated poly(NIPAAm) oligomers and their application to form thermo-reversible polymer–enzyme conjugates

GUOHUA CHEN and ALLAN S. HOFFMAN*

Center for Bioengineering, FL-20, University of Washington, Seattle, WA 98195, USA

Received 18 August 1992; accepted 4 February 1993

Abstract—A thermo-reversible poly(*N*-isopropylacrylamide) poly(NIPAAm) oligomer with a carboxyl functional end group has been synthesized by radical polymerization using β-mercaptopropionic acid as a chain transfer reagent. This polymer has been conjugated to an enzyme, β-D-glucosidase, to form a thermo-reversible water soluble–insoluble polymer–enzyme conjugate. This conjugate can be used for separation, recovery and recycle of an enzyme simply by applying small temperature changes to the reaction medium. In contrast to the random polymer–enzyme conjugates reported in the literature, in this study the enzyme is coupled to each polymer chain by a single end attachment. These preliminary studies show that the conjugated enzyme exhibits very high retention of activity (>90%) compared to the native enzyme and shows improved thermal stability.

Key words: Poly(isopropylacrylamide) oligomer; chain transfer polymerization; β-D-glucosidase; thermo-reversible polymer–enzyme conjugate; soluble–insoluble polymer–enzyme conjugate.

INTRODUCTION

Many water insoluble, immobilized enzyme systems have been investigated because of their ease of recovery from aqueous suspension and their potential for application in fluidized or packed bed bioreactors and chromatographic columns [1, 2]. However, when the enzymes are immobilized in an insoluble form, there is less effective use of the enzyme due to the solid–liquid heterogeneous reaction conditions, which can lead to diffusion-control of substrate into and/or product out of the solid carrier. Furthermore, it is inappropriate to use a solid-phase enzyme when using a solid substrate.

Soluble polymer–enzyme conjugates exhibit many advantages over enzymes immobilized on or within porous solid supports, especially when used for macromolecular or solid substrates [3]. In particular, with reversible soluble–insoluble polymer–enzyme conjugates, the enzymatic reaction occurs in solution, after which the enzyme can be recovered in an insoluble state by small changes of pH [4–7] or temperature [8–11]. Our aim in this study is to prepare thermo-reversible, soluble–insoluble polymer–enzyme conjugates and to apply them to reactions with macromolecular or solid substrates, such as using conjugated polymer–cellulase for conversion of cellulose into glucose.

It is well known that poly(*N*-isopropylacrylamide) poly(NIPAAm) is a thermo-reversible water soluble polymer, whereby aqueous solutions of this polymer exhibit

---

* To whom correspondence should be addressed.

a lower critical solution temperature (LCST) *ca.* 33°C [12]. The reversible phase separation behavior of poly(NIPAAm) has been utilized to make soluble–insoluble antibody [8] and other protein conjugates [13, 14]. However, in all of those studies, the polymers used have been copolymers of NIPAAm with comonomers containing reactive groups. Since the copolymers usually have more than one reactive group in each polymer chain, it is hard to know if the enzyme is coupled with a polymer chain by single or multiple attachment, and also at what point along with the chain the attachment occurs, both of which might significantly influence the activity and properties of the conjugated enzyme. In order to simplify the structure of the conjugated enzyme, we have synthesized a new type of poly(NIPAAm) oligomer having an end-capped carboxyl group, and conjugated it to an enzyme, *β*-D-glucosidase, one of the three enzymes comprising cellulase. This conjugation assures that the enzyme is singly attached to the polymer chain by a single end attachment. Some properties of the conjugated enzyme have been investigated and are reported here.

EXPERIMENTAL

Materials

N-Isopropylacrylamide (NIPAAm, Eastman Kodak, Rochester, NY, USA) was purified by recrystallization from n-hexane and dried in vacuum. 2,2′-Azoisobutyronitrile (AIBN, from J. T. Baker, Phillipsburg, USA) was recrystallized from methanol. *β*-Mercaptopropionic acid (MPA, Aldrich, Milwaukee, WI, USA) was purified by distillation under reduced pressure. All the alcohol solvents were distilled before use. *β*-D-Glucosidase (from almonds), *p*-nitrophenyl *β*-D-glucopyranoside (*p*NPG) and 1-ethyl-3-(3-dimethylaminopropyl) carbodiimide hydrochloride (EDC) were purchased from Sigma, St. Louis, MO, USA.

Preparation of carboxylated poly(NIPAAm)

Poly(NIPAAm) was prepared by the radical polymerization of NIPAAm using AIBN and MPA as initiator and chain transfer reagent, respectively, as shown in Scheme 1. Appropriate quantities of NIPAAm, AIBN, and MPA with an alcohol solvent were placed in a thick-walled polymerization tube and the mixtures were degassed by freezing and evacuating and then thawing (four times). After cooling for the last time, the tubes were evacuated and sealed prior to polymerization. The

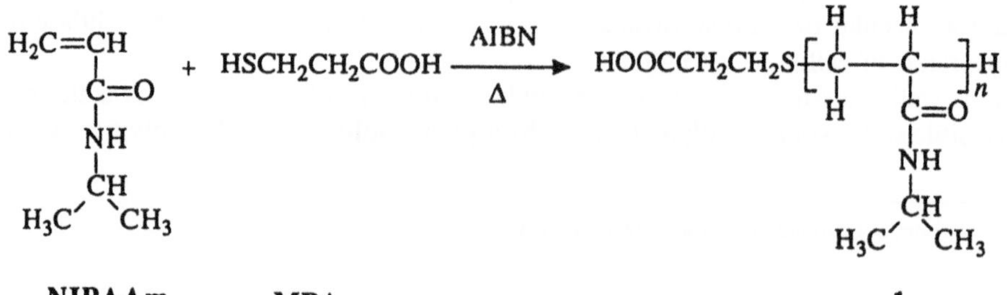
AIBN
Δ
HSCH2CH2COOH
HOOCCH2CH2S

NIPAAm MPA 1

Scheme 1. Synthesis of the carboxylated poly(NIPAAm) oligomer.

tubes were immersed in a water bath at 50 or 60°C for 24 h. The resulting polymer was isolated by precipitation into diethyl ether and weighed to determine yield. The molecular weight of the polymer obtained was determined by titration with 0.01 N NaOH, which detects the carboxyl groups at the end of the polymer molecules.

Lower critical solution temperature (LCST) measurement

The LCSTs of the polymer solutions (2.0 mg/ml) were measured spectrophotometrically by determining the turbidity of the solution at various temperatures. The temperature at 90% light transmittance of the poly(NIPPAAm) solution at 500 nm was defined as the LCST [13].

Conjugation of β-D-glucosidase with poly(NIPAAm)

Conjugation of β-D-glucosidase with the carboxylated poly(NIPAAm) was carried out by activation of the enzyme and polymer mixture with EDC in 0.05 M phosphate buffer (Scheme 2). After gently shaking at 4°C for 16 h, the mixture was heated up to 40°C. The resulting precipitate was recovered by centrifugation (10000 r.p.m. for 15 min at 40°C). The precipitate was redissolved in the same buffer at room temperature and the solution was heated up to 40°C. The resulting precipitate was recovered again by centrifugation. This procedure was repeated twice. The final precipitate was dissolved in 10 ml buffer (pH 6.5) and stored at 4°C for further study on its properties. The enzymatic activity of the solution was tested. The amount of protein bound to poly(NIPAAm) was determined by the BCA protein assay method [15].

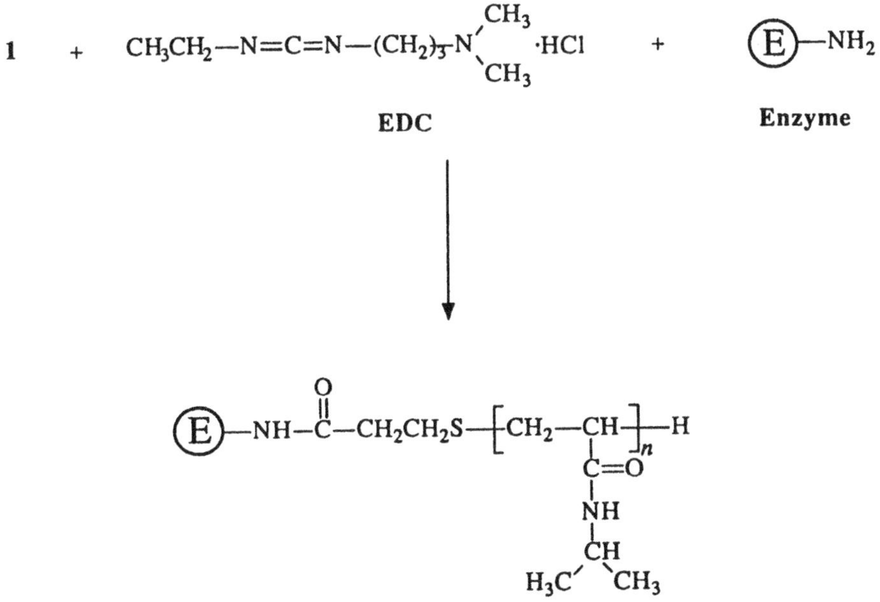
1
EDC
Enzyme

Scheme 2. Coupling of poly(NIPAAm) with β-D-glucosidase.

Enzyme assay

The activities of native and conjugated β-D-glucosidase were determined spectrophotometrically using *p*-nitrophenyl β-D-glucopyranoside (*p*NPG) as the substrate (0.45 mM/l, 0.05 M phosphate buffer, pH 7.0, 25°C, unless otherwise indicated). The enzyme activity was expressed as: one unit of the enzyme produces 1 μM of *p*-nitrophenol per minute at the assay condition.

Kinetic constants

The kinetic parameters for β-D-glucosidase were determined using *p*NPG as the substrate using the concentration range of 0.11–0.68 mM at pH 6.5, 25°C. K_m and V_{max} values were calculated from Lineweaver–Burke (double reciprocal) plots.

Optimum pH

The activities of native and conjugated β-D-glucosidase were measured at different pHs (ranging from 4.5 to 8.5) at 25°C with a *p*NPG concentration of 0.45 mM in 0.05 M phosphate buffer.

The optimum temperature and the thermal inactivation

The optimum temperature of hydrolysis of the substrate *p*NPG by native and conjugated β-D-glucosidase was determined at a variety of temperatures. The enzyme was incubated with *p*NPG at a particular temperature for 10 min with gentle stirring, and the reaction was stopped by adding 0.2 M sodium carbonate solution to adjust the pH to around 9. After cooling to room temperature, the *p*-nitrophenol produced was measured at 405 nm. To examine thermal inactivation, native and conjugated β-D-glucosidase were incubated at 60°C either for 40 min with sampling at 5 min intervals, or at a variety of temperatures for 15 min each, with gentle stirring. After incubation, the samples were kept at room temperature for 2 h before the activity was assayed. The *p*-nitrophenol production was measured at 25°C, pH 7.0. For comparison, controls were based on physical mixtures of enzyme and poly(NIPAAm) using the same concentration of the native enzyme and free poly(NIPAAm) as were in the native enzyme of poly(NIPAAm)-conjugated enzyme experiments.

RESULTS AND DISCUSSION

Polymer synthesis

Synthesis of poly(NIPAAm) is shown in Scheme 1. In the first experiments, four alcohols—methanol, ethanol, iso-propanol, and tert-butanol were used as the solvents. We chose those solvents because it is well known that the chain transfer rates of the polymeric radicals to those alcohol solvents are different, following the order of tert-butanol $\ll$ methanol < ethanol < iso-propanol [16, 17]. One might expect that polymerization in tert-butanol may yield polymers with relatively narrow molecular weight distribution compared to those in the other alcohols since it has the lowest chain transfer rate. Table 1 gives some polymerization data and properties of the polymer. It can be seen that the polymerization in tert-butanol resulted in the highest yield. Also, GPC measurements showed that the polymer has the narrowest

Table 1.
Preparation and properties of carboxylated poly(NIPAAm) oligomers[a]

	Experiment no.			
	1	2	3	4
Solvents	tert-butanol	methanol	ethanol	iso-propanol
W_{NIPAAm} (g)	10.0	10.0	10.0	10.0
W_{AIBN} (g)	0.10	0.10	0.10	0.10
W_{MPA} (g)	0.40	0.40	0.40	0.40
$W_{Polymer}$ (g)	8.31	7.03	5.01	6.06
Yield (w/w%)	83.1	70.3	50.1	60.6
LCSTs (°C)	33.6 ± 0.1	33.3 ± 0.1	33.4 ± 0.0	34.3 ± 0.0
$\bar{M}_n$[b]	4020 ± 20	5030 ± 20	5150 ± 20	4340 ± 10

[a] Each polymerization was carried out at 50°C for 24 h, in 20 ml of solvent.
[b] $\bar{M}_m$ was estimated by end group titration.

molecular weight distribution compared to those obtained in the other alcohol solvents (data are not shown here). Therefore, tert-butanol was chosen as the polymerization solvent and the polymer obtained in this solvent was used in the subsequent conjugation experiments. ^{1}H NMR spectra show the proton peaks from the methylene group in the chain transfer reagent, which confirms the formation of the polymer with the designed structure (Fig. 1). By titration using 0.01 N NaOH solution, which detects the end carboxyl group, the molecular weight was estimated to be 4000–5000.

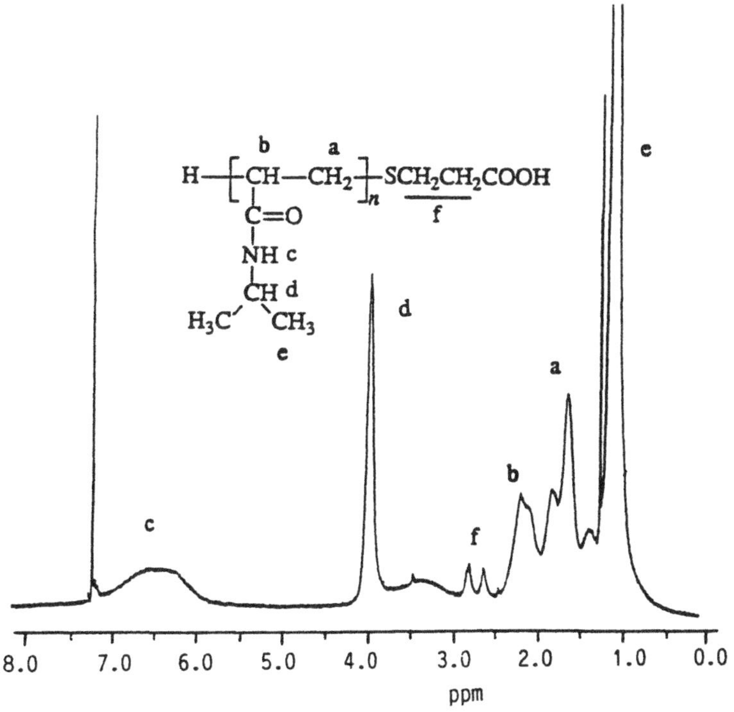
b
a
H—[CH—CH2]n—SCH2CH2COOH
f
C=O
NH c
CH d
H3C CH3
e
e
d
a
b
c
f
8.0
7.0
6.0
5.0
4.0
3.0
2.0
1.0
0.0
ppm

Figure 1. 500 MHz ^{1}H NMR spectrum of carboxylated poly(NIPAAm) (chloroform, 25°C).

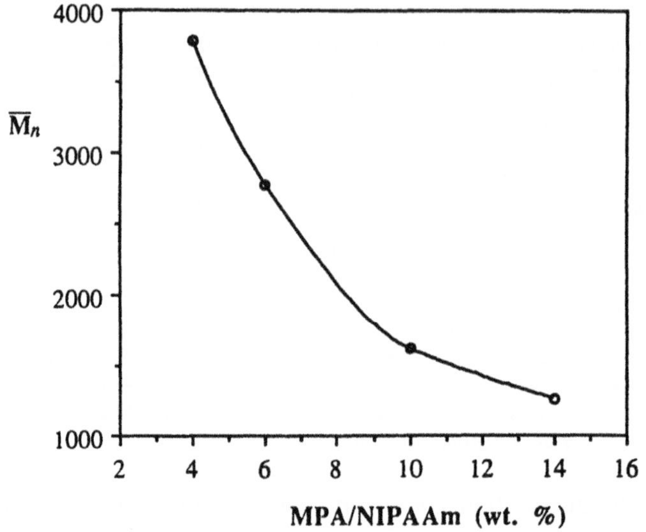
4000
3000
2000
1000
$\overline{M}_n$
2
4
6
8
10
12
14
16
MPA/NIPAAm (wt. %)

Figure 2. Molecular weight dependence of the poly(NIPAAm) oligomer on the ratios of monomer to chain transfer reagent (in tert-butanol, at 60°C for 24 h).

In order to synthesize oligomers with different molecular weights, the polymerizations were carried out in tert-butanol with varying ratios of the monomer, initiator and chain transfer reagent. As shown in Fig. 2 and Table 2, the molecular weight of the oligomer significantly decreases with increase in the ratio of chain transfer reagent to the monomer (Fig. 2), and it is only slightly influenced by changing the ratio of monomer to the initiator (Table 2). It can be concluded that the carboxylated oligomers of poly(NIPAAm) with different molecular weight could be synthesized by simply changing the ratio of chain transfer reagent to the monomer.

From Tables 1 and 2, it can be seen that the oligomers exhibit LCSTs between 33.6 and 34.6°C, slightly higher than the LCST of the homopoly(NIPAAm) (33.0°C), which possibly is due to the relatively hydrophilic end group.

Table 2.
Effect of the ratio of monomer to initiator and to chain transfer reagent on the molecular weight of the oligomers[a]

Experiment no.	NIPAAm : AIBN : MPA (wt. ratio)	$\bar{M}_m$[b]	LCST (°C)
5	100 : 1 : 4	3790 ± 40	33.6
6	100 : 0.5 : 4	3830 ± 40	34.0
7	100 : 1 : 6	2780 ± 30	34.6
8	100 : 1 : 10	1620 ± 40	34.1
9	100 : 2 : 10	2040 ± 30	34.0
10	100 : 1 : 14	1260 ± 30	34.2

[a] Each polymerization was carried out with 10 g of monomer in 40 ml tert-butanol at 60°C for 24 h.
[b] $\bar{M}_n$ was estimated by end group titration.

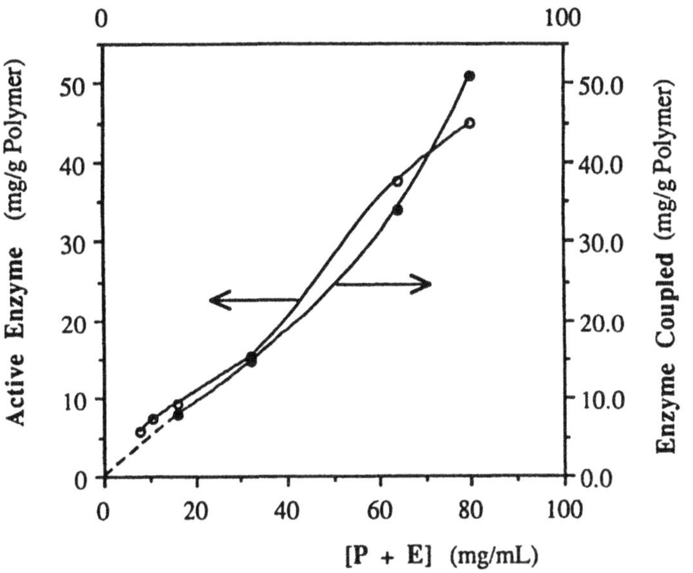
0
100
50
40
30
20
10
0
50.0
40.0
30.0
20.0
10.0
0.0
0
20
40
60
80
100
Active Enzyme (mg/g Polymer)
Enzyme Coupled (mg/g Polymer)
[P + E] (mg/mL)

Figure 3. Effect of the mixture concentration of polymer and enzyme on the total amount of enzyme coupled and the amount of that enzyme which retains its activity (the coupling condition is the same as shown in Table 3, P/E = 1).

Conjugation of β-D-glucosidase with poly(NIPAAm)

Conjugation of β-D-glucosidase with the carboxylated poly(NIPAAm) was carried out by simultaneous activation and coupling with 1-ethyl-3-(3-dimethyl-aminopropyl) carbodiimide hydrochloride (EDC). The polymer chosen for the conjugation was prepared by polymerization in tert-butanol to yield a molecular weight *ca.* 4000 and an LCST around 33.6°C. After coupling, the conjugated enzyme was easily isolated as a precipitate from the reaction medium by heating to *ca.* 40°C and centrifuging. The optimum pH for the coupling was found to be around 6–7. Increase in the ratio of enzyme to polymer increased the amount of conjugated active enzyme. In order to verify that the enzyme was covalently conjugated with

Table 3.
Effect of polymer and enzyme concentration on the coupling reaction[a]

[P + E][b] (mg/ml)	Enzyme coupled (mg/g polymer)	Active enzyme (mg/g polymer)	Retention of activity (%)
8.0	ND	5.63	ND
10.7	ND	7.31	ND
16.0	7.91	9.32	118
32.0	14.81	15.29	103
64.0	34.00	37.41	110
80.0	51.00	44.78	88

[a] Coupling condition: pH, 6.5; W_{EDC}, 40 mg; reaction time, 16 h; temperature, 4°C; W_P/W_E, 1/1.
[b] Total polymer and enzyme concentration in the coupling medium (1/1 wt. ratios); (P = polymer, E = enzyme).

the oligomer, a control mixture of enzyme and polymer was run without addition of activation reagent EDC. After the same treatment as that with EDC, a negligible amount of protein and active enzyme was detected in the precipitate. The most significant variable affecting the coupling reaction is the concentration of the polymer and enzyme, as shown in Fig. 3 and Table 3. It can be seen that the total amount of coupled enzyme, as well as of active, coupled enzyme per gram of polymer significantly increases with increase in the concentration of the polymer and enzyme mixture (P/E = 1) (see Fig. 3), which might be due to the increase of the probability of the reaction between the enzyme and the polymers. The retention of activity for the conjugated enzyme is very high, in some cases even higher than the native one, which may be due to a modification in the microenvironment of the enzyme, as found in the conjugation of poly(ethylene glycol) to trypsin [18]. It is worthwhile noting that in this conjugation experiment, the coupling reaction was run by mixing the polymer, enzyme, and the activation reagent altogether; therefore, it is possible that besides the polymer–enzyme conjugate, enzyme–enzyme conjugates, which could be either water-soluble or water-insoluble, might be formed during the coupling reaction. However, the polymer–enzyme conjugate obtained was collected in its insoluble form in order to remove the unconjugated enzyme, and the isolated precipitate really completely dissolved in cold buffer, indicating that large enzyme aggregates were not present in the precipitate.

Properties of the conjugated β-D-glucosidase

The effect of the pH of the reaction medium on the activity of the native and the conjugated β-D-glucosidase was investigated and the results are shown in Fig. 4. Both native and conjugated β-D-glucosidase are very sensitive to the pH of the reaction medium. However, conjugation did not affect the optimum pH, which is at pH 7.0 for both enzymes.

Conformational ('allosteric') changes of the enzyme may occur on conjugation, causing a change in the affinity between enzyme and substrate, possibly resulting in

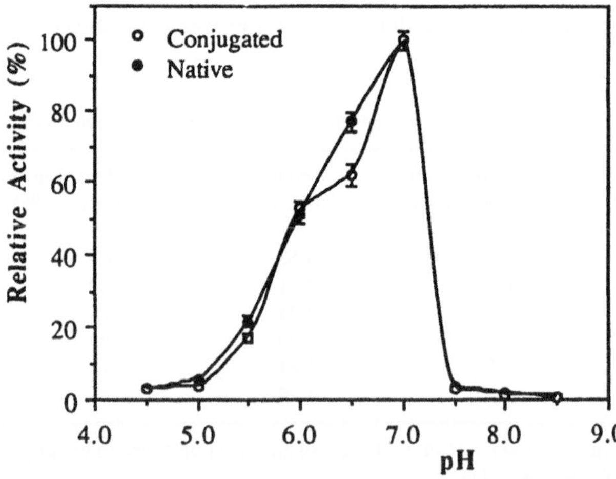
Relative Activity (%)
100
80
60
40
20
0
Conjugated
Native
4.0
5.0
6.0
7.0
8.0
9.0
pH

Figure 4. Effect of pH on the relative activity of native and conjugated β-D-glucosidase at 25°C using *p*NPG as substrate (0.45 mM).

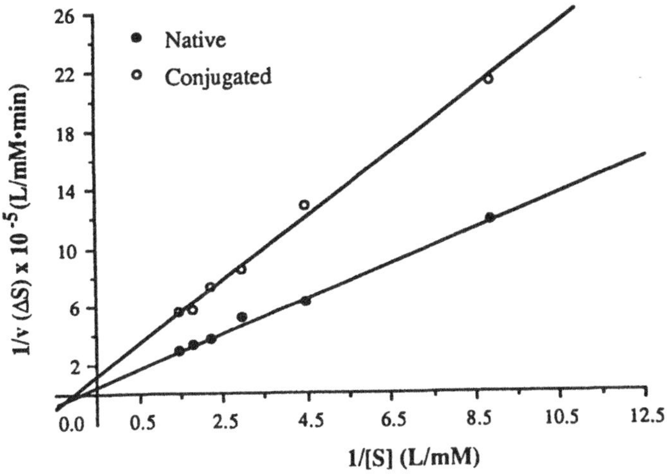
Native
Conjugated
1/v (ΔS) x 10^-5 (L/mM•min)
1/[S] (L/mM)
26
22
18
14
10
6
2
0.0
0.5
2.5
4.5
6.5
8.5
10.5
12.5

Figure 5. Lineweaver–Burk plot for native and conjugated β-D-glucosidase (*p*NPG as substrate at 25°C, pH 6.5).

changes in the enzyme kinetics for the conjugated enzymes. Therefore, determination of the apparent Michaelis–Menten constants (K_m) was performed for the native and the conjugated β-D-glucosidase, using *p*NPG as a substrate at pH 6.5. Figure 5 shows the Lineweaver–Burk plots obtained at various concentrations and the K_m and V_{max} values were calculated by least-squares evaluation as 1.04 ± 0.05 mM, 4.32 ± 0.10 × 10^6 (mM/l-min-mg) for the native enzyme and 0.94 ± 0.05 (mM), 4.63 ± 0.15 × 10^6 (mM/l-min-mg) for the conjugated enzyme, respectively. A slightly lower K_m value (0.94 mM) was found for the conjugated enzyme than the

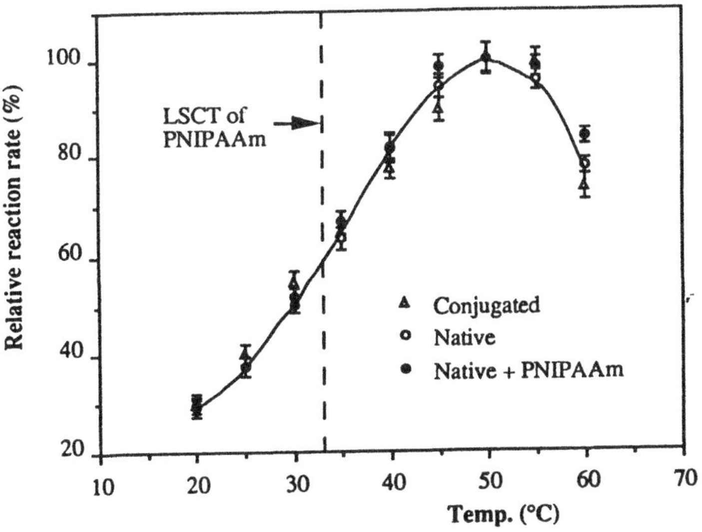
LSCT of
PNIPAAm
Conjugated
Native
Native + PNIPAAm
Relative reaction rate (%)
Temp. (°C)
100
80
60
40
20
10
20
30
40
50
60
70

Figure 6. Effect of temperature on the relative activity of native enzyme, mixture of poly(NIPAAm) with native enzyme and conjugated β-D-glucosidase, using *p*NPG as substrate at pH 7.0. Enzyme activities were determined by reacting the enzymes with *p*NPG at the respective temperatures for 10 min and measuring the production of *p*-nitrophenol at room temperature.

native one (1.04 mM) indicating that the conjugated enzyme may show somewhat higher affinity for the substrate. This increased affinity of the conjugated β-D-glucosidase for the substrate probably resulted from the hydrophilization of the protein surface [18] since below LCST poly(NIPAAm) is very hydrophilic.

The effect of temperature on the rate of *p*NPG hydrolysis is shown in Fig. 6 for native and conjugated enzymes. For comparison, controls were based on physical mixtures with the same amount of native enzyme and free poly(NIPAAm) as for the native or conjugated enzyme studies. It should be noted that above the LCST of poly(NIPAAm) (33°C), the polymer–enzyme conjugate precipitates and the enzymatic reaction occurs heterogeneously. Despite this difference, the temperature dependence of the enzymatic rection is essentially the same as for the native enzyme. This is probably because a small molecular weight substrate was used. The maximal activity was found at 50°C for all three cases. Thus, neither free poly(NIPAAm) nor poly(NIPAAm)-conjugated to the enzyme have a significant effect on enzyme activity compared to free native enzyme alone (see Fig. 6).

Figure 7 shows the effect of heat treatment at a variety of temperatures on the *p*NPG hydrolysis by native enzyme and conjugated enzyme. It is evident that at higher temperature (≥ 50°C) the conjugated β-D-glucosidase is more stable than either the native enzyme or the mixture of native enzyme with poly(NIPAAm). Figure 8 shows the time course for the thermal inactivation of the enzymes at 60°C. Both native β-D-glucosidase and the mixture of native enzyme with poly(NIPAAm) lose *ca.* 80% of their initial activities by a heat treatment at 60°C for 40 min, whereas the conjugated β-D-glucosidase retains 40% of its original activity. It is clear that the conjugate is more thermally stable, indicating that the conjugation of the enzyme with poly(NIPAAm) provides protection against thermal inactivation of the enzyme at temperatures well above the LCST of the polymer. This is different from the well-known stabilization of PEO-conjugated [18] or polysaccharide-conjugated [19] proteins, since the PEO or polysaccharide remains soluble. The protection of

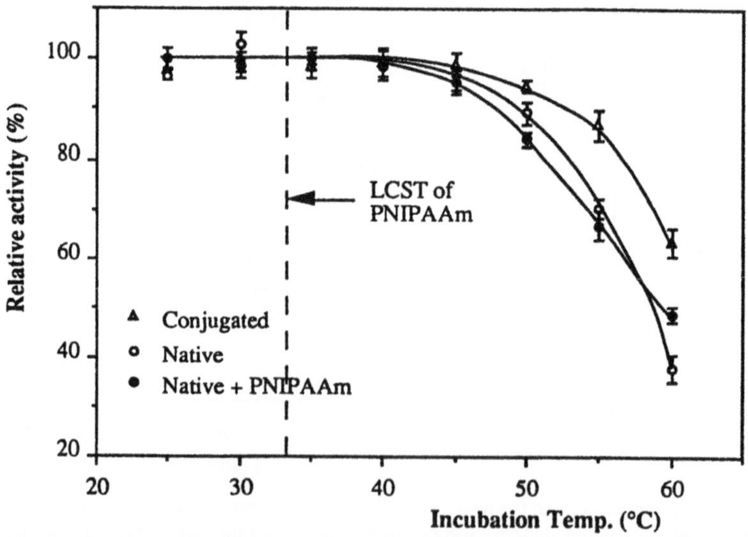
Relative activity (%)
100
80
60
40
20
LCST of
PNIPAAm
Conjugated
Native
Native + PNIPAAm
20
30
40
50
60
Incubation Temp. (°C)

Figure 7. Effect of incubation at various temperatures at pH 6.5 for 15 min on the activity of β-D-glucosidase measured at 25°C and pH 7.0, 120 min after incubation.

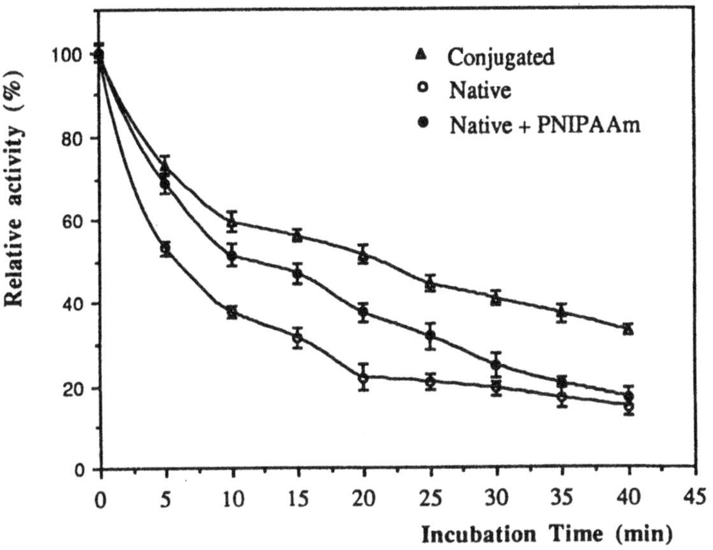
Relative activity (%)
100
80
60
40
20
0
0 5 10 15 20 25 30 35 40 45
Incubation Time (min)
Conjugated
Native
Native + PNIPAAm

Figure 8. Effect of incubation time on β-D-glucosidase inactivation incubated at 60°C and pH 6.5. The activities were measured at 25°C and pH 7.0, 120 min after incubation.

poly(NIPAAm) conjugated enzyme against the thermal inactivation may be due to the reduction of mobility, for the conjugates become insoluble. For immobilized enzyme on solid carriers, many examples have been reported in the literature on the protection against thermal inactivation and this protection has been explained as the reduction of mobility of the enzymes after immobilization on the solid carriers [20–22]. The same explanation may be applicable to the poly(NIPAAm)-β-D-glucosidase conjugates. Therefore, this type of stabilization by a precipitated, conjugated polymer, is worth further investigation.

CONCLUSIONS

Thermally reversible poly(NIPAAm) with carboxyl end groups has been successfully synthesized by radical polymerization using β-mercaptopropionic acid as the chain transfer reagent. The polymer has been utilized for conjugation via its end group to an enzyme, such as β-D-glucosidase. The conjugated β-D-glucosidase retains a high percent of its activity and also shows improved thermal stability over the native enzyme. More extensive study of its use in a recycle enzyme-solid substrate process is under way.

Acknowledgements

The National Science Foundation (NSF) (Grant No. BCS-9101716) and the Washington Technology Center (WTC) are gratefully acknowledged for their financial support of this project.

REFERENCES

1. O. R. Zaborsky, *Immobilized Enzymes*, C.R.C. Press, Cleveland, OH (1973).
2. I. Chibata (Ed.), *Immobilized Enzymes, Research and Development*, Halsted, New York (1978).
3. R. Epton, G. Marr and G. J. Morgan, *Polymer* **18**, 319 (1977).

4. K. Okamura, K. Ikura, M. Yoshikawa, R. Sakaki and H. Chiba, *Agri. Biol. Chem.* **48**, 2435 (1984).
5. A. L. Margolin, V. A. Izumrudov, V. K. Svedas and A. B. Zezin, *Biotechnol. Bioeng.* **24**, 237 (1982).
6. M. Taniguchi, M. Kobayashi and M. Fujii, *Biotechnol. Bioeng.* **34**, 1092 (1989).
7. M. Taniguchi, K. Hoshino, K. Watanabe, K. Sugai and M. Fujii, *Biotechnol. Bioeng.* **39**, 287 (1992).
8. N. Monji and A. S. Hoffman, *Appl. Biochem. Biotechnol.* **14**, 107 (1987).
9. C.-A. Cole, S. M. Schreiner, J. H. Priest, N. Monji and A. S. Hoffman, *ACS Symp. Ser.* **350**, *Am. Chem. Soc.*, Washington D.C., 245 (1987).
10. K. Steinke, and K. D. Vorlop, *DECHEMA Biotechnol. Conf.* **4** (Pt. B, Lect. DECHEMA annu. Meet, Biotechnol. 8th), 889 (1990); CA115: 47742b.
11. A. L. Nguyen and J. H. T. Luong, *Biotechnol. Bioeng.* **34**, 1186 (1989).
12. M. Heskins and J. E. Guillet, *J. Macromol. Sci. Chem.* **A2**, 1441 (1968).
13. J. P. Chen, H. J. Yang and A. S. Hoffman, *Biomaterials* **11**, 625 (1990).
14. J. P. Chen and A. S. Hoffman, *Biomaterials* **11**, 631 (1990).
15. P. K. Smith, R. L. Krohn, G. T. Hermanson, A. K. Mallia, F. H. Gartner, M. D. Provenzano, E. K. Fujimoto, N. M. Geoke, B. J. Olson and D. C. Klenk, *Anal. Biochem.* **150**, 76 (1985).
16. S. N. Bhattacharyya and D. Maldas, *J. Polym. Sci. Polym. Chem. Ed.* **20**, 939 (1982).
17. G. H. Chen, L. van der Does and A. Bantjes, *J. Appl. Polym. Sci.* **45**, 833 (1992).
18. H. F. Gaertner and A. J. Puigserver, *Enzyme Microb. Technol.* **14**, 150 (1992).
19. R. A. K. Srivastava, *Enzyme Microb. Technol.* **13**, 164 (1991).
20. R. Ulbrich, A. Schellenberger and W. Damerau, *Biotechnol. Bioeng.* **28**, 511 (1986).
21. S. Emi, Y. Murase, T. Hayashi and A. Nakajima, *J. Appl. Polym. Sci.* **41**, 2753 (1990).
22. T. Hayashi and Y. Ikada, *Biotechnol. Bioeng.* **36**, 593 (1990).

Synthesis and characterization of a soluble, temperature-sensitive polymer-conjugated enzyme

TAE GWAN PARK* and ALLAN S. HOFFMAN

Center for Bioengineering, FL-20, University of Washington, Seattle, WA 98195, USA

Received 27 July 1992; accepted 29 October 1992

Abstract—The enzyme, alkaline phosphatase, has been conjugated to a temperature-sensitive polymer which exhibits a lower critical solution temperature (LCST). A series of copolymers containing different molar ratios of N-isopropylacrylamide(NIPAAm) and N-acryloxysuccinimide(NAS) were synthesized and then conjugated to the enzyme. These polymer–enzyme conjugates precipitate and flocculate in aqueous solution above the LCST, and redissolve when cooled below that temperature. The kinetics of the conjugated enzymes have been characterized as a function of temperature and compared to free enzyme. The effect of the conjugation degree between polymer and enzyme on the activity of the conjugated enzymes was also investigated.

Key words: Conjugated; enzyme; alkaline phosphatase.

INTRODUCTION

Enzymes have been immobilized in order to stabilize them and also use them in a continuous bioreactor operation. There have been a number of studies on immobilized enzymes based on many different methodologies [1, 2]. Most of them have utilized solid supports or porous materials. On the other hand, the conjugation of enzymes to synthetic polymers has been carried out in order to modify native properties of free enzymes, such as immunogenicity, stability, resistance to proteolytic enzymes, and increased solubility in organic solvents. Hydrophilic water soluble polymers such as poly(ethylene oxide) [3–5], dextrans [6], and poly(N-vinylpyrolidone) [7], have been used to conjugate enzyme molecules for such purposes. Modifications of clinically important enzymes with poly(ethylene oxide) significantly suppress the immune response, and this has attracted much attention recently [8].

The water soluble polymer, poly(N-isopropylacrylamide, NIPAAm), exhibits a lower critical solution temperature (LCST) in aqueous solution [9, 10]. The polymer precipitates when heated above the LCST (32–33°C), but it is redissolved and becomes soluble when cooled below that temperature. This process is fully reversible. Based on the conjugation of this temperature-sensitive polymer to an antibody, a homogeneous immuno-diagnostic assay which facilitates the separation of antigen–antibody complex by raising temperature has been developed [11]. Furthermore, polyNIPAAm-conjugated dye (Cibacron Blue F3G-A) or protein A can separate albumin or immunoglobulin G, respectively, out of solution by thermal affinity precipitation [12, 13].

---

* Present address: School of Pharmacy, Temple University, 3307 North Broad Street, Philadelphia, PA 19140, USA.

In this preliminary study, polyNIPAAm has been covalently coupled to the enzyme, alkaline phosphatase, by conjugating the enzyme to a series of copolymers containing varying molar ratios of NIPAAm and N-acryloxysuccinimide (NAS). NAS has a highly reactive active ester group which reacts with amino groups in the enzyme. The kinetic properties of the conjugated enzymes, such as temperature optimum, pH optimum, and thermal stability, have been studied here and compared to the native alkaline phosphatase. Our main objective is to investigate the effect of the conjugation between the polymer and the enzyme on the activity of the conjugated enzyme, especially below and above the LCST.

MATERIALS AND METHODS

Materials

N-isopropylacrylamide(NIPAAm) was recrystallized with hexane, N-acryloxysuccinimide (NAS) was used without further purification, azobisisobutyronitrile(AIBN) was recrystallized with methanol; all were obtained from Eastman Kodak. The enzyme, alkaline phosphatase (from bovine intestinal mucosa, type VII-S, 1,060 units/mg protein) and the substrate, *p*-nitrophenyl phosphate were obtained from Sigma Co. Coomassie Blue Protein Assay kit was from Pierce Co. All other chemicals were reagent grade.

Synthesis of poly(NIPAAM-co-NAS)

Three activated copolymers were prepared by varying molar ratios of NIPAAm and NAS. Molar ratios of NIPAAm/NAS of 97.5/2.5, 95/5, and 90/10, were dissolved in 120 ml of 50/50 (v/v) of tetrahydrofuran and toluene. The total amount of monomers was 8 g. After removing oxygen by nitrogen purging, 25 mg of AIBN was added as an initiator. The polymerization was carried out at 60°C for 24 h under a slight positive pressure of nitrogen. The polymerization mixture was added dropwise into 300 ml of petroleum ether in order to precipitate poly(NIPAAM-co-NAS). The precipitate was filtered using a glass filter, dried in vacuum, and then stored in desiccator.

Determination of active ester content

N-hydroxysuccinimide, which is liberated by reaction of active ester with isopropylamine, was assayed to estimate the amount of active ester groups in the copolymer backbone. 0.1 g of the NIPAAm/NAS copolymer was reacted with 0.1 g of isopropylamine in 2 ml of DMF overnight, followed by addition of 9 ml of 0.05 M phosphate buffer (pH 7.5). 200 μl of the above solution was pipetted and diluted with 9.8 ml of the same phosphate buffer. The released N-hydroxysuccinimide was then determined by assaying at 259 nm, and the active ester content in the polymer backbone was calculated by assuming 100% reaction yield, as previously described [12].

Determination of molecular weight

Molecular weights of NIPAAm-co-NAS copolymers were determined by measuring intrinsic viscosity. Copolymers were dissolved in THF and their viscosities were measured using a Ubbeholde viscometer at 27°C as a function of copolymer concentration. The number average molecular weight was then calculated, according to Fugishige's equation, $[\eta] = 9.59 \times 10^{-3} M_n^{0.6}$[14].

Conjugation of alkaline phosphatase to NIPAAm/NAS copolymers

One gram of poly(NIPAAm-co-NAS) was dissolved in 5 ml of DMF, which was added dropwise into 50 ml of 0.05 M phosphate/0.05 M NaCl buffer (pH 9) containing 3.1 mg of alkaline phosphatase. The above solution was gently stirred at room temperature for 2 h. In order to selectively separate the polyNIPAAm-conjugated enzyme out of solution, 5 g of NaCl was added and centrifuged at 8000 g to precipitate the conjugated enzyme. It has been shown that the LCST decreases with NaCl concentration [15]. Thus, 10% (w/v) of sodium chloride concentration used in the above step lowered the LCST of the conjugated enzyme below the room temperature, which caused the polyNIPAAm-conjugated enzyme to precipitate. It has also been confirmed that with this NaCl concentration, free enzyme does not precipitate and stays in the supernatant. Therefore, the conjugated enzyme could be separated from the free enzyme by lowering the LCST by addition of sodium chloride. The precipitate was resuspended in 20 ml of 0.05 M Tris/0.05 M NaCl buffer (pH 8) and then dialyzed against 2 l of the same buffer overnight. The conjugated enzyme was stored at 4°C. The amount of conjugated alkaline phosphatase was assayed by a Coomassie blue method [16], using bovine serum albumin as a standard. It was found that other protein assay methods like the BCA method [15] were interfered with by the presence of polyNIPAAm. The three polyNIPAAm-conjugated alkaline phosphatases are denoted as AP-1, AP-2, and AP-3, which signifies that the molar feed ratios of NIPAAm and NAS in the active ester copolymers are 97.5/2.5, 95/5, and 90/10, respectively. The concentrations of alkaline phosphatase in AP-1, AP-2, and AP-3 were 20, 22 and 49 μg/ml, respectively. Since the conjugated enzyme was stored in the solution state, it was not possible to calculate the number of polyNIPAAm chains attached per alkaline phosphatase enzyme molecule.

Determination of LCST for conjugated enzymes

The LCSTs of the three poly NIPAAm-conjugated enzymes were determined by measuring transmittance of 600 nm as a function of temperature. The cuvette contained 0.1% (v/v) of conjugated enzyme in pH 8 buffer. The temperature was controlled by circulating water. As the temperature was raised, the change in transmittance was recorded every 1°C increment. The LCST was defined as the temperature at which 50% change in the transmittance occurs.

Activity assay of conjugated enzyme

Two mM of *p*-nitrophenyl phosphate(pNPP) in 0.05 M Tris/0.05 M NaCl buffer (pH 10) was used as a substrate. To determine the temperature-activity profiles for both free and conjugated enzymes, 5 ml of the above substrate solution was incubated at a particular temperature and 200 μl of the conjugated enzymes or 10 μl of the free enzyme (concentration: 91.6 μg/ml) was added. An initial turnover rate was obtained by incubating the mixture for 3 min and by measuring absorbance at

405 nm. For the conjugated enzymes incubated above the LCST, the enzyme reaction mixture was quickly immersed in an ice–water bath before reading absorbance to make sure that the solution was transparent. This immediate cooling of the cloudy solution below the LCST made it possible to avoid any light scattering interference. The product, *p*-nitrophenol, which is yellow after being cleaved from the colorless substrate, absorbs that wavelength. To determine the pH-activity profiles, 0.05 M Tris/0.05 M NaCl was used to prepare a series of pH buffers. In order to measure the effects of thermal cycling between 30 and 40°C, which are below and above the LCST of the conjugated enzymes, 4 ml of the conjugated enzymes were incubated in a water bath, which was thermally controlled by a thermal cycling microprocessor program. 200 μl samples were taken at various times and activity was assayed at 30°C as described above. Stabilities of the free and conjugated enzymes were determined by incubating them at 50°C and by measuring their activities at 30°C as a function of incubation time.

RESULTS AND DISCUSSION

In previous studies [11–13], polyNIPAAm has been conjugated to antibodies, dye, and protein A and utilized to thermally separate complexes of antigen–antibody, dye–albumin, and protein A-immunoglobulin G. By simply increasing the temperature above the LCST, the complex of the polyNIPAAm-ligand with the target biomolecule could be separated out in aqueous solution. After redissolving the complex in an eluting solution, the polyNIPAAm-ligand could be separately reprecipitated and then recycled. In this case, active ester groups which are highly reactive to primary amino groups in protein molecules, were first introduced in the polymer backbone of polyNIPAAm by copolymerizing NIPAAm and NAS, and then the copolymer was conjugated, via active ester coupling, to alkaline phosphatase using a similar method as reported earlier [12].

Molecular weights of the three poly(NIPAAm-co-NAS) copolymers were determined by viscosity and are listed in Table 1. The molecular weight–viscosity relationship for polyNIPAAm was adopted from Fugishige's work [14]. It was assumed that the dilute solution viscosity of poly(NIPAAm-co-NAS) in THF at 27°C is the same as that of polyNIPAAm, because polyNIPAAm has no LCST property in organic solvents and the molar ratio of NAS in the copolymer backbone is less than 10%. It can be seen in Table 1 that there is no relationship between NAS/NIPAAm feed ratio and molecular weight. The copolymer having the NAS/NIPAAm molar ratio

Table 1.
Characteristics of poly(NIPAAm-co-NAS) copolymers

Active ester polymer	Molar feed ratio, NAS/NIPAAm	Intrinsic viscosity[a] (ml/g)	Molecular weight[b] M_n	Active ester molar ratio in copolymer NAS/NIPAAm
Polymer 1	2.5/97.5	9.94	43,000	1.6/98.4
Polymer 2	5/95	8.03	32,000	3.8/96.2
Polymer 3	10/90	14.94	81,000	9.2/90.8

[a] In tetrahydrofuran at 27°C
[b] Using the equation, $[\eta] = 9.59 \times 10^{-3} M_n^{0.6}$

of 5/95 has the lowest molecular weight among the three copolymers. It can be postulated that any small change in polymerization conditions, such as presence of residual oxygen, may result in differences in molecular weight observed. As expected, the amount of active ester in the copolymers increases with increasing ratio of NAS/NIPAAm.

Since the NAS active ester groups are easily hydrolyzed upon contact with water, it can be expected that some active ester groups will be conjugated to enzyme, and others will be hydrolyzed during the conjugation process to form carboxylic acids. It is well known that as relatively hydrophilic monomers, such as acrylamide and acrylic acid, are copolymerized with NIPAAm, their LCSTs shift to higher temperature [17, 18], while the LCST of the polyNIPAAm is not significantly influenced by its molecular weight and concentration [19]. To minimize the possibility of hydrolysis, copolymer dissolved in DMF was slowly dropped into an excess of enzyme. The LCSTs of the three conjugated enzymes are shown in Fig. 1. It can be seen that AP-1, AP-2, and AP-3 have their LCSTs at 33.1, 36.2, and 36.3°C, respectively. Recalling that homo-polyNIPAAm exhibits its LCST at 32–33°C [10, 11], and noting that AP-1 exhibits no detectable change in the LCST, one might conclude that there was little hydrolysis due to the small amount of hydrolyzable active esters in the AP-1 polymer backbone. There also may be a low level of conjugation of enzyme to this polymer because of its low NAS/NIPAAm ratio. As the amount of active ester groups in poly(NIPAAm-co-NAS) increases, it can be expected that the number of attachments between the copolymer and the enzyme will increase. Both

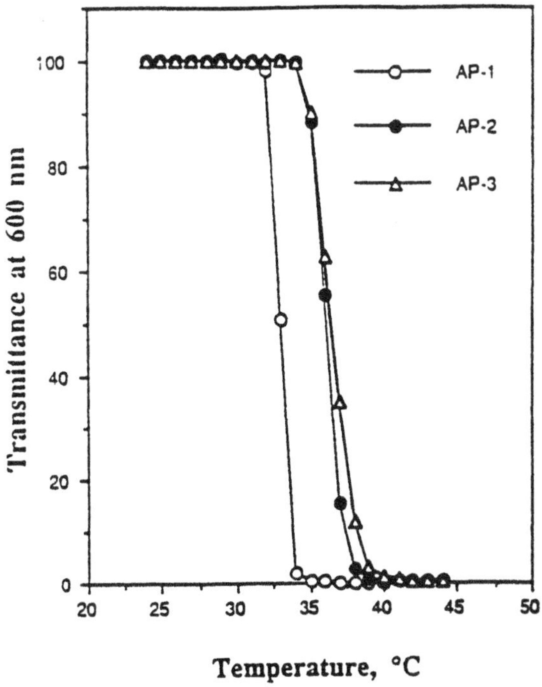
Transmittance at 600 nm
100
80
60
40
20
0
20
25
30
35
40
45
50
Temperature, °C
AP-1
AP-2
AP-3

Figure 1. LCST (lower critical solution temperature)s of polyNIPAAm-conjugated alkaline phosphatases at pH 8, AP-1, AP-2, and AP-3 signifies that molar feed ratios of NIPAAm and NAS are 92.5/2.5, 95/5, and 90/10, respectively.

high level of enzyme conjugation and/or hydrolysis would be expected to raise the LCST a few degrees, as seen in AP-2 and AP-3. Chen and Hoffman [13] have shown that conjugation of polyNIPAAm to proteins raises the LCST a few degrees at most, which is also a possible contributor to the small increases in LCST seen in AP-2 and AP-3. The LCST shifts observed in the AP-2 and AP-3 are due to the increasing amount of hydrolyzed active esters generating carboxylate groups on the polyNIPAAm backbone as well as the enzyme conjugation.

Figure 2 shows the effect of the active ester content in the copolymer on the remaining activity of the conjugated enzymes which was assayed at 55°C and pH 10. It was assumed that the degree of enzyme conjugation was proportional to the active ester content in the polymer chain, because the active ester groups are highly reactive to amino groups in the enzyme. The remaining activity was relative specific activity of the conjugated enzyme compared to that of free enzyme before conjugation. The free enzyme used was not treated in the same procedure (exposure to organic solvent and high ionic strength) as the conjugated enzymes. Thus, native free enzyme activity was used as a control. It is of interest to note that the AP-2 shows the highest activity, compared to the AP-1 and the AP-3. These results suggest that an optimal number of attachments between the polyNIPAAm and the enzyme may play an important role in the activity of the conjugated enzyme. The severe enzyme deactivation observed in the AP-3 might be due to the fact that excessive multi-point attachment elicits a conformational change of the enzyme.

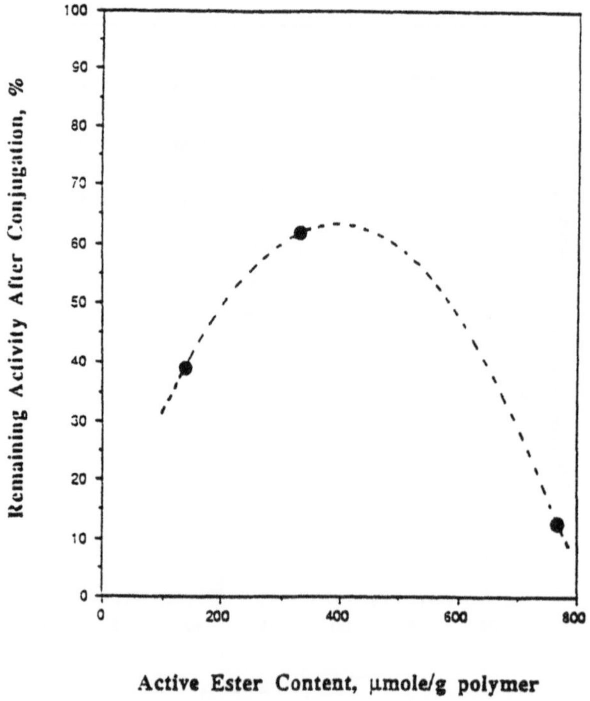
Remaining Activity After Conjugation, %
100
90
80
70
60
50
40
30
20
10
0
0
200
400
600
800
Active Ester Content, µmole/g polymer

Figure 2. Effect of active ester amount in poly(NIPAAm-co-NAS) on the remaining specific activities of the conjugated enzymes after conjugation compared to free enzyme. Specific activity (µmol product/mol·min) was assayed at 55°C and pH 10.

AP-3 may have a relatively dense array of polyNIPAAm molecules on the enzyme surface, which can bind enzyme conjugates to each other above the LCST. In the case of the AP-1 which has fewer attachment sites, it can be postulated that the poly-NIPAAm chain is more mobile and its density around the enzyme may be smaller than that of AP-3. Thus, a significant conformational change of the AP-1 is less likely to happen. A more important parameter to be considered would be micro-environmental difference in the three conjugated enzymes. When a relatively hydrophobic polyNIPAAm is conjugated with the enzyme, it may interact with accessible hydrophobic regions of the enzyme and cause some conformational changes. It could also interact with, and favorably partition, the substrate, *p*-nitrophenyl phosphate, which is somewhat hydrophobic. Thus, it can be said that the AP-3 could have the highest substrate partitioning with the large conformational changes, while the AP-1 should have the lowest partitioning with little, if any, conformational changes. These two effects will act in opposite ways on the activity of the conjugated enzyme which may lead the AP-2 to exhibit a maximum in the retained activity as a function of active ester content in the copolymers. Alternatively, from the observation that the activities increase with decreasing molecular weight of polymer conjugated to the enzyme (AP-2 shows the highest activity with the lowest M_W of 32,000), one may postulate that molecular weights of the conjugated polymers, although their difference is not so large, create the diffusion-limited reaction kinetics because of different microenvironmental viscosity. The conjugation process, which involves the reaction between multi-functional groups in enzymes and active ester polymers, is heterogeneous in nature. Thus, quantitative structural analysis on the conjugated enzymes (for example, measurement of modified isoelectric point and titration of blocked lysine amino groups) will be necessary to support the above hypotheses. Controlled conjugation with different molecular weights or controlled blocking of lysine amino groups with hydrophilic small molecules will be one approach.

A main hypothesis to be tested in this study is that the kinetic behavior of the polyNIPAAm-conjugated enzyme will vary below and above the LCST due to an expansion and collapse of polyNIPAAm chains on the enzyme surface. Increasing temperature may result in microenvironmental change around the active site and/or the flocculation of the conjugated enzyme, both of which should affect the enzyme turnover rate. Activity-temperature profiles of the free and the conjugated enzymes are shown in Fig. 3. It appears that activity profiles of the conjugated enzymes are similar to that of the free enzyme, regardless of different conjugation degrees. Their maximum activities are all at 55°C similar to that of the free enzyme. There also seem to be no discrete discontinuities in the activities below and above the LCST for any of the three conjugated enzymes. However, when the conjugated enzymes are incubated above the LCST, it can be expected that the polyNIPAAm chains will collapse and become hydrophobic, which may increase the partitioning of the substrate molecules near the active site. On the other hand, the collapse of polyNIPAAm chains may alter the enzyme conformation and disrupt the active site as the polymer chains precipitate near the polymer surface. These two opposite effects above the LCST on the activity of the conjugated enzymes may cause such apparent activity continuities below and above the LCST.

In order to more clearly examine the LCST effect on the activity, Arrhenius plots are constructed as shown in Fig. 4. In contrast to the lack of an obvious

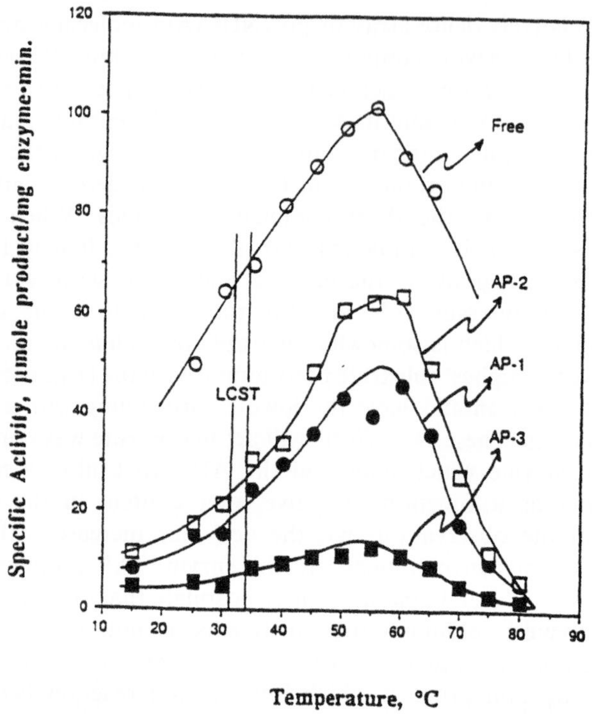
Specific Activity, μmole product/mg enzyme•min.
Temperature, °C
Free
AP-2
AP-1
AP-3
LCST

Figure 3. Temperature-activity profiles of free and conjugated enzymes at pH 10.

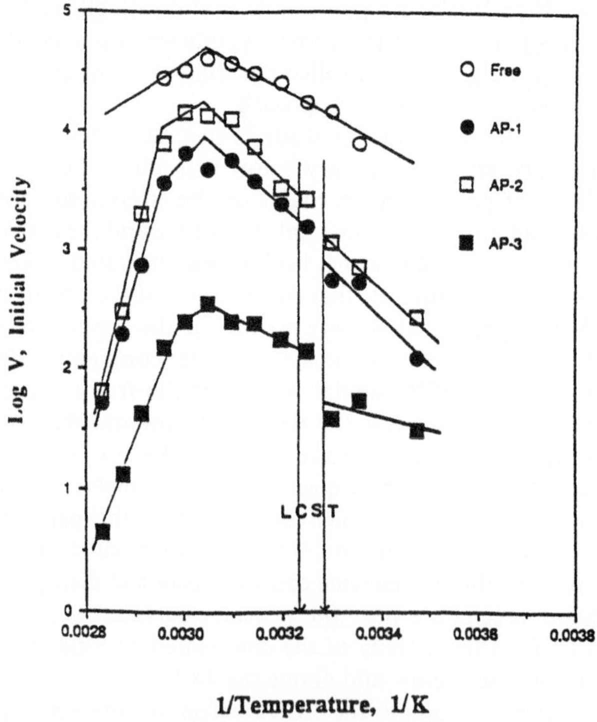
Log V, Initial Velocity
1/Temperature, 1/K
Free
AP-1
AP-2
AP-3
LCST

Figure 4. Arrhenius plots of free and conjugated enzymes.

discontinuity at the LCST, here it can be seen that the activation energies for the conjugated enzymes seem to be different below and above the LCST, which implies that there might be some influence of the LCST on the kinetic behavior of the conjugated enzymes. The AP-3, which may have the most multiple-point attachment sites, appears to exhibit the greatest discontinuity in activation energy. This suggests that the multiple-point attachment of polyNIPAAm on the enzyme may be the most likely to change the conformation of the conjugated enzyme or to affect the partitioning of the substrate near the active site as described above. In previous studies in our laboratory, polyNIPAAm-conjugated asparaginase has also shown such small discontinuities in activation energies when raising the temperature above the LCST.

The pH-activity profiles for the conjugated enzymes at 30°C are shown in Fig. 5. It can be seen that there is also no major change in the pH-activity profiles, and their optimum pH values are the same, implying that active ester copolymers may be coupled to the lysine groups away from the active site. In order to test the reversibility of the enzyme activity below and above the LCST, the conjugated enzymes were thermally cycled between 30°C and 40°C with a heating and cooling rate of 1°C/min. Figure 6 shows that there is no detectable decrease in activities of the conjugated enzymes after many thermal cycles, compared to free enzyme. This result directly indicates that a heat-induced thermal flocculation of the conjugated enzymes does not affect their activity. On the other hand, when the conjugated enzymes were incubated at 50°C, they lost their activities as a function of time, in contrast to the free enzyme which demonstrates no appreciable loss of activity

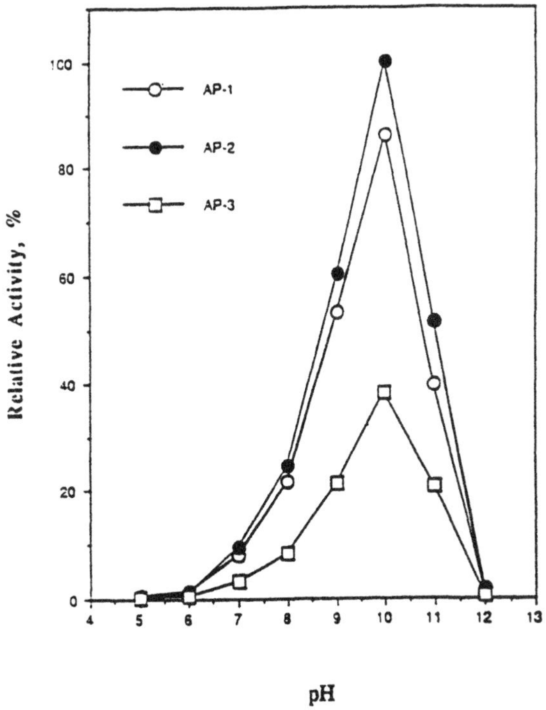
Relative Activity, %
100
80
60
40
20
0
AP-1
AP-2
AP-3
4
5
6
7
8
9
10
11
12
13
pH

Figure 5. pH-activity profiles of conjugated enzymes assayed at 30°C.

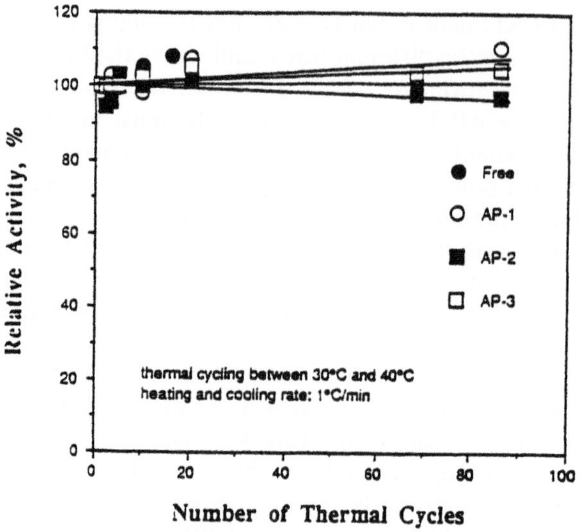
120
100
80
60
40
20
0
Relative Activity, %
0
20
40
60
80
100
Number of Thermal Cycles
Free
AP-1
AP-2
AP-3
thermal cycling between 30°C and 40°C
heating and cooling rate: 1°C/min

Figure 6. Effect of thermal cycling between 30 and 40°C on the activities of free and conjugated enzymes. Activity was assayed at 30°C and pH 10.

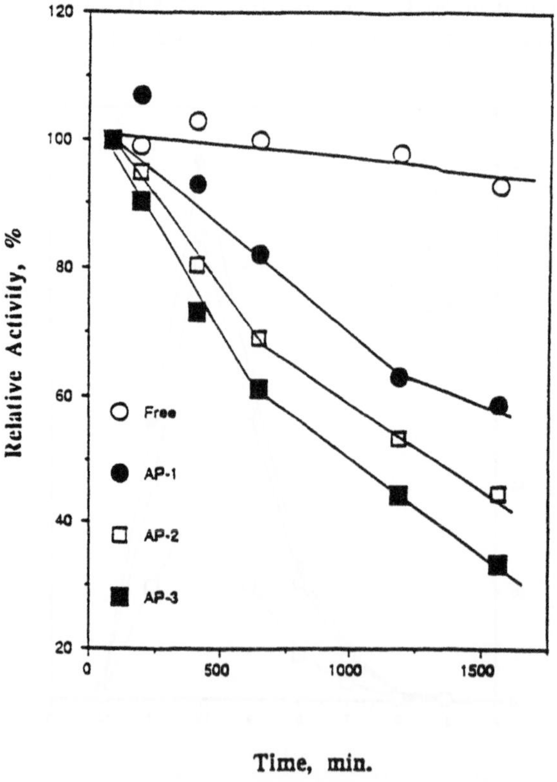
120
100
80
60
40
20
Relative Activity, %
0
500
1000
1500
Time, min.
Free
AP-1
AP-2
AP-3

Figure 7. Stability of free and conjugated enzymes incubated at 50°C. Activity was assayed at 30°C and pH 10.

(Fig. 7). It can also be seen that the stability of the conjugated enzymes is dependent on the number of the conjugation linkage between polyNIPAAm and the enzyme. The AP-3, which has the most possible multi-point attachments, exhibits the lowest stability. One plausible explanation is that the collapsed polyNIPAAm attached to the enzyme at multiple points induces a continuous strain in the conformation of the enzyme, which leads to the destabilization of the conjugated enzymes over the period studied. This result is in contrast to the conventional knowledge that the thermal stability of enzymes is normally enhanced by multi-point attachments between solid supports and the enzymes, in which there are no such strains on the enzyme conformation [20]. In our case, however, the action mechanism of the collapsed polyNIPAAm on the enzyme is quite unique, which may be responsible for that thermal instability observed in the conjugated enzymes.

Based on these kinds of thermo-sensitive biocatalysts, there might be a wide range of applications in the field of biotechnology. For example, one can utilize this new class of technology to recover biocatalysts after a batch reaction by raising temperature, so that water soluble biocatalysts can be recycled for a new batch reaction. Since an LCST polymer-conjugated enzyme will exhibit a temperature-dependent change of hydrophilic/hydrophobicity at the LCST, it is conceivable that one could control the partition behavior of the conjugated enzyme in an aqueous two phase system by adjusting the system temperature. Furthermore, one could try to 'collect' the conjugated enzyme in a homogeneous solution by adding polymeric particles having a hydrophobic surface (e.g. similar to polyNIPAAm), and raising the temperature to approach the LCST. Hydrophobic adsorption of the conjugated enzyme on the surface of the polymer particles will be responsible for that separation. If the temperature is lowered below the LCST, the redissolved polymer-conjugated enzyme could be used for another batch reaction.

In summary, it has been demonstrated that polyNIPAAm can be conjugated to the enzyme, alkaline phosphatase, by using an activated copolymer, poly(NIPAAm-co-NAS), containing highly reactive active ester groups. The activity of the polyNIPAAm-conjugated enzymes depends on the extent of the conjugation. An optimal degree of conjugation is important to achieve the highest activity. It appears that the kinetic behaviors of the conjugated enzymes below and above the LCST are different due to the phase transition of the conjugated polyNIPAAm. Microenvironmental changes near the active site may be responsible for that kinetic behavior.

REFERENCES

1. H. H. Weetall (Ed.), *Immobilized Enzymes, Antigens, Antibodies, and Peptides. Preparation and Characterization*. Marcel Dekker, New York (1975).
2. W. R. Gombotz and A. S. Hoffman, in: *Hydrogels in Medicine and Pharmacy*, p. 95, N. A. Peppas (Ed.). CRC Press, Boca Raton, FL (1986).
3. Y. Inada, T. Yoshimoto, A. Matsushima and Y. Saito, *Trends Biotechnol.* **4**, 68 (1986).
4. Y. Inada, K. Takahashi, T. Yoshimoto, A. Ajima, A. Matsushima and Y. Saito, *Trends Biotechnol.* **4**, 190 (1986).
5. F. M. Veronese, R. Largajolli, E. Boccu', C. A. Benassi, O. Schiavon, *Appl. Biochem. Biotech.* **11**, 141 (1985).
6. B. Chaplin and M. L. Green, *Biotech. Bioeng.* **24**, 2627 (1982).
7. B. Geiger, H. Von Spect and R. Arnon, *Eur. J. Biochem.* **73**, 141 (1977).
8. A. Abuchowski, T. Vanes, C. Palczuk and F. F. Davis, *J. Biol. Chem.* **252**, 3578 (1977).

9. L. D. Taylor and L. D. Cerankowski, *J. Polymer. Sci., Polym. Chem.* **13,** 2551 (1975).
10. A. S. Hoffman, *J. Controlled Release* **4,** 297 (1987).
11. N. Monji and A. S. Hoffman, *Appl. Biochem. Biotech.* **14,** 107 (1987).
12. H. J. Yang, Ph.D. Thesis, University of Washington, Seattle, WA (1989).
13. J. P. Chen and A. S. Hoffman, *Biomaterials* **11,** 625 and 631 (1990).
14. S. Fugishige, *Polymer J.* **19,** 297 (1987).
15. T. G. Park, Ph.D. Thesis, University of Washington, Seattle, WA (1990).
16. Pierce Co. Catalog, p. 214 (1989).
17. L. C. Dong and A. S. Hoffman, *J. Controlled Release* **4,** 223 (1986).
18. T. G. Park and A. S. Hoffman, *Biotech. Bioeng.* **35,** 152 (1990).
19. S. Fugishige, K. Kubota and I. Ando, *J. Phys. Chem.* **93,** 3311 (1989).
20. A. Klibanov, *Analyt. Biochem.* **93,** 1 (1979).

Activated, *N*-substituted acrylamide polymers for antibody coupling: Application to a novel membrane-based immunoassay

NOBUO MONJI*, CAROL-ANN COLE[1] and ALLAN S. HOFFMAN[2]

[1]*Genetic Systems Corporation, 6565 185th Avenue NE, Redmond, WA 98052, USA*
[2]*Center for Bioengineering, FL-20, University of Washington, Seattle, WA 98195, USA*

Received 6 August 1992; accepted 22 February 1993

Abstract—A room-temperature-precipitable, activated terpolymer consisting of *N*-isopropylacrylamide (NIPAAm)/*N*-*n*-butylacrylamide(nBAAm)/*N*-acryloxysuccinimide(NASI) (LCST = 7–13°C) at a monomer feed ratio of 60:40:2.5, respectively, was prepared and conjugated to an antibody. The conjugate was evaluated in a novel cellulose acetate (CA) membrane-based immunoassay which utilizes the especially strong physical attachment of the polymer to CA to bind and concentrate the polymer attached protein onto the membrane. When compared in the CA membrane immunoassay to the antibody–poly(NIPAAm) conjugate prepared via anhydrous copolymerization of NIPAAm and NASI at the monomer feed ratio of 40:1, respectively, the performance of the NIPAAm/nBAAm/NASI terpolymer was superior to that of the NIPAAm/NASI copolymer (LCST = 32°C) when the studies were carried out at room temperature. However, the terpolymer and copolymer gave equivalent performance when the assay mixture was heated to 45°C. These results indicate the importance of the LCST of the polymer component of the Ab–polymer conjugate to its adsorption and binding on the CA membrane.

Key words: Poly *N*-isopropylacrylamide; immunoassay; antibody-polymer conjugate; cellulose acetate membrane; *N*-acryloxysuccinimide; *N*-*n*-butylacrylamide; free radical polymerization; lower critical solution temperature.

INTRODUCTION

Aqueous solutions of poly(*N*-isopropylacrylamide) [poly(NIPAAm)] exhibit a lower critical solution temperature (LCST) around 32°C [1, 2]. Based on its thermoreversible behavior, poly(NIPAAm) has been used in immunoassays [3–5] and other bioseparation processes [6, 7]. Aside from their phase-separating characteristics, we found previously that the conjugate of poly(NIPAAm) and an antibody (Ab) showed preferential binding to a cellulose acetate (CA) membrane when compared to other membrane materials [8]. Based on these findings [8, 9], we developed the membrane affinity concentration immunoassay (MAC-IA). The assay is shown schematically in Fig. 1.

In our previous studies [3, 4, 8], incorporation of Ab into poly(NIPAAm) was carried out by a two-step procedure: first, the conjugation of a vinyl monomer to an Ab, followed by aqueous free radical copolymerization of the monomer-conjugated Ab with NIPAAm (*de novo* copolymerization). Since the *de novo* copolymerization did not provide control and characterization of the size and composition of the polymer, an alternative approach was chosen. In this method, a polymer of known

* To whom correspondence should be addressed.

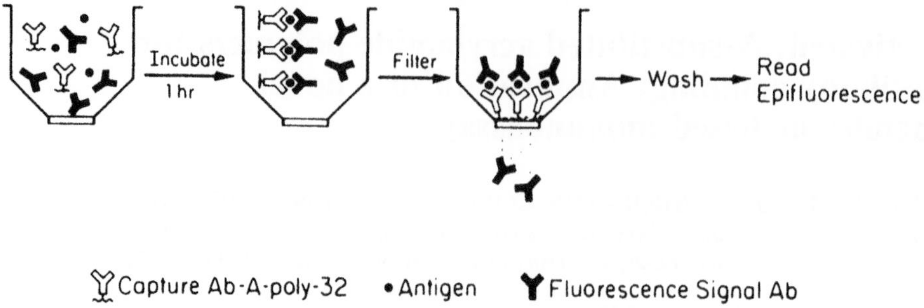
Incubate
1 hr
Filter
Wash
Read
Epifluorescence
Capture Ab-A-poly-32
Antigen
Fluorescence Signal Ab

Figure 1. Assay outline for the membrane affinity concentration immunoassay (MAC-IA).

size and chemical composition was prepared, with at least one active site where an Ab could be coupled covalently. As a result, copolymers of NIPAAm and *N*-acryloxysuccinimide (NASI) having varying amounts of active ester sites were synthesized under anhydrous conditions and characterized [5, 7].

We previously found that, at any particular temperature, the binding efficiency of the Ab–poly(NIPAAm) polymer conjugate to the CA membrane was related to the LCST of the polymer [8]. Enhanced binding occurred above the LCST (i.e. at 45°C). Since most immunoassays are carried out at ambient temperature and in order to perform the MAC-IA more efficiently at ambient temperature, room-temperature-precipitable polymers (LCST < 18°C) were needed. We report here the preparation, characterization and utility in the MAC-IA of two activated polymers, a copolymer designated as Apoly 5, and a room-temperature-precipitable terpolymer designated as Apoly 32.

MATERIALS AND METHODS

Preparation of monomer-conjugated mouse monoclonal antibody to the kappa light chain of human immunoglobulin (MAb_{2H1}) and the aqueous de novo *copolymerization of the monomer-conjugated MAb_{2H1} with NIPAAm*

Conjugation of a monomer to MAb_{2H1}, as well as preparation of the MAb_{2H1} poly(NIPAAm) conjugate through *de novo* copolymerization was carried out by the methods described previously [3, 4].

Synthesis of the copolymer consisting of NIPAAm and NASI

The copolymer consisting of NIPAAm and NASI (Apoly 5) was synthesized under anhydrous conditions at the monomer feed ratio of 40:1, respectively, using the method described previously [5].

Synthesis of room-temperature-precipitable terpolymers consisting of (1) NIPAAm, (2) NASI, and (3) either N*-alkylacrylamide or alkyl acrylate*

The conditions for the terpolymerization were analogous to those described by Pollak [10] with subsequent modification reported by us [5]. A 100-ml two-necked, round-bottomed flask, fitted with a reflux condenser, thermometer, and nitrogen inlet controlled by a Firestone valve, was charged with various ratios of NIPAAm, *N*-alkylacrylamide, or alkyl acrylate, and NASI together with azobis

Table 1.
Comonomers and the reaction percentages used in the terpolymerizations

N-Alkylacrylamide	% Added[a]	Alkylacrylates	% Added[a]
n-Butyl	10, 20, 30, 40	*n*-Butyl	10
t-Butyl	40	*n*-Amyl	10
Diacetone	10, 40	iso-Amyl	10
Isobutoxymethyl	10	Trimethyl hexyl	6, 10
Benzyl	10, 15, 20	*n*-Octyl	6, 10
t-Octyl	10	Hexadecyl	6, 10
Decyl	10		

[a] Mol % of the total monomers added. Mol % of NASI was kept at 2.5% throughout.

(isobutyronitrile) (0.021 g, 0.13 mM) and THF (50 ml). THF was pretreated to control peroxide contamination–deperoxidation by the procedure of Burfield [11]. The mixture was stirred, degassed, heated to 50–55°C internal temperature, maintained under positive nitrogen pressure for 24 h, and allowed to cool to room temperature. The reaction mixture was filtered through a layer of glass wool as the filtrate was stirred into ethyl ether (200 ml). The precipitated product was collected by filtration, washed thoroughly with ethyl ether, and dried (40–45°C) under vacuum to yield 3.9 g. The various monomers and their initial reaction percentages are shown in Table 1. The terpolymer consisting of NIPAAm: *N*-*n*-butylacrylamide (nBAAm): NASI with initial reaction ratio of 60:40:2.5, respectively, was designated Apoly 32.

Conjugation of activated polymers to a murine monoclonal antibody, MAb_{2H1}

Conjugation of MAb_{2H1} to an activated polymer was carried out as follows. An activated copolymer or terpolymer (20 mg) was dissolved in dimethylformamide (DMF, 100 ml). *N*-2-hydroxyethylpiperizine-*N*′-ethanesulfonic acid (HEPES, 0.1 M, 2 ml) buffer, pH 7.5, containing 1.5 mg of MAb_{2H1} was prepared and placed in ice. The DMF solution containing the activated co- or terpolymer was added to the antibody solution, and the mixture was kept in ice with intermittent vortex mixing until the solution was homogeneous. Then it was incubated at 2–8°C for 48 h. The mixture was diluted with distilled water (4 ml) and allowed to come to room temperature before the addition of saturated ammonium sulfate solution (2 ml). The resulting flocculence was centrifuged (approximately 25°C, 1500 × *g*) for 15–20 min. The supernatant was removed; and after addition of ice-cold distilled water (6 ml), the solution was placed in ice to dissolve the pellet. The mixture was returned to room temperature. The steps involving ammonium sulfate precipitation, centrifugation and dissolution of the polymer pellet were repeated two more times. To the final precipitate, distilled water (6 ml) was added and the solution was placed in ice to dissolve.

Hydroxylapatite chromatography was carried out at 4°C. The solubilized precipitate (consisting of copolymers and the MAb–copolymer conjugate) was loaded onto a hydroxylapatite column (1 × 1 cm, packed and equilibrated with distilled water at 4°C). The column was washed with distilled water until the optical density at 214 nm returned to baseline, signifying that the unconjugated copolymer had been washed from the column. The MAb–copolymer conjugate was then eluted with 0.3 M potassium phosphate buffer, pH 6.8. The collected fractions were monitored with a Coomassie blue protein assay reagent (Pierce, Rockford, IL, USA) and the major fractions containing protein were pooled to yield about 0.7 mg of the conjugate as measured by Coomassie blue protein assay.

Preparation of radioiodinated IgG (or M)-Apoly 32

Radioiodination of either mouse immunoglobulin G or M (IgG or IgM) was first carried out using ^{125}I labeled Bolton–Hunter reagent. Benzene, the organic phase of the ^{125}I-Bolton–Hunter reagent (E. I. DuPont de Nemour & Co., Inc., Wilmington, DE, USA) was removed by evaporation using dry nitrogen. IgG or IgM (~25 μg) dissolved in 25 μl of 0.1 M HEPES, pH 7.4, was added and incubated overnight at 4°C. Then the reaction mixture was diluted (200 μl) and chromatographed on a Pharmacia PD-10 column using the same buffer. The first radioactive fractions containing protein were pooled together and used for the subsequent Apoly 32 conjugation. For the preparation of ^{125}I-IgG (or M)–Apoly 32, 1 mg of unlabeled IgG or IgM was added to the radioactive fraction, and conjugated to Apoly 32 using the same procedure as described in the previous section.

Preparation of fluoresceinated MAb_{2C3} (a mouse monoclonal antibody specific for the mu chain of human IgM)

MAb_{2C3} was labeled with the fluorescein isothiocyanate using the method described previously [4].

Preparation of R*-phycoerythrin conjugated MAb_{2C3}*

MAb_{2C3} was labeled with *R*-phycoerythrin using the method described previously [12].

Membrane binding studies and the effect of detergent or common immunoassay additives on the binding of ^{125}I-IgG (or M)–Apoly 32

The binding of the ^{125}I-Ig–Apoly 32 to a membrane was studied using a Blot–Block filtration apparatus (designed primarily for DNA probe assays) which has 96 round bottomless wells, each having a diameter of 6.5 mm and a depth of 15 mm. Each membrane was wetted with water and after blotting to remove excess water, soaked in 1% bovine serum albumin in phosphate buffered saline, pH 7.0, 10 mM (BSA/PBS). The BSA treated membrane was then placed under the Blot–Block apparatus with 4 layers of absorbent paper underneath. Radioactive IgG (or M)–Apoly 32 (20 μl) was added to BSA/PBS (100 μl), and the mixture was added to a well on the Blot–Block apparatus. The solution was drawn through the membrane by capillary action. The membrane was washed three times with 0.05% Tween-20/PBS (100 μl) in the same manner. The membrane was removed from the Blot–Block apparatus. The area where the filtration had taken place was cut out and placed in a polystyrene tube. Radioactivity was counted in a gamma-counter.

For the studies on the effect of detergent on the binding of ^{125}I-IgG (or M)–Apoly 32, radioactive IgG (or M)–Apoly 32 (20 μl) was added to 100 μl of water or saline (0.15 M NaCl) solution containing various amount of detergent. The rest of the filtration procedure was the same as described above.

For the studies on the effect of immunoassay reagent additives, radioactive IgG (or M)–Apoly 32 (20 μl) was added to PBS (100 μl) containing the various additives listed in Table 5. The rest of the assay procedure is as described above.

Nuclear magnetic resonance (NMR) spectroscopy

NMR analysis was performed by J. H. Medley, Bristol—Myers Squibb Co., Syracuse, NY, USA, using a 360 MHz NMR spectrophotometer.

Measurement of LCST

The LCST was measured by visual observation of the temperature at which turbidity first appeared in a solution immersed in a silicone oil bath with the temperature raised at the rate of 3°C/h.

Molecular weight analysis

An absolute molecular weight analysis of activated copolymer was carried out by Modchrom Inc., Mentor, OH, USA using GPC/SEC with differential viscometry as the detection system.

Antigen capture MAC-IA for human IgM

The assay was performed by adding the following reagents: 100 μl of MAb_{2H1}–Apoly 5 or –Apoly 32 conjugate (4.5 μg Ab/assay), fluoresceinated MAb_{2C3} (1.2 μg Ab/assay) or *R*-phycoerythrin conjugated MAb_{2C3} (0.6 μg Ab/assay), and 100 μl of human IgM standard in BSA/PBS. The reaction mixture was incubated for 1 h at room temperature + 15 min at 45°C, or only (using Apoly 32) at room temperature (22°C). A 90 μl aliquot of the reaction mixture was transferred to a Pandex assay plate (Pandex Laboratories, Mundelein, IL, USA) containing a 0.2 μm cellulose acetate membrane which had been blocked with 100 μl of BSA/PBS for 1 h. Filtration of the reaction mixture was carried out by vacuum suction either at 25°C or at 45°C. Another 90 μl of the reaction mixture was added to the same well and vacuum filtered at the respective temperature. The wells were washed twice by filtration with BSA/PBS (100 μl) and two more times with 120 μl of PBS. After the final filtration, the epifluorescence of each well was determined using a Pandex Screen Machine (Pandex Laboratories, Mundelein, IL, USA), with an excitation wavelength of 490 nm and an emission wavelength of 520 nm for fluorescein and 520 and 570 nm, respectively, for *R*-phycoerythrin.

RESULTS

Synthesis of polymers and conjugation to MAbs

NIPAAm and NASI copolymer (Apoly 5). We have prepared an activated copolymer of NIPAAm and NASI under anhydrous conditions with a monomer feed ratio of 40 : 1, respectively, in order to control the size of the polymer conjugated to protein molecules. Absolute molecular weight (M_n) of this polymer is 7000–8000 as we reported previously [4] and consistent with the figure reported by Yang *et al.* [7]. This activated copolymer is designated as Apoly 5. In an aqueous medium (PBS), however, Apoly 5 co-eluted with bovine serum albumin (BSA, M_w 66000) on

Table 2.
Monomer ratios of representative copolymers

	Apoly 2		Apoly 5		Apoly 32		Apoly 32a (Repeat)	
Monomer	T	E	T	E	T	E	T	E
NIPAAm	20	15.5	40	35	24	18.5	24	18.5
NASI	1	1	1	1	1	1	1	1
nBAAm	—	—	—	—	16	10.5	16	11.1

T = theoretical: Molar monomer feed ratios used in the polymerization reactions
E = experimental: Molar ratios found in the isolated polymer [based on the integration of the specific monomer-related peaks in the nuclear magnetic resonance (NMR) spectra]

Sephacryl S-300 gel permeation columns indicating the differences in Stoke's radii between polyacrylamides and proteins under these conditions. In order to examine the extent of copolymerization, Apoly 5 was analyzed by NMR. Results showed 1 mol of NASI was incorporated per 35 mol of NIPAAm (Table 2). Apoly 5 with intact active ester has a LCST of 29–31°C. The hydrolyzed Apoly 5 had a slightly higher LCST (31–34°C). Mouse anti-human kappa MAb_{2H1} was conjugated to Apoly 5 by the method described in the materials and methods.

NIPAAm/N-substituted (or alkyl acrylate) acrylamide/NASI terpolymers. The activated, room-temperature-precipitable terpolymers consisting of NIPAAm, an *N*-substituted acrylamide (or alkyl acrylate) and NASI were synthesized under anhydrous conditions. Table 1 lists the comonomers and feed ratios used in the terpolymerizations. Each of the resulting terpolymers exhibited an LCST below

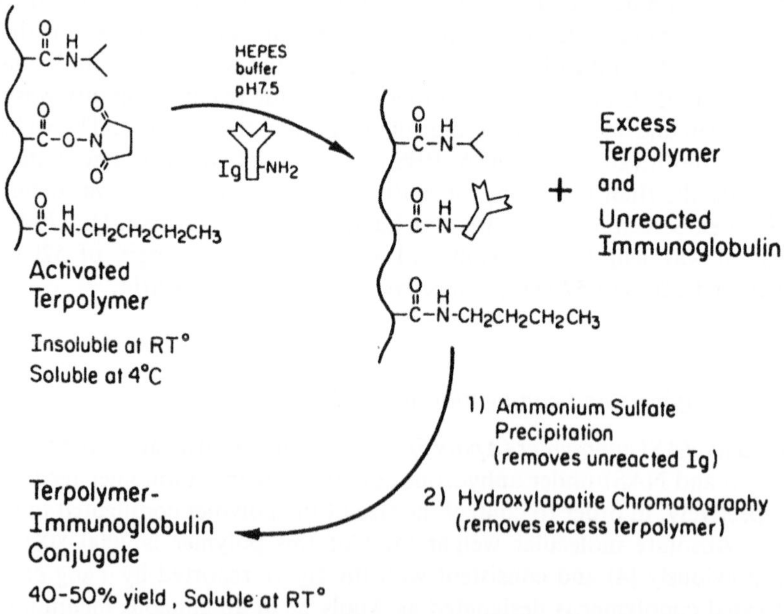
HEPES
buffer
pH7.5
Ig
NH2
Activated
Terpolymer
Insoluble at RT°
Soluble at 4°C
Excess
Terpolymer
and
Unreacted
Immunoglobulin
1) Ammonium Sulfate
Precipitation
(removes unreacted Ig)
2) Hydroxylapatite Chromatography
(removes excess terpolymer)
Terpolymer-
Immunoglobulin
Conjugate
40-50% yield, Soluble at RT°

Figure 2. Conjugation of Apoly 32 to an Ab and the isolation of Ab–Apoly 32 conjugate.

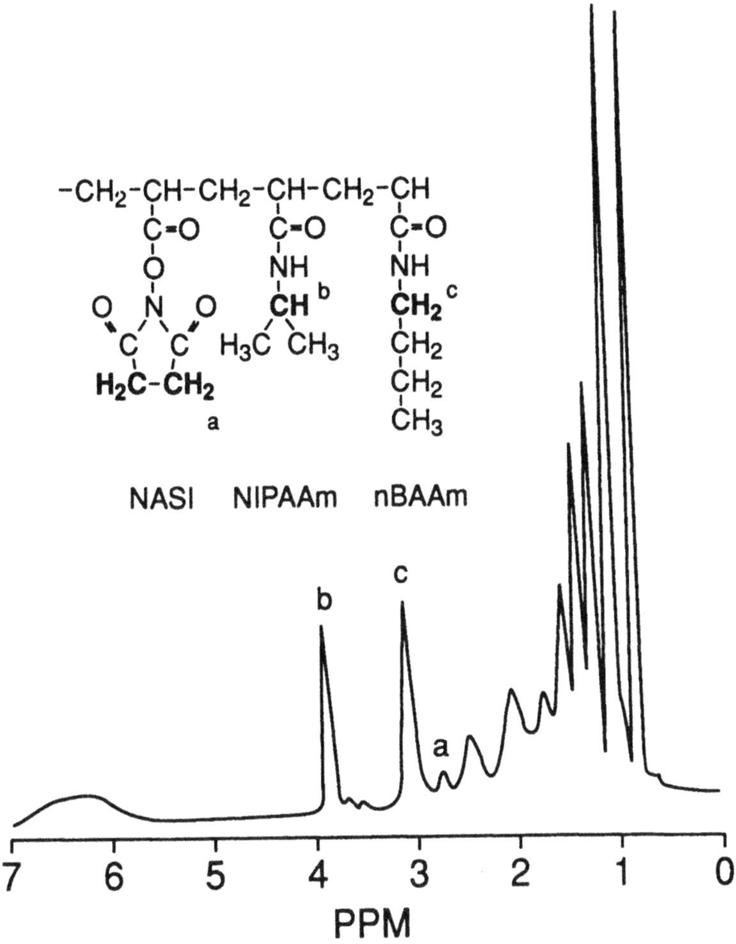

NASI
NIPAAm
nBAAm
a
b
c
7
6
5
4
3
2
1
0
PPM

Figure 3. NMR spectrum (360 MHz) of Apoly 32 (monomer feed ratio of 24:16:1 for NIPAAm/*N*-butylacrylamide/NASI respectively).

Table 3.
Lower critical solution temperature (LCST) of Apoly 32 at various concentrations

% Apoly 32	LCST (°C)[a] in water	in PBS[b]
0.031	12.1	13.1
0.063	10.5	11.4
0.125	8.6	10.3
0.250	7.8	9.0
0.50	7.2	7.8
1.0	6.7	7.1

[a] The LCST was measured by visual observation of the temperature at which turbidity first appeared in a solution immersed in a silicone oil bath with the temperature raised at the rate of 3°C/h^{-1}.
[b] PBS = phosphate buffered saline.

that of the NIPAAm homopolymer. Most were insoluble in an aqueous solution at room temperature (22°C), but dissolved as the temperature was lowered to 4°C. MAb_{2H1} was conjugated to the activated terpolymers by the method described above. When conjugated to an Ab, however, all appeared water soluble (@ 1 mg ml^{-1} Ab concentration) at room temperature by visual inspection. The terpolymer consisting of NIPAAm/nBAAm/NASI (monomer feed ratio of 24:16:1), designated as Apoly 32, consistently gave better positive-to-negative signal ratio than the others on MAC-IA. A brief sketch of the conjugation scheme for Apoly 32 is shown in Fig. 2. Actual copolymerization ratios were examined by NMR and found to be 18:10.5:1 for NIPAAm, nBAAm, and NASI, respectively (Table 2 and Fig. 3). The LCST of Apoly 32 at various concentrations (water or PBS) is shown in Table 3. Depending on the concentration, the phase separation temperature, as measured visually, varied from 7 to 13°C. Apparently the more efficient formation of precipitating aggregates at higher concentration leads to the lower observed LCSTs. Overall the LCST in PBS was slightly higher than that in water.

INTERACTION OF MAb-Apoly 32 WITH VARIOUS MEMBRANE MATERIALS

Binding of ^{125}I-IgM (MAb18C9)-Apoly 32 or ^{125}I-IgG (MAb2H1)-Apoly 32 to various membranes

^{125}I-IgM (MAb18C9) and ^{125}I-IgG (MAb2H1) were conjugated to Apoly 32 by the procedure described above. Five different membrane materials were examined. They were polycarbonate (0.8 μm pore size, from Nuclepore Corp., Pleasanton, CA, USA), mixed esters (0.8 μm), regenerated cellulose (1.0 μm), nylon-66 (1.2 μm), and hydrophilic cellulose acetate (CA, 1.2 μm) (the rest of them were purchased from Schleicher and Schull Co., Keene, NH, USA). As shown in Fig. 4, CA was found

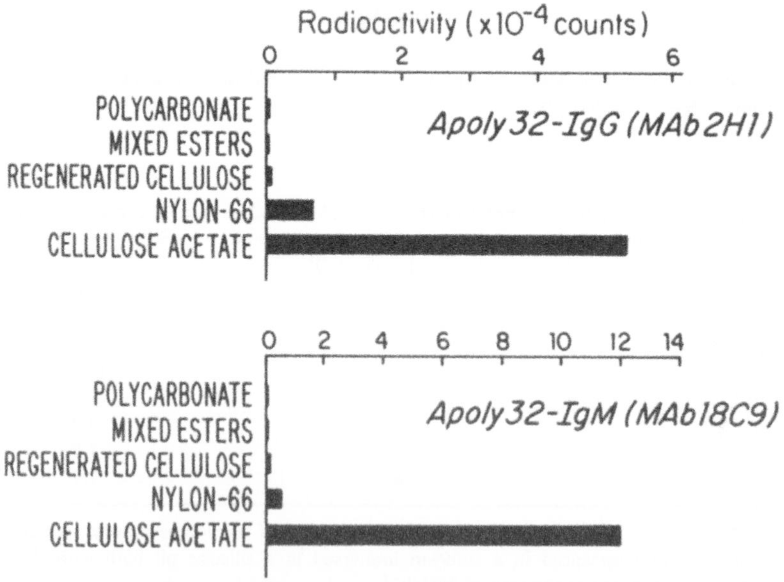
Radioactivity (x10⁻⁴ counts)
0
2
4
6
POLYCARBONATE
MIXED ESTERS
REGENERATED CELLULOSE
NYLON-66
CELLULOSE ACETATE
Apoly 32-IgG (MAb2H1)
0
2
4
6
8
10
12
14
POLYCARBONATE
MIXED ESTERS
REGENERATED CELLULOSE
NYLON-66
CELLULOSE ACETATE
Apoly 32-IgM (MAb18C9)

Figure 4. Binding of IgG(MAb_{2H1})- or IgM(MAb_{2C3})-Apoly 32 to various membranes.

to bind most effectively. This cellulose acetate is claimed (by the manufacturer) to be one of the lowest non-specific protein binding membranes available (Product # ST69). Nylon-66 only binds 10–15% of either ^{125}I-IgM–Apoly 32 and ^{125}I-IgG–Apoly 32 compared to that observed for the CA membrane. The remaining membranes, i.e. polycarbonate, mixed ester, and regenerated cellulose, exhibited negligible binding of both MAb-polymer conjugates. As observed in other NIPAAm copolymer studies [8], the above results indicated the specific nature of the interaction between Ig–Apoly 32 and cellulose acetate.

Effect of detergent on MAb–Apoly 32/CA interaction

The strength of the MAb–polymer binding to the CA membrane (1.2 μm pore size) was examined by the addition of various detergents (Table 4). The surfactants studied were Tween-20, Triton X-100 (TX-100), Nonidet P-40 (NP-40), sodium desoxycholate (DOC), and sodium dodecyl sulfate (SDS). These detergents are commonly used in immunological assays to control various nonspecific interactions.

The results indicated that Tween-20 (0.8%) interfered the least, showing no significant loss in IgG- or IgM–polymer binding. Tween-20 concentrations of greater than 0.8% have not been routinely used in most immunoassays. The most severe binding interference was seen by SDS (0.05–0.8%). With the IgG–Apoly 32 conjugate, as low as 0.0125% SDS caused 80% reduction in the CA binding even in the presence of NaCl. SDS is also known to inhibit phase-separation of pNIPAAm at concentrations $>1\%$ [13]. Listed in order of decreasing ability to interfere with the MAb-polymer/CA interaction, the detergents are: SDS $\gg$ DOC $>$ NP-40 $>$ TX-100 $>$ Tween-20 (see Table 4).

In general, the addition of NaCl (0.15 M) helps to counteract the interference of the detergent (Table 4). Figure 5 shows the dose dependent effect of NaCl on SDS, NP-40, and DOC. The most apparent effect of NaCl was with IgM–Apoly 32

Table 4.
Percentage retention of radioactive conjugates in the presence of various detergents

% Retention	^{125}I-MAb(2C3, IgM)[a] -Apoly 32		^{125}I-MAb(2H1, IgG)[b] -Apoly 32	
Detergent	Water	0.15 M NaCl	Water	0.15 M NaCl
Tween-20 (0.8%)	100	100	98	100
Triton-X100 (0.8%)	89	100	85	100
Nonidet-P40 (0.8%)	63	100	22	92
DOC (0.8%)[c]	59	100	20	62
SDS (0.25%)[d]	3	51	0	0

The pore size of the CA membrane was 1.2 μm
[a] Retention of radioactive conjugate in saline is regarded as 100%. Retention of ^{125}I-MAb(2C3) alone was <3% with and without 0.15 M NaCl
[b] Retention of radioactive conjugate in saline is regarded as 100%. Retention of ^{125}I-MAb(2H1) alone was <0.1% with and without 0.15 M NaCl
[c] DOC = sodium desoxycholate
[d] SDS = sodium dodecyl sulfate

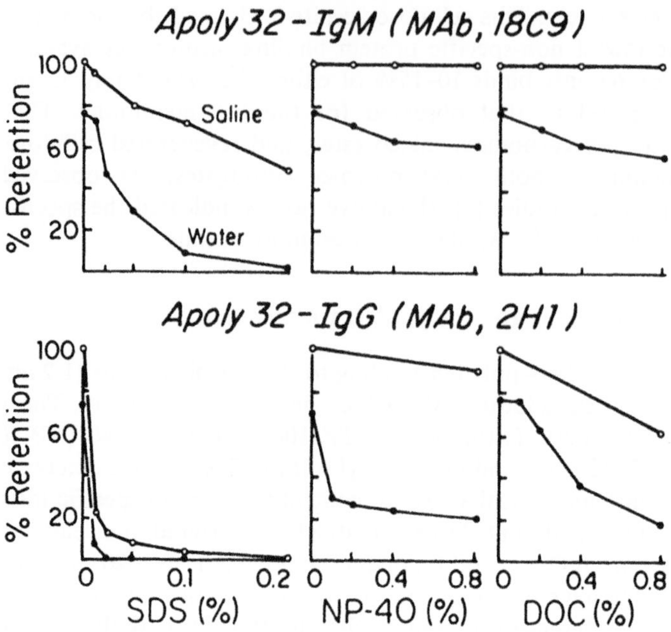
Apoly 32 - IgM (MAb, 18C9)
% Retention
100
60
20
Saline
Water
Apoly 32 - IgG (MAb, 2H1)
% Retention
100
60
20
0
0.1
0.2
0
0.4
0.8
0
0.4
0.8
SDS (%)
NP-40 (%)
DOC (%)

Figure 5. Influence of NaCl on the inhibition of Ab-polymer/CA interaction by SDS, NP-40, and DOC.

conjugate. Without NaCl, SDS (0.1%) caused >90% inhibition of IgM–Apoly 32 conjugate binding to CA; however in the presence of 0.15 M NaCl, only 20% inhibition occurred. These results suggest that: (1) addition of NaCl to the medium causes an increase in hydrophobic bonding; and (2) the IgM–Apoly 32 conjugate appears to be more hydrophobic than the IgG–Apoly 32 conjugate, causing a stronger affinity to CA. These results emphasize the importance of both the physical characteristics of the protein conjugated to the polymer, as well as the polymer itself.

Table 5.
Effect of serum (25%) on the binding of IgG (or M)–Apoly 32 conjugate to CA membrane.

		^{125}I-MAb (2C3, IgM) –Apoly 32		^{125}I-MAb (2H1, IgG) –Apoly 32	
Component serum (25%)		RA[a]	% bound	RA[a]	% bound
PBS only	–	12435	100	24746	100
PBS only	+	12540	100	24704	100
1% BSA/PBS	–	12645	100	24801	100
1% BSA/PBS	+	12588	101	19797	80
Blotto (1 ×)	–	12547	101	24556	99
Blotto (1 ×)	+	10612	85	8661	35

The pore size of the CA membrane used in this study was 1.2 μm.
[a] RA = radioactivity, cpm; average of 3, CV < 5%

Table 6.
MAC-IA standard curve for human IgM

	Epifluorescence	
Human IgM (θg ml^{-1})	MAb (2H1)-poly(NIPAAm)	MAb (2H1)-Apoly 5
0	720	648
0.03	1275	1197
0.06	1796	1721
0.125	3008	2928
0.25	5138	5066
0.50	10020	9962

Effect of other common additives used in immunoassays on MAb–Apoly 32/CA interaction

Besides several detergents mentioned above, the effect of other common media and reagents were studied. They were human serum, BSA, and Blotto (2.5% non-fat dry milk in 0.3 M citrate, pH 7.6; a specimen diluent used by Genetic Systems Corporation). As serum samples are commonly used in immunoassays, 25% serum was chosen as the representative concentration; and the effect on MAb–Apoly 32/CA interaction was tested either with and without addition of 1% BSA or with and without addition of Blotto (Table 5). BSA and non-fat dry milk are two of the most commonly used protein additives to effectively block nonspecific signal binding in immunoassays and serological tests.

Addition of 25% serum alone did not cause interference of either IgG– or IgM–Apoly 32 conjugate binding to CA. Addition of 1% BSA to 25% serum also did not affect the binding. However, the addition of Blotto to the 25% serum caused a 15% reduction of the binding in the IgM–Apoly 32 conjugate and a 65% reduction in the IgG–Apoly 32 binding. Since the effect of Blotto alone was minimal, a lower percentage of non-fat dry milk, in conjunction with 25% serum, was indicated for developing an immunoassay.

Performance of MAb–polymer conjugates in an immunoassay

The MAb–Apoly 5 conjugate was first compared, using the MAC-IA, with a MAb–pNIPAAm conjugate prepared via copolymerization of NIPAAm with the monomer-conjugated MAb. As shown in Table 6 (Standard Curve for Human IgM), both preparations showed comparable performance in the assay, indicating that the CA membrane binding characteristics of the preformed copolymer and the *de novo* polymer were similar.

Comparison of MAb_{2H1}–Apoly 5 and MAb_{2H1}–Apoly 32 was then made, at two different temperatures: room temperature (22°C) and 45°C, using the MAC-IA (Table 7). At 45°C, both capture conjugates gave equivalent signals. However, at room temperature the Apoly 5 conjugate gave only ~50% of the comparable signal produced by the Apoly 32 conjugate at both 22 and 45°C. This increase in signal at room temperature using Apoly 32 was indicative of the enhanced CA/polymer binding above the LCST (7–13°C) of the terpolymer. At room temperature the MAb_{2H1}–Apoly 32 conjugate appeared (by visual observation) to form a

Table 7.
MAC-IA standard curve for human IgM

	Epifluorescence MAb (2H1)–Apoly 5		MAb (2H1)–Apoly 32	
Human IgM (μg ml^{-1})	22°C	45°C	22°C	45°C
0	439	741	680	750
0.03	765	1469	1465	1540
0.06	1091	2095	2092	2165
0.125	1301	3393	3465	3540
0.25	3085	5948	5970	6040
0.50	5810	10340	10470	10540

homogeneous aqueous solution. However, when the terpolymer undergoes the conformational changes occurring above its LCST, this probably results in a more hydrophobic surface and leads to the more efficient binding to the CA membrane. In contrast to the unconjugated terpolymer; the covalently attached, 'hydrophilic' Ab may have 'protected' the conjugated terpolymer from precipitating at room temperature.

The conjugated Apoly 32 exhibited assay sensitivity at room temperature similar to that of the conjugated Apoly 5 at 45°C. When the sensitivity of the MAC-IA was evaluated at room temperature using MAb_{2H1}–Apoly 32 as the capture conjugate, a maximum sensitivity (maximum sensitivity = the ng IgM/assay with a signal value 2.5 times higher than at 0 ng/assay.) of 4 ng IgM/assay was detected when MAb_{2C3}–fluorescein was the signal conjugate. When MAb_{2C3}–*R*–phycoerythrin was the signal conjugate, as little as 0.5 ng IgM/assay was detected (Fig. 6). *R*-Phycoerythrin is a highly fluorescent biomolecule isolated from red sea algae. This simple change in the signal molecule conjugated to the MAb resulted in a >8-fold increase in assay sensitivity.

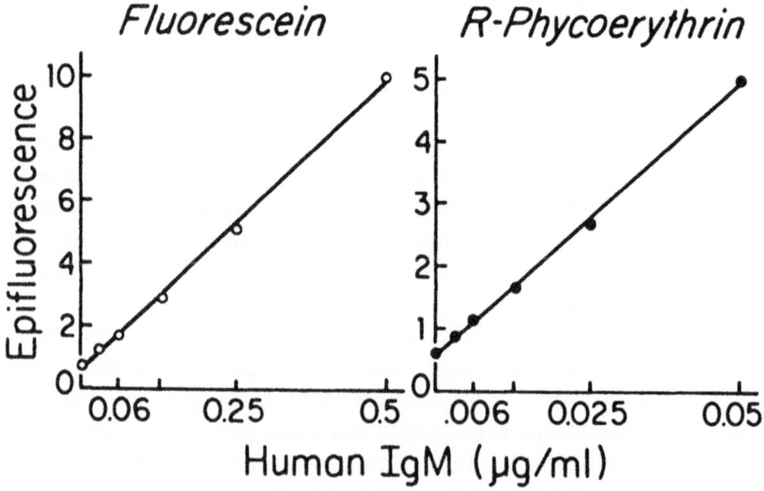
Fluorescein
R-Phycoerythrin
Epifluorescence
10
8
6
4
2
0
5
4
3
2
1
0
0.06
0.25
0.5
.006
0.025
0.05
Human IgM (μg/ml)

Figure 6. Antigen capture MAC-IA for human IgM. The assay procedure is described in Materials and Methods.

DISCUSSION

Priest *et al.* [2] previously reported that copolymerization of higher *N*-alkyl acrylamides with NIPAAm lowers the LCST of the resulting copolymers. Based on their studies, various percentages of *N*-alkyl acrylamides or alkyl acrylates were added as comonomers in an anhydrous free radical polymerization to prepare a room-temperature-precipitable (i.e. LCST $<$ room temperature) activated terpolymer suitable for MAC-IA applications. The present studies demonstrated that it is indeed possible to prepare room-temperature-precipitable polymers that can be conjugated to protein molecules and have enhanced binding capacity for CA at room temperature (22°C) relative to their higher LCST analogues. This development of a conjugate using a low LCST terpolymer to produce a higher assay sensitivity at room temperature is a further improvement of the MAC-IA over systems with homo-poly(NIPAAm) described previously [8, 9].

In order for the present MAC-IA system to be applicable for general immunodiagnostic uses, the interaction between the MAb–Apoly 32 conjugate and the CA membrane needs to be both strong and stable under various conditions, including the presence of common detergents, serum, and other proteins such as BSA and non-fat dry milk, which are commonly used for blocking nonspecific background signals. It was therefore important to study the strength of this conjugate/CA binding under various conditions. The most commonly used detergents in immunoassays are weak nonionic detergents, such as Tween-20 and Triton X-100. Strong anionic detergents, such as SDS, are rarely used in immunoassays because they can cause irreversible damage to proteins even in low concentrations. Up to 0.8% of either Tween-20 or Triton-X100 was acceptable for use in the MAC-IA format without interference with the MAb–Apoly 32/CA interaction. The MAC-IA also was able to withstand the detergent-like effect of BSA and non-fat dry milk even above the typical concentrations (0.1–1%) used in immunoassays. Taken together, these results indicate that the MAC-IA can be applied to many different types of immunoassays without compromising the MAb–Apoly 32/CA interaction due to various assay conditions.

Various solid-phase-based immunoassays have been described in the literature [14, 15]. The solid phase can take the form of a microplate, a molded surface, porous support or microparticles. Although versatile and easy to use, some solid phase immunoassays continue to be plagued by slow reaction kinetics and high nonspecific binding [16]. The kinetics of the binding of a macromolecular antigen to an immobilized antibody are influenced by the presence of an unstirred layer of solution at the surface of the solid phase, the Nernst layer [17, 18], which serves to limit the rate of diffusion of reactants from the bulk solution.

In the present MAC-IA system, the kinetics of the binding of antibody and antigen should be rapid, since all reactants are freely diffusing in solution. Also, because a large bulk solid interface is not present during the immune complex formation, the possibility of nonspecific binding is minimized. The specific signal is captured on the CA membrane only after the immune complexation reaction has taken place in solution, resulting in a very low nonspecific background signal.

As for the application in the immunoassay system, double-antibody-sandwich-assay for the detection of human IgM is presented as an example. The advantages of this human IgM MAC-IA are: (1) fast liquid-phase immune reaction kinetics;

(2) ability to concentrate specific molecules onto a limited, controlled area; (3) no requirement for physical, chemical or biochemical modification of the membrane surface in order to capture the specific signal; and (4) low nonspecific background.

REFERENCES

1. M. Heskins and J. E. Guillet, *J. Macromol. Sci. Chem.* **A2**, 1441 (1968).
2. J. P. Priest, S. L. Murray, R. J. Nelson and A. S. Hoffman, in: *Reversible Polymeric Gels and Related Systems*, p. 255, P. S. Russo (Ed.). ACS Symposium Series 350, American Chemical Society, Washington, DC (1987).
3. N. Monji and A. S. Hoffman, *Appl. Biochem. Biotech.* **14**, 109 (1987).
4. K. Auditore-Hargreaves, R. L. Houghton, N. Monji, J. H. Priest, A. S. Hoffman and R. C. Nowinski, *Clin. Chem.* **33**, 1509 (1988).
5. C.-A. Cole, S. M. Schreiner, J. H. Priest, N. Monji and A. S. Hoffman, in: *Reversible Polymeric Gels and Related Systems*, p. 245, P. S. Russo (Ed.). ACS Symposium Series 350, American Chemical Society, Washington, DC (1987).
6. R. F. S. Freitas and E. L. Cussler, *Chem. Eng. Sci.* **42**, 97 (1987).
7. J. H. Yang, C.-A. Cole, N. Monji and A. S. Hoffman, *Polymer Chem.* **28**, 219 (1990).
8. N. Monji, C.-A. Cole, M. Tam, L. Goldstein, R. C. Nowinski and A. S. Hoffman, *Biochem. Biophys. Res. Comm.* **172**, 652 (1991).
9. N. Monji and C.-A. Cole, U.S. Patent #5,206,178.
10. A. Pollak, H. Blumenfeld, M. Wax, R. L. Baughn and G. M. Whitesides, *J. Am. Chem. Soc.* **56**, 6324 (1980).
11. D. R. Burfield, *J. Org. Chem.* **47**, 3821 (1982).
12. N. Monji and A. Castro, *Rev. Immunoassay Technol.* **1**, 95 (1988).
13. J. Eliassaf, *J. Polym. Sci.* **22**, 873 (1978).
14. J. Kang, P. Kaladas, C. Chang, S. Chen, R. Dondero, A. Frank, S. Huhn, P. Lisi, D. Mochnal, J. Nasser, W. Pottor, E. Schutt, K. Utberg and H. Graham, *Clin. Chem.* **32**, 1682 (1986).
15. J. E. Butler (Ed.), *Immunochemistry of Solid-Phase Immunoassay*. CRC Press, Inc., Boca Raton, FL (1991).
16. G. H. Parsons, *Methods Enzymol.* **73**, 224 (1986).
17. W. Z. Nernst and E. Z. Brunner, *Z. Phys. Chem.* **47**, 52 (1904).
18. E. Katchalski, I. Silman and R. Goldman, *Adv. Enzymol. Relat. Areas Mol. Biol.* **34**, 445 (1976).

Temperature-dependent adsorption/desorption behavior of lower critical solution temperature (LCST) polymers on various substrates

M. MIURA[1*], C.-A. COLE[2], N. MONJI[2] and A. S. HOFFMAN[1]

[1]*Center for Bioengineering, FL-20, University of Washington, Seattle, WA 98195, USA*
[2]*Genetic Systems-Sanofi Diagnostics Pasteur, Inc., 6565 185th Avenue NE, Redmond, WA 98052, USA*

Received 26 July 1992; accepted 15 April 1993

Abstract—We have been studying adsorption and retention (resistance to desorption) behavior of temperature sensitive LCST polymers on different substrates as a function of temperature. According to our studies with Poly 64 (a copolymer of 60% (mol) NIPAAm and 40% (mol) NnBAAm, LCST = 8.5°C in water), the copolymer retention depends on the rinse temperature. When the rinse temperature is above the LCST, the polymer adheres well to most surfaces. On the contrary, at rinse temperatures below the LCST, most of the adsorbed polymer is easily rinsed off. These studies are relevant to our work on the thermally reversible adsorption of LCST polymers conjugated to peptides and proteins, such as affinity ligands, for uses in immunoassays and affinity separations.

The interaction between the LCST polymer and most hydrophobic polymer surfaces is mainly due to hydrophobic interactions, and the critical surface tension (γ_c) and the solubility parameter (δ) of the solid polymer substrate are the most important factors which influence the LCST polymer adsorption and retention. The critical surface tension appears to correlate best with the LCST polymer adsorption levels on different substrates, while the solubility parameter correlates best with the retention of the adsorbed polymer.

According to our preliminary study, *n*-butyl groups probably interact more strongly with the substrates than isopropyl groups because of the greater hydrophobic surface area of the former groups.

Key words: Copolymer; LCST; adsorption; temperature-dependent; N-isopropylacrylamide; N-n-butylacrylamide.

INTRODUCTION

It is known that some water soluble polymers exhibit a lower critical solution temperature (LCST) in aqueous solution [1–7]. For example, poly-*N*-isopropylacrylamide (poly(NIPAAm)) has an LCST around 32°C [2–5]. This polymer dissolves in water at temperatures below the LCST. Above this temperature, the polymer precipitates. This process is reversible. Such stimuli-responsive polymers might be applied in the medical or biotechnological fields for drug delivery systems [8–10] immuno-assays [11–15], and bioseparations [16–21]. For these reasons, many researchers are actively studying LCST polymers [8–15, 22–26, 33].

Very few publications have appeared on the adsorption and retention behavior of an LCST polymer on a solid polymer surface, especially as a function of temperature. In previous work, we have applied the selective and strong physical adsorption of such a polymer, which had been conjugated to an antibody for a membrane-based

* On leave from Asahi Chemical Industry Co. Ltd., Japan.

immuno-assay at room temperature [13–14]. Since the LCST precipitation phenomenon is driven by the release of bound water, causing the polymer to become hydrophobic and to precipitate, it is reasonable to expect that such a polymer will be preferentially retained above its LCST on a hydrophobic polymer substrate, due to hydrophobic interactions, while it should redissolve into the solution below the LCST.

Therefore, we prepared copolymers of *N*-isopropylacrylamide and *N*-*n*-butylacrylamide which have different LCSTs and studied the adsorption and retention behavior of these copolymers and their homopolymers on a range of polar to hydrophobic substrates as a function of temperature.

MATERIALS

We selected five solid polymer substrates for study: hexafluoropropylene-tetrafluoroethylene copolymer (FEP), low density polyethylene (LDPE), polystyrene (PS), polyethyleneterephthalate (PET), and three cellulose acetates (CA).

The Teflon® FEP film was purchased from DuPoint, the LDPE film was supplied by Penn Fibre, and Falcon® 1029 Petri dishes were used as the PS substrate. Thermanox® coverslips (Nunc) were used as the PET substrate, and Eastman® cellulose acetates, CA 432 (D.S. = 2.82), CA 398 (D.S. = 2.46), and CA 320 (D.S. = 1.75), were used as CA substrates. The first numbers following CA show the acetyl content (e.g., the acetyl content of CA 432 is 43.2 wt%, followed by the degree of substitution, D.S.). NIPAAm and NnBAAm were purchased from Eastman Kodak Company and Monomer-Polymer & Dajac Laboratories, Inc., respectively. Dichloromethane (spectrographic grade) was purchased from EM Science, and acetone (spectrographic grade) and *N,N*-dimethylformamide (reagent grade) were obtained from J. T. Baker Chemical. Polyacrylamide, ethanol (HPLC grade) and 1,1,2,2-tetrachloroethane (reagent grade) were purchased from Aldrich Chemical Company, Inc. Water was deionized and glass distilled before use.

EXPERIMENTAL METHODS

Substrate preparation

FEP, LDPE, PS, and PET were cleaned by the following procedures.

FEP, LDPE, and PET. These 8 mm square sheets were cleaned with dichloromethane by sonication for 15 min at room temperature (r.t.). Then the solvent was changed to acetone and the same procedure was repeated. After that, the sheets were rinsed twice with deionized and distilled water by sonication for 15 min at r.t. Then the sheets were dried *in vacuo* overnight.

PS. The 8 mm square sheets were cleaned twice with ethanol by sonication for 15 min at r.t. After that, they were rinsed twice with deionized and distilled water by sonication for 15 min at r.t. and then dried *in vacuo* overnight.

Cellulose acetate substrates. Three different kinds of 1% cellulose acetate solutions were prepared for different CA types.

CA 432, 398, and 320 were dissolved into 1,1,2,2-tetrachloroethane, acetone, and *N,N*-dimethylformamide, respectively. These solutions were filtered through 0.45 μm PTFE membranes.

Clean glass discs were prepared and spin-coated with these solutions. After that, the coated glass discs were dried in a laminar flow hood for 3 h, and dried *in vacuo* overnight.

Water contact angle measurement

Advancing water contact angles on each substrate were measured by a conventional sessile droplet method, using a Rame-Hart goniometer. Three droplets, with a volume between 0.4 and 0.5 μl, were deposited on each surface, and six angles were measured and averaged for each reported value. The angles were all measured with 1 min of droplet deposition on the surface.

LCST polymer preparation

Homopoly-NIPAAm and copolymers of NIPAAm and NnBAAm were synthesized by the following method:

Poly(NIPAAm). NIPAAm (45 mmol) and 2,2′-azobisisobutyronitrile (AIBN, 0.13 mmol) were dissolved in 50 ml of dry tetrahydrofuran. The magnetically stirred solution was degassed, heated to 50°C for 24 h under positive nitrogen pressure, and allowed to cool. The reaction mixture was filtered through a 0.45 μm Teflon filter and the filtrate volume reduced by half. Ether was added with mixing to precipitate the polymer. The precipitate was filtered off, washed with ether, and dried under vacuum to yield 4.3 g of dry product (yield = 84%).

Copolymers of NIPAAm and NnBAAm. These copolymers were prepared by the method described for poly(NIPAAm) except that the same molar amount of the mixture of NIPAAm and NnBAAm was added instead of NIPAAm to the reaction.

Determination of molecular weight

Polymer samples were arranged for molecular weight by GPC. Values ranged from 7000–10 000. Protocol details have been published elsewhere [15].

Determination of LCST

Each LCST polymer was dissolved in water (0.5 mg ml^{-1}) and light transmittance at 600 nm of each solution was measured with increasing temperature (at a rate of 10°C h^{-1}). The temperature at which the transmittance had dropped from 100 to 90% was referred to as the LCST of that polymer.

Adsorption isotherms of LCST polymer (Poly 64) on FEP at 0 and 37°C

Poly 64 solutions were prepared in phosphate-buffered saline (PBS) to the following concentrations: 10, 20, 50, 100, and 200 μg ml^{-1}. FEP substrates were separated into two groups, one for the adsorption at 37°C and the other for adsorption at 0°C (ice bath). First, each FEP film was immersed in bottles containing the Poly 64 solutions at 0°C. Then half of the bottles were transferred from the ice bath to a 37°C water bath. The immersed FEP samples were incubated for 2 h and then removed and rinsed with deionized and distilled water at 37°C for 5 min. After the rinse, each sheet was dried *in vacuo* overnight.

In this experiment, we originally decided to use a 37°C rinse based on the assumption that after the rinse procedure, most of the adsorbed Poly 64 should remain on the FEP surface because 37°C is much higher than the LCST of Poly 64, which is 8.5°C.

Since there is no nitrogen atom in any of the substrates used, the XPS nitrogen signal intensity from the sample surface can be used as an indicator of LCST polymer adsorption and retention. Therefore, the extent of adsorption and retention were estimated using XPS (SSX-100 spectrometer system/Surface Science Instruments, Mountain View, CA) in the NESAC-BIO Surface Analysis Laboratory, University of Washington. The XPS conditions were as follows:

X ray source: Al *K* : Al *Kα*; spot size: 1000 μm; take off angle: 55 deg; quantitative data were collected by detailed scans (20 eV regions at a pass energy of 150 eV). The XPS conditions are the same for all other experiments.

Retention behavior of LCST polymer (Poly 64) on FEP after 0 and 37°C rinses

A 100 μg ml^{-1} solution of Poly 64 (in PBS) was prepared. Each FEP film was immersed in the polymer solution at 0°C, then all the sample bottles containing FEP films were transferred from the ice bath to a 37°C water bath and the samples were incubated for 2 h. After that, the samples were rinsed with deionized and distilled water at either 0 or 37°C, for various times. After the rinse, the samples were dried *in vacuo* overnight.

The surface elemental components were determined by XPS according to the conditions mentioned above.

Temperature dependence of LCST polymer (Poly 64) retention on FEP

A 100 μg ml^{-1} PBS solution of Poly 64 was prepared. Each FEP film was immersed in the polymer solution at 0°C, then all the sample bottles containing FEP films were transferred from the ice bath to a 37°C water bath and the samples were incubated for 2 h. After the incubation, the polymer solution was replaced with deionized and distilled water whose temperature was either 0, 5, 10, 15, 20, or 37°C. Each sample was kept at these conditions for 1 h, then dried *in vacuo* overnight. The surface elemental components were determined by XPS for each film.

Adsorption and retention of LCST polymer (Poly 64) on various substrates

A 100 μg ml^{-1} PBS solution of Poly 64 was prepared. FEP, LDPE, PS, PET, CA 432, CA 398, and CA 320 were used as substrates. Each substrate was immersed into the polymer solution at 0°C, and then all the sample bottles containing each film were transferred from the ice bath to a 37°C water bath. The substrate was incubated for 2 h at 37°C. The samples were split into two groups, one of which was rinsed with deionized and distilled water at 0°C for 1 h while the other was rinsed at 37°C. After that, the samples were dried *in vacuo* overnight. The surface elemental components were determined by XPS for each substrate.

Retention of three LCST polymers and PAAm on FEP

100 μg ml^{-1} PBS solutions of Poly 64, Poly 82, Poly(NIPAAm), and PAAm were prepared. FEP substrate samples were immersed into each polymer solution at 0°C, and then incubated at either 37 or 0°C for 2 h. After that, the samples were rinsed

with deionized and distilled water at either 0 or 37°C for 1 h. The samples were then dried *in vacuo*, and the surface elemental components were determined by XPS.

The surface content of each polymer was calculated by the following equation:

Surface content = $[(N1s\ at\%)_{Obs}/(N1s\ at\%)_{Theo}] \times 100$ where $(N1s\ at\%)_{Obs}$ is the N1s at% determined on FEP after a rinse, and $(N1s\ at\%)_{Theo}$ is the theoretical N1s at% of each polymer.

RESULTS AND DISCUSSION

Characteristics of various substrates

The water contact angles (θ) measured are listed in Table 1 along with literature values or estimated critical surface tensions (γ_c) and solubility parameters (δ) of these substrates.

Table 1.
Water contact angle (θ), critical surface tension (γ_c), and solubility parameters (δ) for for various substrates.

Substrate	θ (deg)	γ_c (dyn cm^{-1})	δ ((cal cm^{-3})$^{0.5}$)
FEP	104.5 ± 1.5	18 [16]	6.5 [20]
LDPE	90.0 ± 1.0	27 [17]	8.1 [19]
PS	81.7 ± 2.2	33 [17]	9.1 [19]
PET	71.3 ± 2.9	40 [17]	10.7 [19]
CA 432	60.3 ± 0.8	—	13.7 [21]
CA 398	59.4 ± 2.6	—	13.5 [21]
CA 320	58.6 ± 1.1	—	13.2 [21]

It can be seen that θ decreases as either γ_c or δ increases, as expected.

LCST of copolymers of NIPAAm and NnBAAm

As described in the Methods section, we prepared different LCST copolymers that have various ratios of NIPAAm to NnBAAm and then measured the LCSTs of these copolymers. The results are shown in Fig. 1. The LCST appears to be proportional to the NIPAAm content in the polymerization mixture. This suggests that the copolymers are mostly random copolymers and not block copolymers or homopolymer blends. (The copolymerization was carried out to 100% completion.) Others have also reported that the more hydrophilic copolymer, the higher LCST it exhibits [3–5]. The LCSTs here reflect the fact that the *n*-butyl group is more hydrophobic than the isopropyl group. For example, Poly 64, for which the ratio of NIPAAm to NnBAAm is 60 to 40, has an LCST in water around 8.5°C while Poly 82, for which the ratio is 80 to 20, has an LCST around 18.5°C.

Adsorption isotherm of LCST polymer on FEP

We decided to use FEP as a substrate since, in principle, fluorocarbon polymers are easy to clean and FEP has a smooth surface in comparison to PTFE. In addition, the fluorine signal is easily detected by XPS and it can indicate whether the adsorbed Poly 64 has totally covered the substrate or not.

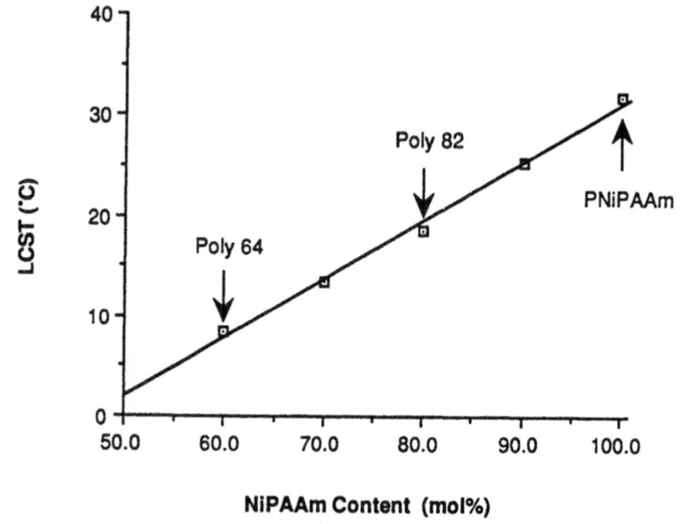
LCST (°C)
40
30
20
10
0
Poly 82
PNiPAAm
Poly 64
50.0
60.0
70.0
80.0
90.0
100.0
NiPAAm Content (mol%)

Figure 1. LCST as a function of NIPAAm content in copolymers of NIPAAm and NnBAAm.

The adsorption results are illustrated in Fig. 2, where the theoretical value of Poly 64 is 11.9% and the experimental value is 11.7% determined by an XPS assay of a glass disc which had been spin-coated with Poly 64. The results show that more polymer adheres to the substrate at 37 than at 0°C, as expected. This is because Poly 64 not only is more hydrophobic at 37 than at 0°C, but is also insoluble at 37°C and soluble at 0°C.

Molecular processes occurring in solution and on the substrate surface

In Fig. 2, both isotherms rise sharply up at low concentrations and appear to start leveling out around 50 μg ml^{-1}. Above this concentration, the adsorption at 37°C is constant. However, at 0°C, it still increases gradually. The difference is due to the LCST phenomenon; since the polymer precipitates at 37°C, the actual polymer

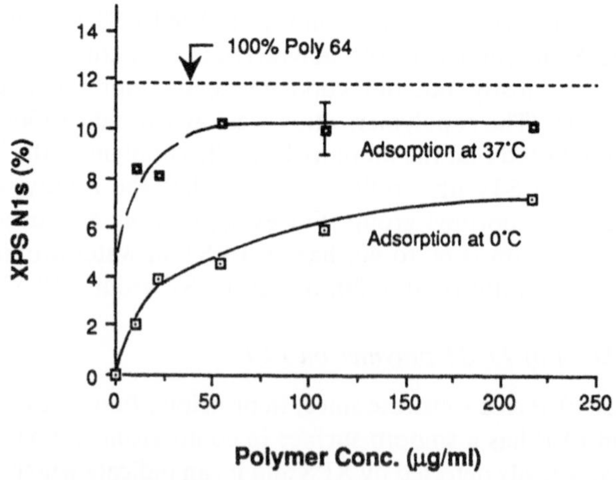
XPS N1s (%)
14
12
10
8
6
4
2
0
100% Poly 64
Adsorption at 37°C
Adsorption at 0°C
0
50
100
150
200
250
Polymer Conc. (µg/ml)

Figure 2. Adsorption isotherms of Poly 64 on FEP at 0 and 37°C.

concentrations in solution at this temperature should approach zero, independent of the initial concentrations. The 'theoretical' % N level for Poly 64 is also shown in Fig. 2 and these data suggest that the surface is almost completely covered by a 50 Å or more layer of Poly 64 at 37°C, but not at 0°C. In order to explain this, it is presumed that there are two adsorption processes occurring as the temperature is raised above the LCST. Below the LCST, Poly 64 molecules are well solvated individually in solution and some of those molecules adsorb on the FEP substrate to reduce the interfacial energy between water and the FEP surface. This is a conventional adsorption process. On the other hand, above the LCST, the precipitation and adsorption processes may both be occurring at once, and if so, these processes could compete with or add to each other. When the temperature is raised through the LCST, the polymer molecules in solution should first collapse individually as they begin to precipitate and some of them may adhere to the exposed FEP or Poly 64-coated FEP surface. In this experiment, the temperature for the 37°C adsorption process is reached by raising the temperature from 0 to 37°C. Since there are Poly 64 molecules on the substrate before the temperature reaches the LCST, when the temperature is raised above the LCST, those adsorbed Poly 64 molecules should also start collapsing (curling up) and this should open up other sites on the surface which other precipitating molecules could occupy. This could also cause more polymer to adhere to the FEP substrate at 37 than at 0°C. The adsorption and adhesion of this precipitated polymer to the FEP surface at 37°C is probably driven by hydrophobic intermolecular forces and is not at all simply a 'gravitational deposition' since there is a wash step before the adsorption level is assayed.

Retention behavior of LCST polymer (Poly 64) on FEP

These experiments were planned in order to test the hypothesis that most of the absorbed LCST polymer should remain after a 37°C rinse (which is above its LCST) but should easily be eluted at 0°C (which is below its LCST). The test concentration was chosen to be 100 μg ml^{-1} because the surfaces should be almost completely covered at 37°C at this condition (Fig. 2).

Figure 3 shows the deposition kinetics for rinses at 37 and 0°C. It can be seen that the poly 64 pre-sorbed at 37°C remains on the surface during a 37°C wash-step even after 48 h, while over 80% of the pre-sorbed polymer is eluted within 20 min during a 0°C rinse. The retention curve at 0°C appears to reach equilibrium between 20 min and 1 h.

These surface retention phenomena reasonably reflect the change in polymer solution characteristics from hydrophilic and soluble below the LCST to hydrophobic and insoluble above the LCST.

Temperature dependency of retention of adsorbed LCST polymer

We also studied the temperature dependency of the retention of Poly 64 after adsorption at 37°C. The result is illustrated in Fig. 4. These results are especially noteworthy because the retention curve exhibits a dramatic change near the LCST. Above or below this region, the retention level is constant. This is clearly due to the LCST phase transition behavior of Poly 64.

According to these results, the LCST polymer maintains almost a monolayer on FEP (note the 100% Poly 64 dashed line) when the rinse temperature is higher than

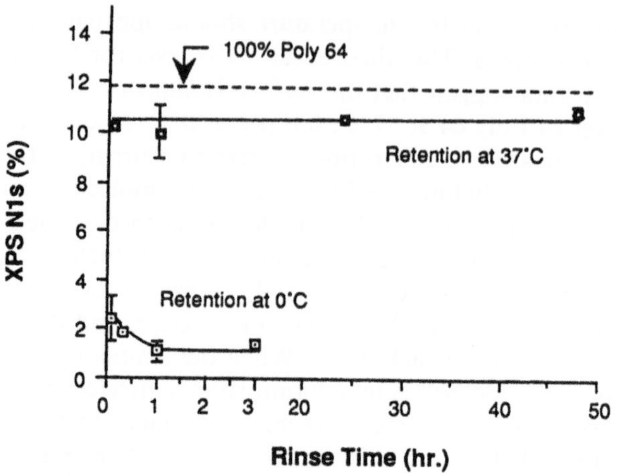
100% Poly 64
Retention at 37°C
Retention at 0°C
XPS N1s (%)
Rinse Time (hr.)

Figure 3. Retention of Poly 64 on FEP after washing in pure water at 0 or 37°C on surfaces which were presorbed by raising the temperature to 37°C.

the LCST. Conversely, if the rinse temperature is below the LCST, most of the adsorbed Poly 64 is desorbed from the surface (Fig. 3 shows the protocol for this experiment). It can be seen that the LCST is the critical temperature for the desorption–retention behavior.

Adsorption and retention of LCST polymer (Poly 64) on various substrates

We are interested in the relationship between LCST polymer composition and its adsorption or retention on a substrate, as a function of the substrate surface characteristics. In this experiment, we have attempted to correlate adsorption and retention data with the water wettability (θ) and solubility parameter (δ) of the substrate. The water contact angle was selected as a measure of substrate hydrophilicity since it should reflect the water–substrate interaction more directly than the surface critical energy of the substrate.

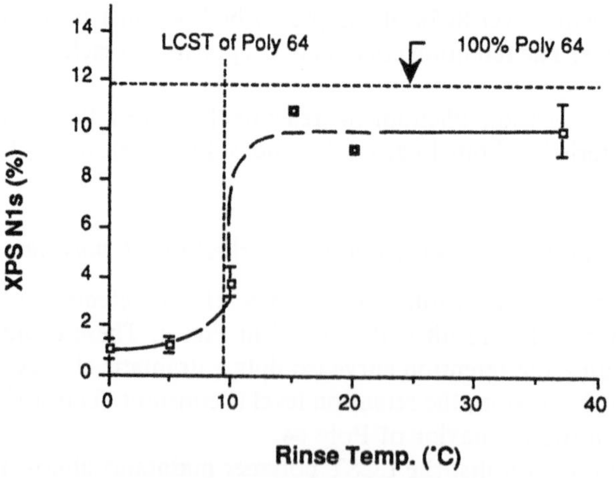
LCST of Poly 64
100% Poly 64
XPS N1s (%)
Rinse Temp. (°C)

Figure 4. Temperature dependency of retention of Poly 64 on FEP.

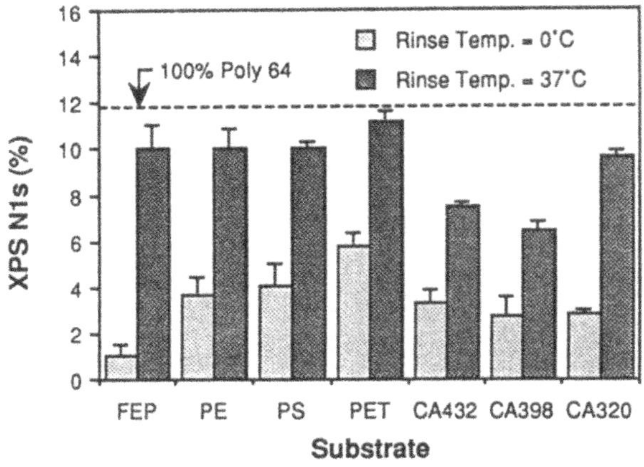
16
14
12
10
8
6
4
2
0
XPS N1s (%)
100% Poly 64
Rinse Temp. = 0°C
Rinse Temp. = 37°C
FEP
PE
PS
PET
CA432
CA398
CA320
Substrate

Figure 5. Retention of Poly 64 on various substrates after adsorption at 37°C.

Figure 5 shows the retention data on different substrates after pre-adsorption at 0°C and rinses at 37°C. The substrates are listed in order of decreasing water contact angles from left to right. The interaction between the LCST polymer and FEP appears to be the weakest of the substrates tested, because FEP has the smallest retention at 0°C. Since our earlier data in Fig. 3 showed that the absorbed polymer was strongly retained on FEP after rinsing with water at 37°C, it is assumed that for all substrates the retention data at 37°C in Fig. 5 are also the same as the adsorption level at 37°C.

The adsorption at 37°C (or retention at 37°C) seems to be independent of the substrate (except perhaps for CA 432 and CA 398 which exhibit lower levels at this condition, the reasons for which are not well understood). However—and perhaps more importantly—the retention at 0°C varies from substrate to substrate. The adsorption of Poly 64 on most of these substrates should be mainly due to hydrophobic interactions. In an aqueous solution, the LCST polymer molecule should be hydrated by bound water molecules and the surfaces of the substrates should be similarly covered with bound water molecules. Then, during the adsorption process, many bound water molecules should be released from the surfaces of both the substrate and the polymer. The greater the hydrophobicity of the substrate, the more water should be bound to its surface. This suggests that since the hydrophobicity of a surface is reflected by the water contact angle on that surface, the water wettability of the substrate may correlate with the results for LCST polymer retention.

Figure 6 plots the adsorption and retention results in Fig. 5 against the water contact angles on the different substrates. It should be emphasized that the water contact angles were measured on the original substrate surfaces, not after the adsorption or rinsing steps.

When the water contact angle is 70 deg and above, the adsorption (37°C rinse) does not change significantly. It can also be seen that the adsorption drops sharply as the contact angle drops to just below 60 deg. This may be due to the increasing polar interaction between water molecules and the CA polymer surfaces, which should reduce the LCST polymer adsorption, since the main driving force for the polymer adsorption is release of bound water followed by hydrophobic (dispersion) interactions.

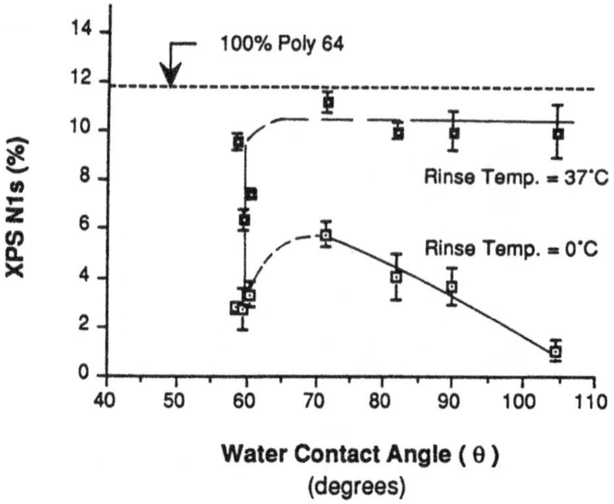
100% Poly 64
Rinse Temp. = 37°C
Rinse Temp. = 0°C
XPS N1s (%)
0
2
4
6
8
10
12
14
40
50
60
70
80
90
100
110
Water Contact Angle (θ)
(degrees)

Figure 6. Retention of Poly 64 vs water contact angle on the original (uncoated) substrate surfaces.

We can also attempt to correlate these data with substrate critical surface tensions [29]. It is well-known that water contact angles agree qualitatively with critical surface tensions. If the critical surface tension of the LCST polymer (as a solid in air) is higher than the critical surface tension of a substrate, their interfacial tensions in water should be the reverse, and therefore the LCST polymer should spread on the substrate surface in order to reduce the interfacial energy between water and the solid polymer surface. Conversely, if the critical surface tension of the LCST polymer (as a solid in air) is lower than the critical surface tension of a substrate (also in air), the polymer should have less tendency to adsorb on it. The contact angle results shown in Fig. 6 suggest that the critical surface tension of Poly 64 is between that of PET (*c.* 40 dyn cm^{-1}) and those of the CAs (not available). It is reasonable (based on thermodynamic arguments) that the adsorption of Poly 64 at 37°C (and thus the retention at 37°C) should decrease with decreasing substrate water contact angle, especially when the surface becomes hydrated and the aqueous interfacial energy drops. The sharp drop, however, is both interesting and unusual.

In comparison to the 37°C data, the retention data at 0°C in Fig. 6 show more complicated behavior. PET, whose water contact angle is 71 deg, shows the highest retention of all substrates after a 0°C rinse. The retention of Poly 64 decreases on substrates with water contact angles above or below *c.* 70 deg. The 0°C retention data most likely reflect all of the possible H-bonding, polar and (mainly) dispersive interactions between the Poly 64 and the different substrates since the Poly 64 should be fully hydrated in a random coil conformation under these conditions. Apparently, PET has a similar balance of polar and (mainly) dispersive groups so as to maximize the adsorption of Poly 64 at 0°C.

Solubility parameters (δ) are commonly used to characterize intermolecular interactions. For example, there is a correlation between solubility parameters and (critical) surface tensions [30, 31]. It predicts that a substrate whose critical surface tension is low should also have a low solubility parameter. Therefore, it is possible that LCST polymer retention can be correlated with solubility parameters.

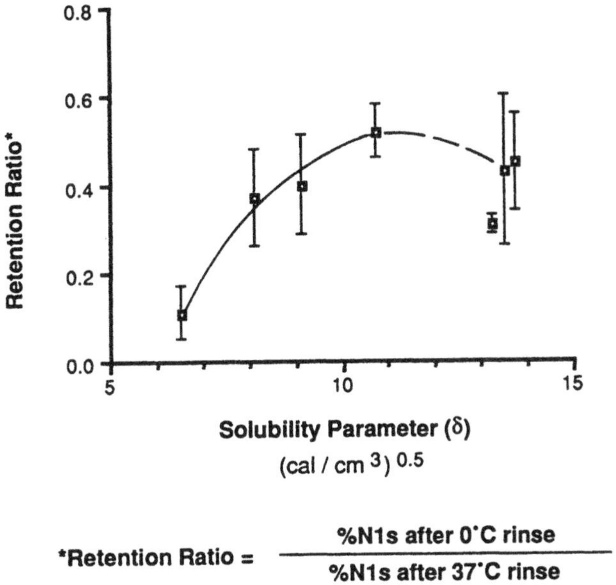
0.8
0.6
0.4
0.2
0.0
Retention Ratio*
5
10
15
Solubility Parameter (δ)
(cal / cm 3) 0.5
*Retention Ratio =
%N1s after 0°C rinse
%N1s after 37°C rinse

Figure 7. Poly 64 retention ratio vs solubility parameters of the original substrate polymers.

On the other hand, since no assumptions are made about water interactions in deriving solubility parameters, and since water molecules play such an important role in the absorption and retention process, it may be erroneous to correlate the adsorption and retention phenomena with the solubility parameter. Nevetheless, since the LCST polymer molecules are already adsorbed in the 37°C retention process, and since there should not be many water molecules at the contact sites because the main driving force for the adsorption is release of bound water, it is reasonable to attempt to correlate the retention of Poly 64 with the solubility parameters.

Figure 7 shows the LCST polymer retention ratios as a function of the solubility parameter. The retention ratio is the ratio of N1*s* content after a 0°C rinse to that after a 37°C rinse. The retention ratio reflects the interaction strength between the polymer and the substrates more accurately than the retention at 0°C, as the latter concerns only the fraction of the adsorbed polymer molecules which bind strongly to the surface at 0°C. The correlation in Fig. 7 shows a possible maximum at a solubility parameter value of around 11. Since the solubility parameters of poly(NIPAAm) and poly(NnBAAm) are 11.18 [32] and 10.32 [32] respectively, Small's method [27] suggests that the solubility parameter of Poly 64 be about 10.8. Therefore, the maximum value agrees fairly closely with the solubility parameter of Poly 64. This is probably due to the optimum match of polar and hydrophobic forces between the adsorbed polymer and the substrate. When the substrate solubility parameter is higher than 11, the interaction between the polymer and the substrate decreases as water–substrate interactions are favored. When δ for the substrate drops below 11, the retention ratio decreases because Poly 64-water interactions may be more favored than Poly 64-substrate interactions, despite the increase in hydrophobically bound water at the substrate surface.

According to our reasoning from these correlations, the critical surface tension influences the extent of contact of the LCST polymer with a substrate surface and the solubility parameter influences the binding strength between them.

Retention of three LCST polymers and PAAm on FEP

In the experiment, we studied the influence of the LCST polymer composition on the retention. The adsorption and retention results at 37°C and 0°C are summarized in Figs 8 and 9, respectively. In these figures, the *x*-axis shows the percent of the theoretical 0°C N1*s* content determined on the FEP substrate after 0 or 37°C, for each LCST polymer.

In contrast with LCST polymers, PAAm is not retained on FEP under all conditions. PAAm is too hydrophilic and water soluble at either 37 or 0°C. Therefore, it is probable that the adsorption of PAAm occurs only weakly during an adsorption process at 37 or 0°C. Even if some adsorption occurs, PAAm should not be retained after either a 37 or 0°C rinse.

It can be seen from Figs 8 and 9 that the retentions of LCST polymers are all higher after a 37°C rinse than after a 0°C rinse, which is logical since all of the polymers have LCSTs between 0 and 37°C. The retention after a 37°C rinse increases with an increase in *n*-butyl content (relative to isopropyl) in both cases (Figs 8 and 9). However, the increases seen in Fig. 8 are more remarkable. This may be due to the balance between polymer–polymer interactions and polymer–substrate interactions. When the temperature is below the LCST, the polar and hydrophobic interactions between the LCST polymer molecules should be relatively weak since the polymer molecules should exist in aqueous solution as solvated coils [25].

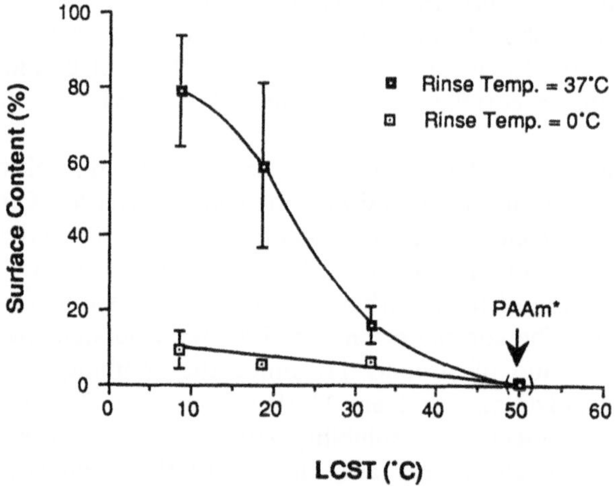
Surface Content (%)
100
80
60
40
20
0
0
10
20
30
40
50
60
LCST (°C)
Rinse Temp. = 37°C
Rinse Temp. = 0°C
PAAm*

* PAAm does not exhibit any LCST in aqueous solution.

$$\text{Surface Content (\%)} = \frac{\text{\%N1s on FEP after a rinse}}{\text{Theoretical \%N1s of each polymer}} \times 100$$

Figure 8. Retention of water soluble polymers on FEP after a 37°C adsorption.

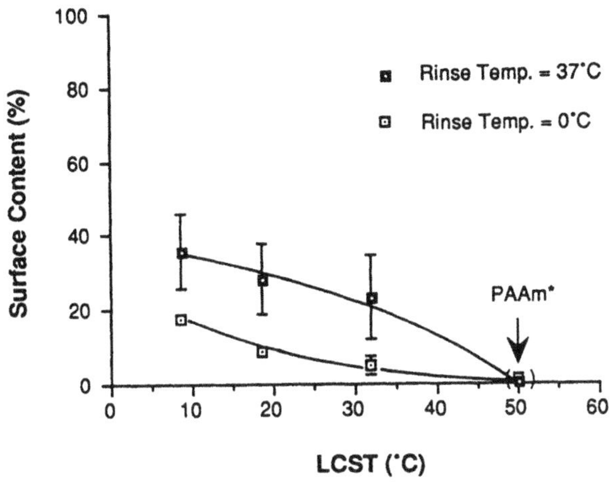
100
80
60
40
20
0
Surface Content (%)
Rinse Temp. = 37°C
Rinse Temp. = 0°C
PAAm*
0
10
20
30
40
50
60
LCST (°C)

* PAAm does not exhibit any LCST in aqueous solution.

$$\text{Surface Content (\%)} = \frac{\text{\%N1s on FEP after a rinse}}{\text{Theoretical \%N1s of each polymer}} \times 100$$

Figure 9. Retention of water soluble polymers on FEP after a 0°C adsorption.

Conversely, when the temperature is above the LCST, the polymer molecules first collapse individually (in dilute solution) and then aggregate due to hydrophobic interactions. Therefore, at a temperature above the LCST, the aggregation will compete with the LCST polymer adsorption on FEP, as described above.

CONCLUSIONS

We have shown that an LCST polymer remains strongly bound to a substrate at a temperature above its LCST, but below the LCST most of the adsorbed polymer molecules are easily rinsed off. We have correlated the unique retention behavior of the LCST polymer with the critical surace tension and solubility parameter of the substrate.

These phenomena may be used to modify a solid polymer surface by a temperature-sensitive adsorption process. For example, one can conjugate a bioactive molecule such as a recognition peptide or an antibody to an LCST polymer, and then adsorb the polymer–biomolecule conjugate on a solid polymer surface above its LCST. In this way the biomolecule can be gently immobilized so as to retain a higher fraction of its activity. We have already done this for IgG and IgM [13, 14]. We have also conjugated a cell recognition peptide such as RGDS (H–Arg–Gly–Asp–Ser–OH) to an LCST polymer, and then adsorbed (above the LCST) the RGDS–LCST polymer conjugate on a solid polymer surface [34]. Anchor-dependent cells such as endothelial cells can then be grown on the surface, and after cultivating the cells, it should be possible to detach those cells from the surface without using proteolytic enzymes, by simply reducing temperature below the LCST [35]. We will publish these studies in detail in a subsequent publication.

Acknowledgements

We gratefully acknowledge the generous use of the National ESCA and Surface Analysis Center for Biomedical Problems (NESAC/BIO) which is supported by the Division of Research Resources, National Institutes of Health, under grant number RR01296. We also thank Dr. David G. Castner, the NESAC/BIO laboratory coordinator and Deborah K. Leach-Scampavia, and Michelle Radeke for their help in many XPS surface analyses. This project was finally supported by Asahi Chemical Industry Co., Ltd. and the Washington Technology Center.

REFERENCES

1. P. Molyneux, *Water-Soluble Synthetic Polymers: Properties and Behavior*, Vol. 1. CRC, Boca Raton, FL (1985).
2. M. Heskins and J. E. Guillet, *J. Macromol. Sci. Chem.* **A2**(8), 1441 (1968).
3. C. K. Chiklis and J. M. Grasshoff, *J. Polym. Sci., Part A-2* **8**(9), 1617 (1970).
4. L. D. Taylor and L. D. Cerankowski, *J. Polym. Sci., Polym. Chem. Ed.* **13**(11), 2551 (1975).
5. J. H. Priest, S. L. Murray, R. J. Nelson and A. S. Hoffman, *ACS Symposium Series* No. 350, 255 (1987).
6. Y. Hirokawa and T. Tanaka, *J. Chem. Phys.* **81**, 6379 (1984).
7. W. C. Wooten, R. B. Blanton and H. W. Coover Jr. *J. Polym. Sci.* **25**, 403 (1957).
8. A. S. Hoffman, A. Afrassiabi and L. C. Dong, *J. Contr. Release* **4**, 213 (1986).
9. A. S. Hoffman, *J. Contr. Rel.* **6**, 297 (1987).
10. Y. H. Bae, T. Okano, R. Hsu and S. W. Kim, *Makromol. Chem. Rapid Commun.* **8**, 481 (1987).
11. N. Monji and A. S. Hoffman, *Appl. Biochem. Biotechnol* **14**, 107 (1987).
12. N. Monji, A. S. Hoffman and J. H. Priest, US Patent 4,780,409 (1988).
13. N. Monji, C.-A. Cole, M. R. Tam, L. Goldstein, A. S. Hoffman and R. C. Nowinski, *Transactions of the 3rd World Biomaterials Congress*, Vol. XI, p. 298, Kyoto, Japan (1988).
14. C.-A. Cole, N. Monji, M. R. Tam, L. Goldstein, A. S. Hoffman and R. C. Nowinski, Transactions of the 3rd World Biomaterials Congress, Vol. XI, p. 299, Kyoto, Japan (1988).
15. C.-A. Cole, S. M. Schreiner, J. H. Priest, N. Monji and A. S. Hoffman, *ACS Symposium Series*, No. 350, 245 (1987).
16. M. Charles, R. W. Coughlin and F. X. Hasselberger, *Biotechnol. Bioeng.* **16**, 1553 (1974).
17. E. Van Leemputten and M. Horisberger, *Biotechnol. Bioeng.* **18**, 587 (1976).
18. K. Okumura, K. Ikura, M. Yoshikawa, R. Sasaki and H. Chiba, *Agric. Biol. Chem.* **48**, 2435 (1984).
19. D. L. Marshall, US Patent 4,530,900 (1985).
20. M. Fujimura, M. T. Mori and T. Tose, *Biotechnol. Bioeng.* **29**, 747 (1987).
21. M. Taniguchi, M. Kobayashi and M. Fujii, *Biotechnol. Bioeng.* **34**, 1092 (1989).
22. R. F. S. Freitas and E. L. Cussler, *Chem. Eng. Sci.* **42**, 978 (1987).
23. L. C. Dong and A. S. Hoffman, *J. Contrs. Release* **4**, 223 (1986).
24. H. J. Yang, C.-A. Cole, N. Monji and A. S. Hoffman, *J. Polym. Sci. A: Polym. Chem.* **28**, 219 (1990).
25. M. K. Bernett and W. A. Zisman, *J. Phys. Chem.* **64**, 1292 (1960).
26. M. Toyama, T. Ito and H. Moriguchi, *J. Appl. Polym. Sci.* **14**, 2039 (1970).
27. P. A. Small, *J. Appl. Chem.* **3**, 71 (1953).
28. A. G. Shvarts, *Colloid J.* **18**, 753 (1956).
29. H. W. Fox and W. A. Zisman, *J. Colloid Sci.* **5**, 514 (1950).
30. J. Hildebrand and R. Scott, *The Solubility of Non-Electrolytes*, 3rd Ed. Reinhold Publishing Corp., New York (1949).
31. L.-H. Lee, *J. Paint Technol.* **42**, 365 (1970).
32. H. Ahmad, *J. Macromol. Sci. Chem.* **A17**(4), 585 (1982).
33. S. Fujishige, K. Kubota and I. Ando, *J. Phys. Chem.* **93**, 3311 (1989).
34. M. Miura, C.-A. Cole, N. Monji and A. S. Hoffman, *Proceedings of the 17th Annual Meeting of the Society for Biomaterials*, p. 130, Scottsdale, AZ (1991).
35. T. Takezawa, Y. Mori and K. Yoshizato, *Bio/Technol.* **8**, 853 (1990).

A polymeric drug delivery system for the simultaneous delivery of drugs activatable by enzymes and/or light

N. L. KRINICK[1], Y. SUN[2], D. JOYNER[2], J. D. SPIKES[3], R. C. STRAIGHT[2] and J. KOPEČEK[1,4]*

[1]*Department of Bioengineering,* [2]*Dixon Utah Laser Institute and Department of Veteran Affairs Research,* [3]*Department of Biology,* [4]*Department of Pharmaceutics and Pharmaceutical Chemistry/CCCD, University of Utah, Salt Lake City, UT 84112, USA*

Received 15 September 1992; accepted 12 January 1993

Abstract—Three water soluble copolymers based on *N*-(2-hydroxypropyl)methacrylamide were prepared. Copolymer I contains adriamycin, a chemotherapeutic agent, attached via enzymatically degradable oligopeptide (glycylphenylalanylleucylglycine; G–F–L–G) side chains. The other two copolymers contained the photosensitizer, meso-chlorin e_6 monoethylene diamine disodium salt (Mce_6). In Copolymer II, the chlorin is attached via the degradable G–F–L–G sequence, and it was bound by the nondegradable glycyl spacer in Copolymer III.

Initially, the copolymers were characterized separately *in vitro* and *in vivo*. Combinations of the copolymer bound chemotherapeutic agent and each of the copolymer bound photosensitizers were then assessed for antitumor effect *in vivo*. Localization/retention studies (A/J mice; Neuro 2A neuroblastoma solid tumor) were performed with the two copolymers containing Mce_6 as well as the free drug. Results of these experiments demonstrated a very different tumor uptake profile for the two copolymers. While the free drug was rapidly cleared from tumor tissue, the copolymer containing Mce_6 attached via the nondegradable bond was retained for an extended period; drug concentrations in the tumor were high even after 5 days. On the other hand, a high concentration of the copolymer containing Mce_6 bound via the degradable sequence was taken up by the tumor, yet its concentration in the tumor was substantially diminished at 48 h after administration. This shows indirect evidence of *in vivo* cleavage of Mce_6 from the copolymer in the lysosomal compartment which is supported by direct evidence of cleavage by cathepsin B (a lysosomal enzyme) *in vitro*.

Antitumor effects were assessed on Neuro 2A neuroblastoma induced in A/J mice for all three copolymers. Photodynamic therapy (PDT) proved the copolymer with Mce_6 bound via the degradable oligopeptide sequence to be a more effective photosensitizer *in vivo* than the other chlorin containing copolymer. The difference in activity was consistent with the results obtained by photophysical analyses in which the free drug had a higher quantum yield of singlet oxygen generation than the polymer bound drug in buffer. The quantum yield of singlet oxygen generation increased with the enzymatic cleavage of the chlorin from the copolymer.

Conditions were subsequently determined for which chemotherapy or PDT would show some antitumor effect, yet be incapable of curing tumors. Finally, combination therapy experiments were performed in which the copolymer bound adriamycin was mixed with either of the copolymer bound chlorin compounds and injected intravenously (*i.v.*) into the tail veins of mice. A time lag was allowed for optimal uptake in the tumor and for the adriamycin to begin to take effect, after which light (650 nm) was applied to activate the photosensitizer. Tumor cures were obtained with the combination therapy that could not be achieved by either chemotherapy or PDT alone.

Key words: N-(2-Hydroxypropyl)methacrylamide copolymer; photodynamic therapy; photosensitizer; chemotherapy; meso chlorin e_6 monoethylene diamine disodium salt; adriamycin; neuroblastoma.

*To whom correspondence should be addressed at Department of Bioengineering, 2480 MEB, University of Utah, Salt Lake City, UT 84112, USA.

ABBREVIATIONS

Adria:	adriamycin
AIBN:	2,2′-azobisisobutyronitrile
CTAB:	cetyltrimethylammonium bromide
DMSO:	dimethyl sulfoxide
F:	L-phenylalanine
FPLC:	fast protein liquid chromatography
G:	glycine
HPMA:	*N*-(2-hydroxypropyl)methacrylamide
i.v.:	intravenous
i.p.:	intraperitoneal
L:	L-leucine
MA:	methacryloyl
Mce_6:	meso-chlorin e_6 monoethylene diamine disodium salt
ONp:	*p*-nitrophenoxy
P:	polymer backbone
PBS:	phosphate buffered saline
THF:	tetrahydrofuran
TRIS:	2-amino-2-hydroxymethyl-1,3-propanediol

INTRODUCTION

Low molecular weight chemotherapeutic agents and photosensitizers are being studied for use in cancer therapy. There are, however, side effects associated with both chemotherapy and photodynamic therapy (PDT). By binding low molecular weight drugs to macromolecular carriers, their side effects may be reduced, and target tissue localization may be enhanced. Synthetic copolymers such as *N*-(2-hydroxypropyl)methacrylamide (HPMA) copolymers are suitable drug carriers [1] because they are biocompatible [2] and can be tailored for specific applications [3, 4].

Adriamycin, a chemotherapeutic agent, shows cumulative dose-dependent cardiomyopathy which limits its long-term use [5, 6]. Binding chemotherapeutic agents to HPMA copolymers reduces anthracycline-related toxicity [7–9]. HPMA copolymers containing anthracycline antibiotics are active only if the drug can be released from the copolymer by the enzymatic degradation of its side chains [10] in the lysosomal compartment so that it can be free to travel to the nucleus, its cellular site of action. There, intercalation with DNA takes place resulting in cell death. On the other hand, a photosensitizer may be active whether or not it is bound to an HPMA copolymer via a degradable bond [11, 12]. However, recent studies indicate that the solution properties of a copolymer bound sensitizer largely influence its photodynamic effects [13].

Photosensitizers are activated to their first excited singlet state by absorbing light of a characteristic wavelength. The first excited singlet state of most photodynamic sensitizers is very short lived and rapidly decays. One decay process called intersystem crossing forms the corresponding excited triplet state. Triplet state species can efficiently give up their energy to molecular oxygen in the environment forming singlet excited oxygen. Singlet oxygen is highly reactive and will oxidize many biomolecules and destroy cell structure and function [14, 15]. PDT has been extensively investigated as a means for tumor treatment. Currently, low molecular weight photosensitizers are being used in clinical trials [16]. Skin photosensitivity, however, remains a problem in which patients injected with low molecular weight photosensitizers must stay in dim light and/or cover their skin for at least 30 days following therapy [17]. Attempts have been made to improve the selectivity of photosensitizers [18–20] by incorporating them into antibody-polymer conjugates with promising results.

In this study, we have investigated the possibility of polymeric combination therapy in which an HPMA copolymer containing a chemotherapeutic agent (adriamycin) and an HPMA copolymer containing a photosensitizer [meso-chlorin e_6 monoethylene diamine disodium salt (Mce_6)] were mixed and administered simultaneously. The rationale behind combination therapy is to achieve a synergistic effect by the combined action of both drugs and to reduce their dose-limiting side effects. Theoretically, lower doses of both drugs are then possible.

MATERIALS AND METHODS

Monomers and chemicals

N-(2-hydroxypropyl)methacrylamide (HPMA) [21], *N*-methacryloylglycine *p*-nitrophenyl ester (MA-G-ONp) [22], and *N*-methacryloylglycylphenylalanylleucylglycine *p*-nitrophenyl ester (MA-G-F-L-G-ONp) [23] were prepared as previously described. AIBN was recrystallized from methanol. Mce_6 was from Porphyrin Products, Logan, UT. Solvents were freshly distilled.

Synthesis of polymers

HPMA copolymer–drug conjugates were synthesized using a two step procedure. First, polymeric precursors containing side-chains terminated in reactive ONp groups were prepared, followed in the second step by binding of the respective drug by aminolysis [22–24].

Preparation of polymer precursors

P-G-ONp. P-G-ONp [25] (Precursor 1) was prepared by radical precipitation copolymerization in acetone of HPMA (2.26 g, 85 mol %) and MA-G-ONp (0.74 g, 15 mol %) using 0.144 g 2,2′-azobisisobutyronitrile (AIBN) with weight percentages of 12.5% monomers, 86.9% acetone, and 0.6% AIBN. The monomers plus AIBN were dissolved in the acetone, filtered, transferred to an ampule, and bubbled with N_2. The ampule was sealed and the mixture polymerized at 50°C for 48 h. The polymer was filtered, washed with acetone and dry ether, and desiccated. The polymer was dissolved in methanol and reprecipitated into acetone, washed with acetone and ether and desiccated. The yield of purified product was 1.52 g. The content of ONp groups determined by spectroscopy ($\varepsilon_{274} = 0.95 \times 10^4$ 1/mol · cm in DMSO) was 10.6 mol %. The weight average molecular weight (17 000) and polydispersity (1.5) were determined after aminolysis with 1-amino-2-propanol by FPLC analysis on a Superose 12 column (10 × 30 cm) calibrated with fractions of polyHPMA (buffer 0.5 M NaCl + 0.05 M TRIS; pH 8).

P-G-F-L-G-ONp. P-G-F-L-G-ONp (Precursor 2) was prepared similarly to Precursor 1 by copolymerization of HPMA and MA-G-F-L-G-ONp (mole ratio 24:1). The copolymer contained 3.7 mol % of ONp groups. $M_w = 21\,000$; $M_w/M_n = 1.6$.

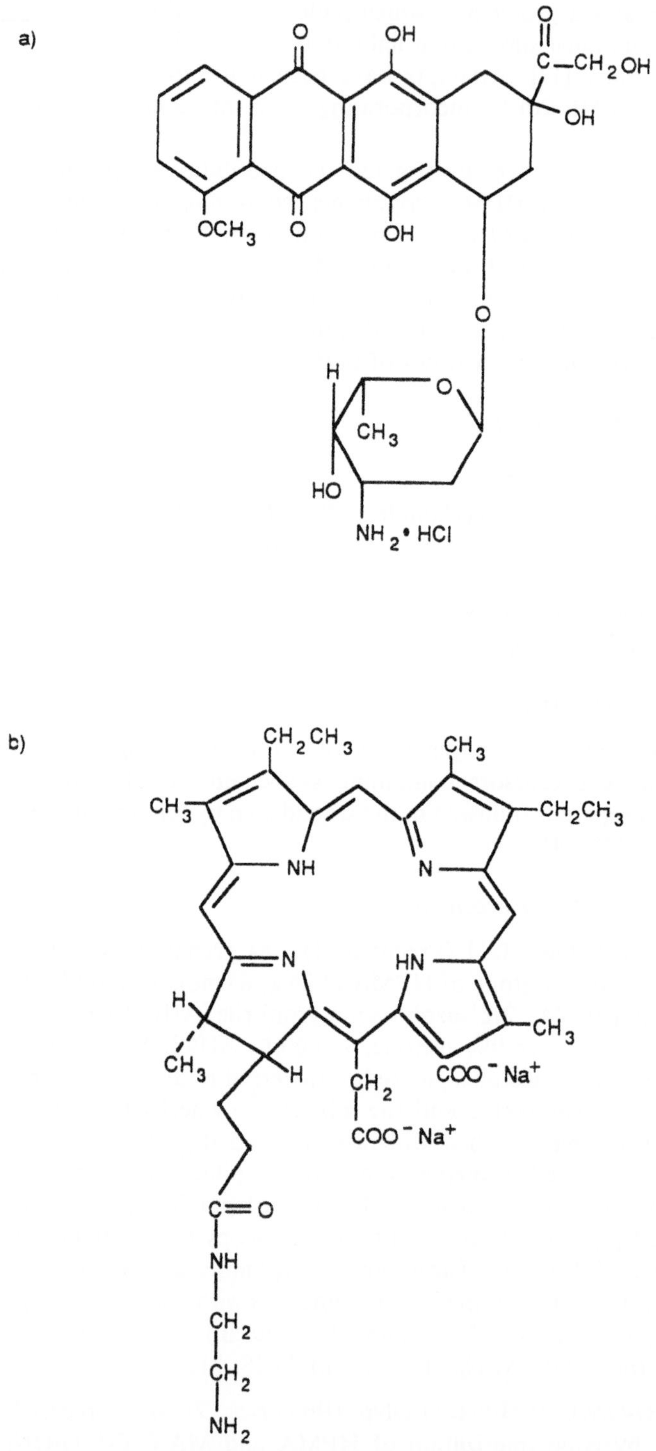
a)
O
OH
O
C—CH2OH
OH
OCH3
O
OH
O
H
O
CH3
HO
NH2• HCl
b)
CH2CH3
CH3
CH3
CH2CH3
NH
N
N
HN
H
CH3
H
CH3
CH2
COO⁻ Na⁺
COO⁻ Na⁺
C=O
NH
CH2
CH2
NH2

Figure 1. Structures of drugs: (a) adriamycin · HCl; and (b) meso-chlorin e_6 monoethylene diamine disodium salt.

Preparation of copolymer bound drugs

The structures of the anticancer drugs, adriamycin and Mce_6 can be found in Fig. 1. The drugs were bound to the copolymer precursors by aminolysis reactions of the aliphatic amino groups of the drugs with active ester functionalities of the copolymer side chains. The structures of Copolymers I, II, and III are shown in Fig. 2 and the synthesis is demonstrated below using Copolymer II as an example.

P-G-F-L-G-adria (Copolymer 1) was synthesized from Precursor 2 as previously described [23, 24]. The adriamycin content ($\varepsilon_{488} = 1.19 \times 10^4$ 1/mol·cm in water) was 7.4 wt %.

P-G-F-L-G-Mce₆ (Copolymer II). P-G-F-L-G-ONp (Precursor 2) (200 mg, 3.7 mol % ONp) was dissolved in 0.75 ml DMSO. Mce_6 (32.9 mg) in a 1.25 times molar excess was dissolved in 0.15 ml DMSO. The Mce_6 solution was added to the polymer mixture (an additional 0.2 ml DMSO added for washing) and stirred for 4 h

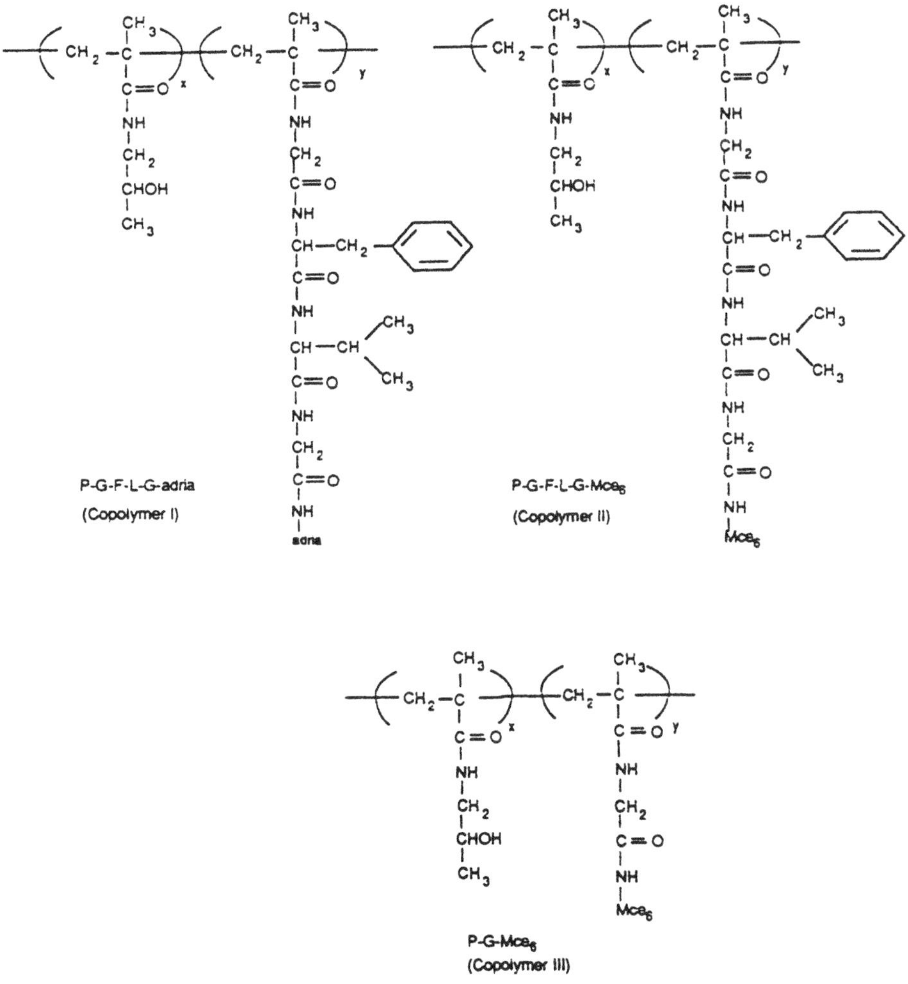
P-G-F-L-G-adria
(Copolymer I)
adria
P-G-F-L-G-Mce6
(Copolymer II)
Mce6
P-G-Mce6
(Copolymer III)

Figure 2. Structures of copolymers used in this study.

at room temperature. 1-amino-2-propanol (6.4 ml) in 3 times excess of the theoretical remaining ONp groups was added and the mixture stirred for 5 additional min. The copolymer was precipitated into a 3:1 acetone:ether mixture, filtered, washed with acetone and ether, and desiccated. The copolymer was then dissolved in 5 ml methanol and applied to an LH-20 column (55 × 3 cm). The copolymer band was collected, evaporated to dryness, dissolved in deionized water, frozen, and lyophilized. The yield of pure product was 168 mg. Copolymer II contained 11.2 wt % Mce_6 as determined spectrophotometrically ($\varepsilon_{394} = 1.58 \times 10^5$ 1/mol · cm in methanol).

P-G-Mce_6 (Copolymer III) was synthesized by aminolysis of Precursor 1 with Mce_6 as previously described [25]. The chlorin content determined spectrophotometrically ($\varepsilon_{394} = 1.58 \times 10^5$ 1/mol · cm in methanol) was 11.2 wt % (2.6 mol %).

Biological testing

Cleavage of Mce_6 from Copolymer II (P-G-F-L-G-Mce_6) with Cathepsin B. Cathepsin B (EC 3.4.22.1) [26], a lysosomal cysteine protease isolated from bovine spleen, was a gift from Dr. M. Baudyš. The concentrations of the enzyme in stock solutions were determined spectrophotometrically using $\varepsilon_{281} = 5.15 \times 10^4$ 1/mol · cm (0.09 M phosphate buffer, pH 6). The activity of the enzyme (0.47 U/mg) was determined using N^{α}-benzoyl-L-arginine *p*-nitroanilide as substrate [27].

Stock solutions of Copolymer II (1.9 mg/ml in 0.09 M phosphate buffer containing 1 mM EDTA; pH 6), enzyme (2.12 mg/ml phosphate buffer), and glutathione (15.36 mg/ml) were prepared. The enzyme stock solution (0.25 ml) plus 0.4 ml more buffer were bubbled with N_2 on ice for 5 min. Glutathione (0.1 ml) was added and the solution was preincubated for 5 min at 37°C to activate the enzyme. Stock solution of copolymer (0.25 ml) was added and the sample was flushed with N_2 and sealed. The final reaction concentrations were: 0.53 mg/ml (1.9×10^{-5} M) of cathepsin B; 5 mM glutathione; 1 mM EDTA; 54.8 μM Mce_6. Six samples were prepared this way and incubated at 37°C in the dark for 4, 8, 12, 24, 49, and 120 h. The reaction mixture at each time interval (0.95 ml plus 1.55 ml water) was applied to a PD-10 column (Sephadex G-25; Pharmacia), equilibrated with water, and 1 ml fractions collected. The P-G-F-L-G-Mce_6 eluted in fractions 1–4 and the free drug in fractions 7–10. One mililiter of 1 *N* NaOH was added after fraction 10 to release free Mce_6 nonspecifically bound to the column. One half milliliter of each fraction was placed into a cuvette and 50 μl of 10% Triton X-100 added. The absorbance at 398 nm was recorded and the percent of cleaved Mce_6 calculated for each sample. After 120 h of incubation, the cathepsin B retained 68% of initial activity as determined using a low molecular weight substrate, N^{α}-benzoyl-L-arginine *p*-nitroanilide.

Quantum yield of oxygen uptake of Copolymer II (P-G-F-L-G-Mce_6) after cleavage with cathepsin B. Three samples of Copolymer II were prepared for enzymatic cleavage. The same stock solution concentrations and reaction mixtures were prepared as in the cleavage experiments described above. For two samples, stock solutions were mixed and the reactions proceeded at 37°C in the dark for 2 days and for 1 week, respectively. The third sample was used as a control. That is, the enzyme and substrate were incubated separately for 2 days and mixed immediately prior to photophysical analysis.

The photochemical quantum yield of oxygen uptake was determined from measurements of a decrease in oxygen concentration (oxygen uptake) with a recording oxygen electrode system. The quantum yield was calculated as the ratio (initial rate of uptake of oxygen molecules)/(initial rate of absorption of photons). The reaction mixtures contained the photosensitizer and furfuryl alcohol as a photooxidizable substrate. Furfuryl alcohol was chosen because it reacts chemically with singlet oxygen with good efficiency (rate constant $1.2 \times 10^8\,\mathrm{l\,M^{-1}}$). In addition, it does not react with hydrogen peroxide or superoxide and most likely does not undergo radical initiated autooxidation [28, 29]. Reaction mixtures were illuminated with a 500 W slide projector provided with a 407 nm interference filter (bandwidth 10 ± 2 nm at 50% peak transmittance; from Corion Corp., Holliston, MA) and the decrease in oxygen concentration was recorded as a function of illumination time. Incident light intensity was measured with a vacuum thermocouple-millimicrovoltmeter calibrated with standard lamps. Incident light irradiance was approximately $2\,\mathrm{mW/cm^2}$. The fraction of light absorbed was determined with a silicon photodiode photometer.

Two hundred fifty microliters of the sample were diluted with 3.75 ml of phosphate buffer (pH 6) and a furfuryl alcohol saturating concentration of 100 mM was added. The solution was air saturated, equipped with an oxygen electrode and the sample was illuminated. The quantum yield of oxygen uptake of the sample was calculated from the quantum yield of singlet oxygen using previously determined values of rose bengal as a standard: 0.375 (quantum yield of oxygen uptake) and 0.75 (quantum yield of singlet oxygen generation) [30]. Errors in quantum yield measurements were ±5–10%.

In vivo *studies*

The Neuro 2A neuroblastoma cell line was used in these studies. It forms a solid, difficult to cure tumor in A/J mice. A/J mice (5–6 weeks old) were routinely injected with approximately 1.5×10^6 viable Neuro 2A neuroblastoma tumor cells in the right costal margin. When tumors became palpable treatment was initiated. Treatment and control groups consisted of five mice per group. The drugs were dissolved in phosphate buffered saline (PBS) or bacteriostatic saline and injected intravenously (*i.v.*) into the tail vein of the mice. Drug doses expressed in mg/kg drug were calculated based on the weight percent of the drug bound to the copolymers.

For the controls and treatment with the chemotherapeutic agent, tumor volumes were followed by measuring the length, width, and height of the tumors with calipers after drug administration. For the photodynamic therapy treatments, an optimal time lag after drug administration was determined from localization/retention experiments (see below) after which light of 650 nm (argon dye laser; Cooper LaserSonics, Inc., Santa Clara, CA) was applied. Several experiments were performed to determine suitable light and drug doses. In the combination chemotherapy and PDT experiments, the drugs were dissolved together and injected into the tail vein. A time lag was allowed for the adriamycin to take effect and for optimal photosensitizer uptake before light was applied.

Chemotherapy: treatment with P-G-F-L-G-adria (Copolymer I). Copolymer bound adriamycin (4.1, 8.2, and 16.4 mg/kg) dissolved in bacteriostatic saline was injected into the tail vein of tumor bearing animals. The treatment day was considered to be day zero and tumor volumes were followed until tumor burdens were indistinguishable from controls. Survivors were followed until day 55 after treatment.

Localization/retention of free and HPMA copolymer bound Mce_6. Localization/retention experiments were performed to compare the body distribution of free Mce_6 with HPMA copolymer bound Mce_6, i.e. with P-G-Mce_6 (Copolymer III) and with P-G-F-L-G-Mce_6 (Copolymer II). Five mg/kg of the respective drug in solution [dissolved in PBS; pH 7.4; the polymeric drug was more soluble in PBS than the free drug. To get the free drug into solution, either the pH had to be raised, or the solution was heated in the dark and cooled to room temperature before injection] was immediately injected into the tail vein of animals with palpable tumors. Animals were sacrificed and tissue samples (tumor, skin, spleen, leg muscle, kidney, abdominal muscle, and liver) removed at various time intervals after injection. Samples were frozen and lyophilized for 2 days. The dried samples were weighed. One ml of water/25 mg dried sample was added. The tissue was then mechanically homogenized and 100 μl of the homogenate transferred to a hydrolysis tube. Fifty percent methylbenzethonium hydroxide in methanol (1 ml) was added to each tube and the tubes were evacuated. The samples were hydrolyzed in a heating block for 1 h at 55°C. After cooling, 2 ml of 50% THF in water was added. After mixing, fluorescence was read (excitation 397 nm, emission 654 nm) and compared to a standard curve for the calculation of Mce_6 concentration in the samples.

Photodynamic therapy. Several PDT experiments were performed to compare the effects of P-G-Mce_6 (Copolymer III) and P-G-F-L-G-Mce_6 (Copolymer II). Different drug and light doses were studied to obtain optimal treatment regimens. Various concentrations of drug were dissolved in PBS and injected into the tail vein of the mice. After a certain time lag for uptake mice were anesthesized with sodium pentobarbitol (stock solution 6.48 mg/ml; 0.013 ml stock solution/g body weight injected) and light (650 nm) (argon-dye laser) applied to the tumors for various time periods.

Combined chemotherapy and PDT. In these experiments either P-G-Mce_6 (Copolymer III) and P-G-F-L-G-adria (Copolymer I) or P-G-F-L-G-Mce_6 (Copolymer II) and P-G-F-L-G-adria (Copolymer I) were dissolved together in PBS and injected *i.v.* into the tail veins of mice with palpable tumors. A 2 day time lag was allowed before light administration (argon-dye laser; 500 mW/cm^2; 5 min) for the combination of Copolymers III + I, whereas 1 or 2 days were allowed for the combination of Copolymers II + I (500 mW/cm^2; 10 min) due to the differences in the localization/retention behaviour of the Copolymers II and III containing the photosensitizer (see results). The day of drug injection was considered to be day zero, the day of treatment.

RESULTS

The structures of all copolymers used in this study: P-G-F-L-G-adria (Copolymer I), P-G-F-L-G-Mce_6 (Copolymer II), P-G-Mce_6 (Copolymer III), can be found in

Table 1.
Physical properties of copolymers

Copolymer	Abbreviated structure of copolymers	M_w/polydispersity of precursor copolymer	Mol% ONp in precursor copolymer	Drug content (wt%)
I	P-G-F-L-G-adria	21,000/1.6	3.7	7.4
II	P-G-F-L-G-Mce_6	21,000/1.6	3.7	11.2
III	P-G-Mce_6	17,000/1.5	10.6	11.2

Fig. 2 and their physical properties in Table 1. It is important to note that purification of conjugates on a column is necessary in order to remove residues of free drugs. Precipitation into a medium which can dissolve the free drug is not sufficient since part of the drug remains associated with the precipitated copolymer.

Enzymatic cleavage of Copolymer II with cathepsin B

The results of the cleavage experiments are shown in Fig. 3. Figure 3 (a) is a graph of percent of Mce_6 cleaved from Copolymer II (P-G-F-L-G-Mce_6) vs time. Figure 3 (b) shows chromatographically the time dependent cleavage. It can be seen that the absorbance for fractions 7–10 (corresponding with the free Mce_6) increases with a concomittant decrease in the absorbance of fractions 1–4 (corresponding with the amount of the decrease in P-G-F-L-G-Mce_6) over time. The exact values of the decrease in concentration of P-G-F-L-G-Mce_6 and the increase in free Mce_6 were calculated and correlated. The recovery of material was calculated for each sample and was within a few percent of 100% for all samples. Copolymer III (P-G-Mce_6) where the drug is bound to the carrier via a nondegradable spacer was studied under the same conditions. No cleavage was observed during incubation with cathepsin B (results not shown).

Photophysical properties after cleavage. In accordance with the results of other photophysical studies showing higher quantum yields of singlet oxygen generation for free Mce_6 compared with P-G-Mce_6 [13], the quantum yield of singlet oxygen generation increased with cleavage of Mce_6 from P-G-F-L-G-Mce_6 (Copolymer II) (Table 2). The 48 h incubation period showed a fivefold increase compared with the control and the 120 h incubation sample had a quantum yield approaching that of free Mce_6.

Table 2.
Quantum yields of singlet oxygen generation from Copolymer II (P-G-F-L-G-Mce_6) after cleavage with cathepsin B

Incubation	$\varphi^1\Delta\hat{g}$
none	0.14
2 days	0.66
1 week	0.71

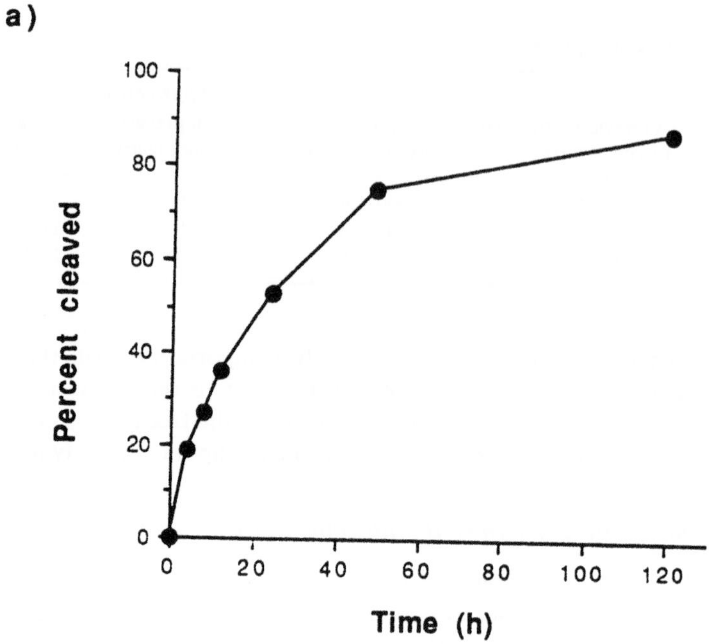
a)
100
80
60
40
20
0
Percent cleaved
0
20
40
60
80
100
120
Time (h)

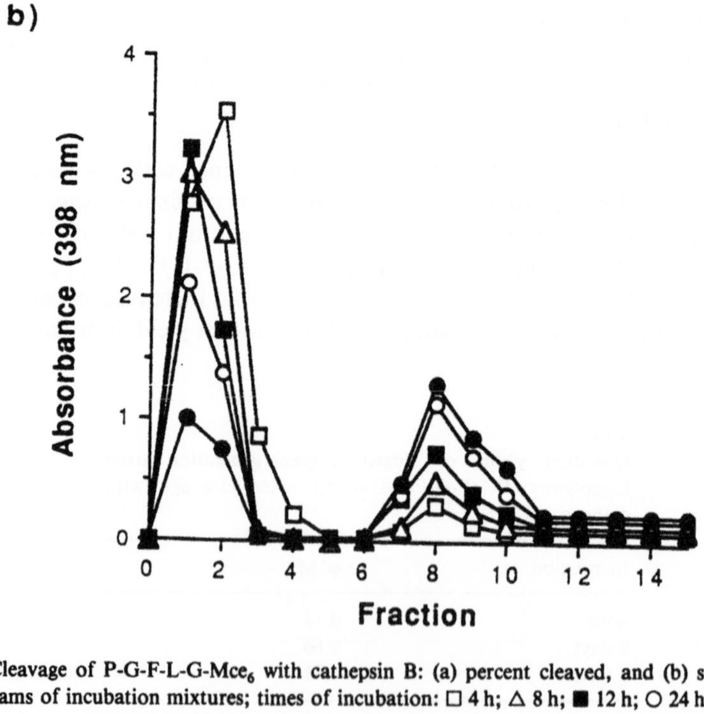
b)
4
3
2
1
0
Absorbance (398 nm)
0
2
4
6
8
10
12
14
Fraction

Figure 3. Cleavage of P-G-F-L-G-Mce$_6$ with cathepsin B: (a) percent cleaved, and (b) size exclusion chromatograms of incubation mixtures; times of incubation: □ 4 h; △ 8 h; ■ 12 h; ○ 24 h; ● 49 h.

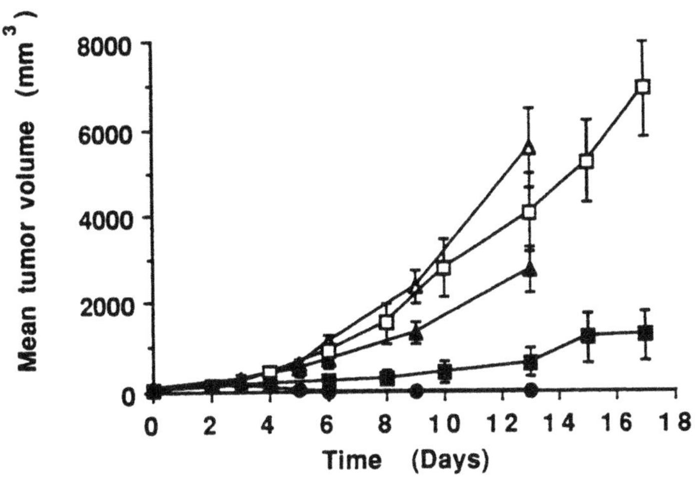

Mean tumor volume (mm 3)
8000
6000
4000
2000
0
0 2 4 6 8 10 12 14 16 18
Time (Days)

Figure 4. Chemotherapy with P-G-F-L-G-adria. Copolymer I was administered *i.v.* into the tail veins of A/J mice bearing Neuro 2A neuroblastoma tumors. Tumor volumes measured wih calipers were followed with time. Error bars represent standard errors. △ Control I; □ control II. Doses of adria: ▲ 4.1 mg/kg; ■ 8.2 mg/kg; ● 16.4 mg/kg.

In vivo *studies*

All data for *in vivo* experiments are represented as the mean value of the number of mice in the group. Error bars represent the standard error of the measurements.

Chemotherapy

Experiments were performed to determine a concentration of P-G-F-L-G-adria (7.4 wt % adriamycin · HCl) (Copolymer I) which would show some tumor reduction effect, yet not cure tumors. A dose dependent study was undertaken (Fig. 4). Doses of P-G-F-L-G-adria were used corresponding to 4.1, 8.2, or 16.4 mg/kg of adria. The 16.4 mg/kg dose was 100% effective in curing tumors. All of the mice in this group remained tumor free until day 55 at which time they were sacrificed. This same dose of free drug was toxic to mice. The 4.1 mg/kg dose group behaved without substantial effect compared with the controls. The 8.2 mg/kg dose had some effect on tumor suppression, but no cures. Tumor growth was suppressed for approximately 5 days. By day 10, tumors were growing at similar rates as the controls. This P-G-F-L-G-adria concentration (8.2 mg/kg) was used in subsequent combination therapy experiments. A control experiment using 8.2 mg/kg P-G-F-L-G-adria plus light (500 mW/cm^2; 650 nm; 10 min) was performed and the results were consistent with chemotherapy alone at this same dose indicating that the light did not affect the activity of the chemotherapeutic agent (results not shown).

Localization/retention of free and HPMA copolymer bound Mce$_6$. Tumor and tissue localization/retention experiments compared the uptake of free meso-chlorin e$_6$ monoethylene diamine disodium salt (Mce$_6$) with that of P-G-Mce$_6$ (11.2 wt % Mce$_6$) (Copolymer III). When tumors were palpable, mice were injected *i.v.* with 5 mg/kg free Mce$_6$ or P-G-Mce$_6$. The free drug reached a maximum concentration in the tumor tussue in 1 h compared with the polymer which was present in high concentration even after 48 h (Fig. 5).

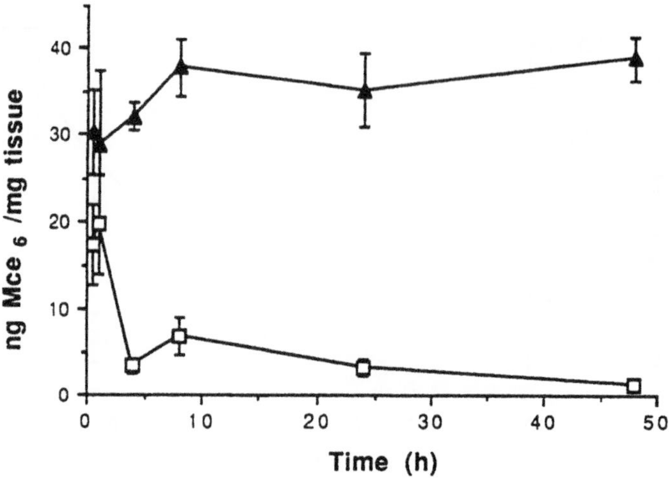
ng Mce 6 /mg tissue
0
10
20
30
40
0
10
20
30
40
50
Time (h)

Figure 5. Localization/retention: Free Mce_6 vs P-G-Mce_6. Five mg/kg of drug were administered *i.v.* into the tail veins of A/J mice bearing Neuro 2A neuroblastoma tumors. Tumor tissue was excised at various times and subjected to an alkaline hydrolysis assay (see experimental). Mce_6 content in the tissue is plotted against time. □ free Mce_6; ▲ P-G-Mce_6.

Another experiment with Copolymer III collected tissue samples 5 days after administration (results not shown). The concentration of Mce_6 in the tumor was still substantial (28 ng/mg tissue), although lower than for the shorter time periods.

The same localization/retention procedure was used for Copolymer II (P-G-F-L-G-Mce_6; 11.2 wt % Mce_6). The results of this study are shown in Fig. 6. In contrast with the results obtained with the noncleavable Copolymer III, the cleavable Copolymer II shows a drastic reduction in Mce_6 content in the tumor by 48 h and

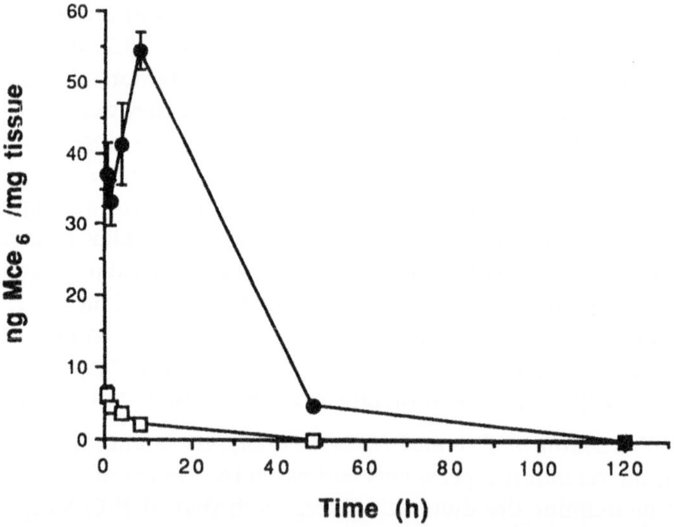
ng Mce 6 /mg tissue
0
10
20
30
40
50
60
0
20
40
60
80
100
120
Time (h)

Figure 6. Localization/retention: Free Mce_6 vs P-G-F-L-G-Mce_6. Five mg/kg of drug were administered *i.v.* into the tail veins of A/J mice bearing Neuro 2A neuroblastoma tumors. Tumor tissue was excised at various times and subjected to an alkaline hydrolysis assay (see experimental). Mce_6 content in the tissue is plotted against time. □ free Mce_6; ● P-G-F-L-G-Mce_6.

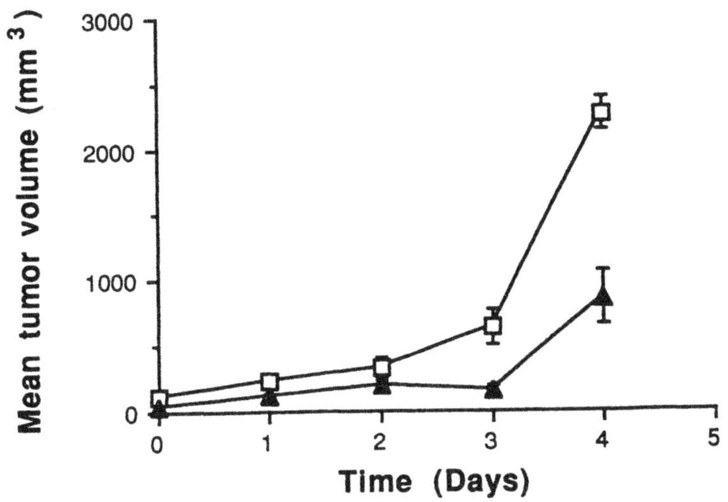
Mean tumor volume (mm 3)
3000
2000
1000
0
0
1
2
3
4
5
Time (Days)

Figure 7. Photodynamic therapy: P-G-Mce$_6$ (Copolymer III; 4 mg/kg) was injected *i.v.* into the tail veins of A/J mice bearing Neuro 2A neuroblastoma tumors. Tumors were irradiated 24 h after drug administration with 650 nm light (500 mW/cm^2; 5 min irradiation). Tumor volumes measured with calipers were followed with time. Error bars represent standard errors. □ Control; ▲ P-G-Mce$_6$.

complete clearance by 120 h. This result provides indirect evidence of Mce$_6$ cleavage from the polymer in tumor cells and shows the body's ability to clear free Mce$_6$.

Photodynamic therapy. A PDT experiment comparing the photodynamic effects of free Mce$_6$ and Copolymer III (P-G-Mce$_6$; 4 mg/kg) was performed. Irradiation (500 mW/cm^2; 5 min) was applied after a 1 or 24 h time lag for tissue uptake. The 1 h uptake time was systemically toxic for both groups upon irradiation (not shown). All animals in both groups died upon light administration. The effects on tumor volume of the 24 h uptake group for Copolymer III are shown in Fig. 7. The free drug which is cleared from the tumor in 24 h behaves the same as the control, while the polymer showed tumor suppression for approximately 3 days. Because of the differences in optimal tumor uptake time between the free Mce$_6$ and P-G-Mce$_6$ (Fig. 5) and insolubility of the free Mce$_6$ at concentrations desirable for the P-G-Mce$_6$, the free drug was no longer studied.

PDT was performed in several experiments comparing the photodynamic effects of P-G-Mce$_6$ (Copolymer III) and P-G-F-L-G-Mce$_6$ (Copolymer II). The cleavable copolymer was more systemically phototoxic in all cases (results not shown). For example, at a concentration of 4 mg/kg Mce$_6$ (24 h uptake) and an irradiation power of 500 mW/cm^2 for 5 min, 60% morbidity resulted in the cleavable group (autopsy showed severe liver and other internal photodynamic damage). All mice in the noncleavable group (Copolymer III) were alive. Qualitative effects indicative of photodynamic action such as bleaching, edema, and black scab formation were more evident in all experiments using the cleavable copolymer compared to the noncleavable one. The best results were achieved when black scab formation was apparent. When this happened, the tumor usually disappeared for a few days; however, it always recurred.

Because of the greater photodynamic effect of the cleavable copolymer (Copolymer II), it was decided to use it instead of P-G-Mce$_6$ (Copolymer III) for all

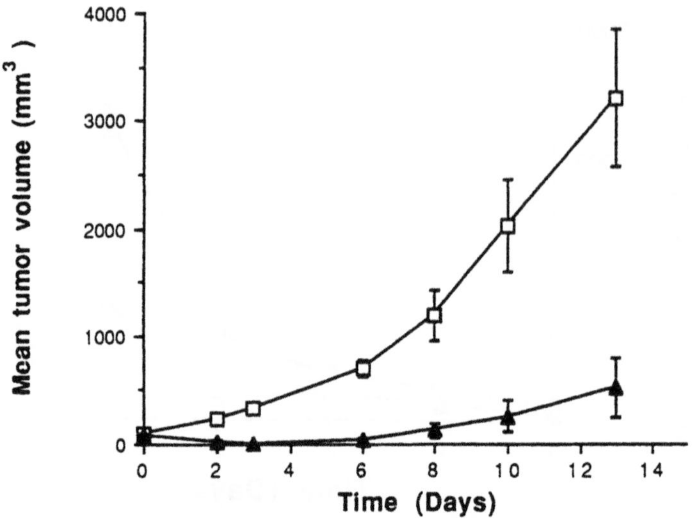
4000
3000
2000
1000
0
Mean tumor volume (mm³)
0
2
4
6
8
10
12
14
Time (Days)

Figure 8. Photodynamic therapy: P-G-F-L-G-Mce_6 (Copolymer II; 2 mg/kg) was injected *i.v.* into the tail veins of A/J mice bearing Neuro 2A neuroblastoma tumors. Tumors were irradiated 24 h after drug administration with 650 nm light (500 mW/cm^2; 10 min irradiation). Tumor volumes measured with calipers were followed with time. Error bars represent standard errors. □ Control; ▲ P-G-F-L-G-Mce_6.

future PDT experiments. The next step was undertaken to optimize the drug and light dose to arrive at parameters yielding maximum effect on tumor growth with minimum systemic phototoxicity. A lower drug dose with an increased light dose proved effective. Doses of Copolymer II (P-G-F-L-G-Mce_6) equivalent to 4, 3.25, 2.5, 2, and 1 mg/kg Mce_6 (24 h uptake) and light of 500 mW/cm^2 for 10 min were compared. Doses from 2.5 to 4 mg/kg Mce_6 (results not shown) consistently gave evidence of systemic phototoxicity and mortality rates between 20 and 100%. The 1 mg/kg Mce_6 dose group was virtually ineffective with the light dose compared with the controls.

The 2 mg/kg Mce_6 dose of Copolymer II (P-G-F-L-G-Mce_6; 24 h uptake) was chosen for study. The light dose (500 mW/cm^2) was varied using 5–20 min irradiation time (results not shown). The 20 min dose group showed 60% lethality while the 5 min group showed little effect. However, substantial effect was achieved for irradiation times of 8 min 20 s, 10 min, and 13 min 20 s (results not shown).

The best tumor effect with the least chance of systemic phototoxicity was shown for a drug dose of 2 mg/kg (Copolymer II; P-G-F-L-G-Mce_6) and a light dose of 10 min with 500 mW/cm^2. The results of this experiment are shown in Fig. 8. Blanching of skin and black scab formation were evident for all members of the treated group. No tumor was detectable for 3 days following treatment after which time tumors started to grow.

Mixed chemotherapy and PDT. Table 3 summarizes the results obtained from combination therapy experiments. Copolymer I (P-G-F-L-G-adria; 8.2 mg/kg adriamycin · HCl) and Copolymer III (P-G-Mce_6; 4 mg/kg) were mixed and injected *i.v.* by tail vein when tumors became palpable. After 48 h tumors were irradiated (500 mW/cm^2; 5 min). The extra time lag (48 h) was allowed to permit adriamycin more time for effect since it was known from the localization/retention experiments that Mce_6 would still be present in high concentration. Cures were

Table 3.
Combination therapy

Samples	Uptake (h)	Irradiation (650 nm)	Long-term survivors
P-G-F-L-G-adria (8.2 mg/kg) + P-G-Mce_6 (4 mg/kg)	48	500 mW/cm^2 5 min	5/15
P-G-F-L-G-adria (8.2 mg/kg) + P-G-F-L-G-Mce_6 (1.5 mg/kg)	48	500 mW/cm^2 10 min	0/5
P-G-F-L-G-adria (8.2 mg/kg) + P-G-F-L-G-Mce_6 (1.5 mg/kg)	24	500 mW/cm^2 10 min	4/5

followed until day 54 at which time the experiment was terminated. The mixed chemotherapy plus PDT was much more effective than either drug alone.

Although a 2 mg/kg dose of Copolymer II (P-G-F-L-G-Mce_6) was found to be effective in the previous PDT experiment (Fig. 8), it was found to be toxic when used in combination with Copolymer I (P-G-F-L-G-adria; 8.2 mg/kg adria). Therefore, the P-G-F-L-G-Mce_6 dose was decreased to 1.5 mg/kg. A combination experiment was performed using 1.5 mg/kg P-G-F-L-G-Mce_6 plus 8.2 mg/kg P-G-F-L-G-adria. Two uptake times (24 and 48 h) and irradiation (500 mW/cm^2/10 min) were investigated. The 48 h uptake group showed significant effect compared with the controls, however, there were no cures (Fig. 9) while the 24 h group (sacrificed at 48 days) showed an 80% long range cure rate (Table 3).

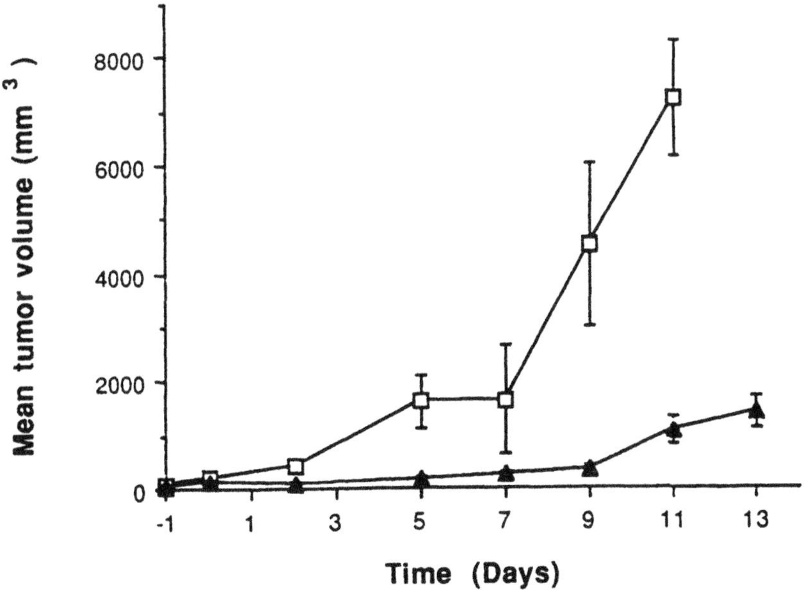
Mean tumor volume (mm 3)
8000
6000
4000
2000
0
-1
1
3
5
7
9
11
13
Time (Days)

Figure 9. Combination therapy: P-G-F-L-G-adria (8.2 mg/kg) and P-G-F-L-G-Mce_6 (1.5 mg/kg) were mixed and injected *i.v.* into the tail vein of A/J mice bearing Neuro 2A neuroblastoma tumors. Tumors were irradiated 48 h after drug administration with 650 nm (500 mW/cm^2; 10 min irradiation). Tumor volumes measured with calipers were followed with time. Error bars represent standard errors. □ Control; ▲ combination therapy.

DISCUSSION

N-(2-hydroxypropyl)methacrylamide copolymers were systematically developed for the controlled release and targeting of antitumor agents [1, 3, 31]. A number of chemotherapeutic agents were attached to HPMA copolymers via oligopeptide side-chains and their anticancer activity studied [1, 4, 31, 32]. A detailed study of the relationship between the structure of oligopeptide sequences and drug/drug model release [27, 31, 33–36] permitted the design of HPMA copolymer–adriamycin and optionally galactosamine (targeting moiety) conjugates [23] which will soon undergo clinical trials. Similar conjugates containing antibodies as targeting moieties were also studied [6].

The design of a polymer–drug conjugate is a complex task. The structure of the carrier itself is important [37]. If nondegradable in the lysosomes, it has to have an optimum molecular weight distribution [38] which is under the threshold of glomerular filtration. Moreover, the structure of the carrier has to be such that the carrier enters cells by fluid phase pinocytosis [37]. Only in this way the capture-recapture cycle can be interrupted [39]. The design of the oligopeptide spacer between the polymer backbone and the drug is governed by the mechanism of drug action. If an anthracycline antibiotic is used it has to be attached via a bond degradable in the lysosomal compartment of the cell. The released drug can cross the lysosomal membrane and intercalate with DNA in the cell nucleus. Following cell death the carrier will be released into the bloodstream and eventually eliminated from the organism. It was observed previously that HPMA copolymer–daunomycin-anti-Thy 1.2 antibody conjugates were active only when the drug was attached via an enzymatically degradable bond [6]. Similar results were obtained when HPMA copolymer conjugates contained galactosamine or fucosylamine as targeting moieties [4, 40, 41]. Based on these results the tetrapeptide G-F-L-G was used as a spacer in Copolymers I and II. This is a sequence stable in blood serum and plasma [42], but readily releasing adriamycin when incubated with lysosomal enzymes [23, 36, 43]. However, it is well known that the rate of drug release depends on the structure of the drug itself. The active site of proteinases usually accommodates two amino acid residues towards the COOH end of the substrate. In polymeric prodrugs, the drug molecule occupies these positions if the oligopeptide spacer is tailor-made to be cleaved between the distal amino acid residue and the drug [37]. It was indeed shown [23] that the structure of the leaving group has a strong influence on the rate of drug release from an HPMA copolymer with a given oligopeptide sequence. Cathepsin B is the most important lysosomal proteinase for the cleavage of oligopeptide side-chains [3]. The release of drug model from oligopeptide side chains of HPMA copolymers by cathepsin B has been studied in detail [27]. To evaluate if a large and rigid molecule (Mce_6) can be released by lysosomal enzymes, Copolymer II was incubated with cathepsin B. The results of the *in vitro* cleavage experiments with P-G-F-L-G-Mce_6 (Fig. 3) indicate that Mce_6 can in fact be cleaved by cathepsin B from HPMA copolymers containing G-F-L-G side chains terminated in Mce_6. In accordance with the previously published results, the rate of cleavage is relatively low (Fig. 3). The low cleavage rate may be caused by the size and shape of the Mce_6 molecule. The rigidity of the macrocycle may slow down the enzyme-substrate complex formation due to energetically unfavorable accommodation into the S_1' and S_2' subsites of the enzyme. No drug was released from Copolymer III where a non-degradable glycine spacer was used.

In vivo localization/retention experiments with the cleavable copolymer II (Fig. 6) show a rapid clearance of Mce_6 from the tumor tissue compared with the noncleavable copolymer III (Fig. 5). This same trend was seen for the abdominal muscle, kidney, spleen, liver, leg muscle, and for skin tissue (results not shown). This lends indirect evidence of cleavage *in vivo*. The rate of cleavage *in vivo* may be regulated by using different oligopeptide spacer arms [3]. Even slower cleavage (3–4 days) may be desirable such that the copolymer not taken up by the tumor is eliminated from the rest of the body before tumor irradiation. In addition, a longer time lag between injection and irradiation would provide the adriamycin (on a spacer arm with a relatively faster release rate than that with Mce_6) more time to take effect.

From the mechanism of Mce_6 action point of view it would appear that it is not important if this photosensitizer is bound via a cleavable bond. Studies with the noncleavable P-G-Mce_6 *in vitro* indicated that it is not necessary for the Mce_6 to be cleaved from the copolymer to have a photodynamic effect [12]. However, the physicochemical properties of the HPMA copolymer–Mce_6 conjugates have to be taken into account. Photophysical experiments have shown that the solution properties of copolymer bound Mce_6 greatly affect its quantum yield of singlet oxygen generation [13]. The P-G-Mce_6 had a much lower yield of singlet oxygen generation in sodium phosphate buffer than did the free drug. Adding detergent (CTAB) to both, enhanced the quantum yield of the free Mce_6 by a small amount and substantially enhanced the quantum yield of singlet oxygen of the P-G-Mce_6. This indicates that the P-G-Mce_6 is much more aggregated in buffer than the free drug, although some monomerization takes place for the free drug when the surfactant is added. Other evidence of aggregation is seen by shorter, broader peaks in the absorbance spectrum and quenching of the fluorescence spectrum of the P-G-Mce_6 in buffer compared with Mce_6 [30]. Perhaps micellar aggregates are formed in aqueous solutions as the hydrophobic Mce_6 molecules are repelled by the water forming a hydrophobic core with the hydrophilic copolymer forming the outer micellar shell in contact with the water. This phenomenon has been shown with other HPMA copolymers containing side chains terminated with hydrophobic molecules [44, 46] as well as with adriamycin–poly(glutamic acid) conjugates [45].

A study in which the quantum yield of oxygen uptake was measured for the P-G-F-L-G-Mce_6 reaction mixture after both 48 h and 1 week incubation periods with cathepsin B showed a marked increase in this value with time of cleavage (Table 2). As the Mce_6 was cleaved from the copolymeric carrier, the photophysical behavior became more and more like the free drug in solution. This may explain the increased antitumor effect of PDT with the cleavable vs noncleavable copolymer *in vivo* (compare Figs. 7 and 8). Perhaps the copolymers are aggregated inside of the lysosomes to a greater extent than the free drug and as the drug is cleaved it becomes less aggregated. The aggregation of HPMA copolymers containing hydrophobic side-chains is strongly dependent on the amount of hydrophobic side-chains per macromolecule [44, 46]. This may also explain why PDT with the noncleavable Copolymer III (P-G-Mce_6) seemed to have less effect on tumor suppression than expected compared to free Mce_6 (results not shown) based on the enhanced localization/retention behavior of the copolymer. Even though more Copolymer III may have localized in the tumor, its PDT effect was not as pronounced as the free drug in the cellular environment. However, a comparison of PDT effect *in vivo* of the free drug vs the P-G-Mce_6 is not very accurate because of the insolubility of the free Mce_6 at the concentrations desirable for effective PDT. In addition, the differences in uptake properties which indicate different uptake times for a maximum concentration of the free vs P-G-Mce_6 in the tumor makes necessary the use of different lag times before tumor irradiation. In addition, the goal of this work was to optimize the delivery of the copolymer bound sensitizer; the PDT properties of the free drug were not investigated further.

Another property that has been shown is a reduction in the quantum yield of photobleaching for Mce_6 bound to the copolymer compared with the free drug [13]. Others have also seen this effect [47]. This is a general phenomenon observed with many light sensitive molecules bound to polymers [48]. This may be either an advantage or a disadvantage. Sometimes photobleaching may be useful in PDT such that light can penetrate deeper and deeper into tumor tissue as the compound photofades after its PDT effect has been exerted. The sensitizer deeper in the tumor can then be activated. Photobleaching can be exploited such that low levels of photosensitizer in surrounding normal tissue can be bleached preserving the normal tissue and enhancing the therapeutic ratio [49]. On the other hand, if the concentration of sensitizers in the tumor is excessive, it will absorb the light and prevent it from penetrating the tumor and activating the photosensitizer deeper in the tumor [50].

Binding of (anticancer) drugs to polymeric carriers changes not only the mechanism of cellular uptake and subcellular trafficking, but also the body distribution of the drug. Consequently, uptake in tissues where the major side effects occur may decrease [37]. Adriamycin (doxorubicin) shows cumulative dose-dependent cardiomyopathy which is its principal dose-limiting side effect. This limits its long-term use [5]. Binding anticancer drugs to HPMA copolymers markedly decreases cardiotoxicity. A 100-fold decrease in the initial adriamycin peak level in the heart was observed when bound to an HPMA copolymer compared with the free drug [51]. Rats which received 4 mg/kg of free adriamycin showed a decrease in cardiac output and died while the same dose of HPMA conjugate caused no decrease in cardiac output [8]. The LD_{50} of the HPMA copolymer-adriamycin conjugate is 5–10 times higher than that observed of free adriamycin [9]. It was also found that a total cumulative dose of 90 mg/kg of adriamycin bound to HPMA copolymers containing a galactosamine targeting moiety showed no overt toxicity in CDF1 mice [9]. If mice were administered 10^5 L1210 leukemia cells (*i.p.* day 0) and 3 doses (days 1, 2, 3) of 30 mg/kg of copolymer bound adriamycin, 17/20 long term survivors resulted. However, only three doses of 10 mg/kg of free adriamycin caused 40/40 toxic deaths.

The higher toxicity of free adriamycin has been indicated in other ways. Decreased toxicity against hematopoietic precursors in bone marrow as measured by the *in vivo* colony-forming unit-spleen assay of P-G-F-L-G-adria compared with free adriamycin was found [2].

We have seen a similar trend in these studies (results not shown). A single dose (16.4 mg/kg) of P-G-F-L-G-adria could cure neuroblastoma tumors (A/J mice) whereas this dose of free adriamycin was toxic to mice. At an even higher dose (20 mg/kg), the free drug caused 100% morbidity. This dose caused initial weight loss in the mice administered the P-G-F-L-G-adria in the first few days after treatment which was quickly regained.

A similar decrease of toxicity was observed with photosensitizers *in vitro* [12]. Using a polymeric carrier for a photosensitizer may decrease side effects such as light ultrasensitivity after treatment because as has been shown in the localization/retention experiments in this study (Figs. 5 and 6), a greater quantity of copolymer bound drug seems to accumulate in the tumor than the free drug even without the incorporation of a targeting moiety. This would allow a lower drug dose to be administered and still have tumor retention of the necessary concentration of sensitizer.

In vivo chemotherapy with Copolymer I (P-G-F-L-G-adria) indicated that Neuro 2A neuroblastoma responded well to HPMA copolymer bound adriamycin (Fig. 4). Photodynamic therapy *in vivo* has shown that the cleavable Copolymer II (P-G-F-L-G-Mce_6) was much more potent than the noncleavable Copolymer III (P-G-Mce_6). Four mg/kg of P-G-Mce_6 (24 h uptake) and a light dose of 500 mW/cm^2 for 5 min was used in PDT experiments; this same dose of P-G-F-L-G-Mce_6 and light was toxic to mice. There was a very fine line between therapeutic effect and toxicity for PDT with the Copolymer II (P-G-F-L-G-Mce_6), however, and a precise combination of drug and light dose was necessary. Toxicity was found to be through liver and other internal injuries to the mice upon light exposure. This would not be the case with a larger animal model (or in the human) in which the light would be administered a distance away from internal organs. However, in the mouse studies this was found to be the limiting factor with implications which will be discussed subsequently.

Optimal conditions for the cleavable Copolymer II (P-G-F-L-G-Mce_6) proved to be a decreased drug dose (2 mg/kg) and an increase light dose (500 mW/cm^2; 10 min) (Fig. 8). Tumor growth was suppressed for a number of days, yet, no cure rate was obtained with this modality alone. The best results were obtained when black scab formation was evident. Photodynamic effect to a lesser extent was observed by a white faded area encircling and covering the tumor after irradiation.

The variable sensitivity of different subpopulations of cells is the main rationale for combination chemotherapy. However the implementation of combination chemotherapy in the clinic has been relatively random. Treatment failure is mainly caused by the presence of cell populations resistant to cytotoxic drugs [52, 53]. There are examples of cancer treatment using PDT in combination with other treatment modalities e.g., the treatment of retinoblastoma using the pharmacologic enhancement of radiotherapy and PDT [54] and the use of adjuvant chemotherapy with PDT in the treatment of advanced gastrointestinal cancers in humans [55].

The only treatment protocol used in our studies capable of producing long term survivors without overt side effects was combined therapy consisting of a mixture of a copolymer containing adriamycin and another containing Mce_6. These results verified our hypothesis that a combination of two copolymers, one containing a chemotherapeutic agent and another containing a photosensitizer, would improve the therapeutic efficacy of either modality used alone in the treatment of a solid tumor.

For both the cleavable (P-G-F-L-G-Mce_6) and noncleavable (P-G-Mce_6) copolymers used, combination therapy in a mixture with P-G-F-L-G-adria proved more effective than PDT or chemotherapy alone with the same doses (Table 3). The best effect was achieved by administering a mixture of 8.2 mg/kg of P-G-F-L-G-adria and 1.5 mg/kg P-G-F-L-G-Mce_6 with a 24 h time lag before irradiation (500 mW/cm^2; 10 min). An 80% cure rate was achieved (Table 3). It is possible that the adriamycin first takes effect by damaging DNA and then the photosensitizer takes effect and kills cells which were not affected by the former. Damage by adriamycin may make the cells more sensitive to PDT. Many more studies are needed to assess the mechanism of action of the combination of the two copolymeric drugs. By combining therapies, it may be possible to overcome the side effects of both drugs. Lower adriamycin doses will reduce the risk of cardiotoxicity and toxicity to nontarget cells; lower doses of Mce_6 could reduce the risk of light ultrasensitivity.

CONCLUSIONS

In conclusion, Mce_6 can be cleaved from an HPMA copolymer containing Mce_6 bound via the G-F-L-G oligopeptide sequence with cathepsin B *in vitro*. Indirect evidence of cleavage *in vivo* was shown by localization/retention experiments. Other indirect evidence for cleavage is the severely reduced therapeutic effect with a 48 h uptake compared with a 24 h uptake for Copolymer II (P-G-F-L-G-Mce_6) in mixed chemotherapy and PDT experiments. This is consistent with results obtained by the localization/retention experiments.

The quantum yield of singlet oxygen generation increases with the cleavage of Mce_6 from Copolymer II with cathepsin B *in vitro*. This is in agreement with the results of PDT experiments comparing the antitumor effects of Copolymer III (P-G-Mce_6) and Copolymer II (P-G-F-L-G-Mce_6); the Copolymer II containing the cleavable side chains proved more efficacious than the one with the glycine spacer.

For both the cleavable P-G-F-L-G-Mce_6 and noncleavable P-G-Mce_6 copolymers, combination therapy in a mixture with P-G-F-L-G-adria is more therapeutically active than PDT or chemotherapy alone. In fact, long term survivors were possible with the combination therapy whereas cure rates could not be obtained with either therapy alone at the dosages used (Table 3).

The copolymer carriers used in this study have several advantages over free drugs including solubility in physiological solutions, increased tumor:normal tissue uptake ratios, and reduced side effects. These experiments give rise to the possibility of incorporating both drugs into the same copolymer to ensure identical body distribution for the two drugs for improved cancer treatment.

Acknowledgements

We are indebted to Dr. P. Kopečková for helpful discussions. The participation of P. Meekins on photophysical analyses, of R. Reed on *in vivo* experiments, and of M. Wolfe on localization/retention experiments is extremely appreciated. We would like to thank Dr. A. Suarato, Farmitalia Carlo Erba, Milano, Italy, for the generous gift of adriamycin, and Dr. M. Baudyš, Institute of Organic Chemistry and Biochemistry, CSAS, Prague, Czech Republic, for the gift of cathepsin B. The Utah Cancer Institute was extremely helpful in providing updated information. The study was supported in part by NIH grant CA 51578 (JK), American Cancer Society grant CH-511 (JDS), Dixon Utah Laser Institute and Medical Free Electron Laser Program (RDS), and by a gift from Insutech, Inc., Salt Lake City, Utah (JK).

REFERENCES

1. J. Kopeček, in: *IUPAC Macromolecules*, p. 305, H. Benoit and P. Rempp (Eds). Pergamon Press, Oxford (1982).
2. B. Říhová, M. Bilej, V. Větvička K. Ulbrich, J. Strohalm, J. Kopeček and R. Duncan, *Biomaterials* **10**, 335 (1988).
3. J. Kopeček, *Biomaterials*, **5**, 19 (1984).
4. J. Kopeček and R. Duncan, *J. Controlled Rel.* **6**, 315 (1987).
5. L. Lenaz and J. A. Page, *Cancer Treatment Rev.* **3**, 111 (1976).
6. B. Říhová, P. Kopečková, J. Strohalm, P. Rossmann, V. Větvička and J. Kopeček, *Clin. Immunol. Immunopathol.* **46**, 100 (1988).
7. L. W. Seymour, K. Ulbrich, J. Strohalm, J. Kopeček and R. Duncan, *Biochem. Pharmacol.* **39**, 1125 (1990).
8. T. K. Yeung, J. W. Hopewell, G. Rezzani, R. H. Simmonds, L. W. Seymour, R. Duncan, O. Bellini, M. Grandi, F. Spreafico, J. Strohalm and K. Ulbrich, *Cancer Chemother. Pharmacol.* **29**, 105 (1991).
9. R. Duncan, L. W. Seymour, K. B. O'Hare, P. A. Flanagan, S. Wedge, I. Hume, K. Ulbrich, J. Strohalm, V. Šubr, F. Spreafico, M. Grandi, M. Ripomonti, M. Farao and A. Suarato, *J. Controlled Rel.* **19**, 331 (1992).
10. J. Cassidy, R. Duncan, G. J. Morrison, J. Strohalm, D. Plocovcá, J. Kopeček and S. B. Kaye, *Biochem. Pharmacol.* **38**, 875 (1989).
11. N. L. Krinick, B. Říhová, K. Ulbrich, J. D. Andrade and J. Kopeček, *Proc. SPIE.* **997**, 70 (1988).
12. N. L. Krinick, B. Říhová, K. Ulbrich, J. Strohalm and J. Kopeček, *Makromol. Chem.* **191**, 839 (1990).
13. J. D. Spikes, N. L. Krinick and J. Kopeček, *Photochem. Photobiol., A: Chem.* **70**, 163 (1993).
14. C. S. Foote, in: *Light Activated Pesticides*, p. 22, J. R. Heitz and K. R. Downum (Eds). ACS Symposium Series 339 (1987).
15. J. D. Spikes and R. C. Straight, in: *Light Activated Pesticides*, p. 98, J. R. Heitz and K. R. Downum (Eds). ACS Symposium Series 339 (1987).
16. T. J. Dougherty, *Semin. Surg. Oncol.* **5**, 6 (1989).
17. T. Dougherty, *J. Invest. Derm.* **77**, 122 (1981).
18. A. R. Oseroff, D. Ohuoha, T. Hasan and J. Bommer, *Proc. Natl. Acad. Sci. USA* **83**, 8744 (1986).
19. F. N. Jiang, S. Jiang, D. Liu, A. Richter and J. G. Levy, *J. Immunol. Methods* **134**, 139 (1990).
20. B. A. Goff, M. Bamberg and T. Hasan, *Cancer Res.* **51**, 4762 (1991).
21. J. Strohalm and J. Kopeček, *Angew. Makromol. Chem.* **70**, 109 (1978).
22. P. Rejmanová, J. Labský and J. Kopeček, *Makromol. Chem.* **178**, 2159 (1977).
23. J. Kopeček, P. Rejmanová, J. Strohalm, K. Ulbrich, B. Říhová, V. Chytrý, J. B. Lloyd, and R. Duncan, U.S. Pat. (5,037,883) (1991).
24. R. Duncan, I. C. Hume, P. Kopečková, K. Ulbrich, J. Strohalm and J. Kopeček, *J. Controlled Rel.* **10**, 51 (1989).
25. N. L. Krinick, M. S. Thesis, University of Utah (1989).
26. J. Pohl, M. Baudyš, V. Tomášek and V. Kostka, *FEBS Lett.* **142**, 23 (1982).
27. P. Rejmanová, J. Kopeček, J. Pohl, M. Baudyš, and V. Kostka, *Makromol. Chem.* **184**, 2009 (1983).
28. M.-T. Maurette, E. Oliveros, P. P. Infelta, K. Ramsteiner and A. M. Braun, *Helv. Chim. Acta* **66**, 722 (1983).
29. W. R. Haag, J. Hoigne, E. Gassma and A. M. Braun, *Chemosphere* **13**, 631 (1984).
30. J. D. Spikes, *Photochem, Photobiol.* **55**, 797 (1992).
31. R. Duncan, *Anti-Cancer Drugs* **3**, 175 (1992).
32. J. Kopeček and R. Duncan, in: *Polymers in Controlled Drug Delivery*, p. 152, L. Illum and S. S. Davis (Eds.) Wright, Bristol (1987).
33. J. Kopeček, P. Rejmanová and V. Chytrý, *Makromol. Chem.* **182**, 1917 (1981).
34. J. Kopeček, P. Rejmanová, R. Duncan and J. B. Lloyd, *Ann. N.Y. Acad. Sci.* **446**, 93 (1985).
35. V. Šubr, J. Kopeček, J. Pohl, M. Baudyš and V. Kostka, *J. Controlled Rel.* **9**, 133 (1988).
36. V. Šubr, J. Strohalm, K. Ulbrich, R. Duncan and I. C. Hume, *J. Controlled Rel.* **18**, 123 (1992).
37. N. L. Krinick and J. Kopeček, in: *Handbook of Experimental Pharmacology, Vol. 100, Targeted Drug Delivery*, p. 105, R. L. Juliano (Ed). Springer, New York (1991).

38. L. W. Seymour, R. Duncan, J. Strohalm and J. Kopeček, *J. Biomed. Mater. Res.* **21**, 1341 (1987).
39. J. B. Lloyd, in: *Drug delivery Systems Fundamentals and Techniques,* p. 95, P. Johnson and J. G. Lloyd-Jones (Eds). Ellis Horwood, Deer Sealed Beach, Florida (1987).
40. R. Duncan, P. Kopečková, J. Strohalm, I. C. Hume, J. B. Lloyd and J. Kopeček, *Br. J. Cancer* **57**, 147 (1988).
41. R. Duncan, I. C. Hume, P. Kopečková, K. Ulbrich, J. Strohalm and J. Kopeček, *J. Controlled Rel.* **10**, 51 (1989).
42. P. Rejmanová, J. Kopeček, R. Duncan and J. B. Lloyd, *Biomaterials* **6**, 45 (1985).
43. R. Duncan, P. Kopečková-Rejmanová, J. Strohalm, I. Hume, H. C. Cable, J. Pohl, J. B. Lloyd and J. Kopeček, *Br. J. Cancer* **55**, 165 (1987).
44. Č. Koňák, P. Kopečková and J. Kopeček, *Macromolecules* **25**, 5451 (1992).
45. M. Nukui, K. Hoes, H. van den Berg and J. Feijen, *Makromol. Chem.* **192**, 2925 (1991).
46. K. Ulbrich, Č. Koňák, Z. Tuzar and J. Kopeček, *Makromol. Chem.* **188**, 1261 (1987).
47. J. S. Bellin, *Photochem. Photobiol.* **8**, 383 (1968).
48. H.-R. Yen, J. Kopeček and J. D. Andrade, *Makromol. Chem.* **190**, 69 (1989).
49. T. S. Mang, T. J. Dougherty, W. R. Potter, D. G. Boyle, S. Somer and J. Moan, *Photochem. Photobiol.* **45**, 501 (1987).
50. S. G. Bown, C. J. Tralau, P. D. Coleridge Smith, D. Akdemir and T. J. Wieman, *Br. J. Cancer* **54**, 43 (1986).
51. L. W. Seymour, K. Ulbrich, J. Strohalm, J. Kopeček and R. Duncan, *Biochem. Pharmacol.* **39**, 1125 (1990).
52. L. M. van Putten, *Cancer Treat. Rev.* **11**, 19 (1984).
53. V. T. DeVita, in: *Cancer Principles and Practice of Oncology*, Updates, Vol. 4, No. 11, p. 1, V. T. DeVita, S. Hellman and S. A. Rosenberg (Eds). Lippincott, Philadelphia (1990).
54. L. White, *Am. J. Pediatr. Hematol. Oncol.* **13**, 189 (1991).
55. M. L. Jin, B. Q. Yang, W. Zhang and P. Ren, *J. Photochem. Photobiol.*-B **7**, 87 (1990).

Formation of poly(glucosyloxyethyl methacrylate)–Concanavalin A complex and its glucose-sensitivity

KATSUHIKO NAKAMAE[1], TAKASHI MIYATA[1], ATSUSHI JIKIHARA[1] and ALLAN S. HOFFMAN[2]

[1]*Department of Chemical Science and Engineering, Faculty of Engineering, Kobe University, Rokko, Nada, Japan*
[2]*Center for Bioengineering, FL-20, University of Washington, Seattle, WA 98195, USA*

Received 6 January 1993; accepted 12 November 1993

Abstract—The complex formation between Concanavalin A (Con A) and a polymer having pendant glucose groups was studied in order to design a glucose-sensitive polymer. The polymer having pendant glucose (poly(glucosyloxyethyl methacrylate) or (poly(GEMA)) forms a complex with Con A in tris HCl buffer (pH = 7.5). The solution then becomes turbid due to the multiple associations between poly(GEMA) and Con A. When free glucose or mannose are added to the turbid solution, the solution becomes transparent again. However, the addition of galactose does not cause the solution to be transparent. This indicates that Con A prefers to form a complex with free glucose or mannose (but not galactose) rather than with the pendant glucose in poly(GEMA). Therefore, the complex between poly(GEMA) and Con A is expected to be glucose- and mannose-sensitive. The apparent dissociation constants of the complexes between saccharide (poly(GEMA), glucose, and mannose) and Con A were also determined by affinity electrophoresis.

Key words: Glucose; Concanavalin A; poly(GEMA); pendant glucose; complex; glucose-sensitivity; affinity electrophoresis.

INTRODUCTION

Stimuli-sensitive polymer gels which swell and shrink sharply and reversibly in response to pH, temperature, electric fields, solvent change, or chemicals are sometimes called 'intelligent' materials. These interesting materials have many potential uses in medicine and biotechnology [1, 2].

For example, many systems have been studied for stimuli-induced release of active agents, such as drugs. Copolymer gels based on poly(*N*-isopropyl acrylamide) (PNIPAAm) and poly(acrylic acid) (PAAc), temperature- and pH-sensitive polymers, respectively, have been studied by Hoffman, Kim, Bae, Okano and coworkers as matrices for environmentally-stimulated drug release [3–6]. Katono *et al.* showed a pulsatile release of drug driven by cyclic temperature-induced formation and dissociation of H-bonds between PAAc and PAAm chains [7]. Nakamae *et al.* introduced pH-sensitive phosphate monomers (phosphato-ethyl methacrylate or Phosmer M) into hydrogels, and discovered an unexpected temperature-dependent response in these interesting gels [8, 9]. Copolymer gels of a pH-sensitive monomer (AAc or Phosmer M) with the temperature-sensitive component, NIPAAm, have been applied to pH- or ion-exchange-controlled drug delivery at 37°C [10, 11].

Hoffman and coworkers proposed covalent immobilization of affinity ligands and enzymes within stimuli-responsive gels for purposes of affinity separations or environmentally-controlled bioreactions [12, 13]. Cells were also immobilized [14].

In the case of immobilized enzymes, when the temperature was raised above the gel collapse temperature (LCST), the enzyme reaction was turned off due to the collapse of the gel pores [12]. When the temperature was cycled above and below the LCST, a significant increase in enzyme–substrate reaction kinetics was also observed, due to the enhanced mass transfer of substrate in and product out of the gel, demonstrating the possibility of controlling both mass transport as well as reactions within gels by means of external stimuli [13]. Monji and Hoffman also conjugated an antibody to soluble PNIPAAm, and developed a novel immunoassay by complexing the antigen and a second, labeled antibody to the polymer–MAb conjugate in solution, and then precipitating the conjugate/complex by raising the temperature above the LCST [15]. This demonstrated the principle of environmentally-induced phase separation of a polymer–ligand conjugate.

Ishihara and coworkers immobilized glucose oxidase (GOD) in a pH-sensitive gel and showed that the conversion of glucose to gluconic acid could stimulate swelling and relase of insulin [16]. This concept was also demonstrated by Horbett and coworkers [17]. Kokufuta and Tanaka similarly showed that the change in solution composition induced by an immobilized enzyme could trigger the phase transition of the gel [18].

Kitano *et al.* demonstrated an increase in permeability of a polyvinyl alcohol–boric acid gel network caused by competitive binding of a free glucose molecule to the boric acid groups [19, 20]. They proposed this gel for use in a glucose-stimulated, feed-back insulin delivery system.

In 1979, Schultz and Sims showed that immobilized concanavalin A (Con A), which binds saccharides, could be used to detect glucose by the competitive binding of glucose to Con A, causing the release of a fluorescent-labeled dextran [21]. The measured increase in free dextran concentration was a direct measure of the glucose concentration [21, 22]. Kim and coworkers applied the complexation of Con A and saccharides to the delivery of insulin [23–26]. They glycosylated insulin (so-called SAPG insulin), complexed it with Con A, and retained the complex within a membrane capsule, which allowed glucose to diffuse in, disrupt the Con A–SAPG insulin complex, and release insulin, which diffuses out. The barrier does not permit the larger molecular weight Con A to diffuse out, so that when the glucose concentration decreases due to the physiologic action of the SAPG insulin outside the microcapsule, the delivery of insulin is shut off. Okahata *et al.* prepared a polymer with pendant sugar groups, which they grafted to a nylon membrane surface, and then complexed with Con A to form a complex network on the surface of the semipermeable nylon membrane [27]. They demonstrated an increase in permeability induced by free glucose, which binds competitively with the Con A, disrupting the network. Tanaka studied a gel system similar to Kim *et al.* [23–26] and Okahata *et al.* [27] to cause ion-induced expansion and collapse of a hydrogel caused by competitive binding of ionized and non-ionic saccharides, respectively, to Con A entrapped within the gel [28].

We have previously investigated a polymer with pendant glucose groups, prepared from the monomer glycosyl-ethyl methacrylate (GEMA). We studied protein adsorption on films of this gel, and found little protein adsorbed [29, 30]. We also studied drug delivery from crosslinked gels prepared from copolymers of HEMA [31] and butyl acrylate (BA) [32] with GEMA. In the latter study, we found that the GEMA–BA gels had excellent adhesive properties, even at 98% relative humidity, and the release rate of vitamin B12 could be controlled by the GEMA content [32].

In the study reported here, we continue our work with polyGEMA and extend it to the development of a novel glucose sensor, based on changes in the light transmission of a polyGEMA–Con A network due to competitive binding of glucose. We report on the binding of three different sugars to Con A and then investigate the behavior of the network complex of poly(GEMA) and Con A in the presence and absence of the different sugars. These studies are also relevant to the development of a glucose-sensitive poly(GEMA)–Con A hydrogel membrane for a glucose-controlled insulin delivery system.

METHODS

Materials

A monomer having a pendant glucose [glucosyloxyethyl methacrylate (GEMA)], was supplied as an aqueous solution by Nippon Fine Chemical Co., Ltd. Lectin concanavalin A (Con A) was purchased from Wako Pure Chemical Industries, Ltd. 2,2′-azobis(2-amidinopropane) dihydrochloride (AIBA) and other analytical-grade reagents were used without further purification and distilled water was used.

Polymerization of GEMA

An aqueous solution of GEMA was mixed with AIBA (2.5 wt% relative to the monomer) to make a 10 wt% solution and was then transferred into a glass tube. The reaction was carried out at 70°C for 20 h under nitrogen atmosphere. The polymer (poly(GEMA)) was isolated by slow precipitation with vigorous stirring into acetone, suction filtering, and repeated washing with acetone and was purified by reprecipitation from water into acetone and dried at 40°C *in vacuo*. The molecular weight of the polymer was determined by gel permeation chromatography (GPC): M_n, M_w, and M_w/M_n of poly(GEMA) were 86000, 310000, and 3.60, respectively.

Transmittance measurements

The resulting poly(GEMA) was dissolved in 0.1 M tris HCl buffer (pH = 7.5) containing 1 mM $MnCl_2$, 1 mM $CaCl_2$, and 0.1 M NaCl to make a solution of 5 wt%. It is well known that Con A contains Mn^{2+} and Ca^{2+}, which play important roles in making a complex of Con A with monosaccharide [33]. In general, therefore, $MnCl_2$ and $CaCl_2$ are added to the aqueous Con A solution [34]. Futhermore, NaCl stabilizes Con A in the aqueous solution and prevents it from forming a complex non-specifically with various saccharides. The desired amounts of Con A were stirred into the poly(GEMA) solution and the desired amounts of glucose or mannose were then added. All solutions were kept at 5°C for 24 h and transmittance of the solutions was then measured with a visible spectrophotomer at 600 nm.

Affinity electrophoresis [35, 36]

The separating gel and poly(acrylamide)(poly(AAm)) gels containing 5 wt% of GEMA as a ligand, were synthesized by redox copolymerization of AAm and

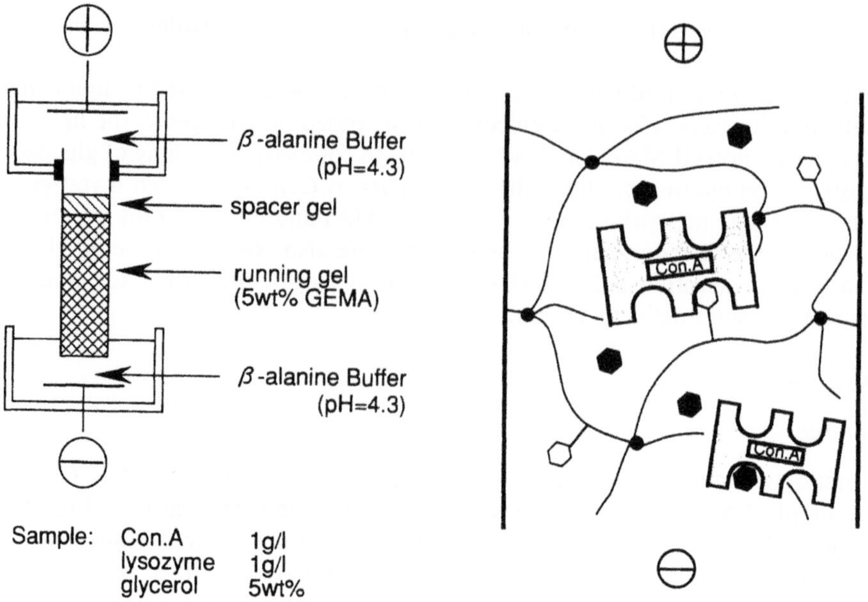
β-alanine Buffer (pH=4.3)
spacer gel
running gel (5wt% GEMA)
β-alanine Buffer (pH=4.3)
Sample: Con.A 1g/l
lysozyme 1g/l
glycerol 5wt%
Con.A
Con.A

Figure 1. Apparatus for affinity electrophoresis.

GEMA using ammonium persulfate and tetramethyl ethylenediamine as initiators. When the gel was washed to remove the monomers, GEMA was not detected from the cleaning solution. Therefore, it is thought that all GEMA was incorporated in the gel. The separating gel as prepared was 5 cm in height, contained 15% gel, and the spacer gel was 1 cm in height. Disc electrophoresis was carried out at 3 mA per tube for 3 h in β-alanine acetate buffer (pH 4.3) at room temperature. The disk electrophoresis apparatus is shown in Fig. 1. One $g\,l^{-1}$ of Con A as a sample, 1 $g\,l^{-1}$ of egg white lysozyme as a reference for calculation of the relative migration distance of Con A, and monosaccharide as an inhibitor in 5 wt% glycerol solution were added to the tube. After electrophoresis, the gel was stained overnight in 0.02% Coomassie Brilliant Blue R 250 in 7% acetic acid, and the migration distances of Con A relative to the lysozyme were measured.

RESULTS AND DISCUSSION

Formation of poly(GEMA)-Con A complex

Figure 2 shows the relationship between transmittance and amounts of Con A added to aqueous poly(GEMA) solutions in the absence and presence of glucose. The concentration of the aqueous poly(GEMA) solution is 5 wt% in 0.1 M tris HCl buffer at pH 7.5. Namely, the solution contains 1.71×10^{-2} mol% of the pendant glucose in GEMA. In the absence of glucose, the transmittance gradually decreased with an increase of Con A added to the poly(GEMA) solution. Under our experimental conditions, Con A is a tetramer [33] and forms a complex with four saccharides, recognizing hydroxyl groups at the 3, 4, and 6-positions [37–39]. The decrease of transmittance by addition of Con A indicates that Con A forms complexes with the pendant glucose groups in poly(GEMA) and that much of the

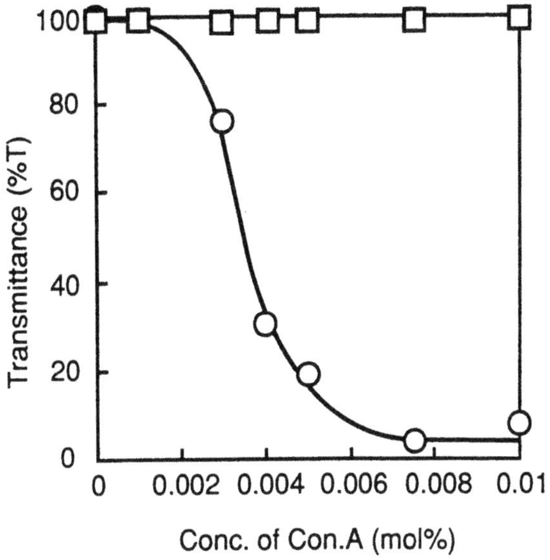
100
80
60
40
20
0
Transmittance (%T)
0
0.002 0.004 0.006 0.008 0.01
Conc. of Con.A (mol%)

Figure 2. Relationship between light transmittance at 600 nm and the amount of Con A added to aqueous poly(GEMA) solution in the absence (○) or presence (□) of 0.1 M glucose. The concentration of the aqueous poly(GEMA) solution is 5 wt% in 0.1 M tris HCl buffer at pH 7.5.

poly(GEMA) is associated with, and precipitated by, Con A. The fact that the transmittance came close to zero sharply at the Con A concentration of approximately 4.0×10^{-3} mol%, which corresponded to quarter concentration of the pendant glucose of GEMA in the solution, supports the supposition that a molecule of Con A can form a complex with four saccharides. On the other hand, when glucose is added to a suspension of poly(GEMA)–Con A complex in water, transmittance rapidly increases to 100%. This result is due to the exchange of free glucose with poly(GEMA) and to the dissociation of the poly(GEMA)–Con A complexes. This implies that Con A associates more preferentially with free glucose than with the pendant glucose in poly(GEMA), which is reasonable in view of the greater steric freedom of free glucose.

Figure 3 shows the relationship between the transmittance and the amounts of glucose, mannose, and galactose added to the suspension of the poly(GEMA)–Con A complex in water. The aqueous solution consisted of 1 wt% poly(GEMA) (3.4×10^{-3} mol% pendant glucose) and 4.0×10^{-3} mol% Con A in 0.1 M tris HCl buffer at pH 7.5. In the case of both glucose and mannose, the transmittance increases from 30 to 100% momentarily when they are added. This is also due to the dissociation of the poly(GEMA)–Con A complex by the addition of monosaccharide. Moreover, mannose was more effective for this action than glucose. The ability of Con A to form a complex with mannose is known to be greater than that with glucose [35, 37–39]. The addition of mannose to the suspension of the poly(GEMA)–Con A complex, therefore, leads to more effective dissociation of the complex than with glucose. On the other hand, the presence of galactose does not result in an increase of transmittance. This is attributable to the well-known small affinity of Con A for galactose. Con A recognizes and forms a complex with the hydroxyl groups at the 3, 4, and 6-positions in the saccharide [37–39]. The

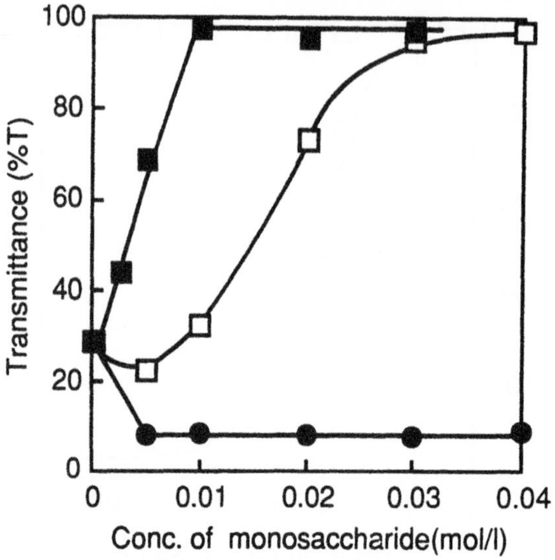
Transmittance (%T)
100
80
60
40
20
0
0
0.01
0.02
0.03
0.04
Conc. of monosaccharide(mol/l)

Figure 3. Effect of the amount of monosaccharide added to an aqueous poly(GEMA)–Con A complex solution on its transmittance. The aqueous solution consists of 1 wt% poly(GEMA) and 0.004 mol% Con A in 0.1 M tris HCl buffer at pH 7.5. The monosaccharides added to the solution are glucose (□), mannose (■), and galactose (●).

orientation of the hydroxyl groups at the 4-position in glucose and mannose is different from that in galactose, as shown in Fig. 4. The difference of hydroxyl groups at the 4-position gives rise to differences in affinity of Con A for glucose, mannose, and galactose. Consequently, the suspension of poly(GEMA)–Con A complex exhibits specific monosaccharide-sensitivity for transmittance.

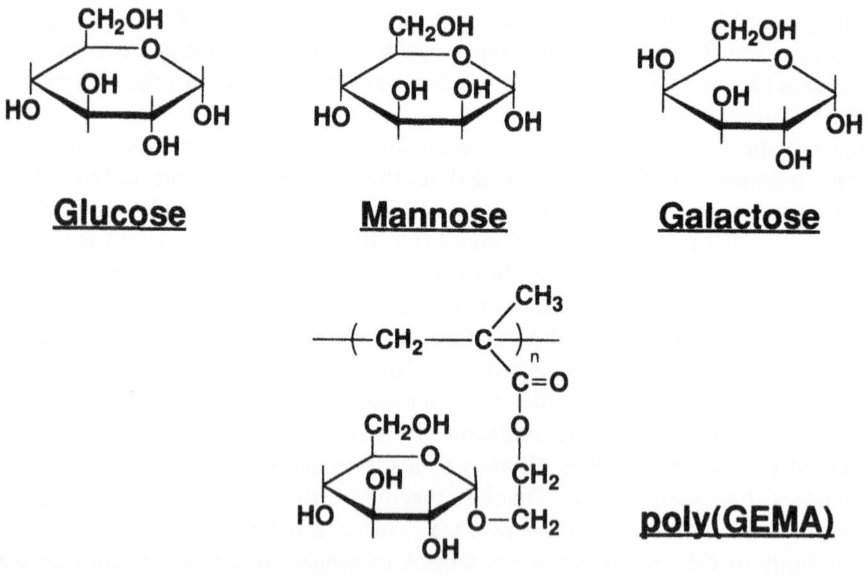
Glucose
Mannose
Galactose
poly(GEMA)

Figure 4. Structure of saccharides used in this study.

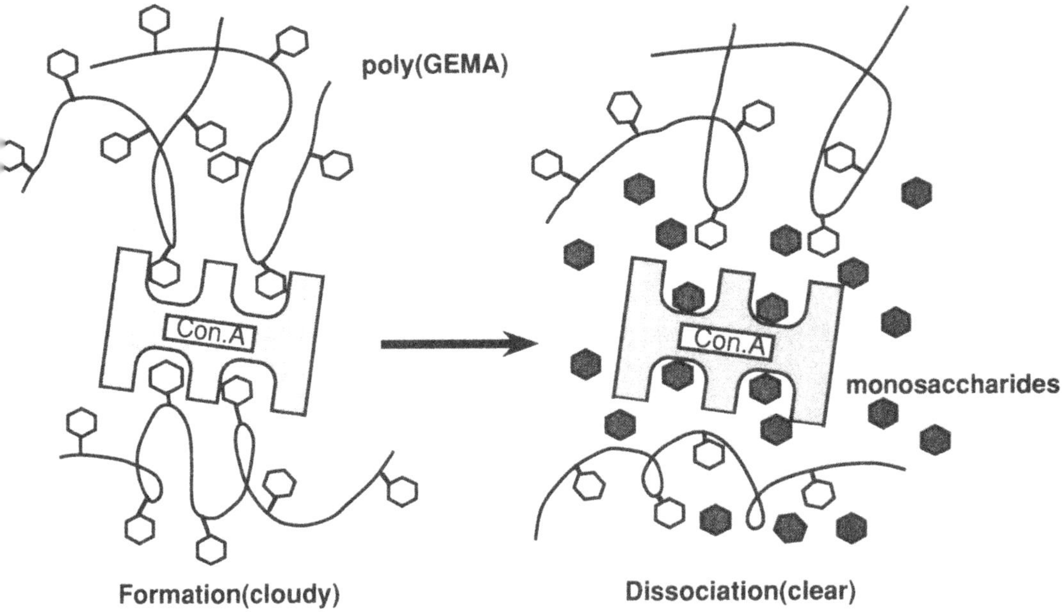
poly(GEMA)
Con.A
Con.A
monosaccharides
Formation(cloudy)
Dissociation(clear)

Figure 5. Schematic of monosaccharide-sensitive poly(GEMA)-Con A complexation.

Interaction of monosaccharides with the poly(GEMA)-Con A complex is illustrated schematically in Fig. 5. Since Con A is in the tetramer state under our conditions, it has four binding sites for saccharides. Therefore, when the pendant glucose groups in poly(GEMA) bind with Con A, the Con A acts as a flocculation agent and causes the poly(GEMA) to precipitate rapidly. When free glucose is added, however, Con A prefers to complex with free glucose rather than with the pendant glucose groups in poly(GEMA) because of the better steric fit, leading to a stronger binding. Consequently, the poly(GEMA)-Con A complex dissociates and redissolves. Thus, the poly(GEMA)-Con A complex exhibits glucose-sensitivity in this process.

Determination of apparent dissociation constants of poly(GEMA)-Con A complex

Figure 6 shows the patterns of inhibition affinity electrophoresis of Con A. In all cases, lysozyme is used as a reference and migrates to the same position. In the absence of monosaccharides, however, the migration distance of Con A in the GEMA-AAm copolymer gel was less than that in a poly(AAm) gel. This is attributable to the interaction of Con A with the pendant glucose in GEMA. On the other hand, the migration distance of Con A increased with an increase in the amount of glucose or mannose added, but is not influenced by the addition of galactose. This result indicates that free glucose or mannose inhibit the formation of the poly(GEMA)-Con A complex, whereas galactose does not. The relationship between the amount of added monosaccharide and the migration distance of Con A (relative to lyoszyme) is shown in Fig. 7. In the presence of glucose, the migration distance increased with an increase in the amount of added glucose and was constant above 0.03 mol l^{-1} glucose. With an increase in the amount of mannose, the

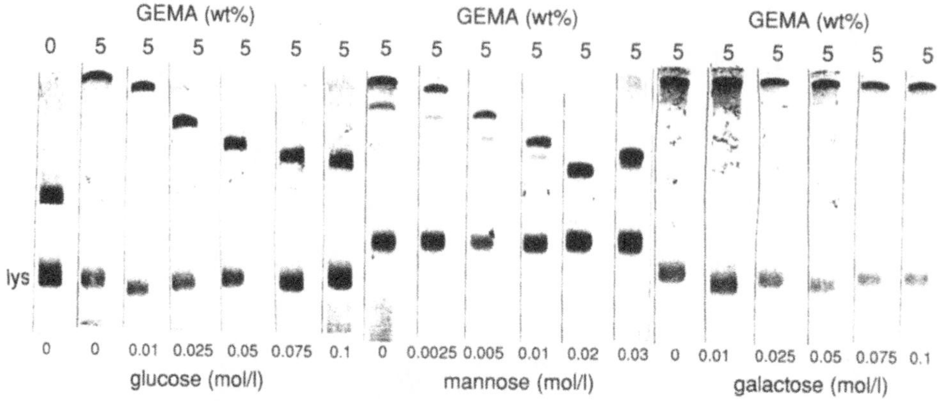
GEMA (wt%)
0 5 5 5 5 5 5
GEMA (wt%)
5 5 5 5 5 5
GEMA (wt%)
5 5 5 5 5 5
lys
0 0 0.01 0.025 0.05 0.075 0.1
glucose (mol/l)
0 0.0025 0.005 0.01 0.02 0.03
mannose (mol/l)
0 0.01 0.025 0.05 0.075 0.1
galactose (mol/l)

Figure 6. Patterns of inhibition affinity electrophoresis of Con A. The electrophoresis was carried out in poly(AAm) and poly(GEMA–AAm) gels containing various amounts of glucose (a), mannose (b), and galactose (c) in acidic buffer at 3 mA per tube for 2 h. Egg white lysozyme was added to the aqueous Con A solutions as a reference.

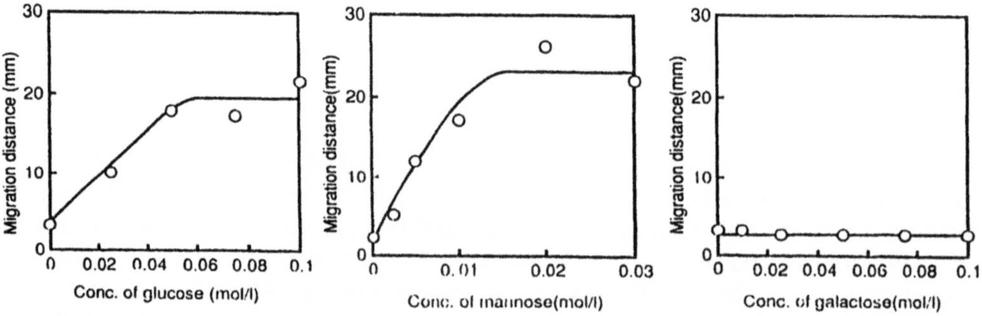
Migration distance (mm)
30
20
10
0
0 0.02 0.04 0.06 0.08 0.1
Conc. of glucose (mol/l)
Migration distance(mm)
0 0.01 0.02 0.03
Conc. of mannose(mol/l)
Migration distance(mm)
0 0.02 0.04 0.06 0.08 0.1
Conc. of galactose(mol/l)

Figure 7. Effect of the amount of added glucose (a), mannose (b) and galactose (c) on the migration distance of Con A relative to lysozyme.

relative migration distance also increased and was constant above 0.01 mol l^{-1} mannose. The presence of galactose did not increase the migration distance of Con A. These results indicate that glucose and mannose act as inhibitors for the complex formation between the pendant glucose group and Con A, but galactose does not. Furthermore, the presence of mannose is more effective for increasing the migration distance than that of glucose. These results are based on the fact that Con A has a stronger affinity for mannose than for glucose and has no affinity for galactose [35, 37–39]. The effect of the presence of free monosaccharide on the migration distance is similar to that on transmittance. Note especially, that the amount of monosaccharide at which the migration distance becomes constant, is approximately the same as that at which transmittance also levels out. This fact supports the conclusion in the previous section that an increase of transmittance by the presence of glucose or mannose is due to the dissociation of the poly(GEMA)–Con A complex.

Dissociation constants of poly(GEMA)–Con A complex can be determined from the migration distance, using the method by Horejsi *et al.* [35]. They calculated dissociation constants of saccharide–Con A complex from electrophoretic mobility of Con A, based on the equation derived by Dunn and Chaiken in affinity chromatography [40].

Con A (C) is equilibrated with GEMA (L) as a ligand in the poly(AAm) gel and free monosaccharide (I) as an inhibitor as follows:

$$C + L = CL \qquad C + I = CI$$

$$K = \frac{[C][L]}{[CL]} \qquad K_i = \frac{[C][I]}{[CI]}$$

where K and K_i are apparent dissociation constants of poly(GEMA)-Con A complex and free monosaccharide-Con A complex, respectively, since Con A has four binding sites. [C] [L], and [I] are the concentrations of Con A, GEMA as a ligand, and free monosaccharide as an inhibitor, respectively.

When x and x_0 are migration distances of Con A in the presence and absence of the free monosaccharide, the following equation is obtained:

$$\frac{x}{x_0 - x} = \frac{K}{[L]}\left(1 + \frac{[I]}{K_i}\right). \qquad (1)$$

When the relationship between $x/(x_0 - x)$ and [I] is plotted at a constant concentration of [L], a straight line should be obtained and is known as the inhibition affinity plot. Its intercepts on the x- and y-axes give K_i and K, respectively. The inhibition affinity plots of glucose-Con A and mannose-Con A in GEMA-AAm gels are shown in Fig. 8 and give straight lines. This demonstrates that the application of Eq. (1) is valid and that the formation of poly(GEMA)-Con A complex is reversible. The apparent dissociation constants obtained by Eq. (1) are presented in Table 1. The dissociation constant of glucose-Con A complex is about six times larger than that of the mannose-Con A complex. This result supports our previous conclusion that the affinity of Con A for mannose is stronger than that for glucose. On the other hand, the dissociation constant of GEMA-Con A complex is 2.07×10^{-3} mol l^{-1}. This value is smaller than that between free glucose and Con A, but is larger than that between free mannose and Con A. This indicates that Con A more preferentially forms a complex with the GEMA group in the gel than with free

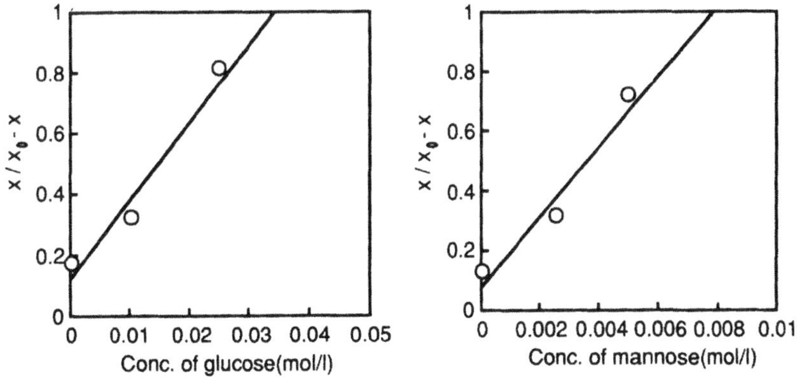
x / x0 - x
0
0.2
0.4
0.6
0.8
1
0 0.01 0.02 0.03 0.04 0.05
Conc. of glucose(mol/l)
x / x0 - x
0
0.2
0.4
0.6
0.8
1
0 0.002 0.004 0.006 0.008 0.01
Conc. of mannose(mol/l)

Figure 8. Affinity plot of the interaction of glucose-Con A and mannose-Con A in poly(GEMA-AAm) gels.

Table 1.
Dissociation constants of complexes of Con A with free monosaccharide and GEMA immobilized as a ligand in a poly(GEMA-AAm) gel.

Free ligand	K_i [M]
D-Glucose	5.15×10^{-3}
D-Mannose	8.87×10^{-4}
Immobilized ligand	K[M]
GEMA	2.07×10^{-3}

glucose and more preferentially forms a complex with free mannose than with the GEMA group. As revealed from the transmittance measurement, however, poly(GEMA)–Con A complex is dissociated by free glucose or mannose, and Con A has stronger affinity for glucose and mannose than poly(GEMA). The difference between the result of the transmittance measurement and dissociation constant from affinity electrophoresis is attributable to the possiblity that the affinity of Con A for GEMA within the poly(GEMA–AAm) gel is different from that for free poly(GEMA) in solution. This is perhaps due to the crosslinked nature of the gel

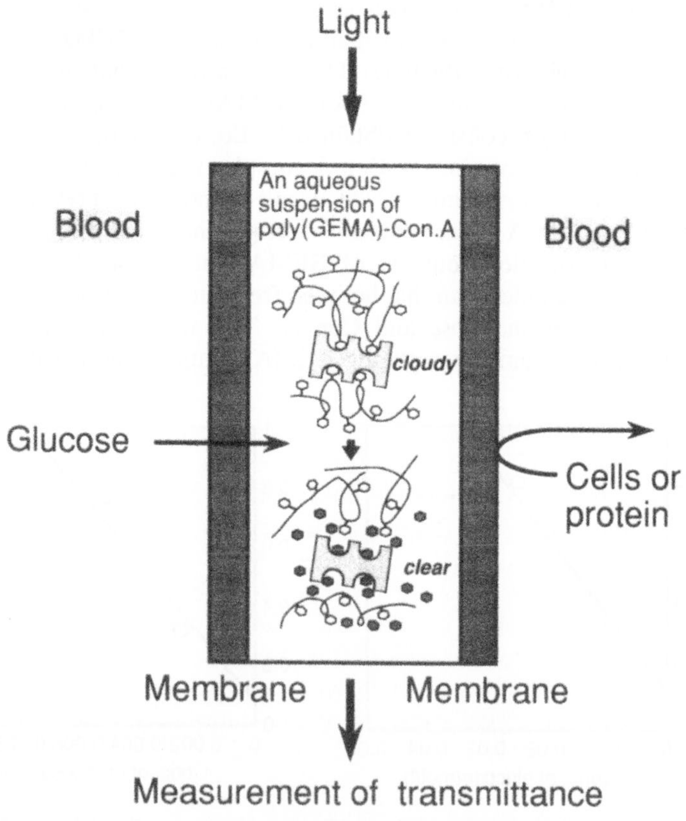
Light
An aqueous
suspension of
poly(GEMA)-Con.A
Blood
Blood
cloudy
Glucose
Cells or
protein
clear
Membrane
Membrane
Measurement of transmittance

Figure 9. Schematic diagram of the glucose sensor using this system.

which limits the conformational movements of both the GEMA units in the gels as well as the Con A, and thereby makes the complex more difficult to disrupt by the monosaccharides. The obtained values of the apparent dissociation constants are similar to those by Horejsi *et al.* [35] and Takeo [36]. Horejsi *et al.* also reported that K values were much lower than K_i values and may be due to the influence of main chains in the gel and the fact that effective or local concentration of immobilized saccharide is different from the total concentration.

It is apparent that poly(GEMA) forms a complex with Con A and precipitates due to multiple associations. The addition of glucose or mannose results in the dissociation of the poly(GEMA)–Con A complex, but that of galactose does not. Therefore, the poly(GEMA)–Con A complex can recognize specific monosaccharides and be specific monosaccharide-sensitive. Those results imply that glucose-sensitive devices can be designed based on the difference in binding affinity of Con A for free glucose vs the pendant glucose groups in poly(GEMA). For example, a conceptional glucose sensor, as shown in Fig. 9, can be built using this system. The sensor consists of an aqueous suspension of poly(GEMA)–Con A complex and an ultrafilter to prevent cells and proteins from passing through. The suspension of poly(GEMA)–Con A complex in the sensor is turbid before the measurement of glucose concentration. When the sensor is dipped in the blood, the transmittance increases with an increase in the glucose concentration because of the dissociation of the complex. The glucose concentration can be determined by the measurement of the transmittance of the solution. Based on this concept, such 'intelligent' materials can also be used for insulin release stimulated by the glucose-sensitivity of the poly(GEMA)–Con A complex. At present, we are applying these glucose-sensitive hydrogels to closed-loop insulin releasing devices.

CONCLUSIONS

The formation of complexes between poly(GEMA) and Con A in tris HCl buffer (pH = 7.5) in the absence and presence of free monosaccharide was investigated by light transmittance measurements and affinity electrophoresis. With increasing Con A concentration, poly(GEMA) is flocculated and the transmittance of an aqueous poly(GEMA) solution decreases. This indicates that Con A acts as a flocculation agent and causes poly(GEMA) to precipitate due to the binding of pendant glucose groups in poly(GEMA) with Con A. When free glucose or mannose is added to the suspension of the poly(GEMA)–Con A complex, the complex dissociates and redissolves due to the exchange of free glucose or mannose with the poly(GEMA). However, the addition of galactose to the suspension does not result in the dissociation of the complex. These results show that dissociation of the poly(GEMA)–Con A complex is different for different monosaccharides and that glucose and mannose are effective for this action. The apparent dissociation constants of poly(GEMA)–Con A complex and free monosaccharide–Con A complex were determined by affinity electrophoresis. The dissociation constant between poly(GEMA) immobilized in an electrophoresis separating gel and Con A is 2.07×10^{-3} mol l^{-1}, which is smaller than that between free glucose and Con A (5.15×10^{-3} mol l^{-1}). This indicates the importance of chain mobility of the poly(GEMA) to the complex affinity with Con A. These results can be applied to a novel glucose sensor, and also potentially to a new closed-loop insulin delivery system.

REFERENCES

1. A. S. Hoffman, *J. Control. Release* **6**, 297 (1987).
2. A. S. Hoffman, *MRS Bull. XVI* **9**, 42 (1991).
3. A. S. Hoffman, A. Afrassiabi and L. C. Dong, *J. Control. Release* **4**, 213 (1986).
4. L. C. Dong and A. S. Hoffman, *J. Control. Release* **13**, 21 (1990).
5. Y. H. Bae, T. Okano, R. Hsu and S. W. Kim, *Makromol. Chem., Rapid Commun.* **8**, 481 (1987).
6. Y. H. Bae, T. Okano and S. W. Kim, *J. Polym. Sci., Part B: Polym. Phys.* **28**, 923 (1990).
7. H. Katono, A. Maruyama, K. Sanui, N. Ogata, T. Okano and Y. Sakurai, *J. Control. Release* **16**, 215 (1991).
8. K. Nakamae, T. Miyata and A. S. Hoffman, *Makromol. Chem.* **193**, 983 (1992).
9. K. Nakamae, T. Miyata and A. S. Hoffman, *Makromol. Chem.* submitted.
10. L. C. Dong and A. S. Hoffman, *J. Control. Release* **15**, 141 (1991).
11. K. Nakamae, T. Nizuka, T. Miyata, T. Uragami, A. S. Hoffman and Y. Kanzaki, Abstract at 15th Annual Japan Biomaterials Soc. Mtg., p. 132. Kobe, Japan.
12. L. C. Dong and A. S. Hoffman, *J. Control. Release* **4**, 223 (1986).
13. T. G. Park and A. S. Hoffman, *J. Biomed. Mater. Res.* **24**, 21 (1990).
14. T. G. Park and A. S. Hoffman, *Biotech. Lett.* **11**, 17 (1989).
15. N. Monji and A. S. Hoffman, *Appl. Biochem. Biotech.* **14**, 107 (1987).
16. K. Ishihara, M. Kobayashi, N. Ishimaru and I. Shinohara, *Polym. J.* **16**, 625 (1984).
17. G. Albin, T. A. Horbett and B. D. Ratner, *J. Control. Release* **2**, 153 (1985).
18. E. Kokufuta and T. Tanaka, *Macromolecules* **24**, 1605 (1991).
19. S. Kitano, K. Kataoka, Y. Koyama, T. Okano and Y. Sakurai, *Makromol. Chem., Rapid Commun.* **12**, 227 (1991).
20. S. Kitano, Y. Koyama, K. Kataoka, T. Okano and Y. Sakurai, *J. Control. Release* **19**, 161 (1992).
21. J. S. Schultz and G. Sims, *Biotechnol. Bioeng. Symp.* **9**, 65 (1979).
22. J. S. Schultz, S. Mansouri and I. Goldstein, *Diabetes Care* **5**, 245 (1982).
23. S. Y. Jeong, S. W. Kim, M. J. D. Eenink and J. Feijen, *J. Control. Release* **1**, 57 (1984).
24. S. Sato, Y. Jeong, J. C. McRea and S. W. Kim, *J. Control. Release* **1**, 67 (1984).
25. S. W. Kim, C. M. Pai, K. Makino, L. A. Seminoff, D. L. Holmberg, J. M. Gleeson, D. E. Wilson and E. J. Mack, *J. Control. Release* **11**, 193 (1990).
26. K. Makino, E. J. Mack, T. Okano and S. W. Kim, *J. Control. Release* **12**, 235 (1990).
27. Y. Okahata, G. Nakamura and Hiroshi Noguchi, *J. Chem. Soc. Perkin Trans.* **II**, 1317 (1987).
28. E. Kokufuta, Y.-Q. Zhang and T. Tanaka, *Nature* **351**, 302 (1991).
29. K. Nakamae, T. Miyata and N. Ootsuki, *J. Biomater. Sci. Polym. Ed.* to be published.
30. K. Nakamae, T. Miyata and N. Ootsuki, *Makromol. Chem.* submitted.
31. K. P. Antonsen, J. L. Bohnert, Y. Nabeshima, M.-S. Sheu, X. S. Wu and A. S. Hoffman, *J. Biomater. Artif. Cells, Immob. Biotech.* **21**, 1 (1993).
32. K. Nakamae, T. Miyata, N. Ootsuki, M. Morizane, S. Ii, T. Nizuka, S. Yamamoto and M. Okumura, *Polym. Prep. Jpn.* **41**, 2421 (1992).
33. G. N. Reeke, Jr., J. W. Becker and G. M. Edelman, *J. Biol. Chem.* **250**, 1525 (1975).
34. Y. Oda, K. Kasai and S. Ishii, *J. Biochem. (Tokyo)* **69**, 285 (1981).
35. V. Horejsi, M. Ticha and J. Kocourek, *Biochim. Biophys. Acta* **499**, 290 (1977).
36. K. Takeo, *Lectins-Biology, Biochemistry, Clinical Biochemistry*, Vol. II. Walter de Gruyter & Co., Berlin, New York (1982).
37. I. J. Goldstein, C. E. Hollerman and E. E. Smith, *Biochemistry* **4**, 876 (1965).
38. B. B. L. Agrawal and I. J. Goldstein, *Biochem. J.* **96**, 23c (1965).
39. R. D. Poretz and I. J. Goldstein, *Biochemistry* **9**, 2890 (1970).
40. B. M. Dunn and I. M. Chaiken, *Biochemistry* **14**, 2343 (1975).

Poly(ethylene oxide) star molecules: Synthesis, characterization, and applications in medicine and biology

EDWARD W. MERRILL

Department of Chemical Engineering, Massachusetts Institute of Technology, Cambridge, MA 02139, USA

Received 17 September 1992; accepted 15 February 1993

Festschrift remark—It was an honor to be invited to honor Allan Hoffman on the occasion of his 60th birthday. It seems like yesterday that Allan and I shared an office in old Building 12 on the M.I.T. campus. At that time Allan had lost a bout with a snow covered mountain while on skis, and consequently moved about awkwardly in our already small room on crutches with a heavy cast. Despite these impediments, he carried on his classroom activities with typical energy. It was even earlier that at M.I.T., under the supervision of Ed Gilliland and myself, he carried out his doctoral research on the radiation grafting of styrene to polyethylene. Both he and I have dabbled in radiation chemistry on and off since then—he far more than I. Neither of us at that epoch had much idea about biomaterials or medical applications of materials—at least I did not—perhaps Allan had it in the back of his mind.

The University of Washington, Seattle, is recognized as one of the great centers of biomedical engineering in the world, both in scope and in depth, and Allan has played a stellar role in the evolution of Seattle's reputation. He has also developed a reputation for outstanding work in biomaterials and is one of the most visible and internationally recognized leaders in the field.

INTRODUCTION

Poly(ethylene oxide) PEO in various forms and under other names: poly(ethylene glycol) PEG, poly(ethylene glycol mono methyl ether) PEGME, etc. is widely recognized as unusual, and perhaps unique, in its lack of interaction with biological matter. For example, the clotting factor fibrinogen is not adsorbed, the complement system is not activated, platelets are not bound, viral particles are not bound, and so on. Various theories to explain these observations have been put forward, without a consensus having been reached. Harris [1] offers a recent resume of the field.

PEO has some unusual physical chemical properties: (1) it is soluble in widely disparate solvents of widely varying cohesive energy densities including water, benzene, chlorinated hydrocarbons and tetrahydrofuran; (2) in organic solvents, especially non polar solvents, it solvates the monovalent metal ions Li^+, K^+, Na^+ and for this reason has been called 'the poor man's crown ether'; and (3) dissolved in ion free water to which a limited quantity of sodium dodecyl sulfate has been added, PEO (François *et al.* [2]) acts like a classical polyelectrolyte and expands upon dilution by pure water so that the relative viscosity *increases* as concentration decreases. This suggests that the twelve carbon alkane chain of SDS wraps around the PEO chain so tenaciously that the sulfate ion is bound as if covalently.

The effect noted in (3) is almost certainly the explanation for the total failure of PEO hydrogels in attempted SDS gel electrophoresis of proteins: the SDS used to denature the proteins also 'decorates' the PEO of the gel, converting it to a polyelectrolyte gel, thereby guaranteeing electroendosmosis when the electric field is turned on. This same effect *may* also account in part for the success of PEO as an 'inert' surface in certain biological applications, especially blood contact. Since blood is a mixture of protein and cellular material containing traces of free fatty acids (FFA) as well as esterified fatty acids (EFA), PEO may adsorb the FFA and EFA and thus make itself appear 'more natural' to the biological environment. This hypothesis ought to be tested.

In any case, PEO is an interesting molecule in its usual form—a linear chain. There are two major issues that arise in its potential biological applications: (1) Is it to be used to cover an otherwise non-specifically binding surface, so as to prevent binding? If the answer is yes, to what density must the PEO chains be packed on the surface to prevent ingress of the biopolymers in the external solution? (2) Is it to be used as a molecular 'leash' for bioactive molecules (peptides, oligosaccharides, antibodies, etc.)? If so, the concentration and availability of terminal *hydrogel* groups is important.

STAR MOLECULES COMPARED TO LINEAR MOLECULES—A PREVIEW

As shown in a number of recent papers, and particularly well by Bauer *et al.* [3] for polyisoprene star molecules, the properties of isolated star molecules diverge increasingly from those of linear counterparts as the number of arms f on the star increases beyond 3 or 4. While with $f = 3$ or 4 the hydrodynamic volume of the star may be about the same as that of a linear molecule of the same molecular weight, as f increases into the tens of arms, the star molecule is more dense that its linear counterpart of equal total molecular weight, by as much as an order of magnitude. This means that in a multiarmed star (e.g., $f = 50$), the concentration of arms within the hydrodynamic volume is up to an order of magnitude greater than in the linear counterpart, and the concentration of terminal groups on the arms is $f/2$ greater than in a linear molecule of the same hydrodynamic volume. This relationship is shown schematically in Fig. 1 and will be discussed further below.

SYNTHESIS OF PEO STAR MOLECULES BY THE CORE-FIRST METHOD

In the synthesis of model stars [3] of polyisoprene it was possible to utilize the arm-first method of anionic synthesis, i.e. the arms are generated from a monofunctional

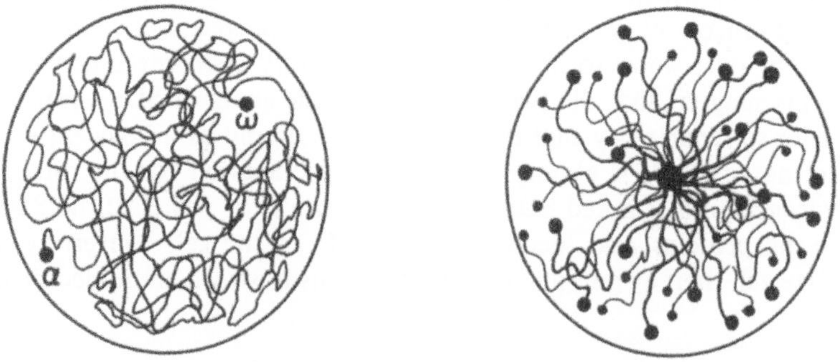

Figure 1. Comparison of linear PEO molecule as a random coil with a PEO star.

initiator (which thereby makes the starting unit a neutral group like butyl), and subsequently divinyl benzene is introduced to the living polymerization in such a mol ratio to the existing living ends that cores of narrow distribution of functionality f are formed, f being determined by the mol ratio experimentally set.

This arm-first procedure cannot be followed if the desired product is PEO star molecules in which the terminals of the arms are hydroxyl groups, since no *mono-functional* anionic initiator would subsequently be convertible to an hydroxyl group. Secondly, whereas living polyisoprene chains will add divinyl benzene, living PEO chains will not. Thus of necessity of 'core-first' anionic polymerization was devised by Rempp and his colleagues [4, 5].

Anionic polymerization of divinyl benzene DVB in tetrahydrofuran is initiated by potassium naphthenide under all the restrictions (purity, etc.) relevant to living anionic polymerizations and at $-40°C$. The mol ratio of DVB to K^+ is maintained in the range of 1.0–2.5, which results in the formation of living cores, each having a number of active carbanions and their corresponding potassium cations.

The longer one waits after initiation the larger the cores become (by association) and the fewer their number per unit volume. Also the higher the DVB/K^+ ratio, the larger is the ultimate average functionality of the cores. The count of carbanions is presumed equal to the count of potassium ions added as naphthenide.

At a certain time ethylene oxide gas is introduced, immediately converting the carbanions to oxanions, i.e.

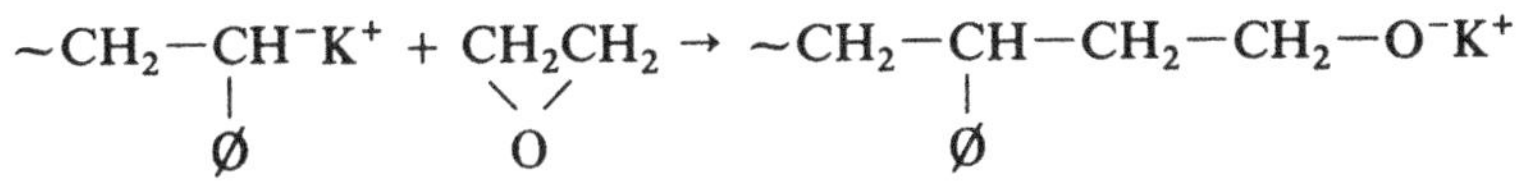

and the *first* of n units of ethylene oxide is thus fixed to the core. The temperature is then raised to $+40°C$ and more ethylene oxide in a predetermined amount is added. The reaction proceeds as the ring-operating polymerization of ethylene oxide. After sufficient time, the reaction is killed with acidified methanol which thereby creates a terminal hydroxyl from the oxanion at the end of each arm.

It is supposed that ring opening polymerization proceeds at nearly the same rate from each original carbanion, and thus that all PEO arms have nearly the same degree of polymerization. By determining the quantity of *unused* ethylene oxide following termination, and subtracting this from the quantity injected, one has the quantity (mols) polymerized. Division of the mols of ethylene oxide *consumed* by the mols of potassium naphthenide used, yields the number of ethylene oxide units added to each original carbanion site, and thus the number average degree of polymerization $\bar{P}_n$ of each arm. The number average should be close to the weight average in such a polymerization. Thus the molecular weight of an arm, M_{arm}, should be simply $44\bar{P}_n$.

The weight average molecular weight M_w of each batch of stars is determined by classical multiangle light scattering, assuming that the refractive index increment for the PEO star is the same as that of linear PEO in the same solvent.

By reason of the fact that the mol ratio of divinyl benzene to potassium naphthenide is of the order of 1.0–2.5, and the mol ratio of ethylene oxide utilized to potassium napthenide is of the order of 100–300, the mol ratio of ethylene oxide to divinyl benzene is of the order of 100, and the weight ratio thus of the order of 33. Thus the final star molecules contains about 3 wt.% DVB and about 97% PEO.

The average functionality is calculated approximately (ignoring the small content of DVB) as:

$$f \approx M_{star}/M_{arm}. \tag{1}$$

Since M_{star} is the weight average, given by a Zimm plot intercept, f must also be the weight average number of arms per star, f_w.

'IMMOBILIZED' PEO STAR MOLECULES

As is the case with linear PEO, a major potential field of biomedical use of PEO star molecules is as immobilized surface layers, which may serve only to prevent non-specific adsorption of biological entities onto the underlying support, *or* in addition to serve as a 'leash' to connect a bioactive ligand to a support [6]. 'Immobilized' suggests more than in fact occurs. PEO chains of any form connected to a surface and in an aqueous environment have molecular mobility.

Radiation cross-linked PEO star hydrogels were studied in an *ex vivo* baboon shunt model [7] in which it has shown that platelets are not deposited over a period of at least 1 h. Other studies in which the radiation dose [8, 9] was varied compared the response of PEO star molecules to linear PEO molecules in aqueous solution. The PEO star hydrogels so formed having a stated e.g., 10 vol.%, initial content of polymer as aqueous solution, subsequently swelled much less than hydrogels created from linear PEO at the same initial volume percent and exposed to the same radiation dose. This was attributed to the incorporation of the DVB cores which serve as multifunctional crosslinks augmenting the tetrafunctional crosslinks introduced by ionizing radiation.

PEO star molecules can be fixed (immobilized) on surfaces by utilizing the hydroxyl at the end of the PEO arm of the star. Surfaces of methacrylate polymers were exposed to ammonia ion plasma, resulting in the implantation of amino groups. PEO star molecules were activated by tresyl chloride in methylene chloride solution by the well established reaction:

$$\mathrm{ARM{-}OCH_2CH_2OH + CF_3CH_2SO_2Cl}$$

$$\rightarrow \mathrm{ARM{-}O{-}CH_2CH_2O\overset{\overset{\displaystyle O}{\|}}{\underset{\underset{\displaystyle O}{\|}}{S}}{-}CH_2CF_3}$$

When subsequently presented to the amino containing methacrylate surfaces at pH 10, the star molecules became attached, presumably as monolayers, by the expected reaction:

$$\mathrm{{-}NH_2 + CF_3CH_2\overset{\overset{\displaystyle O}{\|}}{\underset{\underset{\displaystyle O}{\|}}{S}}{-}O{-}CH_2CH_2OARM}$$

$$\rightarrow \mathrm{{-}\overset{\overset{\displaystyle H}{\|}}{N}CH_2CH_2OARM + CF_3CH_2SO_3H}$$

In preliminary experiments, it was found that the originally hydrophobic methacrylate surface became partially wettable by water following exposure to the ammonia ion plasma. Following exposure to the solution of tresylated PEO star molecules, the surface became completely wettable by water and slippery to the touch, suggesting monolayer attachment of PEO stars.

Subsequent XPS analysis of the surfaces indicated little change in carbon-nitrogen-oxygen content of the surface (the control being methacrylate after ammonia plasma treatment).

It is suspected that this is the consequence of the dehydration necessary prior to XPS analysis. The tresylated star molecules cannot be fixed to the surface in more than a monolayer (they cannot attach to each other).

By way of illustration we will assume that a monolayer of close-contacting, water swollen, spherical PEO star molecules have become attached covalently to a surface. We further assume, for reasons shown below, that a typical molecule might have a diameter D_0 of about 300 Å. We further assume that the volume fraction of PEO star arms within the spherical domain is around 0.05. Then total dehydration of such a surface layer prior to XPS would produce a thin wafer of the same diameter D_0 and height h by Model I (Fig. 2), or as a separate, collapsed sphere by Model II (Fig. 3), of diameter D_f. By the first model, a layer of dehydrated PEO star polymer of 10 Å thickness would result; by the second model the fraction of the surface *uncovered* by retraction of the PEO star to a dense ball would be approximately $1 - D_f^2/D_0^2$, or under the above assumptions, 0.86. Thus if the surface to which the star molecules were grafted is a polymethacrylate, which may show up to 5 at.% nitrogen after plasma treatment, the same elements: carbon and oxygen will appear as in the PEO star coating, and the differences would lie primarily in the detailed carbon spectra: carbonyl, ether, alkane, etc peaks as well as a slight change in the N/C, N/O, C/O ratios.

By either scenario (Model I or Model II), in view of the depth of scanning of conventional wide angle XPS, one would not expect much difference. Angular dependent scattering at increasingly shallow angles, now underway, may elucidate the question, in combination with appropriate label molecules attached to the PEO stars e.g., pentaflurobenzaldehyde.

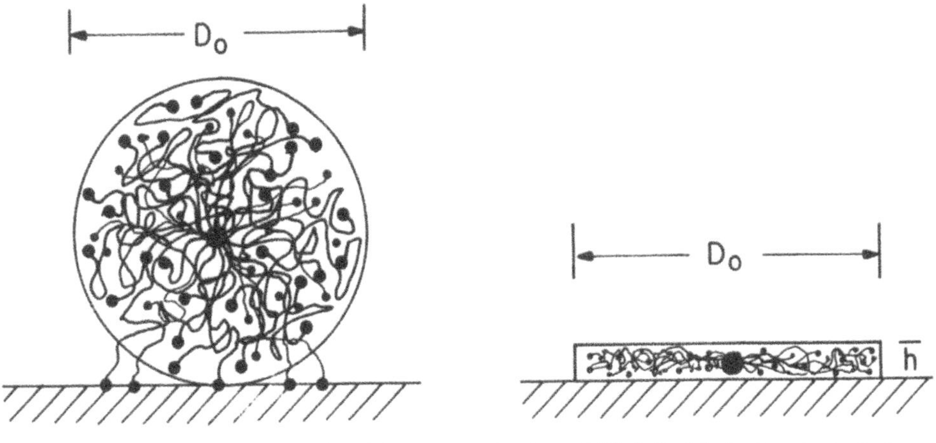
Do
Do
h

Figure 2. Dehydration of an immobilized PEO star molecule model I. Left: fully hydrated spherical star molecule; right: collapsed dehydrated star molecule as cylindrical wafer.

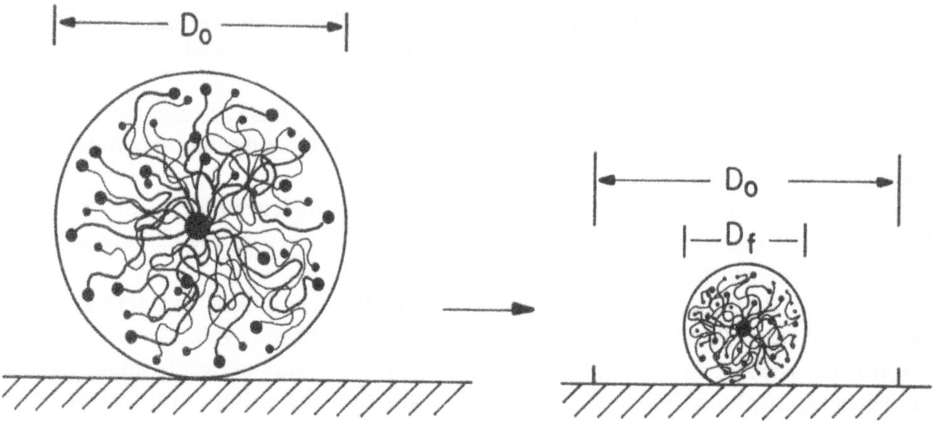
Do
Do
Df

Figure 3. Dehydration of an immobilized PEO star molecule model II. Left: fully hydrated spherical star molecule; right: collapsed dehydrated star molecule as a dense sphere.

Indeed a major purpose of PEO star molecules 'immobilized' on supports is to hold biopolymers such as oligopeptides, enzymes, proteins, heparin—the PEO arms acting as leashes when hydrated. The several biopolymers when attached may assist XPS interpretation.

As to why we made the above assumptions about size and density of PEO star molecules, we turn to the question of molecular volume/mass ratio as a function of the number of arms.

RELATION OF MOLECULAR VOLUME TO MOLECULAR MASS OF ISOLATED MOLECULES IN SOLUTION

The studies of Bauer *et al.* [3] on polyisoprene (PIP) star molecules, in which number of arms f was increased while M_{arm} was held constant, were unusually complete in that three radii were determined for each species: the Einstein radius R_E via intrinsic viscosity, the radius of gyration R_G via angular dependence in light scattering, and the thermodynamic radius R_{th} via concentration dependence in light scattering.

The Einstein equivalent radius R_E is determined from Einstein's equation for the viscosity η_s of a suspension of spheres, of volume fraction ϕ, relative to the viscosity of the continuum η_0

$$\eta_s/\eta_0 = 1 + 5/2\phi. \quad (2)$$

This equation can be rearranged, taking the volume of a single sphere as (4/3) πR_E^3, and the number per unit total volume as C/m or $C/(M/N_A)$, where C = concentration in g/cm^3, m = mass of a sphere, i.e. a single macromolecule, M = molecular weight, and N_A = Avogadro's number:

$$\lim_{C \to 0} \frac{(\eta_s/\eta_0) - 1}{C} = [\eta] = 6.28 \times 10^{24} R_E^3/M \quad (3a)$$

whence:

$$R_E = \left(\frac{[\eta]M}{6.28 \times 10^{24}}\right)^{1/3}. \quad (3b)$$

Since in Eqn (3b), for such a series of PIP star molecules, the total molecular weight is linearly proportional to the number of arms f (M_{arm} = constant), the Einstein radius R_E would vary as $f^{1/3}$ if the intrinsic viscosity were independent of f. Their finding that $R_E \sim f^{0.3}$ was close to such a condition, and by further noting the increasing ratios R_G/R_E and R_{th}/R_E they concluded that as f increased to large values (*c.* 50) the molecule approached increasingly a hard sphere model, and consequently that each arm became increasingly extended from the core, although the arm was of constant molecular weight.

Little information is available on star molecules made by the anionic core-first method (core = DVB) except those of Lutz [private communication] on polystyrene star molecules made analogously (i.e. DVB initiated by potassium naphthenide, to which subsequently styrene monomer was added). Lutz fractionated the batches as synthesized, and then performed light scattering and intrinsic viscosity measurements. As in the PEO star molecules under discussion, the fraction of the PST star which consisted of DVB, not styrene, was of order 0.01–0.03. Thus by knowing the molecular weight of the polystyrene arm (by the styrene/K^+ ratio) and the total molecular weight of the star by light scattering, it was possible to deduce the functionality (number of arms) f for each fraction derived from the initial batch. Lutz found that the intrinsic viscosity of the fraction was nearly constant over a range of $4 \leq f \leq 50$ for three different batches of PST stars, varying in arm molecular weight.

In view of the near constancy of $[\eta]$ observed by Lutz, application of Eqn (3b) leads to the conclusion that $R_E \sim f^{1/3}$, close to the relation observed by Bauer *et al.* [4]. Thus probably the conclusions of Bauer *et al.* with respect to PIP star molecules apply to PST star molecules [3].

What Lutz [10] thereby proved was that by the core first anionic synthesis, a single batch of PST star molecules would consist of species with a wide range of f. The molecular weights of the fractions obtained would therefore range from well below to well above the value $\bar{M}_w$ observed by light scattering on the unfractionated batch, and so the values of f would range from below to above the value $\tilde{f}_w$ calculated by application of Eqn (1).

In contrast to PST star molecules which can be fractionated by classical manipulation of temperature, it was found impossible to obtain fractions of PEO star molecules because crystallization of the PEO intervenes. Thus at present we infer the properties of PEO star molecules made by the core first method by what is known about PST star molecules made by identical starting conditions (DVB/K^+ ratio, temperature, time before addition of styrene, solvent, etc.) and we conclude that in any batch of PEO star molecules, even though arm molecular weight is probably almost constant, there must exist a wide range of the number of arms f.

When a surface containing groups e.g., $-NH_2$ or $-SH$ reactive toward tresylated hydroxyl groups is exposed to a batch of tresylated PEO star molecules, it would be expected that the attachment to the surface would be random, that possibly the greater the number of tresyl groups on a particular star, the greater its probability of attachment, and that after deposition of the largest star, the smaller might eventually be incorporated in the remaining space on the surface. If the objective is to cover a surface to prevent subsequent non-specific binding of constituents from biological milieux, the mixture of stars—varying in their cross-sectional area according to their number of arms f—might produce a more completely covered surface than as if all were of the same size (same f). This has yet to be proved by experiment.

Secondly, one must consider the effective radius of a PEO star molecule in relation to the effective radius of the biological entity to be attached. One way to estimate this is by use of the defined ratio g:

$$g = [\eta]_{\text{star}}/[\eta]_{\text{linear}} \tag{4}$$

wherein $[\eta]$ refers to intrinsic viscosity, respectively of the star molecule and of its linear counterpart of the same total molecular weight. Thus in a series of star molecules in which arm molecular weight M_{arm} is constant, the number of arms f is variable, and the core mass is negligible, the star $[\eta]$ is to be taken as the ratio with $[\eta]$ of a linear chain of molecular weight fM_{arm}.

Lutz [private communication] determined g as a function of f for fractionated PST stars made by the core first method, such as used for PEO stars; and found that g varied from ~ 1 for $f = 5$ to $g \approx 0.2$ for $f = 50$. Utilizing and extrapolating this variation of g with f to apply to PEO stars of arms molecular weight 10 000, $f = 4$ and $f = 100$, and utilizing the known relation for intrinsic viscosity of linear PEO in water at 25°C

$$[\eta]\ \text{cm}^3 g^{-1} = 0.0121 M^{0.784}$$

we estimate the Einstein radii R_E to be respectively about 50 Å for $f = 4$ and 200 for $f = 100$.

Either value is large by comparison with the dimensions of the cell attachment tripeptide RGD or pentapeptide YIGSR and thus either hypothetical species of star might become fully fitted with the peptide, if its terminal hydroxyls had been tresylated, yielding respectively 4 and 100 peptides per star.

In contrast, if the bipolymer to be attached is an antibody of effective radius around 50 Å, it might be possible to fit four antibodies onto the PEO star with $f = 4$, but spatially it would be impossible to fit 100 antibodies onto the star with $f = 100$. The surface area of a star with $R_E = 200$ Å might accommodate about 10–12 antibodies if they are assigned a projected area proportional to the square of the presumed radius (50 Å).

The general implication therefore is that when multifunctional PEO star molecules having tresylated arms and in *the free state* are exposed to biopolymers, the number of biopolymers that become attached may be considerably less than the number of arms, and a particular biopolymer might be attached to the same star by several arms.

When the star molecules are first immobilized on a surface, the number of biopolymers that can be subsequently added would be considerably less than what would be possible if the star molecules were in a free state.

PRACTICAL FRACTIONATION OF PEO STAR MOLECULES MADE BY THE CORE FIRST METHOD

As noted above, classical fractionation is impossible owing to premature crystallization upon concentration. Since, as is evident from PST star molecules prepared by the same route, the star molecules differ in effective volume (and thus effective radius), two routes to fractionation are available: gel permeation chromatography GPC and ultrafiltration. While GPC cannot be called a practical method for large scale separation, ultrafiltration, though imprecise, can be carried out on almost any scale.

Gel permeation chromatography carried out on several batches of PEO star molecules (core-first method) confirms a wide distribution of sizes, spanning a range of elution volumes that correspond to linear PEO calibration standards ranging from 20000 to 500000. Such a distribution of sizes is expected from observation of PST stars.

By tangential flow ultrafiltration, it was possible to remove into the filtrate a fraction of the star molecules in the intial mixture, the fraction depending on the nominal molecular weight cut-off of the membrane. The possibility of ultrafiltration raises interesting questions in respect to star molecules. Linear polymers as random coils in solution can change shape (elongate, 'reptate') and thus pass through ultrafiltration membranes which would retain rigid molecules (e.g., protein) of the same size. As more arms are added to the core of a star, it should be increasingly 'rigid', i.e. less deformable in the flow field of an ultrafiltration membrane pore. Thus two different star molecules having the same effective diameter as judged by GPC elution, one with four longer arms and the other with 100 shorter arms, might be separated by ultrafiltration on the basis of flexibility. These issues are under investigation (Fig. 4).

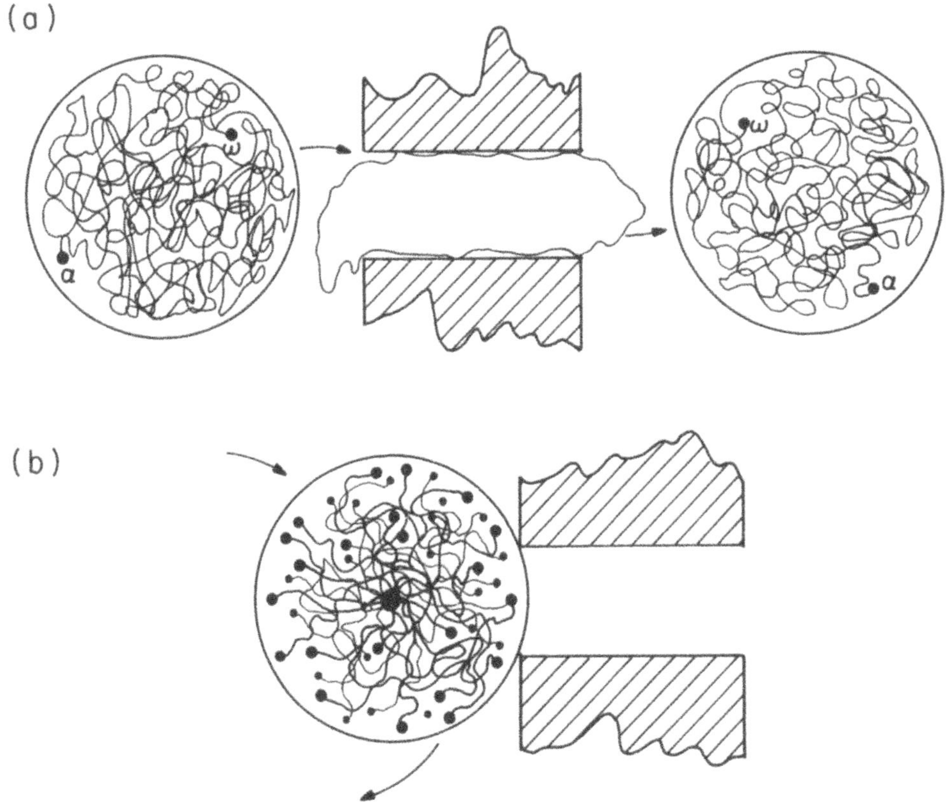
(a)
ω
a
ω
a
(b)

Figure 4. (a) Linear PEO chain shown, or *four-arm* PEO star, readily deforms and moves through pore of ultrafiltration membrane. (b) *Multiarmed* PEO star of same volume is reflected because of inability to deform and enter pore.

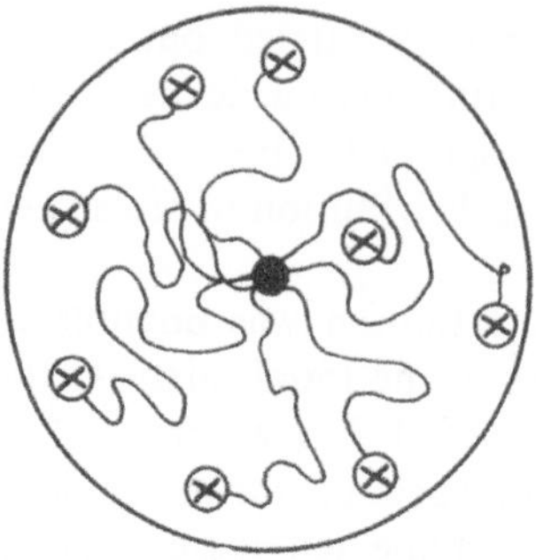

Figure 5. PEO star molecule as carrier of pharmacological agent X (X = oligopeptide for example).

APPLICATIONS FOR NON-IMMOBILIZED PEO STAR MOLECULES

One biomedical application for which prior ultrafiltration might be desirable would be the use of PEO star molecules as carriers of *small* bioactive molecules to be injected into the blood stream (Fig. 5). One would preselect a fraction of PEO stars that could not rapidly be eliminated by the kidney. Bioactive molecules small enough to be eliminated via the kidney would after attachment to PEO star molecules remain much longer in the blood compartment.

A diagnostic application for free PEO star molecules would be as enhancers of antibody-antigen reactions (Fig. 6). Antibodies in general act as bifunctional

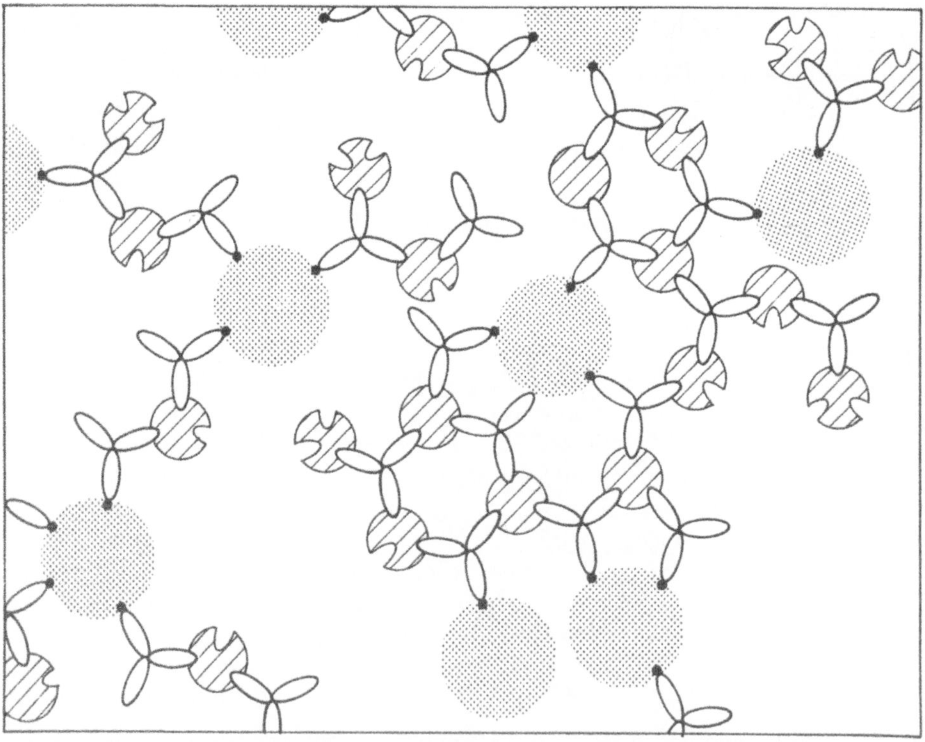

Figure 6. Antibodies attached to PEO star molecules (finely dotted circles) react with trifunctional antigen (diagonally lined circles) to form dense network.

molecules binding to plurifunctional antigens, and thereby causing turbidity, precipitation, etc: signals of diagnostic importance. Presumably the signals would be amplified if the antibodies were first linked to polyfunctional PEO star molecules, at only one point of attachment. This would mean that PEO stars of few but long arms ($M_{arm} \sim 10000, f \sim 4$) would be selected.

Tresylation of the stars followed by exposure to the desired antibody in great excess would result in attachment of the antibodies without cross-linking. (Great excess is required, as we discovered in some experiments with tresylated PEO stars and albumin. The albumin, having several accessible amino groups, coupled with the polyfunctional star molecules creating a gel.)

CONCLUSIONS

This account concerns a new class of star macromolecules, the physical chemical properties of which have yet to be worked out in detail. Numerous opportunities appear to exist for the application of these star molecules as components of biomaterials, and as independent entities in pharmacological and diagnostic products. Consequently this is a very preliminary account.

REFERENCES

1. J. M. Harris (ed.), *Poly(Ethylene Glycol) Chemistry*. Plenum Press, New York (1992).
2. J. François, J. Dayantis and J. Sabbadin. *Eur. Polymer J.* **21**, 165 (1985).
3. B. J. Bauer, L. J. Fetters, W. W. Graessley, N. Hadjichristidis and G. F. Quack, *Macromolecules* **22**, 2337 (1989).
4. P. Lutz and P. Rempp, *Makromol. Chemie* **189**, 1051 (1988).
5. Y. Gnanou, P. Lutz and P. Rempp, *Makromol. Chemie* **189**, 2885 (1988).
6. E. W. Merrill, U.S. Patent 5, 171, 264 (1992).
7. P. Rempp, P. Lutz, E. W. Merrill and A. Sagar, *Polymer Preprints (ACS)* **32**, 687 (1991).
8. E. W. Merrill, P. Rempp, P. Lutz, A. Sagar, R. Connolly, A. D. Callow, K. Gould and K. Ramberg, *Soc. for Biomaterials Proceedings* (1990).
9. P. Rempp, P. Lutz and E. W. Merrill, *Polymer Preprints (ACS)* **31**, 215 (1990).
10. P. Lutz, *private communication* (1991).

Investigations of the architecture of tamarind seed polysaccharide in aqueous solution by different scattering techniques

P. LANG* and K. KAJIWARA

Kyoto Institute of Technology, Matsugasaki, Sakyo-ku, Kyoto 606, Japan

Received 15 June 1992; accepted 29 October 1992

Abstract—The architecture of tamarind seed polysaccharide (TSP) has been investigated by light scattering (LS), small angle X-ray scattering (SAXS) and synchrotron radiation scattering (SRSAXS). The experimental data show that TSP in aqueous solution consists of multi stranded aggregates, with a high degree of particle stiffness. The angular dependence of the scattered intensity is typical for wormlike chains. Data evaluation on the basis of this model yields a statistical Kuhn segment length $l_K = 150$ nm. The cross sectional radius of gyration is estimated as $R_{gcs} = 6.0 \pm 0.5$ Å, which is more than twice the value, published for single stranded polysaccharides. Correspondingly, the experimental value of the linear mass density, measured by LS, is about five times higher than the theoretical value calculated from the primary structure.

Key words: Tamarind seed polysaccharide; light scattering; X-ray scattering; wormlike chain; particle scattering factor.

INTRODUCTION

Polysaccharides are a class of polymers which are applied in very different fields, covering such extremes like food processing on one hand, and tertiary oil exploitation on the other [1]. They also find wide-spread use in the pharmaceutical industry, mainly as pharmacologically inactive additives [2]. In recent years the biomedical significance of polysaccharides has also been discovered. For instance the bacterial polysaccharide schizophyllan [3] shows surprising anti-tumor activity, probably due to its particular structure in solution, a triple helix [4].

Certain properties of polysaccharide solutions which are desired in one field of application may be a severe disadvantage in another. High solution viscosity especially, which is requested in most applications, is a major handicap if such a solution has to be injected intramuscularly or subcutaneously for medical applications. However, the properties of polymer solutions are always closely connected to the architecture of the solute, and thus it is of major interest to know the latter under the conditions of application.

A very powerful method to investigate polymers in solution is the scattering of electromagnetic waves [5, 6], which we have applied to a plant polysaccharide in the present study. Tamarind seed polysaccharide (TSP) is isolated from the seed kernels of *Tamarindus indica*, a tree growing mainly in South East Asia. There is a great

* Present address and to whom correspondence should be addressed: Ivan N. Stranski-Institute for Physical and Theoretical Chemistry, Technical University, Straße des 17 Juni 112, W 1000 Berlin 12, Germany.

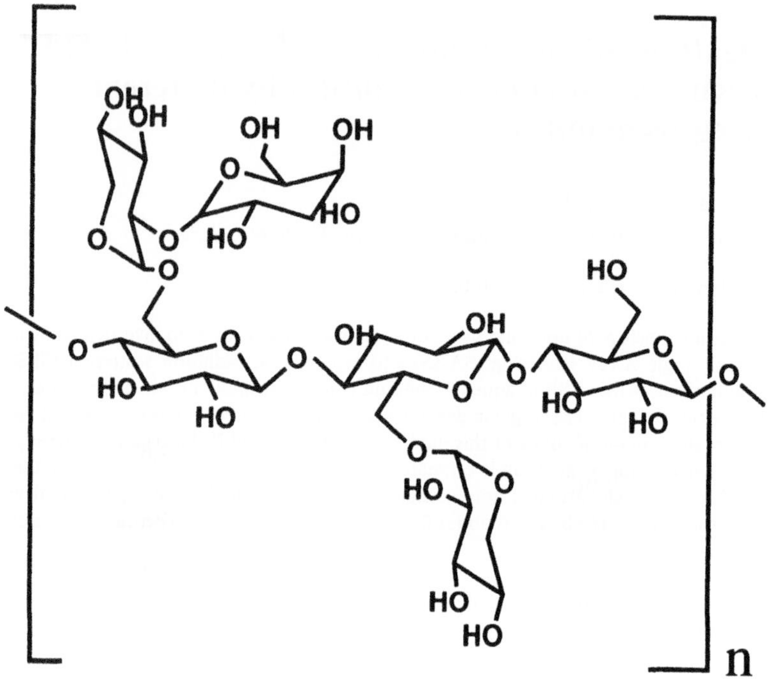

Figure 1. Common elements of the suggested primary structures of TSP. The cellulose type backbone is substituted by xylose and galactoxylose residues. The molar ratio of the residues is subject to discussion.

number of publications [7–10] on the primary structure of this polymer, all agreeing in the type of backbone and the nature of the substituents, but differing in the molar ratios of the residues. There is common agreement about the structural unit, shown in Fig. 1. It consists of a cellulose type spine with α-D-xylose (1 – >6) linked to the glucose residues. A part of xylose residues are substituted in C-2 position by β-D-galactose.

TSP is applied in the Japanese food processing industry because of its ability to enhance the viscosity of aqueous-based systems. This feature is attributed to a high degree of chain persistence of the polymer. However, there was some contradiction in LS and SAXS data, reported recently, as to whether this persistence is an intrinsic property of the polymer chain, due to the bulky side chains [10], or whether the formation of multi stranded aggregates [11] causes the observed particle stiffness.

In the present investigation we have applied both scattering techniques on the very same polymer solutions in order to clarify this problem.

EXPERIMENTAL

All measurements were carried out on a chromatographically purified TSP sample, made available by Dainippon Pharmaceuticals & Co., Osaka, Japan. The quality of the sample was verified by ^{13}C and ^{1}H-NMR, thermogravimetry and elemental analysis, which gave the same results as the elaborate analyses reported in ref. [10]. X-ray scattering data were collected with a commercial Kratky camera, equipped with a Phillips PW1710 diffraction control unit, which provides CuK_{α}-radiation, with a wave length of $\lambda = 1.54$ Å. A range of scattering vectors

$3 \times 10^{-2} < q < 0.55$ Å^{-1} was covered with a resolution of 172 points. This technique is known as small angle X-ray scattering (SAXS). Deconvolution of the smearing, due to finite primary beam profile, was performed, either with a house made software, using statistical procedures described elsewhere [12], or a software based on the program ITP by Glatter [13].

Additional X-ray data were recorded at the beam line 10-C of the Photon Factory in Tsukuba, Japan, using synchrotron radiation. Data collection was performed with a one dimensional position sensitive photon counter, in a range of scattering vectors $5 \times 10^{-3} < q < 0.17$ Å^{-1} with each 256 channels above and below the zero degree level. The number of counts in the corresponding channels in the lower and upper branch were averaged. These data do not require desmearing, since the radiation provided is monochromatized at $\lambda = 1.49$ Å, and the high primary intensity allows for point collimation [14]; data from this source are known as synchrotron radiation SRSAXS data. Static light scattering measurements were carried out on commercial equipment by Sofica, France. A mercury high pressure lamp at $\lambda = 435.8$ nm was used as a light source. Data were collected in so-called VV-geometry, i.e. the polarizer for the incident beam and analyzer for the scattered light were both in vertical position. The scattering intensity was recorded from $\Theta = 22.5$ to 45 deg in steps of 7.5 deg, from $\Theta = 45$ to 135 deg in steps of 15 deg and at 142.5 deg. All solutions were diluted from a stock of 12 g/l TSP in 5 mM NaN_3 solution, which had been prepared 1 week prior to the measurements. The solvent was water, purified by distillation and reverse osmose, to result an electric resistance of 18.1 MΩ. No further care was taken for pH adjustment. Typical concentrations for measurements were 6.0–12.0 g/l for SAXS, 2.0–10.0 g/l for SRSAXS and 0.5–2.0 g/l for light scattering. In the latter case, the solutions were filtered through 0.45 μm filters directly into measuring cells.

THEORETICAL BASICS

The angular dependence of the electromagnetic radiation, scattered by an assembly of macromolecules, is conveniently described by the familiar equation given by Zimm [15]

$$\frac{Kc}{R(q)} = \frac{1}{M_w P(q)}(1 + A_2 c + \ldots) \qquad (1)$$

where c is the solution concentration, q is the scattering vector defined by $q = 4\pi \sin(\Theta/2)/\lambda$, Θ is the scattering angle, $R(q)$ is the ratio of scattered and primary intensity $I(q)/I_0$, K is an apparatus constant, containing the optical contrast factor and calibration constants, M_w is the weight average molar mass of the observed polymer, A_2 is the second osmotic virial coefficient of the solution, and $P(q)$ is the particle scattering factor.

The so-called particle scattering factor is the Fourier transform of the distance distribution of scattering centers $D(r)$ within the observed particle

$$P(q) = \int_0^\infty r^2 D(r) \frac{\sin(qr)}{qr}\, dr \qquad (2)$$

with r being the distance between two scatterers. Thus $P(q)$ is characteristic of the polymer architecture.

To eliminate the influence of inter-particle interference on the angular distribution of the scattered intensity, the data, collected at finite concentrations, have to be extrapolated to infinite dilution. Equation (1) then reads

$$\lim_{c\to 0}\frac{Kc}{R(q)} = \frac{1}{M_w P(q)} \equiv \frac{F}{R(q)}. \tag{3}$$

In general the particle scattering factor may be expanded in a power series of the product of the mean square radius of gyration and the squared scattering vector $\langle S^2\rangle q^2$. For small scattering vectors, i.e. $\langle S^2\rangle q^2 \ll 1$, the series may be truncated after the second term to yield

$$\lim_{c\to 0}\frac{F}{R(q)} = \frac{1}{M_w}\left(1 + \frac{1}{3}\langle S^2\rangle q^2\right). \tag{4}$$

Thus the extrapolation of scattering data to infinite dilution and zero scattering angle provides the weight average molar mass M_w and the radius of gyration $R_g = \langle S^2\rangle^{1/2}$.

Detailed information on the architecture of the observed polymer can be gained from the analyses of the angular dependence of the scattered intensity over the entire range of measured scattering vectors. The particle scattering factor for random coils and rigid rods have been predicted long ago by Debye [16] and Zimm *et al.* [17] respectively:

$$P_{\text{coil}}(q) = \frac{2}{u^4}(u^2 - 1 + e^{-u^2}) \tag{5a}$$

$$P_{\text{rod}}(q) = \frac{1}{x}\int_0^{2x}\frac{\sin w}{w}\,dw - \left(\frac{\sin x}{x}\right)^2 \tag{5b}$$

with

$$x = \frac{Lq}{2} \quad \text{and} \quad u = R_g q$$

where L is the length of the rod. Linear polymer systems, which lie in between these two extremes, are best described by the Kratky–Porod wormlike chain model [18] or by Kuhn's model of a dummy chain [19]. The particle scattering factor of such systems was derived by Koyama [20]:

$$P(q) = \frac{2}{L_c^2}\int_0^{L_c}(L_c - t\,\Phi(q)\,dt \tag{6}$$

where L_c is the contour length of the chain. The term $\Phi(q)$ is a highly complex function of the Kuhn statistical segment length l_K and the second and fourth moment of the distance distribution function. For details the reader is referred to the original paper. The expression for $P(q)$ in Eqn (6) reproduces Eqn (5a) exactly in the case of small scattering vectors. For large values of q it yields the same result as the expression for rigid rods, which asymptotically approaches an inverse linear dependence on the scattering vector:

$$\lim_{q\to\infty} P_{\text{rod}}(q) = \frac{\pi}{2x} = \frac{\pi}{Lq}. \tag{7}$$

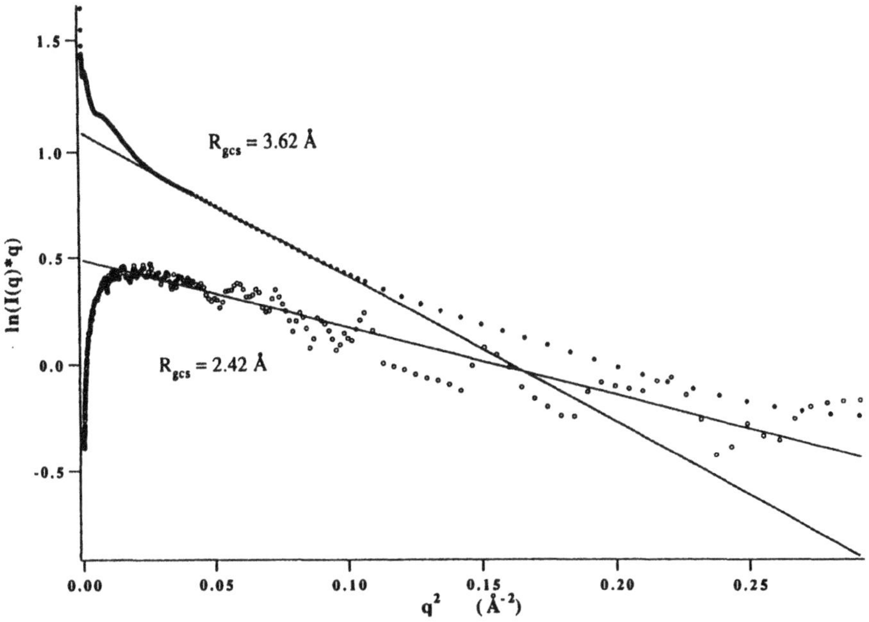
1.5
1.0
0.5
0.0
-0.5
ln(I(q)*q)
R_gcs = 3.62 Å
R_gcs = 2.42 Å
0.00
0.05
0.10
0.15
0.20
0.25
q² (Å⁻²)

Figure 2. Guinier-plot for the cross section. The open circles denote raw data, while the full ones represent the desmeared data. The cross sectional radii of gyration were calculated from the results of the linear regressions, indicated by the straight lines.

RESULTS AND DISCUSSION

The cross sectional radius of gyration R_{gcs} of an extended polymer chain can be extracted from the slope of the linear part of a so-called Guinier plot for the cross section [5, 21], where the product of scattering and scattered intensity $I(q)q$ is plotted versus the squared scattering vector $\mathbf{q}^2$. In Fig. 2 we present such a plot for a polymer solution with a concentration of 11.9 mg/cm^3. For comparison, both the raw and the desmeared data are displayed, and it is immediately recognized that there is a significant difference at low values of the scattering vector. This is due to the experimental artifact, caused by a non-circular profile of the incident beam. In the case of the desmeared data, the cross section radius of gyration was determined by linear extrapolation of a subset of data points, which had been selected by the requirement to yield a correlation better than 99.95%. For the extrapolation of the raw data the same q-range was chosen. The straight lines in Fig. 2 were obtained by this method, and it is obvious that the effect of convolution leads to a considerable underestimation of the cross sectional radius of gyration.

Further, it can be seen from Fig. 3 that there is a distinct increase of the cross sectional radius of gyration with decreasing concentration. Extrapolation of the data, derived from the desmeared scattering curves, to infinite dilution give a value of R_{gcs} = 6.3 Å. The corresponding value, R_{gcs} = 4.04 Å, for the convoluted data is in fair agreement with the data reported by Gidley *et al.* [10].

However, there are two points to be considered: (i) desmearing of SAXS data is never free from arbitrariness; and (ii) we have reported earlier [11] that TSP forms aggregates at random, thus the apparent concentration dependence of the cross

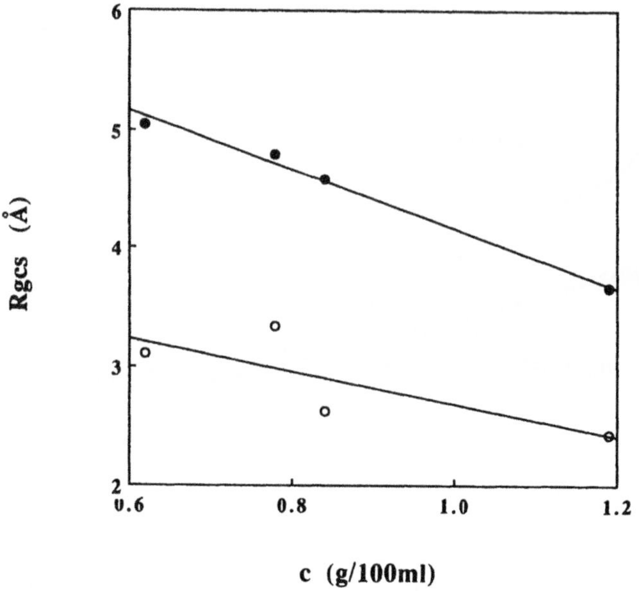
Rgcs (Å)
6
5
4
3
2
0.6
0.8
1.0
1.2
c (g/100ml)

Figure 3. Concentration dependence of the cross sectional radius of gyration from raw (open circles) and desmeared (full circles) SAXS data. The values for R_{gcs} are severely underestimated, if the SAXS data are not desmeared.

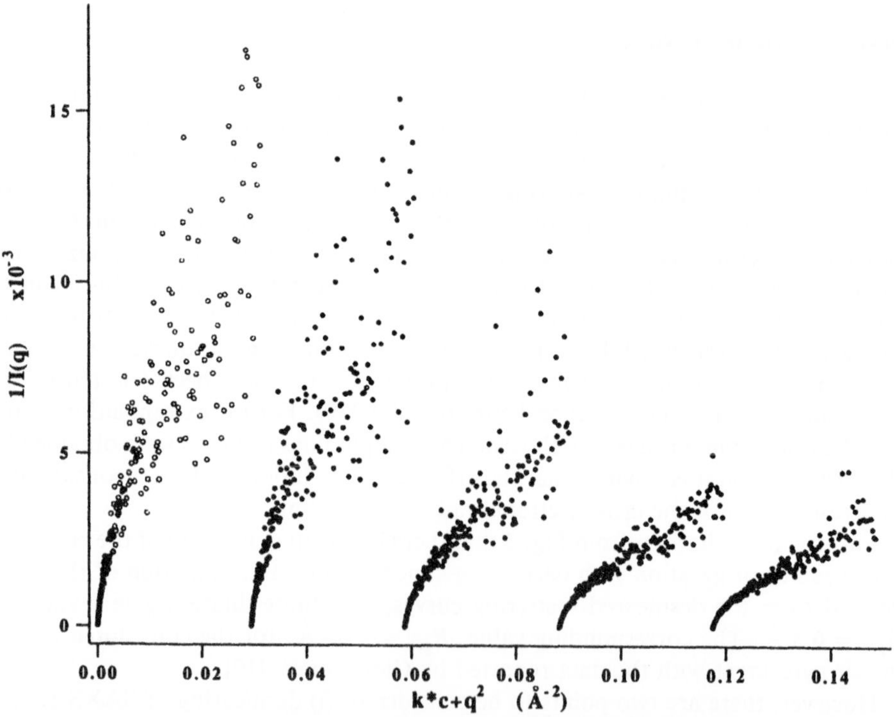
1/I(q) x10^-3
15
10
5
0
0.00
0.02
0.04
0.06
0.08
0.10
0.12
0.14
k*c+q^2 (Å^-2)

Figure 4. Zimm plot of the synchrotron radiation scattering data on a concentration series of TSP solutions. Open circles represent scattering data, extrapolated to infinite dilution at fixed scattering vector.

sectional radius of gyration might as well be incidental. To make sure neither of this was the case we carried out SRSAXS measurements on a series of solutions which had been diluted from a stock. These data, which do not require desmearing (see experimental), were extrapolated in a so called Zimm-plot, which is displayed in Fig. 4. The observed downward curvature of the scattering curves in this representation is typical for polydisperse samples with high particle stiffness. The angular dependence of the scattered intensity will be discussed in more detail below. The cross sectional radius of gyration extracted from Fig. 5 is R_{gcs} = 5.8 Å. This value might have a quite large error, since we did extrapolate to infinite dilution from scattering data, which had been collected at rather high solution concentrations. Thus the assumption of a linear dependence of the scattering intensity on the concentration is not necessarily correct. However, the cross sectional radius of gyration, determined this way, agrees well enough with the value from the desmeared SAXS curves to justify this approach. The average value of the cross sectional radius of gyration $R_{gcs} = 6.0 \pm 0.3$ Å is not compatible with the values of single polysaccharide chains, which are reported [22, 23] to lie in the range of 2.5–3 Å. The observed value for TSP is rather typical for polysaccharides with multi stranded conformations, like Xanthan [24] and the triple helix of Schizophyllan [4], which has a cross sectional radius of gyration of $R_{gcs} = 5.5 \pm 0.5$ Å.

For random coil molecules the particle scattering factor should approach $P(q) \propto u^{-2}$ behavior at large values of the scattering vector, as is evident from

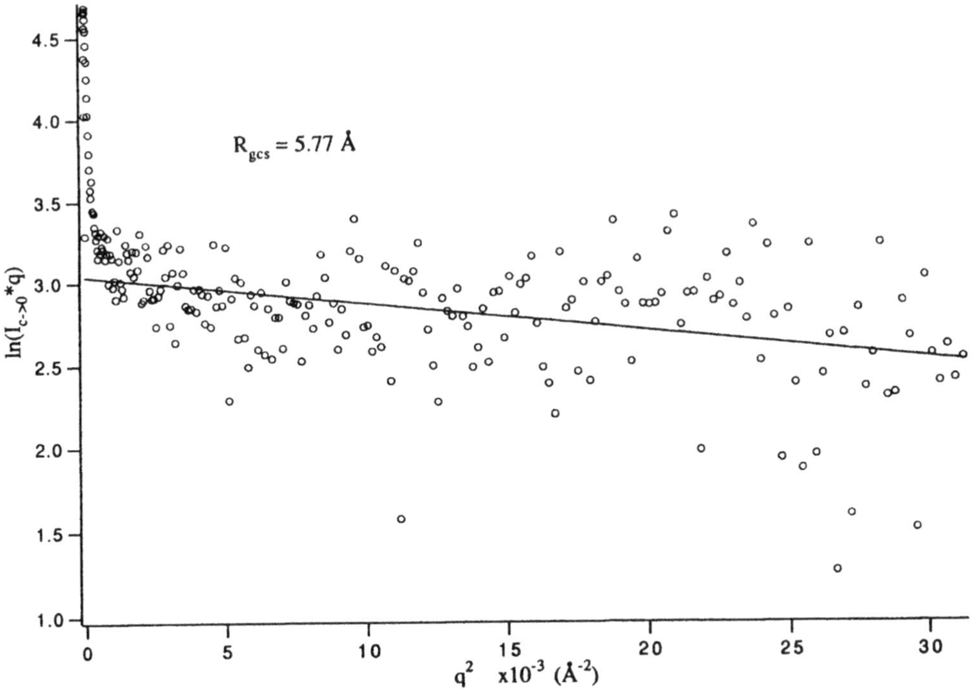

R_gcs = 5.77 Å
4.5
4.0
3.5
3.0
2.5
2.0
1.5
1.0
ln(I_c->0*q)
0
5
10
15
20
25
30
q² x10⁻³ (Å⁻²)

Figure 5. Guinier-plot for the cross section of the SRSAXS curve at zero concentration. The cross sectional radius of gyration was calculated from the result of the linear regression, indicated by the straight line.

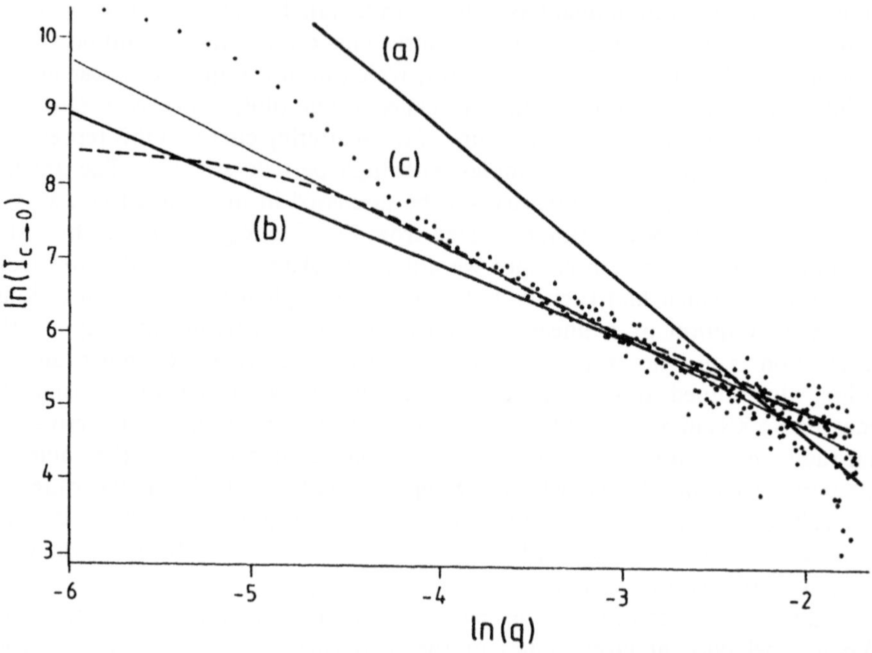

(a)
(c)
(b)
ln(I_{c→0})
ln(q)
10
9
8
7
6
5
4
3
-6
-5
-4
-3
-2

Figure 6. Double logarithmic plot of the scattered intensity from SRSAXS data, extrapolated to infinite dilution. The dots are experimental data, and the dashed line represents the theoretical curve, which was calculated from Eqn (6), using the values listed in Table 1. The full lines are the theoretical predictions for: (a) a random coil, and (b) a rigid rod. The line denoted by (c) represents the final slope of the experimental data $m = -1.25$, which is typical for a wormlike chain.

Eqns (5b) and (7). For rigid rods the corresponding asymptote is $P(q) \propto u^{-1}$ according to Eqn (5b). In Fig. 6 we have plotted the scattered intensity versus the scattering vector in a double logarithmic scale. The final slope of the curve is $m = 1.25$, which is typical for a linear chain with high persistence. A qualitative analysis of the chain stiffness according to Eqns (3) and (6) requires the knowledge of the weight average molar mass and the scattered intensity at zero angle. Measurement of molecular weights is basically possible by X-ray scattering [25], but very delicate, and limited to values of less than about 2×10^4 g/mol. Also the extrapolation of the scattered intensity to zero angle is usually subject to large errors. To cope with these problems, we performed light scattering measurements on TSP. The solutions for the LS experiments had been diluted from the same stock as the ones for the SRSAXS-measurements, to make sure, that we were looking at the same kind of particle. The weight average molar mass and the radius of gyration were determined from a Zimm plot according to Eqns (1), (3) and (4) to result $M_w = 2.5 \pm 0.1 \times 10^6$ g/mol and $R_g = 148 \pm 5$ nm. The errors represent the scattering of the results from different experiments and are well within the limit which is usually considered to be acceptable in light scattering experiments. In terms of experimentally accessible data the particle scattering factor reads:

$$P(q) = \left(\frac{R(q)}{KcM_w}\right)_{c \to 0}. \tag{8}$$

According to Cassasa [26] and Holtzer [27] it is convenient to plot the product $P(q)u$ versus the product of scattering vector and radius of gyration $u = qR_g$. From such a plot, which is displayed in Fig. 7, the following information is available [28]. (i) The height of the plateau at large values of u is a measure for the inverse of the contour length L_c of the chain, as one may see from Eqns (6b) and (7). (ii) The ratio of the height of the maximum to the plateau value is connected to the number of Kuhn segments per contour length N_k. (iii) The position of the maximum is sensitive to the polydispersity of the sample. In the present case the u-range, covered by experimental data, did not extend into the plateau region, therefore we had to fit them with theoretical curves according to Eqn (6) in order to obtain the above mentioned information. Since TSP is very likely to have a non-neglegible polydispersity, the expression given by Koyama was modified for our calculations. We assumed the polymer to have a Schulz-Zimm type [29, 30] distribution of the contour length:

$$W(L_c) = (-\ln p)^{k+1} (L_c^k) \frac{p^{L_c}}{\Gamma(k+1)}$$

with (9)

$$k = \left(\frac{L_w}{L_n} - 1\right)^{-1} \quad \text{and} \quad p = \exp\left\{-\frac{k+1}{L_w}\right\}$$

where L_w and L_n denote the weight and the number average respectively, and Γ is the common gamma-function.

The experimentally accessible z-average of the particle scattering factor then reads:

$$\langle P(q)\rangle_z = \frac{\int_0^\infty W(L_c)L_c P(q)\mathrm{d}L_c}{\int_0^\infty W(L_c)L_c\,\mathrm{d}L_c}. \tag{10}$$

Introducing Eqn (6) into the integral of the numerator in Eqn (10), we obtain a double integral, which we solved numerically by Gaussian quadratures [31]. Varying the independent variables in Eqn (10), the Kuhn segment length l_k, the contour length L_c and the polydispersity parameter $U = M_w/M_n - 1$, we calculated theoretical predictions for the particle scattering factor, and compared them with the experimental data. The curves in Fig. 7 have been calculated this way, and the values for l_k, L_c and U, which resulted in the theoretical curve with the least deviation from the experimental data, are listed in Table 1.

Since the outlined procedure is a three-parameter fit, an additional constraint was introduced, which selected combinations of l_k, L_c and U, which are physically meaningful. Only such combinations of variables were allowed, which would predict the correct result if they were introduced into the Benoit and Doty [32] formula for the radius of gyration of polydisperse [33] wormlike chains:

$$\langle S^2\rangle_z = l_k \frac{k+2}{6y} - \frac{l_k^2}{4} + \frac{l_k^3}{4L_W} - \frac{l_k^4}{8k(k+1)}\left(y_2 - \frac{y^{k+2}}{y + (2/l_k)}\right) \tag{11}$$

where $y = (k+1)/(L_W)$.

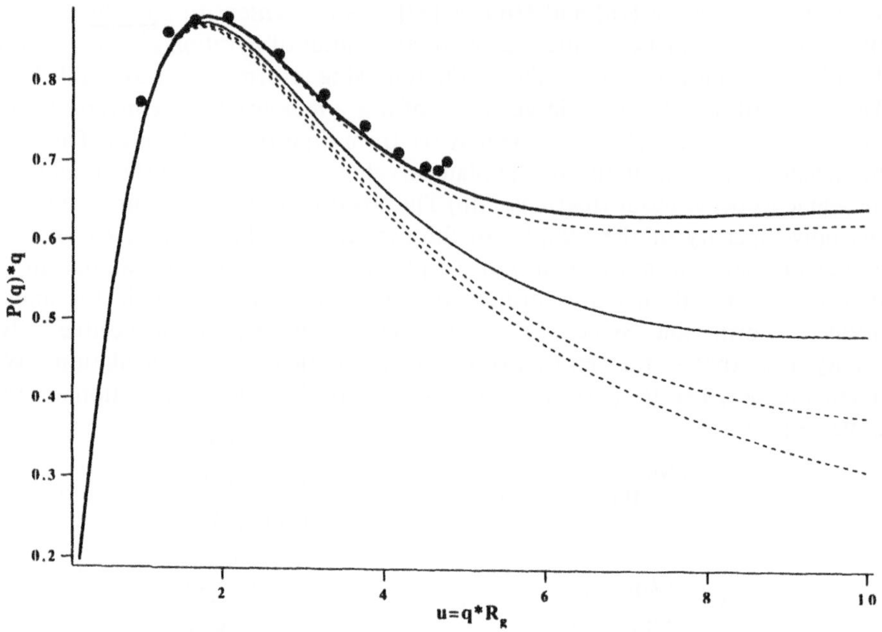
P(q)*q
0.8
0.7
0.6
0.5
0.4
0.3
0.2
2
4
6
8
10
u=q*R_g

Figure 7. Cassasa-Hotzler-plot of the light scattering data, extrapolated to zero concentration. The full circles are experimental values. The dashed lines are theoretical scattering curves, calculated from Eqn (7), using different values for the Kuhn segment and the contour length. The full line represents the best fit to experimental data, the corresponding values for L_c and l_k are listed in Table 1. All curves in this plot are calculated with a polydispersity $U = 2$.

Table 1.
Experimental data on TSP.

$M_w \times 10^{-6}$ (g/mol)	2.5 ± 0.1	R_g (nm)	148 ± 5
L_c (nm)	746.2	l_K (nm)	150
M_l (g/(mol/nm))	3350	$U = M_w/M_n$	2.0
R_{gcs} SAXS[a](Å)	4.04	R_{gcs} SAXS[b](Å)	6.3
R_{gcs} SRSAXS (Å)	5.8		

[a] derived from raw data
[b] derived from desmeared data

According to Holtzer [27], the plateau value of a plot of $P(q)u$ versus u gives the weight average of the contour length L_w, thus it is possible to calculate the absolute value of the mass per unit length M_l, without consideration of polydispersity, from the simple relation

$$M_l = \frac{M_w}{L_w}. \tag{13}$$

The experimental value from the present data is $M_l = 3350\,\mathrm{g\,(mol\,nm)^{-1}}$, which is about five times higher than the value calculated from the primary structure. Even if the latter may vary considerably, depending on the assumed primary structure [7–10], it is obvious that the experimental data are by no means compatible with a single polysaccharide strand. From the present data alone it is not possible to distinguish between a folded single chain and multistranded aggregates. However, we have reported earlier [11] that the linear mass density of different TSP samples varies almost linearly with the molar mass. That means the enhanced mass per unit length is due to lateral aggregation and not to backfolding.

Since the derived value of the linear mass density M_l may seem questionable, we shall add two more points, to prove that our procedure is reasonable.

(i) The dashed curve in Fig. 6 is the theoretical curve, which fitted the LS data in Fig. 7. For comparison, the theoretical predictions for a perfectly rigid rod and a random coil architecture are also displayed in Fig. 6. Although there is a difference in absolute values, which is due to the above mentioned problems of normalizing SRSAXS data, it is evident that the q-dependence of the experimental and the curve, calculated by the Koyama theory, are very similar over a wide range of observed scattering vectors. The deviation in the low q-range is probably due to a remaining concentration range with regard to the SRSAXS data. Thus we may say that the particle scattering factor, which was calculated by use of the data from Table 1, fits the LS data as well as, at least qualitatively, the SRSAXS data.

(ii) From Eqns (11) and (12) it is easy to calculate the exponent ν of the relation between the z-average of the radius of gyration and the weight average molar mass

$$\langle S^2 \rangle_z \sim M_w^{2\nu} \tag{14}$$

Using the values for l_k, L_c and U from Table 1, we derive a value of $\nu_{th} = 0.633$ in the neighborhood of the experimentally determined molar mass. This value is in excellent agreement with the experimental value $\nu_{ex} = 0.63 \pm 0.01$ we have found earlier for linear aggregates of TSP [11].

CONCLUSIONS

The present data support very strongly the model of TSP forming multistranded lateral aggregates in aqueous solution, which has been suggested earlier [11]. Furthermore, this model fits very well with the finding of lateral aggregates in the case of different cellulose derivatives, which has been reported recently [34]. Both, the cross sectional radius of gyration, derived from SAXS and SRS data, and the mass per unit length determined by LS measurements, are in favour of this model. The observed value of the cross section radius of gyration $R_{gcs} = 6.0 \pm 0.3$ Å might be explained by the, surely oversimplifying, model of close lateral packing of tubes of cross section radius $r = 3$ Å. The latter value is slightly higher than the reported value for the cross sectional radius of gyration of single polysaccharide chains [22, 23]. The experimental data are in good agreement with the SAXS [10] and LS [11] results reported earlier, which seemed to be in contradiction concerning the aggregation behavior of TSP. It is obvious that this contradiction results from a severe underestimation of the cross sectional radius of gyration, due to the uncorrected slit smearing of the formerly reported SAXS data. The Kuhn segment length of a single chain of TSP was estimated to be $l_k = 30$ nm [11]; it is therefore very likely that the high particle stiffness, and the related viscosifying capacity of TSP, are not intrinsic features of a single polysaccharide chain, but are rather caused by the formation of multistranded lateral aggregates.

Acknowledgements

We wish to thank the Dainippon Pharmaceuticals Company for the donation of a purified TSP sample. The SRSAXS measurements were performed under the approval of the Photon Factory Program Advisory Committee (Proposal No. 91-217). During the period of this work P.L. was supported financially by the Japan Society for the Promotion of Science (JSPS) and the Alexander von Humboldt Stiftung Institute, to whom grateful acknowledgement is made.

REFERENCES

1. R. L. Davidson (Ed.), *Handbook of Watersoluble Gums and Resins*. MacGraw-Hill, New York (1980).
2. V. Crescenzi, I. C. M. Dea, S. Paoletti, S. S. Stivala and I. W. Sutherland (Eds), *Biomedical and Biotechnological Advances in Industrial Polysaccharides*. Gordon Breach, New York (1989).
3. M. Kinoshita, *Ochanomizu Igaku Zasshin* **34**, 221 (1986).
4. T. Norisuye, T. Yanaka and H. Fujita, *J. Polym. Sci. Polym. Phys. Edn* **18**, 547 (1980).
5. O. Glatter and O. Kratky (eds), *Small Angle X-Ray Scattering*. Academic Press, London (1982).
6. B. Chu, *Laser Light Scattering*. Academic Press, New York (1974).
7. P. Kooimann, *Rec. Trav. Chim. Pays-Bas* **80**, 849 (1961).
8. A. Dali-Youcef, P. LeDizet and J. E. Courtois, *Phytochemistry* **18**, 1949 (1979).
9. H. C. Srivastava and P. P. Singh, *Carbohydr. Res.* **4**, 326 (1967).
10. M. J. Gidley, P. J. Lillford, D. W. Rowland, P. Lang, M. Dentini, V. Crescenzi, M. Edwards, C. Fanutti and J. S. G. Reid, *Carbohydr. Res.* **214**, 299 (1992).
11. P. Lang, Thesis, Albert Ludwigs University of Freiburg, Germany (1991); W. Burchard and P. Lang, *Makromol. Chem.* submitted.
12. Y. Hiragi, H. Urakawa and K. Tanabe, *J. Appl. Phys.* **58**, 5 (1985).
13. O. Glatter, *Acta Physica Austriaca* **47**, 83 (1977); *J. Appl. Cryst.* **10**, 415 (1977).
14. T. Ueki, I. Hiragi, Y. Izumi, H. Tagawa, M. Kataoka, Y. Muroga, T. Matsushita and Y. Amemiya, *Photon Factory Activity Report* V7–V9, V29, and V170–V171 National Laboratory for High Energy Physics Tsukuba, Japan (1982/83).
15. B. H. Zimm, *J. Chem. Phys.* **16**, 1099 (1948).
16. P. Debye, *J. Phys. Colloid. Chem.* **51**, 18 (1947).
17. B. H. Zimm, R. S. Stein and P. Doty, *Polymer Bull.* **1**, 90 (1945).
18. O. Kratky and G. Porod, *Rec. Trav. Chim. Pays-Bas* **68**, 1106 (1949).
19. W. Kuhn, *Kolloid. Z.* **68**, 2 (1934).
20. R. Koyama, *J. Phys. Soc. Japan* **34**, 1029 (1973).
21. A. Guinier, *Ann. Phys.* **12**, 161 (1939).
22. S. K. Gark and S. S. Stivala, *J. Polym. Sci. Polym. Phys. Edn* **16**, 1419 (1978).
23. M. Djabourov, A. H. Clark, D. W. Rowlands and S. B. Ross-Murphy, *Macromolecules* **22**, 180 (1989).
24. T. Coviallo, K. Kajiwara, M. Dentini, W. Burchard and V. Crescenzi, *Macromolecules* **19**, 2826 (1986).
25. O. Kratky, G. Porod and L. Kahovec, *Z. Elektrochem.* **55**, 53 (1951).
26. E. F. Cassasa, *J. Chem. Phys.* **23**, 596 (1955).
27. A. Holtzer, *J. Polym. Sci.* **17**, 432 (1955).
28. M. Schmidt, G. Paradossi and W. Burchard, *Makromol. Chem. Rapid Commun.* **6**, 767 (1985).
29. G. V. Schulz, *Z. Physik. Chem.* **B43**, 25 (1939).
30. B. H. Zimm, *J. Chem. Phys.* **16**, 1099 (1948).
31. W. H. Press, B. P. Flannery, S. A. Teucholsky and W. K. Vettering, *Numerical Recipes, the Art of Scientific Computing*. Cambridge University Press, Cambridge (1986).
32. H. Benoit and P. Doty, *J. Phys. Chem.* **57**, 958 (1953).
33. M. Schmidt, *Macromolecules* **17**, 553 (1984).
34. L. Schulz and W. Burchard, *Das Papier* submitted.

Attempts to stabilize a monoclonal antibody with water soluble synthetic polymers of varying hydrophobicity

K. PICKARD ANTONSEN[1]*, W. R. GOMBOTZ[2]† and A. S. HOFFMAN[1]‡

[1]*Center for Bioengineering, FL-20, University of Washington, Seattle, WA 98195, USA*

[2]*Bristol-Myers Squibb, Pharmaceutical Research Institute, 3005 First Avenue, Seattle, WA 98121, USA*

Received 5 October 1992; accepted 11 August 1993

Abstract—Proteins are subject to a variety of physical and chemical reactions that lead to a loss of activity. These reactions are a particular problem in controlled-release devices, where temperatures and protein concentrations are high. Current approaches to increasing protein stability include the addition of saccharides, amino acids, or polymers. New synthetic polymers may be promising protein stabilizers because properties such as molecular weight and side-chain composition can be controlled.

In this study, the stability of a murine monoclonal antibody, BR96, was evaluated in solution at 37°C. The antibody was incubated in the presence of a series of synthetic polymers that included poly(glucosylethyl methacrylate) (GEMA) and copolymers of *N*-vinylpyrrolidone (NVP) and methyl methacrylate (MMA). Samples were taken periodically up to 30 days. The formation of precipitated antibody in particulate aggregates was measured with a Coulter counter, and the molecular-weight distribution of soluble antibody was measured by size-exclusion chromatography.

Two trends were evident. First, with poly(GEMA) and copolymers of NVP and MMA, protein aggregation increased at higher polymer concentrations. Second, higher molecular weights of the poly(NVP) homopolymer also led to increases in protein aggregation. Effects of polymer hydrophobicity were more complex. A copolymer containing 9 mol% MMA caused immediate protein precipitation, while a copolymer containing 21 mol% MMA did not. The effects of the copolymer containing 21% MMA were strongly concentration dependent. At 1 wt%, the polymer reduced aggregation, but aggregation increased strongly between concentrations of 2 and 3 wt%.

INTRODUCTION

Proteins are intrinsically unstable, and if they are to be employed in applications such as diagnostics, pharmaceuticals, or vaccines, they must be stabilized against the myriad chemical and physical processes that cause them to lose their desired functionality. These processes have received extensive study [1], and in many cases can be controlled by careful selection of buffer conditions, temperature, and physical state (e.g., frozen or lyophilized). In addition, suitable additives, or excipients, are generally included in a protein preparation to enhance its stability.

Many protein inactivation reactions can only take place in the presence of water, or take place at a more rapid rate in an aqueous environment. Thus, lyophilization is one of the most common methods for stabilizing proteins [2]. Nevertheless, solution-state formulations are preferred because: (1) they do not present problems of reconstitution at the site of use; and (2) lyophilization is expensive. Moreover,

---

* Present address: Miles, Inc., Pharmaceutical Division, 4th and Parker Sts., P.O. Box 1986, Berkeley, CA 94701, USA

† Present address: Immunex Corporation, 51 University St., Seattle, WA 98101, USA

‡ To whom correspondence should be addressed.

protein formulations developed for controlled-release devices must take into account the aqueous environment to which such proteins will be exposed once injected, ingested, or implanted [3, 4].

Proteins in solution, or in the hydrated environment of a controlled-release device *in vivo*, may undergo the following inactivation processes: (1) adsorption to the walls of the container; (2) aggregation; (3) intermolecular disulfide bond rearrangement; and (4) denaturation [1, 5]. Each of the first three processes can be initiated by, or cause protein denaturation. Other reactions, such as oxidation or deamidation, may also cause inactivation, but typically do not involve denaturation.

A wide variety of compounds is used to stabilize proteins in solution [6]. These include salts, amino acids, polyhydroxy compounds (e.g., glycerol or sugars), and nonionic detergents. In general, these compounds promote the hydration of protein molecules in solution [7], and may thereby decrease the number of protein–protein contacts that necessarily precede aggregation and disulfide bond interchange. At room temperature, poly(ethylene glycol) also preferentially hydrates proteins. The mechanism in this case is steric exclusion [8]. This enhances stability at low temperatures, including freezing [9]. However, it decreases the thermal stability of some proteins because it binds to nonpolar regions of the protein molecule [7, 8].

The mechanisms by which other water-soluble polymers may stabilize proteins have received less scrutiny. A number of heparin-binding proteins are stabilized by polyanions, but this can be considered a special case, since the stabilization is related to biological activity. Other polymers may reduce surface adsorption. Examples are given in Table 1. For the most part, however, polymers have been used because they were effective in a specific situation, and were not investigated further.

Polymer synthesis is a promising route to new stabilizers because characteristics such as molecular weight, charge, and side-chain composition can be easily manipulated. It is the goal of this study to provide some insight into the properties that might be desirable for stabilizing proteins in solution.

A series of polymers with varying hydrophobicity was prepared in order to examine the effects of hydrophobicity and molecular weight on the aggregation of

Table 1.
Polymers affecting protein stability

Polymer	Protein	Effect	Reference
Poly(ethylene glycol)	dextransucrase		16
	penicillin acylase		17
	hyaluronidase	prevents surface inactivation	18
	thrombin		19
Dextran	dextransucrase		16
Dextran sulfate	tryptase	mimics heparin	20
	fibroblast growth factors	mimics heparin	21
	a murine IgG3 monoclonal antibody	helps solubilize	22
Poly(vinyl alcohol)	hyaluronidase	prevents surface inactivation	18
Poly(*N*-vinyl pyrrolidone)	hyaluronidase	prevents surface inactivation	18
	interferon-β		23

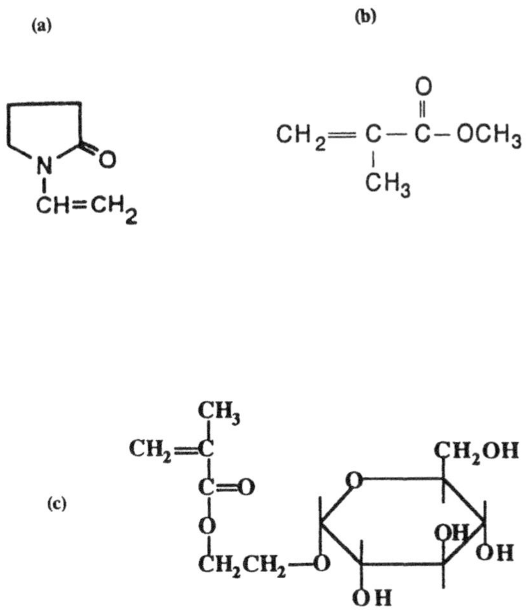
(a)
(b)
(c)

Figure 1. Chemical structure of the monomers used in this study, (a) *N*-vinyl 1-pyrrolidone; (b) methyl methacrylate; (c) GEMA.

a model protein in solution. The polymers studied were: (1) poly(*N*-vinyl pyrrolidone); (2) a series of copolymers of *N*-vinyl pyrrolidone and methyl methacrylate; and (3) poly(glucosylethyl methacrylate), a vinyl polymer with a pendant glucose group. The structures of these monomers are shown in Fig. 1.

A murine monoclonal antibody was selected as a model protein because it has potential use as a therapeutic agent, and because the results of this study may be applicable to the stabilization of other monoclonal antibodies used in therapeutic and diagnostic applications. Moreover, aggregation is a particular issue for therapeutic antibodies because aggregates can lead to the activation of complement [10].

MATERIALS AND METHODS

N-Vinyl 1-pyrrolidone (NVP) and methyl methacrylate (MMA) were obtained from Aldrich Chemical (Milwaukee, WI) and redistilled before use. Azobisisobutyronitrile (AIBN) was also obtained from Aldrich, recrystallized in methanol, and dried under vacuum. Ammonium persulfate was from J. T. Baker (Phillipsburg, NJ) and *N,N,N′,N′*-tetramethylethylenediamine (TEMED) was from Sigma Chemical (St. Louis, MO).

Glucosylethyl methacrylate (GEMA) was a kind gift of Nippon Fine Chemical (Tokyo). Poly(*N*-vinyl pyrrolidone), or PNVP, was obtained from Fluka. K15 (nominal molecular weight 10 000) was practical grade; K25 (nominal molecular weight 24 000) was pharmaceutical grade.

Murine monoclonal antibody BR96 was a gift from Bristol-Myers Squibb, Pharmaceutical Research Institute, Seattle. It binds to a tumor cell-surface marker. Its subclass is IgG_3 [11].

Polymer preparation

Poly(GEMA) The GEMA monomer was supplied as a 52 wt% aqueous solution. It was diluted to 10% with water. Ammonium persulfate (APS) was added at a concentration of 2 gl^{-1}. The solution was sparged with nitrogen for 30 min, and 10 μl TEMED mg^{-1} APS was added to initiate polymerization. The reaction was allowed to run, with stirring, overnight (exposed to room temperature, but without temperature control).

The polymer was precipitated by injecting the reaction solution slowly into rapidly-stirred acetone. The precipitated polymer was then dried under vacuum (without heat) for 30 h. Overall yield was 68%.

Poly(NVP-co-MMA) Both NVP and MMA were redistilled before use. The monomers were water-white liquids at room temperature. Polymerizations were conducted with a total monomer concentration of 200 gl^{-1} in tetrahydrofuran (THF). AIBN was used as an initiator at a concentration of 1 gl^{-1}. The monomer solutions were sparged for 30 min with nitrogen, and then held at 60°C ±5°C for 4 h. The polymer was then precipitated in 5 volumes of diethyl ether, redissolved in THF, and precipitated a second time. The polymer was then dried under vacuum for 24 h. Two polymers were prepared, with initial molar ratios of NVP to MMA of 95:5 and 98:2.

Polymer characterization

The compositions of the NVP–MMA copolymers were measured using NMR spectroscopy. The polymers were dissolved in deuterated chloroform containing tetramethylsilane (TMS) as an internal standard. Scans were taken with a Bruker 500 MHz instrument.

The protons on the backbone methyl group of MMA are found 1 ppm downfield from the TMS and are separate from the signals of other protons. Polymer composition was then estimated by taking the ratio of peak area in this region to the total peak area. If this ratio is r, and the mole fraction of MMA in the polymer is f, then $f = 9r/(3 + r)$,where the constants come from the number of protons in each repeat unit. The results of the measurements are shown in Table 2.

Table 2.
Polymers used in this study

Code	Composition	Nominal molecular weight[a]
	Homopolymers	
K15	Poly(NVP)	10,000
K25	Poly(NVP)	24,000
PG	Poly(GEMA)	—

Code	Initial mole fraction NVP	Initial mole fraction MMA	Mass conversion (%)	Molar percentage MMA in polymer
		Copolymers		
M9	0.98	0.02	15	9
M21	0.95	0.05	32	21

[a] Based on measurements of viscosity.

Aggregation experiments

Frozen antibody solutions were thawed. The buffer was replaced with phosphate-buffered saline (PBS) containing 0.01 M phosphate and 0.15 M NaCl at pH 7.4, by either dialysis or repeated ultrafiltration with buffer exchange using Centricon 30 units (Amicon, Lexington, MA). Antibody concentrations were determined spectrophotometrically at 280 nm using an extinction coefficient of 1.4 ml mg^{-1} cm^{-1} [12].

Concentrated solutions of antibody and the polymers were prepared in PBS. The polymer solutions were added to antibody and vortexed briefly at room temperature to initiate stability studies. Immediately after mixing, solutions containing poly(GEMA) were passed through a 0.22-μm syringe-tip filter (Rainin Instrument, Woburn, MA). All other solutions were filtered through a 0.5-μm Millex FH_4 filter (Millipore, Bedford, MA). Filtered solutions were incubated at 37°C for up to 30 days. Control solutions were also prepared: (1) protein only was held at 4 and 37°C; and (2) polymer only was held at 37°C.

The solutions were periodically sampled and diluted in PBS twice filtered through a 0.22-μm Nylon 66 filter (Gilson, Woburn, MA). The concentration of insoluble aggregates was measured in a Coulter Multisizer (Coulter Electronics, Hialeah, FL) equipped with an orifice of 50 μm. Samples were diluted so that the concentration index was less than 5% [13]. The instrument was capable of counting particles greater than approximately 1 μm in diameter. Both number density and total particle volume were measured; these measurements paralleled each other. A minimum of four measurements was made of each sample; error bars reflect the standard deviation of these measurements. Background counts (of filtered PBS only) were subtracted from the raw values.

Size-exclusion chromatography

After 4 days at 37°C, selected protein solutions were centrifuged to remove large aggregates and then injected onto an HPLC equipped with a TosoHaas (Philadelphia, PA) $TSK3000PW_{XL}$ size-exclusion column. Detection at the outlet was with both absorbance at 280 nm and refractive index, thereby providing an indication of protein elution separate from all constituents. The column was run in 0.1 M potassium phosphate, pH 7.4, at 0.2 ml min^{-1}.

RESULTS

Polymer characterization

The yields and compositions of the polymers used in these studies are reported in Table 2. The compositions of the NVP–MMA copolymers are consistent with literature values of reactivity ratios [14]. These were coded M9 and M21 for the molar percentage of methyl methacrylate in the polymer. All polymers were water-soluble and formed clear solutions.

Aggregation experiments

In an initial experiment, two grades of the homopolymer PNVP were evaluated. Samples of antibody solutions containing 2 and 5 wt% of each polymer were taken after 4, 17, and 30 days at 37°C. Under these circumstances, a measurable amount

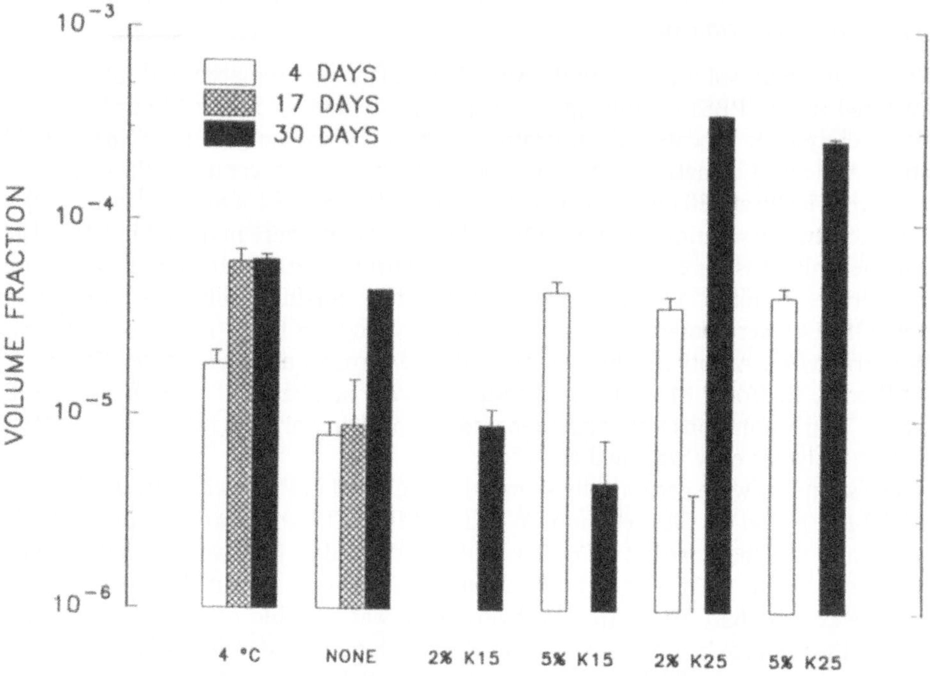
4 DAYS
17 DAYS
30 DAYS
VOLUME FRACTION
10⁻³
10⁻⁴
10⁻⁵
10⁻⁶
4 °C
NONE
2% K15
5% K15
2% K25
5% K25

Figure 2. Aggregation of BR96 in the presence of PNVPs of molecular weights 10,000 (K15) and 24,000 (K25), as measured by the particulate volume fraction. BR96 concentration was 3.2 mg ml^{-1}. Polymer concentrations were 0 ('none'), 2, and 5 wt%. Asymmetry of error bars reflects the logarithmic ordinate. The lower limit (10^6 ml^{-1}) was approximately the lower limit of detection. Some determinations near this limit had means below detection, and their bars are not shown. All samples were incubated at 37°C, except for the 4°C control.

of aggregation took place. The process of aggregation was slow without added PNVP, as shown in Fig. 2, with the particle volume fraction remaining low at 4 days, but increasing with time. Aggregation was more rapid at 4 than at 37°C, but by 30 days, the extent of aggregation was at a similar level.

Small amounts of polymer K15 (2–5%) reduced the extent of aggregation or retarded the aggregation process. However, the higher-molecular-weight polymer, K25, caused increased aggregation (over the control) as soon as 4 days. Higher aggregation was seen up to 30 days.

Measurements of particle volume with control polymer solutions containing only K15 or K25 (results not shown) revealed no detectable particle formation. In fact, with both polymers, particle counts were below background, suggesting that these polymers do not form particulates at all. This is not surprising since PNVP is a highly water-soluble polymer.

Results were substantially different when copolymers of NVP and MMA were added to the antibody. BR96 was incubated with 2 and 5 wt% of copolymers M9 and M21. Copolymer M9, at both concentrations, caused an immediate and visible precipitation reaction upon mixing with the antibody. Further experimentation with this copolymer was discontinued.

No such precipitation was observed with copolymer M21. Solutions of M21 and BR96 were filtered as described in the methods section. Measurements of

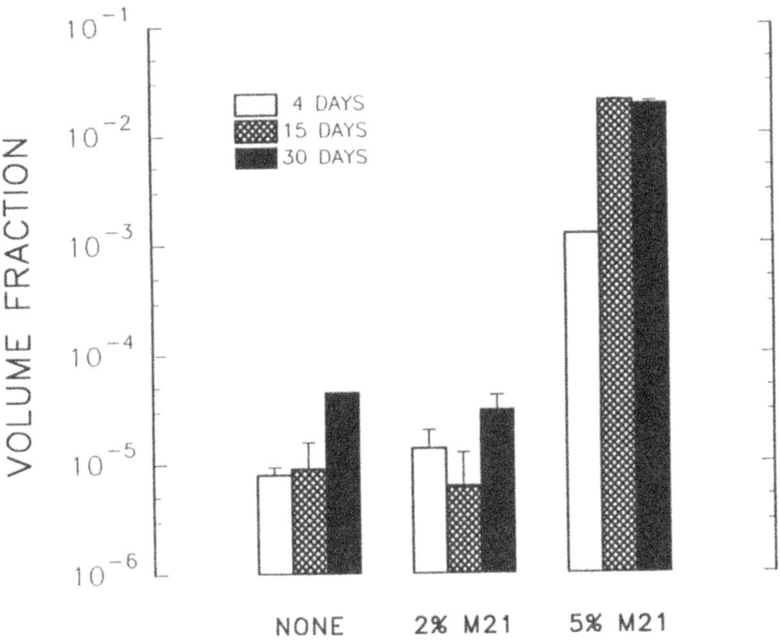

VOLUME FRACTION
10^{-1}
10^{-2}
10^{-3}
10^{-4}
10^{-5}
10^{-6}
4 DAYS
15 DAYS
30 DAYS
NONE
2% M21
5% M21

Figure 3. Aggregation of BR96 in the presence of the M21 copolymer containing 21 mol% MMA and 79% NVP. Experimental conditions were as described for Fig. 2.

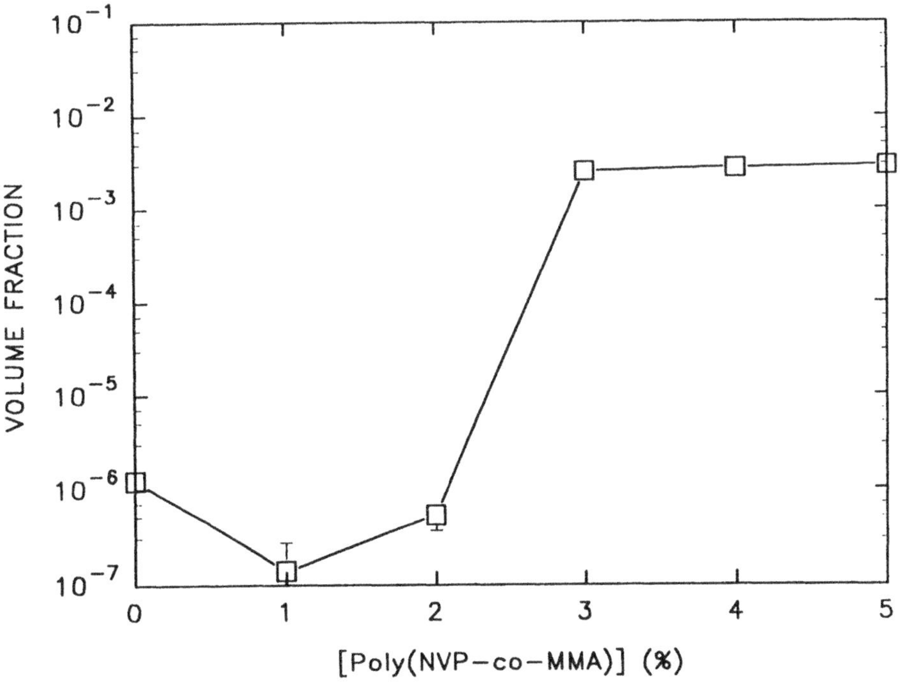

VOLUME FRACTION
10^{-1}
10^{-2}
10^{-3}
10^{-4}
10^{-5}
10^{-6}
10^{-7}
0
1
2
3
4
5
[Poly(NVP-co-MMA)] (%)

Figure 4. Aggregation of BR96 as a function of copolymer M21 concentration after 4 days at 37°C. The reduced lower limit reflects increased sensitivity from adjustment of solution dilution. BR96 concentration was 3.2 mg ml^{-1}.

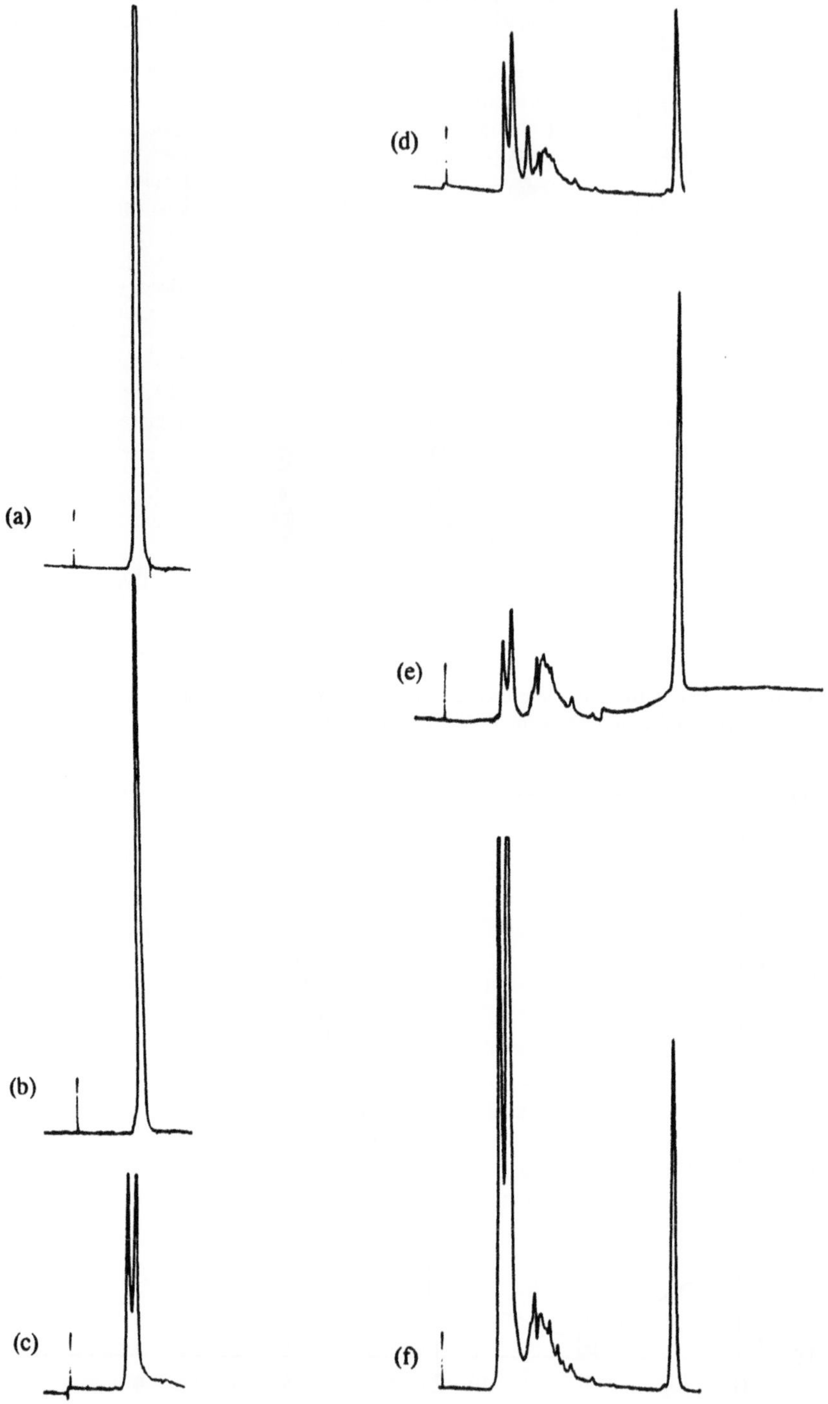
(a)
(b)
(c)
(d)
(e)
(f)

Figure 5. Size exclusion chromatograms of BR96 in solution supernatants containing varying concentrations of copolymer M21. Absorbance at 280 nm is plotted as a function of time at the column outlet. Incubation conditions were described for Fig. 4. Varying peak heights reflect varying protein concentrations in the supernatants and adjustments in instrument sensitivity. (a) 0, (b) 1, (c) 2, (d) 3, (e) 4, (f) 5 wt% M21.

aggregation were carried out at 4, 15, and 30 days (Fig. 3). At a concentration of 2%, copolymer M21 had no significant effect on aggregation. However, at 5%, significant precipitate formation had occurred as early as 4 days and was visible even to the naked eye. There was an increase by a factor of 10^3 in the aggregate volume (over the control).

This experiment suggested that the aggregation phenomenon was sensitive to copolymer concentration. The experiment was repeated using a series of concentrations from 0 to 5 wt% of M21. Particle volume was measured after 4 days at 37°C. At low concentrations (1–2%), copolymer M21 reduced the extent of aggregation, but from 2 to 3% there was a large increase in the amount of aggregation (Fig. 4).

Size-exclusion chromatography was used to probe the behavior of the antibody that remained in solution. The antibody-copolymer solution was centrifuged, and the supernatants injected onto a size-exclusion column. The chromatograms varied with copolymer concentration, as shown in Fig. 5. At low copolymer concentration (0–1%), there was only a single protein peak, eluting at the correct time for IgG (chromatograms a and b). The detector could not discriminate between protein alone and protein coeluting with copolymer. However, the polymer peak (as seen by refractive index) eluted much later.

At 2 wt% M21 (chromatogram c), a second protein peak appeared. This second peak, equal in area to the first, eluted early. At higher polymer concentrations (3–5%), there was no significant protein peak eluting at the normal time; instead, a number of small, late-eluting (after the column void volume) peaks were observed (chromatograms d–f).

The polymer poly(GEMA) provides a contrast to the NVP–MMA copolymers

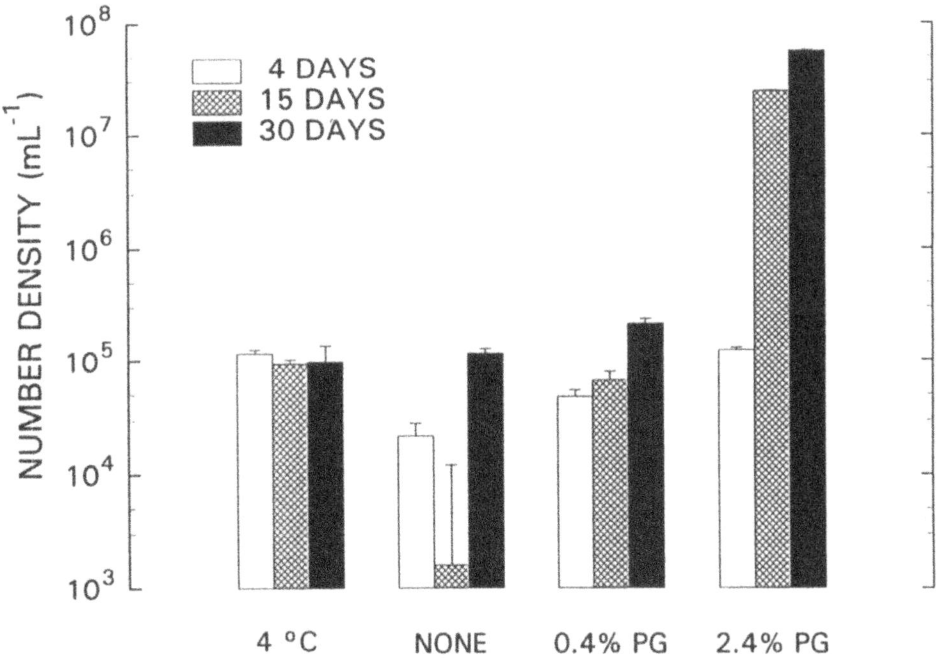
NUMBER DENSITY (mL^{-1})
10^8
10^7
10^6
10^5
10^4
10^3
4 DAYS
15 DAYS
30 DAYS
4 °C
NONE
0.4% PG
2.4% PG

Figure 6. Aggregation of BR96 at 37°C in the presence of poly(GEMA), indicated as PG. BR96 concentration was 3.1 mg ml^{-1}.

because each repeat unit contains both hydrophilic and hydrophobic groups. Experiments were carried out with poly(GEMA) at 0.4 and 2.4 wt%. Control experiments were carried out with BR96 in the presence of 0.4% and 2.4% glucose. Samples were taken after 4, 15, and 30 days at 37°C.

At a poly(GEMA) concentration of 0.4%, BR96 aggregated steadily with time, but even at 30 days, the overall particle count was not significantly greater than with no polymer present (Fig. 6). This was not seen with 2.4% poly(GEMA). After 4 days, little aggregation was observed, but by 15 days, a visible precipitate had formed, with the particle concentration higher by a factor of 10^3. Glucose had no effect on aggregation.

DISCUSSION

The experiments described here show that polymer molecular weight, hydrophobicity, and concentration can all influence the process of protein aggregation in solution. This is a significant and interesting observation because it suggests that copolymers may be synthesized with compositions designed for stabilization of specific proteins. Two rules of thumb emerge from the data: (1) low polymer concentrations are preferred in solution formulations. This was seen most notably with polymers M21 and poly(GEMA), where low concentration had no effect or inhibited aggregation of BR96. (2) Low molecular weight may be preferred to high molecular weight. Polymer K25 caused more aggregation than polymer K15.

One explanation of these observations is that the polymers are merely precipitating the protein, similar to poly(ethylene glycol) [15]. Our experiments suggest that the aggregation has a different origin, and is irreversible, for two reasons. First, all protein solutions were filtered immediately after polymer addition. In general, polymer-induced precipitation is rapid, with equilibrium reached within 30–90 min [15]. Second, protein solutions were all diluted, some as much as 1000-fold, before particle counting. If ordinary protein precipitation were occurring, the particles should have redissolved upon dilution in additional buffer.

The effects of polymer hydrophobicity are more complex than concentration or molecular weight. Copolymer M9, with an MMA content intermediate between copolymer M21 and the PNVP homopolymers, caused immediate protein aggregation, while neither M21 nor the PNVPs did. Nevertheless, the size-exclusion experiments suggest that polymer hydrophobicity does play an important role in causing aggregation and precipitation. At low concentration (1% and below), polymer M21 had no effect on the antibody in solution. Moreover, BR96 aggregation was the lowest in all experiments with this polymer and concentration. At 2%, soluble aggregates were formed. At higher concentrations, the trend was toward precipitated protein rather than toward higher proportions of soluble aggregate. The protein remaining in solution eluted slowly, even later than the void volume, from the size-exclusion column. This means there was an attractive interaction between the column packing and the protein under these conditions. The manufacturer claims that the column packing material has some hydrophobic character. This suggests that the antibody has been unfolded by the polymer, and the exposed hydrophobic residues interact with the column packing. Thus, some hydrophobicity may be desired in a polymer at low concentration, if the hydrophobic groups interact with hydrophobic patches on the protein surface and prevent them from coming into contact. However, at higher concentrations such hydrophobic interactions can help stabilize the unfolded state of the protein, leading to denaturation, increased aggregation and precipitation.

CONCLUSIONS

In some instances, synthetic polymers can be used to provide stable solution formulations for proteins. This study has provided a preliminary examination of variables that may be important in the design of appropriate polymers for use in such applications.

In this study, we found that polymer concentrations must be kept low. Lower molecular weights are preferred. Finally, present dogma is that polymers (indeed, all protein stabilizers) must be hydrophilic. We found that as long as polymer concentration is low (1%), a slightly hydrophobic copolymer (M21) keeps the level of aggregation down to the lowest level seen with any formulation tested.

Acknowledgements

The authors wish to thank the Washington Technology Centers for financial support, Prof. K. Nakamae of Kobe University for providing the GEMA monomer, and Wen Chen of the University of Washington for assistance with particle counting. KPA is the recipient of a National Research Service Award through NIH grant HL07403.

REFERENCES

1. M. C. Manning, K. Patel and R. T. Borchardt, *Pharmaceut. Res.* **6**, 903 (1989).
2. M. J. Pikal, *BioPharm* **3**(8), 18 (1990).
3. M. J. Hageman, in: *Chemical and Physical Pathways of Protein Degradation*, p. 273, T. J. Ahern and M. C. Manning (Eds). Plenum, New York (1992).
4. W. R. Liu, R. Langer and A. M. Klibanov, *Biotechnol. Bioeng.* **37**, 177 (1991).
5. J. Geigert, *J. Parent. Sci. Technol.* **43**, 221 (1989).
6. Y.-C. J. Wang and M. A. Hanson, *J. Parent. Sci. Technol. Suppl.* **42**, S3 (1988).
7. T. Arakawa, R. Bhat and S. N. Timasheff, *Biochemistry* **29**, 1924 (1990).
8. R. Bhat and S. N. Timasheff, *Protein Sci.* **1**, 1133 (1992).
9. J. F. Carpenter and J. H. Crowe, *Cryobiology* **25**, 244 (1988).
10. D. H. Bing, *Am. J. Med.* **76**, 19 (1984).
11. I. Hellström, H. J. Garrigues, U. Garrigues and K. E. Hellström, *Cancer Res.* **50**, 2183 (1990).
12. H. A. Sober, *Handbook of Biochemistry*. Chemical Rubber Co., Cleveland, OH (1970).
13. K. M. Clark and C. E. Glatz, *Biotechnol. Progr.* **3**, 241 (1987).
14. J. Brandrup and E. H. Immergut (Eds.), *Polymer Handbook,* Wiley, New York (1989).
15. D. H. Atha and K. C. Ingham, *J. Biol. Chem.* **256**, 12108 (1981).
16. A. W. Miller and J. F. Robyt, *Biochim. Biophys. Acta* **785**, 89 (1984).
17. E. Andersson and B. Hahn-Hagerdal, *Biochim. Biophys. Acta* **912**, 317 (1984).
18. A. P. Harrison, Biochem. J. **252**, 875 (1988).
19. G. Oshima and K. Nagasawa, *Thromb. Res.* **47**, 59 (1987).
20. L. B. Schwartz, T. R. Bradsord, D. C. Lee and J. F. Chlebowski, *J. Immunol.* **144**, 2304 (1990).
21. M. Tardieu, C. Gamby, T. Avramoglou, J. Jozefonvicz and D. Barritault, *J. Cell. Physiol.* **150**, 194 (1992).
22. W. Jiskoot, A. M. Hoven, A. A. De Koning, M. F. Leerling, C. H. Reubsaet, D. J. Crommelin and E. C. Beuvery, *J. Immunol. Methods* **138**, 181 (1991).
23. S. Cymbalista, United States Patent 4,647,454 (1987).

Part X.A.

Hydrogels

Water and Diffusion in Hydrogels

Review
Do hydrogels contain different classes of water?

WOUTER ROORDA
ALZA Corporation, 950 Page Mill Road, P.O. Box 10950, Palo Alto, CA 94303, USA

Received 13 October 1992; accepted 19 January 1993

Abstract—in the first part of this article calorimetric studies on poly (hydroxy ethyl methacrylate) (PHEMA) are presented. In the past the irregular melting curves in this type of experiment have been interpreted as evidence for the existence of different types of water in these gels. The studies presented here demonstrate that the occurrence of a glass transition in the freezing hydrogels may be responsible for this irregular melting behavior, and that this behavior is not (necessarily) an indication for the existence of different types of water. In the second part results are shown of measurements of the mobility of water in hydrogels, made by relaxation NMR. These results indicate that very rapid interchange occurs between the water molecules, and they support the conclusion that the calorimetric data mentioned above are not indicative for the existence of different classes of water in hydrogels. These results are compared with data from other fields of science, especially from fundamental freeze drying studies, which support the alternative interpretation of the calorimetric measurements.

Key words: Hydrogel(s); differential scanning calorimetry, differential thermal analysis; nuclear magnetic relaxation; poly-(hydroxy ethyl methacrylate).

INTRODUCTION

This article is a summary of work the author had the privilege to be involved in at the laboratories of Professors H. E. Junginger and J. C. Leyte, at the University of Leiden, The Netherlands. The data presented in this article have been reported in earlier publications [1–4], but not in a single, comprehensive overview like the present one. The article seeks to make a contribution to a lively discussion between two schools of thought about water structures in hydrogels. One school favors the opinion that based on calorimetric experiements one can distinguish between thermodynamically different classes of water in hydrogels, a concept that has always had great appeal in areas of pharmaceutical technology. The other school teaches that the irregular behavior of many hydrogels in calorimetric experiments is a consequence of the development of non-equilibrium conditions, and are, by themselves, not an indication for the existence of different classes of water in these systems. Hydrogels are often described as polymeric materials, that swell in water, but do not dissolve, having a water content of about 20% or more. This definition is rather broad, and a wide variety of hydrogels has been studied for many different purposes. This is illustrated in an impressive way by the work of Allan Hoffman, whose group over the years has investigated such aspects as hydrogel synthesis and characterization, thermal reversibility of swelling, protein delivery, hydrogel grafting for hemocompatibility and enzyme immobilization [5–10].

In most of the applications of hydrogels the presence and behavior of water in the gel play a crucial role. A technique that is often used to characterize the behavior of water is differential scanning calorimetry. Phase transitions, induced in the gels at sub-zero temperatures are interpreted in terms of water structure, and the results extrapolated to the situation at room- or body temperature. However, calorimetry is a rather indirect method, measuring the heat effect of a phase transition, induced by temperature changes. For that reason it was decided to compare the results from such calorimetric experiments with those of a spectroscopic technique, measuring molecular mobility.

The model compound selected for these studies was poly-(hydroxy ethyl methacrylate) (PHEMA), a well known and extensively studied hydrogel-forming polymer. When made from well purified monomers, it is essentially uncharged. Even when prepared in the absence of crosslinker, the polymer is not totally soluble in water, but swells to an equilibrium degree of about 40% water per total gel. It is important to note that all conclusions in this article should be strictly limited to this model compound, and that they may not apply to other hydrogel systems, in particular hydrogels with ionizable groups.

MATERIALS AND METHODS

The poly-(hydroxy ethyl methacrylate) (PHEMA) gels used in these studies were prepared from carefully purified HEMA. Special attention was given to removal of crosslinker (ethylene glycol dimethacrylate [EGDMA]), and methacrylic acid. Gels were prepared with 0, 1, 2, and 5% added crosslinker, and will be referred to as 0XL, 1XL, 2XL, and 5XL respectively. All calorimetric experiments were carried out on a Mettler DTA, and NMR experiments on an in-house built NMR spectrometer. For more details see ref. [7–9].

Calorimetry

When frozen hydrogels are thawed in a differential scanning calorimetry (DSC) or differential thermal analysis (DTA) experiment, a more or less complicated pattern of melting phenomena is recorded. A series of typical examples, in this case from PHEMA hydrogels, is represented in Fig 1.

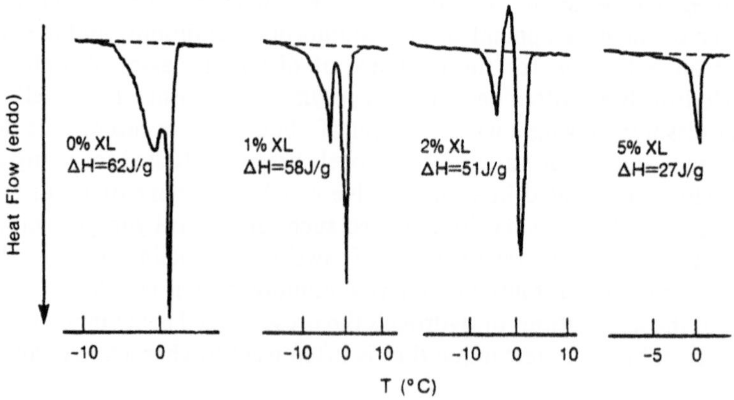
Heat Flow (endo)
0% XL
ΔH=62J/g
1% XL
ΔH=58J/g
2% XL
ΔH=51J/g
5% XL
ΔH=27J/g
-10
0
-10
0
10
-10
0
10
-5
0
T (°C)

Figure 1. DTA melting curves of PHEMA: 0 XL = 41.2% water; 1 XL = 40.0%; 2 XL = 38.3%; 5 XL = 34.3% (melting enthalpy expressed as J/g of total gel).

From these experiments two types of information can be obtained: melting enthalpy, where the total melting enthalpy measured is less than would be expected based on the amount of water in the gel; and shape of the melting curve where the melting curve itself shows a pattern with multiple extremes. These phenomena have been interpreted as an indication for the presence of different classes of water in hydrogels, with different melting points. Terms like bound, semi-bound, interfacial, loosely bound, and free or bulk water are abundant in the literature [11–14]. However, in doing this type of experiment we found that both the total melting enthalpy and the actual shape of the curves were extremely sensitive to the experimental conditions, making interpretation of the data difficult. Therefore, the following set of experiments and calculations was carried out.

Melting enthalpy. Since crystallization in gels is known to be a potentially slow process, an experiment was performed to see whether under the usual conditions of DSC or DTA analysis the freezing of the water had reached its maximum; in other words, whether all the freezable water was actually frozen. The gels were cooled to −25°C to induce crystallization, then kept at −15°C for various periods of time, and finally heated again to room temperature. The total areas under the recorded melting peaks were measured, and are plotted in Fig. 2. The data showed a marked increase in the melting enthalpy with increasing annealing periods, leveling off at about 16 h. This clearly demonstrates that the freezing of water in the hydrogels is indeed a slow process, requiring substantial amounts of time to reach maximum levels. The temperature protocol described above was critical to the experiments. The highest values for the melting enthalpy were recorded after the annealing process at −15°C, but cooling to −25°C was necessary to overcome undercooling phenomena, and to induce crystallization. This experiment was designed to address

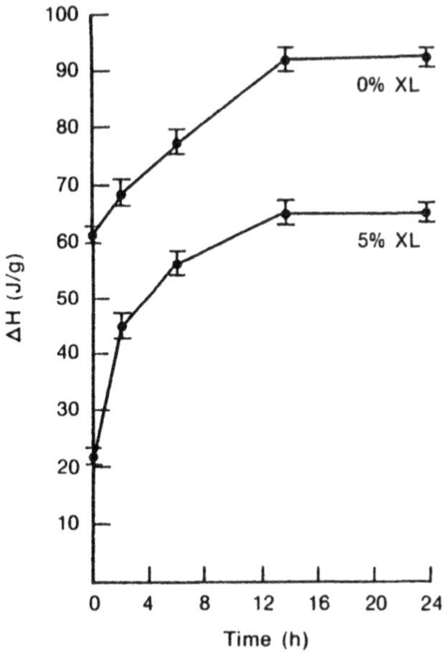
100
90
80
70
60
50
40
30
20
10
ΔH (J/g)
0% XL
5% XL
0
4
8
12
16
20
24
Time (h)

Figure 2. Melting enthalpies as a function of annealing time (compositions as in Fig. 1).

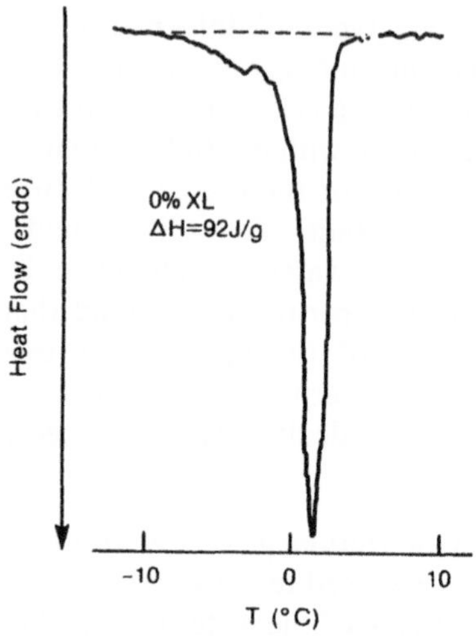
Heat Flow (endo)
0% XL
ΔH=92J/g
-10
0
10
T (°C)

Figure 3. Melting curve for PHEMA after annealing (composition as in Fig. 1).

the problem of the low melting enthalpy of the gels. An unexpected result was the fact that the shape of the melting curves changed significantly: with the exception of a minor shoulder, the curves were recorded with only one single melting peak, as illustrated in Fig. 3. Confirmation of this observation with a different technique will be presented further on under the paragraph on the shape of the melting curve. The fact that under the proper conditions the water in PHEMA gels melts in a single event greatly facilitates interpretation of the data. The amounts of frozen and unfrozen water in the gels can now be calculated from the melting enthalpies. Literature data are available indicating that there is no enthalpic interaction between the freezable water and the polymer in the gel [4], which means that the normal melting enthalpy of 334 J/g can be used for water. It also means that the low melting enthalpy of the gels is due to incomplete freezing of the water in the gels, and not to a lower melting enthalpy of the frozen water. There are several different ways the water content can be expressed, but the most interesting picture emerges when the amount of unfrozen water is calculated relative to the amount of polymer in the gel. The results are given in Table 1 for hydrogels of different cross-linker and water content. A remarkably constant value of about 1.7 molecules of unfreezable water per monomeric unit HEMA is found. The fact that the gel with 14.9% of water does not reach this value is due to the fact that this gel does not contain more than 1.3 molecules of water per monomeric unit.

A potential explanation for these observations was found when the thermomechanical properties of the gels were studied more closely. Dry PHEMA has a glass transition temperature (T_g) of about 120°C, which increases slightly with the crosslinker content. Water has a T_g of about −130°C, and, when absorbed into the polymer, acts as a plasticizer. The T_g of the resulting hydrogel, at different water–polymer ratios, can be calculated by the formula [15]:

$$\frac{1}{T_g(\text{gel})} = \frac{W_1}{T_g(\text{water})} + \frac{W_2}{T_g(\text{polymer})}$$

Table 1.

Cross-linker content (% w/w)	Water content (% w/w)	Water content (mol H_2O/ mol HEMA)	Maximum melting enthalpy (J/g gel)	Non-freezing water content (% w/w)	Non-freezing water content (mol H_2O/ mol HEMA)
0	41.2	5.1	92 ± 2	13.6	1.7
0	27.6	2.8	34 ± 2	17.4	1.7
0	21.0	1.9	10 ± 1	17.8	1.6
0	14.9	1.3	0	14.9	1.3
1	40.0	4.8	86 ± 2	14.4	1.7
2	38.3	4.5	79 ± 2	14.6	1.7
5	34.3	3.8	65 ± 2	15.1	1.6

where W_1 and W_2 are the weight fraction of water and polymer in the gel, and T_g is expressed in K. The calculated and measured values are presented in Fig. 4. From this graph it will be clear that at a water content of 1.7 molecules of water per monomeric unit, the gel has a T_g in the vicinity of 0°C.

The events observed in these experiments are consistent with the following explanation: When a hydrogel is cooled, and water in the gel starts to freeze, a phase separation occurs. Initially the water and the polymer form a single, homogeneous phase. When water starts to freeze, it crystallizes and forms a separate phase of pure, solid ice. While this crystallization proceeds, the ice phase grows, at the expense of water in the hydrogel phase. The hydrogel phase thus loses water. The ice crystals do not contribute to the plasticizing action of water on the gel phase, and consequently the glass transition temperature of the gel phase rises. Water loss from the gel phase continues, until a water content of 1.7 molecules of water per

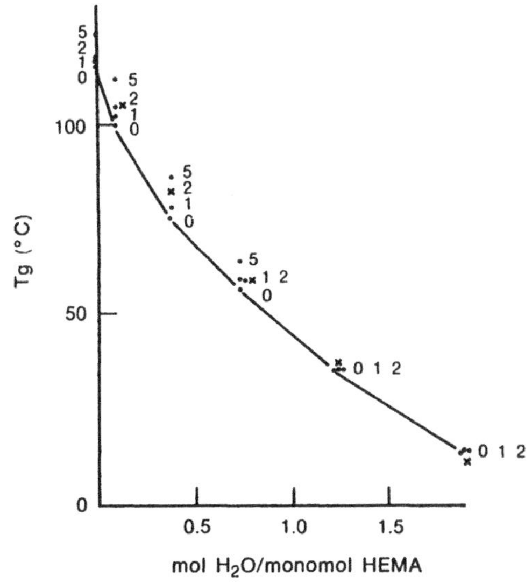

Tg (°C)
100
50
0
0.5
1.0
1.5
mol H2O/monomol HEMA
5 2 1 0
0 1 2

Figure 4. T_g PHEMA vs water content: ● measured data; × calculated data.

monomeric unit is reached. At that particular combination of water content and temperature the gel goes through a glass transition which has dramatic consequences for the whole system. The mobility of the water molecules in the glassy 'gel' drops substantially, and their capability to diffuse to the growing ice crystals is drastically reduced. At the same time the ice crystals are locked in a glassy matrix, which seriously impairs their capability to grow. The overall effect is a virtual stop of the ongoing crystallization process, not because of the presence of different types of water, but because of kinetic factors preventing further crystallization at measurable rates.

Shape of the melting curve. Based on the observation that after proper annealing in a DTA experiment PHEMA gels generate heating curves with a single melting peak an experiment was designed, using an adiabatic calorimeter. Although this type of equipment is not suitable for well controlled, slow cooling rates, it is capable of heating samples at a much slower overall rate than a DSC or DTA, and therefore offers a closer approach of equilibrium during the measurement. Uncrosslinked PHEMA was frozen, and its melting process recorded in the adiabatic calorimeter. The results are shown in Fig. 5, and confirm the earlier observation that under this type of conditions, the melting of PHEMA results in a single melting peak.

The observation of more complicated patterns can be explained by comparing the behavior of the gels with that of pure polymer systems in which crystallization occurs. A common phenomenon in heating partially crystalline polymers is the so-called premelting crystallization, or devitrification. This process shows up as an exothermic peak on a DSC or DTA heating scan, at temperatures slightly below the melting point of the polymer. It is caused by additional crystallization of domains in the polymer that were potentially capable of crystallizing, but were prevented from doing so by a glass transition induced by rapid cooling. A similar process could be responsible for the complicated melting peaks observed in hydrogels. If a devitrification process were to occur simultaneously with the onset of melting in

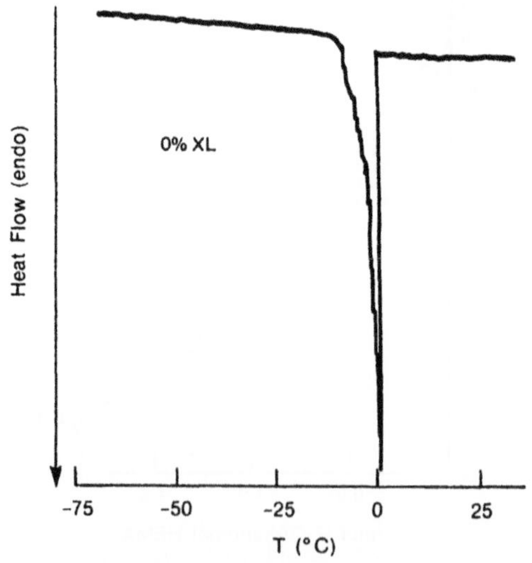
Heat Flow (endo)
0% XL
-75
-50
-25
0
25
T (°C)

Figure 5. Adiabatic melting curve of PHEMA (composition as in Fig. 1).

other regions of the gel, the overlap of the two peaks would result in what appears to be a double melting peak. Similar results have been described for the system poly-(vinyl pyrrolidone)–water [16].

Conclusions. The conclusions from this part of our studies are: The low value for the melting enthalpy of frozen PHEMA gels is due to incomplete freezing of water in the gels. The incompleteness of this process is caused by a glass transition in the system, which kinetically prevents further crystallization of water. The multiple character of melting peaks in DSC or DTA scans of frozen PHEMA gels is due to the occurrence of a devitrification process.

Nuclear magnetic relaxation

A serious limitation of calorimetric experiments is the fact that events occurring in the gel are measured indirectly, in other words, what is measured is their heat effect, and in the case of simultaneous events, the sum of their heat effects. Spectroscopic methods may offer a more direct measurement of events on a molecular, or atomic level. A particularly suitable probe to investigate the behavior of water in hydrogel systems like these is the oxygen isotope ^{17}O. The nuclear magnetic relaxation behavior of ^{17}O in water is inversely related to the rotational mobility of the water molecule. Nuclear magnetic relaxation is closely related to nuclear magnetic resonance, but is used to study molecular mobilities rather than structures [9]. The concept of different classes of water, like free and bound water, would seem to have implications for the mobility of the water molecules, and therefore a number of nuclear magnetic relaxation experiments was carried out, investigating this mobility. Using the oxygen in the water molecule as a probe instead of the hydrogen has a number of advantages, mainly due to the fact that ^{17}O relaxation is these cases is a fast, strictly intra-molecular process, which greatly facilitates interpretation of the data. PHEMA hydrogels were swollen in water with 0.5% of $H_2{}^{17}O$. The nuclear magnetic relaxation rates of the oxygen isotype were measured, and correlated to structural characteristics of the gel. In contrast to the DTA experiments, the NMR studies were done at 25°C. The following results were obtained:

Signal amplitude. The amplitude of the NMR signal is directly proportional to the quantity of labeled nuclei in the sample. A comparison was made of the amplitude of the signal of pure water, enriched with ^{17}O, and of water in the gel. Within an error range of ± 2% the amplitude of the signals was the same, indicating that, within this accuracy all the water in the gels is detected, and no immobilized, or "ice-like" structures of water in the gels existed that are undetectable in this type of NMR experiment.

Magnetic relaxation rates, general. The relaxation rates of the ^{17}O in these systems were in the range of 1000–2000 Hz, indicating timescales for the process in the order of milliseconds. This puts a practical limit on the validity of some of the conclusions of the studies presented here. If different classes of water exist in the gels, and exchange of their water molecules occurs at frequencies higher than those of the relaxation experiments, the water molecules will appear as one class of molecules, with characteristics that are the average of the individual classes. Since millisecond timescales are long in terms of water dynamics, this exchange may very well be present, but is undetected by the NMR methods used. Therefore, the following conclusions of the nuclear magnetic relaxation studies are valid only on timescales of milliseconds or longer, except where indicated otherwise.

Exponentiality relaxation. The relaxation of the ^{17}O nuclei in these experiments is an exponential process, characterized by time constants that are dependent on the mobility of the water molecules. Since all relaxation curves were found to be monoexponential, it was concluded that, on the time scale of the experiments only one type of water exists in these gels in terms of mobility of the water molecules.

Magnetic relaxation rates, correlation with structure of PHEMA gels. A large number of gels was prepared with different water- and crosslinker contents. The relaxation rate of ^{17}O in these gels was compared with its relaxation rate in pure water. This relaxation rate is *inversely* related to the mobility of the water molecules. Since the presence of the polymer *reduces* the mobility of these molecules, the measured relaxation rates in the gels are *higher* than those in water. By dividing the rates in the gels by the rates in water, a Rate Enhancement Factor (REF) was obtained, which served as a quantitative measure of the influence of the polymer on the water mobility. This REF was plotted against the polymer content of the gels, in order to get an impression of the influence of the polymer on the relaxation rate of the ^{17}O, and thus on the mobility of the water molecules.

A similar exercise was carried out for gels swollen in $NaClO_4$ solutions, which results in PHEMA gels with a much higher hydration level than what is obtained with pure water. In this case the relaxation rates in the gels were compared to the rates in $NaClO_4$ solutions with the same composition as the solutions in the gel. (Due to preferential partitioning of $NaClO_4$ into the gel, the solution in the gel has a different composition than the swelling solution in which the gel is immersed.)

The results are shown in Fig. 6. The polymer contents of the gels varied between 20 and 80%, with REFs roughly between 3 and 40. For comparison with later experiments a line was drawn indicating the trend of the data points. No mathematical analysis was made of this trend. An interesting conclusion from this experiment is

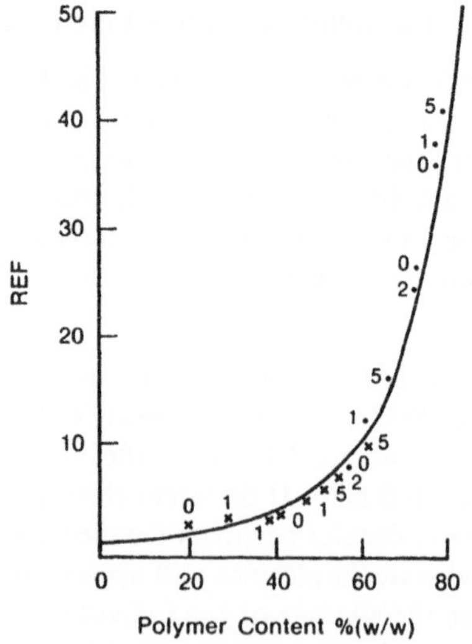
REF
Polymer Content %(w/w)
0
10
20
30
40
50
60
80

Figure 6. REF vs polymer content: ● PHEMA–water; × PHEMA–water–$NaClO_4$

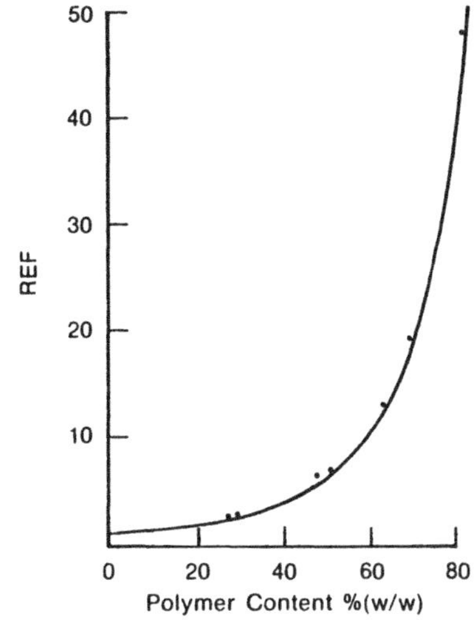

50
40
30
20
10
REF
0
20
40
60
80
Polymer Content %(w/w)

Figure 7. REF vs water content: ● PHEMA–water–$NaClO_4$ *solutions.*

the fact that the trend seems to be completely determined by the polymer content, or degree of swelling, of the gels. The data points contain samples with 0, 1, 2, and 5% crosslinker, but this difference in crosslinker content is not reflected at all in the relaxation behavior of the gel. Likewise, the presence of $NaClO_4$ in the solution seems to be irrelevant to the outcome of this experiment.

Reports in the literature claim that the amount of bound water in these gels increases with the crosslinker content [11], but this is contradicted by the present measurements. Even if rapid exchange of water molecules between bound and free water would make it impossible to distinguish these classes separately, a relative

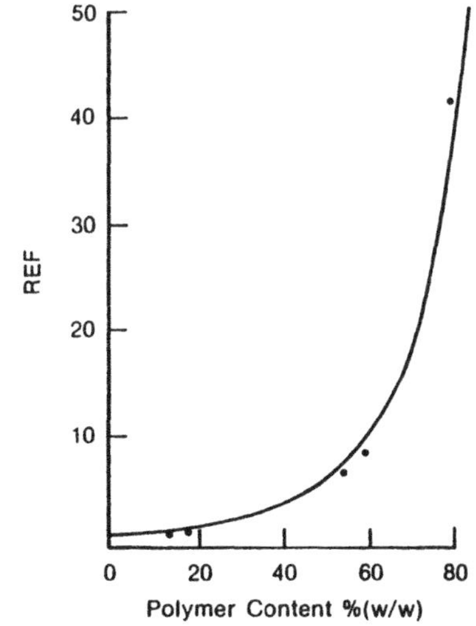

50
40
30
20
10
REF
0
20
40
60
80
Polymer Content %(w/w)

Figure 8. REF vs water content: ● PDHMA–water.

increase in the amount of bound water would change the average mobility of the water molecules, which does not appear to be the case in these experiments. In an experiment not shown here, $NaClO_4$ was replaced by other swelling solutes, like NaI, antipyrine and DMSO, with similar results.

Relaxation rates of PHEMA solutions. PHEMA is not soluble in pure water, but freshly prepared 0XL PHEMA gels are soluble in a number of solvents, including $NaClO_4$ solutions in water.

These PHEMA–$NaClO_4$ solutions were subject to the same measurements as described under IV, and the results plotted in Fig. 7. The drawn line is identical to the one in Fig. 6. Again, all data points follow the same trend, indicating that the mobility of the water in the gels is determined by the degree of swelling, and not by the crosslinker content.

Relaxation rates in an alternative hydrogel. In a final experiment the relaxation of water in an alternative, related type of hydrogel was investigated. The hydrogel in this study was based on poly(dihydroxypropylmethacrylate) (DHPMA). The structure of this polymer is similar to that of PHEMA, but instead of a *monohydroxy*-ethyl side chain, the polymer contains a *dihydroxy*propyl group. Therefore this polymer contains one more potential binding site for water per monomeric unit, in the form of an additional -CHOH group. The REF vs polymer content plot is shown in Fig. 8. Despite the presence of an additional potential binding site for water, the data points follow exactly the same trend as in the PHEMA gels. If truly bound water molecules were present in this type of gel, one would consider the hydroxyl group as a prime candidate for a binding site. One would expect then that introduction of an additional hydroxyl group would lead to a reduction of the average mobility of the water in the gels. Based on these NMR experiments this appears not be the case. This conclusion is not affected by the millisecond time scale of the experiment, since it relates to *average* mobilities.

Conclusions. These experiments led to the following conclusions: At 25°C, on a time scale of milliseconds, and within an error range of ± 2%, rotational mobility measurements on $H_2{}^{17}O$ by nuclear magnetic relaxation detect only one class of water molecules in PHEMA hydrogels. At cross-linker contents between 0 and 5% the rotational mobility of these water molecules is much more determined by the degree of hydration of the gels than by their cross-linker content. Introduction of an additional potential binding site in the form of a -CHOH group in the monomer does not affect the average rotational mobility of the water molecules.

DISCUSSION

The difference in temperature between the two sets of experiments is significant. The NMR experiments were performed at 25°C, and the DTA experiments necessarily at temperatures below zero. It is likely that the exchange rates of water molecules will be lower at sub zero temperatures, and actually, below the glass transition, they are very low indeed. Nevertheless, if different types of water exist, they are in equilibrium at more elevated temperatures, and if some of the water is removed from that equilibrium by a freezing process, the equilibrium will shift. This makes calorimetric experiments inherently difficult to interpret. Moreover, the results of the calorimetric experiments are often extrapolated to higher temperatures, for instance in order to explain diffusion phenomena in hydrogels at room or body temperature. One has to wonder how relevant the low temperature data are for the actual situation at elevated temperatures.

Otherwise the two techniques offer an interesting complementary approach, calorimetry basically measuring properties of the bulk of the material, and NMR offering more insight into events on a molecular level. An important advantage of NMR is the fact that systems can be studied at equilibrium, or in a metastable state like a glass. DTA or DSC studies measure the heat effect of a process, induced in the sample by a temperature change, and therefore essentially are non-equilibrium experiments. If the induced process is significantly slower than the scanning rate of the calorimeter, artifacts may occur, which in the opinion of the author is what has led to the complicated melting peaks of PHEMA and other hydrogels. Careful annealing or the use of more suitable equipment like an adiabatic calorimeter results in single melting peaks of these gels, while the inevitable glass transition in the freezing system can account for the incomplete crystallization of water.

Additional evidence for the role of glass transitions in freezing systems can be found in other branches of science, especially in the field of freeze-drying or lyophilization. The common practice in freeze-drying, by now applied to millions of samples, is to divide the process in two steps: primary drying and secondary drying. The fundamental phenomena underlying these two steps have been described extensively by Mackenzie [17, 18]: The separation between a crystalline and a glassy phase occurs during the initial freezing of the sample to be dried. A solution (normally, but not necessarily, an aqueous one) is frozen, and crystals of pure solvent develop. The remaining solution becomes more concentrated, due to this loss of solvent, and its freezing points drops, with ongoing loss of solvent to the crystalline phase. At some point the descending freezing point curve of the solution intersects with its rising glass transition curve. The glass transition temperature of the solution is rising, because of its increased content of solids. At the intersection the remaining solution solidifies into glassy state, and the crystallization process stops.

During primary drying the solvent crystals are removed, leaving behind an open, glassy structure, which is stable as long as it is glassy. The solvent from this phase is removed during secondary drying. During this process the glass transition temperature of the mixture rises. Raising the temperature of this mixture above the glass transition, or more accurately, above its collapse temperature, destabilizes the porous structure, and the system collapses.

These phenomena have been well studied and documented, and actually recorded on film in a freeze drying microscope. They are practically universally applicable in freeze drying, and occur in gels, polymeric solutions as well as in solutions of low molecular weight compounds like sugars. The events in the freezing part of the procedure apply to hydrogels as well, and this is no reason to assume that the events in freezing a DSC or DTA sample are any different. This is very strong supporting evidence that the freezing of water in a hydrogel is incomplete due to a glass transition. Similar conclusions are reached by the group of Zografi [19, 20], using among others, combinations of vapor pressure measurements and DSC.

Then there is the question of the mobility of the water molecules, and the potential existence of different classes of water with very high exchange rates. The NMR experiments in this article suggest that if different classes of water exist, the exchange of the individual molecules has to take place on a sub-millisecond time scale. In fact, this time scale may be many orders of magnitude smaller. Recent work by Pitt *et al.* [21] shows that ESR measurements on nanosecond timescales do not offer evidence for the existence of separate domains in water swollen polymer systems. For a number of monovalent ions, the residence times of the water molecules in the hydration shells have been calculated using Molecular Dynamics simulations. Even though one would expect this type of ionic interaction to be relatively strong, these residence times were found to be in the pico second range, similar to the translation correlation time of pure water [22].

The results of the studies presented in this article, in combination with the selection of literature data cited above, cast serious doubts on the existence of a class of 'bound' water in hydrogels like PHEMA, the term 'bound' being used in the sense of '(partly) immobilized'. A similar conclusion was reached by Felix Franks in a recent review article on freeze drying (quote): "It is our strong belief, backed up by a growing body of evidence, that the concept of bound water, at least as applied to the process of freeze concentration, has no valid scientific basis" [23].

Therefore, the question posed in the title remains unanswered: Do hydrogels contain different types of water? In view of the evidence presented here, it appears that the question as such is not specific enough. If one includes such aspects as temperature and timescales, some partial answers to the question seem to emerge. First of all and rather obviously, at very low temperatures different types of water are present: frozen and non-frozen water. However, there are no indications as to the consequences of this fact for the situation at room- or body temperature. It should be clear from this article that DSC or DTA experiments on hydrogels cannot be interpreted in terms of different classes of water in the way that this is often done in the literature. Secondly, there is the matter of the timescales of the experiments. If one were able to make observations at infinitely short timescales, in other words, if one were able to stop time, it would be possible to identify each water molecule individually. On the other hand, the NMR experiments described in this article seem to indicate that on millisecond timescales the exchange between the water molecules is so fast that they behave as a single class of water, with properties that are the average of all the water molecules. The cited literature references are indications that this behavior continues to significantly shorter timescales than the millisecond, measured by NMR.

The question that remains unanswered is what happens on a molecular level in hydrogels, at ambient temperatures, and on timescales that are important for e.g. diffusion phenomena. Undoubtedly there are differences between the individual water molecules. Some of them will be closer to hydroxyl groups than others, which will influence their behavior. At the present time it is not clear, however, whether this results in a meaningful distinction between different classes of water, or in a continuum of properties. Finally a word about the term 'bound' water. This term would seem to have implications for the mobility of the water molecules, but disregards for instance the possibility that water structures are disrupted more in terms of orientation than of mobility. The very short residence times of water molecules in the hydration shells of some monovalent ions are interesting in this respect. Perhaps it would be better to substitute the term 'bound' water with the more neutral, but possibly less appealing term 'perturbed water'.

Acknowledgement

The author is especially grateful for the help and guidance of Prof. H. E. Junginger, Prof. J. C. Leyte, Dr. J. A. Bouwstra, Dr. J. de Bleyser, and Ms. M. A. de Vries. The DHPMA, used in the NMR experiments was obtained as a gift from Professor E. Schacht, University of Gent, Belgium.

REFERENCES

1. W. E. Roorda, J. A. Bouwstra, M. A. de Vries and H. E. Junginger, *Biomaterials* **9**, 494 (1988).
2. W. E. Roorda, J. A. Bouwstra, M. A. de Vries and H. E. Junginger, *Pharm. Res.* **5**, 722 (1988).
3. W. E. Roorda, J. de Bleyser, H. E. Junginger and J. C. Leyte, *Biomaterials* **11**, 17 (1990).
4. J. A. Bouwstra, J. C. Van Miltenburg, W. E. Roorda and H. E. Junginger, *Polym. Bull.* **18**, 337 (1987).
5. A. S. Hoffman, *J. Contr. Rel.* **6**, 297 (1987).
6. T. G. Park and A. S. Hoffman, *J. Biomed. Mater. Res.* **24**, 21 (1990).
7. L. C. Dong and A. S. Hoffman, *J. Contr. Rel.* **13**, 21 (1990).
8. L. C. Dong, Q. Yan and A. S. Hoffman, *J. Contr. Rel.* **19**, 171 (1992).
9. A. S. Hoffman, *J. Polym. Sci. Technol.* **8**, 33 (1975).
10. A. S. Hoffman, T. A. Horbett and B. D. Ratner, *Ann. NY Acad. Sci.* **283**, 372 (1977).
11. H. B. Lee, M. S. Jhon and J. D. Andrade, *J. Colloid. Interf. Sci.* **51**, 558 (1976).
12. Y. K. Sung, D. Gregonis, M. S. Jhon and J. D. Andrade, *J. Appl. Polym. Sci* **26**, 3719 (1981).
13. D. G. Pedley and B. J. Tighe, *Br. Polym.* **11**, 130 (1979).
14. Y. Taniguchi and S. Horigome, *J. Appl. Polym. Sci.* **19**, 2743 (1975).
15. M. L. Miller, *The Structure of Polymers*, Reinhold Publishing Company, 291 pp (1966)
16. A. P. MacKenzie and D. H. Rasmussen, in: *Water Structure at the Water-Polymer Interface*, p. 146, H. G. H. Jellinek (Ed.) Plenum Publ. Corp., New York.
17. A. P. MacKenzie, *International Symposium on Freeze Drying of Biological Products. Develop. Biol. Standard.*, **36**, 51. S. Karger, Basel (1977).
18. A. P. MacKenzie, in: *Freeze Drying and Advanced Food Technology*, p. 277, S. A. Goldblith, L. Rey and W. W. Rothmayr (Eds). Academic Press, New York (1975).
19. C. A. Oksanen and G. Zografi, *Pharm. Res.* **7**, 654 (1990).
20. G. Zografi, *Drug Dev. Ind. Pharm.* **14**, 1905 (1988).
21. C. G. Pitt, X. C. Song, R. Sik and C. F. Chignell, *Biomaterials* **12**, 715 (1991).
22. R. W. Impey, P. A. Madden and I. R. MacDonald, *J. Phys. Chem.* **87**, 5071 (1983).
23. F. Franks, *Process Biochem.* **3**, (1989).

Properties controlling the diffusion and release of water-soluble solutes from poly(ethylene oxide) hydrogels 1. Polymer composition

MARION E. McNEILL and NEIL B. GRAHAM*

Department of Pure and Applied Chemistry, University of Strathclyde, Thomas Graham Building, 295 Cathedral Street, Glasgow G1 1XL, UK

Received 16 April 1992; accepted 6 October 1992

Abstract—This study examines the state of water-association with poly(ethylene oxide), as evidenced by diffusivity, in a series of crosslinked polyurethanes made from poly(ethylene glycols) of a range of molecular weights. As a subsidiary underpinning exercise the correlation of diffusivity with water content at relatively high levels of swelling (>45%) using a variety of semi-empirical equations was analyzed. Three water-soluble compounds with similar molecular weights and which exhibit minimal interaction with the polymer, as shown by their partition coefficients, were chosen for this part of the research programme. These were proxyphylline, morphine hydrochloride and caffeine. The best statistical correlations of the data were obtained for plots of: (a) diffusivity against weight percent water; and (b) log diffusivity against the reciprocal of the weight percent of water in the hydrogels. Proxyphylline results for the high levels of swelling compositions were augmented with data from lower swelling compositions and a clear break in the slope of diffusivity against percentage of water in the swollen hydrogel was obtained. This indicated a change in the nature of the diffusion at this point. The probability of this transition point corresponding to a change for diffusion through water bound as trihydrate to diffusion in free water is discussed.

Key words: Water content; trihydrate; free water.

INTRODUCTION

The affinity for water differentiates hydrogels from other polymers. The physical properties of a hydrated hydrogel are quite different from those of the dry hydrogel. Diffusivity, D, in a fully hydrated hydrogel may be several orders of magnitude greater than in the xerogel. It is reasonable to assume that the free volume of water present in the hydrogel is available for diffusion of water-soluble solutes [1]. Across a range of compositions there will be a broad band of permeabilities dependent primarily on the degree of hydration. Since hydrogels are useful carriers for drugs and bioactive molecules correlation between water content and diffusivity could be very important in designing drug delivery systems based on hydrogels.

Although most research and developments of hydrogels for controlled drug delivery systems has been carried out with poly(methacrylates) [2] and poly(vinyl alcohols) [3] the large and versatile family of poly(ethylene oxide) (PEO) hydrogels has already proved useful [4-7] and merits further research and development. A large series of polymers based on poly(ethylene glycols) (PEG) $\bar{M}_n$ 600–8500 with crosslinking ratios of 0.2–8 mol triol per mol PEG has been prepared. All these

* To whom correspondence should be addressed.

hydrophilic polymers can be characterized by their water uptake at a given temperature. Reproducibility in the water uptake between different batches of any one composition is a sensitive test for completion of the reaction between the hydroxyl groups of both the diol and the triol and the stoichiometric equivalence of diisocyanate. Incomplete reaction causes a detectable increase in the water uptake and also loss in weight of the xerogel on drying down. These polymers are normally specified with less than 1% extractable material to consider them suitable as biomaterials.

Three water soluble compounds with similar order molecular weights all exhibiting minimal interaction with the polymer backbone were chosen, namely proxyphylline, morphine hydrochloride and caffeine, in order to relate diffusivity with hydrogel water contents for each solute and then compare the results among the different solutes. A preferred basis for plotting relationships between diffusivity and water content could then be selected and, in the context of this paper, applied to a study of water-content and hydrogel structure.

THEORY

The swelling properties of a crosslinked polymer are controlled by the combination of free energies of mixing between the solvent and polymer chains, and by the elastic response of the network to volume increase due to solvent absorption [8]. At equilibrium the elastic response of the network exactly offsets the swelling potential of the solvent within the network [9].

The analysis by weight of a hydrated polymer may be considered in the simplest interpretation as W% water and P% polymer (Fig. 1). The equilibrium water uptake, EWU, in parts per hundred parts initial dry weight at a given temperature is easily determined by leaving a dry polymer sample (xerogel) in water until the swollen weight remains constant. Then

$$W = \frac{EWU}{100 + EWU}\,\%. \tag{1}$$

W is linked to P by the hydrophilic component of the polymer composition. In the case of PEO-based crosslinked hydrogels the hydrophilic component is the PEO. We have found [10] the water uptake is a direct function of the PEO weight fraction, EO_x of the xerogel. This can be calculated from the stoichiometric formula for the composition. At 37°C

$$\log EWU = 1.54EO_x + 1.01. \tag{2}$$

This empirical equation was found to hold for $EO_x = 0.5$–0.9 irrespective of either PEG $\bar{M}_n$ or the degree of crosslinking. Since water in hydrogels may exist in different forms it is to be expected that each form will have a characteristic diffusity for any specified water soluble compound and that each will make a different contribution to the overall permeability. From studies on crosslinked poly(hydroxyethyl methacrylates) Lee *et al.* [11] examined gels containing 20–50% water by conductivity, thermal expansion and differential scanning calorimetry, DSC, over a temperature range −15 to 25°C and concluded 20% of the hydrated gel water was bound to the polymer network. They also believed there was a transition from a homogeneous to a heterogeneous 2 phase system around 40% water.

Water binding to PEG was examined using DSC by Hager and Macrury [12]. They measured the enthalpy changes for three endotherms in the scans; first polymer melting, second free water melting and third a low temperature melt with a peak at −16°C, related to 2.7 mol water per mol PEG, which they attributed to crystalline eutectic melting. Hey *et al.* [13] also recorded a low temperature endotherm with PEO/water mixtures but they found the peak temperature was not constant. It increased with progressive additions of water. Graham *et al.* [14] reported clear evidence for the existence of a crystalline complex associated with 3 mol of water per ether group. This trihydrate was evident even in the presence of relatively large amounts of free water for PEG $\bar{M}_n \geq 1000$.

The components of a fully hydrated hydrogel are illustrated in a branched diagram (Fig. 1), showing how the hydrogel can be divided into polymer and water components then subdivided into polymer composition and predicted types of water. Fraction EO_x of the polymer P% in the fully swollen hydrogel is PEO EO_p%. If 3 mol of water are associated with each repeat ethylene oxide unit (M_w 44) to form a trihydrate then $3 \times 18/44 = 1.23$ times the percentage of PEO in the fully water swollen hydrogel will be trihydrate water which has an ordered structure and a defined melting endotherm minimum between −16 and −7°C. Free water is also identified in the DSC scan with a minimum 0–2°C, and also frequently a third endotherm minimum between −4 and 1°C attributed to interfacial water which may link the bound and free water states. It is sometimes difficult to separate and quantify the interfacial water from the free water by DSC. Two endothermic minima are as a rule distinguishable but perhaps only 2 deg apart so that they are generally fused together.

The free volume in a polymer may be thought of as the volume fraction of molecular size holes available for diffusion. Free volume theory of diffusion was

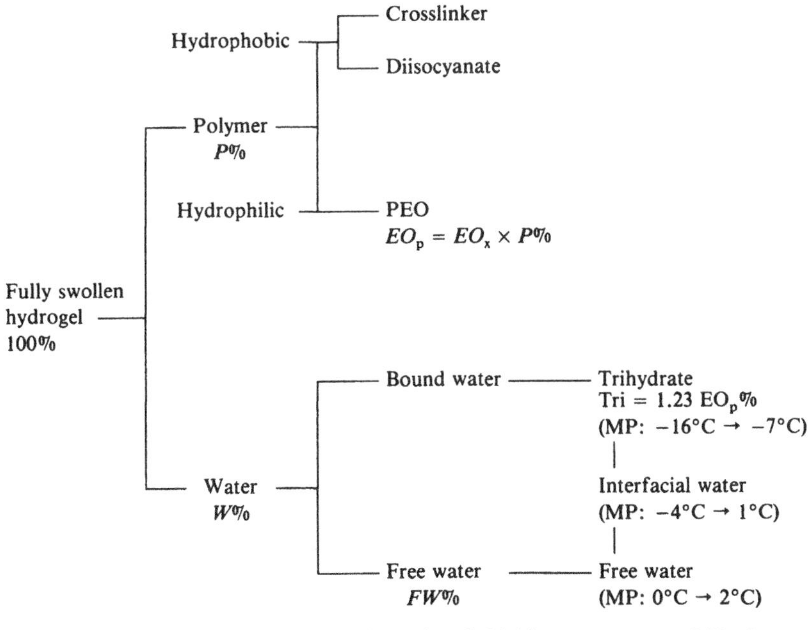
Crosslinker
Hydrophobic
Diisocyanate
Polymer
P%
Hydrophilic
PEO
EO_p = EO_x × P%
Fully swollen
hydrogel
100%
Bound water
Trihydrate
Tri = 1.23 EO_p%
(MP: −16°C → −7°C)
Interfacial water
(MP: −4°C → 1°C)
Water
W%
Free water
FW%
Free water
(MP: 0°C → 2°C)

Figure 1. Structure of a hydrated poly(ethylene oxide) hydrogel. Melting temperatures, MP, given, are the DSC endotherm minima.

applied to the diffusion of sodium chloride in water-swollen membranes by Yasuda *et al.* [15]. They assumed the effective free volume for diffusion was the free volume of water present in the membrane and found

$$\log D_p = \log D_0 - K[(100/W) - 1] \tag{3}$$

where D_p is the diffusion coefficient of the diffusate (in this case sodium chloride), D_0 is the diffusion coefficient in pure water and K is a proportionality constant related to a characteristic volume required to accommodate the permeant molecule.

B. K. Davis [16] studied diffusion in crosslinked water-swollen polyacrylamides and polyvinyl pyrrolidones and found $\log D_p$ decreased linearly with P.

$$\log D_p = \log D_0 - K_s P \tag{4}$$

where K_s is termed a retardation coefficient and increases linearly with molecular weight W_w in accordance with Eqn (5).

$$K_s = 0.022 + 4.4 \times 10^{-7} M_w. \tag{5}$$

Combining Eqns (4) and (5)

$$D_p = D_0 \exp[-(0.05 + 10^{-6} M_w)P]. \tag{6}$$

For compounds with $M_w < 500$ Eqn (6) can be simplified to

$$D_p = D_0 \exp(-0.05P) \tag{7}$$

or

$$\log D_p = \log D_0 - 0.022P. \tag{8}$$

Since $P = 100\text{-}W$ Eqn (8) may be rewritten

$$\log D_p = \log D_0 - 2.2 + 0.022W. \tag{9}$$

Equations describing the release of water soluble solutes uniformly distributed in the matrix of fully hydrated hydrogel monoliths into an aqueous sink have been derived by Crank [17] from Fick's Laws for diffusion. A simplified form of a complex mathematical equation valid for the first 60% of the total release from a thin slab of constant dimensions and constant D_p is

$$\frac{M_t}{M_\infty} = 4\left[\frac{D_p t}{\pi l^2}\right]^{0.5} \tag{10}$$

where M_t is the amount released at time t and M_∞ is the total amount released. l is the slab thickness and D_p is the diffusion coefficient of the solute in the hydrogel [18]. A graph of fractional release vs square root of time should give a linear plot and D_p can be calculated from the gradient. Alternatively D_p can easily be calculated from the half life time, $t_{1/2}$. When $M_t/M_\infty = 0.5$.

$$D_p = \frac{0.0492 l^2}{t_{1/2}} \text{ cm}^2\text{s}^{-1}. \tag{11}$$

EXPERIMENTAL

Preparation of PEO/polyurethane xerogels

A series of PEO/polyurethane polymers based on thoroughly vacuum dried poly(ethylene glycol), PEG, $\bar{M}_n$ 600–8500 (Breox Pharmaceutical Grade, BP

Chemicals) determined precisely by hydroxyl number with various ratios of cross-linker 1,2,6 hexane triol (Aldrich Chemical Co. Ltd.), ranging from 0.3–7 mol per mol PEG and the stoichiometric equivalence of dicyclohexyl methane-4,4′-diisocyanate (Desmodur W, Bayer) was prepared. This is a one-shot method of polymerization. No solvent is used. The polymer was cured in the form of billets which were sliced with a horizontal guillotine into strips 1–1.5 mm thick.

Example of a stoichiometric xerogel formulation

PEO 5830/4HT

Proportions	PEG	:	1,2,6HT	:	Desmodur W
molar	1	:	4	:	7
daltons	5830	:	536	:	1838
fractional	0.7106	:	0.0653	:	0.2240

$$-OH + -NCO \rightarrow -OOC\overset{H}{N}-\text{urethane}$$

∴ 1 mol Desmodur ≡ 1 mol PEG
and 1.5 mol Desmodur ≡ 1 mol 1,2,6HT

EO_x the weight fraction of PEO in a formulation determines the equilibrium water uptake. (See Eqn (2)). In the above example $EO_x = 0.7106$.

Swelling in water

Samples from each composition were weighed, placed in distilled water at 37°C and left to swell to constant weight.

$$\text{Equilibrium water uptake, EWU} = \frac{\text{swollen weight} - \text{dry weight}}{\text{dry weight}} \times 100 \qquad (12)$$

in parts per hundred.

DSC

Samples were cut from the fully water-swollen hydrogels, placed in DSC sample pans, covered, sealed and transferred to a DuPont 9900 Differential Scanning Calorimeter. The temperature was taken down to −40°C then raised at 2°C/min.

Other xerogel samples were placed in contact with water equivalent to 3 mol water per $-CH_2CH_2O-$ group calculated from the formulation. They were left in sealed impermeable bags for several days to equilibrate, then scanned.

Drugs

Morphine hydrochloride, (Vestric) M_w 322. Caffeine, 2,3,7-trimethylxanthine (Aldrich) M_w 194. Proxyphylline, (hydroxypropyl theophylline) (Sigma) M_w 238.

Preparation of hydrogels containing drugs in the matrix

Dimensions of xerogel slices were l_d mm × 9.5 mm × 29 mm with semicircular ends. Thickness l_d was measured with a micrometer to 2 decimal places. The three drugs examined are all water soluble. Polymer slices were weighed and immersed in aqueous solutions at 37°C of morphine hydrochloride 10 mg/ml, caffeine 10 mg/ml and proxyphylline 65 mg/ml and left to equilibrate. On removal, the samples were blotted with a tissue to remove surface solution, weighed, and the swollen lengths and breadths measured. The rubbery texture of the swollen gel makes it difficult to measure the swollen thickness accurately with a pressure sensitive micrometer. However, since water swelling and consequently expansion of these PEO hydrogels is isotropic, the expansion of the two larger dimensions can be used to calculate the swollen thickness, l_s, from the following equation:

$$\text{Linear swelling of factor } f = \frac{\text{swollen length}}{\text{dry length}} \text{ or } \frac{\text{swollen width}}{\text{dry width}}$$

$$l_s = f \times l_d. \tag{13}$$

Analysis of drug concentration

The three drugs used in this study all absorb ultra violet light, allowing solution concentrations to be analyzed in a Cecil 5000 UV Spectrophotometer. Drug solutions were calibrated at 283 nm, morphine hydrochloride, 273 nm, caffeine, and 273 nm, proxyphylline.

Release studies

After weighing, the drug-swollen hydrogel was transferred to an aqueous sink at 37°C to desorb the drug substance. Two slightly different methods of following the increasing concentration of drug in the sink with time were used depending upon the estimated uptake of drug and the intensity of the UV absorbance signal.

(1) A conical flask containing 100 ml distilled water was clamped in a Grant SS40 Thermostat Bath agitated at 60 strokes per minute. Samples of solution were withdrawn periodically, analyzed in a 1 cm quartz cell then returned to the conical flask. Sampling continued at ever increasing time intervals until the absorbance reached a constant value taken as the total release.

(2) A Caleva Tablet Dissolution Apparatus Model 8ST conforming to US Pharmacopoeia XXI with paddles and 6 vessels containing 800 ml was used. The solution was pumped by a Watson Marlow Peristaltic Pump 503U through a flow cell in the Spectrophotometer and the changing absorbance with time was continuously recorded on a chart.

The absorbance values were converted first into concentrations then absolute amounts, M_t, at time, t. M_t vs $t^{0.5}$ provided linear plots for all experiments up to $t_{1/2}$, the half life time when M_t is half the total release, M_∞. Using Eqn (11) the diffusion coefficients were calculated for morphine, caffeine and proxyphylline for a wide range of compositions. The results illustrated are the mean values of duplicates.

RESULTS AND DISCUSSION

We have already reported [10] how the EWU of these PEO-based crosslinked polymers depends on the ethylene oxide content of their composition. This link was further illustrated by plotting the total water content, W, against the ethylene oxide content, EO_p, of the fully hydrated polymer in Fig. 2. The total water is made up

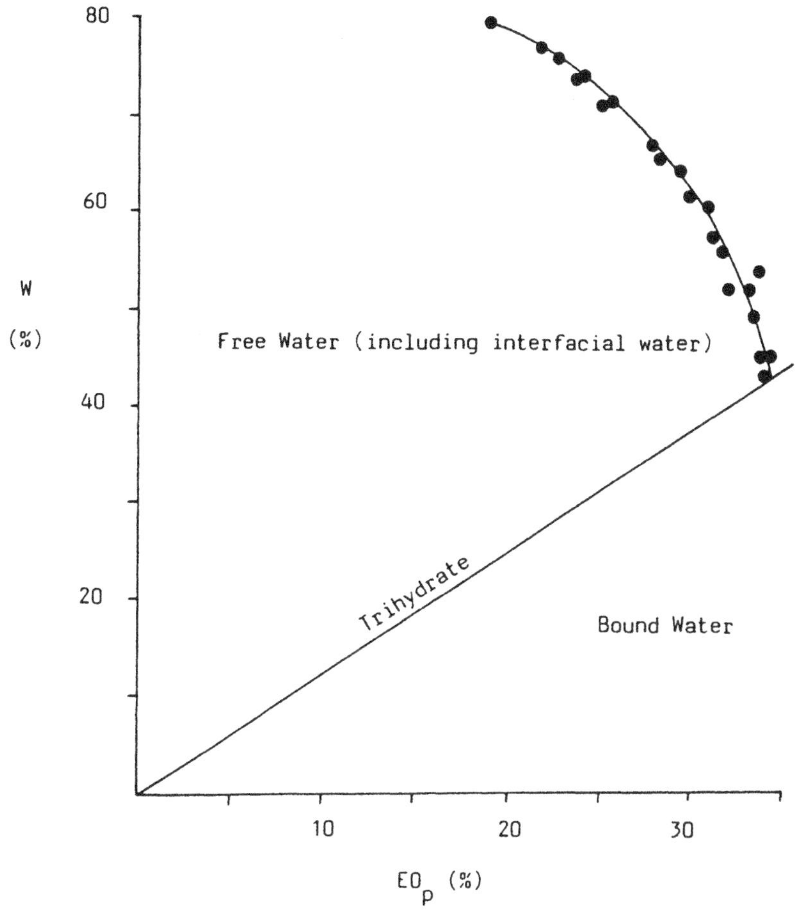
80
60
40
20
W
(%)
Free Water (including interfacial water)
Trihydrate
Bound Water
10
20
30
EO p (%)

Figure 2. Separation of total water content W into bound and free water related to the ethylene oxide content of PEO hydrogels.

of bound, free and interfacial water (Fig. 1). The best computerized fit to the curve through the points shown in Fig. 2 was found to be given by the following truncated polynomial expression.

$$W/(100\text{-}W) = 7.524 - 0.201\ EO_p. \quad (14)$$

When DSC scans were run for a range of swellings from zero to fully swollen they showed a developing endotherm with a minimum temperature of −16°C to −7°C (dependent on composition and degree of hydration) which grew to a maximum size before another endotherm between −4°C and 2°C appeared. This is shown for partly swollen PEO5830/4HT in Fig. 3 and fully swollen in Fig. 4. The low melting endotherm is, we believe, due to the stable complex of 3 mol H_2O per $-CH_2CH_2O-$ as it persists even in the fully hydrated scan (Fig. 4), together with the bifurcated endotherm for interfacial and free water which is difficult to separate. The interfacial water, while recognized, has therefore been included in the free water for subsequent calculations. Graham *et al.* [14] obtained similar results with linear PEG. When trihydrate water equivalent to 1.23 times the ethylene oxide, EO_p, is

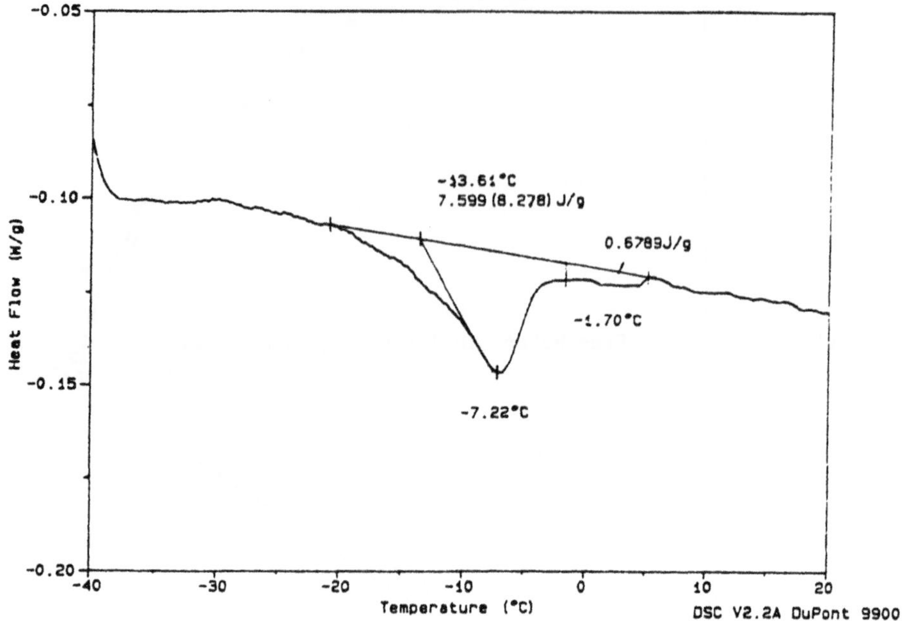
-0.05
-0.10
-0.15
-0.20
Heat Flow (W/g)
-13.61°C
7.599 (8.278) J/g
0.6789J/g
-1.70°C
-7.22°C
-40
-30
-20
-10
0
10
20
Temperature (°C)
DSC V2.2A DuPont 9900

Figure 3. DSC of partly swollen PEO 5830/4HT, *EWU* 64.6 pph. Heating rate 2°C/min.

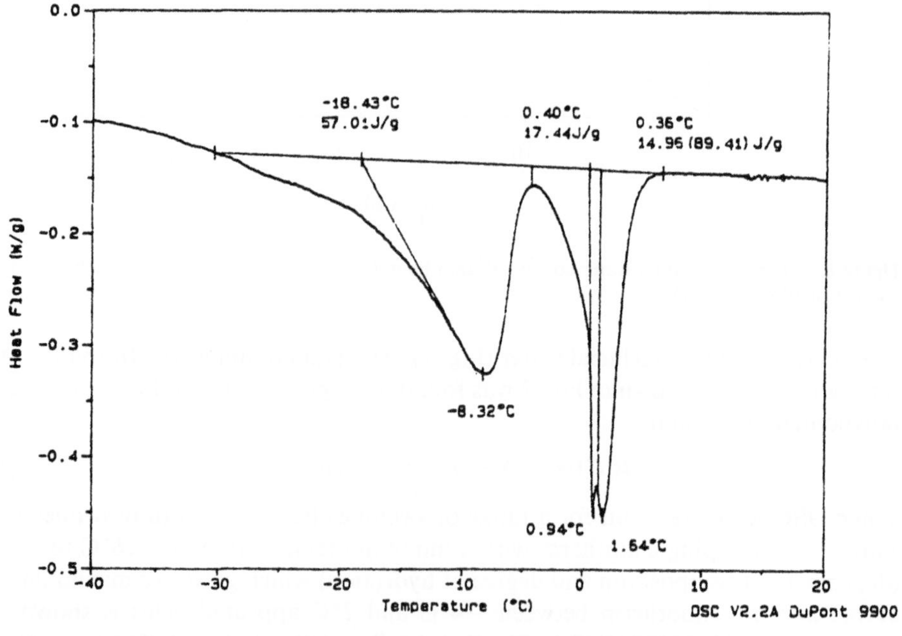
0.0
-0.1
-0.2
-0.3
-0.4
-0.5
Heat Flow (W/g)
-18.43°C
57.01J/g
0.40°C
17.44J/g
0.36°C
14.96 (89.41) J/g
-8.32°C
0.94°C
1.64°C
-40
-30
-20
-10
0
10
20
Temperature (°C)
DSC V2.2A DuPont 9900

Figure 4. DSC of fully swollen PEO 5830/4HT, *EWU* 127 pph. Heating rate 2°C/min.

drawn in Fig. 2 it divides the total water, *W*, into bound and free water, *FW*, and provides an interesting intersection at 42% *W* suggesting there will be no free water in fully hydrated PEO hydrogels containing <42% *W*.

According to the 'free-volume theory' the diffusion coefficient, D_p, of water-soluble compounds in hydrogels decreases exponentially with the inverse of the water content (see Eqn (3)). However according to Davis [16] the diffusion coefficient increases exponentially with the water content (see Eqn 9) for compounds with $M_w < 500$.

We have observed the diffusion coefficient plotted simply against the total water content gives a reasonably good linear fit for a wide variety of water-soluble drugs in PEO-based hydrogels containing 45–80% water.

From Figs 1 and 2 it is clear that water in hydrogels can exist in different forms and one would expect each different type of water to contribute to the overall diffusivity to different degrees with the main contribution coming from the free water form. It may therefore be the case that the best correlation is given by comparing the diffusion coefficient with the free water content.

Assuming water first binds to the polymer as a trihydrate it is possible to calculate the rest of the water as free water (including interfacial water). For example, PEO5830/4HT:

EWU	experimental determination at 37°C	127 pph
W	see Eqn (1)	55.9%
EO_x	calculated from polymer formulation	0.711
P	= 100-W	44.1%
EO_p	= $EO_x \times P$ (see Figure 1)	31.4%
Tri	= trihydrate = $1.23 \times EO_p$	38.6%
FW	= W-Tri	17.3%

In the following figures experimental values of W were used to plot the points rather than values estimated from Eqn (14). The predicted W for the above example 56.0% is almost the same as the measured value of 55.9%. It is possible for 2 compositions prepared using quite different M_ws of PEG and proportions of crosslinker to contain the same percentage of ethylene oxide; then according to Eqn (2) they would be expected to have the same equilibrium water content when hydrated. PEO1610/0.9HT contains 68.6% ethylene oxide, compared with PEO5830/4HT with 71.1% ethylene oxide. Consequently it has a predicted W 2% lower, i.e. 53.8%. The experimental value was 55.0%.

Diffusion coefficient values for morphine, caffeine and proxyphylline were calculated from the results of the release studies described using Eqn (11). The data was entered into a computer program and graphs were drawn by Plotlib using the 'least squares' analysis to fit the best straight line. Log D_p vs $1/W$, log D_p vs W, D_p vs W and D_p vs FW plots have been combined in one figure for comparison, Fig. 5 for morphine hydrochloride, Fig. 6 for caffeine and Fig. 7 for proxyphylline. Regression equations were provided by a computer program [19]; and the coefficient of determination, r^2, was used to critically analyse how close a fit to linearity, i.e. $r^2 = 100\%$ was obtained from the points. The equations were used to calculate the diffusion coefficients with zero water content (in the log D vs $1/W$, $W = 1\%$ rather than 0 was substituted), $W = 100\%$ and $FW = 100\%$ which it was possible might equal the diffusivity in pure water.

The conclusions from the regression equations have been summarized in Tables 1, 2 and 3. The empirical equations given in Figs 5, 6 and 7 were derived from results between 45 and 75% total water content and between 3 and 48% free water (including interfacial water).

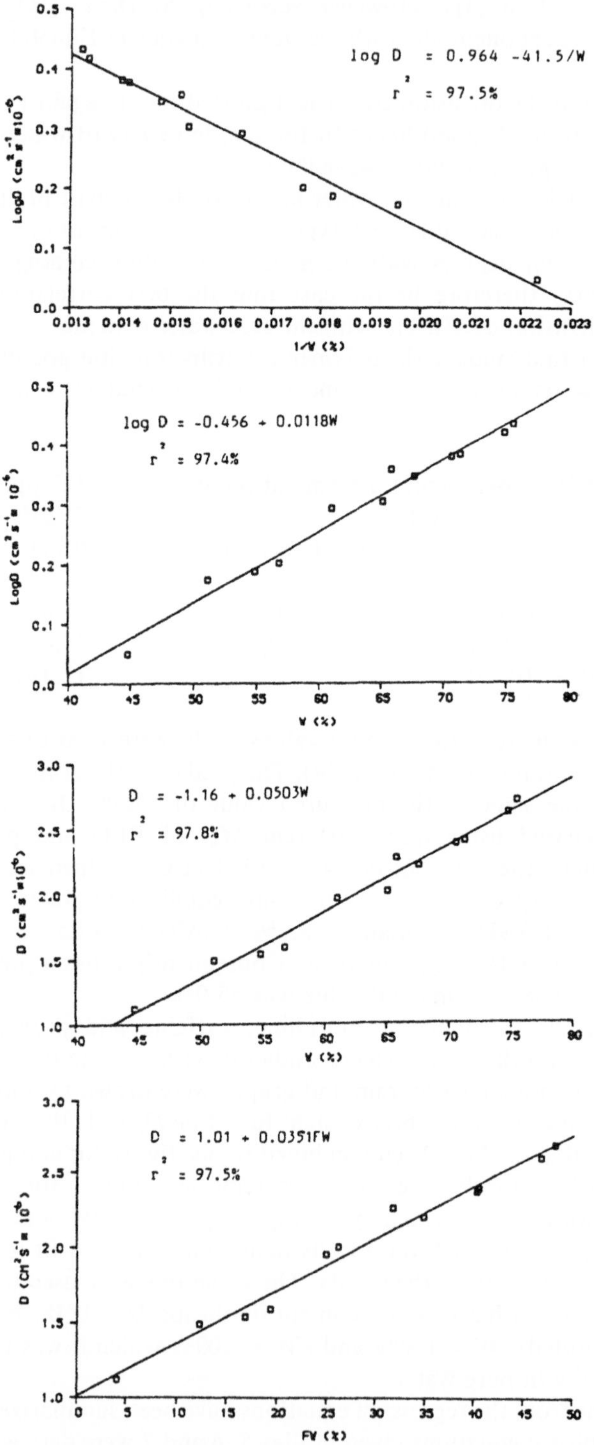
log D = 0.964 -41.5/W
r² = 97.5%
LogD (cm² s⁻¹ × 10⁻⁶)
0.0
0.1
0.2
0.3
0.4
0.5
0.013 0.014 0.015 0.016 0.017 0.018 0.019 0.020 0.021 0.022 0.023
1/W (%)
log D = -0.456 + 0.0118W
r² = 97.4%
LogD (cm² s⁻¹ × 10⁻⁶)
0.0
0.1
0.2
0.3
0.4
0.5
40 45 50 55 60 65 70 75 80
W (%)
D = -1.16 + 0.0503W
r² = 97.8%
D (cm² s⁻¹ × 10⁻⁶)
1.0
1.5
2.0
2.5
3.0
40 45 50 55 60 65 70 75 80
W (%)
D = 1.01 + 0.0351FW
r² = 97.5%
D (CM² S⁻¹ × 10⁻⁶)
1.0
1.5
2.0
2.5
3.0
0 5 10 15 20 25 30 35 40 45 50
FW (%)

Figure 5. Comparison of diffusion coefficient D vs water content W equations and plots for morphine hydrochloride through PEO hydrogels.

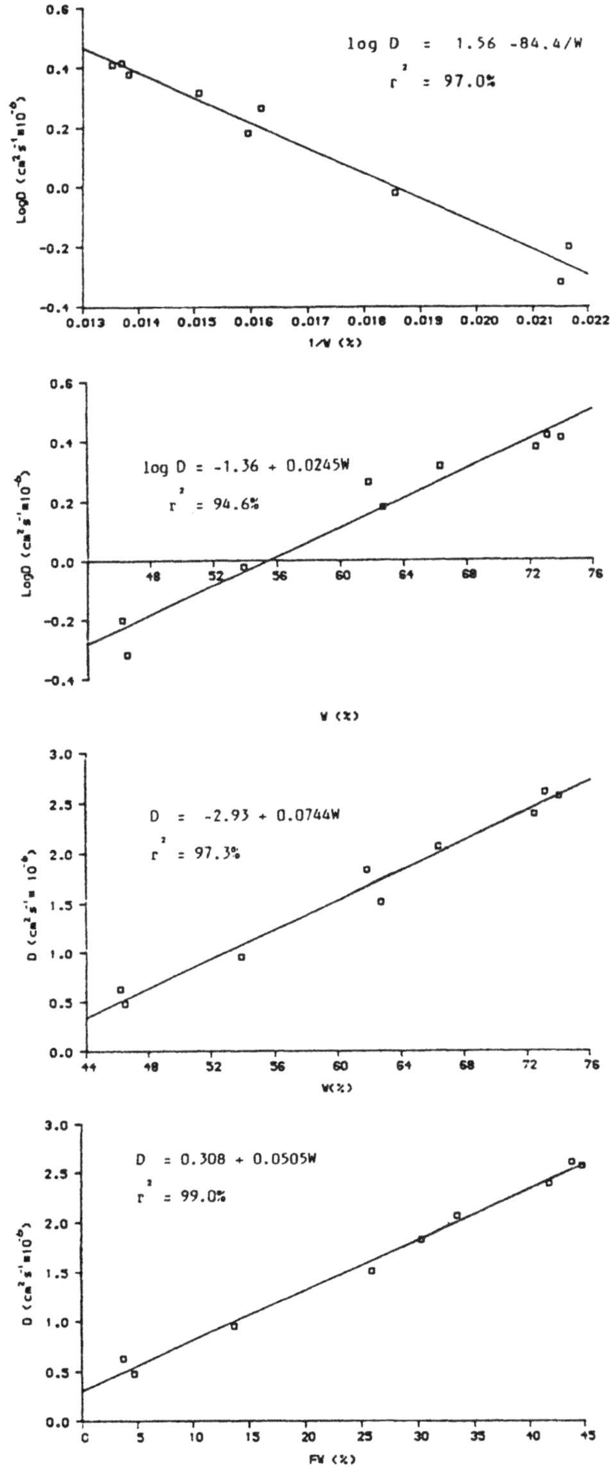
log D = 1.56 -84.4/W
r² = 97.0%
LogD (cm²s⁻¹×10⁻⁶)
1/W (%)
log D = -1.36 + 0.0245W
r² = 94.6%
LogD (cm²s⁻¹×10⁻⁶)
W (%)
D = -2.93 + 0.0744W
r² = 97.3%
D (cm²s⁻¹× 10⁻⁶)
W(%)
D = 0.308 + 0.0505W
r² = 99.0%
D (cm²s⁻¹×10⁻⁶)
FW (%)

Figure 6. Comparison of diffusion coefficient D vs water content W equations and plots for caffeine through PEO hydrogels.

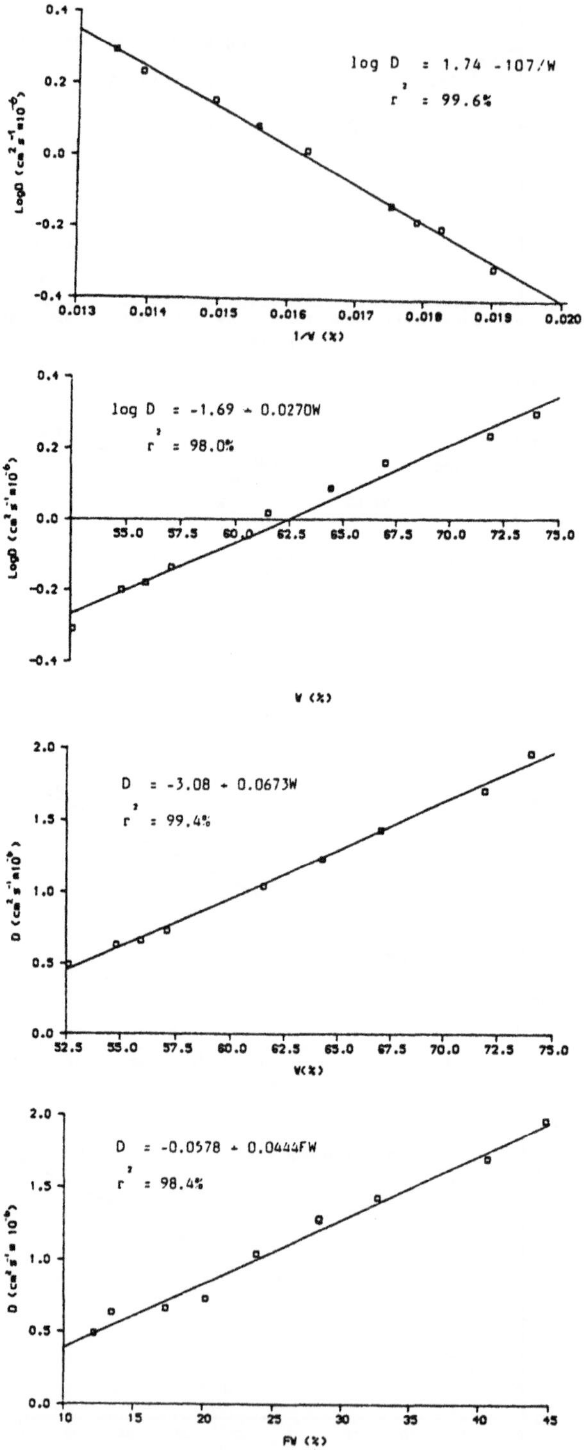
log D = 1.74 -107/W
r² = 99.6%
LogD (cm² s⁻¹ x10⁻⁶)
1/W (%)
0.013
0.014
0.015
0.016
0.017
0.018
0.019
0.020
0.4
0.2
0.0
-0.2
-0.4
log D = -1.69 + 0.0270W
r² = 98.0%
LogD (cm² s⁻¹ x10⁻⁶)
W (%)
55.0
57.5
60.0
62.5
65.0
67.5
70.0
72.5
75.0
D = -3.08 + 0.0673W
r² = 99.4%
D (cm² s⁻¹ x10⁻⁶)
W(%)
52.5
2.0
1.5
1.0
0.5
0.0
D = -0.0578 + 0.0444FW
r² = 98.4%
D (cm² s⁻¹ x 10⁻⁶)
FW (%)
10
15
20
25
30
35
40
45

Figure 7. Comparison of diffusion coefficient D vs water content W equations and plots for proxyphylline through PEO hydrogels.

Table 1.
Comparison of diffusivity equations for morphine hydrochloride through PEO hydrogels. Data derived from regression equations, Fig. 5, $D\,\mathrm{cm^2 s^{-1}} \times 10^{-6}$.

	log D vs $1/W$	log D vs W	D vs W	D vs FW
Regression r^2%	97.5	97.4	97.8	97.5
Predicted D for W or FW = 0%[a]	2.9×10^{-35}	0.35	−1.16	1.01
Predicted D for W or FW = 100%	3.54	5.30	3.87	4.52
Predicted D for W = 65.9%[b] or FW = 31.7%[b]	2.20 ± 0.07	2.10 ± 0.06	2.15 ± 0.05	2.13 ± 0.05
Measured D for W = 65.9% or FW = 31.7%	2.27	2.27	2.27	2.27

[a] For log D vs $1/W$ W = 1%
[b] ±95% C.I.

Table 2.
Comparison of diffusivity equations for caffeine through PEO hydrogels. Data derived from regression equations, Fig. 6, $D\,\mathrm{cm^2\,s^{-1}} \times 10^{-6}$.

	log D vs $1/W$	log D vs W	D vs W	D vs FW
Regression r^2%	97.0	94.6	97.3	99.0
Predicted D for W or FW = 0%[a]	1.4×10^{-77}	0.04	−2.93	0.31
Predicted D for W or FW = 100%	5.20	12.30	4.51	5.36
Predicted D for W = 66.3%[b] or FW = 33.5%[b]	1.99 ± 0.19	1.84 ± 0.23	2.00 ± 0.12	2.00 ± 0.07
Measured D for W = 66.3% or FW = 33.5%	2.06	2.06	2.06	2.06

[a] For log D vs $1/W$ W = 1%
[b] ±95% C.I.

Table 3.
Comparison of diffusivity equations for proxyphylline through PEO hydrogels. Data derived from regression equations, Fig. 7, $D\,\mathrm{cm^2\,s^{-1}} \times 10^{-6}$.

	log D vs $1/W$	log D vs W	D vs W	D vs FW
Regression r^2%	99.6	97.9	99.4	97.4
Predicted D for W or FW = 0%[a]	0	0.02	−3.07	−0.06
Predicted D for W or FW = 100%	4.68	10.72	3.64	4.38
Predicted D for W = 67.0%[b] or FW = 32.6%[b]	1.39 ± 0.04	1.34 ± 0.09	1.42 ± 0.04	1.39 ± 0.09
Measured D for W = 67.0% or FW = 32.6%	1.43	1.43	1.43	1.43

[a] For log D vs $1/W$ W = 1%
[b] ±95% C.I.

The log D vs $1/W$ and log D vs W values for W = 1 or 0% are indeed as low as might be expected for diffusion in a xerogel. The negative values for D vs W are interesting and will be explained further on in this paper together with the $1.01 \times 10^{-6}\,\mathrm{cm^2\,s^{-1}}$ value for morphine from the D vs FW equation.

For the range of solute molecular weights studied the diffusion coefficients in pure water would be expected to be between 4.5 and $6.5 \times 10^{-6}\,\mathrm{cm^2\,s^{-1}}$ [20, 21]. The values obtained for caffeine and proxyphylline from Davis's equation are well outside this range. All the other equations give results within or fairly close to the predicted values from the literature.

The equations were tested by predicting with 95% confidence limits the diffusivity of each solute in water swollen hydrogel compositions which absorb twice their dry weight of water at 37°C. In order to compare each predicted D with a measured D, W values close to 66.7% which tied up with an experimental value were inserted into the regression equation. These results are summarized in Tables 1, 2 and 3. For log D vs $1/W$ all the measured values fall within the predicted 95% confidence limits, although the spread for caffeine is almost the largest in the table. The measured Ds for morphine are higher and outside the limits of the predicted values in all the other proposed equations. However, good agreement for caffeine and proxyphylline was obtained from both D vs W and D vs FW relationships.

The coefficient of determination, r^2, from the regression equations, together with the 95% confidence limits within the range of hydrogels studied, affords a constructive though not rigorous mathematical evaluation of these plots. Using the regression values to compare the different relationships, the mean r^2 for the three solutes log D vs $1/W$ is 98.0% for log D vs W is 96.6% for D vs W is 98.2% and for D vs FW is 98.0%.

The apparent contradiction of log D decreasing linearly with $1/W$ according to the 'free volume' theory and at the same time increasing linearly with W according to Davis is explained by the mathematics of these equations between 45 and 75% W. If log D vs $1/W$ is a perfect fit, i.e. $r^2 = 100\%$ log D vs W will be a shallow convex curve with $r^2 = 97.4\%$. When a straight line is fitted by 'least squares' analysis the computer regression program does not identify a bias of the points. Subjective analysis of all the log D vs W plots shows most of the points in the middle section lie above the straight line but below the line at each end, i.e. there is indeed a bias towards a shallow convex curve rather than a straight line fit to the values. In all cases the r^2 values are weaker for log D vs W than for log D vs $1/W$. Together with the high predicted values for D from extrapolation to 100% W for both caffeine and proxyphylline, the postulated log D vs W linearity is therefore the least favoured relationship.

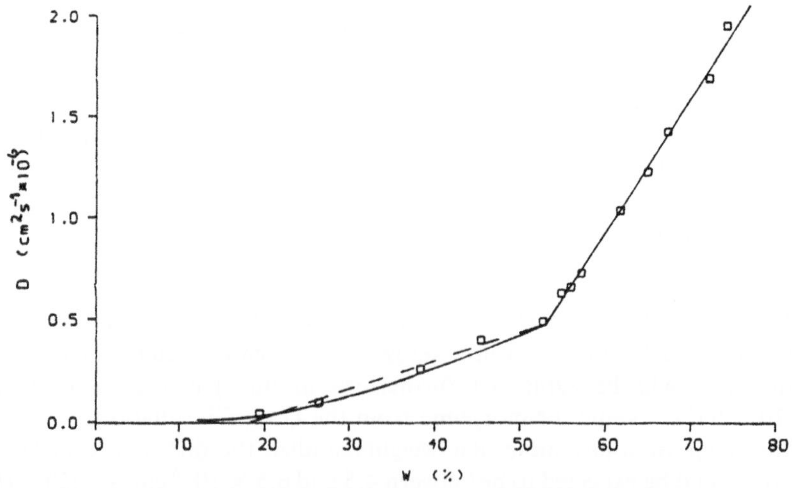
2.0
1.5
1.0
0.5
0.0
D (cm²s⁻¹×10⁻⁶)
0
10
20
30
40
50
60
70
80
W (%)

Figure 8. Proxyphylline diffusivity D vs water content W through PEO hydrogels.

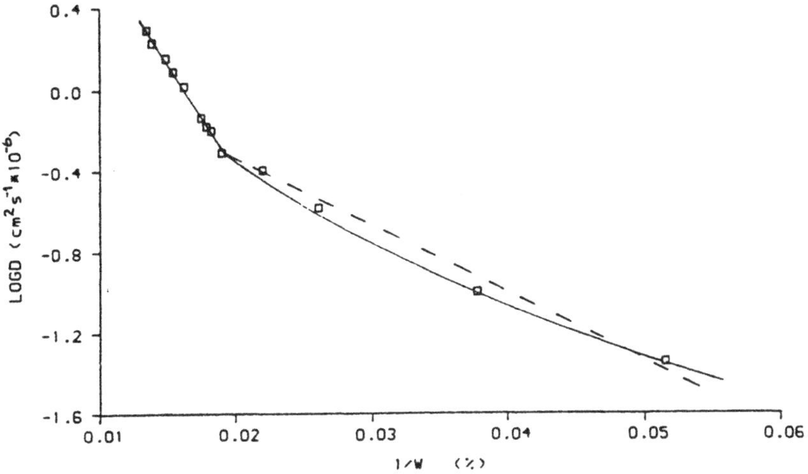
0.4
0.0
-0.4
-0.8
-1.2
-1.6
LOGD (cm2s-1x10-6)
0.01
0.02
0.03
0.04
0.05
0.06
1/W (%)

Figure 9. Log D vs $1/W$ for proxyphylline through PEO hydrogels.

As shown in Fig. 1, water in these hydrogels may be considered to exist in different forms. If water first associates with the PEO as a trihydrate then according to Fig. 2 there may not be any free water in hydrogels with less than 42% *W*. If water-soluble solutes are able to diffuse through this bound water transportation might be expected to be much slower than through the free water channels. In order to test this hypothesis diffusivity results for proxyphylline in three low swelling compositions with <40% water content and no free water and 1 composition with 45.4% *W* and 3.3% *FW* were determined and added to the results shown in Fig. 7 (2nd bottom and top). The extended figures appear as Figs 8 and 9. The compositions, swelling data and diffusion coefficient values for proxyphylline are listed in Table 4.

In Fig. 8 the 'least squares' straight line fit for diffusivity between 52.6 and 74% *W* with $r^2 = 99.4\%$ has been drawn and the *D* values between 19.4 and 52.6% *W* have been linked with a shallow curve (solid line) and also the 'least squares' linear fit,

Table 4.
Diffusion of proxyphylline in PEO hydrogels at 37°C c.f. Figs 1 and 2.

Composition	*EWU* pph	*W* %	EO_x	*P* %	EO_p %	Tri %	*FW* %	*D* $cm^2 s^{-1} \times 10^{-6}$
PEO 600/1HT	24	19.4	0.432	80.6	34.8	36.3	—	0.045
PEO 1610/4HT	36	26.5	0.404	73.5	29.7	36.5	—	0.104
PEO 3330/4HT	62	38.3	0.566	61.7	34.9	43.0	—	0.258
PEO 3330/3HT	83	45.4	0.627	54.6	34.2	42.1	3.3	0.402
PEO 6430/5HT	111	52.6	0.690	47.4	32.7	40.4	12.2	0.493
PEO 3800/2HT	121	54.8	0.742	45.2	33.5	41.3	13.5	0.631
PEO 5830/4HT	127	55.9	0.711	44.1	31.4	38.6	17.3	0.664
PEO 3100/2HT	133	57.1	0.699	42.9	30.0	36.9	20.2	0.734
PEO 3100/1HT	160	61.5	0.797	38.5	30.7	37.7	23.8	1.042
PEO 4360/1HT	203	67.0	0.845	33.0	27.9	34.4	32.6	1.431
PEO 8490/1.2HT	256	71.9	0.904	28.1	25.4	31.3	40.6	1.703
PEO 8490/1HT	285	74.0	0.915	26.0	23.8	29.3	44.7	1.958

$r^2 = 98.3\%$ is shown as a dashed line. The curved line is possibly the more likely fit as it can be extrapolated to the origin. Either way the most interesting feature in this figure is the transition around 50% *W*.

The log *D* vs 1/*W* plot shown in Fig. 7 with $r^2 = 99.6\%$ has been extended to include the data for proxyphylline release from the low swelling compositions in Fig. 9. For these limited number of points it again may be the case that a shallow curve is a better fit (solid line). The 'least squares' linear fit (dashed line) had a regression of 98.7% and again there is a break around 50% *W*.

In a recent study by Domb *et al.* [22] on the permeability of peptides through hydrogel membranes more than 31% hydration was identified as the criterion for the creation of water 'pores' necessary for transport of these larger molecules. Other researchers [23, 24] have also noted a change in gradient in log *D* vs *W* plots between 31 and 40% *W*. A different mechanism termed 'solution-diffusion' was used to describe solute permeability at low hydrations supplemented at hydrations above the significant gradient break point by the more influencial 'pore' diffusion. We noted in the early publication by Yasuda *et al.* [15] on the diffusion of sodium chloride through 26 different compositions of hydrated polymers that no break is shown in the log *D* vs 1/*W* figure, however, only two of their compositions contained more than 41% water and there was a pronounced increased in the diffusion coefficient values for these two, from 1.8 to $8.8 \times 10^{-6}\ cm^2\ s^{-1}$ for an increase of only 61.3 to 65.5% (mean values) water content signifying a change in gradient between low and moderate swelling compositions.

A break in the *D* vs *W* and log *D* vs 1/*W* relationships shown in Figs 8 and 9 indicates a combination of controlling mechanisms dependent on the degree of hydration of the polymers. Either mathematical relationship can validly predict the diffusion coefficient for a stated water content between 50 and 75%, i.e. 1 to 3 parts water per part of polymer. From Fig. 8 the reason for the negative values for *D* extrapolated to $W = 0$ in the *D* vs *W* plots, Figs 5, 6 and 7 may now be explained as follows.

When the 'bound' water believed to be trihydrate is separated from the total water, 'pore' water (including the currently inseparable interfacial water) termed free water, is left. Figures 5, 6 and 7 include plots for diffusivity versus free water. Again a reasonable linearity from the 'least squares' fit has been achieved. This relationship actually provided the best fit to the caffeine data, was almost equal to all the other morphine relationships but less convincing for proxyphylline. The linear extrapolation to zero *FW* for both morphine and caffeine is positive supporting the hypothesis that diffusion can still take place through bound water and contribute to the overall diffusion. In the case of proxyphylline the extrapolation to zero free water was just less than nil although it has already been shown in Figs 8 and 9 that diffusion still occurs when there may be no free water present. When the extra *D* value for the composition containing 45.4% *W*, 3.3% *FW* was included in the regression equation the linear fit extrapolated to $+0.07\ cm^2\ s^{-1} \times 10^6$ although the coefficient of determination was reduced to 96.1%.

CONCLUSIONS

It is clear that diffusion of water-soluble drugs through hydrogels is a function of the water content, *W*. The established 'free volume' theory which predicts log *D* is proportional to $1/W$ and Davis's empirical equation in which log D is proportional to W were compared with a simple proposed relationship in which D is proportional to W. For PEO hydrogels with 50–75% water content containing morphine hydrochloride, caffeine and proxyphylline the D proportionality to W was found to be as good as the log D proportionality to $1/W$ and better than the log D proportionality to W.

Some of the water in PEO hydrogels may be associated with the polymer backbone as a stable trihydrate. When the calculated trihydrate water was subtracted from the total water it was found that the remaining water, termed free water, FW, was also proportional to the diffusivity of the three drugs studied, and when extrapolated to zero FW gave a positive value for D, evidence that diffusion takes place through both bound and free water.

When water swelling was related to the hydrogel composition it appeared $>42\%$ water content was a prerequisite for the presence of free water. This finding was supported by thermal analysis of one composition partly swollen with 39% water and fully swollen with 56% water. The former showed no evidence for free water while the latter clearly showed both forms.

When the D vs W and log D vs $1/W$ plots for proxyphylline were extended to include hydrogels with $<50\%$ water content there was a distinct break in the gradient around 50% W. Solute transport through water swollen PEO hydrogels is apparently a combination of at least two processes, first through the bound water then additionally at a much enhanced rate in hydrogels containing $>50\%$ water through 'pores' of free water.

Acknowledgement

The support of the British Technology Group in this research is gratefully acknowledged.

REFERENCES

1. N. A. Peppas and S. R. Lustig, in: *Hydrogels in Medicine and Pharmacy*, Vol. 1, p. 57, N. A. Peppas (Ed.). CRC Press, Boca Raton, Florida (1986).
2. E. J. Mack, T. Okano and S. W. Kim, in: *Hydrogels in Medicine and Pharmacy*, Vol. 2, p. 65, N. A. Peppas (Ed.). CRC Press, Boca Raton, Florida (1987).
3. M. L. Brannon and N. A. Peppas, *J. Membr. Sci.* **32,** 125 (1987).
4. M. E. McNeill and N. B. Graham, *J. Controlled Release* **1,** 99 (1984).
5. M. P. Embrey, N. B. Graham, M. E. McNeill and K. Hillier, *J. Controlled Release* **3,** 39 (1986).
6. M. E. McNeill, N. B. Graham, in: *Controlled-Release Technology. Pharmaceutical Applications*, p. 157, P. I. Lee and W. R. Good (Eds). ACS Symposium Series 348 (1987).
7. C. D. Hanning, A. P. Vickers, G. Smith, N. B. Graham and M. E. McNeill, *B. J. Anaesthesiol.* **61,** 221 (1988).
8. P. J. Flory and J. Rehner Jr., *J. Chem. Phys.* **II,** 512 (1943).
9. W. R. Good and K. Mueller, in: *Controlled Release of Bioactive Materials*, p. 155, R. W. Baker (Ed.). Academic Press, NY (1980).
10. N. B. Graham and M. E. McNeill, *Makromol. Chem. Macromol. Symp.* **19,** 255 (1988).
11. H. B. Lee, M. S. Jhon and J. D. Andrade, *J. Colloid Interface Sci.* **51,** 225 (1975).
12. S. L. Hager and T. B. Macrury, *J. Appl. Polym. Sci.* **25,** 1559 (1990).
13. M. J. Hey, S. M. Ilett, M. Mortimer and G. Oates, *J. Chem. Soc. Farady Trans.* **86,** 2673 (1990).
14. N. B. Graham, M. Zulfiqar, N. E. Nwachuku and A. Rashid, *Polymer* **30,** 528 (1989).
15. H. Yasuda, C. E. Lamaze and L. D. Ikenberry, *Makromol. Chemie.* **118,** 19 (1968).
16. B. K. Davis, *Proc. Nat. Acad. Sci. USA* **21,** 3120 (1974).

17. J. M. Crank, p. 11, *The Mathematics of Diffusion*, Clarendon Press, Oxford (1975).
18. R. W. Baker and H. K. Lonsdale, in: *Controlled Release of Biologically Active Agents*, p. 15, A. C. Tanquary and R. E. Lacey (Eds). Plenum Press, NY (1973).
19. *Minitab Handbook*, p. 218, Duxbury Press, Boston (1985).
20. S. J. Desai, D. Sing, A. P. Simonelli, W. I. Higuchi, *J. Pharmacol. Sci.* **55**, 1224 (1966).
21. *Handbook of Chemistry and Physics*, p. F62, CRC Press, 60th Edition (1980).
22. A. Domb, G. W. Raymond Davidson III and L. M. Sanders, *J. Controlled Release* **14**, 133 (1990).
23. W. P. O'Neill, in: *Controlled Release Technologies: Methods Theory and Appliations*, Vol 1, p. 166, A. F. Kydonieus (Ed.). CRC Press, Boca Raton, Florida (1980).
24. J. M. Wood, D. Attwood and J. H. Collett, *J. Pharm. Pharmacol.* **34**, 1 (1982).

Properties controlling the diffusion and release of water-soluble solutes from poly(ethylene oxide) hydrogels 2. Dispersion in an initially dry slab

MARION E. McNEILL and NEIL B. GRAHAM*

Department of Pure & Applied Chemistry, University of Strathclyde, Thomas Graham Building, 295 Cathedral Street, Glasgow, G1 1XL, UK.

Received 16 April 1992; accepted 15 December 1992

Abstract—The mechanisms which control the release of dispersed water-soluble drugs from an initially dry hydrogel are complex. The release profile derives from a combination of several contributing factors which may change with time at different rates. It has been possible to isolate controlling factors and investigate their individual contributions to the release kinetics. The hydrogels presented in this paper owe their hydrophilicity to their poly(ethylene oxide) content. They swell and can absorb up to three times their dry weight in water. Having a glass transition temperature (T_g) below body temperature they are essentially different to those studied theoretically or experimentally, by other groups, which have T_g values above body temperature and are initially glassy. A range of diffusates was studied ranging from low water-soluble prostaglandin E_2 to highly water-soluble lithium chloride. Device geometry was restricted to approximations to infinite slabs with more than 85% total surface area over the top and bottom surfaces so that release was predominantly one-dimensional and the controlling variable was thickness. The increase in surface area with time, drug-solubility in the water-swelling matrix and the presence of crystallinity were shown to be important factors governing the profile and level of release rate with time. It was observed that the release profile could be separated into three parts, the most important being the middle section from early in the release until at least the half-life time. This period could be characterized by the exponential time function, t^n. The diffusional exponent, n, is an important indicator of the release mechanism and ranged from 0.79 to 1, i.e. good anomalous to zero order. This is a highly desirable range of values for controlled release devices. The value of n decreases at late-time. The very early-time release can also show a burst or lag effect depending on the diffusate solubility and its loading in the xerogel.

Key words: Xerogel; hydrophilicity; equilibrium water content; dissolution; exponential function; surface area.

INTRODUCTION

In the previous paper [1] we described the mechanism for drug delivery from fully hydrated monolithic slabs of poly(ethylene oxide)-based hydrogels. The amount released at any one time could be determined by judicious selection from a number of variables, but invariably the rate of release diminished with time at a $t^{-0.5}$ proportionality. Starting off with the drug dispersed in a dehydrated hydrogel offers several practical advantages over the hydrated system. When dried down the drug substance is essentially immobile, trapped in the matrix, and contact with penetrating water is necessary to unlock the immobilized solute and trigger the release process.

For a drug such as prostaglandin E_2, which degrades in water, a viscous aqueous gel for vaginal application [2] has to be dispensed in the hospital pharmacy prior to

* To whom correspondence should be addressed.

administration, whereas a dry hydrogel containing dispersed PGE_2 provides a long lasting stable formulation which is immediately available when required [3]. Other advantages in starting with a dry device are the swelling process necessary to activate the release which helps to linearize the decreasing rate normally associated with an initially hydrated device and also the period of effective release is extended. This has been demonstrated for a PGE_2 pessary which provided almost zero order release until the half life time, which was twice as long as the $t_{1/2}$ from an initially fully hydrated pessary [4].

An initially dry hydrogel containing dispersed drug with the added safety that it cannot dump the entrapped medication when administered would in general be of more interest for pharmaceutical formulations than one swollen with or containing a solution of the drug [5].

Although membrane envelope systems can provide constant rate drug delivery it is harder to achieve a constant rate from monolithic matrix-type dispersed formulations [6]. The swelling systems described herein approximate to it. Such swelling systems are difficult to model mathematically but some limiting conditions discussed below have been analysed for hydrogels. Initial attempts to apply 'Bond-Graph' techniques to these rubbery hydrogels are promising [7]. This paper presents phenomenological analysis of release results from swelling matrices.

Release kinetics may employ one of the detailed mathematical models based on the diffusion equations or may be purely empirical e.g., fitting the release curve data to a polynominal expression [8, 9]. The exponential relation $M_t/M_\infty = kt^n$ can frequently be used to describe Fickian and non-Fickian release behaviour of swelling controlled release systems [10] where

$$\frac{M_t}{M_\infty} = kt^{0.5} \qquad n = 0.5 \text{ (Fickian)} \tag{1}$$

$$\frac{M_t}{M_\infty} = kt \qquad n = 1 \text{ Case II transport} \tag{2}$$

$$\frac{M_t}{M_\infty} = k_1t^{0.5} + k_2t \qquad \text{Anomalous} \tag{3}$$

or

$$\frac{M_t}{M_\infty} = kt^n \qquad 0.5 < n < 1 \qquad \text{Anomalous.} \tag{4}$$

Swelling-controlled release of dispersed drugs from initially dry glassy hydrophilic polymers can be achieved by absorption of water resulting in macromolecular relaxation associated with the transition from the glassy to the rubbery state [11]. The glass transition temperature, T_g, is lowered by the water to a temperature below the temperature of the experiment. If the interfacial boundary between the swollen gel and the glassy core advances with constant velocity, the weight gain is directly proportional to time rather than $t^{0.5}$ and there is potential for zero-order release of dispersed drug [12]. When the ratio of the characteristic relaxation time to the characteristic diffusion time, termed the diffusional Deborah Number, D_e, approaches unity, the release approaches zero-order. The Swelling Interface Number, S_w, relates the rate of penetrant uptake to the rate of solute diffusion [13–16]. The latter has to be much faster than the former (i.e. S_w must be small) in order to achieve zero-order release. In a paper by Klier and Peppas [17] swelling of a slab was regarded as being constrained to one-dimensional expansion by the glassy core and only became three-dimensional when the solvent fronts met in the centre.

The release profiles for drugs dispersed in PEO hydrogel slabs can be categorized as anomalous, sometimes approaching zero-order, however, the controlling mechanisms are essentially quite different to the swelling-controlled systems described above. Important differences which may contribute to the release mechanisms are: (1) Most of these PEO xerogels are semi-crystalline, not amorphous. (2) Since their T_g is well below zero they are rubbery not glassy polymers in the dry state. (3) The major portion of the drug content has still to be released at the time when the permeating water fronts from opposite faces meet in the centre. (4) Swelling and expansion from xerogel to fully swollen hydrogel is three-dimensional with a corresponding significant increase in surface area. (5) During release the partly or even fully water swollen regions of the hydrogel may contain both dissolved and dispersed solute, i.e. the Swelling Interface Number is not small.

EXPERIMENTAL

Nomenclature for hydrogel compositions

The preparation of these hydrogels was described in the first paper [1]. In these stoichiometric compositions the average number molecular weight, $\bar{M}_n$, of the PEG determined experimentally by end-group analysis has been used to calculate the quantities required in preparing the hydrogels. This will vary between batches. Thus, PEO8400/1HT and PEO8490/1HT represent different batches of Breox 8000 with 1 mol of crosslinker hexane triol per mol PEG. As a result the properties, in particular the equilibrium water content, may vary slightly between these compositions. Sometimes the proportion of crosslinker was modified in order to maintain the same equilibrium water content between batches of Breox with the same nominal molecular weight but different $\bar{M}_n$ values.

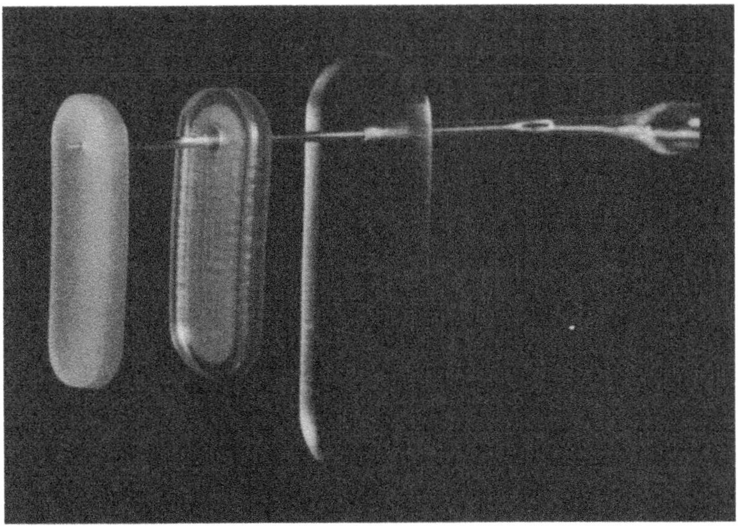

Figure 1. Xerogel, partly swollen hydrogel and fully water swollen hydrogel slab.

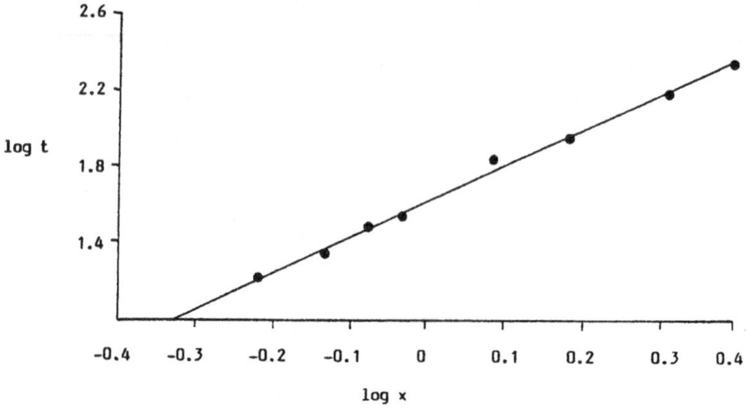
2.6
2.2
1.8
1.4
log t
-0.4
-0.3
-0.2
-0.1
0
0.1
0.2
0.3
0.4
log x

Figure 2. Penetration of water through xerogel PEO8490/1HT at 37°C. Time t mins, distance x mm.

Velocity of water penetration with distance

The presence of crystallinity in these PEO hydrogels provides a useful visual aid so that migration of water through the polymer can readily be followed. Opacity due to the crystalline domains in the xerogel disappears when water molecules penetrate and dissolve the crystallites. Progress of the migrating water front is easily observed (Fig. 1). When the water fronts from opposite sides meet, the hydrogel becomes totally transparent. If the xerogel monolith is l_d mm thick, the path length, x, traversed by the penetrating water front at the moment when the opaque crystalline core disappears is $l_d/2$ mm.

The velocity of penetration of water was studied with composition PEO8490/1HT at 37°C. Slices of xerogel of different thickness l_d (measured by micrometer to 0.01 mm) were placed in water and the time of disappearance of crystallinity, t, recorded against the distance travelled by the water front, x. Log t vs log x values were plotted (Fig. 2).

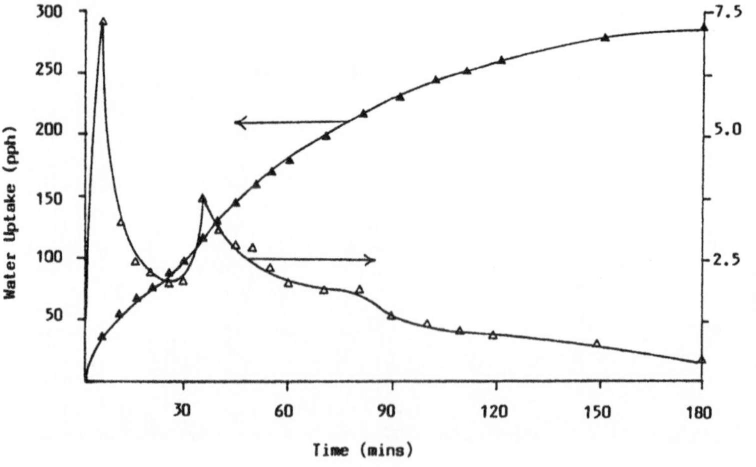
300
250
200
150
100
50
Water Uptake (pph)
7.5
5.0
2.5
Rate of Uptake (pph/min)
30
60
90
120
150
180
Time (mins)

Figure 3. Water uptake of PEO8400/1HT at 37°C, l_d = 2 mm.

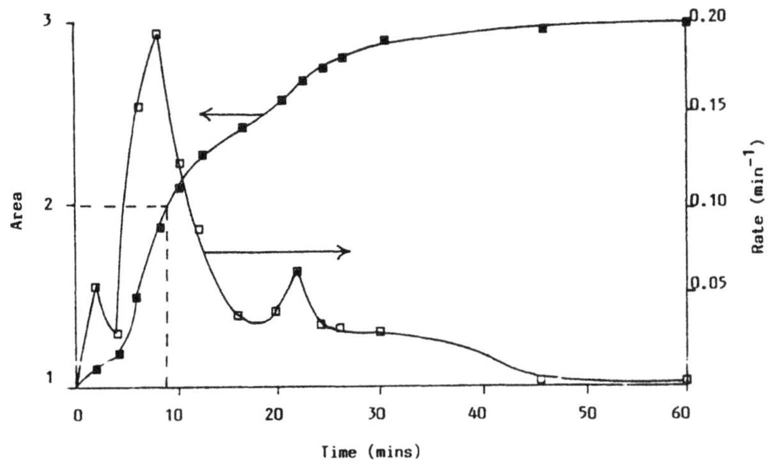
Area
Rate (min^{-1})
Time (mins)
0
10
20
30
40
50
60
1
2
3
0.05
0.10
0.15
0.20

Figure 4. Surface expansion of PEO8400/1HT on swelling in water at 37°C, l_d = 1 mm. Initial area, unity.

Swelling of xerogels in water with time

By weight. Preweighed and identified samples of PEO8400/1HT, 2 ± 0.05 mm thick, were placed in distilled water at 37°C, removed after different periods, blotted dry and weighed. Water uptakes with time as parts per hundred parts initial dry weight (pph) have been plotted in Fig. 3. The incremental rates of water uptake at each recorded time were also calculated and are superimposed in the figure.

By surface area. Swelling of xerogels is accompanied by expansion (see Fig. 1). Slices of PEO8400/1HT 1, 2, and 3 mm thick were placed in distilled water at 37°C, removed after different periods, measured, and returned to the water to continue

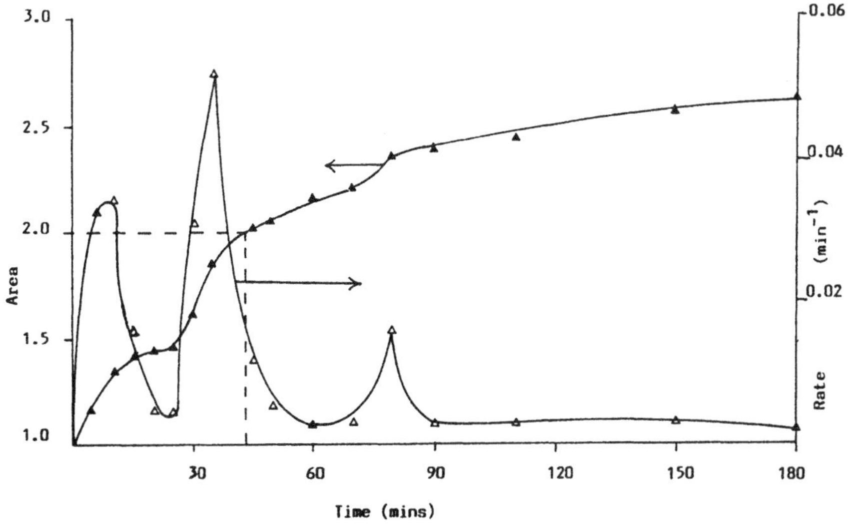
Area
Rate (min^{-1})
Time (mins)
30
60
90
120
150
180
1.0
1.5
2.0
2.5
3.0
0.02
0.04
0.06

Figure 5. Surface expansion of PEO8400/1HT on swelling in water at 37°C, l_d = 2 mm. Initial area, unity.

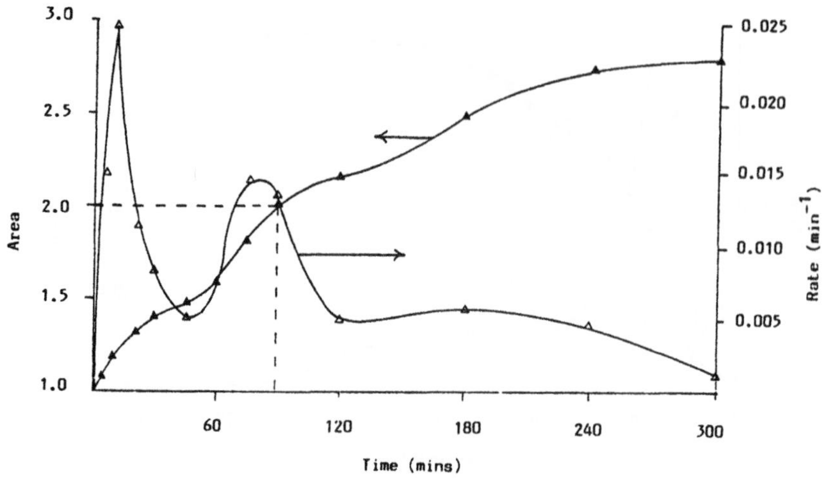
3.0
2.5
2.0
1.5
1.0
Area
0.025
0.020
0.015
0.010
0.005
Rate (min^{-1})
60
120
180
240
300
Time (mins)

Figure 6. Surface expansion of PEO8400/1HT on swelling in water at 37°C, l_d = 3 mm. Initial area, unity.

swelling. The time clock was stopped while the samples were being measured. The surface area was calculated and ratioed to the initial area. The rate of change of surface area for each interval of time was also calculated. The results are illustrated in Figs 4–6.

Composition. The water uptakes with time of two different compositions, 2 mm thick, semi-crystalline PEO8400/1HT and almost amorphous PEO4360/10HT,

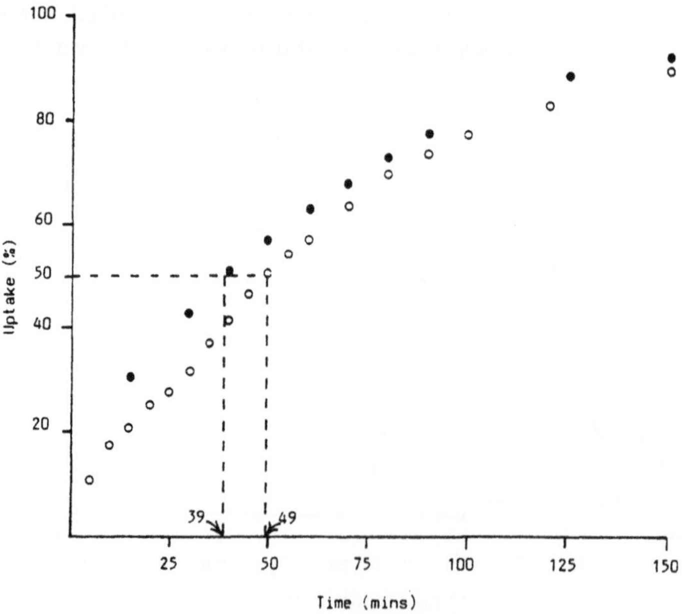
100
80
60
50
40
20
Uptake (%)
39
49
25
50
75
100
125
150
Time (mins)

Figure 7. Swelling of xerogels in water at 37°C. ○ PEO8400/1HT 46.2% PEO crystalline, EWU = 310 pph. ● PEO4360/10HT 4.2% PEO crystalline, EWU = 68 pph.

have been compared in Fig. 7 in order to examine the effect of crystallinity and hydrophilicity on the rate of swelling in water. Since these compositions have very different equilibrium water contents the percentage of the total uptake rather than the absolute amounts of water have been plotted.

Dispersion of an active compound in a xerogel slab

Compounds: Lithium chloride (Fisons), Proxyphylline (Sigma), Salbutamol base (Sigma), Salbutamol sulphate (Sigma), Aspirin (Sigma), and Prostaglandin E_2 (Upjohn).

Aqueous or organic solvent solutions of the active compound were used to impregnate the hydrogel slabs depending upon the solubility of the compound in a good swelling solvent for the polymer and the required dose. The concentration of the swelling solution was calculated from the required total and the equilibrium solvent uptake by the xerogel. The drugs chosen ranged from highly water-soluble lithium chloride to low water-soluble prostaglandin E_2, PGE_2. Proxyphylline is readily water soluble and was selected to examine the effect of drug loading on the release profile and also to compare the profiles from dried-down slabs with initially hydrated slabs [1]. The two forms of salbutamol were used to illustrate the effect of drug solubility on the release profile. Aspirin is a commonly prescribed drug sometimes requiring a sustained release formulation. PGE_2 represents the type of prolonged release which may result from a low water-soluble compound. From the solution uptake the solute uptake was calculated. For all the above compounds the measured total was 100–120% of the estimated total, i.e. none exhibited pronounced preferential absorption by association with the polymer.

Release studies

Slabs of dried-down hydrogel containing dispersed active additive were placed in distilled water. 0.1 *N* sulphuric acid was used for the aspirin release, to suppress hydrolysis to salicylic acid at 37°C. The concentration of released compound was monitored as described in the first paper [1]. Lithium was analysed by atomic absorption spectrophotometry, prostaglandin E_2 was analysed by HPLC and the rest by UV spectrophotometry.

Lithium chloride

Total release M_t from 1, 2 and 3 mm thick slabs of PEO8300/0.75HT was measured until M_t remained constant. The rate of release values were calculated from the incremental release at each recorded time interval.

Proxyphylline

The release of proxyphylline from PEO8400/1.2HT containing 18.3% w/w proxyphylline was measured together with the swelling profile from an identical sample. A correction for loss in weight due to desorbed proxyphylline was made at each time of weighing before plotting the values for water uptake with time.

Salbutamol

The anti-asthmatic drug salbutamol is available as the base (solubility 14 mg ml^{-1}) and the sulphate (freely water soluble). In order to investigate the effect of aqueous solubility on the release profile identical xerogel samples of PEO3800/1HT were impregnated with the same amount, i.e. 17.7 mg of both the base and the sulphate. The freely soluble sulphate was loaded from an aqueous solution. The sparingly soluble base was loaded into the polymer from a methanolic solution. The effect of drug loading (17–63.8 mg) on the release profile was studied for salbutamol base in PEO3800/1HT slabs 1.35 mm thick. The effect of different hydrogel compositions on the release profile and $t_{1/2}$ was studied for salbutamol base.

Aspirin

The higher swelling PEO hydrogel which gave the most constant release rate with salbutamol was selected for studies with aspirin designed to investigate: (i) the effect of widely different drug concentrations on the release profile for a single device thickness; and (ii) the effect of thickness on the profile and $t_{1/2}$ keeping the drug concentration in the hydrogel constant. (i) The release profiles from a low content formulation 14 mg and an identical xerogel with more than ten times the amount of aspirin, 153 mg, were compared. (ii) Xerogel slabs of different thickness, 0.83, 1.32, and 1.77 mm, were swollen in the same solution of aspirin 86.8 mg/g, 1/1, w/w, ethanol/water so that they would have approximately the same concentration of aspirin/polymer when dried down.

Prostaglandin E_2

The effect of slab thickness on the release profiles from 10 mg dispersions of a drug with a low water solubility, PGE_2, was studied.

RESULTS AND DISCUSSION

In the first paper [1] we discussed how the water content of hydrated PEO hydrogels can be related to the PEO content of the polymer composition and how this water content determines to a large extent the permeability of water soluble drugs through the matrix. For fully hydrated hydrogels containing dissolved drug the release into an infinite sink is nearly always Fickian, i.e. $dM_t/dt \propto t^{-0.5}$. However, starting with a dispersion of the drug in a dry hydrogel, the xerogel has first to absorb water which acts as a solvent for the drug followed by diffusion of the solute along the aqueous pathways to the surface to be released to the surrounding fluid.

The mechanism for release would not be expected to be Fickian, more likely anomalous and possibly the often desirable zero-order release. Many parameters may influence the release profile: (1) polymer composition, and hence water uptake; (2) degree of crystallinity; (3) (a) sample dimensions and (b) sample geometry; (4) temperature; (5) molecular weight of solute; (6) water solubility and also hydrophilicity of solute; (7) drug concentration in polymer; (8) association of drug with polymer structure; (9) kinetics of water uptake by hydrogel; and (10) kinetics of surface area increase with swelling.

Item (1) has already been discussed in the first paper on swollen release systems and the effect of composition on release from initially dry gels will now be considered. Items (2), (3a), (6), (7), (9), and (10) will be discussed in this paper. Items (3b) and (8) will be reported in subsequent publications. Results in item (5) were recently presented [18] and will be extended in future publications.

Swelling of xerogels in water

Penetration of water from the surface of a semi-crystalline PEO xerogel towards the interior is readily observed and can be followed by the contrast in light transmission at the boundary of the opaque dry semi-crystalline polymer and the partly water-swollen amorphous region (Fig. 1). A log–log plot for the time, t min, to travel distance x mm provided a good linear fit to the data for PEO8490/1HT at 37°C (Fig. 2) and the following equation:

$$\log t = 1.61 + 1.90 \log x \quad r^2 = 99.1\% \tag{5}$$

where r is the correlation coefficient.

Therefore

$$t = 40.7\, x^{1.90} \tag{6}$$

and

$$x = 0.142 t^{0.53}. \tag{7}$$

Differentiating with respect to time gives the velocity v mm min^{-1} at which water penetrates the xerogel.

$$v = 0.075 t^{-0.47}. \tag{8}$$

It can be seen that the speed of water penetration slows down at a rate approximately proportional to the inverse square root of time.

Water uptake amounts at 37°C in 2 mm thick slabs of PEO8400/1HT over a wide range of periods were plotted (Fig. 3). The final constant swollen weights for 20 samples showed a mean equilibrium water uptake, EWU, of 300 ± 10 pph (parts per hundred parts initial weight). A smooth curve has been drawn through the points. To determine the rate of swelling the incremental weight difference between each recorded weight was used rather than the gradient of the curve. This method of analysis is more sensitive, accurate and less subjective in detecting subtle changes in the swelling pattern. These values have also been joined by a smoothed outline. There are two distinct peaks in the rate plot, at 5 and 35 min and a levelling off period around 80 min. It will be seen that the hydrogel is still swelling, albeit slowly, after 180 min. The water uptake will actually exceed the EWU value of 300 pph then fall back to a steady 300 pph. This overswelling phenomena has been discussed in an earlier paper [19].

The three transitional times are shown even more distinctly by measuring the expanding swollen surface area with time (Fig. 5). The early peak rate between 5 and 10 min is due to the ease of water absorption at a surface in intimate contact with an aqueous environment. The layer of hydrogel close to the surface expands rapidly but is constrained by the adjoining inner dry layer which can only swell by absorbing water from the outer swollen layer. The rate therefore soon falls away as the advancing water fronts from opposite faces approach to meet in the middle. However with time the restraining unswollen core diminishes both in size and intensity leading to a sharp upturn in the rate of water absorption which reaches a peak value at 35 min when the water fronts meet. Just after this happens the last crystals vanish at almost the same time given by Eqn (6) for $x = 1$ mm as 40 min. At this time it is noteworthy (see Fig. 5) that the surface area now measures twice the initial size. The levelling off time around 80 min in the weight measurements shows up in the surface area measurements as a small peak.

Slices of the same polymer composition 1 and 3 mm thick were also swollen in water at 37°C, removed periodically and measured as with the 2 mm thick slice. Again three peaks are discernible in the changing surface area with time plots (Figs 4 and 6), although their relative size and peak times vary between the three thicknesses. For all thicknesses the second peak occurred at the time when the last remnant of unswollen hydrogel disappeared closely followed by 'dissolution' of the last crystals of PEO at times predicted by Eqn (6), i.e. 11 min for a 1 mm thick slice and 88 min for 3 mm. In all cases at this time the surface area is approximately double the initial value.

It is not clear why there is a third peak although it is possible that swelling from the upper surface of the slab is faster than from the under surface which was supported on a glass surface. This non-uniform swelling causes slight curvature of the initially flat surfaces of the slab. The third peak has been observed to occur at the time when the curved slice relaxes and flattens out.

In the first paper [1] we discussed a linear swelling factor, f, which measures the unidimensional expansion of a hydrogel swollen to equilibrium, f is related to the polymer composition. For PEO8400/1HT $f = 1.65$ in water at 37°C therefore $f^2 = 2.72$ which is close to the maximum recorded value of dimensionless area in Figs 4–6 as swelling approaches equilibrium.

To determine how PEO crystalline domains in a xerogel hinder the migration of water through the matrix the swelling profiles for an almost amorphous composition, PEO4360/10HT and semi-crystalline PEO8400/1HT have been compared, Fig. 7. Both slabs were 2 mm thick. The degree of crystallinity in the PEO for the first polymer was 4.2% and more than 10 times this amount, 46.2%, for the second composition. Their equilibrium water uptakes at 37°C were 68 and 310 pph respectively, therefore the water uptakes with time have been plotted as percentage of the EWUs in comparing the swelling profiles. Water uptake from the composition with very little crystallinity is initially faster. After 30 min when water has reached and dissolved the last remnants of crystallinity in the core of the slab of PEO8400/1HT water uptake accelerates to a rate faster than in the PEO4360/10HT slab bringing the degrees of swelling at any one time closer though the overall percentage uptake of the PEO8400/1HT never quite catches up with the PEO4360/10HT. For the same initial size of 2 mm thickness the almost amorphous composition was half swollen after 39 min whereas it took 49 min for the semi-crystalline composition to reach half the equilibrium water uptake.

The effect of increasing surface area on release of drug

The release profiles for lithium chloride, dispersed in slabs 1, 2, and 3 mm thick (Fig. 8), are unique in all showing Super Case II release during the early stages of desorption. This is shown by the rate plots where all thicknesses show a peak rate which diminishes in intensity, becomes broader and occurs later with increasing thickness. The lithium chloride flux peaks tie in with the surface expansion peak times in Figs 4–6 when the last remnant of restraining dry polymer in the centre has just disappeared.

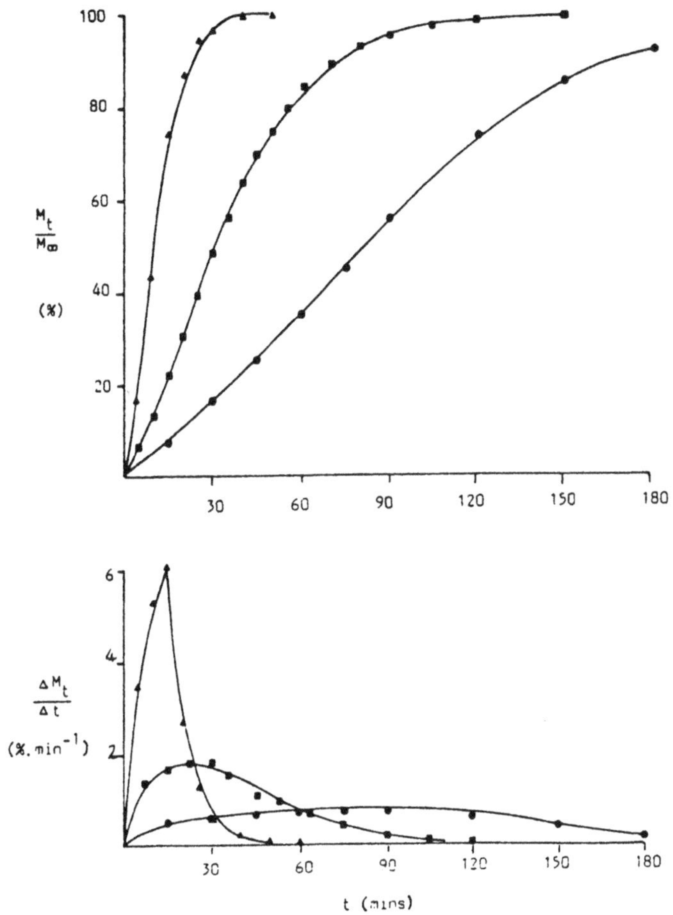
100
80
60
40
20
$\frac{M_t}{M_\infty}$
(%)
30
60
90
120
150
180
6
4
2
$\frac{\Delta M_t}{\Delta t}$
(%.min^{-1})
30
60
90
120
150
180
t (mins)

Figure 8. Release profiles for lithium chloride dispersed in slabs of PEO8300/0.75HT into water at 37°C. ▲ 1 mm; ■ 2 mm; ● 3 mm.

From Fick's Laws of diffusion, flux is directly proportional to surface area for fixed values of other dependent variables. Although the diffusion coefficient, D, depends on the water content [1] and has a fixed value for fully hydrated systems it will not be constant for systems which are swelling while they are releasing active additive. Until the hydrogel is fully swollen there will be a decreasing gradient of diffusivity from the surface to the centre. Release across the surface/sink interface will be partly controlled by the D value in the hydrogel close to the surface [1] which may quickly approach the characteristic D value for the fully hydrated composition [20]. Provided the supply of drug from the interior by diffusion, or dissolution of dispersant in the surface layer can maintain a constant level of concentration of dissolved drug, flux would be expected to increase along with expansion of the surface in accordance with Fick's 1st Law. This has been shown to be the case with lithium chloride dispersed in monolithic slabs of PEO xerogel. The mobility of the small Li ion together with the ease of solution of LiCl probably explains the close similarity between the expansion peak times at 9, 30, and 80 min for 1, 2, and 3 mm thick slabs (see Figs 4–6) and the peak rates for LiCl release at 12, 28, and 90 min respectively (see Fig. 8) and emphasizes the importance of surface expansion with swelling on release profiles.

Comparison of kinetics of swelling and dry release

According to Lee [21, 22] the presence of the active additive will affect the rate of swelling of a hydrogel monolith. This however was not found to be the case for a PEO hydrogel impregnated with 18.3%, w/w proxyphylline. Comparison of the water uptake with a blank xerogel (Fig. 9) shows the swelling profiles to be essentially the same. It therefore appears to be the hydrophilicity of the PEO polymer rather than the properties of the additive which determines the rate of water uptake in this case. Figure 9 also compares water uptake and drug release profiles. Clearly swelling is faster than proxyphylline release in contrast to reported studies [16] on another swelling-controlled system in which transport of the dye Sudan Red was faster than the transport of the solvent *n*-hexane.

After 20 min the polymer slab is already half swollen whereas the drug half life in the slab is 41 min at which time the polymer is 75% swollen. Release up to 70% M_∞ can be seen to be much slower and steadier than water absorption.

When a straight line was drawn between the percentage proxyphylline released at t_5, 10.1% and t_{70}, 72.8%, it can be seen that the release profile is actually a gentle convex curve between these times after an initially faster release (burst) then finally an exponential decrease after 70 min. The log–log plot for the data between 5 and 70 min has a gradient of 0.91 with a correlation coefficient from these values of $r^2 = 99.8\%$. The regression for the linear fit of the log–log plot was obtained from Minitab's correlation and regression program [23] which also provided the 95% confidence interval, C.I., for the log–log gradient for the twelve readings between

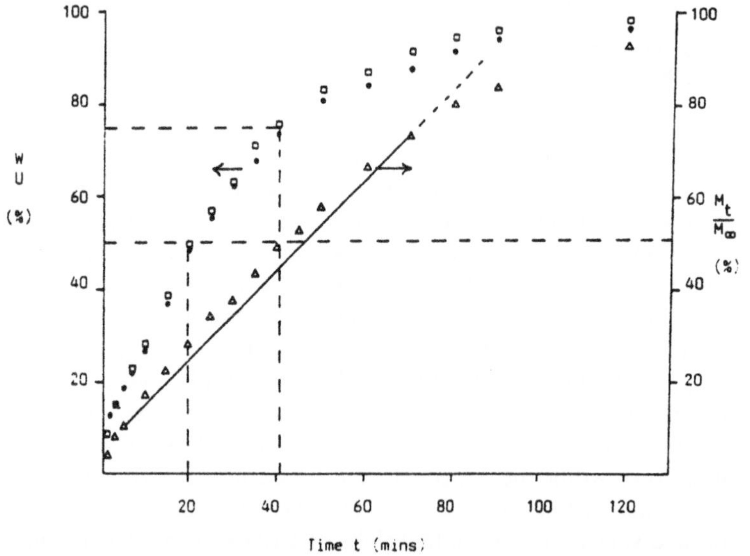
100
80
60
40
20
W
U
(%)
100
80
60 M_t/M_∞ (%)
40
20
20
40
60
80
100
120
Time t (mins)

Figure 9. Release profile for proxyphylline l_d 1.33 mm △. Swelling profile for an identical sample □. Swelling profile for blank ● PEO8400/1.2HT. All at 37°C.

t_5 and t_{70} of 0.91 ± 0.037. From t_5 to t_{70} the percentage release can be calculated at any time from the following derived equation:

$$\frac{100M_t}{M_\infty} = 10.1 + 1.49(t - 5)^{0.91}. \quad (9)$$

This is a specific case of a more general equation, cf. Eqn (4)

$$\frac{100M_t}{M_\infty} = \frac{100M_{t'}}{M_\infty} + k(t - t')^n \quad (10)$$

which allows for a burst effect from surface material lasting until t'. As the exponential function, n, approaches 1 the release approaches zero-order. The constant, k, with dimension time^{-1} is a measure of the rate of release and is useful for comparison in a series of experiments looking at the effect of composition or thickness on the release kinetics.

The concentration of the swelling solution used to impregnate the slabs of polymer was 70 mg/ml. More dilute solutions were not found to affect the release profile; however, increasing the concentration to 400 mg/ml resulted in a white coating of proxyphylline crystals on the surface of the dried-down slice. This produced a massive burst effect when the loaded polymer was placed in water. After 15 min the release steadied to a similar profile to the Fig. 9 plot. There appears to be a limit to the amount of drug which can be trapped by swelling in a xerogel matrix without producing significant surface drug and a burst effect. In the case of proxyphylline this is about 20% dried weight.

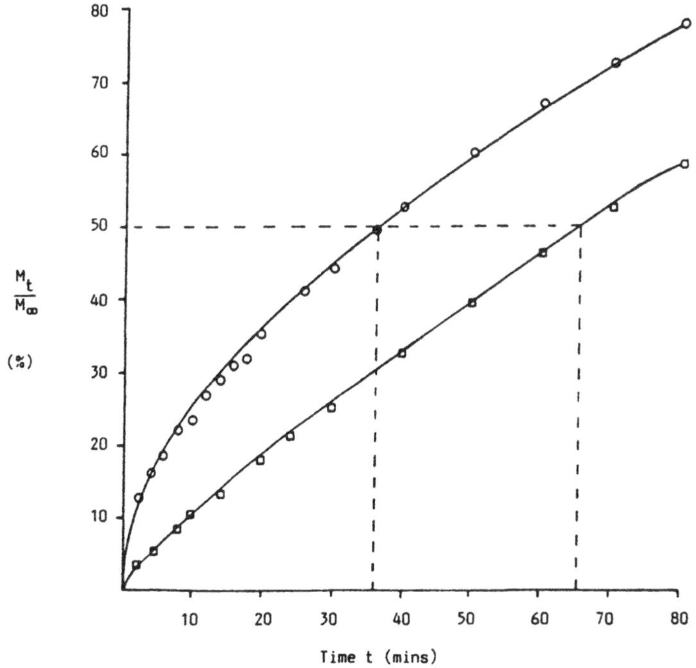

$\frac{M_t}{M_\infty}$ (%)
80
70
60
50
40
30
20
10
10
20
30
40
50
60
70
80
Time t (mins)

Figure 10. Release of salbutamol base at 37°C □ M_∞ = 17.7 mg, l_d = 1.33 mm, $t_{1/2}$ = 65 min, n = 0.95 (10–70 min) and salbutamol sulphate ○ M_∞ = 17.7 mg, l_d = 1.37 mm, $t_{1/2}$ = 36 min, n = 0.89 (10–70 min) from PEO3800/1HT.

The effect of drug solubility in water on the release

The degree of drug dissolution in water is shown in Fig. 10 with release profiles for salbutamol base (solubility 14 mg/ml) and freely water soluble salbutamol sulphate. Both samples contained 17.7 mg drug substance. After 10 min 23.1% sulphate has been released compared to 10.2% from the base form. The following equations were derived from the experimental data.

$$\frac{100M_t}{M_\infty} = 23.1 + 1.41(t - 10)^{0.89} \tag{11}$$

for salbutamol sulphate, and

$$\frac{100M_t}{M_\infty} = 10.2 + 0.87(t - 10)^{0.95} \tag{12}$$

for salbutamol base, both between 10 and 70 min.

The exponential values do not differ appreciably. The main difference in the two formulations can be seen by comparing the half-lives of 65 min for the base and 36 min for the sulphate due to the faster dissolution of the sulphate form of the drug.

The effect of salbutamol concentration on the release profile

We have observed release profiles for proxyphylline are similar up to 20% loading. At a much higher loading proxyphylline was seen concentrated at the surface. We found the same effect with caffeine; a rapid burst was followed by almost zero order release up to 70% total [24]. When the loaded dried slabs were dissected a clear demarcation was seen under the microscope separating the dispersed caffeine in the interior from a concentrated build up at the surface. The effect of salbutamol base loading on the release profile has been studied for this paper and the results from four formulations are summarized in Table 1. After the initial 'burst' taken optionally as 10 minutes, the residual salbutamol in the polymer, $M_\infty - M_{t_{10}}$, was compared with the mean rate of release between 10 and 50 min and a remarkable direct dependency of release rate on the remaining drug content was observed as shown by the last column in Table 1. This at first appears surprising since the solubility of salbutamol base in water has been exceeded for all examples cited. After 10 min, from swelling data the hydrogel should contain 0.2 ml water and 0.4 ml water after 30 min. It might have been expected that amounts exceeding the saturated solution

Table 1.
Release of salbutamol base from PEO3800/1HT at 37°C $l_d = 1.35 \pm 0.02$ mm

M_∞ (mg)	$t_{1/2}$ (mins)	n 95% C.I.	r^2 (%)	$M_{t_{10}}$ (mg)	$M_\infty - M_{t_{10}}$ (mg)	Mean rate at t_{30} (mg min^{-1})	Ratio $\frac{\text{Column 6}}{\text{Column 7}}$
17.7	65	0.95 ± 0.03	99.8	1.9	15.8	0.12	132
35.8	67	0.89 ± 0.01	100.0	4.1	31.7	0.24	132
43.8	67	0.91 ± 0.03	99.8	5.3	38.5	0.29	133
63.8	47	0.85 ± 0.02	99.8	15.8	48.0	0.40	120

Table 2.
Release of salbutamol base from different hydrogels at 37°C. Burst effect until t' $l_d = 1.35 \pm 0.01$ mm. $M_\infty = 20 \pm 3$ mg

Composition	EQU at 37°C (pph)	$t_{1/2}$ (mins)	n from $t' \rightarrow t_{1/2}$ 95% C.I.	t' (mins)	r^2 (%)
PEO3330/2HT	120	98	0.83 ± 0.01	14	99.9
PEO3800/1HT	194	65	0.89 ± 0.01	10	100.0
PEO8490/1HT	304	60	1.00 ± 0.02	0	99.8

value should not have increased the chemical potential, providing a constant rate rather than a concentration dependent rate. However, rate of dissolution depends not only on the amount of solute and the surface area of the diffusate particles but also on the ease of dissolution. Only with a very hydrophilic solute like lithium chloride is a saturated solution likely to be rapidly formed within the confines of a water-swollen polymer network. The greater rate of dissolution of salbutamol sulphate compared with salbutamol base was demonstrated and illustrated in Fig. 10.

The effect of composition on salbutamol base half life

In Table 2 the effect of water content of different polymer compositions on the half-life of salbutamol base is shown. Not only does the $t_{1/2}$ decrease as the characteristic water content of the composition increases but the anomalous exponent approaches 1, i.e. zero order release until the half-life time.

The regression equation which fitted the results for 21.3 mg salbutamol base in PEO8490/1HT between 0 and 90 min (69.4% M_∞) was

$$\frac{100M_t}{M_\infty} = 1.05 + 0.79t \tag{13}$$

$r^2 = 99.8\%$.

It will be seen there is practically no burst effect from this system since the linear fit extrapolates to only 1.05% M_∞.

The effect of low loading and high loading on the release profiles with aspirin and the effect of slab thickness

The release profiles from a low loading of aspirin and another xerogel containing more than ten times the amount of aspirin are compared in Fig. 11. It is interesting to observe there is little difference in the half-lives but a marked difference in the release profiles. The low loading formulation shows almost zero order release up to 70% M_∞. The equation assuming $n = 1$ is

$$\frac{100M_t}{M_\infty} = -2.31 + 0.90t \tag{14}$$

$r^2 = 99.8\%$.

Until the half-life time the points lie below the linear fit then from 50–70% M_∞ are above the line, indicating the first part of the release is actually Super Case II with $n > 1$. This observation is confirmed by the negative intersection of the ordinate in Eqn (14).

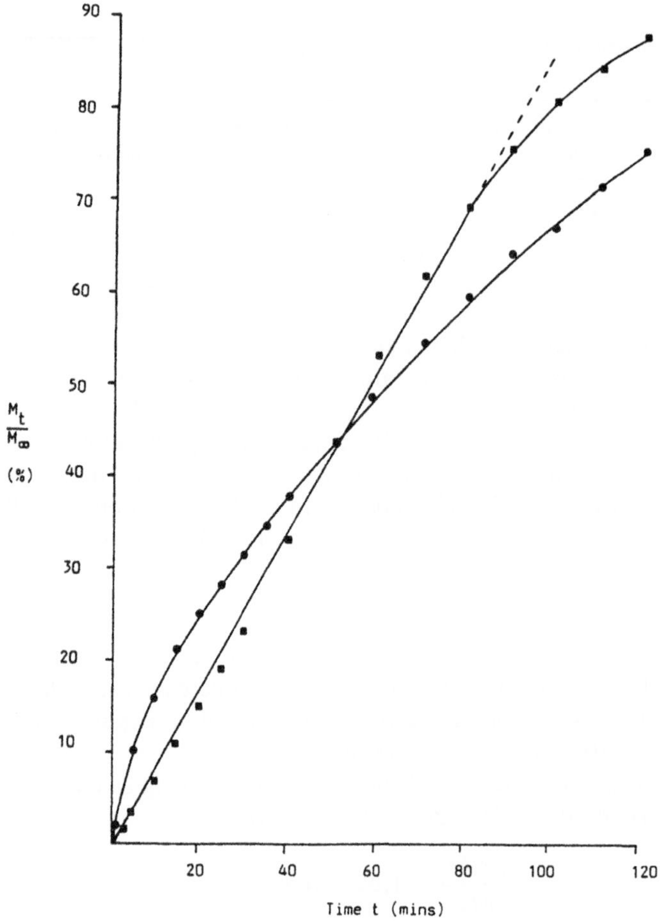
90
80
70
60
50
$\frac{M_t}{M_\infty}$
(%)
40
30
20
10
20
40
60
80
100
120
Time t (mins)

Figure 11. Release of aspirin from PEO8400/1HT at 37°C. ■ l_d = 1.30 mm M_∞ = 14 mg, $t_{1/2}$ = 58 min, n = 1.12 (3–80 min). ● l_d = 1.32 mm M_∞ = 153 mg, $t_{1/2}$ = 61 min, n = 0.80 (10–80 min).

The exponential fit $n = 0.80$ for the high load aspirin device is given in Table 3 along with the data from a thicker and thinner slice all containing the same concentration of aspirin. This table shows how the drug half-life can be controlled by the polymer thickness. The exponential values for the 1.32 and 1.77 mm thick devices are almost identical. In the case of the thin 0.83 mm device the release is practically zero order after 8 min.

Table 3.
Release of aspirin at 37°C from PEO8400/1HT slices of different thicknesses l_d but the same concentration of aspirin in xerogel.

l_d (mm)	M_∞ (mg)	$t_{1/2}$ (mins)	n 95% C.I.	n range (mins)	r^2 (%)
0.83	111	30	0.99 ± 0.06	8 → 44	99.5
1.32[a]	153	61	0.80 ± 0.01	10 → 100	99.9
1.77	200	102	0.79 ± 0.03	10 → 150	99.4

[a] See also Figure 11.

Table 4.
Effect of thickness, l_d, on half-life, $t_{1/2}$, of PGE_2 release at 37°C. $M_\infty = 10 \pm 0.5$ mg (see Fig. 13).

l_d (mm)	$t_{1/2}$ (hours)	Mean rate until $t_{1/2}$ (mg h^{-1})	n until $t_{1/2}$ 95% C.I.	r^2 (%)	k (h^{-1})
0.54	1.1	4.5	0.89 ± 0.06	99.5	51.9
0.63	1.7	2.9	0.94 ± 0.06	99.5	34.5
0.81	2.1	2.4	0.96 ± 0.09	98.6	22.9
1.12	3.8	1.3	0.92 ± 0.05	99.7	16.3
1.34	6.3	0.8	1.01 ± 0.03	99.7	8.2
1.80	8.9	0.6	1.02 ± 0.04	99.6	6.2

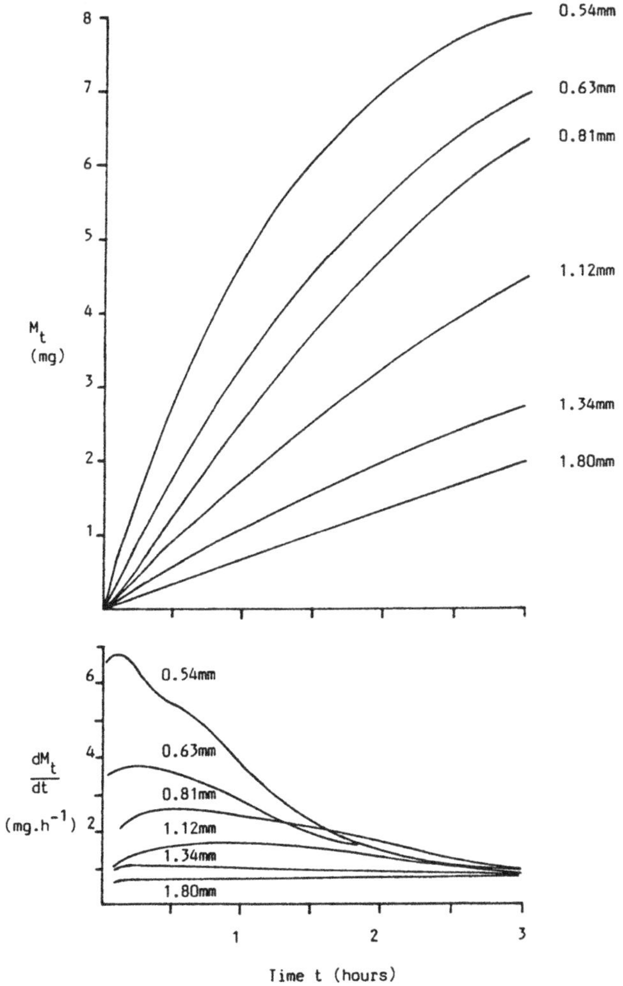

8
7
6
5
4
3
2
1
M_t (mg)
0.54mm
0.63mm
0.81mm
1.12mm
1.34mm
1.80mm
6
4
2
dM_t/dt (mg.h^-1)
0.54mm
0.63mm
0.81mm
1.12mm
1.34mm
1.80mm
1
2
3
Time t (hours)

Figure 12. Comparison of release profiles for 10 mg PGE_2 from PEO8660/1.2HT for different slab thicknesses at 37°C. (i) 0.54 mm; (ii) 0.63 mm; (iii) 0.81 mm; (iv) 1.12 mm; (v) 1.34 mm; (vi) 1.80 mm.

Both the higher loaded samples showed a burst effect from a higher concentration of aspirin next to the surface amounting to about 16% in the first 10 min before setting to $t^{0.8}$ release profiles.

The effect of a low aqueous solubility dispersant on the release profile

Compared with the drugs already discussed prostaglandin E_2 has a low water solubility (1.4 mg/ml at 37°C). Dissolution and desorption would consequently be expected to be considerably slower. In a matrix containing dispersant the rate of dissolution of the suspended drug particles may be so low that dissolution becomes the rate limiting step in the release process [25]. Table 4 lists data resulting from the release of 10 ± 0.5 mg PGE_2 in 6 thicknesses of PEO8660/1.2HT illustrated in Fig. 12. It will be seen that a 1.34 mm device has a $t_{1/2}$ of 6.3 h compared with 1 h for both aspirin (Fig. 11) and salbutamol (Table 1). In all the cases for PGE_2 instead of an initial burst there is a small but perceptible time lag due to the slower dissolution of the PGE_2 followed by a small surge then a period of approximately constant release until the half-life, thereafter a declining rate for the remainder of the PGE_2 release. Since there is no initial burst the release profiles can be characterized by the original unmodified equation for anomalous release.

$$\frac{100M_t}{M_\infty} = kt^n. \qquad \text{(Cf., 4)}$$

In Table 4 it will be seen that n approaches 1 as the slab increases in thickness. The values of k are a measure of the rates of release.

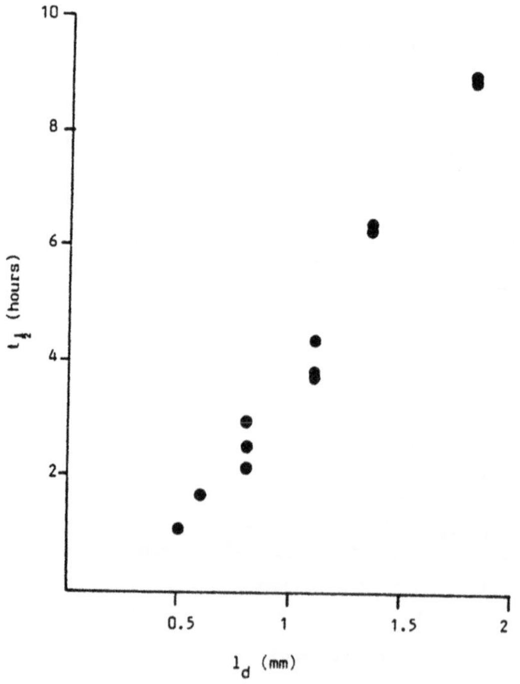
10
8
6
4
2
t½ (hours)
0.5
1
1.5
2
ld (mm)

Figure 13. Half-lives, $t_{1/2}$, of 10 mg PGE_2 in PEO8660/1.2HT vs initial thickness, l_d.

It is worth examining Fig. 12 alongside the previously discussed water swelling profiles for a similar hydrogel. From Fig. 12 it can be predicted that a 1-mm slab will have released only around 20% of a 10-mg dose of PGE_2 after 1 h although Fig. 4 shows that the hydrogel is almost fully swollen at this time. From Fig. 12 a 1.8-mm slab will have released less than 20% of a 10-mg dose of PGE_2 after 3 h whereas Figs 3 and 5 for 2 mm show the hydrogel is approaching its equilibrium water uptake at this time. Thereafter the release continues at a steady rather than declining rate of $t^{-0.5}$, clear evidence for the release of PGE_2 being partially dissolution controlled.

The half-lives for PGE_2 in the hydrogel were plotted against the original slab thickness (Fig. 13). Clearly, the half-life increases with thickness but the points are too scattered to be sure of the best relationship between $t_{1/2}$ and l_d. A clinically useful but only approximate prediction ($r^2 = 97.6\%$) was obtained from the simple $t_{1/2}$ proportionality to l_d.

$$t_{1/2} = 6.3 l_d - 2.5. \tag{15}$$

For example, the half-life for PGE_2 in a 1 mm thickness is predicted with 95% confidence as 3.8 ± 0.3 h.

CONCLUSIONS

Although no single model can predict all the experimental observations of drug release from an initially dry hydrogel placed in an aqueous sink [26] by studying controlling variables separately an elucidation of the mechanisms involved has been achieved.

In the early stages of release dynamic swelling changes the water content from zero to the equilibrium water uptake over a period which can be predicted from the thickness of the slab and the presence of crystallinity in the polymer composition. Transport of solute through the network depends on the degree of water swelling ranging from zero in the xerogel to a maximum which determines the diffusion coefficient, D, for the fully equilibrated hydrogel. It was found that the higher swelling compositions generally showed release profiles nearer to the often desirable zero order but also because of their higher water content and consequently higher D values had shorter half-lives for the drug in the matrix.

It was possible to relate the changing surface area to the drug release profile and again with the higher swelling compositions and consequently with greater increase in area on swelling a flatter profile approaching zero order was frequently achieved.

It was observed that there was no significant change in the release pattern for a significant time even after the hydrogel was fully swollen. This conclusion differs radically from the model presented by Korsmeyer *et al.* [20] who predicted a transition to a continuously decreasing rate once the hydrogel was fully swollen.

The release profile could be considered in three parts, first, and generally briefly, a burst from drug concentrated at the surface. The first part might alternatively be a time lag. In some cases the first and second parts merged imperceptibly and inseparably forming a single section from the beginning. The second, and most important in designing these drug delivery systems, was an anomalous release profile lasting until 50–70% of the total release characterized by the exponential function, t^n. Finally the release rate decreased steadily until all the drug had been desorbed. Values of n ranged from 0.79 to just over 1 (Super Case II transport) depending on drug loading, hydrogel water content and device thickness. These values are higher than the 0.5 → 0.76 values reported by Korsmeyer and Peppas for theophylline [27]. The first controlling variable, drug loading, also includes the properties of the drug substance such as aqueous solubility and ease of solubilization within the restrictions of the hydrogel network. Considerably exceeding the drug solubility by using a better solvent than water for the drug in order to impregnate the slab produces a reservoir of drug particles to draw upon after the hydrogel has reached its fully swollen state. However there was a limit to the amount of drug which the dried-down matrix could contain. Excess amounts then formed a coating of drug on the surface resulting in a burst effect when placed in water.

The safety of a simple system containing trapped drug which can only be activated by contact with water, followed by controlled release at predictable rates approaching steady state release for more than 50% drug content, offers a promising area for drug delivery in areas of the body where there is water available to swell the xerogel.

Acknowledgement

The support of the British Technology Group in this research is gratefully acknowledged.

REFERENCES

1. M. E. McNeill and N. B. Graham, *J. Biomater. Sci. Polym. Edn* **4**(3), 305–322 (1993).
2. I. Z. MacKenzie, F. R. Burnet and M. P. Embrey, *Br. J. Obstet Gynaecol.* **87**, 292 (1980).
3. M. P. Embrey, N. B. Graham, M. E. McNeill and K. Hillier, *J. Controlled Release* **3**, 39 (1986).
4. M. E. McNeill and N. B. Graham, *J. Controlled Release* **1**, 99 (1984).
5. D. A. Wood, in: *Critical Reports on Applied Chemistry*, Vol. 6, *Materials Used in Pharmaceutical Formulations*, A. T. Florence (Ed.). Blackwell Scientific Publications, Oxford. (1984).
6. R. Langer, *Chem. Eng. Commun.* **6**, 1 (1980).
7. R. Paterson, *Swiss Chem.* **3a**, 17, (1988).
8. A. R. Berens and H. B. Hopfenberg, *Polymer* **19**, 489 (1978).
9. N. A. Peppas and R. W. Korsmeyer, in: *Hydrogels in Medicine and Pharmacy*, Vol. III, Ch. 6, p. 109, N. A. Peppas (Ed.). CRC Press, Boca Raton, FL (1988).
10. P. L. Ritger and N. A. Peppas, *J. Controlled Release* **5**, 37 (1987).
11. R. W. Korsmeyer and N. A. Peppas, *Proc. Int. Symp. Contr. Rel. Bioact. Mater.* **8**, 85 (1981).
12. G. W. R. Davidson III and N. A. Peppas, *J. Controlled Release* **3**, 243 (1986).
13. G. W. R. Davidson III and N. A. Peppas, *J. Controlled Release* **3**, 259 (1986).
14. R. W. Korsmeyer, T. L. Rave, N. A. Peppas, R. Gurny, E. Doelker and P. A. Buri, *Proc. Int. Symp. Contr. Rel. Bioact. Mater.* **9**, 65 (1982).
15. P. I. Lee, *J. Controlled Release* **2**, 277 (1985).
16. H. P. Hopfenberg and K. C. Hsu, *Polym. Eng. Sci.* **18**, 1186, (1978).
17. J. Klier and N. A. Peppas, *J. Controlled Release* **7**, 61 (1988).
18. M. E. McNeill and N. B. Graham, *Proc. Int. Symp. Contr. Rel. Bioact. Mater.* **18**, 259 (1991).
19. N. B. Graham and M. E. McNeill, *Biomaterials* **5**, 27 (1984).
20. R. W. Korsmeyer, S. R. Lustig and N. A. Peppas, *J. Polym. Sci. Polym. Phys. Ed.* **24**, 395 (1986).
21. P. I. Lee, *J. Controlled Release* **2**, 277 (1985).
22. P. I. Lee and C. J. Kim, *J. Controlled Release* **16**, 229 (1991).
23. *Minitab Handbook*, Duxbury Press, Boston, 218 (1985).
24. N. B. Graham and M. E. McNeill, *Makromol. Chem. Macromol. Symp.* **19**, 255 (1988).
25. W. E. Roorda, Structure and application of hydrogels as drug delivery systems, PhD thesis, Leiden, p.37 (1988).
26. P. I. Lee, *J. Pharmacol Sci.* **73**, 1344 (1984).
27. R. W. Korsmeyer and N. A. Peppas, *J. Membrane Sci.* **9**, 211 (1981).

Dextran permeation through poly(N-isopropylacrylamide) Hydrogels

LIANG C. DONG[1], ALLAN S. HOFFMAN[2] and QI YAN[2]

[1]*Alza Corp., 950 Page Mill Road, PO Box 10950, Palo Alto, CA 94303, USA*
[2]*Center for Bioengineering FL-20, University of Washington, Seattle, WA 98195, USA*

Received 8 September 1992; accepted 25 February 1993

Abstract—The permeation of macromolecules such as fluoroescein-labeled dextran fractions through thermally reversible hydrogels has been investigated. A permeation model has been formulated, which takes into account hydrogel porosity and tortuosity as well as the combined effect of a geometric restraint for a relatively large solute molecule at a pore entrance and the friction between solute molecules moving through the pores and pore walls. Based on this model, we have estimated the tortuosity and average pore size of a swollen hydrogel, poly(*N*-isopropylacrylamide) [poly(NIPAAm)] and a swollen heterogel, poly(*N*-isopropylacrylamide-co-vinyl-terminated dimethylsiloxane) [poly(NIPAAm-co-VTPDMS)]. The permeation data for dextran molecules up to the size of 43.5 Å in radius show good agreement with the values predicted from the model.

Key words: FITC-dextran, permeation, hydrogel, pore size, poly(N-isopropylacrylamide), diffusion, thermally reversible.

INTRODUCTION

A common factor used to describe and 'correct' diffusion of small solutes through the pores in porous membranes is the ratio of the fractional pore volume, or porosity (ε) to the tortuosity factor (τ) which represents the 'meandering' of the pore and its interconnections across the membrane. If the pores are filled with water, as with hydrogels, the diffusivity of the free solute in water, D_0, is multiplied by ε/τ to yield the 'apparent' diffusivity, D_{app}, across the membrane. For the case of hydrogels, Yasuda *et al.* have proposed a free volume approach which can successfully interpret the solute permeation through hydrogels [1]. Peppas and Reinhart have also developed a model for diffusion of spherical and quasi-spherical solutes through highly swollen membranes [2]. It is always found that the diffusivity (D_p) of a solute through a small pore is less than the free solution diffusivity (D_0) of the solute in water which fills the pores of the hydrogel membranes. The discrepancy has been shown to be attributed to two restriction factors. The hydrodynamic theories of hindered transport [3, 4] have been developed to correct the diffusivity (D_p) of a solute through a pore for these two factors. This finding led to the following Renkin equation for the ratio of D_p/D_0.

$$\frac{D_p}{D_0} = \left(1 - \frac{a}{r_p}\right)^2\left[1 - 2.104\left(\frac{a}{r_p}\right) + 2.09\left(\frac{a}{r_p}\right)^3 - 0.95\left(\frac{a}{r_p}\right)^5\right], \tag{1}$$

where a is the solute radius and r_p is the average pore radius. The first factor is a geometrical restraint, that is, for solute entrance into a pore, the solute molecule must pass through the opening without striking the pore edge. Therefore, the center of the molecule must pass through a circle of radius ($r_p - a$) within the mouth of the pore. The second factor is the friction between the solute moving through a pore and its walls.

In this paper, we have combined these approaches to develop a permeation model which takes into account hydrogel porosity and tortuosity as well as the combined effect of a geometric restraint and the friction between solute molecules moving through the pores and pore walls. We then use this model to correlate data for permeation of dextran macromolecules in two different swollen hydrogels—poly(*N*-isopropylacrylamide) [poly(NIPAAm)] and poly(*N*-isopropylacrylamide-co-vinyl-terminated dimethylsiloxane) [poly(NIPAAm-co-VTPDMS)], one homogeneous and the other heterogeneous containing silicone domains. Also, the pore radius (r_p) of each hydrogel has been estimated by fitting the experimental data from the permeation of low molecular weight dextran to the model. We assumed in this model that the hydrogel pores are cylindrical tunnels, the diffusable solutes (FITC-dextrans) are spherical in shape, and only 'free' (or freezable) water is available for diffusion. Since the hydrogel pores actually are fluctuating free volume voids of irregular shape, the estimated r_p is only an average radius of the pores equivalent to the assumed cylindrical channels.

EXPERIMENTAL

Materials

N-isopropylacrylamide (NIPAAm) (Eastman Kodak Company, Rochester, NY) was purified by recrystallization in a mixture of *n*-hexane and benzene (50/50, vol). Methylene-bis-acrylamide (MBAAm) (electrophoresis grade, Aldrich Chemical Company Inc.), vinyl-terminated polydimethylsiloxane (VTPDMS) (M_w 28 000, Petrach Systems, Bristol, PA, USA) and FITC-dextrans of various molecular weights (Sigma Chemical Co., St. Louis, MO, USA) were all used as received. Solvents used were all of reagent grade (J. T. Baker Chemical).

Membrane preparation

Membrane preparation was described previously [5]. A 30% (w/v) solution in chloroform of NIPAAm for poly(NIPAAm) homogeneous hydrogels or of a 1/1 weight ratio of NIPAAm to VTPDMS for poly(NIPAAm-co-VTPDMS) heterogeneous hydrogels, respectively, were radiation-polymerized between glass plates, the surfaces of which had been silanized with dimethyldichlorosilane solution (2% v/v) in toluene to avoid the adhesion between the polymer and glass plate. Two glass plates were separated by a 1.5 mm Teflon spacer to form a mold. After the solution was purged with nitrogen at 4°C, it was immediately added into the mold, which was then placed in a polyethylene zip-lock bag filled with nitrogen. The polymerization was achieved by exposing the solution to ^{60}Co γ-rays at a dose rate of 0.5 Mrad day^{-1} for 1 day. After polymerization, the gel membrane was cut with a cork borer into 1.5 cm diameter discs, which were subsequently extracted in a Soxhlet extractor with $CHCl_3$ to remove unreacted VTPDMS, dried and then washed and rinsed in deionized water. The NIPAAm homopolymer hydrogel is designated as NS-100 and the NIPAAm-co-VTPDMS copolymer heterogel as NS-50.

Estimation of freezable ('free') water with DSC

Both NS-100 and NS-50 hydrogel membranes to be studied by DSC were carefully blotted with wet filter paper to remove extra surface water after the initially dried hydrogels had been equilibrated in phosphate buffer (pH 7.4) at 25°C. The disc shape samples were used in order to assure good contact between the sample and the sample pan. The DSC measurements were carried out with a Seiko DSC100 differential scanning calorimeter fitted with a liquid nitrogen sub-ambient accessory. The samples were cooled down from room temperature to −30°C at cooling rate of 2–3°C min^{-1}, and then warmed to 30°C at a rate of 1°C min^{-1}. Thermograms were recorded with a SSC5020 Series TA Disk Station (Seiko Instruments Inc.). Enthalpies (ΔH) were calculated from the areas under the melting peak, using a routine program in the instrument. The ΔH values are reproducible to within 2%. The amount of freezing water was calculated from a calibration graph obtained by measuring deionized water at identical conditions, based on the assumption that the heat of water fusion in the hydrogel membranes is the same as that of deionized water. We assume here that only this freezable water is 'free' and available for solute partitioning and diffusion within the aqueous pores. We will also compare this assumption with the assumption that all the gel water is available for permeants (see below).

Permeation of fluorescein-labeled dextran through the gel membranes

Dextran is known for its spherical shape in aqueous solution [6]. However, dextran is not easily analyzed at low concentration. In order to increase the sensitivity and ease of its analysis, fractionated fluorescein-labeled dextrans (FITC-dextrans) have been chosen as model macromolecules for permeation study. Since fluorescein has a very high extinction coefficient and high quantum yield that is almost invariant with environment, even at the concentrations as low as 10–100 ng ml^{-1}, FITC-dextrans can be detected with a fluorimeter. The diffusivities of FITC-dextrans across the NS-100 and NS-50 gel membranes were measured at 25°C with Side-By-Side^{Tm} diffusion cells (Crown Glass Company, Inc., Somerville, NJ, USA).

The hydrogel membranes were equilibrated at 25°C in phosphate buffer (pH 7.4) containing 0.02% (w/v) of sodium azide as a bactericide before being placed between the halves of the permeation cell with an exposed are of 0.64 cm^2. Usually, it takes at least 12 h for the macromolecules to permeate through even a swollen hydrogel membrane to establish a steady state. Therefore, the juncture of the membrane and the halves was wrapped with Parafilm (American Can Company, Greenwich, CT, USA) in order to prevent the periphery of the membrane from drying out, since it could, otherwise, induce strain in the membrane during permeation, and result in breakage of the membrane. Three ml of the buffered water at 25°C were placed into the receptor cell first and then at zero time 3 ml of 200 μg ml^{-1} FITC-dextran solution in the buffered water were added into the donor compartment. A 1 ml aliquot was withdrawn at various times from the receptor compartment and one ml of the fresh buffered water was added to maintain constant volume and zero sink conditions. The preset time intervals depended on the permeation rate. The concentration of the solution was measured with a Perkin-Elmer LS-5B luminescence spectrometer at 485 nm of the excitation wavelength and 515 mm of the emission wavelength, using the buffer solution as a control.

RESULTS AND DISCUSSION

Formulation of permeation model

The apparent diffusivities (D_{app}) of FITC-dextrans with various molecular weights through NS-100 and NS-50 membranes were estimated using the time-lag equation described by Crank [7]. In this expression, the total amount of solute which has diffused across a membrane at time t (M_t) is given by the equation:

$$M_t = \frac{AD(C_0 - C_l)t}{l} + \frac{2lA}{\pi^2}\sum_{n=1}^{n=\infty}\frac{C_0\cos n\pi - C_l}{n^2}\left[1 - \exp\left(\frac{-Dn^2\pi^2 t}{l^2}\right)\right]$$
$$+ \frac{4C_m lA}{\pi^2}\sum_{m=0}^{m=\infty}\frac{1}{(2m+1)^2}\left[1 - \exp\left(\frac{-D(2m+1)^2\pi^2 t}{l^2}\right)\right] \quad (2)$$

where A is the area of the membrane, D is diffusivity, l is membrane thickness, C_0 is the concentration within the membrane at the donor face $x = 0$, C_l is the concentration at the downstream acceptor face $x = l$ and C_m is the initially uniform concentration in the membrane.

For the time lag case, both C_m and C_l are zero and the above equation reduces to

$$M_t = \frac{ADC_0 t}{l} + \frac{2lAC_0}{\pi^2}\sum_{n=1}^{n=\infty}\frac{\cos n\pi}{n^2} - \frac{2lAC_0}{\pi^2}\sum_{n=1}^{n=\infty}\frac{\cos n\pi}{n^2}\exp\left(\frac{-Dn^2\pi^2 t}{l^2}\right)$$
$$= \frac{ADC_0 t}{l} - \frac{lAC_0}{6} - \frac{2lAC_0}{\pi^2}\sum_{n=1}^{n=\infty}\frac{\cos n\pi}{n^2}\exp\left(\frac{-Dn^2\pi^2 t}{l^2}\right). \quad (3)$$

As $t \to \infty$, the exponential terms in Eqn (3) vanish. The result at steady-state is:

$$M_t = \frac{ADC_0}{l}\left(t - \frac{l^2}{6D}\right) \quad (4)$$

If M_t is plotted against time, the slope is:

$$\frac{ADC_0}{l} = \frac{ADKC_s}{l} \quad (5)$$

where K and C_s are the partition coefficient of FITC-dextran into the membrane from the donor compartment and its concentration in the donor compartment, respectively.

Kim *et al.* [8], Wisniewski and Kim [9], and Zentner *et al.* [10, 11] examined the permeation of solutes, both hydrophilic and hydrophobic, through poly(hydroxyethyl methacrylate) and its copolymer hydrogel membranes. These researchers have concluded that permeation of hydrophilic solutes in hydrogels is by a 'pore' mechanism, i.e. that a solute permates through water-filled pores in the gel. For hydrophobic solutes diffusion also occurs predominantly by a 'pore' mechanism in highly swollen hydrogels.

The hydrogel membranes examined in this study are highly swollen, having water contents about 90%. As indicated by DSC results, these hydrogel membranes contain a large fraction of freezable ('free') water. The permeation of FITC-dextrans through the membrane should, therefore, be predominantly by a pore mechanism. If so, the diffusional area and diffusional path length should be the total pore area

(A_p) and average pore length (l_p) instead of the membrane area (A) and thickness (l), respectively. The experimental values for the partition coefficient are usually less than unity, especially for macromolecules. This is due to the geometric restraint. Since our model takes it into account as a factor of $(1 - a/r_p)^2$, the assumption of K, defined as the ratio of the FITC-dextran concentration in water-filled hydrogel pores to its bulk concentration, to be unity seems reasonable.

Thus, the slope of the steady state portion of the plot of M_t vs t becomes

$$\frac{A_p D_p C_s}{l_p} = \frac{\varepsilon A D_p C_s}{\tau l} = \frac{A D_{app} C_s}{l} \tag{6}$$

where D_p is the diffusivity of a solute through the hydrogel pores, ε the porosity and τ the tortuosity. (It will be recognized that $\varepsilon = A_p/A$ and $\tau = l_p/l$.) The 'apparent' diffusivity, D_{app}, as measured in the time-lag experiment, is then seen to be $D_p(\varepsilon/\tau)$.

The porosity (ε) refers to the volume fraction of the hydrogels which is available for diffusion. Since the pore mechanism suggests that the permeation of solutes occurs in free water within the pores, the porosity in this study is considered to be the fraction of free water in the hydrogel membranes. Exclusion of hydrophilic solutes from the bound water regions has been reported [9, 12]. Therefore, it is reasonable to equate the free water fraction to ε in these diffusion equations.

The tortuosity (τ) is defined as the ratio of diffusional path length to the membrane thickness, accounting for the additional distance a molecule must travel due to its tortuous and winding path within the gel membrane pores.

As noted above, it can be seen that the measured 'apparent' diffusion constant is $D_{app} = D_p(\varepsilon/\tau)$. For small solutes this is normally seen as $D_0(\varepsilon/\tau)$, where D_0 is the free solution diffusivity of the permanent, but D_p is substituted for D_0 since the macromolecules studied here are expected to interact with the polymer molecules of the gel pore walls.

Substitution of D_p in Eqn (1) into Eqn (6) gives

$$\frac{D_{app}}{D_0} = \frac{\varepsilon}{\tau}\left(1 - \frac{a}{r_p}\right)^2 \left[1 - 2.104\left(\frac{a}{r_p}\right) + 2.09\left(\frac{a}{r_p}\right)^3 - 0.95\left(\frac{a}{r_p}\right)^5\right]. \tag{7}$$

Equation (7) was employed to estimate the average pore radius (r_p) and tortuosity (τ) for the two hydrogels studied here.

Estimation of free solution diffusivities of FITC-dextrans

To have reliable free solution diffusivities (D_0) of FITC-dextrans with various molecular weights, two independent methods were used to estimate these values. One estimate of D_0 of FITC-dextrans at 25°C is based on D_0 values for dextrans at 20°C which are available from the Polymer Handbook [13]. The diffusivities of dextrans at 25°C can be calculated from those values at 20°C, based on the following equation [14].

$$D_{25} = D_{20}\frac{298}{293}\frac{\eta_{20}}{\eta_{25}} \tag{8}$$

where D_{25} and D_{20} are the free solution diffusivities at 25 and 20°C, respectively, and η_{25} and η_{20} are water viscosities at 25 and 20°C, respectively. η_{20} was reported

Table 1.
Free solution diffusivities of dextrans

M_w of dextran ($M \times 10^{-4}$)	$D_{20} \times 10^7$ [a] ($cm^2 s^{-1}$)	$D_{25} \times 10^7$ [b] ($cm^2 s^{-1}$)
14000	7.50	8.59
36000	4.69	5.37
84000	3.12	3.57
140000	2.40	2.75
240000	1.73	1.98
1990000	0.75	0.86

[a] Obtained from reference [13].
[b] Calculated from Eqn (8).

to be 1.002 (cp) [15] while η_{25} was calculated to be 0.8903 (cp) based on Gragoe's equation [16]. Table 1 shows values obtained from the Polymer Handbook for D_{20} of dextrans with various molecular weights along with calculated values for D_{25}, based on Eqn (8).

The free solution diffusivities for FITC-dextrans with various molecular weights used in this study can be estimated from the following simplified linear equation which is obtained by linear regression of the data in Table 1,

$$\log D_{25} = -4.140 - 0.4687 \log M; \qquad r^2 = 0.9954 \tag{9}$$

where M is dextran molecular weight and r is the correlation coefficient.

The other estimation method is to calculate D_0 by the Stokes–Einstein equation:

$$D_0 = \frac{RT}{6\pi a \eta N} \tag{10}$$

where R is the gas constant, T is the absolute temperature, N is Avogadro's number, η is the viscosity of water, and a is the radius of FITC-dextran. The use of this equation implies the assumption that dextran is a sphere in aqueous solution. The radii of FITC-dextrans may be estimated by interpolation of data given by Stone and Scallan for a series of dextran fractions [6]. Figure 1 shows their data for

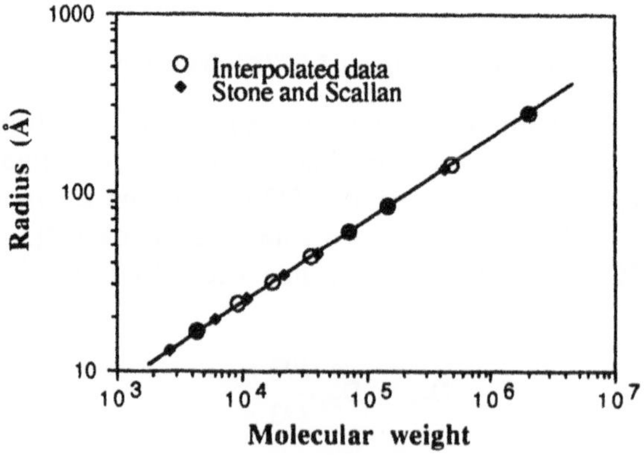
1000
100
10
Radius (Å)
Interpolated data
Stone and Scallan
10^3
10^4
10^5
10^6
10^7
Molecular weight

Figure 1. FITC-dextran molecular radius vs molecular weight.

Table 2.
Estimation of free solution diffusivities (D_0) of FITC-dextrans at 25°C

FITC-dextran code	Molecular weight[a] $\times 10^{-3}$	Molecular radius[b] (Å)	Free solution diffusivity, $D_0 \times 10^6$ [c] ($cm^2\ s^{-1}$)		
			method 1	method 2	average
FD-4	4.4	16.5	1.42	1.49	1.46
FD-10	9.4	23.5	0.995	1.04	1.00
FD-20	17.2	31.0	0.750	0.791	0.771
FD-40	35.6	43.5	0.533	0.564	0.549
FD-70	71.6	59.5	0.384	0.412	0.398
FD-150	152	84.5	0.270	0.290	0.280
FD-500	487	144	0.156	0.170	0.163
FD-2000	2000	280	0.081	0.088	0.085

[a] Molecular weight as stated by Sigma Chemical Co.

[b] Radii of FITC-dextrans were found by interpolation of the data given by Stone and Scallan [6] for a series of dextran fractions (see Fig. 1).

[c] Method 1 based on D_{20} data in Polymer Handbook [13]; Method 2 based on estimation from Stokes-Einstein equation (see text for details).

dextrans and the interpolated data for FITC-dextrans used here. The estimated values of D_0 and data from which they are calculated are listed in Table 2 along with the D_0 values from the first method. The values for free solution diffusivities of FITC-dextrans estimated by these two methods are in good agreement. The average values shown in column 6 of Table 2 have been used for estimating the average pore size and tortuosity of both NS-100 and NS-50 gels in the present work.

Evaluation of apparent diffusivities of FITC-dextrans

Figure 2 is the typical plot of experimental permeation data. The apparent diffusivity of each FITC-dextran fraction was found from the slope of the steady-state portion

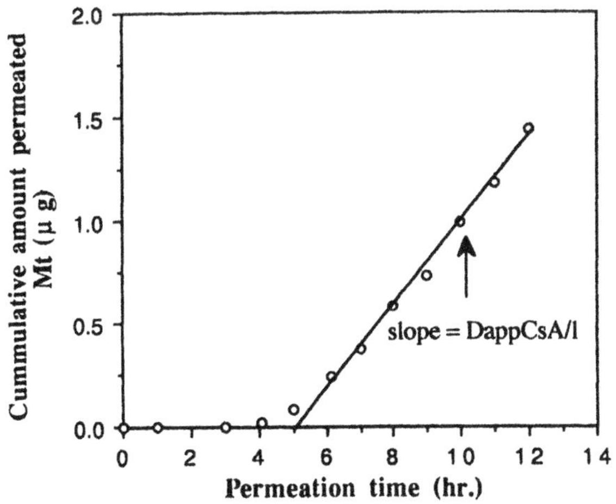
Cummulative amount permeated
Mt (μ g)
2.0
1.5
1.0
0.5
0.0
0
2
4
6
8
10
12
14
Permeation time (hr.)
slope = DappCsA/l

Figure 2. Typical plot of permeation of a FITC-dextran (FD-10) through NS-100 at 25°C.

Table 3.
Apparent diffusivities (D_{app}) of FITC-dextrans through poly(NIPAAm) (NS-100) and poly(NIPAAm-co-VTPDMS) (NS-50) hydrogels

	$D_{app} \times 10^9$ ($cm^2 s^{-1}$)		D_{app}/D_0	
FITC-dextran	NS-100	NS-50	NS-100	NS-50
FD-4	278 (31)	388 (45)	0.190	0.266
FD-10	112 (8)	175 (22)	0.112	0.175
FD-20	134 (14)	114 (60)	0.174	0.148
FD-40	27.8 (4.2)	25.5 (11)	0.051	0.046
FD-70	8.22 (0.02)	56.5 (21)	0.021	0.142
FD-150	20.1 (2.2)	45.8 (2.1)	0.072	0.164
FD-500	8.00 (0.14)	23.3 (23.7)	0.049	0.143
FD-2000	10.4 (0.4)	22.4 (12.6)	0.124	0.264

The numbers in parentheses are standard deviations of three runs.

of the curve. Listed in Table 3 are the apparent diffusivities and relative diffusivity ratios (D_{app}/D_0) of various FITC-dextran fractions permeating through NS-100 and NS-50 hydrogel membranes.

The pore mechanism equation (Eqn (7)) suggests that the apparent diffusivity (D_{app}) should decrease regularly with increase in solute size. This decrease in diffusivity is shown in Fig. 3 which represents the result simulated according to Eqn (7), assuming $\varepsilon/\tau = 1$. It can be seen that the smaller the average pore radius, the more rapidly the apparent diffusivity drops with an increase in the solute size. This drop in diffusivity can be ascribed to the greater hindrance to diffusion of the solute by the molecules making up the smaller pore walls as well as the lower probability for the solute molecules to enter the pores without striking the pore edge. These arguments assume a well-defined average pore, which in reality is probably fluctuating in shape and size with time.

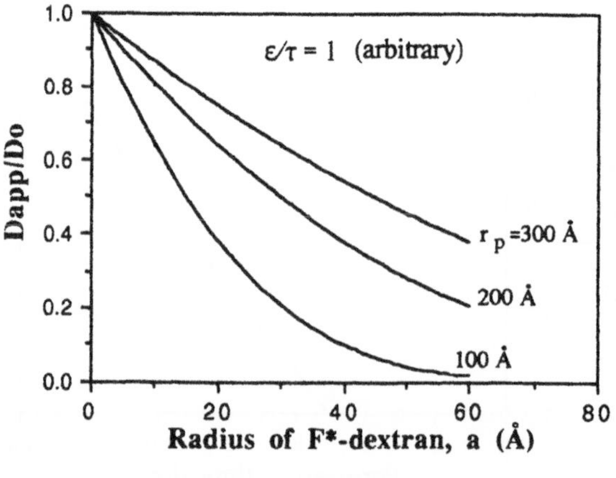
ε/τ = 1 (arbitrary)
Dapp/Do
1.0
0.8
0.6
0.4
0.2
0.0
r p =300 Å
200 Å
100 Å
0
20
40
60
80
Radius of F*-dextran, a (Å)

Figure 3. Simulation (based on Eqn (7)) of apparent diffusivity as a function of FITC-dextran radius for various pore sizes, assuming $\varepsilon/\tau = 1$.

Estimation of ε/τ

According to Eqn (7), ε/τ can be obtained by extrapolating the experimental data points to the solute size equal to zero. If Eqn (7) is converted to a log form and expanded into a series, the higher terms can be truncated when a/r_p is sufficiently small, leading to

$$\log\left(\frac{D_{app}}{D_0}\right) = \log\left(\frac{\varepsilon}{\tau}\right) - 4.104\frac{a}{r_p}. \tag{11}$$

If $\log(D_{app}/D_0)$ is plotted against the solute radius (a), (ε/τ) can be obtained from the intercept of the linear curve. Therefore, the extrapolation was carried out based on Eqn (11), using the experimental data of the four smallest FITC-dextrans, yielding $\varepsilon/\tau = 0.57$.

Separate estimation of porosity (ε) and tortuosity (τ)

The free water fraction of each hydrogel was calculated from the DSC heating thermograms (Fig. 4) for the endothermic water melting peak. The fraction is assumed to be the porosity (ε). The values of the fraction are 0.713 and 0.663 for NS-100 and NS-50, respectively. It follows that the tortuosity is 1.25 for NS-100 and

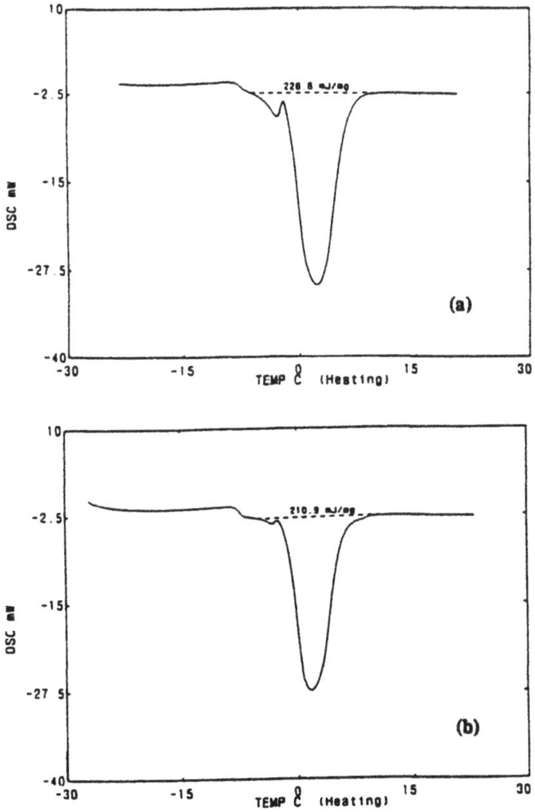

10
-2.5
-15
-27.5
-40
DSC mW
228.8 mJ/mg
(a)
-30
-15
0
15
30
TEMP C (Heating)
10
-2.5
-15
-27.5
-40
DSC mW
210.9 mJ/mg
(b)
-30
-15
0
15
30
TEMP C (Heating)

Figure 4. DSC thermograms of (a) NS-100 and (b) NS-50 hydrogels after equilibration in phosphate buffer (pH 7.4) at 25°C.

1.17 for NS-50. Renkin [3] mentioned that if the pores were oriented randomly with respect to the plane of the membrane, the tortuosity should be 3. Because of connections between the pores, the tortuosity might lie somewhere between 1 and 3. Since both NS-100 and NS-50 are highly swollen, with water contents of 0.90 and 0.89, respectively, the values of tortuosity (τ) close to 1 seem reasonable. (If ε is assumed to be the total water content instead of the freezable water, then τ is 1.58 for NS-100 and 1.56 for NS-50, respectively. These values are also reasonable.

We expected a higher tortuosity value for NS-50 than NS-100, since the NS-50 heterogel contains hydrophobic silicone domains which should be impermeable to FITC-dextrans, making the diffusion pathway more tortuous. The small tortuosity value for NS-50 suggests that the larger pores in NS-50 might more than compensate for the domain contribution.

Estimation of average pore size

In Figure 5, the relative diffusivity ratio is shown as a function of (a/r_p) for the two gels. It can be seen that a good fit of the data to Eqn (7) can be obtained for a/r_p

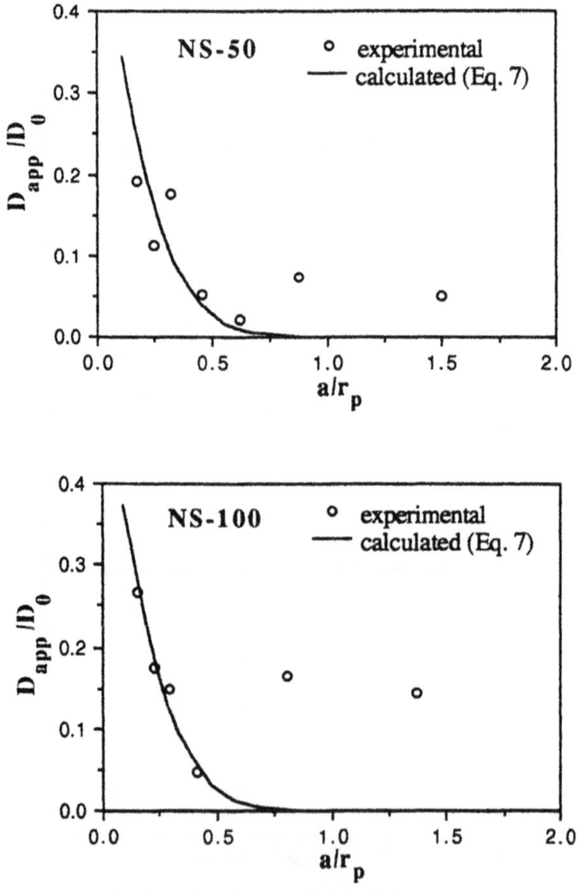
NS-50
experimental
calculated (Eq. 7)
D app /D 0
0.4
0.3
0.2
0.1
0.0
0.0
0.5
1.0
1.5
2.0
a/r p
NS-100
experimental
calculated (Eq. 7)
D app /D 0
0.4
0.3
0.2
0.1
0.0
0.0
0.5
1.0
1.5
2.0
a/r p

Figure 5. Relative dextran diffusivity in hydrogel membrane versus permeant size.

ratios up to *ca.* 0.6, leading to r_p values of 105 Å for NS-50 and 90 Å for NS-100. It has been reported that the Renkin equation applies only to the a/r_p ratio under 0.2 for the case of hard spheres moving through a solvent-filled pores with distinct solid walls [17]. However, it has been demonstrated that this equation can be successfully applied to the ratio of 0.4 for a series of solutes diffusing through mica membranes with uniform, straight pores [18], and up to at least 0.5 for the case of hydrogels with low water content [19]. The further extension of the applicability in this case might be attributed to the high water content of these hydrogels, suggesting more fluctuating hydrogel pores.

The higher average pore size of NS-50 agrees with the smaller value of τ for this gel, as estimated above. Furthermore, this gel (NS-50) shows an extremely fast shrinking rate when heated through its lower critical solution temperature (around 32°C) in comparison with the NS-100 gel [5]. This finding is another reflection of the more open pore structure of the NS-50 silicone domain heterogel.

Deviations at higher (a/r_p) from the calculated curves may be related to a possible change in permeation mechanism of the higher molecular weight molecules from molecular diffusion of random coils to repeating chain segments and/or to the polydispersity of the higher molecular weight FITC-dextran samples.

CONCLUSIONS

A model has been formulated for permeation of macromolecules in hydrogels. The model takes into account hydrogel porosity, tortuosity, and permanent-pore wall interactions. Tortuosity and average pore size of swollen hydrogels can be estimated from this model. Good agreement exists between the permeation data and the model for dextran M_ws up to 35 600. The model should be useful for permeation of small to medium sized proteins in hydrogels as well as for estimation of swollen hydrogel pore size, which is usually very difficult to measure.

Acknowledgement

We would like to gratefully acknowledge the Washington Technology Center for financial support.

REFERENCES

1. H. Yasuda, C. E. Lamaze and L. D. Ikenbery, *Makromolek. Chem.* **118**, 19 (1968).
2. N. A. Peppas and C. T. Reinhart, *J. Membrane Sci.* **15**, 361 (1983).
3. E. M. Renkin, *J. Gen. Physiol.* **38**, 225 (1954).
4. M. G. Davidson and W. M. Deen, *Macromolecules* **21**, 3481 (1988).
5. L. C. Dong and A. S. Hoffman, *J. Contr. Rel.* **13**, 21 (1990).
6. J. E. Stone and A. M. Scallan, *Cellulose Chem. Technol.* **2**, 343 (1968).
7. J. Crank, *Mathematics of Diffusion*. Oxford University Press, London (1956).
8. S. W. Kim, J. R. Cardinal, S. Wisniewski and G. M. Zentner, p. 347, In S. P. Rowland (Ed.), *ACS Symposium Series 127*, A.C.S., Washington DC (1980).
9. S. Wisniewski and S. W. Kim, *J. Membr. Sci.* **6**, 299 (1980).
10. G. M. Zentner, J. R. Cadinal and S. W. Kim, *J. Pharmacol. Sci.* **67**, 1352 (1978).
11. G. M. Zentner, J. R. Feijen and S. Song, *J. Pharmacol. Sci.* **68**, 970 (1979).
12. G. N. Ling, in: *Water and Aqueous Solutions: Structure, Thermodynamics, and Transport Properties*, p. 688, R. A. Horne (Ed.). Wiley-Interscience, New York (1972).
13. J. Brandrup and E. H. Immergut (Eds.), *Polymer Handbook IV-89*, 2nd Edition. John Wiley, New York (1975).

14. A. F. Schick and S. J. Singer, *J. Phys. Chem.* **54**, 1928 (1950).
15. J. F. Swindells, J. R. Coe *et al.*, *J. Res. Nat. Bur. Stand.* **48**, 1 (1952).
16. J. R. Coe and T. B. Godfrey, *J. Appl. Phys.* **15**, 625 (1944).
17. E. L. Cussler, *Diffusion*, p. 189. Cambridge University Press (1984).
18. R. E. Beck and J. S. Schultz, *Science* **170**, 1302 (1970).
19. B. D. Ratner and I. F. Miller, *J. Biomed. Mater. Res.* **7**, 353 (1973).

Poly(vinyl alcohol) hydrogels as soft contact lens material

SUONG-HYU HYON[1], WON-ILL CHA[1], YOSHITO IKADA[1*],
MIHORI KITA[2], YUICHIRO OGURA[2] and YOSHIHITO HONDA[2]

[1] *Research Center for Biomedical Engineering, and* [2] *Department of Ophthalmology, Faculty of Medicine, Kyoto University, 53 Kawahara-cho, Shogoin, Sakyo-ku, Kyoto 606, Japan*

Received 18 September 1992; accepted 12 February 1993

Abstract—A new type of soft contact lens was developed from the poly(vinyl alchol) (PVA) hydrogel prepared by a low temperature crystallization technique using a water–dimethyl sulfoxide mixed solvent. The PVA contact lens materials had a water content of 78% and a tensile strength of 50 kg/cm^2, five times as strong as that of commercial poly(2-hydroxyethyl methacrylate) soft contact lens. The amount of proteins adsorbed to the PVA soft hydrogel material was half to one thirtieth of that of conventional soft contact lenses. Histological and scanning electron microscopic observation of rabbit eyes which had worn the PVA soft contact lens for 12 weeks showed no difference in corneal epithelium and cell arrangement in the corneal epithelium from the non-wearing eyes.

Key words: Poly(vinyl alcohol); hydrogel; soft contact lens; protein adsorption; animal test.

INTRODUCTION

Biomaterials widely applied for soft contact lenses include silicone elastomer [1], copolymer [2] of butyl acrylate (BA), and butyl methacrylate (BMA) in hydrophobic lenses and poly(2-hydroxyethyl methacrylate) (PHEMA) [3], copolymer [4] of methyl methacrylate (MMA), and *N*-vinyl pyrrolidone (VP) in hydrophilic lenses. The hydrophobic materials have disadvantages such as poor wear feeling caused by low water wettability and a remarkable adsorption of lipids. On the other hand, the hydrophilic materials give rise to significant protein adsorption and are very low in mechanical strength, although they give excellent wear feeling and minimum damage to the cornea.

Poly(vinyl alcohol) (PVA) hydrogel has already been proposed as a material for soft contact lenses (SCLs), for instance, by crosslinking with glyoxal or borate [5]. Furthermore, preparation of a PVA soft contact lens through annealing [6] is proposed, but this needs a long and complicated fabrication process, although it has a high water content and a high mechanical strength.

Recently, we have developed a new preparation method of PVA hydrogel, which involves dissolution of PVA in a mixed solvent consisting of water–organic solvent and the subsequent crystallization of PVA at a low, but not extraordinarily low, temperature. This method enabled us to fabricate a transparent hydrogel with high strength and high water content.

The present work was undertaken in an attempt to explore the possibility of using the PVA hydrogel with high water content and high mechanical strength in soft contact lenses.

* To whom correspondence should be addressed.

MATERIALS AND METHODS

PVA hydrogel membrane

Preparation. Commercial PVA with a degree of saponification of 99.5 mol% and a viscosity-average degree of polymerization of 1700 (average $M_w = 74800$) was used to prepare PVA solutions with PVA concentrations from 5 to 30 wt.%. PVA solution was obtained by heating a mixture of PVA powder and a mixed water–dimethyl sulfoxide (DMSO) solvent at 120°C. After the PVA solution was cooled to about 60°C, it was cast on a glass plate and then allowed to stand at −20°C for 24 h to achieve PVA crystallization. The resulting PVA gel film was immersed in water at room temperature for 3 days and then air-dried overnight. The gel was dried further under vacuum overnight to remove the remaining organic solvent, and was again put in water for hydration at 37°C for 2 days. The water content of the hydrated PVA gel was calculated from the equation:

$$\text{water content (\%)} = \frac{(\text{weight of hydrated gel}) - (\text{weight of dried gel})}{(\text{weight of hydrated gel})} \times 100.$$

Physical properties. The PVA membranes (1 mm thickness) were measured for tensile strength and elongation break at 25°C under a tensile speed of 50 mm/min using an Autograph S-100 (load cell: 5 kgf) of Shimadzu Inc. Ltd., Kyoto, Japan, and five measurements were averaged. The membrane samples were cut to dumb-bell test pieces in accordance with JIS specifications.

Light transmittance at 550 nm was measured for the PVA gels of 0.5 mm thickness under immersion in water at 25°C. Oxygen permeability of the hydrogel membranes was measured at 35°C using a conventional gas permeation apparatus applied for polymer films [7].

The proteins used were γ-globulin (IgG), bovine serum albumin (BSA), and lysozyme which was purchased from Sigma Chemical Co. Labeling of proteins with ^{125}I was performed by the chloramine-T method [8]. Protein adsorption was conducted at 37°C in phosphate buffered saline (PBS) at pH 7.4 for 3 h. After adsorption, the SCL was rinsed with 6 ml of PBS and an amount of adsorbed protein was measured with the radioactivity, using an X-ray scintillation counter.

PVA hydrogel soft contact lens

Preparation. A mixed solvent of water–DMSO (80 wt.% DMSO) was added to PVA powder so as to have a PVA concentration of 20 wt.%. The PVA solution was prepared by heating the mixture for 2 h at 120°C. After the PVA solution was cooled to 50–60°C, it was poured into a mould. One surface of the mould was virtually spherical and convex, while the other was spherical and concave. The distance between the central part of the convex and concave surfaces was 0.2 mm. The PVA solution in the mould was allowed to stand for 1 h in a refrigerator kept at −20°C to effect crystallization of PVA. The resulting PVA gel lens was removed from the mould and the organic solvent in the gel lens was exchanged with water by immersing in water for 3 days at room temperature. The base curve and diameter of the PVA contact lens were 8.0 and 14 mm, respectively.

Animal tests. The PVA hydrogel contact lens was applied to albino rabbit eyes continuously for a period of 3 months. After PVA hydrogel contact lens were fitted to the rabbit eyes, one-third of the on-nose part of rabbit eyelid was sutured. Slit-lamp biomicroscopic examination was performed after extended wear for 12 weeks. For each examination, light microscopy was carried out following enucleation of the eyes, which were fixed in 20% formaldehyde and stained with hematoxylin and eosin (HE). At the 12th week after application of the lens, the enucleated eyes were fixed in 2.5% glutaraldehyde and then processed for scanning electron microscopy (SEM) using the standard technique. The central corneal thickness was evaluated *in vivo* using an ultrasonic pachymeter. The measurement was made before insertion of the contact lens and 1, 2, 3, 4, 5, and 6 weeks thereafter. Four readings were made for each eye, and the mean values were recorded and compared with those from the rabbits without PVA lens.

RESULTS

In vitro *evaluation*

The gelation of PVA solutions in mixed solvents consisting of water and a water-miscible organic solvent have been previously reported in detail [9, 10]. Figure 1 represents the effect of DMSO concentration in the mixed solvent on the tensile strength and elongation of the PVA hydrogels crystallized at $-20°C$ for 24 h. Clearly, the mixing ratio of water–DMSO has a significant effect on the tensile properties of the resulting PVA hydrogel membranes. When prepared from a mixed

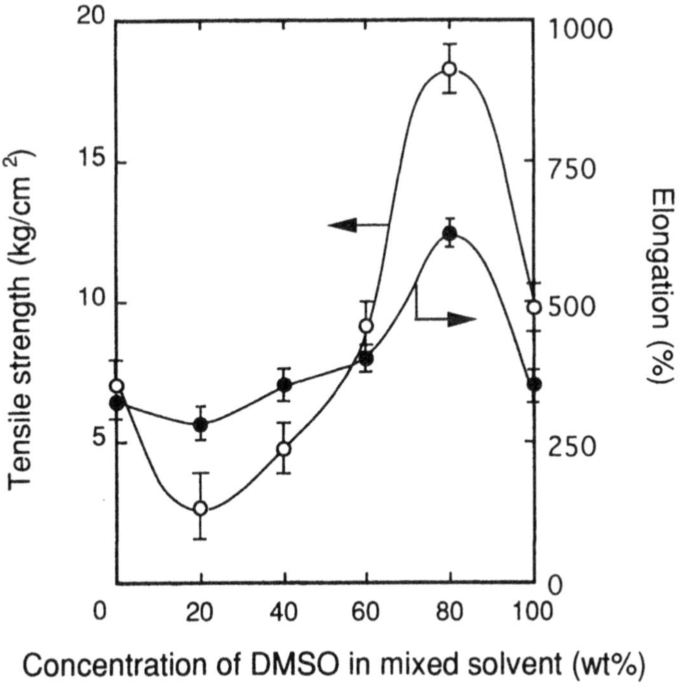

Tensile strength (kg/cm 2)
20
15
10
5
0
20
40
60
80
100
1000
750
500
250
0
Elongation (%)
Concentration of DMSO in mixed solvent (wt%)

Figure 1. Effect of the DMSO concentration on tensile strength and elongation of the PVA hydrogels crystallized from 12 wt.% PVA solution at $-20°C$ for 24 h.

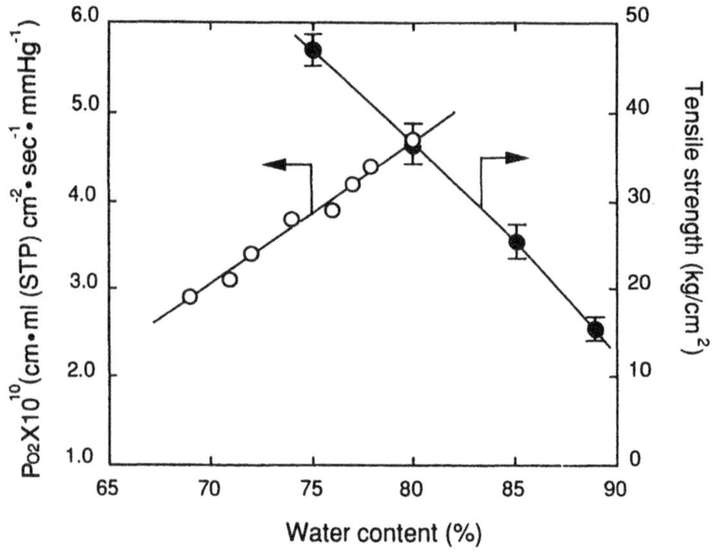

P_{O2}X10^{10}(cm•ml (STP) cm^{-2}• sec^{-1}• mmHg^{-1})
6.0
5.0
4.0
3.0
2.0
1.0
50
40
30
20
10
0
Tensile strength (kg/cm^2)
65
70
75
80
85
90
Water content (%)

Figure 2. Effect of the water content on the oxygen permeability coefficient and tensile strength of PVA hydrogels at 35°C.

water–DMSO solvent at a weight ratio of 2 : 8, the gel membranes showed the highest tensile strength and the highest elongation at break. The mechanical properties of the hydrogels could be further improved by irradiation in a deaerated atmosphere with high energy radiation [11]. Furthermore, the tensile strength of PVA hydrogel membranes could be easily controlled by the water content of PVA hydrogel. The oxygen permeability and tensile strength of PVA hydrogel with different water contents is shown in Fig. 2. As seen in Fig. 2, the oxygen permeability and tensile strength depends greatly on the water content of PVA hydrogels. However, the oxygen permeability exhibits the opposite tendency to tensile strength. The oxygen permeability decreases with lowering water content, while the tensile strength increases with a lower water content.

Table 1 compares physical properties of PVA hydrogel membrane with those of PHEMA and MMA–VP copolymer which are currently being applied clinically as

Table 1.
Comparison of physical properties of PVA hydrogel with other contact lens materials

	PVA	MMA–VP copolymer (Breath-O®)	PHEMA (Hydron®)
Water content (wt.%)	78	78	38
Tensile strength (kg/cm^2)	47	19	10
Elongation (%)	500	160	160
Light transmittance (%)[a]	99–100	99–100	99–100
O_2 permeability[b]	4.4	4.6	1.0

[a] 550 nm, 0.2 mm thick, in water.
[b] 10^{-10} cm·ml(STP)·cm^{-2}·s^{-1}·mmHg^{-1} (35°C).

soft contact lenses. In general, the strength of hydrogel will vary inversely with the water content. However, as Table 1 indicates, PHEMA has relatively low strength although it has a water content as low as 38%. The MMA–VP copolymer is superior in terms of water content, but has only marginally greater strength. In contrast, the PVA hydrogel membrane has a tensile strength approaching 50 kg/cm^2—five times stronger than PHEMA—although the water content is as high as 78%. Oxygen permeability increases simply with the water content, indicating that the oxygen permeabilities of both MMA–VP copolymer and PVA are superior to that of PHEMA.

To study protein adsorption to the PVA hydrogel material, we measured adsorption of three proteins, serum albumin, γ-globulin, and lysozyme. These proteins were used as an artificial tear solution component [12]. The amount of proteins adsorbed to the PVA hydrogel increased with increasing concentration of protein in solutions [13]. Therefore, protein adsorption was conducted at concentrations of 1.0, 3.0, and 1.5 mg/ml for IgG, BSA and lysozyme, respectively. Table 2 gives the results, together with those of commercially available soft contact lenses. Clearly, the PVA hydrogel contact lens exhibits very low protein adsorption, whereas protein adsorption to the MMA–VP copolymer is 10–40 times as high as that of the PVA contact lens. It may be expected that our PVA hydrogel will remain relatively stable and resistant to protein staining over a prolonged period of time.

Table 2.
Protein adsorption to PVA and conventional soft contact lenses (n = 4]

Lens	Water content (wt.%)	IgG ($\mu g/cm^2$)	BSA ($\mu g/cm^2$)	Lysozyme ($\mu g/cm^2$)
PVA	78	0.074 ± 0.006	0.005 ± 0.001	0.195 ± 0.014
MMA–VP copolymer (Breath-O®)	78	0.889 ± 0.011	0.169 ± 0.021	4.991 ± 0.165
PHEMA (Hydron®)	38	0.271 ± 0.013	0.037 ± 0.004	0.230 ± 0.016

In vivo *evaluation*

Many studies have shown that corneal edema results from the overnight wear of rigid contact lenses by humans [14–18]. Figure 3 shows a slit-lamp microscopic photograph of a rabbit eye on which the PVA hydrogel contact lens resided for 3 weeks. This slit-lamp biomicroscopic examination reveals no abnormal findings such as conjunctival hyperemia, corneal edema, corneal neovascularization, and lens pollution during the experimental period. It was usually noticed that the ultrastructural changes of the cornea were related to the wearing duration and the oxygen transmissibility of the contact lens worn [19–22]. Table 3 shows the thickness of cornea which wore the PVA hydrogel contact lens continuously for a period of 3 months, together with that of non-wearing eyes. Normal corneal thickness is about 370 μm. No difference in corneal thickness was noted between the non-wearing and the wearing eyes. The histological section of the rabbits cornea after extended wear for 3, 6, and 9 weeks are shown in Fig. 4. This histological study again shows no

Table 3.
Change in central cornea thickness[a] of rabbit eyes after extended wear of PVA lens (μm)

	Wearing days								
	Initial	1	3	7	14	21	28	35	42
Wearing eye ($n = 4$)	374 ± 11	377 ± 12	376 ± 16	374 ± 15	374 ± 13	378 ± 10	373 ± 17	372 ± 20	376 ± 21
Non-wearing eye ($n = 4$)	373 ± 13	373 ± 10	376 ± 11	372 ± 14	373 ± 10	373 ± 16	—	—	—

[a] Measured with ultrasonic pachymeter 200.

difference in thickness of the corneal epithelium and cell arrangement between the non-wearing and the wearing eyes. SEM photographs of the corneal epithelium of the eyes which wore the PVA contact lens are given in Fig. 5. As is evident, the corneal epithelium of the eyes which wore the PVA lens for 12 weeks shows no abnormal tissue and no difference between the wearing and the non-wearing eye.

DISCUSSION

For the preparation of PVA hydrogel there have been reported a number of methods which do not use any additives while keeping the gel water content high [23–26]. Table 4 lists the reported methods of preparing the PVA hydrogels with high water content and high mechanical strength. However, all of the PVA hydrogels prepared are optically too opaque or translucent to be used as contact lens material. In contrast, our PVA hydrogel prepared by dissolving PVA in a mixed solvent consisting of water and organic solvent, followed by low temperature crystallization of concentrated PVA solution, is optically clear [9, 10]. Testing a variety of organic solvents has disclosed that DMSO gives the most transparent hydrogel with high water content and high mechanical strength. This transparent hydrogel contains neither additives nor chemical crosslinking agents, but retains the water-swollen state through microcrystalline domains formed by hydrogen bonding between the hydroxyl groups of PVA chains.

The water content of the newly developed PVA hydrogel contact lens is 78% ± 1, which is the same as the cornea, and the gel tensile strength is 50 kg/cm^2 which is about three times as high as that of a commercially available soft contact lens. The PVA contact lens may be used for long-term wear because of its high oxygen permeability. Moreover, the lysozyme adsorption test revealed that the protein adsorbed to the PVA lens was of the order of 0.2 μm/cm^2, which is 1/2 to 1/30 that of conventional soft contact lenses. The animal test showed no abnormality after continuous wear for 12 weeks.

Table 4.
Reported preparation methods of PVA hydrogel with high water content and high mechanical strength

1. Freezing of highly concentrated aqueous PVA solution [23].
2. Partial freeze-drying of aqueous PVA solution under vacuum [24].
3. Repeated freezing and defreezing of concentrated aqueous PVA solution [25].
4. Low temperature crystallization of aqueous PVA solution [26].

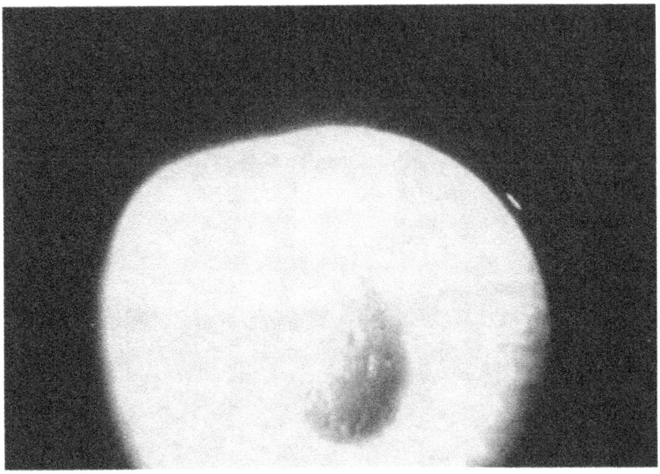

Figure 3. Slit-lamp microscopic photograph of rabbit eye after extended wear of PVA hydrogel contact lens for 3 weeks.

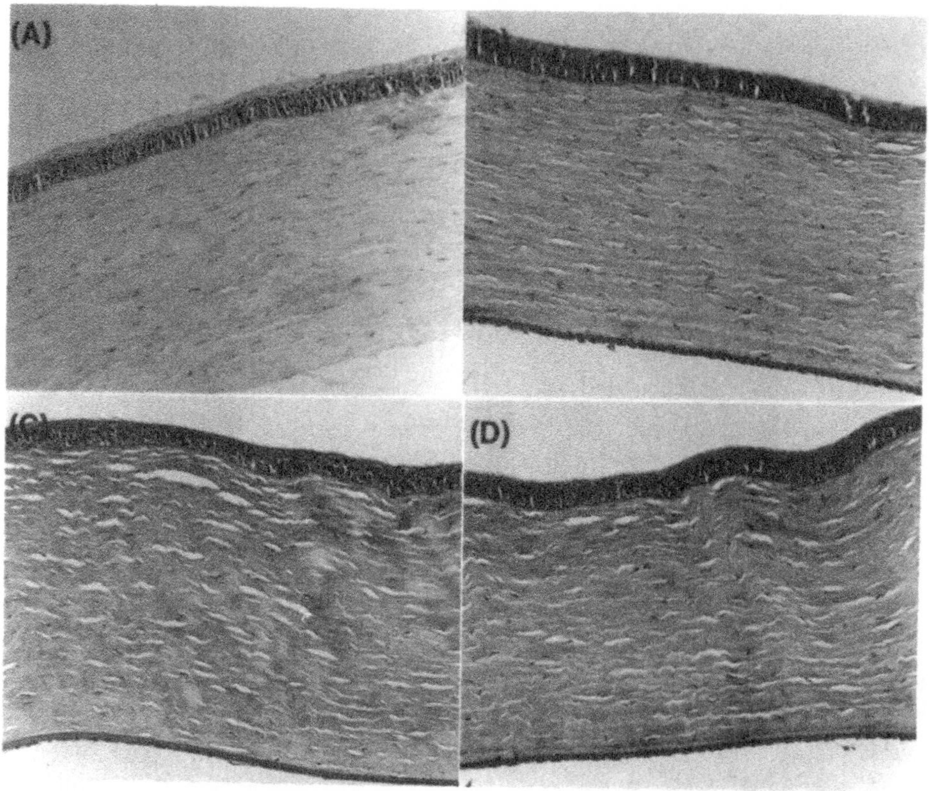
(A)
(C)
(D)

Figure 4. Histological section of rabbit cornea stained with HE after extended wear of PVA hydrogel contact lens: (A) control; (B) 3 weeks wear; (C) 6 weeks wear; and (D) 9 weeks wear.

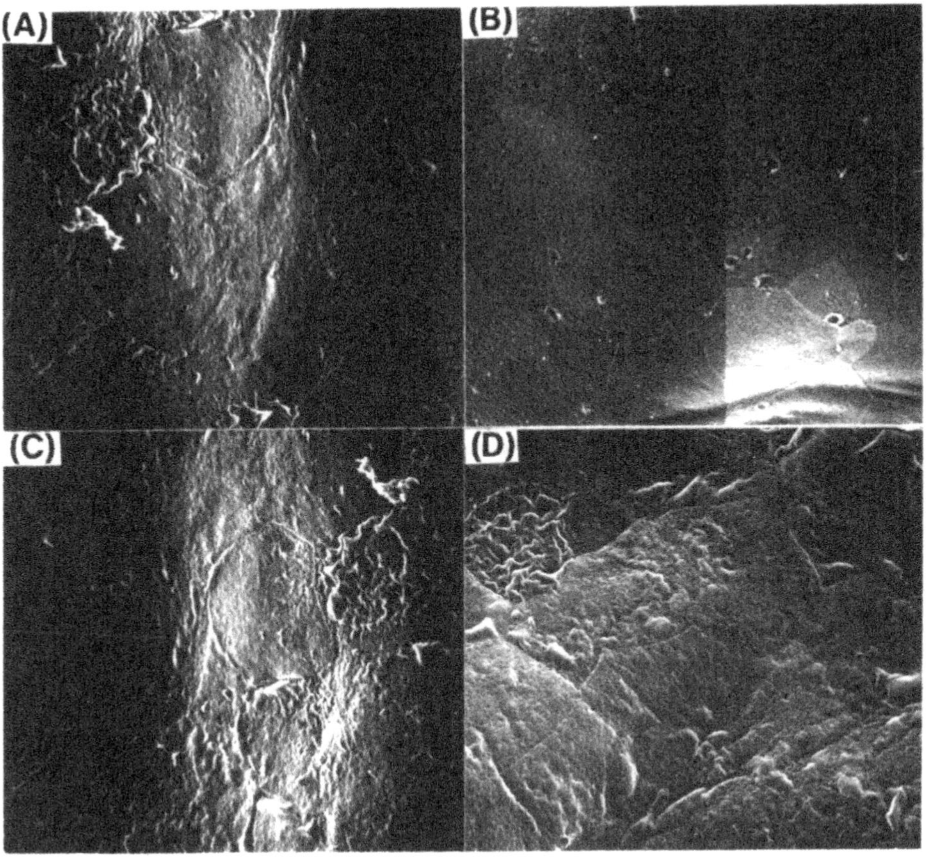
(A)
(B)
(C)
(D)

Figure 5. SEM photographs of rabbit corneal epithelium after extended wear of PVA hydrogel contact lens: (A) control; (B) 6 weeks wear; (C) 9 weeks wear; and (D) 12 weeks wear.

Formation of the PVA hydrogel with excellent transparency and high oxygen permeability as well as high mechanical strength and high water content may be explained in terms of microcrystalline structure of the hydrogel. The solution obtained by dissolving PVA in the mixed solvent consisting of water–DMSO (2 : 8) is unlikely to be molecularly homogeneous. As the temperature of this solution is lowered, the molecular motion of PVA chains must become restricted to some extent and localized regions of high segment density may be produced in solution. Cooling the PVA solution below −20°C may promote formation of small crystalline nuclei and simultaneously crystallization may take place with the lapse of time. The resulting microcrystalline regions must act as physical crosslinks of a three-dimensional network structure, providing a gel with high mechanical strength, high water content, and high transparency.

REFERENCES

1. R. P. Burns, H. Roberts and L. F. Rich, *Am. J. Ophthal.* **71**, 486 (1971).
2. S.-H. Hyon, *Sen-i Gakkaishi* **47**, 156 (1991).
3. N. J. Bailey, in: *Soft Contact Lens*, p. 61, H. E. Kaufman (Ed.). C. V. Mosby Co., St Louis, MO (1972).

4. H. Hosaka, A. Yamada, H. Tanzawa, T. Momose and A. Nakajima, *J. Biomed. Mater. Res.* **14**, 557 (1980).
5. P. I. Lee, U.S. Patent, 4,559,186 (1985).
6. P.I. Lee, U.S. Patent 372,893 (1982).
7. F. C. Greenwood, W. M. Hunter and J. S. Glover, *Biochem. J.* **89**, 114 (1963).
8. S. Hosaka, Y. Adachi and H. Tanzawa, *Maku* **5**, 247 (1980).
9. S.-H. Hyon, W.-I. Cha and Y. Ikada, *Polym. Bull.* **22**, 119 (1989).
10. W.-I. Cha, S.-H. Hyon and Y. Ikada, *Makromol. Chem.* **193**, 1913 (1992).
11. S.-H. Hyon and Y. Ikada, *Proceedings of the 6th Symposium on Radiation Chemistry*, p. 657 (1986).
12. J. L. Bohnert, T. A. Horbett, B. D. Ratner and F. H. Roycet, *Invest. Ophthalmol. Vis. Sci.* **29**, 362 (1988).
13. W.-I. Cha, S.-H. Hyon and Y. Ikada, *Kobunshi Ronbunshu* **48**, 425 (1991).
14. M. D. Sarver, D. A. Baggett, M. G. Harris and K. Louie, *Am. J. Optom. Physiol Optics* **58**, 386 (1981).
15. B. A. Holden, G. W. Mertz and J. J. McNally, *Invest. Ophthalmol. Vis. Sci.* **24**, 218 (1983).
16. M. R. O'Neal, K. A. Polse and M. D. Sarver, *Invest. Ophthalmol. Vis. Sci.* **25**, 837 (1984).
17. B. A. Holden and G. W. Mertz, *Invest. Ophthalmol. Vis. Sci.* **25**, 1161 (1984).
18. E. Kenyon, K. A. Polse and M. R. O'Neal, *CLAO J.* **11**, 119 (1985).
19. J. P. G. Bergmanson and L. W.-F. Chu, *Br. J. Pphthalmol.* **66**, 667 (1982).
20. J. P. G. Bergmanson, C. M. Ruben and L. W.-F. Chu, *Br. J. Opthalmol.* **69**, 373 (1985).
21. H. Ichijima, D. L. Mackeen, H. Hamano, J. V. Jester and H. D. Cavanagh, *CLAO J.* **15**, 290 (1989).
22. H. Ijchijima, W. M. Petroll, J. V. Jester, J. Ohashi and H. D. Cavanagh, *Cornea* **11**, 282 (1992).
23. K. Shibatani, *Polym. J.* **1**, 348 (1970).
24. K. Nishinari, M. Watase, K. Ogino and M. Nambu, *Polym. Commun.* **24**, 345 (1983).
25. M. Watase, *J. Chem. Soc. Japan* No. 9, 1254 (1983).
26. S.-H. Hyon, W.-I. Cha and Y. Ikada, *Kobunshi Ronbunshu* **46**, 673–680 (1989).
27. S.-H. Hyon and Y. Ikada, U.S. Patent 4,874,562 (1989).
28. S.-H. Hyon, W.-I. Cha, Y. Ikada, M. Kita, U. Ogura and Y. Honda, *Proceedings of the Third World Biomaterials Congress*, Vol. 3, 536 (1988).

Polyimide–polyethylene glycol block copolymers: Synthesis, characterization, and initial evaluation as a biomaterial

CHANDRASHEKHAR P. PATHAK†, AMARPREET S. SAWHNEY†, CHRISTOPHER P. QUINN and JEFFREY A. HUBBELL‡

Department of Chemical Engineering, University of Texas at Austin, Austin, TX 78712-1062, USA

Received 8 April 1993; accepted 13 August 1993

Abstract—Block copolyimides with varying amounts of polyethylene glycol (PEG) were synthesized and characterized by copolymerization of diaminodiphenyl ether (DDE), amino terminated PEG, and benzophenone tetracarboxylic acid dianhydride (BTDA). Strong materials were obtained, with enhanced flexibility as compared to the parent DDE–BTDA polyimide homopolymer. Incorporation of PEG led to an increase in water absorption by these copolymers, and hydrophilicity was increased as reflected by a decrease in air-water-polymer contact angle. These materials supported less cell adhesion *in vitro* than the parent polyimide homopolymer. Short term *in vivo* evaluation of these copolymers showed reduced fibrous encapsulation than was observed in the absence of PEG.

Key words: Polyimide; poly(ethylene glycol); copolymer; biocompatibility.

INTRODUCTION

There exists a need for biomaterials that are strong, flexible, and resistant to degradation in the body for use in device components such as heart valve leaflets and pacemaker leads. Polyurethanes are an important class of polymers that, at least in part, have met and continue to meet this need. These polymers are composed of a polyether or polyester soft segment, often a poly(tetramethylene oxide), and a hard segment which usually contains an aromatic component. Elastomeric properties are conferred upon these materials by the phase incompatibility of the hard and soft segments, resulting in materials where the semicrystalline hard segment domains provide physical crosslinks between the relatively flexible soft segment domains [1]. The lowest temperature of operation of these materials is determined by the glass transition temperature of the soft segment, and the highest temperature of operation is determined by the softening temperature of the hard segments.

One important limitation that has been observed with polyurethanes is that they are relatively unstable to enzymatic, hydrolytic, and oxidative degradation within the body. For example, superoxides and peroxides released by inflammatory cells, such as macrophages, polymorphonuclear neutrophils, lymphocytes, and foreign body giant cells, are known to be involved in polyurethane degradation [2–7]. In materials where an ester linkage was present on the polymer backbone along with the urethane linkage, the ester link was identified as the site of chain degradation [8].

† Present address: Focal, Inc., One Kendall Square, Building 600, Cambridge, MA 02139, USA.

‡ To whom correspondence should be addressed.

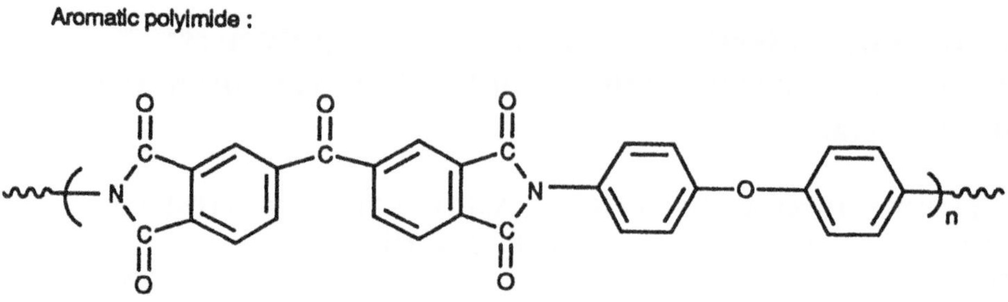
Aromatic polyimide :

Figure 1. Typical structure of aromatic polyimides.

In the present report, we explore the biomaterial properties obtained by substitution of a urethane linkage by a more stable aromatic imide linkage. The absence of an ester linkage along the polymer chain may help in improving the chemical stability of these polymers. Elastomeric properties should be attainable, using the polyimide block as a hard segment, and a polyether block as a soft segment.

Aromatic polyimides have the general structure shown in Fig. 1. These materials have been documented as having excellent mechanical, electrical, and thermal properties and have been widely used in many demanding applications such as in fabrication of thermally stable materials in microelectronics, as a photoresist in photolithography, and as a heat resistant adhesive [9, 10]. Many properties, such as mechanical strength and electrical properties, of polyimides are attributed to the rigid aromatic backbone and semi-ladder structure. The resonance stabilization of aromatic rings gives additional thermal and chemical stability. Recently, the biocompatibility of commercially available polyimides was examined in relation to sensory and neural structures of the cochlea, and was found to be comparable to that of biomaterials such as Silastic and Teflon [11].

With the aim of further improving the biocompatibility of these mechanically resilient and chemically inert materials, to make them more biologically inert, we examined the incorporation of PEG into these materials. Copolymers of a polyether, poly(tetramethylene oxide), and polyimides have been synthesized by Yu *et al.* [12], who found them to have excellent elastomeric properties. They did not, however, report upon the biocompatibility of these materials. PEG has been shown to have excellent biocompatibility due to hydrophilicity, chain mobility and lack of ionic charge [13, 14]. When PEG is immobilized upon a surface, either chemically or physically, it leads to surfaces which are more resistant to protein adsorption than the parent surface [15]. Cell adhesion to a surface is mediated by cell adhesion proteins, such as fibrinogen, fibronectin, and vitronectin, which can be immobilized upon the surface through adsorption [16]. A reduction in protein adsorption makes the surface more cell non-adhesive, which in some cases improves biocompatibility. Thus, PEG-containing materials have been shown, in several types of materials, to be more biologically inert than the parent unmodified materials. For example, Sawhney and Hubbell reported a decrease in pericapsular fibrous overgrowth of alginate-poly(l-lysine) microcapsules by using a graft copolymer of poly(l-lysine) and PEG [17]. Desai and Hubbell reported a decreased intraperitoneal reaction to poly(ethylene terephthalate), which had PEG immobilized on the surface [18]. Liu *et al.* found polyurethanes that had PEG chains immobilized along the backbone, or present as grafted chains, to be more non-thrombogenic than those lacking PEG [19].

We incorporated these biomedically important features of PEG into polyimides by synthesizing block copolymers of PEG and polyimides. The copolymers of polyimides and PEG were characterized by FT-IR and ^{1}H NMR spectroscopy. The increase in hydrophilicity was measured using swelling and contact angles under water. Thermal transitions and stability were evaluated using DSC and TGA. Preliminary evaluation as a biomaterial was carried out using cell culture and short term intraperitoneal implantation in mice.

MATERIALS AND METHODS

Materials

Diaminodiphenyl ether (DDE) (Fluka Ronkonkoma, NY) was recrystallized from ethanol. α, ω-diamine terminated PEG with average molecular weight 2000 (PEG 2K diamine) (amine equivalent = 88 mEq g^{-1}; Scientific Polymer Products Ontario, NY) was dried using azeotropic distillation with benzene. Benzophenone tetracarboxylic acid dianhydride (BTDA) (Fluka) was purified using vacuum sublimation. Dimethyl formamide (DMF) (Aldrich Milwaukee, WI) was purified by vacuum distillation prior to use. All other solvents and chemicals were of reagent grade and were used without purification.

Synthesis of polyamic acids

Several polyamic acids were synthesized, as indicated in Table 1. As an example, the synthesis of a polyamic acid with 50% PEG 2K diamine is detailed below. A 100-ml round bottom flask was flame dried and cooled under argon. Under a dry argon atmosphere, 0.617 g of BTDA, 1 g of PEG 2K diamine, 0.383 g DDE, and 10 ml of DMF were added to this flask. The reaction mixture was stirred under argon for 10 h at room temperature. The polyamic acid solution was precipitated using dry xylene. It was purified by redissolving in DMF and reprecipitating in anhydrous diethyl ether. The product was then dried under vacuum for 24 h at 70°C. The compositions of other polymers synthesized are given in Table 1, based on comonomer feed ratios.

Table 1.
Nomenclature, composition, and appearance of polymers synthesized

Sample no.	% BTDA[a]	% DDE	% PEG	% of diamine as PEG	Appearance
P 62/38/0	62	38	0	0	Hard solid
P 43/23/34	43	23	34	59	Hard, tough solid
P 32/16/52	32	16	52	76	Hard, tough solid
P 26/11/63	26	11	63	85	Tough solid
P 21/6/73	21	6	73	92	Gummy solid
P 12/0/88	12	0	88	100	Gummy solid

[a] Comonomer compositions are shown.

Conversion of polyamic acid into polyimide

A solution of the subject polyamic acid (10–20% w/v in DMF) was cast onto clean, dry glass sheets to give a thin layer of solution. The films were allowed to air-dry overnight and were further dried under vacuum to obtain polyamic acid films. These films were heated under an argon atmosphere to 200°C for 2 h. This resulted in complete conversion of the polyamic acid to a polyimide. The conversion was followed by FT-IR spectroscopy. Figure 2 shows the entire synthetic scheme.

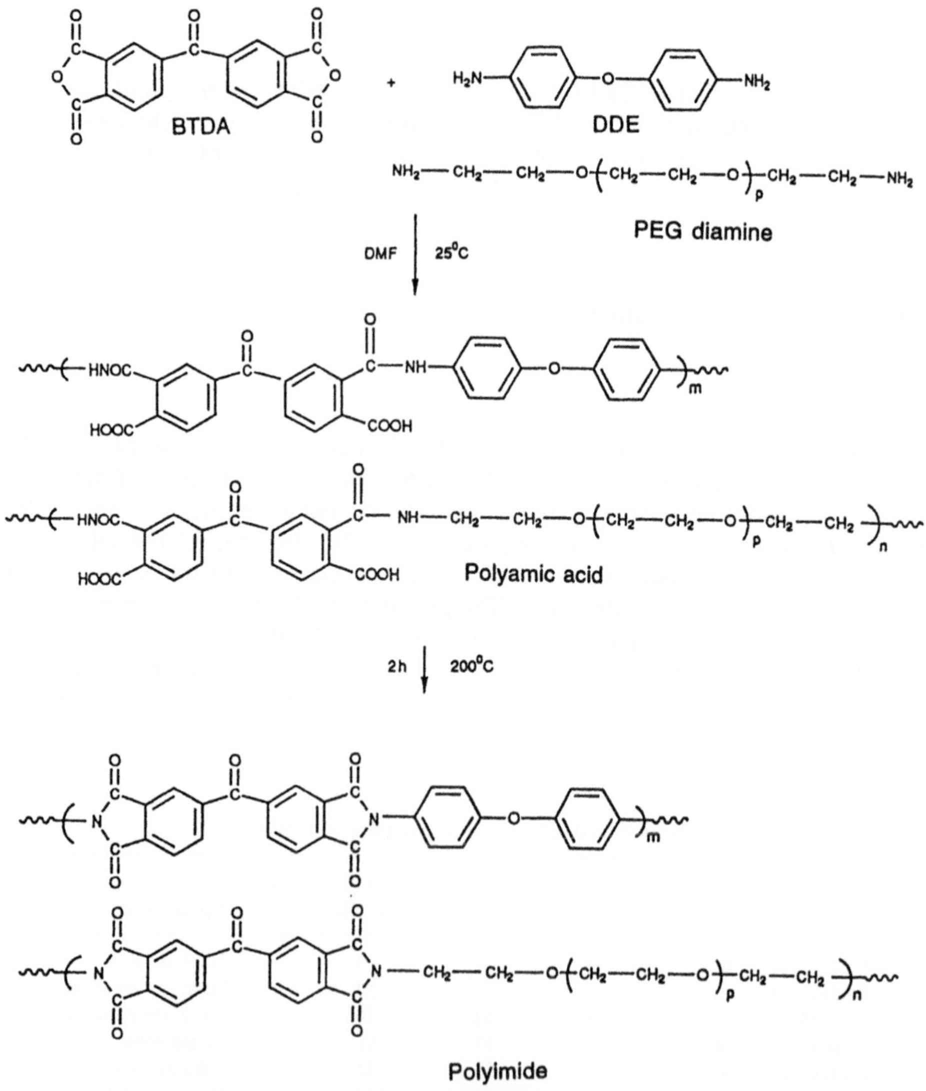
BTDA
DDE
PEG diamine
DMF
25°C
Polyamic acid
2h
200°C
Polyimide

Figure 2. Reaction sequence for the synthesis of polyimide of benzophenone tetracarboxylic acid dianyhydride (BTDA) and diaminodiphenyl ether (DDE) and their copolymers with PEG diamine.

Nomenclature

The various polyimides and their copolymers with PEG were named according to the percentages of BTDA/DDE/PEG contained in the comonomer compositions therein. For example a copolymer containing 43% of BTDA, 23% of DDE, and 34% of PEG comonomer was called P 43/23/34.

Spectral and thermal characterization

FT-IR spectra were measured using a Digilab FTS 15/90 FT-IR spectrophotometer. ^{1}H NMR spectra were measured using a Nicolet NT-360 instrument in deuterated DMSO.

Thermal transitions in the polymers were recorded using a Perkin Elmer DSC-7 system. Samples were equilibrated in deionized water for 48 h prior to analysis. Excess water was carefully swabbed from the surface of these samples and a heating rate of 20°C min^{-1} was used. Thermal stability was determined using a Perkin Elmer TGA-7 system. Samples were prepared as for DSC, and a heating rate of 20°C min^{-1} was used.

Hydration and contact angles

Polymer films of 1 cm × 1 cm in size were hydrated in deionized water for 48 h. The excess water was carefully removed and the hydrated weight of the films was determined (W_1). These films were then dried in a vacuum oven at 60°C for 2 days. At the end of this the weight of the films was again determined (W_2). The percentage of water content of the materials was determined as $(W_1\text{-}W_2)/W_1 \times 100$.

Polymer–air contact angles were measured under deionized water using a stagnant bubble technique. A custom built apparatus was used for these measurements. The air bubble was introduced using a microsyringe on to the inverted surface of the sample film, and the contact angle was calculated from the height and width of the bubble, as measured with a travelling stage and micrometers. The films were hydrated in deionized water for 48 h prior to contact angle measurement.

In vitro *cell adhesion*

The ability of these polymers to resist cell adhesion was evaluated using *in vitro* culture of human foreskin fibroblasts (HFF). HFF cells were obtained from neonatal foreskins and routinely cultured in Dulbecco's Modification of Eagle's Medium supplemented with 10% fetal bovine serum and 1% antibiotic/antimycotic solution at 37°C in a 5% CO_2 environment. The cells were passaged every week and were harvested using trypsin. Cells between the third and twentieth passage were used. Films of homo- and copolyimides of 1 cm × 1 cm in size were sterilized using 70% ethanol. After thorough rinsing in sterile deionized water, HFF cells were seeded at a density of 30 000 cells cm^{-2}. At the end of 24 h, the films were gently rinsed with phosphate buffered saline (PBS, pH 7.4) to remove non-adherent cells. The number of adherent cells and their morphology were recorded using phase contrast microscopy. Five fields, at predetermined locations, were examined at a magnification of × 100 to count the number of adherent cells.

In vivo *biocompatibility*

The ability of these polymers to resist fibrous overgrowth was evaluated by intraperitoneal implantation in mice. *In vivo* biocompatibility of P 62/38/0, P 43/23/34, P 32/16/52, and P 26/11/63 was evaluated. ICR swiss male mice, 6–8 weeks old (Sprague-Dawley) were anesthetized by intraperitoneal injection of 6 mg pentobarbital/100 g body weight. The abdomen was prepared using Betadine and a midline incision was made. Two polyimide films of size 1–2 cm^2 were inserted into the peritoneal cavity through this incision. Two animals were used per material. All the implanted samples were equilibrated in PBS for 48 h before implantation and were sterilized by immersion in 70% ethanol solution. The samples were rinsed thoroughly with sterile PBS before implantation. The animals were sacrificed by carbon dioxide asphyxiation 10 days postoperatively and samples were recovered, fixed in 2% glutaraldehyde, and subjected to histological analysis. Five-micron sections were cut after the films had been embedded in wax and were stained using hematoxylin and eosin. The sections were observed under bright field illumination at a magnification of × 200. Three typical fields were selected and the type and number of cells were enumerated and an average calculated. The histological evaluation was carried out by an observer who was blinded to the identity of the samples.

RESULTS AND DISCUSSION

The comonomer compositions of the various polyimides synthesized and evaluated are shown in Table 1. Sample P 62/38/0, containing no PEG, represents the hard segment model, i.e. the parent DDE–BTDA polyimide homopolymer, where the only diamine is DDE. Sample P 12/0/88 represents the soft segment model, i.e. the polyimide resulting when the only diamine is PEG 2K diamine. The remaining polyimides represent materials with 34, 52, 63, or 73% PEG 2K diamine, where the percentage of the diamine being PEG 2K diamine is 59, 76, 85, or 92%, respectively.

The PEG-containing copolymers that had greater than 73% PEG content by total weight, or greater than 92% of the diamine being PEG 2K diamine, were gummy and could not form films. These two compositions, P 12/0/88 and P 21/6/73, were not used for further studies. The remaining PEG-containing polyimide copolymers exhibited toughness, presumably due to the elastomeric nature of these materials. The homopolymeric polyimide was found to be a brittle hard solid. It lacks an elastomeric nature since the polyether segment used (DDE) has bulky phenyl groups that give it a rigid structure. The presence of the PEG chains lends to free rotation and gives improved toughness to the copolymeric polyimides.

Spectral and thermal characterization

The conversion of the polyamic acid to the polyimide was confirmed by infrared spectroscopy by monitoring the disappearance of the NH stretch at 3400 cm^{-1} and the appearance of an imide CN stretch at 1780 cm^{-1}. The presence of PEG in the copolymers was confirmed by a strong band for the ether stretch at 1110 cm^{-1}. The completion of imidization was confirmed by the absence of a NH bond at

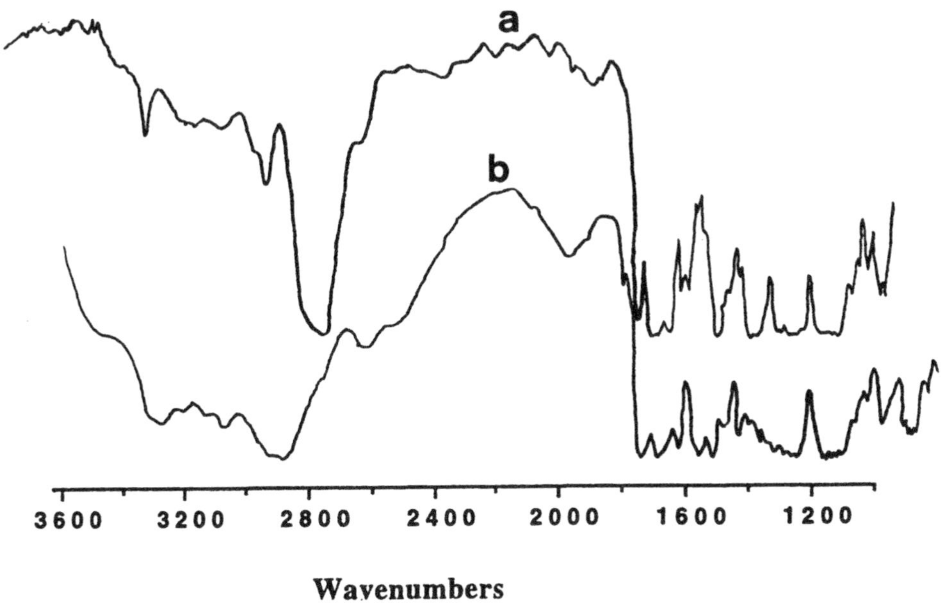
a
b
3600
3200
2800
2400
2000
1600
1200
Wavenumbers

Figure 3. (a) FT-IR spectrum of copolymeric polyimide containing 52% PEG 2K diamine (P 32/16/52). (b) FT-IR spectrum of copolymeric polyamic acid containing 52% PEG 2K diamine (precursor to P 32/16/52) before conversion into the polyimide.

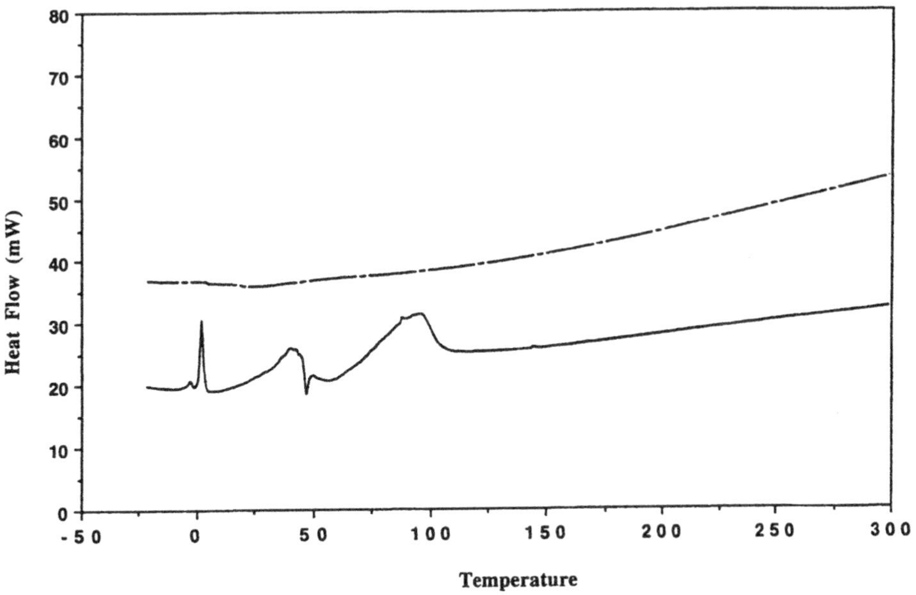
80
70
60
50
40
30
20
10
0
Heat Flow (mW)
-50
0
50
100
150
200
250
300
Temperature

Figure 4. Differential scanning calorimetry thermogram of homopolymeric polyimide (P 62/38/0; upper thermogram) and PEG-containing polyimide with 34% by mass PEG 2K diamine (P 43/23/34; lower thermogram). The samples were equilibrated in water prior to scanning. The endotherm at 6°C is due to water associated with the PEG segments, the endotherm at 60°C is due to the melting of PEG-rich domains, and the endotherm at 95°C is due to water evaporation.

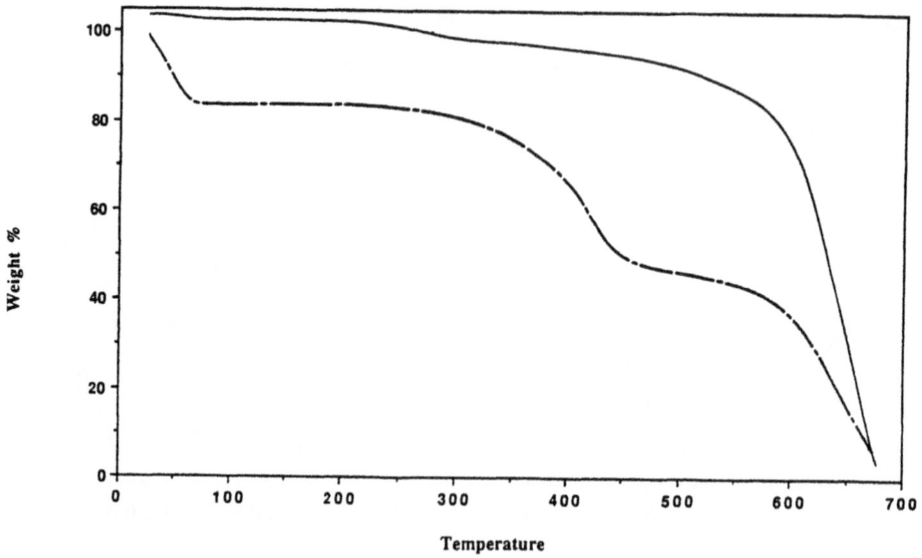
Weight %
100
80
60
40
20
0
0
100
200
300
400
500
600
700
Temperature

Figure 5. Thermogravimetric analysis of homopolymeric polyimide (P 62/38/0; upper thermogram) and PEG-containing polyimide with 34% by mass PEG 2K diamine (P 43/23/34; lower thermogram). The samples were equilibrated in water prior to scanning. The initial weight loss in the PEG-containing polyimide is due to loss of water of hydration. The PEG-containing polyimide was stable to approximately 350°C, while the homopolymer was stable to approximately 550°C.

3400 cm^{-1} in the IR spectrum of P 32/16/52, shown after imidization in Fig. 3a and before imidization in Fig. 3b. Appearance of a CN band 1380 cm^{-1} and an imide carbonyl at 1780 cm^{-1} also confirmed the imidization. The ^1H NMR spectra of P 32/16/52 polyamic acid in DMSO showed a strong peak at 3.4 ppm due to oxyethylene protons, which also confirms the presence of PEG in the copolymers (spectra not shown).

A typical DSC thermogram of samples of homopolymeric polyimide (P 62/38/0) and copolyimide P 43/23/34, which had both been previously equilibrated in water, is shown in Fig. 4. The P 43/23/34 showed three main transitions. The first transition around 6°C was attributed to the free water associated with the oxyethylene segment. A sharp exothermic crystallization transition followed by an endotherm at 60°C was due to melting of PEG segments, demonstrating phase separation. The endotherm around 95°C was due to water evaporation. The PEG-containing polymers did not show any non-PEG polymer melting endotherm up to 300°C. No transitions were observed in the P 62/38/0 polymer up to 300°C.

The thermal stability of these copolymers was determined using thermogravimetric analysis. Typical TGA analyses of P 62/38/0 and P 43/23/34, both of which had been previously equilibrated in water, are shown in Fig. 5. The P 43/23/34 showed 20% weight loss up to 100°C due to loss of water. No further weight loss or decomposition was observed in this copolymer up to 300°C, indicating good thermal stability. However, the presence of PEG in the backbone makes these polymers less thermally stable as compared to the homopolyimide P 62/38/0, which was stable up to 550°C. However, it is still remarkable that these elastomers are stable up to temperatures around 300°C in an air atmosphere.

Table 2.
Extent of hydration and contact angles of polyimide homopolymeric and PEG-containing copolymers

Sample no.	Contact angle[a]	Hydration[a]
P 62/38/0	67.7 ± 4.7[b]	1.64 ± 0.78[b]
P 43/23/34	34.1 ± 2.5[c]	23.9 ± 1.0
P 32/16/52	21.8 ± 2.1	24.5 ± 1.8
P 26/11/63	25.0 ± 4.1	21.4 ± 0.78

[a] Mean ± S.E.M.
[b] $p < 0.001$ compared to the PEG-containing copolymers.
[c] $p < 0.002$ compared to P 32/16/52.

Hydration and contact angles

PEG is water soluble and thus polyimides containing PEG are capable of absorbing water. The extent of water adsorption for this series of copolymers is listed in Table 2. As expected, the degree of hydration increased with incorporation of the hydrophilic PEG component. The homopolymeric polyimide (P 62/38/0) showed minimal water uptake. All the PEG-containing copolymers showed more than 20% uptake of water. However, when compared with polyurethanes containing similar amounts of PEG, polyimide and PEG copolymers absorb considerably less water [19]. This is presumably due to the high rigidity and hydrophobic nature of the aromatic polyimide backbone. This rigidity and hydrophobicity is also presumably responsible for the relatively low dependence of swelling upon the amount of incorporated PEG.

The amount of PEG in the copolymer was controlled by copolymerizing PEG 2K diamine with DDE. Measurement of the contact angles gives an indication of the relative hydrophilicity of these copolymers. A decrease in contact angles under water indicates increasing hydrophilicity of the surface. The results of contact angle measurements are shown in Table 2. The homopolyimide of DDE and BTDA was hydrophobic, and incorporation of PEG blocks led to a decrease in contact angles. PEG is a unique polymer which is soluble in water as well as in several organic solvents. In an aqueous environment, the polymer surface may rearrange to expose its polar components to minimize the interfacial free energy, assuming that mobility is not restricted. Preferential soft segment surface enrichment has been seen to occur by Kajiyama and Takahara [20] in aqueous environments. Aromatic polyimides are rigid rod polymers and have a restricted chain mobility. The PEG segments are more mobile in an aqueous environment and probably preferably orient at the water–polymer interface.

In vitro *cell adhesion*

The attachment and spreading of HFF cells on these homopolymeric and copolymeric films was followed up to 24 h. The number of fibroblasts adherent after 24 h of culture from serum-containing medium is shown in Table 3. All copolymers containing PEG showed lower cell adhesion than the homopolymeric polyimide

Table 3.
Adhesion of human foreskin fibroblasts in serum-containing medium at 24 h

Sample	Number of adherent cells[a]
P 62/38/0	1006 ± 496[b]
P 43/23/34	0 ± 0
P 32/16/52	31 ± 29
P 26/11/63	196 ± 139

[a] Mean ± S.E.M.
[b] $p<0.01$ compared to the PEG containing copolymers.

(P 62/38/0). The P 43/23/34 and P 32/16/52 films showed practically no cell adhesion as compared to P 62/38/0, which showed extensive cell attachment. The cells attached to the P 62/38/0 films were well flattened and well spread. Several cellular processes were evident in these cells. The few cells that were seen attached to the copolymeric films containing PEG were unable to attain a well spread morphology and appeared rounded.

In vivo *biocompatibility*

The measurements of *in vivo* biocompatibility, specifically the ability to resist fibrous encapsulation after intraperitoneal implantation for 7 days in mice, also indicate the effectiveness of PEG segments in the copolymers. The homopolymeric polyimide, P 62/38/0, films showed 65 ± 13% (mean ± SD) of the surface having cellular overgrowth. Histological sections revealed this overgrowth to be four to eight cells thick. Of these, 30% were identified as polymorphonuclear lymphocytes (PMNs) and 65% were identified as macrophages having a foamy, granular cytoplasm. The rest were some unidentified cells and a few foreign body giant cells. No fibroblasts or neutrophils were seen. In the P 43/23/34 polymers, 32 ± 10% of the film surface showed cellular overgrowth, which was observed by histology to be zero to three cells thick. Of these, 70% were PMNs and 25% were macrophages. No signs of a severe inflammatory reaction, such as the presence of neutrophils and foreign body giant cells, were seen. In the P 32/16/52 polymers, 30 ± 9% of the surface showed cellular overgrowth, which was seen to be zero to two cells thick. Of these, 30% were PMN and 60% were macrophages. In the P 26/11/63 films, up to 80% of the surface showed cellular overgrowth, which was seen to be one to four cells thick. Of these, 50% were PMNs and 50% were macrophages. All PEG-containing copolymers showed significant reductions in extent of cellular overgrowth *in vivo*, as measured by the thickness of the adherent cellular layers.

CONCLUSIONS

Copolymers of PEG diamine of molecular weight 2000 g mol^{-1} with a polyimide containing BTDA and DDE were synthesized. Copolymers having less than about 63% PEG by weight were capable of forming good films. The resulting copolymers were found to have a lower thermal stability compared to homopolymeric polyimide, but still exhibited superior heat stability. Thermal analysis showed phase separation into PEG-rich and PEG-poor domains, and elastomeric behavior was observed. The PEG-containing materials were found to absorb about 20% of their weight in water and to be considerably more wettable than the homopolymeric polyimide. The biological response of these materials was found to be reduced, as reflected by a decreased tendency to support cell adhesion *in vitro* and elicit a foreign body response upon short term implantation *in vivo*. These materials may warrant further exploration of usefulness as a biomaterial, to determine if the good thermal stability will result in good stability to degradation *in vivo*.

Acknowledgements

Research support from the NIH (HL-39714) is gratefully acknowledged.

REFERENCES

1. Okkema, T. G. Grasel, R. J. Zdrahala, D. D. Solomon and S. L. Cooper, *J. Biomater. Sci. Polymer Edn.* **1**, 43 (1989).
2. Q. Zhao, R. E. Marchant, J. M. Anderson and A. Hiltner, *Polymer* **28**, 2040 (1987).
3. J. G. Dillon and M. K. Hughes, *Biomaterials* **13**, 240 (1992).
4. C. Batich, J. Williams and R. King, *J. Biomed. Mater. Res.: Appl. Biomater.* **23**, 311 (1989).
5. M. Bouvier, A. S. Chawla and I. Hinberg, *J. Biomed. Mater. Res.* **25**, 773 (1991).
6. B. D. Ratner, K. W. Gladhill and T. A. Horbett, *J. Biomed. Mater. Res.* **22**, 509 (1988).
7. Q. Zhao, N. Topham, J. M. Anderson, A. Hiltner, G. Lodoen and C. R. Payet, *J. Biomed. Mater. Res.* **25**, 177 (1991).
8. C. S. Schollenberger and F. D. Stewart, *J. Elastoplastics* **3**, 28 (1971).
9. K. Mukai, A. Saiki, K. Yamanaka, S. Harada, S. Shoji, *IEEE J. Solid-State Circuits* **SC-13**, 462 (1978).
10. W. H. Morita and S. R. Graves, *Proc. Natl. SAMPLE Symp. Exhib.* **26**, 402 (1981).
11. H. S. Haggerty and H. S. Lusted, *Acta Otolaryngol (Stockh.)* **107**, 13 (1989).
12. X. Yu, S. Chunkang, C. Li and S. Cooper, *J. Appl. Polym. Sci.* **44**, 409 (1992).
13. S. Nagoaka, Y. Mori, H. Takiuchi, K. Yokota, H. Tanzawa and S. Nishiumi, in: *Polymers as Biomaterials*, p. 361, S. W. Shalaby, A. S. Hoffman, B. D. Ratner and T. A. Horbett (Eds). Plenum Press, New York (1984).
14. W. R. Gombotz, W. Guanghui and A. S. Hoffman, *J. Appl. Poly. Sci.* **37**, 91 (1989).
15. N. P. Desai and J. A. Hubbell, *Biomaterials* **12**, 144 (1991).
16. C. A. Buck and A. F. Horwitz, *Ann. Rev. Cell Biol.* **3**, 179 (1987).
17. A. S. Sawhney and J. A. Hubbell, *Biomaterials* **13**, 863 (1992).
18. N. P. Desai and J. A. Hubbell, *Biomaterials* **13**, 505 (1992).
19. S. Q. Liu, Y. Ito and Y. Imanishi, *J. Biomater. Sci. Polymer Edn.* **1**, 111 (1989).
20. T. Kajiyama and A. Takahara, *J. Biomater. Appl.* **6**, 42 (1991).

Evidence for Fickian water transport in initially glassy poly(2-hydroxyethyl methacrylate)

S. H. GEHRKE*, D. BIREN and J. J. HOPKINS

Department of Chemical Engineering, Mail Location 171, University of Cincinnati, Cincinnati, OH 45221-0171, USA

Received 3 July 1992; accepted 1 April 1994

Abstract—Water sorption which is not classically Fickian has been observed in a variety of polymers. Deviation from Fickian kinetics is widely assumed to be caused by rate-limiting polymer relaxation, despite minimal proof of this. To the contrary, the evidence accumulated in this work indicates that water transport in initially glassy poly(2-hydroxyethyl methacrylate) (PHEMA), an important water-swellable biomedical polymer, is controlled by Fickian diffusion. First of all, the fractional water uptake is initially linear and independent of sample thickness when plotted against the square root of time over initial thickness, as expected for a Fickian process. Furthermore, the moving solvent front also advanced with the square root of time. Temperature, polymer thermal history and initial solvent concentration all affected the sorption kinetics of PHEMA in manners consistent with a Fickian process. The invariably Fickian sorption mechanism is believed to be the consequence of the water molecule's small size and affinity for hydrophilic, swellable polymers.

Key words: Anomalous transport; Case II; controlled release; diffusion; hydrogel; poly(2-hydroxyethyl methacrylate); relaxation; sorption; swelling kinetics.

INTRODUCTION

The advantages of controlled drug release systems over conventional dosage forms have made development of controlled release systems one of the fastest growing areas in pharmaceutical research. When a conventional tablet or capsule is taken, the drug concentration in the bloodstream may rise rapidly to a level above the therapeutic range and later drops below this range. Maintaining drug concentrations within the therapeutic range for longer times is the main advantage of controlled release systems. Patient compliance is also improved by reducing dosing frequency.

One of the simplest controlled release systems is the monolithic device of a swellable glassy polymer containing a uniformly dispersed or dissolved drug [1]. The success of these systems is the result of the very low diffusivity of the drug in the dry glassy polymer—effectively entrapping the drug—in contrast to the fairly high diffusivity of the drug in the swollen, rubbery polymer. Also, when fluid penetrates a glassy polymer, a solvent front typically forms which separates the glassy core from the rubbery sheath, and which gradually advances towards the center. Since the drug dissolved in the solvent-swollen rubbery portion will normally diffuse quickly from the device, the overall release rate is typically controlled primarily by the rate of the solvent uptake and the rate of advance of the solvent front. If the sorption occurs by a Fickian diffusion process, the sorption and the front will

* To whom correspondence should be addressed.

advance with the square root of time, and thus a declining rate of release is expected. However, if the solvent is absorbed at a constant rate and the diffusion of the solute is much faster in the swollen polymer than the rate at which the solvent is absorbed, a constant (zero order) release rate of the solute is possible. Since such constant sorption and release rates have been demonstrated using organic solvents, there has been much interest in developing an analogous system suitable for zero-order drug release into aqueous fluid [2, 3].

In this paper we continue our studies into the possibility of developing a hydrophilic polymer which would swell at a constant rate in water [4, 5]. The model polymer studied is poly(2-hydroxyethyl methacrylate) (PHEMA), since this is one of the most common and widely studied polymers for biomedical applications, including monolithic drug release devices. Furthermore, a detailed examination of the swelling behavior of initially glassy PHEMA in water is of general value for controlled drug delivery and other biomedical applications which use swollen PHEMA.

BACKGROUND

Relexation-controlled swelling

The most widely-cited work describing constant swelling rate polymers ascribes the constancy of sorption to slow polymer relaxation [1, 3, 6–11]. Relaxation processes are typically described by an exponential decay function. Since an exponential decay is linear at short times, relaxation-controlled swelling varies linearly with time [7]. The rate of relaxation of polymer chains in response to stress depends most significantly upon whether the polymer is above or below its glass transition temperature (T_g). Below its T_g, a polymer is in the glassy state and relaxation is very slow, while above T_g, the polymer is in the rubbery state and relaxation is very fast. The presence of solvent molecules between the polymer chains usually serves as a molecular lubricant or 'plasticizer', lowering T_g by increasing the mobility of the polymer chains. Thus an initially glassy polymer may become rubbery as a result of either an increase in temperature or in solvent concentration (sorption). However, sorption of solvent by a polymer induces stress within the polymer. Thus relaxation rates of the polymer chains in response to swelling stresses may influence the rate at which solvent molecules can be incorporated within the polymer matrix; i.e. the rate of sorption.

Many models based on the concept of relaxation-controlled sorption have also been published in the literature over the last several decades [1, 3, 6–11]. Here we will briefly review two extremes of relaxation/diffusion modeling of sorption. Berens and Hopfenberg developed a simple empirical model that simply summed contributions of superimposed diffusion and relaxation processes. While effective in fitting data for the sorption of organic vapors by polystyrene and poly(vinyl chloride), its empirical nature limited its ability to interpret the underlying phenomena. One of the most widely cited models which explicitly incorporates relaxation-control in a more fundamental way is that of Thomas and Windle [3, 8]. They developed a model which can explain penetrant transport in glassy polymers using two independent parameters: the diffusivity of the penetrant in the glassy matrix and the viscosity of the solvent-free polymer. They suggested that constant-rate sorption arises from the slow deformation of the glassy matrix as a response to stresses generated by the solvent osmotic pressure. The model is able to predict a broad range of transport phenomena, and the predictions agree well with the features of the methanol sorption in glassy PMMA, but has not been fitted to a wide range of experimental data, including aqueous data. But the idea of combining relaxation and diffusion to interpret sorption of solvents by polymers is well-established in the literature.

Experimental identification of non-Fickian sorption kinetics

The most commonly used method of characterizing the kinetics of solvent sorption by polymers is to fit the first 60% of the sorption curve to the following empirical expression [1, 6, 11]:

$$M_t/M_\infty = kt^n \tag{1}$$

where M_t is the amount of solvent absorbed at time t, M_∞ is the amount of solvent absorbed at equilibrium, k is the empirical rate constant, and n is the transport exponent.

The value of n provides an indication of the transport mechanism controlling the rate of sorption. For a flat sheet, a value of $n = 0.5$ indicates that a Fickian process is rate controlling, while values of n between 0.5 and 1.0 (termed anomalous transport) are assumed to indicate the presence of rate-influencing relaxation processes. A value of $n = 1$ (termed Case II sorption) means that sorption occurs at a constant rate, which is generally assumed to imply that relaxation is the rate controlling mechanism. The term 'non-Fickian' is often applied to any sorption process which deviates from the prediction from the integrated form of Fick's law with constant boundary conditions (which does not necessarily mean that the differential form of Fick's law does not hold). This classification includes not only cases of sorption with $n > 0.5$, but also processes with $n = 0.5$ that lack other hallmarks of classical Fickian behavior, such as a monotonic, asymptotic approach toward equilibrium and dimensional scaling with the square of the characteristic sample dimension.

For a binary diffusion process, Fick's law states that the mass transfer flux is proportional to the concentration gradient; the proportionality constant is called the diffusion coefficient. The diffusion coefficient is not a constant, however, but usually varies with concentration. For a polymer sheet swelling in a solvent under conditions of constant boundary conditions, a polymer/solvent mutual diffusion coefficient D in a polymer-fixed reference frame can be extracted from data using the following equation [4–6, 11–12]:

$$M_t/M_\infty = 1 - \sum_{n=0}^{\infty} \{8/(2n + 1)^2\pi^2\} \exp\{-(2n + 1)^2\pi^2(Dt/L_0^2)\} \tag{2}$$

where L_0 is the initial thickness of the sheet. The appearance of the dimensionless time group (Dt/L_0^2) in this equation indicates that a plot of M_t/M_∞ vs (t/L_0^2) will yield a single master curve that is independent of sample dimension. The thickness dependence can aid in identification of the transport mechanism, since Case II transport will scale with t/L while anomalous transport may have a complex dimensional dependence [3, 6]. A least squares curve fit of Eq. (2) to the data can be used to obtain the value of D; alternatively, the limit of this equation at short times can be used to extract D from kinetic swelling data. The short time limit, valid for $M_t/M_\infty < 0.60$ is simply [11]:

$$M_t/M_\infty = (4/\sqrt{\pi})(Dt/L_0^2)^{0.5}. \tag{3}$$

This equation demonstrates that for a Fickian process, the sorption is initially linear and independent of thickness when M_t/M_∞ is plotted against $\sqrt{t/L_0}$, and the diffusion coefficient can be extracted from the slope.

Thus while a swelling exponent of 0.5 is an indication of Fickian kinetics, other tests are required to confirm it. Similarly, a value of $n \neq 0.5$ does not mean that the process is *not* diffusion controlled, only that the solutions to Fick's law given as Eqs (2) and (3) do not apply to the process at hand. Furthermore, many processes other than relaxation-limitations can cause the value of n extracted from Eq. (1) to be other than 0.5. Processes other than relaxation of polymer chains that can cause apparently non-Fickian sorption include: (1) non-planar geometry [13]; (2) dimensional distortions of polymer geometry due to anisotropic swelling [4–5, 14–15]; (3) external mass transfer resistance to transport of ions which induce polymer swelling [16, 17]; (4) external or surface resistance to solvent transport into the polymer [18, 19]; (5) degradation of crosslinks with time, causing continually increasing polymer swelling [20, 21]; (6) gradual conversion of the polymer to a form having increased affinity to the solvent [22]; and (7) presence of osmotically active solutes in the polymer [23–24].

Thus there are many processes which can cause polymers to swell with transport exponents other than 0.5. Even when these processes can be ruled out as influencing the rate of sorption, additional tests are required to distinguish between diffusion and relaxation-controlled swelling. Such tests include: (1) determination of the dimensional dependence of sorption, as described earlier; (2) measurement of the kinetics of an observable moving solvent front, if any (in a planar geometry, the front should scale with the $\sqrt{t}$ if sorption is diffusion controlled or with t if it is relaxation-controlled); and (3) identification of changes in sorption caused by the addition of plasticizers or changes in temperature, since these affect diffusion- and relaxation-controlled processes differently (though the effects are less quantifiable than tests 1 and 2).

Experimentally, the transport of water in glassy polymers has been studied by a number of research groups over the past several decades; while many of these observe Fickian sorption, a number of papers have been published describing non-Fickian water sorption by polymers. Almost all of these base claims of non-Fickian water sorption upon initially non-linear plots of weight gain against time. 'Relaxation-control' is the usual explanation for the deviation, although possible alternative causes of the deviations as listed above are rarely ruled out, nor are confirming tests such as dimensional dependence typically performed.

The most common source of ambiguity in reports of non-Fickian water transport is tied to the anisotropic dimensional changes observed as the glassy core vanishes for non-spherical geometries [4, 5, 14, 15]. When a glassy polymer is immersed in a good solvent for the polymer, the following process is observed. Soon after immersion, a moving solvent front forms which delineates a swollen rubbery layer from an unswollen glassy core. Due to the resistance of the glass to deformation, the sample expands only in thickness. Once the glassy core vanishes, however, the anisotropic stress which builds up as a result of the anisotropic swelling deforms the sample to a nearly isotropic state relative to the initial state. In other words, the sample becomes thinner with a greater surface area. Both of these changes cause acceleration of the swelling rate. This deformation occurs quickly with respect to swelling, and we have shown previously that this anisotropic stress does not otherwise affect the swelling rate (note that this type of stress relaxation is not the type postulated in the models like that of Thomas and Windle) [5]. Since Fick's law can describe this process once the change in the dimensionality of the problem is noted, and because the stress does not affect molecular transport, our opinion is that this should still be considered 'Fickian'. In other words, these macroscopic distortions are superimposed on top of a rate-limiting diffusion process at the microscopic level.

With respect to the treatment of sorption data, it is important to note that this rearrangement often occurs within the first 60% of the mass uptake. Thus inclusion of data points after this acceleration point in the calculation of the transport exponent will increase its value. Thus, even if the rate-controlling process is Fickian diffusion at the molecular level, a transport exponent greater than 0.5 will be obtained. This effect is particularly significant if only a few points are used to calculate the transport coefficient.

The strongest indications of non-Fickian water sorption are of water vapor in polymers like poly(ethylene-*co*-vinyl alcohol) and regenerated cellulose which absorb less than 15 wt% of water [25–27]. Thus they are not considered hydrogels [6]. In these systems, the surface may only gradually approach equilibrium, especially at low water activities. Thus deviations from Eqs (2) and (3) can arise in the boundary conditions rather than from failure of Fick's law for molecular diffusion. Furthermore, since dual mode theory of vapor diffusion into polymers postulates a hole-filling mechanism followed by a dissolution mechanism, two-stage mechanisms may arise for vapor sorption to low weight gains [7, 28]. These phenomena can be termed 'non-Fickian' in that they deviate from classical behavior, though they may well be driven by random molecular motion at the molecular level that is still essentially 'Fickian'. However, it appears possible that much of the non-classical sorption behavior described in these papers may be the result of simple dimensional rearrangements causing late-stage acceleration described earlier. Even if dimensional changes do not explain the deviations observed with these systems, these effects are unlikely to be significant factors with pharmaceutical hydrogels, since they swell far more than 15% in aqueous solution.

The specific subject of this paper, water sorption by PHEMA and copolymers of PHEMA has been studied previously, but questions persisted about the nature of the transport mechanism. Franson and Peppas studied the sorption of water by copolymers of HEMA with both the more hydrophilic monomer *N*-vinyl pyrrolidone (NVP) and with the more hydrophobic monomer methyl methacrylate (MMA) [29, 30]. In all cases there was no statistically significant deviation from Fickian diffusion. The confidence limits on the transport exponents were quite broad, however, due to the limited number of points obtained during the first 60% of the sorption, and thus firm conclusions about the transport mechanism could not be drawn. Korsmeyer and Peppas also studied the swelling kinetics of P(HEMA-*co*-NVP) in water, but in more detail [15]. They noted the time of the glass–rubber transition and measured the sudden dimensional changes associated with this transition. However, the sorption mechanism still did not appear to deviate significantly from Fickian kinetics (values of n were not reported). Similarly, the water sorption curves for P(HEMA-*co*-MMA) presented by Davidson and Peppas do not appear to be significantly non-Fickian (values of n were not reported) [14]. Robert *et al.* created crosslinked PHEMA spheres (105–610 μm diameter) by suspension polymerization and measured the transport exponent and the velocity of the moving solvent front as a function of crosslinker composition [31]. Their results are the strongest evidence of non-Fickian water sorption in the literature, as the transport exponent did shift from Fickian to anomalous transport as the crosslinking ratio increased above 5 mol%. This result is consistent with relaxation-influenced sorption, since increased crosslinking should increase the relaxation time. Furthermore, the one-dimensional nature of a swelling sphere means that effects of dimensional changes upon disappearance of the glassy core did not compromise calculations of the transport exponent (though the error bars on the data were quite large). But the spherical geometry complicated interpretation of the moving front results; since the transport area decreases as the front progresses, even a diffusion controlled front has a significant linear portion and a late stage acceleration, as Robert *et al.* observed [17].

Our group also saw that increasing the amount of crosslinker to very high levels (up to 30 mol%) in a PHEMA gel increased the water transport exponent from Fickian into the anomalous range (up to $n = 0.68 \pm 0.16$). Considering that the additional crosslinker increased T_g from 96 to 190°C, this change in n is minor [4]. But the other indications were that transport in these systems was Fickian—sorption and moving solvent fronts both advanced with the square root of time for gels of different thicknesses and copolymer ratios. Since the dramatic dimensional changes associated with the loss of the restraining glassy core implied that significant stresses existed within the swelling polymers, modification of swelling rates as a result seemed possible, as this is a stress relaxation effect. Thus we further examined the effects of varying levels of anisotropic stresses in polymers [5]. However, regardless of the magnitude or orientation of anisotropic stress in the swelling sample, the swelling rates always progressed as predicted by Fick's law, including dimensional dependence. In contrast, methanol sorption by PHEMA occurred by an unambiguously anomalous transport mechanism.

Thus further study on the relative importance of diffusion and relaxation on the swelling of glassy polymers in water was warranted. The objective of this paper was to obtain further evidence on this topic beyond measurement of transport exponents and dimensional dependences. To do this, we compared sorption rate, temperature dependence, thermal history (annealing) and residual solvent effects against the expectations of diffusive and relaxation phenomena.

EXPERIMENTAL

Materials

2-hydroxyethyl methacrylate (HEMA) monomer of 97% purity (Aldrich) was further purified by vacuum distillation at 67°C and 3.5 mm Hg. Cuprous chloride was added as an inhibitor to prevent polymerization during distillation. The top and bottom 20% fractions of the distillate were discarded. Purified HEMA monomer was mixed with 0.5 wt% initiator benzoyl peroxide (Aldrich) and deaerated in a desiccator for 30 min to remove oxygen which can inhibit the free radical polymerization. The sample mold consisted of two polypropylene plates separated by a silicon rubber gasket and held together with clamps; the solution was injected into the mold through a hole in one of the polypropylene plates. The mold was then put into a vacuum oven (NAPCO, Model 5831), and polymerization was carried out at 60°C and 25 mm Hg for 24 h. The thickness of the glassy polymer sheet obtained was equal to the thickness of the silicon rubber gasket (1.5 cm). The glassy polymer sheet was cut into smaller square portions with a heated razor blade; the aspect ratio (length : thickness) was kept greater than ten so that one-dimensional transport could be assumed. Certain samples were prepared differently, as described below. Glass transition measurements were made using a Perkin-Elmer differential scanning calorimeter (Model DSC 7) in conjunction with a Perkin-Elmer TAC 7 instrument controller and a Perkin-Elmer 7500 computer. The samples was heated at a rate of 20°C min^{-1} from 30 to 200°C. The glass transition temperatures were then determined from the thermogram using the Perkin-Elmer TAS 7 software package.

Swelling measurements

Sorption experiments were performed at 24°C using a simple blot-and-weigh technique, shown to be reproducible to within ±3%. Samples were removed from the distilled water containers at appropriate intervals and blotted free of surface moisture with filter paper before each weighing. This procedure was repeated until the sample attained constant weight. The solvent front movement was measured by immersing the sample in water and holding it beneath a stereomicroscope (Olympus Model SZH-ILLD) with a clip such that the edge of the sample was seen; the moving solvent front was observed under polarized light and the distance moved by the front with time was measured using a digital Filar micrometer with a processor (LASICO, Model XM).

RESULTS

Solvent uptake phenomena

When a glassy sample of PHEMA is immersed in liquid water, it spontaneously absorbs water. Looking edgewise at a flat sheet of this polymer, a clear, sharp front indicating the extent of water penetration into the polymer is clearly seen. The exact concentration at this front is unknown, but can be assumed to be approximately equal to the concentration which lowers T_g of the polymer to the system temperature, as it is observed that the sample loses all rigidity and becomes completely rubbery at the point when the fronts from opposite sides of the sample meet at the center of the sample.

Figure 1 is a plot of the front position against the square root of time; its linearity is the first indication that the transport of water into the polymer is Fickian. The fractional water uptake (M_t/M_∞) plotted against the square root of time is also linear for over 50% of the sorption interval, as seen in Fig. 2. In fact, these data fit the exact solution of a Fickian process (Eq. (2)) up to the point that the solvent fronts meet at the center of the sample, which makes the polymer entirely rubbery. This deviation results from the dimensional changes which occur at this point: the increase in sample area at the expense of thickness, as described earlier. Both of these changes will enhance the sorption rate of any diffusion process. In a related

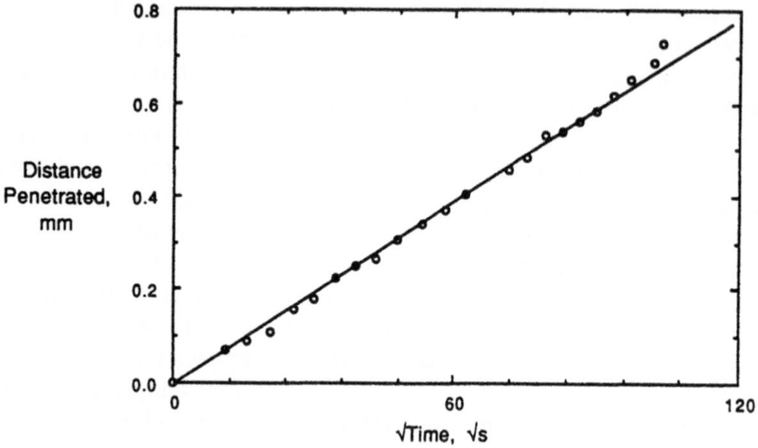
0.8
0.6
Distance Penetrated, mm
0.4
0.2
0.0
0
60
120
√Time, √s

Figure 1. Movement of water penetration front in PHEMA. Linearity against the square root of time is evidence of a Fickian transport mechanism.

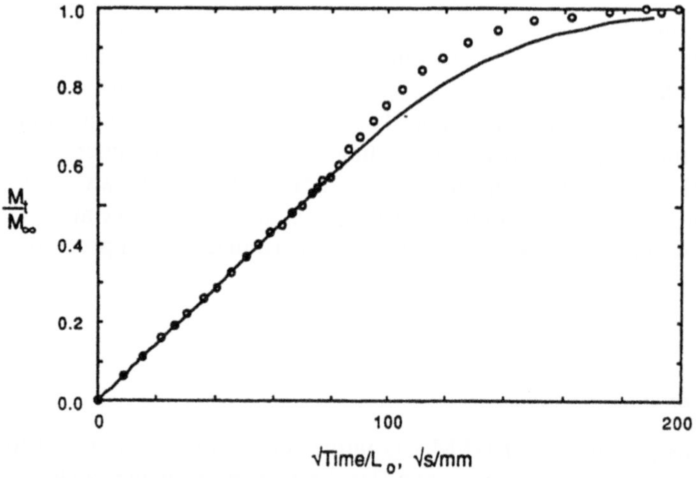
1.0
0.8
0.6
M_t / M_∞
0.4
0.2
0.0
0
100
200
√Time/L_0, √s/mm

Figure 2. Water sorption by glassy PHEMA at 24°C. Transport mechanism is classical diffusion as given by Eq. (2) until the glassy core vanishes; $D = 1.02 \pm 0.07 \times 10^{-7}\ \mathrm{cm^2\ s^{-1}}$.

paper, we have shown that these dimensional changes can be eliminated by preparing glassy samples of PHEMA which contain anisotropic stress fields which counterbalance the swelling stresses [5]. The sorption curves of such samples fit Eq. (2) for all times. In fact, the sorption curves for samples of different thicknesses all fall on the same curve of fractional uptake against square root time over initial thickness [5]. This is compelling evidence that the macroscopic distortions in dimension do not shift the underlying transport mechanism from Fickian diffusion.

Effect of temperature

Relaxation and diffusion processes have different temperature dependencies, and thus the solvent sorption mechanism is expected to change with temperature. Thus we examined the swelling of PHEMA in water over almost the entire temperature range for liquid water at atmospheric pressure, from 4 to 88°C. The polymer volume fraction is nearly constant with temperature at about 0.58, which agrees with the values reported by Peppas and Moynihan [32]. This indicates the value of the Flory-Huggins χ-parameter is also nearly constant, reported to be 0.84 for this system [32]. The large value of the χ-parameter explains why the polymer does not dissolve even without crosslinker: water is not a sufficiently good solvent for PHEMA (also, low levels of HEMA dimers in the monomer cause a certain amount of inadvertent crosslinking).

Figure 3 shows that at all temperatures, polymer samples follow swelling curves comparable to Fig. 2. In other words, all samples match Fick's law up to the point that the sample becomes entirely rubbery. Since the difference between T_g and the ambient temperature declines as the temperature increases, with an increase in temperature a lesser amount of water is needed to make the sample entirely rubbery. This also is seen in Fig. 3, as the samples become entirely rubbery at increasingly smaller values of M_t/M_∞ as the temperature rises. Diffusion coefficients can be obtained from these curves by fitting the portions of the curves prior to the disappearance of the glassy core to Eq. (2); these values are given in Table 1. When the logarithms of the diffusion coefficients are plotted against inverse absolute temperature, as in Fig. 4, a straight line is obtained with a slope corresponding to an activation energy of 6.1 kcal mol^{-1} (29.4 kJ mol^{-1}). This compares favorably to the value of 5.0 kcal mol^{-1} reported for the PHEMA/water system by Wisniewski and Kim [33]. An activation energy of this low magnitude is characteristic of the diffusion of small molecules in polymers [28, 34].

Effect of initial solvent content

The primary difference between the glassy and rubbery state of a polymer is the difference in response to an applied stress; a rubber responds nearly instantaneously, while the response of a glass is time dependent. Thus the sorption

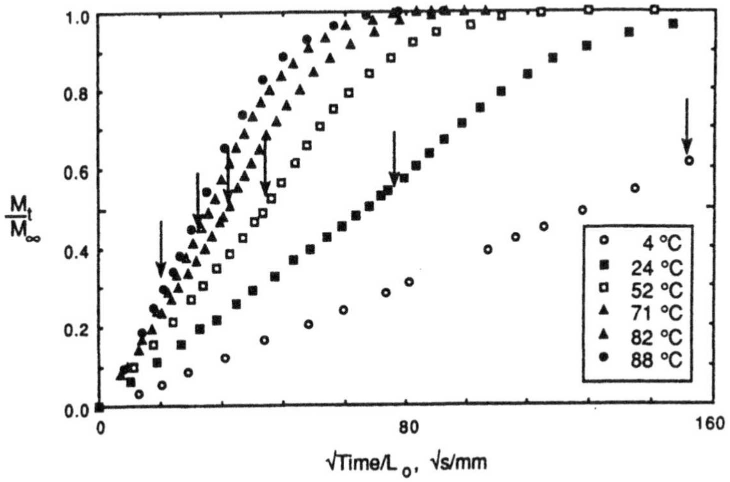

1.0
0.8
0.6
0.4
0.2
0.0
M_t/M_∞
0
80
160
√Time/L₀, √s/mm
4 °C
24 °C
52 °C
71 °C
82 °C
88 °C

Figure 3. Water sorption by glassy PHEMA as a function of temperature. Transport is diffusion controlled over the entire range (arrows indicate the point where the glassy core vanishes).

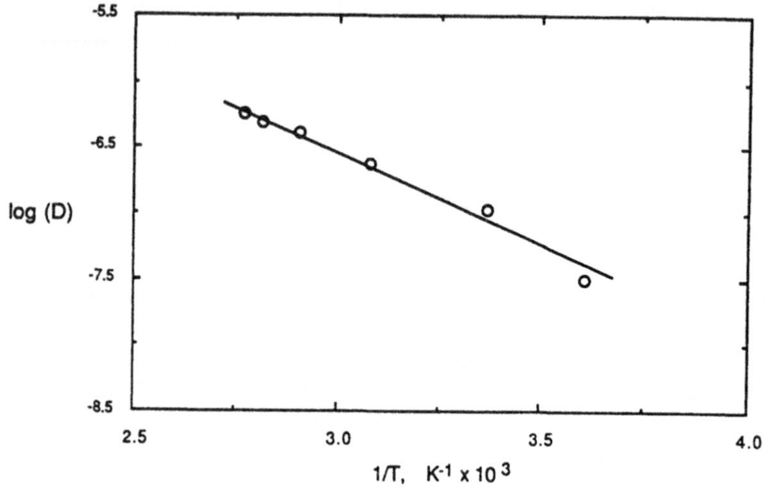
-5.5
-6.5
-7.5
-8.5
log (D)
2.5
3.0
3.5
4.0
1/T, K-1 x 10 3

Figure 4. Arrhenius plot of the diffusion coefficients obtained from Fig. 3. From the slope, the activation energy is 6.1 kcal mol^{-1}.

Table 1.
Water transport parameters in glassy PHEMA at different temperatures (from Fig. 3)

T (°C)	Diffusion coefficient (10^7 cm^2 s^{-1})	Equilibrium swelling ratio Q^a	Transport exponent n	Number of data points used to calculate n	Entirely rubbery at M_t/M_∞
4	0.32 ± 0.02	1.67 ± 0.01	0.55 ± 0.02[b]	12	0.60
24	1.10 ± 0.07	1.57 ± 0.01	0.50 ± 0.01	14	0.55
52	2.43 ± 0.17	1.52 ± 0.01	0.53 ± 0.02	8	0.51
71	4.15 ± 0.25	1.53 ± 0.01	0.52 ± 0.01	8	0.48
82	4.95 ± 0.40	1.52 ± 0.01	0.56 ± 0.03	5	0.42
88	5.64 ± 0.54	1.51 ± 0.01	0.61 ± 0.04[c]	3	0.30[d]

[a] Q = swollen mass/dry mass = (g polymer + g solvent)/(g polymer).
[b] 95% confidence limits.
[c] The slight increase in transport exponents at high temperatures may be due to time elapsed during the weight measurements. Although it takes less than 30 s to take a reading, this amount of time may become significant when the diffusion is fast.
[d] Since this experiment occurs close to T_g, any glassy appearance may have been due to slight cooling when weighing.

of a solvent by rubbery polymers is expected to be relatively fast and diffusion-controlled, since the relaxation rate is virtually instantaneous, while sorption by a glassy polymer should be slower and may be relaxation-controlled, since the relaxation rate is finite [6, 9, 10, 28].

To test this hypothesis, samples of PHEMA were immersed in solvent for a limited period of time, such that sorption was stopped well short of equilibrium. They were then placed in sealed jars for several days to allow the solvent concentration to become uniform throughout the sample. It was found that 33 wt% water and 16.5 wt% of tetrahydrofuran (THF) were sufficient to plasticize PHEMA.

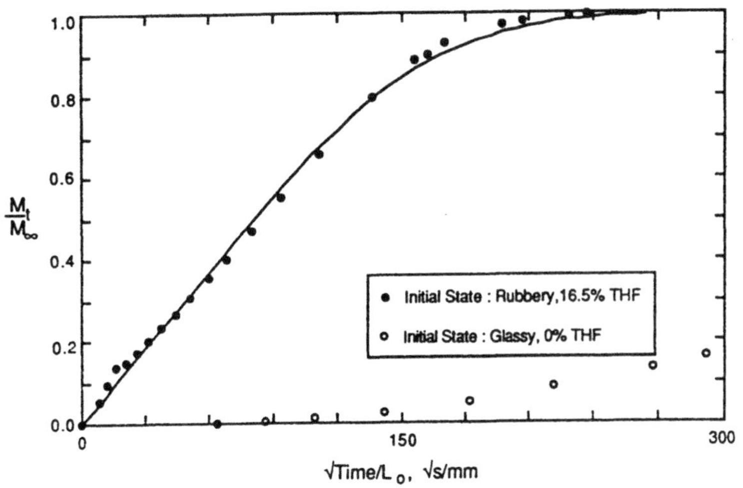
1.0
0.8
0.6
0.4
0.2
0.0
Mt/M∞
0
150
300
√Time/L0, √s/mm
Initial State : Rubbery, 16.5% THF
Initial State : Glassy, 0% THF

Figure 5. Sorption of THF by PHEMA as a function of initial solvent content at at 24°C; the equilibrium swelling ratio is 1.53. The rate increases and the transport mechanism becomes Fickian when the sample is initially rubbery. $D = 7.4 \times 10^{-8}\, cm^2\, s^{-1}$ at 16.5 wt% THF; initial slope at 0% THF is equivalent to $2.0 \times 10^{-12}\, cm^2\, s^{-1}$.

Figure 5 shows that the sorption kinetics of THF varies significantly with initial solvent concentration, consistent with the hypothesis stated above. In the glassy state, THF is absorbed by PHEMA very slowly and by a pronounced non-Fickian transport mechanism. When the polymer initially contains enough solvent to become rubbery, THF transport is classically Fickian, fitting Eq. (2) over the entire sorption curve, and with a diffusion coefficient comparable to that seen for water. Figure 6 compares the sorption rates of PHEMA initially containing 0 wt% water

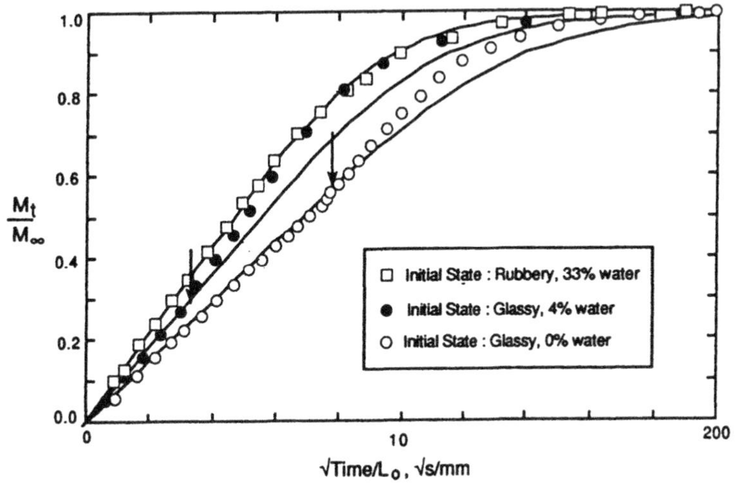
1.0
0.8
0.6
0.4
0.2
0.0
Mt/M∞
0
10
200
√Time/L0, √s/mm
Initial State : Rubbery, 33% water
Initial State : Glassy, 4% water
Initial State : Glassy, 0% water

Figure 6. Sorption of water by PHEMA as a function of initial solvent content at 24°C; the equilibrium swelling ratio is 1.57. Diffusion coefficients and transport mechanisms do not vary significantly with initial solvent concentration of physical state; $D = 2.2 \times 10^{7}\, cm^2\, s^{-1}$ at 33 wt% water; $D = 1.6 \times 10^{-7}\, cm^2\, s^{-1}$ at 4 wt% water; $D = 1.1 \times 10^{-7}\, cm^2\, s^{-1}$ at 0 wt% water. Arrows indicate the point where the glassy core vanishes.

(glassy), and 33 wt% water (rubbery). The initial rubbery sample fits the Fickian equation (Eq. (2)) at all times, as expected. However, its diffusion coefficient is only twice that of the glassy polymer containing no water at all! The glassy sample with 4% water displays intermediate behavior; its early deviation from the Fickian curve indicates the fact that less water must be absorbed to make the sample entirely rubbery. The results of Fig. 6 imply that the water transport in initially glassy PHEMA is similar to the transport mechanism in rubbery polymers, which would be quite surprising if relaxation was the rate-controlling process in glassy PHEMA.

Effect of thermal history

It is well known that the thermal history of glassy polymers can affect their properties by altering polymer chain configurations and the distribution of the free volume within the sample. Since PHEMA is slightly more dense than the liquid monomer (HEMA), bulk polymerization of HEMA at a temperature below the T_g of PHEMA can yield a polymer with more free volume than an annealed sample, due to the resistance of the glassy polymer to the innate tendency of freshly polymerized PHEMA to contract. Furthermore, the rate of cooling after annealing PHEMA should also affect the amount and distribution of free volume. Since the rubbery state has greater free volume than the glassy state, a sample rapidly cooled below T_g should contain more free volume than a slowly cooled sample, since the rapidly cooled polymer chains are frozen into the configurations of the rubbery state. Since solvent transport in polymers occurs through this free volume, samples with differences in free volume distribution might be expected to display differences in sorption behavior [28].

To determine the influence of thermal history on sorption, PHEMA samples were annealed in an oven at 100°C (slightly above T_g) for 24 h. They were cooled back to ambient temperature at two cooling rates. Certain samples were rapidly cooled by plunging them into liquid hexane at ambient temperature. Other samples were

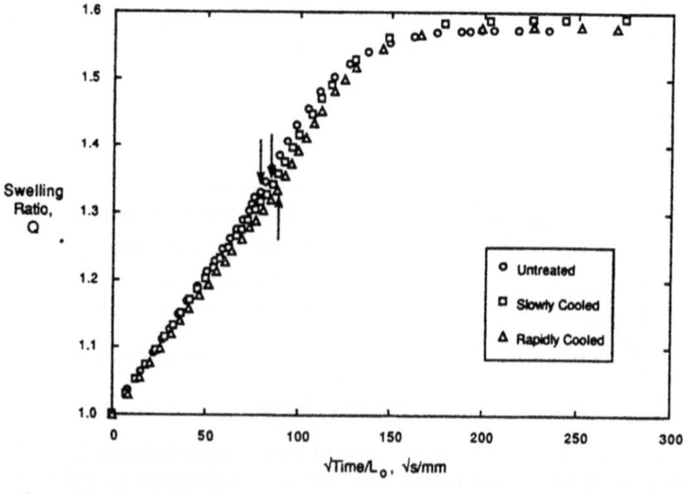
Swelling Ratio, Q
1.0
1.1
1.2
1.3
1.4
1.5
1.6
0
50
100
150
200
250
300
√Time/L₀, √s/mm
Untreated
Slowly Cooled
Rapidly Cooled

Figure 7. Sorption of water by PHEMA as a function of thermal history at 24°C. Thermal history has little effect of the transport behavior of PHEMA/water systems. Arrows indicate the point where the glassy core vanishes.

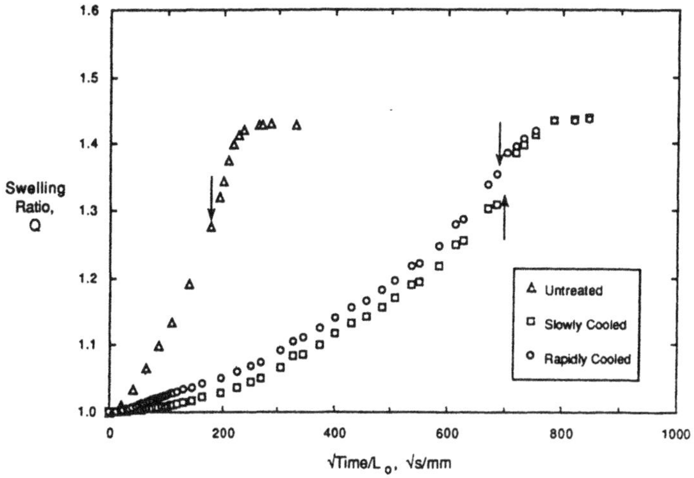
1.6
1.5
1.4
1.3
1.2
1.1
1.0
Swelling Ratio, Q
0
200
400
600
800
1000
√Time/L₀, √s/mm
Untreated
Slowly Cooled
Rapidly Cooled

Figure 8. Sorption of acetone by PHEMA as a function of thermal history at 24°C. Sorption rates in acetone decreased substantially for annealed PHEMA samples. Arrows indicate the point where the glassy core vanishes.

cooled slowly (about 10°C h^{-1}) simply by turning off the oven with the samples inside. The water sorption rates of these annealed samples were then measured at 24°C and compared against the results of Fig. 2.

Figure 7 shows that there is no significant difference in any feature of the three curves. In sharp contrast, Fig. 8 shows that thermal history has a significant effect on the sorption of acetone by PHEMA at 24°C (note also that acetone sorption occurs by a decidedly non-Fickian mechanism). This suggests that while the transport of larger penetrants in glassy polymers is quite sensitive to the free volume distribution, water molecules are too small to be affected.

DISCUSSION

All of the data available indicates that water is absorbed by PHEMA by a Fickian sorption mechanism, in contrast to clearly non-Fickian sorption of THF and acetone. The dimensional changes caused by the disappearance of the glassy core offer the only deviation from classical Fickian behavior, and the evidence indicates that this distortion does not affect the microscopic sorption mechanism. In fact, such deviations would not even be seen for a spherical sample. The transport mechanism is not a function of temperature, and the diffusion coefficients obtained from the portions of the sorption curves prior to the dimensional distortion yield an activation energy consistent with diffusion of small molecules in polymers. The sorption rates and transport mechanisms of large solvents depends strongly upon the initial state of the polymer—whether glassy or rubbery, past history—as expected based on our physical insight into these systems. However, the rates and mechanisms of water sorption are insensitive to the initial state of the polymer sample; sorption is always Fickian with diffusion coefficients on the order of 10^{-7} cm^2 s^{-1}.

Our interpretation of the evidence suggests that the water sorption phenomena observed are the consequences of the water molecule's small size and affinity for hydrophilic polymers. Since the water transport is unaffected by the thermal history of the polymer, and little dependent upon the physical state of the polymer (whether glassy or rubbery), it appears that water molecules can easily penetrate the free volume of the polymer (i.e. water has a low jump activation energy in a hydrophilic polymer). In short, relaxation of the polymer does not influence the sorption rate, because significant deformation of the polymer chains while in the glassy state is not required.

Another effect which encourages Fickian sorption of liquid water by hydrogels is seen in the studies by Berens of organic solvent sorption by poly(vinyl chloride) [34]. Berens found that transport was Fickian in cases where liquid solvents swelled the polymer significantly. He explained this effect in terms of diffusion-limited transport in the thick, swollen rubbery layer which develops upon swelling. Solvent transport in this layer dominates the overall weight gain, regardless of any relaxation processes that might otherwise influence rates at lower solvent activities. Since hydrogels swell significantly in water, by definition, this is another force driving hydrogel swelling to a Fickian limit. However, we note that this conclusion cannot be extended to all solvents, since even in this paper non-Fickian effects are seen with organic solvents that are good swelling agents for the polymer (i.e. THF and acetone in PHEMA). Biren explained this deviation in terms of the Long and Richman model of surface relaxation effects that did not exist for water [18, 35]. Thus the Berens effect is still expected to develop in most hydrogels.

This suggests that the water sorption will generally be Fickian in hydrogels. The reasoning behind this statement is that all glassy polymers have comparable levels of free volume at T_g (≈ 2.5 vol% [37]) and that water will have a favorable interaction with any hydrogel-forming polymer. Thus there seems little basis for expecting substantial differences in water sorption by different hydrogels. This has been our experience with a wide variety of glassy hydrogels besides PHEMA. Thus Case II sorption—relaxation-controlled, constant rate swelling—may be difficult to achieve with a hydrogel, as Peppas' group and our group have found with PHEMA and its copolymers. This is consistent with the literature in that examples of Case II transport of water in swellable polymers are lacking; at best, modest deviations from diffusion-controlled swelling exist. Research focused on molecular level interactions is required to confirm these ideas, however.

This study has implications for the design of swelling-controlled drug delivery systems based on glassy polymers. If water sorption is the rate limiting step and occurs by diffusion, swelling-controlled drug delivery systems will be diffusion-limited and thus display declining rates of drug release, rather than a constant rate as is often desired. This does not mean that constant delivery rate systems cannot be designed, but that to do so will require some other rate-limited or rate-influencing step. For example, non-uniform concentration gradients, chemical reactions, balancing the rate of water uptake with drug diffusion outward, or special geometries are all means of achieving constant release rates in swelling-controlled drug delivery systems [1, 22–24].

CONCLUSIONS

Based on sorption rate measurements as a function of temperature, initial solvent content, physical state (glassy or rubbery) and thermal history, we conclude that water sorption by PHEMA, whether glassy or rubbery, is diffusion-controlled.

Fickian behavior is probably the result of the small molecular size of water and its affinity for the hydrophilic, swellable polymers, and thus is likely to be observed with most water-swellable polymers.

Acknowledgement

This work was supported by grant CBT-8809271 from the National Science Foundation.

REFERENCES

1. N. A. Peppas and R. W. Korsmeyer, in: *Hydrogels in Medicine and Pharmacy Volume III*, p. 109, N. A. Peppas (Ed.). CRC Press, Boca Raton, FL (1987).
2. H. B. Hopfenberg and K. C. Hsu, *Polym. Eng. Sci.* **18**, 1186 (1978).
3. A. H. Windle, in: *Polymer Permeability*, p. 75, J. Comyn (Ed.). Elsevier Applied Science, London (1985).
4. B. Kabra, S. Gehrke, S. T. Hwang and W. Ritschel, *J. Appl. Polym. Sci.* **42**, 2409 (1991).
5. D. Biren, B. Kabra and S. H. Gehrke, *Polymer* **33**, 554 (1992).
6. S. H. Gehrke and P. I. Lee, in: *Specialized Drug Delivery Systems: Manufacturing and Production Technology*, p. 355, P. Type (Ed.). Marcel Dekker, New York (1990).
7. A. R. Berens and H. B. Hopfenberg, *Polymer* **19**, 489 (1978).
8. N. L. Thomas and A. H. Windle, *Polymer* **22**, 627 (1981).
9. C. E. Rogers, in: *Physics and Chemistry of the Organic Solid State. Vol. II*, p. 509, D. Fox, M. M. Labes and A. Weissberger (Eds). Interscience, New York (1965).
10. J. Crank and G. S. Park, in: *Diffusion in Polymers*, p. 1, J. Crank and G. S. Park (Eds). Academic Press, London (1968).
11. J. Crank, *The Mathematics of Diffusion, 2nd Ed.* Clarendon Press, Oxford (1975).
12. B. Kabra, M. K. Akhtar and S. H. Gehrke, *Polymer* **33**, 990 (1992).
13. P. L. Ritger and N. A. Peppas, *J. Contr. Release* **5**, 23 (1987).
14. G. W. R. Davidson III and N. A. Peppas, *J. Contr. Release* **3**, 243 (1986).
15. R. W. Korsmeyer, E. von Meerwall and N. A. Peppas, *J. Polym. Sci.: Poly. Phys. Ed.* **24**, 409 (1986).
16. S. H. Gehrke and E. L. Cussler, *Chem. Eng. Sci.* **44**, 559 (1989).
17. S. H. Gehrke, G. Agrawal and M. C. Yang, in: *Polyelectrolyte Gels, ACS Symposium Series 480*, p. 211, R. S. Harland and R. K. Prud'homme (Eds). American Chemical Society, Washington, D.C. (1992).
18. F. A. Long and D. Richman, *J. Am. Chem. Soc.* **82**, 513 (1960).
19. A. C. Newns, *Nature* **218**, 355 (1968).
20. C. M. Ofner and H. Schott, *J. Pharm. Sci.* **8**, 790 (1986).
21. K. Park, *Biomaterials* **9**, 435 (1988).
22. K. Park, W. S. W. Shalaby and H. Park, *Biodegradable Hydrogels for Drug Delivery*. Technomic, Lancaster, PA (1993).
23. P. I. Lee, *Polymer* **25**, 973 (1984).
24. P. I. Lee, *Polymer (Commun.)* **24**, 45 (1983).
25. A. C. Newns, *Trans. Farad. Soc.* **52**, 1533 (1956).
26. H. B. Hopfenberg, A. Apicella and D. E. Saleeby, *J. Membr. Sci.* **8**, 273 (1981).
27. A. Apicella and H. B. Hopfenberg, *J. Appl. Polym. Sci.* **27**, 1139 (1982).
28. W. R. Vieth, *Diffusion In and Through Polymers*. Hanser, Munich (1991).
29. N. M. Franson and N. A. Peppas, *J. Appl. Polym. Sci.* **28**, 1299 (1983).
30. N. A. Peppas and N. M. Franson, *J. Polym. Sci.: Polym. Phys. Ed.* **21**, 983 (1983).
31. C. C. R. Robert, P. A. Buri and N. A. Peppas, *J. Appl. Polym. Sci.* **30**, 301 (1985).
32. N. A. Peppas and H. J. Moynihan, in: *Hydrogels in Medicine and Pharmacy Volume II*, p. 50, N. A. Peppas (Ed.). CRC Press, Boca Raton, FL (1987).
33. S. Wisniewski and S. W. Kim, *J. Membr. Sci.* **6**, 309 (1980).

34. A. R. Berens, *J. Appl. Polym. Sci.* **37**, 901 (1989).
35. D. Biren, M.S. Thesis, University of Cincinnati, Cincinnati, OH (1990).
36. A. Eisenberg, in: *Physical Properties of Polymers*, p. 55, J. E. Mark, A. Eisenberg, W. W. Graessley, L. Mandelkern and J. L. Koenig (Eds). American Chemical Society, Washington, D.C. (1984).

Part X.B.

Hydrogels

Stimuli-Responsive Hydrogels

Biochemo-mechanical function of urease-loaded gels

ETSUO KOKUFUTA,* YONG-QING ZHANG and TOYOICHI TANAKA
Department of Physics and Center for Materials Science and Engineering, Massachusetts Institute of Technology, Cambridge, MA 02139, USA

Received 5 April 1993; accepted 22 July 1993

Abstract—A gel system is developed that undergoes a reversible volume phase transition in response to a small amount of urea. An *N*-isopropylacrylamide gel in which urease is immobilized by entrapping changes its equilibrium volume discontinuously when urea molecules are hydrolyzed by urease causing a change in pH that alters the osmotic balance of the gel triggering the phase transition. The system demonstrates a method of mechano-biochemical transformation where molecular recognition and biochemical reaction are achieved by an enzyme and the macroscopic amplification of the reaction is carried out by a gel capable of a volume phase transition. The work presented here is dedicated to Professor Allan S. Hoffman to honor his 60th birthday and his pioneering contribution to the science and technology of polymer gels, both as a scientist and as an educator.

Key words: *N*-isopropylacrylamide; acrylic acid; copolymer gel urease; phase transition; biochemical control.

INTRODUCTION

Mechano-chemical or chemo-mechanical systems which exert mechanical work in response to chemical changes have long attracted interest among scientists, medical researchers, and engineers. Katchalsky and coworkers [1–3] were pioneers in the development of such systems using polymer gels and their work has been expanded further by various researchers (e.g., see ref [4]).

Recent findings for discontinuous volume phase transitions in gels have widened the scope of mechano-chemical systems [5–7]. In these systems, changes in the physical or chemical environment trigger the volume phase transition either discontinuously or continuously. The volume change is reversible and can be as large as many thousand times. The discontinuous transition generally requires ionization of polymers to create an internal pressure due to Coulombic charge–charge interaction and counter-ion osmotic pressure. The volume transition can be used to develop artificial muscles, actuators, sensors, controlled delivery systems, and other 'intelligent' devices.

The environmental parameters that trigger a volume transition are not necessarily specific. For example, the phase transition can be induced in copoly(acrylamide/acrylic acid) gels by changing the solvent composition, but the solvent can be alcohol, acetone, or many other nonpolar compounds. The transitions can also be induced in some gels by varying pH, but more or less similar results are obtained for different combinations of acidic and basic solutions so long as the pH values are the same. In contrast, biological reactions are significantly more specific and target

* Present address: Institute of Applied Biochemistry, University of Tsukuba, Tsukuba, Ibaraki, 305 Japan.

dependent. Therefore, by combining the amplifying capability of the gel phase transition and the specificity of an enzymatic reaction, one may be able to develop a 'biochemo-mechanical' system in which biochemical changes catalyzed by the enzymatic reaction are used to trigger the phase transition and to create mechanical energies. The combination of specific sensing and amplification may not only be technologically important, but will also provide a model for simulation of the marvelous sensing mechanisms and energy-conversion available only in biological systems.

In our previous studies [8, 9], two gel systems with a biochemo-mechanical function have been designed and prepared: *N*-isopropylacrylamide gels with an immobilized enzyme (esterase); and a sugar-binding protein (concanavalin A). In the enzyme-loaded gel, however, only the swelling process was controlled and no reversible aspects of the volume changes were explored. Ideally both swelling and shrinking should be reversibly utilized for repeated biochemo-mechanical cycles. In this paper we report on the development of a biochemo-mechanical system using a polymer gel consisting of a network of *N*-isopropylacrylamide-acrylic acid copolymer into which urease was immobilized.

There has been some development of biochemo-mechanical systems, but these have not made use of a discontinuous phase transition.

MATERIALS AND METHODS

The gel consists of a covalently cross-linked copolymer network of *N*-isopropylacrylamide and acrylic acid in which urease is immobilized (Fig. 1). Jack bean urease (or urea amidohydrolase; Sigma, EC 3.3.1.5) was used which had 61 000 U g^{-1} solid of enzyme activity. One unit of the activity liberates 1.0 μmol of ammonia from urea per minute at pH 7.0 and 25°C. The immobilization was achieved by gelling an

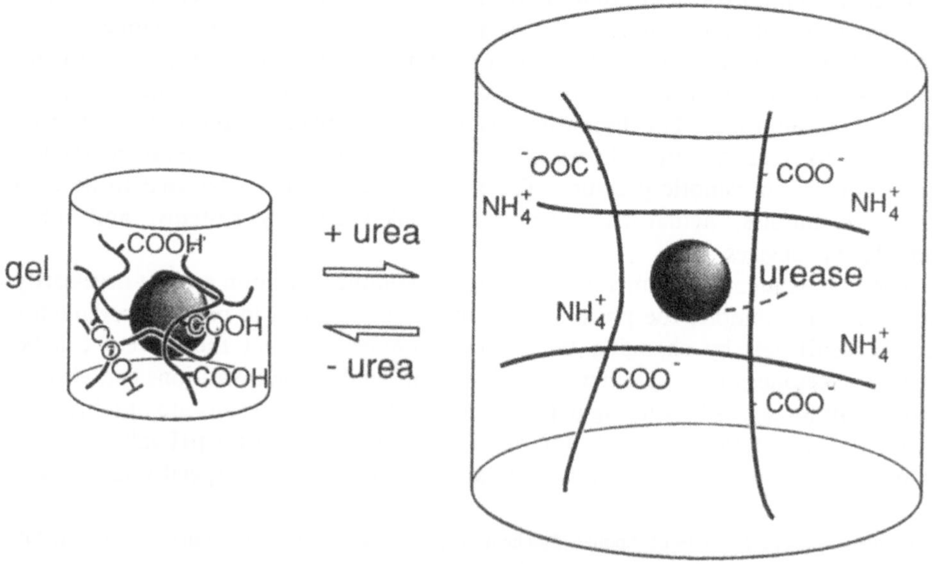
gel
COOH
COOH
COOH
COOH
+ urea
- urea
$^{-}$OOC
COO^{-}
NH_4^+
NH_4^+
urease
NH_4^+
NH_4^+
COO^{-}
COO^{-}

Figure 1. Schematic illustration of a mechano-biochemical system consisting of *N*-isopropylacrylamide, acrylic acid, and immobilized urease. Ammonia, that is enzymatically produced from urea, raises the pH of the gel phase and dissociates a carboxyl group even when the ambient pH is kept at a low value in the acidic range. This brings about the swelling of gel, whereas the gel collapses upon removal of the substrate.

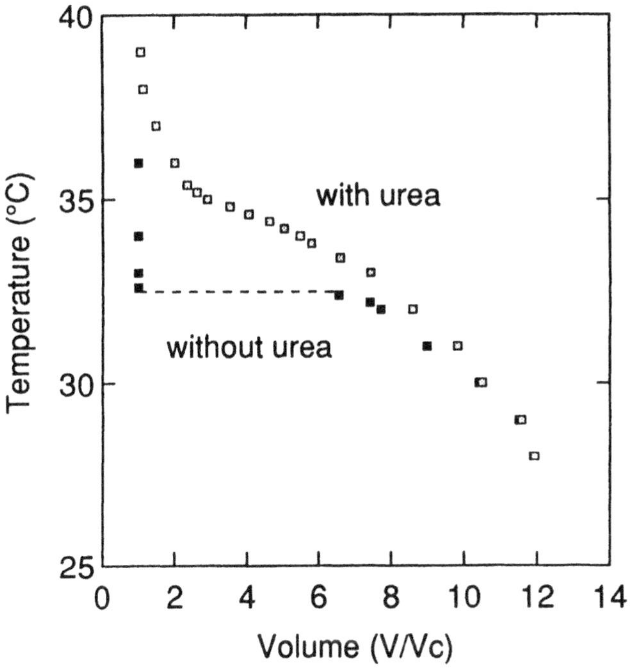

40
35
30
25
0
2
4
6
8
10
12
14
Temperature (°C)
Volume (V/Vc)
with urea
without urea

Figure 2. Swelling curves of urease-loaded gel in a 0.2 M ammonium buffer (pH 4.0) in the presence and absence of urea (1 M) as the substrate. In the absence of urea, hysteresis was observed in the volume changes by heating and cooling procedures, indicating a discontinuous phase transition. The completely collapsed volume for each sample was determined at 50°C.

aqueous solution (1 ml) containing *N*-isopropylacrylamide (75.6 mg), acrylic acid (2.3 mg), *N,N'*-methylenebisacrylamide (1.33 mg), *N,N,N',N'*-tetramethylethylene-diamine (1.85 mg), ammonium persulfate (0.4 mg), and urease (20 mg). The gelation took place at 0°C for 1 h in glass capillaries with 0.13 mm inner diameters. After the gelation was completed, the gels were taken out of the capillaries and thoroughly washed with a NH_4Cl/HCl buffer solution (0.2 M; pH 4.0). All the samples were cut into cylinders of approximately 2 mm length and stored at 3°C before use.

Equilibrium swelling volumes were measured in the buffer solutions described above with and without urea as the substrate (Fig. 2). Gel diameters were determined using a microscope with a calibrated scale, while the temperature was controlled to within 0.1°C between 30 and 40°C using circulating water. The swelling degree is calculated from the equilibrium diameter d_C and that at the complete collapsed diameter d obtained at 40°C as $V/V_C = (d/d_C)^3$.

RESULTS

In the absence of urea, the gel underwent a discontinuous phase transition at 32.4°C. In contrast, the presence of urea brought about a continuous transition, and the transition temperature was shifted to a higher range. The dependence of the volumes of the enzyme-free and enzyme-loaded gels on the urea concentration clearly indicated that the enzyme reaction contributed to the swelling and shrinking of the gel (Fig. 3).

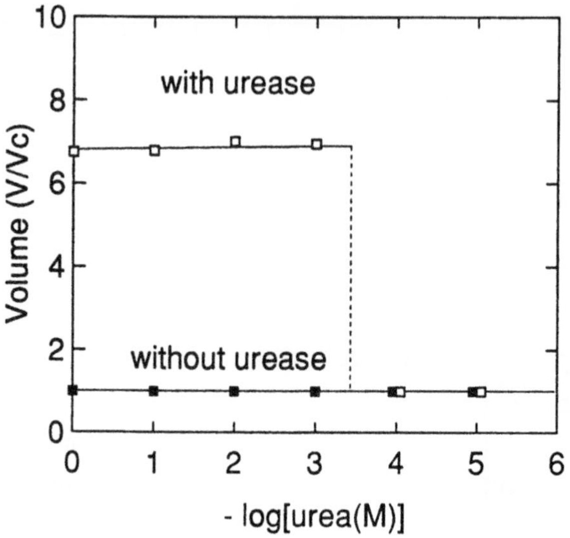
with urease
without urease
Volume (V/Vc)
- log[urea(M)]
0
2
4
6
8
10
0
1
2
3
4
5
6

Figure 3. Effect of the urea concentration on the volume ratios of the gels with and without urease at 33.0°C. The volume ratios were determined from the swelling curves in the same manner described in Fig. 1.

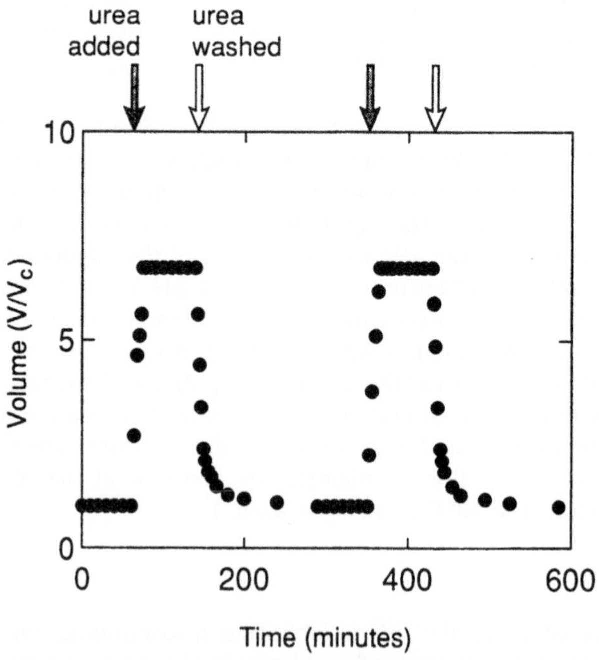
urea
added
urea
washed
Volume (V/Vc)
Time (minutes)
0
5
10
0
200
400
600

Figure 4. Repeated swelling-shrinking control of the urease-loaded gel at 33.0°C by the alternate use of the urea-free and urea-containing 0.2 M ammonium buffer solutions at pH 4.0. The urea concentration was 1 M. The aqueous phase (80 μl) in the cell into which the gel sample was immersed was quickly replaced by 4 ml of another aqueous solution within 2 min. To avoid temperature change during the replacement, both aqueous phases were kept at almost the same temperature.

We attempted to regulate the gel volume reversibly by alternate immersions of the gel in urea-free and urea-containing buffer solutions (Fig. 4). The cylindrical gel in a cell at 33.0°C was shrunken in the urea-free buffer solution. The buffer was then quickly replaced by the urea solution using a syringe. The gel began to swell as soon as the urea was introduced into the cell, and completed within 15 min. The swollen gel collapsed again when the substrate solution was replaced by the buffer. The time needed for a complete collapse was approximately 90 min for the present size of the gels. The swelling and collapsing were repeated several times with satisfactory reproducibility.

DISCUSSION

The mechanism by which the urease-loaded gel reported here undergoes a volume phase transition in response to urea may be understood as follows. The immobilized urease catalyzes the hydrolysis of urea into ammonia and carbon dioxide:

$$(NH_2)_2C{=}O + 3H_2O \Rightarrow CO_2 + 2NH_4OH.$$

The carbon dioxide produced is soluble in water and partially turns into H_2CO_3. The gel phase is saturated with the following species: NH_4OH, NH_4^+, CO_2, H_2CO_3, HCO_3^-, and Cl^-. Some of the components, NH_4OH, NH_4^+, and Cl^-, originate partially from the buffer. The reaction will increase the pH of the gel phase because of an increase in the ammonia concentration in the excess amount of HCO_3^-. The carboxyl groups of the gel are then dissociated and exert an extra osmotic pressure on the networks bringing about an increase in the transition temperature. The observed continuous phase transition of the gel in the substrate solution (see Fig. 1) suggests that a factor other than the counter ion effect causes a volume change, because the ionization of gel networks generally brings about a discontinuous phase transition with an increase in the transition temperature [10, 11]. The gel is swollen when a certain amount of urea is present but collapses in its absence if the temperature is kept within a suitable range. In this system the enzymatically produced chemical energy is converted into mechanical energy through electrostatic work.

As was mentioned above, the time for the gel to shrink in response to urea was approximately 15 min, whereas the time needed for a complete collapse was approximately 90 min for the present size (130 μm diameter). The difference indicates that the swelling and shrinking processes are not in the linear response regime, but in the nonlinear regime. Presumably, small molecules such as urea or hydroxide ions may be quickly washed out when the solvent is replaced. This should happen before the gel shrinking or swelling occur, since the collective diffusion coefficient of gels is typically of the order of 10^{-7} to 10^{-8} cm^2 s^{-1}, whereas the diffusion constants of the small molecules are 10^{-5} cm^2 s^{-1} and a hundred or thousand times faster. We speculate that the slowness may be attributed to less chance for swollen polymers to meet each other to begin collapsing. The more swollen the gel is, the slower the collapse. On the other hand the swelling process requires only the separation of paired segments which does not require polymer collision.

CONCLUSION

The enzymatically driven gel not only demonstrates an artificially constructed biochemo-mechanical system, but it is also available for technological applications.

For example, gels undergoing reversible swelling and shrinking changes in response to urea may serve as a sensor for a quick determination of a level of urea in the blood in clinical laboratories. In addition, they may be used, for example, as a switch to turn on when artificial dialysis of the blood is required. In general, the transition pH can be chosen to be any value in a wide range of pH's as was shown in various polyampholyte gels by varying the chemical composition of the polymer constituents [11]. Thus by choosing a proper chemical composition one can obtain the most suitable phase transition behavior for a particular application in a given biological condition.

Devices which release insulin in response to glucose levels are being developed by several groups using glucose-sensitive membrane systems consisting of hydrogels and glucose-oxidase [12–15]. The latter catalyzes glucose causing a pH change and gel swelling. The mechanism will become sensitive and amplified by using the phase transition behavior [16, 17].

Acknowledgements

The work was supported by NEDO and NSF. The stay of EK in the US was supported by the Japanese Ministry of Education, Science and Culture.

REFERENCES

1. W. Kuhn, B. Hargitay, A. Katchalsky and E. Eisenberg, *Nature* **165,** 514 (1950).
2. I. Z. Steinberg, H. Oplatka and A. Katchalsky, *Nature* **210,** 568 (1966).
3. M. V. Sussman and A. Katchalsky, *Science* **167,** 45 (1970).
4. Y. Osada, *Adv. Polymer Sci.* **82,** 1 (1987).
5. K. Dusek and D. J. Patterson, *Poly. Sci.* **A2,** 1209 (1968).
6. T. Tanaka, D. J. Fillmore, S.-T. Sun, I. Nishio, G. Swislow and A. Shah, *Phys. Rev. Lett.* **45,** 1636 (1980).
7. M. Ilavsky, *Macromolecules* **15,** 782 (1982).
8. E. Kokufuta and T. Tanaka, *Macromolecules* **24,** 1605 (1982).
9. E. Kokufuta, Y.-Q. Zhang and T. Tanaka, *Nature* **351,** 302 (1991).
10. Y. Hirokawa, T. Tanaka and E. S. Matsuo, *J. Chem. Phys.* **81,** 6379 (1984).
11. M. Annaka and T. Tanaka, *Nature* **335,** 430 (1992).
12. R. Siegel, in: *Pulsed and Self-Regulated Drug Delivery,* p. 129, J. Kost (Ed.). CRC Press, Boca Raton, FL (1990).
13. K. Ishihara, M. Kobayashi, N. Ishimaru and I. Shinohara, *Polym J.* **8,** 625 (1984).
14. G. Albin, T. A. Horbett and B. D. Ratner, in: *Pulsed and Self-Regulated Drug Delivery,* p. 159, J. Kost (Ed.). CRC Press, Boca Raton, FL (1990).
15. S. Kitanno, Y. Koyama, K. Kataoka, T. Okano and Y. Sakurai, *J. Controlled Release* **19,** 162 (1992).
16. A. S. Hoffmann, *J. Controlled Release* **6,** 297 (1987).
17. T. Okano, Y. H. Bae, H. Jacobs and S. W. Kim, *J. Controlled Release* **11,** 255 (1990).

Electro-driven chemomechanical polymer gel as an intelligent soft material

HIDENORI OKUZAKI and YOSHIHITO OSADA*

Division of Biological Sciences, Graduate School of Science, Hokkaido University, Sapporo 060, Japan

Received 22 October 1992; accepted 2 March 1993

Abstract—Weakly crosslinked poly(2-acrylamido-2-methylpropanesulfonic acid) (PAMPS) gel was synthesized and the chemomechanical behaviors in the presence of *N*-alkylpyridinium chloride (C_nPyCl $n = 4, 12, 16$) were studied. The principle of this behavior is based upon an electrokinetic molecular assembly reaction of surfactant molecules on the polymer gel caused by both electrostatic and hydrophobic interactions. Under an electric field PAMPS gel underwent significant and quick bending and the response could be controlled effectively by changing the alkyl chain length of the surfactant molecule, the salt concentration, and the current applied. The results allow us to consider that cooperative complex formation between PAMPS gel and C_nPyCl is responsible for this effective chemomechanical behavior.

Key words: Chemomechanical system; energy transducer; artificial muscle; soft actuator; polyelectrolyte gel; surfactant molecule; electrokinetic phenomenon; molecular assembly reaction.

INTRODUCTION

The system which undergoes shape change and produces contractile force in response to environmental stumuli is called a 'chemomechanical system'. This system can transform chemical free energy directly into mechanical work to give isothermal energy conversion and this can be seen in living organisms, for example, in muscle, flagella, and in ciliary movement.

Previously, we have found that any polyelectrolyte gel undergoes shape changes by applying d.c. current [1, 2]. This may be the first model of an electrically activated chemomechanical system working in aerobic as well as in aqueous media. We have also reported that this kind of electro-activated chemomechanical system can be applied to a drug delivery system (DDS) [3] and chemical valve which is able to regulate the permeation of chemicals by changing micropore size in the gel membrane under isometric contraction [4].

However, the contractile velocity of the gel and efficiency of energy conversion of the described system were not sufficient. Thus, we have recently developed a new type of electrically-driven chemomechanical system which shows quick responses with worm-like motility [5]. The principle of motility of this system is based upon an electrokinetic molecular assembly reaction of surfactant molecules on the hydrogel caused by both electrostatic and hydrophobic interactions. However, the detailed mechanism of the motility has not been studied.

The purpose of this paper is to analyze behavior and principles of electrically-driven artificial muscle from both thermodynamic and kinetic viewpoints.

* Author to whom correspondence should be addressed.

EXPERIMENTAL

Materials

2-acrylamido-2-methylpropanesulfonic acid (AMPS) (Nitto Chem. Co., Ltd., Japan) was purified by repeated recrystallization. *N,N'*-methylenebisacrylamide (MBAA) (Tokyo Kasei Co., Ltd., Japan) used as crosslinking agent was recrystallized twice from pure ethanol. Potassium persulfite (Tokyo Kasei Co., Ltd., Japan) which was used as a radical initiator was recrystallized from pure water. *N*-alkylpyridinium chloride (C_nPyCl, $n = 4, 12, 16$) (Tokyo Kasei Co., Ltd., Japan) and sodium sulfate (Tokyo Kasei Co., Ltd., Japan) were used as-received.

Preparation of the gel

A weakly crosslinked poly(2-acrylamido-2-methylpropane sulfonic acid) (PAMPS) gel was prepared by radical polymerization of $1.0\ mol\ l^{-1}$ solution of AMPS monomers in the presence of $0.05\ mol\ l^{-1}$ MBAA and $0.001\ mol\ l^{-1}$ $K_2S_2O_8$. The polymerization was carried out in an ampule at 333 K for 12 h under nitrogen atmosphere. After polymerization, the gel was immersed in a large amount of pure water until it reached its equilibrated size (the degree of swelling was a factor of 45).

Measurement

The configuration of the apparatus in the measurement of chemomechanical behavior (swinging pendulum) is shown in Fig. 1. A sheet of PAMPS gel (20 mm long, 5 mm wide, 1 mm thick, dry weight 0.08 g) was suspended in 10 ml surfactant solution containing a certain amount of sodium sulfate leaving the bottom end free. The ionic strength of the solution was kept constant ($\mu = 0.1$). A pair of parallel

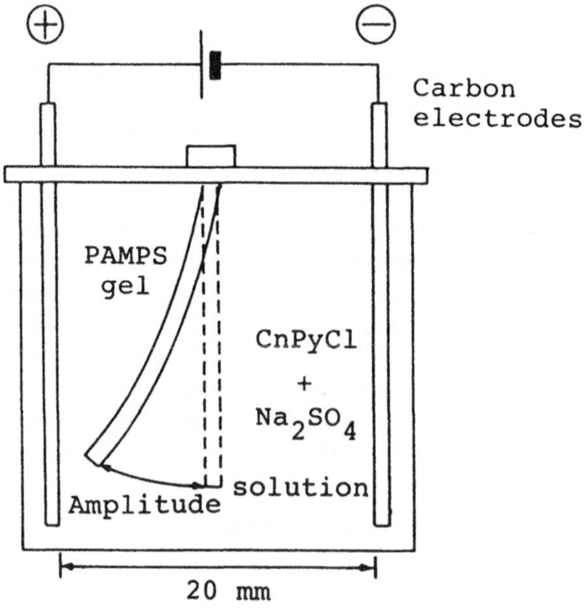
+
−
Carbon
electrodes
PAMPS
gel
CnPyCl
+
Na2SO4
solution
Amplitude
20 mm

Figure 1. Apparatus for measurement of bending and swinging of PAMPS gel under an electric field.

plate carbon electrodes (30 mm long, 10 mm wide, 1 mm thick) was inserted in the solution 20 mm apart and d.c. voltage was applied through the electrodes from an electric source using a potentiometer (HA-501, Hokutodenko, Co., Japan). The chemomechanical response of PAMPS gel under an electric field was observed by video camera and the time profile was analyzed by personal image analyzing system (XL-500, OLYMPUS). The mechanical stress generated by applying voltage was measured using a strain gauge (KYOWA, 120TK-5B). One end of a strip (20 mm long, 5 mm wide, 1 mm thick) was suspended from the strain gauge and hooked the other end of the strip to the bottom of the cell (isometric condition). Tension, not enough to cause stretching by itself, was then applied to the sample. The profile of the complexation of surfactant molecules with the gel was studied spectrophotometrically, i.e. by observing the change in UV adsorption of aqueous solution of surfactant molecules at 259 nm (1×10^{-2} M) containing a piece of cylindrical PAMPS gel (13 mm diameter, 2 mm thick) with time.

RESULTS AND DISCUSSION

Chemomechanical behaviors

When an electric field is applied to a sheet of PAMPS gel suspended in the surfactant solution, the gel showed significant and quick bending toward the anode (Fig. 1). Among the surfactants used, *N*-dodecylpyridinium chloride (C_{12}PyCl) and *N*-hexadecylpyridinium chloride (C_{16}PyCl) showed intensive and steady bending and the magnitude of bending attained 4 ~ 6 mm within 1 s (Fig. 2). On the other hand, *N*-butylpyridinium chloride (C_4PyCl) caused rather weak and slow bending and the direction of which was opposite to those of C_{12}PyCl and C_{16}PyCl (Fig. 2). Similar behavior was obtained by the sodium sulfate solution containing no surfactant molecules. We have described in the previous paper [2] that the electrically-driven bending of polymer gel immersed in sodium sulfate is caused by the incorporation of sodium ions into the sulfonic gel from the side of the gel surface facing the anode. The PAMPS gel swells more extensively when adsorbed with sodium ions due to enhanced electrostatic repulsion of the sulfonate group than with adsorbed proton ions. Therefore, the bending of PAMPS gel towards the cathode in the sodium sulfate solution is associated with anisotropic swelling of the gel at the surface facing the anode. The pH change of the gel occurs at a local level both by a change in the concentration of mobile ions and by the electrode reactions. However, it should be noted that the pH change is not dominant for the shrinkage of the gel because a salt bridge which was used as an electrode instead of carbon or platinum can also induce extensive shrinkage of the gel without any pH change. Conversely, the bending of PAMPS gel toward the anode in solutions of C_{12}PyCl and C_{16}PyCl can be asssociated with the anisotropic assembly reaction of surfactant molecules on the surface of polymer gel facing the anode, whereupon surfactant molecules form micellar-like reaction through hydrophobic interaction of long alkyl chains and cause an effective contraction of the gel.

If the polarity of the electric field is altered, the gel showed repeated pendulum-like swinging in the surfactant solution. The mechanism of such fast electro-activated bending of the gel is connected with the electrokinetic molecular assembly reaction: the positively charged surfactant molecules undergo

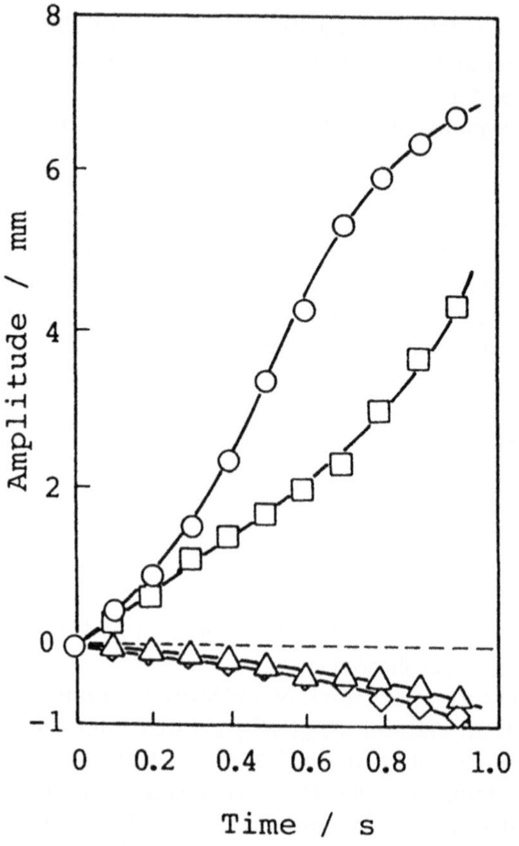
Amplitude / mm
8
6
4
2
0
-1
0
0.2
0.4
0.6
0.8
1.0
Time / s

Figure 2. Electro-driven chemomechanical responses of PAMPS gel in the presence of C_nPyCl under d.c. 10 V. (△) C_4PyCl (1×10^{-2} M) + Na_2SO_4 (3×10^{-2} M); (○) $C_{12}PyCl$ (1×10^{-2} M) + Na_2SO_4 (3×10^{-2} M); (□) $C_{16}PyCl$ (1×10^{-2} M) + Na_2SO_4 (3×10^{-2} M); (◇) Na_2SO_4 (3.33×10^{-2} M).

electrophoretic movement towards the cathode and form the complex with the negatively charged gel, mainly on the side facing the anode of the PAMPS gel and this causes anisotropic contraction resulting in bending towards the anode. When the electric field is reversed, the surfactant molecules adsorbed on the surface of the gel leave and electrophoretically travel away toward the cathode. Instead, new surfactant molecules approach from the opposite side of the gel and form the complex preferentially on that side of the gel and it bends in the opposite direction. Thus, the gel can swing like a pendulum more than 3000 times by changing the polarity every 2 s. The profile is shown in Fig. 3.

Figure 4 shows dependence of amplitude of swinging on electric current. Here, an electric current was controlled by changing the concentration of sodium sulfate while the surfactant concentration was kept constant. It is seen that the swinging amplitude of the polymer gel increases with increased electric current. This result additionally indicates that the chemomechanical behavior described is based on the electrokinetic process, whereupon a movement of ions and subsequent electrochemical reaction at the electrodes plays an important role.

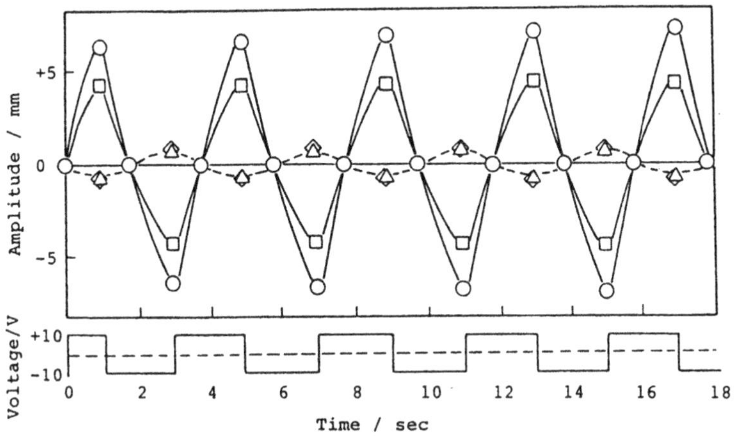
Amplitude / mm
+5
0
-5
Voltage/V
+10
-10
0
2
4
6
8
10
12
14
16
18
Time / sec

Figure 3. Time profiles of electro-driven chemomechanical swinging of PAMPS gel in the presence of C_nPyCl under a. c. 10 V. every 2 s. (△) C_4PyCl (1×10^{-2} M) + Na_2SO_4 (3×10^{-2} M); (○) $C_{12}PyCl$ (1×10^{-2} M) + Na_2SO_4 (3×10^{-2} M); (□) $C_{16}PyCl$ (1×10^{-2} M) + Na_2SO_4 (3×10^{-2} M); (◇) Na_2SO_4 (3.33×10^{-2} M).

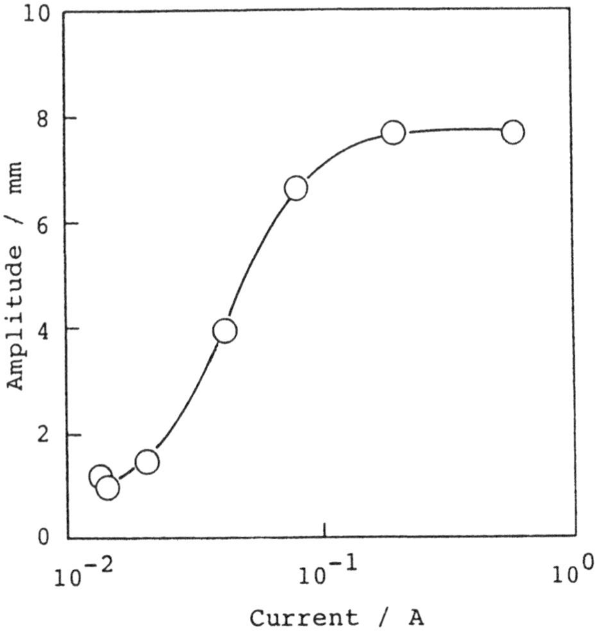
Amplitude / mm
10
8
6
4
2
0
10^-2
10^-1
10^0
Current / A

Figure 4. Dependence of amplitude of swinging on electric current in 1×10^{-2} M $C_{12}PyCl$ solution.

Figure 5 shows the dependence of amplitude of swinging on the surfactant concentration. Here, the surfactant concentration was changed, while ionic strength was kept constant by means of adjusting the sodium sulfate concentration. When the surfactant concentration is below 1×10^{-3} M, the amplitude of swinging is negligibly low, but later the magnitude increased very rapidly with increased surfactant concentration. This result is associated with enhanced complexation of surfactant molecules at the surface of the polymer gel due to an increase in surfactant

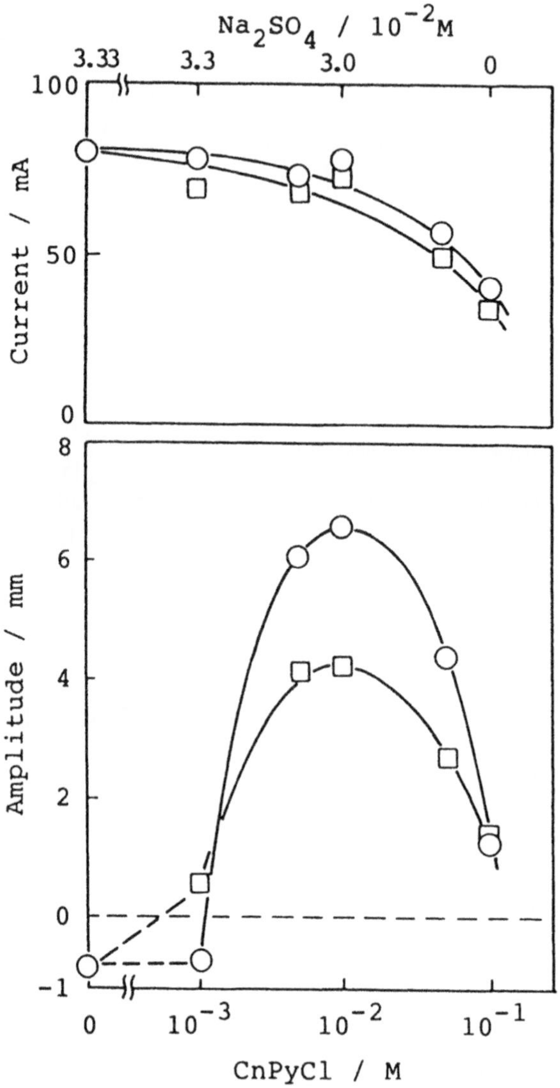

Na_2SO_4 / 10^{-2}M
3.33
3.3
3.0
0
Current / mA
100
50
0
Amplitude / mm
8
6
4
2
0
-1
0
10^{-3}
10^{-2}
10^{-1}
CnPyCl / M

Figure 5. Dependencies of amplitude of swinging and current observed on the surfactant concentration. (O) $C_{12}PyCl + Na_2SO_4$; (□) $C_{16}PyCl + Na_2SO_4$; ionic strength of the medium was kept constant.

concentration. A decrease in the amplitude with further increase of surfactant concentration might be attributed to a decrease in the electric current in this region (increase in surfactant concentration denotes a decrease in concentration of sodium sulfate since the ionic strength of the medium was kept constant). Thus, there exists an optimum surfactant concentration in order to cause effective bending of the polymer gel.

Stability of gel–surfactant molecules complexation

As described above, the principle of the motility of the gel is based on the molecular assembly reaction of cationic surfactant molecules on the negatively-charged hydrogel, whereupon both electrostatic and hydrophobic interaction plays an important role in the effective contraction.

Generally, swelling of the ionic gel is well explained by the difference of osmotic pressure associated with the freely mobile ions between inside and outside the gel. The contraction of the gel in our system is particularly associated with a decrease in osmotic pressure due to the neutralization of negative charges of the gel. As seen in Fig. 5 a marked contraction of the gel is observed only above a certain critical concentration of the surfactant which indicates that the surfactant–gel complexation has a cooperative nature associated with hydrophobic interaction of surfactant molecules.

As is well known, the stability constant (K_u) of the complexation can be calculated as follows:

$$\begin{array}{cccc} \text{Polymer} + & \text{Surfactant} & \overset{K_u}{\rightleftharpoons} & \text{Complex} \\ C_p & C_{so} & & 0 \\ C_p\text{-}(C_{so}\text{-}C_s) & C_s & & C_{so}\text{-}C_s \end{array}$$

$$K_u = (C_{so}\text{-}C_s)/\{C_p\text{-}(C_{so}\text{-}C_s)C_s\}$$
$$= 1/C_{s_{0.5}} \; (\beta = 0.5)$$

where C_p is the concentration of sulfonic acid groups in the gel, C_{so} is an initial concentration of the surfactant, C_s is the surfactant concentration in the surrounding

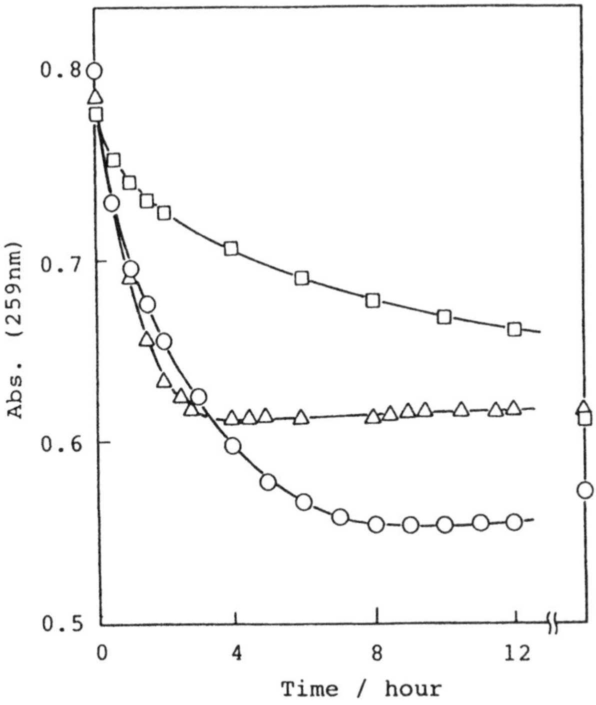

0.8
0.7
0.6
0.5
Abs. (259nm)
0
4
8
12
Time / hour

Figure 6. Time profiles of absorption change in the presence of polymer gel. (△) C_4PyCl (1×10^{-2}M); (○) C_{12}PyCl (1×10^{-2}M); (□) C_{16}PyCl (1×10^{-2}M).

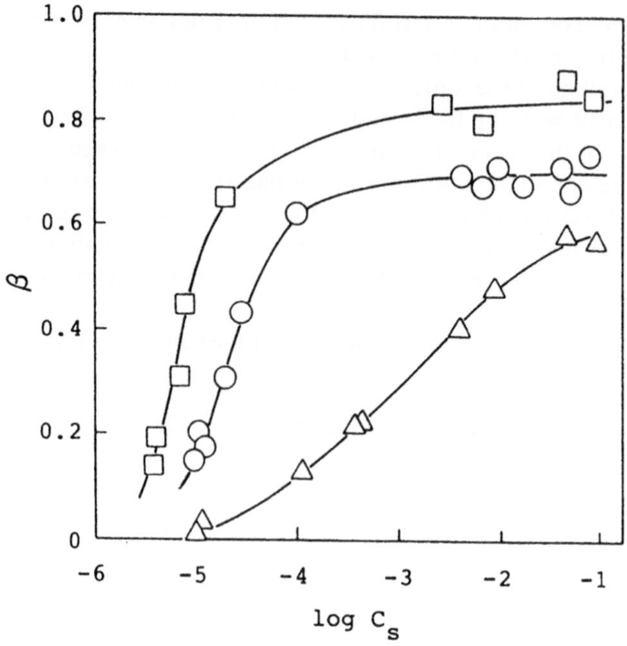

1.0
0.8
0.6
0.4
0.2
0
β
-6
-5
-4
-3
-2
-1
log C_s

Figure 7. Binding isotherms for surfactant molecules into the polymer gel. (△) C_4PyCl; (○) C_{12}PyCl; (□) C_{16}PyCl.

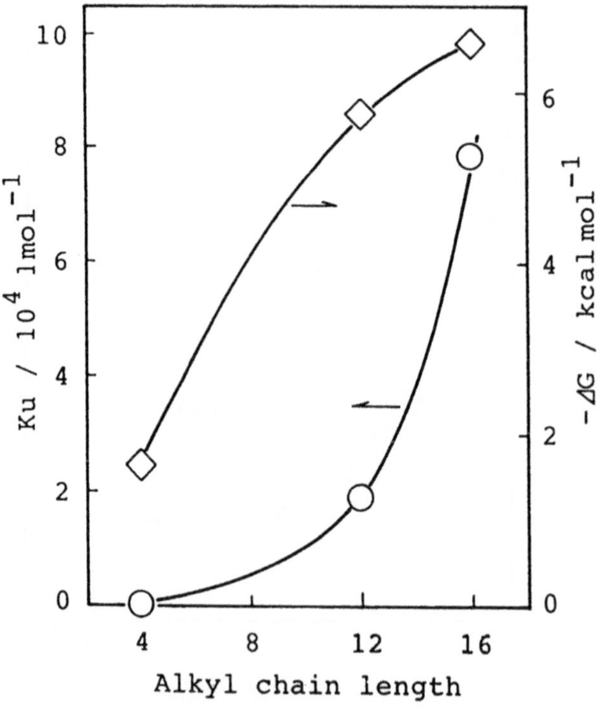

10
8
6
4
2
0
Ku / 10^4 l mol^{-1}
6
4
2
0
-ΔG / kcal mol^{-1}
4
8
12
16
Alkyl chain length

Figure 8. Dependencies of stability constant of the complex (K_u) and free energy change (ΔG) on alkyl chain length of the surfactant molecules.

solution at equilibrium state, and β is the degree of complexation (molar ratio of surfactant adsorbed to the sulfonic acid group in the gel).

Figure 6 shows the dependence of free surfactant concentration when different amounts of surfactant were added to the polymer gel. It is seen that the free surfactant concentration decreases significantly with time due to complexation with the gel. If a degree of complexation (β) derived from the concentration change is plotted as a function of the surfactant concentration (C_s), one obtains Fig. 7. According to reference [6], we can calculate a stability constant of the complex by knowing C_s at $\beta = 0.5$ in Fig. 7. Figure 8 shows the dependencies of the stability constant and free energy change of the complexation obtained by this method for various alkyl chain lengths of surfactant molecules. It is seen that the stability constant (K_u) increases significantly with increasing alkyl chain length of surfactant molecule and the value of C_{16}PyCl is about 5000 times larger than C_4PyCl and more than 4 times larger than C_{12}PyCl.

These results indicate that the hydrophobic interaction of alkyl chain plays an important role in an effective complexation: the positively charged surfactant molecules complex with the sulfonate moiety through an electrostatic salt formation, but the alkyl chains make an additional hydrophobic binding side on the side to give a micellar-like structure in the gel.

Kinetics of complexation of surfactant molecules and contraction behaviors

In the previous section we indicated that the contraction of the gel strongly depends upon the stability of the complex. However, contractile behavior of the gel should also depend on the kinetic aspect of complexation of the surfactant molecules in the gel since complexation of the given case is heterogeneous and occurs by the penetration of surfactant molecules through the polymer network.

Figure 9 shows the time profile of complexation of surfactant molecules and contraction of the gel when no electric field was applied. It is seen that C_4PyCl can penetrate very rapidly into the gel attaining $\beta = 0.6$ within the initial 3 h. However, no significant contraction of the gel occurred (a). C_{12}PyCl can adsorb as rapidly as C_4PyCl and easily attains $\beta = 1.0$. At the same time it gives significant contraction of the gel (b). On the other hand, complexation of C_{16}PyCl occurred very slowly and the velocity of concentration is much slower than C_{12}PyCl (c) (it took 2 days to reach adsorption equilibrium, and more than 6 days to reach equilibrium contraction).

Time profiles of complexation and contraction shown here can easily be explained both by thermodynamic and kinetic viewpoints of the complexation in the polymer networks. In the case of C_4PyCl, surfactant molecules are able to penetrate easily into the swollen network because of its small molecular size and attains equilibrium quickly. This surfactant, however, is not able to cause effective contraction because of its low stability constant. In contrast, C_{16}PyCl penetrates extremely slowly through the network and equilibrium can be established only after 48 h. The contraction of the gel occurs also very slowly and the shrinkage of the gel continued for a considerable period of time after complexation of surfactant molecules has been completed. C_{12}PyCl can penetrate as fast as C_4PyCl and induces rapid and extensive contraction of polymer gel because of its suitable molecular size to penetrate the network and large stability constant sufficient to induce effective contraction.

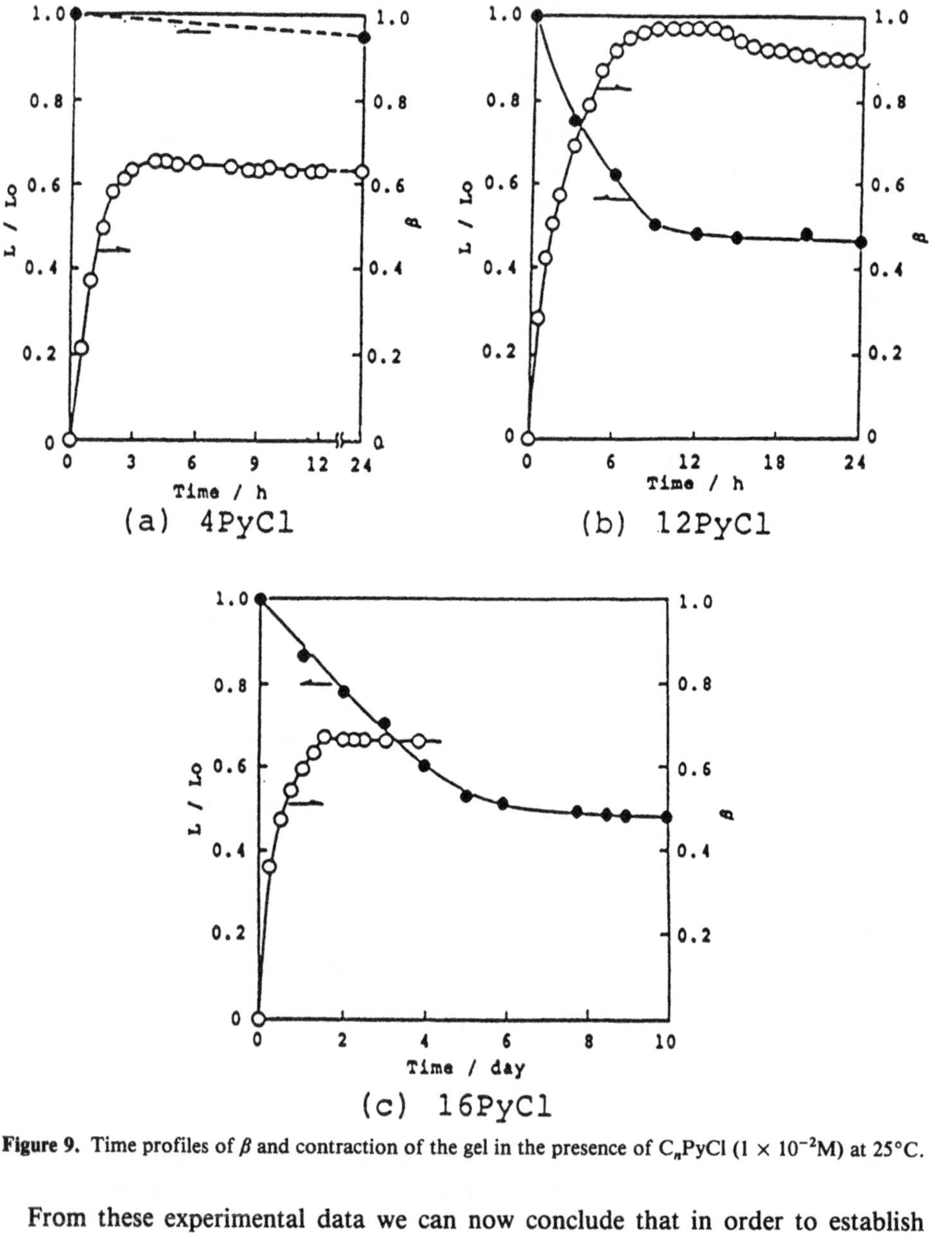
(a) 4PyCl
(b) 12PyCl
(c) 16PyCl
L / Lo
β
Time / h
Time / day

Figure 9. Time profiles of β and contraction of the gel in the presence of C_nPyCl (1×10^{-2}M) at 25°C.

From these experimental data we can now conclude that in order to establish effective chemomechanical behavior, we have to consider both thermodynamic and kinetic viewpoints, and there exists an appropriate molecular size of the surfactant to satisfy both factors.

We have found that the rate of complexation of surfactant molecules could be accelerated by the electric field. As seen in Fig. 10 the rate of complexation increases in proportion to the electric field, whereupon the smaller the molecular size of surfactant, the larger the effect of acceleration. Effective enhancement of the complexation above 4 V indicates that the increase in the rate of complexation is due to enhanced electrophoretic transport of surfactant molecules as described above.

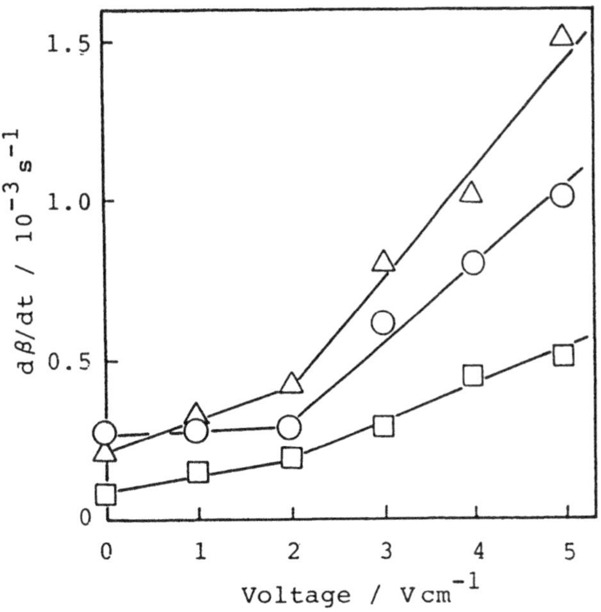
1.5
1.0
0.5
0
dβ/dt / 10^{-3} s^{-1}
0
1
2
3
4
5
Voltage / Vcm^{-1}

Figure 10. Effect of the electric field on the rate of complexation of C_nPyCl with PAMPS gel. (△) C_4PyCl (1×10^{-2}M); (○) $C_{12}PyCl$ (1×10^{-2}M); (□) $C_{16}PyCl$ (1×10^{-2}M).

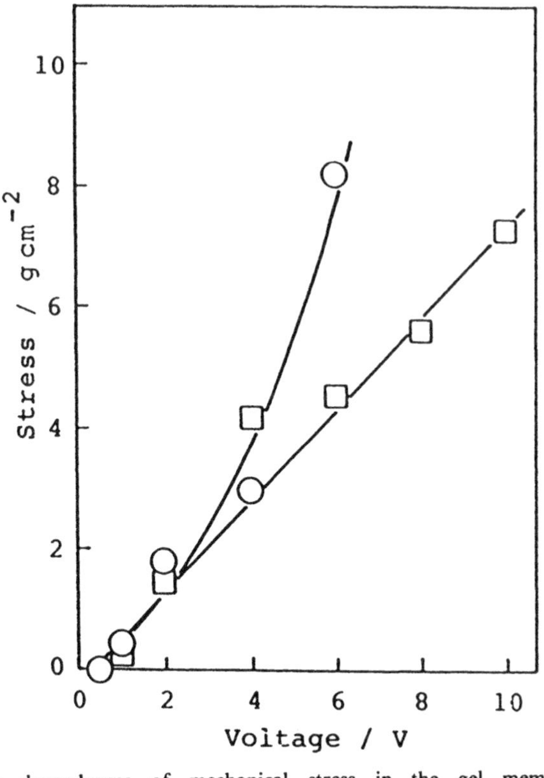
10
8
6
4
2
0
Stress / gcm^{-2}
0
2
4
6
8
10
Voltage / V

Figure 11. Voltage dependences of mechanical stress in the gel membrane. (○) $C_{12}PyCl$ (1×10^{-2}M) + Na_2SO_4 (3×10^{-2}M); (□) $C_{16}PyCl$ (1×10^{-2}M) + Na_2SO_4 (3×10^{-2}M).

Figure 11 shows the voltage dependences of mechanical stress in the strip of polymer membrane. Stress increased largely in proportion to the voltage applied. When the stress becomes larger than 8 g cm^{-2} the strip disconnected and further experiments could not be carried out.

Thus, the mechanism of electro-activated quick bending of the gel is connected with the electrokinetic molecular assembly reaction: the positively charged surfactant molecules undergo electrophoretic movement toward the cathode and form the complex with the negatively charged gel, mainly on a side facing the anode of PAMPS gel and this causes anisotropic contraction to give bending toward the anode. When the electric field is reversed, the surfactant molecules adsorbed on the surface of the gel leave and electrically travel away toward the cathode. Instead, new surfactant molecules approach from the opposite side of the gel and form the complex preferentially on that side of the gel and stretch the gel. Here, the amount of $C_{12}PyCl$ molecules adsorbed in the gel (β) in 2 s is calculated as less than 1×10^{-3}, explaining that the quick and significant swinging of the gel under an electric field shown in Fig. 3 is caused by the complexation primarily at the solution–polymer gel interface.

This kind of moving device of polymer gel may serve as a new type of 'soft-actuator' or 'molecular machine'. Actually, we have developed an electro-driven drug delivery system using PAMPS gel [3]. The principle described can also be applied as permselective 'chemical valve' membrane, where reversible expansion and contraction of the pore are made under an electric field, thereby selectively permeating solute molecules by their size [4]. Unlike in motors and hydrodynamic pumps, the motility of this system is produced by the chemical free energy of molecular assembly reactions at the polymer network, with the electric energy being used to drive the direction and control the state of equilibrium.

Therefore, the chemomechanical muscle produces 'gentle' and 'flexible' action and its movement is unlike the metallic machine systems.

Acknowledgement

This research was supported in part by Grant-in-Aid for Experiment Research 'Electrically Driven Chemomechanical Polymer Gels as Artificial Muscle' from the Ministry of Education, Science and Culture (03555188), Japan.

REFERENCES

1. Y. Osada and M. Hasebe, *Chem. Lett.* **1285** (1985).
2. R. Kishi and Y. Osada, *J. Chem. Soc., Faraday Trans.* **85**, 655 (1989).
3. K. Sawahata, M. Hara, H. Yasunaga and Y. Osada, *J. Contr. Release* **14**, 253 (1990).
4. Y. Osada, *Adv. Mater.* **3**, 107 (1991).
5. Y. Osada, H. Okuzaki and H. Hori, *Nature* **355**, 242 (1992).
6. K. Hayakawa, J. P. Santerre and J. C. T. Kwak, *Macromolecules* **16**, 1642 (1983).

Polyanionic hydrogel as a gastric retentive system

SAU-HUNG S. LEUNG[1], BRIAN K. IRONS[1] and JOSEPH R. ROBINSON[2]

[1]*Columbia Research Laboratories, Madison, WI 53713, USA*
[2]*School of Pharmacy, University of Wisconsin, Madison, WI 53706, USA*

Received 28 July 1992; accepted 11 October 1992

Abstract—Gastric emptying of a polyanionic hydrogel, polycarbophil (PC), from the canine stomach was studied using a duodenal cannulation technique. The basis of the study is to employ a certain quantity of swelling hydrogel that, by virtue of its swollen size and viscosity, converts a fasted stomach to a fed state and resists discharge from the stomach for an extended period of time. Different amounts of PC, in 200 ml water plus buffering agents, were administered orally to fasted canines. The gastric emptying lag time was found to increase with the viscosity of the administered dose. Addition of a base, sodium bicarbonate, to PC increased gastric retention via an increase in its apparent viscosity. The polymer mass is retained in the canine stomach until a sufficient quantity of stomach acid secretion reduces the viscosity of the viscous mass, through protonation of polycarbophil, and, at that point, discharge of the hydrogel will commence. Thus, gastric retention of PC hydrogel in the canine stomach can be prolonged by increasing the apparent viscosity of the hydrogel administered.

Keywords: Gastric retentive; swollen hydrogel; sustained release; stomach; delayed gastric discharge.

INTRODUCTION

For sustained release oral drug delivery systems, prolonged gastrointestinal (GI) residence time of a formulation can affect both duration and extent of drug absorption [1-5]. The GI residence time of a drug delivery system between the mouth and the caecum can be controlled by delaying gastric emptying of the formulation [6-10]. Measurement of gastric emptying of different oral formulations has been studied using several techniques including duodenal cannulation [8, 11, 12], γ-scintigraphy [13-16], radiotelemetry [17, 18], ultrasound and fluoroscopic imaging techniques [19]. Approaches that have been tried, with limited success, to prolong gastric emptying time or GI transit time include either lowering [20] or raising [21] the apparent density of particles, alterations of size [22-26], shape [23, 24], integrity [27] and surface properties of the formulation [28, 29]. The present study describes the use of a viscosity building mucoadhesive hydrogel, polycarbophil, to prolong gastric emptying time in canines. These results may be applied to the design of gastric retentive drug delivery systems.

In man and canine, mixing and propulsion of ingested material is governed by a GI motility pattern. The cyclic pattern of GI motility, which originates in the foregut and propagates to the terminal ileum, can be divided into fasted and fed states (Fig. 1) [30]. The fasted state has four phases; phase I is the quiescent period of minimal or no activity that lasts for 50-60 min [31]; phase II has random motor spikes that last for 20-40 min [32]; phase III is a period of high amplitude contractions at a maximal frequency that lasts for 15-20 min [32], and is commonly known as the 'Housekeeper Wave'; and phase IV is the transitional period between phase III and I [31]. The average length of one complete cycle, commonly known as

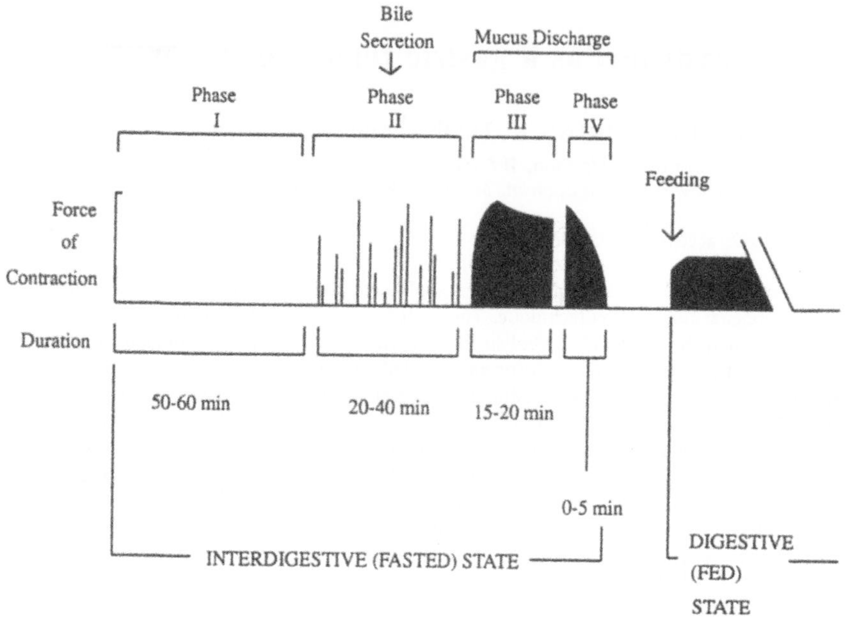
Bile
Secretion
Mucus Discharge
Phase
I
Phase
II
Phase
III
Phase
IV
Feeding
Force
of
Contraction
Duration
50-60 min
20-40 min
15-20 min
0-5 min
INTERDIGESTIVE (FASTED) STATE
DIGESTIVE
(FED)
STATE

Figure 1. A representation of typical motility patterns in the interdigestive (fasted) and digestive (fed) state in the dog. Major secretions associated with fasted motility are also listed.

the interdigestive migrating motor complex (IMMC), ranges from 90–120 min in both man and canine [33]. In contrast, the fed state has continuous contractions, with a force that is approximately half that of the housekeeper wave, which continues until the stomach is emptied [30]. Thus, when the stomach is converted from a fasted to fed state by ingestion of an appropriate amount of digestible and/or non-digestible material, the force of contraction is reduced. Reduction in the force of contraction in the fed state, as compared to the fasted state, may prolong the gastric emptying time of gastric contents. Administration of oral formulations that convert a fasted stomach to a fed state may have a longer gastric emptying time than those formulations that have no effect on GI motility.

Normally, the motility pattern will immediately convert from the fasted to the fed state when a meal is ingested, and the fed state will be maintained until all of the food has left the stomach. A normal meal can sustain the fed state for 3–4 h in humans [34] or up to 8 h in canines depending upon the caloric content of the meal [35]. Liquid materials of 150 ml or more may initiate a fed state, but are emptied rapidly by the stomach, enabling the stomach to return to the fasted state in a short period of time [36]. Thus, converting the stomach into a fed state is by itself not sufficient to keep a formulation in the stomach for an extended period of time. It was found [37] that viscous liquids, e.g. >5000 cps, show a longer lag time before gastric emptying begins. Thus, 150 ml or more of highly viscous hydrogel may substantially prolong gastric emptying.

Polycarbophil (PC), a lightly cross-linked (0.05–1.5% by weight of cross-linking agent, divinyl glycol [38]) polyacrylic acid was chosen to be used as a model hydrogel in this study. Polycarbophil is a known bioadhesive [29, 38–40] that hydrates in aqueous medium and absorbs about 60 times its original weight [41]. At the

Table 1.
Effect of hydration on viscosity of polycarbophil

Ratio $\frac{PC}{DDW}$	$\frac{1}{10}$	$\frac{1}{13.33}$	$\frac{1}{20}$	$\frac{1}{30}$	$\frac{1}{40}$
Viscosity (cps)	204 091	97 164	57 183	28 824	20 456
(S.E.M.)	(31 490)	(2026)	(9895)	(2324)	(2324)

beginning of hydration, a hydrogel with high viscosity develops. As the network further hydrates, the resultant viscosity decreases (Table 1). Thus, the lower the degree of hydration, the higher the viscosity of the hydrogel and the higher will be its resistance to gastric discharge.

Once administered, there are four parameters that influence the degree of hydration and subsequent viscosity of polycarbophil. First, is the volume of available water. When a dose of polycarbophil is administered, the aqueous medium required to hydrate the polymer comes from three different sources: the volume of co-administered fluid, the resting gastric fluid volume, and induced gastric fluid secretions.

The second parameter is pH of the bathing medium. The degree of hydration of polycarbophil is greatly influenced by pH [42]. At a pH greater than 5, the carboxyl groups on the polycarbophil are mostly ionized (pK_a of polyacrylic acid is 4.75 [43]). The charged groups along the polymer chains establish an electrostatic repulsion which tends to expand the polymer network and water is absorbed from the bulk solution following Donnan's equilibrium phenomenon [44]. The apparent volume of equilibrium swelling decreases dramatically when the medium pH drops below 4 [42]. Thus, by varying pH of the stomach content, the degree of hydration of polycarbophil, and hence viscosity, can be varied. There are two types of gastric acid secretion, basal and induced. Basal acid secretion is higher in men than in women, which is probably due to a larger maximal acid secretory capacity in man [45]. The pH of gastric fluid in a resting stomach is around one and can vary greatly between individuals [45]. There are a number of factors that can induce gastric acid secretion, e.g. distention of the stomach, increase in the gastric pH, peptone, gastrin or food. In a 24 h period, the average gastric acid secretion is about 200 mmol [46], with just over two thirds of this amount secreted during the daytime (i.e. 9:00 a.m. to 9:00 p.m.) and the rest at night (i.e. 9:00 p.m. to 9:00 a.m.). The secreted gastric acid affects the luminal pH of the stomach, which can be varied by administration of appropriate amounts of buffer or alkalinizing agents, e.g. sodium bicarbonate.

The third parameter is ionic strength of the bathing medium. Hydration of polycarbophil follows the Donnan's equilibrium phenomenon [44], wherein the amount of water absorbed per gram of polycarbophil decreases with an increase in ionic strength of the bathing medium. Hydration of the polymer network depends on establishing a net osmotic pressure across the polymer network which, in effect, acts as a membrane. The net osmotic pressure decreases as the ionic strength of the external bulk solution increases, with a corresponding decrease in degree of hydration.

The fourth parameter influencing viscosity of the hydrogel is the relative amount of polymer administered and aqueous medium available. The higher the polymer to water ratio, the higher the resultant viscosity and the greater the gastric retention.

MATERIALS AND METHODS

Animals

Adult female beagle dogs, weighing 10–16 kg (Ridglan Research Farm, Mount Horeb, WI), were used in the study. Each dog was prepared with a permanent modified Derlin Thomas duodenal cannula [47, 48]. The duodenal cannula was implanted following the surgical procedure of Reinke *et al.* [49] and the dogs were allowed to recover for at least two weeks before they were used for studies.

Materials and equipment

Polycarbophil (Pharmaceutical grade, Noveon AA-1, previously called Carbopol EX55) was obtained from B. F. Goodrich Co., Brecksville, OH. The diameter of all PC particles were smaller than 53 μm [50], and the pH of 1.0% PC in water is not more than 4.0 [41, 50]. All other chemicals were either pharmaceutical or food grade and were used as received. Modified Derlin Thomas cannulas were manufactured by the University of Wisconsin Physical Plant Machine Shop, Madison, WI. Viscosities were measured at 37°C with rotor speed (n) set at 4 min^{-1} using a Haake Viscometer Rotovisio RV12 (SV rotor and cup), Haake Buchler Instruments, Inc., Saddle Brook, NJ.

Administration of formulations

All hydrogel mixtures were allowed to hydrate overnight before administration. Dry mixtures were administered in 000 gelatin capsules followed by 50 ml of water, the minimum amount of water for the dog to swallow the gelatin capsules. All experiments were done in triplicate.

Prior to each experiment, the dogs were fasted for 16–18 h with a free supply of water. The duodenal discharge collected from the cannula was used to ascertain the phase of gastrointestinal activity. After bile and mucus discharge (indicating phases II and III) ceased, an additional 30 min period of no discharge was allowed in order to make certain that the gastrointestinal motility of the dog was in phase I. At time zero the test formulation was administered in aliquots to the back of the dog's mouth and the animal was allowed to swallow freely. Capsules were administered by placing them at the back of the dog's mouth one at a time followed by 5–10 ml of water. At the end of each experiment, one or two rinsing doses (200 ml) of water were administered to the canine to remove the PC hydrogel still adhering to the stomach and duodenum.

Collection of gastric discharge from cannula

After oral administration, gastric discharge was collected every 15 min from the duodenal cannula. The collected samples were then separately rinsed several times with 1 M NaOH solution to remove the bile pigment. Their hydration volume in 0.1 M pH 2 phosphate buffer was then measured, and the amount of hydrogel in the collected gastric discharge was determined by comparing with the equilibrium hydration volume of known amounts of polycarbophil in 0.1 M pH 2 phosphate buffer.

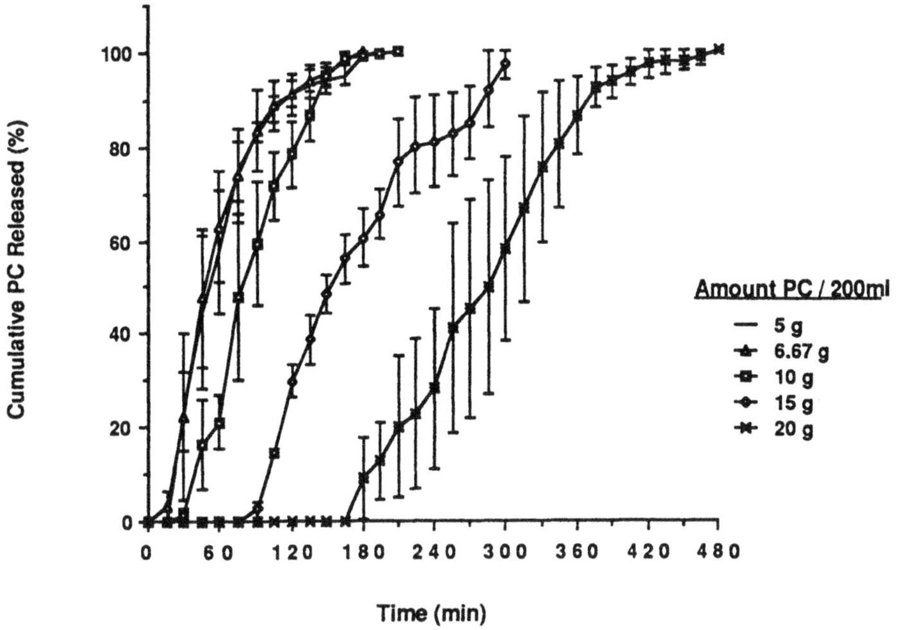
Cumulative PC Released (%)
100
80
60
40
20
0
0 60 120 180 240 300 360 420 480
Time (min)
Amount PC / 200ml
5 g
6.67 g
10 g
15 g
20 g

Figure 2. Gastric release profiles for different amounts of polycarbophil hydrated in 200 ml DDW (mean ± S.E.M.).

RESULTS AND DISCUSSION

The average gastric release profiles, lag times and gastric emptying half-lives for different amounts of polycarbophil hydrated in 200 ml of DDW are shown in Fig. 2 and Table 2. Cumulative percent release of PC equals the cumulative amount of PC collected via the duodenal cannula divided by the total amount of PC collected over time. The total amount of PC collected is around 30–70% of the administered dose. Some of the PC adhered to the stomach and duodenum (1–10%) and some by-passed the duodenal cannula due to cannula design and could therefore not be collected. From Fig. 2, there is a characteristic lag time for gastric release of hydrogel and a gradual release after the lag time. From these studies, the gastric emptying lag time was found to increase with the amount of PC administered. This increase in gastric retention may be due to the combined mechanism of conversion of the canine stomach from a fasted to a fed state and resistance to gastric discharge

Table 2.
Effect of hydrated polycarbophil on gastric emptying of canine.

Amount of PC in 200 ml DDW (g)	Mean gastric emptying lag time ± S.E.M. (min)	Mean gastric emptying half-life ± S.E.M. (min)
5.0	35 ± 5	18 ± 6
6.67	30 ± 9	19 ± 5
10	45 ± 9	23 ± 9
15	95 ± 5	67 ± 9
20	210 ± 30	68 ± 19

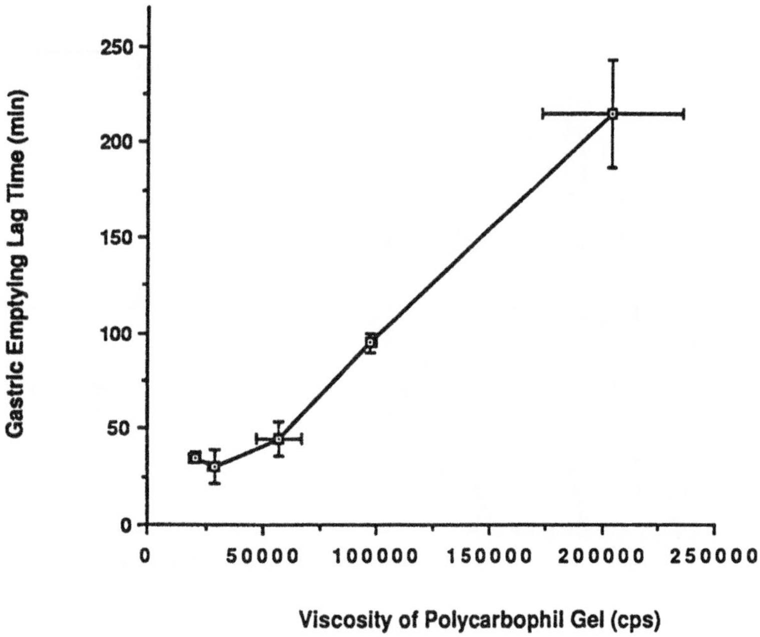
Gastric Emptying Lag Time (min)
250
200
150
100
50
0
0
50000
100000
150000
200000
250000
Viscosity of Polycarbophil Gel (cps)

Figure 3. Effect on polycarbophil hydrogel viscosity on lag time for gastric release (mean ± S.E.M.).

through increased viscosity. When the amount of polycarbophil in 200 ml DDW increased, viscosity of the resultant hydrogel also increased. The effect of polycarbophil hydrogel viscosity on gastric retention is shown in Fig. 3. There is a small increase in lag time for low viscosities; however, the lag time for release was found to increase linearly with viscosity of the hydrogel when the viscosity increased from 50 000 to 200 000 cps. Thus, the higher the viscosity the longer the retention time.

Gastric release of polycarbophil gel was simulated using the STELLA® computer program from High Performance Systems, Inc. STELLA® is a mathematical modeling program that can be used to simulate time-dependent behavior of pharmacodynamic and pharmacokinetic systems [51, 52]. STELLA® has been used

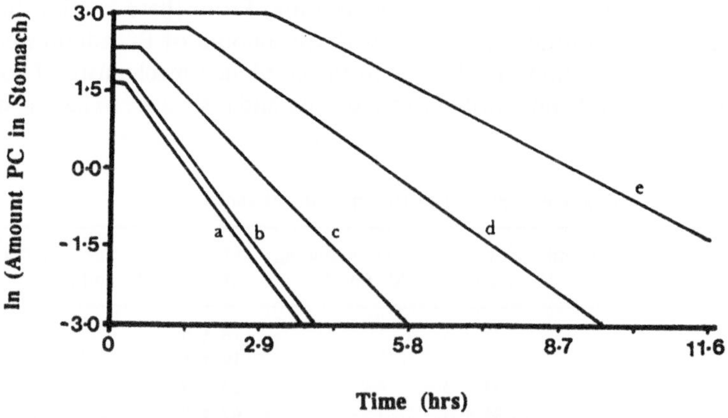
ln (Amount PC in Stomach)
3·0
1·5
0·0
-1·5
-3·0
0
2·9
5·8
8·7
11·6
a
b
c
d
e
Time (hrs)

Figure 4. Computer simulation of gastric release of hydrated polycarbophil. Amount PC hydrated in 200 ml DDW: (a) 5 g; (b) 6.67 g; (c) −10 g; (d) −15 g; and (e) −20 g.

Table 3.
Effect of sodium bicarbonate on pH, ionic strength, viscosity and gastric emptying lag time of 5.0 g PC in 200 ml DDW.

Amount sodium bicarbonate (g)	pH	Ionic strength (mM)	Mean viscosity ± S.E.M. (cps)	Mean gastric emptying lag time ± S.E.M. (mind)
0	3.0	0	22,222 ± 445	35 ± 5
1	4.3	60	106,222 ± 3205	123 ± 3
2	5.0	119	113,778 ± 445	120 ± 15
3	5.6	179	135,111 ± 445	120 ± 23

in pharmacokinetic modeling to predict the behavior of formulations [53–55] and devices [56]. The simulated gastric emptying profiles for different amounts of polycarbophil in 200 ml water correlate very well with the experimental results. The natural logarithm of PC amount versus time is shown in Fig. 4. Gastric release of hydrated PC hydrogel follows first order kinetics, and the gastric emptying rate decreases with an increase in amount of hydrogel. Thus, increasing the amount of hydrogel will increase the resultant viscosity and decrease the rate of gastric release.

One of the objectives of this study is to develop a method to increase the viscosity of polycarbophil hydrogel so that smaller amounts of PC can be used in gastric retentive formulations. Viscosity of the administered polycarbophil can be increased by increasing the amount of polycarbophil, increasing the medium pH, decreasing the water of hydration or by the addition of other viscosity building polymers. Sodium bicarbonate was chosen to increase the medium pH and apparent viscosity of 5 g polycarbophil in 200 ml water. The effect of sodium bicarbonate on the viscosity and gastric emptying lag time of 5 g PC in 200 ml DDW is shown in Table 3. There is a dramatic increase of viscosity from 22 222 to 106 222 cps when sodium bicarbonate is increased from 0 to 1 g. However, when the amount of sodium bicarbonate is further increased to 2 and 3 g, the apparent viscosity increases slightly with the medium pH. This may be due to an increase in ionic strength of the medium (Table 3). This increase in ionic strength decreases the swelling pressure established across the polymer network, which decreases the swelling of the polymer network and resultant apparent viscosity. The gastric emptying lag time follows the same trend, with a great increase when 1 g sodium bicarbonate was added to 5 g polycarbophil in 200 ml water and no increase in gastric emptying lag time above 1 g of sodium bicarbonate. Thus, PC hydrogel administered with an alkalinizing agent results in a longer gastric emptying time than PC administered in the absence of an alkalinizing agent due to its effect on viscosity of the hydrogel.

The effect of hydration of the administered dose on gastric emptying was further studied using dry PC powder. The gastric emptying profiles of encapsulated 3.0, 6.0 and 8.0 g dry polycarbophil powder are shown in Fig. 5. The gastric emptying profiles are similar to that of hydrated PC hydrogel (Fig. 2) with a distinct gastric emptying lag time (Table 4), presumably associated with hydration of this poorly dispersed mass to give a very high, localized, viscosity. The mean gastric emptying lag time and half-life for encapsulated PC powder are significantly greater than hydrated PC hydrogel (Tables 2 and 4). Thus, administration of dry powder

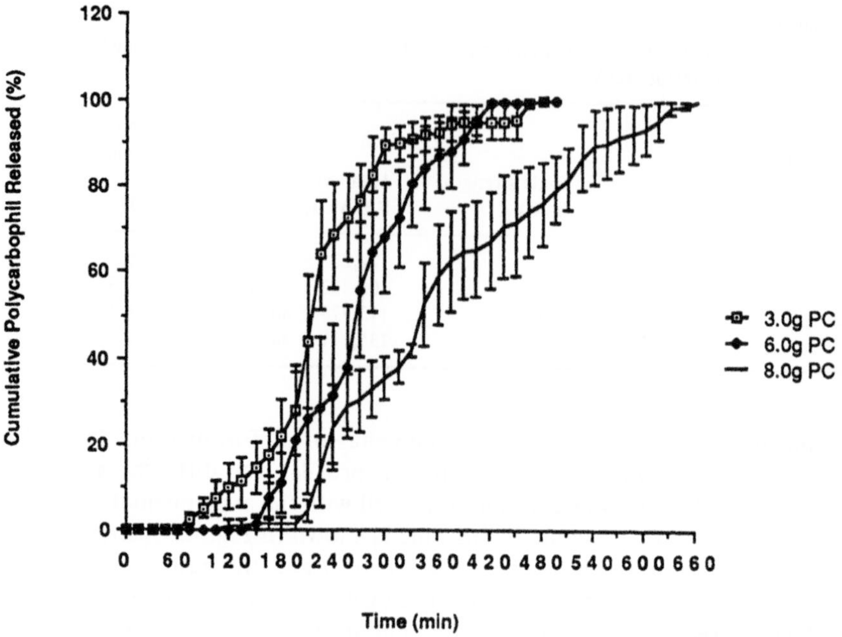
Cumulative Polycarbophil Released (%)
120
100
80
60
40
20
0
0 60 120 180 240 300 360 420 480 540 600 660
Time (min)
3.0g PC
6.0g PC
8.0g PC

Figure 5. Gastric release profiles of encapsulated 3.0, 6.0 and 8.0 g dry polycarbophil powder (mean ± S.E.M.).

substantially prolonged gastric retention. When a known amount of dry polycarbophil is administered orally, it is rapidly hydrated by the resident gastric fluid and coadministered water. At a low degree of hydration, the viscosity is high and the polymer gel resists the contractile force of the gastrointestinal motility and thus remains in the stomach. As more gastric fluid is secreted, the degree of hydration increases, and viscosity decreases. Gastric emptying of hydrogel occurs when the viscosity of the gastric contents is no longer high enough to resist the contractile force of the GI motility. At a high dose of polycarbophil and appropriate hydration (>150 ml), the stomach is converted to a fed state and its contents are retained for a substantial period of time. The time it takes to hydrate a known amount of polycarbophil and reduce its viscosity depends greatly on the degree of hydration of the administered polycarbophil. Thus, the lower the degree of hydration, the longer the gastric emptying lag time.

Table 4.
Effect of encapsulated dry polycarbophil on gastric emptying of canine

Amount PC (g)	Mean gastric emptying lag time ± S.E.M. (min)	Mean gastric emptying half-life ± S.E.M. (min)
3	113 ± 33	111 ± 36
6	186 ± 17	90 ± 29
8	190 ± 36	188 ± 70

CONCLUSION

Ingestion of non-digestible materials of 150 ml or more may convert a fasted stomach to a fed state in the canine. A critical feature of the fed state is the absence of the powerful 'housekeeper wave'. Thus, a useful gastric retentive system can be based on: (a) conversion of the stomach from a fasted to a fed state to eliminate the housekeeper wave and reduce the force of contraction; and (b) raising the viscosity of the stomach contents to resist the GI contractile force and stomach discharge caused by peristaltic waves.

The present paper describes a swellable, bio/mucoadhesive polymer which swells quickly in the canine gastric fluid and attains high viscosity. When different amounts of hydrated PC in 200 ml DDW were administered to previously fasted dogs, the gastric emptying lag time increased with apparent viscosity of the PC hydrogel. When the viscosity of the PC hydrogel was increased with an increase in medium pH via addition of sodium bicarbonate, there was a corresponding increase in gastric retention.

When a known amount of polycarbophil is administered orally, it is hydrated by the gastric fluid and coadministered water. For encapsulated dry PC powder, the viscosity of the hydrogel is high at the beginning of the hydration and is able to resist gastric emptying. As more gastric fluid is secreted, the degree of hydration increases, and viscosity decreases until gastric emptying of hydrogel occurs. Based on the data of the dry, encapsulated PC given with 50 ml of water, it may be that the strategy of using large volumes to convert from the fasted to the fed state as a means of reducing the amplitude of contractions, is less important than the high viscosity generated by the polymer. Thus, the higher the viscosity, the longer the gastric emptying time.

REFERENCES

1. R. Khosla, L. C. Feely and S. S. Davis, *Int. J. Pharm.* **53**, 107 (1989).
2. S. S. Davis, R. Khosla, C. G. Wilson and N. Washington, *Int. J. Pharm.* **35**, 253 (1987).
3. L. C. Feely and S. S. Davis, *Pharm. Res.* **6**, 274 (1989).
4. K. Klokkers-Bethe and W. Fischer, *J. Controlled Release* **15**, 105 (1991).
5. A. J. Coupe, S. S. Davis and I. R. Wilding, *Pharm. Res.* **8**, 360 (1991).
6. R. Khosla and S. S. Davis, *Int. J. Pharm.* **52**, 1 (1989).
7. S. S. Davis, J. G. Hardy and J. W. Fara, *Gut* **27**, 886 (1986).
8. P. Gruber, A. Rubenstein, V. H. K. Li, P. Bass and J. R. Robinson, *J. Pharm. Sci.* **76**, 117 (1987).
9. T. Itoh, T. Higuchi, C. Gardner and L. Caldwell, *J. Pharm. Pharmacol.* **38**, 801 (1986).
10. P. Mojaverian, R. Ferguson, P. Vlasses, M. Rocci, A. Oren, J. Fix, L. Caldwell and C. Gardner, *Gastroenterology* **89**, 392 (1985).
11. P. Sirois, G. Amidon, J. Meyer, J. Doty and J. Dressman, *Am. J. Physiol.* **258**, G65 (1990).
12. J. Meyer, J. Dressman, A. Fink and G. Amidon, *Gastroenterology* **89**, 805 (1985).
13. D. Harris, J. Felt, H. Sharma and D. Taylor, *J. Controlled Release* **12**, 45 (1990).
14. R. Khosla and S. S. Davis, *Int. J. Pharm.* **62**, R9 (1990).
15. S. S. Davis, F. N. Christensen, R. Khosla and L. Feely, *J. Pharm. Pharmacol.* **40**, 205 (1988).
16. F. N. Christensen, S. S. Davis, J. G. Hardy, M. J. Taylor, D. R. Whalley and C. G. Wilson, *J. Pharm. Pharmacol.* **37**, 91 (1985).
17. P. Mojaverian, P. Vlasses, S. Parker and C. Warner, *Clin. Pharmacol. Ther.* **47**, 382 (1990).
18. P. Mojaverian, J. C. Reynolds, A. Ouyang, F. Wirth, P. E. Kellner and P. H. Vlasses, *Pharm. Res.* **8**, 97 (1991).
19. W. S. W. Shalaby, W. E. Blevins and K. Park, *Biomaterials* **13**, 289 (1992).
20. S. Watanake, M. Kayano, Y. Ishino and K. Miyao, U.S. Patent 3 976 764 (1976).
21. H. Bechgaard and K. Ladefoged, *J. Pharm. Pharmac.* **30**, 690 (1978).

22. J. W. Sieg and J. W. Triplett, *J. Pharm. Sci.* **69**, 863 (1980).
23. R. Cargill, L. Caldwell, K. Engle, J. Fix, P. Porter and C. Gardner, *Pharm. Res.* **5**, 533 (1988).
24. H. M. Park, S. M. Chernish, B. Rosenek, R. Brunelle, B. Hargrove and H. Wellman, *Int. J. Pharm.* **35**, 157 (1987).
25. P. W. Carryer, M. L. Brown and J. R. Malagelada, *Gastroenterol.* **82**, 1389 (1982).
26. V. H. K. Li, Gastric emptying of non-digestible solids in the fasted dog. Ph.D. Thesis, University of Wisconsin, Madison, WI (1987).
27. R. Cargill, K. Engle, C. Gardner, P. Porter, R. Sparer and J. Fix, *Pharm. Res.* **6**, 506 (1989).
28. D. Harris, J. Fell, D. Taylor, J. Lynch and H. Sharma, *J. Controlled Release* **12**, 55 (1990).
29. H. S. Ch'ng, H. Park, P. Kelly and J. R. Robinson, *J. Pharm. Sci.* **74**, 399 (1985).
30. C. F. Code and J. M. Marlett, *J. Physiol. Lond.* **246**, 289 (1975).
31. J. P. Miolan and C. Romen, *Am. J. Physiol.* **235**, G366 (1978).
32. R. C. Gill, M. A. Pilto, P. A. Thomas and D. L. Wingate, *Am. J. Physiol.* **249** G655 (1985).
33. Z. Itoh and T. Sekiguchi, *Scand. J. Gastroenterol.* **18** (suppl. 82), 497 (1982).
34. N. Washington, *Antacids and Anti-Reflux Agents*. CRC Press, Boca Raton, FL (1991).
35. I. Dewever, C. Eeckhout, G. Vantrappen and J. Hellemans, *Am. J. Physiol.* **235**, E661 (1978).
36. P. K. Gupta, Processing of liquids and solids by the fasted canine stomach. Ph.D. Thesis, University of Wisconsin, Madison, WI (1990).
37. C. Tasman-Jones, *Scand. J. Gastroenterol.* **73**, 11 (1977).
38. J. R. Robinson, U.S. Patent 4615697, Oct. 7 (1986).
39. S. H. S. Leung and J. R. Robinson, *J. Controlled Release* **5**, 223 (1988).
40. S. H. S. Leung and J. R. Robinson, *J. Controlled Release* **12**, 187 (1990).
41. Pharmacopeial Forum, The United State Pharmacopeial Convention, Inc., Rockville, MD (1990) p. 1149.
42. H. Park and J. R. Robinson, *J. Controlled Release* **2**, 47 (1985).
43. H. L. Greenwald, in: *Handbook of Water-Soluble Gums and Resins*, Chap. 17, R. L. Davidson (Ed.). McGraw-Hill, New York (1980).
44. C. Tanford, *Physical Chemistry of Macromolecules*, John Wiley and Sons, Inc., New York (1967).
45. M. Feldman, in: *Gastrointestinal Disease*, Vol. 1, 4th ed. p. 713, M. H. Sleisenger and J. S. Fordtran (Eds). W. B. Saunders Co., Philadelphia, PA (1989).
46. M. Feldman and C. T. Richardson, *Gastroenterology* **90**, 540 (1986).
47. J. E. Thomas, *Proc. Soc. Biol. Med.* **46**, 260 (1941).
48. R. S. Jones, T. K. Yee and C. F. Michielson, *J. Appl. Physiol.* **30**, 427 (1971).
49. D. A. Reinke, A. H. Rosenbaum and E. R. Bennett, *Am. J. Dig. Dis.* **12**, 113 (1967).
50. Material safety data sheet, number 88033, The B. F. Goodrich Co., Specialty Polymers and Chemical Division, Cleveland, OH (1988).
51. D. K. Bogan, *Science* **246**, 138 (1989).
52. C. Washington, N. Washington and C. Wilson, Pharmacokinetic Modelling using STELLA® on the Apple Mackintosh, Ellis Horwood, Chichester (1990).
53. G. M. Grass and W. T. Morehead, *Pharm. Res.* **9**, 759 (1989).
54. C. G. Wilson, N. Washington, J. L. Greaves, C. Washington, I. R. Wilding, T. Hoadley and E. E. Sims, *Int. J. Pharm.* **72**, 79 (1991).
55. U. Finne and A. Urtti, *Int. J. Pharm.* **84**, 217 (1991).
56. A. Urtti, *Proc. Int. Symp. Controlled Release Bioact. Mater.* **18**, 431 (1991).

Release behavior of bioactive agents from pH-sensitive hydrogels

ATUL R. KHARE and NIKOLAOS A. PEPPAS*
School of Chemical Engineering, Purdue University, West Lafayette, IN 47907-1283, USA

Received 22 June 1992; accepted 18 September 1992

Abstract—Controlled release systems of theophylline, proxyphylline and oxprenolol.HCl exhibiting modulated drug delivery were prepared by using pH-sensitive anionic copolymers of 2-hydroxyethyl methacrylate with acrylic acid or methacrylic acid. Drug release studies were carried out in simulated biological fluids. The initial drug release rates and the drug release mechanisms were dependent upon the pH and ionic strength of the buffer solution as well as its salt composition. Initial drug diffusion coefficients in these swelling-controlled release systems were calculated from the release curves; they were of the order of 10^{-7} cm^2/s and were dependent upon the degree of swelling. The drug release mechanism was non-Fickian in all the dissolution media studied. Lowest release rates were observed for drug release from nonionized polymer networks in agreement with the relationship between ionization, swelling and drug release.

Key words: pH-controlled drug release; non-Fickian diffusion; polyelectrolytes; ionization equilibrium; hydrogels.

INTRODUCTION

In recent years there has been significant interest in novel pulsatile and self-regulated delivery systems for drug peptide and protein release [1, 2]. A selective release of these drugs at a particular site in the body or during a particular time interval offers major advantages over conventional dosage forms [3]. Such advantages are reduction in side-effects, reduction of the amount of drug required to be delivered, and maximum bioavailability [4, 5].

Swelling-controlled release systems exhibiting modulated or pulsatile drug delivery are usually formed by incorporation (loading) of drugs, peptides or proteins in initially dry, hydrophilic polymers that may exhibit significant changes of their dimensional stability and associated diffusional/topological characteristics as a result of changes in the physiological fluid where they have been placed. For example, change of the pH or ionic strength [6–8] can act as stimuli for significant diffusional changes. Hoffman and his collaborators have been pioneers in the utilization of temperature-sensitive gels for the same purpose, as for example in the work of Dong and Hoffman [9]. Other stimuli may include magnetism or ultrasounds [1], electromagnetic irradiation [10] and polymer melting and dissolution. A complete analysis of all these systems is presented elsewhere [1]. A drug which is dispersed within or coated with a polymer can be released depending upon the swelling ratio. This type of 'on–off' control [9, 10] is desirable in insulin and growth factor delivery, or targeting of anticancer agents as well as in other applications.

* To whom all correspondence should be addressed.

Vehicles in such drug delivery systems are polymers of either neutral or ionic nature. A preferred class of drug delivery systems for these applications is that of swellable crosslinked polymers [11, 12]. These are biocompatible, nondegradable matrices with uniformly dispersed or dissolved drug. The drug release is controlled by swelling of the hydrogel when it comes in contact with the physiological fluid.

Penetrant (water) transport into glassy polymer networks leads to formation of a rubbery (gel) phase. The drug from the polymer matrix dissolves into the imbibed water and then diffuses out of the swollen polymer. This swelling rate as well as the diffusional drug release behavior control the overall release rate of the system. The penetrant–polymer, polymer–drug and penetrant–drug interactions play a very important part in the overall release behavior [1, 13].

In previous work [13], we showed that the dynamic swelling of these ionic polymers in different buffer solutions was dependent on pH and ionic strength. Based on these observations, the objectives of this study were to study the diffusional behavior of low molecular weight drugs from pH-sensitive, anionic, initially glassy copolymeric networks of 2-hydroxyethyl methacrylate with acrylic acid and/or methacrylic acid, henceforth designated as P(HEMA-co-AA) and/or P(HEMA-co-MAA), respectively, as functions of pH, ionic strength and the nature of the dissolution medium. In this case, the dissolution media used were simulated physiological fluids such as simulated gastric fluid, a phosphate buffer of pH 7.4 and a glutarate buffer solution of pH 7.

EXPERIMENTAL

Materials

Crosslinked copolymers of 2-hydroxyethyl methacrylate (HEMA, Aldrich Chemical Co., Milwaukee, WI), with methacrylic acid (MAA, Aldrich Chemical Co., Milwaukee, WI) or the acrylic acid (AA, Aldrich Chemical Co., Milwaukee, WI) were synthesized by solution polymerization. HEMA was mixed with appropriate quantities of MAA or AA to form a mixture containing 50 or 80 mol.% HEMA. Quantities of 0.5 wt.% ammonium persulfate and sodium bisulfite were added as redox initiators and 0.45 mol.% ethylene glycol dimethacrylate (EGDMA) was used as a crosslinking agent. Thus, the overall process was a terpolymerization/crosslinking reaction between HEMA, MAA and EGDMA or HEMA, AA and EGDMA. The solution polymerization was carried out in polyethylene vials using deionized water (50 wt.%) as a solvent in a constant temperature water bath at 37°C ± 0.5 for 24 h, followed by gradual cooling to room temperature.

The vials were cut away from the resultant crosslinked polymer samples. Polymer gels were dried at room temperature and thin disks were cut with a lathe. These samples were further dried in a vacuum oven for 1 week at 40°C. The final samples had diameters of 10–12 mm and thicknesses of 0.8–1.0 mm. This range of dimensions (aspect ratio greater than 10:1) was consistent with the requirements of one dimensional penetrant diffusion which will be invoked later in the analysis of swelling and release data.

Drug loading in glassy polymer disks

The drugs used in the release studies were theophylline, proxyphylline, and oxprenolol.HCl (all from Sigma Chemical Co. St. Louis, MO). The drug loading into the polymer discs was achieved by swelling the polymer sample for a week in 25 ml of an aqueous drug solution prepared from deionized water or from a pH 7.4 phosphate buffer of ionic strength I equal to 0.1 M. The drug concentrations used were 3.33 g/l for theophylline, 100 g/l for proxyphylline, and 10 g/l for oxprenolol.HCl. The ensuing swollen polymer gels were dried at room temperature for 2 days to avoid the formation during drying of a surface layer of drug crystallites. The glassy, drug-laden polymer discs were dried further at room temperature under a vacuum of 5 mm Hg for 1 week to remove the remaining water. Drug loading as high as 25 wt.% on a dry basis was achieved by using this technique.

Scanning electron microscopy (SEM) was used to determine the concentration distribution and uniformity of oxprenolol.HCl in polymer samples by measuring the spatial distribution of chlorine concentration in oxprenolol.HCl in thin sections of polymer samples. The drug-loaded polymer disks were first fractured in the center and the exposed surface and cross-sectional areas were coated with aluminum using the standard SEM coater. Then, SEM scans could be taken throughout the surface and cross-sectional areas of the samples to determine the chlorine concentration.

Controlled release studies

Controlled drug release studies were carried out in a dissolution cell (Hansen Research, Northridge, CA) containing 300 ml of dissolution medium at 37°C. The dissolution media used were deionized water, a pH 1.2 simulated gastric fluid without pepsin, a pH 7.4 phosphate buffer of ionic strength 0.1 M, a pH 7 glutarate buffer of I equal to 0.01 M, and a pH 7 glutarate buffer of I equal to 0.1 M. The stirring speed used to maintain uniform drug concentration was 60 rpm. The concentration of the released drug was monitored by using a flow-through cell in a UV-Vis spectrophotometer (model 559, Perkin-Elmer, Norwalk, CT) at 272 nm for proxyphylline, 271 nm for theophylline and 274 nm for oxprenolol.HCl. Solute spectra taken before and after the release experiment showed that there was no change in the physical and chemical structure of the drug molecules studied here.

The drug loading into the polymer sample was determined from the difference between the weights of the drug-loaded polymer samples and their corresponding drug-free samples. After the completion of a release experiment, the polymer samples were also washed several times in water to extract traces of drug from the polymer. By measuring the drug concentration in the extract, the total drug loading was calculated. The drug loading obtained by these two methods differed by only $\leq 2\%$.

RESULTS AND DISCUSSION

General observations

The preparation methods used here were selected because they could provide samples with uniform drug composition and properties as determined by microscopic, swelling and other studies. The samples showed no evidence of stresses or cracks during swelling and did not present any of the unusual instability phenomena during swelling as those reported by Drummond *et al.* [15].

To analyze the drug concentration distribution and uniformity throughout the polymer samples, SEM data were used to determine the chlorine concentration by calculating the ratio of aluminum to chlorine at various points along the cross-section. Typical values of the Al/Cl ratio for a dry sample of P(HEMA-co-MAA) loaded with 4.5 wt.% oxprenolol.HCl are shown in Fig. 2. As the amount of aluminum deposited on the samples was uniform throughout the surface and the cross-section, the results of Fig. 2 indicate that chlorine (and oxprenolol.HCl) was uniformly distributed in the P(HEMA-co-MAA) and P(HEMA-co-AA) polymers containing 80 mol.% HEMA.

To investigate the uniformity of theophylline and proxyphylline concentration in the polymer samples, the latter were crushed and extracted in water for several weeks until there was no change in the drug concentration in the water. It was observed that there was no appreciable change in the drug loading in each piece. From the total drug concentration calculated, the initial drug loading values were determined.

A primary interest in the analysis of the release studies was whether release from ionic, swelling-controlled release systems was diffusion-controlled (Fickian) or could be best described by a coupling of water transport/polymer relaxation (non-Fickian, anomalous transport) as discussed by Davidson and Peppas [13] or Ritger and Peppas [16].

Figure 3 shows typical release data for delivery of oxprenolol.HCL from initially dry P(HEMA-co-MAA) disks containing 80 mol.% HEMA. These systems were immersed in deionized water or a buffered phosphate solution with pH 7.4 at 37°C. It is obvious that the fractional release of oxprenolol is linear with respect to time

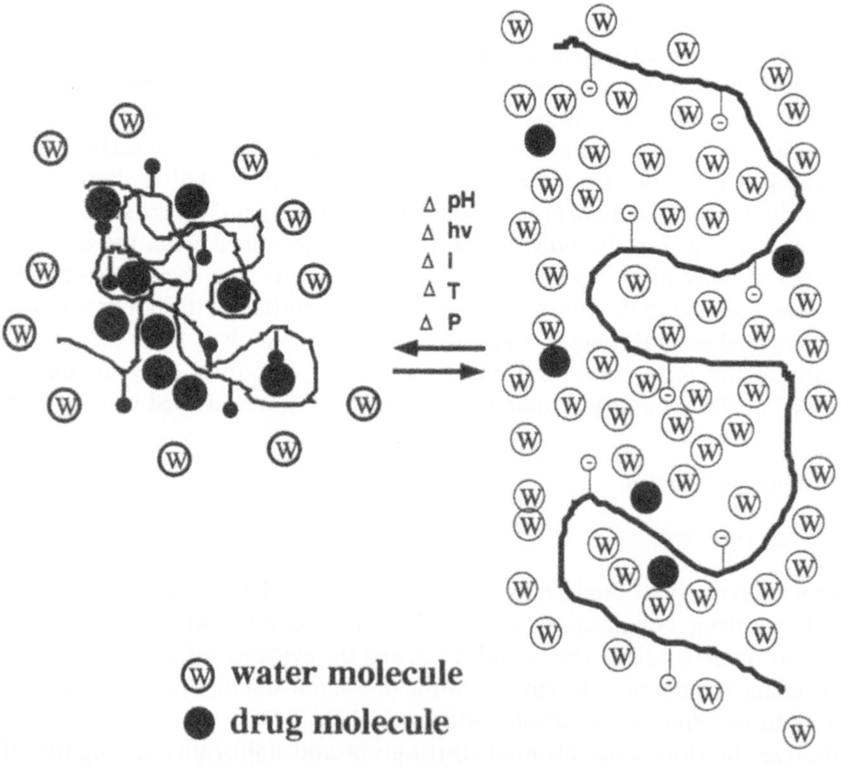
Δ pH
Δ hν
Δ I
Δ T
Δ P
W water molecule
drug molecule

Figure 1. Molecular interpretation of drug transport in pH-sensitive macromolecular systems.

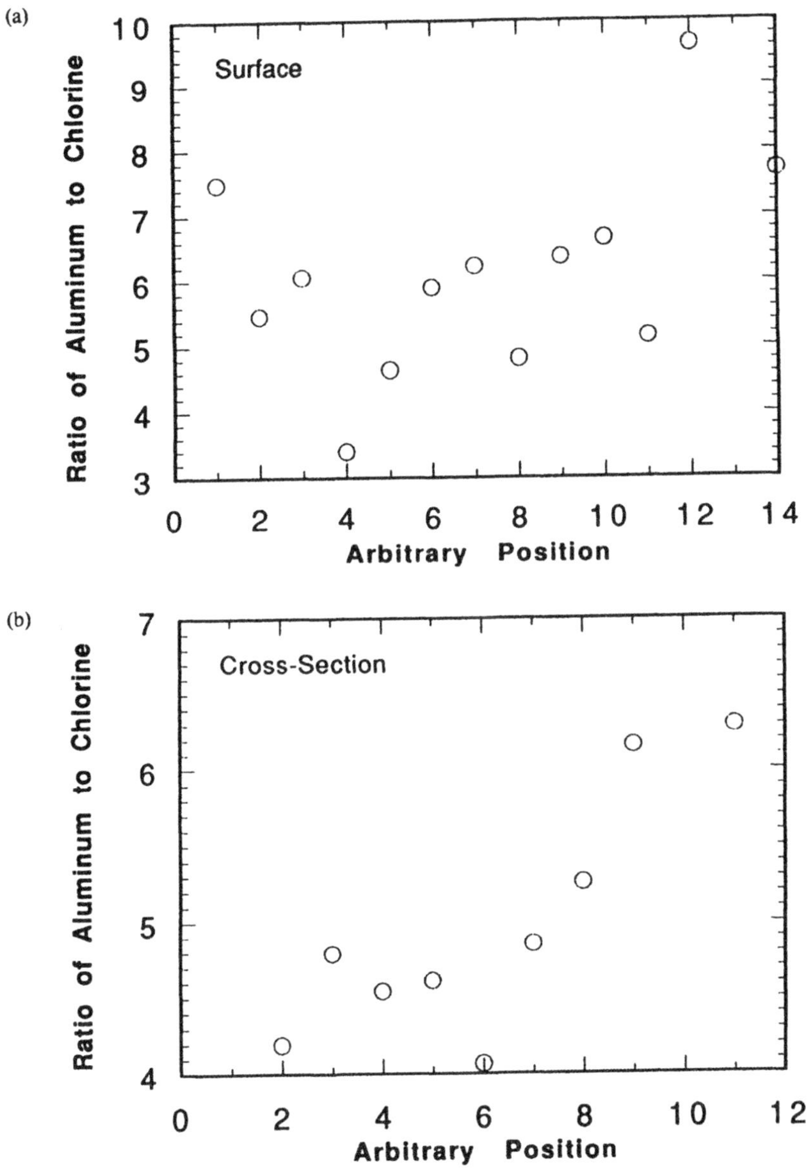
(a)
Surface
Ratio of Aluminum to Chlorine
3
4
5
6
7
8
9
10
0
2
4
6
8
10
12
14
Arbitrary Position
(b)
Cross-Section
Ratio of Aluminum to Chlorine
4
5
6
7
0
2
4
6
8
10
12
Arbitrary Position

Figure 2. Ratio of aluminum to chlorine atoms on (a) the surface and (b) cross-section, versus position in a P(HEMA-MAA) (80 : 20 mol.%) polymer sample containing oxprenolol.HCl. The standard deviation in the ratio was 2 units.

over the first 2 h of release, corresponding to approximately 65% of drug released in water or 32% of drug released in the buffered solution. This linear release behavior indicates that this controlled release system acts as a zero-order release system. The release behavior is clearly non-Fickian and the chain relaxational behavior of the ionic polymer controls the drug release process.

Mathematical verification of this notion can be obtained by analyzing the first 60% of the release data of each curve using the following equation as discussed by

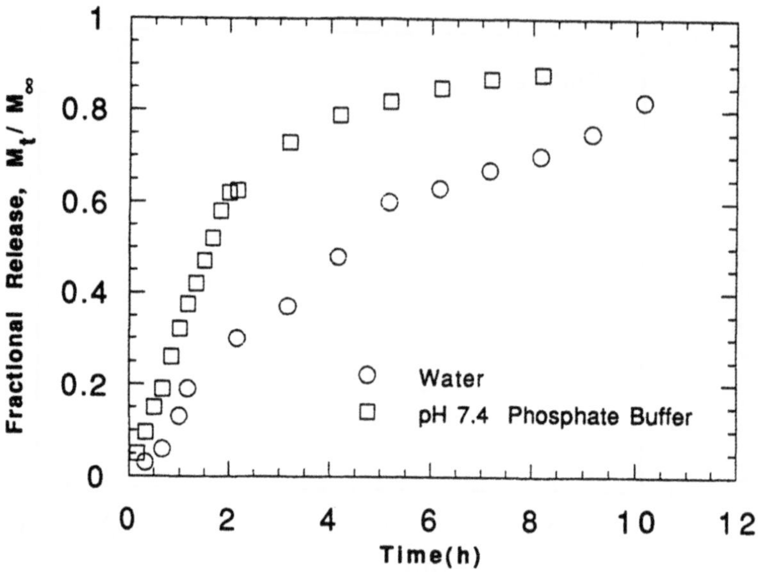
Fractional Release, M_t / M_∞
1
0.8
0.6
0.4
0.2
0
0
2
4
6
8
10
12
Time(h)
Water
pH 7.4 Phosphate Buffer

Figure 3. Fractional oxprenolol.HCl release from P(HEMA-co-MAA) (80 : 20 mol.%) samples with initial drug loading of 4.5 wt.% as a function of time at 37°C.

Ritger and Peppas [16].

$$\frac{M_t}{M_\infty} = kt^n \tag{1}$$

Here, M_t is the amount of drug released up to time t, M_∞ is the amount of drug present in the glassy polymer at time $t = 0$, and k and n are constants. For a planar geometry as used here, an n of 1.0 signifies that the drug release mechanisms is Fickian whereas $n = 1.0$ indicates Case-II or zero order release. During zero order drug release, the drug mechanism is controlled by the relaxation rate of the polymer chains. For $0.5 < n < 1.0$, the release mechanism is anomalous or non-Fickian and a coupling of drug diffusion and polymer chain relaxation control the overall release process.

The determination of the drug diffusion coefficient, D, from the results of these release studies is a rather cumbersome process because of the continuous swelling exhibited by the polymer samples. Exact analysis of swellable systems can be done using the models of Hariharan and Peppas [17]. Two approximate solutions were used here for the calculation of D.

By using release data during the very early portion of the release portion (first 10% of release) where the release is occurring through a sample that is at the initial stage of the swelling process, one can determine the drug diffusion coefficient, D_0, calculated from Eq. (2).

$$\frac{M_t}{M_\infty} = 4\left[\frac{D_0 t}{\pi L_0^2}\right]^{1/2} \tag{2}$$

Here, D_0 is the initial diffusion coefficient and L_0 is the initial thickness of the sample tested. It must be noted that Eq. (2) is the *early time* approximation of the exact solution of Fick's second law [16] under *constant boundary conditions*, i.e. under conditions of no or minimum swelling [16], which may prevail *only at the beginning* of the release process.

Analysis of the release data towards the end of the release process can be done using Eq. (3) which is the *late time* approximation of the exact solution of Fick's second law [16].

$$\frac{M_t}{M_\infty} = 1 - \frac{8}{\pi^2}\exp\left(-\frac{D_g \pi^2 t}{L_g^2}\right) \tag{3}$$

Here, D_g is the drug diffusion coefficient though the equilibrium gel and L_g is the equilibrium swollen gel thickness.

Both Eqs (2) and (3) are approximations as they assume constant thickness during the release period that data are collected (early or late release period).

Drug release from ionic networks

Typical release data of oxprenolol.HCl at 37°C into distilled water and a pH 7.4 phosphate buffer solution from P(HEMA-co-MAA) (80 : 20 mol.%) swelling-controlled release devices are presented in Fig. 3. The drug release rate, calculated from the slope of each curve was high in the pH 7.4 buffer than in water. Since the buffer solution pH of 7.4 is higher than the pK_a of the gel studied, i.e. 5.5 [14], the carboxylic acid groups of the polymer network ionize leading to a high water absorption. Of course, high water uptake or large swelling ratios lead to an increased drug diffusion coefficient in the gel according to the theory of Peppas and Reinhart [18]. On the other side, in distilled water the carboxylic acid groups ionize to a very small extent due to the limited supply of protons, i.e. there is no buffering action in distilled water. In general, decrease in pH reduces the degree of ionization of the gel. Thus, the quantity of water absorbed by the gel is reduced and consequently the drug diffusion coefficient is decreased. This is seen in the results of Table 1. Similar observations were made in the release profiles of theophylline from P(HEMA-co-AA) copolymers containing 20 mol.% AA, as shown in Fig. 4.

The release profiles of oxprenolol.HCl from two HEMA copolymers containing 80% HEMA in a pH 7.4 phosphate buffer are shown in Fig. 5. The drug release rate from MAA-containing copolymers was lower than from AA-containing copolymers; obviously the α-methyl group of the MAA-containing copolymers was responsible for the change. Similarly, the diffusion coefficients of all the drugs were lower in MAA-containing copolymers than in those copolymers containing AA as shown in Table 1.

The diffusional analysis of the release data is presented in Table 1 where the initial and late time approximation values of the drug diffusion coefficients, D_0 and D_g, are summarized for various systems. For each release study, the calculated late time drug diffusion coefficient, D_g, is significantly higher than the initial drug diffusion coefficient, D_0. An order of magnitude difference is typical and indicates that the large swelling occurring through these ionic systems allows fast transport of the incorporated drug. Transport in buffered solutions is seemingly faster than in ionized water but this is only an exhibition to the different ionization levels for the same system in different pH solutions. Ionic strength has the same affect, suppressing the ionization and thus leading to decreased values of D_g and D_0.

Table 1.
Initial and final drug diffusion coefficients calculated from release data from P(HEMA-co-AA) and P(HEMA-co-MAA) samples immersed in different buffers at 37°C[a].

Release system and drug	Initial and final drug diffusion coefficients (cm^2/s)			
	Water	Phosphate buffer pH 7.4, I = 0.1 M	Glutarate buffer pH 7.0, I = 0.01 M	Glutarate buffer pH 7.0, I = 0.1 M
P(HEMA-co-AA) (80:20 mol.%) Oxprenolol.HCl	0.26/3.20	0.33/4.45	0.08/1.26	0.04/0.18
P(HEMA-co-AA) (80:20 mol.%) Theophylline	0.15/3.66	0.36/7.68	0.27/1.99	0.14/3.71
P(HEMA-co-AA) (80:20 mol.%) Proxyphylline	0.13/1.13	0.19/1.61	*	*
P(HEMA-co-MAA) (80:20 mol.%) Oxprenolol.HCl	0.05/0.51	0.14/0.47	0.01/0.75	0.02/0.48
P(HEMA-co-MAA) (80:20 mol.%) Theophylline	0.19/4.84	0.33/6.97	0.03/0.62	0.04/1.40
P(HEMA-co-MAA) (80:20 mol.%) Proxyphylline	0.08/0.64	0.15/1.48	*	*

[a] Initial diffusion coefficients, D_0, were calculated from Eq. (2) whereas final (gel) diffusion coefficients, D_g, were calculated from Eq. (3). Each entry presents D_0 followed by D_d.
* Not available.

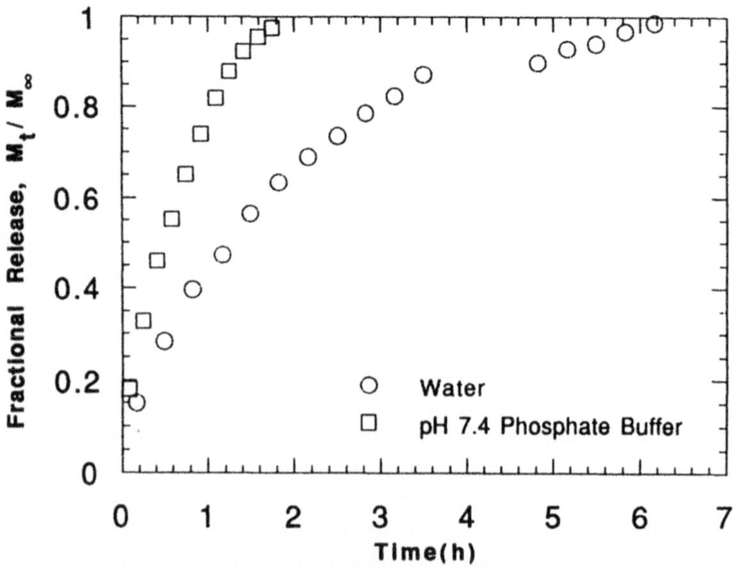

Fractional Release, M_t / M_∞
Time(h)
Water
pH 7.4 Phosphate Buffer
0
0.2
0.4
0.6
0.8
1
0
1
2
3
4
5
6
7

Figure 4. Fractional theophylline release from P(HEMA-co-AA) (80:20 mol.%) samples with initial drug loading of 0.5 wt.% as a function of time at 37°C.

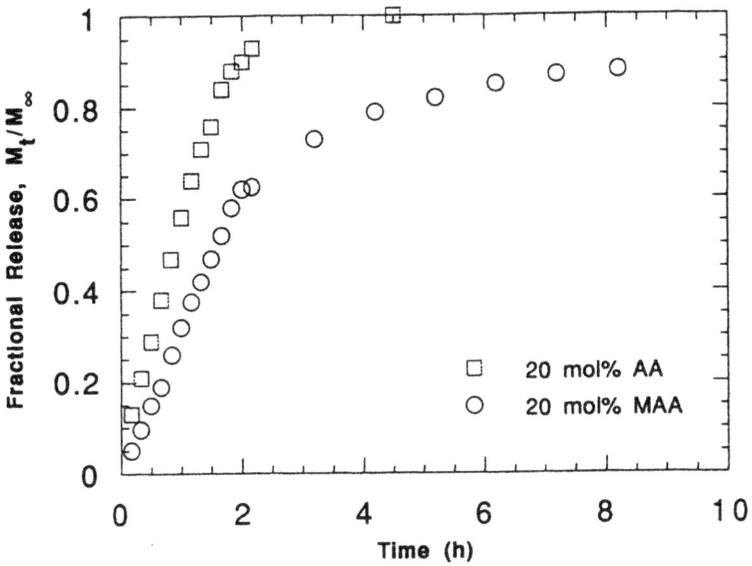
Fractional Release, M_t/M_∞
Time (h)
20 mol% AA
20 mol% MAA
0
0.2
0.4
0.6
0.8
1
0
2
4
6
8
10

Figure 5. Fractional oxprenolol.HCl release from P(HEMA-co-AA) (80 : 20 mol.%) and P(HEMA-co-MAA) (80 : 20 mol.%) samples with initial drug loading of 4.5 wt.% as a function of time at 37°C in pH 7.4 phosphate buffer.

The drug diffusion coefficient values as determined by Eq. (2) and the data of the various figures that follow, were highest in a pH 7.4 phosphate buffer in both AA-containing and MAA-containing polymer systems as shown in Table 1. The drug diffusion coefficients were higher in AA-containing polymers than MAA-containing polymers exhibiting the same ionization. In the case of AA-containing polymers, it was also observed that as the ionic strength increased, the drug diffusion coefficients decreased in pH 7 glutarate buffer system.

The analysis and comparison of the two diffusion coefficients, D_0 and D_g, for each controlled release formulation indicates the need to revise the theory of Peppas and Reinhart [18]. It is clear that the water uptake or swelling ratio is not sufficient to predict diffusion coefficient in ionic systems and that additional information such as degree of polymer ionization, and drug/polymer interactions are equally important.

Figure 6 shows the release behavior of proxyphylline from P(HEMA-co-MAA) in three different dissolution media. The release profiles in pH 1.2 simulated gastric fluid and water were very similar. Since neither fluid has any buffering capacity, the carboxylic acid groups are ionized to a very small extent. Similar observations were made in the release profiles of oxprenolol.HCl and theophylline from these polymers. The rate of release of proxyphylline was highest in the pH 7.4 phosphate buffer for the reasons presented earlier.

Analysis of the release mechanism was performed using Eq. (1). The diffusional exponent n was calculated between 0.5 and 1.0 depending on the ionization of the sample tested. In general, the drug release studied was non-Fickian. The exponent n increased with increasing pH, in agreement with previous studies [6]. The non-Fickian release behavior persisted even in water, something that had not been observed before [14]. The reason for this deviation may be the existence of drug–polymer interactions in these ionic polymers.

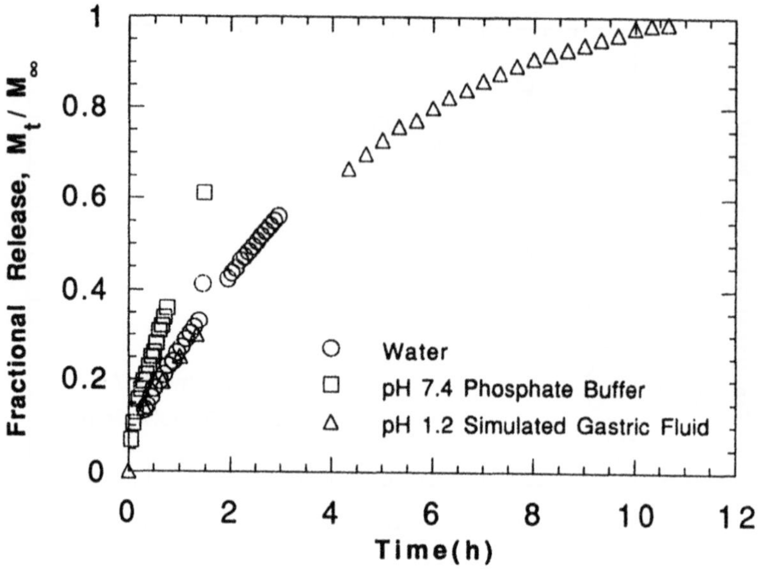

Fractional Release, M_t / M_∞
0
0.2
0.4
0.6
0.8
1
0
2
4
6
8
10
12
Time(h)
Water
pH 7.4 Phosphate Buffer
pH 1.2 Simulated Gastric Fluid

Figure 6. Fractional proxyphylline release from P(HEMA-co-MAA) (80 : 20 mol.%) samples with initial drug loading of 23 wt.% as a function of time at 37°C.

Effect of the drug loading method on drug release

To investigate the effect of the drug loading method on the release behavior of drugs, crosslinked P(HEMA-co-MAA) (50 : 50 mol.%) samples were swollen in a pH 7.4 phosphate buffer containing the desired drug. The ensuing swollen polymers were dried at room temperature and then under vacuum to obtain glassy polymer samples. Figures 7 and 8 show the release profiles of oxprenolol.HCl from

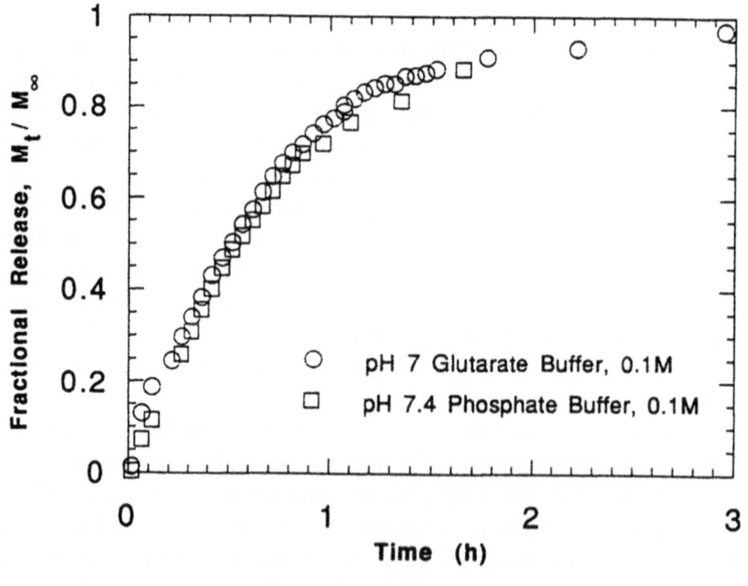

Fractional Release, M_t / M_∞
0
0.2
0.4
0.6
0.8
1
0
1
2
3
Time (h)
pH 7 Glutarate Buffer, 0.1M
pH 7.4 Phosphate Buffer, 0.1M

Figure 7. Fractional oxprenolol.HCl release from P(HEMA-co-MAA) (50 : 50 mol.%) samples with initial drug loading of 29 wt.% as a function of time at 37°C.

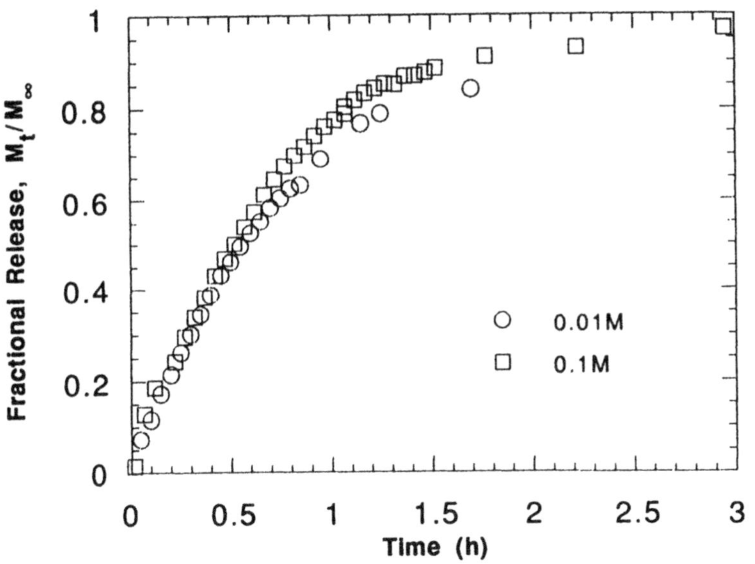
Fractional Release, M_t/M_∞
1
0.8
0.6
0.4
0.2
0
0
0.5
1
1.5
2
2.5
3
Time (h)
0.01M
0.1M

Figure 8. Fractional oxprenolol.HCl release from P(HEMA-co-MAA) (50 : 50 mol.%) samples with initial drug loading of 29 wt.% as a function of time at 37°C in pH 7 glutarate buffer.

P(HEMA-co-MAA) (50 : 50 mol.%) polymers in a pH 7.4 phosphate buffer and in various pH 7 glutarate buffers. There seems to be no major difference in the profiles and the values of the initial drug diffusion coefficients were approximately $D_0 \approx 4.7 \times 10^{-7}$ cm^2/s. This was due to the fact that, the glassy polymer network did not attain their original conformation after imbibition and drying; thus, the polymer remained in its relaxed form. The polymer chains were in their extended conformation and therefore, the network was already in an open conformation at time $t = 0$. It is critical to have the original conformation of the polymer to have a swelling-control over the release behavior of the drugs.

The release profile of oxprenolol.HCl from MAA-containing copolymers in pH 7.4 phosphate buffer is shown in Fig. 9 as a function of drug loading. Higher drug release rate was observed at a higher drug loading level. In general, higher drug loading leads to a higher osmotic pressure differential in the gel; this adds to the water absorption capacity of the gel and leads to higher drug release rates [17].

Effect of ionic strength on drug release behavior

Figure 10 shows the effect of ionic strength on the release behavior of oxprenolol.HCl from P(HEMA-co-AA) (80 : 20 mol.%) in pH 7 glutarate buffer having two ionic strengths. As can be seen, the release rate increased with a decrease in ionic strength due to the higher osmotic swelling pressure which leads to higher swelling [19]. However, as seen from the graph the release profiles were slightly affected by a large change in the ionic strength. Based upon dynamic swelling studies of these gels as a function of ionic strength [14], it was expected that there would be a significant change in the release behavior of the incoporated drug with a change in the ionic strength. The observed release profiles show a very small dependence on ionic strength. This may be due to drug-polymer interactions that are presently examined.

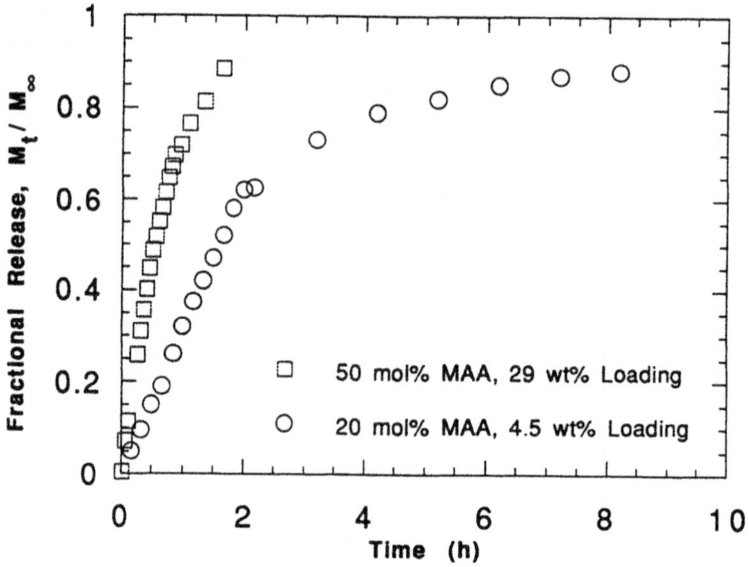
Fractional Release, M_t / M_∞
0
0.2
0.4
0.6
0.8
1
0
2
4
6
8
10
Time (h)
50 mol% MAA, 29 wt% Loading
20 mol% MAA, 4.5 wt% Loading

Figure 9. Fractional oxprenolol.HCl release from P(HEMA-co-MAA) samples as a function of time at 37°C in pH 7.4 phosphate buffer.

Table 2.
Drug diffusional exponent *n* calculated from release data from P(HEMA-co-AA) and P(HEMA-co-MAA) samples different buffers at 37°C.

Release system and drug	Diffusional exponent, *n*			
	Water	Phosphate buffer pH 7.4, I = 0.1 M	Glutarate buffer pH 7.0, I = 0.01 M	Glutarate buffer pH 7.0, I = 0.1 M
P(HEMA-co-AA) (80 : 20 mol.%) Oxprenolol.HCl	0.98	0.72	0.74	0.96
P(HEMA-co-AA) (80 : 20 mol.%) Theophylline	0.52	0.57	0.76	0.63
P(HEMA-co-AA) (80 : 20 mol.%) Proxyphylline	0.64	1.19	*	*
P(HEMA-co-MAA) (80 : 20 mol.%) Oxprenolol.HCl	0.93	0.73	0.85	0.73
P(HEMA-co-MAA) (80 : 20 mol.%) Theophylline	0.87	0.63	0.59	0.63
P(HEMA-co-MAA) (80 : 20 mol.%) Proxyphylline	0.66	0.62	*	*

*Not available.

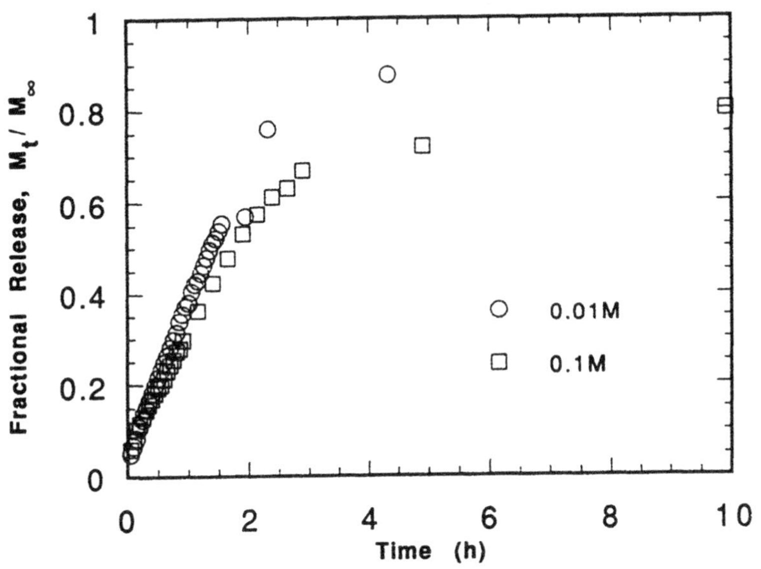
Fractional Release, M_t / M_∞
Time (h)
0.01M
0.1M
0
0.2
0.4
0.6
0.8
1
0
2
4
6
8
10

Figure 10. Fractional oxprenolol.HCl release from P(HEMA-co-AA) (80:20 mol.%) samples with initial drug loading of 4.5 wt.% as a function of time at 37°C in pH 7 glutarate buffer.

Drug transport mechanisms in ionic polymers

The previous studies shed new light onto the mechanisms of drug transport and release from dynamically swelling, ionic polymers. Clearly, the degree of ionization plays an important role in the overall release process. In the early stages of the release, due to limited swelling the drug diffusion coefficient is of the order of 10^{-7} cm^2/s and release is slow and predominantly from the surface layers of the system. As swelling increases, the drug diffusion coefficient can increase by an order of magnitude or more, to 10^{-6} cm^2/s. However, the distance traveled by the drug in the swollen gel is much larger. This balance between 'slow diffusion-small distance' and 'fast diffusion-long distance' leads to a quasi-constant release over a long period of time.

Any additional components that will affect the ionization process will influence the release. The presence of small electrolytes (increase in ionic strength) as well as drug–polymer interactions will affect the release behavior. Any delay in release due to such interactions will be seen as a decrease in the rate from constant (zero-order) release to Fickian release.

Of course, an alternative method of analysis of these transport mechanisms is in terms of the coupled diffusion/relaxation of the system. As discussed before by Brannon-Peppas and Peppas [6], typical values of the Deborah number, *De*, for water transport in crosslinked P(HEMA-co-MAA) samples containing 80 mol.% HEMA and at a pH of 8 are 1.50.

We have previously defined the Swelling Interface number, *Sw*, as an indication and criterion of the non-Fickian release, expressed as the ratio of water transport (with velocity $\bar{v}$) to drug diffusion. Two such *Sw* values can be introduced here, according to Eqs (4) and (5)

$$Sw_0 = \frac{\bar{v}L_0}{D_0} \tag{4}$$

$$Sw_g = \frac{\bar{v}L_0}{D_g}. \tag{5}$$

A typical value of $\bar{v}$ for the systems considered here [6] is $\bar{v} = 7.5\mu m/min = 1.25 \times 10^{-5}$ cm/s. Using the values of $D_0 = 0.5 \times 10^{-6}$ cm^2/s reported for oxprenolol release from P(HEMA-co-MAA) samples in Table 1 and the associated thicknesses of $L_0 = 0.1$ cm and $L_g = 0.3$ cm, respectively, we determined the Swelling Interface numbers as $Sw_0 = 25$ and $Sw_g = 7.35$. Thus, as the swelling started occurring, the Sw values decreased significantly into the region of non-Fickian and Case II behavior.

CONCLUSIONS

The drug release from crosslinked P(HEMA-co-MAA) and P(HEMA-co-AA) samples was found to be a function of pH, ionic strength and the nature of the dissolution medium. The initial rate of drug release was higher in pH 7.4 phosphate buffer than in water or in glutarate buffers. Drug release was also higher from AA-containing copolymers than from MAA-containing copolymers having the same amount of ionizable groups. The drug release mechanism was found to be non-Fickian in all the release studies approaching Case II transport in a number of systems. The initial drug diffusion coefficients were of the order of 10^{-7} cm^2/s, increasing by an order of magnitude towards the end of the swelling process.

Acknowledgements

This research was supported by the National Institute of Health grant number GM43337. This work is dedicated to Professor Allan S. Hoffman of the University of Washington, Seattle on the occasion of his 60th birthday. The insightful research of Professor Allan S. Hoffman, who is also an 'academic brother' of the senior author, has influenced our thinking and research over the past 15 years.

REFERENCES

1. N. A. Peppas, in: *Pulsatile Drug Delivery: Current Applications and Future Trends*, R. Gurny, H. Junginger and N. A. Peppas (Eds), Wischenschaftliche, Frankfurt, in press.
2. V. H. L. Lee (Ed.), *Peptide and Protein Drug Delivery*, Marcel Dekker, New York (1991).
3. J. Kost (Ed.), *Pulsed and Self-Regulated Drug Delivery*, CRC Press, Boca Raton, FL (1990).
4. C. G. Pitt, *Inter. J. Pharm.* **59**, 173 (1990).
5. W. Y. Kuu, R. W. Wood and T. J. Roseman, in: *Treatise on Controlled Drug Delivery*, p. 37, A. Kydonieus (Ed.). Marcel Dekker, New York (1991).
6. L. Brannon-Peppas and N. A. Peppas, *J. Controlled Release* **8**, 267 (1989).
7. J. H. Kou, G. L. Amidon and P. I. Lee, *Pharm. Res.* **5**, 592 (1988).
8. B. A. Firestone and R. A. Siegel, *Macromolecules* **21**, 3254 (1988).
9. L. C. Dong and A. S. Hoffman, *J. Controlled Release* **13**, 21 (1990).
10. I. C. Kwon, Y. H. Bae, T. Okano and S. W. Kim, *J. Controlled Release* **17**, 149 (1991).
11. N. A. Peppas (Ed.), *Hydrogels in Medicine and Pharmacy*, Vol. 1–3, CRC Press, Boca Raton, FL (1987).
12. S. H. Gehrke and P. I. Lee, in: *Specialized Drug Delivery Systems*, p. 333, P. Tyle (Ed.). Marcel Dekker, New York (1990).
13. G. W. R. Davidson III and N. A. Peppas, *J. Controlled Release* **3**, 259 (1986).

14. A. R. Khare and N. A. Peppas, *Macromolecules*, in press.
15. R. Drummond, M. L. Knight, M. L. Brannon and N. A. Peppas, *J. Controlled Release* **7**, 181 (1988).
16. P. L. Ritger and N. A. Peppas, *J. Controlled Release* **5**, 37 (1987).
17. D. Hariharan and N. A. Peppas, *J. Controlled Release*, in press.
18. N. A. Peppas and C. T. Reinhart, *J. Membr. Sci.* **15**, 275 (1984).
19. C. J. Kim and P. I. Lee, *Pharm. Res.* **9**, 10 (1992).

pH, salt, and buffer dependent swelling in ionizable copolymer gels: Tests of the ideal Donnan equilibrium theory

BRUCE A. FIRESTONE* and RONALD A. SIEGEL†

Departments of Pharmacy and Pharmaceutical Chemistry, University of California, San Francisco, CA 94143-0446, USA

Received 9 November 1992; accepted 2 March 1993

Abstract—We present results of equilibrium swelling studies of the ionizable copolymer gel, methyl methacrylate/*N*,*N*-dimethylaminoethyl methacrylate 70/30 mol%, in buffered and unbuffered electrolyte solutions. The experimental conditions were designed to demonstrate the sensitivity of swelling in ionized gels to the electrolyte composition of the external solution. In general, gel swelling as a function of solution ionic strength is shown to be highly nonmonotonic and is particularly sensitive to the valence and concentrations of ions present in solution. A rigorous test of ideal Donnan equilibrium theory shows that the latter is unable to explain all the data in a self-consistent manner. However, a heuristic procedure based on the ideal Donnan theory can predict qualitatively the observed trends. While not quantitative, this heuristic approach provides considerable insight into the mechanisms underlying the swelling behavior under various solution conditions. Possible causes of nonideal behavior are discussed, and some observed specific ion effects are reported and discussed.

Key words: Copolymers; ionizable gels; swelling; Donnan equilibrium; ion exchange.

INTRODUCTION

Ionic gels eventually come to an osmotic equilibrium with surrounding electrolyte solutions. Because net transfer and exchange of solvent (water) and mobile ions occurs between the gel and the external electrolyte solution, changes in the composition of the solution (e.g., pH, ionic strength, ionic valence, cosolvent content) can perturb the osmotic balance and cause profound changes in the equilibrium degree of swelling of the gel. Accounts of these effects in gels have been reported primarily for partially ionized polyacrylamide and poly(*N*-isopropyl acrylamide) gels [1-7]. The presence of ionized groups on the gel network can significantly alter the swelling behavior of the gel compared with an equivalent neutral gel [2], and can cause the gel to undergo discrete or continuous volume transitions with changes in solution composition, temperature or salt content [2, 4].

Several reports detailing the effects of pH and ionic strength on gel swelling have appeared. Typically, the gels chosen for study are relatively hydrophilic, and swelling is carried out in unbuffered solutions of mineral acids and added simple salts. For certain of these systems, in which the density of charge groups fixed to the gel is relatively low, it has been shown that a simple theory based on the ideal Donnan equilibrium can explain the data reasonably well [8-18].

---

* Present address: Allergan Pharmaceuticals, Inc., Pharmaceutical Sciences, 2525 Dupont Drive, Irvine, CA 92713, USA.

† To whom all correspondence should be addressed.

In an earlier study [19], aqueous swelling equilibrium measurements were reported for cationic gels based on *n*-alkyl methacrylates (*n*-AMA, ≥70% mol/mol) copolymerized with *N,N*-dimethylaminoethyl methacrylate (DMA, ≤30% mol/mol) and lightly crosslinked with divinyl benzene (DVB). These gels were immersed in solutions consisting of citrate or phosphate buffer and sodium chloride, the latter being added to set the total ionic strength to desired levels. This study differed in a number of ways from much of the previous work. First, our gels were intrinsically more hydrophobic than those mentioned in the previous paragraph, and swelled negligibly in the neutral and alkaline pH ranges. Gel hydrophobicity was due to the dominant *n*-AMA comonomer. Second, the proportion of ionizable groups was higher than usual. Third, buffered solutions were used.

The use of buffers increases the complexity of the system under study. This probably explains why most physicochemical studies of gel swelling utilize unbuffered solutions, especially when a comparison with theory is to be made. On the other hand, when considering pharmaceutical applications, buffered solutions become important. Buffers are used to stabilize pH and to simulate expected physiologic conditions. Further, it has been shown recently that buffers can speed up gel swelling manyfold compared to unbuffered systems [20, 21].

The question can therefore be asked, whether the ideal Donnan theory is appropriate for the buffered systems, particularly when the gels are relatively hydrophobic and moderately charged. In our previous contribution [19], an indirect comparison was made over a very limited set of data. Although the results were encouraging, it was clear that more extensive data and a more rigorous comparison between theory and experiment are required. The purpose of this paper is to provide a more extensive and rigorous test of ideal Donnan theory.

THEORY

We consider an initially dry gel immersed in an ionic solution which at equilibrium has a hydration level H, where H is defined as the weight fraction of the gel that is water. The exact value of H is assumed to be determined by a balance of three pressures. First, there is the 'network swelling pressure', Π_{net}, which is due to polymer/solvent interactions and the elasticity of the polymer network; this pressure is considered independent of the ionization state of the network and the ionic environment. Second, the effect of the ions is summarized by an 'ion swelling pressure' Π_{ion}, which depends on H and the concentrations $\{C\}$ of mobile ions inside and outside the gel. Finally, an external pressure P_{ext} can be applied to the gel (but not the surrounding medium) in order to increase or reduce H. At swelling equilibrium the combined internal swelling pressures equal the externally applied pressure:

$$P_{ext} = \Pi_{net} + \Pi_{ion}. \tag{1}$$

In this study free swelling was measured, so that $P_{ext} = 0$. We may therefore rewrite Eqn (1) as follows:

$$\Pi_{net}(H) = -\Pi_{ion}(H, \{C\}). \tag{2}$$

In this form we are explicit about the variables which determine the swelling pressures. Viewing Eqn (2), one can conclude that a test for the success of any particular model for the ion swelling pressure Π_{ion} is that different ionic compositions $\{C\}$ that lead to similar *experimental* values of H must have similar *predicted* values for Π_{ion}. Thus, a plot of experimental values of H vs theoretical values of Π_{ion}, in which the experimental values of H are used in calculating Π_{ion}, should show little scatter.

Before proceeding, we note that the model test described above was introduced, although somewhat less explicitly, by Rička and Tanaka [16]. It has the advantage that no assumptions need to be made regarding the relationship between Π_{net} and H. The theories of polymer/solvent interaction (i.e. Flory–Huggins theory) [11] and polymer elasticity [11, 22] could have been used to postulate a theoretical relationship between Π_{net} and H, which would permit direct prediction of H based on $\{C\}$. However, these theories are imperfect [22], and moreover they require independent determination of several parameters. Since our goal is to isolate and test a particular theory of ion swelling pressure, the test embodied in Eqn (2) seems appropriate.

The full derivation of the expression for the ion swelling pressure Π_{ion} based on the ideal Donnan equilibrium theory appears in the appendix. Briefly, let C_i and C_i' represent the concentration of the ith ionic species inside the gel water and in the external solution, respectively, and let z_i be that ion's valence. The ideal Donnan theory relates the internal and external concentrations by

$$C_i = \lambda^{z_i} C_i' \tag{3}$$

where λ is the *Donnan ratio*, given by the only real positive root of the equation

$$(1 - v_p) \sum z_i C_i' \lambda^{z_i} + \frac{\sigma_0 v_p}{1 + \lambda^{-1} 10^{\text{pH-p}K_a}} = 0. \tag{4}$$

In Eqn (4) pK_a is the log ionization constant of the amine groups inside the gel, σ_0 is the concentration (mol l^{-1}) of these groups when the gel is in the dry state, and v_p is the volume fraction of polymer in the gel, calculated as

$$v_p = \frac{1 - H}{1 + (\rho_p - 1)H}$$

where ρ_p is the density of the polymer.

At this point we digress to note that the $(1 - v_p)$ term in Eqn (4) has not appeared in our previous work [12, 18]. It also is absent in Rička and Tanaka's paper, but this is of little consequence since they worked with highly swollen gels for which $v_p \ll 1$. The need for the $(1 - v_p)$ term in models for gel electrostatics was originally recognized by Hill [23]. The justification for this term appears in the appendix.

The computed value of λ from Eqn (4) is used to calculate Π_{ion} using van't Hoff's law

$$\frac{\Pi_{ion}}{RT} = \sum C_i'(\lambda^{z_i} - 1). \tag{5}$$

Thus, our procedure to test the Donnan equilibrium is as follows. The constants σ_0 and ρ_p are characteristics of the gel under study, and the values of C_i' are parameters for a given experiment. When the gel comes to swelling equilibrium, the hydration H is assessed, and sufficient information is now available to solve Eqn (4) for λ. This value is then introduced into Eqn (5) in order to calculate Π_{ion}/RT. A plot of Π_{ion}/RT vs H, for a number of experimental conditions, is then examined for monotonicity and scatter.

As discussed elsewhere [12], the ideal Donnan theory results from some simplifying assumptions: (1) Ideal solution behavior of the ions—all ions retain full osmotic activity in both the external solution phase and the gel interior. This ignores electrostatic interactions between ions (both mobile and fixed) as well as steric and dielectric effects that affect ionic activities in the gel interior. (2) Ionization constants of the gel and other weak electrolytes are constant—the proton binding constants of the gel amine groups and all other weak electrolytes in the gel interior are constant and independent of the extent of swelling and the network charge density. The protons inside the gel are thus assumed to bind to the fixed amines and to mobile buffers in a Langmuirian manner.

EXPERIMENTAL

Materials

Sodium chloride, potassium chloride (Mallinckrodt, Inc.); sodium sulfate, sodium bromide, sodium thiocyanate (J. T. Baker); sodium iodide, sodium nitrate, cesium chloride, lithium chloride (Aldrich), and citric acid (Fisher Scientific) were analytical reagent grade and were used without further purification. Water used in the swelling studies was double distilled and deionized.

Gel preparation and density determination

Copolymer gels consisting of 70% (mol/mol) methyl methacrylate (MMA) and 30% (mol/mol) dimethylaminoethyl methacrylate (DMA), lightly crosslinked by divinyl benzene (DVB) were prepared as thin circular disks with a dry thickness of 0.28 (±0.02) mm and diameter 7 mm according to the method described previously [16, 24, 25].

Dry gel density (ρ_p) was determined by determining dry gel weight in air (W_a) and dry gel weight in propanol (W_p) using a hanging microbalance (Mettler), and then using Archimedes principle, by which

$$\rho_p = \frac{(0.805)W_a}{W_a - W_p}$$

where the factor 0.805 is the density of propanol. Using this method, the dry gel's density was determined to be $\rho_p = 1.16$.

Equilibrium swelling studies

Gel equilibrium swelling was determined in duplicate in 2 l of electrolyte solution at 25°C by a method described previously [16, 24, 25]. The ionic strength dependent swelling properties of the gel were determined in several electrolyte solutions including: (1) simple NaCl solutions; (2) citrate buffers (0.5, 2.1, and 10 mM citrate) at pH 4.0 with excess NaCl added to adjust the ionic strength; (3) 2.1 mM citrate at pH 4.0, 5.0, and 6.0 with added NaCl; and (4) 10 mM citrate buffer pH 4.0 with added Na_2SO_4. Also, swelling was measured in 100 mM solutions of NaCl, KCl, LiCl, CsCl, NaSCN, NaI, $NaNO_3$, and NaBr at 25°C.

RESULTS

We now present results of equilibrium swelling studies of the 70/30 mol% MMA/DMA gel in several types of electrolyte solutions. The experimental conditions were designed to demonstrate the sensitivity of swelling in an ionized gel to the electrolyte composition of the external solution. Swelling results are expressed in terms of equilibrium hydration, H, (g water in gel)/(g swollen gel), and is calculated as $(W_S - W_0)/W_S$, where W_S and W_0 are the sample weights of the swollen gel at equilibrium and in the dry (initial) state, respectively.

Buffer and counterion valence effects

The ionic strength (I) dependent equilibrium swelling behavior of the gel was first studied in three different types of electrolyte solutions, all at a constant pH of 4.0: (1) simple NaCl solution; (2) 10 mM citrate buffer with added NaCl; and (3) 10 mM citrate buffer with added Na_2SO_4. These solution conditions allow the effects of ionic strength and ionic valence on gel swelling to be observed concurrently. For example, the HCl/NaCl solutions contain chloride as the single gel counterion (OH^- can be neglected at pH 4.0), while the citrate/NaCl solutions contain citrate counterions (mono, di, and trianions, the latter in trace amounts at pH 4.0) in addition to chloride. The citrate/Na_2SO_4 solutions contain approximately the same distribution of citrate counterions (the pH is the same) as the citrate/NaCl solutions, but contain the sulfate dianion instead of chloride.

Figure 1 shows the ionic strength dependence of gel equilibrium swelling in each of these solution types. Considering the NaCl results first, it is seen that swelling in NaCl solutions is nonmonotonic, with swelling increasing to a maximum at approximately $I = 7$ mM and decreasing as I is increased further. Gel swelling in

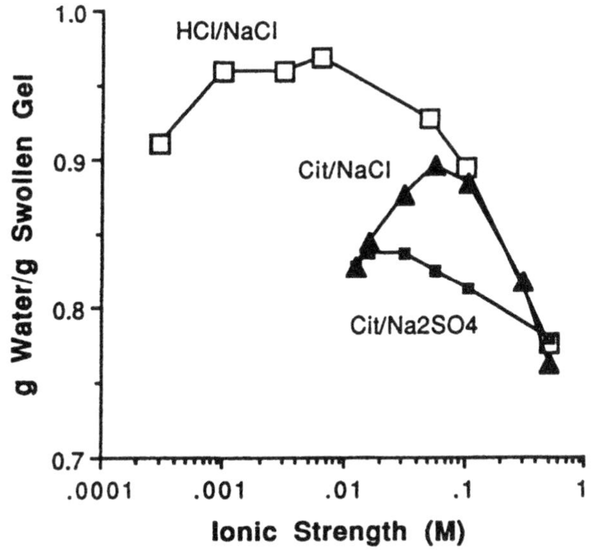
1.0
0.9
0.8
0.7
HCl/NaCl
Cit/NaCl
Cit/Na2SO4
g Water/g Swollen Gel
.0001
.001
.01
.1
1
Ionic Strength (M)

Figure 1. Equilibrium swelling of the MMA/DMA 70/30 mol% gel as a function of ionic strength at pH 4.0 and 25°C in HCl/NaCl solutions (□), 10 mM citrate/NaCl solutions (▲) and 10 mM citrate/Na_2SO_4 (■).

the citrate/NaCl solutions differs significantly from swelling in NaCl solutions in several respects. In the low ionic strength region, swelling is much lower and the position of the maximum in the curves is shifted to higher ionic strength compared with swelling in NaCl solutions. The differences in swelling become less pronounced, and eventually disappear, as solution ionic strength is increased. In citrate/Na_2SO_4 solutions, gel hydration is nearly flat initially as I is increased, but then decreases at higher ionic strength. It should be pointed out that the two citrate curves of Fig. 1 share a common point at I = 12 mM (the lowest ionic strength), which corresponds to 10 mM citrate buffer at pH 4.0 with no added salt. (Note: a number of the data points in Fig. 1 have appeared previously [19]. Figure 1 extends previous results for unbuffered solutions to lower ionic strengths, and also introduces data for citrate/Na_2SO_4 solutions for the first time.)

Buffer concentration effects

Gel swelling is also strongly affected by the total citrate concentration at pH 4.0 as shown in Fig. 2. Solutions contained either 0.5, 2.1, or 10 mM citrate with NaCl added to adjust the total ionic strength to the desired level. The leftmost data-point on each curve corresponds to the minimum ionic strength attainable at each citrate concentration (no added NaCl). Swelling in all cases is highly nonmonotonic and is strongly suppressed by increasing citrate concentration in the low ionic strength region. Furthermore, the observed maxima in the curves shift to higher I with increasing citrate concentration. The differences in swelling become less evident, and finally disappear, as I increases sufficiently past the maximum in each curve, as was observed in Fig. 1.

Effects of pH

A set of swelling curves of similar profile are generated when I is varied at different pHs with constant citrate concentration as indicated in Fig. 3. Citrate/NaCl solutions

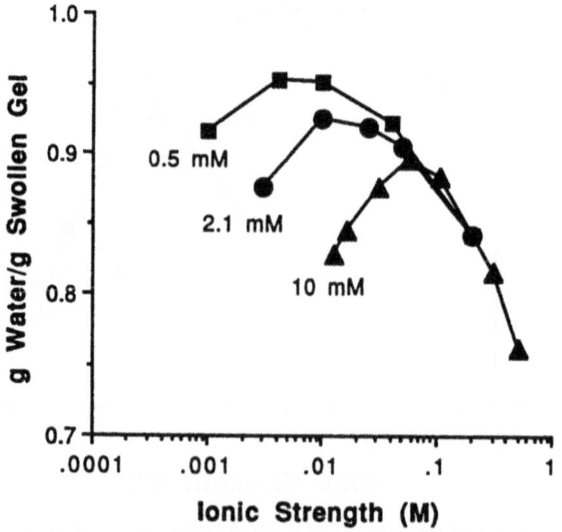

1.0
0.9
0.8
0.7
g Water/g Swollen Gel
0.5 mM
2.1 mM
10 mM
.0001
.001
.01
.1
1
Ionic Strength (M)

Figure 2. Effect of total citrate concentration of citrate/NaCl solutions on gel swelling at pH 4.0 and 25°C in 0.5 mM citrate (■), 2.1 mM citrate (●) and 10 mM citrate (▲).

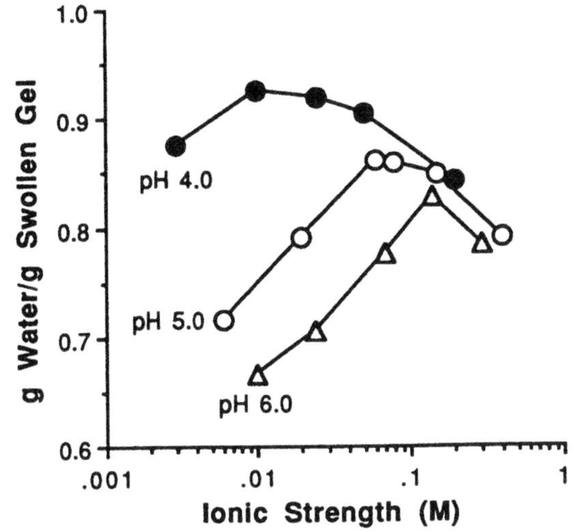
1.0
0.9
0.8
0.7
0.6
g Water/g Swollen Gel
pH 4.0
pH 5.0
pH 6.0
.001
.01
.1
1
Ionic Strength (M)

Figure 3. Effect of pH of 2.1 mM citrate/NaCl solutions on gel swelling at 25°C. pH 4.0 (●), pH 5.0 (○) and pH 6.0 (△).

containing 2.1 mM citrate and various concentrations of NaCl were prepared at solution pHs of 4.0, 5.0, and 6.0. The profiles of the curves are generally similar to those of Fig. 2, with swelling in the low I region being significantly decreased as solution pH increases and with the convergence of the curves at higher I. The position of the maximum in the curves shifts to higher I as pH increases.

Specific ion effects

In order to assess the effects of ion-specific interactions on gel swelling not predicted by the Donnan theory, swelling was measured in a series of four unbuffered solutions containing chloride salts (representing different gel coions) and four sodium salts (representing different gel counterions), each prepared at pH 4.0 and $I = 0.1$ M. Figure 4a indicates that the identity of the gel coion has minimal impact on the gel swelling, whereas swelling is strongly affected by the identity of the counterion as shown in Fig. 4b.

DIRECT TESTS OF IDEAL DONNAN THEORY

The results presented in the previous section highlight the sensitivity of swelling of ionized gels to the electrolye composition of the external solution. In the context of the Donnan theory, the observed swelling behavior can be qualitatively rationalized on the basis of a net mobile ion osmotic pressure (Π_{ion}) between the gel interior and the solution, which can be affected both by gel properties (charge density) and solution properties (electrolyte composition: pH, ionic strength, ionic valence). We now use the data described in the previous section to test the validity of the ideal Donnan theory for the swelling of 70/30 MMA/DMA gels.

Before testing the theory for the buffered systems considered in Figs 1–3, it can be noticed immediately that the ideal Donnan theory embodied in Eqns (4) and (5) fails to predict the specific counterion effects seen in Fig. 4b. In those equations, an ion's effect on swelling is determined only by its concentration and valence. The

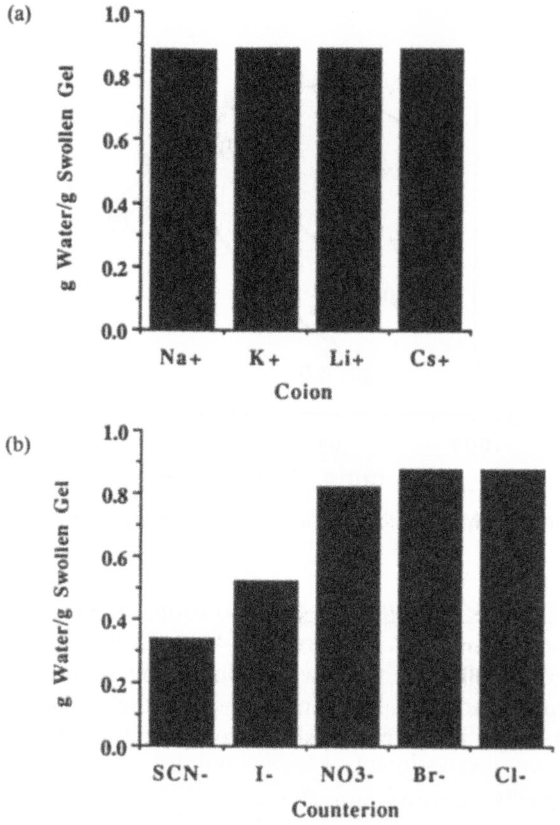
(a)
g Water/g Swollen Gel
1.0
0.8
0.6
0.4
0.2
0.0
Na+
K+
Li+
Cs+
Coion
(b)
g Water/g Swollen Gel
1.0
0.8
0.6
0.4
0.2
0.0
SCN-
I-
NO3-
Br-
Cl-
Counterion

Figure 4. Effect of specific ions on the equilibrium swelling of the MMD/DMA 70/30 mol% gel at pH 4.0 and a total salt concentration of 100 mM. (a) Various gel coions (chloride salts). (b) Various gel counterions (sodium salts). Measurements were made at 25°C.

concentrations and valences of all counterions in Fig. 4b were identical, yet the degree of swelling for SCN^- and I^- was much less than that for Cl^- and Br^-.

We turn now to the buffered systems, for which we will apply the quantitative test of the ideal Donnan theory described previously. Figure 5 is a plot of experimental gel hydration H values, taken from Figs 1–3, vs the ion swelling pressure, Π_{ion}, calculated from Eqns (4) and (5). The concentration of ionizable amine groups in the dry gel was calculated using the formula

$$\sigma_0 = \frac{1000 f \rho_g}{M_{w_{DMA}}}$$

where $M_{w_{DMA}} = 157.2$ is the formula weight of DMA and $f = 0.4$ is the *weight* fraction of DMA in the copolymer, and the dry gel density $\rho_g = 1.16$ (*vide supra*). From these values we obtain $\sigma_0 = 2.95$ equiv l^{-1}.

Symbols in Fig. 5 correspond to those appearing in the original plots. It is clear that, while an overall monotonic decreasing trend in Π_{ion} with increasing H can be inferred, there is considerable scatter. Conditions leading to identical predicted

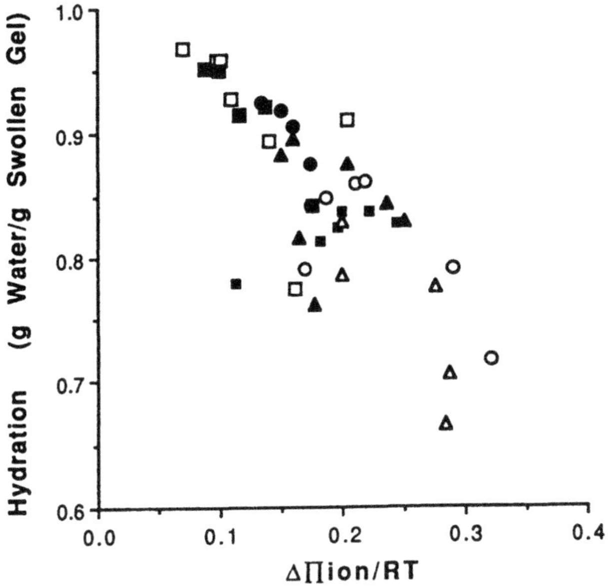
Hydration (g Water/g Swollen Gel)
1.0
0.9
0.8
0.7
0.6
0.0
0.1
0.2
0.3
0.4
ΔΠion/RT

Figure 5. Plot of observed hydration values H vs ideal Donnan osmotic pressure Π_{ion}, as calculated from theory presented in eqns (3)–(5). Data taken from Figs 1–3 with symbols preserved.

Table 1.
Parameter values used in all calculations

Parameter	Definition	Value used
f	wt. fraction of DMA in gel	0.40
$M_{w_{DMA}}$	formula weight of DMA	157.2
ρ_g	density of dry gel	1.16
K_a	ionization constant of gel	$10^{-7.7}$

Π_{ion} can give rise to significantly different degrees of swelling, which violates the criterion for the validity of the ideal Donnan theory. Moreover, the behaviors for some of the investigated conditions, represented by specific symbols in Fig. 5, are erratic. In several cases the relationship between Π_{ion} and H is nonmonotonic for a particular symbol, and in one case it is monotonically *increasing*.

The lack of a well defined monotonic relationship between Π_{ion} and H indicates that the ideal Donnan equilibrium model does not provide a self-consistent explanation of ion effects on swelling pressure. Attempts to explain this observation, plus the specific ion effects described in the previous paragraph, will be deferred till after the next section.

A HEURISTIC PROCEDURE

Despite the quantitative failure of the ideal Donnan theory, the latter is probably the simplest theory that can be applied to the data. We now show, using a procedure similar to that introduced by Grignon and Scallan [17], that the theory can be adapted to provide qualitative explanations of the trends observed in Figs 1–3.

We assume a fictional (though readily implementable [24, 25]) experiment in which the hydration H is fixed at some arbitrary value, H^*, and calculate the resulting Π_{ion} from Eqns (4) and (5). From Eqn (1), we see that the calculated Π_{ion} will equal the excess pressure over ambient that must be applied to the gel to maintain $H = H^*$ (i.e. constant volume), minus the pressure that would have to be applied in the absence of ionic effects. We then postulate the following *heuristic*: any condition leading to higher Π_{ion} at fixed hydration H^* (the fictional case), should lead to increased H in the free swelling (i.e. real) case.

Figures 6a–c represent computed values of Π_{ion} at $H^* = 0.7$, calculated for conditions represented in Figs 1–3, respectively. This value of H^* was chosen because

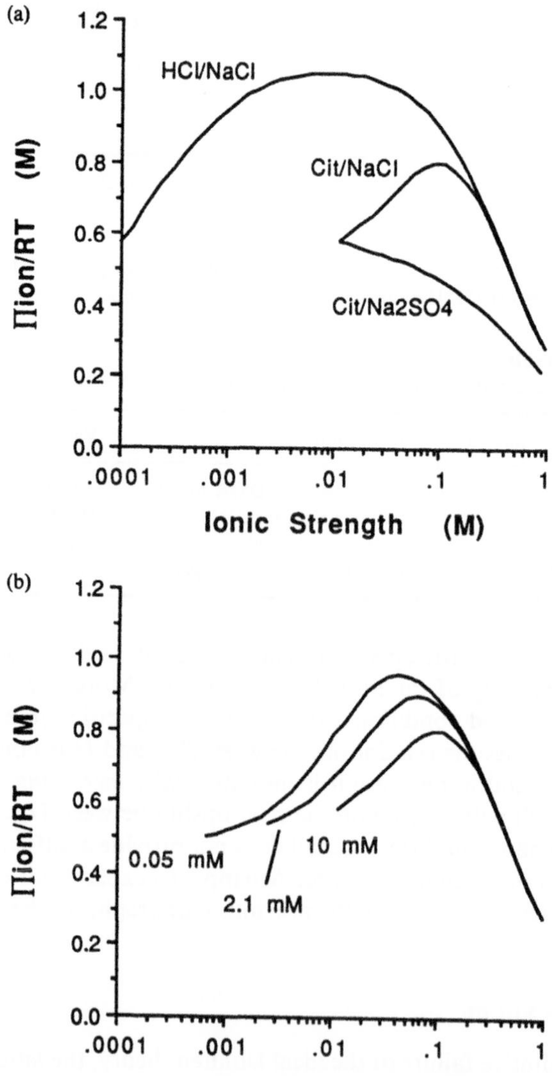
(a)
HCl/NaCl
Cit/NaCl
Cit/Na2SO4
Πion/RT (M)
1.2
1.0
0.8
0.6
0.4
0.2
0.0
.0001
.001
.01
.1
1
Ionic Strength (M)
(b)
0.05 mM
2.1 mM
10 mM
Πion/RT (M)
1.2
1.0
0.8
0.6
0.4
0.2
0.0
.0001
.001
.01
.1
1
Ionic Strength (M)

Figure 6. Ion swelling pressures Π_{ion} calculated according to heuristic model described in text. (a) Curves corresponding to Fig. 1. (b) Curves correspond to Fig. 2. (c) Curves correspond to Fig. 3.

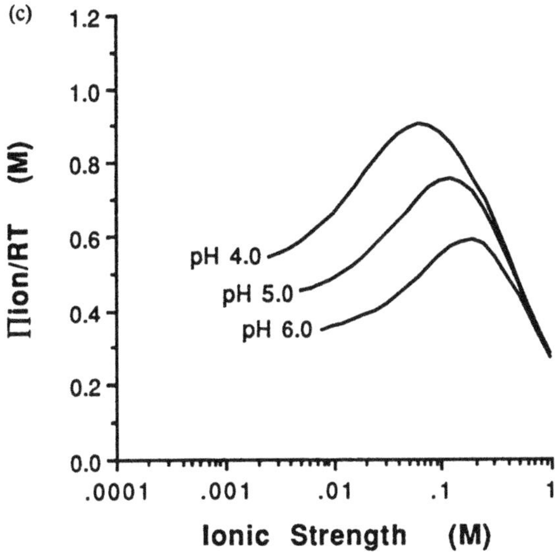
(c)
1.2
1.0
0.8
0.6
0.4
0.2
0.0
Πion/RT (M)
pH 4.0
pH 5.0
pH 6.0
.0001
.001
.01
.1
1
Ionic Strength (M)

Figure 6. *Continued.*

it leads to curves whose peaks are located near those in Fig. 1. Apparently, this procedure does predict reasonably well the existence and relative locations of the swelling peaks, as well as the relative degrees of swelling. It also predicts the generally observed trend that differences in swelling disappear at high ionic strength.

Closer comparison of the curves in Fig. 6 with the corresponding experimental data of Figs 1–3 reveals quantitative discrepancies, however, particularly in the peak positions. For the data of Figs 2 and 3, the peaks are considerably more spread out than is predicted in Fig 6b and c, respectively. This is to be expected, since we have already shown quantitative incompatibilities between the ideal Donnan model and the swelling data. It should also be noted that the position and sharpness of the theoretical peaks depends on the chosen value of H^*. It can be shown that as H^* increases, the peaks shift to lower values of I and tend to broaden (simulations not shown).

We now use the heuristic model to provide explanations for the observed experimental trends. Consider first the curve for unbuffered NaCl solutions at pH 4.0, shown in Fig. 1. The heuristic theory predicts a rise and fall in ion swelling pressure with increasing ionic strength (Fig. 6a), which corresponds to the rise and fall in hydration in the experimental, free swelling case. To explain the initial rise, we calculate the fraction of amine groups that are ionized, ϕ, which is given by the expression

$$\phi = \frac{1}{1 + \lambda^{-1}10^{\mathrm{pH}-\mathrm{p}K_a}} \tag{6}$$

where λ is calculated using Eqn (4). Results are shown in Fig. 7, which displays a rise in ϕ to a plateau value of 1 with increasing ionic strength. In the low ionic strength regime, the availability of gel counterions (chloride in the present case) ultimately limits the extent of gel ionization. The addition of NaCl permits the ionized fraction of the gel to increase, thereby increasing the charge on the gel and thus Π_{ion}. The maximum in Π_{ion} at approximately I = 1 mM (Fig. 6a) corresponds to the complete

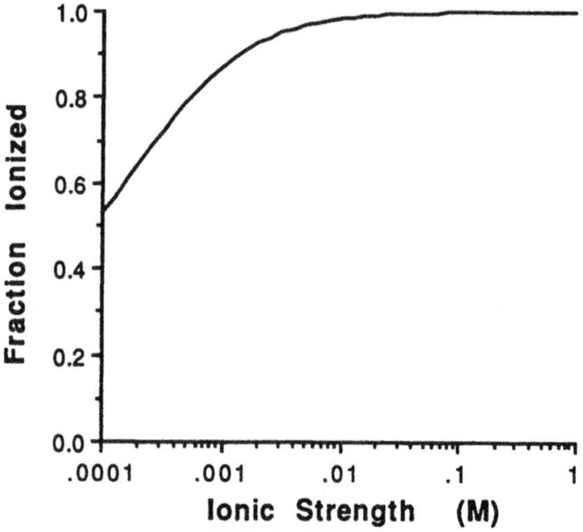
Fraction Ionized
1.0
0.8
0.6
0.4
0.2
0.0
.0001
.001
.01
.1
1
Ionic Strength (M)

Figure 7. Fraction of gel amine groups ionized, ϕ, in NaCl solutions at pH 4.0 computed as a function of ionic strength from the Donnan theory using a Langmuir proton mass action law [Eqn (6)].

ionization of virtually all gel amine groups ($\phi \approx 1$ in Fig. 7). Increasing I past this level simply decreases Π_{ion} due to a reduction in the Donnan ratio towards unity. That this will occur can be seen by comparing the two terms on the right side of Eqn (4). With increasing ion concentration C_i', the first term becomes dominant, and λ must approach unity in order to satisfy the electroneutrality requirement for the outer solution, i.e. $\Sigma z_i C_i' = 0$.

In the citrate/NaCl curve of Fig. 6a, Π_{ion} is significantly reduced at low ionic strengths compared with the NaCl case due to the presence of multivalent citrate species. The multivalent counterions tend to reduce gel osmolarity, since fewer such ions are required inside the gel to provide electroneutrality. Although the predominant citrate counterion in the external solution is the monoanion (at pH 4.0: 76.5% Cit^-, 12.7% Cit^{2-}, and 0.05% Cit^{3-} with remainder unionized), Donnan partitioning enhances the concentration of multivalent ions in the gel phase [see Eqn (1)].

The maximum in the citrate/NaCl curve arises by a different mechanism than in the pure NaCl case. As NaCl is added to increase the ionic strength, isoelectric exchange of chloride from the external solution for citrate anions in the gel phase occurs according to

$$Cit^{z-}{}_{gel} \leftrightharpoons z Cl^-{}_{soln}.$$

This exchange process is illustrated graphically in Fig. 8, in which internal total monovalent (Cit^{1-} plus Cl^-) and divalent (Cit^{2-}) and trivalent (Cit^{3-}) concentrations are plotted as a function of ionic strength. At pH 4.0 there are very few Cit^{3-} ions, so the exchange between Cit^{2-} and Cl^- dominates. The exchange process increases the osmolarity of the gel interior, resulting in an increase in Π_{ion}. The maximum in the citrate/NaCl curve (Fig. 6a) at approximately $I = 100$ mM indicates that the exchange process is virtually complete and the electrolyte composition of the gel resembles that in the NaCl case. This is corroborated in Fig. 8. A further increase in I simply causes a reduction in Π_{ion} in parallel with the NaCl curve (Fig. 6a). The

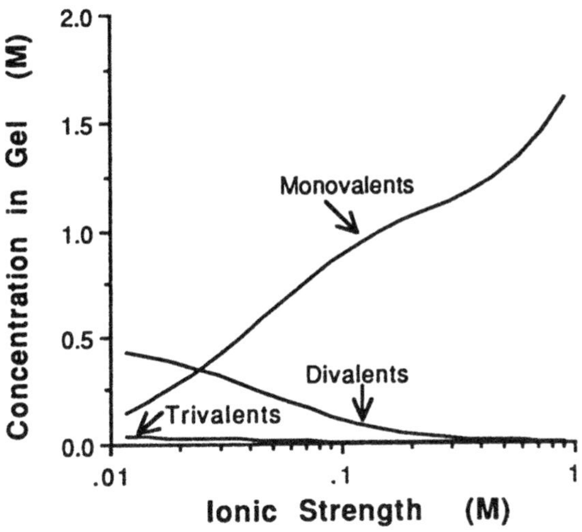
2.0
1.5
1.0
0.5
0.0
.01
.1
1
Concentration in Gel (M)
Ionic Strength (M)
Monovalents
Divalents
Trivalents

Figure 8. Calculated total concentrations of monovalent (Cit^{1-} and Cl^-), divalent (Cit^{2-}) and trivalent (Cit^{3-}) ions in gel as function of ionic strength, according to heuristic theory. Assume pH = 4.0, citrate buffer concentration = 10 mM.

observed swelling in citrate/NaCl solutions of Fig. 1 shows qualitative agreement with the predictions, with the maximum occurring in the same ionic strength range.

For the citrate/Na_2SO_4 case (Fig. 1), Π_{ion} decreases monotonically with increasing ionic strength, without showing a maximum. Addition of Na_2SO_4 results in isoelectric exchange of $SO_4{}^{2-}$ from the solution for citrate anions according to

$$2Cit^{z-}{}_{gel} \leftrightarrows zSO_4{}^{2-}{}_{soln}.$$

In contrast to citrate/chloride exchange, this stoichiometry results in a decrease in the gel osmolarity and thus Π_{ion}. The data of Fig. 1 for citrate/Na_2SO_4 solutions shows an initial small increase in swelling as I is first increased, but then decreases thereafter. This small increase in swelling is not predicted by the theory and could involve nonideal partitioning of the multivalent ions, or the specific ability of multivalent citrate ions to form electrostatic crosslinks with the gel, which are eliminated as exchange with $SO_4{}^{2-}$ occurs.

In the other citrate/NaCl curves of Fig 6b and c (corresponding to the data shown in Figs 2 and 3, respectively), Π_{ion} is shown to be sensitive to the total citrate concentration and to solution pH, respectively. In these sets of curves, Π_{ion} is progressively reduced (at fixed ionic strength) as either the citrate concentration increases at constant pH = 4.0 (Fig. 6b) or the pH increases at constant buffer concentration (Fig. 6c). In both cases the depression of Π_{ion} is attributed to an increase of multivalent citrate species. Also, all curves in Fig. 6b and c show a peak in Π_{ion} which is attributed to the citrate/chloride exchange process described above. With higher concentrations of multivalent citrate ions, more chloride ions must be added to complete the exchange, which explains the peak shifts to higher ionic strengths when either total citrate concentration (Fig. 6b) or pH (Fig. 6c) increase. It is noteworthy that as pH is raised (Fig. 6c), the trivalent counterions become significant, which enhances the effects described here.

DISCUSSION

The direct test

The direct test of the ideal Donnan theory could be used for any other theory of ionic effects in gel swelling, provided such a theory assumes that ion-based swelling forces are additive with the other swelling forces (polymer/solvent interactions, polymer elasticity). In some theoretical frameworks, this additivity will not hold. For example, ionic effects may change the persistence length of the gel's constituent chains, and this may affect the elastic force [26–28]. Alternatively, one may consider that ionizing a group attached to the gel will change the short range forces affecting hydration, in addition to purely electrostatic effects [29]. As a final example, a multivalent counterion may serve to 'crosslink' the network (see below), thus contributing to its elasticity. In any of these cases, the direct test described in this paper will not be applicable.

A direct test of the ideal Donnan theory for polyacrylamide gels, partially substituted by acrylic acid, has been published by Rička and Tanaka [16]. These gels swell much more than those studied here, and at much lower fixed charge densities. Rička and Tanaka found that the ideal Donnan theory makes excellent predictions when the mobile ions are all monovalent, but when divalent and trivalent ions are present the ideal theory overpredicts swelling. No specific monovalent counterion effects were seen. We attribute the theory's greater success in this case to the dilute nature of the gel and the relatively large distance between the fixed charge groups. This reduces the ion activity effects which are discussed in the following paragraphs.

Ion activities

A major simplifying assumption of the ideal Donnan model tested above is that ion activities are equal to their respective concentrations [12, 16]. Given that (a) the external ion concentrations can be well above the range where ideality can be assumed; and (b) the ion concentrations inside the gel are even higher, this assumption must be regarded as an approximation. The simplest activity coefficient correction based on Debye–Hückel theory will also be inapplicable, since this theory is only valid for low concentrations, particularly for multivalent ions [30]. More sophisticated activity coefficient theories might be considered; however, these are often difficult to apply to mixed-ionic media.

The presence of the charged polymer network must also affect the activities of the mobile ion species. Ignoring for a moment the charge on the network, it should be recognized that when the network constitutes a significant volume fraction of the hydrated gel, the network will interfere to some extent with the interactions between mobile ions, yielding different activity coefficients than would be obtained for aqueous solutions with identical ion concentrations. Interference can be due to steric and/or dilectric effects. At present, there is no theory to deal with these effects.

Network fixed charges will also affect ion activities. By virtue of their attachment to the polymer network, these charges cannot be distributed in the same manner as the mobile ions; in turn, the fixed charges will alter the mobile ion distribution within the gel. When the gel is highly charged, as in the present case, counterions are likely to be attracted to the polyelectrolyte chains, leading to a reduced osmotic activity [28, 31].

Thus far we have discussed the effects of the network and fixed charges on ion activities in general terms. More specific hypotheses will be introduced in the following subsections.

Specific ion effects

The invariance of gel swelling in solutions of the various chloride salts (Fig. 4a) is behavior consistent with the expected Donnan exclusion of coions. Under the conditions of Fig. 4a (monovalent salts, pH 4.0, $I = 100$ mM), the gel phase concentration of the coion is calculated from Donnan theory to be about fifteenfold smaller than the counterion concentration ($\lambda \approx 0.26$). This relative exclusion of coions from the gel interior reduces the coion contribution to gel osmolarity. Thus Π_{ion} is determined primarily by the chloride counterion. This results in virtually identical swelling across the series of chloride salts.

On the other hand, gel swelling is strongly affected by the identity of the counterion (anions of the sodium salts), as shown in Fig. 4b, which exemplifies behavior not predicted by the Donnan theory. The counterion effect correlates strongly with the position of the anion in the Hofmeister or lyotropic series [32, 33]. The lyotropic series identifies a certain 'order of effectiveness' of ions that has been found to be remarkably similar in a wide variety of physical and chemical processes, including protein stability [34], swelling of gelatin and collagen [34–37], ion adsorption in gels [33, 38] and in the ion-selective permeabilities of hydrogels [33, 39]. The order of monoanions that has emerged is $SCN^- < I^- < Br^- < NO_3^- < Cl^-$. These specific ion effects are generally regarded as resulting from the water structure perturbing effects of the ions in solution [30, 33]. Relatively small ions and multivalent ions at the right-end of the series are thought to be water structure-makers (kosmotropes). They order and electrostrict surrounding water molecules through ion–dipole interactions to form long range hydration shells. Large monovalent ions at the left-end of the series are thought to be water structure-breakers (chaotropes) which generate relatively weak electrostatic fields capable of perturbing only nearest neighbor water molecules.

Recent studies have shown that the order of elution of anions from a crosslinked dextran gel correlates strongly with the lyotropic series, with the ions at the beginning of the series eluting later than those at the end [38]. This has been interpreted as an indication of the degree to which a given ion is adsorbed to the stationary gel phase (late elution indicating a greater tendency to adsorb) rather than a simple hydrodynamic effect. Since chaotropes such as SCN^- and I^- adsorb to the gel more than kosmotropes such as NO_3^-, Br^-, and Cl^-, it seems that the chaotropes should have a reduced osmotic activity, and hence swelling, compared to the kosmotropes, as observed in Fig. 4b.

Multivalent counterion effects

Most of the interesting trends associated with citrate buffers have been shown to be due to their multivalent nature. It is worthwhile to point out that nonidealities tend to be compounded when multivalent ions are present. For example, the Debye-Hückel theory loses its predictive ability at much lower concentrations for multivalent ions than for monovalents [30]. Similarly, flocculation of colloids is enhanced by multivalent ions in a manner that can't be accounted for by well known analytical theories [40]. Recent statistical mechanical calculations have shown that multivalent counterions lead to attractive forces between colloids, due to so-called 'ion correlations' [41]. While we make no attempt at present to model these effects, we can speculate that they may explain at least some of the nonidealities we have observed.

Another potential effect of the multivalent citrates is a kind of ionic 'crosslinking' or 'bridging'. For example, a single divalent citrate ion may 'bind' transiently to two amine groups on two separate polymer chains, leading to a transient crosslink between those chains. As the citrate ion is rather large, this seems to be a plausible event. Similar effects might be conjectured for trivalent citrate ions.

In our calculations of Eqns (3)–(5) we assume that the citrate buffer ionizes according to its pK_1, pK_2, and pK_3 values given at infinite dilution. It is well known that these pK values will decrease with increasing salt content, due to electrostatic stabilization of the ionic forms of the buffer. Therefore the multivalent components of the buffers will probably be in higher proportions than we have assumed. This effect will be greatest for the trivalent species. This may explain the slight peak in the citrate/Na_2SO_4 experimental curve of Fig. 1: trivalent citrates may be present in sufficient number that their exchange with SO_4^{2-} is significant.

We have considered these shifts in pK values using Debye–Hückel theory [42]. In the resulting simulations we saw no improvement in the self-consistency of the Donnan theory, upon viewing a graph similar to Fig. 5. As might be expected, shifting the pKs downward tends to lower the calculated Π_{ion} curves of Fig. 6 and shifts their peaks to the right. However, this correction of pKs does not significantly improve or degrade the qualitative performance of the Donnan model.

Finally, it should be mentioned that buffer effects on the swelling of cationic gels have been noticed previously. For example the present authors observed, for 70/30 mol% MMA/DMA gels, that citrate buffered solutions lead to lower degrees of swelling than phosphate buffered solutions when buffer concentration is the same and the overall ionic strength, set by addition of NaCl, is also the same. This result is attributed to the higher pKs of phosphate, such that at a given pH the average ionization of citrate is higher than for phosphate [19]. Similar results were obtained by Vasheghani-Farahani *et al.* [28], who studied copolymer gels consisting of acrylamide and the strong cationic salt methacrylamidopropyl triethyl ammonium chloride (MAPTAC). In solutions containing Na_2HPO_4, the swelling of these gels decreases with increasing pH. Since MAPTAC is a strong electrolyte it should not be affected by pH; changes in swelling are attributed to a shift in ionization state of the phosphate ion.

CONCLUSIONS

The ideal Donnan theory, while unable to provide a self-consistent correlation, remains a useful tool for understanding gel behavior, particularly in the presence of buffers of mixed valence. Three primary and distinct processes have been discussed as underlying the ionic strength dependent nonmonotonic swelling behavior of the polybasic 70/30 mol% MMA/DMA gel in aqueous electrolyte solutions. Gel swelling in the dilute salt regime is very sensitive to ionic valence and the availability of counterions from the bulk solution, which may ultimately limit the charge density of the gel. At intermediate salt concentrations, isoelectric ion exchange may occur between ions in the gel and in solution, which can increase swelling depending upon the stoichiometry of exchange. And finally, gel swelling decreases in solutions containing high salt regardless of the ionic composition. These processes have been shown to be qualitatively consistent with the Donnan theory as applied to swelling in ionized, hydrophilic gels.

Gel coions are substantially Donnan excluded and have little effect on gel swelling, while swelling is strongly dependent on the presence of different counterions, indicating nonideal behavior. The effectiveness of the counterions correlates with the lyotropic series.

Though the theory, as implemented here, does not attempt to describe nonideal electrolyte behavior or interactions between mobile electrolytes and fixed charged groups within the gel, we believe it can provide considerable insight into the mechanisms underlying the swelling behavior of ionized gels.

Acknowledgements

This work was funded in part by a predoctoral NIH training grant to BAF (GM07174), NIH grant DK 38035, and a grant from the Whitaker Foundation.

REFERENCES

1. T. Tanaka, *Phys. Rev. Lett.* **40,** 820 (1978).
2. T. Tanaka, D. Fillmore, S.-T. Sun, I. Nishio, G. Swislow and A. Shah, *Phys. Rev. Lett.* **45,** 1636 (1980).
3. M. Ilavsky, *Macromolecules* **15,** 782 (1982).
4. I. Ohmine and T. Tanaka, *J. Chem. Phys.* **77,** 5725 (1982).
5. T. Amiya and T. Tanaka, *Macromolecules* **20,** 1162 (1987).
6. L.-C. Dong and A. S. Hoffman, *J. Contr. Release* **15,** 141 (1991).
7. L. Brannon-Peppas and N. A. Peppas, *Biomaterials* **11,** 635 (1990).
8. F. G. Donnan, *Chem. Rev.* **1,** 73 (1925).
9. J. Th. G. Overbeek, *Prog. Biophys. Biophys. Chem.* **6,** 57 (1956).
10. F. Helfferich, in: *Ion Exchange*, Chapter 5. McGraw-Hill, New York (1962).
11. P. J. Flory, in: *Principles of Polymer Chemistry*, Chapter XIII. Cornell University Press, 1953.
12. R. A. Siegel, in: *Self-Regulating and Pulsed Drug Delivery Systems*, J. Kost (Ed.). CRC Press, Boca Raton, FL, in press.
13. H. R. Proctor, *J. Chem. Soc.* **105,** 313 (1914).
14. H. R. Proctor and J. A. Wilson, *J. Chem. Soc.* **109,** 307 (1916).
15. D. Vermaas and J. J. Hermans, *Rec. Trav. Chim.* **67,** 983 (1948).
16. J. Rička and T. Tanaka, *Macromolecules* **17,** 2916 (1984).
17. J. Grignon and A. M. Scallan, *J. Appl. Polym. Sci.* **25,** 2829 (1980).
18. S. H. Gehrke, G. P. Andrews and E. L. Cussler, *Chem. Eng. Sci.* **41,** 8, 2153 (1986).
19. R. A. Siegel and B. A. Firestone, *Macromolecules* **21,** 3254 (1988).
20. R. A. Siegel, I. Johannes, C. A. Hunt and B. A. Firestone, *Pharmac. Res.* **9,** 76 (1992).
21. L. Y. Chou, H. W. Blanch, J. M. Prausnitz and R. A. Siegel, *J. Appl. Polym. Sci.* **45,** 1411 (1992).
22. J. E. Mark and B. Erman, *Rubberlike Elasticity. A Molecular Primer.* Wiley, New York (1988).
23. T. L. Hill, *J. Chem. Phys.* **20,** 1259 (1952).
24. S. R. Eisenberg and A. J. Grodzinsky, *J. Orthopaed. Res.* **3,** 148 (1985).
25. W. Xie and M. Valleton, *J. Membrane Sci.* **64,** 113 (1991).
26. A. Katchalsky and I. Michaeli, *J. Polym. Sci.* **15,** 69 (1955).
27. J. Hasa, M. Ilavsky and K. Dusek, *J. Polym. Sci.* **13,** 253 (1975).
28. E. Vasheghani-Farahani, J. Vera, D. D. Cooper and M. E. Weber, *Ind. Eng. Chem. Res.* **29,** 554 (1990).
29. D. W. Urry, *Prog. Biophys. Molec. Biol.* **57,** 23 (1992).
30. J. O'M. Bockris and A. K. N. Reddy, *Modern Electrochemistry.* Plenum, New York (1970).

31. A. Katchalsky, *Pure Appl. Chem.* **26**, 327 (1971).
32. F. Hofmeister, *Arch. Exptl. Path. Pharmakol. (Leipzig)* **24**, 247 (1888).
33. K. D. Collins and M. W. Washabaugh, *Q. Rev. Biophys.* **18**, 323 (1985).
34. K. Hamaguchi and E. P. Geiduschek, *J. Am. Chem. Soc.* **84**, 1329 (1962).
35. H. Freundlich and P. S. Gordon, *Trans. Faraday Soc.* **32**, 1415 (1936).
36. E. Heymann, H. G. Bleakley and A. R. Docking, *J. Phys. Chem.* **42**, 353 (1938).
37. A. R. Docking and E. Heymann, *J. Phys. Chem.* **43**, 513 (1939).
38. M. W. Washabaugh and K. D. Collins, *J. Biol. Chem.* **261**, 12477 (1986).
39. C. J. Hamilton, S. M. Murphy, N. D. Atherton and B. J. Tighe, *Polymer* **29**, 1879 (1988).
40. P. C. Hiemenz, *Principles of Colloid and Surface Chemistry*, 2nd Edn. Dekker, New York (1986).
41. R. Kjellander, T. Akesson, B. Jönsson and S. Marcelja, *J. Chem. Phys.* **97**, 1424 (1992).
42. A. Martin, J. Swarbrick and A. Cammarata, *Physical Pharmacy* 3rd Edn. Lea and Febiger, Philadelphia (1983).

APPENDIX

Discussion of Eqns (3)–(5)

The ideal Donnan model is based on four assumptions [9]: (1) the (electro)chemical potential of each mobile species, including water and ions, is the same inside the gel as it is in the outer solution; (2) all mobile species behave ideally, with no effect of the polymer network on the electrochemical behavior of the mobile ions; (3) the ionization equilibrium between protons inside the gel and fixed amine groups is governed by mass action, and the ionization constant is independent of gel hydration; and (4) within the gel, the total charge due to mobile and fixed ions is essentially zero (quasineutrality).

By assumptions (1) and (2), and using the notation of this paper, we have for any mobile ion, i,

$$\mu_i^0 + RT \ln C_i + \bar{v}_i P_{ext} + z_i F\psi = \mu_i^0 + RT \ln C_i' \qquad \text{(A1)}$$

where $\bar{v}_i$ is the partial molar volume of i and ψ is the so-called Donnan potential between the gel and outer solution phases. Ignoring the $\bar{v}_i P_{ext}$ term which can be shown to be very small (see Ref. 9; this term vanishes in the free swelling case where $P_{ext} = 0$), defining $\lambda = e^{-F\psi/RT}$ and rearranging Eqn (A1), we obtain Eqn (3).

Before proceeding, it is necessary to define explicitly the concentrations C_i appearing in Eqn (A1). We take C_i to be the number of moles of i in the gel, divided by the volume of the aqueous component of the gel. This is in contrast to the definition we have employed in past publications [12, 19], in which the total gel (aqueous + polymer) volume was used. The present definition is justified by assumption (2) above. The presumed equality of the standard state chemical potentials in Eqn (A1) reflects the assumption that the ions can interact only with the gel water. The polymer component is thus assumed to have no chemical effect on the escaping tendency of the ions, so that the relevant concentration will be referenced to the water volume. (The $\bar{v}_i P_{ext}$ term represents a very small mechanical effect of the network on the ions which is negligible, as noted above.)

In what follows we will require an expression for the mobile ion densities in the whole gel. These are given by $(1 - v_p)C_i$.

The fixed charge on the gel is due to mass action of protons inside the gel with amine groups (assumption 3). The fixed charge density (moles of fixed charge/gel volume) is then

$$\sigma = \frac{\sigma_0 v_p C_{H^+}}{K_a + C_{H^+}} = \frac{\sigma_0 v_p}{1 + K_a/\lambda C'_{H^+}} = \frac{\sigma_0 v_p}{1 + \lambda^{-1} 10^{pH-pK_a}}. \qquad \text{(A2)}$$

The total charge density, mobile and fixed within the gel is therefore $\sigma + (1 - v_p)\Sigma C_i$ which, with the aid of Eqns (3) and (A2) and the quasineutrality assumption, yields Eqn (4).

Finally, the use of van't Hoff's law, Eqn (5), is based on the assumed ideality of the ions in the aqueous component of the gel phase and the external solution.

Swelling controlled zero order and sigmoidal drug release from thermo-responsive poly(*N*-isopropylacrylamide-co-butyl methacrylate) hydrogel

YUKINARI OKUYAMA[1], RYO YOSHIDA[1], KIYOTAKA SAKAI[1], TERUO OKANO[2]* and YASUHISA SAKURAI[2]

[1] *Department of Chemical Engineering, Waseda University, 3-4-1, Ohkubo, Shinjuku-ku, Tokyo 169, Japan*

[2] *Institute of Biomedical Engineering, Tokyo Women's Medical College, 8-1, Kawada-cho, Shinjuku-ku, Tokyo 162, Japan*

Received 3 August 1992; accepted 28 October 1992

Abstract—Thermo-responsive hydrogels of poly(*N*-isopropylacrylamide-co-butyl methacrylate) (poly-(IPAAm-co-BMA)) are capable of swelling–deswelling changes in response to external temperature. As poly(IPAAm-co-BMA) gels swell larger at a lower temperature, the degree and rate of the swelling could be controlled by temperature without altering the chemical structure. Therefore, drug release profiles were remarkably changed by alternation of temperature. The release profiles of indomethacin from poly(IPAAm-co-BMA) were observed to be zero-order at 20°C. This release profile was explained in terms of a Case-II diffusion mechanism; which indicates relaxation of polymer chains with swelling was rate-determining. In the case of 10°C, release demonstrated a sigmoidal profile. The acceleration of drug release was due to a rapid increase in swelling with disappearance of the glassy core which had constrained swelling. The regulation of the water-uptake process by changing external temperature remarkably affected drug release and resulted in several different release profiles.

Key words: Poly(*N*-isopropylacrylamide); thermo-responsive polymer; swelling kinetics; Case-II diffusion; zero-order release; sigmoidal release.

INTRODUCTION

Recently, new approaches to deliver drugs, in consideration of their chronopharmacology, have been investigated as next generation-type drug therapy to avoid side effects and maximize their effectiveness [1–3]. In particular, systems retaining drugs, which sense an environmental stimuli and respond with controlled drug release, may be termed 'intelligent drug delivery systems' [4, 5]. Pulsatile drug release in response to external stimuli such as specific molecules [6–8], pH [9, 10], electric field [11, 12] and temperature [13] have been achieved using polymeric materials. Kitano *et al.* [8] have synthesized polymer complex between poly(vinyl alcohol) and poly(*N*-vinyl-2-pyrrolidone) with a phenylboronic acid moiety, and applied to pulsatile insulin release in response to glucose concentration changes. Dong and Hoffman [9] have prepared heterogeneous hydrogels with both pH- and temperature-sensitivity from *N*-isopropylacrylamide (IPAAm), acrylic acid and vinyl terminated polydimethylsiloxane (VTPDMS) loaded with indomethacin into the gels, and regulated the release rate of indomethacin from the gels by pH change.

* To whom correspondence should be addressed.

Kwon *et al.* [11] synthesized copolymer gels of 2-acrylamide-2-methylpropane sulfonic acid (AMPS) with *n*-butylmethacrylate (BMA) loaded with a positively charged solute (edrophonium chloride) by an ion-exchange method. The drug was released only when an electric field was applied and 'on-off' drug release could be achieved. The mechanism was explained as an ion exchange between positive solute and hydroxonium ion produced at the anode by water hydrolysis.

We have studied 'on-off' regulation of drug release using temperature-responsive hydrogels [13-21]. Complete 'on-off' drug release in response to temperature has been achieved using poly(*N*-isopropylacrylamide (IPAAm)-co-alkyl methacrylate (RMA)) gels [14]. Polymer chains of poly(IPAAm) hydrate to form expanded structures in aqueous solution at lower temperatures. At higher temperatures, however, the chains form compact structures by dehydration. Crosslinked poly-(IPAAm) gel exhibits remarkable swelling change in aqueous media in the vicinity of 32°C due to the rapid hydration and dehydration alterations of the polymer chains, and the gel swells larger as temperature decreases under 32°C [15]. Therefore, the degree and rate of the swelling could be controlled by temperature without changing the chemical structure of the gels. Furthermore, drug release from the gels could be controlled by alternation of temperature. From this perspective, the release profiles of drugs from poly(IPAAm-co-RMA) gels from the 'on' to the 'off' state have been studied in detail [16-21]. The dense skin layer formed on the gel surface immediately after increasing temperature stops drug release from the polymer matrix and keeps complete 'off' state stably [16, 17]. The density and thickness of the skin layer can be controlled by changing the length of alkyl side chain of RMA [18-20]. It has been demonstrated that the dynamic process of skin formation affects the drug release pattern from the 'on' state to the 'off' state, and the release patterns have been theoretically simulated [21] using a pore model [22].

However, the drug release mechanisms of poly(IPAAm) gels from the 'off' to the 'on' state have not been clarified. In the process from 'off' to the 'on' state, deswollen (dehydrated) gels absorbs water to swell, allowing drug to diffuse within the polymer matrix. Therefore, the swelling kinetics of gel may effect the drug release profiles. Furthermore, the change in temperature difference may modify the degree and rate of swelling. In this study, we have investigated the water-uptake process of poly(IPAAm-co-BMA) gels at decreasing temperatures and the concurrent drug release. The effect of the water-uptake rate on the release rate was discussed from the standpoint of diffusion of water and drug and relaxation of polymer chains with hydration.

EXPERIMENTAL

Synthesis of crosslinked poly(IPAAm-co-BMA)

Crosslinked random copolymer of *N*-isopropylacrylamide (IPAAm) (Eastman Kodak Co., Rochester, NY) with butyl methacrylate (BMA, Tokyo Kasei Kogyo Co. Ltd., Tokyo) (5 wt% in feed composition) were synthesized using 1.5 mol% ethyleneglycol dimethacrylate (EGDMA) (Nakarai Chemicals Ltd., Kyoto) as a crosslinker, *t*-butylperoctanoate (BPO) (Nippon Oil and Fats Co., Ltd., Tukuba, Japan) as an initiator, and distilled 1,4-dioxane (Kanto Chemical Co., Inc., Tokyo) as a diluent. Monomer solution was bubbled with dried nitrogen for 15 min and injected between two Mylar sheets separated by a Teflon gasket (0.5 mm) backed by glass plates. The solution was polymerized at 80°C for 18 h. After cooling to room temperature, the membrane was separated from the Mylar sheets and immersed in 100% methanol for 1 week to remove all unreacted water insoluble compounds. The methanol was changed every other day. The membranes were then soaked in 75/25, 50/50, and 25/75 vol/vol% methanol/distilled water for 1 day each. The final washing was pure water for 1 day.

Swollen membranes were cut into disks (15 mm diameter) using a cork borer and dried ambiently for 1 day and under vacuum for 3 days at room temperature.

Swelling measurements

For swelling kinetics measurements, dried copolymer disks were immersed into phosphate-buffered saline (PBS, pH 7.4) at a fixed temperature (±0.1°C). Copolymer samples were removed from the PBS at specific time points and weighed after being tapped with filter paper to remove excess water on the surface. The swelling ratio is defined as the weight of absorbed water per weight of dried polymer disk (W_{H_2O}/W_P).

Drug loading

Dried poly(IPAAm-co-BMA) gel disks were equilibrated for 3 days in a solution of indomethacin (Sigma Chemical Co., St. Louis, MO) in ethanol–water (80:20, vol/vol%) at 30°C. These swollen disks were dried under vacuum for 1 day at −20°C and for 3 days at room temperature to prevent drug migration to the surface with rapid evaporation of the solvent. Usually, the loaded drug is liable to move to surfaces in the process of drying. This migration of drug, however, was prevented by reducing the temperature to under −20°C in the process of vacuum drying.

Drug release

The device was held using a wire in constant-temperature PBS (pH 7.4, 1 *l*) stirred sufficiently. Samples of media (3 ml) were withdrawn and replaced with the same amount of PBS at specific time points. The concentration of released indomethacin was measured from the absorbance at 265.9 nm by UV spectrophotometer (Model 228, Hitachi Ltd., Tokyo).

RESULTS AND DISCUSSION

The change in swelling kinetics of poly(IPAAm-co-BMA) gel at several external temperatures

The mechanical properties of the gel are improved drastically by a small addition of the hydrophobic comonomer of BMA [13, 14, 16]. Figure 1 shows the equilibrium swelling ratio of poly(IPAAm-co-BMA) (BMA 5 wt%) gel as a function of temperature in PBS. Equilibrium swelling ratio at each temperature was measured as follows. First, dry samples ($n = 3$) were immersed in the highest temperature solution (40°C). At a given time, swollen gels were taken out and the swelling ratio was measured. Swelling reached equilibrium at 40°C in 3 days. After confirming no change in swelling with time, the temperature was lowered to 38°C. In the same manner the equilibrium swelling ratio at each temperature was measured down

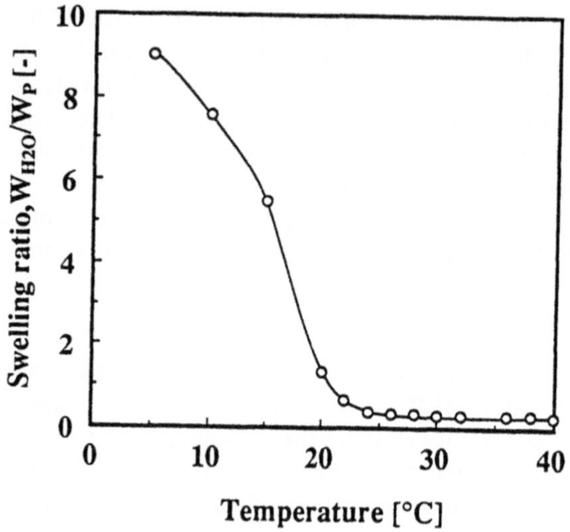
Swelling ratio, W_{H2O}/W_P [-]
10
8
6
4
2
0
0
10
20
30
40
Temperature [°C]

Figure 1. Equilibrium swelling ratio of poly(IPAAm-co-BMA) (BMA 5 wt%) gel (disk shape: 15 mm diameter and 0.5 mm thickness at room temperature) in PBS as a function of temperature.

to 10°C. The plot shows the mean value of the three samples. The deviation is so small that error bars do not have to be indicated. The lower critical solution temperature (LCST) of poly(IPAAm-co-BMA) gel was in the vicinity of 25°C. Although the transition point of poly(IPAAm-co-BMA) occurred at a lower temperature than that of poly(IPAAm) (around 32°C), the gel still maintained high thermosensitivity and showed a remarkable swelling–deswelling change at around 25°C. Figure 2 shows swelling kinetics of poly(IPAAm-co-BMA) gel as a function of time at 40, 30 and 25°C. Initially, water uptake obeyed diffusion theory (i.e. increased in proportion to the square root of time (Fickian diffusion) [23]). As the

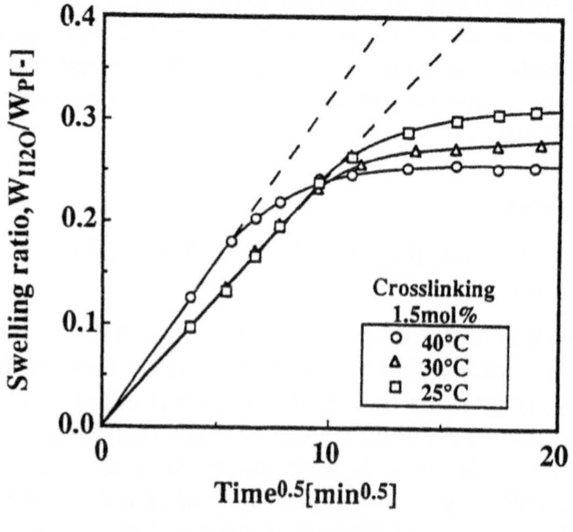
Swelling ratio, W_{H2O}/W_P[-]
0.4
0.3
0.2
0.1
0.0
0
10
20
Time$^{0.5}$[min$^{0.5}$]
Crosslinking
1.5mol%
40°C
30°C
25°C

Figure 2. Swelling kinetics of poly(IPAAm-co-BMA) gel (disk shape; 15 mm diameter and 0.5 mm thickness at room temperature) in PBS from dry state at constant temperature of 25, 30 and 40°C over LCST.

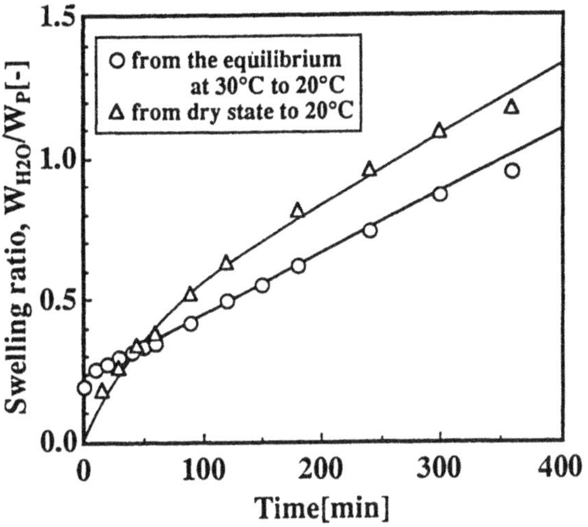

1.5
1.0
0.5
0.0
Swelling ratio, W_{H2O}/W_P[-]
0
100
200
300
400
Time[min]
O from the equilibrium at 30°C to 20°C
Δ from dry state to 20°C

Figure 3. Swelling kinetics of poly(IPAAm-co-BMA) gel (disk shape; 15 mm diameter and 0.5 mm thickness at room temperature) in PBS at constant temperature of 20°C under LCST from dry state and equilibrium swollen state at 30°C.

contraction force is larger than swelling force over LCST, large swelling change yielding a structure change of the polymer network does not occur. Therefore, water uptake was governed only by diffusion of water molecules into the polymer matrix while retaining polymer–polymer interactions.

Figure 3 shows swelling kinetics of the gel at 20°C from dry state and slightly swollen state at 30°C. Swelling from the dry state (triangles) increased linearly after 100 min. On the other hand, swelling from a slightly swollen state (circles) demonstrated a linear increase from time zero. Such linear increase in water-uptake behavior can be explained in terms of Case-II transport mechanism [24]. When polymer chains become hydrated, polymer–polymer interactions are disrupted by water. As the macromolecular relaxation rate with hydration is much slower than the diffusion rate of water in gel, the relaxation process may become rate-limiting for swelling. Then, the polymer matrix is divided into a glassy core and the swollen region separated by an interface as shown in Fig. 4. It has been actually observed that the swelling front moves interior and disappears. These pictures as the evidence of our model will be presented, discussed and compared to swelling kinetics in the next paper. The swelling front proceeds towards the inside of gel at a fixed rate resulting in a constant rate of water-uptake in the case of slab-geometry [25]. During the swelling process, the surface of the poly(IPAAm-co-BMA) gel was homogeneous. This indicates a small difference in swelling between the swelling zone and unswollen zone. The equilibrium swelling ratio of poly(IPAAm-co-BMA) at 20°C was 1.3. As the gel has already reached almost equilibrium swelling state at the time of the disappearance of swelling front, the difference between the swelling ratio when the swelling front disappears (H_2) and the equilibrium swelling ratio (H_3) is small. Therefore, constant water-uptake rate is kept until an equilibrium swelling state is reached.

The plots stands for the mean value of three samples and the standard deviation is so small that the error bar does not have to be indicated. Therefore, the two

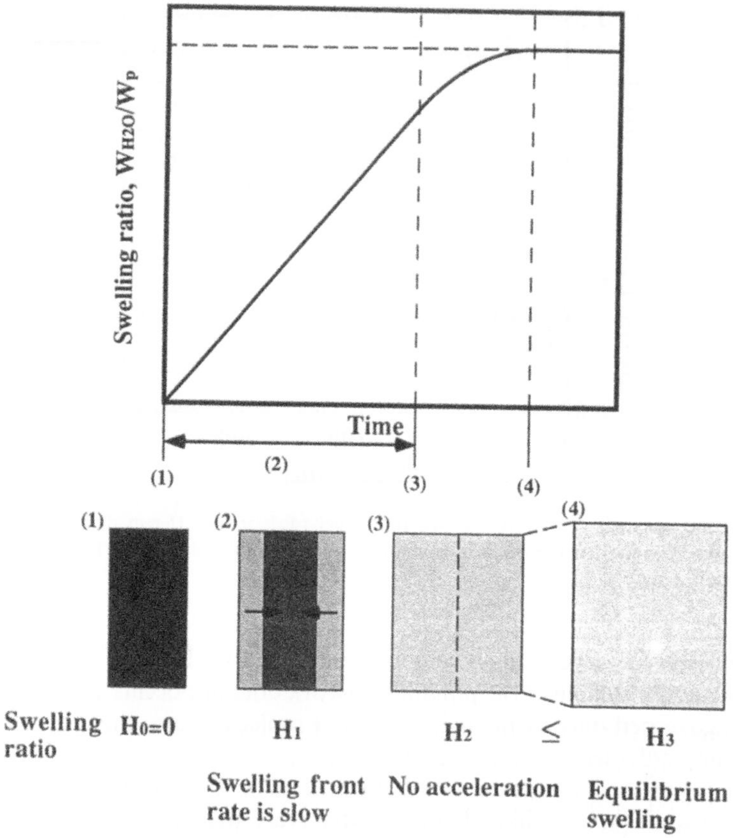
Swelling ratio, WH2O/Wp
Time
(1)
(2)
(3)
(4)
(1)
(2)
(3)
(4)
Swelling ratio
H0=0
H1
H2
≤
H3
Swelling front rate is slow
No acceleration
Equilibrium swelling

Figure 4. Swelling model at 20°C for explaining the linear increase in swelling.

triangle and circle plots have statistically significant difference. The difference may be explained as follows. At 30°C, hydration was proceeded and this process may enhance the dynamic motion of the polymeric matrices. This may cause the hydrophobic aggregated structure by hydrophobic groups which are liable to reorganize the structure to minimize the contact with water. Therefore, the hydrophobic interaction must be disrupted to hydrate from the equilibrium swollen state at 30°C, when the temperature is changed from 30 to 20°C. However, in the case of swelling from dry state, hydrophobic groups at the initial state do not aggregate as strongly as at the equilibrium of 30°C. In the process of swelling, effects to suppress hydration by hydrophobic interaction may be less remarkable than the case of swelling from 30°C. This effect may result in a larger swelling of the triangle plot than the circle plot. The curve of the triangle plot seems to be convex, however, the coefficient of correlation from 90 to 300 min is 0.996 and the curve is appreciably linear. Further, from the results of our previous data [26], the swelling of poly-(IPAAm-co-alkyl methacrylate) at 20°C tends to exhibit linear increase and Case-II model can be applied at 20°C.

Figure 5 shows swelling kinetics of poly(IPAAm-co-BMA) at 10°C. At this temperature, the swelling rate accelerated after 40 min. Then the crease pattern was observed on the gel surface in the process of swelling, as Tanaka *et al.* [27, 28]

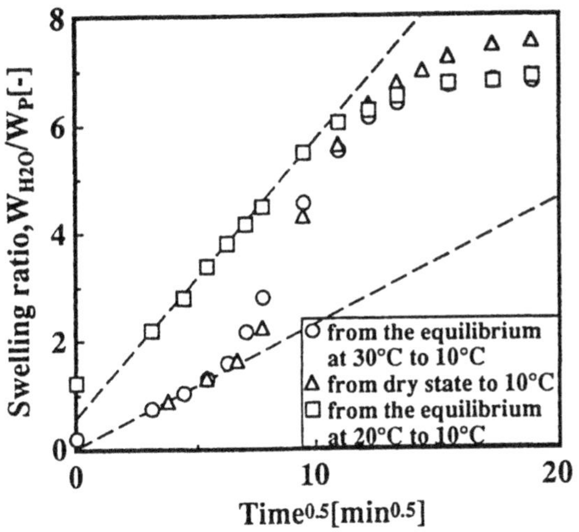

8
6
4
2
0
0
10
20
Swelling ratio, W_{H2O}/W_P[-]
Time$^{0.5}$[min$^{0.5}$]
from the equilibrium at 30°C to 10°C
from dry state to 10°C
from the equilibrium at 20°C to 10°C

Figure 5. Swelling kinetics of poly(IPAAm-co-BMA) gel (disk shape; 15 mm diameter and 0.5 mm thickness at room temperature) in PBS at constant temperature of 10°C under LCST from dry state and equilibrium swollen state at 20 and 30°C.

observed on the surface of poly(IPAAm) gel. This pattern was formed because the swelling of the gel surface was buckled due to mechanical constraint of the glassy core. The mechanism of pattern formation was discussed in detail by Tanaka *et al.* [27, 28]. The swelling curve for the initially dry gel was similar to that for the slightly swollen state at 30°C. For such sigmoidal swelling, Siegel *et al.* [29] have presented a 'swelling front model'. They have observed sigmoidal swelling kinetics for initial glassy poly(methyl methacrylate (MMA)-co-*N*,*N* dimethylaminoethyl methacrylate (DMA)) gels. In the early stage of swelling, the swelling front separates a swollen outer region from a glassy core due to Case-II transport mechanisms. The front proceeds toward the interior as the polymer absorbs water. In this process, the glassy core largely constrains swelling in the direction normal to the front. When the fronts meet at the midplain of polymer core, the glassy region vanishes and the polymer starts to swell largely uni-directionally because the swelling constraint disappears. Therefore, swelling of the gels is accelerated.

Our results can also be explained by the mechanism as shown in Fig. 6. At 10°C, the macromolecular relaxation rate was faster than at 20°C. As the difference in swelling between the swelling region and the unswollen region is larger than that at 20°C, the swollen region stretched the unswollen region and the unswollen region compressed the swelling region. This results in non-isotropic swelling and leads to pattern formation according to heterogeneous swelling structure on the surface [27, 28]. At 10°C, the swelling front proceeds toward the interior at a higher rate than at 20°C, due to the enhanced hydrogen bonding between the polymer and water. Therefore, two swelling fronts meet in the center of the gel and the depression force to swelling presumably disappears. Furthermore, the equilibrium swelling ratio H_3 was larger than the swelling ratio H_2 at that time, allowing for increased swelling. Consequently, water-uptake into poly(IPAAm-co-BMA) gels demonstrates a sigmoidal profile. Now we are trying to measure the velocity of the swelling front by taking a picture of a cross section of the swelling gel and calculating the

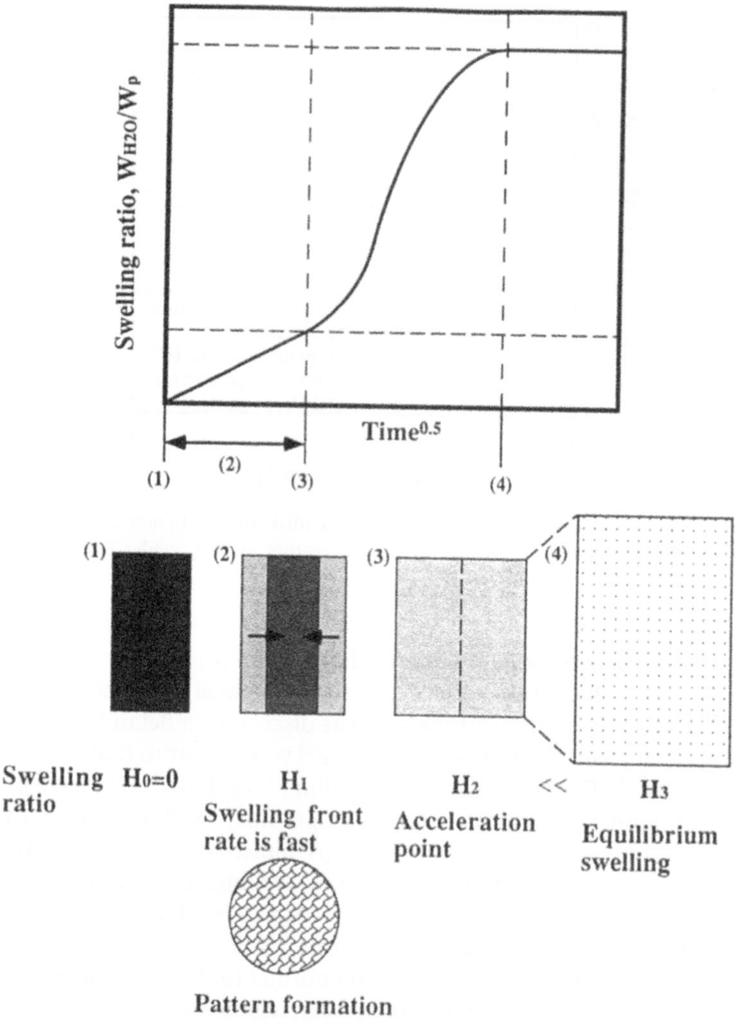
Swelling ratio, W_{H2O}/W_p
Time$^{0.5}$
(1)
(2)
(3)
(4)
(1)
(2)
(3)
(4)
Swelling ratio
H_0=0
H_1
H_2
<<
H_3
Swelling front rate is fast
Acceleration point
Equilibrium swelling
Pattern formation

Figure 6. Swelling model at 10°C for explaining sigmoidal swelling.

mass balance of absorbed water. The swelling front velocity will be obtained and discussed in detail compared to swelling kinetics in the next paper.

Water-uptake of the copolymer from the equilibrium swelling state at 20°C exhibited Fickian swelling behavior (Fig. 5, squares). In the case that polymer chains are completely hydrated at the first stage, relaxation processes do not exist in water-uptake process and swelling is governed by diffusion process.

The change in release pattern of indomethacin from poly(IPAAm-co-BMA) at several external temperatures

If water flux into the device is negligible (that is, device is at the equilibrium swelling state before releasing the drug), the drug may be released by diffusion. The diffusion theory [23] says that 60% of the total released amount of drug is in proportion to the square root of time and the remainder obeys an exponential increase. However,

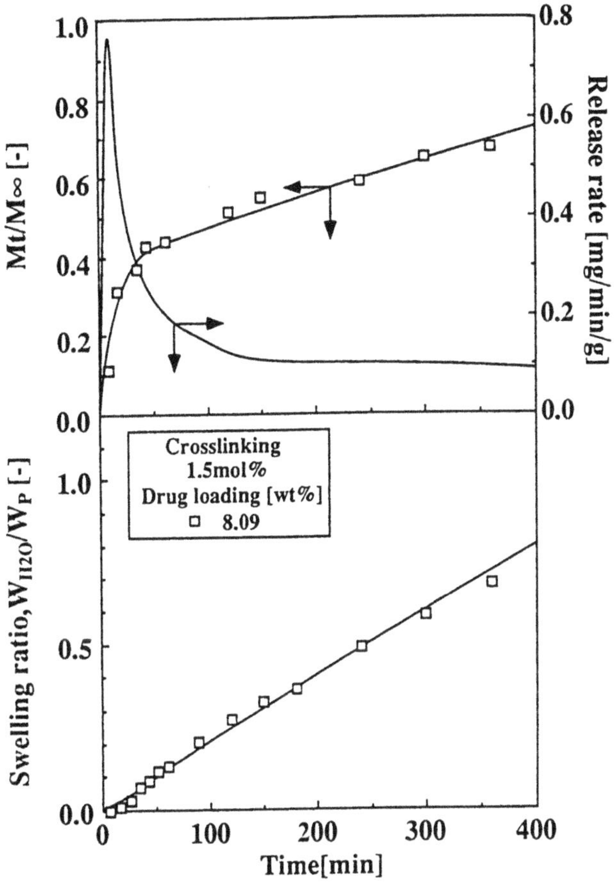

Mt/M∞ [-]
Release rate [mg/min/g]
Swelling ratio, W_H2O/W_P [-]
Crosslinking
1.5mol%
Drug loading [wt%]
8.09
Time[min]
1.0
0.8
0.6
0.4
0.2
0.0
0.5
0
100
200
300
400

Figure 7. Release profiles of indomethacin from poly(IPAAm-co-BMA) gel in PBS and swelling kinetics of the gel during the drug release at 20°C.

drug release from the dehydrated state would proceed with concurrent water uptake. Then, the swelling kinetics of gel may affect the drug release pattern. The influence of water influx into the device on the drug release pattern has been investigated.

The fractional released amount of indomethacin along with release rate from poly(IPAAm-co-BMA) device at 20°C and swelling kinetics during the release experiment are shown in Fig. 7. The release rate curve exhibited a peak at the early stage and reached a constant rate after 100 min. The peak appears to be exerted by the diffusion of indomethacin on the surface (burst effect). The swelling also demonstrated a linear increase from around 100 min. The drug release corresponded to the swelling, which suggests that the drug diffusion rate in the swollen region is much faster than the proceeding rate of the swelling front. The macromolecular relaxation rate of polymer became rate-limiting for drug release, and zero-order release was achieved.

Figure 8 shows the fractional released amount of indomethacin along with the release rate from poly(IPAAm-co-BMA) device at 10°C and swelling kinetics during the release experiment. The release rate at 10°C started with the highest value which seems to be due to burst effect. Since the swelling speed at 10°C is larger than that

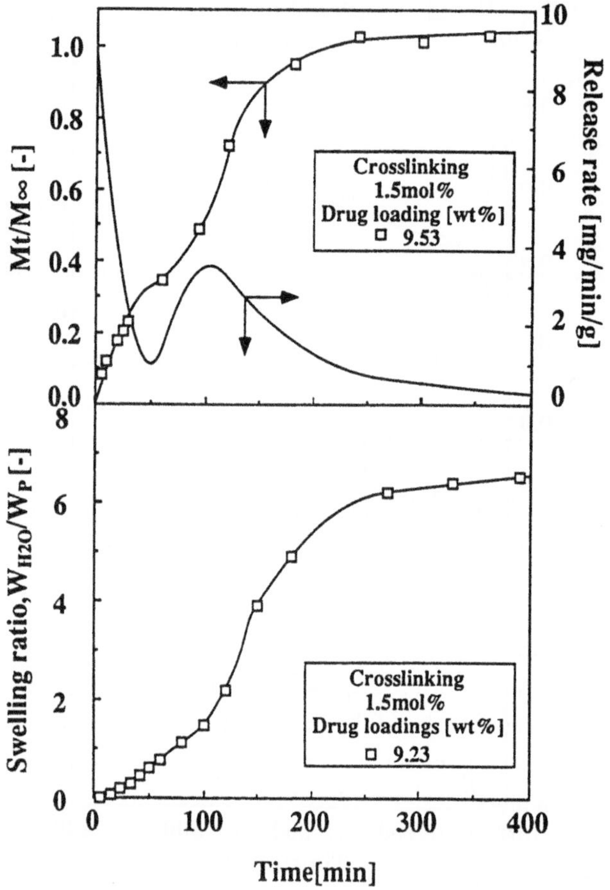

1.0
0.8
0.6
0.4
0.2
0.0
Mt/M∞ [-]
10
8
6
4
2
0
Release rate [mg/min/g]
Crosslinking
1.5mol%
Drug loading [wt%]
9.53
8
6
4
2
0
Swelling ratio, W_{H2O}/W_P [-]
Crosslinking
1.5mol%
Drug loadings [wt%]
9.23
0
100
200
300
400
Time[min]

Figure 8. Release profiles of indomethacin from poly(IPAAm-co-BMA) gel in PBS and swelling kinetics of the gel during the drug release at 10°C.

at 20°C, the burst effect appears immediately after the beginning of drug release. The release of indomethacin at 10°C resulted in a sigmoidal pattern similar to water-uptake. The peak at around 100 min in the release rate curve corresponded to the acceleration of water-uptake. Figure 9 illustrates the mechanism of the drug release at 10°C. The factors governing the release rate are: the change in diffusivity of drug in the device (curve 1); and the change in concentration gradient at the surface of the device (curve 2). While swelling fronts proceed in the device, the swelling ratio of swelling region is almost constant, that is, diffusivity of drug in the device is constant as the core depresses gel-swelling. Thus, a decrease in the concentration gradient only affects the release rate, and it decreases with increasing time at an early stage. After the core disappears, the diffusivity of the drug is enhanced with increasing swelling and the release rate is accelerated. An increase in surface area may also contribute to the rapid release rate. At the late stage of the release, the release rate decreases with decreasing drug concentration in the device. Consequently, the drug release rate has a peak as seen in Fig. 8. Clearly, drug release profiles are influenced by the change in the water-uptake process in response to external temperature change.

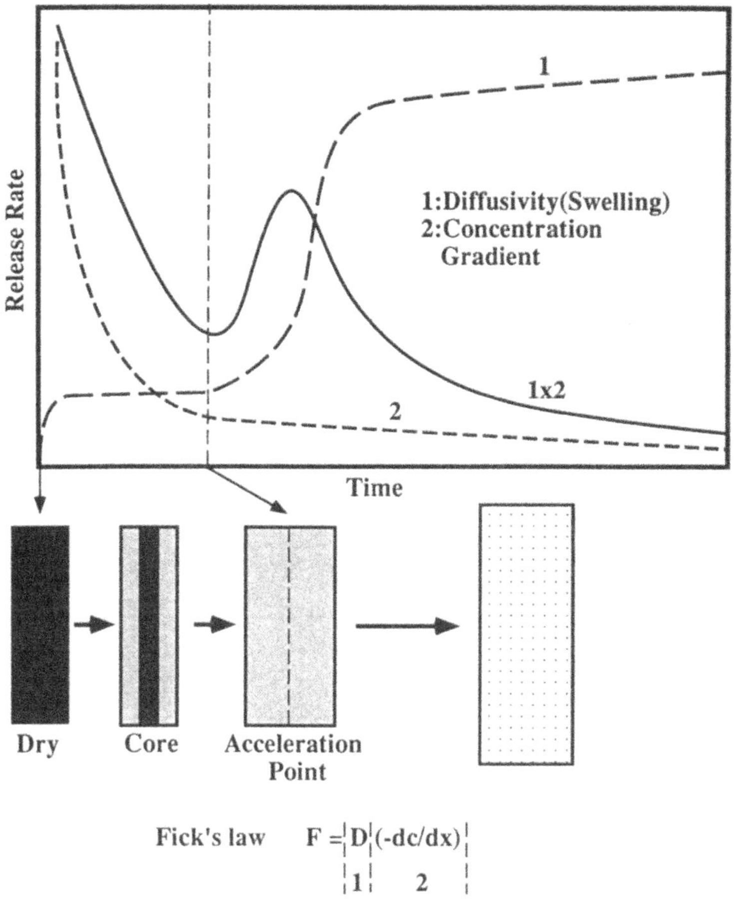
Release Rate
Time
1
2
1x2
1:Diffusivity(Swelling)
2:Concentration
Gradient
Dry
Core
Acceleration
Point
Fick's law
F = D (-dc/dx)
1
2

Figure 9. Concept of mechanism to give the maximum in the drug release profile by changes in diffusivity and concentration gradient at the surface.

CONCLUSIONS

In the swelling process of poly(IPAAm-co-BMA) gel from dehydrated state, swelling profiles are affected by structural changes of polymer chains from glassy state to hydrated state due to disruption of polymer–polymer interactions by water. Over the LCST, swelling is governed by diffusion of water molecules into polymer matrix because relaxation of polymer is negligible. Under the LCST, the gels swell larger with decreasing temperature and the swelling process is affected by relaxation processes. At 20°C, swelling of poly(IPAAm-co-BMA) increased linearly with time due to Case-II diffusion mechanism. At 10°C, the swelling demonstrated a sigmoidal pattern, which was explained in terms of swelling front mechanism. The release of indomethacin from the gel was observed to be a zero-order release profile at 20°C and sigmoidal at 10°C corresponding to water-uptake profiles. The regulation of the water-uptake process of poly(IPAAm-co-BMA) hydrogel by changing external temperature remarkably affected the drug release and resulted in several different release profiles.

Acknowledgements

The authors gratefully acknowledge Dr. Glen S. Kwon, International Center for Biomaterials Science, Tokyo Women's Medical College, for valuable discussions. This research was partially supported by the Ministry of Education, Japan (New Functionally Materials—Design, Preparation and Control).

REFERENCES

1. J. R. Robinson, *Sustained and Controlled Release Drug Delivery Systems.* Marcel Dekker, New York (1978).
2. J. Roseman and S. Z. Mansdorf, *Controlled Release Delivery Systems.* Marcel Dekker, New York (1983).
3. R. W. Baker, *Controlled Release of Biologically Active Agents.* John Wiley & Sons, New York (1987).
4. T. Okano and R. Yoshida, in: *Biomedical Applications of Polymeric Materials*, T. Tsuruta and Y. Kimura (Eds.). CRC Press, in press.
5. R. Yoshida, K. Sakai, T. Okano and Y. Sakurai, in: *Advanced Drug Delivery Reviews, Issues of Modern Hydrogel Delivery Systems*, A. Scranton and N. A. Peppas (Eds.). Elsevier, in press.
6. K. Makino, E. J. Mark, T. Okano and S. W. Kim, *J. Controlled Release* **12,** 235 (1990).
7. T. A. Horbett, B. D. Ratner, J. Kost and M. Singh, in: *Recent Advances in Drug Delivery Systems*, p. 209, J. M. Anderson and S. W. Kim (Eds.). Plenum Press, New York (1984).
8. S. Kitano, Y. Koyama, K. Kataoka, T. Okano, Y. Sakurai, *J. Controlled Release* **19,** 162 (1992).
9. L. C. Dong and A. S. Hoffman, *J. Controlled Release* **15,** 141 (1991).
10. R. A. Siegel, M. Falamarzian, B. A. Firestone and B. C. Moxley, *J. Controlled Release* **8,** 179 (1988).
11. I. C. Kwon, Y. H. Bae, T. Okano and S. W. Kim, *J. Controlled Release* **17,** 149 (1991).
12. I. C. Kwon, Y. H. Bae and S. W. Kim, *Nature* **354,** 291 (1991).
13. T. Okano, Y. H. Bae and S. W. Kim, in: *Pulsed and Self-Regulated Drug Delivery*, p. 17, J. Kost (Ed.). CRC Press, Boca Raton, FL (1990).
14. Y. H. Bae, T. Okano, R. Hsu, S. W. Kim, *Markromol. Chem., Rapid. Commun.* **8,** 481 (1987).
15. Y. H. Bae, T. Okano and S. W. Kim, *J. Polym. Sci., Polym. Phys.* **28,** 923 (1990).
16. T. Okano, Y. H. Bae, H. Jacobs and S. W. Kim, *J. Controlled Release* **11,** 255 (1990).
17. Y. H. Bae, T. Okano and S. W. Kim, *Pharm. Res.* **8,** 531 (1991).
18. R. Yoshida, K. Sakai, T. Okano, Y. Sakurai, Y. H. Bae and S. W. Kim, *J. Biomater. Sci., Polymer Edn* **3,** 155 (1991).
19. R. Yoshida, K. Sakai, T. Okano and Y. Sakurai, *J. Biomater. Sci., Polymer Edn* **3,** 243 (1992).
20. T. Okano, R. Yoshida, K. Sakai and Y. Sakurai, in: *Polymer Gels*, p. 299, D. DeRossi (Ed.). Plenum Press, New York (1991).
21. R. Yoshida, K. Sakai, T. Okano and Y. Sakurai, *Ind. Eng. Chem. Res.* **31,** 2339 (1992).
22. K. Sakai, K. Ozawa, R. Mimura and H. Ohashi, *J. Membrane Sci.* **32,** 3 (1987).
23. J. Crank, *The Mathematics of Diffusion.* Oxford University Press, London (1975).
24. T. Alfrey, Jr., E. F. Gurnee and W. G. Lloyed, *J. Polym. Sci.* **C12,** 249 (1966).
25. D. J. Enscore, H. B. Hopfenberg and V. T. Stannett, *Polymer* **18,** 793 (1977).
26. R. Yoshida, K. Sakai, T. Okano and Y. Sakurai, *Polym. J.* **23,** 1111 (1991).
27. T. Tanaka, *Physica* **140A,** 261 (1986).
28. E. Sato Matsuo and T. Tanaka, *J. Chem. Phys.* **89,** 1695 (1988).
29. R. A. Siegel, in: *Pulsed and Self-Regulated Drug Delivery*, p. 129, J. Kost (Ed.). CRC Press, Boca Raton, FL (1990).

Printed and bound by CPI Group (UK) Ltd, Croydon, CR0 4YY
23/10/2024
01778261-0001